Sustainable Development Goals Series

The **Sustainable Development Goals Series** is Springer Nature's inaugural cross-imprint book series that addresses and supports the United Nations' seventeen Sustainable Development Goals. The series fosters comprehensive research focused on these global targets and endeavours to address some of society's greatest grand challenges. The SDGs are inherently multidisciplinary, and they bring people working across different fields together and working towards a common goal. In this spirit, the Sustainable Development Goals series is the first at Springer Nature to publish books under both the Springer and Palgrave Macmillan imprints, bringing the strengths of our imprints together.

The Sustainable Development Goals Series is organized into eighteen subseries: one subseries based around each of the seventeen respective Sustainable Development Goals, and an eighteenth subseries, "Connecting the Goals", which serves as a home for volumes addressing multiple goals or studying the SDGs as a whole. Each subseries is guided by an expert Subseries Advisor with years or decades of experience studying and addressing core components of their respective Goal.

The SDG Series has a remit as broad as the SDGs themselves, and contributions are welcome from scientists, academics, policymakers, and researchers working in fields related to any of the seventeen goals. If you are interested in contributing a monograph or curated volume to the series, please contact the Publishers: Zachary Romano [Springer; zachary.romano@springer.com] and Rachael Ballard [Palgrave Macmillan; rachael.ballard@palgrave.com].

Eveline M. Ibeagha-Awemu
Sunday O. Peters • Appolinaire Djikeng
John E. O. Rege
Editors

African Livestock Genetic Resources and Sustainable Breeding Strategies

Unlocking a Treasure Trove and Guide for Improved Productivity

Volume 1

Editors
Eveline M. Ibeagha-Awemu
Sherbrooke Research and Development Centre
Agriculture and Agri-Food Canada
Sherbrooke, QC, Canada

Appolinaire Djikeng
International Livestock Research Institute (ILRI)
Nairobi, Kenya

Sunday O. Peters
Department of Animal Science
Berry College
Mount Berry, GA, USA

John E. O. Rege
Emerge Centre for Innovations–Africa
Nairobi, Kenya

ISSN 2523-3084   ISSN 2523-3092 (electronic)
Sustainable Development Goals Series
ISBN 978-3-031-92075-2   ISBN 978-3-031-92076-9 (eBook)
https://doi.org/10.1007/978-3-031-92076-9

International Livestock Research Institute
This work was supported by the International Livestock Research Institute (ILRI) and its funders.

This book is an open access publication.

Color wheel and icons: From https://www.un.org/sustainabledevelopment/
Copyright © 2020 United Nations. Used with the permission of the United Nations.

The content of this publication has not been approved by the United Nations and does not reflect the views of the United Nations or its officials or Member States.

© Crown 2026

Open Access This book is licensed under the terms of the Creative Commons Attribution 4.0 International License (http://creativecommons.org/licenses/by/4.0/), which permits use, sharing, adaptation, distribution and reproduction in any medium or format, as long as you give appropriate credit to the original author(s) and the source, provide a link to the Creative Commons license and indicate if changes were made.
The images or other third party material in this book are included in the book's Creative Commons license, unless indicated otherwise in a credit line to the material. If material is not included in the book's Creative Commons license and your intended use is not permitted by statutory regulation or exceeds the permitted use, you will need to obtain permission directly from the copyright holder.
The use of general descriptive names, registered names, trademarks, service marks, etc. in this publication does not imply, even in the absence of a specific statement, that such names are exempt from the relevant protective laws and regulations and therefore free for general use.
The publisher, the authors and the editors are safe to assume that the advice and information in this book are believed to be true and accurate at the date of publication. Neither the publisher nor the authors or the editors give a warranty, expressed or implied, with respect to the material contained herein or for any errors or omissions that may have been made. The publisher remains neutral with regard to jurisdictional claims in published maps and institutional affiliations.

Editorial Contact: Annette Klaus

This Springer imprint is published by the registered company Springer Nature Switzerland AG
The registered company address is: Gewerbestrasse 11, 6330 Cham, Switzerland

If disposing of this product, please recycle the paper.

Preface

For Africa, the journey to achieving the aspirations of the African Union Agenda 2063 (Goal 1 on quality of life and well-being, and Goal 2 on health and nutrition) and the corresponding ambitions of the United Nation's Sustainable Development Goals (SDGs: SDG1 on Poverty alleviation, SDG2 on ending Hunger, etc.) must consider livestock production and productivity. Understanding the key types, availability, and use options of the livestock genetic resources in Africa is a first step in taking measures toward sustainable increased productivity and resilience in the face of a myriad of current and future challenges. In this connection, members of the executive committee of the African Animal Breeding Network (AABNet, https://www.animalbreeding-africa.org/index.html) started discussions in mid-2019 to explore the landscape and review key drivers and opportunities for establishing breeding goals for adapted and resilient African livestock to meet increasing demand for animal source food, the challenge of climate change, and environmental concerns. As discussions progressed, it became clear that the substantial information gaps on African livestock genetic resources could not be adequately covered, in the depth and breadth we had envisaged, in a single review paper, and the idea to write this book on "African livestock genetic resources and sustainable breeding strategies" was born, and the writing started in earnest in late 2020 involving 83 contributing authors—professionals and experts from animal science, animal breeding, genetics, genomics, and other science disciplines—from 24 countries.

Authored by African livestock professionals representing a wide range of disciplines and focused on an examination of efforts toward understanding indigenous African livestock, their uses and systems of production, improvement challenges, and achievements made to date, this book presents a rich resource, across livestock species, for formulating technical, institutional, and policy interventions going forward and that leverage on rapidly changing technologies and responds to evolving human nutritional needs and other trends.

For thousands of years, natural selection and human activities have generated genetically diverse breeds of domesticated farm animals, which can significantly contribute to the livelihoods of millions of present-day Africans. Indeed, they are a treasure trove. Africa's indigenous livestock are particularly hardy and well-adapted to local production contexts, having developed adaptations to the continent's diverse climatic conditions and environmental pressures. Despite the wealth of desirable genetic traits, some of Africa's iconic and lesser-known livestock are disappearing at an alarming rate. Despite the increasing recognition of the benefits of this diversity, little has been done to understand and optimally harness the full potential of these genetic resources. This book catalogues the main farm animal genetic resources (cattle, sheep, goat, pig, chicken, dromedary, horse, donkey, buffalo, turkey, duck, and geese) and some non-conventional animal genetic resources (frog, rabbit, snail, and honeybee) in Africa and the opportunities that can be leveraged with available technologies and knowledge for achieving rapid genetic gain and improved productivity. While the structure of the chapters differ considerably, they cover five broad areas in the context of animal genetic resources in Africa: their characterization (to understand the unique genetic composition of what we have, where it might have originated from, and its present-day distribution); conservation; use (efforts and approaches to enhance sustainable utilization, focusing on improving production and productivity, and product quality to meet evolving human needs); technological support (available technological developments that support sustainable livestock improvement); and the institutions and institutional arrangements needed to support their characterization, conservation, and sustainable use.

This book is the first to present as one source, the diversity and uniqueness of African farm animal genetic resources, and the possibilities that new technologies present, which will give a better appreciation of the existing resources to a wide range of users. It is a valuable manual for students, professors, researchers, animal science professionals, livestock farmers and farmer organizations, civil society organizations, non-governmental organizations, business professionals, government agencies, including policymakers, the international development community, and anyone who seeks to understand the uniqueness of African livestock genetic resources, production systems, and strategies for sustainable improvement for the African environment.

Sherbrooke, QC, Canada	Eveline M. Ibeagha-Awemu
Mount Berry, GA, USA	Sunday O. Peters
Nairobi, Kenya	Appolinaire Djikeng
Nairobi, Kenya	John E. O. Rege

Acknowledgements

For this book to have its desired impact, it must be freely available to all interested readers. Open access availability of this book was made possible through financial support from the International Livestock Research Institute (ILRI, https://www.ilri.org/) and its funders.

We acknowledge with profound gratitude all the institutions of contributing authors for supporting them during the production of this book.

Eveline M. Ibeagha-Awemu
Sunday O. Peters
Appolinaire Djikeng
John E. O. Rege

Contents of Volume 1

Part I African Livestock Production Systems and Genetic Resources

1. **An Overview of African Livestock Genetic Resources and Improvement Strategies** 3
   Eveline M. Ibeagha-Awemu, Sunday O. Peters, Appolinaire Djikeng, and John E. O. Rege

2. **African Livestock Production Systems: The Past, Present and the Projected Future** 13
   Eveline M. Ibeagha-Awemu, Richard Osei-Amponsah, and Martha N. Bemji

3. **The History, Geography, and Characteristics of Indigenous African Taurine Cattle** 65
   John E. O. Rege, Chi L. Tawah, Isidore Houaga, and Eveline M. Ibeagha-Awemu

4. **The History, Geography, and Characteristics of African Zebu, Zebu–Taurine Derivatives, and Well-Established Exotic Cattle Breeds** 117
   John E. O. Rege, Chi L. Tawah, Donald R. Kugonza, Mizeck G. G. Chagunda, Isidore Houaga, Oluyinka Opoola, and Eveline M. Ibeagha-Awemu

5. **African Goat Genetic Resources, Diversity and Unique Features** 185
   Moses Okpeku, Martha N. Bemji, Isidore Houaga, Khaled Fantazi, Liveness J. Banda, Timothy Gondwe, Sebastine Chenyambuga, Sahar A. Elnahta, Doctor M. N. Mthiyane, Shumuye Belay, Tadelle Dessie, Taiye S. Adewumi, and Oliver Hanotte

6. **African Sheep Genetic Resources, Diversity and Unique Features** 239
   Martha N. Bemji, Semir B. S. Gaouar, Abdelkader Ameur Ameur, Fatima Z. Belharfi, Isidore Houaga, and Anne W. T. Muigai

7 **African Domestic Poultry Genetic Resources, Diversity, and Unique Features** 327
   Tadelle Dessie, Christian K. Tiambo, Liveness J. Banda,
   Raman A. Lawal, Sheila C. Ommeh, Timothy Gondwe,
   Esatu Wondmeneh, Matthew A. Adeleke, and Olivier Hanotte

8 **Pig Genetic Resources of Africa** 357
   Richard Osei-Amponsah and Wilson S. Kaumbata

9 **African Dromedary Genetic Resources, Diversity and Breeding Systems** 395
   Semir B. S. Gaouar, Imane Meghelli, Zoubeyda Kaouadji,
   Félix Meutchieye, and Djalel E. Gherissi

10 **African Donkey Genetic Resources, Diversity and Breeding Strategies** 451
   Martha N. Bemji, Eveline M. Ibeagha-Awemu,
   Abdelkader Ameur Ameur, Albano Beja-Pereira,
   Madani Labbaci, Liveness J. Banda, and Semir B. S. Gaouar

11 **African Horse Genetic Resources and Breeding Strategies for African Input Systems** 497
   Djalel E. Gherissi, Ginette Aumassip-Kadri,
   Mohammed E. A. Benhamadi, Félix Meutchieye,
   Yassine H. Jamali, and Semir B. S. Gaouar

12 **African Water Buffalo Genetic Resources, Diversity, and Unique Features** 593
   Sahar S. E. Ahmed, Amal A. M. Hassan,
   Ibrahim A. H. Barakat, and Hassan M. Al Ashmaoui

13 **Nonconventional Animal Genetic Resources, Diversity, and Unique Features** 619
   Kingsley A. Etchu, Abdelkader Ameur Ameur, M. Chahbar,
   Félix Meutchieye, Annick N. Enangue Njembele,
   Gerald C. Tasse Taboue, and Semir B. S. Gaouar

14 **Contributions of African Livestock Production Systems to Greenhouse Gas Emissions and Global Warming in the Face of Climate Change** 675
   Mizeck G. G. Chagunda, Kingsley A. Etchu,
   Kwamboka Tirimba, and Okeyo Mwai

Part II Opportunities for Improved Utilization of African Livestock Genetic Resources

15 **Defining Breeding Goals and Breeding Strategies for Improving Livestock Under Various Production Systems in Africa: Concept and Brief Overview** 691
   Raphael Mrode, David A. Mbah, and Julie M. K. Ojango

16 **Defining Breeding Goals and Breeding Strategies for Improving the Productivity of Cattle Breeds and Buffaloes in African Production Systems** 705
Raphael Mrode, David A. Mbah, Chi L. Tawah,
Julie M. K. Ojango, Sunday O. Peters,
Eveline M. Ibeagha-Awemu, Oluyinka Opoola,
Isidore Houaga, Richard Osei-Amponsah, Moses Okpeku,
and John E. O. Rege

17 **Breeding Goals and Strategies for Improving Small Ruminant Productivity in African Input Systems** 759
Julie M. K. Ojango, Richard Osei-Amponsah, Isidore Houaga,
Timothy Gondwe, Moses Okpeku, and Donald R. Kugonza

18 **Defining Breeding Goals and Breeding Strategies for Chicken Production Systems in Africa** 785
Sunday O. Peters, Michael O. Ozoje, Adeyemi S. Adenaike,
Blaise A. Hako Touko, Christian K. Tiambo,
Matthew A. Adeleke, Oluyinka Opoola, and Tadelle Dessie

19 **Defining Breeding Goals and Breeding Strategies for Pigs in Various African Production Systems** 819
Donald R. Kugonza and Richard Osei-Amponsah

Contents of Volume 2

Part III   Leveraging Modern Technologies for Improved Utilization of African Livestock Genetic Resources

20 Role of Modern Technologies for Sustainable Genetic Improvement of African Livestock 851
Eveline M. Ibeagha-Awemu, Sunday O. Peters, Martha N. Bemji, Jean M. Feugang, Richard Osei-Amponsah, Daniel Trocmé, and Raphael Mrode

21 The Role of Modern Technologies for Improving the Production Environment of Livestock in Africa 909
Eveline M. Ibeagha-Awemu, Faith A. Omonijo, Martha N. Bemji, Obioha Duranna, Iliya D. Kwoji, Michael O. Ozoje, and Richard Osei-Amponsah

22 Prospects for Utilization of Modern Technologies for Cattle Improvement in Africa 991
Sunday O. Peters, Eveline M. Ibeagha-Awemu, Iliya D. Kwoji, Raphael Mrode, Michael O. Ozoje, David A. Mbah, Isidore Houaga, Moses Okpeku, Peter O. Fayemi, Daniel Trocmé, Oluyinka Opoola, and Matthew A. Adeleke

23 Prospect of Modern Technologies for Poultry Improvement in Africa 1021
Victor E. Olori, Sunday O. Peters, and Matthew A. Adeleke

24 Prospect for Utilization of Modern Technologies for Small Ruminant and Pig Improvement in African Input Systems: Case Studies 1049
Jean M. Feugang, Othman E. -M. Othman, Wilson Nandolo, Donald R. Kugonza, and Richard Osei-Amponsah

Part IV Conservation of African Livestock Genetic Resources

25 Conservation and Management of Animal Genetic Resources in the Context of African Livestock Production Systems: The Case for In Situ and Ex Situ Conservation 1071
Jean M. Feugang, Richard Osei-Amponsah, John E. O. Rege, Khaled Fantazi, Christian K. Tiambo, Felicien Shumbusho, Isidore Houaga, Derradji Harek, Notsile H. Dlamini, and Semir B. S. Gaouar

26 Economic Considerations and Framework of Conservation of African Animal Genetic Resources 1091
Wilson Kaumbata, Maria Wurzinger, and John E. O. Rege

Part V Institutional Arrangements and Enabling Environment for Enhanced Utilization of African Livestock Genetic Resources

27 Capacity Strengthening of Animal Genetic Improvement Education in Africa 1109
Samuel E. Aggrey, Richard Osei-Amponsah, Donald R. Kugonza, Raphael A. Mrode, Romdhane Rekaya, and John E. O. Rege

28 Capacity Building in Livestock Breeding and Genetic Improvement in Achieving UN Sustainable Development Goals for Africa 1123
Christian K. Tiambo, Oluyinka Opoola, Moses Okpeku, Isidore Houaga, Blaise A. Hako Touko, and Tadelle Dessie

29 Harnessing Multi-Country Cooperations, Initiatives, Facilities, and Technologies for Advancing Livestock Genetic Improvement in Africa 1149
Ntanganedzeni O. Mapholi, Cuthbert Banga, Raphael Mrode, Oluyinka Opoola, Isidore Houaga, Lucy T. Nesengani, and Eveline M. Ibeagha-Awemu

30 Policies, Frameworks, Strategies, and Action Plans for Conservation and Sustainable Use of African Animal Genetic Resources 1185
John E. O. Rege and Mure Agbonlahor

31 Resourcing and Institutional Arrangements to Deliver Sustainable Animal Genetic Improvement in Africa 1233
Eveline M. Ibeagha-Awemu, Victor Olori, Ismail Muritala, Olubunmi I. Duduyemi, Mizeck G. G. Chagunda, and John E. O. Rege

Index 1279

List of Figures

Fig. 1.1	Africa's total, rural, and urban population statistics for the years from 2010 to 2021	6
Fig. 2.1	(**a**) Cattle, sheep and goat population trends in Africa and regions from 2005 to 2021. (**b**) Pig and chicken population trends in Africa and regions from 2005 to 2021	17
Fig. 2.2	(**a**) The main agro-ecological zones in Africa (Source: Sebastien (*2009*)). (**b**) Livestock production systems by agro-ecological zones in Africa. (Source: Robinson et al. (*2011*). Available freely through ILRI on Flickr https://www.flickr.com/photos/ilri/14509985147/)	20
Fig. 2.3	Progression of African main livestock production systems according to land availability and level of inputs	24
Fig. 2.4	Pastoralism in West Africa, specifically in the Bamenda Grasslands of Cameroon (**a**) and the Sahel region of Burkina Faso (**b**). (Photos by Dr. Eveline M. Ibeagha-Awemu)	28
Fig. 2.5	Positive synergies between crops, livestock and households in mixed crop–livestock systems	31
Fig. 2.6	Small intensively managed backyard pig farm (**a**) and poultry farm (**b**) in Bambili, Cameroon. (Photos by Azenui B. Abongban)	33
Fig. 2.7	A peri-urban medium-scale, commercial, intensively managed poultry farm in Nkwen, Cameroon. (Photo by Azenui B. Abongban)	35
Fig. 2.8	Large-scale, intensively managed poultry operations: (**a**) broiler, (**b**) layer, (**c**) meat processing and (**d**) chick units at commercial farms in Nigeria	38
Fig. 2.9	Transformational drivers of current and future African high-producing and resilient livestock production systems	53

Fig. 3.1	Origin and migration patterns of domestic cattle in Africa. (Source: Modified from Mwai et al. (*2015*), licensed under CC-BY 4.0)	69
Fig. 3.2	Distribution of cattle breed types in Africa. *Dots indicate the type of cattle in a region (cattle of North Africa and exotic cattle breeds are not shown). (Source: Mwai et al. (*2015*), licensed under CC-BY 4.0)	70
Fig. 3.3	A typical N'Dama bull and herd at Nsukka, Nigeria. (Photos by Dr E.M. Ibeagha-Awemu)	73
Fig. 3.4	A Kuri cattle herd in its Lake Chad environment and typical Kuri animals. (Courtesy: ILRI)	77
Fig. 3.5	Doayo (Namchi) cattle at Wakwa, Cameroon. (Photos by Dr E.M. Ibeagha-Awemu)	85
Fig. 3.6	A Muturu herd (**a**), bull (**b**), and cow with calf (**c**), showing typical black and white colour patterns. (Courtesy: Prof. O. A. Osinowo)	86
Fig. 3.7	A herd of Ghana Shorthorn cattle. (Source: ILRI)	90
Fig. 3.8	A Lagune bull. (Source: Ahozonlin et al. (*2022*), licensed under CC-BY 4.0)	90
Fig. 3.9	Typical Somba (**a**) and Lagune (**b**) cows. (Source: Vanvanhossou et al. (*2021a*), licensed under CC-BY 4.0)	91
Fig. 3.10	A Baoulé cow. (Courtesy: Albert Soudre', Open Source: https://www.eurekalert.org/multimedia/722329)	92
Fig. 3.11	Kapsiki (**a**), Namchi (**b**), Bamenda (**c**), and Bakosi (**d**) cows. (Source: Ojong et al. (*2021*), licensed under CC-BY 4.0)	94
Fig. 3.12	A Sheko cattle herd. (Source: ILRI)	109
Fig. 4.1	A Butana cow with a calf and a Kenana cow. (Source: https://twitter.com/ensembl/status/951859117488005120 - Open Access)	123
Fig. 4.2	A herd of Baggara cattle. (Courtesy: AGTR, ILRI)	123
Fig. 4.3	Typical Karamojong cows with suckling calves. (Courtesy: Dr. Donald Kugonza)	125
Fig. 4.4	A Boran bull (**a**) and herd (**b**) at the Sosian ranch in Kenya and typical Ethiopian Boran bulls in Yabello, Borana, southern Ethiopia. (Courtesy: ILRI)	127
Fig. 4.5	A herd of Small East African Zebu in Tanzania. (Courtesy: Dr. Gladness Mwanga)	130
Fig. 4.6	Angoni bull (**a**) and a cow with a calf (**b**). (Courtesy: Dr. Maria da Gloria Taela)	131
Fig. 4.7	(**a**) A herd of Madagascar Zebu. (Courtesy: Dr. Maria da Gloria Taela). (**b**) A herd of Malawi Zebu. (Courtesy: Dr. Patricia Mayuni)	133

Fig. 4.8	(**a**) Typical Banyo Gudali cow and calf and (**b**) mature bull at Sabga, Cameroon, portraying the typical physical characteristics of the breed. (Courtesy: Dr. Eveline M. Ibeagha-Awemu). (**c**) Sokoto Gudali bull (Nigeria). (Courtesy: Dr. Eveline M. Ibeagha-Awemu). (**d**) Typical Ngaoundere Gudali cow and calf and (**e**) bull (Cameroon). (Courtesy: Dr. Eveline M. Ibeagha-Awemu).	135
Fig. 4.9	Wadara or Shuwa cattle: a bull and a herd	139
Fig. 4.10	Azawak cattle in Niger. (Courtesy: Larry W. Harms, Virginia)	140
Fig. 4.11	Maure cattle. (Courtesy: Dr. Larry W. Harms, Virginia)	141
Fig. 4.12	Gobra cattle. (Courtesy: Dr. Larry W. Harms Virginia)	142
Fig. 4.13	White Fulani bull (**a**) and cow and calf (**b**). (Courtesy: Dr. E. M. Ibeagha-Awemu)	144
Fig. 4.14	Red Fulani herd (**a**) and bull (**b**). (Courtesy: Dr. E. M. Ibeagha-Awemu)	146
Fig. 4.15	Sudanese Fulani cattle. (Courtesy: ILRI)	147
Fig. 4.16	Djelli cattle. (Courtesy Dr. Larry W. Harms, Virginia)	148
Fig. 4.17	A Danakil cow and a Raya Azebo herd. (Courtesy: AGTR, ILRI)	151
Fig. 4.18	Ankole bull and herd showing their characteristic lyre horns. (Courtesy: Dr. Donald Kugonza)	153
Fig. 4.19	An Abigar bull and a herd of Abigar cattle. (Courtesy: AGTR, ILRI)	154
Fig. 4.20	A Landim cattle herd and a heifer in Mozambique. (Courtesy: Maria da Gloria Taela)	155
Fig. 4.21	A typical Nguni bull (**a**) and a cow with a calf (**b**). (AGTR, ILRI)	155
Fig. 4.22	A herd of Tonga cattle. (Courtesy: DAGRIS, ILRI)	157
Fig. 4.23	A herd of Barotse cattle. (Courtesy: Dr. Denis Lembani)	157
Fig. 4.24	A herd of Tswana cattle. (Courtesy: ILRI)	159
Fig. 4.25	A Tuli herd and a bull. (Courtesy: Tuli Cattle Society of Southern Africa)	160
Fig. 4.26	Afrikaner bull and herd. (Courtesy: Afrikaner Cattle Breeders' Society of South Africa)	161
Fig. 4.27	Typical Keteku (**a**) and Djakore (**b**) cattle herds. (Courtesy: AGTR, ILRI)	163
Fig. 4.28	Fogera cow (**a**), bull (**b**), and a cow with a calf (**c**). (Source: Tesfa et al. (*2022*)) (attributed to ILRI)	167
Fig. 4.29	The Ethiopian Horro cattle. (Courtesy: DAGRIS, ILRI)	168
Fig. 4.30	A typical Nganda cow. (Courtesy: Dr. Donald Kugonza)	168
Fig. 4.31	Herds of Tete cattle. (Courtesy: AGTR, ILRI, and Dr. Maria da Gloria Taela)	168
Fig. 4.32	A Bonsmara bull (**a**) and a cow with a calf (**b**). (Courtesy: Bonsmara Cattle Breeders' Society, South Africa)	169

Fig. 4.33	An Mpwapwa bull (**a**) and a cow (**b**). (Courtesy: Aluna Chawala)	169
Fig. 4.34	A herd of Renitelo cattle of Madagascar. (Courtesy: Dr. Maria da Gloria Taela)	171
Fig. 4.35	A herd of Wakwa cattle. (Courtesy: Mr. Lawrence Shang of the TADU Dairy Cooperative, Bui Division, North-West Region of Cameroon)	172
Fig. 4.36	A herd of Drakensberger cows (**a**) and a bull (**b**). (Courtesy: Drakensberger Cattle Breeders' Society)	172
Fig. 5.1	Nubian goat	190
Fig. 5.2	(**a**) Begait does with its kids. (Photo by Shumuye Belay). (**b**) Begait buck. (Photo by Shumuye Belay)	194
Fig. 5.3	(**a**) Kalahari Red bucks at Federal University of Agriculture, Abeokuta (FUNAAB). (Source: Bemji et al. *2014*). (**b**) Kalahari Red doe and kids at FUNAAB. (Photo by O. A. Osinowo)	195
Fig. 5.4	A herd of Abergelle goats. (Photo by Shumuye Belay)	200
Fig. 5.5	A herd of Afar goats. (Anne W.T Muigai)	204
Fig. 5.6	Sudanese hill goat. (Photo: Y. A. Hassan)	205
Fig. 5.7	A Western lowland or Shankela doe. (Photo: ILRI)	207
Fig. 5.8	(**a**) Keffa Central highland goat. (Photo: ILRI). (**b**) Keffa Central highland goat. (Photo: ILRI)	208
Fig. 5.9	Cameroon Grassland Dwarf. (Photo by Jules Fon)	209
Fig. 5.10	West African Dwarf doe at Southwest Nigeria. (Photo by M.N. Bemji)	209
Fig. 5.11	(**a**) A Mubende doe. (Photo by D. Kugonza). (**b**) A Mubende buck. (Photo by D. Kugonza)	210
Fig. 5.12	Small East African goat	211
Fig. 5.13	A herd of Red Sokoto does at the National Animal Production Research Institute (NAPRI), Zaria, Nigeria. (Photo: M.N. Bemji)	212
Fig. 5.14	West African Long-legged at the National Animal Production Research Institute (NAPRI), Zaria, Nigeria. (Photo by J.C. Amabo)	213
Fig. 5.15	Sudanese Desert goat. (Photo: Y.A. Hassan)	214
Fig. 5.16	Angora doe with its kid (Visser and Van Marle-Köster *2018*)	215
Fig. 5.17	(**a**) Damascus buck. (ESGPIP *2008*). (**b**) Damascus doe (ESGPIP *2008, 1971*)	216
Fig. 5.18	Boer goat	222
Fig. 5.19	Original *Bouhezza* cheese from Algeria	229
Fig. 6.1	West African Dwarf Ram at Federal University of Agriculture, Abeokuta, Nigeria. (Photo: Ismaila Muritala)	253
Fig. 6.2	Cameroon Blackbelly Ewe. (Photo: Julius Félix Meutchieye)	254

Fig. 6.3	Cameroon Blackbelly Ram. (Photo: Julius Félix Meutchieye)	255
Fig. 6.4	Balami Ewes at National Animal Production Research Institute, Ahmadu Bello University, Zaria, Nigeria. (Photo: Julius C. Amabo)	255
Fig. 6.5	Uda Ewes at National Animal Production Research Institute, Ahmadu Bello University, Zaria, Nigeria. (Photo: Julius C. Amabo)	256
Fig. 6.6	Yankasa Ewes at National Animal Production Research Institute, Ahmadu Bello University, Zaria, Nigeria. (Photo: Julius C. Amabo)	257
Fig. 6.7	D'man. (From Djaout et al *2017*, licensed under CC-BY 4.0)	258
Fig. 6.8	Bélier Ouled Djellal ram at Biskra. (From Djaout et al. *2017*; licensed under CC-BY 4.0)	262
Fig. 6.9	Peul-Peul sheep in the sylvo-pastoral zone of Senegal. (From Ndiaye et al. *2018*, licensed under CC-BY 4.0)	263
Fig. 6.10	Sudan Desert sheep reared in the semi-desert belt of Sudan. (From Ali et al. *2016*, licensed under CC-BY 4.0)	264
Fig. 6.11	Berbere sheep of the Bouhadjar mountains (El-Tarf), Algeria. (From Djaout et al. *2017*, licensed under CC-BY 4.0)	267
Fig. 6.12	Hamra Ewes in Mecheria City, Nâama Province, Algeria. (From (Djaout et al. *2017*, licensed under CC-BY 4.0)	268
Fig. 6.13	Beni Guil ram of Morocco. (From Bechchari *2009*, published in an open-source report published by the Ministry of Agriculture, Kingdom of Morocco)	270
Fig. 6.14	A Saidi ram. (From Elshazly and Youngs *2019*, licensed under CC-BY 4.0)	272
Fig. 6.15	A Saidi ewe. (From Elshazly and Youngs *2019*, licensed under CC-BY 4.0)	272
Fig. 6.16	Farafra ram (left) and ewe (right) found in the El-Farafra Oasis, Western Egypt. (From Elshazly and Youngs *2019*, licensed under CC-BY 4.0)	273
Fig. 6.17	Afar ram in the Afar State of Ethiopia. (From Getachew *2008*, through ILRI repository; CGSPACE, an open-source repository for Agricultural Research Output)	273
Fig. 6.18	Afar ewe in the Afar State of Ethiopia. (From Getachew *2008*, through ILRI repository; CGSPACE, an open-source repository for Agricultural Research Output)	274
Fig. 6.19	Washera ram (left) and ewe (right) in the Amhara state in Ethiopia. (From Ferede et al. *2014*, licensed under CC-BY 4.0)	275

Fig. 6.20	Horro ewe; Photo: ILRI/Apollo Habtamu. (Available on Flickr for free use under the Creative Commons Attribution Licence. (Image can be viewed here)).	275
Fig. 6.21	Adilo ewe. (Photo: ILRI\Zerihun Sewunet) (Available on Flickr for free use under the Creative Commons Licence (Image can be viewed here))	276
Fig. 6.22	A Bonga ram. (Photo: ILRI/Apollo Habtamu) (available on Flickr for free use under the Creative Commons Licence (Image can be viewed here)).	277
Fig. 6.23	Red Maasai sheep, Photo: ILRI. (Available on Flickr for free use under the Creative Commons Licence (Image can be viewed here))	278
Fig. 6.24	Tanzanian long fat-tailed sheep, Photo: ILRI/Cleaned VC. (Available on Flickr for free use under the Creative Commons Licence. (Image can be viewed here))	279
Fig. 6.25	Indigenous Menz sheep herd in Amhara Region, Ethiopia; Photo: ICARDA\Getachew Mengistu. (Available on Flickr for free use under the Creative Commons Licence. (Image can be viewed here))	286
Fig. 6.26	Menz ram in Menz Area, Ethiopia. (From Ferede et al. *2014*, licensed under CC-BY 4.0)	287
Fig. 6.27	Menz ewe in Menz Area, Ethiopia. (From Ferede et al. *2014*, licensed under CC-BY 4.0)	287
Fig. 6.28	Farta ram in Farta District, Ethiopia. (From Ferede et al. *2014*, licensed under CC-BY 4.0)	289
Fig. 6.29	Farta ewe in Farta District, Ethiopia. (From Ferede et al. *2014*, licensed under CC-BY 4.0)	289
Fig. 6.30	Barki ram. (From Elshazly and Youngs *2019*, licensed under CC-BY 4.0)	291
Fig. 6.31	Barki ewe. (From Elshazly and Youngs *2019*, licensed under CC-BY 4.0)	291
Fig. 6.32	Blackhead Somali sheep in a livestock market in Somaliland, Photo: ILRI. (Available on Flickr for free use under the Creative Commons Licence (Image can be viewed here))	292
Fig. 6.33	Bergui ewes. (Queue Fine de l'Ouest) (Tozeur). (From Khaldi et al. *2011*, licensed under CC-BY 4.0)	296
Fig. 6.34	Rahmani ram. (From Elshazly and Youngs *2019*, licensed under CC-BY 4.0)	296
Fig. 6.35	Rahmani ewe. (From Elshazly and Youngs *2019*, licensed under CC-BY 4.0)	296
Fig. 6.36	Taâdmit ram in Djelfa, Taâdmit. (From Djaout et al. *2017*, licensed under CC-BY 4.0)	300
Fig. 6.37	Noire de Thibar in Zaghouène in Tunisia. (From Djaout et al. *2017*, licensed under CC-BY 4.0)	301
Fig. 6.38	Sicilo Sarde. (From Aloulou et al. *2018*, licensed under CC-BY 4.0)	301

Fig. 6.39	Sheep milk production in Africa (in millions of tonnes), 2009–2019 (FAOSTATS *2021*)	311
Fig. 6.40	Sheep meat production in Africa (in millions of tonnes), 2009–2019 (FAOSTATS *2021*)	312
Fig. 6.41	Sheep greasy wool production in Africa (in millions of tonnes), 2009–2019 (FAOSTATS *2021*)	313
Fig. 7.1	Fayoumi chicken. (Courtesy Dr. Wondmeneh Esatu)	335
Fig. 7.2	Horro chicken. (Courtesy Dr. Wondmeneh Esatu)	338
Fig. 7.3	Konso chicken. (Source: Dana et al. *2010*), Food and Agricultural Organization, reproduced with permission)	339
Fig. 7.4	Tepi chicken. (Courtesy: Dr. Wondmeneh Esatu)	340
Fig. 7.5	Tilili chicken. (Courtesy: Dr. Wondmeneh Esatu)	340
Fig. 7.6	Indigenous Rwandan chickens. (Source: Hirwa et al. 2019, licensed under CC BY 4.0)	341
Fig. 7.7	Ugandan chicken. (Source: Yussif et al. 2023) (Licensed under C-C By 4:0)	341
Fig. 7.8	Ovambo. (Source: Grobbelaar et al. *2010*, Food and Agricultural Organization, reproduced with permission)	341
Fig. 7.9	Potchefstroom Koekoek chicken. (Courtesy: Dr. Wondmeneh Esatu)	342
Fig. 7.10	Venda chicken. (Source: Grobbelaar et al. *2010*, Food and Agricultural Organization, reproduced with permission)	342
Fig. 7.11	Boschveld chicken. (Source: http://boschveld.co.za/)	343
Fig. 8.1	Major centers of livestock domestication based on archaeological and molecular information. From FAO 2007a, licensed under Copyright © FAO 2007	360
Fig. 8.2	Suiforme diversity and phylogenetic relationship. (From Chen et al. *2007*, licensed under CC-BY 4.0). (Source: https://untamedscience.com/family/suidae/)	361
Fig. 8.3	Ashanti Dwarf pig of Ghana	367
Fig. 8.4	Burkina Faso pig. (From Kiendrebeogo et al. *2012*, licensed under CC-BY 4.0)	368
Fig. 8.5	Korhogo pig of Burkina Faso and Cote d'Ivoire. (From Kiendrebeogo et al. *2012*), licensed under CC-BY 4.0)	369
Fig. 8.6	Pigs in local production systems in Liberia. (**a**) Libera native pig; (**b**) crossbred; (**c**) Exotic breed	369
Fig. 8.7	Indigenous pig breeds of Cameroon. (**a**) Mankon long nose; (**b**) Bakosi; (**c**) Bamiléké. (From Motsa'a et al. *2021*, licensed under CC-BY 4.0)	370
Fig. 8.8	Indigenous pig of Uganda. (Ugandan pig). (From Marshall *2022*, licensed under CC-BY 4.0)	371

Fig. 8.9	The Bantu pig of South Africa. (Source: Lilongwe University of Agriculture and Natural Resources 2011, licensed under CC-BY 4.0)	372
Fig. 8.10	The Mukota pig breed of Zimbabwe. (From https://www.petmapz.com/breed/mukota-pig/#, licensed under CC-BY 4.0)	373
Fig. 8.11	The Kolbroek pig breed of South Africa. (From http://farmersweekly.co.za, licensed under CC-BY 4.0)	373
Fig. 8.12	The Duroc pig breed. (Credit: Prof. R. Osei-Amponsah)	375
Fig. 8.13	The Large White pig breed. (Credit: Prof. R. Osei-Amponsah)	376
Fig. 8.14	The Pietrain pig breed. (Credit: Prof. R. Osei-Amponsah)	377
Fig. 8.15	A Hampshire sow with piglets. (From Source: https://agro4africa.com/hampshire-pig-breed/, licensed under CC-BY 4.0)	377
Fig. 8.16	The Landrace pig breed. (Credit: Prof. R. Osei-Amponsah)	377
Fig. 8.17	Extensive pig production practices showing poor housing (**a** and **b**), feeding/watering (**c**), and feed resources	379
Fig. 8.18	A farmer feeding pigs in a small-scale semi-intensive production system. (Credit: Prof. R. Osei-Amponsah)	380
Fig. 8.19	Simplified marketing channel for indigenous pigs from the free-range/scavenging and smallholder semi-intensive systems	384
Fig. 8.20	Simplified marketing channel for pigs from medium and large-scale intensive farms	384
Fig. 9.1	Distribution of large camelids in the world. (From FAOSTAT *2019*, licensed under Open Government License OGLv3.0)	397
Fig. 9.2	Migration map of the historical camelid family. (From Burger et al. *2019*, licensed under CC-BY 4.0) The current distribution of dromedaries and Bactrian camels is presented in red and green colours. The last refugia of the wild two-humped camels in China and Mongolia are shown as dark-green patches. The map was adapted from Mesa Schumacher/Aramco World (https://www.aramcoworld.com/en-US/Articles/November-2018/The-Magnificent-Migration). Reprint permits were granted by AramcoWorld on March 6, 2019	399
Fig. 9.3	Major coat colours found in dromedary populations in Algeria. (Photos by Gaouar S.B.S)	403
Fig. 9.4	Dromedary populations in the southwest of Algeria. (Photos by Gaouar S.B.S)	404

Fig. 9.5	Coat colours as they are called by Chadian dromedary herders. (From Djomtchaigue et al. *2015*, licensed under CC-BY 4.0)	407
Fig. 9.6	Somali dromedary. (From Tilahun et al. *2020*, licensed under CC-BY 4.0)	409
Fig. 9.7	Rendille dromedary. (From Sun *2005*, licensed under CC-BY 4.0). (**a**) A warrior driving camel herds out of the camp for day's herding. (**b**) Transporting wedding goods	409
Fig. 9.8	Turkan young camels with female, in Prosopis thicket at lake, Northern Turkana. (From Carr *2017*, licensed under CC-BY 4.0)	410
Fig. 9.9	Afar dromedary in Afar Region, Northeast Ethiopia (Melkamu et al. *2022*). (From Melkamu et al. *2022*, licensed under CC-BY 4.0)	411
Fig. 9.10	Maghrebi dromedary ecotypes in southern Tunisia according to their tribal affiliation. (From Chniter et al. *2013*, licensed under CC-BY 4.0)	415
Fig. 9.11	Dromedary coat colours nomenclature used by Tunisian dromedary herders. (From Chniter et al. *2013*, licensed under CC-BY 4.0)	416
Fig. 9.12	Female dromedary of Chalfi and Kawa types. (From Chniter et al. *2013*, licensed under CC-BY 4.0)	417
Fig. 9.13	Moroccan dromedary breeds. (From FAOSTAT *2019*, licensed under CC-BY 4.0)	417
Fig. 9.14	Pakistani dromedary. (Photo by: Dr. Asim Faras)	418
Fig. 9.15	Traditional milk production. (From Cherifi et al. *2013*, licensed under CC-BY 4.0)	424
Fig. 9.16	Dromedary slaughter and carcass/meat handling. (From Cherifi et al. *2013*, licensed under CC-BY 4.0)	426
Fig. 10.1	Origin of domestic donkey and routes of dispersion in Africa, Asia and Europe	456
Fig. 10.2	Idabari jenny (**a**) and jack (**b**) at the Federal University of Agriculture, Abeokuta, Nigeria. (Photos by M.N. Bemji)	458
Fig. 10.3	Dunni donkeys at Zaria, Nigeria. (**a**) brown, and (**b**) black types. (Photo by DR. E. M. Ibeagha-Awemu)	460
Fig. 10.4	Fari donkeys at Zaria, Nigeria. (**a**) White with brown patches, and (**b**) pale cream. (Photo by Dr. E. M. Ibeagha-Awemu)	460
Fig. 10.5	Egyptian donkey. (Source Wikipedia: photo Jerome Bon by https://en.wikipedia.org/wiki/List_of_donkey_breeds#/media/File:Egyptian_donkey_(2428011421).jpg, CC-BY-2.0 license)	461

Fig. 10.6 Libyan donkey. (Photo Source: Wikipedia and available at https://en.wikipedia.org/wiki/List_of_donkey_breeds#/media/File:Donkey_in_Wasita_Libya.jpg, CC-BY-4.0 license)... 462

Fig. 10.7 Tunisian. (Source: Wikipedia and available at https://en.wikipedia.org/wiki/List_of_donkey_breeds#/media/File:Barrage_bni_mtir_456789_13.JPG, CC-BY-3.0 license/)... 463

Fig. 10.8 Moroccan. (Source: Wikipedia, photo by Dale Harvey and available at https://en.wikipedia.org/wiki/List_of_donkey_breeds#/media/File:Tamraght-daleharvey-10-donkey.jpg, CC-BY-2.0 license)....................... 463

Fig. 10.9 Kassala donkey carrying grains from a market in Keren, Eritrea. (Source: Wikipedia- photo by David Standley and available at https://en.wikipedia.org/wiki/List_of_donkey_breeds#/media/File:Donkey_Transport_(8383370559).jpg, CC-BY-2.0 license)................. 464

Fig. 10.10 Sudanese Pack. (Source: Wikipedia- photo by David Haberlah and available at https://en.wikipedia.org/wiki/List_of_donkey_breeds#/media/File:Donkey_carrying_Rahal_basket.jpg, CC BY-SA 3.0 license)........ 465

Fig. 10.11 Abyssinian donkeys grazing (**a**), slate-grey Abyssinian adult donkey (**b**), and chestnut-brown Abyssinian jenny and foal (**c**). (Photos by Dr. E. Kefena, Agricultural Transformation Institute, Ethiopia).................... 467

Fig. 10.12 Sinnar donkeys displaying different coat colours: (**a**) white, (**b**) brown, and (**c**) fawn. (Photos by Dr. E. Kefena, Agricultural Transformation Institute, Ethiopia)................................ 468

Fig. 10.13 A herd or drove of Afar donkeys (**a**) and Afar Jenny and foal (**b, c**). (Photos by Dr. E. Kefena, Agricultural Transformation Institute, Ethiopia).................... 469

Fig. 10.14 A herd or drove of Omo donkeys (**a**) and fawn-coloured Omo jack (**b**) and jenny (**c**). (Photos by Dr. E. Kefena, Agricultural Transformation Institute, Ethiopia).......... 471

Fig. 10.15 A Maasai donkey (Source: Wikipedia. Photo by Daryona, https://commons.wikimedia.org/w/index.php?curid=10558134, CC BY-SA 3.0).......... 472

Fig. 10.16 A Somali donkey. (Photo by Dr. E. Kefena, Agricultural Transformation Institute, Ethiopia).................... 472

Fig. 10.17 Two Somali Wild Asses (*Equus africanus somaliensis*) in Marwell Zoo, Hampshire, UK. (Source: Wikipedia- photo by Chris Keating, CC BY-SA 3.0, https://commons.wikimedia.org/w/index.php?curid=15354236)........... 473

Fig. 10.18 Muscat donkey in Tanzania. (Source: Wikipedia- photo by Nevit Dilmen (talk), CC BY-SA 3.0, https://commons.wikimedia.org/w/index.php?curid=11320901)........... 474

Fig. 10.19	Tswana adult donkey showing typical features. (Photo by Dr. E. M. Ibeagha-Awemu)	475
Fig. 10.20	Greyish-brown Tswana donkey ready to transport water at a collection point near Opuwo in Namibia. (© The Donkey Sanctuary 2024)	475
Fig. 10.21	Donkeys grazing in Zaria, Nigeria (**a**) (Photo by Dr. E. M. Ibeagha-Awemu) and in Ethiopia (**b**) (Photo by Dr. E. Kefena)	476
Fig. 10.22	Mule breeding in Ethiopia. Female horse (left) and male donkey (right) ready to mate. (Photo by Dr. E. Kefena, Agricultural Transformation Institute, Ethiopia)	477
Fig. 10.23	Donkeys at a livestock market in Niger. (Source: Wikipedia.com- photo by Vincent van Zeijst, CC BY-SA 4.0, https://commons.wikimedia.org/w/index.php?curid=82422085)	480
Fig. 10.24	Women fetching water to be transported by donkeys at a collection point near Opuwo in Namibia. (©The Donkey Sanctuary 2024)	481
Fig. 10.25	Donkey pulling cart with woman in Somali. (Source: Wikipedia and available at https://en.wikipedia.org/wiki/List_of_donkey_breeds#/media/File:AMISOM_Djiboutian_Contingent_in_Belet_Weyne_11_(8212396061).jpg, CC0 1.0 license)	482
Fig. 10.26	Donkeys ploughing a field at a Fulani settlement in Zaria, Nigeria. (Photo by Dr. E.M. Ibeagha-Awemu)	483
Fig. 10.27	Donkeys transporting humans, goods and animals at Zaria, Nigeria. (Photo by Dr. E.M. Ibeagha-Awemu)	484
Fig. 10.28	Donkey racing during the celebration of Lamu's annual Maulidi festival. (Photos by Zollo Nyambu, https://africaviewfacts.com/facts/lamu-donkey-race-celebrating-tradition-and-community-in-kenya/)	486
Fig. 10.29	The donkey as essential means of transportation of persons and goods in old town of Lamu. (Photo by Alex Malombo, https://africaviewfacts.com/facts/lamu-donkey-race-celebrating-tradition-and-community-in-kenya/)	487
Fig. 10.30	Pictures show the poor slaughter, handling and disposal of donkeys in Nigeria, Kenya and Ghana as deplorable activities associated with the donkey skin trade. (©The Donkey Sanctuary 2024)	489

Fig. 11.1	Barb horses. (**a**) Barb horse from the national stud farm "Chaou-Chaoua" in Tiaret, Algeria (Chikhaoui et al. *2018*) (Image credit: (Chikhaoui et al. *2018*)); (**b**) Moroccan Barb horse used for Fantasia (Talley *2015*) (Image credit: Gwyneth Ursula Jean Talley); (**c**) Barbe horse breed in Mauritania (Photo Courtesy: Ministry of Rural Development, Mauritania) (AU-IBAR *2019*), (**d**) Mounted Barb horse in Nigeria (Image credit: Eveline M. Ibeagha-Awemu)	504
Fig. 11.2	Dongola horse in Eritrea (Ministry of information of Eretria *2021*)	507
Fig. 11.3	Dongola horse (Koçkar *2012*). (Reproduced with permission CC-BY 4.0 license)	507
Fig. 11.4	(**a**) Algerian Arabian Stallion "SAKHI" with his owner Ahmed Feghouli, WAHO 2010 trophy winner (**b**) Diable Du Desert. Breeder: Amraoui Belkasem. Owner: Bouteldja Mohamed, WAHO 2014 trophy winner; (**c**) FOUSHA. Mare, 1999., Breeder: Haras Nationaux, Jumenterie de Tiaret with Mr. Ahmed Bouakkaz, General Director and Head of Stud Book at ONDEEC, and the owner Mr. Zeguaoui Abed, WAHO 2016 trophy winner	508
Fig. 11.5	Typical Selale riding horse (Kefena et al. *2012*)	509
Fig. 11.6	Kafa stallion (Kefena et al. *2012*)	511
Fig. 11.7	Director of Parks Mather insists on A. W.'s being properly mounted on an Abyssinian horse. (Harris *1952*)	512
Fig. 11.8	Pair of Abyssinian horses engaged in cropland cultivation. (Kefena et al. *2012*)	513
Fig. 11.9	Bay Bale horse. (Image credit: Laika ac from UK 2014). Reproduced with permission—CC-BY 4.0 license	513
Fig. 11.10	Borana horse riding. (Sirikoi *2023*)	514
Fig. 11.11	Horro horse in Street Jimma, Ethiopia. (Image credit: Rod Waddington 2014)	516
Fig. 11.12	Kundudo mare with foal. (Paris *2014*)	517
Fig. 11.13	A thin white Kundudo mare standing on a rocky mountain slope with few trees. (Paris *2014*)	517
Fig. 11.14	Gharkawi horse. (Image credit: Sean Woo *2004*)	518
Fig. 11.15	Tawleed horse. (Image Credit—Omdurman-Tuti Island jetty, Sudan)	519
Fig. 11.16	Poneys musey à Gobo (Arthus-Bertrand 2003). (Image credit: Seignobos C). (Arthus-Bertrand *2003*)	520
Fig. 11.17	Coat color diversity of the Kirdi pony in the Sudano-Sahelian region, Kabbia South-West of Chad. (Image credit: authors)	522
Fig. 11.18	Bélédougou horse used for carriage and pulling. (Image credit: Ferdinand Reus) (Claudot-Hawad *1988*)	523

Fig. 11.19 Arewa stallion. (Edward and Bukola *2023*) 525
Fig. 11.20 White Bobo horse harnessed for a wedding procession, in the enclosure of Ouagadougou, Burkina Faso. (Image credit: Amélie Tsaag Valren) 525
Fig. 11.21 M'bayar horse served for light draft in Pikine, Sénégal. (Source: Lejosne *2023*) 527
Fig. 11.22 Fleuve horse breed from Sénégal in West Africa 527
Fig. 11.23 A Foutanke or Narougor horse in Pikine, Sénégal. (Lejosne *2023*) 529
Fig. 11.24 Mogods pony. (Chabchoub et al. *2006*) 531
Fig. 11.25 Cape Boerperd mares and foals at pasture (Van der Merwe and Martin *2002*). (Image credit: Van der Merwe and Martin *2002*) 534
Fig. 11.26 Basuto pony (Van der Merwe and Martin *2002*). (Image credit: Martin J. in Van der Merwe and Martin *2002*) 535
Fig. 11.27 The South African Nooitgedacht stallion named Mac. (FAO *2002*)In 1958, the birth of the stallion Mac marked a pivotal moment in the development of the Nooitgedacht horse breed. Mac exemplified the ideal general-purpose horse sought by project managers, displaying key physical traits such as a strong shoulder, well-muscled back, and robust legs with hard hooves. Additionally, Mac embodied the distinctive characteristics of the Nooitgedacht breed, with his small pointed ears, pronounced brows, and straight to slightly concave profile. This iconic horse left a lasting legacy in the breed's evolution (Photo: Landbouweekblad) 536
Fig. 11.28 A South African Vlaamperd stallion. (SAVBS *2019*) 537
Fig. 11.29 Algerian native Arabian stallion "SAKHI", senior champion stallion of World Arabian Horse Organization (WAHO) 2014 (http://www.waho.org/algeria). 539
Fig. 11.30 Anglo-Arab horse. (Image credit: Personal image) 542
Fig. 11.31 Horse populations from southwestern Ethiopia (Mustefa et al. *2022*). (**a**) Telo stallion; (**b**) Masha stallion; (**c**) Gesha stallion; (**d**) Gesha mare (Mustefa et al. *2022*) ... 544
Fig. 11.32 Kundido feral horses from Ethiopia. (Kefena et al. *2012*)... 548
Fig. 11.33 Namibian wild horses. (Pütz and Schlottmann *2020*) 549
Fig. 11.34 Feeding of the horses with hay. (Pütz and Schlottmann *2020*)......................... 550
Fig. 11.35 Plowing with horses. (Image credit: Author's Own 2022) (Kebede *2023*) 556
Fig. 11.36 (**a**) Barb or Arab-Barb horse-drawn cart for passenger transport; (**b**) Horse-drawn cart with pneumatic wheels and elevated platform used for transporting goods in Tozeur region of Tunisia. (Charles *2022*). 556
Fig. 11.37 Weeding festivity in Chechar city, Khenchla (Algeria) 558

Fig. 11.38 Rider and horse (caparisoned for a wedding ceremony) performing the pace walk. (Baynes-Rock and Teressa *2021*)................................. 559

Fig. 11.39 Awi Agew horse riders during annual festival. (Kebede *2023*).................................... 559

Fig. 11.40 Fulani horse dancing traditionally at the conclusion of the Ramadan. (Image credit: Flickr/Rosemary Lodge) (DLIFLC *2018*).................................. 560

Fig. 11.41 Ndop loincloth during funerals, Cameroun. (Image credit: Foréké Dschang 2017) (Noel *2021*)........ 560

Fig. 11.42 Fantasia in Northern Cameroon. (Image credit: Félix Meutchieye) 562

Fig. 11.43 Horse show at Jida horse market in Ethiopia. (Kefena et al. *2012*)................................ 562

Fig. 11.44 Horse parade to celebrate the 123rd anniversary of Adwa victory as led by prominent Ethiopian musician Hachalu Hundessa. (Kefena et al. *2021*) 563

Fig. 11.45 Durbar festival: an annual equestrian event in Northern Nigeria. (Yusuf et al. *2023*) 563

Fig. 11.46 Lambert Leclezio first Mauritian athlete (Mauritius is a country in eastern Africa) to compete at the FEI Open European Vaulting Championships for Juniors 2011. (FEI *2012*)....................................... 567

Fig. 11.47 Fantasia parade in Mekhatria, at 10 kms north of the seat of the wilaya of Ain Defla, Algeria. (Image credit: Bisker Mohamed) (Bisker *2017*)........................... 569

Fig. 11.48 The Fantasia equestrian parade with warrior inspiration (Blog du cheval) (https://blog.cheval-daventure.com/fr/post/140/la-fantasia-une-tradition-equestre-dinspiration-guerriere) 569

Fig. 11.49 Saddle, breast strap, and bridle on a Barb horse Capital Festival of traditional horseback riding, Rabat, August 2014. (Image credit: Gwyneth Ursula Jean Talley) (Talley *2015*) 570

Fig. 11.50 Lesotho's mountain jockeys race. (Kuwait times *2022*) 571

Fig. 11.51 Tourism activity by riding Nooitgedacht horses in South Africa. (The figure illustrates a scene where a game warden and a tourist, riding Nooitgedacht horses, are approaching a female rhinoceros with a calf visible in the background. These South African horse breeds have demonstrated their exceptional suitability for guiding inexperienced riders during game safaris. (Photo by F.J. van der Merwe) (FAO *2002*))...................................... 572

Fig. 11.52 Etalon Arabe-Barbe alezan attaché devant le domicile de son propriétairefromTozer, Tunisia. (Charles *2022*) 577

List of Figures

Fig. 12.1 Distribution of wild and domestic water buffalo in Africa. (From Furstenburg (*2007*), licensed under CC-BY 4.0, modified by adding distribution of water buffalo) 596

Fig. 12.2 Phylogeny of Bubalus based on complete mitochondrial genomes. The tree was reconstructed with MrBayes software using the 118 mitogenomic haplotypes of Bubalus identified in our alignment of 16,356 nucleotides. The tree was rooted with Syncerus (not shown). Haplogroups showing more than 0.5% of nucleotide divergence are highlighted by different colored rectangles, and those found in the swamp buffalo are named SBa to SBe (the name of the haplotype is followed by a dash and a number when the same haplotype was found in least two individuals). Among the 29 mitogenomes of Egyptian river buffalo, the 21 samples collected from North are indicated in red, whereas the eight samples collected from South are indicated in green. The four mitogenomes of river Buffalo assembled using SRA data available in GenBank are named with the SAMN accessions. For nodes supported by bootstrap percentage (BP) _ 90 in the RAxML analysis (see details in Material and Methods), the two values correspond to the posterior probability (PP calculated using MrBayes software, left of the slash) and BP (right of the slash). An asterisk is used when both MrBayes and RAxML analyses provided maximal support values, i.e., PP ¼ 1 and BP ¼ 100, respectively. No information was provided for nodes supported by BP <90. (From Youssef et al. (*2021*), licensed under C-C BY 4.0)......................... 598

Fig. 12.3 Phylogeny of genus *Bubalus* based on complete mitochondrial genomes. The tree was rooted with African buffalo. (From Hassan et al. (*2022*), licensed under C-C BY 4.0). 600

Fig. 12.4 A phylogenetic tree assembled using the 5′ half of CO1 gene sequence (650-bp) in all the investigated species and the donkey sequence as an outgroup. Accession numbers are given to all samples. (From Oraby et al. (*2011*), licensed under C-C BY 4.0). 601

Fig. 12.5 COI (mitochondrial cytochrome C oxidase subunit 1) barcode-based phylogenetic tree constructed using sequences from four buffalo taxa groups and a sequence of goat (*Capra hircus*) as an outgroup. *Hap*. haplotype. (Hassan et al. *2018*, licensed under C-C BY 4.0).......... 602

Fig. 12.6 Egyptian buffalo types 603

Fig. 13.1 An adult Goliath Frog 622
Fig. 13.2 Nest of Goliath frog 623

Fig. 13.3	Median-joining network for the cyt b mtDNA haplotypes of *C. aspersum*. (Guiller and Madec *2010*)	625
Fig. 13.4	The most common snail varieties in northern Algeria. (Adapted from Bouchiba et al. *2021* and Mezouar et al. *2021*)	626
Fig. 13.5	Distribution of snails according to shell color in Northern Algeria	628
Fig. 13.6	Picture comparing two Achatinadae snails	629
Fig. 13.7	Some species of land snail in South Africa. (Herbert *2010*)	630
Fig. 13.8	Picture comparing two Achatinadae snails: a. *Achatina achatina* snails, b. *Acharchatina marginata* snails	636
Fig. 13.9	Snail nutrition	636
Fig. 13.10	Reproduction of *Archachatina marginata* snails	637
Fig. 13.11	The Kabyle rabbit phenotypes. (Carneiro et al. *2011*)	640
Fig. 13.12	Phenotypes of the white breed. (Gacem et al. *2008*)	640
Fig. 13.13	Phenotypes of the synthetic breed	641
Fig. 13.14	Tadla breed	641
Fig. 13.15	Zemmouri breed	641
Fig. 13.16	Bauscat breed. (El Raffa et al. *2002*)	642
Fig. 13.17	Giza White Breed. (El Raffa et al. *2002*)	642
Fig. 13.18	Chinchilla breed. (Bolet et al. *2012*)	642
Fig. 13.19	Baladi Breed. (El Raffa et al. *2002*)	643
Fig. 13.20	Gabali Breed	643
Fig. 13.21	Phendula	644
Fig. 13.22	The origin of the honeybee, *Apis mellifera,* according to three hypotheses. (Gupta et al. *2014*)	651
Fig. 13.23	Harvesting of wax by African beekeepers. (Mayazine *2019*)	654
Fig. 13.24	a. Traditional log hive for *A. mellifera* bees; b. modern hive; c. transitional Kenyan top bar hive; d. clay pot hive for stingless bees. (Leven et al. *2005*)	655
Fig. 13.25	The clinical signs of American foulbrood *P. larvae* observed in the field. (Onions *1912*)	661
Fig. 14.1	Various methane emission measuring techniques at Mazingira centres of the International Livestock Research Institute (ILRI). Cattle metabolic chamber (**a**), SF6 gas cylinder (**b**), sheep metabolic chamber (**c**) and gas analysis kit (**d**). (Photos by Mizeck Chagunda)	681
Fig. 14.2	Laser methane detector. On the left, a scientist taking methane measurements and on the right, two different models of the laser methane detector. (Photo by A. Ross)	682
Fig. 14.3	Examples of cattle under extensive production systems. Top, Kenya and bottom, Zambia. (Photos by Mizeck Chagunda)	685

Fig. 15.1	A schematic illustration of the main components of a breeding strategy	692
Fig. 15.2	Impact pathway for best practices for selective breeding for improved livestock (Ojango et al. *2018*)	701
Fig. 16.1	Wakwa (Brahman (50%)–Gudali (50%)) bull. (Courtesy of Dr. David Mbah)	718
Fig. 16.2	Holstein (67.5%) Gudali (32.5%) bull. (Courtesy Dr. David Mbah)	724
Fig. 16.3	A 3-year-old Holstein 5/8 Gudali 3/8 Milker. (Courtesy Dr. David Mbah)	724
Fig. 16.4a	Holstein-Gudali programme to produce H 5/8 G3/8 cross-breds [8–12 years]	725
Fig. 16.4b	Holstein-Gudali programme H ¼ G ¾ to produce H 5/8 G 3/8 cross-breds for farmers [8–12 years]	725
Fig. 16.5	A cloned Boran bull and its offspring (calves). Calf 1 died due to low blood glucose while calf 2 survived following supplemental glucose administration. (From Yu et al. (*2016*), licensed under CC-BY 4.0)	746
Fig. 17.1	Strains of Dorper sheep found in Kenya and South Africa	769
Fig. 17.2	Illustration of the FARM Africa farmer-based breeding strategy for goat improvement. Adapted from Ojango et al. (*2010*)	770
Fig. 17.3	Example of cross-breeding to develop a composite breed	771
Fig. 17.4	West African Dwarf Goat in Ghana	776
Fig. 17.5	Map of Kenya indicating the location of main goat projects from 1972 to 2010	777
Fig. 17.6	Galla Goats in Kenya	778
Fig. 18.1	The cycle of a breeding program	791
Fig. 18.2	Basic components for a successful poultry breeding strategy for Africa. (Adapted from Jack Dekkers *1990*)	799
Fig. 18.3	Breeding objective traits in a typical broiler breeding program. (*Source*: Olori (*2012*))	800
Fig. 18.4	Pyramidal structure of poultry breeding. (*Source*: Olori (*2008*))	802
Fig. 18.5	Genomic selection in poultry	809
Fig. 19.1	Frequency of severe heat stress for past and future periods for pigs in Uganda. (From Mutua et al. (*2020b*), licensed under a CC BY 4:0)	826
Fig. 19.2	Risk factors that cause heat stress in pigs. (From Mutua et al. (*2020b*), licensed under a CC BY 4:0)	827
Fig. 20.1	Benefits and beneficiaries of a livestock identification and management system	859

Fig. 20.2   Wearable sensors such as milking collar, leg band and ear tags, temperature and proximity sensors; image-based monitoring systems and internal sensors collect data (**a**) which is analysed (**b**) for informed management decisions. (From Siberski-Cooper and Koltes (Siberski-Cooper and Koltes *2021*), licensed under CC-BY 4.0) 860

Fig. 20.3   iCOW system on mobile phones used for farmer training and advisory applications 867

Fig. 20.4   Genome editing platforms and mechanisms for DSB repair with endogenous DNA. Genome editing nucleases (ZFNs, TALENs and CRISPR/Cas9) induce DSBs at targeted sites. DSBs can be repaired by NHEJ or, in the presence of donor template, by HDR. Gene disruption by targeting the locus with NHEJ leads to the formation of indels. When two DSBs target both sides of the amplification or insertion, a therapeutic deletion of the intervening sequences can be created, leading to NHEJ gene correction. In the presence of a donor-corrected HDR template, HDR gene correction or gene addition induces a DSB at the desired locus. DSB double-stranded break, ZFN zinc-finger nuclease, TALEN transcription activator-like effector nuclease, CRISPR/Cas9 clustered regularly interspaced short palindromic repeat associated 9 nuclease, NHEJ nonhomologous end-joining and HDR homology-directed repair. (From Li et al. (*2020*), licensed under CC-BY 4.0) 877

Fig. 20.5   Diagrammatic representation of CRISPR/Cas9 gene editing using either zygote micromanipulation by microinjection or electroporation, or somatic cell nuclear transfer (SCNT) for the generation of livestock animals for various applications. (From Parisse et al. (*1713*), licensed under CC-BY 4.0) 882

Fig. 20.6   Cellular agriculture uses cells from animals or precision fermentation technologies to produce products of animal origin (meat, eggs, milk) and fermentation products for use as ingredients for the creation of varied products including textiles. 883

Fig. 20.7   Main steps involved in the production of cell-cultured meat 884

Fig. 20.8   Schematic representation of improved farm management practices and technologies that have led to enhanced genetic gain in dairy cattle breeding and improvement in milk production. (From Ibeagha-Awemu and Yu (*2021*), licensed under CC-BY 4.0) 891

Fig. 21.1	A precision livestock farm uses state-of-the-art technologies to continuously collect data on animal (microphones, cameras, sensors) followed by continuous data analysis to provide real time information or data to support management and production decisions as well as resolve health and welfare issues.	916
Fig. 21.2	The one health concept.	919
Fig. 21.3	Agricultural waste management toward a circular bioeconomy. Digestate (10% dilution) resulting from anaerobic digestion of chicken manure was used as nutrients for microalgal (strain *Chlorella vulgaris* CPCC 90) growth resulting in production of biofuel and bioproducts that can be used as fertilizer (for soil treatment), bio-pesticide and as animal feed. (From Rajagopal et al. *2021*), licensed under CC-BY 4.0)	929
Fig. 21.4	Modern techniques for the improvement of livestock health	954
Fig. 21.5	Tie-stall barn (**a**), cubicle barn (**b**), and free walk barn (**c**) dairy cow housing systems	961
Fig. 21.6	A typical portable milking machine.	963
Fig. 21.7	Different parlor milking systems	965
Fig. 21.8	Sample pasteurization system for large-scale use (Tubular Aseptic UHT Pasteurizer with Vacuum Deaerator).	968
Fig. 22.1	IBIS in Burundi: identification, insemination, and consanguinity management. IFAD, Ministry of Livestock, Burundi and Adventiel	1010
Fig. 22.2	Zebuscan, a mobile application for animal traceability in Madagascar 2019, World Bank, Ministry of Agriculture, Livestock and Fisheries, Adventiel	1010
Fig. 22.3	Example of the modular structure of a Livestock Identification and Management System.	1011
Fig. 23.1	A representative SNP Chip	1026
Fig. 23.2	A Barcode Scanner.	1030
Fig. 23.3	A 4G mobile data sim wireless router useful for digital telephony and data transfer	1031
Fig. 23.4	A typical smartphone suitable for communication, data capture, and transfer.	1031
Fig. 23.5	Programmable hand-held terminals that can be used for data recording	1032
Fig. 23.6	A miniature mobile printer that can be used to print instant labels during recording.	1032
Fig. 23.7	A large-capacity portable data storage device (1 TB)	1033

Fig. 28.1	Contribution of animal production to the different Sustainable Development Goals (SDGs). (Source: FAO (*2015*)).	1127
Fig. 28.2	Eight capacity building-related domains to address the causes of genetic erosion in African livestock genetic diversity. (Source: FAO (*2021*)).	1137
Fig. 28.3	The cyber-physical management cycle of smart farming enhanced by cloud-based event and Big Data management. (Source Wolfert et al. (*2014*)).	1141
Fig. 28.4	The data chain of Big Data applications in livestock breeding. APIs Application Programming Interface, PRA Data Collectors/Performance Recording Agents, DA Direct Agreement between firm/farm & the programme, FA Farmer Assistants, MTA Material Transfer Agreement, CRA Collaborative Research Agreement, DSA Data Sharing Agreements, NDA Non-Disclosure Agreement	1141
Fig. 28.5	The collection of technologies that we refer to as advanced technologies can help animal farmers create better outcomes	1142
Fig. 28.6	Key domains of capacity building on machine learning algorithms to create optimal performance in livestock breeding. (Adapted from Neethirajan *2020*).	1143
Fig. 29.1	Organization and core elements of a typical Genome Core facility.	1158
Fig. 29.2	Overview of high-performance computing options showing common clusters (e.g. GPUs, FPGAs, and cloud solutions) and highlighting differences in flexibility, performance, and options for custom design. CPU: Central Processing Units; GPU: Graphics Processing Unit; FPGA: Field Programmable Gate Arrays; HDL: Hardware Description Language; RTL: Register-Transfer Level. (From Lightbody et al. (*2019*), licensed under a CC BY 4:0)	1165
Fig. 29.3	Overview of a high-performance computing core with service centres in other locations	1167
Fig. 29.4	Mapping the five African Regional Animal Genebanks. African regional Genebanks (large triangles-pink, yellow, red, blue, and green colours) showing interconnections between them (blue lines) and countries served by each Genebank (small circles in the colours of their respective regional Genebanks)	1174
Fig. 30.1	The Regional Economic Communities in Africa	1210

List of Figures

Fig. 31.1　Recipients of piglets and feed through The Pig Van Djouke POG (Pass on the Gift) Project. (Photo by NOWEPIFAC).................... 1240

Fig. 31.2　Imported Holstein (**a**) and Jersey (**b**) and Holstein (1995 batch) (**c**) at IRZV Bambui centre, as well as imported hens (**d**). (Photos by HPI, extracted from HPI Cameroon archives and provided by Emmanuel Bassam (HPI Director of Program))....... 1256

Fig. 31.3　Farmers receive Holstein-Friesian cattle at Nseh Mbabu village (**a**), Holstein-Friesian under semi-intensive system of management at Bamendakwe (**b, c**); and HPI staff at a pasture and forage development site at Fonta, Bambui (**d**). (Photos by HPI, culled from HPI Cameroon archives and provided by Emmanuel Bassam (HPI Director of Program))....... 1259

Fig. 31.4　A ram from HPI Project in the far North Region of Cameroon (**a**); Sheep placement in the Mbam Upper Sanaga Valley Integrated Sheep/Goat Project in the Center Region of Cameroon by HPI (**b**); and Sheep placement in the Bui Donga Small Holder Integrated Sheep & Goat Project in the Northwest Region of Cameroon by HPI (**c**). (Photos by HI, extracted from HPI Cameroon archives and provided by Emmanuel Bassam (HPI Director of Program))....... 1260

Fig. 31.5　Boran animals imported from Kenya (**a, b**) and offspring of Boran × Zebu (**c**) distributed to farmers under "*Lake Nyos Livestock Restocking Project*". (Photos by HI, extracted from HPI Cameroon archives and provided by Emmanuel Bassam (HPI Director of Program)) 1261

Fig. 31.6　Livestock farmers showing off their milk products to Cameroon government officials and veterinarians during Heifer International Cameroon's 40th Anniversary Celebration in Bamenda in 2015. (Photo by HPI, extracted from HPI Cameroon archives and provided by Emmanuel Bassam (HPI Director of Program))....... 1262

Fig. 31.7　Collaboration web for successful delivery of sustainable livestock development in Africa...................... 1274

List of Tables

Table 1.1	The number of livestock breeds and breed types by species in Africa relative to other regions of the world	4
Table 1.2	The number of livestock breeds and breed types in Africa and regions	5
Table 2.1	Livestock populations[a] in Africa and regions in the year 2021	16
Table 2.2	Main characteristics of the agro-ecological zones of Africa[a]	21
Table 2.3	Key features of African livestock production systems[a]	25
Table 2.4	Major mixed crop–livestock farming systems of sub-Saharan Africa described by Dixon et al. (*2001*)	32
Table 2.5	Major challenges limiting animal performance under current livestock production systems and possible solutions for the sustainable development of improved livestock productivity	45
Table 3.1	Humpless cattle breeds of Sub-Saharan Africa	71
Table 3.2	Adult linear body measurements by sex and breed of Shorthorn cattle	87
Table 3.3	Special genetic characteristics of some Shorthorn cattle breeds and stabilized crosses	89
Table 3.4	Mean live weight by sex of some Shorthorn cattle breeds and stabilized crosses	96
Table 3.5	Estimates of reproductive parameters of some Shorthorn cattle breeds and stabilized crosses	97
Table 3.6	Estimates of milk production by milking and management system for some Shorthorn cattle breeds	98
Table 3.7	Live weights, carcass weights, and dressing-out percentage of Baoulé, Muturu, and Keteku cattle by sex and age	99
Table 3.8	Birthweight and linear body measurements of Baoulé cattle in Côte d'Ivoire	104
Table 3.9	Best linear unbiased estimates of live weight and body measurements of Namchi and Kapsiki cattle under the natural environment in Cameroon	107

Table 3.10	Population estimates of Shorthorn cattle types of West and Central Africa	110
Table 4.1	Zebu cattle breeds of Eastern and Southern Africa	120
Table 4.2	West and Central African Zebu Cattle	134
Table 4.3	The Sanga cattle breeds of eastern and southern Africa	149
Table 4.4	The pseudo-Sanga of West Africa[a]	161
Table 4.5	Zenga (Zebu–Sanga) cattle, recent derivatives, and synthetic breeds	166
Table 4.6	Summary of exotic cattle breeds used in Africa	175
Table 5.1	Five sub-species of *Capra* (domesticated goat)	188
Table 5.2	African goat genetic resources: breed groups, subgroups and breeds	198
Table 6.1	Position of Africa in the world in terms of sheep production and processed products for the year 2019 (FAOSTATS *2019*)	241
Table 6.2	Sheep production by regions in Africa for the period 2018–2019 (FAOSTATS *2019*)	242
Table 6.3	Classes of sheep breeds in Africa	247
Table 6.4	Threatened conventional sheep breeds in Africa that need conservation (AU-IBAR *2019*)	303
Table 6.5	Trait preference, selection criteria and breeding strategies for genetic improvement of economically important traits in sheep	305
Table 6.6	Heritabilities, expected direct and correlated responses to selection for some reproductive traits in sheep	310
Table 6.7	Global sheep raw skin production (tonnes) 2000–2012 (FAOSTATS *2021*)	314
Table 7.1	Identified characteristics of some African indigenous chicken breeds	333
Table 7.2	Examples of genes in indigenous chicken populations that may be linked to the adaptation to tropical conditions	333
Table 7.3	FAO chicken production systems classification	349
Table 7.4	Examples of chicken genetic improvement programs in Africa	350
Table 8.1	Pig breeds of Africa by country and region	365
Table 8.2	Key outputs, activities, and actors for sustainable utilization and conservation of pig genetic resources in Africa	382
Table 9.1	Number of dromedaries in different African countries (FAOSTAT *2019*)	420
Table 9.2	Evolution of African livestock animal numbers between 1969 and 2019 (FAOSTAT *2019*)	421

Table 10.1	Donkey populations by regions in Africa for the period 2018–2022 (FAOSTAT *2024*)	453
Table 10.2	Distribution of donkey types in Nigeria based on coat colours.	457
Table 11.1	Horse population by African countries (FAO DAD-IS *2021*)	510
Table 11.2	Risk statue of African horse breeds (FAO *2019*)	578
Table 12.1	Genetic diversity of milk traits genes and association with different milk production measures.	605
Table 12.2	Genetic diversity of fertility traits genes associated with different fertility measures	608
Table 13.1	Descriptive analysis of body measurements (height, length, and width of shell and body weight) between different species in Algeria (Bouchiba et al. *2021*)	627
Table 15.1	Summary of main breeding goals by species	696
Table 16.1	Example of breeding goals and breeds developed for meat production using indigenous breeds[a] in different regions of Africa	712
Table 16.2	Breakthrough biotechnological innovations in cattle and buffalo genetics and breeding	742
Table 16.3	Examples of nucleus/community breeding scheme for cattle improvement in West Africa.	748
Table 17.1	Number of sheep in the different regions of Africa from 2000 to 2021[a]	760
Table 17.2	Numbers of goats in the different regions of Africa from 2000 to 2021[a]	761
Table 17.3	Number of different breeds of sheep and goats in the different regions of Africa	761
Table 17.4	Traits and performance indicators associated with different products from sheep and goat	763
Table 17.5	Estimates of genetic parameters for growth traits in some sheep and goat breeds in Africa	765
Table 17.6	Indigenous breeds characterized and developed through planned breeding programmes	767
Table 17.7	Mean performance of various indigenous sheep and goat breeds and their crosses in Africa	768
Table 18.1	Characteristics to be balanced when selecting for dual purpose and tropically adapted chicken breeds	801

Table 19.1	A SWOT analysis of a typical piggery value chain.	821
Table 19.2	Current and futuristic adaptations to heat stress management for pigs (Mutua et al. *2020b*)	827
Table 19.3	Pig production system considering the scale of operation—the Uganda case	829
Table 19.4	Pig breeding companies operating in Africa	838
Table 20.1	Production share of live animals and some products by continent in 2019	854
Table 20.2	Technologies and sequencing systems for next-generation sequencing	869
Table 20.3	Publication of the whole-genome and pangenome sequences of some livestock species	871
Table 20.4	Genome-wide genotyping chips for various livestock species	872
Table 20.5	Quantitative trait loci mapped for various livestock species	873
Table 20.6	Technological developments in gene editing and application in animal production	875
Table 20.7	Summary of gene editing in livestock species	878
Table 20.8	Sample cellular agriculture companies producing a wide range of cultivated- and precision fermentation-based products	885
Table 21.1	Number of poultry and livestock species and the estimated amount of manure produced, and nitrogen and phosphorus excreted in the USA in year 2017	920
Table 21.2	Some chemical preservatives for meat and target organisms.	942
Table 21.3	Practical feed improvement technologies for smallholder livestock production systems and their expected or observed impacts.	947
Table 21.4	Categories of some of the novel feed ingredients or supplements in livestock feed. Adapted from (Chisoro et al. *2023*)	951
Table 22.1	Examples of diagnostic tests available in South Africa.	997
Table 22.2	Examples of gene editing application and transgenes in cattle	1000
Table 24.1	Examples of precision livestock farming technologies	1052
Table 25.1	Constraints of animal selection and crossbreeding in Africa.	1075

Table 26.1	Values of indigenous animal genetic resources	1092
Table 26.2	A summary of the main findings of economic valuation studies that have been undertaken in Africa	1096
Table 28.1	Representativity of top-50 leading agricultural universities and colleges in Africa in 2021	1132
Table 28.2	Objectives and expected impacts of capacity building for sustainable intensification of livestock in Africa	1139
Table 28.3	Examples of Big Data applications/aspects in livestock and fishery smart farming processes	1142
Table 29.1	Documents and standard operating procedures	1159
Table 29.2	Some centres that provide genomic and high-performance computing services in Africa	1168
Table 30.1	Status of National Biosafety Frameworks in SSA countries	1195
Table 30.2	Classification of SSA countries on the basis of enabling environment for applications of biotechnology in livestock	1196
Table 30.3	Strategic Priority Areas (SPA) and Strategic Priorities (SP) of the Global Plan of Action for AnGR	1202
Table 30.4	Summary of achievements of AU-IBAR's Animal Genetic Resources Project	1206
Table 30.5	Summary of status of policies and implementation instruments by African countries	1218
Table 30.6	Number of livestock breeds and breed societies in South Africa	1225
Table 30.7	Cartagena Protocol on Biosafety and Nagoya Protocol on Access and Benefit-sharing: status of ratification by African countries and entry into force	1229
Table 31.1	Sample farmer organizations in Africa	1239
Table 31.2	Sample livestock breed societies in Africa	1242
Table 31.3	Sample professional animal production-based societies in Africa and internationally	1247
Table 31.4	Sample FAO books, reports, and databases on animal genetic resources and their management and improvement strategies	1270

List of Boxes

Box 2.1	Forms of pastoralism	28
Box 2.2	Summary of case studies on the impact of livestock grazing system (LGS) management using a multifunctional approach (Ickowicz et al. 2022)	57
Box 4.1	Summary of the risk status of cattle breeds in Africa	176
Box 15.1	Some Useful Terms in Defining Breeding Goals and Breeding Strategies	692
Box 15.2	The Global Plan of Action (GPA, 2007)	697
Box 15.3	AU-IBAR Animal Genetics Project	698
Box 16.1:	Genomics to Identify Appropriate Cross-Bred Dairy Cattle for Smallholder Farming Systems in East Africa—The Dairy Genetics East Africa Project	713
Box 16.2:	West Africa	713
Box 16.3:	The ADGG Programme (https://portal.adgg.ilri.org/)	714
Box 16.4:	Wakwa, a Synthetic Beef Breed in Cameroon	717
Box 16.5:	Genetic Improvement of Cattle in Ghana	718
Box 16.6:	Genetic Improvement of Cattle for Dairy Production in Cameroon	720
Box 16.7:	Cross-Breeding Azawak Zebu Cattle with Local Sudanese Fulani Cattle for Dairy Production in Burkina Faso	721
Box 16.8:	Genetic Improvement of Cattle for Dairy Production in Ethiopia	722
Box 16.9:	Genetic Improvement of Cattle for Dairy Production in Kenya	723
Box 16.10:	Kenya Sahiwal Breed (Muhuyi et al. 1999)	726
	Definitions See Box 15.1, Chap. 15, Defining Breeding Goals and Breeding Strategies	761
Box 17.1:	Cross Breeding: Dorper Sheep in South Africa (Ramsay et al. 2000; Milne 2000)	767
Box 17.2:	Cross-breeding for smallholder farming systems; *The FARM Africa Goat improvement programme*	769
Box 17.3:	Improved Djallonke Sheep: Côte d'Ivoire Adapted from Yapi-Gnaore (2000)	774

Box 17.4:	D'Man Sheep in Morocco	775
Box 17.5:	The Boer Goat: South Africa (Malan 2000; Visser and van Marle-Köster 2018)	775
Box 17.6:	West African Dwarf WAD Goat	775
Box 17.7:	The Galla Goat: Kenya	776
Advantages of Genomic Evaluation		1025
Box 27.1:	Collaborative Masters in Agricultural and Applied Economics (MAAE)	1118
Box 30.1	Definition of Terms	1189
Box 30.2	Treaty State Descriptors—Ratification, Accession, Approval, and Acceptance	1191
Box 31.1	HPI-Specific Objectives in Cameroon	1257
Box 31.2		1263

Part I

African Livestock Production Systems and Genetic Resources

An Overview of African Livestock Genetic Resources and Improvement Strategies

Eveline M. Ibeagha-Awemu, Sunday O. Peters, Appolinaire Djikeng, and John E. O. Rege

Abstract

This book is the first comprehensive text that presents the rich livestock genetic resources in Africa in one volume. This introductory chapter summarizes the materials covered in the book. It introduces the various topics covered and serves as a guide to topics of interest.

Keywords

African indigenous livestock breeds · Rich genetic diversity · Improved utilization and modern technologies · Institutional arrangements · Conservation of African livestock genetic resources

E. M. Ibeagha-Awemu (✉)
Sherbrooke Research and Development Centre, Agriculture and Agri-Food Canada, Sherbrooke, QC, Canada
e-mail: eveline.ibeagha-awemu@agr.gc.ca

S. O. Peters
Department of Animal Science, Berry College, Mount Berry, GA, USA

A. Djikeng
Internationaly Livestock Research Institute, Nairobi, Kenya

J. E. O. Rege
Emerge Centre for Innovations-Africa, Nairobi, Kenya

Abbreviations

AgGDP	Agricultural gross domestic product
AnGR	Animal genetic resources
AU	African Union
AU-IBAR	African Union-InterAfrican Bureau of Animal Resources
BecA	Biosciences eastern and central Africa
FAO	Food and Agricultural Organization
GDP	Gross domestic product
GHG	Greenhouse gas
ILRI	International Livestock Research Institute
LiDeSA	Livestock development strategy for Africa
MCI	Multicounty cooperations and initiatives
SDG	Sustainable Development Goal
UN	United Nations
UN-SDGs	United Nations Sustainable Development Goals

1.1 Introduction

Animal genetic resources (AnGR), and in particular, those used for food and agriculture, are important components of global biodiversity and

© The Author(s) 2026
E. M. Ibeagha-Awemu et al. (eds.), *African Livestock Genetic Resources and Sustainable Breeding Strategies*, Sustainable Development Goals Series, https://doi.org/10.1007/978-3-031-92076-9_1

key contributors to human survival through the provision of critical components of nutrition and diverse social, cultural, economic, and environmental sustainability needs. An estimated 8800 breeds and strains across animal species are of crucial significance for food and agriculture (FAO 2021, 2023a). Africa has the richest diversity in terms of the number of breeds of cattle, goat, and dromedary. Africa has the second highest number of sheep, chicken, rabbit, and donkey breeds compared to other regions of the world (FAO 2021, 2023a; AU-IBAR 2019) (Table 1.1). Within Africa, cattle are the most dominant in terms of the number of breeds (692), followed by sheep (363), chicken (329), and goat (289) (Table 1.2). This high number of livestock breeds is an invaluable treasure that is critically important to the socioeconomic well-being of Africans. In particular, African indigenous livestock breeds (Table 1.2) exhibit unique and rare adaptive characteristics carefully crafted by years of natural and artificial selection for adaptation to local environmental and production conditions.

Africa's livestock accounts for one-third of the global livestock population (AU-IBAR 2016) and, on average, 40% of the national agricultural GDP, ranging from 10% to 80% in individual countries (Panel, M. M 2020). The demand for animal-source food is projected to increase due to population growth, increased incomes, and urbanization. These demographics point to a trend wherein livestock will be increasingly important in Africa. Being intricately interwoven with the livelihood of farming communities on the continent, livestock has the potential to deliver both the socioeconomic transformation and food security needs of Africa as envisioned in the African Union (AU) Malabo Declaration on Accelerated Agriculture Growth and Transformation, which is a key part of the framework of the AU Agenda 2063, the United Nations Sustainable Development Goals (UN-SDGs), the livestock development strategy for Africa (LiDeSA), and other frameworks and initiatives that cover agriculture and food and nutrition secuirty (AU-IBAR 2015; AU 2015; FAO 2015). This is especially so because, in Africa, livestock farming does not only provide poor people with food, income, traction, and fertilizer but also acts as a catalyst that transforms subsistence farming into income-generating enterprises, allowing poor households to enter the market economy. Indeed, livestock is recognized as being so important to Africa that the Food and Agriculture Organization (FAO) of the United Nations (FAO 2023b) states that the future of African livestock

Table 1.1 The number of livestock breeds and breed types by species in Africa relative to other regions of the world

Species	Africa	Asia	Europe and the Caucasus	Latin America and the Caribbean	Near and Middle East	North America	Southwest Pacific	World
Cattle	692	251	389	141	31	14	32	1550
Goat	289	194	213	36	34	6	11	783
Sheep	363	263	615	64	48	21	38	1412
Chicken	329	318	771	87	32	11	30	1578
Pig	146	227	200	61	1	10	15	660
Horse	134	136	378	78	15	25	25	791
Rabbit	77	16	253	14	7	8	–	375
Buffalo	9	92	8	9	5	1	2	126
Donkey[a]	43	39	48	24	13	6	3	176
Dromedary[b]	63	13	1	–	24	–	2	103
Duck	42	99	100	20	3	1	12	277
Turkey	18	11	42	9	2	8	5	95

Data compiled from various sources (FAO 2021, 2023a; AU-IBAR 2019)
[a]Donkey is also commonly known as ass
[b]Dromedary is also commonly known as camel

Table 1.2 The number of livestock breeds and breed types in Africa and regions

Species	Africa Total	Indigenous	Exotic	East Africa	Central Africa	North Africa	Southern Africa	West Africa
Cattle	692	444	248	167	74	43	275	133
Sheep	363	284	79	99	28	65	109	62
Goat	289	212	77	101	29	33	79	47
Pig	146	50	96	23	24	1	64	34
Chicken	329	109	220	71	33	43	85	97
[a]Camel	63			23	3	23	2	12
[b]Donkey	43			21	2	8	3	9
Buffalo	9			1	3	3	2	–
Horse	134			10	5	20	61	38
Duck	42			3	9	5	17	8
Guinea fowl	41			2	6	1	6	26
Ostrich	8			–	1	1	4	2
Turkey	18			2	2	7	3	4
Rabbit	77			25	8	12	19	13
Total	2254			548	227	265	729	485

Data compiled from various sources (FAO 2021, 2023a; AU-IBAR 2019)
[a]Camel is also commonly known as dromedary
[b]The donkey is also commonly known as ass

will influence the development of the entire continent. Despite the recognition of its potential, livestock resources in Africa are underdeveloped and not in a position to contribute toward meeting the nutrition and economic needs of Africans nor to counter the challenges posed by changing climatic conditions, emerging market demands, and unpredictable upheavals.

The FAO estimates the total African human population at 1.39 billion in 2021, with rural and urban populations of 777 million and 609 million, respectively (FAOSTATS 2022) (Fig. 1.1), and it is projected to reach 2.5 billion by 2050 (UN 2022). A steady annual increase of both the urban and rural populations by 3.7% and 1.8%, respectively, since 2010 is accompanied by an increased demand for animal products. In particular, the livestock sector supports the livelihood of about 65% of Africa's population, especially those living in rural areas. The FAO indicates that the number of undernourished people in Africa rose from 178 million in 2005 to 282 million in 2022, which is 38% of the total number of undernourished people in the world in 2022 (FAO 2023c). The rising number of chronically undernourished people in Africa may pose a challenge to achieving the UN-SDGs of eradicating hunger in Africa by 2050. Livestock is thus critical for Africa, and indigenous AnGR can be the basis of developing a resilient, sustainable, and productive livestock sector, but investments to date have been inadequate, and hence, the livestock sector is underdeveloped, and the potential of indigenous AnGR remains unexploited.

A joint study by the FAO, International Livestock Research Institute (ILRI), and African Union-InterAfrican Bureau of Animal Resources (AU-IBAR) estimated that the meat and milk markets will reach 34.8 and 82.6 million tons by 2050, corresponding to increases of 145% and 155%, respectively, compared to year 2005/2007 (Pica-Ciamarra et al. 2023). As a result, consumer demand will surpass production if concerted actions to increase livestock productivity are not taken. Africa may, therefore, continue to be a prolific importer of livestock products if current national, regional, and continental livestock development strategies are not implemented. The rapid transformation and sustainable development of Africa's livestock sector will require the cooperation of all stakeholders; investments in the development of personnel and the use of innovative technologies such as improved feeding, housing, healthcare, marketing, improved

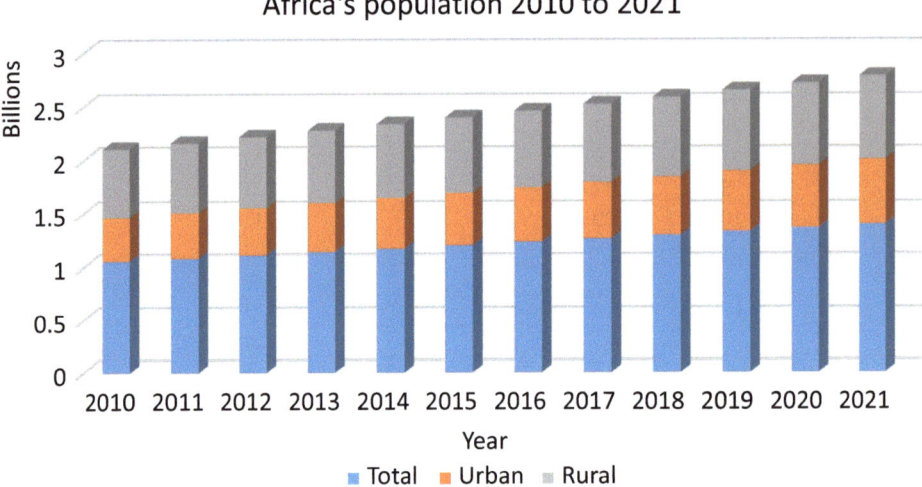

Fig. 1.1 Africa's total, rural, and urban population statistics for the years from 2010 to 2021

animal genetics, and biosecurity; and an evolution of a more business-focused livestock sector. These laudable objectives will, however, require an understanding of every aspect of the livestock genetic resources and production systems for informed targeted interventions. This book chronicles the current facts about the main African livestock genetic resources, specifically cattle, goat, sheep, poultry, pig, dromedary, donkey, horse, buffalo, and nonconventional animal resources (rabbit, snail, frog, and honey bee), addressing five major themes summarized in sections 1.2, 1.3, 1.4, 1.5, and 1.6.

1.2 African Livestock Genetic Resources and Production Systems

The domestication of the present-day African livestock genetic resources has been shaped by a delicate balance between human activities, natural selection, and environmental adaptation (Ibeagha-Awemu et al. 2019) under diverse production systems (Chap. 2). African livestock production systems are heavily dependent on seasonality and the amount of rainfall, which, in part, determines the species of livestock raised in different agroecological zones. According to the population size (number of individuals), chicken (59.37% of the main livestock genetic resources) is the most predominant livestock species in Africa, followed by goat (13.49%), sheep (11.69%), and cattle (10.46%) (FAOSTATS 2022). Chapter 2 details the characteristics, unique features, status, trends, and driving factors of African livestock production systems, which are generally classified as small-scale (pastoral, agropastoral, mixed small holder farming, free range, and urban/peri-urban), medium-scale, and large-scale (pastoral, ranching, commercial intensive and extensive farming, cooperative farming, and state-owned farms). Special emphasis is given to pastoral systems, which play a big role in African livestock production as well as the social, economic, and biophysical aspects of production systems with implications for improvement. Chapter 2 concludes with a discussion of the main challenges limiting livestock production, the possible interventions for livestock development in Africa, and a presentation of the drivers and enabling factors that could shape the future transformation of African livestock production systems to respond to the increasing demand and other trends, which will affect the contexts of both supply and demand of livestock and livestock products.

Knowledge on the historical origins and production characteristics of the main livestock genetic resources in Africa, the breeds, and unique features of taurine cattle (Chap. 3), zebu, zebu-taurine derivatives and established exotic breeds (Chap. 4), goat (Chap. 5), sheep (Chap. 6), poultry (chicken, guinea fowl, and duck) (Chap. 7), pig (Chap. 8), dromedary (Chap. 9), donkey (Chap. 10), horse (Chap. 11), and buffalo (Chap. 12) is essential for their development. These chapters (3–12) chronicle the origins, available knowledge on the breeds/breed types, breed names as used in different countries/regions, physical and production characteristics, special characteristics, and distribution in the African continent and genetic diversity, which together contribute to the reliable prediction of genetic performance of the genetic resources in defined environments so as to provide a basis for their development, taking advantage of their unique attributes. Current uses and geographical distribution (including indicative ecological parameters that define these systems) are also covered. These chapters allude to the role that traditional breeding and indigenous knowledge of livestock keeping and maintenance have played in shaping the diversity in these genetic resources over millennia. Lesser known or underutilized nonconventional animal genetic resources (the Goliath frog, snail, rabbit, and honeybee), which also contribute significantly as food sources and income generators in the continent, are covered in Chap. 13.

Agricultural activities including livestock farming contribute to greenhouse gas (GHG) emissions, with livestock contribution to emissions estimated at about 15% of global methane emissions and 65% of global nitrous oxide emissions, which amount to an estimated 8–11% of global anthropogenic GHG emissions (IPCC 2007). Chapter 14 presents available information on the contributions of African livestock to GHG emissions and global warming in the face of changing climatic conditions. The chapter also examines actions needed to mitigate the impact of animal agriculture on climate change and the role indigenous African genetic resources can increasingly play in enhancing climate adaptation in livestock systems.

1.3 Opportunities for the Improved Utilization of African Livestock Genetic Resources

In comparison to breeds that have been under sustained intentional breeding and selection programs for many decades, African livestock are generally considered underproductive. Such generalizations are not accurate for at least two reasons. First, the majority of African livestock have been subjected neither to sustained selective breeding nor to improved management. Second, productivity measures used have tended to ignore adaptive attributes, which enable indigenous African livestock breeds to survive and produce in extremely challenging environments with near-zero external inputs. Only very few African countries have functional livestock breeding programs that focus on long-term improvement of indigenous breeds. Any livestock breed improvement program starts with defining the breeding goal (Goddard 1998), which will determine the traits to be improved, the direction and rate of genetic change, and the appropriate genetic evaluation system to be implemented. Chapter 15 defines breeding goals and expends the concepts of breeding strategies for improving livestock under various African livestock production systems. It also examines the economic value of breeding goals, the systems for adequate routine data capture, and the opportunities increasingly becoming available to use digital mobile tools for routine data capture. The chapter also presents sample breeding strategies for small ruminants and cattle and highlights the integration of essential elements of breeding strategies in the African context.

Chapters 16 (cattle and buffalo), 17 (goat and sheep), 18 (poultry), and 19 (pig) define breeding goals and breeding strategies for sustainable livestock improvement in African low- to high-input production systems. These chapters cover critical traits and discuss the importance of directional breed formation that responds to demand patterns, the importance of maintaining pure breeds for future use, strategies for gainful exploitation of crossbreeding to improve traits of economic importance, the importance of the male in breed improvement, aspects of adaptation and resilience in African agroecological zones, and the lessons learned and best practices. These chapters also examine past breed improvement efforts and sample results in different countries/regions, lessons learned, and the prospects for future advances.

1.4 Leveraging Modern Technologies for the Improved Utilization of African Livestock Genetic Resources

Technological advances have been instrumental in driving different dimensions of development, and present-day advanced technologies are facilitating breakthrough innovations in medicine and agriculture, including livestock production. It is expected that developments in genomics, phenomics, nutrition, reproduction, health and management technologies will support rapid progress in livestock development, including in developing more productive and resilient animals suitable for the diverse production environments of the continent.

Chapters 20 and 21 chronicle the various technological innovations that have resulted in advances in breeding, nutrition, health, and reproduction in livestock across the globe and the opportunities they present for Africa. Key issues related to the adoption of modern technologies as well as the challenges and opportunities for the application of modern technologies for sustainable livestock production in African livestock production systems are discussed. Using practical examples, the chapters demonstrate that the modern technologies necessary to advance African livestock productivity are available but that these have not been fully harnessed by the continent.

Prospects for the utilization of modern technologies by species are presented in Chaps. 22 (cattle), 23 (poultry), and 24 (goat, sheep, and pig). These chapters focus on specific technologies that have been applied in each species/group and present case studies on their application in Africa's low-, medium-, and high-input livestock systems. The conclusion here is that the technologies with potential are available and can support rapid genetic progress in African livestock. Taken together, the analyses presented in these chapters emphasize that it is imperative that African nations invest in human and institutional capacity to effectively utilize the available and emerging modern technologies in livestock research and development.

1.5 Conservation of African Livestock Genetic Resources

As alluded to above, Africa is home to a high diversity of some of the world's livestock genetic resources (AnGR). Importantly, although the indigenous African livestock genetic resources present unique opportunities for improving production and productivity in the diverse African environments, they are threatened as major genetic erosion continues to happen even before they are adequately characterized. Thus, deliberate interventions are needed to improve the management of AnGR in all situations, including where the AnGR are most at risk (Seré et al. 2008). Conserving Africa's AnGR is an essential initial step toward maintaining AnGR for current and future production objectives and environmental challenges.

Chapter 25 discusses the need, options, and tools for the conservation and management of AnGR. The operational procedures/guidelines of past and current conservation efforts in Africa are also discussed, as well as the opportunities of exploring genomics and other modern technologies in conservation programs. The challenges of applying in situ and ex situ conversation and current conservation efforts are addressed. Issues around policies and regulations governing the conservation and improved management of AnGR and the future opportunities for conserving African livestock genetic resources are also discussed.

Chapter 26 examines the economic considerations, the framework of conservation of livestock genetic resources, and the institutional frameworks for the exchange and utilization of AnGR. The methods and tools for the economic valuation of AnGR and for estimating the economic and social benefits and costs of AnGR conservation are presented. Components of the economic framework for the sustainable use and conservation of AnGR in terms of creating awareness, capacity development, data availability, and access to data, resources, and incentives for the conservation of AnGR are also presented.

1.6 Institutional Arrangements and Enablers for the Enhanced Utilization of African Livestock Genetic Resources

Enhanced utilization of African livestock genetic resources will require the acquisition and mastering of a diverse set of skills and knowledge going well beyond livestock breeding and management. Animal genetic improvement is generally regarded as the science of increasing the genetic merit of a future population through the intentional selection and mating of genetically superior parents. The technical know-how to achieve this requires multilevel education and training that integrates knowledge from multiple disciplines such as quantitative genetics, population genetics, genomics, statistics, molecular biology, biochemistry, computer science, and geography, among others. Chapter 27 presents the current status of animal genetic improvement education in Africa, evaluates graduate education in Animal Breeding and the associated institutions, identifies the gaps in the current curricula, and makes recommendations to strengthen the existing training. Suggested strategies for capacity strengthening include curriculum overhaul for tertiary education, retooling of trainers and researchers, tailored training for farmers, and the facilitation of access to research resources such as sharing arrangements for the use of facilities at national, regional, and international research institutions (e.g., the Biosciences eastern and central Africa (BecA) and shared laboratories at the ILRI campus in Nairobi, Kenya).

Increased livestock productivity with the potential to enhance the livelihoods of rural dwellers is central for the achievement of UN-SDGs (FAO 2015) for Africa. Specifically, the contributions of livestock breeding and genetic improvement and Africa's livestock development strategy aimed at making available animal-sourced foods for the achievement of UN-SDGs for Africa are presented in Chap. 28. Key capacity building areas in livestock breeding and genetic improvement for achieving UN-SDGs for Africa include the improvement and maintenance of livestock genetic diversity, livestock conservation, market-driven capacity development in modern livestock breeding strategies, sustainable intensification of livestock production practices, adaptation of livestock farming practices to changing environmental conditions, tailored breeding for enhanced animal welfare, adaptability and biosafety, smart farming and breeding strategies with the implementation of big data and artificial intelligence, and infrastructural development and tailored policies for technology uptake for sustainable and enhanced livestock development. The analyses of these needs are covered in Chap. 28.

Multicounty cooperation or initiatives (MCIs) are considered powerful approaches to enhance the efficiency and effectiveness of livestock breeding programs. The subcommittee of the international committee for animal recording known as INTERBULL (https://interbull.org/) has effectively deployed multicountry evaluations for dairy bulls across Europe for decades. Chapter 29 examines what would be required, including data recording and analysis methodologies and tools, for African countries to establish robust systems for the genetic evaluation of livestock across multiple countries. In addition to the opportunity to effectively use limited human and physical resources (including data storage and computing equipment) across countries, key benefits of such an approach will be the quantification of how different genotypes interact with the diverse environments and the evaluation of performance under different environmental and management conditions.

Chapter 30 presents an analysis of the policy and legislative environment, including an examination of the status of relevant policy and legislation instruments in African countries, and the engagement of Africa in the development and implementation of global policy instruments and frameworks to ensure that these are relevant for the diverse contexts of Africa. The latter includes coverage of both Nagoya and Cartegena Protocols, which address ownership, sovereignty, access, and benefit sharing of genetic resources. Overall, the chapter attempts to examine the extent to which African countries are paying attention to relevant policies locally and engaging effectively in subregional, continental, and international discourses that are shaping the global policy landscape that has implications for AnGR.

Finally, sustainable livestock production is among the prerequisites for meeting the food needs of Africans and the economic development of the continent, and achieving this *involves the participation and cooperation of many stakeholders*. Chapter 31 concludes this book by presenting the roles of various institutions—public, private, civil society, and farmer organizations—within and between countries, as well as linkages to international players, including research institutions in the North and development partners, which support livestock research and development. The chapter examines the challenges associated with building functional partnerships and concludes by emphasizing the key pillars that underpin effective development interventions: enabling policies, functioning institutional arrangements, access to appropriate technologies, funding and information, and adequate infrastructure and trained personnel.

1.7 Conclusion

This introductory chapter is an overview of the material covered in this book. It is intended to point the reader to materials of interest in the book.

References

AU (2015) Agenda 2063 – the Africa we want. African Union (AU). Available at https://au.int/en/agenda2063/overview. https://au.int/sites/default/files/documents/36204-doc-agenda2063_popular_version_en.pdf. Accessed 21 Sept 2023

AU-IBAR (2015) The livestock development strategy for Africa 2015–2035. African Union-InterAfrican Bureau of Animal Resources (AU-IBAR), Nairobi. Available online at http://repository.au-ibar.org/bitstream/handle/123456789/540/2015-LiDeSA.pdf. Accessed on 15 May 2023

AU-IBAR (2016) Livestock policy landscape in Africa: a review. http://www.au-ibar.org/component/jdownloads/finish/36-vet-gov/2712-livestock-policy-landscape-in-africa-a-review. Accessed 20 Feb 2021

AU-IBAR (2019) The state of farm animal genetic resources in Africa: towards accelerated agricultural growth and transformation by the year 2025. African Union – Interafrican Bureau for Animal Resources (AU-IBAR), Nairobi, 337 pages

FAO (2015) Food and Agricultural Organisation (FAO) of the United Nations, FAO synthesis – livestock and the sustainable development goals global agenda for sustainable livestock. Draft prepared by FAO-AGAL Livestock Information, Sector Analysis and Policy Branch. Available at: http://www.livestockdialogue.org/fileadmin/templates/res_livestock/docs/2016/Panama/FAO-AGAL_synthesis_Panama_Livestock_and_SDGs.pdf. Accessed 8 Sept 2023

FAO (2021) Status and trends of animal genetic resources – 2020 – CGRFA/WG-AnGR-12/23/4/Inf.1. Commission on Genetic Resources for Food and Agriculture. https://www.fao.org/3/cc3705en/cc3705en.pdf. Accessed 8 Sept 2023

FAO (2023a) Data on Animal Diversity Information Services (DAD-IS). https://www.fao.org/dad-is/en/. Accessed 8 Sept 2023

FAO (2023b) Africa sustainable livestock 2050. Food and Agricultural Organization (FAO). https://www.fao.org/in-action/asl2050/en/. Accessed 18 Oct 2023

FAO (2023c) The state of food security and nutrition in the world 2023. Urbanization, agrifood systems transformation and healthy diets across the rural–urban continuum. FAO, IFAD, UNICEF, WFP and WHO, Rome. Available at https://doi.org/10.4060/cc3017en. https://www.fao.org/3/cc3017en/online/state-food-security-and-nutrition-2023/food-security-nutrition-indicators.html. Accessed 20 Aug 2023

FAOSTATS (2022) Food and Agricultural Organization statistics for year 2022. Available at: www.fao.org/faostat/. Accessed 11 Aug 2023

Goddard ME (1998) Consensus and debate in the definition of breeding objectives. J Dairy Sci 81:6–18. https://doi.org/10.3168/jds.S0022-0302(98)70150-X

Ibeagha-Awemu EM, Peters SO, Bemji MN, Adeleke MA, Do DN (2019) Leveraging available resources and stakeholder involvement for improved productivity of African livestock in the era of genomic breeding. Front Genet 10:357. https://doi.org/10.3389/fgene.2019.00357, Review

IPCC (2007) Climate change 2007: the physical science basis. Contribution of Working Group I to the Fourth Assessment Report of the Intergovernmental Panel on Climate Change (IPCC). http://ipcc-wg1.ucar.edu/wg1/docs/WG1AR4_SPM_Approved_05Feb.pd. Accessed 4 Sept 2023

Panel, M. M (2020) Meat, milk and more: policy innovations to shepherd inclusive and sustainable livestock systems in Africa. International Food Policy Research Institute. https://www.ifpri.org/cdmref/p15738coll2/id/133855/filename/134065.pdf. Accessed 23 Feb 2021

Pica-Ciamarra U, Baker D, Morgan N, Ly C, Nouala S (2023) Investing in African livestock: business opportunities in 2030–2050. Livestock Data Innovation in Africa Project, a joint initiative of the World Bank, FAO, ILRI and AU-IBAR. Available at https://openknowledge.worldbank.org/server/api/core/bitstreams/57003a03-3518-5ab3-bb8c-93bce40bb843/content. Accessed 12 Sept 2023

Seré C, van der Zijpp A, Persley G, Rege E (2008) Dynamics of livestock production systems, drivers of change and prospects for animal genetic resources. Anim Genet Resour/Resources génétiques animales/Recursos genéticos animales 42:3–24. https://doi.org/10.1017/S1014233900002510. From Cambridge University Press Cambridge Core

UN (2022) United Nations, Department of Economic and Social Affairs, Population Division. Probabilistic population projections based on the world population prospects 2022. http://population.un.org/wpp/. Accessed 4 Sept 2023

Open Access This chapter is licensed under the terms of the Creative Commons Attribution 4.0 International License (http://creativecommons.org/licenses/by/4.0/), which permits use, sharing, adaptation, distribution and reproduction in any medium or format, as long as you give appropriate credit to the original author(s) and the source, provide a link to the Creative Commons license and indicate if changes were made.

The images or other third party material in this chapter are included in the chapter's Creative Commons license, unless indicated otherwise in a credit line to the material. If material is not included in the chapter's Creative Commons license and your intended use is not permitted by statutory regulation or exceeds the permitted use, you will need to obtain permission directly from the copyright holder.

African Livestock Production Systems: The Past, Present and the Projected Future

Eveline M. Ibeagha-Awemu, Richard Osei-Amponsah, and Martha N. Bemji

Abstract

Africa is home to diverse livestock genetic resources (AnGR), which have contributed and continue to contribute to the livelihoods of Africans and the economic development of the continent (Section 2.1). Livestock farming engages about 65% of Africans, predominantly rural dwellers, and its evolution is shaped by many factors such as agro-ecological conditions, socio-economic variables, land availability, species, goal of production and technological and institutional factors, among others. Consequently, a wide variety of livestock production systems are practised in Africa, even within the same agro-ecological zones or regions. Depending on the agro-ecological zone and the level of investment/technical know-how, the main livestock production systems range from small-scale subsistence to large-scale extensive and intensive systems (Section 2.3). Generally, African livestock production is dependent on rainfall, and while about three-quarters of Africa's agricultural land is classified as grassland, over 70% of livestock productivity takes place in the small-scale production systems characterised by low farm size (animal numbers), limited access to adequate nutrition and healthcare and generally low inputs and outputs (Section 2.4). With the rising human population and demand for livestock products, the current level of livestock productivity is not meeting consumer demand, a situation that will worsen in future years. As a remedy, current livestock production systems must be understood as a first step to putting in place more productive systems. This chapter presents the characteristics of present-day African livestock production systems, their challenges, and factors that could shape future more productive livestock production systems.

E. M. Ibeagha-Awemu (✉)
Sherbrooke Research and Development Centre,
Agriculture and Agri-Food Canada,
Sherbrooke, Quebec, Canada
e-mail: eveline.ibeagha-awemu@agr.gc.ca

R. Osei-Amponsah
Department of Animal Science, School of Agriculture, University of Ghana,
Legon, Accra, Ghana

M. N. Bemji
Department of Animal Breeding and Genetics,
Federal University of Agriculture,
Abeokuta, Ogun State, Nigeria

Keywords

Livestock farming · Ecological zones · Subsistence livestock farming · Technological innovation · Enabling livestock policies · Livestock marketing · Sociocultural values

© The Author(s) 2026
E. M. Ibeagha-Awemu et al. (eds.), *African Livestock Genetic Resources and Sustainable Breeding Strategies*, Sustainable Development Goals Series, https://doi.org/10.1007/978-3-031-92076-9_2

Abbreviations

AfCFTA	African Continental Free Trade Area
AnGR	Animal genetic resources
ASAL	Arid and semi-arid lands
AU-IBAR	African Union-InterAfrican Bureau for Animal Resources
CAADP	Comprehensive Africa Agriculture Development Programme
CSA	Climate-smart agriculture
EAC	East African Community
ECCAS	Economic Community of Central African States
ECOWAS	Economic Community of West African States
FAO	Food and Agricultural Organization
GDP	Gross domestic product
GHG	Greenhouse gas
LGS	Livestock grazing system
LiDESA	Livestock Development Strategy for Africa
SADC	Southern African Development Community
SDG	Sustainable Development Goals
SSA	Sub-Saharan Africa
TLU	Tropical livestock unit
UN	United Nations

2.1 Introduction

Livestock refers to various domesticated animal genetic resources (AnGR) that are of food, cultural and economic value to those who keep them. In Africa, typical animal species that constitute animal agriculture include ruminants (cattle, sheep, goat and buffalo), pseudo-ruminants (camel, rabbit and horse), donkey, pig and various poultry species (chicken, guinea fowl, turkey and duck). Livestock are raised in Africa to provide a broad array of goods and services such as (1) food (meat, milk and eggs), (2) energy for land tilling and transport, (3) manure for fertilising fields, (4) raw materials (hide, skin and hair), (5) employment for smallholder farm families and professionals, (6) social roles (dowry payment, sport and competition, burial and other traditional ceremonies), (7) religious roles (use in religious festivals like Christmas, Ramadan, etc.), (8) status and social symbol, (9) a form of insurance (livestock sold for cash when needed), (10) risk reduction through provision of diversified production, (11) environmental sustainability through nutrient recycling and landscape and biodiversity maintenance and (12) contribution to the circular economy. In addition to livestock playing a central role in the livelihoods of African communities, they contribute significantly to the agricultural gross domestic product of many African countries (10–50%) (AU-IBAR 2019). Livestock is intricately interwoven with the livelihood of farming communities on the continent to the extent that most indigenous breeds are named after tribes or specific production regions. Therefore, it is vital to understand the systems in which the livestock are produced in the face of rapid human population growth and how existing livestock production systems can be sustainably maintained and improved to meet future production goals.

The scale, goal, nature and management of a farming initiative is referred to as a production system. Agricultural production systems are the ways in which farmers make use of available resources to their food and agriculture needs. According to the Food and Agricultural Organization (FAO) (FAO 2020a), livestock production systems encompass all aspects of the use and supply of livestock and livestock products, including their number and distribution, the various production systems, estimates of present and future consumption and production, the people directly implicated in producing livestock, the associated benefits and their economic and environmental impacts. Thus, the ecological system, husbandry and management practices, and levels of input collectively define a livestock production system, and various combinations of these factors depict a rich variety of livestock production systems in Africa, often broadly classified as extensive, semi-intensive and intensive. These systems vary from region to region depending on the resource availability and/or allocation, production objectives, scale of operation, target market and the cultural values of the local communities. The unique features of livestock production systems in Africa include their integration into crop farming, extensive pastoral nature, free range, multiple objectives, heterogeneity of labour, keeping of multiple breeds and crossbreds, and irregularity of feed, veterinary services and markets. The subsistence or smallholder livestock production system predominates and often targets the needs of farm families with no immediate objective of making a profit from the sale of animals and animal products.

A main characteristic of these major systems is their low productivity, which falls short of meeting increasing consumer demands for animal products by an increasing human (Africa) population currently estimated at 1.39 billion and projected to reach 2.5 billion by 2050 (FAOSTATS 2022; UN 2022). To meet the shortfalls in livestock productivity, African governments have resorted to the importation of animal products, a trend that goes up yearly (Ibeagha-Awemu et al. 2019; FAOSTATS 2022). Fuelled by the increasing demand for livestock products driven by urbanisation, increasing standards of living and changing lifestyles, animal production in Africa is slowly evolving from a predominantly traditional subsistence farming relying on large numbers of animals towards intensification and production efficiency. Several factors are critical components of this evolution including the availability of innovative technologies, improved animal genetics, improved management strategies (feeding, housing, healthcare, marketing and biosecurity) and the emergence of more business-focused livestock farmers. As a result, various forms of semi-intensive-to-intensive livestock production systems are emerging across the African landscape.

This chapter presents the main features, status, trends and driving factors of African livestock production systems and the distribution and population trends of commonly farmed animal genetic resources (cattle, sheep, goat, buffalo, camel, donkey, rabbit, horse, pig, chicken, guinea fowl, turkey and duck) and various non-conventional animal genetic resources. The chapter also presents the factors limiting livestock production and possible mitigations, as well as possible enablers that could shape future livestock production systems and directions for more sustainable and environmentally friendly production practices.

2.2 Overview of African Livestock Genetic Resources and Agro-ecological Zones

The livestock genetic resources exploited for food and economic gains on the African continent as well as the agro-ecological zones shaping various livestock production systems will be presented in this section. More detailed descriptions of these AnGRs are found in Chapters 3, 4, 5, 6, 7, 8, 9, 10, 11, 12, and 13.

2.2.1 African Regional Livestock Genetic Resources

The five major regions of the African continent (East, Central, North, South and West) support diverse species of livestock, their introduction or domestication being shaped by a delicate balance between human and natural selection and environmental adaptation (Ibeagha-Awemu et al. 2019). Based on the FAO statistics (FAOSTATS 2022), the populations of the main livestock species in Africa and their rate of annual increases in the past 11 years are shown in Table 2.1 and Fig. 2.1a, b. According to numbers from the year 2021, chicken is the most predominant livestock, constituting about 59.37% of Africa's main livestock genetic resources, followed by goat (13.49%), sheep (11.69%) and cattle (10.46%). Cattle and small ruminants (goat and sheep) are the predominant ruminant species in East and West Africa, accounting for about 67.27% of the continent's ruminant (cattle, sheep, goat and buffalo) livestock. Among the major livestock species, North Africa is the lead producer of chicken (33.91%), followed by West Africa (32.74%); East Africa is the lead producer of cattle (48.05%), followed by West Africa (22.71%); West Africa is the lead producer of sheep (31.14%), followed by North Africa (26.55%); West Africa is the lead producer of goat (37.68%), followed by East Africa (36.84%); and East

Table 2.1 Livestock populations[a] in Africa and regions in the year 2021

	Africa (total)	Annual increase[b] (%)	East Africa	Central Africa	North Africa	Southern Africa	West Africa
Cattle	373,198,697	2.07	179,333,640	51,340,196	40,590,648	17,177,334	84,756,879
Sheep	416,965,073	2.34	103,647,389	48,386,131	110,724,177	24,384,880	129,822,496
Goat	481,063,954	3.00	177,229,471	65,062,755	48,253,665	9,254,419	181,263,644
Pig	43,007,434	3.32	17,649,551	8,139,402	28,914	1,511,304	15,678,263
Camel	34,269,954	2.65	14,018,423	9,401,892	5,852,123	94	4,997,422
Horse	7,412,943	1.99	2,225,032	1,397,710	1,221,493	440,322	2,128,386
Buffalo	1,263,128	−6.70	25	ND[c]	1,263,103	ND	ND
Rabbit	18,364,000	0.34	2,782,000	395,000	8,495,000	125,000	6,567,000
Chicken	2,117,636,000	2.48	380,541,000	145,553,000	718,136,000	180,074,000	693,332,000
Turkey	31,488,000	4.48	2,523,000	3000	28,421,000	541,000	ND
Duck	15,612,000	0.45	8,674,000	85,000	4,063,000	419,000	2,371,000
Geese	19,659,000	1.71	19,371,000	ND	144,000	144,000	ND

[a]Data was extracted from FAOSTATS (2022)
[b]Percent average annual increase in livestock populations in Africa from 2010 to 2021
[c]ND: No data (data not available)

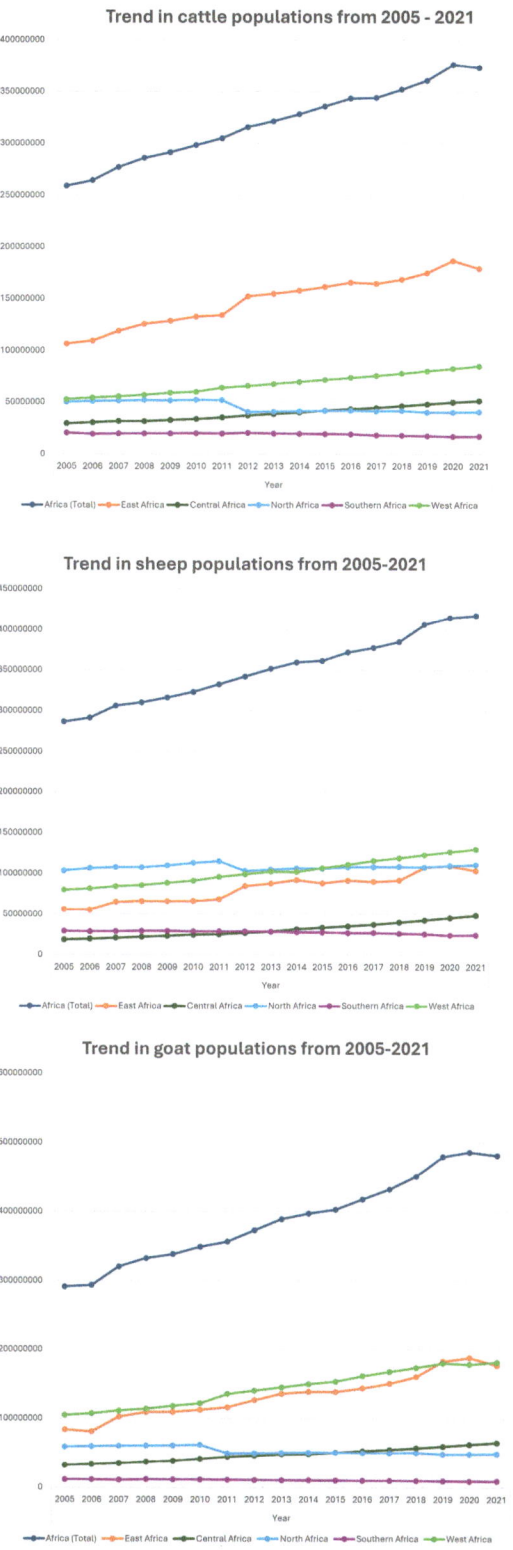

Fig. 2.1 (**a**) Cattle, sheep and goat population trends in Africa and regions from 2005 to 2021. (**b**) Pig and chicken population trends in Africa and regions from 2005 to 2021

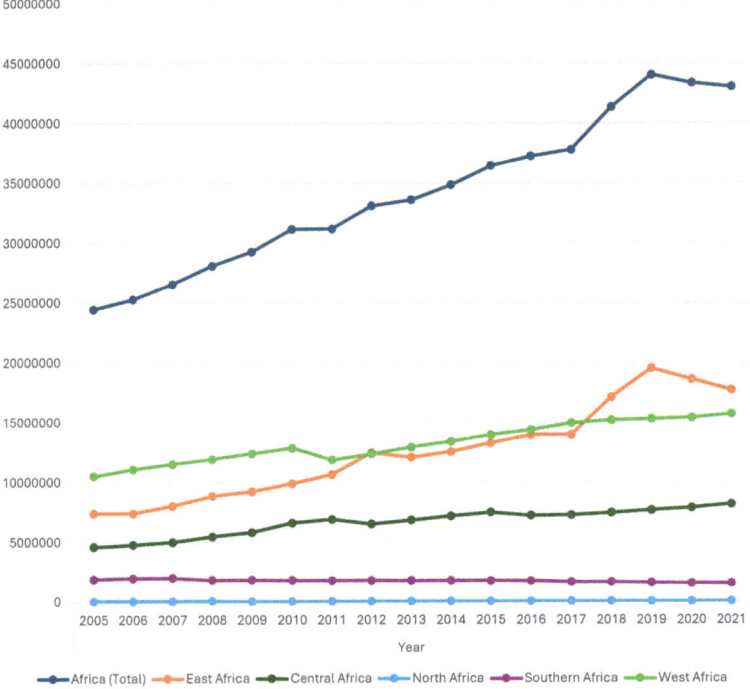

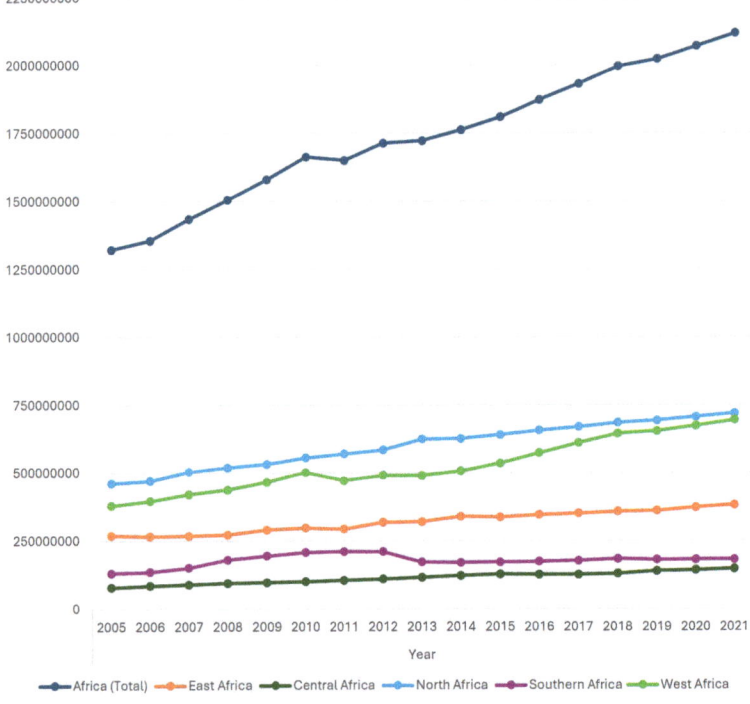

Fig. 2.1 (continued)

Africa is the lead producer of pig (41.04%) followed by West Africa (36.45%). Among the less common domesticated livestock species, North Africa is the topmost producer of buffalo (over 95%), turkey (90%) and rabbit and hare (46.26%). More details on the characteristics and genetic diversity of the major African livestock genetic resources and non-conventional animal genetic resources are presented in Chapters 3, 4, 5, 6, 7, 8, 9, 10, 11, 12, and 13.

African livestock farming is heavily dependent on rainfall, and the livestock populations or their native habitats are shaped by the prevailing agro-ecological zones. Most of the present-day livestock production in Africa is characterised by low inputs and outputs and not being able to meet the increasing demand for livestock products. Due to low livestock productivity, African governments increased beef imports from 482,111 tonnes in 2012 to 612,353 tonnes in 2016 and pork imports from 184,322 tonnes in 2012 to 252,611 tonnes in 2016 (Ibeagha-Awemu et al. 2019). To position the livestock sector to positively contribute to animal protein food supply and the socio-economic development of the continent, it is imperative to ensure sustainability in African livestock production systems since livestock directly contribute to the achievement of all 17 United Nation's Sustainable Development Goals (SDGs) and the goals of the Livestock Development Strategy for Africa (LiDESA) 2015–2035 (AU-IBAR 2015; UN 2015). Since the livestock sector has the potential to enhance the livelihoods of Africa's subsistence farmers, understanding today's livestock production systems will pave a way to putting in place more productive systems.

2.2.2 Agro-ecological Zones of Africa and Livestock Production

Geographical areas displaying similar climatic conditions (e.g. rainfall, temperature, humidity, wind speed, etc.) and vegetation that govern their ability to support rain-fed agriculture including livestock farming are known as agro-ecological zones. Agro-ecological zones in Africa are delineated by various factors including latitude, elevation, highland/lowland (warm or cool based on elevation), temperature, moisture zones (water availability), rainfall (including rainfall amounts), seasonality, duration of rainy season, length of growing season and the species/amount of vegetation, which also influence the livestock species kept by farmers. Based on these factors and considering the major climatic zones of Africa (tropics and sub-tropics), four major agro-ecological zones can be distinguished in Africa: arid, semi-arid, sub-humid and humid (Fig. 2.2a). Within these major agro-ecological zones are found subtropical warm and cool regions as well as tropical warm and cool regions. The main characteristics of African agro-ecological zones are summarised in Table 2.2. The main livestock production systems of the agro-ecological zones determined using factors such as land cover, days of growing period, highland and temperate areas, human population and irrigated areas (Robinson et al. 2011) are depicted in Fig. 2.2b. Notable arid zones of Africa are the Sahara desert in Northern Africa and the Namibian and Kalahari deserts in Southern Africa. As shown in Fig. 2.2b, large areas of sub-Saharan Africa (SSA) are dominated by mixed crop–livestock systems. Moreover, of the four major agro-ecological zones (arid, semi-arid, sub-humid and humid) in SSA (Dixon et al. 2001), arid and semi-arid agro-ecological zones cover 43% of the land area. In West Africa, 70% of the total population lives in the moist sub-humid and humid zones, whereas in East and Southern Africa, only about half the population occupies these areas. According to Otte and Chilonda (2003), production parameters of ruminants in the traditional systems of SSA published in the literature between 1973 and 2000 are generally poor, without marked differences between systems, agro-ecological zones, or sub-regions. Spatial analysis by the authors revealed regional variations in the availability of meat and milk per person, showing that per capita beef and milk supply were highest in the regions with smallholder dairy systems and lowest in the humid zones of Central and West Africa. Three-quarters of agricultural land can be classified as grassland

Fig. 2.2 (**a**) The main agro-ecological zones in Africa (Source: Sebastien (2009)). (**b**) Livestock production systems by agro-ecological zones in Africa. (Source: Robinson et al. (2011). Available freely through ILRI on Flickr https://www.flickr.com/photos/ilri/14509985147/)

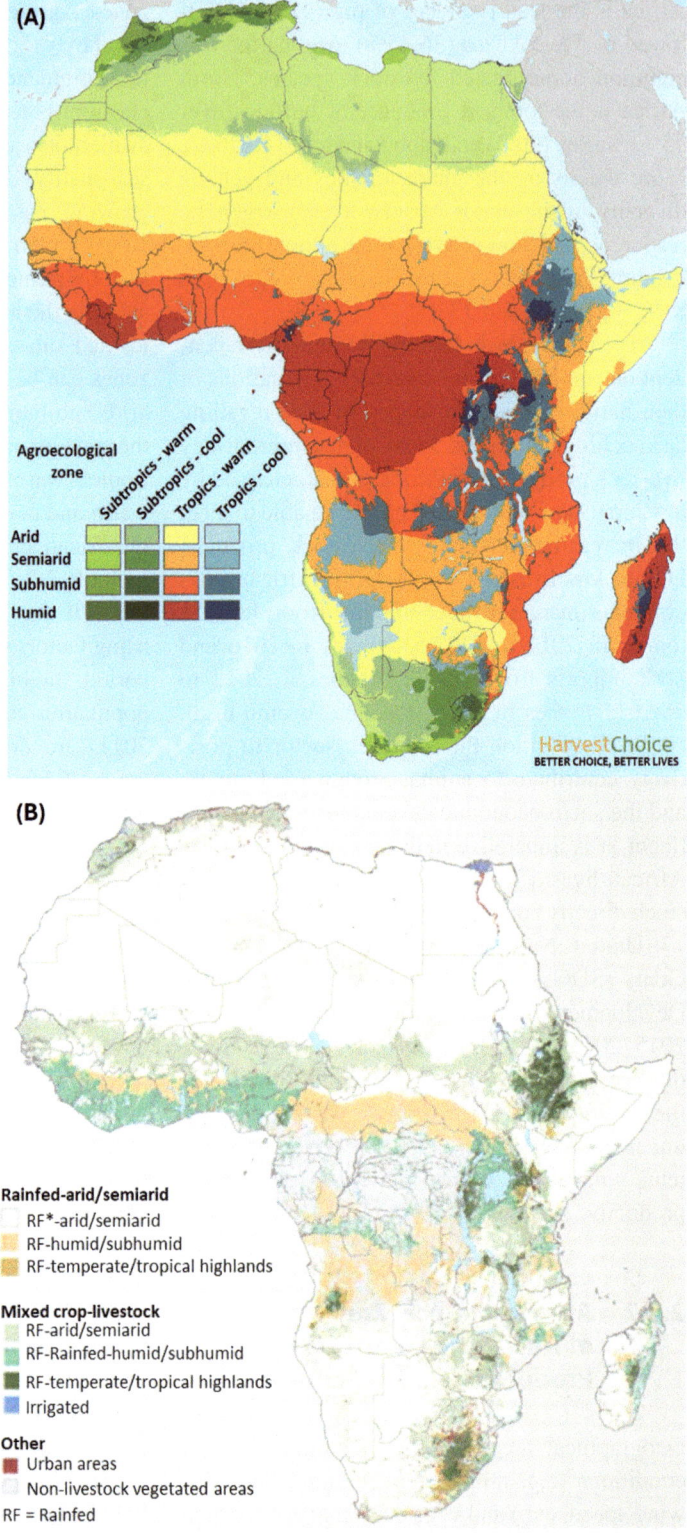

Table 2.2 Main characteristics of the agro-ecological zones of Africa[a]

Region	Factor	Agro-ecological zones			
		Arid	Semi-arid	Sub-humid	Humid
West Africa	Days of growing period/year	30–90	90–240	180–270	240–365
	Mean annual rainfall	Up to 750 mm	750–1250 mm	750–1500 mm	1500–2000 mm
	Mean annual temperature	17–40 °C	17–35 °C	20–35 °C	20–37 °C
	Main livestock species	Sheep, goat, cattle and camel	Sheep, goat, cattle, camel and chicken	Sheep, goat, cattle, camel, chicken and pig	Sheep, goat, cattle, camel, pig and chicken
	Main plant/tree species	Maize, millet, sorghum, rice, groundnut, cowpea, cotton. Kapok, mango, acacia spp., date palm	Millet, sorghum, groundnut, cotton, beans and rice. Tree crops are mango, cashew and kapok	Maize, sorghum, rice, millet, yam, cotton, groundnut, pulses, mango, kapok, cashew, coffee, citrus	The Guinea or derived savannah zone: The crops are mostly maize, yam, rice, millet, sorghum, groundnut and cotton; sugar cane is grown in the wetter parts. The forest zone: The main food crops are maize, yam, cassava, cocoyam plantain, banana, cowpea, cocoa, coffee, coconut, mango, citrus, sugar cane and oil palm
Central Africa	Days of growing period/year	30–90	90–240	180–290	270–365
	Mean annual rainfall (range)	Up to 610 mm	Up to 971 mm	971–1540 mm	1376–3600 mm
	Mean annual temperature	23–35 °C	23.8–31 °C	16–30 °C	16–30 °C
	Main livestock species	Cattle, small ruminants and poultry	Cattle, small ruminants, poultry and pig	Poultry, pig, fisheries and small ruminants	Poultry, pig, fisheries and small ruminants
	Main plant/tree species	Maize, sorghum, groundnut, cotton, muskwari, cowpea and millet	Maize, yam, cassava, sweet potato, rice and cotton	Banana, plantain, cassava, cocoyam, sweet potato, maize, vegetables, groundnut, cocoa, coffee, oil palm, rubber and fruits	Plantain, cassava, banana, maize, cocoyam, sweet potato, cocoa, oil palm, rubber, coffee, coconut, tea, maize and fruits

(continued)

Table 2.2 (continued)

Region	Factor	Agro-ecological zones			
		Arid	Semi-arid	Sub-humid	Humid
North Africa	Days of growing period/year	<60	<150	150–210	210–290
	Mean annual rainfall	<100 mm	100–200 mm	200–350 mm	400–800 mm
	Mean annual temperature (range)	32–45 °C	30–35 °C	20–28 °C	18–25 °C
	Main livestock species	Buffalo, camel, donkey, sheep and goat	Buffalo, camel, donkey, sheep and goat	Goat, sheep, poultry and beekeeping	Goat, sheep, poultry and beekeeping
	Main plant species	Rice, cotton, wheat, maize, sugar cane, soybeans, broad beans, onion, olive, barley, date palm and tomato	Wheat, maize, barley, carrot, tomato, sorghum, rice and saltwort	Potato, sugar cane, olive and date palm	Lemon, lime, grapefruit, olive and date palm
East Africa	Days of growing period/year	100–120	120–140	120–240	200–250
	Mean annual rainfall (range)	200–447 mm	400–800 mm	800–1200 mm	1500–2000 mm
	Mean annual temperature	26–32 °C	23–28 °C	20–27 °C	20–26 °C
	Main livestock species	Cattle, buffalo, sheep, goat, pig and camel	Cattle, sheep, goat, pig and poultry	Cattle, sheep, goat, cattle, pig and poultry	Cattle, sheep, goat, pig and poultry
	Main plant/tree species	Baobab tree, wheat, rice, maize, sweet potato, sorghum and ginger	Cactus, palm tree, ginger, black pepper and acacia	Cassava, potato, maize, ginger, peanut and coconut	Cassava, pear, banana, giant lobelia, tree groundsel, coffee, tea and maize,

(continued)

Table 2.2 (continued)

Region	Factor	Agro-ecological zones			
		Arid	Semi-arid	Sub-humid	Humid
Southern Africa	Days of growing period/year	<60	60–120	120–250	270–330
	Mean annual rainfall (range)	Up to 269.2 mm	Up to 464 mm	650–982 mm	880–1550 mm
	Mean annual temperature	24–31 °C	22–29 °C	20–28 °C	18–25 °C
	Main livestock species	Cattle, goat, sheep, pig and poultry	Dorper sheep, Damara sheep, Boer goat, pig, donkey and poultry	Cattle, goat, sheep and equines	Chicken, cattle, pig, goat, sheep, deer, duck and turkey
	Main plant/tree species	Maize, millet, sorghum, barley, cassava, groundnut, cowpea, cabbage, carrots and date palm	Maize, millet, sorghum, barley, cassava, groundnut, cotton, cowpea, butternut, melon and watermelon	Sweet potato, cotton, cassava and cowpea	Maize, cassava, banana and orange

[a]The data in this table were compiled from the following sources: https://gaez-data-portal-hqfao.hub.arcgis.com/, Fischer et al. (2002), Muimba-Kankolongo (2018) and Vrieling et al. (2013)

across Africa (Otte et al. 2019), with over 70% of livestock production taking place in small-scale production systems characterised by low animal population numbers and generally low inputs and outputs (Ibeagha-Awemu et al. 2019).

2.3 African Livestock Production Systems

In its broadest sense, livestock production systems, which are subsets within the agriculture sector, describe the type of animal genetic resources raised in specific environments using specific management practices and for specific purposes by producers. The path and rate at which livestock production systems evolve depend on many factors such as the interaction of the crop with the livestock, agro-ecological conditions, socio-economic variables (such as days of growing period and distance to markets to acquire inputs and sell produce), technological and institutional factors, land availability, goal of production, intensity of production and product type, as summarised in Fig. 2.3. The level of investments or inputs in terms of feeding, healthcare, housing, innovative technologies and marketing varies between production systems. Consequently, a wide variation in livestock production systems is found in Africa, even within the same agro-ecological zones or regions. Based on climate change, the FAO (2009) classified African livestock production systems into grazing systems, mixed farming systems and industrial systems. Grazing systems were further subdivided into extensive and intensive, while mixed farming systems were broken down into irrigated and rain-fed. Industrial systems include highly intensive cattle, pig and poultry production systems. Large areas of the Horn of Africa, Southern Africa and the Sahel are dominated by grazing systems, and grazing and mixed systems feature in West Africa and some parts of East Africa, with smaller areas in Southern Africa (Simpkin et al. 2020). Recently, the African Union-InterAfrican Bureau for Animal Resources (AU-IBAR) (AU-IBAR 2019) broadly classified African livestock production systems into three main categories, namely (1) agro-pastoral and pastoral systems, (2) mixed crop–livestock systems

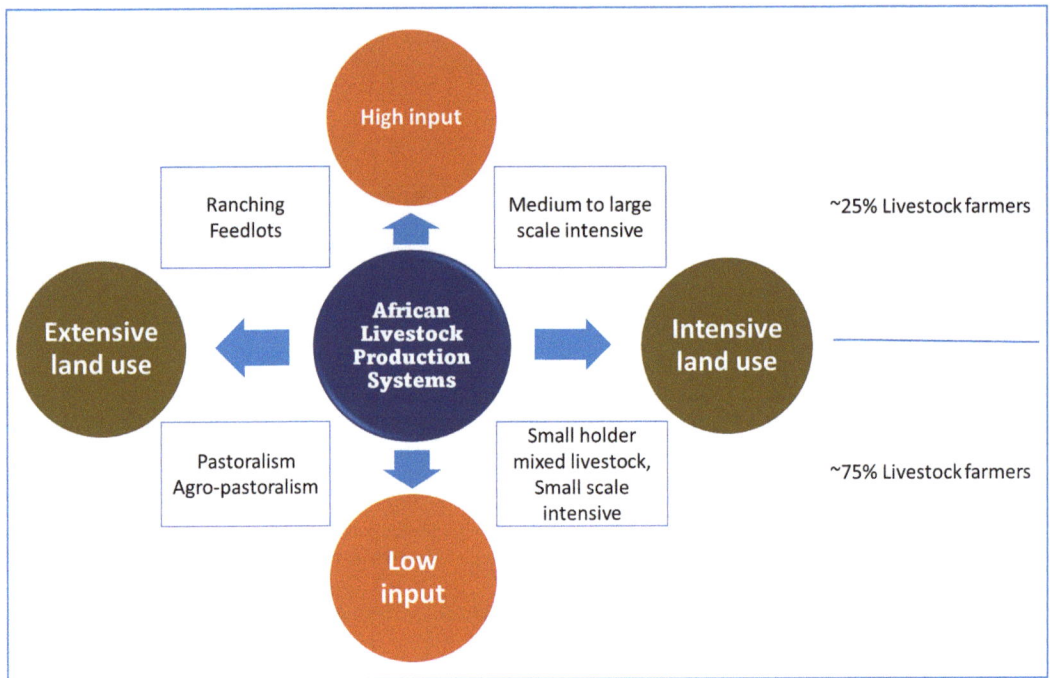

Fig. 2.3 Progression of African main livestock production systems according to land availability and level of inputs

and (3) landless systems. For instance, most ruminants (cattle, sheep and goat) in Africa are reared in production systems that vary from very-low-input, large-scale, extensive pastoral production systems to high-input, small-scale, intensive production systems.

Depending on the agro-ecological zone, different forms of these systems from small-scale (pastoral, agro-pastoral, mixed smallholder farming, free-range and urban/peri-urban) to large-scale (pastoral, ranching, large scale commercial intensive and extensive farming, cooperative farming and state-owned farms) are practised and have various levels of inputs. Subsequent discussion on the various livestock production systems will focus on these major categories.

The main characteristics of African livestock production systems are summarised in Table 2.3.

2.3.1 Small-Scale Livestock Production Systems

These are extensive low-input livestock production systems, which may be broadly categorised as small-to-medium-scale extensive pastoral/agro-pastoral and transhumance systems, mixed smallholder, free-range and urban/peri-urban livestock production systems characterised by low inputs and limited number of animals.

2.3.1.1 Pastoral and Agro-pastoral Systems

Pastoralism, coined from the word pastor, or herder, generally signifies herdsmen, livestock in motion and rangelands. Thus, pastoralism is generally regarded as a finely interwoven interdependent relationship between livestock, local environment or ecology and pastoralists in resource-limited, marginal (climate) and highly variable environmental (temperature, rainfall, pasture, etc.) conditions (Fig. 2.4). Pastoralism represents a multifaceted form of natural resource management that involves the interaction between the natural resource system, the resource users system and the geopolitical system in place (Pratt et al. 1997). The pastoralism livestock system is practised in different forms and scales (Box 2.1), which could be at a small scale (up to 100 heads) or a large scale (more than 100 heads).

Table 2.3 Key features of African livestock production systems[a]

Major category	Production system	Primary livestock types	Source of labour	Characteristics	Input	Output	Utilisation of output	Constraints
Small scale	Pastoral (sedentary)	Cattle, sheep, goat, camel, donkey	Family unit	Few to hundred animals, high mortality	Open communal grazing land shared depending on the herd size, manure	Low production of meat, milk, hide and skin, draft power	Primarily for own consumption and sale	Recurrent drought, conflict, weak governance, disease, lack of veterinary services and feed resources, loss of grazing lands, loss of labour to urban migration
	Transhumance pastoralism	Cattle, sheep, goat, camel, donkey	Family unit	Few to hundred animals, high mortality	Move animals to different pasture lands according to the season	Low production of meat, milk, hide and skin	Primarily for own consumption and sale	Recurrent drought, conflict, weak governance, disease, lack of veterinary services and feed resources, loss of labour to urban migration
	Nomadic pastoralism	Cattle, sheep, goat, camel, donkey	Community	Large community herds	People and livestock settle where pasture is found and migrate to new pastures when necessary	Low production of meat, milk, hide and skin	Primarily for own consumption and sale	Recurrent drought, conflict, weak governance, disease, lack of veterinary services and feed resources

(continued)

Table 2.3 (continued)

Major category	Production system	Primary livestock types	Source of labour	Characteristics	Input	Output	Utilisation of output	Constraints
	Agro-pastoral	Cattle, sheep, goat, camel, poultry	Family members	Settled with few to hundreds of animals but migrate for survival, involved in crop cultivation	Open communal grazing land, crop by-products, manure, labour and seeds	Low production of meat, milk, hide and skin, draft power, crops	Mainly for home consumption, with a small portion sold or battered	Disease, drought, migration of labour to urban areas for jobs, conflict, weak governance, lack of veterinary services and feed resources
	Mixed smallholder farming	Sheep, goat, pig, poultry, cattle	Ranges from family unit to hired labour, depending on the herd size	Small herd size, proximity to markets in urban cities, specialisation in crop and animal farming	Crop by-products, household wastes and forage, limited access to inputs such as commercial feeds, manure, fertiliser, improved seeds and irrigation	Low production of meat, milk, hide and skin, draft power and crops	Home consumption and sale	Disease, drought, loss of labour to urban migration, conflict, weak governance, limited access to veterinary services and feed resources
	Urban/peri-urban	Sheep, goat, pig, poultry, rabbit, cattle	Mainly family unit and hired labour	Small herd size, housed in backyard enclosures	Crop by-products, household waste, commercial feed and pasture	Low production of meat, milk and eggs	Mainly for sale with a limited quantity retained for home consumption and sale	Disease, high cost of drugs and feed and limited water resources and housing

	System	Animals	Labour	Characteristics	Land/Inputs	Products	Purpose	Challenges
Large scale	Pastoral	Cattle	Family unit and hired labour	Large herds of animals (>100), high mobility	Open communal grazing land	High animal numbers, low production of meat, milk, hide and skin	For commercial purpose	Drought, diseases, limited access to veterinary services and land, conflict
	Ranching	Cattle, sheep, goat	Mainly hired herders, depending on the herd size and supported by family members in some instances	Large herd size, high input through fencing, pasture management, rotational grazing, health management, infrastructure	Large rangeland, veterinary services, drugs, vaccines, supplemental feeding, especially during the dry season	Meat, milk and hide	For commercial purpose	Diseases, limited access to veterinary services and land
	Large-scale intensive commercial farming	Cattle, poultry, pig	Hired labour	Access to amenities, infrastructure, organised structure, access to local and export markets	Veterinary services, drugs, vaccines, feed mill, modern equipment	Meat, milk, hide and eggs	For commercial purpose	Diseases and instability in government policies
	Cooperative farming (community breeding)	Cattle, sheep, goat, poultry	Farm families and also hired labour	Limited access to amenities, infrastructure, local and export markets Weak organisational structure	Veterinary services, drugs, vaccines, feeds, few equipment	Meat, milk, hide and eggs	For commercial purposes	Diseases, unstable government policies
	State-owned farms	Cattle, sheep, goat, poultry	Civil servants employed by the state	Financed by government, access to local and export markets	Veterinarians and other livestock professionals, modern equipment and access to government facilities	Meat, milk, hide and poultry	For commercial purpose	Diseases, unstable government policies and bureaucracy

aModified from Ibeagha-Awemu et al. (2019)

Fig. 2.4 Pastoralism in West Africa, specifically in the Bamenda Grasslands of Cameroon (**a**) and the Sahel region of Burkina Faso (**b**). (Photos by Dr. Eveline M. Ibeagha-Awemu)

In Africa, pastoralism is an activity practised by over 22 million people, who belong to a few ethnic groups such as the Bedouins, Berbers, Boran, Fulani, Maasai, Mbororo, Somali and Turkana tribes. Several forms of pastoralism livestock systems are found in Africa (Box 2.1).

The pastoral/agro-pastoral system spans about 20 countries and extends over 4.8 M km² of land, including 0.68 M km² of cropland, in semi-arid regions of North Africa, the Sudano-Sahelian belt stretching across West and Central Africa and East and Southern Africa (Nidumolu et al. 2022). These areas are home to about 99 million people (in 2015) with a mean livestock holding of four tropical livestock units (TLU) and mean harvested area of 3.8 ha per farm (Dixon et al. 2019). Large-scale extensive pastoral and transhumance systems are found mainly in arid and semi-arid areas of the continent where the potential for crop farming is low. Based on the degree of aridity, variations could exist in terms of the amount of rainfall, growing period, type of pastoralism, migration pattern and the types of species reared.

Box 2.1 Forms of pastoralism

Pastoralism: This is the practice of grazing livestock in natural open pastures with little or no further inputs. The main livestock are cattle, sheep and goat.

Transhumance: This is a form of pastoral livestock management system involving the herding of animals to different geographical grazing locations (could be near or distant) over a period, typically determined by seasonal variations.

Nomadic pastoralism: Nomads are communities of people with no fixed settlement, and those who practise pastoralism are referred to as nomadic pastoralists. Nomadic pastoralism is the movement of a community and their livestock from place to place in search of pasture. Animals and nomads will settle where grazing has been found until the location is no longer sustainable, prompting migration.

Agro-pastoralism: Farmers grow crops in addition to keeping livestock under a pastoral system of management.

Silvo-pastoralism: It is a system that integrates livestock (pastoralism), trees and forage production at the same location.

Arid pastoralism: This is the practice of pastoralism around scattered oases in sparsely settled arid areas of Africa. The main livestock are cattle, sheep, goat and camel, and the main crops are date palm, irrigated crops and vegetables.

Characteristics Pastoral livestock system of management is based on grazing natural vegetation, which is often of low nutritional quality, and the availability/distribution depends on the variability and intensity of annual precipitation. Pastoral management is characterised by the adaptation of the feed requirements of the animals to the environment through constant movement in search of grazing or migration. Access to feed resources (grazing areas, crop residues, etc.) and water for livestock in pastoral systems is of critical importance in the arid and semi-arid regions of Africa due to highly variable rainfall amounts, frequency and duration (see Table 2.2). Pastoralists are flexible in the manner of their operations, which gives rise to multiple systems including nomadic, transhumance, agro and arid systems. The nomadic system is very flexible, involving regular seasonal migration of livestock in search of pasture. Nomads do not have a base or home, and wherever pasture is available for their animals is considered home until the pasture is no longer sustainable or available. Generally, there is little variation in daily grazing movements of herds around the base locations (camps, villages and water points) with most radii of dispersion below 6 km but wide variation in the seasonal movements between base locations (Turner and Schlecht 2019). The magnitude of mobility parameters is the highest for transhumance systems moving along latitudinal and elevation gradients. Transhumance, on the other hand, is the regular movement of livestock to specific locations to exploit the seasonal availability of pastures and water. In West Africa for example, animals are moved from the more arid northern areas during the dry season to the humid and sub-humid southern regions in search of pasture and water. Transhumance is beginning to evolve in response to changing cultural and social norms and environmental and social challenges, with constant or more frequent herd mobility to facilitate livestock's year-round access to pastoral resources. Moreover, transhumance is increasingly practised by persons not of pastoral-associated tribes, and there is increasing marked expansion of crop cultivation among traditional pastoralists. Such ongoing changes in pastoral systems may compromise the resilience and sustainability of farming systems (Houessou et al. 2019).

In the agro-pastoral system, farmers grow crops in addition to keeping livestock. The pastoral and agro-pastoral systems are based on the use of natural or semi-natural vegetation. Local practices and knowledge are of great value in these systems to effectively manage a diversity of breeds and their roles and the feeding environment and resources. For example, in the transhumance system, herders move their animals from dry areas during the dry seasons to areas with green pasture for the survival of the animals, and knowledge of the routes and areas with adequate pasture and water is necessary. About 60% of the land area of SSA is covered by rangelands with predominantly native (uncultivated) grasses, grass-like plants, shrubs, woodlands and savannahs suitable for grazing or browsing by domestic and wild animals. The SSA rangelands support a population of more than 100 million pastoralists (Notenbaert et al. 2009). The agro-pastoral areas cover about 19% of the land in SSA, are home to almost 10% of the population and house more than 15 million cattle. One-third of the pastoralists is estimated to be under an agro-pastoral management system involved with livestock and crop production.

Main livestock and products The main livestock species raised under pastoral systems of management are cattle, goat, sheep, camel and donkey. The main products are meat, milk and hide/wool. Although the main reason for keeping livestock under pastoral systems is subsistence, social and cultural functions are also important.

Challenges Challenges related to changes in land use, climate change, fodder availability/quality, grazing reserves, water shortages, disease outbreaks and conflicts with settled farmers are common in pastoral systems. The perennial movement of transhumant cattle from Sahel regions to the coastal regions of West Africa, in particular, in recent times, often leads to conflicts between the livestock herders and resident crop farmers as a result of the shrinkage of land for feeding and watering, lack of well-defined grazing corridors and a general failure of countries to adhere to the Economic Communities of West African States (ECOWAS) protocol on transhumance. These challenges have negative implications on the implementation of strategies for sustainable livestock genetic improvement, being compounded by small herd sizes without proper identification, variability between farms and uncontrolled breeding. Most African government policies towards solving the problems associated with pastoralism are often directed towards resettlement rather than actively addressing the issues directly. Pastoralists are, however, adopting new practices that enhance the productivity of their animals. For example, communication technology such as mobile phones is providing greater access to markets and meteorological information, thereby adding further benefits. With the use of mobile phones, the market price of livestock (cattle, goat, sheep or camel) is easily available, and combined with weather information (forecast), pastoralists/agro-pastoralists are able to be in control of the time to sell their stock or move to specific grazing lands.

To solve the problems associated with overgrazing of rangelands, climate change and pressure on transhumance routes, agro-silvo-pastoral systems of livestock management are gaining importance in Africa. In this system, livestock benefit from grazing pasture and woody perennials and the shade provided by the trees, while the trees benefit from soil fertilisation by animal waste and the improvement of biodiversity.

2.3.1.2 Smallholder Mixed Crop–Livestock Systems

This system constitutes the backbone of animal agriculture in Africa, within which about 70% of livestock is produced. The mixed crop–livestock system includes smallholder extensive subsistence and intensive commercial systems and medium-scale intensive commercial systems. The term 'smallholder' is mainly used to describe farmers or producers with limited resources (resource-poor or subsistence farmers) such as land, labour, capital and feed inputs. In this system, at least 15% of the land is used for crop cultivation, and over 10% of animal feed (dry matter) is derived from crop by-products or stubble and grazing. Based on rainfall, this system is further classified into mixed rain-fed and irrigated systems. The rain-fed system is dominant, whereby animal production is dependent on rainfall to grow the feed/grazing and water the animals. The major species kept within this system are cattle, buffalo, sheep, goat, poultry and pig, while the major crops cultivated are maize, millet, rice, sorghum, pigeon pea, groundnut, soybean, banana, cocoa, coffee, cotton, palm, casava and vegetable crops. This system is common in the tropical highland, semi-arid, humid and sub-humid regions. In mixed irrigated systems, about 10% of the arable land is irrigated for crop production and for watering animals. The mixed irrigated system is practised in semi-arid and arid regions where rainfall is a major limitation for crop cultivation; therefore, irrigation ensures year-round crop production for use in animal feeding and watering. Irrigation, therefore, supports increased fodder production/availability, which lessens feed deficit and makes the intensification and commercialisation of livestock production in urban and peri-urban areas feasible (Dixon et al. 2001; Thornton et al. 2002).

The diversity in mixed systems enables positive synergies between crops and livestock, such as the recycling of animal waste and crop residues, and a multifunctional use of livestock (Fig. 2.5). Fifteen crop–livestock farming systems (Table 2.4) identified in SSA by Dixon et al. (2001) were based on several factors, including (1) the availability of natural resources, which include water, land, grazing area and forest; (2) climate, of which altitude is an important determinant; (3) landscape including slope; (4) farm size and tenure in relation to access to different resources; (5) dominant farm activities and household livelihood patterns (e.g. crops, livestock, trees, aquaculture, hunting and gathering and off-farm activities); (6) technologies and the resulting intensity of production and integration of crops and livestock; and (7) farm management and organisation (e.g. family, corporate, cooperative, etc.). Rain-fed crop production combined with livestock production occupies about 20% of

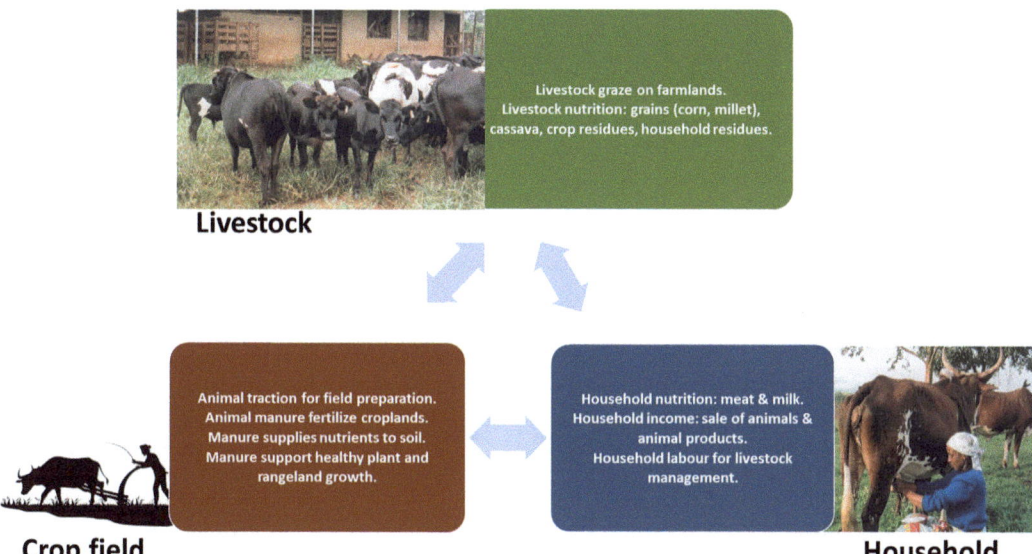

Fig. 2.5 Positive synergies between crops, livestock and households in mixed crop–livestock systems

the land area in SSA (Notenbaert et al. 2009). The cropping systems where cereals dominate occupy most of this area (12%), followed by tree crops (3%) and legumes (2%). Large-scale irrigation is rare in SSA, with only 0.1% of the land area under such management supporting a fraction of the human as well as animal population.

Smallholder Extensive Mixed Crop–Livestock Production Systems

Characteristics The smallholder mixed livestock production systems are extensive subsistence systems with the main production objective of meeting the needs of the farm family with no immediate objective of making a profit out of the sale of animals and animal products. Little or no investment is put into the operations (no hired labour), but it depends on family or close relatives for animal management. There is little or no formal record-keeping, although in some cases, farmers have their own unique ways of identifying individual animals and determining their productivity and market value. For ruminant production, the free-range system is mostly practised with farmers relying on communal grazing lands to feed their stock. Mixed farming is a common feature with two or more livestock species or breeds kept. Animals are fed on various agricultural by-products. The availability and the nutritional plane of agro-industrial by-products and communal pastures vary depending on the season, and this has a bearing on the growth and productivity of livestock. Under this system, pig feeding often relies on household waste and leftovers from restaurants and other food service providers.

Farmers in this system also grow crops, often sharing their attention between crop production and livestock management. For instance, smallholder cattle production in West Africa is extensive and dominated by indigenous cattle breeds with the ability to adapt to the local environment, such as Somba, Lagune, Muturu, N'Dama, Kuri, Gudali, White Fulani, Red Fulani, Ghana Sanga and Borgou, among others. They also have adaptive traits including but not limited to tolerance to ticks and tick-borne diseases, heat and diseases such as trypanosomiasis (Adido and Gicheha 2018). Animals are housed in backyard fencing made from local materials or housed indoors, with little or no investments in nutrition and healthcare. In this system, poultry and pigs are also raised under free-range small-extensive/ extensive scavenging systems with 1 to 50 animals per family. Poultry and pig feed by scaveng-

Table 2.4 Major mixed crop–livestock farming systems of sub-Saharan Africa described by Dixon et al. (2001)

Farming systems	Land area (% of region)	Agriculture population (% of region)	Principal livelihoods
Irrigated	1	2	Rice, cotton, vegetables, rain-fed crops, cattle, poultry
Tree crop/crop	3	6	Cocoa, coffee, oil palm, rubber, yams, maize, off-farm work
Forest-based	11	7	Cassava, maize, beans, cocoyam
Rice-tree crop	1	2	Rice, banana, coffee, maize, cassava, legumes, livestock, off-farm work
Highland perennial	1	8	Banana, plantain, coffee, cassava, sweet potato, beans, cereals, livestock, poultry, off-farm work
Highland temperate mixed	2	7	Wheat barley, teff, peas, lentils, broad beans, potatoes, sheep, goat, livestock, poultry, off-farm work
Root crop	11	11	Yam, cassava, legumes, off-farm work
Cereal-root crop mixed	13	15	Maize, sorghum, millet, cassava, yam, legumes, cattle
Maize mixed	10	15	Maize, tobacco, cotton, cattle, goat, poultry, off-farm work
Large commercial and smallholder	5	4	Maize, pulses, sunflower, cattle, sheep, goat
Agro-pastoral millet/sorghum	8	8	Sorghum, pearl millet, pulses, sesame, cattle, sheep, goat, poultry, off-farm work
Pastoral	14	7	Cattle, camel, sheep, goat, remittances
Sparse (arid)	17	1	Irrigated maize, vegetables, date palm, cattle, off-farm work
Coastal artisanal fishing	2	3	Marine fish, coconut, cashew, banana, yam, fruit, goat, poultry, off-farm work
Urban-based	<1	3	Fruits, vegetables, dairy, cattle, goat, poultry, off-farm work

ing, with little or occasional supplementation with kitchen waste. During the cropping season, pigs are tethered to protect the crops and farmlands from destruction.

Main Species and Products Most indigenous chicken, pig, cattle, sheep and goat breeds are raised in this system. The main products are live animals, milk, meat and eggs.

Challenges The main challenges faced by this system include high mortality rates due to the lack of veterinary attention and medications. Farmers in this system also have limited knowledge of modern animal-farming methods. Consequently, animal productivity is low in the smallholder extensive mixed crop–livestock production system.

Free-Range Livestock Production System

Characteristics Animals roam freely and scavenge for food and water. Supplementary feeding with kitchen waste is offered when it is available. Some form of backyard housing is provided for shelter during the night. No form of disease management is practised in this system. Indigenous chickens are mostly raised under this system, and pig is raised to a lesser extent. Consumer perception is that free ranging provides greater opportunities for livestock to express their natural behaviours, resulting in improve welfare (Campbell et al. 2021). Moreover, free ranging in chicken has been linked to health benefits such as improved plumage condition, reduced pecking and reduced diseases such as footpad dermatitis (Campbell et al. 2021).

Main species and products Livestock species commonly raised under the free-range system are poultry and pig. The main products are live animals, meat and eggs.

Challenges Health risks associated with free-range systems include greater susceptibility to diseases and parasites. Animals are exposed to predation, rainfall, heat stress and other environmental challenges. Productivity in this system is low.

Smallholder Commercial Intensive/Semi-intensive Livestock Production Systems

Characteristics This system is labour-intensive and often uses family labour. Smallholder intensive livestock farming is combined with low-input rain-fed crop production. This system can be found in peri-urban areas and marginalised or isolated areas. Requirements for this system include diverse inputs and labour. Inputs are sourced from local/natural resources such as crops and industrial by-products, grains from own farms and kitchen waste. Figure 2.6 shows small, intensively managed backyard pig and poultry farms. Production is for commercial purpose (local markets), home consumption and non-monetary exchange (e.g. as payment for school fees). Animal waste (manure) is used as the input for crop production.

Producers usually invest in improved management practices such as housing, animal healthcare, improved feeding and the use of improved breeds such as exotic as well as exotic × local crossbreds (chicken, pig and dairy cattle). Herd

Fig. 2.6 Small intensively managed backyard pig farm (**a**) and poultry farm (**b**) in Bambili, Cameroon. (Photos by Azenui B. Abongban)

sizes could be up to 50 (pig) or 200 (poultry), and productivity is higher, and the products are sold in local markets.

Main species and products Cattle, goat, pig, poultry and sheep are raised in these systems and often involve less animal mobility. High-yielding exotics and their crossbreds with indigenous livestock species are also kept in this system. The main products include live animals, meat, milk, eggs, skin and hide.

Challenges Challenges of this system include limited access to information and technologies, poor infrastructure, high cost of inputs (feed, drugs, etc.) or lack of inputs, high disease burden, limited access to markets, limited marketing channels for products or the lack of appropriate marketing channels, limited attention by policymakers, limited research or extension education, environmental factors and climate change.

2.3.1.3 Urban and Peri-urban Livestock Production Systems or Landless Production Systems

The urban and peri-urban livestock systems are generally regarded as landless production systems on small pieces of land, mainly household backyards and yards not demarcated for the keeping of livestock.

In this system, about 10% of the dry matter fed to livestock is farm-produced or kitchen waste, and the average livestock density is 14.6 tropical livestock units (TLU)/km². This system is found in all agro-ecological zones where land is a constraint and livestock feeding depends on external feed resources. In West Africa, it is the most fragmented in terms of spread, covering an area of 52,642 square km (Fernández-Rivera et al. 2004). The system can be further classified as landless small subsistence and smallholder commercial intensive/semi-intensive peri-urban systems. In general, smallholder landless producers keep animals for multiple purposes in backyard settings, while the commercial intensive/semi-intensive peri-urban systems are demand-driven and specialise in meat, egg or milk production.

Peri-urban poultry and pig production are commercially oriented and intensively managed to ensure rapid growth to meet the demands of urban dwellers. Stocking rates could reach 10,000 birds with the provision of adequate housing, feeding, healthcare and waste disposal systems. Figure 2.7 shows a medium-scale, commercially and intensively managed poultry farm in Nkwen, Cameroon. Dairy cattle, goat, pig and rabbit (mainly exotics and crossbreds) are kept under intensive (stall-feeding) or semi-intensive (backyard/tethering) systems, either as livestock-only or with small-scale crop production. In particular, peri-urban dairying (cattle and goat) is growing fast in some countries (e.g. Kenya and Ethiopia) due to rising demands in urban centres. Stall-feeding is also practised, mainly for the festival fattening of sheep and poultry, and in some major livestock markets to recondition stock before reselling them, usually at the same market. According to Fernández-Rivera et al. (Fernández-Rivera et al. 2004), stall feeding is a common and long-standing tradition in West Africa, probably rooted in the fattening of sheep, specifically for the Tabaski Muslim festival. The boundaries of the stall-fed urban and peri-urban livestock systems are defined by sites in the region with a human population density greater than 450 persons/km². In urban centres, crop and livestock production are spatially separated such that all feed-crop residues, cut-and-carry fodder and other supplementary feeds are imported from outside the stall, thereby limiting the benefit of the producer from the usual crop–livestock interaction. Livestock ration depends on the surrounding crop–livestock system; for instance, a peri-urban system located in Niamey (Niger) will likely depend on feeds from pearl millet–cowpea crops, while in Dakar (Senegal), it will depend on feeds from rice and groundnut crops and in Nigeria, on feeds from corn, cassava, cowpea and groundnut haulms. With the current trends in urbanisation, this system is expected to grow in importance and will encourage its practitioners to grow their own feed to ensure a reliable source of nutrition for their livestock.

Fig. 2.7 A peri-urban medium-scale, commercial, intensively managed poultry farm in Nkwen, Cameroon. (Photo by Azenui B. Abongban)

Main Species and Products The common livestock species for these systems are pig and poultry, though cattle and other large stock can be kept. The main products are live animals, meat, milk and skin/hide.

Challenges Major constraints of the urban and peri-urban production systems border on the high cost of inputs, health management, waste disposal and issues of sanitation, particularly in intensive farms near populated cities, with implications on public and environmental health, especially with the expansion of the more market-oriented aspects of the system.

2.3.2 Large-Scale Livestock Production Systems

Large-scale systems include extensive pastoral systems (discussed above), commercial ranching and intensive systems. The commercial ranching and intensive systems are capital-intensive in nature and typically linked with highly controlled environments where land and labour inputs have been maximised through intensification and mechanisation with heavy capital investment. These systems are characterised by high labour inputs in terms of housing, machinery, feeding, genetics, healthcare and labour. Chicken, pig and, to a small extent, dairy animals (cattle and goat) are popular in these systems in Africa, and the main goal of production is cash income. Intensive commercial livestock production systems are mostly located close to urban centres.

2.3.2.1 Large-Scale Extensive Commercial Ranching Livestock System

Large-scale extensive commercial system or ranching represents a relatively small but critical proportion of the production systems on the

African continent (AU-IBAR 2019). Ranching is considered an improvement or modernisation of the pastoral/agro-pastoral systems. In the ranching system, land is owned by individuals, commercially organised groups or government-run institutions. In the southern region of the continent and in some countries of East Africa, robust programmes for selective breeding and improvement of indigenous breeds are managed within these systems. This has greatly contributed to the production of specialised indigenous livestock breeds such as the Boran and Tuli cattle for beef production in East and Southern Africa, local indigenous pig breeds like the Ashanti dwarf pig for pork production in Central and West Africa, the Boer goat for chevron production in Southern Africa, Barki and Beni Guil sheep for mutton in North Africa and chevron production from West African Dwarf goat in West and Central Africa.

Ranching is a labour-extensive undertaking, specialising in the production from one or two livestock species of a marketable commodity, mainly live animals for slaughter for meat, skin and hide but also wool and milk. The function of livestock is mainly to provide cash income. Livestock management is characterised by grazing within fixed boundaries by individual tenure with the intensification of possibilities for feeding and watering.

Fenced feed-lots are incorporated in most ranches to finish livestock to a desired market weight and meat quality faster. Feed-lots can be either dry lots where grain-based rations are used or more extensive, involving grass-fed growing operations. There is limited diversity of livestock within the ranching system, and animals are kept for specific purposes such as meat quality. The ranching system is characterised by high inputs in terms of pasture management, health management, crop production for supplemental feeding and a high level of mechanisation. The forms of ownership may be parastatal, cooperative or private (companies or individuals), and instead of straightforward ownership, there may be lease arrangements. Individual tenure means tenure by the individual ranch system where many management units share the tenure of the land. In present-day Africa, ranching is mostly practised by research institutions (government and private), universities and, to a limited extent, the private sector.

Ranching is increasingly being adopted to address frequent farmer–pastoralist/herder conflicts in many African countries. Recently, Ghana adopted a national cattle ranching policy to establish ranches as solutions to farmer–herder conflicts in the southern region of the country due to transhumance by Fulani herdsmen from the North and Sahelian countries. However, complex land tenure arrangements, identity perceptions of Fulani herders (generally perceived as dangerous, murderous and violent), perceptions of land grab and dynamics of access to animal feed and water sources could make cattle ranching less feasible, thus calling for better understanding of the nature of conflicts and widening public consultation, which will support a ranching policy that is feasible (Ahmed and Kuusaana 2021).

Main Species and Products Livestock species reared in this system include cattle, sheep, goat, camel and buffalo. Locally adapted exotic animals, indigenous breeds and crosses between them are reared for commercial production of meat, milk, hide and skin in the ranching system.

Challenges The main constraints include limited land availability, high costs of inputs (feed, disease management and infrastructure) and changing climatic conditions.

2.3.2.2 Large-Scale Intensive Commercial Livestock Production Systems

These are capital-intensive systems characterised by highly modified environments where land and labour inputs have been replaced with adequate mechanisation and technology (FAO 2018a). In many African countries, poultry and pig are the main species raised in this system. Intensive dairying and beef production (ranching) are also common in some African countries (e.g. South Africa). Poultry farms of this system are large, and some are highly mechanised and extremely productive. With the supply of genetically improved poultry breeds by big companies (e.g. Aviagen, Hendrix Genetics, Hubbard, etc.) producing highly productive poultry strains, intensive commercial poultry production is growing rapidly in many African countries and, with this growth, is the emergence of poultry farmer associations (e.g. Poultry Association of Nigeria [PAN], Ghana National Association of Poultry Farmers, Kenya Poultry Farmers Association, etc.) to support the interest of farmers and find markets for products. For example, membership categories of the Poultry Association of Nigeria include micro-farmers (<2000 birds' capacity), small farmers (2000–10,000 birds), medium farmers (10,000–50,000), large farmers (50,000–200,000) and mega-farmers (>200,000 birds), which testifies to the scale of operations. PAN helps its members maintain a sustainable poultry sector through services like (1) education (poultry education, conducting of and/or assistance in investigative work of a practical and scientific nature and organisation of seminars and courses); (2) advance improvement of poultry production by embracing and coordinating production objectives (egg, broiler, chicks, etc.); (3) coordination of the aims, views and efforts of poultry operators; and (4) protection of the poultry industry from adverse policies and legislation and assistance in obtaining beneficial legislative policies and regulation, among others. Through the activities of the union, farmers are better informed on management practices, housing, healthcare, marketing and acquisition of inputs. Thus, intensive poultry and pig production, which have been the mainstay in high-income countries, are becoming more prevalent in African countries, where they are supplying much-needed animal proteins to feed fast-growing urban populations. These intensive poultry and pig farms are highly efficient as well as extremely productive (Fig. 2.8). For example, Obasanjo Farms (Obasanjo Farms, Ota, Ogun state, Nigeria) operates, among others, a large-scale, intensively managed poultry farm specialising in producing both egg and broiler. Obasanjo Farms is a big player in the Nigerian poultry industry, with a capacity of about 1.3 million meat chickens and 1.26 million laying hens with the ability to produce 20,000 creates of eggs per day. Obasanjo Farms is highly mechanised in its operations, produces its own feed, and processes its broiler birds. Its broiler-processing plant processes up to 2500 chickens per hour. In addition to poultry production, Obasanjo Farms also engages in pig, rabbit, snail, fish and dairy (cattle) production.

Main Livestock and Products The main livestock raised in this system are poultry and pig. The main products are meat, egg and manure.

Challenges Intensive livestock production systems are faced with multifaceted challenges such as the high cost of inputs, limited access to high quality genetics, high cost of drugs and veterinary care, high cost of feeds, unstable electricity and water supply, unstable marketing channels, emergence of antimicrobial resistance due to high use of antimicrobials, high cost of production and environmental impacts through deforestation to set up large farm buildings, disruption of nutrient cycles and pollution with pesticides.

2.3.3 Livestock System Production Highlights of Some African Countries

Within countries, regions and the continent, there is great diversity in the livestock production systems presented above in terms of the level of development and dominance of the different livestock sub-sectors, such as dairying, pastoralism and wool and meat production. For example, pastoralism is mainly practised by certain tribes in East, West and Central Africa. While smallholder dairy production is growing in importance in East Africa, intensive dairy production is common in Southern Africa. In West Africa, smallholder poultry and ruminant production systems are commonplace, and small-intensive to large-scale, intensive poultry and pig production systems are becoming popular. Differences in the production systems for the various species are shaped by the inherent specific geographic and environmental

Fig. 2.8 Large-scale, intensively managed poultry operations: (**a**) broiler, (**b**) layer, (**c**) meat processing and (**d**) chick units at commercial farms in Nigeria

challenges and, to some extent, religious perceptions. For example, pig production is not practised in Northern Africa (except Egypt) for religious reasons. To provide more clarity, the country-specific production systems for poultry (Nigeria), cattle (Kenya) and buffalo (Egypt) are highlighted below.

2.3.3.1 Chicken Production System Highlights of Nigeria

The Nigerian poultry industry is the second largest (after South Africa) in Africa, with over 180 million birds producing about 454 billion tonnes and 3.8 million tonnes of meat and eggs, respectively, with about 80, 60 and 40 million birds raised in extensive, semi-intensive and intensive systems, respectively (Mcdougal 2019). In 2021, its egg production of 14.43 billion eggs was the highest in Africa, followed by South Africa (10.36 billion eggs) and Egypt (10.32 billion eggs) (FAOSTATS 2022). Poultry meat and eggs are unrestricted by any religion or culture in Nigeria, making them the most consumed animal proteins in the country. The poultry sector is the most commercialised livestock sector in Nigeria, accounting for ~8% of Nigeria's gross domestic product (GDP) and ~30% of the total agriculture contribution. It provides about 14 million jobs and accounts for about 58.2% of Nigeria's animal production (Ogunyemi and Orowole 2020; Heise et al. 2015; FAO 2020b).

Three major poultry production systems are practised in Nigeria and include the extensive (free-range), semi-intensive and intensive systems. In 2017, 83.4 M, 58.4 M and 38.4 M birds were found within the extensive, semi-intensive and intensive systems of production, respectively.

(a) *Free-range (extensive) system*: Nearly 50% of chicken production in Nigeria takes place under the free-range or backyard subsistence-oriented system. Birds kept are often of different indigenous breeds and varying ages. Birds are left to scavenge for food and water. Kitchen leftovers and agricultural by-products are provided when available. At night, birds roost outside in trees or nest in bushes close to the farmer's dwelling. In some cases, some form of housing in the form of wooden cages may be provided in the backyards. There is little or no attention in terms of disease management. Productivity is low and characterised by low body weights, longer time to reach maturity and low egg productivity of about 50–65 eggs/hen/year. The average flock size is about 50 birds. Due to the limited or no provision for housing, free-range chickens are exposed to predators, climate and weather stress and high disease burden, with Newcastle disease being the most common disease.

(b) *Semi-intensive system*: The semi-intensive system is practised in family-owned farms, and the main production objective is market-oriented. About one-third of the total chicken population in Nigeria is raised under this system. Most semi-intensive chicken farms are located in peri-urban and urban centres for easy access to markets and inputs. The average flock composition includes indigenous and improved breeds, and the herd size ranges from 50 to about 2000 birds. Management includes the provision of housing, feed, water and healthcare and compliance with basic biosecurity measures. Products (live birds and eggs) are marketed through informal channels, mostly to urban dwellers. Productivity levels are low to medium. The main challenges faced are limited access to feed and healthcare due to low purchasing power, low productivity, disease and environmental perturbations.

(c) *Intensive (commercial/integrated) system*: The intensive poultry sector in Nigeria is commercially oriented, having its own great-grandparent and grandparent hatcheries, layers and broiler farms. About 21% of chickens in Nigeria are raised in commercial and/or integrated farms. The sector has a number of fully, vertically integrated commercial operators, which have adequate capital and technical capacity. Optimal housing, feeding and health management practices ensure high productivity of the system. High biosecurity measures are practised, and productivity levels are high. The Nigerian intensive poultry production landscape consists of three typologies of integrated poultry farms: (1) commercial peri-urban enterprises with flock sizes of between 2000 and 5000 birds, (2) clustered commercial enterprises with flock sizes of over 5000 birds and located in peri-urban areas, and (3) large commercial poultry integrators, which own feed mills, hatcheries and processing facilities, with flock sizes of up to 100,000 birds or more. Some of the large poultry integrators are franchisees or have joint ventures with global poultry operators. Poultry are sold to consumers as dressed and processed chicken in large and small retail outlets. The main problems faced are the high costs of inputs and frequent disease episodes such as Gumboro and highly pathogenic avian influenza.

2.3.3.2 Cattle Production System Highlights of Kenya

Kenya, with a cattle population of 22.9 million heads in 2021 (FAOSTATS 2022), has vibrant dairy and beef cattle sectors. The beef industry is the largest contributor to Kenya's agricultural GDP at about 35%, followed by dairy production (Kosgey et al. 2011). In terms of value and employment, the cattle industry is a significant contributor to the Kenyan economy (Alarcon et al. 2017). The dairy sector in Kenya is the second largest in Africa (after South Africa). In 2021, Kenya produced over six million tonnes of milk, of which 77% was from cattle, 18% from camel and 4.3% from goat (FAOSTATS 2022). Milk is primarily produced by smallholder dairy farmers (about 1.8 M smallholder farmers constituting about 80% of dairy producers), who account for about 80% of the total milk production in the country (KMALF 2010).

Dairy Cattle Production Systems

Three dairy production systems have been identified in Kenya and include large- and small-scale intensive, semi-intensive or semi-grazing system and extensive controlled (10% of farms) and uncontrolled (5% of farms) dairy production systems (FAO 2018b).

Large- and Small-Scale Intensive Dairy Production Systems About 40% of farms practise small-/large-scale intensive dairying. This system is characterised by animal confinement with zero grazing, high levels of health and nutrition management and feed resource planning. The system is dominated by small-scale dairy farms (35%), which also grow crops and predominate in Mount Kenya and central Rift Valley regions. Small-scale intensive dairying is also common in urban and peri-urban centres in the sub-humid and humid regions of Kenya. Flock sizes range from one to 20 for small-scale operations to >20 for large scale operations (Nguluu et al. 2011; Staal et al. 2001). The main breeds used are Holstein, Jersey, Fleckvieh, Ayrshires and Guernsey, and the main product is milk.

Feed for animals is either purchased or grown by producers. Regular health management includes regular tick and internal parasite control and vaccination against diseases such as Rift Valley fever, East Coast fever, black quarter/blackleg, lumpy skin disease, foot and mouth disease and anthrax. Other diseases like brucellosis and mastitis are also common. The type of housing is basic or simple. Milk production is primarily for sale as well as for home consumption. While the large-scale farmers sell their milk to large cooperatives and processors, the small-scale producers sell to middlemen cooperatives (KMALF 2010; Staal et al. 2001). The main constraints faced by producers in this system are the high cost of feed ingredients, poor quality feed, high cost of veterinary drugs, inadequate veterinary services, unfavourable policies limiting peri-urban dairying and limited access to credit. Despite these challenges, intensive dairy production in Kenya has the potential for rapid growth fuelled by the increasing demand for milk and milk products.

Semi-intensive or Semi-grazing Dairy Production System This is the predominant dairy production system (45% of dairy farms) practised in areas of greater land availability in Kenya. In this system, herd management includes grazing animals freely or in paddocks during the day and stall-feeding at night and supplemental feeding with concentrates (grain millings or compounded dairy feeds) during milking (Muia et al. 2011; Nguluu et al. 2011). Thus, feeding is on native and improved pastures, post-harvest grazing and concentrates.

Unlike in the intensive system, animals are raised in mixed herds with other livestock species such as sheep, goat, chicken and donkey, as well as pig (Staal et al. 2001). Semi-intensive dairy farms are predominantly found in areas around Mount Kenya, the central and north Rift Valley, the coastal regions and the western and Nyanza areas where crop production is practised. The semi-grazing herds are small in size (about one to 20 animals), and the main species kept are exotic breeds like Holstein, Ayrshire, Jersey, Guernsey and zebu/Zenga breeds like Sahiwal, Boran and small East African zebu. Reproduction management is mostly by natural mating (70% of farms) because artificial insemination (AI) service is not easily available (Staal et al. 2001). Production is relatively low in this system, averaging ≤6 litres/cow/day or yearly production ranging from 1300 to 4575 kg, depending on the availability of feeds, the breed, the production system and the agro-ecological zone (Staal et al. 2001; Muia et al. 2011; Omore et al. 1999; Mugambi et al. 2015). The milk produced is mostly consumed in homes (~40% farmers) or sold in liquid (unprocessed) form through informal channels.

The main constraints of the semi-intensive dairy production sector include low quality and limited availability of feeds, limited reliable statistical information on milk market outlets, poor infrastructure, limited access to credit, limited technical know-how on husbandry practices, limited access to veterinary services and artificial insemination services and high cost of drugs (Odero-Waitituh 2017).

Extensive dairy production system This system is characterised by pasture-based management and is practised in areas with large farms where controlled grazing is possible and on marginal and communal grazing lands where grazing is not controlled. About 3% of farms with ~35% of the dairy cattle population are involved in extensive dairy production (Omore et al. 1999). Herd sizes range from ≤10 animals (uncontrolled grazing) to about 50 or more animals (controlled grazing).

Controlled grazing involves placing animals on natural and improved pastures in paddocks or strip grazing, as well as supplementation with fodder (of high quality), mineral licks and concentrates. On the other hand, uncontrolled grazing involves free grazing of animals on native or improved pastures with limited supplementation. Housing infrastructure (e.g. hay barns, dips, water troughs and crushes) is mostly provided in the controlled grazing system. Disease management is by vaccination, parasite control (mostly in the controlled grazing system) and the uncontrolled use of acaricides and dewormers, which may increase the development of drug resistance. There is a high prevalence of contagious bovine pleuropneumonia in this system, while tick-borne diseases are the major causes of death (Staal et al. 2001; Tambi et al. 2006). The milk production range of 4–11 l per cow per day is relatively low compared to the intensive system (Lanyasunya et al. 2006). The milk and dairy products from this system are considered to be of high quality (organic) due to the limited use of antimicrobials and are sold in niche markets and at good prices.

Beef Cattle Production Systems

Beef cattle production in Kenya is mainly concentrated in the arid and semi-arid lands (ASALs), which host 70% of the national herd and where beef production from pasture is the main economic activity and main source of household livelihoods (Kahi et al. 2006; MSDNKOAL 2012). Three beef production subsystems are recognised in Kenya: extensive grazing (pastoralism and ranching), semi-intensive grazing (agro-pastoralism) and intensive (feed-lot) systems.

Pastoralism-Extensive Beef Production System The pastoralism system (see Section 2.3.1.1) is a low-input, low-output system where animals depend entirely on grazing native pastures or communal grazing lands for their maintenance and productivity. Other forms of pastoralism like transhumance and nomadism are also practised (Kahi et al. 2006; Ouda et al. 2001). Extensive pastoralists are important beef producers, comprising 34% of beef producers but with the lowest livestock densities of 11 TLU/km^2 or an average herd size of 50 animals (FAO 2018b). The cattle breeds in this system are indigenous breeds (small East African zebu, Boran and Sahiwal) kept mostly in mixed herds with other livestock (sheep and goat). Feeding management is based on grazing on native pastures, which is affected by seasonality in rainfall. Therefore, animals are often moved from one place to another in search of pasture and water, resulting in conflicts over grazing rights and water resources as well as conflicts with large-scale ranch operators (Mwangi et al. 2020; Otieno et al. 2012). Disease management through vaccination is low due to the nature of pastoralism and the climatic and infrastructural conditions of the ASALs (Kimani et al. 2016). Brucellosis and tick infestations are of higher prevalence in pastoralism systems (Kadohira et al. 1997). The annual average beef production in the pastoral system is about 408,000 tonnes (about 70% is from the zebu cattle in the ASALs), and the average yield per animal is estimated at 120 kg (Behnke and Muthami 2011). Pastoralist animals are mainly sold live at animal markets (primary and secondary markets), mostly dominated by traders and middlemen. The beef from pastoralist cattle, which satisfies the bulk of domestic demand, is sold to consumers in urban markets (the most important being Nairobi) (Kahi et al. 2006).

The main challenges of pastoral livestock production are drought, erratic weather patterns affecting the supply of feed and water, wildlife attacks, livestock raids, livestock diseases, poor management of pastures, invasive plants, lack of/weak delivery of extension services, limited veterinary services, high cost of drugs and other inputs, poor market infrastructure, low prices and insecurity and limited prior access to market information (Mwangi et al. 2020; Otieno et al. 2012; Wanyoike et al. 2018; Makokha and Witwer 2013).

Ranching-Extensive Beef Production System
Ranching is not a major beef production system in Kenya, but it is commercially oriented and contributes to most of the beef exports of Kenya (FAO 2018b). Ranching is practised by 11% of farms, which are made up of large land areas and have large herd sizes (average of 150 animals). The majority of ranches are privately owned and are predominantly found in Laikipia and Taita Taveta counties. A few government ranches are spread across the sub-humid and semi-arid zones of Kenya. The main breeds raised are exotic (Hereford, Simmental, Charolaise and Angus), improved Boran and exotic × Zebu crosses (Kosgey et al. 2011; Mwangi et al. 2020; Otieno et al. 2012; Khalil and Sammour 2006). Ranching is fairly labour-intensive, using specialised professional and unskilled labour. Infrastructure for disease control, feeding and water storage are available at most ranches. Feed management includes grazing on natural and improved pastures with some feed supplementation. The main diseases are parasitic (worm and tick infestations), which are controlled through regular deworming and dipping, as well as brucellosis and tuberculosis, which are controlled with antibiotics. The beef from the ranches is sold at local niche and international markets. Most ranches (70%) have access to marketing information, and >75% of them sell live animals to abattoirs, mostly (53%) by regular contract (Otieno et al. 2012).

The main constraints of beef cattle ranching include high feed cost (especially during the dry season), drought, difficulties in targeting and reaching affluent markets, low and fluctuating livestock prices, recurrent conflicts with pastoralists in search of pasture and water, wildlife attacks, disease challenges due to livestock–wildlife interactions, lack of government support and the unavailability/high cost of livestock drugs (Mwangi et al. 2020; FAO 2018b).

Semi-intensive Grazing System or Agro-pastoralism
Agro-pastoralism (see Sect. 2.3.1.2) is a low-input, low-output system where producers keep livestock and also grow crops. The production system is subsistence-oriented and is mostly practised in the semi-arid areas of the country. Generally, agro-pastoralism is practised in more regions of Kenya (coastal, lower eastern, north and south Rift) and on more farms (54%) when compared with other systems. Animal densities are low ranging from 20 TLU/km^2 (lowlands) to 50 TLU/km^2 (highlands), with average herd sizes from ten to 12 animals, with the main animals being crossbreds (exotic × zebu crosses) and pure exotic breeds (Otieno et al. 2012; Mwangi et al. 2020). Animals are predominantly fed through grazing on communal lands or in paddocks (where agro-pastoralists have large amounts of land) and supplemented with crop residues. Disease management is achieved through public veterinary vaccination and deworming services. Occasional tick control and treatment of sick animals with drugs from both formal and informal outlets are practised. Major diseases of livestock in the system are parasitic diseases, brucellosis and tuberculosis (Fèvre et al. 2017). Housing are enclosures known as bomas and are provided at night. Livestock products are live animals (and milk) marketed through channels that comprise middlemen or livestock traders at local markets and with links to abattoirs (Otieno et al. 2012).

The main challenges faced by the agro-pastoral system include seasonal variations in pasture and water availability, which result in fluctuating productivity, risk of mycotoxin contamination of poorly stored crop by-products, high disease burden, high cost of veterinary drugs and limited veterinary services.

Intensive Feed-Lot Beef Production Systems Feed-lot (see Sect. 2.3.2.1) is a commercially oriented beef production system in which animals are kept for a limited period of time (~3 months) for fattening using highly nutritious diets. Only about 1% of farms are considered feed-lots and are operated by individual farmers, livestock traders or corporate companies. Feed-lot operations are both capital- and labour-intensive in terms of infrastructure, labour and inputs, including optimal biosecurity practices and veterinary service practices. The feed-lots are focused on fattening either dairy culls/dairy bull calves or beef breeds. While beef feed-lots could be located anywhere in the country (major commercial feed-lots are found in Nakuru County, Nivasha (keeping up to 3000 animals) and Nyeri County, Kieni (keeping ~500 animals)), dairy cull/bull calf feed-lots are mainly located in the central Rift Valley and Mount Kenya regions (FAO 2018b). Beef breed feed-lots are large and may reach several hundreds of Sahiwal, Boran, small East African zebu and zebu crosses, and specialised beef breeds (Charolaise, Angus and Frisian). Healthcare management includes regular vaccination and deworming and appropriate treatment of sick animals. The primary product is beef, which is sold through formal channels to niche national urban and export markets.

The main constraints are the lack of capital and investment capacities, which limit the expansion of feed-lot systems; animal concentration and issues with dung disposal, which may cause soil, water and air pollution.

2.3.3.3 Buffalo Production System Highlights of Egypt

The bovine sector in Egypt is composed of cattle and buffalo, and it is well-integrated with crop production since the country has limited natural pastures. Female bovines are used for milk production, while males and infertile females are fattened for meat. Bovine (cattle and buffalo) constitute about 23% of Egypt's total agricultural value of 73.5 billion EGP, of which 66% is from meat production and 34% is from milk production (FAO 2018c). The Egyptian buffalo (*Bubalus bubalis*), which belongs to the river type, plays vital roles in the Egyptian economy, serving as a source of food security, an income generator and a source of employment. The Egyptian buffalo supplies the local market with 44% and 39% of its milk and red meat needs, respectively (Fahim et al. 2018).

Egyptian buffaloes are mainly kept in the northern Delta (~57%) and mid-upper (about 43%) regions of Egypt and mostly under traditional mixed crop–livestock farming system (Galal and Elbeltagy 2006; Khalil and Sammour 2006). The main buffalo production systems in Egypt include smallholder mixed crop–livestock/extensive, semi-intensive and intensive production systems.

Smallholder Mixed Crop–Livestock or Extensive Production System This system is characterised by traditional management practices integrating buffalo (bovine) keeping with crop production. It is a low-input, low-output system with limited investments in terms of housing, feeding and health management. Herd sizes range from one to ten, and mainly indigenous breeds of buffalo and cattle are kept, constituting about 33% of the total cattle and buffalo population in Egypt (FAO 2018c). This system is recognised as a major component of rural household livelihood, serving as a source of income and protein in the diet. The key forage crop for feeding animals is Egyptian clover (*berseem*), in addition to corn leaves (*darawa*), hay and straw offered during the summer months. The milk production per animal is low and mostly for home consumption and to feed calves, with excess sold to milk collectors or processed into cheese and ghee and sold at local and urban markets. Another source of income in this system is the sale of surplus calves, unproductive females or cows and bulls, mostly as live animals. Animals in this category are also slaughtered during special occasions like weddings and religious festivals. The animal healthcare management service for this system is during government mass vaccination campaigns against foot and mouth disease, Rift Valley fever and lumpy skin disease.

The main constraints of the extensive production system are poor management and breeding practices, low productivity and profitability, low inputs and outputs, limited access to veterinary drugs/services and extension services, etc.

Semi-intensive Production System It accounts for about 60% of the bovine (cattle and buffalo) population, with herd sizes ranging from ten (cattle) to more than 50 (buffalo), and supplies about 70% of the total domestic milk production and the majority of the meat production (FAO 2018c; Abdel-Salam and Fahim 2018; Abdel-Salam and Fahim 2018). Animal management utilises some modern production and husbandry practices including grazing and the provision of agricultural by-products and animal health services rendered by private veterinary practitioners and regular veterinary services, with constrained access to governmental veterinary services. The main governmental veterinary services received are during mass vaccinations against Rift Valley fever, foot and mouth disease and lumpy skin disease. The milk produced is sold principally as liquid raw milk, while a small percentage is processed into homemade butter, cheese and yoghurt. The milk is principally for home consumption. The surplus milk is sold to milk collectors and distributors for supply to large cities. Some large milk and milk-processing factories also get their supply from this system.

The challenges of the system include varied production practices, a scattered and unorganised farmer community, limited veterinary and extension services, high cost of drugs, limited infrastructure, unregulated value chains, lack of clear marketing channels, limited farmer incentives and high disease burden.

Intensive Buffalo Production System The intensive production system is a high-input, high-output system, accounting for over 7% of the total beef and dairy cattle and buffalo population of the country. This system is made up of large commercial farms with herd sizes from ten to over 1000 animals and accounts for ~4% of the total buffalo and cattle populations, producing ~30% of the marketable milk in the country (FAO 2018c; Abdel-Salam and Fahim 2018). Animals are kept under intensive feeding regimes at commercial farms for the production of regular and/or high-fat milk and at experimental farms for educational (training) and research purposes. The intensive system produces about 84,000 tonnes of meat and 5 million tonnes of milk per year. About 90% of the milk produced is sold to milk processors or milk collection centres, while beef animals are sold to butchers in large cities or directly to slaughterhouses through formal channels. This system enjoys regular access to veterinary and extension services, including vaccinations against endemic and infectious diseases such as brucellosis and government vaccination programmes for Rift Valley disease, foot and mouth disease and lumpy skin disease. The main constraints of the system are the high cost of imported feed ingredients (grains, milling by-products, minerals, vitamins, fats/oils, etc.) and other nutritional supplements.

2.4 Challenges and Interventions for Livestock Development in Africa

Livestock production systems in Africa are faced with numerous challenges including inappropriate policies, poor or low input supply including extension and veterinary services, poor nutrition and limited sources of funding. Additionally, weak institutional and human resource capacities, low private sector participation and conservative farmers often unwilling to try innovations are major limiting factors to developing sustainable animal production practices in Africa. Table 2.5 summarises these major challenges and possible interventions for improved productivity.

2.4.1 Farm Management Practices, Farmer Perceptions and Low Productivity

The current African livestock production is practised to fit local conditions within varied management systems characterised by low/inferior inputs (including genetics) and low outputs and heavy disease burden. Also, most of the extensive and mixed farming systems for many years were not considered serious businesses but secondary activities to crop production handed down

Table 2.5 Major challenges limiting animal performance under current livestock production systems and possible solutions for the sustainable development of improved livestock productivity

Challenge	Possible intervention
Limited knowledge of appropriate production practices that increase animal performance, output and farm incomes	Training of animal owners under the various production systems and other stakeholders including livestock production experts, livestock processors, veterinarians, policy makers and other players along the animal resources value chain on improved management practices and exploitation of livestock for income generation
Poor management practices such as non-identification of animals, variability between farms, uncontrolled breeding, no application of intentional breeding of livestock and high mobility of pastoral flock	Training of livestock keepers on the importance of proper management practices, training on basic animal breeding practices, organising livestock production systems such as cooperative farming and community breeding, formation of national breed societies with regulatory agencies
High disease burden and threat of emerging diseases, limited access to drugs/high cost of drugs and limited veterinary services	Exploration of traditional methods for disease control, training of more veterinarians and provision for free or subsidised veterinary services and medication, increase in funding for research on livestock diseases, breeding for disease resistance and climate change adaptation, and training of farmers on improved management practices
Poor quality and limited availability of feed resources and grazing lands to support improved animal performance	Putting in place grazing land policies that favour access to pasture and pasture development, training of farmers on pasture management and training of animal nutritionists, animal feed producers and extension workers on effective delivery of services to farmers
Frequent clashes between pastoralists and crop farmers	Performance of appropriate consultations between concerned parties to put in place enabling and sustainable solutions to enable access to grazing lands by pastoralist farmers, and putting in place appropriate grazing land policies
Poor production capacity of local breeds under existing production systems and limited information on production characteristics of indigenous breeds	Training of farmers on improved management practices including how to monitor and measure animal performance, undertaking of systematic characterisation of animal productivity in the various production systems for informed decisions on the implementation of improved management strategies, establishment and support of sustained long-term breeding and improvement programmes, systematic characterisation of existing animal genetic resources and conservation through sustainable use and through in situ and ex situ conservation strategies
Lack of appropriate government policies that support gainful exploitation and protection of indigenous livestock	Putting in place government policies that support the gainful exploitation of indigenous livestock breeds; defining the roles of various stakeholders (e.g. NGOs, research, private sector, etc.); strengthening of the linkage between animal research, farmer challenges and extension services; putting in place appropriate incentives and structures for digital animal agriculture by national governments; farmer sensitisation on the use of appropriate technologies to improve farming practices and increase animal performance; resourcing of national agricultural extension systems; private sector involvement to deliver livestock extension services

(continued)

Table 2.5 (continued)

Challenge	Possible intervention
Limited attention and financial support of livestock production and marketing activities by governments	Need for governments to understand the contributions of livestock production to the livelihoods of the people and the national gross domestic product and input in place the necessary funding to support sustainable livestock production, markets and marketing channels; intentional funding of animal research and encouraging the private sector to support government initiatives; concerted efforts required to educate and change the orientation of national policymakers towards funding livestock-related research and increased funding by the government and the private sector
Limited private sector participation in promoting improved livestock production practices	Need for private sector involvement in promoting and funding improved livestock production practices for the benefit of all stakeholders
Lack of appropriate marketing channels/markets for livestock products	Educate farmers on the business aspects of livestock production, establishment of the necessary markets/marketing channels and infrastructure, establishment and support of farmers to run animal markets, modernisation of the livestock trade, putting in place measures to achieve global sanitary and phytosanitary standards, which will stimulate Africa's exports of livestock products
Inappropriate livestock policies, lack of necessary infrastructure for routine recording of production and health traits and limited research facilities, leading to inadequate performance and performance recording	Development of appropriate policies, institutional and legislative frameworks to modernise African livestock production systems; establishment of national/regional recording and improvement schemes to support production practices; upgrade of infrastructure and increase in research funding
Low genetic potential of indigenous livestock breeds and limited information on breed characteristics	Undertaking of massive characterisation (production and genetic potentials) of indigenous livestock breeds, undertaking of intentional breeding programmes for improved productivity of indigenous breeds and training of farmers to do the same, provision of funding to support the characterisation of production systems and the preservation of indigenous breeds for future exploitation, putting in place infrastructure to deliver genomic services and selection
Threats due to climate change, environmental hazards and natural disasters	Educate farmers on production practices that address the effects of climate change, engage farmers in the active preparation and readiness to address environmental hazards and natural disasters that may happen at any time
Limited human capacity for sustainable livestock production practices that support livestock improvement	Continually increase the knowledge base/capacity of farmers and other stakeholders (policymakers, extension workers, veterinarians, students and associated animal production professionals) in animal production and improvement practices
Lack of active farmer groups and breed associations and no linkages across livestock production groups and between universities and research organisations	Building of effective farmer groups and breed associations to harness resources, knowledge sharing for decision-making when needed, government funding to support the establishment of farmer groups and breed associations, and active interactions between farmers, universities and research organizations

(continued)

Table 2.5 (continued)

Challenge	Possible intervention
Limited adoption of improved reproduction and breeding practices	Educate and encourage farmers to adopt improved reproduction and breeding strategies, for example, encouragement of the use of artificial insemination, even in smallholder systems; development and implementation of appropriate selection programmes adapted to each production system; improvement of the provision and affordability of innovative technologies such as artificial insemination, incubation services and genomic selection services

through generations. Therefore, there has been little or no motivation for innovation, nor investment or business plans to help improve the traditional systems. There is little or no monitoring of animal performance, and therefore, there is little data available on which the selection for improvement can be based. Moreover, indiscriminate cross-breeding and poor genotype–environment matching make it difficult to achieve maximum benefits from most breed improvement programmes.

Smallholder farmers are accustomed to expecting aid from governments, policymakers, researchers and international partners to address the issues of low feed quality, limited availability of feed resources and the tackling of animal diseases. However, the practices and attitudes of farmers must change as active solutions must start with them. Adequate feed resources and feed security and disease management are the foundations to the sustainable improvement of animal productivity. Drawing on endogenous knowledge and disease management practices as well as promoting systems that are ecologically sustainable is part of the solution. Integrating crop–livestock production, as seen within the smallholder mixed crop–livestock production system, has been described as a viable agro-ecological system approach that can support feed availability/reduction of feed shortage as well as an effective approach to the management of rangelands (Lenné and Thomas 2006). The mixed crop–livestock production system is commonly practised in West Africa, and more attention from research can find the best synergy between crops/forage crops and animal feeding to make this production system sustainable. In the Republic of Benin, for example, a relationship exists between crop and cattle production in agro-pastoral areas at the household level, whereby cattle herders exploit crop residues from crop farmers after the harvest in exchange for manure (Djenontin et al. 2002). The manure from cattle supplies nutrients to the soil, thereby increasing soil fertility to support healthy plant growth and rangeland health. In Cameroon, the various pastoralist initiatives for rangeland and water resource management include the effective distribution of grazing areas, grazing land management, reorganisation of herd mobility and combination with other adaptive management practices (Moritz 2012). To strengthen the adoption of integrated crop–livestock systems, Sekaran et al. (2021) suggested that government polices need to support capital investments, infrastructure and on-field demonstrations, particularly for small and marginal farmers, and provide greater attention to include small ruminants, protein-rich crops, fruits and vegetables in the system.

2.4.2 Feed Resources' Availability and Quality

Feed availability and quality impact livestock productivity, profitability, environment, nutrition security for humans and animals, animal welfare and ethics and health (animal and human). In semi-intensive-to-intensive systems, feed costs accounts for 70% or more of the total variable costs of production, implying that good-quality feed is an important component of such systems. Proper feeding of high-quality feedstuffs directly translates to improved animal productivity, health, welfare and reproductive performance; increases productive life and profitability and enables higher productivity under a given management regimen and environmental sustainability.

Limited availability and access to quality feed materials continue to be the most important limitations to livestock production in Africa. Inadequate availability and supply of quality feed and water critically limit the efficiency of livestock in terms of production, reproduction, health and welfare and the economic benefits derived from livestock-based products. The mainly extensive and smallholder livestock production systems are shaped by the agro-ecological zones, and the amount of rainfall is a main determinant of the availability and quality of forage for the nutrition of majority of livestock produced on the continent. With increasing drought conditions due to changing climates, feed and water resources are dwindling in amount and quality, calling for tailored interventions. Countries in East Africa identified, a lack of animal feed policy, strategy and institutional framework to support the animal feed sector, as a major constraint hindering subsector growth, livestock productivity, resilience and trade (Opio et al. 2020). The implementation of a regional livestock feed action plan will enhance the efficiency of utilisation of natural resources, opening new avenues for the development of a sustainable bioeconomy in East Africa. Climate change/variability is an important issue in the semi-arid and arid regions of Africa, especially in pastoralists' and agro-pastoralist's food and nutrition security and livelihood outcomes (Opio et al. 2020). A range of actions for sustainable feed production was recently proposed (Balehegn et al. 2020) and includes (i) improving feed productivity by improving biomass production and availability; (ii) enhancing feed quality by focusing on improving nutritional value, palatability, intake and digestibility of low-quality feeds; (iii) maintaining/preserving feed quality by preserving the nutritional quality of feeds during storage for off-season feeding; (iv) enhancing the nutritional status of animals through the supplementation of diets with high-quality nutritious ingredients that supply critical nutrients, and enhancing the digestion and assimilation of feed; and (v) using high-quality analytical and operational technologies to improve feed quality analysis, quality control, marketing, packaging, transportation and feeding.

Generally, informed strategies for the improvement of feed resources, both qualitative and quantitative, tailored to suit the needs of the different livestock species and production systems are prerequisites for sustainable livestock productivity in Africa. For example, access to improved communal rotational grazing pastures and fodder banks, which should be regularly maintained to ensure the quality of feed resources, will enhance sustained livestock production in these systems (Ibeagha-Awemu et al. 2019). Moreover, legislations instituting the development of watersheds, restrictions on indiscriminate burning of grazing land and their use for other purposes, increase grain production for livestock feeding, development of improved pastures and fodders, development of agricultural by-products as feed resources and access to water resources, and feed markets will support a durable livestock sector (Ibeagha-Awemu et al. 2019).

2.4.3 Climate Change

African livestock production systems operate under a wide range of environmental conditions and are increasingly affected by climate change. Global warming and its associated changes in mean climate variables and climate variability affect feed and water resources as well as animal health and production. Climate change also has implications for the processing, storage, transport, retailing and consumption of livestock products. The ability of current livestock systems to support livelihoods and meet the increasing demand for livestock products is thus threatened. Climate change may also impact infectious livestock diseases by changing their spatial distributions, affecting annual and seasonal cycles, altering disease incidence and severity and aiding the susceptibility of livestock to illness. Furthermore, climate change is expected to directly depress animal adaptive response mechanisms, cause heat stress and related welfare issues and compromise the availability of feed crops and quality of forages. For example, most indigenous cattle owned by smallholder farmers are raised entirely on communal rangelands. Communal rangeland productivity is affected by climate change and degradation caused by uncontrolled grazing, encroachment of indigenous woody plants and the spread of invasive alien plants. Therefore, it is important to develop and implement innovative community-based rangeland management and forage conservation technologies (Marshall 2014).

Projected increases in extreme weather events and changes in feed availability, composition and quality of animal diets will affect the availability of animal products and food supply due to increasing human population growth and demand for livestock products. Some of the negative consequences of climate change for ruminant livestock production can be mitigated through the use of adaptive management and breeding approaches. Examples include the use of energy-efficient cooling systems for animals, proper nutrition, water availability and access, use of heat-tolerant (thermotolerant) breeds and genetic selection to improve the production, fertility, health and survival under heat stress conditions (Henry et al. 2018).

2.4.4 Disease Burden

Disease burden and limited access to drugs and veterinary care are major limitations for livestock production in Africa. For example, smallholder pig production in Uganda, like in most countries in SSA, is constrained by poor management and high disease burden, with African swine fever being one of the most important perennial threats responsible for decimating Africa's pig genetic resources. Under poor management and husbandry practices, there is an increased risk of transmission of highly infectious diseases, including African swine fever (Escarcha et al. 2018; Dione et al. 2017). Biosecurity measures are not followed in the majority of livestock enterprises, which calls for the need to raise awareness among farmers about the dangers of disease outbreaks to the farm and society. Given the high cost of imported drugs, the improvement of traditional livestock disease control methods and exploitation requires revisiting and harnessing traditional disease control methods, which in conjunction with improved animal housing, feeding management, and breeding for disease resistance are sustainable solutions to support sustainable animal production and product quality.

Furthermore, the use of antimicrobial products to promote growth and prevent disease in healthy food-producing animals has exacerbated the emergence and spread of resistant microorganisms. Consequently, the sector can only deliver on expectations if, among other measures, the productivity and income of small-scale food producers are improved, sustainable and resilient food systems are promoted, the diversity of genetic resources is maintained, the proper functioning of food markets is ensured, and the use of antimicrobials is reduced through better access to quality veterinary services and good animal husbandry practices (FAO 2018d).

2.4.5 Limited Access to Innovative Technologies

Livestock production systems in Africa are often challenged in many ways regarding the infusion of innovative technologies for improved efficiency. For example, dairy farming is an important source of income and nutrition for many smallholder farmers in East Africa (Mwanga et al. 2019). However, the most challenging critical factors hampering the success of oestrus synchronisation and mass artificial insemination service delivery in dairy cows for smallholders in Ethiopia are the improper selection of cows/heifers, small and scattered herds rendering impossible the insemination of a large number of cows/heifers in 1 day and at a specific place, the absence of a data-recording system, the lack of clearly defined share of responsibilities among stakeholders, poor communication and collaboration among stakeholders, the lack of motivation and skills of artificial insemination technicians, the lack of support and limited availability of inputs, feed shortages, poor heat detection and wrong insemination time by smallholders and lower reproductive performances of both indigenous and crossbred cows, which consequently contribute to limited success of the technologies (Abebe et al. 2021). Genetic diversity losses are often aggravated by focusing on specialised exotic breeds and crossbreds, and production systems, the failure to evaluate indigenous breeds and inappropriate breed replacement, especially in post-disaster rehabilitation programmes that involveg the introduction of breeds that are not adapted to local production environments (Mapiye et al. 2019).

2.4.6 Lack of Clear Breeding Goals and Genetically Improved Livestock

Intentional breeding and selection through setting and implementing clear breeding goals are prerequisites for sustainable animal breed improvement. Unfortunately, this has generally not been the case in the African livestock sector, where breeding is mostly random, and coupled with poor feeding and management practices, livestock productivity is low. When compared with Western breeds, which are under sustained breed improvement programmes and improved management practices, African livestock are erroneously tagged 'unproductive'. To make the transition from 'unproductive' to 'productive', African indigenous livestock too must be intentionally selected and placed under improved management practices. An intentional selection programme will start by setting clear breeding goals and the selection of best-performing individuals for the traits under selection as parents of the next generation, coupled with improved management practices. Taking advantage of improved technologies, like the implementation of genomic selection, will lead to rapid progress in desired traits. Repeated circles of such intentional actions will lead to the development of genetically improved individuals for the traits under selection. Setting the breeding goals to meet today's challenges for sustainable animal breeding should consider the following: (i) market-oriented and non-market-oriented value traits, (ii) the specific needs of farmers, (iii) social and religious needs, and (iv) ecological and environmental adaptation goals (Ibeagha-Awemu et al. 2019). The breeding goals, which must be tailored to fit the conditions of each production system, must be supported by appropriate education, the involvement and cooperation of stakeholders and appropriate government policies and financial input. The setting of appropriate breeding goals of African livestock genetic improvement are discussed in Chapters 15, 16, 17, 18 and 19.

Generally, matching genetics to the production environment is a major challenge. Recording and selective breeding can serve as means of ensuring the identification and/or development of genotypes that fit in current and evolving production environments. Cross-breeding and/or upgrading of African indigenous livestock with exotic breeds have been the commonly rushed strategies implemented to improve the genetics of African livestock for increased productivity. Such introduced genetics, which have been developed over long periods in different environments, do experience genotype by environment interactions, which markedly lower performance after the F1 generation in the African environment where they are introduced. It has been observed that the adoption of 'shortcut approaches' or 'rushed strategies' of utilising exotic genes to improve African livestock has generally not resulted in substantial gains nor sustainable long-term increases in productivity or contributed to poverty alleviation (Ibeagha-Awemu et al. 2019). Moreover, cross-breeding programmes between exotic and indigenous breeds were intended to improve biological and economic efficiency, but their unstructured nature has produced nondescript crossbred individuals, e.g. crossbred cattle that are currently dominant in most smallholder livestock production systems in East Africa, resulting in the erosion of genetic diversity. Careful selection programmes and conservation of indigenous cattle breeds are critical for reversing the unprecedented loss of diversity and ensuring the security of cattle genetic resources for economic, ecological and social benefits for the short and long terms. To improve production through crossing with exotic breeds, careful selection objectives and judicious implementation of cross-breeding with improved indigenous breeds for desired traits in combination with relevant biotechnologies and improved management practices may support the sustainable improvement and use of African indigenous livestock genetic resources. Key strategies to maintain diversity and ensure the sustainable use of indigenous African AnGR should include sustained genetic improvement programmes targeting unique productive and adaptive traits; improved nutrition, healthcare and housing; appropriate infrastructure; integration and empowerment of women and youth; and institution of appropriate marketing channels, capacity building and regulatory and policy frameworks (Mapiye et al. 2019).

2.4.7 Lack of Necessary Infrastructure and Inputs

Access to infrastructure for the general development of livestock production within the African continent is limited, including access to basic equipment for data, storage and traceability, computing infrastructure to deliver genomics services/information, animal housing, pasture development, animal handling and transport, livestock markets, slaughter facilities and health and nutrition management, among others. High cost/limited access to inputs such as veterinary drugs, fertilisers and soil nutrients, among others, are additional factors limiting livestock development in Africa. Making available these infrastructures will support rapid development of the African livestock sector.

2.4.8 Lack of Enabling Government Policies in the Livestock Industry

Successful and sustainable development of the livestock industry in any given country or region depends on the development of clear national, regional and continental policies as well as guiding principles in the conduct of affairs (Ibeagha-Awemu et al. 2019). However, this is not the case in most African countries, which lack clear supply and demand policies that favour local production nor recognise the positive impact of livestock in the livelihood of smallholder farm families and the nutrition and economy of individual countries. Where some sort of policies exist, there is a general lack of political commitment to implement national, regional and continental livestock policies, which are keys to stimulating, developing and financially sustaining the livestock value chain development.

Therefore, the development of enabling livestock policies at all levels of governance (national, regional and continental) is key to realising the full impact of the livestock sector on the continent, and the factors that should be considered for putting in place enabling policies include (1) understanding of the characteristics and challenges of current livestock production systems; (2) broad consultation with all stakeholders (e.g. farmers, consumers, policy makers, professionals, service providers, marketers, etc.); (3) demonstration of the will to address concerns and needs through policy; (4) creation of policy and institutional environments suitable for mainstream livestock farming systems; (5) creation of policies with appropriate incentives for farmers and agribusiness to encourage the implementation of livestock development strategies at all levels; (6) creation of policies that allow access to markets, information, healthcare, nutrition and other agricultural services as these are the backbone to decreasing productivity gaps and ensuring sustainable livestock improvement, food security and improved livelihoods; (7) creation of policies that support increased investments in technological innovations and research such as incorporating transdisciplinary strategies with focus on reducing productivity gaps in all production systems and promotion of novel and effective production systems and uptake of innovative technologies; (8) creation of inclusive policies that develops and strengthens human (paying attention to gender equity) and social capital in suitable ways for each livestock production system; (9) creation of policies to ensure access to grazing lands, cultivated lands and water resources and develop improved pastures and restore degraded lands; (10) creation of policies that support sustainable intensification and diversification within the various production systems with appropriate climate-smart practices to improve livestock productivity in the face of increasing climatic variability and market volatility; and (11) creation of policies that stimulate local production and encourage inter-country and international trade in livestock products. Finally, setting national, regional and continent-wide policies and strategies must be accompanied by the development and implementation of livestock breeding laws that address the technical aspects of animal genetic resource management and harness factors like institutional arrangements, roles and responsibilities of various bodies and decision-making process that influence the implementation of legislation and policies. The policies, frameworks, strategies, and action plans for advancing the African livestock sector are presented in Chapter 30 and the resourcing institutional arrangements are discussed in Chapter 31.

2.5 Future Trends and Drivers of Transformation of African Livestock Production Systems

In general, past livestock production systems in Africa were mostly subsistence smallholder enterprises with the main production objective of satisfying household and/or family needs with little commercial motives. The main animal resources were indigenous breeds, often passed down the family line. In addition, animal farming was part of crop production, with farmers and governments making little investment in housing, feeding, veterinary care and new genetics. However, with the increasing demand for animal protein due to population growth, most farmers are beginning to adopt innovative technologies and improved practices to increase the productivity of their stock.

Present-day livestock production systems in Africa are diverse and dynamic and include (1) those where existing indigenous breeds are currently optimal and likely to remain so, (2) those where non-indigenous breed types are already in common use and (3) systems that are changing by intensification, where the introduction of new breed types represents significant opportunities. These systems will continue to evolve in response to many factors including but not limited to human population growth, increasing demand for livestock products, shift in natural resource availability, climate change, scientific innovation and technological advances, human capacity development and gender equity, livestock business, markets and marketing channels and supporting institutional structures and enabling policies (Fig. 2.9). These drivers present opportunities to support rapid improvements in sustainable livestock production and, consequently, enhance the livelihood of Africans through increased food and nutrition security and environmental sustainability.

2.5.1 Human Population Growth and the Demand for Livestock Products

Increasing global human population growth projected to reach 9.7 billion by 2050 (UN 2022) and increasing incomes and purchasing power will continue to push up the demand for animal proteins and livestock-related products. With more and more people choosing to live in urban areas, it is projected that more than two-thirds of the world population and 60% of Africans will live in urban areas by 2050. Urban dwellers are known

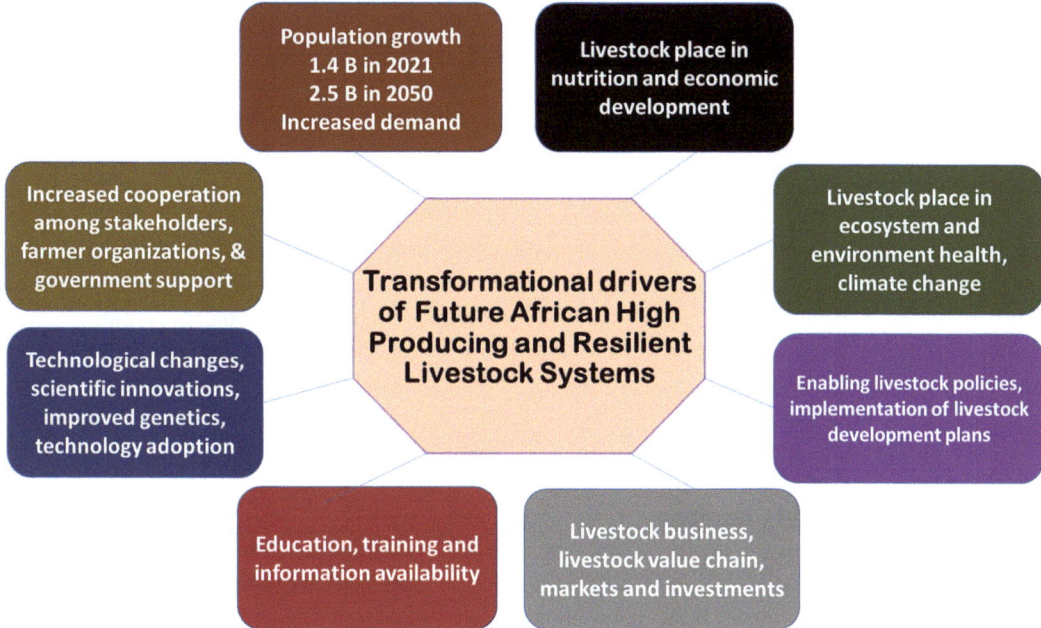

Fig. 2.9 Transformational drivers of current and future African high-producing and resilient livestock production systems

to embrace new habits, including increased consumption of animal-sourced proteins. A recent study estimated that the demand for meat and milk by African urban dwellers may triple by 2050 (Latino et al. 2020). The current growth of peri-urban intensive livestock operations and the growth of intensively managed large poultry and pig farms have been largely driven by the increased demand for animal products by urban dwellers. The current and projected increased demand for animal-sourced foods will only intensify and will add to the existing pressure on ecosystems and biodiversity, and livestock producers will face greater competition for capital, labour, land, water and energy.

In Tanzania, for example, a relatively small number of livestock keepers have adopted exclusively commercial production to meet the growing demand for livestock products by urban populations (de Glanville et al. 2020). Moreover, the commercialisation and intensification of livestock production is strongly promoted by the Tanzanian Government through beef ranching and the establishment of zero-grazing dairy units to support exotic breeds of cattle. In recent times, Maasai pastoralists are increasingly adopting crop production as a response to changing land tenure practices in northern Tanzania. Widespread adoption and subsequent change within traditional pastoral systems could conceivably lead to it becoming broadly indistinguishable (in terms of production) from neighbouring agro-pastoral systems (de Glanville et al. 2020). The ranching system, if popularised and successfully implemented, could help overcome some of the challenges associated with the large-scale extensive pastoral and transhumance systems and lead to improved productivity.

These factors are real opportunities that are set to push the livestock industry towards larger, intensively managed operations, resulting in higher outputs to satisfy the demand. Hence, the increasing demand for livestock products may drive the reformation of existing livestock production systems towards more effective and productive systems.

On the other hand, Bonnet et al. (2020) report that the increased demand and consumption of meat in developed countries may increase the risk of some chronic health diseases, and in particular, increased ruminant production is a significant source of greenhouse gas (GHG) emissions and a major driver of deforestation and worldwide biodiversity loss. Other limitations associated with increased livestock production include environmental pollution (water, soil and air), development of antibiotic resistance, increased incidence of zoonotic diseases, emergence of new disease pathogens and increased animal welfare issues. These potential negative impacts must be considered when developing strategies for increased livestock productivity to satisfy the increasing demand.

2.5.2 Technological Changes and Adoption

Technological advances have been the major drivers of improved livestock productivity, and their massive roles in advancing livestock productivity will only increase. Technological advances in reproduction (e.g. artificial insemination), breeding (e.g. genomic breeding) and feeding (precision feeding) technologies, etc., have left their marks on advancing livestock productivity. Today, these technologies and other emerging technologies (see Chapters 20 and 21) are advancing at a faster pace, and their adoption, including technologies to capture and manage big data, are poised to address the challenges facing low livestock productivity in Africa towards a more productive and rewarding livestock sector. For example, the widespread use of artificial insemination, multiple ovulation and embryo transfer reproductive methods; more efficient statistical tools for genetic merit estimation; the use of molecular genetics and DNA-based methods to study animal genomes for genomic selection to increase the rate of genetic gains for most traits and health management and the possibilities presented by cloning and transgenic livestock production, as well as the possibilities presented by future technologies (Ibeagha-Awemu et al. 2019) are modern tools whose adoption can promote rapid growth of the African livestock sector. In a recent empirical study, smallholder livestock farmers in SSA indicated that the use of digital technologies could enhance their risk mitigation and value chains, which supports the overwhelming need for the use of emerging technologies by smallholder livestock farmers in SSA (Gwaka and Dubihlela 2020). New technologies such as mobile telephones, new strategies and tools for livestock health management and changes in land tenure and land availability may also lead to change within traditionally defined production systems. Therefore, national governments need to invest in these emerging technologies and put in place the appropriate infrastructure to encourage stakeholders to embrace these technologies across Africa.

2.5.3 Education, Training and Information Availability

Africa is home to many higher institutions providing education and training on different aspects of livestock production, transformation and marketing (See Chapters 27 and 28). Although not currently operating at optimal capacity, these institutions have opportunities through inter-country, inter-regional, continent-wide and international collaborations to build capacity and contribute positively towards revamping the livestock sector and realising its full potential. Many international organisations, such as the International Livestock Research Institute (ILRI) and non-governmental and professional organisations on the continent engage in student and farmer training on aspects of livestock production, management and trade. Moreover, open-access availability of scientific information, a concept promoted by many countries and facilitated by technology (e.g. Internet accessibility), means that the knowledge necessary to revamp the African livestock sector is freely available and that harnessing it will result in the rapid growth of the livestock sector. It was stated recently that better collaboration between institutions of higher learning such as universities, research organisations (national, regional, continent-wide and international), farmer groups, non-governmental organisations and national governments is the key to ensuring the flow and sharing of information and knowledge (Ibeagha-Awemu et al. 2019).

2.5.4 Livestock Business, Markets and Investments

Recognition of the livestock sector and its promotion as a viable driver of national economies will garner interest in livestock production as a viable income-generating venture. This requires appropriate government incentives and investment to build a sustainable livestock sector. Most importantly, every livestock enterprise should be market-oriented, and marketing channels should be made available to facilitate trade in livestock products. Also, trade in livestock and livestock products in most African countries is performed through informal channels requiring much effort, and adequate measures need to be put in place for the acquisition of the true livestock marketing statistics within and between African countries.

In recent years, efforts are being made by national governments and regional and continental bodies to promote trade in livestock. For example, the Ethiopian government's strategy of encouraging foreign trade of sheep and goat products has seen the creation of employment opportunities for many of its citizens (Nwogwugwu et al. 2018) as well as the emergence of livestock feed and poultry businesses in West Africa (Konlan et al. 2018). Also, the growing inter-regional trade in livestock between the member states of the Economic Community of West African States (ECOWAS) and the Economic Community of Central African States (ECCAS) and among Southern African Development Community (SADC) and East African Community (EAC), plus Libya and Egypt (Valerio 2020), will continue to grow due to increasing population growth, which will stimulate increased productivity. Moreover, the full implementation of the African Continental Free Trade Area (AfCFTA), which seeks to ease free trade barriers between member states, may help formalise Africa's large informal livestock trade sector and increase trade in livestock products between states and also internationally. Greater involvement of the private sector will drive livestock extension services and stimulate the growth of the sector. Therefore, the increasing African urban demand as well as the global demand for livestock products are potential drivers that can transform Africa's livestock value chains and international trade in livestock products.

2.5.5 Enabling Livestock Policies

Recent time has seen the emergence of various livestock-oriented policies by national governments and regional bodies and continental policies with the goal to achieving rapid growth of the livestock sector.

The AU-IBAR (https://www.au-ibar.org/) has, since its inception in 1951, implemented a mandate to support and coordinate the improved and sustainable utilisation of livestock, fish and wildlife species as viable resources for the sustenance and well-being of Africans. Its main mission of ensuring that animal genetic resources contribute substantially to the reduction of poverty and hunger in Africa has been/is being realised through the implementation of various past, ongoing and future projects. The African Union LiDESA for Africa (2015–2035), developed through an inclusive consultation process involving stakeholders and experts at national, regional and continental levels (AU-IBAR 2015), recognises the central role played by livestock in the lives of Africans. The LiDESA 2015–2035 aims to transform African livestock productivity for accelerated, sustainable and equitable growth and for increased contribution to socio-economic development (see Chapter 30). These goals are in line with ongoing strategies, policy frameworks and guidelines in place in many African countries as well as mirror the frameworks and agenda in individual countries, regional economic communities and throughout the continent. Moreover, livestock are central to achieving all 17 current UN SDGs (FAO 2015), and the careful implementation of these goals in Africa will contribute to sustainable livestock productivity and raising livelihoods. These policy frameworks are, therefore, positive steps towards the sustainable utilisation of African animal genetic resources and present opportunities to stimulate increased livestock productivity.

2.5.6 Livestock's Place in Ecosystem and Environment Health

While the livestock sector is critically important to Africa as it performs vital nutrition and development functions through its contribution to dietary needs, economic growth and environment health, it is also responsible for about 70% of agricultural GHG emissions, overgrazing and associated wide-ranging land degradation and loss of biodiversity. To curb livestock production's negative impacts and ensure sustainability in food systems and the positive role of livestock, a whole system approach to livestock production, climate-smart interventions, carbon-neutral production systems, proper grazing and rangeland management, improved forage-based systems, preservation of natural systems, proper manure management, improved housing, improved animal health management and genetics are increasingly being adopted by many countries.

It is recognised that about 90% of ruminants on the African continent are raised on grazing lands (Robinson et al. 2011), which also support about 30% of global domestic ruminant production and 30% of animal-sourced proteins (Mottet et al. 2018). Globally, about 1.5 billion hectares of land not suitable for growing crops due to poor soil fertility, rainfall and topography are exploited for livestock grazing system (LGS), as well as 54% of total terrestrial landscape (Ickowicz et al. 2022). Most of the unsuitable lands (about 28 M km^2) are found in desert or marginal shrubland areas (ILRI et al. 2021). Many LGSs are characterised by the mobility of livestock and people, as seen in most pastoral systems, and the reliance on natural resources and processes such as native pasture, water source, manure and high human capital input. A global analysis of the land footprint of LGSs, their associated management of livestock and their impacts on the ecosystem dynamics using a multifunctional conceptual model and a holistic approach (production and social, environmental and local development) demonstrated the potential contribution that LGSs may bring to sustainable food systems (Ickowicz et al. 2022). Case studies of LGS impact are summarised in Box 2.2.

Climate-smart livestock strategies and interventions can enhance the productivity of livestock, particularly under low-input or resource-restricted systems, with positive impacts on climate change adaptation, mitigation and environmental resilience (Gaitán et al. 2016). Moreover, climate-smart interventions in various components of the livestock production practice can decrease GHG emissions and the intensities of emissions and increase carbon sequestration, with the consequence of contributing to a system-wide negative carbon balance (Ortiz-Gonzalo et al. 2017).

Agro-forestry systems, generally regarded as carbon-neutral production systems, such as silvopastoral systems, can contribute to reductions in GHG emissions from livestock through the contribution to carbon sequestration and improved feed quality (Feliciano et al. 2018).

Proper grazing and rangeland management can increase biomass production and biodiversity, and well-managed grasslands can store up to 260 tonnes of carbon per ha (Savory and Butterfield 2007; Gebregergs et al. 2019). Moreover, interventions like zero-, moderate- and rotational-grazing can enhance carbon sequestration by grazing lands and increase livestock productivity (Gebregergs et al. 2019). Improved forage-based systems promote increased herbage biomass and livestock productivity (Paul et al. 2020). Integration of forage technologies with food crops was shown to reduce soil loss by ~50% and increase soil organic carbon by 10%, stover yield by 33% and grain yield by 60% (Paul et al. 2020).

Box 2.2 Summary of case studies on the impact of livestock grazing system (LGS) management using a multifunctional approach (Ickowicz et al. 2022)

Case study 1: **Argentina: The Puna high-altitude, dry pastoralism**

    Social indicators: Household members and the number of local organisations in the supply chain.

    Environmental indicators: Plant cover, dry matter production and biodiversity.

    Production indicators: Meat sold in kilograms, diversity of livestock and drought management strategies.

    Local development indicators: Yearly income and number and diversity of marketing channels.

    Impact/results: The approaches for resilience centred on social networks and the diversity of livestock species associated with the local supply chain and household members linked to wealth generation. The system of grazing maintains vegetation condition and diversity, with plant cover regulating soil temperature and water.

Case study 2: **Senegal: Ferlo Rangeland–based dairy milk platform**

    Social indicators: Collaboration between forage producers and social inclusion.

    Environmental indicators: GHG emissions and biomass production.

    Production indicators: Efficiency of milk collection, milk production (litres/day) and biomass flows.

    Local development indicators: Biomass supply network, income from milk and milk value chain development

    Impact/results: Evaluation of three scenarios of dairy intensification identified the trade-offs between inputs and outputs and the environmental and social consequences and assisted in the development of the sector strategy.

Case study 3: **Brazil: Maranhao silvo-pastoral systems**

    Social indicators: Profit and employment.

    Environmental indicators: Carbon balance, animal welfare and biodiversity.

    Production indicators: Amount of meat (kg)/ha/year

    Local development indicators: The number of associated businesses

    Impact/results: Achieved higher profit compared with monoculture, with potential for further gains through payments from additional ecosystem services provided, increased flora and fauna biodiversity, enhanced animal welfare and enhanced soil conservation.

Case study 4: **Mongolia: Bulgan forest steppes conservation coexisting with livestock systems**

    Social indicators: Participating farm families, household income and diverse employment.

    Environmental indicators: Increase in the numbers of existing species and reintroduction of species.

    Production indicators: Livestock production.

    Local development indicators: New business prospects.

    Impact/results: The positive uptake by herder households of conservation-related employment positively alleviated poverty and improved environmental outcomes without affecting existing livestock systems.

Therefore, the implementation of sustainable livestock production systems is key to increasing livestock productivity in Africa.

2.5.7 Adapting Livestock Systems to Mitigate Climate Change Effects

The potential impacts of climate change on current livestock systems worldwide and in Africa, in particular, are a major concern. The consequences of climate change include the production of GHGs such as methane and nitrous oxide. Methane is mostly generated by ruminants during enteric fermentation (a biological process for removing carbon dioxide and hydrogen from the rumen to maintain fermentation) or the anaerobic decomposition of excreted manure. In livestock, high-concentrate-diet-fed livestock are known to produce faecal and urine output rich in nutrients, which leads to environmental pollution. This indicates poor nutrient use efficiency of nitrogen and phosphorus in the agricultural production system. Despite the related adverse environmental effects, agricultural production practices are the primary routes to achieving food security and ending world hunger by improving protein and other nutrient needs in human diets. Nutritional strategies (e.g. feed supplementation) and manure processing (e.g. anaerobic digestion (biogas)) can be used to reduce GHG emissions and recycle nutrients, thereby increasing the nutrient use efficiency of animals. Mostly, the ruminants in Africa are raised for meat production, which is often associated with low feed efficiency and high GHG emission intensity (Herrero et al. 2013) due to slow growth rates. Reducing emissions without lessening the ruminant population requires either good-quality feed supplementation or improved yield. Genetic improvement of indigenous livestock for disease- and heat-tolerant traits will be useful in this regard. Adaptation of livestock to heat stress is an essential practice in sustainable livestock farming because of the higher methane produced from heat-stress-susceptible animals (Hyder et al. 2017). The adaptation to marginal lands and the potential for lower emission intensity could favour small ruminant rearing in Africa.

Therefore, climate-smart agricultural (CSA) practices will ensure the sustainability of Africa's livestock production systems. The CSA practices focuses on three major pillars: (1) sustainable increase in productivity to support the development and equitable increase in farm incomes and food security, (2) increased resilience and (3) reduction or elimination of GHG emissions (mitigation) (Chandra et al. 2018). The CSA practice is a relevant mechanism to support food system transformation in Africa as its role in GHG mitigation is as important as food security and climate change adaptation (Anuga et al. 2020).

2.5.8 Livestock's Role in Responses, Recovery and Rebuilding Measures Against Pandemics

Pandemics and other natural disasters can have profound impacts on nurturing livelihoods and global food security, including livestock production and its sustainability. Evidence from the recent COVID-19 pandemic and past pandemics indicates disruptions to the livestock value chains with dire socio-economic impacts on production, processing and transport (FAO 2020c) (Obese et al. 2021), including reduced access to animal feeds, reduced inputs and services, reduced market access, reduced processing capacity (e.g. meat and dairy processing industries), compromised storage and conservation, constrained informal businesses (e.g. dairy and meat trading in developing countries is mostly informal), constrained national transport (e.g. movement restrictions and transport disruptions), international trade restrictions, modified retailing methods (e.g. online) and reduced consumer purchasing power, demand and public procurement.

To ensure the sustainability of the livestock value chains during pandemics, the FAO proposed the implementation of the actions listed below (FAO 2020c):

1. Tailor measures to protect animal production and markets: Such measures may include (i) the establishment of production safety nets such as new or resupplied feed reserves; (ii) the issuance of special driving permits to allow animal feed distribution in remote areas during pandemics as well as putting in place waivers for agri-food system operations to keep inputs flowing; (iii) empowerment of producer organisations to advance bargaining through collective purchasing and marketing; (iv) active coordination of the supply of inputs for livestock production; (v) promotion and/or funding of the local sourcing and production of feed and supplements for animals; (vi) launch of emergency management procedures and services to provide timely communication to control rumours, guide stakeholders, get feedback and move staff and resources to tackle crisis relief activities such as health management and food inspection; (vii) permission for food markets to stay open and manage physical distancing through public health-conscious rules, procedures, materials and equipment and through the protection of environments (water, land, biodiversity and ecosystems) where diseases are common; and (viii) keeping open borders (within and between countries) for imports and exports of relevance to all areas of the value chain, permission for transboundary livestock movement and allowing for access to essential natural resources for pastoralists.
2. Put in place measures to support processing and retail operations: Such measures may include (i) providing clear guidelines for pandemic control, management and prevention along the supply chains to protect all value chain participants and their families; some of the provisions could be for increased biosecurity, personal hygiene and protective equipment; (ii) increasing funding to support packaging and freezing capacities as well as encouraging enterprises (small and medium) to engage in the production of safe products and those with long shelf lives; (iii) putting in place group slaughterhouses or points and supporting the installation and use of cold chain to reduce slaughtering frequency and improve meat inspection; (iv) ensuring that children participating in school feeding programmes are reached as well as distributing foods rich in animal proteins to enhance nutrition and incomes of smallholder families; and (v) promoting timely collection and distribution of milk to processing companies.
3. Ensure adequate financial measures: Such measures may include (i) funding milk collection centres and factories to enhance the purchasing power for milk supply and processing of products with long shelf lives; (ii) supporting businesses (small- and medium-sized) in the short-term relief of pandemic impacts through temporary tax relief programmes, tax exemptions, emergency loan programmes, direct stimulus payments, loan repayment extensions, grace periods to repay loans, lower interest rates and public investments and subsidies; (iii) putting in place mentoring and training programmes to support enterprises (small- and medium-sized) in assessing and managing the financial impact of pandemics, going digital and finding new markets; and (iv) making available subsidies to agri-food sectors to help them continue activities during lockdown as well as implementing price controls to reduce the effect of inflation on livestock commodities.

2.6 Conclusion and Perspective

African livestock production systems are diverse, dominated by small-scale subsistence livestock farming, and depend largely on rain-fed agriculture. More in-depth characterisation of the existing production systems is, however, needed for informed decisions on their improvement. Current livestock productivity levels are not enough to satisfy the growing consumer demands for livestock products and will continue to remain so if present challenges are not mitigated. Stimulating factors outlined in Section 2.5, like population growth, availability of knowledge and technology and the willingness by national and international bodies to support livestock production, present real opportunities for the rapid evolution of African livestock production value chains of the future. By consideration of the aforementioned stimulating factors, many new livestock production typologies could emerge from demographic challenged, low technical know-how, technology poor, policy starved and environmental change threats, towards more resilient, productive and environmentally friendly livestock production systems. Additionally, national governments and stakeholders by ensuring that appropriate capacity building, land use and animal resource conservation policies backed by appropriate legal and regulatory frameworks are implemented will not only safeguard the rich and unique livestock production systems in Africa but also ensure that its almost 1.5 billion inhabitants derive maximum benefits from them.

References

Abdel-Salam S, Fahim N (2018) Classifying and characterizing buffalo farming systems in the Egyptian Nile Delta using cluster analysis. J Anim Poult Prod 9(1):23–28. https://doi.org/10.21608/jappmu.2018.35763

Abebe B, Alemayehu M, Barreiros JP (2021) Challenges and opportunities on estrus synchronization and mass artificial insemination in dairy cows for smallholders in Ethiopia. Int J Zool 2021:1–6. https://doi.org/10.1155/2021/9914095

Adido S, Gicheha M (2018) The status of cattle genetic resources in West Africa: a review. Adv Anim Vet Sci 7:112–121. https://doi.org/10.17582/journal.aavs/2019/7.2.112.121

Ahmed A, Kuusaana ED (2021) Cattle ranching and farmer-herder conflicts in sub-Saharan Africa: exploring the conditions for successes and failures in Northern Ghana. Afr Secur 14(2):132–155. https://doi.org/10.1080/19392206.2021.1955496. Mbih RA (2020) The politics of farmer–herder conflicts and alternative conflict management in Northwest Cameroon. Afr Geogr Rev 39(4):324–344. https://doi.org/10.1080/19376812.2020.1720755

Alarcon P, Fèvre EM, Murungi MK, Muinde P, Akoko J, Dominguez-Salas P, Kiambi S, Ahmed S, Häsler B, Rushton J (2017) Mapping of beef, sheep and goat food systems in Nairobi – a framework for policy making and the identification of structural vulnerabilities and deficiencies. Agric Syst 152:1–17. https://doi.org/10.1016/j.agsy.2016.12.005

Anuga SW, Chirinda N, Nukpezah D, Ahenkan A, Andrieu N, Gordon C (2020) Towards low carbon agriculture: systematic-narratives of climate-smart agriculture mitigation potential in Africa. Curr Res Environ Sustain 2:100015. https://doi.org/10.1016/j.crsust.2020.100015

AU-IBAR (2015) The livestock development strategy for Africa 2015–2035, Nairobi. Available online at http://repository.au-ibar.org/bitstream/handle/123456789/540/2015-LiDeSA.pdf. Accessed on 15 May 2023

AU-IBAR (2019) The state of farm animal genetic resources in Africa: towards accelerated agricultural growth and transformation by the year 2025. African Union – InterAfrican Bureau for Animal Resources (AU-IBAR), Nairobi, Kenya, 337 pages. http://repository.au-ibar.org/handle/123456789/1243

Balehegn M, Duncan A, Tolera A, Ayantunde AA, Issa S, Karimou M, Zampaligré N, André K, Gnanda I, Varijakshapanicker P (2020) Improving adoption of technologies and interventions for increasing supply of quality livestock feed in low-and middle-income countries. Glob Food Sec 26:100372. https://doi.org/10.1016/j.gfs.2020.100372

Behnke R, Muthami D (2011) The contribution of livestock to the Kenyan economy. IGAD LPI Working Paper 03–11. IGAD Livestock Policy Initiative. https://hdl.handle.net/10568/24972, Addis Ababa

Bonnet C, Bouamra-Mechemache Z, Réquillart V, Treich N (2020) Regulating meat consumption to improve health, the environment and animal welfare. Food Policy 97:101847. https://www.tse-fr.eu/publications/viewpoint-regulating-meat-consumption-improve-health-environment-and-animal-welfare. (Accessed 22 June 2023)

Campbell DLM, Bari MS, Rault J-L (2021) Free-range egg production: its implications for hen welfare. Anim Prod Sci 61(10):848–855. https://doi.org/10.1071/AN19576

Chandra A, McNamara KE, Dargusch P (2018) Climate-smart agriculture: perspectives and framings. Clim Pol 18(4):526–541. https://doi.org/10.1080/14693062.2017.1316968

de Glanville WA, Davis A, Allan KJ, Buza J, Claxton JR, Crump JA, Halliday JEB, Johnson PCD, Kibona TJ, Mmbaga BT et al (2020) Classification and characterisation of livestock production systems in northern Tanzania. PLoS One 15(12):e0229478. https://doi.org/10.1371/journal.pone.0229478.

Dione MM, Akol J, Roesel K, Kungu J, Ouma EA, Wieland B, Pezo D (2017) Risk factors for African swine fever in smallholder pig production systems in Uganda. Transbound Emerg Dis 64(3):872–882. https://doi.org/10.1111/tbed.12452

Dixon J, Gulliver A, Gibbon D (2001) Farming systems and poverty: improving farmers' livelihoods in a changing world. FAO & World Bank, Rome/Washington, DC, p 49. Available at https://www.fao.org/3/ac349e/ac349e.pdf. Accessed on 15 May 2023

Dixon J, Garrity DP, Boffa J-M, Williams TO, Amede T, Auricht C, Lott R, Mburathi G (2019) Farming systems and food security in Africa: priorities for science and policy under global change. Oxon, UK: Routledge - Earthscan. 638p. (Earthscan Food and Agriculture Series). https://doi.org/10.4324/9781315658841

Djenontin JA, Amidou M, Baco NM (2002) Diagnostic sur la gestion du troupeau : gestion des ressources pastorales dans l'Alibori et le Borgou. Actes du colloque, 27–31 mai 2002, Garoua, Cameroun

Escarcha J, Lassa J, Zander K (2018) Livestock under climate change: a systematic review of impacts and adaptation. Climate 6(3). https://doi.org/10.3390/cli6030054

Fahim NH, Abdel-Salam SAM, Mekkawy W, Ismael A, Bakrorcid SA, El Sayed M, Ibrahim MA (2018) Delta and upper Egypt buffalo farming systems: a survey comparison. Egypt J Anim Prod 55(2):95–106. https://doi.org/10.21608/ejap.2018.93242

FAO (2009) State of food insecurity in the world-economic crises, impacts and lessons learnt. FAO, Rome. Available at http://www.fao.org/docrep/012/i0876e/i0876e00.htm. Accessed June 2023

FAO (2015) FAO synthesis – livestock and the sustainable development goals global agenda for sustainable livestock. Draft prepared by FAO-AGAL Livestock Information, Sector Analysis and Policy Branch. Available online at http://www.livestockdialogue.org/fileadmin/templates/res_livestock/docs/2016/Panama/FAO-AGAL_synthesis_Panama_Livestock_and_SDGs.pdf. Accessed 27 Oct 2022

FAO (2018a) Shaping the future of livestock. In: The 10th Global Forum for Food and Agriculture (GFFA). Food and Agricultural Organisation of the United Nation, Berlin. Available online at https://www.fao.org/3/i8384en/i8384en.pdf. Accessed on 15 May 2023

FAO (2018b) ASL2050 – livestock production systems spotlight: cattle and poultry sectors-Kenya. Africa sustainable livestock 2050. FAO, Rome, 12pp. https://www.fao.org/3/i8270en/I8270EN.pdf. (Accessed 28 June 2023)

FAO (2018c) ASL2050 – livestock production systems spotlight: cattle and buffaloes, and poultry sectors-Egypt. Africa sustainable livestock 2050. FAO, Rome, 9pp. https://openknowledge.fao.org/items/184ea12a-4905-4848-8be9-33a1abef74ef. (Accessed 28 June 2023)

FAO (2018d) Transforming the livestock sector through the Sustainable Development Goals. In: World livestock, FAO, Rome, p 222. https://livestockdata.org/publications/transforming-livestock-sector-through-sustainable-development-goals. (Accessed 28 June 2023)

FAO (2020a) Livestock systems [Online]. Food and Agriculture Organization of the United Nations. Available at http://www.fao.org/livestock-systems/en/. Accessed 18 Apr 2023

FAO (2020b) The state of food security and nutrition in the world 2020. The Food and Agriculture Organisation (FAO). Available online: https://www.fao.org/documents/card/en/c/ca9692en. Accessed 11 Aug 2023

FAO (2020c) Mitigating the impacts of COVID-19 on the livestock sector, FAO, Rome. https://doi.org/10.4060/ca8799en. (Accessed 30 June 2023)

FAOSTATS (2022) Food and agricultural organization statistics for year 2022. Available at: www.fao.org/faostat/. Accessed 11 Aug 2023

Feliciano D, Ledo A, Hillier J, Nayak DR (2018) Which agroforestry options give the greatest soil and above ground carbon benefits in different world regions? Agric Ecosyst Environ 254:117–129. https://doi.org/10.1016/j.agee.2017.11.032

Fernández-Rivera S, Okike I, Manyong V (2004) Classification and description of the major farming systems incorporating ruminant livestock. Available at Permanent link to cite or share this item: https://hdl.handle.net/10568/50278. Accessed 15 May 2023

Fèvre EM, de Glanville WA, Thomas LF, Cook EAJ, Kariuki S, Wamae CN (2017) An integrated study of human and animal infectious disease in the Lake Victoria crescent small-holder crop-livestock production system, Kenya. BMC Infect Dis 17(1):457. https://doi.org/10.1186/s12879-017-2559-6

Fischer G, van Velthuizen HT, Shah MM, Nachtergaele FO (2002) Global agro-ecological assessment for agriculture in the 21st century: methodology and results. IIASA research report, iiasa:6667. IIASA, Laxenburg. Available at https://pure.iiasa.ac.at/id/eprint/6667/. Accessed 11 Aug 2023

Gaitán L, Läderach P, Graefe S, Rao I, van der Hoek R (2016) Climate-smart livestock systems: an assessment of carbon stocks and GHG emissions in Nicaragua. PLoS One 11(12):e0167949. https://doi.org/10.1371/journal.pone.0167949.

Galal S, Elbeltagy A (2006) Achievement of research in the field of buffalo production in Egypt. Pages 177-188 in A. Rosati, A. Tewolde, and C. Mosconi (eds) Animal Production and Animal Science Worldwide, WAAP Book of the year 2006, Wageningen Academic, Wageningen, Netherlands, 384 pp.

Gebregergs T, Tessema ZK, Solomon N, Birhane E (2019) Carbon sequestration and soil restoration potential of grazing lands under exclosure management in a semi-arid environment of northern Ethiopia. Ecol Evol 9(11):6468–6479. https://doi.org/10.1002/ece3.5223.

Gwaka L, Dubihlela J (2020) The resilience of smallholder livestock farmers in Sub-Saharan Africa and the risks imbedded in rural livestock systems. Agriculture 10(7). https://doi.org/10.3390/agriculture10070270

Heise H, Crisan A, Theuvsen L (2015) The poultry market in Nigeria: market structures and potential for investment in the market. Int Food Agribus Manag Rev 18(A):1-26. https://doi.org/10.22004/ag.econ.207011

Henry BK, Eckard RJ, Beauchemin KA (2018) Review: adaptation of ruminant livestock production systems to climate changes. Animal 12(s2):s445–s456. https://doi.org/10.1017/s1751731118001301.

Herrero M, Havlík P, Valin H, Notenbaert A, Rufino MC, Thornton PK, Blümmel M, Weiss F, Grace D, Obersteiner M (2013) Biomass use, production, feed efficiencies, and greenhouse gas emissions from global livestock systems. Proc Natl Acad Sci USA 110(52):20888–20893. https://doi.org/10.1073/pnas.1308149110.

Houessou SO, Dossa LH, Diogo RVC, Houinato M, Buerkert A, Schlecht E (2019) Change and continuity in traditional cattle farming systems of West African Coast countries: a case study from Benin. Agric Syst 168:112–122. https://doi.org/10.1016/j.agsy.2018.11.003

Hyder I, Sejian V, Bhatta R, Gaughan JB (2017) Biological role of melatonin during summer season related heat stress in livestock. Biol Rhythm Res 48(2):297–314. https://doi.org/10.1080/09291016.2016.1262999

Ibeagha-Awemu EM, Peters SO, Bemji MN, Adeleke MA, Do DN (2019) Leveraging available resources and stakeholder involvement for improved productivity of African livestock in the era of genomic breeding. Front Genet 10:357. https://doi.org/10.3389/fgene.2019.00357,

Ickowicz A, Hubert B, Blanchard M, Blanfort V, Cesaro J-D, Diaw A, Lasseur J, Thi Thanh Huyen L, Li L, Mauricio RM et al (2022) Multifunctionality and diversity of livestock grazing systems for sustainable food systems throughout the world: are there learning opportunities for Europe? Grass Forage Sci 77(4):282–294. https://doi.org/10.1111/gfs.12588

ILRI, I, FAO, WWF, UNEP, ILC (2021) Rangelands atlas. ILRI, Nairobi. Available online at https://www.rangelandsdata.org/atlas/. Accessed on 15 May 2023

Kadohira M, McDermott JJ, Shoukri MM, Kyule MN (1997) Variations in the prevalence of antibody to brucella infection in cattle by farm, area and district in Kenya. Epidemiol Infect 118(1):35–41. https://doi.org/10.1017/s0950268896007005

Kahi AK, Wasike CB, Rewe TO (2006) Beef production in the arid and semi-arid lands of Kenya: constraints and prospects for research and development. Outlook Agric 35(3):217–225. https://doi.org/10.5367/000000006778536800

Khalil MA, Sammour HB (2006) Economics of some feeding packages application under mixed production system at El-Beheira governorate. In: Proceedings of the 13th Conference of the Egyptian Society of Animal Production, Cairo, 10–11 Dec 2006, vol 43, pp 325–339

Kimani T, Schelling E, Bett B, Ngigi M, Randolph T, Fuhrimann S (2016) Public health benefits from livestock rift valley fever control: a simulation of two epidemics in Kenya. EcoHealth 13(4):729–742. https://doi.org/10.1007/s10393-016-1192-y.

KMALF (2010) Kenya national dairy master plan 2010–2023 Vol. I. Situational analysis. Kenya Ministry of Agriculture, Livestock and Fishers (KMALF). Available online at https://www.kdb.go.ke/download/kenya-national-dairy-master-plan-vol-i/. Accessed 11 Aug 2023

Konlan SP, Ayantunde AA, Addah W, Dei HK, Karbo N (2018) Emerging feed markets for ruminant production in urban and peri-urban areas of Northern Ghana. Trop Anim Health Prod 50(1):169–176. https://doi.org/10.1007/s11250-017-1418-1

Kosgey IS, Mbuku SM, Okeyo AM, Amimo J, Philipsson J, Ojango JM (2011) Institutional and organizational frameworks for dairy and beef cattle recording in Kenya: a review and opportunities for improvement. Anim Genet Resour/Resources génétiques animales/Recursos genéticos animales 48:1–11. https://doi.org/10.1017/S2078633610001220. From Cambridge University Press, Cambridge Core

Lanyasunya TP, Wang HR, Mukisira EA, Abdulrazak SA, Ayako WO (2006) Effect of seasonality on feed availability, quality and herd performance on smallholder farms in Ol-Joro-Orok Location/Nyandarua District, Kenya. Trop Subtrop Agroecosystems 6(2):87–93

Latino LR, Pica-Ciamarra U, Wisser D (2020) Africa: the livestock revolution urbanizes. Glob Food Sec 26:100399. https://doi.org/10.1016/j.gfs.2020.100399

Lenné JM, Thomas D (2006) Integrating crop—livestock research and development in Sub-Saharan Africa: option, imperative or impossible? Outlook Agric 35(3):167–175. https://doi.org/10.5367/000000006778536765. Holt-Giménez E, Altieri MA (2013) Agroecology, food sovereignty, and the new green revolution. Agroecol Sustain Food Syst 37(1):90–102. https://doi.org/10.1080/10440046.2012.716388

Makokha S, Witwer M (2013) Analysis of incentives and disincentives for live cattle in Kenya. Technical notes series. MAFAP, FAO, Rome. https://www.fao.org/in-action/mafap/resources/detail/es/c/396069/

Mapiye C, Chikwanha OC, Chimonyo M, Dzama K (2019) Strategies for sustainable use of indigenous cattle genetic resources in southern Africa. Diversity 11(11):214. https://doi.org/10.3390/d11110214

Marshall K (2014) Optimizing the use of breed types in developing country livestock production systems: a neglected research area. J Anim Breed Genet 131(5):329–340. https://doi.org/10.1111/jbg.12080

Mcdougal T (2019) FAO sustainable livestock Africa 2050 report. Poult World. Available online at https://www.poultryworld.net/home/fao-sustainable-livestock-africa-2050-report/. Accessed 14 Aug 2023

Moritz M (2012) Pastoral intensification in West Africa: implications for sustainability. J R Anthropol Inst 18(2):418–438. https://doi.org/10.1111/j.1467-9655.2012.01750.x. Moritz M, Hamilton IM, Yoak AJ, Scholte P, Cronley J, Maddock P, Pi H (2015) Simple movement rules result in ideal free distribution of mobile pastoralists. Ecol Modell 305:54–63. https://doi.org/10.1016/j.ecolmodel.2015.03.010

Mottet A, Teillard F, Boettcher P, De' Besi G, Besbes B (2018) Review: domestic herbivores and food security: current contribution, trends and challenges for a sustainable development. Animal 12:s188–s198. https://doi.org/10.1017/S1751731118002215

MSDNKOAL (2012) Sessional Paper No. 08 of 2012 on national policy for the sustainable development of Northern Kenya and other arid lands. Ministry of State for Development of Northern Kenya and Other Arid

Lands (MSDNKOAL). http://repository.kippra.or.ke/handle/123456789/1020. Accessed on 10 Aug 2023

Mugambi D, Kimenchu, Mwangi M, Wambugu S, Kairu, Gitunu AMM (2015) Assessment of performance of smallholder dairy farms in Kenya: an econometric approach. J Appl Biosci 85:7891–7899. https://doi.org/10.4314/jab.v85i1.13

Muia JMK, Kariuki JN, Mbugua PN, Gachuiri CK, Lukibisi LB, Ayako WO, Ngunjiri WV (2011) Smallholder dairy production in high altitude Nyandarua milk-shed in Kenya: status, challenges and opportunities. Livest Res Rural Dev 23:Article #108. Retrieved 14 Aug 2023, from http://www.lrrd.org/lrrd23/5/muia23108.htm

Muimba-Kankolongo A (2018) Chapter 2 – climates and agroecologies. In: Muimba-Kankolongo A (ed) Food crop production by smallholder farmers in Southern Africa. Academic Press, Cambridge/San Diego, pp 5–13

Mwanga G, Mujibi FDN, Yonah ZO, Chagunda MGG (2019) Multi-country investigation of factors influencing breeding decisions by smallholder dairy farmers in sub-Saharan Africa. Trop Anim Health Prod 51(2):395–409. https://doi.org/10.1007/s11250-018-1703-7

Mwangi V, Owuor S, Kiteme B, Giger M (2020) Beef production in the rangelands: a comparative assessment between pastoralism and large-scale ranching in Laikipia County, Kenya. Agriculture 10:399. https://doi.org/10.3390/agriculture10090399

Nguluu SN, Njarui DMG, Gatheru M, Wambua JM, Mwangi DM, Keya GA (2011) Feeding management for dairy cattle in smallholder farming systems of semi-arid tropical Kenya. Livest Res Rural Dev 23:Article #111. Retrieved 14 Aug 2023, from http://www.lrrd.org/lrrd23/5/njar23111.htm

Nidumolu U, Gobbett D, Hayman P, Howden M, Dixon J, Vrieling A (2022) Climate change shifts agropastoral-pastoral margins in Africa putting food security and livelihoods at risk. Environ Res Lett 17(9):095003. https://doi.org/10.1088/1748-9326/ac87c1

Notenbaert A, Herrero M, Kruska R, You L, Wood S, Thornton P, Omolo A (2009) Classifying livestock production systems for targeting agricultural research and development in a rapidly changing world. Discussion Paper No. 19. ILRI (International Livestock Research Institute), Nairobi, 41 pp. Available at https://core.ac.uk/download/pdf/132631286.pdf. Accessed on 15 May 2023

Nwogwugwu PC, Lee S-H, Freedom E, Chidozie, Manjula P, Lee J-H (2018) Review on challenges, opportunities and genetic improvement of sheep and goat productivity in Ethiopia. J Anim Breed Genom 2(1):1–8. https://doi.org/10.12972/jabng.20180015

Obese FY, Osei-Amponsah R, Timpong-Jones E, Bekoe E (2021) Impact of COVID-19 on animal production in Ghana. Anim Front 11(1):43–46. https://doi.org/10.1093/af/vfaa056.

Odero-Waitituh JA (2017) Smallholder dairy production in Kenya; a review. Livest Res Rural Dev 29:Article #139. Retrieved 9 Aug 2023, from http://www.lrrd.org/lrrd29/7/atiw29139.html

Ogunyemi OI, Orowole PF (2020) Poultry farmers socio-economic characteristics and production limiting factors in Southwest Nigeria. J Sust Dev Afri 22:151–165

Omore AO, Muriuki H, Kinyanjui M, Owango M, Staal S (1999) The Kenya Dairy sub-sector: a rapid appraisal. Research report of the MoLFD/ KARI/ILRI. Smallholder dairy (Research and Development) Project Report. International Research Institute, Nairobi, 51pp. https://hdl.handle.net/10568/2054

Opio P, Makkar H, Tibbo M, Ahmed S, Sebsibe A, Osman A, Olesambu E, Ferrand C, Munyua S (2020) Regional Animal Feed Action Plan for East Africa: why, what, for whom, how used and benefits. CABI Rev 15(44):1–16. https://doi.org/10.1079/PAVSNNR202015044

Ortiz-Gonzalo D, Vaast P, Oelofse M, de Neergaard A, Albrecht A, Rosenstock TS (2017) Farm-scale greenhouse gas balances, hotspots and uncertainties in smallholder crop-livestock systems in Central Kenya. Agric Ecosyst Environ 248:58–70. https://doi.org/10.1016/j.agee.2017.06.002

Otieno DJ, Hubbard L, Ruto E (2012) Determinants of technical efficiency in beef cattle production in Kenya. 2012 Conference, August 18-24, 2012, Foz do Iguacu, Brazil 125853, International Association of Agricultural Economists (IAAE) Triennial Conference. https://doi.org/10.22004/ag.econ.125853

Otte J, Chilonda P (2003) Classification of cattle and small ruminant production systems in sub-Saharan Africa. Outlook Agric 32(3):183–190. https://doi.org/10.5367/000000003101294451

Otte J, Pica-Ciamarra U, Morzaria S (2019) A comparative overview of the livestock-environment interactions in Asia and Sub-saharan Africa. Front Vet Sci 6:37. https://doi.org/10.3389/fvets.2019.00037.

Ouda JO, Kitilit JK, Indetie D, Irungu KRG (2001) The effects of levels of milking on lactation and growth of pre-weaning calves of grazing Boran cattle. East Afr Agric For J 67(1–2):69–75

Paul BK, Groot JC, Maass BL, Notenbaert AM, Herrero M, Tittonell PA (2020) Improved feeding and forages at a crossroads: farming systems approaches for sustainable livestock development in East Africa. Outlook Agric 49(1):13–20. https://doi.org/10.1177/0030727020906170.

Pratt DJ, Le Gall FG, De Haan C (1997) Investing in pastoralism: sustainable natural resource use in Arid Africa and the Middle East. World Bank Group, Washington, DC. Available at http://documents.worldbank.org/curated/en/453331468774590152/Investing-in-pastoralism-sustainable-natural-resource-use-in-arid-Africa-and-the-Middle-East. Accessed on 15 May 2023

Robinson TP, Thornton PK, Franceschini G, Kruska RL, Chiozza F, Notenbaert A, Cecchi G, Herrero M, Epprecht M, Fritz S, You L, Conchedda G, See L (2011) Global livestock production systems. Rome, Italy: Food and Agriculture Organization of the

United Nations (FAO) and International Livestock Research Institute (ILRI), 152 pp. https://hdl.handle.net/10568/10537

Savory A, Butterfield J (2007) Holistic management: a new framework for decision making; Island press, 1999. FAO. State of the world's forests. Available at http://www.fao.org/3/a0773e/a0773e00.htm. Accessed on 15 May 2023

Sebastian K (2009) 005_africaaez_09.png. Agroecological zones of Africa. https://doi.org/10.7910/DVN/HJYYTI/UX1QND. Harvard Dataverse, V2

Sekaran U, Lai L, Ussiri DAN, Kumar S, Clay S (2021) Role of integrated crop-livestock systems in improving agriculture production and addressing food security – a review. J Agric Food Res 5:100190. https://doi.org/10.1016/j.jafr.2021.100190

Simpkin P, Cramer L, Ericksen P, Thornton P (2020) Current situation and plausible future scenarios for livestock management systems under climate change in Africa. CCAFS Working Paper no. 307. CGIAR Research Program on Climate Change, Agriculture and Food Security (CCAFS), Wageningen. Available at https://www.fao.org/3/i2414e/i2414e.pdf. Accessed on 15 May 2023

Staal S, Owango M, Muriuki H, Kenyanjui M, Lukuyu B, Njoroge L, Njubi D, Baltenweck I, Musembi F, Bwana O, Muriuki K, Gichungu G, Omore A, Thorpe W (2001) Dairy systems characterisation of the greater Nairobi milk shed. Smallholder Dairy (R&D) Project report. ILRI, Nairobi, Kenya. https://hdl.handle.net/10568/1590

Tambi NE, Maina WO, Ndi C (2006) An estimation of the economic impact of contagious bovine pleuropneumonia in Africa. Rev Sci Tech 25(3):999–1011. PMID: 17361766

Thornton PK, Kruska RL, Henninger N, Kristjanson PM, Reid RS, Atieno F, Odero AN, Ndegwa T (2002) Mapping poverty and livestock in the developing world. ILRI, Nairobi. Available at https://hdl.handle.net/10568/915. Accessed on 15 May 2023

Turner MD, Schlecht E (2019) Livestock mobility in sub-Saharan Africa: a critical review. Pastoralism 9(1):13. https://doi.org/10.1186/s13570-019-0150-z

UN (2015) United Nation's sustainable development goals. Available online at: https://sdgs.un.org/goals. Accessed 15 May 2023

UN (2022) World population prospects 2022: summary of results. United Nations Department of Economic and Social Affairs, Population Division. UN DESA/POP/2022/TR/NO. 3. https://www.un.org/development/desa/pd/sites/www.un.org.development.desa.pd/files/wpp2022_summary_of_results.pdf. Accessed 11 Aug 2023

Valerio VC (2020) The structure of livestock trade in West Africa. West African Papers, No. 29. OECD, OECD Publishing, Paris. https://doi.org/10.1787/f8c71341-en. Kurtz JE, Mitik L, Zaki C (2021) African trade in livestock products and value chains. In: Bouët A, Tadesse G, Zaki C (eds) Africa agriculture trade monitor 2021. Chapter 4, Pp. 85-133. Kigali, Rwanda; and Washington, DC: AKADEMIYA2063; and International Food Policy Research Institute (IFPRI). https://hdl.handle.net/10568/143970.

Vrieling A, De Leeuw J, Said MY (2013) Length of growing period over africa: variability and trends from 30 years of NDVI time series. Remote Sensing 5:982–1000. https://doi.org/10.3390/rs5020982

Wanyoike F, Njiru N, Kutu A, Chuchu S, Wamwere-Njoroge GJ, Mtimet N (2018) Kenya Accelerated Value Chain Development Program (AVCD)-livestock component: analysis of livestock and fodder value chains in arid and semi-arid lands in Kenya. Nairobi, Kenya: ILRI. https://hdl.handle.net/10568/91179

Open Access This chapter is licensed under the terms of the Creative Commons Attribution 4.0 International License (http://creativecommons.org/licenses/by/4.0/), which permits use, sharing, adaptation, distribution and reproduction in any medium or format, as long as you give appropriate credit to the original author(s) and the source, provide a link to the Creative Commons license and indicate if changes were made.

The images or other third party material in this chapter are included in the chapter's Creative Commons license, unless indicated otherwise in a credit line to the material. If material is not included in the chapter's Creative Commons license and your intended use is not permitted by statutory regulation or exceeds the permitted use, you will need to obtain permission directly from the copyright holder.

The History, Geography, and Characteristics of Indigenous African Taurine Cattle

3

John E. O. Rege, Chi L. Tawah, Isidore Houaga, and Eveline M. Ibeagha-Awemu

Abstract

This chapter introduces indigenous African cattle—their origins, the history of their introduction, and subsequent dispersal across the continent. The first two sections (Sect. 3.1 on origins and history and Sect. 3.2 on present-day breeds and strains) provide a broad coverage of all African cattle, using a framework that can help trace the roots of the different breeds of African cattle that exist today. Sections 3.3, 3.4, and 3.5 are then dedicated to the African taurine cattle, the cattle considered to have originated and evolved in Africa, consisting of two major groups—the Taurine Longhorns and the Taurine Shorthorns. For each breed or strain within each of these groups, the chapter summarizes the classification; distribution; ecological settings; physical, adaptive, and special genetic characteristics, including selected breed photos; predominant production systems; and production characteristics. Some breeds and strains are similar in various dimensions of the coverage and are thus covered in groups. Moreover, the differential availability of data/information led to the difference in the depth of coverage between breeds and strains. The chapter concludes by presenting a summary of the risk status of the taurine breeds of Africa, especially noting the increasing interbreeding among neighbouring breeds, as well as intentional cross-breeding, and making a case for conservation actions for those at risk of extinction. The chapter also advocates for characterization efforts to enhance understanding of these breeds to inform programmes for their sustainable use. Section 3.5 presents a summary and conclusion.

J. E. O. Rege (✉)
Emerge-Centre for Innovations–Africa,
Nairobi, Kenya
e-mail: ed.rege@emerge-africa.org

C. L. Tawah
African Development Bank (Retired),
Abidjan, Ivory Coast

I. Houaga
The Roslin Institute and Royal (Dick) School of Veterinary Studies, Easter Bush Campus, University of Edinburgh, Midlothian, UK

Centre for Tropical Livestock Genetics and Health (CTLGH), Roslin Institute, Easter Bush Campus, University of Edinburgh, Edinburgh, UK

E. M. Ibeagha-Awemu
Sherbrooke Research and Development Centre, Agriculture and Agri-Food Canada, Sherbrooke, QC, Canada

Keywords

Origins of African cattle · African taurine cattle · Ecological settings · Physical and production characteristics · Genetic characteristics

Acronyms

Alb	Albumin
ARS	Agricultural Research Station (of the University of Ghana)
BC	Before Christ
BoLA	Bovine leukocyte antigen
BP	Before present years
CA	Carbonic anhydrase
CBPP	Contagious bovine pleuropneumonia
CM	Centimetres
CREAT	Centre de recherche et d'élevage
DAD-IS	Domestic Animal Diversity Information System
DAGRIS	Domestic Animal Genetic Resources Information System
DRC	Democratic Republic of Congo
DSB	Double-strand break (of DNA)
FAO	Food and Agriculture Organization
GILMA	Gambian Indigenous Livestock Multipliers' Association
GMT	Greenwich Mean Time
Hb	Haemoglobin
ILCA	International Livestock Centre for Africa
ILRI	International Livestock Research Institute
ITC	The International Trypanotolerance Centre
Kg	Kilograms
MM	Millimetres
ONBS	Open nucleus breeding scheme
PGM	Phosphoglucomutase
RIM	Resource Inventory and Management (Abuja, Nigeria)
Tf	Transferrin
UST	University of Science and Technology (Kumasi, Ghana)

3.1 Origins and History

Archaeologists and biologists agree that there is strong evidence for two distinct domestication events from aurochs: *Bos taurus* in the Near East about 10,500 years ago, and *Bos indicus* in the Indus valley of the Indian subcontinent about 7000 years ago. Archaeological findings, however, have led to the new theory that there was an African centre of domestication in the Sahara from southern Libya and north-western Niger to southern Egypt (MacDonald and Hutton MacDonald 2000) and that this third domestication in Africa (of a domesticate, which has been referred to as *Bos africanus*) happened about 8500 years ago. Further genetic studies have also suggested that the present-day humpless cattle populations are so divergent from similar cattle populations of Europe that separate domestication could, indeed, have occurred in Africa (Bradley and Loftus 2000). This is supported by the genetic evidence provided by Hanotte et al. (2002), which also pointed to an exogenous but minor genetic influence of non-African origin from Europe and/or the Near East in the breeds of North and North-East Africa as well as localized areas of southern Africa. These African taurine cattle were also influenced by a slow genetic introgression by the zebu cattle (*Bos indicus*) of Asian origin. More recent genomic studies (Decker et al. 2014) of 134 modern breeds support the idea of three domestication events but also point towards evidence for later migration waves of animals to and from the three main loci of domestication. A study by Stock and Gifford-Gonzalez (2011) reported that although genetic evidence for African domesticated cattle is not as comprehensive as that for other forms of cattle, the available evidence suggests that domestic cattle in Africa are the result of wild aurochs having been introduced into local domestic *Bos taurus* (or *Bos taurus africanus*) populations. Indeed, African cattle of 5000 BP are distinctly different morphologically from the European humpless *Bos taurus* as well as Asian-originating Zebu (Grigson et al. 2000) and are hence believed to have a separate origin.

While scholars are divided about the likelihood of a third domestication event having occurred in Africa, *Bos* remains have been found at African sites in what is now Egypt, such as Nabta Playa and Bir Kiseiba, as long ago as 9000 BP, implying that they may have been domesticated in Africa. The earliest domesticated cattle in Africa have been found at Bir Kiseiba and Nabta Playa in South Egypt (11,000 BP–6000

BP) and Capeletti in Algeria (about 6500 BP), although early cattle remains have also been found at Wadi el-Arab (8500–6000 BC) and El Barga (6000–5500 BC). According to Wendorf and Schild (1980), the domestication of cattle in Africa happened 9500 BP in the Eastern Sahara near the Jebel Marra massif in north-western Sudan. Marshall and Hildebrand (2002) suggest another location on the eastern part of Tibesti in northern Chad. Thus, there is increasingly convincing genetic and archaeological evidence for domestication within Africa, and it is now generally considered that there are at least three auroch domesticates and that all of them are represented in Africa today.

The first major description of African domestic cattle was prepared by Doutressoulle (1947), while Epstein (1971) did the most comprehensive review, also conjecturing on the chronology of their introduction to the continent. A recension of this was presented by Mason (1984). There are also descriptions of cattle types by Stewart (1937), Gates (1952), Payne (1970), and Fricke (1979). Archaeological material was reviewed by Smith (1980), Muzzolini (1983, 1993), Epstein and Mason (1984), and Clutton-Brock (1989). Blench (1993) presented additional ethnographic and linguistic evidence for the prehistory of African cattle. ILCA (1979a, b, 1992a) and Shaw and Hoste (1987) made an overview of the distribution of the trypanotolerant cattle of West and Central Africa.

Over the past 25–30 years, the global science community, with significant leadership provided by the International Livestock Research Institute (ILRI) scientists, working with local scientists and international partners in Africa and Asia, has undertaken in-depth studies to help establish the origins of and genetic relationships among African cattle.

Rege (1999) and Rege and Tawah (1999) made attempts to do a comprehensive classification of present-day African cattle breeds and strains of various groups, including recent derivatives. This classification formed an important basis for the sampling of biological materials that followed and underpinned the study by Hanotte et al. (2002), which was the first comprehensive continent-wide molecular genetic study of African cattle.

The work by Hanotte et al. (2002) led to the reconstruction of the first genetic history of cattle in Africa, linking livestock to human history and providing a glimpse into the distant past of the peoples and civilizations of Africa (and Asia), and the publication has remained a reference on the history and geography of African cattle to date. The study has inspired several follow-ups to deepen the understanding of the current distribution of cattle genetic diversity and how this relates to the centres of cattle domestication in the Near East and the Indus Valley.

Subsequent autosomal and Y-specific microsatellite analyses have revealed the genetic relationships among the present-day African breeds and have detected little-to-no Asian Zebu or Middle East or European taurine influences among the indigenous West African taurine living within the tsetse fly zone. A major Asian Zebu influence, which probably started around the seventh and eighth century AD, is clearly visible among the cattle populations of Eastern Africa and the Sahel. These populations are the result of interbreeding between the African indigenous taurine and the Zebu of Asian origin. The predominantly taurine genetic background of the southern African Sanga supports the archaeological view of an early arrival of cattle in the southern part of the continent before the major Zebu influence in the Horn of Africa. These studies of the origin and history of present-day African indigenous cattle demonstrate their unique genetic background—both the taurine and the indicine.

Some of the studies that have informed the present-day understanding of the history, geography, and unique genetic attributes of African cattle include Grigson (1991), who provided alternative views on the origins of present-day African Sanga cattle, Frisch et al. (1997), who presented new evidence of the origins of southern African Sanga, and Hanotte et al. (2002), who examined the genetic signatures of the origins of African cattle populations and revealed dispersal history and current geographical distribution of various genetic backgrounds. Kim et al. (2017)

analyzed patterns of African cattle genetic variation by sequencing 48 genomes from five indigenous populations and comparing them to the genomes of 53 commercial taurine breeds. The highest genetic diversity was found among African Zebu and Sanga cattle populations. Kim et al. (2020) undertook a whole-genome sequence analysis of 172 indigenous African cattle from 16 breeds representative of the main cattle groups. The study identified a major taurine × indicine cattle admixture event dated to circa 750–1050 years ago, which supposedly shaped the genome of today's cattle in the Horn of Africa.

It is now well-established that African cattle carry a taurine maternal ancestry originating from the Near East taurine domestication centre(s), while the possible genetic contribution of the now-extinct African auroch (*Bos primigenius opisthonomous*) remains unclear (Decker et al. 2014; Bonfiglio et al. 2012). The pattern of introgression of the Zebu genome across the southern, eastern, and north-western parts of Sub-Saharan Africa has been well-documented using autosomal and Y-chromosome-specific microsatellite loci (Bradley et al. 1994; Hanotte et al. 2000; Freeman et al. 2004).

Taken together, available results strongly suggest that the earliest cattle in Africa were of taurine *B. taurus* type. Subsequent waves of migrations of humped cattle (Zebu or *B. indicus*) then reshaped the genomic landscape of African cattle (Hanotte et al. 2002; Bradley et al. 1994; Freeman et al. 2004). The origins of African and European cattle seem to be more closely related and quite distinct from Indian cattle, the relatively large divergence providing evidence for at least two separate domestication events, and these were presumably of different subspecies of the aurochs, *Bos primigenius* (Loftus et al. 1994). Today, the African continent is uniquely rich in cattle diversity, with the number of unique breeds or strains estimated to be between 145 (Rege 1999) and 150 (Mwai et al. 2015). The genetic distinctiveness between these cattle breeds/strains remains largely unknown, and it may be more appropriate to talk about African cattle ecotypes. The various types are adapted to local environmental conditions (e.g. high temperatures, long periods of drought, and vector-borne diseases), which largely restrict the use of pure exotic breeds of European origin. The origin and migration pattern of cattle in Africa is depicted in Fig. 3.1.

3.2 Present-Day African Cattle Breeds and Strains

Present-day African cattle populations can be classified into four broad categories: the humpless *Bos taurus* (the indigenous African taurine cattle); the humped *Bos indicus* (the Zebu), distributed widely in Africa; *B. taurus* × *B. indicus* derivatives (the Sanga), found mainly in eastern and southern Africa; and Zebu × Sanga types (Zenga), such as the Fogera and Horro of Ethiopia and the Nganda of Uganda. The taurine (humpless) type has two subgroups—Longhorns (*Bos taurus longifrons*) and Shorthorns (*Bos taurus brachyceros*)—both of which are restricted to West and Central Africa. While the Longhorns are represented by two breeds only—the N'Dama and the Kuri—the Shorthorn subgroup has numerous representatives. In addition, Africa is home to a number of recently developed commercial composite breeds that are products of systematic cross-breeding between one or more of the four categories and exotic breeds, primarily of European origin. This forms the fifth breed category. A final category consists of recent and ongoing imports of exotic cattle breeds reputed for dairy, beef, or dual purpose. The classification can, thus, be summarized as follows:

1. The indigenous taurine cattle of Africa
2. The Zebu cattle of Africa
3. The Sanga cattle of Africa
4. The Zenga cattle of Africa
5. African commercial composite cattle breeds
6. Well-established exotic cattle breeds in Africa

These cattle populations (covered in this chapter and Chap. 4) inhabit more than five distinct major agro-ecological zones. Overall, the Zebu cattle (and their crossbred derivatives) are common in the arid and semi-arid northern Sahelo-Sudanian zone as well as in the eastern part of the continent, including the highlands, whereas tau-

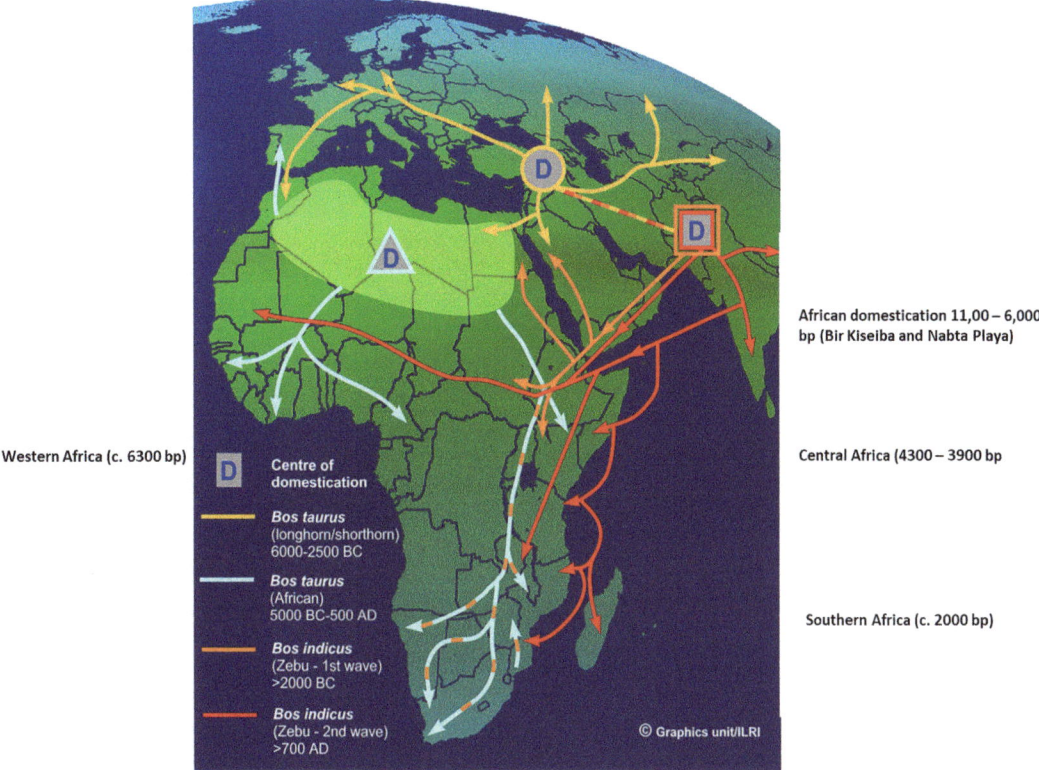

Fig. 3.1 Origin and migration patterns of domestic cattle in Africa. (Source: Modified from Mwai et al. (2015), licensed under CC-BY 4.0)

rine cattle today dominate the sub-humid and humid regions of West Africa, which are heavily infested with tsetse fly. The Sanga are predominant in the eastern African region, around the Great Lakes region, and in the southern part of the continent. The highest genetic diversity is among the African Zebu and Sanga cattle (Kim et al. 2017) (see Chap. 4).

The knowledge of the different categories of cattle on the continent and the uniqueness that they possess is crucial for developing strategies for detailed genetic characterization and have implications for the conservation and use of these genetic resources, as they point to population locations, phenotypes, and uniqueness. Africa is clearly a treasure trove of rich genetic diversity of cattle that are adapted to the diverse eco-zones and production systems covering a wide range of environments—from the harsh fringes of the Sahara Desert in North Africa, the drier areas of the Horn of Africa, through the wet tropical lowlands found along the Congo River, and on to the vast savannahs of Southern Africa. The spatial distribution of cattle breed types in Africa is depicted in Fig. 3.2.

The rest of this chapter presents a description of the indigenous taurine cattle of Africa, with a focus on habitat (ecological settings and production systems), their production and unique genetic characteristics, and their main uses. The next chapter (Chap. 4) undertakes a similar coverage of the remaining cattle groups—the Zebu and the Zebu–taurine derivatives, that is, the Sanga, the Zenga, and commercial composite breeds—and presents a summary of the major well-established exotic cattle breeds introduced into Africa over the past century.

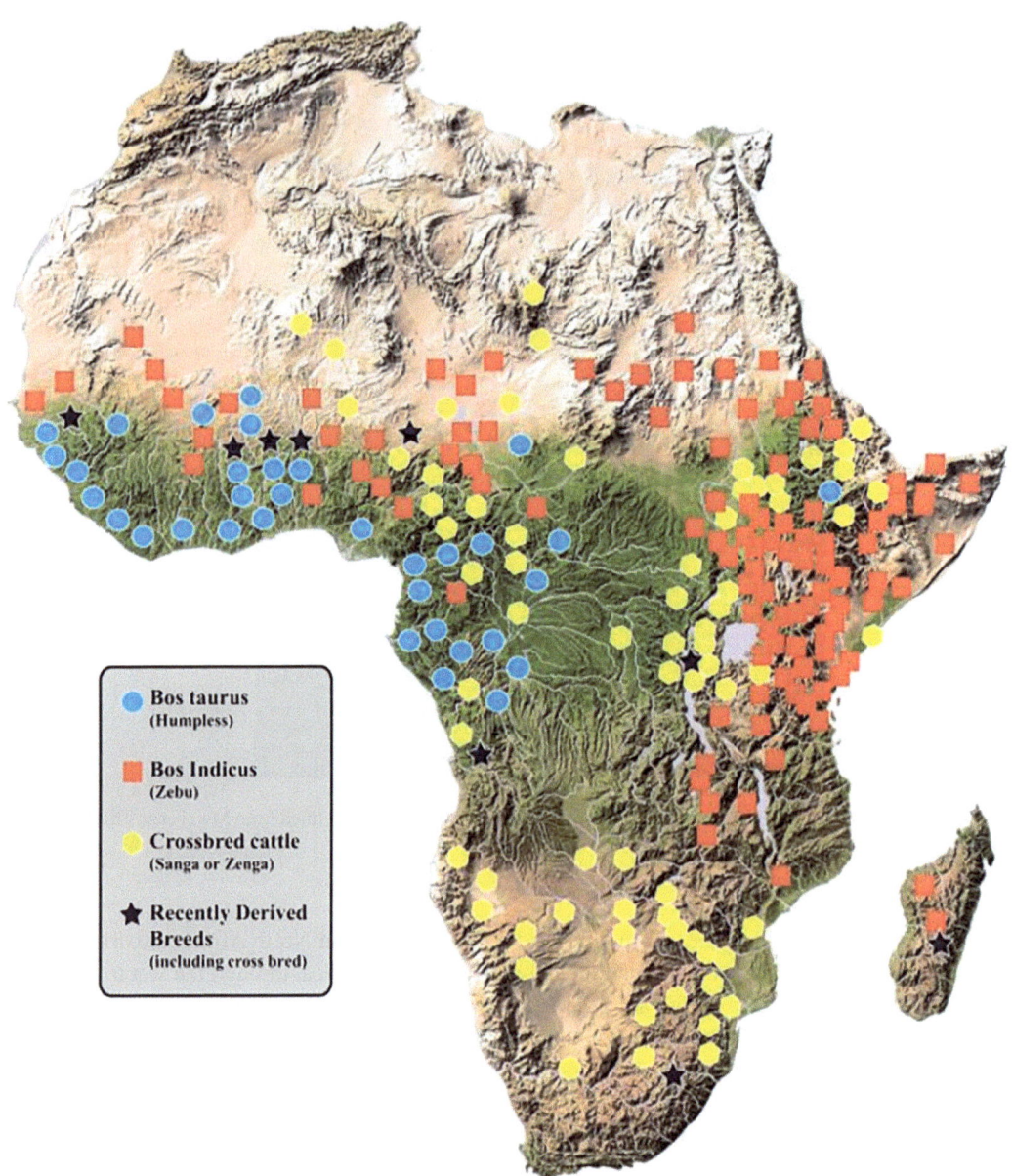

Fig. 3.2 Distribution of cattle breed types in Africa. *Dots indicate the type of cattle in a region (cattle of North Africa and exotic cattle breeds are not shown). (Source: Mwai et al. (2015), licensed under CC-BY 4.0)

3.3 The Indigenous Taurine Cattle of Africa

The taurine (humpless)-type cattle of Africa have two subgroups—the Longhorns (*B. taurus longifrons*) and the Shorthorns (*B. taurus brachyceros*). Both subgroups are restricted in their distribution to West and Central Africa. While the Longhorns are represented by two breeds only—the N'Dama and the Kuri—the Shorthorns are represented by a total of 15 breeds, strains, or ecotypes, all found in West and Central Africa, except one, the Sheko, which is found in southwest Ethiopia (Table 3.1).

Table 3.1 Humpless cattle breeds of Sub-Saharan Africa

Group and breed/strain	Synonyms	Location/country
Humpless Longhorns		
1. N'Dama	Boenca, Boyenca, Fouta Jallon, Fout, Malinke, Mandingo, N'Dama Petite, Fouta Malinke, Futa, N'Dama Peti, Gambian Longhorn	All coastal countries of West and Central Africa plus Mali, Burkina Faso, and Central African Republic
2. Kuri	Baharié, Bare, Borrie, Boundouma, Dongolé, Kouburi, Buduma, White Lake Chad, Boenca, Boyenca	Chad, Niger, Nigeria, Cameroon
Humpless Shorthorns (West and Central Africa)		
Savannah Shorthorns		
3. Ghana Shorthorn	Gold Coast Shorthorn	Ghana
4. Baoulé		Central African Republic, Côte d'Ivoire, Gabon, Burkina Faso
5. Savannah Muturu	Muturu	Nigeria
6. Somba	Atacora (Benin), Mango, Konkomba (Togo)	Benin, Togo
7. Doayo	Namchi, Namshi, Poli	Cameroon
8. Kapsiki	Kirdi	Cameroon
9. Bakosi	Bakuri, Kosi	Cameroon
Dwarf Shorthorns		
10. Ghana Dwarf Muturu	Muturu, Forest Shorthorn	Ghana
11. Lagune	Dahomey, Mayombe (Zaire)	Benin, Togo, Côte d'Ivoire, Congo, Gabon, Zaire
12. Logone	Toupouri	Chad
13. Bakweri	Muturu	Cameroon
14. Lobi	Lobo-Goui, Mere-Lobi	Burkina Faso
15. Liberia Dwarf Muturu	Muturu	Liberia
16. Nigeria Dwarf Muturu	Muturu, Forest Muturu	Nigeria
Humpless Shorthorns (Eastern Africa)		
	Shewa-Ghimira, Goda, Mitzan	Ethiopia

Source: Rege et al. (1994a)

3.3.1 The Longhorn Taurine Cattle

As alluded to above, humpless (*Bos taurus*) Hamitic Longhorn cattle were believed to have descended from the first domesticated cattle populations of the humpless Hamitic Longhorn cattle in the 'Fertile Crescent', possibly in 9000 BC (Payne and Hodges 1997), and are considered to be the first cattle to be introduced into Africa by nomadic people via the land connection with Asia. From here, they spread westwards and southwards. More recently, combining genetic marker analyses with archaeological findings of indigenous cattle from the African continent indicates that the present-day indigenous African humpless cattle were the earliest, with a history dating as far back as 8000 BC (Gifford-Gonzalez and Hanotte 2011). The Longhorns are represented by only two breeds—the N'Dama and the Kuri. The two breeds are quite distinct in their morphology, adaptive and production attributes, and habitats (Rege 1999; Rege and Tawah 1999; Tawah et al. 1997).

Out of a total of ten million trypanotolerant cattle in 1985 in West Africa, the N'Dama was the largest group, accounting for 49.5%, the Savannah Shorthorn was the next largest group with 20%, and this was followed by the Dwarf West African Shorthorn, which accounted for 1.0%. Zebu × N'Dama cross-breeds accounted for 12.6% and Zebu × West African Shorthorn for 16.9% of the trypanotolerant cattle population in the region (ILCA 1992a).

Except for the Kuri, which lives in a very special environment around Lake Chad where tsetse flies have not been recorded, the taurine breeds (both the Longhorns and the Shorthorns) have developed in tsetse-infested areas and have developed varying degrees of trypanotolerance. The N'Dama is the best-known, the most numerous and the most widely spread of the trypanotolerant breeds. Although no recent comprehensive cattle census has been undertaken in West Africa, there is increasing cross-breeding of the N'Dama with Zebu breeds, and other than on-station herds, pure N'Dama population is decreasing at alarming rates on-farm across West Africa.

The N'Dama Cattle

***Other names*:** *Boenca* or *Boyenca* (Guinea-Bissau), *Fouta Jallon, Fouta Longhorn, Fouta, Malinke, Futa, Malinke,* Fouta Malinke, Madingo or *Mandingo* (Liberia), *N'Dama Petite* (Senegal), *Fout, N'Dama Peti, and Gambian Longhorn*

Incorrect names that are sometimes used: *Dama* and *Ndama*

Classification, Distribution, and Ecological Settings

The N'Dama is the most representative '*Bos taurus*' breed in West Africa and is believed to have originated from the Fouta Djallon plateau in Guinea (Conakry). From there, the N'Dama supposedly spread to the Sudanian and Guinean regions. Being trypanotolerant, the breed has been used for large-scale dissemination for grazing savannah across West and Central Africa, especially in the regions infested by the tsetse fly.

Today, the N'Dama is found in most West and Central African countries, namely, Guinea, Guinea-Bissau, The Gambia, Senegal, Sierra Leone, Côted'Ivoire, Mali, Burkina Faso, Togo, Ghana, Nigeria, Central African Republic, Gabon, Cameroon, Congo (Brazzaville), and the Democratic Republic of the Congo (DRC, former Zaire), particularly in the regions infested by tsetse flies (DAGRIS 2022). The estimated population size of the breed in West and Central Africa was about 4.9 million heads (Shaw and Hoste 1987), spread across the countries of West and Central Africa. An experimental population is also found in Kenya (owned by the International Livestock Research Institute).

Originally imported into the Democratic Republic of the Congo (former Zaire) from West Africa starting over seven decades ago, large-scale herds of this breed have been established and improved in Zaire by Compagne J. Van Lanckeic, a private company. The herd size is over 40,000 heads of pure N'Dama. Through a collaboration with ILRI, the ranch has been able to improve the mature weight of the breed by 30–50 kg without affecting inherent hardiness. In addition, large herds of N'Dama are kept exclusively for beef under ranching conditionsin Guinea. The breed has been selected and crossed with Jersey in Côte d'Ivoire since 1955 and has also been crossed with the Sokoto Gudali in Ghana following its original introduction in 1923. The N'Dama has also been crossed with the West African Shorthorn in Togo. Since 1986, the N'Dama has been exported from Senegal to the Virgin Islands, where they have been bred with Red Poll, and a new breed, the Senepol, has been created (Rege and Tawah 1999; Maule 1990).

Physical, Adaptive, and Special Genetic Characteristics

The N'Dama has a compact body that is set on short legs of fine bone. It is of medium size, being 100 cm of height at the shoulder for cows and 120 cm for bulls, and with a short and broad head. The neck is thick and deep, the back is fairly broad, well-fleshed, and straight from withers to tail head, and it has a broad muzzle. The animals have no hump. The typical coat colour is fawn with darker extremities and lighter underside; common solid colour is light-to-dark fawn, grey, dun, light red, chestnut, or red with a black head. The belly and the lower part of the tail may be white. The average horn size is about 60 cm, and the horns are typically curved upwards and outwards or have a lyre shape, although there are different horn shapes and occasional polled individuals, especially in Sierra Leone and Guinea. The dewlap and umbilical folds are poorly developed. An N'Dama cow and herd are shown in Fig. 3.3.

In addition to its well-established tolerance to trypanosomiasis, the N'Dama is also considered

Fig. 3.3 A typical N'Dama bull and herd at Nsukka, Nigeria. (Photos by Dr E.M. Ibeagha-Awemu)

resistant to tick-borne diseases (Mattioli et al. 1993, 1998; Mattioli and Dempfle 1995). Specifically, tick resistance is mainly associated with Zebu breeds. Studies carried out in The Gambia show a higher resistance to ticks and tick-borne diseases in N'Dama than in Gobra (*Bos indicus*) cattle (Mattioli and Dempfle 1995). Tick resistance in N'Dama breed appears to be effective against those species with long hypostome, such as *Amblyomma variegatum* and

Hyalomma spp. In addition, the N'Dama is well-adapted to stressful humid and dry tropical climates. However, the breed is susceptible to rinderpest.

Production Characteristics and Production Systems

To date, the characterization of the N'Dama has predominantly emphasized the resistance to trypanosomiasis—which remains the trait of major interest in the production systems that dominate the habitat of the breed—and, to some extent, growth and milk production. More comprehensive characterization that includes objective measures of both inputs and outputs under local production environments is lacking.

The N'Dama is used in traditional production systems as a multipurpose breed providing milk, meat, manure, and traction. N'Dama oxen are considered good work animals, but cows are not considered very good milk producers. Much of the production data on N'Dama has been collected on government and university research stations and, hence, may not reflect the on-farm performance of the breed.

N'Dama cattle are known for their beef conformation. Average birthweight of 19 and 22 kg was reported in two high and zero trypanosomiasis risk areas, respectively (Feron et al. 1988), while respective bodyweights for mature cows in the two areas were 296 and 331 kg. The average adult weights range from 320 to 360 kg and from 250 to 270 kg for males and females, respectively (Payne 1970; Maule 1990; Starkey 1982, 1984; Portar 1991; Mason 1996). The dressing percentage is around 50%, and the meat has a very good flavour without much fat (Rege and Tawah 1999; Maule 1990).

Based on on-station data collected in a nucleus herd at Yamoussoukro station in Côte d'Ivoire, N'Goran et al. (2016) reported the age at first calving of N'Dama heifers between 20 and 65 months, with an average of 33.2 months. Calving intervals ranged from 300 to 822 days, and the average fertility rate in free mating was 86.8%. Mean birthweight was 17 kg for heifer calves and 19 kg for bull calves. Under ranching conditions involving grazing, salt licking, provision of minerals and dipping, the heifers have been reported to calve at 35–42 months (Maule 1990; Starkey 1984).

Least squares analyses of 668 lactations recorded over a 4-year period by Agyemang et al. (1991) in The Gambia showed a mean lactation length of 420 days, milk offtake of 404.3 kg, fat content of 5.1%, protein content of 3.2%, calf weaning weight of 88.1 kg, and calving interval of 641 days. A productivity index incorporating milk offtake, calf weaning weight, calving rate, and viability gave a mean annual of 140.6 kg of weaner calf plus the weight equivalent to milk offtake per 100 kg of cow metabolic weight. The index was higher than that which had been recorded for larger Zebu (*Bos indicus*) cattle managed under similar production systems elsewhere in Africa. The study concluded that the trypanotolerant N'Dama cattle appeared to be more productive than previously thought and suggested that the breed should be taken more seriously when promoting livestock development, especially in the tsetse-infested areas of Africa where other breeds cannot survive.

It must be noted that partial milking (in consideration of the needs of the calf) is the most frequent practice in the traditional herds of West Africa. In these circumstances, the milk yield (excluding consumption by the calf) is estimated at 70–100 kg per cow per year in most countries. The full lactation yield is about 500–600 kg (Mason 1984; Portar 1991). However, a dairy herd of N'Dama maintained in Sierra Leone from 1944 to 1952 reportedly averaged 540 kg over a lactation period of 283 days in one five-year period, with the highest yields recorded at 1150 kg per lactation (Touchberry 1967).

N'Dama cattle are remarkably productive under the moderate-to-high tsetse-fly-challenge areas (Fall et al. 2003), which characterize large areas of West and Central Africa, and they have been utilized in various production environments to meet the nutritional and livelihood needs of resource-poor farmers. The relatively high number of N'Dama cattle in countries like The Gambia, Guinea, southern Senegal, Guinea-Bissau, Mali, Liberia, Sierra Leone, Congo, and Gabon is directly related to its distinguishable disease tolerance and relatively high adaptability to the local climatic conditions. Moreover, because of its unique characteristics, the breed has been crossed with the Jersey, Red Poll, Sahiwal, Fulani Zebu, Sokoto Gudali, and Simmental (Felius 1995).

An open nucleus breeding scheme (ONBS) for the N'Dama was established in The Gambia by the International Trypanotolerance Centre (ITC) in 1994 with the objectives of enhancing its genetic improvement and facilitating conservation through use (Dempfle and Jaitner 2000). The scheme focused on improving milk and meat production while maintaining trypanotolerance and other adaptive attributes. The involvement in the ONBS of multipliers and their association (Gambian Indigenous Livestock Multipliers' Association (GILMA)) in the dissemination of improved N'Dama bulls as well as the training of livestock assistants, farmers and multipliers, was an important component. Unfortunately, despite the promise demonstrated by this scheme, from 2006, GILMA collapsed as the programme suffered significant funding challenges, including a declining trend in human capacities and extension services. The injection of animals into the nucleus herd declined, and the release of new bulls into farmers' herds essentially stopped.

The Kuri Cattle

Other names: Lake Chad cattle, White Lake Chad, Buduma, Kuburi, Koubouri, Dongolé, Boudouma, Boundouma, Budumu, Budduma, Bare, Baharié, Borrie, and Kouri

While the origin of the Kuri is not certain, there are two theories. One theory considers that it is a pure Hamitic Longhorn, which descended from the ancient Egyptian or Hamitic Longhorn (Sect. 3.1), as depicted in the Egyptian drawings. According to this theory, the Kuri subsequently branched off the stream of the Hamitic Longhorn on its north-westerly passage from Egypt and migrated south-westwards through the Sahara corridor to Lake Chad. The other theory considers the Kuri to be of the Sanga (*Bos indicus* × *Bos taurus*) type that originated from the intermixture of lateral-horned Zebu and the Hamitic Longhorn in upper Egypt and present-day Ethiopia (DAGRIS 2022). The large and unique bulbous horns have facilitated the identification of its presence in rock paintings, which are estimated to be 4000–5000 years old (Epstein and Mason 1984; Baker and Manwell 1980).

Classification, Distribution, and Ecological Settings

Today, Kuri cattle are kept by the Kuri and Buduma ethnic groups. These groups are thought to have descended from the Kanembou tribe, which migrated to the Kanem district from Libya and the French Sudan in historical times.

The Kuri habitat is around the shores and islands of Lake Chad, bordering Chad, northern Cameroon, Nigeria, and Niger (in N'Guigmi Province). Its main habitat is southern Chad and north-eastern Nigeria. Most of the pure Kuri in the lake region are found on the islands of Djibadala, Koremeron, Debala, and Bagabol. Data on population estimates of Kuri cattle are quite outdated. ILCA (1992a) indicated populations of 39,560 heads in Niger and 13,947 in Nigeria. A total of 200,000 heads had been reported earlier by Renard (1972) for the entire Lake Chad Area. However, the restricted and rapidly shrinking habitat of the breed and the ongoing extensive cross-breeding with Zebu breeds in adjoining areas indicate that the breed is under immense pressure and is vulnerable.

The area in which the Kuri cattle are found is between approximately 130 and 160 North latitude and 130 and 170 East longitude and embraces the islands and shores of Lake Chad, both in Bornu Province, Nigeria, and in the Chad territory and N'Guigmi Province of Niger. The Kuri also extends into the Sudan Savannah zone to the west and south of Lake Chad and to the borders of the Sahelian zone in the north-west. The area is flat, with extensive marshes near the shores of Lake Chad, except the area that extends into the Sudan Savannah zone, which is undulating with sand dunes. The average elevation of the area above sea level is approximately 305 metres. The area has the Sudan zone climate with typically a clearly defined wet season extending over 5 months from May to September, while the remainder of the year is dry. During the period from October to February, the days are hot and dry, with temperatures of over 43 °C. While the climate of the waterside areas of the lake is Sahelian, borders to the north have a sub-desert climate.

The diurnal range in both temperature and humidity is, however, considerable, and the nights can be cold and sometimes foggy. During the dry season, dry harmattan winds, carrying fine particles of dust from the desert, blow from the north and north-east and from the east to produce dry fog; during the wet season, monsoon winds blow from the south-west and west. Local secondary winds include daytime breeze from the lake when waterside temperatures are higher than those of the lake and night-time breeze from the land when a reversal in temperatures occurs. Winds are generally stronger in the day than at night. During the rainy months, storms move over the area from the south-west. At this time of the year, although temperatures are lower than in the dry season, the air gets humid, and the sky is frequently overcast.

The vegetation of Lake Chad is typically aquatic or semi-aquatic—a mixture of Sudan and Sahel zone vegetation. The best pastures are on the lowest highlands, which have been levelled by erosion. There are three types of vegetation on these lowlands: Temporarily flooded areas are dominated by *Sporobolus helvolus* and *Sporobolus spicatus*; the sandy bank vegetation is predominantly *Imperata cylindrica, Cynodon dactylon, Paspalindum geminatum,* and *Panicum subalbidum*; and the floating edge vegetation consists of *Phragmites mauritianus, Typha angustifolia, Cyperus papyrus,* and *Pennisetum* spp. Around the shores of the lake, in the marshy area, tall reed grasses of the papyrus type grow profusely, and grass species include *Andropogon* spp. and annual grasses such as *Aristidia mutabilis, Cenchrus biflorus,* and *Eragrostis tremula*. Along the banks of rivers and rivulets that drain into Lake Chad, there is a dense growth of *Commiphora africana, Acacia raddiana, Balanites aegyptiaca, Acacia senegal, Cadaba farinosa,* and *Calotropis procera. Cenchrus echinatus* and *Andropogon* spp. are the most common among the grass types.

Physical, Adaptive, and Special Genetic Characteristics

The Kuri (Fig. 3.4) is a tall animal, distinguished by its enormous horns and the absence of hump. Indeed, standing 15.2 cm taller than the average Zebu cattle in Nigeria, the Kuri is taller than most indigenous cattle of Africa. The height at withers can reach 180 cm in Kuri bulls. Cannon bone length averages 24.8 cm in bulls and 24.5 cm in cows. The head is long, with a straight profile and a wide forehead, to which the prominence of the orbital arches lends a 'degree of concavity'. The horns of the Kuri are its most remarkable feature, and both bulls and cows have horns. They are typically long (70–150 cm), with 'normal' horns being 60–90 cm long. The horns are circular in cross section, are about 20–100 cm in circumference at the base, and are lyre-shaped or crescent-shaped. Sometimes, however, the horns may be short, about 20–30 cm long. These horns may have a surface, which is roughened and ridged, and a cross section, which is flattened, so that their appearance is that of 'enormous ears'. The colouration of the horns is generally light except for the tips, which are black. The horns, although their appearance is massive and lend an aspect of great weight to the head, are not heavy as their structure is fibrous material that is light weight—being made of spongy material with a thin shell. Prunier (1946) had suggested that the bulbous base and spongy interior of Kuri horns was an adaptation for buoyancy. Loose horns can be seen in Kuri herds, and polled animals are not uncommon.

Reported linear body measurements of mature animals are highly variable, perhaps more a reflection of the variability in the definition of 'maturity' in the different studies than the actual variation among Kuri animals. The neck is short and flat. The ears are of medium size and are carried horizontally. The body is long and the top line straight, rising slightly from the withers, which, although thick, does not carry a hump. The hindquarters are of moderate slope. The tips of the dorsal vertebrae show a fused bifid structure. The limbs are long, and the hoofs are large and open. Overall, the body profile of a Kuri animal is straight, long, and disproportionate with

Fig. 3.4 A Kuri cattle herd in its Lake Chad environment and typical Kuri animals. (Courtesy: ILRI)

its cephalic mass. The most common coat colouration is white, but grey shading over the shoulders and the extremities, red, and red and white are sometimes seen. In areas further from Lake Chad, Kuri herds show evidence of a degree of Zebu interbreeding, and small cervico-thoracic humps may be seen as well as an increased proportion of broken and pied coat colours.

There is clear sexual dimorphism in the Kuri, with females having much smaller bodies than males. In addition, the head, tail, ear, and horn measurements of the cows are much smaller than those of the bulls, and both bulls and oxen are taller and broader at the chest than cows. Mature bulls weigh 500–650 kg, oxen weigh 500–750 kg, and cows weigh 360–450 kg.

The Kuri has been living in the hot aquatic milieu of Lake Chad for thousands of years and is well-adapted to this environment. They are excellent swimmers and are able to move around the aquatic habitat and thrive on the natural rangelands on the islands and the shores of the lake. They have a strong preference for fresh pastures. They are reputed for tolerance to insect bites and tend to 'ignore' the swarms of biting insects, which are found in their habitat during the greater part of the year. Their Zebu counterparts in this habitat are sensitive to these insects and have to be moved out of the lake basin before the peak of the biting insects. Kuri animals are, however, intolerant to heat and sunlight and are not able to withstand extended periods of drought. They, thus, tend to spend a significant part of the day immersed in water, with only their nostrils above water. They can graze in water that is as high as their bellies.

Kuri cattle are unable to survive outside the Lake Chad environment, and attempts to introduce them elsewhere have not succeeded, except around the Yobe River, where the environment is similar to that of Lake Chad. This has direct implications for the future survival of the breed since the water levels of Lake Chad are falling and the habitat is changing rapidly and could disappear over time. In addition, while the Kuri seems to be relatively immune to a range of indigenous parasitic diseases, the breed is susceptible to trypanosomiasis and piroplasmosis

(Queval et al. 1971). The Kuri is also reported to be susceptible to rinderpest and contagious bovine pleuropneumonia (CBPP) (Adeniji 1985; RIM 1992). Mortality rates of 27% have been reported in Kuri herds in Nigeria (RIM 1992), and this has been attributed to the CBPP epidemic.

Characterization of the Kuri serum has shown that total globulins are in greater proportion than albumins, which is a possible adaptation of the breed to its aquatic environment and may have implications for its immune responses. However, the relationship with the latter is not clear and requires further investigation.

Haemoglobin (Hb) gene frequency in the Kuri seems to suggest that the breed received its Hb^B through Zebu introgression. On the other hand, the complete absence of Hb^C in Kuri populations may be an indicator of its purity. Like *Bos taurus* breeds, the Kuri differs from the Zebu in their Y-chromosome—which is submetacentric in *Bos taurus* and acrocentric in the Zebu.

Production Characteristics and Production Systems

The Kuri cattle are raised in an unusual production system that integrates livestock production with flood-retreat farming and fishing. There is an abundance of pasture, and crop residues are not major feed sources in the system. The management system is traditional and extensive, with dependence on grazing. Flooding in the low-lying islands during the rainy season forces herds to be moved to the highlands and return to the islands during the dry season, a kind of transhumance. In the dry season, animals in herds of 30–35 heads follow their herdsmen in swimming from island to island in search of water-weeds as feed, a system that has been described as 'aqueous transhumance'. On the other hand, 'upland transhumance' is used to refer to the movement of cattle to the highlands during the wet season (June to September), as practised by the Yedina tribesmen. For both transhumance types, animals are split into transhumant and non-transhumant herds, with the latter comprising mainly lactating cows and calves.

The Kuri is reputed to be a good milk animal, and cows have traditionally been milked by their tribal owners. It is also considered to have great potential as a meat animal, with great fattening ability, but its owners do not use it as such. They are also used as pack animals but only sparingly, and they are not generally considered to be suitable for draft purposes. As a draft animal, the Kuri, on account of its heaviness, lethargic temperament, and slow movement, has poor performance. Even as pack animals, they suffer from the sun and get tired quickly.

Reliable information on Kuri lactation performance as well as milk composition is limited. The daily milk yield of the Kuri is reported to range from 3 to 6 kg over a 6–10-month lactation period. Tawah et al. (1997) summarized some historical lactation performances of the breed, including a record 2400 kg of milk reported in Nigeria over a 314-day lactation period and a 1259 kg lactation yield recorded over a lactation period ranging from 210 to 214 days in Chad. The lactation milk yield of the Kuri increases progressively from the first lactation to the fourth, at which point it plateaus and then declines over subsequent lactations. Tawah et al. (1997) referred to the results of a single study (based on an anonymous source) that estimated the butterfat, solids-not-fat, and lactose contents of Kuri milk at 4.2%, 8.3%, and 48.3%, respectively.

Queval et al. (1971) reported the average birthweight of Kuri at 25.0 and 22.5 kg for male and female calves, respectively. Tawah et al. (1997) have summarized some growth performance data: Yearling weights have been reported at 130 and 125 kg, respectively, for males and females, while corresponding weights at 2 years have been reported at 225 and 200 kg; animals on a 117-day experimental feeding on a diet consisting of fresh Napier (*Pennisetum* spp.) grass, cotton seed, and sodium bicarbonate supplement achieved an average daily weight gain of 0.62–0.65 kg (and consumed 6.9–7.4 kg per kg of weight gain) compared to extensively grazed controls on natural pastures, which gained 0.17 kg/day.

Kuri steers readily achieve 200–250 kg of carcass at 5 or 6 years of age on natural pasture without supplementary feeding and up to 700 kg at slaughter when stall-fed (Renard 1972). A slaughter weight of 500–600 kg at the age of 5 years has also been reported (Tawah et al. 1997). The animals have good beef conformation (Renard 1972), and the meat is tender, juicy, and well-marbled (Renard 1972; Queval et al. 1971). Prime meat has been obtained from animals 4–6 years old in excellent fattening conditions. The most comprehensive slaughter information on the Kuri has been reported by Queval et al. (1971) and Renard (1972) based on the data collected from Fort-Lamy abattoirs in Chad. About 40% of Kuri carcasses weighed over 200 kg. These results have been summarized by Tawah et al. (1997). Kuri hides are reported to be of good quality.

Reproductive traits are the least documented of the Kuri performance traits. Age at first calving has been reported to range from 36 to 48 months (Adeniji 1985), and females have an active productive life of up to 12 years, during which they are able to produce an average of six to eight calves, with 12 calvings not being uncommon. Queval et al. (1971) reported 80% of cows in a Kuri herd producing four to nine or more calves in a lifetime. They breed throughout the year, although the usual breeding season is from July to October. Calving intervals have ranged from 15 to 18 months, and bulls are sexually mature at 3 years of age.

3.3.2 Shorthorn Taurine Cattle

Only limited publications exist on the Shorthorns of West and Central Africa, the earliest available ones being Jeffreys (1953) and Ferguson (1967). Although they are currently termed trypanotolerant, humpless Shorthorn cattle were historically distributed in almost all ecological zones. However, not much is known about the actual movements of Shorthorn-type cattle. Moreover, despite their wide distribution, humpless Shorthorns are not represented in rock paintings to the same degree as Longhorns.

It is speculated that the increasing aridity at the time of their introduction may have made it impossible for the Shorthorns to follow the southwestward migration route across the Sahara. This may explain the route taken westwards along the North African Mediterranean coast. Modern cattle in Egypt exhibit characteristics that resemble those of Shorthorn-type cattle, while existing breeds in the Libyan Arab Jamahiriya, Tunisia, Algeria, and Morocco are almost entirely of the Shorthorn type (Epstein 1971; Payne 1970; Rege et al. 1994a). It is highly likely that, like the Longhorns before them, the Shorthorns split into two routes around Morocco. One stream moved north into present-day France and the British Isles—the Jersey, Guernsey, and Kerry breeds are partly derived from this stock—and the other moved southwards and westwards, eventually coming into the rainforest zone of West Africa where, through exposure to trypanosomiasis, they became tolerant.

The movement of Shorthorn-type cattle southwards along the river Nile is much less documented. Stewart (1937) reasoned that, after entering Egypt from Asia, *B. taurus brachyceros* cattle spread southwards into western Sudan, as well as along the Mediterranean seaboard, as alluded to above. Hartmann (1864) and Keller (1896) recorded sporadic occurrence of humpless Shorthorns in the Sudan among cattle that were, by then, already generally humped. Faulkner and Epstein (1957) referred to a small number of cattle of this type in the Koalib Hills in the Nuba Mountains of the Sudan. This may be the same population that Mills (1953) reported to be tolerant to local forms of trypanosomiasis in a 'tsetse pocket' in the Nuba Mountains. It is possible that cross-breeding with the predominant Zebu-type populations in this area eventually occurred and, hence, Mason and Maule's (1960) conclusion that these cattle were of the Zebu type. Isolated populations of Shorthorns were also reported in East Africa (Stuhlmann et al. 1927). Similar cattle are reportedly found in the island of Socotra off the Horn of Africa and were also reported in the 1920s on the islands of Pemba and Mafia, off the coast of the present-day United Republic of Tanzania (Payne 1964), and also in Madagascar (Dechambre 1951). The Baria of Madagascar, for example, supposedly descended from the crossbreeding between these cattle, Zebus, and a feral small-humped type. These Shorthorn populations may have resulted from movements further south of the introductions through Egypt and/or from a separate introduction through the Horn of Africa.

Doutressoulle (1947) reported that Shorthorns were brought to West Africa by the Berbers from southern Morocco. Domingo's (1976) suggestion that these cattle came through northern Nigeria from Egypt is not supported by Payne (1970) or Oliver (1983). From Morocco, the southern-bound Shorthorns supposedly spread between the Sahara and the Atlantic coast and the Guinea coast and the hinterland of Nigeria. Cameroon formed the extreme eastern and southern limits of their distribution in the region (Epstein 1971).

Whatever the route might have been, short-horned humpless cattle are depicted among the prevailing Longhorns in rock paintings on the Bauchi plateau of Nigeria dating from the second half of the first millennium BC. Epstein (1971) states that the Shorthorns were the most common type of cattle in northern Nigeria before the Fulani invasions of about 1820. It is believed that they migrated from this area to the Atacora mountain regions in Benin, where they were later called Somba (Pagot 1974). Following the mountain chain along Benin, Togo, and Ghana, they supposedly populated the north of Togo and Ghana. Migrations of Akan people from Ghana to settle in Côte d'Ivoire (where they became known as the Baoulé tribe) brought the Shorthorns to the area. Through further migrations from Benin City in Nigeria, the Gulf coast was populated by the Shorthorns, eventually evolving to create the present-day local Shorthorn populations.

The elimination of Shorthorns from Chad, Sudan, and Ethiopia probably took place much earlier and was associated with the primary westward expansion of the Zebu. Traces of Shorthorns are still found in the east and south of Chad. It is believed that the population of Shorthorns in West and Central Africa owes its existence to its tolerance to trypanosomiasis, a competitive advantage over the Zebu. As has been alluded to, the Zebus were introduced much later and spread west and south, stopped only by the western rainforest barrier. Cross-breeding between the Zebu and the taurine types is thought to have formed the Sanga type, which is the predominant type of cattle in the southern African region (Chap. 4). Present-day populations of Shorthorns in the Central African countries have resulted from imports in recent times, during and after the colonial era.

Modern-day populations or breeds of *B. taurus brachyceros* cattle in West and Central Africa include the Ghana Shorthorn, Baoulé, Savannah Muturu, Somba, Namchi (Doayo), Kapsiki, Bakosi, Lagune, Dwarf (Forest) Muturu, Bakweri, and Liberian Dwarf.

Broadly, Shorthorns can be divided into two subgroups according to size and conformation, namely, the larger Savannah type and the smaller Dwarf type. These, in turn, are associated with specific habitats. The larger Savannah type is predominantly found in the Guinean or Sudano-Guinean savannahs from Côte d'Ivoire to Cameroon (ILCA 1979a), while the smaller Dwarf (Forest) Shorthorns are found in lesser numbers in pockets of the coastal and forest regions. Whether the Dwarf Shorthorns are the descendants of a much larger progenitor type—dwarfing having resulted from generations of breeding—under the adverse effects of trypanosomiasis and mineral deficiency (Henderson 1929) or whether the larger Savannah type is a derivative of a Dwarf type—selection for larger size having occurred in the relatively tsetse-free savannahs—is still a matter of discussion.

3.3.2.1 Classification, Distribution, and Ecological Settings

(a) *Muturu, Ghana Shorthorn, Baoulé, and Lagune*

The Muturu cattle are considered the only true native taurine cattle of Nigeria, the taurine Longhorn N'Dama having been imported to upgrade Nigerian Zebu for beef production and conferment of trypanotolerance on the trypano-susceptible (Zebu) breeds (Adebambo 2001). *Muturu* is a Hausa word meaning 'humpless', and in Nigeria and other English-speaking countries in West Africa (Ghana and Liberia), the name Muturu is used for Shorthorn cattle.

However, in Nigeria, no distinction is made between the Savannah and the Dwarf types (ILCA 1979b). The name Kirdi, a general term used by the Fulani for non-Muslim pagans (ILCA 1979b), is sometimes used for Muturu. These populations are found in south-west Nigeria and in the Middle Belt (Adeniji 1985). The past distribution of Muturu ranged across West and Central Africa, particularly in Cameroon, Liberia, Ghana, and Nigeria. Today, the Muturu is increasingly sparsely distributed in the humid forest zone and in a few savannah areas in Nigeria (Rege et al. 1994b). In French-speaking countries, the Savannah Shorthorns are called Baoulé and the Dwarf-type Lagune, although local names or synonyms are also used.

The Muturu breed is considered to have populated most of Nigeria before the arrival of Zebu (*B. indicus*) cattle in the seventeenth century (Felius 1995). Since then, the existence of Muturu has been consistently under threat due to farmer preference for the larger body size of the Zebu breeds. It is also considered that the rinderpest epidemics of the late nineteenth century exterminated many Muturu herds (Blench 1998). Today, the Muturu in Nigeria is mostly found in the forest zone and Guinea savannah of the Middle Belt. The *Dwarf (Forest) Muturu* is found in the rainforest just north of the coastal mangrove swamp, but a few hundred Muturu are reportedly found in the swamps of the states of Cross River, Edo, Delta, Ondo, Ogun, Lagos, and Rivers. The coastal forests, located south of the seventh parallel in the west and below the sixth parallel in the east, have constantly high temperatures (ranging from 26° to 28 °C) and high humidity, with an annual rainfall of between 1800 and 3000 mm. The rainy season lasts for 7–8 months between May and December, interrupted in August by a short dry season that becomes less distinct towards the south and disappears along the coast where there is virtually daily rainfall.

The habitat of the Nigerian *Savannah Muturu* is the 'derived' savannah and the Guinea savannah. The former, which covers a narrow belt between the Guinea savannah to the north and the rainforest to the south, is a climax vegetation created by the destruction of the original forest cover through constant shifting cultivation and population pressures (Olutogun 1976). Located in the higher rainfall belt of the country, the derived savannah represents a better natural grassland. Annual grasses such as *Brachiaria deflexa* and *Digitaria horizontalis* are among the first species to appear after the original forest is cleared. Perennials, including *Pennisetum purpureum*, *Panicum maximum*, *Ctenium newtonii*, *Andropogon tectorum*, and *Hyparrhenia rufa*, appear later. The Guinea savannah receives an average of 1500–2000 mm of rainfall, while the surrounding plateau and plains receive 1000–1500 mm, and the drier west receives 500–1000 mm. It is characterized by tall perennial grasses (1.5–3.0 m high) such as *Andropogon* and *Hyparrhenia* that grow in tussocks and provide wide ground coverage. *Pennisetum* and *Panicum*, among others, form the lower (1.0–1.5 m) layer (Fricke 1979). The type of grass is dictated by soil moisture. Dry soils are dominated by such plant associations as *Loudetia arundinacea*, *Ctenium newtonii*, and *Monocymbium ceresii*, while damp soils support grasses such as *Pennisetum purpureum*, *Chloris*, *Robusta*, and *Brachiaria* spp.

The habitats of both Muturu types (Forest and Savannah) fall within the tsetse belt. *Glossina fusca* is associated with the coastal forest, while the *morsitans* and *palpalis* groups are found in the savannah areas (ILCA 1979b).

In addition to the Muturu distributed in the south-eastern coastal area near Ada and Keta lagoons of Ghana, the country is also home to many Savannah Shorthorn types known locally as **Ghana Shorthorn** and, for which, the broader name West African Shorthorn is sometimes (ambiguously) used. The Ghana Shorthorn is distributed across the country from the Gold Coast to the north. They are particularly concentrated in the North, especially towards the north-western border around Bole, Wa, Lawra, and Tumu. They are also concentrated in the northern part of the Ashanti Region. A high concentration of relatively pure Ghana Shorthorns is also found around Bouna in neighbouring Côte d'Ivoire and near Gaoua in Burkina Faso (ILCA 1979b). In Liberia, typical Dwarf Shorthorns are found in the eastern coastal areas of Maryland and Sinoe counties, while the less characteristic short-horned animals are found in Grand Bassa County and inland in the counties of Grand Gedeh, Bong, and Nimba.

The **Ghana Shorthorn** is found in all vegetation zones, including the humid forest zone. Its habitat is, thus, not limited to the savannah regions alone, despite its classification as a Savannah Shorthorn. The Guinean savannah woodland habitat of the Ghana Shorthorn is found north of the forest zone and stretches down to the south-eastern plains. It comprises a continuous cover of grass and low fire-resistant trees. Typical trees are *Parkia* spp., *Jacaranda* spp., and *Butyrospermum parkii*. There is a single rainy season from March to October in the northern half of the country, during which 1000–1200 mm of rain falls annually. Temperatures are very high, with a mean annual maximum of 34.5 °C in the extreme north. The humid rainforest in the south-west of Ghana receives an average rainfall of between 1210 and 3312 mm annually. The other forests are semi-deciduous, with a bimodal rainfall pattern and mean daily maximum and minimum temperatures of 30° and 21 °C, respectively.

The coastal savannah of Ghana receives an annual rainfall of 800–1000 mm between April and November. During the dry seasons (November and March), maximum daily temperatures can be as high as 32.2 °C. Mean monthly relative humidity is about 70 per cent at 15.00 h(GMT). *Andropogon*, *Brachiaria*, and *Hyparrhenia* spp. are the most common grasses of the coastal savannah. The south-eastern part of this area, near Ada and Keta lagoons, is the habitat of the few Dwarf (Forest) Muturu in Ghana. This is an unusual location for animals whose size is usually associated with a forest habitat. This area forms part of the Accra plains, a savannah region unusual for the coastal belt, which extends eastwards to the Togo border. Tsetse fly infestation in this area is reportedly low (ILCA 1979b).

The Baoulé are distributed in the north and central regions of Côte d'Ivoire (Pagot 1974) and are scattered throughout the forest zone (Shaw and Hoste, 1987). Some can also be found in the eastern parts of the country (Glattleider 1976). Baoulé cattle are grouped into several nucleus herds in the Korhogo (Sinematiali) Region, Bouna, Dabakala, and Central and High Nzi Valley, where they are herded by Peuhls (Fulanis). A study by ILCA (1979b) reported that the breed was still relatively pure towards Bouaké, Dabakala, and Bouna, but the situation is changing fast. In the Bouna Division in the north-east, they are called *Lobi* after a local tribe. Indeed, Baoulé in Burkina Faso are the same as the Lobi or *Méré*, which are found in the southern part of the country. Rege et al. (1994a, b) proposed that the name Méré be restricted in the scientific literature to Zebu crosses with the Savannah Shorthorns in the area. The name is currently used to refer to both purebred Lobi and crosses of Zebu with Lobi (Burkina Faso), with Baoulé (Côte d'Ivoire and Burkina Faso), and with N'Dama (Côte d'Ivoire, Burkina Faso, and Guinea) and is, thus, rather confusing. In Mali, the term Méré (or **Bambara**) is used for stabilized Zebu × N'Dama crosses. Although the name Lobi is widely used for the Savannah Shorthorn cattle in Burkina Faso, these cattle are quite like the Baoulé of Côte d'Ivoire (ILCA 1979b); the Lobi are somewhat smaller than the Baoulé in some areas. In addition, because of the proximity of the populations in the two countries and the cross-border movement of people and livestock, there is extensive interbreeding between Lobi and Baoulé.

Baoulé cattle were imported into the Central African Republic from Côte d'Ivoire and Burkina Faso between 1955 and 1969 and were subsequently distributed under the *métayage* system throughout the country (ILCA 1979b). Small numbers were also imported into Gabon for *métayage* operations in the 1940s (Shaw and Hoste 1987). In 1979, the Centre de recherche et d'élevage (CREAT) station at Avetonou in Togo purchased a herd of 120 Baoulé cattle from the Korhogo and Bouna regions of northern Côte d'Ivoire for trypanotolerance work (Morkramer and Dekpol 1984).

In both Côte d'Ivoire and Burkina Faso, the habitat of the Baoulé has a Sudano-Guinean climate. The climate is tropical and sub-humid with a dry season from November to April, during which the harmattan winds influence the area. The rainy season lasts from May to October, with a peak from July to September. The northern frontiers of Côte d'Ivoire have Sudanian vegetation with occasional islands of dense, dry forest over mostly savannah landscape. Farther south, the vegetation is sub-Sudanian, characterized by dry woodland and savannah. The Bagou River flows through the region from south to north, and its gallery forests form the major habitat of the tsetse fly of the *palpalis* group. In the dry season, *Glossina palpalis* and *G. tachinoides* are mainly confined to the riverine gallery forests. In the rainy season, however, *G. tachinoides* may be detected in forest islands at some distance from watercourses.

All the Baoulé (or *Lobi*) cattle in Burkina Faso are kept under village production systems (Shaw and Hoste 1987), while 99 per cent of Baoulé in the Central African Republic are kept under the *métayage* system (ILCA 1979a). This system is used to introduce cattle husbandry at the village level in regions where cattle rearing has not been a traditional activity. A basic breeding herd consisting of five to ten heifers and a bull is provided under contract to an individual or a small group of villagers by a commercial, governmental, or religious organization that maintains central breeding herds. The central organization undertakes the provision of technical assistance and veterinary inputs. In the event of loss resulting from natural causes, the animals are replaced without charge. The Baoulé cattle in Gabon are kept under the old *métayage* method or are generally left to wander freely with minimal supervision and no external inputs (Shaw and Hoste 1987).

The Lagune As alluded to above, Lagune is the name used for the Dwarf Shorthorn cattle in French-speaking West and Central Africa. These cattle are found mainly in Benin but also in Côte d'Ivoire and Togo, in a zone about 40 km off the coast (Epstein 1971). In Benin, Lagune are found mainly in the southern provinces of Atlantique, Mono, Ouemé, and Zou, an area stretching from the coast to the north of Abomey (Domingo 1976). In Côte d'Ivoire, they are located along the coastal region around Jacqueville, Abidjan, and Sassandra (ILCA 1979b). The population in Togo is mainly located in the maritime and plateau regions. *Logone* cattle (also known as *Toupouri*), found in southern Chad along the banks of the river Logone, are said to be similar to the Lagune (Anonymous 1950). Payne (1970) referred to *Lagone* rather than Logone, but this must be the same population. Whatever the origin of the Chadian population, it is quite likely that, given the habitat, it is genetically different from the coastal Dwarf Shorthorns. In particular, this population is unlikely to be trypanotolerant to the same extent as the Forest Shorthorns.

The introduction of Lagune cattle into the Democratic Republic of the Congo (former Zaire) began in 1904 when 50 heads of cattle were imported from Benin (Mortelmans and Kageruka 1976). Shortly afterwards, the breed was introduced at the Kangu Mission and at the Government Livestock Station at Zambi. Importation continued during the period prior to the First World War, and the animals multiplied and spread throughout Mayombe in Gas-Zaire, Kuilu in Bandundu, and Lisala and Bumba in the Equateur region of Zaire. These animals came to be known as Mayombe or Dahomey. Lagune cattle were later imported into Gabon from Zaire in 1945, 1948, and 1958 (ILCA 1979b). Some Lagune were also imported by Owendo Farms in Gabon from Benin in 1948 (Anonymous 1950). A few Baoulé cattle were imported into Gabon from Côte d'Ivoire around 1956 as well. These Lagune and Baoulé cattle in Gabon are now indistinguishable from each other and are commonly referred to as Baoulé (ILCA 1979b). They are found mainly in the north of Gabon as well as in the Ngounie and Nyanga regions.

In general, Dwarf (Forest) Shorthorns, including the Lagune, have a restricted ecological zone around the coastal forests of Côte d'Ivoire, Togo and Benin, the Atacora Mountains of Benin, and the riverine areas of Chad (for 'Logone'), as well as the tropical forests of Gabon, Zaire and the Congo, and the Guinea savannah of Zaire and the

Congo. The coastal forests are characterized by a Guinean or Sudano-Guinean climate with an annual rainfall ranging from 1200 mm in Benin to 2400 mm in Côte d'Ivoire. Two rainy seasons occur from March to July and from September to November. The most common tsetse fly species in these habitats are *G. palpalis*, *G. fusca*, *G. medicorum*, *G. pallicera pallicera*, and *G. longipalpis*. Trypanosomes commonly found are *T. vivax*, *T. brucei*, and *T. congolense*.

The humid tropical forests have an annual rainfall varying from 1600 mm to 3000 mm. In Gabon, this habitat is infested with *G. palpalis*, *G. tabaniforms*, and *G. haningtoni*. In the Guinea savannahs of the Congo and Zaire, rainfall varies from 1200 mm in the Congo to 2000 mm at the equator in Zaire. Gas-Zaire, Bandundu, and Equateur are the most heavily infested regions in Zaire. The main tsetse species here are *G. fuscipes* and *G. palpalis*. Common trypanosomes are *T. vivax* and *T. congolense*. Although the Logone breed is trypanotolerant, no tsetse flies are present in its habitat in the riverine areas of Chad (Landais 1980).

(b) **Bakweri, Kapsiki, Doayo, Somba, and Bakosi**

A small population of the Dwarf Muturu is found in Cameroon at the foot of Mount Cameroon, between Buea and Victoria in the South-West Region, and is known locally as **Bakweri**. Cameroon also has three localized subpopulations of the Savannah Shorthorn type, known as **Kapsiki, Doayo,** and **Bakosi**.

Kapsiki cattle (also called **Kirdi**, as are the Nigerian Muturu) have historically been found around the Mandara Hills of Cameroon along the Nigerian border in the Margui-Wandala Division of the Far North Region between Mokolo and Bourrah, as well as around Mogode at the border with Nigeria. There are also a few Kapsiki herds on the Nigerian side of the border. The Doayo people keep a Savannah Shorthorn type of cattle in a small area in the north-western foothills of the Poli Mountains in the Bénoué Division of the North Region. These cattle are locally known as **Doayo, Namchi** (**Namshi**), or **Poli**. Figure 3.5 shows a herd of Doayo (Namchi) cattle in Cameroon.

The **Bakosi** cattle are kept by a tribe of the same name located in south-western Cameroon, west of Nkongsamba on the border between the South-West and Littoral Regions. However, the cattle-rearing areas of the Bakosi people are limited to the north-eastern part of the boundary of the Bakosi tribe on the western slopes of Mount Manengouba in the Banghemu subdivision of the South-West Region. It has been reported (Epstein 1971) that the location of the Shorthorn cattle, otherwise mainly coastal, in the highlands of Cameroon, is as a result of the fact that their owners were forced to retreat from the fertile plains of the Diamaré and the Adamawa regions at the time of the Fula migration at the beginning of the twentieth century.

In Cameroon, the habitat of the Bakweri is around the foot of Mount Cameroon, between Buea and Victoria (present-day Limbe) in the South-West Region. This area has a humid forest climate with an annual rainfall between 1500 and 4000 mm. The habitat is similar to that of the Muturu in the forest areas of Nigeria and Liberia. The dominant tsetse species are *G. tabaniformis*, *G. nigrofusca*, *G. pallicera*, and *G. caligenia* (ILCA 1979b).

The name **Somba** is used for the Savannah Shorthorn populations in Benin (where the name **Atacora** is also used) and Togo (where they are also known as **Mango**). The Somba tribe inhabits the area around the Atacora Mountains in Benin. This mountain range has enabled the Somba breed to be maintained pure in its original habitat. Somba cattle are particularly associated with the district of Boukombé, as well as with Natitingou and Tanguiete. The Somba are also found in Togo, where they represent over 50% of the national herd. They are raised by the Tamberma people, who are believed to have the same ethnic origin as the Somba of Benin (Avegan 1984). These cattle populations are concentrated in the north of Plateau Region; in Oti Division south of the Savannah Region, where they are known as Mango cattle; and in the Central and Kara regions, where they are called **Konkomba**.

Somba cattle are principally found in the Sudano-Guinean savannahs, while the Kapsiki and Namchi inhabit the Sahelo-Sudanian zone,

Fig. 3.5 Doayo (Namchi) cattle at Wakwa, Cameroon. (Photos by Dr E.M. Ibeagha-Awemu)

with the Kapsiki being predominantly on the Sahelian side and the Doayo (Namchi) in the more Sudanian climate. The habitat of the Kapsiki cattle in the Mandara Mountains is at an altitude of 600–1000 m. The rainy season is traditionally between June and October, with the maximum rainfall occurring in August. Rainfall ranges between 500 and 850 mm, while minimum and maximum temperatures have historically averaged 8–10 °C and 32 °C, respectively. The dry season lasts for 7 months. It must be noted that these climatic conditions are significantly changing due to climatic variability.

The vegetation is savannah interspersed with such trees as *Isoberlina* spp., *Adansonia digitata*, *Boswellia dalzielii*, *Combretum* spp., *Daniellia oliveri*, and *Ficus populifolia*. The most common grasses are *Hyparrhenia hirta*, *Andropogon gayanus*, *Pennisetum pedicellatum*, *Cymbopogon giganteus*, *Rhynchelytrum repens*, *Thelepogon elegans*, and *Aristida* spp. The habitat of the Kapsiki breed is considered to be tsetse-free. The Bakosi area in Banghem is derived savannah merging into humid forest. Annual rainfall ranges from 2000 to 3000 mm. The level of tsetse fly infestation in the area is not fully known, but it is unlikely that there are any tsetse flies in the higher altitude areas, and varying degrees of infestation may possibly be found in the forests.

3.3.2.2 Physical, Adaptive, and Special Genetic Characteristics

(a) **Muturu, Ghana Shorthorn, Baoulé, and Lagune**

The **Muturu** is a small and compact animal, very similar to the Lagune, but well-muscled and with a good beef conformation. Indeed, the Muturu are the smallest cattle breed of the Nigeria cattle and the smallest among the Shorthorns in West Africa, with a reported range in wither heights varying from 71 to 100 cm (Aboagye et al. 1994)—from extremely small Dwarf (Forest) Muturu (<89 cm tall at withers) to small Savannah Muturu (>89 cm tall at withers). The Nigerian Dwarf Muturu has been described as a smaller and purer *B. brachyceros* than the other Shorthorns in the region. Like the Lagune, the legs are thin and short, with the height to chest ranging from 37 to 49 cm. Sexual dimorphism is quite pronounced. Contrary to expectations, the males are generally smaller than the females, although they have well-developed and robust hind legs. The hump is absent and the dewlap is extremely small, typical of the Shorthorns. The Muturu head is long and relatively large compared to its body. The face is triangular in profile and slightly concave, with a flat and wide forehead. The nose is straight, and the muzzle is large and black. Typical Muturu animals are depicted in Fig. 3.6.

The horns of the Muturu are small and short, especially in the females, and thick at the base in the males. They can attain lengths of about 7.8 cm, with a basal girth of 16.6 cm and a basal core circumference of 13.2 cm. The horns are oriented outwards and forwards and laterally to the exterior, forming a 'C' in the females. Rudimentary horns also exist. Polled animals are also common. The ears are small and laterally attached below the horns. The neck is of average length and flabby and appears to be wider longitudinally. It is thin in the females and sturdy in

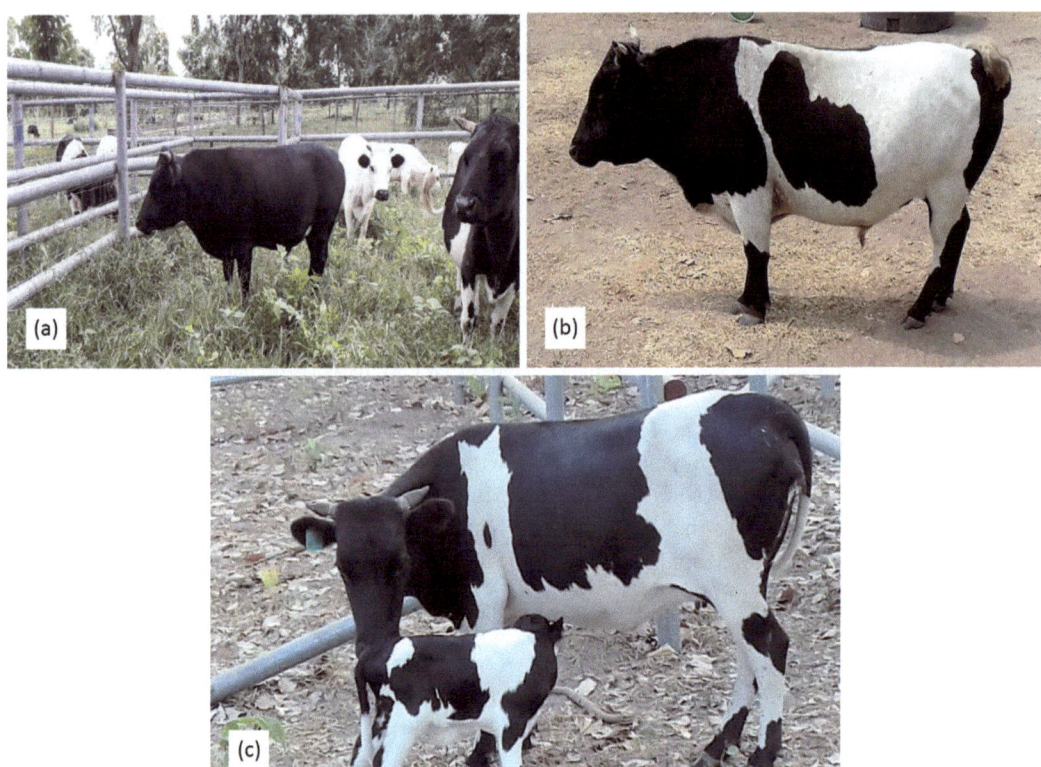

Fig. 3.6 A Muturu herd (**a**), bull (**b**), and cow with calf (**c**), showing typical black and white colour patterns. (Courtesy: Prof. O. A. Osinowo)

the males. The chest is wide and deep, and the back is well-muscled. The top line is inclined slightly forwards in the females and is horizontal in the males. The udders of the cows are poorly developed.

The Muturu has a multiplicity of coat colours, with plain black and black and white being the most common and typical of the Shorthorns (Maule 1990; Mason 1988). However, coats that are brownish black or dark brown with white markings have also been reported. Black or darker shade coats are common among the Forest type, while black and white or lighter shade coats are typical of the Savannah type.

Estimates of adult linear body measurements in Table 3.2 indicate a wide variation across breeds/strains and environments, especially in the females. It must be noted that substantial differences exist in body size measurements between the two ecotypes—Forest (Dwarf) and Savannah—in the region.

Muturu cattle have acquired hardiness and tolerance to tsetse flies, tick-borne diseases, and trypanosomiasis, as well as the ability to maintain an excellent body condition on grazing or browsing alone. Despite their reputed hardiness, Muturu cattle are extremely susceptible to rinderpest (Ferguson 1967). It is the well-recognized ability of the Muturu to thrive in the presence of the trypanosome challenge that has drawn much attention to the breed. Studies comparing the trypanotolerance of the Muturu, N'Dama, and Zebu under different combinations of trypanosome species have concluded that the Muturu was intermediate in its ability to withstand mixed infections. Results from a more recent study (Tijjani et al. 2019), which examined genomic regions in Muturu, which are under common signatures of selection with those in the more widely studied taurine Longhorn, the N'Dama cattle, including for adaptive immunity and heat tolerance, suggest shared mechanisms of adaptation

Table 3.2 Adult linear body measurements by sex and breed of Shorthorn cattle

Trait	Sex	Ghana Shorthorn	Baoulé	Kapsiki	Namchi	Lagune	Muturu
Body length (*cm*)	Male	143.0	120.0–121.2	114.0–142.0	125.0–145.0	121.4	86.5
	Female	111.4–122.7	108.6–112.3	115.0–132.0	110.0–130.0	108.6–119.7	93.5–107.0
Withers height (*cm*)	Male	115.5	100.1–105.7	105.0–116.2	100.0–110.0[a]	99.1	95.0
	Female	99.0–109.0	90.0–100.0	100.0–109.0	97.0–106.0	85.0–101.4	88.0–95.4
Heart girth (*cm*)	Male	169.0	140.5–140.7	135.0–148.0	140.0–159.0[a]	137.3	125.0
	Female	139.0–147.1	128.4–132.6	130.0–140.7	125.0–140.0	122.6–137.5	127.0–130.0

Source: Aboagye et al. (1994)
[a]Steers

to environmental challenges for these two taurine cattle. These results, while providing insights into the candidate genes under selection in Muturu, are forming a basis for the identification of genes and their polymorphisms linked to the unique tropical adaptive traits, including the trypanotolerance of the Muturu.

Muturu cattle have been reported to be sensitive to solar radiation and, therefore, are very shade-dependent. This behaviour is related to their low rate of cutaneous evaporation compared with that of the N'Dama and the Zebu (Amakiri and Mordi 1975). Therefore, an appropriate husbandry system for the Muturu should provide shade.

The hardiness of the Muturu is exemplified by its low mortality rates compared with those of other Shorthorns under similar management conditions (Aboagye et al. 1994). Its cumulative calf mortality rate at 1 year of age has been relatively low compared to those of other Shorthorns, both on-station and on-farm. Few studies have reported on the haemoglobin (Hb), albumin, vitamin D–binding protein, carbonic anhydrase, and transferrin (Tf) polymorphisms in the Muturu (Table 3.3). The Muturu has a different sequence of Hb than that of other Shorthorns and the N'Dama, an indication that they may be genetically different from the other breeds. However, Baoulé and Muturu cattle, with similar transferrin alleles, are different from the N'Dama, which also has the Tf^e allele (Braend and Khanna 1968). The presence of another blood factor—Z'—in the Zebu and not the taurine breeds (Muturu and N'Dama) is further evidence of different ancestries of these cattle types. These alleles differentiate the humpless Shorthorns and Longhorns from each other as well as from the humped Zebu.

The ***Ghana Shorthorn*** (Fig. 3.7) is a small-sized animal with good beef conformation. Its head and neck are both long, but the neck is thin and the forehead flat. It has short and thin horns, like those of the Lagune, averaging 20.3 cm in length. Unlike the Lagune and Muturu, however, polled Ghana Shorthorns are rare. The top line is concave, with a rising rump. The hump is absent, and the dewlap is poorly developed, typical of *Bos taurus* cattle. Ghana Shorthorns vary considerably in colour markings: black and white animals are common, in addition to other colour patterns such as solid black, white, and mottled black and white. The Ghana Shorthorn is considered to be tolerant to trypanosomiasis and tick-borne diseases and is adapted to the harsh hot, humid tropics. Ghana Shorthorns have also been shown to have good heat tolerance with substantial individual variation (Kahoun 1971).

The ***Lagune*** is the smallest of the West and Central African Shorthorns. Indeed, standing at 80–105 cm at the withers, it is thought to be one of the smallest cattle breeds in the world. Lagune cattle are compact, with a straight top line. The legs are fine and short, about 47 cm to the chest, with a cannon bone length of about 20 cm. Like the rest of the Shorthorns, it is humpless, with a poorly developed dewlap. The head is massive, with a rectilinear facial profile, and it has a conspicuous poll and protruding eyes. The head varies in length from 27 to 39 cm and appears longer than that of the other Shorthorns. However, the length of the body compares favourably with that of the other Shorthorns. The forehead is flat or slightly concave, with prominent orbits. The horns are imperfect, often thin, or flat and, sometimes, loose or absent. The horns are short, huge at the base, and round and pointed at the extremities. They measure between 14 and 24 cm in length. The surface of the horns is coarse, and they are lighter in colour at the base and darker at the extremities. They are cylindrical in the males and oval-shaped and obliquely forwards in the females. Drooping horns and polled animals are common. The neck is short, thick, and bulky in the bulls and thin in the cows. The rump is narrow in the cows and muscular in the bulls, with the tail ending in a 10-cm switch. The back line slopes down from the rump to the withers more steeply (about 3–5 cm) than in the typical Savannah Shorthorn. The udders of the cows are very poorly developed. The coat colour is usually plain black, black with white spots, or black and white. Even though coats tend to be pure black among the Dwarf Shorthorns, they are especially so among the Lagune. Red or red and white animals are rare, while fawn, dark grey, or spotted light grey individuals are frequently encountered.

Table 3.3 Special genetic characteristics of some Shorthorn cattle breeds and stabilized crosses

	Breeds						
	Somba[b]	Lagune[b]	Borgou[b]	Boulé[c]	Muturu[d]	Namchi	N'Dama
Phenotypic frequency of erythrocyte factors							
A	1.00	0.93	0.98	–	–	–	–
B	0.56	0.64	0.23	–	–	–	–
C	0.69	0.64	0.45	–	–	–	–
FV	0.49	0.34	0.19	–	–	–	–
FF	0.44	0.47	0.56	–	–	–	–
W	0.07	0.19	0.24	–	–	–	–
J	0.42	0.47	0.63	–	–	–	–
L	0.46	0.68	0.60	–	–	–	–
S	0.69	0.90	0.87	–	–	–	–
Z	0.91	0.93	0.87	–	–	–	–
R'	0.08	0.19	0.15	–	–	–	–
T'	0.49	0.44	0.82	–	–	–	–
Allelic frequency							
Haemoglobin							
A	0.99	0.92	0.90	0.96	0.72–1.00	0.95	0.83
B	0.01	0.08	0.10	0.04	–	0.05	0.17
D	–	–	–	–	0.28	–	–
Transferrin							
A	–	–	–	0.14 (0.02)	0.22–0.53	0.40	0.36
B					–	0.02	0.04
D	–	–	–	0.81 (0.02)	0.47–0.78	.383	0.56
E	–	–	–	0.05 (0.01)	–	0.13	0.04
F						0.07	
Phosphoglucomutase							
A	–	–	0.61 (0.10)	–	–	–	–
B	–	–	0.39 (0.10)	–	–	–	–
Albumin							
A	–	–	–	0.25	1.00	0.55	0.96
B	–	–	–	0.75	–	0.42	0.04
G					–	0.03	–
Carbonic anhydrase							
S	–	–	–	0.99	1.00	1.00	0.96
F	–	–	–	0.01	–	–	–
Z					–	–	0.04
Vitamin D–binding protein							
A					0.97	0.62	0.88
B					0.03	0.37	0.12
C						0.02	

Source: Aboagye et al. (1994), Ibeagha-Awemu et al. (2004)

The patchy red, brown, or red and white animals found in Côte d'Ivoire are believed to be an admixture with the N'Dama. The eyelids, the mucosae, and the hoofs are black, and the muzzle is black or brown and thick (ILCA 1979b; Maule 1990; Adeniji 1985; Mortelmans and Kageruka 1976; Agbemelo 1983). A typical Lagune bull is shown in Fig. 3.8, while Fig. 3.9 shows a Somba and a Lagune cow.

In Togo, Leclercq (1970) and Agbemelo (1983) reported that animals from the maritime region in the coastal savannah have significantly longer (116–121 vs 110 cm) and deeper (133–138 vs 129 cm) bodies and are taller (96–101 vs

Fig. 3.7 A herd of Ghana Shorthorn cattle. (Source: ILRI)

Fig. 3.8 A Lagune bull. (Source: Ahozonlin et al. (2022), licensed under CC-BY 4.0)

93 cm) than those from Assahoun in the humid forest zone. These trends may have resulted from differences in nutrition resources, the prevalence of parasitic diseases, climatic conditions, and other contextual differences. However, similar values have been reported for animals under coconut plantations and village conditions in the maritime region. Much lower measurements than these have been reported for herds in villages and palm plantations in Côte d'Ivoire (ILCA 1979b). The ratio of heart girth to height at withers of the Lagune (1.36–1.53) (Aboagye et al. 1994) is comparable to those of the European beef breeds, such as the Limousin and the Charolais, and indicates that the Lagune is more compact than the Savannah Shorthorns. Figures for young Lagune cattle show that at 3–4 years of age, they are similar in ratio of heart girth to height at withers with the mature Zebus—the Gudali and the Fulani—in the region (Aboagye et al. 1994). Domingo (1976) reported similar ratios for Lagune cattle from 1.3 to 1.5 and from 3.0 to 3.5 years of age. The Lagune and Somba are similar in morphology, particularly in terms of height at withers. A recent study has reported high morphological diversity in the Lagune cattle of South Benin

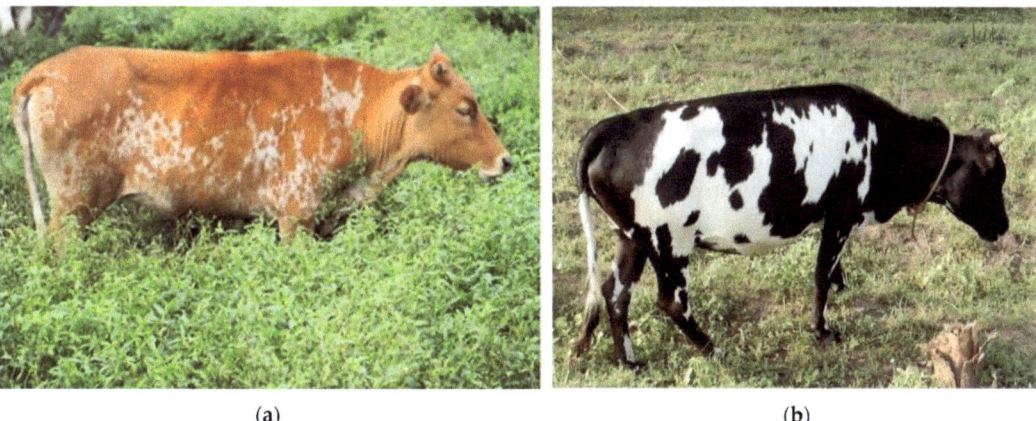

(a) (b)

Fig. 3.9 Typical Somba (**a**) and Lagune (**b**) cows. (Source: Vanvanhossou et al. (2021a), licensed under CC-BY 4.0)

based on body measurements (Ahozonlin et al. 2020).

Lagune cattle, like other Shorthorns, have been described as hardy, trypanotolerant, and adapted to the hot, humid tropical conditions typical of where they live (Maule 1990; Adeniji 1985). Their habitat is characterized by poor feed resources; a high prevalence of parasitic diseases, including trypanosomiasis; and very harsh climatic conditions. Even though these attributes have not been objectively quantified, the fact that these animals live and produce in the tsetse belt indicates that they have a certain degree of tolerance. Moreover, in a case study of Lagune and Borgou (a cross between the White Fulani and the Somba or the Lagune) herds located about 50 km apart but similarly managed, with the Borgou under light tsetse challenge at M'Betecoucou ranch and the Lagune under medium challenge at the Samiondji station, ILCA (1979a) reported that only 51% of positive cases of trypanosomiasis were diagnosed among the Lagune cows over a two-year period compared with 86% for the Borgou cows, with the former showing only 4.2% positively diagnosed animals per month compared with 11.5% for the Borgou.

The highest Lagune calf mortality rate (24%) was recorded on the Samiondji station, which was reportedly the result of a medium tsetse challenge (ILCA 1979a), followed by the *métayage* operations (15%) in Benin (Aboagye et al. 1994). The much higher abortion rate (8.6%) under Samiondji station conditions in Benin (ILCA 1979a) compared with that (4.0%) under conditions of the maritime region in Togo (Shaw and Hoste 1987) was considered an indication of the effect of trypanosomiasis. Mortality has ranged from 9.3 to 37.9% in the males and from 13.4 to 31.3% in the females (Agbemelo 1983).

Besides the morphological similarity alluded above, Lagune and Somba cattle have also been shown to have similar erythrocyte factors, except in the L and S systems (Table 3.3), further suggesting that these breeds are identical, separated only by a geographical barrier. Likewise, haemoglobin gene frequencies are similar in both breeds as well as in the Baoulé. The smaller size of the Lagune is obviously a consequence of natural selection in an environment characterized by poor nutrition, a high prevalence of parasitic diseases, and very harsh climatic conditions.

The *Baoulé* cattle vary in size from small to medium. They are straight and compactly built, with a good general profile. Baoulé cattle are short-legged, with a height at sternum of 46.8–48.5 cm. The Baoulé body conformation points to good beef animal qualities; however, they are not stocky. Udders and teats are retracted and degenerate. The Baoulé has a massive head with a straight profile. The forehead is large and flat in the males but slightly narrow between the horns and orbits in the females. The muzzle is large and black, and the ears are small, laterally located and upwardly oriented in some and horizontally attached in others. The horns are short and sturdy, with a solid base, although they vary in form (crown, half-moon, or hook) in the bulls. In the females, the horns have a slightly different shape and are smaller and oriented obliquely forwards and upwards. The neck is thick and very strong in the males but short and thin in the females. Baoulé cattle, like all taurine cattle, are humpless, with a retracted and degenerate dewlap.

As in the Ghana Shorthorn, the Baoulé (see Fig. 3.10) has a multiplicity of coat colour patterns, although solid black or black and white flecking is typical. Fawn or pied-fawn animals are rare. Various shades of colours with irregular spots, such as small black mottles on a predominantly white coat, occur in some Baoulé populations. Cases of red-pied, grey and yellow, or yellow-pied Baoulés are not uncommon. In addition, Baoulé animals with red backs are common in some locations. Healthy Baoulé cattle have short, fine, and glossy hair (Maule 1990; Morkramer and Dekpol 1984).

The Baoulé is considered tolerant to trypanosomiasis, resistant to tick-borne diseases, and hardy to the hot, humid tropics. Trypanotolerance in the Baoulé (like the N'Dama) has been much more extensively studied (Akol et al. 1986; Maehl et al. 1987; Pinder et al. 1987, 1988; Duvallet et al. 1988) than that of the other Shorthorns. Results from studies have indicated individual variations in the natural resistance to trypanosomiasis.

Table 3.3 shows the allelic frequencies of the transferrin (Tf), haemoglobin (Hb), erythrocyte phosphoglucomutase (PGM), albumin (Alb), and carbonic anhydrase (CA) polymorphisms in Baoulé cattle. Transferrin polymorphisms are common in livestock. Except for alleles Tf^B and Tf^f, which are also found in the Zebu, only alleles Tf^A, Tf^D, and Tf^E are found in the Baoulé and N'Dama. The Baoulé, like the Somba and Lagune, has a different sequence of Hb alleles from the Muturu and the N'Dama. These humpless cattle generally lack Hb^B and Hb^C, which are present in the humped Zebu. Comparatively, the Baoulé and most other Shorthorns have predominantly Hb^A and traces of Hb^B, except the Muturu, which have predominantly Hb^A and traces of Hb^D (Bangham and Blumberg 1958; FAO 1976). Y chromosomes in the Baoulé and N'Dama are sub-metacentric as in the European breeds, while those of the Zebu and Zebu derivatives are acrocentric (Pagot 1985). The structure of this chromosome can, therefore, be useful in detecting the purity of these breeds and in tracing their evolutionary history.

(b) ***Somba, Kapsiki, Namchi, Bamenda*, and *Bakosi***

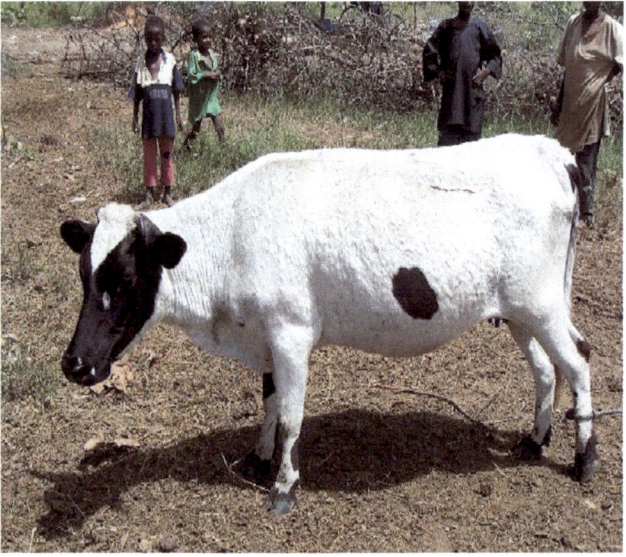

Fig. 3.10 A Baoulé cow. (Courtesy: Albert Soudre', Open Source: https://www.eurekalert.org/multimedia/722329)

The *Somba, Kapsiki, Namchi, Bamenda, and Bakosi* have generally been described as small Savannah Shorthorns, although variations exist within and among them. In general, these breeds have a much-reduced dewlap and umbilical folds, typical of Bos taurus cattle. Of these breeds, only the physical appearance of the Somba has been well-documented.

The *Somba* is not readily differentiable from the Lagune morphologically. It is a stocky animal with good beef conformation. It has a straight and compact body profile with a cylindrical trunk. The top line is straight but inclined forward in the females, although slightly convex in multiparous cows. It is horizontal but with a slight concavity near the back in the males. The Somba head is long and narrow, with prominent orbital arcades, presenting a concavity to its large forehead. The head droops on a short and thick neck in the males. In contrast, the neck is thin in the females. The horns are short and thin, arching above the head, and light at the base and darker towards the extremities. They are circular in cross section in the males and oval in the females and have an outward and forward orientation towards the extremities. Somba cattle with polled and drooping horns also exist. The ears are short and horizontally attached. The limbs are small and short, and the tail is long, extending from the pin bone to the hock. The muzzle and the area around the eyes of the Somba are dark, and its skin is darkly pigmented. Its coat is usually dark, either uniformly black, black and white, red and white, or pied, generally with dark extremities. Grey and brown coats are rare in the Somba.

The Somba has an average height at withers of 90–100 cm, heart girth of 130–137 cm, and body length from pin bone to shoulder of 105–120 cm at maturity (at least 5 years of age) (ILCA 1979b; Maule 1990; Domingo 1976).

The *Kapsiki* (Fig. 3.11) has horns of medium length, averaging 20–40 cm. It has short and glossy hair and supple skin. While there is a variety of colour markings in the Savannah Shorthorns, those of the Kapsiki are even more varied. Pied-black is the predominant coat colour in Kapsiki. Other colour markings include solid black, pied-red, solid white, fawn, pied-brown, red, dark brown, wheat grey, and mottled pied-black. The coat colour of the B**amenda** and *Namchi*, on the other hand, is uniformly black, black and white, or black with white spots, and in some cases, it may be brown or spotted brown (Fig. 3.10). The coat colour of the *Bakosi* varies from black to white, with more than half of the population either brown or black.

Adult linear body measurements of the Kapsiki have been elaborately studied under village conditions (Aboagye et al. 1994) and are considered useful in estimating body weights of young stock, especially under village conditions where weighing is often not done. Reported linear body measurements attest to the small size of these animals. In general, the males are taller and deeper-bodied than the females. Tawah and Mbah (1989) reported adult height at withers of 112–116 cm and 104–109 cm, respectively, for the male and female Kapsiki under station conditions in Cameroon. An estimate of 110 cm was reported by ILCA (1979b) for height at withers in mature *Bakosi* animals. Adult linear body measurements for the Kapsiki and Namchi cows are quite comparable. In general, the Baoulé, Somba, Kapsiki, Namchi, Bamenda, and Bakosi are very similar in size at maturity. However, they are shorter than the Ghana Shorthorn but taller than the Muturu.

Like the other Shorthorns in the region, Somba cattle are believed to have acquired natural tolerance to trypanosomiasis. Although somewhat tolerant to tick bites, they have been reported to be susceptible to infectious diseases such as rinderpest, foot-and-mouth disease, contagious bovine pleuropneumonia, and digestive tract infections of the young (Avegan 1984). The Kapsiki cattle are also believed to be adapted to the Sudano and Sudano-Guinean climate of their habitat. The trypanotolerance of Kapsiki cattle is doubtful as these animals have lived in an environment now believed to be free of tsetse flies. Indeed, Achukwi et al. (1997) demonstrated that the Kapsiki cattle have more Zebu introgression than the Namchi cattle, the latter being similar to the N'Dama cattle in trypanotolerance, while the Kapsiki are similar to the Gudali (Zebu) in trypano-susceptibility. Meanwhile, the Doayo (Namchi)

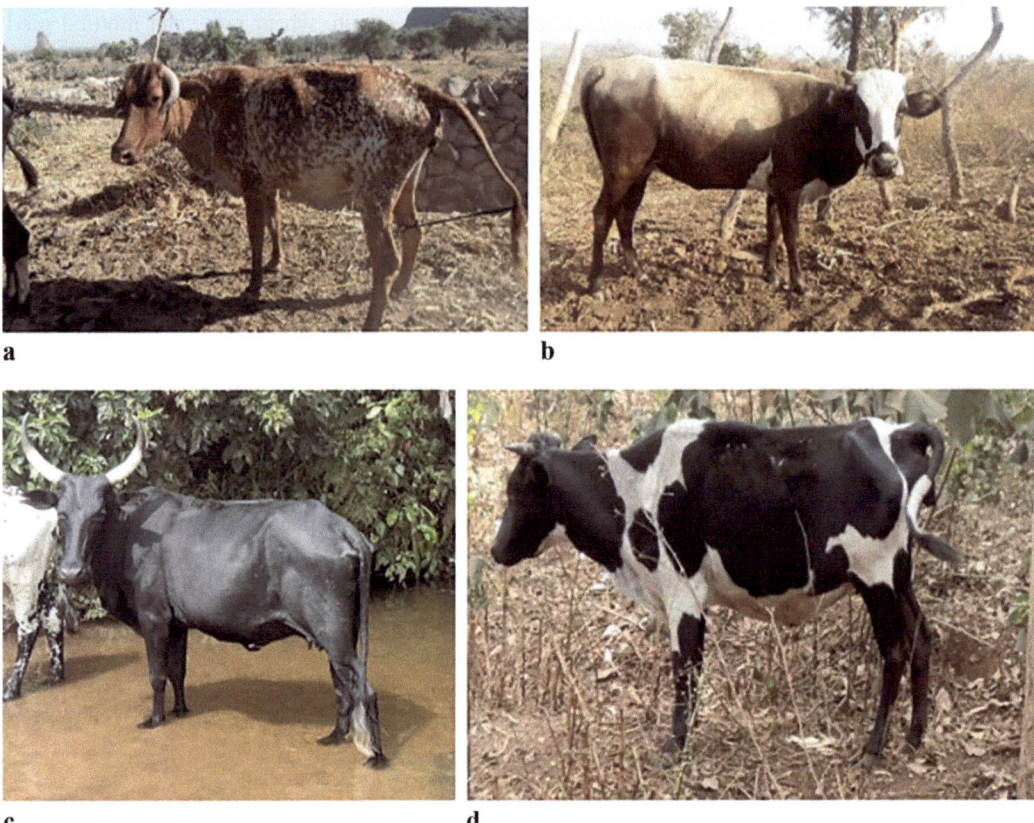

Fig. 3.11 Kapsiki (**a**), Namchi (**b**), Bamenda (**c**), and Bakosi (**d**) cows. (Source: Ojong et al. (2021), licensed under CC-BY 4.0)

is reported to be trypanotolerant, following challenges with T. congolense and Karankasso 83/CRTA/57 and under natural vector challenge (Achukwi et al. 1997, 2009).

The cumulative calf mortality rate to 1 year under improved conditions has been reported to average 12.9% for the Somba (ILCA 1979a) and 6.0 per cent for the Kapsiki (Tawah and Mbah 1989), which was much lower than the estimated 56.8% for the Kapsiki (Dineur et al. 1982) under village conditions. Moreover, the adult mortality rate has averaged as low as 1.9% for the Somba (ILCA 1979a) and as high as 5.9% for the Kapsiki (Tawah and Mbah 1989) under station conditions.

There is limited information on the genetic characteristics of these breeds in the literature. Phenotypic frequencies of the erythrocyte factors in Table 3.3 indicate that, except for the L and S systems, the Somba and the Lagune are genetically similar breeds. They also have similar haemoglobin gene frequencies. Vanvanhossou et al. (2021b) have identified the sub-region of bovine leukocyte antigen (BoLA) class IIb (including *DSB* and *BoLA-DYA*) in Somba cattle, which is uncommon in other African breeds. These results suggest the need for further investigations to understand its association with specific adaptation to endemic diseases.

3.3.2.3 Production Characteristics and Production Systems

The West and Central African humpless Shorthorns have reasonable levels of production, despite the hostile nature of their production environments. However, competition from the generally larger and supposedly more productive Zebu cattle restricts the ecological habitat of trypanotolerant cattle to the tsetse-infested zones where the Zebu cannot survive. Indeed, the single most important factor determining the distribution of trypanotolerant cattle in the region and, to an extent, their production is the distribution of tsetse flies. This, in turn, is closely related to geographical relief, hydrography, climate, and vegetation. ILCA (1979a) presented a summary description of some of the features of the West and Central African regions where trypanotolerant livestock are found. The northern limit of the area is the 750-mm isohyet, above which is the dry, tsetse-free Sahelian zone. Between the 750-mm and 1500-mm isohyets is a transition climate that, in terms of rainfall and vegetation, can be divided into a Sahelo-Sudanian zone up to about 1250 mm and a Sudano-Guinean zone from 1250 to 1500 mm. Farther south, the humid tropical climate can be subdivided, mainly in terms of vegetation, into the Guinea savannah and forest zones. These zones differ in both amount and distribution of rainfall.

Superimposed on this broad zonal delineation are the effects of such local relief features as inland basins, river valleys, and mountains or highland areas. Between the coastal regions and the basins of the Niger River, Western Congo, and Lake Chad, prominent massifs are found. Local highland areas include the Fouta Djallon Ranges (1000–1500 m above sea level), the Guinea Dorsal (800–1752 m above mean sea level), the Togolese Mountains (600–920 m), the Jos Plateau in Nigeria (800–1690 m), the Adamawa Mountains in Cameroon (900–3008 m), and the Crystal Mountains in Gabon (600–1000 m). In addition to the coastal rainforest, the river basins, ramified by a network of rivers and streams, are also covered by dense rainforest.

The effects of these local relief features on local ecology and, hence, on livestock production can be considerable. For example, the Sudan savannah is characterized by high temperatures, low relative humidity of about 8 per cent in dry months, generally dry conditions with rainfall averaging 635 mm annually, an extended dry season of 5 or 6 months, and a single rainy season. The Atacora Mountains, however, have an annual rainfall of more than 1300 mm. In addition, the natural vegetation here is open woodland with grassland comprising *Andropogon*, *Hyparrhenia*, and *Imperata* spp. or woodland savannah with species such as *Isoberlina doka* and *Monotes kerstingii*. Similarly, the Namchi area, home of the Namchi, or Doayo, breed at the foothills of Poli Mountain in Cameroon, receives more than 1000 mm of rain over a period of only about 3 months. As has been stated, the distribution of trypanotolerant livestock is closely linked with that of the tsetse fly and is influenced by the degree of disease challenge presented by varying tsetse fly population densities and the ability of a species to transmit the disease. Tables 3.4, 3.5, 3.6, and 3.7 summarize some production characteristics of the Shorthorn cattle. These are then discussed in the sections that follow under each breed or group.

(a) ***Muturu, Ghana Shorthorn, Baoulé, and Lagune***

The Muturu are generally not milked (Fricke 1979; Ngere 1990) since their yield is just sufficient for their calves. Throughout most of southern Nigeria, however, milk is extracted from Muturu by native doctors for medicinal preparations. The animals and their hides are used mainly for ritual sacrifices and ceremonies (Fricke 1979), particularly funerals (ILCA 1979b). When a pagan dies, for example, one or more Muturu oxen are sacrificed, and the corpse is rolled up in the hides of the slaughtered animals, while the meat forms part of the ceremonial feast. Muturu are commonly kept for prestige or dowry purposes and are also used for draught in some areas (Adeniji 1985; Domingo 1976). They are, however, not widely used as work animals. Occasionally, Muturu cattle are slaughtered specifically to provide meat. Some villages own only a few heads, which scavenge around the homesteads.

Because of the wide distribution of the Muturu, it is difficult to make general statements about management practices, which range from permanent confinement with stall-feeding through to tethering and to year-round grazing. Grandin (1980) gave a detailed description of the Muturu-keeping systems found among the Egun in the 1970s, but significant changes have

Table 3.4 Mean live weight by sex of some Shorthorn cattle breeds and stabilized crosses

Trait	Sex	Ghana Shorthorn	Baoulé	Lagune	Muturu	Borgou
Birthweight *(kg)*	Male	19–20	9–16	10	11–16	16–17
	Female	17–19	9–14	10–12	8–16	15–16
Weight at 3 months *(kg)*	Male	47–53	36–42	–	27–52	
	Female	46–47	37–38	–	28–52	
Weight at 6 months *(kg)*	Male	75–84	51–61	47–49	49–94	86–91
	Female	68–101	56–62	47–53	47–87	66–72
Weight at 9 months *(kg)*	Male	–	64–82	–	81–119	
	Female	–	60–85	–	61–117	
Weight at 12 months *(kg)*	Male	109–135	78–94	34–83	90–122	79–130
	Female	82–134	70–98	45–87	78–115	79–117
Weight at 18 months *(kg)*	Male	134	86–130	–	105–174	164
	Female	147	78–124	–	90–127	152
Weight at 24 months *(kg)*	Male	164–178	112–162	–	114	200
	Female	136–182	116–146	164	136	207
Weight at 36 months *(kg)*	Male	193–256	170–213	–	196	226
	Female	173–193	151–166	153–221	177	157–226
Weight at 48 months *(kg)*	Male	–	–	–	–	265
	Female	–	–	158–253	–	227
Mature weight *(kg)*	Male	190–380	191–298	–	141–255	193–330
	Female	163–273	150–240	165–262	140–270	181–295

Source: Rege et al. (1994c)

occurred in these systems since then. Akinwumi and Ikpi (1985) also made a comprehensive presentation of the production systems of the Muturu in Nigeria. The Resource Inventory and Management Limited presented a detailed coverage of management systems of the Muturu in different regions of Nigeria (RIM 1992). These reports indicate a very wide variation in the systems of herding, feeding, and housing. Availability of labour and access to grazing land seem to be the overriding determinants of the type of husbandry system used. On the one hand, expanding cultivation in many areas, especially in Igboland, has either tended to discourage the keeping of Muturu or forced owners to turn to stall-feeding or tethering systems. On the other hand, some of the northern Muturu populations, such as those in the north-east and on the Dimmuk escarpment, have ready access to mountain pastures that are of no value as arable land and can, therefore, be grazed by cattle with little supervision without risk of damage to crops.

In Igboland, an estimated 45 per cent of the herds are taken out to graze, 45 per cent are tethered, and 10 per cent are permanently stall-fed. The animals are kept in enclosures surrounded by hedges (*oka-efi*) with built-in sheds for shelter during the night or in bad weather. Traditionally, these cattle were sacred (*juju* cattle) in this region and were considered the property of local deities or were dedicated to the shrine. With the decline in tradition and the destruction of shrines, however, many villages are without a bull, which is one of the threats to the survival of these cattle populations.

The Koma people of the Antlantika Mountains in Gongola State keep semi-feral Muturu, which are neither tended nor supplemented, on the slopes outside their villages. Individual animals required for sacrifice have to be hunted down, although by-laws were passed in the 1970s to restrict the hunting of Muturu to year-round cut-and-carry systems in Yorubaland. This has not worked very well, mainly because of the labour-intensiveness of stall-feeding. As a result, both free-range and stall-feeding are practised in Yorubaland.

Traditions and beliefs relating to the Muturu on the Jos Plateau appear to reflect the 'wild' nature of the animal (RIM 1992). Not only are

Table 3.5 Estimates of reproductive parameters of some Shorthorn cattle breeds and stabilized crosses

	Ghana Shorthorn		Baoulé			Lagune			Muturu		Keteku	Bergou
	ARS (Legon)	UST (Kpong)	Improved Bouaké (Côte d'Ivoire)	CREAT (Togo)	Village Burkina Faso	Village Benin	Togo	Improved Benin	Improved	Village	Ranch Nigeria	Village Benin
Age at first calving (months)												
Mean	42.4 ± 6.5 (57)[a]	37.1 ± 1.4 (11)	25.7 ± 1.3 (63)	–	56.0 (271)	–	–	29.1 (17)	28.1 ± 7.7 (15)	–	43.6 ± 4.4 (567)	48.5 ± 2.4 (378)
Range	34.8–47.9	35.9–38.5	24.0–25.7	37.0–43.0	34.0–56.0	36.0–48.0	42.0–60.0	24.0–42.0	18.1–43.5	48.0–60.0	38.0–47.0	42.7–50.0
Calving interval (months)												
Mean	15.7 ± 0.9 (181)	17.0 ± 0.8 (130)	14.0 (234)	18.6 ± 5.8 (81)	17.0 (448)	24.3	24.0	13.6 (60)	12.4 ± 0.6 (93)	–	17.7 ± 1.4 (1087)	19.1 ± 1.7 (685)
Range	14.8–16.7	15.4–18.6	–	11.0–36.5	–	–	–	12.0–13.6	10.2–14.8	18.0–24.0	16.4–19.1	15.3–20.0
Calving rate (%)												
Mean	71.0	66.0 ± 6.8	–	–	57.7	–	–	–	–	–	–	65.4 ± 13.1
Range	–	60.2–73.8	74.0–85.0	32.0–61.0	57.0–71.0	34.0–45.0	42.0–49.0	58.0–70.0	–	–	–	64.4–66.9

Source: Rege et al. (1994c)
[a]Figures in brackets are the number of observations

Table 3.6 Estimates of milk production by milking and management system for some Shorthorn cattle breeds

Milking system	Trait	Mean	Ghana Shorthorn Station	Ghana Shorthorn Station Range	Northern Côte d'Ivoire Village	Baoulé Village	Station Southern Côte d'Ivoire	Somba Village Togo	Lagune Village Togo
Weigh-suckle-weigh	Lactation milk yield (kg)	656.9 ± 183.6 (20)[a]	384–774	–	–	318 ± 69	356.0 ± 78.3	228.5 ± 47.6	–
	Lactation length (days)	261 ± 53 (20)	–	–	182–295	180	210	143 ± 13	–
	Daily milk yield (kg)	2.5	0.9–9.0	–	–	–	1.7	1.8 ± 0.7	–
	Butterfat (%)	1.3 ± 0.4 (9)	0.3–3.9	–	–	–	–	6.0 ± 0.4	–
	Total solids (%)	10.6 + 0.4 (9)	9.3–13.8	–	–	–	–	–	–
	Solids-not-fat (%)	9.3 ± 0.2 (9)	9.0–9.5	–	–	–	–	–	–
	Proteins (%)	3.2 ± 0.2 (9)	2.1–4.3	–	–	–	–	–	–
	Ash (%)	0.75 ± 0.02 (9)	0.64–1.10	–	–	–	–	–	–
Direct milking	Lactation milk yield (kg)	–	–	130–150	215.0 ± 29,4	–	–	295.3 (23)
	Lactation length (days)	–	–	210–122	285 ± 33	–	–	–	225
	Daily milk yield (kg)	–	–	0.36–0.70	0.8	–	–	–	1.4–1.7
	Butterfat (%)	–	–	–	5.0	–	–	–	6.2

Source: Rege at al (1994c)
[a]Figures in brackets are the number of observations

Table 3.7 Live weights, carcass weights, and dressing-out percentage of Baoulé, Muturu, and Keteku cattle by sex and age

Breed	Sex	Age (months)	n[a]	Live weight (kg)	Carcass weight (kg)	Dressing-out percentage
Baoulé	Bull	18	–	130.0	65.0	50.0
		24	23	157.0	79.5	57.0
		36	20	–	102.7	–
		48	27	–	104.5	–
		60	26	–	103.0 ± 11.6	–
	Cow	18	–	112.0	56.0	50.0
		24	12	146.0	72.7	50.0
		36	7	–	105.7	–
		48	21	–	98.1	–
		60	64	–	94.0 ± 3.8	–
Muturu	Cow	12–24	6	109.0 (92–121)[b]	–	–
		24–36	11	145.0 (122–174)	–	–
		36–48	14	167.0 (135–202)	–	–
		>48	11	179.0 (153–221)	–	–
Borgou	Bull	–	24	265.0	137.0	52.0
	Cow	–	8	227.0	117.0	52.0
Keteku	Steer	12–24	12	159.0 ± 17.8	–	–
		24–36	12	235.0 ± 21.9	127.0	46–51

Source: Rege et al. (1994c)
[a]Number of observations
[b]Figures in brackets are ranges

the animals not milked, but it is also considered taboo to tie them up. In addition, ticks are not removed as they are considered to be harmless. In this area, cattle are confined during the night in pens set close to the compound. Manure from these pens is collected at the beginning of the rain to fertilize the fields.

Most Muturu in southern Nigeria are in the hands of private owners, whether individuals, the community, or traditional chiefs and royal families. The animals receive little care and largely survive by scavenging around the village. They are usually seen grazing and sleeping in the fields of community schools. In some areas, cattle ownership was historically limited to traditional rulers and highly placed chiefs from whom permission had to be sought before an animal was killed.

The Muturu village production systems can be classified into three broad groups: free-range, household (compound), and communal (cooperative). In addition to these, a few government stations such as Ado Ekiti, Imala, and Odeda keep Muturu (Akinwumi and Ikpi 1985). The free-range system has been associated with the 'wildness' of the Muturu (Fricke 1979; Akinwumi and Ikpi 1985). Household production systems are characterized by the provision of a shed and stall-feeding. Common feeds used in this system include cassava leaves, chopped tubers, yam peels, maize on the cob, fresh maize leaves, kola nut pods, and salt.

The communal production system is common in the states of Ondo and Rivers in Nigeria. In Rivers State, the usual system of management is to allow the cattle to range freely throughout the year. The semi-feral animals are not milked. They receive little veterinary care and are not given any supplementary feed. Little attention is paid to their breeding. These factors, plus the importation of Zebus from the north, make the disappearance of Muturu from Rivers State almost inevitable (RIM 1992). More organized communal systems also exist, however, where owners pool their animals, often uniquely tagged, and engage one or two Fulani herders. A tract of land is set aside within the village for housing. Local government authorities occasionally provide a kraal, within which improved pastures are developed and made available at a fee to the cattle owners. This system was adopted decades ago in response to the persistent agitation of crop farmers over the destruction of their crops and has had differing results in different states. In Ondo State, the acceptance of this system has essentially ensured the survival of Muturu, while in the adjacent Bendel State, several communities were forced to slaughter their cattle because of constant pressure from crop farmers. This was considered a causal factor for the decline of Muturu populations in this state between 1977 and 1984.

The Muturu in Liberia are also kept under village conditions although commercial production was initiated on the rubber plantations of Liberian Agricultural Company and Firestone, as well as on private commercial farms. Under village conditions, these animals are rarely herded and receive very little care. In some cases, they are tethered to avoid crop damage (ILCA 1979b).

Growth and live weights Birthweights of the Muturu have ranged from 8.3 kg for calves born to heifers under 24 months of age to 12.1 kg for calves of heifers at least 24 months old under the environment of light tsetse challenge in the derived savannah. Heavier birthweights, averaging 16 kg, have been reported in high-altitude tsetse-free zones in the northern Guinea savannah. Weights at 6 and 12 months of age of 87 and 119 kg, respectively, have been reported at the Upper Ogun Ranch, while corresponding figures of 56–58 kg and 91–92 kg have been reported at Ado Ekiti and 37–39 kg and 61–71 kg at Vom in Nigeria. This trend is obviously the result of more intensive animal management, including routine prophylactic treatments and proper feeding, at the Upper Ogun Ranch than at the Vom Institute, given the pattern of tsetse challenge in these areas. This demonstrates that the effect of trypanosomiasis can be reduced through adequate feeding and regular prophylactic dosing.

Carcass characteristics Carcass weights have ranged from 90 to 100 kg, with dressing-out percentages of 46–53 per cent. Despite differences in live weight, dressing-out percentages of Muturu and Zebu cattle in the region are comparable, and the meat quality of the Muturu has been judged to be excellent (Ezekwe and Machebe 2005).

Reproductive characteristics Muturu cattle are considered to have good fertility, with the capacity to produce one viable calf annually. Furthermore, they mature earlier than the Zebu in the region. The Muturu compares favourably with the Baoulé and the Lagune and is better than the Ghana Shorthorn in age at sexual maturity under improved management. Also, the intervals between consecutive carvings were shorter under improved management (10–15 months) than under traditional village (18–24 months) systems. As with other Shorthorn and Zebu cattle in the region, heifers tend to have their first calves at an older age under village conditions than under improved conditions.

The age of Muturu bulls at first service under village conditions has been estimated at 4–5 years (ILCA 1979b). Even though these bulls are capable of year-round breeding, some studies have indicated seasonal changes in the characteristics of their ejaculates (Igboeli et al. 1987), including volume (1.8 and 2.3 ml in the dry and wet seasons, respectively), motility (36.2 and 37.7 per cent), and morphologically normal sperm (70.0 and 79.1 per cent). In general, semen characteristics of the Muturu are inferior to those of the Gudali and the White Fulani under similar conditions. In addition to producing low-quality semen, Muturu bulls have been reported to be generally shy and nervous (Nwakalor et al. 1979). Moreover, the low motility and sperm concentration, coupled with low amounts of fructose in the ejaculates (Igboeli et al. 1987), are major impediments to the processing of Muturu semen for artificial insemination purposes.

Milk production characteristics Since Muturu cows are normally not milked, data on their milk production are not well-documented. Nonetheless, lactation milk yields for lactation lengths of 120–216 days have been reported to range from 127 to 421 kg (Fricke 1979; Olaloku 1976). Ezekwe and Machebe (2005) confirmed the low milk production of the Muturu and reported a butterfat content of 4.7 per cent and that the other milk constituents were within the values reported for tropical breeds.

Lagune cattle are mainly used for beef (Adeniji 1985; Domingo 1976; Mortelmans and Kageruka 1976), but, like the Muturu, they are poor milk producers. They are not used for traction. In Gabon, a multiplication herd of 53 heads of Lagune cattle was established at the Estuaire Region with the main objective of manure production (ILCA 1979b).

The management and production systems under which Lagune cattle are raised are mainly traditional, even though some animals are kept under improved conditions (Adeniji 1985). In Benin, Lagune animals graze freely on natural pastures, under palm trees and coconut trees, and on fallow land. There are no provisions for feed supplementation or for veterinary care. On the contrary, under improved conditions, such as at the Samiondji station, animals graze natural or sown pastures in fenced paddocks and receive veterinary attention.

Generally, Lagune cattle graze and roam freely around the villages (Agbemelo 1983). Under these conditions, the husbandry systems are no different from the free-range system under which most southern Nigerian Muturu populations are kept. Unlike free-range Muturu, however, Lagune cattle are reported to be docile. In Côte d'Ivoire, Lagune cattle are kept in areas where farmers do not traditionally keep cattle, and they are rarely herded. Lagune cattle in Zaire (the Democratic Republic of the Congo) also roam freely in small herds of three to six, living on forest undergrowth and fallow land. They seek grazing everywhere and sometimes traverse long distances. This is like the situation in the Congo (Congo-Brazaville), where the animals are kept exclusively in village herds. As alluded to above, Lagune cattle imported into Gabon from Zaire were introduced at the village level through *métayage* operations (ILCA 1979b). Here, they were kept in herds of seven to 12, roaming freely during the day and penned at night. They were seldom given mineral supplements and were not sprayed against ticks. These animals were originally intended for Gabonese farmers, but some were sold to immigrants from Cameroon, who kept them under the same conditions for commercial purposes.

Growth and live weights Birthweights of Lagune have averaged 10–12 kg under village conditions, while growth rates and yearling and mature weights varied substantially. Rege et al. (1994c) reported that the mature weight of Lagune range from 180 to 280 kg in males and from 165 to 262 kg in females.

Carcass characteristics Dressing-out percentages for the Lagune have ranged from 48 to 54 per cent. Despite differences in carcass weight between the Lagune and Somba cattle, their dressing-out percentages are quite similar under village conditions.

Reproductive characteristics Age at first calving of the Lagune has ranged from 24 months under improved village conditions to 60 months under traditional village conditions. Mean calving intervals have varied from as short as 12 months under station conditions to as long as 24 months under village conditions, while calving percentages have ranged from 34 per cent under traditional management to 70 per cent under improved management. Most of the Lagune cows (74 per cent) tend to calve at least every 2 years under village conditions in Togo, with as few as 13 per cent calving every 3 years and the rest calving annually (Agbemelo 1983). The calving rate is related to the tsetse challenge, although the Lagune performs better under a medium challenge than do the Borgou under a light tsetse challenge (ILCA 1979a). Lagune females are known to calve without difficulty (Agbemelo 1983).

Milk production characteristics The lactation yield reported for the Lagune breed by Agbemelo (1983) under village conditions was higher than the available estimates for the Baoulé and the Somba under similar conditions. Daily milk yields ranging from 1.5 to 2.0 kg over lactation lengths of 120–225 days have been reported under village conditions. Milk butterfat in these animals has been estimated at 6.2 per cent, which is like that of the Zebu, Baoulé, and Somba cattle but higher than that of the Ghana Shorthorn and Holstein Friesian cattle.

Ghana Shorthorns are poor milkers (Maule 1990) and so are generally not milked. Montsma (1960) concluded that the milk yield of Ghana Shorthorns was too low to warrant the sale of fresh milk without adverse effects on the growth and general well-being of their calves. Moreover, when milked in the absence of the calf, they go dry within a few weeks (Ngere et al. 1975). However, the breed has good conformation for meat production (Domingo 1976) and yields reasonable beef carcasses (Ngere et al. 1975). They are also used for draught.

Over 99 per cent of Ghana Shorthorns are kept in village herds (Shaw and Hoste 1987). Most cattle owners in Ghana are resident farmers, but town dwellers are increasingly becoming absentee cattle owners using hired herders, usually the Fulani. In such situations, the herders milk the animals, which forms a large part of their salary. Cattle are grazed on communally owned land. In many cases, especially in the main livestock region in the north, animals of several households are herded together and tended by village children, partly because the animals have good temperament (ILCA 1979b). Animals are penned at night. In the south-east, a herd may be allowed to graze on a farmer's land in exchange for the manure produced. In northern Ghana, herd sizes are usually small, with a typical family owning five to 20 heads of cattle. A high percentage of oxen, up to 10 per cent of a herd, are kept in areas where they are used for draught.

Growth and live weights Birthweights of the Ghana Shorthorns have ranged from 17 to 20 kg (see Table 3.4). Growth rates in the literature reviewed varied considerably, as did weaning and mature weights. In general, the performance of the Ghana Shorthorn (the Savannah type) has been better than that of the Dwarf (Forest) Shorthorn type. The wide range (64–154 kg) of weaning weights in Ghana Shorthorns may be the result of the variation (7–9 months) in weaning age, in addition to differences in management. Supplementation studies have shown that proper feeding can improve the growth performance of the animals. Mature weights on-station ranged from 163 to 380 kg and were, on average, higher than those on-farm, which varied from 170 to 200 kg. The variation in mature weights may also be associated with the lack of precise age at maturity in these studies.

Carcass characteristics Slaughter and carcass weights of the Ghana Shorthorn averaged 250 and 125 kg in the males and 192 and 102 kg in the females, respectively. Corresponding dressing-out percentages were 50 and 53 per cent. An on-station study of a sample of 32 Dwarf Shorthorns (Appiah 1988) in the humid forest zone produced slaughter and carcass weights of 142 and 67 kg, respectively, and yielded a dressing-out percentage of 47.3 per cent.

Reproductive characteristics The mean age at first calving of Ghana Shorthorn was better on-station than on-farm (Shaw and Hoste 1987; Cockcroft 1977). This is clearly the effect of poor nutrition in village herds. Likewise, the mean calving interval ranged from 15.4 to 18.6 months for the Dwarf (Forest) Shorthorn and from 9.3 to 30.8 months for the Savannah-type Shorthorn. Ngere and Cameron (1972) reported a much shorter calving interval of 13.8 months for Ghana Shorthorns under an improved production system. The annual calving rate, defined as a proportion of the number of calves dropped to the number of cows mated per year, has been estimated to be slightly higher for the Ghana (Savannah) Shorthorn than for the Forest type. This is obviously the result of the different tsetse challenges in the two agro-ecological zones. Under improved conditions, the calving rate has ranged from 60 to 70 per cent for Ghana Shorthorn cows.

Milk production characteristics The total lactation milk yield of Ghana Shorthorn has been estimated using the weigh-suckle-weigh method, and results have shown that improving the level of concentrate feeding from late pregnancy through lactation increases the milk yield of the Ghana Shorthorns, up to as much as 1002 kg in 252 days. Rege et al. (1994c) reported a variation in lactation performance of cows from a milk yield of 5 kg produced over a period of 9 days to 3059 kg over 454 days. Normally, Ghana Shorthorn cows lactate for no more than 10 months, with a daily milk yield of at least 0.9 kg. However, this variation points to the potential for genetic improvement. In the absence of the calf, Ghana Shorthorn cows, like Zebu cows, tend to dry off within a few weeks of lactation. There is only limited data available on the milk composition of Ghana Shorthorns. Compared with the Holstein Friesian (4 per cent), Zebus (4–5 per cent), and some of the other Shorthorns (5–6 per cent), the Ghana Shorthorns' milk is low in butterfat (1.3 per cent) and proteins (3.2 per cent) (DAGRIS, https://dagris.info/ko/node/8359).

The Baoulé Unlike the Ghana Shorthorns, milking is practised in some Baoulé herds in northern Côte d'Ivoire (Godet et al. 1981). Here, herders usually receive a salary and all or part of the milk from the herds. Among the Lobi people of Bouna in the Northern Region of Côte d'Ivoire, milking is rarer than elsewhere—less than 50 per cent of the herds were reportedly milked, and only 25 per cent of these were milked regularly (Godet 1976). As has been noted, sales for cash represent an important cattle function in these production systems. In Baoulé herds, there appears to be a high offtake of young males between 2 and 4 years of age. Results from a survey carried out in Dabakala, Korhogo, and Bouna in Côte d'Ivoire (Poivey and Seitz 1977) revealed that herds consisted of 31 per cent males and 69 per cent females, with virtually no steers and an insignificant proportion of males over 2 years of age. Some herds had as many as six adult (mature) bulls. These obviously function as the breeding bulls at the village level in communal pastures. About 99 per cent of the cattle are kept under village conditions (Shaw and Hoste 1987), and the rest are kept in research stations and ranches, most of which are government-owned. In the north of Côte d'Ivoire, Baoulé cattle from one village are often herded together by hired Fulani herders who also milk the animals. Calves are separated from their dams in the evenings on return from grazing. In the mornings, they are allowed to suckle to stimulate milk let-down, facilitating milk offtake. Generally, a maximum of three teats of the udder are milked, after which calves are allowed to consume the remainder of the milk (Godet et al. 1981).

Growth and live weights Soro et al. (2015) undertook a comprehensive study of the morphometrics of Baoulé cattle in Côte d'Ivoire, including adult birthweights (Table 3.8). Birthweights of the Baoulé have ranged from 9 kg under village conditions to 16 kg under improved conditions, and growth rates and weaning and mature weights have varied substantially. Mature weights have ranged from 160 to 300 kg in males and from 150 to 240 kg in females (Rege et al. 1994c). Supplementary feeding yields substantial improvements in the growth performance of the Baoulé (Kouakou 1984; Tiemoko et al. 1990).

Carcass characteristics Only limited information is available on the carcass traits of the Baoulé. Carcass weights have ranged from 90 to 200 kg. Despite live weight differences, carcass weights of the Baoulé, the N'Dama, and the Sahelian Zebus are comparable. Dressing-out percentages are moderate (50 per cent), with percentages ranging from 53.2 to 54.8 for Baoulé cattle under station conditions in Côte d'Ivoire. A study by Tidori et al. (1975) reported that Baoulé

Table 3.8 Birthweight and linear body measurements of Baoulé cattle in Côte d'Ivoire

Parameters	Male ($n = 44$)	Female ($n = 121$)
Body weight (kg)	184 ± 39.9[a]	191 ± 36.5[a]
Heart girth (cm)	128 ± 10.3[a]	130 ± 9.65[a]
Height at withers (cm)	94.1 ± 5.25[a]	94.9 ± 5.48[a]
Horn length (cm)	14.8 ± 7.93[a]	15.6 ± 8.59[a]
Ear length (cm)	15.0 ± 1.37[a]	14.9 ± 1.51[a]
Body length (cm)	120 ± 7.43[a]	122 ± 10.1[a]
Muzzle circumference (cm)	38.0 ± 2.68[a]	38.8 ± 4.24[a]
Pelvic width (cm)	36.1 ± 2.65[a]	36.2 ± 3.60[a]
Head length (cm)	37.6 ± 2.48[a]	38.4 ± 2.98[a]

Source: Soro et al. (2015)
n number of animals measured
[a]Means of males and females were not significantly different ($P < 0.05$)

carcasses have a poor finish, with a 0.5-per cent fat index (defined as the weight of kidney fat as a percentage of hot carcass weight), against the recommended 2 per cent. Thus, Baoulé meat can be classified as poorly marbled.

Reproductive characteristics Age at first calving of the Baoulé has ranged from 24 months on-station to 56 months on-farm, while calving intervals have ranged from 11 to 37 months under station conditions. Chicoteau (1989) reported results from experiments carried out in Burkina Faso using both hormonal and behavioural criteria, which demonstrated that Baoulé females reach puberty at an average age and weight of 14 months and 125 kg, respectively. Despite differences in weight, Baoulé and European heifers are quite comparable in age at puberty. Moreover, Baoulé cattle reach puberty much earlier and are reportedly more fertile than Zebus and other Shorthorns in the region (Tidori et al. 1975). Calving rates have ranged from 32 to 85 per cent under station conditions, although a calving rate as high as 85 per cent was reported (Shaw and Hoste 1987) from a sample of 44 *métayage* herds in the Bossenbélé area of the Central African Republic. Given the difficulty of deriving a true calving rate in village herds, where immature heifers are usually grouped together with breeding females, some researchers have previously proposed that the term 'fecundity rate', defined as the number of calves dropped per 100 breeding-age females per year (Glattleider 1976), be used as a fertility measure in village herds. The fecundity rate of Baoulé has ranged from as low as 28 per cent in village herds in the Central African Republic to 85 per cent in some station herds in Côte d'Ivoire. Godet et al. (1981) pointed out that the extraction of milk for human consumption, in addition to affecting calf growth (Hoste et al. 1983), has a depressing effect on the fertility of the Baoulé.

Information on post-partum behaviour, sexual function, and cyclicity in cattle can be used to improve herd fertility. Although such studies are generally lacking for the West African Shorthorns reviewed here, a few have been made on the Baoulé (Chicoteau 1989; Djabakou and Grundler 1988). For example, it has been reported that the duration of uterine involution is between 30 and 34 days post-partum in lactating Baoulé cows (Chicoteau 1989; Hoste et al. 1983). This has a significant influence on the post-partum resumption of ovarian activity, the ability of the cow to breed again, the length of the calving interval, and the calving percentage. Moreover, post-partum resumption of cyclicity, an important determinant of the duration of the calving interval, takes about 57 days in lactating Baoulé cows (Chicoteau 1989), although it can be longer than 160 days (Meyer and Yesso 1988). Tidori et al. (1975) reported that sexual cyclicity in the Baoulé cow is significantly influenced by season, which is clearly a reflection of the effect of seasons on the nutritional status of animals in this pasture-dependent systems.

Male reproduction has been better documented for the Baoulé than for the other Shorthorn breeds in the region. Chicoteau (1989) reported the average age and weight at puberty as being 18 months and 155 kg, respectively, taken from a sample of 15 young Baoulé bulls in Burkina Faso. Despite the differences in their weights, Baoulé and White Fulani bulls reach puberty at comparable ages. Baoulé bulls were reported to have attained about 61 per cent of their mature body weight at puberty. Chicoteau (Tiemoko et al. 1990) reported that, although young Baoulé bulls ejaculate first at 17 months of age, they produce first motile spermatozoa only after 30 months of age. The semen quality of Baoulé bulls is mediocre, which is indicated by their pubertal semen characteristics: 0.85 per cent motility, 0.8 ml sperm volume, $6.0 \times 10^7/mm^3$ sperm concentration, 40 per cent live sperm, and 85 per cent abnormal sperm. Chicoteau (1989) reported scrotal circumferences of 23 and 27 cm in pubertal and adult Baoulé bulls, respectively. This measure is used for the early evaluation of bull soundness and fertility.

Milk production characteristics Milk yield estimates under station conditions have been lower in the Baoulé than in the Ghana Shorthorn but generally higher than those of the Somba under village conditions. Milk yields have ranged from 118 to 387 kg over lactation periods of 279–341 days under village conditions. The month or season of calving seems to have a more significant effect on milk yields of the Baoulé than parity (Hoste et al. 1983), which is a further indication that nutrition is perhaps the most limiting factor. Information on the composition of Baoulé milk is completely lacking, except for an estimate made in a village herd in Bouaké of a milk butterfat content of 4.2–5.1 per cent (Glattleider 1976).

(b) *Somba, Kapsiki, Namchi, and Bakosi*

Kapsiki, Doayo (Namchi), and Bakosi cattle play an important role in the traditional life of the Kapsiki, Namchi, and Bakosi peoples. These animals are neither milked (ILCA 1979b) nor exploited commercially and are used in much the same way as the Muturu cattle in Nigeria for dowry, special feasts, and rituals. The Kapsiki cattle in Cameroon are also associated with burial ceremonies and, like the Muturu in Nigeria, play an important part in complex ritual rites. The Kapsiki are used in some areas as draught animals. Intricate entrustment relations have also been developed among the Kapsiki people to ensure even distribution of the cattle throughout the community (van Beek 1978). However, the role of cattle in religious ceremonies is declining as Islam becomes widespread in the area.

The herd size of Kapsiki cattle is quite variable. Households generally own five to ten heads and, sometimes, fewer (ILCA 1979b). Dineur and Thys (1986) reported an average herd size of 7 ± 2 heads with a range of 2 to 30 heads from a sample of 33 herds. These are brought together in collective herds and tended by children. This herding practice is also employed for the Namchi and Bakosi. At night, animals are kept in pens close to the compound. Reproduction among the Kapsiki is said to be controlled, a practice that has been suggested to be responsible for the maintenance to date of the breed in a relatively pure form (Dineur and Thys 1986).

Bakosi cattle in Cameroon are also neither herded nor milked and are believed to be getting smaller and less fertile as a result of isolation and inbreeding (ILCA 1979b). One animal may have several owners. The few Dwarf Shorthorns in Cameroon—the Bakweri—are kept on palm plantations where they graze the legume *Pueraria phaseoloides* (kudzu), which grows under the palm trees (ILCA 1979b).

Somba is mainly a beef animal kept for social functions (Maule 1990) although it is milked in some areas (Avegan 1984). Because of its small size, however, the Somba is not used for draught. They are usually tended by hired Fulani herders or by children. Herd sizes vary from 10 to 100 animals and can even reach 300 heads. The animals are grazed exclusively on pastures, although traders or functionaries who may own these cattle provide salt-licks, agro-industrial by-products, and crop residues on the pastures (Avegan 1984). In the Somba area in Togo, for example, more than 80 per cent of the residents practise mixed crop–livestock farming. They cultivate food crops, mainly maize, yams, millet, and rice; cash crops, such as cotton, groundnuts, coffee, and cocoa; and horticultural crops, including pineapples, pears, bananas, sugar cane, oranges, and tangerines. Residues from these crops are the main sources of supplementary livestock feed.

Although Togo is not an important livestock country (Somoko-Balantpli and Freitas 1978) and livestock rearing is a minor, almost marginal, activity, the Kontomba of Central Region and the Kabré of Kara Region, the two major regions of Somba cattle, seem to place more importance on livestock. Husbandry here differs from that of the Fulani in the north. After harvest, animals are allowed to roam freely, unattended, until the next cropping season.

Growth and live weights Birthweights of Somba cattle under village conditions have ranged from 9 to 12 kg for the males and from 7 to 9 kg for the females. These are lower than the average of 21.4 kg reported under station conditions by Grell et al. (1982). The mature weights of the Somba have ranged from 145 to 175 kg for cows, from 200 to 210 kg for bulls, and from 185 to 200 kg for oxen under village conditions. Mature weights varying from 220 to 260 kg have been reported on-station. Mature weights of the Namchi have averaged from 180 to 210 kg for bulls and 150 kg for cows under village conditions. These estimates were based on a rather limited sample of animals and are, therefore, only indicative. Table 3.9 shows the best linear unbiased estimates of live weight, height at withers, and heart girth of limited samples of Namchi and Kapsiki cattle under their natural environments in Cameroon.

Carcass characteristics Information on carcass traits is limited for the Somba and completely lacking for the Kapsiki, Namchi, and Bakosi. Available dressing-out percentages of the Somba have ranged from 48 to 52 per cent under village conditions. These compare well with those of other humpless Shorthorns and Zebus in the region.

Reproductive characteristics Age at first calving under village conditions has averaged 36–48 months for the Somba, 48 months for the Kapsiki, and 36 months for the Namchi, while calving intervals have averaged 12–14 months for the Somba, 18–24 months for the Bakosi, and 12 months for the Namchi under village conditions. Calving interval on-station has been estimated at 15 months for the Kapsiki (Tawah and Mbah 1989). The Namchi appear to mature earlier and have shorter calving intervals than the Kapsiki and the Bakosi in Cameroon. However, Kapsiki's environment is more stressful, both in terms of climate and nutrition, than that of the Namchi. Generally, calving intervals have been much shorter for these breeds than for the other Shorthorns and Zebus in the region, even under village conditions. Calving rates have averaged 60 per cent for the Somba on-station and ranged from 70 per cent on-farm (Dineur et al. 1982) to 84 per cent on-station (Ojong et al. 2021) for the Kapsiki.

In general, most of these breeds have good fertility and, with adequate feeding, calving can occur every year. They have excellent longevity (ILCA 1979b), with the productive lifetime of the Somba estimated to be five to seven lactations (Avegan 1984). Young Somba bulls are sexually mature at 2.5 years of age and remain sexually active for about 12–14 years. However, ILCA (1979b) pointed out that isolation and dwindling populations, with resultant inbreeding, tend to impair the fertility of most of these breeds, especially the Bakosi.

Milk production characteristics Available estimates of milk production in the Somba indicate a range of 1–2 kg of milk per day for 150–180 days of lactation. As with the other breeds, parity influences the lactation milk yields of the Somba. Their lactation milk offtake increased from 176 kg, for a 121-day lactation in the first parity, to 280 kg, for a 152-day lactation in the fourth parity and then declined to 214 kg for a 147-day lactation in the fifth parity. Butterfat content in Somba milk has ranged from 5.5 to 6.5 per cent and is, thus, higher than that of the Ghana Shorthorn and European dairy breeds but comparable to that of the other Shorthorns and Zebus in the region.

Table 3.9 Best linear unbiased estimates of live weight and body measurements of Namchi and Kapsiki cattle under the natural environment in Cameroon

Breed	Sex	Age (years)	N[a]	Live weight (kg)	Height at withers (cm)	Heart girth (cm)
Namchi		0–3		158.1 ± 10,4	95.3 ± 1.9	126.2 ± 2.5
		4–7		197.6 ± 5.9	102.9 ± 1.1	138.7 ± 1.4
		8–11		183.8 ± 10.8	102.6 ± 2.0	138.3 ± 2.6
		>11		145.7 ± 20.1	101.1 ± 3.6	134.4 ± 4.8
Kapsiki		0–3		167.7 ± 25.1	107.2 ± 4.5	131.0 ± 6.1
		4–7		218.5 ± 18.3	113.6 ± 3.3	144.7 ± 4.4
		8–11		186.0 ± 17.8	111.4 ± 3.2	138.3 ± 4.3
		>11		168.8 ± 23.9	109.1 ± 4.3	132.2 ± 5.8
Namchi	Male		45	171.3 ± 7.4	103.7 ± 1.3	134.5 ± 1.8
	Female		75	171.3 ± 5.6	97.2 ± 1.0	134.6 ± 1.4
Kapsiki	Male		3	221.7 ± 33.8	116.5 ± 6.1	145.4 ± 8.2
	Female		84	148.9 ± 6.3	104.2 ± 1.1	127.7 ± 8.2
Namchi			122	171.3 ± 5.5	100.5 ± 1.0	134.5 ± 1.3
Kapsiki			85	185.3 ± 17.5	110.3 ± 3.2	136.5 ± 4.2

Source: Ebangi et al. (2011)
[a]Sample size

3.3.3 Other Shorthorn Cattle Breeds of West and Central Africa

Other less known Shorthorn cattle breeds include the **Manjaca** and the **Gambian Dwarf**. A few Manjaca cattle are reportedly found towards the coast in the Cacheu Region of Guinea-Bissau and also on the islands. Da Costa (1933) had earlier described the Manjaca of the Brames Region of the former Portuguese Guinea as '*small, well-proportioned and of typical Shorthorn conformation, distinguished by the black colour of the coat*'. The breed was reportedly disappearing in the 1970s (ILCA 1979b) because of crossing and replacement by other breeds. The subsequent FAO/ILCA studies (Shaw and Hoste 1987; ILCA 1992b), however, made no mention of the Manjaca. This breed has probably been wiped out through interbreeding. In The Gambia, the remnants of the Shorthorns are still found south of the Gambia River, but these populations are being absorbed by the N'Dama, principally owing to the economic superiority of the latter in this environment. Whether purebred or crossed with Zebu, N'Dama is the predominant type in these locations. Equatorial Guinea has about 300 heads of cattle, a small proportion of which are West African Shorthorns of the Savannah type, thought to have originated from the neighbouring Cameroon (Shaw and Hoste 1987). There is a small population of Shorthorns in eastern Chad, known locally as *Taurin de l'Est* (Eastern Taurine).

3.3.4 The Sheko Cattle of Eastern Africa

Classification, Distribution, and Ecological Settings

Sheko cattle are humpless (taurine) cattle of the Shorthorn group. They were first reported in 1929 and later in 1982 in the Shewa Ghimira of the former Kefa region of Ethiopia. They are also referred to as Gida or Mitzan cattle. They inhabit the humid Bench Maji zone in Sheko and Bench districts of the south-western part of Ethiopia, adjacent to the Sudanese border. The Sheko are believed to be the last remnants of the original humpless Shorthorn (*Bos taurus*) cattle in eastern Africa and are the only indigenous humpless cattle found in eastern Africa today. However, although the Sheko are classified as an African taurine breed, a phylogenetic study by Hanotte et al. (2000) showed a high frequency of indicine allele (90%) and low taurine allele frequency (10%) in Sheko males. More recently, Bahbahani et al. (2018) have also reported that the genomes of Sheko cattle are an admixture of Asian Zebu and African taurine ancestries. The same study suggested that natural selection took place before admixture. The breed is considered to be trypanotolerant with a high potential for dairy production. A small population of these animals that existed in the Nuba Mountains of the Sudan, not far from the present habitat, seems to have been crossed with Zebus.

The Sheko has been traditionally kept by small numbers of local farmers in the warm and humid Sheko and Bench districts under mixed crop–livestock farming systems.

Physical and Adaptive Characteristics
Sheko cattle (Fig. 3.12) have small and compact bodies, without hump or with only small-to-medium-sized hump. They have small horns, but the majority are polled. They have prominent eyes with folded eyelids and a gently sloping rump. Their coat colour is most often brown or multicoloured with black and white. However, brown colour predominates in a plain (75%), patchy (15%), or spotted (9%) pattern. The udders are well-shaped and balanced. Adult height at withers has been reported to be 105 cm, body length to be 102 cm, heart girth to be 136.7 cm, and live weight to be 179 kg (DAGRIS 2022). The breed is well-adapted to live in this environment and produce and reproduce under the tsetse challenge and is known for its milk production and traction capacity.

Production Characteristics
The Sheko is used for milk, meat, and draught. Milk yield is about 2 litres per day, but the persistence is low, and it drops quickly to 1 litre per day in a six-to-eight-month lactation period. Lactation milk yield is 420 litres over an average lactation period of 210 days. The average age at puberty for male and female populations is 41.6 and 42.1 months, respectively, and the average age at first calving is 36–48 months, while the calving interval is 15.6 months, and the average reproductive lifespan of the cow is 14.7 years (Taye et al. 2008).

To inform the design of genomic breeding programmes, further studies of the breed using a larger sample size with full genome sequence data are needed to confirm the candidate regions under selection that were reported by Bahbahani et al. (2018) to better characterize the underlying haplotypes.

The Sheko is considered endangered mainly because of interbreeding with the neighbouring Zebu and Sanga (specifically the Abigar) and the scarcity of pure Sheko breeding bulls (Takele 2005), which force farmers to use Zebu bulls. Population estimates have been highly variable due to the difficulty in identifying pure Sheko animals. Figures as high as 31,000 heads have been reported. Taye et al. (2008) placed the population of Sheko at only 4040 heads. Moreover, the heterozygosity levels in the Sheko population are lower than expected, and the total number of alleles found in Sheko cattle was the least compared to other contemporary Ethiopian breeds, pointing to high levels of inbreeding in the Sheko breed (Dadi et al. 2008). Molecular genetic evidence has shown that about 90 per cent of Sheko bulls have indicine rather than taurine allele, indicating the extent of introgression of Zebu genes (Hanotte et al. 2000) in the breed.

3.4 Risk Status of Taurine Breeds

As alluded to in Sect. 3.2, of the Taurine Longhorns, the Kuri has a limited ecological habitat—a special environment around Lake Chad where tsetse flies are absent—while the other Taurine Longhorn breed, the N'Dama, and the Shorthorn breeds and strains have evolved in tsetse-infested areas and have developed varying degrees of trypanotolerance. In this regard, the N'Dama is the best-known, the most numerous, and the most widely spread of the trypanotolerant breeds. Population estimates of pure N'Dama cattle across countries in West and Central Africa between 1977 and 1985 have ranged between 7.6 and 9.8 million heads. Thus, this breed is certainly not at risk of extinction.

The Kuri population has previously been estimated at 120,000 in Chad, which hosts about 75% of the breed, and 2350 in Cameroon (DAD-IS: Domestic Animal Diversity Information System 2005). However, based on the meta-analysis of figures in the literature, Rege and Tawah (1999) reported a lower figure of 110,000 for the total population in the Lake Chad

Fig. 3.12 A Sheko cattle herd. (Source: ILRI)

region, covering Chad, Cameroon, and Nigeria. The population of the breed is declining rapidly, with the major drivers being drought, protracted civil conflicts in the region, its inability to thrive outside the lake habitat that is shrinking due to retreating waters of the lake, and extensive crossbreeding with the Zebus that graze on the shores of the lake.

Regarding the Shorthorns, Rege (1999) provided population statistics of the various strains of Shorthorn cattle and their crossbred derivatives (Table 3.10). The table shows that the stabilized crosses—Borgou, Méré, Ghana Sanga, and Keteku—have such large and expanding populations that they are not in any danger. Indeed, they are expanding at the expense of the 'parental breeds', e.g. the Ghana Shorthorn and the Ghanian Dwarf Shorthorn. With an estimated population of only 100–200 heads, the latter is at serious risk of extinction—classified as 'Critical'. The Bakweri (800–1300 head) is also at serious risk, as are the Savannah Shorthorns Bakosi (1000–1300 head), Kapsiki (3000–5000), and, to a lesser extent, Doayo (5000–7500). Among the 'pure' Shorthorns, the Ghana Shorthorn, The Baoulé, and the Somba have the largest populations, while for the 'stabilized' crosses, the Méré, Borgou, and Ghana Sanga have the largest (and increasing) populations. Priority actions for the 'stabilized' crosses include systematic characterization and breed improvement focused on the stabilization of the populations towards distinct breeds or strains. Quantitative data should be collected systematically on their performance relative to that of the parental types.

3.5 Summary and Conclusion

An attempt has been made to collate much of the information presented here from available mainstream and grey literature. However, there are major gaps in knowledge in many of the cattle breeds covered. There is a dire need for up-to-date information, especially comparative data, on many characteristics of African taurine cattle, including their trypanotolerance trait and other adaptive and special genetic attributes. Studies on the relationships between adaptive qualities and production traits in these breeds are particularly limited, as well as comparative breed data on physical characteristics. The reproductive performance of Shorthorns is characterized by late puberty, late maturity, and long calving intervals, resulting in low fecundity rates. However, results obtained in some locations within the region indicate that the improvement of these traits is possible.

Cattle rearing in humid West Africa was nearly impossible in the past owing to the prevalence of trypanosomiasis. However, in recent times, with population pressure, deforestation, crop cultiva-

Table 3.10 Population estimates of Shorthorn cattle types of West and Central Africa

Breed	Synonyms/composition	Approximate population
Savannah Shorthorns		
Ghana Shorthorn	Gold Coast Shorthorn	600,000–800,000
Baoulé[a]	–	745,000–870,000
Savannah Muturu	Muturu	50,000–80,000
Somba	Atacora (Benin), Mango, Konkomba (Togo)	215,000–220,000
Doayo	Namchi, Namshi, Poli	5000–7500
Kapsiki	Kirdi	3000–5000
Bakosi	Bakuri, Kosi	1000–1300
Dwarf Shorthorns		
Lagune	Dahomey, Mayombe (Zaire)	46,000–72,000
Logone	Toupouri	NA
Nigerian Dwarf Muturu	Muturu	25,000–40,000
Ghanaian Dwarf Muturu	Muturu, Forest Shorthorn	100–200
Bakweri	Muturu	800–1300
Liberian Dwarf	Muturu	5000–12,000
'Stabilized' crosses		
Borgou	Zebu × Somba or Lagune	380,000–580,000
Méré	Zebu × Lobi or Baoulé (and N'Dama)	600,000–700,000
Ghana Sanga[b]	Ghana Shorthorn × Zebu	120,000–125,000
Keteku	Muturu × White Fulani, 'Borgu', Kaiama	100,000–300,000
Biu	Dwarf Muturu × White Fulani	NA

Source: Rege (1999)
NA not available
[a]Includes the Lobi of Burkina Faso but excludes the crossbred types (Méré)
[b]Occasionally N'Dama × Zebu crosses; Zebu used is principally White Fulani but occasionally Sokoto Gudali

tion, and tsetse control measures, the challenge has been reduced. Consequently, there has been an influx of transhumant cattle keepers who visit the zone for dry season grazing and return to the safer sub-humid and semi-arid zones in the wet season. An increasing number of them are settling in the humid zone and are adopting crop–livestock mixed farming. There is also a tendency among some local crop farmers to adopt livestock in the farming systems. Consequently, new (crop–livestock) farming systems are emerging in the zone. The overall impact of these changes has been the increasing interbreeding of the original taurine cattle with Zebus and the emergence of a 'cocktail' of *taurine–indicus* intermediates. This trend is demonstrated, for example, by the changing population of the Muturu cattle: The size of Muturu animals increases as one moves northwards from the deep coastal belt, and this suggests that the animals are being extensively crossed with Zebus and that this practice is more common in tsetse-free areas. Indeed, a review of population estimates from Nigeria in the 1990s already suggested that the Muturu population had declined from about 0.4 million heads in 1960 to somewhere between 50,000 and 80,000 in the late 1980s (ILCA 1992b). This trend is responsible for some reports in the literature that the population of the N'Dama and some Taurine Shorthorns has been increasing (ILCA 1992a).

Owing to their relatively small sizes, the taurine breeds cannot compete with more productive breeds, such as the Zebu, under conditions in which their adaptive attributes do not confer advantage—e.g. under low tsetse challenge. A real effort must be made to improve their performance in traits in which they are weak to exploit their superiority in adaptive attributes. However, their value is in their adaptive attributes, and this is what should underpin their development, conservation, and use, especially for smallholder farmers who are constrained by limited landholdings and limited access to inputs.

References

Aboagye GS, Tawah CL, Rege JEO (1994) Shorthorn cattle of West and Central Africa III. Physical, adaptive, and special genetic characteristics. World Anim Rev 78:22–32

Achukwi MD, Tanya VN, Hill EW, Bradley DG, Meghen C, Sauveroche B, Banser JT, Ndoki JN (1997) Susceptibility of the Namchi and Kapsiki cattle of Cameroon to trypanosome infection. Tropl Anim Health Prod 29(4):219–226

Achukwi MD, Ibeagha-Awemu EM, Musongong GA, Erhardt G (2009) Doayo (Namchi) Bos taurus cattle with low zebu attributes are trypanotolerant under natural vector challenge. Online J Veterin Res 13(1):94–105

Adebambo OA (2001) The muturu: a rare sacred breed of cattle in Nigeria. Anim Genet Resour 31:27–36. https://doi.org/10.1017/S1014233900001450

Adeniji KO (1985) Review of endangered cattle breeds of Africa. In: Animal genetic resources in Africa: high potential and endangered livestock. 2nd OAU Expert Committee Meeting on Animal Genetic Resources in Africa, 24-28 November 1983, Bulawayo, Zimbabwe. Nairobi, Kenya, OAU/STRC/IBAR, pp 20–32

Agbemelo TK (1983) Contribution à l'étude des races bovines autochtones du Togo: la race des Lagunes. Institut polytechnique rural de Katibougou, Bamako, Mali, p 97. (Thesis)

Agyemang K, Dwinger RH, Grieve AS, Bah ML (1991) Milk production characteristics and productivity of N'Dama cattle kept under village management in The Gambia. J Dairy Sci 74:1599–1608. https://doi.org/10.3168/jds.S0022-0302(91)78322-7

Ahozonlin MC, Dossa LH, Dahouda M, Gbangboche AB (2020) Morphological divergence in the West African shorthorn Lagune cattle populations from Benin. Tropl Anim Health Prod 52:803–814. https://doi.org/10.1007/s11250-019-02071-1

Ahozonlin MC, Gbangboche AB, Dossa LH (2022) Current knowledge on the Lagune Cattle Breed in Benin: a state of the art review. Ruminants 2(2):271–281. https://doi.org/10.3390/ruminants2020018

Akinwumi JA, Ikpi AE (1985) Trypanotolerant cattle production in southern Nigeria. Report to International Livestock Centre for Africa (ILCA) Humid Zone Programme, Ibadan, Nigeria. ILCA, Addis Ababa, Ethiopia, 31

Akol GWO, Authie E, Pinder M, Moloo SK, Roelants GE, Murray M (1986) Susceptibility and immune responses of Zebu and taurine cattle of West Africa to infection with *Trypanosoma congolense* transmitted by *Glossina morsitans centralis*. Vet Immunol Immunopathol 11(4):361–373. https://doi.org/10.1016/0165-2427(86)90038-3

Amakiri SF, Mordi R (1975) The rate of cutaneous evaporation in some tropical and temperate breeds of cattle in Nigeria. Anim Prod 20(1):63–68. https://doi.org/10.1017/S0003356100035017

Anonymous (1950) Rapport sur l'élevage en Afrique équatoriale française. Rev Elev Med Vet Pays Trop 4:185–192

Appiah P (1988) A comparative study on productivity and temperament of N'Dama cattle under two management systems and N'Dama and West African Shorthorn under the same management system. Department of Animal Science, University of Science and Technology, Kumasi, Ghana. (B.Sc. dissertation)

Avegan DK (1984) Etude des aptitudes de la race bovine Somba: importance de son élevage au Togo (Thesis), p 73

Bahbahani H, Afana A, Wragg D (2018) Genomic signatures of adaptive introgression and environmental adaptation in the Sheko cattle of southwest Ethiopia. PLoS One 13(8):e0202479. https://doi.org/10.1371/journal.pone.0202479

Baker CMA, Manwell C (1980) Chemical classification of cattle. 1. Breed groups. Anim Blood Groups Biochem Genet 11:127–150. https://doi.org/10.1111/j.1365-2052.1980.tb01503.x

Bangham AD, Blumberg BS (1958) Distribution of electrophoretically different haemoglobins among some cattle breeds of Europe and Africa. Nature 181:1551–1552. https://doi.org/10.1038/1811551a0

Blench RM (1993) Ethnographic and linguistic evidence for the prehistory of African ruminant livestock, horses and ponies. In: Blench R (ed) The Archaeology of Africa, pp 71–103

Blench RM (1998) Le West African Shorthorn au Nigeria. Cameroun: Des taurins et des hommes, pp 249–292. Google Scholar

Bonfiglio S, Ginja C, De Gaetano A, Achilli A, Olivieri A, Colli L, Tesfaye K, Agha SH, Gama LT, Cattanaro F, Penedo MCT, Ajmone-Marsan P, Torroni A, Ferretti L (2012) Origin and spread of *Bos taurus*: new clues from mitochondrial genomes belonging to haplogroup T1. PLoS One 7(6):e38601. [PMC free article] [PubMed] [Google Scholar]

Bradley DG, Loftus R (2000) Two Eves for *taurus*? Bovine mitochondrial DNA and African cattle domestication. In: Blench RM, MacDonald KC (eds) The origins and development of African Livestock: archaeology, genetics, linguistics and ethnography. Univ. College London Press, London, pp 244–258

Bradley DG, Machugh DG, Loftus RT, Sow RS, Hoste CH, Cunningham EP (1994) Zebu-taurine variation in Y chromosome DNA: a sensitive assay for introgression in West African trypanotolerant cattle populations. Anim Genet 25:7–12. https://doi.org/10.1111/j.1365-2052.1994.tb00048.x

Braend M, Khanna ND (1968) Haemoglobin and transferrin types of some West African cattle. Anim Prod 10(2):129–134. https://doi.org/10.1017/S0003356100026076

Chicoteau P (1989) Adaptation physiologique de la fonction sexuelle des bovine Baoulé au milieu tropical sud-soudanien. Université Paris XII, p 174. (Thesis)

Clutton-Brock J (1989) Five thousand years of livestock in Britain. Biol J Linn Soc 38(1):31–37. https://doi.org/10.1111/j.1095-8312.1989.tb01560.x

Cockcroft FL (1977) Agricultural development planning project: Ghana Meat Development Project. A UNDP FAO Consultancy Report, Rome, Italy, 192

Da Costa AM (1933) L'élevage et les services vétérinaires dans les domaines portugais d'outre-mer. Les Colonies Portugaises, Lisbon

Dadi H, Tibbo M, Takahashi Y, Nomura K, Hanada H, Amano T (2008) Microsatellite analysis reveals high genetic diversity but low genetic structure in Ethiopian inssssssssdigenous cattle populations. Anim Genet 39:425–431. https://doi.org/10.1111/j.1365-2052.2008.01748.x

DAD-IS: Domestic Animal Diversity Information System (2005) The Food and Agriculture Organization of the United Nations (FAO), Rome, Italy. http://dad.fao.org/en/home.htm

DAGRIS (2022) Domestic animal genetic resources information system. International Livestock Research Institute (ILRI), Nairobi, Kenya and Addis Ababa, Ethiopia. http://dagris.ilri.cgiar.org

Dechambre E (1951) Origines des animaux domestiques de Madagascar. Terre et Vie 98:187–196

Decker JE, McKay SD, Rolf MM, Kim JW, Alcalá AM et al (2014) Worldwide patterns of ancestry, divergence, and admixture in domesticated cattle. PLoS Genet 10:e1004254. https://doi.org/10.1371/journal.pgen.1004254

Dempfle L, Jaitner J (2000) Case study about the N'Dama breeding programme at the International Trypanotolerance Centre in The Gambia. In: Galal S, Boyazoglu J, Hammond K (eds) workshop on developing breeding strategies for lower input animal production environments. ICAR technical Series 3, pp 347–354

Dineur B, Thys E (1986) Les Kapsiki: race taurine de l'extrême-nord camerounais. I. Introduction et barymétrie [Kapsiki: a breed of cow from the far north of Cameroon. I. Introduction and barymetry]. Rev Elev Med Vet Pays Trop 39(3-4):435–42. French. PMID: 3659497

Dineur B, Oumate O, Thys E (1982) Les taurins Kapsiki: race bovine des monts du Mandara (Nord Cameroun). In: Actes du colloque sur les productions animales tropicales au bénéfice de l'homme. Institut de médecine tropicale Prince Léopold, Antwerp, Belgium, pp 181–188

Djabakou K, Grundler G (1988) Post-partum behaviour of the trypanotolerant females of the N'Dama and Baoulé breeds at CREAT: preliminary results. In: Report of the first workshop on the reproduction of Trypanotolerant Livestock in West and Central Africa, 7-11 March 1988, Addis Ababa, Ethiopia. FAO, Rome, pp 21–22

Domingo PM (1976) Contribution à l'étude de la population bovine des Etats du golfe du Bénin. Ecole inter-Etats des sciences et de médecine vétérinaires de Dakar, 148

Doutressoulle G (1947) L'élevage en Afrique occidentale française. Larose, Paris, p 298

Duvallet G, Ouedraoga A, Pinder M, van Melick A (1988) Observations following the cyclical infection with *Trypanosoma congolense* of previously uninfected Baoulé and Zebu cattle. In: Proc. meeting on Livestock production in Tsetse-affected areas of Africa. ILRAD/ILCA, Nairobi, Kenya/Addis Ababa, Ethiopia, pp 318–325

Ebangi AL, Achukwi MD, Messine O, Abba D (2011) Characterization of Doayo and Kapsiki taurine cattle breeds of Cameroon in their natural environment. Tropl Anim Health Prod 43(6):1117–1122

Epstein H (1971) The origin of the domestic animals of Africa, vol I. Africana Publishing Corporation, New York, pp 327–556

Epstein H, Mason IL (1984) Cattle. In: Mason IL (ed) Evolution of domesticated animals. Longman, New York, pp 6–27

Ezekwe AG, Machebe NS (2005) Milk yield and composition of Muturu cattle under the semi-intensive system of management. Nigerian J Anim Prod 32(2):287–292. https://doi.org/10.51791/njap.v32i2.1309

Fall A, Bosso A, Corr N, Agyemang K (2003) Genetic improvement programme of trypanotolerant cattle, sheep and goats in The Gambia: an overview of the features and achievements. In: PROCORDEL national conference. International Trypanotolerant Centre, The Gambia, pp 18–28

FAO (1976) Rapport de la première Consultation d'experts sur la recherche concernant la trypanotolérance, Rome, p 42

Faulkner DE, Epstein H (1957) The indigenous Cattle of the British Dependent Territories in Africa. In: Colonial Advisory Council of Agriculture, Animal Health and Forestry (ed) With material on certain other African Countries, vol 5. HMSO Publication, 185pp

Felius M (1995) Cattle breeds: an encyclopedia. Misset, C. bv, Doetinchem (The Netherlands), p 799

Ferguson W (1967) Muturu cattle of Western Nigeria. J West Afr Sci Assoc 12:37–44

Feron A, Sheria M, Mulungo M, Pelo M, Kakiese O, d'Ieteren GDM, Durkin J, Itty P, Maehl JHH, Nagda SM, Rarieya JM, Thorpe W, Trail JCM, Paling RW (1988) Productivity of ranch N'Dama cattle under trypanosomiasis risk. In: Livestock production in tsetse affected areas of Africa. International Laboratory for Research on Animal Diseases/International Livestock Centre for Africa, Nairobi, pp 246–250

Freeman AR, Meghen CM, Machugh DE, Loftus RT, Achukwi MD et al (2004) Admixture and diversity in West African cattle populations. Mol Ecol 13:3477–3487

Fricke W (1979) Cattle husbandry in Nigeria: a study of its ecological conditions and social-geographical differentiations. Heidelberger Geographischen Arbeiten. Geographisches Institut der Universität Heidelberg, p 330

Frisch JE, Drinkwater R, Harrison B, Johnson S (1997) Classification of the southern African sanga and east African shorthorned zebu. Anim Genet 28(2):77–83. https://doi.org/10.1111/j.1365-2052.1997.00088.x

Gates GM (1952) Breeds of cattle found in Nigeria. Farm Forest 11:19–43

Gifford-Gonzalez D, Hanotte O (2011) Domesticating animals in Africa: implications of genetic and archeological findings. J World Prehist 24(1):1–23

Glattleider DL (1976) Caractérisation des races locales de Côte d'Ivoire. Rapport préliminaire. CRZ No. 14 Zoot. Op. 3.02, Ministère de la recherche scientifique. Centre de recherches zootechniques de Bouaké-Minankro, Côte d'Ivoire, p 38

Godet G (1976) Rapport annuel de la cellule d'appui (section alimentation et zootechnie). SODEPRA, p 20

Godet G, Landais E, Poivey JP, Agabriel J, Mawudo W (1981) La traite et la production laitière dans les troupeaux villageois sédentaires au nord de la Côte d'Ivoire. Rev Elev Med Vet Pays Trop 34(1):63–71

Grandin BE (1980) Small cows, big money: wealth and dwarf cattle production in southwestern Nigeria. Stanford University, Stanford, CA, USA. (PhD thesis).

Grell H, Schlote W, Morkramer G, Johnson B (1982) L'évolution pondérale des N'Dama, race locale et croisements. CREAT (Centre de recherche et d'élevage), Avetonou, Togo

Grigson C (1991) An African origin for African Cattle? some archaeological evidence. Afr Archaeol Rev 9:119–144. https://www.jstor.org/stable/i25130531

Grigson C (2000) Bos africanus (Brehm)? Notes on the archaeozoology of the native cattle of Africa. In: Blench RM, MacDonald KC (eds) The origins and development of African Livestock: archaeology, genetics, linguistics, and ethnography. UCL Press, London, pp 38–60

Hanotte O, Tawah CL, Bradley DG, Okomo M, Verjee Y, Ochieng JW, Rege JEO (2000) Geographic distribution and frequency of a taurine (*Bos taurus*) and an indicine Zebu (*Bos indicus*) Y specific allele amongst sub-Saharan African cattle breeds. Mol Ecol 9:87–396

Hanotte O, Bradley DG, Ochieng JW, Verjee Y, Hill EW, Rege JEO (2002) African pastoralism: genetic imprints of origins and migrations. Science 296:336–339

Hartmann R (1864) Die Haussaugetiere der Nillander. Ann. Landwirtschaft. 22 Jahrg., 43 und 44. Berlin

Henderson WW (1929) Annual report of the Veterinary Department (Nigeria) for 1928, Lagos, Nigeria

Hoste C, Cloe L, Deslandes P, Poivey JP (1983) Etude de la production laitière et de la croissance des vèaux et estimations des productions laitières. Rev Elev Med Vet Pays Trop 36(2):197–205

Ibeagha-Awemu EM, Jäger S, Erhardt G (2004) Polymorphisms in blood proteins of *Bos indicus* and *Bos taurus* cattle breeds of Cameroon and Nigeria, and description of new albumin variants. Biochem Genet 42(5-6):181–197. https://doi.org/10.1023/b:bigi.0000026633.10296.e5

Igboeli G, Nwakalor LN, Orji BI, Onuora GI (1987) Seasonal variation in the semen characteristics of Muturu (*Bos brachyceros*) bulls. Anim Reprod Sci 14(1):31–38

ILCA (1979a) Trypanotolerant livestock in West and Central Apical Vol. 1: general studies. International Livestock Centre for Africa (ILCA) Mono. No 2. Addis Ababa, Ethiopia, p 148

ILCA (1979b) Trypanotolerant livestock in West and Central Apical Vol. 2: country studies. International Livestock Centre for Africa (ILCA) Mono. No 2, Addis Ababa, Ethiopia, p 303

ILCA (1992a) Trypanotolerant Livestock in West and Central Africa Volume 3- a decade's results. ILCA Monograph No. 2. The International Livestock Centre For Africa, Addis Ababa Ethiopia. https://cgspace.cgiar.org/bitstream/handle/10568/4210/ILCA_Monograph_2(3).pdf?sequence=2&isAllowed=y

ILCA (1992b) Annual report and programme highlights. The International Livestock Centre for Africa, Addis Ababa, Ethiopia. https://cgspace.cgiar.org/bitstream/handle/10568/2719/ILCAr92.pdf?sequence=1&isAllowed=y

Jeffreys MDW (1953) *Bos brachyceros* or dwarf cattle. Vet Rec 65:410–415

Kahoun J (1971) Heat tolerance in West African cattle. Ghana J Agric Sci 11(1):19–26

Keller C (1896) Das afrikanische Zeburind und seine Beziehungen zum europäischen Brachyceros-Rind. Vierteljahresschr Naturforsch Ges Zürich XLl:2

Kim J, Hanotte O, Mwai OA, Dessie T, Bashir S, Diallo B, Agaba M, Kim K, Kwak W, Sung S, Seo M, Jeong H, Kwon T, Taye M, Song KD, Lim D, Cho S, Lee HJ, Yoon D, Oh SJ, Kemp S, Lee HK, Kim H (2017) The genome landscape of indigenous African cattle. Genome Biol 18(1):34. https://doi.org/10.1186/s13059-017-1153-y. PMID: 28219390

Kim K, Kwon T, Dessie T, Yoo D, Mwai OA, Jang J, Sung S, Lee S, Salim B, Jung J, Jeong H, Tarekegn GM, Tijjani A, Lim D, Cho S, Oh SJ, Lee HK, Kim J, Jeong C, Kemp S, Hanotte O, Kim H (2020) The mosaic genome of indigenous African cattle as a unique genetic resource for African pastoralism. Nat Genet 52:1099–1110. https://doi.org/10.1038/s41588-020-0694-2

Kouakou YM (1984) Comportements pondéraux des veaux de races Baoulé et N'Dama sous leurs mères ainsi que des vaches sur deux cultures fourragères: Panicum maximum et Brachiaria ruziziensis. Note technique n° 1O/84 Zoot./C.E. Idessa. Centre d'élevage de Bouaké, Côte d'Ivoire, 28

Landais E (1980) L'élevage bovin dans les zones tropicales du sud du Tchad. In: Premier colloque international: recherches sur l'élevage bovin en zone tropicale humide, pp 589–599

Leclercq P (1970) L'élevage bovin dans la région maritime du Togo. IEMVT, Maisons-Alfort, p 115

Loftus RT, MacHugh DE, Bradley DG, Sharp PM, Cunningham P (1994) Evidence for two independent domestications of cattle. Proc Natl Acad Sci U S A 91:2757–2761

MacDonald KC, Hutton MacDonald R (2000) The origins and development of domesticated animals in arid West Africa. In: Blench R, MacDonald K (eds) The origins and development of African Livestock: archaeology, genetics, linguistics and ethnography. UCL Press, London, pp 127–162

Maehl HM, Coulibaly L, Defly A, D'Ieteren GDM, Dumont P, Feron A, Grundler G, Itty P, Jeannin P, Leak SGA, Morkramer G, Mulungo M, Nagda SM, Ordner G, Paling RW, Rarieya JM, Schuetterle A, Sheria M, Thorpe W, Trail JCM, Yangari G (1987, March) Health and performance of trypanotolerant cattle breeds exposed to quantified trypanosomiasis risk at five sites within the African Trypanotolerant Livestock Network. Meeting International Scientific Council for Trypanosomiasis Research and Control, Lomé, Togo, p 4

Marshall F, Hildebrand E (2002) Cattle before crops: the beginnings of food production in Africa. J World Prehist 16:99–143

Mason IL (ed) (1984) Evolution of domesticated animals. Longman, London, UK, p 452

Mason IL (1988) A world dictionary of Livestock breeds, types and varieties, 3rd edn. CAB International, The Cambrian News Ltd, UK, p 348

Mason IL (1996) A world dictionary of livestock breeds, types, and varieties. CAB International

Mason IL, Maule JP (1960) The indigenous livestock of eastern and Southern Africa. Commonwealth Agricultural Bureaux. Farnham Royal, Bucks, p 29

Mattioli RC, Dempfle L (1995) Recent acquisitions on tick and tick-borne disease resistance in N'dama (Bos taurus) and Gobra zebu (Bos indicus) cattle. Parassitologia 37(1):63–67

Mattioli RC, Bah M, Faye J, Kora S, Cassama M (1993) A comparison of field tick infestation on N'Dama, Zebu and N'Dama × Zebu crossbred cattle. Vet Parasitol 47:139–148

Mattioli RC, Jaitner J, Clifford DJ, Pandey VS, Verhulst A (1998) Trypanosome infections and tick infestations: susceptibility in N'Dama, Gobra zebu and Gobra x N'Dama crossbred cattle exposed to natural challenge and maintained under high and low surveillance of trypanosome infections. Acta Trop 71(1):57–71. https://doi.org/10.1016/s0001-706x(98)00051-5

Maule JP (1990) The cattle of the tropics. Centre for Tropical Veterinary Medecine, University of Edinburgh, Edinburgh, UK

Meyer C, Yesso P (1988) A study of cyclicity variations in N'Dama and Baoulé cows on station. In: Report of the First Workshop on the Reproduction of Trypanotolerant Livestock in West and Central Africa, 7-11 March, Addis Ababa, Ethiopia, pp 15–16

Mills HD (1953) *Bos brachyceros* in Africa. Vet Rec 65:587–588

Montsma G (1960) Observations of milk yield and calf growth and conversion rate on three types of cattle in Ghana. Trop Agric (Trinidad) 37:293–302

Morkramer GL, Dekpol K (1984) Characteristics of a Baoulé herd. In: Trypanotolerance and animal production. Pub. No. 3, pp 52–60

Mortelmans J, Kageruka P (1976) Trypanotolerant cattle breeds in Zaire. World Anim Rev 19:14–17

Muzzolini A (1983) L'art rupestre du Sahara central: classification et chronologie. Le boeuf dans la préhistoire africaine. Thèse 3e cycle, Université de Provence

Muzzolini A (1993) The emergence of a food producing economy in the Sahara. In: Shaw T, Sinclair P, Andah B, Okpoko A (eds) The archaeology of Africa: food, metals and towns. Routledge, London, pp 227–239

Mwai O, Hanotte O, Kwon YJ, Cho S (2015) African Indigenous Cattle: unique genetic resources in a rapidly changing world. Asian-Austr J Anim Sci 28(7):911–921

N'Goran KE, Zakpa LG, Dago DN, Lallié HDMN, Sokouri DP, Doumbia L (2016) Production and reproduction parameters analysis of N'Dama Cattle Breed in the Dairy Station of Yamoussoukro (SLY), in the Savannah Zone, in Côte d'Ivoire. World J Res Rev 3(6):15–20

Ngere LO (1990) Endangered livestock breeds in West Africa. In: Animal genetic resources: a global programme for sustainable development. FAO Animal Production and Health, pp 189–196. Paper No. 80

Ngere LO, Cameron CW (1972) Crossbreeding for increased beef production. 1. Performance of crosses between local breeds and either Santa Gertrudis or Red Poll. Ghana J Agric Sci 5(1):43–49

Ngere LO, Hagan R, Oppong ENW, Loosli JK (1975) Milking potential of the West African Shorthorn cow. Ghana J Agric Sci 8:31–35

Nwakalor LN, Igboeli G, Orji BI (1979) Sexual behaviour in Muturu, N'Dama and Holstein-Friesian bulls in a humid tropical environment. World Rev Anim Prod 15(3):35–46

Ojong ET, Oben PM, Hako TBA, Etchu KA, Motsa'a JS, Wozerou NN, Keambou TC (2021) Socio-economic, technical characteristics and challenges in indigenous (taurine type cattle) beef production in Cameroon. Appl Anim Husb Rural Dev 14:9–21

Olaloku EA (1976) Milk production in West Africa: objectives and research approaches. J Assoc Adv Agric Sci Afr 3:5–13. https://hdl.handle.net/10568/67006

Oliver J (1983) Beef cattle in Zimbabwe. Zimb J Agric Res 21:1–17

Olutogun OL (1976) Reproductive performance and growth of N'Dama and Keteku cattle under ranching conditions in the Guinea Savannah of Nigeria. Ph.D. Thesis. Univ. Ibadan, Nigeria, p 126

Pagot JR (1974) Les races trypanotolérantes: colloque sur les moyens de lutte contre les trypanosomes et leur vecteurs, Paris

Pagot J (1985) L'Elevage en Pays Tropicaux. Editions Maisonneuve, G. P. et Larousse, Paris, pp 380–383

Payne WJA (1964) The origin of domestic cattle in Africa. Empire J Exp Agric 32(126):97–113

Payne WJA (1970) Cattle production in the tropics, vol I. Tropical Agricultural Series, Longman Group Ltd, Western Printing Services Ltd, Bristol, Great Britain, p 336

Payne WJA, Hodges J (1997) Tropical cattle: origins, breeds and breeding policies. Blackwell Science Ltd.

Pinder M, Fumoux F, van Melick A, Roelants GE (1987) The role of antibody in natural resistance to African trypanosomiasis. Veterin Immunol Immunopathol 17(1-4):325–332

Pinder M, Bauer J, van Melick A, Fumoux F (1988) Immune responses of trypanosusceptible cattle after cyclic infection with *Trypanosoma congolense*. Veterin Immunol Immunopathol 18(3):245–257

Poivey YP, Seitz JL (1977) Enquête sur les ressources génétiques bovines de Côte d'Ivoire et mise au point d'un système de contrôle du troupeau. Rapport annuel 1976. Centre de recherches zootechniques de Bouaké-Minankro, Côte d'Ivoire, 29

Portar V (1991) Cattle - A handbook to the breeds of the world. A&C Black, Christopher Helm Publishers, London, UK

Prunier R (1946) Les bovins du Lac Tchad (with annotated English translation). Farm Forest 7(2):123

Queval R, Petit JP, Hascoet MC (1971) Analyse des hémoglobines du Zébu arabe (*Bos indicus*). Rev Elev Revue D'élevage et de Médecine Vétérinaire Des Pays Tropicaux 24(1):47–51

Rege JEO (1999) The state of African cattle genetic resources I. Classification framework and identification of threatened and extinct breeds. Anim Genet Resour Inf 25:1–25

Rege JEO, Tawah CL (1999) The state of African cattle genetic resources II. Geographical distribution, characteristics and uses of present-day breeds and strains. Anim Genet Resour Inf 26:1–25

Rege JEO, Aboagye GS, Tawah CL (1994a) Shorthorn cattle of West and Central Africa I. Origin, distribution, classification, and population statistics. World Anim Rev 78:2–13

Rege JEO, Aboagye GS, Tawah CL (1994b) Shorthorn cattle of West and Central Africa II. Ecological settings, utility, management, and production systems. World Anim Rev 78:14–21

Rege J, Aboagye G, Tawah CL (1994c) Shorthorn cattle of West and Central Africa. IV. Production characteristics. World Anim Rev 78:33–48. Google Scholar

Renard P (1972) Avant project pour la Sauvegarde de la race bovine Kouri et l'extension de son elevage a l' ensemble du perimetre du Lac Tchad. Commission du Bassin du Lac Tchad. No. 04.14. Fort-Lamy (Tchad), p 24.

RIM (1992) Nigerian Livestock Resources (6 vols). Report by Resource Inventory and Management Limited (RIM) to FDLPCS, Abuja, Nigeria

Shaw APM, Hoste CH (1987) Trypanotolerant cattle and livestock development in West and Central Africa. FAO Animal Production and Health. Paper No. 67/1

Smith AB (1980) Domesticated cattle in the Sahara and their introduction into West Africa. In: Williams MAJ, Faure H (eds) The Sahara and the Nile. A.A. Balkema, Rotterdam, pp 489–503

Somoko-Balantpli M, Freitas MA (1978) Considération sur l'élevage, la production et les industries animales au Togo. In: Réunion sous-régionale OUA/IBAR de l'Afrique de l'Ouest. juillet, Lomé, Togo, pp 10–14

Soro B, Sokouri PD, Dayo G-K, N'guetta ASP, Yapi-Gnaoré CV (2015) Morphometric and physical characteristics of Baoulé cattle in the "Pays Lobi" of Côte d'Ivoire. Livestock Res Rural Dev 27(7):Article #124. Retrieved November 11, 2022, from http://www.lrrd.org/lrrd27/7/soko27124.html

Starkey PH (1982) N'Dama cattle as draught animals in Sierra Leone. FAO, World Anim Rev 42:19–26

Starkey PH (1984) N'Dama cattle – a productive trypanotolerant breed. World Anim Rev 50:2–15

Stewart JL (1937) The cattle of the Gold Coast. Vet Rec 49(41):1289–1297

Stuhlmann F, Tagebücher D, von Emin Pascha (1927) Vol IV. Braunschweig, Berlin, Hamburg

Takele T (2005) On-farm phenotypic characterization of Sheko breed of cattle and their habitat in Bench Maji Zone, Ethiopia. MSc Thesis. Alemaya University, Ethiopia, p 105

Tawah CL, Mbah DA (1989) Cattle breed evaluation and improvement in Cameroon: a review of the situation. Institute of Animal Research, Ngaoundéré, p 29

Tawah CL, Rege JEO, Aboagye GS (1997) A close look at a rare African breed - the Kuri cattle of Lake Chad Basin: origin, distribution, production and adaptive characteristics. S Afr J Anim Sci 27(2):31–40

Taye T, Ayalew W, Hegde BP (2008) On-farm characterization of Sheko breed of cattle in southwestern Ethiopia. Ethiopian J Anim Prod 7(1):80–105. https://agris.fao.org/agris-search/search.do?recordID=QM2007000127

Tidori E, Serres H, Richard D, Adjusziogul J (1975) Etude d'une population taurine de race Baoulé en Côte d'Ivoire. Rev Elev Med Vet Pays Trop 28(4):499–511

Tiemoko Y, Bouchel D, Kouao Brou J (1990) Effet de différents niveaux de complémentation d'une ration de fourrage vert (*Panicum maximum*) par de la graine de coton mélassée sur la croissance de taurillons Baoulé en post-sevrage. Rev Elev Med Vet Pays Trop 43(4):529–534

Tijjani A, Utsunomiya YT, Ezekwe AG, Nashiru O, Hanotte O (2019) Genome sequence analysis reveals selection signatures in endangered Trypanotolerant West African Muturu Cattle. Front Genet 10:442. https://doi.org/10.3389/fgene.2019.00442

Touchberry RW (1967) A study of the N'Dama cattle at Musaia Animal Husbandry Station in Sierra Leone. Published by University of Illinois, Agricultural Experiment Station, , Bulletin 724. University of Illinois, Urbana (USA), p 40

van Beek WEA (1978) Bierbrouwers in de bergen. Mededeling No. 12, Utrecht, the Netherlands, ICAU

Vanvanhossou SFU, Dossa LH, König S (2021a) Sustainable management of animal genetic resources to improve low-input livestock production: insights into local Beninese cattle populations. Sustainability 13:9874. https://doi.org/10.3390/su13179874

Vanvanhossou SFU, Yin T, Scheper C, Fries R, Dossa LH, König S (2021b) Unraveling admixture, inbreeding, and recent selection signatures in West African Indigenous Cattle populations in Benin. Front Genet 12:657282. https://doi.org/10.3389/fgene.2021.657282

Wendorf F, Schild R (1980) Prehistory of the eastern Sahara. Academic Press, New York

Open Access   This chapter is licensed under the terms of the Creative Commons Attribution 4.0 International License (http://creativecommons.org/licenses/by/4.0/), which permits use, sharing, adaptation, distribution and reproduction in any medium or format, as long as you give appropriate credit to the original author(s) and the source, provide a link to the Creative Commons license and indicate if changes were made.

The images or other third party material in this chapter are included in the chapter's Creative Commons license, unless indicated otherwise in a credit line to the material. If material is not included in the chapter's Creative Commons license and your intended use is not permitted by statutory regulation or exceeds the permitted use, you will need to obtain permission directly from the copyright holder.

4

The History, Geography, and Characteristics of African Zebu, Zebu–Taurine Derivatives, and Well-Established Exotic Cattle Breeds

John E. O. Rege, Chi L. Tawah, Donald R. Kugonza, Mizeck G. G. Chagunda, Isidore Houaga, Oluyinka Opoola, and Eveline M. Ibeagha-Awemu

Abstract

The origins and history of African cattle are covered in the introduction to chapter 3. This chapter covers the African Zebu, Zebu–taurine derivatives, and well-established exotic cattle breeds on the continent. The coverage of breeds is in terms of ecological settings, physical and production characteristics, and adaptive and special genetic characteristics. The African Zebu cattle are classified under the following categories: the Large East African Zebu (LEAZ) and the Small East African Zebu (SEAZ), the Southern African Zebu, and the Shorthorn and the Longhorn Zebu of West and Central Africa. The dominance of the Zebu-type cattle in Eastern Africa and the Horn and the relative absence of distinct Taurine cattle in the sub-region are explained by the two waves of cattle introduction into the continent and their subsequent spread from the point of entry. The other group, the Sanga—breeds and strains derived from interbreeding between African Zebu and Taurine cattle—are found in East and Southern Africa. The 'Sanga' of West Africa are relatively recent, with the process still actively underway and derived breeds still not fully stabilized. They are referred to in this chapter as *'Pseudo-Sanga'*. The chapter also covers another group of breeds known as the Zenga, which are Zebu–Sanga derivatives. These are mainly found in East Africa but also have rep-

J. E. O. Rege (✉)
Emerge Centre for Innovations–Africa (ECI-Africa), Nairobi, Kenya
e-mail: ed.rege@emerge-africa.org

C. L. Tawah
African Development Bank (Retired), Abidjan, Ivory Coast

D. R. Kugonza
School of Agricultural Sciences, College of Agricultural and Environmental Sciences, Makerere University, Kampala, Uganda

M. G. G. Chagunda
Department of Animal Breeding and Husbandry in the Tropics and Subtropics, University of Hohenheim, Stuttgart, Germany

I. Houaga · O. Opoola
The Roslin Institute and Royal (Dick) School of Veterinary Studies, Easter Bush Campus, University of Edinburgh, Midlothian, UK

Centre for Tropical Livestock Genetics and Health (CTLGH), The Roslin Institute, Easter Bush Campus, University of Edinburgh, Midlothian, UK

E. M. Ibeagha-Awemu
Sherbrooke Research and Development Centre, Agriculture and Agri-Food Canada, Sherbrooke, QC, Canada

© The Author(s) 2026
E. M. Ibeagha-Awemu et al. (eds.), *African Livestock Genetic Resources and Sustainable Breeding Strategies*, Sustainable Development Goals Series, https://doi.org/10.1007/978-3-031-92076-9_4

resentative breeds in Southern (Mozambique) and Central (the Democratic Republic of the Congo (DRC)) Africa. Also covered in this chapter are the commercial composite breeds developed to combine local adaptation and productivity traits (milk and/or beef) through the incorporation of one or more specialized exotic breeds as sources of the productivity genes. The chapter also summarizes the major well-established exotic cattle breeds on the continent—dairy, beef, and dual-purpose—widely used in crossbreeding for milk and/or beef production. Several Zebu breeds and their derivatives are at risk of extinction, some have become extinct, and many are critically endangered. While no comprehensive analysis of the risk status of African cattle breeds has been done since 1999, it is clear that many more breeds are at escalated risk levels. An updated comprehensive survey is overdue, even as steps are being taken to save the most threatened breeds.

Keywords

Zebu cattle · Zebu–Taurine derivatives · Exotic cattle breeds · Ecological settings · Physical and production characteristics · Genetic characteristics

Acronyms

AD	Anno Domini
BC	Before Christ
Cm	Centimetres
DAD-IS	Domestic Animal Diversity Information System
DAGRIS	Domestic Animal Genetic Resources Information System
EASZ	East African Shorthorn Zebu
ECF	East Coast fever
FAO	The Food and Agriculture Organization of the United Nations
FDLPCS	Federal Department of Livestock and Pest Control Services (Nigeria)
ILCA	International Livestock Centre for Africa
ILRI	International Livestock Research Institute
IRZ	Institute of Animal Research (Cameroon)
ISBN	International Standard Book Number
KEASZ	East African Shorthorn Zebu (Kenya Cluster)
Kg	Kilograms
LEAZ	Large East African Zebu
mtDNA	Mitochondrial deoxyribonucleic acid
OAU	Organization of African Unity
QTLs	Quantitative trait loci
RIM	Resource Inventory and Management Limited (Nigeria)
SEASZ	Small East African Shorthorn Zebu
SEAZ	Small East African Zebu
SNP	Single nucleotide polymorphism
SSA	Sub-Saharan Africa
TSZ	Tanzania Shorthorn Zebu
UBOS	Uganda Bureau of Statistics

4.1 The Zebu Cattle of Africa

African *Bos indicus* cattle (Zebu), also known as humped cattle, are known to have descended from the introduction of Zebu into Africa from Asia. As presented in detail in Chap. 3, the domestication of the Zebu is thought to have taken place more than 8000 years ago in South Asia, and the centre of origin of Zebu cattle is believed to be India—from where they spread to Africa and south-east Asia.

Zebu cattle are characterized by a fatty hump on their shoulders, drooping ears, and a large dewlap. The Zebu is represented by some 75 breeds in Africa, making it the largest single cattle type (Rege 1999). Approximately 61 of the Zebu breeds in Africa are found in eastern Africa and neighbouring countries in southern–central Africa, while the rest are found in West Africa. Faulkner and Epstein (1957) coined the term 'East African Zebu', to embrace the substantial variation in the morphology of the Shorthorn

Zebu in the region. The term is used today to collectively describe the 'Shorthorn Zebu' of both eastern and southern Africa (Rege and Tawah 1999).

4.1.1 The Zebu of Eastern Africa

Based on body size, the East African Zebu cattle can be further divided into two subgroups—the 'Small' and the 'Large'. The term 'Small East African Zebu' (SEAZ) portrays the small frame of the animals in this subgroup and helps avoid confusion with other types of East African cattle (Mason and Maule 1960). Represented by a total of 49 breeds, the Small East African Zebu (SEAZ) are the majority. The Zebu of eastern and southern Africa are well-adapted to high temperatures and are farmed throughout these regions, both as pure Zebu and as hybrids with Taurine cattle. They are used as draught oxen, as well as for milk and meat production. They also produce by-products such as hides and dung used for fuel and manure.

The Large East African Zebu (LEAZ) is represented by 13 breeds or strains, which are restricted to the relatively drier parts of Sudan, Eritrea, Ethiopia, Somalia, Kenya, Tanzania, and Uganda. The isolations over time that have been imposed by physical/ecological restrictions and/or tribal/cultural boundaries have been responsible for the genetic differentiations that have led to the emergence of different breeds and strains (Rege and Tawah 1999). However, variations in nomenclature associated with different tribes and ecologies do not necessarily imply genetic differences. Thus, breeds or strains that have a common ancestry can be classified further, according to whether they occupy the same geographical area (e.g. a country) and/or a defined ecological zone within one or more countries. Rege and Tawah (1999) referred to these two classifications as clusters and groups, respectively (Table 4.1).

4.1.1.1 The Large East African Shorthorn Zebu

Butana and Kenana Cattle
Origin, distribution, and population statistics: The Butana and Kenana cattle together contribute the larger portion of the 44 million cattle population in Sudan and South Sudan. The rest of the cattle in the area are either Baggara or Nilotic cattle. The Butana cattle breed descended from the first Zebu introduction into Africa. The breed largely remained with nomadic pastoralist tribes (the Batahin and Shukria), especially in the Butana plains in central Sudan, east of Khartoum, in the Butana region—the acacia scrub and desert area lying between the Blue Nile and Atbara rivers. However, some of the Butana strains, for example, the Dongola and Shendi strains, are kept by settled farmers. The breed has spread to Gizera in central Sudan and along River Nile in the northern region. They are referred to as Dar El Reih cattle across the White Nile in the northern part of Darfur and Kordofan.

The Butana is not considered to be at risk of extinction. Their population size was estimated at one million heads in 1994 (DAD-IS 2005). Rege (1999) reported a lower figure of 258,000 based on a compilation of reports of 'pure' Butana over a period ending in 1991. The breed has been subjected to extensive crossbreeding with exotic (European) cattle since the 1950s and has also been exposed to cycles of droughts. Thus, the breed is declining in numbers, and the actual population size is most likely lower than what is available in the literature. Pure breeding of Butana is still practised on-station (e.g. at the Atbara Research Station).

Ecological settings, management, and production systems: Butana cattle are kept by the Batahin and Shukria ethnic groups that mainly practise nomadic pastoralism and the Dongola and Shendi ethnic groups that are typical agro-pastoralists. They are kept in the Butana plains of central Sudan between the River Nile tributaries of Atbara and Blue Nile, an area that is arid (Rege and Tawah 1999), receiving about 300 mm of rainfall a year. The Kenana cattle, on the other hand, are reared by semi-nomadic pastoralist Fung and White Nile ethnic groups, mainly in the Blue Nile province that receives between 300 and 800 mm of rainfall in one half of the year, the other being dry entirely (Bahbahani et al. 2018). Overall, the main system of production of Kenana is the range system, which includes the pastoral

Table 4.1 Zebu cattle breeds of Eastern and Southern Africa

	Group and breed/strain	Location/country		Group and breed/strain	Location/country
1.	***Large East African Zebu***		*2.4*	*2.4 The Kenya cluster*	
1	Ethiopian Boran (Borana)	Ethiopia	35	Kikuyu	Kenya
2	Kenya (Improved) Boran)	Kenya, Tanzania, Zambia	36	Taita	Kenya
3	Unimproved Boran (Borana)	Kenya	37	Giriama	Kenya
4	Orma Boran	Kenya	38	Duruma	Kenya
5	Somali Boran	Somalia	39	Kamba	Kenya
6	Karamojong Zebu	Uganda	40	Maasai	Kenya
7	Toposa	Sudan	41	Winam (Kavirondo)	Kenya
8	Murle	Sudan, Ethiopia	42	Nandi	Kenya
9	Butana	Sudan	43	Samburu	Kenya
10	Kenana	Sudan	44	Watende	Kenya
11	Baggara	Sudan	2.5	***The Teso group***	
12	Barka	Eritrea	45	Teso	Kenya, Uganda
13	Turkana	Kenya	46	Usuk	Uganda
2.	***Small East African Zebu***		47	Kyoga	Uganda
2.1	***Abyssinian short-horned Zebu***		48	Serere	Uganda
14	Arsi	Ethiopia	2.6	***The Tanzania cluster***	
15	Adwa	Ethiopia	2.6.1	***Tanganyika short-horned group***	
16	Ambo	Ethiopia	49	Iringa Red	Tanzania
17	Bale	Ethiopia	50	Maasai	Tanzania
18	Goffa	Ethiopia	51	Ugogo Grey	Tanzania (mainland)
19	Guraghe	Ethiopia	52	Mkalama Dun	Tanzania
20	Harar	Ethiopia	53	Singida White	Tanzania
21	Jem-Jem	Ethiopia	54	Pare	Tanzania
22	Smada	Ethiopia	55	Tarime (Shashi)	Tanzania
23	Mursi	Ethiopia	56	Chagga (Wachagga)	Tanzania
24	Hammer	Ethiopia	2.6.2	***Zanzibar group***	
25	Jijiga	Ethiopia	57	Zanzibar Zebu	Tanzania (Zanzibar & Pemba)
26	Ogaden	Ethiopia, Somalia	3	Tanzania (Zanzibar and Pemba)	
2.2	***The cluster of Southern Sudan and vicinity***		*3.1*	***The Angoni group***	
27	Lugware	Uganda, DRC	1	Agoni	Zambia
28	Mongolla	Sudan	2	Malawi Zebu	Malawi
29	Nkedi	Uganda	3	Angonia (Angone)	Mozambique
30	Nuba Mountain Zebu	Eritrea	3.2	***The Madagascar group***	
2.3	***The Somali group***		*4*	Madagascar Zebu	Madagascar
31	Garre (Gherra)	Somalia	5	Baria	Madagascar
32	Gasara	Somalia			
33	North Somali Zebu	Somalia			
34	Baherie	Eritrea			

Source: Rege (1999)

and semi-pastoral types. Nomadism in Kenana is relatively mild.

Physical, adaptive, and special genetic characteristics: The phenotype of Butana cattle presents a characteristic deep red coat colour, while Kenana cattle phenotypically possess a light blue-grey fur coat that turns from brown-red coat colour in calves up to 6 months of age. Kenana cattle have a shading of nearly black on the head, neck, hump, hindquarters, and legs (Yousif and El-Moula 2006). They have short black horns that rarely grow beyond 30–45 cm, which are relatively shorter in males than in females; loose horns are common. Males possess very large and prominent cervico-thoracic hump and dewlap, just as is typical in other Zebu cattle (Rege et al. 2001).

Like other Zebu cattle, the Butana cattle are known to be superior to the humpless cattle in regulating body temperature—hence, they have lower body water requirements. Their hardened hooves and lighter bones enable them to endure long migrations in their acacia scrub and desert habitat. These adaptive attributes are considered to have facilitated their importation and spread by Indian and Arabian merchants across the Red Sea to the drier agro-ecological regions of the Horn of Africa.

The average cattle herd size among typical smallholder farmers of Butana cattle is six animals, while for Kenana, it is ten animals (Mohammed et al. 2014). Prevalent diseases in the Kenana and Butana cattle-rearing areas include brucellosis, foot and mouth disease, theileriosis, and trypanosomosis. The production environment of the Butana and Kenana cattle is characterized by high temperature, low feed quality and quantity, and a high disease and parasite challenge (Musa et al. 2008). Considering some major production traits, these cattle compare very well with 50% crossbred dairy *Bos taurus* × *B. indicus* cattle in Sudan, as well as with some of the best cattle breeds in other tropical countries (Musa et al. 2008). Breed improvement options that target crossbreeding should ensure that *Bos taurus* genetics are kept to a very moderate level, if at all. The possibility of developing a community-based breeding programme for Butana cattle as a dairy breed focusing on optimizing milk yield, growth performance, and lactation length is real, considering that over 66% of farmers who rear these cattle are willing to exchange breeding bulls and establish farmers' associations (Omer et al. 2021). At the mitochondrial DNA (mtDNA) level, only taurine mtDNA sequences have been identified, with high diversity (Salim et al. 2014).

Production characteristics: Butana cattle have typical dairy characteristics with an established history of active selection for milk production. Although the breed is raised as a dual-purpose breed, it is known for its good milk production in its harsh habitat. Milk yield per lactation ranges from 1200 to 1800 kg in an average lactation of 240 days. Under on-farm conditions, the average milk offtake is 1.5 kg per cow per day for durations often exceeding a year, and the milk is consumed fresh or used to make ghee and butter, as the milk butterfat content is quite high (4.5% for Butana and 5.5% for Kenana cattle) (Rege and Tawah 1999). Under on-station conditions, their production averages 1662 kg for Butana in a 268-day lactation length (Musa et al. 2005, 2006) and 2264 ± 131 litres in a mean lactation length of 283 ± 11 days (Medani 2003). For the same conditions, the Kenana cattle breed has a mean yield of 1400–2100 kg for a 198–257-day lactation (Musa et al. 2005, 2006) and 1836 ± 186 litres in a mean lactation length of 308 ± 6 days (Medani 2003). Contemporary characterization studies, however, reported that the productive performance of Butana and Kenana cattle is similar at an average of 1415 kg of milk per lactation of 251 days (Lutfi et al. 2005), with a maximum record for an individual cow of 4530 kg in a lactation (Saeed et al. 1987). An average lactation yield of 3626.7 kg has been reported in a meta-analysis (DAGRIS 2022). These characteristics lend to the classification of these cattle as Africa's indigenous Zebu-type dairy cattle. An assessment of the environmental and genetic factors affecting the milk production of the Butana cattle based on single-record milk yield on-station reported a heritability of 0.278 ± 0.232 and a repeatability of 0.415 ± 0.037 for milk yield (Alim 1962).

Ageeb and Hillers (1991) (Yousif and El-Moula 2006; Wilson 2018) have reported some growth and reproduction parameters. Age at first calving in Kenana cattle averages 4.5 years under traditional production systems, but under improved management, it reduces to 3.0–3.5 years (Wilson 2018). The average birthweight of Kenana cattle is 22.7 kg (range 21–26 kg in males and 20–24 kg in females), rising to 56.7 kg at 6 months and adult bulls and cows attaining 300–500 kg (average 370 kg) and 250–350 kg (average 320 kg), respectively. Calving intervals measured on-station averaged 18 months with a lifetime production of 4.02 calves (mean) and 12 calves (maximum), while on-farm, under traditional systems, cows are culled after three calvings, with an average lifetime production of 2.9 calves (Wilson 2018). Earlier studies had reported a mean calving interval of 416 days (13.8 months) (Alim 1962). The rather low production level on-farm could be partly explained by the high age at first calving. Repeatability for calving interval has been estimated at 0.111 ± 0.039, 0.415 ± 0.043 for lactation period, and 0.221 ± 0.037 for dry periods. The average calving rate was 59%, measured over four decades ago, and was distinctly higher (60%) in migratory than in sedentary (40%) production systems.

The two breeds (Butana and Kenana) produce an estimated 7.4 million tons of milk annually (Wilson 2018). They also contribute an estimated 67.2% of red meat. Butana cattle additionally provide draught power and insurance and meet the sociocultural needs of the rural communities that rear the breed (Omer et al. 2021).

Figure 4.1 presents a Butana cow with a calf and a Kenana cow.

Baggara Cattle

Distribution and population statistics: Like their neighbours, Butana and Kenana, the Baggara cattle descended from the first Zebu introduction into Africa. The Baggara are reared in the west, central, and southern Darfur; central and southern Kordofan; Nuba Mountains; and the Sulien Baggara sub-region of Sudan. They are named after the Baggara people of western Sudan and central Chad, who raise these cattle. *Baggara* means cattle people in the local Shuwa Arabic language. Baggara cattle have common grazing lands and migratory routes with small Nilotic and large Fulani cattle, which results in indiscriminate crossbreeding, leading to admixture, with unknown breed composition and high variability in body conformations. Baggara cattle are estimated to constitute 33% of the total 44 million national cattle herd of Sudan.

Physical characteristics: They are medium-sized, smaller, and thinner than the Boran breed of Kenya and Ethiopia and have varied coat colours, horn shapes, and conformation. They have cervico-thoracic humps, which are relatively large, particularly in males, and their dewlaps are well-developed and prominent. The cattle in Darfur possess the largest horns. Baggara cattle are smaller than the Boran breed of Kenya and Ethiopia. Baggara cattle are horned with cervico-thoracic humps and have a large and prominent dewlap. Figure 4.2 shows a herd of Baggara cattle.

Production characteristics and uses: The breed is primarily used for beef and is the major beef animal in Sudan. Baggara cows typically produce 1.2–2.5 kg of milk per day, with a total 305-day adjusted lactation milk yield of 366 kg. Under favourable conditions and good management, 70% of the cows become pregnant, compared to 45% of cows under unfavourable conditions. About 25% of the cows that conceive normally fail to maintain the foetus to full term and abort, usually towards the end of the pregnancy period. Malnutrition, arising from substantial seasonal feed shortages, and stresses from diseases have been suggested as the main causes of foetal loss. There have been concerted breed improvement programmes of the Baggara by the Sudanese Department of Agriculture.

Adaptive characteristics: The Baggara are characterized by their adaptive characteristics and high performance in hot and dry agro-ecosystems such as in the dry Sahel. The related Butana and Kenana breeds of the Nile Valley, known for their good dairy attributes, need much more feed and water than the Baggara.

Fig. 4.1 A Butana cow with a calf and a Kenana cow. (Source: https://twitter.com/ensembl/status/951859117488005120 - Open Access)

Fig. 4.2 A herd of Baggara cattle. (Courtesy: AGTR, ILRI)

The Karamojong Zebu

Distribution and population statistics: The Karamojong cattle is the only LEAZ type found in Uganda. They were estimated to be 510,000 heads in 1999, and the population has most likely increased (even doubled) with the control of cattle rustling and increased livestock management infrastructure in the Karamoja region, animal disease management, the establishment of livestock water reservoirs, and pasture improvement efforts. These cattle are kept by the Jie, Pian, Matheniko, Tepeth, Dodoth, Nyaakwae, Bakora, and Pokot tribes, who are collectively classified as the Karamojong people, numbering about 1.2 m people (UBOS., The Uganda Demographic and Health Survey (UDHS), Uganda Bureau of Statistics (UBOS) 2011), and are nomadic pastoralists who roam large areas in search of water and pasture for their livestock. They occupy the entire North-Eastern region of the Republic of Uganda, incorporating the districts of Abim, Amudat, Kaabong, Karenga, Kotido, Moroto, Nabilatuk, Nakapiripirit, and Napak, an area of 27,900 square kilometres (Loquang 2003).

Ecological settings, *utility, and management:* The Karamojong tribe that maintains this breed is the most influential tribe among the Karamoja-Nilotic people in the north-eastern region of Uganda, and their herds are maintained in the original homeland. These cattle are kept for their multipurpose roles that include the provision of milk, meat, and blood (all used as food); payment of dowry; the provision of hides that are used for shelter, beddings, and making of traditional sandals; the provision of dung that is used for building homesteads and as manure; and the use as a source of power for cultivation, transport, and haulage (Loquang 2003). The skin of calves is used as clothing, though this practice is slowly dying out.

Physical, adaptive, and special genetic characteristics: Karamojong cattle are a large breed with coat colours varying from red, light fawn, or tan (with or without white patching), roan, grey to white. Their horns vary widely in length and shape, and they possess a large and dome-shaped hump. The dewlap, umbilical fold, and sheath are larger in the Karamojong type than in the Toposa, and their udder is moderately well-developed. The cattle are similar in conformation to the Boran cattle—deep body, long legs, and well-developed pyramidal hump. Figure 4.3 shows typical Karamojong cows.

The Karamojong cattle have several unique characteristics that include being drought-tolerant, disease-resistant, and generally a 'hardy breed'—e.g. they do not require a lot of forage, can survive for many days without drinking water, and can easily graze on rough terrain. In addition, their meat is considered tastier and more tender than that of the other cattle breeds in the area. The cattle-keeping community practises selective breeding with particular focus on the breeding of cows and bulls that are hardy and can trek for long distances in search of pasture and water, are aggressive and can ward off predator attack, are disease-tolerant, and do not experience calving difficulty (Rege and Tawah 1999). Selective breeding is made possible by applying strict selection criteria that target animals of well-known lineage, with parents known to perform well, from a lineage of animals that are disease-tolerant, and with offspring that remain healthy right from birth to maturity. Consequently, these cattle are well-adapted to their very dry climate and semi-nomadic pastoral system.

Production characteristics: Karamojong cattle are largely reared for milk and meat (Rege 1999). In addition, they are used for ploughing. The mature live weight of these cattle is 320–490 kg in males and 300–410 kg in females. A study by Behnke and Arasio (2019) reported that the cattle in Karamoja make very substantial economic contributions to Uganda's national livestock economy—accounting for 39% of the national cow milk value and 27% of the national cattle offtake value.

The Boran/Borana Cattle

Origin, distribution, classification, and population statistics: The Boran is a Zebu breed maintained by the Borana pastoralists of southern Ethiopia and contiguous areas of Kenya and Somalia. It is considered that all the Boran cattle were derived from the native Shorthorn Zebu cattle of the cattle-

 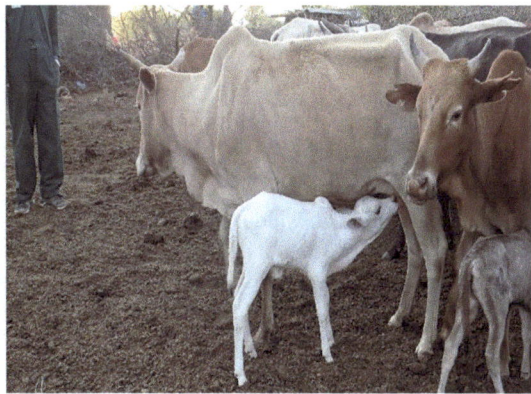

Fig. 4.3 Typical Karamojong cows with suckling calves. (Courtesy: Dr. Donald Kugonza)

keeping tribes of southern Ethiopia—home of the Borana people. It is estimated that their ancestors arrived in the Horn of Africa about 1300–1500 years ago from south-west Asia. Large numbers of these animals then migrated from the Liben Plateau of southern Ethiopia to Somalia, where they are named as the Awai cattle, and to Kenya, where they are known as the Tanaland Boran and Orma Boran. In the 1920s, European ranchers in Kenya purchased the Tanaland Boran cattle and, through selection, developed the Improved Boran. The present-day Boran cattle populations in eastern Africa can be classified under five strains (Rege 1999): Ethiopian Boran (Borana), Unimproved Boran (Boran), Orma Boran, Somali Boran (Awai), and the Improved or Kenya Boran (commercial breed).

Today, the Ethiopian Boran (Borana) cattle are mainly found on the Borana Plateau, from the Liban Plateau to the extreme south, where they number about 1,896,000 heads. The Improved Boran is essentially a commercial breed kept in the commercial ranches in Kenya, has been selected for beef production, has 580,570 animals, and has been exported to Tanzania, Uganda, the Democratic Republic of Congo (DRC), and Zambia. Over time, the development of the Improved Boran has incorporated the Orma Boran, the Borana of Ethiopia, and the Somali Boran. The Kenyan Boran Cattle Breeders' Society has been in existence since the early 1950s and has been spearheading the selective breeding of the Boran in the country. In more recent times, the Boran breed was introduced to Australia, the United States, Brazil, and Mexico. As of 2008, there were approximately 454 beef ranches in Kenya, under five categories: group ranches, private company ranches, cooperative ranches, public company ranches, and government ranches. The Boran breed is a common feature across these ranch categories in Kenya.

The Kenyan Unimproved Boran cattle are kept in north-east Kenya, where there are 1,882,000 animals, largely owned by pastoral herders, while the Orma Boran cattle breed is kept in the North-east of Kenya, with a total of 547,000 heads.

Ecological settings, *management, and production systems*: The habitat of the Borana or Boran cattle in Ethiopia, Kenya, and Somalia is typically the arid and semi-arid lands characterized by recurrent drought, degraded rangelands, and an increasing reduction in access to traditional grazing lands. The region has a savannah landscape, marked by gently sloping lowlands and floodplains vegetated predominantly with grass and bush land. Pasture is the main (and often the only) source of feed for livestock in these pastoral systems. Severe grazing shortage is common. Severe livestock losses and reduced livestock productivity are also common during dry periods. This situation is being exacerbated by climate change and invasive weeds such as *Ipomoea* spp., *Prosopis* spp., and *Parthenium* spp., which are also accelerating the degradation of the rangeland ecosystem in these areas.

Annual mean temperatures in the Boran habitat vary between 19 and 24 °C. Rainfall in the Borana lowlands is bimodal, with long rains expected between March and May (59% of the rainfall) and short rains (27%) between October and November, but these have changed significantly recently. Mean annual rainfall between the years 1988 and 2001 was 412–566 mm. The scarcity of surface water is widespread, and livestock are trekked long distances in search of both water and pasture. Over the past three decades, there has been an increasing incidence of recurrent droughts, reduced rainfall, and a decreasing availability of grazing and water resources. However, the feed situation is significantly better on the ranches—typically with improved pastures, fodder crops, and other supplementary feeds, as well as water sources. The woodlands of Borana rangelands are characterized by species from the genera *Combretum* and *Terminalia*, whereas the bushlands and thickets, which cover major parts of the Borana lowlands, are dominated by *Acacia* and *Commiphora* species. Pastures are predominantly short-lived perennial or annual grasses with scattered eucalyptus trees. They occur on level-to-undulating plains. Soils are variable-depth loams, sometimes calcareous, and the surface may be stony.

Physical, adaptive, and special genetic characteristics: The Boran cattle breed has the ability to survive, produce, and reproduce under high ambient temperatures, low-quality forage resources, long watering intervals, and tick infestation. As such, the breed is versatile and well-adapted to arid and semi-arid environments. Boran cows are efficient converters of pasture forage into body fat deposits, which are mobilized during periods of feed scarcity and lactation, minimizing the loss of body condition during lactation or slight droughts (Hagos 2016).

A phylogenetic study (Hanotte et al. 2000) indicated that the Boran has a 100% indicine Y allele. However, a subsequent study (Hanotte et al. 2003) analyzed the autosomal loci of the Boran and reported the proportions as 64% *Bos indicus*, 24% European *Bos taurus,* and 12% African *Bos taurus*. The European-Near East taurine background in the Boran is of some antiquity and possibly recent crossbreeding with exotic European taurine, as well as an African indigenous taurine background, which is not found in any Asian Zebu crosses such as Sahiwal or Brahman.

The Ethiopian Borana is a medium-sized, deep-chested, and long-legged cattle breed. Its coat colour is white, light grey, or fawn with black or dark brown shading on the neck, head, shoulders, and hindquarters (Rege et al. 2001). Its horns are thick at the base, short and erect; the hump is well-developed and hangs over to one side in males, as is typical of Boran cattle, and cows have well-built udders. The Kenyan Improved Boran, on the other hand, possesses a white coat colour with black spots and resembles other types of Boran but mainly differs from the Unimproved Boran by having a straight top line and well-developed hindquarters. Ethiopian Borana may also have other colours ranging from brown to red, though these are rare in occurrence. The breed is medium in size with a short head, small ears, loose dewlap, and a large hump above the shoulders. They can be horned or polled. They vary in height at the withers from 114 cm to 147 cm. Their skin is loose, thick, and extremely pliable—an attribute that is linked to enhanced insect repellence. The skin is also dark-pigmented with fine, short hair, which is considered to provide heat tolerance.

The Orma Boran cattle typically have a white or fawn coat colour with unpigmented skin. It is mostly used for milk and meat. The Orma are well-adapted to local conditions and parasites. They are also known for their fertility, early maturation (compared to other Zebu breeds), hardiness, and docility. Studies have indicated that the Orma Boran may have a degree of resistance to African trypanosomosis and can be productive under tsetse challenge with minimum drug intervention (Dolan 1997; Bett et al. 2004). Figure 4.4 presents a Kenya Boran bull (a) and herd (b) in a ranch and Ethiopian Boran bulls (c).

A recent review (Bayssa et al. 2021) has attributed a higher resistance to biting insects and tick infestation in the Boran to the following attributes of the breed: (a) a highly sensitive and motile skin with a thick, well-developed

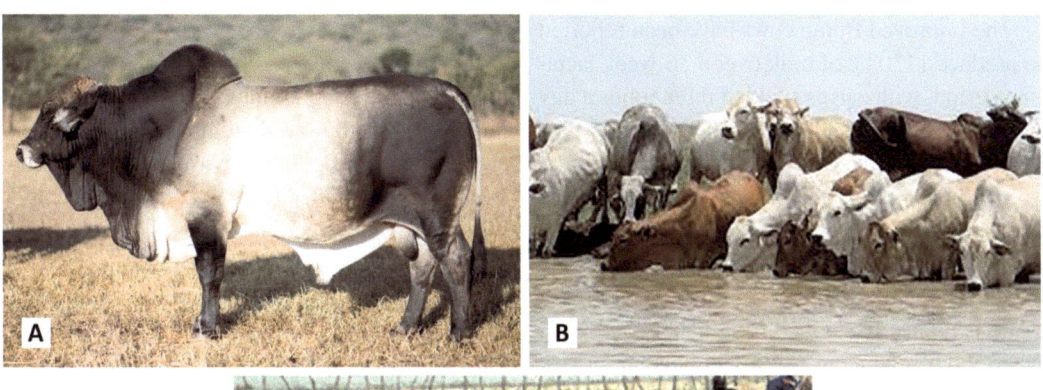

Fig. 4.4 A Boran bull (**a**) and herd (**b**) at the Sosian ranch in Kenya and typical Ethiopian Boran bulls in Yabello, Borana, southern Ethiopia. (Courtesy: ILRI)

layer of subcutaneous tissue, which causes the muscles beneath the skin to contract and move in reaction to insects—thus enabling them to shake off external pests; (b) short hair coat, which makes it difficult for insects to attach onto the hides of the Boran cattle; (c) a waxy and oily secretion from the skins of the Boran, which makes them less desirable hosts for ticks and flies; and (d) prominent, protective eyebrows and long eyelashes, which protect their eyes from bright sunlight, dust, and other irritants that predispose cattle to pinkeye infection. Other studies have attributed the ability of the Boran to thrive well in hot and dry environments to the high number of sweat glands, which allow it to withstand high ambient temperatures (DAGRIS 2022).

Production characteristics: Boran cattle are used mainly for milk and meat. The Improved Boran is primarily reared for meat, milk, and work (Rege et al. 2001), but the breeding focus is on beef traits. The average birthweight of the Ethiopian Boran is 25 kg for male calves and 23 kg for females, while weight at maturity is 300–385 kg (male) and 300–350 kg (females). The Improved Kenyan Boran differs from other Boran sub-types due to its larger size and well-developed hindquarters. The prominent sizes of the Kenyan Improved Boran make them the heaviest in the region, at 550–850 kg in mature males and 400–550 kg in mature females. The Kenyan Unimproved Boran is much smaller than the Improved Boran and attains a mature weight of 255–395 kg (males) and 250–355 kg (females). The mature males of the Somali Boran weigh 295–410 kg, and females weigh 230–355 kg, while the male Orma Boran range in size from 225 to 395 kg and the females from 250 to 355 kg. Besides the well-developed beef conformation, the Boran also exhibits good carcass quality characteristics: The depth of eye muscle, marbling, an even fat cover, and the ratio of hind to forequarter stand out in the Boran compared to other local breeds and make well-finished Boran steers the favourites of butchers and hoteliers in Kenya.

The Improved Boran cows have been reported to produce 1130 kg of milk over a 36-week lactation period, with calves suckled three times a day (DAGRIS 2022). Under a more favourable production environment, Boran cows produce up to 1657 kg of milk over a 252-day lactation period (Zander et al. 2009). Overall, although their production potential is less compared with exotic and crossbreeds, the level of production of Boran cattle is relatively stable during harsh conditions where high-producing exotic animals are at risk (Abdurehman 2019).

4.1.1.2 The Small East African Shorthorn Zebu

The Small East African Zebu (SEAZ) is the main type of cattle populating East Africa (Rege et al. 2001). The SEAZ are preferred by the majority of local smallholder farmers over the pure exotic highly productive taurine breeds due to their superior adaptability to the local environment, which is characterized by a warm climate (20–23 °C), high humidity (60–80%), and high disease challenges (e.g. *Theileria parva, Ehrlichia ruminantium*, and *Haemonchus placei*) (de Clare Bronsvoort et al. 2013). These cattle show a degree of resistance to *Rhipicephalus appendiculatus* tick and the vector of East Coast fever (ECF) protozoan parasite *T. parva* (Latif et al. 1991; Latif and Pegram 1992) as well as tolerance to poor forage and water scarcity (Western and Finch 1986). A mortality rate of 16% has been observed in an SEAZ population of western Kenya under the traditional management system (de Clare Bronsvoort et al. 2013; Thumbi et al. 2013). The mortality was mainly attributed to ECF, haemonchosis, and heartwater.

The genetic structure of these cattle has been investigated by Mbole-Kariuki et al. (2014) using mid-density genome-wide single nucleotide polymorphism (SNP) data. The study confirmed the long-held position by many researchers that the East African Shorthorn Zebu (EASZ) is a stabilized admixed population of Zebu and African taurine ancestries. The average genome proportions reported by this study were 0.84 ± 0.009 and 0.16 ± 0.009, respectively. However, some of the EASZ animals showed recent European taurine introgression, likely linked to artificial insemination programmes aiming to improve indigenous cattle productivity (Mbole-Kariuki et al. 2014). This crossing with European taurine cattle has been linked to the increased vulnerability of EASZ to infectious diseases (Murray et al. 2013).

A recent high-density genome-wide SNP analysis of the EASZ from Kenya (KEASZ) and Zebu × Taurine–admixed cattle populations from Uganda and Nigeria (Bahbahani et al. 2017) found that 25 of the regions were shared between KEASZ and Uganda cattle, and seven regions were shared across the KEASZ, Ugandan, and Nigerian cattle. The identification of common candidate regions allowed the mapping of 18 regions that intersect with genes and quantitative trait loci (QTLs) associated with reproduction and environmental stress (immunity and heat stress) and was interpreted to suggest that the genomes of the Zebu × Taurine–admixed cattle have been uniquely selected to maximize the hybrid vigour, in terms of both reproduction and survivability.

The ***Teso Zebu*** group of Uganda consists of two strains, the *Kyoga* and the *Usuk*. The Kyoga inhabits the Lango and Kaberamaido areas, is larger than the Nkedi, and has deeper chest and shorter legs. The forehead is broad, and ears are pendulous. The Usuk is found around the northeast and North Teso areas adjacent to the Karamoja and is also larger than the Nkedi (Rege and Tawah 1999). The Nkedi cattle occupy Bukedi, Lango, and much of South-Eastern Uganda (Pallisa, Iganga, Kamuli, and Tororo areas). In Kenya, the Teso Zebu is reared by the Teso ethnic group in the Busia district of Western Kenya, near the border with Uganda.

The ***Highland Zebu*** cattle of Kenya (also known as Kikuyu Zebu) occupy the highlands of central Kenya, where they have been heavily crossed with purebred exotic European breeds. The breed is currently endangered. The ***Lowland or Coastal Zebu***, on the other hand, include four strains named according to the tribes that keep them, namely Kamba, Taita-Taveta, Giriama, and Duruma. The Maasai Zebu have their homeland in Southern Kenya and the adjoining areas of northern Tanzania, particularly in the Moshi, Upase, and Usambara areas. Due to the centuries-old traditional practice of cattle raids, present-day cattle of the Maasai are very heterogeneous, having some similarities with the adjacent Kenyan Zebu breeds (Rege and Tawah 1999). In Tanzania, the Maasai Zebu is found distributed from the northern to the southern parts of the country. Genetically, Maasai cattle are clustered with the Nandi, Taveta, Turkana, and Teso Zebu cattle (Rege et al. 2001).

The **Kavirondo Zebu** are kept by the Luo and Luhya ethnic groups that occupy the Nyanza and Western provinces/regions of Kenya in the Lake Victoria Basin. A small cluster of SEAZ cattle called the **Lugware** includes ecotypes called the Bahu, Lagware, and Lugwaret. The Lugware ecotypes are kept in the Democratic Republic of Congo in the Kibali-Ituri district, in the northwestern part of Uganda, west of River Nile.

The Tanganyika Shorthorn group is made of the **Iringa Red** that is reared in the Singida region of Central Tanzania and the **Maasai Grey** that thrives in Moshi, Upase, and Usambara regions of Northern Tanzania. The **Ugogo Grey**, **Mkalama Dun**, **Singida White**, and **Pare** cattle occupy central Tanzania, while the **Tarime** and **Chagga** are in the Mara and Usambara regions, respectively.

Ecological settings, *utility, management, and production systems*: Poor nutrition is one of the major problems affecting livestock productivity in the habitats of the Zebu cattle. This is normally reflected in reduced live weights, perpetual low animal productivity, late age at first parturition, increased parturition intervals, prolonged non-productive life, and high mortalities. The production systems in most habitats of the SEAZ are typified by natural pastures as the major feed resource, and its utilization is through continuous grazing.

Tethering is common during the cropping season in some locations—e.g. in Kitui County of Kenya. This system is often adopted in response to the shortage of herding labour when farming activities are at their peak. Tethered animals can obtain sufficient access to pasture during this period of the year while also ensuring control of animals that could otherwise stray onto other people's farms. Feed conservation is commonly practised in the form of crop residues (cereal straws) and hay from native pastures. These conserved feeds are fed to animals during the dry season, work oxen during the cropping season, and in-calf and lactating cows. Supplementation of the indigenous breeds using high energy and protein concentrates is rarely practised in these systems. Equally rare is the provision of mineral licks as part of cattle diets. However, animals exploit various natural salt deposits that exist in the grazing fields. These practices are similar to those prevalent in other regions of Africa—e.g. in Nigeria (Pullan 1979; de Jode et al. 1992) and Botswana (Trail et al. 1977).

Physical, adaptive, and special genetic characteristics: The typical SEAZ is a relatively small, dual-purpose animal and is stocky in appearance. They have no dominant coat colour, but black, black and white, brown, brown and white, and a 'dalmatian' pattern are common. The horns are short and thick at the base and curve slightly outwards. The musculo-fatty hump is prominent and thoracic in position. The ears are medium-sized and directed outwards. The skin is of medium thickness, and the hair is short and smooth. The hooves are medium and hard. Horns are short and lyre-shaped, and the hump is well-developed with a varying posterior overhang. The dewlap is of moderate size. The Lowland Zebu varies in coat colour from black to red, and there are significant variations in horn shape, horn presence, and hump size. The Lugware are small-sized, hardy, and fine-boned, with coat colours varying from black to grey and from brown to white. Their skin is pigmented, and the horns are short and grow sideways and upwards in a crescent shape, while the hump is large, as is typical of Zebu cattle, and tends to hang over to one side at the rear. The dewlap is moderate in size, and the umbilical fold is very small. Figure 4.5 shows an SEASZ in Tanzania.

Production characteristics. The Taita-Taveta Zebu weight ranges are 194–405 kg in mature males and 125–360 kg in mature females (Rege et al. 2001). The bodyweights of the adult Maasai

Fig. 4.5 A herd of Small East African Zebu in Tanzania. (Courtesy: Dr. Gladness Mwanga)

Zebu range between 300 and 445 kg in mature males and between 275 and 385 kg in mature females. The Kavirondo Zebu cattle have a black coat colour, though white and a combination of both are also seen. The mature weights of these cattle are 215–419 kg (males) and 195–365 kg (Females) (Rege et al. 2001). Mwacharo and Rege (2002) reported average milk production per day of 1.6 litres in Kenya, with significant differences being observed between locations (counties) and, hence, Zebu strains, at the start and peak of lactation.

4.1.2 The Southern African Zebu

The Zebu breeds of southern Africa are in two groups, *the Angoni group* and the ***Madagascar group***. The former includes *Angoni (*of Zambia*)*; *Angonia* or *Angone* (Mozambique); and ***Malawi Zebu,*** while the latter has two breeds, the ***Baria*** and the ***Madagascar Zebu***. A brief description of these breeds is presented below.

The Angoni
Distribution and ecological settings: The Angoni belongs to an SEAZ group of cattle believed to have been introduced to Africa by the Indian and Arabian merchants. The Angoni group of cattle is believed to be the descendant of the original Zebu introduced through the Horn of Africa. These cattle were maintained by the Angoni people who descended from the Nguni tribe in South Africa. The Angoni people are supposed to have lost their original Sanga cattle and restored them with Zebu as they wandered as far north as Lake Tanganyika and subsequently settled on the Angoni plateau of eastern Zambia and the adjoining parts of Malawi between 1850 and 1870. These cattle became known as Angoni cattle. The Angoni people later spread southwards, reaching north-western Mozambique. The cattle breeds that emerged in this process are known today as *Malawi Zebu* in Malawi, *Angoni* in Zambia, and *Angone* or *Angonia* in Mozambique (Rege and Tawah 1999). Angoni cattle are mostly found in the Eastern province of Zambia, parts of Isoka district in Muchinga province, the adjoining areas of Malawi between the Lungwa River in the west and Lake Malawi in the east, and into Tanzania. They also spread southwards, reaching north-western Mozambique. In Mozambique, they are distributed north of the Zambezi River, in the plateau of Tete in the Tete district, in a small area of Angonia to the extreme north-east on the Malawi border.

Physical and production characteristics *and* uses*:* The coat colour of the Angoni in Zambia is highly variable—red, brown, black, red or black and white, or brindle. Similar to Malawi Zebu, the horns of the Zambian Angoni are short and thick and lateral rather than upright. The hump and dewlap are well-developed. The Angoni has a compact body frame that is generally larger than the Tonga but smaller than the Barotse. The body size varies from medium to large, with males weighing up to 730 kg. The Mozambique *Angone* is much smaller than the Zambian Angoni and resembles the Malawi Zebu more than it does the Zambian Angoni. The large size of males and the large compact body make them ideal for draught power. Nearly all the males and, in some instances, even females are used as work animals. The females have characteristic large udders that make the breed ideal for crossing with exotic breeds for milk production. Typical Angoni bull and cow are shown in Fig. 4.6.

With an estimated population of 797,000 heads in Malawi, 300,000 in Zambia, and 65,000 in Mozambique, the Angoni is not yet at risk. However, its continued interbreeding with neighbouring Sanga populations may put it at risk over time.

The Madagascar Zebu

Distribution and ecological settings: The *Madagascar Zebu*, also called *Omby Malagasy*, is believed to have originated from the interbreeding of the Indo-Pakistani Zebu cattle and the African Sanga cattle. Indonesian immigrants are known to have brought Indo-Pakistani Zebus to Madagascar from India and Arabia, while Hamitic and Bantu people from East Africa introduced humpless and Sanga cattle. East African Zebus were also imported to the island from 1506 onwards, and the interbreeding among these cattle produced a stabilized indigenous breed of the Madagascar Zebu today (Rege and Tawah 1999). More recent studies have found a higher indicine ancestry in the Madagascar Zebu populations as compared to other African continental Zebus and have attributed this to a second pulse of Indian Zebu introgression into an already admixed African taurine × Zebu population of likely East African origin that traces back to the twelfth century (Hanotte 2002; Magnier et al. 2021).

Physical characteristics: The animals have thoracic humps and are short-horned. The hump is large and pyramidal. Some animals, especially in the west of the island, do have long horns. The coat colour of the Madagascar Zebu cattle is very much a result of active selection by different tribes for particular colours and patterns. Indeed, there are 140 words in the Malagasy language that describe these variations (Felius 1995).

Adaptive and special genetic characteristics: Madagascar Zebu is considered to be well-adapted to the conditions on the islands where it is found. In particular, they are generally regarded as more resistant to heat stress and tick-borne diseases than imported European breeds, such as African Zebu and N'Dama breeds.

The Malawi Zebu

The Malawi Zebu are found throughout Malawi and are the dominant cattle breed in Malawi, con-

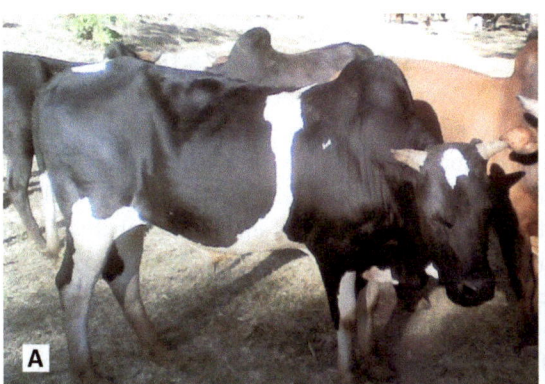
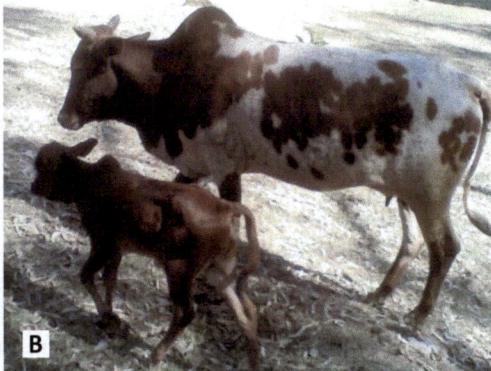

Fig. 4.6 Angoni bull (**a**) and a cow with a calf (**b**). (Courtesy: Dr. Maria da Gloria Taela)

stituting the largest proportion of the 800,000-strong national cattle herd. They are generally similar to the Angoni of Zambia. However, because its hump is more cervico-thoracic and it has larger ears and less developed dewlap (similar to the Nguni), it is considered to have some Sanga blood, from the opportunistic interbreeding with the Nguni Sanga inhabiting the same area. Consequently, the populations in central and southern Malawi are slightly different in conformation from the other Angoni. This has led to the classification of the Malawi Zebu into *North Malawi Angoni* and *South Malawi Angoni* (Felius 1995).

Herds of Madagascar and Malawi Zebu breeds are shown in Fig. 4.7a, b, respectively.

Baria Cattle

The Baria cattle are found in the western region of Madagascar in the Plateau de Bemaraha. It is a small, wild, and free-roaming population. They have not been sufficiently characterized, and their origin is uncertain. They have been referred to in different reports as Sanga, Zebu, and Taurine. However, the Madagascar official reports consistently classify the Baria as Zebu. Their limited use and neglect have caused a decline in the population of the breed. With a population estimated at about 500, the breed is considered critically endangered (Rege 1999).

4.1.3 West and Central African Zebu

Most of the West and Central African Zebu cattle are reared in the Sahel belt, which serves as a geographic and ecoclimatic bridge between the Sahel and the Savannah and Forest belts (Ouedraogo et al. 2021). They have historically inhabited the dry savannah and Sahelian zones, while the taurine (humpless) breeds have been widely distributed across the moist savannah and sub-humid coastal forest belts (Rege and Tawah 1999). The latter zones are infested with tsetse flies, which are the vectors of Nagana or trypanosomosis in cattle. However, the distribution of these populations has been shifting towards the coastal belt because of frequent droughts and insecurity in the Sahel. Consequently, many pastoralists have migrated with their animals into the humid and sub-humid areas (Boutrais 2007; Traore et al. 2015) where their animals have interbred with the humpless cattle. This crossbreeding between the *Bos taurus* and *Bos indicus* cattle has resulted in a new group called the West African (*Pseudo*) Sanga cattle (e.g. Borgou, Djakore, Ketekou, Ghana Sanga, and Pabli) (Sect. 4.2.3).

Rege (1999) classified the West and Central African Zebu cattle into the Gudali (represented by the Adamawa, Sokoto, and Ngaoundere Gudali) and the Fulani (represented by the lyre-shaped White Fulani and the long lyre-shaped Red M'bororo types, the Awazak, the Shuwa, and the Maure). Table 4.2 summarizes the West and Central African Zebu cattle breeds.

The Zebu of West and Central Africa are raised in a range of production systems—from on-station (intensive, mostly research stations) and ranches (public or privately owned) to the majority, which are on-farm mostly extensive, including pastoral and agropastoral production systems. The extensive system is dominant and primarily comprises smallholder producers with indigenous cattle as their dominant stock because of their adaptability to the prevailing ecological conditions. These animals contribute to the economies of this sub-region in significant ways in terms of food, income, employment, culture, and ecosystem stability. The two types, Shorthorns and Longhorns, are described in the sections below.

4.1.3.1 West and Central African Shorthorn Zebu Cattle

The West and Central African Shorthorn Zebus are similar in origin, physical appearance, and conformation to the East African and Indo-Pakistan Shorthorn Zebus (Payne 1970). They are classified as being among the Sahelian Zebu breeds that are believed to have descended from the Indo-Pakistan Zebu, which entered the Horn of Africa by way of the Persian Gulf and south Arabia. The Arabian invaders spread the Zebu to the south and west over the continent from 669 BC (DAGRIS 2022).

Fig. 4.7 (**a**) A herd of Madagascar Zebu. (Courtesy: Dr. Maria da Gloria Taela). (**b**) A herd of Malawi Zebu. (Courtesy: Dr. Patricia Mayuni)

Grouped under this category are the Azawak (found in Mali, Niger, and Nigeria), Djelli (Niger), Gudali (Cameroon and Nigeria), Maure (Mali, Mauritania, and Senegal), Tuareg (Mali), and Shuwa (Cameroon, Chad, and Nigeria). According to Gates (1952), these breeds have a common ancestry. They are known for their physical endurance that gives them the ability to survive in arid and semi-arid conditions. For example, the Gudali cattle are heavy and stocky animals that exhibit effective adaptability characteristics in both the humid and wet/dry and the arid conditions of the West and Central African sub-region. They are preferred for their meat production and draught power, while the White Fulani are preferred for both their milk and meat potential and trekking ability. The Shuwa cattle have similar endurance characteristics as well as high bodyweight at maturity (Rege and Tawah 1999) and are mainly reared for their dairy potential.

The Gudali Cattle
Classification, distribution, and population statistics: The Gudali cattle represents a large group of Shorthorn Zebus in West and Central Africa and comprises the Adamawa, Ngaoundere, and Sokoto subgroups, which are found principally in Cameroon, Nigeria, and the Central African Republic, with a small population in Ghana (Rege and Tawah 1999; Tawah and Rege 1996b). They are traditionally reared by Fulani and Hausa

Table 4.2 West and Central African Zebu Cattle

Breed name		Synonym
West and Central African Shorthorn Zebu		
1.1 Gudali	Adamawa Gudali	Banyo, Boyo
	Sokoto Gudali	Bokoloji, Godali, Sokoto
	Ngaoundere Gudali	Adamawa Fulani, Cameroon Fulani, Ngaundere
	Yola Gudali	Tattabareji, Tattabereji
1.2 Wadara or Shuwa cattle		Arab Shuwa, Arab Choa, Wadera
1.3 Azawak cattle		Adar (Nigeria), Shanun Adar (in Hausa), Azawa, Azawal, Azawaje (in Fulani), Arab (in Niger), Azaouak, Azouak, Darmeghou, Tagana, Tagama (Niger), Tuareg (in Burkina Faso), Touareg
1.4 Maure cattle		Arab, Mauritanian, Moor, Moorish
West and Central African Longhorn Zebu		
2.1 Gobra cattle		Senegalese Fulani, Senegal Zebu, Toronke
2.2 White Fulani		Bunaji, Yakanaji, Akou, Aku, White Bororo, Fellata, White Kano
2.3 Red Fulani cattle		Abori, Bodadi, Brahaza, Djafoun (Cameroon), M'Bororo, Mbororo, Red Bororo, Fellata (Chad and Ethiopia), Fogha, Gabassae, Gadehe, Hanagamba, Kreda (Chad), Ogha, Rahaji, Rahaza, Rahau, Red Longhorn, Wodabe, Wodaabe
2.4 Sudanese Fulani cattle		Peul Fulani, Bogoro Fulani, Fulani Nam, White Umboroa, Misse, Sego, Baouro, Bourgou, Toronke, Sambourou, Seno, Kaarta, Fellata, Bororo
2.5 Djelli cattle		Nigerian Fulani, Diali, Djali, Djeli, Jali, Jeli, Jalli, Peuhl, Tuareg, Tuoaureg

pastoralists. *Gudali* is a Hausa word for *'short-horned and short-legged animals'*. It is generally used to embrace a large group of Shorthorn Zebus, which are also referred to as *Fulbe* or *Peuhl* Zebu in West and Central Africa. Different names are used to designate these cattle populations (Table 4.2). These names are either based on their place of origin (e.g. Adamawa Gudali, Banyo Gudali, Ngaoundere Gudali, Sokoto Gudali, and Yola Gudali) or the owning tribe (e.g. Fulani Gudali, Fulbe Gudali, Peul, and Poulfoulo). In some cases, however, their dominant coat colour markings are used to designate them. For example, *Tattabareji*, a Fulani word for speckled coat colour, is used to refer to the Yola Gudali.

The *Adamawa Gudali* is primarily found in the Adamawa mountain ranges stretching from Nigeria to Cameroon. There are two sub-types: the **Banyo** or Boyo type found in the Banyo and Bamenda highland regions of Cameroon and the Mambilla highland regions of Nigeria (Taraba State) often have a white face and red eye patches or a red and white combination and, often, patches on their heads (Fig. 4.8a, b), and the **Yola** type are red, black, and white, often in patches. The Banyo type has rather large horns because of either Bunaji or Rahaji introgression (which one specifically is uncertain), while the Yola Gudali is considered to have a Muturu admixture (Gates 1952), and it has progressively been diluted since the 1950s, such that the Yola type is no longer recognized as a distinct variety by local herders (Blench 1999). Both the Kanuri and Fulani pastoralists in Nigeria own the Adamawa Gudali cattle. It is rare for these pastoralists to have herds of only Adamawa Gudali; they are often mixed with the Bunaji, Rahaji, and Wadara cattle (Blench 1999).

The *Sokoto Gudali* is commonly found in Nigeria, although small populations are also found in Burkina Faso, Ghana, Mali, Niger, and Northern Benin (Tawah and Rege 1996b). The population in Ghana is about 10,000, and most of them have been used for crossbreeding with the Ghana Shorthorn. In Nigeria, the breed used to be mainly found in the north-western region but is now widely distributed across the country. About 90% of the Sokoto Gudali are owned and managed by Fulani and Hausa pastoralists and transhumant herders, who graze their cattle on communally owned grazing lands and browse especially in the dry season (Tawah and Rege 1996b).

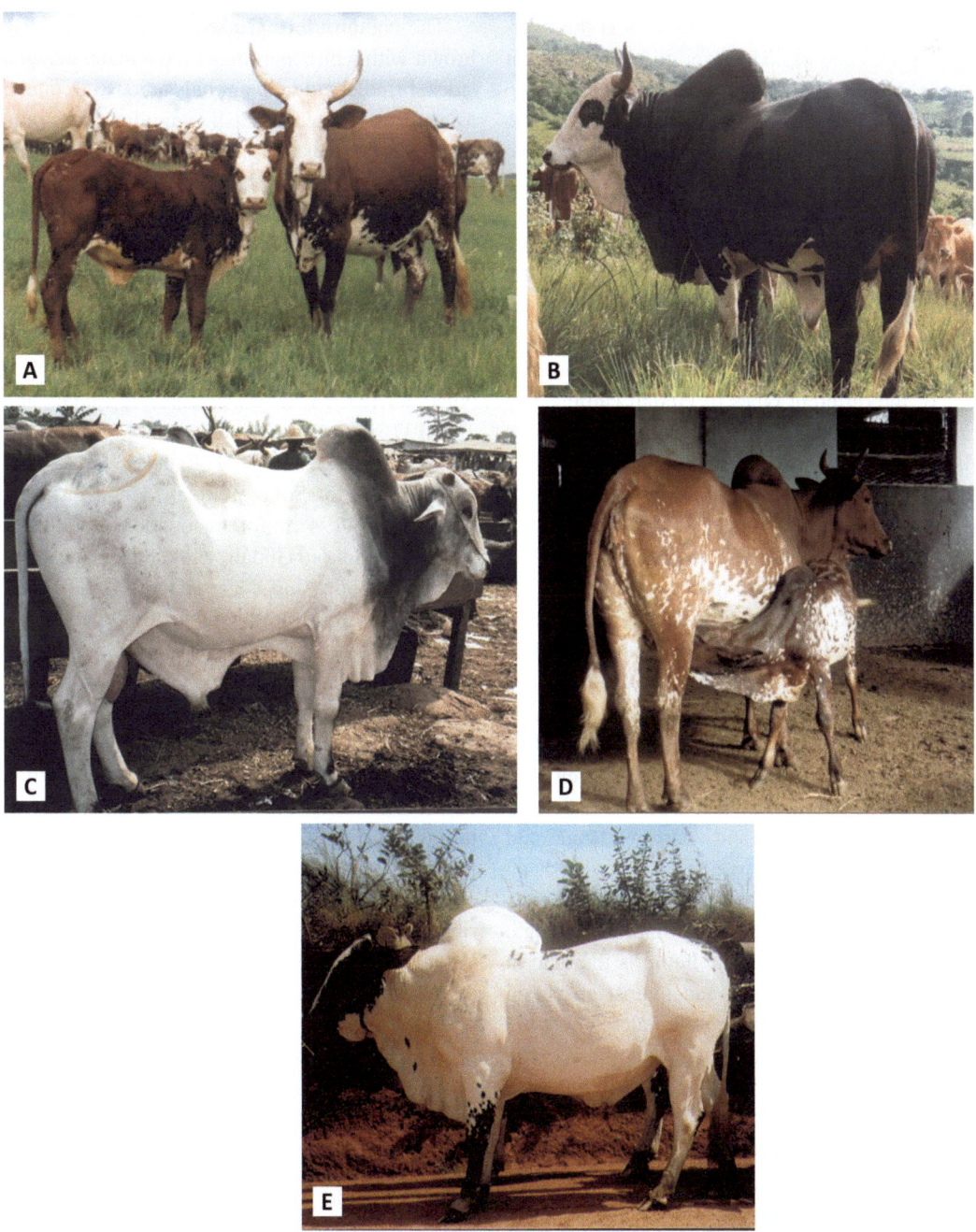

Fig. 4.8 (**a**) Typical Banyo Gudali cow and calf and (**b**) mature bull at Sabga, Cameroon, portraying the typical physical characteristics of the breed. (Courtesy: Dr. Eveline M. Ibeagha-Awemu). (**c**) Sokoto Gudali bull (Nigeria). (Courtesy: Dr. Eveline M. Ibeagha-Awemu). (**d**) Typical Ngaoundere Gudali cow and calf and (**e**) bull (Cameroon). (Courtesy: Dr. Eveline M. Ibeagha-Awemu)

The *Ngaoundere Gudali* is mainly found in the Adamawa region of Cameroon, with some populations also in the Central African Republic.

Accurate estimates of the population sizes of the various Gudali breeds/strains are not available because of their frequent movements inside and outside their ecological zones and the overlap in their ecotypes. However, the population of the Sokoto Gudali has previously been estimated at about 4.35 million in Nigeria, the Banyo Gudali at about 140,000 in Cameroon, the Ngaoundere Gudali at about one million in Cameroon and 400,000 in the Central African Republic, and the Adamawa Gudali at about three million in Cameroon and Nigeria (Rege 1999; Tawah and Rege 1996b).

Physical characteristics: Tawah and Rege (1996b) have described the physical characteristics of the different types of Gudali cattle. They are similar in conformation, size, and origin to the Large East African Shorthorn Zebu and closely resemble the Boran and Kenana cattle (Sect. 4.1.1).

The *Adamawa Gudali* resemble the White Fulani (Bunaji) in conformation and are medium- to large-sized, with medium-length, thick, and crescent-shaped horns. Their large and pendulous hump is the main characteristic that differentiates them from the White Fulani. The coat colour varies from pied, white, black, and brownish yellow to red and white, black and white, and grey with brown spots. They have a well-developed dewlap and a short-to-long and narrow head and muzzle.

The *Sokoto Gudali* have long and broad ears, well-developed humps, and a pronounced sheath in the male (Fig. 4.8c). The hump is thoracic in position. Horns are short and almost absent in males but tend to be longer in females. There are large variations in coat colour in the breed. The most common coat colour is black and white with a light underside. However, uniform cream, light grey, or dun is also frequent. Colour variations exist and include grey and white or light dun with dark shading over the head, neck, and shoulders. The tail switch is black. The head is long and wide between the eyes and across the forehead, with a straight or slightly convex facial profile. Their dewlaps and skin folds are highly developed. The hair is short, and the skin is thick and pigmented. Ears are long, large, and convex, sometimes pendulous, although not to the same degree as in some of the Indo-Pakistan Zebus such as the Sahiwal.

The **Ngaoundere Gudali** are mostly reddish brown with white markings on the underside and sides; brindle and roan animals are also frequent, as well as spotted reddish individuals in white and brown coats (Fig. 4.8d, e). The dewlap, sheath, and umbilical folds are poorly to moderately developed. The horns are short-to-medium in length and crescent-shaped. The hump of the Ngaoundere Gudali is very large and pendulous, generally hanging over on one side and having the appearance of being broken.

Height at withers for males is 133–136 cm for Ngaoundere Gudali, 122 cm for Yola Gudali, and 130–137 cm for Sokoto Gudali, while for females, it is 123 cm for Ngaoundere Gudali and 117–131 cm for Sokoto Gudali. Heart girth for males is 190–194 cm for Ngaoundere Gudali and 190 cm for Sokoto Gudali, while for females, it is about 170 cm for Ngaoundere Gudali and 144–166 cm for Sokoto Gudali. Body length for males is 158–179 cm for Ngaoundere Gudali, 88 cm for Yola Gudali, and about 154 cm for Sokoto Gudali, while for females, it is 145 cm for Ngaoundere Gudali, 76 cm for Yola Gudali, and 124–145 cm for Sokoto Gudali. Despite the differences in conformation and other physical features, what the Gudali group has in common includes the long, well-balanced, and relatively compact, deep, and wide body and well-sprung ribs. They have deeper bodies than the White Fulani and present a close-to-the-ground appearance.

Production systems, husbandry practices, and production characteristics: Gudali cattle are raised under different traditional husbandry practices in different habitats and production systems. These include communal grazing and short- and long-distance transhumance pastoral systems. Long-distance transhumance occurs in the dry season when herders move southwards in search of pasture and water. When the rains return, there is a reverse movement while ensuring that the tsetse-infested and flooded areas are avoided. During transhumance, lactating cows are left with the offspring in the homestead and cared for semi-intensively (grass and occasional concentrates offered in addition to field grazing). Other than the castration of unwanted bulls, which is a common practice, breeding is generally uncontrolled. The Gudali are also raised in ranches, on research stations, and on government-owned farms in the sub-region. The ranches may be state-owned or privately owned.

As alluded to above, Gudali cattle are heavy and stocky animals. They are well-known for their meat and milk qualities (Yakubu et al. 2010) and for their draught power. They are used for ploughing and carting because of their sturdiness and docility despite their slow and sluggish nature. Bullocks are usually out to work at 3 or 4 years of age. Gudali cattle have been shown to respond well to intensive feeding (Lhoste and Dumas 1972; Olayiwole et al. 1981) and are, thus, believed to be the most promising beef animals in the sub-region (Leclercq 1976; Olutogun 1976; Pagot 1985).

Tawah and Rege (1996b) and Rege (1999) have reported large variations in weight at maturity between and within the Gudali sub-types as follows: male mature weights are 400–563 kg for Ngaoundere Gudali, 400–408 kg for Banyo Gudali, 495–660 kg for Sokoto Gudali, and 350–352 kg for Adamawa Gudali. Female mature weights are 330–408 kg for Ngaoundere Gudali, 360–363 kg for Banyo Gudali, 240–355 kg for Sokoto Gudali, and 335–336 kg for Yola Gudali. Mature weights of oxen (castrated bulls) are 350–499 for Ngaoundere Gudali, 453 kg for Banyo Gudali, and 509–662 kg for Sokoto Gudali. Management practices are important contributors to the differences in mature weights. The Adamawa Gudali is considered a multipurpose breed, used as a draught animal and for beef and milk production. It fattens easily on grazing and is very docile. They respond easily to the use as draught animals and, for this reason, are favoured by locals for ploughing. However, when they become too large to pull a plough effectively, they are fattened and sold as meat animals (Babayemi et al. 2014).

Gudali cattle are also good milkers; the udders in the female are well-developed with good teats, and hence, they are regarded as an indigenous dairy breed. The lactation length for the milking cows is about 228 days, with a daily milk yield averaging about 7.5 kg. Sokoto Gudali cows produce an average of 1500 kg of milk per lactation. They produce more milk than White Fulani cows under similar research station conditions (Alphonsus et al. 2012). The age at first calving ranges from 38.6 to 49.5 months, and the calving interval ranges from 360 to 537 days (Tawah and Rege 1996b). The long interval between calvings is because of the ovarian inactivity in the post-partum period, coupled with high rates of abortions and stillbirths (ranging from 4.7% to 5.9%) in the breed.

Adaptive and special genetic characteristics: Gudali cattle exhibit adaptability to both humid, wet, dry, and arid conditions and are also known for their hardiness in the arid environments in the north. Sokoto Gudali are reputed for browsing, and herds are usually seen lopping trees towards the end of the dry season (RIM 1992). Buvanendran et al. (1992) reported rectal temperatures of 38–39 °C and 38.5–39.1 °C, respiration rates of 19.9–38.3, and 24.9–42.6 flank movements per minute, respectively, for Sokoto Gudali in the cool and hot seasons in Nigeria. Using these values, corresponding coefficients of adaptability to heat stress for these animals were estimated at 1.77–2.69 and 2.09–2.87. A higher value of this coefficient indicates poor adaptability to heat stress. Also recorded was a sweating rate of 54.7–93.3 and 170.8–224.1 g per m^2 per hour. These results show that Sokoto Gudali cattle are less adapted to the conditions of the Guinea savannah than White Fulani cattle (Buvanendran et al. 1992; Ngere 1985).

Microsatellite, milk genes, and milk and blood protein polymorphism studies of the Zebu and taurine cattle of Western and Eastern Africa (Braend and Khanna 1968; Braend 1971; Queval et al. 1971; Ibeagha-Awemu et al. 2004) have revealed that the Gudali and other Zebu breeds in the region may share the same evolutionary history with the Indo-Pakistan and Eastern African Zebus. Similarly, there is evidence that the Zebu cattle are distinct from the Shorthorn and Longhorn Taurine cattle of West and Central Africa (Ibeagha-Awemu et al. 2004). The Y chromosome is acrocentric in these Zebus and sub-metacentric in the taurine Baoulé and N'Dama cattle of West Africa and the European taurine breeds (Popescu 1980). Moreover, Braend and Khanna (1968) observed that blood factor Z' is common in the Gudali and Fulani Zebus but absent in the Muturu and N'Dama cattle. Thus, there is a clear genetic distinction between the Taurine and Zebu cattle. These polymorphic systems can also be used as genetic markers for investigating the genetic relationships between these populations and their genetic purity.

The Wadara or Shuwa Cattle

Distribution and population statistics: The Wadara or Shuwa cattle are typical short-horned Zebu and are also known by various synonyms (Table 4.2). They are found throughout Chad (except the south-west), in north-eastern Nigeria, and in the extreme north region of Cameroon. The Wadara cattle in Nigeria are considered the indigenous cattle of Borno State and are often referred to as *'our cattle'* by the Koyam, Shuwa, and related pastoralists. Wadara cattle are considered to have some humpless Shorthorn gene introgression, although they are phenotypically predominantly Zebu. They represent 6.6% of the Nigerian herd (RIM 1992). They are also frequently referred to as 'Shuwa' in the literature, after the Shuwa Arabs who also keep them. A related breed, known as *Ambala*, is found throughout the Lake Chad region. The Ambala are said to often migrate into Nigeria from Chad. Their population is estimated at 4.55 million (Rege 1999).

Physical characteristics: The Wadara are small but well-muscled, medium-sized, lightly to deeply built animals with rounded ribs. Their coat colour is usually dark red or plain red, black, pied or brown, and sometimes, black and patchy. They are short-horned and have a small erect hump. Their height at withers is 135–140 cm in males and 125–128 cm in females (Rege 1999). The *Ambala* have a white coat or are dappled. Figure 4.9 shows mature male and female Wadara (Shuwa) cattle portraying the typical physical characteristics of the breed.

Production characteristics and uses: The Wadara are good dairy animals, with milk yields of 450–1820 kg in a 240–370-day lactation period. Consequently, like the Kenana and Butana cattle of eastern Africa, they are mainly reared for their dairy potential (Belemsaga et al. 2005; Kubkomawa 2017). Mature weights are 350–475 kg for males and 250–300 kg for females (Rege 1999). These animals are also used for riding and as pack animals. The Ambala ecotype is a large-sized animal and has a much larger mature weight of about 800 kg.

Adaptive characteristics: The Wadara have similar endurance characteristics to the other members of the Shorthorn Bos indicus family of the West and Central African Zebu population (Rege and Tawah 1999).

The Azawak Cattle

Distribution and population statistics: Azawak cattle are indigenous to the Azaouak region situated at the borders between Niger and Mali. *Azaouak* means sandy land without any vegetation in Tamajeq. Azawak cattle are raised by nomadic Tuareg and Arab inhabitants of the Azaouak valley, which is a vast, windy depression, stretching 3°–7° E longitude and 15°–20° N latitude and covering the arid eastern Mali and extending from the Niger River through southern Niger to parts of northern Nigeria. A small population of Azawak cattle is found in Nigeria throughout the year and constitutes about 0.7% of the national herd, while the majority are seasonally transhumant. They are generally found on the border to the North and West of Sokoto and North-West of Borgu. They are also dotted along the frontier from Sokoto to Katsina (Meghen et al. 1999). Azawak cattle are also referred to by different synonyms (Table 4.2). They are the most important cattle population in Niger, representing about 33% of the national cattle herd, and are found throughout the country. They are also found in Benin, Burkina Faso, the Central African Republic, and Chad. Their overall population is estimated at about 506,000 (Rege 1999). They have been used in crossbreeding with the Sudanese Fulani Zebu in Burkina Faso.

Physical characteristics: The Azawak are lightly built, average in size, stocky-shaped, and compact and are, therefore, suitable for beef production. Their coat colour is usually a mixture of

Fig. 4.9 Wadara or Shuwa cattle: a bull and a herd

red and white, black and white, fawn with white patches, or fawn red. On traditional farms, white, grey, and black are common, while on experiment stations, mahogany is common. The Azawak cattle in the Niger Republic are commonly red, while those found in Nigeria are usually light fawn, though they can also be white, brown, pied, and black. The dewlap is well-developed. The hump is large and narrow (12–16 cm thick). They have medium-length horns, which are short and thick at the base in males and lyre-shaped in females. Ears are 21 cm long in cows. The skin is flaccid with average thickness. The height at withers for males is 120–135 cm, and for females, it is 110–130 cm (Rege 1999). Figure 4.10 shows mature male and female Azawak cattle portraying the typical physical characteristics of the breed.

Production characteristics and uses: The Azawak are suitable for beef and draught. The average mature bodyweight of males is 350–500 kg and 250–410 kg for females in traditional systems (Rege 1999). However, males can attain mature weights of up to 600 kg on-station. They are early-maturing and aesthetic animals. Birthweight averages 23–24 kg for males and 21–22 kg for females. Carcass yield ranges from 48 to 52%. Their daily milk yield averages 1.5 kg over an average lactation length of 290 days (Traore et al. 2015), but they are able to produce between 5 and 15 litres of milk daily under favourable conditions. The age at first calving ranges from 35 to 40 months, with a calving rate of 70–75% and a calving interval of 11–13 months. Their average mature weight of 345 kg and high carcass yield account for the breed's suitability for meat production. Azawak cattle have been under selection at the Toukounous Sahelian Experiment Station in the Republic of Niger since the 1950s.

Adaptive characteristics: The Azawak are hardy and well-adapted to drought conditions and the general harsh climatic conditions of Niger and the surrounding environments. They have similar tolerance and endurance qualities as the Maure cattle. Their physical endurance, coupled with their ability to walk long distances, makes them highly suitable for pastoral farming (Traore et al. 2015).

The Maure Cattle

Distribution and population statistics: Maure cattle, also known as Moorish cattle and other synonyms (Table 4.2), are found in the Sahel region of Mauritania and the adjoining regions of Mali. They are owned by nomadic Arabs and Berbers. The humps are more prominent in the eastern part of their distribution range, suggesting possible Fulani influence. These animals are also in Côte d'Ivoire, Mali, and Senegal. Their population is estimated at about 673,000 (Rege 1999).

Physical characteristics: The coat colour of the Maure is fawn, red, or black in plain or patchy pattern. However, the most common coat colour is black or black and white, although dark red

Fig. 4.10 Azawak cattle in Niger. (Courtesy: Larry W. Harms, Virginia)

coats are common in the east. The Maure cattle are loosely built and tend to be leggier than the Azawak cattle. The height at withers ranges from 125 to 140 cm in males and from 110 to 130 cm in females (Rege 1999). Figure 4.11 shows a mature Maure cow portraying the typical physical characteristics of the breed.

Production characteristics and uses: Maure cattle have a large and stocky body size, making them suitable for beef production. Their mature bodyweight is about 250–700 kg for males and 250–350 kg for females (Rege 1999). Cows are valued for their milk yield, which averages between 471 and 650 kg over a lactation period of between 210 and 270 days. Milk yield per day is about 8.0 kg at peak lactation. They are also used as pack and riding animals.

Adaptive characteristics: Maure cattle have a strong muscular body, which makes them suitable for draught power (Penda et al. 2014). They also show a high tolerance for high temperatures and low-humidity conditions and, hence, their preference by farmers in the arid parts of Mali. Moreover, they are adaptable to varying temperatures and are, thus, able to survive in the high humidity conditions of Côte d'Ivoire and Senegal (Santoze and Gicheha 2019).

4.1.3.2 West and Central African Longhorn Zebu

The longhorn *Bos indicus*–type of the West and Central African Zebu cattle, also known as the Fulani type, includes the **Gobra, White Fulani, Red Fulani, and Djelli** cattle. Their origins are not known. Tawah and Rege (1996a) have summarized existing theories about their genetic constitution, including evidence suggesting that they are an admixture of *Bos taurus* and *Bos indicus* ancestry. This theory is supported by a report of more than 7% African taurine and more than 10% European taurine genetic influence in the White Fulani and Red Fulani populations of Cameroon and Nigeria, based on the data on 225 alleles at 28 loci (Ibeagha-Awemu et al. 2004). Indeed,

Fig. 4.11 Maure cattle. (Courtesy: Dr. Larry W. Harms, Virginia)

while the Fulani people are thought to be conservative pastoralists, their livestock management practices suggest that they do cross and change cattle breeds to adapt to new ecological or sociopolitical conditions (Boutrais 2007).

The Fulani cattle are a unique group in West and Central Africa, which differ from the typical Shorthorn Zebu of Central, Eastern, and West Africa by the presence of long horns and from the cervico-thoracic-humped Sanga cattle by the presence of a thoracic, or sometimes, intermediate hump. The Fulani family has been classified into two subgroups: the lyre-horned subgroup consisting of the Gobra (Senegalese Fulani), the Sudanese Fulani, and the White Fulani, also known as Bunaji in Nigeria and Akou in Cameroon, and the long lyre-horned subgroup represented principally by the Red Fulani, also known as Rahaji in Nigeria (Payne 1970; Mason 1988). Rege and Tawah (1999) have described the Djelli (or Diali) as a strain of the Fulani family found on the floodplains and adjacent valleys of the Niger River in the Niger Republic and south-west Nigeria. Blench (1999) described the Djelli as exotic to Nigeria. Recent reports show that the Fulani cattle are characterized by a high degree of genetic diversity (Ibeagha-Awemu et al. 2004; Ibeagha-Awemu and Erhardt 2006; Nourezzine et al. 2019). Moreover, there is a high genetic connectedness between the Red and White Fulani (Ibeagha-Awemu and Erhardt 2006) as well as the Kuri cattle (Epstein 1971). Individual members of the Fulani family are described in the following paragraphs.

The Gobra Cattle

Distribution and population statistics: The Gobra is the dominant breed in the Sahelian belt of Mali and Senegal and in the adjoining regions of Mauritania. They are referred to as the Toronke Cattle in Mali. Their population is estimated at about 1.3 million (Rege 1999). The Gobra cattle are also known by different names such as the Senegalese Fulani, Senegal Zebu, or Toronke (Table 4.2).

Physical characteristics: Gobra cattle have either lyre-shaped or crescent-shaped horns, which are sometimes loose and can reach 70–80 cm in length. Their dewlap is well-developed. Their hump is prominent, especially in the adult males. They have thick and loose skin. Most of the cattle are grey with dark patches, but animals with white backs and black speckles are not uncommon. The Gobra and White Fulani cattle are predominantly white in coat colour and are much larger than the Sudanese Fulani, whose coat colour is quite variable but usually spotted light grey. The Gobra cattle have an average height at withers of 130–144 cm for males and 124–140 cm for females. They have a heart girth

Fig. 4.12 Gobra cattle. (Courtesy: Dr. Larry W. Harms Virginia)

of 171 cm and a body length of between 70 and 127 cm. Figure 4.12 shows mature male and female Gobra Zebu cattle portraying the typical physical characteristics of the breed.

Production characteristics and uses: Gobra cattle possess an average weight at maturity of between 300 and 578 kg for males and between 250 and 450 kg for females (Rege 1999). Sexual dimorphism is marked between males and females, with significant differences. The age at first calving is about 44.8 months, and the interval between calvings is about 15.5 months. They are mainly kept for meat production because of their high mature bodyweight, but they can also produce about 675 kg of milk over a 330-day lactation period (Missohou et al. 1997). The bulls and oxen are used for riding and as pack animals.

Adaptive characteristics: Because of their physical endurance in the arid and semi-arid conditions, Gobra cattle are preferred in pastoral production systems.

The White Fulani Cattle

Fulani is the Hausa word for the pastoral peoples of Nigeria belonging to the 'Fulbe' migratory ethnic group. Pastoralism, as a livelihood, is coming under increased pressure across Africa, including among the Fulbe owners of Fulani cattle, due to changing socio-economic and environmental conditions. In the past (especially, the 1950s and earlier), there was a symbiotic relationship between pastoralists (practising transhumance) and the more sedentary crop farmers. During the dry season, pastoralists migrated to the pasture-rich areas with minimal cropping in the southern parts of the Guinea savannah zone. During the rainy seasons, pastoralists would migrate away from the tsetse-infested forest areas towards the northern Sudan savannah zone, supplying dairy products to the local farming community and avoiding trypanosomosis. The farming community supplied pastoralists with grain, and after the harvest, cattle were permitted to graze on crop residues in the fields, leaving behind valuable manure and the fields ready for the next cropping season.

Today, population pressure has resulted in deforestation and increased settlement in the Guinea savannah and rainforest regions. At the same time, the control of tsetse and trypanosomosis has enabled cattle and crop farmers to inhabit these zones all year round. This has increased the competition for resources, including land, with grazing areas and transhumance routes coming under pressure and crop land being increasingly limited. With pastoralists having limited or no rights to land, there are intense conflicts when livestock damage crops. The situation is getting worse with climate change and its impacts on the reliability of rains and, hence, the availability and access to water and pasture.

Distribution and population statistics: The White Fulani cattle are the most numerous and widespread of all the Nigerian cattle breeds, representing about 37.2% of the national cattle population (FDLPCS 1992; Alphonsus et al. 2012; Kubkomawa 2017), and are distributed in an area stretching from the South-western states, across the middle belt to Sokoto, Katsina, and Kano states. In Cameroon, where their distribution stretches from the Adamawa, Far North, to the North-west Regions of the country, they represent 33% of the national cattle herd and are only second to the Red Fulani (Tawah and Rege 1996a; Messine et al. 1995), and in Ghana, the White Fulani constitute about 80% of all Ghanaian Zebus (Aboagye 1995). They are also represented in small populations in the Benin Republic, Chad, Niger Republic, and Sudan. They are known by different names (Table 4.2) and are found in both tsetse and non-tsetse-infested agro-ecological zones of West and Central Africa. The population of the breed is estimated at about 9.67 million (Rege 1999).

Physical characteristics: The White Fulani are medium-to-large-sized animals with a well-balanced body of good depth and width. Height at withers is 130–152 cm for adult males and 118–138 cm for adult females. Heart girth is 193 cm for males and 145–161 cm for females, while body length is 152 cm for males and 117–137 cm for females. They are of good length and have a marked slope from the hook to the pin bones, which tend to be narrow in some animals. The barrel is well-sprung and of good capacity, and the top line is strong and slopes gently from the hump to a somewhat high sacrum. The hump is well-developed in the males than in the females and sometimes tends to hang over at the back in the former, like that of the Gudali. The hump is either thoracically or cervico-thoracically placed on the backline and is musculo-fatty. The tail is thin and long, with the tail switch almost touching the ground. The White Fulani have light and hard bones and a well-developed and large dewlap, especially in the bulls. The head is long and of good proportion and is straight or slightly dished (concave). The neck is strong and deep, providing an upright carriage for the head. Figure 4.13 shows mature male and female White Fulani cattle portraying the typical physical characteristics of the breed.

The White Fulani have lyre-shaped horns, which are slender, well-proportioned, and of medium-to-long length, measuring about 80–107 cm. The horns are round in cross section and are curved outwards and upwards immediately above the head. Most of the White Fulani animals have horns with an outward twist at the tips, giving them the characteristic lyre shape. The ears are of medium size, erect, and set horizontally, with the inner parts having black points showing to the front. The udders are well-developed and of good shape and are strongly attached, and the teats are well-positioned and are of medium-to-large size.

White Fulani cattle have a predominantly white coat colour on a black skin with black ears, eyes, muzzle, hooves, horn tips, and tail tip. There are a few cases with black coat colour mixed with dark patches or red and white coat colours. Sometimes, red marks can be found on the ears, feet, and sides. Black flecking on the sides and limbs is quite common, while red markings are frequent. In some cases, variations in coat colour include combinations of black and white on a black skin or red and white on a white skin (Ogunsiji 1974). The skin is loose and pigmented with soft hair.

Production characteristics and uses: The White Fulani cattle are predominantly found in the tropical climate characterized by two well-defined seasons—the wet and dry seasons—and two prevailing wind systems—the south-west rain-bearing wind from the Gulf of Guinea and the dry north-easterly dust-laden wind (the harmattan). The vegetation varies from closed forest in the derived savannah to light forest and open woodlands in the Guinea savannah. Most trees in the Guinea savannah are fire-tolerant, and the major grass species have been described by (Tawah and Rege 1996a), with *Hyparrhenia* spp. being the most common. Luxuriant growth of many tall grasses, such as the Gamba grass (*Andropogon tectorum* and *A. gayanus*) and

Fig. 4.13 White Fulani bull (**a**) and cow and calf (**b**). (Courtesy: Dr. E. M. Ibeagha-Awemu)

Guinea grass (*Panicum maximum*), is characteristic of the derived and Guinea savannah zones.

The White Fulani are well-known for their dual-purpose characteristics and are mainly reared for their beef and dairy production (Rege and Tawah 1999). Their traditional owners rear them primarily for milk. Their dairy potential is much better than that of most Zebu cattle in the region but comparable to that of the Kenana cattle in Sudan (Osman 1984). White Fulani are also good beef animals and fatten quite well in feedlots and on natural pastures. Indeed, much of the beef consumed throughout Nigeria is from the White Fulani (Payne and Wilson 1999; Alphonsus et al. 2012). White Fulani bulls are good draught animals by virtue of their docility, tractability, conformation, and body size.

White Fulani cattle attain mature bodyweights of between 425 and 665 kg for males and between 250 and 380 kg for females in improved systems. Mature weight for females in traditional systems has been reported at 270–310 kg. Birthweights average between 18 and 20 kg in traditional systems and between 21 and 24 kg in improved systems. They reach sexual maturity late and have long intervals between calvings and short lactation lengths compared to the Sokoto Gudali (Kubkomawa 2017). The milk yield varies considerably from 720 to 2950 kg for a variable lactation length of between 196 and 427 days (Mrode 1988). The age at first calving ranges from 26 to 45 months, but in Fulani pastoralist herds, it can be as late as 60 months (Aganga et al. 1986).

Adaptive and special genetic characteristics: The White Fulani cattle are tolerant of intestinal helminths and are more resistant to dermatophilosis than the Muturu and N'Dama cattle (Nwufoh and Amakiri 1981). They are also known for their genetic predisposition for hardiness, heat tolerance, and adaptation to local conditions (Alphonsus et al. 2012). For example (Amakiri and Mordi 1975), reported that White Fulani cattle are more heat-tolerant than N'Dama and Sokoto Gudali in Nigeria, which is reflected in their low respiration rate and heat tolerance index. White Fulani cattle tend to pant and salivate less under heat stress (Buvanendran et al. 1992) and sweat much more profusely than Gudali, N'Dama, and Muturu cattle when exposed to similar high ambient temperatures (Amakiri and Onwuka 1980). Igono and Aliu (1982) described the White Fulani as the least stressed animal in the hot climates of Nigeria. They are superior to all the other breeds of Zebu in resisting diseases, with the ability to thrive under a variety of conditions (Blench 1993; Meghen et al. 1999). Their leggy appearance, which is from the shallowness of their body and lack of width, has been associated with their ability to trek long distances (Oyenuga 1967; Capitaine 1972). This attribute has been described as an adaptation to their nomadic lifestyle.

The Red Fulani Cattle

Distribution and population statistics: The Red Fulani cattle, also known by other names (Table 4.2), including *Rahaji* in Nigeria, are predominantly found in the arid and semi-arid regions of West and Central Africa. They are distinguished by their chestnut-to-deep mahogany coat colour and are the most striking of the Fulani family. They are found in northern Cameroon, southern Niger, and central and northern Nigeria. They are specific to the Bororo Fulani people of Cameroon, Niger, and Nigeria.

The Red Fulani is the second most numerous of the cattle populations in Nigeria, representing about 22% of the national herd, and the most numerous of the cattle populations in Cameroon, representing about 37% of the national herd (FDLPCS 1992; Messine et al. 1995). They are mainly located in the Far Northern States of Nigeria, although they are significantly absent in Borno State, where the Wadara cattle are predominant, and in the Adamawa, Far North, North, and North-west Regions of Cameroon. The Fulani pastoralists consider the Red Fulani as an extremely prestigious breed, and they are well-liked because of their beautiful appearance, high intelligence, and obedience to commands. Rege (1999) reported an estimated population of 4.92 million.

Physical characteristics: The Red Fulani cattle is one of the largest breeds in size (second only to the Kuri) in the region. They are distinguished by their deep burgundy-coloured coat and pendulous ears and horns that are gigantic—long (up to 63 cm) and thick. The hump is thoracically or cervico-thoracically located, well-developed, muscular in structure, and generally pyramidal in shape. Their dewlap, umbilical folds, and sheath are highly developed. Figure 4.14 shows Red Fulani cattle herds portraying the typical physical characteristics of the breed.

Production characteristics: The mature weight of the Red Fulani is between 400 and 450 kg for males and between 255 and 410 kg for females (Rege 1999). Their age at first calving ranges from 48 to 53 months, with the calving interval averaging between 537 and 585 days. The average milk yield per lactation has been estimated at 900 kg. In Cameroon, the daily milk production has been reported to range from one to two litres per cow, with a lactation length of 140–180 days (IRZ 1987).

Adaptive characteristics: The Red Fulani cattle are adapted to the arid and semi-arid regions and are susceptible to diseases in the humid areas (such as trypanosomosis) and poor nutrition. They are raised in conditions of low rainfall (3–5 months of rain, followed by a long dry season). They graze well on mature grasses, shrubs, and tree leaves and are well-noted for their herd instincts, walking gait, and ability to trek long distances (Mason 1996). They are also able to withstand heat, ticks, and insect bites.

The Sudanese Fulani Cattle

Distribution and population statistics: The Sudanese Fulani cattle, also known by other names (Table 4.2), are similar in origin to the Gobra of Senegal, with which they merge in transition zones. The Sudanese Fulani cattle are mainly owned by the nomadic Fulani people who occupy the belt between the Sahara and the coastal rainforest zones from the west of River Senegal to the east of Lake Chad, including parts of western Senegal, southern Mauritania, in and around the floodplains of Niger, Chad, northern Nigeria, and Cameroon (Rege and Tawah 1999). They are said to interbreed with the Azawak cattle in eastern Mali during seasonal migrations. The breed is found all over southern and southwestern Mali, along the floodplains of the Niger River from Segon to Timbuktu. They are also found in Senegal. In both Senegal and Mali, they are known as the Toronke cattle. In Burkina Faso and Côte d'Ivoire, interbreeding with the humpless cattle is common. Sudanese Fulani have an estimated population of about 5.62 million (Rege 1999).

Physical characteristics: The horns of Sudanese Fulani cattle are either lyre-shaped or crescent-shaped in appearance. Most have grey coat colours with dark patches, but animals with white backs and black speckles are also common. The height at withers ranges from 120 to 138 cm in males, and from 115 to 126 cm in females

Fig. 4.14 Red Fulani herd (**a**) and bull (**b**). (Courtesy: Dr. E. M. Ibeagha-Awemu)

(Rege 1999). The Sudanese Fulani cattle are usually spotted light grey, although quite variable in coat colour. Figure 4.15 shows mature Sudanese Fulani cattle portraying the typical physical characteristics of the breed.

Production systems, husbandry practices, and production characteristics: The major habitat of the Sudanese Fulani, the Niger delta, is in the semi-arid zone and has rainfall ranging from 200 to 600 mm per annum. The vegetation is strongly influenced by the annual flooding of River Niger. The floodplain pastures consist mostly of tall grass formations, which give way to rain-fed shrubland and browse trees in areas characterized by late and irregular floods. The vegetation varies according to the flood regime and consists of high-quality *bourgou* pastures (*Echinochloa stagnina*) and perennial Gramineae, such as *Vetiveria nigritiana, Andropogon gayanus,* and *Oryza longistaminata*.

The herds begin transhumance with the first rains in July, moving from the dry delta to the Sahelian uplands in the north-west, where they remain from August to October, grazing rain-fed pastures (Wagenaar et al. 1986). They return to the delta in November/December when the flood has subsided and follow its withdrawal to the north until the next rains. Due to the sharp climatic contrasts between the single rainy season of 2–4 months and the long dry season, there are large seasonal variations in the quantity and quality of the range resources available in the Sahel.

Birthweights of 17 and 21 kg have been reported in the Sudanese Fulani herds in traditional systems and on the ranch, respectively. The corresponding weaning weights were 55 and 79 kg. Mature weights are between 280 and 345 kg for males and between 248 and 300 kg for females (Rege 1999). Live bodyweight of cows of 1–2 years of age has been reported at 210 kg, and the bodyweight of mature animals over 6 years old reached up to 500 kg, with an average of 384.7 kg (Hamza et al. 2021).

Adaptive characteristics: The Sudanese Fulani cattle are able to trek long distances of up to 300 km in search of grazing and water and are adapted to the arid and semi-arid conditions.

The Djelli Cattle

Origin, distribution, and population statistics: The origins of the Djelli cattle, also known by different names (Table 4.2), are contentious. Some believe that they were derived from the Dwarf West African Shorthorn, primarily a cross between the Shorthorn Zebu and *Bos taurus*. It is postulated that the Djelli cattle have a stronger *Bos taurus* ancestry (genetic background) than any other West African Zebu cattle. They are found in the river basin of Niger, throughout north-eastern Burkina Faso, the south-west corner of Niger, the neighbouring areas of northern Benin, north-western Nigeria around Lake Chad, and Cameroon. However, they are predominantly found in Niger, mainly in pastoral systems (Rege and Tawah 1999).

Physical characteristics: Djelli cattle are of medium build with lyre-shaped horns. They have short, slender limbs. Their chest is full and deep. Neckline and shoulders are short, while dewlap and hump are very marked. The back dips slightly

Fig. 4.15 Sudanese Fulani cattle. (Courtesy: ILRI)

behind the hump and the waistline is nicely rounded. The skin is supple and slightly thin. The head is long and has medium lyre-shaped horns. The coat colour is black and white in a speckled or patchy pattern, although the dominant colour is white. Some animals have piebald coats (i.e. dirty white with black spots). The Djelli cattle population in Nigeria has been influenced by its neighbouring Sokoto Gudali cattle because of the seasonal migrations of Fulani pastoralists. Figure 4.16 shows mature male and female Djelli cattle portraying the typical physical characteristics of the breed.

Production characteristics: They are used primarily for meat and animal traction and have good fattening capacity. They have a mature bodyweight of 300 kg and a height at withers of between 115 and 130 cm. The carcass yield ranges from 48 to 50%. The cows produce about 2–3 litres of milk per day on average.

Adaptive characteristics: Like the other members of the Fulani family, the Djelli cattle are believed to be highly tolerant of diseases and exhibit cross-environmental adaptability. They are known to be able to survive in the arid and semi-arid conditions in Niger and its surroundings.

4.2 The Sanga Cattle

The original African cattle are divided into two major categories, namely humpless (*Bos taurus*) and humped (*Bos indicus*) (Chap. 3). The latter category is subdivided into Zebu proper and Zebu crossbred types. The Sanga is an intermediate type of cattle, which is a cross between *Bos taurus* and *Bos indicus*. *Sanga*, an Ethiopian word meaning 'bull' or 'ox', relates to the origin and centre of dispersal of this group of cattle breeds (DAGRIS 2022). One school of thought about the origin of the Sanga posits that the Sanga likely evolved in Ethiopia from the interbreeding between the original longhorn humpless cattle with cervico-thoracic-humped (Zebu). These cattle were then spread southwards by Hamitic (Bantu) tribes migrating from north-eastern Africa about 600 years ago (Rege and Tawah 1999) into western and southern Uganda and to Rwanda-Burundi and adjacent areas, including the present-day Democratic Republic of the Congo. This theory is buttressed by the fact that food production and the keeping of cattle seem to have begun in the highland and Rift Valley regions of East Africa in the first millennium and have derived from peoples who were probably southern Kushites from Ethiopia.

The Sanga are humped, but the hump is cervico-thoracic rather than thoracic. The Sanga is nowadays considered a separate group of cattle. There are 30 Sanga cattle breeds and strains on the African continent. They are discussed here by geographical regions—eastern, southern, western, and central Africa. Rege (1999) categorized the Sanga of eastern Africa into three groups: Nilotic Sanga of Sudan and Ethiopia, the Abyssinian Sanga of Ethiopia and Eritrea, and the Ankole group of Burundi, Democratic

Fig. 4.16 Djelli cattle. (Courtesy Dr. Larry W. Harms, Virginia)

Republic of Congo, Rwanda, Tanzania, and Uganda. The Sanga cattle in southern Africa are classified into six clusters, namely, the Shona of Zimbabwe, the Nguni group of Angola and Zambia, the Ovambo, the Setswana group, and the Afrikaner group of Southern Africa. Table 4.3 summarizes the classification of the present-day Sanga breeds/strains of eastern Africa and southern Africa. The following sections summarize the Sanga of eastern, southern, and West Africa.

4.2.1 The Sanga of Eastern Africa

The Abyssinian Group of Sanga Cattle

The *Danakil* and *Raya Azebo* cattle breeds are known members of this group.

Distribution and population statistics: The Danakil (also known as Adal, Afar, or Keriyu) cattle are kept by the Afar people of north-eastern Ethiopia in the steppe and semi-desert areas of the Afar region of Ethiopia, south-eastern part of Eritrea (along the southern Red Sea coast), and northern part of Djibouti. The Danakil depression—which the breed is named after—lies in the central part of the distribution area. The breed has also spread along the Awash River to parts of the Western Hararghe and Eastern Shewa zones. The related breed known as Raya Azebo is kept by the adjacent Raya and Keriyu tribes in Eastern Shoa, South Wollo, and Western Hararghe. The population of the Danakil has been estimated at 680,500 (DAD-IS 2005).

Physical characteristics: The Danakil cattle have a straight profile; long, thin legs; and long horns, which measure 1 metre or more in length and are lyre- or crescent-shaped. In some cases, the two horns loop to form a complete circle. The animals have a straight back but with a sloping rump. Their average height at withers is 125–145 cm in males and 120–125 cm in females. Danakil humps are small and cervico-thoracic, and the dewlaps are small. The navel flap and preputial sheath are also small. Hind legs are bowed with weak hocks and thin thighs. Their coat colour is usually ash grey, cream, or light chestnut. Figure 4.17 presents pictures of typical Danakil and Raya Azebo cattle.

Production characteristics: The Danakil cattle are kept for milk, meat, and work but mainly for milk production. Male animals weigh 250–380 kg, while females weigh 200–305 kg. Their meat is very fibrous. Their milk production averages 200–300 kg in a lactation period of 160–225 days.

Adaptive or special genetic characteristics: The Danakil are able to cope with extended periods of drought. They are browsers and are able to walk long distances in search of the few scattered patches of coarse grasses that may be available during these periods.

The Ankole Group

Distribution and population statistics: These cattle are found in eastern and central Africa and comprise the *Bahima, Bashi, Kigezi, Nsagalla,*

Table 4.3 The Sanga cattle breeds of eastern and southern Africa

Group and breed/strain	Location/country	Estimated population[a]	Mature weight (kg)		Withers height (cm)		Main users (in the order of importance)
			Male	Female	Male	Female	
1. The Sanga of Eastern Africa							
1.1 The Nilotic group							
1. Abigar	Ethiopia	548,600					Milk, meat, work
2. Aliab Dinka	Sudan	NA					Milk, meat
3. Aweil Dinka	Sudan	NA					Milk, meat
4. Nuer	Sudan	NA					Milk, meat
5. Shilluk	Sudan	NA					Milk, meat
1.2 The Abyssinian Sanga							
6. Danakil	Ethiopia, Eritrea	680,500	250–380	200–305	130–145	120–125	Milk, meat, work
7. Raya Azebo	Ethiopia	521,000					Work, milk, meat
1.3 The Ankole group							
8. Watusi	Uganda, Rwanda, Burundi, Tanzania, DRC	1,600,000	350–425	290–350	132–135	110–127	Milk, meat, work
9. Bahima	Uganda, DRC	*					Milk, meat, work
10. Kigezi	Uganda	*	220–380	200–330	112–132	108–120	Milk, meat, work
11. Bashi	Democratic Republic of Congo (DRC)	*	235–400	220–340	115–135	110–124	Milk, meat, work
12. Ruzizi	DRC, Rwanda, Burundi	*					Milk, meat, work
2. The Sanga of Southern Africa							
2.1 The Shona group							
13. Mashona	Zimbabwe	500,000	350–635	260–410			Meat, work
2.2 The Nguni group							
14. Nguni	South Africa, Swaziland	2,156,000	400–680	225–450			Meat, work, milk

(continued)

Table 4.3 (continued)

Group and breed/strain	Location/country	Estimated population[a]	Mature weight (kg) Male	Mature weight (kg) Female	Withers height (cm) Male	Withers height (cm) Female	Main users (in the order of importance)
15. Nkone	Zimbabwe	400		300–450			Meat, milk
16. Pedi	South Africa	400					Meat, work, milk
17. Shangani	South Africa	600					Meat, work, milk
18. Landim	Mozambique	536,400					Meat, work, milk
2.3 The Zambia/Angola cluster							
19. Tonga	Zambia	993,000	485–530	310–495	118–130		Milk, meat, work
20. Porto Amboim	Namibia	NA	400–530	350–425	120–140	118–129	Meat
2.4 The Ovambo and south-western cluster							
21. Ovambo	Namibia	NA					Meat, work, milk
22. Kaokoveld	Namibia	NA					Meat, work, milk
23. Okavango	Namibia	NA					Meat, work, milk
24. Caprivi	Namibia	NA					Meat, work, milk
25. Humbi	Angola	NA					Meat, work, milk
2.5 The Setswana group							
26. Barotse	Zambia, Angola	793,000	255–700	240–455	120–137	114–129	Milk, meat, work
27. Damara (Herero)	Namibia	NA					Milk, work, meat
28. Tswana	Botswana	1,395,000	310–520	290–420	140	131	Meat, work, milk
29. Tuli	Zimbabwe	3300	450–820	360–570			Meat
2.6 The Afrikaner group							
30. Afrikaner	South Africa	302,000	450–950	360–555	130–150	128–140	Meat, work

Source: Rege (1999)
NA = Not Available
[a]Latest available estimate; if multiple in the same year, the highest estimate used; combines estimates from different countries, if applicable

Fig. 4.17 A Danakil cow and a Raya Azebo herd. (Courtesy: AGTR, ILRI)

Ruzizi, and *Watusi* (FAO 2010). Ankole cattle are generally classified in the wider group of the Sanga cattle. It is considered that Ankole cattle, with their characteristic long horns, further evolved from the original Sanga around the present-day Uganda, developing further into the local strains/sub-populations (Felius 1995). This position is also supported by Banyankole mythology that Ankole cattle were domesticated in their current location by the Bachwezi, a semi-mythical people (Infield 2003). The central and eastern African 'Ankole cattle region' is also generally referred to as the inter-lacustrine or Great Lakes region, as it is bordered by Lakes Victoria, Kyoga, Albert, George, Kivu, and Tanganyika. The Bahima and Nsagalla strains are considered by some to be the same and should not be separated (Bett et al. 2013). These strain names were derived from the tribal names of their traditional owners. The *Bahima* are kept by the Bahima race of the Banyankole tribe of Uganda, the *Kigezi* are kept by the Banyakigezi/Bakiga tribe, the *Bashi* are kept by the Bashi people in the Democratic Republic of Congo, and the *Watusi* are reared by the Tutsi people of Rwanda. A particular lineage of the Watusi called Inyambo is descended from the Tutsi kings who ruled the Rwanda-Urundi kingdom and is currently reared in Eastern and Northern Rwanda. The *Ruzizi* is found in the Ruzizi Valley between Lake Tanganyika and Lake Kivu. In Uganda, the Ankole is the prevalent type of cattle, found in the western, central, and south-western parts of the country (Nakimbugwe and Muchunguzi 2003). The multiplicity of strains notwithstanding, the Ankole are mainly owned by the Bahima pastoral communities (Kajura 2001) in the districts that form the Ankole region (Lamwaka 2006) and other districts of the cattle corridor.

Physical characteristics: Ankole cattle typically have a rusty dark red coat colour, but dun and black coat colours are not uncommon. The coat colour is usually solid, but they can also be spotted or speckled. Both males and females have large horns, which spread up to 2.4 metres long and are considered the largest horns of any cattle in the world. The body is of medium size, the neck is relatively short, and the legs are weak.

Production characteristics and uses: The Ankole is used for meat and milk production, draught power, and many cultural functions such as funerals, weddings, or religious sacrifices (Infield 2003; Kugonza et al. 2011). Cows have excellent maternal abilities. Ankole meat has low fat content compared to that of other breeds in the region and specialized beef breeds. The skin was traditionally used for making cultural regalia like drums, stools, sandals, and clothing. Indeed, the Ankole cattle hold a central role in the culture and lives of the Bahima (Bonte 1991; Schoenbrun 1993), and their economic contribution is critical to the livelihoods of these communities (Infield 2002). The height of mature animals ranges from 165 to 198 cm. Birthweights range from 4.7 to 6.8 kg. Mature males weigh 450–720 kg, and females weigh 405–540 kg. Milk yield is about

500 litres over a 240-day lactation period (DAGRIS 2022). At 7%, the butterfat content of Ankole milk is quite high. In areas of land scarcity (Burundi, Rwanda, and western Uganda), a clear trend from pure Ankole cattle towards crossbred animals can be observed, and recent studies (Wurzinger et al. 2006; Kugonza et al. 2011) show that these cattle are increasingly under threat from crossbreeding with taurine breeds.

Adaptive or special genetic characteristics: The Ankole are hardy and efficient grazers, with the ability to thrive on rough forage. An attribute commonly mentioned by Ankole keepers is their (general) disease resistance.

Ankole were first introduced to the United States in the 1960s from a small seed stock in European zoos. Today, fewer than 1000 purebred Ankole animals are registered in the United States. Figure 4.18 shows typical Ankole animals with their characteristic horns.

The Nilotic Group

This group is composed of the *Abigar, Aliab Dinka, Aweil Dinka, Nuer*, and *Shilluk* cattle breeds.

The Abigar

Distribution and population statistics: The home of Abigar cattle is the lowlands of south-west Ethiopia, around the border with Sudan, along the White Nile, in the area covering the Akobo area of Gambella. It is considered a sub-type of the Nuer cattle (Rege 1999) and is reared and maintained by the Nuer tribes of Gambella Regional State. Abigar is the dominant cattle type in the region and plays a major role in the socio-cultural and economic life of pastoral and agro-pastoral communities that depend on it. The breed has a population of 548,600 heads.

Physical characteristics: The Abigar cattle possess a wide array of coat colour patterns, with white/cream being the most dominant. Mean heart girth circumference is 142.5 cm in females and 148.9 cm in males, and the height at withers is 114 and 118.3 cm, respectively. Figure 4.19 shows an Abigar and a herd.

Production characteristics and uses: Abigar cattle are mainly reared for milk and meat and as a source of farm power. The daily milk offtake per cow is 1.3–3 litres (Minuye et al. 2018), the mean lactation length is 8–9 months, and the lifetime calf crop production is 7.4–10 calves (Mureja 2002). The breed is the major source of milk and milk products for the people in the Gambella Region. The average age at puberty has been reported to be 36.2 months, while the age at first calving and calving interval were 42.5 and 14.1 months, respectively, under extensive management (Minuye et al. 2018). Overall, the breed is a productive breed, considering its harsh environment—disease prevalence and frequent droughts—and the extensive management. Abigar animals are docile and easy to handle.

Adaptive or special genetic characteristics: Abigar is known for its tolerance to tsetse fly challenge, high heat load, and periodic flooding.

The *Aliab Dinka* breed is found in south-east Sudan and the *Aweil Dinka*—the smaller-sized of the two—in the north-west territory of Sudan (Rege 1999). The Aliab Dinka cattle are owned by a tribe of the same name—a subdivision of the Dinka people of South Sudan. They traditionally lived in an area west of the upper White Nile River. The cattle have white, grey, or light fawn coat colours, large bodies, long horns, and small cervico-thoracic humps.

The *Shilluk* breed neighbours with the other strains and is more diffuse. These cattle are owned by the agro-pastoralist, Nilotic-speaking Shilluk people, who form part of the larger Luo ethnic group of South Sudan. They live on both banks of River Nile, around the city of Malakal, and between Lake No in the south and Kosti in the north, extending on the east bank of the Nile as far as Anakdier in the east. The area has flat-lying plains with moderate rainfall. Vegetation is made up of thick, tall grass and a few trees and shrubs.

4.2.2 The Sanga of Southern Africa

This group of Sanga cattle is further subdivided into six breed clusters or groups, namely, the

Fig. 4.18 Ankole bull and herd showing their characteristic lyre horns. (Courtesy: Dr. Donald Kugonza)

Shona subgroup, the *Nguni* cluster, the Zambia/Angola cluster, the *Ovambo* and south-western cluster, the *Setswana* group, and the *Afrikaner* group. The Afrikaner group has only one breed, the Afrikaner breed, which is located primarily in South Africa with a population of 302,000 heads.

The Shona cluster is made of the Mashona cattle breed that is mostly reared in east and central Zimbabwe and estimated at 500,000 heads. The mature weight for Mashona bulls ranges between 350 and 635 kg, while that of mature cows has been estimated at 260–410 kg (Rege 1999) and 275–350 kg (Khombe 2002). The median calving interval of Mashona cows is 548 days (Hall 1998). The cattle are reared for meat and as work animals. The dominant coat colour is black, with secondary colours, including red and various shades of brown. Generally, these cattle are closely related to the other Sanga in the sub-region—that is, Nkone, Tuli, and Nguni (Khombe 2002).

The Nguni cluster: The Nguni group of cattle is considered by some as the 'true Sanga cattle of southern Africa', historically owned by the Bantu (the Nguni people represent a south-eastern Bantu people) on the coast. These cattle descended from the original Sanga that were introduced into southern Africa when the Khoikhoi (Hottentots) and their Sanga cattle first crossed the Zambezi River about 700 AD. The Nguni cattle in South Africa and the Kingdom of Eswatini number about 2.2 million heads. The Nguni cluster is made of four breeds or strains, namely *Landim, Nguni, Nkone, Pedi*, and *Shangan*. The Nguni cattle have a mean weight in male cattle of 400–680 kg and 225–450 kg in females. The cattle are reared for meat, milk, and work (Rege 1999).

The *'Landim'* strain is found in the eastern part of the Kingdom of Eswatini and the Zululand territory of South Africa. In Mozambique, the Landim cattle are kept in areas of the plains of Gaza, in the Inhambane coastal area, and in Maputo Province. Small herds are also found on the northern shores of the Limpopo River near Zimbabwe, where they interbreed with the Mashona cattle (Alberro 1983). These cattle survive on marginal land and are tolerant of local diseases and adapted to a hot environment. Landim cattle have a wide array of coat colours and patterns but stand out as black, white, dark brown, brown, and white (Kotze et al. 2000). The breed is threatened by extensive crossbreeding and replacement with exotic cattle (Bessa et al. 2009). A Landim herd and a heifer are shown in Fig. 4.20, while a typical Nguni bull and cow are shown in Fig. 4.21.

The Zambia/Angola cluster includes the *Tonga* (Zambia) and *Porto Amboim (*Namibia*)*, while another cluster called the *'Ovambo and south-western cluster'* includes the *Ovambo, Kaokoveld, Okavango, and Caprivi* strains; all of Namibia; and the *Humbi* strain found in Angola.

The Tonga

Distribution and ecological settings: The Tonga breed is found in the Southern Province of Zambia, on the Zambezi Plateau, between the

Fig. 4.19 An Abigar bull and a herd of Abigar cattle. (Courtesy: AGTR, ILRI)

Kafue and Zambezi rivers and Northern Zimbabwe, and the valley areas surrounding Lake Kariba. The Tonga breed is named after its owners, the Bantu-speaking Tonga people who inhabit the southern portion of Zambia (and the neighbouring areas of northern Zimbabwe and Botswana). Estimated at more than one million in the early twenty-first century, the Tonga people are concentrated along the Zambezi Escarpment and along the shores of Lake Kariba.

Butonga, the core Tonga area, lies between 16° and 18° S and between 26° and 29° E and is bounded on the north by the Kafue and Sanyati rivers, in Zambia and Zimbabwe, respectively, and on the south by the Zambezi and Gwai rivers. It includes the southern Zambian plateau, which rises to more than 1000 metres above sea level. The Middle Zambezi Valley is known as Gwembe Valley. Mean annual rainfall ranges from 80 cm on the escarpment to 40 cm in northern Gwembe Valley. Drought years are frequent, and rains are expected to start by mid-November, tapering off through March to April, as the cold, dry season sets in. Light frost may be experienced in the June–July period, before the hot, dry season sets in abruptly about late August, with temperatures that can reach 45 °C in northern Gwembe.

The Tonga cattle breed is believed to have been in Zambia even before the Bantu migration from East Africa. Due to the continuous interbreeding with exotic breeds, the original Tonga breed is now concentrated around the valleys of the Zambezi River, where the conditions are unfavourable for the other breeds of cattle (Mwaanga and Parés-Casanova 2017). The breed is made up of a mixture of various ecotypes found in different districts of the province. Because of this variation, Tonga is yet to be registered as a distinct breed with the Herd Book Society of Zambia. Due to its adaptability to the local environment, the Tonga breed is widely used in crossbreeding with exotic breeds to produce crossbred beef and dairy animals. The habitat of the Tonga is no longer limited to the Southern Province of Zambia. It has continued to migrate northwards as the Tonga people move in search of new farming areas. This represents a major threat for the breed through the uptake of commercial cattle breeding, interbreeding with other cattle populations in their new locations, and the impact of East Coast fever in the province. As more people acquire high-producing exotic bulls at the expense of local Tonga bulls that are deemed inferior, the indigenous Tonga is gradually being replaced through upgrading. Rege (1999) reported a population estimate of Tonga cattle at 993,000.

Physical characteristics: The differences between the ecotypes found in the different districts make the breed quite heterogeneous phenotypically. Generally, Tonga cattle have a small body size and short-to-medium-sized horns that spread outwards from the head (Musimuko 2014). The coat colour is variable, ranging from pure black to white, but brown tends to dominate in most herds, especially those found on the pla-

Fig. 4.20 A Landim cattle herd and a heifer in Mozambique. (Courtesy: Maria da Gloria Taela)

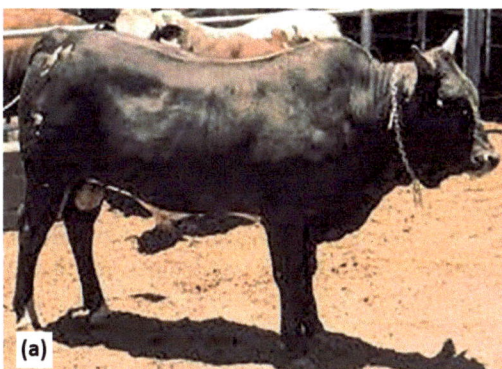 

Fig. 4.21 A typical Nguni bull (**a**) and a cow with a calf (**b**). (AGTR, ILRI)

teau areas of Sinazongwe and Maamba. The ecotypes found in the valleys tend to be relatively larger than those found on the plateau. One ecotype found on the shores of Lake Kariba in Kalomo District is locally referred to as the Mazambezi and is considerably larger than the upland strains. A herd of Tonga cattle is shown in Fig. 4.22.

Uses and production characteristics: The Tonga is mostly kept for meat, although, as is common among pastoral communities, the Tonga people will rarely slaughter an animal for food or sale unless there is a funeral or wedding in the family. Because of its small body stature, the Tonga breed is not well-adapted as a work animal. However, nearly all farmsteads have a set of working animals that are used for ploughing and the transportation of produce to and from markets.

Adaptive or special genetic characteristics: Due to its adaptability to the local production environment, the breed is widely used in crossbreeding with exotic breeds to produce crossbred beef and milk animals.

The members of **the Setswana group** are the *Barotse, Damara (Herero), Tswana*, and the *Tuli* cattle. These are summarized below.

The Barotse Cattle

Distribution and population statistics: The homeland of the Barotse cattle is Barotseland of the West Province of Zambia and the Kalahari and Namib Deserts in the neighbouring Angola, Namibia, and Botswana. The territory is characterized by the floodplain of the Zambezi, extending into Angola on the west. The breed is kept principally by the Lozi people living on the Zambezi floodplain and is also known as Lozi,

Rowzi, Rozi, or Baila. The Barotse population appears to have expanded after the rinderpest epidemic (Felius 1995). There are a number of strains of the Barotse from the various districts: the Shang'ombo strain—more closely related to populations in Angola—tends to be relatively larger than the Kalabo and Lukulu types. The Barotse population has been estimated at 793,000 (Rege 1999), and the population in Zambia has been estimated to be less than 100,000 (DAD-IS 2005). The majority of this breed is found in Angola. High mortality caused by contagious bovine pleuropneumonia (CBPP) and the effects of indiscriminate crossbreeding and drought are the major risk factors for the breed.

Physical characteristics: The Barotse has typical Sanga conformation with long horns and legs; they are coarse-boned and have a large, long appearance due to their long legs (Felius 1995). The main distinguishing features of Barotse cattle include the large, long but relatively slender body frame with long lyre-shaped horns that spread and curve backwards (Musimuko 2014). Despite the large body frame, it is rare to see a fat, well-rounded Barotse animal.

The coat colours are usually brown, dark red, fawn, or black, sometimes mixed with white. White or brown patches on the sides and under the belly are not uncommon. Pure white colour has not been reported. The Barotse hump is small, often barely visible in the cow. The height at withers of mature male and female animals are 120–137 and 114–129 cm, respectively (Mason and Maule 1960). The hump is small in males and barely visible in cows. A herd of Barotse is shown in Fig. 4.23.

Production characteristics and uses: Barotse cattle are used as beef and milk animals, and the sour milk made from Barotse cow milk (the *Muzilili*) is known for its special taste and flavour. However, beef is the most important product. The animals are docile with good temperament, and hence, the bulls are good work animals. Under good management, they weigh between 485 kg (for females) and 700 kg (for bulls), compared to weights of 400 kg for females and 580 kg for males under average management. The age at first calving is about 44 months, while the lactation milk yield is 1077 litres in an average lactation length of 302 days. A conception rate of 67% and a calving interval of 620 days have been reported (Thorpe et al. 1981).

The Baila

There is debate on whether the Baila of the Kafue floodplains in the Southern Province of Zambia is a separate breed or not, and despite having distinct features, the Baila is generally considered to be an intermediate crossbreed between the Barotse and Tonga cattle breeds. It is not among the recognized breeds of cattle in Zambia. This could be because it only accounts for about 1% of the cattle population in the country (Kaluba 1984). They are mainly found along the Kafue floodplains in the Mumbwa and Namwala districts of Zambia. The animals have a relatively large bone frame. Their large horns are similar in shape but shorter than those of the Barotse breed and longer than those of Tonga. They have a clearly defined cervical hump. It is a meat breed. The animals have good heat tolerance and disease resistance abilities (Mwaanga and Parés-Casanova 2017).

Tswana or Setswana Cattle

Distribution and population statistics: It is considered that the Tswana people settled with their livestock in the Ngami region of Botswana early in the nineteenth century and that, over time, their cattle displaced the original Ngami cattle (Rege and Tawah 1999). Today, Botswana is the epicentre of the Setswana cattle, and the most typical Tswana cattle occur mainly in the north and north-west of Botswana in Ngamiland and along the Boteti River. Tswana-type cattle extend into the south-western Zimbabwe. In South Africa, they were originally found in the Northern Cape of Transvaal. The population of Tswana cattle in Botswana was reported at 1.4 million heads by Rege (1999). No estimate is available on the population size of the breed in South Africa and Namibia. The purity of the Tswana breed is declining due to crossbreeding with the Brahman and exotic (European) breeds (DAD-IS 2005). Other names used locally for the breed include Bechuana, BataWana, Mangwato, Sechuana, Seshaga, and Sengologa.

Fig. 4.22 A herd of Tonga cattle. (Courtesy: DAGRIS, ILRI)

Fig. 4.23 A herd of Barotse cattle. (Courtesy: Dr. Denis Lembani)

Physical characteristics: Today, the Tswana cattle are quite phenotypically heterogeneous. Typically, they have long, lyre-shaped horns, but dehorning is a common practice, especially on commercial farms. The Tswana are well-built and well-fleshed, with moderate to long legs. They are multicoloured with colour patterns of black, brown, grey, and white, the typical ones being red and red and white. A herd of Tswana cattle is shown in Fig. 4.24.

Production characteristics and uses: Tswana cattle are mainly used for beef, milk, and draught power. Their beef is considered to be of high quality. They are fertile, and the females are known for their ease of calving, and the quality of hides is good. Calving percentage is 70–81%. The average age at first calving is 36–48 months, and the calving interval is 18–24 months in traditional herds. The Tswana outperforms other South African breeds including Tuli and Afrikaner in growth traits, but the breed has lower growth rates than the Bonsmara and its crossbreeds. The mature weight of male animals is 310–520 kg and that of females is 290–420 kg (DAGRIS 2022). The average daily milk yield of a Tswana cow is 1.4 kg. They are used for crossbreeding with the exotic dairy and beef bulls for improved milk and meat production. Calf mortality is about 7.5%.

Adaptive or special genetic characteristics: The breed is known to have a high level of tick and heat tolerance and is adapted to dry and hot environments.

Damara or Herero Cattle The Himba, a Herero-speaking people, settled during the sixteenth century in North-western Namibia with their Sanga cattle. Today, this area is known as Kaokoland. The Himba have a nomadic lifestyle and roam freely in search of water and grazing for their livestock. Although the Herero primarily reside in Namibia, there are also significant populations in Botswana and Angola. Not much has been done to document the unique characteristics of the Herero/Damara cattle that differentiate them from the neighbouring Sanga cattle.

The Tuli

Distribution and population statistics: Originally called the *Amabowe* cattle of Ngwato Setswana type in the south-west corner of Zimbabwe, an area stretching from Tuli to Plumtree, the rinderpest epidemic of 1896 reduced the number of these cattle substantially. Much of what was left was crossed with the Afrikaner cattle. Remnant herds of the original population were restricted in the Tuli area; hence, the present-day name of the breed. Today, the Tuli is distributed in several countries in the region, especially in the neighbouring Botswana and South Africa. It has been used in forming the synthetic breed *Akouma* in Gabon. It has also been exported from Zimbabwe to Australia, the United States, and Canada (Rege and Tawah 1999; Mpofu 2002). The Tuli cattle population in the national commercial herds is about 10,000 in Zimbabwe and 2339 in South Africa (DAD-IS 2005). There are seven stud farms in Zimbabwe, and excess bulls from these farms are used in many commercial farms in the country. There is no estimate available on the number of Tuli cattle held by smallholder producers. However, the existence of a breed society and sustained efforts to improve the breed and popularize it internationally have significantly reduced the chances that the breed could be endangered.

Physical characteristics: The Tuli is a medium-sized 'pure African Sanga' breed and is one of the few that has benefitted from organized breeding: The Tuli breed society was formed in 1961, and several stud breeders are operational in Zimbabwe (Mpofu 2002). The Tuli is a heavy-boned animal with long, wide, spreading horns and long, strong legs. The coat colour is mostly red, red and white, and golden brown. The Tuli have been selected for their golden brown colour and with selection preference for polled animals, and selection against shyness and poor breeding. A Tuli herd and bull are shown in Fig. 4.25.

Production characteristics: The Tuli breed are characterized by a high calving percentage of 70.6%, mortality between birth to 2 years estimated at 7.5%, and a weight at 18 months of age of 284 kg. The Tuli cattle are good beef animals. Weights at birth, at 90 days old, at weaning, at the age of 18 months, and at 5.5 years were 31.9 kg, 105.8 kg, 189.7 kg, 284.1 kg, and 411.1 kg, respectively (Ward and Tawonezvi 1983).

Adaptive or special genetic characteristics: Today's Tuli cattle are characterized by high fertility, early maturity, hardiness, adaptability, ease of calving, good mothering ability, good carcass characteristics, and docility (Assan 2011).

The Afrikaner

Distribution and population statistics: The Afrikaner cattle are believed to have been brought to the present-day South Africa by the Khoikhoi people during the first century AD. The Hottentots owned large herds of Sanga-type cattle when the Dutch established the Cape Colony in 1652. The oxen drew the wagons, which carried Boer farmers and families on the Great Trek of 1835–1836 from the Cape of Good Hope to the Orange Free State, Natal, and the Transvaal to escape British rule. The present-day breed was developed from these original Hottentot cattle in Cape Province. Today, the breed is the most popular indigenous breed in South Africa (Maule 1990), being bred throughout the country, the main ones being North-east Cape, Orange Free State, Transvaal, and West Natal. The Afrikaner cattle have been exported to several countries in and outside Africa, including Namibia, Botswana, Swaziland, Zimbabwe, Zambia, Malawi, the Democratic Republic of the Congo, Australia, and the United States (Rege and Tawah 1999).

Physical characteristics: The Afrikaner coat colour ranges from dark to light red. The hump is

Fig. 4.24 A herd of Tswana cattle. (Courtesy: ILRI)

prominent but not large and is cervico-thoracic. The long spreading horns leave the head in a downward and backward direction and then, at maturity, bend gracefully forwards, upwards, and backwards. Through breeding, a polled type of Afrikaner has also been developed. Typical Afrikaner bulls are shown in Fig. 4.26.

Production characteristics and uses: The breed was developed as a draught animal (or *trek ox*), and it is only in the past half century or so that its beef potential has been exploited. It is among the largest breeds in Africa, with birthweights of male animals being about 33 kg. Adult bulls weigh 450–950 kg, with an average of 745 kg, and adult cows weigh 360–555 kg (average 525 kg). Weights of up to 1100 kg have been recorded in show animals (Rege and Tawah 1999). The age at first calving is about 36 months, and the calving interval is about 445 days. Calving percentages are highly variable and are lower than those of other indigenous breeds, e.g. Tswana, Tuli, Angoni, and Mashona.

Adaptive and special genetic characteristics: The legs of the Afrikaner are slightly sickle-shaped. The animals have good resistance to tick-borne diseases and are well-adapted to the local hot, arid conditions: the sweat glands in their skin are reportedly more active than those of taurine cattle, and this makes them more tolerant of heat than the European breeds.

4.2.3 The Pseudo-Sanga of West Africa

A distinction must be made between the general cattle group 'Sanga', which is used to mean the long-term, established Zebu × Taurine derivatives represented by several breeds in eastern and southern Africa (Sects. 4.2.1 and 4.2.2 above) and the more recent West African Zebu × Taurine crossbreeds described in this section. Even though we have classified the West African recently stabilized Zebu × Taurine crosses as Sanga here, to avoid attendant confusion, the use of *pseudo-Sanga* in reference to the more recent Zebu × humpless crossbred populations is advised. Pseudo-Sanga is used to connote two things: the first is to emphasize the heterogeneity of the genotype compositions of these populations, which are constantly changing as the crossbreeding processes continue, and the second is to distinguish these dynamically evolving populations from the Sanga of eastern and southern Africa, which are more stabilized and some of which are registered as unique, commercial breeds. Table 4.4 summarizes the breeds/strains of the present-day *pseudo*-Sanga cattle populations of West Africa, indicating the parental breeds from which they were (and are being) derived, and the countries with major populations of the breeds/strains (Rege 1999). The major

Fig. 4.25 A Tuli herd and a bull. (Courtesy: Tuli Cattle Society of Southern Africa)

groups/clusters of the pseudo-Sanga of West Africa are described below.

The Evolving Pseudo-Sanga Strains

The Sanga of West Africa include the ***Borgou, Méré, Ghana Sanga, and Keteku*** (or ***Borgu***). Earlier reports (e.g. (Doutressoulle 1947; Flamigni 1951)) used the term ***Borgu*** to refer to all forms of humpless × Zebu cattle crossbreeds in this region. In reference to crossbreeds of this type (Doutressoulle 1947), stated that '*…these same animals are encountered in the corresponding regions of the neighbouring Togo, Nigeria, Ivory Coast (**Méré-Lobi**), with small differences due to environment and* selection'. The same report states that the Borgu is known by different names such as Borgowa, Kettije, Ketaku, Keteku, Ketari, and Kaiama. Epstein (1971) states that '*… the general conformation of the body as well as the occasional occurrence of relatively long horns in the Borgu indicate the influence of White Fulani and possibly N'Dama Longhorn blood*' and that '*… in Western Nigeria, this type of cattle, called Keteku, is derived from a mixture of the humpless Dwarf Muturu of the south and the Zebu of the north*'.

The definition of Keteku has become more problematic in recent years, with an increasing proportion of Zebu blood in 'Keteku' herds. As Fulani pastoral herds push even further south and increasingly inhabit regions previously restricted to trypanotolerant stock, more Zebu stock are brought into these herds from village Zebu herds. For example, the 'Biu', a Zebu × Savannah Muturu cross found near Biu in southern Borno (e.g. (Gates 1952)), has effectively become submerged in the local Zebu gene pool. Thus, the application of the name Keteku to an individual animal may reflect as much the owner's cultural background as its actual genetic composition. The population size of Keteku in Nigeria was 180,000 heads (ILCA 1979a). Keteku are significantly less common than previously thought, and their distribution is quite different. It is unlikely that there are as many as 100,000 of all types. The distribution and productivity of **Keteku** cattle have been studied in more detail (Blench et al. 1998).

Epstein (1971) also recognizes the ***Biu*** of the Bornu State of Nigeria as a type similar to the Borgu but smaller. Gates (1952) states that the Biu is the result of White Fulani crossbreeding with the humpless Dwarf Shorthorn. Mason (1988) and Maule (1990) recognize the Biu as a distinct breed; however, Mason (1988) and ILCA (1979a) observed that the Biu has been absorbed by Zebu. The general consensus is that the Biu, when it existed, differed from the Keteku as it is known today: the Keteku is a broader mixture of Zebu (mainly White Fulani) crosses with N'Dama and/or Shorthorns, while the Biu was more specifically defined as a cross derived from White Fulani and Dwarf Muturu. Reference has also been made to the characteristic long horns and white with black colour markings of the Biu (Epstein 1971; ILCA 1979a). Today, large num-

Fig. 4.26 Afrikaner bull and herd. (Courtesy: Afrikaner Cattle Breeders' Society of South Africa)

Table 4.4 The pseudo-Sanga of West Africa[a]

Breed/strain	Parental breeds/strains	Country	Population[b]
Borgou	Zebu × Somba or Lagune	Benin, Togo	380,000-580,000
Méré	Zebu × Lobi or Baoulé (and N'Dama)	Burkina Faso, Côte d'Ivoire	600,000-700,000
Ghana Sanga[c]	Ghana Shorthorn × Zebu	Ghana	120,000-125,000
Keteku	Muturu × White Fulani, 'Borgu', Kaiama	Nigeria	100,000-300,000
Biu	Dwarf Muturu × White Fulani	Nigeria	NA

[a]'Pseudo-Sanga' is suggested here to distinguish the West African group, which is more recent, still highly variable, and under development, from the more established Sanga breeds of eastern and southern Africa, which are largely recognized as distinct breeds
[b]From Rege (1999)
[c]Occasionally N'Dama × Zebu crosses; Zebu used is principally White Fulani but occasionally Sokoto Gudali

bers of Keteku are found in Kwara State of Nigeria, where they are thought to be an extension of the Borgou population of Benin.

ILCA (1979b) points out the difficulty of identifying a standard type of the Borgou breed in Togo since all intermediate types between the humpless breeds and Zebu are commonly grouped under this name. However, the same report defines the Borgou in Benin as a cross between Zebu (mainly White Fulani) and West African Shorthorn (Somba or Lagune). The *Méré*, like the Togolese **Borgou**, refers to a whole range of crossbreeds between Zebu and humpless breeds (including the N'Dama) in both Burkina Faso and Côte d'Ivoire. In Burkina Faso, the Méré is found mainly in a belt from the west to the east, including Bobo Dioulasso and Koudougou, which widens parallel to the southern border, south of Koupela and Fada N'Gourma. In Côte d'Ivoire, there is little information about the Méré crossbreeds or their numbers, although crossbreeding is extensively practised in the North. Figure 4.27 shows a typical Keteku cattle herd.

The Ghana Sanga is a cross between Zebu and Ghana Shorthorn. The Zebu breed involved is commonly White Fulani, but occasionally, Sokoto Gudali is used. Similarly, the N'Dama occasionally replaces the Ghana Shorthorn in the cross. The term is also used to include similar crosses involving the Dwarf Shorthorn. The Ghana Sanga resembles the Shorthorn type more closely than it does the Zebu, but, like the Zebu, it is found in the drier areas of the country towards the northern border and on the Accra Plains, extending into the Volta Region.

In addition to the above stabilized crosses, there are other Zebu × Taurine (mostly N'Dama) crosses in the region. These include the **Djakore** or **Djokore** of Senegal, **Bambara** (also known as *Méré*) of Mali, and various gradations of Zebu × N'Dama crosses in Guinea, the Democratic Republic of the Congo, The Gambia, and Gabon. In the Central African Republic, these crosses are also called Bambara. Maule (1990) defines the

Bambara as a *'variety of N'Dama with Zebu blood'*. Other than in Mali and the Central African Republic, the name **Borgou** is widely used to refer to various strains of Zebu × humpless crossbreeds in French-speaking countries. In Ghana (Sanga) and Nigeria (Keteku and Biu), more specific names are used. In all cases, however, care needs to be taken to ensure that, as much as possible, a population ('breed' or 'strain') in question is accurately defined in terms of its composition and location.

Production systems: As alluded to above, there are four broad types of Zebu × Taurine (mostly Shorthorn) crossbreeds: the *Méré* in Côte d'Ivoire and Burkina Faso, the *Ghana Sanga* in Ghana, the *Borgou* in Togo and Benin, and the *Keteku* (or *Borgu*) in Nigeria. The common environment of these populations is the intermediate Sudanian zone between the Zebu-dominated north and the tsetse-infested south, which provides the greatest opportunity for interbreeding between the two groups. This is an important agricultural area, producing groundnuts, cotton, and sorghum. Farming is intensive, and the livestock density has traditionally been high. Livestock are confined to fallow land or areas unsuitable for cultivation during the rainy season. In the dry season, animals are left to graze freely, and crop residues, mainly cereal straws and groundnut haulms, make important contributions to their diet. Fodder is not usually grown, but there is an abundance of groundnut stalks and standing hay. When the cattle owners or herders are Fulani, the animals are managed in a similar way to any Fulani herd. Draught animals are widely used to pull carts and ploughs in many of these areas, such as Sine Salown in Senegal, southern Mali, southern Burkina Faso, and the province of Borgou in Benin. Cattle are also used for manuring fields by letting animals feed on crop residues in the field. In some areas, animals are tethered in the fields. Herds are often large, and the proportion of males in the herds is high (28–32%).

Physical characteristics: These populations, being unstabilized crosses in which the injection of new genetics continues, exhibit considerable variations in size, conformation, and colour markings. The Keteku and the Borgou are generally very similar, although their sizes usually vary with the amount of Zebu breeding. In Nigeria, for example, the Keteku is larger and taller in the northern Guinea savannah than it is in the southern Guinea savannah. Likewise, in Benin, the Borgou varies in size from the small 'true' Borgou of the south to the large Borgou–Zebu of the north. In body conformation, the Keteku is of fair depth, although it is inclined to lack width at the chest and to be flat over the ribs. The Borgou has an elliptic and compact body with a straight profile, while the Ghana Sanga and Borgou–Zebu have similar conformations as that of the Zebu (Domingo 1976). The head of the Keteku is generally well-proportioned, straight, and 'dished' in profile, while that of the Ghana Sanga has a long, straight, and convex profile. The head of the Borgou is generally triangular and thick at the extremities, but it is long with a flat forehead among the small Borgou. Horns are of medium length in the Méré, but in the Ghana Sanga, they are long and U-shaped, as in the Bunaji. The necks of both the Borgou and Ghana Sanga are short and sturdy. The neck of the Borgou is thick and robust in the males and skinny in the females, and the neck of the small Borgou is very muscular around the shoulders as well. The top line in the Keteku slopes up to the sacrum, with the rump tending to slope steeply between the hooks and the pin bones. The top line of the Borgou is generally not straight, and the rump is short. The top line of the small Borgou is straight, with a long and narrow back. In contrast, the back of the Ghana Sanga is short and concave, like that of the Borgou–Zebu, with an elevated rump. The hump is rudimentary and usually inconspicuous in these populations. It is cervically positioned in the Keteku, but with increasing White Fulani breeding, the hump tends to be larger and cervico-thoracically positioned. The dewlap and umbilical folds are retracted and poorly developed in the Keteku, although they are better developed in the Borgou–Zebu, depending on the degree of Zebu breeding. The chest of the Borgou is particularly deep in the males, and it is very solid in the Borgou–Zebu, which has a short and robust body. The hindquarters of the Keteku are poorly developed,

Fig. 4.27 Typical Keteku (**a**) and Djakore (**b**) cattle herds. (Courtesy: AGTR, ILRI)

with the thighs lacking in width and fullness, although it has a broad hip (40 cm in males and 39 cm in females). The tail-head is set high, and the legs are averagely long, with light and fine bones. The small Borgou has short legs, and the females have well-developed udders. Coats are similar in colour markings and patterns in the Keteku, Borgou, and Ghana Sanga, but considerable variation exists within each of these populations, with pure white being the dominant colour, while black is the most common in the Ghana Sanga and white with black spots and black markings in the Keteku and Borgou. Solid black or black and white or spotted grey or fawn coats are also evident in the Borgou. A variety of other colours, including black, black and white, faded red, red and white, or blue tinge, also exist in the Keteku. Méré cattle have mostly typical pure black coats, although black and white or brown and white coats are also common.

Adaptive and special genetic characteristics: The Keteku is more susceptible to trypanosomosis and dermatophilosis than the N'Dama and the Muturu in the same locality, but, like the Ghana Shorthorn, it is more tolerant than the White Fulani or the Gudali. Borgou cattle, like the Keteku, are adapted to the harsh conditions of production and are tolerant of trypanosomosis. Some studies (e.g. Doko (1993) and Dehoux and Hounsou-Ve (1993)) have demonstrated that the Borgou has trypanotolerant attributes that are similar to those of the Lagune and N'Dama but with a higher degree of within-breed variation.

Production characteristics: Birthweights of Borgou calves are slightly lower than those of the Keteku under ranch conditions. Birthweights of the Méré calves on-station in Burkina Faso have averaged 18 and 17 kg, respectively, in males and females. Corresponding weights at weaning (about 8 months of age) have averaged 128 and 112 kg. The Keteku cattle raised on a ranch in Nigeria with heavy tsetse challenge had a higher growth performance than the Borgou cattle raised on a ranch in Benin with light tsetse challenge. This points to a greater tolerance to trypanosomosis of Keteku than Borgou cattle. Mature weights have also varied considerably across breed types. Typical mature weights have ranged from 190 to 280 kg for the Borgou cattle in Benin, from 295 to 330 kg for the Keteku cattle in Nigeria, and from 260 kg for the Méré cattle in Côte d'Ivoire to 320 kg for the Ghana Sanga cattle (ILCA 1979a). In these systems with limited animal recording, the considerable variation in mature weights may reflect the lack of knowledge of precise age at maturity. However, in general, these stabilized crosses are heavier at maturity than their humpless parental stock but lighter than their humped parental stock. The age at first calving in the Borgou cattle in Benin has ranged from 43 months in sedentary village herds to 46 months in transhumant herds, averaging 43.5 months, and has varied from 38 to 47 months for the Keteku cattle under ranch conditions in Nigeria and from 48 to 60 months for the Méré cattle under village conditions in Burkina Faso. The mean calving intervals for the Borgou cattle in Benin have ranged from 15 months under sedentary husbandry to 18 months under transhumant husbandry and averaged 16 months. Corresponding figures for the fecundity rate were 64 and 67 per cent, with an average of 65 per cent. Surprisingly, reproductive performances under these different systems of production reported in the literature have tended to be quite similar, which does not support the suggestion by Wilson (1988) that transhumant herds are relatively better nourished and, therefore, are expected to perform better than sedentary herds. The mean calving intervals have varied from 15 to 20 months in Borgou cattle and from 18 to 24 months in Méré cattle under village conditions. For Keteku cattle, they have ranged from 16 to 19 months under ranch conditions. The calving rate in the Borgou cattle in Benin has ranged from 33 to 75 per cent under ranching management at M'Betecoucou, a location characterized by light tsetse challenge (ILCA 1979a, b; Lopez 1985), and from 40 to 67 per cent under village conditions (Dehoux and Hounsou-Ve 1993). These differences may partly be the result of variations in the levels of trypanosomosis challenge. The bimodal calving frequency in village herds in Benin (ILCA 1979b) is an indication that the breeding in Borgou herds is seasonal in character, which is a direct effect of nutrition on the fertility in traditional village herds. In addition (Dehoux and Hounsou-Ve 1993), reported that about 45 per cent of the Borgou herds in Benin have only one young breeding bull. Since young bulls can only breed 30 cows per year at most, the herd fertility is bound to suffer in these systems where herd sizes are large. Other factors contributing to the poor fertility of Borgou cows in the village herds in Benin are prolonged suckling, trypanosomosis, brucellosis, and other infectious diseases.

Population sizes: While this chapter does not put emphasis on the population sizes of the breeds covered, the dynamism in population sizes of West African cattle breeds, especially the rapid interbreeding between Zebu and Taurine types and the consequences of this, requires special mention. Population sizes and trends are discussed in Chap. 3. As pointed out in Sect. 3.4 of that chapter, the stabilized crosses require systematic characterization, with a focus on gathering both phenotype and genotype data on the production and adaptive traits of these new 'breeds' or 'strains' relative to those of the parental types, as part of a process to determine the attributes to define the end point of the envisioned new breeds and breeding plans towards genetic stabilization.

4.3 The Zenga Cattle

The 'Zenga', a term introduced by Rege (1999), refers to the cattle derived from the interbreeding between Sanga and Zebu cattle. The Zenga is heavier than the Zebu but lighter than the Sanga (Kajura 2001). Exclusively located in eastern Africa, some members are found in predominantly the Zebu habitat and others in the Sanga habitat. Indeed, the location of Zenga forms a natural division between the Zebu country in the north and the predominantly Sanga country in the south of the African continent. The members of the Zenga, a total of eight, are the Arado, Fogera, and Horro (of Ethiopia); Jiddu (southern Somalia); Alur, also called Nioka (Nyoka) or Blukwa cattle (the Democratic Republic of Congo), derived from crossing Ankole cattle with Small East African Zebu in the area (that is, Lugware and Nkedi); Nganda (Uganda); Sukuma (Tanzania); and Bovines of Tete (Mozambique). Table 4.5 presents a summary of the classification of the present-day Zenga breeds/strains of eastern Africa. They are briefly described below.

Arado Cattle

The Arado cattle are red- and black-coated and are small-bodied and hardy. They are the most common cattle variety in the north Ethiopian highlands, where the breed is reared mainly for draught power. Closely related to the Arado are the Abergele cattle, which are the smallest breed of north Ethiopia in the Abergele lowlands and at the south-western lower slopes of Dogu'a Tembien district. The Abergele ecotype is known for its adaptation to the hotter and drier lowlands, is tolerant of diseases and parasites, and can cope with feed shortages during long dry periods. Another closely related Zenga ecotype is the Begayt, which is currently being crossed with the Arado to increase the milk production of the Arado.

Fogera Cattle
The Fogera breed (Fig. 4.28) is found in the Fogera Plains, which lie to the east of Lake Tana, covering the area around the town of Woreta, between Bahir Dar and Gondar in Ethiopia. The area is a flat, open plain, across which the Rib River flows into Lake Tana, with the Gumera River forming the southern boundary. Among the typical characteristics of the Fogera cattle are the broad hooves, which allow it to move more easily in the marshes of the Fogera Plain. The breed is highly tolerant of heat stress and solar radiation and is also known for its tolerance to high altitudes, parasite and disease infestation, fly burden, wet soils, and swampy areas. The Fogera is considered at risk, and conservation interventions are being taken.

Horro Cattle
Also known as Wallega or Wollega cattle, the habitat of the Horro breed (Fig. 4.29) is the Horro Guduru Welega Zone of western Ethiopia. The population of cattle in the Horro area was estimated to be around 47,700 heads in 2017 (Mekonnen et al. 2021). The breed is mainly used as a draught animal and for milk production. The Horro have a large body size, small-to-medium hump size, small dewlap, fine skin, and uniform brown colour, which is lighter around the muzzle. The adaptation to their environment and the relative resistance to trypanosomosis disease make the Horro an important breed in its habitat.

Nganda Cattle
The Nganda, also known as Sese Island cattle, is a Sanga–Zebu intermediate, which was developed as a result of interbreeding the East African Shorthorn Zebu and the Ankole and the Alur cattle (Rege and Tawah 1999). It occupies Central Uganda and parts of Western Uganda, south of Lake Albert (Kajura 2001). The Nganda was developed mainly from the interbreeding of the Small East African Zebu in Uganda, mainly the Nkedi, and the Ankole Sanga (mainly the Baima but also Watusi in recent times). Indeed, the Ankole cattle have been mixed with the adjacent Bukedi Zebu over the past three centuries (Rege and Tawah 1999), and the interbreeding continues. Although it is not currently on the red or yellow list of threatened breeds, the Nganda deserves conservation attention. The only on-station open nucleus breeding herd for the breed in Uganda was disbanded and is being considered for reconstitution.

The Nganda breed habitat is the Buganda region in central Uganda and, hence, the name Nganda. In the area north of Lake Victoria and between Lake Kyoga (Busoga region), it is referred to as Nsoga type, and in the areas around Lake Albert (Bunyoro region), it is called the Nyoro type. This breed of cattle is also found in the tsetse-infested areas south of Lake Albert and east of Semiliki River. The bulls have characteristic wrinkles around the eyes.

The special characteristics of Nganda cattle are horns that are typically long and round in cross section and white and thick at the base; a hump that is small and cervico-thoracic in location; and the dewlap that varies but is generally moderate in size, sometimes appearing a little fleshy but has few folds. The coat colours are variable, with red, black, and brown being more common.

Nganda are used for milk and work and as meat animals (Rege and Tawah 1999). The adult body length of male Nganda cattle ranges between 283 and 416 cm, that of the female range between 300 and 338 cm, and the height at withers in male Nganda averages 116 cm and 115–122 cm in females. The Nganda cattle are hardy and have high reproductive rates, even under poor nutrition. On average, they produce a calf every year (DAGRIS 2022). A typical Nganda cow is shown in Fig. 4.30.

Sukuma Cattle
Sukuma cattle are a product of the interbreeding of various ecotypes of Tanganyika Zebu with the Ankole from adjacent areas of Rwanda, Burundi, and southern Uganda. The current habitat is lim-

Table 4.5 Zenga (Zebu–Sanga) cattle, recent derivatives, and synthetic breeds

Group and breed/strain	Location/country	Mature weight (kg) Male	Mature weight (kg) Female	Withers height (cm) Male	Withers height (cm) Female	Main uses (in the order of importance)
1. Zenga (Zebu × Sanga) cattle						
1. Arado	Eritrea, Ethiopia	205–430	192–350	117–144	96–126	Work, milk, meat
2. Fogera	Ethiopia			110–145	100–121	Work, milk, meat
3. Horro	Ethiopia	320–480	210–400			Work, milk, meat
4. Jiddu	Somalia	340–590	325–430	109–133	108–124	Milk, meat
5. Alur (Nioka, Blukwa)	DRC	290–550	225–380	110–131	108–122	Milk, meat, work
6. Nganda	Uganda	280–420	200–340	115–124	115–122	Milk, meat, work
7. Sukuma (Tinde)	Tanzania	230–410	210–370	94–132	93–127	Milk, meat
8. Bovines of Tete	Mozambique					Meat, milk, work
2. Recent derivatives						
1. Borgou	Benin, Togo	190–330	180–295			Meat, work, manure
2. Méré	Burkina Faso, Côte d'Ivoire					Meat, work, manure
3. Ghana Sanga	Ghana					Milk, meat, work, manure
4. Keteku	Nigeria					Meat, work, manure
5. Djakore	Senegal					Milk, meat, work, manure
6. Biu	Nigeria					Meat, work, manure
7. Basuto	Lesotho					Work, meat, milk
8. Barra do Caunzo	Angola	370–620	360–500			Meat, work, milk
9. Rana (Omby Rana)	Madagascar					Milk, meat
3. Systematic synthetics/composites						
1. Drankensberger	South Africa, Swaziland					Meat, work, milk
2. Bonsmara	South Africa					Meat
3. Renitelo	Madagascar	420–655	305–450			Meat, work
4. Manjan'i Boina	Madagascar					Milk
5. Mpwapwa	Tanzania	360–620	290–455			Milk
6. Wakwa	Cameroon					Meat

Source: Rege (1999)

ited to Tanzania's Sukumaland—south of Lake Victoria on the eastern border of Mwanza and the north-eastern parts of Shinyanga. The area covers Seke, Lalago, Nyalikangu, and the upper Semu River. The horns of Sukuma cattle are medium-to-long and crescent-shaped. They have small humps, small dewlap with thin skin, and a coat colour that varies between red, grey roan, light dun, and solid red. The animals are mainly used for milk and meat but also for work (Rege and Tawah 1999).

Bovines of Tete

Bovines of Tete (Fig. 4.31) were developed from the interbreeding of the Angonia Zebu and the Landim Sanga in Mozambique. Their habitat is the coastal area of northern Mozambique between Lake Malawi and the Indian Ocean. They are considered to have some level of trypanotolerance. They are primarily used for meat but also for milk and work (Rege and Tawah 1999). They are morphologically similar to the Landim but are smaller in size and have larger humps and lateral black horns. They can produce carcasses weighing 85–110 kg.

4.4 Commercial Composite Breeds

Sub-Saharan Africa (SSA) is home to at least five commercial composite breeds with varying proportions of exotic blood. Unfortunately, only one of them—the Bonsmara of South Africa—is secure in terms of numbers and the existence of programmes for continuing genetic improvement. The other four are the Mpwapwa of Tanzania (dual-purpose); the Mangan'i Boina of Madagascar (a dairy breed); the Renitelo, also of Madagascar (beef); and the Wakwa of Cameroon (beef). These breeds are summarized below (see also Rege and Tawah 1999).

The Bonsmara

The Bonsmara breed (Fig. 4.32) is a composite breed of cattle known for high-quality beef. Originating in South Africa as a scientific experiment of Professor Jan Bonsma, the breed was created from the mating between local Afrikaner cows and the European Shorthorn and Hereford bulls, with the final breed composition being approximately 5/8 Afrikaner, 3/16 Shorthorn, and 3/16 Hereford. The outcome of this scientific experiment is a breed that incorporates the hardiness of the Afrikaner and the beef traits of the European Hereford and Shorthorn cattle. The breed's name was coined from the researcher's name, Bonsma, and the research station, Mara. The breed was recognized in 1964, and the Bonsmara Cattle Breeders' Society of South Africa was founded the same year, and breed improvement programmes have been operating since then. The breed was registered in 1972. The Bonsmara has become the biggest of all beef and dual-purpose breeds in South Africa through concerted research undertaken to improve the breed. Bonsmara cows have excellent mothering ability under diverse environmental conditions and wean calves that are suitable for finishing on pasture or in feed-lots and have excellent meat qualities. The animals are well-adapted to warm bushveld and subtropical areas. They are also reported to be tolerant of ticks.

Fig. 4.28 Fogera cow (**a**), bull (**b**), and a cow with a calf (**c**). (Source: Tesfa et al. (2022)) (attributed to ILRI)

Fig. 4.29 The Ethiopian Horro cattle. (Courtesy: DAGRIS, ILRI)

Fig. 4.30 A typical Nganda cow. (Courtesy: Dr. Donald Kugonza)

Fig. 4.31 Herds of Tete cattle. (Courtesy: AGTR, ILRI, and Dr. Maria da Gloria Taela)

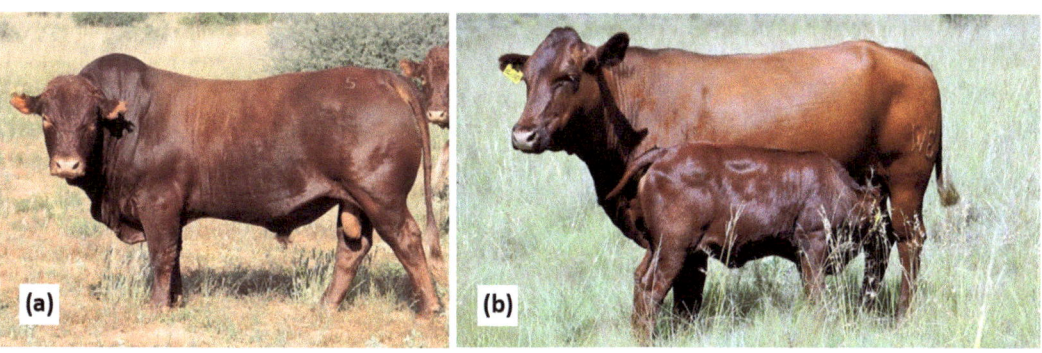

Fig. 4.32 A Bonsmara bull (**a**) and a cow with a calf (**b**). (Courtesy: Bonsmara Cattle Breeders' Society, South Africa)

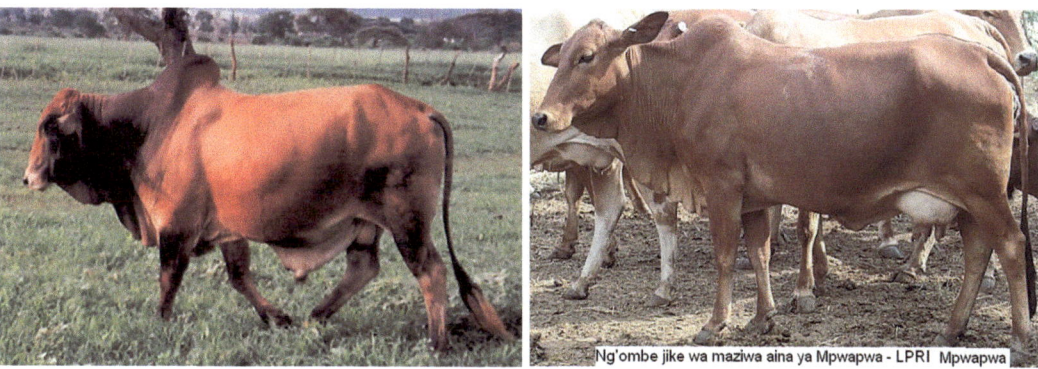

Fig. 4.33 An Mpwapwa bull (**a**) and a cow (**b**). (Courtesy: Aluna Chawala)

The Bonsmara has been exported to countries in and outside Africa where it has become well-established. Today Bonsmara is found in Zimbabwe, Namibia, Zambia, Swaziland, Botswana, Angola, Mozambique, Rwanda, the Democratic Republic of Congo, Australia, Argentina, Paraguay, Brazil, Argentina, and the United States, where its attributes under similar conditions are highly valued.

The Mpwapwa

The *Mpwapwa* breed (Fig. 4.33) was developed as a composite dual-purpose breed for milk and meat production in medium- and low-input production environments (Chawala et al. 2017). The breed was developed through a crossbreeding programme involving East African Zebu, Indian Zebu, and European dairy breeds, mainly Ayrshire. This crossbreeding programme was initiated in the 1920s.

The initial programme was halted in 1940, and a new programme was started in 1944, with the aim of creating an animal capable of producing 2300 kg of milk in a 305-day lactation, having a calving interval of 380 days, and steers attaining a 250-kg carcass weight at 4 years of age (Syrstad 1990; Bwire et al. 2005). To achieve this, additional genetics were introduced, including the Kenya Boran (*Bos indicus*), the Jersey, and Guernsey in addition to the Ayrshire as sources of European (*Bos taurus*) dairy genetics. The composite was declared a breed in 1958 (Kiwuwa and Kyomo 1970; Rushalaza et al. 1993), and a nucleus herd of Mpwapwa was established in 1958 to drive the breeding programme (Syrstad 1990). By the mid-1960s, at the Mpwapwa Livestock Breeding Station, the breed contained approximately 20% Tanganyika Zebu, 10% Boran, 5% Ankole, 55% Red Sindhi and Sahiwal, and 10% Ayrshire (Syrstad 1990). The breed development continued, and in 1971, the breed composition was estimated to be 32% Red Sindhi, 30% Sahiwal, 19% Tanganyika Zebu, 10% Boran, and 9% Ayrshire and Shorthorn. Then the breed was selected towards Sahiwal, and by 1988, the Sahiwal content had reached 75% (Katyega 1987).

Unfortunately, the breeding programme has not been consistent throughout this period, and genotype stability (fixed genotype composition) has not been achieved. This phase was further complicated by factors such as disease outbreaks, droughts, inadequate nutrition, and a lack of enabling political and economic environments, all worsened by technical staff who were inadequate in driving the programme. For example, in the 1990s, the Mpwapwa cattle population declined significantly because of mortalities caused by East Coast fever (ECF), which led to the near extinction of the breed (Bwire et al. 2005). Consequently, breeding targets were not achieved, and new genetic lines, especially Sahiwal (*B. indicus*) from Kenya, were introduced.

There has never been more than 1000 pure Mpwapwa cows at any one time, and there has been little dissemination outside the original station (Wilson 2021). The Food and Agriculture Organization (FAO) of the United Nations has classified the Mpwapwa breed as being at risk of extinction (Syrstad 1990; Rege 1999), even before it has really been fully established. However, multiplication efforts from 1999 onwards supposedly increased the population of the breed owing to farmers' increased demand for improved cows and bulls for draught power. The efforts involved the multiplication of the pure Mpwapwa breed and the intensification of the increase in the number of these animals by back-crossing the available pure Mpwapwa with Boran and Sahiwal. The underlying consideration and assumption driving the breed choices and mating plans were that the Mpwapwa breed production potential was on par with Sahiwal, while the Boran cows were known to be better milkers than the Tanzania Shorthorn Zebu (TSZ) (Bwire et al. 2005).

However, a major challenge has been how to scale up a breeding programme to cover the needs of a wider breeder and consumer community—generating a large number of animals and systematically distributing these to farmers (Chawala et al. 2017). The target of 2300 kg of milk per 305-day lactation has not been realized, while the goal of 230 kg carcass per steer at the age of 4 years was realized in 1985. On-farm breed performance studies in the 1970s–1980s showed that the Mpwapwa cows were yielding four times more milk and had higher growth rates than the TSZ cows (Rushalaza et al. 1993), and the breed has been widely accepted by the community owing to its milk and meat yield, disease tolerance/resistance, and the ability of the bulls to be used for draught power (Komwihangilo et al. 2009).

There has not been a recent population census or estimate. By 1997, the estimate was 1000–1500 heads, and other than the animals on government stations, there were no 'purebreds' left on farms (Rege 1999). The Mpwapwa is clearly an example of a composite breed with a potential for high productivity and is liked by farmers. The failure for the breed to emerge as a major breed being used widely in Tanzania and neighbouring countries is principally an institutional failure to put in place a functional breeding programme to stabilize the breed and scale its use. If no steps are taken to address these, the Mpwapwa will disappear.

The Renitelo Breed

The Renitelo breed (Fig. 4.34) is a three-way composite beef cattle breed created at the Kianjasoa station (in Madagascar) by crossing Malagasy Zebu (25%) × Limousine (50%) × Afrikander (25%). The development of this breed was initiated in 1951, building on related cross-breeding work, which started in 1946, with the objective to produce a beef cattle breed suited for the typical Madagascar conditions. Although the breed was recognized in 1965, it was not registered. In 1992, the government took steps to revitalize the work on the breed. Interest in the breed had waned in the preceding period. Despite its

Fig. 4.34 A herd of Renitelo cattle of Madagascar. (Courtesy: Dr. Maria da Gloria Taela)

many valuable traits (e.g. its adaptation to a wide range of conditions in Madagascar, high growth rate, and draught power), the breed suffers from high sensitivity to the skin disease dermatophilosis (previously called streptothrichosis). The breed is currently at high risk of extinction.

The Mangan'i Boina Breed

The *Mangan'i Boina* (also known as *Manjan'i Boina*) is found in the Mahajanga province of Madagascar. It is a dairy type of breed created for the hot coastal climate of the Mahajanga province. It is considered to consist of (unknown proportions) the local Madagascar Zebu and the imported dual-purpose (dairy/beef)-type *Brune des Alpes* from France. Breed development was started in the 1980s, and it is still under formation. There were only 200 animals in 1998. The breed has not been registered, and without any concerted focus on its development, it will likely disappear.

The Wakwa Breed

The *Wakwa* breed (Fig. 4.35) is a composite cattle breed developed from crossing the American Brahman as the paternal line and the local Gudali of Cameroon as the dam line. The resulting F1 generation, although considered then to be susceptible to streptothricosis, proved to be most adaptable to the Cameroon environment. Through subsequent generations of inter se mating, the cross (named *Wakwa*) turned out to be tolerant of streptothricosis and was found to be a suitable beef animal for the local environment.

The Drakensberger

Although there is literature that refers to the Drakensberger of South Africa (Fig. 4.36) as a composite breed, it has a long history, which does not compare to the more recent composite breeds in this section. It dates back to the indigenous cattle possessed by the Khoi and Bantu tribes of the Western Cape of South Africa in the seventeenth century. Jan van Riebeeck supposedly crossbred the indigenous black cattle in 1665 with eight imported Gröningen bulls, which became known as 'Vaderlanders' in the early 1700s. Thus, the Drakensberger, also known as the Uys cattle (after the Uys family that played a significant role in the early days of the breeding), is considered to be the world's first synthetic or composite cattle breed, although there is no data on the proportionate genetic content of different cattle breeds from which it originated.

Fig. 4.35 A herd of Wakwa cattle. (Courtesy: Mr. Lawrence Shang of the TADU Dairy Cooperative, Bui Division, North-West Region of Cameroon)

Fig. 4.36 A herd of Drakensberger cows (**a**) and a bull (**b**). (Courtesy: Drakensberger Cattle Breeders' Society)

4.5 Well-Established Exotic Breeds in Africa

Although livestock have been part of the culture in SSA for a long time, the commercialization of animals based on specific traits is relatively new—a twentieth-century phenomenon. For a long time, individual herds of cattle were all that counted. The oldest archaeological evidence of domestic cattle dates back to between 6000 and 5000 BC. The Kushite Empire (300 BC–350 AD) used cattle for draught power to drive irrigation. However, the first recorded use of cattle in trading is reported in 1250–1450 AD in Great Zimbabwe, where cattle were traded together with ivory, gold, and salt. Further, cattle are documented to have been traded in the Zulu nation (1816–1897 AD). Recorded literature indicates that milk was a free commodity, representing good hospitality and community sharing. The indigenous breeds, which, by today's definition, are not dairy cattle breeds catered to these needs. With the exception of a few countries, the main dairy products consumed were liquid milk, sour milk, and butter.

Following the establishment of the first herd books in the late 1700s, most European cattle populations have developed into clearly defined breed types, and the selection for specific traits (e.g. with a focus on dairy, beef, or dual purpose) has been accompanied by an increase in the uniformity of colour and other external characteristics within each breed (Cunningham and Syrstad 2005). This specialization of breeds has, among other things, led to a significant reduction in the number of breeds in Europe and North America.

For example, today, more than half of the dairy cows in Europe and North America are of one breed type, the Holstein-Friesian.

The main driver for the utilization of popular and well-established exotic breeds in Africa has been the commercialization of livestock products stimulated by the post-Second World War globalization and driven by European settler farmers in the period before and following the political independence of African states. Subsequently, as the demand for animal products (meat, milk, and other dairy products) increased, so did the need for high-producing breeds. This has led to the importation of European cattle breeds, which, initially, were meant for the so-called high management farms and ranches. This trend has substantially increased post-independence, supported by international development agencies, and is increasingly driven by international private livestock breeding companies. In some countries, the intentional importation of tropical dairy (or dual-purpose) breeds such as Sahiwal or Brahman was done to serve the needs of poorly managed or 'low-to-medium-input' (smallholder) farms. This restriction policy for the European dairy breeds had some negative consequences for livestock development. For example, the introduction of Sahiwal from India and Pakistan during the colonial era essentially replaced efforts for the within-breed genetic improvement of indigenous breeds for smallholders in the highlands of Kenya. On the other hand, the unintended consequence of lifting the restriction policy for the European dairy breeds for the smallholder farms was the rapid introduction of the 'not-fit-for-purpose' European dairy breeds into poorly managed farms.

Exotic dairy breeds that are dominant in Africa today include Friesian, Jersey, Ayrshire, Holstein-Friesian, and Brown Swiss. Also widespread are Guernsey, Simmental/Fleckvieh, Red Dane, Dairy Shorthorn, and Normande, among others. Today, the dominant exotic beef breeds in Africa include Angus, Hereford, Charolais, Limousin, Brahman, and Gelbvieh.

Of the three main strategies for using exotic germplasm—breed replacement, continuous crossbreeding systems (e.g. rotational or criss-cross crossbreeding), and the creation of synthetic or composite breeds—the one widely witnessed across the continent is breed replacement, in which some farmers, usually large- and medium-scale, remove or reduce indigenous breeds on their farms to focus on one or more specific pure exotic breeds as a basis of their farming enterprise. As alluded to above, Africa has had a few initiatives aimed at developing composite breeds, but these have not been consistently successful. With regard to crossbreeding, while the practice is widespread, it is opportunistic and haphazard. Consequently, the continent has a 'fruit salad' of crosses of all kinds. For example, in eastern Africa, where the Holstein-Friesian has been widely used, there is a wide range of Friesian genes across and within smallholder farms. This would not be a problem if genetic composition was consistently a choice by farmers. Unfortunately, it is more a reflection of the absence of programmes to systematically develop appropriate breed combinations to supply farmers with the genetics they need. Where crossbreeding programmes have been tried, and there have been many across SSA, they have been short-term donor-funded research or development 'projects', and they have generally failed to consider the needs of farmers, which include production traits (e.g. milk, meat, and traction), adaptive attributes, and other more nuanced considerations such as product taste, animal colour, etc.

Overall, crossbreeding with exotic dairy cattle in Africa has been more widespread and has taken root in agro-ecological zones, which are similar to the temperate origins of these breeds. This can be seen in eastern Africa (the highlands of Kenya, Ethiopia, Uganda, and Tanzania) and South Africa.

The Holstein-Friesian (a combination of the original European Friesian, the North American Holstein, and intermediates between the two) dominates the global dairy industry and is also the leading exotic dairy breed in terms of numbers in pure and crossbred forms across the African continent. Most countries, which have considered specialized dairy animals, have almost invariably tried this breed, with varying levels of success. Lessons learnt, especially in relation to its large body size and a lack of adaptability to the myriad production environmental and production system challenges—the inadequacy and low quality of feeds, heat, disease, etc.—have, in many countries, informed the introduction of smaller, 'less fragile' breeds such as the Jersey. Indeed, the Jersey is probably the second most widespread exotic cattle breed on the continent presently—with an estimated 34 African countries reported to have imported the Jersey at some point in the recent past or are actively doing so at the present time (Opoola et al. 2022).

Crossbreeding for milk and beef production in Africa using some of the major exotic breeds has been a subject of various studies (e.g. Wilson (2018), Cunningham and Syrstad (2005), Gregory et al. (1984), Wilson (2009), and Moyo et al. (1994)).

Table 4.6 presents a summary of some of the main exotic cattle breeds in Africa.

Although some of the exotic breeds are being used as pure breeds in Africa, all exotic breeds that have been introduced into Africa have been used in some form of crossbreeding with indigenous breeds. Indeed, their use in crossbreeding is more widespread than their use as pure breeds.

While some of the exotic breeds have been successfully utilized, there are significant yield gaps, that is, gaps between what is possible given the genetic composition of the animals and what is realized. This is particularly evident in the dairy sector. Some of the reasons for these gaps include a mismatch between the exotic genetics and the local smallholder production systems. Partly, this has been due to the fact that breeding goals and long-term breeding strategies for smallholder farmers have not been defined clearly. Although indigenous livestock breeds may not have levels of production (milk yield, growth rate, number of eggs, etc.) as high as their exotic counterparts under certain production systems, the indigenous breeds possess valuable traits such as tolerance and resistance to disease, high fertility, good maternal ability, unique product qualities, longevity, and adaptation to harsh environments and poor-quality feeds, which have not been appropriately exploited in the breeding programmes aimed at improving productivity. Overall, the use of exotic breeds in Africa has been consistently successful only under high levels of management in medium-to-large farms. However, in eastern Africa, especially Kenya, the successful use of exotic crossbreeds in smallholder dairy systems is a demonstration that smallholders can utilize these breeds when conditions, both production and market, are right.

4.6 Summary and Conclusion

Of the four main categories of indigenous African cattle—the humpless (*Bos taurus* or Taurine) cattle (discussed in Chap. 3), the humped (*Bos indicus* or the Zebu), the Sanga (Taurine × Zebu) hybrids, and the Zenga (Sanga × Zebu hybrids)—the majority of cattle breeds and strains on the continent are Zebu.

The Zebu and its derivatives (Sanga and Zenga types) are well-adapted to hot and dry environmental conditions and are known to be more resistant to tick infestations compared to Bos taurus cattle—both indigenous to Africa and exotic. Consequently, the Zebu are widely distributed in Africa, inhabiting the production systems in the expansive arid and semi-arid areas across the continent. They generally have larger body sizes and higher overall productivity levels in tsetse-free areas of the continent.

The population of the Zebu is highest in the East and the Horn of Africa region—in both highlands and lowlands, including the large arid and semi-arid land masses—compared to other regions. This distribution is, in large part, a legacy of the history of cattle introduction (the Zebu having entered the continent primarily through the Horn of Africa) as well as human migration patterns on the African continent. The two waves of cattle introduction into the continent and their subsequent migration led to a dominance of the Zebu-type cattle in Eastern Africa and explain the relative absence of distinct Taurine cattle in the sub-region. Indeed, there is only one remaining Taurine breed in the sub-region, the Sheko of

Table 4.6 Summary of exotic cattle breeds used in Africa

Exotic breed	Primarily imported from (country)	Country of established use (in Africa)	Focal trait(s)
Holstein-Friesian[a]	Denmark, the United Kingdom, the United States, New Zealand, Canada, Sweden, Australia	Burkina Faso, Burundi, Cameroon, Kenya, Côte d'Ivoire, DRC, Egypt, Ethiopia, Malawi, Nigeria, Rwanda, Senegal, South Africa, Sudan, Tanzania, The Gambia, Uganda, Zambia, Zimbabwe	Dairy
Jersey	The United Kingdom (Island of Jersey), New Zealand, the United States	Angola, Botswana, Burkina Faso, Burundi, Cameroon, Chad, Côte d'Ivoire, Democratic Republic of the Congo, Egypt, Eritrea, Eswatini, Ethiopia, The Gambia, Guinea, Ghana, Lesotho, Kenya, Liberia, Libya, Madagascar, Malawi, Mozambique, Namibia, Nigeria, Rwanda, Senegal, Seychelles, Somalia, South Africa, Sudan, Tanzania, Uganda, Zambia and Zimbabwe	Dairy
Ayrshire	The United Kingdom, Canada	Kenya, South Africa, Tanzania	Dairy
Guernsey	The United Kingdom (Island of Guernsey)	South Africa, Kenya	Dairy
Tarentaise	France	Burkina Faso, Egypt, Algeria, Morocco, Tunisia, Zimbabwe	Dual purpose (Dairy and beef)
Montbéliarde	France	Niger, Burkina Faso, Cameroon, Côte d'Ivoire, Senegal, Sudan, The Gambia, Morocco, Zimbabwe	Beef
Charolais	France	Botswana, Kenya, Namibia, Rwanda, South Africa, Zambia, Zimbabwe	Beef
Brown Swiss	The United States	Nigeria, Kenya, South Africa	Dual purpose (Dairy, beef)
Simmental	Switzerland	Cameroon, Kenya, Namibia, Nigeria, South Africa, Zambia, Zimbabwe	Dual purpose (Dairy and beef, and also draught power)
Danish Red	Denmark	Tanzania, Zimbabwe	Dairy
Fleckvieh	Austria, Germany	Kenya, Namibia, South Africa	Dual purpose (Dairy, beef)
Gelbvieh	Germany	South Africa	Dual purpose (Dairy, beef)
Pie Rouge des Plaines	France	Tunisia	Dairy
Hereford	Originally, the United Kingdom (Herefordshire in England); more recent imports come from secondary sources (e.g. the United States, Canada, and Australia)	Kenya, South Africa, Zambia, and Zimbabwe	Beef
Angus	Scotland (original home) and secondary sources the United States, Canada, Argentina, New Zealand, and Australia	Angus: Kenya, Namibia, South Africa, and Botswana Brangus (the United States and Canada): Developed from Brahman and Angus and notable for its resistance to heat; is present in South Africa and a few other African countries, including Namibia and Kenya	Beef

(continued)

Table 4.6 (continued)

Exotic breed	Primarily imported from (country)	Country of established use (in Africa)	Focal trait(s)
Charolais		Kenya and South Africa	Beef
Sahiwal	India and Pakistan	Burundi, Kenya, Sierra Leone, Nigeria, Somalia, and, in smaller numbers, in several other countries in Africa	Dual purpose
Brahman	The United States	Botswana, Namibia, South Africa, Swaziland, Zambia, Zimbabwe, and, to a smaller extent, a few other African countries	Beef

[a]Refers to European Friesian, North American Holstein, and intermediates between the two

---

Box 4.1 Summary of the risk status of cattle breeds in Africa

Africa is home to a total of 145 cattle breeds/strains comprising two Taurine Longhorns, 15 Taurine Shorthorns, 75 Zebu (*Bos indicus*), 30 Sanga, eight Zenga (Zebu–Sanga), nine breeds that are derived from the interbreeding of indigenous breeds/strains located in close proximity to each other, and six systematically created composite breeds. Out of the 145 breeds identified from the survey, 47 (about 32%) were considered to be at risk of extinction. The risk categories used were: Critical (most severe), Endangered, Vulnerable, and Rare (least severe). Of the breeds identified to be at risk of extinction, six were in the Rare category, ten were Vulnerable, another ten were Endangered, and 15 were in the Critical category. A total of 22 breeds (about 13%) previously recognized on the continent have become extinct in the past century. This number excludes some populations, which have lost their individual identity due to admixtures involving two or more originally distinct breeds.

Source: Rege (1999).

South-west Ethiopia and the eastern parts of South Sudan. Box 4.1 is a summary of the findings by Rege (1999) with regard to the risk status of cattle breeds in Africa. The situation has changed significantly, and many more breeds are at escalated risk levels than they were back in 1999. An updated comprehensive survey is overdue.

Although across the continent, the population of the Zebu as a group remains large and the overall genetic diversity high, there are threats, increasingly placing several specific strains or breeds (of Zebu and its stabilized derivatives) at risk of extinction, with some already extinct and others facing extinction. Crossbreeding presents the highest risk. As a result, several existing Zebu breeds and their derivatives are critically endangered, having fewer than 1000 animals. Four of these are Small East African Zebu (Mkalama Dun, Pare, Chagga, and Baria), and three are South African Sanga (Nkone, Pedi, and Shangan).

Many early African societies, especially nomadic groups, depended almost entirely on livestock, with cattle playing a central role. Pastoralism remains an important system on the continent today, as exemplified by the Fulani, Maasai, and the Tuareg communities. Across the continent, cattle remain major sociocultural assets, playing important economic and sociocultural roles, as well as major sources of animal protein (milk and meat), draft power, and manure that contribute to crop agriculture.

African indigenous breeds are considered to be poor performers relative to exotic breeds in terms of outputs of products (e.g. meat or milk) per animal and unit of time. While this is true on the surface, it does not consider the adaptive attributes of the indigenous breeds and the fact that the more 'productive' exotic breeds can hardly survive in the (most dominant) production environments in which the indigenous breeds are able to not only survive but also produce and support the livelihoods of their owners. Moreover, while commercial exotic cattle breeds, provided with enabling production environments, excel because they have been continuously selected for the specific productivity traits, some indigenous African breeds are known to inherently have either good dairy or beef potential or both. But only limited effort has been made to fully exploit these potentials. Outstanding examples of Zebu cattle and their derivatives covered in this chapter include the dual-purpose group (exhibiting good dairy and beef production attributes), comprising the White Fulani cattle (West African Zebu) and Fogera cattle (Zenga); the dairy types, including the Kenana and Butana cattle of Sudan, which are among the best milk-producing Zebu breeds globally; and the beef types, which are exemplified by the Boran or Borana, which are now globally recognized and sought after. The potential for within-breed improvement among indigenous African breeds is, indeed, huge.

Moreover, while intensive and consistent artificial selection and management interventions have resulted in marked productivity improvements among cattle (and livestock, more broadly) in the global North, this has come at a cost: reduced genetic diversity. The immediate consequence is expressed in such performance traits as fertility, which suffers significantly in inbred animals (an extreme case of lack of diversity). The long-term consequence is more dire: loss of options for addressing yet unknown production challenges and/or human requirements of the future. In contrast, most African cattle breeds have not been selected consistently for productivity gains. The main (natural and artificial) selection focus has been on survival in the largely unpredictable, harsh, and changing environmental conditions. The major threat African breeds have faced has been genetic dilution, historically, through interbreeding among neighbouring breeds, and more recently, through haphazardly and inconsistently executed, widespread crossbreeding. Although artificial selection programmes have been tried, these have tended to be short-term 'projects', and there are only limited cases of sustained efforts on within-breed genetic improvements.

The narrative that indigenous African breeds have inherently low productivity has led to the shift in attention to genetic 'upgrading' with and replacement by exotic breeds from the global North. Today, the continent has, in addition to the four main indigenous cattle categories, a fifth as well as a sixth category. The fifth category consists of the composite breeds developed through systematic crossbreeding involving one or more indigenous breeds with one or more imported commercial breeds. Crossbreeding and the formation of composite breeds is intended to create fit-for-purpose breeds for specific production contexts. The sixth category includes the pure exotic breeds that are directly imported, maintained, and used for commercial production. Some of the exotic breeds are thriving in specific locations of the continent under commercial production systems for dairy, beef, or dual purpose. On the other hand, efforts to create composite breeds have faltered, in major part because they were driven by the public sector—government research—efforts to create genotypes that should (at least from technical considerations) provide a desirable combination of adaptation and productivity traits. Unless the private sector is involved in the design and implementation of these breeding programmes and the needs and capabilities of smallholders are considered, such programmes will continue to be restricted in existence only on research stations and government ranches.

References

Abdurehman A (2019) Physiological and anatomical adaptation characteristics of Borana cattle to pastoralist lowland environments. Asian J Biol Sci 12:364–372

Aboagye GS (1995) White Fulani. Ghana Country Report. http://dad.fao.org/

Aganga AA, Oyejola BA, Aganga AO, Yaakugh IDI (1986) Reproductive performance of White Fulani cows. Thai J Agric Sci 19:225–229

Ageeb AG, Hillers JK (1991) Production, and reproduction characteristics of Butana and Kenana cattle of The Sudan. World Anim Rev 67:49–56

Alberro M (1983) The indigenous cattle of Mozambique. World Anim Rev 48:12–17

Alim KA (1962) An assessment of the environmental and genetic factors affecting milk production of Butana cattle. J Dairy Sci 45(2):242–247. https://doi.org/10.3168/jds.S0022-0302(62)89372-2

Alphonsus C, Akpa GN, Barje PP, Finangwai HI, Adamu BD (2012) Comparative evaluation of linear udder and body conformation traits of Bunaji and Friesian x Bunaji cows. World J Life Sci Med Res 2(4):134–140

Amakiri SF, Mordi R (1975) The rate of cutaneous evaporation in some tropical and temperate breeds of cattle in Nigeria. *Anim Prod* 20:63–68

Amakiri SF, Onwuka SK (1980) Quantitative studies of sweating rate in some cattle breeds in a humid tropical environment. Anim Prod 30:383–388

Assan N (2011) Indigenous cattle breeding studies in Zimbabwe. Ed. Lambert Academic Publishing Company, Saarbrucken. ISBN: 978-3-8465-1585-3

Babayemi OJ, Abu OA, Opakunbi A (2014) Integrated animal husbandry for schools and colleges, 1st edn. Positive Press, Ibadan, Nigeria, pp 20–122

Bahbahani H, Tijjani A, Mukasa C, Wragg D, Almathen F, Nash O, Akpa GN, Mbole-Kariuki M, Malla S, Woolhouse M, Sonstegard T, Van Tassell C, Blythe M, Huson H, Hanotte O (2017) Signatures of selection for environmental adaptation and Zebu × Taurine hybrid fitness in East African Shorthorn Zebu. Front Genet 8:68. https://doi.org/10.3389/fgene.2017.00068

Bahbahani H, Salim B, Almathen F, Al Enezi F, Mwacharo JM, Hanotte O (2018) Signatures of positive selection in African Butana and Kenana dairy zebu cattle. PLoS One 13(1):e0190446. https://doi.org/10.1371/journal.pone.0190446

Bayssa M, Yigrem S, Betsha S, Tolera A (2021) Production, reproduction, and some adaptation characteristics of Boran cattle breed under changing climate: a systematic review and meta-analysis. PLoS One 16(5):e0244836. https://doi.org/10.1371/journal.pone.0244836

Behnke RH, Arasio RL (2019) The productivity and economic value of livestock in Karamoja subregion, Uganda. Karamoja Resilience Support Unit. USAID/Uganda, Irish Aid and UKaid, Kampala. https://karamojaresilience.org/publications/item/the--productivity-and-economic-value-of-livestock-in-karamoja-sub-regionuganda

Belemsaga DM, Lombo Y, Thevenon S, Sylla S (2005) Inventory analysis of West Africa cattle breeds. In: Applications of gene-based technologies for improving animal production and health in developing countries. Springer, Dordrecht, pp 167–173

Bessa I, Pinheiro I, Matola M, Dzama K, Rocha A, Alexandrino P (2009) Genetic diversity and relationships among indigenous Mozambican cattle breeds. S Afr J Anim Sci 39(1):61–72

Bett B, Orenge C, Irungu P, Munga LK (2004) Epidemiological factors that influence time-to-treatment of trypanosomosis in Orma Boran cattle raised at Galana Ranch, Kenya. Vet Parasitol 120(1–2):43–53

Bett RC, Okeyo AM, Malmfors B, Johansson K, Agaba M, Kugonza DR, Bhuiyan AFK, Filho AEV, Mariante AS, Mujibi FD, Philipsson J (2013) Cattle breeds: extinction or quasi-extant? Resources 2(3):335–357

Blench RM (1993) Ethnographic and linguistic evidence for the prehistory of African ruminant livestock, horses and ponies. In: The archaeology of African food, metals and towns, pp 71–103

Blench R (1999) Traditional livestock breeds: geographic distribution and dynamics in relation to the ecology of West Africa. Working Paper 122, vol 1-69. Overseas Development Institute, Portland House, London, UK

Blench RM, De Jode A, Gherzi E, Di Domenico C (1998) Keteku and N'Dama crossbred cattle in Nigeria: history, distribution and productivity. In: Seignobos C, Thys E (eds) Des Taurins au Cameroun et Nigeria, pp 293–310

Bonte P (1991) To increase cows, God created the king: the function of cattle in inter-lacustrine societies. In: Galty JD, Bonte P (eds) Herders, warriors and traders: pastoralism in Africa. Westview Press, Boulder, pp 62–86

Boutrais J (2007) The Fulani and cattle breeds: crossbreeding and heritage strategies. Africa 77(1):18–36. https://doi.org/10.3366/afr.2007.77.1.18

Braend M (1971) Haemoglobin variants in cattle. Anim Blood Groups Biochem Genet 2:15–21

Braend M, Khanna ND (1968) Haemoglobin and transferrin types of some West African cattle. Anim Prod 10(2):129–134

Buvanendran V, Adamu AM, Abubakar BY (1992) Heat tolerance of Zebu and Friesian-Zebu crosses in the Guinea Savanna Zone of Nigeria. Trop Agric 69(4):394–396

Bwire JMN, Mkonyi JI, Masao D (2005) Development and multiplication of Mpwapwa breed cattle for meat and milk production in Tanzania. In: Proceedings of hundred years of livestock services in Tanzania 1905–2005. Ministry of Water and Livestock Development, Tanzania, Mpwapwa, pp 96–105

Capitaine P (1972) Projet de Développement de l'Elevage au Ghana. Etudes de factibilité de quatre ranches. Etude Zootechnique. IEMVT, France, p 103

Chawala AR, Banos G, Komwihangilo DM, Peters A, Chagunda MGG (2017) Phenotypic and genetic parameters for selected production and reproduction traits of Mpwapwa cattle in low-input production systems. S Afr J Anim Sci 47:307–319. https://doi.org/10.4314/sajas.v47i3.7

Cunningham E, Syrstad O (2005) Crossbreeding of Bos indicus and Bos taurus for milk production in the tropics. In: Types and breeds of tropical and temperate cattle. FAO (ISBN 92–5), Rome, Italy, p 1987. http://www.fao.org/docrep/009/t0095e/T0095E04.htm

DAD-IS (2005) Domestic animal genetic diversity information system. The Food and Agriculture Organization of the United Nations. FAO (FAO), Rome, Italy. https://www.fao.org/dad-is/en

DAGRIS (2022) Domestic animal genetic resources information system. International Livestock Research Institute (ILRI), Nairobi, Kenya and Addis Ababa, Ethiopia. http://dagris.ilri.cgiar.org

de Clare Bronsvoort BM, Thumbi SM, Poole EJ, Kiara H, Auguet OT, Handel IG, Jennings A, Conradie I, Mbole-Kariuki MN, Toye PG, Hanotte O, Coetzer JAW, Woolhouse MEJ (2013) Design and descriptive epidemiology of the Infectious Diseases of East African Livestock (IDEAL) project, a longitudinal calf cohort study in western Kenya. BMC Vet Res 9:171. http://www.biomedcentral.com/1746-6148/9/171

de Jode A, Reynolds L, Matthewman RW (1992) Cattle production systems in the derived Savannah and

Southern Guinea Savannah regions of Oyo State, Southern Nigeria. Trop Anim Health Prod 24:90–96. https://doi.org/10.1007/BF02356951

Dehoux JP, Hounsou-Ve G (1993) Productivité de la race bovine Borgou selon les systèmes d'élevages traditionnels au Nord-Est du Benin. Ticks in a changing world. Revue mondiale de Zootechnie 74/75:36–48

Doko AS (1993) Etude sur la trypanosomiase et la trypanotolérance bovines au Bénin. Mémoire MSc. Institut de médecine tropicale, Anvers, Belgique

Dolan RB (1997) The Orma Boran: a trypanotolerant East African breed. In: FAO *Revue Mondiale de Zootechnie*. Revista Mundial de Zootecnia (FAO)

Domingo PM (1976) Contribution à l'étude de la population bovine des Etats du Golfe du Benin. DVM thesis. Ecole Inter-Etats des Sciences et de Médecine Vétérinaires de Dakar, Senegal, p 148

Doutressoulle G (1947) L'élevage en Afrique occidentale française. Larose, Paris, p 298

Epstein H (1971) The origin of the domestic animals of Africa. Volume In Africana Publishing Corporation, New York, pp 327–556

FAO (2010) FAOSTAT data 2010., http://faostat.fao.org/faostat. Accessed 12 July 2010

Faulkner DE, Epstein H (1957) The indigenous cattle of the British dependent territories in Africa. Colonial Advisory Council of Agriculture, Animal Health and Forestry, HMSO Publication No. 5, p 186

FDLPCS (1992) Nigerian livestock resources, vol. II., national synthesis. Federal Department of Livestock and Pest Control Services (FDLPCS), Abuja, Nigeria

Felius M (1995) Cattle breeds: an encyclopaedia. Misset uitgeverij bv. Doetinchem, The Netherlands, p 690

Flamigni A (1951) Le gros bé'tail au Mayumbe. Bull, agric. Congo beige 42:91–106

Gates GM (1952) Breeds of cattle found in Nigeria. Farm Forest 11:19–43

Gregory KE, Trail JCM, Sandford J, Durkin J (1984) Crossbreeding cattle in beef production programmes in Kenya. Trop Anim Health Prod 16:181–186

Hagos B (2016) Ethiopian cattle genetic resources and unique characteristics under a rapidly changing production environment. A review. Int J Sci Res 6(12):1959–1968

Hall S (1998) Traditional livestock in semi-arid north eastern Zimbabwe: Mashona Cattle. Trop Anim Health Prod 30:351–360. https://doi.org/10.1023/A:1005192604460

Hamza AE, Fathi AM, Mohammed OE, Shuiep ES (2021) Phenotypic characterization and morphometrical measurements of Sudanese Red Fulani cattle (Kuri) in South Darfur, Sudan. EC Veterin Sci 6(8):15–25

Hanotte O (2002) African pastoralism: genetic imprints of origins and migrations. Science 296:336–339. https://doi.org/10.1126/science.1069878

Hanotte O, Tawah CL, Bradley DG, Okomo M, Verjee Y, Ochieng JW, Rege JEO (2000) Geographic distribution and frequency of a taurine (Bos taurus) and an indicine Zebu (Bos indicus) Y specific allele amongst sub-Saharan African cattle breeds. Mol Ecol 9:387–396

Hanotte O, Ronin Y, Agaba M, Nilsson P, Gelhaus A, Horstmann R, Sugimoto Y, Kemp S, Gibson J, Korol A, Soller M (2003) Mapping of quantitative trait loci controlling trypanotolerance in a cross of tolerant West African N'Dama and susceptible East African Boran cattle. Proc Natl Acad Sci 100:7443–7448

Ibeagha-Awemu EM, Erhardt G (2006) An evaluation of genetic diversity indices of the Red Bororo and White Fulani cattle breeds with different molecular markers and their implications for current and future improvement options. Trop Anim Health Prod 38:431–441

Ibeagha-Awemu EM, Jann OC, Weimann C, Erhardt G (2004) Genetic diversity, introgression and relationships among West/Central African cattle breeds. Genet Sel Evol 36:673–690

Igono MO, Aliu YO (1982) Adaptive thermal responses of the Zebu, half and three-quarter Friesian-Zebu crosses to the hot-dry and hot-humid seasons. Bull Anim Hlth Prod Afr 30(4):329–333

ILCA (1979a) Trypanotolerant livestock in west and central apical Vol. 1: general studies. International Livestock Centre for Africa (ILCA) Mono. No 2, Addis Ababa, Ethiopia, p 148

ILCA (1979b) Trypanotolerant livestock in west and central apical Vol. 2: country studies. International Livestock Centre for Africa (ILCA) Mono. No 2, Addis Ababa, Ethiopia, p 303

Infield M (2002) The culture of conservation: exclusive landscapes, beautiful cows and conflict over lake Mburo National Park, Uganda, PhD thesis. School of Development Studies, University of East Anglia, Norwich, UK, p 318

Infield M (2003) Conserving nature and the nature of conservation – national parks as cultural entities. Policy Issues (special issue for the World Parks Congress) 12:64–70. www.IUCN.org/themes/ceesp/publications/newsletter

IRZ (1987) Summary of research results. Institute of Animal Research (IRZ), Bambui, Cameroon, pp 27–38

Kajura S (2001) Beef production. In: agriculture in Uganda. In: Mukiibi JK (ed) Livestock and fisheries, vol 4. Fountain Publishers /CTA/NARO, pp 1–17

Kaluba EM (1984) Milk production systems in Zambia. In: The potential for small scale milk production in eastern and southern Africa. IDRC manuscript reports, vol 98e, pp 103–109

Katyega IMJ (1987) Mpwapwa cattle of Tanzania. Anim Genet Resour/Resources Génétiques Animales/Recursos Genéticos Animales 6:25–28. https://doi.org/10.1017/S1014233900000237. ftp://ftp.fao.org/docrep/fao/012/s6460t/s6460t00.pdf

Khombe CT (2002) Genetic improvement of indigenous cattle breeds in Zimbabwe: a case study of the Mashona Group Breeding Scheme. Animal genetics training resources case study. The International Livestock Research Institute (ILRI), Nairobi, Kenya. http://agtr.ilri.cgiar.org/

Kiwuwa GH, Kyomo ML (1970) Milk composition and yield characteristics of Mpwapwa cattle. E magAfr Agr Forest J 36:290

Komwihangilo DM, Mkonyi JI, Masao DF, Moto E, Mahiza AMO, Mnzava V (2009) Performance and challenges in the management of improved cattle in agro-pastoral systems of Central Tanzania. Livestock Res Rural Dev 21:75. http://www.lrrd.org/lrrd21/5/komw21075.htm

Kotze A, Harun M, Otto F, Van der Bank, F.H (2000) Genetic relationships between three indigenous cattle breeds in Mozambique. S Afr J Anim Sci 30(2):92–97

Kubkomawa HI (2017) Indigenous breeds of cattle, their productivity, economic and cultural values in Sub-Saharan Africa: a review. Int J Res Stud Agric Sci 3:27–43

Kugonza DR, Nabasirye M, Mpairwe D, Hanotte O, Okeyo AM (2011) Productivity and morphology of Ankole cattle in three livestock production systems in Uganda. Anim Genet Resour 48:13–22

Lamwaka S (2006) Ugandan pastoralists hit by market reforms. PANOS features, PANOS, London. www.panos.org.uk/newsfeatures /featuredetails.asp

Latif AA, Pegram RG (1992) Naturally acquired host resistance in tick control in Africa. Insect Sci Appl 13:505–513

Latif AA, Punyua DK, Capstick PB, Nokoe S, Walker AR, Fletcher JD (1991) Histopathology of attachment sites of Amblyomma variegatum and *Rhipicephalus appendiculatus* on Zebu cattle of varying resistance to ticks. Vet Parasitol 38(2–3):205–213

Leclercq P (1976) Principales races d'animaux domestiques des zones tropicales. Institut d'Elevage et de Médecine vetérinaire des Pays tropicaux (IEMVT). ENS/II-35, France, pp 35–36

Lhoste F, Dumas R (1972) Embouche intensive des Zébus de l'Adamaoua. I. Comparaisons de différents systémes d'alimentation (1970). Rev Elev Méd vét Pays trop 25(2):259–280

Lopez G (1985) Technical assistance mission to CEBV. Final Report FAO Project GCP/RAF/191/ITA, Ouagadougou, Burkina Faso

Loquang TM (2003) Karamojong, Uganda. In: Köhler-Rollefson I, Wanyama J (eds) The Karen Commitment. In Proceedings of a conference of indigenous livestock breeding communities on animal genetic resources. German NGO Forum on Environment and Development, Bonn, pp 24–29

Lutfi MAM, Mohamed-Khair AA, Peters KJ, Zumbach B, Kamal EAG (2005) The reproductive and milk performance merit of Butana cattle in Sudan. Archiv fur Tierzucht 48:445–459

Magnier J, Druet T, Naves M, Ouvrard M, Raoul S, Janelle J, Moazami-Goudarzi K, Lesnoff M, Tillard E, Gautier M, Flori L (2021) The genetic history of Mayotte and Madagascar cattle breeds mirrors the complex pattern of human exchanges in Western Indian Ocean. G3. https://doi.org/10.1101/2021.10.08.463737

Mason IL (1988) A world dictionary of livestock breeds, types and varieties, 3rd edn. CAB International: The Cambrian News Ltd, UK, p 348

Mason IL (1996) A world dictionary of livestock breeds, types and varieties. CAB International

Mason IL, Maule JP (1960) The indigenous livestock of eastern and Southern Africa. Commonwealth Agricultural Bureaux. Farnham Royal, Bucks, England, p 29

Maule JP (1990) The cattle of the tropics. Centre for Tropical Vet. Med/University of Edinburgh, Edinburgh, UK

Mbole-Kariuki MN, Sonstegard T, Orth A, Thumbi SM, de Bronsvoort BMC, Kiara H, Toye P, Conradie I, Jennings A, Coetzer K, Woolhouse MJJ, Hanotte O, Tapio M (2014) Genome-wide analysis reveals the ancient and recent admixture history of East African Shorthorn Zebu from Western Kenya. Heredity 113:297–305

Medani MA (2003) Animal resources and animal production in Sudan, 2nd edn. University of Khartoum Printing Press, Sudan

Meghen C, MacHugh DE, Sauveroche B, Kana G, Bradley DG (1999) Characterization of the Kuri Cattle of Lake Chad using molecular genetic techniques. In: Blench RM, MacDonald KC (eds) The origin and development of African livestock. University College Press, London, pp 28–86

Mekonnen A, Haille A, Dessie T, Mekasha Y (2021) On farm characterization of Horro cattle breed production systems in western Oromia, Ethiopia. Livest Res Rural Dev 24:6. https://www.lrrd.cipav.org.co/lrrd24/6/meko24100.htm

Messine O, Tanya VN, Mbah DA, Tawah CL (1995) Ressources génétiques animales du Cameroun passé, présent et avenir: le cas des ruminants. Anim Genet Resour Inf 16:51–69

Minuye N, Abebe G, Dessie T (2018) On-farm description and status of Nuer (Abigar) cattle breed in Gambella Regional State, Ethiopia. Int J Biodivers Conserv 10(6):292–302. https://doi.org/10.5897/IJBC2017.1168

Missohou A, Bankole AA, Niang AT, Ragounandea G, Talaki E, Bitar I (1997) Le Zébu Gobra: Caractères ethniques et performances zootechniques. Anim Genet Resour Inf 22:53–60

Mohammed SA, Rahamtalla S, Ahmed SS, Ishag IA, Dousa BM, Ahmed MKA (2014) Smallholder dairy production systems of Kenana and Butana cattle in Sudan. Wayamba J Anim Sci 6:992–997

Moyo S, Rege JEO, Swanepoel FJC (1994) Evaluation of indigenous, exotic and crossbred cattle for beef production in a semi-arid environment. In: Smith C, Gavora JS, Benkel B, Chesmais J, Fairfull W, Gibson JP, Kennedy BW, Burnside EB (eds) Proceedings of the 5th world congress on genetics applied to livestock production (WCGALP) and poultry breeding; avian biotechnology; behaviour genetics; reproductive biotechnology; immunogenetics and disease resistance genetics; breeding in the tropics and extreme environments, pp 344–347

Mpofu N (2002) The multiplication of Africa's indigenous cattle breeds internationally: the story of the Tuli and Boran breeds. Animal Genetics Training Resource (AGTR) case study. The International Livestock Research Institute (ILRI), Nairobi, Kenya

Mrode RA (1988) Lactation performance of the White Fulani cattle in Southern Nigeria. Trop Anim Health Prod 20:149–154. https://doi.org/10.1007/BF02240083

Mureja S (2002) A survey on cattle management and utilization in Gambella region. In: Livestock in food security - roles and contributions. In proceedings of the 9th conference of the Ethiopian Society of Animal Production (ESAP), pp 91–98

Murray GGR, Woolhouse MEJ, Tapio M, Mbole-Kariuki MN, Sonstegard TS, Thumbi SM, Jennings AE, van Wyk IC, Chase-Topping M, Kiara H, Toye P, Coetzer K, de Clare Bronsvoort BM, Hanotte O (2013) Genetic susceptibility to infectious disease in East African Shorthorn Zebu: a genome-wide analysis of the effect of heterozygosity and exotic introgression. BMC Evol Biol 13:246. https://doi.org/10.1186/1471-2148-13-246

Musa L, Ahmed M, Peters K, Zumbach B, Gubartalla K (2005) The reproductive and milk performance merit of Butana cattle in Sudan. Arch Tierz Dummerstorf 48(5):445–459

Musa LM-A, Peters KJ, Ahmed M-KA (2006) On farm characterization of Butana and Kenana cattle breed production systems in Sudan. Livest Res Rural Dev 18(12):1–15

Musa LM-A, Bett RC, Ahmed MKA, Peters KJ (2008) Breeding options for dairy cattle improvement in The Sudan. Outlook Agric 37(4):289–295. https://doi.org/10.5367/000000008787167781

Musimuko E (2014) Genetic diversity and estimation of genetic parameters for economically important traits in Zambian cattle. M.Phil thesis. University of Adelaide, School of Animal and Veterinary Science, Adelaide, Australia

Mwaanga ES, Parés-Casanova PM (2017) In: Michael B (ed) The Zambian Indigenous Cattle - a review of their origins and phenotype. Pub. Reach Publishers. ISBN 13, 9789982703550

Mwacharo JM, Rege JEO (2002) On-farm characterization of the indigenous small East African Shorthorn Zebu cattle (SEAZ) in the Southeast rangelands of Kenya. Anim Genet Resour Inf 32:73–86. https://doi.org/10.1017/S1014233900001577

Nakimbugwe H, Muchunguzi C (2003) Bahima, Uganda. In: The Karen Commitment. In Proceedings of a conference of indigenous livestock breeding communities on animal genetic resources. Karen, Nairobi, Kenya, pp 34–40

Ngere LO (1985) The White Fulani (Bunaji) of Nigeria. In: 2nd OAU Expert Committee meeting proceedings of animal genetic resources in Africa: high potential and endangered livestock, pp 67–77

Nourezzine A, Duksi F, Tsvetkova AD, Ulybina EA, Gins MS, Yacer RN, Klenovitsky AA, Nikishov AA, Amirshoev F, Digha J, Gladyr EA (2019) Genetic characterization of White Fulani in Nigeria: a comparative study. J Adv Vet Anim Res 6(4):474–480

Nwufoh KJ, Amakiri SK (1981) The normal skin bacterial flora of some cattle breeds in Nigeria. Bull Anim Hlth Prod Afr 29:103–105

Ogunsiji O (1974) Length and level of preparum feeding on the milk of W. Fulani cows in Ibadan. Ph.D. thesis. Univ. Ibadan, Nigeria, p 356

Olayiwole MB, Buvanendran V, Fulani IJ, Ikhatua JU (1981) Intensive fattening of indigenous breeds of cattle in Nigeria. World Rev Anim Prod 17(2):71–77

Olutogun O (1976) Reproductive performance and growth of N'Dama and Keteku cattle under ranching conditions in the Guinea Savannah of Nigeria. Univ. Ibadan, Nigeria, Ph.D. Thesis, p 126

Omer EAM, Addo S, Roessler R, Schäler J, Hinrichs D (2021) Exploration of production conditions: a step towards the development of a community-based breeding program for Butana cattle. Trop Anim Health Prod 53:9. https://doi.org/10.1007/s11250-020-02459-4

Opoola O, Shumbusho F, Hambrook D, Thomson S, Dai H, Chagunda MGG, Capper JL, Moran D, Mrode R, Djikeng A (2022) From a documented past of the Jersey breed in Africa to a profit index linked future. Front Genet:13. https://doi.org/10.3389/fgene.2022.881445

Osman AH (1984) Sudanese indigenous cattle breeds and the strategy for their conservation and improvement. FAO Anim Prod Health 44(1):58–66

Ouedraogo D, Soudre A, Yougbare B, Ouedraogo-Kone S, Zoma-Traore B, Khayatzadeh N, Traore A, Sanou M, Meszaros G, Burger PA, Mwai OA, Wurzinger M, Solkner J (2021) Genetic improvement of local cattle breeds in West Africa: a review of breeding programs. Sustain For 13:2125

Oyenuga VA (1967) Agriculture in Nigeria: An Introduction. FAO, Rome, Italy, pp 240–253

Pagot J (1985) L'Elevage en Pays Tropicaux. Editions Maisonneuve, G. P. et Larousse, Paris, France, pp 380–383

Payne WJA (1970) Cattle production in the tropics. Volume I. Tropical agricultural series. Longman Group Ltd, Western Printing Services Ltd, Bristol, p 336

Payne WJA, Wilson RT (1999) Animal husbandry in the tropics, 5th edn. Blackwell Science, Oxford, UK

Penda NN, Adama S, Saliou N, Mbacke S, Jerome SG (2014) Phenotypical characterization of Senegalese local cattle breeds using multivariate analysis. J Anim Vet Adv 13(20):1150–1159

Popescu CP (1980) Etude cytogénétique d'un lot de bovins Africains. Résultats préliminaires. Premier Colloque International: Recherches sur l'Elevage Bovin en Zone Tropicale Humide, 18–22 Avril, Bouaké. Côte d Ivoire, pp 821–822

Pullan NB (1979) Productivity of White Fulani cattle on the Jos plateau, Nigeria I. Herd structures and reproductive performance. Trop Anim Health Prod 10:229–236

Queval R, Petit JP, Hascoet MC (1971) Analyse des hémoglobines du Zébu arabe (Bos indicus). Rev Elev Méd vét Pays trop 24(1):47–51

Rege JEO (1999) The state of African cattle genetic resources. I. Classification framework and identification of threatened and extinct breeds. FAO/UNEP Anim Genet Resour Inf Bullet 25:1–25

Rege JEO, Tawah CL (1999) The state of African cattle genetic resources II. Geographical distributions, characteristics and uses of present-day breeds and strains. FAO/UNEP Anim Genet Resour Inf Bullet 26:1–25

Rege JEO, Kahi AK, Okomo-Adhiambo M, Mwacharo J, Hanotte O (2001) Zebu cattle of Kenya: uses, performance, farmer preferences, measures of genetic diversity and options for improved use. Animal Genetic Resources Research 1. ILRI (International Livestock Research Institute), Nairobi, Kenya, p 103

RIM (1992) Nigerian livestock resources (6 vols). Report by Resource Inventory and Management Limited (RIM) to FDLPCS, Abuja, Nigeria

Rushalaza VG, Kasonta JS, Nkungu DR, Jumbe BH (1993) On-farm performance evaluation of the dual purpose Mpwapwa breed cattle in semi-arid central Tanzania. In: Proceedings of the 20th scientific conference of the Tanzania Society of Animal Production, pp 62–70

Saeed AM, Ward FA, Light D, Durkin JW, Wilson RT (1987) Characterization of Kenana cattle at Um Banein, Sudan. ILCA research report, No. 16. The International Livestock Centre for Africa (now ILRI), Addis Ababa, Ethiopia

Salim B, Taha KM, Hanotte O, Mwacharo JM (2014) Historical demographic profiles and genetic variation of the East African Butana and Kenana indigenous dairy zebu cattle. Anim Genet 45:782–790. https://doi.org/10.1111/age.12225

Santoze A, Gicheha M (2019) The status of cattle genetic resources in West Africa: *a review*. Adv Anim Veteri Sci 7(2):112–121. https://doi.org/10.17582/journal.aavs/2019/7.2.112.121

Schoenbrun DL (1993) Cattle herds and banana gardens: the historical geography of the Western Great Lakes region, ca AD 800-1500. Afr Archaeol Rev 11:39–72. https://doi.org/10.1007/BF01118142

Syrstad O (1990) Mpwapwa cattle. An Indo -Euro-African synthesis. Trop Anim Health Prod 22:17–22. https://doi.org/10.1007/BF02243492

Tawah CL, Rege JEO (1996a) White Fulani cattle of West and Central Africa. Anim Genet Resour Inf 17:127–146. https://doi.org/10.1017/S101423390000064X

Tawah CL, Rege JEO (1996b) Gudali cattle of West and Central Africa. Anim Genet Resour Inf 17:147–164. https://doi.org/10.1017/S1014233900003448

Tesfa A, Bimerew T, Tilahune M, Kassahun D, Kebede A, Mengesha W (2022) Evaluation of the breeding practices and population trend of the Fogera cattle breed in Ethiopia. Front Anim Sci 3:998628. https://doi.org/10.3389/fanim.2022.998628

Thorpe W, Cruickshank DKR, Thompson R (1981) Genetic and environmental influences on beef cattle production in Zambia. 4. Weaner production from purebred and reciprocally crossbred dams. Anim Prod 33(2):165–177. https://doi.org/10.1017/S0003356100024004

Thumbi SM, Bronsvoort MBM, Kiara H, Toye PG, Poole P, Ndila M, Conradie I, Jennings A, Handel IG, Coetzer JAW, Steyl J, Hanotte O, Woolhouse MEJ (2013) Mortality in East African shorthorn zebu cattle under one year: predictors of infectious-disease mortality. BMC Vet Res 9:175. http://www.biomedcentral.com/1746-6148/9/175

Trail JCM, Buck NG, Light D, Rennie TW, Rutherford A, Miller M, Pratchett D, Capper BS (1977) Productivity of Africander, Tswana, Tuli and crossbred beef cattle in Botswana. Anim Sci 24:57–62. https://doi.org/10.1017/S0003356100039209

Traore A, Koudande DO, Fernandez I, Soudre A, Granda V, Alvarez I, Diarra S, Diarra F, Kabore A, Sanou M (2015) Geographical assessment of body measurements and qualitative traits in West Africa cattle. Trop Anim Health Prod 47(8):1505–1513. https://doi.org/10.1007/s11250-015-0891-7

UBOS., The Uganda Demographic and Health Survey (UDHS), Uganda Bureau of Statistics (UBOS) (2011) Government of Uganda. https://microdata.worldbank.org/index.php/catalog/1539

Wagenaar KT, Diallo A, Sayers AR (1986) Productivity of transhumant Fulani cattle in the inner Niger delta of Mali. ILCA research report No. 13. The International Livestock Centre for Africa (ILCA), Addis Ababa, Ethiopia

Ward HK, Tawonezvi HPR (1983) Production traits of Mashona, Nkone and Tuli cattle and of some beef breeds exotic to Zimbabwe. Animal genetic resources in Africa. In: 2nd OAU Expert Committee meeting on animal genetic resources in Africa. OAU/STRC/IBAR publication, Nairobi, Kenya, pp 86–95

Western D, Finch V (1986) Cattle and pastoralism: survival and production in arid lands. Hum Ecol 14(1):198. https://link.springer.com/content/pdf/10.1007/bf00889211.pdf

Wilson RT (1988) La production animale au Mali central: études à long terme sur les bovine et les petite ruminants dans le système agropastoral. Rapport de recherche. no 14. ILCA, Addis Ababa, Ethiopia

Wilson RT (2009) Fit for purpose – the right animal in the right place. Trop Anim Health Prod 41:1081–1090. https://doi.org/10.1007/s11250-008-9274-7

Wilson RT (2018) Crossbreeding of cattle in Africa. J Agric Environ Sci 7(1):16–31. https://doi.org/10.15640/jaes.v7n1a3

Wilson RT (2021) When is a "breed" not a breed: the myth of the Mpwapwa cattle of Tanzania. Trop Anim Health Prod 53:233. https://doi.org/10.1007/s11250-021-02669-4

Wurzinger M, Ndumu D, Baumung R, Drucker AG, Okeyo AM, Semambo DK, Sölkner J (2006) Assessing stated preferences through the use of choice experiments: valuing (re)production versus aesthetics in the breeding goals of Ugandan Ankole cattle breeders. In: Proceedings of 8th world congress on genetics applied to livestock production (WCGALP). Belo, Horizonte

Yakubu A, Idahor KO, Haruna HS, Wheto M, Amusan S (2010) Multivariate analysis of phenotypic differentiation in Bunaji and Sokoto Gudali cattle. Acta Agric Slov 96:75–80

Yousif IA, El-Moula AAF (2006) Characterization of Kenana cattle breed and its production environment. Anim Genet Resour 38:47–56. https://doi.org/10.1017/s1014233900002042

Zander KK, Drucker AG, Holm-Muller K (2009) Costing the conservation of animal genetic resources: the case of Borana cattle in Ethiopia and Kenya. J Arid Environ 73:550–556. https://doi.org/10.1016/j.jaridenv.2008.11.003

Open Access This chapter is licensed under the terms of the Creative Commons Attribution 4.0 International License (http://creativecommons.org/licenses/by/4.0/), which permits use, sharing, adaptation, distribution and reproduction in any medium or format, as long as you give appropriate credit to the original author(s) and the source, provide a link to the Creative Commons license and indicate if changes were made.

The images or other third party material in this chapter are included in the chapter's Creative Commons license, unless indicated otherwise in a credit line to the material. If material is not included in the chapter's Creative Commons license and your intended use is not permitted by statutory regulation or exceeds the permitted use, you will need to obtain permission directly from the copyright holder.

African Goat Genetic Resources, Diversity and Unique Features

5

Moses Okpeku, Martha N. Bemji, Isidore Houaga, Khaled Fantazi, Liveness J. Banda, Timothy Gondwe, Sebastine Chenyambuga, Sahar A. Elnahta, Doctor M. N. Mthiyane, Shumuye Belay, Tadelle Dessie, Taiye S. Adewumi, and Oliver Hanotte

Abstract

This chapter introduces indigenous African goats, exploring their origins of domestication, dispersal and distribution across the African continent. The first two sections provide a general overview (Sects. 5.1 and 5.2). The following three sections (Sects. 5.3, 5.4, and 5.5) are devoted to the history of domestication, dispersal and distribution, tracing the journey from the centres of domestication to

M. Okpeku (✉)
Discipline of genetics, School of Life Science, University of KwaZulu-Natal, Durban, South Africa
e-mail: OkpekuM@ukzn.ac.za

M. N. Bemji
Department of Animal Breeding and Genetics, Federal University of Agriculture Abeokuta, Adzho, Nigeria

I. Houaga
The Roslin Institute, University of Edinburgh, Edinburgh, UK

Centre for Tropical Livestock Genetics and Health (CTLGH), The Roslin Institute, University of Edinburgh, Edinburgh, UK

K. Fantazi
Research Division of Animal Production, Algerian National Institute of Agronomic Research (INRAA), Algiers, Algeria

L. J. Banda · T. Gondwe
Department of Animal Science, Bunda College, Lilongwe University of Agriculture and Natural Resources, Lilongwe, Malawi

S. Chenyambuga
Sokoine University of Agriculture, Morogoro, Tanzania

S. A. Elnahta
Biotechnology Institute, National Research Centre, Giza, Egypt

D. M. N. Mthiyane
Department of Animal Science, School of Agricultural Sciences, Faculty of Natural and Agricultural Sciences, North-West University, (Mafikeng Campus), Mmabatho, South Africa

Food Security and Safety Focus area, Faculty of Natural and Agricultural Sciences, North-West University, Mmabatho, South Africa

S. Belay
Tigray Agricultural Research Institute, Tigray, Ethiopia

T. Dessie
International Livestock Research Institute, Livestock Genetics, Addis Ababa, Ethiopia

T. S. Adewumi
Department of Parasitology and Entomology, Federal University of Agriculture Abeokuta, Adzho, Nigeria

O. Hanotte
International Livestock Research Institute, Livestock Genetics, Addis Ababa, Ethiopia

Faculty of Medicine & Health Sciences, Nottingham, UK

© The Author(s) 2026
E. M. Ibeagha-Awemu et al. (eds.), *African Livestock Genetic Resources and Sustainable Breeding Strategies*, Sustainable Development Goals Series, https://doi.org/10.1007/978-3-031-92076-9_5

their present-day locations. Section 5.5 focuses on the characterisation of available genetic diversity, categorising goats into three major groups: indigenous, exotic and composite breeds. Each subsection examines the origin, distribution, classification, physical characteristics, adaptive characteristics and production characteristics of named breeds. Section 5.6 discusses the characterisation of genetic diversity in African goats. It highlights the role of modern molecular markers in the characterisation of genetic diversity and the utilisation of known gene regions to understand adaptation, physiology and production functions in goats. Section 5.7 presents an overview of indigenous knowledge on the utilisation of various goat products. The chapter concludes with a future outlook (Sect. 5.8) on African goat genetic resources, utilisation and potential conservation strategies for the present and future.

Keywords

Africa · goat · genetic resources · diversity · characterisation

5.1 Introduction

Africa is home to the largest animal diversity on earth, with sub-Saharan Africa accounting for the majority of this diversity (UNEP-WCMC, IUCN 2016; Chapman et al. 2022). Goats are a ubiquitous animal genetic resource found across the globe (Stella et al. 2018), and Africa is the second largest home to goats in the world, accounting for 36.4% of live animals, with total goat meat and milk production estimated at 1,082,120.07 tonnes (23.4%) and 3,674,644.96 tonnes (23.1%), respectively (FAOSTAT 2019). Nigeria, Sudan, Chad and Kenya are key producers of goats on the continent, bringing them into the rank of ten top producers in the world (FAOSTAT 2019).

African goats are referred to as 'the poor man's cow' (Steele 2006) because of their important roles in rural agricultural systems. They feature conspicuously in the economy of most nations of the world, particularly in the rural poor agricultural systems where they are kept for food and their roles in the sociocultural and financial security of their keepers (FARM-Africa 1994). Their small stature has many advantages, which include easy acquisition, rearing and maintenance, cheaper to raise than cows with nearly the same economic return, and exploitation of marginal lands.

A total of 289 goat breeds/populations exist in Africa (AU-IBAR 2019), the majority of which are not characterised. Eastern Africa has the highest number of goat breeds (101), followed by Southern Africa (79), Western Africa (47), Northern Africa (33) and the Central African region (29). The highest number of goat breeds is found in Ethiopia (26), followed by Tanzania (17), Sudan (12) and Kenya (10). Cameroon and Chad have seven and eight goat breeds, respectively. Nigeria and Mali have the highest number, at nine breeds each in West Africa, while the number of breeds ranged from seven to nine in North Africa, comprising Egypt (9), Algeria (8), Mauritania (8) and Morocco (7).

African goats have a long and intertwining history with their human keepers' culture and traditions: featuring in tales and stories, used for religious worship, marriage contracting and dowry payment, and much more. Apart from these, goats are kept majorly for the meat, a savoured delicacy, relished in many African communities that goes well with many traditional dishes and has no taboo attached (Oluwatayo and Oluwatayo 2012). Almost all parts of the goat are useful. Major economic values come from the meat (protein), hide is used for quality leather works, the horns, hooves and bones are also sourced after for art and craft materials. Their fur finds great use in the fabric industry, while faecal droppings and intestinal contents are regarded as a valuable source of manure exploited by the rural poor farmers as a source of cheap fertiliser for use on farmlands.

Goats found with local residents are referred to as **indigenous goat types**; these are particularly adapted to the regions where they are found and have a history as old as the local inhabitants. Non-resident goat types imported from elsewhere within the continent or from outside the continent into a region are regarded as **exotic goat types;**

these are often imported to cross the indigenous type (for varied reasons), and the hybridised generation is referred to as **locally developed goat types** (Dekkiche 1987; Takoucht 1998).

Indigenous African goats represent a strategic genetic resource with unique features and diversity (Monua et al. 2020; Magoro et al. 2022). Their unique features include: **Adaptability**; the ability to thrive and survive on marginalised resources (Daramola and Adeloye 2009). This includes the evolution to thrive in a range of environments, from arid and semi-arid to regions of extreme humidity (Monua et al. 2020). This also includes the evolution of a digestive system that enables them to extract the maximum amount of nutrients from their food and disease tolerance, which enhances their survivability in harsh and hostile environments in Africa (Daramola and Adeloye 2009; Berihulay et al. 2019; Akinmoladun et al. 2020). In terms of **genetic diversity**, there are many different breeds of indigenous African goats, each with its own unique characteristics. The evolution of a diversity of coat types and colours and varied morphological characteristics, which make them fit well into the specific locations where they are found and the culture and traditions of their owner make them remarkable (Berihulay et al. 2019) specimens. Different breeds of goats are subsequently discussed, taking into consideration the following: *origin and distribution*, p*hysical characteristics, adaptive and special genetic characteristics* and t*ypical production systems and production characteristics.* See Chap. 2 for comprehensive information on production systems of some named breeds to reflect their geography, adaptation and uses where they are found. The named goat breeds discussed are listed in Table 5.2.

5.2 Origin and Domestication of African Goats

Animal genetic resources represent actual genetic material and/or their potential value. Goats are small ruminants belonging to the tribe called *Caprini,* which is further divided into two genera, *Capra* and *Hemitragus*. The domesticated goat originates from the *Capra* genus with 60 chromosomes and includes five species (Table 5.1), as reported by Steele (2006).

The development of the domestic goats of the world has long been a subject of serious debate. It was general knowledge that the domesticated goats of the world were derived from the ancient wild species, but there was no strong evidence of the true ancestor of the domesticated goat among the known wild goat types: *Capra aegagrus, Capra ibex, Capra Caucasica, Capra pyrenaica*, and *Capra falconeri* (Table 5.1). However, recent research suggests the Bezoar (*Capra aegagrus*), native to the montane forests of the Caucasus and the Zagros Mountains and also found in Eastern Europe, the Middle East and Asia (Eckart 2024) is the most likely ancestor of domestic goats—*Capra hircus* (Lin et al. 2013). A DNA study by Colli et al. (2015) places Bezoar goats (*Capra aegagrus*) as the wild ancestor of domestic goats, with multiple domestication centres contributing to the existence of the domesticated goats (Daly et al. 2018). The author and the team emphasise the need for ancient genomics to shed more light on the pattern of domestication, which is mosaic. They suggested that although this domestication took place in different places, with different people, the same wild goat (Bezoar goat) was linked to the domestication of African goats. Another study by Daly et al. (2018) pointed out the weak phylogenetic relationships between the other four wild goats with *Capra hircus,* discarding their contribution to the domestic goat development.

A signature of population expansion in bezoars of the C haplogroup suggested an early domestication centre on the Central Iranian Plateau (Yazd and Kerman Provinces) and in the Southern Zagros (Fars Province), possibly corresponding to the management of wild flocks. However, the contribution of this centre to the domestication of the goat population of this era is rather low (1.4%). A second domestication centre covering a large area in Eastern Anatolia, and possibly in Northern and Central Zagros, was proposed (Naderi et al. 2008). This last domestication centre (around modern-day Iran) (Alberto et al. 2018) is taken as the likely origin of almost all domestic goats today; this is consistent with

Table 5.1 Five sub-species of *Capra* (domesticated goat)

S/N	Species of goat	Centre of origin
1	*Capra hircus* (Bezoar)	West Asia
2	*Capra ibex* (Ibex)	Central Asia, Near East, Alps
3	*Capra Caucassica* (Tur)	West Caucasia
4	*Capra pyrenaica* (Spanish Ibex)	Pyrenees
5	*Capra falconeri* (Markhor)	Afghanistan/Pakistan

archaeological data identifying Eastern Anatolia as an important domestication centre (Daly et al. 2018; Naderi et al. 2008; Alberto et al. 2018; Naderi et al. 2007; Zheng et al. 2020; Daly et al. 2021). Goat domestication was probably achieved in present-day Iran and Iraq by 10,000 BC, from where goats entered into North Africa around 7000–6000 BC (Mason 1984; Shackleton 1997; Zeder 2008) and dispersed southward until they reached the coastal regions of the South.

True goats do not occur wild in Africa. African goats were introduced from the centre of domestication through Southeast Asia trade routes. Although the exact origin and path of arrival in Africa have been long debated, goats found in Africa are believed to have originated from the Zagros Mountains, from where they migrated as companion animals with the humans (Zheng et al. 2020; Joubert 1973).

5.3 Dispersal and Distribution of Goats Across Africa

The dispersal and distribution of goats across Africa from archaeological and genetic findings provide insights into their migration and genetic history, suggesting that goats entered Africa from the Middle East around 7000 years ago, accompanying human migration (Zheng et al. 2020). Genetic studies also identified multiple waves of migration and mixing with North African goats as humans migrated, shedding light on the complex dispersal patterns of goats across the continent (Zheng et al. 2020; Chynoweth et al. 2015; Pereira et al. 2009; Amills et al. 2017; Cooper 2021). Following domestication, goat dispersal followed human movement and trade routes as companions (Pereira et al. 2009; Amills et al. 2017).

They were originally kept for food and carried along because of the ease of transportation and their quick adaptation. The preference for goat meat and milk by human beings as food helped in the fast dispersion across Africa and the rapid spread to Central Sahara and the Ethiopian highland between 6500 and 5000 years ago (Amills et al. 2017; Cooper 2021). The quick dispersal and distribution of goats across Africa was aided by their ability to adapt quickly and survive on meagre feed, where other larger ruminants failed. Their dispersal across Central Africa might have been slowed down by trypanosomiasis, transmitted by tsetse fly (Amills et al. 2017). However, the population that survived are believed to have continued to travel with their human companions and finally reached the southern parts of Africa about 2000 years before the present time. Today, goats are found everywhere, from the high-temperature, arid northern region to the marshy swamps and salty coastal region in the south of the continent (Zheng et al. 2020; Amills et al. 2017).

5.4 African Goat Genetic Resources

Across Africa, from the arid north to the ocean bank in the south, across plains, grassland and safaris, even mountains, dense forests and marsh swamps on the continent have some form of goat inhabiting them. The African goat genetics resources represent a great diversity described differently **by size** into small, medium and large types; **by production** into dairy and meat types; by hair length as long and short hair types; by location into Northern, Western, Central, Eastern and Southern types. On the account of historical origin, they have been broadly grouped into

indigenous, **exotic** and **developed** breed types. Among the indigenous class, five breed groups have been recognised, which include (i) Lop-eared, (ii) Short-Eared Small-Horned and (iii) Short-Eared Twisted-horned. This complex diversity has also been described phenotypically using known morphology, but physical resemblance is limited, so molecular characterisation has also been explored in these goats to understand their genetic architecture for immediate and future use and conservation. This chapter summarises the different modes of classification of these complex diversities (Sect. 5.6) and presents some important value chains derived from the African goats.

5.5 Characterisation of Genetic Diversity in African Goats

The characterisation of animal genetic resources provides valuable information for classification, use and prioritisation for conservation. Such information includes phenotype, genotype and history of use and distribution. Genetic and historical information curated on the goat varies from place to place in Africa and is dependent on countries, national priorities and goals for animal genetic resources. In order to synchronise the different characterisations used in literature, we have arranged the different groups by origin into indigenous, exotic and developed types and, under these, described some common types by phenotype, production and physiology. Later in the chapter, we considered molecular characterisation and indigenous knowledge on some value chain products.

Research indicates that goats did not enter sub-Saharan Africa until about 4000 years ago, with genetic markers suggesting that the dispersal of goats into North Africa may have been the result of maritime trading, with male goats being introduced from Southwest Asia along the North African coast (Chynoweth et al. 2015).

The article 'A Brief History of Goat Domestication' by Tamsin Cooper emphasises the social nature and adaptability of goats, which enabled their domestication and subsequent spread across different regions. It highlights the importance of protecting local and heritage goat breeds to maintain the diversity of the worldwide gene pool, crucial for the survival of goats in a changing future (Cooper 2021).

In summary, the dispersal and distribution of goats across Africa have been influenced by human migration and trade, as evidenced by archaeological and genetic studies. The genetic history of goats provides valuable insights into their complex migration patterns and the need to preserve genetic diversity for the sustainability of goat populations.

5.5.1 Indigenous Goat Breeds

Indigenous or local goats represent goat populations developed in a particular region over time and are adapted to the prevailing local environment. They are an active part of the culture and tradition of the people and locality where they are found. Across all regions of Africa, there are unique goat types associated with the people and their cultural history. These classes of goats can best be described by regions as northern, western, eastern, central and southern in origin. Some of these also bear regional names, like the West African dwarf goats found in the western, central and eastern regions of the continent. Their northern counterparts are sometimes referred to as northern or Arab goats. Within each region, the diversity exhibited by these goats has necessitated the use of unique breed names as well. The majority of the native goats are generally characterised by small stature, multicoloured coats, and hardiness, with unique adaptation features to regions where they are found. They can be very prolific, with twins and triplets being common among some breeds. They are particularly close to and found very useful applications in local, smallholder agriculture (Monua et al. 2020; Okpeku et al. 2011a; Tarekegn et al. 2019).

5.5.1.1 Lop-Eared Goats
This group of goats is characterised by the long drooping (lop) ears and includes the Zaraibi of Egypt and the **Nubian** of the Sudan. Similar

types of goats are heavily represented in the region of North Africa. The ancestral stock might have evolved either in India, subsequent to the Indus Valley civilisation or west of India, possibly Iran, from where it spread to Syria and Egypt in the west. It is also possible that this goat type evolved from the screw-horned goats common throughout the ancient world, from India in the east to Libya in the west, judging from the screw horns common in the different breeds found in different parts of Africa. The so-called Nubian goat probably does not in fact originate from Nubia (the area of southern Egypt and northern Sudan) and certainly not from Ethiopia, but bears the convex profile, a common characteristic of goats found in the Middle East and India (Mason 1984). It is therefore possible that the different breeds found in Africa are close relatives or descendants of those found in the Middle East and Asia. The lopped-ear goats share common features which have been extensively documented in the literature. However, there are a couple of unique features that separate breeds from region to region. It has been argued that these features were perhaps developed at a later stage after domestication in response and adaptation to the prevailing environments where they are found. This adaptation feature is perhaps the reason why goats are so ubiquitous and found everywhere.

5.5.1.2 Nubian Goats

Origin, Distribution and Classification: Nubian goats (Fig. 5.1) are said to have originated in Africa, especially in the Sudan region. However, they have long been popular in India and the Middle East as well. The Lop-eared Nubian goat, also known as the Anglo-Nubian goat, was developed in the nineteenth century as a result of extensive crossbreeding between imported Nubian goats and a variety of native British goats. They are found commonly in the Middle East, North Africa and more than sixty other countries (Porter et al. 2016). They are believed to have been brought from France, through the eastern Mediterranean to the regions between Egypt and Sudan along the Nile, referred to as

Fig. 5.1 Nubian goat

Nubia (Britannica 2016), this is where the word 'Nubian' was first used. Despite their existence in several Asian, African and Central/South American nations, this species is widespread and not at risk. However, the lack of suitable, unrelated breeding partners puts available small, isolated groups in danger.

Physical Characteristics: The Nubian is frequently referred to as a royal animal with long, broad ears that droop, huge almond-shaped eyes, a prominent forehead, a convex 'roman' nose, a tall, flat-sided body, long legs and a short, shiny coat, all contributing to the Nubian's unique appearance. The goat's short coat might be spotted, parti-coloured, or any other solid colour, including golden brown, reddish-brown, black and dark grey, with the most common colour being a blend of brown and white (Porter et al. 2016). White facial stripes might be a sign of crossbreeding with Swiss-born goats. Bucks average 36 in. (90 cm), and does 32 in. (80 cm). Mature weights are usually a minimum of 174 lb. (79 kg) for both sexes and a maximum of 309 lb. (140 kg) and 243 lb. (110 kg) for bucks and does, respectively (Porter et al. 2016; Transboundary breed summary 2024).

Adaptive Characteristics: Nubians are able to adapt to hot temperatures due to their wide ears and flat flanks. They do not, however, fare well in humid environments. They have great fecundity and can reproduce all year long.

Production Characteristics: Compared to most other goat breeds, the Nubian has milk with a greater fat content. They produce 6.6 lb (3.9 kg) of milk on average every day, or 1920 lb (871 kg) over a period of 305 days, containing 4.8% butterfat and 3.5% protein. The majority of Nubians have genes that result in high levels of alpha s1-casein synthesis, a crucial protein for cheesemaking and a significant goat milk advantage (Rahmatalla et al. 2022). They are also produced mainly for meat and milk (Kholif et al. 2020). Crossbreeding with native stock is very common in Latin American, Asian and African nations to increase milk or meat output. These are unique products of economic benefit derived from Nubian goats.

5.5.1.3 Sinai Goats

Origin and Distribution: Lop-eared Sinai goats are domestic goats that originated in the Sinai Peninsula, a triangular landmass located between the Mediterranean Sea and the Red Sea, and are found specifically in the southern part of the Sinai Peninsula, in Egypt (Pereira et al. 2009; Monteiro et al. 2021).

Physical Characteristics: Lop-eared Sinai goats are characterised by their drooping or lop ears, which is a distinct feature of the breed (Monteiro et al. 2021). The breed has a medium-sized body with a well-developed chest and a straight or slightly convex profile, with a short and dense coat that comes in various colours, including white, black, brown and a combination of these (Monteiro et al. 2021).

Adaptive and Special Genetic Characteristics: These goat types are adapted to the arid and harsh environment of the Sinai Peninsula, characterised by high temperatures, low rainfall and limited vegetation (Monteiro et al. 2021). They are known for their ability to browse on a wide range of plant species, including shrubs, trees and herbs, which allows them to survive their arid habitats. The breed is also known for its resistance to various diseases and parasites, a valuable trait in the challenging environment of the Sinai Peninsula (Monteiro et al. 2021; Nguluma et al. 2021; Kardjadj and Ben-Mahdi 2017).

Typical Production Systems and Production Characteristics: The production systems for the lop-eared Sinai goats are extensive, with the goats being raised in semi-nomadic or nomadic herds (Monteiro et al. 2021). They are raised mainly for the meat and milk, which contribute importantly to the local community diet, particularly cheese, which is highly valued in the region. They also contribute meat as protein (Monteiro et al. 2021).

5.5.1.4 Swazi/Nguni

Origin and Distribution: Lop-eared Swazi/Nguni goats are indigenous to South Africa, specifically the Swaziland and Nguni regions. They are part of the Indigenous Veld Goats (IVG) group, which includes pure-bred indigenous ecotypes represented by the IVG-Association (Visser and Snyman 2023).

Physical Characteristics: Lop-eared Swazi/Nguni goats are medium- to small-framed, well-proportioned, with strong, fine legs (FAOSTAT 2019). They have lop ears that hang down (Visser and Snyman 2023). They have a variety of coat colours, including black, white and multicoloured, and possess morphological variations between and within populations, leading to some categorisation based on quantitative data (Nguluma et al. 2021).

Adaptive and Special Genetic Characteristics: The breed is referred to as an untapped genetic resource (Visser and Snyman 2023) that is highly adapted to the South African region. Adaptation to harsh environments in this breed may be linked to the morphological appearance and/or physical attributes, including lop ears (Onasanya et al. 2021).

Typical Production Systems and Production Characteristics: Indigenous goat production in the southern Africa region is typically an extensive system with goats often allowed to scavenge for food and water.

5.5.1.5 Berber

Origin and Distribution: Lop-eared Berber goats are a breed of domestic goats that are native to the Amazigh communities of Morocco. These goats were adopted by the Berbers for subsistence agriculture, and they have been an integral part of the local farming systems for many years (Hossaini-Hilaii and Benlamlih 1995).

Physical Characteristics: This goat type is known for its distinctive drooping ears, which hang on the side of its head. The breed has a compact and sturdy body with a deep chest and straight back with varied coat colours, including white, black, brown or a combination of these colours. Both males and females of this breed have horns, which are usually twisted and grow outward and backward (Benjelloun et al. 2011; Boujenane et al. 2016).

Adaptive and Special Genetic Characteristics: Lop-eared Berber goats are well adapted to the mountainous and arid regions of Morocco, where they are raised. These goats have excellent adaptability to difficult environments, which contributes to their success in various production systems. They are also among a few goat breeds that are well adapted to climbing trees (Benjelloun et al. 2015, 2016).

Typical Production Systems and Production Characteristics: Lop-eared Berber goats are primarily used for subsistence agriculture by the Amazigh communities in Morocco. They are raised for meat and milk production mainly and also for their hides and fibre. Their production systems are not well documented in the literature, but it is basically similar to the extensive production system common with most indigenous African goats. In addition, local farmers use them for bush control, as they are very effective in this role (Hobart 2022).

5.5.1.6 Benadir

Origin and Distribution: Lop-eared Benadir goats are a breed of goat that originated in Somalia. They are commonly found in the Benadir region of southern Somalia as well as in the neighbouring regions of Kenya and Ethiopia (Fereja 2016).

Physical Characteristics: Lop-eared Benadir goats are medium-sized goats, with males weighing between 45 and 55 kg and females between 30 and 40 kg (Fereja 2016). They have long legs and drooping ears that hang over their shoulders; their coat colour is typically black or brown, although some individuals may have white markings on their faces or legs (Fereja 2016).

Adaptive and Special Genetic Characteristics: There is limited information available on the adaptive genetic characteristics of this breed. However, as a breed that has evolved in a harsh, arid environment, it is likely that they possess traits that allow them to thrive in such conditions, such as drought tolerance and disease resistance.

Typical Production Systems and Production Characteristics: Lop-eared Benadir goats are primarily raised for meat production (Fereja 2016). They are well-suited for the arid, semi-arid and sub-humid regions of East Africa, where they are often kept in pastoral production systems characterised by extensive grazing on natural pastures, with little or no supplementary feeding (Fereja 2016). According to Nguluma et al. (2021), indigenous goat production in East Africa faces several challenges. The authors also reported morphological variations in the goat populations from this region, suggesting the possibility of selective breeding to improve productivity. Sufficient understanding of their genetics and production characteristics is important before embarking on improvement work.

5.5.1.7 Wahati

Origin and Distribution: Lop-eared Wahati are indigenous to Tanzania (Nguluma et al. 2021). They are found in different agroecological zones of Tanzania, including Pare, Gogo, Maasai, Tanga, Sukuma, Lindi, Ujiji, Pwani, Fipa, Songwe and Newala (Nguluma et al. 2021).

Physical Characteristics: Morphological variations between and within goat populations based on quantitative data have been reported for the breed (Nguluma et al. 2021). High twinning has also been observed in Ujiji and Lindi goats and low for Sukuma. The dominant coat colour is white in Pare, Gogo, Maasai and Tanga. Other coat colour patterns are mixed black and white for Sukuma and reddish-brown for Pwani and Maasai (Nguluma et al. 2021).

Adaptive and Special Genetic Characteristics: There is a paucity of information on adaptive and special genetics for this population in the literature, suggesting the need for research into this.

Typical Production Systems and Production Characteristics: This goat breed is often raised in village goat production systems in Tanzania, where feed shortage, prevalence of diseases and water scarcity are the major goat production constraints (Nguluma et al. 2021). Common diseases in the breed population include contagious caprine pleuralpneumonia and helminthiasis (Nguluma et al. 2021).

5.5.1.8 Begait

Synonyms: Barka, Hassan, Bellenay and Beni-Amer

Origins, Distribution and Classification: Lop-eared Begait or Barka goats descended from the pasang (*Capra aegagrus*), which is probably native to Asia (Nigatu 1994). These goat breeds are grouped under the Nubian goat family (FARM-Africa 1996). They are among indigenous goat types and distributed in the northwestern lowlands of Ethiopia (west and northwest Tigray), Eritrea's Barka Valley and eastern Sudan. More than 110,000 Begait goat populations are estimated to be found in the northwestern and western zones of the Tigray region only. The goat producers in Ethiopia sometimes call it 'Barka' for the Begait goats because they have very similar phenotypic characteristics, though they have some unique features in each breed (eg in coat colour difference and neck length). The name 'Barka' is commonly used in Eritrea and Begait in Ethiopia. The breed is also named Bellanay and Beni-Amer because they are kept by these two ethnic groups in Eritrea. All in all, this breed is grouped under the Nubian family and distributed in three African countries.

Physical Characteristics: Body size, horn shape, ear shape and coat types are key physical characteristics used in classifying Ethiopian goat families and breeds (Gizaw et al. 2010). Begait goats are a dual-purpose goat population mainly used for milk and meat production. They have a large body size, horns with backwards orientation, heavy and drooping ears, a concave face profile, white (43%) and white with black (31%) coat colours, fast growth, a large udder and teat, and have no wattles. Under an extensive production system, the mean body weight of a mature buck is 45 kg and 38 kg for does; height: male 74 cm and female 68 cm (Berihulay et al. 2019).

Adaptive and Special Genetic Characteristics: Adaptive features in this meat goat breed include environmental and production adaptability, reproductive and growth rates, and carcass (Luginbuhl et al. 1998) and milk characteristics. It is adaptable to arid and semi-arid environments. The SCN9A and CAB39L candidate genes are identified in Begait goats, which are associated with body size and weight, and cell adhesion receptors, and growth factor receptors, respectively (Berihulay et al. 2019).

Typical Production Systems, Husbandry Practices and Production Characteristics Goat production systems in Ethiopia vary depending on the region and the breed of goat. Almost all goat production systems in Ethiopia have been classified as traditional, with the majority of goats being raised under extensive management systems (Fereja 2016). Natural pasture, browse species, crop residue and crop aftermath are the main feed resources for Begait goats. Communal grazing is the most abundant feed source for these goats, but grazing drastically reduces in the dry season. They do not have a permanent shelter because they migrate from place to place in search of food and water, especially during the dry season. However, farmers in urban areas used permanent houses. Goats and sheep are managed or housed together. Average weight at birth, 3, 6, 9 and 12 months is 2.8 ± 0.04, 11.1 ± 0.43, 16.5 ± 0.49, 19.6 ± 0.37 and 24.1 ± 0.48 kg, respectively, in the semi-intensive production system and 2.6 ± 0.03, 10.3 ± 0.46, 14.8 ± 0.57, 17.1 ± 0.49 and 20.6 ± 0.54 kg in the extensive production system. Litter size of the Begait goat varies from 1 to 3, with the mean of 1.53 ± 0.03 in semi-intensive and 1.51 ± 0.03 in extensive production systems. Begait does (Fig. 5.2a) supplemented with vetch hay provide 48 kg of milk within 71 days of lactation length (Berhane and

 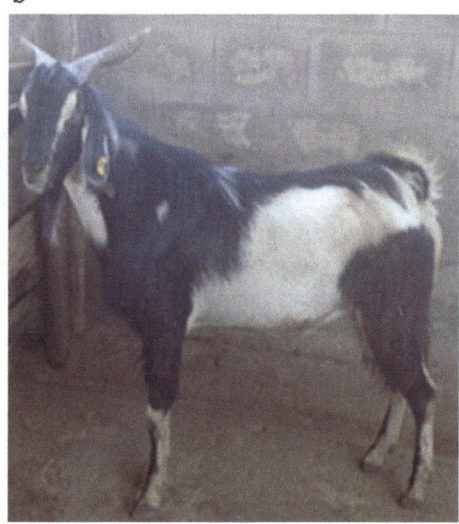

Fig. 5.2 (**a**) Begait does with its kids. (Photo by Shumuye Belay). (**b**) Begait buck. (Photo by Shumuye Belay)

Eik 2006), bucks (Fig. 5.2b) are also maintained on supplements in harsh weather.

5.5.1.9 Baladi

Origin and Distribution: Lop-eared Baladi goats are indigenous goats found mainly in the Nile Delta region and the northern coast of Egypt (Hassanane et al. 2010).

Physical Characteristics: Lop-eared Baladi goats have long ears that hang down to their shoulders. They have a medium-sized body with a narrow chest and straight back. Their coat colour is usually black or brown, but can also be white or spotted (Hassanane et al. 2010).

Adaptive and Special Genetic Characteristics: There is limited information in literature on the adaptive and special genetic characteristics of Baladi goats, making them a population that is poorly researched and poorly exploited. However, they are a hardy breed that is well-suited to the local environment, recognised as an important source of food and income for the small-scale farmers in Egypt (Hassanane et al. 2010).

Typical Production Systems and Production Characteristics: Lop-eared Baladi goats are raised for their meat, milk and skin. They are usually raised in small herds by small-scale farmers, are well adapted to the hot and dry climate of Egypt and survive well on low-quality feed. This breed is also resistant to many of the goat diseases prevalent in the region and classified as average milk producers, producing around 1.5 litres of milk per day (Hassanane et al. 2010).

5.5.1.10 Zaraibi

Origin and Distribution: The lop-eared Zarabi goats are part of the indigenous Nubian goats of Egypt, found in the northeast of the Nile Delta (DAGRIS 2019).

Physical Characteristics: The breed is characterised by elongated drooping ears, a convex facial profile with diverse coat colours that include brown, reddish-brown and spotted (DAGRIS 2019).

Adaptive and Special Genetic Characteristics: Information about their genetic uniqueness is not documented in literature, but we believe they share a similar genetic profile with other breeds found in the Nile Delta region.

Typical Production Systems and Production Characteristics: The breed is primarily a dairy goat and secondarily for meat. The meat is savoury, with a taste that is described as intermediate between beef and lamb (DAGRIS 2019).

5.5.1.11 Zaghawa

Origin and Distribution: Lop-eared Zaghawa goats are indigenous breeds found in the Zaghawa region of Sudan (Jnied et al. 2012).

Physical Characteristics Lop-eared Zaghawa goats have lop ears, which is a distinguishing feature of the breed. They have a medium-sized body with a compact muscular build (Nguluma et al. 2021).

Adaptive and Special Genetic Characteristics: There is a dearth of information on adaptive and special genetic features of the breed.

Typical Production Systems and Production Characteristics: Lop-eared Zaghawa goats are raised in traditional production systems where they are used for milk and meat production; they are known for their high milk yield and good meat quality (Jnied et al. 2012). However, there is a paucity of information on other production characteristics, creating room for research and production exploits with the breed.

5.5.1.12 Kalahari Red

Origin and Distribution: Kalahari Red (Fig. 5.3a, b), a commercial goat breed native to South Africa. Two lines were identified (Campbell 2003; FAO 1991). One was developed from red-head Boer goats, and another was developed from unimproved indigenous goats (Native South African and Namibian landraces). The name of the Kalahari Red goats is derived from their red coat and the *Kalahari Desert,* which spans the borders of Botswana, South Africa and Namibia. Some populations of the Kalahari Red goats are being developed in Australia, Brazil and the United States (Roy's Farm 2021), while some were imported from South Africa to Nigeria (Bemji et al. 2014).

Physical Characteristics: Kalahari Red is a large and tall goat breed (54.05 cm) with a long (69.8 cm), deep body (Pieters et al. 2009), a medium-to-large frame and strong legs. It has solid body colour shades from light to dark red-brown (Fig. 5.1a). White patches sometimes occur in offspring, as their ancestors had white and brown coats. Short, glossy hair bears little undercoat during winter. The pigmented skin is loose and supple. The horns present in both sexes are dark, round, curving backwards behind broad pendulous ears, soft brown eyes and a slightly Roman nose as found in the Boer goat (Fig. 5.18). Does have full and properly attached udders and teats. Multiple, divided, or extra non-functional teats may occur (FAO 1991; Roy's Farm 2021; Sanni et al. 2018).

Adaptive and Special Genetic Characteristics: Kalahari Red goats are extremely hardy and well adapted to free-ranging in arid and semi-arid

Fig. 5.3 (**a**) Kalahari Red bucks at Federal University of Agriculture, Abeokuta (FUNAAB). (Source: Bemji et al. 2014). (**b**) Kalahari Red doe and kids at FUNAAB. (Photo by O. A. Osinowo)

savanna in South Africa and the Kalahari Desert. They have strong herding instincts for protection, and their red coat provides camouflage from predators within the red soil of their native land (Roy's Farm 2021). Pigmented skin provides resistance to high radiation, enabling them to endure heat and continue to forage in hot weather. They are known to tolerate all climates (Bemji et al. 2006, 2012, 2018, 2021).

Typical Production Systems and Production Characteristics: Kalahari Red goats are managed on free-range system mainly for meat and skins. Does are calm, gentle and excellent mothers (Fig. 5.1b), both in their care of young and their protective instincts of flocking well and hiding their kids. Age at first breeding is 6 months. Does are fertile and prolific, bearing twins of equal weight, with peak fertility occurring in fall. They can breed several times a year, raising three litters over 2 years. Birth weight: 2.5 kg; weight at 6 months: 30 kg; mature live weight: 115 kg for males and 75 kg for females (FAO 1991; Roy's Farm 2021) were reported in South Africa. Kids grow fast (about 1.5 kg per week) and yield tender, delicious, low-fat meat (Roy's Farm 2021). In Nigeria, estimates of 151 days (gestation length), 2.14 (litter size) and 3.04–3.38 kg for birth weight of kids from does aged 2 to 2½ years were reported (Oderinwale et al. 2017). Omotosho et al. (2020) obtained 2.30 kg for birth weight of kids from younger does, 8.88 kg for 3-month weaning weight, while daily milk offtake and milk yield were estimated at 26.27 ml and 695.52 ml, respectively (Umejesi et al. 2016).

5.5.1.13 Kil

Origin and Distribution: Information on Lop-eared Kil goats is limited in the literature. However, available information linked the Kil goat to the Tanzania region.

Physical Characteristics: The physical characteristics of the lop-eared Kil goats are also not well documented. However, available information suggests diversity in phenotypic and genetic characteristics, especially in coat colour and ear length (FAO 2012).

Adaptive and Special Genetic Characteristics: Although there is no specific information in literature in this regard, there are suggestions that there is a strong physiological and genetic adaptability in the population to harsh environments (Silanikove and Koluman 2015).

Typical Production Systems and Production Characteristics: Information available suggests that the production indices in this goat breed are similar to those of indigenous goat breeds in sub-Saharan Africa.

5.5.1.14 Barki

Origin and Distribution: Lop-eared Barki are indigenous to Egypt, raised primarily around the Nile Delta region, particularly in the Dakahlia Governorate (DAGRIS 2023).

Physical Characteristics: The breed is referred to as the smallest Egyptian goat breed; it is of medium, compact body size with a straight profile and possesses long, drooping ears and a white or light brown coat colour (DAGRIS 2023).

Adaptive and Special Genetic Characteristics: Lop-eared Barki goats are well adapted to the hot and dry climate of the Nile Delta region. They are resistant to many common goat diseases in the area, such as contagious caprine pleural pneumonia and helminthiasis (Nguluma et al. 2021). They are also known for their high twining rate. Morphological variation exists between and within these goat populations based on qualitative data (Nguluma et al. 2021).

Typical Production Systems and Production Characteristics: Lop-eared Barki goats are raised for meat. They are usually kept in small flocks that are left to graze freely on the natural vegetation (4). The production system for these goats is not well characterised, but feed shortage, prevalence of diseases and water scarcity are the major constraints to their production in the area.

5.5.2 Short-Eared Small-Horned Goat

This goat type was introduced into North Africa (around Egypt) from the east about 3500 BC. Although their dispersal route across Africa is poorly documented, they are also found in regions including: East Africa, the equatorial west, Central Africa and the humid hot climate of West and Central Africa. Goats of this type tend to be dwarfs, believed to be a response to natural selection on thermoregulation under the unfavourable humid and hot climate. In addition, the West African Dwarf goat has short, bowed legs attributed to achondroplasia. The distribution of this goat type extends southwards through central Africa as far as Zaire, Angola and the north of Namibia. They are also found in the western African countries of Nigeria, Ghana and Cameroon.

In Africa, breeds are often named after the main ethnic group keeping them, e.g. the Boer goat, etc. However, some of the names date from the colonial era and can take on a pejorative meaning when used in modern times, e.g. the Galla goat in Kenya. The breeds can also be named after the region in which they are found, which is sometimes synonymous with ethnicity. Types can also be referred to in a more neutral manner by naming them after an important identifying characteristic, e.g. West African Dwarf goat, the Long-eared Somali goat. The names given to the clusters reflect geographical location as much as possible, rather than ethnic group (FARM-Africa 1996).

There are many goat breeds that are classified as short-eared small-horned (Table 5.2). Three families have been identified, and these are the Small East African (SEA), Somali and Ethiopian Rift Valley families. Examples of Small East African breeds are Western Highland, Western Lowland, Central Highland, Keffa, Sudanese Hill Goat, Small East African, Mubende, Rwanda–Burundi and Kigezi (FARM-Africa 1996). They are small-horned goats with short, fine hair and variable coat colour, with some males having a short mane extending down the back and flanks. They are found in a wide range of arid and semi-arid areas and are raised in agropastoral systems where they mostly graze freely and are tethered during crop production seasons. They are mainly used for meat and skin production. FARM-Africa (1996) reported that these goats descended from the Somali Arab goats found in Somalia. The breeds in this family are distributed in various agroecological zones in East and Southern Africa.

According to FARM-Africa (1996), Somali goats comprise the short-eared Somali, the long-eared Somali and the Somali Arab breeds, while the Ethiopian Rift Valley family includes several breeds such as Arsi-Bale, Afar, Woyto-Guji, Abergelle and Worre. The breeds are believed to have originated from a Rift Valley goat breed that migrated to Ethiopia from Yemen and Saudi Arabia.

5.5.2.1 Abergelle

Synonyms: Abergale, Abergalle.

Origin and Distribution: Descendant of the Rift Valley goat type from southwest Asia and is related to Afar and Worre goat breeds. The breed is found in Ethiopia and Eritrea. The goats are kept by the Agew and Tigray ethnic groups (Gizaw et al. 2010; Yami and Merkel 2008).

Physical Characteristics: Abergelle (Fig. 5.4) is stocky, compact and well-built, with mean height at shoulders estimated at 71.4 and 65 cm for adult bucks and does, respectively. The goats have a straight to concave facial profile. Both males and females have horns, and in most cases, the horns in males are much bigger and spiral-shaped. Most goats have plain and patchy colour, and the reddish-brown colour is the most common. Spotted coat colours are also common. The hair is short and smooth in both sexes, and males have beards and ruffs. Wattles are almost entirely absent (Alemu et al. 2020).

Adaptive and Special Genetic Characteristics: The Abergelle goat is an indigenous goat population found in Ethiopia. Apart from a few resistance genes, little is known of the genetic and adaptation traits of this breed. Though not well characterised, Abergelle goats contribute to their productivity and suitability to the agroecological zones of Ethiopia, where they are found (Berihulay et al. 2019). Research has identified genetic signatures of selection in its genome (Berihulay et al. 2019) associated with adaptation to the environment and may be linked to specific traits such as size, colour, horn shape and reproductive and productive characteristics (Bertolini et al. 2018). These genetic adaptations make the Abergelle goat well-suited to walking long dis-

Table 5.2 African goat genetic resources: breed groups, subgroups and breeds

Breed group	Breed	Breed name synonyms	Location
Lop-eared goats	Nubian	Beladi, Bledi, Hassen, Langae, Sciucria, Shukria and the Sudanese Nubian	Sudan
	Sinai	Bedouin, Baladi (literally means local)	Egypt
	Swazi/Nguni	Herero, Tswana	Eswatini, South Africa, Mozambique and some places in Zimbabwe
	Berber	Algerian Berber, Libyan Mahalli, Moroccan Berber, Tunisian Berber	Algeria, Morocco, Tunisia, Libya
	Benadir	Bimal, Cherre, Tuni	South Somalia
	Wahati	Baladi (literally means local)	Egypt
	Begait	Barka, Bellenay, Beni-Amer, Hassan	Ethiopia, Eritrea and eastern Sudan
	Baladi	Egyptian; Bedouin	Egypt
	Zaraibi	Egyptian Nubian; Nuba; Theban	Egypt
	Zaghawa		Chad, Sudan
	Kalahari Red		South Africa
	Pafuri		Mozambique
	Kil	Kara Keci (Black goat)	
	Barki	Sahrawi (literally means desert)	Egypt
Short-eared small-horned goats	Abergelle	Abergale, Abergalle	Ethiopia
	Landim	Landim	Mozambique
	Mzabite	Nubian, Algerian Red, Touggourt, M'zab	Algeria
	Mudugh	Modugh	Algeria, Kenya
	Ndebele	Bantu, Matabeleland, Matebele	Zimbabwe Madagascar
	Somali	Abgal, Bimal, Denghier, Deghiyer, Dighi Yer, Dighier, Galla Habab, Maria, Modugh, Mudugh, Ogaden, Issa-Somali	Somalia, Kenya, East Africa
	Long-eared Somali	Large-White Somali, Galla, Digodi, Melebo, Boran Somali, Benadir, Gigwain	Somalia, Ethiopia
	Afar	Abyssinian short-eared, Adal, Assaorta, Denakil	Ethiopia, Eritrea, Djibouti
	Zimbabwe small goat	Small East African, Mashona	Zimbabwe, Kenya
			Congo, Angola, Cameroon, Nigeria
	South Sudan (dwarf)	Dinka, Ingessana, Latuka-Bari, Montain, Nuba Nilotic, Southern Sudanese, Toposa, Yei	Sudan
	Madagascar		Madagascar
	Sudanese Hill Goat		
	Boran	Galla, Somali, Modugh/Mudugh, Short-eared Somali, Dighier/Deghiyer/Deghi Yer	Ethiopia, Kenya, Somalia, East Africa
	Masai	Congo, Chaga, Arusha, Small East Africa	Kenya, Tanzania, East Africa
	Rwanda and Burundi	Chèvre Locale, Chèvre Commune, Long-Hair, Short Hair	Rwanda, Buundi, D R Congo

(continued)

5 African Goat Genetic Resources, Diversity and Unique Features

Table 5.2 (continued)

Breed group	Breed	Breed name synonyms	Location
	Western Lowland	Shankila, Gumuz/Gumez	Ethiopia
	West African Dwarf	African Dwarf, African Pygmy, Angola Dwarf, Cameroon Dwarf, Chevre de Fouta Djallon, Chevre Naine, Chevre de Casamance, Chevre guineene, Congo Dwarf, Forest Dwarf, Ghana Dwarf, Ghana Forest, West African Dwarf, Grassland Dwarf, Guinean Dwarf, Nigerian Dwarf	Benin, Cameroon, Chad, Congo, Cote d'Ivoire, DRC, Equatorial Guinea, Gabon, Ghana, Guinea Bissau, Liberia, Nigeria, Sierra Leone and Togo
	Small East African/Mubende		
	Malawian	Small East African	Malawi
	Congo Dwarf/Angola	Bahu	Uganda, D R Congo
Short-eared twisted-horned	Red Sokoto	Kano Brown (Kyasuwa), Katsina Light Brown, Mambilla, Bornu White (Buduma (Chad)), Damagaran dapple-grey, Maradi in Niger	Nigeria, Niger, Chad
	Barguirmi Maradi		
	Sahelian	Sahel, West African Long-legged, Desert, Sudan, Fulani (Peul, Peulh), Voltaique, Maure, Touareg, Arabian, Makatia (Algeria)	Chad, Burkina Faso, Mali, Togo
	Sudan Desert		Sudan
Exotic goats	Angora	Ankara-Kecisi, Tiftik-Kecisi, Mohair goat, Sybokke (South Africa)	Lesotho, South Africa, Kenya and Egypt
	Damascus	Aleppo, Baladi, Chami, Damascene, Halep or Shami, Qahr	Egypt, Algeria
	Toggenburg	Swiss Alpine, Togg, Toggenburger Ziege, Chèvre du Toggenbourg, Capra del Toggenburgo	Sub-Saharan African countries
	Saanen	Chèvre de Gessenay (French) or French Saanen, Saanenziege (German), Capra di Saanen (Italian), Israeli Saanen, British Saanen, Russian white, Sable Saanen (New Zealand),	Sub-Saharan African countries
	Anglo-Nubian	Nubian (USA), Nubia (North-eastern Africa)	Sub-Saharan African countries
	Alpine	Alpine polychrome, American Alpine or American Oberhasli, French Alpine, German Alpine or German Fawn, British Alpine, Swiss Chamoisee	Sub-Saharan African countries
Composite goat breeds	Pafuri		Southwest Mozambique
	Tanzanian Blended goat		Tanzania
	Vogan		Togo
	Arabi	Arabia and Sahelian	Algeria
	Makatia		Algeria
	Boer	Africander, South African Common, Boer goat, Kalahari, Boerbok	South and East Africa

Fig. 5.4 A herd of Abergelle goats. (Photo by Shumuye Belay)

tances, drought and heat stress tolerance as well as developed resistance to many common diseases in their environment, making them more resilient to the environmental conditions and production systems in Ethiopia than some imported breeds found in the same location.

Typical Production Systems, Husbandry Practices and Production Characteristics: Mixed farming and agropastoral systems and free grazing during all seasons, supplemented with crop residues in the dry season. Some localised transhumance is practised, moving flocks to graze along the banks. Abergelle goats are milked for domestic consumption. Their skin is also used to make aprons, containers, etc. According to Birhanie et al (2018), the daily milk yield of progenies of selected bucks (372 ± 14.8 ml) was significantly lower than that of base flock does (408 ± 6.72 ml). Moreover, there was significant variation of daily milk yield between the goat populations in the study districts (404 ± 14.2 ml in Tanqua-Abergele and 375 ± 7.10 ml in Ziquala) and seasons (403 ± 11.2 ml in wet and 376 ± 8.49 ml in dry). The overall mean weights at birth, 3 months, 6 months and 9 months were 2.28 ± 0.02 kg, 7.40 ± 0.09 kg, 9.48 ± 0.15 kg and 11.38 ± 0.19 kg, respectively. Comparatively, the average birth weight for progenies of selected bucks was 2.39 ± 0.02 kg, while 2.17 ± 0.02 kg was estimated for progenies of base flocks (Jembere et al. 2020). A study on carcass characteristics and sensory analysis of Abergelle goats and Abergelle cross-bred goats fed hay supplemented with a concentrate mixture showed that genotype had a significant effect on carcass parameters such as skin, foreleg and hind leg weight. However, the study found no significant difference in non-edible offal components between the two groups (Alemu et al. 2020). Abergelle goats have a high productivity rate, with a mean observed and expected heterozygosity value of 0.56, which indicates a close relatedness among the populations (Hassen et al. 2012).

5.5.2.2 Landim

Synonyms: Mozambique ('Landim' = Portuguese 'Landrace'), Small East African.

Origin and Distribution: Landrace in Portuguese resembles Mozambique Small East African goats and possibly has some early incorporation of European blood. The breed is found in southern Mozambique, south of the Limpopo and in Tete Province.

Physical Characteristics: Landim is fairly large-sized, 65 cm. Weight: male 50 kg; female 35–40 kg. Head is fairly heavy, concave profile in females, and slightly convex in males. Horns are present in both sexes: 96 percent of all animals but 31 percent have only rudimentary horns or scurs; heavier in males than females, but grow upwards and backwards in both sexes; length 11.3 cm, males 2.3 cm longer than females. Ears are medium-long, carried erect or horizontal, 13.3 cm. Toggles are present in 7 percent of both sexes. All males have beards, but only 12 percent of females. Male has short, stiff mane extending down the back line. Neck is short and thick. Chest is well developed, girth measurement exceeding withers height considerably at all ages. Back is short. Croup is fairly long but sloping. Legs are medium in length. Colour variable is commonly dark brown (36.3%), black (23.6%), pied (23.6%), white (4.5%), yellow (4.5%) and several combinations of colours. Coat is usually short and fine.

Typical Production Systems: The main product from Landim is meat, the production system being a sedentary agropastoral and the cultivation of annual subsistence crops being the dominant farm enterprise.

5.5.2.3 Mzabite

Synonyms: Algerian Red, Touggourt, M'zab, Mozabite

Origin and Distribution: The M'zab goat, or red goat of the oases, or goat of Touggourt, is mainly found within the M'zab oases of the Ghardaia region in Algeria.

Physical Characteristics: They are medium in size, 64–67 cm high at the withers (67.1 ± 0.2 cm for males and 64.2 ± 0.2 cm for females). The coat has three colours: chamois, which is dominant, white and black. The black goat has a regular line on the spine, which sometimes has two extensions on the shoulders. The belly may be spotted black and white. The length of the hair varies from 3 to 11 cm for males and 3 to 7 cm for female (Fantazi 2004).

Adaptive and Special Genetic Characteristic: M'zabite adapts well to warm oasis regions and is resistant to certain diseases such as scrapie. Fantazi et al (2018) detected several polymorphisms in PRNP, particularly in the more ancient African *Kabyle Dwarf*, the oldest breed in North Africa. The resistant allele K222 was present at a high level in South Italian breeds and at a very low frequency in *Kabyle Dwarf,* whereas a frequency similar to other Mediterranean countries was detected in M'zabite. The overall results showed how polymorphisms in PRNP of goat populations from different areas of North Africa and the Mediterranean basin can differ in terms of variability and relative frequencies.

Production Characteristics: M'zabite is mainly raised for milk and meat. Milk yield could reach 2–3 litres per day. Its production per lactation is 460 kg for a period of 180 days. It is also a good meat breed since its prolificacy can reach 180% (Fantazi et al. 2017).

5.5.2.4 Mudugh

Synonym: Modugh

Origin and Distribution: The Short-eared Mudugh goat is an indigenous breed of Somalia and is also found in Kenya and Ethiopia (Gebreyesus et al. 2012; Muigai et al. 2016). The breed is named after the Mudugh region in Somalia, where it is commonly found (Gebreyesus et al. 2012).

Physical Characteristics: The Mudugh goat has a short, smooth coat that is mainly white (76%) with brown (9%), black (7%) and grey (7%) occasionally in spotted patterns (DAGRIS 2023; Cooper 2022). The breed has short, forward-pointing ears. They have a slim but well-muscled frame, with long legs and neck, a straight facial profile, short spiral horns and a tail typically carried high and curved. Polled animals are common. The height to withers is 24–28 inches (61–70 cm) for the Short-eared Somali and 27–30 inches (69–76 cm) for Long-eared Somali. The weight of the breed ranges from 55 to 121 lbs. (25–55 kg), with Long-eared Somali tending to be larger than the Short-eared varieties (Cooper 2022).

Adaptive and Special Genetic Characteristics: The Short-eared Mudugh goat is well adapted to the harsh environment of Somalia and other arid and semi-arid regions (Gebreyesus et al. 2012; Muigai et al. 2016). They are known for their ability to walk long distances, drought and heat stress tolerance (Gebreyesus et al. 2012). The breed has developed resistance to many common diseases in their environment, making them more resilient than some imported breeds (Muigai et al. 2016; Cooper 2022).

Typical Production Systems and Uses: The Short-eared Mudugh goat is often raised in small-scale, family based production systems. They are used for meat, milk and other products (DAGRIS 2023; Muigai et al. 2016). The breed is an important source of income and food for many people in Somalia, Kenya and Ethiopia (Gebreyesus et al. 2012; Muigai et al. 2016). The breed is also culturally significant and plays an important role in the traditions of many African communities (Gebreyesus et al. 2012; Muigai et al. 2016) where they are found.

5.5.2.5 Ndebele

Synonyms: Bantu, Matabeleland, Matebele

Origin and Distribution: The Short-eared Ndebele goat is an indigenous breed of Zimbabwe (Gebreyesus et al. 2012). It is named after the Ndebele people, who are one of the major ethnic groups in Zimbabwe (DAGRIS 2023).

Physical Characteristics: The Short-eared Ndebele goat has a short, smooth coat that is usually black or brown (DAGRIS 2023; Gebreyesus et al. 2012). The breed has short, forward-pointing ears. They have a slim but well-muscled frame, with long legs and neck, straight facial profiles, short spiral horns and a tail typically carried high and curved. The height to withers is 24–28 inches (61–70 cm). The weight of the breed ranges from 55 to 121 lbs (25–55 kg) (DAGRIS 2023).

Adaptive and Special Genetic Characteristics: There is limited information available on the adaptive and special genetic characteristics of the Short-eared Ndebele goat, which makes this breed a poorly explored breed that requires characterisation as a priority.

Typical Production Systems: The Short-eared Ndebele goat is often raised in small-scale, family based production systems. They are used for meat, milk and other products. The breed is an important source of income and food for many people in Zimbabwe (DAGRIS 2023; Gebreyesus et al. 2012). The breed is also culturally significant and plays an important role in the traditions of the Ndebele people (DAGRIS 2023).

5.5.2.6 Somali

Synonyms: Denghier or Deghiyer, Abgal, Bimal, Dighi Yer, Dighier, Galla Habab, Maria, Modugh, Mudugh, Ogaden, Issa-Somali (FARM-Africa 1996).

Origin and Distribution: Probably related to the Arab goats in Somalia, which were introduced directly from Arabia. The related type is Long-eared Somali. *The breed is found in* northern and eastern parts of Ogaden (Jijiga, Degeh Bur and Werder), where they are kept by the Isaaq and Mijertein Somali clans and Dire Dawa, Issa and Gurgura (FARM-Africa 1996).

Physical Characteristics: The Short-eared Somali goat is smaller than the Long-eared Somali type. It has a straight facial profile, and most males bear straight (46%) upward-pointing (64%) horns. Females appear to bear more curved horns (50%), most of which point upwards (55%), but 27% are orientated backwards and 12% are lateral-pointing. Polled goats are found in 5% of males and 7% of females. There is a low incidence (6%) of spiral horns in both sexes. The Short-eared Somali goat has a short, smooth coat which is mainly white (76%) with brown (9%), black (7%) and grey (7%) occasionally in spotted patterns (12%). No ruffs were observed in either sex, but beards are present in 79% of males and 14% of females. Wattles were found in 5% of all goats. Males have a height at the withers of 64.9 cm and an average weight of 32.8 Kg against 61.8 cm in height and 27.8 Kg of weight for females (FARM-Africa 1996).

Typical Production Systems, Husbandry Practices and Production Characteristics: The animals are herded during the day by family members, usually with sheep. During the dry season, goats are watered infrequently, every 5–8 days. The goats return to the thorn enclosures at night. The average number of kids born per breeding female was 2.4. Somali goats mainly give birth to single kids (97.5%) with very few (2.5%) twins. Somali goat owners reported a tradition of selecting against twinning, and past selection appears to have been very effective (FARM-Africa 1996).

Somali goats are widely milked (Hanna ear), and milk is consumed fresh and also made into butter. The Somalis use butter for both food and medicinal purposes. Meat may be eaten fresh or preserved by cutting into slices, frying in butter or animal fat and keeping in a container (odka). Preserved in this way, the meat can keep for up to 5 years. Odka may be given to the groom by the parents of the bride. Meat may also be preserved by air-drying strips (solei). Fresh meat may be roasted and eaten with rice (wesla). Goat skins are widely used as sitting or sleeping mats and prayer mats (harek or okedi). The Somali make water containers (karbit) and use goat skins to churn butter. The Somali also use strips of goat skin for tying firewood and constructing their houses (FARM-Africa 1996).

Genetic Characteristic: Short gastrointestinal parasites and respiratory problems, including occasional outbreaks of contagious caprine pleuropneumonia (CCPP).

5.5.2.7 Long-Eared Somali

Synonyms: Large-white Somali, Degheir, Galla, Digodi and Melebo, Boran Somali, Benadir, Gigwain.

Origins and Distribution: Long-eared Somali goats are also known as *Large-white Somali*, *Degheir*, *Galla*, *Digodi* and *Melebo*. They are probably related to the descent of the Arab goats in Somalia, introduced directly from Arabia. Long-eared Somali goats are distributed throughout the Ogaden, lowlands of Bale, Borana and southern Sidamo. They are kept by the Hawia, Ogaden, Rare Bare, Digodi clans of the Somali ethnic group and the Boran, Gabra, and Geri ethnic groups (Alemu 2015).

Physical Characteristics: Large; white; short hair. Other features: predominantly straight face; horns are curved (41% in males, 46% in females) and pointed backwards in 38% of males and upwards in 48% of females; 13% of horns in both sexes have a lateral orientation; polledness is 19% in males and 8% in females; colour is plain white (92%), brown (4%), black (3%) and grey (1%); spotted coat pattern is observed in 21% of males; ruffs occur in 21% of males but never in females; beards in 66% of males and 7% of females; wattles in 6% of males and 3% of females (Abegaz and Awgichew 2009).

Adaptive and Special Genetic Characteristic:
Typical Production Systems, Husbandry Practices and Production Characteristics: Somali goats are extensively milked by the Somali and Boran pastoralists. Goat meat is preferred to mutton in most areas where the Long-eared Somali goat is kept.

5.5.2.8 Afar

Synonyms: Adal, Danakil

Origin and Distribution: The Short-eared Afar goat is an indigenous breed of Ethiopia (Solomon et al. 2008, 2014). It is found in very hot or arid regions of Ethiopia, where the temperature sometimes climbs to about 50 °C with low rainfall (150–300 mm). It is a valuable breed to the people of the Afar Region of Ethiopia.

Physical Characteristics: The Short-eared Afar goat has a short, smooth coat that is mainly white with brown or black markings. The breed has short, forward-pointing ears. They have a slim but well-muscled frame, with long legs and neck (Fig. 5.5), straight facial profile, short spiral horns and tail typically carried high and curved. The height to withers is 24–28 inches (61–70 cm). The weight of the breed ranges from 55 to 121 lbs (25–55 kg) (Solomon et al. 2008, 2014).

Adaptive and Special Genetic Characteristics: The Short-eared Afar goat is well adapted to the harsh environment of the Afar Region of Ethiopia. They are known for their ability to walk long distances, drought and heat stress tolerance. The breed has developed resistance to many common diseases in their environment, making them more resilient than some imported breeds (Solomon et al. 2008, 2014).

Typical Production Systems: The Short-eared Afar goat is often raised in small-scale, family based production systems. They are used for meat, milk and other products. The breed is an important source of income and food for many people in the Afar Region of Ethiopia. The breed is also culturally significant and plays an important role in the traditions of the Afar people

Fig. 5.5 A herd of Afar goats. (Anne W.T Muigai)

(Solomon et al. 2008, 2014). Its unique characteristics make it well-suited to its environment and an important part of local cultures and economies.

5.5.2.9 Zimbabwe Small Goat

Synonyms: Mashona, Small East African, Sebei, Karamoja, Tanzania, Zambian

Origin and Distribution: Found mainly in Manicaland and Mashonaland provinces, Masvingo and parts of the Midlands provinces in Zimbabwe and a wide and diverse range of environments in Kenya, Tanzania, Uganda and Zambia.

Physical Characteristics: A generalised and diverse group of goats with variable type, conformation and size of body. Adult goats stand on average 60 cm at the withers and weigh 25 kg; they are hardy animals generally used for meat rather than milk. Both sexes are horned, and horns are of variable size, ranging from 3 to 25 cm in length; shape of ears is also variable, but typical are prick ears of moderate size; wattles are fairly common; the coat hair is short, fine and smooth; in some males the shoulders, back and flanks are covered with longer hairs; and the coat is variable in colour type and pattern (Epstein 1971).

It is a small, compact and hardy goat. The goat has short ears and they are held horizontally. It appears in multiple colours. Tassels or toggles or wattles hanging under the neck are common.

Production Characteristics: Mature weights of the Mashona goat range from 25 kg to 35 kg. Kid birth weight is about 2.4 kg, with weaning weights ranging from 10 to 12 kg. Fertility is about 67.2%, and litter size is from 1.1 to 1.3. Twining rates range from 14% to 30%. The gestation period is about 5 months on average.

5.5.2.10 South Sudan Dwarf

Origin and Distribution: The Short-eared South Sudan Dwarf goat is an indigenous breed of South Sudan (FARM-Africa 1996). It is found in the southern part of the country, a region with a tropical climate and high rainfall.

Physical Characteristics: The Short-eared South Sudan Dwarf goat has a short, smooth coat that is usually black or brown. The breed has short, forward-pointing ears. They have a slim but well-muscled frame, with long legs and neck, a straight facial profile, short spiral horns and a tail typically carried high and curved. The height to withers is 20–24 inches (50–60 cm). The weight of the breed ranges from 44 to 66 lbs (20–30 kg) (FARM-Africa 1996).

Typical Production Systems: The Short-eared South Sudan Dwarf goat is often raised in small-scale, family based production systems. They are used for meat, milk and other products. The breed is an important source of income and food for many people in South Sudan. The breed is also culturally significant and plays an important role in the traditions of many African communities. Its unique characteristics make it well-suited to its environment and an important part of local cultures and economies.

5.5.2.11 Madagascar

Origin and Distribution: The Short-eared Madagascar goat is an indigenous breed of Madagascar. It is found throughout the island of Madagascar (Ngere 1987).

Physical Characteristics: The Short-eared Madagascar goat has a short, smooth coat that is usually black or brown. The breed has short, forward-pointing ears. They have a slim but well-muscled frame, with long legs and neck, a straight facial profile, short spiral horns and a tail typically carried high and curved.

Typical Production Systems and Uses: The Short-eared Madagascar goat is often raised in small-scale, family based production systems. They are used for meat, milk and other products. The breed is an important source of income and food for many people in Madagascar. The breed is also culturally significant and plays an important role in the traditions of many Malagasy communities (Ngere 1987).

5.5.2.12 Sudanese Hill Goat

Synonyms: Sudan desert, Goat Southern Sudanese, Southern Sudan, Nuba Mountain, Ingessana, Latuka-Bari, Nilotic, Yei, Toposa, Dinka Dinka, Ingessana, Latuka-Bari, Montain, Nuba Nilotic, Southern Sudanese, Toposa, Yei.

Distribution: The goats are mainly found in South Sudan.

Physical Characteristics: In general, the Sudanese Hill goat has mixed coat colours (Fig. 5.6). Black or white with or without spots or patches were more frequent (40%), followed by black colour (31.6%), brown colour (18.33%) and white colour (10.00%). Indigenous Sudanese desert goats have a straight rump shape. About 97% of the studied goats were horned. Most had no wattle (100%) and were bearded (71.6%), 8.33% were toggled. The average body length is 56.98 cm. Horn length is 11.61 cm, which is an important self-defensive mechanism to enable them to thrive in the harsh surroundings in which they are raised. The average chest girth was 68.45 cm and longer ears of 19.28 cm (Warsame and Turyasingura 2022).

Adaptive and Special Genetic Characteristics: The breed is adapted to arid zones, extending to hyper-arid zones during transhumance and nomadic migration. Black-coloured animals are better adapted to seasonal cold weather or cold nights because the dark pigment allows them to absorb heat faster than other coat colours.

Typical Production Systems and Uses: The breed is mainly raised for meat and milk under traditional agropastoral and pastoral systems.

5.5.2.13 Boran

Synonyms: Galla; Somali, Modugh/Mudugh, Short-eared Somali, Dighier/Deghiyer/Deghi Yer, Denghier, Abgal, Ogaden, Issa-Somali, Habab, Maria, Bimal.

Origin and Distribution: The Boran is part of the Small East African group, Northern Kenya, southern Somalia and parts of southern and southeastern Ethiopia.

Physical Characteristics: The Boran is of small size (60 cm at the withers). The male weighs 30–40 kg; the female, 25–30 kg. Benadir is slightly larger. Head is fine, the muzzle is narrow, and the facial profile is convex. Horns are small, usually slender, with no marked twist, in

Fig. 5.6 Sudanese hill goat. (Photo: Y. A. Hassan)

about 97% of animals. Ears are short to medium, pricked sideways and slightly forwards and upwards. Toggles in about 5% of both sexes. Neck is medium length. Chest narrow, girth exceeding withers height by about 10%. Back is fairly long and slightly dipped. Croup is sloping, and legs are long.

Colour is brilliant white (>70% in central Somalia). Some Ethiopian goats from the Ogaden have black spotting or solid black on the head and forepart of the neck; some varieties have a black dorsal stripe. Hair is short, shiny and smooth with thin skin.

Adaptive and Special Genetic Characteristics: The Short-eared Boran goat is well adapted to the arid and semi-arid environments of Kenya and Ethiopia. They are known for their ability to walk long distances, drought and heat stress tolerance. The breed has developed resistance to many common diseases in its environment, making it more resilient than imported breeds.

Typical Production Systems and Husbandry Practices: The management systems are pastoral and agropastoral. In the Kenya Coast, about 20 per cent of farms keep cattle, but about 60 per cent have small ruminants, mostly goats. In Central Somalia, Animals are herded during the day and penned in thorn enclosures at night, kids separated from adults. Kids are allowed to suckle twice a day after milk for human consumption has been taken off. Bucks run continuously with does to ensure kidding (and milk supply for humans) all year round.

5.5.2.14 Maasai

Synonyms: Congo, Chaga, Arusha, Small East Africa

Origin and Distribution: The majority of goats in the Maasai steppe zone of Northern Tanzania are strains of the indigenous breed of Small East African goats. The Maasai steppe agroecological zone in Hai district has an altitude range of 900–1873 metres above sea level, with an average annual rainfall of 521 mm and 23.3 °C of average annual temperature (FARM-Africa 1996).

Physical Characteristics: Body weight ranged from 25.83 kg to 30.34 kg for goats aged below 2 years and above 3 years, respectively. The majority of goats had straight hairs, with beards present in 80% of bucks and 9.1% of does. Ears are short and erect, horns are present in more than 88% of the studied goat population, and about 48% of the horns were lateral and 18.1% straight. Facial profile was mainly concave, back was straight and wattles were absent (FARM-Africa 1996).

Adaptive and Special Genetic Characteristics: The breed is adapted to arid zones, extending to hyper-arid zones. Black-coloured animals are better adapted to seasonal cold weather or cold nights because the dark pigment allows them to absorb heat faster than other coat colours.

Typical Production Systems and Production Characteristics: The breed is mainly raised for meat and milk under traditional agropastoral and pastoral systems.

Twinning rate is 8.7%, bucks and does reach sexual maturity at 11 months of age, and does have their first kids at 16.7 months old. Kidding interval was 7.7 months, lactation length was 82.3 days (2.9 months), and pre-weaning kid survival rate was 77.1% and 79.9% for the dry and rainy seasons, respectively (FARM-Africa 1996).

5.5.2.15 Rwanda Goat/Burundi Goat

Synonyms: Chèvre rwandaise; chèvre burundaise.

Origin and Distribution: The Rwanda and Burundi goats are part of the Small East African group. The 'types' from Rwanda, Burundi and eastern Zaire are, for all practical purposes, indistinguishable. The goats are found in Rwanda, Burundi and Kivu province of Zaire and extend into southern Uganda and the extreme west of Tanzania. The ecological zones include sub-humid east-central African highlands from 1200 m to 2500 m altitude in the 800 mm to 1500 mm rainfall zone. The estimated total number of goats and sheep (of which probably 75 per cent were goats) in Burundi was 1,313,000 in 1984; total goats in Rwanda were 940,000 in 1983, according to an administrative census (for tax purposes), but a sample agricultural survey at the same time estimated 2.2 million goats (FARM-Africa 1996).

Physical Characteristics: The breed is of small size, 64 cm (60–67 cm). The male weighs 35 kg while the female is 27 kg. Horns are present in both sexes: curving outwards and backwards in males, up to 20 cm in length; females have lighter and scimitar-shaped horns; polled animals are very rare. Ears short to medium length, pricked forward and upward. Toggles are present in both sexes (14%). Most males and some females are bearded. Some males have a topknot and a mane along the whole length of the spine is almost universal in this sex. Neck is fine and medium length. Chest reasonably well rounded, girth measurement 20–25% greater than withers height. Withers level with the sacrum. Back is short and straight. Legs normally proportioned in relation to the body, front cannon bone circumference about 7–8 cm. Udder is rounded and small with short teats. Colour is very variable, whole blacks common, but many multicoloured animals. Coat is fine and short, but very few males have long hair on the hind legs (FARM-Africa 1996).

Typical Production Systems, Husbandry Practices and Production Characteristics: Agropastoral and agricultural systems are practised, verging on pastoral in Ankole/Bahima areas of eastern and lower areas of Rwanda.

Generalised flock structure for meat production has females 82.6% (65.7% weaned); males 17.4% (3.7% weaned); 51.3% of all goats in the traditional system have milk teeth only (FARM-Africa 1996).

Multiple births are very common; 61.1% single, 37.1% twin, 1.8% triplet in traditional system in north of Burundi; 54.6, 42.5, 2.8 and 0.2% for single, twin, triplet and quadruplet, respectively in large-scale traditional study in three countries combined; 41.4% single, 58.6% multiple under station management in southeast Rwanda. Litter size was 1.44 in traditional system; 1.75 on station, not differing significantly with season, but larger litters at older parities were noted. Annual reproductive rate was 1.86 on the station. For lifetime production, most females do not exceed 5 parturitions, but up to 12 recorded; average of 2.39 parturitions for 1340 does in large-scale traditional system survey. Age at first kidding was 640 ± 27.8 days on station in Rwanda, where does born as twins, kidded more than 3 months later than those born as singles and females out of older dams kidded younger than those out of junior dams. Kidding interval was 343 ± 13.8 days on station, but this is largely due to an imposed breeding season (FARM-Africa 1996).

5.5.2.16 Western Lowland Goat

Synonyms: Shankila/Shankela, Gumuz/Gumez

Origin and Distribution: Western Lowland (Fig. 5.7) goats are also known as Shankela and Gumez locally. They are derived from the mixing of types in the past. They are distributed in the western lowlands bordering Sudan in Gojam (Metekel), Wellega (Assosa) and Illubabor (Gambela). Most closely related to the Central (Fig. 5.8a, b) and Western Highland goats in Ethiopia.

Indigenous Ethiopian goat genetic resources have been classified phenotypically into 11 types. These are Abergalle, Arsi-Bale, Afar, Central Highland, Gumez, Hararghe Highland, Keffa, Long-eared Somali, North-West Highland, Short-eared Somali and Woyto-Guji (Tesfaye 2004).

Physical Characteristics: The Gumuz goat (34.7 kg) population was significantly heavier than Central Abergelle (27.9 kg) and Abergelle

Fig. 5.7 A Western lowland or Shankela doe. (Photo: ILRI)

 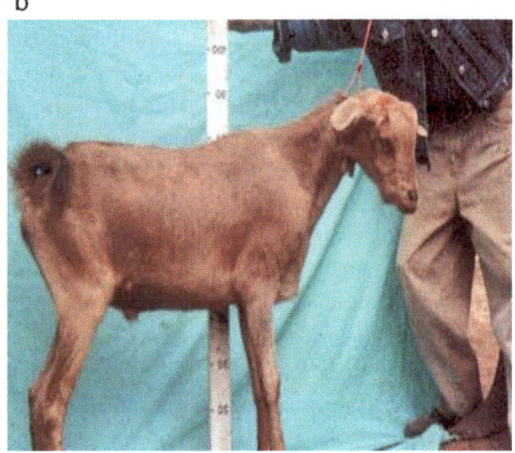

Fig. 5.8 (**a**) Keffa Central highland goat. (Photo: ILRI). (**b**) Keffa Central highland goat. (Photo: ILRI)

goat (28.1 kg) ecotypes, with an overall mean of 31.14 kg. The Begia-Medir goat, with an average height of 71 cm at withers, was significantly taller than all other goat ecotypes, while Gumuz goats were the shortest with 65 cm height at withers (Hassen et al. 2012). The goat has a straight facial profile (100%), short, smooth coat (81%), with 16% having a relatively coarse coat. The main colours are white (42%) and fawn (38%), with some black (9%) and grey (11%), occurring mainly in patches (73%). Most male goats have straight horns (85%) orientated backwards (77%). There are 12% polled males in the population. A ruff is present in 96% of all males and a beard in 70% of males. Wattles are present in 12% of all male goats.

Adaptive and Special Genetic Characteristics: The breed is adapted to arid zones, extending to hyper-arid zones. All the Ethiopian goat populations are very closely related to each other.

Typical Production Systems and Production Characteristics: Pastoral and agropastoral groups in the area of distribution use goat milk extensively. Goat meat is widely eaten. The horn is used as a musical instrument (zoombara) in Assosa. The average number of kids born per breeding female is 3.5. Western Lowland goats are remarkably prolific, with 56% giving single births, 41% twin births and 3% triplets. Quadruplets were also reported (Tesfaye 2004).

5.5.2.17 West African Dwarf (WAD)

Origin and Distribution: Synonyms for the breed include Djallonke, African Pygmy, 'Chèvre de Fouta Djallon', 'chèvre guinéenne', 'chèvre naine', Guinean Dwarf, Cameroon Dwarf or pygmy, Kirdi (southern Chad and northern Cameroon), Ghana Dwarf, Ghana Forest, Forest Dwarf, Nigerian Dwarf, Cameroon Grassland Dwarf, Congo Dwarf and Chèvre de Casamance (Wilson 1991; DAGRIS 2007). The breed is an achondroplastic dwarf with a lack of ossification at the cartilage joints, which probably evolved in response to the conditions of the humid forest zone by selection of recessive genes for dwarfism (Wilson 1991). With the exception of the Central African Republic, they are found throughout the Atlantic coastline of 15 West and Central African countries (Guinea Bissau, Guinea, Liberia, Sierra Leone, Cote d'Ivoire, Ghana, Togo, Benin, Nigeria, Cameroon, Congo, DRC, Equatorial Guinea, Gabon and Chad). According to the latter author, they constitute the only goat breed (100%) in Liberia, Sierra Leone, Congo, Equatorial Guinea and Gabon. For subtypes, many are recognised, usually by the name of the country of their location and the type of habitat, such as Cameroon grassland (Fig. 5.7), Ghana forest, Cote d'Ivoire dwarf and Congo Dwarf. Slightly larger goats than the typical WAD, such as the Mossi of Burkina Faso, goats of southern Mali and the Kirdi of southern Chad and northern

Cameroon, are also sometimes included in the main type.

Physical Characteristics: WAD goat is very short (30–50 cm), with an average height at the withers estimated at 44.74 ± 0.64 (Fajemilehin and Salako 2008). Coat colour is variable (Figs. 5.9 and 5.10) depending on region (Okpeku et al. 2011a; Wilson 1991). Dark brown with black points is common, but black, white, red, pied and mixed colours also occur. In addition to three basic colours (black, brown and white), larger face, agouti and spotting patterns with over 22 combinations have been reported in Nigeria (Bemji et al. 2012). Review by Wilson (1991) showed that the coat of WAD goat has stiff and short hair with longer hair and varying degrees of waviness in some subtypes. Horns are present in both sexes, curling outwards and backwards in males and fairly strong, while light, sharp and pointing upwards and backwards in females. Ears are short to medium length, narrow and carried horizontally. Toggles may be present in both sexes. Males are bearded with a weak mane, while females occasionally have beards, and the degree of bearding varies depending on sub-type. Neck is strong and fairly long, chest broad and deep, and the udder is small and well-shaped.

Adaptive and Special Genetic Characteristic: WAD goat is hardy, trypanotolerant and resistant to gastrointestinal nematodes (Chiejina and Behnke 2011). They are well adapted to humid rainforest zones. Goats found in Africa can be broadly categorised into three namely; indigenous/native, exotic and composite (locally developed hybrid) types. Indigenous goats are hardy and well adapted to prevailing agroecological and environmental challenges in the regions of Africa where they are found.

Typical Production Systems and Production Characteristics: WAD is mainly raised for meat under agricultural and urban systems, while the agropastoral production system is rare. In southwest Nigeria, three types of management systems (free roaming, confined for part of the year and confined the whole year) are practised. Reproductive characteristics as reviewed by Wilson (1991) include age at first kidding: 12–18 months; Kidding interval: 210–290 days; multiple births are many, twins are very common, triplets are common with occasional quadruplets. Average litter size is 1.40–1.85 and can be more than 2 depending on parity; Oestrus cycle is 16–25 days, with heat lasting 16.4–40.0 hours; Gestation period: 142–149 days; birth weight: 1.04–1.62 kg; weights at 3, 9 and 12 months were 4.6 kg, 6.0 kg and 9.5 kg, respectively; average daily gain from birth to 90 days, 90–150 days and 150–365 days were 35 g, 20 g and 16 g respectively; mature liveweight ranged from 20 to 25 kg for male and 18 to 22 kg for female. In Southwest Nigeria, mature weights were 26.3 kg for white goats, while brown and black goats weighed 23.40 and 22.24 Kg, respectively (Ebozoje and Ikeobi 1998) Dressing percentage was 63% for a live weight of 23.5 kg; Carcass composition:

Fig. 5.9 Cameroon Grassland Dwarf. (Photo by Jules Fon)

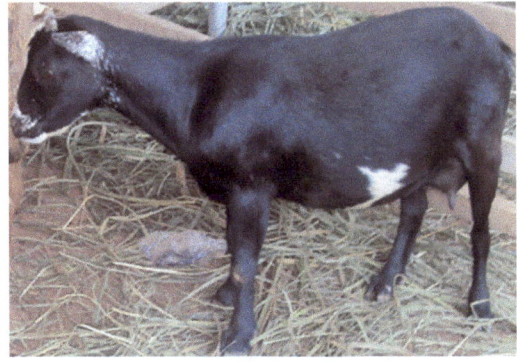

Fig. 5.10 West African Dwarf doe at Southwest Nigeria. (Photo by M.N. Bemji)

meat/bone ratio of 0.41. The carcass has more fat than dwarf sheep in the same environment (Wilson 1991). In the humid zone of Nigeria, the daily milk offtake for West African Dwarf goats was reported to be approximately 14.58 ml, with values influenced by parity and lactation stage, peaking at 36 ml on the eighth day before declining (Olalere & Bemji, 2024). Wilson (1991) reported a lactation length of 126 days with peak yield of 710 g at about 40 days; Milk composition was DM: 19.2%; fat: 8.3%; protein: 5.1%; lactose: 4.5%; energy: 123 Kcal/100 g.

5.5.2.18 Small East African/Mubende

Synonyms: Mashona

Origin and Distribution: The Mubende goat is a breed found in the sub-humid highland areas of Uganda and Zimbabwe, known for its unique black or black-and-white hair coat (Fig. 5.11a, b) in both sexes, high-quality skin and valuable meat production. The main location in Zimbabwe is in the southeast of the country; this goat is replaced by the Ndebele type (Wilson 1991). According to Onzima et al. (2018), the Small East African exists in Uganda, and in a study, reported high genetic variability of the Ugandan goat populations with sufficient genetic potential for further improvement of the breeds for heritable economic traits. The Ugandan indigenous goats are weakly differentiated, consisting of two breeds forming more uniform clusters (Kigezi and Small East African), two breeds clearly cross-bred (Karamojong and Sebei), and Mubende showing signs of gene flow from all these goat populations.

Typical Production Systems, Husbandry Practices and Production Characteristics: From a flock of traditionally managed Mashona Small East African goats monitored over a 6-year period, kids born in the hot dry season were significantly heavier at 60, 90 and 180 d (6.4, 8.0 and 10.9 vs 5.7, 6.7 and 9.2 kg, respectively) than kids born in the hot wet season. There was a steady decline in growth rates from age 15 d (60.2 g/d) to weaning age of 180 d (41.7 g/d). Overall, 19.4% of the 294 kids born during the study period exited the flocks before age 180 d. Major causes of exits were lost kids and predation, which together accounted for 61% of exits by kids. Seasonal effects on exits were mainly at ages below 30 d, where 10.3% of kids born in the hot dry season exited compared to 4.3% and 4.6% for the hot wet and cool dry seasons. Kidding intervals were 321 and 259 days. Due to the slow growth rates and long kidding intervals, the flock productivity in terms of weaned live kid weight (kg) per doe per year was low (Ndlovu and Simela 1996).

5.5.2.19 Malawian (Small Goat)

Synonym: Small East African

Origin and Distribution: This is a small breed under the Small East African goats. It is found in Malawi in various agroecological zones ranging

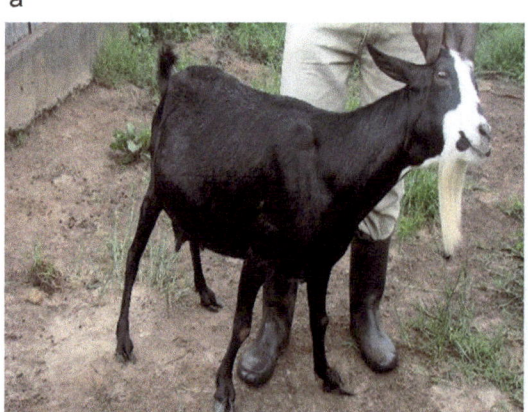

Fig. 5.11 (**a**) A Mubende doe. (Photo by D. Kugonza). (**b**) A Mubende buck. (Photo by D. Kugonza)

from semi-arid to sub-humid uplands. They are raised in agricultural systems with an average flock size of nine and mainly for meat production. They are mostly tethered during the day in cropping seasons and left on free range after harvest. They are housed at night in different types of housing. Flock structure is related to meat production.

Physical Characteristics: They are small in size, with a length of 62 cm and a weight of about 29.1 kg for females. The head is fine with a facial profile that is straight or dished. They have horns in both males and females, with pricked ears. The neck is fine and fairly long, and the chest is fairly well developed. The girth measurement is greater than the withers height by about 20%. They have a rump that is higher than the withers. The most common colours are black, black and brown, brown and red and white with a very wide variation. Some have 'Badger' and reverse badger face markings, and the hair is generally short and fine.

Typical Production Systems, Husbandry Practices and Production Characteristics: Age at first kidding was 15.6 months in the village system; 17.5 months on the ranch; 451 days (14.8 months) on the development project. Kidding interval was 44.9 weeks in the village; 35.2 weeks on the ranch; 254 days on the development project. Multiple births are common. Birth weight was 1.76 kg. Weight at 280 days was 12.2 ± 2.4 kg on the development project. Average daily gain from birth to 280 days was 36 ± 8 g on the development project. Post-partum weights were 29.5 kg in the traditional system and 28.5 kg on the ranch.

5.5.2.20 Small East Africa/Congo Dwarf/Angola

Synonym: Bahu

Origin and Distribution: The Congo Dwarf/Angola Dwarf goat is found in the eastern part of the Congo on the Upper Uele and Upper Ituri rivers, as well as near Lake Kivu. The Congo dwarf is classified under the subgroup of Small East African Meat goat. It occupies the eastern part of Congo on the Upper Uele and Upper Ituri rivers, the vicinity of Lake Kivu in Burundi, and the area north and west of Lake Tanganyika (Epstein 1971).

Special Characteristics: The short-coated Congo Dwarf goat stands at only 40–50 cm (Fig. 5.12) at the withers, but in the Savannah, it may reach a height of 60–70 cm, suggesting that its size is dependent on the environment rather than on race. It is generally characterised by a black-and-white colour pattern, less frequently by a brown or black coat; a short head with a prominent forehead; occasionally polled or furnished with short scimitar-like or twisted horns; short erect ears; absence of a beard, occasional presence of throat lappets, a large chest and a plump; very compact and well-covered body standing on short, straight or crooked legs. The milking qualities are negligible (Epstein 1971; DAGRIS 2007).

5.5.3 Short-Eared Twisted-Horned Goats

The earliest domestic goats of Africa belonged to a small generalised type (DAGRIS 2007). Ancient Egyptians received these dwarf goats and passed them to the south up the Nile Valley and across to Lake Chad before they received the specialised screw-horned goats. The centre of evolution of the screw-horned goats is not very clear. Since Egypt and India are peripheral areas of its range, the centre of evolution is speculated

Fig. 5.12 Small East African goat

to be in present-day Iran and Iraq. It is also possible that not all ancient screw-horned goats were derived from the same source. Selection is believed to have affected the forms of ears and horns and probably led to the dominance of long drooping ears. These ancient screw-horn goats have fairly long legs and short hair, and their colour was red, black, fawn, white or pied. From the northern region of Africa, the screw-horned, lop-eared type later spread to the west and south. Today, their distribution is limited to the Sahel, from western Sudan in the east, across Mali and Niger to Mauritania and Senegal in the west. Examples of goat breeds in this group include: Red Sokoto, Barguirmi Maradi, Sahalian, Sudan Desert and Sukria.

Fig. 5.13 A herd of Red Sokoto does at the National Animal Production Research Institute (NAPRI), Zaria, Nigeria. (Photo: M.N. Bemji)

5.5.3.1 Red Sokoto

Origin and Distribution: Synonyms include Sokoto Red, Kano Brown (Kyasuwa), Katsina Light Brown, Mambilla, Bornu White, Buduma (Chad), Damagaran dapple-grey and Maradi in Niger (Chèvre rousse de Maradi) (DAGRIS 2007). The breed is one of the Savanna goats with a fairly small size, suggesting the possibility of crosses with dwarf goats before selection in their current habitat. The hypothesis that a fusion occurred is supported by the relatively high prolificity of the Sokoto goat (Wilson 1991). Subtypes outside the main centres of its distribution have varied colours, such as Kano Brown, Bornu White, etc. Red Sokoto breed is mainly found in southern Niger and northern Nigeria between latitudes 12°N and 14°N and longitudes 4°N and 10°E. In Nigeria, it is common in Sokoto and Kano States.

Physical Characteristics: Red Sokoto breed is of small size, about 60 cm height at the withers (60–65 cm for males and 54–65 cm for females), but Bornu White is up to 80 cm (Okpeku et al. 2011a; Wilson 1991). Literature reviewed by Wilson (1991) shows that coat colour is deep red in Sokoto (Fig. 5.13) but lighter and occasionally almost chestnut in Maradi. Males are darker than females and may have a black back stripe. Bornu White has occasional black or brown spots on the nose, ears and around the eyes. Head is fine, the forehead is prominent, and the profile is short and straight or slightly dished. Males and females are horned (short to medium in length and somewhat heavier in males), ears are short, medium in width and carried horizontally. They are longer (semi-pendulous) in Niger and in Bornu White. Beard has profuse hair in males but is absent in females. Forehead is longer, bushier and darker in males than in females. Neck is short, thin and very mobile. Chest is rounded, legs are short, strong and well-muscled in fore and hind limbs. Tail hairs are black. Coat has fine and short hair, but males may have longer and wavier hair. Udder has good conformation, well rounded with well-spaced teats.

Adaptive Characteristic: The breed is adapted to semi-arid areas with a single rainfall season of 4–6 months duration.

Typical Production System and Production Characteristics: Red Sokoto is mainly kept by the sedentary agropastoral Hausa tribe. Flock sizes are small (<10 head), the breed is kept for meat, skins and milk. Production characteristics reviewed by Wilson (1991) include age at first oestrus (157 ± 5.92 days); Oestrus cycle (15–30 days); heat lasts for 24–120 hours in Kano Brown; Gestation period of 153 days (range 142–165) in Niger; age at first kidding (243–882 days); Kidding interval (240–343 days). Multiple births are common; 32.6% for single, 58.8% for twin, 7.2% for triplet and 1.8% for quadruplet at Shika (Nigeria); Litter size is 1.17–2.0 depending on parity; Annual reproductive rate was 1.50–2.00; 1.67 in Niger. Repeatability of litter size was 0.28 ± 0.07 (first to second kiddings) and heritability at first kidding (0.08 ± 0.02) by dam-daughter regression at Shika; Birth weight (1.5–2.0 kg); Weight at 12 weeks (7.6 kg) for singles and 5.7 kg for twins on station in Nigeria; Average daily gain of 7–120 days (males 63 g and females 55 g); Daily milk yield of 500–1000 g/d (545 g/d in 12-week lactation, does with twins produced 50 kg, better than singles by 20%). Daily milk offtake (14.09 ml) and milk yield (281.98 ml) were quite low within the humid zone of southwest Nigeria, where the breed is not well adapted (Bemji et al. 2021). For milk composition, total solids (18.2%); fat (4.7–7.8%), protein (3.8–4.7%); lactose (4.7%); energy (22.2 KJ/g dried milk) and 381 KJ/100 g whole milk (all values except lactose were higher in colostrum than in milk); dressing percentage was 45–50% (44.7–48.6%) for 20.5–25.0 kg live weight for Bornu White and 43.7–48.1% at 19.8–24.2 kg live weight for Red Sokoto at Maiduguri); young male castrates have 54–55%. For skins, average dry weight of skins from Nigeria and Niger is about 420 g (extra light is 250 g and heavy 625 g). Useful tanning area is 3–7 ft^2 (0.28–0.65 m^2). Red Sokoto skins are of excellent quality, known as 'Morocco' in the tannery trade. They are in demand for the fancy goods trade, particularly for gloves, high-quality shoes, patent leather and suede clothes. Skins of Kano goats are heavier and weigh 430–460 g (Wilson 1991).

5.5.3.2 Barguirmi Maradi

Origin and Distribution: The Barguirmi Maradi goat constitutes part of the West African Twisted horn goats, otherwise known as Zinder Brown, Chevre reusse de Maradi. The breed is the same as Red Sokoto goat of Nigeria and mainly found in Niger (DAGRIS 2007). They are restricted in Maradi and Tessoua Departments, while their subtypes extend to the west and east.

Physical characteristics: Phenotypically, Maradi goats of Niger are similar to Red Sokoto of Nigeria they are intermediate in size between the Sahel goat and the Djallonke (West African Dwarf), having approximately 65 cm height at withers and weigh 20–25 kg at maturity.

5.5.3.3 Sahelian

Synonyms: Synonyms for this breed include Sahel, West African Long-legged, Fulani (Peul, Peulh), Voltaique, Maure (Mauritania and northern Mali), Touareg, Arabian, Makatia (Algeria and Arab (Chad) (Wilson 1991; DAGRIS 2007).

Origin and Distribution: The breed (Fig. 5.14), according to Wilson (1991), is of the Savanna group, which includes several Saharan types from Egypt, Libya, Tunisia, Algeria and Morocco. For subtypes, local names are given, such as Gorane, Peul (Chad), Voltaïque (Burkina Faso), Nioro, Niafounké (Mali). Crosses with dwarf goats resulted in Vogan in Togo. The breed is distributed north of 12°N parallel from central Chad in the east to the Atlantic coast in the west and well into the southern Sahara and has migrated further south due to drought.

Physical Characteristics: The breed is large, with a withers height of 70–85 cm (80–85 cm for males and 70–75 cm for females) (Wilson 1991). Colour is very variable and depends on sub-type. Sahel type is mostly red or red pied, but grey, white and black are numerous (Fig. 5.14); Touareg is red pied or black pied, Maure is red pied or fawn pied, Voltaique has white (predominant) and other colour combinations. Coat is

Fig. 5.14 West African Long-legged at the National Animal Production Research Institute (NAPRI), Zaria, Nigeria. (Photo by J.C. Amabo)

short with stiff hair, except the mane, which could be 15 cm long. The head is small, fine and triangular in shape with narrow nostrils, fine lips, and large or less large supraorbital processes. Horns are present in both sexes (99.9% in Mali) but absent in Touareg. Horns in males are strong, spiral and up to 40 cm in length, sometimes horizontal, while in females they are 13 cm long, finer, scimitar-shaped, curving out and then in. Ears are long (21 cm) and wide or shorter (11 cm), pendent or semi-pendulous. Toggles and beards are common, manes are common in males, neck is long and thin and chest is narrow and shallow, with girth circumference slightly exceeding or about withers height. Back is straight with a prominent backbone. Legs are long and spindly, lightly boned with sickle hocks. Udder is elongated and divided into two halves with large teats.

Adaptive and Special Genetic Characteristics: Sahel goats are heat tolerant, long-legged and adapted for long-distance foraging in semi-arid and arid zones. Few are found in higher rainfall areas, but are not trypanotolerant and do not survive for long in forest savanna (Wilson 1991).

Typical Production Systems and Production Characteristics: The breed is managed under agropastoral, pastoral and urban settings for meat, milk and skins. Production characteristics as reviewed by Wilson (1991) include age at first kidding (275–1104 days); Kidding interval (234–291 days); Gestation period (148 days); Multiple births are fairly common with moderate litter size of 1.19–1.24 (1.34 for does of fourth and higher parities); Yearly reproductive rate was 1.49 in Mali, increasing with parity; Lifetime production of about 5 parturitions or less; Birth weight (2.2–3.2 kg); Weight for age at 90 days (7.7 kg); 365 days (20.2 kg); Adult weight was 31–50 kg (male 40 kg; female 27 kg); Daily milk yield (600–1500 g/d); Lactation length was 134.7 ± 5.6 (54–155 days); Milk composition: DM (12.7%); fat (3.9%); lactose (4.6%); density (1.030); Dressing percentage: castrates (49.4%) for 26.7 kg live weight, entire males (48%) for 27.4 kg and females (48.1%) for 27.8 kg. Estimates varied depending on environmental conditions.

5.5.3.4 Sudan Desert

Origin and Distribution: The breed is alternatively called Sudanese Desert, originating as Savanna type, similar to West African Long-Legged (Wilson 1991). It is common in dry areas of the Republic of Sudan, north of 12°N, while north of 10°N in Darfur, western Kordofan and also in some parts of Eritrea and westwards in Chad. Intermediate types with small forest goats are found at the southern limit in Sudan.

Physical Characteristics: The breed has a large size, with 65–85 cm height at the withers (Wilson 1991). Coat colour varies, including white (Fig. 5.15), black, greys (common), and mixed (black back stripe in dark colours and grey in light colours). Coat is short and fine. Some, especially Zaghawa, have longer hair which may cover the whole body or be limited to hindquarters and legs. Head is fine with a flat forehead and straight or a little dished profile. Horns are present in 95% of both sexes. They are long (35 cm), large, twisted and flattened in cross-section in males, projecting outwards or backwards, while in females they are finer, curving upwards and backwards. Ears are lopped and medium to very long (12–20 cm). Toggles are present in 15% of both sexes. Both sexes have beards, which are very bushy in males. Males may have a mane to the shoulders or extending the whole length of

Fig. 5.15 Sudanese Desert goat. (Photo: Y.A. Hassan)

the back, but sometimes present in females. Neck is short, chest shallow and pinched, back is short and straight, croup is very weak and sharply sloping with a low tail. The legs are long and poorly boned.

Adaptive Characteristic: The breed is adapted to semi-arid zones, extending to hyper-arid zones during transhumance and nomadic migration.

Typical Production Systems and Production Characteristics: The breed is mainly raised for meat and milk under traditional agropastoral and pastoral systems. The meat is more tender and juicier than many domestic breeds found in the region (Fadlelmoula et al. 2014). Production characteristics as reviewed by Wilson (1991) include: Age at first kidding (290 days); Kidding interval is 238 ± 41 days (Southern Darfur traditional system); 9 months (Kordofan); Multiple births (numerous) with a litter size of 1.57 (1.30 in primiparous does and 1.68 in multiparous); lifetime production is 9–10 kids (2.41 annually) in Southern Darfur and 4–7 in Kordofan; Birth weight (2.13 kg), single weighed 2.27 kg (male 2.30 kg, female 2.25 kg); twin weighed 2.05 kg (male 2.03 kg, female 2.07 kg); triplet weighed 1.82 kg (male 1.88 kg, female 1.73 kg); weight for age at 4 weeks (4.8 kg), 13 weeks (9.9 kg), Average daily gain from birth to 13 weeks (86.9 g), mature weight for male (40–60 kg); female (32.7 ± 5.22 kg); dressing percentage for entire males is 48.2% (44.2–52.8%) at live weight of 34.7 kg and castrates are 51.2% (46.6–33.1%) at live weight of 35.8 kg in Southern Darfur traditional system.

5.5.4 Exotic Goat Breeds

In general, goats not found resident in a location as described above, but imported from another region or from outside of the continent, are generally referred to as exotic goat breeds. European goats imported with the aim of so-called upgrading or improvement of indigenous goats fall into this group. Majority of these are characterised by a long and thin neck and may have horns. Their sizes usually vary from 85 to 90 cm. The body is narrow with a sharp back, with a short and sloping rump, and with high limbs. may have horns or be polled. These improvement programmes, if carried out correctly, should produce unique breeds like the ones listed in the sections for the **developed goats**, with higher production and productivity. However, when done indiscriminately and randomly without genetic records and proper monitoring, they become the source of genetic dilution and erosion of valuable local gene pool.

5.5.4.1 Angora

Synonyms: Ankara-Kecisi, Tiftik-Kecisi, Mohair goat, Sybokke (South Africa)

Origin and Distribution: Angora goat (Fig. 5.16) originated in Turkey (Özgecan and Ertuğrul 2012), from where the animals were exported to Europe in the sixteenth century and in 1838 to South Africa to establish a mohair industry without sacrificing body weight (Visser and Van Marle-Köster 2014). However, currently, this breed is widespread in many countries worldwide, including Kenya, Lesotho and Egypt from African countries (Visser and Van Marle-Köster 2014).

Physical Characteristics: Angora or Ankara goats are known for their mohair production, but they are also used for meat. They have medium-sized, horned, heavy and drooping ears. Have different coat colours but are dominated by white,

Fig. 5.16 Angora doe with its kid (Visser and Van Marle-Köster 2018)

and some goats also have black, brown, or grey coat colours. They have uniform fleece regarding the length and fineness, with good lustre, solid style and good character. Male and female Angora goats can produce up to 5.4 and 3.6 kg of skirted fibre per year, respectively. Hair length ranges from 13 to 25 cm. Mature buck weighs 45 kg and does weighs 35 kg, height: male 61 cm and female 51 cm (Özgecan and Ertuğrul 2012).

Adaptive and Special Genetic Characteristic: Angora goats are well-suited to harsh and semi-arid environments. They have high genetic diversity within populations (H_E), ranging from 0.371 in the South African population to 0.397 in the Argentina population (Visser et al. 2016). Hence, goats that are kept geographically in different countries are genetically distinct due to country-specific selection and breeding strategies. Candidate genes such as keratin-associated proteins (KAP1–KAP9) and intermediate filament proteins (KRT genes) play a functional role in the expression of fibre quality (Visser and Van Marle-Köster 2014). Marker-assisted selection for mohair traits (e.g. fleece weight, fineness, length of the fibre, etc.) using these genes and growth traits in combination with accurate estimated breeding value information could lead to a rapid genetic improvement for the Angora goat breeds.

Typical Production Systems, Husbandry Practices and Production Characteristics: Angora goats are good browsers and love grazing and eating green grasses, corns, plants, etc. They also generally eat a lot of roughage that includes weeds, bark, leaves, or woody shrubs. In South Africa, Angora goats were primarily farmed under extensive production systems, but currently, the breed is also managed under an intensive production system for the purpose of mohair production (Visser and Van Marle-Köster 2014).

Limitation and Challenges Breed Use and Classification: In the past 50 years, limited new genetic material has been available due to restrictions on imports and exports in countries like South Africa. Introduction of a new gene pool of the same breed but from other countries could improve their genetic diversity and productivity.

5.5.4.2 Damascus

Synonyms: Aleppo, Baladi, Chami, Damascene, Halep or Shami

Origin and Distribution: Damascus goat (Fig. 5.17a, b) is a native breed of the Middle East (Cyprus, Syria and Lebanon). It is believed that these goats were taken by the British to Cyprus, the place where the goat got the name Damascus. It is developed through crossbreeding of Nubian (African origin) and Jamnapari goats of India and belongs to the Nubian group (ESGPIP 1971). This breed is also introduced to African countries like Egypt and Algeria.

Fig. 5.17 (**a**) Damascus buck. (ESGPIP 2008). (**b**) Damascus doe (ESGPIP 2008, 1971)

Physical Characteristics: Damascus goat breeds are used as a multipurpose breed. They are prolific, and a doe can give birth to two kids per doe on average, but three and four kids per doe are also common. Birth weight ranges from 3.5 to 5.5 kg depending on the birth type, season and sex. At an early age, kids have a normal face. But at a later stage, they change from beautiful to a monster face. They are also known as noble creatures in the Middle East and have a long neck, legs and hair, a large body size, long (27–32 cm) and drooping ears and a small head. They have a unique look, with an almost scrunched-up face. Their head has a convex bridge, polled (81.2%), coat colour varies from brown (42%), grey (27%) and black (21%), which resulted from continuous selection of the preferred coat colour (Halima et al. 2016). An average body weight of a mature buck and doe is 75 ± 5 kg and 65 ± 5 kg, respectively. Height at wither is 80 cm for males and 74 cm for females, with body circumference ranging from 97 cm to 99 cm.

Adaptive and Special Genetic Characteristic: The most desirable feature of Damascus or Shami goat is its adaptative to high temperatures and suitablity for breeding in lowland areas (Tatar et al. 2019).

Typical Production Systems, Husbandry Practices and Production Characteristics: The breed is considered one of the best dual-purpose (meat and milk) breeds of the Middle East countries under a semi-intensive or intensive production system. It is not successfully raised under an extensive grazing system. Hence, it requires an improved management and feeding environment to express its full genetic potential. They are good milk producers and could produce 1.9 litres of milk per day on average. They can reach puberty within 5–10 months, depending on the feed availability and other management factors, including veterinary service and housing.

Limitation and Challenges Breed Use and Classification: Some Damascus goats have long protruding lower jaw that makes them difficult to graze and nurse.

5.5.4.3 Toggenburg

Synonyms: Swiss Alpine, Togg, Toggenburger Ziege, Chèvre du Toggenbourg, Capra del Toggenburgo

Origin and Distribution: Toggenburg goat was developed in Switzerland for the first time (Bosman et al. 2015). They are developed through crossing of Appenzell and Chamoisee goats (ESGPIP 1971). Currently, it is distributed worldwide. It is a dairy goat type.

Physical Characteristics: Toggenburg goats are a compact, robust, dairy-type animal (that is, with a dished or straight facial line and wedge-shaped body). They have short or medium in length, soft, fine and lying flat hair. Their coat colour is solid, varying from light fawn to dark chocolate with no preference for any shade. They are usually slightly smaller than other Alpine breeds. Adult bucks and does weigh 75 kg and 55 kg, respectively. Height at the wither ranges from 66 to 91 cm (mean 85 cm) and 63.5 to 91 cm (mean 75 cm) for adult males and females, respectively.

Adaptive and Special Genetic Characteristic: The Toggenburg are adaptable to different climatic conditions but best performed in cooler climates and less suited to tropical climates (Burren et al. 2016). It is better adapted than other exotic dairy breeds, such as Saanens or Anglo-Nubians.

Typical Production Systems, Husbandry Practices and Production Characteristics: The Toggenburg goats are adapted to grazing in rough pastures or an alpine environment (Burren et al. 2016). They are good milk producers and could produce from 1 to 3 litres of milk per day, depending on feed quality and accessibility. Jackson et al. (2014) also reported that Toggenburg goat breeds are better milk producers in sub-humid areas (0.3 litres of milk per day better) than semi-arid areas. This has mainly happened due to the availability of feed resources both in quality and quantity. However, they are also suitable for stall feeding. Their milk yield and flavour are also better if they can range extensively on a variety of forages.

Limitation and Challenges Breed Use and Classification: Handling these goats requires experience and expertise. They are active and aggressive. Hence, they are tough to handle, and they can even break out of their fences.

5.5.4.4 Saanen

Synonyms: It has different names in different countries. Example: Chèvre de Gessenay (French) or French Saanen, Saanenziege (German), Capra di Saanen (Italian), Israeli Saanen, British Saanen, Russian white, Sable Saanen (New Zealand),

Origin and Distribution: Saanen goat is one of the most important dairy goat breeds, originating in Switzerland, from the Saanen Valley (De Vasconcelos et al. 2021). They are distributed in more than 80 countries of the world (FAO 2007). These goat breeds are used to develop other dairy goat subtypes such as German, British, Dutch, French, Israeli, Russian, Australian and American Saanen breeds (ESGPIP 1971).

Physical Characteristics: They are characterised as all-white or all-cream coat colour, short hair with occasional black spots on the udder, ears and nose, a straight face, short and pointed ears (ESGPIP 1971). Horns are sabre-shaped and pointed backwards. Adult males weigh 85 kg and females 60 kg; height is 90 cm for males and 80 cm for females (Pesmen and Yardimci 2008).

Adaptive and Special Genetic Characteristic: The Saanen goats are adaptable to temperate and wet environment (Devendra and Haenlein 2016). They are sensitive to excessive sunlight and perform best in cooler conditions. The provision of shade is essential.

Typical Production Systems, Husbandry Practices and Production Characteristics: Saanen goats are mainly managed under an intensive production system. Saanen goats can produce an average milk yield of 700 kg within 260–280 days of lactation length. But they can produce more than 1000 kg of milk yield in some productive does, depending on the husbandry practice taking place from country to country. Because of their high-level milk production and low level of milk fat content, they are compared with Holstein Friesian dairy cattle breeds. The milk fat content ranges from 3% to 4%. However, this milk composition and yield can be affected by environmental factors such as feed, season, climatic conditions, etc.

5.5.4.5 Anglo-Nubian

Synonyms: Nubian (USA), Nubia (North-eastern Africa)

Origin and Distribution: The Anglo-Nubian goat originated in the United Kingdom (FAO 2007). It is all-purpose breed known for meat, milk and hide production. The breed was developed through the crossing of a British goat with bucks of Indian (Jamnapari) and African (Zaraiby goat from Egypt breeds with Nubian and Damascus ancestry) (FAO 2007). The breed is distributed in more than sixty countries.

Physical Characteristics: Anglo-Nubians are characterised by large body size, pride, grace, long legs, either polled or small horns, a convex facial profile and pendulous ears. They have different coat colours but commonly black, chestnut, cream, fawn or white with solid and pattern coat colour. They are prolific, docile and friendly in nature. Adult males weigh up to 140 kg (average 79 kg) and females up to 110 kg (average 61 kg); height is 90 cm for males and 80 cm for females. Birth and weaning weight of Anglo-Nubian kids are recorded as 2.54 kg and 13.12 kg, respectively (Lele et al. 2018). However, birth and weaning weight can be affected by season and year of birth, sex of the kid, type of birth and parity.

Adaptive and Special Genetic Characteristic: Anglo-Nubian goats are well adapted to hot climatic conditions and tolerate frequent weather changes. They have been used in grading-up programmes in many tropical countries to increase the milk and meat production of local breeds.

Typical Production Systems, Husbandry Practices and Production Characteristics: Anglo-Nubian goats have an extended breeding season, so they can produce milk year-round. They provide an average milk yield of 260 ± 129.11 kg within 191 days of lactation length (Lele et al. 2018). The author also added that the minimum and maximum daily milk yield varies from 0.24 to 3 litres. These goats are not heavy milk producers, but they produce quality milk with an average butterfat content of ~5% and protein content of ~4%.

5.5.4.6 Alpine

Synonyms: Alpine polychrome, American Alpine or American Oberhasli, French Alpine, German Alpine or German Fawn, British Alpine, Swiss Chamoisee, etc.

Origin and Distribution: Alpine goats are mountain breeds of Switzerland and are a good dairy goat type (Devendra and Haenlein 2016).

Physical Characteristics: The Alpine goat breeds have different coat colours, which leads to the formation of distinctly separate breeds. Their coat colour can range from white or grey to brown and black. They have medium-to-large size, hair is medium to short, alert, graceful, erect ears, face is straight and prolific goats (1–3 kids at a time). Mature Alpine male goat weighs 77 kg and female 6 1 kg; height at withers 81 cm for male and 76 cm for female.

Adaptive and Special Genetic Characteristic: The Alpine goats are hardy and adaptable and succeed in any climate conditions while maintaining good health and excellent production.

Typical Production Systems, Husbandry Practices and Production Characteristics: The Alpine goat can be managed under extensive, semi-intensive and intensive production systems, but their production performance is higher in intensive production compared to the two production systems. This breed is extremely well known in the dairy industry for their docile temperament, high-quality milk production and long lactation length. On-farm research conducted in Kenya indicated that an adult Alpine dam aged 5–6 years can provide ~3 litres of milk per day, whereas 2–4.9 years old gives 1.7–2.7 litre (Monica et al. 2014). These goats were supplemented with concentrate feed and minerals. Therefore, milk yield is affected by the age of the dam and feeding practice. They also reach sexual maturation within 4–5 months for buck kids and 5–6 months for doe kids. However, doe kids should not be bred until they are at least 34–36 kg in weight.

5.5.4.7 Status of the Exotic Goat Breeds in Africa

In Africa, most of the goat breeds/populations are dual-purpose breeds, mainly kept for meat production and at medium to low levels for milk production. Even though there are some dairy- or meat-type goat breeds, no genetic improvements have been made so far and managed under an extensive production system. As a result, their production performance, in both meat and milk, is low. To improve their production performance in a short time at a low cost and reduce poverty, exotic goat breeds, mainly dairy goats, were introduced to African countries by non-governmental organisations and some Universities and research institutions. However, in most African countries, except in Kenya (in some breeds: e.g. Saanen) and South Africa, they were unsuccessful and unsustainable. The main reasons were: these exotic breeds were imported and introduced to beneficiaries without any clear justification, lack of adequate planning, such as a structured breeding programme and the development of useful technologies, and efficient management of high-producing animals, such as the use of concentrates, veterinary service, housing, etc. Beneficiaries did not clearly understand the nature of the exotic breeds and lacked participation in decision-making. A harsh environment is also another factor for the unsuccessfulness of the exotic goat breeds in Africa. Indiscriminate crossbreeding also poses a threat to the local genetic resources. Hence, livestock professionals and politicians do not have a common sense of understanding and do not fully participate in the genetic improvement programme of the local goat breeds and their sustainability.

5.5.5 Composite Goat Breeds

The need for specific products in goats and market demands for quality and improved quantity of certain products motivate goat farmers to develop novel goat types through crossing different local breeds and/or crossing local with imported breeds (so-called upgrading). While this has been responsible for the production of novel goat types of important value, indiscriminate, unguided ventures only result in undue dilution and erosion of local gene pools. In the following section, we mention some successful, well-guided goat types developed in Africa and their uses.

5.5.5.1 Arabi

Origin and Distribution: The Arabi goats are also known as Arabia and Sahelian. It is a variety of Sahelian goats. There are two subtypes of Arabi goats, a sedentary and a transhumant type. They are found distributed in the high plateau and Saharan areas of Algeria.

Physical Characteristics: The sedentary type has longer hair (14–21 cm) compared to the transhumant type (10–17 cm). Predominant colours of the sedentary type are black, pied black and brown. Other coat colours include black, grey or chestnut. They have long ears and are either horned or polled.

Adaptive/Special Characteristics: This breed is highly susceptible to trypanosomiasis and can be reared only in areas not affected by this disease. It is a very hardy breed.

Productive Characteristics: Body weights at birth and adult age for males/females were recorded to be 2.5/2.0 and 60/40 kg, respectively. The average kidding interval and litter size are reported to be 80% and 1.3 kids per kidding, respectively. The females produce 0.5–1.0 litres of milk per day and a total of 220 litres over a lactation period of 120 days. The breed is mainly used for milk production. The population is considered safe.

5.5.5.2 Pafuri

Origin and Distribution: The Pafuri lop-eared goats found in South Africa are a sub-type of the Xhosa goat breed (Magoro et al. 2022). The Pafuri resulted from the crossing of imported Boer males on indigenous Landim females. The Boers were introduced from the northern Transvaal of South Africa in 1928. The breed is restricted to a small area known as Pafuri in southwest Mozambique, near the border with South Africa and Zimbabwe. This area has a semi-arid to arid subtropical climate with annual unimodal rainfall of less than 400 mm. The production system is agropastoral to transhumant.

Physical Characteristics: This goat breed is characterised by long, drooping ears, with a medium-sized body with a straight back and deep chest, and varied coat colour, but brown or black with white markings are typically common (Magoro et al. 2022). Large body size, with males weighing up to 60 kg and females up to 43 kg; convex facial profile; horns well developed and diverging in males, but smaller and scimitar-shaped in females; ears are lopped or semi-lopped with rounded tips; both males and females are bearded; coat colour is variable, and the coat hair is short.

Adaptive and Special Genetic Characteristics: Adaptive traits of these indigenous goats, including the Xhosa breed, are important for climate change and affordable maintenance (Magoro et al. 2022). Studies indicated healthy and high levels of genetic diversity in these indigenous goat populations of South Africa. However, there is a need for further research to identify genes associated with production and adaptive traits in indigenous goats (Baenyi et al. 2020).

Typical Production Systems and Production Characteristics: Goat production plays an important role in the livelihood of small farming communities in South Africa. The production potential of the indigenous goat ecotype in South Africa is high, and they are well adapted to local conditions (Magoro et al. 2022). Among sub-Saharan countries, the production system is more diverse in South Africa than most, with both extensive and intensive systems (Baenyi et al. 2020) in goat production. The milk production of this indigenous ecotype, like in most sub-Saharan goat types, is low with potential for improvement through selective breeding and improved management practices (Baenyi et al. 2020). There is also a need for conservation and sustainable utilisation of indigenous goat ecotypes, particularly in South Africa, where improvement emphasis is more on cattle than on goat (Magoro et al. 2022).

Productive Characteristics: Body weights at birth, yearling age, 18 months and adult age were recorded to be 2.4, 16.7, 25.8 and 43–60 kg, respectively. The age at first kidding and litter size are reported to be 35.1 months and 1.1 kids per kidding, respectively. The females produce 0.3–0.4 litres of milk per day and a total of 50 litres per lactation period. According to a report in 1991, the population in Mozambique was less than 100 animals. The breed is thus considered critically at risk.

5.5.5.3 Tanzanian Blended Goat

Origin and Distribution: The Tanzanian Blended Goat is also known by the common names Blended and Malya blended. The breed was developed and stabilised as a composite breed in the late 1960s at the Malya Livestock Research Centre in Tanzania. It is constituted of 55% Kamaria, 30% Boer of South Africa and 15% of indigenous stock.

Physical Characteristics: They have large pendulous ears and resemble the Angol-Nubian breed.

Productive Characteristics: The breed is kept mainly for meat production. Body weights at birth and weaning age were recorded to be 2.6 and 12.2 kg, respectively. The average age at first kidding, kidding interval, twinning rate and litter size are reported to be 25.5–27.3 months, 261 days, 29% and 1.3 kids per kidding, respectively. The females produce 153.3 litres of milk with a butterfat content of 3% over a lactation period of 215.6 days. At slaughter, the males produce a carcass with a dressing percentage of 48%.

5.5.5.4 Vogan

The Vogan is reported to be a cross of the Djallonke with the Sahel goat from southern Togo and neighbouring Benin in the vicinity of Vogan, Aneho and Tabligbo districts. The area has a sub-humid to humid climate with a long unimodal rainfall. The production system is principally urban. The goat is characterised by horns that are triangular in males, loosely spiralled and 30–40 cm long; only 5.6% of them carry small horns; long neck; coat colour is variable, with red and black pied animals being the commonest; the hair is short.

5.5.5.5 Makatia

Origin and Distribution: This breed may be the result of a crossing between the Charkia (an Oriental exotic breed) and the long-haired Arabi goat from the Ouled Nail region. It is a variety of Sahelian goats. Found mostly in the high plateau in the northern regions of Algeria.

Physical Appearance: Large-sized animals; long and drooping ears; grey, beige, white or brown coat colour; toggles and beards common.

Adaptive/Special Characteristics: This breed is highly susceptible to trypanosomiasis and can be reared only in areas not affected by this disease. However, it is a very hardy breed.

Productive Characteristics: The breed is mainly used for milk production. Body weights at birth, yearling and adult age for males/females were recorded to be 3.0/2.5, 18–25, and 50/34 kg, respectively. The average kidding percentage and litter size are reported to be 90% and 1.5 kids per kidding, respectively. The females produce 1–2 litres of milk per day for a period of 210 days.

5.5.5.6 Boer

Origin and Distribution: The Boer goat (Fig. 5.18) is also known by other common/local names, including Africander, South African Common, Boer goat Kalahari and Boerbok. The breed was developed by Dutch settlers at the Cape mainly out of the indigenous Bantu stocks through selection for improved meat production since the 1920s although the name Boer was used in the mid-nineteenth century. Subsequently, goat breeds of European and Indian origin were used to upgrade the breed standard. The lop ears and often convex facial profile of the Boer may have arisen from Indian breeds (e.g. Jamnapari) imported during the eighteenth and nineteenth centuries. The breed society for the 'improved Boer' was established in South Africa in 1959 to improve the breed, to establish breed standards and to develop training, inspection, promotion and marketing capacity. These animals were kept in thornveld areas for bush control.

South Africa is the main location. The breed has been exported to Lesotho, Swaziland, Botswana, Mozambique, Zimbabwe and Namibia. Small populations were also imported

Fig. 5.18 Boer goat

Productive Characteristics: Body weights at birth, weaning and adult age were recorded to be 3.2, 24.9 and 50.7 kg, respectively. The age at first kidding, kidding percentage, kidding interval, twinning percentage and litter size are reported to be 15 months, 82.4%, 357.5 days, 65.9% and 1.6 kids per kidding, respectively. The females produce 2.0 litres of milk per day with a fat percentage of 5.7% and a total of 118.4 litres over a lactation period of 102 days.

Boer goat is primarily used for meat production. The population in South Africa in 1991 was estimated at 5000,000.

5.5.5.7 South African Savannah Goat

Origin and Distribution: The Savannah goat breed was developed in South Africa in the 1950s by breeding Boer goats for their meat quality (Chapman et al. 2022; UNEP-WCMC, IUCN 2016). Although the real origin of this breed is murky, it is generally attributed to a ranch in South Africa near the Vaal River (FARM-Africa 1994). Since the 1990s, Savannah goats have been imported to the United State (FARM-Africa 1994).

Physical Characteristics: Savana goats are large-framed, well-muscled breed primarily used for meat (FARM-Africa 1994). They typically have white coats, as the dominant coat allele. However, their skin, horn and hooves have black pigmentation to protect from the sun (FARM-Africa 1994).

Adaptive and Special Genetic Characteristics: Savannah goats are bred for their white coat, black skin and ability to withstand harsh savannah conditions like direct sunlight, changing temperatures and drought (Chapman et al. 2022). They eat plants that other flock and herd animals will not and have good reproductive traits: high fertility, mothering ability and milk production, a high twinning rate, and their kids have a good growth rate (FARM-Africa 1994).

Typical Production Systems and Production Characteristics: The Savannah goats are used in diverse goat production systems in many African countries.

to Kenya, Burundi, Tanzania and recently, Ethiopia. The breed is also distributed throughout the world in 43 other countries. The total population might be about 5 million, about 2.2 million of which are 'improved'. The improved Boer performs best in arid areas of South Africa, where the production systems are agropastoral, pastoral and ranching.

Physical Appearance: A red-headed, large-framed goat with a white body; droopy ears; strong and moderate-length horns, which are placed moderately apart with a gradual backwards curve; a beard is present in males (Fig. 5.18).

Adaptive/Special Characteristics: Boer goats are the most sought-after meat goat breed in the world. They are good browsers that can be used to control bush and weeds; have excellent-quality skin; low-fat, tender and tasty meat at a young age; also produce good quality white cashmere in small quantities. The breed is highly fertile with multiple births, including triplets and quadruplets, and is a non-seasonal breeder. It has an exceptional ability to withstand and resist diseases such as blue tongue, prussic acid poisoning and, to a lesser extent, enterotoxaemia (pulpy kidney). It is also less susceptible to infections caused by internal parasites as the breed prefers to graze above the ground.

5.6 Characterisation of Genetic Diversity in African Goats

5.6.1 Early Molecular Genetic Studies on African Goat Populations

Domestication gave rise to a great diversity in goats across the world, with Africa being home to the largest diversity (Stella et al. 2018; Paim et al. 2019; Colli et al. 2018; Selionova et al. 2021; Michailidou et al. 2019). Molecular and archaeological evidence suggests that African goats have a genetic affinity with Asian goat populations (Paim et al. 2019; Colli et al. 2018). African goats, however, have higher diversity, occasioned by their evolution of resistance to diverse climatic and environmental features that characterise the continent and the presence of very few improvement programmes tailored toward speciality production on the continent (Paim et al. 2019). Diversity in different populations was classified using different indices (Colli et al. 2018; Michailidou et al. 2019). Genotypic information housed in the DNA is unique not only for populations but also for individuals. The need to verify phenotypic classification with molecular data is borne from the fact that many local breeds bare very close semblance. Thus, molecular characterisation of goats and other livestock to determine what is original to, or imported to, a location is very important (FAO 1986b, 2012). Furthermore, molecular genetic tools are also very useful in the determination of the maternal and paternal origin of an animal, tracing poached and stolen animals and much more (FAO 1986a, b).

Molecular genetic characterisation studies involve the study of genetics at the molecular basis, leveraging genetic information housed in deoxyribonucleic acid (DNA) of living organisms. Microsatellite markers and mitochondrial DNA (mtDNA) markers were among the first sets of molecular markers used in goat genetic studies in Africa. Prior to this, blood protein markers, including haemoglobin and serum proteins, were used. Multiple maternal origins of the mitochondrial DNA (mtDNA) hypervariable region (Luikart et al. 2001) is among the foundational information that first classified global goat populations by lineages into A and B, based on unique haplotype grouping. Lineage B was found to be mostly Asian, and a few individuals in lineage were assigned Sub-Saharan African, including Namibia (Naderi et al. 2007) and South Africa (Awotunde et al. 2015). Over the years, more studies using the molecular characterisation approach have been carried out in different parts of the world, including Africa, to delineate indigenous from imported goat types using phylogenetic trees.

Phylogenetic study conducted to determine the relationship between the West African Dwarf, Red Sokoto and Karahari Red suggested *Capra aegagrus* as their wild ancestor (Awotunde et al. 2015). Using the same techniques, indigenous Nigerian goats have been regrouped into demographic types as northern and southern goat breeds rather than the initial phenotypic typing into three breeds (Ahmed et al. 2017). In another study by Tarekegn et al. (2019), shared haplotypes were observed between goats in Cameroon, Mozambique, Namibia, Zimbabwe, Kenya and Ethiopia, with an indication of an initial dispersion of goats to Cameroon and central Africa from northeast Africa following the Nile Delta, with autosomal genomic DNA analysis corroborating a common origin for Moroccan, Ethiopian and Iranian goats. Molecular characterisation has even improved beyond mere phylogenetic assay of selected regions of the DNA. In recent times, with the reduction in the cost of sequencing and availability of the technology in Africa, though not easily found everywhere, molecular characterisation has taken the form of genomic exploration as well.

5.6.2 Genomic Characterisation of African Goat Breeds

High-throughput sequencing platforms and chip sequencing have been carried out on some African goat breeds (Monua et al. 2020; Okpeku et al. 2011b, 2017), shedding more light on the genetic architecture and population structure, determining the evolutionary genetic basis for adaptation to diverse environments, and providing insight into functionally important genetic variants (Andersson and Georges 2004) within and across breeds, Contributing to a better understanding of population genetic structure in goats (Monua et al. 2020; Okpeku et al. 2011a, b, 2016a; Sanni et al. 2018), coat colour genetics (Adefenwa et al. 2013), genetics of immune response (Yakubu et al. 2013, 2016, 2017a, b; Ajayi et al. 2014; Okpeku et al. 2016b), genetics basis of muscle formation in myo-statin genes (Sanni et al. 2018), evolution of thermotolerance of disease resistance (Okpeku et al. 2016b) and the evolutionary process leading to domestication (Zheng et al. 2020).

Genomic studies have provided evidence that supports a high level of genetic diversity but weak phylogeographic structure (low genetic differentiation between breeds) in native African goats, attributed to extensive intermixing, confirming earlier reports based on mtDNA (Awotunde et al. 2015; Tarekegn et al. 2019) and microsatellite markers (Okpeku et al. 2011a; Nguluma et al. 2018) studies. Genome-wide autosomal markers using a caprine 50K single nucleotide polymorphism (SNP) chip (Monua et al. 2020; Tarekegn et al. 2019) and being used to validate reports of goat genetics determined with mtDNA haplo-grouping. This goes to strengthen claims of genetic relatedness with Asian goats and shared ancestry with ancient domesticated goats (Zheng et al. 2020). Naderi's A haplotype is now confirmed to be the predominant type found in all regions of the continent, B has been identified in South Africa and Namibia goats, while G is reported to be limited to some East African (Kenya, Somalia, Sudan and Ethiopia) and Egyptian goats. This provision of confirmed high genetic diversity has implications for further improvement through selection within populations (Tarekegn et al. 2019).

Incorporation of genetics and genomics studies in breed improvement and breed development programmes is also becoming key in goat production systems, not only in the advanced world but also in Africa. Although a very limited number of studies exist that reported mutations in candidate genes affecting economically important traits in some African goat breeds with potential for marker-assisted selection (MAS) toward genetic improvement. These association studies, however, require further validation before implementation of MAS, given that economic traits are under the influence of many loci.

Mutations in genes affecting litter was reported implicating four SNPs, including: inhibin alpha (*INHA*) gene as g.−65C>G (5'UTR), g.2518G>A (a novel variant in exon 2), g.3041A>G (exon 3) and g.3234C>T (3'UTR) identified in Kalahari Red (KR), Red Sokoto (RS) and West African Dwarf (WAD) goats (Isa et al. 2017). MAS breeding of the WAD and KR goats suggest moderate genetic diversity at g.3041A>G locus, and does having CT genotype at g.3234 C>T locus significantly ($P = 0.01$) impacted larger litter size compared with CC in WAD goats, while SNP genotype at other loci were not significantly associated with litter size in goats.

In like vein, gonadotropin-releasing hormone receptor genes investigated as candidate genes for litter size in WAD (Bemji et al. 2018), Khalahari Red goats and Red Sokoto goats (Isa et al. 2019) suggest that the WAD population possesses novel SNPs (g.-29T>G in 5'UTR region), g.48G>A and g.209T>G in exon 1 that promote amino acid change from methionine to arginine. Moderate diversity and strong linkage disequilibrium was also found to exist at the SNP loci with significant association between allele G at g.-29T>G with higher mean litter size for homozygous (GG) mutant does compared with heterozygotes (GT) or homozygotes (TT), while the relationship between SNPs at the two loci detected in exon 1 and litter size was not significant. For Khalahari Red goat and Red Sokoto goat populations were also observed to possess three SNPs (g.-29T>G, g.48 G>A and g.209 T>G) detected in Red Sokoto and one (g.48 G>A) in Khalahari Red goats, all mutations occurring within 5'UTR and exon one of the gene. The g.209T>G was non-synonymous with an amino acid change from methionine to arginine at position 70 of the GnRHR polypeptide.

Milk protein polymorphism at the casein (*CSN3*) gene locus studied in Red Sokoto goats by Bemji et al. (2006) revealed three alleles: A (45.3%), B (52.3%), and M (2.3%) or variants and five phenotypes at the haemoglobin locus in the study of susceptibilities to helminth infestation. While these alleles were found similar in the studied populations, they differed in their expected proportions, suggesting differential susceptibilities to helminth infestation (Wilson 1991).

Several other polymorphisms in the prion protein gene *(PRNP)* of goat populations have been shown in North Africa, cases of Algeria cited in Fantazi et al. (2018), particularly in the more ancient African *Naine de Kabylie*, the oldest breed in North Africa. The polymorphism analysed in four main Algerian goat breeds (Naine de Kabylie, Arbia, Mzabite, Makatia) suggests that the stress-resistant allele K222 was present at a very low frequency in *Naine de Kabylie*, whereas a frequency similar to other Mediterranean countries was detected in Mzabite. The overall results showed how polymorphisms in *PRNP* of goat populations from different areas of the Mediterranean basin can differ in terms of variability and relative frequencies. Genomic information is thus changing the breeding programme in Africa from indiscriminate gambling processes to purpose-driven programmes.

reared in Egypt: Baladi, Barki, Damascus, and Zaraibi. Their results revealed that the CSN1S1 gene exhibited considerable allelic diversity among these breeds, indicating genetic variation that could influence milk production traits. Three of the genotypes were homozygous (A/A (4.5%), B/B (6.8%) and D/D (2.3%)), and the other two were heterozygous (A/C (40.9%) and B/D (45.5%)). Egyptian goat breeds also carried CSN1S2 gene A, B and F alleles, while C, D, E and O alleles were not present. The frequency of homozygous and heterozygous genotypes varied significantly, and polymorphism of gene loci was also documented. Polymorphism associated with milk production parameters was reported by Eid et al. (2018).

In another study, Nowier and Ramadanl (2020) reported a significant association of G/A SNP with the lactoferrin gene, protein, solids-not-fat and total solid content of goat milk, suggesting that these could be useful markers in MAS programmes to improve goat milk composition. These results advance our understanding of the genetic basis for milk composition and would help in planning improvement, management and breeding programmes for dairy goats.

5.6.4 Association of Molecular Markers with Growth Trait

Growth hormone (GH) is the main regulator of animal growth and is encoded by the GH gene that exhibits active gene variants improving growth. The investigation of genetic polymorphisms in two functional genes (myostatin (MSTN) and prolactin (PRL)) was reported by Abdel-Aziem et al. (2018) in three goat breeds using Sanger sequence and restriction fragment length polymorphism (RFLP). In the study, the 337 bp MSTN gene profile displayed five SNPs. The MSTN-HaeIII/PCR-RFLP complex was used in genotyping goats for growth traits, revealing three different genotypes designated as AA, AB and BB, but polymorphism was limited to some breeds. The study also demonstrated polymorphism in the PRL-Eco24I/PCR-RFLP complex across all breeds, suggesting the latter as a

5.6.3 Association of Molecular Markers with Milk Production Traits

Milk production is an essential component of livestock production. High quantity and quality of milk production is usually desired to provide nourishment for young kids and exploited for producing nourishment in meals of the keepers, excess is sold fresh or processed as dairy products sold support the keepers' economy. Genes encoding major protein and whey protein of milk are reported to be candidate genes that can be exploited in MAS to improve the milk productivity in farm animals. Ahmed and Othman (2009) genotyped four milk protein genes—CSN1S1, CSN1S2, KCN, and β-LG—in four goat breeds

better marker for genetic differentiation between goat breeds relative to MSTN-HaeIII/PCR-RFLP.

Growth hormone (GH) gene that promotes goat muscles, bone formation, regulating fat content and other important traits in goats was molecularly analysed by Mahrous et al. (2018) in a study that identified three loci of GH gene (GH1, GH2 and GH6) in three goat breeds using polymerase chain reaction–restriction fragment length polymorphism (PCR-RFLP) and gene sequencing analyses. In another study, El-Halawany et al. (2017) documented SNP polymorphism in the exon-3 region of the GH gene, confirming previous study findings, and concluded that candidate growth genes could be used as a marker in the selection of goats with high growth value traits.

5.6.5 Association of Molecular Markers with Reproduction and Immune Response Trait

The association of molecular markers with reproduction and immune response traits in African goats has been studied in various research. A study that assessed the acquired immune response of different goat breeds, including South African Indigenous goats (Nguni), Saanen goats, and cross-bred goats, found that all three goat breeds exhibited an acquired immune response, but the Nguni response was significantly higher than that of the Saanen goats (Gopalraj et al. 2013). Another study on Innate Immune Response to Rift Valley Fever Virus (RVFV) found that the virus infects dendritic cells (DCs) and monocytes, inhibiting the IFN-α response and allowing rapid replication (Nfon et al. 2012). A study by Chiejina and Behnke (2011) on the Nigerian West African Dwarf goat breed found that they had a strong resistance and resilience to gastrointestinal nematode infections. The researchers believe that this trait is genetically determined and could be a breed characteristic, particularly regarding haemonchotolerance. Some other research on the impact of heat stress on goats found that heat stress results in a decrease in growth, milk and meat production and immune response in goats. However, the exact immune-regulatory mechanisms are still underutilised in goats, especially in South African goats (Visser and Snyman 2023). These researchers, associating molecular markers with reproduction and immune response traits in goats, found that various breeds exhibited different patterns of gene expression related to traits such as faecal egg count, packed cell volume, immunoglobulin A (IgA), immunoglobulin G (IgG), interleukin family genes and interferon gamma (IFN-γ) (Visser and Snyman 2023). Despite these, the general opinion is that further research is needed to better understand the underlying mechanisms and potential applications of these molecular markers in goat breeding and health management.

A study conducted by Metawi et al. (2019) analysed the polymorphism of two genes, namely growth differentiation factor 9 (GDF9) and follicle-stimulating hormone (FSHβ), in goat breeds known for high prolificacy. These genes are associated with fecundity, which implies that they can serve as molecular markers for the selection of goats with high litter size. The study suggests that these markers can be used in goat improvement programmes to enhance their breeding for high prolificacy. Several other candidate genes have been identified that are associated with reproduction traits in African goats, including fecundity and prolificacy. Some of these candidate genes include: Adenylate Cyclase 1 (ADCY1) involved in energetic metabolism, cellular mitosis and growth hormone releasing. This gene has been described in West African Dwarf goats and Cameroon goats (Salgado Pardo et al. 2022; Zonaed Siddiki et al. 2020). Heat Shock Protein 4 (HSF4), Vacuolar Protein Sorting 13A (VPS13A) and Nucleobindsin E1(NBEA) are candidate genes associated with adaptation to desert regions, dry weather and milk traits in goats (Amiri Ghanatsaman et al. 2023); these genes have been described in Asian goats raised in arid regions similar to most of sub-Saharan Africa's production environment with promising prospects in African goat relatives. Interleukin 7 (IL7), Interleukin 5 (IL5), Interleukin 23 subunit alpha (IL23A), and Leucine Rich Repeat and

Fibronectin Type III Domain Containing (5 LRFN5) (Amiri Ghanatsaman et al. 2023). These genes are associated with immune response traits, litter size traits and production function in indigenous African goats (Wang et al. 2022). These candidate genes provide insights into the genetic potential of African goats for fecundity and prolificacy. However, more research is needed to better understand the specific roles of these genes in goat reproduction and to develop strategies for improving reproductive performance in African goats.

5.6.6 Candidate Genes Associated with Immune Response/Adaptive Traits

Analysis of immune response genes to understand genetic adaptation and response to evolutionary forces in African goat evolution was reported in many studies in connection with indigenous African goat breeds. Polymorphisms of the MHC class II DRB gene in the Egyptian (Ahmed and Othman 2006; Ahmed et al. 2018) and Nigerian goats (Yakubu et al. 2017b) using an amplified fragment of 285-bp of the gene revealed various allele frequency patterns of the MHC class II DRB. Furthermore, genetic diversity of 2-decyl-4-quinazolinyl amine exon2 (DQA2 exon2) gene (Yakubu et al. 2017a; Ahmed et al. 2018) and DQB (Yakubu et al. 2013), Interleukin-2 (IL-2) gene (Yakubu et al. 2016), agouti signalling protein (ASIP) gene (Adefenwa et al. 2013) among different African goat populations from different agro-climatic areas have also been studied. These reports strengthen the knowledge that African goats possess adaptive genetic properties for disease resistance (Okpeku et al. 2016a), particularly those that are regionally prevalent where they are found. These further strengthen reported diversity in native African goats with unlimited potential for genetic manipulations for improved meat, dairy and other economic traits.

5.7 Indigenous Knowledge, Unique Features and Products from African Goats

Traditional knowledge and indigenous goat production in Africa play a crucial role in the economy, food security and livelihoods of local communities. By leveraging these resources and promoting sustainable practices, it is possible to enhance the productivity and value addition of goat production on the continent. Some of the key values include:

Meat and Milk Production: Goats raised for milk production in sub-Saharan Africa are comprised of indigenous and cross-bred breeds. Goat meat and milk are essential sources of protein for pastoralist communities in the region (Kahi and Wasike 2019).

Natural Resource Management: Traditional methods of Rwandan goat production and management involve allowing goats to graze in open areas, which contributes to sustainable land use and conservation of natural resources (Khowa et al. 2023).

Livelihood Opportunities: Goats serve as an economic safety net for pastoralist communities, providing income through the sale of animals, meat and milk products. Goat products are also used for various purposes, such as blood for food in some pastoralist communities, skins for the leather industry and manure for fertiliser (Kahi and Wasike 2019).

Cultural Significance: Goat meat and milk are traditionally accepted in some cultures within countries and are usually promoted as livelihood programmes. Urban goat production exists but is scanty, and commercial production is limited (Kahi and Wasike 2019). These products, which have strong cultural, economic and medicinal values, reflect social and economic dynamism and employ a variety of methods, with a majority of the workforce being females. Despite their roots in the African culinary tradition, these products have evolved to play significant roles in not only the food and economy of the producers but also reflect the region and sometimes tribes from where they are made, with special reference to traditionally made butter, curd, cheese and yoghurt. Additionally, other values derived from goats in different cultures confirm their roles in improving household food security, income and sociocultural roles among rural communities in Africa (Pollott and Wilson 2009).

Community-Based Breeding Programmes: Community-based breeding programmes for improved goat production with a focus on the improvement of observable phenotypes, production and productivity have been implemented in some countries, leveraging traditional knowledge and indigenous goat production to enhance productivity and value addition (Monau et al. 2020). Generally, goat breeding is often associated with rural and pastoral agriculture, with a focus on meat and dairy production to support household needs and contribute to farmers' economy on a small scale. See Chap. 17, which addresses utilisation dimensions through the development of appropriate breeding goals and strategies.

Market Opportunities: Sheep and goats contribute to supplying markets with food and non-food products, offering export-earning and import-saving opportunities for local communities (Pollott and Wilson 2009). The manufacturing of these products employs indigenous knowledge, passed down from generation to generation (CNL 2019). Processing and preservation employ techniques that vary by regions and territories. The use of goatskin water bags or pottery in some regions confirms specificity in manufacturing, linked to traditional practices (Khoualdi 2017). For instance, salting, draining and refining of *Bouhezza* cheese are done simultaneously in a permeable goatskin water bag or 'Chekoua' with the incorporation of red chilli powder; thus, the production lasts several weeks to several months with a strong acidulous taste characterised by this cheese (Medfouni and Benidir 2018). These are some value chain products from goat rearing apart from the skin sold for money and processed into leather for bags, shoes and belts, among others. The Red Sokoto goat skin was particularly famous for making quality leather. Horns and bones from goats are have also been known to be burnt and processed into China plates and other products for household uses. Ethiopian hides and skins have a good reputation in the international market. The international leather market for its unique natural substance of fitness, cleanness, compactness of texture, producers thickness, flexibility and strength. Specifically, the goat variety around the Bati area in Ethiopia is known for its superior quality. According to Made (2017) 'Bati-genuine' is associated with the highest quality class goatskins in the world. The particular characteristics of Ethiopian bati-genuine goat skins are their thickness, high flexibility and clean inner surface, and they are known worldwide for being an excellent raw material for producing high-quality suede leather.

Dietary Contributions: Meat and milk products from small ruminants, including goats, contribute to dietary diversity and provide essential nutrients for human consumption (Pollott and Wilson 2009). Milk is perishable, so it is processed into products often determined by the traditional and cultural needs of locales.

Butter from goat milk is known as *S'men* in some parts of North Africa. It is made by churning plain goat milk, cultured milk or cream in mixers until a butter lump is formed and then cooled and separated from the buttermilk. These either contribute to household meals, are sold for money or offered as a gift.

Fermented milk products include yoghurt, *Raib* (curdled milk), and *D'han* (fermented curdled milk), obtained at the end product of curdling milk from goat and other ruminants. These are often served as dairy beverages chilled and contribute to family incomes as well.

Cheese, known variously as *Djben*, *Klila* (Hallel 2001) in Northern Africa, called *Wara* by the Yoruba-speaking tribes in Nigeria, West Africa, and other various names across the continent, is among the list of traditional milk products prepared from raw milk (Ould Abeid et al. 2013), consumed within 10 or 15 days of its preparation. There is also *Bouhazza* (Fig. 5.19), a traditional cheese, refined, soft and spicy, unmolded, and very common in the East of Algeria (Aissaoui-Zitoun et al. 2012). *Michouna* is a fresh cheese made from the milk of cows or goats or the mixture of both (Derouiche and Zidoune 2015). *Klila* is a traditional cheese, eaten fresh or dried, prepared from the milk of a cow, sheep or goat or a mixture of the three. Its empirical manufacturing process is still in force, and characterised by spontaneous fermentation of the original Flores of milk at room temperature. After clotting and draining, the product is consumed fresh or after drying for several weeks (Boubekri and Otha 1996). *Medeghissa* is a traditional Algerian cheese known to the Chaouia under its original name, *Imdeghest*. It is a melted cheese, prepared by cooking *Klila* semi-dry in the whole milk of a cow or goat, on a low heat (Khoualdi 2017).

Benefits to the Farming System Goats contribute to sustainable farming systems by consuming local feeds, utilising marginal lands and providing manure for fertiliser (Pollott and Wilson 2009).

Fig. 5.19 Original *Bouhezza* cheese from Algeria

Preservation of Local Breeds Indigenous goat production contributes to the conservation of local breeds, which are adapted to specific environments and ecological niches, providing genetic resources for future improvement and development of African goat breeds (Monau et al. 2020).

In Africa, several value chains have been identified in goat-rearing communities across the continent based on their development potential and their relevance to the targeted populations, particularly in terms of the inclusion of a large number of small enterprises and beneficiaries (Abdelfettah and Fantazi 2020), major among these are milk and dairy products, known to support significant local demands but face serious competition from imported products such as powdered milk.

5.8 Conclusion and Future Perspectives

5.8.1 Conclusion

This chapter discussed the African goat genetic resources in terms of its importance, origin, geographical distribution, introgression between and among populations, and the nature of phenotype, molecular and genomic characteristics.

Archaeological and molecular studies suggest the Middle East and the Asian region as centres of domestication for goats. Their dispersal was mainly through human migration and trade, leading to the creation of sub-species based on five key ecological regions of Africa. Through localisation, adaptation to farming systems, climate and human needs, and introgression, a total of 289 goat breeds currently exist in Africa. Eastern and Southern Africa dominate with 180 breeds, while Western, Northern and Central Africa share the rest.

Morphological and genomic studies show that Africa has the richest diversity of goats, each locally adapted to specific locations and niches, hence described as local breeds. Apart from these, a few breeds have been developed into composites and have been used to improve local breeds in other countries, hence termed exotic breeds, led by Boer and Angora goat breeds. Saanen, Togenberg and Alpine have been introduced in a few countries to promote goat milk production. Within other countries such as South Africa, Mozambique, Tanzania, Togo, Algeria and Morocco, some attempts have been made to develop locally adapted breeds.

Regardless of region and ecological zones, African goats are commonly utilised for meat, a source of income, and for sociocultural roles among the majority of rural communities in Africa. Goat milk is traditionally accepted in some cultures within countries and usually promoted as livelihood programmes. Urban goat production exists but is scanty and so is commercial production. In commercial settings and research and academic institutions, goat meat and milk values chains produce biltong and goat cheese, respectively.

African goats also share similarities in production and reproductive traits, with local breeds maturing at between 30 and 40 kg, while exotic breeds reach up to 60 kg live weights. Twining is common. Trait variations exist within and between populations, meaning there is a high likelihood of improving African goats using crossing and line breeding strategies.

Being a poor man's cow, African goat offers a better livelihood and commercialisation tool that can enhance the livelihood and equity of wealth distribution among the majority of rural communities in Africa. Their commonality in roles provides a clue that similar strategies can be employed to improve their productivity and value addition and similarly drive the commercialisation of African goat.

5.8.2 Future Perspectives

Indigenous goats of Africa's unique breeds have great potential to strengthen agriculture, tradition, culture and the economic structure of the communities where they exist. Despite their valuable features and products, the majority of these breeds are still underutilised and understudied. More so, available diversity makes them promising genetic resources for developing improved African goat breeds that are valuable locally and for international exports.

To fully utilise these genetic resources, it is crucial to prioritise and intensify research on the genetic characterisation of indigenous goat breeds to understand their genetic architecture, particularly at the molecular level. This will enable the design of effective breeding programmes for improvement.

Prioritising research on the genetic characterisation of indigenous goat breeds and promoting public-private partnerships can help unlock the potential of these unique breeds and contribute to the sustainable development of African communities. To support research, cushion breeding and provide an enabling environment for breeding, public-private partnerships are recommended. Indigenous livestock breeds are valuable assets of the communities where they are found. As such, future interventions for improvement should be carried out in partnership with locals who will be the owners and end-users of future improved products.

References

Abdel-Aziem SH, Mahrous KF, Abd El-Hafez MA, Abdel Mordy M (2018) Genetic variability of myostatin and prolactin genes in popular goat breeds in Egypt. J Genet Eng Biotechnol 16:89–97

Abdelfettah M, Fantazi K (2020) The annual implementation plan of Live2Africa's activities in the sub-region North Africa, case of Algeria. In: Paper presented in the regional consultative workshop on the annual implementation plan of Live2Africa activities in North-Africa, 28th –30th January 2020, Cairo, Egypt

Abegaz S, Awgichew K (2009) Estimation of weight and age of sheep and goats. In: AlemuYami KA, Gipson TA, Merkel RC (eds) Ethiopia Sheep and Goat Productivity Improvement Program (ESGPIP) technical bulletin No 23. USAID, Prairie View A and M research Foundation, MoA and American Institute for goat research. Branna Printing Enterprise, Addis Ababa, Ethiopia, p 11. Accessed 2024/03/10

Adefenwa MA, Agaviezor BO, Peters SO, Oboh BO, Wheto M, Adekoya KO, Okpeku M, Ikeobi CON, Williams GO, Singh M, De Donato M, Imumorin IG (2013) Identification of single nucleotide polymorphisms in the agouti signaling protein (ASIP) gene in some goat breeds in tropical and temperate climates. Mol Biol Rep 40(7):4447–4457

Ahmed S, Othman EO (2006) A PCR-RFLP method for the analysis of Egyptian goat MHC Class II DRB gene. Biotechnology 5(1):58–61

Ahmed S, Othman O (2009) Genotyping analysis of milk protein genes in different G--oat breeds reared in Egypt. J Genet Eng Biotechnol 7(1):1–6

Ahmed S, Grobler J, Madisha T, Kotzé A (2017) Mitochondrial D-loop sequences reveal a mixture of endemism and immigration in Egyptian goat populations. Mitochondrial DNA A DNA Mapp Seq Anal 28(5):711–716. https://doi.org/10.3109/24701394.2016.1174225

Ahmed S, Abdel-Rahman S, Grobler P, Kotzé A (2018) Allelic diversity of DQA2 exon 2 gene in Egyptian goat populations. Indian J Anim Res 52(8):1101–1106

Aissaoui-Zitoun O, Pediliggieri C, Benatallah L, Lortal S, Licitra G, Zidoune MN, Carpino S (2012) *Bouhezza*, a traditional Algerian raw milk cheese, made and ripened in goat skin bags. J Food Agric Environ 10(2):289–295

Ajayi OO, Adefenwa MA, Agaviezor BO, Ikeobi CON, Peters SO, Wheto M, Okpeku M, Yakubu A, De Donato M, Imumorin IG (2014) A novel *TaqI* polymorphic site in the coding region of ovine *TNXB* gene in the MHC Class III region influences morphostructural and physiological indices. Biochem Genet 52(1–2):1–14

Akinmoladun OF, Fon FN, Mpendulo CT, Okoh O (2020) Performance, heat tolerance response, and blood metabolites of water-restricted Xhosa goats supplemented with vitamin C. Transl Anim Sci 4(2):txaa044. https://doi.org/10.1093/tas/txaa044

Alberto FJ, Boyer F, Orozco-terWengel P, Streeter I, Servin B, de Villemereuil P, Benjelloun B, Librado P, Biscarini F, Colli L, Barbato M, Zamani W, Alberti A, Engelen S, Stella A, Joost S, Ajmone-Marsan P, Negrini R, Orlando L, Rezaei HR, Naderi S, Clarke L, Flicek P, Wincker P, Coissac E, Kijas J, Tosser-Klopp G, Chikhi A, Bruford MW, Taberlet P, Pompanon F (2018) Convergent genomic signatures of domestication in sheep and goats. Nat Commun 9(1):813. https://doi.org/10.1038/s41467-018-03206-y

Alemu A (2015) ON-FARM phenotypic characterization and performance evaluation of Abergelle and central-Hihgh-land goat breeds as an input for designing community-based breeding program. MSc Thesis. Accessed 9 March 2024

Alemu T, Alemu D, Alemu T (2020) Carcass characteristics and sensory analysis of Abergelle goat breed and Abergelle crossbred goat fed hay supplemented with concentrate mixture. Transl Anim Sci 4(3):txaa149. https://doi.org/10.1093/tas/txaa149

Amills M, Capote J, Tosser-Klopp G (2017) Goat domestication and breeding: a jigsaw of historical, biological and molecular data with missing pieces. Anim Genet 48(6):631–644. https://doi.org/10.1111/age.12598

Amiri Ghanatsaman Z, Ayatolahi Mehrgardi A, Asadollahpour Nanaei H, Esmailizadeh A (2023) Comparative genomic analysis uncovers candidate genes related with milk production and adaptive traits in goat breeds. Sci Rep 13:8722. https://doi.org/10.1038/s41598-023-35973-0

Andersson L, Georges M (2004) Domestic animal genomics: deciphering the genetics of complex traits. Natl Res Gen 5(3), 202–212. https://doi.org/10.1038/nrg1294

AU-IBAR (2019) The state of farm animal genetic resources in Africa: towards accelerated agricultural growth and transformation by the year 2025. African Union - Interafrican Bureau for Animal Resources (AU-IBAR), Nairobi, Kenya, p 337. https://www.au-ibar.org/sites/default/files/2020-10/gi_20191107_state_farm_animal_genetic_resources_africa_full_book_en.pdf

Awotunde EO, Bemji MN, Olowofeso O, James IJ, Ajayi OO, Adebambo AO (2015) Mitochondrial DNA sequence analyses and phylogenetic relationships among two Nigerian goat breeds and the South African Kalahari Red. Anim Biotechnol 26:180–187. https://doi.org/10.1080/10495398.2014.977907

Baenyi PS, Owino JJ, Keambou TC, Bwihangane BA, Karume K, Mekuriaw TG, Winyo OJ (2020) Production systems, genetic diversity and genes associated with prolificacy and milk production in indigenous goats of sub-Saharan Africa: a review. Open J Anim Sci 10:735–749. https://doi.org/10.4236/ojas.2020.104048

Bemji MN, Ibeagha-Awemu EM, Osinowo OA, Erhardt G (2006) Casein (CSN3) genetic variability of the Nigerian Red Sokoto goat. Nig J Genet 20:1–8

Bemji MN, Ogundiyi AI, Adebambo OA, Dipeolu MA, Onagbesan OM, James IJ, Osinowo OA (2012) Prevalence of coat colour phenotypes and its influence on mange infestation of West African Dwarf goat. Bull Anim Prod Health for Africa 60(4):331–338

Bemji MN, Awotunde EO, Olowofeso O, James I, Oduguwa BO, Okwelum N, Osinowo O (2014) Maintenance of mtDNA diversity in Kalahari Red goat of South Africa imported to Nigeria. Anim Gen Res 55:39–46. https://doi.org/10.1017/S2078633614000289

Bemji MN, Isa AM, Ibeagha-Awemu EM, Wheto M (2018) Polymorphisms of caprine GnRHR gene and their association with litter size in West African Dwarf goats. Mol Biol Rep 45(1):63–69. https://doi.org/10.1007/s11033-017-4141-0

Bemji MN, Osinowo OA, Ozoje MO, Adebambo A, Aina A (2021) Live weight changes during lactation and its relationship with milk off-take and yield in West African Dwarf and Red Sokoto goats intensively managed within the humid zone of Nigeria. Nig J Anim Prod 33(1):145–150. https://doi.org/10.51791/njap.v33i1.2225

Benjelloun B, Pompanon F, Ben-Bati M, Chentouf M, Ibnelbachyr M, El-Amiri B, Rioux D, Boulanouar B, Pierre T (2011) Mitochondrial DNA polymorphism in Moroccan goats. Small Rumin Res 98:201–205. https://doi.org/10.1016/j.smallrumres.2011.03.041

Benjelloun B, Alberto FJ, Streeter I, Boyer F, Coissac E, Stucki S, BenBati M, Ibnelbachyr M, Chentouf M, Bechchari A, Leempoel K, Alberti A, Engelen S, Chikhi A, Clarke L, Flicek P, Joost S, Taberlet P, Pompanon F, NextGen Consortium (2015) Characterizing neutral genomic diversity and selection signatures in indigenous populations of Moroccan goats (Capra hircus) using WGS data. Front Genet 6:107. https://doi.org/10.3389/fgene.2015.00107

Benjelloun B, Benbati M, Chikhi A, Boulanouar B (2016) Performances de croissance et de viabilité pré-sevrage des races parentales et des moutons croisés «D'man x Boujaâd» de générations F1, F2 et F3. Options Méditerranéennes. The value chains of Mediterranean sheep and goat products. Organisation of the industry, marketing strategies, feeding and production systems Zaragoza. CIHEAM 115:407–411

Berhane G, Eik LO (2006) Effect of vetch (Vicia sativa) hay supplementation on performance of Begait and Abergelle goats in northern Ethiopia I. Milk yield and composition. Small Rumin Res 64(64):225–232. https://doi.org/10.1016/j.smallrumres.2005.04.021

Berihulay H, Li Y, Gebrekidan B, Gebreselassie G, Liu X, Jiang L, Ma Y (2019) Whole genome resequencing reveals selection signatures associated with important traits in Ethiopian indigenous goat populations. Front Genet 10:1–12. https://doi.org/10.3389/fgene.2019.01190

Bertolini F, Servin B, Talenti A, Rochat E, Kim ES, Oget C, Palhière I, Crisà A, Catillo G, Steri R, Amills M, Colli L, Marras G, Milanesi M, Nicolazzi E, Rosen BD, Van-Tassell CP, Guldbrandtsen B, Sonstegard TS, Tosser-Klopp G, Stella A, Rothschild MF, Joost S, Crepaldi P, AdaptMap consortium (2018) Signatures of selection and environmental adaptation across the goat genome post-domestication. Genet Sel Evol 50(1):57. https://doi.org/10.1186/s12711-018-0421-y

Birhanie M, Alemayehu K, Mekuriaw G (2018) Performance evaluation of Abergelle goat under community based breeding program in selected districts, Northern Ethiopia. Livest Res Rural Dev 30(4) http://www.lrrd.org/lrrd30/4/msmi30064.html.u

Bosman L, Van Marle-köster E, Visser C (2015) Genetic diversity of South African dairy goats for genetic management and improvement. Small Rumin Res 123:224–231. https://doi.org/10.1016/j.smallrumres.2014.12.003

Boubekri K, Otha Y (1996) Identification of lactic acid bacteria from Algerian traditional cheese, El-Klila. J Sci Food Agric 70:501–505

Boujenane I, Derqaoui L, Nouamane G (2016) Morphological differentiation between two Moroccan goat breeds. J Livestock Sci Technol 4:31–38

Britannica T (2016, April 22) Editors of Encyclopaedia. "Nubian." Encyclopedia Britannica, https://www.britannica.com/animal/Nubian-goat

Burren A, Neuditschko M, Frischknecht M, Reber I, Menzi F (2016) Genetic diversity analyses reveal first insights into breed-specific selection signatures within Swiss goat breeds. Anim Genet:1–13. https://doi.org/10.1111/age.12476

Campbell QP (2003) Origin and adaptation of South African indigenous goats. S Afr J Anim Sci 4:18–22

Chapman CA, Abernathy K, Chapman LJ, Downs C, Effiom EO, Gogarten JF, Golooba M, Kalbitzer U, Lawes MJ, Mekonnen A, Omeja P, Razafindratsima O, Sheil D, Tabor GM, Tumwesigye C, Sarkar D (2022) The future of sub-Saharan Africa's biodiversity in the face of climate and societal change. Front Ecol Evol 10:790552. https://doi.org/10.3389/fevo.2022.790552

Chiejina SN, Behnke JM (2011) The unique resistance and resilience of the Nigerian West African Dwarf goat to gastrointestinal nematode infections. Parasites Vectors 4:12. http://www.parasitesandvectors.com/content/4/1/12

Chynoweth MW, Lepczyk CA, Litton CM, Hess SC, Kellner JR, Cordell S (2015) Home range use and movement patterns of non-native feral goats in a tropical Island montane dry landscape. PLoS One 10(3):e0119231. https://doi.org/10.1371/journal.pone.0119231

CNL (2019) Specification cheese Bouhezza. National Committee for the Labeling Of Agricultural Products or of Agricultural Origin, CNL, p 25

Colli L, Lancioni H, Cardinali I, Olivieri A, Capodiferro MR, Pellecchia M, Rzepus M, Zamani W, Naderi S, Gandini F, Vahidi SM, Agha S, Randi E, Battaglia V, Sardina MT, Portolano B, Rezaei HR, Lymberakis P, Boyer F, Coissac E, Pompanon F, Taberlet P, Ajmone MP, Achilli A (2015) Whole mitochondrial genomes unveil the impact of domestication on goat matrilineal variability. BMC Genomics 16(1):1–12

Colli L, Milanesi M, Talenti A, Bertolini F, Chen M, Crisà A, Daly KG, Del Corvo M, Guldbrandtsen B, Lenstra JA, Rosen BD, Vajana E, Catillo G, Joost S, Nicolazzi EL, Rochat E, Rothschild MF, Servin B, Sonstegard TS, Steri R, Van Tassell CP, Ajmone-Marsan P, Crepaldi P, Stella A, AdaptMap Consortium (2018) Genome-wide SNP profiling of worldwide goat populations reveals strong partitioning of diversity and highlights post-domestication migration routes. Genet Sel Evol 50(1):58. https://doi.org/10.1186/s12711-018-0422-x

Cooper T (2021) Brief history of goat domestication. The livestock conservancy. Accessed 2023/09/24 from https://livestockconservancy.org/2022/10/11/a-brief-history-of-goat-domestication/

Cooper T (2022) Breed Profile: Somali Goat. Backyard Goats - Countryside. https://backyardgoats.iamcountryside.com/goat-breeds/somali-goat-breed-profile/. Accessed 2024/03/10

DAGRIS (2007) In: Kempo S, Mamo Y, Astrat B, Dessie T (eds) Domestic animal genetic resources information system. International Livestock Research Institute, Addis Ababa, Ethiopia. http://dagris.ilri.cgiar.org/regions/64/breeds?order=title&sort=asc&page=2. Accessed 2024/03/10

DAGRIS (2019) Domestic animal genetic resources information system. International Livestock Research Institute (ILRI) Online. http://dagris.ilri.cgiar.org/

DAGRIS (2023) Short-eared Somali all Traits. https://dagris.info/breed/2521/trait_type/all?order=title_1&page=1&sort=asc. Accessed 2024/03/10

Daly KG, Maisano-Delser P, Mullin VE, Scheu A, Mattiangeli V, Teasdale MD, Hare AJ, Burger J, Verdugo MP, Collins MJ, Kehati R, Erek CM, Bar-Oz G, Pompanon F, Cumer T, Çakırlar C, Mohaseb AF, Decruyenaere D, Davoudi H, Çevik Ö, Rollefson G, Vigne JD, Khazaeli R, Fathi H, Doost SB, Rahimi SR, Vahdati AA, Sauer EW, Azizi KH, Maziar S, Gasparian B, Pinhasi R, Martin L, Orton D, Arbuckle B, Bradley DG (2018) Ancient goat genomes reveal mosaic domestication in the Fertile Crescent. Science 361(6397):85–88. https://doi.org/10.1126/science.aas9411

Daly KG, Mattiangeli V, Hare AJ, Davoudi H, Fathi H, Doost SB, Amiri S, Khazaeli R, Decruyenaere D, Nokandeh J, Richter T, Darabi H, Mortensen P, Pantos A, Yeomans L, Bangsgaard P, Mashkour M, Zeder MA, Bradley DG (2021) Herded and hunted goat genomes from the dawn of domestication in the Zagros Mountains. Proc Natl Acad Sci 118(25):e2100901118

Daramola JO, Adeloye AA (2009) Physiological adaptation to the humid tropics with special reference to the West African Dwarf (WAD) goat. Trop Anim Health Prod 41(7):1005–1016. https://doi.org/10.1007/s11250-008-9267-6

De Vasconcelos AM, Osterno JJ, Rogério MCP, Façanha DAE, Landim AV, Pinheiro AA, Silveira RMF, Ferreira JB (2021) Adaptive profile of Saanen goats in tropical conditions. Biol Rhythm Res 52:748–758. https://doi.org/10.1080/09291016.2019.1603691

Dekkiche Y (1987) Study of the zootechnical parameters of an improved goat breed (Alpine) and two local populations (Makatia and Arbia). In: Intensive breeding in a steppe zone (LAGHOUAT). Thesis Ing. INA El-Harrach, p 120

Derouiche M, Zidoune MN (2015) Characterization of a traditional cheese, Michouna from the region of Tebessa, Algeria. Livest Res Rural Dev 27:11

Devendra C, Haenlein GFW (2016) Goat breeds. Encycl Dairy Sci Third Ed 1:77–97. https://doi.org/10.1016/b978-0-08-100596-5.00622-3

Ebozoje MO, Ikeobi CON (1998) Colour variation and reproduction in the West African dwarf (WAD) goats. Small Rumin Res 27(2):125–130. https://doi.org/10.1016/S0921-4488(97)00045-X

Eckart E FORESTS AND FORESTRY i. In Persia, Encyclopædia Iranica, X/1, pp 86–90, available online at http://www.iranicaonline.org/articles/forests-and-forestry-i. Accessed on 3 March 2024

Eid JI, Teleb DF, Mohamed SA, El-Ghor AA (2018) DGAT1 polymorphism in Egyptian Zaraibi goat breed and their association with milk yield and composition. J Basic Appl Zool 81:38

El-Halawany NK, Abd-El-Razek FM, El-Sayed YA, Afaf El-Werdany A, Abd-El- AS, Al-Tohamy AF, Abdel-Shafy H (2017) Genetic polymorphisms in Exon-3 region of growth hormone gene in the Egyptian goat breeds. Acad J Biolog Sci 9(2):1–8

Epstein H (1971) The origin of the domestic animals of Africa. Africana Publishing Corporation, New York

ESGPIP (1971) Sheep and goat production handbook for Ethiopia. Ethiop. Sheep goat product. Improv. Program. In: Alemu Yami RC, Merkel Epstein H (eds) The origin of the domestic animals of Africa. Africana Publishing Corporation, New York. 2008. Accessed 2024/03/10

ESGPIP (2008) Sheep and goat production handbook for Ethiopia. Addis Ababa, Ethiopia

Fadlelmoula AA, Yousif I, Ismail A (2014) Genetic and phenotypic parameter estimates of morphometric traits in Sudan desert goats. Online J Vet Res 15(2):106–111

Fajemilehin O, Salako A (2008) Body measurement characteristics of the West African Dwarf (WAD) goat in deciduous forest zone of Southwestern Nigeria. Afr J Biotechnol 7:2521–2526

Fantazi K (2004) Contribution to the study of the genetic polymorphism of Algerian goats, case of the Valley of Oued Righ (Touggourt). Magister thesis. INA El-Harrach, Algeria, pp 85–95

Fantazi K, Tolone M, Amato B, Sahraoui H, Vincenzo di Marco LP, La Giglia M, Gaouar SBS, Vitale M (2017) Characterization of morphological traits in Algerian indigenous goats by multivariate analysis. Genet Biodiv J 1:20–30

Fantazi K, Migliore S, Kdidi S, Racinaro L, Tefiel H, Boukhari R, Federico G, Vincenzo Di Marco LP, Gaouar S, Bechir S, Vitale M (2018) Analysis of differences in prion protein gene (PRNP) polymorphisms between Algerian and Southern Italy's goats. Ital J Anim Sci 17(3):578–585. https://doi.org/10.1080/1828051X.2017.1420430

FAO (1986a) Animal genetic resource data banks. 2. Descriptor lists for cattle, buffalo, pigs, sheep and goats. Animal Production and Health Paper No. 59/2, Rome

FAO (1986b) Animal genetic resource data banks. 1. Computer systems study for regional data banks. Animal Production and Health Paper No. 59/1, Rome

FAO (1991) Small ruminant production and the small ruminant genetic resource in tropical. Animal production and Health Paper. Available at: https://www.fao.org/3/t0376e/T0376E10.htm. Accessed 2024/03/09

FAO (2007) In: Rischkowsky B, Pilling D (eds) The state of the world's animal genetic resources for food and agriculture, Rome

FAO (2012) Regional Workshop on "Characterization and Value Addition to local Breeds and their Products". North Africa and Near-East, 19–21 November, Rabat, Morocco

FAOSTAT (2019) Food and Agricultural Organization of the United Nations Statistics. http://www.fao.org/faostat/en/#data/QP/visualize. Accessed 15 July 2021

FARM-Africa (1994) Dairy goat development programme annual report, Addis Ababa, Ethiopia

FARM-Africa (1996) Goat types of Ethiopia and Eritrea. Physical description and management systems. Published jointly by FARM-Africa, London, UK, and ILRI (International Livestock Research Institute), Nairobi, Kenya, p 76

Fereja G (2016) Characterization of African goat production and productivities: the case of Ethiopia: a review. Glob J Sci Front Res 16:1–11. https://journalofscience.org/index.php/GJSFR/artcle/view/1829

Gebreyesus G, Haile A, Dessie T (2012) Participatory characterization of the Short-eared Somali goat and its production environment around Dire Dawa, Ethiopia. Livest Res Rural Dev 24(10) http://www.lrrd.org/lrrd24/10/gebr2410.html

Gizaw S, Tegegne A, Gebremedhin B, Hoekstra D (2010) Sheep and goat production and marketing systems in Ethiopia: characteristics and strategies for improvement. IPMS working paper 23. ILRI, Nairobi, Kenya

Gopalraj JB, Clarke FC, Donkin EF (2013) Assessment of acquired immune response to Rhipicephalus appendiculatus tick infestation in different goat breeds. Onderstepoort J Vet Res 80(1):614. https://doi.org/10.4102/ojvr.v80i1.614

Halima H, Barbara R, Adnan T, Ghassen J, Aynalem H, Michael B, Samir L (2016) Morphological and molecular genetic diversity of Syrian indigenous goat populations. Afr J Biotechnol 15:745–758. https://doi.org/10.5897/ajb2015.15062

Hallel A (2001) Traditional Algerian cheese. What future? Agroline Rev 14:43–47

Hassanane MS, El-Kholy AF, El-Rahman ARA, Somida MR (2010) Genetic variations of two Egyptian goat breeds using microsatellite markers. Egypt J Anim Prod 47(2):93–105

Hassen H, Lababidi S, Rischkowsky B, Baum M, Tibbo M (2012) Molecular characterization of Ethiopian indigenous goat populations. Trop Anim Health Prod 44(6):1239–1246. https://doi.org/10.1007/s11250-011-0064-2

Hobart E (2022) The real story behind Morocco's tree-climbing goats. National Geographic. https://www.nationalgeographic.com/animals/article/moroccos-tree-climbing-goats. Accessed 2024/03/10

Hossaini-Hilaii J, Benlamlih S (1995) La chèvre Noire Marocaine capacités d'adaptation aux conditions arides. Anim Genet Resour 15:43–48

Isa AM, Bemji MN, Wheto M, Williams TJ, Ibeagha-Awemu EM (2017) Mutations ininhibin alpha gene and their association with litter size in Kalahari Red and Nigerian goats. Livest Sci 203: 106–109. https://doi.org/10.1016/j.livsci.2017.07.012

Isa AM, Bemji MN, Wheto M, Williams TJ, Ibeagha-Awemu EM (2019) Characterization of gonadotropin releasing hormone receptor gene in Sokoto and Kalahari Red goats. Anim Reprod Sci 208:106–109. https://doi.org/10.1016/j.anireprosci.2019.106109

Jackson M, Chenyambuga SW, Ndemanisho EE, Komwihangilo DM (2014) Production performance of Toggenburg dairy goats in semi-arid and sub-humid areas of Tanzania. Livest Res Rural Dev 26:1–7

Jembere T, Haile A, Dessie T, Kebede K, Okeyo AM, Rischkowsky B (2020) Productivity of Abergelle, Central Highland and Woyto-Guji goat breeds in Ethiopia. Livest Res Rural Dev 32:Article #125. http://www.lrrd.org/lrrd32/8/tjbak32125.html

Jnied AM, Rashed MA, Anous MR (2012) Molecular genetic characterization for two Egyptian goat breeds. Egypt J Genet Cytol 41:285–296

Joubert DM (1973) Goats in the animal agriculture of Southern Africa. Z Tlerzuchtungsblol 90:245

Kahi AK, Wasike CB (2019) Dairy goat production in sub-Saharan Africa: current status, constraints and prospects for research and development. Asian Australas J Anim Sci 32(8):1266–1274. https://doi.org/10.5713/ajas.19.0377. Epub 2019 Jul 1

Kardjadj M, Ben-Mahdi M (2017) Major infectious diseases with impact on goat production in north African countries. https://doi.org/10.1007/978-3-319-71855-2_13

Kholif AE, Gouda GA, Hamdon HA (2020) Performance and milk composition of Nubian goats as affected by increasing level of Nanochloropsis oculata microalgae. Animals 10:2453

Khoualdi G (2017) Characterization of traditional Algerian cheese "Medeghissa". Thesis Master University of the Brothers Mentouri Constantine 1, p 141

Khowa AA, Tsvuura Z, Slotow R, Kraai M (2023) The utilisation of domestic goats in rural and peri-urban areas of KwaZulu-Natal, South Africa. Trop Anim Health Prod 55(3):204. https://doi.org/10.1007/s11250-023-03587-3

Lele Z, Feitong W, Xing L (2018) Genetic and phenotypic trends in milk production traits of Anglo Nubian goats from selected farms in The Philippines. Philipp J Vet Anim Sci 12:10–13

Lin BZ, Odahara S, Ishida M et al (2013) Molecular phylogeography and genetic diversity of East Asian goats. Anim Genet 44(1):79–85

Luginbuhl JM, Harvey T, Green J, Poore M, Mueller J (1998) Use of goats as biological agents for the renovation of pastures in the Appalachian region of the United States. Agrofor Syst 44:241–252. https://doi.org/10.1023/A:1006250728166

Luikart G, Gielly L, Excoffier L, Vigne JD, Bouvet J, Taberlet P (2001) Multiple maternal origins and weak phylogeographic structure in domestic goats. Proc Natl Acad Sci USA 98(10):5927–5932. https://doi.org/10.1073/pnas.091591198

Made B (2017) Major factors affecting hide and skin production, quality and the tanning industry in Ethiopia. Adv Biol Res 11(3):116–125

Magoro AM, Mtileni B, Hadebe K, Zwane A (2022) Assessment of genetic diversity and conservation in South African indigenous goat ecotypes: a review. Animals (Basel) 12(23):3353. https://doi.org/10.3390/ani12233353

Mahrous KF, Abdel-Aziem SH, Abdel-Hafez MAM, Abdel-Mordy M, Rushdi HE (2018) Polymorphism of growth hormone gene in three goat breeds in Egypt. Bull Natl Res Cent 42:35. https://doi.org/10.1186/s42269-018-0035-0

Mason IL (1984) Goat. In: Mason IL (ed) Evolution of domesticated animals. Longman, London, pp 85–99

Medfouni S, Benidir K (2018) Characterization of traditional Algerian goat's *Bouhezza* cheese and determination of its shelf life during refrigeration. Thesis Master University Larbi Ben M'hidi d'Oum El-Bouaghi, p 71

Metawi HR, Rashed MA, Anous MR, Khattab AS, Heba Abd El Halim H, Mohamed L (2019) Polymorphisms GDF9 and FSHβ genes and its association with litter size in Egyptian goat breeds. https://www.tropentag.de/2019/abstracts/full/296.pdf

Michailidou S, Tsangaris GT, Tzora A, Skoufos I, Banos G, Argiriou A, Arsenos G (2019) Analysis of genome-wide DNA arrays reveals the genomic population structure and diversity in autochthonous Greek goat breeds. PLoS One 14:e0226179. https://doi.org/10.1371/journal.pone.0226179

Monau P, Raphaka K, Zvinorova-Chimboza P, Gondwe T (2020) Sustainable utilization of indigenous goats in Southern Africa. Diversity 12(1):20. https://doi.org/10.3390/d12010020

Monica M, Beatrice M, Anselimo M, Simon M (2014) Factors affecting Kenya Alpine Dairy goat milk production in Nyeri region. J Food Res 3:160. https://doi.org/10.5539/jfr.v3n6p160

Monteiro A, Santos S, Gonçalves P (2021) Precision agriculture for crop and livestock farming—brief review. Animals 11:2345. https://doi.org/10.3390/ani11082345

Monua PI, Visser C, Muchadeyi FC, Okpeku M, Nsoso SJ, Van Marie-Köster E (2020) Population structure of indigenous southern African goats based on Illumina Goat50K SNP panel. Trop Anim Health Prod:1–11. https://doi.org/10.1007/s11250-019-02190-9

Muigai A, Matete G, Aden HH, Tapio M, Mwai O, Marshall K (2016) The indigenous farm genetic resources of Somalia: preliminary phenotypic and genotypic characterization of cattle, sheep and goats. ILRI project report. International Livestock Research Institute (ILRI), Nairobi, Kenya

Naderi S, Rezaei HR, Taberlet P, Zundel S, Rafat SA, Naghash HR, Mohamed AA, El-Barody MAA, Ertugrul O, Pompanon F (2007) Large-scale mitochondrial DNA analysis of the domestic goat reveals six Haplogroups with high diversity. PLoS One 2(10):e1012. https://doi.org/10.1371/journal.pone.0001012

Naderi S, Rezaei HR, Pompanon F, Blum MGB, Negrini R, Naghash HR, Balkız Ö, Mashkour M, Gaggiotti OE, Ajmone-Marsan P, Kence A, Vigne JD, Taberlet P (2008) The goat domestication process inferred from large-scale mitochondrial DNA analysis of wild and domestic individuals. Proc Natl Acad Sci 105(46):17659–17664. https://doi.org/10.1073/pnas.0804782105 DAGRIS

Ndlovu LR, Simela L (1996) Effect of season of birth and sex of kid on the production of live weaned single born kids in smallholder East African goat flocks in North East Zimbabwe. Small Rumin Res 22(1):1–6. https://doi.org/10.1016/0921-4488(95)00844-6

Nfon CK, Marszal P, Zhang S, Weingartl HM (2012) Innate immune response to Rift Valley fever virus in goats. PLoS Negl Trop Dis 6(4):e1623. https://doi.org/10.1371/journal.pntd.0001623

Ngere LO (1987) Principles for Indigenous Animal Improvement in the Tropics - African Experiences with Sheep and Goats. Animal genetic resources Strategies for improved use and conservation. Food and Agriculture Organization (FAO) Animal Production and Health Paper 66. https://www.fao.org/3/ah806e/ah806e.pdf. Accessed 2024/03/10

Nguluma AS, Huang Y, Zhao Y et al (2018) Assessment of genetic variation among four populations of Small East African goats using microsatellite markers. S Afr J Anim Sci 48:117–127. https://doi.org/10.4314/sajas.v48i1.14

Nguluma A, Kyallo M, Tarekegn GM, Loina R, Nziku Z, Chenyambuga S, Pelle R (2021) Mitochondrial DNA D-loop sequence analysis reveals high variation and multiple maternal origins of indigenous Tanzanian goat populations. Ecol Evol 11(22):15961–15971. https://doi.org/10.1002/ece3.8265

Nigatu A (1994) Characterization of indigenous goat types of Eritrea, Northern and Western Ethiopia. MSc thesis. Alemaya University of Agriculture, Alemaya, Ethiopia, p 136

Nowier AM, Ramadan SI (2020) Association of β-casein gene polymorphism with milk composition traits of Egyptian Maghrebi camels (*Camelus dromedarius*). Arch Anim Breed 63(2):493–500. https://doi.org/10.5194/aab-63-493-2020

Oderinwale OA, Oluwatosin BO, Sowande OS, Bemji MN, Amosu SD, Sanusi GO (2017) Concentrate supplementations of grazing pregnant Kalahari Red goats: effects on pregnancy variables, reproductive performance, birth types and weight of kids. Trop Anim Health Prod 49(6):1125–1133. https://doi.org/10.1007/s11250-017-1303-y

Okpeku M, Yakubu A, Peters SO, Ozoje MO, Ikeobi CON, Adebambo OA, Imumorin O (2011a) Application of multivariate principal component analysis to morphological characterization of indigenous goats in southern Nigeria. Acta Argiculturae Slovenica 98(2):101–109. https://doi.org/10.2478/v10014-011-0026-4

Okpeku M, Ozoje MO, Adebambo OA, Agaviezor BO, O'Neill MJ, Imumorin IG (2011b) Preliminary analysis of microsatellite-based genetic diversity of goats in southern Nigeria. Anim Genet Resour 49:33–41

Okpeku M, Peters SO, Imumorin IG, Caires KC, Sharma VK, Wheto M, Tamang R, Adenaike SA, Ozoje MO, Thanganraj K (2016a) Mitochondrial DNA hypervariable region 1 diversity in Nigerian goats. Anim Genet Resour 59:47–54. https://doi.org/10.1017/S2078633616000266

Okpeku M, Esmailizadeh A, Adeola AC, Shu L, Zhang Y, Wang Y, Sanni TM, Imumorin IG, Peters SO, Zhang J, Dong Y, Wang W (2016b) Genetic variation of goat interferon regulatory factor 3 gene gene and its implication in goat evolution. PLoS One 11(9):e0161962. https://doi.org/10.1371/journal.pone.0161962

Okpeku M, Peters SO, Imumorin IG et al (2017) Mitochondrial DNA hypervariable region 1 diversity in Nigerian goats. Anim Gen Res 59:47–54. https://doi.org/10.1017/S2078633616000102

Olalere MO, Bemji MN (2024) Milk Offtake In West African Dwarf goats during early Lactation. Nigerian J Anim Prod 821–825. https://doi.org/10.51791/njap.vi.5940

Oluwatayo IB, Oluwatayo TB (2012) Small Ruminants as a source of financial security: a case study of women in rural Southwest Nigeria. Institute for Money, Technology and Financial Inclusion (IMTFI), Working Paper 2012–2. https://www.imtfi.uci.edu/files/blog_working_papers/2012-2_oluwatayo.pdf

Omotosho BO, Bemji MN, Bamisile K, Ozoje MO, Wheto M, Lawal A, Oluwatosin B, Sowande OS, James I, Osinowo O (2020) Comparative study of growth patterns of Kalahari Red goats and West African Dwarf goats reared in Southwest. Nig J Anim Prod 47(5):213–226. https://doi.org/10.51791/njap.v47i5.1334

Onasanya GO, Msalya GM, Thiruvenkadan AK, Sreekumar C, Tirumurugaan GK, Fafiolu AO, Adeleke MA, Yakubu A, Ikeobi CON, Okpeku M (2021) Heterozygous single-nucleotide polymorphism genotypes at heat shock protein 70 gene potentially influence Thermo-tolerance among four Zebu breeds of Nigeria. Front Genet 12:642213. https://doi.org/10.3389/fgene.2021.642213

Onzima RB, Upadhyay MR, Mukiibi R, Kanis E, Groenen MAM, Crooijmans RPMA (2018) Genome-wide population structure and admixture analysis reveals weak differentiation among Ugandan goat breeds. Anim Genet 49:59–70

Ould Abeid A, Berkani M, Quasmaoui A, Ouhssine M, Mennane Z (2013) Microbiological and physicochemical quality of fresh cheese (Jben) from Rabat and Salé. Rev Microbiol Ind San Environn 2:162–174

Özgecan KA, Ertuğrul O (2012) Assessment of genetic diversity, genetic relationship and bottleneck using microsatellites in some native Turkish goat breeds. Small Rumin Res 105:53–60. https://doi.org/10.1016/j.smallrumres.2011.12.005

Paim TP, Faria DA, Hay EH, Lanari R, Chaverri-Esquivel L, Cascante MI, Jimenez-Alfaro E, Mendez A, Faco O, de Moraes Silva K, Mezzadra CA, Mariante A, Rezende-Paiva S, Blackburn HD (2019) New world populations are a genetically diverse reservoir for future use. Sci Rep-UK 9:1476. https://doi.org/10.1038/s41598-019-38812-3

Pereira F, Queirós S, Gusmão L, Nijman IJ, Cuppen E, Lenstra JA, Econogene Consortium, Davis SJM, Nejmeddine F, Amorim A (2009) Tracing the history of goat pastoralism: new clues from mitochondrial and Y chromosome DNA in North Africa. Mol Biol Evol 26(12):2765–2773. https://doi.org/10.1093/molbev/msp200

Pesmen G, Yardimci M (2008) Estimating the live weight using some body measurements in Saanen goats. Arch Zootech 11:30–40

Pieters A, Marle-Koster EV, Visser C, Kotze A (2009) South African developed meat type goats: a forgotten animal genetic resource? Anim Gen Res Info 44:33–43

Pollott G, Wilson RT (2009) Sheep and goats for diverse products and profits. FAO Diversification Booklet N°9, p 54. http://www.fao.org/3/i0524e/i0524e00.htm

Porter V, Lawrence A, Hall SJG, Sponenberg DP (2016) Mason's world Encyclopedia of livestock breeds and breeding, 6th edn. CABI, Wallingford. ISBN 9781780647944

Rahmatalla SA, Arends D, Brockmann GA (2022) Review: genetic and protein variants of milk caseins in goats. Front Genet 13:995349. https://doi.org/10.3389/fgene.2022.995349

Roy's Farm (2021) Kalahari Red Goat: characteristics, uses & full breed information. https://www.roysfarm.com/kalahari-red-goat/. Accessed 2024/03/09

Salgado Pardo JI, Delgado Bermejo JV, González Ariza A, León Jurado JM, Marín Navas C, Iglesias Pastrana C, Martínez Martínez MDA, Navas González FJ (2022) Candidate genes and their expressions involved in the regulation of milk and meat production and quality in goats (*Capra hircus*). Animals (Basel) 12(8):988. https://doi.org/10.3390/ani12080988

Sanni MT, Okpeku M, Onasanya GO, Adeleke MA, Wheto M, Adenaike AS, Oluwatosin BO, Adebambo OA, Ikeobi CON (2018) Genetic morphometry in Nigerian and South African Kalahari Red goat breeds. Agric Trop Subtrop 51(2):51–61. https://doi.org/10.2478/ats-2018-0006. ISSN: 1801-0571

Selionova MI, Trukhachev VI, Aybazov AMM, Stolpovsky YA, Zinovieva NA (2021) Genetic markers of goats (review). Agric Biol 56:1031–1048. https://doi.org/10.15389/agrobiology.2021.6.1031eng

Shackleton DM (1997) Classification adopted for the Caprinae survey. Wild Sheep and goats and their relatives: status Survey and Conservation Action Plan for Caprinae, 9–14

Silanikove N, Koluman ND (2015) Impact of climate change on the dairy industry intemperate zones: predications on the overall negative impact and on the positive role of dairy goats in adaptation to earth warming. Small Rumin Res 123:27–34. https://doi.org/10.1016/j.smallrumres.2014.11.005

Solomon A, Kassahun A, Alemu Y, Girma A, Sileshi Z, Adane H (2008) Records and record keeping. In: Yami A, Merkel RC (eds) Sheep and goat production handbook for Ethiopia, pp 360–366. Accessed 2024/03/10

Solomon AK, Mwai O, Grum G, Haile A, Rischkowsky BA, Solomon G, Dessie T (2014) Review of goat research and development projects in Ethiopia. ILRI project report. International Livestock Research Institute, Nairobi, Kenya. Accessed 2024/03/10

Steele M (2006) Goats. Tropical Agriculturalist. CTA. Macmillan Publishers Limited, p 152

Stella A, Nicolazzi EL, Van Tassell CP, Rothschild MF, Colli L, Rosen BD, Sonstegard TS, Crepaldi P, Tosser-Klopp G, Joost S, AdaptMap Consortium (2018) AdaptMap: exploring goat diversity and adaptation.

Genet Sel Evol 50(1):61. https://doi.org/10.1186/s12711-018-0427-5

Takoucht A (1998) Attempt to identify the visible genetic variability of goat populations in the M'ZAB Valley and the AHAGGAR Mountains, Thesis Ing. Agronomic Institute Blida, Algeria, p 52

Tarekegn GM, Wouobeng P, Jaures KS, Mrode R, Edea Z, Liu B, Zhang W, Mwai OA, Dessie T, Tesfaye K, Strandberg E, Berglund B, Mutai C, Osama S, Wolde AT, Birungi J, Djikeng A, Meutchieye F (2019) Genome-wide diversity and demographic dynamics of Cameroon goats and their divergence from east African, north African, and Asian conspecifics. PLoS One 14(4):e0214843. https://doi.org/10.1371/journal.pone.0214843

Tatar AM, Tuncer SS, Şireli HD (2019) Comparison of yield characteristics of Damascus and Kilis goats in dry climatic conditions. Aust J Vet Sci 51:61–66. https://doi.org/10.4067/S0719-81322019000200061

Tesfaye AT (2004) Genetic Characterization of Indigenous Goat Populations of Ethiopia using Microsatellite DNA Markers. A PhD thesis submitted to the National Dairy Research Institute, Deemed University, Karnal-Haryana, India. Accessed 2024/03/10

Transboundary breed summary: Goat: Anglo Nubian. Domestic Animal Diversity Information System of the Food and Agriculture Organization of the United Nations. Accessed March 2024

Umejesi SI, Bemji MN, Ozoje MO, James IJ (2016) Milk off-take of West African Dwarf and Kalahari Red goats reared in south-west Nigeria. Nig J Genet 30:101–113

UNEP-WCMC.; IUCN (2016) Protected planet report. UNEP-WCMC and IUCN, Cambridge UK and Gland, Switzerland. 2016-051.pdf (iucn.org)

Visser C, Snyman MA (2023) Incorporating new technologies in breeding plans for South African goats in harsh environments. *Anim Front* 13(5):53–59. https://doi.org/10.1093/af/vfad040

Visser C, Van Marle-Köster E (2014) Strategies for the genetic improvement of South African Small ruminant production and the small ruminant genetic resource in tropical Africa. FAO Animal Production and Health Paper, no. 88

Visser C, Van Marle-Köster E (2018) The development and genetic improvement of South African goats. INTECH. https://doi.org/10.5772/intechopen.70065

Visser C, Lashmar S, Van Marle Koster E, Poli M, Allain D (2016) Data from: genetic diversity and population structure in South African, French and Argentinian Angora Goats from genome-wide SNP data. PLoS One 11:1–15. https://doi.org/10.5061/dryad.p1b90

Wang JJ, Li ZD, Zheng LQ, Teng Z, Shen W, Lei CZ (2022) Genome-wide detection of selective signals for fecundity traits in goats (Capra hircus). Gene 818. https://doi.org/10.1016/j.gene.2022.146221

Warsame AAA, Turyasingura B (2022) Phenotypic characterization of Sudanese desert goat. Sci J Anim Sci 10:716–725. https://doi.org/10.14196/sjas.v10i1.1672

Wilson RT (1991) Small ruminant production and the small ruminant genetic resource in tropical Africa. FAO animal production and health paper 88. Bartridge Partners, Umberleigh, North Devon, UK. http://www.fao.org/3/t0376e/T0376E00.htm. Accessed 2024/03/10

Yakubu A, Salako AE, De Donato M, Peters SO, Adefenwa MA, Okpeku M, Wheto M, Agaviezor BO, Ajayi OO, Onasanya GO, Sanni TM, Takeet MI, Kundayo OJ, Ilori BM, Amusan SA, Imumorin IG (2013) Genetic diversity in exon 2 at the major histocompatibility complex DQB1 locus in Nigerian goats. Biochem Genet 51(11–12):954–966

Yakubu A, Salako AE, De Donato M, Takeet MI, Peters SO, Wheto M, Okpeku M, Imumorin IG (2016) Interleukin-2 (IL-2) gene polymorphism and association with heat tolerance in Nigerian goats. Small Rumin Res 141:127–134

Yakubu A, Salako AE, De Donato M, Takeet MI, Peters SO, Okpeku M, Wheto M, Imumorin IG (2017a) Nucleotide sequence variability analysis of major histocompatibility complex class II DQA1 gene in Nigerian goats. Genetika 49(3):865–874

Yakubu A, Salako AE, De Donato M, Peters SO, Takeet MI, Wheto M, Okpeku M, Imumorin IG (2017b) Association of SNP variants of MHC class II DRB gene with thermo-physiological traits in tropical goats. Trop Anim Health Prod 49:323–336. https://doi.org/10.1007/s11250-016-1196-1

Yami A, Merkel RC (2008) Sheep and goat production hand book for Ethiopia, Ethiopian Sheep and Goat Productivity Improvement Program (ESGPIP). Ministry of Agriculture and Rural Development, Ethiopia

Zeder MA (2008) Animal domestication in the Zagros: an update and directions for future research. MOM Éditions 49(1):243–277

Zheng Z, Wang X, Li M, Li Y, Yang Z, Wang X, Pan X, Gong M, Zhang Y, Guo Y, Wang Y, Liu J, Cai Y, Chen Q, Okpeku M, Colli L, Cai D, Wang K, Huang S, Sonstegard TS, Esmailizadeh A, Zhang W, Zhang T, Xu Y, Xu N, Yang Y, Han J, Chen L, Lesur J, Daly KG, Bradley DG, Heller R, Zhang G, Wang W, Chen Y, Jiang Y (2020) The origin of domestication genes in goats. Sci Adv 6(21):eaaz5216. https://doi.org/10.1126/sciadv.aaz5216

Zonaed Siddiki AM, Miah G, Islam MS, Kumkum M, Rumi MH, Baten A, Hossain MA (2020) Goat genomic resources: the search for genes associated with its economic traits. Int J Genom 2020:5940205. https://doi.org/10.1155/2020/5940205

Open Access This chapter is licensed under the terms of the Creative Commons Attribution 4.0 International License (http://creativecommons.org/licenses/by/4.0/), which permits use, sharing, adaptation, distribution and reproduction in any medium or format, as long as you give appropriate credit to the original author(s) and the source, provide a link to the Creative Commons license and indicate if changes were made.

The images or other third party material in this chapter are included in the chapter's Creative Commons license, unless indicated otherwise in a credit line to the material. If material is not included in the chapter's Creative Commons license and your intended use is not permitted by statutory regulation or exceeds the permitted use, you will need to obtain permission directly from the copyright holder.

African Sheep Genetic Resources, Diversity and Unique Features

6

Martha N. Bemji, Semir B. S. Gaouar, Abdelkader Ameur Ameur, Fatima Z. Belharfi, Isidore Houaga, and Anne W. T. Muigai

Abstract

Africa is host to over 363 breeds of sheep, the majority of which have not been characterised. The total sheep population is 27.1% of the global sheep genetic resources, placing Africa in the second position worldwide. This Chapter introduces indigenous African sheep breeds, their origins, their history of domestication, dispersion in Africa and their role on the continent (Sects. 6.3 and 6.4), thus providing a broad coverage of African sheep, using a framework that could trace the roots of current breed groups. Section 6.5 is devoted to a discussion on different sheep breeds of Africa, which are classified into two major classes based on tail morphology (thin-tailed and fat-tailed or fat-rumped) and further subdivided into seven subtypes (thin-tailed hair sheep, thin-tailed coarse wool sheep, fat-tailed hair sheep, fat-tailed coarse wool sheep, fat-rumped hair sheep, introduced breeds and developed or composite sheep breeds). For each breed, the chapter summarises the origin, distribution and classification, physical characteristics, including some breed photos, adaptive and special genetic characteristics, typical production systems and production characteristics. The depth of discussion is dif-

M. N. Bemji (✉)
Department of Animal Breeding and Genetics, Federal University of Agriculture,
Abeokuta, Ogun State, Nigeria
e-mail: bemjimn@funaab.edu.ng

S. B. S. Gaouar · A. Ameur Ameur
Laboratory of Applied Genetics in Agronomy, Ecology and Public Health (GenApAgiE), Faculty SNV/STU, Abou Bekr Belkaid University, Tlemcen, Algeria

F. Z. Belharfi
Laboratory of Applied Genetics in Agronomy, Ecology and Public Health (GenApAgiE), Faculty SNV/STU, Abou Bekr Belkaid University, Tlemcen, Algeria

Laboratory of Physiopathology and Biochemistry of Nutrition (PpBioNut), Faculty SNV/STU, Abou Bekr Belkaid University, Tlemcen, Algeria

I. Houaga
Centre for Tropical Livestock Genetics and Health (CTLGH), The Roslin Institute, University of Edinburgh, Easter Bush Campus, Roslin, UK

A. W. T. Muigai (✉)
School of Biological Sciences, Jomo Kenyatta University of Agriculture and Technology, Nairobi, Kenya

National Defence University-Kenya, Nakuru, Kenya
e-mail: awtmuigai@ndu.ac.ke

© The Author(s) 2026
E. M. Ibeagha-Awemu et al. (eds.), *African Livestock Genetic Resources and Sustainable Breeding Strategies*, Sustainable Development Goals Series, https://doi.org/10.1007/978-3-031-92076-9_6

ferent between breeds due to the differential availability of data or information. Sections 6.6, 6.7, 6.8, 6.9 and 6.10 further appraised the current status of sheep diversity in Africa, conservation of sheep genetic resources, sheep breeding goals and strategies for sustainable sheep genetic improvement, responses of existing sheep to selection, structure and organisation of the sheep sector in Africa and sheep production capacity in terms of meat, milk, wool and skin. In Sects. 6.11 and 6.12, the chapter concludes with an overview of challenges to sustainable genetic improvement in the sheep industry and future prospects.

Keywords

Sheep · Genetic resources · Characterisation · Selection · Breeding · Adaptive traits · Africa

Abbreviations

AU-IBAR	African Union InterAfrican Bureau for Animal Resources
BP	Before Present
CBBP	Community-Based Breeding Programmes
DNA	Deoxyribonucleic Acid
FAO	Food and Agricultural Organisation
GDP	Gross Domestic Product
GPA	Global Plan of Action
mtDNA	Mitochondrial DNA
PCA	Principal Component Analysis
RFLP	Restriction Fragment Length Polymorphism
RGA	Return with Goal Accomplished
SNP	Single Nucleotide Polymorphism
TWW	Total Weaning Weight
WAD	West African Dwarf

6.1 Introduction

Africa is endowed with enormous sheep genetic resources, which is the second largest mammalian species, with over 363 breeds, most of them originating from Southern Africa (109), followed by East Africa (99), North Africa (65), West Africa (62) and Central Africa (28) (AU-IBAR 2019). However, only a limited number of the breeds have been characterised. As indicated in Table 6.1, this huge genetic resource positions the continent as the second largest producer in the world in terms of percentage of live animals (27.1%), meat (mutton) (18.2%), fresh skin (16.2%) and edible offal (21.2%), while ranking third in total fresh milk production valued at 2,040,376 tonnes (21.9%) (FAOSTATS 2019). The total African sheep genetic resources for the year 2019 were estimated at 407,652,676. As outlined in Chap. 2, with more current data, West Africa ranks highest in sheep production, followed by North Africa, East Africa and Central Africa, while the least production comes from Southern Africa, corroborating the earlier trend between 2018 and 2019 (Table 6.2).

In the most vulnerable arid and semi-arid zones in sub-Saharan Africa (see Chap. 2 for a detailed description of the main agro-ecological zones in Africa), sheep are among the most frequently owned animals (Seo 2021) that perform numerous roles in many smallholder farm families who depend on sheep production for their livelihood. Almost two-thirds of rural households keep livestock in sub-Saharan Africa (FAO 2014). The minority of livestock keepers in Africa, who are pastoralists, migratory or sedentary, or agro-pastoralists, keep relatively large herds, while the majority keep an average of 5.5 animals, with half of these also keeping other livestock species such as cattle or goats and a few sheep (FAO 2014). Studies have shown that sheep farming offers several advantages over goat farming. The average annual rates of return on capital for the production of the dwarf sheep and goats in Southwest Nigeria was 55% for sheep compared to a lower estimate of 34% for goats (Upton 1985), with a sensitivity analysis further showing that at high levels of mortality, the risk of loss appeared lower for sheep.

Outstanding advantages that make the sheep industry pivotal in reducing poverty, which is endemic in Africa, include the sheep products, including meat, milk, skins, fibre, horn and offal, by-products such as manure used as fertiliser,

Table 6.1 Position of Africa in the world in terms of sheep production and processed products for the year 2019 (FAOSTATS 2019)

Region	Percentage (%) of live animals	Annual production in tonnes (percentage in parentheses)						
		Meat	Milk (fresh)	Cheese	Butter and ghee	Skins (fresh)	Edible offal	Fat
Africa	27.1	1,521,992.15 (18.2)	2,040,376 (21.9)	44,104.8 (6.6)	1813.04 (3.1)	258,117.12 (16.2)	314,171.85 (21.6)	47,117.27 (9.6)
Europe	12.7	1,298,721.81 (15.6)	3,021,721.5 (32.4)	370,372.04 (55.4)	–	248,380.31 (15.6)	187,601.19 (12.9)	84,655.38 (18.2)
America	8	413,302.92 (5)	87,977.62 (0.9)	6947.96 (1)	–	83,319.96 (5.2)	70,732.12 (4.9)	17,703.77 (3.6)
Asia	40.6	3,959,530.23 (47.5)	4,171,529.88 (44.8)	247,454.56 (37)	56,709.84 (96.1)	770,6.606.5 (48.3)	736,551.62 (50.6)	239,953.96 (48.7)
Oceania	11.6	1,150,728.81 (13.8)	–	–	–	233,734.38 (14.7)	145,254.46 (10)	103,319 (21)
African countries among top 10 producers in the world	Sudan (former), Sudan, Nigeria	Sudan, Sudan (former), Algeria	Sudan (former), Sudan	Niger	Morocco, Tunisia	Sudan (former), Sudan	Sudan (former), Sudan, South Africa	Sudan (former)

Table 6.2 Sheep production by regions in Africa for the period 2018–2019 (FAOSTATS 2019)

Area	Sheep number (2018)	Sheep number (2019)	Annual increase (sheep number)	Annual increase, %
Africa	389,389,503	407,652,676	18,263,173	4.69
East Africa	95,583,003	106,110,387	10,527,384	11.01
Central Africa	39,584,348	42,236,002	2,651,654	6.70
North Africa	108,297,080	110,981,099	2,684,019	2.48
Southern Africa	26,507,377	25,525,622	−981,755	−3.70
West Africa	119,417,695	122,799,566	3,381,871	2.83

dung used as fuel and for biogas production. The indirect benefits of keeping sheep include their use for weed control and for intangible benefits such as their use as a store for capital, income generation, reduction of risk from crop failure, fulfilling social, cultural and religious obligations, providing status or prestige and use in sport, culture and recreation (Pollott and Wilson 2009). Growth in meat demand in sub-Saharan Africa between 2013 and 2015 indicates that the sheep industry grew impressively, and sheep meat consumption was sustained at a similar rate, with imports accounting for a very small share of additional consumption compared to meat from other livestock species (OECD and FAO 2016). Consumption preferences reflected cultural and religious preferences. Growth was significantly faster in Eastern Africa compared with other regions in sub-Saharan Africa, where sheep are extensively raised on pasture. This trend has been maintained, being supported by more current statistics (Wilson 1991). Major producing countries are Sudan and Ethiopia (East Africa), Nigeria, Mali and Niger (West Africa), Algeria, Morocco, Tunisia and Egypt (North Africa), South Africa (Southern Africa) and Chad (Central Africa).

The two major classes of sheep, recognised as thin-tailed and fat-tailed, are further subdivided into four types (Wilson 1991) based on tail morphology and coat type. They include thin-tailed with hair, thin-tailed with wool, fat-tailed and fat-rumped. As indicated in a review by Muigai and Hanotte (2013), fat-tailed sheep constitute the majority in most parts of North Africa (from Egypt to Tunisia) and in Eastern and Southern Africa (from Eritrea to South Africa). Thin-tailed sheep are mainly found in North Africa (Morocco to Egypt), Sudan and West Africa (Senegal to Nigeria). Two subtypes of thin-tailed sheep (the African long-legged and the tropical dwarf sheep) have been identified. The former is widely distributed in the arid and semi-arid zones of the Sahel and West Africa, while the latter is mainly found in the sub-humid and humid zones of West and Central Africa.

Diversity studies with molecular markers, for example, microsatellites (Gaouar 2002; Gizaw et al. 2007; Gaouar et al. 2016; Ameur Ameur et al. 2018; Nigussie et al. 2020), mitochondrial DNA (Agaviezor et al. 2012; Brahi et al. 2015; Ghernouti et al. 2017) and genome-wide SNP chip (Gaouar et al. 2017; Edea et al. 2017) analyses revealed high levels of genetic admixture between African breeds of sheep, particularly among populations within the same geographic area, leading to higher within-breed than between-breed genetic diversity. This could be largely attributed to uncontrolled breeding, which often characterises most production systems.

The following topics will be discussed in this Chapter: (1) a brief history of the introduction and dispersal of African sheep, (2) the importance and uses of sheep genetic resources in Africa, (3) phenotypic and genetic characterisation of sheep breeds with emphasis on unique attributes of adaptation and productivity, (4) the current status of sheep diversity and conservation, (5) responses of existing sheep breeds to selection and their limitations, (6) sheep breeding goals and strategies and (7) structures and organisations of the sheep sector in Africa.

6.2 Structures and Organisation of the Sheep Sector in Africa

African sheep production is a subset of the larger livestock production systems that have been comprehensively discussed in Chap. 2 (Seré and Steinfeld 1996; FAO 2007). In Africa, however, there is no single classification that can be used to map the livestock production systems across the continent (Herrero et al. 2010; Robinson et al. 2014). Sheep production largely depends on rainfed pastures that are dependent on seasonal rainfall patterns and existing environmental conditions. As a result of this, the production of sheep in Africa can broadly be described as follows: (i) agropastoral and pastoral systems, (ii) mixed crop-livestock systems and (iii) landless systems, which have been discussed in Chap. 2.

The 'Livestock Revolution' discourse (Delgado et al. 1999), which points to growth in demand for livestock and their products (meat and milk) due to rising incomes and urbanisation in developing countries including Africa, has so far perhaps largely been confined to emerging countries, and it is difficult to predict a comparable explosion in the consumption of animal products in sub-Saharan Africa in the coming decades. No doubt increases in human population, urbanisation and changes in consumption patterns will result in changes to consumption of livestock and livestock products, with the FAO predicting a doubling of per capita meat consumption in Africa by 2050 (from 11 to 22 kg/year/capita). There are also indications that Africa, over time, will become a net importer of meat and animal products, with imports accounting for between 12% and 15% of the meat and meat products being consumed on the continent.

6.3 Brief History of Domestication, Introduction and Dispersion of Sheep in Africa

6.3.1 The Domestication of Sheep

The first animal to be domesticated was the dog (Clutton-Brock 1997) followed by sheep and goats, which are the earliest bovids to have been domesticated around 10,500 BP (Zeder 2011). Present-day sheep, *Ovis aries*, evolved in Eurasia (Uerpmann 1987; Ryder 1983) in the early Pleistocene period around 2.5 million years ago (Ryder 1983) from its wild progenitor, the *Ovis orientalis*, which was confined to the Taurus/Zagros Arc region (Zeder 2017). Domestication of sheep, goats, cattle, yak, water buffalo and perhaps pigs was not a one-time event but rather a process that began with the animal first being hunted by man in regions that were remote from human settlements, followed by their being managed as game and finally to their being fully domesticated (Zeder 2011; Larson and Burger 2013). Domestication involved bringing the animals into close proximity with humans, which led to their being managed and bred extensively by man (Zeder 2011; Larson and Burger 2013).

Mitochondrial DNA sequence analysis is a molecular tool that can be used to infer the origin of species and has been widely used to reconstruct the patterns of domestication (Groeneveld et al. 2010). It has been revealed that at least three distinct lineages of the wild sheep, *Ovis orientalis*, were involved in the domestication process (Pedrosa et al. 2005), probably in the eastern Taurus and northwestern Zagros region, which has also been associated with goat domestication (Demirci et al. 2013). These lineages do not, however, provide evidence of multiple and independent domestication processes, but instead show mitochondrial introgression was a result of continuous admixture between wild and initial domesticated animals outside the domestication regions (Larson and Burger 2013). This mouflon–domestic sheep cross-breeding is thought to have been either naturally occurring and/or human-mediated (Lauvergne et al. 1977; Schröder et al. 2016; Barbato et al. 2017). Evidence for the matings between feral and domestic sheep has been provided by microsatellite marker analysis (Schröder et al. 2016; Lorenzini et al. 2011; Guerrini et al. 2015).

6.3.2 Dispersal of Sheep Out of Their Centre of Domestication into Anatolia and Europe

While man was domesticating animals, he was also domesticating plants, resulting in the expansion of human populations and the development of communities that were engaged in cultivation and husbandry of domesticates (Bocquet-Appel and Bar-Yosef 2008). These populations then migrated away from the centres of domestication (eastern Taurus and northwestern Zagros region), taking their domesticated animal and plant species with them. Managed sheep that were in various stages of domestication appeared in the Zagros around ca. 9000 (Zeder 2008) and in the Levant slightly before this time (Horwitz and Ducos 1998). The earliest evidence of sheep populations outside the Levant came from Cyprus, dating to ca. 10,500 BP, reaching the other islands of Corsica and Cyprus by ca. 6000–7000 (Zilhão 2001; Vigne et al. 2014) having been transported across the eastern Mediterranean in boats. Morphologically wild but managed sheep appeared in Central Anatolia in ca. 10,000. There is debate on whether these sheep were the result of independent domestication of local wild sheep or whether they were managed sheep dispersed from the east (Buitenhuis 1997; Vigne et al. 1999). Nevertheless, there is evidence of managed caprines by early animal keepers as early as ca. 10,400–10,100 (Stiner et al. 2014). By 9500 BP, there were sheep in south-central Anatolia (Russell et al. 2005) ,where sheep were 2–3 times more prevalent than goat (Çakirlar 2012a, b). From Anatolia, domestic livestock, including sheep, was dispersed into Europe following two pathways, the first was across the Mediterranean Basin, beginning in Cyprus (10,500 ka) and moving onto Crete (9000 BP (Efstratiou et al. 2004)) and both Corsica and Sardinia (7700 BP), and the other was dubbed the 'Danubian dispersal', through the river valleys of south-eastern and central Europe and beyond, with sheep being reported in southern Istria at the head of the Adriatic in 7700 BP, southern Italy 8000 BP (Skeates 2003), southern France 7700 BP (La Guilaine 2006; Guilaine et al. 2007; Rowley-Conwy 2013), Iberian Peninsula (7700–7500 BP) and Portugal (7400–7300 BP) (Zilhão 2001; Zilhão 2003).

6.3.3 The Dispersal of Sheep into Africa

Dispersal of sheep into Africa is not as well documented as done for the other continents. This is partly due to the initial focus of early African archaeology on early and modern humans and their dispersal, but also because the environmental conditions experienced in Africa (such as adverse soil chemistry) made the preservation of faunal remains more difficult. Although the case for an African centre of domestication for cattle has been made (Gautier 1987; Wendorf and Schild 1998; Muzzolini 2000), the lack of strong genetic evidence to support this has made it likely that all domestic livestock were introduced into Africa.

There are two possible routes for the introduction of domestic sheep into Africa. The first follows the southerly migration of sheep through the Negev of southern Palestine, across the Sinai Peninsula to the Red Sea Coast (Close and Wendorf 1992; Garcea 2004). Evidence for this route is provided by sheep remains recovered from the Nile Delta, the eastern Sahara and the Red Sea Hills, dated between 7000 and 7500 BP (Gautier 1987; Close 1992; Vermeersch et al. 1994). These are the earliest remains of sheep in Africa, and they are likely to have originated from the Sinai Peninsula, where sheep finds are dated *c.* 7000 BP (Oren 1979).

The second dispersal route of sheep into Africa is across the Mediterranean Basin, which possibly led to the introduction of sheep into the Libyan Cyrenaica, northern Algeria and Mediterranean Morocco between 7000 and 6500 BP (Higgs 1967; Gilman 1975; Roubert and La Carter 1984; Klein and Scot 1986; Pereira et al. 2006). The livestock keepers seem to have settled along the northern African Mediterranean coastline but eventually migrated southwards, settling in the central Nile Valley (6000 BP) and the more hospitable regions within the Sahara (Close 1992).

Livestock herding spread quickly but incompletely across the Sahara, with herders migrating southwards and eventually westwards into West Africa as climatic conditions in the Sahara deteriorated during the sixth millennium BP (Drake et al. 2011), reaching western Niger and the West African Sahel at approximately 6000 BP and 5400 BP, respectively (MacDonald and MacDonald 2000). At approximately the same period herders following the River Nile migrated south from the Nile Delta to better-watered regions such as the Khartoum Nile in southern Sudan (Gautier 1987; Clark 1976; Peters 1986), reaching the Lake Turkana Basin between 4500 and 3500 BP (MacDonald and MacDonald 2000; Barthelme 1985), thus introducing sheep into East Africa. In addition to this, the direct introduction of sheep into East Africa could also have occurred via trade routes that occurred between the Arabian Peninsula and present-day Eritrea (Marshall 2000; Boivin and Fuller 2009). The picture of the dispersal of sheep into Central Africa has not clearly emerged, partly due to the low number of excavations occurring in this region, but also the poor state of faunal finds. Here, the earliest dates for caprines occur at approximately 1800–1000 BP (Van Neer 2000).

The sheep of southern Africa came from further north, with different routes and chronologies being proposed (Elphick 1985; Bousman 1998; Russell 2004). When the first Europeans arrived on the southwestern coast of southern Africa, they found indigenous Khoe speakers with sheep (Elphick 1985; Stow 1905; Cooke 1965). Questions on the route these sheep took and the identity of their keepers have been seriously debated. New evidence is pointing to the fact that the earliest sheep arrived in southern Africa, with cattle, approximately 2000 years ago via two separate routes (Sadr 2015). One route located in the extreme western part of the continent involved hunter-gatherers migrating southwards along the Atlantic seaboard, reaching the most southern tip of Africa (Sadr 2015). Around the same time, a smaller group of hunter-gatherers, with sheep and cattle, reached the middle Limpopo River Basin, migrated westwards into the Kalahari Drainage Basin as far as Lake Ngami and migrated further to the north and west of Lake Ngami (Sadr 2015). Genetic evidence for the southerly migration of pastoralists from the East African nation of Tanzania into southern Africa has been reported (Henn et al. 2008, 2011). These early pastoralists who came from East Africa are believed to be Khoe-Kwadi speakers. The introduction of livestock into the South African region is thought to be the result of small-scale infiltrations of pastoralists with their animals (Sadr 2015).

6.4 Importance and Use of Sheep Genetic Resources in Africa

6.4.1 Role in Agriculture

The indigenous sheep of Africa play various roles in African communities and form an integral part of their day-to-day activities. Sheep and sheep products are direct contributors to food and nutrition security, providing meat, milk and other products (Table 6.1). Sheep and other livestock species are direct contributors to the agricultural economies of African countries, contributing up to 40% to the GDP, excluding the contributions made by draught power and manure (Winrock International 1992). For example, in 2009, the net annual sheep offtake rate reported in Uganda, South Sudan and Sudan was USD 10 million, USD 1.5 million and USD 4.6 million, respectively (Behnke and Osman 2012; Kosgey et al. 2008). The sales of fresh sheep milk were also reported to generate USD 810,000 and USD 2.5 million in South Sudan and Sudan, respectively, in the year 2009 (Behnke and Osman 2012). Incomes, especially those from exports, are also generated from the sales of live animals, leather and leather products.

Sheep are an important resource for African women and children who, in many regions, are their owners and managers. Studies conducted in Kenya, Tanzania and Mozambique demonstrated that after chickens, sheep were the most important species owned, followed by goats, although the proportion of animals owned was generally lower than what was owned by men (Njuki and Mburu 2013).

Few African farmers have access to commercial fertilisers and the few who do find the cost prohibitive. Manure and urine from indigenous sheep are essential components of African agricultural production. In addition to this, dried dung is used as a source of fuel. However, the exact contribution of manure from sheep and other livestock is not known.

The pastoralists who keep the majority of the sheep in Africa inhabit some of the most fragile ecosystems in the world that continuously face challenges of recurrent droughts and threats of desertification. However, these pastoralists have inhabited these lands for generations and thus know how to manage these resources sustainably by migrating out of ecosystems when the water and grass resources diminish and returning when they have had a chance to be replenished. Their utilisation of these lands that would otherwise not be utilised is an important benefit and contribution. The pastoralists contribute significantly to the economies of these countries by providing food to their families and the surrounding villages and towns. Unfortunately, this important contribution by pastoralists is seldom recognised by governments and administrators (Blench 2001), especially because of the concerns that have been raised on the role that overgrazing plays in environmental degradation (Foley et al. 2005).

6.4.2 Sociocultural Role of Sheep Across Africa

The primary reason for keeping sheep in Africa is for sustenance and for providing an income at the household level. However, sheep also perform important cultural roles and are used as part of African expression at different ceremonies, including weddings, funerals and day-to-day living. Sheep are important in the day-to-day lives of their owners, conferring status and power, and are used in the development and nurturing of relationships between members of a community. Many African cultures sacrifice sheep and other livestock as part of their religious or customary practices.

6.5 Sheep Breeds of Africa and Their Adaptive Features

There are an estimated 1.3 billion head of sheep in the world, one-third of these (approximately 418 million head) are found in Africa (FAOSTATS 2021). Illustrations of sheep from ancient Egypt suggest that the sheep present during the pre-dynastic and early dynastic periods were thin-tailed. These sheep are depicted with a straight head profile, long, thin legs, a long, thin tail that reached the hocks and were often black, white or pied. However, they were later replaced with fat-tailed sheep. Sheep illustrations after the 12th Dynasty in ancient Egypt show fat-tailed sheep (Ryder 1983). It is considered that these sheep easily replaced the thin-tailed breeds because of their coarse wool and fat tails, which were a valuable storage organ to pastoralists (Ryder 1983).

Thin-tailed sheep of large size are found in the northern dry tropics, while small or dwarf sheep are in humid areas of west to central Africa. Fat-tailed sheep are predominant in east Africa, extending to Mozambique, while fat-rumped sheep are most common in traditional systems in north-east Africa, but have also spread to commercial systems in Zimbabwe and other countries within the southern region.

Nine subtypes of sheep, as presented in Table 6.3, can be distinguished in sub-Saharan Africa (Gatenby 2006) comprising a variety of breeds and a large number of varieties or strains. Although there is no consistency in the classification of sheep breeds identified by different authors in Table 6.3 (Wilson 1991; Gatenby 2006; DAGRIS 2007), five of the types, namely, the thin-tailed hair, thin-tailed coarse wool, fat-tailed hair, fat-tailed coarse wool and fat-rumped hair sheep, are indigenous to Africa. Other classes include exotic or introduced sheep breeds and developed breeds through cross-breeding for specific purposes, which will all be considered in subsequent discussions.

6.5.1 Thin-Tailed Hair Sheep

The present-day thin-tailed hair sheep of Africa are mainly found in West Africa, extending from Sénégal to Nigeria, northwards towards Morocco and eastwards to Sudan. These sheep closely resemble the sheep that were found in ancient Egypt and are considered the most primitive on the continent (Ryder 1983). They were the first to migrate into Africa from the centre of domestication through the Isthmus of Suez and diffused

Table 6.3 Classes of sheep breeds in Africa

S/N	Breed type/breed	Synonym(s)	Location
1	**Thin-tailed hair sheep**		
(i)	*Dwarf type*		
	West African Dwarf	Djallonké, Fouta Djallon, Forest sheep, Mouton Guinéen, Mouton Nain d'Afrique occidentale, Kumasi, Ghana Dwarf, Nigerian Dwarf, West African Maned, Blackbelly, Mayo-Kebbi, Kirdi (southern Chad/northern Cameroon) Cameroon Dwarf and Congo Dwarf	Southern Senegal to Chad, Cameroon, Gabon and Congo
	South Sudanese	Sudanese Maned, Southern Sudan, Nilotic, Nuba Maned, Mongalla	Southern Sudan
(ii)	*African Long-legged type*		
	Balami	Fulani, Borno white	Senegal to Cameroon
	Uda	Oudah; Peul; Bali-Bali; Zaghawa (Chad); Bororo and Fellata (Western Sudan)	Senegal to Cameroon, central Chad to western Sudan
	Yankasa	White Fulani, Y'ankasa, Hausa	Senegal to Cameroon
	D'man	D'men, Damana, Demnan, Demmane, Demane and Tafilalet	South-eastern Morocco, southwestern Algeria and southern Tunisia
	Ouled Djellal	Arab white sheep	Algeria
	Ara Ara		Central and South-central Niger
	Peul-Peul	Fulani (English), Foulbé	Central Senegal
	Sudan Desert Sheep	Arab, Northern Sudanese, Desert Sudanese, Ashgur, Shugor, Butana, Beja, Drashiani, Dubasi, Gash, Kababish, Watish, Baqqara, Barka (Shukria), Hamale	Northern Sudan, Western Eritrea, Chad
	Maure, Black Maure	Moor, Mauritania, Arab	Mauritania (Hodh region), northern Mali, western Niger flood zone, northern Senegal
	Touareg	Grand Targui (Ara-Ara, Argooradji), Petit Targui, Tuareg	North-east Mali, Niger, Timbuctoo, south Algeria
	Angola Long-Legged	Goitred, Zenu	Northern and Eastern Angola
	Baluba		Katanga, Zaire
	Dongola (Coarse wool?)		Northern Sudan
	Zaire Long-legged, Congo Long-legged Baluba	Congo Long-legged, Sudanese	North-east Zaire
	Touabire	Landoum, White Maure, White Arab, short hair Maure	
	Angola Maned	Coquo	

(continued)

Table 6.3 (continued)

S/N	Breed type/breed	Synonym(s)	Location
	Terguia-Sidaou		
	Rembi		
	South Sudanese/Southern Sudan	Sudanese Maned, Southern Sudan, Nilotic, Nuba Maned, Mongalla	
	Sicilo-Sarde		
	Boujaad		
	Fellata	Zaghawa	
	Toronké		
	Mossi		
	Atlas Mountain		
2	**Thin-tailed Coarse wool sheep**		
	Macina	Koundoum, Goundou (Niger), Tillabery Sheep (Niger)	Central Delta of Niger and Mali
	Berbere	Berbere (Morocco, Algeria), Ait-Barka (Morocco), Ati-Mohad, Ait Haddidou, Aknoul, Marmoucha	Northern Algeria, Mountains of Morocco
	Hamra	Beni-Ighil	Algeria, Morocco (High Atlas)
	Darâa		Algeria
	Beni Ahsen		Northwest Morocco
	Timahdite	Zaian, El Hammam-Azrou	Morocco
	Sardi		Morocco
	Beni Guil	Beni Guil, Petit Oranais, Hamra, Hamyan, Harcha, Race des Plateaus de l'Est, Beniguil, Daghma, Tounfite, Tounsint	Northeastern Morocco and Western Algeria
	Tadla		Morocco
	Arrit		Northern Eritrea
	Saidi		Upper Egypt
	Farafra (*Developed*)		Western Egypt
	Dongola	North Riverain Wooled, Riverain Wooled	Northern Sudan
	Ati Barka		
	Dane Zaila	Zaila	
	Karayaka		
	Zemmour		Tunisia
	Aknoul		
	Turkish Merino		
	Algerian Arab	Race Arabe; Raimbi; Rembi; Ouled Jellal; Hodna; Laghouat; Chellala; Taguine; Saida; Djebel-Amour; Sidi-Aissa (Algeria)	Algeria
	Zel	Mazandarani, Chiva	
	Doukkala	Doukkala-Abda, Doukkalide	Ksar es Souk, Morocco
	Hadina	Hadine, Toubou	
	Kivircik		
	Ait Mohad		

(continued)

Table 6.3 (continued)

S/N	Breed type/breed	Synonym(s)	Location
	Hemsin		
	Gokceada (Imroz)		
	PIrlak		
3	**Fat-tailed hair sheep**		
(i)	*East African fat-tailed*		
	Afar	Adal, Danakil and Elle	Eastern Ethiopia
	Washera	Agew, Dangilla	Amhara state in Ethiopia
	Horro	Abyssinian, Ethiopian, Bonga, Wollaga	Western and southern Ethiopia
	Adilo		Ethiopia
	Bonga	Gesha, Menit	South western Ethiopia
	Red Maasai	Maasai, Tanganyika Short-tailed, Kipsigis, Luo, Nandi, Samburu	Tanzania, Kenya, Uganda
	Mondombes	Herero, Macubias	
	Malawi	Nyasaland	Malawi
	Norduz		
	Hottentot	Namaqua, Cape	
(ii)	*African long fat-tailed*		
	The Tanzanian long-tailed	Tanganyika Long-legged, Ugogo, Gogo, Rwanda, Burundi	Tanzania, Rwanda, Burundi, Democratic Republic of Congo, western Uganda
	The Rwanda-Urudai		Rwanda, Burundi, Democratic Republic of Congo, southwestern Uganda, northwest Tanzania
	The Ronderib Afrikaner	Cape fat-tailed, Namaqua Africander, Ronderib Africander, Blinkhaar, Steekhaar, Nama	South Africa and Namibia
	Damara		South Africa, Botswana, Namibia, Mozambique, Zimbabwe
	Namaqua Afrikaner, Transvaal		South Africa
	Zulu	Nguni	KwaZulu-Natal Province of South Africa
	Pedi	Bapedi	Limpopo Province, South Africa
	Swazi	Bapedi, Zulu	Rural Eswatini
	Landim	Nguni, (Landim is Landrace in Portuguese)	Mozambique (south of the Limpopo river)
	Sabi	Rhodesian, Zimbabwie, Mashona	Zimbabwe
	Tswana		Botswana, Southwest Zimbabwe
	Madagascar	Malagasy, Malgache	Madagasy Republic
	Candir		

(continued)

Table 6.3 (continued)

S/N	Breed type/breed	Synonym(s)	Location
	Arabi		
	Mehrabani		
	Kurdi-Khorasan		
	Sakiz	Chios, Cesme	
	Gray Shiraz	Kaboudeh Shiraz, Gray Karakul	
	Chal	Shal	
	Kesber		
	Karadi		
	Moghani		
	Sanjabi		
	Tuj	Kars, Cildri, Kesik, Tushin, Tushinski	
	Haraki		
	Dalagh	Semi fat-tailed Turkmen, Atabai	
	Karakas		
	Gicik		
	Hamdani		
	Sangsari		
	Lori	Lori-Bakhtiyari, Arabi	
	Ghezel	Kizil	
	Arabi		
	Candir		
	Haraki		
	Farta		
	Odemis		
	Barbarine		
	Daglic		
	Kallekouhi		
4	**Fat-tailed coarse wool**		
	Menz	Ethiopian Highland, Abyssinian, Legagora, Shoa	Central Ethiopia
	Tikur	Lasta, Tikur Tekur, Tukur, Tucur	Amhara state, Ethiopia
	Arsi-Bale	Ethiopian Highland, Abyssinian, Arusi	Eastern and south-central Ethiopia
	Wello	Wollo	South Wollo (Amhara state), Ethiopia
	Farta		South Wollo (Amhara state), Ethiopia
	Semien		Ethiopia (Simien Mountains)
	Sekota	Tigray highland, Abergelle	Central zone (Tigray state), Ethiopia
	Ossimi	Ausimi, Osemi, Ausemy, Meraisi	Lower Egypt
	Barki	Arab, Barqi, Bedouin, Dernawi (from the town of Derna), Maryuti, Libyan, Mariouti	North-West Egypt

(continued)

Table 6.3 (continued)

S/N	Breed type/breed	Synonym(s)	Location
	Barbary	Libyan Barbary, Tunisian Barbarin, Algerian Barbarin	North Africa (Libya, Tunisia, Algeria, Egypt)
	Somali Arab		Coast of Somalia
	Fellahi	Fellah (literally meaning peasant), Baladi (literally meaning local or village), Sharkawi.	Nile delta, Egypt
	Ethiopian Highland	Abyssinian, Akale Guzay (Shimenzana), Arsi-Bale, Menz (Legagora, Rashaidi, Tukur (Lasta)	
	Kurassi		
	Rahmani		
	Ghimi		
	Dangila		
	Akale-Guzay	Shimenzana, Rashaidi	
	Wooled Persian	Russian Persianer, Persian Red	
5	**Fat-rumped hair sheep**		
	Blackhead Somali	Berbera Blackhead, Blackhead Ogaden, Toposa (Sudan), Murle, Turkana, Gabbra, Boran, Adali (Afar blackhead), East African Blackheaded Persian (Uganda)	Somalia, Eastern Ethiopia, Northern Kenya
	Blackhead Persian (Developed)	South African Persian, Swartkoppersie (Afrikaans)	South Africa (for meat)
	East African Blackhead Persian	Tanganyika Long-legged, Ugogo	Western Uganda and North-West Tanzania
6	**Introduced sheep breeds**		
	Karakul (*Fat-tailed fur*)	Astrakhan, Bukhara, Arabi, Duzabi, Kambar, Shirazi, Sur; Persian Lamb	Namibia, South Africa, Botswana, Mozambique, Angola
	South African mutton Merino	German Mutton Merino	South Africa
	Barbados Black Belly		
	Bergui	La race ovine Queue Fine de l'Ouest	Algeria, Tunisia
	Rahmani	Rahmany	Beheira, Lower Egypt
7	**Developed or composite breeds**		
	Dorper (*Fat-rumped*)	White Dorper, Dorsian, Dorsie (Afrikaans)	South Africa, Namibia, Zimbabwe, Zambia, Angola, Botswana, Kenya, Burundi, Malawi, Mauritius
	Afrino		Southern Free, Eastern Cape and Northern Cape provinces of South Africa
	Van Rooy	Van Rooy White Persian, Van Rooy Persie (Afrikans)	Orange Free State, South Africa (for meat)
	Vandor		Central and western South Africa
	Meat Master		South Africa

(continued)

Table 6.3 (continued)

S/N	Breed type/breed	Synonym(s)	Location
	Dohne Merino	Dohne, El Dohne Merino	Eastern Cape, Free State and Western Cape Provinces of South Africa, Namibia
	Dormer	Dorman	Gauteng and Free States of South Africa
	Taadmit	Queue Fine de l'Ouest (Thin-tailed of the West, in Tunisia)	Central Algerian steppes, Tunisia, Morocco
	Thibar	Noire de Thibar, Black Merino	Algeria, Tunisia
	Tunisian Milk	Sardinian, Silico-Sarde	Tunisia
	Sidi Tibet cross		
	Sohagi		Southern Egypt
	Vogan		Togo (for meat)
	Blackhead Persian (Fat-rumped)?		South Africa (for meat)?
	Wiltiper		Zimbabwe (for meat)
	Nungua Blackhead		Ghana (for meat)
	Warle		
	Kazakh fine wool		
	Walrich Mutton Merino		
	Kanzi		
	Okouma		
	Ingessana		
	Maenit		
	White Wooled Mountain		
	Permer		
	Abidi (*Fat-tailed coarse wool*)	Ebeidi, Ibidi	
	Meidob		
	Sanabawi		
	Bezuidenhout Africander		
	Aboudeleik		
	South Moroccan	Rehmane-Sraghna, Rehmana-Srarhna, Zemrane	South Morocco
	Toronke		
	Fallahi		
	Jaffna	Deshiya Betaluwa, Lanka	
	Letelle Merino		
	Zoulay		
	Sidi Tabet		

Wilson (1991), Gatenby (2006), DAGRIS (2007)

overland into north and west Africa (Muigai and Hanotte 2013).

They are divided into two subtypes, the tropical dwarf and the African long-legged (Rege et al. 1996). The tropical dwarf sheep are mainly found in sub-humid and humid zones of west Africa, Sudan and South Sudan (Table 6.3). They are characterised by a straight nose and erect ears. These sheep are small in size, with an approximate height of 40–60 cm and weighing an average of 20 kg. The African long-legged sub-type is kept in large migratory flocks mainly in dry, arid and semi-arid areas that range from Mauritania on the west coast of Africa extending to Mali, Niger, Chad and Sudan in the east (Gatenby 2006) (Table 6.3). These sheep are characterised by a convex nose, pendulous ears and exceptionally long legs that give them a height of over 60 cm, which is an adaptation to the dry environments and the pastoralist lifestyle of their owners (Wilson 1991; Ryder 1983). All breeds classified in this group have a hairy coat, except the Macina breed, which has a coarse wool coat. The litter size is small, where twins are not common and triplets are very rare. The long-legged sheep breeds of Africa, namely, the Fulani, Maure and Tuareg, are named after the pastoralists who are their owners and managers. In West Africa, the sheep of the nomadic Fulani, also known in French as Peul or Foulbe include Balami (Bornu), Uda, Yankasa, Samburu, Torunke and some other strains (Gatenby 2006). Different breeds of thin-tailed hair sheep will be appraised in subsequent sections.

6.5.1.1 West African Dwarf and Cameroon Blackbelly Sheep

West African Dwarf Sheep

Distribution and Classification: The West African Dwarf (WAD) Sheep (Fig. 6.1) is also known as Djallonké, Fouta Djallon, Forest sheep, Mouton Guinéen, Mouton Nain d'Afrique occidentale, Kumasi, Ghana Dwarf, Nigerian Dwarf, West African Maned, Blackbelly, Mayo-Kebbi, Kirdi (southern Chad/northern Cameroon), Cameroon Dwarf and Congo Dwarf (Wilson 1991; DAGRIS 2007). With respect to subtypes, the Kirdi (Kirdimi or Massa) in northern Cameroon and southwest Chad is a black variant of WAD sheep, also known as Poulfouli in the far north of Cameroon and several other local or regional names in West Africa from southern Senegal to Chad, Cameroon, Gabon and Congo.

Physical Characteristics: The WAD sheep (Fig. 6.1) is a very small and compact sheep breed of 40–60 cm at withers (Wilson 1991). Apart from the Kirdi, which has been selected for a black coat, other members of the breed have varied coat colours ranging from brown, white, black or spotted black or brown on a white coat (Adu and Ngere 1979; Bemji et al. 2012). Other characteristics as reviewed by Wilson (1991) include short and stiff hair (males may have a heavy mane and apron of long hair), well-developed horns in males which are wide at the base, curving backwards, outwards and forward, fine and short horns when present in females, a strong and broad head with a flat forehead (profile slightly bulging in males), a wide muzzle and less prominent eyes, short ears (10 cm), narrow and pendent or semi-pendent, neck is long and chest is fairly deep with chest circumference 20% greater than the withers' height.

Adaptive and Special Genetic Characteristics: The WAD sheep is a very strong and hardy breed that has been described as being trypanotolerant and resistant to roundworms (Gatenby 2006). The breed is adapted to the humid rain forest

Fig. 6.1 West African Dwarf Ram at Federal University of Agriculture, Abeokuta, Nigeria. (Photo: Ismaila Muritala)

zone of Central and West Africa, also extending into the derived savanna and sub-humid regions.

Typical Production Systems and Production Characteristics: The WAD sheep is an exclusive meat breed owned by many ethnic groups in various agricultural systems, including traditional smallholder, migratory (nomadic and transhuman) and modern (ranching and finishing units) (Wilson 1991; Njuki and Mburu 2013). Estimates for reproductive traits reviewed by Wilson in 1991 (Wilson 1991) include: A*ge at first oestrus:* 250 days (206–322), *oestrus cycle:* 17.4 days (16–19) with heat lasting for 36 h (12–60 h), *gestation length*: 148.1–150.2 days on station in Côte d'Ivoire, age at first lambing ranged from 350 to 572 days depending on management system, *lambing interval*: 244–663 days depending on management system and parity of ewe, 241–275 days under controlled breeding systems, multiple births are numerous, twins common, triplets rarer with average litter size ranging from 1.15 to 1.5 depending on environmental conditions, f*ertility* (number of females giving birth/females mated): 90–96%, *fecundity (*lambs/100 ewes/year): 221% for WAD and 231% for Kirdi (north Cameroon), b*irth weight:* 1.2–2.5 kg, w*eight for age:* 1 month–4.7 kg, 3 months–9.6 kg, 12 months–18.0 kg, adult weight: male 25–30 kg, female 20–25 kg, *dressing percentage:* 43.0% at 20.2 kg live weight (Cameroon), 43.7% at 19.2 kg (Nigeria), *lactation length:* 117 days in Mali; *lactation yield*: 87 kg in Mali for supplemented ewes; maximum of 810 g/d and 1200 g/d with lactation yield of 57.4 ± 16.60 kg and 86.4 ± 29.21 kg, respectively for single and twin bearing ewes in Togo, *milk composition:* dry matter 16.5%, fat 6.0%, ash 0.8% and protein 5.4%. Heritabilities of 0.19 were estimated for age at first conception, 0.46 (dam-daughter pairs) for lambing interval and 0.26 for litter size in Cameroon. For repeatability, inter- and intra-dam variances of 0.11 for lambing interval were obtained in Senegal; estimates for weights (intra- and inter-ewe variances) were respectively 0.22 ± 0.07, 0.18 ± 0.07 and 0.24 ± 0.08 at birth, 60 and 120 days for lambs, while 0.57 ± 0.05 was estimated for ewe post-partum weight. Phenotypic correlations between ewe weight and lamb weight at birth, 60 and 120 days were, respectively, 0.20, 0.38 and 0.39.

Cameroon Blackbelly Sheep

Distribution and Classification: The Cameroon Blackbelly sheep (Fig. 6.2) is also known as 'Cameroon' or 'Cameroon Dwarf'. It is a sheep breed in Cameroon classified under the West African Dwarf group of breeds. The sheep was initially bred in West Africa for meat production but has also been domesticated in Europe, where they are used as grazers or raised as companion animals because they do not have to be shaved (https://rainwaterrunoff.com/the-cameroon-sheep-breed/). Current investigation on genetic diversity indicates that the Cameroon Blackbelly sheep are genetically distinct from other African sheep populations and are associated with a single ancestry (Wiener et al. 2022). Genes found in genomic regions of low diversity suggest biological functions that could be under strong selection, which include immune function, fertility and pigmentation.

Physical characteristics: The Blackbelly sheep has a height at withers estimated at 60–70 cm. The coat colour is brown with a black belly and black markings on the head and legs (Fig. 6.2). It has a hair coat and, in winter, grows a fine undercoat which it sheds in springtime. The females are polled, while the males have spiral horns (Fig. 6.2), a mane and a throat ruff (Rain

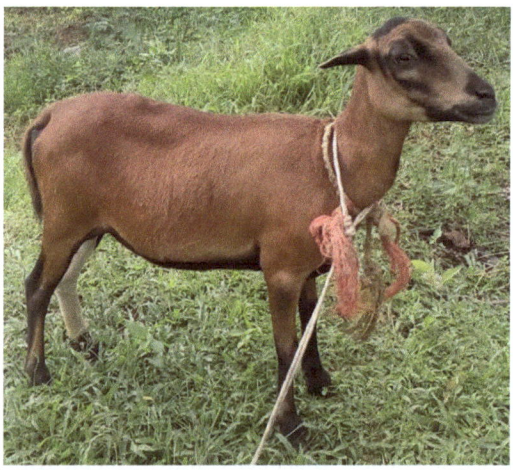

Fig. 6.2 Cameroon Blackbelly Ewe. (Photo: Julius Félix Meutchieye)

Water Run Off 2021) (https://rainwaterrunoff.com/the-cameroon-sheep-breed/).

Adaptive and Special Genetic Characteristics: The Cameroon Dwarf is a hardy sheep with characteristics similar to the WAD sheep.

Typical Production Systems and Production Characteristics: The Blackbelly sheep are mostly grazers. They have high prolificacy, with lambing occurring three times in 2 years, and are also noted for early maturity. Twins or triplets are common with commercial sheep breeds, but ewes can only cope with twins as the number of teats is limited to two. The gestation period is 147–50 days, weaning (3–5 months), mature live weight (30–40 Kg) and life span (12–16 years) (Njwe and Manjeli 1992). Range for average daily milk yield estimated by the indirect method of weighing lambs before and after suckling during a period of 12 weeks was 0.31–0.50 kg for ewes suckling single lambs, and maximum weekly milk yield was 3.52 kg at the 4th week of lactation, after which production declined (Njwe and Manjeli 1992) (Fig. 6.3).

6.5.1.2 African Long-Legged Sheep

Balami

Distribution and Classification: The Balami sheep (Fig. 6.4) is also known as Bornu white, Bornu being the location in northeastern Nigeria where the breed is predominant. It is part of the West African Sahel (long-legged group). For *subtypes or races*, Balami is considered to be the same as the 'Bouli' or white Uda variety in Niger (Wilson 1991).

Physical Characteristics: Balami is a large sheep breed with a white and hairy coat and a distinct convex head with a dull depression (Fig. 6.4) (Adu and Ngere 1979). It has a dewlap-like fold of skin and a white mane, the ear is large and droopy, and the tail is thin and long. Horns are present in rams, while ewes are polled.

Adaptive Characteristics: The breed is long-legged, hardy and well-adapted to forage within the semi-arid and drier sub-humid ecological zone.

Typical Production System and Production Characteristics: The main management system for Balami is agropastoral. The breed is heavier than Uda sheep, fast-growing and can attain 18 kg weaning weight in 12 weeks. Yearling weights of 35–40 kg for ewes and 45–60 kg for rams were reported (Adu and Ngere 1979). The mean litter size increased from 1.02 at first parity

Fig. 6.3 Cameroon Blackbelly Ram. (Photo: Julius Félix Meutchieye)

Fig. 6.4 Balami Ewes at National Animal Production Research Institute, Ahmadu Bello University, Zaria, Nigeria. (Photo: Julius C. Amabo)

to 1.6 at 54 months old, and birth weights were estimated at 3.5 kg for singles and 3.1 kg for twins (Adu and Ngere 1979; Buvanendran and Adu 2021). Estimated daily average gains for single males were reported at 142 g (101–190) (DAGRIS 2007), twins at 108 g (67–167), single females at 137 g (76–198 g) and 103 g (69–133 g) for twins at Katsina in Nigeria. The first lambing was at 18 months old and subsequent lambings at approximately 9-month intervals, with peak lambing taking place during the rainy season. Ewes lambing in the wet season had a shorter lambing interval by 50 days than ewes that lambed in the dry season. Lamb mortality to 10 weeks old was 43%, with lower survival rates for twin lambs, wet season-born lambs and lambs from primiparous ewes. Repeatabilities of lambing interval, litter size, birth and 10-week weight were 0.091, 0.087, 0.052 and 0.273, respectively.

Uda Sheep

Origin, Distribution and Classification: The Uda (Fig. 6.5) is the most numerous of the Fulani sheep (Gatenby 2006). They are alternatively known as Oudah, Peul, Bali-Bali, Zaghawa (Chad), Bororo and Fellata (Western Sudan). They originated as part of the West African Sahel (long-legged group). In Niger, the pied or Oudah bicolour and white or Bouli are considered colour variants. In Nigeria, the white variety is known as Balami and has been given a full breed status. In Chad and Northern Cameroon, the pied variety is known as Foulbé, while the white variety is known as Waïla. The Bali-Bali is occasionally seen as a separate type and is not considered a synonym in Niger. The Uda breed occurs throughout the Sahelo-Sudan vegetation zone of Nigeria (Adu and Ngere 1979), Southern Niger, Central Chad to Western Sudan and Cameroon (Wilson 1991).

Physical Characteristics: The breed is large and long-lopped with average measurements of 85 cm at the withers, with males measuring 75–85 cm, while females are 65–75 cm. The face profile is convex, the tail is long and thin. The body is covered with a coat of short coarse hair that is coloured entirely black or brown, beginning from the anterior forehead to the limb-abdominal girth line and the white posterior (Fig. 6.5). Ears are long, large and pendulous. Rams have large spiral horns while ewes are polled (Wilson 1991; Njuki and Mburu 2013).

Adaptive Characteristics: The breed is long-legged and adapted to semi-arid monomodal rainfall lowlands to arid zones. It thrives in hot and dry environments, with a poor survival rate beyond this ecological zone. The Uda is well adapted to extensive grazing and well-known for its trekking abilities (Adu and Ngere 1979).

Typical Production Systems and Production Characteristics: The breed is mostly raised for meat under the agropastoral and pastoral transhumant management systems. Mature live weights range from 45 to 65 kg for rams and 35 to 45 kg for ewes (Wilson 1991; Adu and Ngere 1979). Other production characteristics reviewed by the latter authors include *first oestrus:* 351 days in Niger; *Oestrus cycle:* 16.8 ± 0.06 days (heat lasts 33.2 ± 3.1 h); *Ovulation rate:* 1.3; *gestation period:* 154.8 days; *age at first lambing:* 15–17 months on station (Niger), 448 days (314–662) in Cameroon; *lambing interval:* 270 days (Shika, Nigeria); average litter size: 1.1; *birth weight:* 3.5 kg on station (Niger); males 3.9 kg, females 3.5 kg (Shika, Nigeria); *weight*

Fig. 6.5 Uda Ewes at National Animal Production Research Institute, Ahmadu Bello University, Zaria, Nigeria. (Photo: Julius C. Amabo)

for age: on station in Niger, weights at 30, 90 and 180 days increased from 7.3, 13.6 and 20.4 kg in 1981 to 11.0, 18.9 and 27.0 kg in 1987, probably due to better management; *average daily gain from birth to 18 kg*: single males 150 g (114–192), twin males 110 g (75–174 g), single females 140 g (79–170 g), twin females 113 g (101–142 g) in Katsina (Nigeria); *Milk yield* on station in Niger was estimated at 60 litres in 150 days. The *dressing percentage for meat* was 48–50% on station in Niger and 38–41% at 13–14 months with a carcass weight of 11.5 kg at Shika (Nigeria).

Yankasa

Distribution and Classification: Yankasa sheep (Fig. 6.6) is part of the West African Sahel (long-legged group), distributed from Senegal to Cameroon, with the possibility of some admixture among sheep from farther south (Wilson 1991). Yankasa is widely distributed, making up approximately 60% of the Nigerian flock in Northern Guinea and Sudan Savanna, where it is locally called white Fulani or 'Hausa'.

Physical Characteristics: Yankasa is intermediate in size between Uda and West African Dwarf sheep (Adu and Ngere 1979), 68 cm at withers, coat is white with black patches around the muzzle, ear, eyes and occasionally the feet, ears are relatively short and semi-pendent, rams have curved horns and heavy hairy white mane while ewes are polled and some may have tassels on the neck (Fig. 6.6).

Adaptive Characteristics: They are adapted to sub-humid and semi-arid regions, extending to the Northern Guinea savanna in the south.

Typical Production Systems, Husbandry Practices and Production Characteristics: The Yankasa is maintained mainly in agropastoral and pastoral management systems. In the Fulani agropastoral system near Zaria (northern Nigeria), 70% of families keep an average of 12.5 sheep with a flock structure of 76% females, which are allowed to graze with cattle. Reproductive characteristics reviewed by Wilson (1991) include: *Age at first oestrus*: 238 days, *weight at first oestrus*: 18.4 ± 0.4 kg, *oestrus cycle*: 18.1 ± 1.7 days; *heat:* 25 ± 2.2 h and *ovu-*

Fig. 6.6 Yankasa Ewes at National Animal Production Research Institute, Ahmadu Bello University, Zaria, Nigeria. (Photo: Julius C. Amabo)

lation rate of 1.36 ± 0.34. Other estimates include 575 ± 9.6 days for *age at first lambing*, 249.8 ± 2.7 days for *lambing interval* (Bemji et al. 2001a), 1.22 for litter size (Bemji et al. 2001b), 2.48–2.91 kg *birth weight* (Bemji et al. 1996; Afolayan et al. 2006), 31 kg at maturity (25–40 kg for ewes and 35–50 kg for rams), and 42% (*dressing percentage*) at 30 kg live weight (Review by Wilson; 99 (Wilson 1991)). *Heritability* estimates of 0.07 (*litter size*), 0.41 ± 0.08 (*birth weight*), 0.12 ± 0.07 (*litter birth weight*), 0.99 ± 0.38 (*age at first lambing*) and 0.73 ± 0.26 (*lambing interval*) (Bemji et al. 1996, 2001a, c).

D'man

Origin, Distribution, Classification and Population Statistics: The D'man (Fig. 6.7) is also known as the D'men, Damana, Demnan, Demmane, Demane and Tafilalet (DAGRIS 2007). As reviewed by Boubekeur in 2015 (Boubekeur et al. 2015), D'man is distributed within the pre-Saharan regions of Southern Maghreb, around the oases of Errachidia, Ouarzazate, Dades and Draa (South-eastern Morocco), oases of Saoura, Touat and Gourara (Southwestern Algeria) and Tozeur and Kebili in Southern Tunisia. Over 617,000 D'man sheep are raised in the oases of South Morocco (Boujenane et al. 2013). The breed is classified among the endangered breeds.

The origin of this breed has always been disputed, and a study using 30 microsatellites (Gaouar et al. 2015) concluded that this breed may be the result of a cross between the Sid Aoun race and a white race from the North of Algeria. In addition, there is a homonymy of two populations evolving in the Algerian southwest, one at the level of the Saoura and the other in the region of Adrar, both bearing the same name 'D'men'. However, they may constitute different breeds that may need further studies.

Physical Characteristics: D'man is a large goat breed with 63.9 cm (females) and 72.3 cm (males) height at wither, coat colour varies (white, black and brown with their combinations), black or black in combination with other colours are predominant (Boubekeur et al. 2015). Most animals have a white spot on the head, both sexes are polled, the tail is thin, long (42.9 cm) with a white tip, and the ear length is 13.6 cm (Fig. 6.7).

Adaptive Characteristics: D'man is mainly a sheep breed of the Oasis in Morocco, Algeria and Tunisia but tolerates Saharan conditions.

Typical Production Systems and Production Characteristics: The breed is mainly raised for meat and milk under pastoral and family farming systems and in a few breeding farms. It has high reproductive performance, including high prolificacy of 200%, early puberty (6 months) and a long breeding season (from May to February), although the growth rate is low (140 g/day) (Boujenane et al. 2013). Conception rate, litter size at birth, litter size at weaning, litter weight at birth, litter weight at weaning, gestation length and ewe's body weight averaged 94.4%, 2.38 lambs, 2.03 lambs, 6.23 kg, 39.6 kg, 149.9 days and 44.8 kg, respectively.

Ouled Djellal

Distribution: Ouled Djellal, also known as the Arab white sheep (Fig. 6.8), is a dominant sheep breed in Algeria, accounting for more than 63% of the Algerian sheep populations (Gaouar et al. 2017). Due to high productivity, the breed is frequently used for cross-breeding with other local Algerian sheep breeds. This has resulted in genetic admixture with Rembi and Taâdmit breeds (Gaouar et al. 2015).

Physical Characteristics: Ouled Djellal is a large, long-legged sheep, with a withers height of 80.13 cm and 91.40 cm for females and males, respectively. The coat colour is predominantly off-white or pale-yellow, while the head and limbs are either white or white speckled with fawn (Harkat et al. 2015). The males are horned, with the horns being small (budded) or spiral with a lateral orientation. The majority of females (92.21%) are polled. Where horns are present in females, they are either small or curved. The ear length varies and can be either short, medium or long. The ear shape can either be semi-horizontal or drooping, the latter being the characteristic in the majority. The forehead profile is slightly hooked in the majority of sheep. The tail length varies between short, medium and long (Fig. 6.8).

Adaptive Characteristic: The breed is adapted to arid climates.

Typical Production Systems and Production Characteristics: Ouled Djellal is managed under sedentary or semi-sedentary systems with the use of one or two seasonal pastures and a transhumant farming system. The breed is mainly raised for meat and wool. Mature body weight is 61.27 kg for females and 95.87 kg for males (Harkat et al. 2015).

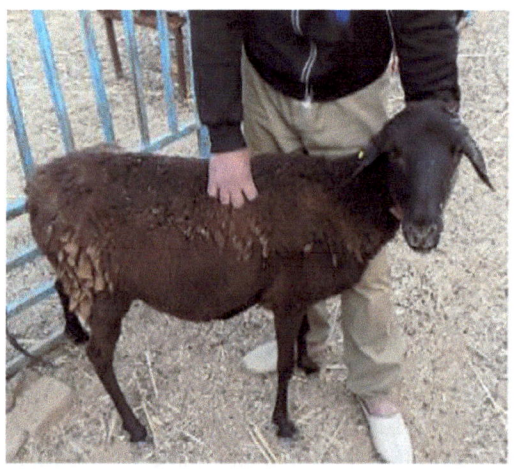

Fig. 6.7 D'man. (From Djaout et al 2017, licensed under CC-BY 4.0)

Ara Ara

Distribution and Classification: The Ara Ara belong to the West African Sahel or long-legged breeds of sheep distributed in central and south-central Niger.

Physical Characteristics: The breed is similar to the Uda sheep in physical characteristics.

Adaptive Characteristics: The Ara Ara is adapted to semi-arid conditions.

Typical Production Systems and Production Characteristics: The breed is managed under agropastoral and pastoral systems for meat. Litter size is 1.18 (84.6% single, 13.0%, 2.4% triplet). Multiple births are common for arid zone sheep. Birth weight is 2.7 kg with single males, twin males, single females and twin females weighing 3.0 kg, 2.4 kg, 2.7 kg and 2.3 kg, respectively, on the Maradi station (Wilson 1991).

Peul-Peul

Distribution and Classification: Fulani (English) and Foulbé are synonyms for the breed, which constitutes part of the West African Sahel or long-legged sheep found in central Senegal (Wilson 1991).

Physical Characteristics: Peul-Peul (Fig. 6.9) is a medium-sized sheep breed of about 65–75 cm at the wither and a live weight of 30–50 kg at maturity. The coat colour varies; however, most animals are white with black or red spots or entirely red. The coat is stiff and short. Loosely spiralled horns are present in males, while females are polled.

Adaptive Characteristics: The sheep are adapted to semi-arid to sub-humid zones within the Sahel and Sudano-Sahel regions.

Typical Production Systems and Production Characteristics: Peul-Peul is managed under pastoral and agropastoral systems for meat. Reproductive includes *Age at first lambing*: >24 months; *Lambing interval*: 12 months in Senegal's traditional system and 7.3 months for sheep raised on station; *Birth weight*: male 3.1 kg and female 2.9 kg (for dams provided supplementary feeding). *Average daily gains*: 0–40 days: 124.4 g, 40–180 days: 110.6 g and the *dressing percentage* was estimated at 50%.

Sudan Desert Sheep

Origin, Distribution and Classification: Sudan Desert sheep (Fig. 6.10) are the most common sheep in Sudan and are found in all parts of the country except the southern regions. They are alternatively called Arab, Northern Sudanese, Desert Sudanese, Sudanese Desert (Wilson 1991; Gatenby 2006). Numerous tribal types are recognised in Sudan, which include Dubasi, Watish, Kababish, Baqqara, Shugor, Ashgur, Butana, Beja, Drashiani, Gash, Barka (Shukria) and Hamale (Wilson 1991; DAGRIS 2007). They are found at 10 °N of Sudan, extending eastwards into Eritrea and westwards into Chad (Wilson 1991; Gatenby 2006). Other populations include Gezira and perhaps Barka and Wollega, found in Ethiopia. A crossbred ecotype (Sudan Desert x Southern Sudan) was recognised in the central belt of Sudan, extending eastwards into Eritrea and westwards into Chad within different ecological regions (arid, semi-arid and riverine).

Physical Characteristics: Sudan Desert sheep is a large breed (Fig. 6.10) of up to 80 cm or more at the withers, depending on the tribal type, with adults weighing 60 kg (male) and 50 kg (female). As reviewed by (Wilson 1991), the head is strong, the forehead broad and flat with a convex profile (the Dubasi have a markedly convex profile) with eyes set high. Sheep from eastern Sudan are polled, but 5% of males and females from the west are horned (spiral type), with lengths of up to 60 cm. Ears are medium to long (12–18 cm) and drooping. Toggles are present in both sexes in about 10% of sheep. The chest is well developed with a fairly heavy and long neck. The withers are prominent and broad in most types, with legs long, sometimes very long, and lightly fleshed. The sacrum is usually higher than the withers. The group is fairly well developed. The tail is thin and long, with variable amounts of fat at the base or extending down depending on the tribal type. The tail length can be 60% of the withers height to greater than the withers height trailing the ground.

The coat is fine to coarse and short to long with variable coat colours depending on the tribal type (Fig. 6.10). The Shugor are moderately large, with coat colour ranging from light to dark brown and sometimes patches of wool beneath the hair. They are found along the White Nile and areas to the west. Herds of Shugor sheep are predominant in the western part of Gezira. They travel more than the Dubasi sheep, which are the prototype sheep of the northern part of the Gezira area and are found in greater numbers in the villages of Dubaseen tribes (hence Dubasi) in El-Fawar, Umbusha, Selaim, Kab El-Gidad and north of Khartoum, where they are called Butana. Watish sheep are smaller than Shugor and Dubasi.

Adaptive Characteristics: Sudan Desert sheep are hardy and adapted to arid, semi-arid and rive rain areas.

Typical Production Systems and Production Characteristics: Sudan Desert sheep are mainly raised in large flock sizes for meat, milk and skins under pastoral, agropastoral and urban systems. Flock structure favours meat or milk production, constituting 77.8% females (55.8% breeding females), males 22.2% (4.2% >15 months). Reproductive parameters as reviewed by Wilson (1991) include *Age at first lambing*: 13–15 months in the Southern Darfur traditional system; *Lambing interval*: 275–426 days; *Litter size*: 1.14–1.22 depending on the type, *Annual reproductive rate*: 1.11 lambs per ewe on station; 1.50; *Oestrus cycle*: 17–21 days; *Duration of heat*: 25 h; *Gestation period*: 147–166 days; *Birth weight*: males 4.1 ± 0.63 kg, females 3.9 ± 0.66 kg (Shugor and Dubasi heavier than Watish); *Weight for age*: 30 days—7.4 kg, 90 days—14 kg; *Average daily gain*: 0–30 days: 136 g, 30–120 days: 110 g; *Lactation length*: 188 days on station; *lactation yield*: 137 kg; *Milk composition*: estimates were reported for total solids (12.4%) and fat (4.28%); *Dressing percentage*: 49.1% and 46.0%, respectively, at 32.3 kg and 34.9 kg live weights on low and high fibre diets, while 52.7% was recorded for an empty body weight of 30 kg.

Black Maure

Origin, Distribution and Classification: Synonyms include Moor, Mauritania and Arab (DAGRIS 2007). The breed has been reported to have originated as part of the West African Sahel (long-legged group), with the likelihood that it descended from Rio de Oro sheep of northern Mauritania, which in turn descended from the Maghreb type in Morocco (Wilson 1991). Subtype includes the Zaghawa (Arid Upland) of northwestern Darfur and eastern Chad. Black Maure is owned by Moor tribes in southern Mauritania (Hodh region), northern Mali in Nara, Nioro and Niono areas and eastwards to the western border of the Niger flood zone, and northern Senegal.

Physical Characteristics: The Black Maure is a large sheep breed of 75–90 cm at the withers (male 80 cm, female 74 cm) with a mature weight of 45 kg (male) and 32 kg (female) (Wilson 1991). The coat colour is black with a white tip to the tail. The breed has a strong head (forehead flat), a convex profile in males but less in females and well-developed tear glands. Females are polled, while horns are well developed in males, triangular in shape, spiral outwards from the head and up to 30 cm long. Horns have less pronounced ribs in Zhagawa. Ears are long (20 cm in Maure), medium (12 cm in Zhagawa), broad and pendulous. Toggles are long and present in a few of both sexes. They have a long and thin neck, a narrow and shallow chest and a long and dipped back. The tail is long (36 cm), thin and falls below the hocks. The coat is long with coarse, stiff black hair in loose ripples on a softer undercoat.

Adaptive Characteristics: The Black Maure is adapted to arid areas with low or high seasonal rainfall. The areas include Northern Sahel and Saharo-Sahelian zones with annual grasses (Aristida, Cenchrus), some perennials and scattered acacia scrub.

Typical Production Systems and Production Characteristics: The main husbandry system for Black Maure is pastoral. The breed is mainly raised for hair, meat, milk, skins and pelts. Reproductive and production characteristics as reviewed by Wilson (1991) include *Age at first lambing*: 12–18 months; *Lambing interval*: 8–10 months; *Litter size*: 1.03 (multiple births of 2–4%); *Annual reproductive rate*: 1.4; *Gestation length*: 153 days (Sudanese Zaghawa); *Birth weight*: 3 kg; *Weight for age*: 5 months (16.0 kg), 12 months (24.0 kg) and 36 months (32.4 kg);

Hair yield: multiple clips yield (200 g per year); *Fibre length* (3.3–10 cm) and *Fibre diameter* (26.6 um); *Dressing %*: 35–40 in West Africa and 39–42 in Chad.

Touareg

Origin, Distribution and Classification: Synonyms for Touareg sheep include Grand Targui (Ara-Ara, Argooradji), Petit Targui and Tuareg (DAGRIS 2007). The breed originated as part of the West African Sahel (long-legged group). A smaller 'Touareg' sheep (sub-type) is sometimes distinguished in Gourma (Mali). The breed is distributed North-east of Mali from the top of the Niger bend and Timbuctoo, north to the Adrar n'Iforas at about 19°N and eastwards to Niger, from Niamey and Dosso to the Aïr region in the north (Wilson 1991). This breed is also very popular in the great south of Algeria, where it is very appreciated (Djaout et al. 2017).

Physical Characteristics: The Touareg sheep is a large animal measuring 75–80 cm at the withers and weighing 40–60 kg. The coat colour could be white and spotted or red. The coat is of short, stiff hair. A review by Wilson (1991) shows that the head is strong with a prominent forehead and convex profile. The muzzle is narrow and the eyes are less prominent. Females are polled, while horns in males are strongly ribbed and spiralling backwards. Ears are 15 cm in length, and toggles are present in both sexes. They have a long neck with fat folds at the nape in males. Withers is prominent, and the chest is narrow and shallow, with a girth circumference hardly exceeding withers height. The back is straight and long. Rump is less sloping compared to neighbouring Maure and Toronké types. Legs are long with flat thighs. The tail is of medium length, reaching the hocks. The udder is well-shaped with long, well-spaced teats (Wilson 1991).

Adaptive Characteristics: The Touareg sheep is adapted to semi-arid and arid to extremely arid conditions.

Typical Production Systems and Production Characteristics: The Touareg sheep is raised for meat and milk and is managed under pastoral, transhumant, nomadic and agropastoral systems to a lesser extent. Flock size is fairly large, at about 50–100 and mostly raised for meat production in Mali and for milk production in Niger. Reproductive and production traits, as reviewed by Wilson (1991), include *Age at first lambing*: 2 years; *Lambing interval*: ≥365 days; *Litter size*: 1.03, *Annual reproductive rate*: 0.59; *Lifetime production*: 4% of ewes produce more than 4 lambs; *Milk yield*: 200–400 g/d (dry season); 400–600 g/d (wet season); *Dressing percentage*: 46%.

6.5.2 Thin-Tailed Coarse Wool Sheep of Africa

Thin-tailed coarse wool sheep have been given the collective name Maghreb by Epstein in 1971 (Epstein 1971) and are distributed in the northern African mountainous regions of Algeria, Morocco and Tunisia, along the central Niger River delta and in east African countries of Eritrea and South Sudan. They probably evolved from the original thin-tailed hair sheep populations (Devendra and McLeroy 1982). Examples of these breeds are the Macina, Barbarin, Toubour and Dané Zaïla. Their bodies are covered with a coarse carpet-type wool, which is often white or coloured, from which the pastoralists use to make blankets and coarse cloth. They range in size from small to medium, with an average withers height of 60–80 cm, and have a heavy head, convex face profile and pendulous ears. Rams have large spiral horns that either curve backwards or slightly upwards. Ewes can be polled or have small horns (Rege et al. 1996; Epstein 1971).

6.5.2.1 Macina

Origin, Distribution and Classification: Synonyms or subtypes of Macina include Koundoum, Goundou (in Niger) and Tillabery Sheep (in Niger) (Wilson 1991; DAGRIS 2007). The breed is mainly found in the central Delta of Niger and Mali, with Macina being the Fulani name for the Niger River flood zone. According to Wilson (1991), they possibly descended from the wooled thin-tailed sheep from North Africa and migrated to the present area with the Moors and Moroccans during the conquest of Timbuctoo in the fifteenth and sixteenth centuries. It was previously mistakenly believed to have descended from the Karakul breed or from various crosses of Merion with Syrian or Barbary sheep. Three other types of 'wool' sheep in West Africa include Hadina (large black sheep kept by the Toubou in the extreme east of Niger); Dané Zaïla (very small white sheep from the same area kept by Arabs) and wool sheep of West Kanem found in the north and west of Mao in Chad. The Macina breed is confined to the floodplain of River Niger in central Mali (± 50,000 km^2) along a narrow band on both sides of the river to Niamey (Wilson 1991; Gatenby 2006).

Physical Characteristics: The Macina sheep is a medium-sized breed averaging 60–80 cm at the withers. The adult average weight is 40 kg in males and 30 kg in females. The forehead is broad and straight with a straight profile or slightly convex in males. There is no interorbital depression, and supraorbital processes are pronounced. The nose is narrow with a longer upper jaw compared to the lower one. Horns are well developed with deep grooves in males (65%) and a classic spiral 'ram's horn' shape. A small percentage (0.5%) have multiple horns, although in the early twentieth century, and 4.0% had multiple horns. In the present-day population, 8% of females have weak horns or scurs. The ears are medium (12 cm long), wide and pendulous. Toggles, which were initially absent in the original stock, are found in 15% of animals. The neck is short, and the chest is shallow and narrow. The withers are prominent, while the back is straight. The croup is tucked and thinly fleshed. Legs are long with light flesh. The tail is thin and descends below the hocks. The coat colour is commonly white, diversely spotted with black or red colours, especially around the eyes and ears. The body, forehead, knees and hocks are covered with a coarse wool coat that is mixed with hair, but the underside is bare. Goundoun is similar in most of these features.

Adaptive Characteristics: The Macina sheep thrives within humid areas associated with the seasonal flooding cycle of the Niger River ecological zone, which forms its main distribution area.

Typical Production Systems and Production Characteristics: The Macina sheep is kept by Fulani, who are the main inhabitants of the Niger River floodplain area. They keep the Macina mainly for wool, milk and meat production under pastoral and agropastoral systems. Transhumant management involves short treks outside the flood zone when flooding is at its peak between July and October. Flock sizes are generally large. Reproductive and production parameters as reviewed by Wilson (1991) include *Age at first lambing*: 371–766 days; *Lambing interval*: 170–485 days; *Multiple births*: twins: 3.1%, triplets: very rare; *Litter size*: 1.03; *Birth weight*: 2.7 ± 0.62 kg; *Weight for age*: 90 days—10.3 kg, 365 days—24.4 kg; *Mature weight*: males—60 kg; *Lactation length*: 135 days (range 85–165); *Lactation yield*: 50 kg; *Wool yield*: 2 clips per year with total of 685 ± 42.8 g (males 836 ± 52.5 g, females 534 ± 65.0 g); *Fibre length*: 4.6 cm; Fibre diameter: 39 um; about 10 crimps per 100 mm; resistance is 14 g with very little grease. The fibre is used for blankets and coarse cloaks. *Dressing %* is 40 with low-fat meat.

6.5.2.2 Berbere

Origin and Distribution: Azoulai Berber wool and A'arbia, Zoulay are synonyms of the Berbere sheep (Fig. 6.11). The Berbere breed is the oldest Algerian sheep breed. It is distributed across the mountainous chain of northern Algeria in Bouhadjar, Souk Ahras, El-Tarf, Annaba, along

Fig. 6.8 Bélier Ouled Djellal ram at Biskra. (From Djaout et al. 2017; licensed under CC-BY 4.0)

the Algerian-Tunisian borders, Tlemcen, Souk Ahras, Maghnia, Jijel (Collo), Edough, Ouarsenis and the mountains of Tiaret. In Algeria, it is known as A'arbia (Djaout et al. 2017). It is also found in the upper Moulouya Valley of Morocco. In Morocco, the Berbere breed origins are believed to be from the Tousint and Berber breeds (Mason 1996). It is called '*A'arbia*' by the breeders because it was believed to have originated from the Arabian Peninsula.

The Berbere sheep are reared in small flocks that do not usually exceed 20 heads per breeder. The current distribution area of Berbere sheep is much smaller than what was reported by Chellig (1992). Recent studies have shown that in the mountainous region of Jijel, the Berbere sheep was replaced by the Ouled Djellal and Hamra breeds as a result of a great reduction in the artisanal activity, which uses Berbere's wool for the manufacture of burnous, other traditional clothing and leather goods. The reduction in this traditional activity has led breeders to replace this breed with the Hamra breed and, to a greater extent, with the Ouled Djellel breed (Gaouar and de La Biodiversité 2009). Berbere sheep have been classified as an endangered breed.

Physical Characteristics: The Berbere sheep is a small animal with bright, shiny white wool (Azoulai). The Berbere sheep is described as a robust animal. The coat colour could be white (Fig. 6.11), brown and sometimes black or a mixture of brown, white or black and white. The head is short and concave. Ears are medium, thin and horizontal. The wool is white and long, sometimes mixed with brown or black, not crimped, with an open, drooping fleece.

Adaptive and Special Genetic Characteristics: Berbere ewes are good mothers. The sheep are milked and mainly used for consumption at the household level. Breeders prefer this breed for its hardiness and due to the fact that it is resistant to parasites and adapted to the extreme cold. Meat quality is poor (Djaout et al. 2017).

6.5.2.3 Hamra

Distribution, Classification and Population Statistics: The Hamra (Fig. 6.12) is an indigenous Algerian breed, known locally as '*Beni-Ighil*' in the Moroccan High Atlas (Morocco), where it is reared by the *Beni-Ighil* tribe, by which it is named (Gaouar et al. 2015). In Algeria, however, it is known as '*Deghma*' because of its dark red colour. The Hamra's natural habitat is from Chott Chergui to the Moroccan border (Chellig 1992).

The population was 3.2 million head at the beginning of the 1990s (Chellig 1992), with the number increasing to 500 million head in 2003 (Feliachi 2003). However, the population has decreased to under 100 million in recent years (Ayachi et al. 2017). This decrease has been mainly attributed to the massive introduction of the Ouled Djellal breed by breeders, resulting in the replacement of the Hamra breed in the main distribution areas (Gaouar et al. 2015). Hamra sheep is currently located in the western steppe, up to the Moroccan border, at the level of the Wilayas of Saïda, El Bayed and Tlemcen (Djaout et al. 2017).

Fig. 6.9 Peul-Peul sheep in the sylvo-pastoral zone of Senegal. (From Ndiaye et al. 2018, licensed under CC-BY 4.0)

Fig. 6.10 Sudan Desert sheep reared in the semi-desert belt of Sudan. (From Ali et al. 2016, licensed under CC-BY 4.0)

Physical Characteristics: Hamra sheep is medium-sized and has a very close phenotype to the Moroccan *Beni-Iguil* breed (Benoudifa 1990), with which it shares the same origin (Gaouar et al. 2015; Chellig 1992). The body is covered with white coarse wool, with a fineness of 74.17 ± 49.44 μm (Belharfi et al. 2017a). The rams have medium-sized horns with black streaks (Djaout et al. 2017). The head is dark coloured, with three main colours being reported, namely, dark mahogany to an almost black, dark mahogany and light mahogany (Fig. 6.12) (Djaout et al. 2017).

Adaptive Characteristics: It is very popular for its hardiness, but above all, for the flavour and finesse of its meat.

Production Characteristics: The Hamra breed (also referred to as *Oranie* in the French market) has the ideal conformation of a meat-producing sheep and bears a remarkably fine frame. The medium size is the main reason why this breed is being replaced by the Ouled Djellal breed. The Hamra ewe is characterised by good milking ability, with milk production under the medium category and with a maximum production of between 50 and 60 kg during 4–5 months of lactation (Chellig 1992).

6.5.2.4 Darâa

Distribution: The Darâa is an indigenous Algerian breed distributed throughout Algerian territory. Although currently experiencing low numbers, it is found in flocks of mixed breeds and races. It closely resembles the French breed *Noire Velay* (Daniel 2000), the Tunisian breed (Noire de Thibar) and a variety of the D'man breed described by Boukhliq (2002).

Physical Characteristics: The Darâa breed has a completely black head and legs. The body is covered with wool, characterised as closed or semi-closed, brown in colour and measuring 45.00 ± 19.115 μm (Belharfi et al. 2017a). The head is short and thin, the muzzle is straight, the limbs are thin, and the tail is medium or long; horns are absent in females, but males can be horned (Djaout et al. 2017).

6.5.2.5 Beni-Ahsen

Distribution: The Beni Ahsen is a Moroccan sheep breed cherished for its wool traits. It has the weightiest fleece and finest wool compared to other local breeds. Studies have indicated that the Beni Ahsen may be an ancestor of the Merino and specifically the Iberian Merino (Kandoussi et al. 2022).

Physical Characteristics: The Beni Ahsen is a thin-tailed breed of 80–100 cm and 70 cm height at the withers for males and females, respectively. The face is brown with strong head conformation. Males are horned while females are polled. The fleece is white, neck long with a pronounced fold and dewlap. Shoulders are high with an appearance of a hanging rump. Studies from the late 1970s through mid-1980s recorded 3.54 kg and 3.43 kg as average birth weights for males and females, respectively, and 18.6 kg and 17.0 kg at 90 days for males and females, respectively (Boujenane 2002). A review by the author showed that the weighted average for adult ewes (≥2 years) was 44.8 kg, with adult rams averaging 60–80 kg.

Breeding Practices and Production Characteristics: Beni-Ahsen ewes are seasonally bred between May and August to lamb between

January and February (Boujenane 2002). As reviewed by the authors, ewes lamb for the first time at 24.9 months with a gestation length of 149.9 days. Studies from the mid-1970s to the mid-1980s reviewed by the authors recorded average fertility (85.6%), prolificacy (104%) and lamb survival rate from birth to 90 days (93%), with 81% of lambings concentrated between November and February.

Lactation performance as reviewed by Boujenane (2002) showed that average milk yields of 70.9 kg and 64.4 kg were reported for two different 10-week lactation studies, while 57.4 kg was obtained using weight differences (WD) of lambs before and after suckling during a 12-week lactation. A higher average yield of 113.7 kg was obtained using oxytocin. A 14-week lactation study realised 81 kg under the WD method, while 132 kg was obtained using the oxytocin method. Dry matter, fat and protein contents ranged from 22.1 to 22.3%, 9.9 to 10.6% and 5.0 to 5.5%. Beni Ahsen sheep are treasured for their wool quality and quantity. Average fleece weight and fibre fineness score were, respectively, 2.7 kg and 52.9 kg.

6.5.2.6 Timahdite

Location: The Timahdite is one of the main indigenous Moroccan sheep breeds. It is the country's best meat breed and is valued for its good conformation, ease of fattening, carcass yield and high milk production (Boujenane 2002). The breed is very hardy with excellent adaptation.

Physical Characteristics: The Timahdite has a brown face, white coarse fleece and white legs. Height at withers is estimated at 50–55 cm for females and 60–70 cm for males (Boujenane 2002). The tail is thin, and horns are present in rams, but ewes are polled.

Breeding Practices and Production Characteristics: Timadite ewes are seasonally bred between May and September to lamb between January and February (Boujenane 2002). Review of reproductive traits by the author revealed that litter size at birth ranged from one to three lambs. Lambings of single litters represented 93.3%, twin litters 6.5% and triplets 0.1%; age at first lambing: 48 months; average gestation length: 148 days; fertility: 82.4%; prolificacy: 104%; lamb survival from birth to 90 days: 84.5%; body weights: 3.31 kg vs 3.25 kg for males and females at birth, 17.6 kg vs 16.4 kg at 90 days, 37.7 kg vs 36.8 kg at 1 year and 57.2 kg vs 40.6 kg for adult rams and ewes (≥2 years); average daily gain: 171 g; feed conversion: 6.2 kg feed/kg weight gain; average slaughter weight: 29.8 kg at 221 days of age; average carcass weight: 14.4 kg and dressing %: 48.1.

Lactation yield estimated by the weight difference of lambs before and after suckling was 84.3 kg. Dry matter, fat and protein contents of milk from oxytocin-treated ewes averaged 24%, 11.7% and 5.1%, respectively, with a higher average milk yield of 123.4 kg for a 12-week lactation period. Fleece weight averaged 2 kg, fibre diameter was 31.6 mm and staple length was 8.75 cm.

6.5.2.7 Sardi

Location: The Sardi sheep is a prominent breed reared in Morocco. It is highly regarded nationally, and the male is highly sought after for the Sheep Sacrifice Festival (Aid Adha). In Morocco, the main local breeds, Timahdite, Sardi, Boujaad and Beni Guil, show similar performances, particularly concerning characters of economic interest.

Physical Characteristics: The Sardi sheep is characterised by a white head that is devoid of wool. Black spots are present around the eyes, muzzle and the tips of the ears and legs. The belly and the limbs are devoid of wool. The muzzle is straight in the ewe, broad and slightly arched in the ram. The horns (absent in females) are well developed and open in males. The fleece is closed without spots or jars with short and tired wicks. The adult weight in males varies from 70 to 100 kg and in females from 45 to 60 kg. Adult height at withers varies from 80 to 90 cm in males and 55–65 cm in females.

Adaptive Characteristics: The Sardi breed is rustic and well adapted to poor pastures, especially those of the Central West plateaus.

Breeding Practices and Production Characteristics: The breed has low prolificacy (estimated at <120%). To increase its productivity, the breed is often crossed with other breeds. Development of efficient cross-breeding schemes, such as industrial or double-decker cross-breeding involving the Sardi breed, particularly outside its cradle area, would help to cope with growing demand from more Moroccan consumers, especially in large urban centres where consumers demand quality meat with less fat (Boujenane 2002).

A 12-week lactation yield of Sardi ewes obtained by weighing lambs before and after suckling averaged 59.3 kg, while two 8-week lactation studies had a weighted average of 55.4 kg (Boujenane 2002). The average milk yield at 10 weeks of lactation by the oxytocin method was 68.5 kg. Suckling in combination with hand milking prolonged the lactation period to 20 weeks with an average milk yield, dry matter content, fat content and protein content of 112.7 kg, 16.9%, 4.6% and 5.8%, respectively. Average fleece weight was estimated at 2 kg from four studies, while one of the studies recorded 6.43 cm for staple length (Boujenane 2002).

6.5.2.8 Beni Guil

Distribution, Classification and Population: The Beni Guil (Fig. 6.13) is one of the main Moroccan local breeds included in the National Programme for Genetic Improvement since 1980. The Beni Guil breed is found located around the Oriental plateaus, namely, Tendrara, Bouarfa and Figuig. Its name comes from the Beni Guil tribes who own this breed. Its population was estimated at 1.2 million in Morocco in 2018.

Physical Characteristics: The Beni Guil (Fig. 6.13) is a medium-sized sheep of 60–65 and 40–50 cm height at the withers for males and females, respectively (Boujenane 2002). Phenotypically, it has white fleece, brown limbs, a brown large face and belly (Belhaj et al. 2021), resembling Beni-Ahsen. It has a rectangular shape, a short neck and a fleece of medium fineness. The breed is thin-tailed, and rams have fairly well-developed spiral horns (Boujenane 2002).

Adaptive Characteristics: The Beni Guil breed adapts well to the poor vegetation of the Oriental plateaus (Tendrara, Bouarfa and Figuig) and can acclimatise in other regions of Morocco.

Breeding Practices and Production Characteristics: Beni guil ewes are seasonally bred between May and September to lamb between January and February (Boujenane 2002). Review of reproductive traits reported by the author includes age at first lambing: 25.3 months; gestation length: 149.3 days; mean fertility: 85.9%; prolificacy: 109%; 90-day litter weight: 16.5 kg; single litter: 94.1%; twin litter: 5.9%; lamb survival from birth to 90 days: 94.3%; average birth weights: 3.32 kg for males and 3.14 kg for females; 90-day weight: 16.1 kg and 14.8 kg, for males and females, respectively; adult body weights: 70–90 kg and 45–50 kg, respectively, for males and females; average daily gain and feed conversion for lambs: 215 g and 5.5 kg feed/kg weight-gain, respectively; average slaughter weight at 160 days: 29.1 kg; carcass weight: 13.6 kg and dressing percentage: 47.7%.

Beni Guil ewe milk yields at 8, 12 and 20 weeks of lactation were estimated at 40.3 kg, 52.3 kg and 97.9 kg, respectively, while mean dry matter, fat and protein contents at 12 and 20 weeks were 19.5 g, 16.4 g and 5.4 g, and 8.4 g, 4.3 g and 5.6 g, respectively (Sefiani 1980; Battar 1983; Boujenane et al. 1996). These studies used the method of weighing lambs before and after suckling to estimate yield. Various studies have shown an average fleece weight of 1.70 kg. An average staple length of 7.12 cm has been reported (Abdelali 1988).

6.5.2.9 Tadla

Distribution and Classification: The Tadla is a Moroccan sheep breed that is maintained in its pure state on the plateaus of Kasba-Tadla, Oued-Zem and El Borouj. The breed is part of the concave, arched, sublongilinear, eumetric type, with white fleece and pigmented ends.

Physical Characteristics: It is a fairly large sheep, ranging between 85 and 90 cm in height at the withers. The breed has a strong frame (9.5 cm around the barrel) and a heavy head bearing large and powerful horns that sometimes weigh up to 500 g each. The neckline is long, thick and has a slight dewlap, and the top is quite wide. The thigh is muscular and slightly flat. The adult average weight for rams ranges between 75 kg and 85 kg. Lambs at 18–20 months old weigh between 50 kg and 55 kg. Birth weights as high as 6 kg have been reported. The cream-white fleece is closed, and the locks are square and tight. The sheep produces a pale yellowish-white coloured wool that is long, fine and wavy strands with an average staple length of between 7 cm and 9.5 cm and an average shearing weight of 1.5–1.6 kg and sometimes reaches 3.8 kg (Ben Lakhal 1995).

The Béni-Meskine sheep, a variety of the Tadla breed, is distinguished from the Tadla breed by having a less fine fleece and a better conformation. This breed is distinguished from the Sardi (previously described) by its white pigmented head with black around the nose and around the eyes.

6.5.2.10 Arrit

Origin and Distribution: The Arrit breed, an indigenous breed from Eritrea, is thought to be an intermediate between the Sudan Desert and Ethiopian Highland sheep reared by the Arab tribes located in the Keren region in Northern Eritrea (Mason and Maule 1960). It phenotypically resembles the Dongola sheep of Sudan. It is partially hairy with some coarse wool and has a seasonal accumulation of fat at the base of its tail, suggesting its origin may be related to the woolled thin-tailed sheep breeds of North Africa and the Sahara, as well as the fat-tailed sheep of the East African highlands.

Physical Characteristics: The breed is typically polled; however, some rams have open spiral horns. The breed is medium-sized with semi-pendulous ears. The coat is white. The facial profile is nearly straight, and the tail is thin, long and straight, extending to below the hocks.

6.5.2.11 Saidi

Location: The Saidi sheep is found in Upper Egypt, especially in the Assiut governorate.

Physical Characteristics: The body is black (Fig. 6.14) or dark brown (Fig. 6.15), but sometimes creamy white and black and white individuals are found. The head is large with a curved

Fig. 6.11 Berbere sheep of the Bouhadjar mountains (El-Tarf), Algeria. (From Djaout et al. 2017, licensed under CC-BY 4.0)

nose. The ears are medium-sized. The head is covered with wool. The neck is long, and both sexes are polled. Saidi has a long, cylindrical, fat tail.

Production Characteristics: Mature body weights for rams and ewes average 50 kg and 35–45 kg, respectively. Birth weights range from 2.5 to 3.1 kg, while weaning weights average 13 kg. The mean average daily weight gain from birth to weaning is 104 g, and the post-weaning average daily growth rate is 88 g. Fertility rates for ewes are from 4.5% to 7.0%, and average mortality rates from birth to weaning range from 10% to 18.2% (Elshazly and Youngs 2019).

6.5.2.12 Farafra

Origin and Distribution: The Farafra sheep (or Farafra Oasis sheep) (Fig. 6.16) is an indigenous Egyptian sheep found in the El-Farafra Oasis located in the desert region of Western Egypt, where average temperatures range between 39.5 °C and 41 °C and rainfall is low. It is thought to have originated by crossing Ossimi and Desert sheep breeds such as the Barki breed.

Physical Characteristics: The Farafra sheep breed has a body covered with white fleece (Elshazly and Youngs 2019). The head is brown in colour (Fig. 6.16), both sexes are polled, the tail is long and cylindrical, the average weight at weaning is 15 kg, ewe mean fertility rate is 90%, and the percentage of twin births ranges from 30% to 60% (Elshazly and Youngs 2019).

6.5.2.13 Dongola

Distribution: Synonyms for the Dongola breed include North Riverine wooled sheep and Riverain Wooled. It is a breed from a town in Northern Sudan, located on the banks of the River Nile, from where it gets its name. The breed is kept by the Sudanese Dongola Arabs, renowned as great merchants and travellers.

Physical Characteristics: The Dongola sheep are small sheep, with an average withers' height of 57–65 cm and an average adult weight of 13–32 kg. They are among the most fertile Sudanese sheep. The breed is covered with a top coat made up of long, sparse hairs and a woolly undercoat, similar to that found in the Karakul (discussed in Sect. 6.4.6.1) (Maisonneuve et Larose 1993).

6.5.3 Fat-Tailed Hair Sheep of Africa

The present-day African fat-tailed sheep are the most populous sheep with distribution across North Africa from Egypt to Algeria and from East Africa to Southern Africa. The fat-tailed sheep are divided into two subtypes; the first group comprises the East African fat-tailed breeds made up of the East African short-fat-tailed sheep and East African long fat-tailed sheep, and the second group comprises the African long-fat-tailed breeds (Rege et al. 1996).

The African long-legged sub-type of the African long fat-tailed is found mainly in arid and semi-arid zones of West Africa, Chad, central Africa and the eastern African country of South Sudan. These sheep are characterised by a convex nose, pendulous ears and very long legs that give the sheep a height of over 60 cm, an adaptation to the dry environments they are found in and the pastoralist lifestyle of their owners (Wilson 1991; Ryder 1983).

6.5.3.1 East African Fat-Tailed Sheep

Afar

Origin and Distribution: Synonyms for the Afar breed include Adal, Danakil and Elle. The Afar sheep is believed to have entered the African continent through the Horn of Africa from the Arabian peninsula (Muigai and Hanotte 2013). They are found mainly in northern Ethiopia, in the Afar State that borders Tigray, Amhara, eastern and western Harerghe and eastern Shoa regions of Oromia and are kept mainly by the Afar, Amhara and Tigray communities. The environment where these sheep are reared can be described as arid and semi-arid in mid-highland (1200–1900 m) to lowland regions (elevations less than 1000 m). The sheep are therefore reared in pastoral/agropastoral production systems (Gizaw et al. 2008a).

Physical Characteristics: The Afar is a medium sheep (Fig. 6.17), with mature weight

Fig. 6.12 Hamra Ewes in Mecheria City, Nâama Province, Algeria. (From (Djaout et al. 2017, licensed under CC-BY 4.0)

ranging from 30 to 35 kg. The sheep is characterised by a wide (16 ± 9 cm) fat-tail, which is long (19.1 ± cm) and has a slight 'S' shape that reaches below the hock (Gizaw et al. 2013). The large tail size qualifies its characterisation as a fat-rump sheep. However, it has been argued by Gizaw and others (Gizaw 2008) that they are fat-tailed sheep and ought to be classified as such. The body of the Afar sheep is covered with short hair fibre (3.2 ± 3 cm long) with a predominant cream or white colour (Figs. 6.17 and 6.18). The average measurements in females are: withers height—63.6 ± 8 cm; body length—58.3 ± 8 cm, heart girth—73.2 ± 6 cm and substernal height—35.6 ± 6 cm. The head is small, with a slightly convex profile in males, while females bear a straight profile. Pads of fat are observed on the nose and behind the poll. The ears are rudimentary in 80% of the animals; however, in a few animals, short ears of approximately 3.8 ± 4 cm are observed. Both sexes are polled, and about 5% of the animals bear toggles. The neck is short, often with a prominent dewlap. The chest is shallow and narrow. The tail head is higher than the withers (Wilson 1991; Gizaw 2008).

Adaptive and Special Genetic Characteristics: The Afar sheep breed is a hardy animal, adapted to the drought-prone arid and semi-arid regions of the Afar region of Ethiopia. Rainfall in these regions is erratic, with annual precipitation patterns ranging from 300 to 700 mm. The breed has developed adaptations to the harsh environmental conditions, including high temperatures, poor-quality feed and shortage of water. The breed has long legs that enable it to trek long distances in search of water.

Typical Production Systems and Production Characteristics: The Afar sheep is kept under pastoral and agropastoral systems mainly for meat production. The farmers, however, also milk the sheep and consume the milk at the household level. The sheep are milked an average of two times a day, producing an average of 224 ml of milk per day. The lactation average length is 3.8 months with a lactation yield of 25.5 litres per animal (Gizaw et al. 2013). The Afar lambs are once a year in traditional farming systems with an average litter size of between 1.03 and 1.14. Birth weight is 2.40 ± 0.6 kg and has an approximate weight of 11.9 ± 0.07 kg and 16.8 kg at three and 6 months, respectively. Adult males have a mature weight of 54 kg. The animals are kept for meat, milk and hides (DAGRIS 2007; Gizaw 2008). The agro-pastoralists and the pastoralists in the Afar region of Ethiopia mainly keep the Afar as a source of income and to provide milk for the household. A further reason for keeping the Afar aside meat is as a source of savings (Gebre et al. 2018). During emergencies, rams under the age of 6 months are the first choice for agro-pastoralists and pastoralists to take to the market for sale. Due to high demand, young rams were easily disposed of. Culling and sales of females are rarely practised, even for animals with unwanted coat colour, because most females in flocks are retained for breeding (Gebre et al. 2018).

Washera

Distribution, Classification and Population Statistics: The Washera breed (Fig. 6.19), which is also known as Agrew and Dangilla, is classified as an alpine short fat-tail sheep. They inhabit the West and East Gojam and Agew Awi zones of Amhara State in Ethiopia. They are approximately 1,227,200 head and inhabit terrain that can be described as wet and warm mid-highlands with altitudes of 1600–2000 m above sea level. They are kept by the Amhara and Agaw communities in (Gizaw et al. 2008a).

Physical Characteristics: Washera sheep have a short, fat tail (Fig. 6.19) with long ears that are approximately 10.6 ± 1 cm long. The body is large, with a withers height of 69.4 ± 3 cm for adult females and a body length of 66.7 ± 5 cm. The heart girth is 74.1 ± 6 cm and the substernal height is 38.6 ± 3 cm.

Adaptive and Special Genetic Characteristics: The Washera breed is tolerant to gastrointestinal helminth infections (Getachew et al. 2015).

Typical Production Systems and Production Characteristics: The Washera sheep are kept in the highlands by mixed crop-livestock farmers growing cereals. The animals are kept mainly for meat, manure and skin production. The production system is a semi-intensive to low-input production system (Gizaw et al. 2008a). The sheep are prolific animals with an average litter size of 1.8. Adult females have an approximate weight of 32.8 ± 9 kg (Gizaw et al. 2008a).

Horro

Origin, Distribution, Classification and Population Statistics: Synonyms for the Horro breed (Fig. 6.20) include Abyssinian, Ethiopian, Bonga and Wollaga. The Horro breed is part of the Ethiopian group of sheep whose origin is believed to be Arabia via the Arabian Peninsula (Muigai and Hanotte 2013). The breed is classified as a long, fat-tailed hair breed. The sheep is the most widely distributed in the western and southern parts of Ethiopia, from the humid mid-highlands (1600 m) to the cool wet highlands (2991 m) in regions with high agricultural production potential. Relative extinction probabilities calculated for Horro indicate that the breed is relatively safe and not at risk of extinction. In 2008, 3,409,300 head were reported (Gizaw et al. 2008a). The breed is kept by the Oromo, Benishangul and Gambella communities found in Ethiopia.

Physical Characteristics: The Horro is a large breed (Fig. 6.20), with a withers height of 70.6 ± 6 cm and a substernal height of 38.1 ± 4 cm. The average female body length is 71.6 ± 6 cm with a heart girth of 76.9 ± 8 cm. Horro sheep have a long fat-tail that extends to below the hock (Fig. 6.20). The tail (average length 35.6 ± 6 cm, width 9.9 ± 3 cm) is straight in 51.4% of the population and coiled or twisted with a tapering at the end in 48.6% of the population. The majority of the population is brown-coloured or fawn, but black-, white- or fawn-coloured animals are also found. Ears are present and average 10.8 ± 2 cm. The body is covered with short, smooth hair that is on average 2.6 ± 1 cm. The females' belly is usually lighter in colour, especially in adult ewes, and both sexes are polled (Gizaw et al. 2008a).

Adaptive and Special Genetic Characteristics: The breed is well adapted to the regions where it is reared, producing good-quality mutton. Its habitat is located on the fringe of the trypanosome-infested environment, although the degree of trypanotolerance in Horro is not known (Gizaw et al. 2008a).

Typical Production Systems and Production Characteristics: The Horro breed is reared by sedentary mixed crop-livestock farmers mainly for meat (Edea et al. 2017). The greatest asset of this breed is its large size, which makes it popular among households that keep animals for meat. The live weights recorded in yearling males and females are 31.2 ± 1.70 kg and 26.5 ± 0.31 kg, respectively. Birth weights average 2.7–3.0 kg; weaning weights 11.8–13.5 kg; and 6-month weights are 15.4 kg–17.9 kg. The Horro breed is among the topmost prolific of the Ethiopian breeds, with a twinning rate of 1.29 ± 0.08. On average, households keep approximately 12 heads per household (Gizaw et al. 2008a).

Adilo

Distribution and Population Statistics: The Adilo sheep (Fig. 6.21) is described as a highland long fat-tailed sheep reared in the warm and wet mid-highland regions of the Southern State and northern Borena districts of Ethiopia. The sheep were approximately 407,000 heads in 2008 (Gizaw et al. 2008a).

Physical Characteristics: The Adilo is a large sheep (Fig. 6.21) that averages 28.1 ± 5 kg in mature weight, withers height of 65.5 ± 4 cm, substernal height of 35.8 ± 5 cm, body length of

Fig. 6.13 Beni Guil ram of Morocco. (From Bechchari 2009, published in an open-source report published by the Ministry of Agriculture, Kingdom of Morocco)

62.1 ± 5 cm and a heart girth of 71.8 ± 6 cm. It has a long and wide fat-tail that reaches the hocks, with an average length of 28.1 ± 1 cm and a width of 6.7 ± 3 cm. The males have short horns, while the majority of the ewes are polled, with 18.4% bearing horns. The predominant body colour is brown (94.3%) (Fig. 6.21), brown with white patches (32%), black (19%) and black with brown patches (9%). The body is covered with short hair (4.4 ± 2 cm long). The ears are long and average 11.7 ± 5 cm (Gizaw et al. 2008a).

Adaptive Characteristics: The Adilo sheep are adapted to the wet highlands of Ethiopia, which range between 1800 and 2000 m and produce high-quality mutton under these conditions.

Typical Production System and Production Characteristics: The Adilo sheep are reared by small-scale crop-livestock farmers on natural pastures and crop residues. Breeding is not controlled, and rams are left free in communal grazing areas. The sheep are kept in small flocks, which are tethered to a tree if they are one or two animals or left free during the day. During the cropping season, the animals are tethered in grazing areas. At night, they are housed in small enclosures. The average mean birth weight of Adilo lambs is 2.30 ± 0.03 kg and the mean average weight at 150 days is 15.7 ± 0.20 kg.

Bonga

Distribution and Population Statistics: The Bonga (Fig. 6.22) sheep, whose synonyms are Gesha and Menit, are native to the Kaffa region in Southwestern Ethiopia and in the Keffa, Sheka and Bench zones of the Southern State of Ethiopia (Gizaw et al. 2008a; Mamiru et al. 2018). Approximately 517,500 heads were reported in 2008 (Gizaw et al. 2008a).

Physical Characteristics: The Bonga sheep are large sheep (Fig. 6.22) with an average adult body weight of 34.2 ± 8 kg and a withers height of 66.7 ± 6 cm. The heart girth is 73.5 ± 7 cm, with a substernal height of 36.4 ± 4 cm. Bonga sheep has a long fat tail (25.9 ± 9 cm in length and 8.1 ± 3 cm in width) that hangs just at the hocks or below the hocks and tapers towards the end in a straight (67% of the population) or twisted manner (33% of the population). Most of the sheep are brown in colour (57.9%), though some white (10.5%) or black and white (0.9%) individuals can be found. The body is long (69.4 ± 5 cm) and is covered with short hair (~ 2.9 ± 1 cm in length). The ears are short (9.8 ± 2 cm in length), and both sexes are polled. Wattles are present in only 5.1% of the population (Gizaw et al. 2008a).

Adaptive Characteristics: The Bonga sheep are adapted to the humid mid-highland zone that ranges between 1200 m and 2500 m (Edea 2008).

Typical Production Systems, Husbandry Practices and Production Characteristics: The Bonga sheep are predominantly kept for meat production (Mamiru et al. 2018). The age at first lambing is 14.9 ± 3.1 months, with an average lambing interval of 8.9 ± 2.1 months. The average litter size is 1.40 with an average reproductive life span of 7.9 ± 3.1 year (Zewdu 2008).

Red Maasai

Distribution and Population Statistics: The Red Maasai sheep (Fig. 6.23) is commonly referred to as Maasai. It is predominantly owned by the Maasai ethnic group of East Africa, though in the past it was also kept by other Eastern African ethnic groups, such as the Nandi and the Abaluhya ethnic groups of Kenya (Wilson 1991). The Red Maasai sheep are distributed throughout northern Tanzania and south-central Kenya (Wilson 1991). Indiscriminate cross-breeding with breeds such as the Dorper has resulted in dilution of the breed. However, the breed has not been reported as threatened to date.

Physical Characteristics: The Red Maasai sheep is a large animal with an average height of 72 cm and 66 cm in males and females, respectively. The tail is long and fat, especially in sheep that are in good condition. The coat is red-brown with relatively long, smooth, coarse hair that can be up to 4 cm long (Fig. 6.23). Most animals have a woolly undercoat. The profile is convex in males and straight in females. The forehead is short and broad, and both sexes have horns that are up to 27 cm in length. The ears are semi-pendulous and are medium in length, 11–15 cm, although vestigial ears are present in approxi-

Fig. 6.14 A Saidi ram. (From Elshazly and Youngs 2019, licensed under CC-BY 4.0)

Fig. 6.15 A Saidi ewe. (From Elshazly and Youngs 2019, licensed under CC-BY 4.0)

mately 7% of the population. Toggles are present in approximately 15% of the population, and their size and position vary. The neck is short with a pronounced dewlap. The chest is narrow and shallow. The brisket is pronounced with a dewlap that carries fat. The average heart girth is 75.1 cm and 63.3 cm for males and females, respectively. The legs and back are short with a sloping croup (Wilson 1991; DAGRIS 2007).

Adaptive and Special Genetic Characteristics: The breed is adapted to arid and semi-arid regions. It is resistant to gastrointestinal nematodes and shows higher resistance than the Dorper, Blackhead Somali and Romney Marsh breeds (Wilson 1988; Baker et al. 2002). It is also resistant to bluetongue virus infection (Davies 1987) and has survived under high trypanosome challenge (Wilson 1991).

Typical Production Systems, Husbandry Practices and Production Characteristics: The sheep is raised within pastoral and agropastoral production systems and herded in mixed herds usually with goats and more recently with the Dorper breed. The animals move around during the day with their keepers and are contained within makeshift enclosures made from thorny shrubs. In these pastoral production systems, the flock structure is made up of 65% females and 31.5% males (Wilson 1991). First lambing is reported at 549 ± 112.1 days on a group ranch in Kenya, which is quite late and can be explained by the fact that the Maasai community uses leather aprons on males to control breeding. Multiple births are rare. The litter size is approximately 1.05 ($n = 1009$) in a group ranch in Kenya with an annual reproductive rate of 1.22. Birth weights are approximately 2.7 kg with an average daily gain of 73 g between 0 and 150 days in the pastoral production system. Mature weights are 21–47 kg and 32.55 ± 4.55 kg in males and females, respectively. The breed is kept mainly for meat, skin and lard (Wilson 1991).

6.5.3.2 African Long Fat-Tailed Sheep

The African long fat-tailed sheep comprises the East African long fat-tailed, which includes the Tanzanian long-tailed and Rwanda-Uruadi breeds, and the southern African long fat-tailed breeds, which comprise, among others, the Afrikander, Damara, Namaqua Afrikaner, Nguni, Tswana and Pedi breeds (Table 6.3) (Rege et al. 1996).

The Tanzanian Long-Tailed Sheep

Distribution: The Tanzanian long-tailed sheep are also known as Tanganyika long-legged, Ugogo, Gogo, Rwanda, Burundi and are widely distributed in the arid and semi-arid regions of Tanzania, Democratic Republic of Congo, Rwanda, Burundi and western Uganda.

Physical Characteristics: The Tanzanian long-tailed sheep are medium-sized sheep (Fig. 6.24) with an average adult weight of

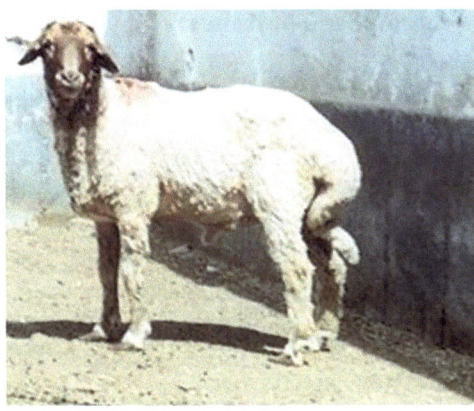 

Fig. 6.16 Farafra ram (left) and ewe (right) found in the El-Farafra Oasis, Western Egypt. (From Elshazly and Youngs 2019, licensed under CC-BY 4.0)

25–40 kg and an average height of 55 cm–65 cm. They have a thick, long, fat tail that tapers at the end. The head has a convex facial profile and is broad and short. Males have fat pads on the nose. Both males and females are polled. The ears are pendulous, hanging towards the front and are medium-sized in length. The body colour is black with white patches or brown and white patches. The body is covered with short or medium-length hair, with males having a mane and cape at the shoulders and a long apron of hair from the throat down the chest to the brisket (Wilson 1991).

Adaptive Characteristics: The Tanzanian long-tailed sheep are adapted to the dry conditions in which they are found. The thick, long, fat tail makes these sheep cope with seasons where there are critical food shortages (Devendra and McLeroy 1982).

Typical Production System and Production Characteristics: The sheep are maintained by small-scale crop-livestock farmers in mixed herds consisting of goats and cattle in an extensive livestock production system where the sheep are grazed in communal grazing land and kept in outdoor Kraal enclosures during the night (Nziku et al. 2016). The sheep are mainly reared for meat and income generation at the household level (Devendra and McLeroy 1982). Age at first lambing averaged 714 days on a station in Rwanda with an average lambing interval of 413 days (n = 863). Birth weights averaged 2.6 ± 0.02 kg on the station in Burundi (Wilson 1991).

Fig. 6.17 Afar ram in the Afar State of Ethiopia. (From Getachew 2008, through ILRI repository; CGSPACE, an open-source repository for Agricultural Research Output)

The Rwanda-Urudai

Origin, Distribution and Classification: The Rwanda-Urudai sheep are part of the East African long-fat tailed group. These sheep are distributed throughout Rwanda, Burundi, the Democratic Republic of Congo, southwestern Uganda and north-west Tanzania. In Rwanda, it is taboo to consume sheep meat, though the hides are used in the homesteads. This contributed to the rapid decline in the numbers of this breed and the breed is now at risk of extinction (Wilson 1991; Ndayambaje 2016).

Physical Characteristics: The Rwanda-Urudai sheep are small-bodied sheep, averaging 55–65 cm, with an adult average weight of 45 kg (male) and 35 kg (female). The head is convex, the forehead is broad and short. Males have pads of fat behind the nose. Both sexes are polled. Ears are medium-sized and pendulous, with occasional vestigial ears. The tail is long and fat with a tapering end. The neck is strong and long. The chest is well-rounded with an average circumference of 72 cm. The legs are long and not very fleshy. The sheep are predominantly white and black in colour. The body is covered with long hair that is either stiff, wavy or fine. Males may have a mane and cape over the withers and shoulders and an apron of long hair from the throat down the chest to the brisket (Wilson 1991).

Typical Production System and Production Characteristics: The Rwanda-Urudai sheep are kept in mixed herds comprising cattle and sheep by crop-livestock farmers in high-altitude regions. The age of first lambing is 714 ± 18.4 days, and the lambing interval is 406 ± 7.5 days, with an average of 1.14 litter size in the traditional system and 1.43 on station in Rwanda. Multiple births are not common. For lifetime production, average parturitions are 2.29 in traditional systems, few ewes exceed 5 parturitions on station in 643 ewes (Wilson 1991). Average birth weights are 2.5 ± 0.56–2.6 ± 0.02 kg on station farms, with average weights of 6.3 kg and 31 kg in 30 days and 365 days, respectively (Wilson 1991).

Fig. 6.18 Afar ewe in the Afar State of Ethiopia. (From Getachew 2008, through ILRI repository; CGSPACE, an open-source repository for Agricultural Research Output)

The Ronderib Afrikaner

Origin, Distribution, Classification and Population Statistics: Like most indigenous South African sheep, the Afrikander originated from the Middle East and migrated into east Africa before migrating in a southerly direction with the KhoiKhoin people. The sheep first arrived in South Africa and eventually migrated to the west coast at the southernmost part of the African continent. Ronderib Afrikaner has a multicoloured coat and it is also known as Cape fat-tail, Afrikaner or Africaner. However, Cape Dutch farmers successively selected against the multicoloured coat, resulting in single-coloured animals that are prevalent today. The name 'Ronderib' is a reference to the cross-section of the ribs that are oval in shape (Bisschop et al. 1954). Two varieties are recognised, the Blinkhaar Ronderib Afrikaner, with soft shiny white hair and the Steekhaar Ronderib Afrikaner, which has coarse hair. The breed is distributed in South Africa and Namibia. The Steekhaar Ronderib Afrikaner is believed to be extinct (Epstein 1960; Snyman 2014a).

Physical Characteristics: Ronderib Afrikaner is a large fat-tailed sheep with adult live weights of 60–80 kg and 35–60 kg for rams and ewes, respectively (Epstein 1960; Ramsay et al. 2001). The fat tail is large, having a broad upper part with the same width as the rump hanging down to about three inches above the hocks, weighing between 2 and 3 kg and can go up to 5 kg (Snyman 2014a; Ramsay et al. 2001). The animals have large heads with a wide muzzle. The neck is long and narrow, with a prominent crown composed of fat and muscle tissue at the junction of the head and neck. The dewlap is well developed and extends to the brisket, which is prominent. The chest is well developed and capacious. The animal is tall with a withers height of 62.8 cm and 66.8 cm for rams and ewes, respectively and a height at hook bones of 69.4 cm and 68.4 cm for rams and ewes, respectively. The length of the body is 66.5 cm and 68.2 cm for rams and ewes, respectively. The heart girth is 99.7 cm and 89.2 cm for rams and ewes, respectively. The animals have long, thin legs. The convex forehead and Chaffron are separated by a slight depression

Fig. 6.19 Washera ram (left) and ewe (right) in the Amhara state in Ethiopia. (From Ferede et al. 2014, licensed under CC-BY 4.0)

in the interorbital region. The face is long, the eyes large, with ears that are fairly long, narrow, pointed and slightly pendulous. The rams have moderately sized horns that are spiral with distinct transverse wrinkles and grow backwards in a spiral that is close to the head. Only a small portion of the rams are polled and these animals have a more rounded head. Ewes are generally polled. The length of the rump is 23.5 cm and 22.6 cm in rams and ewes, respectively (Epstein 1960). The face and legs are covered with short, glossy, soft hair, and the body is covered with a glossy fleece of fine, smooth cream wool, approximately 3–6 cm long. The body is generally white, while the face is brown or reddish brown in colour (Epstein 1960; Snyman 2014a; Ramsay et al. 2001).

Adaptive and Special Genetic Characteristics: The Ronderib Afrikaner is a hardy sheep that can survive extremely adverse dry conditions. It produces a heavy carcass with flesh that has been described as stringy and coarse (Epstein 1960). It is well adapted to the dry, arid and semi-arid regions of southern Africa. It is also heat-tolerant and tolerant to parasites. The breed has long legs that enable it to walk long distances. It has a good foraging ability (Epstein 1960; Snyman 2014a).

Production Characteristics: The breed has high fertility, with the ram being active all year round. The ewes are good mothers. The age at first lambing is 8 months. The purebred animals are slow-growing and are best slaughtered at

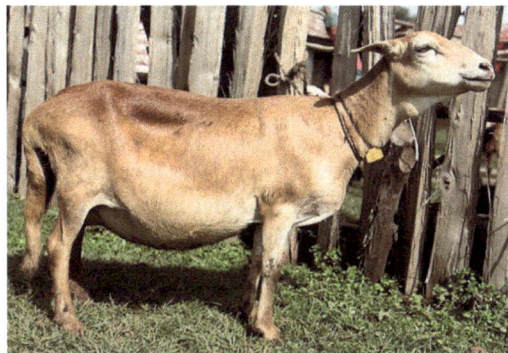

Fig. 6.20 Horro ewe; Photo: ILRI/Apollo Habtamu. (Available on Flickr for free use under the Creative Commons Attribution Licence. (Image can be viewed here))

10–12 months when they produce a good-grade carcass. The breed is kept for meat, which is high in fat. The fat-tail is usually in high demand for use as an additive in game meat and to make sausages. The Blinkhaar Ronderib Afrikaner has been used to develop several South African breeds, including Vandoor, Van Rooy and Afrino (Epstein 1960; Snyman 2014a).

Damara

Origin, Distribution and Classification: The Damara breed has its origins in the Middle East and migrated first into east Africa and then migrated southwards into the southern African region, into present-day Namibia and Angola. Classified as an African long fat-tail sheep, these sheep are native to the Damaraland region, a region known as 'The Koakoveld' in northern Namibia, where they were herded by the Himba and Tjimba ethnic groups for many years with little external influence. Himba have been hailed as one of the most successful pastoralist communities to date (Ramsay et al. 2001; Du Toit 2008). The breed is distributed in South Africa, Botswana, Namibia, Mozambique and Zimbabwe. The Damara breed is stable and thriving with several commercial farmers in South Africa and Namibia (Du Toit 2008).

Physical Characteristics: The Damara sheep are large animals with long legs. Both sexes have spiral horns that stand away from the head. The head has a strong profile, predominantly Roman-shaped. The tail is long and fat, reaching below the hock and gradually tapering down to a thin and pointed end. The eyes are large, bright and brown in colour. The teeth are hard and strong. The body is covered with short hair, though it grows longer during the winter and sheds during the warmer months. The body is long, oval and fairly deep. The body colour is predominantly brown, though white, black, grey and spotted multicoloured animals are also found. The top line is concave, dips to the neck, hollow, rises again, over the shoulder, dips into the back and rises again over well-developed loins. The legs are long, strong with well-shaped hooves. The dewlap is present and prominent in females (Snyman 2014b).

Adaptive and Special Genetic Characteristics: Damara sheep are adapted to the arid and semi-arid regions of southern Africa and can go for long periods without water. They can forage on grass, bush and shrub and survive and produce under very stressful natural conditions. The sheep have a high temperature tolerance and are very resistant to several sheep diseases, especially heartwater disease (Cowdriosis), parasites, intestinal worms, flies and blowflies. This enables the breed to be farmed where water and forage are scarce. The ewes have a strong maternal instinct and make good mothers, and low mortality rates of 2–4% are reported. The herd has high flocking instincts, with the mothers always gathering in a group that encircles the lambs, thus protecting them from wild animals. The sheep also easily adapt to other environments (Ramsay et al. 2001; Almeida 2011).

Production Characteristics: These sheep require minimum care and give high returns from the sale of meat and skins. The Damara sheep has a high fertility of 89–95%. The rams are active throughout the year, the females have a long productive life, and lambings occur on average every eight months (Du Toit 2008). The lambing intervals are short, and ewes can be mated throughout the year. The age at first lambing ranges between 14.6 and 17.3 months. Birth weights are 4–4.5 kg and 3–4.2 kg for males and females, respectively, with twinning rates reported at 35%. The weight at 100 days is 25 kg and 22 kg, with a mature weight of 60 kg and 45 kg for males and females, respectively. The weights reported at six and 12 months for both sexes are 32.3 kg and 46.6 kg. The Damara are kept for meat, the fat tail and

Fig. 6.21 Adilo ewe. (Photo: ILRI\Zerihun Sewunet) (Available on Flickr for free use under the Creative Commons Licence (Image can be viewed here))

skin (Schoeman 2000). The meat has a small amount of fat (1–2 mm) and is considered to be quite tasty, juicy with a lot of flavour. The tail is fatty, high in quality and is an ideal ingredient in sausage. The skins are used to produce very high-quality leather (Ramsay et al. 2001; Almeida 2011).

Namaqua Afrikaner

Origin, Distribution and Population Statistics: The Namaqua has its origins in the Middle East and migrated southwards into east Africa and then into southern Africa with the Khoisan people. The Namaqua breed is an endangered sheep breed with an estimate of only 2000 animals in 2014. Indiscriminate cross-breeding is the main reason why the animal is endangered. Two flocks of 100 ewes are currently being maintained in government-managed stations in South Africa (Snyman 2014c).

Physical Characteristics: The Namaqua breed is a tall animal with a narrow body and long, lean legs that enable it to walk long distances in search of water. The tail is fatty, weighing between 8 kg and 12 kg and stores up to 38% of the animal's fat. The tail is also long, reaching up to the hocks and bearing a distinctive twist at the end. The body is white with a black or red head. The hooves and horns are black. The body is covered with short, smooth and shiny hair that grows during the winter and sheds during the warmer months (Ramsay et al. 2001; Snyman 2014c).

Adaptive and Special Genetic Characteristics: The Namaqua sheep is a hardy and prolific animal with high reproductive performance even when reared in severe and stressful environments. The Namaqua ewes can mobilise their fat reserves to wean heavy lambs even under extreme drought conditions. These ewes have excellent mothering instincts and defend their lambs fiercely against predators. The breed is adapted to extremely hot, dry desert conditions with intense infrared radiation during the day and extremely low temperatures during the night. Namaqua sheep are able to walk long distances in search of water and can reproduce under low-quality grazing conditions.

Production Characteristics: The Namaqua is a breed that requires little care. The average birth weights for rams and ewes, respectively, are 4.55 ± 0.33 kg and 4.27 ± 0.03 kg. The 100-day weaning weight is 26.1 ± 0.2 kg and 24.7 ± 0.2 kg for rams and ewes, respectively. The average weight gain from birth to 100 days is 215.4 ± 1.6 g/day and 204.8 ± 1.7 g/day, respectively, for rams and ewes, while between 100 days and 12 months, it is reported at 51.9 ± 0.2 kg (rams) and 44.0 ± 0.2 kg (ewes). The weight at 18 months is 58.7 ± 0.3 kg (rams) and 50.4 ± 0.03 kg (ewes). Conception rates of $86.0 \pm 2.0\%$ and fecundity rates of $156.4 \pm 1.5\%$ were recorded for a flock maintained on a private farm. The breed is farmed for its meat and skin. The skins produce high-quality leather, which is graded as Glover skins and is used to make high-value leather products such as jackets and gloves. The fat tail is prized in the meat processing industry, where it is used as a meat additive. The meat is, however, considered to be quite fatty and thus the carcass is not considered to be high-grade meat (Ramsay et al. 2001; Snyman et al. 1993).

Nguni

The Nguni sheep constitute four ecotypes, namely the Zulu, Pedi, Swazi and Landim sheep (Ryder 1983). They originated in the Middle East, from where they migrated into east Africa. They are believed to have migrated southwards

Fig. 6.22 A Bonga ram. (Photo: ILRI/Apollo Habtamu) (available on Flickr for free use under the Creative Commons Licence (Image can be viewed here))

Fig. 6.23 Red Maasai sheep, Photo: ILRI. (Available on Flickr for free use under the Creative Commons Licence (Image can be viewed here))

into the southern African region with the Nguni people, following the eastern coastline of South Africa. The Nguni people separated and dispersed in different directions, each taking sheep along with them. One group, the Zulu people (after their leader 'Shaka Zulu'), settled in the plains of KwaZulu-Natal and then dispersed further south in search of suitable crop-growing areas (Ramsay et al. 2001). Their sheep are known as the Zulu sheep. Another group of sheep migrated with the Bapedi people to Sekhukhuneland, North East South Africa, in present-day Limpopo and Mpumalanga regions and became the present-day Pedi sheep (Ryder 1983). The third group of sheep migrated into present-day Eswatini (Swaziland) and are known as the Swazi breed (Kunene et al. 2007). Each of these four ecotypes is discussed below.

Zulu

Distribution and Population Statistics: The Zulu sheep are distributed throughout KwaZulu-Natal Province of South Africa, with the northern part of the province having larger numbers of sheep than the southern regions (Kunene et al. 2009; Mavule 2013). Despite its rich history and deep association with the Zulu, the breed is threatened with extinction (Kunene et al. 2009) due to replacement and indiscriminate cross-breeding with non-indigenous breeds (Mavule et al. 2013). Farmers' perceptions of its small size have led to perceptions that the breed is not favoured by consumers, leading to its threatened status (Mavule 2013).

Physical Characteristics: The Zulu sheep are medium- to small-bodied, average live weights for rams and ewes are 30.88 kg and 27.20 kg, respectively (Mavule et al. 2016). The average adult heart girth for rams and ewes is 73.62 cm and 71.75 cm, respectively. The sheep can have two types of tail phenotypes, a long, fat, carrot-shaped tail found in 39.5% of the population or a long, thin tail averaging 33.7 cm. Their body is multicoloured, with dark brown (26.6%) and dark brown with white patches (23.9%) being their predominant colour, followed by brown with white patches (17.5%) (Mavule et al. 2016). Solid brown, fawn, black and white coloured sheep are also found, though in smaller numbers (Mavule et al. 2016). The majority (73%) of the animals are polled, while only a small number (26.9%) are horned. The ears are medium-sized (28.55–30.09 cm), with a horizontal (70.3%) or dropping (28.9%) orientation. Vestigial ears were found in 19.7% of the population. The body is covered with wool.

Adaptive and Special Genetic Characteristics: The Zulu sheep are hardy and able to walk long distances in search of forage and water (Ramsay et al. 2001). They are highly adapted to the harsh climate that includes hot and humid conditions, resistant to gastrointestinal parasites and tick-borne diseases and have strong foraging ability (Ramsay et al. 2001; Nyamukanza et al. 2010).

Typical Production System and Production Characteristics: These sheep are kept by the Zulu people, who are mixed crop-livestock farmers (Kunene and Fossey 2006). The farmers also

Fig. 6.24 Tanzanian long fat-tailed sheep, Photo: ILRI/Cleaned VC. (Available on Flickr for free use under the Creative Commons Licence. (Image can be viewed here))

keep other livestock such as cattle, goats and chickens (Mavule et al. 2013). The sheep provide meat and meat products to the household and a source of income (Kunene and Fossey 2006). The sheep have low maintenance and are able to utilise low-grade farm by-products that are high in forage (Kunene et al. 2007).

Pedi

Distribution: The Pedi sheep are located in the Limpopo Province (Mpumalanga region) of South Africa.

Physical Characteristics: The Pedi sheep are small, with adult weights of 40 kg and 30 kg for rams and ewes, respectively. They are polled and have a long, thin, carrot-shaped tail that reaches between the hocks and hooves (Ryder 1983). Spiral tails are found in some animals, but this phenotype has been selected against by the breeders (Snyman 2014d). The body is small (withers height 62.94 ± 0.97 cm; thorax depth 32.22 ± 0.22 cm; body length 62.87 ± 0.32 cm) and shallow (heart girth 82.20 ± 0.34 cm) with long legs. The head is small (width 11.79 ± 0.15 cm), and the ears are also small (8.73 ± 0.25 cm). The rump is wide 16.42 ± 0.26 cm and long 20.98 ± 0.33 cm. The sheep are multicoloured, with most of the animals having a white body with a reddish-brown head due to selection for this phenotype (Snyman 2014d). Some animals' coat colours are brown, black with or without white or red/brown patches (Snyman 2014d).

Adaptive and Special Genetic Characteristics: The Pedi breed is tolerant to ticks and tick-borne diseases. Like the other Nguni sheep, the Pedi sheep are very hardy, rarely sick nor needing to be vaccinated. They are resistant to gastrointestinal helminths, Heartwater, Pulpy Kidney and Blue Tongue diseases. The sheep do not need to be dewormed (Thaba Manzi Ranch 2021). They are also good foragers and are able to walk long distances in search of forage and water (Ramsay et al. 2001).

Typical Production Systems, Husbandry Practices and Production Characteristics: The Pedi sheep are reared for their meat, hides and fat tails. Their meat is tender and very tasty and lean (Ramsay et al. 2001). The fatty tail is valued for its additive value, especially in processed meats such as sausages (Thaba Manzi Ranch 2021). The hides produce leather that is of superior quality and grade. The Pedi sheep require very low maintenance but are slow-growing, taking up to 12 months to reach a slaughter weight of 18 kg (30 kg live weight) (Thaba Manzi Ranch 2021). The Pedi sheep are very fertile sheep, producing two lambs per year and are very good mothers that defend their lambs at all costs (Ramsay et al. 2001).

Swazi

Distribution: The Swazi sheep are predominantly found in rural Eswatini (Baker and Rege 1994; Lebbie and Ramsay 1999).

Physical Characteristics: The Swazi sheep are predominantly black, though multicoloured varieties with brown, reddish brown and white are found (Wilson 1991). The body is covered with hair that grows long along the back and rib cage (Wilson 1991). They are small sheep with an average body weight of 30.41 ± 0.41 kg. The other body measurements are 82.20 cm ± 0.34 (heart girth), 61.87 cm ± 0.36 (wither height), 30.28 cm ± 0.21 (thorax depth), 64.39 cm ± 0.31 (body length), 10.87 cm ± 0.14 (head width), 9.63 cm ± 0.24 (ear length), 15.51 cm ± 0.25 (rump width) and 21.73 cm ± 0.30 (rump length) (Gwala et al. 2015).

Adaptive and Special Genetic Characteristics: The Swazi sheep, like the Nguni, are adapted to the hot-humid and hot-dry climatic conditions of Eswatini. They are hardy and resistant to gastrointestinal parasites and local sheep diseases. They have developed a high tolerance for ticks and survive in environments that are endemic for heartwater, ticks and fleas (Ramsay et al. 2001).

Typical Production System: The sheep are kept in communal grazing lands by small-scale crop-livestock farmers (Ramsay et al. 2001).

Landim

Distribution: The Landim sheep are part of the Nguni group of sheep and are found in Mozambique, south of the Limpopo River (Wilson 1991).

Physical Characteristics: The Landim breed is a small animal with an adult withers height of 61–65 cm, adult body weight of 55 kg (males) and 35 kg (females) and adult heart girth of 80.7 cm ± 4.9. The sheep have a thorax depth of 30.19 ± 0.23 cm, a body length of 63.11 ± 0.32 cm and a rump width and length of 14.4 ± 0.25 cm and 21.96 ± 0.34 cm, respectively. The average head width is 12.23 ± 0.14 cm, with an average ear length of 10.13 ± 0.24 cm (Gwala et al. 2015). The breed has a long fat-tail (average 35.6 cm) reaching after the hocks but above the hooves. The body is covered with short (4 cm) coarse hair. The legs are long and thin. The head is convex, and the forehead is broad and short. Males have pads of fat on the face. The neck is long, and the back is long and straight. Both sexes are polled, with ears short (12.9 cm) and pendulous. Vestigial ears are present in 12% of the population (Wilson 1991).

Adaptive and Special Genetic Characteristics: The Landim breed is adapted to arid and semi-arid and the sub-humid tropics of Africa (Wilson 1991). They survive in dry tsetse-infested regions where zebu cattle cannot survive (McKinnon and Rocha 1985).

Typical Production System and Production Characteristics: The Landim breed is kept by small-scale crop-livestock farmers of Mozambique mainly for meat. The age at first lambing is approximately 805 ± 151.7 days, with a lambing interval of 412 ± 161.3 days. Twins are common and the litter size averages 1.41. The annual reproductive rate is 1.40 (Gwala et al. 2015).

Sabi

Distribution: The Sabi sheep are part of the African long-fat-tailed sheep and are the most common sheep found in Zimbabwe. They are distributed in the arid and semi-arid regions of Zimbabwe.

Physical Characteristics: The Sabi sheep is a large animal with an average withers height of 60–70 cm, with adult weights of 40 kg in females and 50 kg in males. The head is strong, with a prominent forehead, convex in males and a poorly developed dewlap. They often have pads of fat behind the nostrils in males. The males can be horned (often with one twist) or polled, while females are often polled. Ears are short and horizontal or downward facing, the neck is short and thick, brisket is well developed, the chest is pinched and the withers are level with the tail head, legs are long and fleshy, the tail is long, often reaching the hocks and tapering to a point, and the tail is thin with no fat deposits. The body colour is dark or light brown, and black-coloured sheep are also found. The body is covered with short, coarse hair (Wilson 1991).

Adaptive and Special Genetic Characteristics: The Sabi breed is hardy and adapted to dry and arid regions with production rates higher than that of the Doper breed reared in the same environment (Matika 2001). The Sabi breed is resistant to gastrointestinal parasites (Matika et al. 2003; McKenzie 1987). When comparisons were made between the Sabi and Dorper breeds, the Sabi was found to be more resistant than the Dorper (Matika et al. 2003).

Typical Production Systems and Production Characteristics: The breed is kept in agropastoral, pastoral and ranching production systems for meat production. The Sabi breed is fertile, with a 19% twinning rate. The birth weights average 2.7 kg on a government farm, and lambs weigh 19.8 kg at 100 days, with an average daily weight gain of 130.7 ± 0.09 g for 0–140 days (Wilson 1991). The Sabi breed are slow-growing, and females conceive at 10 months with a body weight of approximately 18–20 kg (Devendra and McLeroy 1982), while males reach sexual maturity at 169 days with a weight of 21 kg (Arrowsmith and Ward 1983).

Tswana

Origin, Distribution, Classification and Population Statistics: The Tswana sheep are part of the Nguni group of sheep. Currently, they are found in Zimbabwe and Botswana. The breed is not at risk of extinction, with approximately 250,000 heads reported in Zimbabwe in 2018 (Molotsi et al. 2020).

Physical Characteristics: The Tswana sheep is a medium-sized sheep with an adult weight of 44.3 ± 2.6 kg in males. Height at withers is 45.0 ± 3.9 kg (males) and 63.2 ± 0.5 kg (females). The body's adult heart girth measures 65.1 cm ± 2.0 (males) and 65.8 cm ± 3.0 (females). The average body length is 57.7 ± 2.8 cm (male) and 57.0 ± 0.8 cm (females), with a shoulder width of 17.1 ± 1.1 cm (males) and 14.7 ± 0.3 cm (females). The ears are long (11.4 ± 0.8 cm (males) and 11.5 ± 0.2 cm (females)) and horizontal. The Tswana sheep have short horns with an average length of 0.5 cm (males) and 3.8 cm ± 2.3 (females) (Nsoso et al. 2004). The coat colour is predominantly white or white with multicoloured patches. The tail is short, fat, with an upward-pointing tip at the end. Wattles are present (Bolowe et al. 2021).

Adaptive Characteristics: The Tswana breed is a very hardy breed and able to produce in very dry regions.

Production Characteristics: Twin births are common, with a kidding rate of 1.21 per ewe and an average flock lambing rate of 80% on a government farm. The lambs have a weaning weight of 17 kg and an average weight of 26 kg in 12 months (Senyatso and Masilo 1996).

6.5.4 Fat-Tailed Coarse Wool Sheep of Africa

6.5.4.1 Menz

Origin and Distribution: The Menz sheep, also known as Legegora, Shoa, Abyssinian, or Ethiopian highland sheep, is believed to be a descendant of the Arabian sheep crossed with the narrow Bab-el-Mandeb straits located at the mouth of the Red Sea. It is classified as a short-fat-tailed sheep. It is one of the few woolled sheep found in Ethiopia and is mainly kept by the Amhara community. It is widely distributed from the cool alpine mountainous climate in northern Ethiopia to the arid and semi-arid lands located in the lowland regions of Ethiopia (Fig. 6.25) (Tibbo et al. 2004).

Physical Characteristics: The Menz is a small sheep with short legs. The body is covered with a long fleece of coarse, wavy wool that is 8–11 cm long. The tail is fat with an approximate length of 17–23 cm and a width of 8–11 cm, the end of which is twisted and stops halfway to the hocks. Most are black or dark brown in colour, although light brown- and roan-coloured animals are also found (Fig. 6.25). White or brown spots on the head, neck and legs are common. The animals have an open face with a straight profile and short semi-pendulous ears that are 7–11 cm long. Approximately 12% of the population has rudimentary ears. The neck is short and thin. The back is short but straight. The rams have long twisted horns (Fig. 6.26) while ewes are mostly polled (Fig. 6.27). The body is compact and inclines forward slightly. Dewlap and wattles are absent. The average height at withers is 64–65 cm for males and 58–59 cm for females. The heart girth is 66–70 cm, and the average body length is 59–63 cm. Both the males and females have a nervous temperament (Gizaw et al. 2008a; Galall 1983).

Adaptive and Special Genetic Characteristics: The Menz sheep inhabit cool subalpine mountainous regions that are approximately 2500–3000 m above sea level. These regions have a severe cold climate, with temperatures ranging between 7.6 °C and 22.1 °C with strong cold winds and frosts mainly experienced between November and January. Functional analyses of the genes revealed that PPP1R12A, RELN, PARP2 and DNAH9 were involved in biological processes associated with adaptation to extreme altitudes, including influencing the development of the respiratory system and ensuring a smooth signalling pathway (Edea et al. 2019). The Menz sheep is tolerant to gastrointestinal helminth infections (Getachew et al. 2015).

Typical Production Systems and Production Characteristics: The Menz sheep are mainly kept to generate income for the households. This is mainly done through the sale of wool and meat in that order. Some of the meat is consumed at the household level and adjudged to be quite tasty. The sheep are also valued for manure production (Getachew et al. 2010). Flock sizes are generally small- to medium-sized, with up to 23 animals owned at the household level (Wilson 1991). The ewes reach first oestrus at around 350 ± 12 days at a mean weight of approximately 22–29 ± 0.6 kg and first lambing at 523 ± 13 days (Mukasa-Mugerwa and Lahlou-Kassi 1995) with a lambing interval of 255 days and 270 days in crop-livestock and pastoral production systems, respectively (Gizaw et al. 2013). Males reach sexual maturity at approximately 347 and 210 days in crop-livestock and pastoral production systems, respectively (Gizaw et al. 2013). Lambs recorded an average live weight at birth of 2.3 ± 0.1 kg and 9.8 ± 0.9 kg at weaning, with 14.3% being twins (Mukasa-Mugerwa and Lahlou-Kassi 1995).

6.5.4.2 Tikur

Origin, Distribution, Classification and Population Statistics: The Tikur, also known as Tekur, Tukur, or Tucur, like other Ethiopian sheep, is thought to have descended from Asia, entering Africa through the Horn of Africa (Muigai and Hanotte 2013). They are fat-tailed sheep kept by the Amhara community in the sub-alpine highland region of the North Wollo zone of Amhara state, Ethiopia (Gizaw et al. 2008a). The sheep numbers reported in 2008 were 525,300 heads, with a possibility that the numbers are declining due to decreased market demand for black-coloured sheep and sheep wool, resulting in selection against this coat colour in multicoloured, heterogenous populations (Gizaw et al. 2008a).

Physical Characteristics: The Tikur sheep is classified as a small fat-tail sheep with a woolly undercoat. The coat colour is predominantly black (60% of the population). The sheep has the following average measurements: body weight 25.4 ± 6 kg; withers height 64.1 ± 6 cm; body length 63.6 ± 6 cm; heart girth 69.7 ± 6 cm; substernal height 35.9 ± 5 cm. The ears are short,

6.8 ± 5 cm, and the tail measures 17.3 ± 6 cm long and 8.9 ± 3 cm wide. The coat is woolly with a fibre length of 7.4 ± 3 cm (Gizaw et al. 2008a; Ethiopia Sheep and Goat productivity Improvement Program (ESGPIP) 2008).

Adaptive Characteristics: The sheep is adapted to cold alpine and semi-alpine environmental conditions with feed shortages and is able to produce wool under these extreme conditions (Devendra and McLeroy 1982).

Typical Production System and Production Characteristics: The Tikur sheep is reared by small-scale mixed crop livestock farmers in barley growing regions of Ethiopia, an environment characterised as subalpine that is above 3000 m above sea level (Gizaw et al. 2008a). The sheep are reared under a traditional/free grazing management system (Weldeyesus 2020), mainly for meat and wool. The average litter size is 1.08 with a lamb survival rate of 95% (Devendra and McLeroy 1982). The average weaning age is 5.1 months (Mohammed et al. 2015).

6.5.4.3 Arsi-Bale

Origin, Distribution and Population Statistics: The Arsi-Bale sheep is believed to have its origins in the Arabian Peninsula (Muigai and Hanotte 2013). They are widely distributed in the cool, wet and highlands of eastern and south-central Ethiopia, ranging from 2000 to 3000 m. The climate in these regions ranges from semi-arid to sub-humid with annual rainfalls above 1500 mm. The sheep is reared by the Oromo communities, with 6,345,100 heads reported in 2008 (Gizaw et al. 2008a).

Physical Characteristics: The Arsi-Bale sheep is large in size with a withers height of 64.1 ± 6 cm and a substernal height of 35.3 ± 4 cm. The animals have a long, fat-tail, measuring 28.4 ± 6 cm in length and 6.2 ± 3 cm in width and have a twisted end. The heart girth is 73.3 ± 6 cm. The body is 62.3 ± 8 cm long and covered with coarse, wavy wool that is 4.2 ± 1 cm long. The colour is brown with white patches or black with white patches. Males and 52% of females are horned (Gizaw et al. 2008a).

Adaptive and Special Genetic Characteristics: These sheep inhabit the wet, cool highland areas with extreme temperatures. Candidate gene analyses showed that the Arsi-Bale sheep has genes that are associated with adaptations to extremely cold conditions (Edea et al. 2019) under which they are able to produce (Devendra and McLeroy 1982).

Typical Production Systems: The Arsi-Bale are raised in a range of production systems (agro-pastoral to agricultural and urban). Sheep are kept for meat, manure and hides. In the highland regions where farmers grow barley, the sheep are also kept in low semi-intensive production systems and are used for cereal cropping. In the perennial cash crop production system, the sheep are kept in low- to semi-intensive production systems where they are allowed to roam freely. The farmers in the wetter lowland regions keep large numbers of Arsi-Bale sheep while also practising cereal, sesame and cotton farming in semi-intensive low production systems (Devendra and McLeroy 1982; Gizaw et al. 2008a).

6.5.4.4 Wello

Location and Population Statistics: The Wello or Wollo sheep is an indigenous highland sheep found in the south Wollo administrative zone of Amhara state and is mainly reared by the Amhara community (Amare et al. 2018). A population size of 1,395,900 was reported in 2008 (Gizaw 2008).

Physical Characteristics: Wello sheep is characterised by a short fat tail, approximately 20.4 ± 6 cm in length, 7.2 ± 2 cm in width, which is twisted and turned up at the end. The animal has a small body with an average female body weight of 21.7 ± 5 kg, height, substernal height and length of 62.7 ± 5 cm, 34.3 ± 4 cm and 61.2 ± 5 cm, respectively. The heart girth for adult females is 67.9 ± 5 cm. The sheep has long hair (average 7.9 ± 4 cm in length) with a woolly undercoat. Most of the animals are either black, white or brown in colour, with white, black or brown patches. The males are horned (Gizaw et al. 2008a).

Adaptive and Special Genetic Characteristics: The breed performs well in dry regions and resists local diseases well (Ethiopia Sheep and Goat productivity Improvement Program

(ESGPIP) 2008). It is able to produce wool under extreme dry conditions (Ethiopia Sheep and Goat productivity Improvement Program (ESGPIP) 2008).

Typical Production Systems, Husbandry Practices and Production Characteristics: Sheep in this region are maintained in a mixed crop-livestock production system, with farmers releasing their animals to graze in communal and private grazing areas during the day, and at night they are kept in an enclosure together with the other livestock species owned at the household level (Amare et al. 2018). Breeding practice in the traditional systems is mainly random mating.

6.5.4.5 Farta

Origin, Distribution, Classification and Population Statistics: The Farta sheep (Figs. 6.28 and 6.29) is an indigenous sheep found in the South Wollo zone of Amahara State, Ethiopia, and is classified as a subalpine short-fat-tailed sheep. It is found in alpine and subalpine regions of Ethiopia, ranging between 2000 and 4000 m above sea level (Gizaw et al. 2008a). In 2008, 555,600 heads were reported. The Farta sheep is mainly reared in the cereal-growing regions in South Wollo of Amhara State in Ethiopia (Gizaw et al. 2008a).

Physical Characteristics: The tail is short, fat, measuring 22.9 ± 8 cm long and 9.6 ± 2 cm wide and reaches above the hocks. The average adult body weight is 28.3 ± 7 cm with a withers height of 67.9 ± 5 cm and a body length of 65.7 ± 7 cm. The average heart girth measures 72.0 ± 7 cm, and the substernal height is 37.3 ± 4 cm. The ears are long, measuring 9.9 ± 3 cm. The body is covered with a woolly undercoat with an average fibre length of 7.5 ± 3 cm. The body colour is predominantly white (37.5%) (Fig. 6.28), though brown (27.5%) and black with brown belly (15%), white and brown (Fig. 6.28), and brown and white patches are also found. Males are horned while females are polled (Fig. 6.29) (Gizaw et al. 2008a).

Adaptive Characteristics: The Farta sheep is reared in the highland regions of Ethiopia, where the environment is harsh and dry. Feed shortages are common. The breed is fully adapted to these conditions (Bimerow et al. 2011).

Typical Production System: The Farta breed is maintained in small-scale sheep production systems that can be further described as a low-input production system that involves mainly free range grazing (Gizaw et al. 2008a).

6.5.4.6 Semien

Distribution: The Semien sheep are found in the Simien Mountains region, which is rich in alpine mountainous ecology. The specific districts where the sheep are found include the South Gondar zone, Gondar zuria, Belesa and Dembia districts of Ethiopia and are reared by the Amhara community (Gizaw et al. 2008a; Melaku et al. 2019).

Physical Characteristics: The Simien sheep are the largest of the highland sheep. They have a short, fat tail measuring 12.8 ± 6 cm long and 9.6 ± 2 cm in width. The coat is woolly with a well-developed undercoat. The wool fibre is 8.2 ± 3 cm in length. The body colour is plain brown, plain white or brown with white patches or white with brown patches. The ears are medium in length, 8.3 ± 5 cm. The average body measurements are as follows: body weight -26.9 ± 4 kg, withers height -66.6 ± 6 cm, body length -64.7 ± 6 cm, heart girth -73.2 ± 6 cm, substernal height -35.9 ± 5 cm. The sheep are horned, with the males having long spiral horns and the females having short horns.

Adaptive Characteristics: The sheep are adapted to extremely cold, high mountainous regions (Melaku et al. 2019).

Typical Production Systems, Husbandry Practices and Production Characteristics: The Semien sheep are maintained in a smallholder production system where they graze on natural pastures in communal grazing areas where breeding is uncontrolled (Melaku et al. 2019). The Simien sheep are reared in barley-growing regions and are allowed to graze the crop stubble, thus clearing the farms (Melaku et al. 2019). Overgrazing of the sheep on the mountainous grasslands has resulted in their erosion, rendering them unproductive (Melaku et al. 2019).

6.5.4.7 Sekota

Distribution and Population Statistics: The Sekota sheep, also known as Tigray Highland and Abergelle, are found in the Central Zone of Tigray State in Ethiopia and are reared by Agew, Tigray and Amhara communities (Gizaw et al. 2008a).

Physical Characteristics: The Sekota sheep is a large sheep, with a long, fat tail, averaging 19.9 ± 8 cm long and 9.5 ± 3 cm wide. Sekota sheep have an average body weight of 26.6 ± 7 kg; withers height of 62.3 ± 6 cm; body length of 62.2 ± 6 cm; heart girth measurement of 69.9 ± 5 cm and substernal height of 35.9 ± 5 cm. The ears are medium in length, average of 8.3 ± 5 cm, semi-pendulous or rudimentary (Gizaw et al. 2008a).

Adaptive Characteristics: The Sekota sheep are adapted to the dry, lowland, semi-arid regions of Tigray state, which is 1500 metres above sea level, and are able to produce in these adverse conditions (Gizaw et al. 2008a).

Typical Production System and Production Characteristics: The Sekota sheep are reared in a lowland mixed crop-livestock production system in grasslands that are reported as degraded. The mean birth weight is 2.73 ± 0.06 kg, and the lambs weigh an average of 11.9 ± 0.21 kg at 3 months. The mean average weight gain from birth to weaning is 101 ± 2.66 gm. The breed is mainly kept for meat (Yiheyis et al. 2012).

6.5.4.8 Ossimi

Origin and Distribution: The Ossimi breed's origin is the Giza Governorate of Egypt, and it gets its name from a village near Cairo called Ossim. It is also known as Ausimi and Meraisi. It is the most popular sheep breed in the Nile and the Nile Delta. The breed is also found in the Middle East. The breed numbers are growing at the expense of the other local breeds and are considered to be the most productive breed in Middle Egypt. However, in Southern Egypt, it is not considered to be as productive.

Physical Characteristics: The Ossimi sheep has distinct white coarse wool which covers the entire body, although the neck, legs and abdomen have little wool. The head is brown or dark brown in colour and is convex in shape. The neck is short, often brown in colour or can have white patches. The sheep are medium-sized with a narrow, long body and long, thin legs. The males have curved, medium-sized horns that are black in colour with a downward-facing curvature, but females are polled. The average withers height is 75 cm and 72 cm for females and males, respectively. The average mature body weights are 45 kg (females) and 60 kg (males). The Ossimi have a fatty, cylindrical tail that weighs an average of 2.5–4.0% of the animal's weight (Elshazly and Youngs 2019).

Adaptive Characteristics: The Ossimi sheep are hardy and adapt easily to various environments found along the Nile Valley.

Typical Production Systems, Husbandry Practices and Production Characteristics: The Ossimi sheep is primarily kept for wool, which is mainly used in the carpet industry. The fleece weighs between 1.4 kg and 2.8 kg. The Ossimi is also a milk-producing sheep and has an average lactation length of 84 days and produces an average of 65 kg milk. The milk has an average fat content of 6.3%. The average daily weight gain is 100 g. The average mature carcass weight is 16 kg with an average dressing percentage of 46%. Average birth weights recorded for the breed are 3.2 and 2.9 kg for males and females, respectively. A twinning rate of 10–20% has been recorded with an average conception rate of 6.1%. The carcass produces good-quality meat, with most of the fat being concentrated around the tail region.

6.5.4.9 Barki

Origin, Distribution, Classification and Population Statistics: The Barki sheep (Figs. 6.30 and 6.31), often also called Desert sheep, are distributed in the coastal Mediterranean zone of Northern Africa. The Barki sheep gets its name from a province in Lybia known as Barka (Elshazly and Youngs 2019).

Physical Characteristics: The Barki sheep are small- to medium-sized (Figs. 6.30 and 6.31). The body is covered with white coarse wool, softer in texture when compared to the wool produced by the Ossimi or Rhamani sheep breeds.

The head is black, brown or white and bears small semi-pendulous ears. Males have horns which curve backwards (Fig. 6.30), while the females are polled (Fig. 6.31). The legs are long, thin and are covered with minimum wool, which enables the sheep to walk long distances. The sheep has a fatty tail, which is wide at the top and narrow at the bottom and often reaches the hocks.

6.5.5 Fat-Rumped Hair Sheep of Africa

The African fat-rump hair breeds are divided into the East African fat-rumped breeds that comprise the Blackhead Somali and the South African fat-rumped breeds, for example, the South African Blackhead Persian (Rege et al. 1996). They are widely distributed across the hot, dry desert regions of East Africa in Uganda, Somalia and South Africa. The Blackhead Persian of South Africa is documented to have arrived in 1868 on a ship from the Persian Gulf, with a high probability of the sheep having been loaded in Aden or off the Somali coastline (Mason and Maule 1960).

6.5.5.1 Blackhead Somali

Distribution: The Blackhead Somali (Fig. 6.32), also known as Blackhead Ogaden, is a fat-rumped tail sheep (Gizaw et al. 2008a) that is indigenous to the Ogaden area of the Somali region. It is widely distributed in the Horn of Africa in semi-arid and arid regions and is reared by pastoralists located in Ethiopia, Somalia and Kenya. There is a high demand for this breed in the Middle East, with live exports of sheep and goats from Somalia reported at 4.6 million in 2014 (FAO et al. 2017).

Physical Characteristics: Blackhead Somali sheep has a short-fat tail with a stumpy appendage, 14.7 ± 6 cm long and 14 ± 9 cm wide. The body is white in colour except for the neck and head that are black (Fig. 6.32) with an average body length of 59.9 ± 9 cm. The body is covered with short hair (4.0 ± 3 cm long). The weight for mature females is 27.9 ± 9 kg with a heart girth of 71.5 ± 6 cm. The animal is large, with a withers height of 63.3 ± 6 cm and a substernal height of 35.1 ± 6 cm. The sheep has a convex face, which is prominent in the males. The ears are outward and forward dropping and 9.6 ± 4 cm in length. Both sexes are polled, and the dewlap is well developed.

Adaptive and Special Genetic Characteristics: The Blackhead Somali sheep is a hardy sheep

Fig. 6.25 Indigenous Menz sheep herd in Amhara Region, Ethiopia; Photo: ICARDA\Getachew Mengistu. (Available on Flickr for free use under the Creative Commons Licence. (Image can be viewed here))

Fig. 6.26 Menz ram in Menz Area, Ethiopia. (From Ferede et al. 2014, licensed under CC-BY 4.0)

Fig. 6.27 Menz ewe in Menz Area, Ethiopia. (From Ferede et al. 2014, licensed under CC-BY 4.0)

that is well adapted to heat, feed and water shortages experienced in semi-arid and arid environments (Gizaw et al. 2008b). The breed is adapted to extreme conditions where they continue to produce all year round (Rocha et al. 1990).

Typical Production System and Production Characteristics: This breed is mainly reared for meat by pastoralists. The age at first lambing is reported at 666.73 days (DAGRIS 2007). The mean lambing interval of the Blackhead Somali reared in the Kenyan Maasai pastoral system is 312 days (Peacock 1996).

6.5.5.2 Blackhead Persian

Origin and Distribution: Synonyms for the breed include Persian and Swartkoppersie (Afrikaans). The Blackhead Persian sheep of South Africa has its origins in the Horn of Africa. This breed was founded from one ram and three ewes that were obtained from a shipwreck off the Cape coast in 1869. They have a white body with a black head, but animals with red heads are also common. Red speckled animals, together with blackheads, are still found in the population. Somali traders along the eastern coast are thought to have introduced more animals. Three distinct varieties are recognised: the Blackheaded Persian, the Redheaded Persian and the speckled (skilder) Persian (Ramsay et al. 2001). This breed has been used to develop the Dorper breed by crossing it with the Dorset Horn (which was developed by crossing Spanish and English breeds in the sixteenth century). The Black Head Persian is not at risk.

Physical Characteristics: The Blackhead Persian is a small- to medium-sized sheep whose body is covered with smooth white hair. The rump and base of the tail have fat deposits, giving rise to a stumpy fat rump. The head and neck are coloured black. The ears are long and pendulous, and both sexes are polled. The average mature weight of rams and ewes is 68 kg and 52 kg, respectively, while the birth weights are 2.6 kg and 2.55 kg for males and females, respectively (Ramsay et al. 2001).

Adaptive and Special Genetic Characteristics: The Blackhead Persian sheep is fertile, extremely hardy and well adapted to the arid and semi-arid climates in South Africa. The sheep is resistant to helminths. They are able to walk for long distances and survive for long periods without water.

Typical Production System and Production Characteristics: The Blackhead Persian sheep is mainly bred for meat under extensive veld conditions, although previously the fat stored in the tail region was highly valued for cooking. Although the body is covered with a mixture of wool and hair, they are not sheared. The age at first breeding is 7 months. The Blackhead Persian ewe have good mothering qualities, producing large quantities of milk and suckling their young for approximately 84 days. The ewe produces approximately 50 kg of milk during a lactation period. The milk is of good quality and contains 5.9 percent fat. The average litter size is 1.08. The lambs grow very fast and are ready for slaughtering at an average of 4 months of age. The rams fare well in cross-breeding programmes. The breed is described as having an even temperament. It is fertile and usually gives lambs without complications. The fat tail is used as an additive to processed and game meats. The hide is used to produce high-quality leather that is known as Cape glover. The ewes are bred to Dorset Horn sheep to produce Dorper sheep breeds (Ramsay et al. 2001).

6.5.5.3 East African Blackhead Persian

Distribution and Classification: The East African Blackhead Persian is part of the fat-rumped hair sheep of Africa. It is found distributed in southern Uganda, west of Lake Victoria, regions around Mbarara in Uganda, northwestern parts of Uganda around Moroto and south and east into Sukumaland in Tanzania. The ecological region inhabited by the sheep can be described as semi-arid and sub-humid in medium- to low-altitude areas.

Physical Characteristics: The animal is small in size, with an adult female weight of 25 kg. The head has a finer shape when compared with other fat-tailed breeds, with a long neck and fat pads present on the nose and behind the ears. Both sexes are polled. The ears are short (5–8 cm), pendulous or vestigial. The chest can be described as pinched. The withers are level with the sacrum, or slightly lower. The back is short, straight or dipped. The brisket is well developed, but the dewlap is not very prominent except in very fatty sheep. The tail is prominent, with a large, fat tail. The body is white with a black head. The coat is short and coarse wool.

Typical Production Systems and Production Characteristics: The sheep is managed in pastoral and agropastoral production systems. Age at first lambing is 532 ± 8.1 days with a lambing interval of 255 days ±2.3. Twinning frequency is reported at 18%, with the sheep producing an average of 7 lambs in 5–6 years. The average birth weight is 2.5 ± 0.02 kg, and the average weight at 2 months is 10.1 ± 0.07 kg, while at 5 months it is 15.5 kg and 14.2 kg for males and females, respectively. The average daily weight gain between birth and 2 months is 123 g and 95 g for single births and multiple births, respectively.

6.5.6 Introduced Breeds

6.5.6.1 Karakul

Origin, Distribution and Population Statistics: The Karakul breed, also known as Astrakhan, is a fat-tailed fur sheep and is native to Central Asia and has its origins from the steppes of Turkistan (Mason 1996; Snyman 2014e). The Karakul were first imported into Namibia in 1907, with 12 sheep (2 rams and 10 ewes). Approximately 750 Karakul sheep were imported at the beginning of the 1900s (Wilson 1991; Ryder 1983; Snyman 2014e). The Karakul expanded rapidly in the 1930s, spreading into the surrounding regions of Namibia, South Africa, and Botswana, with a few reaching Angola (Viljoen 1981). By the 1970, it was the predominant breed in Namibia, with five million head constituting 95% of the Namibian flock, with exports of up to five million pelts out of South Africa and Namibia in 1979 (Rawlingson 1994). However, the African Karakul population has seen a decline in numbers as a result of various factors, including a rapid decline in Karakul stud breeders due to declining wool prices/demand for Swakara pelts and changing climatic conditions that have led to successive droughts (Snyman 2014e; Rawlingson 1994). The breed is found in South Africa, Botswana, Namibia, Mozambique and Angola. The population was estimated at 200,000 animals in Namibia and approximately 40,000 animals in South Africa. Total estimates for Africa are between 160,120 and 240,784 animals (Schoeman 2000; Molotsi et al. 2020; Snyman 2014e).

Physical Characteristics: The Karakul breed is a medium-sized sheep. The pelt is mainly black, but grey, white and brown are also found. The body is covered with a mixture of coarse and fine fibres that range in colour from black to grey. The tail is wide and fatty. The head is long and narrow, with slight indentations between the eyes. The ears are long and point downwards. The rams are predominantly horned, while the ewes are polled (Snyman 2014e).

Fig. 6.28 Farta ram in Farta District, Ethiopia. (From Ferede et al. 2014, licensed under CC-BY 4.0)

Fig. 6.29 Farta ewe in Farta District, Ethiopia. (From Ferede et al. 2014, licensed under CC-BY 4.0)

Adaptive Characteristics: The breed is well adapted to the semi-arid and arid areas and has thrived in the dry regions of southern Africa where it was introduced (Snyman 2014e).

Production Characteristics: The breed is kept for wool, pelt, meat (which is lean) and milk. The milk is creamy and used for butter and cheese production. The wool is used to make a large variety of wool and pelt products (Snyman 2014e).

6.5.6.2 South African Mutton Merino

Origin, Distribution, Classification and Population Statistics: The South African Mutton Merino, also known as German Mutton Merino, was first imported from Germany into South Africa in 1932. The goal was to develop a sheep breed that would utilise the winter pastures by having lambs born in autumn, rearing them quickly so that they could be marketed before the grazing dried up during the dry summer months. It is not a true landrace and has been developed through selection and used in the development of other South African breeds, namely Afrino, Dohne Merino, Dormer and Vandor. Although the current numbers of the South African Merino are not known, the breed is not threatened (Ramsay et al. 2001; Snyman 2014f).

Physical Characteristics: The South African Mutton Merino is a large animal with a mature weight of 125 kg and 70 kg for rams and ewes, respectively. The average birth weight is 4.1 kg and 3.8 kg for rams and ewes, respectively. The 100-day weight is 32 kg and 29 kg for rams and ewes, respectively. The breed produces good-quality wool with an average fibre diameter of 21–23 microns. The wool production averages at 4.5 kg and 3.4 kg per ram and ewe, respectively (Ramsay et al. 2001; Snyman 2014f).

Adaptive Characteristics: The South African Merino is adapted to the dry, arid and semi-arid regions. It also does well in regions that are not dry (Snyman 2014f).

Husbandry Practices and Production Characteristics: The South African Mutton Merino is kept for both mutton and wool. The wool provides a good source of income. The ewes have high fertility and good mothering ability and produce multiple births. The rams are suitable for cross-breeding. The breed grows fast and has a high feed conversion rate. The animals are non-selective when it comes to grazing. The carcass produced is of high quality, with exceptional conformation and tender meat. This breed has been used in the development of other breeds, such as the Dormer and Dohne Merino.

6.5.6.3 Barbados Blackbelly

Origin: Barbados Blackbelly is widely accepted as an African sheep breed. Although there can be little doubt that the Blackbelly has African ancestry, there is compelling historical evidence that the Barbados Blackbelly, as a breed, originated and evolved on the island of Barbados, following the colonisation of Barbados by the English in 1627. The evidence reviewed supports the view that the Barbados Blackbelly evolved from crosses of African hair sheep and European woolled breeds. For Barbados Blackbelly, the parent stocks necessary for the evolution of a highly prolific, tropically adapted breed were established in Barbados in the first quarter-century of colonisation.

Physical Characteristics: Mature ram body weights range between 47.6 kg and 56.6 kg, while adult ewes weigh approximately 45 kg. The coat bears all shades of brown, tan or yellow, with deeply contrasting black underparts that extend down the inside of the legs. The animals bear black parts on the nose, forehead and on the inside of the ears. Rams have a neckpiece of thick hair, which extends down the neck to the brisket and may cover part of the shoulder in some animals. The head is medium-sized with good conformation down to the neck and body. Ears point forward from the side (not droopy). Both sexes are polled. The neck is medium in length, slender in comparison to the body size. Rams are heavier and have well-set shoulders. The body is fairly deep with well-sprung ribs. The back and loins are medium-sized, with deficient hindquarters that are very similar to what is observed in goats. The rump is quite steep from the hips to the pin bones, and the tail is set very low. The legs are fairly long and thin and are generally well set. When observed from a distance, the Barbados Blackbelly sheep resembles a small deer or antelope in appearance. The sheep are very active and lively and are alert at all times. Compared to the more recognised mutton sheep breeds, the Barbados Blackbelly sheep are less 'squatty' and are slower growing.

Adaptive and Special Genetic Characteristics: Barbados Blackbelly sheep has a rare attribute of adaptation to a wide range of environments. In cooler climates, they develop protective wool undercoats during the fall and winter that shed in the spring. Barbados Blackbelly is resistant to intestinal parasites.

Production Characteristics: Barbados Blackbelly ewes have high prolificacy due to high reproductive efficiency, which accounts for their average of two lambs per litter and an average lambing interval of 8–9 months. Average lambing rate ranges between 1.50 and 2.30 per ewe. Body weights of yearling ewes are variable due partly to the tendency to breed ewes that are less than 12 months old. However, most weigh between 36 kg and 40 kg. Carcass studies of 5- to 7-month-old rams sent to slaughter show that the Barbados Blackbelly breed has much less body fat compared to other sheep breeds. The flavour of the meat has been described as excellent and milder compared to other breeds. Estimated daily average gains from 5- to 7-month single rams range between 181 g and 204 g when fed on rations of alfalfa hay and wheat with mineral supplement. These gains are perhaps 60–70% of the normal expected gain for wool sheep on similar rations. Crosses of Barbados Blackbelly with the Dorset Horn and Suffolk have resulted in a pronounced increase in growth rates of progeny.

6.5.6.4 Bergui

Origin, Distribution and Population Statistics: The Bergui breed (Fig. 6.33), also known as La race ovine Queue Fine de l'Ouest, was introduced by the Romans (Trouette 1929), who were known to be great lovers of wool, in the fifth century from Taranto in Italy, where it exists to date (Snoussi 2003). It is found in Algeria and Tunisia, occupying the western regions of Tunisia. Their numbers in Tunisia increased steadily over the years, reaching 32% of Tunisian sheep breeds by 2013.

Physical Characteristics: It is a wool breed with a fine tail. The breed is taller than the Barbarine and it has a straight dorsal line and a long, thin tail (Fig. 6.33).

Typical Production System: The breed is reared for milk and meat in mainly extensive production systems (Khaldi et al. 2011).

6.5.6.5 Rahmani

Origin and Distribution: The Rhamany breed (Fig. 6.34), also known as Rahmany, originated in northern Syria and northern Turkey and was

Fig. 6.31 Barki ewe. (From Elshazly and Youngs 2019, licensed under CC-BY 4.0)

Fig. 6.30 Barki ram. (From Elshazly and Youngs 2019, licensed under CC-BY 4.0)

introduced into Egypt in the ninth century. The original stock is the Red Karman from Turkey. The breed is named after Rahmania, a village in the Beheira Governorate in the north of the Delta. The breed is distributed in the central and northern Egyptian delta regions.

Physical Characteristics: The breed is the largest of the Egyptian breeds and is characterised by its red wool. The animal has small ears. The males have large horns (Fig. 6.34) and the females are polled (Fig. 6.35). The average wither height is 43 cm for females and 47 cm for males. The average mature weight is 62 kg for males and 50 kg for females. The breed has a fat tail and often has no earlobes (Elshazly and Youngs 2019).

Adaptive and Special Genetic Characteristics: The breed is believed to have some resistance to gastrointestinal parasites.

Production Characteristics: The Rahmani is mainly reared for wool and milk production to a lesser degree. The wool produced is of high quality, often used in the manufacture of carpets. Mature animals produce an average of 3 kg wool. The average lactation period is 85 days, with an average production of 60 kg milk with 3.2% fat. The daily weight gain for the breed is 145 g with an average dressing percentage of 52%. The average birth weight for lambs is 1.2 kg for females and 1.22 kg for males, with a relatively high twinning rate. The animals breed all year round. The weaning weight is 18–21 kg; yearling weight is 38.6–42.2 kg; twinning rate is 12–40%; conception rate is 41.5–57.8%; and weaning rate is 74–78%.

6.5.7 Developed or Composite Sheep Breeds of Africa

Commercial composites are breeds that have been developed through the crossing of indigenous African sheep breeds with varying proportions of exotic breeds. Commercial farmers in Africa, who were mainly of European origin, considered the indigenous African breeds generally not adequate for commercial production systems geared mainly towards wool and meat export. As a result, they imported European breeds. However, these European breeds were poorly adapted to the dry and tropical African conditions and suffered seasonal weight loss, parasitism, diseases and were attacked by predators (Almeida 2011). In order to mitigate these challenges, European breeds were crossed with indigenous African breeds, mainly in South Africa, resulting in new composite breeds, namely Dorper, Afrino, Van Rooy, Vandor and Meatmaster (Ramsay et al. 2001; Almeida 2011).

6.5.7.1 Dorper

Origin and Distribution: Synonyms for the breed include White Dorper, Dorsian, Dorsie (Afrikaans). The need to produce a high-grade

Fig. 6.32 Blackhead Somali sheep in a livestock market in Somaliland, Photo: ILRI. (Available on Flickr for free use under the Creative Commons Licence (Image can be viewed here))

mutton sheep breed, able to withstand dry South African conditions and produce fast-growing lambs, led to the crossing of Dorset Horn rams and Blackhead Persian ewes between 1933 and 1946 at the Grootfontein Research Station (Milne 2000). The results were half-cross Dorset Horn x Blackhead Persian, from which selections were made to form the foundation flock for the current breed. The breed was formally recognised in 1950 when a breed society was established (Ramsay et al. 2001; Milne 2000). The breed's hardiness and ability to produce in extremely dry conditions led to its popularity and rapid spread (Marais and Schoeman 1990). It is currently widely distributed in South Africa, has spread to neighbouring countries of Namibia, Zimbabwe, Zambia, Angola and Botswana and has been exported to Kenya, Burundi, Malawi, Mauritius, Palestine, Saudi Arabia and Australia (Milne 2000). The current numbers are not known; however, in 1997, the South African flock was approximately seven million heads (South African Department of Agriculture 1997).

Dorper sheep were imported to Germany in 1995 at a time when the wool market was in decline because they did not need to be sheared. Australian Dorper breeders are currently exporting their animals to Vietnam and India. Additionally, the Dorper was crossed with the Damara, a South African fat-tailed breed, resulting in the Damper breed. Damper rams are crossed with Merino females to produce meat animals in Australia and then sold to the Middle East for slaughter (South African Department of Agriculture 1997).

Physical Characteristics: The Dorper breed is a large animal with a black head and white body. It has an average adult body weight of 52.84 ± 0.30 kg, an average withers height of 64.43 ± 0.32 cm and a heart girth of 84.23 ± 0.32 cm. The average body length is 71.14 ± 0.30 cm with a rear rump width of 20.71 ± 0.24 cm and a rump length of 24.29 ± 0.32 cm. The average thorax depth is 36.54 ± 0.21 cm. The ears are long with an average of 11.43 ± 0.023 cm. The head is large with a width of 13.83 ± 0.11 cm (Gwala et al. 2015). The body is covered with a mixture of wool and hair, while the bare parts of the body are covered with white hair; shearing is not necessary.

Adaptive and Special Genetic Characteristics: The breed is very well adapted to arid and semi-arid environments and is able to produce well in dry conditions. It is able to survive dehydration and quickly replenishes its body weight when water becomes available (Degen and Kam 1992). The Dorper's thick skin offers protection against harsh weather conditions (Buvanendran and Adu 2021). It has a higher browsing ability when compared to Merino breeds (Du Toit 2008). The breed has high resistance to diseases, including fly strike and fleece rot, though it has been found to be resistant to gastrointestinal helminths (Baker et al. 1999, 2002).

Typical Production Systems and Production Characteristics: The Dorper breed is adapted for both veld and feedlot production (Ramsay et al. 2001). The breed matures early, with Dorset ewes reaching maturity at 213 days and a live weight of 39 kg, with an average first-time lambing at 346 days and overall high lamb survival rates (Greeff et al. 1988). The Dorper ewes make good mothers and produce a large quantity of milk, averaging 1.22 kg of milk per day over approximately 77 days (Bonsma 1944). Fifteen years of production data taken in an arid region of Kenya reported Dorper ewes had a twinning rate of 10%, lambing (fertility) rate of 78%, average birth weight of 3.6 kg and pre-weaning lamb growth and mortality rates of 178 kg per day and 14%, respectively (King'oku et al. 1975). The breed produces high-quality mutton that is described as tender and juicy (Ramsay et al. 2001). There is high demand for their skin, which is used to make high-quality leather products.

6.5.7.2 Afrino

Origin, Distribution and Population Statistics: The Afrino sheep is a dual-purpose meat and white wool-producing sheep found in the Southern Free, Eastern Cape and Northern Cape provinces of South Africa. It was developed in the late 1960s, specifically to meet the needs of commercial farmers who needed a sheep breed able to produce high-quality white wool similar to that of the Merino breed and, at the same time, produce high-grade mutton while being hardy and adapted to grazing in extensive production systems (Snyman 2014g). The result was the Afrino, which was developed from a combination of 25% Merino, 25% Ronderib Afrikaner and 50% South African Mutton Merino (Snyman 2014g). An Afrino breeders' society was established in 1980. The risk status of the breed is unknown. However, reported numbers in the period 1990–2018 are between 1000 and 3141 heads (Molotsi et al. 2020).

Physical Characteristics: The Afrino sheep is large, and its body is covered with white wool. The average mature weight is 80 kg (males) and 60 kg (females). The head is large, with strong upper and lower jaws and prominent nostrils, and the face is covered with soft cream wool. The average body length is 60–80 cm. The rams may have small horns while the ewes are polled. The chest is deep and wide, and the hindquarters are wide across the hips and pin bones. The legs are muscular and well-fleshed. The legs are of average length (Ramsay et al. 2001; Snyman 2014g; Afrino Sheep Breeders Society n.d.).

Adaptive and Special Genetic Characteristics: The breed is adapted to the arid and semi-arid veld and is able to produce in extremely harsh environments. The ewes make excellent mothers, leading to high survival lamb rates (Ramsay et al. 2001; Snyman 2014g).

Production Characteristics: The average age of first lambing is 10–17 months. Average birth weight is 4.9 kg and 4.6 kg for rams and ewes, respectively. The weaning weight at 120 days is 32 kg and 29.5 kg for rams and ewes, respectively, while corresponding weights at 9 months are 49 kg and 43.5 kg. The sheep are kept mainly for meat (80%) and wool (20%) production. The average carcass weight is 19.8 kg. The sheep produces Merino-type wool with an average diameter of between 18 and 22 microns. Ewes that raise lambs produce between 2.5 kg and 3 kg wool per year (Snyman 2014g; Afrino Sheep Breeders Society n.d.).

6.5.7.3 Van Rooy

Origin and Distribution: Synonyms for the breed include Van Rooy White Persian, Van Rooy Persie (Afrikans) and White Persian. This breed was developed in 1906 by a South African farmer known as JC van Rooy, who first mated a Ronderib Afrikaner ram to a French Merino *de Rambouillet* ewe. In order to improve carcass characteristics, he introduced a polled

Wensleydale ram. The breed is mainly reared in South Africa. The breed is not under threat and numbers are not known (Almeida 2011; Snyman 2014h).

Physical Characteristics: The Van Rooy is a medium- to large-framed animal with a predominant fat rump tail that has a huge oval from which a little tail hangs vertically. Both sexes are polled. They have a characteristic dewlap from the jaw to the brisket with a prominent chest and brisket. The body is covered with a mixture of hair and wool (Ramsay et al. 2001; Snyman 2014h).

Adaptive and Special Genetic Characteristics: The breed is adapted to the arid and semi-arid savannah, especially the dry north and northwestern Bushveld regions of South Africa around the Orange River and in Gauteng. It is heat and drought-tolerant and can cope with extreme hot and cold temperatures during the day and at night (Ramsay et al. 2001; Snyman 2014h).

Production Characteristics: The Van Rooy is a minimum care breed that is kept mainly for meat and skins. The breed is fertile and ewes can be mated throughout the year, making the breed suitable for accelerated lambing systems (Ramsay et al. 2001; Snyman 2014h).

6.5.7.4 Vandor

Origin, Distribution and Population Statistics: The Vandor sheep was developed by a South African farmer, Mr. CJ Van Vuuren, in 1944, when he crossed Van Rooy and Dorset Horn and later in 1958 with a German Merino in order to get a composite that would produce high-quality wool and meat. A Vandor breeders' society was established in 1968 (Ramsay et al. 2001; Almeida 2011). Presently, the breed is found in central and western South Africa.

Physical Characteristics: The Vandor is a large animal with an average weight of 100 kg for rams and 90 kg for ewes. It has a distinct face with large ears. The body colour is mainly white. The body is covered with wool.

Adaptive and Special Genetic Characteristics: It is especially adapted to the northern Cape Province of South Africa, where the natural environment consists of grassveld and busveld (Ramsay et al. 2001).

Production Characteristics: The Vandor rams are fertile (2 rams per 100 ewes), and in normal conditions, the lambing rate is 150%. The lambs are born small but grow at a very fast rate of 300–400 g per day. The ewes have good mothering instincts, resulting in low mortalities. The rams weigh an average of 40–45 kg in 100–120 days. The sheep are kept for meat and wool production (Ramsay et al. 2001; Snyman 2014h).

6.5.7.5 Meatmaster

Origin: The Meatmaster is a composite breed that was developed in South Africa to survive the harsh, dry, arid and semi-arid environments and produce with low inputs and minimum care. The breed was the result of crossing Damara with several sheep breeds that included Ile de France, Van Rooy, South African Merino, Dorper, Namaqua and Ronderib (Peters et al. 2010).

Physical Characteristics: The breed is small- to medium-framed with a mature weight of 60–90 kg (rams) and 35–50 kg (ewes). The legs are well-placed, enabling it to walk long distances. Both sexes are polled. The body is multicoloured and covered with short shiny hair. The tail is long, thin and carrot-shaped and reaches the hock.

Adaptive and Special Genetic Characteristics: The Meatmaster is adapted to bushveld and savannah environments. It is tick-resistant and tolerant of tick diseases (Ramsay et al. 2001).

Typical Production System and Production Characteristics: The Meatmaster is reared mainly in an extensive production system. The breed matures early, lambs easily and the ewes make excellent mothers, resulting in low mortality rates. Birth weights average 3.2 kg and 2.8 kg for rams and ewes, respectively. The 100-day weaning weight is 27 kg and the animals weigh an average of 47 kg at 270 days. The adult weaning weight is 185 kg for males (Ramsay et al. 2001; Snyman 2014h).

6.5.7.6 Dohne Merino

Origin and Distribution: Synonyms for the Dohne Merino include Dohne and El Dohne Merino. Dohne Merino sheep was developed at the Dohne Agricultural Research Station in Eastern Cape, South Africa (Snyman 2014i). The goal was to develop a hardy sheep able to produce mutton and wool in the harsh Sourveld environment. The Dohne Merino was developed by crossing a local merino with the South African Mutton Merino. Presently, the sheep are farmed in the Eastern Cape, Free State and Western Cape Provinces of South Africa and are also distributed in Namibia. The total number of Dohne Merino sheep is not known (Snyman 2014i).

Physical Characteristics: The Dohne Merino sheep are medium- to large-sized animals with an average live body weight of 80–100 kg (rams) and 50–65 kg (ewes). The coat colour is white and is covered with white wool. The Dohne Merino sheep are polled.

Adaptive Characteristics: The Dohne Merino sheep are hardy and well-adapted to the arid and semi-arid environments.

Typical Production Systems and Production Characteristics: The Dohne are suitable for both extensive and intensive production systems. Merino is a dual-purpose sheep that produces a high-quality carcass. The lambs do not accumulate fat at an early age, and hence, they can be marketed at a later stage. The breed has good fertility, producing lambs with a birth weight of 5.2 kg and 4.9 kg for rams and ewes, respectively. The lambs have a weight of 29.9 kg (rams), 31.1 kg (ewes) at 100 days, 46.3 kg (rams), 37.8 kg (ewes) at 8 months and 60.3 kg (rams), 48.6 kg (ewes) at 12 months. Dohne produces a very high-quality wool with an average fleece weight of 3–5.5 kg with a fibre diameter of 17.21 micron (Ramsay et al. 2001).

6.5.7.7 Dormer

Origin and Distribution: The Dormer breed was developed from crossing Dorset Horn rams and German Merino ewes in 1927. The name Dormer comes from an abbreviation of Dorset-Merino (Dormer Sheep Breeders' Society of South Africa 2022). The goal was to develop a mutton breed that could adapt to the cold and rainy winter conditions experienced in the Western Cape Province of South Africa. A Dormer Breed Society was established in 1965, and the breed was formally recognised in 1970. Presently, it is widely distributed in Gauteng and the Free States of South Africa (Ramsay et al. 2001; Snyman 2014j).

Physical Characteristics: The Dormer is a mutton sheep breed whose body is muscular and covered with white wool. The average adult weight is 75 kg. Both sexes are hornless, with a smooth body and good meat conformation.

Adaptive and Special Genetic Characteristics: The Dormer is adapted to temperate climates and does well in extreme cold and rainy conditions. It is also resistant to the Muellerius lungworm.

Typical Production System and Production Characteristics: The Dormer breed is raised in natural and planted pastures under an extensive production system for meat and wool. The lambs have a good feed conversion rate. The rams can be used in cross-breeding programmes to improve meat production. The breed is very fertile, and multiple births occur frequently. Lambs are born small but grow very rapidly. Birth weights average 3.9 kg and a weaning weight of 29 kg. The age at first lambing is 18 months. The breed matures early and is ready for the market after 3 months, with a live weight of 38.57 kg (rams) and 34.57 kg (ewes) and a carcass weight of 16–22 kg. The sheep produce fleece with an average weight of 3.5–5 kg with a diameter of 27 microns. The wool is strong, free of kemp and coloured fibres. The breed qualities make it a favourite choice of sires used in cross-breeding (Du Toit 2008; Sagne 1950). The efficiency of the Dormer has improved markedly, especially over the past decade, mainly due to performance testing that is compulsory for all South African breeders (Dormer Sheep Breeders' Society of South Africa 2022).

6.5.7.8 Tâadmith

Origin, Distribution and Population Statistics: The Tâadmith breed (Fig. 6.36) originated from a cross between the Eastern Merino and an indigenous breed from the Djelfa region of Algeria (Jores D'Arces 1947; Sagne 1950). The breed was developed in the 1860s at a national research station at Taâdmit, hence its name. The main objective of its development was to improve the

Fig. 6.33 Bergui ewes. (Queue Fine de l'Ouest) (Tozeur). (From Khaldi et al. 2011, licensed under CC-BY 4.0)

Fig. 6.34 Rahmani ram. (From Elshazly and Youngs 2019, licensed under CC-BY 4.0)

Fig. 6.35 Rahmani ewe. (From Elshazly and Youngs 2019, licensed under CC-BY 4.0)

wool quality of the Ouled-Djellal breed (Chellig 1992). The Tâadmit breed has a wide distribution in the central Algerian steppes. Currently, there are only a few hundred animals at the Djelfa Wilaya, especially in the Taâdmit region, due to gradual replacement by the Ouled-Djellal breed. A nucleus herd is currently being maintained at the Institut National de la Recherche Agronomique Research Station at Hmadna (Wilaya of Relizane).

Physical Characteristics: The Tâadmith breed is characterised by a white head with a hooked profile and a long body, with males having large and bulky horns (Fig. 6.36). The animal is leggy, the fleece is extended, covering the forehead, sometimes extending down to the hocks. The wool is superfine to fine. The tail is long and thin.

6.5.7.9 Noire de Thibar

Origin and Distribution: Noire de Thibar (Fig. 6.37) is a Tunisian breed distributed in the basins of Medjerda and Béja. It is a composite breed developed by a monk from the Netherlands called Novat, in the wet and rainy northwestern region of Tunisia. He began the development of this breed in 1908 through the reciprocal crossing of Merinos d'Arles from southern France with the Algerian thin-tail breed (Queue Fine de l'Ouest) (Khaldi 1984). The goal was to produce animals with a uniform black colour that were naturally able to avoid consuming the poisonous St Johnswort (*Hypericum perfoliatum*) plant. In 1945, the breed was officially recognised, and a flock book opened under the name Race ovine Noire de Thibar, where approximately 3.5% of Tunisian sheep livestock are recorded (Chalh et al. 2007).

Physical Characteristics: The Noire de Thibar breed has a uniform black coat (Fig. 6.37), a cylindrical body and a good conformation. The head of the animal is black and has an elongated, concave shape. The sheep are polled in both sexes. According to Mason (1967), rams have a mature average weight of 80–90 kg and a mature height of 60–65 cm. The skin and mucous membranes are pigmented. The entire body is covered with a medium to fine fleece, which weighs 4–5 kg for rams and 2–3 kg for ewes. The skin and mucous membranes are black in colour, leaving only the head, throat, inner side of the tail and perineum exposed (Meyer et al. 2015).

Production Characteristics: The breed has a prolificacy rate of 130–134% with a fertility rate of 68–88% (Djemali et al. 1995). The seasonal anoestrus has been reported to be less among the Tunisian sheep breeds (Gouhis 1989), therefore allowing the breeders the opportunity to adapt their breeding to any season.

6.5.7.10 Sicilo Sarde

Origin, Distribution and Population Statistics: The Sicilo Sarde (Fig. 6.38), also known as Sicilian-Sardinian, is a composite breed developed from crossing an exotic Sardinian and Sicilian sheep. The original stocks of Sardinian and Sicilian sheep were introduced into Tunisia by Italian settlers before the 1900s. By 1949, there were 40,000 Sicilo Sarde sheep in the Tunisian districts of Mateur, Bizerta, Beja, Tunis and Grombalia. The Sicilian sheep had greater influence on the composite Tunisian milk sheep until further large importations of the Sardinian sheep were made in 1948. As a consequence of this, the Sicilio sheep are also known as Sicilian-Sardinian (Mason 1967).

The Sicilo Sarde sheep was not threatened until the 1990s. Since then, the numbers of this breed have fallen sharply from 219,000 heads to 37,000, 20,000 and 8500 heads recorded in the years 1995, 2005 and 2000, respectively (Ouhichi 2014). These sheep are currently being maintained in government-owned farms and private breeders, especially around the Beja region of Tunisia. The breed enjoys considerable socio-economic interest for rural Tunisia and has been prioritised for conservation.

Physical Characteristics: The Sicilo Sarde breed is a medium breed with a height at withers of 70–80 cm, and average adult weights of 45 kg and 65–75 kg for female and male, respectively. The coat colour is variable with white (Fig. 6.38), black or spotted animals. The fleece shape is heterogeneous and is described as jarred. The head is elongated with a straight face profile, which bears a slight curvature, and the muzzle is straight. The Sicilo Sarde sheep has a long, thin tail, and its legs are long and thin. Males have longer horns than females. The neck is long, and the ears are small and horizontal. The rump is long and the abdomen rounded (Rafik et al. 2018). The udder is well developed with straight teats (Khaldi and Farid 1981).

Typical Production Systems and Production Characteristics: The Sicilo Sarde breed is reared mainly for milk and meat under semi-intensive and extensive production systems. The semi-intensive production system is used in government farms and by private large-scale farmers with large flock sizes of between 200 and 300 ewes. The Sicilo Sarde is raised on pasture consisting of natural rangelands, stubble and barley, supplemented with hay, straw and concentrates during the dry periods or when animals are lactating (Rouissi et al. 2001).

In the extensive production system, the animals are reared under free-range grazing of pastures with high supplementation with concentrates (Rekik et al. 2005). Natural matings occur during the spring, which results in most lambings occurring between August and November and peaking between September and October (Djemali et al. 1995; Rouissi et al. 2001).

The Sicilo Sarde sheep is milked for household consumption. The milk is also used to make cheese. The cheese industry has, however, grown and has evolved into a traditional cottage industry, making specialised cheeses using milk from this breed (Khaldi and Farid 1981).

6.5.7.11 Sidi Tabet Cross

Origin, Physical and Production Characteristics: The Sidi Tabet Cross is a composite of Portuguese

Black Merino and Thibar breeds. The breed has a black coat and is polled in both sexes. It is specifically reared for wool production. The wool produced is of medium type (DAD-IS n.d.).

6.5.7.12 Sohagi

Distribution: The Sohagi sheep breed is widely distributed in the Egyptian Governorate of Sohag in southern Egypt.

Physical Characteristics: The body is covered with white wool. The Sohagi has a small head, covered with dark brown wool, with creamy black rings around the eyes. The neck is long, and the animal has long legs. The rams are horned, but females are polled (Elshazly and Youngs 2019).

6.6 Current Status of Sheep Diversity in Africa

6.6.1 Morpho-Biometric Studies and Population Characterisation

Morphological traits are important tools in livestock population characterisation. Furthermore, morphological characters are used by farmers as an indicator of the growth and reproduction performance of their animals. Traits such as the heart girth and wither height measurements are good growth indicators, and measuring the scrotal circumference gives a good indication of ram fertility, hence an indication of how a ram's daughters would perform reproductively (Portes et al. 2018). However, the use of morphological traits is often misunderstood and may be erroneously linked to practices that do not conform to cultural traditions, including superstition.

Morpho-biometric traits are therefore important tools in the characterisation of sheep populations and are complementary to the molecular markers used in population genetic diversity studies. African sheep breeds have been largely characterised using morphological characters such as rump width, rump length, tail length, wither height, heart girth, paunch girth, rump height, ear length, foreleg length, rear-leg length, body length, shoulder width, neck circumference, head length, head width, horn length and hock length and also the size of udder and quality of wool, amongst others.

These morphometric characters have been used to characterise the Nigerian Yankasa, Balami, Uda, and West African Dwarf sheep (WAD) breeds (Yunusa et al. 2013) where the results showed that the tail length was the most discriminating trait in the four populations, and the longest genetic distance was found between WAD and Uda sheep, while the closest breeds were Balami and Uda (Yunusa et al. 2013). A morphometric study of WAD sheep in Guinea Bissau showed phenotypic diversity in the studied population and three genetic groups, namely, large, intermediate and small (Dayo et al. 2022).

Similarly, Belharfi et al (Belharfi et al. 2017a). used 13 biometric traits to characterise six sheep breeds from western Algeria, namely, the Ouled Djellal, Hamra, Barbarine, Srandi, Daraa and Tazegzawt breeds, and the results showed distinct differences between males and females. Principal component analyses (PCA) of the measurements revealed three distinct clades of sheep, with one clade representing most of the breeds, an indication that the sheep breeds from this region have undergone dilution due to a high rate of gene flow between breeds. These results were later confirmed using genome-wide analysis (Gaouar et al. 2016).

Morphometric measurements analysed from 280 adult WAD sheep (77 males, 203 females) from the Sudano-Guinean zone of Cameroon showed high variability with distinct characteristics that can be used to distinguish the sheep (Deribe et al. 2021). Three genetic types with high intra-genetic variability clustered into two sub-groups of the local sheep population. The study concluded that selection is the most appropriate method to improve the sheep population.

A total of 22 phenotypic measurements taken from 288 Blackbelly sheep from three Central African countries, Cameroun, Congo Brazzaville and Gabon, revealed phenotypic dimorphism between the Blackbelly sheep populations in the three countries, with the tallest and heaviest animals from Congo Brazzaville (head width = 62.217 ± 5.288 cm and body weight = 27.44 ± 6.08 kg), the longest from Gabon (TBL = 84.69 ± 8.70) while the largest were from Cameroon (CW = 15.01 ± 2.531 and rump width = 14.716 ± 2.351) (Meka et al. 2021). Further analyses using PCA revealed that six components could account for 73.1% of the observed variability in sheep body measurements. The study revealed a perfect correlation between the thoracic circumference and body weight, suggesting that the thoracic circumference could be used as an indirect indicator of body weight (Meka et al. 2021).

The morphological diversity of three fat-tailed and three thin-tailed indigenous sheep breeds of Ethiopia was determined based on live body weight and linear body measurements and revealed significant variations between breeds (Deribe et al. 2021). Multivariate analyses clustered the studied sheep breeds into distinct populations, with Mahalanobis distance showing significant differences between breeds.

Mavule and others (2016) examined eight Zulu sheep populations in KwaZulu-Natal sampled from Empangeni, Escourt, Eshowe, Jozini, Msinga, Mtubatuba, Nongoma and Nqutu rural communities, with findings that dark-brown coat colour, either in solid form or white patched, was most frequent. Analyses of linear body measurements and qualitative traits showed significant variations, with an indication that sheep from Nqutu had the highest estimates in most morphometric variables, while sheep from Empangeni had the lowest. Discriminant analysis identified rump width, head width, heart girth, thorax depth and tail length as the most distinguishing variables amongst sheep populations. Hierarchical cluster analysis revealed two major groups, one formed by populations from Empangeni, Mtubatuba and Nongoma, and the other by Jozini, Msinga and Eshowe populations. Estcourt and Nqutu were separated from these groups as individual entities. About 62% of individual sheep could be correctly identified with populations from which they were sampled. Nqutu had the highest percentage (88.9%) of correct assignments, whilst Mtubatuba had the lowest (46.7%).

High assignment errors were ascribed to gene flow between sheep populations.

Udder measurements have also been effectively used to characterise sheep breeds. James et al (2009). evaluated udder and teat morphometrics, including udder and teat length, width, circumference, distances between teats, udder and teat shape of 143 extensively managed WAD sheep in southwestern Nigeria and reported that age, pregnancy and lactation status greatly influenced the teat length and udder circumference. Udder and teat dimensions significantly increase with age, while the smallest udder and teat dimensions were found in lactating and dry or non-lactating animals.

Analyses of sheep wool quality and dimensions undertaken in western Algeria on a total of 60 sheep sampled from the Ouled Djellal, Hamra, Barbarine, Srandi, Daraa and Rembi breeds by Belharfi et al. (2017b). showed that sex did not affect wool finesse in the breeds. The study also reported significant differences in the average wool fibre diameter among the breeds studied.

6.6.2 Biochemical Characterisation of African Sheep

The use of protein polymorphisms allowed for the characterisation of livestock species before the advent of molecular genomics. The technique uses the principle of separating proteins according to their electric charge and molecular weight.

Due to the degeneration of the genetic code, approximately three out of four mutations can cause an amino acid to be substituted for another during protein synthesis. In addition, only a third of these substitutions, on average, can change the net charge of the protein and are therefore detectable by electrophoresis. This, therefore, makes it possible to detect at the protein level about a quarter of the mutations existing in the coding part of the corresponding gene (Grosclaude 1988). The immunological and biochemical characters have a fairly simple mode of hereditary transmission that conforms to the laws of Mendelian genetics. Some of the best-known traits used for characterisation are blood groups

Fig. 6.36 Taâdmit ram in Djelfa, Taâdmit. (From Djaout et al. 2017, licensed under CC-BY 4.0)

and blood proteins. They are genetically independent of each other; their analysis makes it possible to better characterise the hereditary attributes of breeds compared to studies based on morphological characters.

Eight polymorphic loci were used to characterise five Swiss sheep breeds (Blanc des Alpes, Brun-noire du pays, Oxford, Nez noir du Valais and sheep from Hauts-Grisons) (Delacrétaz-Wolff 1997) and differences in frequencies obtained for blood specificities were tested and found to be statistically significant, thus allowing these sheep to be characterised using blood groups.

In sheep, 80 proteins were identified, 14 of which are polymorphic. The serum proteins include transferrin, albumin, haemoglobin, amylase I and carbonic anhydrase II. Of these, transferrin is the most polymorphic serum protein in sheep, with 12 clearly recognised variants (Delacrétaz-Wolff 1997) including transferrin H and K, which were found in 6 sheep breeds from the former Czechoslovakia (Stratil 2009). Nigussie et al (2016). investigated protein (haemoglobin and serum albumin) polymorphism and reported three phenotypes for both loci in indigenous sheep breeds of eastern Ethiopia. The frequency of HbA estimated at 0.77 was higher in Hararghe highland sheep, while HbB was higher in Afar (0.7) and Black head Somali (0.67), with heterozygosity (H_E) ranging from 0.42 ± 0.09 in Hararghe highland sheep to 0.48 ± 0.04 in Afar.

The closest genetic relationship was between Afar and Black head Somali ($D = 0.01$), while Afar and Hararghe highland sheep were more distant apart ($D = 0.22$). The genetic relationship in the three indigenous Ethiopian breeds was associated with their adaptation to the environment/agroecology. It was demonstrated that the occurrence of errors in sheep pedigrees can be identified by the use of protein polymorphic traits as genetic markers (Wang and Foote 1990). The authors carried out a sheep parentage test using four polymorphic proteins (transferrin, serum arylesterase, haemoglobin and a red-blood-cell lysate protein) and observed that a minimum of 9.2% of the progeny had errors in parentage records. The efficiency of protein genetic markers in solving uncertain parentage cases was 67.2%. An estimate of 58.3% among all cases were solved using genetic incompatibility with transferrin alone.

In cattle, the synthesis of 1000 publications on protein polymorphism by Baker and Manwell (1980) made it possible to establish the phylogenetic tree of 196 breeds. For this study, ten polymorphic loci were identified, including the five previously mentioned blood proteins and five milk proteins. In order to understand the role milk proteins play in determining the quality of milk and cheese, a number of studies have been done using milk from livestock such as cattle and goat milk, though not as many studies have been done in sheep. However, a few significant studies have been done on sheep (Stratil 2009).

Other significant work utilising blood proteins for characterisation conducted on African sheep includes work done on Algerian sheep (Sargent et al. 1999), West African Dwarf sheep breeds (Osamede and Adebowale 2018), on Kenyan breeds and (Sargent et al. 1999) on South African breeds (Sargent et al. 1999).

6.6.3 Genetic Diversity and Population Structure of African Sheep

A new dawn arose for the characterisation of African sheep populations with the discovery of

Fig. 6.37 Noire de Thibar in Zaghouène in Tunisia. (From Djaout et al. 2017, licensed under CC-BY 4.0)

Fig. 6.38 Sicilo Sarde. (From Aloulou et al. 2018, licensed under CC-BY 4.0)

gel electrophoresis, which allowed for protein and later genetic polymorphisms to be separated. The first application of gel electrophoresis was performed by Poulik and Smithies (1958) and quickly spread in the early 1960s, giving rise to a profusion of studies, especially on blood proteins, including the erythrocyte enzyme characterisation work published in 1975 by McDermid and others (McDermid et al. 2009). These foundational studies made it possible to extend the list of markers, including blood polymorphisms, to approximately 30–40 markers per species. The number, however, was still too small to allow for chromosome mapping of breeds. This ultimately led to the development of techniques that were more sensitive, such as those detecting polymorphisms at the DNA level. Among the first such techniques to be developed was the Restriction Fragment Length Polymorphism (RFLP), which was quickly followed by mitochondrial DNA, microsatellites and ultimately genomic sequencing.

Restriction Fragment Length Polymorphisms have been used to characterise African sheep breeds. Investigations on three mutations (Saleh Ahmed et al. 2020) in three candidate genes found in six Egyptian breeds showed that all the breeds did not carry the Fec-B mutation, while the mutations of FecXG and Fec-GH were detected in Rahmani and Rahmani X Barki crosses that are associated with high twinning rates found in the crosses. The characterisation of four Nigerian sheep breeds (Abubakar et al. 2020) using RFLP, namely, Balami, Yankasa, Ouda and WAD, revealed high genetic polymorphisms in the loci studied, high gene flow between the populations, high within-population genetic variances, and suggested that these populations need greater attention to prevent further genetic erosion.

Mitochondrial DNA (MtDNA) sequencing has been extensively used to study domestication of sheep, centres of origin, migration routes followed by the sheep and including studies on how the populations have changed over time. Several studies using MtDNA on African sheep include that by Ghernouti et al (2017). to determine the origin of three Algerian sheep breeds (Ouled Djellal, Rembi and Berbère). The study showed that 87% of the Algerian breeds belong to the haplotype B, with the rest belonging to haplotype C and that the Algerian sheep arrived through three different migration routes, with the Ouled Djellal being the last breed to arrive in Algeria from the Middle East. Similarly, work on Kenyan sheep (Kandoussi et al. 2020) showed that the sheep breeds belonged to haplotype B and had not mixed with sheep from West Africa, while those from Ethiopia belonged to haplotype A and B, with more sheep having haplotype B than A. Phylogeographic analyses of the African sheep breeds identified ancestral haplotypes that are likely to have been introduced to Africa and are now widely distributed throughout the continent (Poulik and Smithies 1958). The genetic relationship between Moroccan and Iberian sheep using mtDNA sequencing was determined (Kandoussi et al. 2020). This study showed that the first founder Moroccan sheep populations were Iberian in origin (79%), with 21% originating from a territory between the Middle East and the African continent, with population expansion dates estimated at between 7100 and 8600 years BP, respectively. This suggests that neolithisation was introduced through a double influence, from Iberia and from another route, maybe Oriental or sub-Saharan.

In order to investigate the origin of Cameroonian WAD sheep (Djallonke), 16 sequences of a 989 bp fragment of the MtDNA D-loop region in West Cameroon were analysed with previously published sequences from African, European, Japanese and Chinese sheep genotypes (Justin et al. 2022). The phylogenetic analyses revealed that the Cameroon sheep belong to haplotype B and all had a common origin with the African, European and Asian breeds. There is a likelihood that African breeds were domesticated through a series of events.

Wanjala et al. (2021) recently conducted a meta-analysis on 399 previously sequenced African indigenous sheep breeds and showed that overall African sheep populations have undergone selective signatures for adaptation. The results indicated heavy selective pressure, mainly driven by adverse production environments being experienced by East African sheep populations. Analyses of the West and North African sheep populations showed that these populations recently expanded from small populations to the present-day populations (Wanjala et al. 2021) with genetic diversity in different populations mainly being attributed to the effects of breed and location (Aboul-Naga et al. 2023).

DNA microarray studies on African sheep populations have been undertaken on Algerian sheep breeds (Gaouar et al. 2016), with an indication that the populations have been influenced by African and European (especially Italian) breeds. Some of the breeds, especially the Berber and Rembi, and to a lesser extent the Barbarine, have undergone extreme dilution of their genetic diversity due to uncontrolled mating with the Ouled-Djellal breed. The information obtained from this study should be used to inform conservation and population management strategies.

DNA sequencing is an optimal method for the determination of species identity and gene discovery. Until recently, obtaining high-quality genomic sequences for all species was laborious and time consuming (Satam et al. 2023). Several African sheep breeds have been characterised using genomic techniques to determine genetic diversity and breed status (Djaout et al. 2017; Ouhrouch et al. 2021). Genome-wide analysis as well as single nucleotide polymorphism (SNP) genotyping of African sheep breeds have provided information that could guide breeding, conservation and management strategies for the indigenous African sheep populations (Gaouar et al. 2017; Edea et al. 2019; Álvarez et al. 2020; Belabdi et al. 2019; Abied et al. 2021).

Whole genome sequencing has also revealed associations with genes of interest, including genes involved with adaptations of sheep to their environments, disease resistance, phenotypic trait selection among many others (Álvarez et al. 2020; Mastrangelo et al. 2019; Yaro et al. 2019; Wiener et al. 2021), providing key information that can be applied to management, breeding and sheep development programmes. A recent study (Aboul-Naga et al. 2023) identified candidate genes and variants associated with heat stress adaptation in three Egyptian sheep breeds (Saidi, Wahati and Barki). The SNPs associated with heat tolerance were located in MYO5A, PRKG1, GSTCD and RTN1 genes. Association between genetic and phenotypic variations also showed that OAR1_18300122.1, located in ST3GAL3, had the greatest positive effect on heat tolerance. Genome-wide association analysis further identified SNPs associated with heat tolerance in the PLCB1, STEAP3, KSR2, UNC13C, PEBP4 and GPAT2 genes.

6.7 Conservation of Sheep Genetic Resources

Following domestication and the dispersal of sheep throughout Africa, the livestock keepers of Africa have, over the millennia, selected and bred animals to meet their particular needs (Gifford-Gonzalez and Hanotte 2011). This has resulted in a large number of indigenous sheep breeds that have developed a large and diverse range of genes enabling the species to inhabit and produce in a wide range of environments, including environments that are prone to disease, drought and are considered extreme in nature.

Some of the African unique genetic resources are, however, at risk of becoming extinct (Table 6.4) or being subjected to dilution due to indiscriminate cross-breeding among local breeds and between exotics and local breeds. This has led to the continuous genetic erosion and loss of genetic diversity, leading to the loss of breeds carrying unique adaptive traits before these resources have been fully documented and characterised.

The indigenous genetic resources of Africa can make the continent self-sufficient in the production of meat and meat products; hence, there is a great urgency to ensure that they are conserved. Conservation of this unique genetic resource is also important for the African livestock keepers who own and manage these genetic resources and depend on them for their livelihood (Mwai et al. 2015). This would only be possible with concerted efforts.

Two basic strategies for the conservation of indigenous genetic resources can be adopted, in situ and ex situ conservation. In situ conservation involves the conservation of biological elements, including species in the environment where they are naturally found (their natural ecological and livestock production environments or in the wild) (see Sect. 25.3 of Chap. 25 for in situ conservation methods), while ex situ conservation refers to the conservation of biological elements, including species outside the environment where they are naturally found (Maxted 2013). See Sect. 25.4 of Chap. 25 for a detailed discussion on ex situ conservation methods.

Table 6.4 Threatened conventional sheep breeds in Africa that need conservation (AU-IBAR 2019)

Sheep breed	Country
Mouton Kirdi à poils ras	Cameroon
Hamari	Ethiopia
Djallonké	Nigeria, Gambia and Liberia
Red Maasai Sheep	Kenya and Tanzania
East African long-tailed	Kenya and Tanzania
Mouton à laine du Macina	Mali
Mouton Maure à poil long	Mauritania
Mouton Maure poil ras (Ladem)	Mauritania
Landim	Mozambique
Hamari	Sudan
Blackhead Persian	Tanzania
Urambo	Tanzania

The Global Plan of Action (GPA) for Animal Genetic Resources, and specifically The Strategic Priority 10, recognises the need to establish national conservation policies and establish and strengthen in situ and ex situ national conservation programmes, including those found in Africa.

6.7.1 Sustainable Use of Sheep Genetic Resources

More than 6379 documented breeds/populations comprising approximately 30 livestock species have been developed over the 12,000 years since they were first domesticated (Scherf 2000; Gibson et al. 2012). These breeds have developed adaptations that allow them to be productive in a wide range of situations, including some of the most hostile natural environments ever inhabited by humans. These naturally evolving genetic characteristics provide a consistent set of enduring solutions for disease resistance, survival and yield. It is estimated that 35% of mammalian breeds and 63% of avian breeds are threatened with extinction and one breed is lost every week. Although not clearly established, threats to animal genetic resources in developing countries appear to be increasing rapidly, mainly due to rapid changes in production systems and the extensive use of interbreeding. Emerging threats, such as the implementation of slaughter policies to prevent the spread of diseases in highly traded livestock such as foot-and-mouth disease, and zoonoses such as bovine spongiform encephalopathy and avian influenza, may exacerbate the disease risk of extinction of races in developed countries and also in developing countries. The effective conservation of livestock genetic resources, whether in situ or ex situ, involves the mobilisation of significant social and economic resources over long periods of time. These resources are often available in developed countries, where there is often a high proportion of rare breeds of livestock that are already efficiently preserved. However, the majority of genetic diversity in livestock is found in developing countries, where resources for conservation are scarce. Therefore, it seems likely that difficult choices will have to be made about what to conserve with the limited resources available. Information on genetic diversity is useful in optimising conservation and strategies for the use of agricultural genetic resources. Ideally, it is important to ensure that all existing genetic variations remain available for future use in the most cost-effective manner. In practice, there will often be insufficient resources to maintain the full genetic diversity of a given species. Even when resources are sufficient, we do not have complete knowledge of all the genetic functional variations of a species. Thus, ensuring the conservation of almost 100% of the functional variations would imply the creation of an inefficient process of retaining many more individuals or populations (breeds, memberships) than would be necessary if we had complete information (Gibson et al. 2012).

Since African indigenous sheep breeds play an important role in the livelihood of communal farmers, their utilisation should be approached taking into consideration the socio-ecological system within which they are found. Many factors hinder the sustainable utilisation of sheep genetic resources in rural communities and need to be addressed. These include lack of structured breeding programmes, poor record keeping, poor feed resources, disease and health challenges, as well as lack of technical skills and knowledge (Molotsi et al. 2020). Genomic breeding (see Chap. 20) can be applied to improve areas of environmental sustainability of African smallholder sheep farming systems but must target specific production environments, challenges, as well as the opportunities of smallholder production (Molotsi et al. 2020).

In general, an animal resource management programme is more important and successful when it integrates the triangular holder 'Return with Goal Accomplished' (RGA)-decision makers-researchers (this type of model is very well applied in Morocco, in particular via the national association of sheep and goats, ANOC). However, our experience in Africa leads us to adopt most often a method with two actors, holder RGA-researchers, but this action must go through an important effort of sensitisation of the breeders on the risks of the extinction of the races and the efficiency of African researchers (this type of model is in the process of being applied, in particular in Algeria via the genetics training team at the University of Tlemcen) (Gaouar 2002; Gaouar and de La Biodiversité 2009). The lack of a state animal identification policy in most African countries is a major problem in the management and development of domestic animals.

6.8 Sheep Breeding Goals and Strategies for Sustainable Genetic Improvement

The aim of any breed improvement project is to improve animal traits with a futuristic goal of ending up with an improved population with the best animal phenotype. For sheep, the goals mainly include increased production of meat, milk, wool and resistance to parasites for increased revenue to the farmer (see Chaps. 15 and 17 for a detailed discussion on breeding goals and strategies).

In Africa, the majority of sheep breeds are reared under traditional breeding systems characterised by a lack of improved modern breeding technologies (de Aguiar et al. 2020). There is a need to consider differences in objectives of production, farmers' trait preferences as well as breeding practices in various systems of production before designing sustainable breeding strategies to improve productivity (Nigussie et al. 2013) (see Sect. 17.2.1 of Chap. 17). Preferences for traits and order of importance vary depending on the environmental conditions of the region and farming systems (Table 6.5). Breeding objectives that incorporate adaptive traits, reproduction and production could be derived by combining highly favoured and measurable characteristics that are heritable (Molotsi et al. 2020). From an extensive review on relevant genetic traits to sustain smallholder sheep farming systems in South Africa, the authors emphasised growth, carcass and reproduction traits for high input as well as high output farming systems within high rainfall areas; survival, disease or pathogen resistance, adaptability and reproduction could be more important in low input and low output farming systems in drier arid regions, corroborating findings (de Aguiar et al. 2020) on community-based breeding programmes in Brazil. Therefore, selection criteria at the smallholder level should include estimated breeding values of traits that are relevant to the breeding objective as well as preferences of farmers (Amare et al. 2018; Abebe et al. 2020).

Preferred traits in defining breeding objectives under crop-livestock mixed systems were increasing meat production (improving growth rate and conformation), while improving milk yield and meat production were favoured in pastoral and agropastoral systems (Nigussie et al. 2013). General appearance, fast growth and colour were the most preferred characteristics in selecting rams for breeding. For ewes, general appearance and traits related to mothering ability were favoured. Tail type and adaptation to heat stress and feed shortage were preferred in the pastoral production systems compared to agropastoral and mixed crop-livestock systems.

Table 6.5 Trait preference, selection criteria and breeding strategies for genetic improvement of economically important traits in sheep

Preferences by farmers in order of importance	Breeding objectives	Selection criteria	Breeding practices/strategies	Sheep breed/country
Traits: good mothering ability, coat colour, large body size, short lambing interval, rapid growth rate *Purpose of sheep farming*: Income generation, meat and manure	Production, reproduction and adaptive sheep traits	Body size, brown coat colour, growth rate, horn, ear size and tail type	Selection of rams and ewes within the sheep flock using diverse criteria (body size, coat colour, etc.); culling of sheep due to small body size, unfavourable coat colour, old age, fertility problems for both male and female sheep and poor mothering ability of ewes. Although pedigree-based records are lacking, random mating with communal rams is considered to reduce the effect of inbreeding (Abebe et al. 2020)	Northwest highlands of Ethiopia (Gifford-Gonzalez and Hanotte 2011)

(continued)

Table 6.5 (continued)

Preferences by farmers in order of importance	Breeding objectives	Selection criteria	Breeding practices/strategies	Sheep breed/country
Growth rate, body size, marketing value	Twin birth type, fast age at first lambing, short lambing interval and fast weaning rate for ewes and rams	Growth rate, body size, weaning rate, body conformation, age at first lambing, lambing rate, drought resistance, colour type, disease resistance, hair type, marketing value, twine rate, horn type, tail type and temperament traits	Village community-based control sheep cross-breeding with communal selected ram use; Private-controlled, mixed-type, village communities-based controlled breeding and random mating practices All breeding strategies and strata had their own source of ram breed type, which could be obtained from their own flock, donor project, purchased from local markets and government supply from the nucleus farm (Amare et al. 2018)	Wollo and Washera highland sheep, Ethiopia (Snowder and Fogarty 2008)
Income generation, breeding, meat production, manure	–	Appearance/body size, colour, fast growth and pedigree	Natural mating, uncontrolled mating due to communal grazing and watering point and lack of a sufficient number of rams (Haile et al. 2015)	North Shoa, Ethiopia (Olivier et al. 2001)
No studies have been done in South Africa to identify traits' preference by smallholder sheep farmers	–	1. Reproduction: wet-dry, fertility (ewes lambing per ewe joined), number of lambs born/ewe joined, number of lambs weaned/ewe joined, Lamb survival; 2. disease and pathogen resistance: faecal egg count, tick resistance; 3. growth traits: birth weight, adult weight; 4. feed conversion ratio; 5. Carcass weight	Application of traditional breeding and genomics will enhance sustainable genetic improvement in low-input farming systems and commercial intensive sheep farming (Molotsi et al. 2020)	South Africa (McKinnon and Rocha 1985)

(continued)

Table 6.5 (continued)

Preferences by farmers in order of importance	Breeding objectives	Selection criteria	Breeding practices/strategies	Sheep breed/country
Cultural/family traditions, food source for family, savings, personal fulfilment and income generation	Adaptation of the animals to the climate and management conditions of the region, the resistance to diseases and productive performance	Adaptability, disease resistance, host resistance of worms, growth rate, fertility, longevity, body conformation, body size and age at first lambing	Community-based breeding programmes which take into account the desires and needs of the local farmers (de Aguiar et al. 2020)	Brazil (Maxted 2013)

The most economically important trait in sheep is reproduction, which is considered to be complex due to the fact that its expression is closely tied to other traits (Molotsi et al. 2020; Snowder 2008). Reproduction is measured by component traits, including the number of lambs born per ewe's lifetime, or number of lambs weaned per ewe lifetime, litter size, conception rate, lamb survival and mothering ability (Snowder and Fogarty 2008). Reproductive performance has been defined as the total weight of lambs weaned per ewe joined (Olivier et al. 2001), which could be improved genetically by direct or indirect selection. As a composite trait, genetic improvement can occur by selecting individual component traits, some combinations of individual component traits, or direct selection for the composite trait. It has been considered that rather than selecting for a single component trait, selection for a composite trait, for example, litter weight weaned, will improve reproductive efficiency under most production and environmental systems (Snowder 2008). Furthermore, selection for litter weight could lead to a balanced biological composite trait with favourable responses in component traits like fertility, number of lambs at birth, lamb survival, lactation and lamb growth. Despite the fact that reproduction and survival traits are generally lowly heritable in livestock, genetic progress was possible for over 20 years of selection in South African sheep.

Strategies for genetic improvement could follow three pathways (Tibbo et al. 2006). They include (i) selection within breed, (ii) selection between breeds and (iii) cross-breeding (see Sect. 17.3 of Chap. 17). Due to the low productivity of indigenous African sheep breeds, strategies involving cross-breeding in different regions to upgrade performance have generally been unsuccessful due to poor adaptability of crossbreds with high exotic blood (Gizaw et al. 2013). The exception is Dorper sheep of South Africa, which is the most successful sheep breed in Africa developed through cross-breeding (Sect. 17.3.1 of Chap. 17). Indiscriminate cross-breeding has proven to be unsuitable in many production areas (Gibson 2007). Current strategies for breeding of sheep involve indiscriminate mating and cross-breeding, village-controlled breeding and privately owned controlled breeding (Amare et al. 2018). Smallholder sheep producers achieve their breeding objectives by introducing new superior rams from highly performing indigenous breeds for cross-breeding with local breeds.

6.8.1 Pure Breeding

Selection and mating within a breed based on measurable productivity traits such as growth, survival and litter size (Tibbo et al. 2006) constitutes pure breeding. Considerable genetic improvement is possible when within-breed genetic variation for traits of interest is substantial. To achieve a successful within-breed improvement programme, there is a need to define overall breeding objectives, characterise production systems, identify suitable breeds, identify the breeding goal traits and derive goal values for each of the traits. Breeding goal traits should have (i) large genetic variability; (ii) easily and cheaply measurable, or must have high genetic correlation with a trait (indicator trait) that is easily measurable, has high heritability or can be measured earlier in life than the goal trait it represents; and (iii) desirable economic value, either as a marketable commodity or as a means of reducing production costs (Tibbo et al. 2006).

To guarantee long-term genetic improvement, indigenous genetic diversity conservation, protection of ecosystem health, as well as improvement of livelihood, a participatory programme has been proposed (Tibbo et al. 2006). The authors proposed an open-nucleus breeding system with flocks on government ranches at topmost of a three-tier system of flocks. High-performing breeding ewes are brought in from sub-nucleus herds for pure-breeding to nucleus flocks in the ranches. Superior rams from the ranches are selected and distributed to participating farmers in sub-nucleus flocks for mating. Afterwards, village flocks receive selected superior rams from the sub-nucleus herds. To ensure success, the programme should be managed by a nationally authorised Animal Genetic Resources Institution, collaborating with research institutions and overseeing all activities related to the programme.

As a case study in Kenya demonstrated (Zonabend König et al. 2017) pure breeding of Red Maasai under adverse environmental conditions improved meat production and livelihoods. With a well-designed nucleus breeding scheme, considerable genetic gain in live weight and carcass weight at the commercial level was realised, although there was a lag period until the improvement in any trait reached the commercial tier. However, the selection of female replacements in the commercial tier would reduce the lag but requires resources for extended sheep recording. The predictable gain in a low heritability trait, such as survival, is low. Cross-breeding of Red Maasai with Dorper was not recommended for harsh environmental conditions, except that Dorper is used as a terminal sire breed for crossing with Red Maasai ewes in less harsh environments. See Sect. 17.4 of Chap. 17 for examples of other indigenous sheep breeds (Djallonke and D'man) improved through targeted selection.

It has also been demonstrated in the dairy industry within the European setting that the most effective selection scheme for local dairy sheep breeds is based on pyramidal management of the population with breeders of nucleus flocks at the top, where pedigree and official milk recording, artificial insemination, controlled natural mating and breeding value estimation are carried out to generate genetic progress (Gibson et al. 2012). Genetic improvement realised is transferred to commercial flocks through artificial insemination or the use of superior rams for natural mating.

6.8.2 Community-Based Breeding Programme

Livestock improvement programmes in Africa have not always achieved high levels of success in smallholder settings, especially those in low-input production systems and with small flocks of sheep and goats. In these setups, farmers have limited access to breeding stock and rely mainly on traditional community breeding practices. The community-based breeding programme (CBBP) was developed as a more suitable alternative for such smallholder settings and has seen greater success in some African countries, especially in Ethiopia, where it has been widely implemented (Gizaw et al. 2013; Haile et al. 2019; Weldemariam and Mezgebe 2021). The CBBP's success comes from the involvement of farmers in the design of breeding programmes from the project inception stage, through to implementation. In this model, decisions are made by the farmers and they actively take part in the breeding improvement programmes, and their views, choices, and decisions are taken into consideration and incorporated. The farmers are organised into groups and receive technical input and support from extension workers and government- and institutional-based breeders who transmit sound breeding methods. The males used for breeding are selected from the community, and thus, local and indigenous genetic resources have a great influence on the breeding outcomes. This approach has seen small but significant improvements in the productivity of indigenous breeds, without undermining the breeds' resilience, genetic diversity and with interventions that are within the

6.9 Responses of Existing Sheep Breeds to Selection

Phenotypic changes under selection of sheep are ongoing, resulting in a variety of current breeds adapted to various environments for the production of meat, milk and fine wool. In a broader sense, genome-wide analysis of the world's sheep breeds revealed strong selection signals with 31 regions comprising genes for coat pigmentation, skeletal morphology, body size, growth and reproduction, and adaptation to diverse environments (Kijas et al. 2012; Fariello et al. 2014). The strongest selection signal occurred in response to breeding for polledness (Kijas et al. 2012). Some indigenous sheep breeds of Africa (Djallonke and D'man) have been improved for some economic traits through targeted selection (see Sect. 17.4 of Chap. 17).

Limited number of studies have been reported on the evaluation of direct and correlated selection responses for economically important traits in African sheep breeds. A study involving Merino sheep of South Africa showed that direct selection for total weaning weight (Table 6.6) would be the best selection criterion to improve reproductive performance without increasing litter size at birth (Olivier et al. 2001). This is important if an increase in the number of lambs is not desirable, where lambing and weaning percentages are high.

6.9.1 Responses to Selection for Meat, Milk and Wool Traits

6.9.1.1 Responses to Selection for Meat Traits

Litter weight weaned is an important trait to target when the aim is to increase the overall reproductive rate in sheep because it has a positive and direct impact on the income of a farming enterprise (Yaro et al. 2019; Olivier et al. 2001). It can be used as a biological index for selection (Yaro et al. 2019). Although the heritability estimates are low (0.1–0.19) for litter weight at birth (Bemji et al. 1996; Boujenane et al. 2013) and at weaning (Boujenane et al. 2013; Olivier et al. 2001), intense selection can result in a favourable selection response given their large phenotypic variances (Snowder 2008). Long-term selection for litter weight weaned would result in a balanced biological system within the environment and production system selected. Although a small negative genetic correlation may exist between litter weight weaned and grease fleece weight, there is no other antagonistic genetic correlation with litter weight weaned.

Selection responses for weaning weight and total weaning weight (TWW) were expressed as the average gain (kg) per generation, and for the number of lambs born and the number of lambs weaned as the average gain in the number of lambs born or weaned per generation (Olivier et al. 2001). Expected direct selection responses by two populations of Merino sheep were higher for TWW compared to the other reproductive traits studied (Table 6.6). Furthermore, correlated selection responses per generation based on TWW were lower for all traits compared to expected direct selection responses.

Reports indicate that the Dormer direct breeding value for weaning weight increased by 0.12% during a period of 17 years, while post-weaning weights improved by 0.32% per annum, and the average predicted direct breeding values of birth weight decreased by 0.055% (Zishiri 2009). The Ile de France sheep breed saw an increase in predicted breeding value at birth (0.025% per annum), averaged direct breeding values for pre-weaning weight and weaning weight increased at an annual rate of 0.23% and 1.21%, respectively. In the Merino Land sheep, the predicted direct breeding value for birth weights decreased by 0.04% per annum, and pre-weaning and weaning weights increased by 0.36% and 0.10%, respectively. For D'man sheep, estimated annual genetic progress was very small, being 0.0009 lambs for litter size at birth and − 0.0078 kg for ewe's body weight (Boujenane et al. 2013).

Table 6.6 Heritabilities, expected direct and correlated responses to selection for some reproductive traits in sheep

Sheep Breed	Country	Trait	Heritability	Direct selection response per generation	Correlated response with total weaning weight	References
Grootfontein Merino	South Africa	Number of lambs born	0.23	0.41	0.33	Olivier et al. (2001)
Carnarvon Merino	South Africa	Number of lambs born	0.19	0.27	0.25	Olivier et al. (2001)
Grootfontein Merino	South Africa	Number of lambs weaned	0.17	0.31	0.33	Olivier et al. (2001)
Carnarvon Merino	South Africa	Number of lambs weaned	0.16	0.24	0.26	Olivier et al. (2001)
Grootfontein Merino	South Africa	Total weaning weight	0.19	9.03	–	Olivier et al. (2001)
Carnarvon Merino	South Africa	Total weaning weight	0.21	6.37	–	Olivier et al. (2001)
Grootfontein Merino	South Africa	Weaning weight	0.21	1.16	0.70	Olivier et al. (2001)
Carnarvon Merino	South Africa	Weaning weight	0.30	1.36	0.89	Olivier et al. (2001)
Sabi	Zimbabwe	Weaning weight	0.38	0.14[a]	–	Assan et al. (2002)
Sabi	Zimbabwe	Birth weight	0.27	0.80[a]	0.26	Assan et al. (2002)
Yankassa	Nigeria	Birth weight	0.41 ± 0.08	–	–	Bemji et al. (1996)
Yankassa	Nigeria	Litter weight at birth	0.12 ± 0.07	–	–	Bemji et al. (1996)
D'man	Algeria	Litter weight at birth	0.1	–	–	Boujenane et al. (2013)
D'man	Algeria	Total weaning weight	0.1	–	–	Boujenane et al. (2013)
D'man	Algeria	Litter size at birth		0.0009[a] lambs		Boujenane et al. (2013)

[a]Actual estimates

6.9.1.2 Responses to Selection for Milk Production

Milk production is part of the objective of sheep production in pastoral and agropastoral systems, but not common in the mixed crop-livestock farming (Nigussie et al. 2013). The expected gain in low heritability traits like milk yield and survival is low (Zonabend König et al. 2017). The low heritability for milk yield in African sheep populations is likely due to the fact that sustained direct selection for milk yield has not been implemented for most African breeds. Within the European setting, realised genetic gains of most breeds show an increasing trend of milk production as a result of the development of management techniques (Carta et al. 2009). However, in semi-extensive systems where grazing represents an important portion of feeding, this trend is lower and irregular because of annual variations of herbage availability to which these systems are very sensitive (Carta et al. 2009).

6.9.1.3 Responses to Selection for Wool Traits

Genetic trends observed in Karakul sheep (Greeff et al. 1993) indicated that selection for pelt traits can be relatively effective based on realised heritabilities of 0.20, 0.34, 0.63 and 0.56 for pattern, hair quality, hair length and curl development, respectively. Corresponding estimates for mean annual response to selection were 2.52%, 1.31%, 2.96% and 7.1%. The control flock did not remain genetically stable, as pattern and hair quality had small positive changes, whereas curl development and hair length had small negative changes. According to the authors, responses to selection compared favourably with characteristics in other sheep breeds. Annual responses of between 1 and 3% were obtained for hair quality, hair length and pattern. A response of more than 7% per year was reported for selection for a decrease in curl development. The relatively high responses indicated that pelt traits can be precisely estimated.

6.9.2 Main Environmental Selection Factors and Limitations in Responses

Estimates of heritabilities for animal genetic and permanent environmental and maternal genetic effects are small as a result of substantial effects of environmental factors (season of breeding, year, age of ewe, breed of ewe and hormone treatment) on reproductive traits and to non-normal distributions of traits (Rosati et al. 2002). As further reviewed by the authors, low heritability of reproductive traits is likely due to a larger proportional influence of environmental factors, little genetic variability for fertility, litter size, lamb survival and lambing frequency and other reproductive traits.

Designing efficient selection and breeding strategies for genetic improvement and appropriate genetic evaluation of local breeds involves precise estimates of genetic parameters (Boujenane et al. 2013; Safari et al. 2005), which include heritabilities, repeatabilities, genetic and phenotypic correlations. Lack of records on the animals or lack of continuity in data submission constitutes the most serious challenge to genetic parameters' estimation and genetic progress (Zishiri 2009). Estimates are often biased due to a few records or inconsistencies in data structure.

6.10 Sheep Production Capacity of Africa in Terms of Milk, Meat, Wool and Skin

6.10.1 Sheep Milk Production

Sheep milk production (Fig. 6.39) and consumption are important to many African nations. Sheep milk provides a wide spectrum of nutrients, including proteins and vitamins and has been found to be richer in fat content, solids and minerals when compared to cow milk, making it ideal for cheese production. Sheep milk provides

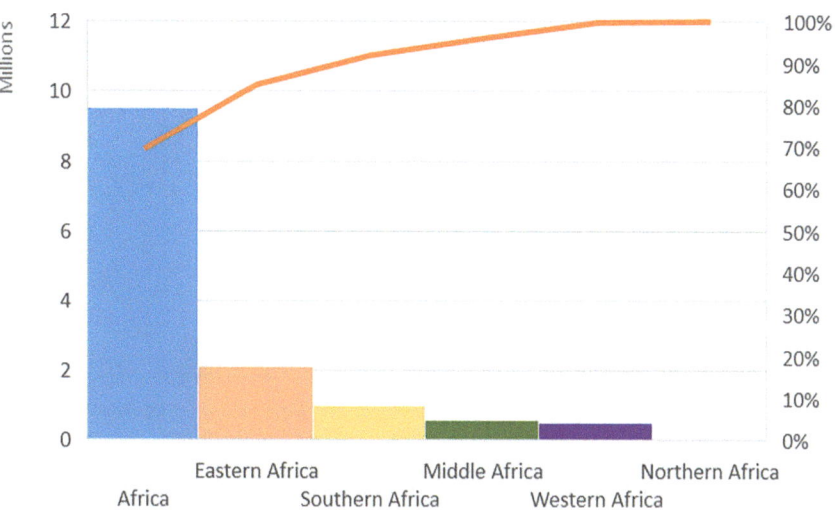

Fig. 6.39 Sheep milk production in Africa (in millions of tonnes), 2009–2019 (FAOSTATS 2021)

an important resource to households in terms of nutrition, especially in arid and semi-arid regions. It is also used to make a variety of products, including cheese, yoghurt and butter, which provide additional sources of income for households. There are relatively few sheep breeds that have been developed for dairy, and the most productive breeds are the Lacaune, East Friesian, Sarda, Assaf, Chios and the Awassi breeds (Skapetas and Kalaitzidou 2017).

6.10.2 Meat Production

Africa produced approximately 1.6 million tonnes of sheep meat (Fig. 6.32), generating approximately USD 8.3 million in 2019 (FAOSTATS 2019). Indigenous sheep produced approximately 1.4 million tonnes in 2019. Sheep production in Africa has seen a steady increase from 614,915 tonnes produced in 1970 (FAOSTATS 2019). North African countries are the largest sheep meat producers in Africa (Fig. 6.40). Specifically, Algeria is the largest sheep meat producer in Africa (331,967 tonnes in 2019), followed by Sudan (265,000 tonnes in 2019) and Morocco (178,770 tonnes in 2019) (FAOSTATS 2021).

6.10.3 Wool Production

Wool production was of major importance throughout human history (Ryder 2007) and was the first product to be involved in the international trade of goods. In addition to being used to make clothes, wool is used to make carpets, blankets, house rugs, felt and many other products. While African wool production only represents 3% of world fibre production, sheep fibre and wool are very important to the economies of many African nations. Africa produces a total of 203,000 tonnes of greasy wool, with the North African countries being the largest producers (Fig. 6.41) (FAOSTATS 2019, 2021). Morocco is the largest greasy wool producer at 64,000 tonnes in 2019, followed by South Africa, Algeria and Ethiopia (42,000, 34,700 28,500 and tonnes in 2019, respectively) (FAOSTATS 2019).

Greasy wool production (Fig. 6.41) provides farmers with alternative sources of income, a trade and a way of life for many communities. The gradual increased use of synthetic fibres has resulted in a decrease in the demand for wool, which has caused a drastic impact on sheep breeds kept primarily for wool production. Sheep in Africa are primarily kept for meat, fat, milk and skin, and as a result of this, production of wool

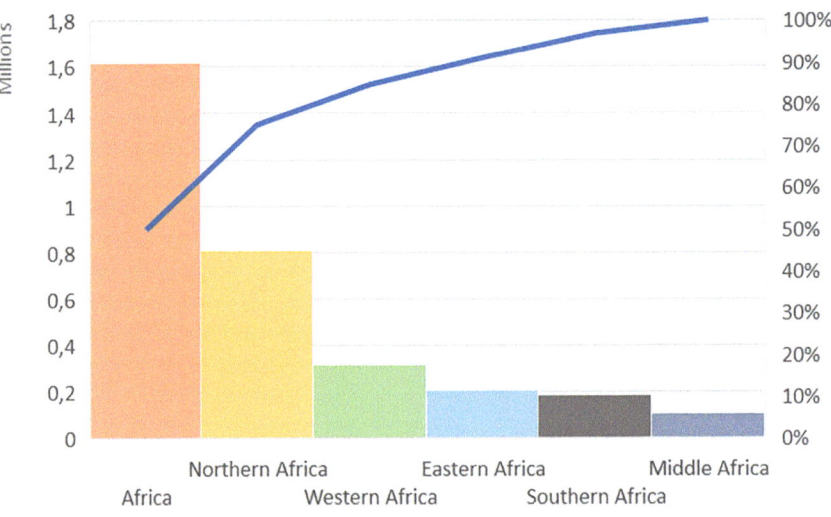

Fig. 6.40 Sheep meat production in Africa (in millions of tonnes), 2009–2019 (FAOSTATS 2021)

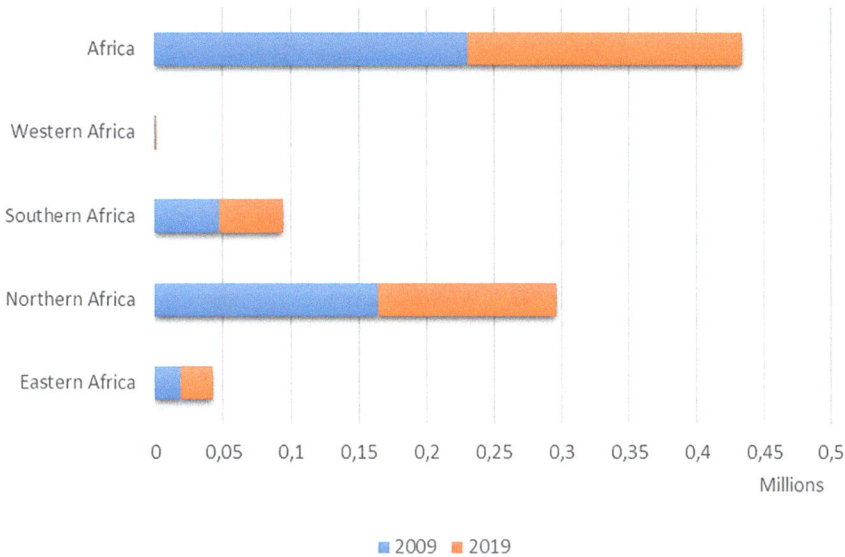

Fig. 6.41 Sheep greasy wool production in Africa (in millions of tonnes), 2009–2019 (FAOSTATS 2021)

from sheep has been undermined, with many African nations not maximising wool farming. Wool farming is also a profitable venture and can increase sheep farmers' income by at least 20%.

6.10.4 Skin Production

Raw skins and hides are by-products of sheep that are reared and slaughtered for meat. The raw skins and hides are direct inputs to the leather industry. Africa produced approximately 282,323 tonnes of raw skins in 2012 (Table 6.7), most of which is exported, thus earning many countries much-needed foreign exchange. Ethiopia, for example, earned USD 139 million in 2010 from exports of finished leather products (Central Statistical Authority 2011).

Sheepskin produces high-quality leather that is used in the production of many leather goods, including shoes, gloves and clothing. The sheep fleece is also used to make a wide range of products, including fibre, string, soft wool lining and coverings (Skapetas and Kalaitzidou 2017).

6.11 Challenges to Sustainable Genetic Improvement in the Sheep Industry

Several technical and non-technical constraints militate against successful genetic improvements and sustainability in the sheep sector (Pollott and Wilson 2009; OECD and FAO 2016; Tibbo et al. 2006; Wiener 1994), which are briefly highlighted below.

- Indifferent attitude of the population to low genetic potentials of african indigenous sheep breeds: Although the low production potential of African indigenous sheep is not a constraint to genetic improvement, an indifferent attitude by the population limits proper planning (poor participation of livestock farmers and other stakeholders) in the design for implementation of more appropriate genetic improvement programmes.
- Limited information on the characterisation and conservation of most African sheep genetic resources.

Table 6.7 Global sheep raw skin production (tonnes) 2000–2012 (FAOSTATS 2021)

s	Year 2000	2012	Change (%) 2000–2012		Contribution 2012
Asia	851,791	1,049,474	23.2	11.7	
Africa	200,064	282,323	41.1	3.15	
Oceania	251,136	7,223,508	2776	80.7	
Europe	376,886	299,085	−20.6	3.34	
EU (28)	236,966	206,245	−13	2.30	
Americas	92,311	93,160	0.92	1.04	
World	1,772,189	8,947,550	404.89	100	

- Uncontrolled cross-breeding is the foremost cause of genetic erosion of sheep breeds in Africa (Leroy et al. 2016).
- Lack of skilled manpower: The design and implementation of a sound breeding improvement programme requires trained experts, which are often lacking.
- Highly mobile small flock sizes, single-sire flocks, lack of animal identification, lack of performance and pedigree recording due to the low level of literacy of farmers.
- Inadequate feed supplies and poor animal nutrition.
- Widespread distribution of livestock diseases.
- Poor Infrastructural Facilities (transportation, marketing and implementation of artificial insemination): Many countries in sub-Saharan Africa are characterised by poor road network systems, which make transportation of animals and semen from one location to another difficult. It is also difficult to successfully implement artificial insemination, which involves the production, collection, storage, distribution and effective use of semen. The marketing of agricultural products is also limited by poor infrastructure.
- Non-technical constraints include the absence of government-driven clear livestock development policies and strategies, as well as incentives or credit facilities for farmers.

6.12 Conclusion and Future Prospects

The sheep industry remains one of Africa's most promising small ruminant sectors, with huge genetic resources of over 363 breeds. A limited number of these breeds have been characterised using phenotypic (qualitative and quantitative) traits, as well as molecular markers, which demonstrated evidence of high within-breed genetic diversity. Growth in demand for meat from sheep is almost entirely met by supply from indigenous production, with limited import compared to other livestock species. Despite substantial challenges to genetic improvement, the sector has huge potential to improve food security and encourage economic development in most countries in sub-Saharan Africa. To drive sustainable breeding and genetic improvement programmes, there is an urgent need to intensify efforts on characterising more African sheep breeds, conserving the existing diversity, developing human resources, encouraging cooperative farming among farmers, and forming breed societies. Given the enabling environment along with a simple and well-designed nucleus breeding programme, possibilities exist for improving the sheep genetic potential for more efficient production of meat, milk, wool and skins using traditional breeding (see Chap. 17) and recent genomic techniques, as well as relevant modern technologies as outlined in Chap. 21. This will also involve re-orientation of the role of governments and other relevant institutions to take appropriate responsibilities.

Acknowledgements The authors would sincerely like to acknowledge in a very special way the African sheep keepers, managers, breeders and processors who, through their utilisation of the sheep in Africa, have enhanced their conservation. They are also sincerely thankful to individuals who provided photographs of various African sheep breeds included in this chapter. Citations were provided for photographs obtained from the literature, and sources of information were acknowledged using references.

References

Abdelali A (1988) Contribution à l'estimation Des Facteurs de Correction Des Effets Du Milieu Pour Les Caractères Lainiers; Mémoire de 3ème Cycle Agronomie. Institut Agronomique et Vétérinaire Hassan II, Rabat

Abebe AS, Alemayehu K, Johansson AM, Gizaw S (2020) Breeding practices and trait preferences of smallholder farmers for indigenous sheep in the northwest highlands of Ethiopia: inputs to design a breeding program. PLoS One 15(5):e0233040. https://doi.org/10.1371/journal.pone.0233040

Abied A, Ahbara AM, Berihulay H, Xu L, Islam R, El-Hag FM, Rekik M, Haile A, Han J-L, Ma Y, Zhao Q, Mwacharo JM (2021) Genome divergence and dynamics in the thin-Tailed Desert sheep from Sudan. Front Genet 12:659507. https://doi.org/10.3389/fgene.2021.659507

Aboul-Naga AM, Alsamman AM, Nassar AE, Mousa KH, Osman M, Abdelsabour TH, Mohamed LG, Elshafie MH (2023) Investigating genetic diversity and population structure of Egyptian goats across four breeds and seven regions. Small Rumin Res 226:107017. https://doi.org/10.1016/j.smallrumres.2023.107017

Abubakar GR, Ezewudo EA, Egena SSA, Usman A (2020) Genetic diversity of Nigerian indigenous sheep breeds at the βLactoglobulin gene locus. Genet Biodivers J 4(2):40–49

Adu IF, Ngere LO (1979) The indigenous sheep of Nigeria. World Rev Anim Product 15(3):51–62

Afolayan RA, Adeyinka IA, Lakpini CAM (2006) The estimation of live weight from body measurements in Yankasa sheep. Czeh J Anim Sci 51(8):343–348

Afrino Sheep Breeders Society (n.d.) Afrino Breed Standards. http://afrino.org.za/

Agaviezor BO, Peters SO, Adefenwa MA, Yakubu A, Adebambo OA, Ozoje MO, Ikeobi CO, Wheto M, Ajayi OO, Amusan SA, Ekundayo OJ, Sanni TM, Okpeku M, Onasanya GO, De Donato M, Ilori BM, Kizilkaya K, Imumorin IG (2012) Morphological and microsatellite DNA diversity of Nigerian indigenous sheep. J Animal Sci Biotechnol 3(1):38. https://doi.org/10.1186/2049-1891-3-38

Ali AS, Ibrahim MT, Mohammed MM, Elobied AA, Lühken G (2016) Growth differentiation factor 9 gene variants in Sudanese Desert sheep ecotypes. S Afr J Anim Sci 46(4):373–379. https://doi.org/10.4314/sajas.v46i4.5

Almeida AM (2011) The Damara in the context of southern Africa fat-tailed sheep breeds. Trop Anim Health Prod 43(7):1427–1441. https://doi.org/10.1007/s11250-011-9868-3

Aloulou R, Marnet P-G, M'Sadak Y (2018) Revue Des Connaissances Sur La Micro-Filière Ovine Laitière En Tunisie: État Des Lieux et Perspectives de Relance de La Race Sicilo-Sarde. Biotechnol Agron Soc Environ 22(3):188–198

Álvarez I, Fernández I, Traoré A, Pérez-Pardal L, Menéndez-Arias NA, Goyache F (2020) Genomic scan of selective sweeps in Djallonké (West African Dwarf) sheep shed light on adaptation to harsh environments. Sci Rep 10(1):2824. https://doi.org/10.1038/s41598-020-59839-x

Amare T, Goshu G, Tamir B (2018) Flock composition, breeding strategies and farmers' traits of interest evaluation of Wollo Highland sheep and their F1 crosses. J Anim Sci Technol 60(1):14. https://doi.org/10.1186/s40781-018-0173-9

Ameur Ameur A, Ata N, Benyoucef MT, Djaout A, Azzi N, Yilmaz O, Cemal İ, Gaouar SBS (2018) New genetic identification and characterisation of 12 Algerian sheep breeds by microsatellite markers. Ital J Anim Sci 17(1):38–48. https://doi.org/10.1080/1828051X.2017.1335182

Arrowsmith SP, Ward HK (1983) Indigenous sheep selection programme and productivity of indigenous sheep and goats, Annual report 1980/81; Division of livestock and pastures. Department of Research and Specialist Services, Harare, pp 92–95

Assan N, Makuza S, Mhlanga F, Mabuku O (2002) Genetic evaluation and selection response of birth weight and weaning weight in indigenous Sabi Sheep. Asian-Aust J Anim Sci 15(12):1690–1694

AU-IBAR (2019) The State of Farm Animal Genetic Resources in Africa: towards accelerated agricultural growth and transformation by the year 2025. African Union – Inter-African Bureau for Animal Resources (AU-IBAR), Nairobi

Ayachi A, Belkhadem S, Gaouar SBS (2017) Preservation and valorization of the Hamra Sheep Breed. Genet Biodivers J (GABJ) 1(1):19–25

Baenyi P, Meutchieye F, Ayagirwe BR, Bwihangane BA, Karume K, Mushagalusa NG, Ngoula F (2018) Biodiversity of indigenous Djallonke sheep (Ovis Areas) in Sudano Guinean Region in Cameroon. Gen Biodiv J 2(2):1–10

Baker CM, Manwell C (1980) Hemical classification of cattle. 1. Breed groups. Animal blood groups. Biochem Genet 11:127–150

Baker RL, Rege JEO (1994) Genetic resistance to disease and other stresses in improvement of ruminant livestock in the tropics. In: Proceedings of the fifth world Congress on genetics applied to livestock production, vol 20, pp 405–441

Baker RL, Mwamachi DM, Audho JO, Aduda EO, Thorpe W (1999) Genetic resistance to gastro-intestinal nematode parasites in red Maasai, Dorper and red Maasai X Dorper ewes in the sub-humid tropics. Anim Sci (UK) 69(2):335–344

Baker RL, Mugambi JM, Audho JO, Carles AB, Thorpe W (2002) Comparison of red Maasai and Dorper sheep for resistance to gastro-intestinal nematode parasites: productivity and efficiency in a humid and a semi-arid Environment in Kenya. In: Institut National de la Recherche Agronomique, Paris (France), pp 639–642

Barbato M, Hailer F, Orozco-terWengel P, Kijas J, Mereu P, Cabras P, Mazza R, Pirastru M, Bruford MW (2017) Genomic signatures of adaptive introgression from European mouflon into domestic sheep. Sci Rep 7(1):7623. https://doi.org/10.1038/s41598-017-07382-7

Barthelme JW (1985) Fisher-hunters and Neolithic pastoralists in East Turkana, Kenya, British archaeological reports, International Series 254. BAR Publishing, Oxford

Battar M (1983) Utilisation Des Béliers Sardi et Beni Guil Sur Les Brebis Timahdit, Beni Guil, Sardiet Beni Guil. Etude de La Production Laitière Des Brebis En Relation Avec La Croissance de Leursagneaux.; Mémoire de 3ème Cycle Agronomie; Institut Agronomique et Vétérinaire Hassan II: Rabat, Morocco

Bechchari A (2009) Characterization of the BG breed and its link with the terroir. Ministry of Agriculture, Fisheries, Rural Development, Water and Forests, Kindom of Morocco

Behnke R, Osman HM (2012) The contribution of livestock to the Sudanese economy. IGAD LPI working paper No. 01-12

Belabdi I, Ouhrouch A, Lafri M, Gaouar SBS, Ciani E, Benali AR, Ould Ouelhadj H, Haddioui A, Pompanon F, Blanquet V, Taurisson-Mouret D, Harkat S, Lenstra JA, Benjelloun B, Da Silva A (2019) Genetic homogenization of indigenous sheep breeds in Northwest Africa. Sci Rep 9(1):7920. https://doi.org/10.1038/s41598-019-44137-y

Belhaj K, Mansouri F, Tikent A, Taaifi Y, Boukharta M, Serghini HC, Elamrani A (2021) Effect of age and breed on carcass and meat quality characteristics of Beni-Guil and Ouled-Djellal sheep breeds. Sci World J 2021:1–8. https://doi.org/10.1155/2021/5536793

Belharfi FZ, Djaout A, Ameur A, Gaouar SBS (2017a) Barymetric characterization of Algerian sheep breeds in Western Algeria. Genet Biodivers J 1(2):31–41

Belharfi FZ, Djaout A, AmeurAmeur A, Sahraoui H, Gaouar SBS (2017b) Comparative study of wool quality in sheep breeds in Western Algeria. Genet Biodivers J 2(1):19–25

Bemji MN, Ogungimi AA, Ode AJ, Okediji SR, Akinwunmi AT, Sanyaolu TO, Salawudeen BS, Kelani BA, Ogunsola AO, Agunbiade MO, Adenaike AS, Ogundiyi AI (2012) Prevalence of coat colour phenotypes in west African dwarf sheep reared by small holder farmers in South Western Nigeria. AJEA 2(4):587–596. https://doi.org/10.9734/AJEA/2012/1544

Bemji M, Osinowo O, Ehoche O, Aduku A (1996) Birth weight and litter birth weight in Yankasa sheep: environmental factors and heritability estimates. Niger Soc Anim Product 23(10):5–11

Bemji MN, Osinowo OA, Ehoche OW, Aduku AO (2001a) Environmental factors and heritability estimates of lambing interval and age at first lambing in Yankasa sheep under controlled mating system. ASSET Ser A 1(2):37–44

Bemji MN, Osinowo OA, Ehoche OW, Aduku AO (2001b) Litter size in Yankasa sheep: environmental factors, additive correction factors and heritability estimates. ASSET Ser A 1(1):109–144

Bemji MN, Osinowo OA, Ehoche OW, Aduku AO (2001c) Litter size in Yankasa sheep: environmental factors, additive adjustment factors and heritability estimate. ASSET Ser A 1(1):107–114

Ben Lakhal M (1995) D'abÀttage Des Asneffi*. Tmnghdit, Sardi, Mérinos 7:1

Benoudifa M (1990) Production et Composition Du Lait Des Brebis Timahdit. Influence Du Numéro Delactation et Du Sexe de l'agneau; Mémoire de Fin d'Etudes, Ecole Nationale d'Agriculture; Meknès, Morocco

Bimerow T, Yitayew A, Taye M, Mekuriaw S (2011) Morphological characteristics of Farta sheep in Amhara Region, Ethiopia. Online J Anim Feed Res 1(16):299–305

Bisschop, J. H. R.; Strake, J. S.; Lategan, A. W.; Meyer, W. D. The indigenous sheep of Hottentot origin in South Africa. Mimeograph 1954

Blench RM (2001) Pastoralism in the new millennium, FAO: animal health and production series, no 150. Rome, Food and Agriculture Organisation

Bocquet-Appel J-P, Bar-Yosef O (2008) The Neolithic demographic transition and its consequences. Springer Science+Business Media, New York

Boivin N, Fuller DQ (2009) Shell Middens, ships and seeds: exploring coastal subsistence, maritime trade and the dispersal of domesticates in and around the ancient Arabian Peninsula. J World Prehist 22(2):113–180. https://doi.org/10.1007/s10963-009-9018-2

Bolowe MA, Thutwa K, Kgwatalala PM, Monau PI, Malejane C (2021) On-farm phenotypic characterization of indigenous Tswana sheep population in selected districts of southern Botswana. Afr J Agric Res 17(10):1268–1280

Bonsma FN (1944) Milk production studies with sheep, Part XIX Farm. S. Afr, pp 311–328

Boubekeur A, Benyoucef M, Lounassi M, Slimani A, Amiali M (2015) Phenotypic characteristics of Algerian D'man sheep breed in Adrar oases. Livest Res Rural Dev:27 (7)

Boujenane I (2002) Development of the DS synthetic breed of sheep in Morocco. Small Rumin Res 45(1):61–66. https://doi.org/10.1016/S0921-4488(02)00115-3

Boujenane I, Berrada D, Mihi S, Jamai M (1996) Production Laitière Des Brebis de Races Timahdite, Sardi et Béni Guil En Race Pure et En Croisement. Revue Marocaine des Sciences Agronomiques et Vétérinaires 16(3):11–18

Boujenane I, Chikhi A, Sylla M, Ibnelbachyr M (2013) Estimation of genetic parameters and genetic gains for reproductive traits and body weight of D'man Ewes. Small Rumin Res 113(1):40–46. https://doi.org/10.1016/j.smallrumres.2013.02.009

Boukhliq R (2002) Cours En Ligne Sur La Reproduction Ovine.; Cours 1.; Institut Agronomique et Vétérinaire Hassan II.: IAV Hassan II –2002

Bousman CB (1998) The chronological evidence for the introduction of domestic stock into Southern Africa. Afr Archaeol Rev 15(2):133–150

Brahi OHD, Xiang H, Chen X, Farougou S, Zhao X (2015) Mitogenome revealed multiple Postdomestication genetic mixtures of west African sheep. J Anim Breed Genet 132(5):399–405. https://doi.org/10.1111/jbg.12144

Buitenhuis H (1997) A msıklı Höyü k: a "Protodomestication" site. Anthropozoologica 25(26):655–662

Buvanendran V, Adu IF (2021) Balami sheep: performance in north-eastern Nigeria. Niger J Anim Prod 17:29–35

Çakirlar C (2012a) Evolution of Animal Husbandry in Neolithic Central-West Anatolia: The Zooarchaeological Record from Ulucak Höyü k (ca. 7,040–5,660 Cal. BC, Izmir, Turkey). Anatol Stud 61:1–33

Çakirlar C (2012b) Neolithic Dairy Technology at the European-Anatolian Frontier: Implications of Archaeozoological Evidence from Ulucak Höyük, I zmir, Turkey, ca. 7,000–5,700 Cal. BCA Anthropozoologica 47(2):77–98

Carta A, Casu S, Salaris S (2009) Current state of genetic improvement in dairy sheep. J Dairy Sci 92(12):5814–5833. https://doi.org/10.3168/jds.2009-2479

Central Statistical Authority (2011) Agricultural Sample Survey, 2010/11; Report on livestock and livestock characteristics (Private peasant holdings) Volume II, Statistical Bulletin 505; Addis Aba, Ethiopia

Chalh A, El Gazzah M, Djemali M'N, Noureddine C (2007) Genetic and phenotypic characterization of the Tunisian noire De Thibar lambs on their growth traits. J Biol Sci 7(8):1347–1353. https://doi.org/10.3923/jbs.2007.1347.1353

Chellig R (1992) Les Races Ovines Algériennes. Office des Publications Universitaires, Ben-Aknoun/Alger

Clark JD (1976) The domestication process in Sub-Saharan Africa with Special Reference to Ethiopia. In: Proceedings of the 9th congress of the U.I.S.P.P. UISPP: Nice, pp 56–115

Cloete JM (1978) The Namaqua Afrikaner. Karoo Agric 1(1):41–43

Close AE (1992) Holocene occupation of the eastern Sahara. In: Klees F, Kuper R (eds) New light on the northeast African past. Koln, Heinrich Barth Institute, pp 155–183

Close AE, Wendorf F (1992) The beginnings of food production in the eastern Sahara. In: Gebauer AB, Price TD (eds) Transitions to agriculture in prehistory. Prehistory Press, Madison, pp 63–72

Clutton-Brock J (1997) Origins of the dog: domestication and early history. In: Serpell J (ed) Domestic dog: its evolution, behavior and interactions with people. Cambridge University Press, Cambridge, pp 2–19

Cooke CK (1965) Evidence of human migrations from the rock art of Southern Rhodesia. Africa 35:263–285

DAD-IS. Domestic Animal Diversity Information System

DAGRIS (2007) In: Kempo S, Mamo Y, Astrat B, Dessie T (eds) Domestic animal genetic resources information system. International Livestock Research Institute, Addis Ababa. http://dagris.ilri.cgiar.org

Daniel B (2000) Racesovines et Caprines Françaises; Edition, France Agricole

Davies FG (1987) Recent advances in virology at Kabete relevant to small ruminant production. In: Proceedings of the 6th KVA/small ruminant CRSP workshop, pp 206–210

Dayo G-K, Houaga I, Somda MB, Linguelegue A, Ira M, Konkobo M, Djassi B, Gomes J, Sangare M, Cassama B, Yapi-Gnaore CV (2022) Morphological and microsatellite DNA diversity of Djallonké sheep in Guinea-Bissau. BMC Genom Data 23(1):3. https://doi.org/10.1186/s12863-021-01009-7

de Aguiar AL, da Silva RR, Alves SM, da Silva LP, de Morais OR, Lobo RNB (2020) Breeding objectives and selection criteria of a participatory community-based breeding programme of goats and sheep. Trop Anim Health Prod 52(4):1933–1943. https://doi.org/10.1007/s11250-020-02209-6

Degen AA, Kam M (1992) Body mass loss and body fluid shifts during dehydration in Dorper sheep. J Agric Sci 119(3):419–422. https://doi.org/10.1017/S0021859600012260

Dekhili M (2014) A morphometric Study of sheep reared in North-East Algerian. Archivos de zootecnia 63(244):623–631

Delacrétaz-Wolff AS (1997) Etudes Génétiques et Sérologiques Des Systèmes de Groupes Sanguins Du Mouton. Doctoral dissertation, ETH, Zurich

Delgado C, Rosegrant MW, Steinfeld H, Ehui SK, Courbois CB (1999) The coming livestock revolution. Choices 14(316):2016–7248

Demirci S, Koban Baştanlar E, Dağtaş ND, Pişkin E, Engin A, Özer F, Yüncü E, Doğan ŞA, Togan İ (2013) Mitochondrial DNA diversity of modern, ancient and wild sheep (Ovis Gmelinii Anatolica) from Turkey: new insights on the evolutionary history of sheep. PLoS One 8(12):e81952. https://doi.org/10.1371/journal.pone.0081952

Deribe B, Beyene D, Dagne K, Getachew T, Gizaw S, Abebe A (2021) Morphological diversity of northeastern fat-tailed and northwestern thin-tailed indigenous sheep breeds of Ethiopia. Heliyon 7(7):e07472. https://doi.org/10.1016/j.heliyon.2021.e07472

Devendra C, McLeroy GB (1982) Goat and sheep production in the tropics, Intermediate tropical agriculture series. Longman, London

Djaout A, Afri-Bouzebda F, Chekal F, El-Bouyahiaoui R, Rabhi A, Boubekeur AA, Benidir M, Gaouar SBS (2017) Etat de La Biodiversité Des «races» Ovines Algériennes. Genet Biodivers J 14(1):11–18

Djemali M, Ben M'Sallem I, Boraoui R (1995) Effet Du Mois, Mode et Âge d'agnelage Sur La Production Laitière Des Brebis Sicilo-Sardes En Tunisie. In: Caja G, Djemali M, Gabiña D, Nefzaoui A (eds) L'élevage ovin en zones arides et semi-arides. Cahiers Optiones: Méditerranéennes, p 6

Dormer Sheep Breeders' Society of South Africa. Dormer Sheep Breeders' Society of South Africa History. http://www.dormer.co.za/p2/dormer-sa/dormer-sheep-breeders'-society-of-south-africa-history.html. Accessed 15 Jan 2022

Drake NA, Blench RM, Armitage SJ, Bristow CS, White KH (2011) Ancient watercourses and biogeography of the Sahara explain the peopling of the desert. Proc Natl Acad Sci 108(2):458–462. https://doi.org/10.1073/pnas.1012231108

Du Toit DJ (2008) The indigenous livestock of southern Africa. http://www.damaras.com/newsletters/du-toit.pdf. Accessed 10 Sept 2021

Edea Z (2008) Characterization of Bonga and Horro indigenous sheep breeds of smallholders for designing community based breeding strategies in Ethiopia. MSc thesis, School of Graduate Studies of Haramaya University, Dire Dawa, Ethiopia

Edea Z, Dessie T, Dadi H, Do K-T, Kim K-S (2017) Genetic diversity and population structure of Ethiopian sheep populations revealed by high-density SNP markers. Front Genet 8:218. https://doi.org/10.3389/fgene.2017.00218

Edea Z, Dadi H, Dessie T, Kim K-S (2019) Genomic signatures of high-altitude adaptation in Ethiopian sheep populations. Genes Genom 41(8):973–981. https://doi.org/10.1007/s13258-019-00820-y

Efstratiou NA, Karetsou E, Banou E, Margomenou D (2004) The Neolithic Settlementof Knossos: new light on an old picture. In: Cadogan G, Hatzaki E, Vasilakis E (eds) Knossos: palace, city, state (British School at Athens Studies), vol 12. British School at Athens, London, pp 43–49

Elphick R (1985) Khoikhoi and the founding of White South Africa. Ravan Press, Johannesburg

Elshazly AG, Youngs CR (2019) Feasibility of utilizing advanced reproductive Technologies for Sheep Breeding in Egypt. Part 1. Genetic and nutritional resources. Egypt J Sheep Goat Sci 14(1):39–52

Epstein H (1960) History and origin of the Ronderib and Namaqua Afrikaner sheep. Z Tierzuecht Zuechtungsbiol 74(1–4):267–292. https://doi.org/10.1111/j.1439-0388.1960.tb00132.x

Epstein H (1971) The origin of the domestic animals of Africa. Africana Publishing Corporation, New York

Ethiopia Sheep and Goat productivity Improvement Program (ESGPIP) (2008) Sheep and goat production handbook for Ethiopia. ESGPIP

FAO (2007) The state of the world's animal genetic resources for food and agriculture. Rome, FAO

FAO (2014) Business and Livelihoods in African Livestock Investments to Overcome Information Gaps. Livestock Data Innovation in Africa; World Bank Report Number 86093-AF

FAO, IFAD, UNICEF, WFP, WHO (2017) The state of food security and nutrition in the world 2017. Building Resilience for Peace and Food Security. http://www.fao.org/3/a-I7695e.pdf

FAOSTATS (2019) Food and Agriculture Organization of the United Nations

FAOSTATS (2021) Food and Agriculture Organization of the United Nations

Fariello M-I, Servin B, Tosser-Klopp G, Rupp R, Moreno C, International Sheep Genomics Consortium, Cristobal MS, Boitard S (2014) Selection signatures in worldwide sheep populations. PLoS One 9(8):e103813. https://doi.org/10.1371/journal.pone.0103813

Feliachi K (2003) Rapport National Sur Les Ressources Génétiques Animales: Algérie. Commission nationale AnGR. Direction Générale de l'INRAA, p 46

Ferede Y, Amane A, Mazengia H, Mekuriaw S (2014) Prevalence of major sheep diseases and analysis of mortality in selected model sheep villages of South Gondar administrative zone. Ethiop Ethiop Vet J 18(2):83–97

Foley JA, DeFries R, Asner GP, Barford C, Bonan G, Carpenter SR, Chapin FS, Coe MT, Daily GC, Gibbs HK, Helkowski JH, Holloway T, Howard EA, Kucharik CJ, Monfreda C, Patz JA, Prentice IC, Ramankutty N, Snyder PK (2005) Global consequences of land use. Science 309(5734):570–574. https://doi.org/10.1126/science.1111772

Galall ESE (1983) Sheep germ plasm in Ethiopia. Anim Genet Resour Inf 1:5–12. https://doi.org/10.1017/S1014233900000018

Gaouar SBS (2002) Contribution à l'étude de La Variabilité Génétique Des Races Ovines Par l'utilisation Des Microsatellites: Caractérisation de Deux Races Ovine Algériennes Hamra et Ouled-Djellal. Thèse de Magistère, spécialité génétique, option: polymorphisme génétique, nstitut des sciences de la nature, Université d'Es-Sénia, Oran

Gaouar SBS, de La Biodiversité E (2009) Analyse de La Variabilité Génétique Des Races Ovines Algériennes & de Leurs Relations Phylogénétiques Par l'utilisation de Microsatellites. PhD thesis, Université d'Oran1-Ahmed Ben Bella

Gaouar SBS, Da Silva A, Ciani E, Kdidi S, Aouissat M, Dhimi L, Lafri M, Maftah A, Mehtar N (2015) Admixture and local breed marginalization threaten

Algerian sheep diversity. PLoS One 10(4):e0122667. https://doi.org/10.1371/journal.pone.0122667

Gaouar SBS, Kdidi S, Ouragh L (2016) Estimating population structure and genetic diversity of five Moroccan sheep breeds by microsatellite markers. Small Rumin Res 144:23–27. https://doi.org/10.1016/j.smallrumres.2016.07.021

Gaouar SBS, Lafri M, Djaout A, El-Bouyahiaoui R, Bouri A, Bouchatal A, Maftah A, Ciani E, Da Silva AB (2017) Genome-wide analysis highlights genetic dilution in Algerian sheep. Heredity 118(3):293–301. https://doi.org/10.1038/hdy.2016.86

Garcea E (2004) An alternative way towards food production: the perspectives from the Libyan Sahara. J World Prehist 18:18 (2)

Gatenby RM (2006) Sheep production in the tropics. Longman Publisher, London

Gautier A (1987) Prehistoric men and cattle in North Africa: a dearth of data and a surfeit of models. In: Close AE (ed) Prehistory of arid North Africa. Southern Methodist University Press, Dallas, pp 163–187

Gebre KT, Yfter KA, Teweldemedhn TG, Gebremariam T (2018) Production objectives, selection criteria and breeding practices of Afar sheep in Abaala, Afar region. Ethiopia J Drylands 8(2):834–845

Getachew T (2008) Characterization of Menz and Afar indigenous sheep breeds of smallholders and pastoralists for designing community-based breeding strategies in Ethiopia. MSc Animal Genetics and Breeding, Haramaya University, Haramaya, Ethiopia

Getachew T, Haile A, Tibbo M, Sharma AK, Souml J, Wurzinger M (2010) Herd management and breeding practices of sheep owners in a mixed crop-livestock and a pastoral system of Ethiopia. Afr J Agric Res 5(8):685–691

Getachew T, Alemu B, Sölkner J, Gizaw S, Haile A, Gosheme S, Notter DR (2015) Relative resistance of Menz and Washera sheep breeds to artificial infection with Haemonchus Contortus in the highlands of Ethiopia. Trop Anim Health Prod 47(5):961–968. https://doi.org/10.1007/s11250-015-0815-6

Ghernouti N, Bodinier M, Ranebi D, Maftah A, Petit D, Gaouar SBS (2017) Control region of mtDNA identifies three migration events of sheep breeds in Algeria. Small Rumin Res 155:66–71. https://doi.org/10.1016/j.smallrumres.2017.09.003

Gibson J (2007) Red Maasai Sheep -accelerating threats. In: Rischkowsky B, Pilling D (eds) The state of the world's animal genetic resources for food and agriculture. FAO, p 444

Gibson J, Ayalew W, Hanotte O (2012) Les Mesures de La Diversité En Tant Qu'apports à La Prise de Décisions Liées à La Conservation Des Ressources Génétiques Du Bétail. In: Jarvis DI, Padoch C, Cooper HD (eds) Gestion de la biodiversité dans les écosystèmes agricoles. Bioversity Internationa

Gifford-Gonzalez D, Hanotte O (2011) Domesticating animals in Africa: implications of genetic and archaeological findings. J World Prehist 24(1):1–23. https://doi.org/10.1007/s10963-010-9042-2

Gilman A (1975) A later prehistory of Tangier, Morocco, American School of Prehistoric Research, Peabody Museum Bulletin 29. Harvard University, Cambridge

Gizaw S (2008) Sheep resources of Ethiopia: genetic diversity and breeding strategy. Wageningen University, Wageningen

Gizaw S, Van Arendonk JAM, Komen H, Windig JJ, Hanotte O (2007) Population structure, genetic variation and morphological diversity in indigenous sheep of Ethiopia: genetic and morphological diversity in Ethiopian sheep. Anim Genet 38(6):621–628. https://doi.org/10.1111/j.1365-2052.2007.01659.x

Gizaw S, Komen H, Hanotte O, Arendonk JA (2008a) Indigenous sheep resources of Ethiopia: types, production systems and farmers preferences. Anim Genet Resour/Resources génétiques animales/Recursos genéticos animales 43:25–39

Gizaw S, Komen H, Windig JJ, Hanotte O, van Arendonk JAM (2008b) Conservation priorities for Ethiopian sheep breeds combining threat status, breed merits and contributions to genetic diversity. Genet Sel Evol 40(4):433–447. https://doi.org/10.1051/gse:2008012

Gizaw S, Getachew T, Edea Z, Mirkena T, Duguma G, Tibbo M, Rischkowsky B, Mwai O, Dessie T, Wurzinger M, Solkner J, Haile A (2013) Characterization of indigenous breeding strategies of the sheep farming communities of Ethiopia: a basis for designing community-based breeding programs, ICARDA working paper. ICARDA, Aleppo, 47pp

Gouhis F (1989) PMSG injection and breed influences on Ewes reproductive parameters. Engineering Specialization Thesis, Central Library of I.N.A.T. Memoire de fin d'etudes du Cycle de Specialisation, Bibliotheque Centrale de l'INAT. Tunis, Tunisia

Greeff JC, Hofmeyr JH, Wyma GW, Van Deventer JFPJ (1988) Preliminary results on Heterosis and breed transmitted effects in respect of fertility and survival rate of Romanov and Dorper crossbreds. In: Proceedings of the 3rd World. Congress on sheep and beef cattle breeding, Paris, vol 2, pp 19–23

Greeff JE, Faure AS, Minnaar GJ, Schoeman SJ (1993) Genetic trends of selection for pelt traits in karakul sheep. Direct responses. S Afr J Anim Sci 23:5–6

Groeneveld LF, Lenstra JA, Eding H, Toro MA, Scherf B, Pilling D, Negrini R, Finlay EK, Jianlin H, Groeneveld E, Weigend S, The GLOBALDIV Consortium (2010) Genetic diversity in farm animals – a review. Anim Genet 41:6–31. https://doi.org/10.1111/j.1365-2052.2010.02038.x

Grosclaude (1988) Le Polymorphisme Génétique Des Principales Lactoprotéines Bovines. Relations Avec La Quantité, La Composition et Les Aptitudes Fromagères Du Lait. Prod Anim 1(1):5–17

Guerrini M, Forcina G, Panayides P, Lorenzini R, Garel M, Anayiotos P, Kassinis N, Barbanera F (2015) Molecular DNA identity of the mouflon of Cyprus (Ovis Orientalis Ophion, Bovidae): near eastern origin and divergence from Western Mediterranean conspecific populations. Syst Biodivers 13(5):472–483. https://doi.org/10.1080/14772000.2015.1046409

Guilaine J, Manen CA, Vigne J-D (2007) Pont de Roque-Haute: Nouveaux Regards Sur La Néolithisation de La France Méditerranéenne. Centre de recherche sur la préhistoire de la Méditerranée, Toulouse

Gwala PE, Kunene NW, Bezuidenhout CC, Mavule BS (2015) Genetic and phenotypic variation among four Nguni sheep breeds using random amplified polymorphic DNA (RAPD) and morphological features. Trop Anim Health Prod 47(7):1313–1319. https://doi.org/10.1007/s11250-015-0865-9

Haile D, Gizaw S, Kebede K (2015) Selection criteria and breeding practice of sheep in mixed crop livestock farming system of North Shoa. Ethiop J Biol Agric Healthc 5(21):168–174

Haile A, Gizaw S, Getachew T, Mueller JP, Amer P, Rekik M, Rischkowsky B (2019) Community-based breeding Programmes are a viable solution for Ethiopian small ruminant genetic improvement but require public and private investments. J Anim Breed Genet 136(5):319–328. https://doi.org/10.1111/jbg.12401

Harkat S, Laoun A, Benali R (2015) Phenotypic characterization of the major sheep breed in Algeria. Revue Méd Vét 166(5–6):138–147

Henn BM, Gignoux C, Lin AA, Oefner PJ, Shen P, Scozzari R, Cruciani F, Tishkoff SA, Mountain JL, Underhill PA (2008) Y-Chromosomal evidence of a pastoralist migration through Tanzania to southern Africa. Proc Natl Acad Sci 105(31):10693–10698. https://doi.org/10.1073/pnas.0801184105

Henn BM, Gignoux CR, Jobin M, Granka JM, Macpherson JM, Kidd JM, Rodriguez-Botigue L, Ramachandran S, Hon L, Brisbin A, Lin AA, Underhill PA, Comas D, Kidd KK, Norman PJ, Parham P, Bustamante CD, Mountain JL, Feldman MW (2011) Hunter-gatherer genomic diversity suggests a southern African origin for modern humans. Proc Natl Acad Sci 108(13):5154–5162. https://doi.org/10.1073/pnas.1017511108

Herrero M, Thornton PK, Notenbaert AM, Wood S, Msangi S, Freeman HA, Bossio D, Dixon J, Peters M, van de Steeg J, Lynam J, Rao PP, Macmillan S, Gerard B, McDermott J, Seré C, Rosegrant M (2010) Smart Investments in Sustainable Food Production: revisiting mixed crop-livestock systems. Science 327(5967):822–825. https://doi.org/10.1126/science.1183725

Higgs ES (1967) Environment and chronology – the evidence from Mammalian Fauna. In: McBurney CBM (ed) The Haua Fteah (Cyrenaica) and the Stone Age of the South-East Mediterranean. Cambridge University Press, Cambridge, pp 149–164

Horwitz LK, Ducos P (1998) An investigation into the origins of domestic sheep in the Southern Levant. In: Buitenhuis H, Bartosiewicz L, Choyke AM (eds) Archaeozoology of the Near East III. ARC Publications, Groningen, pp 80–95

James IJ, Osinowo OA, Adegbasa OI (2009) Evaluation of udder traits of west African dwarf (WAD) goats and sheep in Ogun state, Nigeria. J Agric Sci Environ 9:75–87

Jores D'Arces P (1947) L'élevage En Algérie: Amélioration et Développement. Éditions Guianchain, Alger, p 93

Justin N, Arthur MZI, Rodrigue AB, Félix M (2022) Genetic diversity of the Cameroon Western highlands' Djallonke sheep assessed by mitochondrial d-loop. GABJ 6(1):72–79. https://doi.org/10.46325/gabj.v6i1.200

Kandoussi A, Boujenane I, Piro M, Petit D (2020) mtDNA genetic characterization of an isolated sheep breed in south of Moroccan atlas. Small Rumin Res 193:106250. https://doi.org/10.1016/j.smallrumres.2020.106250

Kandoussi A, Boujenane I, Piro M, Petit DP (2022) Genetic diversity and population structure of Moroccan Beni Ahsen: is this endangered ovine breed one of the ancestors of merino? Ruminants 2(2):201–211. https://doi.org/10.3390/ruminants2020013

Khaldi G (1984) Research on sheep and goats in Tunisia. In: FAO expert consultation on small ruminant Research and Development in the Near East; Tunis, Tunisia

Khaldi G, Farid M (1981) Encyclopédie Des Productions Animales Dans Le Monde Arabe. Cas de La République Tunisienne. Arab Center for the Studies of Arid zones and Dry lands (ACSAD), Damas

Khaldi Z, Haddad B, Souid S, Rouissi H, Ben Gara A, Rekik B (2011) Caracterisation Phenotypique de La Population Ovine Du Sud Ouest de La Tunisie. Anim Genet Resour 49:1–8. https://doi.org/10.1017/S2078633611000361

Kijas JW, Lenstra JA, Hayes B, Boitard S, Porto Neto LR, San Cristobal M, Servin B, McCulloch R, Whan V, Gietzen K, Paiva S, Barendse W, Ciani E, Raadsma H, McEwan J, Dalrymple B, Other members of the International Sheep Genomics Consortium (2012) Genome-wide analysis of the world's sheep breeds reveals high levels of historic mixture and strong recent selection. PLoS Biol 10(2):e1001258. https://doi.org/10.1371/journal.pbio.1001258

King'oku JM, N'Thome J, Ogutu EM, Rakozi C (1975) Fifteen years production data on Dorper Sheep at Katumani Research Station, Sheep and Goat Development Project Technical Note No. 12. Ministry of Agriculture, Kenya, Machakos, p 12

Klein RG, Scot K (1986) Re-analysis of faunal assemblages from the Haua Fteah and other late quaternary archaeological sites in Cyrenaican Libya. J Archaeol Sci 13:515–542

Kosgey IS, Rowlands GJ, van Arendonk JAM, Baker RL (2008) Small ruminant production in smallholder and pastoral/extensive farming Systems in Kenya. Small Rumin Res 77(1):11–24. https://doi.org/10.1016/j.smallrumres.2008.02.005

Kunene NW, Fossey AA (2006) A survey on livestock production in some traditional areas of northern Kwazulu Natal in South Africa. Livest Res Rural Dev 18:30–33

Kunene NW, Nesamvuni EA, Fossey AA (2007) Characterization of Zulu (Nguni) sheep using linear body measurements and some environmental factors affecting these measurements. S Afr J Anim Sci 37:11–20

Kunene NW, Bezuidenhout CC, Nsahlai IV (2009) Genetic and phenotypic diversity in Zulu sheep populations: implications for exploitation and conservation. Small Rumin Res 84(1–3):100–107. https://doi.org/10.1016/j.smallrumres.2009.06.012

La Guilaine J (2006) Néolithisation de La Méditerranée Occidentale. In: Guilaine J, Grifoni R, Helgouach JL (eds) The Neolithic of the Near East and Europe. A.B.A.CA.O. Edizione, Forlí, pp 53–68

Larson G, Burger J (2013) A population genetics view of animal domestication. Trends Genet 29(4):197–205. https://doi.org/10.1016/j.tig.2013.01.003

Lauvergne J-J, Denis B, Théret M (1977) Hybridation Entre Un Mouflon de Corse (Ovis Ammon Musimon Schreber, 1872) et Des Brebis de Divers Génotypes: Gènes Pour Lacoloration Pigmentaire. Ann Genet Sel Anim 9:151–161

Lebbie SHB, Ramsay K (1999) A perspective on conservation and Management of Small Ruminant Genetic Resources in the Sub-Saharan Africa. Small Rumin Res 34(3):231–247. https://doi.org/10.1016/S0921-4488(99)00076-0

Leroy G, Baumung R, Boettcher P, Scherf B, Hoffmann I (2016) Sustainability of crossbreeding in developing countries; definitely not like crossing a meadow.... Animal 10(2):262–273. https://doi.org/10.1017/S175173111500213X

Lorenzini R, Cabras P, Fanelli R, Carboni GL (2011) Wildlife molecular forensics: identification of the Sardinian mouflon using STR profiling and the Bayesian assignment test. Forensic Sci Int Genet 5(4):345–349. https://doi.org/10.1016/j.fsigen.2011.01.012

MacDonald KC, MacDonald RH (2000) The origins and development of domesticated animals in arid West Africa. In: Blench RM, MacDonald KC (eds) The origins and development of African livestock: archaeology, genetics, linguistics, and ethnography. Routledge, London/New York, pp 127–162

Maisonneuve et Larose (1993) Le Mouton. La Croissance Des Jeunes Agneaux, vol Tome I

Mamiru M, Banerjee S, Haile A (2018) Selection practices of Bonga sheep reared in Southern Ethiopia. Proc Zool Soc 71(2):164–169. https://doi.org/10.1007/s12595-017-0207-1

Marais PG, Schoeman A (1990) Geographic distribution of Dorper sheep in the republic. Dorper News 46:4–5

Marshall F (2000) The origins and spread of domestic animals in East Africa. In: Blench RM, MacDonald KC (eds) In the origins and development of African livestock: archaeology, genetics, linguistics, and ethnography. Routledge, London/New York, pp 119–221

Mason IL (1967) Sheep breeds in Mediterranean; commonwealth agricultural Bureax. Farnham Royal

Mason IL (1996) A world dictionary of livestock breeds types and varieties. CAB International, Wallingford

Mason IL, Maule JP (1960) The Indigenous Livestock of Eastern and Southern Africa, Technical Communication no. 14 of the Commonwealth Bureau of Animal Breeding and Genetics, Edinburgh. Commonwealth Agricultural Bureaux, Farnham Royal

Mastrangelo S, Bahbahani H, Moioli B, Ahbara A, Al Abri M, Almathen F, da Silva A, Belabdi I, Portolano B, Mwacharo JM, Hanotte O, Pilla F, Ciani E (2019) Novel and known signals of selection for fat deposition in domestic sheep breeds from Africa and Eurasia. PLoS One 14(6):e0209632. https://doi.org/10.1371/journal.pone.0209632

Matika O (2001) A genetic evaluation of the Matopos Sabi Sheep Flock in Zimbabwe. PhD thesis, The University of the Orange Free State, Bloemfontein, South Africa

Matika O, Nyoni S, van Wyk JB, Erasmus GJ, Baker RL (2003) Resistance of Sabi and Dorper ewes to gastrointestinal nematode infections in an African semi-arid Environment. Small Rumin Res 47(2):95–102. https://doi.org/10.1016/S0921-4488(02)00251-1

Mavule B (2013) Phenotypic Characterisationof Zulu sheep: implications for conservation and improvement. University of Zululand

Mavule B, Muchenje V, Kunene NW (2013) Characterizationof Zulu sheep production system: implications for conservation and improvement. Sci Res Essays 8:1226–1238

Mavule BS, Sarti FM, Lasagna E, Kunene NW (2016) Morphological differentiation amongst Zulu sheep populations in KwaZulu-Natal, South Africa, as revealed by multivariate analysis. Small Rumin Res 140:50–56. https://doi.org/10.1016/j.smallrumres.2016.06.001

Maxted N (2013) In situ, ex situ conservation. In: Encyclopedia of biodiversity. Elsevier, pp 313–323. https://doi.org/10.1016/B978-0-12-384719-5.00049-6

McDermid EM, Agar NS, Char CK (2009) Electrophoretic variation of red cell enzyme Systems in Farm Animals. Anim Blood Groups Biochem Genet 6(3):127–174. https://doi.org/10.1111/j.1365-2052.1975.tb01361.x

McKenzie RL (1987) The tolerance of indigenous sheep in Zimbabwe to Haemonchus Contortus. Zimb Vet J 17:20–23

McKinnon D, Rocha A (1985) Reproduction, mortality and growth of indigenous sheep and goats in Mozambique, C.P. 25. Institute of Animal Reproduction and Breeding, Matola Mozambique

Meka Z, Martin A, Tadakeng Y, Meutchieye F, Fonteh F (2021) Biometric assessment of Blackbelly sheep in Central Africa. Genet Biodivers J 5(2):149–163

Melaku S, Kidane A, Abegaz S, Tarekegn A, Tesfa A (2019) Phenotypic characterization of Simien sheep in Simien Mountain region, Ethiopia. International Journal of Agriculture and Biosciences 8(4):178–185

Meyer C, Kreft H, Guralnick R, Jetz W (2015) Global priorities for an effective information basis of Biodiversity distributions. Nat Commun 6(1):1–8

Milne C (2000) The history of the Dorper sheep. Small Rumin Res 36(2):99–102. https://doi.org/10.1016/S0921-4488(99)00154-6

Mohammed T, Kebede K, Mekasha Y, Abera B (2015) On-farm phenotypic characterization of native sheep types in North Wollo Zone, Northern Ethiopia. Direct Res J Agric Food Sci 3(3):48–56

Molotsi AH, Dube B, Cloete SWP (2020) The current status of indigenous ovine genetic resources in southern Africa and future sustainable utilisation to improve livelihoods. Diversity 2020(12):14

Muigai AWT, Hanotte O (2013) The origin of African sheep: archaeological and genetic perspectives. Afr Archaeol Rev 30(1):39–50. https://doi.org/10.1007/s10437-013-9129-0

Mukasa-Mugerwa E, Lahlou-Kassi A (1995) Reproductive performance and productivity of Menz sheep in the Ethiopian highlands. Small Rumin Res 17(2):167–177. https://doi.org/10.1016/0921-4488(95)00663-6

Muzzolini A (2000) Livestock in Saharan Rock Art. In: Blench RM, MacDonald KC (eds) The origins and development of African livestock. Archaeology, genetics, linguistics and ethnography. University College London Press, London, pp 87–110

Mwai O, Hanotte O, Kwon Y-J, Cho S (2015) Invited review – African indigenous cattle: unique genetic resources in a rapidly changing world. Asian Australas J Anim Sci 28(7):911–921. https://doi.org/10.5713/ajas.15.0002R

Ndayambaje JD (2016) The state of Rwanda's Biodiversity for food and agriculture. FAO Country reports

Ndiaye B, Diouf MN, Ciss M, Wane M, Diop M, Sembène M (2018) Morphologie et pratiques d'élevage du mouton peul-peul du sénégal. Int J Adv Res 6(5):727–738

Nigussie H, Mekasha Y, Kebede K, Abegaz S, Kumar Pal S (2013) Production objectives, breeding practices and selection criteria of indigenous sheep in eastern Ethiopia. Livest Res Rural Dev 25(127):1–6

Nigussie H, Pal SK, Diriba S, Mekasha Y, Kebede K, Abegaz S (2016) Phenotypic variation and protein polymorphism of indigenous sheep breeds in eastern Ethiopia. Livest Res Rural Dev 28(139)

Nigussie H, Mwacharo JM, Osama S, Agaba M, Mekasha Y, Kebede K, Abegaz S, Pal SK (2019) Genetic diversity and matrilineal genetic origin of fat-Rumped sheep in Ethiopia. Trop Anim Health Prod 51(6):1393–1404. https://doi.org/10.1007/s11250-019-01827-z

Nigussie H, Mwacharo JM, Osama S, Agaba M, Mekasha Y, Kebede K, Abegaz S, Pal SK (2020) Correction to: genetic diversity and matrilineal genetic origin of fat-Rumped sheep in Ethiopia. Trop Anim Health Prod 52(6):3933–3933. https://doi.org/10.1007/s11250-020-02427-y

Njuki J, Mburu S (2013) Gender and ownership of livestock assets. In: Njuki J, Sanginga PC (eds) Women, livestock and markets: bridging the gender gap in Eastern and Southern Africa. ILRI and IDRC

Njwe RM, Manjeli Y (1992) Milk yield of Cameroon Dwarf Blackbelly Sheep. In: Small ruminant research and development in Africa, African small ruminant research networo. ILCA, Nairobi, pp 527–532

Nsoso SJ, Podisi B, Otsogile E, Mokhutshwane BS, Ahmadu B (2004) Phenotypic characterization of indigenous Tswana goats and sheep breeds in Botswana: continuous traits. Trop Anim Health Prod 36(8):789–800. https://doi.org/10.1023/B:TROP.0000045979.52357.61

Nyamukanza CC, Scogings PF, Mbatha KR, Kunene NW (2010) Forage-sheep relationships in commonly managed moist Thornveld in Zululand, Kwazulu-Natal, South Africa. Afr J Range Forage Sci 27:11–19

Nziku ZC, Katule A, Chenyambuga SW, Mruttu H (2016) The reasons, herd characteristics and Management of Indigenous Gogo Sheep in Central Tanzania. J Vet Sci 2(2):8–12

OECD and FAO (2016) OECD-FAO agricultural outlook 2016–2025; OECD-FAO Agricultural Outlook. OECD. https://doi.org/10.1787/19428846-en

Olivier WJ, Snyman MA, Olivier JJ, van Wyk JB, Erasmus GJ (2001) Irect and correlated responses to selection for total weight of lamb weaned in merino sheep. S Afr J Anim Sci 31(2):115–121

Oren ED (1979) Die Landbrüke Zwischen Asien Und Afrika. Archäologie Der Nordsinai Bis Zur Klassischen Periode. In: Rothenberg B (ed) Sinai, Pharaonen, Bergluete, Pilger und Soldaten. Kümmerly und Frey, Bern, pp 181–192

Osamede OH, Adebowale SE (2018) Genetic structure of indigenous sheep breeds in Nigeria based on electrophoretic polymorphous Systems of Transferrin and Haemoglobin. Afr J Biotechnol 17(12):380–388

Ouhichi R (2014) Dispositifs d'appui à l'innovation et Au Développement Territorial En Tunisie: Cas de La Brebis Sicilo-Sarde—Tunisie. In: 2e Séminaire Méditerranéen LACTIMED; Zahlé, Liban

Ouhrouch A, Boitard S, Boyer F, Servin B, Da Silva A, Pompanon F, Haddioui A, Benjelloun B (2021) Genomic uniqueness of local sheep breeds from Morocco. Front Genet 12:723599. https://doi.org/10.3389/fgene.2021.723599

Peacock C (1996) Improving goat production in the tropics, Manual for development work. Arm Africa and Oxfam

Pedrosa S, Uzun M, Arranz J-J, Gutiérrez-Gil B, San Primitivo F, Bayón Y (2005) Evidence of three maternal lineages in near eastern sheep supporting multiple domestication events. Proc R Soc B 272(1577):2211–2217. https://doi.org/10.1098/rspb.2005.3204

Pereira F, Davis SJM, Pereira L, McEvoy B, Bradley DG, Amorim A (2006) Genetic signatures of a Mediterranean influence in Iberian Peninsula sheep husbandry. Mol Biol Evol 23(7):1420–1426. https://doi.org/10.1093/molbev/msl007

Peters, J. A revision of the faunal remains from two central Sudanese sites: Khartoum Hospital and Esh Shaheinab.

5th international congress of Archaeozoologia, Bordeaux; 1986; pp. 11–35

Peters FW, Kotze A, van der Bank FH, Soma P, Grobler JP (2010) Genetic profile of the locally developed Meatmaster sheep breed in South Africa based on microsatellite analysis. Small Rumin Res 90(1–3):101–108. https://doi.org/10.1016/j.smallrumres.2010.02.005

Pollott G, Wilson RT (2009) Sheep and goats for diverse products and profits; FAO Diversification booklet number 9

Portes JV, Somavilla AL, Grion AL, Dias LT, Teixeira RA (2018) Short communication: genetic parameters for post-weaning visual scores and reproductive traits in Suffolk Sheep. Span J Agric Res 16(1):e04SC01. https://doi.org/10.5424/sjar/2018161-11612

Poulik MD, Smithies O (1958) Comparison and combination of the starch-gel and filter-paper electrophoretic methods applied to human sera: two-dimensional electrophoresis. Biochem J 68(4):636–643. https://doi.org/10.1042/bj0680636

Rafik A, Marnet P-G, M'Sadak Y (2018) Revue Des Connaissances Sur La Micro-Filière Ovine Laitière En Tunisie: État Des Lieux et Perspectives de Relance de La Race Sicilo-Sarde. Biotechnology, Agronomy, Society and Environment, p 3

Rain Water Run Off. The Cameroon sheep breed. https://rainwaterrunoff.com/the-cameroon-sheep-breed/. Accessed 5 June 2021

Ramsay KH, Harris L, Kotzé A (2001) Landrace breeds: South Africa's indigenous and locally developed farm animals. Farm Animal Conservation Trust, Pretoria

Rawlingson J (1994) The meat industry of Namibia, 1835 to 1994. Macmillan, Gamsberg

Rege JEO, Yapi-Gnaore CV, Tawah CL (1996) The indigenous domestic ruminant genetic resources of Africa. In: Proceedings of the second all africa conference on animal agriculture, Pretoria, South Africa

Rekik M, Aloulou R, Ben Hammoud M (2005) Small Ruminant Breeds of Tunisia. In: Iniguez L (ed) Characterization of small ruminant breeds in West Asia and North Africa, vol 2. International Center for Agricultural Research in the Dry Areas

Resende A, Gonçalves J, Muigai AWT, Pereira F (2016) Mitochondrial DNA variation of domestic sheep (Ovis Aries) in Kenya. Anim Genet 47(3):377–381. https://doi.org/10.1111/age.12412

Robinson TP, Wint GRW, Conchedda G, Van Boeckel TP, Ercoli V, Palamara E, Cinardi G, D'Aietti L, Hay SI, Gilbert M (2014) Mapping the global distribution of livestock. PLoS One 9(5):e96084. https://doi.org/10.1371/journal.pone.0096084

Rocha A, McKinnon D, Wilson RT (1990) Comparative performance of Landim and blackhead Persian sheep in Mozambique. Small Rumin Res 3(6):527–538. https://doi.org/10.1016/0921-4488(90)90048-B

Rosati A, Mousa E, Van Vleck LD, Young LD (2002) Genetic parameters of reproductive traits in sheep. Small Rumin Res 43(1):65–74. https://doi.org/10.1016/S0921-4488(01)00256-5

Roubert CA, La Carter PI (1984) Domestication Au Maghreb: État de La Question. In: Krzyzaniak L, Kobusiewica M (eds) Origin and early development of food-producing cultures in North-Eastern Africa. Polish Academy of Sciences and Poznan Archaeological Museum, Poznan, pp 437–451

Rouissi H, Souissi NB, Dridi S, Chaieb K, Tlili S, Ridene J (2001) Performances Zootechniques de La Race Ovine Sicilo-Sarde En Tunisie. In: Rubino R, Morand-Fehr P (eds) Production systems and product quality in sheep and goats, Série A, 46. CIHEAM, Options Méditerranéennes, Zaragoza, pp 231–236

Rowley-Conwy P (2013) North of the frontier: early domestic animals in Northern Europe. In: Colledge S, Conolloy J, Dobney K, Manning K, Shennan S (eds) The origins and spread of domestic animals in Southwest Asia and Europe. Left Coast Press, Walnut Creek, pp 283–312

Russell TM (2004) The spatial analysis of radiocarbon databases: the spread of the first farmers in Europe and of the fat-tailed sheep in southern Africa. Archaeopress, Oxford

Russell N, Martin L, Buitenhuis H (2005) Cattle domestication at Çatalhöyük revisited. Curr Anthropol 46(5):S101–S108

Ryder ML (1983) Sheep & man. Duckworth, London

Ryder ML (2007) Sheep & [and] man. Duckworth, London

Sadr K (2015) Livestock first reached southern Africa in two separate events. PLoS One 10(8):e0134215. https://doi.org/10.1371/journal.pone.0134215

Safari E, Fogarty NM, Gilmour AR (2005) A review of genetic parameter estimates for Wool, growth, meat and reproduction traits in sheep. Livest Prod Sci 92(3):271–289. https://doi.org/10.1016/j.livprodsci.2004.09.003

Sagne J (1950) L'Algérie Pastorale. Ses Origines, Sa formation, Son Passé, Son Présent, Son Avenir. Imprimerie Fontana: 27

Saleh Ahmed A, Hammoud MH, Dabour NA, Hafez EE, Sharaby MA (2020) BMPR-1B, BMP-15 and GDF-9 genes structure and their relationship with litter size in six sheep breeds reared in Egypt. BMC Res Notes 13(1):215. https://doi.org/10.1186/s13104-020-05047-9

Sargent J, van der Bank FH, Kotze A (1999) Genetic variation in blood proteins within and between 19 sheep breeds from Southern Africa. S Afr J Anim Sci 29(3):245–257

Satam H, Joshi K, Mangrolia U, Waghoo S, Zaidi G, Rawool S, Thakare RP, Banday S, Mishra AK, Das G, Malonia SK (2023) Next-generation sequencing technology: current trends and advancements. Biology 12(7):997. https://doi.org/10.3390/biology12070997

Scherf BD (2000) World watch list for domestic animal diversity, 3rd edn. Food and Agriculture Organization of the United Nations, Rome

Schoeman SJ (2000) A Comparative assessment of Dorper sheep in different production environments and systems. Small Rumin Res 36(2):137–146. https://doi.org/10.1016/S0921-4488(99)00157-1

Schröder O, Lieckfeldt D, Lutz W, Rudloff C, Frölich K, Ludwig A (2016) Limited hybridization between domestic sheep and the European mouflon in Western Germany. Eur J Wildl Res 62(3):307–314. https://doi.org/10.1007/s10344-016-1003-3

Sefiani M (1980) Studies on Sardi Ewes and Beni Guil Ewes about the Milk productivity and the milking ability (in Morocco)

Senyatso EK, Masilo BS (1996) Animal genetic resources in Botswana. Anim Genet Resour 17:51–60

Seo SN (2021) Friends of animals: a story of sheep and goats in the Sahel in Sub-Saharan Africa. In: Climate change and economics. Springer International Publishing, Cham, pp 27–41. https://doi.org/10.1007/978-3-030-66680-4_2

Seré C, Steinfeld H (1996) World livestock production systems. Current status, issues and trends. Rome, FAO

Skapetas B, Kalaitzidou M (2017) Current status and perspectives of sheep sector in the world. Livest Res Rural Dev 29(2):21

Skeates R (2003) Radiocarbon dating and interpretations of the Mesolithic-Neolithic transition in Italy. In: Ammerman AJ, Biagi P (eds) The widening harvest: the Neolithic transition in Europe looking Back, looking forward. Archaeological Institute of America, Boston, pp 157–185

Snoussi S (2003) Situation de l'élevage Ovin En Tunisie et Rôle de La Recherche Réflexions Sur Le Développement d'une Approche Système. Cahiers Agricultures 12(6):419–428

Snowder GD (2008) Genetic improvement of overall reproductive success in sheep: a review. 16

Snowder G, Fogarty N (2008) Composite trait selection to improve reproduction and ewe productivity: a review. Anim Prod Sci 49:9–16. https://doi.org/10.1071/EA08184

Snyman MA (2014a) South African Sheep Breeds: Ronderib Afrikaner, Info-pack ref. 2014/027. Grootfontein Agricultural Development Institute

Snyman MA (2014b) South African Sheep Breeds: Damara, Info-pack ref. 2014/014. Grootfontein Agricultural Development Institute

Snyman MA (2014c) South African Sheep Breeds: Namaqua Afrikaner, Info-pack ref. 2014/023. Grootfontein Agricultural Development Institute

Snyman MA (2014d) South African Sheep Breeds: Pedi, Info-pack ref. 2014/025. Grootfontein Agricultural Development Institute

Snyman MA (2014e) South African Sheep Breeds: Karakul Sheep, Info-pack ref. 2014/019. Grootfontein Agricultural Development Institute

Snyman MA (2014f) South African Sheep Breeds: South African Merino, Info-pack ref. 2014/028. Grootfontein Agricultural Development Institute

Snyman MA (2014g) South African Sheep Breeds: Afrino, Info-pack ref. 2014/013. Grootfontein Agricultural Development Institute

Snyman MA (2014h) South African Sheep Breeds: Van Rooy, Info-pack ref. 2014/030. Grootfontein Agricultural Development Institute

Snyman MA (2014i) South African Sheep Breeds: Dohne Merino, Info-pack ref. 2014/015. Grootfontein Agricultural Development Institute

Snyman MA (2014j) South African Sheep Breeds: Dormer, Info-pack ref. 2014/016. Grootfontein Agricultural Development Institute

Snyman MA, Olivier JJ, Cloete JAN (1993) Productive and reproductive performance of Namaqua Afrikaner sheep. Karoo Agric 5(2):21–24

South African Department of Agriculture (1997) Abstract of Agricultural Statistics. Directorate of Agricultural Information Services Private Bag X144, Pretoria 0001, South Africa

Stiner MC, Buitenhuis H, Duru G, Kuhn SL, Mentzer SM, Munro ND, Pollath N, Quade J, Tsartsidou G, Ozbaaran M (2014) A forager-herder trade-off, from broad-spectrum hunting to sheep Management at A Kl Hoyuk, Turkey. Proc Natl Acad Sci 111(23):8404–8409. https://doi.org/10.1073/pnas.1322723111

Stow GW (1905) The native races of South Africa. Swan Sonnenschein

Stratil A (2009) Two new sheep transferrin variants and the effect of neuraminidase. Anim Blood Groups Biochem Genet 4(3):153–159. https://doi.org/10.1111/j.1365-2052.1973.tb01291.x

Thaba Manzi Ranch. Thaba Manzi Pedis. Thaba Manzi Pedis. http://www.pedisheep.co.za/breed.htm. Accessed 11 Sept 2021

Tibbo M, Ayalew W, Awgichew K, Ermias E, Rege JEO (2004) On-Station characterisation of indigenous Menz and Horro sheep breeds in the central highlands of Ethiopia. Anim Genet Resour Inf 35:61–74. https://doi.org/10.1017/S1014233900001814

Tibbo M, Philipsson J, Ayalew W (2006) Sustainable sheep breeding programmes in the tropics: a framework for Ethiopia. University of Bonn, Tropentag

Trouette M (1929) Les «races» d'Algérie. In Congrès du mouton, pp 299–302

Uerpmann H-P (1987) The ancient distribution of ungulate mammals in the Middle East: Fauna and Archaeolog. In: Sites in Southwest Asia and Northeast Africa; Beihefte zum Tübinger Atlas des Vorderen Orients Reihe A, Naturwissenschaften. Reichert, Wiesbaden

Upton M (1985) Returns from small ruminant production in south West Nigeria. Agric Syst 17:65–83

Van Neer W (2000) Domestic animals from archaeological sites in Central and West- Central Africa. In: Blench RM, MacDonald KC (eds) The origins and development of African livestock: archaeology, genetics, linguistics, and ethnography. Routledge, London/New York, pp 163–190

Vermeersch PM, Van Peer W, Moeyersons J, Van Neer W (1994) Sodmein cave site. Red Mountains Sahara 6:32–42

Vigne J-D, Buitenhuis H, Davis S (1999) Les Premiers Pas de La Domestication Animale à l'Ouest de l'Euphrate: Chypre et l'Anatolie Centrale. Paléorient 25(2):49–62

Vigne J, Zarro A, Cucchi T, Carrère I, Briois F, Guilaine J (2014) The transportation of mammals to Cyprus sheds light on early voyaging and Boats in the Mediterranean. Eurasian Prehistory 10:157–176

Viljoen JJ (1981) Die Geskiedenis van Die Karakoelboerdery in Suidwes-Afrika 1907–1950. Randse Afrikaanse University, Johannesburg

Wang S, Foote WC (1990) Protein polymorphism in sheep pedigree testing. Theriogenology 34(6):1079–1085. https://doi.org/10.1016/S0093-691X(05)80007-X

Wanjala G, Bagi Z, Kusza S (2021) Meta-analysis of mitochondrial DNA control region diversity to shed light on phylogenetic relationship and demographic history of African sheep (Ovis Aries) breeds. Biology 10(8):762. https://doi.org/10.3390/biology10080762

Wanyangu SW, Mwendia CMT, Bain RK, Mugambi JM, Stevenson P (1993) Production potentials of the red Maasai and black-headed Somali sheep under pastoral management conditions without trypanosome and helminth control. Small Ruminant Collaborative Research Support Programme, Nairobi, pp 167–180

Weldemariam B, Mezgebe G (2021) Community based small ruminant breeding programs in Ethiopia: Progress and challenges. Small Rumin Res 196:106264. https://doi.org/10.1016/j.smallrumres.2020.106264

Weldeyesus GB (2020) Phenotypic variations of indigenous sheep breed ecotypes of Ethiopia: a review. J Cell Anim Biol 14(1):6–20. https://doi.org/10.5897/JCAB2019.0467

Wendorf F, Schild R (1998) Nabta playa and its role in northeastern African prehistory. J Anthropol Archaeol 17:97–123

Wiener G (1994) Animal breeding. Macmillan published in co-operation with the CTA, Basingstoke

Wiener P, Robert C, Ahbara A, Salavati M, Abebe A, Kebede A, Wragg D, Friedrich J, Vasoya D, Hume DA, Djikeng A, Watson M, Prendergast JGD, Hanotte O, Mwacharo JM, Clark EL (2021) Whole-genome sequence data suggest environmental adaptation of Ethiopian sheep populations. Genome Biol Evol 13(3):evab014. https://doi.org/10.1093/gbe/evab014

Wiener P, Salavati M, Djikeng A, Van Tassell CP, Rosen BD, Spangler GL, Simo G, Tanya V, Meutchieye F, Clark EL (2022) 412. Genetic diversity of the Cameroon Blackbelly Sheep, an Indigenous Sheep from West Africa. In: Proceedings of 12th World Congress on Genetics Applied to Livestock Production (WCGALP). Wageningen Academic Publishers, Rotterdam, pp 1717–1720. https://doi.org/10.3920/978-90-8686-940-4_412

Wilson RT (1988) Small ruminant production Systems in Tropical Africa. Small Rumin Res 1(4):305–325. https://doi.org/10.1016/0921-4488(88)90058-2

Wilson RT (1991) Small ruminant production and the small ruminant genetic resource in tropical Africa, Food and Agriculture Organisation animal production and health paper 88. Rome, Food and Agriculture Organisation

Winrock International (1992) Assessment of animal agriculture in Sub-Saharan Africa. Winrock International, Mprrilton, p 20

Yaro M, Munyard KA, Morgan E, Allcock RJN, Stear MJ, Groth DM (2019) Analysis of pooled genome sequences from Djallonke and Sahelian sheep of Ghana reveals co-localisation of regions of reduced heterozygosity with candidate genes for disease resistance and adaptation to a tropical environment. BMC Genomics 20(1):816. https://doi.org/10.1186/s12864-019-6198-8

Yiheyis A, Tegegn F, Melekot MH, Taye M (2012) Pre-weaning growth performance of Sekota sheep breed in Waghimra zone, Ethiopia. Online J Anim Feed Res 2(4):340–343

Yunusa AJ, Salako AE, Oladejo OA (2013) Morphometric characterization of Nigerian Indigenoussheep using multifactorial discriminant analysis. Int J Biodivers Conserv 5(10):661–665

Zeder MA (2008) Animal domestication in the zagros: an update and directions for future research. In: Vila E, Gourichon L, Choyke A, Buitenhuis H (eds) Archaeozoology of the Near East VIII. Travaux de la Maison de l'Orient et de la Méditerranée (TMO), Lyon, pp 243–278

Zeder MA (2011) The origins of agriculture in the near east. Curr Anthropol 52(S4):S221–S235. https://doi.org/10.1086/659307

Zeder MA (2017) Out of the fertile crescent: the dispersal of domestic livestock through Europe and Africa. In: Boivin N, Crassard R, Petraglia M (eds) Human dispersal and species movement: from prehistory to the present. Cambridge University Press, Cambridge, pp 261–303. https://doi.org/10.1017/9781316686942.012

Zewdu E (2008) Characterization of Bonga and Horro indigenous sheep breeds of smallholders for designing community based breeding strategies in Ethiopia. MSc, School of graduate studies, Haramaya University

Zilhão J (2001) Radiocarbon evidence for maritime Pioneer colonization at the origins of farming in West Mediterranean Europe. Proc Natl Acad Sci 98(24):14180–14185. https://doi.org/10.1073/pnas.241522898

Zilhão J (2003) The Neolithic transition in Portugal and the role of Demic diffusion in the spread of agriculture across West Mediterranean Europe. In: Ammerman AJ, Biagi P (eds) The widening harvest: the Neolithic transition in Europe looking Back, looking forward. Archaeological Institute of America, Boston

Zishiri OT (2009) Genetic analyses of south African terminal sire sheep breeds. M.Sc. thesis, Department of Animal Sciences, Faculty of AgriSciences, Stellenbosch University, 114

Zonabend König E, Strandberg E, Ojango JMK, Mirkena T, Okeyo AM, Philipsson J (2017) Purebreeding of red Maasai and crossbreeding with Dorper sheep in different environments in Kenya. J Anim Breed Genet 134(6):531–544. https://doi.org/10.1111/jbg.12260

Open Access This chapter is licensed under the terms of the Creative Commons Attribution 4.0 International License (http://creativecommons.org/licenses/by/4.0/), which permits use, sharing, adaptation, distribution and reproduction in any medium or format, as long as you give appropriate credit to the original author(s) and the source, provide a link to the Creative Commons license and indicate if changes were made.

The images or other third party material in this chapter are included in the chapter's Creative Commons license, unless indicated otherwise in a credit line to the material. If material is not included in the chapter's Creative Commons license and your intended use is not permitted by statutory regulation or exceeds the permitted use, you will need to obtain permission directly from the copyright holder.

7. African Domestic Poultry Genetic Resources, Diversity, and Unique Features

Tadelle Dessie, Christian K. Tiambo,
Liveness J. Banda, Raman A. Lawal,
Sheila C. Ommeh, Timothy Gondwe,
Esatu Wondmeneh, Matthew A. Adeleke,
and Olivier Hanotte

Abstract

This chapter provides a detailed account of African domestic poultry genetic resources, their diversity, and unique features. It aims to bring together information to provide a comprehensive picture of the topic. Section 7.1 introduces African poultry genetic resources, their importance, and previous interventions conducted to better utilize this important resource. Section 7.2 deals with the origin of poultry genetic resources and their dispersal into Africa. The introduction of domesticated chickens into Africa is not well documented.

Although indigenous birds have several adaptive traits and associated genes that are important in the tropics, the real value of indigenous chicken breeds is often underestimated. They have a poor appearance, relatively low productivity, and are consequently considered to have low value compared to their improved commercial counterparts. Chicken genetic resources are endangered and under-conserved. However, there has been recent emphasis on local chicken breed development and conservation, which has gained attention. The section demonstrates how chickens have become a key part of the African poultry production system and how they can remain vital and instrumental, considering the current climate change. Sections 7.3, 7.4, 7.5 and 7.6 describe the available chicken and duck resources in different parts of Africa and out-

T. Dessie (✉) · E. Wondmeneh
International Livestock Research Institute (ILRI), Addis Ababa, Ethiopia
e-mail: t.dessie@cgiar.org

C. K. Tiambo
Centre for Tropical Livestock Genetics and Health (CTLGH) – ILRI, Nairobi, Kenya

L. J. Banda · T. Gondwe
Lilongwe University of Agriculture and Natural Resources, Lilongwe, Malawi

R. A. Lawal
The Jackson Laboratory, Bar Harbor, ME, USA

S. C. Ommeh
Institute for Biotechnology Research (IBR), Jomo Kenyatta University of Agriculture and Technology (JKUAT), Nairobi, Kenya

M. A. Adeleke
Discipline of Genetics, School of Life Sciences, University of KwaZulu-Natal, Durban, South Africa

O. Hanotte
International Livestock Research Institute (ILRI), Addis Ababa, Ethiopia

Centre for Tropical Livestock Genetics and Health (CTLGH) – ILRI, Addis Ababa, Ethiopia

The University of Nottingham, School of Life Sciences, Nottingham, UK

© The Authors(s) 2026
E. M. Ibeagha-Awemu et al. (eds.), *African Livestock Genetic Resources and Sustainable Breeding Strategies*, Sustainable Development Goals Series, https://doi.org/10.1007/978-3-031-92076-9_7

line their unique features and population estimates. Considering the population size of these genetic resources, conservation efforts to support their utilization are addressed in Sect. 7.8. Furthermore, the importance of conserving poultry genetic resources using germplasm cryobanking (cryoconservation) is described. Section 7.10 describes genetic improvement attempts in Africa involving indigenous chicken populations. The information appears to be limited to a few examples, as not many improvement programs are ongoing, and information on some programs is not available. Section 7.11 describes the major constraints of poultry production in Africa, followed by the economic contributions in Sect. 7.12. The last chapter implies the way forward in the utilization of poultry genetic resources in Africa.

Keywords

Poultry · Genetic resources · Cryoconservation · Africa

Abbreviations

AU-IBAR	The African Union—InterAfrican Bureau for Animal Resources
CTLGH	The Centre for Tropical Livestock Genetics and Health
DAGRIS	Domestic Animal Genetic Resources Information System
DNA	Deoxyribonucleic acid
ESCs	Embryonic Stem Cells
FAO	Food and Agriculture Organization of the United Nations
FAO DAD-IS	FAO – Domestic Animal Diversity Information System
GDP	Gross Domestic Product
ILRI	International Livestock Research Institute
iPSCs	Induced Pluripotent Stem Cells
KALRO	Kenya Agricultural and Livestock Research Organization
NARS	National Agricultural Research Systems
MSCs	Mesenchymal Stem Cells
PGC	Primordial Germ Cell
SFRB	Scavenging Feed Resource Base

7.1 Introduction

7.1.1 Overview of African Poultry Genetic Resources

Poultry genetic resources have existed for thousands of years and many generations. They are found in a diversity of environments and production systems, and they were subjected to breeding improvement and natural selection (FAO 2014). The chickens, once domesticated, have adapted to specific environmental and farming conditions, resulting in within-population unique gene combinations. When direct measures of genetic diversity are unavailable, the number of breeds and ecotypes may provide a first indication of the farm species' genetic diversity. Breeds are commonly classified as indigenous or exotic, with indigenous populations often present in low-input–low-output production systems and exotic ones referring to commercial lines and imported fancy breeds. The poultry genetic resources, the entire chicken populations, exotic and commercial, contain genetic material of present or future value. The diversity of poultry genetic resources reflects a long history of natural and human selection as well as crossbreeding among not only domestic chickens but also, as recently discovered, with wild *Gallus* populations outside the center(s) of domestication (Lawal et al. 2020; Wang et al. 2020; Lawal and Olivier 2021).

Beyond their potential use as inputs to improve productivity and other traits of interest, understanding the resources and designing interventions to conserve and utilize species, breeds, and ecotypes is vital to benefit the households that depend on them. The poultry genetic resources are a starting material for breeding activities to improve the traits of interest. Conserving genetic resources is a means of safeguarding the resources further to be utilized as food and other purposes considered necessary by the users. While in situ in vivo, we may call it conservation through utilization. Here, the utilized genetic resources are a warranty for their conservation. Conservation efforts should ensure broad-based genetic diversity to ensure availability for use by present and future generations. This can only be possible by knowing the within-species diversity and keeping important information in an orderly manner in breed databases such as the (DAGRIS 2021; DAD-IS n.d.).

Domestic chickens first appeared in Africa many centuries ago and are now an established part of the African agricultural landscape (Alders and Pym 2009). There are about 126 poultry genetic resources regarded as varieties or ecotypes in Africa (http://dagris.ilri.cgiar.org/). Most of them are hardly studied to show their actual worth, but some studies have revealed their potential as better-performing and adapted chicken populations. There has still been little effort in African countries to conserve the local chicken breeds or lines (Manyelo et al. 2020). Although African poultry genetic resources are resistant to prevalent disease, they are also known for low productivity (Dessie 2003). Poultry genetic resources in Africa are admired for their tolerance to common poultry diseases and uncertainty in feed quality and availability. They exist with minimal input supply (Desta and Wakeyo 2012). However, any improvement in the productivity of local chickens will require close attention to nutritional, breeding, and health aspects. Smallholders in Africa are the custodians of poultry genetic resources, and they kept them under a traditional scavenging system (Magothe et al. 2012a; Desta et al. 2013). Interventions to introduce high-performing chicken strains into Africa have been largely unsuccessful (Tadelle et al. 2003; Wondmeneh et al. 2016). Poultry keepers in African villages cannot afford the high input requirements (housing/shelter, commercial diets, and strict disease control/vaccination programs) associated with more genetically efficient breeds (Tadelle et al. 2003).

7.1.2 Importance of African Poultry Genetic Resources

Although indigenous birds have several adaptive traits and genes with utility in the tropics (Horst 1989a), the real value of indigenous chicken breeds is often underestimated, primarily due to their poor appearance, relatively low productivity, and alleged low "commercial" values. African poultry genetic resources are neglected, and relatively little attention has been given to them in the research and development agendas. Hodges (1990) stated that developing countries, in most cases, opt for high-performing commercial breeds from developed countries to increase animal productivity through crossbreeding or, if conditions allow, by breed substitution without adequately investigating the production system and potential of the indigenous birds. Although poultry (both meat and eggs) is an essential source of food and a means of investment that is important to the welfare of women and children in traditional and low-input systems, an alarming 34% of all chickens in Africa are at risk of being lost (FAO 2000).

7.1.3 Initiatives to Utilize Indigenous Poultry Genetic Resources

Emphasis on local chicken breed development and conservation has gained attention through several pioneering programs led by national institutions. For example, independent indigenous chicken improvement programs have been established involving promising local chicken breeds in Kenya (KALRO 1 and KALRO 2) (Miyumo et al. 2023), Ethiopia (Horro and Tilili) (Mulugeta et al. 2020), Tanzania (Horasi) Esatu 2022 (a) (Esatu et al. 2022), Nigeria (Noiler and Funaab-alpha) (Bamidele et al. 2019), and Cameroon (IRZ (Van Marle-Köster and Casey 2001)). They aim to improve egg production, growth, survival, and age of the first egg. These chicken strains were identified through previous characterization and evaluation studies and possessed qualities appreciated by the poultry producers and consumers. Five of these programs were established through the support of the ILRI-led African Chicken Genetic Gains (ACGG) and Tropical Poultry Genetic Solution (TPGS) programs, while the programs in Nigeria and Cameroon were started earlier. Seven of these programs are applying mass selection supported by BLUP (Best Linear Unbiased Prediction) techniques to select the best-performing chickens based on the estimated breeding values. Further, there is the potential in the future of applying genomic selection with the characterization of the breeds at the genome level. This is currently being explored through collaboration between the newly established Center for Tropical Livestock Genetics and Health (CTLGH) and national partners. The CTLGH poultry program also includes the identification of local ecotypes and ex situ PGC (Primordial Germ Cell) conservation initiatives (see Sect. 7.8).

7.2 The Origin and Purpose of African Poultry in Relation to African Farming Systems: A Historical and Production Systems Perspective

7.2.1 Origin of Domestic Chicken and Its Dispersal to Africa

Domestic chickens are assigned to the genus *Gallus,* which includes four extant wild species. Several studies support Asia as the continental origin of chicken (Wang et al. 2020; Fumihito et al. 1994; Peters et al. 2022; Clutton-Brock 1993) with the specific geographic domestication center likely in Southeast Asia (Wang et al. 2020; Fumihito et al. 1994). One of the red junglefowl subspecies, i.e., *Gallus gallus spadiceus,* is the primary ancestor from which the chicken was domesticated about 8000 years ago (Lawal et al. 2020; Fumihito et al. 1994; Peters et al. 2022). However, subsequent genetic introgression after domestication from different *Gallus* species (*G. lafayettii, G. sonneratii, G. varius*) and the other four *Gallus* subspecies (*G. g. bankiva, G. g. gallus, G. g. jabouillei, G. g. murghi*) have been reported[2,325]. The consequence of these introgression episodes may explain the observed high genetic diversity seen today in modern chickens and, perhaps, also their rapid adaptation to new environments. From Southeast Asia, the domestic chicken migrated, as human livestock commensal, to other parts of the world, adapting to different environmental conditions.

The introduction of domesticated chickens into Africa is not well documented. Clutton-Brock (1993) and Mwacharo et al. (2013) have summarized the archaeological findings on the domestic fowl in Africa. It is worth emphasizing here that Egypt played a key role in chicken introduction to the African continent. First, its connection to the Fertile Crescent presented an important trading network between Africa and South Asia. Second, the early civilization of Egypt attracted international diplomatic relationships with the Persians and with the North of the Mediterranean Sea, especially during the Egyptian Greek and Roman reigns. Egypt was therefore an important route of entry and dispersion of chicken to the continent, either through the Mediterranean coast or overland.

Domestic chicken bones have been found in Ethiopia as early as 1055–825 BC (D'Andrea et al. 2011). In East Africa, they have been found in two Iron Age sites in Mozambique and South Africa from the eighth century. In South Africa (Plug 1996), mentions that they were uncommonly found in early Iron Age sites. In West Africa, they have been excavated from the Iron Age site of Jenne-Jalo in Mali, dating from 500 to 800 AD.

Also, multiple lines of evidence support different timescales and patterns of chicken dispersion to the African continent (Adebambo et al. 2010; Muchadeyi et al. 2008; Mwacharo et al. 2011; Razafindraibe et al. 2008). Two migration waves and entry points have been proposed. (i) An overland entry following possibly the Berbers and the Phoenician expansion along the North of Africa and across the Sahara (Kiple and Ornelas 2000), with most domestic chickens on the North, West, East, and South parts of the continent likely originating from this wave. This wave might be characterized by a single mitochondrial DNA haplogroup present in all African countries but also commonly found on the Indian subcontinent (Adebambo et al. 2010; Muchadeyi et al. 2008; Mwacharo et al. 2011; Hassaballah et al. 2015). (ii) A more recent arrival of chickens to Africa likely followed the Indian Ocean maritime trading routes along East Africa, the Red Sea, India, and Southeast Asia coastal areas (Fuller and Boivin 2009).

Today, chickens play significant roles in the cultural life of rural people. In Africa, different indigenous chicken types are valued for cultural, social, and religious functions (Sonaiya 1999). Basic features with sociocultural significance are the color, sex, and comb type of the bird, and these are often related to the spirituality of their owners(Sonaiya 1999; Tadelle 1996).

7.2.2 Domestic Chickens as a Key Part of the African Farming Systems

At the onset of domestication, chickens were likely kept for cultural and recreational purposes, including cockfighting. In Africa, the precise time of transition from a cultural to a food-producing animal is unknown, but it may have followed urbanization and/or cultural habit changes as observed in Europe around 1000 AD (Loog et al. 2017). Since then, chickens have remained a key part of the African farmyard, and globally, they have become the most important poultry species (FAO 2014). It is the most common domestic animal on the African continent, and it is now kept primarily for meat and eggs and mostly under the custody of women and children.

There are two major poultry production systems in Africa: commercial (intensive) and indigenous village chickens (extensive). In Africa, the cost of raising commercial chickens is high. As a result, most poor African farmers keep indigenous village chickens within a small-scale farming system. In Tanzania, indigenous chicken represents 94% of the chicken population (R.I.U 2012), a proportion similar across the continent. Like their wild ancestors, indigenous village chickens are natural scavengers. They are raised under an extensive system of production where they forage on free range and under natural environmental conditions. Though the productivity of the extensively raised village chickens is less than the commercial chickens, the overall input is minimal, especially for housing, disease control, management, and supplementary feeding (Alders et al. 2018). Thus, it is economically efficient to be kept by farmers. Indigenous chickens represent an important genetic resource for breed improvement programs in Africa. They are characterized by high genetic diversity, adaptation to common poultry diseases, and poor-quality feed (Bettridge et al. 2018; Lyimo et al. 2014). Hence, they represent a key resource to enhancing food security, nutrition improvement, and economic growth (Alders and Pym 2009; Melesse 2014).

7.2.3 Challenges with Small-Scale Backyard Farming

Small-scale farming systems are faced with multiple risks, including a high mortality rate associated with predator attacks and disease outbreaks. A sudden change in the weather cycle affecting the availability of feed resources due to droughts can also increase mortality (Ayanlade et al. 2017; Debela et al. 2015). As part of their survival instinct, indigenous village chickens mostly roost in trees, high fences, or other elevated structures. However, deforestation, urbanization, and climate-induced fire outbreaks constitute a current threat to them. The lack of an organized marketing system, capital, and, in the event of a disease outbreak, the lack of adequate vaccination programs are key challenges facing backyard farming. All these pose risks to productivity, economic growth, and general rural development.

7.2.4 The Impact of Climate Change on Poultry Farming in Africa

Smallholder farmers significantly contribute to the food supply in Africa (Kamara et al. 2019). Their daily survival is also dependent on the livestock they keep, both for food and income. Though Africa, as well as poultry farming, contributes the least to greenhouse gases, compared to other continents and ruminant livestock species (Abioja and Abiona 2020), the impact of climate change is expected to be severe on the continent, disrupting natural ecosystems (Ngaira 2007), causing famine, starvation, and death. For instance, a warmer climate will lead to drought, creating competition for water, forages, heat stress, and disease outbreaks. Heat stress, for instance, has a direct impact on growth, productivity, and mortality rate of poultry (Liverpool-Tasie et al. 2019; Tankson et al. 2001). Though this may be severe, indigenous village chickens can serve as a key model to mitigate against the impact of climate crises on the continent through adequate breeding improvement programs. Their genomes have adaptively evolved to better cope with extreme climatic variations, as shown, for example, by a recent study including 245 Ethiopian indigenous village chickens from different agro-ecologies (Gheyas et al. 2011).

7.3 The Current State of Knowledge on Poultry Genetic Resources: Introduction and Distribution in Africa

According to (AU-IBAR 2019), a diversity of both local and exotic poultry breeds is found on the continent. Chickens have the largest number of "breeds" (329), followed by ducks (41) and guinea fowls (40); other species, like domestic ostriches and turkeys, are also present but in much lower numbers. Local breeds follow the path of smallholding in rural communities, while exotic breeds are for commercial production, and they are common in peri-urban areas. The increased proportion of exotic breeds observed across the continent reflects the demand for livestock products following the population size increase. Except for dual-purpose breeds that are used in crossbreeding programs, exotic strains are usually raised in confinement following an all-in-all-out system. In this respect, they do not pose a threat to local breeds. According to (DAD-IS n.d.), West and Southern Africa host the largest number of described "breeds," followed by East and Central Africa, with a lower number of breeds reported for North Africa. A high number of plumage and morphological phenotypes may be found within the indigenous chicken population, some of which have been associated with specific breed types (Table 7.1). Also, there are genes in indigenous chicken populations that may be associated with adaptation to tropical conditions (Table 7.2).

7.4 Examples of African Indigenous Chicken Breeds

Most poultry in Africa are chickens, ducks, and other species. Africa is home to various local and exotic chickens. The various chicken and duck breeds were introduced from the Southeast Asian region and adapted to the African conditions for thousands of years. In this and the next section, we present information on chicken breeds according to the regions of Africa. It should be noted that information regarding the indigenous chickens of Africa is limited. Most of the information is only available (DAD-IS n.d.).

7.4.1 Chicken Breeds in North Africa

A few indigenous chicken breeds in North Africa are characterized and studied. Although most indigenous chicken breeds are nondescript, some show morphological differences; they may be considered breeds or ecotypes (Negash et al. 2023). Such breeds as Fayoumi (Egypt), which is adapted to North Africa's harsh and dry conditions, can be used in future breed improvement programs. Descriptions of significant North African indigenous chickens are given below.

Table 7.1 Identified characteristics of some African indigenous chicken breeds

Country	Local name	Identifiable characteristics	Mature male weight (kg.)	Mature Female weight (kg.)	References
Burkina Faso (DAD-IS n.d.)	Cou nu, Joub-kole	Na: naked neck F: frizzle	1.5	1.2	DAD-IS (n.d.)
Chad (DAD-IS n.d.)	Chicken of Moulkou	P: pea comb	1.5	1	DAD-IS (n.d.)
Chad (DAD-IS n.d.)	Dijded	P: pea comb	1.5–2.0	1.0–1.5	DAD-IS (n.d.)
Ghana (DAD-IS n.d.)	Local Ghanaian	Na: naked neck, F: frizzle, P: pea comb	1.2	1.1	DAD-IS (n.d.)
Lesotho (DAD-IS n.d.)	Basotho	P: pea comb	1.8	1.6	DAD-IS (n.d.)
South Africa (DAD-IS n.d.)	Kaalnekke	Na: naked neck	–	–	DAD-IS (n.d.)
Swaziland (DAD-IS n.d.)	Inkhukhu	Na: naked neck	2.1	1.6	DAD-IS (n.d.)
Sudan (DAD-IS n.d.)	Large balady	Na: naked neck	–	–	DAD-IS (n.d.)
Ethiopia (DAD-IS n.d.)	Melata	Na: naked neck	–	–	DAD-IS (n.d.)
Cameroon (DAD-IS n.d.)	Sudano–Sahelian, Forest zone chicken, Northwest/West chicken	Na: naked neck F: frizzle C: Crested S: feathered shanks, T: feathered tarsus, N: normal feathered	1.4–1.7	1.2–1.3	

Table 7.2 Examples of genes in indigenous chicken populations that may be linked to the adaptation to tropical conditions

Gene	Mode of inheritance	Direct effects	Indirect effects
Dw: Dwarf	Recessive, sex-linked, multiple allele	Reduction of body size between 30 and 10% from the normal size	Reduced metabolism, improved fitness, disease tolerance
Na: Naked neck	Incomplete dominant	Loss of neck feathers, reduction of pteryla width, reduction of secondary feathers	Improved ability for convection, reduced embryonic livability (hatchability), improved adult fitness
F: Frizzle	Incomplete dominant	Curling of feathers, reduced feathering	Decreased fitness under temperate conditions, improved ability for convection
K: Slow feathering	Dominant, sex-linked, multiple allelic	Delay of feathering	Reduced protein requirement, reduced fat deposition during juvenile life, increased heat loss during early growth
Id: Non-inhibitor	Recessive, sex-linked, multiple allelic	Dermal melanin deposition in the skin and shanks	Improved ability for radiation from shanks and skin
Fm: Fibro-melanosis	Dominant with multifactorial modifiers	Melanin deposition all over the body, sheaths of muscles and nerves, and blood vessel walls	Protection of skin against UV radiation, improved radiation from the skin, increased packed cell volume, and plasma protein
P: Pea comb	Dominant	Change of skin structure, compact comb size, reduction of pterylae width, development of breast ridges	Decreased frequency of breast blisters, sex limited, improvement of late juvenile growth

Source: Horst (1989b)

7.4.1.1 Beheri

Distribution and Population Statistics: The breed originated in the Northern region and the Delta of Egypt with an estimated population size of 800,000 (DAD-IS n.d.).

Physical Characteristics: Detailed description of the breed is unavailable; however, the average body weight of mature Beheri chicken weighs 1.7 and 1.3 kg for the cocks and hens, respectively.

Adaptive and Special Genetic Characteristics: Beheri chickens are resistant to leukosis and spirochetosis, have high fertility (more than 95%), and are early maturing for cocks (4.5 months) and hens (5 months).

Typical Production Systems, Husbandry Practices, and Production Characteristics: Information regarding the production system, husbandry practices, and production characteristics is unavailable. However, being an adaptive local chicken, the Beheri chicken would suit low-input and low-output production conditions.

7.4.1.2 Dandarawi

Distribution and Population Statistics: Dandarawy is an indigenous chicken from Egypt, genetically the closest to the other Egyptian ecotypes (Sabry et al. 2021). According to (DAD-IS n.d.), Dandarawi is a locally adapted breed with an estimated population of 20,000.

Physical Characteristics: The cocks are heavier than the hens, with an average body weight of 1.7 kg (cocks) and 1.2 kg (hens). Hens are pinkish with white breast feathers and a Colombian pattern, with black markings on the neck, wing, and tail.

Adaptive and Special Genetic Characteristics: Like Beheri chickens, they have high fertility (95%). Hens are also good layers with 185 eggs/hen/year. They mature early at the average age of 4.5–5 months for the cocks and hens, respectively.

Typical Production Systems, Husbandry Practices, and Production Characteristics: Information regarding the production system is unavailable; however, being early maturing and good layers, they must be suitable for environments where ideal management conditions are challenging.

7.4.1.3 Dokki-4

Distribution and Population Statistics: Dokki-4 is a cross between Fayoumi and Barred Plymouth Rock with an estimated population of 20,000 (DAD-IS n.d.). It is developed in Egypt.

Physical Characteristics: Being crossbred, the average body weight for cocks and hens is 2.4 and 1.7 kg, respectively. Adult birds have blue plumage; neck feathers are silvery-white and red ear lobes; day-old chicks have grayish-brown stripes.

Adaptive and Special Genetic Characteristics: No unique, adaptive, bad, or special genetic characteristics were reported. However, Dokki-4 is a crossbred expected to perform better with many adaptation qualities from Fayoumi.

Typical Production Systems, Husbandry Practices, and Production Characteristics: Dokki-4, although not reported, is expected to suit the smallholder to semi-commercial production systems.

7.4.1.4 Fayoumi (Fig. 7.1)

Distribution and Population Statistics: Fayoumi breed originated in the Fayoum Province of Egypt (Hossaryl and Galal 1994).

Physical Characteristics: The cocks and hens weigh 2.2 and 1.6 kg, respectively. Fayoumi chickens resemble the silver Campine breed in body shape and plumage color.

Adaptive and Special Genetic Characteristics: Fayoumi chickens mature early cocks (5 months) and hens (5.5 months). They are fertile (95%) and produce 205 eggs/hen/annually on average. They lay eggs of good shell quality and a high percentage of yolk/albumin. They also resist such diseases as spirochetosis and leukosis.

Typical Production Systems, Husbandry Practices, and Production Characteristics: Fayoumi chickens are widespread in the tropics, where the temperature is high. Reports

Fig. 7.1 Fayoumi chicken. (Courtesy Dr. Wondmeneh Esatu)

show their suitability for smallholder poultry production systems.

7.4.1.5 Gimmizah

Distribution and Population Statistics: Gimmizah is a crossbred between the White Plymouth Rock and Dokki-4 (Galal 2007). It is considered a local chicken of Egypt.

Physical Characteristics: The average body weight for the cocks and hens is 2.2 kg and 1.6 kg, respectively.

Adaptive and Special Genetic Characteristics: Identification of chick sexing is possible for this breed at day old. They reach sexual maturity at 5.5 and 6 months for the cocks and hens, respectively. They lay, on average, 195 eggs annually. The breed resembles the Barred Plymouth Rock in its shape and color. The feathers are light gray and are barred with a darker gray. The Gimmizah showed an improvement in the local breeds in their productive and reproductive abilities.

Typical Production Systems, Husbandry Practices, and Production Characteristics: Detailed information regarding the production system is scarce.

7.4.1.6 Alaraby

Distribution and Population Statistics: Alaraby is distributed in the West and South of Libya. According to (DAD-IS n.d.), it has an estimated population of 50,000.

Physical Characteristics: They have yellow skin color, single comb type, red and black plumage color, gray shank, and feet. Hens lay cream, white, or pale grayish eggs. The average body weight of mature chickens is 1.9 kg for cocks and 1.9 kg for hens.

Adaptive and Special Genetic Characteristics: Alaraby chickens resist heat and diseases and adapt to dry and poor environments. The age at the first egg for hens is 6.6 months, and they lay 138 eggs annually.

Typical Production Systems, Husbandry Practices, and Production Characteristics: The Alaraby chickens are a good option for smallholder production conditions and can survive by scavenging available feed resources.

7.4.1.7 Moroccan Beldi

Distribution and Population Statistics: Moroccan Beldi (Beldi meaning "native" in Arabic) is an indigenous breed population with an estimated population of one million. The Beldi chicken is raised in rural areas of the Khenifra Region of Morocco.

Physical Characteristics: Beldi chickens have large phenotypic variability. They appear in black, brown, gray, and white colors, pure or mixed. Sexual maturity is 154 days for roosters and 168 days for hens in chickens. The age at first egg averaged 5.8 months for hens. The number of eggs laid per hen per year was 78 for the hens with a hatchability rate of 78%. The chickens weigh 1.9 kg (male) and 1.5 kg (female).

Adaptive and Special Genetic Characteristics: No adaptive and unique genetic characteristics were reported. However, being a chicken from drier areas, they may possess the ability to withstand harsh environments.

Typical Production Systems, Husbandry Practices, and Production Characteristics: The Moroccan Beldi can be a good chicken for drier areas of the tropics, where feed supply can be challenging.

7.4.1.8 Large Sudanese Baladi

Distribution and Population Statistics: Sudanese native chicken breed is one of the main ecotypes (Yousif and Eltayeb 2011).

Physical Characteristics: The average body weight for cocks and hens is 1.9 kg and 1.4 kg, respectively.

Adaptive and Special Genetic Characteristics: Large Sudanese Baladi is a hardy breed, able to thrive under very harsh conditions; on average, hens mature late (8 months) and lay 50 eggs annually.

Typical Production Systems, Husbandry Practices, and Production Characteristics: Large Sudanese Baladi chickens can be kept under smallholder production conditions in drier areas.

7.4.1.9 Sudanese Bare Neck Baladi

Distribution and Population Statistics: Estimated population of 1,000,000 (DAD-IS n.d.).

Physical Characteristics: The chickens are naked neck, with an average body weight of 1.1 kg in hens, and they lay on average 106 eggs annually and are an important local chicken in Sudan (Binda et al. 2012).

Adaptive and Special Genetic Characteristics: Being naked neck, they exhibit adaptation to heat stress.

Typical Production Systems, Husbandry Practices, and Production Characteristics: They can produce in areas of high heat stress.

7.4.2 West and Central Africa

Various colors and conformations phenotypically characterize local chickens in West Africa. Despite this phenotypic diversity, West African local chickens have many traits in common with wild chicken populations (Guisso et al. 2022). Their local adaptation calls for conservation and utilization strategies.

7.4.2.1 Naine

Distribution and Population Statistics: Naine is an indigenous breed from Burkina Faso with a small population of 10,000 (DAD-IS n.d.).

Physical Characteristics: They are lightweight, with an average of 1.1 kg cocks and 0.8 kg hens. Sometimes they exhibit frizzle feathers.

Adaptive and Special Genetic Characteristics: The average age of maturity for cocks and hens is seven and 6 months, respectively. The hens lay 60 eggs per year.

Typical Production Systems, Husbandry Practices, and Production Characteristics: With frizzle feathers and light body weight, Naine is suitable for harsh hot environments where feed supply is challenging.

7.4.2.2 Naked Neck

Distribution and Population Statistics: This is an indigenous breed, southern and central Burkina Faso, with an estimated population of 10,000 (DAD-IS n.d.).

Physical Characteristics: The average body weight for cocks and hens is 1.5 kg and 1.2 kg, respectively. The flock exhibits both naked-necked and frizzled feathers. The average age of maturity for both cocks and hens is 6 months. Hens produce 60 eggs annually.

Adaptive and Special Genetic Characteristics: They are resistant and tolerant to some diseases like mycoplasmosis, pseudo avian plaque, and pasteurellosis. They show remarkable thermotolerance.

Typical Production Systems, Husbandry Practices, and Production Characteristics: They are suitable for harsher and drier areas of the tropics.

7.4.2.3 Souche Kondé

Distribution and Population Statistics: The origin of this strain is not known, but it is widely available in southern Burkina Faso.

Physical Characteristics: Both the cocks and hens are heavy, with the average weight for males and females of 2.5 kg and 1.5 kg, respectively. They mature at 6 months. Hens produce 60 eggs annually.

Adaptive and Special Genetic Characteristics: No unique qualities were reported.

Typical Production Systems, Husbandry Practices, and Production Characteristics: Although information regarding the production system is unavailable, they are a good candidate for a smallholder production system.

7.4.2.4 Ghanian Fowl

Distribution and Population Statistics: This is an indigenous chicken widely present in Ghana; the estimated population size is about four million (DAD-IS n.d.).

Physical Characteristics: Most of the local chickens are usually feathered (96%), with a small proportion of naked neck (2%) and frizzled (2%). The average body weight for cocks and hens is 1.2 kg and 1.1 kg, respectively.

Adaptive and Special Genetic Characteristics: Local chickens are disease-tolerant and excellent scavengers. The hens are broody all year round, start egg laying at 5.5 months, and lay 150 eggs annually.

Typical Production Systems, Husbandry Practices, and Production Characteristics: As good scavengers, disease-tolerant, and good mothers, the local chickens of Ghana are suitable for smallholder production conditions.

7.4.2.5 Poule De Benna

Distribution and Population Statistics: This is an indigenous chicken found in the Region Du Benna of Guinea. According to (DAD-IS n.d.), the population is estimated to be 10,000.

Physical Characteristics: Poule De Benna is a big chicken with long shanks, a naked neck, and black eyes. Both cocks and hens weigh, on average, 2.3 kg.

Adaptive and Special Genetic Characteristics: Poule De Benna chickens are resistant to Newcastle disease. The hens are very broody, start egg laying at 6.5 months, and produce 80 eggs per year.

Typical Production Systems, Husbandry Practices, and Production Characteristics: Poule De Benna are suitable for smallholder poultry production systems.

7.4.2.6 Poule du Guinea

Distribution and Population Statistics: An indigenous chicken from Guinea with a population of 15 million (DAD-IS n.d.).

Physical Characteristics: These chickens are variable in size but short beaks and single combs are dominant; they are multicolored, but red, white, and black are dominant. The average body weight of mature chickens is 1.5 kg for cocks and 1 kg for hens.

Adaptive and Special Genetic Characteristics: Hens have a robust brooding instinct and produce, on average, 60 eggs per annum.

Typical Production Systems, Husbandry Practices, and Production Characteristics: These chickens are suitable for the village poultry production system.

7.4.2.7 Nigerian Normal Feathered Chicken

Distribution and Population Statistics: Might have originated from the junglefowl of Java and the Indonesian islands. It is distributed throughout Nigeria.

Physical Characteristics: The hens, on average, weigh 1.07 kg. The hens start egg laying at 25 weeks. With five clutches, they produce 180 eggs per year, weighing 23–38 g.

Adaptive and Special Genetic Characteristics: No unique characteristics were reported.

Typical Production Systems, Husbandry Practices, and Production Characteristics: As most indigenous chickens, the chickens can be kept in village management conditions.

7.4.2.8 Poule du Sénégal

Distribution and Population Statistics: This is an indigenous breed distributed throughout Senegal. According to (DAD-IS n.d.), the estimated population size is about 29 million.

Physical Characteristics: Poule du Sénégal are small, rustic birds usually uni-colored (light brown or white). The Senegal chicken varies in color, but the most frequent colors are brown (13.8%), white (12.4%), yellowish (8.4%), white and brown (8.4%), and light brown (7.8%) (Missohou et al. 1998). The cocks weigh 1.5 kg, the hens 1 kg. Hens lay 60 eggs annually.

Adaptive and Special Genetic Characteristics: Small size, rustic bird, unicolored. Hens have 3–4 egg laying cycles per year, laying between 6 and 15 eggs per cycle.

Typical Production Systems, Husbandry Practices, and Production Characteristics: Chickens are reared by all ethnic groups in Senegal.

7.4.2.9 Cameroon Chicken

Distribution and Population Statistics: Cameroon national poultry flock was estimated at 35 million, 70% being indigenous chickens (Fotsa et al. 2011).

Physical Characteristics: Males average 1.4–1.7 kg, while females average 1.2–1.3 kg.

Adaptive and Special Genetic Characteristics: Across all agro-ecological zones, the presence of genes for naked neck, frizzle, normal feathered, feathered shanks, crested, and rose comb is indicated.

Typical Production Systems, Husbandry Practices, and Production Characteristics: In all the agro-ecological zones, family/backyard production systems are the norm.

7.4.3 Eastern Africa

Local chicken in East Africa shows extensive phenotypic diversity. Chickens appeared in the region relatively late, around ~700 BC, with their geographic origin remaining unknown, with the Indian subcontinent, Southeast Asia, and North Africa as possible primary centers of origin (R.I.U 2012). Local chickens in East Africa are found across the entire region and in all agro-ecological zones; they show large within-population phenotypic diversity in plumage color, feather morphology, pattern, skin color, comb type, etc. (Msoffe et al. 2001; Dana et al. 2010).. Large maternal genetic diversity in the region could potentially support genetic improvement programs (Mwacharo et al. 2013).

Fig. 7.2 Horro chicken. (Courtesy Dr. Wondmeneh Esatu)

7.4.3.1 Horro (Fig. 7.2)

Distribution and Population Statistics: Indigenous breed, Oromia regional state in western Ethiopia, Horro is an indigenous chicken type named after the geographic region of origin, located in the western part of Ethiopia near the Blue Nile gorge. There are about 30,000 chickens restricted to this original environment (DAD-IS).

Physical Characteristics: Horro chickens are predominantly brown, and hens lay about 178 eggs annually that weigh 52–58 grams. They also have crests.

Adaptive and Special Genetic Characteristics: Horro chickens are hardy and resist diseases.

Typical Production Systems, Husbandry Practices, and Production Characteristics: Horro chickens are suitable for smallholder poultry production systems. They are good scavengers and excellent mothers.

Fig. 7.3 Konso chicken. (Source: Dana et al. 2010), Food and Agricultural Organization, reproduced with permission)

7.4.3.2 Konso (Fig. 7.3)

Distribution and Population Statistics: Konso chickens are indigenous chickens found in the Southern Region of Ethiopia.

Physical Characteristics: Adult cocks weigh about 1.4 kg and hens 1.0 kg; most cocks have red body plumage, whereas brown, zigrima, and black are the prominent plumage colors in hens. Both white and yellow skin colors exist. The shape of the head is mainly flat (DAD-IS n.d.).

Adaptive and Special Genetic Characteristics: No unique features were reported.

Typical Production Systems, Husbandry Practices, and Production Characteristics: Reared under scavenging management. Konso chickens are suitable for smallholder/village poultry production conditions.

7.4.3.3 Mandura

Distribution and Population Statistics: The breed is available in the western part of Ethiopia in Benshangul Gumuz regional state, with a small population size of 21,000 (DAD-IS n.d.).

Physical Characteristics: Brown is the most predominant plumage, and complete red is typical of cocks but absent in hens. Almost all chickens have normal feather distribution. The average shank length of adult males is 8.4 cm, and that of females is 7.1 cm; adult cocks weigh about 1.6 kg, and hens 1.4 kg.

Adaptive and Special Genetic Characteristics: No unique features were reported.

Typical Production Systems, Husbandry Practices, and Production Characteristics: As the breed adapted to the drier areas of western Ethiopia, where feed supply is a challenge.

7.4.3.4 Tepi (Fig. 7.4)

Distribution and Population Statistics is an indigenous breed adapted to the southwestern parts of Ethiopia.

Physical Characteristics: The population of the breed is estimated to be about 50,000 (DAD-IS n.d.).

Adaptive and Special Genetic Characteristics: No unique features were reported.

Typical Production Systems, Husbandry Practices, and Production Characteristics: The breed can suit village or smallholder poultry production conditions.

7.4.3.5 Tililli (Fig. 7.5)

Distribution and Population Statistics: An indigenous breed that originated from and is widely spread in the Sekela areas of Ethiopia near the source of the Blue Nile (DAD-IS n.d.).

Physical Characteristics: This ecotype is known for its better growth and egg production (1.2 kg and 147 eggs/hen/year).

Adaptive and Special Genetic Characteristics: No unique features were reported.

Typical Production Systems, Husbandry Practices, and Characteristics: Suit smallholder poultry production systems.

7.4.3.6 Kenyan Chicken

Distribution and Population Statistics: Kenyan chicken is an indigenous breed adapted to rural and peri-urban areas of the country, with an estimated population of about 33 million (DAD-IS n.d.).

Fig. 7.4 Tepi chicken. (Courtesy: Dr. Wondmeneh Esatu)

Fig. 7.5 Tilili chicken. (Courtesy: Dr. Wondmeneh Esatu)

Physical Characteristics: Average body weight for cocks and hens is 2.2 kg and 1.6 kg, respectively. No unique pattern within the feather, yellow skin color, single comb type, various plumage colors, various shank and foot colors, cream white to pale grayish eggshell color.

Adaptive and Special Genetic Characteristics: No unique features were reported.

Typical Production Systems, Husbandry Practices, and Characteristics: They are suitable for smallholder poultry production systems.

7.4.3.7 Rwandan Chicken

Distribution and Population Statistics: Indigenous breed from Rwanda.

Physical Characteristics: Rwandan indigenous chickens have four comb types (strawberry—most dominant, whiteness of earlobes, rounded ear lobe shape, and curved beaks), varied beak color (green, black, yellow, and brown), evenly distributed feathers (most common), and naked-neck phenotypes (this was rare, occurring in about 0.2% of chickens), brown eyes, yellow-colored thick skins, and yellow-colored shanks (Fig. 7.6). The chickens are small and rustic and produce 50 eggs annually.

Adaptive and Special Genetic Characteristics: No unique features have been reported.

Typical Production Systems, Husbandry Practices, and Characteristics: The breed can suit village or smallholder poultry production condition.

7.4.3.8 Ugandan Chicken (Fig. 7.7)

Distribution and Population Statistics: This is an indigenous chicken breed widely distributed in the eastern and northern regions of Uganda. The population is estimated to be more than 22 million (DAD-IS n.d.).

Physical Characteristics: Both the cocks and hens are heavy, with an average body weight of 4 and 2 kg for females and males, respectively. The hens show five clutches of 10–12 eggs.

Adaptive and Special Genetic Characteristics: No unique features were reported.

Typical Production Systems, Husbandry Practices, and Characteristics: The breed can suit village or smallholder poultry production condition.

Fig. 7.6 Indigenous Rwandan chickens. (Source: Hirwa et al. 2019, licensed under CC BY 4.0)

Fig. 7.7 Ugandan chicken. (Source: Yussif et al. 2023) (Licensed under C-C By 4:0)

Fig. 7.8 Ovambo. (Source: Grobbelaar et al. 2010, Food and Agricultural Organization, reproduced with permission)

7.4.4 Southern Africa

South African domestic chickens are a significant bird genetic resource, and more conservation efforts are being made to save these unique genotypes. South African chickens are the result of multiple introductions and dispersion, with their high genetic diversity potentially originating from China, Southeast Asia, and the Indian subcontinent (Mtileni et al. 2011).

7.4.4.1 Ovambo (Fig. 7.8)

Distribution and Population Statistics: The population is small and limited to 10,000.
Physical Characteristics: Dark in color and small in body size.
Adaptive and Special Genetic Characteristics: Very aggressive and agile and can catch and eat mice and young rats. Can fly and roost on the top of trees to avoid predators.
Typical Production Systems, Husbandry Practices, and Production Characteristics: Scavenging and semi-scavenging chicken production system.

7.4.4.2 Potchefstroom Koekoek (Fig. 7.9)

Distribution and Population Statistics: Northwest Province and Gauteng Province of South Africa (DAD-IS n.d.).

Fig. 7.9 Potchefstroom Koekoek chicken. (Courtesy: Dr. Wondmeneh Esatu)

Fig. 7.10 Venda chicken. (Source: Grobbelaar et al. 2010, Food and Agricultural Organization, reproduced with permission)

Physical Characteristics: Average height for cocks and hens 2.51 kg and 1.7 cm, respectively; light and dark gray with a white barred pattern within the feather (barred, sex-linked), skin color (yellow), comb type (single), plumage color (gray and white), shank and foot color (yellow), eggshell color (light brown). The average age at the first egg for a hen is 4.27 months, body weight at the hatch for both sexes is 35 g, and annual egg production is 196 eggs.

Adaptive and Special Genetic Characteristics: The breed has a specific resistance or tolerance to poultry diseases, adaptability to cold and heat, wet and drought, high egg production, brooding well, and the ability to hatch their offspring.

Typical Production Systems, Husbandry Practices, and Production Characteristics: Scavenging and semi-scavenging chicken production system.

7.4.4.3 Venda (Fig. 7.10)

Distribution and Population Statistics: Discovered in Limpopo province, Vhembe, and Capricorn district, and other parts of Gauteng, Mpumalanga, Free State Province, of South Africa, with a small population of 10,000 (DAD-IS).

Physical Characteristics: The average body weight for cocks and hens is 2 and 1.9 kg, respectively. They have a single comb type and lay cream white to a pale grayish egg.

Adaptive and Special Genetic Characteristics: Ability to fly away from predators and to roost in the tree, resistance or tolerance to diseases and parasites, adaptability to the high temperate area and harsh environment, good quality egg production, brooding, and good mothering instincts. Hens lay, on average, 153 eggs annually (Grobbelaar et al. 2010).

7.4.4.4 Boschveld (Fig. 7.11)

Typical Production Systems, Husbandry Practices, and Production Characteristics: The breed suits village or backyard production conditions. They are good scavengers too.

7.5 Local and Exotic Duck Breeds in Africa: Characteristics, Distribution, and Features

7.5.1 Indigenous and Exotic Duck Breeds of North Africa

Duck breeds in North Africa are limited in numbers and almost entirely imported from abroad. They are less popular as compared to chickens

Fig. 7.11 Boschveld chicken. (Source: http://boschveld.co.za/)

that are widely kept. Below, the description of duck strains is given.

7.5.1.1 Domiati

Distribution: Domiati is widely distributed in the north of Egypt (DAD-IS n.d.).

Physical Characteristics: Females are small.

Adaptive and Special Genetic Characteristics: No unique features were reported.

Typical Production Systems, Husbandry Practices, and Production Characteristics: Information regarding the production system is not available.

7.5.1.2 Pekin

Distribution and Population Statistics: It is an imported breed into the lower Egypt and delta regions (DAD-IS n.d.).

Physical Characteristics: The plumage is creamy white; the legs and feet are a yellowish orange. The beak is yellow, short, and almost straight.

Adaptive and Special Genetic Characteristics: Disease resistance, good adaptation to local conditions, and known for its hardiness.

Typical Production Systems, Husbandry Practices, and Production Characteristics: Raised at the commercial and village levels.

7.5.2 West and Central Africa

7.5.2.1 Ghanaian Duck

Distribution and Population Statistics: This is an indigenous breed with a country-wide distribution. The estimated population size is 200,000 (DAD-IS n.d.).

Physical Characteristics: Average body weight for cocks and hens is 2.4 and 1.7 kg, respectively.

Adaptive and Special Genetic Characteristics: Resistant to various diseases, the average age of maturity for hens is 6 months, and they lay about 60 eggs per year.

Typical Production Systems, Husbandry Practices, and Production Characteristics: They are raised in smallholder production conditions.

7.5.3 Eastern Africa

7.5.3.1 Albet

Distribution and Population Statistics: Gambella and Benishangul regions of Ethiopia (DAD-IS n.d.).

Physical Characteristics: White and black colors are dominant, and they lay 57 eggs annually.

Adaptive and Special Genetic Characteristics: No unique features were reported.

Typical Production Systems, Husbandry Practices, and Production Characteristics: They can be raised in smallholder production conditions.

7.5.3.2 Ugandan Duck

Distribution and Population Statistics: Crossbreeds of the original Muscovy ducks imported from Britain (DAD-IS n.d.). They are distributed throughout Uganda, especially in urban and peri-urban areas, with an estimated population of about a million.

Physical Characteristics: Large bird with an average body weight for cocks and hens is 10 and 6 kg, respectively.

Adaptive and Special Genetic Characteristics: No unique features were reported.

Typical Production Systems, Husbandry Practices, and Production Characteristics: They are raised scavenging feed resources available in the household and backyards.

7.5.4 Southern Africa

7.5.4.1 Aylesbury

Distribution and Population Statistics: Aylesbury is imported and locally adapted to all agro-ecological zones of Zambia (DAD-IS n.d.).

Physical Characteristics: The Aylesbury duck is a breed of domesticated duck, bred mainly for its meat and appearance. It is a large duck with pure white plumage, orange legs, and feet. On average, the body weight of cocks and hens is 4.5 and 4 kg, respectively. They are unicolored (white).

Adaptive and Special Genetic Characteristics: No unique features were reported.

Typical Production Systems, Husbandry Practices, and Production Characteristics: Can be raised under smallholder production conditions.

7.5.4.2 Madada

Distribution and Population Statistics: This is an indigenous breed from Zambia with an estimated population of 100,000 (DAD-IS n.d.).

Physical Characteristics: Average weight for cocks and hens is 2.5 and 1.8 kg, respectively.

Adaptive and Special Genetic Characteristics: This breed resists most duck diseases. The hens mature at the age of 6 months and produce 40 eggs per year.

Typical Production Systems, Husbandry Practices, and Production Characteristics: Can be raised under smallholder production conditions.

7.6 Exotic Chicken Breeds in Africa

Introduction Although not well developed, the poultry sector in Eastern African countries has grown from a backyard poultry-keeping operation to a more commercial-oriented system (Vernooji et al. 2018). However, the introduction of exotic genetics into smallholder-intensive chicken production has yet to prove competitive in sub-Saharan Africa because of management issues and high costs of feed, veterinary, and energy inputs (Tabler et al. 2023). Africa's leading countries in chicken meat production are Algeria, Egypt, Morocco, South Africa, and Nigeria (DAD-IS n.d.), which mainly depend on chicken strains. The broiler chicken strains include Arbor Acres, Cobb 500, Cobb 700, Ross, and Ross Indian River. The layer chicken breeds are Hy-Line Brown, Hy-Line Silver Brown, ISA Brown, ISA White, Lohmann Brown, Lohmann Silver, and Lohmann White. Breeds such as white and brown Leghorns, Rhode Island Red, New Hampshire, Cornish, Australorp, and Light Sussex were crossed with local chickens to improve the genetic potential of indigenous breeds (Dana et al. 2011).

Physical Characteristics The exotic African chicken breeds originated from combining different worldwide chicken strains. Most of the inputs to develop are dual-purpose chicken strains, such as Rhode Island Red and white leghorn. Mostly, the broilers are white in color and heavy in body weight, while the layers are either white in the white egg layer or brown in the case of the brown egg layer. Layers are generally small to medium-sized.

Adaptive and Special Genetic Characteristics The commercial chickens, layers, and broilers are meant for improved management conditions where good management and proper feeding, health, and housing are provided. Such strains have been less efficient in Africa as the breeds lack adaptation traits for harsh environments. There are efficient feed converters (broilers) or layers (laying hens).

Typical Production Systems, Husbandry Practices, and Production Characteristics The standard production system for such chicken strains and commercial management conditions; however, farmers in Africa keep them with less sophisticated management conditions. The broiler birds attain 2–2.5 kg live weight at 6–7 weeks and are ready for the market, while layer birds reach 16 weeks before the pullets start laying eggs. The average industry egg production is 230–250 eggs/layer/year (Kusi et al. 2015). Most of the exotic chicken breeds introduced into Africa are commercially mainly kept and managed by medium to big farms.

7.7 The Importance and Role of Poultry for Smallholders

Most of the poultry population in Africa exists in villages with low-input, low-output systems (Guèye 2000; Gilbert et al. 2015). Although large-scale commercially produced poultry products are available, the indigenous chicken remains vital (Wondmeneh 2015), as an example of food sovereignty, where communities prefer a sustainable production system that supplies healthy, culturally appropriate food (Patel 2009; Wong et al. 2017). African consumers prefer eggs and chicken meat from indigenous stocks to those derived from commercial flocks of imported ones, and they, therefore, fetch premium prices (Gueye 2009). Rural poultry supply 70–90% of poultry products in Africa (Alabi et al. 2006; Branckaert and Guèye 2000; Kitalyi 1998; Mack et al. 2005); income from poultry products is often the primary source of income for female-headed households. In contrast, male-headed families usually have multiple income sources and even take over the poultry business when they realize financial feasibility (Aklilu et al. 2008; Muchadeyi et al. 2004). Small-scale poultry production systems have been integrated with human livelihoods for thousands of years, enhancing the rural poor's diet, income, and food and nutrition security (Alders and Pym 2009). The potential contributions and impacts of extensive, small-scale scavenging poultry production systems in rural, resource-poor areas differ significantly from more intensive systems in urbanized settings (Wong et al. 2017). Poultry in Africa has several roles. Some of the roles have direct economic implications, while others have more social implications. Below are some detailed important roles of poultry production in Africa, namely, contribution to food security, income generation and employment, GDP contribution, and sociocultural.

(a) *Role in Food Security*: Food security remains a severe challenge for many households in Africa (Silvestri et al. 2015). Mild to moderate protein-energy malnutrition is common throughout the developing world, which contributes to poor growth, diminished mental development, and *illness* in children (Neumann et al. 2002). Animal source foods address malnutrition by supplying the essential nutrients that are lacking in plant-sourced foods, such as micronutrients like iron, zinc, vitamin B-12, riboflavin, and conjugated linoleic acids (Ndlovu 2010). Farmers in rural areas have better access to poultry products as compared to others. Poultry contributes 20–32% of total animal protein intake (Tadelle et al. 2003; Kitalyi 1998).

(b) *Roles in Income Generation and Employment*: Increasingly, with the impact of climate change, including on crop productivity, farmers need to look for alternative sources of income (Vermeulen et al. 2012). Chickens can adapt to drier areas with the potential to cope with feed shortages. A scavenging chicken produces eggs with little or no supplementation. Few eggs can be collected and sold to the local markets to support the family income, to meet some basic needs, such as buying clothes for children or for meeting some financial obligations of learners in school. Women have better access to the income from chicken, although men take over with better prospects of income from poultry. Poultry can serve as an income-generating activity, providing sustainable employment for rural households (Pica-Ciamarra and Otte 2010). The better the level of inputs used, the higher the production will become. The need for better interventions, such as tropically adapted and yet productive chicken strains and supporting packages, might be the first step to transforming the existing system.

(c) *Roles in GDP Contribution*: The economic contribution of poultry at a household level cannot be overlooked. The sector employs 80–85% of the population and contributes 40% to the total GDP. Rural chicken in Ethiopia represents a significant part of the national economy in general and the rural economy and contributes to 98.5% and 99.2% of the national egg and chicken meat production, respectively (Asresie et al. 2015). Apart from that, poultry in villages serves as disposable cash at any need and is named poor men's "ATM" (Pal et al. 2020). The ability of the live product to be 'stored' for a more extended period is a unique characteristic. Regardless of its enormous impact, the actual contribution of the village poultry to the national GDP is barely quantified. Considering the size of the population keeping village chickens, the contribution to the national GDP is significant.

(d) *Roles in Sociocultural Issues*: Apart from the monetary and food terms, poultry in Africa has other functions. The sociocultural importance of poultry is age-old. Keeping poultry by village communities throughout Africa has been practiced for many generations. More than 85% of rural families in sub-Saharan Africa support some poultry species. For example, approximately three chickens for every two people (Gueye 2009). In addition to providing farmers with eggs and meat for their home consumption, poultry products' sale (or swap) enables poultry keepers to obtain money to spend on their own and family specific needs. Moreover, poultry is symbolic in many social activities and religious ceremonies (Akinola and Essien 2011). In Zimbabwe, the preferred taste of chicken meat is available and reserved for special guests or ceremonial gatherings (weddings or funerals) (Muchadeyi et al. 2004). In Kenya, chickens are helpful in several social, cultural, and spiritual activities such as entertainment, gifts, funeral rites, and spiritual cleansing (Magothe et al. 2012b).

7.8 Indigenous Poultry Breeds Conservation Efforts and Biobanking in Africa

A speedy and constant decline in animal species populations has been recognized in recent years (IUCN 2010). The chicken genetic resources are the most endangered and under-conserved (Hoffmann 2009). At the global level, more than 1600 local chicken breeds have been identified; 33% are considered endangered to critical breeds, and another 40% have unknown risk status (FAO 2007). The 126 indigenous breeds recognized for Africa are most often generalized under the name of local chicken despite their great variability and specific traits (DAGRIS 2021). These breeds contain vast ranges of phenotypic and genetic diversity resulting from the diversity of environmental and human selection pressures. Regrettably, many of these local breeds are classified as at risk due to the introduction and adoption of exotic breeds, acceptance of intensive chicken production systems, changes in the environment, disease conditions, and adverse development policies. The conservation of African indigenous poultry breeds is vital in light of their rapid loss through commercial dilution and breed replacement. All varieties of these breeds are important candidates for conservation. While in situ in vivo conservation strategies are highly relevant to conserving diversity, other complementary conservation strategies, such as germplasm cryopreservation, are needed not only to conserve unique adaptations and gene combinations but also to be able to respond to the needs of future generations. To reduce this genetic erosion, it is crucial to improve knowledge of local breeds and production systems, improve planning, and raise awareness of the threat at the policy level. New innovations in genetic preservation technologies for chickens are also needed. All African indigenous chicken populations with economic potential, scientific use, and cultural or aesthetic interest should be part of the conservation efforts.

7.8.1 Conservation Efforts of Indigenous Poultry Breeds in Africa

Conservation of poultry genetic resources engages numerous activities, including strategies, management, planning, policies, and effective actions, intended to ascertain that the diversity of the genetic resource is maintained to contribute to current and future agricultural and food production (Rege and Gibson 2003). Conservation of genetic resources encompasses characterization, identification, monitoring, and utilization to ensure management for the best short-term use and longer-term availability (Moyo 1995). The economic framework for the sustainable use and conservation of animal genetic resources, including poultry, is outlined in Chap. 25.

7.8.1.1 In Situ Conservation Programs
In African countries, there has been little effort to conserve the local chicken breeds or lines (Manyelo et al. 2010). It is alleged that 33% of indigenous chicken breeds are facing extinction (FAO 2007; Manyelo et al. 2010). Most poultry genetic resources are conserved in situ in the living populations, which are facing. The challenges of epidemics and climate (Assan 2015). In situ conservation programs for indigenous chickens in Africa mostly involve various individual smallholder farmers and farmers associations. Unfortunately, this category of custodians is not benefiting from enough support for their contribution to the conservation effort.

7.8.1.2 Ex Situ Conservation Programs
In vivo ex situ conservation of chickens is mainly practiced by breeding and dissemination centers, held by national research institutions, universities, and industrial poultry farms, as important reservoirs of indigenous chicken biodiversity through the collection of frozen semen. These are generally targeted poultry populations collected from their natural environment and transferred to breeding centers for research or development purposes. Ex situ protection programs may have some sustainability challenges in Africa. Most often, they may be coupled to cryopreservation and reproduction technologies for the conservation and dissemination. These introduce the aspects of biotechnology applications and the need for gene banks as biorepositories where representative samples of genetic resources are preserved. This should be supported by data banks where metadata related to the attributes of breeds are stored in a systematic way. The technological advancement of poultry reproductive technologies and precision breeding provides insight into ex situ conservation as it facilitates the capture of the entire genetic makeup of the populations.

7.8.2 Poultry Germplasm Biobanking in Africa

African poultry germplasm is a live information source for all the genes present in the African indigenous poultry breeds, which can be conserved for long periods and regenerated whenever required in the future. The conservation of these genetic resources through cryopreservation, referred to as biobanking, is an important component for the conservation and revival of rare or endangered species. Germplasm cryobanking, or cryoconservation, is the most successful method to conserve the genetic traits of endangered and commercially valuable species. Germplasm cryobanking involves the harvesting and freezing of gametes, embryos, gonadal tissues, somatic tissues, or PGCs of species threatened with extinction. There is considerable diversity in the cryobiological conditions among cell types and tissues of each species. Research done by scientists from the Centre for Tropical Livestock Genetics and Health (CTLGH) at the Roslin Institute, University of Edinburgh, has focused on developing techniques for the conservation and recovery of the genetic material of chickens. This is an attractive complement to the indispensable maintenance of genetic diversity through living animals. However, research is needed to effectively apply these techniques in a multidisciplinary program aimed at the preservation of the numerous subpopulations of animals that exist within a given species. Previously, it has not been possible to biobank chicken genetic material, but the recent innovation at the Centre for Tropical Livestock Genetics and Health (CTLGH) and International Livestock Research Institute (ILRI) using Primordial Germ Cells (PGCs) has changed that. There is now a way forward to preserve the future biodiversity of African poultry breeds.

7.8.3 Biobanking Using Stem Cell Techniques

Cryopreservation of stem cells is valuable to offer storage of high cell numbers, fast transport, and to preserve cells for long periods. Due to the increased level of endangered poultry breeds, it is important to preserve genetic material for future applications. Conservation of poultry breeds and genetic lines poses challenges. Most poultry genetic resources are maintained in situ in living populations. However, this conservation approach of genetic resources always carries the risk of losing the population owing to pathogen outbreaks, genetic problems, breeding cessation, and/or natural disasters. Biobanking, whereby sperm, eggs, or zygotes are conserved, is the ex situ alternative. Common in mammalian species, the approach does not work well for avian species in females due to the large amount of lipid deposited in the oocyte (Petitte 2006; Whyte and McGrew 2015). Other genetic preservation and propagation techniques, such as cloning using somatic cell nuclear transfer, are not possible because embryo transfer cannot be done in avian species (Kjelland et al. 2014). Cryobanking of germplasm in birds has been mainly limited so far to the use of semen.

7.8.3.1 Primordial Germ Cells (PGCs) Technique

As an alternative, avian primordial germ cells (precursor cells for gametes, which temporally circulate in the vasculature during early development) can be incorporated into the gonads (Yasuda et al. 1992) and differentiated into functional gametes following transplantation to recipient embryos (Tajima et al. 1993; Ono et al. 1998). An avian PGC transplantation technique has been established and is readily available now at ILRI. To date, several techniques for PGC manipulation, including purification, cryopreservation, depletion, and long-term culture, have been developed in chickens. PGC transplantation combined with recent advanced PGC manipulation techniques has enabled ex situ conservation of poultry genetic resources in their complete form. In collaboration with AU-IBAR, the poultry PGC technologies for conservation have been introduced to African NARS scientists and are ready for deployment.

7.8.3.2 Induced Pluripotent Stem Cells (iPSCs) and Mesenchymal Stem Cells (MSCs) Techniques

Induced pluripotent stem cells (iPSCs) are somatic cells reprogrammed by ectopic expression of transcription factors or small molecule treatment, which resemble embryonic stem cells (ESCs) (Zhang et al. 2014). They represent a great promise for conserving and generating genetically improved and locally adapted poultry genetic resources. Numerous reports emphasize the importance of chicken iPSCs as well as MSCs for their self-renewal potential and multilineage differentiation as well as current knowledge concerning their usefulness for conservation, various biological properties and health studies, and the use of MSCs as a feeder layer or as a very promising tool for immunomodulatory cell therapy in immune-mediated diseases (Zhao et al. 2016). Hence, stem cells beyond conservation purposes provide a useful model in the field of chicken biological studies.

7.9 Chicken Production Systems

Chickens in Africa are in different management and production systems. Most classifications follow the breed type, input and output level, mortality rate, type of producer, production purpose, length of broodiness, growth rate, and number of chickens reared. There are four poultry production systems in the world, as identified (FAO 2014; Alders et al. 2018). For producers to benefit from the poultry production systems and continue to ensure positive and sustainable contributions, production and marketing should be tailored to local conditions and important value chain nodes while maintaining genetic diversity (Table 7.3).

7.10 Chicken Breed Improvement Programs in Africa

Breeding programs for chickens started several decades ago in the tropics. Most attempts were to improve the production performance of local chickens with better adaptation, but with poor production potential. The most common approach was to consider crossbreeding in the form of a cockerel exchange scheme and the direct intro-

Table 7.3 FAO chicken production systems classification

Criteria	Small-extensive scavenging	Extensive scavenging	Semi-intensive	Small-scale intensive
Production/farming system	Mixed, poultry, and crops, often landless	Mixed, livestock, and crops	Usually poultry only	Poultry only
Other livestock raised	Rarely	Usually	Sometimes	No
Flock size	1–5 adult birds	5–50 adult birds	50–200 adult birds	>200 broilers >100 layers
Poultry breeds	Local	Local or crossbred	Commercial, crossbred, or local	Commercial
Source of new chicks	Natural incubation	Natural incubation	Commercial day-old chicks or natural incubation	Commercial day-old chicks or pullets
Feed source	Scavenging; almost no supplementation	Scavenging; occasional supplementation	Scavenging; regular supplementation	Commercial balanced ration
Poultry housing	Seldom; usually made from local materials or kept in the house	Sometimes; usually made from local materials	Yes; conventional materials; houses of variable quality	Yes; conventional materials; good-quality houses
Access to veterinary services and veterinary pharmaceuticals	Rarely	Sometimes	Yes	Yes
Mortality	Very high, >70%	Very high, >70%	Medium to high, 20–>50%	Low to medium <20%
Access to reliable electricity supply	No	No	Yes	Yes
Existence of conventional cold chain	No	Rarely	Yes	Yes
Access to urban markets	Rarely	No, or indirect	Yes	Yes
Products	Live birds, meat	Live birds, meat, eggs	Live birds, meat, eggs	Live birds, meat, eggs
Time devoted each day to poultry management	<30 min	<1 h	>1 h	>1 h

Source: FAO (2014), Alders et al. (2018)

Table 7.4 Examples of chicken genetic improvement programs in Africa

Country	Species	Blood level	Use
Ethiopia	Horro (Mulugeta et al. 2020)	Pure local	Dual
	Tilili (Esatu et al. 2022)	Pure local	Dual
	DZ-white (Mulugeta et al. 2020)	Composite	Dual
Kenya	KALRO 1 (Miyumo et al. 2023)	Composite	Dual
	KALRO 2 (Miyumo et al. 2023)	Composite	Dual
Nigeria	Noiler (Bamidele et al. 2019)	Composite	Dual
	Funnab Alpha (Bamidele et al. 2019)	Pure local	Dual
	Shika-Brown (Bamidele et al. 2019)	–	Layer
	Fulani (Bamidele et al. 2019)	Pure local	
South Africa	Boschveld (Van Marle-Köster and Casey 2001)	Composite	Dual
	Koekoek (Van Marle-Köster and Casey 2001)	Composite	Dual
Cameroon	Northwest (IRZ 1986)	Local and composite	Dual

duction of exotic chicken strains. As exotic chickens could not survive in African settings and required continuous importation, countries have focused on developing their chicken strains using locally available chickens. The need for better management practices is also a problem. In recent years, new breeding programs (see Sect. 7.1.3) are now bearing fruit. The table below shows some of the successful chicken breed improvement programs in Africa. It is good to note that more programs have yet to be reported (Table 7.4).

7.11 Challenges and Opportunities of Chicken Production in Africa

Poultry production in tropical countries is based on the traditional scavenging system. The share of family poultry to the total poultry population in developing countries in Africa is estimated to reach 70–80% (Sonaiya 1990; Gueye 1998; Sonaiya et al. 1999).

7.11.1 Diseases and Predations

Estimates show that about 60% of chicks die before 8 weeks. Incidence of mortality continues at significant rates even during adult stages. A considerable proportion of loss in adult birds is associated with diseases, although losses due to predators also contributed substantially (Gueye 1998). The significant challenges related to an infection at the producer household level include farmers' inadequate knowledge and awareness of better health care practices and insufficient veterinary services.

7.11.2 Improved Genetics

Poor productivity of local breeds and lack of access to productive and adaptable chicken breeds are the other most critical challenges to increasing the economic contribution of the sub-sector. Most of the chickens kept by smallholder farmers are unimproved indigenous flocks with slow growth rates and poor egg productivity (Sonaiya et al. 1999). Attempts to increase the productivity of the sub-sector are mainly focused on introducing high-yielding exotic chickens to replace indigenous stocks, which have generally failed due to the failure of imported breeds to adapt to local conditions. On the other hand, efforts to develop productive species adaptive to local circumstances have borne fruit in recent years (Wondmeneh 2015). Crossbreeding selected local breeds with selected exotic breeds could produce adapted and productive breeds.

7.11.3 Feeds and Nutrition

The backyard or village chicken production system, the leading practice in Africa, mainly depends on a scavenging feed resource base (SFRB) that is minimally supplemented with household feed scraps. The availability and quality of the scavenging feed resources are highly seasonal, and the carrying capacity of the SFRB can only support a minimal number of chickens (Sonaiya et al. 2002). Similarly, feed, accounting for about 70% of the production cost of commercial chicken meat and egg production, could be supplied more and of better quality. Apart from that, the price of imported items such as premixes and locally produced central feed raw materials such as maize and soyabean is exceptionally high due to the inadequate supply of these grains (Negash 2020).

7.11.4 Training and Extension Support

Lack of awareness and knowledge of smallholder producers on modern chicken production practices is another critical challenge to backyard chicken production.

7.11.5 Institutional Constraints

The development of the chicken sector should be viewed in the context of addressing the entire value chain. Engaging and mobilizing the multiple arrays of actors, both public and private, involved in the value chain is critical to address the many issues involved in developing the sector, such as marketing, delivery of inputs, provision of services and credit facilities, and building capacities of research and relevant development institutions, etc.

7.12 Economic Contribution of Chickens in Africa

Village chickens make substantial contributions to household food security throughout the developing world. Indigenous chickens serve as an investment and source of security for households in addition to their use as sources of meat and eggs for consumption and income (Muchadeyi et al. 2007). Chicken production can be regarded as an investment, contributing to the welfare of women and children in traditional, low-input farming systems in the tropics. According to Ref. (Dana et al. 2010), an average flock of five adult chickens helped women in Central Tanzania to have an additional income equivalent to 10% of the average annual income. In the Niger Delta family, poultry husbandry contributes 35% of household women's income, representing about 25% of the Nigerian minimum wage and 50% of the per capita income (Alabi et al. 2006). Village poultry can be used as an effective means of empowering women and as a tool for poverty alleviation. Experiences in many other developing countries have been reported (Kitalyi 1998).

7.13 Implication and Ways Forward

Poultry production in Africa and beyond is an essential source of food, income, and employment. The African poultry sector is so underdeveloped that the demand for poultry products still needs to be met. Previous interventions by governments and nongovernmental organizations have not been fully applied. The introduction of exotic blood into the harsh environment cannot be successful. The problems involve the threat to the existence of the local chicken that the farmers heavily depend on. Future interventions should work on the balance: the need for more meat, eggs, and income on one side and the utilization of local and exotic chicken in appropriate breed combinations coupled with animal recording.

The plans should start with knowing what Africa has and strategically approaching the problem.

7.14 Conclusion

An extensive indigenous poultry conservation and biobanking program will not only support research and development to prevent problems with inbreeding and preserve at-risk poultry breeds but also reduce the large number of live animals needed to be kept for research across the world. It could also have an important role within poultry breeding companies to maintain important parental lines of mainstream poultry breeds used in commercial poultry production without the need to keep large populations of live birds. It is important to note that more work needs to be undertaken, and it should be government policy to encourage the conservation of these breeds to avoid extinction. Therefore, conservation strategies and genetic improvement can be developed and implemented simultaneously in a coordinated approach. This strategy will allow new progeny to be developed and studied while the original breed is preserved.

References

Abioja M, Abiona J (2020) Impacts of climate change to poultry production in Africa: adaptation options for broiler chickens. In: African handbook of climate change adaptation, pp 1–22

Adebambo AO, Mobegi VA, Mwacharo JM et al (2010) Lack of phylogeographic structure in Nigerian village chickens revealed by mitochondrial DNA D-loop sequence analysis. Int J Poult Sci 9(5):503–507

Akinola LAF, Essien A (2011) Relevance of rural poultry production in developing countries with special reference to Africa. Worlds Poult Sci J 67(4):697–705

Aklilu HA, Udo HMJ, Almekinders CJM (2008) How resource-poor households value and access poultry: village keeping in Tigray. Ethiop Agric Syst 96:175–183

Alabi RA, Esobhawan RA, Aruna MB (2006) Econometric determination of the contribution of family poultry to women's income in Niger-Delta. Nigeria J Cent Eur Agric 7(4):753–760

Alders R, Pym R (2009) Village poultry: still important to millions, eight thousand years after domestication. Worlds Poult Sci J 65(2):181–190

Alders RG, Dumas SE, Rukambile E et al (2018) Family poultry: multiple roles, systems, challenges, and options for sustainable contributions to household nutrition security through a planetary health lens. Matern Child Nutr 14:e12668

Asresie A, Eshetu M, Adigrat E (2015) Traditional chicken production system and marketing in Ethiopia: a review. J Mark Consum Res 8:27–34

Assan N (2015) Prospects for indigenous chicken gen Zimbabwe. Agric Adv 4:49–56

AU-IBAR (2019) The State of farm animal genetic resources in Africa. AU-IBAR Publication

Ayanlade A, Radeny M, Morton JF (2017) Comparing smallholder farmers' perception of climate change with meteorological data: a case study from southwestern Nigeria. Weather Clim Extremes 15:24–33

Bamidele O, Sonaiya EB, Adebambo OA et al (2019) On-station performance evaluation of improved tropically adapted chicken breeds for smallholder poultry production systems in Nigeria. Trop Anim Health Prod 52:1541–1548

Bettridge JM, Psifidi A, Terfa ZG et al (2018) The role of local adaptation in sustainable village chicken production. Nat Sustain 1(10):574–582. https://doi.org/10.1038/s41893-018-0150-9

Binda BD, Yousif IA, Elamin KM, Eltayeb HE (2012) A comparison of performance among exotic meat strains and local chicken ecotypes under Sudan conditions. Int J Poult Sci 11(8):500

Branckaert RDS, Guèye EF (2000) FAO's programme for support to family poultry production. In: Dolberg F, Petersen PH (eds) Proceedings of a workshop on poultry as a tool in poverty eradication and promotion of gender equality, Tune, Denmark, pp 244–256. https://doi.org/10.1017/s0043933909000117

Clutton-Brock J (1993) The spread of domestic animals in Africa. In: Shaw T, Sinclair P, Andah B, Okpko A (eds) The archaeology of Africa: food, metals and towns. Routledge Press, London, pp 61–70

D'Andrea AC, Richards MP, Pavlish LA et al (2011) Stable isotopic analysis of human and animal diets from two pre-Aksumite/Proto-Aksumite archaeological sites in northern Ethiopia. J Archaeol Sci 38(2):367–374

DAD-IS (n.d.) https://www.fao.org/dad-is

DAGRIS. http://dagris.ilri.cgiar.org/species. Consulted 09 Sept 2021

Dana N, Dessie T, Van der Waaij LH et al (2010) Morphological features of indigenous chicken populations of Ethiopia. Anim Genet Resour/Resources génétiques animales/Recursos genéticos animales 46:11–23

Dana N, Megens HJ, Crooijmans RP et al (2011) East Asian contributions to Dutch traditional and western commercial chickens inferred from mtDNA analysis. Anim Genet 42(2):125–133

Debela N, Mohammed C, Bridle K et al (2015) Perception of climate change and its impact by smallholders in pastoral/agropastoral systems of Borana, South Ethiopia. SpringerPlus 4(1):1–12

Dessie T (2003) Phenotypic and genetic characterization of local chicken ecotypes in Ethiopia, Doctoral dissertation, Humboldt-Universität zu Berlin

Desta TT, Wakeyo O (2012) Uses and flock management practices of scavenging chickens in Wolaita zone of Southern Ethiopia. Trop Anim Health Prod 44(3):537–544

Desta TT, Dessie T, Bettridge J et al (2013) Signature of artificial selection and ecological landscape on morphological structures of Ethiopian village chickens. Anim Genet Resour/Resources génétiques animales/Recursos genéticos animales 52:17–29

Esatu W, Kassa B, Lulie B, Kebede A, Genetu G, Yismaw K, Jember Z, Yeheyis L, Girma M, Dessie T (2022) A guide to setting up a selective indigenous chicken improvement program: the Tilili breed in Ethiopia, ILRI Manual 57. ILRI(b), Nairobi

FAO (2000) Production yearbook, vol 53. FAO, Rome

FAO (2007) In: Pilling D, Rischkowsky B (eds) The State of the world's animal genetic resources for food and agriculture – in brief. FAO

FAO (2014) Decision tools for family poultry development, FAO animal production and health guidelines, no. 16. FAO, Rome

Fotsa JC, Kamdem DP, Bordas A, Tixier-Boichard M, Rognon X (2011) Assessment of the genetic diversity of Cameroon indigenous chickens by the use of microsatellites. Livest Res Rural Dev 23:5

Fuller DQ, Boivin N (2009) Crops, cattle, and commensals across the Indian Ocean. Current and potential archaeobiological evidence. Etudes Ocean Indien 42–43:13–46

Fumihito A, Miyake T, Sumi S et al (1994) One subspecies of the red junglefowl (Gallus gallus gallus) suffices as the matriarchic ancestor of all domestic breeds. Proc Natl Acad Sci 91(26):12505–12509

Galal S (2007) Farm animal genetic resources in Egypt: factsheet. Egypt J Anim Product 44(1):1–23

Gheyas AA, Vallejo-Trujillo A, Kebede A et al (2011) Integrated environmental and genomic analysis reveals the drivers of local adaptation in African indigenous chickens. Mol Biol Evol 38(10):4268–4285

Gilbert M, Conchedda G, Van Boeckel TP et al (2015) Income disparities and the global distribution of intensively farmed chicken and pigs. PLoS One 10(7):e0133381

Grobbelaar JAN, Sutherland B, Molalakgotla NM (2010) Egg production potentials of certain indigenous chicken breeds in South Africa. Anim Genet Resour 46:25–32. https://doi.org/10.1017/S2078633610000664

Gueye EF (1998) Village egg and fowl meat production in Africa. Worlds Poult Sci J 54:73–87

Guèye EF (2000) Women and family poultry production in rural Africa. Dev Pract 10(1):98–102

Gueye EF (2009) Small-scale family poultry production: the role of networks in information dissemination to family poultry farmers. Worlds Poult Sci J 65(6):115–124

Guisso TA, Moula N, Issa S et al (2022) Phenotypic characterization of local chickens in West Africa. Syst Rev Poult 1(4):207–219

Hassaballah K, Zeuh VA, Lawal R et al (2015) Diversity and origin of Indigenous Village chickens (Gallus gallus) from Chad, Central Africa. Adv Biosci Biotechnol 6:592–600. https://doi.org/10.4236/abb.2015.69062

Hodges J (1990) Conservation of animal genetic resource in developing countries. In: Alderson L (ed) Genetic conservation of domestic livestock. C.A.B. International, pp 128–145

Hoffmann I (2009) Open questions on poultry genetic diversity. In: Proceedings of the 6th European poultry genetic symposium, Poland, pp 61–73

Horst P (1989a) Native fowls as a reservoir of Genomes and major genes with Direct and Indirect effect on the adaptability and their potential for tropically oriented breeding plans. Arch Geflugelk 53(3):93–101

Horst P (1989b) Native fowl as a reservoir for genomes and major genes with direct and indirect effect on the adaptability and their potential for tropically oriented breeding plans. Archiv für Guflügelkunde 53(3):93–101

Hossaryl MA, Galal ESE (1994) Improvement, and adaptation of the Fayoumi chicken. Anim Genet Resour/Resources génétiques animales/Recursos genéticos animales 14:33–39

IRZ (1986) Annual Report 1985–1986

IUCN Species survival commission, 2010

Kamara A, Conteh A, Rhodes ER, Cooke RA (2019) The relevance of smallholder farming to African agricultural growth and development. Afr J Food Agric Nutr Dev 19(1):14043–14065

Kiple KF, Ornelas K (2000) The Cambridge world history of food, vol 2. Cambridge University Press

Kitalyi AJ (1998) Village chicken production systems in rural Africa: household food security and gender issues. Food and Agriculture Organization (FAO), Rome

Kjelland ME, Romo S, Kraemer D (2014) Avian cloning: adaptation of a technique for enucleation of the avian ovum. Avian Biol Res 7(3):131–138

Kusi LY, Agbeblewu S, Anim IK et al (2015) The challenges and prospects of the commercial poultry industry in Ghana: a synthesis of literature. Int J Manag Sci 5:476–489

Lawal RA, Olivier H (2021) Domestic chicken diversity: origin, distribution, and adaptation. Anim Genet 52(4):385–394

Lawal RA, Martin SH, Vanmechelen K et al (2020) The wild species genome ancestry of domestic chickens. BMC Biol 18(1):1–18

Liverpool-Tasie LSO, Sanou A, Tambo JA (2019) Climate change adaptation among poultry farmers: evidence from Nigeria. Clim Chang 157(3):527–544

Loog L, Thomas MG, Barnett R, Allen R, Sykes N, Paxinos PD et al (2017) Inferring allele frequency trajectories from ancient DNA indicates that selection on a chicken gene coincided with changes in medieval husbandry practices. Mol Biol Evol 34(8):1981–1990

Lyimo C, Weigend S, Msoffe P et al (2014) Global diversity and genetic contributions of chicken populations from African, Asian, and European regions. Anim Genet 45(6):836–848. https://doi.org/10.1111/age.12230

Mack S, Hoffman D, Otte J (2005) The contribution of poultry to rural development. Worlds Poult Sci J 61(1):7–14

Magothe TM, Okeno TO, Muhuyi WB et al (2012a) Indigenous chicken production in Kenya: I. Status World's Poult Sci J 68(1):119–132

Magothe TM, Okeno TO, Muhuyi WB, Kahi AK (2012b) Indigenous chicken production in Kenya: I. Current status. World's Poult Sci J 68(1):119–132

Manyelo TG, Selaledi L, Hassan ZM, Mabelebele M (2010) Local chicken breeds of Africa: their description, uses and conservation methods. Animals 10(12):2257

Manyelo TG, Selaledi L, Hassan ZM et al (2020) Local chicken breeds of Africa: their description, uses and conservation methods. Animals 10(12):2257

Melesse A (2014) Significance of scavenging chicken production in the rural community of Africa for enhanced food security. Worlds Poult Sci J 70(3):593–606

Missohou A, Sow R, Ngwe-Assoumou C (1998) Caractéristiques morphobiométriques de la poule du Sénégal. Anim Genet Resour/Resources Génétiques Animales/Recursos Genéticos Animales 24:63–69

Miyumo S, Wasike CB, Ilatsia ED, Bennewitz J, Chagunda MG (2023) Genetic and non-genetic factors influencing KLH binding natural antibodies and specific antibody response to Newcastle disease in Kenyan chicken populations. J Anim Breed Genet 140(1):106–120

Moyo S (1995) Evaluation of breeds for beef production in Zimbabwe. In: Dzama K, Ngwerume FN, Bhebhe E (eds) Proceedings of the international symposium on livestock production through animal breeding and genetics. University of Zimbabwe, Harare, pp 122–129

Msoffe PLM, Minga UM, Olsen JE, Yongolo MGS, Juul-Madesen HR, Gwakisa PS, Mtambo MMA (2001) Phenotypes including immunocompetence in scavenging local chicken ecotypes in Tanzania. Trop Anim Health Prod 33:341–354

Mtileni B, Muchadeyi FC, Maiwashe A et al (2011) Diversity and origin of South African chickens. Poult Sci 90(10):2189–2194

Muchadeyi F, Sibanda S, Kusina N et al (2004) The village chicken production system in Rushinga District of Zimbabwe. Livest Res Rural Dev 16:6

Muchadeyi FC, Wollny CBA, Eding H et al (2007) Variation in village chicken production systems among agro-ecological zones of Zimbabwe. Trop Anim Health Prod 39:453–461

Muchadeyi F, Eding H, Simianer H et al (2008) Mitochondrial DNA D-loop sequences suggest a southeast Asian and Indian origin of Zimbabwean village chickens. Anim Genet 39(6):615–622

Mulugeta S, Goshu G, Esatu W (2020) Growth performance of DZ-white and improved Horro chicken breeds under different agro-ecological zones of Ethiopia. J Livest Sci 11:45

Mwacharo JM, Bjørnstad G, Mobegi V et al (2011) Mitochondrial DNA reveals multiple introductions of domestic chicken in East Africa. Mol Phylogenet Evol 58(2):374–382

Mwacharo JM, Bjørnstad G, Jianlin H et al (2013) The history of African village chickens: an archaeological and molecular perspective. Afr Archaeol Rev 30(1):97–114

Ndlovu L (2010) The role of foods of animal origin in human nutrition and health. In: The role of livestock in developing communities: enhancing multifunctionality, p 77

Negash D (2020) Evaluation of commercial animal feed quality and manufacturing status in Ethiopia. Acta Sci Nutr Health 4(2):1–13

Negash F, Abegaz S, Tadesse Y, Jembere T, Esatu W, Dessie T (2023) Evaluation of growth performance and feed efficiency in reciprocal crosses of Fayoumi with three exotic chicken breeds. Acta Agric Scand Sect A Anim Sci 72:1–14

Neumann C, Harris DM, Rogers LM (2002) Contribution of animal source foods in improving diet quality and function in children in the developing world. Nutr Res 22(1–2):193–220

Ngaira JKW (2007) Impact of climate change on agriculture in Africa by 2030. Sci Res Essays 2(7):238–243

Ono T, Matsumoto T, Arisawa Y (1998) Production of donor-derived offspring by transfer of primordial germ cells in Japanese quail. Exp Anim 47(4):215–219

Pal S, Prakash B, Singh Y (2020) Review on backyard poultry farming: resource utilization for better livelihood of the rural population. Int J Curr Microbiol App Sci 9(05):2361–2371

Patel R (2009) Food sovereignty. J Peasant Stud 36(3):663–706

Peters J, Lebrasseur O, Irving-Pease EK, Paxinos PD, Best J, Smallman R, Callou C, Gardeisen A, Trixl S, Frantz L, Sykes N (2022) The biocultural origins and dispersal of domestic chickens. Proc Natl Acad Sci 119(24):e2121978119

Petitte JN (2006) Avian germplasm preservation: embryonic stem cells or primordial germ cells? J Poult Sci 85(2):237–242. https://doi.org/10.1093/ps/85.2.237

Pica-Ciamarra U, Otte J (2010) Poultry, food security and poverty in India: looking beyond the farm-gate. Worlds Poult Sci J 66(2):309–320

Plug I (1996) Domestic animals during the early Iron Age in South Africa. In: Pwiti G, Saper R (eds) Aspects of African archaeology. University of Zimbabwe

R.I.U. (2012) Taking poultry subsector to scale: a call for commercial expansion of the indigenous poultry industry (Tanzania policy brief, no. 1) Available at: https://assets.publishing.service.gov.uk/media/57a08a9f40f0b649740006b2/riu040412-tz-policy-brief1.pdf. Accessed 6 Oct 2021

Razafindraibe H, Mobegi VA, Ommeh SC et al (2008) Mitochondrial DNA origin of indigenous malagasy chicken. Ann N Y Acad Sci 1149:77–79. https://doi.org/10.1196/annals.1428.047

Rege JEO, Gibson JP (2003) Animal genetic resources and economic development: issues in relation to economic valuation. Ecol Econ 45(3):319–330

Sabry A, Ramadan S, Hassan MM, Mohamed AA, Mohammedein A, Inoue-Murayama M (2021) Assessment of genetic diversity among Egyptian and Saudi chicken ecotypes and local Egyptian chicken breeds using microsatellite markers. J Environ Biol 42(1):33–39

Silvestri S, Sabine D, Patti K et al (2015) Households and food security: lessons from food secure households in East Africa. Agric Food Secur 4(1):1–15

Sonaiya EB (1990) The context and prospects for development of small holder rural poultry production in Africa. In: Proceedings, CTA seminar on smallholder rural poultry production, Thessaloniki, Greece, vol 1, pp 35–52

Sonaiya EB (1999) Culture and family poultry development. J Inst Cult Stud 7:1–10

Sonaiya EB, Branckaert RDS, Guèye EF (1999) The scope and effect of family poultry research and development. Research and Development options for family poultry. First INFPD/FAO Electronic Conference on Family Poultry. http://www.faoext02.fao.org/

Sonaiya EB, Dazogbo JS, Olukosi OA (2002) Further assessment of scavenging feed resource base. Characteristics and parameters of family poultry production in Africa. Publication of FAO/IAEA Co-ordinated

Tabler T, Ayres V, Thornton T, Wells J, Moon J (2023) Department of Animal Science

Tadelle D (1996) Studies on village poultry production systems in the central high lands of Ethiopia. M.Sc. thesis, Swedish University of Agricultural Sciences

Tadelle D, Kijora C, Peters KJ (2003) Indigenous chicken ecotypes in Ethiopia: growth and feed utilization potentials. Int J Poult Sci 2(2):144–152

Tajima A, Naito M, Yasuda Y, Kuwana T (1993) Production of germ line chimera by transfer of primordial germ cells in the domestic chicken (Gallus domesticus). Theriogenology 40:509–519

Tankson J, Vizzier-Thaxton Y, Thaxton J et al (2001) Stress and nutritional quality of broilers. Poult Sci 80(9):1384–1389

Van Marle-Köster E, Casey NH (2001) Phenotypic characterization of native chicken lines in South Africa. Anim Genet Resour/Resources génétiques animales/Recursos genéticos animales 29:71–78

Vermeulen SJ, Aggarwal PK, Ainslie A et al (2012) Options for support to agriculture and food security under climate change. Environ Sci Pol 15(1):136–144

Vernooji A, Masaki MN, Meijer-Willems D (2018) Regionalization in poultry development in Eastern Africa, Report 1121. Wageningen Livestock Research, Wageningen

Wang MS, Thakur M, Peng MS et al (2020) 863 genomes reveal the origin and domestication of chicken. Cell Res 30(8):693–701

Whyte JBE, McGrew MJ (2015) Chapter 12: Increased sustainability in poultry production: new tools and resources for genetic management. In: Sustainable poultry production in Europe. CABI Publishing, p 214. https://doi.org/10.1079/9781780645308.0000. https://www.cabi.org/vetmedresource/ebook/20163136485

Wondmeneh E (2015) Genetic improvement in indigenous chicken of Ethiopia. PhD thesis, Wageningen University

Wondmeneh E, Van der Waaij EH, Udo HMJ et al (2016) Village poultry production system: perception of farmers and simulation of impacts of interventions. Afr J Agric Res 11(24):2075–2081

Wong JT, de Bruyn J, Bagnol B et al (2017) Small-scale poultry and food security in resource-poor settings: a review. Glob Food Sec 15:43–52

Yasuda Y, Tajima A, Fujimoto T et al (1992) A method to obtain avian germ-line chimaeras using isolated primordial germ cells. J Reprod Fertil 96:521–528

Yousif IA, Eltayeb NM (2011) Performance of Sudanese native dwarf, and bare neck chicken raised under improved traditional production system. Agric Biol J N Am 2(5):860–866

Zhang Y, Wei C, Zhang P, Li X, Liu T, Pu Y et al (2014) Efficient reprogramming of naïve-like induced pluripotent stem cells from porcine adipose-derived stem cells with a feeder-independent and serum-free system. PLoS One 9(1):e85089

Zhao Q, Ren H, Han Z (2016) Mesenchymal stem cells: immunomodulatory capability and clinical potential in immune diseases. J Cell Immunother 2(1):3–20

Open Access   This chapter is licensed under the terms of the Creative Commons Attribution 4.0 International License (http://creativecommons.org/licenses/by/4.0/), which permits use, sharing, adaptation, distribution and reproduction in any medium or format, as long as you give appropriate credit to the original author(s) and the source, provide a link to the Creative Commons license and indicate if changes were made.

The images or other third party material in this chapter are included in the chapter's Creative Commons license, unless indicated otherwise in a credit line to the material. If material is not included in the chapter's Creative Commons license and your intended use is not permitted by statutory regulation or exceeds the permitted use, you will need to obtain permission directly from the copyright holder.

Pig Genetic Resources of Africa

Richard Osei-Amponsah and Wilson S. Kaumbata

Abstract

The African pig genetic resource is valuable for food security, income generation, and livelihood sustenance, as well as for sociocultural and religious activities. This chapter highlights the unique adaptive and resilient attributes of the indigenous pig breeds found in various regions of Africa and makes a case for their sustainable improvement, use, and conservation. The first section (Sect. 8.1) highlights the advantages of pigs over other livestock species and the key roles pigs play in enhancing the livelihoods of both rural and urban farmers across the African Continent. The origins, domestication, and classification of pigs, including dispersal to various parts of Africa and their subsequent development into distinct western, eastern, and southern African pig breeds, are detailed in Sects. 8.2, 8.3, and 8.4. Distribution of pigs to North African countries is scanty due to sociocultural and religious restrictions. Sections 8.5 and 8.6 provide information on various pig production systems practiced in rural and urban settings, including the status of existence and implementation of various pig improvement programs across the continent. The sections further highlight status and challenges relating to the conservation of indigenous pigs, considering the proliferation of exotic breeds and prevalence of uncontrolled mating and unregulated crossbreeding, particularly in rural settings where the free-range pig production system dominates. The chapter concludes by presenting general challenges faced by pig farmers in Africa, the existence of opportunities, and possible solutions, including strategies that can be implemented to enhance pig production and profitability in smallholder farms.

R. Osei-Amponsah (✉)
Department of Animal Science, School of Agriculture, College of Basic and Applied Sciences, University of Ghana, Legon, Accra, Ghana
e-mail: ROsei-Amponsah@ug.edu.gh

W. S. Kaumbata
Department of Animal Science, Lilongwe University of Agriculture and Natural Resources, Bunda College, Lilongwe, Malawi

Keywords

Sus scrofa · Kolbroek · Busia pig · Mukota pig

Abbreviations

ADP	Ashanti Dwarf Pig
AI	Artificial Insemination
ASF	African Swine Fever
AnGR	Animal Genetic Resources
AU-IBAR	African Union Inter-African Bureau of Animal Resources
GPA	Global Plan of Action
FAO	Food and Agriculture Organization of the United Nations
FST	Wright Fixation index (Average inbreeding coefficient of subpopulations relative to the total population)
SNPs	Single Nucleotide Polymorphisms

8.1 Introduction

The pig industry in Africa is a major source of meat, income, employment, and livelihood support. The pig (*Sus scrofa*) is among the few animals with the ability to convert byproducts that would have gone waste from the agricultural industry and households into high-quality animal proteins. Pig production is important to many smallholder livestock farmers, particularly in sub-Saharan Africa. Their relatively short generation interval, high fecundity, early maturity, prolificacy, ability to survive in a wide range of environments and on a variety of feed resources, feed conversion efficiency, as well as their relatively good carcass production relative to ruminants, are comparative advantages for pig producers. Pigs contribute to the animal protein food security of many households in Africa and are a source of wealth creation, particularly for women and the youth. In rural settings where banking services are unavailable, pigs play the role of an inflation-proof cash reserve used by pig farmers during emergencies or when extra cash is needed to meet critical household needs. In most African communities, pigs also play valuable roles in various sociocultural practices, including their use as gifts, payment for dowries, and pork for special dishes during festivities.

Unfortunately, the development of the African pig sector has, over the years, been hindered by several constraints; namely: inadequate and poor nutrition, poor animal healthcare, relatively low production potential of the indigenous pig, and unfavorable policy and institutional environment. A sustainable pig industry, however, will require national governments to invest in the capacity of stakeholders in the pig industry and promote multistakeholder public-private partnerships at all levels (FAO et al. 2017). An important requirement to achieve this objective is the development and implementation of appropriate livestock policies to support stakeholder activity along the pig value chain. Furthermore, knowledge of pig farmers on biosecurity should be regularly updated to help prevent infectious disease outbreaks, including African Swine Fever (ASF). This will motivate them to adopt appropriate, practical, and affordable technologies for sustainable biosecurity in small-scale free-ranging pig production system (FAO et al. 2017). Over the years, a lack of adequate support in pig farming has constrained the growth of the industry and negatively impacted local producers. It is therefore important for national governments and other stakeholders to increase investments in this regard to improve farmers' livelihoods and food security. In line with these, the objective of this chapter is to synthesize available information on *Sus* genetic resources in Africa (indigenous and locally adapted exotic breeds), pig production systems, sustainable use, and conservation of pig genetic resources for wealth creation and food security, as well as the challenges of the pig value chain and the opportunities they present for sustainable improvement of the African pig genetic resources.

8.2 Origin, Domestication, and Classification of African Pigs

8.2.1 Origin and Domestication of African Pig

Pigs are believed to have been first domesticated roughly over 40,000 years ago, making them one of the earliest animals domesticated by humans. The origins of African domestic pig breeds are unclear, poorly documented, and shrouded in controversy due to insufficient archaeological and genetic evidence to establish sound hypotheses about how, when, and where the *Sus* species were founded (FAO 2007a; Amills et al. 2013; Ramirez et al. 2009; Weka et al. 2020) on the continent. Available archaeological evidence from the Middle East indicates that pigs were domesticated as early as 9000 years ago, when most livestock were utilized by nomadic people, while swine were utilized by settled farming communities (Weka et al. 2020; Motsa'a et al. 2021). Other research findings (Amills et al. 2013; Halimani et al. 2020) indicate that indigenous African pigs originated from the Iberian Peninsula between the third and seventh centuries, with most of the present-day populations having similar phenotypic characteristics in all African countries, but referred to by different local and often tribal names. For example, the African indigenous pig is known in South Africa as Kolbroek, in Mali as Somo, in Kenya as Busia pig, in Cameroon as Bakosi, in Nigeria as the Local/Black pig, in Ghana as the Ashanti Dwarf pig, in Togo as Bush pig, in Zimbabwe as Mukota pig, and in Mozambique at the Landim (Amills et al. 2013; Halimani et al. 2020).

It is also believed that the African pig was domesticated following the introduction of Asian and/or European pigs through commercial trade routes (Sowls 1984). This was further substantiated by genetic studies on the local populations of African indigenous pig breeds. One study by Ramirez and colleagues (2009) showed that North and West African pig breeds have significant European ancestry, which may be in line with Portuguese exploration in the fifteenth century. Other archaeological findings posit that the contemporary Eastern African pig breeds' ancestry could have either been by direct introgression with Far Eastern breeds or through a European intermediary, since the earlier British breeds were shown to have carried some Eastern alleles (Weka et al. 2020). As highlighted in Fig. 8.1, archaeological records and molecular studies strongly support Europe (Region 3), the Middle (Region 5), and the Far East (Regions 9 and 10) as the major centers of ancient pig domestication.

The ancestry of the indigenous African pigs probably originated from these centers, because this is yet to be confirmed by scientific evidence.

8.2.2 Classification of Pigs

Pigs are omnivores, and their food range is close to that of humans. In fact, between pigs and humans, there are so many significant similarities that pigs are used as experimental models for human medical research (Sykes 2022) and for the production of organs for human health purposes (xenotransplantation) (Montgomery et al. 2022). The pig is one of the oldest domesticated animals, and the majority of the breeds are known to have descended from the Eurasian Wild Boar (*Sus scrofa*) (Weka et al. 2020; Larson et al. 2005). An independent domestication of both European and Asian subspecies of the wild boar has been clearly established (Giuffra et al. 2000). Pigs are classified as follows (Giuffra et al. 2000):

Kingdom	*Animalia*
Phylum	*Chordata*
Sub-phylum	*Craniata* (Vertebrata)
Class	*Mammalia*
Sub-class	*Theria* (Viviparous)
Infra-class	*Eutheria* (Placenta)
Order	*Ungulata* (hoofed mammals)
Sub-order	*Artiodactyla* (even toed)
Sub-division	*Sunia*
Family	*Suidae* (domesticated pigs)
Genus	*Sus*

Pigs are the species that make up the family *Suidae* are known to have originated in Old World Europe, Africa, and Asia. New World pigs, such as the peccaries, differ from true pigs in certain skeletal and dental features and belong to a different family (*Tayassuidae*). A peccary is a medium-sized, pig-like ungulate found throughout Central and South America, Trinidad in the Caribbean, and in the southern deserts of America. The phylogenetic relationships of 19 species in the family Suidae are depicted in Fig. 8.2 and include the following (Larson et al. 2005; Chen et al. 2007):

Fig. 8.1 Major centers of livestock domestication based on archaeological and molecular information. From FAO 2007a, licensed under CC-BY 4.0. Copyright © FAO 2007.

Key: 1: turkey; 2: guinea pig, llama, alpaca; 3: pig, rabbit; 4: cattle, donkey; 5: cattle, pig, goat, sheep, Bactrian camel; 6: cattle, goat, chicken, river buffalo; 7: horse; 8: yak; 9: pig, swamp buffalo, chicken; 10: chicken, pig, Bali cattle; 11: dromedary; 12: reindeer.

- *Babyrousa babyrussa*: Golden Babirusa
- *Babyrousa celebensis*: Sulawesi Babirusa
- *Babyrousa togeanensis*: Togian Babirusa
- *Hylochoerus meinertzhageni*: Giant Forest Hog
- *Phacochoerus aethiopicus*: Cape, Somali, or Desert Warthog
- *Phacochoerus africanus*: Common Warthog
- *Porcula salvania*: Pygmy Hog
- *Potamochoerus larvatus*: Bushpig
- *Potamochoerus porcus*: Red River Hog
- *Sus ahoenobarbus*: Palawan Bearded Pig
- *Sus barbatus*: Bearded Pig
- *Sus bucculentus*: Vietnamese Warty Pig
- *Sus cebifrons*: Visayan Warty Pig
- *Sus celebensis*: Celebes Warty Pig
- *Sus heureni*: Flores Warty Pig
- *Sus oliveri*: Mindoro Warty Pig
- *Sus philippensis*: Philippine Warty Pig
- *Sus scrofa*: Wild Boar, Domestic Pig
- *Sus verrucosus*: Javan Warty Pig

8.2.3 Dispersal and Genetic Diversity of African Pigs

Pigs are omnivores and have survived in a variety of habitats, including grasslands, rainforests, wetlands, scrublands, temperate forests, and savannas, everywhere in the world except Antarctica, Northern Africa, and far northern Eurasia. It is generally acknowledged that livestock species were incorporated into diversified subsistence economies that utilized a range of domestic, managed, and free-living plant and animal resources. Thus, the movement of both cattle and swine down the eastern and western arms of the Fertile Crescent was slow, with no domestication evidence detected in the southernmost reaches of the Fertile Crescent until about 2000 years after their initial domestication. The poor performance of pigs in this dispersal story, usually coming in either a distant fourth after cattle and caprines in the pastoral economy of Anatolia and Europe, or, as in Africa, generally absent, may also have some relationship to their biological attributes. Higher water requirements

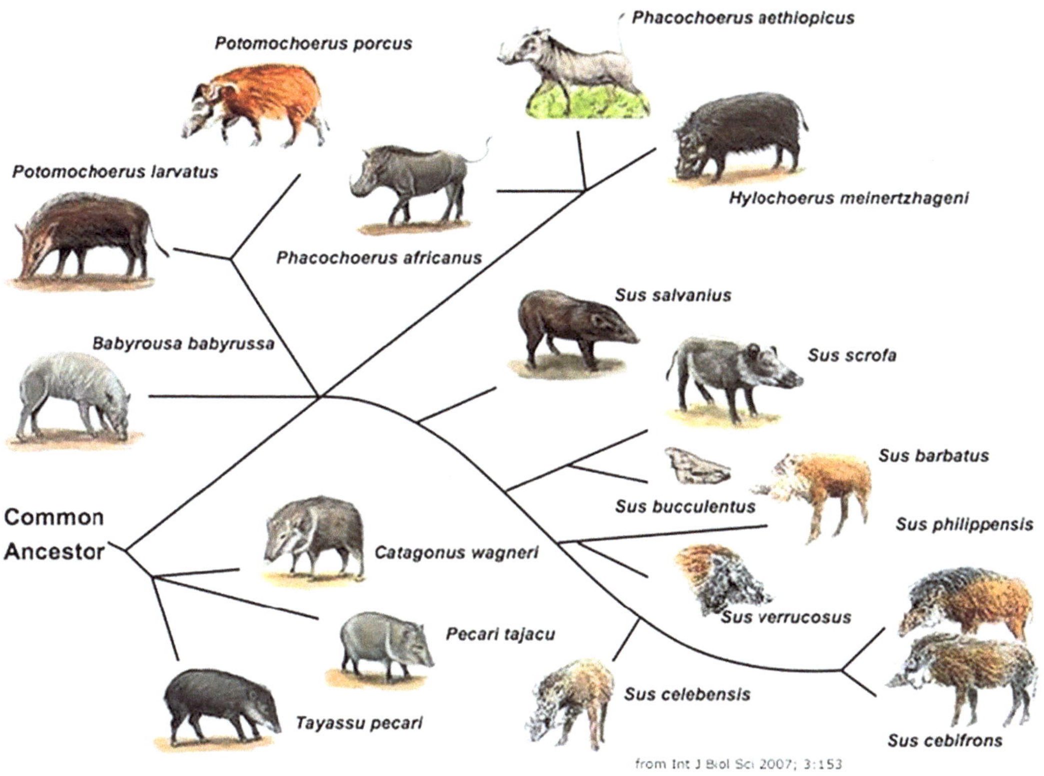

Fig. 8.2 Suiforme diversity and phylogenetic relationship. (From Chen et al. 2007, licensed under CC-BY 4.0). (Source: https://untamedscience.com/family/suidae/)

and the fact that pigs are temperamentally ill-suited to long-distance transhumance (Zeder 2017) may well have played an important role in shaping their dispersal story.

The genetic diversity of African pigs indicates how close or how different the various Sus genetic resources on the continent are. In general, the relationship between the African indigenous pigs and their ancestors has not been elucidated or studied extensively like that of the European and Far Eastern pig breeds (Amills et al. 2010; Ramirez et al. 2009). undertook the first genetic survey of a number of pig breeds distributed in Western (Nigeria and Benin) and Eastern (Kenya and Zimbabwe) Africa but did not find any close relationship between the Near Eastern wild boars and African pigs, which is opposed to reported close relationships between Near Eastern and European *S. scrofa* populations and Near Eastern wild boars (Weka et al. 2020; Larson et al. 2005). Resolution of the genetic make-up of the African pig breeds has also revealed substantial differences between Western and Eastern African populations. The West African pigs share some alleles that are abundant in European breeds, while eastern and southern African breeds harbor Far Eastern alleles at very high frequencies (Ramirez et al. 2009). Two alternative scenarios might explain the presence of Far Eastern alleles in East African pig breeds. First, they might have been introduced through either a European intermediary, given that British pig breeds were strongly admixed with Chinese genotypes in the eighteenth–nineteenth centuries, and second, there might have been a direct genetic introgression with Far Eastern breeds. These findings are probably because 79% of African pigs analyzed with microsatellites were originally from East Africa, where the Far Eastern introgression has been particularly strong (Ramirez et al. 2009; Halimani et al. 2020). In Western Africa, pigs do not display Far Eastern alleles, suggesting that they descended from the admixture of indigenous populations and exotic breeds with European ancestry (i.e., Iberian pigs brought by the Portuguese in the fifteenth century as well as swine introduced by the European colonizers in the nineteenth–twentieth centuries). In strong contrast, Eastern African pigs harbor Far Eastern alleles at significant frequencies, evidencing that they have a mixed European/Chinese origin (Ramirez et al. 2009; Amills et al. 2010). At the same time, archaeological records have shown that the indigenous African pig breeds, typified by the ADP of Ghana, Kolbroek of South Africa, and the Bakosi of Cameroon and Gabon, probably originated from the Iberian region (Weka et al. 2020; Halimani et al. 2020).

In reference to the first strategic priority area of the Global Plan of Action (GPA) on Animal Genetic Resources (AnGR) (FAO 2007b), not much research has been done on the characterization of African indigenous pig breeds. While the African Union Inter-African Bureau for Animal Resources (AU-IBAR) has developed a tool and portal for characterization of the continent's animal genetic resources, there has been very little commitment by most national governments to allocate resources for activities directed toward boosting both human and institutional capacity for sustainable use and conservation of Africa's rich AnGR (AUIBAR 2019). Available data on pig characterization depicts inadequate research coverage of the populations and countries, or largely fragmented studies that are not well coordinated. There is thus an urgent need for adequate characterization of indigenous African pig breeds for informed decisions on mapping their distributions, population status, diversity, and sustainable utilization (Giuffra et al. 2000; Bester and Küsel 1998).

8.3 The Importance and Use of Pig Genetic Resources in Africa

Madmizure et al. (2013) list some of the uses of pigs as provision of meat, income generation, sociocultural values, savings and investment, family pride/status, and manure. Indeed, African indigenous pig breeds are of great agricultural and food security importance, uniquely adapted to the existing production systems and environments, and possess resilient traits for their survival. Although pigs are raised as homestead animals to provide food for the household, most pig farmers also raise pigs mainly as a source of income (Aryee et al. 2019). For instance, one study in Central Ethiopia found that farmers did not slaughter pigs for home consumption but mainly kept them as a source of income (Gebrehawariat et al. 2017). The majority of pig farmers interviewed in Ghana kept crossbred pigs with income as their main motivation (Aryee et al. 2019). Pig production in Africa is thus of economic, nutritional, and sociocultural importance to livelihoods (Aryee et al. 2019). Regardless of the social and religious restrictions that exist in certain African communities, pig farming is playing an important role in tackling poverty, malnutrition, and a deficit in animal protein, thanks to its numerous advantages. Due to its low production cost, fast growth rate, and short cycle, pig farming has been recommended as an alternative source of cheap, high-quality dietary protein for the escalating human population (Motsa'a et al. 2021).

Indigenous pigs utilize feedstuffs of low nutritive value, are thermotolerant, are resistant to certain diseases and parasites, and can survive and reproduce in low-input and resource-limited systems. However, due to their slow growth rates and small body sizes, the productivity of indigenous pig breeds is low and not able to cope with increased pork demand in most African countries. To satisfy the surging demand for pork and pork products, national governments have often resorted to the importation of high-producing exotic pig breeds, mostly from Europe and America. Dotché et al. (2019) provide extensive information on the reproductive performance of local breeds of pigs, such as age at puberty, gestation length, litter size at birth, litter size at weaning, birth weight, weaning weight, and age at weaning, recorded in some African countries. It should be stressed that these performance parameters are affected by many environmental factors, including feeding, the breeding program, and

greatly improved by crossbreeding with exotic breeds using artificial insemination.

These exotic pig breeds have existed alongside the local African breeds for decades and have been naturalized in different environments of Africa. However, the commercial exotic and the native African pig breeds and their crossbreds exist in distinct production systems and play key roles in African animal agriculture. The commercial exotic breeds dominate in medium and large-scale intensive farms. The medium-scale intensive farms are commonly found in and around urban and peri-urban environs of major cities with good markets for pork and pork products. Different actors play various roles along the pig value chain, creating employment and livelihoods for millions of people in the process. Processing, sale, and consumption of pork is commonplace in many African countries such as Burkina Faso, Ghana, Nigeria, Uganda, Kenya, Tanzania, and Cameroon. Compared to chicken meat and beef, pork is relatively cheaper for most consumers.

Not surprisingly, there has been a fast growth in the number of actors in the pig value chain, with a resultant growth in the pig sector estimated at between 5% and 10% annually (FAO et al. 2017). Pig production has been an attractive venture to an increasing number of smallholder rural farmers in Africa. It is estimated that over 70% of the pig output in African countries comes from smallholder rural pig farms. The indigenous African pig breeds and their crossbreds with exotic breeds dominate in these farms. The relatively short production cycle of the *Sus* species, their prolificacy and ability to survive on a wide range of feeds, as well as high carcass are key motivators for their production. In a rural setting where banking services are unavailable, pigs play the role of an inflation-proof cash reserve to be used during emergencies or when extra cash is needed to meet household needs (Aryee et al. 2019).

8.4 Pig Breeds of Africa

Africa boasts a rich diversity of indigenous and locally adapted exotic pig genetic resources and their crosses, which have been a great source of animal protein, support of livelihoods, wealth, and sociocultural values over the years. African indigenous pig breeds display varying phenotypic and adaptive characteristics (Osei-Amponsah et al. 2017). The AU-IBAR (2019) describes indigenous African pigs as small, dark-colored animals with small ears, a short forehead, a straight tail, and an elongated snout. The body is often narrow, carried on relatively long legs. Coat color is variable and sometimes covered with long, coarse hairs and a distinct mane along the spine. They may vary in size, but adult pigs rarely weigh above 60 kg. The pigs are generally adapted to low-input systems practiced in many rural areas of Africa and other developing countries. Generally, the indigenous African pig breed has relatively good adaptive traits, such as tolerance to endemic diseases and survival under poor management, including heat stress, and it is also able to handle fibrous feeds much better than exotic breeds. Sample indigenous pig breeds include Kolbroek (South Africa), Somo (Mali), Busia (Kenya), Bakosi (Cameroon and Gabon), West African Dwarf pig (Nigeria), ADP (Ghana), Bush pig (Togo), Mukota (Zimbabwe), and Tswana (Botswana) (Halimani et al. 2020). In some countries, the indigenous pigs have no specific names but are generally referred to as local, indigenous, or forest pigs, and these may also refer to crossbreds or admixed populations. However, the breed status of most *Sus* breeds on the continent remains unknown, and literature on most of the indigenous breeds remains largely unavailable. Therefore, in the subsequent sections, we document as much available information as possible on indigenous and locally adapted exotic pig breeds on the African continent with emphasis on their unique adaptive traits and the production environments that promote their conservation and sustainable use.

8.4.1 Geographical Distribution of African Pig Breeds

In terms of numbers, a total of 43,007,434 pigs are found in Africa, with 28,914 in North Africa, 15,678,263 in West Africa, 8,139,402 in Central Africa, 17,649,551 in Eastern Africa, and 1,511,304 in Southern Africa in the year 2021 (FAOSTAT 2022). In terms of number of breeds, Southern Africa has the highest number of pig breeds (64), followed by Western Africa (34), Central Africa (24), and Eastern Africa (23), with Northern Africa reporting just a single breed from Egypt due to sociocultural and religious reasons (AUIBAR 2019). According to the Food and Agricultural Organization (FAO) Domestic Animal Diversity Information System (DAD-IS) database (Food and Agriculture Organisation of the United Nations (FAO) 2020) (https://www.fao.org/dad-is/en/) and FAO country reports, the following *Sus* Genetic Resources are commonly found in various regions of Africa (Table 8.1).

8.4.2 Pig Breeds of North Africa

Although pigs are not common in North Africa, some archaeological data indicate that pigs were bred in Ancient Egypt during the fourth century and dispersed to North and Sub-Saharan Africa through the Nile corridor and subsequently to West, East, and Southern Africa. However, the pig population in North African countries is believed to have been substantially affected by the growth, domination, and expansion of Islam. Although Islam forbids Muslims to eat pork, Muslims living in multireligious communities have tolerated pigs kept by others and have at times utilized pork for magical purposes (Epstein 1971). On the other hand, where Islam is the dominant religion, pig production is forbidden, the cause of the disappearance of the species in most parts of Northern Africa.

Additionally, some religions in ancient Ethiopia and Egypt with strong links to Christian Old Testament teachings prohibited their followers from raising pigs, the reason for the relatively low pig populations in these parts of Africa. Additionally, research on African pigs was for many years also not common on the agenda of the International Livestock Research Institute (ILRI), as well as the fact that the *Sus* species compete with humans for food, and thus, feeding them may be expensive. Collectively, these facts, among others, have also contributed to the scant information on pig breeding and production in Africa. It is believed that pigs were herded, raised, and occasionally eaten throughout Egypt from the Predynastic period into the Late Period and Graeco-Roman times (Parsons 2017). The domestic pig of ancient Egypt descended from an indigenous ancestor, *Sus scrofa*, the Wild Boar. Once abundant in the country with a fairly extensive range throughout the Nile Valley, in the Delta, the Faiyum, and the Wadi Natrun, the pigs only became locally extinct around the turn of the twentieth century, mainly as a result of overhunting, loss of their prime habitat, and the spread of Islam. Indeed, pig remains have been found throughout Egypt at sites such as Hierakonpolis, Maadi, Abydos, and Armant, near graves of the lower social class, indicating that pork was an important component of their diet, particularly during the Predynastic period. If there was a prohibition against eating pigs among the upper classes, there was none against raising them. Pigs' teeth, together with other ingredients, are on record to have been mixed and bandaged onto infected parts of the body to expel exudations, perhaps a reference to pus or eczema. The literature also indicates that pig viscera, including the brain, was an ingredient in a cure to combat cancer (Parsons 2017).

8.4.2.1 Baladi Pig Breed

The FAO's DAD-IS (Food and Agriculture Organisation of the United Nations (FAO) 2020) records only one pig breed in Egypt, called Baladi, which means land or soil. The Baladi is an indigenous pig breed of Egypt (Galal 2007), with a population estimate of more than 300,000 heads. However, very little information is available or has been published on the Baladi pig. Pig farming and consumption is concentrated in Egypt's Coptic Christian minority, estimated at 10% of the population. Pig farms were sited near

Table 8.1 Pig breeds of Africa by country and region

Region	Country	Indigenous breeds (synonyms)	Exotic breeds
North Africa	Egypt	Baladi	–
West Africa	Benin	Porc de Benin	Landrace, Large White/Yorkshire, Meishan
	Burkina Faso	Korogho; Porc de Burkina Faso	Landrace, Large White
	Côte d'Ivoire	Korhogo; Porc de Côte d'Ivoire	Landrace, Large White
	Ghana	Ashanti Dwarf Pig (Ashanti Black Forest Pig)	Landrace, Large White
	Guinea	Guinea pig	Landrace, Large White
	Guinea-Bissau	Guinea-Bissau pig	Duroc, Landrace, Large White
	Liberia	Liberia Native pig (Mba, Saynee, Gbocho, Que Bayee, Black pig, and Pepee)	Hampshire, Landrace, Large White/Yolkshire
	Mali	Korogho, Somo	Large White, Race Chinoise
	Niger	Porc Africain	
	Nigeria	West African Dwarf pig (Nigerian pig)	Hampshire, Landrace, Large White/Yolkshire
	Senegal	Porc de Senegal	Large White
	Sierra Leone		Berkshire, Duroc, Landrace, Large White
	Togo	Porc De Dapaong, Porc de Togo	Landrace, Large White
East Africa	Burundi	Porc de Burundi	Landrace, Large White
	Kenya	Kenyan pig	Hampshire, Landrace, Large White
	Madagascar	Danoise, Porc de Madagascar	Landrace, Large White
	Rwanda	Rwandan pig (Porc indigene rwandais)	Berkshire, Landrace, Large Black, Large White/Yorkshire, Middle Large, Piétrain
	Tanzania	Tanzanian pig	Hampshire, Landrace, Large White, Saddleback
	Uganda	Ugandan pig	Landrace, Large White, Wessex Saddleback
Central Africa	Cameroon	Bakosi, Bamiléké, Kousseri, Mankon Long Nose	Duroc, Landrace, Large White, Hampshire
	Central African Republic	Porc central africain	Duroc, Landrace, Large White
	Chad	Porc tchadien	
	Congo	Porc africain, Porc Congolese	Large White/Yorkshire
	Democratic Republic of the Congo	Bandundu, Kasaï	Duroc, Landrace, Large Black, Large White; Pietrain
	Gabon	Porc Africain; Bakosi	Duroc, Landrace, Large White

(continued)

Table 8.1 (continued)

Region	Country	Indigenous breeds (synonyms)	Exotic breeds
	Sao Tome and Principe	Sao Tome pig	
Southern Africa	Angola	Ganda, Jambona Munhanda, Porco do Bengo, Porco do Kunene	Duroc, Landrace, Large White
	Botswana	Tswana, Warthogs	Duroc, landrace, large white
	Eswatini	Black China, Spotted Swazi	Swazi Large Black
	Lesotho	Basotho	Duroc, Landrace, Large White
	Malawi	Malawian pig	Landrace, Large White, Camborough
	Mozambique	Landim	
	Namibia	Namibian pig	Large White, Topigs
	South Africa	Bantu, Kolbroek, QM Hamline, South African pig, Robuster, Welsh, Windsnyer	American Hampshire, Chester White, Duroc, Large Black, Large White; Pietrain, Landrace
	Zambia	Lusitu, Nsenga	Landrace, Large White, Camborough
	Zimbabwe	Dalland, Mukota	Duroc, Landrace, Large White

garbage pits at the outskirts of cities and were often exposed to poor hygienic production conditions. Local pig breeds like the Baladi were reared in slum areas by rubbish collectors who fed the pigs mainly on leftover feed from the rubbish dumps, and in this way, the pigs help in the disposal of organic waste. Meat from the Baladi is sold in special shops in larger cities to pork consumers (Galal 2007).

8.4.3 Pig Breeds of West Africa

West Africa is home to several indigenous and locally adapted exotic pig genotypes as well as many crossbreds between indigenous and exotic genotypes with different levels of admixture. Like in other parts of Africa, indigenous pig genetic resources are known by various names in the region, such as Somo (Mali), *Porc de Togo* (Togo), West African Dwarf pig (Nigeria), and Ashanti Dwarf Pig (Ghana), also known as the Ashanti Black Forest pig, among others. The origin of indigenous pigs of West Africa is controversial, but recent work on molecular genetic characterization suggests that they harbor alleles similar to pigs in the Middle East and the Far East. West African pig breeds may thus have contributed to the Eastern pig germplasm through trade routes across the Indian Ocean (Amills et al. 2013; Ramirez et al. 2009; Dotché et al. 2019; Osei-Amponsah et al. 2017; Agbokounou et al. 2016).

8.4.3.1 Ashanti Dwarf Pig

The Ashanti Dwarf Pig (ADP), also known as the Ashanti Black Forest Dwarf Pig, is an indigenous breed of Ghana raised at the subsistence level in mixed farming systems. Unlike many local pig breeds in Africa, the ADP has been characterized both at the phenotypic and molecular levels (Osei-Amponsah et al. 2017; Adjei et al. 2015a, b; Aryee 2023). Findings of genetic characterization studies indicate that the Ashanti Dwarf pig has both European and Asian ancestry, with both mitochondrial data and coat color suggesting a stronger contribution from Asian genetics in the north of Ghana (Osei-Amponsah et al. 2017). Black coat color is the most predominant within the ADP, with black MC1R alleles of both Asian and European origin present. European alleles for spotting are present at a low frequency and may account for the occurrence of spotted piglets in some APD litters (Osei-Amponsah et al. 2017). On a global level, the ADP cluster closely with European pigs of commercial origin, but FST (average inbreeding coefficient of subpopulations relative to the total population) analyses identified loci for ADP-specific traits (Osei-Amponsah et al. 2017). Adjei et al (2015a).

reported wide variations in morphological attributes of the ADP population in Ghana. In general, animals classified by owners as ADPs presented a concave head profile (85.9%), black coat color (67.5%), a plain coat color pattern, erect ears (84.7%), projecting backward (52.2%), and a short, cylindrical snout. Although the majority had short and straight body hair type, others had long and dense, or long and curly coats. A few ADPs (15%) had a straight back line, but the majority were swaybacked (85%). Other indigenous pigs in the survey had semi-lop ears, straight head profiles, white and black coat color types, patchy and spotted coat color patterns, and long and cylindrical snouts. It is possible these differences in phenotypes occur owing to genotypic differences in the indigenous swine gene pool, reflecting their admixture, which could potentially dilute highly beneficial adaptive genes. Porter et al. (2016) confirmed some of these findings, indicating that the ADP is usually black (Fig. 8.3) but may also be black pied, brown, or white and occasionally gray. It has a long head, a long tapering snout, upright ears, and quite long legs. Weka et al (2020). describe the ADP as hardy, able to survive under poor management, mostly scavenging for their food, and able to digest high fibrous matter; they are well adapted to resist heat stroke as well as other harsh environmental conditions and are less susceptible to many local diseases and have parasites and one good mothering ability (Weka et al. 2020).

The ADP is tolerant to most common diseases and stressors and can survive under poor management and extreme environmental conditions (Osei-Amponsah et al. 2017; Barnes and Fleischer 1998). Indigenous pigs in general have low production potential but are well adapted to extensive, low-input production systems. On the other hand, exotic breeds require intensive husbandry practices (adequate feeding, improved housing, and regular veterinary care), which many small-scale farmers cannot afford. Indigenous pig breeds like the ADP thus offer a viable alternative for many small-scale local farmers to increase livestock production (Osei-Amponsah et al. 2017). Average mature adult

Fig. 8.3 Ashanti Dwarf pig of Ghana

weighs around 40–60 kg, with females reaching an average mature age of 7.5 months and average litter size between 5 and 8 (Weka et al. 2020; Porter et al. 2016). Crosses of ADPs with exotic breeds have generated animals with relatively high growth rates, but their ability to forage on fibrous feed and resistance to diseases are reduced. Thus, despite its relatively low cost of production, the ADP has, over the years, faced the threat of high gene introgression from exotic breeds such as the Large White, Landrace, and Duroc (Osei-Amponsah et al. 2017, 2021).

8.4.3.2 Burkina Faso Pig

In many countries in Africa, the pig breeds are not inventoried and well-defined as breeds. They are regarded as "indigenous or local," or "none improved" pig breeds. The indigenous pig breed of Burkina Faso or Burkina Faso pig (Fig. 8.4) is small in body size, rustic/local of white, black, or mapie and white coat, with very hard silks, and is of great heterogeneity in production. The live weight of adult pigs varies between 40 and 60 kg (Kiendrebeogo et al. 2012). Like most of the local pigs in Africa, information on the phenotypic and molecular characterization of the local pig of Burkina Faso is scant, and there is a need for more research on this breed.

8.4.3.3 Korhogo Pig

The Korhogo breed (pig of Korhogo) is commonly found in Burkina Faso and Cote d'Ivoire.

The Korhogo pig breed started with a cross between the Craonnaise (West French White) pig breed and indigenous female pigs. The best-performing descendants were further crossed with boars of Yorkshire (Large White). The stabilization of these final genotypes in the village of Korhogo in the Republic of Côte d'Ivoire produced the Korhogo pig breed (Fig. 8.5). It is considered a rustic/local and prolific breed, found in all pig production systems in and around Bobo-Dioulasso's area (Kiendrebeogo et al. 2012) of Burkina Faso. The Korhogo breed, like other improved pig breeds such as the Large White, is superior in terms of the number of piglets per litter in all the production systems (Kiendrebeogo et al. 2012) compared to indigenous pig breeds. Information on phenotypic and molecular characterization of the breed is lacking, and this research gap needs to be filled.

8.4.3.4 Liberia Native Pig

The dominant body coat color pattern of the Liberia native pig is solid/uniform/plain (80%), spotted (11%), and patchy/pied (8%) (Karnuah et al. 2018) (Fig. 8.6). In terms of coat color, the pigs were predominantly white (79%) with some pigs being fawn (9%) or black (7%) (Karnuah et al. 2018). The head profile of the Liberia native pig is mostly straight (83%) or concave (16%), and the tails are straight (66%) or curly/kinked (34%) with straight (81%) or swayed back (19%) backline (Karnuah et al. 2018). Information on the molecular characterization of Liberia's local pigs is lacking, and this research gap needs to be filled.

8.4.4 Pig Breeds of Central Africa

As shown in Table 8.1, the Central African region is home to several pig breeds due to the popularity of pig farming in Cameroon, the Central African Republic, Chad, Congo, the Democratic Republic of Congo, and Gabon. Despite the popularity of pigs in this region, information on their phenotypic and molecular characterization remains scarce.

8.4.4.1 Bakosi Pig

A popular indigenous pig breed of Cameroon is the Bakosi (Fig. 8.7), a variety of the West African pig (Iberian in origin), which is black and often with random and quite extensive white patches, and it is like the forest pig of Nigeria and Ghana in size and performance. Phenotypically, the Bakosi pig is of a small size and has a long head with a short, almost flat forehead, a straight muzzle, and an elongated snout. The ears are small, thick, and horizontal or erect. The neck is short, the back slightly convex, and quite long. The croup is sloping, and the ham is sparse. The limbs are slender, not very muscular, with sows having, on average, five pairs of teats. The bristles are long, and the coat color can be black, gray, and red piebald (Mason and Maule 1960). In general, Ghomsi et al. (2022) reports the following mor-

Fig. 8.4 Burkina Faso pig. (From Kiendrebeogo et al. 2012, licensed under CC-BY 4.0)

Fig. 8.5 Korhogo pig of Burkina Faso and Cote d'Ivoire. (From Kiendrebeogo et al. 2012), licensed under CC-BY 4.0)

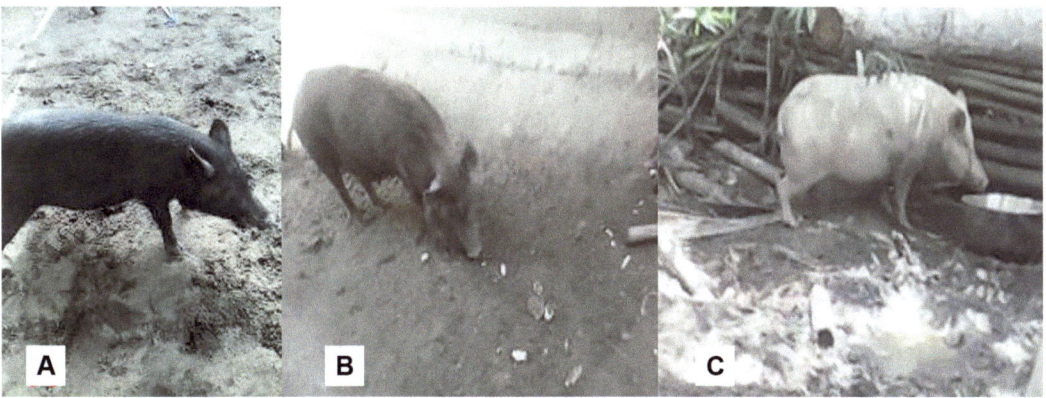

Fig. 8.6 Pigs in local production systems in Liberia. (**a**) Libera native pig; (**b**) crossbred; (**c**) Exotic breed

phological attributes of indigenous pigs of Cameroon: coat color—black, white, or black and white coat color (68%), plain coat pattern (82%), straight hair (85%), smooth skin (78%), concave head profiles (61%), long and thin snouts (92%), prick ears (82%) projected either upwards (54%) or backward (39%), straight backline (85%), and tails (68%).

Apart from the Bakosi pig, other indigenous pig breeds found in Cameroon include Bamiléké, Kousseri, Bakim, and Mankon Long Nose (the names are linked to different people or places) (Porter et al. 2016; Audrey et al. 2021) (Fig. 8.7). The Mankon Long Nose is said to be more prolific than the others, and it is under a pig improvement program by the Agricultural Research Institute for Development, Mankon Center (https://irad.cm/). Bankim pigs have multiple coat colors and mainly (65%) front-oriented erect ears, teat number ranging from 5 to 7, similar to most indigenous pigs of Africa (Audrey et al. 2021). Like other breeds, information on the molecular characterization of local pig breeds of Cameroon is lacking, and this research gap needs to be filled.

8.4.5 Pig Breeds of East Africa

The genetic diversity of pigs in the East African region has not been fully characterized, and their breed composition has been the subject of considerable speculation (Mujibi et al. 2018). Genetic characterization of indigenous East African pigs in relation to disease susceptibility is important in understanding the genetic deter-

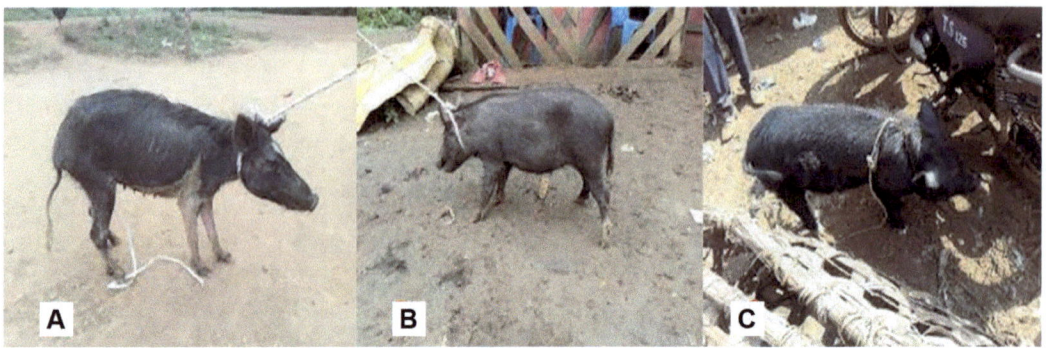

Fig. 8.7 Indigenous pig breeds of Cameroon. (**a**) Mankon long nose; (**b**) Bakosi; (**c**) Bamiléké. (From Motsa'a et al. 2021, licensed under CC-BY 4.0)

minants of disease tolerance, which may impact the design of appropriate control strategies. Limited studies indicate that East African pigs have a complex ancestry, with haplotypes from Asian, Far Eastern, and European pigs all present in certain populations (Mujibi et al. 2018). The application of genome-wide single nucleotide polymorphism (SNP) markers in characterizing the genetic diversity of domestic pigs revealed that local domestic pigs in Kenya are a nonhomogeneous group. The Busia population consisted of admixed pigs of various breeds, while the Homabay population represents pigs that are homogenous and whose composition is significantly different from that of the international commercial breeds (Mujibi et al. 2018).

8.4.5.1 Ugandan Pig

Pig production in Uganda is widespread and appears to be increasing at a high rate. There is limited information on the types of specific breeds and breeding practices in different pig production systems. Pigs in Uganda can be white, black, or black and white (Tatwangire 2014). This contrasts with the black color of local pigs, which are considered to be indigenous. There are no commercial breeding services for pigs in Uganda, and the use of artificial insemination (AI) in pigs remains limited (Tatwangire 2014). The Ugandan pig, or local pig of Uganda (Fig. 8.8), is mainly black in color with some spotting or white patches in crossbreds between local and exotic breeds, resulting from various pig improvement interventions (Babigumira et al. 2021). The body length is medium (60–80 cm), with a big head, prominent face, and tapering snout. The ears are medium to large-sized, tails are long (17–20 cm) (Marshall 2022). In terms of performance, sows have good mothering ability, with average litter sizes of 6–8 piglets, but intensively managed sows have superior litter sizes up to 15 piglets. The Ugandan pig is hardy, can cope with low-quality feed, and high disease burden, especially mange and worms. However, their carcasses are often very fatty. A mature sow weighs between 70 kg and 120 kg, while a mature boar weighs between 100 kg and 150 kg (Marshall 2022). Crossbred sows produced from crosses between the Ugandan pig and exotic breeds such as the Large White or Landrace are suitable for small-scale producers that are not able to provide high standards of housing, healthcare, and feed (Marshall 2022).

8.4.5.2 Tanzanian Pig

According to Mason and Maule (1960) the Tanzanian indigenous pigs are believed to be descendants of earlier introductions by missionaries during the colonial period (Mbaga et al. 2005). More than 90% of these local pigs are found under traditional systems and are an integral part of smallholder farming systems (Mbaga et al. 2003). Pig farming in Tanzania is mainly carried out by smallholder farmers, involving over 500,000 rural households, representing about 22.4% of agricultural households. Most pigs are kept in high-altitude areas, with about 54% of the pigs in the Southern highlands of

Fig. 8.8 Indigenous pig of Uganda. (Ugandan pig). (From Marshall 2022, licensed under CC-BY 4.0)

Tanzania (SHT) regions of Mbeya, Iringa, Rukwa, and Ruvuma. However, a lack of systematic breeding plans, poor husbandry practices, genetic drift, and possibly mutation have led to pigs of varied phenotypes. Coat color of the pigs varies, but are predominantly white (28%), black and white (24%), and solid black (19.8%). The majority (78%) of pigs have droopy ears and are significantly ($p<0.01$) heavier with a longer trunk and increased body length when compared with those with erect ears. Other features include a long and straight face and a short-curled tail (Mbaga et al. 2005). A study undertaken in the Mbeya region showed that the predominant management system practiced is free-ranging and occasionally tethering. The average mature body weights for boars and sows are 57.4 kg and 54 kg, respectively. The overall mean birth weight of 0.9 kg, weaning weight of 10.8 kg, litter size at birth of 6.6 piglets, and litter size at weaning of 4.3 piglets have been reported for the Tanzanian pig (Lekule and Kyvsgaard 2003). Like other local pig breeds, little research work on the molecular characterization of these breeds has been undertaken to identify and provide more information on genes responsible for their adaptation to relatively harsher production environments.

8.4.6 Pig Breeds of Southern Africa

According to FAO's DAD-IS (Food and Agriculture Organisation of the United Nations (FAO) 2020), all the Southern African Development Community (SADC) countries, except the Comoros Islands, have three major international pig breeds (Large White, Landrace, and Duroc) and several indigenous pig breeds. The main attributes of the indigenous breeds are hardiness, foraging ability, heat tolerance, high fertility, good mothering ability, good-quality meat, tolerance to endemic diseases and parasites, and adapted to low management levels (Halimani et al. 2020).

8.4.6.1 Bantu Pig Breed

The Bantu pig (Fig. 8.9) is mostly found in southern Africa. It is generally considered to have been bred from imported pigs from Asia and Europe and is usually small or medium-sized at maturity. Studies at the University of Pretoria found Traces of the Poland China breed and the English breed "Gloucester Old Spot" in numerous litters produced by one male Bantu pig and several Bantu sows. The Bantu is thought to have been developed from early importations of swine from Europe and Asia. The standard body color of the Bantu Pigs is white with black spots, but reddish-tan pigs are also common in the breed. Reddish tan Bantu pigs are born with longitudinal lines similar to the native bush pig, suggesting past breeding with bush pigs (Porter et al. 2016). Bantu Pigs are regarded as very social animals that coexist very well with others once a herd is formed, though the introduction of new females, gilt or sow, can cause males in the herd to become quite agitated (Porter et al. 2016). Bantu pigs produce, on average, 6–10 piglets in a litter with two to three litters per year. The average weight of a Bantu piglet at birth is about 0.9 kg, and the body weight of an average adult Bantu pig ranges from 95.3 to 99.8 kg. Little research work on the molecular characterization of the Bantu breed has been undertaken to identify genes responsible for their adaptation and production traits.

8.4.6.2 Mukota Pig Breed

Mukota is an indigenous breed found primarily in Zimbabwe. It is also known as the Rhodesian or Zimbabwe Indigenous pig. The Mukota pig predominates in Zimbabwe, and it is also found in some parts of Mozambique and Zambia. Mukota pigs (Fig. 8.10) are believed to have been introduced in the seventeenth century through trade between Europe and China. There are basically two classes of Mukota pigs. One class is short, fat, and has a short snout resembling the Chinese Lard pig, while the other resembles the Windsnyer (Wind cutter) in being long-nosed with a razor back. They are, however, both black in color without any shades or spots. Although the Mukota indigenous pigs are adapted to the local environment, their numbers are declining largely due to livestock production policies that prefer the use of fast-growing imported breeds (Chimonyo and Dzama 2007). Mukota pigs are well known to be hardy and well adapted to the harsh tropical environment in terms of heat stress, disease resistance, and inadequate nutrition. Mukota pigs have a selective advantage to survive and reproduce on a low plane of nutrition. Mukota pigs can

Fig. 8.9 The Bantu pig of South Africa. (Source: Lilongwe University of Agriculture and Natural Resources 2011, licensed under CC-BY 4.0)

easily survive under very unhygienic conditions, testifying to their high disease resistance. These indigenous Zimbabwean pigs show moderate parasite tolerance, greater than the imported or western varieties. Mukota pigs have been shown to be less susceptible to *Ascaris suum*, and their meat is organoleptically more acceptable to the rural people than meat from European breeds. Mukota pigs, which have been demonstrated to adapt to survive under rural low-input production systems, have been shown to exhibit relatively low growth rates, which could be an advantage in that they do not require large amounts of concentrate feeds (Chimonyo and Dzama 2007). Studies provide evidence that Mukota are better able than imported pigs to utilize agricultural byproducts, such as maize cobs. It has been demonstrated that Mukota pigs are able to utilize fibrous feeds and high-tannin red sorghum better than European breeds such as the Large White (Chimonyo and Dzama 2007).

8.4.6.3 The Kolbroek Pig

The Kolbroek (Fig. 8.11) is a synthetic breed developed from a variety of breeds in South Africa. Development of the modern Kolbroek began in 1996 at the Agricultural Research Council (ARC), South Africa, under the leadership of Dr. Danie Visser. He and his team started by refining the nearly forgotten old Kolbroek breed and ultimately succeeded in producing a unique, indigenous pig breed. The modern Kolbroek consists of, among others, Windsnyer, Sandveld Red, Tamworth, and Great White genetics. The Kolbroek is a domestic free-range pig found in South Africa (Halimani et al. 2010).

Kolbroek pigs come in a variety of colors, including black with white blotches and brown with gray blotches. The body frame is small and short compared with other pig breeds in South Africa. Distinctive characteristics are a pot belly and a concave face. Weka et al (2020), reports that Kolbroek pigs are short, with prickled ears, a short snout, and a squashed face. A mature Kolbroek weighs about 150 kg, and fares extraordinarily well on a high-fiber diet. The average purebred Kolbroek weaner weighs approximately 8 kg at between 8 and 10 weeks of age. Kolbroeks

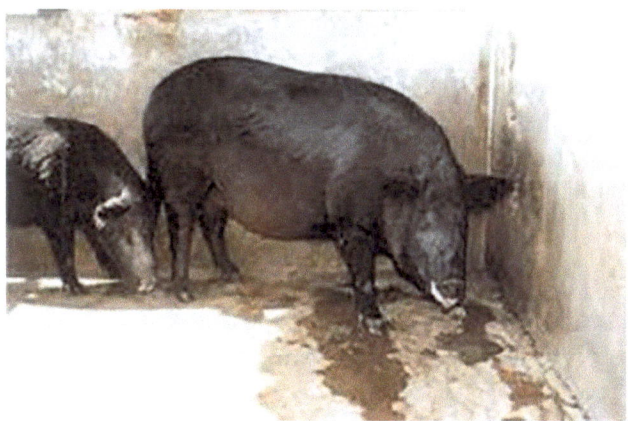

Fig. 8.10 The Mukota pig breed of Zimbabwe. (From https://www.petmapz.com/breed/mukota-pig/#, licensed under CC-BY 4.0)

Fig. 8.11 The Kolbroek pig breed of South Africa. (From http://farmersweekly.co.za, licensed under CC-BY 4.0)

fatten up very rapidly, with four-month-olds presenting up to a 35 mm back fat layer over the ideal of 25 mm. Kolbroek meat is naturally marbled, and the carcasses of young pigs are ideal for manufacturing boerewors, salami, prosciutto hams, and a variety of sausages. Like other indigenous pig breeds, it is known for its hardiness, foraging ability, heat tolerance, high fertility, good mothering ability, good quality meat, tolerance to endemic diseases and parasites, and adaptation to low management levels.

8.4.6.4 Indigenous Pigs of Zambia

Like other African countries, pig production remains crucial to the livelihood of farmers in Zambia, and thus, there are policies to support the conservation of local animal genetic resources, including pigs. The indigenous pig breeds of Zambia are the Lusitu and Nsenga, and they originate from Gwembe and Petauke districts, respectively (Abigaba et al. 2022). These two breeds are more adapted to local conditions, constitute 65% of the national pig flock, and are reared by the majority of traditional farmers (65%) in Zambia (Abigaba et al. 2022). Although studies on the characterization of indigenous pigs to a range of harsh rearing conditions (Abigaba et al. 2022) are scanty, when asked about unique traits they prefer in indigenous pigs, 25.9% and 23.8% of farmers indicated disease resistance and fertility as their preferred traits, respectively. Other important traits include meat quality and foraging ability. For instance, the Nsenga pig is one of the breeds that farmers want conserved because of its good litter size (8–10), and can thus help them increase income at a faster rate (Abigaba et al. 2022). Like most local pig breeds in Africa, there is little or no information on the phenotypic and molecular characterization of the Lusitu and Nsenga pig breeds of Zambia, an urgent research priority that should be considered by stakeholders.

8.4.6.5 Landim Pig

The Landim is an indigenous pig of Mozambique. In Mozambique, livestock plays an important role in the agriculture sector, due to its contribution to socioeconomic development and poverty reduction. Given the variations in climate, soil fertility, rainfall pattern, and altitude among the different areas, the country is divided into ten agroecological regions, each one with its own characteristic production systems and livestock breeds. Livestock is mostly composed of indigenous breeds: Landim cattle, goats, pigs, and chicken, Angoni and Bovino de Tete cattle breeds, and Pafuri goat. Most native breeds are named as Landim, which in the South of Mozambique means indigenous or local (Cumbula and Taela 2020). The Landim pig is a small black pig kept by smallholder pig producers with 2–10 pigs per family on average (Chilundo et al. 2017). Unfortunately, little or no phenotypic and molecular characterization has been carried out on the Landim pig, and this needs to be urgently addressed by stakeholders to generate important data for its sustainable use and conservation.

8.4.6.6 Indigenous Pig Breeds of Angola

The *Porco do Bengo*, *Porco do Kunene*, and *Jambona Munhanda* are the main indigenous pig breeds of Angola. Traditional management practices include feeding pigs with scraps and crop residues. In peri-urban and rural areas, pigs forage on rubbish dumps. Pigs are a crucial source of meat and income in Angola. Like most local pig breeds in Africa, the *Jambona Munhanda* has social and cultural importance, while the other breeds have economic importance. Unfortunately, little research on the phenotypic and molecular characterization of the indigenous pig breeds of Angola has been carried out, and this needs to be addressed by stakeholders to provide valuable information for their sustainable use and conservation (Osei-Amponsah et al. 2017). The exotic breeds introduced and farmed in intensive systems include Large White, Landrace, and Duroc, which have continually been imported over the last years (Porter et al. 2016).

8.4.7 Exotic Pig Breeds and Pig Production in Africa

The major commercial exotic pig breeds that have influenced African Animal Agriculture over the years include the Large White, Landrace, Duroc, Hampshire, and Pietrain (FAO 2007a, 2015; Weka et al. 2020; AUIBAR 2019; Goraga et al. 2017). Although the use of these exotic breeds of pigs is popular in all regions of Africa (Table 8.1) on account of their relatively high production, they require more investments in housing, feeding, and health care. Consequently, most exotic breeds are found in medium-high production commercial systems where farmers invest in feed provision and health care. At the same time, local farmers are increasing crossbreeding their local stocks with these exotic genetics. Often, the crossbreeding is indiscriminate, leading to the dilution of adaptive characteristics in the crossbreds and often low production performance due to poor feeding and veterinary care. There is a need for improved capacity building for stakeholders, particularly pig farmers, to enable them to benefit from the improved genetics. Rwanda's modern pig husbandry practices usually involve imported exotic breeds kept in confinement and fed with concentrate feed, provided with clean water, vaccines, and other important inputs. Over 70% of pig enterprises in Rwanda are rearing exotic breeds, and the most popular pig breeds are Large White, Landrace, Pietrain, and Duroc, followed by indigenous and some crossbreeds between these exotic and indigenous ones (Kruger 2015). Hirwa et al. (2022) reported that 41 percent (41%) of households kept crossbred pigs, local (19.7%), Landrace (18.7%), Large White (12.1%), Pietrain (8.1%), and other breeds (0.3%). In most African countries, pig producers decide to breed any type of breed according to their availability, but also for production purposes. Commercial farmers mainly use improved breeds, while traditional farmers solely rely on the indigenous or crossbreeds. However, like in other countries, the lack of breeding centers (seed stock nucleus) poses a serious problem, since the majority of pigs in the country are highly inbred (Shapiro et al. 2017).

In the following sections, we provide summaries of the key features of the most popular exotic pig breeds in Africa.

8.4.7.1 Duroc

The Duroc pig breed (Fig. 8.12) originated from the eastern United States, and the foundation stock was originally called the Duroc-Jersey because of the red hogs that went into their foundation. One source of the red or reddish-brown hogs found in the United States was reputed to be those that came from the Guinea coast of Africa. Durocs have considerable variation in color, ranging from a very light golden, almost yellow color, to a very dark red that approaches mahogany. The red is a very practical color that suits pork producers, and since it is a solid color, there is no concern about fancy points of proper markings. Durocs also have a medium length and a slight dish of the face with drooping. Indeed, Durocs constitute an important component of commercial semi-intensive and intensive pig management systems in many African countries on account of their superior growth, reproductive, and relatively good carcass traits (AUIBAR 2019; FAO 2015; Lichaba 2022). In the Republic of Congo, an average litter size per sow of 9 ± 0.8 head, a litter weight per sow of 25 ± 8 kg, and a breeding age of 8.3 ± 0.7 months have been recorded (Ognika et al. 2021). The average live weight of piglets is 1.32 ± 0.4 kg at birth, with average daily gains ADGs of 230 ± 0.5 g in the 2nd week, 370 ± 0.3 g in the 4th week, and 650 ± 0.8 g in the 16th week, respectively (Ognika et al. 2021).

8.4.7.2 Large White

The Large White was one of the original founder breeds of the British Pig Association, and the first herd book was published in 1884. The Large White got to South Africa in the 1890s and, as shown in Fig. 8.13, they are distinguished by their erect ears and slightly dished faces. They are long-bodied with excellent hams and fine white hair, and, as their name suggests, they are characterized by large size. Large Whites are valued for their bacon production.

The Large White has proved itself as a rugged and hardy breed that can withstand variations in climate and other environmental factors. Their ability to cross with and improve other breeds has given them a leading role in commercial pig production systems and breeding pyramids in many African countries (Aryee et al. 2019; Adjei et al. 2015b; Chimonyo and Dzama 2007; Lichaba 2022; Kanengoni et al. 2015). They and their descendants, the Yorkshire, are to be found in practically all crossbreeding and rotational breeding programs using two or more breeds. Modern breeding programs have developed separate sire and dam lines to produce purebred Large White terminal sires that excel in growth rate and lean meat percentage and are incorporated in most terminal sire breeding programs. In many African countries, including South Africa, Ghana, Kenya, Rwanda, and Nigeria, improved production performance has been obtained using the Large White in intensive management systems or crossing the breed with local sows (AUIBAR 2019; Osei-Amponsah et al. 2017; Kanengoni et al. 2015; Dube et al. 2012; Mbuthia et al. 2015). However, studies on blood antibody titers, protein, and globulin levels indicate that the Nigerian local breed, for instance, is superior to crosses

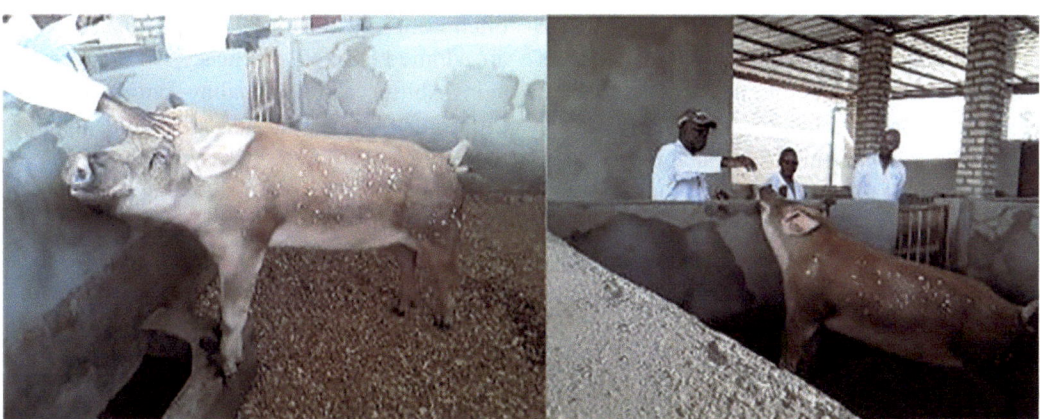

Fig. 8.12 The Duroc pig breed. (Credit: Prof. R. Osei-Amponsah)

Fig. 8.13 The Large White pig breed. (Credit: Prof. R. Osei-Amponsah)

between the Large White and Landrace (Abonyi et al. 2018). In the semi-intensive system, farmers in Kenya mainly kept Large White and Landrace and their crossbreeds on account of their faster growth rates and high mature weights. Additionally, the Large White in particular was preferred as the sire line in most commercial pig production enterprises due to high prolificacy and mothering ability (Mbuthia et al. 2015).

8.4.7.3 Pietrain

The Pietrain pig breed's birthplace is Pietrain, a village in Belgium. The breed is of medium size and is white with black spots. Around the black spots, there are characteristic rings of light pigmentation that carry white hair and erect ears. The Pietrain (Fig. 8.14) is very popular as a terminal sire in two of Europe's largest pig-producing countries, Germany and Spain. The Pietrain is renowned for its very high yield of lean meat, but this is often associated with the presence of the halothane gene for Porcine Stress Syndrome. For this reason, the use of purebred Pietrain in British pig production is relatively rare, and it is most commonly found in crossbred and synthetic terminal sire lines (https://www.thepigsite.com). Like the Large White, Landrace, Duroc, and Hampshire, the Pietrain is a major commercial exotic pig breed that has, over the years, been used in various pig improvement programs in Africa with varying degrees of success (FAO 2007a, 2015; Weka et al. 2020; AUIBAR 2019; Goraga et al. 2017; Hirwa et al. 2022).

8.4.7.4 Hampshire

The Hampshire pig breed (Fig. 8.15) was developed in the United States of America and is now one of the world's most important breeds. At a meeting of American breeders in 1890, the breed was renamed the "Hampshire," as the original pigs were imported from a farm in Hampshire, Wessex, UK. A breed society was established at the same time, and herd book recording can be traced for more than 100 years. The Hampshire is used extensively as the sire of crossbred pigs for the pork and manufacturing markets in many countries. It has the reputation of being the leanest of the North American Breeds, and most carcass competitions in North America are won by Hampshires and Hampshire crosses. In Africa, Hampshire has been popular in semi-intensive and intensive pig production systems on account of their lean carcasses, docility, and good mothering ability of its sows (AUIBAR 2019). Although often criticized for their large size, Hampshires are admired for their prolificacy, hardy vigor, foraging ability, and outstanding carcass qualities (http://afs.okstate.edu/breeds/swine/hampshire/index.html).

8.4.7.5 Landrace

Landraces have white skin, free from black hair, are lop-eared pigs (Fig. 8.16) with a long middle, light forequarters, and excellent ham development. The major faults with the original Landrace were leg weakness, splay legs, and nervous disorders such as porcine stress syndrome (PSS), which still occur in some strains. Landraces are

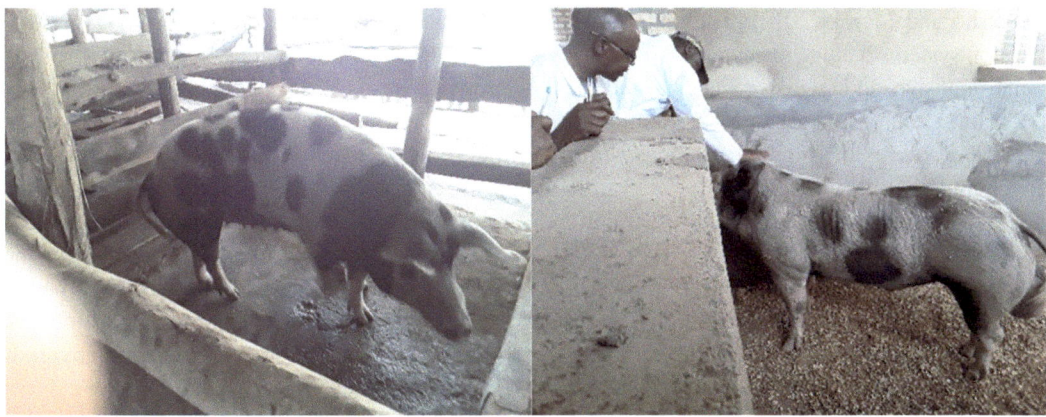

Fig. 8.14 The Pietrain pig breed. (Credit: Prof. R. Osei-Amponsah)

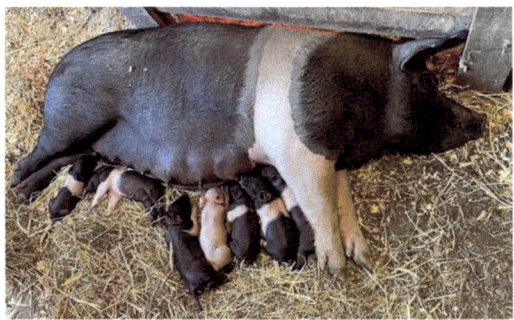

Fig. 8.15 A Hampshire sow with piglets. (From Source: https://agro4africa.com/hampshire-pig-breed/, licensed under CC-BY 4.0)

Fig. 8.16 The Landrace pig breed. (Credit: Prof. R. Osei-Amponsah)

fast-growing and have a high percentage of carcass weight in the ham and loins. The sows are prolific breeders that farrow large pigs and like the Large White and the Duroc, have been used in many semi-intensive and intensive pig production systems in Africa (AUIBAR 2019). While this has helped improve the growth and reproductive performance of crossbred pigs, this poses a threat in terms of dilution of the adaptive traits of indigenous pigs (Osei-Amponsah et al. 2017, 2021). A comparative analysis of growth and reproduction performance between the Landrace, Large White, and local pigs in Uganda (Greve 2015) indicated an average litter size of 12, 11, and 8, respectively. The weaning to service interval (days) was also significantly better for the Landrace (38.3) and the Large White (47.1) compared to the local genotype (Greve 2015). In terms of pre-weaning mortality, however, the local breed was significantly ($p \leq 0.05$) superior, averaging 1.2 compared to 1.8 and 1.7 of the Landrace and the Large White (Greve 2015).

8.5 Sustainable Use and Conservation of Pigs in Africa

8.5.1 Pig Production Systems in Africa

Pigs are raised in a variety of production systems in Africa. The pig production systems are usually classified based on herd size, level of inputs in the form of housing, feeding, veterinary services, and purpose of production (Adjei et al. 2015b;

Montsho and Moreki 2012). The main pig production systems include the extensive/free-range, semi-intensive, and intensive systems (see Chap. 2). In most countries, the extensive traditional pig rearing system is the most predominant, mostly involving local breeds kept in the backyard and fed on kitchen waste and agricultural byproducts.

8.5.1.1 Extensive/Free-Range/Scavenging System

The free-range scavenging system is a low-input–low-output production system. This means that there is little or no investment in housing, nutrition, and healthcare (Fig. 8.17) (Adjei et al. 2015b). The indigenous African breeds dominate in this system since they can survive and reproduce under low-quality feed and limited healthcare conditions. In this system, pigs freely roam around the homesteads, surrounding areas, and in garbage dump sites (FAO et al. 2017), scavenging and hunting for their own feed. For instance, most pig farmers in Cameroon (75%) are engaged in the traditional system of production, where mostly the local breeds of pigs (Bakweri, Bamileke, Mankon Long Nose, hybrids (local and exotic)) are reared with limited input (Ebwanga et al. 2021). The animals are left to fend for themselves and can be seen roaming the villages (Ebwanga et al. 2021; Pondja et al. 2015). Like many African countries, pig production in most areas of Ghana, Nigeria, Uganda, Tanzania, and Mozambique are dominated by traditional production systems and practices, characterized by small herds, low level of biosecurity and productivity, low market off-takes, and poor food safety, which pose substantial limitations on the public health and economic viability of smallholder farmers and public (Adjei et al. 2015b; Pondja et al. 2015; Kimbi et al. 2015). A study to characterize pig production systems in Ghana reported that free range and semi-intensive systems of producing local pigs were common, possibly due to the fact that the majority of the farmers who kept local pigs were involved in crop cultivation or fishing (Adjei et al. 2015b). Predominantly, local pig farmers practiced extensive/scavenging (56%) with very few practicing an intensive system of production (19%) (Adjei et al. 2015b). In most countries, during the crop growing seasons, pigs are tethered to protect the crops. Herd sizes range from two to ten sows. At night, pigs are housed in enclosures surrounded by a fence of sticks or in simple grass-thatched roof houses to provide shelter and protection from cold, rain, and predators (Fig. 8.17). The scavenged feeding materials are usually characterized by high fiber content and low nutritive value. Sometimes, kitchen refuse or crop byproducts such as rice or maize bran, banana or cassava peels, cooked pumpkins, etc., are offered to the pigs as supplements. The inadequate nutrition, coupled with poor healthcare and the low genetic potential of the indigenous breeds, are the reasons for their low productivity. In most cases, the pigs are not kept for meat for the household or as a regular source of cash income, but as an inflation-proof savings account or insurance policy. This entails that pigs are only sold when there is an emergency requiring some cash or when extra cash is needed to meet some household needs, such as the purchase of seeds or fertilizer and food during lean periods, payment of school fees and medical bills, and cash needs during festive seasons or funeral emergencies (Kaumbata 2012). Even though local pigs are reared under low-input systems, they have demonstrated resilience in adapting to low-input systems and have been instrumental in contributing to the improved livelihoods of millions of resource-poor farmers in rural areas. This presents an opportunity to exploit this adaptive potential by implementing genetic and nutrition improvement programs to enhance their productivity.

8.5.1.2 Small-Scale Semi-intensive System

This system is characterized by investment in improved husbandry practices such as construction of standard pig houses, healthcare, improved feeding, and acquisition of improved breeds (Fig. 8.18). Herd sizes are usually more than 5 sows but not up to 25 sows (Sharma et al. 1999), and in most cases, exotic boars are used in crossbreeding to take advantage of hybrid vigor. Pigs

Fig. 8.17 Extensive pig production practices showing poor housing (**a** and **b**), feeding/watering (**c**), and feed resources

may be released occasionally for grazing, but most of the time they are kept in enclosures. As in a study reported from the Democratic Republic of Congo, the pigs are mainly fed locally available feedstuffs such as maize/rice bran, green vegetables, kitchen leftovers, molasses, institutional/industrial byproducts, and often supplemented with commercial feeds (Akilimali et al. 2017). In Rwanda, the majority of pig producers are smallholder farmers who rear pigs in their backyards and house them in semi-permanent structures. Semi-intensive production system is also practiced in the country, where improved breeds are kept, and some commercial pig production management principles are applied. In Cameroon, about 20% of pig farmers practice the semi-intensive system of production, often in small family setups, where pens are constructed behind homes and on rented land (Ebwanga et al. 2021).

8.5.1.3 Medium- and Large-Scale Intensive Production System

These systems aim to produce meat for the market efficiently and profitably, usually with larger numbers of pigs. There is significant investment in the form of housing, feeding, healthcare, and marketing. Herd sizes are relatively large, 25–75 sows (Sharma et al. 1999) or more, and mostly exotic breeds are used. Breeding stock is usually imported or acquired locally from reputable sources such as research, training and religious institutions, and registered breeders. The pigs are kept in concrete-floored houses partitioned to accommodate different pig classes separately. The pigs are fed commercial feeds throughout the production cycle. Biosecurity measures are strictly observed to prevent disease and parasite outbreaks. In most cases, producers are certain about the market or where to sell the pigs. Meat processors may sign contracts with the pig producers to supply pigs of specific attributes at regular intervals. In Liberia, pig farming is a predominantly commercial activity with over 75% of farmers in commercial production systems, with peasant farming comprising about 20% and breeding and multiplication making up only 4% (Karnuah et al. 2018). Pigs in Liberia are mostly kept in housing, with 71% of pig farmers providing permanent structures and 24% providing

Fig. 8.18 A farmer feeding pigs in a small-scale semi-intensive production system. (Credit: Prof. R. Osei-Amponsah)

sheds. Only 5% of pig farmers provided no housing for their stock. The main reason for the high level of housing provision is the fact that most farmers are keeping exotic breeds such as the Landrace, Large White, and Hampshire in intensive systems. As expected, therefore, most pigs in Liberia are continuously confined (82%), with a little over a tenth of pig farmers (11%) never confining their animals (Karnuah et al. 2018).

8.5.2 Pig Improvement Programs

Pig production is a growing animal resource sector with the potential to contribute to protein food security, wealth creation. and sociocultural values while contributing to sustainable use and conservation of the available *Sus* resources. Reports on the existence of organized indigenous pig conservation and improvement breeding programs in African countries are lacking. As earlier described, the indigenous pigs are commonly kept under free-range scavenging production systems. The proliferation of exotic breeds in many African countries led to unplanned crossbreeding, particularly in semi-intensive production systems, to improve the productivity of the indigenous pig breeds. The crossbred pigs usually find their way into the free-range production system, where uncontrolled mating is rampant. This poses a serious threat to the indigenous pig genetic resources, which possess valuable adaptive characteristics suitable for low-input systems. Despite the absence of systematic within-breed genetic improvement programs for indigenous pig breeds and the prevalence of unplanned crossbreeding, there is high potential for success for such programs because superior genes can be spotted in a few animals within the African indigenous pig breeds.

The exotic pig breeds in Africa represent a valuable genetic resource that is significantly contributing to the development of the pig industry in many African countries. For the exotic pig genetic resources to benefit smallholder farmers in rural areas, careful and proper breeding strategies and designs are needed. The within-breed genetic improvement program for local pigs at the community level, coupled with systematic and controlled crossbreeding, can be a powerful tool to improve pig productivity on smallholder farms. The prolificacy and fast-growing nature of pigs and the adaptive characteristics of indigenous breeds are potent factors that have not been exploited yet to the benefit of smallholder farmers. Such programs can represent the best in situ conservation strategies for indigenous genetic resources that are currently contributing to food production (FAO 2007a). In situ conservation allows the animals to keep adapting to changes in their environment while performing other important roles, such as ecosystem services. Sustainable conservation, improvement, and utilization of the African indigenous pig genetic resources require a thorough characterization of the attributes and possible uses for the breeds and the development of niche markets for their products. Utilization of indigenous pig genotypes in communal areas has the potential to increase food security, reduce poverty, and improve the livelihoods of resource-poor farmers (Halimani et al. 2020). The Convention on Biological Diversity (CBD) defines sustainable use as the use of components of biological diversity in a way and at a rate that does not lead to the long-term decline in biological diversity, thereby maintaining its potential to meet the needs and aspirations of present and future generations. In designing community-based breeding programs, breeding goals of smallholders, which are often multifaceted compared to the commercial pig farmers who focus on a few traits of economic importance, such as fast growth rates, larger carcasses, and disease tolerance, should be considered. Goals for smallholders include: color pattern, temperament, mothering ability, foraging behavior, adaptability, and the ability to survive and reproduce in low-input systems (Halimani et al. 2020). There is also a need to incorporate indigenous knowledge in the breeding programs. Several possible schemes have been proposed, including sire rotation or loan schemes, nucleus-based programs run by the public sector and linked to community-level multipliers, and other community-based programs where selection is done at the community level. There is an opportunity to use recent advances in technology, especially assisted reproductive technologies and genomics, to quicken the process without a loss of diversity. Possible interventions suggested to achieve sustainable conservation and utilization of pig genetic resources in Africa (Halimani et al. 2010, 2020) are summarized in Table 8.2.

8.5.3 Conservation of African Pig Genetic Resources

Detailed information on the significance of conservation of Africa's AnGR in the context of the GPA on AnGR, including past and current initiatives as well as emerging technologies and the role of stakeholders, has been presented in Chap. 25. In this section, we throw more light on specific conservation interventions, challenges, and opportunities in the conservation of pigs. In Africa, the numbers, breeds, population genetic structures, attributes, and risk status of local pig genetic resources are understudied. Available research findings indicate that local pigs are tolerant to parasites that are endemic in their production environment. They also have a better chance of surviving various disease outbreaks and have a higher capacity to utilize fibrous and poor-quality feed resources compared to exotic breeds (Halimani et al. 2010). Sustainable use and conservation of pig genetic resources is the surest way to ensure that they remain functional parts of the prevailing systems and play their useful roles of contributing to protein food security, supporting livelihoods, and providing employment to many in Africa. Smallholder farmers in Africa have, for millennia, played key roles in conserving and using native breeds to support their livelihoods. They have maintained the rich biodiversity and the associated local knowledge, in addition to the crucial role in sustaining livelihoods and other environmental benefits. Future conservation interventions should, therefore, aim to support these efforts in order to maintain the existing genetic wealth (AUIBAR 2019).

Two main methods are available for conserving AnGR in Africa, namely in vivo conservation of animals (which includes in situ and ex situ) and in vitro conservation of germplasm material (see Chap. 25). In terms of in situ conservation, smallholder farmers in Africa have, for millennia, played key roles in conserving and using native breeds to support their livelihoods. They have maintained the rich biodiversity and the associated local knowledge in addition to the crucial role in sustaining livelihoods and other environmental benefits (AUIBAR 2019). However, conserving their pig genetic resources solely in situ places them at a huge risk from artificial and natural disasters. African countries should therefore improve their human and institutional capacity to take advantage of ex situ technologies, including ex situ in vivo or cryopreservation. Already, the AU-IBAR and its stakeholders have established regional gene banks in Africa to provide a backup facility where countries can cryo-

Table 8.2 Key outputs, activities, and actors for sustainable utilization and conservation of pig genetic resources in Africa

Variable	Expected outputs	Activities	Responsibility
Food and nutrition	Number of pig farmers increased	Conduct a needs assessment study, followed by various awareness campaigns. Facilitate information dissemination among farmers and between farmers and extension through [a]ICTs	Researchers, extension, government, farmers, [b]NGOs
	Reproductive efficiency enhanced and mortality rates reduced	Improve animal health and housing management. Implement farmer capacity-building programs	Researchers, extension, government, farmers, NGOs
	Environmental and public health improved	Strengthen veterinary services. Establishing biosecurity structures to control zoonotic diseases Improve product quality and quantity, and timing, as well as addressing price and policy issues	Researchers, extension, government, farmers, NGOs
Income generation	Lucrative pig market identified	Conduct qualitative and quantitative value chain analysis. Create niche markets. Establish processing facilities	Researchers, extension, government, farmers, NGO
	Pig producer cooperatives established	Incentive group farming and contract farming. Facilitate credit support for the farmer groups in production. Create small farmer abattoirs	Researchers, extension, government, farmers, NGOs

Source: Adapted from Halimani et al. (2010, 2020)
[a]*NICT* Information Communication Technology, [b]*NGOs* nongovernmental Organizations

preserve their unique porcine germplasm (AUIBAR 2019). As a result of the predominant free-range/extensive production systems in Africa, understanding the effects of admixture on important adaptive and economic traits of local pig breeds will be critical for developing sustainable conservation programs to prevent the decline of these genetic resources (Osei-Amponsah et al. 2017). Data on phenotypic characteristics and production systems of pigs are essential in developing breeding, management, and conservation programs, but these remain unavailable in most countries of Africa (Karnuah et al. 2018).

As signatories to the Global Plan of Action (GPA) on Animal Genetic Resources (FAO 2007b), African countries have committed to developing national conservation policies, establishing or strengthening in situ and ex situ conservation programs, developing and implementing regional and global long-term conservation strategies, and developing approaches and technical standards for conservation. Although the commonest conservation methods to date have been the continuous keeping of animals in their natural environments (in situ conservation), events over the last four decades indicate clearly that Africa needs to embrace and use other forms of conservation of its animals. For instance, there have been many natural and man-made disasters that have wiped out entire herds, leading to the need for total restocking. Additionally, the negative effects of climate change on animal production, disease emergence, epidemics and pandemics, increasing demand for animal protein, and other uncertainties indicate the risk faced by Africa's animal genetic resources. In view of these, other means of conservation, including keeping animals on government farms, research station farms, or national parks (ex situ in vivo conservation), and cryopreservation (ex situ in vitro conservation) should be embraced. The reasons for conserving local pig genetic resources vary from current utility to the ability to meet future challenges in a dynamic environment. Unfortunately, there is a big policy gap in most African countries with regard to the conservation of AnGR in general, and it is imperative that policymakers be made aware of the value of AnGR. The costs of conservation can be met by increasing the market value of neglected breeds so that they eventually become self-sustaining (Halimani et al. 2010).

8.6 The Economics of Pig Production in Africa

In many African countries, pigs and pig products are marketed through formal and informal marketing channels. The indigenous and the crossbred pigs from the free-range/scavenging and semi-intensive systems are usually marketed through informal channels, while the pigs from medium and large-scale intensive pig farms are marketed through formal marketing channels. The free-range/scavenging and semi-intensive systems usually supply the local consumers in peri-urban and rural areas through a simple marketing channel presented in Fig. 8.19. The marketing channels are dominated by male intermediaries who collect and assemble pigs from producers' homesteads and rural markets at low prices. The pigs are then resold to butchers at high market margins. The animals are then slaughtered, and the meat is supplied to meat kiosks, restaurants, bars, and pork joints. The challenges associated with the informal marketing channels and the possible interventions are detailed under Sect. 8.7.3.

The medium- and large-scale intensive pig farms usually supply consumers in big cities and urban areas. The market system is well developed ann organized and requires a regular supply of large numbers of pigs with specific attributes. Producers are certain about market availability, and production is sometimes based on contract farming, which specifies the number of pigs required at specific intervals. Pricing is based on the cost of production and the desired level of profits; hence, producers are the price makers. Figure 8.20 shows the simplified marketing channel for pigs from medium- and large-scale intensive pig farms.

8.7 Challenges of African Pig Production

African pig farmers, especially smallholder farmers in peri-urban and rural areas, face many technical and management problems that limit the development of the pig industry in Africa. For instance, farmers in both subsistence and commercial pig production systems in Southern Africa indicated diseases and parasites, followed by feed shortages, inbreeding, and abortions as major constraints for pig production (Madzimure et al. 2013). Additionally, like the case of Botswana, several challenges face commercial pig operations in Africa, including high feed costs, inadequate slaughtering facilities, unorganized marketing, poor breeding stock, transboundary diseases, and inadequate extension service (Montsho and Moreki 2012; Shyaka et al. 2022). A major constraint in conserving pig genetic resources in Southern Africa is the lack of market participation of the majority of pig farmers, who keep small herds mainly for subsistence. The major barriers to market participation are production constraints, information asymmetry, underdeveloped markets and support infrastructure, limited finance and other resources, and inadequate knowledge (Halimani et al. 2020). These challenges have significant negative impacts, ranging from poor productivity to the entry of unsafe products in the pig value chain, which poses serious health risks to consumers. The challenges can be grouped as follows:

(i) Prevalence of parasites, diseases, and poor disease control measures
(ii) Poor pig nutrition, high feed costs, and feed availability
(iii) Unavailability of systematic breeding programs and indiscriminate crossbreeding
(iv) Unorganized and low-value marketing system
(v) Erratic meat inspection and inadequate hygiene slaughter facilities
(vi) Inadequate extension and veterinary services

Overcoming the challenges requires concerted and coordinated efforts by various stakeholders and players in the pig value chains (Gebrehawariat et al. 2017; Shyaka et al. 2022; Ibeagha-Awemu et al. 2019).

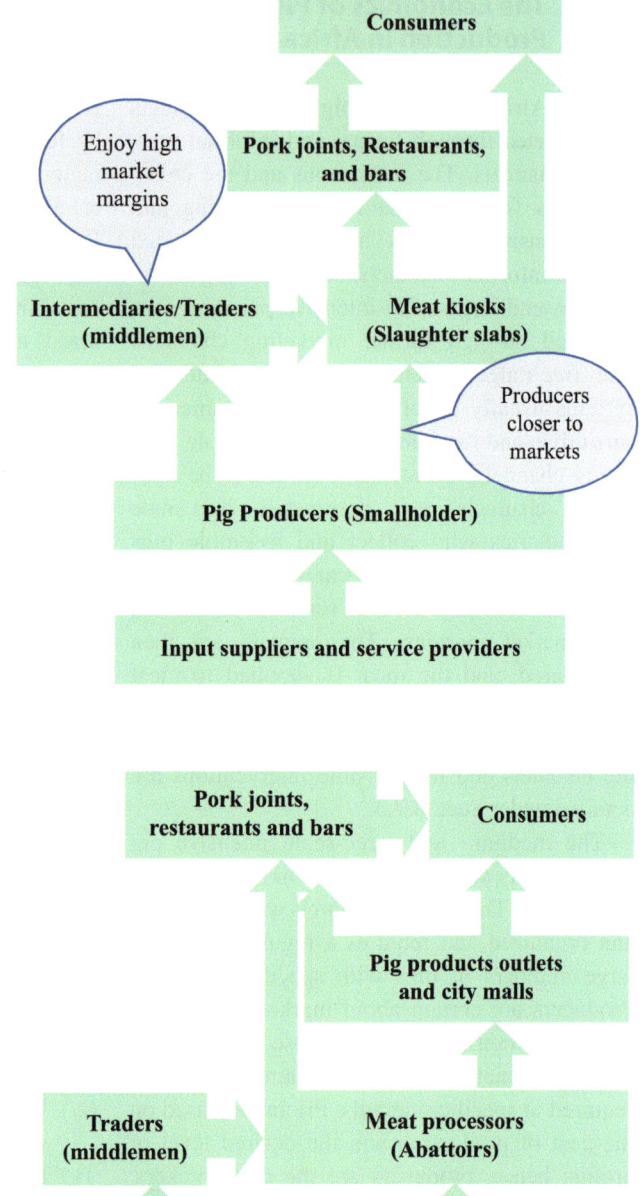

Fig. 8.19 Simplified marketing channel for indigenous pigs from the free-range/scavenging and smallholder semi-intensive systems

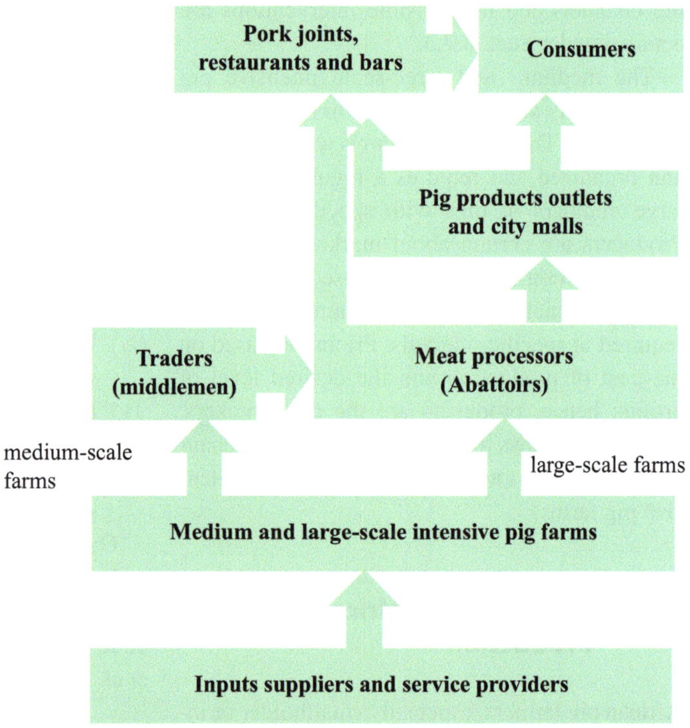

Fig. 8.20 Simplified marketing channel for pigs from medium and large-scale intensive farms

8.7.1 Prevalence of Parasites and Diseases and Poor Disease Control Measures

Disease and parasite infestation is a big challenge for pig production in many African countries. The most prevalent and endemic disease is African swine fever (ASF), a viral disease that spreads rapidly and is associated with high morbidity and mortality (Ebwanga et al. 2021). Other prevalent infectious diseases include swine erysipelas, brucellosis, exudative dermatitis (greasy pig), respiratory diseases, swine dysentery, mastitis, and porcine parvovirus (Weka et al. 2020; Dione et al. 2017). African Swine Fever is a deadly disease affecting the pig industry in Africa. Unlike most livestock diseases, to date, vaccine development to help control ASF remains elusive. Attempts aimed at preventing and controlling ASF have been complex, expensive, and unsuccessful and will require more innovative strategies (FAO et al. 2017) and funding. Additionally, parasitic diseases such as Helminthosis (Strongylid parasites, *Strongyloides ransomi, Ascaris suum, Metastrongylus* sp., *Trichuris suis, Taenia solium*), protozoa (coccidiosis and trypanosomiasis), and ectoparasitism also erodes the economic gains due to reduced weight gain and litter size, poor growth rates, condemnation of carcass at slaughter, and sometimes death (Weka et al. 2020).

Disease control measures in most smallholder pig farms in Africa are generally poor (Dione et al. 2017). Vaccination is a major focus of disease prevention and herd-health management in pig production. Biosecurity measures are absent in most pig farms, and insufficient confinement/quarantine of newly introduced breeding stock, leading to disease introduction and circulation from one farm to another farm (Shyaka et al. 2022). The free-range production system predominant in most peri-urban and rural areas of Africa makes it almost impossible to bring disease outbreaks and parasite infestation under control. Additionally, public and private veterinary services, as well as pig farmers, lack funding and compensation and insurance policies, making investment in commercial pig enterprises riskier and increasingly unattractive. High cost of veterinary drugs and services, and proliferation of fake, expired, and ineffective drugs frustrate disease and parasite control efforts (FAO et al. 2017; Tatwangire 2014).

8.7.2 Poor Pig Nutrition and High Feed Costs

Pig feeding in Africa is usually dictated by the prevalent production system, and challenges related to pig feeding also differ by the production system practiced. The free-range scavenging system is a low-input–low-output production system (Adjei et al. 2015b). This means that there is little or no investment in housing, nutrition, and healthcare. Pigs roam around the homesteads in search of feed. Feed availability and optimal nutritive value of the feedstuffs are seasonal. The scavenging feed resources are abundant during the rainy season and at harvest, and the nutritive value is optimal during the same periods. In the dry season, feed becomes scarce and is usually characterized by high fiber content and low nutritive value. Sometimes the pigs are supplemented with crop byproducts (cooked pumpkins, rice/maize bran, etc.), kitchen waste, and leftover human food. In the smallholder semi-intensive and medium and large-scale intensive pig farms, pigs are fed with commercial feeds. Due to the rising costs of feed ingredients, the commercial feeds are increasingly becoming expensive (Madzimure et al. 2013, 2014). These high prices are due to a lack of raw materials and feed ingredients on local markets, as corn, for example, is used for human food, and expensive imported feed ingredients. Although the establishment of local feed mills has been promoted as a strategy to improve the availability of animal feed, there is a need for research and development of alternative, cheap, and innovative sources of animal protein feeds, such as insects (Shyaka et al. 2022). Some farmers are now practicing what is known as the blended feeding approach, especially in smallholder semi-intensive systems, where, in addition to the commercial feeds, animals are also supplemented with locally available feedstuffs such as vegetables, crop byproducts, kitchen/institutional food remains, and industrial byproducts. This feeding approach has resulted in significant reductions in feed costs and increased profitability of pig enterprises.

8.7.3 Unorganized and Low-Value Marketing System for Smallholder Rural Farmers

Pig market structure in Africa varies according to country, region, and production system. Markets for indigenous pigs from low-input systems are mostly informal and exploitative. In many areas, farmers are not organized into formal organizations, road infrastructures are poor, and access to essential services (such as veterinary, credit/loans, etc.) is a challenge. This puts them into a weak negotiation position, increases the transaction costs, and reduces their competitiveness when dealing with supply chains that are becoming formalized and upgraded.

The informal market structure is usually dominated by pig brokers and traders (intermediaries/middlemen) who assemble live animals from farmers' homesteads and local markets. At these selling points, the pigs are sold based on physical appraisals of size, body conformation, and health status, and not on a live weight basis. The prices are usually arrived at based on how skillful the buyer is in price negotiation and how desperate the farmer is for money, to meet the need at hand. The brokers are highly knowledgeable about pig price trends and act as key persons connecting rural producers to different slaughter slabs/houses and other higher-level traders (Shyaka et al. 2022). In Rwanda, for instance, live pig markets constitute a major avenue for stakeholders to buy and sell pigs. Middlemen or brokers play a key role in this pig trade: setting animal prices, examining pigs to ensure that they are free of disease, helping supply pigs to various abattoirs and for export (Shyaka et al. 2022). For instance, lingual palpation helps identify cysticercosis infection, with affected pigs sold at reduced prices. Some of the risks identified, among others, include a lack of biosecurity and unsanitary conditions, particularly at abattoirs (Shyaka et al. 2022).

A major constraint in conserving pig genetic resources in southern Africa has been reported as the lack of market participation of most pig farmers who keep small herds mainly for subsistence, due mainly to production constraints, information irregularity, underdeveloped markets and support infrastructure, limited finance, and inadequate knowledge (Halimani et al. 2020). Pig farmers, especially in rural Africa, are poorly organized and therefore unable to utilize the advantages of collective marketing and high bargaining power. This limits farmers' efforts to upgrade into various pig-related market exchanges at different nodes of the value chain, creates inefficiencies, and opens doors for exploitation and poor-quality products (Tatwangire 2014). Absence of market information puts the producers at a disadvantaged position, which turns them to be price takers instead of price makers. Formal pig markets exist but require a regular supply of a large number of pigs with specific attributes. These are usually accessed by farmers in intensive and semi-intensive systems because the herd sizes in these systems are relatively large and uniform, hence can meet a regular supply of pigs. Therefore, there is a need to empower and build the capacity of smallholder rural pig producers so that they become active participants in the pig value chain. This can be done through capacity building and the establishment and operationalization of community-based pig producer cooperatives. Functional pig producer cooperatives are instrumental in consolidating smallholders' bargaining power, increasing access and management of lucrative markets, enjoying the benefits of economies of scale, enhancing access to credits/loans and livestock extension services, and having a voice on issues that affect their well-being.

8.7.4 Unavailability of Systematic Breeding Programs and Indiscriminate Crossbreeding

Past and ongoing pig breeding programs in Africa, the breeding companies involved, as well as molecular and assisted reproductive innovations, are detailed out later in this book (refer to Sect. 19.6 of Chap. 19). Unlike the imported exotic breeds, most indigenous African pig breeds have no distinct and fixed characteristics that can qualify them to be classified into distinct breeds. The indigenous pig populations are dominated by crossbreds with no pedigree information for traceability. The free-range production system predominant in smallholder pig farms and the absence of record keeping make it difficult to trace the parentage and genetic composition of offspring. Small-scale intensive and large-scale commercial farms import breeding stock or obtain them from local breeders and institutions. Smallholder rural producers acquire breeding stock from neighbors/friends and sometimes from research, training, and religious institutions. Selection of breeding stock is often based on reproductive and growth performance, such as large litter size, conception rate, weaning to conception interval, mothering ability and growth rate, feed conversion efficiency, weaning, and slaughter weights, respectively. Inadequate and unreliable availability of breeding stock and absence of planned breeding schemes for smallholder pig farmers have resulted in high levels of inbreeding, thus leading to small litter sizes, poor growth rates, and small animals (Tatwangire 2014; Shyaka et al. 2022).

The uncontrolled exchanges and hazardous crossing between pig breeds constitute some risks of either the extinction of the indigenous breeds, or the problem of consanguinity, or even the threat of erosion of genetic diversity (Kiendrebeogo et al. 2012). There is a risk of losing the indigenous pig biodiversity because of the race to breed for high production capacity of pigs via crossing with imported exotic breeds. In some countries, indigenous breeds have been replaced with exotic breeds such as Large White, Landrace, Hampshire, and Duroc to help meet the demand of the market system (Weka et al. 2020). Indiscriminate crossbreeding of indigenous pigs with imported pigs should be discouraged because it dilutes the ability of indigenous pigs to resist disease challenges, hence threatening the genetic resource. In addition, indigenous pigs survive well under resource-limited, low-input systems. The conservation of indigenous pigs that will be bred and sold to rural farmers for sustainable rural development is recommended (Madzimure et al. 2013). Indiscriminate crossbreeding, replacement of indigenous pigs with imported pigs, and lack of clearly defined policies on conservation of indigenous genotypes threaten their continued existence (Halimani et al. 2020) and chance to contribute to the development of future breeds. Furthermore, the lack of the requisite human and institutional capacity for sustainable use, improvement, and conservation of pig genetic resources in Africa remains a major challenge and needs appropriate stakeholder intervention and support, as well articulated later on in Chaps. 27 and 28 of this book.

8.7.5 Erratic Meat Inspection and Inadequate Standard Slaughter Facilities

Livestock products, including pork and other pig products, pose a potential public health risk if poorly handled and processed. The supply of safe pork and other pig products relies on sound parasite and disease control programs and strictly regulated pig slaughter and carcass handling and processing systems. Reports show that there is poor and inadequate infrastructure development in many African countries to guarantee safe and hygienic handling of pork and other pig products (Shyaka et al. 2022). While formal pig abattoirs exist in many urban and peri-urban areas for the slaughter and processing of animals from medium and large commercial farms, the majority of pigs from the informal markets are slaughtered at informal slaughter facilities. Reports (FAO et al. 2017; Tatwangire 2014; Shyaka et al. 2022) on the prevalence and existence of informal, unregulated, and unhygienic pig slaughter slabs are not uncommon on the continent. In many areas, sufficient formal and regulated slaughterhouses are not available to handle the supply of pigs from smallholder rural farms.

Veterinary and public health officials in many African countries are entrusted and mandated with the responsibility of enforcing sanitary standards and regulations that improve the quality and safety of pork and other meat products. However, in many instances, meat inspection is rarely carried out in farms, and individuals practice backyard slaughter on market days and during festive seasons. Unfortunately, even at certified slaughter areas where routine veterinary inspections are carried out, poor and unhygienic conditions are prevalent. In terms of traceability, no records are kept on the source of animals, the number of animals slaughtered at a particular period, carcasses approved for consumption, organs, or carcasses condemned, as well as the final destination of slaughtered animals (Shyaka et al. 2022). Sometimes, there is a lack of clean water, dysfunctional refrigeration facilities, and unhygienic trucks transporting carcasses, leading to increased potential for contamination. Although abattoirs could be good locations for passive disease surveillance, the absence of records defeats this important function. The level of awareness among pig producers, processors, veterinarians, and public health officials on issues of pork safety is still low and a big concern for pork consumers. There is poor enforcement of regulations and safety standards even when relevant policies are clear; there is a lack of inspectors to conduct ante-mortem, visual inspection of meat organs and lymph nodes, and further testing of pork in the laboratories.

8.7.6 Poor Access to Livestock Extension, Veterinary, and Other Services

The problem of inadequate and often poor human and institutional capacity for sustainable use, production, breed improvement, and conservation of Africa's AnGR and the urgent need for stakeholder intervention have been highlighted in many sections of this book. Chapters 27 and 28 call for improvement in human and institutional capacity, including the provision and maintenance of appropriate infrastructure (see Sects. 27.7 and 28.2.4). For instance, livestock extension and veterinary services are instrumental for enhancing pig husbandry practices and market access. Producers need relevant and timely information on feeding, breeding, healthcare management, and market availability, as well as access to inputs. Unfortunately, in most regions of Africa, personnel to provide such services to pig farmers are usually inadequate or are not available in many areas. In many countries, qualified and experienced veterinarians are being replaced by less qualified and inexperienced para-veterinarians and community animal health workers. In other instances, public veterinarians have limited their services only to cattle production, leaving the other livestock production systems to para-veterinarians who, in most cases, are not effectively supervised (Shyaka et al. 2022). Other reports indicate the prevalence of high extension officer–farmer ratios, such as one officer to 2300 farmers, as opposed to the recommended ratio of 1 officer to 400 farmers (Benor et al. 1984). Hence, appropriate interventions are needed by governments and other relevant stakeholders to create enabling environments for the pig industry to grow and thrive. There is a need to enhance livestock extension and veterinary service delivery by making sure that the agriculture and public health sectors are adequately staffed with well-trained and skilled personnel to provide appropriate veterinary services in support of animal production. In addition to the provision of animal health services, extension officers are also a useful link between farmers and researchers and play critical roles in farmer capacity building through technical and leadership trainings and facilitation of farmer-to-farmer learning through exchange visits, on-farm demonstrations, field days, and agricultural/animal shows (Kaumbata et al. 2021).

Providing adequate access to formal financial services for smallholder pig farmers in most developing countries is a challenge because commercial banks do not consider them as viable clients, or the service centers are located very far away from their communities. In certain livestock improvement initiatives, the village saving and credit scheme (VSCS) has been included to help address this challenge (Kaumbata 2012). Such schemes will help pig farmers access credit at affordable interest rates. Others, including veterinarians, community animal health workers, and pig assemblers and traders, can also benefit from such initiatives.

8.8 Opportunities for Sustainable Pig Production in Africa

Pig production offers wealth creation opportunities to actors on the pig value chain. Smallholder farmers benefit from quick returns for manure used in crop production. Several opportunities exist that, if well exploited and utilized, can enhance the growth of the pig industry in Africa. In the first place, demand for animal products, including pig products, in developing countries will continue to rise during the next three decades due to human population growth, urbanization, per capita income growth, and changing eating habits (Delgado 2005). The rising demand for animal products creates ready markets for pig products and direct income growth opportunities for pig producers, including smallholder rural farmers. In terms of consumer health concerns, meat quality and sensory properties have become important aspects of meat production and consumption in Africa (Hoffman et al. 2005). There is an urgent need for the pig industry to respond appropriately by improving acceptance of pork, particularly with reference to consumer demand for leaner pork.

Secondly, African indigenous pigs demonstrate resilience in surviving and reproducing under poor nutrition and healthcare management conditions. Indigenous pigs have, for millennia, proved to be instrumental in enhancing the livelihoods of many rural farmers because of their ability to survive and reproduce under conditions of poor nutrition, housing, disease, and parasite control. Despite demonstrating these important adaptive traits, the indigenous pigs have been ignored because they are apparently regarded as less productive without subjecting them to appropriate improvement programs. These adaptive traits need a thorough study on how they can be improved and used for the benefit of the pig industry (Ibeagha-Awemu et al. 2019). At the same time, uncontrolled exchanges and hazardous crossing between breeds constitute some risks of either the extinction of the indigenous breeds, or the problem of consanguinity, or even the threat of erosion of genetic diversity. Genetic characterization of indigenous African pig breeds, for instance, should help develop more sustainable breeding programs, particularly in low-input breeding systems (Kiendrebeogo et al. 2012).

The community-based within-breed genetic improvement program for indigenous pigs, coupled with systematic and controlled crossbreeding, can be a powerful tool to exploit heterosis for adaptive, reproductive, and productive traits for the improvement of pig productivity in smallholder farms. Indigenous and crossbred pigs are known for their ability to convert low-quality feedstuffs into valuable outputs. This presents an opportunity to use abundant green forages, crop residues, kitchen and institutional remains of human food, and industrial byproducts for feeding the pigs for the production of valuable pig products. Finally, the availability of imported exotic pig genetic resources in many African countries entails that producers have breeds suitable for existing production systems. Hence, producers can use indigenous, exotic breeds, or their crossbreds depending on consumer preferences and available niche markets.

8.9 Implications and the Way Forward

The rich diversity of pig genetic resources in Africa, their unique adaptive traits, and socioeconomic, cultural, and economic values, as well as their roles in ensuring protein food security, cannot be overemphasized. Africa, however, needs to ensure the sustainable use and conservation of these valuable animal genetic resources by promoting the many opportunities they offer while addressing various challenges associated with various production systems (Ibeagha-Awemu et al. 2019). There is high potential for using indigenous pigs in subsistence-oriented production systems and for crossbreeding of indigenous pigs with imported breeds in market-oriented systems (Madzimure et al. 2013).

Most importantly, all stakeholders along the pig value chain need to recognize their unique roles in achieving this noble objective and ensuring that pigs remain important components of the animal agriculture sector in Africa. Sustainable development of the pig sector in Africa hinges on active participation of all stakeholders working together to address constraints to pig production and marketing (Ibeagha-Awemu et al. 2019). Constraints in pig production call for integrated, sustainable, practical policies on health, genetics, feeds, and management practices, as well as the organization of producers. In order to meet the demand for proteins by the growing human population in Africa, pig production presents immense opportunities for stakeholders (FAO et al. 2017).

For a sustainable utilization of pig genetic resources, therefore, we need to:
(i) Improve on the capacity of pig farmers and producer organizations to be able to adopt emerging technologies in pig production and health.
(ii) Promote technical and managerial skills among all stakeholders.
(iii) Establish pig breeding centers for breeding pigs and artificial insemination, and train farmers, technicians, and extension agents in AI and on the importance of record keeping, and provide the infrastructure for implementation.
(iv) Improve the capacity of farmers for biosecurity and disease control.
(v) Encourage feeding formulation based using locally available feed resources and subsidized imported feed ingredients.
(vi) Utilize animal biotechnology in the characterization and genetic improvement of local pigs.
(vii) Add value to local pig genetic resources through niche markets to ensure their sustainable use and conservation.

Investments in research, development, and promotion of alternate feed sources, superior genetics, and veterinary services are also needed. Regular training and supply of information on biosecurity, as well as provision of needed resources, will help in disease prevention and control programs, including epizootic and zoonotic diseases. Investments are also needed to support value chain actors through the establishment of community-based producer cooperatives, strengthening market information, and construction of more standard slaughterhouses designed to ensure food safety and public health (Shyaka et al. 2022). All these actions require commitment from national governments in terms of appropriate policies and funding to assist pig farmers and other stakeholders to improve production. Additionally, effective regular education and cheap technologies that help improve biosecurity, especially for the small-scale free-ranging pigs, should be encouraged. There should be more incentives for investment in pig farming to help improve the livelihoods of smallholder farmers and consequently improve food security (FAO et al. 2017). Recognizing the challenges and the opportunities that exist in the pig sector, the AU-IBAR, the FAO, and the International Livestock Research Institute (ILRI) have been collaborating since March 2013 to develop an Africa-wide strategy for the prevention and control of African Swine Fever (ASF). To this effect, a Taskforce drawn from the three organizations was set up and has worked extensively on the strategy (FAO et al. 2017).

Sustainable management and conservation of Africa's pig genetic resources will no doubt contribute to achieving the 2030 Sustainable Development Goals (SDGs), particularly Goal 2, which aims to end hunger and ensure access by all people to safe, nutritious, and sufficient food all year round; and Goal 8, which promotes sustained, inclusive, and sustainable economic growth, full and productive employment, and decent work for all. In this chapter, we have highlighted the rich pig genetic resources available in Africa, including the role of stakeholders in the value chain, and made a case for their sustainable use and conservation. Most of the challenges confronting pig production in Africa also provide many opportunities for bringing on board the private sector and modern technologies for improved production and greater efficiency.

There is a need for African national governments to collaborate with the private sector to encourage investments in the pig industry while empowering farmers through their producer cooperatives to access credit and providing them with regular training on various aspects of pig production, including the use of appropriate technology. Governments need to invest in the characterization, inventory, and monitoring of local pig breeds while improving on human and institutional capacity to ensure their sustainable use and conservation. Moreover, stakeholders should invest in research to develop a vaccine for ASF, which has the potential to totally wipe out all pigs on the African continent. Until this objective is achieved, strict adherence to recommended biosecurity protocols, intensive and more controlled pig production, as well as vibrant national agricultural livestock extension services, should be encouraged. Finally, in situ and ex situ in vivo conservation of Africa's pig genetic resources should be considered a priority. An important route to ensure sustainable use and conservation of Africa's pig genetic resources is stakeholder awareness creation on the value of pigs and capacity building of players along the value chain to enable them to take advantage of existing and emerging technologies in their operations.

8.10 Conclusions

In this chapter, we have provided information on the rich pig genetic resources of Africa to motivate stakeholders to take appropriate actions for their sustainable use and conservation.

African indigenous pig breeds and their crosses that are well adapted in Africa are characterized by small body size (adult around 60 kg), relatively poor growth rate and performance, sturdy, small litter size, difficult husbandry conditions, and can survive on low-quality feed, including crop residues, household wastes, and grass. Clearly, a lot needs to be done in terms of phenotypic and molecular characterization of indigenous pig breeds of Africa to provide valuable information on their adaptive, production, carcass quality, and disease resistance profiles and make a case for their conservation. To ensure optimum production of indigenous pig breeds, it is important that stakeholders improve their feed resources using locally available products with higher protein and energy contents to allow for faster growth and improved carcass quality.

Due to an expected ever-growing demand for animal products in Africa, fueled by human population growth, urbanization, increased per capita incomes, and changing eating habits, many African countries are likely to see more large-scale investment in pig production. The number of smallholder pig producers in rural and peri-urban areas will also continue to increase. Growth in domestic and regional demand for pork is likely to remain higher than growth in pig production. Therefore, the business-as-usual scenario of pig production will not satisfy the local demand for pork and other pig products in the coming years, unless measures are taken to promote improvements in the pig value chains, particularly for the smallholder pig producers.

Pig production will continue to contribute to poverty reduction in various ways, including an increase in food supply, a source of income, and a means for capital accumulation, employment opportunities, and supply inputs and services for crop production. Pigs are also among the few animals able to convert the otherwise waste byproducts of agroindustry into quality animal protein for human consumers, and thus, pig production provides meat and will therefore remain an increasing income earner for smallholder livestock farmers in Africa. As a species, the pig's relatively short production cycle, high prolificacy, ability to survive on a wide range of feedstuffs, and high carcass yield will continue to make it valuable to African animal agriculture.

Challenges faced by smallholder pig producers have been highlighted in this chapter for immediate attention by stakeholders. At the same time, opportunities exist that, if well exploited and utilized, can enhance the rapid growth of the pig industry and overall development of animal agriculture in Africa. Finally, the need for concerted and coordinated efforts by all players in the pig value chain to capitalize on the existing opportunities to elicit growth, development, and improvement of the valuable pig genetic resources of Africa for food security and improved livelihoods cannot be overemphasized.

References

Abigaba R, Sianangama PC, Nyanga PH, Mwenya WN, Mwaanga ES (2022) Traditional farmers' pig trait preferences and awareness levels toward reproductive biotechnology application in Zambia. J Adv Vet Anim Res 9(2):255

Abonyi F, Arinzechukwu U, Eze D, Eze J, Machebe N (2018) Comparative evaluation of growth performance, serum biochemical profile and immunological response of the Nigerian indigenous and large white x landrace crossbred pigs. Niger Vet J 39(1):81–91

Adjei OD, Osei-Amponsah R, Ahunu BK (2015a) Morphological characterization of local pigs in Ghana. Bull Anim Hlth Prod Afr (AnGR Special Edition):299–304

Adjei OD, Osei-Amponsah R, Ahunu BK (2015b) Characterization of pig production systems in Ghana. Bull Anim Hlth Prod Afr (AnGR Special Edition):337–342

Agbokounou AM, Ahounou GS, Youssao IAK, Mensah GA, Koutinhouin B, Hornick J-L (2016) Caractéristiques de l'élevage du porc local d'Afrique. J Anim Plant Sci 30(1):4701–4713

Akilimali J, Wasso D, Baenyi P, Bajope J (2017) Characterization of smallholder swine production systems in three agro-ecological zones of South Kivu (Democratic Republic of Congo). J Appl Biosci 120:12086–12097

Amills M, Clop A, Ramírez O, Pérez-Enciso M (2010) Origin and genetic diversity of pig breeds. In: Encyclopedia of life sciences (ELS) evolution and diversity of life. Wiley

Amills M, Ramírez O, Galman-Omitogun O, Clop A (2013) Domestic pigs in Africa. Afr Archaeol Rev 30(1):73–82. https://doi.org/10.1007/s10437-012-9111-2

Aryee S (2023) Genome analysis of the Ashanti Dwarf pig of Ghana. PhD thesis, University of Cambridge, Cambridge

Aryee S, Osei-Amponsah R, Owusu Adjei D, Ahunu BK, Skinner B, Sargent C, Affara N (2019) Production practices of local pig farmers in Ghana. Int J Livestock Prod 10:175

Audrey TS, Manjeli Y, Meutchieye F (2021) The Bankim pigs: a native Cameroonian breed assessed by biometric features. Genet Biodivers J 5(1):101–111

AUIBAR (2019) The state of farm animal genetic resources in Africa. African union Interafrican Bureau for Animal Resources (AUIBAR), Nairobi

Babigumira BM, Sölkner J, Mészáros G, Pfeiffer C, Lewis CR, Ouma E, Wurzinger M, Marshall K (2021) A mix of old British and modern European breeds: genomic prediction of breed composition of smallholder pigs in Uganda. Front Genet 12:676047

Barnes AR, Fleischer JE (1998) Growth and carcass characteristics of the indigenous Ashanti Dwarf Pig. Ghana J Agric Sci 31:217–221

Benor D, Baxter MW, Harrison JQ (1984) Agricultural extension: the training and visit system. The World Bank, Washington, DC

Bester J, Küsel U (1998) Early domesticated animals in South Africa. In: Proceedings of the 4th global conference on conservation of domestic animal genetic resources, Nepal

Chen K, Baxter T, William M, Muir WM, Groenen MA, Schook LB (2007) Genetic resources, genome mapping and evolutionary genomics of the pig (Sus scrofa). Int J Biol Sci 3(3):153–165

Chilundo AG, Mukaratirwa S, Pondja A, Afonso S, Miambo R, Johansen MV (2017) Prevalence and risk factors of endo-and ectoparasitic infections in smallholder pigs in Angónia district, Mozambique. Vet Parasitol Region Stud Rep 7:1–8

Chimonyo M, Dzama K (2007) Estimation of genetic parameters for growth performance and carcass traits in Mukota pigs. Animal 1(3):317–323

Cumbula D, Taela M (2020) Animal genetic resources (AnGR) in Mozambique. In: IOP conference series: earth and environmental science, vol 482. IOP Publishing, p 012045

Delgado C (2005) Rising demand for meat and milk in developing countries: implications for grasslands-based livestock production. In: Grassland: a global resource, p 2939

Dione MM, Akol J, Roesel K, Kungu J, Ouma EA, Wieland B, Pezo D (2017) Risk factors for African swine fever in smallholder pig production systems in Uganda. Transbound Emerg Dis 64(3):872–882. https://doi.org/10.1111/tbed.12452. From NLM Medline

Dotché IO, Thilmant P, Gabriel A, Bonou GA, Dahouda M, Dehoux J-P, Mensah GA, Farougou S, Youssao IAK, Benoît G et al (2019) Reproductive performances of local pigs in west African countries: a review. J Adv Vet Res 10(1):49–55

Dube B, Mulugeta SD, Dzama K (2012) Estimation of genetic and phenotypic parameters for sow productivity traits in South African Large White pigs. S Afr J Anim Sci 42(4):389–397

Ebwanga EJ, Ghogomu SM, Paeshuyse J (2021) African swine fever in Cameroon: a review. Pathogens 10(4):421

Epstein H (1971) The origin of the domestic animals of Africa

FAO (2007a) The state of the world's animal genetic resources for food and agriculture. Commission on Genetic Resources for Food and Agriculture Food and Agriculture Organization of the United Nations, Rome

FAO (2007b) Global plan of action on animal genetic resources and the Interlaken declaration commission on genetic resources for food and agriculture. Food and Agriculture Organization of the United Nations, FAO, Rome

FAO (2015) The second report on the state of the world's animal genetic resources for food and agriculture. Food and Agriculture Organisation of the United Nations (FAO), Rome. http://www.fao.org/3/a-i4787e/index.html

FAO, AU-IBAR, ILRI (2017) Regional strategy for the control of Africa Swine Fever in Africa. The Food and Agriculture Organization of the United Nations (FAO) and African Union-Interafrican Bureau for Animal Resources (AUIBAR) and International Livestock Research Institute (ILRI)

FAOSTAT (2022) In: FAOSTAT, (FAO) (ed) Food and Agriculture Organization of the United Nations FAOSTAT database. FAO, Rome. Available at: www.fao.org/faostat/. Accessed 11 Aug 2023

Food and Agriculture Organisation of the United Nations (FAO) (2020) Domestic Animal Diversity Information System (DAD-IS). www.fao.org/dad-is/en/. Accessed 15 Aug 2023

Galal S (2007) Farm animal genetic resources in Egypt: factsheet. Egypt J Anim Prod 44(1):1–23

Gebrehawariat E, Animut G, Urge M, Mekasha Y (2017) Husbandry practices, farmers' perception and constraints of pig farming in Bishoftu and Holeta areas, Central Ethiopia. East Afr J Vet Anim Sci 1(1):31–34

Ghomsi OSM, Wirnkar CN, Etchu KA, Fontanesi L, Bilong CFB, Moundipa PF (2022) Morphological and serum biochemical characterizations of local pig populations from three different agro-ecological areas of Cameroon. Int J Livestock Prod 13(2):33–42

Giuffra E, Kijas JMH, Amarger V, Carlborg O, Jeon J-T, Andersson L (2000) The origin of the domestic pig: independent domestication and subsequent introgression. Genetics 154:1785–1791

Goraga Z, Adamu M, Ali S, Guteta A, Mengesha M, Lima GJ (2017) Swine production, productivity and breeding practices in Ethiopia. In: Cupido M (ed) An insight into the livelihood of small-scale pig farmers in the Western Cape, South Africa. Stellenbosch University, Stellenbosch

Greve D (2015) Analysis of performance, management practices and challenges to intensive pig farming in peri-urban Kampala, Uganda. Int J Livestock Prod 6(1):1–7

Halimani T, Muchadeyi F, Chimonyo M, Dzama K (2010) Pig genetic resource conservation: the southern African perspective. Ecol Econ 69(5):944–951

Halimani TE, Mapiye O, Marandure T, Januarie D, Imbayarwo-Chikosi VE, Dzama K (2020) Domestic free-range pig genetic resources in southern africa: Progress and prospects. Diversity 12(2):68

Hirwa CDA, Mutabazi J, Nsabimana JDD, Dusengemungu L, Kayitesi A, Semahoro F, Uwimana G, Nyabinwa P, Kugonza DR (2022) Challenges and opportunities of smallholder pig production systems in Rwanda. Trop Anim Health Prod 54(5):305

Hoffman L, Styger E, Muller M, Brand T (2005) Sensory, physical and chemical quality characteristics of bacon derived from south African indigenous and commercial pig breeds. S Afr J Anim Sci 6:36–48

Ibeagha-Awemu EM, Peters SO, Bemji MN, Adeleke MA, Do DN (2019) Leveraging available resources and stakeholder involvement for improved productivity of African livestock in the era of genomic breeding. Front Genet 10:357

Kanengoni A, Chimonyo M, Ndimba B, Dzama K (2015) Feed preference, nutrient digestibility and colon volatile fatty acid production in growing South African Windsnyer-type indigenous pigs and Large White× Landrace crosses fed diets containing ensiled maize cobs. Livest Sci 171:28–35

Karnuah AB, Osei-Amponsah R, Dunga G, Wennah A, Wiles WT, Boettcher P (2018) Phenotypic characterization of pigs and their production system in Liberia. Int J Livestock Prod 9(7):175–183

Kaumbata W (2012) Evaluation of genetic and non-genetic performance of pigs and profitability of piggery stud breeding among smallholder farmers in Manjawira Extension Planning Area in Ntcheu District. Masters thesis, University of Malawi, Lilongwe, Malawi

Kaumbata W, Nakimbugwe H, Nandolo W, Banda LJ, Mészáros G, Gondwe T, Woodward-Greene MJ, Rosen BD, Van Tassell CP, Sölkner J et al (2021) Experiences from the implementation of community-based goat breeding programs in Malawi and Uganda: a potential approach for conservation and improvement of indigenous small ruminants in smallholder farms. Sustain For 13(3). https://doi.org/10.3390/su13031494

Kiendrebeogo T, Logtene YM, Kondombo SR, Kabore-Zoungrana CY (2012) Characterization and importance of pig breeds in the pork industry of the zone of Bobo-Dioulasso (Burkina Faso, West Africa). Int J Biol Chem Sci 6(4). https://doi.org/10.4314/ijbcs.v6i4.13

Kimbi E, Lekule F, Mlangwa J, Mejer H, Thamsborg S (2015) Smallholder pigs production systems in Tanzania. J Agric Sci Technol A 5:47–60

Kruger DA (2015) Genetic analyses of progeny performance from local and imported boar semen used in the south African pig industry. University of Pretoria

Larson G, Dobney K, Albarella U, Fang M, Matisoo-Smith E, Robins J, Lowden S, Finlayson H, Brand T, Willerslev E (2005) Worldwide phylogeography of wild boar reveals multiple centers of pig domestication. Science 307(5715):1618–1621

Lekule FP, Kyvsgaard NC (2003) Improving pig husbandry in tropical resource-poor communities and its potential to reduce risk of porcine cysticercosis. Acta Trop 87(1):111–117. https://doi.org/10.1016/s0001-706x(03)00026-3. From NLM Medline

Lichaba M (2022) The contribution of small-scale commercial piggery farming to farmers' livelihoods in Teyateyaneng Urban Council. National University of Lesotho, Berea

Madzimure J, Chimonyo M, Zander KK, Dzama K (2013) Potential for using indigenous pigs in subsistence-oriented and market-oriented small-scale farming systems of southern Africa. Trop Anim Health Prod

45(1):135–142. https://doi.org/10.1007/s11250-012-0184-3. From NLM Medline

Madzimure J, Bovula N, Ngorora GP, Tada O, Kagande SM, Bakare AG, Chimonyo M (2014) Market opportunities and constraints confronting resource-poor pig farmers in South Africa's Eastern Cape Province. J Ind Distrib Bus 5(2):29–35. https://doi.org/10.13106/jidb.2014.vol5.no2.29

Marshall, K. Local/Ugandese breed of pig: pig breed factsheet for Uganda. 2022

Mason IL, Maule JP (1960) The indigenous livestock of eastern and southern Africa

Mbaga S, Moses Lyimo C, Kifaro G, Kimbi E (2003) Indigenous pigs of the Southern highlands of Tanzania

Mbaga SH, Lymo CM, Kifaro GC, Lekule FP (2005) Phenotypic characterization and production performance of local pigs under village settings in the Southern Highland zone. Tanzania Anim Genet Resour Inf 37:83–90. https://doi.org/10.1017/S1014233900001991. From Cambridge University Press Cambridge Core

Mbuthia JM, Rewe TO, Kahi AK (2015) Analysis of pig breeding management and trait preferences in smallholder production systems in Kenya. Anim Genet Resour/Resources génétiques animales/Recursos genéticos animales 56:111–117

Montgomery RA, Stern JM, Lonze BE, Tatapudi VS, Mangiola M, Wu M, Weldon E, Lawson N, Deterville C, Dieter RA (2022) Results of two cases of pig-to-human kidney xenotransplantation. N Engl J Med 386(20):1889–1898

Montsho T, Moreki JC (2012) Challenges in commercial pig production in Botswana. J Agric Technol 8(4):1161–1170

Motsa'a JS, Defang HF, Hako TBA, Ojong ET, Mube KH, Nguekem CL, Tagning ZPD, Mouchili M, Keambou TC (2021) Socioeconomic and productive characteristics of indigenous pig farming in Cameroon. Appl Anim Husb Rural Dev 14:32

Mujibi FD, Okoth E, Cheruiyot EK, Onzere C, Bishop RP, Fevre EM, Thomas L, Masembe C, Plastow G, Rothschild M (2018) Genetic diversity, breed composition and admixture of Kenyan domestic pigs. PLoS One 13(1):e0190080. https://doi.org/10.1371/journal.pone.0190080. From NLM Medline

Ognika AJ, Akouango P, Gnanga ARE, Ngatse SD (2021) Comportement reproductif des porcs de race Duroc en République du Congo. Int J Biol Chem Sci 15(5):1853–1862

Osei-Amponsah R, Skinner BM, Adjei DO, Bauer J, Larson G, Affara NA, Sargent CA (2017) Origin and phylogenetic status of the local Ashanti dwarf pig (ADP) of Ghana based on genetic analysis. BMC Genomics 18(1):193. https://doi.org/10.1186/s12864-017-3536-6. From NLM Medline

Osei-Amponsah R, Kanlisi R, Adjei O, Naazie A (2021) A review on the impact of gene flow on diversity of animal genetic resources in West Africa. Ghana J Anim Sci 12(1):26–36

Parsons M (2017) Pigs in ancient Egypt

Pondja A, Neves L, Mlangwa J, Afonso S, Fafetine J, Willingham AL III, Thamsborg SM, Johansen MV (2015) Incidence of porcine cysticercosis in Angónia District, Mozambique. Prev Vet Med 118(4):493–497

Porter V, Alderson L, Hall SJ, Sponenberg DP (2016) Mason's world encyclopedia of livestock breeds and breeding, 2 volume pack. Cabi

Ramirez O, Ojeda A, Tomas A, Gallardo D, Huang LS, Folch JM, Clop A, Sanchez A, Badaoui B, Hanotte O et al (2009) Integrating Y-chromosome, mitochondrial, and autosomal data to analyze the origin of pig breeds. Mol Biol Evol 26(9):2061–2072. https://doi.org/10.1093/molbev/msp118. From NLM Medline

Shapiro BI, Gebru G, Desta S, Nigussie K (2017) Rwanda livestock master plan

Sharma KR, Leung P, Zaleski HM (1999) Technical, allocative and economic efficiencies in swine production in Hawaii: a comparison of parametric and non-parametric approaches. Agric Econ 20(1):23–35

Shyaka A, Quinnell RJ, Rujeni N, Fevre EM (2022) Using a value chain approach to map the pig production system in Rwanda, its governance, and sanitary risks. Front Vet Sci 8:720553. https://doi.org/10.3389/fvets.2021.720553. From NLM PubMed-not-MEDLINE

Sowls LK (1984) The peccaries. University of Arizona Press

Sykes M (2022) Developing pig-to-human organ transplants. Science 378(6616):135–136

Tatwangire A (2014) Uganda smallholder pigs value chain development: situation analysis and trends. International Livestock Research Institute (ILRI)

Weka R, Bwala D, Adedeji Y, Ifende I, Davou A, Ogo N, Luka P (2020) Tracing the domestic pigs in Africa. IntechOpen

Zeder MA (2017) Out of the fertile cresent: The dispersal of domestic livestock through Europe and Africa. In: Out of the Fertile Crescent: the dispersal of domestic livestock through Europe and Africa

Open Access This chapter is licensed under the terms of the Creative Commons Attribution 4.0 International License (http://creativecommons.org/licenses/by/4.0/), which permits use, sharing, adaptation, distribution and reproduction in any medium or format, as long as you give appropriate credit to the original author(s) and the source, provide a link to the Creative Commons license and indicate if changes were made.

The images or other third party material in this chapter are included in the chapter's Creative Commons license, unless indicated otherwise in a credit line to the material. If material is not included in the chapter's Creative Commons license and your intended use is not permitted by statutory regulation or exceeds the permitted use, you will need to obtain permission directly from the copyright holder.

African Dromedary Genetic Resources, Diversity and Breeding Systems

9

Semir B. S. Gaouar, Imane Meghelli, Zoubeyda Kaouadji, Félix Meutchieye, and Djalel E. Gherissi

Abstract

This chapter thoroughly investigates the multifaceted world of dromedaries in Africa, employing a structured approach. Initially, Sect. 9.1 delves into the origins and evolution of dromedaries on the continent, including the domestication process and their historical roots in Africa. Subsequent sections meticulously dissect the presence of dromedaries across various African countries, delineating their physiognomy, classification and unique characteristics, exemplified by detailed examination of populations in Algeria, Mali, Egypt, Niger, Nigeria, Mauritania, Chad, Kenya, Somalia, Ethiopia, Eritrea, Djibouti, Sudan, Libya, Tunisia, Morocco and other countries. Section 9.2 shifts focus to the organisational structures within the dromedary sector, exploring infrastructure, support systems and regulatory frameworks, while Sect. 9.3 scrutinises the structure and systems of dromedary husbandry, encompassing geographical distribution, breeding systems and production methodologies. Section 9.4 further elaborates on dromedary production objectives, covering primary production objectives, pharmacopoeia aspects, sociocultural roles and modern uses, including racing, trekking and other activities. Environmental implications, reproduction, prevalent diseases and biodiversity conservation efforts are comprehensively addressed in Sects. 9.5, 9.6, 9.7, and 9.8, paralleling discussions on dromedary diseases and biodiversity conservation programmes. Finally, Sect. 9.9 offers a concluding perspective, advocating for conservation measures and highlighting future directions for the sustainable management of dromedary populations in Africa.

S. B. S. Gaouar (✉) · I. Meghelli
Applied Genetics in Agriculture, Ecology and Public Health (GenApAgiE), Faculty Life and Natural Sciences, Earth and Universe Sciences, University of Tlemcen, Chetouane, Algeria

Z. Kaouadji
Applied Genetics in Agriculture, Ecology and Public Health (GenApAgiE), Faculty Life and Natural Sciences, Earth and Universe Sciences, University of Tlemcen, Chetouane, Algeria

Faculty of nature and life, University of Sidi Bel Abbes, Sidi-Bel-Abbès, Algeria

F. Meutchieye
Faculty of Agronomy and Agricultural Sciences, Department of Animal Science, Biotechnology and Bioinformatics Research Unit, University of Dschang, Dschang, Cameroon

D. E. Gherissi
Laboratory of Biotechnologies, Animal Productions and Health, University of Souk Ahras, Ahras, Algeria

Keywords

Breeds · Conservation · Dromedary · Camel · Genetics · Production

Abbreviations

AD	Anno Domini or Common Era
ATP	Adenosine triphosphate
BCE	Before the Common Era
BRET	Bioluminescence resonance energy transfer
BSE	bovine spongiform encephalopathy
CA.RA.VA.N	ttoward a Camel tRAnsnational VAlue chaiN
CATT/T. Evansi	card agglutination test
C. Bactrianus	Camelus Bactrianus
C. dromedarius	Camelus dromedarius
CBD	Convention on Biological Diversity
CJD	Creutzfeldt-Jakob disease
CPrD	dromedary prion disease
CIRAD	Centre de coopération internationale en recherche agronomique pour le développement
Cm	centimetres
DAD-IS	Domestic Animal Diversity Information System
DNA	Deoxyribonucleic acid
ERK1/2	signal-regulated kinases 1 and 2
FAO	Food and Agriculture Organization
FAOSTAT	Food and Agriculture Organization Corporate Statistical Database
Fig	fig.
F1	Filial one
FST	The fixation index
HIV/AIDS	human immunodeficiency *virus*/acquired immune deficiency syndrome
hIR	human insulin receptor
ICSI	intracytoplasmic sperm injection
ICO	International Camel Organization
ILRI	International Livestock Research Institute
ISOCARD	International Society for Camelid Research and Development
ITS1-PCR	internal transcribed spacer 1-Polymerase Chain Reaction
Kg	kilograms
L	Litre
m	Metres
Ma	million d'années or million years
MANOVA	Multivariate analysis of variance
Mdh	malate dehydrogenase
ME	malicisoenzymes
mtDNA	mitochondrial Deoxyribonucleic acid
nDNA	nuclear Deoxyribonucleic acid
NGO	*Non-Governmental Organization*
Nov	November
NRCC	National Research Camel Center
ONDEC	National Office for the Development of Equine and Camel Breeding
OIE	World Organization for Animal Health founded as OIE
PCR	Polymerase Chain Reaction
PRNP	The prion protein gene
PrPSc	Pathognomonic neurodegeneration and disease-specific prion protein
SD	Standard deviation
SNP	single nucleotide polymorphism
STDM	standard trypanosome detection methods
STR	Short Tandem Repeat
SUCA	Sudan Dromedary Association
T. evansi	Trypanosomaevansi
T. simiae	Trypanosomasimiae
TYR	Tyrosinase

UAE	United Arab Emirates
UNESCO	United Nations Educational, Scientific and Cultural Organization
Yr	Year

9.1 Origins and Evolution of Dromedary in Africa

9.1.1 Domesticating Dromedaries

Dromedary domestication occurred somewhat late in human history, most likely at the transition between the second and first millennia BC. With the emergence of molecular markers, archaeologists have pieced together the domestication process of the dromedary. A significant reduction in the size of bone remains a growing association with human settlements and unambiguous artistic depictions in the fossil record from 2000 years ago. Dromedaries thus only cohabited with other domestic animals for a little over a millennium, at a time when their range was probably limited to the coastal mangrove habitats of the Arabian Peninsula, where their physiological salt requirements could be supported (Orlando 2016).

Today, dromedaries live mostly in the desert areas: North Africa, the Sahara and Sahel Regions, the Horn of Africa, throughout the Arabian Peninsula and the Levant (Fig. 9.1). A feral population also developed in Australia following their introduction in the nineteenth century AD (Orlando 2016).

Almathen et al. (2016) used spatial models of genetic diversity to decipher the history of dromedary domestication (Almathen et al. 2016). Because dromedary reference genomes have only recently become available, Almathen et al. (2016) used classic population genetic tools to characterise 1083 animals with 17 nuclear and mitochondrial microsatellites (Almathen et al. 2016). There were no phylogenetic patterns in mitochondrial DNA (mtDNA) data, as the two main haplogroups (HA and HB) were found throughout their range. As a result, it appears to be impossible to trace the central geographical location of the dromedary's domestication genetically from modern individuals, since a source of domestication should have the highest levels of genetic diversity. Dromedary movements along transcontinental trade routes may have eroded pre-existing phylogeographic models resulting from their initial domestication (Almathen et al. 2016). Alternatively, due to the animal's high dispersal abilities, it is possible that

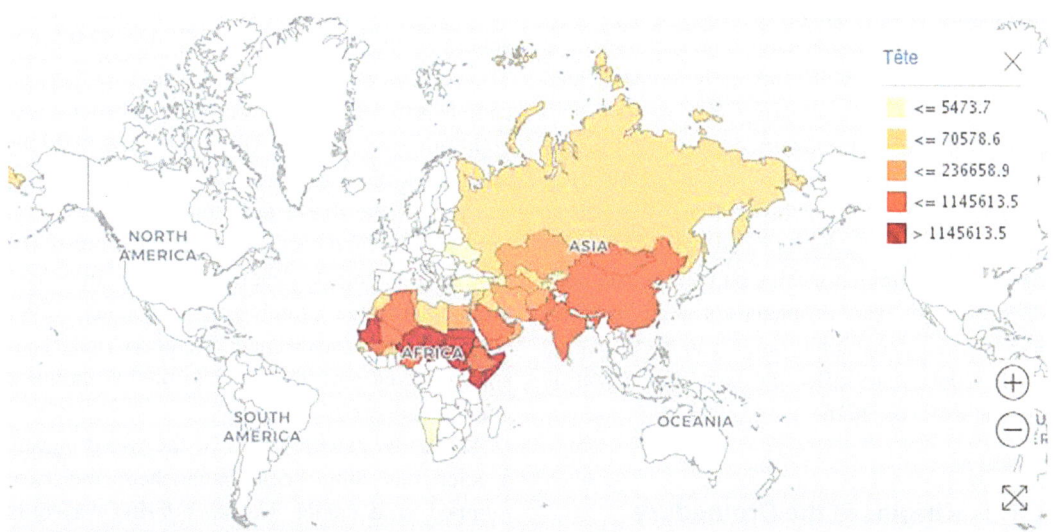

Fig. 9.1 Distribution of large camelids in the world. (From FAOSTAT 2019, licensed under Open Government License OGLv3.0)

such a population structure never existed in nature. In modern populations, detected haplotypes were found to be still segregating. Eight ancient specimens found were wild and dated from 5000 to 500 BC. Three more new haplotypes were discovered, indicating that the domestication process only captured a portion of the pre-existing diversity in the wild state. Data from ancient DNA recovered from wild individuals restricted to the United Arab Emirates (UAE), Middle East region, was unfortunately inconclusive with regards to the presence or absence of structure in Arabian Peninsula populations prior to domestication (Orlando 2016). The authors were able to calibrate the mtDNA clock to date significant population changes because the age of the old mtDNA sequences could be inferred from their archaeological context. This method takes advantage of the well-known benefit of old DNA data and time series in general (Orlando 2016). The authors were able to detect a massive population expansion about 600 years ago using a statistical software package that uses time-structured datasets to co-estimate the mtDNA clock and past demographic profiles. This date is consistent with the rise of the Ottoman Empire and the increasing use of dromedaries over long distances in an empire that extended from the coastal regions of North Africa to Arabia (Orlando 2016).

On the other note, Burger (2016) successfully identified south-eastern Arabia as the region of first domestication by analysing 7000-year-old bone DNA from wild and early domesticated dromedaries and comparing the results with genetic profiles of modern dromedary populations around the world (Burger 2016). The authors came to the conclusion that this domestication was followed by repeated crossing of wild camels with populations that had previously been domesticated. The dromedary's wild ancestor had a small geographic range and went extinct about 2000 years after the first domestication (Almathen et al. 2016).

9.1.2 Origins of the Dromedary in Africa

Modern camels belong to the order of Artiodactyla (even-toed ungulates), suborder Tylopoda, and the family of Camelidae, consisting of the tribes Camelini (Old World camels) and Lamini (New World camels), which diverged 16.3 (9.4–25.3) million years ago (Mya) (Wu et al. 2014).

Similar to other large mammals, the earliest-known ancestors of the camelid family, Protylopus, originated in the North American savannah during the Eocene (~45 Ma), with a size smaller than a domestic goat. At least 20 genera of camelids, such as Megacamelus and Procamelus, developed and disappeared again over the following million years (Honey et al. 1998; Rybczynski et al. 2013), until the ancestors of Old World camels reached Eurasia via the Bering land bridge around 6.5–7.5 Mya. Fossils of Paracamelus and other giant camels have been recorded in Asia (Kozhamkulova 1986; Flynn 1997), Europe (e.g. Spain (Pickford et al. 1995)), Northern Africa (*Camelus thomasi* (Peters 1998)) and the Arabian Peninsula (e.g. Syria (Martini et al. 2015)). The progenitors of the New World camels entered South America around 3 Mya (Rybczynski et al. 2013; Prothero and Schoch 2002) (Fig. 9.2).

The ancestors of dromedaries and Bactrian camels first appeared on the North American continent about 40 Mya. Camels (Lamini) on the New World level can be distinguished from dromedaries (Dromedary) on the Old World level. Dromedaries migrated to the eastern hemisphere via the Bering land bridge (the Old World). The split between one-humped and two-humped camels is thought to have occurred between 5 and eight million years ago. This contradicts previous theories, which suggest that the wild Bactrian camel was the direct and common ancestor of dromedaries and Bactrians. Furthermore, the observed prenatal development of the one-humped Arabian camels has refuted an earlier suggestion that the dromedary embryo undergoes similar developmental stages as the two-humped camels (Burger 2016).

While the genetic status of the two domestic species has long been established, the two-humped wild camel, *Camelus ferus,* was only recently recognised as a separate species based on molecular genetic data (Burger 2016). The demographic history of Old World camels derived from genomic drafts reveals the three species had

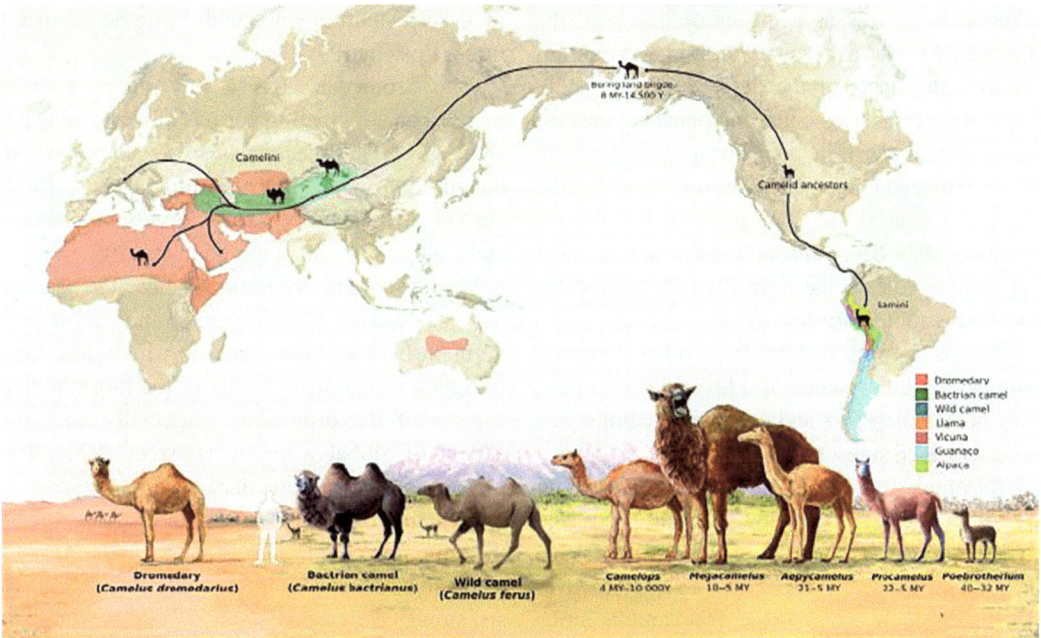

Fig. 9.2 Migration map of the historical camelid family. (From Burger et al. 2019, licensed under CC-BY 4.0)The current distribution of dromedaries and Bactrian camels is presented in red and green colours. The last refugia of the wild two-humped camels in China and Mongolia are shown as dark-green patches. The map was adapted from Mesa Schumacher/AramcoWorld (https://www.aramcoworld.com/en-US/Articles/November-2018/The-Magnificent-Migration). Reprint permits were granted by AramcoWorld on March 6, 2019

independent evolution over the past 100,000 years, with severe bottlenecks occurring during the last Ice Age and in the recent past. Wild two-humped camels are also excluded as direct ancestors of modern Bactrian domestic camels due to this long-term divergence (Almathen et al. 2016).

The dromedary, also known as the one-humped camel (*Camelus dromaderius*), represented a historic turning point for the human populations of the Arabian Peninsula, the Levant, North Africa, the Saharo-Sahelian region and beyond. According to Doutresoulle (1947), the dromedary was introduced into Africa through regular contact with the Arab world and established itself above the 14th north parallel, where the lack of glossines and horseflies, as well as the abundance of salty lakes, suited it perfectly (Doutresoulle 1947). The dromedary has expanded its range in Africa over the last 600 years, now being found in Senegal, Mauritania, Nigeria, Niger, Chad, Cameroon, the Magreb, Sudan and the Horn of Africa. The trans-Saharan salt trade was a significant activity that justified the continued presence of large herds of dromedaries in the desert. Dromedary rearing is very old in some countries, such as Niger, and is associated with typical sociological groups, such as pastoralists or nomads (Chaibou 2005). The introduction of the dromedary through contact with Arabs is true for almost the entire Sudano-Saharan deserts. Other introductions in Kenya and Namibia are believed to be more recent.

9.1.3 Dromedary in Africa

9.1.3.1 Physiognomy and Description of Dromedary Populations in Africa

The dromedary is the best-adapted pseudoruminant animal to the desert environment, characterised by scorching heat and extreme desiccation. Thus, its anatomy and physiology are different from other animals that live in a more aimable environment. In addition to its role as a means of transport, the dromedary serves as a source of food and comfort for the inhabitants of the desert (Wilson 1999). They are a versatile, essential component of the desert ecosystem and a source of cheap animal protein for the indigenous population because they have the ability to convert the rare plant and cellulosic resources of the desert into milk and meat for human consumption, hence the interest in the domestication of this animal (Epstein 1971).

Camels are very distinct from other domestic animals by the presence of a broad head, a thin, long neck, small ears and wide, protruding eyes, wide and deep sinuses. The dromedary presents a lateral blind sinus sac, which is not observed in any other species. This allows the nasal passage to recover water at the time of exhalation. It also has an upperlip that is divided and very sensitive; a broad and low hang in the gearlap and its limbs are very powerful. Dromedary body varies in length from 220 to 250 cm and weight from 400 to 1100 kg, depending on the population (breed), with a life expectancy of 25 years (Wilson 1999). Doutresoulle (1947) identified in French West Africa a dromedary population with a height at withers of 170 cm for pack animals, 180–190 cm for loads and more than 200 cm for riding (Doutresoulle 1947).

One of the anatomical elements that clearly distinguish the dromedary from ruminants and other herbivorous species is the nature of its wide and elastic legs, which are well-suited for sandy soils. The hump is mostly comprised of adipose tissue of soft consistency that vary in volume with the nutritional status of the animal. The digestive anatomy of the dromedary differs from that of other animals in terms of structure and function. It is particularly suited to the natural plant resources of desert environments, including thorny plant species inaccessible to other herbivores (Djomtchaigue et al. 2015). The animal has glands behind the head that are used for perspiration. The skin is supple and covered with hair. The female has four quarters at the level of the udder, while the males' testicles are positioned high behind the thighs (as in cats and dogs). These morphological and anatomical features explain the adaptation capacity of the dromedary in desert environments with extreme temperatures (Epstein 1971).

The dromedary has a physiology entirely focused on anticipating periods of food and water shortages. The mechanisms of adaptation to dehydration are complex in this species. These could be evaluated by a very large set of parameters which, when combined, give this unique behaviour among domestic species. The dromedary's resistance to dehydration and heat is proverbial. In fact, this feature, which has long intrigued researchers, is one of the most studied aspects of the dromedary, especially with the advent of global warming (Epstein 1971). The physiology of the dromedary is tuned for survival in difficult conditions. In a state of dehydration, the animal is able to save body water by mechanisms that reduce water losses (decrease in diuresis, stop sweating, decrease in basic metabolism, change in body temperature) while maintaining a homeostasis vital to its survival, both by limiting the variation in the concentration of vital parameters and by ensuring maximum excretion of metabolic waste. This is demonstrated by the emission of highly concentrated urine. However, the excretion of elements requiring large quantities of water (glucose, urea in particular) is strictly controlled. These adaptation mechanisms, which are the hallmark of the dromedary, also explain why it is one of the domestic species that has not left its natural home range (Bengoumi and Faye 2002).

Because dromedaries generally bear the names of the tribes that keep them, we can find several populations in different tribes with not much recognisable difference. Modern classifications have made little progress beyond these concepts, as little effort has been made in conventional breeding and genetic selection, or to attribute quantitative production parameters that are now so important in other species for the description of the breed. The current quantitative approach uses six morphological and biological characteristics, such as habitat, function and geographic distribution, physical size based on linear measurements, ease of treatment and speed of weight gain as parameters in the description of the breed. In Saudi Arabia, the most commonly used classification is based on coat colour, as is done on dromedary populations in Algeria and other North African countries (Cherifi et al. 2013; Harek et al. 2017). The colour variants vary from one region to another, but it is not yet clear whether there are differences in production between or among colour types. While light-coloured animals from Sudan are particularly appreciated for their speed, pastoral camel keepers have an overall assessment of their animals based on morphological, phaneroptic and metric parameters, in particular height at the withers (Doutresoulle 1947; Djomtchaigue et al. 2015). It is possible that, with a well-designed and funded programme, the constraints of low fertility and the difficulty of breeding dromedaries can be significantly reduced. There are sufficient variations in body weight at maturity, conformation characteristics, working capacity and production potential to make genetic improvement possible, and improved husbandry and health management can reduce camel calves' mortality. However, studies conducted by Almathen et al. (2016) on global populations and Cherifi et al. (2017) on North African populations show that the populations studied are very close genetically and that no structuring into populations could be clearly observed (Almathen et al. 2016; Cherifi et al. 2017).

9.1.3.2 Classification of Indigenous Dromedary Populations

Globally, high variability is observed among dromedary populations, although great confusion is reported. Chapter 1 of this book states that the total number of African dromedary camel breeds is 63, distributed respectively as follows: 23 breeds in Eastern Africa, 3 breeds in Central Africa, 23 breeds in Northern Africa, 2 breeds in Southern Africa and 12 breeds in Western Africa. Indeed, several breed names exist, but more than one breed name could represent the same animal genotype. In an earlier report (Blanc and Ennesser 1989), 48 breeds of dromedaries were described worldwide and classified into eight subgroups according to their physical attributes, but this classification was only based on the general conformation of the animals.

Dromedaries were also classified into two major classes: ride and pack dromedaries. This classification was suitable to satisfy the needs of caravans, transport and military purposes during the first half of the twentieth century. A new system of classification was proposed based on the fact that the dromedary is a major component of the agropastoral systems in vast pastoral areas in Asia and Africa. Dromedaries have also been classified into four major classes: meat, dairy, dual-purpose and race dromedaries (Wardeh 2004). However, the keepers of dromedaries or owning tribes have their own system of classification or nomenclature, which is based, in general, on the owning tribe, colour and region (Cherifi et al. 2013; Harek et al. 2017).

There is great confusion in the description of the biodiversity of camels, as mentioned by many authors and attested by results using molecular genetics tools. Thus, the different phenotypes described in Kenya, Tunisia, Chad, Egypt and Algeria have little or no significant genetic differences between them (Cherifi et al. 2017; Djomtchaigue et al. 2015; Jianlin et al. 2000; Chniter et al. 2013; Harek et al. 2015). In Saudi Arabia, nine dromedary phenotypes were described, but after molecular genetics analysis, only three subtypes were identified (Almathen 2014; Faye and Porphyre 2011). The camel phenotypes described in these countries are presented below. However, additional molecular genetic analyses will help in classifying the populations into distinct breeds or genotypes.

9.1.3.3 Algerian Dromedary Populations

In Algeria, dromedary populations are named according to ethnicity or geographical area. Within each ethnicity, the keepers name the populations according to coat colour (Cherifi et al. 2013; Harek et al. 2017). Seven camel populations named according to coat colour in Algeria are shown in Fig. 9.3a–g and described below.

Hamra

The Hamra has a dark red or brown uniform coat colour (Fig. 9.3a). This type of coat is very appreciated and highly valued by farmers. The brown uniform coat is preferred by some farmers for reasons including aesthetics, religion and sometimes commercial (for cloth making). Other farmers consider the dromedary with the red coat as best suited to the environment, especially in terms of disease resistance and drought tolerance. Hamra's coat is very appreciated because of its hair quality, which makes it suitable for the manufacture of many traditional items. The Hamra accounts for about 30% of the dromedary population in Algeria and about 47% of the population in Tunisia in 2009 (Ould Ahmed 2009). The difference in proportion between the two regions is probably due to the decline of the craft in Algeria.

Safra

The Safra Camel has a yellowish uniform coat colour (Fig. 9.3b). It constitutes about 27.5% of the dromedary population. This coat colour is among the most favoured by farmers (Ould Ahmed 2009).

Chegra

Chegra dromedary has a clear reddish coat (Fig. 9.3c). It is found across the southwest of Algeria, constituting 11% of the dromedary population. This coat is not very popular amongst farmers.

Beydha

Beydha dromedary has a white or very light grey coat colour on its entire body (Fig. 9.3d). Beydha is characterised by its strong and sometimes aggressive behaviour and its susceptibility to diseases.

Zarga

The coat colour of Zarga is completely black (Fig. 9.3e). This phenotype is less preferred when compared to the other phenotypes. Individuals with this coat are not susceptible to disease. The polymorphism of the Zarga (black) coat colour occurs at a frequency of 7.5% in the study area compared to a frequency of 6% in Tunisia (Ould Ahmed 2009). These frequencies suggest that the trait is recessive in nature and is favoured by natural selection.

Hajla

Hajla is characterised by white colour on the head and legs and a light yellowish colour on the body (Fig. 9.3f). Animals with this coat colour are highly appreciated aesthetically and considered by the farmers to attract or draw luck. It is, however, very rare, with a low frequency (9.5%) on farms, which may be attributed to natural selection drivers.

Zarwala (Azargaf)

The coat of Zarwala is a mixture of blue, white and black (crossbred) (Fig. 9.3g). It is less appreciated by farmers. The dromedaries of this colour are characterised by a severe form of deafness. Theories suggest that white spots are linked to the domestication process and are sometimes associated with health disorders (Holl et al. 2017).

The significant variations in the colour of the coat (Hamra, Safra, Beydha, Chegra, Zarga, Hajla and Zarwal) of the Algerian camel populations indicate an admixture of different genetic backgrounds. The high frequency of the red colour appears to be a criterion for adaptation to the arid environmental conditions of Algeria. This hypothesis needs to be carefully studied in order to establish a relationship between adaptation and the mechanisms of the dromedary's coat colour variations in Algeria.

Many socio-geographical factors are used to distinguish between different dromedary populations in Algeria. The notations are linked to the geographical area or tribe to which the animal belongs. Indeed, the dromedary populations in Algeria are also named after five geographic locations (Fig. 9.4a).

Targui

Targui, as the name indicates, is an animal raised by the Tuareg. It is an animal with strong muscular legs and a little bump (Fig. 9.4a). It is considered an excellent animal for racing. The coat is

Fig. 9.3 Major coat colours found in dromedary populations in Algeria. (Photos by Gaouar S.B.S)

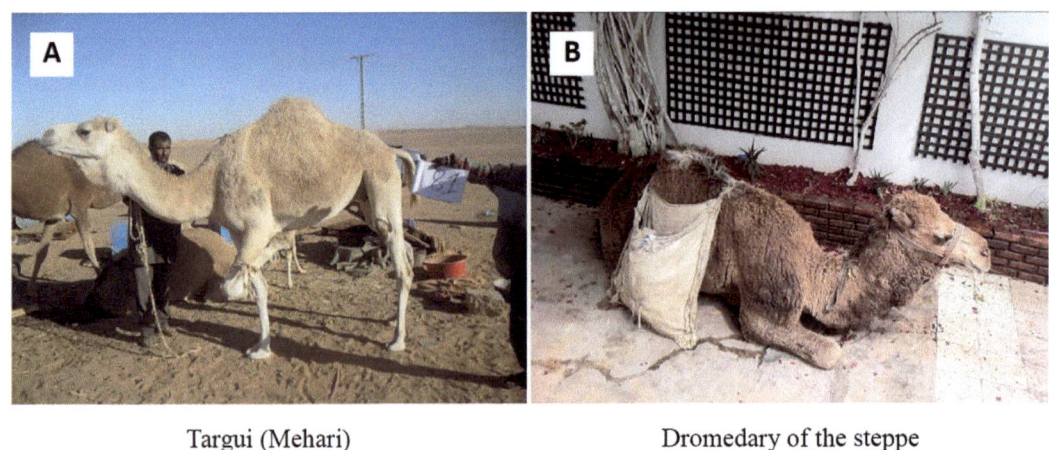

Fig. 9.4 Dromedary populations in the southwest of Algeria. (Photos by Gaouar S.B.S)

usually light beige. The Targui is mainly used for tourism, especially in the regions around the Hoggar and Tassili areas. It is sometimes used for cross-breeding with other populations, depending on the desired characteristics.

Dromedary of the Steppe (Naili)

The animal raised by the tribe of Ouled Neil (Djelfa Wilaya) is found in the Saharan borders and especially at the edge of the Steppe and the Sahara. The population is in decline. It is a relatively small dromedary and a poor carrier (Fig. 9.4b). But it is very appreciated for the quality of fur, especially for making traditional items.

Rguibi

Rguibi or Mniae is the local type raised in the area of Bechar and Tindouf by the Rguibet and Mniae tribes of Yemeni origin. It has multiple abilities with medium production of meat and milk and excellent adaptation to different ecoclimatic conditions. Rguibi is characterised by various size variants (light to heavy) and fur (medium length) (Fig. 9.4c). This population is found in the extreme south of Ain Safra, south of Algeria, and the extreme northeast of Mauritania and northwest of Mali (Wardeh 2004).

Azawak

Azawak is the riding dromedary for the Tuareg from Niger and Mali. It is suitable for racing. It is slender-looking, light-coloured (light brown) and hairless (Fig. 9.4d). There are many different types depending on the nuances of the coat; it was introduced by the Tuareg farmers, and it is found throughout the southwest of Algeria (up to Timyaouine Ain Safra).

Ouled Sidi Cheikh

Ouled Sidi Cheikh (Fig. 9.4e) is found in the highland regions of Algeria, and it is less appreciated by the farmers in the south. It is adapted to a climate marked by a dry summer and a very cold winter.

Chaambi

Chaambi is characterised by a short stature, corpulence and robustness in saddled transport. It is the animal of the peasant, the auxiliary of the oasis dwellers. It is commonly used for agricultural work and appreciated for its meat. It is found throughout North Africa but mainly in Tunisia and eastern Algeria (CIRAD 2022).

Sahraoui

Sahraoui (Fig. 9.4f) was obtained from a cross between Chaambi and Ouled Sidi Cheikh. Its territory extends from all the Grand Erg Occidental of the Sahara (north of Adrar). It is a breed with lactation performance under intenisf or semi intenisf systems (Chergui et al. 2023, 2024).

Ayat Al-Khabbash

Ayat Al-Khabbash are well-developed pack animals that respond to fattening. They are found throughout North Africa, but mainly in Tunisia and eastern Algeria, and are raised by the tribes in Southern Algeria (Wardeh 2004). Unfortunately, this population seems to have become extinct (Cherifi et al. 2013).

9.1.3.4 Mali Dromedary Population

Traoré et al. (2014) identified two classification systems of dromedaries by Gurma and Hausa herders (Traoré et al. 2014). The Hausas classify them into two types (the Tilabayaten and the Talmorokit) on the basis of the overall conformation, production ability (milk, speed, draught) and colour. The Tilabayaten is described as white or light grey, with a slender body shape. Its milk production is reported to be higher than that of the Talmorokit. This type has a dark coat and is relatively smaller. Its heavier conformation makes it suitable for the transportation of goods. Nine types are described by the Gurma herders. Seven are distinguished and named on the basis of coat colour, similar to the classification in Algeria described earlier. It includes the Azargaf, characterised by its piebald colour and blue eyes. The two remaining types are the Adignas, which means 'trustable animal', and the Awinag, which is characterised by its totally white colour and a vision defect.

No definite inventory of dromedary breeds or types in Mali is available. The types cited in Hausa are all included in the list of 12 types reported by Ouologuem et al. (2004). The Abzaw and Azargaf types might correspond to the Abzin and Azaghaf types described by Chaibou (2005) in Niger (Chaibou 2005). This proposition is based on the similarity in the names used as well as the respective coat colours of these types.

9.1.3.5 Egypt Dromedary Population

Four dromedary breeds, Sudani, Maghrabi, Falahi (or Fellahi or Baladi) and Mowalled, are recognised in Egypt (Cherifi et al. 2017). Nowier et al. (2020) also cited Somali as a breed reared in Egypt (Nowier et al. 2020). The Al-Delta strain is found in the Egyptian Delta, used mainly for agricultural draft and transport (Nowier et al. 2020). This Delta dromedary is not a distinctive breed in the real sense but a general-purpose dromedary that was developed with a considerable intake of green feed and water. It attains a large size with well-developed muscles, which allows it to carry heavy loads and be suitable for various agricultural functions (Wardeh 2004).

Maghrabi

Maghrabi is found in the coastal regions of Northern Africa that extend from Egypt to Morocco. The Maghrebi generally refer to several strains of dromedaries which vary in size, body conformation and colour. It is believed to be a mixture of the Sudanese, Egyptian, Libyan and Tunisian dromedaries. The Maghrebi dromedary is medium in size with a small pointed hump. Besides pack use, the Maghrebi dromedary is used for all kinds of agricultural, industrial and draft purposes. It generally responds to feeding and might gain about 700–1000 grams per day during the first year under intensive conditions (Wardeh 2004).

Al-Fellahi or Fellahi

Al-Fellahi dromedary breed is kept by farmers along the Nile in Upper (Southern) Egypt, from which it derived its name. Al-Fellahi dromedary is characterised by its heavy bones and muscles, big size and its slow pace. It is generally used in agricultural work and transport. It also responds to fattening (Wardeh 2004).

9.1.3.6 Niger and Nigeria Dromedary Populations

Doutresoulle (1947) associated various morphotypes of dromedaries with Niger and Nigeria. However, there is not enough information for their rigorous classification (Doutresoulle 1947). Niger has one of the fastest-growing dromedary population, doubling in the last few decades in regions where erratic rainfalls has led to the regression of other domestic livestock species (Chaibou 2005). Abdussamad et al. (2011) have classified different dromedary ecotypes based on coat colour under the Azawak dromedary breed (Abdussamad et al. 2011). Though the dark-brown ecotype was the most preferred, their study indicated that there is likely a survival strategy that sustains pastoral life in a fragile ecosystem, which largely depends on the complementary performance of an assortment of dromedary ecotypes at different seasons of the year.

Azawak

The Azawak dromedary is prevalent in semi-arid northeastern Nigeria. It is an important pack, ride, dairy and meat animal. The dromedaries were made up of predominantly mixed breeds originating from Eastern and Northern Africa regions (Jaji et al. 2017).

9.1.3.7 Mauritania Dromedary Population

Al-Jandaweel (Gandoil)

The Al-Jandaweel dromedary is found in Southern Mauritania along the Senegal River. These Al-Jandaweel dromedaries are not typical lowland reverine types, though they are suspected to have diverse origins. They are dark-coloured, heavy animals mainly used for draft. The feed availability in the Senegal River valley might have resulted in the development of this heavy dromedary (Wardeh 2004).

Mehari (Mehri)

The Mehari is a thoroughbred race dromedary that descended from the Murrah dromedaries in the Sultanate of Oman. They are beautiful, light-coloured and very famous race dromedaries in the Sahara. The colonial French troops formed the Meharist corps based on this breed. They used the Mehari dromedaries for riding and divided them into three categories according to height at the withers. Dromedaries that were 190–200 cm high were used by officers, while those 185 cm high were ridden by the ranks and files. Shorter dromedaries that were at least 180 cm at the withers tall were used as pack animals (Wardeh 2004).

9.1.3.8 Chad Dromedary Population

According to Martin et al. (1996), dromedaries are mostly raised in the Manga and Kanem regions in the Lake Chad Basin, east of Niger and in western Chad (Martin et al. 1996). The two other countries sharing Lake Chad, Cameroon and Nigeria, have few dromedaries. Around Lake Chad, numerous changes in production systems consequent on the recent climatic changes are happening. Thus, the relative development of dromedary production and the increases in the dromedary products' trade have resulted in a broader distribution of various types. It is possible, however, to classify the dromaderies found in this region into four major groups. The dromedary population in Chad is close to 2.5 million heads. Breeds are classified based on their coat colours, which is also not very defendable in a genetic context (Djomtchaigue et al. 2015). Although there has been no genetic characterisation of these groups, they have distinct morphological characters and occupy well-defined geographical areas (Fig. 9.5).

Tibesti

Tibesti breed was probably introduced by Toubou pastoralists from Borku and Tibesti. This is a mountain breed with a withers height of 1.75–1.85 m used as a pack and riding animal. The coat colour varies from grey to dark brown, and the Tibesti dromedary is found mainly in the Kanem region of Chad.

Manga

Manga dromedary breed is 1.85–2.00 m high at the withers and found mostly in the Chadian plains. It weighs 550 kg, well-muscled but not very hardy. The colour is fawn or reddish, and this breed is kept mainly for meat, especially in the Bornu region of Nigeria and in the northern zone of Lake Chad. Another breed called the Komadougou has been described, which seems

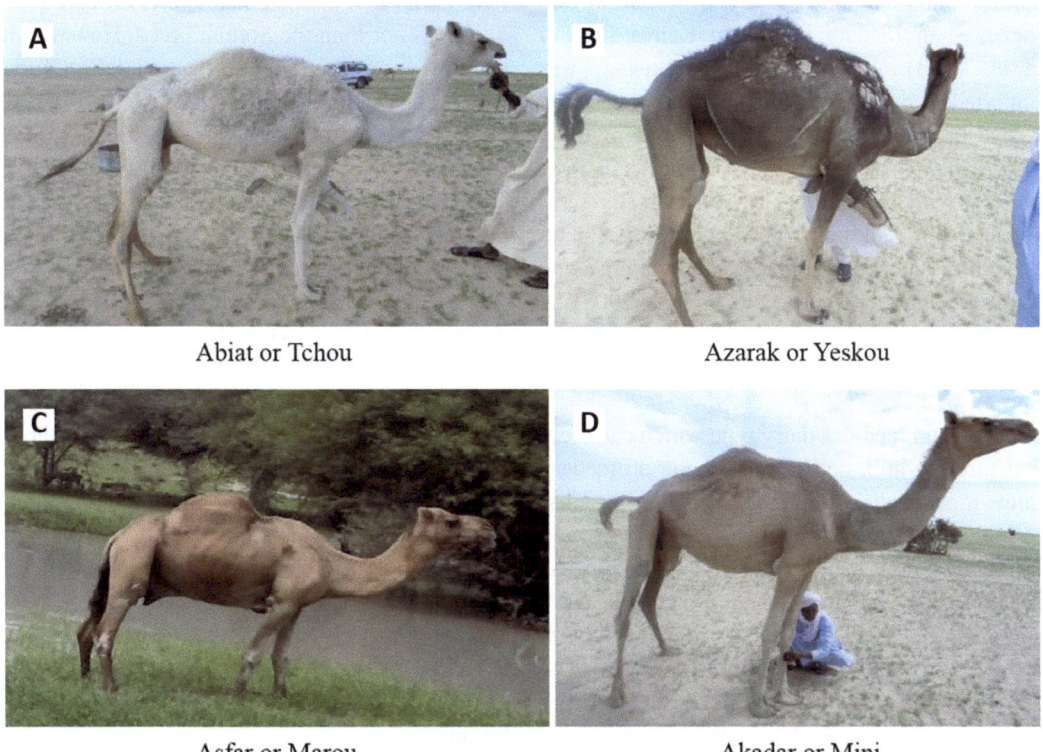

Fig. 9.5 Coat colours as they are called by Chadian dromedary herders. (From Djomtchaigue et al. 2015, licensed under CC-BY 4.0)

similar to the Manga. This breed is considered to be a meat and transport animal found amongst the Fulani people in Nigeria.

Touareg

The Touareg breed originates from the Aïr but is also present in the Lake Chad Basin. It is described as a mountain type with 1.85–1.90 m high at the withers, weighs about 370 kg and is very light or pied in colour. This light breed is mainly used as a saddle animal in the search for grazing areas and for social visits.

Arab

This breed is common in the mountains and plains. It is very large, weighing about 450–500 kg. The colour is pied or greyish, and it is used as a pack animal. It is found in the Arab settlement area of the Convention Basin. The typical coat colours for breeds are indicated in Fig. 9.4.

9.1.3.9 Kenya Dromedary Population

The Somali, Turkana and Rendille/Gabbra are the three dromedary breeds in Kenya (Mburu et al. 2003). A fourth called Pakistani was imported from Pakistan into Laikipia ranches in Kenya in the early 1990s. Only a few of these pure Pakistan camels now exist; instead, their crosses with Somali or Turkana breeds have spread from Laikipia to Samburu, East Pokot, Kajiado, Northern Tanzania, Mandera and Marsabit districts (Kuria and Walaga 2008). A description of these breeds is presented below.

Somali

The Somali breed is a dairy type which can yield 3–5 litres of milk daily. Milking the dromedary three times daily guarantees more milk. First calving occurs between 4 and 5 years of age, and lactation could last 1–1.5 years post-calving. The coat colour of Somali is cream/brown (Fig. 9.6), and it weighs about 450–850 kg at maturity with an average standing height of 2 metres. The Somali requires about 8–12 h of feeding time per day, depending on feed availability. It is more comfortable while feeding on shrubs because of its height. Early maturity and high milk production are two important advantages of this breed, while poor adaptability to rough terrain and rocky hillsides due to its large size is its main disadvantage. It also suffers more with low feed availability (Kuria and Walaga 2008).

The Somali breed has four subtypes, namely, Hoor, Siftarr, Aidimo and Gelab, which differ in physical appearance, production and adaptability characteristics. While the Hoor is the highest milk-producing subtype, it is the least hardy. Gelab, on the other hand, is the smallest in body size and the least milk-producing, but the hardiest.

Rendille/Gabbra

The Rendille breed produces an average of 1–3 litres of milk per day with a lactation length of 1–1.5 years, and first calving occurs between 5 and 6 years of age. The coat colour is cream or brown (Fig. 9.7) with a mature live bodyweight around 550 kg. The average standing height is 1.8 metres. Its feed requirement is less than the Somali breed, needing 8–10 h of grazing per day. This breed does better under poor pasture conditions and rough terrain and tolerates drought conditions better than the Somali breed. However, it has a lower milk yield and matures later (Kuria and Walaga 2008).

Turkana

The Turkana breed produces about 1–2.5 litres of milk per day, which is lower than what the Somali and Rendille/Gabbra breeds produce. The first calving occurs between 5 and 6 years of age, and it lactates for about 1 year after calving. The body colour is mainly greyish/dark (Fig. 9.8). It is a late-maturing breed, attaining a mature live bodyweight around 250–500 kg and an average standing height of 1.7 metres. It requires about 7 h of grazing. It is the hardiest of all the breeds, agile and able to climb steep hills because of its small body size. It is able to cope better with feed scarcity.

9.1.3.10 Somalia Dromedary Population

Somali

Somali dromedary breed is commonly found throughout Somalia. The adult males average 650 kg and the females 575 kg. The average lactation yield is 1650 kg (Fig. 9.6) (DAD-IS 2021).

Fig. 9.6 Somali dromedary. (From Tilahun et al. 2020, licensed under CC-BY 4.0)

Fig. 9.7 Rendille dromedary. (From Sun 2005, licensed under CC-BY 4.0). (**a**) A warrior driving camel herds out of the camp for day's herding. (**b**) Transporting wedding goods

Afar

The Afar breed is also commonly found throughout Somalia (Fig. 9.9) (DAD-IS 2021).

Sifdar

Sifdar breed is found in the lower Shabbily river belt in Somalia. The breed is tall, light and with grey to reddish for coat colour. It reaches matu-

Fig. 9.8 Turkan young camels with female, in Prosopis thicket at lake, Northern Turkana. (From Carr 2017, licensed under CC-BY 4.0)

rity at 5–6 years but is able to reproduce at 6–7 years. This breed slowly loses weight during the dry season and is relatively resistant to biting flies. Sifdar dromedaries produce about 4 kg of milk daily (2–6 kg) and about 1000 kg over a period of 6–10 months (Wardeh 2004).

Eyddimo

The Eyddimo dromedary breed is distributed in the Bay, Gedo and Juga regions of Somalia. They are tall, heavily built, with a relatively bigger hump, long necked, white-coloured and having Roman nose. Eyddimo dromedaries have a slow rate of weight gain during the wet season and slow weight loss during the dry season. They mature at 7–8 years of age, produce about 1000 kg of milk during 6–10 months (4 kg/day) and are very susceptible to biting flies (Wardeh 2004).

9.1.3.11 Ethiopia Dromedary Population

According to Legesse et al. (2018), there are eight breeds in Ethiopia as Jigjiga, Issa, Hoor, Ayden, Liben, Borena, Kerreyu and Afar (Legesse et al. 2018). Wilson (2020) confirmed that the genetic dromedary resources are also generally referred to by the name of the ethnic group owning them and the colour of their coat in Ethiopia and Eretria (Wilson 2020).

9.1.3.12 Eretria Dromedary Population

According to Dioli (2006), the classification of Eritrean dromedaries is based on dromedary-owning tribes, coat colour, region/location, and cultural or social functions (Dioli 2006). However, dromedary breeds have not yet been evaluated based on their performance (Wilson 2020). The four regional populations found in Eritrea are as follows: Gash-Barca, Anseba, Northern Red Sea and Southern Red Sea. The dromedaries of the western lowlands of Eritrea have much in common with those of the Sudan, while those of the eastern lowlands are identical to the dromedaries of the Afar region in Ethiopia, for this breed is transboundary (Gebrehiwet 2021).

9.1.3.13 Djibouti Dromedary Population

Djibouti is a country enclosed between Eritrea, Ethiopia and Somalia, so the breed description follows the same situation as cited for these countries and cited by Dioli (2006) and Wilson (2020).

9.1.3.14 Sudan Dromedary Population

Sudanese dromedary ecotypes include the Shanbali, Kenani, Maalia, Maganeen, Butana, Kabashi and Lahwee, and they have similar

Fig. 9.9 Afar dromedary in Afar Region, Northeast Ethiopia (Melkamu et al. 2022). (From Melkamu et al. 2022, licensed under CC-BY 4.0)

morphological characteristics (Ishag et al. 2011). They have either grey, brown or yellow coat colours. They have a large body conformation and a well-developed large hump with medium-sized udder and teats. Pack dromedaries are the ecotypes that make up the majority of this heavy group. Despite being classified as 'pack dromedary', the Rashaidi dromedary animals have different phenotypic characteristics from the above. They have dark grey or pinkish red coat colour; they have a light weight, with a shorter shoulder height than earlier ecotypes. The Anafi and Bishari breeds, on the other hand, are classified as riding and racing dromedaries, and they have similar characteristics (white, yellowish coat colour, light weight). The Arabi dromedary (Shanbali, Kenani, Maalia, Maganeen and Lahwee dromedary) has better performance and higher milk production and is considered a dual-use dromedary (meat and dairy). Wardeh (2004) described some breeds not cited by Ishag et al. (2011). These breeds are Shallageea and Aririt.

Shallageea

The Shallageea dromedaries are found along the Red Sea coast in Northeastern Sudan. They are tough and eat mostly 'salty Adlib' or Shrubby Seablite (*Suaeda fruticosa*) as well as other mangrove plant species leaves and fruits. The Shallageea are experts at walking into the sea and nipping the lemon-like fruits from mangrove bushes. They frequently wade so far into the water that only the head and hump are visible above the surface. The Shallageea are excellent dairy types, especially during the ad lib fruiting season in November, March and July. One dromedary can give up to 6 kg of milk during early night milking. It may give another 3.5 kg 3 h later. The morning milk could then amount to 6–7 kg. With three milkings per day, each animal may give a daily yield of 15–18 kg. During a rainy period and excellent pasture, an animal can produce 18–21 kg of milk daily (Wardeh 2004).

Aririt

The Aririt dromedary breed is found in the western vast plateau of the Red Sea hills of Aulib in Sudan. This breed is endurant; it can cover long distances at a steady pace without water. The Aririt is also a fair dairy type. Under good conditions, animals may give up to 2.5–5 kg of milk at midday and evening milking and 3.5–5 kg in the morning. The volume of milk production depends on the feeding resources made available (Wardeh 2004).

Kenani Dromedary

The Kenani breed is also known as the Rufaa dromedary. Rufaa, Agilieen, Dighame and Kenana tribes own the breed, which is found in Sinnar and Blue Nile states. Kenani dromedaries have average barrel girth, chest girth, height at shoulder, and body weight, respectively, of 2.48 ± 0.23 m, 2.07 ± 0.10 m, 1.95 ± 0.08 m, and 492 ± 71.0 kg, respectively. The dominant coat colours of these dromedaries are dark brown, grey and yellowish. They are distinguished by long hair that covers the entire body, particularly on the hump and neck. The hump is well-developed in these dromedaries, and it is located in the middle of the back. The udder and teats are medium to large in size, and the milk vein is well-developed. Dromedaries are usually found in the south and north of Sinnar state during the dry seasons (winter and summer).

Rashaidi Dromedary (Rachaida)

The Rashaida nomadic tribe migrated from Saudi Arabia to Sudan and bred the dromedaries in Eastern Sudan (Gadaref and Kassala states). Dominant coat colours are dark grey and pinkish red. These dromedaries' average barrel girth, chest girth, shoulder height and body weight are 2.56 ± 0.23 m, 1.94 ± 0.08 m, 1.76 ± 0.07 m and 423 ± 65.9 kg, respectively. The animals from this type are distinguished by their shorter height at the shoulders, heavier weight and exceptional ability to survive in drought conditions (harsh environment). They have short to medium hair and a small hump, as well as a large and well-developed udder and teats. These dromedaries can be found in Al-showak and Gabat Al-feel (Gadaref state) during the dry season, and they migrate north from New Halfa town (Kassala state) in the wet season.

Lahwee Dromedary

This Lahwee is kept by the Lahween tribe in Gadaref state. This breed is mostly found in Al-showak and Gabat Al-feel areas but migrates to Al-soubag area in the wet season. The average barrel girth, chest girth, height at shoulder and body weight of the Lahwee dromedary are 2.51 ± 0.20 m, 1.98 ± 0.08 m, 1.84 ± 0.07 m and 444 ± 50.7 kg, respectively. The coat colours common in this breed are brown, red and yellowish. The hair is of medium length, the hump is centrally located with an erect or side-bending orientation with a medium size udder and teats.

Anafi Dromedary

The Anafi dromedary can be found in Gadaref state in Eastern Sudan amongst the Rashaida and Lahween tribes. In Gezira state in central Sudan, Anafi is by the Shukria Bataheen and Ahameda tribes, and in Sinnar state in central Sudan, Anafi is also found among the Rufaa and Kenana tribes. This breed is kept in small herds and mixed with other types of dromedaries. The barrel girth, chest girth, height at shoulder and body weight of an Anafi dromedary are 2.39 m, 1.95 m, 1.84 m and 416 kg, respectively. Anafi predominantly have white coat colour, but yellowish animals are also found. It has short and soft hair with a small erect hump in the middle of their backs. The animals have small-sized udders and teats.

Bishari Dromedary

The Bishari breed is mostly found in Eastern Sudan. It is kept by the Bishareen, Amarar, Beni Amir and Hadendowa tribes. Their average barrel girth, chest girth, shoulder height and body weight are 2.36 m, 1.94 m, 1.84 m and 410 kg, respectively. Their coat colour is white or yellowish with short hair and a concave face. The hump is small to medium in size, erect, and sits in the middle of the back. Bishari dromedary udders and teats are characterised by their small size.

Kabbashi Dromedary

The Kabbashi breed is found in North Kordofan state, where it is kept by the Kababish tribe. It has an average barrel girth, chest girth, height at shoulder and body weight of 2.25 m, 1.98 m, 1.91 m and 451 kg, respectively. The primary coat colours of this breed are grey, red and yellow. The hair is medium to long in size, the hump is small, erect and located in the middle of the back, and it has medium-sized udders and teats. The Kabbashi dromedary shares characteristics with the

Kawahla and Hamar dromedaries. During the dry season, the owners of these dromedaries live near Al-obied town, and during the wet season, they relocate to the north (Soudari area).

Maganeen Dromedary
The Maganeen dromedary breed is reared in Northern Kordofan state. The barrel girth, chest girth, shoulder height and body weight of Maganeen dromedaries are 2.42 m, 2.08 m, 1.92 m and 473 kg, respectively. Coat colour is mostly grey, red, or yellow, with straight, short to medium hair length. The Maganeen dromedaries have a large erect hump in the middle of their back with large udders and teats. During the dry season, some of the dromedary's owners migrate to Southern Kordofan, while others migrate north to Gabel Al-ain. They return to Al-mazroob during the wet season.

Shanbali Dromedary
The Shanbali dromedary is found in North and South Kordofan. This breed is reared by the Shanabela and Awamera tribes, who travel between the two states seasonally. They have an average barrel girth of 2.6 m, chest girth of 2.06 m, height at shoulder 1.91 m and body weight of 506 kg. This breed is known for its heavy weight. The most common coat colours are brown, red, grey or yellow with straight, long hair. The well-developed hump is erect or bent sideways. The owners of these dromedaries migrate to Southern Kordofan state near Kadugli during the dry season and return to Al-mazroob during the wet season.

Malian Dromedary
Malian dromedary is located in North and South Kordofan states, which are its cradle and where it is kept by the Maalia and Maagela tribes. During the dry season, these tribes live in South Kordofan (near Kadugli and Daleng towns) and migrate to North Kordofan (Al-mazroob area) during the wet season. The average barrel girth, chest girth, shoulder height and body weight are 2.29 m, 2.12 m, 2.00 m and 479 kg, respectively, in this breed. The height of the shoulders of this breed is quite high compared to other breeds. The coat colour varies from red to grey to yellow, with straight, long hair. It has a large well-developed hump located in the middle of the back. It has large udders and teats.

Butana Dromedary
The Butana dromedary breed is found in the Butana chains (east of Gezira state). These dromedaries have different names based on the owner's tribe. The Butana breed is reared by the Shukria, Ahameda, Magareba and Bataheen tribes. The live body measurements, average barrel girth, chest girth, height at shoulders and live body weight of the Butana dromedary are 2.40 m, 1.94 m, 1.88 m and 446 kg, respectively. Coat colour is mostly red or grey with short to medium hair length. The medium-sized hump is either erect of bent sideways or located in the middle of the back. In the dry season (November to June), the Butana dromedary is found near Tambool town and in the Al-soubag area during the wet season.

9.1.3.15 Libya Dromedary Population
According to Abdalmulaa et al. (2019), three different breeds, Fakhreya, Sirtaweya and Mahari, exist in Libya (Abdalmulaa et al. 2019).

The Sirtawi (Sirtaweya) breed is found mainly in the Sirt area in the middle coastal zone in Libya. The coat colours in this breed vary from light to dark brown. This breed has a medium size with a poorly developed hump and a fairly developed udder. Selected females in some private farms under intensive feeding conditions produce about 3000–4000 kg of milk over a period of 305 days (Wardeh 2004).

The Fakhreya dromedary breed is well known for dairy production (3500 kg/year) under natural grazing conditions in the southern and western areas of Benghazi in Libya (Wardeh 2004).

The OuladBouSayf, Orfella and Fezzan types are variants of the Mehari and its crosses that are found in the western Libyan Oases. They are medium in size, with coat colours close to light, and mainly used for riding, packing purposes and dairy production.

The Tibisti breed is found in southern areas of Libya and northern Chad. They are small riding dromedaries suitable for stony deserts (Wardeh 2004).

9.1.3.16 Tunisia Dromedary Population

Nabul

This Nabul dromedary breed is found in Nabul area at the coastal zone Northern Tunisia and are heavier than the other coastal dromedaries and have larger humps (Wardeh 2004).

Al-Qerawan

Al-Qerawan breed is found in the middle of Tunisia, in the Qerawan district. Its size is of average body conformation, well-built, with a developed hump and muscles, and it is prone to fattening. This dromedary has been extensively trained to perform draft work in agriculture and transportation (Wardeh 2004). According to tribal affiliation (Fig. 9.10), Chniter et al. (2013) identified five groups of Maghrebi dromedaries in southern Tunisia (Chniter et al. 2013): the Gueoudi at Ouled Gharib tribe from Kébili, the Guiloufi in the Beni Guilouf tribe from Kébili, the Merzougui in the Marazigues tribe from Kébili, the Ourdhaoui Tataouine in Oudarna tribe and the Ourdhaoui Médenine in the Tawazins tribe (Chniter et al. 2013). Nine body measurements (body length, neck length, thoracic girth, abdominal circumference, height at the hump, height at the withers, width between shoulders, length of the anterior limb and length of the tool) were taken from a total of 304 female dromedaries (aged 6 years) to identify these groups. The height at the hump and the height at the withers, the thoracic girth and the abdominal circumference, and the body length and the neck length all showed a positive correlation. The tail length, on the other hand, had a negative relationship with the abdominal circumference and height at the withers.

When compared to the Ourdhaoui Médenine dromedary group, the Gueoudi, Guiloufi and Merzougui dromedary groups had longer bodies and larger abdominal circumferences. In comparison to Ourdhaoui Médenine, the Guiloufi, Merzougui and Ourdhaoui Tataouine dromedary groups had the highest thoracic girth and height at the withers. In comparison to the Gueoudi, Guiloufi and Merzougui dromedary groups, the Ourdhaoui Tataouine and Ourdhaoui Médenine dromedary groups had the longest tails. The five groups were clustered together, allowing three main classes of Maghrebi dromedaries to be described: large, medium and small. Six vernacular names were identified based on coat colour (Fig. 9.11): Chagra, Chaâla, Safra, Hajla, Hamra and Zarga. The quality of the hair was mostly rough and thick, that way the female dromedary was named Nagga Chalfi. One small proportion was attributed to the females who had sleek and heavy hair, which was named Nagga Khawar (Fig. 9.12).

9.1.3.17 Morocco Dromedary Population

Dromedary classification in Morocco is mainly based on morphological characteristics (Fig. 9.13), but they are poorly differentiated based on genetic analyses (Piro et al. 2020). Three populations were thus differentiated on this basis in the southern region, and they are the Marmouri, Khouari, Guerzni and Harcha. A multivariate analysis based on morphological measurements found that the Guerzni and Marmouri populations were similar and slightly different from the Khouari population (Boujenane et al. 2019). There is almost no information available on Moroccan dromedary populations. However, some surveys conducted in the south of the country indicate that the Marmouri population is appreciated by keepers for its suitable dairy yield, while the Guerzni population is preferred for its meat production.

9.1.3.18 Other Countries

The Al-Jandaweel (Gandoil) dromedary found in Senegal is similar to the breed of the same name in Mauritania described earlier. The dromedary population in Burkina Faso is similar to that found in Mali. The population found in Namibia is believed to have been introduced by the British Army and is not a natural population of this region. A few herds of dromedaries exist in Botswana, South Africa and the Canary Islands, where they are generally used for touristic activities.

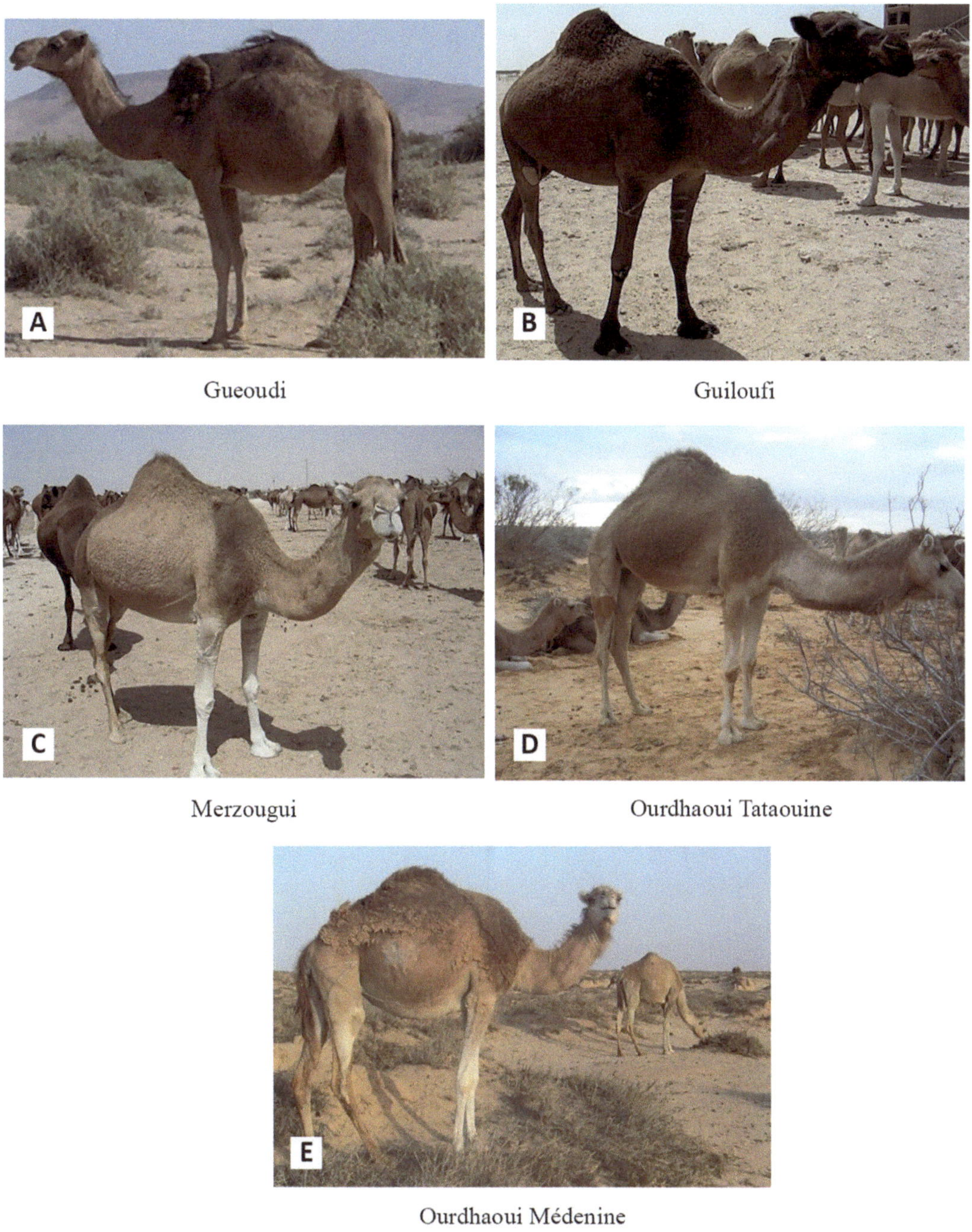

9 African Dromedary Genetic Resources, Diversity and Breeding Systems

Fig. 9.10 Maghrebi dromedary ecotypes in southern Tunisia according to their tribal affiliation. (From Chniter et al. 2013, licensed under CC-BY 4.0)

Fig. 9.11 Dromedary coat colours nomenclature used by Tunisian dromedary herders. (From Chniter et al. 2013, licensed under CC-BY 4.0)

9.1.4 Introduced Dromedary Populations

An example of the dromedary breed introduction can be found in Kenya and Namibia. In Kenya, pure Pakistan dromedary exists and their crosses with Somali or Turkana breeds have since spread to the Samburu, East Pokot, Kajiado, Northern Tanzania, Mandera and Marsabit districts (Kuria and Walaga 2008).

Under ranch conditions, the Pakistani breed produces 4–7 litres of milk daily, making it one

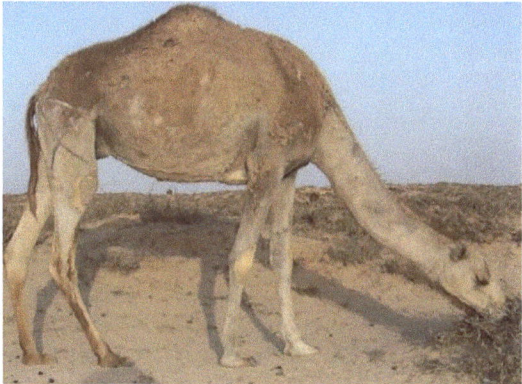

Fig. 9.12 Female dromedary of Chalfi and Kawa types. (From Chniter et al. 2013, licensed under CC-BY 4.0)

Fig. 9.13 Moroccan dromedary breeds. (From FAOSTAT 2019, licensed under CC-BY 4.0)

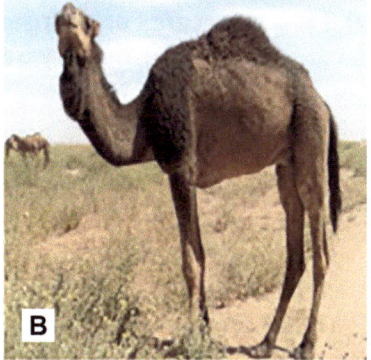

of the best animals for dromedary milk production. The coat is mostly of chocolate colour (Fig. 9.14). Mature body weight ranges from 400 to 700 kg, with a drooping lower lip. The first calving occurs between 4 and 5 years. The average standing height is 1.9 m. This breed is a heavy feeder, insufficiently tested on rough terrain and less hardy (Kuria and Walaga 2008).

Fig. 9.14 Pakistani dromedary. (Photo by: Dr. Asim Faras)

9.2 Structures and Organisation of the Dromedary Sector in Africa

9.2.1 Infrastructure and Support Structures for the Dromedary Industry in Africa

Veterinary Services

Veterinary services for dromedaries are provided by both the government and the private sector. Generally, the central government veterinary services under the Ministry of Agriculture (or Livestock) and the Chief Veterinary Officer are responsible for policies, regulations, guidelines, food safety and oversight of public bodies. The provincial (or regional or county) veterinary services under local or regional governments are responsible for preventive animal disease control. The requirements on the health status of the dromedary depend on the World Organization for Animal Health (WOAH) recommendations for a number of animal diseases in the respective regions. So, the strategic objectives could differ from one country to another. For example, in Algeria, the fight against brucellosis is based on testing and slaughtering programmes. However, obtaining freedom from disease status from WOAH is a serious challenge for several African countries surrounded by neighbours where many animal diseases, such as foot-and-mouth disease, brucellosis, tuberculosis, etc., persist, given the frequent, albeit unofficial, inter-country trading of animals and livestock products and movement of animals for grazing across boundaries.

Breeding Services

Breeding services are intended for the selection of genetically superior animals and control of the reproductive process to increase the presence of genes from these superior animals in the overall population (see Chap. 15 for specific breeding goals and alludes to species-specific breeding goals). However, no African country has been able to develop this kind of service for the improvement of the dromedary species. Dromedaries could then be considered neglected in this aspect.

Research Institutions

There are research centres in different African countries interested in research on the dromedaries' species. These include the 'Desert Research Center' in Egypt, the 'Institute of Arid Regions' in Tunisia, the 'Arid Regions Research Centre' and the 'National Institute for Agronomic Research' in Algeria, 'Agricultural and Livestock Research' and the International Livestock Research Institute (ILRI) in Kenya, the 'Institute of Development Research' in Ethiopia and then the 'Institute for Agricultural Research' in Senegal. The universities and other higher education institutions in veterinary and animal sciences also participate in research on dromedaries (for example, the University of Tlemcen and the agricultural school in Ouargla in Algeria and the

University of Ain Shams in Egypt). These research centres contribute to a better knowledge of local dromedary populations; they also participate in the development of local breed conservation techniques or activities and the improvement of their productivity. Globally, there is the NRCC (National Research Camel Center) in India, the Camel Research Center in Inner Mongolia (China), and several public and private centres in the Gulf countries (such as the camel breeding centre in Dubai or the Royal Dromedary Corps in Oman), especially in Saudi universities.

In terms of organisations, there is the International Society for Camelid Research and Development (ISOCARD), a federation of scientific researchers on camelids (camellodists) and scientific and professional associations in the field. The objectives of these Societies are centred on camelids: give an international scientific status to the camelids' sciences, promote the science and practice of camelids, promote the contributions of camelid scientists to the development of camelids' breeding, promote scientific publications in the field of camelids, set high standards of education and training on camelids, promote standards of health and well-being among camelids, organise the International Camelids Conference every 3 years, encourage the exchange of information on the interest of camelids between the members and the various networks and organisations involved, and finally establish and maintain relations with other organisations. The International Camel Organization (ICO) is a non-profit organisation founded by Sheikh Fahd Bin Falah Bin Hathlin, based in Riyadh, on 21st March 2019. Many African countries are currently among the 105 Member States from the six continents of the world. The organisation aims at developing and serving for the recognition of all dromedary-related legacies.

Many national associations or organisations are interested in keeping and the use of dromedaries. In Algeria, there is the National Office for the Development of Equine and Camel Breeding (ONDEC) and the association, the National Interprofessional Council of Camel Breeding. In Sudan, there is the Sudan Dromedary Association (SUCA) and in Kenya, the Kenya Dromedary Association. This kind of association and federation can be found in many countries in Asia and Europe. Examples include the Dromedary Club (Saudi Arabia), Wild Dromedary Protection Foundation in Magnolia, Dromedary Association of Pakistan, French Federation of the Breeders of Camelids and the equivalent European Federation of the camelids. Collaboration between the African, Asian, European and Australian authorities, institutions and organisations could promote policy and strategy development processes for the development of dromedary genetic resources and their utilisation.

9.2.2 Regulation and Legal Bases of the Dromedary Industry

The regulatory instruments that govern the dromedary's keeping conditions, welfare and health are generally the same as those that govern all farmed animals. The texts are more or less country-specific, but they all follow FAO and WOAH recommendations. For instance, FAO published a report in 2011 based on a national consultation on the analysis of constraints affecting the Mauritanian dromedary value chain, with a focus on animal and zoonotic diseases (FAO 2011). This report identified four main types of constraints to the development of the Mauritanian dromedary sector: lack of support for the sector's development, lack of data and knowledge about dromedaries, lack of infrastructure and animal health facilities. Regulatory short comings are highlighted in particular for genetic resource preservation, strengthening the application of the pastoral code, improving epidemiological surveillance of dromedary diseases, the appropriate use of veterinary products, the safety of dromedary food and the processing and marketing facilities of dromedary products (an analysis of broader policy, strategies and action plans for animal genetic resources (AnGR) in Africa including dromedary camels are provided in Chap. 30).

A priority is to improve on local regulations concerning dromedary keeping, its welfare and the uniqueness of its production. For a better monitoring process, there is a need to extend and harmonise this regulation to the entire sector.

9.3 Structure and System of Dromedary Husbandry in Africa

9.3.1 Geographical Distribution of Farms and Numbers of Dromedary Populations

A detailed review of the world camelid population was recently undertaken by Faye (2020) (Faye et al. 2000). The review showed that dromedary herds have been officially declared in 46 countries worldwide. There are 19 African countries (Table 9.1), 25 Asian countries and 1 European country among them (Ukraine). Dromedaries are only found in Africa, the Near and Middle East, and South Asia, while Bactrians are only found in Central Asia. Kazakhstan is among the few countries where the two species are found. It should be kept in mind that the former Soviet Union included all of Central Asia

Table 9.1 Number of dromedaries in different African countries (FAOSTAT 2019)

Rang	Country	Heads
1	Chad	8,276,416
2	Somalia	7,243,792
3	Sudan	4,895,000
4	Kenya	4,721,900
5	Niger	1,834,943
6	Mauritania	1,500,973
7	Ethiopia	1,281,468
8	Mali	1,241,093
9	Algeria	416,519
10	Eritrea	388,152
11	Nigeria	289,794
12	Tunisia	237,516
13	Egypt	119,885
14	Djibouti	70,894
15	Libya	66,667
16	Morocco	60,808
17	Burkina Faso	20,345
18	Senegal	5030
19	Namibia	93

and that former Ethiopia included Eritrea, and therefore these national entities only represented 38 countries in 1961. Namibia is a relatively new dromedary country, having emerged only in the 2000s. Due to a lack of readily available livestock census, 70% of the data from African countries and 42% of the data from Asian countries are based on estimates. According to the most recent data available in 2022, there were 41,772,353 registered camels worldwide (FAOSTAT 2022). In Africa, there were 34,269,954 with an annual increase rate of about 2.56% (see Chap. 1: African livestock production systems: The past, present and the projected future). However, the population of dromedaries in Africa is not evenly distributed; 14,018,423 heads are in Eastern Africa, 9,401,892 in Central Africa, 5,852,123 in Northern Africa, 4,997,422 in Western Africa and about 94 in Southern Africa. Around 60% of the world's dromedary population is in the Horn of Africa's countries (Somalia, Sudan, Ethiopia, Kenya and Djibouti) (FAOSTAT 2019).

Somalia and Chad alone hold 47.5% (7,243,792 heads) of the dromedaries' population of the African continent (FAOSTAT 2019) and they are part of only eight countries on the African continent with a dromedary population of over one million heads. The other countries include Sudan, Kenya, Niger, Mauritania, Ethiopia and Mali (FAOSTAT 2019). The dromedary population is probably underestimated, especially in the Shelia countries (Mauritania, Mali, Niger, Chad, Sudan and Ethiopia), where the number of heads has been readjusted after exhaustive censuses. For example, in Chad, the number of heads has increased from 800,000 to 1.3 million in a few short years (Faye et al. 2013). In 1961, the date of the first data available, 60% of the 38 national entities reported official data. The total dromedary population at that time was estimated at 12,926,638 heads. The annual growth of the world dromedary population is estimated at 3.07%, but with contrasting national rates varying between −1.95% (Kyrgyzstan, calculated from independence in 1992) and + 45.3% (Oman, calculated from 1961). Overall, 35% of national entities recorded a negative growth rate. An average growth of more than 10–15% cannot be maintained without the importation of live animals because the natural growth of a camel herd can be estimated at about 5–10% (Bonnet 1996; Adamou 2008). In general, the evolution of the dromedary population of the African continent is constant between 1961 and 2019. Thus, the dromedary population multiplied by 1.57 between 1961 and 1980, by 1.33 between 1980 and 2000 and by 1.80 between 2000 and 2019 (FAOSTAT 2019). According to Faye et al. (2013), some African countries are experiencing rapid growth during the recent years such as Algeria, Chad, Mali, Mauritania, Ethiopia and Eritrea, while some have been experiencing steady growth since 1961, such as Burkina Faso, Djibouti, Egypt, Kenya, Niger, Nigeria, Pakistan, Somalia, Sudan, Tunisia and the Western Sahara. Others have stable dromedary numbers, such as Libya and Senegal, and those with a very large decline in dromedary numbers, such as Morocco.

The Algerian dromedary population increased from 286,670 heads in 2007 to 416,519 heads in 2019 (FAOSTAT 2019). This increases the results from several dromedary development programmes established by the Algerian state. In fact, the dromedary livestock industry experienced a considerable boom from the year 2000, following the introduction of the birth premium, which is a kind of financial incentive granted to farmers for any birth of a new dromedary by the Ministry of Agriculture (Bedda et al. 2019).

Over the past 50 years, camelids have grown rapidly, after goats and buffaloes, as shown in Table 9.2. Besides the various data gaps of camelids increased from 16,494,169 in 1969 to 37,509,691 heads in 2019.

Table 9.2 Evolution of African livestock animal numbers between 1969 and 2019 (FAOSTAT 2019)

	2019	1969
Camelids	37,509,691	16,494,169
Donkeys	50,583,572	38,342,749
Horses	59,041,725	60,100,684
Buffalo	204,342,419	106,224,223
Goats	1,094,068,295	377,651,477
Sheep	1,238,719,591	1,080,656,610
Cattle	1,511,021,075	1,069,858,656

9.3.2 Dromedary Breeding Systems in Africa

Like in all domestic species, breeding activities rely on the capacity to record and manage data. Dromedaries are not really on major livestock mainstreams, giving very narrow room for comprehensive breeding schemes. Nevertheless, based on available information, breeding strategies are selection and cross-breeding and very rarely interspecific crossing. Mating systems are controlled for small herds and random when numerous herds meet during nomadic movements.

9.3.2.1 Selection

Selection pressure on the dromedary is low. Apart from keeping animals for a specific purpose (conditioning, riding, milking), few programmes based on selection according to the performance of their offspring have been developed, with the exception of race animals.

Traditional dromedary breeding systems are extensive, based on the use of natural resources and herd mobility. The biological peculiarity of the dromedary is its very low fecundity. Age at sexual maturity is about 3 years, gestation length is long (13 months), and calving interval is generally about 2 years. In addition, calf mortality is high (could reach 20% and even more), hence the reproductive rate is low, though the longevity of the dromedary could compensate for it. Despite these limitations, there is potential for significant productivity gains through selection and breeding, hence scope for the improvement of dromedary meat and dairy productivity in intensified production systems (Breulmann et al. 2007).

Generally, the intensity of selection for females is low, based mostly on reproduction, mothering ability and conformation. This is because almost all breeding females are required for the maintenance and growth of the herd. Conversely, a high selection intensity can be placed on the males. Once a male has been selected, it can be used continuously for 2–12 years, with an average of 7 years. The ideal age for sexual activity for a male when it can be used for breeding is between 6 and 12 years old.

At younger ages, the weakening of the rut does not allow it to mate efficiently, while at older ages, its fertility declines and the male becomes more selective; thus, some females become abandoned without a mate. However, females remain well fertile over 25 years old and can produce 7–9 calves in their lifetime. Most breeders, however, start by eliminating the least fertile females at the age of 15. The breeding objectives include many aspects other than commercial meat and dairy production. They can include aesthetic preferences (coat colour and colour pattern) and behavioural aspects, such as good maternal instinct, ability to live in herds and ability to travel long distances (Doutresoulle 1947; Chaibou 2005; Djomtchaigue et al. 2015). The dromedary's ability to withstand extreme weather conditions (drought, etc.), coating is another important advantage that affects high productivity. Breeders practice phenotypic selection to identify the best, mostly among male candidates. The selection of breeding males is based on available information on their parents (family selection) and themselves (individual selection). A farmer's survey on criteria for the choice of the future sire indicated that male selection is based on information from its dam. The features considered include the genetic type, size of the body and hump, milk yield, health status, disease resistance and drought tolerance. Some criteria also considered include fertility, duration of lactation, offspring quality and coat colour. In selecting a male, breeders also take into account certain qualities of the sire, such as genetic type, the large size with a large bump (conformation), the quality of its progeny and health status with disease resistance, tolerance to drought and fertility. The individual characters considered in the selection of candidates include large size, conformation, beauty (slender appearance and fine hair), docility (young dromedaries that show early aggression are often removed), disease resistance, drought tolerance and coat colour (Cherifi et al. 2017).

The development of modern mini-dairy, particularly in Mauritania and Algeria, is currently helping to direct selection towards the most milk-producing animals. The creation of interprofessional organisations and breeders' associations is amplifying this trend (Gherissi and Gaouar 2022a).

9.3.2.2 Cross-Breeding

This breeding strategy, which consists of mixing at least two distinct gene pools, has been described by Doutresoulle (1947). It was a strategy used to upgrade or create genotypes destined for racing or disease resistance during the colonial era in French-speaking West Africa.

9.3.2.3 Interspecific Crossing

Hybridisation between *C. dromedarius* and *C. bactrianus* was thought to be impossible for centuries. However, a lot of the research on cross-breeding between the dromedary and Bactrian varieties has been undertaken in the former Soviet Union, particularly in Kazakhstan. Such hybrids have long been known in Turkey, northern Iran and Afghanistan (Khan et al. 2003). Planned hybridisation was very important in Turkey in the second decade of the last century, when up to 8000 female dromedaries were imported from Syria and further south to be mounted by Bactrian males. F1 male hybrids with normal spermatogenesis are almost always fertile. When bred with males of either species, the hybrid female produces offspring that look like the male parents (dromedary or Bactrian). In terms of size, hardiness, endurance and longevity, the F1 is heterogeneous with a longer single hump that is less developed than in the dromedary. Some Bactrian characteristics, such as the beard and hairy legs, are preserved (Khan et al. 2003).

The dromedary crosses with the Bactrian to produce a strong draught animal with a wool yield similar to the Bactrian. The hybrid's dairy production and milk fat content are midway between the yields of the two parental types. Other crosses, particularly those resulting from F1 hybrids, appear to produce weak, poorly conformed animals that are difficult to maintain. (Khan et al. 2003) successfully crossed the camel and the llama in Dubai (Khan et al. 2003). 'Camella' was the name given to the offspring (López 2010).

9.3.3 Dromedary Production Systems in Africa

9.3.3.1 Nomadic System

Nomadic pastoralists do not have permanent fixed habitats and the whole family follows the movements of the herd, sometimes over long distances (hundreds of kilometres) (Chergui et al. 2023). They move at the head of the herd, of which they are not always owners but guardians. They lead a seemingly unstable existence: they move constantly, and the movements are traditionally done on donkeys. Dromedaries move from one water point to another, using rivers, streams and wells. As a result of this mobility, nomads practice little or no agricultural activity (Cherifi et al. 2013).

H'mil (Free Range)

In the great Sahara region, where great distances allow families to isolate themselves in the vastness, dromedaries are often left with complete freedom (Chergui et al. 2023). They know the well where they can find the shepherd who provides water, and so they return quite often when they are thirsty (Radfar and Aminzadeh Gowhari 2013; Habte et al. 2021; Chergui et al. 2023). This system is practised mainly in poor seasonal conditions (e.g. no rainfall, etc.) where dromedaries are left to freely search for water and pasture. The disadvantage is that products of the dromedary, such as milk, hair, etc., cannot be exploited during this period, and there is also a high risk of disease spread within the herds, an increase in mortality rate due to road accidents and loss of dromedary calves that are not marked. This system also results in difficulties with dress coating (Gauthier-Pilters 1977). It is a free-range extensive production system practised mostly when the year is presumed unfavourable due to harsh weather, and hence a lack of pasture. It is a system characterised by movements towards water points and the search for pasture that thrives with little rain. The majority of breeders who practice this system are sedentary, giving themselves to various activities, including trade, throughout the year. The control and recovery of dromedaries is done near water points in the summer season (Bensemaoune et al. 2018; Chergui et al. 2023).

9.3.3.2 Sedentary System

'Sedentarisation' is sometimes used to describe a process of evolution and adaptation of nomadic populations which reduce the amplitude of their movements and include agricultural practices in their activities (Kaufmann 1998; Chergui et al. 2023). The sedentary production system is characterised by movement of herds over long distances during the day, with return to the village at evening (Bourbouze 2006). Currently, in Africa, very few sedentary herds exist (Chergui et al. 2023). The few that do exist are found in peri-urban regions, and most of them have a dairy vocation (Gherissi and Gaouar 2022b).

9.4 Dromedary Production Objectives

Dromedary rearing has long been a vital part of the social and economic lives of people living in arid and desert areas of Africa and Asia. Dromedaries are multifunctional animals used in African pastoral production systems to provide milk, meat, blood, hides and skins, transportation, barter trade (sale and exchange) and social and cultural functions (Kuria and Walaga 2008). The image of the dromedary is a symbol of mankind's survival in the desert, and it is associated with the history of the great nomadic civilisations of dry and hot regions characterised by long periods of unfavourable weather, often exceeding 8 months, and rare and weak precipitations (Ramet 2013). As a result, in recent years, this animal has become the animal of choice for responding to the phenomenon of global warming.

9.4.1 Primary Production Objectives

9.4.1.1 Milk Production

Dromedaries have an outstanding milk production, especially in harsh environments, when

 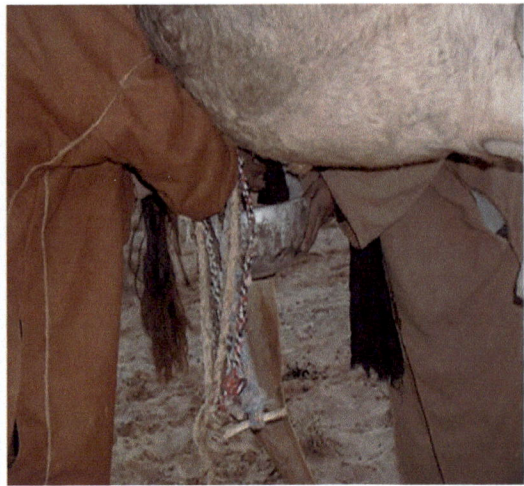

Fig. 9.15 Traditional milk production. (From Cherifi et al. 2013, licensed under CC-BY 4.0)

compared to cattle and small ruminants under similar harsh environments. Their lactation persists well into the dry seasons and rarely ceases even during extended dry periods (Chergui et al. 2024). Generally, the females are milked twice a day, morning and evening, with the presence of the dromedary calf that acts as a stimulus for milk letdown in the udder (Fig. 9.15). Dromedary milk production is difficult to quantify due to the mobility of herders and irregular milking. The frequency of milking varies from once a day to twice a day, depending on the individual breeder (Nega and Tefera 2012; Kebede et al. 2015). The milk is taken mostly for home consumption and sometimes for sale. Daily milk yield ranges from 5 to 10 litres, depending on breed and production system. For one of the most famous Algerian dairy camel, the Sharaoui camel breed, Chergui et al. (2024) reported the following lactation performances: an average daily milk (DMY), fat (DFY) and protein (DPY) yield were 6.77 ± 0.82 kg/day, 4.15 ± 0.91% and 4.49 ± 0.20%, respectively (Chergui et al. 2024). The mean of total milk yield (TMY) was 2696.39 ± 343.86 kg during a mean lactation length (LL) of 398.38 ± 20.65 days. There is no significant difference between intensive and semi-intensive breeding systems for TMY, DMY and LL. However, the total amount of fat was significantly higher in the intensive system, and the DPY content was significantly higher in the semi-intensive system (Chergui et al. 2024).

The development of peri-urban farms around could constitute the core of dromedary dairy farming in the future. The use of dromedary in dairy production is becoming more rational in Mauritania, Niger and Chad (Chaibou 2005; Djomtchaigue et al. 2015; Faye et al. 2000). It is expected that dromedary milk production, which was previously practised mainly for the survival of indigenous peoples in arid regions, will grow with the application of technologies to become a key alternative source of milk and dairy products for the general population. In this regard, commercial dromedary milk is to be increased (Fig. 9.15) (Cherifi et al. 2013).

In some pastoralist tribes, such as the Afar and Somali, women are not allowed to milk dromedaries because they believe lactating dromedaries do not let down enough milk for women (Kebede et al. 2015; Seifu 2009; Simenew et al. 2013). However, amongst the Rashaida tribe, dromedary milking is practised by women. This is also the case in Kenya, Somalia and most parts of Sudan (Eisa and Mustafa 2011). Smoking the milking vessels with local herbs is thought to enhance the flavour, taste and quality of the milk while also extending its shelf life (Seifu 2009; Wanjala et al. 2016; Gebremichael et al. 2019). Smoking milk-handling containers inhibited microbial growth and could be used for sanitation and preservation of raw dromedary milk in arid and semi-arid areas where cold chain facilities for milk preservation are unavailable (Gebremichael et al. 2019). The estimated average daily milk yield per dromedary in Ethiopia under pastoralist milking practices ranged from 2 to 6 litres, according to (Gebremichael et al. 2019). The Algerian dromedary's female can produce 3.96 ± 1.24 litres of milk per day on average. Bakheit et al. (2008) conducted a comparative study of dromedary milk production in Sudanese traditional and semi-intensive dromedary farms, finding that the average daily milk yield for the semi-intensive and traditional systems, respectively, was 6.85 ± 1.32 and 3.14 ± 0.66 liters (Bakheit et al. 2008). AM et al. (2018) found a similar result in Egyptian dromedary herds, finding that daily or total milk yield was significantly ($P = 0.001$) higher for dromedaries under improved farming systems than for those under traditional pastoral systems (26.12 vs 13.41 litres of milk, respectively). With improved feed, water and husbandry, lactating dromedaries can yield up to 20 litres per day (Kuria and Walaga 2008).

Benefits of Dromedary Milk

The dromedary milk has a high nutritional value, often discussed in medical literature. It is slightly saltier than cows' milk, three times as rich in Vitamin C and is known to be rich in iron, unsaturated fatty acids and Vitamin B. Surveyed farmers believe it is suitable for treating diseases such as diabetes and liver problems. It may play a role in reducing coronary heart disease, and some recommend it to HIV/AIDS patients. Research is beginning to demonstrate the scientific basis for these properties (Cherifi et al. 2013). Dromedary's milk is thus preferred over other livestock species' milk because of its taste, nutritive value, and health properties, and it is perceived that dromedary milk prevents thirst even when walking long distances. It is a natural and essential food item in areas where there is water scarcity.

By virtue of its nutritional qualities, dromedary's milk constitutes the essential basis of the diet of nomads, as well as a remedy against many ailments (Viateau 1998). While the virtues of dromedary milk have been prophesied over 14 centuries ago, scientists are only now just beginning to test prophetic medicine, widely brought up to date by the Sahrawi, Tuareg and Berbers. The value of dromedary milk has been confirmed or is in the process of being confirmed by modern science. Scientists are also beginning to discover the preventive and curative medicinal properties of dromedary milk and its value in the treatment of many diseases. It is believed that it contains antibiotics of a very rare type and substances that can fight diseases that modern medicine cannot yet cure (Konuspayeva et al. 2004).

The therapeutic benefits of dromedary milk are numerous, with diverse roles in autoimmune diseases, metabolic disorders, tuberculosis, hepatitis, cancer, rickets, diabetes, Crohn's disease, cirrhosis, autism and rotavirus diarrhea (Zibaee 2015). Previous research on the Arabian dromedary (*Camelus dromedarius*) revealed that its milk has beneficial effects in a variety of human diseases, including significant hypoglycaemic activity. The cellular and molecular mechanisms underlying such effects, however, are unknown. As a result, Abdulrahman et al. (2016) theorised that dromedary milk might act on the human insulin receptor (hIR) and related intracellular signalling pathways (Abdulrahman et al. 2016). Using bioluminescence resonance energy transfer (BRET) technology, they investigated the effect of dromedary milk on the activation of hIR transiently expressed in human embryonic kidney 293 (HEK293) cells. BRET was used to examine the physical interaction between hIR and insulin receptor signalling proteins (IRS1) and growth factor receptor-bound protein 2 in live cells and in real time (Grb2). In the absence of insulin stimulation, dromedary milk did not result in an increase in the BRET signal between hIR and IRS1 or Grb2. It did, however, potentiate the maximal insulin-promoted BRET signal between hIR and Grb2, but not between IRS1 and Grb2. Dromedary milk appears to have a differential impact on downstream signalling because it significantly activated ERK1/2 and potentiated insulin-induced ERK1/2 activation but not Akt activation. The peptide/protein nature of the active component in dromedary milk is suggested by preliminary fractionation. For the first time, this study shows that dromedary milk has an allosteric effect on insulin receptor conformation and activation, with different effects on intracellular signalling. These findings should contribute to a better understanding of the hypoglycaemic activity of dromedary milk with potential therapeutic applications.

9.4.1.2 Meat Production

Meat is a preferred source of animal proteins because of its richness in amino acids (Geay et al. 2002). The dromedary has a high carcass yield and meat with dietary qualities that are appreciated and consumed on a large scale in the Algerian Sahara (Faye et al. 2013; Ould El Hadj et al. 2002). It is for this reason that dromedaries, adapted to the extensive conditions of the arid zones, are increasingly involved in intensive production systems (Faye and Porphyre 2011; Ould El Hadj et al. 2002; Salter 2013; Smith et al. 2013).

The contribution of camelids to the world production of red meat is very marginal due to the size of the camels' population: it represents 0.13% of the meat produced in the world and 0.45% of red meat. However, because a high proportion of camels are confined to arid countries, their contribution to meat production appears to be higher locally. In East Africa, dromedary meat represents 4.1% of red meat produced, 4.8% in North Africa, 2.9% in West Africa, and 3.7% in the Near East. Everywhere else, it represents less than 1% of total red meat. In Northern Africa, the total production of dromedary meat increased between 1961 and 2011, from 21,600 to 63,143 tonnes, i.e. three times in 50 years, and therefore in proportions not commensurate with the increase in milk production (Faye et al. 2014). While the dromedary population of North Africa represents only 3.5% of the world's dromedary population, dromedary meat production is equivalent to 15.4% of world's dromedary meat production. Dromedary meat represents 4.8% of red meat produced in North Africa, but it varies between 1.1% in Morocco (66.9% in the Saharan provinces), 1.2% in Tunisia, 1.3% in Algeria, 4.6% in Egypt and 9% in Libya (Faye et al. 2014). This shows the place of this product in the Maghreb countries. Dromedary meat is widely consumed in certain regions of Africa, such as Somalia, Mauritania and Western Sahara. Per capita rate of consumption is 0.14 kg/year in Algeria, accounting for 4.2% of the total red meat consumed (Faye et al. 2014). The growth observed in dromedary meat consumption is linked to an increase in the slaughter rate in this growing population and to a significant increase in the weight of the carcasses. This is evidenced by the situation in Chad between 1991 and 1996, where production increased from 244,708 tonnes in 1991 to 755,232 tonnes in 1996, an increase of 67.59% in 6 years (Mbaïgaou 1998).

Surveyed breeders confirm that production is primarily driven by the sales of livestock and that meat production is the true purpose and source of income for the farmer in Northern Africa (Fig. 9.16) (Cherifi et al. 2013). Due to the low

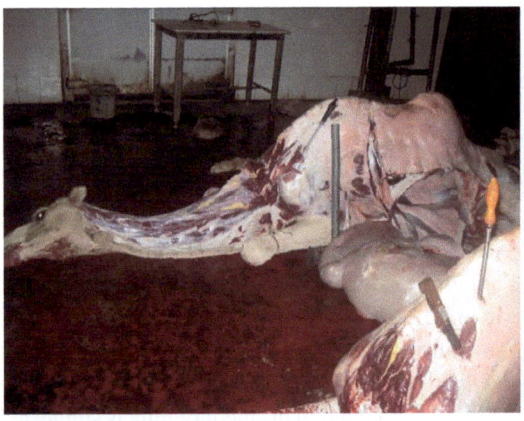 

Fig. 9.16 Dromedary slaughter and carcass/meat handling. (From Cherifi et al. 2013, licensed under CC-BY 4.0)

reproductive performance of dromedaries compared to cattle and small ruminants, dromedary meat production is less important in the eastern African region. The pastoral communities' main source of meat is small ruminants, but dromedaries are occasionally slaughtered for meat. Pastoralists claim that dromedaries'meat is of higher quality because of its nutritional value and flavour. The majority of dromedaries are slaughtered at home for personal consumption, with any surplus meat sold to a butcher (Kuria and Walaga 2008).

The Virtues of Dromedary Meat
In dromedary keeping, meat is the first source of animal proteins, thanks to its richness in amino acids, it is among the noble proteins (Geay et al. 2002). The dromedary has a high carcass yield and meat with dietary qualities that is appreciated and consumed on a large scale in the Sahara (Faye et al. 2013; Ould El Hadj et al. 2002) and where available. Thus, dromedaries are increasingly included in intensive production systems (Ould El Hadj et al. 2002; Salter 2013; Smith et al. 2013; Faye et al. 2011).

9.4.1.3 Wool Production
The wool 'Oubar' is considered a secondary product of dromedary production. Farmers use the wool for the manufacture of Barnous, tents and dromedary chest caches. Herders are not always aware of the market value of the dromedary wool and its usefulness in the manufacture of handicrafts (carpets, clothing, decorative items, etc.) with great export potential. A cottage industry based on these high-value products will have a high commercial value (Cherifi et al. 2013). In countries where tourism is well-developed, products developed from the dromedary wool and skin could have a significant commercial outcome.

9.4.1.4 Bone and Blood
Dromedaries also provide blood, which is mixed with milk to create a diet component for young children, including the herders (Kuria and Walaga 2008). They also serve for social and cultural purposes, facilitating social transactions such as gifts, loans to relatives and friends and the provision of food during ceremonies. Dromedaries are sold and exchanged only during droughts or when pastoralists require large sums of money for health or school fees. The importance of the dromedary in the Samburu community is explained by these unique and strategic uses of dromedaries and related products. Despite all of the advantages of dromedary production in East Africa's pastoral areas, dromedaries still face challenges in their natural habitat, including diseases, drought and predation, which expose the pastoralists to risks of losing their source of livelihood (Kuria and Walaga 2008).

Dromedary bones were once used as tent poles when wood was scarce (Lasnami 1986). Ibn Khaldoun (1377) reported that during the drought, after prolonged fasting, nomadic inhabitants collect bleached dromedary bones, pulverise them and mix them with water to form a paste for consumption. While the Muslim religion prohibits the use of blood as food, certain nomadic populations of southern Ethiopia and northern Kenya take 5–7 litres of blood, 2 or 3 times a year, from each animal to drink fresh or mixed with milk (Faye and Bengoumi 2002).

9.4.1.5 Other Products
The development of by-products from wool and leather remains below its potential. In recent years, some progress has been achieved through projects aimed at promoting technical innovation in these sectors. For example, dromedary wool was traditionally only used to make burnous-type coats or the winter gown (wazra) of Libyan and southern Tunisian dromedary drivers. However, in this sector, which is largely linked to craft processing, the automatic separation of the type of fibres (separation of jars hair from better-quality fibres) and innovations in finished products for customers seeking more refined dromedary wool fabrics have enabled advances in this sector under the impetus of the procamel project in Tunisia and Egypt. Similarly, the leather and skins sector, long neglected due to the low quality of the product, resulting from poor slaughter procedure (quality of collection), quality of treatment in tannery and skin diseases, is currently undergo-

ing some development in Tunisia and Egypt. This work aims at creating specifications for the collection and technological innovation to obtain finished products of high quality suitable for export markets. In Tunisia, collaboration between the producers and the Technical Centre for Leather and Footwear, which works to find quality products, is promoting the development of this by-product (Faye 2014).

9.4.1.6 Pharmacopoeia Aspects of Dromedary Urine

Almost everything obtained from the dromedary is useful in one way or another. Some keepers claim that they use everything, even the dung and saliva, which is exploited for its baking soda content (Mammeri and Khir 2018).

The traditional pharmacopoeia in many countries uses dromedary urine as a therapeutic ingredient. The consumption of dromedary urine, either alone or mixed with dromedary milk, has been traditionally used for health (Alkhamees and Alsanad 2017). Young nomadic girls and women use the dromedary urine collected as 'champing'. According to the nomads, this strengthens the hair and makes the hair red like using henna (Lasnami 1986). Studies have demonstrated the existence of antimicrobial activity against pathogenic bacteria in urine associated with dromedary milk (Benkerroum et al. 2004) and lactoferrin activity against hepatitis C virus infection and cancer cells (Esmail et al. 2008). Since ancient times, dromedary milk and urine have been used as medicines in certain parts of Asia and Africa. Dromedary milk, alone or in combination with dromedary urine, has been shown to be effective in the treatment of a variety of clinical conditions, including diabetes, cancer, food allergies, autism, viral hepatitis and a variety of other viral, bacterial and parasitic infections, according to significant evidence derived from laboratory and limited clinical studies. Furthermore, a number of potential cardiovascular benefits of dromedary milk and urine, particularly their antiplatelet and fibrinolytic actions, have been demonstrated (Abdel Galil and Alhaider 2016).

9.4.2 Use of the Dromedary in Africa

Among the Tuareg and Saharawi, the dromedary has always been considered a prestigious animal and is exhibited during cultural events. Indeed, the animal partakes in many religious (Enid El Adwa), sporting and customary ceremonies. The dromedary has long been used as a means of transport for pilgrimages in Tuareg and Saharawi and for funeral ceremonies that require the use of the dromedary as a means of transport. The place of dromedaries is central in the culture of the Sahara as a source of wealth, prestige and serves as dowry (at weddings); thus, at weddings, the parents of the bride or groom may gift 2–4 dromedaries to their children. To accompany the new bride to her future husband, a procession is organised on the backs of the most beautiful dromedaries.

The dromedary is also used among the Tuareg as a means of compensation in dispute resolution. It is used as a piece of embellishment, especially for honoured guests, and constitutes a treasure for the tribes where they are regarded as 'friends and companions of road and transhumances of the Sahara'. Thus, among all these ethnic groups, the greatest honour that can be done to a distinguished guest or a relative is to offer him a meal based on meat from the dromedary.

9.4.2.1 Coatage

Coatage refers to all the accessories necessary for an activity that the dromedary will carry out. The coatings for the pack saddle start from the age of 4–5 years, but the full load is assigned to the dromedary only from 6 to 8 years, not to disturb the growth and avoid joint deformations. The career of a 'carrier' dromedary can last 12 years. Coatings for agricultural activities begin at the age of 3 years, when a ring is placed in the nostrils of the animal. Only one man is needed to drive the dromedary when it is well coated. Depending on the husbandry conditions, the productive life of the dromedary at work varies from 6 to 20 years. The coatage for saddle riding starts from the age of 3 years, but only the adult animal, around 6 years, is actually used (Lhoste et al. 2010).

During the entire coating period, which is a period of stress, animals must be properly fed and watered; an additional 3 kg of bran cereals per day is recommended (Lhoste et al. 2010).

The dromedary coatings includes a basic coat, which lasts from 6 to 8 weeks, during which the animal learns to sit, lie down, move forward and backwards, turn and carry a load on its back. It takes about 2 more weeks to coat wagons at the pull, a little more if a coat is combined with the pulling of agricultural tools. The dromedaries are guided by a rope passed through the nasal ring. They are led to walk and trot with animals accustomed to being loaded, to move with a load and to obey the instructions given through the rope. Straight line walking is acquired quickly (Lhoste et al. 2010).

For a heavy work period, frequent breaks are advised. The length of the ploughing lines can also be limited to about 50 metres to allow the animals to catch their breath at the end of the furrow. Never allow animals to stop during a skate for no apparent reason. It is risky to 'overload' a hitch even once, as animals can become suspicious and reticent (Lhoste et al. 2010).

9.4.2.2 Agricultural Work and Importance of the Dromedary in Breeding

The use of the dromedary in agriculture is traditional both in oases (water exhort) and in agricultural regions of medium mountains (ploughing, sowing), as a single draught animal (Tunisia, Egypt) or paired with a donkey or mule (Morocco). Its coupling to various wagons for the transport of property or people is much less common than in other countries such as India or Pakistan. As a transport animal, it is still the pack which is the most common, for example, for the transport of wood (Algeria) or dates (Tunisia, Morocco) (Faye et al. 2014).

The productivity and integration of the dromedaries into a generally poor, arid environment is remarkable. This coping capacity is quite superior to cattle, especially exotic breeds recently introduced into Africa. The dromedary, nicknamed 'ship of the desert', is an excellent animal for pack and saddle. It is used to till the soil, draw the water from the wells and turn the grain wheels. The dromedary is a great walker and feeds on almost all the plants, herbaceous and woody species in the pasture area. However, it takes small amounts at a time and is considered to be one of the pseudo-ruminants that does the least overgrazing, unlike, for example, goats. When the pasture is used by other animals, it can, thanks to its large size, still find forage where other species could not reach: 70% of the dromedary feed comes from the same thorny shrubs, while 80% of cattle feed comes from grass. It has the ability to graze up to 80 km from a water point and to remain without drinking for several days, if the water content of the plants allows it. This is not the case for cattle. The resistance of dromedaries to climatic rigour is remarkable. During the great droughts of the past years, cattle losses were of the order of 80% in some regions, while the dromedary population suffered only a relatively minute loss, not exceeding 20%. According to a recent study carried out in Sahelian areas, the dromedary herd seems to revolve around a numerically constant nucleus of adults aged 5–12 years old. This partly explains why the dromedary population can recover fairly quickly after high mortality. Adapted physiology of the dromedary to survive in the most arid environments comes from a number of physiological originalities. In addition, unlike other warm-blooded animals, the dromedary does not waste its water reserves during the hot hours of the day to cool down by sweating (Bengoumi and Faye 2002).

In Eritrea, dromedaries are frequently used as pack animals, for riding, for ploughing and for driving oil mills known as Assara. In addition to providing transport during migrations, they are regularly used for carrying fuelwood, trade goods and, most importantly, water for household consumption (Gebrehiwet 2021).

9.4.3 Sociocultural Role of Dromedary Livestock

9.4.3.1 Saddle Dromedary

There is no definition of the ideal physical feature of the dromedary. Generally speaking, the perfect dromedary should be sturdy, heavy, muscular and well-boned. Several factors determine the load to be transported by the saddle dromedary: age, size, breed and distance to travel. The other factors involved are: the terrain, the pace and the forage covering the road. The animal must be educated between 2 and 5 years, and it is only when it reaches its full maturity that it must be required to transport complete loads, proportional to its strength. The load of the animal must be balanced, and the best system is therefore to divide the luggage into two equal parts which are suspended on both sides of the hump.

9.4.3.2 Saddle and Racing Dromedaries

The saddle and racing dromedary is an essential means of communication in remote arid and semi-arid regions of Africa. It enables police patrols, nomadic guards, extension workers and nomadic populations themselves to move. It is likely that this mount role will not be completely abandoned by the pastoral people even in the face of the rampant proliferation of off-road vehicles. The use of saddle dromedaries as war animals probably dates back to 190 BC (Tennews 2015). The ideal saddle dromedary is thin; it has long legs and a strong bone structure. An adult male saddle dromedary must give an impression of male strength, boldness and symmetry. A short head with a high forehead, a buzzed nose and a rather deep knout with firm lips are desirable characteristics. As an example, the dromedary of Targui in Algeria and in the countries of the Sahel is used as a saddle animal, unlike other breeds, which are used for the pack and the production of meat.

Four speeds have been identified to describe the different gaits at which the saddle dromedary moves: the walk, the small trot, the large trot and the amble. Normal walking speed is about 4 km per hour, while the small trot usually allows the animal to move at 8–12 km/hour on flat ground. At the high trot, the speed described for North African and Arab dromedaries is 14–19 km per hour (Sharp 2016). According to Philipponneau (2014), the dromedary can run very fast. In racing, it reaches 70 km/h, which is faster than a horse (Philipponneau 2014).

9.4.3.3 Sociocultural Uses

In cultural ritual ceremonies, the dromedary is used in various ways (pack animal, saddle animal, race animal, etc.). In these cases, the richly adorned animal can be used in the company of many others for demonstrations, parades or even breeding. For example, for the Chadian ethnic group called Kréda, during the first wedding ceremony: 'ossana', for 3 days, the son-in-law and his companions come to camp next to the *Ferrick* (traditional housing in Chad) of the in-laws. They make the turn three times before getting off their steed (dromedary), richly adorned in the middle of *youyous* (cries of joy emitted by the women). Although the dromedary is not directly involved in religious ceremonies, it contributes indirectly by providing the needs of the cameleers. They can sell one or more dromedaries to meet their food needs during the Ramadan period (sacred fasting), dress the family for religious holidays, or buy very good rams for the feast of Eid al-Adha. They are sometimes sacrificed on the occasion of this feast in place of the sheep.

In marriages of some Berber tribes in the south of Tunisia, as in the society Touareg or Afar, the dromedary is used to pay the dowry (Morin 2006; Cabalion 2013). Among the Fili, the dowry is higher, and if the marriage is endogamous (between clan members), 10–30 sheep, a dromedary and money are required for the dowry. However, if the marriage is exogamous, i.e. outside the clan, the dowry can exceed 40 sheep (in addition to the dromedary and money) (Waefelaer 1982). Among the Tuareg nobles, the service that legitimises marriage (often called '*dot*' in French) and which is intended for the father of the bride must consist exclusively of dromedaries. Only people of low extraction will give the equivalent in the form of cattle or, worse still, in small ruminants. In some parts of sub-Saharan Africa, the parent-in-law helps the young couple to settle by giving them a dromedary and a well-harnessed horse, a loan of 30 cows with a bull and a donation of about ten cows. Similarly, the newborn receives from his parents, grandparents, 7 days after the birth, a gift of some heads of dromedaries, cows, etc., which already constitute his small patrimony. These show how dromedaries play a leading role not only in marriage but also in the start of the independent life of the new households.

9.4.4 Modern Uses of Dromedary

Because it is rich in vitamins C and contains iron and vitamins A, B1, B2 and B12, dromedary milk is an excellent raw material for cosmetic products. For example, organic dromedary milk soap reduces stains on the face, regenerates cells, deeply moisturises and nourishes the skin and makes it softer. Dromedary fat is also believed by some to be a remedy against rheumatism, osteoarthritis, asthma, hair loss and the treatment of respiratory problems (Agrimaroc 2021).

In Mongolia, a Swiss corporation once helped to develop and promote a line of cosmetic products based on dromedary milk. They introduced a programme to train farmers to produce and supply clean, good-quality milk as raw material to the processing industry. The 'Dromedary act' brand products were made available in stores in the capital, Ulan-Bator and for export (Lhoste et al. 2010).

According to the Swiss corporation, the added value obtained from the dairy products of the Bactrian dromedary increases the income of the producers by 80% on average.

The industrial exploitation of dromedary products (wool and milk) is clearly a way to improve the place of this species in Mongolian agriculture (Agrimaroc 2021). This activity and launch phase in Algeria via the collaboration that was born through the CA.RA.VA.N. The objective was the implementation of a modern dromedary selection system project involving the dromedary owners and the students of the University of Tlemcen through start-up projects for the cosmetic valorisation of dromedary milk and its fat (Lakermi and Sabrou 2021).

A technical innovation developed in India transforms dromedary faeces into paper. The dromedary faecal excretions are characterised by a low nitrogen composition, which makes its manure unsuitable as fertiliser. However, faecal material is rich in indigestible fibres that can be exploited after an adequate treatment in the form of pulp. This process was previously tested on elephant droppings, which are also rich in fibre; however, its development from dromedary droppings is a first. This is happening in India and more precisely in Rajasthan, a region of the country where the dromedary husbandry remains important and where the raw material is not lacking. Products such as diaries, notebooks or postcards made from the fibres of dromedary droppings were presented at the annual Pushkar Fair by the NGO '*Lokhit Pashy-Palak Sansthan*' (LPPS), which extolled this ecological product by pointing out the multi-use valuation permitted by dromedary keeping in the country (CIRAD 2009).

Nanobodies, or single-domain antibodies, have found their way into research, diagnostics, and therapy over the last two decades, according to Beghein and Gettemans (2017) in their review paper (Beghein and Gettemans 2017). These antigen-binding fragments, derived from camelids' heavy chain-only antibodies, have unique properties that make them preferable to conventional antibodies or fragments of antibodies in certain research areas. The authors evaluated nanobodies' current status as research tools in a variety of fundamental research areas. They focused on recent advances in super-resolution microscopy and discussed the use of nanobodies as detection reagents in fluorescence microscopy. Secondly, the use of nanobody technology in studying protein–protein interactions is discussed, with a focus on potential applications in mass spectrometry. It has also been recently reported that nanobodies have applications in targeted protein degradation in a review which discusses their potential value in studying protein function (Beghein and Gettemans 2017). They highlighted state-of-the-art engineering strategies that could expand nanobody versatility and suggest future applications of the technology in selected fundamental research areas throughout the review.

9.4.5 Dromedary Racing

Dromedary racing was registered in 2020 on the Representative List of the Intangible Cultural Heritage of Humanity. Dromedary racing, a popular social practice and festive heritage associated with dromedaries in the communities concerned. The preparation of a dromedary race includes several stages. Selected dromedaries, according to their type, origin and age, receive a special diet. They are trained on the racetrack in groups to participate in races. Dromedary breeds are organised on designated routes under the supervision of specialised committees in the communities. The demand for special breeds of racing dromedaries, with higher endurance and resilience, is quite high. For each race, there are usually between fifteen and twenty dromedaries competing, and the distance to be travelled is determined according to the age of the animals. Traditions, customs and principles recognised by the communities govern the breeds and practices of the associated communities. In addition, a preparatory committee is responsible for verifying the origin of each dromedary. The transmission of knowledge and know-how is achieved through the joint efforts of community representatives, government agencies, specialised centres, and the federation of breeds and clubs. Children and youth gradually acquire the knowledge and skills associated with practice through observation, simulation and oral expression. The dromedary race is a fundamental aspect of their nomadic way of life as well as a source of inspiration and creativity in poetry and song. Its importance and continuity in Bedouin society are linked to the predominant role of dromedaries in desert areas (UNESCO 2020).

The development of new techniques to measure oxidative stress and endurance in racing dromedaries has very high potential (Wilson 1999). To do this, studies on the mitochondrial DNA (mtDNA) of dromedaries are being undertaken. Indeed, this DNA is 16,643 bp in size and encodes 23 subunits of the electron transfer chain associated with the production of cellular adenosine triphosphate (ATP) by oxidative phosphorylation (Cui et al. 2007). Cells that require a lot of energy in terms of ATP, such as neurons and muscles, maintain high mtDNA copy numbers (Dickinson, 2013). Polyacrylamide gel electrophoresis was used to analyse malate dehydrogenase (Mdh) and malic (ME) isoenzymes in Arabian camels for energy production and running. The results indicated the need for mitochondrial Mdh-2 for energy production in the racing breeds and cytosolic Mdh-1 for lipogenesis and energy production in both racing and non-racing breeds (Pickford et al. 1995). Likewise, the ratio of mitochondrial DNA (mtDNA) to nuclear DNA (nDNA) is also considered to be a good predictor of the metabolic state of the tissue. This ratio was found to be high in racing dromedaries compared to dairy dromedaries (Soman and Tinson 2016). Several regions have declared the dromedary as their flag animal and hold events nationwide to promote and showcase dromedary keeping among local residents.

In the countries of the Gulf, in particular the United Arab Emirates, the dromedary race is a real institution in a similar way to the thoroughbred horses destined for the same activity in the West. The racing stables are maintained with great care, the feeding of the animals, their training, biosecurity and welfare or their protection or their selection fall in all cases under the principles of intensification, the production of these farms being the sporting performance (Faye 1997; Seboussi et al. 2004). The intensification of breeding of racing dromedaries increasingly requires the use of biotechnologies applied to breeding. The perfect knowledge of the physiological peculiarities is indispensable to optimise the interventions in the dromedary species, which is not the least if compared to those of the horse, considered the noble species of the northern countries of the Mediterranean basin (Zarrouk et al. 2003).

In Europe, especially in France, Jockeys compete in the French Dromedary Racing Championship on the racecourse of La Chartre-sur-le-Loir (Sarthe).

In Africa, for example, in Niger, in the region of Ingall, where one of the main racing dromedary breeds of the Sahara is found, breeds are organised for competition in the 'Cure Salée festival', a great celebration of Saharan pastoralism. Although dromedary breeds are not as popular as they are in the Gulf countries, they are nevertheless organised on the occasion of festivities such as the marathon of Douz (Tunisia), the festival of Marrakech (Morocco) or the fantasia of Ouargla (Algeria). The ascent of the dromedary as a saddle animal is also regularly practised in most countries of the Maghreb, especially among the Tuareg (Algeria, Libya) or Saharawi populations (Morocco) (Faye 2014).

9.4.6 Dromedary Trekking and Tourism

Ecotourism has a low impact and is often an alternative to traditional commercial (mass) tourism that involves visiting or encroaching on fragile, pristine and relatively undisturbed natural areas. Environmental conservation includes ecotourism, which requires taking into account the needs of the native communities so as to help them improve their livelihoods. Dromedary keepers know and benefit from the interest maintained in the field of leisure and tourism, in particular touristic activity in the desert, although this activity has decreased in some countries, such as Algeria or Libya, for security reasons. In other contexts, dromedaries are part of the decor for touristic places or sites (dromedary rides at the foot of the pyramids of Giza in Egypt or on the island of Djerba in Tunisia, or in Essaouira in Morocco, to give only a few emblematic examples) (Faye 2014).

The dromedaries hold a place of choice in all aspects of the social life of dromedary drivers (festivals and games). The occasions of local festivals are where the fantasia is omnipresent, attracting a large audience at the same time. In Morocco, for example, owners supply dromedaries or their services to tourists to rent. They may also rent their whole herd to hotel owners or intermediaries who are in contact with tourist agencies, which allows keepers and their families to earn income and meet their daily needs. Moreover, they do not define themselves as keepers but rather as 'organisers of dromedary treks'. During the high tourist season, they keep their dromedaries in enclosures in the villages. In the low tourist season, they are released, and keepers will take care of their watering. The development of touristic activity has consequently led to the emergence of a new husbandry system (Faye 2014).

9.4.7 Other Uses of the Dromedary

The dromedaries are, unfortunately, also used in public fights. These are organised based on the fury of the male in the period of rut; the fighting animals are muzzled to avoid possible injuries. The dromedary is an element of the return to the source, an unfailing companion. It is therefore an important element in the touristic walks in the Sahara in a desolate universe and in complete break with modernity. Coated animals are bred for the needs of the film industry in the studios of the Atlas in Ouarzazate, Morocco (Camps et al. 1996).

A Case Study of Tsabong Ecotourism Park
A study with the aim of assessing herd size, composition, dromedary types, revenue sources, the role of dromedaries in ecotourism and benefits to communities using the Purposive Sampling Technique was conducted at Tsabong Ecotourism Dromedary Park (Botswana) by Seifu et al. (2018). A semi-structured questionnaire was developed and used to interview key informants. Field observations were also carried out to assess the conditions of the dromedary herd and facilities. The results showed that dromedaries in the Park were used for riding by tourists. Respondents mentioned that the dromedary riding safari was the main touristic activity provided by the Park. According to key informants, the

Park's income was derived from both dromedary- and non-dromedary-related activities. Dromedary riding, trekking, wedding ceremonies, photographs and entrance fees were dromedary-related income sources. Informants' experiences showed an increase in the trend of tourists visiting the Park between 2015 and 2017. The results showed that the Park hosted more local visitors than foreigners, with locals accounting for 91% of visits. Improved promotion is required to attract international tourists to the Park. Dromedary racing seems to be one of the potential business areas to attract tourists to Botswana. Such efforts would improve incomes and livelihoods in the local communities and strengthen a sustainable tourism attraction.

9.5 Dromedary and Environment

The dromedary thrives in the harsh and very restrictive desert conditions of the Saharan ecosystem, where it not only survives but also manages to reproduce and even produce meat and milk. It is the only domestic species capable of valuing this vast, inhospitable environment, thanks to different parameters of adaptation (Chehma and Faye 2009). Faye and Tisserand (1989) cited in Faye and Tisserand (1989), showed that the dromedary requires less water per unit of dry matter ingested and digests more plant walls and less dietary nitrogen than sheep (Faye and Tisserand 1989). It does not require lush forages, but only highly salted forages that are well adapted to the arid areas (Chehma 2005).

The dromedaries are capable of consuming several types of food, some of which are rejected by ruminants. The plant species consumed by the dromedary vary from legumes, grasses, forage trees, and herbaceous to woody plants. Its feed ration comprises 90% of woody forage in the dry season and about 50% in the rainy season (Faye and Tisserand 1989). It eats very thorny plants also for the bitter taste. This makes the dromedary an animal suitable for mixed grazing with other domestic species, where it does not compete with them for feeding resources (Gauthier-Pilters 1977). The technique of feed harvesting, by "brushing" the branches by pressing them laterally in his mouth, allows him to unstuck, without difficulty, the branches of the most thorny woody (Richard 1995).

Although this animal is selective, it does not cause degradation in rangelands. On the contrary, it contributes to the conservation of its extremely fragile ecosystem. It also plays an important role in the regeneration of certain species of plants that germinate only through the digestive tract of this animal (Longo et al. 2007). Unlike small ruminants and cattle that cause intense overgrazing around water points, it behaves solitarily on the rangeland, avoiding overload that causes the degradation of vegetation and the environment. It is able to stabilise its annual nutrient intake, despite temporal fluctuations caused by climate irregularity in its environment (Chehma and Faye 2009).

In general, the dromedary has a set of adaptation mechanisms that make it the animal of choice for the Saharan areas. This choice is more necessary in view of the need to satisfy the nutritional needs of the populations living in these regions, to restore the balance of these ecosystems and the maintenance of certain plants that contribute to the stabilisation of dunes (Faye and Tisserand 1989; Trabelsi et al. 2012).

9.6 Reproduction

During the rutting season, the dromedary becomes aggressive towards other dromedaries and towards its owner and can be extremely dangerous. A horny male grinds his teeth, whips his tail, wags his head and neck, foams at the mouth and urinates frequently, splashing urine all around. Rutting is a period of high sexual activity for a limited time, presumably controlled by the level of testosterone (Gherissi et al. 2017). The heat can be induced by treatment with gonadotropins and better nutrition (El-Bahrawy et al. 2015). Dromedaries are seasonal breeders, but the activity of males is more affected than that of females. With regard to males, the seasonal breeding behaviour of the male dromedary is called 'rut'. It is usually expressed for the first time around when the animal is aged 3 years (Gherissi et al. 2020b). Males in heat become more aggressive, and the size of the testes increases. In the presence of a female or other male in sight, they stand with their legs apart, pour urine on the back with the tail, foam at the mouth while making gurgling noises and overtaking their Dulla. The pollen glands become active and exude a dark, acid secretion, which seems to attract females (El-Bahrawy et al. 2015).

Females, separated from the males, will emit a gurgling sound, almost similar to that of the male in heat, when ready to mate. She will get as close as possible to the male as far as the situation allows. For controlled mating, the traditional procedure is to take the female to a sandy location and lay her down with her head slightly down the hill, if possible. About 50% of females ready for mating spontaneously lie down when approaching the male. Others, especially young animals, may be forcibly restrained in a supine position with the front legs tied together around the animal's neck. In a free mating situation, the male will dominate while battling the female on the ground. Coitus takes place with the female lying down. The male covers the female by placing one foreleg on either side of the female's hump, her pedestal on her dorsal midline behind her hump, and squats back and down in the coital position. Depending on the relative sizes of male and female, the male's foreleg may or may not reach the ground (Khan et al. 2003).

According to Cherifi et al. (2013) the reproduction and weaning for dromedaries take place during the period of low temperatures and heavy rains and when the grass is of high quality (Cherifi et al. 2013). The breeding season extends from November to April in the south-western Algeria, but it is really intense from December to February (Gherissi et al. 2018). The farmers have developed reproductive strategies enabling them to optimise the genetic quality of their animals according to the constraints imposed upon them by the environment. The management of reproduction is characterised by close cross in the same herd that reduces genetic variation and increases inbreeding. As the breeding season approaches, breeders who do not have a breeding male borrow one from other keepers and introduce it to their flock for mating. This practice is free of charge and rather is part of the good social relationship among neighbouring breeders.

The herd of females in hot weather is constantly guarded by the breeding males, sometimes in groups, in order to more accurately determine when they come into heat. According to the farmers, one breeding male is enough for a herd of 20–50 females. For those with large herds, the presence of 2 or 3 breeding males is common.

Weaning of dromedaries is practised between 8 and 18 months, and an average age of about 12 months. This, however, depends on the mother and her diet. The smaller calves tend to stay with their mother for over a year, especially if the mother is not pregnant during the second year. Voluntary withdrawal of dromedaries is rare in these regions, as in most cases, calves are weaned by their mothers very often at the age of 1 year.

Gherissi et al. (2020a) conducted a survey which was aimed at evaluating dromedary herds' fertility and fecundity under Algerian extreme arid conditions (Gherissi et al. 2020a). Progeny history testing data obtained from 14 dromedary herds (78 females and 20 males) was analysed and compared with standard objectives and thresholds. The age at first rut, the first oestrus, and the first male and female mating (months ± SD) were 37.2 ± 16.29 months, 31.07 ± 8.97 months, 42.6 ± 14.28 months and 35.52 ± 8.55 months, respectively. The birth-to-conception interval, open days, age at first calving and calving interval were 40.35 ± 9.41 months, 340 ± 203 days, 51.05 ± 9.59 months and 22.32 ± 5.63 months (Gherissi et al. 2020a). The mean male-to-female ratio was 1:40 (Gherissi et al. 2020a). Pregnancy diagnosis was performed 21.81 ± 16.4 (days) post-mating, and the duration of pregnancy was on average 12.80 ± 0.30 months (Gherissi et al. 2020a). The mean herd's annual fertility was $56.2 \pm 6.6\%$; the mean culling age of males per herd was 15.30 ± 2.47 years, whereas females were culled at 23.31 ± 5.64 years with a mean number of 5.23 ± 2.91 lactations (Gherissi et al. 2020a). However, all the considered herds showed annual fertility out of threshold. The lack of a significant strategy to improve age at first calving, calving interval and selective use of dromedary bulls is likely to affect fertility and productivity of Algerian dromedary herds. Such a negative trend could hamper the genetic improvement of autochthonous dromedary ecotypes and compromise the dromedary sector and the ecosystem services provided by local cameleers (Gherissi et al. 2020a).

It is becoming obvious that the classical system of reproduction cannot respond to the increased importance of dromedary breeding in Africa (and around the world) and especially in the deserts and in newly reclaimed areas. To overcome this, the Artificial Insemination and Embryo Transfer Laboratory, Mariout Research Station, Alexandria, Egypt, has conducted several studies for over a decade to improve the reproductive performance of dromedaries by developing assisted reproduction techniques (Rateb et al. 2020). Gherissi and Lamraoui (2021) have highlighted different pharmacological and biological strategies for the management of dromedary reproduction (Gherissi and Lamraoui 2021). These authors indicated different methods used to improve the cryopreservation capacity of sperm by enriching diluents with additives as well as developing procedures to remove viscosity from dromedary sperm intended for fresh, short-term and long-term conservation (Gherissi and Lamraoui 2021). Finally, a comparison of the conception rate of female dromedaries according to different sperm conservation methods showed wide variability due to conservation time, method, semen quantity, dilution method, extenders and post-throw insemination (Gherissi and Lamraoui 2021).

With regard to females, new techniques and modalities for the induction of multiple synchronised ovulations have been extended, while the practical detection of oestrus in the field has, for the first time, been appropriately determined. In addition, the use of sophisticated molecular genetic techniques to identify the potential fertility of males has been attempted. The development of an in vitro fertilisation procedure and the production of unconventional embryos by intracytoplasmic sperm injection (ICSI) have also been described.

9.7 Some Dromedary Diseases in Africa

For generations, nomadic pastoralists have learnt to manage their herds' health, and particularly their dromedaries, because of the great financial value of these and the capital they represent. They thus acquired a very detailed knowledge of the signs of disease in this species, which they classify and name according to specific systems. The pathologies reported by keepers as being the most worrying in their opinion are gastrointestinal verminosis, dromedary's diarrhoea, tick infestations, smallpox, sarcoptic mange and bronchopneumonia. The presence of misidentified entities is also reported (Antoine-Moussiaux et al. 2006). As a result, we report here two diseases (Prion disease and trypanosomiasis) not reported by many owners, but with a high risk of spreading rapidly in Africa. Indeed, the rush of humans for energy has made the Sahara a dumping ground for household waste, especially some of which are toxic for the dromedary (Babelhadj et al. 2018). These are believed to be the first cause of the onset of prion disease. Also, the development of Saharian agriculture has led governments to create water points, which for the most part, attracted the tsetse fly, the main vector of trypanosomiasis.

9.7.1 Prion Diseases

Human Creutzfeldt–Jakob disease (CJD), scrapies in small ruminants and bovine spongiform encephalopathy (BSE) are all fatal and transmissible neurodegenerative diseases caused by prions. Following the outbreak of BSE and related human infections in the United Kingdom in 1996, animal prions sparked widespread concerns. As a result of these concerns, a large number of countries have begun to implement surveillance strategies in order to prevent the disease. Thanks to collaboration between Algerian and Italian researchers, a prion disease in dromedaries (*Camelus dromedarius*) was discovered in Algeria. In the years 2015–2016, 3.1% of dromedaries brought to the Ouargla slaughterhouse showed signs of prion disease. Pathognomonic neurodegeneration and disease-specific prion protein (PrPSc) were found in brain tissue from three symptomatic animals, confirming the diagnosis. Prions have been discovered in a variety of animals. The presence of prions in lymphoid tissue suggests that the disease is infectious. The biochemical profile of PrPSc differed from that of BSE and scrapies. The discovery of this prion disease in a widely farmed species necessitates the immediate implementation of surveillance and risk assessments for human and animal health (Babelhadj et al. 2018). The lack of effective preventive and therapeutic approaches against prion diseases represents a serious problem for their management. This is especially true for animal prion diseases, which behave like infectious and contagious diseases. In the case of scrapies, the well-known role of variations in the prion protein gene (PRNP) in resistance/susceptibility represents an opportunity that has been exploited to select populations genetically resistant to the disease and therefore the possibility of eradicating this disease. Kaouadji et al. (2020) studied a dromedary prion disease (CPrD) in Algeria while investigating the possible existence of genetic determinants useful for a selection of the most resistant animals (Kaouadji et al. 2020). In this work, researchers studied the variability of PRNP in 232 animals from six populations of dromedaries (Azawad, Hybrid, Naili, Rguibi, Sahraoui and Targui) in Algeria. A Gly69Ser mutation was observed in a single animal from the Targui population and a Gly134Glu polymorphism in the Azawad, Hybrid and Rguibi populations, with a frequency of the 134 Glu allele of 2.6%, 7.7% and 7.1%, respectively. Although this work highlights a low variability of PRNP in Algerian dromedaries, as a possible indication of a recent evolutionary history of CPrD, it also provides evidence for PRNP variants whose role in resistance/susceptibility to prion diseases deserves to be deepened via a case-control. Dromedary keepers currently have no other solution than culling animals suffering from this disease. The investigation by neurologists in the region of Ourgala, where the disease was detected in Algeria, did not reveal the presence of any patient with CJD, showing a non-zoonotic pattern in this case.

9.7.2 Dromedary Trypanosomiasis

In Somalia, which has the world's largest population of dromedaries (*Camelus dromedarius*), dromedary trypanosomiasis, or surra, is a major concern. As a result of the civil war in Somalia, the country's educational, research, economic and social structures were destroyed, and the country scored very low on most humanitarian indicators. Previous studies on the detection of Trypanosoma species in Somali dromedaries, which used standard trypanosome detection methods, were only conducted in the 1990s (STDM). Due to a lack of current knowledge on dromedary trypanosomiasis in Somalia, the goal of this study was to determine the prevalence of *Trypanosoma* spp. in three Somali districts. A total of 182 *C. dromedarius* blood samples from nomadic and dairy farms were tested using STDM, serological (CATT/*T. evansi*) and PCR methods.

A total of 125/182 (68.7%, 95% CI: 61.4–75.3%) dromedaries were seropositive for *T. evansi* by CATT/*T. evansi*. Dromedaries reared in a nomadic system were more likely to be seropositive for *T. evansi* than those reared in a dairy system (OR: 5.6, 95% CI: 2.1–15.2, P = 0.0001). Five out of 182 dromedaries (2.7%, 95% CI: 0.9–6.3%) tested positive for *Trypanosoma* sp. by ITS1-PCR. Sequencing of the ITS1 region of the *Trypanosoma* species detected here revealed that the dromedaries were infected with *T. evansi* and *T. simiae*. These researchers conclude that *Trypanosoma evansi* is widespread in dromedaries in the Banadir region of Somalia, especially in nomadic herds. This study is the first to confirm *T. evansi* and *T. simiae* infections in Somali herds by DNA sequencing. These data highlight the need to implement adequate control measures aimed at reducing the impact on dromedary production in the country (Hassan-Kadle et al. 2019).

By another approach, Nowier et al. (2020) aimed in their studies at estimating firstly the polymorphism of the tyrosinase (TYR) gene in four breeds of Egyptian dromedaries (Nowier et al. 2020), on the other hand, to estimate the body and udder measurements of the most important dromedary breeds in Egypt (Maghrebi) and thirdly, to study the possible association between the polymorphism of the TYR gene and measurements of the body and udder of the Maghrebi breed (Nowier et al. 2020). Hair samples were taken from 124 female dromedaries belonging to four Egyptian breeds: Maghrebi ($n = 70$), Sudani ($n = 17$), Somali ($n = 25$) and Falahy ($n = 12$) for the extraction of DNA. Fourteen body measurement traits and twelve udder traits were evaluated for 35 female Maghrebian dromedaries. The non-synonymous C/T SNPs of the TYR gene were genotyped using the restriction enzyme DdeI. Among the four dromedary breeds studied, the heterozygous CT genotype recorded the highest frequency (45%). Our results showed that the C/T SNP of the TYR gene had a significant association with neck length (NL), height at withers (HW) and chest circumference (TG), with the highest values observed for individuals carrying the CT genotype. In addition, this study showed that the SNP C/T of the TYR gene had a significant association with teat separation (TS) and teat distance (TFD) traits, with the highest values observed for dromedaries carrying CC and CT genotypes, respectively. The information from this study may be useful in designing an appropriate breeding and breeding strategy for optimal use and improvement of the genetic resources of the Egyptian dromedary.

Currently, an important collaborative work between Algeria (University of Tlemcen) and Egypt (University of Ain Shems) is launched on the genetic investigation concerning the detection of *Trypanosoma* species using molecular tools and bioinformatics. Preliminary results indicate that a significant prevalence of this agent exists in these areas with a great diversity of species. Currently, keepers lacking medication rub used engine oil on wounded skin caused by the disease.

9.8 State of Diversity of Dromedary in Africa

The invention of agriculture (horticulture) and animal husbandry provides mankind with greater food security, and the population has made a leap forward. The second leap that humanity made was due to another revolution; the Industrial Revolution. In a few hundred years, major cities will have occupied considerable areas on the planet. These concrete jungles replaced other real ones and devoured vast territories and rural habitats. The speed at which they have spread has left little time for evolutionary adaptations. These sudden changes have put an end to many species, and some are undoubtedly disappearing even before their existence has been realised. Each species is a library of information acquired by evolution over hundreds of thousands, if not millions, of years. We are losing entire libraries. Unfortunately, despite the effects of biodiversity loss (less food security, climate change, etc.), humanity is not aware of what is lost in terms of wealth (Gaouar 2009).

Preserving this heritage, therefore, becomes a necessity in order to keep a genetic reservoir capable of responding to future global constraints. This strategic goal can only be achieved through the mandatory passage of existing biodiversity knowledge. In Africa, more than anywhere else, the need for this stage is felt even more. To do this, several levels and techniques of biodiversity analysis were undertaken by the researchers on dromedary genetic resources.

9.8.1 Morpho-Biometric Studies

According to Gaouar (2009), this method, commonly used to characterise and compare breeds of livestock (goats, sheep, cattle, camels and equines), is a good start for a better knowledge of domestic breeds in Africa (Gaouar 2009). Indeed, there is a lack of classical means and studies to identify and exploit the dromedary's biological potential. There is a great need for standardised approaches. These include the identification of genotypes adapted to different conditions (feed availability, heat resistance, etc.) as well as the analysis of phenotypic characteristics (morpho-biometrics) to establish an available preliminary database that will contribute positively to the better use and promotion of the dromedary resource.

The phenotypic characters were the first traits used for the selection and definition of breeds, it is the result of the animals features observation (coat colour, leg conformation, morphology of the head and body, etc.), or morphological measurements (height, weight, body width, etc.), or performance measurements (growth rate, dairy production, hair production, etc.).

In addition, the similarity between morphological traits can be considered a reflection of a genetic identity, but this may be due to different alleles or genes at different loci. A difference in phenotypic expression can be explained by a real difference at the genetic level, but it can also be due to a variability of expressivity or penetrance of the genetic and/or climatic environment (Gaouar 2002).

These considerations limit the use of morphological characters as elements for the reconstruction of the phylogeny of breeds and their characterisation (Gaouar 2002).

Alhajeri et al. (2021) and Gherissi et al. (2022) propose a new approach to data collection for more interesting morphometric and precise morphometric data (Alhajeri et al. 2021; Gherissi et al. 2022). In broad distribution, animals exhibit large phenotypic variation, which so far has mostly been examined using traditional distance-based morphometric approaches. The narrow sampling of breeds and geographic locations, as well as the relatively few and ambiguously defined morphometric measurements, which often do not cover all functionally important traits, are the main pitfalls of previous studies. So, these researchers offer some suggestions for standardising morphometric data collection, as well as an overview of more advanced methods for capturing morphometric data in dromedaries. This would make data collection much easier in these animals, which can be difficult to work with at times. Moreover, its applicability to breed identification and delimitation was studied for a variety of purposes. Finally, they suggest possible applications for these morphometric studies, particularly in the context of developing breeds specialised for specific purposes such as meat, milk and leather production.

Several papers on African dromedary breeds using the morphometric approach have been published. Chniter et al. (2013) identified five groups of Maghrebi camels in southern Tunisia based on tribal affiliation (Chniter et al. 2013): the *Ourdhaoui Médenine* in the Tawazins tribe, the *Ourdhaoui Tataouine* in the Oudarna tribe, the *Guiloufi* in the Beni Guilouf tribe from Kébili, the *Gueoudi* in the Ouled Gharib tribe from Kébili. The nine body measurements taken include the body length, length of the neck, thoracic girth, abdominal circumference, height at the hump, height at the withers, width between shoulders and height at the withers, the length of the anterior limb and the length of the tail. The height at the hump and the height at the withers, the thoracic girth and the abdominal circumference, and the length of the body and the length of the neck all showed a positive correlation. The length of the tail, on the other hand, had a negative relationship with the abdominal circumference and withers height. When compared to the Ourdhaoui Médenine dromedary group, the Gueoudi, Guiloufi and Merzougui dromedary groups had longer bodies and larger abdominal circumferences. In comparison to Ourdhaoui Médenine, the Guiloufi, Merzougui and Ourdhaoui Tataouine dromedary groups had the highest thoracic girth and height at the withers. In comparison to the Gueoudi, Guiloufi and Merzougui dromedary groups, the Ourdhaoui Médenine and Ourdhaoui Tataouine dromedary groups had the longest tails. The five groups were clustered together, allowing three main classes of Maghrebi dromedaries to be described: large, medium and small size Maghrebi camels. Six vernacular names were identified based on coat colour: Chagra, Chaâla, Safra, Hajla, Hamra and Zarga. Because the hair was mostly rough and thick, the female dromedary was given the name Nagga Chalfi. Females with sleek and heavy hair, known as Nagga Khawar, were given a small portion of the population.

Boujenane et al. (2019) looked at 14 body measurements from 132 adult female dromedaries from three different populations (Boujenane et al. 2019). Guerzni (60), Khouari (28) and Marmouri (44) were reared in 38 herds across eight Southern Moroccan provinces in order to identify homogeneous groups based on conformation. Chest girth (CG), hump girth (HG), height at withers (HW), body length (BL), fore limb length (FLL), chest width (CW), chest depth (CD), fore hoof circumference (FHC), head length (HL), distance between eyes (DE), ear length (EL), neck length (NL), neck circumference (NC), and tail length (TL) were the measurements. Multivariate analyses were used to compare the three populations based on their mean body measurements. Only HG, HW, BL and FLL were found to be significantly influenced by the population. Furthermore, the MANOVA revealed that the Guerzni and Marmouri populations differed significantly, whereas the Khouari population did not differ significantly from the Guerzni or Marmouri populations. Out of 14 variables, discriminant analysis revealed that BL and FLL were the most discriminant, resulting in two significant canonical variables (CAN1 and CNA2). CAN1 could best distinguish the Khouari population from Guerzni and Marmouri, and CAN2 could best distinguish Guerzni from Marmouri. According to the discriminant analysis, 46.7% of Guerzni, 60.7% of Khouari and 40.9% of Marmouri animals were correctly classified in their original population. Two Moroccan dromedary groups emerged from the clustering of the three populations: Guerzni and Marmouri in the first and Khouari in the second.

Meghelli et al. (2020) recently conducted a study using body measurements to identify two different dromedary breeds reared in Algeria (Meghelli et al. 2020). A total of 115 animals from the Steppe ($n = 55$) and Sahraoui ($n = 60$) dromedary breeds made up the study's animal material. Body measurements like head length, neck length, neck girth, tail length, distance between eyes, distance between ears, body length, withers height, chest girth and live weight were determined, as well as eye and coat colours.

The least squares means for head length, neck length, neck girth, tail length, distance between eyes, distance between ears, body length, withers height, chest girth and live weight for Steppe and Sahraoui dromedary breeds, respectively, were 48.2, 116.9, 65.7, 55.6, 24.1, 22.5, 152.2, 184.5, 141.2 cm and 217.2 kg for Steppe and 48.1, 101.2, 56.2, 51.2. The Steppe dromedary breed had a distribution of brown and black eye colours of 58.2% and 41.8%, respectively, while all of the Sahraoui dromedaries studied had brown eyes. Coffee, dark coffee and red colours account for 1.8%, 83.6% and 14.6% of the Steppe dromedary's body colour, respectively, and 98.3%, 1.7% and 0.0% for the Sahraoui dromedary, respectively. As a result, this study concluded that the height at withers and chest girth could better estimate the body weight in the two breeds of dromedaries at different ages (Meghelli et al. 2020). In another study, Legesse et al. (2018) examined seventeen morphometric variables to determine intraspecific variation among eight pastoralist-designated breeds of dromedaries (Legesse et al. 2018).

9.8.2 Immunogenetic or Biochemical Studies

Knowledge of the variability of biochemical markers makes it possible to develop hypotheses about the origin of animal populations and to determine the relations between breeds of the same population or species. Immunogenetic markers include blood groups' polymorphism and blood proteins (Martins et al. 2002). This type of marker has the advantage of giving more precise information on the physiological differences of the studied populations and therefore on their zootechnical potential. Protein polymorphisms were the first markers to be used in livestock.

The biochemical and haematological blood profile in Libyan dromedaries was described by Abdalmulaa et al. (2019), and the effects of breed, gender and age on these values were evaluated (Abdalmulaa et al. 2019). The levels of enzymes, metabolites, electrolytes and haematological indices were measured in blood samples taken from 24 male and 42 female apparently healthy dromedaries (Abdalmulaa et al. 2019). Male dromedaries had higher levels of aspartate aminotransferase (AST), Lactate dehydrogenase (LDH), Amylase (AMS), total proteins, globulin and phosphorus (Ph) in their blood than female dromedaries, which had higher levels of glucose, albumin/globulin (A/G) ratio, urea, Iron (Fe), Calcium (Ca) and Packed Cell volume (PCV) in their blood, Haemoglobin (Hb), erythrocyte osmotic fragility, Mean Corpuscular Volume (MCV), the number of neutrophils and monocytes, as well as the mean corpuscular volume (MCV) and mean corpuscular haemoglobin (MCH). Many haematological and biochemical analyses reveal significant sex differences between male and female Libyan dromedaries in this study (Abdalmulaa et al. 2019).

9.8.3 Molecular Studies in the Dromedary

Recent major developments in molecular genetics have provided new and powerful tools, called molecular markers, to assess the origins of farmed animal species and their geographic distribution as well as their diversity (Duran et al. 2009).

Nuclear markers, which increasingly number in ranges from Restriction Fragment Length Polymorphism (RFLP), Amplified Fragment Length Polymorphism (AFLP), to microsatellites or 'Single Tandem repeat': tandem repeated sequences (STR), Mitochondrial DNA polymorphisms (mtDNA) and recently single nucleotide polymorphisms (SNPs), have become the markers of choice for studying the genetic structures and evolutionary history of organisms (Fitak et al. 2016).

Molecular marker technologies offer the most advanced and likely most effective way to understand the fundamentals of genetic diversity. They are effective and precise tools for identifying and measuring genetic variation quickly and thoroughly.

Genetic diversity and relationships among dromedary (*Camelus dromedarius*) populations have been poorly studied. Yet, understanding the genetic diversity and structure of dromedary populations is critical for long-term herd management and the implementation of breeding programmes in this species. Mburu et al. (2003) used 14 microsatellite loci to study four recognised Kenyan dromedary breeds (Somali, Turkana, Rendille, Gabbra) as well as dromedaries from Pakistan and the Arabian Peninsula (Saudi Arabia, United Arab Emirates) (Mburu et al. 2003). Phylogenetic analysis indicated that Kenyan dromedaries are distinct from Arabian and Pakistani populations. The Kenyan dromedaries were less diverse relative to non-Kenyan populations based on the expected heterozygosity and allelic diversity values. The Kenyan dromedaries were poorly differentiated (average $F_{ST} = 0.009$), with only one to two loci separating the Gabbra, Rendille and Turkana populations studied ($P < 0.05$), with the exception of the Somali population. Individual assignments were performed using the maximum likelihood method. When using an allocation stringency of a log of the odds ratio > 2, a correct breed assignment of only 39–48% was observed for Kenyan dromedaries. The findings refute the current classification of the native Kenyan dromedary into four distinct breeds based on socio-geographical factors. Instead, these findings point to only two genetic entities: Somali type and a group that includes Gabbra, Rendille and Turkana populations.

In the same context, Cherifi et al. (2017) characterised a total of 331 dromedaries from Northern Africa, representative of six populations and thirteen Algerian and Egyptian geographic regions, using 20 STR markers (Cherifi et al. 2017). The nineteen polymorphic loci displayed an average of 9.79 ± 5.31 alleles, ranging from 2 (CVRL8) to 24 (CVRL1D). Average: He was 0.647 ± 0.173. Eleven loci deviated significantly from Hardy-Weinberg proportions ($P < 0.05$), due to an excess of homozygous genotypes in all cases except one (CMS18). Distribution of genetic diversity along a weak geographic gradient, as suggested by network analysis, was not supported by either unsupervised or supervised Bayesian clustering. Traditional extensive/nomadic herding practices, together with the historical use as a long-range beast of burden and its peculiar evolutionary history, with domestication likely occurring from a bottlenecked and geographically confined wild progenitor, may explain the observed genetic patterns.

Additionally, DNA sequences from the mitochondrial cytochrome-b gene and genotyping of six nuclear microsatellite loci were examined to assess genetic diversity and phylogenetic relationships of Ethiopian dromedaries. Examination of 525 individuals revealed significant morphometric differentiation in Afar as compared with the remaining seven breeds. Analysis of cytochrome-b sequences failed to recover monophyletic groups associated with pastoralist-recognised breeds. Analysis of six microsatellite loci from 104 individuals depicted no resolution of distinct genetic lineages in accordance with geographical or designated breeds. Overall, the separation of two ecotypes based on the morphometric data was supported; however, genetic analysis of cytochrome-b and microsatellite data failed to support any unique genetic lineage or statistically significant population structure.

In the other context of gene identification, Holl et al. (2017) aimed at identifying the genetic origin of the white spotting phenotype in dromedary (Holl et al. 2017). This study includes animals from Africa and Asia. While the typical Arabian dromedary is characterised by a single coloured coat, there are rare populations with white coat spotting patterns. White coat spotting patterns are found in virtually all domesticated species but are rare in wild species. Theories suggest that white spotting is linked to the domestication process and is occasionally associated with health disorders. Though mutations have been found in a diverse array of species, fewer than 30 genes have been associated with coat spotting patterns, thus providing a key set of candidate genes for the Arabian dromedary. We obtained 26 spotted dromedaries and 24 solid controls for candidate gene analysis. One spotted and eight solid dromedaries were whole-genome sequenced as part of a separate project. The spotted dromedary was heterozygous for a frameshift deletion in KIT (c.1842delG, named KITW1 for White spotting 1), whereas all other dromedaries were wildtype (KIT+/-KIT+). No additional mutations unique to the spotted dromedary were detected in the EDNRB, EDN3, SOX10, KITLG, PDGFRA, MITF and PAX3 candidate white coat spotting genes. Sanger sequencing of the study population identified an additional five KITW1/KIT+ spotted dromedaries. The frameshift results in a premature stop codon five amino acids downstream, thus terminating KIT at the tyrosine kinase domain. An additional 13 spotted dromedaries tested KIT+/KIT+, but due to phenotypic differences when compared to the KITW1/KIT+ dromedaries, they likely represent an independent mutation. Our study suggests that there are at least two causes of white spotting in the Arabian dromedary, the newly described KITW1 allele and an uncharacterised mutation.

Whole-genome-wide identification and annotation of genetic variations in dromedaries is in its first steps. In their study, Khalkhali-Evrigh et al. (2018) have identified a total of 4,727,238 single nucleotide polymorphisms (SNPs) and 692,908 indels (insertions and deletions) were found by mapping raw reads to the dromedary reference assembly (Gen Bank Accession: GCA_000767585.1) (Khalkhali-Evrigh et al. 2018). *In-silico* functional annotation of the discovered variants under study samples revealed that most SNPs (2,305,738; 48.78%) and indels (339,756; 49.03%) were located in intergenic regions. A comparison of the identified SNPs in Iranian dromedaries with those of the African dromedary (Bio Project Accession: PRJNA269274) indicated that they had 993,474 SNPs in common. The authors found 15,168 non-synonymous SNPs in the shared variants of the studied dromedaries that could affect gene function and protein structure. Some other genetic studies concerning molecular disease analyses are reported in the previous Chapter (see Chap. 7).

9.8.4 Conservation Programmes of Dromedary Genetic Resources in Africa

The Convention on Biological Diversity (CBD) plays a crucial role in ensuring the sustainable management of animal genetic resources (AnGRs), including the dromedary camel, to meet future needs in various sectors such as food production, fibre, fertiliser and traction. These resources provide essential inputs for agriculture and contribute to the resilience of farming systems, enabling farmers to adapt to changing production conditions effectively. Chapter 25, focusing on the Conservation and Management of AnGR, delves into the fundamental aspects of managing and conserving these resources. It discusses strategies and approaches tailored to different conditions under which animals or 'materials' are managed, emphasising the importance of biodiversity conservation and sustainable use. Additionally, Chap. 30 addresses policy frameworks, strategies and plans essential for safeguarding AnGRs, aligning with the objectives outlined in the CBD. By integrating conservation efforts into policy development and management strategies, stakeholders can ensure the long-term viability of dromedary camels and other AnGRs while promoting agricultural sustainability and resilience in the face of evolving production challenges.

Conservation efforts for African dromedary camels necessitate tailored strategies to address the continent's unique challenges and diverse ecosystems. Genetic resource management initiatives focus on developing breeding programmes specifically designed for African dromedary camel populations, aiming to both preserve and enhance genetic diversity. This involves establishing comprehensive databases and genetic repositories within African countries to catalogue and safeguard valuable genetic information. Concurrently, selective breeding programmes prioritise traits endemic to African dromedaries while mitigating the risk of inbreeding, thus safeguarding the resilience of these populations. Conservation breeding centres, strategically located across African countries, serve as vital hubs for controlled breeding efforts, ensuring the preservation of rare genetic variants and preventing genetic bottlenecks. Community engagement lies at the heart of conservation endeavours, with initiatives targeting local herders to promote sustainable husbandry practices that support genetic diversity. By educating communities about the importance of maintaining diverse breeding stocks and preserving traditional breeding techniques, conservation efforts are deeply integrated into local cultures. Genetic monitoring programmes, specific to African dromedary camel populations, facilitate ongoing assessment of genetic diversity levels and identify emerging threats. This research-driven approach enables the development of contextually relevant conservation strategies, addressing the intricacies of African ecosystems. Through international collaboration, African countries leverage shared expertise and resources to bolster conservation efforts, ensuring that genetic management practices are standardised and effectively implemented across diverse African regions.

9.9 Conclusion and Future Perspectives

Large camelids are undeniably under renewed interest in many fields of research around the world. The scientific community is getting organised to give their research recognition commensurate with the challenges to which these species are called upon to respond (CIRAD 2022). Many projects have been carried out on this animal, especially after the ecologists gave the alert of an unavoidable and global climate change. Dromedaries are found throughout Northern and Eastern Africa, the Middle East and parts of Asia, where they represent an extremely important animal species for millions of people who live in remote ecosystems (arid areas) of the planet. Possibly originated in Arabia or Somalia, the dromedary camel is therefore sometimes referred to as the Arabian or East African camel. After extensive bibliographic research, the present Chapter summarised the mosaic of camel breeds or genepools in Africa. It is characterised by a panoply of ecotypes, landraces and varieties, which bear various names that often refer to their origin and owning tribe(s), their distribution area, their coat colour, their characteristics and their use… but several of them belong to the same animal type, as attested by molecular genetics analysis, showing the great confusion in the description of African dromedary biodiversity in many African countries and little or no genetic differences between them. In this manuscript, we have processed the classification notion of African indigenous dromedary populations (in 18 African countries) and the introduced dromedary populations in order to provide exhaustive information allowing us to distinguish the ethnozoo-technical particularities of each group. The concepts of dromedary selection and purebred breeding, cross-breeding and interspecific cross-breeding in camel species were taken up to discuss the prospects of these breeding practices on the African continent. In addition, the manuscript reminds us of the morpho-biometric, immunogenetic, biochemical and molecular studies undertaken on the scale of African dromedary populations to expose the state-of-the-art on the biodiversity of dromedaries in Africa. On the other hand, the manuscript also contains a synthesis of the main facilities and organisations of the dromedary sector in Africa and describes dromedary breeding systems in Africa by highlighting the diversity and production potential, pharmacopoeia aspects of dromedary products, the sociocultural roles and the traditional uses of the dromedary in African communities, the environmental roles of this species and the modern uses of dromedary species, and its contribution in trekking, racing and tourism… Finally, through this chapter, we have been able to formulate the essentials of in situ and in vitro conservation practices and technologies for the camel population of African countries.

Unfortunately, our investigations show that most of the African countries where the dromedary can make a difference are affected by many challenges, namely, the lack of advanced technologies in the field of breeding and proper policies to take full advantage of this animal. The policy maker-researcher-breeder relationship should then be strengthened in African countries to advance the development of this almost native rich animal genetic resource.

References

Abdalmulaa AM, Benashourb FM, Shmelac ME, Alnagard FA, Abograrae IM, Buker AO (2019) Blood profileinnormal one humped dromedary (*Camelus dromedarius*) Dromadaries in Libya. Part 3: effect of sex variationon biochemical and hematological blood profile. Int J Sci: Basic Appl Res (IJSBAR) 48(1):9–24. https://www.gssrr.org/index.php/JournalOfBasicAndApplied/article/view/10172/4425

Abdel Galil MAG, Alhaider AA (2016) The unique medicinal properties of dromedaryproducts: a review of the scientific evidence. J Taibah Univ Med Sci 11(2):98–102. https://doi.org/10.1016/j.jtumed.2015.12.007

Abdulrahman AO, Ismael MA, Al-Hosaini K, Rame C, Al-Senaidy AM, Dupont J, Ayoub MA (2016) Differential effects of camel milk on insulin receptors ignaling—toward understanding the insulin-like properties of camel milk. Front Endocrinol 7(4). https://doi.org/10.3389/fendo.2016.00004

Abdussamad AM, Holtz W, Gauly M, Suleiman MS, Bello MB (2011) Reproduction and breeding in dromedary dromadaries: insights from pastoralists in some selected villages of the Nigeria-Niger corridor. Livest Res Rural Dev 23(8). http://www.lrrd.org/lrrd23/8/abdu23178.htm

Adamou A (2008) L'élevage du dromadaire en Algérie: quel type pour quel avenir? Sécheresse 19(4):253–260 afs.okstate.edu/breeds/other/camel, consulted 28.10.2021

Agrimaroc (2021) Lait de chamelle complément alimentaire et cosmétique. https://www.dromedary-idee.com/les-vertus-des-produits-dromedaryins-en-cosmetique/, 15/08/2021

Alhajeri BH, Alhaddad H, Alaqeely R, Alaskar H, Dashti Z, Maraqa T (2021) Dromedary breed morphometrics: current methods and possibilities. Trans R Soc S Aust 145(1):90–111. https://doi.org/10.1080/03721426.2021.1889347

Alkhamees OA, Alsanad SM (2017) A review of the therapeutic characteristics of camel urine. Afr J Tradit Complement Altern Med 14(6):120–126. https://doi.org/10.21010/ajtcam.v14i6.12

Almathen F (2014) Genetic diversity and demographic history of dromedary dromedary (Camelus dromedarius). Thesis, The University of Nottingham, Nottingham

Almathen F, Charruau P, Mohandesan E, Mwacharo JM, Orozco-terWengel P, Pitt D, Abdussamad AM, Uerpmann M, Uerpmann H-P, De Cupere B, Magee P, Alnaqeeb MA, Salim B, Raziq A, Dessie T, Abdelhadi OM, Banabazi MH, Al-Eknah M, Walzer C, Faye B,

Hofreiter M, Peters J, Hanotte O, Burger PA (2016) Ancient and modern DNA reveal dynamics of domestication and cross-continental dispersal of the dromedary. Proc Natl Acad Sci 113(24):6707–6712. https://doi.org/10.1073/pnas.1519508113

AM N, TH M, AM AES (2018) Some Studies on Milk Production and its Composition In Maghrebi She-Camel Under Farming And Traditional Pastoral Systems In Egypt. Assiut Veterinary Medical Journal. 64(156):51–63

Antoine-Moussiaux N, Faye B, Vias G (2006) Connaissances ethnovétérinaires des pathologies camélines dominantes chez les Touaregs de la région d'Agadez (Niger). In: Centre de Coopération Internationale en Recherche Agronomique pour le Développement (CIRAD), Paris, p 12

Babelhadj B, Di Bari MA, Pirisinu L, Chiappini B, Gaouar SBS, Riccardi G, Marcon S, Agrimi U, Nonno R, Vaccari G (2018) Prion disease in Dromedarycamel, Algeria. EID 24(6). https://doi.org/10.3201/eid2406.172007

Bakheit SA, Majid AM, Abu-Nikhila AM (2008) Dromadaries (Camelus dromedarius) under pastoral systems in North Kordofan, Sudan: seasonal and parity effects on milk composition. J Camelid Sci 1:32–36. https://api.semanticscholar.org/CorpusID:130198626

Bedda H, Adamou A, Bouammar B, Babelhadj B (2019) Le déclin des systèmes de production dromedaryins dans le Sahara septentrional algérien – cas de la cuvette de Ouargla, le M'zab et le Ziban. Livest Res Rural Dev 31. http://www.lrrd.org/lrrd31/3/bedda31044.html

Beghein E, Gettemans J (2017) Nano body technology: a versatile toolkit for microscopic imaging, protein-protein interaction analysis, and protein function exploration. Front Immunol 8:771. https://doi.org/10.3389/fimmu.2017.00771

Bengoumi M, Faye B (2002) Adaptation du dromadaire à la déshydratation. Sécheresse 13:121–129

Benkerroum N, Mekkaoui M, Bennani N (2004) Antimicrobial activity of dromedary's milk against pathogenic strains of Escherichia coli and Listeria monocytogenes. Int J Dairy Technol 57:39–43. https://doi.org/10.1111/j.1471-0307.2004.00127.x

Bensemaoune Y, Beziou S, Senoussi A, Chehma A (2018) Le système d'élevage du dromadaire dans la région de Ghardaïa: situation et perspectives. Revue des Bio Ressources 8(2)

Blanc CP, Ennesser Y (1989) Approche zoogéographique de la différenciation intraspécifique chez le dromadaire Camelusdromedarius Linné, 1766 (Mammalia: eameliadae). Revue Elev Med Vet Pays Trop 42(4):573–587. https://doi.org/10.19182/remvt.8766

Bonnet D (1996) La notion de négligence sociale à propos de la malnutrition de l'enfant. Présentation du numéro spécial «La malnutrition de l'enfant: fait culturel, effet de la pauvreté ou du changement sociale?». Sciences sociales et santé 14(1):5–16

Boujenane I, El Khattaby N, Laghouaouta H, Badaoui B, Piro M (2019) Morphological diversity of female dromedary (*Camelus dromedarius*) populations in Morocco. Trop Anim Health Prod 51(6):1367–1373. https://doi.org/10.1007/s11250-019-01813-5

Bourbouze A (2006) Systèmes d'élevage et production animale dans les steppes du nord de l'Afrique: une relecture de la société pastorale du Maghreb. Science et changementsplanétaires Sécheresse 17:31–39

Breulmann M, Boer B, Wernery U, Wernery R, El Shaer H, Alhadrami G, Gallacher D, Peacock J, Chaudhary SA, Brown G, Norton J (2007) The dromedary, from tradition to modern times. Unesco Doha Publ, Doha, 44p

Burger PA (2016) The history of Old World dromedaryids in the light of molecular genetics. Trop Anim Health Prod 48(5):905–913. https://doi.org/10.1007/s11250-016-1032-7

Burger PA, Ciani E, Faye B (2019) Old world camels in a modern world – a balancing act between conservation and genetic improvement. Anim Genet 50(6):598–612. https://doi.org/10.1111/age.12858

Cabalion S (2013) Le «système domesticatoire» touareg (Tagaraygarayt, Niger). vol 1 et 2. Thèse en anthropologie sociale et ethnologie, EHESS, Paris (France), 515 p

Camps G, Peyron M, Chaker S (1996) Dromadaire. Encyclopédie Berbère

Carr CJ (2017) Turkana survival systems at Lake Turkana: vulnerability to collapse from Omo Basin development. In: River Basin development and human rights in Eastern Africa – a policy crossroads. Springer, Cham. https://doi.org/10.1007/978-3-319-50469-8_9

Chaibou M (2005) Productivité zootechnique du désert: le cas dubassin laitier d'Agadez au Niger (Thèse de doctorat). Montpellier II France, 310p. http://dromedaryides.cirad.fr/fr/science/pdf/these_chaibou.pdf

Chehma A (2005) Etude floristique et nutritive des parcours camelin du Sahara septentrional algérien. Cas des régions de Ouargla et Ghardaïa. Thèse Doctorat Université Badji Mokhtar Annaba, 178 p

Chehma A, Faye B (2009) Stratégies de valorisation des ressources alimentaires de l'écosystème saharien par le dromadaire. In: Séminaire International sur la Protection et Préservation des Ecosystèmes Sahariens, Ouargla, Algérie

Chergui M, Titaouine M, Gherissi DE (2023) Descriptive typology and structural analysis of camel farms in the region of El Oued, Algeria. Genet Biodivers J 7(2):75–93. https://doi.org/10.46325/gabj.v7i2.360

Chergui M, Gherissi DE, Titaouine M, Kaouadji Z, Harek D, Koutti S, Boumaraf H, Gaouar SBS (2024) Lactation traits and reproductive performances of Sahraoui female camel in two breeding systems at Algerian Sahara. Trop Anim Health Prod 56(2):70. https://doi.org/10.1007/s11250-024-03902-6

Cherifi YA, Gaouar SBS, Moussi N, Tabet Aoul N, Saïdi-Mehtar N (2013) Study of dromedary in a biodiversity in southwestern of Algeria. J Life Sci 7(4):416–427

Cherifi YA, Gaouar SBS, Guastamacchia R, El-Bahrawy KA, Abushady AMA, Sharaf AA (2017) Weak Genetic Structure in Northern African dromedary camels reflects their unique evolutionary history.

PLoS One 12:e0168672. https://doi.org/10.1371/journal.pone.0168672

Chniter M, Hammadi M, Khorchani T, Krit R, Benwahada A, Ben Hamouda M (2013) Classification of Maghrebi dromadaries (*Camelus dromedarius*) according to their tribal affiliation and body traits in southern Tunisia. Emir J Food Agric 25(8):625–634. https://doi.org/10.9755/ejfa.v25i8.16096

CIRAD (2009) Du papier avec de la crotte de dromadaire: une innovation technique mise au point en Inde. http://dromedaryides.cirad.fr/fr/curieux/papcrotte.html, mai 2021

CIRAD (2022) http://camelides.cirad.fr/fr/actualites/archives/dossier_mois13_16.html.mars

Cui P, Ji R, Ding F, Qi D, Gao H, Meng H, Yu J, Hu S, Zhang H (2007) A complete mitochondrial genomesequence of the wildtwo-humpedcamel (*Camelus bactrianusferus*): an evolutionaryhistory of camelidae. BMC Genomics 8:241. https://doi.org/10.1186/1471-2164-8-241

DAD-IS (2021) Domestic animal diversity information system (DAD-IS). FAO

Dickinson A (2013) The role of mitochondrial DNA in tumorigenesis (Doctoral dissertation, University of Warwick)

Dioli M (2006) Studies on dromadaries of Eritrea: a review. The International Scientific Congress on Dromadaries

Djomtchaigue HB, Meutchieye F, Manjeli Y (2015) Caractéristiques phénotypiques des dromadaires de la région de bahr-el gazal au Tchad. Bull Anim Health Product Afr (Special issue African Animal Genetics Resources):51–64

Doutresoulle G (1947) Les chameaux (cinquième partie). In: L'élevage en Afrique occidentale française. Larose, Paris, pp 271–277

Duran C, Appleby N, Edwards D, Batley J (2009) Molecular genetic markers: discovery, applications, data storage and visualisation. Curr Bioinforma 4:16–27. https://doi.org/10.2174/157489309787158198

Eisa MO, Mustafa AB (2011) Production systems and dairy production of Sudan dromedary (*Camelus dromedarius*): a review. Middle East J Sci Res 7(2):132–135. https://doi.org/10.1002/qua.20166

El-Bahrawy KA, Khalifa MA, Rateb SA (2015) Recent advances in dromedary reproduction: an Egyptian field experience. Emir J Food Agric 27(4):350–354. https://doi.org/10.9755/ejfa.v27i4.19907

Epstein H (1971) The origin of the domesticated animals of Africa. Africana Publ. Corp, New York/London/Munich. pp 1:1–573, 2: 1–719

Esmail MEF, Ashraf T, El-Wahab AA, Bakry MH et al (2008) Potential activity of dromedary milkamylase and lactoferrin against hepatitis C virus infectivity in HepG2 and lymphocytes. Hepatitismonthly 2008:101–109. https://brieflands.com/articles/hepatmon-70042.pdf

FAO (2011) Rapport sur la consultation nationale pour l'analyse des contraintes affectant la filière dromedaryine mauritanienne, avec un accent sur les maladies animales et zoonotiques. http://faostat.fao.org. Accessed 17 Oct 2021

FAOSTAT (2019) Dromedary milk production in 2019, Livestock primary/Regions/World list/Production Quantity: "Milk, whole fresh dromedary". UN Food and Agriculture Organization, Corporate Statistical Database (FAOSTAT). Retrieved 28 February August 2021

FAOSTAT (2022) Food and Agricultural Organisation of the United Nations. FAOSTAT statistical database. http://www.fao.org/faostat/en/#data/QC

Faye B (1997) Guide de l'élevage du dromadaire. Editions SANOFI, Libourne

Faye B (2014) The dromedary today: assets and potentials. Anthropozoologica 49:167–176. https://doi.org/10.5252/az2014n2a01

Faye B (2020) How many large camelids in the world? A synthetic analysis of the world camel demographic changes. Pastoralism 10(1):25

Faye B, Bengoumi M (2002) Adaptability of the dromedary camel to the mineral under-nutrition. In: Proceedings of the 27th world veterinary congress, no 142, Tunis, September 25–29, p 73

Faye B, Porphyre V (2011) Le dromadaire et le cochon: deux visions opposées de l'élevage? Nat Sci Soc 19:365–374. https://www.cairn.info/revue-natures-sciences-societes-2011-4-page-365.htm

Faye B, Tisserand JL (1989) Problèmes de la détermination de la valeur alimentaire des fourrages prélevés par le dromadaire. Séminaire sur la nutrition et l'alimentation du dromadaire, Ouargla, Algérie. Options méditerranéennes: Séries séminaires 2:61–65. https://om.ciheam.org/article.php?IDPDF=CI000428

Faye B, Bonnet P, Charbonnier G, Marti A (2000) Etat des recherches sur le dromadaire à partir de l'analyse bibliométriques des publications. Cas particulier des recherches sur le chamelon. Rev Elev Med Pays Trop 53:125–131

Faye B, Abdelhadi OMA, Ahmed AI, Bakheit SA (2011) Dromedary in Sudan: future prospects. Livest Res Rural Dev 23(10). http://www.lrrd.org/lrrd23/10/faye23219.htm

Faye B, Abdelhadi O, Raiymbek G, Kadim I, Hocquette JF (2013) La production de la viande de chameau: état des connaissances, situation actuelle et perspectives. Prod Anim 26:247–258

Faye B, Jaouad M, Bhrawi K, Senoussi A, Bengoumi M (2014) Elevage camelin en Afrique du Nord: état des lieux et perspectives. Rev Elev Med Pays Trop 67(4):213–221. https://doi.org/10.19182/remvt.20563

Fitak RR, Mohandesan E, Corander J, Burger PA (2016) The de novo genomeassembly and annotation of a female domesticdromedary of North Africanorigin. Mol Ecol Resour 16:314–324. https://doi.org/10.1111/1755-0998.12443

Flynn LJ (1997) Late Neogene mammalian events in North China. Actes du Congre's Biochro M'97. Meomoires et Travaux EPHE, Institut Montpellier 21, 183–192

Gaouar SBS (2002) Contribution à l'étude moléculaire de la variabilité e génétique: caractérisation de deux races ovines algériennes Thèse de Magister, Université des sciences et de technologie d'Oran (USTO)

Gaouar SBS (2009) Etude de la biodiversité: Analyse de la variabilité génétique des races ovines algériennes et de leurs relations phylogénétiques par l'utilisation des microsatellites, Thèse de Doctorat, Université des sciences et de technologie d'Oran (USTO)

Gauthier-Pilters H (1977) Contribution à l'étude de l'écophysiologie du dromadaire en été dans son milieu naturel (Moyenne et Haute Mauritanie), Extrait du bulletin de l'IFAN, série A, n°2

Geay YBD, Hocquette JF, Culioll J (2002) Valeur diététique et qualités sensorielles des viandes des ruminants. Incidence de l'alimentation des animaux. INRA Prod Anim 15(01):37–52. https://doi.org/10.20870/productions-animales.2002.15.1.3686

Gebrehiwet T (2021) The dromedary in Eritrea: an all-purpose animal ENE LE DROMADAIRE. https://modaina.com

Gebremichael B, Girmay S, Gebrun M (2019) Dromedary milk production and marketing: Pastoral areas of Afar, Ethiopia. Pastoralism 9:16. https://doi.org/10.1186/s13570-019-0147-7

Gherissi DE, Gaouar SBS (2022a) Dromedary milk quantitative and qualitative assessments. Archivos de Zootecnia 71(274):120–122. https://doi.org/10.21071/az.v71i274.5658

Gherissi DE, Gaouar SBS (2022b) Camel diversity survey in El Oued region (South East Algeria). Archivos de Zootecnia 71(274):124–126. https://doi.org/10.21071/az.v71i274.5659

Gherissi DE, Lamraoui R (2021) Reproduction management and artificial insemination in dromedary Camel. In: Yata VK et al (eds) Sustainable agriculture reviews 54. https://doi.org/10.1007/978-3-030-76529-3_2

Gherissi DE, Afri-Bouzebda F, Bouzebda Z (2017) Seasonal changes in the testicular morphology and interstitial tissue histomorphometry of Sahraoui camel under Algerian extreme arid conditions. Biol Rhythm Res 49(2):1744–4179. https://doi.org/10.1080/09291016.2017.1357331

Gherissi DE, Bouzebda Afri F, Bouzebda Z, Bonnet X (2018) Are female camels capital breeders? Influence of seasons, age, and body condition on reproduction in an extremely arid region. Mamm Biol 93:124–134. https://doi.org/10.1016/j.mambio.2018.10.002

Gherissi DE, Monaco D, Bouzebda Z, Afri Bouzebda F, Gaouar SBS, Ciani E (2020a) Dromedary herds' reproductive performance in Algeria: objectives and thresholds in extreme arid conditions. J Saudi Soc Agric Sci 9(7):482–491. https://doi.org/10.1016/j.jssas.2020.09.002

Gherissi DE, Boukhili M, Gherissi A (2020b) Genital histomorphometrical evaluation and survey on reproductive traits of male camel (*Camelus dromedarius*) in relation to the pubertal age under extreme arid conditions. Asian J Agric Biol 8(4):436–446. https://doi.org/10.35495/ajab.2019.12.591

Gherissi DE, Lamraoui R, Chacha F, Gaouar SBS (2022) Accuracy of image analysis for linear zoometric measurements in dromedary camels. Trop Anim Health Prod 54:232

Habte M, Eshetu M, Maryo M, Andualem D, Legesse A, Admassu B (2021) The influence of weather conditions on body temperature, milk composition and yields of the free-ranging dromedary camels in Southeastern rangelands of Ethiopia. Cogent Food Agric 7(1):1930932. https://doi.org/10.1080/23311932.2021.1930932

Harek D, Berber N, Cherifi YA, Yakhlef H, Bouhadad R, Arbouche F, Sahel H, Djellout NE, Saidi-Mehtar N, Gaouar SBS (2015) Genetic diversity and relationships in saharan local breeds of Camelus dromedarius as inferred by microsatellite markers. J Camel Pract Res 22(1):1–9. https://doi.org/10.5958/2277-8934.2015.00001.6

Harek D, Ikhlef H, Bouhadad R, Cherifi YA, El Mokhefi M, Boukhtala S, Gaouar SBS, Arbouche F (2017) Genetic diversity status of dromedary's resources (Camelus dromedarius. Linnaeus, 1758) in Algeria. Genet Biodivers J 2(1):1–22. https://journals.univ-tlemcen.dz/GABJ/index.php/GABJ/article/view/87

Hassan-Kadle AA, Ibrahim AM, Nyingilili HS, Yusuf AA, Vieira TSWJ, Vieira RFC (2019) Parasitological, serological and molecular survey of dromedary trypanosomiasis in Somalia. Parasit Vectors 12:598. https://doi.org/10.1186/s13071-019-3853-5

Holl H, Isaza R, Mohamoud Y, Ahmed A, Almathen F, Youcef C, Gaouar S, Antczak DF, Brooks S (2017) A frameshift mutation in kit is associated with white spotting in the arabian dromedary. Genes 8:102. http://www.mdpi.com/2073-4425/8/3/102/pdf

Honey JG, Harrison JA, Prothero DR, Stevens MS (1998) Camelidae. In: Janis CM, Scott KM, Jacobs LL (eds) Evolution of Tertiary Mammals of North America: volume 1, terrestrial carnivores, ungulates, and ungulate like mammals. Cambridge University Press, Cambridge, pp 439–461

Ishag IA, Eisa MO, Ahmed MKA (2011) Phenotypic characteristics of Sudanese dromedaries (*Camelus dromedarius*). Livest Res Rural Dev 23(4). http://www.lrrd.org/lrrd23/4/isha23099.htm

Jaji AZ, Elelu N, Mahre MB, Jaji K, Ghali Mohammed LI, Audu Likita M, Kigir ES, Onwuama KT, Saidu AS (2017) Herd growth parameters and constraints of dromedary rearing in Northeastern Nigeria. Pastoralism 7:16. https://doi.org/10.1186/s13570-017-0089-x

Jianlin H, Mburu D, Ochieng J, Kaufmann B, Rege JE, Hanotte O (2000) Application of new world dromedaryidae microsatellite primers for amplication of polymorphic loci in Old World dromedaryids. Anim Genet 31:404–419. https://doi.org/10.1046/j.1365-2052.2000.00683.x

Kaouadji Z, Meghelli I, Cherifi Y, Babelhadj B, Gaouar SBS, Conte M, Capocefalo A, Agrimi U, Chiappini B, Vaccari G (2020) Variability of the prion protein gene (PRNP) in Algerian dromedary populations. Anim Gene:17–18. https://doi.org/10.1016/j.angen.2020.200106

Kaufmann B (1998) Analysis of pastoral dromedary husbandry in northern Kenya, Hohenheim tropical. Margraf, Verlag, p 194

Kebede S, Animut G, Zemedu L (2015) The contribution of dromedary milk to pastoralist livelihoods in Ethiopia an economic assessment in Somali Regional State. Drylands and Pastoralism. ISBN: 9781784311513. https://www.iied.org/sites/default/files/pdfs/migrate/10122IIED.pdf

Khalkhali-Evrigh R, Hafezian SH, Hedayat-Evrigh N, Farhadi A, Bakhtiarizadeh MR (2018) Genetic variants analysis of three dromedary camels using whole genome sequencing data. PLoS One 13:e0204028. https://doi.org/10.1371/journal.pone.0204028

Khan BB, Iqbal A, Muhammad R (2003) Production and management of dromadaries- Part – I. Department of Livestock Management University of Agriculture Faisalabad, p 302

Konuspayeva G, Loiseau G, Faye B (2004) La plus-value «santé» du lait de chamelle cru et fermenté: l'expérience du Kazakhstan. Rencontres Recherches Ruminants 11:47–50

Kozhamkulova BS (1986) The late Cenozoic two-humped (Bactrian) camels of Asia. Quartarplaontolgie 6:93–97

Kuria T, Walaga HK (2008) Camel Breeds In Kenya. https://www.kalro.org/fileadmin/publications/brochuresII/Came_breeds_in_Kenya.pdf

Lakermi Y, Sabrou S (2021) Contribution à l'étude phénotypique et moléculaire de la population dromedaryine locale TARGUI dans wilaya d'Adrar. Mémoire de Master génétique. Université de Tlemcen, année universitaire

Lasnami K (1986) Le dromadaire en Algérie, perspectives d'avenir. Magister. INA El Harrach, Alger, Algeria, 185 p

Legesse YW, Dunn CD, Mauldin MR, Ordonez-Garza N, Rowden GR, Gebre YM, Kurtu MY, Mohammed Ali S, Whibesilassie WD, Ballou M, Tefera M, Perry G, Bradley RD (2018) Morphometric and genetic variation in 8 breeds of Ethiopian dromadaries (*Camelusdromedarius*). J Anim Sci 96(12):4925–4934. https://doi.org/10.1093/jas/sky351

Lhoste P, Havard M, Vall É (2010) La traction animale, Collection: Agricultures tropicales en poche, Janvier

Longo HF, Siboukeur O, Chehma A (2007) Aspects nutritionnels des pâturages les plus appréciés par Camelusdromedarius en Algérie. Cahiers Agric 16(6):477–483. https://doi.org/10.1684/agr.2007.0144

López PJW (2010) Inducción de ovulación con plasma seminal o análogo de GnRH (acetato de buselerina) y su efecto sobre la tasa de concepción en alpacas (Vicugna pacos), inseminadas con semen fresco. UNMSM-Tesis Universidad Nacional Mayor de San Marcos. https://hdl.handle.net/20.500.12672/976

Mammeri A, Khir A (2018) Survey on health and therapeutic virtues of she-dromedary milk among customers and staffs of she-dromedary milk dairies in the region of M'Sila, Algeria. Renc. Rech Ruminants, p 24

Martin A, Bonnet P, Bourzat D, Lancelot R, Souvenir P (1996) Importance de l'élevage et sa place dans l'économie des pays de la Commission du Bassin du Lac Tchad. In: Atlas d'élevage du bassin du Lac Tchad. CTA, Wageningen, pp 79–96. http://agritrop.cirad.fr/388414/1/Importance%20de%20l%27%C3%A9levage.pdf

Martini P, Costeur L, Le Tonsorer J-M, Schmid P (2015) Pleistocene camelids from the Syrian Desert: the diversity in ElKowm. Anthropologie 119:687–693

Martins LO, Soares CM, Pereira MM, Teixeira M, Costa T, Jones GH, Henriques AO (2002) Molecular and biochemical characterization of a highly stable bacterial laccase that occurs as a structural component of the Bacillus subtilis endospore coat. J Biol Chem 277:18849–18859. https://doi.org/10.1074/jbc.M200827200

Mbaïgaou M (1998) Etude de l'impact scio-économique du Dromadaire (*Camelusdromedarius*) au Tchad. Thèse d'état pour l'obtention du diplôme de docteur vétérinaire. Ecole Inter-Etats des Sciences et Médecine Vétérinairesde Dakar (Sénégal), 136p

Mburu DN, Ochieng JW, Kuria SG, Jianlin H, Kaufmann B, Rege JE, Hanotte O (2003) Genetic diversity and relationships of indigenous Kenyan dromedary (*Camelus dromedarius*) populations: implications for their classification. Anim Genet 34(1):26–32. https://doi.org/10.1046/j.1365-2052.2003.00937.x

Meghelli I, Kaouadji Z, Yilmaz O, Cemal İ, Karaca O, Gaouar SBS (2020) Morphometric characterization and estimating body weight of two Algerian dromedary breeds using morphometric measurements. Trop Anim Health Prod 52(5):2505–2512

Melkamu S, Angesom H, Teshager D, Mulat A (2022) Major trade sensitive diseases and problems in the afar's dromedary camels market chain and its impact on livelihood of the pastoral community in afar region, north east Ethiopia. J Anim Health Behav 6:160. https://doi.org/10.37421/ahbs.2022.6.160

Morin D (2006) Le chameau Afar: entre diététique et éthique. In: L'Homme et l'animal dans l'Est de l'Afrique. Publ. Les Éthiopisants Associés, Paris, p 252

Nega S, Tefera M (2012) Production and productivity of the dromedary. In: Tefera M, Abebe G (eds) Dromedary in Ethiopia. Ethiopian Veterinary Association, Addis Ababa, pp 76–92

Nowier AM, El-Metwaly HA, Ramadan SI (2020) Genetic variability of tyrosinase gene in Egyptian dromedary breeds and its association with udder and body measurements traits in Maghrebi dromedary breed. Gene Rep 18:100569. https://doi.org/10.1016/j.genrep.2019.100569

Orlando L (2016) Retour aux racines et itinéraires de la domestication du dromadaire. Proc Natl Acad Sci 113(24):6588–6590

Ould Ahmed M (2009) Caractérisation de la population des dromadaires en Tunisie, PhD thesis, Université du 7 Novembre de Carthage

Ould El Hadj MD, Bouzgag B, Bouras A, Et Moussaoui S (2002) Etude comparative de quelques caractéristiques chimiques et physico-chimiques de la viande du dromadaire chez des individus du type "sahraoui", de différents âges. Première journée sur la recherche cameline. Université de Ouargla Algérie, pp 4–6

Ouologuem B, Moussa M, Coulibaly MD (2004) Etudeet amélioration du système d'élevage camelin, Rapport final derecherche, 10esession Comité de Programme, IER, 45 p

Peters J (1998) Camelus thomasi Pomel, 1893, a possible ancestorof the one-humped camel? Mamm Biol 63:372–376

Philipponneau O (2014) L'actu du jour Des courses de... dromadaires! 9 août. p 1. www.1jour1actu.com

Pickford M, Morales J, Soria D (1995) Fossil camels from the Upper Miocene of Europe: implications for biogeography andfaunal change. Geobios 28:641–650. https://doi.org/10.1016/S0016-6995(95)80217-7

Piro M, Mabsoute FE, El Khattaby N, Laghouaouta H, Boujenane I (2020) Genetic variability of dromedary populations based on microsatellite markers. Animal 14(12):2452–2462. https://doi.org/10.1017/S1751731120001573

Prothero DR, Schoch RM (2002) Horns, tusks and flippers – the evolution of hoofed mammals. Johns Hopkins University Press, Baltimore, pp 45–55

Radfar MH, Aminzadeh Gowhari M (2013) Common gastrointestinal parasites of indigenous camels (Camelus dromedarius) with traditional husbandry management (free-ranging system) in central deserts of Iran. J Parasit Dis 37:225–230. https://doi.org/10.1007/s12639-012-0170-8

Ramet RP (2013) La technologie des fromages au lait de dromadaire (Camelus dromedarius). Etude FAO production et santé animale

Rateb SA, El-Bahrawy KA, Khalifa MA, Abd El-Hamid IS, Zaghloul AA, Kamel AM, El-Hassanein EE (2020) Recent achievements for improving reproductive efficiency of dromedary camels in Egypt. Egypt J Anim Prod 57(Suppl):61–65. https://journals.ekb.eg/article_98190_261d7ad7ee2c048b3597b3f905dc865e.pdf

Richard D (1995) Le dromadaire et son élevage, Institut d'Elevage et de Médecine vétérinaire des pays Tropicaux. Ed Maisons-Alfort, Paris, p 161

Rybczynski N, Gosse JC, Harington CR, Wogelius RA, Hidy AJ, Buckley M (2013) Mid-Pliocene warm-period deposits in thehigh Arctic yield insight into camel evolution. Nat Commun 4:1550. https://doi.org/10.1038/ncomms2516

Salter AM (2013) Impact of consumption of animal products on cardiovascular disease, diabetes, and cancer in developed countries. Anim Front 3(1):20–27. https://doi.org/10.2527/af.2013-0004

Seboussi R, Faye B, Alhadrami G (2004) Facteurs de variation de quelques élémentstrace (sélénium, cuivre, zinc) et d'enzymes témoins de la souffrance musculaire dans le sérum du dromadaire (Camelusdromedarius) aux Emirats arabes unis. Revue Elev Méd vét Pays trop 57:87–94

Seifu E (2009) Analysis on the contributions of and constraints to dromedary production in Shinile and Jijiga zones, Eastern Ethiopia. J Agric Environ Int 103(3):213–224. https://doi.org/10.12895/jaeid.20093.33

Seifu E, Angassa A, Boitumelo WS (2018) Community-based dromedary ecotourism in Botswana: current status and future perspectives. J Camelid Sci 11:33–48. http://www.isocard.net/en/journal33

Sharp NC (2016) Animal athletes: a performance review. Feature. http://veterinaryrecord.bmj.com/

Simenew K, Dejen T, Tesfaye S, Fekadu R, Tesfu K, Fufa D (2013) Characterization of dromedary production system in Afar pastoralists, North East Ethiopia. Asian J Agric Sci 5(2):16–24. https://doi.org/10.19026/AJAS.5.2579

Smith J, Sones K, Grace D, MacMillan S, Tarawali S, Herrero M (2013) Beyond milk, meat, and eggs: role of livestock in food and nutrition security. Anim Front 3(1):6–13. https://doi.org/10.2527/af.2013-0002

Soman SS, Tinson A (2016) Development and evaluation of a simple and effective real time PCR assay for mitochondrial quantification in racing camels. Mol Cell Probes 30:326–330. https://doi.org/10.1016/j.mcp.2016.07.006

Sun X (2005) Dynamics of continuity and change in pastoral subsistence among the Rendille in northern Kenya: with special reference to livestock management and responses to socio-economic change. Afr Study Monogr Suppl. 31:1–94. https://web.archive.org/web/20200321041256id_/, http://jambo.africa.kyoto-u.ac.jp/kiroku/asm_suppl/abstracts/pdf/ASM_s31/ASM31_finalprint.pdf

Tennews A (2015) Bsf dromedary sangram's last amble down rajpath this republic day parade. https://tennews.in/

Tilahun S, Serda B, Baissa M, Mohammed AS (2020) Camel research articles abstracts-Part III breeding. Somali Pastoral and Agro Pastoral Research Institute (SoRPARI), p 114. ISBN: 978-99944-79-62-7. http://publication.eiar.gov.et:8080/xmlui/handle/123456789/3357

Trabelsi H, Senoussi A, Chehma A (2012) Etude de la dissémination des graines des plantes spontanées dans les fèces du dromadaire dans le Sahara septentrional Algérien. Sécheresse 23(2):94–101. https://doi.org/10.1684/sec.2012.0338

Traoré B, Moula N, Toure A, Ouologuem B, Leroy P, Antoine-Moussiaux N (2014) Characterisation of dromedarybreeding practices in the Ansongo region. Trop Anim Health Prod 46:1303–1312. https://doi.org/10.1007/s11250-014-0644-z

UNESCO (2020) La course de dromadaires, pratique sociale et patrimoine festif associés aux dromadaires. https://ich.unesco.org/fr/RL/la-course-de-dromadaires-pratique-sociale-et-patrimoine-festif-associs-aux-dromadaires-01576. Consulted 06.10.2021

Viateau E (1998) La composition du lait de chamelle et ses vertus médicinales. DESS productions animales en régions chaudes. CIRAD. https://catalogue-bibliotheques.cirad.fr/cgi-bin/koha/opac-detail.pl?biblionumber=17260

Waefelaer C (1982) Rites du mariage dans une tribu Berbère en Tunisie: Les Frechich, vol 32(1). Institut de Sociologie de l'Université de Bruxelles, pp 187–208. https://www.jstor.org/stable/41803026

Wanjala NW, Matofari JW, Nduko JM (2016) Antimicrobial effect of smoking milk handling containers' inner surfaces as a preservation method in pastoral systems in Kenya. Pastoralism 6(17). https://doi.org/10.1186/s13570-016-0064-y

Wardeh MF (2004) The dromedary applied research and development network (CARDN). J Dromedary Sci 1(1):1–135. http://www.fao.org/docs/eims/upload/211081/CARDN.pdf

Wilson GR (1999) Australian camel racing. Rural Industries Research, Canberra

Wilson RT (2020) The one-humped dromedary in Eritrea and Ethiopia: the critical review of the literature and a bibliography. J Camel Pract Res 27(3):229–262. https://doi.org/10.5958/2277-8934.2020.00034.X

Wu H, Guang X, Al-Fageeh MB, Cao J, Pan S, Zhou S, Zhang L, Abutarboush MH, Xing Y, Xie Z, Alshanqeeti AS, Zhang Y, Yao Q, Al-Shomrani BM, Zhang D, Li J, Manee MM, Yang Z, Yang L, Liu Y, Zhang J, Altammami MA, Wang S, Yu L, Zhang W, Liu S, Ba L, Liu C, Yang X, Meng F, Wang S, Li L, Li E, Li X, Wu K, Zhang S, Wang J, Yin Y, Yang H, Al-Swailem AM, Wang J (2014) Camelid genomes reveal evolution and adaptation to desert environments. Nat Commun 5:5188. https://doi.org/10.1038/ncomms6188

Zarrouk A, Souilem O, Beckers JF (2003) Actualités sur la reproduction chez la femelle dromadaire (*Camelus dromedarius*). Revue Elev Méd vét Pays trop 56:95–102

Zibaee S (2015) Nutritional and therapeutic characteristics of dromedary milk in children: a systematic review. Electron Physician 7:1523. https://doi.org/10.19082/2F1523

Open Access This chapter is licensed under the terms of the Creative Commons Attribution 4.0 International License (http://creativecommons.org/licenses/by/4.0/), which permits use, sharing, adaptation, distribution and reproduction in any medium or format, as long as you give appropriate credit to the original author(s) and the source, provide a link to the Creative Commons license and indicate if changes were made.

The images or other third party material in this chapter are included in the chapter's Creative Commons license, unless indicated otherwise in a credit line to the material. If material is not included in the chapter's Creative Commons license and your intended use is not permitted by statutory regulation or exceeds the permitted use, you will need to obtain permission directly from the copyright holder.

African Donkey Genetic Resources, Diversity and Breeding Strategies

10

Martha N. Bemji, Eveline M. Ibeagha-Awemu, Abdelkader Ameur Ameur, Albano Beja-Pereira, Madani Labbaci, Liveness J. Banda, and Semir B. S. Gaouar

Abstract

The African continent is a leading producer of donkeys in the world, with major concentrations in the arid and semi-arid environments.

Notwithstanding, the donkey remains the least studied among major livestock species despite its enormous support to low- and middle-income communities. The first two sections of this Chapter (Sections 10.1 and 10.2) provide a comprehensive overview of current statistics of donkey production by regions in Africa, as well as the centre of origin of domestication and possible routes of dispersion in Africa. Section 10.3 is dedicated to discussion on 27 different African donkey breeds or types, most of which are named after the country or region where they are raised. For each breed or type, the chapter summarises: origin, distribution, population statistics if available, physical characteristics including some breed photos, adaptive and resilience features, typical production systems and production characteristics. The depth of discussion is different between breeds due to the differential availability of data or information. Sections 10.4, 10.5, 10.6 and 10.7 further considered donkey breeding and production systems in Africa, diversity and population structure, conservation and important roles that donkeys play in

M. N. Bemji (✉)
Department of Animal Breeding and Genetics, Federal University of Agriculture, Abeokuta, Ogun State, Nigeria
bemjimn@funaab.edu.ng

E. M. Ibeagha-Awemu (✉)
Sherbrooke Research and Development Centre, Agriculture and Agri-Food Canada, Sherbrooke, Quebec, Canada
eveline.ibeagha-awemu@agr.gc.ca

A. Ameur Ameur · M. Labbaci
Applied Genetics in Agriculture, Ecology and Public Health Laboratory (GenApAgiE), Faculty SNV/STU, University of Tlemcen, Chetouane, Algeria

A. Beja-Pereira
DGAOT, Faculty of Sciences, Universidade do Porto, Porto, Portugal

CIBIO, Centro de Investigação em Biodiversidade e Recursos Genéticos, InBIO Laboratório Associado, Campus de Vairão, Universidade do Porto, Vairão, Portugal

BIOPOLIS Program in Genomics, Biodiversity and Land Planning, Campus de Vairão, Vairão, Portugal

L. J. Banda
Lilongwe University of Agriculture and Natural Resources (LUANAR), Bunda College, Lilongwe, Malawi

S. B. S. Gaouar
Laboratory of Applied Genetics in Agronomy, Ecology and Public Health (GenApAgiE), Faculty SNV/STU, Abou Bekr Belkaid University, Tlemcen, Algeria

© The Author(s) 2026
E. M. Ibeagha-Awemu et al. (eds.), *African Livestock Genetic Resources and Sustainable Breeding Strategies*, Sustainable Development Goals Series, https://doi.org/10.1007/978-3-031-92076-9_10

Africa. In Sections 10.8 and 10.9, the chapter concludes with an overview of challenges in the donkey enterprise, including major threats to the donkey populations in Africa and future prospects.

Keywords

Donkey · Genetic resources · Diversity · Uses · Africa

Abbreviations

AU-IBAR	African Union-Inter-African Bureau for Animal Resources
BCE	Before the Christian era
DAD-IS	Domestic Animal Diversity and Information System
FAOSTAT	Food and Agricultural Organization of the United Nations Statistics
ISAG	International Society for Animal Genetics
mtDNA	Mitochondrial deoxyribonucleic acid
PCR	Polymerase chain reaction

10.1 Introduction

The African wild ass or wild donkey (*Equus asinus africanus*) is a wild member of the horse family, Equidae, and is considered an ancestor of the domestic donkey. Donkeys are commonly raised in developing countries, especially in rural communities, for their multiple roles. The male donkey is called a jack, the female a jenny or jennet and the young a foal. The donkey can survive under poor management and environmental conditions with unique characteristics (tough, obedient and docile), which allow women and children to handle them (Orhan et al. 2012; Yilmaz et al. 2012a). The donkey population in Africa constitutes about 70% of the African equine population (FAOSTAT 2024) and is mainly found in the arid and semi-arid areas, offering a reliable, environmentally friendly and renewable source of draught power to millions of poor communities. In terms of world head counts (FAOSTAT 2024), Africa ranks first in donkey production, estimated at 32,686,779 (63.24%), followed by Asia (12,391,888; 23.97%), Americas (6,499,519, 12.57%), and Europe (100,208; 0.19%), while Oceania is least with 8879 donkeys (0.02%). The donkey is among the predominant species (cattle, sheep, goat and camel) managed under large-scale extensive production systems (pastoral and transhumance), which underscores its importance in the lives of the African people (AU-IBAR 2019). See Chapter 2 for a detailed description of the livestock production systems and agro-ecological zones in Africa. Based on FAOSTAT (2024), the semi-arid areas of the continent, particularly East Africa, produce most (34.10%) of the donkeys, while the remaining 66% of production is shared among North Africa (29.89%), West Africa (22.96%), Central Africa (11.50%) and Southern Africa (1.54%). The populations of donkey have been fluctuating across the African continent over the last 5 years (Table 10.1) with less than 1% average annual increase (East, Central and West Africa) or decrease (North and Southern Africa).

The contributions of donkey to livelihoods are substantial and more complex than earlier understood and documented (Maggs et al. 2021). In terms of socio-economic and cultural importance, donkeys are renowned for the following: 1. They are highly treasured by their owners since they provide a pathway out of poverty. A study in Ghana where owners described donkeys as priceless (Maggs et al. 2021) showed that donkey ownership confers up to six different income benefits (accounting for about 30–60% of their income), including (1) income generated from renting out donkeys; 2. Donkey reduce the physical hard work of land ploughing; 3. They assist in domestic chores such as transportation of goods to and from the market, transportation of water from water points/streams and firewood, etc.; 4. Donkey help in transportation of people and goods and contribute to animating the weekly markets between villages (Sow et al. 2012); 5. Donkey is a source of milk and meat. Annual production of donkey meat in Burkina Faso was esti-

Table 10.1 Donkey populations by regions in Africa for the period 2018–2022 (FAOSTAT 2024)

	Donkey number by year					Average annual increase	
Region	2018	2019	2020	2021	2022	Number	%
Africa	31,299,133	31,714,794	33,128,431	32,866,771	32,686,779	346,911.50	0.04
East Africa	10,837,847	11,165,307	11,966,839	11,228,224	11,145,584	76,934.25	0.03
Central Africa	3,441,686	3,665,918	3,904,920	4,159,743	3,759,486	79,450.00	0.10
North Africa	10,004,458	9,801,855	9,971,827	9,826,663	9,770,732	−58,431.50	−0.02
Southern Africa	552,966	548,628	520,656	517,425	504,551	−12,103.75	−0.08
West Africa	6,462,175	6,533,087	6,764,189	7,134,716	7,506,426	261,062.75	0.15

mated at 800 tonnes in 2008 (Sow et al. 2012); and 6. There is an increasing demand for donkey skin, which is used for the production of a traditional Chinese medicine (ejiao), thereby threatening the global donkey population (Maggs et al. 2021).

Despite the fact that the donkey is important to the subsistence of many communities in the arid and semi-arid regions of most developing countries, they have received very limited or no attention from research and development agencies compared to other domesticated species. In a few African countries (Ethiopia, Nigeria, Cameroon, Algeria, Lesotho, etc.), phenotypic characterisation of indigenous donkey breeds has been reported (see Sect. 10.3). There is therefore an urgent need to intensify research efforts on donkey phenotypic and genetic characterisation, which will facilitate classification of donkey breeds/types and boost utilisation for improved productivity and conservation in Africa. This chapter presents an overview of donkey production in Africa with a focus on origins, history of introduction and dispersion, breeds/types and diversity, breeding and production systems, various uses, challenges encountered in the donkey enterprise and future perspectives.

10.2 Origin of the Domestic Donkey and Dispersion in Africa

The donkey, or ass, is a member of the equine family. The equine family includes the horse, mule, donkey and zebra. The domestic donkey is believed to originate from the African wild ass or African wild donkey (*Equus africanus*). The domestic donkey (*Equus africanus*) is often regarded as the most neglected and least studied animal species in the world. However, the domestication of donkeys radically changed the ancient transport systems in Africa and Asia, allowing the overland movement of people and goods and influencing the organisation of the first cities and pastoral societies (Rossel et al. 2008). The original motive for domesticating the donkey is unknown, and it is not certain that it would necessarily reflect its common usage today as a means of transport for people and goods. It may have been domesticated for its meat and milk, with its use for portage a later development (Blench 2012).

The African wild ass or African wild donkey is a wild member of the horse family, Equidae, and considered the ancestor of the domestic donkey. It lived in the deserts and other arid areas of the Horn of Africa, Eritrea, Ethiopia and Somalia. Today, only seven species of equids remain, and many of them face some sort of risk or are threatened with extinction (IUCN 2022). The seven extant equid species are characterised by their rapid chromosomal divergence and recent speciation (Librado and Orlando 2021; Orlando et al. 2009). Of the seven living species of the genus Equus, only two wild species were domesticated, namely, the Botai wild horse (*Equus caballus*) and the African wild ass (*Equus africanus*) (Beja-Pereira et al. 2004; Kimura et al. 2011, 2013; Librado et al. 2021).

The reconstruction of the origin and spread of the domestic donkey was initially contentious due to the limited archaeological record of early donkeys (Kimura et al. 2013; Mitchell 2018). Archaeological evidence of donkeys was found in North Africa, from the Moroccan Atlas Mountains to Somalia in Eastern Africa, such as at Egypt's El Omari (4800–4500 BCE) and Maadi (4000–3500 BCE), carvings on Libyan palettes depicting lines of walking animals (donkey, cattle and sheep) and archaeological remains in Abydos, Egypt (3200–3000 BCE) with suggestions of domestication in a region extending from the Northeastern Sahara, the Atbara River, the Nile Valley, and the Red Sea Hills of the Sahara Desert to Eritrea in the Horn of Africa (Rossel et al. 2008; Beja-Pereira et al. 2004; Kimura et al. 2011; Librado et al. 2021; Todd et al. 2022; Levtzion and Hopkins 1981). These narratives suggest that the donkey was domesticated in Africa by pastoralists to help with mobility and transportation of goods in the difficult landscape of the Sahara around 5000–7000 years ago BCE (Beja-Pereira et al. 2004; Mitchell 2018; Xia et al. 2019). Other evidence based on DNA sequence variation and patterns of mitochondrial DNA (Beja-Pereira et al. 2004; Todd et al. 2022; Wang et al. 2020) supports the African origins of the donkey. However, other regions outside of Africa are also proposed as alternative domestication centres, such as Ash Shuman in Yemen (ass remains dated to ~6500 BCE) and Mesopotamia (iconographic, textual, and zooarchaeological evidence dated to the fourth and third millennia BCE) (Todd et al. 2022).

The ancestors of the donkey are generally considered the two extant wild African ass subspecies (Nubian wild ass- *E. africanus africanus* and the Somali wild ass- *E. africanus somaliensis*) and the Asian wild half-asses (*E. hemiones* and *E. kiang*). Reports on mitochondrial DNA data from ancient archaeological and historic museum samples of modern-day donkeys revealed two distinct mtDNA haplogroups, Clade I and II, proposing the Nubian wild ass (*Equus africanus africanus*) as the ancestor of Clade I and either the extinct Atlas wild ass (*Equus africanus atlanticus*), which was found in North Africa, or an undescribed subspecies found outside of Africa as the ancestor of Clade II (Beja-Pereira et al. 2004; Kimura et al. 2011). Moreover, data on mtDNA control regions from both Asian wild half-asses and the two extant wild African ass subspecies (the Nubian wild ass, and the Somali wild ass) clearly eliminated the possibility of the Asiatic half-asses as progenitors of domestic donkeys (Beja-Pereira et al. 2004). While the Nubian wild ass was considered the ancestor of Clade I, the Somali wild ass had substantial mitochondrial divergence from the Nubian wild ass and domestic donkeys, raising new questions concerning the second ancestor for the donkey. With no distinct phylogeographic structure, information on wild ancestors and geographical origins for the domestic mitochondrial clades could not be obtained from the haplogroups. The results of Kimura and colleagues (2011), therefore, exposed the complexity of the domestication process with implications for whether a single maternally inherited marker could completely capture the ancestry of the domestic donkey. To resolve these controversies surrounding the origin and dispersal of domestic donkey, Todd and colleagues (2022) recently reconstructed a comprehensive genome panel of 207 modern and 31 ancient donkeys, as well as 15 wild equids, to clarify their domestication history, and found a strong phylogeographic structure in modern donkeys that supports a single domestication event in Africa ~5000 BCE. This was followed by further expansions on the continent and Eurasia and ultimately returning to Africa. The findings by Todd and colleagues (2022) validated early domestication in Africa that spread at an even rate into the Arabian Peninsula and Eurasia and flowed back into Nubia and Maghreb. Modern donkeys from the Horn of Africa and Kenya are considered to be descendants of the earliest donkeys. Modern donkeys in a phylogenetic reconstruction were grouped according to sampling locations into Clade A (African donkeys) and Clade B (mostly non-African donkeys) (Todd et al. 2022). Within Clade A, donkeys from the Horn of Africa (Ethiopia and Somalia) plus Kenya were separated from those from West Africa (Ghana, Mauritania, Nigeria and Senegal). Clade B also included donkeys from Nubia (Egypt and Sudan) that showed affinities to the Levant (Syria) and Anatolia (Turkey), as well as donkeys from Maghreb (Tunisia), with closer genetic closeness to European sub-populations, suggesting gene flow into Africa from donkeys that were native to Anatolia and the Levant, but not to the Arabian Peninsula. The authors concluded that the domestication centre of modern-day donkeys is the Horn of Africa and Kenya in about 5000 BCE, followed by expansion in continental Africa and Eurasia and ultimate return to Africa (Todd et al. 2022). Figure 10.1 depicts this theory of the African centre of donkey domestication and spread to other parts of the world and back to Africa.

The spread of the domestic donkey out of its domestication centre can be divided into three key phases: before European contact, during European contact and the subsequent era. The study by Todd and colleagues (2022) narrowed the domestication of donkeys to a unique African location (~5000 BCE) and subsequently spread into Eurasia from ~2500 BCE, while Central and Eastern Asian sub-populations differentiated ~2000–1000 BCE. After early domestication, African donkeys further differentiated in the West and the Horn of Africa plus Kenya but also received streams of genetic ancestry from Western Europe as well as a region encompassing the Levant, Anatolia and Mesopotamia. Donkeys spread by land from farmer to farmer and slowly diffused during the first phase. For the second phase, donkeys were spread by carriage in ships during the colonial era, through projects, state institutions, and rapid diffusion of animals. Earlier information on the donkey in the West African Sahel from Arabic material described the region and referred to the donkey as already domesticated (Levtzion and Hopkins 1981; Lewicki 1974). The diffusion of the donkey in pre-European contact times seems to have been strictly by land, mostly across the Sahara (Fig. 10.1), spreading gradually from country to country. There is a reference to 'Muscat' donkeys in Tanzania in the 1950s (Mason and Maule 1960), describing light-coloured donkeys associated with the Arab people, which may have been brought from the Gulf region or from Egypt, where they have a long tradition of use. In the past, donkeys diffused principally from farmer to farmer or were sold by occupationally specialised pastoralists, as in West Africa. During the colonial era, donkeys were brought into Africa mostly by sea, which served as a means of inland transportation by colonial masters and draft power for agricultural purposes. However, in recent times, the spread of the donkey is through national agricultural programmes, aid agencies and acquisition by farmers to support agricultural work and transportation. Most importantly, the donkey has been recommended for traction in regions with light, sandy soils, and the industrial manufacture of axles for donkey carts is being made available to farmers to boost farming activities. The informal diffusion of donkeys continues even today. The clearing of the savannah forest South of the Sahel and the consequent decline in tsetse challenge have permitted donkeys to spread southwards. Donkeys can survive on low-quality diets and can find food in the peri-urban wastelands surrounding many African towns. With deforestation and land degradation leading to decreased biodiversity, donkeys can also feed on the shrubs that persist under these conditions (Blench 2012).

Although the species name of modern donkeys is commonly written as *Equus asinus*, recent evidence pointing to Africa as the centre of donkey domestication implies that *Equus africanus* is the correct nomenclature. The name *Equus asinus* was introduced by Asinus Linnaeus (1758), referring to donkey populations in Turkey (Eurasia), which as a subspecies of the African wild ancestor, is referred to as *Equus africanus* subsp. *asinus* (https://www.gbif.org/species/113279747). The scientific classification of the present-day domestic donkey is:

Kingdom	Animalia
Phylum	Chordata
Class	Mammalia
Order	Perissodactyla
Family	Equidae
Genus	Equus
Species	*E. africanus*
Subspecies	*Equus africanus* subsp. *asinus*

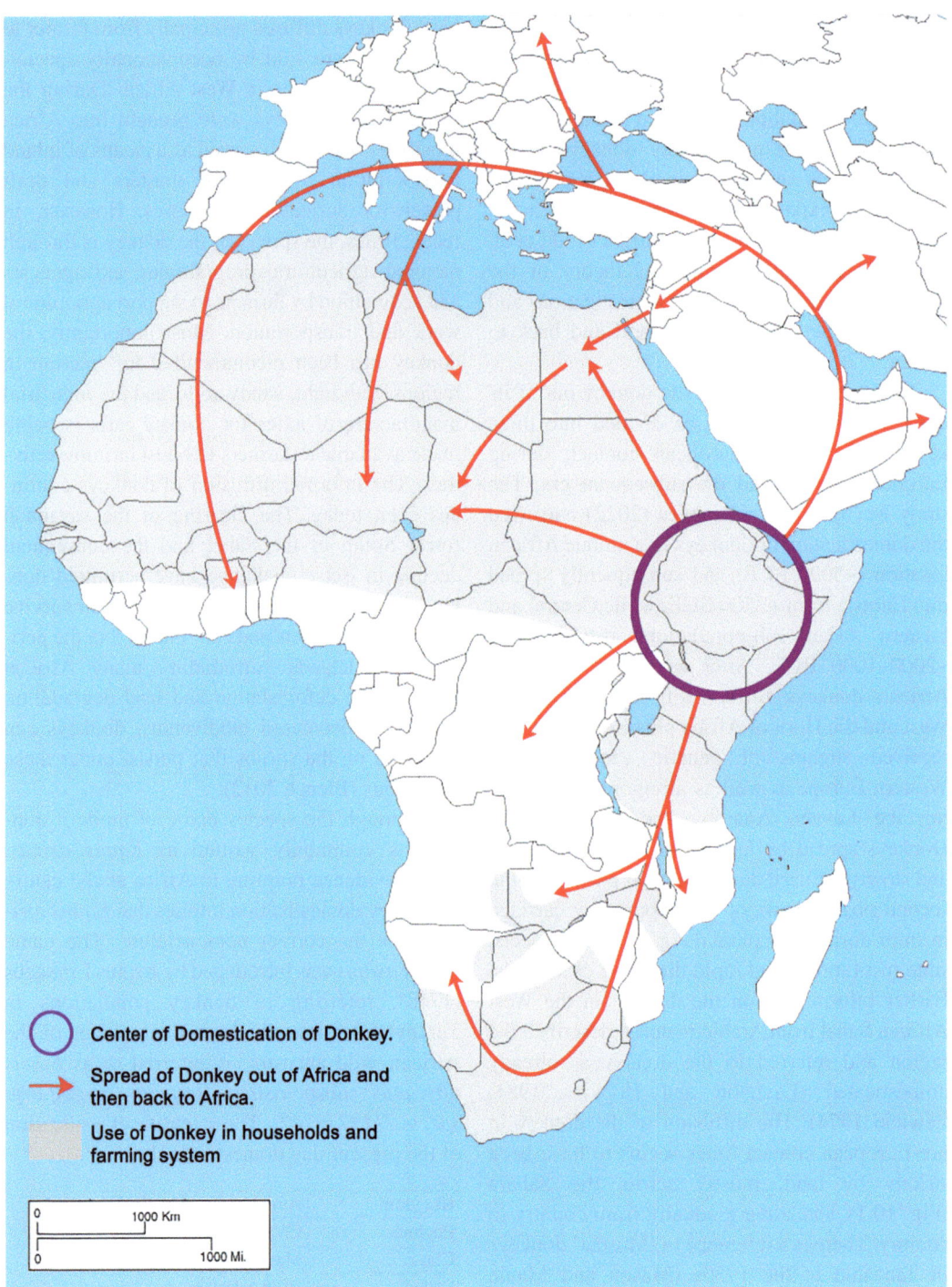

Fig. 10.1 Origin of domestic donkey and routes of dispersion in Africa, Asia and Europe

10.3 Donkey Breeds and Types in Africa

The donkey breeds of Africa are of different sizes and colours and have acquired versatile abilities to adapt to the increasingly changing and dry environments of the arid and semi-arid regions of Africa. In general, the donkey breeds of Africa possess the following characteristics: (1) The breeds vary in body size, a characteristic that helps them to adapt to limited feed resources and water, and high ambient temperatures of its diverse range of environments; (2) have ability to walk long distances and to navigate a wide range of difficult terrains and harsh environments such as sandy, rocky and mountainous terrains; (3) have ability to carry heavy loads and plough the soil, thus making them valuable assets to their owners; (4) they have a working life of about 12–15 years and bring profitable earnings to their owners; (5) they have a digestive system adapted to extract nutrients from poor quality feeds; (6) they have large ears that helps them to cool and regulate their body temperatures ensuring survival in very hot environments; (7) they have varied coat colours that helps in heat deflection and protection against predators; (8) they have ability to graze easily and throughout the day thus ensuring acquisition of available feed resources; (9) they have a loud bray for communication with other donkeys; (10) they are docile, friendly and easy to handle by women and children; and (11) they are a source of enhanced security and well-being to their owners.

10.3.1 Donkey Breeds and Types in West and Central Africa

Donkeys are more prevalent in the arid and semi-arid regions and mostly owned by pastoralists and settled farmers who use them as a source of traction power or means of transport. Four major types or breeds have been reported (Blench et al. 1992) based on coat colour in Nigeria. The names in Hausa are: Auraki (rust or red), Duni (dark brown to black), Fari (pale cream to white) and Idabari (grey to light-medium brown). Khaleel et al. (2020). categorised donkeys in Kano (Nigeria) into seven different coat colour types (Table 10.2), with the majority of individuals having light grey colour (30.6%). Multi-coloured donkeys having light brown with dark stripes, locally called L'âne du Miankala, are also widely distributed in Mali. Information on the presence of donkey is limited to a few countries within the two sub-regions, which include Burkina Faso, Côte d'Ivoire, Ghana, Mali, Mauritania, Niger, Nigeria, Senegal and Togo (West Africa); Cameroon, Chad and the Central African Republic within the Central African sub-region. The grey donkey is the most prominent breed or type among other colours, based on limited reports from several countries highlighted below.

Table 10.2 Distribution of donkey types in Nigeria based on coat colours

Colour	Hausa name	Percentage (%)
Light grey		30.6
Light Brown	Idabari	16.7
Dark brown	Duni	13.9
Dark grey		13.9
Black	Duni	11.1
White with brown	Idabari-fari	8.3
White	Fari	5.5

Source: Khaleel et al. (2020)

10.3.1.1 Idabari

Synonyms: Other names by which Idabari (grey or light-medium brown donkeys) are known in Africa include L'âne du Gourma, L'âne du Miankala, L'âne du Sahel, L'âne du Yatenga, Native of North Africa (Mali); Native of North Africa, Âne commun d'Afrique (Mauritania, Senegal); Anes, l'Ane Africaine (Niger, Togo), and L'Ane Africaine or Rifaï (Chad).

Distribution and Population Statistics: Idabari are the most predominant among different colour variants in the West African sub-region, notably in Nigeria (Khaleel et al. 2020), Mali, Mauritania, Niger and Togo; as well as Cameroon (Nguekeng et al. 2017), Central African Republic and Chad within the Central African sub-region. Donkeys were widespread in Nigeria, but the number drastically reduced during the oil boom in the mid-1980s when they were being replaced by

 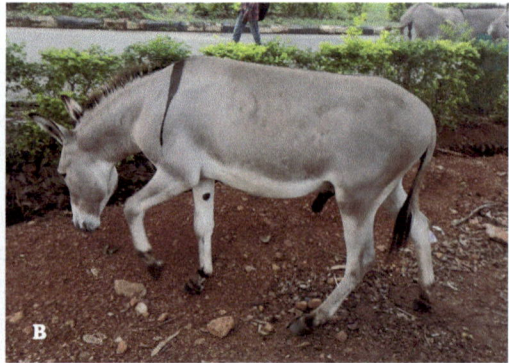

Fig. 10.2 Idabari jenny (**a**) and jack (**b**) at the Federal University of Agriculture, Abeokuta, Nigeria. (Photos by M.N. Bemji)

motorised transport (Blench et al. 1992). However, following economic depression, the demand for donkeys increased substantially. The populations of donkeys were estimated at 1,009,615 in Burkina Faso, with an annual growth rate of 2% (Sow et al. 2012); 250,000 for Native of North Africa and Âne commun d'Afrique (2006 census) in Mali and Mauritania; and 3086 (2006) in Togo (DAD-IS 2024). The donkeys in Chad were imported from Egypt in 1921. They are found in arid and semi-arid areas of Chad and the Sahel and are considered to be at risk, with a population size of 349,982 in 2000 (DAD-IS 2024). According to Ebangi and Vall (1998), the donkey population in Cameroon increased from 700 to 21,000 between 1978 and 1997.

Physical Characteristics: The donkeys within the sub-region are generally of small to medium size, with height at withers being 90 cm (males) and 110 cm (females) for donkeys in Chad. Data from grey and white donkey populations in Cameroon (Nguekeng et al. 2017) showed that average adult liveweight was 122.11 ± 1.24 kg. Females were heavier (124.72 ± 1.65 kg) than males (120.44 ± 1.79 kg). In an earlier study within the same region, the average liveweights of adult (6–13 years old) female and male donkeys were 132.5 ± 2.3 and 123.2 ± 2.1 kg (Ebangi and Vall 1998), respectively. The coat colour of Idabari is predominantly made up of shades of grey (94%) with a black band on the shoulder region (shoulder cross-stripe) and shades of white around the nostrils, the ventral regions and the inner portion of the legs (Fig. 10.2). A study from three localities (Bogo, Gazawa and Kaélé) in the Sudano-Sahelian Zone of Cameroon (Nguekeng et al. 2017) showed that donkeys had predominantly grey coat colour (80.02%) while those with white coat colour (Fari) (4.85%) were least represented, corroborating an earlier study by Ebanji and Vall (1998). The majority of the donkeys displayed the St. Andrew's cross (84.50%) and not zebra lines (76.20%). For morphological characteristics, estimates were 98.05 ± 6.88 cm (height at withers), 108.31 ± 6.82 cm (thoracic), 101.37 ± 13.07 cm (trunk length), 40.47 ± 8.79 cm (neck length), 25.09 ± 1.24 cm (right ear length) and 25.43 ± 1.07 cm for length of the left ear. Data on phenotypic characterisation of donkeys in Nigeria with different colours (Red, Black, White, Brown and Brown-white) by John et al. (2017). showed that majority of adult donkeys had smooth and short hair type (90%), brown coat colour (88.57%), thick skin type (86.07%), solid coat colour pattern (99.64%), pendent tail shape (50.36%), black eye colour (94.64%) and straight head profile (92.86%). Most of the distributions were similar for young donkeys. Donkeys in northwestern Nigeria were mostly of solid coat colour pattern, with short and smooth hair type, brown coat colour, thick skin type, pendent tail shape, black eye colour and straight head profile.

Adaptive and Resilience Features: Donkeys are monogastric herbivores; they eat a wide variety of feeds (roughages) and utilise cellulose and hemicellulose efficiently; hence, they are easy to manage and can practically survive on poor-quality feeds (Aganga et al. 2000). They thrive under adverse climatic conditions and can tolerate considerable heat and dehydration (Smith and Pearson 2005). They also tolerate some tropical diseases and parasites.

Typical Production Systems and Production Characteristics: Fulbe pastoralists mainly use donkeys as pack animals as they migrate from one grazing area to another. During transhumance, donkeys carry tents, camp utensils, children, women, old people, poultry in cages, lambs and kids to new locations. Donkeys are tethered at night, but during the day, they are untethered and allowed to scavenge for grazing in the vicinity of the camp (Blench et al. 1992). In the Sudano-Sahelian zone of Cameroon, the donkeys usually fend for themselves on open natural pastures during the dry season, with little or no health care. During the farming period (rainy season), they are brought back to graze around the houses, and little health care is offered. Varying amounts of sorghum grains and cotton seed cake are given as feed supplements (Ebangi and Vall 1998). At the National Animal Production Research Institute, Zaria, Nigeria, donkeys are semi-intensively managed with grazing being supplemented with straw and concentrates (Sheriff Yusuf et al. 2020).

Reproductive parameters of donkeys reported in Nigeria include the mean age of breeding (96 ± 29 SD months), mean age at first foaling (57 ± 17 SD months), foaling interval (26 months), and mean number of previous parities (2.1) (Blench et al. 1992). Length of the oestrous cycle is about 24 days, while the length of oestrus is 6–7 days, with oestrus occurring throughout the year in tropical climates (Fielding 1988). Gestation period is about 1 year. Body condition of breeding females may not depreciate to the extent of completely inhibiting fertility, with the conception pattern reflecting the ability of donkeys to thrive on the poorest of diets. Free mating of donkeys is allowed in Nigeria when they are herded, but there is restricted access of females to males when jennies are used for work. A mortality rate of 7% was reported in Nigeria (Blench et al. 1992). Most males are sold soon after weaning, with few kept for breeding and work. The pastoral system in Niger and the rural system in Nigeria have a similar donkey herd structure, with 30% males, 70% females and about 50% breeding females, while the urban system in Mali has more males (85%), fewer females (15%) and breeding females (11%).

Local donkeys are highly regarded in Niger and the extreme north of Togo (in the savanna area), and they play very prominent roles such as: animal therapy, draught power, fuel (manure), meat, pack/baggage, pelt or fur, skin/hides, transport, vegetation management, waste recycling of non-human edible feed, weed control and biomass residue management, dressage, racing, cultural and religious ceremonies, and riding (children). Donkey racing is found countrywide in Mali.

10.3.1.2 Auraki, Duni and Fari Donkey Types

The Auraki (rust or red), Duni (dark brown to black; Fig. 10.3), and Fari (pale cream to white; Fig. 10.4) donkey types constitute the minority in the countries of the West and Central African sub-regions, where limited information has been provided on characterisation. Apart from reports on separate estimates for their frequencies in mixed populations (see Sect. 10.3.1), they were not separated from their Idabari counterparts in the analysis for quantitative variables. More phenotypic and molecular characterisations are required to establish the differences among various donkey types for improved utilisation and conservation.

10.3.2 Donkey Breeds and Types in North Africa

The North African sub-region is the second largest producer (29.89%) of donkeys in Africa (Table 10.1). Scanty information on the donkey is limited to a few countries (Algeria, Egypt, Libya, Morocco, Tunisia and Sudan), in which the breeds are discussed.

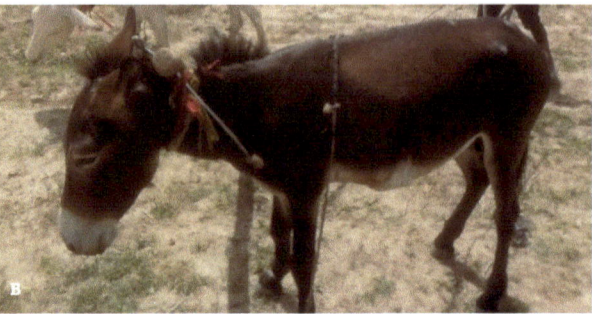

Fig. 10.3 Dunni donkeys at Zaria, Nigeria. (**a**) brown, and (**b**) black types. (Photo by DR. E. M. Ibeagha-Awemu)

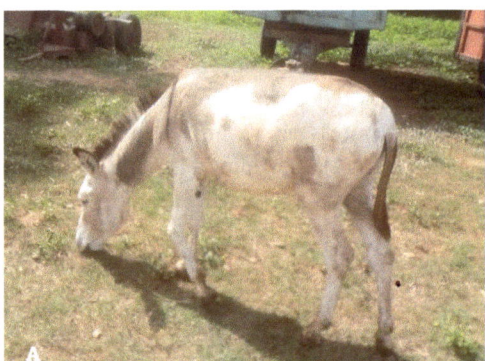

Fig. 10.4 Fari donkeys at Zaria, Nigeria. (**a**) White with brown patches, and (**b**) pale cream. (Photo by Dr. E. M. Ibeagha-Awemu)

10.3.2.1 Algerian Ass

Synonyms: The Algerian ass is also known as the Algerian Ardi or the African donkey.

Distribution and Population Statistics: The breed is native to Algeria. A review (Hannani et al. 2020) showed that the donkey population in Algeria reduced from 315,000 in 1961 to 136,000 in 2013.

Physical Characteristics: The Algerian Ass is medium in size with an average wither height of about 105.3 cm for donkeys in the Kabylie area of Algeria (Ayad et al. 2019), 110.5 cm for extreme eastern Algerian domestic donkey (Hannani et al. 2020), while estimates of 121.92 cm at the shoulder for males and 111.76 cm for females have also been reported (GO 2024). Body weights estimated from linear body measurements using two different equations were 144.3 ± 23.9 and 171.5 ± 28.8 kg, respectively (Ayad et al. 2019). The donkeys have a short, thick coat which varies in colour (grey, black and white). Frequencies for colour variants were estimated at 59.5% for various shades of brown, 27% grey and 13.5% black in the Kabylie area of Algeria (Ayad et al. 2019). In another study, bay and greyish colours were dominant, with a prevalence of 61.5% and 38.5% respectively, for extreme eastern Algerian domestic donkey (Hannani et al. 2020). The head, nose and eyes contour colours were mainly grey with 52.3%, 58.5% and 50.8% respectively. The donkeys were longilinear, or a rectilinear profile, compact with a massive body. Linear body measurements reported by Ayad et al. (2019). were: head length (46.3 cm), ear length (23.6 cm), neck length (48.9 cm), chest width (24.7 cm), back length (61.7 cm), body length (109.6 cm), hips width (32.5 cm), umbilical circumference (142.8 cm), back height (107.8 cm), height at the rump (108.7 cm), thoracic circumference

Fig. 10.5 Egyptian donkey. (Source Wikipedia: photo Jerome Bon by https://en.wikipedia.org/wiki/List_of_donkey_breeds#/media/File:Egyptian_donkey_(2428011421).jpg, CC-BY-2.0 license)

(110.2 cm), chest depth (49.6 cm), front leg length (77.2 cm), cannon circumference (14.8 cm), cannon length (20 cm) and cannon height (30.3 cm).

Adaptive and Resilience Features: The Algerian donkey is hardy and well adapted to the harsh desert climate of North Africa. It is well known for its strength and endurance and is used for different purposes, such as transportation, farming and as a pack animal.

Production Characteristics: The animals are good for meat production (Hannani et al. 2020), milk, and as a source of work for transportation, ploughing and other farming activities.

10.3.2.2 Egyptian

Synonyms: Baladi or white Egyptian of Lower Egypt; Hassawi (widespread); and Saidi (White Egyptian of Upper Egypt) (Yilmaz et al. 2012b; Orhan et al. 2012). Donkeys in ancient Egypt were called Aa-hemet or Eeyore in Hieroglyphic (Attia 2017).

Distribution and Population Statistics: The Egypt Baladi is found in Lower Egypt (northern Egypt), Hassawi is distributed countrywide, while Saidi is found in Upper Egypt (southern Egypt). Based on a review (Attia 2017), there are approximately three million working donkeys in Egypt. Despite their importance to local communities, the Baladi donkey population is facing threats from habitat loss, competition from other livestock species, and crossbreeding with other donkey breeds.

Physical Characteristics: Baladi donkeys (Fig. 10.5) are typically medium to large in size, with height ranging from 122 to 142.25 cm (Yilmaz et al. 2012b; Orhan et al. 2012); 142.24 to 152.4 cm at the shoulder (GO 2024). Donkeys in Egypt generally vary in size depending on breed, with body weights ranging between 80 and 480 kg and height at withers between 79 and 160 cm (Attia 2017). The donkey in ancient Egypt was about 1.2 metres high and had a body weight of about 272.16 kg, much larger than the modern donkey. There are variations in Baladi coat colour (white, grey, black and brown).

Adaptive and Resilience Features: The Egyptian Baladi donkey has renowned strength and endurance and is versatile due to its muscular build and strong legs. It is also known for its

Fig. 10.6 Libyan donkey. (Photo Source: Wikipedia and available at https://en.wikipedia.org/wiki/List_of_donkey_breeds#/media/File:Donkey_in_Wasita_Libya.jpg, CC-BY-4.0 license)

smooth gait and good temperaments and can adapt to a wide range of environments and uses (GO 2024).

Typical Production Systems and Production Characteristics: Donkeys are allowed to roam in herds, watched over by a herdsman. A jenny has a gestation period of about 12 months and gives birth to a single foal (twin births are rare). The donkey was the most important transporter of loads and people in ancient Egypt. It is also used for agriculture and as pack animals. Working donkeys in the poorest countries have a life expectancy of 12–15 years and may live between 30 and 50 years in wealthy countries (Attia 2017).

10.3.2.3 Libyan

Distribution: The Libyan donkey breed is native to Libya.

Physical Characteristics: Libyan donkey is a medium-sized breed, with height at withers ranging from 121.92 to 142.24 cm and body weight between 181.44 and 249.48 kg (GO 2024). The smaller variety is bay with a pale belly, and the larger variety has a dark bay or grey coat. They have a variety of colours, including grey, black, and brown (Fig. 10.6).

Adaptive and Resilience Features: The Libyan donkey is highly versatile with a strong and sturdy body, hard-working and intelligent. It is also well adapted to different terrains and climates, making it a preferred choice for farmers in the region (GO 2024).

Production Characteristics: The breed is used for a variety of tasks, including ploughing fields, carrying heavy loads, and serving as a pack animal. It is also considered an important part of Libya's cultural heritage.

10.3.2.4 Tunisian

Synonym: Anne Tunisien.

Distribution and Population Statistics: The Tunisian donkey is native to Tunisia and the surrounding regions. The population estimated at 240,200 in 2021 is not considered to be at risk (DAD-IS 2024).

Physical Characteristics: The Tunisian donkey is a medium-sized breed with height at withers estimated between 121.92 and 142.24 cm (GO 2024). It has a distinctive appearance with a short, rounded head and large, floppy ears (Fig. 10.7). The build is sturdy and muscular, with well-defined withers. Coat colour varies, including grey, black, and brown.

Adaptive and Resilience Features: It is very hardy, intelligent and has a good temperament. With regards to personality, the donkey breed is very friendly and has a curious nature, eager to please their owner and easy to handle and train

Fig. 10.7 Tunisian. (Source: Wikipedia and available at https://en.wikipedia.org/wiki/List_of_donkey_breeds#/media/File:Barrage_bni_mtir_456789_13.JPG, CC-BY-3.0 license/)

Fig. 10.8 Moroccan. (Source: Wikipedia, photo by Dale Harvey and available at https://en.wikipedia.org/wiki/List_of_donkey_breeds#/media/File:Tamraght-daleharvey-10-donkey.jpg, CC-BY-2.0 license)

Production Characteristics: Tunisian donkey is well-suited for work as pack animals and for transportation. Despite the increasing use of modern machinery and transportation in recent years, the Tunisian donkey still plays a critical role in the daily lives of many rural communities in the region (GO 2024).

10.3.2.5 Moroccan

Synonyms: The Barb or the Atlas donkey.

Distribution: The breed is native to Morocco.

Physical Characteristics: The Moroccan donkey is a medium-sized breed with height at withers estimated between 121.92 and 142.24 cm and body weight between 272.16 and 362.87 kg (GO 2024). It has a variety of colours, including grey, black, and brown (Fig. 10.8).

Adaptive and Resilience Features: The Moroccan donkey is very strong with a sturdy build, versatile, hard-working and patient in nature. It is well adapted to different terrains and climates, hence, a preferred choice for farmers in the region (GO 2024).

Production Characteristics: The breed does a variety of tasks (carrying heavy loads, ploughing fields and serving as a companion animal), as (GO 2024). It can adapt to a variety of environments, making it well-suited for work in a range of conditions in arid and semi-arid regions.

Fig. 10.9 Kassala donkey carrying grains from a market in Keren, Eritrea. (Source: Wikipedia- photo by David Standley and available at https://en.wikipedia.org/wiki/List_of_donkey_breeds#/media/File:Donkey_Transport_(8383370559).jpg, CC-BY-2.0 license)

well as an important part of Morocco's cultural heritage.

10.3.2.6 Dongola

Synonym: Dongolawi.

Distribution and Population Statistics: Dongola donkey is found in the northern regions of Sudan, including the Dongola region for which the breed is named. The population was estimated at 175,004 in 2020 (DAD-IS 2024). Despite its reputation in local communities, the population is facing threats from habitat loss and competition from other livestock species, thereby necessitating conservation efforts to preserve the breed.

Physical Characteristics: The breed is medium to large in size (larger than the Sudanese Pack), with height at withers ranging from 142.24 to 152.4 cm (GO 2024). Coat colour is variable, including black, dark brown or reddish-grey, occasionally pale grey or white with a white belly, muzzle and internal side of the thighs and arms.

Adaptive and Resilience Features: The Dongola donkey breed is noted for its strength and endurance, making it suitable for hot and arid environments. It has a muscular build and strong legs and is also renowned for its smooth gait and good temperament.

Production Characteristics: Dongola is highly appreciated in the region for its use as a pack animal and for transportation.

10.3.2.7 Kassala

Distribution: It is a breed of donkey found in the Kassala region of Sudan.

Physical Characteristics: The breed (Fig. 10.9) is medium size, with height at withers ranging from 121.92 to 142.2 cm and weighing between 181.44 and 249.48 kg (GO 2024). Coat colour is variable, including black, grey and brown.

Adaptive and Resilience Features: Kassala is known for its strong and sturdy body, endurance, versatility and high adaptability to different terrains and climates. It is hard-working and intelligent in nature.

Production Characteristics: Kassala is used for a variety of tasks, including ploughing fields, carrying heavy loads, and serving as a pack animal. It is also an important part of Sudan's cultural heritage.

10.3.2.8 Sennar

Distribution: The breed is commonly found in the Sennar region of Sudan.

Physical Characteristics: The breed is medium to large in size, has a sturdy build, strong

Fig. 10.10 Sudanese Pack. (Source: Wikipedia- photo by David Haberlah and available at https://en.wikipedia.org/wiki/List_of_donkey_breeds#/media/File:Donkey_carrying_Rahal_basket.jpg, CC BY-SA 3.0 license)

legs, with muscular and well-defined withers. It has a distinctive appearance, with a long sloping back, long legs, and large floppy ears. Coat colour varies, including grey, black, brown and white (DAD-IS 2024).

Adaptive and Resilience Features: The breed is very hardy and can adapt to harsh environments, making it well-suited for work in the hot, arid conditions of the Sennar region.

Production Characteristics: Sennar donkey is mainly used for ploughing fields, carrying heavy loads, and for milk and meat (DAD-IS 2024).

10.3.2.9 Sudanese Pack

The Sudanese Pack (Fig. 10.10) is a domestic donkey breed that is native to Northern Sudan. The breed is considered to be at risk, with an estimated population of 7609 in 2018 (DAD-IS 2024).

The Toposa donkey is found in Southeastern Sudan and the surrounding regions. It is not considered to be at risk, with a population of 570,664 in 2020 (DAD-IS 2024). In terms of physical characteristics, adaptive/resilience features and production characteristics, both breeds (Sudanese Pack and Toposa) are similar to the Tunisian donkey (see Sect. 10.3.2.4).

10.3.2.10 Sudanese Riding

Synonyms: Riffawi, Shindawi Riding Ass.

The Sudanese riding donkey has various sub-types or sub-breeds found in northern Sudan and northwestern Eritrea. They include: 1. Atabai, which is medium in size and has a height at the withers estimated at 102–112 cm, with varying coat colours (grey, white, reddish or dark brown, often without a cross); 2. Kassal is larger and has a reddish coat with a cross; 3. Massawa is almost white (Yilmaz et al. 2012b; Orhan et al. 2012). They are locally adapted and not at risk, with a population of 4,139,217 in 2020 (DAD-IS 2024); 4. Etbai (Elgash) is a smaller variety of Sudanese Riding found in the Bija Tribal Country, the Red Sea Mountains and the coastal area, north of Port Sudan in northwestern Sudan. The coat colour is grey with black or dark bilateral stripes running vertically from the wither to the tip of the shoulder. With regard to special qualities, Etbai is reported to be a trotter with a difficult character. The population was estimated at 1,133,719 in 2020 (DAD-IS 2024).

10.3.3 Donkey Breeds and Types in East Africa

The East African region possesses the largest donkey population on the African continent (Table 10.1), and in particular, Ethiopia has the largest population of donkeys in the world

(10.79 M, 10.02 M and 9.93 M heads in 2020, 2021 and 2022, respectively), while Sudan recorded the second largest populations in 2020 (7.63 M heads), 2021 (7.64 M heads) and 2022 (7.65 M heads) (FAOSTAT 2024). Ethiopia's strategic location in the Horn of Africa, which serves as a possible domestication centre as well as a major point of entry, and its ecological attributes are thought to have a major influence on its donkey populations in terms of number, geographical distribution, conformational differences, coat colour patterns and body sizes. While the donkey is amongst the most neglected livestock species on the African continent, the Ethiopian donkeys have received the most research attention (Kefena et al. 2011; Getachew et al. 2023; Wassie et al. 2023; Geiger et al. 2023; Eriso et al. 2023; Hassen et al. 2022; Kefena et al. 2021; Geiger et al. 2020), with several breeds reported: Abyssinian, Afar, Hararghe, Ogaden, Omo, Hamer and Sinnar. Other donkey breeds found in East Africa are Somali, Dongolawi, Maasai and Muscat. The donkeys of East Africa play important economic roles. They are mostly exploited for draught power and transport; they are prominent in festivals and entertainment and used in ceremonies such as gifts during weddings and food at funerals (Geiger et al. 2020, 2023; Admassu and Shiferaw 2011; Asteraye et al. 2024; Geiger and Hovorka 2015).

Donkey ownership has been reported to improve household income and livelihoods (Admassu and Shiferaw 2011; Asteraye et al. 2024; Curran and Smith 2005). In particular, ownership of a working-donkey is a key aspect of women empowerment through assistance with daily work burden (donkey used to fetch water; collect firewood for home use and sale; transport farm produce; transport grain to the mill house; transport items purchased at market back to the household, and till the fields, etc.) and improved economic status and standing.

In the peri-urban and urban areas of Ethiopia, working donkeys are common among youths, who use them for income generation by offering merchandise transportation services.

10.3.3.1 Abyssinian

Synonym: Ethiopian.

Distribution: The Abyssinian is widely distributed throughout Ethiopia, mostly along both sides of the Ethiopian highlands and in South Africa.

Physical Characteristics: Abyssinian is characterised by a small body size or dwarf, predominantly brown, occasionally chestnut-brown, slate-grey and hairy, has the primitive equine stripes with a clear dorsal stripe along the backline from the poll area to the tail and a shoulder stripe that runs across the withers area to make a cross; its ventral side, muzzle and nostril points are white in colour (Fig. 10.11). They are hardy and heavily working animals. Height at withers is 94 cm for both males and females.

Adaptive and Resilience Features: Abyssinian is a compact animal shown to be morphologically distinct from other donkey populations in Ethiopia (Kefena et al. 2011). The small body size is probably an adaptation to the highland ecology of its predominant habitat. Its main uses include draught power, carting, manure (fuel) and transport. It is also important for habitat provision and biodiversity transport through pollination and seed dispersal, weed control and biomass residue management. The gestation period of the Abyssinian ranges from 11 to 13 months, with a lifespan of 30–40 years. The composition of Abyssinian milk is closer to that of human breast milk in its protein, lactose, vitamin C, zinc, pH and density contents, while its calcium, iron and magnesium contents are higher than those of human breast milk (Tadesse et al. 2014).

10.3.3.2 Sinnar

Distribution: Sinnar's geographical niche is the northwestern lowlands along the Ethio-Sudanese border (from Humera to Assosa).

Physical Characteristics: Sinnar is the most peculiar donkey in Ethiopia with unique morphological characteristics. It is considered the tallest of the Ethiopian donkey breeds, with height at withers of 100–114 cm; average matured body weight of 126 kg for males and 129 kg for females (Kefena

Fig. 10.11 Abyssinian donkeys grazing (**a**), slate-grey Abyssinian adult donkey (**b**), and chestnut-brown Abyssinian jenny and foal (**c**). (Photos by Dr. E. Kefena, Agricultural Transformation Institute, Ethiopia)

et al. 2011; Mustefa et al. 2020). Sinnar has variable coat colours such as white, black, fawn, leopard and brown, with white and leopard coat colours being the most common, and most have white abdominal colour (Fig. 10.12). The leopard coat-coloured Sinnars have a ridge on their ribs. Morphologically, it is the second most distinct donkey breed in Ethiopia (Kefena et al. 2011).

Adaptive and Resilience Features: Sinnar is well adapted to the desert region of the country. It finds use as a working animal (draught power), carting, manure (fuel), transport and riding. It preserves biodiversity through seed dispersal and weed control. Sinnar jack, with its unique and peculiar phenotypic characteristics, is used by mule breeders to produce mules that command a premium price in Ethiopia (Kefena et al. 2011).

10.3.3.3 Afar

Distribution: Afar is mostly found in the Afar Plain as well as neighbouring western Hararghe mid and lowlands of Ethiopia.

Physical Characteristics: Afar has varied coat colours, including grey, greyish-red and light-red animals (Fig. 10.13). Leg stripes are common (Fig. 10.13b). Height at withers is 104 cm (Kefena et al. 2011). Afar is used to a small extent for draft power and manure (fuel). Its main use is in biodiversity maintenance through weed control, pollination and seed dispersal, and biomass residue management (Fig. 10.13a). Morphologically, Afar, Harar and Ogaden are very similar (Kefena et al. 2011).

Adaptive and Resilience Features: Afar is a free-ranging animal and is less used for work.

Fig. 10.12 Sinnar donkeys displaying different coat colours: (**a**) white, (**b**) brown, and (**c**) fawn. (Photos by Dr. E. Kefena, Agricultural Transformation Institute, Ethiopia)

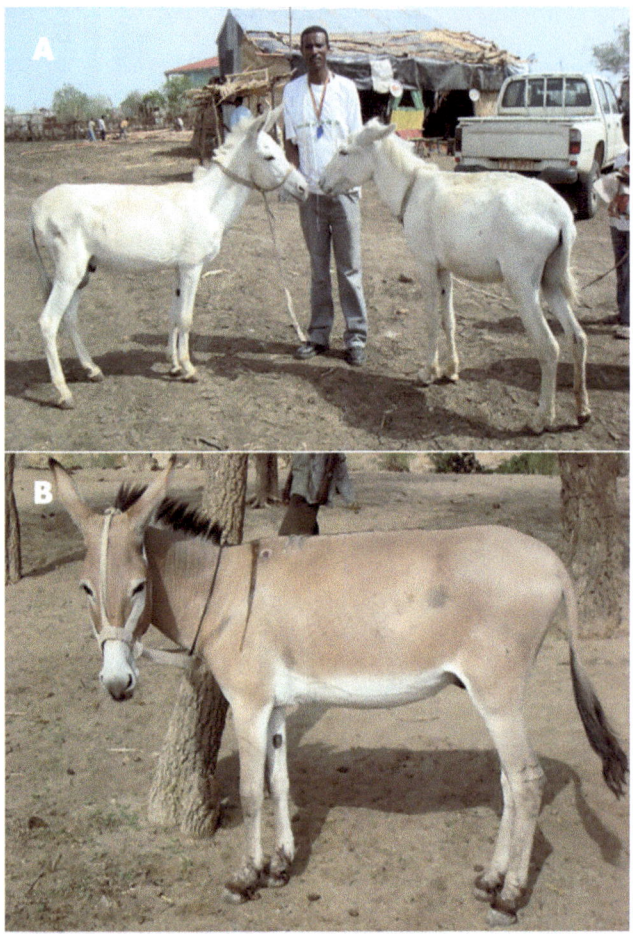
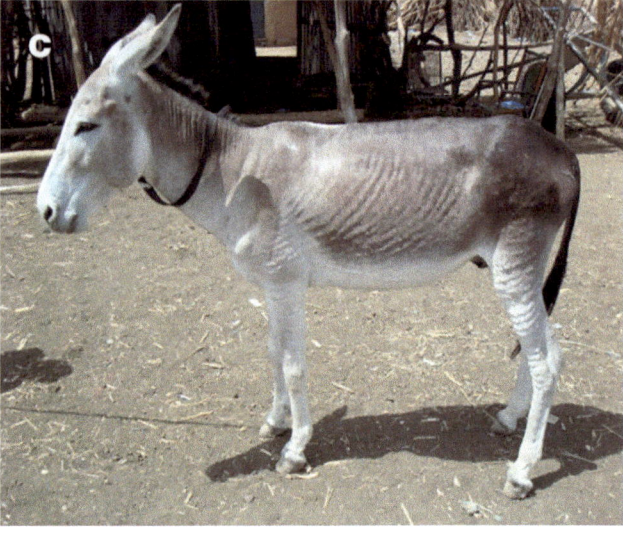

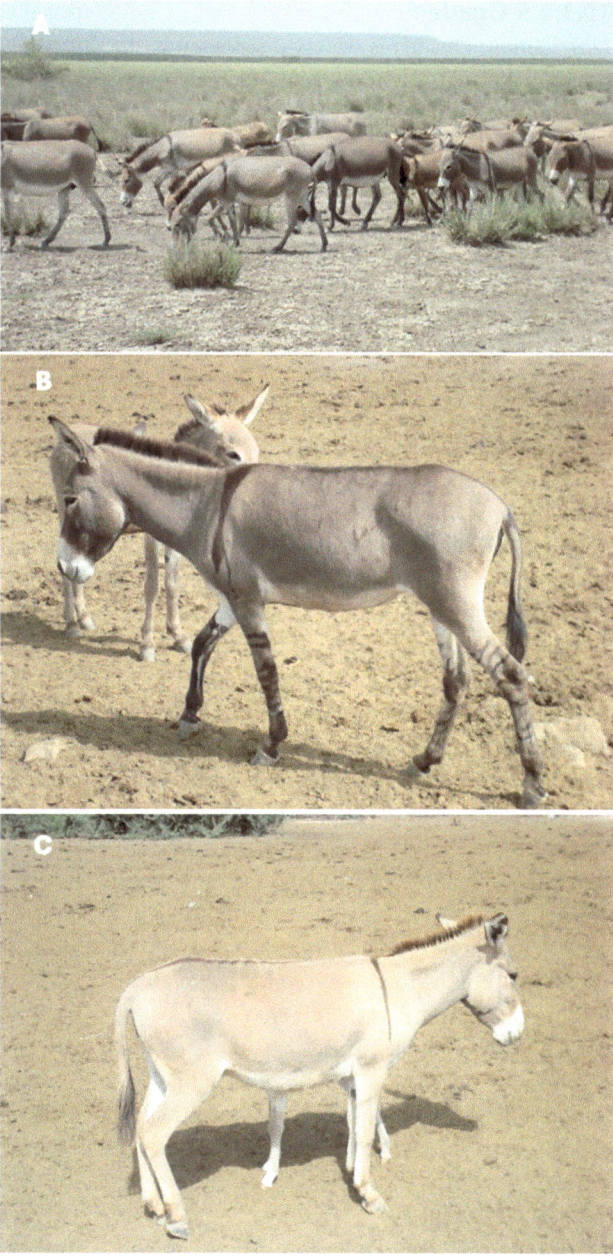

Fig. 10.13 A herd or drove of Afar donkeys (**a**) and Afar Jenny and foal (**b, c**). (Photos by Dr. E. Kefena, Agricultural Transformation Institute, Ethiopia)

10.3.3.4 Hararghe

Distribution: It is found in the Oromia/Hararghe highlands of Eastern Ethiopia.

Physical Characteristics: Hararghe has varied coat colours, with common colours being grey and brown; leg strips are common. It is a heavy working animal. Other functions include carting, draught power, manure (fuel), transport and biodiversity maintenance. Height at withers is 10.3 cm (Kefena et al. 2011).

Adaptive and Resilience Features: Hararghe is adapted to the Hararghe highland and shares phenotypic similarity with both Ogaden lowland and Afar donkey populations, but is obviously distinct morphologically (Kefena et al. 2011).

10.3.3.5 Ogaden

Distribution: Ogaden is widely distributed in the pastoral and agro-pastoral areas in the Somali Regional State (Somali/Ogaden plain) of Ethiopia.

Physical Characteristics: The coat colour is mostly grey and greyish red, and leg stripes are found in some individuals. Ogaden is used for carting, manure (fuel), weed control and biodiversity maintenance.

Adaptive and Resilience Features: Ogaden is a heavily built animal and is relatively used for work.

10.3.3.6 Omo

Distribution: Omo is predominantly found in the South Omo pastoral lowlands around Lake Turkana and Hamerin, Gnyangatom and Dasenech districts in Southern Ethiopia.

Physical Characteristics: This breed is peculiar as it has predominantly spotted dark-brown, fawn and light grey with shiny skin (Fig. 10.14). The fawn coat-coloured donkeys are mostly found in Hammer, Benatsemay and Dasenect districts (Getachew et al. 2023). It is a source of manure (fuel) and it maintains biodiversity through pollination, seed dispersal, and weed control. Omo served as a source of proteins for some tribal groups, and in recent times, served occasionally as a food source. Mean height at withers range from 104 to 110 cm (Kefena et al. 2011; Getachew et al. 2023).

Adaptive and Resilience Features: Omo is a fatty and heavily built donkey. It is free-ranging and is less used for work.

10.3.3.7 Maasai

Distribution and Population Statistics: Synonym: Masai. The Maasai donkey is found in the pastoral areas of southern Kenya and northern Tanzania and is reared by the Maasai pastoralists. In 2018, its population was estimated at about 600,000 in Tanzania, so the breed is not at risk.

Physical Characteristics: The Maasai donkey is a small animal with an average body weight of about 110 kg. It has a greyish-brownish (grey-dun) colour. The ventral side of the body, muzzle and nostril points are white, and it has the traditional equine dorsal stripe line from the poll area to the tail (Fig. 10.15). The average body measurements of Maasai donkeys were estimated at 156 kg (body weight), 100 cm (height at withers), 113 cm (body length) and 114 cm (heart girth) (Gichure et al. 2020a). It is mainly used by Maasai women to transport goods to the market and for draught power. It is not used as a food source by the traditional Maasai community. The Maasai is a valuable asset for semi-nomadic communities who depend on the donkey to transport movable goods to a new location. Culturally, it is used for racing and as a touristic attraction (such as donkey riding).

Adaptive and Resilience Features: The Maasai is well adapted to the Maasai system of free-range grazing.

10.3.3.8 Somali

Distribution and Population Statistics: The Somali donkey is native to the Horn of Africa, especially modern-day Somalia, northern Kenya, Djibouti and Yemen. Its population is declining, and it is considered an 'at risk breed'.

Physical Characteristics: Somali donkey is a small to medium-sized animal, with the withers height from 100 to 120 cm. It is slender and agile built, with long, slim legs, long, slender neck and large, floppy ears (Fig. 10.16). Its coat colour is in a variety of shades, including grey, brown, black and white. For centuries, the Somali donkey has been used for transportation, ploughing and herding. Because of its agility, speed, and endurance, the Somali is used as a riding and pack animal. Culturally, it is used for racing and for tourism as a riding animal.

Adaptive and Resilience Features: The Somali donkey is agile, runs fast and endures heavy loads and long distances. It is well adapted to the arid conditions of Somalia.

10.3.3.9 Somali Wild Ass

Distribution and Population Statistics: The Somali wild ass, *Equus africanus somaliensis*, is a variety of the African Wild Ass (the other variety is the Nubian wild ass). It is the smallest of the wild equids, a group made up of horses, asses and zebras. The Somali wild ass is found in

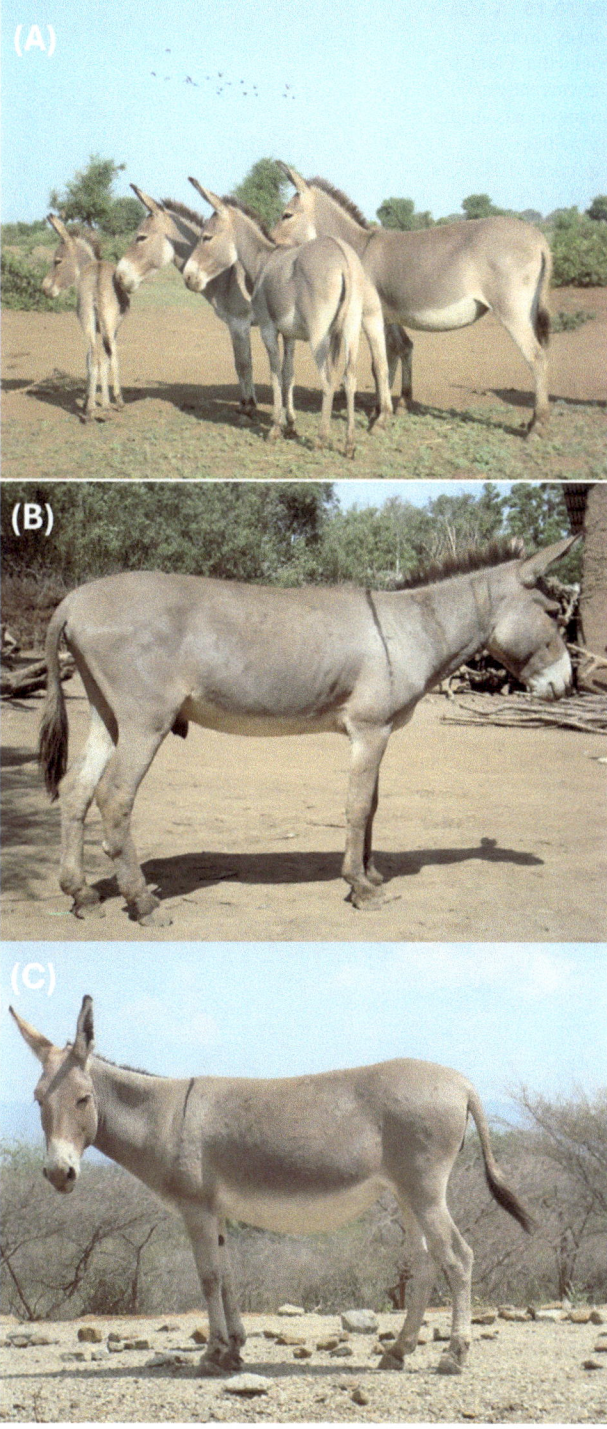

Fig. 10.14 A herd or drove of Omo donkeys (**a**) and fawn-coloured Omo jack (**b**) and jenny (**c**). (Photos by Dr. E. Kefena, Agricultural Transformation Institute, Ethiopia)

Somalia, the Afar region of Ethiopia, and the southern Red Sea region of Eritrea and Djibouti. It is a sleek, graceful and majestic animal that differs from the horse and zebra by its small size, larger ears, stiff mane, tufted tail and louder bray. Domesticated donkeys descended from the Somali and Nubian wild asses, as their images have been found painted in caves by ancient peo-

Fig. 10.15 A Maasai donkey (Source: Wikipedia. Photo by Daryona, https://commons.wikimedia.org/w/index.php?curid=10558134, CC BY-SA 3.0)

Fig. 10.16 A Somali donkey. (Photo by Dr. E. Kefena, Agricultural Transformation Institute, Ethiopia)

ple of North Africa (https://animals.sandiegozoo.org/animals/somali-wild-ass). Today, a few hundred animals are found in zoos in many countries, such as San Diego Zoo (California, USA), Franklin Park Zoo (Boston, Massachusetts, USA), Dallas Zoo (Texas, USA), St. Louis Zoo (Missouri, USA).

A population of 300 heads was reported in Somalia in 2006 (DAD-IS 2024) and likely less than 1000 Somali wild asses are found in the wild. Thus, it has been listed in the Red List of endangered species by the International Union for Conservation of Nature (https://www.iucn.org/), meaning that it is critically endangered and face the risk of extinction.

Physical Characteristics: The Somali wild ass has a reddish-grey coat, white on the ventral side of the body, muzzle and nostril points, a black-

Fig. 10.17 Two Somali Wild Asses (*Equus africanus somaliensis*) in Marwell Zoo, Hampshire, UK. (Source: Wikipedia-photo by Chris Keating, CC BY-SA 3.0, https://commons.wikimedia.org/w/index.php?curid=15354236)

and-grey upright (spiky) mane, and unique black and white stripes on their legs like their zebra relative (Fig. 10.17). In fact, it is the only donkey type with striped legs. It seldom possesses a dorsal stripe and shoulder cross-stripe common to most donkey breeds. It also has small, narrow hooves that facilitate fast and safe movement through its rocky habitat. The average height at withers for adults ranges from 120 to 150 cm, and the average adult weight is 270 kg. The gestation length is 11–12 months, and the birth weight of foals ranges from 20 to 25 kg.

Adaptive and Resilience Features: Somali wild ass is built to survive conditions of extreme heat, extreme high temperatures and very little water. Its large ears help with heat dissipation and to pick up sounds. The Somali wild ass feeds on arid and semi-arid scrubs, grass, bark and tough desert plants. Its habitat is hilly and stony desert lands.

10.3.3.10 Muscat

Distribution: The Muscat donkey is found in Tanzania and the Sebei and Karamoja regions of Uganda. It is believed to have descended from the Egyptian or Arabian donkeys.

Physical Characteristics: Visibly, the Muscat is larger and paler than the Maasai donkey (Fig. 10.18). Its main use is for draught power, transport, riding by children and as pack animals.

It is small to medium-sized in build (compact and sturdy). Its appearance is distinctive with long ears and a convex profile. The coat colour is varied, including grey, black and brown. They are of good temperament. The muscat is grazed with cattle, and breeding is random throughout the year.

10.3.4 Donkey Breeds and Types of Southern Africa

Prominent breeds in the Southern African region include Tswana, Abyssinian and Namibian donkey. A recent study reported donkeys with different coat colours, including dark brown, light brown, white, grey and charcoal in South Africa's Limpopo province (Maswana et al. 2022).

10.3.4.1 Tswana

Distribution and Population Statistics: Synonym: The Tswana is also known as Botswana ass. Tswana is native to Botswana, and it is widely distributed in the country. It is also found in neighbouring regions. The population of Tswana was reported as 351,000 heads in 2015 (DAD-IS 2024). The animal is therefore not at risk.

Physical Characteristics: Tswana is a medium-sized breed with an average adult weight of 250 kg. Tswana has distinct features

Fig. 10.18 Muscat donkey in Tanzania. (Source: Wikipedia-photo by Nevit Dilmen (talk), CC BY-SA 3.0, https://commons.wikimedia.org/w/index.php?curid=11320901)

such as a short-rounded head, large, floppy and erect ears, muscularly built with well-defined withers, has varied coat colours—grey, black, brown and white (less frequent), and the belly, muzzle, nostril and eye points are white in colour (Fig. 10.19). Its main use is for draught power, transportation (Fig. 10.20), racing, riding for sports and riding by children. Tswana is also consumed, and its use as a food source is increasing in recent years.

Adaptive and Resilience Features: Tswana is well adapted to the arid and semi-arid regions of Botswana. Its resilient features make it particularly useful for draught power when compared to cattle or horse, such as (1) ability to survive drought better due to their small size and ability to survive on low-quantity/quality feed materials and lower water requirements; (2) cheaper to purchase and maintain; (3) better body condition for ploughing; (4) easier to train, handle and manage; (6) higher survival rate in areas infested with tsetse flies (cause trypanosomiasis) and African horse sickness; (7) little or no gender restriction on their use; (8) have longer working life if well managed; and (9) can be an avenue for poverty alleviation in rural areas (Nsoso et al. 2008; Nengomasha et al. 1999).

10.3.4.2 Namibian

Distribution and Population Statistics: Synonyms: okasino, donki, tonki, ndongi and donkie. The Namibian donkeys are believed to have been brought to Namibia from 1656 by White colonial settlers who used the animal as pack and draught power for on-farm activities and for pulling carts and wagons, and later spread to the rest of the country from the early nineteenth century (Mwenya and Keib 1997). The Namibian donkey population has remained steady over the years, with an estimated 120,500 heads in the northern communal areas in 1997 and about 120,000 heads in 2014 (DAD-IS 2024; Mwenya and Keib 1997).

Physical Characteristics: The breed is medium-sized, and the coat colour is varied, including greyish-brown, brown, etc. The donkey is used for transportation, draught power, meat, riding (work), transport and riding by children. Donkey meat is a delicacy in some localities in Namibia.

Fig. 10.19 Tswana adult donkey showing typical features. (Photo by Dr. E. M. Ibeagha-Awemu)

Fig. 10.20 Greyish-brown Tswana donkey ready to transport water at a collection point near Opuwo in Namibia. (© The Donkey Sanctuary 2024)

Adaptive and Resilience Features: The breed is locally adapted to the arid regions of Namibia.

10.4 Donkey Breeding and Production Systems in Africa

Donkey husbandry varies based on its purpose, herd size, usage, and environment. Efficient use of donkeys depends on many factors, including socio-economic and cultural factors. In extensive systems (Fig. 10.21), donkeys graze freely throughout the day, similar to feral donkeys in the wild. In pastoral systems, donkeys are herded with other livestock like cattle, goat and sheep. They can survive on poor diets without showing clinical signs of anaemia or metabolic disorders and are resistant to various infectious pathogens and pests common in the tropics. Unlike horses, donkeys are less prone to colic and azoturia, likely due to physiological adaptations. In *mixed and small-scale systems*, donkeys are often an additional livestock for smallholder rural dwell-

Fig. 10.21 Donkeys grazing in Zaria, Nigeria (**a**) (Photo by Dr. E. M. Ibeagha-Awemu) and in Ethiopia (**b**) (Photo by Dr. E. Kefena)

ers, primarily used for draught and transport purposes in free-range systems, which are widespread in sub-Saharan Africa. In *semi-intensive systems*, donkeys graze periodically and receive some supplemental feeding. They spend part of the day in the field and the night in a stable, fence, or barn, as seen in farms integrating donkeys with other livestock, such as dairy production in Kenya, using paddocks and fences for pasture management. See Chap. 2 for a detailed discussion on African livestock production systems.

Donkey management in Africa involves inbreeding and introgression from divergent lineages, although the inbreeding detected in African donkey populations is limited, being attributed to free-range local management practices that allow for continued interbreeding with wild and feral subpopulations (Todd et al. 2022). Admixture modelling suggests ongoing introgression from African wild asses into modern donkeys from Africa and the Southern Arabian Peninsula. Donkeys are bred throughout the world and often play an important role in agricultural activities, particularly in poorer regions and the least accessible. Traditionally, jacks are considered challenging to breed in domestic conditions, whether for natural breeding or semen collection using either jennies or mares. Donkey's natural sexual behaviour significantly differs from that of other domestic animals. This presents challenges for in-hand donkey breeding, particularly on mule studs where normally only jacks and mares are kept for breeding. In Africa, as elsewhere, donkey breeding is mostly random,

Fig. 10.22 Mule breeding in Ethiopia. Female horse (left) and male donkey (right) ready to mate. (Photo by Dr. E. Kefena, Agricultural Transformation Institute, Ethiopia)

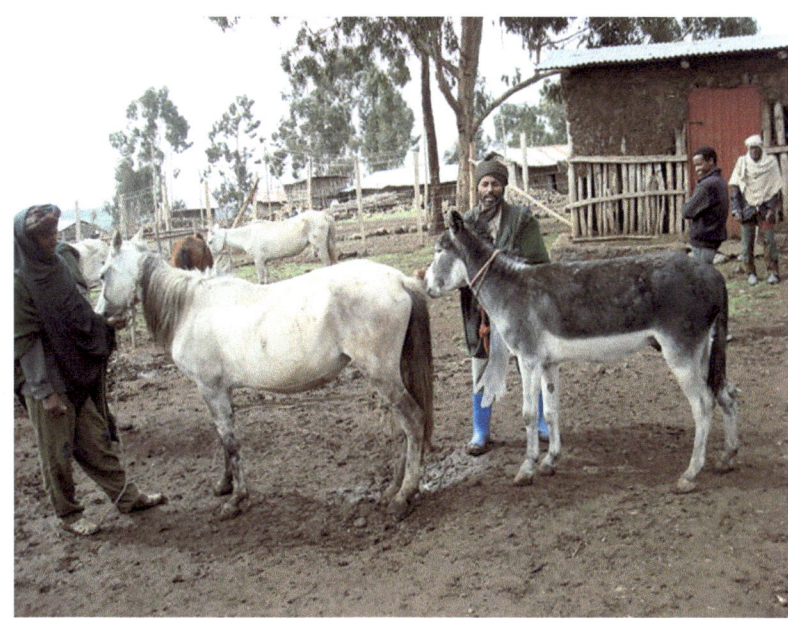

and very little care is taken in the selection of mating pairs. Since importing of exotic breeds is practically not feasible, the risk of breed outcrossing is negligible. Limited data on donkey reproduction indicate relatively long gestation periods of about 11.5–12.5 months or 353–371 days and varying conception rates. Donkeys grow slowly, taking up to 24 months to reach maturity.

While there is limited intentional breeding of the donkey in Africa, mule breeding has been practised in many parts of the world, including Africa (Fig. 10.22), for centuries. The mule is a hybrid cross of a male donkey (jack) with a female horse (mare). The mule crossbreed of the donkey with 62 chromosomes and the horse with 64 chromosomes has 63 chromosomes, and it is infertile. The mule has been deliberately bred by man since ancient times, such as in Mesopotamia and Anatolia in the third century BC, and it was known in Egypt before 3000 BC (Babb 2017; Yilmaz and Wilson 2013). During the Roman Empire, the mule was the most frequently used beast of burden (Devriese 2012). It is believed that the mule combines valuable characteristics from the donkey and horse, making it a good work animal. The mule has the calm, stoic nature of its donkey father and a more energetic, flighty reaction of its horse mother. Unlike the horse, the mule is less likely to panic in dangerous situations and is able to carry more weight than donkeys. The role of the mule as a beast of burden is founded on these great attributes, which make it a great means of transport in remote mountain and desert environments, where roads and other transport infrastructures have failed to penetrate (Debarbieux et al. 2014; Giampiccoli 2017). These attributes prompted Guénon (Guénon 1999) to refer to the mule as the ship of the mountains, while the camel is the ship of the desert. Today, the mountain tourism industry employs mules across the world, from the Alps to the Andes and the Himalayas to the High Atlas (Giampiccoli 2017; Cousquer and Allison 2012; Blakeway and Cousquer 2018; Watson et al. 2023; Cousquer et al. 2023).

10.5 Donkey Diversity in Africa

In Africa, few studies were carried out in the field of conservation using molecular tools on donkeys. In addition, most of the studies were carried out within the framework of foreign collaborations with research laboratories or service providers. As a result, most characterisation studies on African donkey populations focused on morphometry.

10.5.1 Morpho-Biometric Studies and Population Characterisation

During the last decade, a sparse number of phenotypical studies have been conducted to characterise the donkey populations of some regions of Africa. Morphological measurements are combined in indices to characterise the anatomic parts or whole body, coat colour variations, head and body shape, as discussed under Sect. 10.3 (donkey breeds and types). Often, this type of data is statistically analysed using multivariable descriptive analysis such as discriminatory, factorial, and clustering. Morphometric differences are then used to estimate genetic relationships between populations (Ayad et al. 2019; Kefena et al. 2011; Khaleel et al. 2020; Apagu 2016; Labbaci et al. 2018). There are no specific guidelines regarding donkey measurements, and many studies have adapted those used in horse or even other livestock species, and commonly, these studies use the methods previously reported as a reference. The major objective of this type of study is usually to establish a breed studbook (Folch and Jordana 1997) to guide breeding and conservation (Navas et al. 2017). In some studies, live weight estimations and the body condition are also used to characterise the donkey populations, e.g. (Labbaci et al. 2018) using as reference Pearson and Ouassat (2000), which also provide a very useful tool about animal welfare.

10.5.2 Genetic Diversity and Population Structure of African Donkeys

Short tandem repeats (STRs) or microsatellites have been the most popular genetic markers used for studies of genetic diversity and breeding management (e.g. paternity testing) in the donkey and other livestock species. Nonetheless, only a couple of studies were made on African donkey populations using these markers (Rosenbom et al. 2015). Examining the genetic diversity and population genetic structure of Ethiopian donkey populations with 12 equine microsatellite markers, Kefena and colleagues (2021) reported higher diversity within populations than between populations and a grouping of the populations into highland (Abyssinian, Afar and Hararghe) and lowland (Ogaden, Omo and Sinnar) genetic lineages, which agrees well with the traditional donkey classification system in Ethiopia. Using mitochondrial DNA (mtDNA) sequence polymorphism, past population demographic and spatial expansion was evident in the Ethiopian donkey populations, with Sinnar being considerably diverged from the other donkey breeds or populations (Kefena et al. 2014). Meanwhile, an Ethiopian donkey haplogroup formed the centre of a median-joining network of donkey breeds from Ethiopia, China and worldwide, which suggests Ethiopia as a possible centre of diversities for domestic donkeys in the Horn of Africa (Kefena et al. 2014).

Globally, low genetic differentiations in domestic donkeys (Rossel et al. 2008; Beja-Pereira et al. 2004; Kimura et al. 2011; Marshall 2007) are attributed to two factors: (1) no large mammals have been added to the stock of successful domesticates since the initial domestication event, implying not enough time to build up or acquire novel genetic variation within donkey populations, and (2) the fact that donkeys being initially tamed for transport probably facilitated the exchange of breeding individuals among populations across large geographic areas, which led to higher inter-population gene flow than intra-population gene flow (Marshall 2007).

With regards to the use of microsatellites markers for parentage assessment, the International Society for Animal Genetics (ISAG) recommended a set of 13-dinucleotide microsatellites loci isolated from the horse genome (AHT4, ASB23, HMS02, HMS03, HMS06, HMS07, HMS18, HTG7, HTG10, TKY297, TKY312, TKY337 and TKY343 (International Society for Animal Genetics 2017)), while the

FAO recommended 30-dinucleotide microsatellites also isolated from the horse genome, divided into six multiplex groups (AHT04, AHT05, ASB02, ASB17, ASB23, COR007, COR018, COR022, COR058, COR069, COR071, COR082, HMS02, HMS03, HMS06, HMS07, HMS20, HMS45, HTG06, HTG07, HTG10, LEX33, LEX34, LEX54, LEX63, LEX68, LEX73, NVHEQ054, SGCV28, VHL209). These represent all the autosomal chromosomes (with the exception of chromosome 6) for the genetic characterisation of donkey populations (FAO 2011). Both panels share markers and have been used in an extensive number of population genetic studies on donkey populations, e.g. (Rosenbom et al. 2015; Colli et al. 2013; Ivankovic et al. 2002; Jordana et al. 2017; Stanisic et al. 2017; Yatkin et al. 2020; Zeng et al. 2019). It is important to note that cross-species use of microsatellites has the potential to cause ascertainment bias due to the impact of the mutational process on the length of alleles, e.g. see (Li and Kimmel 2013; Dang et al. 2020). As the donkey genome is now available with good quality (Renaud et al. 2018), there is a need to map donkey-specific microsatellites.

A report on mtDNA data from ancient archaeological and historic museum samples of existing donkeys (*Equus asinus*) in Northeast Africa revealed two distinct mtDNA haplogroups, namely the Nubian wild ass (*Equus africanus africanus*) and Somali wild ass (*Equus africanus somaliensis*), with no geographical sub-structuring (Kimura et al. 2011). It is more obvious from a recent comprehensive study on the genome panel of 207 modern and 31 ancient donkeys and 15 wild equids by Todd et al. (2022). that current sub-populations from Kenya and the Horn of Africa best represent the descendants of the earliest donkeys. A phylogenetic reconstruction grouped modern donkeys according to sampling locations into Clade A for African donkeys and Clade B for mostly non-African donkeys (Todd et al. 2022). Further structure within Clade A separated donkeys from the Horn of Africa (Ethiopia and Somalia), plus Kenya from those from West Africa (Ghana, Mauritania, Nigeria and Senegal). Clade B also included donkeys from Nubia (Egypt and Sudan). The whole Y-chromosome sequencing of the donkey was recently reported, and the first analysis has shown low variability and discordant paternal and maternal histories (Wang et al. 2020). Nonetheless, the study lacked a fair representation of donkeys from Africa.

10.6 Conservation of the Donkey Genetic Resources in Africa

The subject of conservation of donkey genetic resources in Africa has received very limited attention in the face of growing concerns bordering on critically endangered Somali Wild Ass (AU-IBAR 2019), which is facing the risk of extinction. Other populations at risk include the Idabari of Chad and Sahel, as well as the Somali donkey. Most donkey breeds or types do not have information on population statistics that could be used to assess risk status.

In the past, donkeys in developing countries were most valued as working animals. However, their roles and economic values have changed greatly over the years. The socio-economic values of donkeys have not been fully harnessed, and the demand for donkey hides in China has triggered a global trade of donkeys. Sub-Saharan Africa exports about 1.8 million donkey hides to China annually (www.donkeysforafrica.org). Such an unprecedented rise in donkey trade is affecting the less protected and the world's poorest communities, which are still heavily dependent on donkeys as work animals (Köhle 2018).

Although the African nations urgently need to design and implement conservation plans for the donkey populations (see Chapter 25 for detailed discussion on conservation techniques), it is not feasible to conserve populations with limited information on characterisation. Therefore, more characterisation of local populations is urgently needed; otherwise, any conservation programme will be shortsighted and inefficient.

10.7 Uses of Donkey in Africa

10.7.1 Sociocultural Role of Donkey

In recent years some studies have revisited the importance of the donkey in African societies at several levels; one proposed the evaluation of its role: as a generator of income, the owner's social status, reducing vulnerability and encouraging resilience, husbandry, and gender dynamics (divisions of labour between men and women typically manifest in the allocation of different responsibilities based on societal norms and expectations), and differences between rural and urban societies (Geiger et al. 2020). This and other studies on the benefits and production challenges of working donkeys in smallholder farming systems (Gichure et al. 2020b) show a growing interest in evaluating the role of this species across Africa. For example, the variations in views and practices between urban and rural settings suggest that assessing the socio-economic value of donkeys in different locations within the same area or country is critical, rather than assuming that similar views are held between compatriots.

It must be recognised that the donkeys are increasingly at risk from external threats like environmental changes, diseases and health issues. The recent rapid emergence of the donkey skin trade to meet global demand for the raw materials to make the traditional Chinese medicine (*ejiao*) is a risk factor to African donkey populations. The trade has resulted in escalating donkey theft in many African countries and has wide consequences for donkey welfare and the livelihoods of some of those who depend upon them. Future studies on the implications of this global trade for the socio-economic value of working donkeys within their communities are needed.

10.7.2 The Role of the Donkey in Household Management and Income Generation

The donkey plays a significant role in supporting household chores and in generating income (Fig. 10.23). In rural areas, donkeys are often used in farming and for transportation; they pull ploughs and carts, deliver goods to market, and collect water from wells (Fig. 10.24). In urban areas, they are mainly used in construction, transport of people and goods, and garbage collection (Angara et al. 2011a; Maggs et al. 2021; Smith

Fig. 10.23 Donkeys at a livestock market in Niger. (Source: Wikipedia.com- photo by Vincent van Zeijst, CC BY-SA 4.0, https://commons.wikimedia.org/w/index.php?curid=82422085)

Fig. 10.24 Women fetching water to be transported by donkeys at a collection point near Opuwo in Namibia. (©The Donkey Sanctuary 2024)

2004; Valette 2014). In Ethiopia, 56% of households keep donkey for pack services (for income generation and household use), 26% for cart service (income generation), 14% for pack use by household and 4% exclusively to rent for breeding or petty trade to generate income (Admassu and Shiferaw 2011). In the Hulet Eju Enese district of Ethiopia in particular, the donkey is mainly used for pack (100%) and cart (34%) services and mule for riding (83%), pack (57%) and cart (34%) services, as well as renting out to other users (Ayalew et al. 2018). In Algeria, the Casbah neighbourhoods of Algiers (capital of Algeria) are mostly inaccessible by car, so the donkey is the sole alternative means of transportation of goods and people. By participating in work, donkeys boost the economic capacity of a region so much that in Ethiopia, there is a saying: 'If you don't have a donkey, then you are a donkey'. It has been reported that the income derived from the use of equines (including donkeys) in Ethiopia accounted for 14% of total income compared to 13% from other livestock (Admassu and Shiferaw 2011).

Millions of people across the world spend several hours every day collecting safe, clean water from near and distant locations. The donkey plays a major role in transporting water for household use, farming and livestock feeding. The donkey reduces the labour and time required to perform these tasks. The extra income generated through working animals also allows people to save money, reinvest in growth, and fund access to education. Donkeys' ability to transport

Fig. 10.25 Donkey pulling cart with woman in Somali. (Source: Wikipedia and available at https://en.wikipedia.org/wiki/List_of_donkey_breeds#/media/File:AMISOM_Djiboutian_Contingent_in_Belet_Weyne_11_(8212396061).jpg, CC0 1.0 license)

goods increases the potential for wider access to quality nutrition through local food markets. The donkey is therefore a valuable asset to many owners. Besides working, donkeys increase productivity by reducing the time and labour in the field or on-farm transportation, they enable farmers to reach inaccessible markets or areas. This allows farmers to sell their crops while allowing the community to access more diverse foods.

Owning a working donkey or animal is a major source of empowerment for women who rely on it to do tasks they would otherwise have to perform themselves, from collecting water and tilling the land to transporting goods to and from the market and from the fields (Figs. 10.24 and 10.25). By enabling women to be economically active, they also increase their economic and community status. This economic capability can prevent the worst form of destitution for lone women, whether working in rural or urban settings. Since donkeys assist in vital chores, parents can give children the adequate attention they need at home.

10.7.3 Role of the Donkey in Agriculture

In small-scale farming systems, the donkey has advantages over machinery due to its low cost of maintenance and environmental safety. Traction capability differs between animals, mainly depending on their body weight. For different species, however, the proportion of power provided based on body weight can vary. For bullocks, it is 12% of body weight, for buffaloes also 12%, for camels 18%, whereas the traction capability of the donkey is 24% of its body weight, which is superior to that of bullocks (12% of body weight), buffaloes (12% of body weigh), and camels (18% of body weight). Thus, the draft ability of donkeys is therefore superior to that of other draft animals. An earlier study showed that a pair of donkeys can generate a draught force equivalent to 15–20% of their body weight with no abnormal sign of fatigue. The average weight of a pair of donkeys is about 250 kg, and they can produce adequate force to do tillage work in light soil (Orhan et al. 2012; Bekele et al. 2001). For these and other reasons, donkeys are of crucial economic importance in many developing countries. Since ancient times, donkeys have been used in Africa, Asia, and Europe to transport people and goods, as well as to pull carts and light ploughs. Although they are not as fast as horses, donkeys are more cost-effective to own and maintain. They have been employed for farm tillage (Fig. 10.26), threshing of grains, drawing of water and milling, for example (Jones 2010).

Fig. 10.26 Donkeys ploughing a field at a Fulani settlement in Zaria, Nigeria. (Photo by Dr. E.M. Ibeagha-Awemu)

Donkeys are an undervalued power source in a large part of the world. Their potential to work is very high, and their contribution to any household or even national economy is considerable (Valette 2014). Generally, the buying and selling price of donkeys is far below their true value, which should be calculated based on the work they give over the 14 years they can work, if well cared for. If a donkey works 6 h a day, 4 days a week, over many years, it will have given about 15,000 hours of work (Orhan et al. 2012; Bekele et al. 2001). The low price of donkeys, therefore, reflects the distorted perception of their role. However, growing donkey slaughter across Africa for their skin has triggered the regulation of the trade in donkey skin in some African countries, leading to a better perception of its value.

10.7.4 Role of Donkey in Transportation

The use of donkey is closely related to road infrastructure and the price of rural transport. It is very common in Khartoum to see working donkeys either pulling carts (91.1%) or utilised as pack animals (8.9%) (Angara et al. 2011b). At the city centre, transportation of building and domestic materials by the donkey is most prominent, while transportation of water and people is common in remote areas (Angara et al. 2011b). In Nigeria, the oil boom era led to massive importation of small 4-wheel drive trucks (i.e., pickup trucks), and these became the preferred means of transportation of farm produce to the market. The prices of both vehicles and fuel were so low that many farmers sold their donkeys, and donkey keepers in the semi-arid regions turned to other enterprises. However, once the recession set in at the end of the 1980s, the economics of motorised rural transport became more doubtful, and farmers became anxious to acquire donkeys again. Having receded in Nigeria, the donkey is once again spreading (Blench 1992).

An individual donkey can carry more than one-third of its body weight (i.e., between 40 and 80 kg), twice as much as the quantity that human beings can carry (Fig. 10.27), relieving women of much time and energy-consuming work (Valette 2014). Donkeys can be used to transport people, young and vulnerable animals, water, firewood, manure, and farm inputs and produce, as depicted in Fig. 10.25. Mules, too, can make excellent pack as well as ride animals. Being larger and stronger, they can carry more but are more expensive to maintain than a donkey (Pearson et al.

Fig. 10.27 Donkeys transporting humans, goods and animals at Zaria, Nigeria. (Photo by Dr. E.M. Ibeagha-Awemu)

2003). In urban and peri-urban regions of Ethiopia, youths generate income by using donkey and other working animals to transport merchandise (Metaferia et al. 2011; Saville et al. 2020). In 2016, the United Nations recognised working equids as working livestock and considered them critical to the livelihoods and resilience of millions of families in low- and middle-income countries (Saville et al. 2020).

10.7.5 The Donkey in Household Nutrition

For centuries, donkey meat and milk have been consumed in many parts of the world. While donkey meat is forbidden in some communities in Africa, it is regarded as an affordable source of protein in others. In West Africa, it is commonplace to trade donkeys for their meat when they are sick, exhausted or old. In 2017, the demand for donkey meat and hide increased following trade restrictions on donkey hides by many African countries (ScienceX 2017). In Turkana County of Northern Kenya, donkey milk, meat and blood are widely consumed as a source of protein (Twerda et al. 1997). There is a general conception that portrays donkey meat as dangerous for human consumption, resulting in its general lack of acceptance as a source of protein in many communities. Thus, the consumption of meat and milk from donkey has not been fully adopted due to a lack of knowledge and legislation regarding production and benefits (Hassan et al. 2022). For example, in Kenya, most licences offered to legalise slaughterhouses for donkeys are export-only licences, which can be attributed to the taboo around the consumption of donkey meat locally (Hassan et al. 2022; Chege 2016). Europe in recent years has seen an increase in the consumption of donkey meat and milk, including an increase in the number of donkey dairies in Italy, Serbia and Turkey (McLean and Navas Gonzalez 2018). For example, a Serbian donkey dairy farmer produced cheese (named pule) from donkey milk, which has been tagged the world's most expensive cheese, as it sold for up to £880/kg (Kriel 2017). Meat from donkey is also a desired product in Denmark, Germany, France and China. The rise in the donkey meat and milk industry is helping to increase the populations of rare breeds in Europe, especially those facing extinction (Kriel 2017).

In recent years, interest in donkey products (meat, milk and hide) has been increasing due to research data portraying the importance of these products. Donkey meat contains low fat and cholesterol contents, a favourable fatty acid profile, and is rich in iron (Polidori et al. 2008). The concentration of minerals in donkey milk is higher than in human milk, and it is considered safer and more similar to human milk than milk from other livestock, and it is recommended as an alternative to infant milk (Polidori et al. 2008; Fantuz et al. 2012). Donkey milk is considered to have properties such as antidiabetic, antioxidant, antimicrobial and antiproliferative properties that could contribute to the management of diseases (Hassan et al. 2022). Donkey milk is also used as a cosmetic, and it is rich in vitamins and polyunsaturated fatty acids.

10.7.6 Use of the Donkey in Festivals, Sports (Racing) and Leisure

In addition to its prominent use as a means of transportation and draught power, the donkey also plays prominent roles in festivals and entertainment, with ceremonial roles in funerals and religious services such as weddings (Geiger et al. 2020; Admassu and Shiferaw 2011; De Klerk et al. 2020). Donkey race is a major attraction at Lamu's annual Maulidi festival. At over 700 years old, Lamu, a quaint island town off the coast of Kenya, is amongst the oldest inhabited Swahili townships on the planet and is Kenya's oldest continuously occupied settlement. In its heyday, it served as one of Africa's most important international trading hubs. Donkeys have been cemented in Lamu's history from its early days and were increasingly important during the time of the trade. To preserve the traditional culture of the community and due to its narrow streets and sandy pathways, the government banned the use of all cars (even bicycles), leaving the donkey as the only form of transport within the old town. One of the few motor vehicles allowed in the old town is an ambulance for donkeys. Today, donkey racing forms a part of Lamu culture. During the celebration of its annual Maulidi festival, Lamu's inhabitants and visitors gather to watch thrilling donkey races (Figs. 10.28). Lamu's donkeys, which also serve as companions and as an essential means of transportation of persons and goods (Fig. 10.29), take centre stage in exciting competitions to win prizes for their owners. This festival is a celebration of Lamu culture and history and includes dhow (sailboat) racing, which is also important in Lamu's history. Although there are donkey racing events in several destinations or locations, the race through Lamu's old town is a sight that cannot be rivalled (Figs. 10.28).

10.7.7 Use of the Donkey in Medicine and Onotherapy

Onos is the Greek word for donkey, and onotherapy is the term for therapy with the help of donkeys. With onotherapy, the physical and behavioural characteristics of the donkey can be valued (small in size, soft to the touch, affectionate and patient, with slow movements). Because of these characteristics, the donkey can offer valuable services to people affected by a disability or discomfort, with results that can be noticed quickly and can be documented (Karatosidi et al. 2013). The activities of 'Mediation with the Donkey' propose a methodology of reappropriation of man and animal. It allows rediscovering the cultural warmth of relationships through which man acquires a greater awareness of the identity of the donkey and the sense of their shared roles in history (Karatosidi et al. 2013). Today, in the context of culture, there is a continuing attempt to reconcile man with the environment in which he lives, especially with animals. The rediscovery of the relationship between man and animal is a journey through time to rediscover the harmonious relationship they had in the past. It is a practice that uses horseback riding as a therapeutic 'tool' and it is a combination of education and rehabilitation techniques, aiming to succeed on a sensory, motor, emotional or behavioural injury of man (Borioni et al. 2012). Some activities cover a therapeutic purpose because they affect the general health of a person. The achievement of a state of psychological wellbeing also influences the general health of an individual. Positive moods significantly affect behaviour, mental processes, and the expression of emotions and good feelings in humans. In this sense, the work done during onotherapy aims to achieve a higher state of wellbeing (Borioni et al. 2012).

Donkey hide is a highly sought commodity for the manufacture of a traditional Chinese medicine known as 'ejiao'. Ejiao is therefore believed to have a variety of health benefits, such as treatment of reproductive problems and improvement of blood circulation (Kriel 2017; DS 2017).

Fig. 10.28 Donkey racing during the celebration of Lamu's annual Maulidi festival. (Photos by Zollo Nyambu, https://africaviewfacts.com/facts/lamu-donkey-race-celebrating-tradition-and-community-in-kenya/)

Fig. 10.29 The donkey as essential means of transportation of persons and goods in old town of Lamu. (Photo by Alex Malombo, https://africaviewfacts.com/facts/lamu-donkey-race-celebrating-tradition-and-community-in-kenya/)

10.8 Challenges of the Donkey Enterprise in Africa

10.8.1 Limited Information on Donkey Keeping

Donkeys have altered human history as vital beasts of burden for long-distance movement, especially across semi-arid and upland environments, but remain inadequately studied despite globally expanding and providing key support to low- to middle-income communities (Todd et al. 2022). Although widespread in the Sahelian region of Africa, they are not frequently considered like other species by governments and private agencies for research and development. Thus, there is limited research attention and documentation of breed, breed characteristics and production characteristics.

10.8.2 Feed and Water Scarcity

The main constraints facing donkey farmers are lack of good quality feed and access to clean water, especially within the arid and semi-arid environments where most donkeys are raised. In Ethiopia, donkeys in most cases are left to scavenge for food with no form of supplemental feeding provided (Gebreab et al. 2004). This is especially challenging for working donkeys, which cannot scavenge low-quality grasses to meet their energy requirements for work. In Godey town in Ethiopia, donkeys provide cart services, transport various items weighing more than 300 kg at a time and work for about 8 hours a day and 6 days per week (Hassen et al. 2022). Adequate nutrition is essential for meeting the energy requirements for growth, work and reproduction.

10.8.3 Poor Breeding Management

Donkey breeding is generally characterised by uncontrolled breeding with no human-directed selection or assisted breeding. In this system, animals mate randomly, a practice that does not promote the improvement of desirable traits. It was observed that donkeys breed during the early rainy season, which might be due to the availability of feed resources during the rainy season.

10.8.4 Diseases and Lack of Veterinary Facilities

Although donkey appear to stay healthy on varied and often poor-quality diets and with low management inputs, they are still plagued with numerous health challenges. Nine commonly encountered diseases of donkeys in northern Nigeria have been described (Blench et al. 1992). Colics (Anomari), ulcerative lymphangitis, inflammation of the tongue, interdigital dermatitis, ear sore, fistulous withers, unknown causes of lameness and a nervous system disorder called locally as 'chinkai chinkai' are frequently reported diseases affecting donkeys in some villages and towns in northern Nigeria (Bale et al. 2003). In Ethiopia, the main health problems of donkeys include back sores, respiratory problems (coughing and nasal discharge as common symptoms), bite wounds, lameness, hoof overgrowth, foot diseases, eye problems, digestive problems and high parasite burden (Hassen et al. 2022; Feleke et al. 2015). Donkeys are also susceptible to common zoonotic diseases and transboundary animal diseases such as African horse sickness, the eastern, western and Venezuelan equine encephalomyelitis, equine infectious anaemia, Equine Herpesvirus, Equine Influenza, Equine Viral Arteritis, Japanese encephalitis, West Nile Fever and rabies.

10.8.5 Poor Animal Welfare

Animal welfare in relation to donkey production is compromised in Africa due to limiting factors such as poverty and limited knowledge on donkey production and management. Often, donkeys are not given due attention because of misplaced love for them and the general notion that they are worthless. Donkeys are appropriate and economical pack animals, especially under smallholder farming systems in regions with low levels of road construction, road networks, and rugged terrain. Moreover, donkeys are used under difficult conditions such as intense heat, difficult topography, dehydration, limited water supply, poor nutrition, wounds on different body parts, hoof problems, poor housing and management practices, working for long hours, walking long distances usually carrying heavy loads, and suffering poor handling during loading and poor harnessing devices, all of which have negative impacts on the welfare and quality of life of donkeys. Paying attention to the welfare of donkeys is critically important to ensure their health and survival, which translates to enhanced livelihoods of the people who depend on them.

10.8.6 Poor Housing

Donkeys under the pastoral and smallholder production systems are mostly allowed to roam freely with no housing provided or housed in an open area by tethering around the household (Hassen et al. 2022; Tuaruka and Agbolosu 2019).

10.8.7 Policy Issues and Poor Funding

Even though the donkey contributes significantly in providing work and transportation services and in enhancing the livelihood of low-income

Fig. 10.30 Pictures show the poor slaughter, handling and disposal of donkeys in Nigeria, Kenya and Ghana as deplorable activities associated with the donkey skin trade. (©The Donkey Sanctuary 2024)

families, there are no government policies on its use and management, and no attention is given to its promotion and protection by the majority of African countries.

10.8.8 The Donkey Skin Trade

In addition to the many roles of the donkey, donkey hide products have recently gained importance for their use in a traditional Chinese medicine called 'ejiao'. The donkey hides are processed by boiling to produce a firm gel, which is dissolved in alcohol or hot water and used in beverages and beauty products. However, increased demand for the supply of donkey hides for ejiao production in China is a threat to the donkey population worldwide (Maggs et al. 2021, 2023), as over 1.8 million donkey hides are exported to China annually (www.donkeysforafrica.org). Africa's donkey population is currently in decline due to this high demand for donkey skin by China, especially in North and Southern Africa (Table 10.1). According to the Donkey Sanctuary, about 5.9 million donkeys are slaughtered yearly for their skin, using slaughter methods that are unregulated, inhumane and unsanitary, and poor handling leading to the deaths of large numbers of donkeys on their way to slaughterhouses (DS 2024). Also, the Donkey Sanctuary indicates that the trade and shipping channels used by donkey skin traders are controlled by criminals involved in wildlife, drug and arms trafficking and other illegal activities (e.g. corruption, money laundering and passport fraud). Furthermore, the handling, slaughter and transportation of untreated skins and improper disposal of donkey carcasses (Fig. 10.30) is a potential risk for the spread of infectious diseases such as anthrax and exposure to shipping workers or abattoir workers (DS 2022).

The devastating consequences of the donkey skin trade to countries and the local populations in affected areas include: (1) stolen livelihoods of women, children and the communities who rely on donkeys for their social and economic wellbeing; (2) horrific treatment of donkeys; (3) decline in donkey populations and corresponding threat of the extinction of the species; (4) potential for the spread of zoonotic diseases such as anthrax; (5) the rate of donkey extraction is disproportionately higher than the rate of donkey multiplication through random breeding; (6) fuel commercial interest, theft and maltreatment of donkey owners; (7) misuse of government issued licences to donkey slaughter houses. The PADCO (2022) further highlighted the following: (8) negative impact on adolescent girls, and women who are already compromised by pregnancy and small children, and have to continue the burden of performing the tasks that the disappearing donkeys do, including carrying water, firewood and farm produce, with the resulting negative impacts on their health and development; (9) negative impact of the shortage of donkeys on persons with disability and the elderly which may be on their health, mobility and independence [i.e., increased dependency level]; (10) the reduction of donkey populations has a negative impact on the few surviving donkeys since they will be overworked, with a negative impact on their welfare; (11) promotion of rising insecurity within communities through donkey theft incidences; (12) increased rates of illegal cross border smuggling of donkeys; and (13) increased rates of poverty.

Consequently, the donkey skin trade has been termed a cruel and largely unregulated international trade in donkey skin by the Donkey Sanctuary, affecting African countries and the African Union (DS 2022). In 2022, the Donkey Sanctuary referred to the Donkey Skin Trade as 'A Ticking Time Bomb' (DS 2022). To curtail the impact of the trade in donkey skin on the communities that depend on the donkey, many African countries banned the trade in donkey skin by 2017, which did not have much effect. The most effective action started with the historic Dar es Salaam declaration during a Pan African Donkey Conference in Tanzania in December 2022, to an African-wide ban on the trade in donkey skin by the African Union at its 37th Summit on February 18, 2024, in Ethiopia (PADCO 2022; AU 2024).

10.9 Conclusion and Future Perspectives

10.9.1 Understand the Donkey

Intensify research efforts to characterise and document every aspect of donkey growth, development, health, reproduction, nutrition, welfare and genetics. This will generate essential baseline data that will guide efforts tailored to improving the productivity, welfare and disease management of donkeys. There is a need to enhance community awareness of the declining populations of donkeys and the role of the communities in the conservation, utilisation, and sustainable development of the donkey. Capacity building to deliver extension services to donkey owners and organisations is vital.

10.9.2 Structured Breeding

There is a need to implement structured donkey breeding and sustainable farming systems. This will require a breeding stock of high quality, which is attainable through phenotypic and genotypic characterisation of donkeys, quality nutrition and healthcare. It is imperative to develop breeding programmes in agricultural stations or within community breeding programmes to supply farming communities with stronger and healthier breeding animals, particularly to solve the rapid decline in donkey populations caused by the trade in donkey skin.

10.9.3 Economic Impact of Donkey

Current knowledge about the socio-economic value of donkeys as working livestock and as a means of transportation, as well as the impact of the donkey skin trade, is limited, highlighting the need for comprehensive research. Moreover, coordinated research on the socio-economic contributions of donkeys and other equids to livelihoods, poverty alleviation, national GDP and economic growth is vital, as it will enhance the value of the species as well as attract interest to it.

10.9.4 Donkey Conservation

To effectively conserve donkeys, it is crucial to first understand their production characteristics, nutritional requirements, disease management, growth and health attributes, amongst others. Conservation plans should be developed with local community involvement, reflecting their needs and the essential role of donkeys in their livelihoods. International cooperation is also necessary to regulate the global donkey trade, establish stricter trade regulations, and promote the breeding of the donkey to satisfy the need for ejiao production. Preserving donkey genetic resources requires urgent and multifaceted approaches, including scientific research, policy development, and community engagement. By working together, it is possible to balance the demand for donkey skins with the need to conserve this vital genetic resource.

10.9.5 Donkey Welfare and Other Issues

These can be achieved by implementing the following recommendations of PADCO (2022): (1) coordinate and guide implementation of the Animal Welfare Strategy for Africa (AWSA); (2) enhance collaboration, cooperation, coordination, partnership and promote information sharing with stakeholder and resourcing organisations; (3) support the formulation of Africa's common position on animal welfare; (4) advocate for political commitment for compliance with animal welfare decisions by the African Union Commission (AUC) and Member States; (5) ensure the inclusion of working animals (Donkeys, Horses and Camels), in animal resource policies, strategies, programmes and projects at national and regional levels; (6) advocate for the development and inclusion of working animals, including donkeys and their health care in the curricula of animal health courses and professional programmes; (7) generating data on donkey populations, productivity, health care and trade in Africa to inform policy makers; (8) institute a coordinated Pan African Conference action on the exploitation and utilisation of the donkey for the wellbeing of rural communities in the continent.

10.9.6 Legislation to Protect Africa's Donkey Populations

Develop national and continent-wide policies tailored to protect the donkey, as well as strategies and programmes for the development of donkey and other equines and the donkey trade within the animal resources sector of Africa.

In conclusion, donkey serve vital roles in the livelihoods of local communities by alleviating poverty, creating job opportunities, assisting in household chores, being a source of nutrition, serving as mobile companion for pastoralists and is a means of transportation of goods and persons through terrains where other modes of transport are not feasible. The donkey, therefore, deserves attention and protection and should be included in the livestock policy planning by governments and other relevant livestock agencies.

References

Admassu B, Shiferaw Y (2011) Donkeys, horses and mules—their contribution to people's livelihoods in Ethiopia. The Brooke, Addis Ababa. Available at https://www.thebrooke.org/sites/default/files/Ethiopia-livelihoods-2020-01.pdf. Accesses 10 Sept 2024

Aganga AA, Letso M, Aganga AO (2000) Feeding donkeys. Livest Res Rural Dev 12:11. http://www.lrrd.org/lrrd12/2/agan122.htm. Accessed 18 September 2024.

Angara T-E, Ibrahim A, Ismail A (2011a) The use of donkeys for transport: the case of Khartoum State, Sudan. WIT Trans Ecol Environ 150:651–660. https://doi.org/10.2495/SDP110541

Angara TE, Ibrahim AM, Ismail A (2011b) The use of donkeys for transport: the case of Khartoum state, Sudan. In: Sustainable development and planning V. WIT transactions on ecology and the environment, vol 150, pp 651–660. https://doi.org/10.2495/SDP110541

Apagu JP (2016) Some characterization of donkeys (Equus asinus) in northwestern nigeria using morphological and morphometric measures. Ahmadu Bello University, Zaria

Asteraye GB, Pinchbeck G, Knight-Jones T, Saville K, Temesgen W, Hailemariam A, Rushton J (2024) Population, distribution, biomass, and economic value of Equids in Ethiopia. PLoS One 19(3):e0295388. https://doi.org/10.1371/journal.pone.0295388. From NLM

Attia VI (2017) Donkeys in Ancient Egypt. Available at https://www.academia.edu/34228771/Donkeys_in_Ancient_Egypt_Donkeys_in_Ancient_Egypt. Accessed 6 Aug 2024

AU (2024) 37th Africa Union Sumit, Addis Ababa, Ethiopia, 15–18 February, 2024. Available at https://au.int/en/summit/37. Accessed 6 Sept 2024

AU-IBAR (2019) The state of farm animal genetic resources in Africa: towards accelerated agricultural growth and transformation by the year 2025. African Union—Interafrican Bureau for Animal Resources (AU-IBAR), Nairobi, 337 p. http://repository.auibar.org/handle/123456789/1243. Accessed 24 June 2024

Ayad A, Aissanou S, Amis K, Latreche A, Iguer-Ouada M (2019) Morphological characteristics of donkeys (Equus asinus) in Kabylie area, Algeria. Slovakia J Anim Sci 52:53–62

Ayalew H, Melekot MH, Taye M (2018) Monitoring of husbandry practices and harnessing of working equines in Hulet Eju Enese District, East Gojjam, Amhara Regional State, Ethiopia. Appl J Hyg 7(1):1–9. https://doi.org/10.5829/idosi.ajh.2018.01.09

Babb D (2017) History of the Mule. In American Mule Museum. Available at http://www.mulemuseum.org/history-of-the-mule.html. Accessed 8 Sept 2024

Bale OOJ, Lakpini CAM, Mohammed AK, Amodu JT, Chiezey UF, Ahmed HU, Achazie AA, Otchere EO (2003) An appraisal study of donkey in three Northern states of Nigeria. Niger J Anim Prod 30:203–208. https://doi.org/10.51791/njap.v30i2.1486

Beja-Pereira A, England PR, Ferrand N, Jordan S, Bakhiet AO, Abdalla MA, Mashkour M, Jordana J, Taberlet P, Luikart G (2004) African origins of the domestic donkey. Science 304(5678):1781–1781. https://doi.org/10.1126/science.1096008

Bekele Z, Geza M, Sisaye A, Ibro A, Bullo T (2001) Draught characteristics of a pair of working donkeys in the Rift Valley of Ethiopia. Draught Anim News 35:2–5

Blakeway S, Cousquer G (2018) Donkeys and mules and tourism. Tourism and animal welfare. CAB International, Wallingford/Oxfordshire, pp 126–131

Blench R (1992) Nigerian national livestock resource survey. In: Resource inventory and management. Federal Department of Livestock and Pest Control Services, Abuja, Nigeria

Blench RM (2010) In Wild asses and donkeys in Africa: interdisciplinary evidence for their biogeography, history and current use. In: Emory E (ed) Proceedings of the Donkey Conference at School of Oriental and African Studies (SOAS), London, UK, 9 May 2012. University of London: School of Oriental and African Studies, London

Blench R (2012) Wild asses and donkeys in Africa: interdisciplinary evidence for their biogeography, history and current use. In: Emory E (ed) Proceedings of the Donkey conference, SOAS, School of Oriental and African Studies. University of London, London, pp 8–9

Blench RM, de Jode A, Gherzi E (1992) Donkeys in Nigeria: history, distribution and productivity. In: Starkey P, Fielding D (eds) Donkeys, people and development. A resource book of the Animal Traction Network for Eastern and Southern Africa (ATNESA). ACP-EU Technical Centre for Agricultural and Rural Cooperation (CTA), Wageningen, 244p. ISBN 92-9081-219-2

Borioni N, Marinaro P, Celestini S, Del Sole F, Magro R, Zoppi D, Mattei F, Dall'Armi V, Mazzarella F, Cesario A et al (2012) Effect of equestrian therapy and onotherapy in physical and psycho-social performances of adults with intellectual disability: a preliminary study of evaluation tools based on the ICF classification. Disabil Rehabil 34(4):279–287. https://doi.org/10.3109/09638288.2011.605919

Chege N (2016) Government approves Sh300 Million Donkey Slaughterhouse. Kenyans.co.ke. Available at https://www.kenyans.co.ke/news/government-approves-sh300-million-donkey-slaughterhouse. Accessed 5 Sept 2024

Colli L, Perrotta G, Negrini R, Bomba L, Bigi D, Zambonelli P, Verini Supplizi A, Liotta L, Ajmone-Marsan P (2013) Detecting population structure and recent demographic history in endangered livestock breeds: the case of the Italian autochthonous donkeys. Anim Genet 44(1):69–78. https://doi.org/10.1111/j.1365-2052.2012.02356.x

Cousquer G, Allison P (2012) Ethical responsibilities towards expedition pack animals. Ann Tour Res 12:1839–1858. https://doi.org/10.1016/j.annals.2012.05.001

Cousquer G, Alyakine H, Lindsay-McGee V (2023) The history and welfare of working mules in the valleys of the Toubkal massif, in the High Atlas of Morocco. Front Vet Sci 10. https://doi.org/10.3389/fvets.2023.1256501

Curran MM, Smith DG (2005) The impact of donkey ownership on the livelihoods of female Peri-urban dwellers in Ethiopia. Trop Anim Health Prod 37(1):67–86. https://doi.org/10.1007/s11250-005-9009-y

DAD-IS (2024) Breed data sheet. Domestic Animal and Diversity Information Sys (DAD-IS), Food and Agricultural Organization of the United Nations. Available at https://www.fao.org/dad-is/browse-by-country-and-species/en/. Accessed 22 July 2024

Dang W, Shang S, Zhang X, Yu Y, Irwin DM, Wang Z, Zhang S (2020) A novel 13-plex STR typing system for individual identification and parentage testing of donkeys (Equus asinus). Equine Vet J 52(2):290–297. https://doi.org/10.1111/evj.13158. From NLM

De Klerk JN, Quan M, Grewar JD (2020) Socio-economic impacts of working horses in urban and peri-urban areas of the Cape Flats, South Africa. J S Afr Vet Assoc 91:e1–e11. https://doi.org/10.4102/jsava.v91i0.2009.

Debarbieux B, Oiry Varacc M, Rudaz G, Maselli D, Koehler T (2014) Sustainable Mountain development series hopes, fears and realities tourism in mountain regions. Available at: https://boris.unibe.ch/63699/1/Tourism_in_Mountain_Regions_EN.pdf. Accessed 8 Sept 2024

Devriese L (2012) From mules, horses and livestock to companion animals: a linguisticetymological approach to veterinary history, mirroring animal and (mainly) human welfare. Vlaam Diergeneeskd Tijdschr 81:237–246

DS (2017) Under the skin: the emerging trade in donkey skins and its implications for donkey welfare and livelihoods. The Donkey Sanctuary. Available at https://www.thedonkeysanctuary.org.uk/end-the-donkey-skin-trade. Accessed 5 Sept 2024

DS (2022) The global trade in donkey skin: a ticking time bomb. The Donkey Sanctuary Report 2. Available at https://www.thedonkeysanctuary.org.uk/sites/default/files/2022-11/report-2-the-global-trade-in-donkey-skins-a-ticking-time-bomb-2022.pdf. Accessed 2 Sept 2024

DS (2024) The Donkey skin trade: stop the slaughter, end the skin trade. The Donkey Sanctuary. Available at https://www.thedonkeysanctuary.org.uk/end-the-donkey-skin-trade. Accessed 5 Sept 2024

Ebangi AL, Vall E (1998) Phenotypic characterization of draft donkeys within the Sudano-Sahelian zone of Cameroon. Revue d'Elevage et de Médecine Vétérinaire des Pays Tropicaux 51(4):327–334. https://doi.org/10.19182/REMVT.9617

Eriso M, Mekuriya M, Abose Y (2023) Wounds in working donkeys: prevalence, causes, and risk factors at Duna Woreda, Hadiya Zone, Ethiopia. J Vet Anim Res 5(1):102

Fantuz F, Ferraro S, Todini L, Piloni R, Mariani P, Salimei E (2012) Donkey milk concentration of calcium, phosphorus, potassium, sodium and magnesium. Int Dairy J 24(2):143–145. https://doi.org/10.1016/j.idairyj.2011.10.013

FAO (2011) Molecular genetic characterization of animal genetic resources, vol 9. FAO Animal Production and Health Guidelines, Rome, Italy. https://www.fao.org/4/i2413e/i2413e00.htm. Accessed 5 September 2024.

FAOSTAT (2024) Food and Agricultural Organization of the United Nations Statistics. http://www.fao.org/faostat/en/#data/QP/visualize, 2022. Accessed 10 July 2024

Feleke A, Getachew K, Tamrat G (2015) Assessment of the production status, utilization and Management of Donkeys in Gena Bossa Woreda, Dawuro Zone, Southern Ethiopia. Food Sci Qual Manag 45:1–6

Fielding D (1988) Reproductive characteristics of the jenny donkey—Equus asinus: a review. Trop Anim Health Prod 20:160–166. https://doi.org/10.1007/BF02240085.

Folch P, Jordana J (1997) Characterization, reference ranges and the influence of gender on morphological parameters of the endangered Catalonian donkey breed. J Equine Vet Sci 17(2):102–111. https://doi.org/10.1016/S0737-0806(97)80347-4

Gebreab F, Wold AG, Kelemu F, Ibro A, Yilma K (2004) Donkey utilisation and management in Ethiopia. In: Starkey P, Fielding D (eds) Donkeys, people and development. A resource book of the Animal Traction Network for Eastern and Southern Africa (ATNESA). ACP-EU Technical Centre for Agricultural and Rural Cooperation (CTA), Wageningen

Geiger M, Hovorka AJ (2015) Animal performativity: exploring the lives of donkeys in Botswana. Environ Plann D: Soc Space 33:1098–1117

Geiger M, Hockenhull J, Buller H, Tefera Engida G, Getachew M, Burden FA, Whay HR (2020) Understanding the attitudes of communities to the social, economic, and cultural importance of working donkeys in rural, Peri-urban, and Urban Areas of Ethiopia. Front Vet Sci 7:60. https://doi.org/10.3389/fvets.2020.00060

Geiger M, Hockenhull J, Buller H, Engida GT, Jemal Kedir M, Goshu L, Getachew M, Banerjee A, Burden FA, Whay HR (2023) Being with donkeys: insights into the valuing and wellbeing of donkeys in Central Ethiopia. Soc Anim. https://doi.org/10.1163/15685306-bja10134

Getachew TB, Kassa AH, Megersa AG (2023) Phenotypic characterization of donkey population in South Omo Zone, Southern Ethiopia. Heliyon 9(8). https://doi.org/10.1016/j.heliyon.2023.e18662. Acccessed 2024/07/03

Giampiccoli A (2017) The effect of the use of mules in tourism: a historical perspective. Afr J Hosp Tour Leis 6(3). Available at: https://www.ajhtl.com/uploads/7/1/6/3/7163688/article_23__vol_6__3__2017.pdf

Gichure M, Onono J, Wahome R, Gathura P (2020a) Assessment of phenotypic characteristics and work suitability for working donkeys in the central highlands in Kenya. Vet Med Int 2020(1):8816983. https://doi.org/10.1155/2020/8816983. Acccessed 26 July 2024

Gichure M, Onono J, Wahome R, Gathura P (2020b) Analysis of the benefits and production challenges of working donkeys in smallholder farming systems in Kenya. Vet World 13(11):2346–2352. https://doi.org/10.14202/vetworld.2020.2346-2352

GO (2024) Livestock of the World: Donkeys, Global Orange Inc © 2007. Available at https://www.livestockoftheworld.com/Donkeys/Default.asp. Accessed 5 Aug 2024

Guénon A (1999) La grande histoire du mulet. Le Poiré-sur-Vie, France. Editions du Vieux Crayon (Original work published 1899). Available at https://maultier.info/Artikel/la_grande_histoire.pdf. Accessed 8 Sept 2024

Hannani H, Bouzebda Z, Bouzebda-Afri F, Hannani A, Khemis MDEH (2020) Morphometric characteristics of the extreme eastern Algerian domestic donkey (Equus asinus). Folia Veterinaria 64(1):66–76. https://doi.org/10.2478/fv-2020-0009

Hassan ZM, Manyelo TG, Nemukondeni N, Sebola AN, Selaledi L, Mabelebele M (2022) The possibility of including donkey meat and Milk in the food chain: a southern African scenario. Animals 12(9). https://doi.org/10.3390/ani12091073. From NLM

Hassen G, Abdimahad K, Welday K, Ma'alin A, Mahamed A, Omer A (2022) Management practices, utilization and challenges of donkey in Godey Town, Somali Regional State, Ethiopia. Open J Anim Sci 12:616–628. https://doi.org/10.4236/ojas.2022.124044

International Society for Animal Genetics (2017) Equine genetics and thoroughbred parentage testing standardisation workshop. In: 36th International Society for Animal Genetics Conference, Dublin, Ireland. https://www.isag.us/Docs/EquineGenParentage2023.pdf. Accessed 8 September 2024.

IUCN (2022) The International Union for Conservation of Nature's (IUCN) Red List of Threatened Species. Version 2021-2. 2021. https://www.iucnredlist.org. Accessed 11 Nov 2023

Ivankovic A, Kavar T, Caput P, Mioc B, Pavic V, Dovc P (2002) Genetic diversity of three donkey populations in the Croatian coastal region. Anim Genet 33:169–177. https://doi.org/10.1046/j.1365-2052.2002.00879.x

John PA, Iyiola-Tunji AO, Akpa GN, Iriso BV, Millam JJ, Mallam I (2017) Distributions of qualitative traits among weaner, young and adult donkeys in Northwest Nigeria. J Anim Prod Res 29(2):231–237

Jones PA (2010) Donkeys for development. ATNESA and Agricultural Research Council of South Africa

Jordana J, Goyache F, Ferrando A, Fernández I, Miró J, Loarca A, López ORM, Canelón JL, Stemmer A, Aguirre L et al (2017) Contributions to diversity rather than basic measures of genetic diversity characterise the spreading of donkey throughout the American continent. Livest Sci 197:1–7. https://doi.org/10.1016/j.livsci.2016.12.014

Karatosidi D, Marsico G, Tarricone S (2013) Modern use of donkeys. Iran J Appl Anim Sci 3(1):13–17

Kefena E, Beja-Pereira A, Han JL, Haile A, Mohammed YK, Dessie T (2011) Eco-geographical structuring and morphological diversities in Ethiopian donkey populations. Livest Sci 141(2):232–241. https://doi.org/10.1016/j.livsci.2011.06.011

Kefena E, Dessie T, Tegegne A, Beja-Pereira A, Yusuf Kurtu M, Rosenbom S, Han JL (2014) Genetic diversity and matrilineal genetic signature of native Ethiopian donkeys (Equus asinus) inferred from mitochondrial DNA sequence polymorphism. Livest Sci 167:73–79. https://doi.org/10.1016/j.livsci.2014.06.006

Kefena E, Rosenbom S, Beja-Pereira A, Kurtu MY, Han JL, Dessie T (2021) Genetic diversity and population genetic structure in native Ethiopian donkeys (Equus asinus) inferred from equine microsatellite markers. Trop Anim Health Prod 53(3):334. https://doi.org/10.1007/s11250-021-02776-2

Khaleel AG, Lawal LA, Nasir M, Hassan AM, Abdu MI, Salisu N, Kamarudin AS (2020) Morphometric characterization of donkeys (Equus asinus) in D/Kudu Kano State for selective breeding and genetic conservation. J Agrobiotechnol 11(2):12–21. https://doi.org/10.37231/jab.2020.11.2.216

Kimura B, Marshall FB, Chen S, Rosenbom S, Moehlman PD, Tuross N, Sabin RC, Peters J, Barich B, Yohannes H et al (2011) Ancient DNA from Nubian and Somali wild ass provides insights into donkey ancestry and domestication. Proce Biol Sci 278(1702):50–57. https://doi.org/10.1098/rspb.2010.0708. From NLM

Kimura B, Marshall F, Beja-Pereira A, Mulligan C (2013) Donkey domestication. Afr Archaeol Rev 30(1):83–95. https://doi.org/10.1007/s10437-012-9126-8

Köhle N (2018) Feasting on donkey skin. In: China story yearbook 2017: prosperity. ANU Press

Kriel G (2017) Donkey production: great economic opportunity or potential minefield? Farmer's Weekly 2017(17044):36–40. https://doi.org/10.10520/EJC-aafee36ca

Labbaci M, Djaout A, Benyarou M, Ameur Ameur A, Gaouar SBS (2018) Morphometric characterization and typology of donkey farming (Equus asinus) in the wilaya of Tlemcen. Genet Biodivers J 2(1):60–72

Levtzion N, Hopkins JFP (1981) Corpus of early Arabic sources for west African history. Cambridge University Press, Cambridge

Lewicki T (1974) West African food in the Middle Ages. Cambridge University Press, Cambridge

Li B, Kimmel M (2013) Factors influencing ascertainment bias of microsatellite allele sizes: impact on estimates of mutation rates. Genetics 195(2):563–572. https://doi.org/10.1534/genetics.113.154161

Librado P, Orlando L (2021) Genomics and the evolutionary history of Equids. Annu Rev Anim Biosci 9:81–101. https://doi.org/10.1146/annurev-animal-061220-023118

Librado P, Khan N, Fages A, Kusliy MA, Suchan T, Tonasso-Calvière L, Schiavinato S, Alioglu D, Fromentier A, Perdereau A et al (2021) The origins and spread of domestic horses from the Western Eurasian steppes. Nature 598(7882):634–640. https://doi.org/10.1038/s41586-021-04018-9

Maggs HC, Ainslie A, Bennett RM (2021) Donkey ownership provides a range of income benefits to the livelihoods of rural households in northern Ghana. Animals 11:3154

Maggs HC, Ainslie A, Bennett RM (2023) The value of donkeys to livelihood provision in northern Ghana. PLoS One 18: e0274337. https://doi.org/10.1371/journal.pone.0274337

Marshall F (2007) African pastoral perspectives on domestication of the donkey: a first synthesis. In: Denham TP, Jriarte J, Vrydaghs L (eds) Rethinking agriculture: archaeological and ethno archaeological perspectives, One world archaeology series. Left Coast Press, Walnut Creek, pp 371–407

Mason IL, Maule JP (1960) The indigenous livestock of eastern and southern Africa, Technical communication no. 14 of the Commonwealth Bureau of Animal Breeding and Genetics. Farnham Royal: Commonwealth Agricultural Bureaux, Edinburgh

Maswana M, Mugwabana TJ, Tyasi TL (2022) Evaluation of breeding practices and morphological characterization of donkeys in Blouberg Local Municipality, Limpopo province: implication for the design of community-based breeding programme. PLoS One 17(12):e0278400. https://doi.org/10.1371/journal.pone.0278400

McLean AK, Navas Gonzalez FJ (2018) Can scientists influence donkey welfare? Historical perspective and a contemporary view. J Equine Vet Sci 65:25–32. https://doi.org/10.1016/j.jevs.2018.03.008

Metaferia F, Cherenet T, Gelan A, Abnet F, Tesfay A, Abdi J, Gulilat W (2011) A review to improve estimation of livestock contribution to the national GDP, IGAD LPI working papers. Ministry of Finance and Economic Development, Ministry of Agriculture, Addis Ababa

Mitchell P (2018) The donkey in human history: an archaeological perspective. Oxford University Press. https://doi.org/10.1093/oso/9780198749233.001.0001

Mustefa A, Assefa A, Misganaw M, Getachew F, Abegaz S, Hailu A, Emshaw Y (2020) Phenotypic characterization of donkeys in Benishangul Gumuz National Regional State. Online J Anim Feed Res 10(1):25–35. https://doi.org/10.36380/scil.2020.ojafr4

Mwenya E, Keib G (1997) History and utilisation of donkeys in Namibia. In: Starkey P, Fielding D (eds) Donkeys, people and development. A resource book of the Animal Traction Network for Eastern and Southern Africa (ATNESA). ACP-EU Technical Centre for Agricultural and Rural Cooperation (CTA), Wageningen, 244p. ISBN 92-9081-219-2

Navas F, Jordana J, León J, Barba C, Delgado J (2017) A model to infer the demographic structure evolution of endangered donkey populations. Animal 11((12):2129–2138. https://doi.org/10.1017/S1751731117000969

Nengomasha EM, Pearson RA, Wold AG (1999) Empowering people through donkey power into the next millennium farmers. In: Kaumbutho PG, Pearson RA, Simalenga TE (eds) Proceedings of the workshop of the animal Traction Network for Eastern and Southern Africa (ATNESA) held on 20–24 September 1999, Mpumalanga, South Africa, pp 22–31

Nguekeng CL, Tiambo CK, Fonteh AF, Manjeli Y (2017) Phenotypic diversity of the domestic donkey (Equus africanus Asinus) in the Sudano-Sahelian Zone of Cameroon. Bull Anim Health Prod Afr 65(4)

Nsoso SJ, Thutwa K, Monkhei M (2008) A survey of donkey farmers in Bobonong village of Central district in Botswana. Botswana J Agri Appl Sci 5(1):81–88. ISSN 1815-5574

Orhan Y, Boztepe S, Ertuğrul M (2012) The domesticated donkey: III – Economic importance, uncommon usages, reproduction traits, genetics, nutrition and health care. Can J Appl Sci 2:139–157

Orlando L, Metcalf JL, Alberdi MT, Telles-Antunes M, Bonjean D, Otte M, Martin F, Eisenmann V, Mashkour M, Morello F et al (2009) Revising the recent evolutionary history of equids using ancient DNA. Proc Natl Acad Sci 106(51):21754–21759. https://doi.org/10.1073/pnas.0903672106

PADCO (2022) Donkeys in Africa Now and in the Future; Towards a common position on conserving donkeys from exploitation for the skin trade amongst the AU Members. Report of the Pan African Donkey Conference 2022, 1st–2nd December 2022, Dares Salaam, Tanzania. Available at https://panafricandonkeyconference.org/wp-content/uploads/2023/04/REPORT_PADCO-2022_PAN-AFRICAN-DONKEY-CONFERENCE_opt.pdf. Accessed 6 Sept 2024

Pearson RA, Ouassat M (2000) A guide to body condition scoring and live weight estimation of Donkeys. Centre for Tropical Veterinary Medicine, University of Edinburgh

Pearson RA, Simalenga TE, Krecek R (2003) Harnessing and hitching donkeys, horses and mules for work; Center for tropical veterinary medicine. University of Edinburgh; and Department of Agriculture and Rural Engineering. University of Venda for Science and Technology. South Africa, South Africa

Polidori P, Vincenzetti S, Cavallucci C, Beghelli D (2008) Quality of donkey meat and carcass characteristics. Meat Sci 80(4):1222–1224. https://doi.org/10.1016/j.meatsci.2008.05.027

Renaud G, Petersen B, Seguin-Orlando A, Bertelsen MF, Waller A, Newton R, Paillot R, Bryant N, Vaudin M, Librado P et al (2018) Improved de novo genomic assembly for the domestic donkey. Sci Adv 4(4):eaaq0392. https://doi.org/10.1126/sciadv.aaq0392

Rosenbom S, Costa V, Al-Araimi N, Kefena E, Abdel-Moneim AS, Abdalla MA, Bakhiet A, Beja-Pereira A (2015) Genetic diversity of donkey populations from the putative centers of domestication. Anim Genet 46(1):30–36. https://doi.org/10.1111/age.12256

Rossel S, Marshall F, Peters J, Pilgram T, Adams MD, O'Connor D (2008) Domestication of the donkey: timing, processes, and indicators. Proc Natl Acad Sci USA 105(10):3715–3720. https://doi.org/10.1073/pnas.0709692105

Saville K, Bambara C, Marry A, Perry B (2020) Invisible livestock'–On the central roles of working horses, donkeys and mules on the smallholder farms that feed the world. International Livestock Research Institute. Available at https://www.ilri.org/news/invisible-livestock-central-roles-working-horses-donkeys-and-mules-smallholder-farms-feed. Accessed 30 July 2024

ScienceX (2017) A donkey's tale: Nigeria becomes key hide export hub (2017, December 19). Available at https://phys.org/news/2017-12-donkey-tale-nigeria-key-export.html. Accessed 10 Sept 2024

Sheriff Yusuf I, James Enam S, Hassan R, Bello TK-D, Jocktong CM (2020) The possible effect of gender and age on haematological and some biochemical parameters in donkey foals (Equus asinus) in Zaria, Nigeria. Veterinarski Arhiv 90:377–383. https://doi.org/10.24099/vet.arhiv.0696

Smith D (2004) Use and Managment of Donkeys by Poor Societies in Peri-Urban Areas of Ethiopia; R7350. Center for Tropical Veterinary Medicine, Edinburg

Smith DG, Pearson RA (2005) Review of the factors affecting the survival of donkeys in semi-arid regions of sub-saharan africa. Trop Anim Health Prod 37:1–19. https://doi.org/10.1007/s11250-005-9002-5

Sow A, Kalandi KM, Ndiaye NP, Bathily A, Sawadogo GJ (2012) Clinical biochemical parameters of Burkinabese local donkeys' breeds. Int Res J Biochem Bioinformatics 2(4):84–89

Stanisic LJ, Aleksic JM, Dimitrijevic V, Simeunovic P, Glavinic U, Stevanovic J, Stanimirovic Z (2017) New insights into the origin and the genetic status of the Balkan donkey from Serbia. Anim Genet 48(5):580–590. https://doi.org/10.1111/age.12589

Tadesse T, Lemma A, Retta N (2014) Investigation into the nutritional content and microbiological property of Abyssinian Donkey's Milk. Ethiop Vet J 18(1):73–82

Todd ET, Tonasso-Calvière L, Chauvey L, Schiavinato S, Fages A, Seguin-Orlando A, Clavel P, Khan N, Pérez Pardal L, Patterson Rosa L et al (2022) The genomic history and global expansion of domestic donkeys. Science 377(6611):1172–1180. https://doi.org/10.1126/science.abo3503. From NLM

Tuaruka L, Agbolosu A (2019) Assessing Donkey Production and Management in Bunkpurugu/Yunyoo District in the Northern Region of Ghana. J Anim Husb Dairy Sci 3:1–5

Twerda M, Fielding D, Field C (1997) Role and management of donkeys in Samburu and Turkana pastoralist societies in northern Kenya. Trop Anim Health Prod 29(1):48–54. https://doi.org/10.1007/bf02632348. From NLM

Valette D (2014) Invisible helpers: women's views on the contributions of working donkeys, horses and mules to their lives. The Brook, London. https://www.thebrooke.org/sites/default/files/Brooke%20News/Invisible%20Helpers%20Report.pdf

Wang C, Li H, Guo Y, Huang J, Sun Y, Min J, Wang J, Fang X, Zhao Z, Wang S et al (2020) Donkey genomes provide new insights into domestication and selection for coat color. Nat Commun 11(1):6014. https://doi.org/10.1038/s41467-020-19813-7

Wassie AM, Getachew TB, Kassa AH, Megersa AG, Ayele T (2023) Donkey production systems and breeding practices in selected districts of South Omo Zone, Southern Ethiopia. Rangel J 45(3):97–108

Watson T, Kubasiewicz LM, Nye C, Thapa S, Chamberlain N, Burden FA (2023) The welfare and access to veterinary health services of mules working the mountain trails in the Gorkha region, Nepal. Austral J Vet Sci 55:9–22. https://doi.org/10.4067/S0719-81322023000100009

Xia X, Yu J, Zhao X, Yao Y, Zeng L, Ahmed Z, Shen S, Dang R, Lei C (2019) Genetic diversity and maternal origin of Northeast African and South American donkey populations. Anim Genet 50(3):266–270. https://doi.org/10.1111/age.12774

Yatkin S, Ozdil F, Unal EO, Genc S, Kaplan S, Gurcan EK, Arat S, Soysal MI (2020) Genetic Characterization of Native Donkey (Equus asinus) Populations of Turkey Using Microsatellite Markers. Animals (Basel) 10(6). https://doi.org/10.3390/ani10061093

Yilmaz O, Wilson RT (2013) The domestic livestock resources of Turkey: status, use and some physical characteristics of mules. J Equine Sci 23(4):47–52. https://doi.org/10.1294/jes.23.47

Yilmaz O, Wilson RT (2012) The domestic livestock resources of Turkey: status, use and some physical characteristics of mules. J Equine Sci 23(4):47–52

Yilmaz O, Boztepe S, Ertuğrul M (2012a) The domesticated Donkey: I—species characteristics. Can J Appl Sci 4(2):339–353

Yilmaz O, Boztepe S, Ertuğrul M(2012b) Domesticated donkeys—part II: types and breeds. Can J App Sci 2012; 2(2): 267–286

Zeng L, Dang R, Dong H, Li F, Chen H, Lei C (2019) Genetic diversity and relationships of Chinese donkeys using microsatellite markers. Arch Anim Breed 62(1):181–187. https://doi.org/10.5194/aab-62-181-2019

Open Access This chapter is licensed under the terms of the Creative Commons Attribution 4.0 International License (http://creativecommons.org/licenses/by/4.0/), which permits use, sharing, adaptation, distribution and reproduction in any medium or format, as long as you give appropriate credit to the original author(s) and the source, provide a link to the Creative Commons license and indicate if changes were made.

The images or other third party material in this chapter are included in the chapter's Creative Commons license, unless indicated otherwise in a credit line to the material. If material is not included in the chapter's Creative Commons license and your intended use is not permitted by statutory regulation or exceeds the permitted use, you will need to obtain permission directly from the copyright holder.

11 African Horse Genetic Resources and Breeding Strategies for African Input Systems

Djalel E. Gherissi, Ginette Aumassip-Kadri, Mohammed E. A. Benhamadi, Félix Meutchieye, Yassine H. Jamali, and Semir B. S. Gaouar

Abstact

This chapter presents a synthesis of current knowledge on the horse in Africa in its different facets. The first part of the chapter deals with the origin and domestication of the horse and the ways in which the ancestors of this animal were tamed in Africa (Sect. 11.1 on origins and evolution of horses in Africa). The second part presents a description of the equine sector in Africa (Sect. 11.2) while exploring local and exogenous equine resources of Africa (Sect. 11.2.1 describes major African horses' genetic groups, while Sect. 11.2.2 presents physiognomy and description of the horse populations in Africa, Sect. 11.2.3 on horse production systems in Africa, and Sect. 11.2.3 on the feral "wild" horses of Africa). The third part (Sect. 11.3) describes the main structures and organizations that manage the equine sector in Africa. The fourth section provides the traditional use (Sect. 11.4.1) and equestrian sports (Sect. 11.4.2), model competition and paces (Sect. 11.4.3), equestrian games (Sect. 11.4.4), traditional sports activities (Sect. 11.4.4), and other

In Memoriam: Ginnette Aumassip Kadri (1928–2025)
This chapter is dedicated to the memory of Ginnette Aumassip Kadri, who passed away at the age of 97, on January 7th, 2025, after contributing to the preparation of this chapter and this book. A pioneer in the study of Saharan prehistory and rock art—particularly in Algeria—she was also a passionate advocate for the preservation and recognition of the Barb horse. She was the director of the Laboratory for Research on African Prehistory at Centre National de Recherches Préhistoriques, Anthropologiques, Historiques, Alger, Algeria under the auspices of Centre National de la Recherche Scientifique (CNRS), Paris, France. She taught African prehistory at various universities. Her lifelong commitment to research, cultural heritage, and the defense of this noble breed continued until her final days.
May she rest in peace.

D. E. Gherissi (✉)
Laboratory of Animal Production, Biotechnology, and Health. Institute of Agricultural and Veterinary Sciences, University of Souk Ahras, Souk Ahras, Algeria
e-mail: d.gherissi@univ-soukahras.dz

G. Aumassip-Kadri
Centre National de Recherches Préhistoriques, Anthropologiques Historiques, Alger, Algiers, Algeria

Centre national de la recherche scientifique (CNRS), Paris, France

M. E. A. Benhamadi · S. B. S. Gaouar
Department of Applied Geneticsin Agriculture, Ecology and Public Health (GenApAgiE), Faculty of Natural and Life Sciences, Earth and Universe Sciences, University of Tlemcen, Chetouane, Algeria

F. Meutchieye
Biotechnology and Bioinformatics Research Unit, Department of Animal Science, Faculty of Agronomy and Agricultural Sciences, University of Dschang, Dschang, Cameroon

Y. H. Jamali
Actes Sud, Rabat, Morocco

© The Author(s) 2026
E. M. Ibeagha-Awemu et al. (eds.), *African Livestock Genetic Resources and Sustainable Breeding Strategies*, Sustainable Development Goals Series, https://doi.org/10.1007/978-3-031-92076-9_11

service perspectives such as leisure and equestrian tourism, used in army and security forces and in hippophagea (Sects. 11.4.6, 11.4.7 and 11.4.8). The fifth part presents an inventory of research on the biodiversity of the African equine population (Sect. 11.5), emphasizing the main tools used in this context, namely: morpho-biometric (Sect. 11.5.1), immunological (Sect. 11.5.2), biochemical, and molecular biology (Sect. 11.5.3) studies. Finally, the last section presents the current situation of African horse production and reproduction performance and the conservation strategies of the local equine resources (Sect. 11.6). Section 11.7 presents a summary and conclusion.

Keywords

African Confederation of Equestrian Sports · caballus · Barb horse · Husbandry practices

Abbreviations

ACES	African Confederation of Equestrian Sports
Al	Albumin
ALAT	Alanine Aminotransferase
ANECB	National Association of Barb Horse Breeders
ANEPCRP	National Association of Breeders and Owners of Purebred Horses
ANTE	National Association for EquestrianTourism
AQHA	American Quarter Horse Association
ARMECPSA	Association Royale Marocaine des Eleveurs de Chevaux Pur Sang Arabes
ASAT	Aspartate Aminotransferase
BC	Before Christ
CABI	Centre for Agriculture and Bioscience International
CAC GALOP	The Horse Friends Club
CAMGALLOP	Cameroon Horse Racing Cup
CCC	Consultative Committee for Races
CCE	Complete Equestrian Competition
CCI	International Complete Competition
CEC	Complete Equestrian Competition
CIC	International Combined Competition
CIRAD	Centre de coopération Internationale en Recherche Agronomique pour le Développement
CNIAEB	Centre National d'Insémination Artificielle Equine de Bouznika
CNIAG	Centre National de l'Insémination Artificielle et Génétique
CNRTL	National Center for Textual and Lexical Resources
DAD-IS	Domestic Animal Diversity Information System
EEF	Egyptian Equestrian Federation
EFAS	European Association of Animal Production
Es	Basic Esterase
FADE	Angolan Federation of Equestrian Sports
FAO	Food and Agriculture Organization
FE	Fédération équestre
FEI	International Equestrian Federation
FIHB	International Horseball Federation
FIP	International Polo Federation
FNARC	Fédération nationale d'amélioration de la race chevaline
FPS	Friesch Paarden Stamboek
FPSSA	Friesian Horse Studbook of Southern Africa
Gc	GC component
GWAS	Genome-Wide Association Studies
ILCA	International Livestock Centre for Africa
ILRAD	International Laboratory for Research on Animal Diseases
INRA	National Institute of Agricultural Research
ISAG-FAO	International Panel on Animal Genetic Diversity
ISRA	Institut Sénégalais de Recherches Agricoles
JO	Jeux Olympiques
LAGEV	Laboratoire d'Analyses Génétiques et Vétérinaires
LNERV	Laboratoire National de l'Elevage et de Recherches Vétérinaires

METN	Ministry of Environment and Tourism Namibia	TBA	Thoroughbred Breeders Association
NHA	National horse racing authority	Tf	Transferrin
OEFCCA	Oromia Environment, Forest and Climate Change Authority	THI	Temperature Humidity Index
		UNDP	United Nations Development Program
OIE	Organisation mondiale de la santé animale	WAHO	World Arabian Horse Organization
OMCB	Organisation Mondiale du Cheval Barbe	WBFSH	World Breeding Federation for Sport Horses
ONDEEC	Office National de Développement des Élevages Équidés et Camelins	Xk	postbumin
ORC Course	Orientation and Regularity		
OSULBD	Oklahoma State University Livestock Breeds' Database		
Pi	prealbumin		
PMU	Pari Mutual Urbain		
PMUM	Pari Mutuel Urbain Marocain		
RIs	Reference Intervals		
SADA 2006	South Africa Department of Agriculture		
SADA	South Africa Department of Agriculture		
SAEF	South African Equestrian Federation		
SAQHA	South African Quarter Horse Association		
SAS	South African Saddle Horse		
SASB	South African Stud Book and Livestock Improvement Association (SASB)		
SASBA	South African Stud Book Association		
SASBA	The South African Stud Book and Livestock Improvement Association		
SAVBS	South African Vlaamperd Breeders Society		
SAWB	Warm Blood South African		
SCB	Sudan Country Breed		
SCHPM	Société des Courses Hippiques et du Pari Mutuel		
SCHT	Company of Horse Races		
SECF	Société d'Encouragement à l'Elevage du Cheval Français		
SOREC	Royal Society of Encouragement of the Horse		
SPANA	Society of Animals and Nature Protection		

11.1 Origins and Evolution of Horses in Africa

11.1.1 Origins of the Horse in Africa

The species *caballus*, the horse, did not take its current form until the Pleistocene, a million years ago. It is the culmination of many descendants dating back to the Eocene (55 to 34 million years ago) and which, four to five million years ago, gave rise to the genus *Equus* in North America. Two to three million years later, the horse was reported to have crossed Beringia and reached Eurasia, then Africa, where its presence was, for some two million years, in the form of a zebra, *Equus stenonis*.

The basic form was unearthed in 1825 in the gypsum of Paris by Cuvier and named *Paleotherium*; discovered in 1841 in London, it was called *Hyracotherium;* and then in 1876, it was called America *Eohippus*. About the size of a dog, it had 44 teeth, three fingers on the forelimbs, and four on the hind legs, all ending in pads. All the stages of its development could be followed in North America, where a long, continuous transformation has made it a prototype of evolution. It disappeared around 10,000 BC while it remained elsewhere. Over time, the multiplicity of discoveries has only led to a bushel of species, without altering the pattern.

11.1.2 The African Niche

For a long time, it was a widely held belief that no wild horse inhabited the African continent, although the genus *Equus* was seen throughout the Quaternary period in various forms; some in the ancestry of zebras and others as donkeys. The *caballus* form was not found in Africa until 1983, when it was simultaneously discovered in the suburbs of Algiers (Bagtache and Hadjouis 1983) and the region of Tiaret (Chaid-Saoudi 1984–1986). A widely accepted hypothesis was that of Le Couteulx de Canteleu (1880), that a domesticated horse was introduced into Egypt by the Hyksos around 1500 BC, and then the Dongola diffused to the south, and the Barb to the west. The discovery of Tassili paintings in the 1950s did not support this proposition, as they depicted a square model with a convex profile, a low-set tail, a model quite distinct from the Egyptian representations of the same era, which featured an elongated body, concave profile, and slender limbs. Linguistics further reinforces this observation by revealing different roots for the word "horse" in Egyptian and North African languages (Galand-Pernet 2006).

The identification of an African wild horse through fundamental modification of the data has led many hippatrists to consider it as the ancestor of the Barb horse and possibly even the Dongola breed; the absence of wild horses was indeed the sole and insurmountable obstacle to the concept of a localized Barb horse domestication.

11.1.3 Domesticating Horses

Numerous studies concerning the species *Equus caballus* have given rise to multiple hypotheses about the domestication site of the horse. However, pinpointing the exact location of domestication remains a challenge. Based on archaeological and more recent genetic data, there is speculation that horse domestication occurred in Eurasia. The archaeological evidence is sourced from three key aspects: skeletal modifications, geographical distribution, and the human-horse relationship. Genetic data, on the other hand, stem from observations of decreasing genetic diversity and increasing variability in coat forms, particularly more pronounced since the third millennium (Epstein 1971; Arne et al. 2009). Linguistic studies, such as Lehmann (1992), associate domestication with the eighth millennium, bringing the notion of a taming period preceding domestication closer to the eighth millennium (Lehmann 1992). Well-established indications of unique connections between humans and horses from the Pontic steppe date back to around 4500 BC (Anthony 2007), these dates are quite late in the history of animal domestication.

In contemporary times, the horse population is declining in all countries due to modernization and mechanization, which have greatly diminished its roles as a warhorse, means of transportation, and racing animal. Its current distribution across the African continent is uneven and hindered by diseases like trypanosomiasis (or nagana). Most equines today are concentrated in the northern and northeastern parts of Africa, where breeds like Barb and Dongola have been documented since the second millennium BC. The introduction of horses to the southern region is recent and closely tied to European presence there (Anthony 1996).

11.1.4 Domestication Hotbeds

Two predominant questions about the horse have been posed: Is *Equus caballus* a distinct species or a subspecies known as *Equus ferus caballus*? Additionally, where and when did horse domestication occur? While no consensus exists regarding the answer to the first question, the second question has generated diverse hypotheses that have replaced earlier propositions (San 1869; Franck 1875). If horse domestication happened in the cradle of civilization in Central Asia, then only one location is established, leading to a single ancestral origin for the horse that subsequently dispersed (Edwards 1973).

Consequently, horse domestication is attributed to the SrednyStog culture, dating back to 4000 BC, primarily due to a notably high concentration of horse bones, predominantly young males, and the discovery of objects that could be interpreted as horse bits in Dereivka, Ukraine (Anthony 1996). More ancient leg relics of domesticated horses are found in Khvalynsk, Russia, where horse remains accompany those of domestic cattle, sheep, and goats in graves dating from 4800 to 4400 BC (Anthony 2007). Other potential centers, such as the Pontic Steppes, Anatolia (Benecke 2006; Arbuckle 2012), Hungary-Romania, Portugal, and Spain (Digard 2002) were initially suggested, but subsequently discarded.

A particular point of interest emerges from the Botai culture that flourished around 3500 BC in Kazakhstan, particularly Krasni Yar and its environs (Outram et al. 2009). These sites contained corrals with horse dung, and pottery preserved traces of mare's milk. Recent research reveals that the Botai horse is not the ancestor of today's domesticated horse; rather, it gives rise to the Przewalski horse, a breed that was tamed rather than a true wild horse (Gaunitz et al. 2018).

An African cradle?

Traditionally, it has been widely accepted that the introduction of the horse into Africa in a domesticated state occurred during the eighteenth or fifteenth century BC, coinciding with the Hyksos invasion of Egypt. Subsequently, the horse spread southward, giving rise to breeds like the Dongola and the Barb, which in turn contributed to the ancestry of most other African horse breeds. However, a challenge to this narrative arose in the 1950s with the discovery of cave paintings depicting horses that differed from those associated with the Hyksos. Gautier (1952) and other researchers examining this discrepancy in North Africa concluded that the absence of wild horses not only hindered, but definitively ruled out, the notion of a local origin (Gautier 1952). This obstacle raised questions about the unique bond between the Maghreb people and their horses.

In 1983, the identification of a wild horse species, *Equus algericus*, at two sites—one near Algiers (Bagtache and Hadjouis 1983) and the other close to Tiaret (Chaid-Saoudi 1984–1986)—along with its presence at various sites spanning from 40,000 to 8000 years ago (Bouzid 1991; Hadjouis 1993), revived the issue and brought forgotten discoveries back to the forefront (Thomas 1879; Reygasse 1922). The consideration of a local origin addresses the quandary posed by artistic representations and human behavior. This perspective aligns with the findings of Lira et al. (2010), who suggest the possibility of an independent domestication occurring in the Iberian Peninsula or North Africa because of specific haplotypes found on the peninsula (Lira et al. 2010). This view is also supported by paleontologist Chaid-Saoudi (2006), who noted a shared "geographical and paleontological area between Southern Europe and the Maghreb" (Chaid-Saoudi 1984–1986).

Jansen et al. (2002) observed a striking similarity in phenotype between Barb and Iberian horses (Jansen et al. 2002). Variants of Barb horses, identified by Ouragh (2006), are exclusively shared with populations having historical connections, distinguishing them from other breeds (Ouragh 2006). Oelke (undated) isolated a D1 genotype (Iberian / Barb), which diverged from the A6 mother genotype around 500,000 BC. These converging lines of evidence point toward a local domestication of the Barb horse, completely independent of other horse breeds.

11.1.5 The Reasons for Domestication

It is widely accepted that domestication arises from material needs, but its development necessitates a favorable cultural and social context. In the case of horse domestication, it signifies a novel requirement that aligns with a process closely associated with the impacts of climate change. Numerous rationales have been put forward for this phenomenon. Chiefly, it revolves around the notion of a "food reserve," encompassing both meat and milk. Archaeological evidence from the Botai culture highlights that horses constituted a significant portion of their meat consumption. However, during the presumed period of domestication, the steppe's grasses reached towering heights of up to two meters, making the hunting of numerous horses impractical. It is postulated that humans would have captured some of these horses to utilize them as mounts for protecting their herds. Thus, aside from providing sustenance, horses would have become invaluable hunting companions.

Religious considerations also emerge as a motivating factor, manifested through horse burials and the practice of hippophagia. These practices may be intricately tied to shamanic rituals aimed at assimilating the virtues of the horse. Digard (2004) highlights the allure of "fiery, fast, and powerful" game, alongside the innate human desire for ownership and possession (Digard 2004).

In Africa, the earliest evidence of domesticated horses comes from pastoral regions that underwent desertification. As herds required significantly larger territories to thrive, the question arises as to whether horse domestication was instrumental in safeguarding these herds. The prospect of domestication within this same region is not implausible. In the TassiliAjjers region, an even older, painting dating possibility to the eighth millennium depicts an arrangement of diminutive equines with small ears. However, representations from this era have primarily focused on humans, with only four animal species, including the giraffe, antelope, and mouflon, bearing indications of domestication or attempts at domestication. The quest remains ongoing to identify a site analogous to Dereivka or Krasni Yar that could definitively address this question.

11.1.6 Number of Horses in Africa

As of 2021, according to FAO data, there were 60,193,503 horses worldwide, distributed across the five continents (FAO DAD-IS 2021). In Africa, there were 7,412,943 (FAO DAD-IS 2021) with an annual increase rate of about 2% (see Chap. 1: African livestock production systems: The past, present and the projected future). The African continent accounts for 21.31% of the total global horse population, resulting in a lower ratio of 5.3 horses per 1000 people compared to the global ratio of 7.7 horses per 1000 people. South America boasts the highest horse population, while Oceania has the lowest.

Horse populations in African countries vary widely, influenced by factors such as economy, culture, and their diverse uses. Ethiopia has one of the continent's largest horse populations, primarily used for transportation and agricultural labor. Chad and Mali also heavily rely on horses for farming and transportation needs. Senegal houses a significant number of horses utilized for various purposes, including agricultural work. Nigeria has a growing equine sector with a substantial horse population.

In North African countries, there is a higher concentration of horses due to longstanding equestrian traditions. Egypt and Sudan have a rich history of using horses, particularly in agriculture and transportation. In South Africa, a well-established equestrian industry supports a significant horse population, used for various purposes, including sports and leisure. On the other hand, some regions may have smaller horse populations, and these horses remain integral to their respective countries, playing significant roles in various aspects of daily life.

11.2 Equine Sector in Africa

11.2.1 The Major Horse Genetic Groups of Africa

At present, the African equine population is primarily characterized by two predominant groups: the Barb and the Dongola breeds. The north coastal area is characterized by the Arabian horse type. The western variation of it is called the Barb, and it is distinguished also by blood polymorphisms (Ouragh 2006). Most African breeds are derived from this Barb type and from Dongola varieties adapted to the north-east and central African environments. The Dongola practically disappeared in its homeland, and residual populations can be found only in Sudan and in the western lowland of Eritrea. Emerging during the second millennium, these two populations have served as the foundation for numerous breeds or lineages.

The Chap. 1 of this book states that the total number hose breeds is 134, distributed respectively as follows: 10 breeds in the Eastern Africa, 5 breeds in Central Africa, 20 breeds in Northern Africa, 61 breeds in Southern Africa and 38 breeds in Western Africa, The United Nations Development Program (UNDP) has identified a total of 98 distinct breeds resulting from the geographic isolation of specific groups of horses. This adaptability is a key trait of horses, as they possess remarkable plasticity in their morphology, enabling them to readily conform to their environment. However, this adaptability encounters a subtle hindrance in the form of the tsetse fly, which poses a significant challenge. Covering over a third of the continent's land area, the presence of the tsetse fly and its effects are influenced by fluctuating climate conditions (Devisse 1995).

11.2.1.1 Barb Horse

Origin and Breed Distribution

The Barb horse is notably distributed across North Africa (Algeria, Libya, Morocco, Tunisia), where it originated through the domestication of the wild form, *Equus caballus algericus* (Fig. 11.1).

The Barb horse likely serves as the ancestor of several small horse breeds across different regions. In the southeast of Mauritania, particularly in Tagant, it is referred to as the *Hodh* or the horse of Hodh. In Mali, it is known by various names such as *Banamba, Bélédougou*, and possibly *Barouéli* or *Gueniekalari*, and even referred to as the horse of *Nioro* (Ndiaye 1978). Additionally, similar small horse breeds include the *Bobo* in Burkina Faso and the *Bhirum* or *Pagan* in Nigeria. In Senegal, the *M'Bayar*, which was introduced through trans-Saharan trade during the Middle Ages, potentially has roots dating back to the eighth-seventh centuries BC. The *M'Par* or *Cayorrose*, situated between the Senegal and Saloum rivers, represents three small horse varieties closely resembling each other (Dupuy 1995). The *Koto-koli*, a pony related to the *Barb*, is the smallest breed in West Africa and can be found in Burkina Faso, Nigeria, and Togo. In northern Benin, the *Berba* is a specific type of *Koto-koli* named after the ethnic group that breeds it. Some specialists suggest that the *Mogods* or *Nefza* pony in Tunisia exhibits Barb influence, a notion that is partially supported by research on blood protein polymorphism (Chabchoub et al. 2006).

The Barb horse serves as the progenitor of various breeds or sub-breeds, with one of the most prominent being the Arab-Barb (Jamali 2020). This lineage emerged in Algeria at the end of the nineteenth century through the crossbreeding of local Barb broodmares with Arab stallions imported from the Near East. Arab-Barbs were raised in Algeria as well as in Morocco, Tunisia, and Chad (Husser 2017; Jamali 2020). In their region of origin, an open studbook recognizes any horse with Arab lineage ranging from 25% to 75% as Arab-Barb, resulting in a diverse array of models.

Physical Characteristics

The Barb horse from Algeria exhibits specific physical characteristics that highlight its unique attributes. This breed stands at an average height of 152.2 cm at the withers and 150.1 cm at the rump, with a total length of 160 cm and a scapular-iliac length of 118.2 cm. Notably, the Barb horse showcases dimensions such as an ilium length of 37.4 cm, thigh width of 45.8 cm, and shoulder width of 54.4 cm. Circumference measurements include 175.5 cm at the chest, 40.4 cm at the forearm, 30.1 cm at the knee, 28.6 cm at the ball, 19.8 cm at the anterior cannon, and 20.6 cm at the posterior cannon (Benhamadi et al. 2017). These physical attributes are further indicated with various indices such as the body profile index of 0.95, signifying its well-proportioned build, and the compactness index of 2.68, reflecting its overall compactness. The body mass index is 1.51, while the Dactylorthracic index is 0.11. These indices indicate the breed's distinctive features. Finally, the relative body index is 0.91, showcasing its relative harmony (Benhamadi et al. 2017). The Barb horse breed possesses an average body weight of 409.9 kg (Benhamadi et al. 2017). This comprehensive physical description underscores the uniqueness and individuality of the Barb horse breed, making it appreciated within the equestrian world. Reported results confirm that the Barb horses from Tiaret and Laghouat regions in Algeria adhere to the international

Fig. 11.1 Barb horses. (**a**) Barb horse from the national stud farm "Chaou-Chaoua" in Tiaret, Algeria (Chikhaoui et al. 2018) (Image credit: (Chikhaoui et al. 2018)); (**b**) Moroccan Barb horse used for Fantasia (Talley 2015) (Image credit: Gwyneth Ursula Jean Talley); (**c**) Barbe horse breed in Mauritania (Photo Courtesy: Ministry of Rural Development, Mauritania) (AU-IBAR 2019), (**d**) Mounted Barb horse in Nigeria (Image credit: Eveline M. Ibeagha-Awemu)

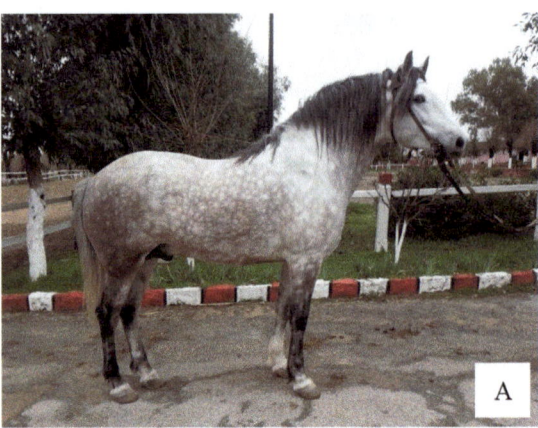
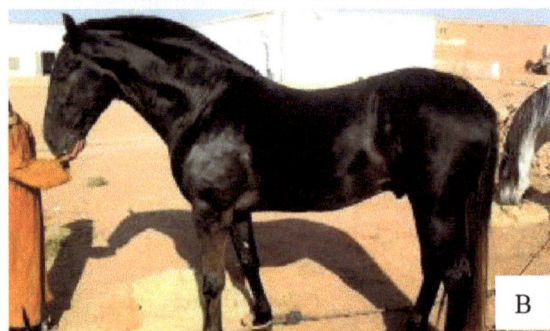

standard for the Barb horse breed (Benhamadi et al. 2017).

Although the Barb horse possesses attributes such as a prominent forehead, broad chest, and sturdy limbs associated with fast-moving saddle horses, it does not excel in speed races when compared to the Thoroughbred. This is primarily due to its elongated features, including a long shoulder, back, kidney, and relatively shorter forelock. However, the Barb horse holds its own in endurance races, especially over long distances, given its morphological similarities to the Thoroughbred Arab. In fact, it can be a significant competitor to the latter in endurance raids, with some enthusiasts even considering the Arab-Barb hybrid as the best choice for this challenging discipline (Benhamadi et al. 2017).

Moreover, the Barb horse's wide limbs and robust joints contribute to its "safe foot," rating, making it an excellent candidate for novices of horseback riding. It is also noted for treks in rugged and remote mountainous terrains. The Barb horse's adaptability to various equestrian disciplines underscores its versatility and value within the world of equine sports and recreational riding (Benhamadi et al. 2017; Jamali 2020).

Husbandry Practices

Barb horse husbandry is a harmonious blend of cultural reverence, adaptability, climate management, endurance focus, and conservation commitment. These practices ensure that the Barb horse remains a symbol of North African culture and a breed celebrated for its exceptional traits and adaptability across various roles. The husbandry practices for the Barb horse in North Africa are distinctively characterized by the keepers' unwavering commitment to heritage preservation and the cultural significance of the horse. Breeders and owners take immense pride in maintaining the breed's genetic purity, thus ensuring that the Barb horse's unique traits, such as agility, endurance, and a docile temperament, are passed down through generations. This dedication is practical and deeply rooted in the cultural fabric of the region, making Barb horse care a symbol of tradition and cultural heritage.

What sets Barb horse husbandry apart is the breed's exceptional versatility and adaptability. These horses excel in diverse roles, from equestrian sports (conventional and nonconventional sport as indicated in the section "Use of the Horse in Africa" below) to agricultural work and livestock herding. Consideration of North Africa's unique arid environment demand specialized care practices suscah as the provision of shelter from extreme weather conditions. However, the hallmark of Barb horse husbandry is the unwavering focus on its endurance. These horses are renowned for their stamina, and their care and training are geared toward enhancing their endurance capabilities, especially in long-distance riding and racing. Furthermore, active participation in conservation efforts marks the Barb horse husbandry practices as a proactive and preservation-oriented.

Production Systems

The husbandry practices for Barb horses in North Africa combine traditional wisdom with modern care standards. These practices not only ensure the health and performance of these remarkable horses but also safeguard their cultural relevance and significance in the region. The production systems for Barb horses in North Africa encompass a wide range of activities and goals, reflecting the breed's adaptability and historical significance in the region. One prevalent production system is rooted in tradition, where Barb horses are employed in herding livestock and assisting with agricultural tasks. These horses play essential roles in rural communities, embodying cultural heritage and providing indispensable support in day-to-day activities.Another prominent production system revolves around equestrian sports and racing, particularly endurance riding. Breeders and owners focus on conditioning and training Barb horses to excel in competitive events. This production system emphasizes physical fitness and stamina development, allowing these horses to showcase their remarkable endurance capabilities. In regions with picturesque landscapes and challenging terrains, Barb horses are employed in tourism and recreational riding. These systems aim to produce reliable and sure-footed horses that can navigate rugged environments, offering tourists and adventurers memorable experiences.

For those dedicated to preserving the purity and unique traits of the Barb horse, breeding and conservation programs take center stage. Selective breeding ensures that the breed's distinct characteristics, such as agility and docile temperament, are perpetuated. These programs often include genetic testing and pedigree recording to maintain the breed's integrity. Some production systems involve crossbreeding Barb horses with other breeds to enhance specific attributes or create new breeds tailored to particular purposes. Improvement programs may focus on refining performance capabilities or adapting the Barb horse to different climates and roles. Traditional training methods, passed down through generations, are still relevant in some production systems. These methods emphasize the development of a horse's innate abilities and temperament, aligning with the breed's historical roles and showcasing its versatility. The diverse production systems for Barb horses in North Africa reflect their adaptability and rich history. From traditional practices and cultural heritage to competitive sports, conservation efforts, and tourism, these systems cater to various purposes, highlighting the breed's enduring significance and its ability to thrive in a variety of roles.

11.2.1.2 Dongola

Origin and Breed Distribution

The origin of the Dongola breed is still uncertain. It owes its name to the province of Dongola where the hyksos horse is said to have established itself, a hypothesis which needs to be reconsidered in the light of current knowledge. When John Burckhardt traveled to the Nubian region in the 1880s, he found that the Dongola horse was famous throughout Sudan, Ethiopia, and the rest of the Near East. He stated that the breed was originally from Arabia, and it was one of the finest he had seen: "the horses possess all the superior beauty of Arabian horses, but they are larger." Burckhardt also noted that a prime stallion was worth five to ten slaves and that the Mamluks were particularly interested in Dongola horses (Epstein 1971; O'Fahey 1980).

The Dongola is however very close to the Barb horse. Its presence in Eastern Africa or its wildness suggests a very old introduction (Fisher 1972) (Figs. 11.2 and 11.3). In addition to its involvement is creating of the *Arewa* varieties of horses, the Dongola developed under various names such as the *Kreikibat* in the south-east of Mauritania, the *Todori,Noungou* or *Fada*, *Hausa* in Niger, under the names of *Bobo Oulé, Boulsa, Boussougha, Boussouma, Dori* in Burkina Faso as well as *Bahr el Ghazal, Kréda,* and *Ganaston* in Chad. The *Kirdimi* is perhaps a type of Dongola or the product of a cross with the Barb.

Physical Characteristics

The Dongola has a convex profile, a rather large head, a flat chest, and long and thin legs and croup, which detracts from its attractiveness. Standing at about 152–155 cm at the withers, the Dongola does not have an imposing presence. Its back is long, and its loins are rather poorly attached. Many Dongola horses have white markings on their faces and lower legs. In addition, the typical Dongola horse has a black or chestnut-colored coat. However, the most common color is a deep, reddish bay.

H. Epstein maintains that the "original breeding center of the Dongola begins that noble race of horses justly celebrated all over the world.... What figure the Nubian breed would make in point of fleetness is very doubtful, their make being so entirely different from that of the Arabian; but of beautiful and symmetrical parts, great size and strength, the most agile, nervous, and elastic movements, great endurance of fatigue, docility of temper, and seeming attachment to man, beyond any other domestic animal, can promise anything for a stallion, the Nubian is, above all comparison, the most eligible in the world" (Heidorn 1997).

Husbandry Practices

The original Dongola horse breed is on the brink of extinction due to interbreeding with the Arab, Barb, and Arab-Barb crossbreeds. Some attribute

Fig. 11.2 Dongola horses in Eritrea (Ministry of information of Eretria 2021)

Fig. 11.3 Dongola horse (Koçkar 2012). (Reproduced with permission CC-BY 4.0 license)

the decline in their population to inadequate management practices. In Sudan, stallions are given preference over mares, leading to the mating of strong stallions with weak mares, which results in offspring of inferior quality. Today, the Dongola horse is primarily used as a riding horse, typically standing at a height of 15–15.2 hands. They are often showcased in festivals and ceremonies, adorned in traditional attire (Mason 1996; Porter et al. 2016; Hendricks 1995).

11.2.1.3 Other Breeds

In addition to the indigenous horse breeds, various foreign breeds have been introduced over time, leading to the formation of their distinct lines and sub-breeds. The Arabian breed is believed to have been introduced to Egypt in the fourth century from the Arabian Peninsula (Olsen 2017). Having gained prevalence in the Maghreb, the Arab horse was brought to Algeria in 1870 and subsequently spread to Tunisia, Morocco, and several other Western countries.

In Algeria, a unique population of Arab horses was developed exclusively in Tiaret for over a century (Fig. 11.4). This population is notable for not having a male lineage for nearly a hundred years; all stallions originated from the breed's homeland. Until 1978, the breed consisted solely of "ASSIL" horses, maintaining its original lineage (Fig. 11.5). Algeria was awarded the WAHO trophy for the first time in 2010. The trophy was awarded to the excellent stallion Diable Du Desert after he won the most prestigious race in the country, the "Grand Prix du President de la Republique." (Fig. 11.4a). After that, Sakhi stallion was awarded Algeria's WAHO trophy for the year 2014 at the age of 6 years. This was in recognition of his winning the title of Senior Champion Stallion at the Algerian "Salon Du Cheval 2014" National Arabian Horse Show (Fig. 11.4b). In 2016 WAHO trophy was awarded to the 17-year-old mare, Fousha (Fig. 11.4c), bred at the Tiaret State Stud (Algeria), a lovely mare, with excellent conformation and type, descended on her

Fig. 11.4 (**a**) Algerian Arabian Stallion "SAKHI" with his owner Ahmed Feghouli, WAHO 2010 trophy winner (**b**) Diable Du Desert. Breeder: Amraoui Belkasem. Owner: Bouteldja Mohamed, WAHO 2014 trophy winner; (**c**) FOUSHA. Mare, 1999., Breeder: Haras Nationaux, Jumenterie de Tiaret with Mr. Ahmed Bouakkaz, General Director and Head of Stud Book at ONDEEC, and the owner Mr. Zeguaoui Abed, WAHO 2016 trophy winner

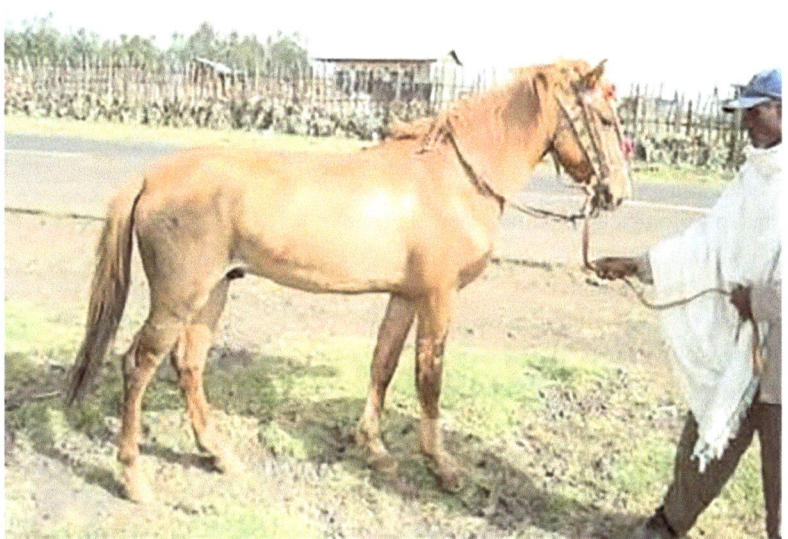

Fig. 11.5 Typical Selale riding horse (Kefena et al. 2012)

dam line from Orkana [Fari II (GB) / Karillon (GB)], who was imported to Algeria from Sweden in 1978. Fousha was awarded the WAHO trophy in respect of her winning success during the major "Salon du Cheval de Tiaret" show, held in September 2016.

The Thoroughbred breed is found in various regions, including South Africa, Algeria, Kenya, Lesotho, Morocco, Tunisia, and, to a lesser extent, Cameroon, Ghana, Senegal, and Sudan. In Sudan, crossbreeding Thoroughbreds with local and Arabian horses resulted in the Sudanese breed. Furthermore, crossing Thoroughbreds with Sudanese horses gave rise to the Tawleed breed.

In areas where trypanosomiasis prevails, the horse populations consist primarily of foreign breeds or those with recent foreign ancestry. This is observed in countries like South Africa, Zambia, and Kenya. In South Africa, the Basuto breed was created in the nineteenth century through a combination of Arabian, Thoroughbred, and Cape horses. It subsequently evolved in the challenging environment of Lesotho. Influenced by Barb and Indonesian horse breeds, the Basuto contributed to the development of the Nooitgedacht breed in Botswana, alongside the Tswana and Boerperd breeds. The Boerperd breed from South Africa is a revival of the historical Boer or Cape Horse, an almost extinct breed established in the seventeenth century through the blending of various imported breeds. The modern Boerperd breed emerged from two varieties: Cape Boer and South African Boer. These breeds are the result of incorporating Flemish, Hackney, and Cleveland Bay bloodlines. The Vlaamperd breed, bred in the Cape region, originated from crossings between Frisian, Thoroughbred, Hackney, and Cape horse breeds. While rare, the Vlaamperd is highly protected due to its uniqueness and limited population.

11.2.2 Physiognomy and Description of the Horse Populations in Africa

11.2.2.1 Indigenous Horse Breeds

The aboriginal horse breeds in Africa are the Dongola in Eastern Africa and the Barb in North Africa. Both breeds are believed to have spread to the Sahelian region, undergoing changes and interactions with each other and with local populations. These preexisting equine herds probably arrived during the Caballine period around nineteen centuries BC. Additionally, some Arabian horses accompanied the latest invasions. This theoretical pattern still requires confirmation or affirmation through genetic research over a larger timescale (including fossils) and larger spatial scale, encompassing the entirety of Eastern, Northern, and Western Africa.

While transitioning from drylands to tropical zones, horses underwent certain morphological and physiological modifications that were initially considered as degeneration (Marcenac 1969; Doutressoulle 1947). It is questionable, however, whether an increase in hardiness or tolerance to trypanosomiasis is truly a degeneration. History and geographic constraints have shaped uniform equine populations. At times, humans have selected the finest individuals for hunting, racing, warfare, and raids. The culmination of this multifactorial process resulted in the creation of superbly gifted horse breeds, such as the Dongola, Barb, Arabe-barb, or a combination of these breeds.

It is worth noting that the borders of African countries do not always align with the ranges of these breeds. For example, the Dongola and the Barb are cross-border breeds. The numbers of local indigenous horses in each country are presented in Table 11.1:

Selale or Oromo

Origin and breed distribution: The Selale, also known as Oromo, is an indigenous horse breed of Ethiopia commonly found in the highlands of the central plateau, in the northern region of Shewa. The most typical form of this breed originates from the Wichale and Jida districts. There is no official census of their population.

Characteristics: Selale horses are well-conformed saddle horses bred by the Oromo ethnic group. They are said to be influenced by some English stallions imported by Emperor Haile Silassie in the 1930s and 1940s. Morphologically similar to the Abyssinian horse (Kefena et al. 2012). The reference measurements recorded in the DAD-IS databases are an average of 1.296 m for females and 1.337 m for males (Kefena et al. 2012). The Selale horses are taller and have a more elongated body and

Table 11.1 Horse population by African countries (FAO DAD-IS 2021)

Order	Area	Value (heads)
1.	Ethiopia	2,192,443
2.	Chad	1,379,167
3.	Sudan	794,045
4.	Mali	607,786
5.	Senegal	582,927
6.	Sierra Leone	445,118
7.	South Africa	328,220
8.	Niger	260,861
9.	Morocco	190,000
10.	Nigeria	107,357
11.	Egypt	84,387
12.	Mauritania	66,383
13.	Tunisia	57,118
14.	Algeria	50,231
15.	Libya	45,712
16.	Lesotho	43,372
17.	Namibia	42,868
18.	Burkina Faso	42,331
19.	Zimbabwe	28,294
20.	Botswana	24,138
21.	Cameroon	17,145
22.	Guinea	3583
23.	Ghana	3312
24.	Guinea-Bissau	2482
25.	Gambia	2431
26.	Kenya	2183
27.	Togo	2072
28.	Eswatini	1724
29.	Benin	1161
30.	Angola	1021
31.	Somalia	893
32.	Cabo Verde	581
33.	Malawi	554
34.	Madagascar	514
35.	Sao Tome and Principe	309
36.	Mauritius	150
37.	Congo	68
38.	Democratic Republic of the Congo	–
39.	South Sudan	–

limbs compared to other Ethiopian horses (Fig. 11.5). These horses have a symmetrical conformation and are perceived as elegant, representing the typical saddle horses of Ethiopia. Their topline is straight, and their legs are slender. The stallions have a well-developed crest, which symbolizes the Ethiopian saddle horse (Kefena et al. 2012).

Uses: Indigenous horses in Ethiopia are extensively used for transport, traction, and agricultural work such as ploughing.

Kafa

Origin and breed distribution: Tropical forest of Sheka and Keffa zones, from the rainforests of Sheka and Keffa in the southern nations.

Description and Morphology: The Kafa is a well-conformed and robust breed. The average recorded height in the DAD-IS database is approximately 1.26 m for mares and 1.33 m for males. The Kafa is among the largest horse breeds (Fig. 11.6) in Ethiopia and exhibits the characteristics of a "heavy forest horse" type. In terms of morphology, the Kafa closely resembles the Horro and the Bale but is notably different from the Selale. It has a higher wither height, a longer body length, and a larger chest circumference compared to other Ethiopian horses. The chest and shoulders are deep, and the body is robust with well-rounded ribs. The mane is often long and typically falls on the right side of the neck. There is no studbook or population estimate for this breed (DAD-IS 2010). These horses typically live outdoors throughout the year and rarely have access to shelters. A subsequent genetic study in 2017 classified the Kafa in the same genetic group as the Selale.

Uses: The Kafa horse is the largest among the Ethiopian horse breeds and is primarily employed for draft purposes in hard areas.

Abyssinian

Origin and breed distribution: The Abyssinian is a horse breed originating from northern Ethiopia. The breed is also known as "Galla," but this name is considered offensive by the Oromos (Mason 1996; Porter et al. 2016). It is often confused with the Oromo horse. The Abyssinian horses are believed to have arrived in Ethiopia through the Red Sea coast. They were exported to England in the 1860s (Rousseau 2014). The Abyssinian horses are distributed in the North and north central parts of Ethiopia, particularly in the Semien Mountains in the area of Gondar, in the Amhara region (Kefena et al. 2012).

Description and morphology: The DAD-IS database reports an average height of 1.25 m for females and 1.29 m for males (DAD-IS 2010) (Fig. 11.7). The height can vary significantly. The Abyssinian is closely related to the Arab-Barb but has irregular morphology, often characterized by a bulging belly and frequently a hollow back

Fig. 11.6 Kafa stallion (Kefena et al. 2012)

Fig. 11.7 Director of Parks Mather insists on A. W.'s being properly mounted on an Abyssinian horse. (Harris 1952)

(Rousseau 2014). Their chest circumference is often smaller compared to other Ethiopian horse breeds, and their overall size is generally more compact (DAD-IS 2010). This is attributed to the intensive use of the breed for work (Kefena et al. 2012). These horses come in various coat colors. A distinctive feature is their dense coat and a mane that grows in a disheveled manner. Many of them have green eyes due to the presence of a specific gene (Mason 1996; Porter et al. 2016; DAD-IS 2010). In terms of characterization, the Abyssinian differs morphologically from the Bale but is very similar to the Selale. These two latter breeds belong to the same group (Kefena et al. 2012). There is no official studbook for this breed.

Uses: The Abyssinian horse is used for a wide range of tasks, particularly for pulling and transportation, including plowing and hauling loads. They serve as the primary source of animal traction in their native region. Despite their small size, Abyssinian horses are valued for their strength and ability to live and work in mountainous areas (Fig. 11.8) (Kefena et al. 2012; DAD-IS 2010).

Bale

Origin and breed distribution: The Bale horse is a horse breed native to the eastern part of Ethiopia. The characterization of this breed is relatively recent, dating back to 2012 (Kefena et al. 2012). Little known, Ethiopian horses were often collectively referred to as "Abyssinians" in written sources without distinction. A characterization study focusing on 45 stallions and 55 mares of a distinct breed specific to the Bale region was published in 2012 (Kefena et al. 2012). There is no studbook for this breed (DAD-IS 2010).

Description and morphology: The morphology of this breed is considered less advantageous, characterized by a hollow back and a bulging belly.

The reference measurements recorded in the DAD-IS database show an average height of 1.255 m for females and 1.292 m for males (DAD-IS 2010). These horses are described as typical Ethiopian ponies. Their conformation is considered "flawed," characterized by a bulging belly, a short and hollow back, and a coarse body (Kefena et al. 2012). They have

Fig. 11.8 Pair of Abyssinian horses engaged in cropland cultivation. (Kefena et al. 2012)

Fig. 11.9 Bay Bale horse. (Image credit: Laika ac from UK 2014). Reproduced with permission—CC-BY 4.0 license

deep chests and shoulders, long necks, and tails compared to other Ethiopian horse breeds (Fig. 11.9). In comparison to other Ethiopian horse breeds, the Bale has deeper chests and shoulders, a longer neck, and a shorter back (Kefena et al. 2012). Among the other Ethiopian horse breeds, the Bale is closely related to the Horro but is very different from the Abyssinian and the Selale. It falls into the same group as the Horro and the Kafa (Kefena et al. 2012).

Uses: These horses are used for draft work and transportation, especially for pack carrying (DAD-IS 2010).

Borana

Origin and breed distribution: They are raised by Borana Oromo pastoralists, particularly in the area of the city of Megga in southern Ethiopia. The Borana-Oromo who breed them are organized in the *Gadaa* system and consider horses as equal if not superior to humans (Coppock 1994). There is no known census of the population (Helland 1980; Mulualem et al. 2015). The breed was recently described by Epstein in 1971, who referred to it as an "African pony (Kefena et al. 2012). They are exclusively saddle horses with a significant cultural importance and are subject to rituals.

Description and morphology: There is no studbook for this breed. It remains to be fully characterized. Its coat is always bay, and it exhibits great resistance to drought. The general conformation of the Borana breed is typical of lowland Ethiopian horses (Fig. 11.10). There is no specific size characterization due to the difficulty in measuring these animals (Kefena et al. 2012; DAD-IS 2010). The coat of the Borana horse is uniformly bay.

Uses: The breed is particularly hardy and resilient to the recurring droughts that affect its region. The Borana horse is primarily used for mounted transportation, but according to oral tradition in its breeding area, it is not permitted to be harnessed (DAD-IS 2010).

These horses hold significant cultural value through the *Gadaa* system, a code of conduct toward animals, which mandates, for example, that horses have priority access to water sources before any person or other animal. Additionally, during the dry season, these horses are regularly led to water points. Male horses receive a burial ceremony similar to those organized for humans upon their death.

Wilwal

Origin and breed distribution: It is concentrated in the east of Ethiopia at the Djidjga region (the traditional Ogaden). The origin of the breed is unknown, particularly due to the absence of written sources (Kefena et al. 2012; Coppock 1994). These horses are locally known by the names Wilwal, Ogaden, Dirdaa'u, or "Somali" horses. According to oral tradition, these horses are believed to have arrived in Ethiopia from the neighboring Somalia, originating from the Nugaal Valley (Kefena et al. 2012). The name "Wilwal" is said to come from the surname of Governor Wilwal Farah Hersi, who was well-

Fig. 11.10 Borana horse riding. (Sirikoi 2023)

known in the region for using horses of this breed in fighting against British colonists (Kefena et al. 2012). However, this theory is not confirmed by demographic and historical studies, which have yet to be conducted (Kefena et al. 2012). There is no stud book for this breed (Coppock 1994). Nowadays, this horse breed can be found in the cities of Jijiga, Dhik, Kebribeya, and Waju. Some of these horses of this breed live in the wild in Aware, near the border with Somalia. There is no inventory of the population in the DAD-IS database (Kefena et al. 2012; Coppock 1994).

Description and morphology: It is a draft horse with a regular conformation (Kefena et al. 2012), rather large and elegant (Coppock 1994). The breed could not be characterized by Kefena et al. (2012) in terms of size because the animals refused to stay still (Kefena et al. 2012).

Uses: It is a beautiful, well-conformed horse, used as saddle horses and cart horses (Kefena et al. 2012).

Horro

Origin and breed distribution: It is a breed of small draft horses originating from the western part of the central plateau of Ethiopia. These horses are found in the Horro district, located in the Oromia region. The breed is considered to be widespread (Kefena et al. 2012), but the DAD-IS database does not provide any population figures. There is no studbook for this breed, and it is not mentioned in the 2016 edition of the CAB International encyclopedia (Mason 1996; Porter et al. 2016; DAD-IS 2010).

Description and morphology: The Horro breed was recently characterized through the doctoral thesis and study by Effa Delesa (Kefena et al. 2012), which included a total of 95 horses from this population identified within eight Ethiopian domestic horse breeds. Most of these horses are considered ponies: the average recorded height in the DAD-IS database is 1.25 m on average for mares and 1.28 m for males (Kefena et al. 2012). In terms of morphology, the Bale and Kafa breeds are very similar to the Horro; these three breeds also belong to the same group, which also includes the Abyssinian, Ogaden, and Selale (Kefena et al. 2012). The morphology is irregular, with these horses being characterized by a defective physical appearance (Fig. 11.11) (Kefena et al. 2012). The chest circumference is similar to that of the Selale and Bale breeds. The genetic diversity of this breed is good.

Uses: It is used for traction, transportation, and plowing, which makes it an extremely useful horse despite its unsound conformation. It is especially used for carrying heavy loads (Kefena et al. 2012).

Kundido

Origin and breed distribution: This population of horses gets its name from the Mont Kundudo, where they have been living in the wild for a long time (Helland 1980; Mulualem et al. 2015). Very little historical information is available because there are no written sources (Kefena et al. 2012; Paris 2014). According to oral tradition collected from the oldest local residents, these horses have been known for over 200 years. It is said that Emperor Haile Selassie I captured one of these horses with the help of his uncle at the age of ten (Kefena et al. 2012; Viganò 2014). One oral hypothesis about the origin of these horses suggests that their ancestors could be military mounts that remained in the area after the war between Adal and Ethiopia in the sixteenth century. It is possible that a small group of 10–15 horses survived despite the past presence of lions and leopards, but there is no tangible evidence to support this theory (Viganò 2014).

The relatively close genetic distance between the wild horses of Kundudo and domestic horses of the Abyssinian breed suggests that the Kundudo horses may be a subpopulation of Abyssinian horses that returned to the wild in the recent past, possibly during the military events of the sixteenth century (Kefena et al. 2012).

The horses of Mount Kundudo were rediscovered in the early twenty-first century during an exploration of specific ecozones for Ethiopian horses. In 2008, researchers found a single mare of about 11 years in the region, showing no signs of domestication. They collected her DNA and nicknamed her "Basra (Viganò 2014)." In total, 18 horses were recorded in the area in October

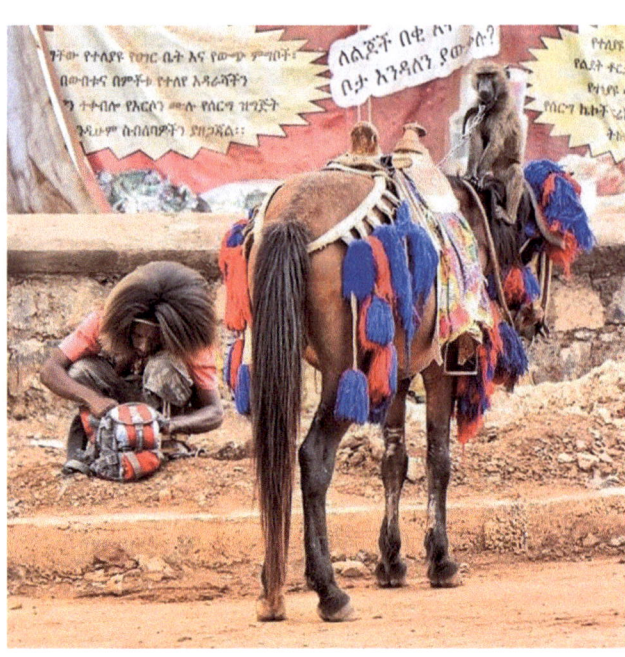

Fig. 11.11 Horro horse in Street Jimma, Ethiopia. (Image credit: Rod Waddington 2014)

2010, but in 2013, an expedition by the Ethiopian Biodiversity Institute could only count 11 horses (Kefena et al. 2012; Viganò 2014).

There is no studbook for this breed. Some of these horses are undergoing a process of redomestication, as they are captured annually by a local farmer for use in work (Mason 1996; Porter et al. 2016; Kefena et al. 2012; DAD-IS 2010).

The DAD-IS database does not provide any figures, but it is important to note that the population of these horses is extremely small. They are primarily located in the region of Mount Kundudo in eastern Ethiopia, where they are considered rare (Kefena et al. 2012; DAD-IS 2010). These horses graze on an area of 13 hectares and drink from a water source located at the summit of the mountain, which never dries up even during the hot season. They represent the last wild horses in East Africa (Viganò 2014). Their survival is seriously threatened due to their rarity, inbreeding, poor breeding practices, and demand for their foals (Kefena et al. 2012; DAD-IS 2010). Since 2011, this animal population has been classified as critically endangered (Paris 2014). The Ethiopian Biodiversity Institute has implemented conservation measures, including collecting and freezing semen from stallions (Paris 2014). The significance of this breed in the context of its connection to Ethiopian history has been emphasized (Kefena et al. 2012).

In 2016, CAB International cited the breed as "virtually extinct," as the last count in 2013 recorded only 11 horses, a number far too small to ensure the long-term survival of this population (Mason 1996; Porter et al. 2016).

Description and morphology: The morphology of the Kundudo horses is described as defective (Figs. 11.12 and 11.13), including irregular shapes, a short back with a sloping topline, and a potbelly (Kefena et al. 2012; DAD-IS 2010). However, the small number of horses studied does not allow for consistent morphological data to be deduced.

The Kundudo horses are one of the eight horse breeds identified in Ethiopia. Among these, the Borana breed, which is the most genetically distant from the Kundudo, shows the greatest genetic divergence (Kefena et al. 2012). The Kundudo is assigned its own gene cluster. The closest Ethiopian breed to the Kundudo is the Abyssinian (Kefena et al. 2012). The population also exhibits low genetic diversity, the lowest among all the studied Ethiopian horse populations. It is likely that these horses have experienced genetic drift due to a small number of founding individuals, isolation over an extended period, and no cross-

Fig. 11.12 Kundudo mare with foal. (Paris 2014)

Fig. 11.13 A thin white Kundudo mare standing on a rocky mountain slope with few trees. (Paris 2014)

Uses: Even though these horses have returned to a wild state and constitute one of the three "feral" horse populations in Africa, some of them are regularly captured by a local farmer who puts them to work during the harvest season before releasing them (Viganò 2014). They are reputed to be of limited utility for this purpose, with low pulling power, and they tend to resist the work required. However, this farmer has likely captured or sold some foals (Viganò 2014).

Studies recommend gradually transforming the grazing area at the summit of Mount Kundudo into a tourist attraction, following the model of the Namib Desert horses, capitalizing on the tourist potential of observing wild horses (Kefena et al. 2012). The Namib Desert horses are indeed globally known, albeit erroneously, as the only "wild" horses in Africa (Kefena et al. 2012).

Gharkaoui

Origin and breed distribution: The breed is native to the regions of Kordofan and South Darfur in Sudan. The breed is known by many names (listed in the DAD-IS database), including "Sudanese Western Pony," "Darfur Pony," Kordofani, Mayray, Reziegi Taaishi, and Messeri (DAD-IS 2010). It was formed in the early twen-

breeding with horses of external origin. However, their genetic heritage is not exceptional or unique compared to other similar equine populations (Kefena et al. 2012).

tieth century and likely descended from the Barb horse, with crossbreeding with Arabian and Thoroughbred horses occurring during the twentieth century, especially in the 1950s and 1960s, under the influence of the Sudanese government at the time (Rousseau 2014). It is widely distributed in southern Darfur and southwestern Kordofanalso in southern Chad (Rousseau 2014).

The only population survey published, dating back to 1994, reported a population of 8000 to 10,000 individuals, with a declining trend. A study conducted by Uppsala University, published in August 2010 for the FAO, indicated that the Western Sudan Pony is an African local horse breed that is not threatened with extinction. There is no more recent information available about this breed.

Description and morphology: Gharkaoui is the smaller of the two commonly found horse types in Darfur (O'Fahey 1980) (Fig. 11.14). According to the Delachaux guide, it stands at 1.40–1.45 m (Rousseau 2014). DAD-IS reports an average height of 1.40 m, with a weight of 400 to 450 kg (DAD-IS 2010). CAB International distinguishes the Gharbui, which has an average height of 1.47 m, from the Darfur Pony (Mason 1996; Porter et al. 2016).

As a riding horse, it resembles the Barb horse, with a convex profile head, short and muscular neck, and long, slender limbs (Rousseau 2014).

It usually has a bay, chestnut, or gray coat with white markings (Rousseau 2014). According to CAB International, the Gharbui typically carries primitive markings (stripes) (Mason 1996; Porter et al. 2016). Mares typically nurse their foals until they are 7 or 8 months old (Rousseau 2014).

Uses: It is locally ridden, including in equestrian sports, and used for light traction (Rousseau 2014).

Tawleed

Origin and breed distribution: The creation of the Tawleed breed is a result of Sudan's colonial history before its independence, with the influence of Britain and Egypt. During the British era, equine breeding was significantly influenced by imports of English horses. The Tawleed breed originated from crosses made with the local Sudan Country-Bred (SCB) horse, along with various breeds of foreign origin, especially the English Thoroughbred (DAD-IS 2010; Rousseau 2014). The initial goal was to produce a riding horse (Mason 1996; Porter et al. 2016).

The importation of English and European horses to Sudan increased in the mid-twentieth

Fig. 11.14 Gharkawi horse. (Image credit: Sean Woo 2004)

century, initially to facilitate polo for colonists. After Sudan gained independence in 1956, these horses were left behind and acquired by local people. Tawleed horses gained a great reputation in Sudan, but many crosses were made with other local breeds.

The Tawleed is considered a local breed specific to Sudan, raised only in the Khartoum region (DAD-IS 2010; Meyer 2019). The study conducted by Uppsala University, published in August 2010 for the FAO, lists the Tawleed as an African local horse breed with an unknown level of threat (Khadka 2010). The threat level for this breed in Sudan is also not provided in the DAD-IS database (DAD-IS 2010), which does not contain any population data. It is likely a rare breed, with perhaps around a hundred individuals (Porter 2020).

Description and morphology: The Tawleed is a light and harmonious-looking blood horse compared to other local African breeds (Fig. 11.15). It has a straight head profile. The breed lacks resistance to the climatic conditions of its breeding region, including drought, food shortages, and the absence of shade (DAD-IS 2010; Rousseau 2014; Porter 2020). The coat can be black, bay, chestnut, or gray (DAD-IS 2010). There is no selective breeding, breed association, or studbook management for Tawleed (DAD-IS 2010).

Uses: It is primarily used for equestrian sports, particularly racing in the Khartoum region. The influence of the Thoroughbred has given it significant racing speed (Rousseau 2014).

The Bahr el-Ghazal

Origin and breed distribution: The breed is also known as "Kréda" (Cadéac et al. 1914) or "Ganaston" (Mason 1996; Porter et al. 2016; DAD-IS 2010). It is generally considered a particularly pure type or lineage of the West African Dongola, specific to the Bahr el Ghazal region in Chad. Most sources describe it as a variety of the West African Dongola (Mason 1996; Porter et al. 2016; DAD-IS 2010; Hendricks 2007). The breed enjoyed a strong reputation during the first half of the twentieth century in Chad and neighboring countries.

Horses do not adapt well to humid regions due to the presence of tsetse flies. As a result, breeding is more common in sub-desert regions such as Kanem, Bahr el Ghazal, and Khozzam.

The study conducted by Uppsala University, published in August 2010 for the FAO, lists the Bahr el-Ghazal as an African local horse breed that is not threatened with extinction. The threat level for the breed is not provided in the DAD-IS database, which records a population of fewer than 10,000 individuals in 1994.

Description and morphology: Measurements provided in the DAD-IS database indicate an average height of 1.50 m and a weight of 350–450 kg. A 2004 FAO study indicates a weight range of 350–400 kg (DAD-IS 2010). These horses are known for their light conformation, being elegantly and gracefully built (Batello et al. 2004).

Their coat is typically dark, often bay or black, with extensive white markings on the legs, a broad blaze on the head, and white under the belly (Mason 1996; Porter et al. 2016). They often have heterochromia, where their eyes have different colors. The breed is resilient and energetic.

Fig. 11.15 Tawleed horse. (Image Credit—Omdurman-Tuti Island jetty, Sudan)

Uses: It is a hot-blooded horse, particularly renowned for racing and for hunting.

Logone Pony (Kirdipony)

Origin and breed distribution: The DAD-IS database reports a multitude of French names for these animals: *Poney Kirdi, Logone, Mbai, Pagan, Poney Hoho, Poney de la Kabia, Sara,* and *Lakka*. Other confirmed names include Cameroonian Pony, Mousseye Pony, and M'baye (Mason 1996; Porter et al. 2016; DAD-IS 2010; Gouraud 2002).

The origins of the Logone pony are not well-documented (Rousseau 2014), but it is found in Cameroon and Chad, in West Africa (Mason 1996; Porter et al. 2016; DAD-IS 2010). The breed may have descended from the Barb horse and could be related to the Nigerian breed, with natural selection allowing it to adapt to its harsh environment. It was used for war and hunting, and its geographic isolation preserved it from crossbreeding with other horse breeds (Rousseau 2014). There are many descriptions of these small horses owned by the Marba-Musey people living in the floodplains along the Logone River, in southwestern Chad and northern Cameroon (DAD-IS 2010).

In 1987, 7500 Logone ponies were recorded in Chad, with a declining trend (DAD-IS 2010).

These ponies mainly inhabit the vicinity of the Logone River. The Logone pony is registered as a local African breed in southwestern Chad, along both banks of the Logone River (DAD-IS 2010). It is also found in northern Cameroon (Mason 1996; Porter et al. 2016). As of 2007, it is not considered endangered according to the FAO (Khadka 2010). However, in Cameroon, the breed is considered a relic of the past and is endangered (as of 2003) (FAO 2016 2003). Efforts by Cameroonian authorities aim to preserve it. Its resistance to tsetse flies may enable its breeding in other African regions.

Description and morphology: The average size recorded for Chadian ponies in the DAD-IS database is 1.24 m for females and 1.26 m for males, with an average weight of 175 kg (DAD-IS 2010). CAB International's dictionary indicates a median height of 1.22 m (Mason 1996; Porter et al. 2016). The Delachaux guide provides an average height of 1.25 m at the withers (Rousseau 2014). These ponies exhibit Barb-type characteristics (Mason 1996; Porter et al. 2016), being small and sturdy, with a moderately concave profile, lightly pronounced brow ridges, and small ears (Rousseau 2014) (Fig. 11.16). They have a short, thick neck and a long back. Their legs are short but strong, with small hooves. They have a shaggy mane and a tail set low (Rousseau 2014).

The most common coat colors, according to CABI, are bay and gray (Mason 1996; Porter et al. 2016), while the Delachaux guide indicates bay as the most frequent, followed by chestnut, with some roan and dun coats (Rousseau 2014). This breed is likely resistant to African trypanosomiasis and is known for its extreme hardiness and robustness, allowing it to carry heavy loads (Mason 1996; Porter et al. 2016; DAD-IS 2010). They are reputed to have gentle temperaments and light gaits (Rousseau 2014).

Uses: These ponies primarily serve as mounts or for various agricultural tasks, and could potentially make good mounts for children (Rousseau 2014). They are used both for riding and as draft animals, and are highly respected by the Musey people. It is central to many rituals resembling

Fig. 11.16 Poneys musey à Gobo (Arthus-Bertrand 2003). (Image credit: Seignobos C). (Arthus-Bertrand 2003)

rites of passage. It has been used as a form of currency and is almost treated as an equal of a human being. Owners of these ponies generally take great care of them. Upon their death, it is not uncommon for owners to bury and mourn them (Mason 1996; Porter et al. 2016; Rousseau 2014). Traditional healing practices were used for these ponies without resorting to surgery, although this practice has declined. The consumption of horse meat is unthinkable for the Musey people (Rousseau 2014). They have developed natural resistance to trypanosomiasis, a disease transmitted by tsetse flies that infest the Logone River (Fig. 11.17).

The Seno-Gondo or Bandiagara

The Bandiagara horse, originating from Mali and Niger in Africa, is a medium-sized horse breed known for its docile nature. This breed, characterized by its Barb-type lineage and delicate yet charming facial features, has shown remarkable adaptability to the challenging climatic conditions of its region. With an average height ranging from 142.24 to nearly 152.4 cm, these horses possess distinctive physical attributes, including short and thick necks, ample chests and shoulders, long backs with rounded croups, low-set tails, robust yet sinewy legs, and hard, small hooves, making it able to fight in rocky terrains (Dehoux et al. 1996). Traditional coat colors for the Bandiagara horse include grey and bay. In terms of temperament, they are recognized for their spirited and bold nature, making them well-suited for purposes such as transportation and cart hauling (http://www.theequinest.com/breeds/).

The Banamba or Beledougou

Origin and breed distribution: The Bélédougou horse, also known as the Banamba horse (DAD-IS 2010; Doutressoulle 1952), is a horse breed originating from Mali and belonging to the group of West African Barb horses. It is traditionally raised in a prosperous region by the Marka people (Doutressoulle 1952). In her essay on the Bambaras in 1954, Pâques (1954) described the Bélédougou horse as "a Barb transformed by climate and man," with the breeding center located in the Touba canton, 90 km northeast of Koulikoro (Pâques 1954). Bélédougou is a local African horse breed with an unknown level of threat (Khadka 2010). The level of threat to the breed is also not provided in the DAD-IS database, which does not offer any population data (DAD-IS 2010).

Description and morphology: The Bélédougou horses exhibit the typical characteristics of West African Barb horses (Mason 1996; Porter et al. 2016). The southern Bélédougou horse was described as generally small (1.34–1.44 m) to well-built, narrow, and unattractive (Challamel 1904; Pierre 1905). The Bélédougou are medium-sized horses, approximately 1.45 m to 1.50 m (Doutressoulle 1952). They are known to be more muscular than horses from desert regions (Dechambre 1921).

Uses: They are used for riding, light draft work, and pulling (TheHorseGuide.com 2010) (Fig. 11.18).

The Nara Horse or Hodh Horse

Origin and breed distribution: The Mechdouf tribe, who raise these horses, trade some of them as far as the banks of the Senegal River, where they influence the local breed, the River Horse. The Mechdoufs are the most famous horse breeders among the Moorish tribes at the beginning of the colonization of Africa (Ndiaye 1978). They regularly descend to the banks of the Senegal River to sell their surplus horses, usually foals. They keep their stallions and broodmares for reproduction (Ndiaye 1978).

This horse breed is also known as the "Barb of Hodh." The Hodh horse is presumed to be descended from the Barb breed (Mauny 1965), which was widely distributed throughout West Africa and the Sahel (Elfasi and Hrbek 1988). According to the tradition of Mechdouf breeders, their so-called "purebred" horses are descended from the four mares taken as loot from the Arabs, with the first one being scabby (Doutressoulle 1947). During the colonization of French West Africa, the Hodh horse was described as one of the best horse models in all of French West Africa (Doutressoulle 1952).

The DAD-IS database and the CAB International encyclopedia classify it as a regional

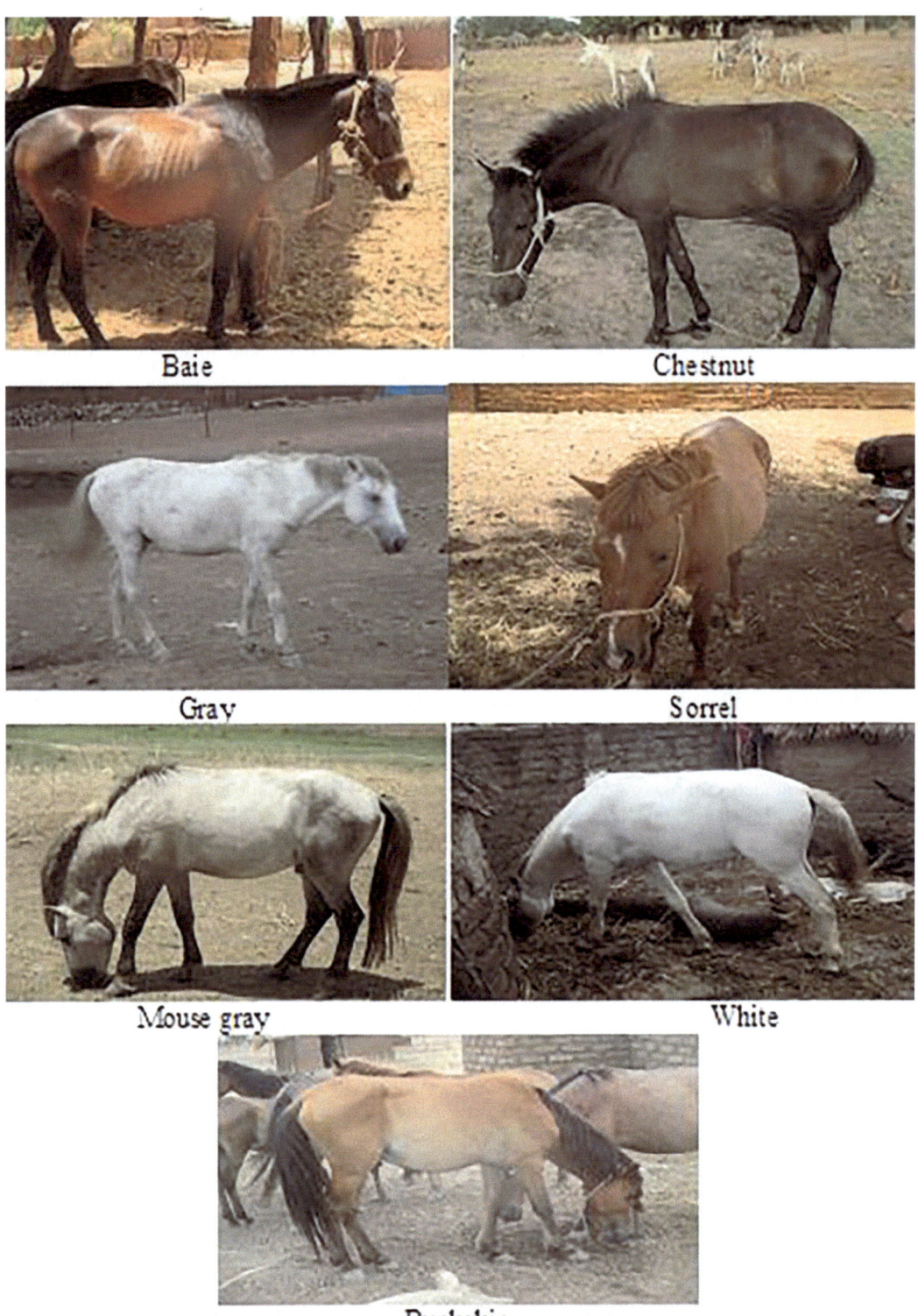

Fig. 11.17 Coat color diversity of the Kirdi pony in the Sudano-Sahelian region, Kabbia South-West of Chad. (Image credit: authors)

Fig. 11.18 Bélédougou horse used for carriage and pulling. (Image credit: Ferdinand Reus) (Claudot-Hawad 1988)

breed in Mauritania and Mali. In Mauritania, the Hodh horse is specific to the region of the same name, where it is raised by the Mechdoufs. The extinction threat level is unknown. The Hodh horse is a transborder local African breed with an unknown status (DAD-IS 2010).

Description and morphology: Hodh horse exhibits the Arab-Barb type (Doutressoulle 1952). However, CIRAD classifies it with the Barb (Meyer 2019), as does the DAD-IS database and the CAB International encyclopedia (Mason 1996; Porter et al. 2016; DAD-IS 2010), which group it among the "West African Barb" breeds. It stands at 1.45–1.48 m and weighs 325–375 kg (Doutressoulle 1952). Its coat is always gray (DAD-IS 2010).

A variety called Kreikibat is described as highly appreciated by the Moors for its gaits and endurance. This breed is perfectly adapted to the Sahelian North and is very sensitive to humidity (Doutressoulle 1952).

Uses: The Hodh horse breed is primarily ridden. The Mechdouf merchants' horses are partly responsible for the Senegalese horse breed called the River Horse (Ndiaye 1978), due to their trade along the banks of the Senegal River. Indeed, the expression *"Fas u nar u gor,"* which designates the River Horse, is very close to the Barb model in terms of conformation and size, and it means "Moorish stallion" (Ndiaye 1978).

Arewa

Origin and breed distribution: The Arewa (Fig. 11.2), a splendid horse from West Africa, is prevalent along the Niger River. It is believed to have descended from a Barb-Dongola cross. This breed is known by various names linked to the ethnic groups that possess it, including variants such as *Bagazan, Ganja, Manga, Gobir, Ader, Djerma* in the central Niger region, Nigeria, and Bornu in the East (DAD-IS 2010). In Mali, it takes on names like *Bandiagara* or *Gondo*, and in Burkina Faso, it is known as *Liptako* or *Djiligoldgi, Mossi*, and even Yagha in the North. In Chad, the Kadara results from a Barb-pony cross. The River or Narugor in Senegal originated from the crossing of local ponies with the Hodh breed. In Western Sudan, the Gharkawi or Kordofani breed emerged from crossbreeding Barb, Arabian, and Thoroughbred horses. The Foutanké or Fouta breed is a result of the Fleuve-M'Bayar cross (DAD-IS 2010).

The breed originates from a mixture of Barb and Dongola horse breeds. It developed during the years when Niger was a French colony, between 1922 and 1960. The French introduced Barb horses, which were better adapted to the hot climate than French horse breeds, to the region. They were crossbred with Dongola horses from Sudan and Eritrea. With the assistance of French breeders, a breed was developed to suit the work demanded in the hot and dry climate of Niger (DAD-IS 2010). The breed has spread between Tombouctou and Say, as well as in central Niger. It is quite widespread, with over 200,000 of these horses recorded in Niger in 2017 (Dehoux et al. 1996). In 1951, Ian Lauder Mason characterized livestock and domestic horse breeds in West Africa, describing a breed of horses raised by the Djerma (Mason 1951).

In 2007, the FAO had no data on the potential threat level to the Djerma. The study conducted by Uppsala University, published in August 2010 for the FAO, reports the "Djerma" as a local African horse breed with an unknown threat level (Khadka 2010). The breed is indicated as not being endangered in the DAD-IS database (Dehoux et al. 1996). In 2018, 253,172 Aréwa horses were recorded throughout Niger, with this population growing (Dehoux et al. 1996).

Description and morphology: The Arewa horse belongs to the group of West African Barb or West African Dongola horses, often considered a variety of Dongola (Fig. 11.19). The build is light and fine. The head profile is convex (Rousseau 2014; Mason 1951). It is a mixture of Barb and degenerate Dongola, measuring 1.40 m to 1.48 m (Mason 1951). The Delachaux guide indicates an average height of 1.40 m (Rousseau 2014).

In 1949, a French colonial monograph mentions "the Djerma horse, raised along the Niger," described as "a small, common, and poorly balanced animal (Anonymous1949). The withers are marked. The croup is slightly sloping with a low-set tail (Rousseau 2014). The legs are slender, ending in small hooves. The mane and tail are fine, smooth, and not very full (Rousseau 2014). The coat is generally dark, bay, chestnut, or black, with white leg markings, but it can also be gray (Rousseau 2014; Mason 1951). These horses are particularly heat-resistant and frugal. They are strong, despite appearing weak due to their external appearance (Rousseau 2014).

Uses: They are used for all daily tasks, including riding, harness work, labor, leisure, and equestrian sports (Rousseau 2014).

Bobo

Origin and breed distribution: The breed probably owes its name to the Bobo-Dioulasso region. It is also known as "Bobodi" or (Dehoux et al. 1996) "Boboli" (Meyer 2019). These horses are relatively lesser-known. The Delachaux and Niestlé guide (2014), titled "Tous les chevaux du monde" and presented as comprehensive, does not mention the Bobo, Min(i)anka, or Koniakar breeds (Rousseau 2014). However, more reputable sources, such as the DAD-IS database and the CAB International encyclopedia, have entries for the Bobo breed (Mason 1996; Porter et al. 2016). Hendricks (2007) mentions the existence of this breed under the name "Bobo" but does not provide a description due to a lack of information (Hendricks 2007). The CIRAD animal science dictionary identifies Bobo ponies (Burkina Faso), Min(i)anka, and Koniakar (Mali) as belonging to the same type in a similar ecological niche, often interbred (Meyer 2019).

A group of Burkinabian researchers reports the presence of horse breeders in the urban area of Bobo-Dioulasso (Sanou et al. 2016), who are supplied with forage collected from natural environments to feed their animals. Horses are common on the streets of Ouagadougou (Manson et al. 2011).

This breed is listed as a local African horse breed with an unknown level of threat (Khadka 2010). The level of threat to the breed is also not provided in the DAD-IS database, which lacks population data (Dehoux et al. 1996).

Description and morphology: The Bobo ponies exhibit the West African "Barb pony" type (Mason 1996; Porter et al. 2016) (Fig. 11.20). The animals are reputed to be well-built, with CIRAD attributing a height of 1.20–1.35 m for Minianka ponies from Mali, weighing around 200 kg. The Koniakar from the Kayes region is reported to be 1.35–1.40 m tall, weighing approximately 250 kg (Meyer 2019). These ponies are likely resistant to Trypanosomiasis (Meyer 2019).

Uses: These ponies are used for riding.

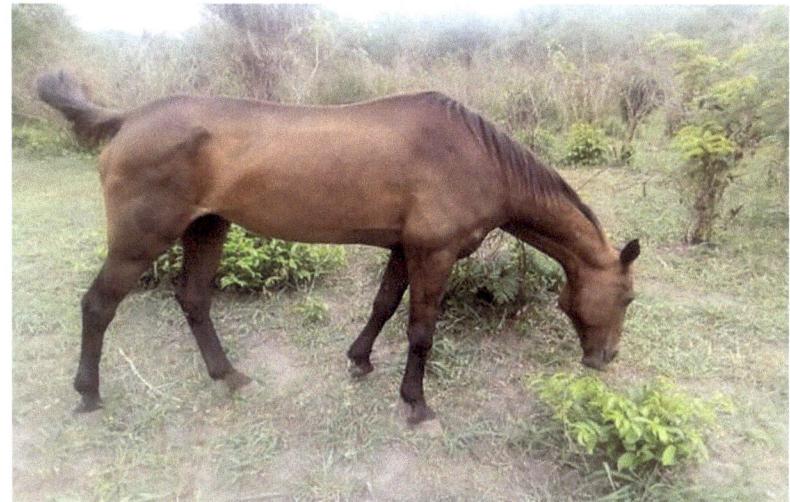

Fig. 11.19 Arewa stallion. (Edward and Bukola 2023)

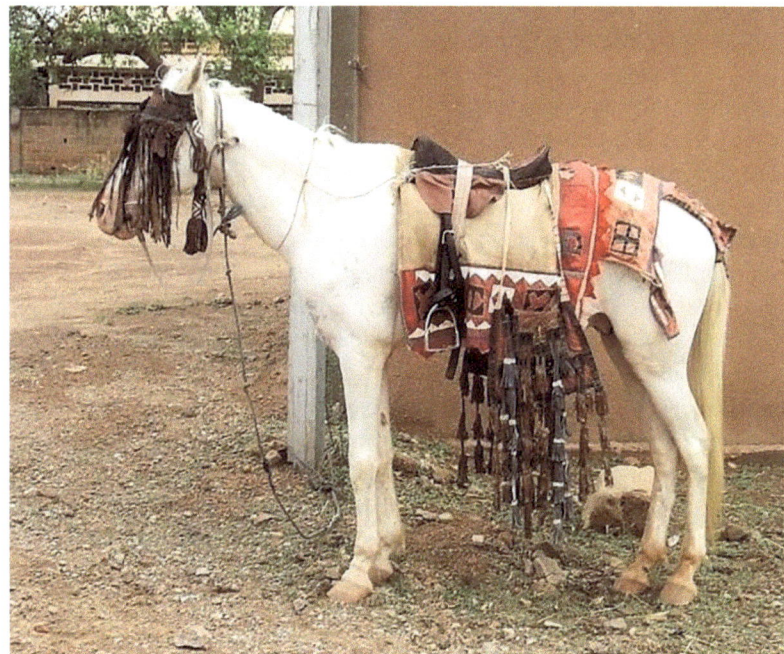

Fig. 11.20 White Bobo horse harnessed for a wedding procession, in the enclosure of Ouagadougou, Burkina Faso. (Image credit: Amélie Tsaag Valren)

The M'bayar Pony

Origin and breed distribution: The name of this breed can also be spelled as *Mbayar* or *M'bayar* (Rousseau 2014). Due to limited documentation on the origins of Senegalese horses (Rischkowsky and Pilling 2007), it is postulated that the *M'Bayar* is an ancient indigenous breed, possibly descending from Barb horses that came from the Maghreb region (ISRA 2003). Historically, this breed was primarily used for the military cavalry of ancient traditional kingdoms. In the twentieth century, the M'Bayar was influenced by Barb horses imported from Mali, Mauritania, and Morocco, as well as Arabian and Anglo-Arabian horses from the Dahra Zootechnical Research Center (ISRA 2003).

In 1996, Senegal had approximately 400,000 horses of all breeds, the largest horse population among all West African countries (ISRA 2003). The M'Bayar is the most common of the four

Senegalese horse breeds, with the others being *M'Par, Fleuve,* and *Foutanké* (ISRA 2003; https://horsebreedslist.com/fleuve/). It is considered a locally adapted breed in Senegal, particularly in the Baol region, and its population size is unknown. There is no available data regarding the threat level to the *M'Bayar* breed. In both 2007 and 2010, FAO did not have information related to the conservation of this breed (DAD-IS 2010; Khadka 2010). The M'Bayar is commonly found in the Diourbel and the Wolof regions of Cayor, Baol, Waalo, and Djolof. In the Sine Saloum region, herds of these horses live semi-freely in the plains (Rousseau 2014).

Description and morphology: The average height of the M'Bayar ranges from 1.34 m to 1.44 m (ISRA 2003), with the Delachaux guide indicating a range of 1.37 m to 1.45 m (Rousseau 2014). DAD-IS database records an average height of 1.37 m among males (Dehoux et al. 1996). This size places it closer to a pony. CAB International's encyclopedia (2016) classifies this breed as part of the West African pony group (Mason 1996; Porter et al. 2016). It is a compact and sturdy small horse with a broad chest and a short neck. Its legs are robust but often exhibit cow hocks and other characteristics considered as faults (Rousseau 2014; ISRA 2003).

The coat color is usually solid, with bay being the predominant color, followed by chestnut. Gray and roan coats are also common (DAD-IS 2010; Rousseau 2014; ISRA 2003). The M'Bayar is known for its calm and docile temperament, hardiness, and endurance (ISRA 2003). It is highly resistant to the climatic conditions of the Sahel, which is characterized by dry and hot weather (Rousseau 2014). The breeding of M'Bayar horses in Senegal is managed by the Senegalese Agricultural Research Institute (ISRA) (https://horsebreedslist.com/fleuve/).

Uses: In Senegal, the M'Bayar is used for riding (Mason 1996; Porter et al. 2016), as well as for draft work, such as pulling carriages or agricultural plowing (DAD-IS 2010) (Fig. 11.21). It plays a significant role in the daily life of Senegalese people, both in rural areas and major cities like Dakar, Saint-Louis, Kaolack, etc. It is also used for horse racing in racetracks (ISRA 2003). The M'Bayar horse is often crossed with Arabian Thoroughbreds by breeders in Senegalese studs to produce crosses for horse racing. The M'Bayar has contributed to the development of the Foutanké breed through crossbreeding with the Fleuve breed (Mason 1996; Porter et al. 2016).

The "Fleuve"

Origin and breed distribution: According to the DAD-IS database and the Delachaux guide, these horses are known as "Naru Gor" (DAD-IS 2010) or "narugor" (Rousseau 2014) in Wolof. However, the CAB International dictionary (2016) indicates that "narugor" is the Wolof name for the "Foutanké" breed (Mason 1996; Porter et al. 2016). The name "Cheval du Fleuve," in French, derives from the word "Fleuve," meaning "river." The breed originates from a cross between the *Barb* horses from Hodh or Kayes and the local Senegalese pony population (Mason 1996; Porter et al. 2016; Dehoux et al. 1996). However, recent crossbreeding among Senegalese horses has made the characterization of breeds difficult (Rousseau 2014).

The breed is specific to Senegal, where it is considered "locally adapted." According to Khadka (2010), the "Fleuve" is a local African breed, but the level of threat to its survival is unknown (Khadka 2010). The DAD-IS database from 2018 also lacks information on population numbers or threat levels (Dehoux et al. 1996).

Description and morphology: The average height of the "Fleuve" is 1.40 m, with a median weight of 325 kilograms (Dehoux et al. 1996). Others suggest the height ranges from 1.40 m to 1.50 m at the withers (Rousseau 2014). They are lightweight riding horses (Dehoux et al. 1996). They are very close to the "Foutanké" but slightly finer (Ndiaye 1978). They have a relatively fine head, a rather narrow chest, a croup with a sloping shape, and slender, long limbs (Rousseau 2014) (Fig. 11.22). Their coat comes in various shades of gray (Rousseau 2014; Dehoux et al. 1996). These horses are known for being lively and energetic, well-suited to the local climate (Rousseau 2014). The management of the breed is overseen by ISRA and *Laboratoire National de*

Fig. 11.21 M'bayar horse served for light draft in Pikine, Sénégal. (Source: Lejosne 2023)

Fig. 11.22 Fleuve horse breed from Sénégal in West Africa

l'Elevage et de Recherches Vétérinaires (LNERV) (Dehoux et al. 1996; https://horsebreedslist.com/fleuve/).

Uses: These horses are used for riding and driving (Rousseau 2014).

Foutanke

Origin and breed distribution: Foutanké is the local name of the breed, also known as "Narougor" according to CAB International (Mason 1996; Porter et al. 2016). There is no written documentation regarding the history of horse breeding in Senegal (Rischkowsky and Pilling 2007). The Foutanké is the result of a cross between a stallion from the Fleuve breed and an M'Bayar mare (Ndiaye 1978; Mason 1996; Porter et al. 2016; Rousseau 2014; Lejosne 2023). CAB International classifies it among the Barb-type ponies originating from West Africa, in the group of "West African Barbs" (Mason 1996; Porter et al. 2016). The Foutankébreed is specific to Senegal (Rousseau 2014). The Foutanké is raised in the Sine-Saloum region (Doutressoulle 1947). In 2007, the FAO had no data on the potential threat level to the Foutanké (https://experiencingnaija.com/what-did-you-know-about-hausa-fulani-wedding/). Khadka (2010) reported the Fouta as a local African breed with an unknown threat level (Khadka 2010). Its threat level is also unknown in the DAD-IS database from 2018 (Dehoux et al. 1996).

Description and morphology: According to Larrat (1947) there are three Senegalese horse breeds: M'Bayar, Foutanké, and M'par. The Foutanké was described as resulting from an attempt to transform the M'Bayar through crossbreeding to increase its size; the results of these crossbreeding are not always very successful, with many horses from these crosses being narrow or greyhound-like (Rischkowsky and Pilling 2007). According to him, the best Foutankés measure between 1.38 m and 1.43 m. According to Rousseau (2014), the Foutanké measures from 1.35 m to 1.50 m at the withers (Rousseau 2014). Measurements taken by Larrat (1947) gave an average height of 1.41 m, describing it as a proportionate horse (Rischkowsky and Pilling 2007). It closely resembles the horse of the Fleuve but is coarser. The most common coat colors are gray and chestnut (Rousseau 2014). These horses are known for being rugged and resilient (Fig. 11.23). The Foutanké is one of the four recognized horse reeds in Senegal, based on Larrat's description (Rischkowsky and Pilling 2007), but extensive crossbreeding among Senegalese horses makes the concept of a "pure breed" difficult to apply (https://horsebreedslist.com/fleuve/).

Uses: These horses are used for riding, driving, and agricultural work (Rousseau 2014). They are particularly sought after for racing (ISRA 2003) (Fig. 11.23).

The M'Par

Origin and breed distribution: The origins of horses in Senegal are not documented. The M'Par could be an indigenous breed with ancient origins in the region (Rousseau 2014). Others, like Larrat (1947), view the horses in Senegal as descendants of Barb horses originating from North Africa (Rischkowsky and Pilling 2007). Regardless, its original homeland is the Cayor region in Senegal, West Africa, which is why it is also known as the "Cayor pony". It is described as a locally adapted horse breed in Senegal, specific to the former Cayor region, with unknown population numbers (Dehoux et al. 1996). The region of origin actually admitted is Louga and Thiès. In 2007, the FAO had no data to estimate the conservation status of the M'Par breed, but agreed it is likely a local breed (https://experiencingnaija.com/what-did-you-know-about-hausa-fulani-wedding/).

Description and morphology: The M'Par is the smallest of the four Senegalese horse breeds (Rischkowsky and Pilling 2007), with the others being the M'Bayar, the Fleuve, and the Foutanké (https://horsebreedslist.com/fleuve/). The average height of males is approximately 1.30 m (Dehoux et al. 1996), with a general range of 1.25–1.35 m (Ndiaye 1978; Rousseau 2014). In terms of morphology, the M'Par resembles the Barb horse (Mason 1996; Porter et al. 2016), although it is generally considered to have poor conformation, including a heavy head, a long back, thin legs, a flat chest, and often defective leg conformation (Rischkowsky and Pilling 2007). This morphology may result from breeding and feeding conditions, as better-nourished horses tend to have more harmonious conformation (Rousseau 2014).

The coat of the M'Par is typically bay, in various shades, or chestnut, but it can vary. In terms of temperament and care, the M'Par is known for exceptional endurance and hardiness (Dehoux et al. 1996; Rischkowsky and Pilling 2007).

Uses: Horses play a significant role in the social and economic life of Senegal. The M'Par is used as a draft horse, particularly in agriculture (Ndiaye 1978) (Rousseau 2014);. Due to its small size, it can only pull light carts and carriages (Ndiaye 1978). It could also be used as a teaching horse for young riders (Rousseau 2014).

The Bagzan

Origin, naming, and breed distribution: The Bagzan is also known by the names "Berrezem", "Baguezzan" (Doutressoulle 1947), "Baguezane," or "Kinaboutout," (Epstein 1971). The name "Bagzan" refers to both this specific horse breed

Fig. 11.23 A Foutanke or Narougor horse in Pikine, Sénégal. (Lejosne 2023)

and the winged horse in Tuareg mythology (Bernus 2002). The Bagzan is considered mythical and legendary, and only a good Tuareg is deemed worthy of riding it, as opposed to the common and less valuable horse, known as the efàkre (Baroin and Boutrais 1999; Bernus 1995). The DAD-IS database includes the "Bagazan" in the group of the national horse breed of Niger, known as Aréwa (Benecke 2006). However, it strongly resembles the Arabian horse (Dehoux et al. 1996).

Bagzan horse breed has its origins in the Bagzane Mountains of Africadate back to the 1850s. It explores the legendary origins of the breed, including stories of a stallion from Istanbul and Arabian horses brought to the mountains (Haddad et al. 2014). There are various oral traditions surrounding the breed's creation, including its use by the Kel Gress tribe in the eighteenth century. Despite its legendary status, the Bagzan breed faced population challenges due to historical conflicts in the region. This breed is known for its elegance, endurance, and ability to navigate rocky terrain, making it highly regarded among the Tuareg people (Hamami 1989). The text also mentions a French military officer's documentation of these features and a colonial veterinarian's description of the breed in 1947 (Doutressoulle 1947).

Description and morphology: The Bagzan is a small, square-built horse, standing at approximately 1.40 m, known for its exceptional elegance (Doutressoulle 1947). It has a short and square head with a straight profile, a broad forehead, expressive eyes, and short, mobile ears (Doutressoulle 1947). The neck is slightly arched and covered with a luxurious mane, with occasionally wavy strands (Anthony 1996). The back is short and straight, providing good support (Doutressoulle 1947). The croup is round and horizontal, resembling an apple, with a tail set high. The chest is broad, the ribs are rounded, and the flanks are short. The limbs are slender, even delicate, with a cannon circumference of less than 18 cm, but they are sturdy (Anthony 1996). The hooves are tough (Doutressoulle 1947).

The coat of the Bagzan is always gray (Epstein 1971; Doutressoulle 1947). This breed is renowned for its simplicity, speed, endurance, and impressive, elongated gaits (Doutressoulle 1947). It is said to be able to go without water for three days. The Tuareg people feed these horses with millet and camel milk (Epstein 1971).

Uses: The Bagzan has significant use among the Tuaregs, especially for long journeys across rocky terrains (Haddad et al. 2014). However, its ownership is reserved for the wealthiest Tuaregs due to the need to provide it with supplementary nutrition (Doutressoulle 1947). The Bagzan is mentioned in poems and narratives where it symbolizes nobility (Haddad et al. 2014). Traditional battle accounts attribute numerous feats to its supernatural speed and the ability to safely bring back its wounded master (Haddad et al. 2014). It is also said to have the ability to understand human language, to weep, and to refuse food if insulted. These traditional poems always mention the horse by name. Bagzan horses are traditionally buried after their death, and the burial sites are known to Tuareg riders (Haddad et al. 2014).

The Mogods Pony

Origin, naming, and breed distribution: The origins of the Mogods pony remain unclear, although research is ongoing (Chabchoub et al. 2006). According to one hypothesis, the Mogods pony may be a variant of the Barb, Arab-Barb, or Arab horse. Another hypothesis considered it to be a breed distinct from the Barb and the Arabian horses (Sebbag 2002). For others, it could be a primitive indigenous breed whose owners, fleeing invasions, took refuge in the mountains. It is believed that the breed became adapted to mountainous regions during this time. The breed is native to the extreme northwest of Tunisia (Chabchoub et al. 2006; Dehoux et al. 1996). The Mogods region is its cradle. Mogod ponies can also be found in the regions of Amdoun, Héllil Mogod, and Nefza (Chabchoub et al. 2006). These areas are characterized by rugged terrain, cork oak forests, and summer drought, all of which have shaped the pony's morphology and character.

The breeding of this pony breed flourished in the twentieth century particularly in the early twentieth century. The Mogods pony was highly prized for equestrian sports, such as polo. According to the Bulletin of Agriculture and Trade of Tunisia, published in 1903, approximately 500 ponies were exported annually to Italy, Malta, France, and England. It is reported that Lord Mountbatten, the last Viceroy of India, purchased a batch for his polo team. The Tunisian administration decided in August 1902 to manage this breed through a specific studbook. However, World War I put an end to these efforts. The production of the pony significantly declined after the end of World War II, mainly due to the mechanization of agriculture (Fondation Nationale d'Amélioration de la Race Chevaline (FNARC) 1988). To prevent the breed's extinction, Tunisian authorities entrusted the management of this breed to the National Foundation for the Improvement of Horse Breeds, a public institution under the Ministry of Agriculture and National Federation for the Improvement of the Horse Breed (FNARC) (SADA 2006). Its population was estimated at around one thousand individuals in 2002 (Chabchoub et al. 2006; Fondation Nationale d'Amélioration de la Race Chevaline(FNARC) 1988). There is no recorded population data in the DAD-IS database (Dehoux et al. 1996); the breed requires an inventory and conservation measures (Rousseau 2014).

Description and morphology: The Mogods pony measures between 1.20 m and 1.45 m according to reported studies (Chabchoub et al. 2006; Rousseau 2014). It closely resembles the Barb horse with a well-proportioned body and pronounced muscles (Rousseau 2014). It has a short and strong neck, a straight back, and short and lean limbs. Its chest is wide and deep, and it has an expressive head. The pony has very regular conformation and hard hooves, eliminating the need for shoeing (Fondation Nationale d'Amélioration de la Race Chevaline(FNARC) 1988) (Fig. 11.24).

Some studies suggest that the Mogod pony's morphology is more similar to that of a horse rather than a pony (Fondation Nationale d'Amélioration de la Race Chevaline(FNARC) 1988). Its eyes, withers, croup, and loins strongly resemble those of the Barb horse. Some of its morphological characteristics do not correspond to those of a pony: it is taller than long, with a cannon bone length greater than that of the Barb. The most common coat colors are bay, chestnut, and gray. Primitive markings, such as stripes, are possible. The Mogods pony is a robust pony accustomed to harsh living conditions. It is a mountain breed, very sure-footed and with a

Fig. 11.24 Mogods pony. (Chabchoub et al. 2006)

calm yet energetic temperament. It possesses great working capacity and is often handled by children or the elderly.

Uses: The Mogods pony is highly versatile. In its original mountainous region, it is used as a pack animal for carrying water or goods. It is also employed as a means of transportation, often ridden bareback and without a bit (Chabchoub et al. 2006) (Fig. 11.24). However, due to increasing mechanization observed even in the most remote rural areas, it is becoming less commonly used for these purposes.

With the renewed interest it has gained, a return to recreational or sports use, as it existed in the early twentieth century, is possible. The Mogods pony could be a suitable partner for activities such as driving, trekking, and more. Because of its calm and docile nature, it can serve as an excellent teaching pony (Chabchoub et al. 2006).

The Cape Horse or Boerperd

Origin, naming, and breed distribution: The Cape Horse is also known as Hantam and Boer (DAD-IS 2010). It was one of the first horse breeds developed in South Africa (Cothran and Van Dyk 1998), accompanying European colonization in its history and development (Mason 1996; Porter et al. 2016). It never had a stud book and originally formed a unique strain with the Basuto (DAD-IS 2010), both descended from horses imported from the Indonesian islands (Mason 1996; Porter et al. 2016).

The Cape Horse resulted from multiple influences of successive waves of horse importations to South Africa. Initially, there were Oriental horses between 1652 and 1778, likely of Arab-Barb and Persian origin (Mason 1996; Porter et al. 2016), followed by Criollos in 1778, Thoroughbreds from 1782 to 1860, and Hackneys from 1860 to 1891 (DAD-IS 2010). The blending of these imported breeds formed the Cape Horse in the early 1800s. Starting in the 1820s, it was mainly influenced by the importation of Arabian and Thoroughbred horses (Mason 1996; Porter et al. 2016). Despite the loss of many Cape horses during the Great Trek, the Cape Horse breed persisted locally in its original form (Cothran and Van Dyk 1998). Breeding horses allowed white European settlers to differentiate themselves from indigenous people, reaffirming their European identity (Bankoff and Swart 2008).

The Cape Horse nearly went extinct in the early twentieth century due to crossbreeding, epidemics in the late nineteenth century, and the Anglo-Boer Wars of 1880 and 1899 (Mason 1996; Porter et al. 2016). During the Second Anglo-Boer War, the British captured or eliminated Cape Horses because they were naturally resistant to African equine diseases and could serve as weapons of war Wessels (1991). Only a handful of these horses remained in the Cape region, where they were eventually crossed with other breeds. The South African Department of Agriculture estimated the population of the Cape Horse to be around 1900 (Cothran and Van Dyk 1998).

The South African horse breed nomenclature becomes even more perplexing due to the activities of a group of breeders primarily based in the Cape Province. Beginning in the 1950s, these breeders established the SA Riding Horse Breeders' Society, with the aim of enhancing the existing Boerperd breed by introducing American Saddlebred and Morgan stallions (Kaplan 1974). Over the years, they diligently pursued recognition from both the Department of Agriculture and the South African Stud Book Association. Their efforts eventually bore fruit in May 1984 when the Cape Boerperd Breeders' Association was officially recognized by the Minister for Agriculture, and they received a certificate of incorporation from the South African Stud Book Association (Mason 1996; Porter et al. 2016; Cothran and Van Dyk 1998; FAO 2002).

For 130 years, the Cape Horse was widely exported throughout the British colonial empire, mostly for military use (Cothran and Van Dyk 1998). Like other horse breeds developed by settlers, it embodied local identity and accompanied the rise of nationalism (Bankoff and Swart 2008). In comparison to the English Thoroughbred, the Cape Horse was not considered a high-status mount but represented the pride of white South African colonists (Bankoff and Swart 2008). It is listed as an extinct local South African and African horse breed (Khadka 2010; Dehoux et al. 1996).

Description and morphology: The Cape Horse has a light build. While its exact morphology is not known, it likely does not resemble a pony (Hendricks 2007). Natural selection for adaptation to the climate and human selective breeding, favoring working animals with a calm disposition, resulted in a uniform breed with its characteristics (Cothran and Van Dyk 1998). The Cape Horse is known for its gentleness (Cothran and Van Dyk 1998).

Uses: The Cape Horse served as a working mount in farming (Dehoux et al. 1996) and a military mount when needed (Mason 1996; Porter et al. 2016). It was the first horse breed introduced to Australia. The Cape Horse also contributed to the development of the Boer du Cap, South African Boer, Basuto, Nooitgedacht, Namaqua (Dehoux et al. 1996), and Carrossier du Cap breeds (Kreuz 1958). It gained a strong reputation during the Great Trek, accompanying Boer farmers' colonization, earning it the name Boerperd (Cothran and Van Dyk 1998).

Boer du Capor Kaapse boerperd

Origin, naming, and breed distribution: The correct spelling of the name of this horse breed uses an initial capital letter in the word "Boer," as this breed is named after the Boer ethnic group (Giammatteo 2019).

Derived from the Cape Horse (which is probably not a pony) (Hendricks 2007), the Boer du Cap is an "improved" version of the latter, originally selected for agricultural work (Mason 1996; Porter et al. 2016). Its ancestors were crossed with Hackneys and Saddlebreds (Rousseau 2014). During the second half of the nineteenth century, it was influenced by the Frison du Hackney, Norfolk Trotter, and the Cleveland Bay (Hendricks 2007).

An association of breeders, the Kaapse Boerperd Breeders Society of South Africa, was established in 1948 (FAO 2002). Disagreements among South African breeders led to a split, creating two associations with different breeding objectives. The creation of the Boerperd Breeders' Society of South Africa resulted in crossbreeding the remaining population of historical Boer ponies with Saddlebreds, Hackneys, Thoroughbreds, and Arabians (Hendricks 2007). In 1993, recognizing the risk of losing the breed's identity, breeders reintroduced eight crossbred stallions to reinforce the distinctive characteristics of the Boer du Cap (FAO 2016 2003). The breed remains under selection, with its studbook only being closed in 1999.

The breed is native to South Africa. Rousseau (2014) and Boerperd Breeder's Society of South

Africa indicate a population of 1924 individuals in 2013 (Rousseau 2014).

Description and morphology: Rousseau (2014) reported the height as ranging from 1.42 m to 1.62 m with a long head (Rousseau 2014). It was described as dry, finely chiseled, with a straight profile, large-spaced eyes, moderately long ears, and a refined throat latch (Hendricks 2007). The neck is carried high, long, and slightly arched, with a sloping and muscular shoulder (Rousseau 2014; Hendricks 2007). The back is short, the body is deep, and the legs are strong, with wide joints (Mason 1996; Porter et al. 2016). The coat is generally bay (Mason 1996; Porter et al. 2016). The breed can exhibit additional gaits (Rousseau 2014). These horses are known for their endurance and sure-footedness (FAO 2002).

Uses: Widely associated with agricultural work and military actions of the Afrikaners, it is now extinct in its original form due to extensive crossbreeding (Fig. 11.25). It has contributed to the development of breeds like the Carrossier du Cap, Boer du Cap, South African Boer, Nooitgedacht, and Namaqua.

Boer Sud-Africain or Suid-Afrikaanse Boerperd

Origin, naming, and breed distribution: In Afrikaans, "boerperd" means "farmer's horse," but it can also be a reference to the Afrikaners. This breed descends from the first horses introduced to South Africa in 1652 from Java, likely of Arab-Barb type (Mason 1996; Porter et al. 2016). Later, Mongolian horses influenced this population (Mason 1996; Porter et al. 2016). The Boer shares common origins with the Basuto, as both were selected from the Cape Horse during the nineteenth century. However, the Boer receives bloodline contributions from Flemish horses, Hackneys, and Cleveland Bays. Its living conditions are not as harsh as those of the Basuto; it is a larger breed with a more powerful conformation. During the Second Anglo-Boer War, the great mobility of this small horse helped the Boers move and resist the British Empire for 3 years.

The Kaapse Boerperd Breeders Society of South Africa, an association of breeders, was established in 1948 (FAO 2002). The breed saw its first declines in the 1950s. A new association, the Boerperd Society of South Africa, was formed in 1973. It changed its name to become the Historiese Boerperd Breeders Society in 1977, then the South African Boerperd in 1998 (SASB 2003). The Historiese Boerperd was officially recognized by the South African Department of Agriculture in 1996 (SASB 2003). This association is a member of the SA Stud Book Association (Promerová et al. 2014). South African Boerperds live in isolated herds in the Southeastern Transvaal, Northern Natal, Eastern Free State, and Northeastern Cape Province.

Description and morphology: It typically measures between 1.30 m and 1.50 m. The most common coat colors are black, chestnut, bay, gray, roan, dun, and palomino. The South African Boerperd has been the subject of a study to determine the presence of the DMRT3 gene mutation responsible for additional gaits. The study of 20 individuals detected the presence of this mutation in 15% of them, without confirmation of the existence of horses with additional gaits within the breed. However, there is no specific distinction made between South African Boerperd and Boer du Cap in this context (Promerová et al. 2014).

Uses: The Boer is the origin of the creation of the Calvinia horse.

The Basuto Pony

Origin, naming, and breed distribution: The Basotho or Basuto is a pony breed originating from Lesotho (formerly Basutoland) and South Africa. The breed was developed in Lesotho, a landlocked country in South Africa (Mason 1996; Porter et al. 2016). Initially, its history was closely linked to that of the Cape Horse (Dehoux et al. 1996), which, in turn, had diverse origins, including Javanese, European, and possibly Mongolian horses (Mason 1996; Porter et al. 2016). These animals were repeatedly crossbred in the eighteenth century with Arabians, Thoroughbreds, and Persian horses (Mason 1996; Porter et al. 2016). The Basuto's origins date back to approximately 1825 when it was

Fig. 11.25 Cape Boerperd mares and foals at pasture (Van der Merwe and Martin 2002). (Image credit: Van der Merwe and Martin 2002)

introduced to Basutoland by the Zulu chief Msosesha (Dehoux et al. 1996). The Basuto pony has maintained its small number and has adapted to its environment, situated at an altitude of about 1400 m. This breed is currently endangered. The Basuto pony was listed as distinct in the DAD-IS (2006) list. It classified the Basotho as an extinct local African breed (Khadka 2010). Furthermore, the 2012 edition of Equine Science categorizes it among the relatively unknown pony breeds at an international level (Parker 2012).

Description and morphology: The Basuto pony has an average height of 1.42 m (Rousseau 2014), and is generally less than 1.47 m (Mason 1996; Porter et al. 2016). It is slender yet agile, exceptionally resilient, robust, enduring, and brave (Fig. 11.26). This breed can carry heavy loads over long distances. Its upbringing has made it rugged and thrifty, with extremely surefooted hooves. The coat of the Basuto pony is often chestnut or bay, occasionally gray. White markings are infrequent and limited. It possesses a refined head, carried by a long and often arched neck (swan neck), sparse manes, powerful hindquarters, short and sturdy limbs, and tough hooves.

The Basuto pony has been the subject of a study aimed at determining the presence of the DMRT3 gene mutation responsible for additional gaits. The study, which involved 30 subjects, found the presence of this mutation in 6.7% of them, although there was no confirmation of the existence of horses with extra gaits within the breed (Promerová et al. 2014).

Uses: The Basuto pony primarily serves as a working horse (Dehoux et al. 1996), but it is also used for pack and light draft work (Mason 1996; Porter et al. 2016). This pony has participated in various wars, including the Second Boer War, where it gained recognition as a fast and agile mount used by Boer farmers. Today, it is sought after as a leisure and sport riding horse, popular for hiking, polo, and occasionally racing. It has also given rise to the Nooitgedacht breed.

The Nooitgedacht

Origin, naming, and breed distribution: The Nooitgedacht (Afrikaans: *Nooitgedachter*) is a breed of small riding horses originating from Southern Africa. It is a rare breed primarily used for leisure riding. In Afrikaans, these horses are called *Nooitgedachter* or *Nooitgedachtperd* (DAD-IS 2010). The breed's origins can be traced back to the characterization of South African horses, initiated in the late 1940s (NHBS 2021). In December 1951, the Basuto pony was selected by the South African Department of Agriculture for its qualities of adaptability and hardiness (NHBS 2021). The Nooitgedacht is a breed artificially created at the Nooitgedacht (which translates to "never imagined") experimental farm in the southeastern Transvaal from the Basuto Pony

Fig. 11.26 Basuto pony (Van der Merwe and Martin 2002). (Image credit: Martin J. in Van der Merwe and Martin 2002)

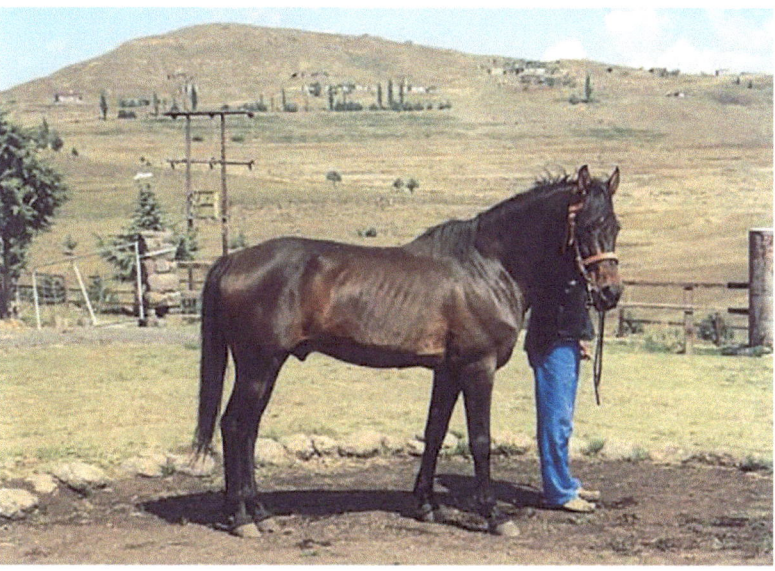

(Rebts 2017). Crossbreeding was employed to create an animal suitable for local climate conditions. The founding stallion of the Nooitgedacht, Vonk, is originally a Basuto and is the ancestor of all current Nooitgedacht horses (Rebts 2017). Crossbreeding with Arab horses, part-bred Arab horses, and Boer ponies was also practiced. A significant influence on the breed's development was the Nooitgedacht stallion named Mac, born in 1958.

An association for the breed was established in 1968, but Nooitgedacht was only officially recognized in South Africa 8 years later, in 1976. In March 1977, the Nooitgedacht farm organized a public animal sale, contributing to the breed's spread. The breed's selection was entirely entrusted to the breeders' association, which had 60 members by 1979 (Rebts 2017).

With only 400 individuals registered, the breed is considered rare. A study conducted by Uppsala University in August 2010, in collaboration with the FAO, categorizes the Nooitgedacht as a regional transboundary breed of Africa in danger of extinction.

Description and morphology: The Nooitgedacht is an elegant yet robust small riding horse with quality bone structure (Fig. 11.27). The height typically ranges from 1.38 m to 1.63 m. They have strong hooves that generally do not require shoeing. Their back is short, and they have well-formed shoulders (Rebts 2017). They typically have gray, bay, chestnut, or roan coats, with piebald and skewbald patterns being prohibited. Nooitgedacht horses are hardy and low-maintenance. They are intelligent and have excellent temperaments, showing affection toward people (Rebts 2017). A study on the presence of the DMRT3 gene mutation based on 14 subjects found the mutation in 21.4% of the horses. This confirms the existence of horses with additional gaits within the breed (Promerová et al. 2014).

Uses: These horses are appreciated for their endurance as riding horses and are used for show jumping, gymkhana, and leisure riding. They excel in polo and are often employed in reserves for patrolling in dangerous areas (Rebts 2017).

The Vlaamperd

Origin, naming, and breed distribution: The Afrikaans name "Vlaamperd" translates to "Flemish Horse" in French (Mason 1996; Porter et al. 2016; Dehoux et al. 1996). It is essential not to confuse this breed with the European Flemish horse (Vlaams Paard) (Rousseau 2014), as they have different origins and uses. In South Africa, the term "Flemish" was applied to any horse imported from the Netherlands or Flanders during a period when the export of Dutch horses was

Fig. 11.27 The South African Nooitgedacht stallion named Mac. (FAO 2002)In 1958, the birth of the stallion Mac marked a pivotal moment in the development of the Nooitgedacht horse breed. Mac exemplified the ideal general-purpose horse sought by project managers, displaying key physical traits such as a strong shoulder, well-muscled back, and robust legs with hard hooves. Additionally, Mac embodied the distinctive characteristics of the Nooitgedacht breed, with his small pointed ears, pronounced brows, and straight to slightly concave profile. This iconic horse left a lasting legacy in the breed's evolution (Photo: Landbouweekblad)

prohibited, regardless of the actual genetic origin of the animals (Mason 1996; Porter et al. 2016).

The South African Vlaamperd is the result of crossbreeding with Frisian horses imported into the country in the early twentieth century. These Frisians were crossed with the local horse population, primarily composed of the Cape Horse (Mason 1996; Porter et al. 2016), as well as Thoroughbreds, Hackneys, and Cleveland Bays (Dehoux et al. 1996). An Oriental Frisian stallion known as Kemp also had a significant influence on the modern Vlaamperd (SAVBS 2019).

The breed's studbook was established in 1983 (Dehoux et al. 1996), coinciding with the founding of the SA Vlaamperd Breeders Society in Bloemfontein (SAVBS 2019). By 1992, there were only 165 registered Vlaamperd horses, including 65 mares in the studbook (Dehoux et al. 1996). In 1999, the population had likely increased, estimated to be between 100 and 1000 individuals (Dehoux et al. 1996).

The Vlaamperd is considered a local South African breed (Dehoux et al. 1996), predominantly found in the Western Cape region (Mason 1996; Porter et al. 2016; Rousseau 2014). In the past, it was close to extinction (Mason 1996; Porter et al. 2016; Hendricks 2007) but was recognized as one of the locally developed exotic horse breeds by the South African Department of Agriculture in June 2006 (Cothran and Van Dyk 1998). The Vlaamperd is not mentioned in the study conducted by Rupak Khadka from Uppsala University, published in August 2010 for the FAO (Khadka 2010). According to the Delachaux guide, the breed still exists but in very limited numbers, with approximately 200 individuals in 2013 (Rousseau 2014).

Description and morphology: The Vlaamperd's height ranges from approximately 1.47–1.57 m (Mason 1996; Porter et al. 2016), with an average of 1.54 m (Dehoux et al. 1996). It has a lightweight (Mason 1996; Porter et al. 2016; Dehoux et al. 1996), medium-sized build and is morphologically similar to the Frisian horse that influenced its development, albeit slightly finer (Rousseau 2014). The head is moderately long with a straight profile while the neck is carried high and features a characteristic arch. The croup is round, and the legs are long (Rousseau 2014). The breed's mane and tail are abundant, as are the feathering on its lower legs (Rousseau 2014) (Fig. 11.28).

The Vlaamperd's coat is typically black (Mason 1996; Porter et al. 2016; Dehoux et al. 1996), although a dark bay is also accepted for mares (Rousseau 2014; SAVBS 2019). Some horses may exhibit bay-brown shading in their mane (Mason 1996; Porter et al. 2016) (Fig. 11.28).

The Vlaamperd is known for its power, cooperative nature, and elevated gaits (Rousseau 2014) (Fig. 11.28). The breed's selection is overseen nationally by the South African Stud Book and Livestock Improvement Association (SASBLIA) and is particularly supported by the Suid Afrikaanse Vlaamperdtelersgenootskap (Dehoux et al. 1996; SAVBS 2019).

Uses: The Vlaamperd is primarily used for harness (Dehoux et al. 1996), where its spectacular gaits make it highly successful in various forms of carriage competitions (Rousseau 2014; SAVBS 2019). However, it can also be ridden with success and has gained popularity in dressage (Mason 1996; Porter et al. 2016).

The Calvinia

Origin, naming, and breed distribution: The Calvinia horse was created by European settlers in South Africa's Calvinia region, using the Cape Horse, Thoroughbred, Hackney, and Cleveland Bay (Mason 1996; Porter et al. 2016; Dehoux et al. 1996), after the Great Trek. A stud farm was indeed established in the town of Calvinia (Hendricks 2007), supplying the surrounding regions with selected specimens of Cape Horses, known as "Hantam horses". It is one of the two notable horse breeds selected at that time by Boer farmers, along with the Cape Trotter (Kreuz 1958), resulting from the last influences on the local strain of the Cape Horse (Mason 1996; Porter et al. 2016).

Its extinction date is not specified (Mason 1996; Porter et al. 2016; Cothran and Van Dyk 1998), but the last representatives of this breed are likely the ancestors of the current Vlaamperd, whose breed association was established in 1983 (http://www.theequinest.com/breeds/Calvinia/).

The Calvinia horse is native to the Western Cape (Porter 2020). The fifth edition of the CAB International dictionary (2002) and the South African Department of Agriculture (2006) indicate that it is an extinct breed, as does DAD-IS,

Fig. 11.28 A South African Vlaamperd stallion. (SAVBS 2019)

which classifies it as a local extinct breed of South Africa (Porter 2020; Cothran and Van Dyk 1998). Similarly, the study conducted by Khadka (2010) describes the Calvinia as an extinct local African horse breed (Khadka 2010). An English-language website states in 2019 that the breed would be "incredibly rare," if it still exists (http://www.theequinest.com/breeds/Calvinia/).

Description and morphology: It exhibits the characteristics of a carriage driving horse. The morphology is light (Dehoux et al. 1996). The coat is generally bay or black. The temperament is reputed to be good and docile.

Uses: There is no specified uses of this horse breed in the DAD-IS database (Dehoux et al. 1996). The constituent crosses and subsequent selection of the Vlaamperd suggest that it was used for carriage driving.

11.2.2.2 Introduced (Exotic) Horse Breeds

Arabian Horse
Due to a performance-based definition rather than one based on morphology, the term "Arabian horse" has been confusing, especially in North Africa. All lightweight horses are given the name "Arabian horse," including Barb horses and Oriental horses. The confusion did not entirely disapper with the creation of studbooks specific to the Arabian horse. It was not until its introduction to racing in the 1970s that a formal distinction was established.

The Arabian horse is present in North Africa, in all countries bordering the Mediterranean Sea, as well as in Mauritania, Senegal, Mali to the west, and Sudan to the east. It is also found in the southern regions of South Africa, Lesotho, Namibia, and Botswana. It has a significant presence in other countries, such as Chad. Its production is overseen by the World Arabian Horse Organization (WAHO), an organization established in 1970, responsible for validating studbooks. The primary aims of Arabian horse breeding are the improvement of local breeds, flat racing, and the enhancement of endurance and walking. Horse shows are held in Egypt, Morocco, and various countries in the south. Some shows are reserved exclusively for the Straight Egyptian type. The most recent of these, the Straight Egyptian World Championship Arabian Horse Show, was held on October 16th–17th, 2021, in Milan. Arabian horses are not used for work.

The Arabian horse can be primarily categorized into two types: the *Saglawi* line with a concave profile, long and curved neck, and usually a bay coat; and the *Muniqi* type, elongated with a straight profile, long back and neck, long legs, and angular morphology. Most are bred on mixed farms with only a few head each. Specialized royal or private stud farms exist in Morocco. In Tunisia and Algeria, only national stud farms like El Batar or Chaouchaoua, with a production of approximately thirty head per year, engage in Arabian horse breeding (Fig. 11.29). Egypt is the only country with studs exclusively specialized in Arabian horses. One of these is EzZahraa Stud Farm with a herd of about one hundred mares. It breeds the Egyptian type of Arabian horses and has become the world's largest Arabian horse stud.

The proven presence of the Arabian horse in Africa dates back to the beginning of the fourteenth century when it was introduced into Egypt by an admirer, Sultan Malik al-Nâsir ibn Sayf al-DînQalawûn. In other countries, its introduction did not truly occur until the late nineteenth century. Horses established in Tunisia by the Phoenicians, which were able to breed, could not be assimilated into the pure Arab race due to its nonexistence at the time. Horses introduced to the Maghreb in the eighth to eleventh centuries through Arab conquest were in small numbers and were met with resistance from local populations who were deeply attached to their own lineages. Consequently, these introductions did not lead to any specific breeding efforts. Scholars Aumassip-Kadri and Souidi agree that, in the absence of such breeding, any introduction would have necessarily been diluted by local stock.

In Egypt, the state has been heavily involved in the breeding of Arabian horses since the seventeenth century. The number of horses was significantly increased at the beginning of the nineteenth century by Mehemed Ali, a devoted admirer of

Fig. 11.29 Algerian native Arabian stallion "SAKHI", senior champion stallion of World Arabian Horse Organization (WAHO) 2014 (http://www.waho.org/algeria)

the breed, particularly after receiving a strong tribe herd of horses from the Nedj during the conquest of that region. His grandson, Abbas I Hilmi, further enriched these numbers with the best Bedouin horses from the Arabian Peninsula. Although the farms were disbanded after his death, proportions were acquired by Ali Pasha Sherif and Lady Blunt, the granddaughter of Lord Byron. They formed the foundation of the English Crabbet stud, and were also used to establish a stud at Sheikh-Obeyd near Cairo. Through astute selection over a few generations, Egyptian breeders created and standardized the "Egyptian Arab" type, known for its dry conformation and distinct expressions, particularly in the head. Presently, breeding efforts are concentrated in northern Egypt, where numerous studs are dedicated to this type, including EzZahraa, founded in 1908 by the Royal Agricultural Society under the chairmanship of Kamel Eddine Hussein. Egyptian breeding activities mainly occur in Lower Egypt and the Nile delta, especially its eastern region and around Cairo. A studbook specifically for the Arabian horse was inaugurated in 1969. It is worth noting that Tahaoui's horse population in Egypt is considered sufficiently purebred Arabian, though not officially recognized.

In the Maghreb countries, the introduction of the Arabian horse occurred in the late nineteenth century. Its presence was driven by the French desire to establish a local horse breed and improve the Barb, which they had access to but which was of low quality. This effort is closely linked to the Pompadour stud farm in southwestern France, where Napoleon initiated the breeding of Arabian horses following his expedition to Egypt in 1798–1801. Little data is available for Libya, where an Arab studbook was established in 1959. Parentage verification was carried out in Morocco by LAGEV in 2011 and 500 Arabian horses were registered.

After years of sporadic and unsystematic breeding, organized efforts were eventually implemented. Currently, much of the workforce has been imported from Tunisia and Algeria and is used in flat racing and endurance events. Arabian horse breeding began in Tunisia in 1881, with the importation of the first four mares from the Pompadour mare farm to the Sidi Thabet breeding establishment. Descendants of these mares were used to create various breeds, resulting in a population of approximately 5000 individuals by 2015. Some stallions introduced from the East and the South-West of France have given rise to the notions of "eastern" and "western" types, with the latter being larger and displaying superior racing performance. These types were separated in 1950 by a beylical decree that established a state monopoly on their importation. Subsequently, national stud farms initiated substantial breeding reforms.

Under the guidance of Dr. Hosny Khaled, a stringent control system was established for mating and births, requiring them to occur within the enclosures of various national stud farm establishments. In order to prevent inbreeding, Egyptian type Arabian horses were imported in 1970 and 1980, despite their initially modest performance, resulting in the development of a model with international courier qualities.

Under pressure from breeders, several French stallions were imported in 1999, which has led to a dilution of the Tunisian breed's distinctiveness. The El Batar stud now has the responsibility of preserving the breed, managed by the FNARC, established in 1988 as a successor to the original national stud farms founded in 1913. The breeding activities of Arabian horses are primarily concentrated in the northern regions, including Tunis, Sidi Bouzid, Kairouan, and Sfax, as well as the South. A dedicated Arabian horse studbook has been in place since 1956. Arabian horse products are used in crossbreeding with the Barb, as well as in flat and endurance races.

The Arabian horse in Algeria was mainly used as a saddle horse. The first introductions date back to the mid-nineteenth century. During this period, horses from Pompadour and Syria were brough in and primarily bred in the mare farm of Tiaret. This effort led to the creation of a lineage known as Tiaret- line Asil, which held the unique distinction of not having a male line for around 70 years, and it produced horses of remarkable quality. To prevent inbreeding, stallions were imported from Europe (Sweden and England) in 1978. However, by 1990, importation was opened to the private sector, resulting in a disorderly proliferation of parentage, which posed a threat to the unique characteristics of the lineage. In Algeria, Arabian horses are predominantly used in flat races, managed by the Horse Racing and Pari-Mutuel Society (SCHPM), and in endurance races overseen by the Equestrian Federation (FE). They also participate in various competitions, including the Tiaret Horse Show, and engage in traditional activities. The current population is estimated to be around 1000 individuals.

Breeding primarily takes place in the Tiaret and Sétif regions. An Algerian studbook specific to the Arabian horse, established in 1932, was adopted in 1982 and is managed by ONDEEC. This organization issues certificate of approval and authorization for breeding stallions, registers each birth, and provides descriptive booklets and registration cards.

In Morocco, the introduction of the Arabian horse took place at the beginning of the twentieth century, leading to the establishment of the Temara and Meknes mare farms in 1906. An Arab stud-book was created in 1982. During the 1980s and 1990s, several stud farms were established. Morocco now boasts 150 traditional Arabian Thoroughbred farms, albeit with a small number of horses. Since 1985, the creation of the Royal Moroccan Association of Thoroughbred Horse Breeders (ARMECPSA) by Hassan II has promoted the breed by allowing crossings only with the most renowned strains.In Mauritania, the Arabian horse is known as a local variant, the Hodh, primarily used as a saddle horse. In Mali, it exists in the Adrar des Iforas region in a local variant. In Senegal, private breeders introduced the Arabian horse around 1950 to enhance local breeds. Additionally, the *"Centre de Recherches Zootechniques de Dahra"* contributed to its introduction from 1959 to 1968. While these populations are heavily influenced by the Arabian horse, they are not officially recognized by WAHO or any studbook.

In Sudan, the introduction of the Arabian horse took place in the twentieth century, driven by a desire to improve local breeds. It played a role in the creation of the Tawleed breed, which has become a local racehorse.In the Southern African countries, the Arabian horse was not introduced until 1902, and its breeding only gained momentum since 1945. This development began with the importation of stallions from England, the USA, and Egypt. A breed introduced in the seventeenth century disappeared due to crossbreeding with Barb, Thoroughbreds, and Persians, resulting in the Cape Horse, which faced disappearance in harsh climates and evolved into the Basuto pony. Breeding of the Arabian horse in these countries is exclusively private, serving purposes such as racing and crossbreeding with Thoroughbreds and Boerperds.

Thoroughbred Horse

The fascination with horses also reached England, where kings, starting from the sixteenth century, began selecting high-speed horses for racing. To achieve this, they crossbred their local horses with other breeds such as the Barb or the Spanish. After two centuries of experimenting with the local stables, British breeders turned to the horses that English travelers had encountered during their journeys to the East in the late seventeenth and early eighteenth centuries, namely, Arabian horses—known for their speed and robustness. Three stallions gave rise to the entire breeding stock of the future Thoroughbred horses: the Byerley Turk, which originated the lineage of the legendary Eclipse; the Darley Arabian, who fathered the first great racehorse, Flying Childers; and the Godolphin Arabian, ancestor of the famous Matchem and Eclipse.

Exclusively reserved for flat races, Thoroughbreds preceded Arabian horses in African countries. In the countries of Southern Africa, they were imported in large numbers at the end of the eighteenth century and the beginning of the nineteenth century, following the establishment of races. South Africa is the second-largest breeding country in the world, boasting approximately 300 registered active breeders. Around the turn of the millennium, South African Thoroughbred horses managed to win championship races in Dubai, Hong Kong, the USA, Singapore, Australia, and Bahrain, as well as top races in Britain and France. South African breeders have long sought international bloodlines to enhance the local breed.

The Thoroughbred is bred in northern Libya at Al Shaab Stud, the largest Thoroughbred stud in the country. In Algeria, the Laghouat region produces the finest specimens. In Tunisia, breeding is concentrated in the northern part of the country, specifically in the governorates of Ariana, Manouba, and Bizerte. The English Thoroughbred made its appearance in stud farms in Morocco from 1968, specifically in El Jadida. By 1973, there were 160 thoroughbreds racing, out of which 90 were born and bred in Morocco. In these countries, the Anglo-Arab (Fig. 10.12), or more generally, the Anglo-Arab-Barb obtained by crossing the English Thoroughbred with the Arab or Arab-Barb, made it possible to obtain a faster horse than the Arab-Barb. Races serve as its main outlet.

SelleFrançais

The SelleFrançais stands as one of the world's most esteemed breeds of horses for show jumping. The SelleFrançais Studbook secured second place in the 2020 World Breeding Federation for Sport Horses (WBFSH) ranking for show jumping and eventing. A blend of various sport horse breeds, including the Anglo-Normand, TrotteurFrançais, Thoroughbred, and Anglo-Arab, contributes to the SelleFrançais' makeup (Fig. 11.30). It achieved official breed status in 1958.

The Selle Français Studbook orchestrated several endeavors that bolstered breeding abroad. For example, organizing an international SelleFrançais championship in Algeria marked a significant show jumping event on the African continent. Since 2011, the Selle Français has been bred by the Hocine El Mansour stud farm in Mostaganem in western Algeria. It boasts a SelleFrançais breeding core comprising broodmares and performers from SF confirmed in France, notably from the Forêts de Fabrice in Paris. After assessing initial generations, Jean Louis Lenoury, President of SF Judges, affirmed that all signs indicate this breed's excellent adaptation to the country's climatic conditions.

The Moroccan Saddle, designed for equestrian sport and particularly show jumping, stands at a withers height of 1.40–1.72 m. It was developed through crosses between the Hanoverian local breed, for size. The Selle Français for mass, and the Thoroughbred for energy. Presently, this studbook remains open and engaged in the selection process, particularly in Sidi Berni, embracing substantial crosses. Roughly 100 births take place annually across three principal farms situated in Rabat and Agadir, Morocco. As an off-

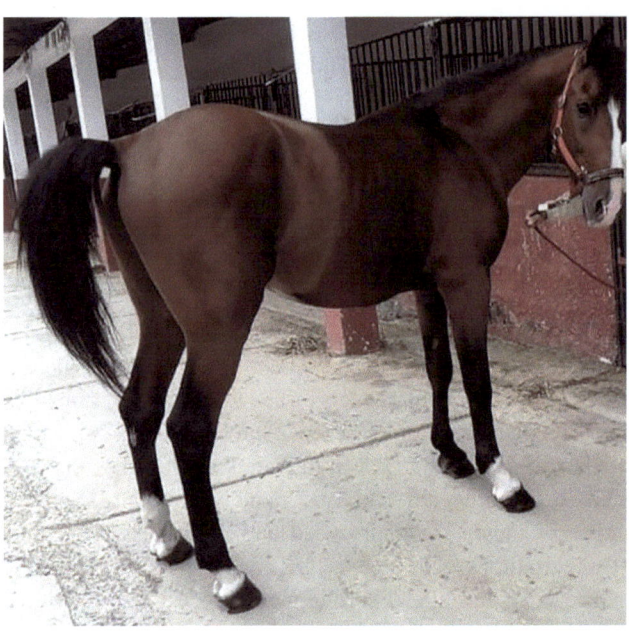

Fig. 11.30 Anglo-Arab horse. (Image credit: Personal image)

shoot of the French studbook, Morocco plays a vital role in approving the breed's studbook.

An 8-year-old steed, "Mister d'Eclipse," procured as a foal by Jean-Charles Grand Montagne and held in high regard by him, has recently come under the Moroccan saddle. The horse is intended for OuaddarAbdelkebir, a talented Moroccan athlete who qualified for the Tokyo Olympics with "Mister d'Eclipse" and was triumphant in the foreign horse final at Fontainebleau at the age of 6. Notably, the SelleFrançaisImperialed'Euskadi, born in Bidart under the care of Marc Hiriart, found its place with Mohammed VI and is set to join the Moroccan national riding team soon (Fourcade 2021).

Trotteur Français

Initially, the TrotteurFrançais breed emerged from a crossing between the Anglo-Normand and other trotting breeds. This breed primarily engages in sulky races but also in mounted trotting races and various other equestrian disciplines. The TrotteurFrançais was meticulously cultivated for harness racing in nineteenth-century in Normandy. Its foundation stock comprised predominantly of Anglo-Normand, Standard bred, Norfolk Trotter, Thoroughbred, and Hackney horses. Despite the Standard bred's lateral gait, the French Trotter retained its distinctive diagonal trot. Officially recognized in 1922, the breed maintained an open studbook until 1937. TrotteurFrançais horses typically exhibit a longer maturation period but often enjoy extended racing careers (Szathmary 2021).

Algeria and South Africa stand as collaborative partners of the *Société d'Encouragement à l'Elevage du Cheval Français* (SECF), widely known as the *Société du Cheval Français*. In Algeria, 38 annual French trotting horse races take place. There is a dedicated hippodrome and a total of 78 training horses owned by 30 different proprietors. The status of the breed in South Africa remains undisclosed. Morocco introduced the half-blood trotter during its protectorate in 1974, yet only about thirty half-blood trotting horses participated in races, predominantly owned by European breeders across the 75 horse farms. The trotter failed to captivate the attention of Moroccan breeders or studs and gradually faded from Morocco's equestrian landscape (Barakat 1974).

Percheron

The Percheron horse breed emerged in the aftermath of confrontations with the Muslim army between Tours and Poitiers in central France. Historical accounts suggest that the creation of the Percheron involved the integration of horses from the Arabian Desert among those used by the French. These Arabian horses were captured following the successful repulsion of the invasion by Muslims and other eastern nations. In subsequent breeding efforts, another Arab influence likely played a role. Notably, two Arabian stallions, Le Pin Godolphin and Gallipoly, were reportedly utilized to aid the breed's recovery. The Arabian horse's pivotal role in shaping the Percheron's development is noteworthy.

Percheron horses arrived in South Africa shortly before the Anglo-Boer War, and were used in the vineyards and wheat farms of the Western Cape. Their robustness and manageability led to their use in urban areas for heavy hauling tasks, surpassing the strength of oxen. The initial entries into the South African Studbook consisted of two French-imported stallions named Hammer and Jones. Subsequent imports, often orchestrated by the government, continued the breed's expansion, with breeding taking place in state-owned farms and educational institutions such as Potchefstroom, Grootfontein, and Elsenburg near Stellenbosch. A noteworthy endeavor included the crossbreeding of Percheron horses with Spanish donkey stallions at the Onderstepoort Veterinary Institute, where they were employed for pulling ploughing, planting, and harvesting alfalfa until approximately 1970.

In contemporary times, a multitude of private breeders have emerged, elevating the breed's status. The Percheron breed now enjoys recognition at equestrian events such as the Horse Show and other major exhibitions across Southern African countries (Mac 2015).

Quarter Horse

The Quarter Horse, originally known as the American Quarter Running Horse, descended from the first Spanish horses, with Arab blood, and then an English strain. It is a horse with an explosive character in middle-distance races, and is also known for its gentle and calm temperament as well as its versatility. The South African Quarter Horse Association (SAQHA) was started in the 1970s with the importation of American Quarter Horses by Gary Player. More anecdotally, quarter horses were imported as part of a western riding project.

Friesian Horse

The Friesian Horse breed originates from Friesland in the Netherlands (Douma 1994). The inception of the Friesian breed's lineage can be traced back to the first herd book for Friesians, namely "Het Friesch Paarden Stamboek" (FPS), which was established in 1879 in Roordahuizum, Friesland with the aim of improving the breed through pedigree recording and the implementation of stud registers (Douma 1994). During 1906, two Friesian stallions were imported to South Africa, followed by several imports of mares and stallions from the Netherlands. In 1980, the Friesian Horse Breeders' Society of South Africa was established, consisting of 12 members (Campher et al. 1998) and an additional breeders' organization was formed in 1989, namely, the Friesian Horse Studbook of Southern Africa (FPSSA).

Friesian horses became popular in South Africa for recreation but were also used in crossbreeding programs with Hackneys, Boerperds, and other breeds in an effort to improve draught horses (Campher et al. 1998). In South Africa, Friesian horses are currently selected mostly on the grounds of their aesthetic value rather than for functional efficiency. Breeders make substantial financial investments, and therefore, there is sufficient reason to evaluate the genetic potential as well as traits for inclusion in selection programs.

11.2.2.3 Other Variants

There are many other variants with over forty other names, some of which may encompass the same population; the distinctions lie primarily in the breeding grounds and the languages spoken by the breeders. Currently, there are no clues regarding their origins, except for the Somali, an esteemed pony believed to have descended from

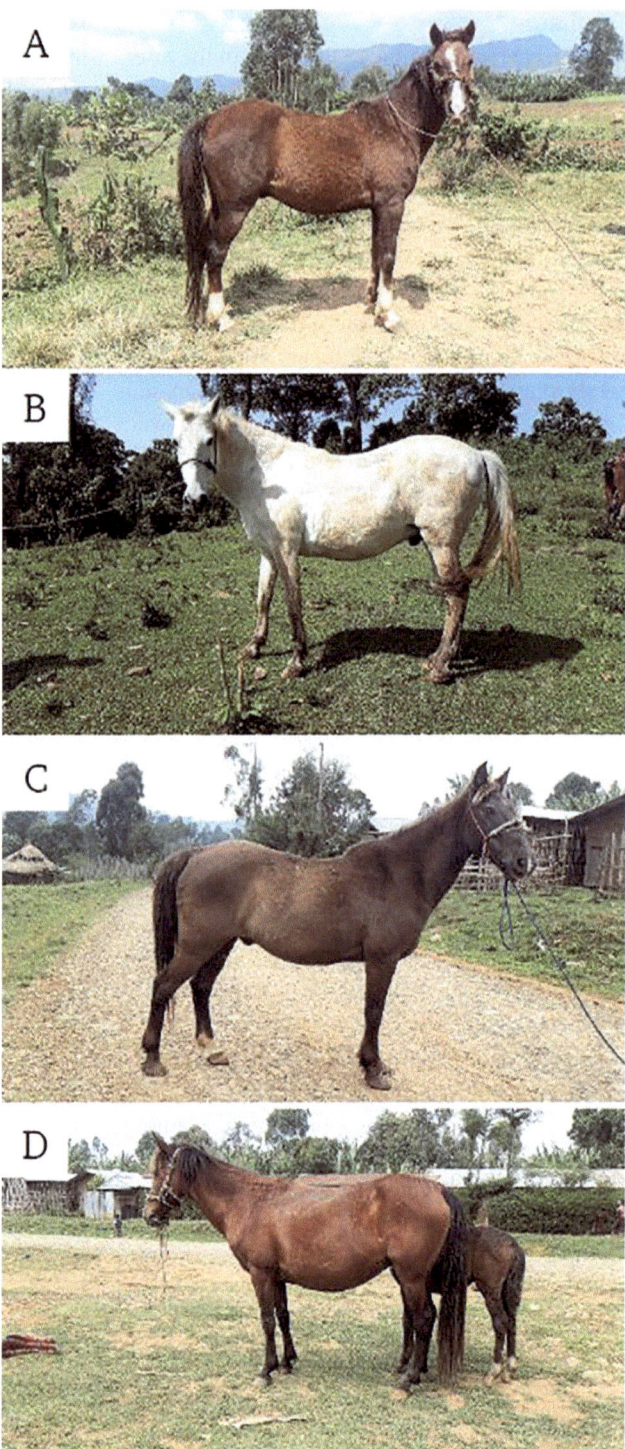

Fig. 11.31 Horse populations from southwestern Ethiopia (Mustefa et al. 2022). (**a**) Telo stallion; (**b**) Masha stallion; (**c**) Gesha stallion; (**d**) Gesha mare (Mustefa et al. 2022)

the Arabs. However, for others like the Berma or Materi found in West Africa (Fig. 11.31), there is little information about their origins. The same lack of origin knowledge applies to breeds such as Gala, Baladi, Egyptian Horse, Dombi, Djerma and Songhoï from Mali, Musey from Cameroon and Chad, Sulebawa from Nigeria and Minianka pony from Burkina Faso.

In addition, there are a few new breeds or foreign breeds of unknown official descent. The Percheron was found in Burkina Faso. The Breton, a sturdy draft horse, found its way into North Africa, particularly Algeria, to aid in plowing and heavy hauling tasks. They were introduced in significant numbers in the nineteenth century and appear to have no official descendants. In Morocco, they were also used for crossbreeding with the Barb or Arab-Barb, a medium-weight horse not well-suited for strenuous labor. By 1974, only the *Haras de Meknes* in Morocco and a few private individuals continued to maintain draft stallions. Their primary focus was on producing females, which served as the foundation for mule breeding.

Morocco created the Moroccan Saddle with the opening of a studbook in 1985. It is a sport horse, directed toward jumping. It is the product of an initial cross between local Arab-Barb with Anglo-mares. Arab and Anglo-Arab-Barb and Hanoverian stallions, the mares from this first cross were recrossed with Thoroughbred and French saddle stallions and then selected on test. South Africa has recently created two breeds of sport horses. These are the South African Saddle Horse (SAS) and the South African Warm Blood (SAWB). A pet horse called the South African Miniature was derived from a cross between the Shetland and small Arab horses.

Among the African horse populations, there are some that have developed a tolerance to trypanosomiasis, such as the Koto-koli in West Africa, Minianka (Bobo) in Burkina Faso, Koniankan and Minianka in Mali, Musey in Cameroon and Chad, and Logone in Chad and Somalia. These populations are small in size, which raises the question of a relationship with the parasite. The small sizes of these populations are probably linked to their challenging environments.

These remarks make it possible to hypothesize that the West African ponies originate from a reduction or maintenance of the size of the first horses introduced into these regions. They probably measured from 1.10 to 1.30 m at the withers. They came from a slide of the horse from north to south from the second millennium BC, when a domestic horse of the Barb type occupied the central Sahara. In the light of history, it appears that the equine population of Africa rests on a background of Barb and Dongola horses (while waiting to get to know the latter better). Everything suggests that there are different names for the same horse breed arising because they are raised in different regions, or because they belong to "tribes" with different expressions. It is also possible that their isolation in different ecological niches for quite a long time led them to evolve specific characteristics. Behind these names probably exists a wide variety of adaptations that can only be confirmed using molecular approaches.

Over time, stud farms shifted away from mule production to the Barb and Arab-Barb horses, which were better adapted to the natural conditions of the country. The use of French draft horses faced challenges in acclimatizing to local conditions in Algeria. At present, draft horses are no longer present in Morocco (Husser 2017). In Algeria, rare Breton specimens can be found in the western regions of the country. The Arab-Barb horses from Tchad are slightly smaller than the North African Arab-Barb: mares are 1.42 m at the withers on average, while the stallions are 1.47 m. Other horse breeds, including the Anglo-Hispano-Arab, are bred and employed by the military. Argentinian horses, known for their prowess in the sport of polo (Polo Argentino), are also utilized for polo games (Husser 2017).

11.2.3 Horse Production Systems in Africa

Breeding stands as the cornerstone for both the quantitative and qualitative advancement of African equine herds. Its diversity mirrors the horses' utilization and hinges on available natural and human resources dedicated to animal management. The breeding landscape is characterized by the coexistence of a modern system alongside a dominant traditional system. Within rural farming contexts, almost all horses are reared through the traditional system. In these regions, the equine population's size is closely linked to the agricultural land area cultivated by the farmer. Predominantly, local breeds are the most utilized. Conversely, modern production system finds their foothold in urban and peri-urban areas, where a fusion of local and exotic breeds is kept. Notably, exotic breeds are most generally kept in stud farms. Numerous African nations are grappling with burgeoning demographic growth and the encroaching desertification witnessed in recent years. Consequently, this context has intensified challenges, including disputes over cropland and grazing territories. Thus, there is an increasing recognition of the need to transition gradually from traditional nomadic farming practices toward sedentary or modern farming approaches.

11.2.3.1 Traditional Production System

The horse holds a special place within the African family, being considered an integral part of the family unit. Unlike the donkey, which may receive less care, the horse is often granted preferential treatment. It is not merely a luxury animal but rather a creature of importance. To prevent conflicts between stallions, they are often housed individually in stables composed of a wooden rod palisade topped with a straw roof, allowing proper ventilation (Ndiaye 1978). Traditional stables typically feature a fine sand floor that is frequently replaced. Alternatively, stallions may be tethered to trees, utilizing branches for shade during the day. In such cases, the ground is sandy and regularly cleaned, although non-renewable.

Mares in the traditional system roam freely during the dry season and graze in pastures when it rains and are tethered to stakes or trees in the evenings. They are provided with water and feed, commonly consisting of straw and agricultural crop residues. Stallions are introduced to the pasture only at the start of the breeding season. The primary feed for horses in this system is of plant origin, including fodder, grains, crop residues, and by-products from the food industry. Concentrated feed is rarely employed in this system. Mares are introduced to stallions for mating only during estrus. Foals are usually born at the beginning of the mating season or in the middle of winter (spring in North Africa) (Bazarusanga 1995). During the free grazing period, horses maintain good condition, but they tend to lose weight in the dry season due to a decline in natural pasture quality (Djimadoum 1994).

Traditional herding revolves around free grazing, encompassing two main forms: nomadic and sedentary herding. In nomadic herding, horses, sometimes alongside other livestock, are moved seasonally across expansive regions to locate natural pasture. Sedentary herding, on the other hand, involves raising a smaller number of horses within limited forage areas like fallow lands or village fields. This approach allows for more intensive land use, facilitates livestock management, and enhances product utilization, given the proximity of the livestock to villages. In the traditional rural setting, few animals are housed. Typically, animals are tethered near trees, utilizing the foliage as makeshift roofing. In some cases, they are tethered to open stakes without any cover. Available shelters are basic structures, often constructed with thatched roofs supported by four poles. These shelters provide minimal protection from wind, rain, and dust. The floors are often bare, and bedding is usually absent. Regular cleaning of shelters and makeshift structures is practiced, and the collected droppings are utilized as organic fertilizer in fields, particularly facilitated by ownership of carts (Diouf 1997).

11.2.3.2 Modern Breeding System

The modern breeding system is practiced in urban and peri-urban areas, where horses are accommodated in contemporary stables. This system predominantly involves improved breeds of horses, demanding heightened attention from breeders. Accommodation design often hinges on the specific horse breeds being raised. Modern stables, constructed entirely of concrete walls on suitable grounds, offer optimal ventilation and adequate shade for the horses' well-being (Amiot

1982). Strategic orientation of openings minimizes exposure to winds (Marcenac 1969). These stables might feature automatic feeders and drinkers to enhance efficiency and prevent moisture in the bedding. Straw can be employed to safeguard against bed sores, with stable upkeep undertaken by grooms. Within this breeding system, two types of stables exist: common stables and individual stables. Common stables, situated in single or multiple buildings, may have horses placed in a single row facing the wall and opposite the box entrance, or in two rows placed rump to rump. In stables common to multiple buildings, stalls are replaced by interconnected boxes with contiguous bays. Individual stables, also referred to as breeding stables or boxes, are frequently used for sport horses to ensure improved rest (Ndiaye 1978; Marcenac 1969; Ménard et al. 2002).

Irrespective of stable type or horse breed, high-quality nutrition is essential to meet maintenance and workload demands. While extensive research has been conducted in the field, limited attention has been directed toward the feeding and rationing of African horse breeds in various contexts. This has led to certain differences among breeders concerning feeding practices. This modern breeding method is applied within stud farms and private farms. Stud farms serve as hubs to optimize logistics, pool resources and personnel, and enhance service quality. They facilitate the availability of breeding stations, providing support and services to breeders. Private horse farms range in size, spanning from family-owned farms to large-scale operations. Common challenges in horse breeding, regardless of the operational system, include issues such as animal theft and diseases arising from inadequate breeding conditions (infections, deficiencies, allergies, etc.). Additionally, challenging environmental conditions characterized by high temperatures, humidity, and drought pose significant difficulties. Low socioeconomic status among breeders and limited technical expertise further hinder the progress of equine breeding in Africa. The transmission of contagious and transboundary diseases also remains a significant concern.

11.2.3.3 Horse Production on Multispecies Farms

Horses have shown an ability to adapt when raised with other animal species. Multispecies grazing is a practice observed across various parts of Africa, where horses coexist with different species (camel, sheep, and goat) in a herd (Gherissi and Gaouar 2022; Chergui et al. 2023). The primary consideration here is maintaining uniform pastures. However, there are challenges in managing and optimizing this system, largely because of insufficient comprehension of grazing principles and animal behavior. Multispecies grazing can take two forms: the preservation of animal diversity on pastures either over time or space. For instance, horses may share grazing grounds with goats or cattle, or even partake in a rotational grazing scheme with other species. This necessitates understanding the daily forage requirements of each species, the projected pasture forage yield, and the variations in forage preferences among the different co-sharing species. A comprehensive scientific understanding of animal diversity within African multi-species farms employing horses for various functions is deficient (Dumont et al. 2020; Martin et al. 2020).

Proper horse husbandry demands a minimum daily ration of 50 kg of fodder, potentially supplemented by barley or oats at a ratio of 0.5 kg of grain per 1 kg of fodder. It is imperative to note that exceeding a substitution rate of 60% could lead to colic due to the horse's elongated intestine. Additionally, around twenty liters of clean water are required each day. In a pasture setting, horses require an average of one hectare per horse, subdivided into plots to facilitate vegetation rejuvenation. The presence of cattle can enhance this regeneration process as their consumption patterns differ from those of horses, and their waste contributes to soil enrichment.

The scientific community has recognized certain advantages regarding the environmental sustainability of horse-related breeding. Horses play a role in upholding pasture plant diversity (Loucougaray et al. 2004) and controlling invasive species (Ménard et al. 2002; Loucougaray et al. 2004; Loiseau and Martin-Rosset 1988). However, a significant challenge associated with multispecies grazing lies in the transmission of nematodes between different species (Dumont et al. 2020; Forteau et al. 2020). Due to these factors, it becomes imperative to conduct comprehensive studies across diverse African regions to ascertain optimal strategies for managing horses within breeding farms alongside various animal species. Such strategies will depend notably on the specific resources available in each country.

11.2.3.4 The Feral (Wild) Horses of Africa

One of the most enigmatic creatures on the African continent is the wild horse of the Namib Desert (Fig. 11.32). For over a century, it has held the status of an exotic species, yet its origins remain shrouded in mystery, giving rise to various theories. Some suggest that these horses may have arrived alongside diamond prospectors, while others propose the intriguing notion that a ship bound for Australia became marooned, inadvertently releasing a cohort of Thoroughbreds (Pinguet 2020). However, the most plausible explanation revolves around the idea that the Namib Desert horse descended from South African military horses that interbred with the Shagya Arabian lineage (Marchant 2018).

In stark contrast, the newly discovered Ethiopian Kundido Wild Horses derive their name from the mountainous terrain where they roam freely (Fig. 11.32). Due to the absence of documented records concerning this equine population, their historical background remains veiled in uncertainty. Insights gathered from focused discussions with local elders, accompanied by a collection of anecdotal narratives (Kefena et al. 2012) shed light on the notion that these horses wandered the expanse of Mount Kundido for an undisclosed time span. These elders within the region propose an intriguing connection between the emergence of Kundido horses in their feral state and the historically significant Ethiopian Muslim-Christian conflict spanning from 1528 to 1560. During this epoch, Ahmed Gragn, the leader of the Muslim forces, clashed with AtseLebnaDengel, commanding the Christian army. It is believed that Kundido mountain, endowed with strategic importance, served as a pivotal military outpost, potentially allowing one of the warring chiefs to leverage its vantage point to impede the advancement of their adversary. Presently, scant substantial evidence exists to substantiate these intriguing conjectures (Kefena et al. 2012) (Fig. 11.33).

Fig. 11.32 Kundido feral horses from Ethiopia. (Kefena et al. 2012)

Fig. 11.33 Namibian wild horses. (Pütz and Schlottmann 2020)

Borana horses in Ethiopia constitute another intriguing wild population, primarily concentrated around the town of Megga within the Borana region. These equines graze across the pastures of southern Ethiopia and are commonly utilized by esteemed Borana Oromo breeders, standing as quintessential representatives of Ethiopian lowland horses. Despite the challenges posed by recurrent severe droughts in the area, these horses exhibit a resilient and robust breed that has adeptly acclimatized to the water scarcity and demanding conditions prevalent in southern Ethiopia (Kefena et al. 2012).

Moving eastward to the Somali lowlands of eastern Ethiopia, we encounter the Somali or Wilwal horses. Among local pastoralists, they are also referred to as "Dirdaa'u" or "Wilwal" horses. Another distinct population of horses resides exclusively in the wild within a region called Aware, positioned on the border between Somalia and Ethiopia. Regrettably, a lack of documented origins hinders our understanding of the Wilwal horses' heritage. Local informants speculate that they may have migrated into Ethiopia from the Nugal Valley in Somalia. Wilwal horses stand out for their robust physique, sleek form, and notable elegance, often characterized by their impressive height (Kefena et al. 2012).

While there exist substantial tracts of suitable habitat for *E. ferus* in sub-Saharan Africa, Australia, and New Zealand, the prospects for rewilding this species in these regions are less evident. This is due to the fact that *E. ferus* is not indigenous to any of these areas (Naundrup and Svenning 2015). The presence of ecologically and socio-behaviorally similar species, such as zebras (*E. quagga, E. zebra, and E. grevyi*), likely plays a role in excluding *E. ferus* from sub-Saharan Africa. These zebras' striped skin provides a defense against tsetse flies, potentially minimizing the risk of trypanosomiasis, a disease transmitted by these flies (Caro et al. 2014).

Notably rugged and adaptive, these wild horses have demonstrated a remarkable ability to thrive in challenging terrain and environments that could prove inhospitable for numerous other species. During droughts, they can endure up to 30 h without water in the summer and up to 72 h in the winter. Extensive research has revealed that over the span of a century living in these arid conditions, the Namib Desert horse population has evolved physiological mechanisms that significantly enhance their water conservation capabilities (Marchant 2018; Hall et al. 2018). In the late 1990s, a severe drought led to distressing scenes of emaciated and ailing horses, prompting public outcry and the emergence of private initiatives, notably the Namibia Wild Horses Foundation (NWHF). These efforts exerted pressure on the Ministry of Environment and Tourism,

Fig. 11.34 Feeding of the horses with hay. (Pütz and Schlottmann 2020)

eventually leading to the legalization of hay feeding for horses during drought periods (Pütz and Schlottmann 2020) (Fig. 11.34).

11.3 Structures and Organizations of the Equine Sector in Africa

11.3.1 Regulations and Legal Bases of the Horse Industry

Horse husbandry in Africa boasts a rich historical legacy and tends to flourish within communities that hold horse riding in high esteem. While a handful of countries, Tunisia for instance, have established and closely monitored stud farms, particularly for breeding Arabian horses intended for horse racing, the broader regulatory landscape remains significantly under-documented. Regulations primarily revolve around the procurement and ownership of horses, with less focus on their management. The multifaceted uses of horses are generally ingrained in traditional and religious customs.

Hippophagia, the consumption of horse meat, is strictly regulated and prohibited in African countries where strong Muslim beliefs prevail (Benkheira 2000). Islam and specific Christian sects, along with traditionalist groups, vehemently oppose this practice (Blench 2000). Conversely, due to societal taboos, hippophagia remains inadequately documented and regulated. In Cameroon, certain communities, particularly in urban centers, display an avid interest in horse meat (Nyock 2014), often driven by elitism. The availability of horse meat is occasionally facilitated through cross-border trade. A parallel scenario unfolds in Nigeria, where a thriving horse trade is evident, likely echoed across the West Coast of Africa (RIM 1992). Similarly, in Chad, due to limited legislation in this area, no comprehensive data is available regarding the contribution of equine meat to the butchery trade, despite its existence. This discrepancy can be attributed, in part, to the fact that horse breeding's primary objective is not meat production but rather encompasses agriculture, transportation, traditional rituals, racing, and even battle. The influence of cultural taboos further compounds this lack of documentation, exposing the species to inadequately regulated practices across most countries (Ndiaye 1978).

11.3.2 Official Equestrian Institutions and Authorities

11.3.2.1 Institutional Equestrian Authorities

Stud farms are present across various African countries, contributing to the advancement of horse breeding. In Algeria, the national Chouchaoua Stud is located in Tiaret, alongside the National Center for Artificial Insemination and Genetic Improvement (CNIAAG), both equipped with frozen semen production centers. Morocco hosts five significant studs: Meknès, Oujda, Marrakech, El Jadida, and the Bouznika studs, which features a frozen semen production center under the CNIAEB (Haras Régional de Bouznika, 2006). Tunisia is home to the El Batan and Sidi Thabet national studs. In some instances, mare yards dedicated to breeding and brood mare production are established. The fundamental role of these studs is to promote and enhance the equine sector, boosting horse numbers and elevating their value. A core focus is placed on both quantitative and qualitative development of the equine population. Premium stallion coverings are monetized, and breeding competitions are held to recognize top-performing horses, guiding breeders in qualitative progress. Additionally, studs engage in research to enhance equine behavior, health, and fertility. Genetic, infectious, and hereditary disease studies, as well as advancements in breeding methods like natural reproduction, artificial insemination, and embryo transfer, are supported by the studs.

Collaborative partnerships extend beyond studs, involving organizations and associations in each country. For instance, in Algeria, the National Office for Equine and Camel Breeding Development (ONDEEC) operates under the agricultural ministry. ONDEEC oversees users, breeders, and their associations, contributing to breed preservation, development, promotion, and studbook maintenance. It allocates a portion of the Pari Mutuel Urbain stakes to fund its activities. Morocco features the Royal Society of Encouragement of the Horse (SOREC), facilitating efficient management and acting as a link between horse breeding and racing. SOREC supervises codified horse racing societies and regional horse racing societies, housing the Consultative Committee for Races (CCC) and Pari Mutuel Urbain Marocain (PMUM).

Stallion depots and breeding stations play a pivotal role, concentrating on breeding elite individuals on stud farms and then distributing them to locations across the region for mare coverings. This gradual enhancement of the entire herd according to breed-specific criteria aims to benefit breeders and access stud farm services. These stations can host local breeding competitions, attracting breeders, public interest, and serving as training grounds. Algeria possesses two national breeding depots and a diminishing number of breeding stations, while Morocco boasts 53 breeding stations and an Equine Breeding Promotion Center.

The realm of horse racing involves organizations responsible for overseeing, organizing, and supporting horse races. South Africa features the National Horse Racing Authority and Racing Association, while Algeria operates the Horse Races and Mutual Betting Society (Decree No. 87–17 of January 13, 1987). Morocco's Consultative Committee of Races (CCC) and PMUM, along with Tunisia's Company of Horse Races (SCHT 2012), contribute to this vibrant sector.

Horses are also significant in the military domain, engaging in homeland security, equestrian sports, and parades. Notably, the Royal Armed Forces' equestrian center in Morocco and the Republican Guard's horse breeding center in Algeria underscore the military sector's investment in equines.

11.3.2.2 Socio-professional Organizations and Associations

The Equestrian Federations of African countries are associative organizations that play a crucial role in regulating various sectors, particularly in the realm of sports. Their primary objective is to promote, guide, and coordinate modern equestrian sports activities involving horses or ponies. They hold a vested interest in the development of traditional sports like Fantasia. Furthermore, these federations aim to foster an appreciation for equestrian activities among young individuals and actively participate in controlling their instruction. Their role extends to the governance and organization of equestrian competitions. Operating autonomously, these federations also fulfill their public service mission within their disciplines, adhering to both country-specific regulations and those established by the International Equestrian Federation (FEI), to which they are affiliated.

The African Equestrian Federation (FEA) was established in Cairo in 2002, prompted by the President of the Egyptian Equestrian Federation. The founding countries include Botswana, Congo, Egypt, Mauritius, Niger, Senegal, Sudan, and Zimbabwe.

The African Confederation of Equestrian Sports (ACES) stands among the five continental associations acknowledged by the FEI, the global governing body for equestrian sport. Presently, twenty African nations possess national organizations recognized by both the FEI and ACES. The Egyptian Equestrian Federation (EEF) holds the distinction of being the oldest member, joining the network in 1946, while the most recent addition is the Angolan Federation of Equestrian Sports (FADE), which joined the network in 2014.

In Morocco, the SOREC (Société Royale d'Encouragement du Cheval) plays a pivotal role by managing the entire equine sector using modern methods, characterized by efficient and transparent management. SOREC serves as the vital link between horse breeding and horse racing. It oversees codified horse racing societies and regional horse racing societies, with two key bodies based at its headquarters: the Consultative Committee of Races (CCC) and the Moroccan Pari Mutuel (PMUM).

Within the professional equine framework, there are various associations, including breed associations, that aim to spur the development of breeds and equestrian activities specific to each region. These associations propose development programs, undertake initiatives to promote particular breeds, and establish sectors devoted to those breeds. For instance, in Morocco, significant associations include the National Association of Barb Horse Breeders (ANECB), the National Association for Equestrian Tourism (ANTE), and the Moroccan Royal Association of Arab Purebred Horse Breeders (ARMECPSA).

In Algeria, the infrastructure for breed promotion units is not yet fully established, and breeders and owners of recognized purebred horses lack well-structured specific associations. The National Association of Breeders and Owners of Purebred Horses (ANEPCRP) is the most well-known to date, with accreditation from the Ministry of the Interior (accreditation 29 of 25 April 1993, Ministry of the Interior). The Adjr Foundation for the Promotion of the Horse is particularly interested in the enhancement of the Barb breed. In Algeria, there are 350 traditional equestrian associations spread across the entire national territory, responsible for promoting the Fantasia tradition. The direct involvement of breeders in these associations is highly encouraging.

In South Africa, the Thoroughbred Breeders Association (TBA) comprises 193 registered Thoroughbred studs, 509 registered breeders, and approximately 13,500 Thoroughbred horses. The South African Jockeys' Association, a national body representing licensed jockeys, is recognized by all key players in the South African racing industry and abroad (Pienaar and Du Toit 2008). The National Horse Racing Authority (NHA) in South Africa acts as the regulatory authority for the Sport of Thoroughbred horse racing. In this country, horse racing is overseen by two organizations: Phumelela Gaming and Leisure Limited and Gold Circle.

Both governmental and non-governmental organizations and associations are involved in the conservation of certain African horse breeds and the development of equestrian activities. These include the World Organization of the Barb Horse, which coordinates national breeders' associations, manages stud books, establishes breed standards, and enforces regulations related to the reproduction of the Barb breed and its derivatives. The headquarters of this organization are located in Algiers. Presently, it has eight member countries, namely Algeria, Morocco,

Tunisia, France, Germany, Belgium, Switzerland, and Luxembourg (OMCB 2012).

The World Horse Welfare Organization (WHWO) operates predominantly in South Africa, collaborating with the Cart Horse Protection Association in Cape Flat. It is in the process of establishing a network of horse welfare organizations across the Western Cape. The organization's mission is to support and enhance the horse-human partnership in all its forms (WHWO 2021). The Society for Animals and Nature Protection in Morocco, affiliated with SPANA of Great Britain, participates in various training courses for veterinarians and farriers, as well as influencing the evolution of equine sector legislation in collaboration with local authorities (SPANA 2008). The American Fondouk in Morocco provides free treatment to working animals and other animals owned by economically disadvantaged individuals. This organization collaborates with American veterinarians, Moroccan technicians, and workers, treating 50–100 animals daily, totaling over 18,000 animals annually (York and Ulrich 2009).

11.3.3 Infrastructure and Support Structures for the Horse Industry in Africa

11.3.3.1 Training Institutions for Horse Tradesmen

Cavalry training institutions are specialized in providing training for technicians in various equestrian trades. Examples of such institutions include the Royal Cavalry School of Témara in Morocco 137, the Equestrian Training Institute in Johannesburg, South Africa, and the Equestre Center d'OuledFayet in Algeria.

11.3.3.2 Research Institutions

On the African continent, there are a total of 77 university training establishments dedicated to veterinary medicine (OIE 2021). These institutions, along with those specialized in natural and life sciences, as well as scientific research establishments, are responsible for educating veterinary doctors and researchers with a focus on equine health and management. These trained professionals are involved in stud farms where they manage animal husbandry, breeding, and equine health. They also engage in scientific research covering various aspects related to the health, conservation, and enhancement of local horse breeds, as well as those intended for sports and cultural activities specific to each region.

Notable examples of these institutions include the National High Veterinary School and the National Institute of Agronomic Research. Their services primarily encompass the identification and verification of origins, genetic characterization of equine populations, prevention of hemolytic disease in newborn foals, detection of genetic and contagious diseases, and gene typing. Research is conducted at horse stud farms in collaboration with a network of partners, including veterinary faculties, universities, and horse breeding organizations. These establishments also undertake additional initiatives to enhance the health, behavior, and fertility of horses.

11.4 Use of the Horse in Africa

11.4.1 Traditional Uses of the Horse

The horse was initially domesticated and subsequently spread by humans who employed it in various activities such as agriculture, hunting, transportation, sport, warfare, and monitoring of other domestic animals. This domestication of horses played a pivotal role in the advancement of trade and the emergence of civilizations across expansive territories. Often referred to as "the noblest conquest of man," the horse holds a distinct position among all animals due to its profound impact on human history and progress. Its contributions extend to sociocultural and economic advancements.

The breeding of horses, including ponies, proves to be a highly lucrative endeavor in specific regions of the Sahel, yielding significant "invisible" income from horse marketing (Blench 2000). This activity not only underscores the economic viability of horse production but also highlights its broader implications for the local economy and society.

11.4.1.1 Agricultural Work

In sub-Saharan Africa, the majority of farmers operate mixed production systems on small plots of land, usually smaller than four hectares (Pearson and Vall 1998). For these farmers, animal energy offers an alternative to manual labor for both cultivating food crops (such as cereals in semi-arid zones and tubers and plantains in wetlands) and growing cash crops (like rice, cotton, peanuts, coffee, banana, and cocoa) (Starkey 1994). These animals also have roles in crop-related tasks and water pumping. When equipped appropriately, they rotate wheels to crush grains and power water pumps. In this manner, an individual can pump 3600 liters of water from a depth of 10 m in just 20 min (Oudman 2004). Animals are employed as draft animals to pull heavy loads and even for plowing (Cynthia 2011). The animals used in these settings are characterized by their dense bone structure and robust legs. Draft horses, in particular, are known for their sturdy build, pulling strength, and docile nature. Generally weighing between 700 and 1000 kg, they stand at a height of approximately 160–180 cm at the withers.

The advantage of horses and donkeys lies in their overall faster movement compared to work oxen. As a result, equines are selected for tasks that require lower traction forces, such as transport, sowing, and weeding on light sandy soils. For instance, in West African countries like Senegal, Mali, and Niger, direct sowing of crops like groundnuts, millet, and sorghum is widely practiced using a single donkey or a small horse. This approach allows swift seeding over large areas to enhance crop success, particularly in light sandy soils of semi-arid regions with uncertain rainfall (Pearson and Vall 1998). Despite this, animal energy is used in only about 15% of cultivated land in sub-Saharan Africa (Jahnke and Sievers 1981; FAO 1987). In Senegal, for instance, where mechanization is absent, all agricultural activities rely heavily on animal energy provided by equines (Ly et al. 1998).

Draft horses play a crucial role in both agricultural production and transportation across sub-Saharan Africa 229. According to Ndiaye (1978), an adult horse can cover a daily agricultural area of 3.5 hectares, while a pair of oxen can cover eight hectares (Ndiaye 1978). Similarly, a young horse can manage 2.5 hectares per day, whereas a pair of oxen is limited to three hectares. Furthermore, horse-drawn carts are widely used to transport agricultural inputs (such as fertilizers, pesticides, and seeds) and agricultural products (harvests, animal products), facilitating distribution and marketing.

Analyzing the profitability of a work horse, a horse with dimensions of 1.35 m at the withers and a chest circumference of 1.45 m is estimated to have a power of $D = 22.5\ C2$ based on Dechahbre's formula for European work horses. However, for the horses studied by Ndiaye (1978) in Senegal, a value of $D = 20\ C2$ was adopted, corresponding to $D = 42$ kg m/s. These horses work for a minimum of 5 h per day (Ndiaye 1978). Nonetheless, work capacity encompasses more than just the animal's power. Tolerance to heat is crucial and directly affects the work output, especially in hot climates where exertion generates additional heat stress. Work potential is a combination of pulling power, speed, and endurance. Some of these traits can be genetically enhanced (Rouvier and Wiener 2009). Equines generally work faster than cattle. When the labor force increases, both speed and work duration tend to decrease. Finally, heavier horses exhibit greater endurance during intense pulling tasks. Horses display rapid recovery and a minimal response to the Temperature Humidity Index (THI). The THI-induced increase was around 1.0–2.0 °C in donkeys, 0.5–1.8 °C in zebus, and 0.5–1.2 °C in horses. Additionally, horses exhibit high energy expenditure compared to maintenance—about 2.24–2.50 times their maintenance level (Perez et al. 1996).

Genetic studies have enabled the estimation of the heritability of traction force: $h2 = 0.20$, indicating relatively low heritability. The Mbayar horse in Senegal is considered a foundational breed for draft horse production due to its conformation and capabilities. In North Africa, the draft horse, typically of Arab-Barb breed or unrecognized origin, is utilized for agricultural tasks like plowing, water transportation, and threshing. However, other draft animals like donkeys and mules pose a threat to this horse population, as the growing ratio of mules to horses implies a shift in production toward mules (Bouazzaoui 1998).

In rural areas, the agricultural draft horse and horse-drawn traction continue to play a crucial role in trade, marketing agricultural products, transporting construction materials, and collecting household waste (Starkey 1994; Akpo 2004; Bathily et al. 2014). Notably, the use of horses in agricultural work is prevalent in developing countries, accounting for 25% of plowed areas in Africa (Bathily et al. 2014). In urban settings, a distinction is made between horse carts, mainly used for transporting materials and goods, and cabs, employed for people transportation (Bathily et al. 2014).

Additionally, organic matter resulting from animal excreta (feces and urine) mixed with litter (such as straw and fern) after composting serves as a fertilizer in agriculture (Oudman 2004).

11.4.1.2 Transportation

The utilization of animal traction for transport was initially introduced in African coastal and inland ports between the seventeenth and nineteenth centuries (Starkey 1994). As favorable social, economic, and ecological conditions emerged, this technology gradually spread across the African continent, facilitated by traders, colonizers, missionaries, and administrative authorities (Bathily et al. 2014). Despite the advent of agricultural mechanization and the increasing dominance of draft oxen, the horse continues to serve as a work companion for peasants in certain African countries. It is estimated that working equines support approximately 600 million people worldwide, especially in impoverished and marginalized communities (Valette 2015). In rural Africa, working equines stand as the primary means of transport and traction, upon which rely the production chains of various sectors such as construction, agriculture, and the transportation of people and goods (Figs. 10.15 and 10.16).

Transport horses are employed either as riding mounts or harnessed animals. In Africa, you can find elegant and lightweight horse breeds from the northern regions, as well as robust breeds from the eastern areas. Additionally, there are pony-type equines, which are essentially small horses at adult age (measuring less than 1.49 m at the withers). Ponies typically originate from sub-arid and arid regions where food might be scarce, and temperatures can be extreme (Porter 2020). There are various types of ponies that are easily managed for riding purposes. These include the Logone Pony, Middle Logone Pony, Hoho Pony, Kabbia Pony, M'Par Pony, Cayor Pony, and others. These stocky and robust animals are small (weighing around 200–250 kg), yet resilient and enduring (Nyock 2014).

Horses that are suitable for riding are relatively light and better suited for the purpose compared to those intended for harnessed transport, which require greater robustness. During the twentieth century, horses were somewhat neglected due to the industrialization and mechanization of societies in developed countries and those that followed suit. In recent times, the use of horses for transportation is declining sharply while their use in performing many other tasks is increasing.

In developing African nations, particularly in rural regions with challenging conditions, horses continue to contribute significantly in the transportation of construction materials, goods, and water in places often inaccessible to motor vehicles. They are also employed for people transportationand for the collection of household waste or water in large cities in South Tunisia (Figs. 11.35 and 11.36), Dakar, and Cairo.

11.4.1.3 Production of Mullets and Hinnys

One of the valuable contributions that horses can make in agriculture is the production of hybrid animals that exhibit enhanced robustness and resilience to the challenging conditions of dry areas. The crossbreeding of different equine species, such as horses and donkeys, results in hybrids that possess distinctive characteristics. However, these hybrids are predominantly sterile

Fig. 11.35 Plowing with horses. (Image credit: Author's Own 2022) (Kebede 2023)

Fig. 11.36 (**a**) Barb or Arab-Barb horse-drawn cart for passenger transport; (**b**) Horse-drawn cart with pneumatic wheels and elevated platform used for transporting goods in Tozeur region of Tunisia. (Charles 2022)

due to their chromosomes being precisely intermediate between those of the two parental species (Petrus 2003). Specifically, they contain 64 chromosomes from the horse, 62 from the donkey, and 63 from the mule and hinny (Trujillo et al. 1962). Despite their occasional sterility (Jussiau et al. 2013), mules and hinnys inherit the strength of horses along with the robustness and hardiness of donkeys (Petrus 2003).

Mules exhibit a blend of the morphological traits of their two parent species. They possess a size that falls between that of a donkey and a mare, while their physical characteristics resemble those of a donkey. Hinnys, similarly, resemble mules in appearance but tend to be larger. Hinnys lean toward the horse in terms of conformation and are capable of consistent daily work, contributing to the improvement of living conditions for many individuals in rural areas. On the contrary, quantifying the productivity of the hinnys, being a nonfood agricultural species, is a more complex endeavor. These hybrid equine species represent a fusion of the advantageous traits from their parent species, creating animals that are uniquely suited to particular tasks and environments, such as the demanding conditions of arid regions.

11.4.1.4 Hunting

To provide a sense of the endurance and swiftness of Sahelian saddle horses, it is essential to recall instances from colonial times when references document successful gazelle and antelope hunting on horseback. During these hunting expeditions, hunters managed to capture gazelles and antelopes due to the exceptional stamina of their horses. This echoes historical traditions seen in earlier centuries with breeds like the Arabian horse in the Middle East and the Barb horse in North Africa. Consequently, Sahelian saddle horses stand on an equal footing with renowned historical breeds. It is highly likely that the last deserts where riders on horseback pursued and caught gazelles and antelopes were not in the Middle East or North Africa, but in the Sahel region (Belin de Launay 1877; Gillet 1964). This showcases the impressive capabilities of Sahelian saddle horses, highlighting their endurance and speed in a manner reminiscent of these distinguished historical equine breeds.

11.4.1.5 Cattle Herding

Horses have played a significant role for generations in various parts of Africa in assisting with the herding of other animals, including cattle, sheep, and goats (Oudman 2004; Eulmi et al. 2023). Numerous pastoralist communities have integrated horses into their practices for sedentary, transhumant, or nomadic livestock management. A prominent example is observed among the Fulani people in the Sahel region. In extensive livestock production, the use of horses for controlling and managing cattle herds is most prevalent in South Africa. In this country, the American Quarter Horse is commonly employed as a working cow horse by South African cowboys. These ranchers require a small, agile, and highly nimble horse capable of maneuvering effectively in challenging conditions and displaying assertiveness around livestock, while being safe for a young child to ride. The horsemanship displayed by South African cowboys is held in high regard, and an increasing number of riders seek them out for lessons. Despite the challenges posed by the African climate, these cowboys are driven by a deep love for their horses (AQHA: American Quarter Horse Association, Undated).

11.4.1.6 Family Celebrations

In societies rich with equestrian traditions, the horse holds a central role in expressions of joy and celebration. As such, the horse is perceived as a symbol of reward due to its prestige and status in society. Its presence, particularly when employed as a mount or escort during marriage ceremonies (Figs. 11.37, 11.38, 11.39 and 11.40), holds great cultural significance (Westermarck 2003). In certain communities, the horse even plays a part in the dowry process. For instance, within the Grassfields communities of Cameroon, the use of the ponytail is a widespread practice in funeral rituals and traditional ceremonies, and symbolizes victory among these peoples (Nyock 2014) (Fig. 11.41).

However, customs surrounding horses vary widely, not only from one country to another but

Fig. 11.37 Weeding festivity in Chechar city, Khenchla (Algeria)

Fig. 11.38 Rider and horse (caparisoned for a wedding ceremony) performing the pace walk. (Baynes-Rock and Teressa 2021)

Fig. 11.39 Awi Agew horse riders during an annual festival. (Kebede 2023)

also between tribes and even within the same tribe (Figs. 11.37, 11.38, 11.39 and 11.40). In specific regions of northern Africa, the father of a young girl might desire one or more mares as part of what he seeks from the family of the man proposing marriage to his daughter. Likewise, the groom's family might offer horses to the bride's family. In some instances, the bride could return to her new home on horseback, especially if the distance to be covered is considerable. The horse chosen for this purpose could be either a mare or a stallion. They believe that using a mare would ensure the bride's fertility, while employing a stallion would increase the likelihood of her bearing male children. In either case, the animal the bride rides is often believed to have a magical influence on her (https://www.aucoeurdeschevaux.com/f/fou-tanke-131/). It is important to note that customs are far from uniform, varying not just between countries and tribes but also within the same tribe. These practices underscore the diverse ways in which horses are woven into the fabric of cultural celebrations, rites, and rituals, reflect-

Fig. 11.40 Fulani horses performing a traditional dance at the conclusion of Ramadan. (Image credit: Flickr/Rosemary Lodge) (DLIFLC 2018)

Fig. 11.41 Ndop loincloth during funerals, Cameroun. (Image credit: Foréké Dschang 2017) (Noel 2021)

ing their profound significance across different societies.

In any case, the saddled horse should not be taken without a rider. For the journey, a brother, a cousin of the groom, or a young boy could ride the mare. An empty saddle is considered a bad omen. When moving the procession to the bride's family house, the mare is sometimes led by hand. The groom's sister or sisters could grasp the tail, while two other people walk on each side with their hands on the stirrups. Upon the convoy's arrival at the bride's village, it might be pelted with stones, mimicking an enemy troop, before receiving a cheerful welcome. This reception likely symbolizes resistance to the young girl's departure (https://www.aucoeurdeschevaux.com/f/foutanke-131/).

The circumcision ceremony in certain communities sees the youth treated like a newlywed. Just like a groom, he is paraded with grandeur and accompanied by music prior to the ceremony, passing through the main marabouts of the city (El-Alami 1971). Traditionally, this parade takes place on horseback, symbolizing virility, power, and pride. However, due to the decline in horse population, rural migration, and mechanization, this tradition has become less frequent, but it remains more prevalent among the working classes. Fantasias are occasionally organized during these events.

11.4.1.7 Pilgrimage to Mecca
The horse is regarded as a prestigious animal, affording its owner a position of honor in societies with an equestrian tradition. As a means of upholding specific traditions, particularly in the Maghreb region, it is not unthinkable that even making the pilgrimage to Mecca on horseback is deemed a noble endeavor.

11.4.1.8 Parades
The horse has maintained a connection with authority since ancient times. Despite the advancement of transportation industries, the horse remains linked to activities involving demonstrations or displays of prestige and power. In the United Kingdom, as well as in most Western countries and parts of Africa, ceremonial cavalry units exist (Figs. 11.42, 11.43 and 11.44). Some public appearances occur exclusively during national holidays in the political capital.

The horse is a prominent presence in parades on numerous occasions, including religious festivals, national celebrations, sporting events, and cultural festivities. In South Africa, the Durban July gathering, a horse event that has attracted horse enthusiasts and luxury aficionados for over 120 years, takes place annually (Fig. 11.45). This event includes a horse parade and races, with wagers surpassing US$30 million in 2016. In the North Cameroon region, select religious leaders use prestigious horses, particularly as symbols of authority, especially those with white markings on all four legs.

In Northern Nigeria, an area steeped in the history of ancient Islamic kingdoms, the annual Durbar Festival is celebrated in states such as Kano, Katsina, Zaria, and Sokoto. Initially associated with wartime allegiance, the festival features a display of horsemanship in front of the Emir to demonstrate the readiness and loyalty of his regiments. The Durbar showcases vibrantly attired horsemen in colorful traditional attire and turbans, led by local leaders, musicians, and traditional performers. These horsemen proceed by regiments to the Emir's palace, punctuated by the sound of muskets, delighting the palace audience as they pay homage to the Emir. This vibrant event, known for marking significant Islamic holidays like Eid-al-Fitr and Eid-al-Adha, is a testament to the region's rich cultural heritage (Okoro 2017). In April 2021, a grand historical procession took place involving the transfer of 22 mummies from the Egyptian Museum in Tahrir to the National Museum of Egyptian Civilization. This skillfully orchestrated show included 150 horses adorned in Pharaonic attire, pulling chariots accompanied by extras in period costumes. Drums and symphonic music provided a dramatic backdrop to the spectacle.

Across various African nations, festivals dedicated to horses are regularly organized. Examples include the national horse festival in Tiaret, Algeria; the horse show in Jadida, Morocco; the horse festival in Parakou, Benin; and the West African Horse Show (SOAC).

Fig. 11.42 Fantasia in Northern Cameroon. (Image credit: Félix Meutchieye)

Fig. 11.43 Horse show at Jida horse market in Ethiopia. (Kefena et al. 2012)

These events serve as opportunities for riders to participate in parades, while folklore troupes follow suit in a jubilant atmosphere characterized by vibrant colors and cultural and traditional inspirations. Similar occasions occur in other African regions, such as Gaani in Nikki (northeastern Benin, near the Nigerian border). In Burkina Faso, the Peuls of the Barani department, situated close to the Malian border, continue the tradition of organizing equestrian gatherings during the Tabaski (Aïd-el-Kebir) holiday. These gatherings often involve parades of lavishly adorned horses, accompanied by riders in traditional garb, constituting opening and closing ceremonies (Adapted from the original text).

Fig. 11.44 Horse parade to celebrate the 123rd anniversary of Adwa victory and led by a prominent Ethiopian musician, Hachalu Hundessa. (Kefena et al. 2021)

Fig. 11.45 Durbar festival: an annual equestrian event in Northern Nigeria. (Yusuf et al. 2023)

11.4.2 Equestrian Sports

Horse shows are hosted in most countries as platforms to promote the discipline. Governed by regulations established by Equestrian Federations, these shows encompass a range of disciplines including dressage, show jumping, endurance competitions, and the complete equestrian competition (CEC), which combines dressage, cross-country (merging jumping and endurance), and jumping obstacles. Equestrian sports find expression through equestrian centers, each accompanied by a multitude of equestrian clubs and farms. The concentration of equestrian centers is notable along the Atlantic and Mediterranean coastlines. Countries such as Algeria, Morocco, Nigeria, South Africa, and Tunisia boast more than fifteen centers affiliated with their respective Equestrian Federations. Despite the development and organization of equine sports activities in other continents, the same progress is yet to be fully achieved in Africa. This is primarily due to the relative sketchiness of regulations in the region, even though the significance of the sport and its associated opportunities is not to be underestimated.

11.4.2.1 Horse Racing

Horse racing stands out as the most prevalent sporting activity on the African continent, although its development remains limited, except in South Africa. Only a handful of African countries lack a hippodrome, with many also hosting informal races on unfenced grounds. Senegal boasts three enclosed hippodromes and sixteen municipal ones. Algeria and South Africa both feature eight equipped hippodromes, while Morocco has seven. These venues cater for flat races, spanning distances from 1000 to 2400 m, and are guided by racing codes. Saddle horses and those intended for equestrian sports typically exhibit warm-blooded characteristics. Morphologically, they are often slender at the extremities, with a withers height ranging from 140 cm to 175 cm and a live weight between 380 kg and 600 kg (Cynthia 2011).

The Thoroughbred, Arabian horse, Dongola, and, to a lesser extent, the Arab-Barb horse, are prominent choices for flat racing. Various other horse breeds are esteemed in different African countries, such as Arewa and Foutanké in West Africa, Koto-Koli in Nigeria and neighboring nations, and Gondo in Mali. The Sudan Countrybreed, a result of crossing local horses with English and Arabian Thoroughbreds, and the Tawleed, obtained by crossing local horses with English Thoroughbreds, are highly regarded for racing. Morocco also engages in Anglo-Arab races, while Algeria hosts trotting races featuring Half-bloods. In several countries, horse racing, in part or whole, supports the equine sector through entities like the Pari Mutuel Urbain (PMU). Given its popularity, the racing sector significantly contributes to herd development and enhancement, largely due to its role as a promoter of peace, moral support, and social cohesion. Periodic competitions occur between local and cross-border teams, as seen in the cases of Cameroon, Nigeria, Niger, and Chad. In Cameroon's Adamawa region, horse breeding primarily centers around racing, supported by horse racing associations like CACGALLOP (The Horse Friends Club) and CAMGALLOP (Cameroon Horse Racing Cup) (Nyock 2014; Mohamadou 2007). The finest Thoroughbreds utilized in European horse racing often hail from Tunisia.

In South Africa, horse racing enjoys immense popularity among the affluent and well-known individuals of the country. Starting from humble beginnings in the late 1700s, horse racing has grown into a multi-million-rand pastime, engaging thousands of participants and garnering widespread support. Within South Africa, three renowned horse races occur in its principal cities. Johannesburg hosts The Summer Cup, boasting a first prize of R1.2 million in prize money. Cape Town is the venue for The J&B Met, held at Kenilworth Race Course, while Durban hosts the prestigious event known as The Durban July, a tradition that dates back to 1897. South Africa also has a number of world-class venues for hosting horse races. In the Eastern Cape, there are the Arlington and Fairview racecourses. The Free State is home to the Bloemfontein and Vaal racecourses, and Gauteng is recognized for its Newmarket and Turffontein racecourses.

11.4.2.2 Endurance Raids

Endurance raids are a practiced discipline that occurs outside of equestrian centers. They involve covering long distances (ranging from 10 to 160 km) over the course of one or more days along a marked route, with the option of following either an imposed or free speed. The well-being and health of the horse are taken into consideration through veterinary checkpoints placed along the route. TREC encompasses multiple equestrian practices, including orientation and regularity courses (ORC), courses across various terrains (PTV), and pace control and presentation activities that focus on harness utilization.

Endurance riding, characterized by its extreme distances and breathtaking landscapes, is both intense and tactical. The sport places substantial demands on mental and physical strength, offering a test not only of the body but also of the mind, with the well-being of both the athlete and the equine being of paramount importance (FEI Endurance World Championship 2021). Ilyasse El Messaoudi from Morocco holds the highest African rank in the FEI endurance open riders' world ranking, standing at 173rd. The Tunisian team secured a seventh position in the World

Endurance Championship for Young Riders and Juniors 2017. In the most recent edition of 2021, Tunisia participated with five athletes (FEI Endurance World Championship 2021). The most challenging endurance race globally is organized by Rocket Horse and traverses the stunning landscapes of the South African coast. Covering a total distance of 350 km, the race spans five days. Each rider is supported by a team of three experienced horses, and only one team can emerge victorious. This event has attracted riders from various countries, including Australia, Canada, Germany, Guatemala, Mexico, South Africa, the UAE, the UK, and the USA. In 2018, Briton Rosie Riall secured victory (Pinguet 2019).

Endurance horses must possess specific morphological and locomotor attributes that enable them to excel in long-distance races. Different breeds offer distinct advantages and disadvantages based on factors such as the breed type, duration, and climatic conditions (Ancelet 2014). Some attribute the Barb horse with the greatest endurance among oriental equines. Its unique characteristic is having a fifth lumbar vertebra, which reduces strain on the muscles supporting its spine (Angelik 2014). In endurance horse racing between 2003 and 2013, Arabian horses represented an average of 52.7% of the top 5 ranked, Arabian half-breeds accounted for 13.7%, and Anglo-Arabians contributed 8.9% of the total races.

The performance of endurance horses is significantly influenced by their genotype, which explains the diversity of physiological responses across various horse breeds. A comparative study of three breeds revealed that Andalusian horses exhibited a notably lower aerobic working capacity compared to Arabian and Anglo-Arabian horses (Castejón et al. 1994; Castejón-Riber et al. 2012). The same study demonstrated that at speeds of 15, 20, and 25 km/h, the amount of lactic acid produced during standardized field exercise was lower in Arabian and Anglo-Arabian horses compared to Andalusians (Castejón et al. 1994).

11.4.2.3 Show Jumping

Show jumping stands as the most prevalent equestrian practice, manifested through "runs" conducted within horse clubs. These runs encompass guiding horses at various gaits and navigating "show jumps." The heights of these jumps range from 0.40 to 0.90 m for ponies and 0.65 to 1.15 m for horses, with a record pinnacle of 2.47 m. Additionally, long jumps extend from 2.70 to 4.20 m, with the record stretching to 8.40 m. North African countries have participated five times in the last 18 Mediterranean Games. Notably, show jumping has been the exclusive event in nine out of the last ten editions. The most recent edition in Barcelona in 2018 featured participation from Algeria, Egypt, Libya, Morocco, and Tunisia. In the team rankings, Egypt secured the fourth position, Morocco the seventh, Tunisia the 10th, and Algeria the 12th out of 22 competing nations. Within the same competition, the highest-ranking African athlete in the individual standings was an Egyptian, attaining the third place overall (FEI Endurance World Championship 2021).

This equestrian pursuit maintained a presence in the nineteenth edition planned for Oran, Algeria. Moreover, show jumping singularly graced the equestrian events of the 12th All-Africa Games held in Rabat, Morocco, in 2019. During this edition, Algeria, Egypt, and Morocco dominated the team rankings. Noteworthy individual performances included the top three athletes from Morocco, the fourth from Algeria, and the fifth from Egypt. The FEI General Assembly, convened by the South African Equestrian Federation (SAEF), took place in November 2020. The SAEF was honored with the FEI Group IX Africa Cup, a distinction earned through the meticulously organized Penbritte Equestrian Centre event in Johannesburg from May 15th to 19th, 2019. The event showcased disciplines such as Jumping, Dressage, Endurance, and Vaulting, attracting participants from 14 African countries. Notably, Lisa Williams from South Africa emerged as the sole African contender eligible for the Longines FEI Jumping

World Cup™ Final 2019 in Gothenburg, riding her remarkable horse, Discovery Campbell (Norman 2019). A monumental achievement was witnessed at the 2018 Youth Olympic Games in Buenos Aires, Argentina, as young athletes from Egypt, Mauritius, South Africa, Zambia, and Zimbabwe combined forces to secure a groundbreaking bronze medal in the continental Jumping event (FEI Fédération equestre international 2018).

11.4.2.4 Complete Equestrian Competition

The Complete Equestrian Competition (CCE) is a multifaceted discipline that amalgamates three distinct events: dressage, cross-country, and show jumping. This discipline stands as one of the seven approved by the FEI and has held the status of an Olympic discipline since 1912. Its origins trace back to the military's requirement for resilient, robust, and manageable horses. At the pinnacle of international events in this domain is the International Complete Competition (CCI). This competition stands apart from the International Combined Competition (CIC), which remains a national event accessible to other federations. The differentiation primarily hinges on the level of the events and the intricacies of the cross-country segment. The cross-country event itself unfolds in four phases: a road phase encompassing 3–6 km at a trotting pace, a steeplechase course spanning 3.1 km with obstacles of various forms, a road phase once more, and finally, the cross-country phase.

11.4.2.5 Cross-country

Cross country serves a dual role within the equestrian realm—it can either be an integral component of eventing or stand alone as an independent test. Its core objective lies in showcasing the remarkable amalgamation of speed, endurance, and agility inherent in the horse-rider partnership. This timed event unfolds across a meticulously designed course, replete with 12–35 fixed, natural obstacles characterized by diverse profiles. The terrain itself exhibits an array of variations, further accentuating the challenge. Due to its inherently perilous nature, this discipline necessitates specialized equipment to ensure the safety of both horse and rider.

11.4.2.6 Dressage

One of the foundational stages in the training of jockeys and riders involves immersing themselves in the realm of dressage, which in itself is a distinct sport. The exercises encompassed by "dressage" serve as a platform for genuine choreography, culminating in captivating equestrian performances. The essence of dressage lies in cultivating a saddle horse's docility, instilling regularity in its gait, and enhancing its overall rideability. This discipline reaches its pinnacle in events known as equestrian shows. At its core, dressage strives to imbue any horse with qualities of obedience, consistency in gait, and an enjoyable riding experience. On the grandest stage, dressage finds its place in the Olympic Games. During this advanced level of competition, horses are capable of executing remarkable maneuvers such as the "passage," which involves a condensed trot, and the "piaffe," an elegant trot executed in place.

In South Africa, a distinct equestrian discipline known as "reining" flourishes. This style falls under the umbrella of western riding and encompasses intricate maneuvers such as backing up, pivots, sharp stops, and sliding stops. Even the tiniest of ponies hold a special role, serving as an ideal choice for children's mounts while remaining equally popular among adults within the team. Over the years, this discipline has introduced numerous generations to the joys of horseback riding, creating a lasting bond between riders and their equine partners.

11.4.3 Model Competition and Pace

Competitions centered on the evaluation of models and gaits, often referred to as "shows" for Arabian horses, provide a platform for assessing the horse's physical structure and behavior across its three gaits. Another facet of equestrian competitions involves driving, encompassing a trio of distinct tests: dressage, the marathon (a timed challenge conducted across diverse terrains), and maneuverability exercises.

11.4.4 Equestrian Games

11.4.4.1 Vaulting

Vaulting originated more than a century ago in the Tyrolean Mountains. It shares a historical similarity with the North African fantasia, both of which serve as preparatory exercises for war. Like other traditional equestrian activities, vaulting occasionally shares similarities with show jumping (INRA 2014). Despite its limited documentation in Africa, vaulting does find a unique practice ground in southern Tunisia. Here, practitioners display impressive feats such as standing on the saddle, maintaining unstable positions while clinging to the horse, relying on a lone stirrup for support, galloping upright on the horse, and even resting on the horse's head. These exercises evoke a sense of competition akin to those performed by the Cossack horsemen. Historically, vaulting was a widespread activity in various African nations during the era when horses were utilized for warfare. It played a role in preparing riders for combat. Over time, these horseback acrobatics transitioned into captivating acts within circus performances. Vestiges of this practice persist in different countries in the form of equestrian games.

An example of such a game is tent pegging, which is practiced in Sudan and South Africa. This traditional game also has roots in the Algerian Highlands, although it is gradually fading there. Tent pegging involves retrieving an object, often transformed into a target for a second rider, using a knife or by hand. In Bou Saada, a notable trial during the enthronement of young riders involved delicately lifting a bowl of honey or melted butter without spilling a single drop. The last African cup of vaulting was held in South Africa with the participation of three nations: Mauritius, Zimbabwe, and South Africa (Fig. 11.46).

Fig. 11.46 Lambert Leclezio is the first Mauritian athlete to compete at the FEI Open European Vaulting Championships for Juniors in 2011. (FEI 2012)

11.4.4.2 Polo

Polo, an ancient game hailing from Central Asia and likely originating with the Scythians, is believed to have evolved in the African continent as well. It has gained substantial popularity in South Africa, where the South African Polo Association was established in 1905, marking one of the country's oldest sports organizations. Beyond South Africa, polo is actively practiced in Egypt, Kenya, Morocco, Nigeria, Tunisia, Zambia, Zimbabwe, and Uganda, all proud members of the International Polo Federation (FIP). Established in 1982 with its headquarters in Montevideo, Uruguay, the FIP is entrusted with the management and global development of polo. Notably, it achieved official recognition by the International Olympic Committee in 1998 (FIP Federation of International Polo 2014). The sport's prestigious World Cup, hosted every 3 years since 1989, has drawn international attention. The most recent edition took place in Australia in 2017, while the last African country to participate was South Africa in 2008 in Chile.

Polo unfolds on a field measuring 275 × 145 m and is structured into 4–8 periods, each lasting 7 min and 30 s to 8 min. The game assembles two teams, each composed of four players, who employ long-handled mallets to maneuver a ball into the opposing team's goals. In the realm of polo, horses sport shaved manes and braided tails to prevent entanglement with the mallet. Their legs are safeguarded by specialized bands, while a martingale restricts head movements. Players don helmets often equipped with grids and knee pads for safety.

In the Algerian Highlands, a similar game named "taoud" or "gaous" was historically played. This version featured two teams of ten riders on a 30-meter pitch, with goals marked by stakes on either side. Players contested possession of a wooden ball using a curved stick or a palm tree stalk, striving to propel it into the opposing goal.

11.4.4.3 Horseball

Adapted from the Argentinian sport of Pato, horseball is played on a field spanning 60–70 × 20–30 m. The game involves a ball equipped with six leather handles, which must be strategically thrown into a vertically positioned circle measuring 1 m in diameter and situated at a height of 4 m. Matches are divided into two 10-min periods, featuring teams composed of 4 players each. For the protection of the horses, gaiters shield their limbs, bells are worn on their feet, and a martingale limits head movements. Stirrups are connected by a strap to prevent saddle rotation, and players can opt to wear knee pads for additional safety. Horseball can be likened to a fusion of rugby and horse basketball, where two teams engage in competitive play, aiming to score as many goals as possible by throwing the ball through high hoops. The landscape of horseball competitions is characterized by its amateur nature, yet its popularity is on the rise, attracting an increasing number of enthusiasts. France is a dominant force in international horseball competition. The sport has also reached North Africa, Canada, the western United States, the Middle East, and select countries in the East.

Horseball remains a developing amateur sport. It has found a presence in Algeria, South Africa, and Togo. Among these countries, Algeria has joined the International Horseball Federation (FIHB) and holds the 12th position among the federation's 17 member countries. The FIHB's world ranking includes eight Algerians, with Mr. Abderaim Sad ranked 855th among the 1502 members. Despite this, the sport suffers from a lack of championships, visibility, publicity, and funding. Matches often occur within private clubs and the Republican Guard. However, the sport's inherent appeal and dynamics could position it as a key element for sporting development. In other sub-Saharan African regions, horseball tournaments are organized. For example, in September 2019, a mini-promotional tournament was hosted by the Hymane Club de Lomé in Togo. The aim was to introduce the local population to this engaging equestrian discipline. South Africa also participated in the grand horseball—pony-games—polo tournament held in Lamotte-Beuvron, France, between the 18th and 20th of May, 2013.

11.4.5 Traditional Sports Activities

11.4.5.1 Fantasia (Tbourida)

The horse holds a significant place in traditional African societies, as evidenced by practices like fantasia, a captivating equestrian art that has historical and cultural significance. Fantasia originated as a practice to bolster the combat capabilities of soldiers, resembling the North African fantasia, which served as a war preparation exercise (Figs. 11.47, 11.48 and 11.49). However, fantasia has also evolved into a popular form of entertainment, often performed during festivals like Ouaada or Moussem (veneration of the wali) and other holidays.

In the Maghrebian dialect, the term for fantasia is commonly referred to as "fantazya" or "tburida," derived from the root word "BRD," meaning "powder." In the Maghreb region, these shows are known as "laâbel-baroud" or "laʿb al-bārūd" (meaning "game of powder") and "laâbel-khayl" or "laʿb al- ḵayl" (meaning "game of horses"). These names have historical roots, and the term "fantasia" arose due to misinterpretation. The term "fantasia" is believed to have been misinterpreted by the French painter Eugène

11 African Horse Genetic Resources and Breeding Strategies for African Input Systems

Fig. 11.47 Fantasia parade in Mekhatria, at 10 kms north of the seat of the wilaya of Ain Defla, Algeria. (Image credit: Bisker Mohamed) (Bisker 2017)

Fig. 11.48 The Fantasia equestrian parade with warrior inspiration (Blog du cheval) (https://blog.cheval-daventure.com/fr/post/140/la-fantasia-une-tradition-equestre-dinspiration-guerriere)

Fig. 11.49 Saddle, breast strap, and bridle on a Barb horse at Capital Festival of traditional horseback riding, Rabat, August 2014. (Image credit: Gwyneth Ursula Jean Talley) (Talley 2015)

Delacroix, who erroneously associated it with "equestrian show." The true origin of the word can be traced to classical Arabic, where confusion between two words built on the same trilateral verbal root, ḵayāl (meaning "imagination") and ḵayyāl (meaning "riders"), contributed to this misconception. The practice of fantasia has deep cultural ties in Algeria and Morocco, known respectively as "El Goum" and "Tbourida." Fantasia is categorized into various forms, such as Temerad and Guelba in Algeria, and Nassirya, Charqaouia, Kheyyatiya, and Heyyaniya in Morocco. There's also individual fantasia, played mostly in eastern Algeria. The essence of fantasia is rooted in its historical connection to Berber traditions of horse riding, particularly the use of Barb horses or "war horses."

The core of fantasia involves riders on richly harnessed mounts simulating a cavalry charge, firing black powder rifles in coordinated salvos. The artistry of the game is often likened to a display of strength, courage, and symbolic confrontation. This tradition represents a modern continuation of Arab-Turco-Berber horse-riding skills from North Africa. Fantasia showcases the bond between horse and rider, as Barb or Arab-Barb horses are trained meticulously for the sport. The riders and their horses execute choreographed figures with precision and coordination, accompanied by the thunderous rhythm of their galloping hooves and synchronized gunshots (Pierre 1905).

In Algeria, approximately 350 traditional equestrian associations continue the practice of fantasia, while in Morocco, around a thousand troops and nearly 15,000 horses partake in competitions organized by national stud farms during religious festivals and other significant occasions. Though the practices of fantasia differ in various regions, their common thread is the celebration of equestrian skills, history, and culture. This vibrant tradition continues to thrive, deeply rooted in the heritage of North Africa.

11.4.5.2 Traditional Horse Racing

Horse racing is a widespread practice that takes place in various regions, often serving as a platform to showcase the prowess of both the horse and the rider. Races can be as straightforward as impromptu gatherings where riders aim to demonstrate the superiority of their mounts or their own skills. Many of these races occur on open grounds, without extensive preparation, and the participation of horses is often organized through informal communication networks (Fig. 11.50). However, certain horse racing events hold traditional and celebrated statuses, such as the Sahel Festival in Amdjarass, located in eastern Chad. The Sahel Festival is an example of a renowned and time-honored event that draws participants

Fig. 11.50 Lesotho's mountain jockeys race. (Kuwait times 2022)

and spectators alike. While the dynamics of horse racing can differ across locations and occasions, the overarching essence remains consistent—a display of the swiftness and strength of horses alongside the riding proficiency of their jockeys. Whether it is a casual gathering or a prominent festival, horse racing remains a captivating and enduring expression of equestrian skill and tradition.

11.4.5.3 Horse Dancing

Among the Peuls and Wolofs of Senegal, horses held a distinct role as part of the dowry demanded from the groom. This practice might be linked to the perceived protective power of horses over the family, possibly explaining the tradition of hanging horseshoes at the entrances of homes. The "Fassu Kalifa" horse, known for its nobility, was highly sought after by both customary and religious leaders. This horse type symbolized prestige and authority, embodying a sense of nobility that resonated with traditional and spiritual figures. Additionally, a specific category of horses called "Pekh" was bred and trained exclusively for dance performances. These horses executed choreographed routines, a tradition that had been perpetuated since the time of the Cayor Empire. Victories and celebrations were marked by these captivating displays known as "fantasies." These dancing horses continue to participate in customary ceremonies, adding an artistic and cultural element to the events.

In northern Cameroon, horses maintain their significance as symbols of prestige and authority. Religious leaders, in particular, breed horses for their esteemed reputation. The coloration of the horse, particularly chestnut with four white or light gray balzanes (sock-like markings on the legs), adds to its importance. These horses are meticulously selected and trained to perform a choreography that originated with the arrival of Adama, the General of Conqueror Ousman Dan Fodio. This choreography serves to commemorate victories and achievements. This tradition is upheld across various lamidates (chiefdoms) and sultanates in Cameroon and Nigeria, underlining the enduring connection between horses and authority in this region (Mohamadou 2007).

11.4.6 Leisure Riding and Equestrian Tourism

Equestrian clubs and centers across Africa are embracing the development of circuits designed for horseback hiking or safaris, often in collaboration with tourism development agencies. Equestrian safaris offer a unique and immersive experience in countries such as Botswana, Kenya, Namibia, Tanzania, South Africa, and Uganda

(Fig. 11.51). These safaris typically involve a group of 8–10 experienced riders who embark on multi-day journeys through natural landscapes, surrounded by the magnificent African wildlife (Fig. 11.51). The safaris can either be itinerant, with camps shifting regularly, or they can involve stays in lodges or established camps. The accessibility of the horse safari varies, catering to experienced riders, intermediate level riders, families, and visitors accompanied by guides.

Certain countries, including Egypt, Ethiopia, Madagascar, Morocco, Mozambique, Senegal, and Zimbabwe, opt for equestrian hikes that span from 1 to 6 days, allowing participants to explore tourist-friendly regions and engage with local communities. Depending on the location, riders can savor the beauty of nature during bivouacs or enjoy stops at comfortable accommodations. Each day, baggage and supplies are transported to the next stop using vehicles or pack animals. It is crucial to ensure proper care and maintenance of the horses during these hikes. Some expeditions venture into remote and challenging terrains, involving steep paths and basic amenities such as portable toilets.

It is worth noting that circus riding is not a native tradition in Africa, and the horses typically used for these activities often come from European contexts. The growing popularity of equestrian tourism, hikes, and safaris contributes to a deeper connection between people, nature, and the rich cultural heritage of the African continent (Fig. 11.51).

11.4.7 Horse in the Army and Security Forces

The use of escort cavalry has been popularized by Middle Eastern kingdoms, initially as a military authority and later as a political symbol. This practice is notably visible in African political capitals, as well as in the remnants of Sultanates and Lamidates. The escort cavalry, which is often a part of the armed forces in certain countries, plays a vital role in escorting national dignitaries during protocol missions. This includes occasions such as visits from foreign heads of state, presidents, and national holidays.

Furthermore, the cavalry service of national security agencies also fulfills traditional roles in maintaining order. This involves tasks like overseeing sporting events, both national and international competitions, festivals, and responding to

Fig. 11.51 Tourism activity by riding Nooitgedacht horses in South Africa. (The figure illustrates a scene where a game warden and a tourist, riding Nooitgedacht horses, are approaching a female rhinoceros with a calf visible in the background. These South African horse breeds have demonstrated their exceptional suitability for guiding inexperienced riders during game safaris. (Photo by F.J. van der Merwe) (FAO 2002))

situations of civil unrest. The cavalry unit is also entrusted with urban patrols and the surveillance of diverse territories, encompassing forests, coasts, and more. These police officers are versatile and may be involved in missions beyond horseback, such as static guard duty or the transportation of prisoners. This multifaceted role highlights the significance of cavalry units in ensuring security and upholding public order in various contexts across the African continent.

11.4.8 Hippophagea (Meat and Milk)

The consumption of horse meat in Africa is very low and has been decreasing over time. Hippophagy, the practice of consuming horse meat, dates back to prehistoric times, although it was not widespread. In the Maghreb, the consumption of meat and milk products from horses and donkeys was known. However, due to concerns about its potential as a vector for diseases like salmonellosis and trichinellosis, there is a certain level of mistrust and various taboos associated with it—some rooted in religious beliefs and others in cultural norms. This aversion is linked to the close relationship humans have with the horse.

In fact, the closer an animal is to humans, the more it can substitute for them in rituals and sacrificial meals. Conversely, consuming animals too closely associated with humans can be seen as sacrilegious. In ancient times, horse meat consumption was considered barbaric, and the Catholic Church opposed it. The Church even prohibited it in the eighth century, possibly to counter the practices of Nordic religions. Horse meat consumption is indirectly prohibited in Judaism, as the Torah only permits the consumption of mammals with cloven hooves. According to the hadith, Islamic tradition allows the consumption of horse meat.

Supporters of hippophagy praise its qualities, extolling its exceptional meat that has historically been recommended for treating anemia. However, opponents, including animal protection associations like the Brigitte Bardot Foundation, focus on the conditions of animal transport and slaughter. They do not take a stance on consumption. Tasty horse meat has low moisture content (2–4%), is rich in iron (4–5 mg per 100 g), minerals like zinc and selenium, and vitamins (B3, B6, and B12). It contains 60–70% unsaturated fatty acids, including some that the body cannot synthesize (linoleic acid and alpha-linolenic acid), which seem to aid in preventing cardiovascular diseases.

In Africa, horse meat often comes from injured or culled animals that are no longer fit for their initial roles. Senegal is the only African country with information on horse meat production. In 2009, it produced an estimated 9500 tons, ranking 13th globally in production 180. Cameroon is known for valuing horse intestines and ascribing medicinal properties to "fired forelimbs," especially around Mount Manengouba. While horse consumption is rare, it has motivated some to breed horses (Mohamadou 2007). Mare's milk is comparable to human milk, with low lipids and high lactose. It is rich in vitamin C and contains half the lipids of human or cow's milk. Production varies by breed, averaging around 20 liters, requiring three milkings due to small udder volume. It is consumed as a drink because it does not curdle. Mare's milk and koumis, a traditional fermented mare's milk drink, are popular in Central Asia. They are believed to treat liver, stomach, and heart issues. They are also used in cosmetic products, particularly to prevent wrinkles due to their tightening effect. Ancient civilizations recognized these benefits with figures like Herodotus and Cleopatra praising them.

Breeding horses solely for meat production is generally not profitable due to their energy needs. However, interest in milk production is increasing, driven by cosmetic uses. Notably, donkey production is valuable for the Chinese market; donkey meat, known for its aphrodisiac properties, and donkey skin, used to make ejiao or "donkey skin gelatin," are sought after. This has led to a decline in donkey populations in Africa and export bans.

11.5 African Horse Diversity

11.5.1 Morpho-Biometric Characterization

According to Gaouar (2009), this method is widely used to characterize and compare breeds of farm animals (goats, sheep, cattle, camels, and horses) (Gaouar 2009). The phenotypic traits were the first traits used for the selection and the definition of the breeds. These traits result from a simple observation of the animal's characteristics, such as coat color, leg color, head and body morphology, and more. They can also include morphological measures like height, weight, and body width, as well as performance indicators such as growth rate, milk production, and hair production (Gaouar 2009).

Furthermore, the similarity of morphological characteristics can reflect genetic identity, although it might arise from different alleles or genes located at different loci. Differences in phenotypic expression can stem from actual genetic differences, but they can also be attributed to variations in expression or penetration levels (Gaouar 2002). Penetration refers to the frequency of expression of a dominant trait in relation to the number of known carriers of the gene in the population. Expressiveness pertains to the degree of intensity of a hereditary trait. These considerations limit the use of morphological characters in reconstructing breed phylogeny and characterization (Gaouar 2002). However, it is useful as initial characterization and understanding African horse heritage due to limited detailed genetic characterization research on horses in Africa. These tools are of significant importance and are gaining momentum, particularly with the emergence of methods like Genome-Wide Association Studies (GWAS).

Most studies on the morphometric characterization of horse breeds on the African continent have focused on the Barb breed in North Africa. Chabchoub et al. (2006) conducted a study on 23 morphological parameters of 41 pure Barb horses from Tunisia (Chabchoub et al. 2006). Subsequently, a study in Algeria by Rahal et al. (2009) delved into the morphometric characterization of 18 quantitative parameters and four semi-qualitative parameters measured on 35 Barb horses in the Tiaret region (Rahal et al. 2009). Another evaluation was carried out on 151 pure and presumed (uncertain origins) Barb and Arab-Barb horses in Algeria, revealing that the Barb horses studied adhere to the international standard (Guedaoura 2012). Similar results were obtained by Benhamadi et al. (2017, 2019) in their study of 58 Barb horses in Northwestern Algeria (Benhamadi et al. 2017). This research concluded that the Algerian Barb horse possesses eumetric characteristics, is mid-linear with a body index of 0.955 (a square horse), boasts a median size of 152.5 cm, and a chest circumference of 175.5 cm (Benhamadi et al. 2017). Boujenane et al. (2008) studied morphometric characteristics of Arab-Barb horses in Morocco (Boujenane et al. 2008). A similar study in Algeria, Mebarki et al. (2018) found higher height at the withers and lower chest circumference than similar measures reported by Boujenane et al. (2008) in Morocco. These results could help establish a standard for the North African Arab-Barb horse, potentially leading to the creation of a closed studbook.

While research on other African horse breeds is less common, there was a study in Senegal 150 focused on the morpho-biometric parameters of traction horses, revealing similarities with earlier studies on indigenous breeds (Rischkowsky and Pilling 2007). Kefena et al. (2012) studied Ethiopian horse populations (Kefena et al. 2012), examining 17 selected variables on 503 horses from five of the eight communities in Ethiopia. Their results revealed three major breed groups and five distinct horse populations. A morphological study was undertaken comparing Arab horses with Desert Egyptian, Polish, and Polish-Egyptian ancestry, within the Arabian breed in Egypt (Machmoum et al. 2019). The results indicated significant morphological differences between the three lines, with the Desert Arabian breed and Polish Arabian line showcasing distinct profiles, while the Straight Egyptian profile fell in between.

11.5.2 Immunogenetic and Biochemical Studies

Biochemical parameters encompass the concentrations of molecules that offer insights into an organism's condition. Of these, the most widely utilized pertain to hepatic functions: alanine aminotransferase (ALAT) and aspartate aminotransferase (ASAT); serum proteins like albumin and globulins; milk proteins; and blood group proteins. Notably, blood group proteins have been extensively employed in equine breed characterization, serving as a crucial tool in parentage control until 2000. In resource-constrained African countries, they present an attractive alternative to molecular tools, especially with their confidence rate exceeding 98%.

It is important to consider breed variations when establishing reference intervals (RIs) for blood parameters and interpreting blood tests (Witkowska-Piłaszewicz et al. 2021). Horse hemotyping entails the analysis of red blood cells and genetic serum markers, constituting a vital tool in paternity and/or maternity verification. Its applications extend to identification, diagnosis, and the management of neonatal isoerythrolysis, as well as selecting blood transfusion donors (Bowling and Clark 1985). Cothran and Van Dyk (1998) examined genetic variability across seven blood groups and ten biochemical genetic loci in the Nooitgedacht, Boerperd, and Pony Basuto South African horses (Cothran and Van Dyk 1998). They subsequently extended this investigation to wild horse populations in Namibia (Cothran et al. 2001). The studies unveiled intermediate heterozygosity among domestic South African horses, with the Boerperd displaying the highest and the Basuto Pony the lowest heterozygosity (Cothran and Van Dyk 1998). All three breeds demonstrated greater genetic similarity with each other than with other domestic horse breeds (Cothran and Van Dyk 1998).

When compared to international breeds, South African horse breeds exhibited closer genetic ties to the Thoroughbred, Holstein, Trakehner, Hanovarian, and North American breeds like the Saddlebred, Standardbred, and Morgan Horse (Cothran and Van Dyk 1998). In contrast, genetic variability in feral Namibian horses was notably lower compared to domestic horse breeds, likely due to recent small population and founding sizes. These horses exhibited the highest genetic similarity to Arabian-type horses, although their conformation did not closely resemble that type (Cothran and Van Dyk 1998). Podliachouk and Quéval (Podliachouk and Queval 1961) studied 59 horses in Chad, looking at the distribution of erythrocyte factors (Podliachouk and Queval 1961). They concluded that Chad's horse population primarily stems from the Aryan and Mongolian breeds (Podliachouk and Queval 1961). They exhibited varying levels of purity, mixed with environmental and farming system influences. Four types were differentiated: Arab, Barb, Dongola, and Kirdi Pony from Logone (Podliachouk and Queval 1961). Despite such heterogeneity, the horse population in Chad remains relatively homogeneous even within regions, with the various types frequently intermixed. The study found that the frequency of various globular factors in Chadian horses closely resembled that observed in Garches (France), indicating comparable distributions. Chabchoub et al. (2006) conducted a study on the Mogod pony in Tunisia, a breed adapted for mountainous terrain in the northwest of the country (Chabchoub et al. 2006). Electrophoretic analysis of six protein systems in 47 Mogod ponies revealed allele frequencies and highlighted genetic differences between this breed and the Arabian, Barb, and Arab-Barb Thoroughbred horses.

There are relatively few studies on the biochemical parameters of heavy horses in Africa. Bathily et al. (2014) studied traction horses in Senegal, analyzing biochemical parameters of 97 male horses across different age groups and body condition scores using commercial kits (Bathily et al. 2014). Variations were observed in parameters such as phosphorus, calcium, cholesterol, urea, AST, and creatinine, with significant differences based on region and age. In Algeria and Morocco, studies on Barb horses and Arab-Barb breeds found that hematological indices fell within the reference range for warm-blooded horses (Chikhaoui et al. 2018; Talbaoui 1986). Notably, differences in hematological parameters between these groups have been reported, including lower hematocrit in cold-blooded horses and higher erythrogram values in warm-blooded horses (Weiss and Wardrop 2010).

11.5.3 Molecular Studies of the Diversity and Structure of Horse Populations

Very few studies have been conducted in African countries to gain insights into the genetic structure of horse populations on the continent. Most of these employed the international panel of 17 microsatellites recommended by the FAO-ISAG International Panel on Animal Genetic Diversity Advisory Group for global equine breed identification and origin control. This situation presents an intriguing opportunity for data compilation across various African nations. In Algeria, Benhamadi (2020) used these markers to characterize eight horse breeds (Benhamadi 2020). Their study revealed a significant loss of genetic variability due to uncontrolled crossbreeding in populations of Algerian horses, such as the Barb, Barb of unknown origin, Arab-Barb, Arab, Thoroughbred, French Trotter, and Algerian Saddle horse (Benhamadi 2020). A similar alarming situation was expressed by Jamali (2020) regarding the genetic diversity of the Barb horse (Jamali 2020). They found that horse breeds in Algeria could be categorized into three major groups, each warranting separate preservation efforts to maintain their genetic diversity.

Berber et al. (2014) undertook a molecular characterization and genetic differentiation of five horse breeds in Algeria (Barb, Arab-Barb, Arab, Thoroughbred, and French Trotter) using 14 microsatellite markers (Berber et al. 2014). Their research indicated clear genetic differentiation between indigenous Algerian horses (Barb, Arab-Barb) and other breeds. The Barb and Arab-Barb breeds showed close genetic relatedness. Guedaoura (2012) characterized the Barb horse and its main variations by analyzing molecular polymorphism with eleven microsatellite markers (Guedaoura 2012). They reported that the Barb breed exhibits high allelic richness (57 alleles in total), an average heterozygosity rate (H) of 0.6651, and an inbreeding coefficient (F) of -0.0620 (Guedaoura 2012). The study calculated probabilities of identity and exclusion, revealing that the probability of identity (PI) is 1.98×10^{-9}, and the probability of exclusion (PE) is 96.4% with only one known parent and 99.99% with both parents known. Analysis of phylogenetic relationships indicated that the Garrano Portuguese pony is genetically closest to the Algerian Barb horse, sharing the maximum number of alleles in common (Guedaoura 2012).

In Tunisia, Haddad et al.74 conducted a molecular characterization with 13 microsatellites on indigenous horse breeds (Sebbag 2002). Their study highlighted a high rate of homozygosity and inbreeding within the analyzed Barb population, emphasizing the urgent need for genetic conservation and management. The molecular characterization of Tunisian Barb horses confirmed the existence of a genetic subdivision into two ecotypes, namely the Central East and Central West ecotypes. The variation previously attributed to environmental factors such as terrain and diet was found to have a genetic origin. Another study evaluated genetic diversity using ten microsatellites on 63 individuals from the pony population of Mogods in Bazina and Nefza, Northwestern Tunisia. The researchers demonstrated the presence of two genetically distinct ecotypes, one specific to the Bazina region and the other to the Nefza region (Sebbag 2002).

Jemmali et al. (2015) investigated the genetic diversity of the Barb and Arab Barb breeds in Tunisia, amplifying 17 microsatellites in a study involving (FAO 2019) horses (Jemmali et al. 2015). The research revealed slight heterozygosity deficiencies in the two populations. The overall population exhibited less heterozygosity compared to each population separately. Principal component analysis, dendrogram results, and factorial correspondence analysis all indicated overlapping genetic makeup, confirming a shared genetic basis between the two breeds (Fig. 11.52) (Piro et al. 2019).

A comparative analysis and genetic comparison of Barb horses from Morocco and Tunisia

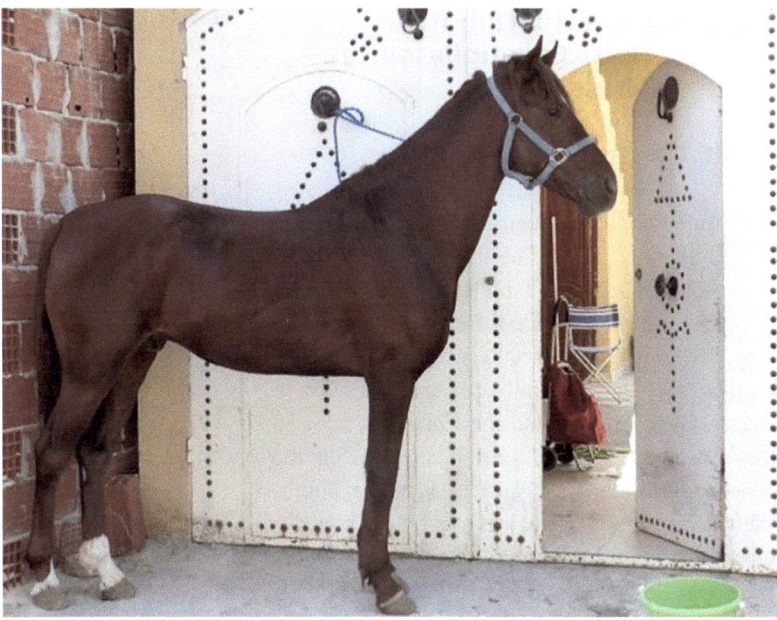

Fig. 11.52 A chestnut Arab-Barb stallion tied up in front of its owner's home in Tozer, Tunisia. (Charles 2022)

was carried out by Piro et al. (2019). This study establishes phylogenetic relationships between the Barb horse and other existing breeds in Morocco using 17 microsatellites. They concluded that the Barb populations from both countries are heterogeneous and almost identical, with similar average heterozygosity rates (Piro et al. 2019). Comparison of the Barb horse's genetic structure in this study with other breeds (Thoroughbred, Arabian, and Arab-Barb) highlighted specific alleles and strong heterozygosity, showing maximum genetic distance from the Thoroughbred. The phylogenetic tree indicated a significant clustering of the two Barb populations and distinct separation from the other three breeds (Thoroughbred, Arabian, and Arab-Barb) (Piro et al. 2019).

In South Africa, Breytenbach et al. (2020) investigated mutations associated with gait and height in SA Boerperd horses, analyzing the genetic diversity using 17 microsatellite markers (Breytenbach et al. 2020). The study found that observed heterozygosity remained consistent with levels from 15 years ago, while inbreeding levels decreased from 8.4% to 3.2%. Despite this, the population structure analysis indicated the contributions of two distinct founder populations to the current breed formation. Compared to nine European breeds, SA Boerperd horses exhibited above-average heterozygosity and numerous private alleles (17.6%) (Breytenbach et al. 2020).

In a recent study in Ethiopia, molecular biology tools were employed to analyze mitochondrial DNA polymorphism in indigenous equine populations (Kefena et al. 2021). The study revealed poor genetic differentiation among the studied animals. The authors also attempted to trace the matrilineal genetic roots of native Ethiopian horses by analyzing haplotypes and reference sequences from the GenBank (Kefena et al. 2021). The network analysis indicated separate clusters formed by haplotypes from native Ethiopian horses, suggesting distinct genetic origins. This study encourages further exploration of the genetic heritage of native African horses by analyzing additional mtDNA sequences from the GenBank for African and Eurasian horses (Kefena et al. 2021).

11.6 Conservation Programs of Horse Genetic Resources in Africa

11.6.1 Current Situation

The equine population census in Africa was conducted in only 37 out of the total 54 African countries (FAO 2019) (Table 11.2). Only 46% of African countries provide information on equine breeds while reporting rates are higher in other regions, such as 100% in North America, 93% in Asia, 91% in Europe and the Caucasus, 65% in Latin America and the Caribbean, 58% in the Near and Middle East, and 23% in the South-West Pacific (FAO 2006a). Horse breeds contribute significantly to the total number of mammal breeds worldwide (14%), even though their contribution in terms of population size is comparatively (FAO 2006a). This indicates a considerable genetic diversity within this species. The African equine population represents 6% of the global equine population, with 7% of the ethnological diversity compared to reported breeds worldwide (FAO 2006a). This corresponds to 34 local (indigenous) breeds out of the 570 breeds reported globally.

It is important to note that the reported number of horse breeds in Africa and worldwide can vary depending on the organizations and methodologies used for studying equine diversity. Different sources report different numbers. For instance, in June 2010, the Oklahoma State University Livestock Breeds' Database 202 listed 217 horse breeds worldwide. Earlier, Hall and Raune (1993) reported 427 breeds (Hall and Raune 1993) while the European Association of Animal Production (EAAP) genetic data bank contained 707 entries representing 110 horse breeds in 1992 (Simon 1992), Mason's breeding dictionary included 592 horse breeds including breed varieties (Mason 1988), and a study by Koenen et al. (2004) indicated 527 horse breeds worldwide (Koenen et al. 2004). In Africa, there are 48 local breeds, 9 regional transboundary

Table 11.2 Risk statue of African horse breeds (FAO 2019)

N°	Breed	Country	Situation
1.	Basotho Pony	Lesotho and South Africa	Extinct
2.	Calvinia	South Africa	Extinct
3.	Cape Harness	South Africa	Extinct
4.	Cape Horse	South Africa	Extinct
5.	Namaqua Horse	South Africa	Extinct
6.	Nefza Pony	Tunisa	Extinct
1.	English WB	South Africa	Critical
2.	Horse (Uganda)	Uganda	Critical
1.	Nooitgedacht	South Africa	Endangered
2.	Pony European WB	South Africa	Endangered
3.	Horse (Tanzania)	Tanzania	Endangered
4.	Namib horse	Namiba	Endangered
5.	SA Miniature horse	South Africa	Endangered
6.	Somali Pony	Somalia	Endangered
7.	Vlampeerd	South Africa	Endangered
1.	El-Ghazal	Tchad	Not at-risk
2.	Bandiagara	Mali, Niger	Not at-risk
3.	Dongola	West Africa (Chad, Eritrea, Mali, Sudan)	Not at-risk
4.	Egyptian	Egypt	Not at-risk
5.	Locale		Not at-risk
6.	Logone pony	Cameroun	Not at-risk
7.	Sudan Country breed	Sudan	Not at-risk
8.	West African Barb	West African	Not at-risk
9.	Western Sudan Pony	Sudan	Not at-risk
1.	Abyssinan	Ethiopia	Unknown
2.	Beledougou	Mali	Unknown
3.	Bhirum Pony	Nigeria	Unknown
4.	Bobo	Burkina Faso	Unknown
5.	Boer	South Africa	Unknown
6.	Bornu	Nigeria	Unknown
7.	Chadian	Tchad	Unknown
8.	Cheval de Nioro	Mali	Unknown
9.	Djerma	Niger	Unknown
10.	Dombi	Mali	Unknown

(continued)

Table 11.2 (continued)

11.	Fleuve	Senegal	Unknown
12.	Fouta	Senegal	Unknown
13.	Hausa	Mali, Niger, Nigeria	Unknown
14.	Hodh	Mali, Mauritania	Unknown
15.	Koto-Koli Pony	Benin, Togo	Unknown
16.	Arabe-Barb	Algeria, Mali, Mauritania, Morocco, Senegal, Tunisia	Unknown
17.	M'bayar	Senegal	Unknown
18.	Mogods pony	Tunisia	Unknown
19.	Mossi	Burkina Faso	Unknown
20.	M'ParPoney	Senegal	Unknown
21.	Rancher	South Africa	Unknown
22.	SA Sporting Horse	South Africa	Unknown
23.	SA WB	South Africa	Unknown
24.	Sahel	Mali	Unknown
25.	Songhoi	Mali	Unknown
26.	Sulebawa	Nigeria	Unknown
27.	Tawleed	Sudan	Unknown
28.	Torodi	Niger	Unknown
29.	Tsawana	Botswana	Unknown
30.	West African Dongola	West Africa	Unknown
31.	West African Pony	West Africa	Unknown
32.	Yagha	Burkina Faso	Unknown

breeds, and 26 international transboundary horse breeds reported by the FAO in 2006 (FAO 2006a). Notably, the number of African countries reporting the presence of Arab, Thoroughbred, Arab-Barb, Barb, and Dongola horses was 8, 7, 6, 6, and 3, respectively (DAD-IS 2008).

Globally, there are 272 horse breeds with an unknown risk status, 52 breeds in critical situation, 10 in continued critical situation, 95 in danger, 24 in continued danger, 246 not at risk, and 87 missing. FAO statistics (2006b) show that 384 breeds of African mammalian species, including horses (with population size unreported for over two-thirds of these populations), are in an unknown risk situation (32 breeds) (FAO 2006b). Regarding horses, 6 breeds have been officially declared extinct in Africa 208. Among African international transboundary horse breeds, the Barb breed is highly represented, and the Dongola breed is well-known among regional transboundary African breeds. The Nooitgedacht breed is an example of an African regional transboundary breed that is endangered (Rebts 2017). With only 400 identified individuals, it qualifies as a rare breed (Khadka 2010).

The lack of comprehensive knowledge about the current status of various African horse breeds creates challenges for coordinated efforts in the establishment of conservation priorities. There are several crucial reasons for classifying the risk status of breeds, including genetic uniqueness (Mendelsohn 2003), high endangerment levels (Gandini et al. 2004), economic, cultural, scientific, and ecological value, as well as the efficient allocation of resources (Simianer et al. 2003). However, the future prospects of breeds, whether equine or of other species, are heavily influenced by their current and potential roles within husbandry systems. As circumstances change, certain breeds might fall into decline and become threatened with extinction, unless alternative strategies are implemented (Hiemstra et al. 2006). Various conservation methods offer advantages beyond traditional technologies and enhance their efficacy in the characterization, preservation, and utilization of equine genetic resources. While a detailed discussion of classic selection methods is beyond the scope here, it is undeniable that such methods have been and will continue to be extremely valuable within the African equine sector.

11.6.2 Reproduction Management

Equine genetic resources serve as the fundamental foundation for advancing the equine sector and ensuring the sustainability of natural equine populations. Efficient reproduction lies at the core of conserving, enhancing, and developing horse herds. In the wild, horses adapt by developing various breeding strategies to ensure optimal foaling during periods of abundant forage and warm climate (Cilek 2009). However, domestication has had a substantial impact on reproductive performance, as selective pressures on fertility have been limited, resulting in domestic horses having comparatively lower reproductive efficiency, as is the case in other species such as the dromedary and (Eulmi et al. 2023; Gherissi et al. 2020).

The timing of the breeding season varies from one country to another, typically spanning 4–5 months. For instance, the breeding season starts on August 15th in African countries located in the southern hemisphere and February 15th for those in the northern hemisphere (Mota and Regitano 2012). In Morocco, the breeding season officially begins at the start of February and continues until the end of June (SOREC 2020), while generally, in North Africa, it extends from February 15th to June 15th (Aouane et al. 2019). In South Africa, the official mating season for horses commences on August 1st. While Stallions can mate year-round, mares only ovulate during the summer months.

Schulman et al. (2003) conducted a study using 25 years of recorded reproduction data from South African Thoroughbred horses (Schulman et al. 2003). They reported an average annual live foal production rate ranging from 52.3% to 62.0%. The percentages of slips and stillborn foals ranged from 2.5% to 3.9% and 7.5% to 9.4%, respectively. The "barren and no return" category showed a percentage of 29.0% to 35.1% (Schulman et al. 2003). The relatively low fertility rate observed in African horses has been noted across different breeds, countries, and with different breeding purposes (Gómez et al. 2020). Factors contributing to this include hormonal dysfunction, genital infections in mares, parasitic infections, and inadequate management prior to the breeding season (Mota and Regitano 2012; Sullivan et al. 1975). The extended reproductive cycle, combined with a long prepubertal period, extended gestation period, high generation interval and low fertility rate, along with occurrences of genital pathologies, abortions, and neonatal mortality, collectively contribute to the limited numerical productivity of horse herds, particularly in extensive breeding systems. Considering that reproductive traits generally have low estimates of heritability and are selected in horses through indirect methods (Taveira and Mota 2007), achieving genetic improvements in this aspect is neither rapid nor straightforward, especially in species like horses with prolonged intergenerational intervals.

11.6.3 Governmental Conservation Programs

11.6.3.1 Characterization Programs

Characterization serves as a fundamental prerequisite for determining which genetic resources need to be conserved and in what order of priority. It plays a pivotal role in optimizing the utilization of resources, especially in situations of limited availability. A prime example is the extensive study carried out by Kefena et al. (2012) on the morphological and phenotypic characterization of the Ethiopian equine population (Kefena et al. 2012). This study not only established a valuable database containing essential information on the geographical distribution of Ethiopian horses but also laid the groundwork for managing and conserving subsequent generations of the five indigenous horse populations. However, certain challenges, such as the elusive behavior of wild horses, their relatively small numbers, and other factors, impeded the morphological characterization of certain wild horse breeds like Borana, Kundido, and Somali/Wilwal. The Somali/Wilwal breed, for instance, is classified as endangered (FAO DAD-IS 2021).

In this context, the lack of widespread utilization of biotechnologies, including molecular markers and "omics" technologies, in the characterization process across African countries presents a significant hurdle in advancing the conservation of the continent's equine genetic heritage. These advanced biotechnologies have the potential to provide valuable insights into the genetic diversity, population structure, and unique traits of horse breeds, which can be critical for their conservation and sustainable management. Efforts to integrate such biotechnological tools into the conservation strategies can enhance the understanding of equine genetics, aid in making informed conservation decisions, and contribute to the long-term viability of these valuable genetic resources in Africa.

11.6.3.2 Conservation Programs

Issues around conservation of AnGR, including approaches, specific challenges in Africa, and needed actions, are described in Chap. 25. In the realm of conservation of African horse genetic resources, two primary strategies are employed: in situ conservation and ex situ conservation. In situ conservation involves the preservation of species within their natural habitats, allowing for ongoing evolution and adaptation. There is a notable scarcity of conservation initiatives for endangered equine populations across Africa. For example, the Kundido wild horses in Ethiopia faced the challenge of domestication due to poor knowledge of horse management practices, inbreeding, and high market demand for newborn foals. To address this, collaborative efforts between Haramaya University, the Ethiopian Biodiversity Institute, and the Oromia Environment, Forest, and Climate Change Authority (OEFCCA) were initiated. This partnership aims to research the Kundido mountain horse species, design sustainable conservation strategies, including ex situ conservation. It also hopes to develop ecotourism opportunities to generate income from genetic resources. Such collaborations are pivotal in increasing the wild Kundido horse population and boosting tourism income.

In the Namib Naukluft Park, there is an intriguing program that studies wild horses in conjunction with the conservation of plants and the interaction between these wild animals and their environment. Predation by hyenas poses a significant threat to foals, leading to management strategies such as the construction of sanctuaries, managing water and food, as well as maintaining a diverse prey population to deter hyenas. Despite these efforts, challenges remain, including a decline in foal survival rates, possibly leading to reproductive challenges in the future.

Understanding and managing wild horse behavior, their response to human care, and the dynamics of their interactions with their environment are crucial aspects of effective conservation. The complexities in managing wild horses are exemplified by conflicts between scientists, advocates, opponents, and stakeholders. In North America, Boyce et al. (2021) highlighted how scientists conducting research may find themselves caught between various interests and stakeholders, affecting their ability to provide informed decision-making (Boyce et al. 2021).

Ex situ conservation involves preserving genetic resources outside their natural habitat. This can be achieved through cryopreservation, where genetic material is stored at extremely low temperatures. Cryopreservation offers numerous benefits, including cost-effectiveness, long-term conservation, reduced risk of loss, minimal space requirement, and the retention of genetic material for future improvement. However, its systematic use is often limited in developing African countries due to the lack of specialized facilities, unreliable energy supply, and difficulties in obtaining liquid nitrogen. Despite these challenges, efforts are being made to establish cryopreservation practices in countries like Senegal, Morocco, and Algeria. Ultimately, both in situ and ex situ conservation strategies, when properly implemented and collaboratively managed, can play vital roles in preserving equine genetic diversity and ensuring the sustainable future of these valuable genetic resources in Africa.

11.7 Conclusion and Future Perspectives

The chapter reviews the origin, number, qualities, characteristics, and importance of equine livestock in Africa. The quality spectrum ranges from the hardiness exhibited by all indigenous strains and breeds to the disease resistance or tolerance observed in certain populations. It also extends to their adaptation to endemic pathologies. The stamina and saddle ability found in the noblest breeds underscore their significance.

These multifaceted attributes of horses in Africa serve to reinforce the recommendations by the FAO that there is a need to maintain a high degree of genetic diversity within domestic species. For example, given the current era of global warming, it can be envisaged that genes from the African equine populations adapted to the hot environment could in the future be introgressed into their European horses, although the appropriateness of such action could be in question.

At present, these populations stand as a testament to adaptation to the harsh and challenging environments of their habitats. They represent a valuable asset, contributing significantly to agricultural, cultural, and transport activities. Consequently, they merit thorough investigation, preservation, and enhancement. The following strategies may be implemented for the resilience and sustainability of African horse genetic resources.

11.7.1 Genetic Markers Studies

A good knowledge and understanding of breed-specific markers in African horses is critical. It has helped, for example, in consolidating populations that were previously assumed to be distinct. Genetic markers play a pivotal role in facilitating parentage verification. Construction of phylogenetic tree for breeds within and across continents helps give a clear understanding of breed relationships and origin. This visual representation illuminates the intricate interconnections between breeds and underscores the impact of human migrations on their evolution. The high cost and advanced complexity of such investigation should no longer present barriers, given the increasing accessibility within educational and research establishments and institutions to funds and human resources. The execution of such research endeavors would not only contribute to the objective classification of diverse breeds but also serve to oversee the establishment of comprehensive studbooks.

11.7.2 Biochemical Studies

In Africa, where many countries lack the necessary infrastructure for molecular studies, investigations into erythrocyte markers can have significant utility. The extensive diversity of blood groups among horses affords an accuracy of approximately 98%. Nonetheless, it is worth noting that other biochemical attributes, particularly those associated with mare's milk, also hold promise for insightful exploration and examination.

11.7.3 Morphometric Studies

An accurate description of a population necessitates an extensive array of measurements. Such comprehensive studies facilitate the development of precise and intricate breed standards. Morphometric analyses rely on straightforward and easily manageable tools such as measuring boards, tape measures, and cameras. This renders conducting a morphometric study involving tens or even hundreds of horses a feasible endeavor for students working on their theses. The cumulative effort of numerous student studies would collectively offer an intricate portrayal of the equine population within each African country.

11.7.4 Anthropological Studies

Horses hold a prominent place in rituals and beliefs across cultures, serving as the protagonists of numerous tales and proverbs. Their esteemed symbolism also renders them integral to celebrations and ceremonies. The realm of traditional medicine, primarily rooted in medicinal plants, remains prevalent. The upbringing of young colts receives meticulous attention, evident in practices like bathing foals on their third day of life or engaging in superstitious acts such as splitting an ear during birth to ward off repeated foal mortalities.

Traditional knowledge of physical attributes crucial for horse selection holds significant value and needs to be preserved and passed on to the younger generations. Attributes such as body proportions, the angles of radius bones, or distinctive tufts of hair sometimes constitute an informal standard, or at the very least, merit inclusion within one. Yet, this indigenous empirical knowledge is often disregarded by researchers and stakeholders tasked with establishing standards. The perception that the concept of standards and morphological criteria is absent contradicts the rich cultural heritage encapsulated within oral traditions.

Morphological criteria, like the width between the jaws, can be used in scientific validation and incorporation into breed standards. If they appear unrelated, they still warrant documentation as a repository of traditional wisdom. This cultural legacy is deserving of rediscovery, scholarly assessment, and preservation to perpetuate the memory of the continent, enrich our understanding of these breeds, and honor the wisdom of indigenous traditions.

Through genetic, biochemical, morphometric, and anthropological studies, opportunities arise to explore rearing methods, dietary practices, equine healthcare, and more. The conservation of traditional harnesses and saddlery not only upholds the identity of equestrian cultures but also safeguards local arts and crafts, potentially offering economic prospects through tourism. Innovations within the technical sphere can distinctly benefit draft horses. For instance, adoption of a modern horse pull collar over the less efficient breast band harness for plowing or cart pulling could considerably diminish power loss and unnecessary energy expenditure. The creation of affordable horse pull collars by local artisans and their gradual popularization could substantially advance the well-being of millions of draft horses and their owners throughout Africa.

Venturing beyond the equestrian realm, mention must be made of the seed drill for direct sowing, a pivotal agricultural tool whose significance aligns with the expansive tracts cultivated by horses, mules, and donkeys in Africa. This ecologically friendly implementation drastically reduces time and physical exertion for both humans and equines. Its dissemination across Africa would serve multiple objectives, including enhancing animal welfare. Exploring local blacksmithing techniques offers potential for improvements without incurring additional costs, utilizing existing methods and tools. This technical enhancement, albeit partial, holds promise for broader implementation, yielding substantial economic gains due to the extensive target population.

11.7.5 Genetic Improvement

Emphasizing the intrinsic strengths of African breeds is crucial in preserving their unique genetic heritage and promoting sustainable breeding practices. Rather than attempting to mold African breeds into unsuitable sports through crossbreeding, focus should be on leveraging their inherent strengths. African breeds excel naturally in disciplines such as endurance riding, TREC, working equitation, and driving. Respecting these strengths safeguards their genetic integrity and ensures their continued relevance in diverse equestrian activities. Principles around breeding goals and strategies, and broad issues/challenges in Africa, as described in Chap. 15 (on breeding goals and strategies).

Breeding programs aimed at enhancing equine traits prioritize performance and utility attributes such as speed, endurance, and temperament. Research indicates the heritability of these traits, suggesting potential improvements through selective breeding practices (McGowan et al. 2002). Moreover, conformation and health play crucial roles in equine welfare and performance, with studies linking certain conformation traits to performance outcomes (Koenen et al. 2004). Additionally, adaptability to diverse environmental conditions is vital for African horse breeds. Studies have identified genomic regions associated with environmental adaptation, indicating opportunities for selective breeding for environmental resilience (Librado et al. 2017). Furthermore, acknowledging the cultural significance of indigenous horse breeds, research on ethnozoology and cultural aspects of horse breeding provides insights into traditional knowledge guiding breeding practices (Madimabe 2016).

Breeding purebred horses in Africa faces numerous challenges, including limited breeding stock, inadequate infrastructure, economic constraints, lack of awareness, and environmental factors. To overcome these obstacles, collaborative efforts are crucial, focusing on investing in education, infrastructure development, genetic conservation, and sustainable breeding practices. By addressing these issues, the success and sustainability of purebred breeding can be ensured, thereby fostering the growth of the equine industry in Africa.

In the context of preserving genetic diversity and maximizing the utility of equine populations, rethinking crossbreeding approaches is paramount. Rather than viewing crossbreeding as a one-size-fits-all solution, it should be approached as an experimental process, rigorously assessing adaptability and performance improvement. It is imperative to avoid repetitive crossbreeding with European breeds, which can deplete local genetic pools and diminish the hardiness of indigenous breeds. Tailoring crossbreeding for regional needs is essential to develop versatile equine populations that meet specific regional requirements. By exploring crossbreeding between Barb horses and local African breeds, it is possible to combine desirable traits from both parent breeds to benefit agriculture, transportation, cultural practices, and recreational activities.

Drawing from principles surrounding breeding goals and strategies, within-breed selection programs prioritize enhancing performance traits and genetic conservation. Additionally, crossbreeding initiatives seek to introduce desirable traits and enhance genetic diversity. Research on African horses has investigated the impacts of crossbreeding on performance and adaptability, underscoring the importance of managing hybrid vigor and maintaining breed purity. To inform future endeavors, studying successful crossbreeding initiatives, such as those involving Barb horses, provides valuable insights into enhancing versatility and performance across various disciplines. Furthermore, preserving distinct indigenous populations is crucial for maintaining genetic diversity and safeguarding cultural heritage.

Despite progress, several challenges persist. Limited access to veterinary services and breeding technologies hampers systematic breeding efforts in African countries. Indigenous horse breeds are particularly vulnerable to threats to genetic diversity, including inbreeding and population decline, necessitating the implementation of conservation programs. Moreover, economic constraints pose significant barriers, requiring sustainable funding mechanisms and capacity-building initiatives to support breeding programs. Addressing these challenges will be instrumental in advancing the equine industry in Africa and ensuring the preservation of its diverse equine heritage.

Capacity-building initiatives, including training programs for breeders and veterinarians, can enhance breeding skills and improve animal health management (AHST n.d.). Investments in veterinary infrastructure and genetic laboratories are crucial to support breeding programs, with organizations like the African Union's Inter-African Bureau for Animal Resources advocating for infrastructure development to enhance livestock production (AU-IBAR n.d.). Collaboration between governments, research institutions, and breed associations is essential for knowledge sharing and resource mobilization, as demonstrated by initiatives like the African Union's Pan-African Animal Genetic Resource (n.d.).

11.7.6 Development of African Horse Breeding

Formulating a comprehensive development strategy for the equine sector in Africa requires the establishment of regional African horse networks. These networks should be designed to champion the interests of these animals, promote local equine populations, and devise solutions for the diverse local and regional challenges associated with breeding and utilization.

The creation of studbooks for local breeds is a crucial step, with the aim of establishing studbooks at local, national, and transnational levels for African horse breeds. Such an initiative would confer visibility and formalize the identity of these populations. Additionally, it would facilitate the formation of local associations, which play a pivotal role in breed development and recognition by showcasing their contributions to traditional sports, traction, transportation activities, and equitherapy.

An effective valuation strategy for African horses must encompass the promotion of equestrian disciplines that align with their inherent strengths. While flat races generate excitement and financial activity, there is a risk that crossbred animals might surpass local breeds, even going so far as to assume their names. Equestrian disciplines like endurance racing, wherein breeds like the Nara horse, the Baguezzan, and the Bahr El Ghazal have the potential to excel, should be a focal point. These breeds exhibit qualities honed over centuries of raids and hunting expeditions. An added advantage of endurance racing is that it requires minimal infrastructure, making organization relatively straightforward. Preserving and enhancing local breeds and equestrian heritage can be achieved through traditional equestrian activities such as fantasia, long-distance races, Mata (a North Moroccan variation of Buzkashi), guugsi, djerid, and more. It is essential to empower practitioners of these disciplines, as they naturally possess the authority for judging competitions or codification due to their intimate connection with tradition.

The establishment of inter-professional groups within the equine sector, alongside awareness programs, is vital. Collaborating with professional equine organizations, these groups should collectively propose medium and long-term development strategies. Priorities must be identified for each equine population according to their aptitudes, within the context of sustainable development programs tailored to each region (traditional activities, tourism, transport, and traditional sport). These organizations should create a comprehensive strategy outlining their missions, objectives, tasks, and contributions to the public. Guided by this strategy, the equine sector should be represented in broader development initiatives of regional, continental, or international significance.

Encouraging investment in horse breeding necessitates the implementation of national aid strategies, including breeding contracts, bonuses for special or endangered breeds, stallion approval incentives, and breeding competitions. Such efforts are instrumental in propelling the equine sector's growth and success.

References

Akpo Y (2004) Contribution à l'identification des métiers liés du cheval dans la région de Dakar et comparaison avec la situation au Maroc. Thèse Méd Vét Dakar:TD04-11

Amiot R (1982) Le cheval. Ed. Presses Universitaires de France, Paris, p 125

Ancelet E (2014) Les raids d'endurance équestre. Éditions Crépin-Leblond, p 294

Angelik B (2014) Le Barbe, un cheval taillé pour l'endurance. Animogen le 24 aout. http://www.animogen.com/2014/

Anthony DW (1996) Bridling horse power: the domestication of the horse. In: Olsen S (ed) Horses through time. Carnegie Museum of Natural History, Boulder, p 63

Anthony DW (2007) The horse, the wheel, and language: how bronze age riders from the Eurasian steppes shaped the modern world. Princeton University Press, Princeton

Aouane N, Nasri A, Bekara MA, Metref AK, Kaidi R (2019) Retrospective study of the reproductive performance of Barb and Thoroughbred stallions in Algeria. Vet World 12(7):1132–1139

Arbuckle BS (2012) Animals and inequality in chalcolithic Central Anatolia. J Anthropol Archaeol 31:302–313

Arne L, Pruvost M, Reissmann M, Benecke N, Brockmann GA, Castaños P, Cieslak M, Lippold S, Llorente L, Malaspinas AS, Slatkin M, Hofreiter M (2009) Coat color variation at the beginning of horse domestication. Science 324(24):485

Arthus-Bertrand Y (2003) Chevaux. Éditions du Chêne, Paris; dans: Éric CARDINALE et Christian SEIGNOBOS 2004. Le poney musey et les pratiques vétérinaires (région de Gobo, Nord-Cameroun). Anthropozoologica, 39(1):43–60

AU-IBAR (2019) The state of farm animal genetic resources in Africa. AU-IBAR Publication

Bagtache B, Hadjouis D (1983) Deux nouvelles espèces d'Equus (Mammalia, Perissodactyla) dans le gisement atérien des Phacochères (Alger). Libyca 30:165–186

Bankoff G, Swart S (2008) Breed of empire: the Inventionof the horse in Southeast Asia and Southern Africa 1500–1950. National Institute of Agrobiological Sciences, Copenhague. ISBN 87-7694-014-4, OCLC 753966176

Barakat A (1974) Organisation et rôle des Haras au Maroc. Thèse Méd Vét, Toulouse, 92: 91

Baroin C, Boutrais J (1999) L'homme et l'animal dans le bassin du lac Tchad: actes du colloque du réseau Méga-Tchad, Orléans, 15–17 octobre 1997. Editions IRD, Institut de recherche pour le développement, Paris, pp 415–416. ISBN 2-7099-1436-0 et 9782709914369

Batello C, Marzot M, Touré AH (2004) The future is an ancient lake: traditional knowledge, biodiversity and genetic resources for food and agriculture in Lake Chad Basin ecosystems. Food and Agriculture

Organization, pp 90–91. ISBN 92-5-105064-3 et 9789251050644

Bathily A, Sow A, Kalandi M, Sawadogo GJ (2014) Paramètres morphobiométriques et biochimiques des chevaux de traction au Sénégal. Int J Biol Chem Sci 8(6):2731–2739

Baynes-Rock M, Teressa T (2021) Shared identity of horses and men in Oromia. Ethiop Soc Anim 30(3):297–315. https://doi.org/10.1163/15685306-12341603

Bazarusanga T (1995) Contribution à l'étude de l'épidémiosurveillance de la peste équine au Sénégal: enquêtes sérologiques dans les zones de Rufisque, Kaffrine et Dahra. Thèse: Med Vet: Dakar, p 33

Belin de Launay J (1877) Voyage dans le Soudan occidental, abrégé par Reliure inconnue – 1 janvier 1877. Hachette, p 307

Benecke N (2006) On the beginning of horse husbandry in the southern Balkan Peninsula-the horse bones from Kırklareli -Kanligeçit (Turkish Thrace). In: Mashkour M (ed) Equids in time and space. Honour of Véra Eisenmann. Oxbow Books, Oxford

Benhamadi ME (2020) Caractérisation et nouvelle identification de la race Barbe et Selle Algérien. Thèse de Doctorat. Université de Tlemcen

Benhamadi M, Mezouar K, Benyarou M, Bouhandasse A, Gaouar SBS (2017) Morphometric characterization of the equine Barbe breed in Northwest of Algeria. Genet Biodivers J 1(2):48–65

Benkheira MH (2000) Islam et interdits alimentaires: juguler l'animalité. PUF, Paris, pp 91–99

Berber N, Gaouar S, Leroy G, Kdidi S, Tabet Aouel N, Mehtar NS (2014) Molecular characterization and differentiation of five horse breeds raised in Algeria using polymorphic microsatellite markers. J Anim Breed Genet 131(5):387–394. https://doi.org/10.1111/jbg.12092

Bernus E (1995) Le cheval Bagzan des Touaregs: Pégase ou Bucéphale?. Cavalieri dell'Africa: storia, iconografia, simbolismo: atti del ciclo di incontri: organizzato dal Centro studi archeologia africana di Milano, febbraio-giugno, pp 75–86

Bernus E (2002) Les touaregs, Paris, Vents de sable, coll. « Initiation aux cultures nomades » p 171. ISBN 2-913252-70-2 et 9782913252707

Bisker M (2017) Fantasia, Algeria. https://www.flickr.com/photos/bisker/35832087770

Blench RM (2000) A history of donkeys and mules in Africa. In: Blench RM, MacDonald KC (eds) The origin and development of African livestock. University College Press, London, pp 339–354

Bouazzaoui I (1998) Le cheval au Maroc: Elevage et maladies. Thèse Méd Vét Rabat 34:224

Boujenane I, Touati I, Machmoum M (2008) Mensurations corporelles des chevaux Arabe-Barbes au Maroc. Rev Med Vet 159:144–149

Bouzid S (1991) Etude paléontologique des Equidés quaternaires de certains gisements marocains. Thèse de 3ème Cycle, Université Mohammed V, Rabat, p 126

Bowling TA, Clark RS (1985) Blood group and proteinpolymorphismgenefrequencies for seven-breeds of horses in the United States. Anim Blood Groups Biochem Genet 16(2):93–108. https://doi.org/10.1111/j.1365-2052.1985.tb01458.x

Boyce PN, Hennig JD, Brook RK, McLoughlin PD (2021) Causes and consequences of lags in basic and applied research into feral wildlife ecology: the case for feral horses. Basic Appl Ecol 53:154–163. https://doi.org/10.1016/j.baae.2021.03.011

Breytenbach N, Grobler JP, Bindeman H (2020) Performance trait analysis and genetic diversity of the SA Boerperd. S Afr J Anim Sci 50(1):99–108. https://doi.org/10.4314/sajas.v50i1.11

Campher JP, Hunlun C, Van Zyl GJ (1998) South African livestock breeding. Biennal Report. South African Stud Book and Livestock Improvement Associon, p 172

Caro D, Davis SJ, Bastianoni S, Caldeira K (2014) Global and regional trends in greenhouse gas emissions from livestock. Clim Chang 126:203–216. https://doi.org/10.1007/s10584-014-1197-x

Castejón F, Rubio D, Tovar P, Riber C (1994) A comparative study of aerobic capacity and fitness in three different horse breeds (Andalusian, Arabian and Anglo Arabian). Zentralbl Veterinarmed A 41(9):645–652. https://doi.org/10.1111/j.1439-0442.1994.tb00132.x

Castejón-Riber C, Muñoz A, Trigo P, Riber C, Santisteban R, Castejón F (2012) Comparative Ergoespirometric adaptations to a treadmill exercise test in untrained show Andalusian and Arabian horses. Vet Res Commun 36(1):41–46. https://doi.org/10.1007/s11259-011-9510-x

Chabchoub A, Mbarki Z, Lasfar F, Landolsi F, Turki I, Ouragh L (2006) Polymorphisme protéique sanguin chez le poney de Mogod de Tunisie. Revue Élev Méd vét Pays Trop 59(1–4):91–95. https://doi.org/10.19182/remvt.9960

Chaid-Saoudi Y (1984–1986) Etude Systématique du Genre Equus (Mammalia. Perissodactyla) de l'Epipaléolithique de Columnata (Algérie Occidentale). Libyca XXXII–XXXIV:175–208

Challamel A (1904) Bulletin du Jardin colonial et des jardins d'essai des colonies françaises, Nogent-sur-Marne (France). Jardin Colonial 3:179

Charles A (2022) Les Équidés à Tozeur (Jérid tunisien): nouveaux usages et nouvelles identités?. Les dynamiques identitaires dans les sociétés plurielles de peuplement composite, ffhal-03825098f

Chergui M, Titaouine M, Gherissi DE (2023) Descriptive typology and structural analysis of camel farms in the region of El Oued, Algeria. Genet Biodivers J 7(2):75–93. https://doi.org/10.46325/gabj.v7i2.360

Chikhaoui M, Smail F, Adda F (2018) Blood hematological values of barb horses in Algeria. Open. Vet J 8(3):330–334. https://doi.org/10.4314/2Fovj.v8i3.13

Cilek S (2009) The survey of reproductive success in Arabian horse breeding from 1976-2007 at Anadolu State Farm in Turkey. Int J Animal Vet Adv:8389396. https://medwelljournals.com/abstract/?doi=javaa.2009.389.396

Claudot-Hawad H (1988) Caravane de la soif, Edisud, 2eèdi, p 43. ISBN 2-85744-197-5 et 9782857441977

Coppock DL (1994) The Borana Plateau of Southern Ethiopia: Synthesis of Pastoral Research, Development, and Change, 1980–91. International Livestock Centre for Africa et ILRAD, 5, 9. (DAD-IS Système d'Information sur la Diversité des Animaux Domestiques – DAD-IS 2012.)

Cothran EG, Van Dyk E (1998) Genetic analysis of three South African horse breeds. J S Afr Vet Assoc 69(4):120–125. https://doi.org/10.4102/jsava.v69i4.839

Cothran EG, Van Dykb E, Van Der Merwe FJ (2001) Genetic variation in the feral horses of the Namib Desert. Namibia Tydskr S Afr Vet Ver 72(1):18–22. https://doi.org/10.4102/jsava.v72i1.603

Cynthia M (2011) The Amboseli elephants: a long-term perspective on a long-lived mammal. University of Chicago Press, Chicago. ISBN 9780226542232

DAD-IS (2008) Système d'Information sur la Diversité des Animaux Domestiques, géré et développé par la FAO

DAD-IS (2010) Domestic Animal Diversity Information System (DAD-IS), Food and Agriculture Organization of the United Nations. http://www.fao.org/dadis/Kafa/Ethiopia(Horse)

Dechambre P (1921) Traité de zootechnie, Les équides, Librairie agricole de la maison rustique, 2ème édition, vol 2

Dehoux JP, Dieng A, Buldgen A (1996) Le cheval Mbayar dans la partie centrale du bassin arachidier sénégalais. Anim Genet Resour Anim 20:35–54

Devisse J (1995) Le cheval dans l'Histoire de l'Afrique. Cavalieri dell'Africa, G. Pezzolied., Actes Rencontre Cavalieri dell'Africa, storia, iconografia, simbolismo, Centro studiarcheologiaafricana, Milano, pp 27–47

Digard JP (2002) Rose-Marie Arbogast, Benoît Clavel, Sébastien Lepetz, Patrice Méniel et Jean-Hervé Yvinec, Archéologie du cheval. Des origines à la période moderne en France. Paris, Éditions Errance, 2002, 128 p., bibl., ill. (Collection des Hespérides), Études rurales, pp 163–164. https://doi.org/10.4000/etudesrurales.128

Digard JP (2004) Une histoire du cheval. Art, techniques, société. Arles, Actes Sud, 232 p

Diouf MN (1997) Le rôle du cheval dans les exploitations du Sud Sine Saloum Rapports d'activités. ISRA, Dakar. 164 p

Djimadoum J (1994) Dominantes pathologiques chez les chevaux de trait urbain dans la région de Dakar Thèse: Med Vet: Dakar, p 60

DLIFLC (2018) Defense Language Institute Foreign Language Center. Cultural orientation: HAUSA. https://fieldsupport.dliflc.edu/products/hausa/hs_co/website/hausa.pdf

Douma AKW (1994) Paarden van eigen bodem: Het Friese paard in kort bestek. Koninklijke Vereniging "Het Friesch Paarden-Stamboek", Drachten. 9–21, 1994

Doutressoulle G (1947) L'Élevage en Afrique Occidentale Française. Éditions Larose

Doutressoulle G (1952) L'élevage au Soudan français, son économie. Alger Imbert 2:374

Dumont B, Puillet L, Martin G, Savietto D, Aubin J, Ingrand S, Niderkorn V, Steinmetz L, Thomas M (2020) Incorporating diversity into animal production systems can increase their performance and strengthen their resilience. Front Sustain Food Syst 4:109. https://doi.org/10.3389/fsufs.2020.00109

Dupuy Ch (1995) Bovins montés et chevaux, puis chevaux montés dans l'art rupestre de l'Adrar des Iforas (Mali). Cavalieri dell'Africa, G. Pezzolied., Actes Rencontre Cavalieri dell'Africa, storia, iconografia, simbolismo, Centro studiarcheologiaafricana, Milano

Edward OO, Bukola AA (2023) Ulcerative lymphangitis in a 12 year old Stallion in Ibadan, Nigeria: a case study. J Appl Vet Sci 8(2):62–66. https://doi.org/10.21608/javs.2023.142098.1204

Edwards GB (1973) The Arabian: war horse to show horse. Rich Publishing

El-Alami M (1971) Le protocole et les usages au Maroc des origines à nos jours. Dar El Kitab, Casablanca, p 186

Elfasi M, Hrbek I (1988) International Scientific Committee for the Drafting of a General History of Africa, Africa from the Seventh to the Eleventh Century, vol. 3 de Africa from the Seventh to Eleventh Century. "General history of Africa", UNESCO, coll., Paris/London/Berkeley. p 133. ISBN 92-3-101709-8 et 9789231017094

Epstein H (1971) The origin of the domestic animals in Africa, vol II. Meier & Holmes Ltd, New York/Londres/Munich, p 429

Eulmi H, Deghnouche K, Gherissi DE (2023) Dairy cattle breeding practices, production and constraints in arid and semi-arid Algerian bioclimatic environments. Int J Environ Stud 81:1238. https://doi.org/10.1080/00207233.2023.2228616

FAO (1987) FAO African agriculture: the next 25 years. FAO, Rome

FAO (2002) Animal genetic resources information. Bulletin 32:64

FAO (2006a) État des ressources zoogénétiques pour l'alimentation et l'agriculture dans le monde. FAO

FAO (2006b) L'état de la biodiversité de l'agriculture dans le secteur de l'élevage. Bulletin de la FAO

FAO stat Base de données sur l'alimentation et l'agriculture 2009

FAO (2019) Enquête sur les races de la FAO. http://www.fao.org/dad-is/en/

FAO 2016 (2003) Comité Consultatif National du Cameroun, «Rapport national sur les ressources zoo-génétiques des animaux d'élevage du Cameroun» FAO, Rome, p 49. ISBN 9789251057629. ftp://ftp.fao.org/docrep/fao/010/a1250e/annexes/CountryReports/Cameroon.pdf

FAO DAD-IS (2021) Système d'Information sur la Diversité des Animaux Domestiques

FEI (2012) FEI sports forum shaping the future. Mauritian vaulter continues on pat h to success. FEI Insight, London, p 8

FEI Endurance World Championship (2021). https://ewc2021.com/en/

FEI Fédération equestre international (2018) African equestrian is already looking at how it can build on the success achieved at this year's Youth Olympic Games. https://www.fei.org/stories/lifestyle/horse-human/africa-equestrian-solidarity

FIP Federation of International Polo (2014). https://www.fippolo.com/

Fisher HJ (1972) He Swalloweth the ground with Fiercenes and rage. The horse in the Central Sudan: I. Its introduction. J Afr Hist 13(3):367–388. https://www.jstor.org/stable/180584

Fondation Nationale d'Amélioration de la Race Chevaline(FNARC) (1988) Créé en 1988 par la loi n°82–88 du 11 Juillet

Forteau L, Dumont B, Sallé G, Bigot G, Fleurance G (2020) Horses grazing with cattle have reduced strongyle egg count due to the dilution effect and increased reliance on macrocyclic lactones in mixed farms. Animal 14:1076–1082. https://doi.org/10.1017/s1751731119002738

Fourcade V (2021) Bidart: le pur-sang « Impérial d'Euskadi » acheté par le roi du Maroc. https://www.sudouest.fr/pyrenees-atlantiques/pays-basque/bidart-le-pur-sang-imperial-d-euskadi-achete-par-le-roi-du-maroc-5958267.php

Franck L (1875) Beiträgezur Rassenkundeunsererpferde. LandwirtchaftlicheJahrbuch

Galand-Pernet P (2006) Cheval-image et cheval-mot. Problèmes d'étymologie. Problèmes d'intercompréhension entre préhistoire et linguistique. Mélanges sahariens en l'honneur d'Alfred Muzzolini. Cahiers de l'AARS, N° 10, Août: 59–78

Gandini GC, Ollivier L, Danell B, Distl O, Georgoudis A, Groeneveld E, Martyniuk E, van Arendonk JAM, Woolliams JA (2004) Criteria to assess the degree of endangerment of livestock breeds in Europe. Livest Prod Sci 91(1–2):173–182. https://doi.org/10.1016/j.livprodsci.2004.08.001

Gaouar SBS (2002) Contribution à l'étude moléculaire de la variabilité e génétique: caractérisation de deux races ovines algériennes Thèse de Magister, Université des sciences et de technologie d'Oran (USTO)

Gaouar SBS (2009) Étude de la biodiversité: Analyse de la variabilité génétique des races ovines algériennes et de leurs relations phylogénétiques par l'utilisation des microsatellites, Thèse de Doctorat, Université des sciences et de technologie d'Oran (USTO)

Gaunitz C, Fages A, Hanghøj K, Albrechtsen A, Khan N, Schubert M, Seguin-Orlando A, Owens IJ, Felkel S, Bignon-Lau O, de Barros Damgaard P, Mittnik A, Mohaseb AF, Davoudi H, Alquraishi S, Alfarhan AH, Al-Rasheid KAS, Crubézy E, Benecke N, Olsen S, Brown D, Anthony D, Massy K, Pitulko V, Kasparov A, Brem G, Hofreiter M, Mukhtarova G, Baimukhanov N, Lõugas L, Onar V, Stockhammer PW, Krause J, Boldgiv B, Undrakhbold S, Erdenebaatar D, Lepetz S, Mashkour M, Ludwig A, Wallner B, Merz V, Merz I, Zaibert V, Willerslev E, Librado P, Outram AK, Orlando L (2018) Ancient genomesrevisit the ancestry of domestic and Przewalski'shorses. Science 360(6384):111–114. https://doi.org/10.1126/science.aao3297

Gautier EF (1952) Le passé de l'Afrique du Nord. Les siècles obscurs. Payot, Paris

Gherissi DE, Gaouar SBS (2022) Camel diversity survey in El Oued region (South East Algeria). Archivos de Zootecnia 71(274):124–126. https://doi.org/10.21071/az.v71i274.5659

Gherissi DE, Monaco D, Bouzebda Z, Afri Bouzebda F, Gaouar SBS, Ciani E (2020) Camel herds' reproductive performance in Algeria: objectives and thresholds in extreme arid conditions. J Saudi Soc Agric Sci 19(7):482–491. https://doi.org/10.1016/j.jssas.2020.09.002

Giammatteo G (2019) How to Capitalize Anything «24 Horse breeds», Inferno Publishing Company, p 366. ISBN 978-0-9850302-9-2 et 0-9850302-9-1

Gillet H (1964) Pâturages et faune sauvage dans le nord Tchad (rapport de mission). J Agric Tradit Bot Appl 11(5–7):155–176. https://www.persee.fr/doc/jatba_0021-7662_1964_num_11_5_2767

Gómez MD, Sánchez MJ, Bartolomé E, Cervantes I, Poyato-Bonilla J, Demyda-Peyrás S, Valera M (2020) Phenotypic and genetic analysis of reproductive traits in horse populations with different breeding purposes. Animal 14(7):1351–1361. https://doi.org/10.1017/S1751731120000087

Gouraud JL (2002) L'Afrique par monts et par chevaux. Éditions Belin, novembre 2002, p 175. ISBN 2-7011-3418-8, LCCN 2004541387

Guedaoura S (2012) Identification et caractérisation des principales races équines en Algérie: Le Barbe et ses dérivés. Université Chadli Bendjedid El Tarf, Doctorat

Haddad MM, Jemmali B, Bedhiaf A, Bedhiaf S, Djemali M (2014) Diversité génétique de la population des Poneys des Mogod du Nord- Ouest de la Tunisie. J New Sci 3:22–28. https://www.jnsciences.org/agri-biotech/11-volume-3/14-diversite-genetique-de-la-population-des-poneys-des-mogods-du-nord-ouest-de-la-tunisie.html

Hadjouis D (1993) Répartition paléogéographique et biostratigraphique de Equus algericus. Anthropologie 97:135–141. https://djillali-hadjouis.fr/articles_pdf/R%C3%A9partition_pal%C3%A9og%C3%A9ographique_et_biostrati_de_Equus_algericus.pdf

Hall SJG, Raune J (1993) Livestock breeds and their conservation: a global review. Conserv Biol 7(4):815–825. https://doi.org/10.1046/j.1523-1739.1993.740815.x

Hall LK, Larsen RT, Knight RN, McMillan BR (2018) Feral horses influence both spatial and temporal patterns of water use by native ungulates in a semi-arid environment. Ecosphere 9(1):1–15. https://doi.org/10.1002/ecs2.2096

Hamami DM (1989) Au carrefour du Soudan et de la Berbérie: le sultanat touareg de l'Ayar. Études Nigériennes, Niamey 55:133–134

Harris AW (1952) Our forty-Dollar horse, and other reminiscences. Published by Privately Printed, Chicago

Heidorn LA (1997) The horses of Kush. J Near East Stud 56(2):105–114. https://www.jstor.org/stable/545752

Helland J (1980) Social Organization and Water Control Among the Borana of Southern Ethiopia, ILRI (aka ILCA and ILRAD), p 9

Hendricks BL (1995) International encyclopedia of horse breeds. University of Oklahoma Press, Norman, p 158

Hendricks BL (2007) International Encyclopedia of Horse Breeds, Norman, 2nd edn. University of Oklahoma Press, p 107. ISBN 0-8061-3884-X, OCLC 154690199

Hiemstra SJ, Drucker AG, Tvedt MW, Louwaars N, Oldenbroek JK, Awgichew K, Abegaz Kebede S, Bhat PN, da Silva Mariante A (2006) Exchange, use and conservation of farm animal genetic resources. Identification of policy and regulatory options. Centre for Genetic Resources, the Netherlands (CGN), Wageningen University and Research Centre

Husser B (2017) Le cheval barbe dans son destin franco-algérien (1542–1914). Thèse diplôme d'archiviste-paléographe: Histoire moderne et contemporaine: Paris, Ecole nationale des chartes

INRA (2014) Institut National de la Recherche Agronomique

ISRA (2003) Institut Sénégalais de Recherches Agricoles, «Rapport national sur l'état des ressources zoogénétiques au Sénégal» Rome, FAO. ISBN 9789251057629

Jahnke HE, Sievers M (1981) Agricultural mechanisation and the demand for agricultural machinery and equipment in Africa to the year 2000. UNIDO/FAO, Vienna/Rome. (Mimeograph)

Jamali YH (2020) Le Cheval Barbe. Actes Sud, p 272. ISBN-10: 2330131119

Jansen T, Forster P, Levine MA, Oelke H, Hurles M, Renfrew C, Weber J, Olek K (2002) Mitochondrial DNA and the origins of the domestic horse. In: 13° Congress of the Europeean Anthropological Association. Zagreb, 30 Août-3 Septembre 2002

Jemmali B, Haddad MM, Ouled Ahmed H, Lasfer F, Ben Aoun B, Ezzar S, Kribi S, Gtari S, Ezzaouia MH, Rekik B (2015) Investigation de la diversité génétique des races Barbe et Arabe Barbe en Tunisie. J New Sci Agric Biotechnol 21(1):948–956

Jussiau R, Papet A, Rigal J, Zanchi E (2013) Amélioration génétique des animaux d'élevage. Educagri Editions, p 365

Kaplan LZ (1974) American saddle horses in South Africa. Cape and Transvaal Printers, Cape Town

Kebede AA (2023) A short history of Awi Agew horse culture, Northwestern Ethiopia. Cogent Arts Hum 10(1). https://doi.org/10.1080/23311983.2023.2231705

Kefena E, Dessie T, Han JL, Kurtu MY, Rosenbom S, Beja-Pereira A (2012) Morphological diversities and ecozones of Ethiopian horse populations. Anim Genet Resour 50:1–12. https://doi.org/10.1017/S2078633612000021

Kefena E, Rosenbom S, Han J, Dessie T, Beja-Pereira A (2021) Genetic diversities and historical dynamics of native Ethiopian horse populations (Equus caballus) inferred from mitochondrial DNA polymorphisms. Genes 12(2):155. https://doi.org/10.3390/genes12020155

Khadka R (2010) Global horse population with respect to breeds and risk status, Uppsala. Faculty of Veterinary Medicine and Animal Science – Department of Animal Breeding and Genetics, pp 62–95

Koçkar MT (2012) Horse breeds and distribution. Thesis Eskişehir Osmangazi University Mahmudiye Vocational School Instructor Eskişehir, p 248

Koenen EPC, Aldridge LI, Philipsson J (2004) An overview of breeding objectives for warmblood sport horses. Livest Prod Sci 88(1–2):77–84. https://doi.org/10.1016/j.livprodsci.2003.10.011

Kreuz W (1958) Types and breeds of Horses in the Union of South Africa. Proceedings of the First Congress of the South African Genetic Society. Publikasies van die Universiteit van Pretoria, South African Genetic Society, 1 à 4(7): 53

Kuwait times (2022) Lesotho's mountain jockeys race in the mist. Lifestyle: Features Monday, October 17, 2022, p 13. https://storage.com/pdf/2022/oct/17/p13.pdf

Larrat R (1947) L'élevage du cheval au Sénégal. Rev Elev Med Vet Pays Trop 1(4):257–265

Le Couteulx de Canteleu, J-E (1880) Les chevaux de Dongola, Bulletin de la Société nationale d'acclimatation et de protection de la nature de France. 3e série, VII

Lehmann WP (1992) Linguistic and archeological data for handbooks of protolanguages » dans Proto Indoeuropeans Studies in honor of M. Gimbutas, Skomal et al editions

Lejosne P (2023) La santé des équidés de travail, un paramètre incontournable dans l'approche global health de certaines régions d'Afrique. Thèse Présentée à l'Université Claude Bernard Lyon 1 (Médecine – Pharmacie) pour obtenir le titre de Docteur Vétérinaire. https://bibnum.univ-lyon1.fr/nuxeo/nxfile/default/f7e77733-c4e7-4de3-9585-987893e265d3/file:content/THv_2023LYO1V016.pdf

Lira J, Linderholm A, Olaria C, Brandström Durling M, Gilbert MT, Ellegren H, Willerslev E, Lidén K, Arsuaga JL, Götherström A (2010) Ancient DNA reveals traces of Iberian Neolithic and Bronze Age lineages in modern Iberian horses. Mol Ecol 19:64–78. https://doi.org/10.1111/j.1365-294x.2009.04430.x

Loiseau P, Martin-Rosset W (1988) Evolution à long terme d'une lande de montagne pâturée par des bovins et des chevaux. I. Conditions expérimentales et évolution botanique. Agronomie 8(10):873–880. https://doi.org/10.1051/agro:19881006

Loucougaray G, Bonis A, Bouzillé J-B (2004) Effects of grazing by horses and/or cattle on the diversity of coastal grasslands in western France. Biol Conserv 116:59–71. https://doi.org/10.1016/S0006-3207(03)00177-0

Ly C, Fall B, Camara B, Ndiaye CM (1998) Le transport hippomobile urbain au Sénégal – Situation et importance économique dans la ville de Thiès. Revue Élev Méd Vét Pays Trop 51(2):165–172. https://doi.org/10.19182/remvt.9643

Mac O (2015) Know your horse breed: Percheron. Farmer's Weekly Magazine June 27, 2015

Machmoum M, Mezrour L, Boujenane I, Piro M (2019) Morphologie comparative des chevaux arabes de lignées égyptienne, polonaise et Desert Breed. Rev Mar Sci Agron Vét 7(1):158–166

Manson K, Knight J, Harvey G (2011) Burkina Faso, Bradt Travel Guides, coll. « Bradt Guides », 41. ISBN 978-1-84162-352-8 et 1-84162-352-0

Marcenac LN (1969) Encyclopédie du cheval: 2éme édition librairie Maloine S A. OG, 1, p 1248

Marchant L (2018) The Fascinating Feral Horses of the Namib Desert. Rhino Africa. https://blog.rhinoafrica.com/2018/08/24/fascinating-feral-horses-namib-desert/

Martin G, Barth K, Benoit M, Brock C, Destruel M, Dumont B, Grillot M, Hübner S, Magne M-A, Moerman M, Mosnier C, Parsons D, Ronchi B, Schanz L, Steinmetz L, Werne S, Winckler C, Primi R (2020) Potential of multi-species livestock farming to improve the sustainability of livestock farms: a review. Agric Syst 181:102821. https://doi.org/10.1016/j.agsy.2020.102821

Mason IL (1951) The classification of West African livestock. Commonwealth Agricultural Bureaux, p 26

Mason IL (1988) A world dictionary of livestock breeds, types and varieties, 3rd edn. CAB International, Wallingford

Mason IL (1996) A world dictionary of livestock breeds, types and varieties, 4th edn. C.A.B International, p 273

Mauny R (1965) Tableau géographique de l'Ouest africain au moyen âge d'après les sources écrites, la tradition et l'archéologie. Revue belge de Philologie et d'Histoire 43(1):145–148

Mebarki M, Kaidi R, Benhenia K (2018) Morphometric description of Algerian Arab-Barb horse. Rev Med Vet 169(7–9):185–190

Ménard C, Duncan P, Fleurance G, Georges J-Y, Lila M (2002) Comparative foraging and nutrition of horses and cattle in European wetlands. J Appl Ecol 39:120–133. https://doi.org/10.1046/j.1365-2664.2002.00693.x

Mendelsohn R (2003) The challenge of conserving indigenous domesticated animals. Ecol Econ 45(3):501–510. https://doi.org/10.1016/S0921-8009(03)00100-9

Meyer C (2019) Cheval Barbe, sur dico-sciences-animales.cirad.fr, Montpellier. Dictionnaire des Sciences Animales

McGowan CM, Fordham T, Christley RM (2002) Incidence and risk factors for exertional rhabdomyolysis in thoroughbred racehorses in the United Kingdom. Vet Rec. 151:623–626. https://doi.org/10.1136/vr.151.21.623

Ministry of information of Eretria (2021) A Monthly Newsletter prepared by the Ministry of Agriculture. https://shabait.com/2021/01/08/ministry-of-agriculture-using-artificial-insemination-to-improve-horse-breeds/

Mohamadou M (2007) Contribution à l'identification des métiers du cheval au Cameroun. Université Cheikh Anta Diop, Dakar

Mota MDS, Regitano LCA (2012) Some peculiarities of horse breeding, livestock production, Khalid Javed. IntechOpen. https://doi.org/10.5772/50519

Mulualem T, Molla M, Getachew M (2015) Assessment of livestock genetic resource diversity in Ethiopia: An implication for conservation. J Genet Environ Resour Conserv 3(2):150–163

Mustefa A, Engdawork A, Sinke S, Hailu A (2022) Phenotypic characterization of Gesha horses insouthwestern Ethiopia. Genet Resour 3(5):36–50. https://doi.org/10.46265/genresj.KPIL8781

Naundrup PJ, Svenning J-C (2015) A geographic assessment of the global scope for rewilding with wild-living horses (Equus ferus). PLoS One 10:e0132359. https://doi.org/10.1371/journal.pone.0132359

Ndiaye M (1978) Contribution à l'étude de l'élevage du cheval au Sénégal Thèse: Med. Vet Dakar, 15, 183 p

NHBS (2021) The Nooitgedacht Horse Breeders Society. History of the Nooitgedachter (co.za)

Noel KB (2021) Symbolism of funerals in the Bamiléké people of Cameroun facing the chalenge of globalisation. EPH-Int J Hum Soc Sci 6(2). https://doi.org/10.53555/eijhss.v6i2.103

Norman J (2019) FédérationInternationaleequestre: South Africa's equestrian goals. https://www.fei.org/stories/lifestyle/horse-human/south-africa-equestrian-goals

Nyock RAF (2014) Caractéristiques socio-économiques et techniques de l'élevage équin dans la région du nord-ouest Cameroun. Mémoire d'Ingénieur Agronome, FASA, Université de Dschang

O'Fahey RS (1980) State and Society in DārFūr. C. Hurst & Co. Publishers, p 96. ISBN 0-905838-04-1 et 9780905838045

OIE (2021) Manuel des tests de diagnostic et des vaccins pour les animaux terrestres. https://pastel.archives-ouvertes.fr/pastel-00947028/document

Okoro E (2017) Inside the Durbar festival in Northern Nigeria. African Voices (CNN)

Olsen SL (2017) Insight on the ancient Arabian horse from north Arabian petroglyphs. Int J Archaeol Soc Sci Arabian Peninsula 8. https://journals.openedition.org/cy/3282#article-3282

OMCB (2012) Organisation Mondiale du Cheval Barbe

Oudman L (2004) Donkeys for traction and tillage, 2nd edn. Agromisa Foundation, Wageningen, 84p

Ouragh L (2006) Le cheval barbe: caractéristiques génétiques et phylogénétiques. Le cheval Barbe Rev de l'OMCB:63–64

Outram AK, Bendrey R, Olsen SL, Kasparov A, Zaibert V, Thorpe N, Evershed RP (2009) The earliest horse harnessing and milking. Sci Univ Exeter 323:1332–1335

Pâques V (1954) Les Bambaras, Presses Universitaires de France, coll. «Monographies ethnologiques africaines», 7, 36

Paris M (2014) The Kundudo feral horses of Ethiopia. Institute for Breeding Rare and Endangered African Mammals

Parker R (2012) Equine Science, Cengage Learning. 4e éd., 608, p 63. ISBN 1-111-13877-X

Pearson RA, Vall E (1998) Performances et conduite des animaux de trait en Afrique sub-saharienne: une synthèse. Revue ÉlevMédvét Pays Trop 51(2):155–163. http://revues.cirad.fr/index.php/REMVT/index

Perez R, Valenzuela S, Merino V, Cabezas I, García M, Bou R, Ortiz P (1996) Energetic requirements and physiological adaptation of draught horses to ploughing work. Anim Sci 63:343–351. https://doi.org/10.1017/S1357729800014909

Petrus I-C (2003) Les hybrides interspécifiques chez les équidés, Thèse de doctorat, Ecole Nationale Vétérinaire, Maison Alfort, 144 p

Pienaar JJ, Du Toit ASA (2008) Management of intellectual capital in the South African horseracing industry. South S Afr J Inf Manag 3(10):1–13. http://www.sajim.co.za/default.asp?to=peer3vol10nr2

Pierre (1905) Etude sur le Soudan agricole: Recueil d'hygiène et de médecine vétérinaires militaires. T VI. p 626

Pinguet E (2019) Venez participer à la course d'endurance la plus dure du monde. Cheval Magasine

Pinguet E (2020) Les chevaux sauvages de Namibie sont menacés. Cheval Magazine

Piro M, Alyakine H, Ezzaouia M, Lasfar F, Ouled Ahmed H, Ouragh L (2019) Analyse génétique and relations phylogénétiques du cheval Barbe par l'utilisation des microsatellites. Rev Mar Sci Agron Vét 7(1):149–157

Podliachouk L, Queval R (1961) Les groupes sanguins des Équidés du Tchad. Ann Inst Pasteur 14:133–136

Porter V (2020) Mason's world dictionary of livestock breeds, types and varieties, CABI publishing series, 6th edn. CABI, coll, p 360. ISBN 1789241537

Porter V, Alderson L, Hall SJG, Sponenberg DP (2016) Mason's world encyclopedia of livestock breeds and breeding, vol 1, 4th edn. CAB International, 107 p. ISBN 1-84593-466-0, OCLC 948839453

Promerová M, Andersson LS, Juras R, Penedo MCT (2014) Worldwide frequency distribution of the 'G aitkeeper' mutation in the DMRT 3 gene. Anim Genet 45(2):274–282. https://doi.org/10.1111/age.12120. (SASBA 2016. SA Stud Book Association)

Pütz R, Schlottmann A (2020) Contested conservation – neglected corporeality: the case of the Namib wild horses. Geogr Helv 75:93–106. https://doi.org/10.5194/gh-75-93-2020

Rahal K, Guedioura A, Oumouna M (2009) Paramètres morphométriques du cheval barbe de Chaouchaoua.
Rev Méd Vét 160:586–589. https://www.revmedvet.com/2009/RMV160_586_589.pdf

Rebts MV (2017) Le Nooitgedacht, une jeune race. Cheval Mag 543:59–60

Reygasse M (1922) Etudes de palethnologie maghrébine (2e série). Recueil des notices et mémoires de la Société archéologique, historique, et géographique. de Constantine. LIII, pp 159–204

RIM (1992) Nigerian National Livestock Resource Survey. (IV vols). Report by Resource Inventory and Management Limited (RIM) to FDL&PCS, Abuja, Nigeria

Rischkowsky B, Pilling D (2007) List of breeds documented in the global databank for animal genetic resources, annex to the state of the world's animal genetic resources for food and agriculture. FAO, Rome, p 101. ISBN 9789251057629

Rousseau É (2014) (ill. Yann Le Bris), Tous les chevaux du monde. Delachaux et Niestlé, « Gharwawi », 412-413. ISBN 2-603-01865-5.

Rouvier R, Wiener G (2009) L'amélioration génétique animale. Editions QUAE GIE; 1er edition. ISBN-13: 978-2759203000.P220

SADA (2006) South Africa Department of Agriculture 2006 South African Country Report on Farm Animal Resources, South Africa Department of Agriculture. https://www.fao.org/agriculture/animal-production-and-health/en/

San A (1869) C.R. Ac. Sc., 6 déc., LXIX:, 1204–1207.

Sanou KF, Ouédraogo S, Nacro S, Ouédraogo M, Kaboré-Zoungrana C (2016) Durabilité de l'offre et valeur nutritive des fourrages commercialisés en zone urbaine de Bobo-Dioulasso, Burkina Faso. Cahiers Agric 25(1):15002

SASB (2003) The South African Stud Book and Livestock Improvement Association (SASB)

SAVBS (2019) South African Vlaamperd Breeders Society, Vlaamperd Horses in South Africa. http://www.savlaamperdtelersgenootskap.co.za/p2/sa--vlaamperd/vlaamperde-in-suid-afrika.html

SCHT (2012) Société des Courses Hippiques de Tunisie. https://www.schippique.tn/webstecourse/

Schulman ML, Marlow CHB, Nurton JP (2003) A survey of reproductive success in South African Thoroughbred horse breeding from 1975 to 1999. J S Afr Vet Ass 74:17–19. https://doi.org/10.4102/jsava.v74i1.492

Sean Woo (2004) General counsel to Sen. Brownback, or John Scandling, chief of staff to Rep. Wolf, per description on p. 11 of the report. Public domain as work of U.S. federal employees

Sebbag S (2002) Contribution à l'étude des caractéristiques morphologiques du poney de Mogod en Tunisie (thèse de doctorat). École nationale de médecine vétérinaire de Sidi Thabet, Sidi Thabet, p 92

Simianer H, Marti SB, Gibson J, Hanotte O, Rege JEO (2003) An approach to the optimal allocation of conservation funds to minimize loss of genetic diversity between livestock breeds. Ecol Econ 45:377–392. https://doi.org/10.1016/S0921-8009(03)00092-2

Simon DL (1992) Summary of breeds in the EAAP animal genetic data bank. Hannover Livest Prod Sci 32:92–93. https://doi.org/10.4236/ojas.2012.23025

Sirikoi (2023) Additional activities 2023, horse riding. https://content.wetu.com/Resources//21571/additional_activities_20231.pdf

SOREC (2020) Société Royale d'Encouragement du Cheval: Procédures Administratives- Reproduction. https://haras.sorec.ma/univers-des-haras-nationaux/proceduresadministratives/reproduction

SPANA (2008) Society for the Protection of Animals Abroad. https://spana.org/about-us/our-work/morocco/

Starkey PH (1994) A world-wide view of animal traction highlighting some of the key issues in eastern and southern Africa. In: Starkey P, Mwenya E, Stares J (eds) Improving animal traction technology. Proceedings of the 1st work animal traction network for Eastern and Southern Africa, Lusaka, Zambia, January 18–23, 1994. CTA, Wageningen, pp 6–81

Sullivan JJ, Turner PC, Self LC, Gutteridge HB, Bartlett DE (1975) Survey of reproductive efficiency in the Quarter-Horse and Thoroughbred. J Reprod Fert Suppl:23315318

Szathmary H (2021) Common Native French Horse Breeds (History, Facts & Information). horseyhooves.com, part of the Hopnetic network

Talbaoui EM (1986) Study of the hemoglobin and the biochemical serum profile of the Arabe-Barbe horse. These (DoctoratVeterinaire). 94 tables; 76 graphs; 141 ref.; Summaries (Ar, En, Fr). Availability: CND, BP 826, Haut Agdal, Rabat – Morocco

Talley GHJ (2015) Fantasia: performing traditional equestrianism as heritage tourism in Morocco. A thesis for the degree of master of arts in anthropology. University of California Los Angeles, p 72. https://escholarship.org/content/qt6s92s4ft/qt6s92s4ft_noSplash_97a480e7b49d8b1c47b74fde758e61e2.pdf

Taveira RZ, Mota MDS (2007) Genetic and quantitative evaluation of breeding traits in Thoroughbred mares. Rev Electron de Vet VIII(5):1–11. https://www.redalyc.org/pdf/636/63612669009.pdf

TheHorseGuide.com. Archived from the original on 2010-12-25

Thomas P (1879) Note sur quelques Equidés fossiles des environs de Constantine. Bull Soc Algér de Climatologie 16:3–4

Trujillo JM, Steniu C, Christian LC, Ohno S (1962) Chromosomes of the horse, the donkey, and the mule. Chromosoma (Berl) 13:243–248

Valette H (2015) Invisible workers: the economic contributions of working donkeys. Horses and Mules to Livelihoods, The Brooke

Van der Merwe FJ, Martin J (2002) Four southern African horse breeds. Anim Genet Resour Inf 32:57–72. https://doi.org/10.1017/S1014233900001565

Viganò M (2014) Kondudo Feral Horses, essentials. http://www.gursum.com/menu/kundudo-whildhorse.pdf

Weiss D, Wardrop KJ (2010) ématologie vétérinaire de Schalm, 6e édn, Wiley-Blackwell

Wessels (1991) Die Anglo-Boere-Oorlog 1899–1902. 'nOorsig van die militêreverloop van die stryd, Bloemfontein, Oorlog Museum. cité par Van Dyk et Cothran en 1998, p 120

Westermarck E (2003) Les cérémonies du mariage au Maroc, 2nd edn. Du jasmin, Clichy, 393p

WHWO (2021) World Horse Welfare Organization. https://www.worldhorsewelfare.org

Witkowska-Piłaszewicz O, Cywińska A, Michlik-Połczyńska K, Czopowicz M, Strzelec K, Biazik A, Parzeniecka-Jaworska M, Crisman M, Witkowski L (2021) Variations des paramètreshématologiques et biochimiques chez les poneyssains. BMC Vet Res 17:38. https://doi.org/10.1186/s12917-020-02741-5

York JC, Ulrich C (2009) Maroc: Prendre soin des animaux, avec l'American Fondouk. Global Voices. https://fr.globalvoices.org/2009/05/12/8845/

Yusuf DA, Ahmed A, Zhu J, Usman AM, Gajale MS, Zhang S, Jialong J, Hussain JU, Zakari AT, Yusuf AA (2023) Quest for an innovative methodology for retrofitting urban built heritage: an assessment of some historic buildings in Kano metropolis, Nigeria. Buildings 13:1899. https://doi.org/10.3390/buildings13081899

Open Access This chapter is licensed under the terms of the Creative Commons Attribution 4.0 International License (http://creativecommons.org/licenses/by/4.0/), which permits use, sharing, adaptation, distribution and reproduction in any medium or format, as long as you give appropriate credit to the original author(s) and the source, provide a link to the Creative Commons license and indicate if changes were made.

The images or other third party material in this chapter are included in the chapter's Creative Commons license, unless indicated otherwise in a credit line to the material. If material is not included in the chapter's Creative Commons license and your intended use is not permitted by statutory regulation or exceeds the permitted use, you will need to obtain permission directly from the copyright holder.

African Water Buffalo Genetic Resources, Diversity, and Unique Features

12

Sahar S. E. Ahmed, Amal A. M. Hassan, Ibrahim A. H. Barakat, and Hassan M. Al Ashmaoui

Abstract

This chapter presents the buffalo genetic resources in Africa. The first section (Sect. 12.1) discusses the origin and distribution of African buffalo with emphasis on the domestic water buffalo types (swamp and river buffalo). The African water buffalo phylogeny is presented in Sect. 12.2, while the various approaches to mapping the genes of buffalo and genetic diversity characterization are presented in Sects. 12.3 and 12.4, respectively. The production and husbandry systems of African buffalo and genetic resources conservation are discussed in Sects. 12.5 and 12.6. This chapter concludes with a presentation of the challenges facing buffalo production and future perspectives on the development of sustainable genetic resources management for the African buffalo.

Keywords

River buffalo · Gene mapping · Genetic diversity · Biodiversity conservation

Abbreviations

MAS	Marker-assisted selection
QTL	Quantitative trait loci
DNA	Deoxyribonucleic acid
RNA	Ribonucleic acid
rRNA	Ribosomal RNA
mtDNA	Mitochondrial DNA
SNPs	Single-nucleotide polymorphisms
CO1	Cytochrome C oxidase subunit 1
PCR-RAPD	Random amplified polymorphic DNA
BBU1 map	Chromosome 1
LG1	Linkage group 1
LG2	Linkage group 2
RH map	Radiation map
STSs	Sequence-tagged sites
GWAS	Genome-wide association studies
CAPN1	Micromolar calcium-activated neutral protease
LEP	Leptin
PCR	Polymerase chain reaction
RFLP	Restriction fragment length polymorphism
IGF2	Insulin-like growth factor 2

S. S. E. Ahmed (✉) · A. A. M. Hassan
I. A. H. Barakat · H. M. Al Ashmaoui
Department of Cell Biology, Biotechnology Research Institute, National Research Centre, Giza, Egypt

GH	Growth hormone	INHBA	Inhibin beta-A
GHR	Growth hormone receptor	STAT5A	Signal transducer and activator of transcription5
GHRH	Growth hormone releasing hormone	LHR	Luteinizing hormone receptor
SSCP	Single-strand conformation polymorphism	ERα	Estrogen receptor-α
IGF2	Insulin-like growth factor 2	CYP19A1	Cytochrome P450 aromatase
ADG	Average daily gain	BMP15	Bone Morphogenetic Protein 15
Q	Glutamine		
aa	Amino acid	FSHB	FSH beta-subunit
H	Histidine	LF	Lactoferrin
MY	Milk yield	LEP	Leptin
FY	Milk fat yield	GnRHR	Gonadotropin releasing hormone receptor
PY	Milk protein yield		
FP	Milk fat percentage	PIT-1or POU1F1	Pituitary-Specific Transcription Factor Gene
PP	Milk protein percentage	G	
LP	Lactose percentage	F	Furcation
LY	Lactose yield	MDMs	Molecular detection methods
ADMYD	Average daily milk yield deviation	GD	Genetic diversity
LL	Lactation length	TMA	Trait Measuring Association
k-casein *(CSN3)*	Kappa-casein		
ABCG2	ATP Binding Cassette Subfamily G Member 2	AN	Accession number
		(MHC) subclass IIα	The BoLA-DRB3 gene is one of the major histocompatibility complex
OLR1 or *Ox-LDL* receptor 1 or *LOX1*	Oxidized low-density lipoprotein receptor 1	FMD	foot and mouth disease
PPARGC1A or PGC1A	Proliferator-activated receptor gamma, coactivator 1 alpha	IL8RB	The interleukin 8 Receptor Gene
SCD1	Stearoyl-CoA desaturase 1	NCBI	National Center for Biotechnology Information
PRL	Prolactin		
PRL promoter	Prolactin promoter	D-loop	Mitochondria non-coding control region
PIT-1 or *POU1F1*	Pituitary-specific transcription factor	H	Heavy strand
Text	Microsatellite markers	L	Rich in guanines and a light strand
β-LG	β-Lactoglobulin		
FSHR	Follicle-stimulating hormone receptor	tRNA	Transfer RNA
		NADH	Nicotinamide adenine dinucleotide
IGF-1	Insulin-like growth factor1	COX	Cytochrome c oxidase
IGF-1R	Insulin-like growth factor1- receptor	ATP	Adenosine triphosphate

CYTB   Cytochrome b gene
PCGs   Protein-coding genes
ARTs   Assisted reproductive technologies
AI   Artificial insemination
IVEP   In vitro embryo production

12.1 Introduction

The buffalo belongs to the Bovidae family (bovids), and two genera are recognized, *Syncerus* and *Bubalus*. The genus *Syncerus* is the wild African buffalo (*Syncerus caffer*) while the genus Bubalus is made up of four species, namely *Bubalus arnee* (water buffalo wild type), *Bubalus bubalis* (water buffalo domestic type), *Bubalus mindorensis* (tamaraw), *Bubalus depressicornis* (lowland anoa), *and Bubalus quarlesi* (mountain anoa) (Huffman 2006; Owak and Paradiso 1983).

The African buffalo wild type is a large animal with large horns for both sexes. Their bodies are covered with thick reddish hair at birth, while by age the mature animals are dark brown or black. According to the chromosomal numbers, Di Berardino and Iannuzzi (1981) divided the African buffalo into two groups: *Syncerus caffer nanus* (2n = 54) and *Syncerus caffer caffer*.

The African buffalo distribution patterns are influenced by season, topography, and herd sizes and are divided into forest buffalo, West African savanna buffalo, Central African buffalo, and southern savanna buffalo (Furstenburg 2007, Fig. 12.1). Some of the African nations with buffalo populations are Ethiopia, Somalia, Zambia, Zimbabwe, Namibia, Botswana, Mozambique, South Africa, Kenya, and Tanzania. African buffalo are formidable fighters and have never been successfully domesticated. Although in the past, the population of African buffalo was in the tens of millions, it now stands at about 900,000, mostly found in the savannas of Eastern Africa. One of the reasons for this decline in numbers is hunting, both for food and for sport. The African buffalo is not considered to be in danger of extinction as long as it remains protected in parks and reserves, although habitat loss continues (Huffman 2006; ICUN 2016). In Asia, dwarf buffalo populations known as mountain anoa (*Bubalus quarlesi*) and lowland anoa (*Bubalus depressicornis*), which are similar in appearance and size (weight 150–300 kg) have been classified as endangered, and the population continues to decrease (ICUN 2016). Both populations are found on the island of Sulawesi in Indonesia and the nearby island of Butung. The tamaraw (*Bubalus mindorensis*) is native to the island of Mindoro in the Philippines. It is believed to have once also thrived on the greater island of Luzon. The tamaraw buffaloes are smaller than other buffaloes, have more hair and are dark brown to grayish black in color. They can be aggressive and have been known to attack humans (Huffman 2006). Tamaraw has been considered highly endangered.

The wild water buffalo (*Bubalus arnee*) is one of the largest members of the Bovidae family in the Indian subcontinent. It was historically distributed across Europe to South and Southeast Asia; they are currently restricted to India, Nepal, Bhutan, Thailand, Cambodia, and Myanmar. With an estimated current global population size of <4000 and a rapidly declining population trend, they are considered as "endangered" by the IUCN (Kaul et al. 2019).

Asiatic buffalo (*Bubalus bubalis*) is a domesticated water buffalo that is assumed to have originated from the wild buffalo (*Bubalus arnee*) that was found in the northeastern region of India (Fahimuddin 1975), and is considered an important animal that affects the agricultural economy. Its worldwide population is approximately 207 million heads (FAO 2019; Rehman et al. 2021). Within the Asian region, approximately 74.80% of buffaloes are found in South Asia, 12.80% in East Asia, and only 8.40% in South-East Asia (Hamid et al. 2017).

There are two types of domestic water buffalo: swamp buffalo, which inhabits Northeast India, Bangladesh, China, and Southeast Asian countries, and river buffalo, which inhabits the Indian subcontinent, South Asia, and the Mediterranean area (Egypt, Italy, and the Balkans). The two buf-

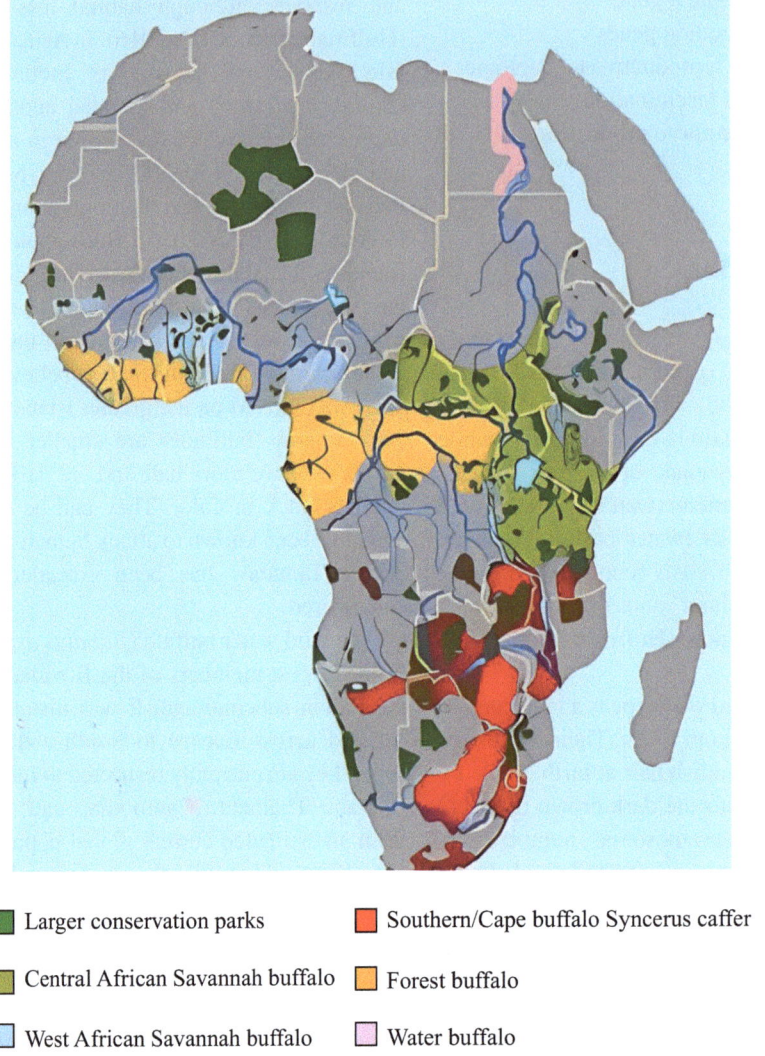

Fig. 12.1 Distribution of wild and domestic water buffalo in Africa. (From Furstenburg (2007), licensed under CC-BY 4.0, modified by adding distribution of water buffalo)

falo types were introduced into South America and Australia (Cockrill 1981).

The river buffalo is black with horns showing a double curvature (at first, they are directed downward and backward, and then curl upward in a spiral) whereas the swamp buffalo is generally dark gray with white chevrons on the throat, white socks, and semi-circular horns that always remain approximately in the same plane as the forehead (Macgregor 1941; Zhang et al. 2020). The river buffalo population constitutes approximately 70% of the world's domestic water buffalo population (FAO 2015). River and swamp buffaloes differ in behavioral traits and purposes with different geographical distributions as well (Cockrill 1974; Borghese 2011). The river buffalo has been selected mainly for milk and meat production with several recognized breeds, while the swamp buffalo has primarily been used for draught power (Borghese 2011; Kierstein et al. 2004; Yindee et al. 2010). The two types also differ in the number of chromosomes; while the river buffalo has 25 pairs of chromosomes (n = 50), the swamp buffalo has 24 chromosome pairs (n = 48) due to a tandem fusion translocation between river buffalo chromosomes 4 and swamp buffalo chromosome 1. All the chromosomal pairs, including the sex chromosomes, are acrocentric in the two types except for the first five chromosomal pairs, which are bi-armed (Kierstein et al. 2004; Iannuzzi and Di Meo 2009; Michelizzi et al. 2010).

The natural history of domesticated mammals (Clutton-Brock 1996), molecular data, including single-nucleotide polymorphisms (SNPs) and mitochondrial DNA (mtDNA) (Kumar et al. 2007), and archaeological evidence (Colli et al. 2018) have demonstrated the westward migration of the river buffalo from its domestication epicenter to areas as far as the Balkans, Italy, and Egypt. Colli et al. (2018) suggested two independent events of migration that are well matched with population genetic diversity: the first being the proto-Mediterranean gene pool of Italy that came through the Balkans and the second the proto-Middle Eastern gene pool toward the Caspian Sea and Mesopotamia, later extended to Turkey and northern Africa (only in Egypt).

The data of migration routes suggested that the combined effects of multiple migration events which occurred at different stages of the post-domestication history shaped the genetic structure of each water buffalo population in different areas of the world where the buffalo in Italy, Balkan, and Turkey, have unique characteristics that differ from Egyptian water buffalo, which suggests that the Egyptian buffalo is an intermediate breed between the Indian (Eastern) and European (Italian) buffaloes (Michelizzi et al. 2010; Şahin et al. 2014, 2019).

12.2 African Water Buffalo Phylogeny

12.2.1 Maternal Origin

Mitochondrial DNA (mtDNA) polymorphism has a tremendous usage in the study of genetic variation and evolution of various species. This is facilitated by the speed and ease of genotyping a large number of individuals and by the complete absence of genetic recombination in mtDNA, which is inherited through the maternal lineage only. Two of the studies (Ramadan and El-Hefnawi 2008; Hassan et al. 2009) focused on the mitochondrial D-loop region, which is the most variable part of mtDNA due to a higher substitution rate than in the rest of the mitochondrial genome. The results of Hassan et al. (2009) showed the mixed origin of the two Egyptian populations (the Northern and the Southern) and probably indicated the presence of many ancient maternal lineages for the Egyptian buffalo with high diversity in mtDNA sequences. The results also suggested that the Egyptian buffalo did not derive from a unique breed imported from Mesopotamia or India, but rather from multiple migrants. The two studies, however, confirm that the Egyptian buffalo has a common origin with river buffalo in other regions of the world (Hassan et al. 2009; Ramadan and El-Hefnawi 2008).

Because the genetic analysis of the whole mitochondrial genome (mitogenome) of livestock, rather than partial mtDNA sequences, has become an important and indispensable part in breed characterization (Ślaska et al. 2014), mitogenome sequencing was applied in population genetic and taxonomic studies in Egyptian buffalo. Youssef et al. (2021) sequenced and analyzed the mitogenome of 29 Egyptian Northern and Southern river buffalo populations and deposited the mitogenome sequences in the GenBank under the accession numbers: MT237604- MT237632. The authors also made a comparison between the Egyptian buffalo mitogenome and the mitogenome sequences of Bubalus (6 river buffaloes, 112 swamp buffaloes, and the anoa (*Bubalus depressicornis*)) to study their phylogenetic relationships. The phylogenetic analyses supported the paraphyly of the water buffalo species since the anoa (*B. depressicornis*), endemic to Indonesia, was found to be the sister group of the river buffalo, with the swamp buffalo at the outside. This result suggested that river buffalo and swamp buffalo belong to distinct species (Fig. 12.2). With the recent availability of the wild water buffalo (*Bubalus arnee*) mitogenome sequence (Pacha et al. 2021; Hassan et al. 2022), Hassan et al. (2022) reconstructed the phylogenetic relationship between the members of the genus *Bubalus*, including the wild, river, and swamp water buffalo types, and reported close genetic architecture

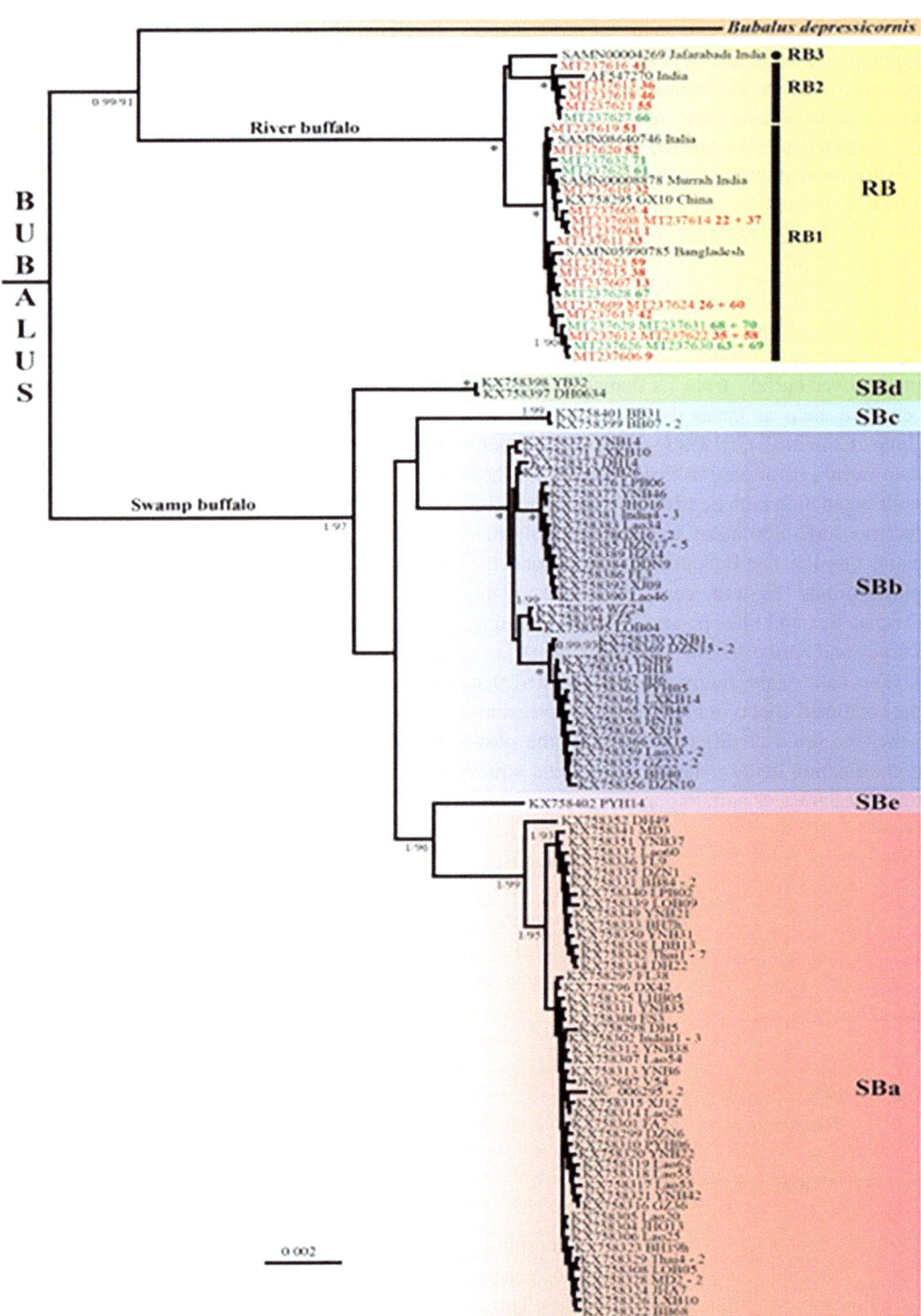

Fig. 12.2 Phylogeny of Bubalus based on complete mitochondrial genomes. The tree was reconstructed with MrBayes software using the 118 mitogenomic haplotypes of Bubalus identified in our alignment of 16,356 nucleotides. The tree was rooted with Syncerus (not shown). Haplogroups showing more than 0.5% of nucleotide divergence are highlighted by different colored rectangles, and those found in the swamp buffalo are named SBa to SBe (the name of the haplotype is followed by a dash and a number when the same haplotype was found in

between the river and wild water buffalo types. The paraphyly of water buffalo species, including *lowland anoa* and *mountain anoa*, river and wild water buffalo, are in a separate clade, leaving swamp buffalo on the outside. The authors confirmed the hypothesis that the river buffalo and the swamp buffalo are two separate species and that the river buffalo is a descendant of wild water buffalo (*Bubalus arnee*) (Fig. 12.3).

12.2.2 DNA Barcoding

DNA barcoding as a genetic tool focuses on the delineation of species rather than their relationships (Hajibabaei et al. 2007). It differs from molecular phylogeny in that the main goal is not to determine classification but to identify an unknown sample in terms of a known classification (Kress et al. 2005). It seeks to assemble a standardized reference library. DNA barcoding is based on the premise that a short standardized sequence can distinguish individuals of a species because genetic variation between species exceeds that within species (Hebert et al. 2003, 2004). The 5′ half (648-bp region) of the mitochondrial cytochrome C oxidase subunit 1 (CO1) gene was proposed as the standard marker for DNA barcoding in animal species (BOLD 2023). Enrichment of barcode databases with COI barcode sequences in different animal taxa has become important for the identification of animal source in food samples to prevent commercial fraud. The 5′ half of the mitochondrial COI gene was used as a DNA barcode for differentiation between different buffalo species and other bovid species, including the Egyptian buffalo. One study differentiated between river buffalo, swamp buffalo, sheep, goat, and cattle (Oraby et al. 2011).

The barcode-based phylogenetic tree assembled using the 5′ half of CO1 gene sequence (650-bp) successfully delineated the investigated species (Fig. 12.4). While another study using the same segment differentiated successfully between the closely related members of the different buffalo species (river buffalo, swamp buffalo, lowland anoa, and African buffalo) (Fig. 12.5) (Hassan et al. 2018).

12.2.3 Genetic Distance Among African Water Buffalo Populations

The Egyptian buffalo population admixture revealed high gene flow between populations, prompting the conclusion that the Egyptian buffaloes belong to one breed (El-Kholy et al. 2007). To detect the genetic variation among Northern and Southern Egyptian buffalo populations, the polymerase chain reaction random amplified polymorphic DNA (PCR-RAPD) and microsatellite markers were used. The results revealed that the populations have the same origin and belong to one breed without significant genetic differences between them (Othman et al. 2012). Other studies used the microsatellite loci to determine the genetic differences between Egyptian buffalo populations from different agroclimatic regions. The results recorded a high level of inbreeding in all populations, a low level of genetic differentiation and reduction of heterozygosity among the different buffalo populations (Abou-Bakr and Attia 2012; Merdan et al. 2020). Although several studies confirmed that the Egyptian buffalo is one breed, it has many local names according to the governorate in which they are raised, such as Al-Buhairi, Al-Menoufi, and Al-Saeedi (Fig. 12.6).

Fig. 12.2 (continued) least two individuals). Among the 29 mitogenomes of Egyptian river buffalo, the 21 samples collected from North are indicated in red, whereas the eight samples collected from South are indicated in green. The four mitogenomes of river Buffalo assembled using SRA data available in GenBank are named with the SAMN accessions. For nodes supported by bootstrap percentage (BP) _ 90 in the RAxML analysis (see details in Material and Methods), the two values correspond to the posterior probability (PP calculated using MrBayes software, left of the slash) and BP (right of the slash). An asterisk is used when both MrBayes and RAxML analyses provided maximal support values, i.e., PP ¼ 1 and BP ¼ 100, respectively. No information was provided for nodes supported by BP <90. (From Youssef et al. (2021), licensed under C-C BY 4.0)

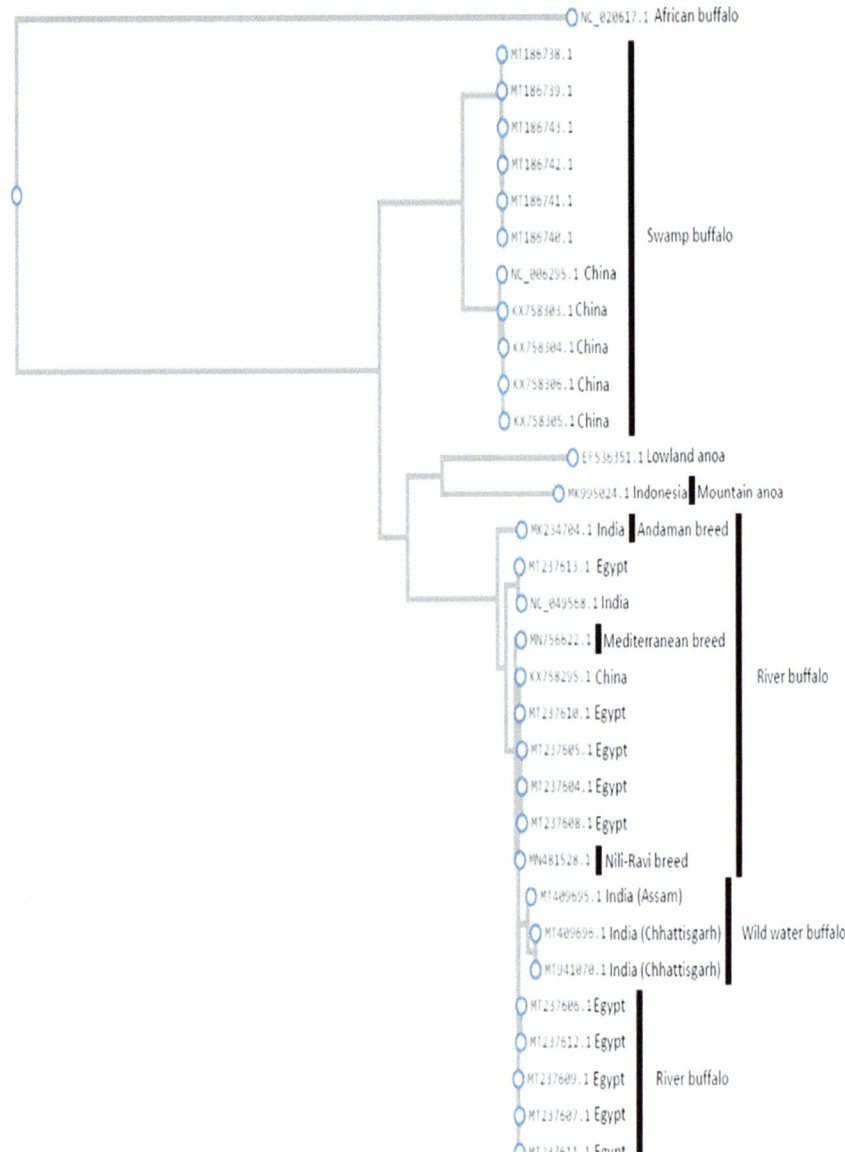

Fig. 12.3 Phylogeny of genus *Bubalus* based on complete mitochondrial genomes. The tree was rooted with African buffalo. (From Hassan et al. (2022), licensed under C-C BY 4.0).

12.3 Buffalo Gene Mapping

Gene mapping is an important tool for mapping disease genes or quantitative trait loci (QTL) used for genetic improvement.

12.3.1 Physical Gene Map

The buffalo genome was established as a first step to identifying loci controlling physiological and quantitative traits. A plethora of research

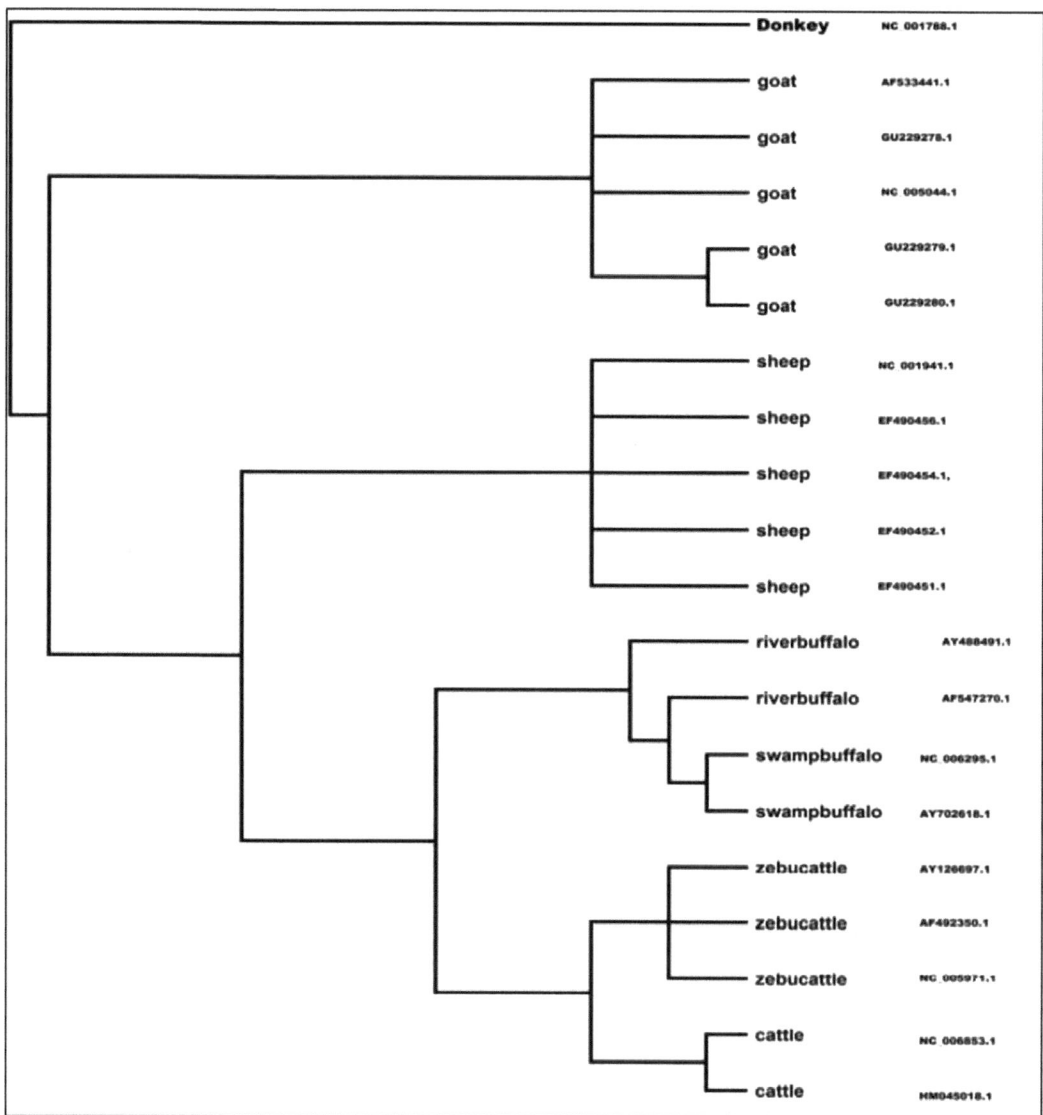

Fig. 12.4 A phylogenetic tree assembled using the 5′ half of CO1 gene sequence (650-bp) in all the investigated species and the donkey sequence as an outgroup. Accession numbers are given to all samples. (From Oraby et al. (2011), licensed under C-C BY 4.0)

studies around the world have investigated and assign genes of buffalo traits on different chromosomes of the Egyptian buffalo (de Hondt et al. 1997; Nahas et al. 1993, 1996, 1997, 1999; El Nahas et al. 1996a, b, 1997, 2001; Hassanane et al. 1993, 1994; Oraby et al. 1998; Othman and Nahas 1999; Ramadan et al. 2000; Mahfouz 2007). The physical cytogenetic buffalo map provided information for 388 gene loci, most of them having homologs mapped in other species (Di Meo et al. 2008) and/also assignment of the genes' syntenic groups to chromosomes (El Nahas et al. 1996a).

12.3.2 Radiation Map

The radiation map (RH map) of river buffalo was established based on cattle-derived markers, with aim to provide a more extensive resource for

Fig. 12.5 COI (mitochondrial cytochrome C oxidase subunit 1) barcode-based phylogenetic tree constructed using sequences from four buffalo taxa groups and a sequence of goat (*Capra hircus*) as an outgroup. *Hap.* haplotype. (Hassan et al. 2018, licensed under C-C BY 4.0).

Fig. 12.6 Egyptian buffalo types

positional candidate mapping of genes associated with complex traits, obtain information for genome-wide scans and detect chromosomal regions contributing to economically important traits and for large-scale physical mapping of the river buffalo genome. However, despite the important information provided by the RH map, the buffalo RH map is still based on limited studies.

The construction of the RH map for river buffalo chromosome 1 (BBU1 map) presented two linkage groups which included linkage group 1 (LG1; a total of 58 markers (39 coding genes, 15 microsatellites and four ESTs) spanning the entire short arm and extending across much of the long arm, with 47 markers placed as framework and 11 markers placed in bins) and linkage group 2 (LG2; 11 markers (nine coding genes and two microsatellites) covering the remaining portion of the BBU1 long arm with 9 framework markers and two placed in bins) (Miziara et al. 2007). The first-generation whole genome radiation map (RH map) for buffalo was developed by Amaral et al. (2008). The results of their study detected 3990 markers on the BBURH5000 panel, of which 3072 were cattle-derived SNPs. The remaining 918 were classified as cattle sequence tagged site (STS), including coding genes, ESTs, and microsatellites. RH mapping analysis by El Nahas et al. (2014) reported that the BBU7 RH5000 map is composed of 22 RH markers (19 genes and three sequence-tagged sites (STSs). Nine genes (ADRB3, ATP2C1, COPB2, CRYGS, P2RY1, SLC5A3, SLC20A2, SST, and ZDHHC2) and one microsatellite (CSSM043) were added by Stafuzza et al. (2015) to the set of markers previously mapped on BBU1.

12.4 Genetic Diversity in Africa Buffaloes

Advanced molecular technologies such as whole-genome sequencing, genome-wide association study (GWAS), gene expression profiling, next-generation sequencing (RNA and DNA), and genome-wide CHIP-seq scanning are used to detect genetic variants and study gene regulation, gene functioning, and single nucleotide polymorphism (SNP) required to differentiate molecular phenotypes and predict breeding values to screen superior mates to produce improved progeny and sustainable management of genetic resources (See Chap. 20).

12.4.1 Marker-Assisted Selection and Quantitative Trait Loci

Quantitative trait loci (QTL) are specific regions on chromosomes that harbor genes controlling reproduction, production, and disease resistance traits. QTL analysis is an important tool to correlate between two types of information, phenotypic data (trait measurements) and genotypic data (molecular markers) that are used to carry-out marker-assisted selection (MAS) of desirable traits required for inclusion in a breeding program and sustainable management of the animal genetic resources (See Chap. 20).

12.4.1.1 Growth Trait
Growth trait is an important economic trait in farm animals. Studying the genes associated with growth is essential for the identification and inclusion of the desirable genotypes in breeding strategies. However, few studies on growth traits have been conducted on buffalo.

The micromolar calcium-activated neutral protease (*CAPN1*) gene is reported as the primary enzyme in the postmortem tenderization process, and the Leptin (*LEP*) gene is considered a candidate gene for performance, carcass, and meat quality traits in beef cattle. Using the method of polymerase chain reaction (*PCR*)—restriction fragment length polymorphism (*RFLP*), Othman et al. (2011) reported genetic diversity in these genes in the buffalo population, whereby tested buffaloes were genotyped as CC for *CAPN1* and AA for *LEP* gene. Genetic characterization of growth hormone and its receptor genes was studied using the *PCR-RFLP* method. The result showed that all tested buffaloes are genotyped as LL for growth hormone (*GH*) and GG for growth hormone receptor (*GHR*) gene, with GenBank accession numbers JN809916 and JN809917, respectively (Othman et al. 2012). Insulin-like growth factor 2 (*IGF2*) is considered a marker for growth traits in farm animals. *PCR-SSCP* (single-strand conformation polymorphism) results recorded a novel single-nucleotide polymorphism (SNP), C287A in exon 10 of the IGF2 gene, with SSCP frequency patterns of 54% (AA), 32.5% (AC), and 13.5% (CC). Association analyses revealed that the AA individuals had a higher average daily gain (ADG) than individuals of CC and AC genotypes from birth to 9 months of age (Abo-Al-Ela et al. 2014). The hypothalamic hormone, growth hormone-releasing hormone (*GHRH*) and its receptor are candidate genes for growth and milk traits. The *GHRH* acts as a stimulator of pituitary growth hormone (*GH*) synthesis and secretion. It has an influence on increasing milk yield, protein, and fat content. The genetic diversity of buffalo *GHRH* and *GHRHR* genes based on *PCR-RFLP* and sequence alignment revealed that all tested buffalo are of genotype AA (GenBank accession number is KC295414) (Othman et al. 2015).

12.4.1.2 Milk Traits
Genetic characterization and association of genetic variants with milk traits (e.g., milk secretion, protein, and fat) is important in breeding programs for milk production (Togashi et al. 2019). Table 12.1 shows the genetic diversity of candidate milk trait genes of kappa-casein (*k-casein (CSN3)*), ATP Binding Cassette Subfamily G Member 2 (*ABCG2*), oxidized low-density lipoprotein receptor 1 (*OLR1* or *Ox-LDL* receptor 1 or *LOX1*), peroxisome, proliferator-activated receptor gamma, coactivator 1 alpha (*PPARGC1A* or PGC1A), stearoyl-CoA desaturase 1 (*SCD1*), prolactin (*PRL*), prolactin promoter (*PRL* promoter), pituitary-specific

Table 12.1 Genetic diversity of milk traits genes and association with different milk production measures

Gene	Function	Molecular detection method	Genetic diversity	Trait	Accession number	Marker-trait association	Reference
k-CN (CSN3)	Casein gene family	PCR-RFLP	Monomorphic	–	–	–	Othman (2005), Othman et al. (2011), Gouda et al. (2013), El Nahas et al. (2013), and Al-Shawa et al. (2020)
		PCR-nucleotide sequencing	Two polymorphic variants in exon IV at codon 135 and 136	–	–	–	
PRL	Role in mammary gland development, lactogenesis, and expression of milk protein genes	PCR-RFLP	Monomorphic	–	–	–	Othman et al. (2011)
PRL promoter		PCR-SSCP and nucleotide sequencing	Two genotypes	MY, FP, PP, LP, and LL	–	no significant association with milk traits production and composition	El-Magd et al. (2015)
PIT-1 (POU1F1)	Role in pituitary development and hormone secretion in mammals	PCR-RFLP	Monomorphic	–	–	–	Othman et al. (2011)
Microsatellite markers (BM1706, BMS711, BM143, BM415, BM6438, ETH131, BM1443, ETH2)	Highly reproducible and specific	PCR	Polymorphic	ADMYD, FP, PP, FY, PY	–	BM415 has a significant influence on protein yield	Rushdi et al. (2017)
ABCG2	It plays a protective role	PCR-RFLP	Two polymorphisms	–	–	–	Hassanane et al. (2016)
OLR1	It plays a role in lipid metabolism	PCR-RFLP	Two polymorphisms	–	–	–	Hassanane et al. (2016)
SCD1	Its expression affects the fatty acid composition of membrane phospholipids	PCR-RFLP	Two polymorphisms	–	–	–	Hassanane et al. (2016)

(continued)

Table 12.1 (continued)

Gene	Function	Molecular detection method	Genetic diversity	Trait	Accession number	Marker-trait association	Reference
PPARGC1A (P616L; T19C, A968C)	It regulates a whole range of nuclear hormone receptors and other transcription factors, affecting adipogenesis, adipocyte differentiation, and mitochondrial biogenesis	PCR-RFLP	Two polymorphisms	–	–	–	Hassanane et al. (2016)
β-LG	It is the major whey protein of ruminant species. Is an important source of bioactive peptides	PCR-RFLP	Monomorphic	–	–	–	Al-Shawa et al. (2020)
GHR (4, 5, 6, and 8 exons)	Receptor for pituitary gland growth hormone involved in regulating body growth.	PCR-SSCP and nucleotide sequencing	Two SNPs: 380G>A and 836T>A	MY, PP, and FP	KC107765	Two SNPs were associated with higher performance of MY, FP, and PP for AA haplotype	El-Komy et al. (2020)
Candidate milk traits genes	A Molecular tool used to scan markers across the complete sets of DNA, or genomes	GWAS	Different SNPs across the genome	LP and LY	–	32 significant SNPs distributed on 18 chromosomes for LP	Awad et al. (2020)

transcription factor (*PIT-1* or *POU1F1*), microsatellite markers, β-Lactoglobulin (*β-LG*), and growth hormone receptor (*GHR*) in Egyptian buffalo.

12.4.1.3 Fertility Traits

Fertility may be considered as two traits, which are inherent fertility and expressed fertility. Inherent fertility is referred to as the genetic potential for reproduction performance and is not directly measurable (Ahmed et al. 2011). Candidate genes were investigated to detect the molecular characteristics of genes associated with fertility traits in buffalo. PCR-RFLP, SSCP, and sequencing techniques were used to detect the genetic variations in candidate genes of reproduction traits. The fertility measures were calving interval, ovarian activity, repeat breeder, anestrus, and semen traits. Table 12.2 shows the genetic variations of the candidate fertility traits genes follicle-stimulating hormone receptor (*FSHR*), insulin-like growth factor1 (*IGF-1*), insulin-like growth factor1-receptor (*IGF-1R*), inhibin beta-A (*INHBA*), signal transducer and activator of transcription5 (*STAT5A*), luteinizing hormone receptor (*LHR*), estrogen receptor-α (*ERα*), cytochrome P450 aromatase (*CYP19A1*) gene, bone morphogenetic protein 15 (*BMP15*), FSH beta-subunit (*FSHB*) lactoferrin (*LF*) and leptin (*LEP*). For semen traits, the genes are gonadotropin-releasing hormone receptor (*GnRHR*) gene and pituitary-specific transcription factor (*PIT-1or POU1F1*) gene.

12.4.1.4 Immune Traits

The immune system is regulated by several thousand genes (8–9% of the genome), indicating its high genetic priority as a critical fitness characteristic that supports species survival. Identifying and raising livestock with an inherent ability to produce superior immune responses can reduce disease incidence, improve milk quality, and increase farm profitability. However, few studies have been conducted on Egyptian buffalo immune traits.

The *BoLA-DRB3* gene is one of the major histocompatibility complex (*MHC*) subclass IIα genes. Due to its role in the development of an immune response and its highly polymorphic nature, it is considered the most important candidate gene in disease resistance. Genetic characterization of *BoLA-DRB3* exon 2 by *PCR-RFLP* method (284-bp product) revealed five genotypes, including three homozygous genotypes AA (6%), BB (34%) and DD (4%), and two heterozygous genotypes AB (36%) and BD (20%) in tested animals (Othman and Ahmed 2010). Another study correlated the polymorphism of the *BoLA-DRB3* gene (exon 2) with foot and mouth disease (*FMD*) resistance. Specifically, the genotype AA may be associated with *FMD*-resistance, especially *FMDV* serotype O, compared to the AC genotype (Othman et al. 2018).

Mastitis is an infectious disease of the mammary gland that leads to reduced milk production and changes in milk composition. The interleukin 8 receptor gene (*IL8RB*) is a genetic marker in inflammatory responses during mastitis, while complement component (C3) plays a major role as a central molecule of the complement cascade involved in the killing of microorganisms, either directly or in cooperation with phagocytic cells. The CXCR2 sequence of IL8RB 124-bp covering a part of intron 1, exon 2 and a part of 3'UTR (Genbank accession #KY399457.1) was analyzed to detect the genetic structure associated with mastitis. The result indicated a potential association between SNP C/A at c.127 and mastitis (El Nahas et al. 2017). Six novel SNPs were detected in buffalo C3 cDNA sequences, and the C>A substitution (ss: 1752816097) in exon 27, together with the milk somatic cell score, are considered indicators of mastitis (El-Halawany et al. 2017a).

12.4.2 Whole-Genome Sequencing and Characterization

Whole-genome sequencing and genome-wide association studies (GWAS) provide an opportunity for the analysis of genotype–phenotype relationships in livestock and identification of selective sweeps associated with performance traits and adaptation. A total of five annotated buffalo genome assemblies have been deposited

Table 12.2 Genetic diversity of fertility traits genes associated with different fertility measures

Gene	Function	Genetic diversity	Trait	Accession number	Marker-trait association	Reference
FSHR	It is the binding receptors for FSH hormone, expressed by granulose cells and during the luteal phase in the secretory endometrium of the uterus	Three polymorphic patterns in exon 10	Calving interval	—	Pattern 2 was recorded with high fertility	Ahmed et al. (2011)
			Repeat breeder		The genotype GG recorded with repeat breeder animals	Fouda et al. (2021)
IGF-1	Necessary for the process of follicular development and ovulation	Two polymorphic patterns	Calving interval	—	Does not affect fertility	Ahmed et al. (2011)
IGF-1R	Critical for the normal development of the mammary gland tissue during pregnancy and lactation	Four polymorphic patterns	Calving interval		Does not affect fertility	Ahmed et al. (2011)
INHBA	Involved in the recruitment and development of ovarian follicles during folliculogenesis	Monomorphic	Calving interval		Does not affect fertility	Ahmed et al. (2011)
STAT5A	Responsible for the initiation and maintenance of pregnancy in ruminants	Three polymorphic patterns	Calving interval		Pattern 2 was recorded with high fertility	Ahmed et al. (2011)
LHR	Regulates gonadal function, including steroidogenesis and gametogenesis	Monomorphic in exon 11	—	JQ885687	Not detected	Othman and Abdel-Samad (2013) and Sosa et al. (2016)
ERα	Estrogen receptors play important roles in reproduction, development of the mammary gland, growth and differentiation of cells	Two polymorphic alleles in exon C	—	JQ308795 JQ308796	Not detected	Othman and Abdel-Samad (2013)
Cyp19	Key enzyme in estrogen biosynthesis	Three polymorphic patterns in exon 2	ovarian activity	KJ551928 KJ551929	The substitution of nucleotide T by C at position 72 in exon 2 of Cyp19	Othman et al. (2014)

Gene	Function	Polymorphism	Trait	Accession	Finding	Reference
CYP19A1	Regulates estrogen biosynthesis	Four single nucleotide polymorphisms (SNPs) in exon 10 and 2 SNPs in exon 5	Anestrus		SNP-anestrus association analyses revealed that genotypes (CC, AA and GG) and alleles (C, A and G) of the −135T>C, c.559G>A and c.1394A>G SNPs respectively were high risk for anestrus	El-Bayomi et al. (2018)
BMP 15.1 BMP 15.2	Play crucial roles in ovarian follicular development and ovulation rate.	Monomorphic—in 5UTR & exon 1	repeat breeders	KU043381	Does not affect fertility	El-Bayomi et al. (2016)
FSHB	Essential for regulation of gametogenesis and follicular growth	Two polymorphic patterns		M83753.1		Sosa et al. (2017)
LF	Has wide biological roles	Two polymorphic alleles in intron 6	- Repeat breeder - Anoestrum	KX228134 KX228137 KX228135 KX228136		Eldebaky et al. (2020)
LEP	Involved in the metabolism and growth of animals and plays a role in regulation of the reproductive hormones GnRH/LH	Two polymorphisms in intron 2	Normal and abnormal ovary and uterus	AH013754.2	Does not affect fertility	El-Debaky et al. (2020)
GnRHR	Stimulate the synthesis and release of both follicle-stimulating hormone and luteinizing hormone	Two polymorphic patterns. Two SNPs as a transversion base substitution mutation at positions 20 (T/A) and 193 (A/T)	Semen trait	EU621854.1	Pattern I was seen in good motility, with incidence of 100%. Pattern II appeared only in poor semen motility.	Mahmoud et al. (2021)
		Three polymorphic patterns with 5 single nucleotide polymorphisms including one nucleotide insertion, one nucleotide deletion and 3 nucleotides substitutions between the three patterns	Repeat Breeder Anoestrum	NM_001290857.1	Pattern III reported infertile animals showed insertion was at position 4, deletion was at position 189, and the 3 nucleotides substitutions were at positions 204 (T/G), and 207 (A/T)	Sosa et al. (2016)
PIT-1 (POU1F1)	Stimulate expression of growth hormone, prolactin, thyroid hormone (TSH) β-subunit and gonadotropin-releasing hormone receptor genes	Monomorphic	Semen traits	–	Does not affect fertility	Hasanain et al. (2016)

in the National Center for Biotechnology Information (NCBI) database. The nucleotide sequences submitted to GenBank include 538 microsatellites, 490 minisatellites, 503 satellite sequences, and 273,061 whole-genome sequences. A total of 1,921,573 nucleotide and 34,831 water buffalo gene sequences, including 471 mitochondrial sequences, have been deposited in the GenBank database (Hassan et al. 2022).

The GWAS detected and annotated 436 SNPs along with 34 indels in 38 fertility-related candidate genes in Murrah buffaloes (Surya et al. 2019). For buffalo milk production traits, four GWASs have been conducted so far, including two studies performed on 1018 individual Italian Mediterranean buffaloes (de Camargo et al. 2015; Liu et al. 2018) and two on Egyptian and Brazilian buffaloes (de Camargo et al. 2015; El-Halawany et al. 2017b).

A partial genome assembly of the Egyptian water buffalo has been achieved by El-Khishin et al. (2020). A total of 21,128 genes were identified in the partial assembly including milk virgin-related genes; milk pregnancy-related genes, milk lactation related genes, milk involution-related genes and milk mastitis-related genes.

Recently, the mitogenome of 29 buffaloes from the North and South of Egypt has been sequenced and analyzed by Youssef et al. (2021). The mitogenome of the Egyptian buffalo is a circular double-stranded DNA sequence that is 16,357–16,359 base pairs in length. Similar to other mammalian mitogenomes, the two strands of buffalo mtDNA are a heavy strand (H) rich in guanines and a light strand (L) rich in adenines. The overall base composition of the mitogenome is A: 33.1%, C: 26.6%, G: 13.9%, and T: 26.4%. The genome contains 37 genes in addition to a non-coding control region (D-loop), which is the only region that has different lengths in the 29 investigated buffalo mitogenomes (926–928 bp). The 37 genes comprise 22 transfer RNA (tRNA) genes, 2 ribosomal RNA (rRNA) genes and 13 protein-coding genes. The 22 tRNA genes are tRNA-Phe, tRNA-Val, tRNA-Leu1, tRNA-Ile, tRNA-Gln, tRNA-Met, tRNA-Trp, tRNA-Ala, tRNA-Asn, tRNA-Cys, tRNA-Tyr, tRNA-Ser1, tRNA-Asp, tRNA-Lys, tRNA-Gly, tRNA-Arg, tRNA-His, tRNA-Ser2, tRNA-Leu2, tRNA-Glu, tRNA-Thr, and tRNA-Pro. The two rRNA genes encode for 12S and 16S rRNAs. The 13 protein-coding genes (PCGs) are composed of seven genes: ND1, 2, 3, 4, 4L, 5, and 6 that code for 7 reduced nicotinamide adenine dinucleotide (NADH) dehydrogenase subunits, 3 genes: COX1, 2, and 3 that code for 3 cytochrome c oxidase subunits, 2 genes: ATP 6 and 8 that code for 2 adenosine triphosphate (ATP) synthetase subunits, and a cytochrome b gene (CYTB) that encodes one subunit of cytochrome c reductase. Twenty-eight genes are located on the mitochondrial DNA heavy strand and 9 genes (ND6, tRNA-Gln, tRNA-Ala, tRNA-Ans, tRNA-Cys, tRNA-Try, tRNA-Ser1, tRNA-Glu, and tRNA-Pro) are located on the light strand. The gene arrangement in the Egyptian river buffalo mitogenome is conserved with other mammalian species. All protein-coding genes of the mtDNA have the start codon ATR (ATG or ATA), except ND4L (GTG) and ND2 (ATW = ATT or ATA). Further investigations revealed that eight of the PCGs (ND1, COX1, COX2, ATP8, ATP6, ND4L, ND5, ND6) use the TAA stop codon, whereas ND2 and CYTB use the TGA and AGA stop codons, respectively. However, the stop codons are incomplete in Cox3, ND3, and ND4. The authors detected 78 nucleotide variations along the whole mitochondrial sequence. Forty-four out of the 78 nucleotide variations were detected in the PCGs while the rest of the variations (34 nucleotide variations) were distributed as follows: 3 in the tRNA genes, 2 in the rRNA genes, and 29 in the D-loop region.

12.5 Production Systems, Husbandry Practices, and Production Characteristics

Domesticated water buffalo is an important animal for many countries throughout the world, as it contributes to human food security and social and cultural aspects (Desta 2012; Mohammed 2018). In developing countries, they play a major part in agricultural economies as important sources of milk, meat and meat products, and skin (El-Debaky et al. 2020). After dairy cattle, the water buffalo is the second most important species in terms of milk production in the world (Coroian et al. 2013). Buffalo milk has a higher content of protein, fat, lactose, and calories than cattle milk (Hamid et al. 2017). However, it only accounts for 13% of global milk consumption due to their small animal population (Liu et al. 2018). Buffalo meat is lean with lower cholesterol and fat contents and with comparatively better taste than beef (Hassan et al. 2022). Furthermore, when compared to cows, buffaloes are better at transferring low-quality forages with low digestible characteristics to milk or meat (Ibrahim 2012; El-Debaky et al. 2020; Li et al. 2020) and have a high ability to adapt and survive in a variety of habitats with varying topography, climate, and vegetation (Li et al. 2020). Buffalo also contributes a significant amount of leather to the global market (FAOSTAT 2014). Its manure is utilized as a fuel and heating source, as well as a fertilizer to enrich the soil and reduce the usage of chemical fertilizers (El-Debaky et al. 2020).

With all these distinctive qualities, buffaloes in Egypt occupy a prominent place in the life of Egyptian farmers as an important source of red meat, milk, and a long productive life (Attia et al. 2014; El-Magd 2015; FAO 2019). The Egyptian customer prefers buffalo milk because of its white color, great quality (in rural areas where unpasteurized milk is consumed), and high fat percentage (El-Magd 2015; Al-Hosary et al. 2015). The Egyptian buffalo population is around 3.5 million heads (FAO 2019; Di Stasio and Brugiapaglia 2021). The production of buffalo accounts for 44% and 36% of Egypt's total milk and meat production, respectively (FAO 2019).

Buffalo breeding management in Egypt is mainly two systems: smallholder and pri-urban production systems (See Chap. 2). The smallholder production system is traditional, characterized by one to five heads of buffalo, and consists of keeping buffaloes indoors at night and confined in fenced areas during the day. The input and output are relatively low. Buffalo milk from this system is used either for family consumption or sold to neighbors. The peri-urban production system is usually near cities and with an average herd size of nine heads per household. The system is characterized by higher milk production, longer lactation length, and better feeding. Almost all buffaloes are traditionally managed, hand milked and naturally bred. Intensive meat and milk production systems characterized by having more than 50 heads of buffaloes per farm and with better management and mechanical milking systems occur at a limited scale (Arefaine and Kashwa 2015).

The average annual red meat production of the Egyptian buffalo reached 495,000 tons, contributing 45% to overall meat produced; the average slaughter weight is 500 kg, at the age of 18–24 months; carcass yield is 51% and overall growth rate is 700 g/day (Ibrahim 2012). The birth weight of calves ranged from 22 to 36 Kg, while the weight after puberty is 200–250 Kg (Perera et al. 2005). The total milk yield is between 1266 ± 229 and 2607 ± 500, milk/Kg 6.5 ± 1.2 and 10.2 ± 1.62, lactation period/day 198.11 ± 37.8 and 258.1 ± 37.51 (Abdel-Salam and Fahim 2018).

Buffaloes are polyestrous, meaning they can breed at any time of the year. The Egyptian buffalo reproductive features are distinguished by the first calving age at 31.98 ± 4.1 to 32.5 ± 3.8 (month), calving interval at 391.9 ± 44.9 to 404.5 ± 55.7 (day), estrus cycle length 17–26 (day), length of gestation averaged 310 days, while the period from calving to the first service ranged from 49.1 ± 23.7 to 59.6 ± 29.1 (day) (Perera et al. 2005; Abdel-Salam and Fahim 2018).

12.6 Genetic Resources Conservation

Advanced biotechnology techniques, molecular genetic tools, ex situ and in situ conservation tools are used to achieve the goals of sustainable livestock management strategies (Baruselli et al. 2013).

12.6.1 Biotechnology Tools

Biotechnology is widely acknowledged as having a role to play in resolving the production restrictions of small-scale or resource-poor farmers, who account for over 70% of food production in developing countries. A wide range of assisted reproductive technologies (ARTs), from artificial insemination to advanced transgenesis and genome editing, have been effectively employed to improve livestock productivity and value. Recent advancements like recombinant DNA technology and associated procedures, monoclonal antibody techniques, embryo manipulation technologies (Baruselli et al. 2007; Hoshino et al. 2009; Selokar et al. 2014) have been introduced as well. Also, artificial insemination (AI), in vitro embryo production, sex preselection, and stem cell technology are examples of revolutionary ARTs that have boosted the production potential of domesticated animals (See Chap. 20).

Artificial insemination (AI) is the only ART tool used for improving the Egyptian buffalo. It is however only used in 1% of the medium to large herds farms. There are six AI stations owned by the government and one by the university, which house a total of 70 bulls. Artificial insemination is still performed at the research level; usually only one semen dose is offered at each estrus, conception at the first estrus being 30% (Abdel-Salam and Fahim 2018; Mohammed 2018).

12.6.2 In Situ and Ex Situ Conservation

Conservation of genetic resources is based on keeping the genetic variation as gene combinations in a reversible form and keeping specific genes of interest. In situ conservation is the maintenance of the breed within the livestock production system and its environment through the enhancement of its production characteristics. This approach needs extensive infrastructure and management, which involves the maintenance of the genetic variation through continued use by livestock keepers. Successful in situ conservation usually requires a performance recording and development of breeding programs with special emphasis on maintaining the genetic diversity within the breed by the livestock keepers, along with the willingness to invest in technologies that improve productivity, such as AI, veterinary services, or supplemental feeding (FAO 2013) (See Chap. 25).

Ex situ in vitro conservation is the preservation of genetic material in haploid form (semen and oocytes), diploid (embryos), or DNA sequences. The Convention on Biological Diversity recommended that ex situ conservation be complementary to in situ conservation for farm animal genetic resources (Prentice and Anzar 2011). Few studies have been reported for cryopreservation of semen, oocytes, and embryos to optimize the methodology in vitro. Successful cryopreservation of Egyptian buffalo ovaries using in situ oocyte cryopreservation was achieved by Abd-Allah (2009). Another study by El-Sayed et al. (El-sayed et al. 2017) proved that the cryotop is an effective tool for cryopreservation of in vitro produced Egyptian buffalo blastocysts.

12.7 Challenges of Buffalo Production and Opportunities

The improvement of the productivity of the Egyptian buffalo, preservation of its genetic resources and its sustainable management face many problems that include poor nutrition, inadequate housing, lack of appropriate veterinary care, and poor management and breeding methods. Additionally, AI, milk recording, genetic screening, and milk marketing systems are not available. Furthermore, most buffaloes at smallholder farmers are bred spontaneously with untested bulls, i.e., bulls that have not been selected for improved production and have not had their reproductive disorders investigated. Furthermore, there are no connections between recording, genetic assessment, and AI institutions, therefore, there are no buffalo genetic enhancement initiatives in Egypt. Despite a major objective to retain and protect native breed genetic materials, the risky use of imported frozen semen without awareness and measures could impair the genetic efficiency of native buffalo breeds.

12.8 Implications and the Way Forward

Advancements of biotechnology and molecular genetics combined with improved breeding programs should help advance improvements in production, environmental adaptation, and disease resistance for this economically important livestock. Therefore, future development plans for the water buffalo should focus upon:

1. Use advanced biotechnologies in buffalo breeding programs, as well as select and use superior males and females in breeding programs.
2. Create banks for oocytes, sperms, and embryos produced in vitro from animals with superior genetic traits in order to preserve the buffaloes' genetic material.
3. Focus on research data that correlates the phenotype (trait measurements) and genotype (molecular markers).
4. Develop MAS of desirable traits in breeding programs.
5. Sustainable management of the buffalo genetic resources.
6. Develop community-based breeding programs for buffalo breeders.
7. Increase efforts by state departments and scientific and research institutions on improving buffalo production in Egypt.
8. Create animal production projects and enact supporting laws and legislations that prevent the slaughter of genetically superior animals from the buffalo population.

References

Abd-Allah SM (2009) Successful cryopreservation of buffalo ovaries using in situ oocyte cryopreservation. Vet Ital 45(4):507–512. From NLM

Abdel-Salam S, Fahim N (2018) Classifying and characterizing Buffalo farming systems in the Egyptian Nile Delta using cluster analysis. J Anim Poult Prod 9(1):23–28. https://doi.org/10.21608/jappmu.2018.35763

Abo-Al-Ela HG, El-Magd MA, El-Nahas AF, Mansour AA (2014) Association of a novel SNP in exon 10 of the IGF2 gene with growth traits in Egyptian water buffalo (Bubalus bubalis). Trop Anim Health Prod 46(6):947–952. https://doi.org/10.1007/s11250-014-0588-3 From NLM

Abou-Bakr S, Attia MH (2012) Genetic characteristics of Egyptian Buffalo using DNA microsatellite markers. Egypt J Anim Prod 49:121

Ahmed S, Abd el eziz KB, Hassan NA, Mabrouk DM (2011) Genetic polymorphism of some genes related to reproductive trait and their association with calving interval period in Egyptian buffalo. Genomics Quant Genetics 2011(3):1–8

Al-Hosary AA, Ahmed LSE-D, Seitzer U (2015) First report of molecular identification and characterization of Theileria spp. from water buffaloes (Bubalus bubalis) in Egypt. Adv Anim Vet Sci 3:629–633

Al-Shawa M, El-Zarei MF, Ghazy A, Ayoub MA, Merdan SM, Mokhtar SA (2020) Polymorphism, allelic and genotypic frequencies of κ-casein and β-LG genes in Egyptian buffaloes. J Agric Sci 12(4):2020. https://ssrn.com/abstract=3770611

Amaral ME, Grant JR, Riggs PK, Stafuzza NB, Filho EA, Goldammer T, Weikard R, Brunner RM, Kochan KJ, Greco AJ et al (2008) A first generation whole genome RH map of the river buffalo with comparison to domestic cattle. BMC Genomics 9:631. https://doi.org/10.1186/1471-2164-9-631 From NLM

Arefaine H, Kashwa M (2015) A review on strategies for sustainable Buffalo milk production in Egypt. J Biol Agric Healthc 5:63–67

Attia M, Abou-Bakr Mahmoud S, Nigm A (2014) Genetic differentiation and relationship among Egyptian Nile delta located Buffalo using microsatellite markers. Egypt J Anim Prod 51(2):71–77

Awad MAA, Abou-Bakr Mahmoud S, El-Regalaty H, El-Assal SE-D, Abdel-Shafy H (2020) Determination of potential candidate genes associated with milk lactose in Egyptian Buffalo. World Vet J 10:35–42. https://doi.org/10.36380/scil.2020.wvj5

Baruselli PS, Gimenes LU, Carvalho NAT d, Sá Filho MF d, Ferraz ML, Barnabé RC (2007) O estado atual da biotecnologia reprodutiva em bubalinos: perspectiva de aplicação comercial. Revista Brasileira de Reprodução Animal 31(3):285–292

Baruselli PS, Soares JG, Gimenes LU, Monteiro BM, Olazarri MJ, Carvalho NAT d (2013) Control of Buffalo follicular dynamics for artificial insemination, superovulation and in vitro embryo production. Buffalo Bull 32:160–176

BOLD (2023) The DNA barcode of life data system (BOLD). https://www.boldsystems.org

Borghese A (2011) Situation and perspectives of buffalo in the world, Europe and Macedonia. Macedonian J Anim Sci 1(2):281–296

Clutton-Brock J (1996) A natural history of domesticated mammals. Cambridge University Press, Cambridge

Cockrill WR (1974) The husbandry and health of the domestic Buffalo. FAO, Rome

Cockrill WR (1981) The water buffalo: a review. Br Vet J 137(1):8–16. https://doi.org/10.1016/s0007-1935(17)31782-7 From NLM

Colli L, Milanesi M, Vajana E, Iamartino D, Bomba L, Puglisi F, Del Corvo M, Nicolazzi EL, Ahmed SSE, Herrera JRV et al (2018) New insights on water Buffalo genomic diversity and post-domestication migration routes from medium density SNP Chip data. Front Genet 9:53. https://doi.org/10.3389/fgene.2018.00053 From NLM

Coroian A, Erler S, Matea CT, Mireşan V, Răducu C, Bele C, Coroian CO (2013) Seasonal changes of buffalo colostrum: physicochemical parameters, fatty acids and cholesterol variation. Chem Cent J 7(1):40. https://doi.org/10.1186/1752-153x-7-40 From NLM

de Camargo GM, Aspilcueta-Borquis RR, Fortes MR, Porto-Neto R, Cardoso DF, Santos DJ, Lehnert SA, Reverter A, Moore SS, Tonhati H (2015) Prospecting major genes in dairy buffaloes. BMC Genomics 16:872. https://doi.org/10.1186/s12864-015-1986-2 From NLM

de Hondt HA, Gallagher D, Oraby H, Othman OE, Bosma AA, Womack JE, El Nahas SM (1997) Gene mapping in the river buffalo (Bubalus bubalis L.): five syntenic groups. J Anim Breed Gen 114(1–6):79–85. https://doi.org/10.1111/j.1439-0388.1997.tb00494.x. Accessed 05 Mar 2024

Desta TT (2012) Introduction of domestic buffalo (Bubalus bubalis) into Ethiopia would be feasible. Renew Agric Food Syst 27(4):305–313. https://doi.org/10.1017/S1742170511000366. From Cambridge University Press Cambridge Core

Di Berardino D, Iannuzzi L (1981) Chromosome banding homologies in Swamp and Murrah buffalo. J Hered 72(3):183–188. https://doi.org/10.1093/oxfordjournals.jhered.a109469 From NLM

Di Meo GP, Perucatti A, Floriot S, Hayes H, Schibler L, Incarnato D, Di Berardino D, Williams J, Cribiu E, Eggen A et al (2008) An extended river buffalo (Bubalus bubalis, 2n = 50) cytogenetic map: assignment of 68 autosomal loci by FISH-mapping and R-banding and comparison with human chromosomes. Chromosome Res Int J Mol Supramol Evolut Aspects Chromosome Biol 16(6):827–837. https://doi.org/10.1007/s10577-008-1229-3 From NLM

Di Stasio L, Brugiapaglia A (2021) Current knowledge on river Buffalo meat: a critical analysis. Animals 11(7) https://doi.org/10.3390/ani11072111 From NLM

El Nahas SM, Oraby HA, de Hondt HA, Medhat AM, Zahran MM, Mahfouz ER, Karim AM (1996a) Synteny mapping in river buffalo. Mamm Genome 7(11):831–834. https://doi.org/10.1007/s003359900245 From NLM

el Nahas SM, Ramadan HA, Abou-Mossallem AA, Kurucz E, Vilmos P, Ando I (1996b) Assignment of genes coding for leukocyte surface molecules to river buffalo chromosomes. Vet Immunol Immunopathol 52(4):435–443. https://doi.org/10.1016/0165-2427(96)05597-3 From NLM

El Nahas SE, Oraby HO, Othman OE (1997) Gene mapping of river buffalo by somatic cell hybridization. Egypt J Genet Cytol 27:171–179

El Nahas SM, de Hondt HA, Womack JE (2001) Current status of the river buffalo (Bubalus bubalis L.) gene map. J Hered 92(3):221–225. https://doi.org/10.1093/jhered/92.3.221 From NLM

El Nahas SM, Bibars MA, Taha DA (2013) Genetic characterization of Egyptian buffalo CSN3 gene. J Genetic Eng Biotechnol 11(2):123–127. https://doi.org/10.1016/j.jgeb.2013.08.003

El Nahas SM, Abou Mossallam AA, Mahfouz ER, Bibars MA, Sabry N, Seif El-Din S, Amaral ME, Womack JE (2014) Radiation hybrid map of buffalo chromosome 7 detects a telomeric inversion compared to cattle chromosome 6. Anim Genet 45(5):762–763. https://doi.org/10.1111/age.12196 From NLM

El Nahas SM, El Kasas AH, Abou Mossallem AA, Abdelhamid MI, Warda M (2017) A study on IL8RB gene polymorphism as a potential immunocompromised adherent in exaggeration of parenteral and mammo-crine oxidative stress during mastitis in buffalo. J Adv Res 8(6):617–625. https://doi.org/10.1016/j.jare.2017.07.002 From NLM

El Nahta S (1996) Physical mapping in the buffalo: a first step towards the improvement of production and reproduction. PhD dissertation, Cairo University, Cairo

El-Bayomi KM, El Araby IE, Ayman A (2016) Screening for single nucleotide polymorphisms in BMP 15 gene in Egyptian buffaloes. Alexandria J Vet Sci 49(2):157–162

El-Bayomi KM, Saleh AA, Awad A, El-Tarabany MS, El-Qaliouby HS, Afifi M, El-Komy S, Essawi WM, Almadaly EA, El-Magd MA (2018) Association of CYP19A1 gene polymorphisms with anoestrus in water buffaloes. Reprod Fertil Dev 30(3):487–497. https://doi.org/10.1071/rd16528 From NLM

Eldebaky H, El-Razik K, Mahmoud K, Mahmoud M, Kandiel M, Ahmed Y, Sosa A (2020) Sequence based polymorphism of Lactoferrin gene in relation to fertility in Egyptian buffaloes. Int J Vet Sci 9:97

El-Debaky H, Mahmoud KGM, EL-Razik KAA, Sosa ASA, Kandiel MMM, Ahmed YF (2020) PCR-SSCP and sequencing analysis for studying leptin gene polymorphism and its association with reproductive status of Egyptian buffalo. Egypt J Vet Sci 51(1):11–21

El-Halawany N, Abd-El-Monsif SA, Al-Tohamy Ahmed FM, Hegazy L, Abdel-Shafy H, Abdel-Latif MA, Ghazi YA, Neuhoff C, Salilew-Wondim D, Schellander K (2017a) Complement component 3: characterization and association with mastitis resistance in Egyptian water buffalo and cattle. J Genet 96(1):65–73. https://doi.org/10.1007/s12041-017-0740-8 From NLM

El-Halawany N, Abdel-Shafy H, Shawky A-E-MA, Abdel-Latif MA, Al-Tohamy AFM, Abd El-Moneim OM (2017b) Genome-wide association study for milk production in Egyptian buffalo. Livest Sci 198:10–16. https://doi.org/10.1016/j.livsci.2017.01.019

El-Khishin DA, Ageez A, Saad ME, Ibrahim A, Shokrof M, Hassan LR, Abouelhoda MI (2020) Sequencing and assembly of the Egyptian buffalo genome. PLoS One 15(8):e0237087. https://doi.org/10.1371/journal.pone.0237087 From NLM

El-Kholy AF, Hassan HZ, Amin AMS, Hassanane MS (2007) Genetic diversity in Egyptian buffalo using microsatellite markers. Arab J Biotechnol 10(2):219–232

El-Komy SM, Saleh AA, Abdel-Hamid TM, El-Magd MA (2020) Association of GHR polymorphisms with milk production in buffaloes. Animals 10(7):1203. https://doi.org/10.3390/ani10071203 From NLM

El-Magd M (2015) Effect of SNPs in prolactin promoter on milk traits in Egyptian Buffalo. Adv Dairy Res:03. https://doi.org/10.4172/2329-888X.1000128

El-sayed A, Gad A, AboelEla M, Ashour G (2017) The effeciency of cryotop in vitrification of in vitro produced Egyptian Buffalo (Bubalus Bubalis) embryos. Journal of Animal and Poultry Production 8:97–100. https://doi.org/10.21608/jappmu.2017.45784

Fahimuddin M (1975) Domestic water Buffalo. In: Domestic water Buffalo. Oxford and IBH Publishing Co, New Delhi

FAO (2013) In vivo conservation of animal genetic. https://www.fao.org/3/i3327e/i3327e.pdf

FAO (2015) The second report on the State of the World's Animal Genetic Resources for Food and Agriculture (Ed. by B.D. Scherf and D. Pilling). FAO Commission on Genetic Resources for Food and Agriculture Assessments, Rome. Available at http://www.fao.org/publications/sowangr/en/

FAO (2019) Live animals data, FAO, UN, Rome

FAOSTAT (2014) FAO statistics division. FAO, Rome. www.fao.org

Fouda M, Hemeda S, El-Bayomi K, El-Araby I, Hendam B, Ateya A (2021) Genetic polymorphisms in FSHR/ ALUI and ESRα /BGlI loci and their association with repeat breeder incidence in buffalo. J Hellenic Vet Med Soc 72(2):2869–2878. https://doi.org/10.12681/jhvms.27525. Accessed 23 Feb 2024

Furstenburg D (2007) Gemsbok. SA Game Hunt 13(4):6–11

Gouda E, Galal M, Ahmed S (2013) Genetic variants and allele frequencies of kappa casein in Egyptian cattle and Buffalo using PCR-RFLP. J Agric Sci:5. https://doi.org/10.5539/jas.v5n2p197

Hajibabaei M, Singer GA, Hebert PD, Hickey DA (2007) DNA barcoding: how it complements taxonomy, molecular phylogenetics and population genetics. Trends Genet 23(4):167–172. https://doi.org/10.1016/j.tig.2007.02.001 From NLM

Hamid MA, Zaman MA, Rahman A, Hossain KM (2017) Buffalo genetic resources and their conservation in Bangladesh. Res J Vet Sci 10:1–13. https://scialert.net/abstract/?doi=rjvs.2017.1.13

Hasanain MH, Mahmoud KGM, El-Menoufy AA, Sakr AM, Ahmed YF, Othman OE (2016) Semen characteristics and genotyping of pituitary-specific transcription factor gene in Buffalo using PCR-RFLP. Egypt J Vet Sci 47(1):13–26. https://doi.org/10.21608/ejvs.2016.1079

Hassan AA, El Nahas SM, Kumar S, Godithala PS, Roushdy K (2009) Mitochondrial D-loop nucleotide sequences of Egyptian river buffalo: variation and phylogeny studies. Livest Sci 125(1):37–42. https://doi.org/10.1016/j.livsci.2009.03.001

Hassan AAM, Balabel EA, Oraby HAS, Darwish SA (2018) Buffalo species identification and delineation using genetic barcoding markers. J Genet Eng Biotechnol 16(2):499–505. https://doi.org/10.1016/j.jgeb.2018.07.006 From NLM

Hassan AMA, Youssef NA, El-Ghor AA, El Nahas SM (2022) Mitogenome analyses of water Buffalo: closeness of the genetic architecture of river Buffalo and wild Buffalo (Bubalus arnee) excludes swamp Buffalo. Egypt J Vet Sci 53(1):43–47. https://doi.org/10.21608/ejvs.2021.90888.1268

Hassanane MS, Gu F, Chowdhary BP, Andersson L, Gustavsson I (1993) In situ hybridization mapping of the immunoglobulin gamma heavy chain (IGHG) gene to chromosome 20q23-q25 in river buffaloes. Hereditas 118(3):285–288. https://doi.org/10.1111/j.1601-5223.1993.t01-1-00285.x From NLM

Hassanane MS, Chowdhary BP, Gu F, Andersson L, Gustavsson I (1994) Mapping of the interferon gamma (IFNG) gene in river and swamp buffaloes by in situ hybridization. Hereditas 120(1):29–33. https://doi.org/10.1111/j.1601-5223.1994.00029.x From NLM

Hassanane M, Ali N, Abo B, Zeinab, Abdel-Hamid G, Hassan N (2016) Genetic polymorphism for some quantitative trait genes in Egyptian cattle and Buffalo. J Pharm Biol Sci 4(3):74–84

Hebert PD, Cywinska A, Ball SL, deWaard JR (2003) Biological identifications through DNA barcodes. Proc Biol Sci 270(1512):313–321. https://doi.org/10.1098/rspb.2002.2218 From NLM

Hebert PD, Stoeckle MY, Zemlak TS, Francis CM (2004) Identification of birds through DNA barcodes. PLoS Biol 2(10):e312. https://doi.org/10.1371/journal.pbio.0020312 From NLM

Hoshino Y, Hayashi N, Taniguchi S, Kobayashi N, Sakai K, Otani T, Iritani A, Saeki K (2009) Resurrection of a bull by cloning from organs frozen without cryoprotectant in a −80 degrees C freezer for a decade. PLoS One 4(1):e4142. https://doi.org/10.1371/journal.pone.0004142 From NLM

Huffman B (2006) Ungulates of the world: species fact sheets. Available at https://www.ultimateungulate.com/Ungulates.html. Accessed on 5 Sept 2023

Iannuzzi L, Di Meo G (2009) Water buffalo. In: Genome mapping and genomics in domestic animals. Springer, Berlin/Heidelberg, pp 19–31

Ibrahim MAM (2012) Effect of enrollment in milk recording systems on improving milk production of Egyptian Buffalo. J Anim Poult Prod 3(1):39–46. https://doi.org/10.21608/jappmu.2012.82765

ICUN (2016) International union for conservation of nature. https://www.iucn.org/content/red-list-bangladesh-volume-1-summary

Kaul R, Williams AC, Rithe K, Steinmetz R, Mishra R (2019) Bubulus arnee. The IUCN red list of threatened species. https://doi.org/10.2305/IUCN.UK.2019-1.RLTS.T3129A46364616.en

Kierstein G, Vallinoto M, Silva A, Schneider MP, Iannuzzi L, Brenig B (2004) Analysis of mitochondrial D-loop region casts new light on domestic water buffalo (Bubalus bubalis) phylogeny. Mol Phylogenet Evol 30(2):308–324. https://doi.org/10.1016/s1055-7903(03)00221-5 From NLM

Kress WJ, Wurdack KJ, Zimmer EA, Weigt LA, Janzen DH (2005) Use of DNA barcodes to identify flowering plants. Proc Natl Acad Sci USA 102(23):8369–8374. https://doi.org/10.1073/pnas.0503123102 From NLM

Kumar S, Nagarajan M, Sandhu JS, Kumar N, Behl V (2007) Phylogeography and domestication of Indian river buffalo. BMC Evol Biol 7:186. https://doi.org/10.1186/1471-2148-7-186 From NLM

Li M, Hassan FU, Guo Y, Tang Z, Liang X, Xie F, Peng L, Yang C (2020) Seasonal dynamics of physiological, oxidative and metabolic responses in non-lactating Nili-Ravi buffaloes under hot and humid climate. Front Vet Sci 7:622. https://doi.org/10.3389/fvets.2020.00622 From NLM

Liu JJ, Liang AX, Campanile G, Plastow G, Zhang C, Wang Z, Salzano A, Gasparrini B, Cassandro M, Yang LG (2018) Genome-wide association studies to identify quantitative trait loci affecting milk production traits in water buffalo. J Dairy Sci 101(1):433–444. https://doi.org/10.3168/jds.2017-13246 From NLM

Macgregor R (1941) The domestic buffalo. Vet Rec 53(31):443–450

Mahfouz E (2007) Sequence analysis of a comparative anchored tagged sequence (Biglycan) and its assignment to the X chromosome of river Buffalo. J Biol Sci:7. https://doi.org/10.3923/jbs.2007.231.238

Mahmoud KG, Sakr AM, Ibrahim SR, Sosa AS, Hasanain MH, Nawito MF (2021) GnRHR gene polymorphism and its correlation with semen quality in Buffalo bulls (Bubalus bubalis). Iraqi J Vet Sci 35(2):381–386. https://doi.org/10.33899/ijvs.2020.126886.1407

Merdan S, El-Zarei M, Ghazy A, Al-Shawa Z, Ayoub M, Mokhtar S (2020) Genetic diversity analysis of five Egyptian Buffalo populations using microsatellite markers. J Agric Sci 12:271–282. https://doi.org/10.5539/jas.v12n4p271

Michelizzi VN, Dodson MV, Pan Z, Amaral ME, Michal JJ, McLean DJ, Womack JE, Jiang Z (2010) Water buffalo genome science comes of age. Int J Biol Sci 6(4):333–349. https://doi.org/10.7150/ijbs.6.333 From NLM

Miziara MN, Goldammer T, Stafuzza NB, Ianella P, Agarwala R, Schaffer AA, Elliott JS, Riggs PK, Womack JE, Amaral ME (2007) A radiation hybrid map of river buffalo (Bubalus bubalis) chromosome 1 (BBU1). Cytogenet Genome Res 119(1-2):100–104. https://doi.org/10.1159/000109625 From NLM

Mohammed KM (2018) Application of advanced reproductive biotechnologies for buffalo improvement with focusing on Egyptian buffaloes. Asian Pac J Reprod 7(5):193

Nahas SM, Hondt HA, Othman OS, Bosma AA, Haan NA (1993) Assignment of genes to chromosome 4 of the river buffalo with a panel of buffalo-hamster hybrid cells. J Anim Breed Genetics = Zeitschrift fur Tierzuchtung und Zuchtungsbiologie 110(1–6):182–185. https://doi.org/10.1111/j.1439-0388.1993.tb00730.x From NLM

Nahas SME, Oraby HO, Othman OE, Hondt HA d, Bosma AA, Womack JE (1997) Use of molecular markers for the identification of river Buffalo chromosomes: chromosome one. J Anim Breed Gen 114(1–6):451–455. https://doi.org/10.1111/j.1439-0388.1997.tb00531.x

Nahas BSME, Hondt HA, Soussa SF, Ghor AE, Hassan AA (1999) Assignment of new loci to river buffalo chromosomes confirms the nature of chromosomes 4 and 5. J Anim Breed Gen 116(1):21–28. https://doi.org/10.1111/j.1439-0388.1999.00167.x

Oraby HA, El Nahas SM, de Hondt HA, El Ghor A, Samad MFA (1998) Assignment of PCR markers to river buffalo chromosomes. Genet Sel Evol 30(1):71. https://doi.org/10.1186/1297-9686-30-1-71

Oraby HA, Hassan AA, El Nahas SM (2011) Differentiation between four Bovidae species using two transfer RNAs. J Appl Biosci 44:2972–2981

Othman O (2005) The identification of kappa-casein genotyping in Egyptian river buffalo using PCR-RFLP. Arab J Biotech 8:265–274

Othman OE, Abdel-Samad MF (2013) RFL polymorphism of three fertility genes in Egyptian buffalo. J Appl Biol Sci 7(2):94–101

Othman EO, Ahmed S (2010) Genetic polymorphism of BoLA-DRB3 Exon 2 in Egyptian Buffalo. Gene Genomes Genomics 4(1):70–73

Othman BOE, Nahas SME (1999) Synteny assignment of four genes and two microsatellite markers in river buffalo (Bubalus bubalis L.). J Anim Breed Gen 116(2):161–168. https://doi.org/10.1046/j.1439-0388.1999.00177.x. Accessed 28 Feb 2024

Othman OE-M, Zayed FA, Gawead AAE, El-Rahman MRA (2011) Genetic polymorphism of two genes associated with carcass trait in Egyptian buffaloes. J Genetic Eng Biotechnol 9:15–20

Othman O, Abdel-Samad M, Abo N, Maaty E, Sewify K (2012) Evaluation of DNA polymorphism in Egyptian Buffalo growth hormone and its receptor genes. J Appl Biol Sci 6:37–42

Othman O, Ahmed W, Balabel E, Zaabal M, Khadrawy H, Hanafy E (2014) Genetic polymorphism of Cyp19 gene and its association with ovarian activity in Egyptian buffaloes. Global Veterinaria 12:768–773. https://doi.org/10.5829/idosi.gv.2014.12.06.83270

Othman O, Abdel-Samad M, El-Maaty N, Sewify K (2015) Genotyping and nucleotide sequences of growth hormone releasing hormone and its receptor genes in Egyptian Buffalo. Br Biotechnol J 5:62–71. https://doi.org/10.9734/BBJ/2015/11619

Othman OE, Khodary MG, El-Deeb AH, Hussein HA (2018) Five BoLA-DRB3 genotypes detected in Egyptian buffalo infected with foot and mouth disease virus serotype O. J Genet Eng Biotechnol 16(2):513–518. https://doi.org/10.1016/j.jgeb.2018.02.009 From NLM

Owak RM, Paradiso JL (1983) Walker's mammals of the world. Johns Hopkins University Press., ISBN 0801825253., Baltimore

Pacha AS, Nigam P, Pandav B, Mondol S (2021) Sequencing and annotation of the endangered wild buffalo (Bubalus arnee) mitogenome for taxonomic assessment. Mol Biol Rep 48(2):1995–2003. https://doi.org/10.1007/s11033-021-06165-8 From NLM

Perera BM, Abeygunawardena H, Vale WG et al (2005) Buffalo. In: livestock and wealth creation improving the husbandry of animals kept by poor people in developing countries. In: Owen E, Kitayi A, Jayasuriya N, Smith T (eds) Livestock production program, natural resources international limited, Nottingham, pp 451–471

Prentice JR, Anzar M (2011) Cryopreservation of mammalian oocyte for conservation of animal genetics. Vet Med Int 2011:146405. https://doi.org/10.4061/2011/146405

Ramadan HAI, El-Hefnawi MM (2008) Phylogenetic analysis and comparison between cow and buffalo (including Egyptian buffaloes) mitochondrial displacement-loop regions. DNA Seq 19(4):401–410. https://doi.org/10.1080/19401730802351004

Ramadan H, Othman O, Samad M, Abou Mossallam A, El Nahas S (2000) Somatic cell hybrids characterization by monoclonal antibodies and assignment of CD71 to the Q arm of buffalo and cattle chromosome one. J Egypt Ger Soc Zool 31:133–142

Rehman SU, Hassan FU, Luo X, Li Z, Liu Q (2021) Whole-genome sequencing and characterization of Buffalo genetic resources: recent advances and future challenges. Animals 11(3) https://doi.org/10.3390/ani11030904 From NLM

Rushdi HE, Moghaieb R, Abdel-Shafy H, Ibrahim M (2017) Association between microsatellite markers and milk production traits in Egyptian buffaloes. Czeh J Anim Sci 62:384–391. https://doi.org/10.17221/80/2016-CJAS

Şahin A, Yıldırım A, Ulutaş Z (2014) Some physicochemical characteristics of raw milk of Anatolian Buffaloes. Italian J Food Sci 26:398–404

Şahin A, Yıldırım A, Ulutaş Z (2019) The effects of storage temperature and storage time on the somatic cell count of Anatolian Buffaloes. Buffalo Bull 38:299–309

Selokar NL, Saini M, Palta P, Chauhan MS, Manik R, Singla SK (2014) Hope for restoration of dead valuable bulls through cloning using donor somatic cells isolated from cryopreserved semen. PLoS One 9(3):e90755. https://doi.org/10.1371/journal.pone.0090755 From NLM

Ślaska B, Makarevič A, Surdyka M, Nisztuk S (2014) Application aspects of animal and human genomics. Acta Sci Pol Zootechnica 13(2):3–18

Sosa A, Mahmoud K, Eldebaky H, Kandiel M, El-Roos MEA, Nawito M (2016) Single nucleotide polymorphisms of GnRHR gene and its relationship with reproductive performance in Egyptian buffaloes. Egypt J Vet Sci 47:41–50. https://doi.org/10.21608/ejvs.2016.1081

Sosa A, Mahmoud K, Kandiel M, Eldebaky H, Nawito M, Abou-El-Roos M, Mahmoud M (2017) Genetic characterization of FSH beta-subunit gene and its association with buffalo fertility. Asian Pac J Reprod 6:193–196. https://doi.org/10.4103/2305-0500.215928

Stafuzza NB, Naressi BC, Yang E, Cai JJ, Amaral-Trusty ME (2015) A framework radiation hybrid map of buffalo chromosome 1 ordering scaffolds from buffalo genome sequence assembly. Genetics Mol Res 14(4):13096–13104. https://doi.org/10.4238/2015.October.26.5 From NLM

Surya T, Vineeth MR, Sivalingam J, Tantia MS, Dixit SP, Niranjan SK, Gupta ID (2019) Genomewide identification and annotation of SNPs in Bubalus bubalis. Genomics 111(6):1695–1698. https://doi.org/10.1016/j.ygeno.2018.11.021 From NLM

Togashi K, Osawa T, Adachi K, Kurogi K, Tokunaka K, Yasumori T, Takahashi T, Moribe K (2019) Selection on milk production and conformation traits during the last two decades in Japan. Asian Australas J Anim Sci 32(2):183–191. https://doi.org/10.5713/ajas.18.0259 From NLM

Yindee M, Vlamings BH, Wajjwalku W, Techakumphu M, Lohachit C, Sirivaidyapong S, Thitaram C, Amarasinghe AA, Alexander PA, Colenbrander B et al (2010) Y-chromosomal variation confirms independent domestications of swamp and river buffalo. Anim Genet 41(4):433–435. https://doi.org/10.1111/j.1365-2052.2010.02020.x From NLM

Youssef NA, Curaudeau M, El Nahas SM, Hassan AAM, Hassanin A (2021) Haplotype diversity in the mitochondrial genome of the Egyptian river buffalo (Bubalus bubalis). Mitochondrial DNA B Resour 6(1):145–147. https://doi.org/10.1080/23802359.2020.1852622 From NLM

Zhang Y, Colli L, Barker JSF (2020) Asian water buffalo: domestication, history and genetics. Anim Genet 51(2):177–191. https://doi.org/10.1111/age.12911 From NLM

Open Access This chapter is licensed under the terms of the Creative Commons Attribution 4.0 International License (http://creativecommons.org/licenses/by/4.0/), which permits use, sharing, adaptation, distribution and reproduction in any medium or format, as long as you give appropriate credit to the original author(s) and the source, provide a link to the Creative Commons license and indicate if changes were made.

The images or other third party material in this chapter are included in the chapter's Creative Commons license, unless indicated otherwise in a credit line to the material. If material is not included in the chapter's Creative Commons license and your intended use is not permitted by statutory regulation or exceeds the permitted use, you will need to obtain permission directly from the copyright holder.

Nonconventional Animal Genetic Resources, Diversity, and Unique Features

13

Kingsley A. Etchu, Abdelkader Ameur Ameur,
M. Chahbar, Félix Meutchieye,
Annick N. Enangue Njembele,
Gerald C. Tasse Taboue, and Semir B. S. Gaouar

Abstract

The African continent abounds in rich faunal diversity, which should contribute significantly to human nutrition. Before the introduction and establishment of conventional livestock and farming systems, Africans fed on animals found in their environment. Here is a non-exhaustive list of some edible animals: gastropods (marine or terrestrial snails), amphibians, small mammals (grasscutters, porcupines, rabbits, etc.), and insects and their larvae (caterpillars, grasshoppers, palm weevil larvae, etc.). The development of conventional animal husbandry systems has resulted in the reduction or even the abandonment of the consumption of certain types of animals now called nonconventional livestock. However, the conventional livestock system struggles to meet the animal protein needs of the African population. In order to fill this shortage in animal proteins, Africans have turned to and are relearning to consume abandoned traditional animal resources. In this chapter, four species, namely the frog, land snail, rabbit, and honeybee, are highlighted.

K. A. Etchu (✉)
Institute of Agricultural Research for Development, Yaoundé, Cameroon

A. Ameur Ameur
Applied Genetic in Agriculture, Ecology and Public Health Laboratory, Faculty of Natural and Life Sciences and Earth and Universe Sciences, University of Tlemcen, Tlemcen, Algeria

M. Chahbar · S. B. S. Gaouar
Laboratory of Agronomy and Environment, Department of Biology, Faculty of Sciences and Technology, University of Tissemsilt, Tissemsilt, Algeria

F. Meutchieye
Biotechnology and Bio Informatics Research Unit, Department of Animal Science, Faculty of Agronomy and Agricultural Sciences, The University of Dschang, Dschang, Cameroon

A. N. Enangue Njembele
Specialized Research Station for Marine Ecosystems, Institute of Agricultural Research for Development, Kribi, Cameroon

G. C. Tasse Taboue
Multipurpose Research Station, Institute of Agricultural Research for Development, Bangangté, Cameroon

Keywords

Nonconventional · Genetic resources · Goliath frog · Snails · Rabbit · Honeybee

Abbreviations and Acronyms

ABPV	Acute bee paralysis virus
ADG	Average daily gains
AMOVA	Analysis of molecular variance
BC	Before Christ
BEA	Biotic element analysis
BL	Body length

© The Author(s) 2026
E. M. Ibeagha-Awemu et al. (eds.), *African Livestock Genetic Resources and Sustainable Breeding Strategies*, Sustainable Development Goals Series, https://doi.org/10.1007/978-3-031-92076-9_13

BQCV	Black queen cell virus	NZW	New Zealand white
BWT	Body weight	OIE	International Office of Epizootics
CAH	Hierarchical classification	OVGP1	Oviduct-specific glycoprotein
CCR6	Chemokine receptor 6	PAE	Parsimony analysis of endemicity
CNRS	Centre national de la recherche scientifique	PCA	Principal component analysis
COI	Cytochrome oxidase subunit I	PCR-RFLP	Polymerase chain reaction restriction fragment length polymorphism
COII	Cytochrome oxidase subunit II		
CVA	Canonical variable analysis	PCR	Polymerase chain reaction
Cyt b	Cytochrome b	QTL	Quantitative trait loci
Dectin-1	Dendritic-cell associated C-type lectin 1	RAPD	Random amplified polymorphic DNA
DFA	Discriminant function analysis	RB	Red Baladi
DNA	Deoxyribonucleic acid	RNA	Ribonucleic acid
DWV	Deformed queen virus	SB	Sudan Baladi
EL	Ear Length	SEA	Southeast Africa
EMBL	European Molecular Biology Laboratory	SG	Sinai Gabali
		SNP	Single nucleotide polymorphisms
FAOSTAT	Food and Agriculture Organization of the United Nations Statistics Division	STAT3	Signal transducer and activator of transcription 3
		TIMP1	Metallopeptidase inhibitor 1
GH1	Growth hormone 1	TL	Tail length
GHR	Growth hormone receptor	TLR4	Toll like receptor 4
HG	Heart girth	UPGMA	Unweighted pairs with arithmetic means
HL	Head length		
IAPV	Israeli acute paralysis virus	USA	United States of America
IGF2	Insulin-like growth factor 2	USD	United States Dollar
INRA	Institut national de la recherche agronomique	VDV1	Varroa destructor virus
		VHD	Viral hemorrhagic disease
JAK1	Janus kinase 1	XAF	Central Africa CFA
LBM	Linear body measurements		
LW	Live weights		
MCA	Multiple correspondence factor analysis		
MINFOF	Ministère des Forêts et de la Faune		
MPA	Maputaland-Pondoland-Albany		
MSTN	Myostatin		
mtDNA	Mitochondrial DNA		
MyD88	Myeloid differentiating factor 88		
NCBI	National Center for Biotechnology Information		
NLRP12	NLR Family Pyrin Domain Containing 12		
NMSA	Natal Museum, Pietermaritzburg, South Africa		
NOD2	Nucleotide-binding oligomerization domain 2		

13.1 Introduction

Conventional livestock are animals raised in an agricultural system in order to produce meat, eggs, milk, and other products used for human consumption. Therefore, animals that do not follow the above definition are called nonconventional livestock. Animal protein consumption in Africa is below the WHO recommendation. The conventional livestock system continues to endeavor to meet the animal protein needs of the African population. Africans have turned to or are relearning to consume abandoned traditional animal resources. In this chapter, four species are highlighted: frog, land snail, rabbit, and honeybee.

13.2 Frog in Africa: The Goliath Frog

13.2.1 Introduction

Amphibians are known for their dual lifestyles, living in water and on dry land. They have medicinal properties and are used in cultural settings (Gonwouo and Rödel 2008; Mohneke et al. 2010; Efenakpo et al. 2016). Amphibians constitute a source of protein and other nutrients in human and livestock diets (Gonwouo and Rödel 2008; Karamoko et al. 2011; Oduntan et al. 2012). Amphibians also have cultural values, and their use in cultural rites such as weddings has been documented (Gonwouo and Rödel 2008). They have been used to cure human infertility by the Bakossi people of Cameroon (Gonwouo and Rödel 2008). In sub-Saharan Africa, various amphibians are consumed by humans. These are directly harvested from the field. The rate at which they are naturally harvested has led to consideration of captive breeding. Some markets are dedicated to frogs in West Africa, where they are sold as dried or smoked stock (Mohneke et al. 2010). In Central Africa, for instance, frogs are harvested for home consumption and either sold on the roadsides or on market days.

The Goliath frog (*Conraua goliath*), known as the largest living frog on earth, is commonly called the giant slippery frog or the goliath bullfrog. It is a species of frog in the family Conrauidae.

The population of Goliath frogs in the wild in Cameroon is decreasing because of accelerated habitat loss caused by deforestation, hunting, and overharvesting from the wild for the pet trade and consumption. Despite its protected status in Cameroon (class A species: not allowed for hunting (MINFOF 2020)), this frog is systematically collected from the field mainly for consumption. An adult Goliath frog has an average snout–vent length of 22–32 cm for males with an average weight ranging from 1.5–2.7 kg and 15–32 cm snout–vent length for females with an average weight of 0.6–3.3 kg (Krüger 1912; Sabater-Pi 1985; Channing and Rödel 2019).

Adult *Conraua goliath* has no tail, and the head is large and flat with a mouth capable of opening widely. The body is massive with long and muscular hind limbs that extend accordingly to provide the necessary force during jumping. The body is arranged in order to provide the necessary power for saltatory locomotion. There is a strong pectoral girdle and a vertebral column arranged for the forelimbs to absorb the shock of landing (see Fig. 13.1).

13.2.2 Semi-domestic Frog Genetic Resources

Goliath frogs are indigenous to Cameroon and mainland Equatorial Guinea. This species belongs to the genus *Conraua*, which has eight species, namely, *Conraua goliath, C. robusta crassipes, C. beccarii, C. alleni, C. derooi, C. sagyimase, and C. kamancamarai*. The genus *Conraua* is the sole member of the Conrauaidae family, considered as a sister family of the Petropedidae. Genetic diversity of the Goliath frog in Cameroon, highlighted from a study of two mitochondrial genes (Cytochrome Oxidase subunit 1 (COI) and 16S ribosomal RNA gene (16S)), revealed a demarcation between populations, potentially due to geographic distance (Nguiffo et al. 2019a). This species is facing threats from overhunting, climate change, conversion of its natural habitat for agriculture, and construction of large infrastructures such as dams. It is, therefore, important to consider captive breeding in order to save this species.

13.2.3 Housing and Environmental Adaptation

The following must be considered when caring for amphibians, especially the Goliath frogs: (1) enclosures, (2) water (sources and quality), (3) environmental conditions (light, temperature, and humidity), (4) food, and (5) veterinary care.

Goliath frog lifecycle stages require different-sized enclosures. Tadpoles can easily be kept in a plastic box of 40 cm × 50 cm × 30 cm. However,

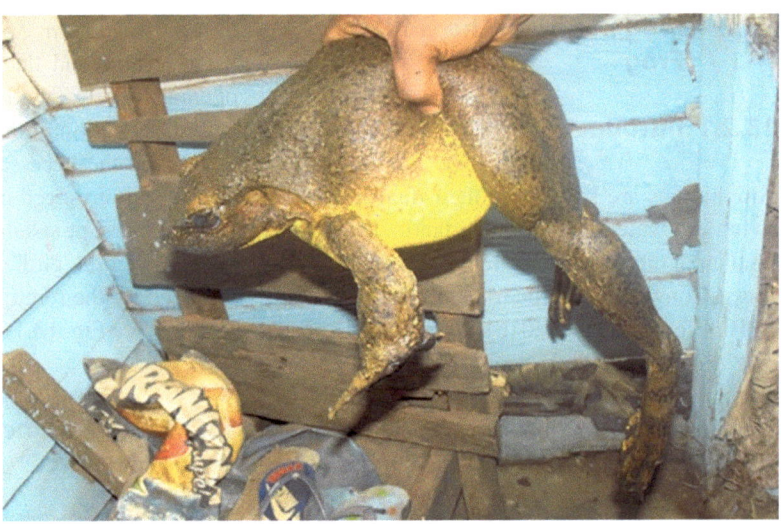

Fig. 13.1 An adult Goliath Frog

juvenile and adult frogs need extra space. Very small enclosures might lead to various traumas such as scoured rostrums, lethargy (due to lack of exercise), and sometimes death. The core idea for making a tank for frog keeping is to use materials made from glass or plastic that are easy to clean and disinfect. The frog tank should be cleaned, disinfected, dried, and fresh water added before receiving any animal.

Water remains one of the most important considerations for amphibian husbandry. Amphibians rely on water to keep their skin moist and permeable for respiration. Water is vital for amphibian reproduction, for eggs and tadpoles to dwell. Water quality is, therefore, essential for amphibian husbandry, and its temperature, pH, ammonia, and nitrite should be tested and maintained to ensure suitable conditions for all lifecycle stages (Odum and Zippel 2008; Wright and Whitaker 2001; Browne et al. 2007).

13.2.4 Reproduction

Female amphibians release eggs that are then fertilized by the male. Goliath frogs are known to build nests for reproduction, and to date, three different types have been documented (Schäfer et al. 2019). Nests (See Fig. 13.2) are known to harbor various cohorts of tadpoles at different ages. Goliath frogs are known to breed during the dry season when the water level is low (Sabater-Pi 1985; Schäfer et al. 2019). Locals have pointed out that a drop in water level in water bodies harboring the Goliath frog during the dry season acts as a signal for reproduction. However, chemical signals are associated with reproduction, like water parameters required for mating.

13.2.5 Health and Care

Goliath frogs are known to harbor parasites such as helminths and microfilariae. Three major groups of helminths have been documented (Nematoda, Trematoda, and Pentastomids) compared to a single microfilaria (Nguiffo et al. 2015, 2019b). The deadly fungus, *Batrachochytrium dendrobatidis,* responsible for the collapse of several populations of amphibians worldwide, is known to be present across the geographical distribution of the Goliath frog (Miller et al. 2018). However, there is no report on the mortality of the Goliath frog due to this fungus. To avoid the escape and movement of populations into the colony that might potentially spread pathogens into the farm, biosecurity and restrictions would be needed, and this may include homemade or natural barriers. Additionally, a limitation of visits to a single colony per outing is encouraged.

Fig. 13.2 Nest of Goliath frog

13.2.6 Core Challenges and Perspectives

Amphibians have been used for various purposes. Their population is also experiencing a decline in Africa due to disease and overexploitation. Captive breeding programs might be the solution to provide sustainable use of this resource. However, due to a loathing of amphibians in several ethnic and economic constraints, sustainable breeding of amphibians has not been a priority. There is an absence of coordination of action as frog hunters are essentially isolated and have no major interest in frog keeping. Additionally, land issues remain a challenge, especially for captive breeding programs involving large amphibians such as the Goliath frog. The lack of funding remains a major problem in frog farming. This is exacerbated by the lack of economic attractiveness of the practice. There is still a long way to go for amphibian husbandry efforts in Central Africa. Indeed, to be effective in addressing the numerous challenges in amphibian husbandry, the development of multidisciplinary research programs in combination with increasing awareness of local communities is necessary. Through education, the development of better management tools for key iconic species like Goliath frogs is needed to fully integrate the needs of local populations. Enhancing local expertise to handle amphibian husbandry efforts and increase awareness is of vital importance. Combined with these, long-term monitoring programs are needed for wild amphibian populations to document the status and trends in these populations and relate trends to threats and the need for action. Actions need to be informed by scientific data on population structure, movement of individuals, health, and climate change. Such information will serve as a basis for current and future actions. These might range from potential translocation efforts to the prioritization of areas with high biological and genetic diversity. The engagement of various stakeholders would ultimately serve as a medium for the exchange of ideas, boost collaboration, encourage in-country students and researchers to conduct research in order to improve the husbandry approaches, and propose specific models and site management plans. Public engagement (brochures, posters) and the involvement of media in publicity campaigns in conservation education programs targeting schools and local communities at the national level should be considered.

13.3 Snails

13.3.1 Introduction

The phylum of mollusks, by the number of its species (more than one hundred thousand), is second only to arthropods and is, therefore, one of the most important and most varied groups. A mollusk is an invertebrate of a large phylum that includes snails, slugs, mussels, and octopuses. Mollusks are triploblastic metazoans with fundamentally bilateral symmetry. They live in aquatic or damp habitats, and most of them have an external calcareous shell. Their body is soft, unsegmented, and consists of three fundamental parts: a head, a foot, and a visceral mass (Maissiat et al. 2011). Meglitsch (1974)reports that mollusks have seven classes, namely: Monoplacophores, Aplacophores, Polyplacophores, Scaphopods, Lamellibranchs (Bivalves), Gastropods, and Cephalopods. Gastropods account for about 80% of mollusk species: they represent the major part of the seven existing classes (Belanger 2009). During their organogenesis, these mollusks undergo very profound anatomical modifications, which upset the anatomical relationships of their organs; these are mainly flexion and torsion movements, to which is added a winding (Maissiat et al. 2011). Originally, all gastropods have a shell and gills, and are aquatic. During their evolution, some species have lost all or part of these characters. Species that have lost their shells are called "slugs" (Kerney et al. 1999), while species with shells, such as the snail, are gastropod mollusks. There are 103,000 species of gastropods; the oldest known fossils date back to the Cambrian, 510 million years old (Lecointre and Le Guyader 2001). Despite their great biodiversity, their evolutionary, geological, ecological, and economic values, terrestrial gastropods are not well known, both from a point of view of biology and of the distribution of species, and most of the data comes from old studies (Karas 2009).

13.3.2 Systematic Position of the Snails

The systematic position of the snail is as follows:
Kingdom: Animalia
Phylum: Mollusca
Class: Gastropoda
Subclass: Pulmonata
Order: Stylomatophora
Superfamily: Helicacea
Family: Helicidae
Subfamily: *Helicina*
Genus: *Helix*
Species: *Aspersa* (and many other species)

13.3.3 Common Snail Species in Africa

Among the species widely sampled in North Africa, from the West of Morocco to the East of Tunisia, and to explain the historical processes influencing their current distribution, the land snail, *Cornu aspersum* or *Helix aspersa*, provides an excellent biological model to understand phylogeographic patterns across North Africa and the surrounding regions of the Western Mediterranean basin. This species is a model for evaluating hypotheses leading to population differentiation.

This land snail, formerly known as *Helix aspersa* (Müller, 1774), is native to the Mediterranean countries. It includes a set of forms and subspecies endemic to North Africa that were described at the beginning of the twentieth century on the basis of the characteristics of the shells (Taylor 1913). The most common, *Cornu aspersa. Aspersum* (syn. *Helix aspersa. aspersa*) has become very abundant in all habitats disturbed by humans in regions with Mediterranean, temperate, and even subtropical climates. To reconstruct the biogeographic history of this invasive form in the Western Mediterranean, the variation in the spatial patterns of shellfish, genital, and molecular characters was previously estimated by investigating more than a hundred populations representative of the distribution area of the *aspersum* (Western Mediterranean and European coasts).

Regardless of the set of populations and/or markers used, the combination of all the different types of data leads to a clear picture of the geographic structure. Indeed, (i) two groups of anatomically and biochemically divergent populations are in the West and East, and the separation that occurs in Kabylia (Algeria) is invariably observed throughout the North African coastal region; (ii) almost all European populations are grouped with those of the Western part of North Africa, with genetic distances smaller than those between the Western and Eastern parts of North Africa (Guiller and Madec 2010) (see Fig. 13.3).

In Africa, generally, the information about snail species (geographical distribution, inventory, characterization, etc.) is scant. Three studies have examined the African snail species through inventory and characterization. According to Oke and Alohan (2006), 46 taxa are recorded from Okomu National Park, Southwestern Nigeria, of which 23 could be identified to species level, while 23 taxa seem to be of unclear status or may be new species. The molluscan fauna is dominated by two families, the *Streptaxidae* (15 species) and the *Subulinidae* (12 species), constituting over 50% of the total number of species. The *Streptaxidae* was the most dominant family, constituting 33% of the total number of species collected and 35% of the total number of individuals. Two common species, namely *Archachatina marginata* and *Achatina fulica*, are listed in Cameroon (Tsayo et al. 2020). In Algeria, the common species are *Otala punctata*, *Helix aspersa*, *aperta*, and *pomatia* (Bouchiba et al. 2021; Mezouar et al. 2021; Skendraoui 2015). Moreover, these are the same dominant species in neighboring countries like Tunisia and Morocco (MAPM 2017). The genus *Eremina (Helicidae)* comprises a few species of land

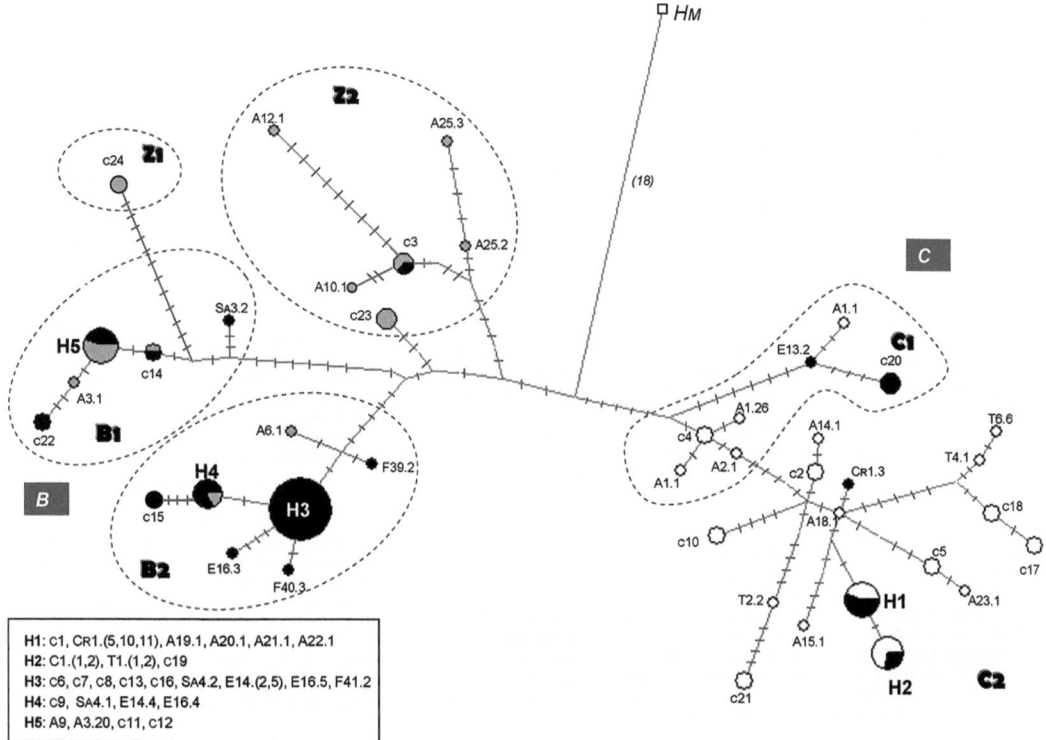

Fig. 13.3 Median-joining network for the cyt b mtDNA haplotypes of *C. aspersum*. (Guiller and Madec 2010)

snails living mainly in desert or semidesert habitats of North Africa and the Near East. Four Moroccan species are recognized as *E. dillwyniana, E. vermiculosa, E. duroi,* and *E. inexspectata*. The two taxa of the land snails, *Eremina d. desertorum* and *Eremina desertorum irregularis,* are differentiated by shell size and shape and are separated by a narrow hybrid zone west of Alexandria, Egypt (Ali et al. 2016).

13.3.4 Geographical Distribution of Common Snail Species in Africa

13.3.4.1 Snails of North Africa

Prevalent Species in North Africa

Fourteen species of land snails in Algeria show high phenotypic diversity of the shell (Fig. 13.4) (Bouchiba et al. 2021; Mezouar et al. 2021). These studies, conducted in 27 regions in Algeria and including 1090 snails, indicated that the district of El Tarf presents more species variability compared to other districts. The species found in El Tarf include *Helix pomatia, Helix aspersa, Pseudotachea splendida, Gllandia annularis, Cryptomphalus aspersus, Cepea hortensis, Cepeanemoralis, Helicellaitemorala,* and *Cepea sylvatica*. The rest of the districts presented few varieties such as Constantine (*Helix aspersa, Arianta arbustrum, Helix pomatia*), Algiers (*Otala punctata, Theba pisana*), Tlemcen (*Theba pisana, Helix pomatia, Pseudotachea splendida, Otala punctata*), Mila, Annaba and Setif (*Helix aperta*), Oran, Sidi Bel Abbès, Ain Témouchent, Saida, Mostaganem, Mascara (*Otala punctata*), and Naama (*Sphincterochila boissieri*).

In Tunisia, out of 400 species of snails reported, only six are the object of significant commercial interest: the Bourgogne snail (*Helix pomatia*), the small gray snail *(Helix aspersa aspersa or muller)*, the big gray snail (*Helix Aspersa Maxima*), the tapado or Attupatelli snail (*Helix aperta*), the mourguette (*Eobania vermiculata*), and the Turkish snail (*Helix lucorum*). Breeding is carried out either outdoors, indoors, or both indoors and outdoors (See: Sect. 13.2.6. Snail production systems*)*.

Also, land snails represent an important natural resource for Morocco through their diversity, productivity, and marketability. In Morocco, the

Fig. 13.4 The most common snail varieties in northern Algeria. (Adapted from Bouchiba et al. 2021 and Mezouar et al. 2021)

main species exploited are *Helix aspersa aspersa*, *Helix aspersa maxima*, *Otala lactea*, *Eobania vermiculata*, and *Eupharipha pisana*. However, the national specialty is the big gray, which is only found in Morocco and Algeria (MAPM 2017). Finally, in Egypt, the land snails are widespread in the Northern Governorate, Alexandria, El-Beheira, Kafr El-Shikh, and Domietta (Eshra 2013). Nine terrestrial snail species in the Upper Egypt region have been identified by the Assiut Governorate. All the species recorded belong to the order *Pulmonata* from eight families. Two species of land snails are recorded: *Monacha obstracta* (Montagu) and *Eobania vermiculata* (Muller) in the Sohag Governorate, in Egypt (Desoky et al. 2015).

External Morphology

Four characteristics are selected (shell "height, length, width, and body weight") to define the body measurements of the Algerian land snail populations. In fact, there is a remarkable variation in the weight of the land snail between the 15 species (Table 13.1). A small species, such as *Cepea sylvatica* and *Helicella itala*, with a weight that does not exceed 2.5 g, was found. On the other hand, *Otala punctata* and *Helix pomatia* are relatively heavy species with an average weight of around 10 g. The rest of the species have average weights that are sometimes similar (Table 13.1). Nevertheless, the shell morphology measurement in the species shows a very homogeneous variation between them, except *Otala punctata*, which shows quite some variability. Hence, the height, length, and width of the shell present a highly significant difference between the studied species with an overall average of 1.99 ± 0.56 cm, 3.35 ± 0.93 cm, and 2.67 ± 0.68 cm, respectively (Table 13.1).

Regarding the qualitative traits, two characters were taken into consideration, which dominate the phenotype (the flesh and the shell color). The results of shell and flesh color show that the Algerian land snail exhibits five different flesh colors: White, brown, yellow, black, and beige, either united or mixed (Bouchiba et al. 2021; Mezouar et al. 2021). The results also show that the El Tarf district has more shell color variability than the other districts, with seven and six shell phenotypes. Then, Constantine and Annaba with four colors, followed by Algiers and Mila, which include three shell colors. In parallel, the other districts are characterized by a single shell color (brown and white) like Naama, Setif, Oran, Sidi

Table 13.1 Descriptive analysis of body measurements (height, length, and width of shell and body weight) between different species in Algeria (Bouchiba et al. 2021)

Species	N	Height shell Mean ± SE	Length shell Mean ± SE	Width shell Mean ± SE	body weight (g) Mean ± SE
Helix aperta	116	1.56 ± 0.27	2.46 ± 0.32	1.94 ± 0.53	4.33 ± 1.60
Helix pomatia	55	2.47 ± 0.91	3.52 ± 0.86	2.61 ± 0.56	9.34 ± 3.46
Helix aspersa	51	1.89 ± 0.40	2.76 ± 0.56	2.19 ± 0.50	7.07 ± 2.69
Arianta arbustrum	10	1.73 ± 0.24	2.68 ± 0.37	1.95 ± 0.24	7.40 ± 2.63
Otala punctata	292	2.13 ± 0.40	4.25 ± 0.45	3.34 ± 0.31	10.37 ± 3.11
Theba pisana	109	1.75 ± 0.57	2.60 ± 0.52	2.16 ± 0.41	3.99 ± 1.10
Pseudotachia splendida	35	1.67 ± 0.32	2.72 ± 0.33	2.28 ± 0.25	5.56 ± 2.69
Helicella itala	16	1.57 ± 0.27	2.35 ± 0.15	2.14 ± 0.27	2.35 ± 0.25
Eobania vermiculata	3	1.88 ± 0.11	2.95 ± 0.32	2.68 ± 0.16	3.26 ± 0.15
Cepea sylvatica	4	1.49 ± 0.09	2.32 ± 0.04	2.12 ± 0.15	2.12 ± 0.09
Cepea nemoralis	1	/	/	/	/
Cepea hortensis	1	/	/	/	/
Cryptomphalus aspersus	17	2.85 ± 0.33	3.41 ± 0.39	2.90 ± 0.31	5.58 ± 2.29
Gllandia annularis	8	3.02 ± 0.62	3.59 ± 0.86	2.88 ± 0.46	5.37 ± 3.99
Sphincterochila boissieri	33	2.33 ± 0.36	2.85 ± 0.11	2.50 ± 0.59	5.57 ± 1.52
Total	751	1.99 ± 0.56	3.35 ± 0.93	2.67 ± 0.68	7.34 ± 3.82

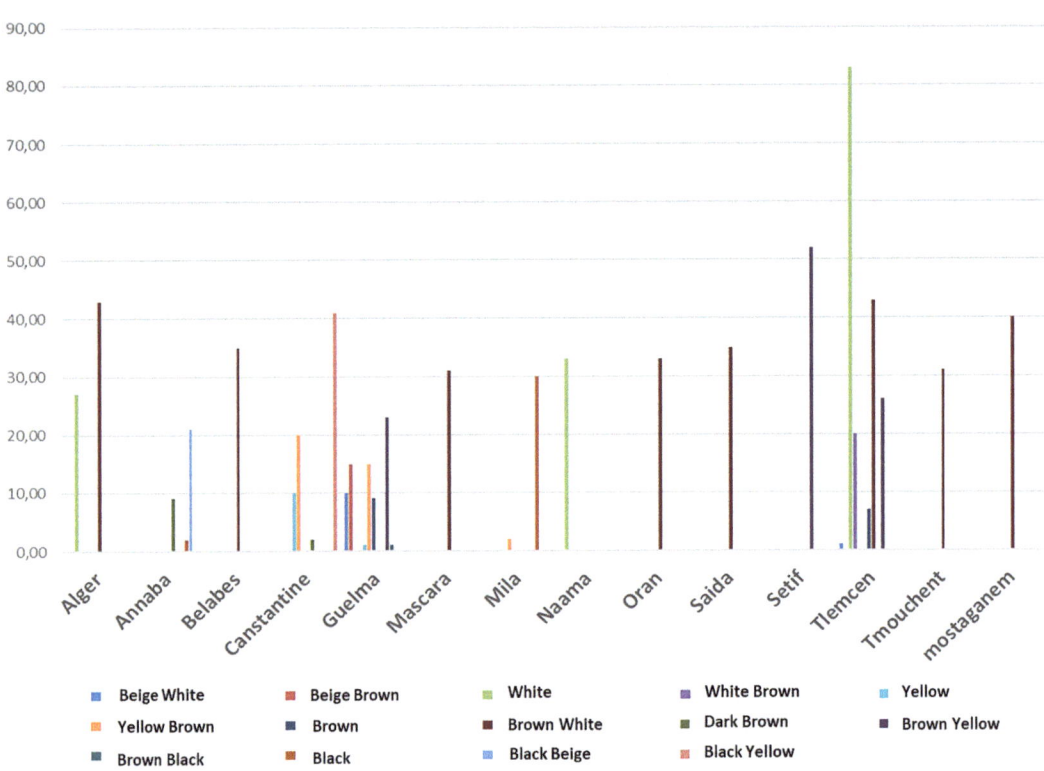

Fig. 13.5 Distribution of snails according to shell color in Northern Algeria

Bel Abbès, Mascara, Saida, Timouchent, and Mostaganem (Fig. 13.5). Generally, the presence of the majority of species at the El Tarf district and their different morphologies may be due to the geographical position; effectively, this part of Algeria is the most humid and is part of a humid climate stage. In addition, the sociocultural state of the citizens of El Tarf is different compared to the other districts visited; they are not interested in snails, neither for national marketing nor for self-consumption, which explains the presence of several species.

13.3.4.2 Snails of Central and West Africa

Prevalent Species: The Giant African Snail

Giant African snail is one of the largest land snails in the world, reaching up to 19 cm in length (Peterson 1957). In part because of its polyphagous diet, it has been recognized as one of the world's most damaging pests and is considered one of the top 100 invasive species by the World Conservation Union, IUCN (ISSG 2003).

According to Tsayo et al. (2020), only two predominant species, namely *Archachatina marginata* and *Achatina fulica*, were identified out of a total of 693 adult giant land snails studied in six localities (Odza, Mbankolo, Biyem-assi, Nyom, Nkolbisson, and Simbok) in the Mfoundi Division of Cameroon. These two species are the most widely harvested edible snails in the Central and West African forest zones.

External Morphology of the African Giant Snail

According to Tsayo et al. (2020), the diverse colors of the shell opening observed in the edible snail populations of the equatorial region indicate that there is genetic variability. The colors are influenced by many factors, including genetics, the action of the environmental conditions and diet, and the different interactions between them. In fact, the African land snail in Cameroon shows a black coloration with yellowish stripes and a red tip, observed in *Archachatina marginata*, in contrast to other observations made by Nkwedem et al. (2019) in Moungo (Cameroon) and those of Mbetid-Bessane (2006) in the Central African Republic, which had shown that the black color with whitish stripes was the most observed. Nevertheless, the brown color with white stripes and white tip observed in *Achatina fulica* is also mentioned by Nkwedem et al. (2019) in Moungo (Cameroon), Hana et al. (2016) in Nigeria, and Mbetid-Bessane (2006) in the Central African Republic.

Furthermore, the results obtained by Okon et al. (2012) about a comparative study between two species of giant land snail, *Archachatina marginata* and *Achatina fulica*, revealed that there are significant correlations between shell length and shell "mouth" width and between shell "mouth" length and shell "mouth" width for *A. fulica* snails. Thus, these quantitative or phenotypic traits of the two species could be chosen to differentiate as well as characterize growing snails in the Niger Delta region of Nigeria (Fig. 13.6).

13.3.4.3 Snails in South Africa

Origin and Distribution
Herbert (2010) examined in detail the exotic terrestrial mollusk fauna of South Africa: 34 species were considered to have been introduced into the country, of which 28 became established and 13 invasive. The author concluded that the introductions continue at a rate of about two species every 10 years, with no signs of stabilizing. He also highlighted that the agricultural and horticultural industries are the main contributors to the introduction and spread of exotic species. The composition of this exotic fauna shows considerable similarity to that known from Southern Australia, reflecting the regions' similar colonial history and climatic correspondence with the regions of origin in Western Europe and the Mediterranean. Each species was discussed based on its distinctive characteristics, habitat preferences, date of introduction and first record, native range and global distribution, distribution in Southern Africa, pest status, and similarity to native species. In addition, some attention has been paid to potentially harmful species whose occurrence in South Africa is not yet known but which represent a significant risk of future introduction. The author also reports the presence for the first time in South Africa of the species: *Discus rotundatus, Hawaiia minuscula, Vitrea contracta,* and *Aegopinella nitidula* (Fig. 13.7).

Characteristics of Prevalent snail Populations in South Africa
Perera et al. (2021) analyzed the zoogeography of the terrestrial malacofauna of Southeast Africa (SEA), proposing the first digital regionalization based on mollusks for the area. They also dis-

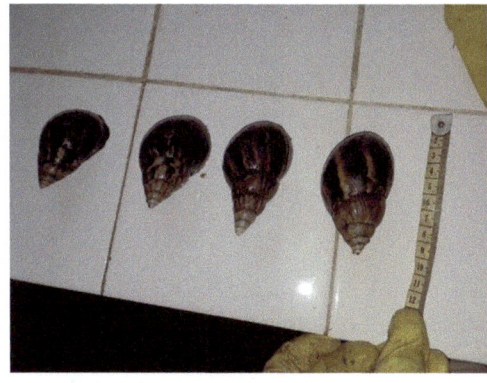

a *Achatina fulica*

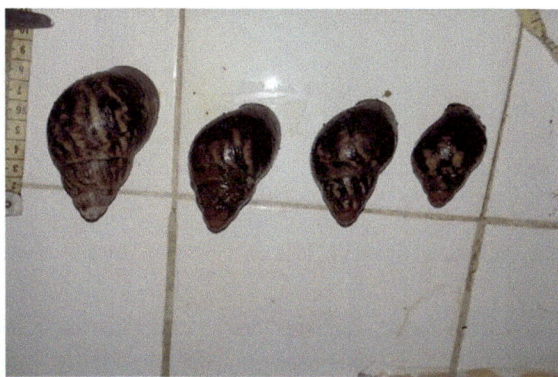

b *Archachatina marginata*

Fig. 13.6 Picture comparing two Achatinadae snails

Eobania vermiculata (Müller, 1774), South End cemetery, Port Elizabeth (NMSA V8627 [2000]).

Cochlicella barbara (Linnaeus, 1758), Constantia, Cape Peninsula (NMSA V8326 [2000]).

Cornu aspersum (Müller, 1774), scalariform freak, Cape Town, W. Cape (NMSA V7391 [1999]).

Oxychilus draparnaudi (Beck, 1837), Kirstenbosch, Cape Peninsula (NMSA W5705 [2007]).

Theba pisana (Müller, 1774), crawling animal, Cape Agulhas (NMSA V7750 [2000]).

Fig. 13.7 Some species of land snail in South Africa. (Herbert 2010)

cussed the models and centers of endemism of terrestrial snails, thus assessing the importance and delineation of the Maputaland-Pondoland-Albany (MPA) biodiversity hotspot for their conservation. An incidence matrix compiled for relatively well-collected lines of terrestrial snails and slugs (73 taxa in twelve genera) in 40 *a priori* operational geographic units was subjected to (a) a phenetic agglomerated hierarchical clustering using the method of groups of unweighted pairs with arithmetic means (UPGMA), and (b) endemicity parsimony analysis (PAE) and biotic element analysis (BEA). Responding to the primary objective of the study, the UPGMA dendrogram provided hierarchical regionalization and identified five centers of mollusk endemism for SEA, while the PAE confirmed six areas of endemism, also supported by the BEA. The regionalization covers a zoogeographic province similar to the MPA hotspot, but with a notable extension westward to Knysna (toward Cape Town). The MPA province, centers, and areas of endemism, biotic elements, as well as spatial patterns of species richness and endemism, support the MPA hotspot but suggest further extensions resulting in greater land snail endemism. The MPA region (also with a northward extension in Sky Islands, Southpansberg, and Wolkberg) is similar to that observed for vertebrates. The large MPA region provides a more robustly defined region of conservation concern, with centers of endemism serving as local conservation priorities.

13.3.5 Importance of Snails

13.3.5.1 Snails as a Dietary Resource

The snail constitutes food that is highly appreciated for its tasty, tender flesh and is very rich in amino acids, mineral salts, and particularly iron. Snail is an important alternative source of animal protein in rural communities. The strong pressure to collect snails due to the growing demand as a food source, as well as the destruction of their biotope by humans and natural predators, is a factor that contributes to a reduction in snail stocks in the natural environment. Faced with this situation, snail breeding, as an activity, appears necessary to compensate for the seasonal shortages of snails and to ensure the sustainability of the species (Bouchiba et al. 2021; Karamoko et al. 2011).

13.3.5.2 Soil Mineral Source

In agronomy and pedology, mineralization is the decomposition of the organic part of a soil material, which also contains a mineral part. The snail contributes to the mineralization of organic substances and the formation of humus, which benefits cultures and is the first link in the chain of decomposition of organic matter (Stiévenart et al. 1990).

13.3.5.3 Use in Traditional Medicine

Edible snails are widely used in traditional medicine. These snails are generally farmed and have no toxicity, like inedible and wild snails. In fact, for the treatment of respiratory infections, it is recommended to use a mixture prepared with thyme and snail, which are cooked inside argan oil. This mixture is left to cool and then filtered; thereafter, it is administered in drops orally (Radi 2003). In Ghana, the bluish liquid remaining in the shell after the flesh has been extracted is recommended to combat ulcers and asthma. Also, a recent study showed that the glandular substances present in the flesh of edible snails caused the clumping of certain bacteria, a phenomenon that can help fight a variety of diseases, including whooping cough (Cobbinah et al. 2008).

13.3.5.4 Ecotoxicological Interest

Ecotoxicology is the study of the interactions between the various chemical products widespread in the environment and the biotope. Hence, it is essential to assess the quality of biomass and the ecosystem. The snail *Helix aspersa* was shown to be a powerful bioindicator and bioaccumulator, since it was used to monitor the ecotoxicity of the environment (Cobbinah et al. 2008).

As a result of their permanent contact with the soil surface, land snails are more exposed to several pollutants found in the soil. These contaminants could be absorbed through various routes, such as respiratory, digestive, and transcutaneous (Gomot-de Vaufleury and Pihan 2000). It was shown that *Helix aspersa* snails can accumulate heavy metals such as lead in their shells (Beeby and Richmond 2011). The reproduction and growth of snails are indicators to evaluate the soil pollution by pesticides (Russell et al. 1981). The *genus Helix*, raised under favorable conditions, can be used as a bioindicator of terrestrial pollution, or it can also be used as a test organism for the toxicity of metals (Russell et al. 1981). It is an element capable of transferring pollutants from the soil to plants and predators (Beeby 1985).

13.3.6 Snail Production Systems

Different systems of snail production have been experimented with over the years. Three snail production systems are currently in use:

13.3.6.1 Outside Farming

Outside, snail farming is conducted completely outdoors, such as open parks, where the snails are located in areas surrounded by snail-proof fences. The enclosures contain the vegetation chosen for its aptitude for feeding snails. Snails are left to breed, reproduce, and grow almost unattended until harvest. This system was adopted by the Italians, and almost all snails produced for the market are bred in this way. Italian snail farms designated very large areas for this system (2.5 hectares or more). The amount of labor and material is reduced to a minimum, and this is more advantageous than other types of snail farming described below. Unfortunately, snails are only collected during the few months of the growing season, and so, the amount of product available on the market is limited (Murphy 2001).

13.3.6.2 Indoor Farming

This system integrates air-conditioned buildings in which the snails spend their time under defined parameters to facilitate the optimal environment for snail reproduction and feeding: all the nutritional needs necessary for each stage of snail development are furnished. This level requires a high initial capital investment. It needs to be constantly monitored by requiring the presence of human supervision seven days a week. The indoor method has been the subject of much research in France for 20 years. This type of snail farming is the result of winters in the Northern Hemisphere that can be quite severe and could curtail snail production. Growing snails indoors has enabled the farmer to produce snails year-round and, therefore, supply markets with fresh produce on demand (Murphy 2001).

13.3.6.3 Mixed (Indoor and Outdoor) Farming

This system incorporates particular management practices of the two types of snail farming above. Reproduction and the nursery environment use the indoor method, while the growth phase takes place outdoors. This system allows the snail farmer to cultivate and produce snails year-round and, therefore, has the product available to the market at all times. This type of snail farming is now widely preserved in France and has been practiced for some time. The indoor/outdoor method is suitable for Australia as the climate along the southern end of the coast is quite mild, and extreme temperatures are not of great concern as they are in Europe. The breeding greenhouse allows snail growers to control the environment to such an extent that the temperature of the outside air does not interfere with the breeding program. The same goes for the nursery greenhouse, where young snails can feed and develop without the overall conditions interrupting the desired growth rate. The growing area is completely covered with a shade cloth that protects the snails from the sun, wind, and hail. The method of raising and managing snails indoors and outdoors gives the Australian grower the best of both systems developed and proven overseas for many years (Murphy 2001).

13.3.7 Genetic Resources of Snails

The first studies in the field of mollusk cytology were carried out at the end of the nineteenth century. Due to inferior optical equipment and methods, many of the first reports were found to be inaccurate (Patterson and Burch 1978).

13.3.7.1 Karyotype

The karyotypes of 76 species from eight archeogastropod families were studied using the hot dry chromosome method of Kligerman and Bloom (1977) and a computer-assisted analysis system (Burch 1968). The number of haploid chromosomes (N) varies from 9 to 21 in the 76 gastropod species. The Patellidae (5 species, N = 9), *Acmaeidae* (15 species, N = 10), and *Neritidae*

(21 species, N = 12; 1 species, N = 9, 10; and 1 species, N = 11, 12 and 14) have a relatively small number of chromosomes, while the *Trochidae* (13 species), *Turbinidae* (3 species), *Stomatellidae* (1 species), and *Helicinidae* (3 species) have a greater number of chromosomes (N = 18), and a *trochid Granatalyrata* (N = 20 or 21). Chromosomal morphologies have been studied in 42 species. More than half of the karyotypic complements are meta- and/or sub-metacentric in each of them, while no family is characterized by complements consisting only of meta and/or sub-metacentrics. The lengths of the chromosomes at the mitotic metaphase are mainly 1–4 micrometers. The total length of diploid chromosomes was measured for 34 species and shows an increase as the number of chromosomes increases (Nakamura 1986). Until now, the systematics of lower taxonomic categories have remained an enigma, both from the practical point of view of identifying specimens and of understanding the mechanisms of speciation and evolution (Burch 1968).

13.3.7.2 Shell Polymorphism in the Hedge Snail (*Cepaea nemoralis*)

On the same site, it is common to observe snails of similar sizes and shapes but differing in the characteristics of their shells. Their colors, as well as the bands that adorn them, can vary widely. *Cepaea nemoralis* is a species that includes individuals whose shell is yellow (it can sometimes be very dark and brown) and others whose shell is pink (likely to drift to purple). Schematically, the color of the shell is controlled by a gene exhibiting two alleles, denoted j and j +, j + being dominant and j recessive. Individuals with two j (j / j) alleles have a yellow shell, individuals with two j+ + (j+/j +) alleles or one j allele and one j+(j +/ j) allele have a pink shell.

Some shells are devoid of bands, while others are decorated with longitudinal brown bands. The absence or presence of bands is determined by a gene with two alleles, b and b+. The b+ allele is responsible for the absence of bands and is dominant over the b allele responsible for the presence of bands. Individuals without bands have two b + (b + / b +) alleles or one b + allele and one b (b + / b) allele, and individuals with five bands have two b (b / b) alleles. Taking into account the color on the one hand and the presence of stripes on the other hand, there are 32 combinations, of which some are common and others are very rare. The possibilities are more numerous, the bands adorning the shell sometimes being fused, discontinuous, or weakly pigmented. The populations of hedge snails, therefore, appear to be polymorphic, with various morphological types coexisting on the same site. In fact, from one observation site to another, the proportions of the different phenotypes are variable, suggesting variations in allele frequencies.

13.3.7.3 Genetics of Body Color Variation of Land Snails

The main factor in variability in color is the amount, density, and distribution of melanin pigmentation. Melanin is dark brown/purple/black in color, which is intensified by dense compaction of melanin granules in the cells of the upper layers of the skin. A mutation in the tyrosine enzyme gene produces a protein with reduced functionality, thereby resulting in reduced melanin production. When the color of the pigments is apparent, the DNA sequence is read (translated into RNA) and translated into the color pigments (protein). At this time, several elements can appear in different ways: The gene code for the color of the pigments can be completely or partially deleted. The gene may be inactive because the regulatory sequences are missing or dysfunctional. If a snail has two copies of an active color gene, the snail is called a wild-type homozygote for the color gene. If a snail has two non-active copies of a color gene, the snail is called a color gene knockout homozygote. If a snail has an active gene copy and a nonfunctional copy, the snail is heterozygous for that color gene. The color variations in the snail are the result of mutations in several genes responsible for the pigmentation of the shell and the body. A major black pigment determines whether the body is dark or albino (colorless). Snails that are not lacking in pigment still vary in the amount of pigment they have. Many genes influence the expression of the major genes, so the intensity of expression can vary, even for normal traits (Afiane 2019).

13.3.7.4 Genetic Structure of Land Snail Populations

The results obtained by Woogeng et al. (2017), where they used mtDNA as a genetic marker, show that the isolation by distance plays a significant role in determining the genetic structure of extant indigenous land snails in Cameroon, in line with the limited ability for dispersal in this species. The *Archachatina marginata* population, nevertheless, displays a diverse gene pool despite its exploitation in recent years. In contrast, *Achatina fulica* shows much less diversity and structure, which is most likely the signature of a bottleneck and weak genome dynamism.

In China, Zhou et al. (2007) proposed a study of the species *Camaena cicatricosa*, which is an important and harmful land snail found in Southern China. It not only damages crops, leading to reductions in yield and quality, but also spreads a zoonotic food-borne parasitic disease and causes substantial damage to human and animal health (Zhou et al. 2007). Based on mitochondrial gene (COI and 16srRNA, mtDNA) and internal transcribed spacer sequences (ITS2), the results of the study revealed significant fixation indices of genetic differentiation and high gene flows between most populations. Furthermore, the phylogenetic trees of haplotypes indicated nonobvious genetic structure. Similar results were obtained based on the synonymous and non-synonymous sites of 347 sequences of the COI gene studied.

Powelliphanta lignaria, a genus of large carnivorous land snail, is endemic to New Zealand and displays phenotypic variation within comparatively small geographic distances. Daly et al. (2019) used information from the data combination of mitochondrial sequence and other microsatellite loci to study the genetic structure of *Powelliphanta lignaria*. The level and structure of genetic diversity provided a picture of a naturally fragmented lineage restricted to a particular ecological zone.

- *Microsatellite*

Jaksch et al. (2017) developed microsatellite markers in the *Clausiliidae* family of snails. The *Clausiliidae* family, known under the name "door" snails, are gastropods with a very high diversity in shell morphology. For the best genetic diversity approach, 13 microsatellite loci with a tetranucleotide repeat were isolated and tested in three geographically close Montenegrina populations.

Furthermore, the study of Shweiger et al. (2004) suggests that the metapopulation structure of such species in fragmented landscapes depends on both the landscape features and the shape of the dispersal function. In fact, four trinucleotide and one tetranucleotide microsatellite loci previously published were analyzed (Davison 1999). The loci Cne1, Cne11, Cne15, and Cne6 were developed in a multiplex PCR to investigate the impact of habitat fragmentation on the spatial genetic structure of an organism with limited dispersal ability of the land snail, *Cepaea nemoralis* (L.). The findings of both genetic and morphological patterns showed spatial structuring at two scales. At the local scale, a deme-structured subdivision of a continuous subpopulation corresponded to the limited active dispersal ability. At the mesoscale, rare dispersal events, most likely driven by passive displacement, were suggested to lead to metapopulation persistence in a fragmented landscape.

- *DNA Sequencing*

The giant African snail, *Achatina fulica* (NCBI:txid 6530), is a gastropod species originating from East Africa. It is the largest terrestrial mollusk, with a voracious appetite, strong environmental adaptability, and a high growth and reproduction rate (Schreurs 1963). The chromosome sequence of *Achatina fulica* will provide the research community with a valuable resource for population genetics and environmental adaptation studies for the species.

13.3.8 African Giant Snails

Snails' consumption by humans has been going on for centuries. In Africa, particularly in West Africa, the consumption of snail meat is part of the culture. Several nutritional benefits were conferred to snail meat compared to the meat of conventional livestock. However, their breeding is not yet widespread and is still done in the traditional way, which is collection in forests and fields.

13.3.8.1 Snails' Genetic Resources

Snails are invertebrate animals belonging to the group of mollusks due to their shell. They are classified in the group called gastropods since they have one shell. The shell covers and protects the soft part of the body (Stiévenart et al. 1990). The body consists of three parts that are not well defined:

- *The head* has two pairs of retractable tentacles, the longer pair of the tentacles possesses the eyes at their top and the shorter pair has a sensitive function and a mouth located below the head.
- *A long, muscular foot,* whose separation is not clear enough from the head (cephalopod), allows the animal to move.
- *A bumpy visceral mass* is hidden under the shell and composed of its vital organs, important in the functions of digestion, reproduction, and respiration.

In sub-Saharan Africa, particularly in the humid forest areas, the well-known and consumed snails are from the Achatinidae family. Commonly named African giant snails because of their weight, which can reach 250 g as adults for some species. A group of giant land snails native to Western, Eastern, and Southern parts of Africa possess lungs (*Pulmonate*) with a long conical shell. The *Achatinadae* group is also subdivided into 14 subgroups, including the Achatina and Archachatina subgroups.

A description of the differences between the *Achatina* and *Archachatina* subgroups was made by Hodasi in 1984 (Stiévenart et al. 1990). According to their general characterization, the *Achatina achatina* group has a relatively oval shell with a regular spiral and a narrow top, while the *Acharchatina marginata* group has a wide-topped shell in a dome shape (see Fig. 13.8).

13.3.8.2 Housing, Environmental, and Adaptation Factors

Heat and humidity are the two main factors important for the survival of *Achatinadea* snails. This explains their geographical distribution in humid regions. Therefore, they can be found in a wide variety of environments such as agricultural land, coastlines, forests, and peri-urban areas. *Achatinadae* snails are animals whose metabolism varies with their environment (poikilotherms). They are very active in humid climatic conditions and are observed generally during the rainy season. During the dry season, when the temperatures are usually high, *Achatinadae* snails live reclusively in their shells that they close with a thin white calcium layer. They are predominantly nocturnal animals, but could have limited activities during the day.

13.3.8.3 Nutrition, Feeds, and Feeding

African giant snails are mostly vegetarians (see Fig. 13.9). However, empty snail shells can be observed inside their cages, suggesting that they may be preying on dying animals. In general, they consume leaves, fruits, and roots (Cobbinah et al. 2008; Agbogidi and Okonta 2011). Based on field observation, mature adults are mostly found around the flowers and leaves of fruit trees (oil palms and plantains). The young are found inside fallen fruits (papaya, tomato). According to snail collectors, snails seem to be affectionate to rotting cassava roots, and are used as bait for them. For breeding needs, many plant species have been identified for their palatability or feed efficiency (Cobbinah et al. 2008). Also, calcium is required for the proper development of their shell, which is obtained in the soil in their natural habitats. In captive breeding, calcium is usually sourced from eggshells or chalk (Cobbinah et al. 2008). A good source of minerals, especially phosphorus, is necessary for shell development (Aktaş et al. 2019).

13.3.8.4 Reproduction, Management, and Breeding Strategies

There are differences in the reproductive tract between *Achatina achatina* and *Archachatina marginata*. For *Achatina achatina*, the

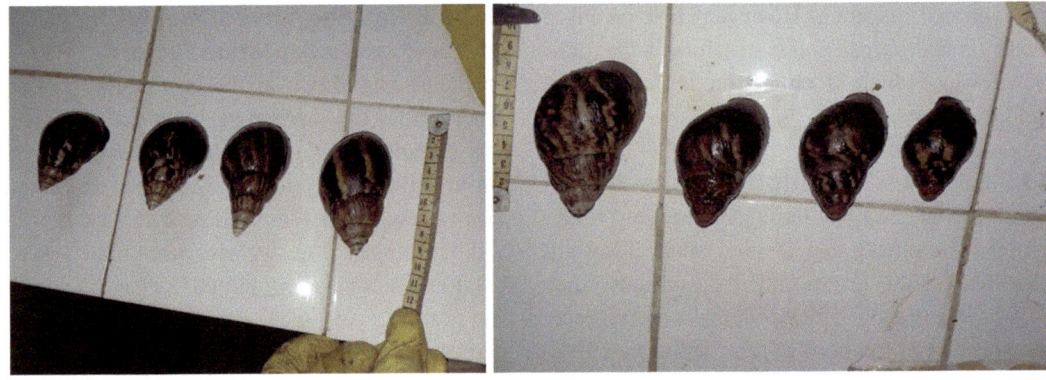

a *Achatina fulica*   b *Archachatina marginata*

Fig. 13.8 Picture comparing two Achatinadae snails: a. *Achatina achatina* snails, b. *Acharchatina marginata* snails

a. Young snails consuming rotten papaya

b. Snail eating banana tree flowers

c. Snail eating watermelon in a husbandry

Fig. 13.9 Snail nutrition

reproductive system comprises an elongated tubular vagina and a short sperm duct (Abiona et al. 2012). For *Archachatina marginata,* the reproductive system comprises a short tubular vagina and a very long and tapered sperm duct (Abiona et al. 2012). There are also differences in reproduction between the two species. According to Hodasi (1979), *Achatina achatina* is self-fertilizing. He concluded based on his observations that there is almost no mating between two snails, even if they can be seen very close to each other (Abiona et al. 2012). This species lays eggs during the rainy season. Laying of 30–300 eggs takes place in holes previously dug by the animal of about 4 cm deep, during the night. The eggs are likely oval and have a yellow color. They are small, 8–9 mm long and 6–7 mm wide. Eggs hatch about two to three weeks after laying (Cobbinah et al. 2008). The eggs of *Achatina* snails are very fragile. As a result, the neonatal mortality rate is high (Hardouin et al. 1995). The juvenile phase is 1–2 months after hatching until the stage of sexual maturity (14–20 months), which corresponds for this species to the adult phase. An adult can live on average 5–6 years. However, some can live up to 9–10 years (Cobbinah et al. 2008; Hodasi 1979).

For *Acharchatina marginata*, cross-fertilization is required for eggs to be viable (Cobbinah et al. 2008) (see Fig. 13.10). The eggs are relatively large, from 12 mm in length to 7 mm in width, and their average weight at breeding is 4.8 g. Similar to *Achatina achatina,* their eggs are yellow (see Fig. 13.10). The number of eggs per laying is about 4–18. Eggs are laid in the soil at a depth of 10 cm (Radi 2003; Hodasi 1979; Hardouin et al. 1995). Hatching of eggs occurs about four weeks after laying (see Fig. 13.10). The juvenile period is about 15 months. The period during which the shell will thicken and grow. *Archachatina* snails reach sexual maturity around 10–12 months, when they reach 100–125 g (Cobbinah et al. 2008). They can live beyond 10 years (Stiévenart et al. 1990).

13.3.8.5 Health, Production Systems, and Value Chains

The giant snail farming system in Africa was inspired by the major snail-consuming countries of Nigeria, Ghana, and Cote d'Ivoire. Indeed, three production systems are applicable: extensive, intensive, and semi-intensive (Stiévenart et al. 1990; Cobbinah et al. 2008; Hardouin et al. 1995).

- *Extensive System*: the animals are placed outdoors in an environment similar to their natural environment. A vegetable garden with plants eaten by the snails is chosen (papaya, plantain, sweet potato, etc.). Then that area is demarcated with a fence to prevent them from escaping.
- *Semi-Intensive System*: the animals are put in a place similar to the extensive system, but their food is controlled, and snails are supplemented with industrial food that can increase their growth. In addition, the eggs are separated from the adults. They are left to hatch and grow for 2 months before reintroducing them.

Fig. 13.10 Reproduction of *Archachatina marginata* snails

a Mating of snails

b Eggs

c Hatching of eggs

- *Intensive System*: is a completely closed system where the environment is controlled. The enclosures can be pits constructed with bricks and covered on top with wire mesh to prevent animals from escaping. A pen can also be a pit on the ground covered with a wire mesh. Wooden boxes can be used as well as old car tires superimposed together, or even old basins.

Regardless of the type of system chosen by the breeder, to optimize the growth of animals and have a better yield, the following conditions are expected to provide optimal productivity: flat land, shady, humid (above 85%), loamy soil rich in organic material, and moderate environmental temperature (25–28 °C).

Snails have several predators such as rodents (mice, rats and shrews), frogs and toads, birds (crows, ducks, and turkeys), reptiles (lizards, snakes), insects (beetles, millepedes, and centipedes), and slug (Cobbinah et al. 2008; Nyameasem and Borketey-La 2014). Snails are susceptible to fungal diseases mainly caused by *Fusarium spp.*, also known as pink egg disease. The eggs turn red-reddish brown when affected by the disease. This disease is mainly due to poor maintenance of pens and rotten food that is not removed (Cobbinah et al. 2008; Aktaş et al. 2019). Parasites such as *Alluaudihella Flavicornis* and *Angiostoma aspersae* are disease vectors for snails in their wild and farming environments. Both parasites lay their eggs on the snail, which hatch into worms that cause death (Aktaş et al. 2019). Bacterial diseases caused by *Pseudomonas* spp. Mostly, *Pseudomonas aeruginosa* causes intestinal infections in snails, which disrupts their growth (Aktaş et al. 2019).

The snail business involves several actors that include field gatherers, breeders, dealers, cleaners, and traders of cooked meats (Mbétid-Bessane 2006; Nzouankeu 2015). Snail farmers have two ways of marketing their animals. The new producers use them as breeders to begin their breeding, which is eventually sold to resellers in the markets. Snails of the dealers are intended for the catering of households. The consumption of snail meat that was once reserved for certain ethnic groups, especially in the South-West region (Kaldjob et al. 2019) in Cameroon, is increasing in the major cities of Cameroon (Nzouankeu 2015). Indeed, most of the snail meat is sold by itinerant traders in grilled form, ready to be consumed. It will also be found in restaurants prepared in various local sauces or simply in the family. Snail meat is certainly valued because of its many proven nutritional qualities (low in fat, high in protein, high in essential amino acids, and rich in vitamins and minerals) when compared to meat from conventional farm animals (beef, chicken, and pork) (Babalola and Akinsoyinu 2009; Eneji et al. 2008; Fagbuaro et al. 2006; Malik et al. 2011; Srilatha et al. 2013). Indeed, snail slime is very popular in cosmetics and pharmacology. According to some studies, it has, among other things, healing, anti-inflammatory, and moisturizing properties (Castro et al. 2007; Chinaka et al. 2021; Laneri et al. 2019). The shell is used in animal feed, particularly in the diet of laying hens (Houndonougbo et al. 2012).

13.3.9 Core Challenges and Perspectives

The diverse use of snails shows that traditional collection in the field can no longer cope with the growing demand (Joshi and Pandey 2019). Breeding would then be an ideal solution to overcome this problem. Indeed, there are several advantages to embarking on snail farming. There is a growing demand for snails because of their several advantages. First, snail meat has several nutritional benefits. Second, snails have other applications, especially in other sectors such as pharmacology, cosmetics, and animal feed. Snail farming is less expensive than conventional farming because the breeding and animal feeding do not require much effort. Furthermore, the snail business is an economic source of income for several actors in the sector. However, breeders encounter problems in getting this sector of activity going. This situation could be explained by several factors. First, there is a lack of funding and training to support farmers. Snail breeding is climate dependent; production is reduced in the dry season since animals either enter aestivation or die. In addition, snails' reproduction is not yet very well controlled, and the percentage of eggs hatching is very low compared to the number of eggs laid. Finally, there is also a problem of predator invasion that reduces snail production (Woogeng et al. 2017; Baruwa 2012; Ndah et al. 2017). To solve these problems, government and private investors should provide more financing to the sector in order to enhance its potential and to allow breeders to be well-trained. The outcomes will be more snails for the local demand and for the export of byproducts of this mollusk. Indeed, snail slime could be supplied to the pharmaceutical and cosmetic industries. Snail meat could be exported to other countries where snails are consumed as well.

13.4 Rabbit

13.4.1 Origin and Domestication of Rabbit

Historically, the rabbit is a mammal that is native to Southern Europe and North Africa. It may have been discovered by the Phoenicians when they came into contact with the Spanish around the year 1000 BC (Lebas 2002). The oldest fossils of the genus are dated around six million years ago and were found in Andalusia. It is the only domesticated mammal whose paleontological origin is the *Oryctolagus cuniculus* and is located in Western Europe (Lebas 2002).

The rabbit has truly entered the era of genomics with the sequencing of its genome, which has paved the way for work to better understand the evolutionary history of this species (Garreau and Gunia 2018). Indeed, studies have shown the existence of two strongly divergent lineages, called A and B, which are partly related to the subspecies *O. c. algirus* and *O. c. cuniculus* (Rogel-Gaillard et al. 2009). These two lineages share a common ancestor dating back two million years. The two lineages are still geographically separated today: the wild populations of the southwest of the Iberian Peninsula belong to group A, while those of Northern Spain, France, and Tunisia are in group B, as all are domestic breeds. The Pyrenees range seems to have constituted an important genetic barrier since migration from the north of Spain to the north of the Pyrenees resulted in a loss of variability of around 12% (Carneiro et al. 2011). The most recent genetic study confirmed that all domestic rabbit breeds originated from the domestication of wild rabbits of the subspecies *O. c. cuniculus* that were present in France in the Middle Ages. Although the genetic variability of rabbit breeds remains high compared to other domestic mammals, the process of domestication and breed creation has been accompanied by a further genetic depletion of around 21% (Carneiro et al. 2011).

13.4.2 Rabbit Genetic Resources in Africa

13.4.2.1 North African Rabbit Breeds

Algeria Rabbit Breeds

Rabbit farming has existed in Algeria for a very long time (Ait Tahar and Fettal 1990). But according to Berchiche and Kadi (2002), in Algeria, there was no study on indigenous rabbit populations before 1990. In the nineteenth century, the invasion and the arrival of European populations led to the development of rational units in the Maghreb, but this sector did not appear in Algeria until the early 1980s (Colin and Lebas 1996). Three indigenous rabbit breeds exist in Algeria, namely: *Kabyle breed*, Synthetic breed, and White population.

Kabyle Rabbit Breed

Belonging to the local population from the Tizi Ouzou region (Fig. 13.11), this breed is characterized by an average adult weight of 2.8 kg. This value classifies this population in the group of light breeds (Zerrouki et al. 2001, 2004), a body of medium length (arched type), descending in a gradual curve from the base of the ears to the base of the tail, and of good height, carried on limbs of medium length. Its posterior part is well developed with well-filled loins; the tail is straight. The head is convex with erect ears. Its

Fig. 13.11 The Kabyle rabbit phenotypes. (Carneiro et al. 2011)

Fig. 13.12 Phenotypes of the white breed. (Gacem et al. 2008)

coat is soft, presenting several color phenotypes, a consequence of the contribution of imported breeds such as Fauve de Bourgogne, White New Zealand, and Californian (Berchiche and Kadi 2002). This breed has a good adaptation to local climatic conditions and is used mainly in meat production, but its prolificacy and adult weight are too low to be used as such in meat-producing farms. The numerical productivity recorded in the females of this population is of the order of 25–30 weaned rabbits/female/year (Berchiche and Kadi 2002; Zerrouki et al. 2005).

The White Breed

It is a population of "commercial strains" (See Fig. 13.12), which was imported from France by Algeria (1985–1986), characterized by a dominant albino phenotype and produced by a state cooperative. According to Zerrouki et al. (2004), young individuals are weaned at 30–35 days old with an average litter size at birth of 7.1 ± 2.4, of which 6.7 ± 2.8 are born alive; litter size at weaning is 5.8 ± 2.4; stillbirth proportion is 7.3% of total kits born, and birth to weaning mortality is 15.8% of kits born alive; prolificacy is similar to that observed for the indigenous population, but kit's mortality is lower; the average live weight at mating is 15% higher than the indigenous population (3.34 vs. 2.9 kg).

The Synthetic Population

The synthetic population is called "ITELV 2006" (Fig. 13.13), created in 2003 by a cross between the females of the indigenous population and the males of the strain INRA 2666, to improve the potential of rabbits intended for meat production in Algeria. It is of medium size but heavier and more productive; its coat is characterized by several color phenotypes: brown, white, black, gray, and sometimes mixed (spotted) (Gacem and Bolet 2005; Gacem et al. 2008; Bolet et al. 2012).

Moroccan Rabbit Breeds

Rabbit activity in Morocco is relatively recent. The main objective of rabbit breeding by Moroccan breeders is the production of meat for 99.85% of them. Due to the traditional and secondary nature of breeding, 91.1% do not practice any registration. Most animals used on farms (87.7% of cases) consist of purebred rabbits or indigenous populations.

Tadla Rabbit Breed (See Fig. 13.14)

The Tadla breed is of small size and raised for meat production. Tadla rabbits are of medium body length. Chest circumference is relatively similar to that of Zemmouri rabbits, but the loin appears to be relatively narrow in width, which may impair meat production. The hips are strong,

   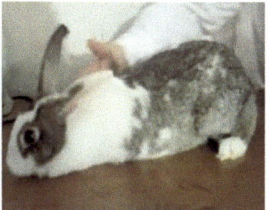

Fig. 13.13 Phenotypes of the synthetic breed

Fig. 13.14 Tadla breed

Fig. 13.15 Zemmouri breed

with a convex head, gray eyes, and erect ears. In addition, the length of the feet and legs is about 22.5 and 12.8 cm, respectively, and the tail is curly. Furthermore, the Tadla rabbits are well adapted to hot weather (40 °C in summer). They have resistance to some diseases (fungal infections) but are more sensitive to viral hemorrhagic disease (VHD) than other breeds.

Zemmouri Rabbits Breed (Fig. 13.15)

At the beginning of the invasion of Morocco by France and Spain (1912), some exotic rabbit breeds were brought by the European religious people. These rabbits were naturally mated to the native stock of the indigenous population in Azrou (Middle Atlas Mountains) and Temara (Rabat region). After several generations, the resulting populations were adapted to local environmental conditions. Then, the inter-mating of the populations of both areas probably gave origin to the Zemmouri population, which is now quite stable regarding phenotypical traits. The Zemmouri breed is of medium size and raised for meat production, the skin being used for traditional manufacture. Regarding its large body length, Zemmouri rabbits seem to have a good ability for meat production; the hips are quite strong; the loin, which is quite large, carries a good amount of meat; the chest circumference balances in good shape with the rest of the body with a convex head, maroon eyes and erect ears, feet and legs: medium in length and a straight tail. Regarding sexual maturity, Zemmouri rabbits seem to be sexually mature earlier than the exotic breeds known in Morocco. This trait needs to be further studied. However, the body weight at first mating for both sexes is within the normal range (80–85% of their expected adult weight). Further, the conception rate in Zemmouri rabbits is relatively lower than that seen in exotic improved breeds (Californian and New Zealand White). Also, the litter sizes at birth and at weaning, and the litter weights at birth and at weaning are still low in comparison to improved breeds.

Fig. 13.16 Bauscat breed. (El Raffa et al. 2002)

Fig. 13.17 Giza White Breed. (El Raffa et al. 2002)

Fig. 13.18 Chinchilla breed. (Bolet et al. 2012)

Egyptian Rabbit Breeds

- The Bauscat Rabbits (Fig. 13.16)

This breed is a medium-sized breed and is used mainly for meat production. This breed has well-rounded hips with a well-filled loin. The ribs are carried forward to combine with the shoulders, which balance with the rest of the body. The shoulders blend smoothly into the midsection, and the midsection smoothly extends into the hindquarters. The body is of medium length with good depth. The top bodyline rises in a gradual curve from the base of the ears to the center of the hips and then falls in a smooth curve downward to the base of the tail. The sides taper slightly from the hindquarters toward the shoulders. The back is markedly convex ventrally without being potbellied. The skin is smooth (El Raffa et al. 2002).

- The Giza White Rabbits (Fig. 13.17)

In 1932, a native stock of rabbits (Baladi rabbits) was bred by the Animal Breeding Department, Cairo University, Giza, Egypt, in an attempt to form a breed of uniform characteristics, El-Khishin. These rabbits were of different colors and sizes. Colors were isolated, and black and albino colors were segregating. In 1937, systematic breeding took place with the objective of obtaining an albino type of rabbit with a faster rate of growth and a larger litter size, which is presently known as the Giza White breed. Closed breeding in the albino population was performed for several years. This is a medium-sized breed and is used mainly for meat production. The body is of medium length with good depth and soft, silky fur. In addition, the top bodyline rises in a gradual curve from the base of the ears to the center of the hips and then falls in a smooth curve downward to the base of the tail. The sides taper slightly from the hindquarters toward the shoulders. The back is markedly convex ventrally without being potbellied. The skin is smooth. Giza White rabbits are late in their sexual maturity, and the conception rate of Giza White rabbits is relatively higher than that of standard breeds raised in Egypt.

- Chinchilla Breed (Fig. 13.18)

This is a medium-sized breed and is used mainly for meat production. In fact, the real Chinchilla has a dark slate blue undercolor at base; the intermediate portion is pearl—and should be as light as possible. The top edge has a very narrow black band, above this is a very light

Fig. 13.19 Baladi Breed. (El Raffa et al. 2002)

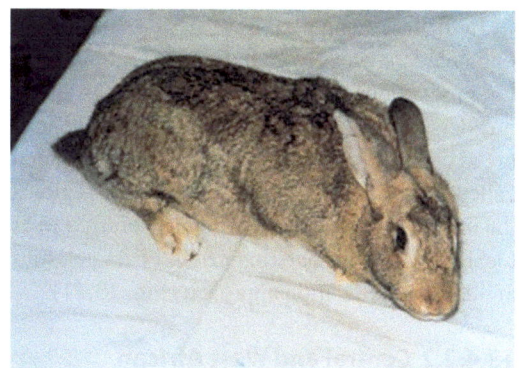

Fig. 13.20 Gabali Breed

band brightly ticked with black hairs, either wavy or even ticked to make the beautiful chinchilla surface color. The body of this breed has well-rounded hips with a well-fitted loin and hips. The body is of medium length, with a rather compact and chubby body. The top bodyline rises in a gradual curve from the base of the ears to the center of the hips and then falls in a smooth curve downward to the base of the tail. The back has a slight, gradual arch starting at the ear base. The skin is smooth (El Raffa et al. 2002).

- Baladi Breed (Fig. 13.19)

The origin of the Baladi breed is a result of several crossbreedings practiced between indigenous (native) rabbits and Flemish Giant (G) in stations of the Poultry Breeding Section, Ministry of Agriculture (Galal and Khalil 1994). The breeding plan is used for producing three native strains of Baladi Red (R), Baladi White (W), and Baladi Black (B). This is a medium-sized breed and is used mainly for meat production. The body of the three strains of Baladi rabbit has well-rounded hips with a well-filled loin. The ribs are carried forward to combine with the shoulders, which balance with the rest of the body. The shoulders blend smoothly into the midsection, and the midsection smoothly extends into the hindquarters. The body is of medium length with good depth. The top bodyline rises in a gradual curve from the base of the ears to the center of the hips and then falls in a smooth curve downward to the base of the tail. The sides taper slightly from the hindquarters toward the shoulders. The back is markedly convex ventrally without being potbellied. The skin is smooth. The Baladi breed has a medium to high conception rate since this rate ranges from 42% to 85% (El Raffa et al. 2002).

- Gabali Breed (Fig. 13.20)

This breed is from Sinai and the Eastern and Western (in the north coast belt) deserts of Egypt. They are raised by the Bedouins for their food. They were also raised in some western Giza Governorate areas by individual persons. It is a medium-sized breed and is used mainly for meat. The color is yellowish-brown with black hairs spread all over the body, with soft fur. Black hairs are intense on the tail, the body has well-rounded hips and well-fitted loins and ribs, extending forward to combine with the shoulders. The shoulders blend smoothly into the midsection, which extends into the hindquarters, with medium body length and a good depth. The top bodyline shows a gradual curve to the base of the tail. The sides taper slightly from the hindquarters to the shoulders, and the back is markedly ventrally convex. The animals are not potbellied. The skin is smooth. Regarding reproductive performance, the average age at first kindling in Gabali rabbits is about 7 months, which is later than 6.2 and 6.3 months reported by El Raffa et al. (2002) on New Zealand White and Californian rabbits, respectively. Moreover, the average performance of Gabali rabbits for litter size traits is within the average range (8–12 rabbits) (El Raffa et al. 2002).

In the following regions of Africa, it is rare to find breeders who use an indigenous rabbit breed for production, except for South Africa, where breeders use exotic breeds and an indigenous breed called Phendula, and Sudan, where a breed called Baladi Sudan is used. In the other regions, the indigenous breeds are generally referred to as local breeds, with heterogeneous characteristics and no standardization approach (Fig. 13.21).

13.4.2.2 Central and West African Rabbit Breeds

In Central and West Africa, the rabbit breeds kept are exclusively exotic breeds such as: Dutch rabbit, New Zealand White, Cottontail rabbit, Hyla rabbit, Chinchilla rabbit, California rabbit, Giant Flemish rabbit, Rex rabbit, and Satin rabbit.

13.4.2.3 East African Rabbit Breeds

In many countries in East Africa, French Ear Lop, California White, Flemish Giant, New Zealand White, and some local crossbreeds (crosses between indigenous and exotic rabbit breeds) are kept.

(a) Sudan

Sudan Baladi is an indigenous breed of Sudan (El Raffa et al. 2002). Researchers compared feedlot performance, carcass and non-carcass traits of a temperate rabbit breed, New Zealand white (NZW), and Sudan Baladi (SB). Thirteen individuals of each breed were used. The results revealed that breeds significantly ($P < 0.05$) affected final live weight and weight gain. The traits recorded 1.8 vs. 1.4 kg for live weight and 17.4 vs. 10.7 g/day for weight gain for NZW and SB, respectively. For carcass traits, slaughter weight, hot carcass weight, cold carcass weight, and one-side carcass weight showed significant ($P < 0.05$) breed differences. They estimated 1672 vs. 1375 g for slaughter weight, 940 vs. 749 g for hot carcass weight, 882 vs. 693 g for cold carcass weight, and 413 vs. 324 g for one-sided carcass weight, respectively, for NZW and SB breeds. The meat/bone ratio was significantly ($P < 0.05$) higher in SB (5.5%) than in NZW (4.4%). Most of the non-carcass traits studied showed no significant breed differences. It was, therefore, concluded that the temperate breed was superior to the local breed in meat yield and weight gain.

(b) Kenya

Mailu et al. reported that the common rabbit breeds used for commercial production in Kenya include the French Ear Lop, California White, Flemish Giant, New Zealand White, and some local crossbreeds (Mailu et al. 2013).

13.4.2.4 Southern African Rabbit Breeds

Most countries of Southern Africa use imported rabbit breeds. These breeds generally include Flemish Giant, Grey Giant Rabbit, Soviet Chinchilla Rabbit, New Zealand White Rabbit, New Zealand Black Rabbit, California Rabbit, and Dutch and Angora Rabbits.

13.4.2.5 South Africa

In South Africa, the rabbit breeds are used for meat and skin production (Steenkamp 2021). The majority of breeders used exotic breeds, as well as a local breed called Phendula. The Phendula is a local landrace breed developed in 2002 and was recognized officially by the South African Rabbit Judges Council in 2014. The Phendula was selected with the objective of producing a hardy, adaptable breed that is suited to the South African climate and that incorporated genetics from Chinchilla Giganta, Flemish Giant, New Zealand

Fig. 13.21 Phendula

Red, and New Zealand Black rabbits (recognized breeds in South Africa), as well as those from hardy, nondescript rabbits in the area.

13.4.3 Breeding System

The rabbit rearing system is a system whose productivity depends mainly on the reproductive performance of the doe (fertility and prolificacy) and on the growth and health of the rabbits (Castellini et al. 2003). Rabbit meat is obtained under four farming systems. Thus, Colin and Lebas (1996) have described three types of rabbit farming: traditional, intermediate, and commercial. Another so-called organic production system has appeared in recent years to meet consumer demands (Lebas-Fraczak 2009).

13.4.3.1 Traditional Cuniculture
Composed of small farms that keep about 8–20 females. Located in rural areas or on the outskirts of towns (Saidj et al. 2013), the animals used are indigenous breeds; they are housed in old, recovered premises and sometimes in traditional buildings specially designed for this breeding. The diet is almost exclusively based on grass and domestic byproducts (plants and table scraps), sometimes supplemented with bran (Berchiche and Kadi 2002), which is common to several regions in the world (Finzi 2011). This system provides a significant protein intake for the local human population. In addition, it can recover a large amount of household waste and unusable byproducts. Rabbits from traditional farms are characterized by modest zootechnical performance. These animals are increasingly rare on the market due to the disappearance of traditional breeding (Lebas 2002).

13.4.3.2 Intermediate Rabbit Culture
Composed of medium-sized farms (8–100 females) for both food and commercial purposes, using semi-intensive methods. The diet is farm-type, supplemented with products bought off the farm, and a large proportion of rabbits are produced and marketed. This type of farming is found in peri-urban as well as urban areas (Lebas 2002).

13.4.3.3 Rational Cuniculture
Made up of large farms (more than 100 females) using rational marketing-oriented techniques, the promotion of this breeding is initiated by the exploitation of hybrid breeders of the Hyplus type, introduced from France or Belgium (Berchiche et al. 2000), but their adaptation has often proven difficult due to climatic conditions and local diet (Berchiche 1992).

Rabbits are housed in cages inside closed, lighted, and ventilated buildings; they are heated in winter and cooled in summer (Lebas 2000). These farms are grouped into cooperatives, themselves supervised by various technical institutes (Colin and Lebas 1995).

13.4.3.4 Organic Cuniculture
Currently, consumers in Europe are increasingly demanding meat from an organic production method (impact on the environment and on health). Thus, the organic meat market has grown (Lebas 2008). The organic rabbit production system implements most of the agroecological principles. Rabbits, generally of rustic breed, are reared in the open air in mobile cages on unfertilized multispecies meadows. The cages are moved every day to provide fresh grass to the animals, which limits contact with their droppings and thus reduces parasite infestation (coccidia). In addition to pasture, the animal feed is mainly composed of dry fodder and a mixture of cereals and protein crops grown in combination, possibly supplemented by commercial organic whole granulated feed. This type of rabbit production system is generally small in size (around 40–60 breeding females) and driven by an extensive reproductive rate (80–90 days between kindling); this makes the system much less productive (20 rabbits/female/year); consequently, its economic viability is only allowed by a higher wind price (Lebas 2000; Fortun-Lamothe et al. 2013).

13.4.4 Genetic Characterization of African Rabbits

Generally, genetic variability can be defined, at a given locus, as the diversity of alleles encountered. For a set of loci, it is defined as the diversity of alleles and their combinations. In absolute terms, genetic variability can be defined over the entire genome, but in the absence of a large number of loci easily studied on large numbers. There are different approaches for defining genetic variability: phenotypic (morphologic), biochemical (isoenzyme, blood groups, and karyotype), and molecular (mtDNA, SNP, SSR, ARFL, and RABD) information, which gives access to this or that part of the genetic variability.

Chromosomes are the support of the genes that carry genetic information. According to cytological studies, the rabbit genome consists of two sex chromosomes, XX (for a female) and XY (for a male), and 21 pairs of autosomes ($2n = 42$). The rabbit karyotype was first described in 1926 by T.S. Painter. Mapping genes and markers involves determining their position along the chromosomes. A comparative chromosome map between humans and rabbits has been published by Korstanje (Korstanje 2000), this map is based on the results of reciprocal chromosome painting. The rabbit karyotype was refined in 2002 (Hayes et al. 2002), which resulted in the fine localization of 250 new genes distributed on all chromosomes (Rogel-Gaillard et al. 2009). A chromosomal map of microsatellite markers was also produced (Chantry-Darmon et al. 2005), which made it possible to construct a genetic map with all of the linking groups between markers positioned and oriented on the chromosomes (Chantry-Darmon et al. 2005).

Genetic markers must have at least four qualities for genetic mapping methods to be optimal: be polymorphic, codominant, neutral, and homogeneously distributed on the genome (Falconer and Mackay 1996).

In rabbits, as in mice and rats, the size of the animal as well as the ease and speed of reproducing made it possible to easily position morphological and biochemical markers on the first linkage groups. In 2000, the rabbit linkage map was made up of 37 loci including 13 morphological markers, 17 immunological markers and seven biochemical markers, forming nine linking groups, of which four are assigned: linkage group I to chromosome OCU1, the group of linkage VII to chromosome OCU12, linker group VIII to chromosome OCU2, and linker group IX to chromosome OCU17 (Korstanje 2000). Ebegbulem (2012) studied in humid tropical southern Nigeria the bodyweight (BWT) and linear body measurements (LBM) of 21 growers and 21 breeders, crossbreed rabbits. The LBMs were: head length (HL), body length (BL), heart girth (HG), ear length (EL), and tail length (TL). The breeders showed significantly higher mean values in BWT and all LBMs than the growers, except in TL. Sex, however, did not have any significant influence on BWT and LBMs between the two groups of rabbits, though the bucks showed slightly higher numerical values than the does. The correlation matrix showed high, positive, and significant values among most of the traits studied ($p < 0.01$). The highest coefficient was between BWT and HG ($r = 0.848$). Results of the regression coefficients showed that HG was the best predictor of BWT, contributing 93% of the total variability. In their study, North et al. examined the meat production of the New Zealand White (NZW) and Phendula rabbit breeds (North et al. 2019). The live weights (LW) and average daily gains (ADG) of 80 (44 males, 36 females) NZW and 40 (22 males, 18 females) Phendula rabbits, housed in single-sex groups of three, were recorded from weaning (5 weeks) until slaughter (11 weeks). The slaughter weight, carcass, organ, and carcass portion weights were recorded for 10 male rabbits and 10 females of each breed, and the physical and proximate chemical quality of the loin meat was determined. The breeds differed for the reference carcass (RC) yield (NZW: $85.3 \pm 0.14\%$; Phendula: $84.9 \pm 0.24\%$) and the proportions of the low-value fore (NZW: $38.6 \pm 0.26\%$; Phendula: $37.6 \pm 0.28\%$) and high-value intermediate (NZW: $19.6 \pm 0.16\%$; Phendula: $20.4 \pm 0.28\%$) parts. Females had greater LW and ADG at 11 weeks old and reduced dressing percentages but greater RC yields owing to lighter heads and red offal. Females also had smaller proportions of the fore part. Meat quality did not differ between the breeds or sexes. All breeds are defined according to the international standards based on external appearance. Quantitative traits were measured on 70 rabbits belonging to different agroecological localities (Steppic, Mountainous, and Saharan). A comparison was carried out in order to identify differences and similarities between males and females. The barometric characterization revealed a significant phenotypic diversity in the population studied. Multiple correspondence factor analysis (MCA) and ascending hierarchical classification (CAH) identified three distinguished classes. In general, the local rabbit has an average weight of 1.97 kg, an average body length of 28.8 cm, and a chest measurement of 26.1 cm. Identification of the local rabbit population in western Algeria is ongoing, specifically in the two wilayas of Tlemcen and Sidi Bel Abbès. This study is based on a phenotypic approach via morphometric tools using 15 quantitative characters, including (lt, LT, lo, Lo, Do, Dy, Pc, lp1, lp2, lc, ltc, lQ, lD, LD, TP) measured on 93 rabbits belonging to the following breeds: Synthetic, local, Half-Giant, Giant, White Population, Berber, Butterfly, Havana Butterfly, and Lion's Head. Moreover, eight qualitative characters (Cp, Cy, CT, Cn, Co, CQ, Cm1, Cm2) were also assessed. The findings between sexes were discussed, of which the barometric characterization revealed a clear phenotypic diversity between the breeds studied, with very highly significant differences ($p < 0.001$) for most of the body measurements used. On the other hand, the biometric measurements of the rabbit breeds studied did not vary with sex; the results showed that there were no significant differences at $p < 0.001$ between males and females for all the characteristics studied. Multiple correspondence factor analysis and the ascending hierarchical classification (CAH) identified four classes in the studied population.

Van Haerigen et al. constructed a genetic map, comprising 103 AFLP markers distributed in 12 linkage groups and covering 583 cm (van Haeringen et al. 2002). This AFLP map was subsequently enriched and includes 400 markers distributed over 24 linkage groups, 14 of which are assigned to rabbit chromosomes. However, even if this map allowed the detection of QTL of interest for the study of atherosclerosis (Hayes et al. 2002), it cannot be used for the identification of candidate genes, because the AFLP markers do not allow an integration of genetic and cytogenetic maps and do not provide comparative mapping information.

Ragoju et al. used RAPD markers to assess the genetic variability among three rabbit breeds: Soviet Chinchilla, White Giant, and Grey Giant. They reported no significant genetic variability among and within these three rabbit breeds (Rangoju et al. 2007).

Before 2001, only 31 microsatellites were characterized in rabbits, as few laboratories were interested in the subject. The British group of G. Hewitt developed nine microsatellites by screening a genomic DNA plasmid library with radioactive probes (Surridge et al. 1997). A French team from the CNRS, interested in the genetic diversity of wild rabbit populations, has published nine microsatellites: six isolated from a plasmid library of genomic DNA (Sat5, Sat7, Sat8, Sat12, Sat13, and Sat16) and three provided through the EMBL database (Sat2, Sat3 and Sat4) (Mougel et al. 1997). The Dutch group from Utrecht identified 184 microsatellite sequences in the EMBL database and determined primers for 13 of them (Van Haeringen et al. 1996).

In Tunisia, the genetic status of 12 Tunisian indigenous rabbit populations has been tested by using 36 microsatellite loci. A total of 264 rabbits from the villages of the Tozeur and Kebili regions were studied. This study is the first detailed analysis of the genetic diversity of Tunisian indigenous rabbit populations, and the data generated provides valuable information about the genetic structure and the priorities for their conservation (Ben Larbi et al. 2014).

In Egypt, three native rabbit breeds—Gabali, Baladi Red, and Baladi Black—in addition to the New Zealand White, were genotyped using 12 microsatellite markers. The results of this study confirmed the applicability and efficiency of this microsatellite panel for assessing genetic diversity and setting the conservation priorities for Egyptian local rabbits (Badr et al. 2019). Furthermore, the selected cytochrome c oxidase subunit I (COX1) gene and four microsatellite markers were used to investigate the population genetic structure of Red Baladi (RB) and Sinai Gabali (SG) rabbits as the Egyptian native rabbit breeds. Both microsatellite and the COI gene as a DNA barcode proved to be exceedingly successful in distinguishing between breeds under investigation (Fortun-Lamothe et al. 2013).

In Nigeria, seven microsatellite markers were used to determine genetic variation amongst the rabbit breeds (New Zealand White, New Zealand Red, Californian White, and Chinchilla). The genetic characterization, as revealed by these microsatellite markers, showed that the four rabbit breeds have more within-breed variation than between-breed variation. Also, the low values of genetic differentiation regarding the inbreeding estimates indicated a relatively high outbreeding among the four rabbit breeds.

In Kenya, Owuor et al. (2019) evaluated the population structure of domesticated rabbits by using 263 base pairs of the mtDNA D-loop region. The rabbit population analyzed in this research was from the Western region (Kakamega, Vihiga, and Bungoma), the Central region (Laikipia and Nyandarua), and the Eastern region (Kitui, Machakos, and Makueni). In this study, the researchers found five unique haplotypes in the mtDNA D-loop region. The results suggested that the Kenyan domesticated rabbits may have originated from Europe. The integration of exotic breeds into breeding programs could have contributed to the low genetic diversity.

Due to its position in the phylogenetic tree of mammals and for its role as a model species in biomedical research, the rabbit was selected for the "Mammalian Genome Project" (Lindblad-Toh et al. 2011). This project aims to identify sequences conserved between species in order to improve the annotation of the human genome and to produce a complete atlas of functional DNA sequences. So, for the first time, the genome of a New Zealand rabbit was completely sequenced at low density (OryCun1.0) by the "Broad Institute" (Boston, USA). Many laboratories at the international level working on rabbits joined forces to write a "white paper," which called for deeper sequencing. The rabbit genome was then sequenced a second time, more densely (OryCun2.0) (Carneiro et al. 2014), and included the sequencing of clones of large DNA fragments (BAC) provided by INRA (France). Annotation of the genome, that is, the location and description of the biological function of genes, was carried out by the Ensembl genomic platform using in particular the sequencing of RNA strands (RNA-Seq) of tissues of rabbits provided by INRA and the human orthologous annotation (similar genes shared by the two species) (http://www.ensembl.org/Oryctolagus_cuniculus/Info/Annotation).

Genome annotation identified 19,203 protein-coding genes, 3375 noncoding genes, and a total of 24,964 transcripts. The University of Uppsala (Sweden) and the University of Porto (Portugal) have sequenced six breeds and wild French and Spanish rabbits, belonging to the two subspecies *O. c. algirus* and *O. c. cuniculus* (Carneiro et al. 2014). This sequencing made it possible to identify 51 million markers of the single nucleotide polymorphism (SNP) type, that is to say, regions of the genome that present differences between individuals for a single DNA base, as well as 5.6 million insertion/deletion polymorphisms were detected. SNPs represent the most abundant source of variation in the genome. Genotyping is the discipline that aims to determine the nature of the alleles (variations) of the markers of each individual. The standardization and mechanization of genotyping techniques now make it possible to quickly know the sequence of several tens or hundreds of thousands of SNP markers. As part of the European project "A Collaborative European Network on Rabbit Genome Biology – RGBNet," a commercial A y metrix chip, supporting 200,000 SNP markers, has been developed, thus offering the possibility of carrying out high-throughput genotyping of rabbits. It was released on the market in 2016.

The goal of genetic association and linkage studies is to identify chromosomal regions (QTL for "Quantitative Trait Loci") involved in the expression of traits of interest. They are based on the statistical demonstration of a link between the genotype of certain markers and the phenotypic value of the characters. The search for QTL is one of the stages in the search for causal mutations. Only one QTL research study has been published to date for rabbits (Ben Larbi et al. 2014). The most significant QTLs were identified on chromosome 7 for growth traits, on chromosome 9 for bone weight, and on chromosome 12 for water loss on wiping. Due to the small number of markers used, the regions of the genome highlighted were too large to be able to precisely identify the genes involved in the development of these characters.

The candidate gene approach consists of studying the association between the alleles of certain genes chosen, on the basis of prior knowledge, and the expression of phenotypes (color, growth, reproduction, feed efficiency, resistance to disease). A variation in the gene can indeed affect one of the key physiological functions in the development of the phenotype. The first step in this approach is to select a list of genes whose biological functions are known from studies carried out in the target species or other species. The second step is to identify variations in the DNA sequence of these genes. These variations are identified by genome sequencing. The third step is based on association studies, which will make it possible to statistically verify the link between variations in the gene and the phenotype in the populations studied. This strategy has the advantage of finding potentially interesting polymorphisms (variations in genes) very quickly and inexpensively. Its drawback is that it is limited to genes for which we know a priori neither the value of their effect on the characters of interest nor their variability in the population studied. Numerous candidate gene studies have been conducted in rabbits (Miller et al. 2014). They can be divided into three categories corresponding to the groups of traits studied: i) growth and meat production traits, ii) female reproductive traits, and iii) disease resistance traits.

13.4.4.1 Growth and Carcass Traits of African Rabbits

Due to the economic importance of growth and meat production, most studies of candidate genes in rabbits have focused on this group of traits (Owuor et al. 2019; Lindblad-Toh et al. 2011; Carneiro et al. 2014; Sternstein et al. 2015; Miller et al. 2014). They investigated the most commonly recorded and selected growth trait in commercial meat production populations, 70-day weight. Other authors (Fontanesi et al. 2011) have studied growth, carcass, and meat quality traits measured at different ages. The myostatin gene (MSTN) has been extensively studied in rabbits because of the well-known effects of mutations in this gene on the development of muscle tissue. The effects of hypermuscularity linked to variations in this gene, MSTN, have so far been found in rabbits in the breeds and lines studied (Peng et al. 2013).

The genes involved in the functioning of the somatotropic axis, and in particular the gene for growth hormone (GH1), growth hormone receptor (GHR), insulin-like growth factor (IGF2), and the melanocortin receptor gene, can have significant effects on growth or production traits. These effects have been described in rabbits (Fontanesi et al. 2012a, 2013; Fontanesi 2016). Most of these studies showed that the most favorable allele of these genes is also the most common in the commercial populations studied. This observation demonstrates the effect of selection, which tends to increase the frequency of favorable alleles by selecting at each generation the best animals (carriers of favorable genes) to reproduce them and constitute the following generations.

Reproductive Traits

The candidate gene studies for reproductive traits are all based on a single experimental divergent population selected for uterine capacity (maximum number of fetuses that a rabbit can carry when ovulation rate is not a limiting factor) for ten generations (Fontanesi et al. 2012b). An association between the polymorphism of three genes (OVGP1, PGR, and TIMP1) and some reproductive characteristics, including embryonic implantation and litter size, has been demonstrated in this experimental population (Peiró et al. 2008).

Disease Resistance Traits

Candidate gene studies have been carried out by Chinese teams from the Sichuan Agricultural University in Chengdu on various local and imported breeds for traits of resistance to digestive disorders. These teams have demonstrated the involvement of various genes encoding receptors participating in the recognition of pathogens: "Toll like receptor 4" (Zhang 2011) (TLR4), "Nucleotide-binding oligomerization domain 2" (Zhang et al. 2013) (NOD2), and "Dendritic-cell associated C-type lectin 1" (Peng et al. 2013) (Dectin-1). They also identified the following genes: "Myeloid differentiating factor 88" (MyD88) encodes a protein that has a central role in the activation of innate and adaptive immunity (Chen et al. 2013); NLR Family Pyrin Domain Containing 12 (NLRP12), which plays a role in the regulation of inflammation; Janus kinase (JAK1)–an intracellular signaling molecule, "Signal transducer and activator of transcription 3" (STAT3)–a factor for transcription and differentiation of immune cells, and Chemokine receptor 6 (CCR6)–a receptor controlling the movement of leukocytes.

13.5 The Honeybee in Africa

The bee is an important insect that has been exploited for its honey, wax, royal jelly, pollen, propolis, and even venom. Bee products are known not only for their economic importance but also for their beneficial properties on human health. Bees pollinate (fertilization, production of fruits and seeds) and, thus, maintain the sustainability of the plants and the ecosystem. Beekeeping is a lucrative activity that provides a sustainable source of employment and additional income for beekeepers and artisans making beehives (Fotso et al. 2021). A multitude of subspecies and ecotypes of honeybees are reported. Beekeepers used mostly traditional methods, that is, fire, to harvest honey produced by wild colonies. Fire could be a threat to honeybees as the beehives and colonies are often destroyed (Tchoumboue et al. 2001). The deforestation, breeding, urbanization, the use of chemicals by farmers, and locust control are also factors that are reducing the number and diversity of these insects.

13.5.1 Generality on the Honeybee

13.5.1.1 The Bee Species of the Genus *Apis*

The bee is an eusocial flying insect, belonging to the animal kingdom, the phylum Arthropoda, the insect class, and the order Hymenoptera (Peiró et al. 2008). It has been around for over 70 million years (Linksvayer et al. 2012). There are about ten recognized bee species belonging to the genus *Apis* (Engel 1999; Arias and Sheppard 2005). All of these species are represented by three distinct groups according to phylogenetic analyses of nuclear DNA and mitochondrial (Arias and Sheppard 2005). The species of the first group are dwarf bees (*Apis florea* and *Apis andreniformis*), the second group species are giant bees (*Apis dorsata, Apis laboriosa, Apis bingham,* and *Apis nigrocincta*), and those of the third group are nesting bees (*Apis mellifera, Apis cerana, Apis koschevnikovi,* and *Apis nulensis*) (Arias and Sheppard 2005; Raffiudin and Crozier 2007). Of all bee species in the genus *Apis*, only two species, namely *Apis mellifera* (in Africa and Europe) and *Apis cerana* (in Asia), are domesticated by humans (Koeniger 1976). According to Ruttner et al. *(*Ruttner and Volprecht 1983*)*, *Apis mellifera* and *Apis cerana* are in the first stages of speciation.

13.5.1.2 The Origins of the Honeybee *Apis mellifera*

Apis mellifera is native to Africa, Europe, and the Middle East. It was introduced by humans to America, Australia, and other countries around the world (Gupta et al. 2014). Three hypotheses (Fig. 13.22) are proposed regarding the origin of the honeybee *Apis mellifera* (Gupta et al. 2014). The first hypothesis (i), based on morphometric analyses, suggests an expansion from the Middle East, involving the colonization of Europe, toward two trajectories, one Eastern and another Western (Mogga and Ruttner 1988). The second hypothesis (ii), based on a molecular study of mitochondrial DNA, suggests an expansion from the Middle East, which does not involve Western colonization of Europe (Garnery et al. 1992). The third hypothesis (iii) suggests an expansion from

13 Nonconventional Animal Genetic Resources, Diversity, and Unique Features

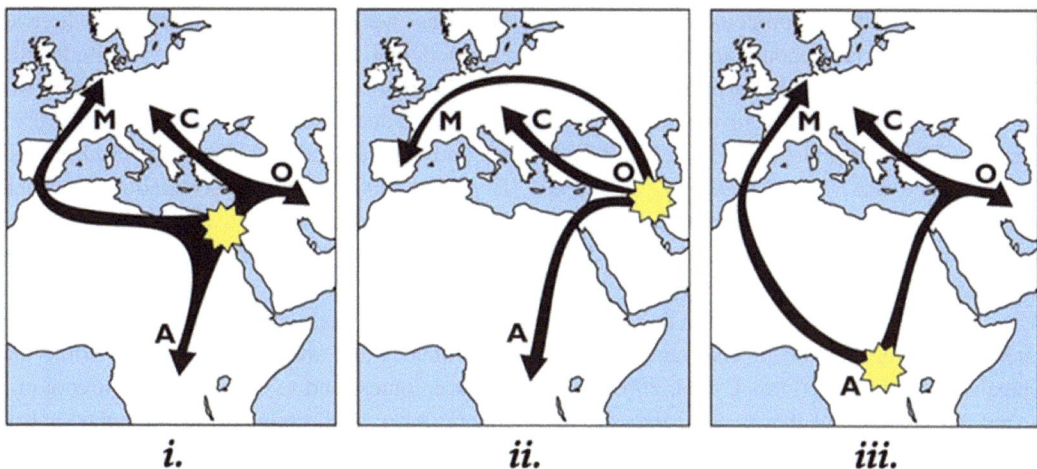

Fig. 13.22 The origin of the honeybee, *Apis mellifera*, according to three hypotheses. (Gupta et al. 2014)

Africa toward both trajectories, east and west (Wilson 1971), which was confirmed by the analyses of 1136 SNPs (Sinha et al. 2006).

13.5.1.3 The Subspecies of *Apis mellifera* in the World

The subspecies of *Apis mellifera* in the world are grouped at the level of five evolutionary branches (A (African), C (Carnica), M (Mellifera), O (Oriental), and Y (Yemenetica) (Garnery et al. 1992; Ruttner et al. 1978). Several studies have characterized the subspecies of *Apis mellifera* based on their morphological and genetic differences (Ruttner et al. 1978; Miguel et al. 2007; Miguel et al. 2011).

13.5.1.4 Geographical distribution of *Apis mellifera* in Africa

The honeybee *Apis mellifera* is found in all African countries (Mogga and Ruttner 1988; Chahbar et al. 2016; Haddad et al. 2015). According to Ruttner et al. (Ruttner et al. 1978), all individuals of the *Apis mellifera* species draw a very clear, lying "Y" (Fig. 13.22i). In the three branches of the "Y," the studied samples are arranged in a characteristic order; in the trunk of the "Y" are the races of Africa south of the Sahara, in one branch the races from the eastern part of the Mediterranean region, and in the other those from the western part (branches A, C, and M). These data are justified by mitochondrial DNA analysis (Garnery et al. 1992; Cornuet and Garnery 1991; Meixner et al. 2011). Specialists distinguish at least 29 subspecies of *Apis mellifera* in Africa (Ruttner et al. 1978; Meixner et al. 2011).

The most dominant races are *Apis mellifera scutellata* and *Apis mellifera adansonii*. For *Apis mellifera scutellata*, they are found in Eastern Africa from Ethiopia to Southern Africa, passing through Burundi, Kenya, Tanzania, and Zimbabwe (Gupta et al. 2014; Hepburn and Radloff 1998). As for *Apis mellifera adansonii*, it covers Benin, Cameroon, Chad, Congo Basin, Nigeria, and Senegal (Ruttner 2013; Amakpe 2010). The two subspecies, *Apis mellifera scutellata* and *Apis mellifera adansonii*, in the presence of the third race *Apis mellifera capensis*, are recorded in South Africa and Namibia (Hepburn and Radloff 1998; Radloff and Hepburn 1997). In Southeast Africa, Mozambique, Zambia, and Zimbabwe, members of honeybee populations of three subspecies, those of *Apis mellifera litorea*, *Apis mellifera scutellata*, and *Apis mellifera adansonii* (Hepburn and Radloff 1998). For the North African regions, five subspecies of *Apis mellifera* are exploited, namely *Apis mellifera intermissa*, *Apis mellifera sahariensis*, *Apis mellifera major*, *Apis mellifera lamarckii* (formerly called *fasciata*), and *Apis mellifera iberica* (Koudjil 1990; Garnery et al. 1993, 1995). The tellian bee, *Apis mellifera intermissa*, is found in the north of Libya to Morocco, passing through Tunisia and Algeria (Hepburn and Radloff 1998). *Apis mellifera sahariensis* is exploited in the desert of Algeria, particularly in the southwest of Morocco (Chahbar et al. 2016; Haccour 1960). In Egypt, the race *Apis mellifera lamarckii* (Sheppard et al. 1996) was introduced in the nineteenth century (Schiff and Sheppard 1993; Schiff and Sheppard 1995). In Northeast Africa, *A. m. jemenitica* is being identified from Mali to Sudan (Amakpe 2010). In the same way, in Somalia, Franck et al. (Franck et al. 2001) mentioned the presence of the *A. m. litorea* species. Contrarily, there are no native *Apis mellifera* in the Pacific and Caribbean regions where the European lineage bees have been imported (Paterson 2006).

13.5.1.5 The Influence of Environmental Factors on the Geographical Distribution of Races and Bee Characteristics

The high variability between the different African races of *Apis mellifera* is mainly due to climatic factors (Hepburn and Radloff 1998). Therefore, the biogeographic distribution of the races of *Apis mellifera* varied depending on the ecosystem (El-Niweiri and Moritz 2008). All the colonies in the regions of Burundi, Kenya, Malawi, and Tanzania belong to localities that are associated with a mountain system, and they are considered members of the same race, *Apis mellifera monticola* (Hepburn and Radloff 1998). Following the Mogga and Ruttner (Mogga and Ruttner 1988) classification system, the low-altitude bees are members of the race *Apis mellifera scutellata*, and those at high altitudes are members of *Apis mellifera monticola*. According to Ruttner (2013), the bee populations from high altitudes differ from those of low altitudes in size and pigmentation. For example, within the same breed, *Apis mellifera monticola*, two colors of pigmentation can be observed. In Kenya and Tanzania, the race *Apis mellifera monticola* is black in color; meanwhile, it is yellow in Malawi (Miguel et al. 2007). A significant correlation was found between the altitude and the size of bees, as well as their pigmentation. It was reported that for each increase in altitude, the bees become larger with a lighter color. On the contrary, a decrease in altitude results in the bees becoming small and darker in color (Radloff and Hepburn 2000). A strong correlation has been found between some adaptive traits and environmental factors in bee races of tropical Africa (Gupta et al. 2014). According to Paterson (Paterson 2006), strong colonies are generally in hot climates and are more aggressive, while the weak colonies tend to be quite mild in cool climates. Also, anthropogenic factors play a key role in this character. In areas where inhabitants search for honey, bee colonies have become more aggressive, as the milder ones are destroyed by these inhabitants (Paterson 2006).

13.5.2 Importance of Beekeeping

13.5.2.1 Ecological and Economic Importance of the Honeybee

The pollination of flowering plants is carried out by several animals and insects, such as birds and bats. Among these pollinators, insects and honeybees, *Apis mellifera* (Allsopp 1993), are the most important. *Apis mellifera* is considered one of the most important pollinating animals (Morse and Calderon 2000). Its role is essential for plants and crops in agriculture (Mouton 2011), forestry, and other ecosystems (Allsopp 1993). In 2005, the total economic value of pollination worldwide was estimated at 153 billion euros. But it is almost impossible to give any real monetary value to the pollination service because there are non-entomophilic plants, which do not need pollinators to produce fruits; however, they benefit from the visits by the pollinators, which ensure the production of larger, more symmetrical, and sweeter fruits. In Algeria, it has been shown that the pollination of bean (Benachour et al. 2007), cucumber (Benachour and Louadi 2011), and other plants belonging to different natural ecosystems (Bendifallah et al. 2012) by a large diversity of wild and domestic bees (e.g., *Apis mellifera intermissa*) greatly enhances their productivity. In Southern Africa, the two bee species,

Apis mellifera scutellata and *Apis mellifera capensis*, play a very important role in the pollination service of forests, as well as urban (Masehela 2017) and agricultural plants, such as industrial apples (Mouton 2011). In other African countries, the evidence by researchers of the pollination of flowering plants via the honeybee *Apis mellifera* has been well established, notably in some African countries. However, the supply and demand of honeybee *Apis mellifera* colonies for crop pollination is a very responsive service in several countries, however, this is not the case in the majority of African countries (Masehela 2017). On the contrary, in several African countries, beekeepers pay farmers to allow them to place the hives on their farms during the honey season.

13.5.2.2 Bee Relationship—Bioindication of the Environment

Bees also serve as biological indicators (Skorbiłowicz et al. 2018; Goretti et al. 2020). In fact, bees signal the presence of chemical degradation in the environment in which they live through two indicators: the high degree of mortality and the residues that can be found on the bodies of bees or in bee products (phytosanitary substances used in agriculture, antiparasitics and pollutants such as heavy metals and radionuclides) (Calatayud-Vernich et al. 2016).

In addition, many ethological and morphological characteristics also make bees a good ecological indicator. Indeed, most of their bodies are covered with hairs that capture the substances they encounter in flight. It is, therefore, highly sensitive to most antiparasitics, which may be present in the environment (Sabatini 2005). But it is especially at the level of bee products that the presence of toxic products (especially phytosanitary) is felt most and are, therefore, bioindicators of choice compared to environmental pollution (Skorbiłowicz et al. 2018; van der Steen 2016). Losfeld et al. (2014) showed that the rates of environmental contamination by heavy metals are higher in bee products than those detected in the body of the bee because this insect is capable of filtering and purifying nectar containing these pollutants during the honey production (Conti et al. 2018; Borsuk et al. 2021).

13.5.2.3 Beehive Products and Their Uses

Other bee-related products are jelly, wax, propolis, and venom, which are of economic importance (Oyerinde et al. 2012; Abou-Shaara 2014). Moreover, it is important to note that many communities consume the insect itself (especially the larvae) as a source of animal protein (Bamidele et al. 2021). The ancient Egyptians and Romans consumed bees for over 70 million years (Linksvayer et al. 2012).

In Africa, beekeepers do not have the same beekeeping production goals, but honey remains the top priority in animal husbandry. For example, in Algeria, beekeepers do not produce enough beeswax, and the state remains dependent on the importation of wax mixed with paraffin (Mayazine 2019). All Cameroonian beekeepers (100%) collect honey, (69.9%) wax (see Fig. 13.23), (44.2%) propolis, (15.9%) pollen, and only (3.5%) harvest royal jelly (Meutchieye et al. 2016). In Senegal and Rwanda, beekeepers do not know the production of wax; unfortunately, a very large part of the wax is discarded because it is considered waste on the one hand, and on the other, the method of extracting honey consists of burning the wax (Mayazine 2019).

Honey is a highly prized natural product because of its nutritional properties and therapeutic applications. Two types of honey are produced naturally, depending on their origins: that of flower nectar and the other of honeydew (Bogdanov 2016). Honey is used by humans for two reasons: nutritional and medicinal (Oyerinde et al. 2012). Honey of jujube, *Ziziphus lotus*, is part of the popular traditional food and medicinal practices of the inhabitants of the greater Maghreb, but its exploitation is recent and dates from the 1990s (Mayazine 2019). The pH of this type of honey varies between 6 and 7, which provides an environment that promotes the activity of diastase, an enzyme beneficial to human health. It has an interesting microbial activity, which is due to its richness in polyphenols, flavonoids, tannins, and alkaloids, which are

Fig. 13.23 Harvesting of wax by African beekeepers. (Mayazine 2019)

molecules with medicinal properties (Mayazine 2019).

The study of Benlacheheb et al. in Algeria showed that wax has a very interesting anti-inflammatory effect, especially when mixed with plant extracts (Benlacheheb Ahmed and Benchiko 2021). More, wax is used by Senegalese craftsmen and traditional healers (traditional doctors) to plug holes as well as to make seeders, basins, earrings, and medicinal balms (Mayazine 2019).

Another bee product, royal jelly, which is rarely produced by African beekeepers (Conti et al. 2018), is very beneficial for human health and has been reported to have favorable outcomes with people suffering from Alzheimer's disease (Antonelli and Donell 2019).

Propolis has shown a beneficial effect on human and animal health (Rahman et al. 2010). All breeds of bees collect plant resins from their environment and deposit them in their hives as propolis, although some colonies deposit relatively little compared to others. Propolis has significant antimicrobial activity, but may be higher if multiple resin sources are available, where one resin can compensate for a lack of function in another, "functional balance" resin. Functional redundancy and functional balance have been demonstrated during resin collection in wild bees. This implies that the composition of propolis varies depending on the botanical origin of the resin. Generally, it is composed of a resinous substance at 50%, wax at 30%, aromatic substances at 10%, pollen at 5%, and other substances at 5% (Aygun et al. 2012). Propolis also contains active substances, namely enzymes, phenolic compounds, and flavonoids (Meutchieye et al. 2016; Bogdanov 2016). For human health, propolis can be used to fight diseases caused by *Escherichia coli*, *Salmonella*, and *Staphylococcus aureus* (Mayazine 2019). Propolis has very effective effects against diseases and enemies of the honeybee *Apis mellifera*, such as American foulbrood and ascospherosis (Wilson et al. 2015). Therefore, propolis has also been used to develop the performance of laying hens at a rate of 30 g / l of water or 5 g / kg of feed (Aygun et al. 2012). This dose can also stimulate the immune system of the laying hen (Çetin et al. 2010).

13.5.3 The Types of Beehives Used in Africa

13.5.3.1 Traditional Beehives (See Fig. 13.24a)

Multiple types of beehives are used in Africa (Mayazine 2019; Shenkute et al. 2012); some are modern and others are traditional (Berkani 2007a; Gratzer et al. 2021). But in most African countries, traditional beehives are the most commonly used (Akinwande and Badejo 2009). However, traditional beehives do not allow beekeepers to visit their bee colonies regularly. Traditional beekeeping is transmitted from one generation to another only through contact with parents and neighbors, and vocational training in bee breeding institutions is totally absent (Paterson 2006; Betayene 2008). For the

Fig. 13.24 a. Traditional log hive for *A. mellifera* bees; b. modern hive; c. transitional Kenyan top bar hive; d. clay pot hive for stingless bees. (Leven et al. 2005)

construction of beehives, traditional beekeepers use natural materials (Fischer 2013). Most often, beehives are created from tree trunks, cork, clay, or straw (Malaka and Fasasi 2002). The traditional beehive most used in Africa is the hive with bars, which is also called the top bar hive (Paterson 2006; Lonyo III 1996).

Many African countries have used the bar hive, which is also called the Kenyan hive, used in particular in Botswana, Cameroon, Egypt, Ghana, Kenya, Nigeria, Tanzania, and Zambia (Betayene 2008; Leven et al. 2005). The bar hive is a trough-shaped box tongue, the two sidewalls forming approximately an angle of 115° with the top (see Fig. 13.24c). It is possible to build the hive with straight sticks carefully assembled with wire (Leven et al. 2005). It is painted or varnished in forest areas and covered by soil and a mixture of cow dung and clay in areas of high temperatures (Betayene 2008). Beehives with movable frames are the most productive, but they should be made from good-quality materials at a reasonable cost and should only be used in areas where theft is not a problem (Garnery et al. 1993). For areas where theft and honey hunting prevail, the hive could be made from cement and mud brick (Betayene 2008; Leven et al. 2005).

In Egypt, Kenya, and Tanzania, the bar hives used are not well known in the other African countries. The numbers of movable frames vary between 22 and 27 wooden bars or slats (Lonyo III 1996). The width of the bar is the only standard to be most rigorously respected, which varies between 32 and 33 mm and depends on the size of the bee to be raised (Betayene 2008). Another type of top bar beehive is called the Tanzanian top bar beehive, particularly used in Botswana and Tanzania (Leven et al. 2005). The side walls are straight for ease of construction. This type of beehive can only be used for colonies of bees that do not attach their combs to the walls (Betayene 2008).

The East African trough hives used in Tanzania, Uganda, Zambia, and other East African countries, the production of honey takes its place next to the brood chamber and not above (Lonyo III 1996). Honey is at the back of the hive, while the brood is in the front. Therefore, the honey harvest is done from the combs located at the back of this traditional hive to avoid damaging the brood. A cylindrical hive of one meter long and 40 cm in diameter corresponds to a volume of 125l (Leven et al. 2005).

The David hive is particularly widespread in the Southern Sahel, Sudan, Senegal, and Ethiopia (Shenkute et al. 2012; Sahle et al. 2018). The hive is built from natural materials (Gratzer et al. 2021). Beekeepers use wood, banana fibers, straw, clay, and thin branches (Gratzer et al. 2021) (see Fig. 13.24d). In North Africa, particularly in Algeria and Morocco, traditional beehives are made from the ferrule trunk and cork, while in the South, beekeepers use Alfa foliage (Behidj 2012). The honey extraction in traditional beekeeping is carried out after pressing it by hand (Leven et al. 2005).

13.5.3.2 Modern Beehives (See Fig. 13.24b)

In the nineteenth century, beekeeping saw considerable progress in the beehives' improvement, of which different models were created, notably those of Langstroth, Dadant, and Layens. In modern African beekeeping, only two types of beehives are chosen by African beekeepers, namely those called Dadant and Langstroth (Berkani 2007a). The Dadant hive is used on a large scale in various African countries such as Algeria, Ethiopia, Madagascar, Morocco, Senegal, and Tunisia. The Langstroth beehive was created by the American Lorenzo Lorraine Langstroth (1810–1895), who is considered the father of American beekeeping. Currently, this type is found everywhere in Africa, particularly in the countries of the greater Maghreb.

13.5.4 The State of Beekeeping in Africa

13.5.4.1 North Africa and the Big Maghreb

North Africa is the cradle of beekeeping (Berkani 2007a). Currently, in the Maghreb, beekeeping is identical to that of Mediterranean countries (Berkani 2007a). The best producers of honey in North Africa are Egypt and Morocco (Berkani 2007a). Although Algeria experienced a significant increase in honey production, the state was forced to fill the honey deficit for the needs of the population through imports. In Algeria, in the nineteenth century, during the French invasion, indigenous populations practiced beekeeping using traditional hives, but at the same time, there were modern hives operated only by European farmers. From the 1970s, in Algeria, three models of beehives were used, namely Langstroth, Dadant, and the traditional hive (Berkani 1985). In traditional Algerian beekeeping, the foliage of Alfa, the trunk of the ferrule, and the bark of cork oak are used by beekeepers for the construction of beehives (Behidj 2012). In Algeria, orientation of beekeepers and guides in beekeeping is provided by the technical institute of breeding (formerly called the institute of small animals) (Berkani 2007a). Beekeeping is practiced across the country (Chahbar et al. 2016). The tellian *Apis mellifera intermissa* and the Saharan *Apis mellifera sahariensis* are commonly used (Hepburn and Radloff 1998). The tellienne is answered throughout the territory with the exception of a few regions in the Southwest where the Saharan *Apis mellifera sahariensis* is used (Haccour 1960; Chahbar 2013). For Libya, beekeeping was practiced only traditionally, and the use of modern hives was maintained from the 1950s (Brittan 1956). In Morocco, researchers report the existence of three races of bees: *Apis mellifera intermissa, Apis mellifera major,* and *Apis mellifera sahariensis*. The most common race in the Gharb region is *Apis mellifera intermissa*. This Moroccan region is considered the best in terms of honey production. In Egypt, honey production is important for two seasons: the first is that of clover in June, and the second is that of cotton in August (Moustafa 2001). Mainly, three types of honey are produced in Egypt: that of citrus in the spring, that of clover in the summer, and cotton in the fall (Moustafa 2001). Thus, the Egyptian flora is very rich in the number of species for the production of honey and pollen (Abou-Shaara 2014). The breed of bee exploited is *Apis mellifera lamarckii*, which has benefited from several breeding programs to increase domestic production. *Apis mellifera lamarckii* is a breed that is resistant to varroasis as well as to other parasitic mites (Moustafa 2001). Bees have also been introduced along other evolutionary routes in North African countries such as Morocco (Franck et al. 2001), Sudan (El-Niweiri and Moritz 2008), and Libya (Shaibi et al. 2009). In Sudan, analysis of mitochondrial DNA shows that Sudanese honeybee populations are members of several subspecies belonging to the three evolutionary branches, in this case A, O, and C (El-Niweiri et al. 2008). The biogeographic distribution of the races of *Apis mellifera* extant in Sudan varies depending on the ecosystem. For the humid savannah and forest ecosystems, the lineage A (*Apis mellifera adansonii* and *Apis mellifera Scutellata*) has been recorded. Desert, semidesert, and dry savannah honeybees are members of three subspecies, namely *Apis mellifera lamarckii*, *Apis mellifera syriaca*, belonging to the O lineage, and *A. m. carnica*, which belongs to lineage C. The race *Apis mellifera carnica* is an imported breed (El-Niweiri and Moritz 2008). The most common subspecies in Sudan is *Apis mellifera jemenitica* (Garnery et al. 1995). Sudanese beekeeping is practiced in the presence of two farming methods, modern and traditional (El-Niweiri et al. 2008). Beekeepers seek to improve their beekeeping herds by importing other breeds from other African countries, notably Egypt (El-Niweiri and Moritz 2008). Algeria has not escaped the phenomenon of genetic pollution caused by the uncontrolled and unofficial importation of foreign races. Indeed, for several years, Algerian beekeepers unofficially introduced foreign breeds (*Apis mellifera ligustica* and *buckfast*) to improve honey production.

13.5.4.2 Central Africa

Beekeepers in Central Africa have attempted several times, but without success, to introduce milder European bees. Colonies, after replacing their queens, begin to grow, but then they become weak and die (Paterson 2006). This may be due to several factors, including disease and the phenomenon of adaptation. Historically, the *Varroa destructor* mite is the natural parasite of the Asian bee *Apis cerana* (Anderson and Trueman 2000), but with the importation of queens of *Apis cerana* by European beekeepers, the European bee *Apis mellifera* was parasitized by this plague (Anderson and Trueman 2000; Beetsma et al. 1999), which caused enormous damage to the bees (Bailey 1991). According to another hypothesis, the workers of the introduced European bee continue to work throughout the day, even when the temperature is too high, and end up exhausting themselves, unlike the workers of the local bee, who only come out when it is less hot (Paterson 2006). Eucalyptus honey is known in most of the Central African countries, notably in Cameroon, Guinea, and Mozambique (D'Albore and Piatti 2004). In Cameroon, the breeds exploited are *Apis mellifera adansonii*, *Apis mellifera jemenitica,* and *Apis mellifera monticola* (Hepburn and Radloff 1998). Nevertheless, *Apis mellifera adansonii* is the most common in Cameroon (Fotso et al. 2021), and traditional beekeeping is the most practiced (Mayazine 2019). The expansion of apiaries is carried out by two methods: those of trapping and recovering the swarms (Betayene 2008). In Cameroon, the forests cover more than 46% of the national territory, which corresponds to about 23 million hectares being characterized by a very rich biodiversity (Mayazine 2019). In this country, beekeeping is mostly done by peasants, and it is mainly carried out by retirees and elderly people (Betayene 2008). The beekeeping sector in most of the Central African countries remains underexploited and marginalized (Betayene 2008). Village beekeeping is based on honey hunting (Mayazine 2019), whereby colonies are destroyed during fire harvesting. Therefore, the harvested honey is dirty, very low in quantity, and of poor quality (Betayene 2008). In central African countries, two breeding methods are followed, namely honey hunting and the Kenyan beehives. For the production of honey, in some regions, three types of honey are produced, namely white honey comes from the Oku within the Northwest region, brown honey from forest areas, and black honey from Sahelian areas (Mayazine 2019). White honey is the most abundant, white in color, which is due to the plant species *Schefflera abyssinica, Nuxia congesta,* and *Prunus africana* (Mayazine 2019), but the national production of honey has remained very low, with poor quality.

13.5.4.3 East Africa

In East Africa, the best honey-producing countries are Ethiopia, Madagascar, and Tanzania (Mamo 2016; Berkani 2007b), where breeding and production of wax are practiced. In terms of honey production in the East African countries, not only the honey that has been the subject of this breeding, but also the production of wax is important (FAOSTAT 2020). East African countries export important quantities of beeswax (Moustafa 2001; FAOSTAT 2020). In Ethiopia, people have practiced beekeeping for about 5,000 years (Gezahegne 2001). To date, beekeeping in this country is based mainly on traditional techniques (Gratzer et al. 2021; Yirga and Teferi 2010). For the construction of beehives, beekeepers use natural materials (Akinwande and Badejo 2009). Usually, beehives are constructed from wood, straw, and clay (Aygun et al. 2012). Traditional beekeeping is considered among the oldest agricultural activities in Ethiopia and is part of the country's current agricultural economy (Fikru 2015). Despite the traditional breeding techniques used, Ethiopia has become an important beekeeping country (Gratzer et al. 2021), and beekeeping has great potential to contribute to the development of the country's economy (Bareke et al. 2018; Fikadu 2019). Currently, this country is considered among the best producers of honey and wax in Africa (FAOSTAT 2020).

However, this activity is practiced only by men, as it is considered a tillage forestry activity, and the harvest of bee products is done at night (Shenkute et al. 2012). The task of Ethiopian women in beekeeping remains limited to marketing (Kebede et al. 2018). Honey obtained from traditional beehives is called raw honey, and it is a mixture of honey, wax, bee parts, and pollen (Shenkute et al. 2012). The average honey production per hive has been estimated at 5–8 kg and 15–20 kg for traditional and modern hives, respectively (Beyene et al. 2016; Gemechis 2016).

In order to enlarge their apiaries, beekeepers purchase colonies from local markets as well as hunt bees from naturally existing swarms or from the swarming of their colonies (Gebretinsae and Tesfay 2014; Hailu and Tadesse 2016). Unfortunately, no beekeeping improvement program has been launched in Ethiopia (Gratzer et al. 2021).

The subspecies of *Apis mellifera* being used in Ethiopia belongs to the branch Y (Franck et al. 2001; Boardman et al. 2020). Their introduction traces queens belonging to other evolutionary branches from northern Ethiopia (Hailu and Tadesse 2016). Beekeepers mainly exploit three subspecies, namely *Apis mellifera jemenitica*, *Apis mellifera scutellata*, and *Apis mellifera monticola* (Franck et al. 2001; Boardman et al. 2020). However, three other subspecies have been mentioned, namely *Apis mellifera bandasii*, *Apis mellifera sudanensis,* and *Apis mellifera woyi-gambell (*Radloff and Hepburn 2000*)*. However, Meixner et al. (Meixner et al. 2011) grouped all Ethiopian honeybee breeds into a single *Apis mellifera simensis*. This subspecies is endemic to the volcanic dome system of Ethiopia (Meixner et al. 2011).

In Tanzania, several breeds of bees exist: *Apis mellifera scutellata, Apis mellifera monticola,* and *Apis mellifera littorea (*Hepburn and Radloff 1998*)*. It has been reported that the mountain systems in Tanzania are dominated by *Apis mellifera monticola* with black pigmentation (Hepburn and Radloff 1998). However, the breed most used by Tanzanian beekeepers is *A. m. scutellata* (Gupta et al. 2014). Given the zootechnical performance of *Apis mellifera monticola*, its softness, productivity, and adaptive capacity to low temperatures, beekeepers belonging to the low-altitude sectors have introduced this native race to the mountain above 2500 m altitude (Paterson 2006).

13.5.4.4 West Africa

Beekeeping in West African countries is done by two farming methods, the traditional and the modern. In Nigeria, it is based mainly on the traditional method (Akinwande and Badejo 2009). Beekeepers use natural materials, including tree trunks and wood, for the construction of beehives (Malaka and Fasasi 2002). Generally, two methods are followed in breeding: the top bar method, whose hive is rectangular and horizontal, with a length that is more than twice that of the Langstroth hive, and has removable frames in numbers that exceed 20 executives. For the second method, beekeepers use Langstroth hives. There are no professional beekeepers, and the average number of hives per beekeeper varies between 20 and 27. More than 51 melliferous species belonging to more than 32 botanical families are exploited to produce honey. The tiger lotus honey, *Nymphaea lotus,* is the most abundant. The subspecies present in Nigeria is *Apis mellifera adansonii* (Amakpe 2010). This breed is known for its honey productivity and pollination activity (Ikediobi et al. 1995), but also for its aggressiveness (Fletcher 1978). In Benin, the subspecies exploited by beekeepers is *Apis mellifera adansonii (*Amakpe 2010*)*. This subspecies occurs in two forms. The first has a relatively small size, is yellow in color, with the forest as its natural habitat (Paraïso et al. 2011). However, the second bee is black, large, and habituated to the savanna (Paraïso et al. 2011). The forest type is more productive than the Savannah type. The subspecies *Apis mellifera adansonii* is characterized by high diversity (Amakpe 2010). About 60% of honey production is obtained by the honey hunting technique (Paraïso et al. 2011). This technique has caused bee populations to decline because the bee products are harvested after the bee colonies are destroyed by fire (Yeates 1978).

13.5.4.5 Southern Africa

Beekeeping in Southern Africa is traditional, based on the use of beehives made from barks and logs. It is part of the local economy of the Savannah in South-Central Africa (Fischer 2013). However, the so-called traditional farming method does not allow frequent visits to the colonies. Beekeeping in Southern Africa is carried out by men, but in several regions, women are engaged in the processing and sale of bee products (Fischer 2013; Clauss 1991), except in a few areas of Zambia, where women participate in the search for honey (Clauss 1991).

13.5.4.6 South Africa

Two breeds of honeybees are exploited in South Africa, namely *Apis mellifera capensis* (Cape honeybee) and *Apis mellifera scutellata* (African honeybee/savannah) (Hepburn and Radloff 1998). The two subspecies differ in their geographic distribution and several other morphological and behavioral traits (Johannsmeier 2001). However, in 1990, transhumance of the Cape bee *Apis mellifera capensis* was carried out by South African beekeepers in the Limpopo province in South Africa, which is the natural biotope of the African Savannah bee *Apis mellifera scutellata* (Allsopp 1993). These beekeeping practices, transhumance, have led to a decline in populations of the Savannah bee. This depopulation is known as the "capensis problem" (Allsopp 1993). The capensis problem is caused by the ability of *Apis mellifera capensis* workers to produce females by the phenomenon of thelytokous parthenogenesis (Onions 1912; Anderson 1963). In the other subspecies of *Apis mellifera*, in the absence of the queen, the workers produce males by the phenomenon of arrhenotokous parthenogenesis. As a result, the workers of *Apis mellifera capensis* parasitize the colonies of *Apis mellifera scutellata* by producing pseudo-queens that end in a reproductive dominance of *A. m. capensis* and a decline in *Apis mellifera scutellata* (Pirk et al. 2014). The dominant honeybee breed in South Africa is *Apis mellifera capensis;* however, the number of colonies of *Apis mellifera scutellata* has declined (Dietemann et al. 2009). South African beekeeping is very advanced, with bees kept in modern Langstroth-type hives (Johannsmeier 2001). The majority of beekeepers practice transhumance to seek resources in honey plants, and the most abundant type of honey is Eucalyptus (Johannsmeier 2001; Hutton-Squire 2014). More than 58 plant species have been the subject of honey production in this country (Masehela 2017). Professional beekeepers own up to 7000 hives per beekeeper, one commercial beekeeper manages around 1,000 hives, and a developing beekeeper has a maximum of 500 hives (NAMC Report 2008). Despite the advanced technologies in breeding and the production of good-quality honey, the local market is under threat from cheap imports of poor-quality honey (NAMC Report 2008).

13.5.5 Main Risk Factors Threatening the Honeybee

In Africa, the honeybee *Apis mellifera* is threatened by several diseases: viruses (Loucif-Ayad et al. 2015), parasites (Chahbar et al. 2016), fungal diseases (Chahbar 2017), and bacteria, including the American foulbrood *Paenibacillus larvae* (Haddad et al. 2015). In addition, it is also threatened by a neurological disease, which is in fact due to the consumption of neonicotinoids. Neonicotinoid insecticides include imidacloprid, thiacloprid, dinotefuran, thiamethoxam, nitenpyram, clothianidin, and acetamiprid. Thiamethoxams are pesticides used for coating seeds.

13.5.5.1 Nosemosis

Nosema apis and *Nosema ceranae*, downgraded from protozoa and classified as fungi (Adl et al. 2005), are parasites of the European bee *Apis mellifera* and the Asian bee *Apis ceranae*, respectively. Nosemosis, caused by these parasites, is widespread throughout the world; it is the major factor in economic losses in beekeeping. Recently, in 2017, a new species of *Nosema* was detected in Uganda, namely, *Nosema neumanni*. Nosemosis is a cosmopolitan disease, which has been detected in almost all African countries. In Algeria, *Nosema* spores are detected in both races, the tellienne *Apis mellifera intermissa* and the Saharan *Apis mellifera sahariensis*. However, when an experimental infection was performed, the Tell bee exhibited greater resistance than the Saharan bee (Chahbar 2017). In Kenya, nosemosis was only recently detected (Muli et al. 2014), but a few years after its introduction, different levels of infection were noted in different agro-ecological zones. This plague has also been detected in South Africa and Zimbabwe. *Nosema ceranae* was reported in only two African countries, in Benin and Algeria, in *Apis mellifera adansonii* and *Apis mellifera intermissa*, respectively (Paraïso et al. 2011); however, in some African countries, in Ghana and Sudan, nosemosis was not detected (Abdal Aziz and El-Niweiri 2013).

13.5.5.2 American Foulbrood

Among the factors threatening the survival of honeybees, American foulbrood, *Paenibacillus larvae,* is a devastating disease of colonies at a fairly high rate, preventing the development of beekeeping (Haddad et al. 2015). This disease is caused by a bacterium, *Paenibacillus larvae,* which occurs as a Gram-positive rod. This plague has been declared by the International Office of Epizootics (OIE) as a notifiable contagious disease. It is a cosmopolitan disease (Haddad et al. 2015) but found mainly in temperate countries and/or subtropical environments (Gregorc and Poklukar 2003). However, in most African countries, this disease is found with low levels of infection (Paraïso et al. 2011). After 2006, American foulbrood was detected in several African countries, namely Kenya, Senegal, South Africa, Tanzania, and Uganda, but not before (Allsopp 1993). In South Africa, the first appearance of this disease was presumed between 2006 and 2008 (Veldtman and Allsopp 2011). Algeria, which has the highest beekeeping production, has also recorded a high frequency of transhumance and high levels of infection (Chahbar 2017). However, some colonies infected with this disease did not develop clinical signs, namely the shooting mass in the hexagonal capped cells, which can be observed in the field by the match test (Fig. 13.25a), and the appearance of a mosaic brood and collapsed gills (Fig. 13.25b).

13.5.5.3 Varroasis

Varroasis, induced by the agent *Varroa destructor,* is a disease listed by the OIE (WOAH 2023). Discovered at the beginning of the twentieth

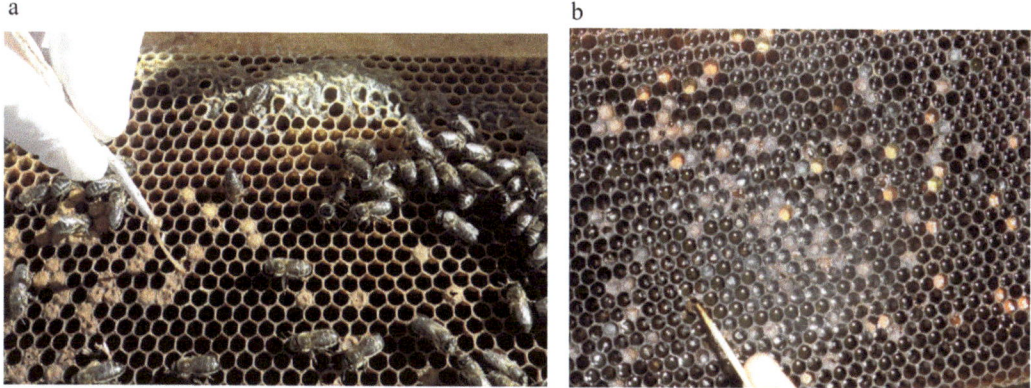

Fig. 13.25 The clinical signs of American foulbrood *P. larvae* observed in the field. (Onions 1912)

century in Indonesia, on the bee *Apis cerana*, the *Varroa jacobsoni* Oudemans mite develops rapidly on its new host, *Apis mellifera*. It was later renamed *Varroa destructor* (Anderson and Trueman 2000). The development cycle of this ectoparasite takes place mainly in the brood and lasts about eight days. Adult females invade brood cells a few hours before operculation (Beetsma et al. 1999). About 60 h after the operculation, the female mite lays her first egg, which will give birth to a male, and the subsequent eggs will give birth to females. Thus, the capping time of the worker brood of *Apis mellifera* generally allows for the production of three mature females of the parasite, and that of the male brood, the production of five mature females. The parasite feeds on the hemolymph of nymphs and adults. Infestation by *Varroa destructor* is extremely damaging to honeybee colonies. The main deleterious effects are caused by breeding females, which, by feeding on the hemolymph of larvae, nymphs, and workers, weaken them, and this affects the entire colony. This mite is also a vector of other pathogens, in particular viral agents.

In Africa, the *Varroa destructor* has been identified in Algeria, Egypt, Morocco, and Tunisia (Haddad et al. 2015). By comparing the genome of *Apis mellifera intermissa* with those of other subspecies, it appears that the latter has significant adaptation potential and has evolved for resistance to high temperatures and against infestations of the *Varroa destructor* parasite (Haddad et al. 2015). In Libya, the arrival of this formidable enemy of the bee was not until 1976 (Crane 1979). In Kenya, *V. destructor* was probably introduced between the years 1998 and 2009, and then from the year 2014, it spread to all regions of the country except a few regions in the North (Muli et al. 2014). In Nigeria, the detection of varroasis in bee colonies was not until 2012 (Akinwade et al. 2012). However, reports from Kenya and Nigeria showed that the presence of varroasis did not have a negative impact on bee health (Muli et al. 2014; Akinwade et al. 2012). *V. destructor* has been found in Benin, Ethiopia, Senegal, Niger and Zimbabwe (Paraïso et al. 2011; Scholmke and Schmolke 2003). In South Africa, *V. destructor* was first detected in 1997 and has caused significant bee colony mortalities (Pirk et al. 2014).

13.5.5.4 Viruses

There are more than 25 viruses affecting honey bees in Africa (Loucif-Ayad et al. 2015), most of which are RNA viruses. The most common viruses are deformed wing virus (DWV), black queen cell virus (BQCV), chronic bee paralysis virus (CBPV), Kashmir bee virus (KBV), sacbrood virus (SBV), Varroa destructor virus (VdV1), acute bee paralysis virus (ABPV), and Israeli acute paralysis virus (IAPV). Some could be present in colonies without causing visible signs. Thus, many colonies continue to appear healthy, even when several viruses are simultaneously present. However, in some cases, these viruses weaken the colonies. These viruses are mainly transmitted by the mite *V. destructor*. In Ethiopia, 25 viruses have been detected. In Algeria, six viruses were detected with different prevalence. CBPV was the most prevalent virus, followed by BQCV, DWV, IAPV, ABPV, and SBV (Loucif-Ayad et al. 2015). The two viruses, BQCV and DWV, are the most common in Uganda. In Kenya, the presence of viruses such as BQCV, DWV, and ABPV is associated with that of Varroa (Muli et al. 2014). In Kenya, viruses were not detected in northern areas where *V. destructor* was absent (Muli et al. 2014). In addition, the DWV virus was systematically present on the *Varroa* mites taken from each apiary. It is recognized that the spread of *V. destructor* has resulted, among other things, in altering the prevalence of bee viruses and their impact on colony health. Of all the bee viruses, only two can cause fatalities and characteristic clinical signs. These two viruses are sacciform brood virus (SBV) and chronic bee paralysis virus (CBPV), which are also known as black disease. Sacciform brood virus (SBV) kills larvae, described as sac-forming, due to the accumulation of fluid between the seed coat and the body. Chronic paralysis virus (CBPV) is the only viral disease of adult bees causing visible symptoms in flight steps and in front of hives. This disease can lead to abnormal wing and body tremors of infected bees, inability to fly, and death within days of the onset of symptoms.

13.5.6 The Fight Against Bee Diseases Using Natural Products

African researchers and beekeepers have tried to replace chemicals used to fight honeybee diseases with other natural products. For example, to control *Varroa* mite, oxalic acid and formic acid are used successfully. In addition to acids, plant extracts have been used, including neem oil (Qayyoum et al. 2013) and thymol powder. Other natural products that have been successfully used include steaming rice vinegar, which inhibits fungal spore germination. In contrast, treatments for ascospherosis are not recommended with antibiotics (Chahbar 2017). These treatments have negative side effects on brood, adult bees, and honey. Another method in the confirmation phase is the use of plant resins by bees, in the form of propolis (Wilson et al. 2015). Transhumance in regions characterized by abundant coniferous plants has very high effects against bee diseases, such as American foulbrood and ascospherosis (Chahbar 2017). It, therefore, becomes very important to test the activity of propolis against multiple pathogens by considering the functional role of propolis in bee health (Wilson et al. 2015). However, the main mechanism of resistance to ascospherosis is the hygienic behavior of adult bees.

13.5.7 Characterization of Honeybees

13.5.7.1 Morphometric Approach

A typological characterization of the local bee conducted by Khedim et al. (Khedim et al. 2021) was carried out by means of a questionnaire in Algeria. These researchers also carried out a classical morphological characterization at the level of several localities in the Algerian Northwest. For this purpose, 15 worker bee specimens were sampled at the level of 30 hives per beekeeping farm (15 morphometric parameters were analyzed). At the same time, they carried out a geomorphometric characterization using the geometry of the wings, which highlights the geotechnical potential for the production of bee populations. The analyses revealed a biogeographic variation within the honeybee colonies in Algeria. These results, thus, made it possible to discriminate among local populations. Assielou et al. (2016) used morphometry to identify the possible breeds and ecotypes of honeybees in central Côte d'Ivoire. Samples of 30 workers per hive were collected in 2–3 hives in beekeeping apiaries located in N'Guessankro, Soungassou, Kouassikouassikro, Lengbe-kouassikro de Yobouekro, and the "Institut National Polytechnique" Félix Houphouet-Boigny in Yamoussoukro. A total of 18 morphometric parameters were measured using a binocular magnifying glass and a microscope. Analyses revealed biogeographic variation within honeybee colonies in Côte d'Ivoire, allowing them to be subdivided into local populations.

A study conducted by Bamidele et al. (2021) evaluated the body allometry of honeybees, *Apis mellifera adansonii* Latreille (Hymenoptera, Apidae), from the rainforest, the Guinean Savannah, and the derived Savannah in Nigeria. Honeybee samples were dissected, and morphological parameters were measured using the PhotoScan method. They showed significantly higher mean body weight, total body length, and mouth width ($P < 0.05$) in honeybees from the rainforest zone. The thorax and abdomen lengths of honeybees from the tropical rainforest and Guinean savanna areas showed a significant linear relationship with the total body length of honeybees. In addition, the lengths of the abdomen, thorax, head, antenna, proboscis, pollen basket, jawbones, hind limbs, and fore and hind wings of honeybees in the three ecological zones exhibited a negative allometric growth pattern with total body length. The length–weight relationship, however, revealed a positive allometric growth pattern in honeybees from the tropical forest zone and negative allometry in those from the Guinean Savannah and derived savanna zones. Thus, these researchers deduced that the length–weight relationship could be a more sensitive tool for evaluating the allometric model of honeybees from different environments.

Fotso et al. (Fotso et al. 2021) conducted a study in the Western Highlands of Cameroon to better understand the biodiversity of honeybees for their preservation and genetic improvement. Samples of 420 worker bees belonging to *Apis mellifera* subspecies were obtained from 14 localities in the study zone. Body measurements in mm were: length (10.98 ± 0.06), abdomen length (6.71 ± 0.03), width of the yellow band on the second (2nd) abdominal tergite (1.48 ± 0.01), width of fourth (4th) abdominal tergite (1.32 ± 0.01), length of cover hair on the fifth (5th) abdominal tergite (0.29 ± 0.00), antenna length (4.27 ± 0.02), proboscis length (4.39 ± 0.11), length of nervure A (0.60 ± 0.00), length of nervure B (0.25 ± 0.00), discoidal shift (−0.13 ± 0.01), anterior right wing length (9.31 ± 0.03), anterior right wing width (3.14 ± 0.01), posterior right wing length (6.38 ± 0.03), and posterior right wing width (1.69 ± 0.01). A correlation coefficient ($r = 0.72$; $p < 0.01$) was obtained between the length of the posterior right wing and the length of the anterior right wing. The cubital index is 2.36 ± 0.04. Populations of honeybees studied consist of three genetic types/ecotypes; the cubital types 2 and 3 were not different. The observed biodiversity suggests that honeybees constitute a natural resource with genetic variability needed for preservation and genetic improvement in Cameroon.

Radloff and Hepburn (2000) studied morphometric characters and sting pheromones of worker honeybees, *Apis mellifera* Linnaeus, to delineate the honeybee populations of the Horn of Africa. Four discrete and statistically homogeneous populations were identified: *Apis jemenitica* Ruttner, *Apis mellifera bandasii* Mogga, *Apis mellifera sudanensis* Rashad in Ethiopia, and an unclassified group in southwestern Somalia. There are two endemic subspecies of Western honeybees (*Apis mellifera* L.) in the Republic of South Africa (RSA), *A. m. capensis* and *A. m. scutellata*. They have traditionally been identified using morphometric characteristics, but the geometric morphometric data of honeybee wings is easier to collect, making them perhaps a useful alternative for identifying these subspecies. Bustamante et al. (2020) compared the precision of morphometric and geometric morphometric methods using linear discriminant and classification and regression tree analyses. The authors found that using the geometric shape data of the fore and hind wings resulted in lower classification accuracy (73.7%) than using models derived from the full set of standard morphometric data (precision at 97%) in the cross-validation. The color of the tergite and the average number of ovarioles were the most important characteristics to distinguish the two subspecies. Finally, these researchers used Kreiger's interpolation to construct maps illustrating the probable distributions of *Apis mellifera capensis* and *Apis mellifera scutellata* in the Republic of South Africa.

13.5.7.2 Biochemical Approach

Proteins and isozymes were used to assess the biochemical characterization of three subspecies of worker bees, *Apis mellifera* L. (Egyptian, Italian, and Carniolan subspecies). Low percentages of polymorphism were recorded in different protein profiles, ranging from 18% to 42%. The isoenzyme systems also recorded low percentages of polymorphism, with the exception of peroxidase (67%). The dendrogram separated the Egyptian subspecies from the other two subspecies with a significant distance of 0.25. The Italian and Carniolan subspecies were grouped into a single group with a genetic distance of 0.01 between them (El-Bermawy et al. 2012).

13.5.7.3 Molecular Approach

The study of Loucif-Ayad et al. (2015) evaluated the genetic diversity of honeybees (*Apis mellifera*) in Algeria (North Africa), using the mtDNA COI-COII molecular marker (Cytochrome Oxidase I and II). In total, 582 beekeeping workers were sampled in 22 regions of the country. Polymerase chain reaction restriction fragment length polymorphism (PCR-RFLP) analysis of mtDNA samples distinguished evolutionary bee lineages and mtDNA haplotypes from each region. These data revealed the presence of three different bee lineages among the populations studied, including the African (A), North Mediterranean (C), and West Mediterranean (M) lineages. Eight different mtDNA haplotypes were recorded at different frequencies (A1, A2, A8, A9, A10, A13, C7, and M4). For the first time, these results identified a weak genetic introgression (3.1%) of non-local mtDNA haplotypes (C7 and M4) among Algerian local bees, most likely due to the importation of foreign bees.

Awodiran et al. (2021) studied the genetic diversity of four populations of honeybees, *Apis mellifera*, from two vegetation zones in Southwestern tropical forest and the North-central derived savanna of Nigeria using 15 morphometric characters and 5 microsatellite loci. Discriminant function analysis of morphometric data revealed considerable variation in morphological characters between sampled localities, while principal component analysis and canonical variable analysis produced clusters overlapping populations sampled, indicating a lack of separation between different populations. Genetic diversity (F_{ST}) revealed poor differentiation between the populations, suggesting that geographic distance was not a barrier to gene flow between populations. All four populations had a heterozygous deficiency, suggesting the presence of inbreeding within the populations. Analysis of molecular variance showed that 91% of the total molecular variance existed within populations while 9% existed between populations. These researchers suggested that there is an apparent loss of genetic diversity in populations of *Apis mellifera* studied in the two vegetation zones of Nigeria.

Fuller et al. (2015) used a new suite of integrated web-based programs to examine population dynamics and selection signatures across the genome using several well-established assays, including FST, pN/pS, and McDonald-Kreitman. These researchers applied these techniques to study populations of honeybees (*Apis mellifera*) in East Africa. In Kenya, there are several descriptions of the subspecies of *Apis mellifera*, which these researchers believe are located in distinct ecological regions. Whole genome sequencing of 11 worker bees from apiaries throughout Kenya was performed using 3.6 million SNPs. The dense coverage allowed several computational procedures to be applied to study population structure and evolutionary relationships between them and to detect signs of adaptive evolution across the genome. Although there is considerable gene flow among the sampled populations, there were clear distinctions between populations in the Northern desert region and those in the temperate savanna region. The study identified several genes showing genetic population models compatible with positive selection within African bee populations and between these populations and the populations of European *Apis mellifera* or Asian *Apis florea*.

Honeybees, *Apis mellifera* subspecies, of the East African populations were also assessed by Kibogo (2017) for phylogenetic relationships using mitochondrial markers. *Apis mellifera cutellata, Apis mellifera litorea,* and *Apis mellifera monticola* were shown to have a close genetic relationship and are largely indistinguishable. However, based on altitude, three groups were documented: lowland, medium altitude, and mountain bees. Populations of *Apis mellifera* from Kenya, Uganda, Tanzania, and Madagascar were also genotyped with eight microsatellite loci. The mean number of alleles and the mean observed heterozygosity were high at 8.201 ± 2.5035 and 0.781 ± 0.00346, respectively, for all populations. The average genetic diversity per locus and per population was high at 0.783 for all populations. A low overall mean population differentiation value of 0.056 ± 0.04 signified moderate levels of genetic structuring and differentiation between populations. The probability values for the genetic assignment of individuals to populations were less than 0.5, indicating that the populations are not isolated and that there was substantial hybridization. The results of clustering, mixing, genetic diversity attributed to variance, and phylogenetics showed that the dynamics of the honeybee subspecies in East Africa are associated with a relatively stable population demographic structure, especially in unfragmented habitats, natural forests, and mountainous regions. In these regions, the results suggest a living demographic historical pattern of their existence characterized by recent evolutionary expansions and no bottlenecks. Association studies have shown that, on average, colonies store more nectar (69%) than pollen (31%). There is a positive correlation between sting and pln (0.184458), suggesting that defensive colonies feed more on nectar than less defensive colonies. The search for candidate genes identified three genes, each associated with foraging (GB46589, GB44258, GB44259) and tingling behavior (GB48999, GB49000, GB55730).

Eimanifar et al. (2020) recently performed whole mitogenome sequencing using SNPs to study the genetic diversity of *Apis mellifera capensis* and *Apis mellifera cutellata* with data from the Republic of South Africa (N = 813 bees from 75 apiaries) using 19 DNA microsatellite loci. The populations averaged between 9.2 and 11.3 alleles per locus, with unbiased heterozygosity values ranging from 0.81 to 0.86 per population. Bayesian clustering analyses revealed two distinct evolutionary units, although the results did not match those of previous morphometric and molecular analyses. This suggests that the microsatellites these researchers tested were not sufficient for identifying subspecies, especially for Cape bees and hybrid bees. Nevertheless, the microsatellite data show considerable genetic diversity within the two populations and a larger than expected hybridization zone between the natural distributions of *Apis mellifera capensis* and Apis *mellifera cutellata*.

13.6 Conclusion and Future Perspectives

The colossal amphibian, Goliath Frog (*Conraua goliath*), occupies a unique place in the animal kingdom due to its enormous size. This is not only important ecologically, but also culturally in certain African regions. Preserving its habitat and genetic diversity is crucial. Various species of snails, such as the giant African snail (Achatina species), have economic value for their meat and shells. Sustainable management of snail populations is essential to avoid overexploitation. Although not unique to Africa, rabbits provide valuable genetic diversity for food production. Local breeds must be preserved and managed for their potential contribution to nutrition and livelihoods. African bee subspecies are essential pollinators for agriculture. Protecting the genetic diversity of these bees is essential for food security and ecosystem health.

The main challenges of nonconventional animal genetic resources include: (1) habitat loss/destruction due to urbanization, agriculture and deforestation, which threatens the survival of these species; (2) overexploitation, which impacts their population growth; (3) climate change, which disrupts habitats, potentially pushing these species into new areas or reduction of available habitats; and (4) invasive or non-native species, which can outcompete native species, thereby affecting genetic diversity.

Opportunities for the preservation of nonconventional animal genetic resources include tailored breeding strategies to increase productivity as well as generate other products that can be used by humans, such as shell and slime from snails and the wax from bees. African countries can collaborate on conservation efforts, creating protected areas and implementing sustainable harvesting practices. Further research into the genetics and biology of these species can guide conservation strategies. Economic diversification through exploiting these unique species for ecotourism or sustainable agriculture can provide economic incentives for conservation. Education of farmers through raising awareness among local communities of the value of these species can encourage their conservation. Establishing gene banks for these species can ensure that their genetic diversity is preserved for future use.

In conclusion, unconventional animal genetic resources in Africa, including the Goliath frog, snails, rabbits, and bees, represent both ecological and economic assets. Protecting their diversity and unique features requires concerted conservation efforts, sustainable practices, and increased awareness of their importance among local communities and policymakers. These resources have the potential to contribute to food security, livelihoods, and biodiversity conservation in Africa and beyond.

References

Abdal Aziz S, El-Niweiri MAA (2013) Survey of Nosema Apis (ZANDER), Acarapis Woodi (RENNIE) ,and Varroa Destructor in Sudan. 9th COLOSS conference, book of abstracts, Ukraine, 27–29 SEPTEMBER 2013. 2013, p 33

Abiona J, Osinowo O, Eruvbetine D, Abioja M, Smith O, Daramola J, Ladokun A, James I, Abe O, Onagbesan O (2012) Accessory reproductive organs dimension of two species of Giant African land snails Archachatina Marginata and Achatina Achatina at three Liveweight ranges. J Agric Sci 4(2):179

Abou-Shaara HF (2014) Recycling behaviour and wisdom in the beehive. Bee World 91(1):12–13

Adl SM, Simpson AG, Farmer MA, Andersen RA, Anderson OR, Barta JR, Bowser SS, Brugerolle G, Fensome RA, Fredericq S (2005) The new higher level classification of eukaryotes with emphasis on the taxonomy of Protists. J Eukaryot Microbiol 52(5):399–451

Afiane A (2019) Effet de l'environnement Sur La Caractérisation Des Coquilles de l'escargot Helixaspersa Dans Les Régions Mascara et Mostaganem

Agbogidi OM, Okonta B (2011) Reducing poverty through snail farming in Nigeria. Small 10:15–15

Ait Tahar N, Fettal M (1990) Témoignage Sur La Production et l'élevage Du Lapin En Algérie, 2ème Conférence Sur La Production et La Génétique Du Lapin Dans La Région Méditerranéenne; Z Qagazig

Akinwade KL, Badejo MA, Ogbogu SS (2012) Incidence of the Korea haplotype of Varroa destructor

Akinwande K, Badejo M (2009) Improving honey production in worker bees (Apis Mellifera Adansoni L.) hymenoptera: Apidae through artificial modification of their feeding activities. Afr J Food Agric Nutr Dev 9(7)

Aktaş D, Şereflişan H, Alkaya A (2019) Disease in land snail culture. In International conference on environment, technology and management

Ali RF, Neiber MT, Walther F, Hausdorf B (2016) Morphological and genetic differentiation of E Remina Desertorum (G Astropoda, P Ulmonata, H Elicidae) in E Gypt. Zool Scr 45(1):48–61

Allsopp M (1993) Summarized overview of the Capensis problem. Afr Bee J 65(6):127–136

Amakpe F (2010) The biodiversity of the honey bees (Apis Mellifera Adansonii) in the district of Djidja, Republic of Benin. Int J Environ Cult Econ Soc Sustain 6(6):90–104

Anderson R (1963) The laying worker in the cape honeybee, Apis Mellifera Capensis. J Apic Res 2(2):85–92

Anderson D, Trueman J (2000) Varroa Jacobsoni (Acari: Varroidae) is more than one species. Exp Appl Acarol 24:165–189

Antonelli M, Donell DI (2019) Potential efficacy of Royal Jelly supplementation in Alzheimer's disease: A systematic review. Communication orale, IX Congresso Nazionale SINut., n° 2, BOLOGNA 30 May–1 June 2019, p P68

Arias MC, Sheppard WS (2005) Phylogenetic relationships of honey bees (hymenoptera: Apinae: Apini) inferred from nuclear and mitochondrial DNA sequence data. Mol Phylogenet Evol 37(1):25–35

Assielou BA, Wandan EN, Abo K, Iritie M (2016) Caractérisation Morphométrique Des Abeilles Mellifères Élevées Dans Le Centre de La Côte d'Ivoire. Eur Sci J 15:155–170

Awodiran MO, Amoo TE, Kehinde TO (2021) Genetic diversity of four populations of honey bee, Apis Mellifera (Linnaeus, 1758) from two vegetation zones in Nigeria. J Entomol Nematol 13(1):1–11

Aygun A, Sert D, Copur G (2012) Effects of Propolis on eggshell microbial activity, hatchability, and Chick performance in Japanese quail (Coturnix Coturnix Japonica) eggs. Poult Sci 91(4):1018–1025

Babalola OO, Akinsoyinu AO (2009) Proximate composition and mineral profile of snail meat from different breeds of land snail in Nigeria. Pak J Nutr 8(12):1842–1844

Badr O, El-Shawaf I, Khalil M, Refaat M, Ramadan S (2019) Molecular genetic diversity and conservation priorities of Egyptian rabbit breeds. World Rabbit Sci 27(3):135–141

Bailey JE (1991) Toward a science of metabolic engineering. Science 252(5013):1668–1675. https://doi.org/10.1126/science.2047876

Bamidele JA, Idowu AB, Ademolu K, Osipitan AA (2021) Nutritional composition of Apis Mellifera Adansonii L.(hymenoptera: Apidae) from three ecological zones of Nigeria. J Apic Res 60(3):445–456

Bareke T, Addi A, Wakjira K (2018) Role and economic benefits of honey bees' pollination on fruit yield of wild apple (malus Sylvestris (L.) mill.) in central highlands of Ethiopia. Bee World 95(4):113–116

Baruwa IO (2012) Economics of raising African Giant land snail (Archachatina Marginata) in Osun state, Nigeria. Int J Agric Econ Rural Dev 5(1):46–54

Beeby A (1985) The role of Helix Aspersa as a major herbivore in the transfer of Lead through a polluted ecosystem. J Appl Ecol:267–275

Beeby A, Richmond L (2011) Magnesium and the deposition of lead in the shell of three populations of the garden snail Cantareus Aspersus. Environ Pollut 159(6):1667–1672

Beetsma J, Boot WJ, Calis J (1999) Invasion behaviour of Varroa Jacobsoni oud.: from bees into brood cells. Apidologie 30(2–3):125–140

Behidj K (2012) La Compétitivité de La Filière Apicole Algérienne

Belanger, D. Utilisation de La Faune Macrobenthique Comme Bioindicateur de La Qualité de l'environnement Marin Côtier. 2009

Ben Larbi M, San-Cristobal M, Chantry-Darmon C, Bolet G (2014) Population structure in Tunisian indigenous rabbit ascertained using molecular information. World Rabbit Sci 22(3):223–230

Benachour K, Louadi K (2011) Comportement de Butinage Des Abeilles (Hymenoptera: Apoidea) Sur Les Fleurs Mâles et Femelles Du Concombre (Cucumis Sativus L.) (Cucurbitaceae) En Région de Constantine (Algérie), vol 47. Taylor & Francis, pp 63–70

Benachour K, Louadi K, Terzo M (2007) Rôle Des Abeilles Sauvages et Domestiques (Hymenoptera: Apoidea) Dans La Pollinisation de La Fève (Vicia Faba L. Var. Major) (Fabaceae) En Région de Constantine (Algérie), vol 43. Taylor & Francis, pp 213–219

Bendifallah L, Doumandji S, Louadi K, Iserbyt S (2012) Geographical variation in diversity of pollinator bees at natural ecosystem (Algeria). Int J Sci Adv Technol 2(11):26–31

Benlacheheb Ahmed A, Benchiko B (2021) Evaluation de l'effet Anti Inflammatoire d'un Produit Naturel Supplémentaire Par Un Produit de La Ruche. Mém. Master 2 En Biochim. Appliquée Dép. SNV L'institut Sci. Technol. Cent. Univ. Tissemsilt Algér

Berchiche M (1992) Systèmes de Production de Viande de Lapin Au Maghreb. Sémin. Approfondi Sur Systèmes Prod. Viande Lapin CIHEAM Saragossa Spain, pp 24–26

Berchiche M, Kadi SA (2002) The Kabyle rabbits (Algeria). Options Méditérr

Berchiche M, Zerrouki N, Lebas F (2000) Reproduction performances of local Algerian does raised in rational conditions. In: Proceeding of 7th world rabbit congress, Valencia, pp 43–49

Berkani M (1985) Comparaison de Deux Types de Ruches: Dadant et Langstroth Dans Les Littoral Est et Algérois

Berkani ML (2007a) Etude Des Paramètres de Développement de l'Apiculture Algérienne

Berkani, M. L. Etude Des Paramètres de Développement de l'Apiculture Algérienne. 2007b

Betayene D (2008) Manuel de Formation Apicole: Abeille- Environnement- Développement, Débuter En Apiculture,. Ed Rev Cent Pour L'Environnement Dév Cameroun, 44

Beyene T, Abi DCG, Mekonen Wolda Tsadik M, Zeway E (2016) Evaluation of transitional and modern hives for honey production in the mid Rift Valley of Ethiopia. Bull Anim Health Prod Afr 64(1):157–165

Boardman L, Eimanifar A, Kimball R, Braun E, Fuchs S, Grünewald B, Ellis JD (2020) The mitochondrial genome of Apis Mellifera Simensis (Hymenoptera: Apidae), an Ethiopian honey bee. Mitochondrial DNA Part B 5(1):9–10

Bogdanov S (2016) Honeys types book of honey Chapter 6

Bolet G, Zerrouki N, Gacem M, Brun J-M, Lebas F (2012) Genetic parameters and trends for litter and growth traits in a synthetic line of rabbits created in Algeria, pp 195–199

Borsuk G, Sulborska A, Stawiarz E, Olszewski K, Wiącek D, Ramzi N, Nawrocka A, Jędryczka M (2021) Capacity of honeybees to remove heavy metals from nectar and excrete the contaminants from their bodies. Apidologie:1–14

Bouchiba I, Ameur AA, Gaouar SBS (2021) Geometrical morphology characterization and taxa identification of land snail Populationsin Algeria. Genet Biodivers J 5(3):140–148

Brittan O (1956) Introduction of modern beekeeping to Cyrenacia (Libya). Bee Craft 37(12):145–146

Browne RK, Odum RA, Herman T, Zippel K (2007) Facility design and associated services for the study of amphibians. ILAR J 48(3):188–202

Burch JB (1968) A tissue culture technique for Caryotype analyses of Pulmonate land snails. Venus Jpn J Malacol 27(1):20–27

Bustamante T, Baiser B, Ellis JD (2020) Comparing classical and geometric morphometric methods to discriminate between the south African honey bee subspecies Apis Mellifera Scutellata and Apis Mellifera Capensis (Hymenoptera: Apidae). Apidologie 51:123–136

Calatayud-Vernich P, Calatayud F, Simó E, Suarez-Varela MM, Picó Y (2016) Influence of pesticide use in fruit orchards during blooming on honeybee mortality in 4 experimental apiaries. Sci Total Environ 541:33–41

Carneiro M, Afonso S, Geraldes A, Garreau H, Bolet G, Boucher S, Tircazes A, Queney G, Nachman MWMW, Ferrand N, van der Maesen LJG (2011) The genetic structure of domestic rabbits. Mol Biol Evol 28(6):1801–1816

Carneiro M, Rubin C-J, Di Palma F, Albert FW, Alföldi J, Barrio AM, Pielberg G, Rafati N, Sayyab S, Turner-Maier J (2014) Rabbit genome analysis reveals a polygenic basis for phenotypic change during domestication. Science 345(6200):1074–1079

Castellini C, Dal Bosco A, Mugnai C (2003) Comparison of different reproduction protocols for rabbit does: effect of litter size and mating interval. Livest Prod Sci 83(2–3):131–139

Castro E, Linares M, Robles A (2007) Study on gamma irradiation of snail slime for cosmetic applications

Çetin E, Silici S, Çetin N, Güçlü B (2010) Effects of diets containing different concentrations of Propolis on hematological and immunological variables in laying hens. Poult Sci 89(8):1703–1708

Chahbar N (2013) Evaluation de La Toxicité d'un Produit Phytopharmaceutique Sur Les Abeilles Domestiques Locales (Apis Mellifera Intermissa et Apis Mellifera Sahariensis) et Diversité Génétique

Chahbar M (2017) Principales Maladies et Ennemis de l'abeille Domestique Apis Mellifera L., 1758 En Algérie

Chahbar M, Tefiel H, Adidou-Chahbar N, Doumandji-Mitiche B, Gaouar S (2016) First spatial distribution of Nosemosis (Nosema Sp) Infected Local Bee, Apis Mellifera Intermissa L. in Algeria

Channing A, Rödel M-O (2019) Field guide to the frogs & other amphibians of Africa. Penguin Random House South Africa

Chantry-Darmon C, Urien C, Hayes H, Bertaud M, Chadi-Taourit S, Chardon P, Vaiman D, Rogel-Gaillard C (2005) Construction of a cytogenetically anchored microsatellite map in rabbit. Mamm Genome 16(6):442–459

Chen S, Zhang W, Zhang G, Peng J, Zhao X, Lai S (2013) Case–control study and MRNA expression analysis reveal the MyD88 gene is associated with digestive disorders in rabbit. Anim Genet 44(6):703–710

Chinaka NC, Chuku LC, George G, Oraezu C, Umahi G, Orinya OF (2021) Snail slime: evaluation of anti-inflammatory, phytochemical and antioxidant properties. J Complement Altern Med Res:8–13

Clauss B (1991) Bees and beekeeping in the North Western Province of Zambia

Cobbinah JR, Vink A, Onwuka B (2008) Snail farming: production, processing and marketing. Agromisa/CTA

Colin M, Lebas F (1995) Le Lapin Dans Le Monde. Lempdes, A. éditeur, Ed

Colin M, Lebas F (1996) Rabbit meat production in the world. A proposal for every country. In: 6. World rabbit congress. Association Scientifique Française de Cuniculture

Conti ME, Canepari S, Finoia MG, Mele G, Astolfi ML (2018) Characterization of Italian multifloral honeys on the basis of their mineral content and some typical quality parameters. J Food Compos Anal 74:102–113

Cornuet J, Garnery L (1991) Mitochondrial DNA variability in honeybees and its Phylogeographic implications. Apidologie 22(6):627–642

Crane E (1979) Fresh news on the Varroa mite. Bee World 60(8)

D'Albore, Piatti (2004)

Daly EE, Walker KJ, Morgan-Richards M, Trewick SA (2019) Spatial genetics of a high elevation lineage of Rhytididae land snails in New Zealand: the Powelliphanta Kawatiri complex. Molluscan Res 39(3):280–289

Davison A (1999) Isolation and characterization of long compound microsatellite repeat loci in the land snail, Cepaea Nemoralis L. (Mollusca, Gastropoda, Pulmonata). Mol Ecol 8(10):1760–1761

Desoky ASSSS, Sallam AAA, Abd El-Rahman TMMMM (2015) First record of two species from land snails, Monacha Obstracta and Eobania Vermiculata in Sohag governorate, Egypt. Direct Res J Agric Food Sci 3(11):206–210

Dietemann V, Pirk CWW, Crewe R (2009) Is there a need for conservation of honeybees in Africa? Apidologie 40(3):285–295

Ebegbulem V (2012) Body conformation characteristics of domestic rabbits in humid tropical southern Nigeria. J Agric Vet Sci 4:65–70

Efenakpo OD, Ayodele IA, Ijeomah HM (2016) Assessment of frog meat utilisation in Ibadan, Oyo state, Nigeria. J Res For Wildl Environ 8(3):31–43

Eimanifar A, Pieplow JT, Asem A, Ellis JD (2020) Genetic diversity and population structure of two subspecies of Western honey bees (Apis Mellifera L.) in the Republic of South Africa as revealed by microsatellite genotyping. PeerJ 8:e8280

El Raffa A, Kosba M, Khalil M, Baselga M (2002) The chinchilla rabbits (Egypt). Opt Méditerranéennes Sér B Etudes Rech 38:79–82

El-Bermawy SM, Ahmed KS, Al-Gohary HZ, Bayomy AM (2012) Biochemical and molecular characterization for three subspecies of honey bee worker, Apis Mellifera L. (Hymenoptera: Apidae) in Egypt. Egypt Acad J Biol Sci Entomol 5(2):103–115

El-Niweiri MA, Moritz RF (2008) Mitochondrial discrimination of honeybees (Apis Mellifera) of Sudan. Apidologie 39(5):566–573

El-Niweiri M, El-Sarrag M, Satti A (2008) Survey of diseases and parasites of honey bees (Apis Mellifera L.) in Sudan. Sudan J Basic Sci 14:141–159

Eneji CA, Ogogo AU, Emmanuel-Ikpeme CA, Okon OE (2008) Nutritional assessment of some Nigerian land and water snail species. Ethiop J Environ Stud Manag 1(2):56–60

Engel MS (1999) The taxonomy of recent and fossil honey bees (Hymenoptera: Apidae; Apis)

Eshra E (2013) Survey and distribution of terrestrial snails in fruit orchards and ornamental plants at Alexandria and El-Beheira governorates, Egypt. Alex Sci Exch J, April–June 34:242–248

Fagbuaro O, Oso JA, Edward JB, Ogunleye RF (2006) Nutritional status of four species of Giant land snails in Nigeria. J Zhejiang Univ Sci B 7(9):686–689

Falconer DS, Mackay TFC (1996) Introduction to quantitative genetics. Ltd., L. G., Ed, London

FAOSTAT (2020) Food and agriculture Organization of the United Nations. Rome

Fikadu Z (2019) The contribution of managed honey bees to crop pollination, food security, and economic stability: case of Ethiopia. Open Agric J 13(1)

Fikru S (2015) Review of honey bee and honey Production in Ethiopia. J Anim Sci Adv 5(10):1413–1421

Finzi A (2011) No Title. Integrated back yard system. http://www.fao.org/ag/AG/AGAInfo/subjects/documents/ibys/default.htm. Accessed 21 Mar 2020

Fischer FU (2013) L'elevage d'abeilles Dans l'economie de Base de La Savane Arboree de Miombo Au Centre de l'afrique Australede Miombo Au Centre de l'afrique Australe. Reseau For. Pour Dev. Rural Doc. RDFN Numéro 15c Lusaka Zamb. 1–9

Fletcher DJ (1978) The African bee, Apis Mellifera Adansonii, in Africa. Annu Rev Entomol 23(1):151–171

Fontanesi L (2016) The rabbit in the genomics era: applications and perspectives in rabbit biology and breeding, pp 15–18

Fontanesi L, Scotti E, Frabetti A, Fornasini D, Picconi A, Russo V (2011) Identification of polymorphisms in the rabbit (Oryctolagus Cuniculus) Myostatin (MSTN) gene and Association analysis with finishing weight in a commercial rabbit population. Anim Genet 42(3):339–339

Fontanesi L, Dall'Olio S, Spaccapaniccia E, Scotti E, Fornasini D, Frabetti A, Russo V (2012a) A single nucleotide polymorphism in the rabbit growth hormone (GH1) gene is associated with market weight in a commercial rabbit population. Livest Sci 147(1–3):84–88

Fontanesi L, Martelli P, Scotti E, Russo V, Rogel-Gaillard C, Casadio R, Vernesi C (2012b) Exploring copy number variation in the rabbit (Oryctolagus Cuniculus) genome by Array comparative genome hybridization. Genomics 100(4):245–251

Fontanesi L, Scotti E, Colombo M, Allain D, Deretz S, Dall'Olio S, Russo V, Oulmouden A (2013) Investigation of the Premelanosome protein (PMEL or SILV) gene and Identification of polymorphism excluding it as the determinant of the dilute locus in domestic rabbits (Oryctolagus Cuniculus). Arch Anim Breed 56(1):42–49

Fortun-Lamothe L, Thomas M, Tichit M, Jouven M, García EG, Dourmad J-Y, Dumont B (2013) Agro-Écologie et Écologie Industrielle: Deux Voies Complémentaires Pour Les Systèmes d'élevage de Demain. Applications Potentielles Aux Systèmes Cunicoles, pp 121–131

Fotso PRK, Costa-Maia FM, Keambou CT, Baenyi SP, Defang HF, Manjeli Y, Teguia A (2021) Diversity of honey bee Apis Mellifera subspecies (hymenoptera: Apoidae) from the Western highlands of Cameroon based on morpho-biometry. Genet Biodivers J 5(3):1–11

Franck P, Garnery L, Loiseau A, Oldroyd B, Hepburn H, Solignac M, Cornuet J-M (2001) Genetic diversity of the honeybee in Africa: microsatellite and mitochondrial data. Heredity 86(4):420–430

Fuller ZL, Niño EL, Patch HM, Bedoya-Reina OC, Baumgarten T, Muli E, Mumoki F, Ratan A, McGraw J, Frazier M (2015) Genome-wide analysis of signatures of selection in populations of African honey bees (Apis Mellifera) using new web-based tools. BMC Genomics 16(1):1–18

Gacem M, Bolet G (2005) Création d'une Lignée Issue Du Croisement Entre Une Population Locale et Une Souche Européenne Pour Améliorer La Production Cunicole En Algérie. In: Proc 11èmes Journ. Rech. Cunic. 29–30 Novemb. 2005 Paris Fr. 15, 18

Gacem M, Zerrouki N, Lebas F, Bolet G (2008) Strategy for developing rabbit meat production in Algeria: creation and selection of a synthetic strain. In: 9th World Rabbit Congress-June 10–13, 2008, Verona, Italy, pp 85–89

Galal E, Khalil M (1994) Development of rabbit industry in Egypt. Cah. Opt Mediterr. CIHEAM

Garnery L, Cornuet J, Solignac M (1992) Evolutionary history of the honey bee Apis Mellifera inferred from mitochondrial DNA analysis. Mol Ecol 1(3):145–154

Garnery L, Solignac M, Celebrano G, Cornuet J-M (1993) A simple test using restricted PCR-amplified mitochondrial DNA to study the genetic structure of Apis Mellifera L. Experientia 49:1016–1021

Garnery L, Mosshine E, Oldroyd B, Cornuet J (1995) Mitochondrial DNA variation in Moroccan and Spanish honey bee populations. Mol Ecol 4(4):465–472

Garreau H, Gunia M (2018) La Génomique Du Lapin: Avancées, Applications et Perspectives. INRA Prod Anim 31(1):13–22

Gebretinsae T, Tesfay Y (2014) Honeybee colony marketing practices in Werieleke District of the Tigray region, Ethiopia. Bee World 91(2):30–35

Gemechis LY (2016) Honey production and marketing in Ethiopia. ABJNA 7(5):248–253

Gezahegne T (2001) Marketing of honey and beeswax in Ethiopia: past, present and perspective features, pp 3–4

Gomot-de Vaufleury A, Pihan F (2000) Growing snails used as sentinels to evaluate terrestrial environment contamination by trace elements. Chemosphere 40(3):275–284

Gonwouo NL, Rödel MO (2008) The importance of frogs to the livelihood of the Bakossi people around Mount Manengouba, Cameroon, with special consideration of the hairy frog, Trichobatrachus Robustus. Salamandra 44(1):23–34

Goretti E, Pallottini M, Rossi R, La Porta G, Gardi T, Goga BC, Elia A, Galletti M, Moroni B, Petroselli C (2020) Heavy metal bioaccumulation in honey bee matrix, an indicator to assess the contamination level in terrestrial environments. Environ Pollut 256:113388

Gratzer K, Wakjira K, Fiedler S, Brodschneider R (2021) Challenges and perspectives for beekeeping in Ethiopia. A review. Agron Sustain Dev 41(4):1–15

Gregorc A, Poklukar J (2003) Rotenone and oxalic acid as alternative Acaricidal treatments for Varroa destructor in honeybee colonies. Vet Parasitol 111(4):351–360

Guiller A, Madec L (2010) Historical biogeography of the land snail Cornu Aspersum: A new scenario inferred from haplotype distribution in the Western Mediterranean Basin. BMC Evol Biol 10(1):1–20

Gupta RK, Reybroeck W, van Veen JW, Gupta A (2014) Beekeeping for poverty alleviation and livelihood security

Haccour P (1960) Recherche Sur La Race d'abeille Saharienne Au Maroc. CR Soc Sci Nat Phys Extr Belg Apic 25:13–18

Haddad NJ, Loucif-Ayad W, Adjlane N, Saini D, Manchiganti R, Krishnamurthy V, AlShagoor B, Batainh AM, Mugasimangalam R (2015) Draft genome sequence of the Algerian bee Apis Mellifera Intermissa. Genomics Data 4:24–25

Hailu TG, Tadesse A (2016) Queen rearing and colony multiplication for promoting beekeeping in Tigray, Ethiopia. Elixir Ent 92:39257–39259

Hannah EE, Essien AE, Peace AO (2016) Phenotypic characterization of two snails breeds Archachatina Marginata and Achatina Fulica in Calabra, Cross River state. Asian J Appl Sci 4(2):534

Hardouin J, Stiévenart C, Codjia JTC (1995) L'achatiniculture. Rev Mond Zootech 2:29–39

Hayes H, Rogel-Gaillard C, Zijlstra C, De Haan NA, Urien C, Bourgeaux N, Bertaud M, Bosma AA (2002) Establishment of an R-banded rabbit karyotype nomenclature by FISH localization of 23 chromosome-specific genes on both G-and R-banded chromosomes. Cytogenet Genome Res 98(2–3):199–205

Hepburn HR, Radloff SE (1998) Honeybees of Africa

Herbert DG (2010) The introduced terrestrial Mollusca of South Africa

Hodasi JKM (1979) Life-history studies of Achatina (Achatina) Achatina (Linné). J Molluscan Stud 45(3):328–339

Houndonougbo MF, Chrysostome C, Odoulami RC, Codjia JTC (2012) Snail Shell as an efficient mineral feedstuff for layer hens: effects and optimum rate. Livest Res Rural Dev 24(9):162

Hutton-Squire JP (2014) Historical relationship of the honeybee (Apis Mellifera) and its forage; and the current state of beekeeping within South Africa

Ikediobi CO, Obi VC, Achoba IA (1995) Beekeeping and Honey Production in Nigeria. Niger Field 50:59–70

ISSG (2003) Global invasive species database; invasive species specialist group. IUCN, Auckland. www.issg.org

Jaksch K, Kruckenhauser L, Haring E, Fehér Z (2017) First establishment of microsatellite markers in Clausiliid snails (Mollusca: Gastropoda: Clausiliidae). BMC Res Notes 10(1):1–5

Johannsmeier MF (2001) Beekeeping History. Handbook No.14, Pretoria,. Beekeep. South Afr. Third Ed. Johannsmeier MF Ed Plant Prot. Res. Inst, 1–8

Joshi N, Pandey S (2019) Meat demand-snailed it: a comprehensive review on snail rearing, to meet the meat demand in future India

Kaldjob MC, Enangue NA, Siri BN, Etchu K (2019) Socio-economic perception of snail meat consumption in Fako division, south-west region Cameroon. Int J Livest Prod 10(5):143–150. https://doi.org/10.5897/ijlp2018.0543

Karamoko M, Memel J-D, Kouassi Kouadio D, Otchoumou A (2011) Influence de La Densité Animale Sur La Croissance et La Reproduction de l'escargot Limicolariaflammea (Müller) En Conditions d'elevage. Acta Zool Mex 27(2):393–406

Karas F (2009) Gastéropodes Terrestres, Invertébrés Continentaux Des Pays de La Loire. Gretia:379–387

Kebede H, Lemma T, Dugassa G (2018) Assessment on the authenticity of imported honey in Ethiopia. J Nutr Health Food Eng 8(6):442–445

Kerney MP, Cameron RAD, Bertrand A (1999) Guide Des Escargots et Limaces d'Europe: Identification et Biologie de plus de 300 Espèces. Delachaux et Niestlé

Khedim R, Halfaoui Y, Mediouni R, Gaouar SBS (2021) Diversity of honey bee Apis Mellifera from the North-West of Algeria based on Morpho and geo-biometry. Gen Biodiv. J: 6 (2): In press

Kibogo HG (2017) Molecular characterisation of honey bees, Apis Mellifera subspecies in East Africa

Kligerman A, Bloom S (1977) Rapid chromosome preparations from solid tissues of fishes. Wsq Womens Stud Q 34:266–269

Koeniger N (1976) Interspecific competition between Apis Florea and Apis Mellifera in the tropics. Bee World 57(3):110–112

Korstanje R (2000) Development of a genetic and comparative map of the rabbit. Tool QTL Mapp

Koudjil M (1990) Etude de La Race Locale Apis Mellifeca Intermissa et Essai d'élevage de Reine Dans La Région de Chlef

Krüger B (1912) Zur Kenntnis Des Grössten Lebeden Frosches, Rana Goliath. Blgr Bl Für Aquar- Terr-Kunde 23(24):383

Laneri S, Lorenzo RD, Sacchi A, Dini I (2019) Dosage of bioactive molecules in the Nutricosmeceutical helix Aspersa Muller mucus and formulation of new cosmetic cream with moisturizing effect. Nat Prod Commun 14(8):1934578X19868606

Lebas F (2000) Les Techniques d'élevage Au 7 Ème Congrès Mondial de Cuniculture. ASFC

Lebas F (2002) La Biologie Du Lapin. INRA, Paris

Lebas F (2008) Méthodes et Techniques d'élevage Du Lapin. Historique de La Domestication et Des Méthodes d'élevage. https://www.researchgate.net/publication/272015465_Methodes_et_techniques_d'elevage_du_lapin_Historique_de_la_domestication_et_des_methodes_d'elevage

Lebas-Fraczak L (2009) Description «communicative» Des Déterminants Français En Vue de La

Didactisation. Rech En Didact Lang Cult Cah Acedle 6:6–2

Lecointre G, Le Guyader H (2001) Classification Phylogénétique Du Vivant; Belin: 2ème édition. Paris

Leven L, Boot W-J, Mutsaers M, Segeren P, Velthuis H (2005) L'apiculture Dans Les Zones Tropicales. Agromisa/CTA

Lindblad-Toh K, Garber M, Zuk O, Lin MF, Parker BJ, Washietl S, Kheradpour P, Ernst J, Jordan G, Mauceli E (2011) A high-resolution map of Human evolutionary constraint using 29 mammals. Nature 478(7370):476–482

Linksvayer TA, Fewell JH, Gadau J, Laubichler MD (2012) Developmental evolution in social insects: regulatory networks from genes to societies. J Exp Zoolog B Mol Dev Evol 318(3):159–169

Lonyo D III (1996) La Ruche KTB et Sa fabrication. Fiche Tech, Sucré-Villages Molga

Losfeld G, Saunier J-B, Grison C (2014) Minor and trace-elements in apiary products from a historical Mining District (les Malines, France). Food Chem 146:455–459

Loucif-Ayad W, Achou M, Legout H, Alburaki M, Garnery L (2015) Genetic assessment of Algerian honeybee populations by microsatellite markers. Apidologie 46:392–402

Mailu S, Wanyoike M, Serem J (2013) Rabbit breed characteristics, farmer objectives and preferences in Kenya: a correspondence analysis

Maissiat J, Baeher J-C, Picaud J-L (2011) Biologie Animale, DUNOD, ED

Malaka S, Fasasi K (2002) Beekeeping in Lagos and its environment. Occas Publ Entomol Soc Niger 34:92–97

Malik AA, Aremu A, Bayode GB, Ibrahim BA (2011) A nutritional and organoleptic assessment of the meat of the Giant African land snail (Archachatina Maginata Swaison) compared to the meat of other livestock. Livest Res Rural Dev 23(3):60

Mamo YS (2016) The honey industry in Comesa: opportunities and challenges. Bull Anim Health Prod Afr 64(1):207–216

MAPM (2017) Ministère de l'Agriculture et de La Pêche Maritime. Prem Journ Natl. L'héliciculture Rabat

Masehela TS (2017) An assessment of different beekeeping practices in South Africa based on their needs (bee forage use), services (pollination services) and threats (hive theft and vandalism)

Mayazine (2019) L'arbre et l'abeille Au Nord et Eu SudEd. Maya Beekeep Dev Liège, 5, n°3, 35 P

Mbetid-Bessane E (2006) Analyse de La Filière Des Escargots Comestibles Dans La Région de l'Equateur En République Centrafricaine. Tropicultura 24(2):115–119

Mbétid-Bessane E (2006) Analyse de La Filière Des Escargots Comestibles Dans La Région de l'Equateur En République Centrafricaine. Tropicultura 24(2):115–119

Meglitsch PA (1974) Zoologie Des Invertébrés, Tome 2, Des Vers Aux Arthropodes (Annélides, Mollusques, Chélicérates). ED Dion, Paris

Meixner MD, Leta MA, Koeniger N, Fuchs S (2011) The honey bees of Ethiopia represent a new subspecies of Apis Mellifera—Apis Mellifera Simensis n. Ssp. Apidologie 42:425–437

Meutchieye F, Kamogne F, Zango P, Youbissi A, Tchoumboue J (2016) Technical and socioeconomic assessment of honey Production in Cameroon Western highlands. Bull Anim Health Prod Afr 64(1):237–245

Mezouar M, Rayah A, Benkhlifa I, Ameur AA (2021) Caractérisation Morpho-Géométrique et Phénotypique Des Escargots Terrestres Dans l'ouest Algérien. Mémoire de licence génétique, Université de Tlemcen

Miguel I, Iriondo M, Garnery L, Sheppard WS, Estonba A (2007) Gene flow within the M evolutionary lineage of Apis Mellifera: role of the Pyrenees, isolation by distance and post-glacial re-colonization routes in the Western Europe. Apidologie 38(2):141–155

Miguel I, Baylac M, Iriondo M, Manzano C, Garnery L, Estonba A (2011) Both geometric morphometric and microsatellite data consistently support the differentiation of the Apis Mellifera M evolutionary branch. Apidologie 42:150–161

Miller I, Rogel-Gaillard C, Spina D, Fontanesi L, de Almeida AM (2014) The rabbit as an experimental and production animal: from genomics to proteomics. Curr Protein Pept Sci 15(2):134–145

Miller CA, Tasse Taboue GC, Ekane MM, Robak M, Sesink Clee PR, Richards-Zawacki C, Fokam EB, Fuashi NA, Anthony NM (2018) Distribution modeling and lineage diversity of the Chytrid fungus Batrachochytrium Dendrobatidis (Bd) in a Central African amphibian hotspot. PLoS One 13(6):e0199288

MINFOF (2020). Ministère Des Forêts et de La Faune, Arrêté No 0053/MINFOF Du 01 Avril 2020 Fixant Les Modalités de Répartition Des Espèces Animales En Classe de Protection

Mogga J, Ruttner F (1988) Apis Florea in Africa: source of the founder population. Bee World 69(3):100–103

Mohneke M, Onadeko AB, Rödel MO (2009) Exploitation of frogs–a review with a focus on West Africa. Salamandra 45(4):193–202

Morse RA, Calderon NW (2000) The value of honey bee pollination in the United States. Bee Cult 128(18):1–15

Mougel F, Mounolou J, Monnerot M (1997) Nine polymorphic microsatellite loci in the rabbit, Oryctolagus Cuniculus. Anim Genet 28(1):58–59

Moustafa (2001)

Mouton M (2011) Significance of direct and indirect pollination ecosystem services to the apple industry in the Western cape of South Africa

Muli E, Patch H, Frazier M, Frazier J, Torto B, Baumgarten T, Kilonzo J, Kimani JN, Mumoki F, Masiga D (2014) Evaluation of the distribution and impacts of parasites, pathogens, and pesticides on honey bee (Apis Mellifera) populations in East Africa. PLoS One 9(4):e94459

Murphy B (2001) Breeding and growing snails commercially in Australia. RIRDC

Nakamura HK (1986) Chromosomes of Archaeogastropoda (Mollusca: Prosobranchia), with some remarks on their Cytotaxonomy and phylogeny. Publ Seto Mar Biol Lab 31(3–6):191–267

NAMC Report (2008) 3rd Draft Report. Sect. 7 Comm. Investig. South Afr. Beekeep. Ind. Pretoria South Afr

Ndah NR, Lucha CF-B, Chia EL, Andrew EE, Yengo T, Anye DN (2017) Assessment of snail farming from selected villages in the Mount Cameroon range, south west region of Cameroon. Asian Res J Agric 6(4):1–11. https://doi.org/10.9734/arja/2017/35113

Nguiffo DN, Wabo JP, Mpoame M (2015) Gastrointestinal helminths of goliath frogs (Conraua Goliath) from the localities of Loum, Yabassi and Nkondjock in the Littoral region of Cameroon. Glob Ecol Conserv 4:146–149. https://doi.org/10.1016/j.gecco.2015.06.009

Nguiffo DN, Mpoame M, Wondji CS (2019a) Genetic diversity and population structure of goliath frogs (Conraua Goliath) from Cameroon. Mitochondrial DNA Part A 30(4):657–663

Nguiffo DN, Wondji CS, Wabo JP, Mpoame M (2019b) Microfilariae infestation of goliath frogs (Conraua Goliath) from Cameroon. PLoS One 14(5):e0217539

Nkwendem DG, Kana JR, Meutchieye F (2019) Genetic diversity of edible snails' population (Archachatina Marginata and Achatina Fulica) Expoited in the coastal region of Cameroon. Cameroon J Exp Biol 13(1):49–55

North MK, Dalle Zotte A, Hoffman LC (2019) The effects of dietary quercetin supplementation on the meat quality and volatile profile of rabbit meat during chilled storage. Meat Sci 158:107905

Nyameasem JK, Borketey-La EB (2014) Pest incidence, mortality, aestivation, feed intake and growth in west African Giant snails (Achatina Achatina) reared under different housing Systems

Nzouankeu AM (2015) Cameroun: Ces Escargots Qui Valent de l'or; Agency, A., Ed

Odum RA, Zippel KC (2008) Amphibian water quality: approaches to an essential environmental parameter. Int Zoo Yearb 42(1):40–52

Oduntan OO, Soaga JA, Jenyo-Oni A (2012) Comparison of edible frog (Rana Esculenta) and other bush meat types: proximate composition, social status and acceptability. E3 J Environ Res Manag 3(7):124–128

Oke OC, Alohan FI (2006) The land snail diversity in a square kilometre of tropical rainforest in Okomu Natioal Park, Edo state, Nigeria. Afr Sci 7(3):134–142

Okon B, Ibom AL, Ettah EH, Udoh HU (2012) Comparative differentiation of morphometric traits and body weight prediction of Giant African land snails with four whorls in Niger Delta region of Nigeria. J Agric Sci Tor 10(4):205–211

Onions G (1912) South African fertile-worker bees. Agric J Union South Afr 3(5):720

Owuor MA, Icely J, Newton A (2019) Community perceptions of the status and threats facing mangroves of Mida Creek, Kenya: implications for community based management. Ocean Coast Manag 175:172–179

Oyerinde AA, Dike MC, Bamaiyi LJ, Adamu RS (2012) Comparative studies of Behaviourial variations of Apis Mellifera L. species in Nigeria. Glob J Sci Front Res Agric Vet Sci 12(7):1–6

Paraïso A, Cornelissen B, Viniwanou N (2011) Varroa destructor infestation of honey bee (Apis Mellifera Adansonii) colonies in Benin. J Apic Res 50(4):321–322

Paterson PD (2006) L'apiculture. Ed Quæ CTA Press. Agron. Gembloux Paris Fr, 158 P

Patterson, M. C.; Burch, J. B. Chapitre 4. Chromosomes Des Mollusques Pulmonés. In Fetter V. & Peake J. (éd.) Pulmonates. Vol. 2A. Systématique, évolution et écologie.; Academic San Francisco, 1978; pp. 172–217

Peiró R, Merchan M, Santacreu MA, Argente MJ, García ML, Folch JM, Blasco A (2008) Identification of single-nucleotide polymorphism in the progesterone receptor gene and its association with reproductive traits in rabbits. Genetics 180(3):1699–1705

Peng J, Zhang G-W, Zhang W-X, Liu Y-F, Yang Y, Lai S-J (2013) Rapid genotyping of MSTN gene polymorphism using high-resolution melting for association study in rabbits. Asian Australas J Anim Sci 26(1):30

Perera SJ, Herbert DG, Procheş Ş, Ramdhani S (2021) Land snail biogeography and endemism in southeastern Africa: implications for the Maputaland-Pondoland-Albany biodiversity hotspot. PLoS One 16(3):e0248040

Peterson GD (1957) Studies on the control of the Giant African snail on Guam. Hilgardia 26:643–658

Pirk CW, Human H, Crewe RM, VanEngelsdorp D (2014) A survey of managed honey bee Colony losses in the Republic of South Africa–2009 to 2011. J Apic Res 53(1):35–42

Qayyoum MA, Khan BS, Bashir MH (2013) Efficacy of plant extracts against honey bee mite, Varroa destructor (Acari: Varroidae). World J Zool 8(2):212–216

Radi N (2003) L'arganier Arbre Di Sud-Ouest Marocain, En Péril, à Protéger. Thèse de docteur en pharmacie, Université de NANTES

Radloff S, Hepburn HR (1997) Multivariate analysis of honeybees, Apis Mellifera Linnaeus (hymenoptera: Apidae), of the horn of Africa. Afr Entomol 5(1):57–64

Radloff S, Hepburn R (2000) Population structure and morphometric variance of the Apis Mellifera Scutellata Group of Honeybees in Africa. Genet Mol Biol 23:305–316

Raffiudin R, Crozier RH (2007) Phylogenetic analysis of honey bee behavioral evolution. Mol Phylogenet Evol 43(2):543–552

Rahman MM, Richardson A, Sofian-Azirun M (2010) Antibacterial activity of Propolis and honey against staphylococcus aureus and Escherichia Coli. Afr J Microbiol Res 4(18):1872–1878

Rangoju P, Kumar S, Kolte A, Gulyani R, Singh V (2007) Assessment of genetic variability among rabbit breeds by random amplified polymorphic DNA (RAPD)-PCR. World Rabbit Sci 15(1):03–08

Rogel-Gaillard C, Ferrand N, Hayes H (2009) Rabbit. In: Genome mapping and genomics in domestic animals. Springer, pp 165–230

Russell LK, DeHaven JI, Botts RP (1981) Toxic effects of cadmium on the Gardensnail (Helix Aspersa). Bull Environ Contam Toxicol

Ruttner F (2013) Biogeography and taxonomy of honeybees. Springer Science & Business Media

Ruttner F, Volprecht M (1983) Experimental analysis of reproductive interspecies isolation of Apis Mellifera L. and Apis Cerana Fabr. Apidologie 14(4):309–327

Ruttner F, Tassencourt L, Louveaux J (1978) Biometrical-statistical analysis of the geographic variability of Apis Mellifera LI material and methods. Apidologie 9(4):363–381

Sabater-Pi J (1985) Contribution to the biology of the Giant frog (Conraua Goliath, Boulenger). Amphib-Reptil 6(2):143–153

Sabatini A (2005) L'abeille Bio-Indicateur: L'abeille, Sentinelle de l'environnement. Biodiversité 5(108):1216

Sahle H, Enbiyale G, Negash A, Neges T (2018) Assessment of honey production system, constraints and opportunities in Ethiopia. Pharm Pharmacol Int J 6(1):42–47

Saidj D, Aliouat S, Arabi F, Kirouani S, Merzem K, Merzoud S, Merzoud I, Ainbaziz H (2013) La Cuniculture Fermière En Algérie: Une Source de Viande Non Négligeable Pour Les Familles Rurales. Livest Res Rural Dev 25(8)

Schäfer M, Tsekané SJ, Tchassem FAM, Drakulić S, Kameni M, Gonwouo NL, Rödel M-O (2019) Goliath frogs build nests for spawning–the reason for their gigantism? J Nat Hist 53(21–22):1263–1276

Schiff N, Sheppard W (1993) Mitochondrial DNA evidence for the 19th century introduction of African honey bees into the United States. Experientia 49:530–532

Schiff NM, Sheppard WS (1995) Genetic analysis of commercial honey bees (hymenoptera: Apidae) from the southeastern United States. J Econ Entomol 88(5):1216–1220

Scholmke W, Schmolke M (2003) Varroa has arrived in Zimbabwe. Bees Dev J 68:13

Schreurs J (1963) Investigations on the biology, ecology and control of Giant African snail 290 in West New Guinea. Manokwari Agricultural Research Station

Schweiger O, Frenzel M, Durka W (2004) Spatial genetic structure in a metapopulation of the land snail Cepaea Nemoralis (Gastropoda: Helicidae). Mol Ecol 13(12):3645–3655

Shaibi T, Fuchs S, Moritz RF (2009) Morphological study of honeybees (Apis Mellifera) from Libya. Apidologie 40(2):97–105

Shenkute A, Getachew Y, Assefa D, Adgaba N, Ganga G, Abebe W (2012) Honey Production Systems (Apis Mellifera L.) in Kaffa, Sheka and Bench-Maji Zones of Ethiopia. J Agric Ext Rural Dev 4(19):528–541

Sheppard W, Rinderer T, Meixner M, Yoo H, Stelzer J, Schiff N, Kamel S, Krell A (1996) Hin Fl variation in mitochondrial DNA of old world honey bee subspecies. J Hered 87(1):35–40

Sinha S, Ling X, Whitfield CW, Zhai C, Robinson GE (2006) Genome scan for cis-regulatory DNA motifs associated with social behavior in honey bees. Proc Natl Acad Sci 103(44):16352–16357

Skendraoui F (2015) Inventaire Qualitatif et Quantitatif Des Gastéropodes Terrestres Au Niveau de Deux Stations de Tizi-Ouzou (Makouda et Drâa Ben Khedda). Thèse de Master en Agronomie, Université Mouloud Mammeri de Tizi-Ouzou

Skorbiłowicz E, Skorbiłowicz M, Cieśluk I (2018) Bees as bioindicators of environmental pollution with metals in an urban area. J Ecol Eng 19(3):229–234

Srilatha G, Chamundeeswari K, Ramamoorthy K, Sankar G, Varadharajan D (2013) Proximate, amino acid, fatty acid and mineral analysis of clam, Meretrix Casta (Chemnitz) from Cuddalore and Parangipettai coast, South East Coast of India. J Mar Biol Oceanogr 02(02)

Steenkamp L-A (2021) A classification framework for carbon tax revenue use. Clim Pol 21(7):897–911

Sternstein I, Reissmann M, Maj D, Bieniek J, Brockmann GA (2015) A comprehensive linkage map and QTL map for carcass traits in a cross between Giant Grey and New Zealand white rabbits. BMC Genet 16(1):1–12

Stiévenart C, Prince SD, Onwueme IC, Hardouin J, Delhove GE (1990) Les Escargots Géants Africains. Technical Centre for Agricultural and Rural Cooperation

Surridge AK, Bell DJ, Rico C, Hewitt GM (1997) Polymorphic microsatellite loci in the European rabbit (Oryctolagus Cuniculus) are also amplified in other lagomorph species. Anim Genet 28(4):302–305

Taylor J (1913) Monograph of the land and freshwater Mollusca of the British Isles; Brothers T, Ed.; Leeds

Tchoumboue J, Tchouamo I, Pinta J, Njia M (2001) Caractéristiques Socioéconomiques et Techniques de l'apiculture Dans Les Hautes Terres de l'Ouest Du Cameroun. Tropicultura 19(3):141–146

Tsayo TS, Meutchieye F, Etchu K, Nkwendem DMG, Dongmo DF, Ngoula F (2020) Phenotypic characteristics of native edible snails Achatina Fulica and Archachatina Marginata in equatorial region of Cameroon. Genet Biodivers J 5(1):147–158

van der Steen JJ (2016) Beehold: the colony of the honeybee (Apis Mellifera L.) as a bio-sampler for pollutants and plant pathogens

Van Haeringen WA, Den Bieman M, van Zutphen LF, Van Lith HA (1996) Polymorphic microsatellite DNA markers in the rabbit (Oryctolagus Cuniculus). J Exp Anim Sci 38(2):49–57

van Haeringen WA, Bieman MGD, Lankhorst AE, Van Lith HA, van Zutphen LFM (2002) Application of AFLP markers for QTL mapping in the rabbit. Genome 45(5):914–921

Veldtman R, Allsopp M (2011) The price of being important: the effective management of honeybees as a critical ecosystem service

Wilson EO (1971) The insect societies. Harvard University Press [Distributed by …], Cambridge, MA

Wilson M, Brinkman D, Spivak M, Gardner G, Cohen J (2015) Regional variation in composition and antimicrobial activity of US Propolis against Paenibacillus larvae and Ascosphaera Apis. J Invertebr Pathol 124:44–50

WOAH (2023). Diseases of Bees. https://www.woah.org/en/disease/diseases-of-bees/

Woogeng IN, Coetzer WG, Etchu KA, Ndamukong KJN, Grobler JP (2017) Current patterns of genetic diversity in indigenous and introduced species of land snails in Cameroon reflect isolation by distance, limited founder size and known evolutionary relationships. Mitochondrial DNA Part B 2(2):375–380

Wright KM, Whitaker BR (2001) Amphibian medicine and captive husbandry. Krieger Publishing Company

Yeates MNDP (1978) Bibliography of trop. APIC Part II:11

Yirga G, Teferi M (2010) Participatory technology and constraints assessment to improve the livelihood of beekeepers in Tigray region, northern Ethiopia. Momona Ethiop J Sci 2(1)

Zerrouki N, Berchiche M, Bolet G, Lebas F (2001) Caractérisation d'une Population Locale de Lapins En Algérie: Performances de Reproduction Des Femelles. In: 9èmes journées de la recherche cunicole (Paris, 28–29 novembre 2001), pp 163–166

Zerrouki N, Bolet G, Berchiche M, Lebas F (2004) Breeding performance of local Kabylian rabbits does in Algeria. In: Proceeding of 8th world rabbit congress, Puebla, Mexico, pp 7–10

Zerrouki N, Bolet G, Berchiche M, Lebas F (2005) Evaluation of breeding performance of a local Algerian rabbit population raised in the Tizi-Ouzou area (Kabylia). World Rabbit Sci 13:29–37

Zhang J (2011) Comparison studies of the structural stability of rabbit prion protein with Human and mouse prion proteins. J Theor Biol 269(1):88–95

Zhang S, Li H, Tao H, Li H, Cho S, Hua Y, Chen J, Chen S, Li Y (2013) Delayed early passive motion is harmless to shoulder rotator cuff healing in a rabbit model. Am J Sports Med 41(8):1885–1892

Zhou WC et al (2007) The intermediate host of Angiostrongylus Cantonensis molluscan. Chin J Zoon 23:401–408

Open Access This chapter is licensed under the terms of the Creative Commons Attribution 4.0 International License (http://creativecommons.org/licenses/by/4.0/), which permits use, sharing, adaptation, distribution and reproduction in any medium or format, as long as you give appropriate credit to the original author(s) and the source, provide a link to the Creative Commons license and indicate if changes were made.

The images or other third party material in this chapter are included in the chapter's Creative Commons license, unless indicated otherwise in a credit line to the material. If material is not included in the chapter's Creative Commons license and your intended use is not permitted by statutory regulation or exceeds the permitted use, you will need to obtain permission directly from the copyright holder.

Contributions of African Livestock Production Systems to Greenhouse Gas Emissions and Global Warming in the Face of Climate Change

14

Mizeck G. G. Chagunda, Kingsley A. Etchu, Kwamboka Tirimba, and Okeyo Mwai

Abstract

Greenhouse gases (GHG) have the ability to trap radiant energy from the sun in the atmosphere. The accumulation of GHGs in the atmosphere significantly contributes to climate change. Methane is 21 times more potent than CO_2 as a greenhouse gas (GHG), however, this is one area where there is a lack of robust data to reflect the amount of GHG different breeds of cattle in Africa produce per day. With an increasing global population placing pressure on land resources, coupled with environmental degradation of existing agricultural land, climate change mitigation and adaptation becomes even a more urgent issue to deal with. Livestock contributes to approximately 15% of global methane (CH_4) emissions and 65% of global nitrous oxide (N_2O) emissions, contributing to an estimated 8–11% of global anthropogenic GHG emissions. The production of GHG from ruminants and their impact on climate change are a major concern worldwide. Although the amounts of GHG from African livestock systems may not be as huge as from other production systems in the world, the inefficiencies in African livestock production systems are a call for concern and deserve quantification. Increasing the efficiency of ruminant production is not only vital in reducing the GHG intensity, but also important for food production to meet increasing demands. Strategies to reduce GHG emissions should include the effects of different feeding systems, the level of intensification and genetics/breeding for reduced GHG emissions. Increasing productivity per animal, that is, efficiency, will reduce GHG emissions per kg of livestock product that is needed to improve livestock production without compromising the environment.

M. G. G. Chagunda (✉) · K. Tirimba
Department of Animal Breeding and Husbandry in the Tropics and Subtropics, University of Hohenheim, Stuttgart, Germany

Centre for Tropical Livestock Genetics and Health (CTLGH), University of Edinburgh, Edinburgh, United Kingdom
e-mail: mizeck.chagunda@uni-hohenheim.de

K. A. Etchu
Division of Animal Production and Fisheries, IRAD Head Office, Yaoundé, Cameroon

O. Mwai
International Livestock Research Institute (ILRI), Nairobi, Kenya

Keywords

Ruminants · Greenhouse gas emissions · Mitigation · Climate change

14.1 Introduction

Greenhouse gases have the ability to trap radiant energy from the sun in the atmosphere. This alters the Earth's average near-surface air temperature. As observed by the Intergovernmental Panel on Climate Change (IPCC), warming of the climate system is unequivocal (Intergovernmental Panel on Climate Change (IPCC) 2007). Climate change is widely recognised as a serious potential threat to the world's environment. However, the greenhouse effect of the Earth's atmosphere is a natural phenomenon, without which the Earth's temperature would be much lower, whereby atmospheric concentrations of water vapour and carbon dioxide (CO_2) trap infrared radiation. The IPCC observes that most of the increase in global average temperatures since the mid-twentieth century is due to an increase in anthropogenic greenhouse gas (GHG) concentrations in the atmosphere (Intergovernmental Panel on Climate Change (IPCC) 2007). The main GHGs emitted globally are carbon dioxide (CO_2), methane (CH_4) and nitrous oxide (N_2O). Greenhouse gas emissions have been escalating since the Industrial Revolution (Intergovernmental Panel on Climate Change (IPCC) 2007). The atmospheric concentration of CO_2 has risen to its highest level in at least 800,000 years (Lüthi et al. 2008). If unchecked, future climate implications of this rise could include further increases in mean annual temperature and a reduction in mean precipitation (Lahn 2021). At the present rate, the IPCC projects a global mean temperature rise of 0.2 °C for each of the next two decades. In the longer term, projections range between a 2 and 4 °C rise by 2100 compared to the end of the twentieth century, depending on global GHG emissions scenarios (Intergovernmental Panel on Climate Change (IPCC) 2021).

The recent Living Planet Report stated that if GHG emissions continue at or above current rates, the natural resilience and natural adaptability of many ecosystems are likely to be exceeded (Lupinacci and Happel-Parkins 2016), with far-reaching consequences and an uncertain future for global biodiversity, agriculture, and livestock production. With an increasing global population placing pressure on land resources, coupled with environmental degradation of existing agricultural land, the number of people per hectare of arable land has been steadily increasing. The United Nations Environment Programme predicts that 25% of the world's food production may become lost due to environmental breakdown by 2050, largely as a result of climate change and agricultural practices, including overgrazing. Collectively, these issues point to an increasing demand on current grazing land and forage production areas, particularly where crop production for direct human consumption is viable, emphasising the need for improved efficiency and adaptation of livestock systems in future climates (Eckard et al. 2018). The Intergovernmental Panel on Climate Change (IPCC) reports that livestock contributes approximately 15% of global CH_4 emissions and 65% of global N_2O emissions, contributing an estimated 8–11% of global anthropogenic GHG emissions. The energy and nitrogen losses associated with these GHG also represent two of the most significant inefficiencies in livestock production systems. Losses of nitrogen and energy are greater in ruminant production systems (e.g. cattle, sheep, goats) than in livestock with a simple single-chambered stomach (e.g. pigs and poultry), with greater gains in breeding for improved feed and energy conversion efficiency having been made in the latter (Eckard et al. 2018).

14.2 Climate Change and Food Security

Climate change is among the biggest threats to food security and thus to human existence. The adverse effects of increased frequency of sporadic rainfall, severe floods and droughts seriously undermine the achievement of food security and sustainable livelihoods of communities in middle-income and low-income countries, where agriculture is the backbone of the economy. Agricultural systems are more vulnerable to climate change as they are dependent on nature. The impacts of climate change on food systems are more prevalent in the global south, where most agricultural activities are carried out by smallholder farmers. Consequently, their nutritional security is limited as they rely heavily on their own production to meet their food security and economic needs. The agricultural sector is currently faced with the challenge of feeding a growing population predicted to peak at 9.2 billion people by 2050 (Alexandratos et al. 2006), whilst meeting social and environmental obligations to reduce greenhouse gas (GHG) emissions. Whilst having an overview of the global climate change issues, this chapter focuses on the contributions of African livestock production systems to greenhouse gas emissions and hence global warming.

14.3 The Greenhouse Gases

Since the beginning of industrialisation around 1750, the atmospheric concentration of CO_2 has increased from around 280 parts-per-million (ppm) to around 379 ppm, with over half of that increase arising since 1970 (Parry 2007). Atmospheric CH_4 has increased from 715 to 1774 parts-per-billion (ppb) and N_2O from 270 to 319 ppb over the same period. These different GHGs vary in their capacity to reflect or trap energy in the atmosphere, referred to as their global warming potential (GWP). Traditionally, this has been a comparison of each GHG with a standardised 100-year period and expressed in units of kilogrammes of CO_2 equivalent (kg CO_2e). In this method, the GWP of CH_4 and N_2O are estimated to be 25 and 298 times greater, respectively, than that of CO_2 (Eggleston et al. 2006). However, this approach has been modified to account for two things. First, methane degrades in about ten years from the time when it is emitted into the atmosphere. Second, enteric CH_4 essentially is methane that is produced from carbohydrates from feed whose carbon got into the feed through the process called photosynthesis from the atmosphere when the plant was growing. This then indicates that the carbon in enteric CH_4 is recycled carbon in CH_4 continuing it's life as it gets involved in the biogenic cycle (Chen et al. 2019). Traditionally, biogenic methanogenesis is the formation of methane under strictly anoxic conditions by microbes from the domain Archaea (Chen et al. 2019). In the past decade, there has been growing evidence that CH_4 associated with animal agriculture is not new carbon in the atmosphere (Liu et al. 2021). Further, Liu et al (2021) argue that in the natural biogenic carbon cycle, plants assimilate CO_2 from the atmosphere during photosynthesis and store it as carbohydrates (e.g. cellulose or starch). Ruminant animals consume the plants and convert some of the carbon contained in plant carbohydrates into CH_4, which is then exhaled or belched out into the atmosphere. Therefore, biogenic carbon is "recycled carbon" and not new and additional to our atmosphere, though the warming effects during its atmospheric presence should still be recognised (Liu et al. 2021).

Apart from livestock, GHGs have other sources. For example, multiple processes emit CO_2: carbon dioxide enters the atmosphere through burning fossil fuels (coal, natural gas, and oil), solid waste, trees and other biological materials, and also as a result of certain chemical reactions (e.g. manufacture of cement). Methane is emitted during the production and transport of coal, natural gas, and oil. Methane emissions also result from livestock and other agricultural practices, land use and the decay of organic waste in municipal solid waste landfills. Nitrous oxide is emitted during agricultural land use, industrial activities, combustion of fossil fuels and solid waste, as well as during the treatment of wastewater. Nitrous oxide emissions account for about 10% of global GHG emissions, with about 90% derived from agricultural practices (Henry and Eckard 2009). Other powerful GHGs such as hydrofluorocarbons, perfluorocarbons, sulphur hexafluoride and nitrogen trifluoride are emitted from a variety of industrial processes. Fluorinated gases are sometimes used as substitutes for stratospheric ozone-depleting substances such as chlorofluorocarbons, hydrochlorofluorocarbons, and halons. These gases are typically emitted in smaller quantities but are said to be potent GHGs. They are sometimes referred to as high global warming potential (GWP) gases (Lee 2007).

Agricultural activities contribute substantially to anthropogenic GHG emissions (Weiske et al. 2006). About 49% of CH_4 and 63% of N_2O emissions can be attributed to agricultural production (Weiske et al. 2006). The nutrition of animals has environmental consequences. Nutrition has important effects on the environment as a consequence of the processes of digestion (CH_4 and phosphorus) or a combination of digestion and metabolism N. Methane is produced by fermentation in the gut and the degradation of manure. Ruminants contribute substantially to emissions of GHG. Large emissions occur from enteric fermentation in ruminants and also from anaerobic storage of manure (Hensen et al. 2006). Nitrous oxide arises from the degradation of animal wastes and from all parts of the agricultural nitrogen cycle, and hence, there may be diverse sources of N_2O emissions from livestock farms (Hensen et al. 2006). Methane and nitrates are important GHGs from livestock production. Quantification of GHGs in livestock production is a subject of great interest. Together, enteric fermentation and manure represent some 80% of agricultural CH_4 emissions and about 30–40% of the total anthropogenic methane emissions (FAO 2006). The following are the major GHGs from livestock.

14.3.1 Carbon Dioxide

The major source of CO_2 is animal respiration, followed by less significant emissions from manure storage and the barn floor. Carbon dioxide is sequestered when it is absorbed by plants as part of the biological carbon cycle and emits CO_2 through plant and soil respiration. It should be noted that the plants capture more CO_2 through photosynthesis than is emitted through respiration. Furthermore, the simple sugars produced in the first stage of carbohydrate digestion in the rumen are rarely detectable in the rumen liquor because they are immediately taken up and metabolised intracellularly by the microorganisms. Consequently, carbohydrate is converted to pyruvate that is linked with the major end products of rumen carbohydrate digestion, which are acetic, propionic and butyric acids, CO_2 and CH_4 (Henry and Eckard 2009; McDonald et al. 2011).

14.3.2 Methane

Methane is a by-product of the anaerobic digestive process (enteric fermentation). In the first stage of digestion, the forage is acted on by the varied population of microorganisms, including bacteria, fungi, and protozoa, in the fore stomach. This process releases hydrogen while producing volatile fatty acids and microbial cells containing energy and essential proteins to be made available for the growth of the animal. In ruminants, the hydrogen is removed through the action of a group of microbes called methanogenic archaea (methanogens) that gain their energy through combining CO_2 with H_2 to form CH_4. Hence, CH_4 emissions provide a mechanism for preventing hydrogen from building up in the rumen with resultant adverse effects on animal productivity, and therefore strategies to reduce CH_4 emissions must also provide for an alternative pathway to remove hydrogen (Henry and Eckard 2009; McDonald et al. 2011; Rotz et al. 2016).

So far, it has been estimated that 70% of the CH_4 in the Earth's atmosphere is a result of human activity, and agriculture accounts for 60% of this. Enteric fermentation makes up around 80% of the agricultural production of CH_4, and around 20% arises from animal wastes. Rumen and hindgut fermentation are similar processes. During rumen fermentation of organic matter, hydrogen is produced, and methanogenic bacteria use this with CO_2 to produce CH_4 and water. The CH_4 leaves the rumen by eructation, and this represents a loss of 6–10% of the gross energy of fermented foods. However, the methanogenic bacteria use hydrogen as an energy source for growth and thereby make a small contribution to the microbial matter to be digested in the small intestine. The removal of hydrogen has a beneficial effect on the fermentation of plant cell wall carbohydrates. Hydrogen is also used by other bacteria to synthesise propionic and butyric acids, and the pathway of use depends on the diet and the pH of the rumen contents (McDonald et al. 2011). Within the agriculture sector, the production of CH_4 is the primary concern. Methane emissions account for 40–45% of GHG emissions from ruminant livestock, with over 90% of these emissions arising from enteric fermentation. Consequently, enteric CH_4 emissions are by far the single most important emission source that can be targeted for mitigation within the ruminant production cycle (Meale et al. 2012a).

14.3.3 Nitrous Oxide

Nitrous oxide is produced from nitrogen fertilisers, urine deposited by livestock on soils and from manure and effluent during storage and treatment. Of the dietary nitrogen consumed by ruminants, less than 30% is utilised for production, with more than 60% being lost from the grazing system (Henry and Eckard 2009; McDonald et al. 2011). Nitrates coming from livestock represent the feed nitrogen component that is not utilised by livestock and has the polluting potential of underground water systems (Jonker et al. 2002). It is important, therefore, to account for the effects of genotype and feed (production system) when one considers the polluting effect of livestock (Ross et al. 2014).

The global challenge associated with lowering GHG emissions from livestock production is that most of the livestock are in developing countries (72% of all cattle and 67% of all sheep and goats) and remote rangeland systems of the world, where mitigation practices are unlikely to be a high priority or where regular intervention is not practical (Eckard et al. 2018). Although livestock production has a significant role to play in global GHG emissions, numerous technical options have been identified to mitigate these emissions (Gerber et al. 2013).

14.4 The Contribution of African Livestock Production Systems to GHG Emissions

In Africa, livestock are kept for multiple purposes such as milk, meat, hides, horns, fertiliser and income generation, and hence are important both socially and economically in households. Although in Africa total emissions from livestock are still lower than in the member states of the Organisation for Economic Cooperation and Development (OECD), the emissions intensities per unit of animal product produced are very high, which is a cause for concern given the rapid growth projected for the sector (Herrero et al. 2013). Inefficient production systems lead to high GHG emissions intensity, measured as the amount of GHG per unit of product (meat, milk, calories, protein). For example, the intensity of CH_4 emissions (g/kg milk) is much larger in Africa compared to the rest of the world (Beauchemin et al. 2009). The mean milk production in Africa ranged from 108 to 3368 kg/cow per year (FAOSTAT 2018), with about half of the countries producing below 500 kg of milk/cow per year (FAOSTAT 2018). Consequently, African countries contribute to about 10% of enteric CH_4 emissions from dairy cattle worldwide despite producing only 3.9% of the world's milk (FAOSTAT 2018).

Husbandry and management practices play a big role in African livestock GHG emissions. Recent reports from the North West Region of Cameroon stipulated a 98% CH_4 contribution to total GHG emissions from cows kept in groups overnight (Ngwabie et al. 2019). Mali witnessed a doubling (increased by 102%) of GHG emissions from the agricultural sector (58%) and manure left on pasture (38%) from 1990 to 2014 (Galford et al. 2020). In the Northwest region of Cameroon, cattle herding poses several challenges and has marked effects on the biodiversity of the Savannah system. For instance, overgrazing is progressively accounting for the replacement of pasture by weeds to the detriment of feed for cattle, and the amount of GHG emitted by cattle wastes keeps growing with a corresponding increase in cattle population (Gideon et al. 2017). Assessing GHG emissions from outdoor cattle sleeping areas in Cameroon indicated that mitigation strategies should be geared more towards CH_4 in the sleeping areas during the wet season (Ngwabie et al. 2019). The GHG emission factors estimated in this study were the first of their kind in Cameroon and can be used as a basis for planning management practices that mitigate emissions.

Apart from cattle, one of the contributing sectors to African livestock GHG emissions is the small ruminant sector. For example, small ruminants are a major source of CH_4 emissions in the South African agricultural sector. The sheep industry contributed an estimated 167 Gg of CH_4 in 2010, and the goat industry 40.7 Gg, with a combined 15.6% of South Africa's total livestock CH_4 emissions in 2010. The commercial sheep industry contributed an estimated 91% of sheep emissions, whereas 56% of goat CH_4 emissions originated from the emerging/communal sector (Du Toit et al. 2013).

14.5 Current Situation in GHG Emission Work in Africa

The GHG of focus in African livestock production systems has been CH_4. While there has been research progress on enteric CH_4 emissions from SSA in recent years, there are still relatively few studies and several major data gaps for important categories (Graham et al. 2022). A further gap in GHG work is the generally insufficient finance for climate-related goals, lack of research capacity and poor research infrastructure (López-Ballesteros et al. 2018). In order to fill in this gap, the International Livestock Research Institute (ILRI) at its Nairobi campus are undertaking various studies to quantify CH_4 from different livestock production systems. This work is being undertaken at the Mazingira centre using metabolic chambers, SF6 technique and other gas analysis techniques Fig. 14.1).

14.6 Why Should Lack of Quantification of GHG Emissions Be a Cause for Concern in African Livestock Production Systems?

The production of GHG from ruminants and their impact on climate change are a major concern worldwide. Although the amounts of GHG from African livestock systems may not be as huge as from other production systems in the world, the inefficiencies in African livestock production systems are a cause for concern and deserve quantification. Quantification is a good starting point for GHG mitigation. Methane is one of the three main GHGs, together with CO_2 and N_2O. The contribution of these three gases to the different activities involved in livestock farming has been estimated using the life cycle assessment method. It has been reported that enteric CH_4 is the most important GHG emitted (50–60%), at the farm scale, in ruminant production systems. Decreasing enteric CH_4 emissions from ruminants without altering animal production is desirable both as a strategy to reduce global GHG emissions and as a means of improving feed conversion efficiency. However, before determining any mitigation strategies, it is important to quantify the GHG of concern. Different methods are used to measure and estimate GHGs from livestock. Examples of such methods are respiration calorimetry chambers, isotopic techniques, tracer techniques (sulphur hexafluoride (SF_6)) and mass balance/micrometeorological techniques (Russell et al. 2007). In the respirometer/calorimeter method, air is typically passed through a chamber containing one or more animals. The volume of gas passed through and the inlet and outlet concentrations of oxygen, carbon dioxide and methane are usually measured (Murray et al. 1999; Kelly et al. 1994). The tracer technique involves placing a source of SF6 in the rumen with a known release rate and measuring the amount of release of SF6 released from the source over a period. The concentration of methane is calculated from the

Fig. 14.1 Various methane emission measuring techniques at Mazingira centres of the International Livestock Research Institute (ILRI). Cattle metabolic chamber (**a**), SF6 gas cylinder (**b**), sheep metabolic chamber (**c**) and gas analysis kit (**d**). (Photos by Mizeck Chagunda)

ratio of CH_4 to SF6 in the expired breath of the test animals, factored by the relative molecular weights of SF6 and CH_4, and the known SF6 release rate (Johnson et al. 1994; McGinn et al. 2006). Mass balance and micro-meteorological techniques range from the use of aircraft to sample CH_4 concentrations at different levels in the atmosphere to paddock scale applications (Lassey et al. 2000; Chagunda et al. 2009). Other methods that do not just concentrate on CH_4 are methods such as the use of Flux towers. Flux towers measure GHG emissions (CH_4 and CO_2 fluxes) over landscapes. As such, this method may be appropriate for grazing animals and rangeland management. With a rapid increase in livestock numbers across some landscapes resulting in an increase in GHG emissions, changes in micro-climate, and carbon sink, such methods would help in mapping hot-spots over landscapes. These methods also use the simulation of production and utilisation of rangeland modelling to pinpoint and quantify GHG gas emissions, nutrients, and carbon fluxes under varying rangeland management scenarios involving different numbers of livestock. Other methods include spot measurements such as the use of the proprietary, laser methane detector (LMD), which measures the CH_4 concentration from the expired breath without any disturbance to the normal activity of the animals (McGinn et al. 2006; Crowcon Detection Instruments 2006). The LMD is a hand-held gas detector for remote measurements of column density for methane-containing gases is manufactured by Tokyo Gas Engineering (Fig. 14.2). The LMD has been widely used in detection applications such as gas transmission networks, landfill sites and other areas where methane leakage or build-up is a risk (Yates et al. 2000). Since the first publication on the feasibility of the use of the LMD in livestock (Chagunda, et al. 2009), different research groups and livestock practitioners have adopted the use of the LMD. In the African context, this presents a practical, easy-to-use, and yet cost-effective proxy for methane quantification in both extensive grazing systems and remotely located and yet intensive smallholder farms.

Fig. 14.2 Laser methane detector. On the left, a scientist taking methane measurements and on the right, two different models of the laser methane detector. (Photo by A. Ross)

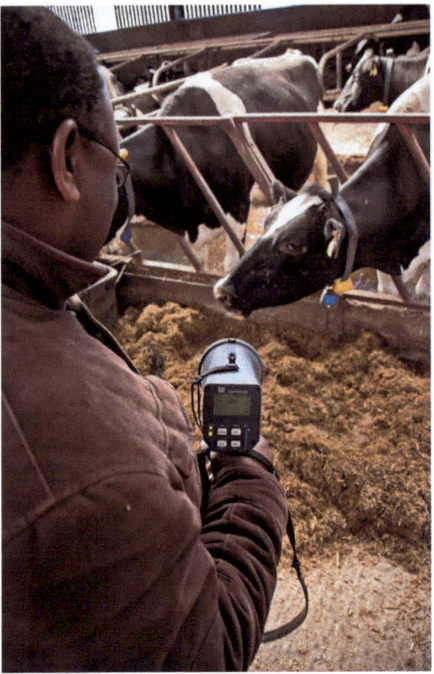

Enteric methane production per cow per day can also be predicted using equations that use total dry matter (DM) intake, DM intake from concentrate component of the feed (CDMI), and the neutral detergent fibre (NDF) in the feed (Martin et al. 2010), that is, CH_4 (MJ/day) = 1.36 + 1.21 × DMI − 0.825 × CDMI + 12.8 × NDF. Other methods include assessing the rumen microbiome diversity to predict the GHG emission of the host animal. Rumen microbe diversity can be modelled utilising information from the host genotype, microbial flora, forage, and the environment to predict GHG emissions. The GHG estimating equations can also be used in a larger system of equations, such as the IPCC methodologies that apply life cycle analysis (LCA). This scientific method has been used in previous empirical studies to calculate the carbon footprints of products and processes. The transferability of the LCA across various industries and regions, as well as its flexible methodological structure, makes this method particularly suited to the many variables and complexities such as mixed farming systems, various land and herd sizes and different feeding systems.

14.7 Mitigation Strategies for GHG in the African Context

Methane mitigation in ruminants is possible through various strategies. According to (Wood and Knipmeyer 1998), the feeding management approach is the most developed; other strategies (biotechnologies, additives) are promising, but the diversity and plasticity of functions of the rumen bacterial and methanogenic communities may be a limiting factor for their successful application (Wood and Knipmeyer 1998). One most promising and cost-effective way to reduce GHG emissions from livestock is to improve the efficiency of the production system (Manson and Leaver 1989). In dairy cattle, for example, an efficient cow will have lower GHG emissions relative to the amount of milk and meat it produces. Although highly productive livestock usually emit more GHG per head in absolute terms than less productive ones, they also produce more products (Manson and Leaver 1989). As a result, the ratio of GHG emissions per product produced is lower than in less productive livestock. Management efficiencies such as reducing the number of unproductive animals and increasing reproductive efficiency to enable fewer breeders to provide the same number of offspring per year will reduce methane emissions as well as increase profitability.

Increasing the efficiency of ruminant production is not only vital in reducing the GHG intensity, but also important for food production to meet increasing demands. Despite the social and economic importance of ruminant products, their contribution to GHG emissions is often criticised. Intensification of production systems may not necessarily lead to increased efficiencies. It is important to identify the changes and the trajectory of change in the production systems that lead to increased efficiency. Sustainable intensification aims to achieve this by identifying specific problems and challenges that come with intensification. In this process, it is important to have a detailed characterisation of the production systems and the biophysical conditions in which the animals are raised. In this characterisation, direct cattle-based measurements to assess animal health and welfare parameters are used. This is important because there is a link between health and welfare, productivity and GHG emissions. Examples of such measurements are: scoring systems developed to assess lameness (Cook 2007), claw length and cleanliness (Wildman et al. 1982), body condition (Rutherford et al. 2008), hock and knee lesions (Gibbons et al. 2009), behaviour (Gibbons et al. 2010), as well as other direct health and welfare changes (Sørensen et al. 2001) in dairy cattle. In addition, animal-based systems of classification have the advantages of describing relevant and significant aspects of what matters from the point of view of the animal, express change over time, are capable of being influenced by decisions taken by the individual farmer and can be measured in a relatively cheap and easy manner (Chagunda et al. 2010). An effective disease-management plan will improve productivity of the herd and result in a reduction in CH_4 emissions per unit of product (Henry and Eckard 2009).

Good feed quality and maintaining animal health will improve the fertility of the herd and increase the weaning rate, with flow-on effects to lower total CH_4 emissions from the herd (Henry and Eckard 2009). Feed is a key limiting factor and often the most expensive input in livestock production. However, poorly utilised feed is also a source of pollution, including GHGs. Generally, in dairy cattle as an example, high forage systems have higher CH_4 production per kg of milk than low forage systems. However, low forage systems have higher N loss than high forage systems. The N loss is mainly a reflection of unutilised protein that is lost through urine and faeces (Smith et al. 2013). Hence, animals with increased productivity in terms of milk yield and body energy reserves are associated with a decrease in CH_4 emissions per kg of milk but an increase in N loss. This highlights the important dynamics of emission burden and pollution potential of different livestock feeding systems. Obviously, feeding should not be considered in isolation. Integrating feed and forage, and pest management with improved cattle health and genetics can lead to significant enhancements in cattle production, up to 240% (Herrero et al. 2013). In mixed crop–cattle systems, which often have the potential to intensify, the most important contributors to feed resources are forages, crop residues and rangelands (Herrero et al. 2013), while in pastoral and agro-pastoral systems, grazing of rangelands is the principal, often the only, source of feed. Feed shortages, in terms of both quantity and nutritional quality, either seasonally or, in the case of dryland pastoral systems, interannually, are widespread, slowing the sector's growth and periodically causing severe losses. Improved feed and forage management practices, such as better management of rangeland, increased cut-and-carry of pasture resources, improved feed utilisation of crop residues and other agricultural by-products, have considerable underexploited potential to improve cattle productivity, while also contributing to the resilience of agro-ecosystems and environmental sustainability (Rao et al. 2015; McAllister and Newbold 2008). The digestibility and quality of feed are major determinants of energy available for animal growth and, therefore, of the performance of ruminants and of CH_4 production. The efficiency of nutrient utilisation by microbial organisms in the rumen controls the fermentation process, which in turn affects the activity of methanogens relative to other microbial species (Henry and Eckard 2009). Improvements in livestock feeding could be achieved by the following:

- Feeding forages with lower fibre and higher soluble carbohydrates, changing from C4 tropical grasses to mostly temperate C3 species, or grazing less mature pastures (Beauchemin et al. 2009).
- Addition of grain to a forage diet increases starch and reduces fibre intake, reduces rumen pH and promotes the production of propionate in the rumen (Wattiaux 2023). Propionate production tends to reduce methanogenesis in the rumen. Methane emissions are also commonly lower with higher proportions of forage legumes in the diet, partly due to lower fibre content, faster rate of passage and, in some cases, the presence of condensed tannins (Beauchemin et al. 2009). Plant breeding, therefore, offers some potential to improve the efficiency of digestion while reducing CH_4 production.
- Aguirre-Villegas and Larson (2017) showed that the adoption of best nutritional, cropping, and manure management practices may substantially reduce on-farm GHG emissions in all dairy systems. Modelling and assessment methods must be improved to better account for interactions among system components and co-products of milk production.

Apart from the feed management strategy, there are other strategies, such as waste management strategies. An example is highlighted by (Galford et al. 2020), who demonstrated that a GHG ($CH_4 + N_2O$) emission rate of 241 mg CO_2e m^{-2} min^{-1} indicates that CH_4 contributes 83% of the total, even though it is stored in a pile. Stringent waste management strategies, such as anaerobic digestion for biogas production or the use of covers over waste heaps, are needed in abattoirs to mitigate GHG emissions. In addition, long-term measurements are needed to cover seasonal variations in outdoor waste storage in different abattoirs.

Mitigation strategies such as genetic selection, use of chemicals and vaccines, and the capture of CH_4 have been proposed, yet dietary manipulation is considered the most promising strategy for the abatement of CH_4 from ruminant production systems (Meale et al. 2012b). According to Meale et al. (2012b), there have been several reviews published on different CH_4 abatement strategies for ruminant livestock (Boadi et al. 2004; Balehegn et al. 2021; Wood and Knipmeyer 1998). Selection for genetic lines of sheep and cattle that have lower CH_4 emissions (both in absolute terms and as a function of productivity) has the potential to be an effective long-term and economically sound approach to reducing CH_4 emissions from livestock. Genetic approaches are suited to extensive grazing systems, where management and husbandry interventions will continue to be impractical, at least in the near future (Fig. 14.3). In more intensive livestock systems, genetic improvements can be combined with management approaches. One of the innovations being implemented includes climate-smart farming practices (CSFPs). A study in Malawi show that the practices, when adopted by farmers, reduced the adverse impacts of climate change (drought and flooding) experienced by farming communities (Jones et al. 2014).

It should be noted that the mitigation strategies provide opportunities to reduce GHG emissions from ruminants, while maintaining (and potentially increasing) productivity. The degree of abatement achievable is mostly not large hence not suitable to deliver the level of reduction desirable to mitigate the threat of dangerous climate change. Significant levels of abatement will require additional innovation from intensive research and development over some years, and this will be achieved only with dedicated resources. The challenge would be demanding for extensive production systems, where few options currently exist for practical intervention considering African livestock production systems. In all this, however, there has to be emphasis on the importance and need to link national and regional policies to global strategies to reduce emissions of greenhouse gases and enhance carbon sequestration potential (Mgalula et al. 2021).

Fig. 14.3 Examples of cattle under extensive production systems. Top, Kenya and bottom, Zambia. (Photos by Mizeck Chagunda)

14.8 Conclusion

Climate change is among the biggest threats to food security and thus to human existence. The adverse effects of increased frequency of sporadic rainfall, severe floods and droughts seriously undermine the achievement of food security and sustainable livelihoods of communities in middle-income and low-income countries, where agriculture is the backbone of the economy. Agricultural systems are very vulnerable to climate change as they are dependent on nature. The impacts of climate change on food systems are more prevalent in the global south, where most agricultural activities are carried out by smallholder farmers. Consequently, their nutritional security is limited as they rely heavily on their own production to meet their food security and economic needs. While African livestock are responsible for the livelihoods of millions of farmers across the continent, they are also responsible for contributing to GHG emissions, which cause climate change. The majority of the GHG is CH_4. Due to a lack of data, the estimates of GHGs in SSA are associated with a huge level of uncertainty. This provides a huge opportunity to act.

14.9 Future Perspectives

Filling data gaps on GHG emissions will require additional studies using both direct and indirect methods. For CH_4 emissions, respiration chambers are considered the "gold standard" for direct measurements. These, however, are expensive to run and need highly trained technicians. The use of proxies that can be used in the animals' natural environment can assist in not only collecting the much-needed data but also account for animal x environmental interactions. This provides a huge opportunity to act.

References

Aguirre-Villegas HA, Larson RA (2017) Evaluating greenhouse gas emissions from dairy manure management practices using survey data and lifecycle tools. Journal of cleaner production 143:169–179.

Alexandratos N, Gerold Bödeker JB, Schmidhuber J, Broca S, Shetty P, Maria Grazia Ottaviani M, FAO (2006) World agriculture: Towards 2030/2050, Interim Report. Food and Agriculture Organization of the United Nations, Rome

Balehegn M, Kebreab E, Tolera A, Hunt S, Erickson P, Crane TA, Adesogan AT (2021) Livestock sustainability research in Africa with a focus on the environment. Anim Front 11(4):47–56

Beauchemin KA, McAllister TA, McGinn SM (2009) Dietary mitigation of enteric methane from cattle. CAB Rev Perspect Agric Vet Sci Nutr Nat Resour 4:1–18

Boadi D, Benchaar C, Chiquette J, Massé D (2004) Mitigation strategies to reduce enteric methane emissions from dairy cows: update review. Can J Anim Sci 84:319–335

Chagunda MGG, Ross D, Roberts DJ (2009) On the use of a laser methane detector in dairy cows. Comput Electron Agric 68(2):157–160

Chagunda MG, Flockhart JF, Roberts DJ (2010) The effect of forage quality on predicted enteric methane production from dairy cows. Int J Agric Sustain 8(4):250–256

Chen L, Zuo L, Jiang Z, Jiang S, Liu K, Tan J, Zhang L (2019) Mechanisms of shale gas adsorption: evidence from thermodynamics and kinetics study of methane adsorption on shale. Chem Eng J 361:559–570

Cook NB (2007) A toolbox for assessing cow, udder, and teat hygiene. In: Proceedings of the 46th annual meetings of the National Mastitis Council, San Antonio, TX (National Mastitis Council, Madison, Wisconsin, USA), pp 31–43

Crowcon Detection Instruments (2006) New-hand-held-device-detects-methane-up-to-150-metres-away. http://www.crowcon.com

Du Toit CJL, Van Niekerk WA, Meissner HH (2013) Direct greenhouse gas emissions of the South African small stock sectors. South Afr J Anim Sci 43(3):340–361

Eckard RJ, Henry BK, Beauchemin KA (2018) Review: adaptation of ruminant livestock production systems to climate changes | animal | Cambridge Core [WWW Document]. URL https://www.cambridge.org/core/journals/animal/article/review-adaptation-of-ruminant-livestock-production-systems-to-climate-changes/CF8B075D4B8D54DDFE0E60A3CBE4A53C. Accessed 5.25.21

Eggleston, H.S., Buendia, L., Miwa, K., Ngara, T. and Tanabe, K., 2006. 2006 IPCC guidelines for national greenhouse gas inventories

FAO (2006) Livestock's long shadow: environmental issues and options. Food and Agriculture Organization, Rome

FAOSTAT (2018) www.fao.org

Galford GL, Peña O, Sullivan AK, Nash J, Gurwick N, Pirolli G, Richards M, White J, Wollenberg E (2020) Agricultural development addresses food loss and waste while reducing greenhouse gas emissions. Sci Total Environ 699:134318

Gerber PJ, Benjamin H, Makkar HP (2013) Mitigation of greenhouse gas emission in livestock production. Food and Agriculture Organization of the United Nations

Gibbons J, Lawrence AB, Haskell MJ (2009) Responsiveness of dairy cows to human approach and novel objects. Appl Anim Behav Sci 116:163–173

Gibbons JM, Kawonga B, Gondwe TN, Chagunda MGG, Roberts DJ (2010) Measuring welfare of dairy cattle in Malawi—challenges, constraints and opportunities. In: The 5th all Africa conference on animal agriculture, Addis Ababa, Ethiopia. All Africa Society of Animal Agriculture

Gideon T, Biswas W, Pritchard D (2017) Sustainability assessment of cattle herding in the north west region of Cameroon, Central Africa. J Dev Agric Econ 9:289–302

Graham MW, Butterbach-Bahl K, du Doit CL, Korir D, Leitner S, Merbold L, Mwape A, Ndung'u PW, Pelster DE, Rufino MC, van der Weerden T (2022) Research progress on greenhouse gas emissions from livestock in sub-Saharan Africa falls short of national inventory ambitions. Front Soil Sci 2:927452

Henry B, Eckard R (2009) Greenhouse gas emissions in livestock production systems. TG Trop Grassl 43:232

Hensen A, Groot TT, van der Bulk WCM, Vermeulen AT, Olesen JE, Schelde K (2006) Dairy farm CH4 and N2O emissions, from one square metre to the full farm scale. Agric Ecosyst Environ 112:146–152

Herrero M, Havlík P, Valin H, Notenbaert A, Rufino MC, Thornton PK, Blümmel M, Weiss F, Grace D, Obersteiner M (2013) Biomass use, production, feed efficiencies, and greenhouse gas emissions from global livestock systems. Proc Natl Acad Sci 110(52):20888–20893

Intergovernmental Panel on Climate Change (IPCC) (2007) Climate change 2007: the physical science basis. Contribution of Working Group I to the Fourth Assessment Report of the Intergovernmental Panel on Climate Change. http://ipcc-wg1.ucar.edu/wg1/docs/WG1AR4_SPM_Approved_05Feb.pd

Intergovernmental Panel on Climate Change (IPCC) (2021) Summary for policymakers. In: Climate change 2021: the physical science basis. contribution of working group I to the sixth assessment report of the intergovernmental panel on climate change

Johnson K, Huyler M, Westberg H, Lamb B, Zimmerman P (1994) Measurement of methane emissions from ruminant livestock using an SF6 tracer technique. Environ Sci Technol 28:359–362

Jones AD, Shrinivas A, Bezner-Kerr R (2014) Farm production diversity is associated with greater household dietary diversity in Malawi: findings from nationally representative data. Food Policy 46. https://doi.org/10.1016/j.foodpol.2014.02.001

Jonker JS, Kohn RA, High J (2002) Dairy herd management practices that impact nitrogen utilization efficiency. J Dairy Sci 85:1218–1226

Kelly JM, Kerrigan LP, Milligan LP, McBride BW (1994) Development of a mobile, open circuit indirect calorimetry system. Can J Anim Sci 74:65–72

Lahn B (2021) Changing climate change: the carbon budget and the modifying-work of the IPCC. Soc Stud Sci 51(1):3–27

Lassey KR, Gimson NR, Wratt DS, Brailsford GW, Bromley AM (2000) Verifying agricultural emissions of methane. In: Van Ham J, Baede APM, Meyer LA, Ybema R (eds) Non-CO$_2$ greenhouse gases: scientific understanding, control and implementation. Kluwer Academic Publishers, Dordrecht, pp 107–114

Lee H (2007) Intergovernmental panel on climate change

Liu S, Proudman J, Mitloehner FM (2021) Rethinking methane from animal agriculture. CABI Agric Biosci 2(1):1–13

López-Ballesteros A, Beck J, Bombelli A, Grieco E, Lorencová EK, Merbold L et al (2018) Towards a feasible and representative pan-African research infrastructure network for GHG observations. Environ Res Lett 13(8):085003. https://doi.org/10.1088/1748-9326/aad66c

Lupinacci J, Happel-Parkins A (2016) In the living planet report 2014 by the WWF (formally known as the 4 World Wildlife Fund), researchers introduce a new index that considers "10,380". In: The educational significance of human and non-human animal interactions: blurring the species line. Springer, p 13

Lüthi D, Le Floch M, Bereiter B, Blunier T, Barnola JM, Siegenthaler U, Raynaud D, Jouzel J, Fischer H, Kawamura K, Stocker TF (2008) High-resolution carbon dioxide concentration record 650,000–800,000 years before present. Nature 453(7193):379–382

Manson FJ, Leaver JD (1989) The effect of concentrate: silage ratio and of hoof trimming on lameness in dairy cattle. Anim Sci 49(1):15–22

Martin C, Morgavi D, Doreau M (2010) Methane mitigation in ruminants: from microbe to the farm scale. Animal 4:351–365

McAllister TA, Newbold CJ (2008) Redirecting rumen fermentation to reduce methanogenesis. Aust J Exp Agric 48:7–13

McDonald P, Edwards RA, Greenhalgh JFD, Morgan CA, Sinclair LA, Wilkinson RG (2011) Animal nutrition, 7th edn. Essex, Pearson Education Limited

McGinn SM, Beauchemin KA, Iwaasa AD, McAllister TA (2006) Assessment of the sulfur hexafluoride (SF6) tracer technique for measuring enteric methane emissions from cattle. J Environ Qual 35:1686–1691

Meale S, Mcallister T, Beauchemin K, Harstad O, Chaves A (2012a) Strategies to reduce greenhouse gases from ruminant livestock. Acta Agric Scand 62. https://doi.org/10.1080/09064702.2013.770916

Meale SJ, Chaves AV, Baah J, McAllister TA (2012b) Methane production of different forages in in vitro ruminal fermentation. Asian Australas J Anim Sci 25(1):86

Mgalula ME, Wasonga OV, Hülsebusch C et al (2021) Greenhouse gas emissions and carbon sink potential in Eastern Africa rangeland ecosystems: a review. Pastoralism 11:19. https://doi.org/10.1186/s13570-021-00201-9

Murray PJ, Moss A, Lockyer DR, Jarvis SC (1999) A comparison of systems for measuring methane emissions from sheep. J Agric Sci 133(4):439–444

Ngwabie NM, Wirlen YL, Yinda GS, VanderZaag AC (2019) Quantifying greenhouse gas emissions from municipal solid waste dumpsites in Cameroon. Waste Manag 87:947–953

Parry ML (ed) (2007) Climate change 2007-impacts, adaptation and vulnerability: working group II contribution to the fourth assessment report of the IPCC, vol 4. Cambridge University Press

Rao IM, Peters M, Castro A, Schultze-Kraft R, White D, Fisher M, Miles JW, Lascano Aguilar CE, Blümmel M, Bungenstab DJ, Tapasco J (2015) LivestockPlus: the sustainable intensification of forage-based agricultural systems to improve livelihoods and ecosystem services in the tropics. CIAT Publication

Ross SA, Chagunda MGG, Topp CF, Ennos R (2014) Effect of cattle genotype and feeding regime on greenhouse gas emissions intensity in high producing dairy cows. Livest Sci 170:158–171

Rotz CA, Skinner RH, Stoner AM, Hayhoe K (2016) Evaluating greenhouse gas mitigation and climate change adaptation in dairy production using farm simulation. Trans ASABE 59(6):1771–1781

Russell JM, Barnett JW, Desilets E, Bertrand S (2007) Mitigation strategies to reduce GHG emissions from the dairy industry. In: Reduction of greenhouse gas emissions at farm and manufacturing levels, Int. Dairy Fed. Bull. No. 422. International Dairy Federation

Rutherford KMD, Langford FM, Jack MC, Sherwood L, Lawrence AB, Haskell MJ (2008) Hock injury prevalence and associated rick factors on organic and non-organic dairy farms in the United Kingdom. J Dairy Sci 91:2265–2274

Smith P, Haberl H, Popp A, Erb KH, Lauk C, Harper R, Tubiello FN, de Siqueira Pinto A, Jafari M, Sohi S, Masera O (2013) How much land-based greenhouse gas mitigation can be achieved without compromising food security and environmental goals? Glob Chang Biol 19(8):2285–2302

Sørensen JT, Sandøe P, Halberg N (2001) Animal welfare as one among several values to be considered at farm level: the idea of an ethical account for livestock farming. Acta Agriculturae Scandinavica, Sect A Anim Sci Suppl 30:11–16

Wattiaux MA (2023) Sustainability of dairy systems through the lenses of the sustainable development goals. Front Anim Sci 4:1135381

Weiske A, Vabitsch A, Olesen JE, Schelde K, Michel J, Friedrich R, Kaltschmitt M (2006) Mitigation of greenhouse gas emissions in European conventional and organic dairy farms. Agric Ecosyst Environ 112:221–232

Wildman EE, Jones GM, Wagner RL, Boman HF, Troutt J, Lesch TN (1982) A dairy cow body condition scoring system and its relationship to selected production characteristics. J Dairy Sci 65:495–501

Wood C, Knipmeyer CK (1998) Global climate change and environmental stewardship by ruminant livestock producers. National Council for Agricultural Education, University of Missouri

Yates CM, Cammell SB, France J, Beever DE (2000) Prediction of methane emissions from dairy cows using multiple regression analysis. Proc Br Soc Anim Sci 2000:94

Open Access This chapter is licensed under the terms of the Creative Commons Attribution 4.0 International License (http://creativecommons.org/licenses/by/4.0/), which permits use, sharing, adaptation, distribution and reproduction in any medium or format, as long as you give appropriate credit to the original author(s) and the source, provide a link to the Creative Commons license and indicate if changes were made.

The images or other third party material in this chapter are included in the chapter's Creative Commons license, unless indicated otherwise in a credit line to the material. If material is not included in the chapter's Creative Commons license and your intended use is not permitted by statutory regulation or exceeds the permitted use, you will need to obtain permission directly from the copyright holder.

Part II

Opportunities for Improved Utilization of African Livestock Genetic Resources

Defining Breeding Goals and Breeding Strategies for Improving Livestock Under Various Production Systems in Africa: Concept and Brief Overview

Raphael Mrode, David A. Mbah, and Julie M. K. Ojango

Abstract

This chapter presents an overview of the concept and functions of breeding goals in livestock improvement programs. The first section (Sect. 15.1 Breeding Goals in Livestock Improvement Programs) provides an overview of the definition and functions of breeding goals, critical elements required for their development, and variations in their adoption across different species. The section also outlines global and continental frameworks and intervention initiatives aimed to facilitate the wide adoption of breeding goals for the diverse species and breeds found in Africa. The second section (Sect. 15.2 Breeding Strategies and Livestock Improvement) presents the concept and options related to breeding strategies to realize the defined breeding goals. The section highlights essential elements for successful breeding strategies to increase productivity and incomes including livestock performance data, farmer participation, appropriate dissemination of outcomes, and continuous evaluation. Examples of breeding strategies for cattle and small ruminants are presented to highlight the integration of some of the essential elements of breeding strategies in the African context. The last section of this chapter (Sect. 15.3 Future Perspective) stresses the futuristic importance of breeding goals and strategies for livestock production in line with changing environments and global demands for products.

R. Mrode (✉)
International Livestock Research Institute, Nairobi, Kenya

Scotland Rural College, Edinburgh, Scotland, UK
e-mail: raphael.mrode@sruc.ac.uk

D. A. Mbah
Cameroon Academy of Sciences, Yaounde, Cameroon

J. M. K. Ojango
International Livestock Research Institute, Nairobi, Kenya
e-mail: j.ojango@cgiar.org

Keywords

Breeding goals for low input systems · Sociocultural values · Digital tools · Breeding strategies in Africa

15.1 Breeding Goals in Livestock Improvement Programs

15.1.1 Fundamental Role of Breeding Goals

A breeding goal consists of the traits to be improved and the relative emphasis to be placed on each trait (Box 15.1). Hazel (Hazel 1943) defined a breeding goal as a linear combination of traits in which the coefficient for each trait represents its effect on profit when all other traits are held constant. Breeding goals are futuristic in the sense that they define what we want the population to be in the future and therefore occupy a central role in any breed improvement program. The definition of breeding objectives is the first step (see Fig. 15.1) in the design of any breeding program (Goddard 1998) and it can result in large inefficiency if inappropriate objectives are defined resulting in the application of too much selection pressure on the wrong traits.

The major aim of defining breeding goals is to ensure the profitability of farmers in the production system. Therefore, breeding goals must be relevant to the farming system targeted with the aim of delivering economic and social benefits. It may be added that this should be achieved without causing detrimental effects to the animal's health and welfare or the environment. Usually, profitability to the farmers depends on more than one trait, hence, in most cases breeding goals consist of more than one trait.

Breeding goals are fundamental in terms of determining the direction and rate of genetic change (Box 15.1); the traits to be collected, the genetic evaluation system, and appropriate dissemination for selected animals. In short, in any breeding program, breeding goals enable us to decide on the choice of individual animals as parents, the choice of breeds or crosses, and the evaluation of investment in breeding programs in terms of the economic value of the predicted genetic change (Goddard 1998; Hutu et al. 2020).

> **Box 15.1 Some Useful Terms in Defining Breeding Goals and Breeding Strategies**
>
> 1. *A breeding goal is the specification of the traits to be improved including the emphasis given to each trait.*
> 2. *A breeding goal gives the direction in which we want to improve the population. A notion of what* constitutes the

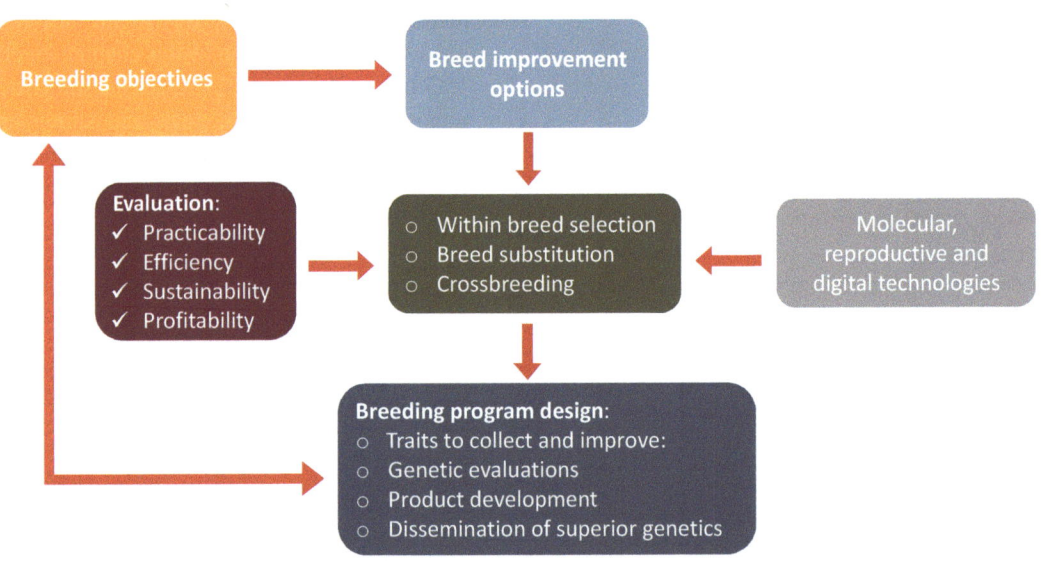

Fig. 15.1 A schematic illustration of the main components of a breeding strategy

best or desired animal phenotype for the population (equivalent to the "vision" of the breeding strategy)
3. The combination of traits in a breeding goal each with their relative weight or emphasis is referred to as a selection index.
4. Breeding strategies: Refer to different ways of using parental generation(s) to generate animals of *desired types. Examples include. Within-breed selection, line breeding, crossbreeding, line crossing, etc*
5. A breeding program is a setup designed to generate the next and subsequent generations of animals that (ultimately) deliver the defined breeding objectives.
6. Genetic evaluation system refers to the procedures and models used in the estimation of the breeding values of animals in a population
7. Rate of genetic change is the rate of change in the estimated breeding value of a population per unit of time such as year or generation

15.1.2 Factors that Underpin the Definition of Breeding Goals in Practice and Challenges of the African Production System

(i) The needs (economic, social, and cultural) of farmers and the production system: In many farming systems in Africa, livestock deliver more than direct economic benefits to families as they contribute to other important societal and cultural values including the provision of traction, provide a means of asset building, provide insurance (ability to sell animals to raise cash to pay medical and educational expenses), and convert crop residue to manure used either as biofuel or as fertilizer. These factors should be considered in defining a breeding goal, since productivity may not necessarily be the key trait but a wide range of other traits related to adaptation, disease resistance, and survival could be very important. Some of these traits could be more difficult to measure or put values on, therefore farmers' involvement is required to appropriately include these traits in the breeding goal.

In many societies in Africa, gender dynamics are strong determinants of livestock ownership patterns, management practices, trait preferences, and breed adoption (Njuki and Sanginga 2013). The need to integrate gender differences in livestock improvement programs has been advocated for sheep and goat production under pastoral systems (Kariuki et al. 2021) as it is evident that gender dynamics can influence breeding outcomes.

In a recent study involving the analysis of 68 research works (Chawala et al. 2020) involving smallholder and medium to large dairy farms, the top four main groups of trait preferences identified were (a) productivity in terms of milk yield per lactation, fat and protein percentages, and lactation length; (b) disease resistance and reproduction, age at first service and calving, days open, number of services per conception, calving interval, and survival; (c) body conformation and temperament and (d) growth rate of animals. However, from the cluster analysis, they reported that trait preferences differed between production systems. While production traits have a similar emphasis on both smallholder and large-scale dairy systems, smallholder dairy farmers put an equal emphasis on reproduction and disease resistance, and then on growth and body conformation. However, large-scale farmers gave equal emphasis to reproduction and growth and then to body conformation and temperament.

(ii) The traits considered important for profitability and of value to the farmer (Philipsson et al. 2011) should easily be measured in a large proportion of the population to be improved either directly or through other proxy-related traits. Another requirement for traits in the breeding goal is that these should be collected with very good precision, at a low cost, and be heritable. In most African settings, this posed a real challenge prior to the advent and transformational use of information and communications technologies (ICT) from the year 2010. Production systems in Africa lack routine methods and systems for collecting performance data. Since 2010 however, innovative approaches for data collection involving the use of digital tools such as mobile applications with data flowing to a database have started to emerge in Africa (Oyieng et al. 2020) and these are discussed in detail in Chap. 16. However, capturing precise information on traits with societal and cultural values which are rather subjective, and the value associated with them poses a challenge. Some studies have employed discrete choice experiment methodology or focus group discussion to gauge the value of traits preferred by farmers (Wurzinger et al. 2008; Leroy et al. 2017; Onzima et al. 2018; Chawala et al. 2019).

(iii) Determining the appropriate emphasis or weight to be given to each trait in the breeding goal. When the traits in a breeding goal are each assigned a weight, it is usually referred to as a selection index (Box 15.1). For example, if the intention is to set a breeding goal to improve growth rate and final weight in goats for example, then a selection index (I) to achieve this goal could be:

I = wt_1(growth rate) + wt_2 (final weight); where wt_i are the weights or relative emphasis on each trait

The relative emphasis or weights can be derived solely on the basis of economic returns to the farmer. Such an index is termed an economic index. In simple terms, the economic weight of a trait in the index could be defined as the effect of a marginal (one unit, for instance, one genetic standard deviation) change in the genetic level of the trait on the breeding objective, keeping all other traits in the index constant. Economic weights can be derived using several methods including:

(a) *Simple accounting* method, where the economic index (A) is the difference between the extra return (g) and the cost (c) from a one-unit increase in the mean of trait: A = (g—c). Note that g and c are evaluated as marginal increases of the trait value above its current value and attempts must be made to avoid double counting of information when dealing with several correlated traits in the index. For example, if computing an economic value for milk yield, the mean of correlated traits such as fat or protein yield must be kept constant.

(b) *Profit functions* have been used extensively and involve the use of an equation to compute the difference between income and expenses of a given system using all herd sources of income and expenses derived from several biological and economic parameters (Ponzoni and Newman 1989; Goddard 1998). With this approach, the economic value of a trait is derived as the first partial derivative of the profit function evaluated at the current population mean for all traits.

(c) *Bioeconomic model* which involves the use of several equations to model inputs or cost and outputs (revenue) for the full-cycle production system utilizing relevant biological and economic parameters. Some examples: bioeconomic models for Brazilian Aberdeen Angus (Campos et al. 2014), sheep (Bohan et al. 2019), dairy cattle (Mbah and Hargove 1982; Schmidtmann 2021), and goats (Bett et al. 2007; Gizaw et al. 2018). The application of the bioeconomic models usually involves several steps: (1) Initially running the model using the means of all traits for the current population, including the trait of interest. The average profit per animal is computed under this scenario (scenario I) (2). Running the model with the mean of the trait of interest increased by a defined amount while the means for all other traits were kept constant. The average profit per animal is then computed under this scenario (scenario II) (3). The economic weight for the trait of interest can be computed as the difference in the average profit of the animal in scenario II and scenario I, divided by the amount of change in scenario II.

However, for societal or cultural traits, the emphasis to be placed on such traits could be determined through focus group meetings or choice experiments as mentioned earlier. Weights for these traits are an outcome of farmers' preferences and predetermined changes with no economic consideration. Such breeding goals are usually termed desired gain indexes.

15.1.3 Types of Breeding Goals (General Objectives) and Issues Relevant to Different Livestock Species

While livestock breeding goals (general objectives) may differ from species-to-species relative to country/region, they share the same concepts discussed in Sects. 15.1.1 and 15.1.2. Generally, the main principle guiding the definition of breeding goals for different species relates to the primary products that define profitability, social, and cultural values in those species.

Thus, as outlined below for different species and summarized in Table 15.1, the breeding goals vary depending on the products derived from them, with the main emphasis on increasing their productivity for food and income.

Beef Cattle The main breeding goal is increased production of beef for food and income.

The associated strategies to meet the specific objective within the goal are different and include the following:
(a) Increased meat production (increased body weight) and meat quality through selection within specialized breeds.
(b) Increased meat production (increase in body weight) through planned crossbreeding of specialized local breeds with specialized exotic breeds.

Dairy Cattle The main breeding goal is increased production of milk for food and income.

Different strategies to meet the specific objectives within the goal include the following:
(a) Increased milk yield and components such as fat and protein yield milk through selection in open/close nucleus of specialized local breeds.
(b) Increased milk yield and components such as fat and protein yield milk through crossbreeding of specialized local breeds with prioritized exotic breeds.

Further reading: Tebong (1985), Tawah et al. (1993, 1994, 1999), Ebangi et al. (1999), Djoko et al. (2003), Ema et al. (2018), Pingpoh et al. (2019), Ouedrago et al. (2020).

Sheep The main breeding goals include increased production of mutton/lamb, milk, and revenue.

The strategies involved include the following:
(a) Increased meat production (growth-related traits) or milk production or more lambs (litter size) through selection (open/close nucleus) of local breed(s) for increased mutton/lamb production.
(b) Increased meat production (growth-related traits) or milk production or more lambs (litter size) through crossbreeding of prioritized local breed(s) with prioritized exotic breed(s) for increased production of mutton/lamb.

Goat The main breeding goals include increased meat and milk production and meeting sociocultural needs.

The strategies involved include the following:
(a) Increased meat production (growth-related traits) or milk production through selection (open/close nucleus) of local prioritized breed(s) for increased goat meat.
(b) Increased meat production (growth-related traits) through crossbreeding of prioritized local breed(s) with prioritized exotic breed(s) for increased goat meat.

Further reading (sheep and goats): Kosgey (2004), Yapi-Gnaore et al. (1997), Bosso et al. (2007), Ojango et al. (2021), Djoko (2019).

Table 15.1 Summary of main breeding goals by species

Specie	Main breeding goals	Sources
Cattle Beef Dairy	Improved growth rate for meat Increased milk production and improved fertility	Tebong (1985), Tawah et al. (1993, 1994), Ebangi et al. (1999), Djoko et al. (2003), Ema et al. (2018), Pingpoh et al. (2019)
Sheep Goats	Improved growth rate for mutton, increased milk, resistance to parasites Improved growth rate for meat, increased milk production	Kosgey (2004), Yapi-Gnaore et al. (1997), Bosso et al. (2007), Ojango et al. (2021), Djoko (2019)
Camels	Improved rate of growth for meat and increased milk production	Morton (1984)
Pigs	Improved growth rate for pork	Dotche et al. (2020), Verhulst (1993)
Poultry	Improved growth rate, increased eggs production	IRZ (1986), Hans (2012)
Rabbits	Improved growth for meat	Nwakalor and Ngo Ndjon (2000), Ngo Ndjon and Nwakalor (1999)
Horses	Race performance, trotting, pacing	IRZ (1986), Amasou and van Vleck (2000), Pretorius (2003), Tennah et al. (2014)
Donkeys	Traction/work, transport	Blench (2000). Ebangi and Vall (1998)

Camels The main breeding goals include increased meat production and milk yield for food and revenue.

Strategies involved include the following:
(a) Selection of local breeds for increased camel meat production (growth-related traits) and milk yield.
Further reading: Morton (1984).

Pig The main breeding goals include increased production of pork for food and revenue.

Strategies involved include the following:
(a) Increased pork production through the selection of prioritized local breeds for growth-related traits.
(b) Increased pork production through crossbreeding of local breeds with exotic breeds for improvement in growth-related traits.
Further reading: Dotche et al. (2020), Verhulst (1993).

Poultry The main breeding goals include increased production of table birds for meat and egg production.

Strategies involved include the following:
(a) Improvement in body weight or increased egg production through selection in specialized local breeds.
(b) Improvement in body weight or increased egg production through crossbreeding of specialized local and exotic breeds.
Further reading: IRZ (1986), Hans (2012).

Rabbits The main breeding goals include increased production of rabbit meat for food and revenue.

Strategies involved include the following:
(a) Improvement in growth-related traits through selection of local breeds.
(b) Improvement in growth-related traits through crossbreeding of specialized local and exotic breeds.
Further reading: Nwakalor and Ngo Ndjon (2000), Ngo Ndjon and Nwakalor (1999).

Horse The main breeding goal is improvement in race performance, trotting, and pacing.

Strategies involved include the following:

(a) Selection of local breeds for increased racing ability or trotting /pacing ability.
(b) Crossbreeding of specialized local and exotic breeds for racing ability or trotting /pacing ability.

Further reading: IRZ (1986), Amason and van Vleck (2000)., Pretorius (2003), Tennah et al. (2014).

Donkeys The main breeding goals include increased traction(work) ability and transportation ability.

Strategies involved include the following:
(a) Improvement in traction ability or transportation ability through selection in local breeds.

Further reading: Blench (2000), Ebangi and Vall (1998).

Two breeding strategies that are common to all the species listed above include the following:
(a) Evaluation of adaptation of selected genotypes to biotic and abiotic stresses
(b) Socioeconomic evaluation of selected genotypes

15.1.4 Adopting Breeding Goals in Livestock Improvement at the Continental Level

To initiate change, breed improvement programs in the continent of Africa have been driven by governments and bilateral organizations seeking to change the productivity within the existing systems. Through these programs, some basic structures were put in place at the individual country level. Glaring gaps in information on practices and breeding strategies for indigenous genetic resources across the continent however made it difficult for many African countries to provide information on their livestock populations for the "State of the World Animal Genetic Resources" report prepared by the FAO (2007). The publication of the FAO report in 2007 led to the international community developing the Global Plan of Action (GPA) for Animal Genetic Resources (AnGR) *and the* Interlaken Declaration on AnGR. The Global Plan of Action (Box 15.2) emphasized the need for improved productivity and long-term breeding strategies if the use of indigenous livestock breeds is going to be sustainable. Details on the GPA are outlined in Sect. 30.2.4 (Chap. 30). Countries in Africa followed up with a concerted effort to characterize animal genetic resources in the continent and the practices adopted in their management.

> **Box 15.2 The Global Plan of Action (GPA, 2007)**
> *Strategic areas*
>
> Characterization, Inventory and Monitoring.
> Sustainable Use and Development.
> Conservation.
> Policies, Institutions and Capacity-building.
>
> *Aims*
>
> Support and increase the overall effectiveness of national, regional and global efforts for the sustainable use, development and conservation of animal genetic resources,
> Contribute to the development of a comprehensive framework for the management of agricultural biodiversity.
> Facilitate international cooperation and the mobilization of resources.

A second report on the "State of the World Animal Genetic Resources" was published in 2015 (FAO 2015). To broaden the scope and adoption of breed improvement, the African-Union Inter-African Bureau for Animal Resources (AU-IBAR) developed an overarching policy framework, and published information on farm animal genetic resources in Africa (AU-IBAR 2019), (See Box 15.3).

Box 15.3 AU-IBAR Animal Genetics Project

To fill the gap in information on the existing animal genetic resources on the African Continent that was required as a basic building block for developing sustainable breeding programs, in 2013 AU-IBAR led a continental effort to characterize and develop an inventory of Animal Genetic Resources (AnGR) in Africa, outlined in detail at Sect. 30.3.1 (Chap. 30). In addition to instituting more targeted breeding management interventions, the project aimed to build capacity for effective formulation and implementation of policies and strategies for the management of AnGR.

Through the project, reports submitted to the FAO by 44 African states on genetic resources and breeding programs in place in 2015 were reviewed and summarized in "The State of Farm Animal Genetic Resources in Africa (AU-IBAR 2019).

Additionally, a series of policy briefs on breeding program options for the continent were published and availed. The brief on Crossbreeding in Africa (AU-IBAR 2020) revealed that among 42 of the African Union countries in 2015, crossbreeding for improvement of animals was instituted in:

- 30.2% of all breeding improvement programs for cattle.
- 10.6% of all breeding improvement programs for sheep.
- 32.6% of all breeding improvement programs for goats.

Results from the study indicated that the main reasons for crossbreeding among local and exotic breeds were for general improvement in meat and milk production. Often the local breeds were deemed to have poor meat production, and/or milk production.

Most of the systems utilizing crossbreeding among sheep and goats sought to produce dual-purpose animals.

15.1.5 Summary of Essential Elements of an Appropriate Breeding Goal for Low Input Systems

A critical element for the successful implementation of appropriate breeding goals in low input systems is that they must have a farmer focus. Goals must ensure profitability for farmers in the production system, delivering economic, cultural, and social benefits to the farmers. In addition, the participation of farmers in identifying the appropriate traits in the goal is essential for adoption and effectiveness. Farmer involvement has been attributed as one of the major factors for the success of the community breeding programs for small ruminants implemented in Ethiopia (Haile et al. 2019). It must be added that breeding goals should have no detrimental effects on animals' health and welfare or the environment. These goals should result in outputs for which there is regional and national demand; in other words, there should be market opportunities. Therefore, a good understanding of the future consumers' demands or good foresight of both farmer and consumer needs and future market trends is fundamental. Finally, such goals should be linked with or rooted in government policies (e.g., national livestock development policy) to promote government buy-in and widespread adoption.

15.2 Breeding Strategies and Livestock Improvement

Breeding strategies involve the design and evaluation of different breeding improvement options including the integration of new technologies, for the purpose of maximizing particular breeding objectives or performance of existing livestock. At the national level, the livestock policy of some countries may include animal breeding strategy components that govern the development and genetic improvement of their animal genetic resources. In general, the overall purpose of breeding strategies is to produce future generations of animals that will produce more efficiently under future production systems.

Therefore, the fundamental components of a breeding strategy involve a clearly defined breeding objective, and evaluation of different breed improvement options to address this objective in terms of practicability, efficiency, sustainability, and profitability to the farmer (Fig. 15.1). For example, let us assume that we wish to improve milk yield in smallholder systems in Southern Kenya. To realize this goal, we can consider several strategies: (1) *within breed selection* of the dairy cattle in Southern Kenya, (2) *importation* of some other dairy breed from another African country or a developed country in another continent, or (3) undertake *crossbreeding* between the existing indigenous breed and an imported breed. We can evaluate these various options in terms of cost (efficiency), practical issues, sustainability (cost and impact on the local breeds, environmental impact), and farmers' profitability. After this evaluation stage, the next steps will involve the detailed design of an actual breed improvement program in terms of defining an appropriate selection index for the breeding objective, the capture of relevant phenotypes, genetic evaluation systems, and selection of the superior genetics based on the breeding objectives as parents of the next generation.

In large farms, the superior stock selected could be used within farms and the implementation of the improvement program will consider issues such as mating ratios and controlling of inbreeding. In some large beef systems, the use of reproductive technologies such as artificial insemination (AI), multiple ovulation and embryo transfer (MOET) are employed. However, in Africa, where smallholder farmers play an important role in the livestock sector, accounting for about 80% of the productivity (Ojango et al. 2018), the delivery of superior genetics to the farmers becomes more complicated and may involve working with national artificial inseminations centers (NAIC) and enrolling superior bulls for AI. This may be coupled with other reproductive technologies such as synchronization to enhance the success of AI.

Reviewing the major components of breeding strategies in Fig. 15.1, some of the bottlenecks for the successful conceptualization and implementation of breeding strategies in developing countries include (i) the definition of breeding objectives that do not account for the farmer's perspective in addition to the production system. Dekkers and Gibson (Dekkers and Gibson 1998) rightly indicated that breeding objectives and selection criteria may never be used in practice if those definitions do not take into account the perceptions and wishes of the breeders for whom they are designed; (ii) narrow breeding options often focused on breed replacement and/or indiscriminate crossbreeding of native with exotic breeds with no plan on how to maintain suitable exotic blood levels and no selection in the local breed (Wurzinger et al. 2014), and (iii) the lack of systematic systems for performance data capture for improved farm management and implementation of genetic improvement programs. However, advances in modern digital, genetic, and reproductivity technologies are providing opportunities to overcome some of these bottlenecks, and in subsequent chapters (18 and 19), these are outlined in more detail in different livestock species.

15.2.1 Example Breeding Strategies Adopted for Targeted Outcomes

It has become more evident that the sustainability of breeding programs in different communities in Africa is highly influenced by socioeconomic factors that affect access to livestock resources, and the use of the resultant products by different gender groups. This also affects the adoption of interventions in the form of breeding management practices. In Kenya's mixed crop-livestock production systems, the introduction of improved livestock breeds with markets for dairy products leads to an increased demand for women's labor resulting from zero-grazing strategies, and men's appropriation of milk incomes—traditionally under women's control (Wangui 2008). Civil unrest and global events such as the Covid-19 pandemic from 2020–2021 also affect breeding practices and choice of breeds in different regions of the continent. For example, the national unrest in Rwanda in 1994 which resulted in loss of livelihoods also resulted in huge losses in existing livestock breeds. To restore and resettle the population and improve livelihoods and household nutrition, the "Girinka" program was initiated through which one dairy cow was provided per family to improve milk production (Ezeanya and Kennedy 2017).

15.2.1.1 Example Strategy Adopted for Dairy Production in Africa

Crossbreeding of indigenous tropical livestock breeds with exotic more temperate breeds is the most widely adopted strategy across the African continent for dairy production. It is however notable that despite using the practice for more than 100 years to improve milk production in cattle, a stabilized crossbred with a known proportion of *Bos taurus* and *Bos indicus* genes has not yet been developed for the low input farming systems of Africa. In a review of the achievements, challenges, and opportunities of using crossbreeding to increase milk production in cattle (Galukande et al. 2013), it was noted that milk production by local animals was more than doubled when they had 50% Bos *taurus* blood. The greatest challenges reported in adopting crossbreeding include uncoordinated crossing among animals, limited understanding of how to use crossbreeding to achieve the long-term breeding goal by the farming communities, lack of national breeding policies, and a lack of consistent records on the performance of crossbred animals to guide selection decisions (Djoko et al. 2003; Roschinsky et al. 2015). At the end of intervention efforts to introduce new breeds into populations, the livestock keepers are unable to follow through on the breeding for a stabilized breed combination in line with their goal to increase milk production.

15.2.1.2 Examples from Sheep and Goat Production in Different Systems

Sheep and goat breeding programs in Africa present a unique diversity in approaches, from breeding programs designed to supply animals for smallholder farming systems in which few animals are reared on small land holdings under controlled management to produce milk to meet household requirements (e.g., FARM Africa Goat program, Heifer project international) to interventions involving the use of improved breeding males of exotic genotypes to enhance growth rates under larger scale operations (Chap. 17). The most common type of breeding programs implemented for sheep and goats in developing countries since the year 2000 are Community-Based Breeding Programs (CBBP) (Mueller et al. 2015; Weldemariam and Mezgebe 2021) that take into consideration and build on indigenous knowledge of communities on breeding practices, resulting in greater appreciation and use of indigenous breeds.

15.2.1.3 Breeding Strategies for Pastoral Production Systems

One way of demonstrating that a breeding program will result in change is to adopt impact pathway models. The impact pathway presented in Fig. 15.2 was adopted to introduce the most optimal breeding practices for sheep and goat production under pastoral systems in Kenya (Ojango et al. 2018). Through the progressive community interactions outlined in which critical socioeconomic aspects related to gendered ownership and use of livestock were taken into consideration (Kariuki et al. 2021), pastoral communities in the arid lands of Kenya have embraced improved breeding management strategies for their small ruminants (Ojango et al. 2021).

15.2.2 Summary of Essential Elements for a Successful Breeding Strategy

A successful breeding strategy(specific objective) would be characterized by the following essential elements: (a) two-way triangular free flow of information among four key partners (research (ministry/institute), development (livestock ministry) and producers (livestock farmers, and consumers); (b) dissemination of results gen-

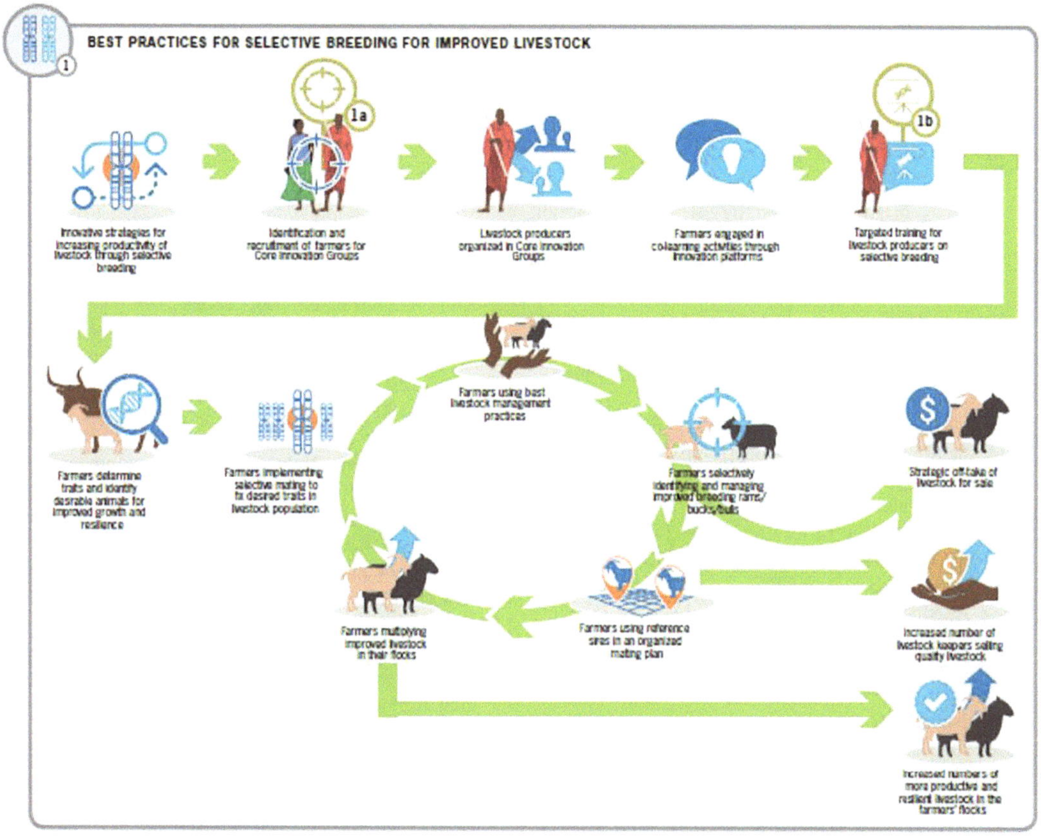

Fig. 15.2 Impact pathway for best practices for selective breeding for improved livestock (Ojango et al. 2018)

erated by the strategy; (c) farmers' participation at all stages of the strategy including design and implementation; (d) continuous evaluation and feedback to farmers researchers and development partners, and (e) increased productivity leading to improved revenue for farmers and fulfillment of several sociocultural roles.

Further reading: IRZ (1986), Tebong (1985).

15.3 Future Perspectives

Emerging markets, changes in consumer attitude, and advances in technology are critical in the design of breeding goals and strategies so that future trends can be accommodated. Notable changes in climate that have a huge impact on livestock productivity and consumers' desire for more healthy diets and preferences for livestock raised in management systems with more welfare considerations are now apparent. These factors will have to be considered in the definition of future breeding goals. Breeding strategies for the mitigation of climate change and the adaptation of livestock systems to accommodate climate change will play central roles in the design of future breeding goals in addition to the incorporation of emerging reproduction, molecular, and digital technologies. Changes in the design of breeding strategies to accommodate some of these new trends are discussed in subsequent chapters on breeding goals for different livestock species.

Acknowledgments The authors acknowledge the contributions of Prof. Ed Rege to the brief definitions in Box 15.1 in this chapter.

References

Amasou T, van Vleck, LD (2000) Genetic improvement of the horse.. Faculty Papers and Publications in Animal Science https://core.acuk/download/pdf/17215923/pdf

AU-IBAR. (2019) The state of farm animal genetic resources in Africa. Edited by AU-IBAR. Nairobi, Kenya

AU-IBAR (2020) Crossbreeding: panacea or curse to African animal genetic resources improvement? Policy Brief No 5. https://www.au-ibar.org/sites/default/files/2020-11/pb_20201116_crossbreeding_panacea_curse_african_animal_genetic_resources_improvement_en.pdf

Bett RC, Kosgey IS, Bebe BO et al (2007) Breeding goals for the Kenya dual purpose goat. II. Estimation of economic values for production and functional traits. Trop Animal Health Product 39:467–475

Blench R (2000) The history and spread of donkeys in Africa. In: Starkey P, Fielding D (eds) Donkeys, people and development, p 240

Bohan A, Shaloo L, Creighton P et al (2019) Deriving economic values for national sheep breeding objectives using a bioeconomic model. Livestock Science. Elsevier. https://www.elsevier.com/open-access/userlicence/1.0

Bosso NA, Cisse MF, van der Waaij EH et al (2007) Genetic and phenotypic parameters of body weight in West African Dwarf goat and djalonke sheep. Small R Res 67(2–3):271–278

Campos GS, Braccini Neto J, Oaigen RP, Cardoso FF, Cobuci JA, Kern EL, Campos LT, Bertoli CD, McManus CM (2014) Bioeconomic model and selection indices in Aberdeen Angus cattle. J Anim Breed Genet 131:305–312. https://doi.org/10.1111/jbg.12069

Chawala AR, Banos G, Peters A et al (2019) Farmer-preferred traits in smallholder dairy farming systems in Tanzania. Trop Anim Health Prod 51:1337–1344. https://doi.org/10.1007/s11250-018-01796-9

Chawala AR, Mwai AO, Peters A, Banos G, Chagunda MGG (2020) Towards a better understanding of breeding objectives and production performance of dairy cattle in sub-Saharan Africa: a systematic review and meta-analysis. CAB Rev 15(007):1–15. https://doi.org/10.1079/PAVSNNR202015007

Dekkers JCM, Gibson JP (1998) Applying breeding objectives to dairy cattle improvement. J Dairy Sci 81:19–35

Djoko TD (2019) Determination des principaux parameters pour le controle de la reproduction chez la chevre naine(Capra hirus) de la zone deforest humid du Cameroun. These doctorat/PhD en Biologie des Organismes Animale: Option Physiologie Animal, Universite de Yaounde Yaounde I

Djoko TD, Mbah DA, Mbanya J et al (2003) Crossbreeding cattle for dairy production in the tropics: effects of genetic and environmental factors on improved genotypes on the Cameroon western plateau. Revue Elev Med Vet Pays tropicaux 56(1–2):63–72

Dotche IO, Bonou AG, Dahouda M et al (2020) Reproductive performances of local pigs in West African countries: a review. J Adv Veterin Res 10:49–55

Ebangi AL, Vall E (1998) Phenotypic characterization of draft donkeys within the Sudano-sahelian zone of Cameroun. Rev Elev Med Vet Pays Tropicaux 51(4):327–334

Ebangi AL, Erasmus GJ, Mbah DA et al (1999) Genetic parameter estimates for growth traits in pure-bred Gudali and two-breed synthetic Wakwa beef cattle a tropical environment. J Cam Acad Sci 6:118–130

Ema P. J. N., Lassila L., Missouhou A. et al(2018) African. Food, Nutrition and Development 18(3): 13572–13587

Ezeanya C, Kennedy A (2017) Integrating clean energy use in national poverty reduction strategies: opportunities and challenges in Rwanda's Girinka programme. Oxford University Press

FAO (2007) Global plan of action for animal genetic resources and the Interlaken declaration. Commission on Genetic Resources for Food and Agriculture, Food and Agriculture Organization of the United Nations, Rome, Italy

FAO (2015) In: Scherf BD, Pilling D (eds) The second report on the state of the world's animal genetic resources for food and agriculture. FAO Commissnion on Genetic Resources for Food and Agriculture Assessments, Rome. http://www.fao.org/3/a-i4787e/index.html

Galukande E, Mulindwa H, Wurzinger M, R. et al (2013) Cross-breeding cattle for Milk production in the tropics: achievements, challenges and opportunities. Animal Genetic Resourc 52:111–125. https://doi.org/10.1017/S2078633612000471

Gizaw S, Abebe A, Bisrat A et al (2018) Defining smallholders' sheep breeding objectives using farmers trait preferences versus bio-economic modelling. Livest Sci 214:120–128

Goddard ME (1998) Consensus and debate in the definition of breeding objectives. Dairy Sci 81:6–18

Haile A, Gizaw S, Getachew T et al (2019) Community-based breeding programmes are a viable solution for Ethiopian small ruminant genetic improvement but require public and private investments. J Anim Breed Genet 136:319–328

Hans L. (NL) (2012) Chickens in Africa. Aviculture Europe. www.aviculture-europe.nl/nummers/12E04A10.pdf

Hazel LN (1943) The genetic basis for constructing selection indexes. Genetics 28:476–490

Hutu I, Oldenbroek K, Van der Waaij L (2020) Animal breeding and husbandry chapter 3. Breeding goals. Agroprint, Timişoara, Wageningen, pp 170–191

IRZ.(1986) Annual report 1985–1986. Institute of Animal Research, Cameroon

Kariuki J, Galie A, Birner R et al (2021) Does the gender of farmers matter for improving small ruminant productivity? A Kenyan case study. Small Ruminant Res 206:106574

Kosgey IS (2004) Breeding objectives and breeding strategies for small ruminants in the tropics. PhD Thesis, Animal Breeding and Genetics Group, Wageningen University

Leroy G, Baumung R, Notter D, Verrier E, Wurzinger M, Scherf B (2017) Stakeholder involvement and the management of animal genetic resources across the world. Livest Sci 198:120–128

Mbah DA, Hargove GL (1982) Genetic and economic implications of selection for milk protein. J Dairy Sci 65:632–637

Morton RH (1984) Carmels for meat and milk production in Su-sahara Africa. J Dairy Sci 67:1548–1553

Mueller JP, Rischkowsky B, Haile A et al (2015) Community-based livestock breeding programmes: essentials and examples. J Anim Breed Genet 132:155–168

Ngo Ndjon M, Nwakalor LN (1999) Heterosis in maternal traits from crossbreeding of exotic and local breeds of rabbits. Trop J Anim Sci 1(2):9–15

Njuki J, Sanginga PC (2013) Gender and livestock: key issues, challenges and opportunities. In: Women, livestock ownership and markets. Bridging the gender gap in Eastern and Southern Africa. Routledge/International Devlopment Research Centre, New York/Ottawa

Nwakalo LN, Ngo Ndjon M (2000) Crossbreeding exotic and local breeds of rabbits in Cameroon: breed of sire and breed of dam effects. Agro-Sci 1:94–99

Ojango J, Oyieng E, Milia D et al (2018) Best practices for selective breeding for improved livestock productivity. ILRI, Nairobi

Ojango JMK, Muigai AW, Oyieng EP et al (2021) Developing community-based breeding programs to improve productivity of sheep and goats in Turkana, Isiolo and Marsabit counties of Kenya, Nairobi, Kenya

Onzima RB, Gizaw S, Kugonza DR et al (2018) Production system and participatory identification of breeding objective traits for indigenous goat breeds of Uganda. Small Rumin Res 163:51–59

Ouedrago D, Soudre A, Yougbare B et al (2020) Breeding objectives in three local cattle breed production systems in Burkina Faso with the implication for the desgn of breeding programs. Livest Sci 232:1–3

Oyieng E, Ojango JM, Dooso R et al (2020) A manual to guide collection of data using the ADGG ODK tools. ILRI, Nairobi

Philipsson J, Rege JEO, Zonabend E et al (2011) Sustainable breeding programmes for tropical low – and medium farming systems. In: Ojango JM, Malmfors B, Okeyo AM (eds) Animal genetics training resource. Version 3, 2011. International Livestock Research Institute/Swedish University of Agricultural Sciences, Nairobi/Uppsala

Pingpoh DP, Mbah DA, Tawah CL (2019) Profitability of agricultural research: the case of genetic improvement for milk production in Cameroon. J Cam Acad Sci 15(1):3–8

Ponzoni RW, Newman S (1989) Developing breeding objectives for Australian beef cattle breeding. Anim Prod 49:35–47

Pretorius SM (2003) Evaluation of the selection and breeding of Friesian horses in Southern Africa. MSc. Dissertation, University of Pretoria. http://hdl.handle.net/2263/26913

Roschinsky R, Kluszczynska M, Sölkner J et al (2015) Smallholder experiences with dairy cattle crossbreeding in the tropics: from introduction to impact. Animal 9(1):150–157. https://doi.org/10.1017/S1751731114002079

Schmidtmann C (2021) Genomic characterization of Red dairy cattle in Northern Europe and the optimization of their breeding programs. PhD dissertation zur Erlangung des Doktorgrades der Agrar- und Ernährungswissenschaftlichen Fakultät der Christian-Albrechts-Universität zu Kiel. https://www.tierzucht.uni-kiel.de/de/forschung/dissertationen-1/Dissertation_Schmidtmann.pdf#page=81

Tawah CL, Mbah DA, Rege JEO et al (1993) Genetic evaluation of birth and weaning weight of pure-bred and two-breed synthetic beef cattle populations under selection in Cameroon. I: genetic and phenotypic parameters. Anim Sci 57:73–79

Tawah CL, Rege JEO, Mbah DA et al (1994) Genetic evaluation of birth and weaning weight of pure-bred and two-breed synthetic beef cattle populations under selection in Cameroon.II: genetic and phenotypic trends. Anim Sci 58(1):25–34

Tawah CL, Mbah DA, Messine O et al (1999) Crossbreeding cattle for dairy production in the tropics: effects of genetic and environmental factors on the performance of improved genotypes on the Cameroon highlands. Anim Sci (UK) 69(1):59–67

Tebong ED (1985) Activites de recherché zootechniques: sommaires des resultants de recherché zootechniques. Wakwa, Cameroon, p 49

Tennah S, Farmir F, Kafidi N et al (2014) Selective breeding of Arabian and Thoroughbred race horses in Algeria: perceptions, objectives and practices of owners-breeders. R Bras Zootec 43(4):188–196

Verhulst A (1993) Lessons from field experiences in the development of monogastric animal production. In: FAO Technical Papers: Strategies for sustainable animal agriculture in developing countries. www.fao.org/3/to582E27.htm. AU-IBAR, 2019. The state of farm animal genetic resources in Africa, AU-IBAR (ed), (Nairobi, Kenya)

Wangui E (2008) Development interventions, changing livelihoods, and the making of female Maasai pastoralists. Agric Hum Values 25:365–378

Weldemariam B, Mezgebe G (2021) Community based small ruminant breeding programs in Ethiopia: Progress and challenges. Small Rumin Res 196:106264. (Elsevier B.V.)

Wurzinger M, Ndumu D, Okeyo AM et al (2008) The lifestyle and herding practices of Bahima pastoralists in Uganda. Afr J Agric Res 3:542–548

Wurzinger M, Mirkena T, Sölkner J (2014) Animal breeding strategies in Africa: current issues and the way forward. J Anim Breed Genet 131:327. https://doi.org/10.1111/jbg.12116

Yapi-Gnaore CV, Rege JEO, Oya A et al (1997) Analysis of an open nucleus bredding programme for Djallonke sheep in Ivory Coast. 2. Response to selection on body weight. Anim Sci 64:301–307

Open Access This chapter is licensed under the terms of the Creative Commons Attribution 4.0 International License (http://creativecommons.org/licenses/by/4.0/), which permits use, sharing, adaptation, distribution and reproduction in any medium or format, as long as you give appropriate credit to the original author(s) and the source, provide a link to the Creative Commons license and indicate if changes were made.

The images or other third party material in this chapter are included in the chapter's Creative Commons license, unless indicated otherwise in a credit line to the material. If material is not included in the chapter's Creative Commons license and your intended use is not permitted by statutory regulation or exceeds the permitted use, you will need to obtain permission directly from the copyright holder.

16

Defining Breeding Goals and Breeding Strategies for Improving the Productivity of Cattle Breeds and Buffaloes in African Production Systems

Raphael Mrode, David A. Mbah, Chi L. Tawah,
Julie M. K. Ojango, Sunday O. Peters,
Eveline M. Ibeagha-Awemu, Oluyinka Opoola,
Isidore Houaga, Richard Osei-Amponsah,
Moses Okpeku, and John E. O. Rege

Abstract

The previous chapter discussed the fundamental concepts and principles that underpin the definition of breeding goals and strategies. In this chapter, some of the peculiar aspects of breeding goals and strategies for the improvement of cattle and buffaloes in Africa, in addition to lessons drawn from past cattle improvement projects, are examined. The chapter is broadly presented in

R. Mrode (✉)
International Livestock Research Institute,
Nairobi, Kenya

Scotland Rural College, Edinburgh, UK
e-mail: raphael.mrode@sruc.ac.uk

D. A. Mbah
Academy of Sciences, Yaounde, Cameroon

C. L. Tawah
African Development Bank,
BP, Abidjan, Cote d'Ivoire

J. M. K. Ojango
International Livestock Research Institute,
Nairobi, Kenya
e-mail: j.ojango@cgiar.org

S. O. Peters
Department of Animal Science, Berry College,
Mount Berry, GA, USA
e-mail: speters@Berry.edu

E. M. Ibeagha-Awemu
Sherbrooke Research and Development Centre,
Agriculture and Agri-Food Canada,
Sherbrooke, QC, Canada
e-mail: Eveline.Ibeagha-Awemu@agr.gc.ca

O. Opoola
The Centre for Tropical Livestock Genetics and
Health (CTLGH), University of Edinburgh,
Edinburgh, UK
e-mail: Oluyinka.Opoola@ctlgh.org

I. Houaga
University of Edinburgh, Edinburgh, UK

R. Osei-Amponsah
Department of Animal Science, School of
Agriculture, University of Ghana, Accra, Ghana
e-mail: ROsei-Amponsah@ug.edu.gh

M. Okpeku
Discipline of Genetics, School of Life Sciences,
University of Kwa-Zulu Natal, Durban, South Africa
e-mail: OkpekuM@ukzn.ac.za

J. E. O. Rege
Emerge-Centre for Innovations Africa (ECI-Africa),
Nairobi, Kenya
e-mail: ed.rege@emerge-africa.org

© The Author(s) 2026
E. M. Ibeagha-Awemu et al. (eds.), *African Livestock Genetic Resources and Sustainable Breeding Strategies*, Sustainable Development Goals Series, https://doi.org/10.1007/978-3-031-92076-9_16

four sections. Section 16.1 deals with the concept of breeding goals in the context of the most dominant production system in Africa and examines evolving trends in breeding goals that incorporate health and welfare considerations. This is followed by Sect. 16.2 which presents the fundamental role of market situations in the design of breeding goals and the influence of some regional institutions and policies on breeding programmes. Breeding goals for beef and dairy cattle production and some lessons from the design of breeding goals from previous breeding programmes were then outlined. Section 16.3 considers breeding strategies aimed at promoting sustainable animal production systems in African diverse environments. It then delves into the different breeding strategies including breed substitution and cross-breeding, selection approaches for cattle improvement, nucleus or group breeding schemes, community-based livestock breeding schemes and unplanned selection programmes. Section 16.3.2 is focused on emerging molecular and reproductive technologies in breeding strategies relevant to genetic improvement in various cattle and buffalo production systems in Africa. It deep dived into the different technologies as follows: molecular and genomic technologies for breeding strategies, reproductive technologies for African cattle and buffalo breeding strategies, nucleus breeding schemes for assisted reproduction in cattle and buffaloes, in vitro fertilization, somatic cell nuclear transfer and multiple ovulation and embryo transfer. The chapter concludes with some final remarks and future perspectives.

Keywords

Breed substitution · Cultural values · Farmers participation · Climate change · Reproductive and molecular technologies

Abbreviations

ADG	Average Daily Gain
ADGG	Africa Dairy Genetic Gains
AfDB	African Development Bank
AgGDP	Agricultural Gross Domestic Product
AI	Artificial Insemination
ART	Assisted Reproductive Technology
ASARECA	Association for Strengthening Agricultural Research in East and Central Africa
AU–IBAR	African Union Inter-African Bureaux of Animal Resources
BLUP	Best Linear Unbiased Prediction
CBBP	Community Breeding Programs
CCARDESA	Centre for Coordination of Agricultural Research and Development for Southern Africa
CIRAD	Centre International de Recherche Agricole pour le Developpement
CH4	Methane
CORAF	West and Central African Council for Agricultural Research and Development
DAGRIS	Domestic Animal Genetic Resources Information
DFID	Department of Foreign and International Development
DNA	Deoxyribonucleic Acid
DGEA	Dairy Genetics East Africa
EBV	Estimated Breeding Value
ET	Embryo Transfer
ETA	Estimated Transmitting Ability
FAO	Food and Agriculture Organization
GBLUP	Bayesian or weighted
GHG	Greenhouse Gases
GIC	Common Initiative Groups
GIZ	German Technical Cooperation
HD	High density

ICT	Information and Communication Technologies	
IEMVT	Institut d'Elevage et Medicine veterinaire des Pays Tropicaux	
ILRI	International Livestock Research Institute	
IRAD	Institute of Agricultural Research for Development	
ITC	International Trypanotolerance Centre	
MAAIF	Ministry of Agriculture, Animal Industry and Fisheries	
MAS	Marker Assisted Selection	
MINEPIA	Ministry of Livestock, Fisheries and Animal Industries	
MOET	Multiple Ovulation and Embryo Transfer	
NAGRC & DB	National Animal Genetic Resources Centre and Data Bank	
NBS	Nuclear Breeding Systems	
PCR	Polymerase Chain Reaction	
QTL	Quantitative Trait Loci	
SCNT	Somatic Cell Nuclear Transfer	
SNP	Single Nucleotide Polymorphism	
SOGELAIT	Dairy Production Company	
TDCS	Tadu Dairy Cooperative Society	
TALIRO	Tanzania Livestock Research Organization	
UK	United Kingdom	
USA	United States of America	

16.1 Overview of Breeding Goals in the Context of Current Production System in Africa

In Chap. 15, the fundamental concepts and principles that underpin the definition of breeding goals and strategies were defined and discussed. In this chapter, some of the peculiar aspects of breeding goals and strategies for the improvement of cattle and buffaloes in Africa are examined in addition to lessons drawn from past cattle improvement projects.

Production traits such as milk, fat and protein yields, and most other traits in the breeding goal for dairy cattle and buffaloes are only measurable in the cows and are, thus, termed as sex-limited traits. However, most of the genetic improvement in the dairy system comes from the selection of bulls. In general, genetic improvement in the dairy systems is through the four pathways: bulls to breed bulls (bb), bulls to breed cows (bc), cows to breed bulls (cb) and cows to breed cows (cc), with the proportion of genetic gains from the four pathways being 39%, 27%, 32% and 2%, respectively (Schmidt and Van Vleck 1974). Thus 66% (bb + bc) of the genetic improvement comes from the selection of bulls.

In developing countries of Africa, farms are predominately managed by smallholders, with 80% of land holdings being smaller than ten hectares (Lowder et al. 2021). Thus, smallholder dairy farming has a huge potential to substantially contribute to sustainable household livelihoods through economic well-being, household food security and nutritional stability. To achieve these potentials, it is important that appropriate breeding strategies are implemented in line with the production systems (Chagunda et al. 2016). A detailed description of the various production systems is presented in Sect. 2.3 of Chap. 2. Therefore, traits in the breeding goals for these systems must be relevant, taking into account the environment and the level of management. Such traits must be easy to measure either directly or through other proxy-related traits in large number of daughters and also early in life to achieve efficient rate of genetic improvement. The top four main groups of trait preferences identified for dairy cattle from a summary of 68 published research work (Chawala et al. 2020) for small and medium large farms in Africa are (a) productivity in terms of milk yield per lactation, fat and protein percentages and lactation length; (b) disease resistance and reproduction, age at first service and calving, days open, number of services per conception, calving interval and survival; (c) body conformation and temperament and (d) growth rate of animals. Usually, traits in the breeding goals should be measurable with very good precision, at low cost and are heritable.

From the list of goal traits identified by Chawala et al (2020), milk yield, age at first caving and lactation length meet these various criteria to a high degree, but other traits such as fat and protein yields, disease traits such as mastitis or its indirect predictor such as somatic cell count are more challenging in an African context as they are not regularly measured.

In general, most African countries lack routine systems for individual animal identification and data collection even for easy-to-measure traits such as milk yield. The advent of information and communications technology (ICT) has resulted in the piloting of innovative approaches for data collection involving the use of digital tools such as mobile apps with data flowing to centralized databases in some African countries (Gebreyohanes et al. 2021). In addition, through data platforms like the Africa Dairy Genetic Gains (ADGG), the recording of fat and protein percent, and somatic cell counts in milk have been initiated in Kenya, Tanzania and Ethiopia in smallholder dairy systems. The ADGG project highlights the important role of digital tools in capturing performance traits in the production systems in Africa and hence efficient design of breeding goals.

Current African indigenous cattle populations have mainly dual-purpose functions, producing both milk and beef or beef production and draught purpose. The low productivity of the indigenous animals, especially for milk and related traits, has created the perception that faster genetic progress in dairy productivity can be achieved by importing exotic breeds. This has resulted in most national governments not developing any comprehensive policy on genetic improvement for dairy cattle. The characteristics of the resultant breeding schemes are that these are usually unplanned or not designed with any focus, and are therefore unsustainable. The imported animals are usually not adapted to the production systems and environmental stressors in Africa, leading to a high rate of mortality due to diseases, lack of adequate feed resources and inadequate management of these animals. In a majority of the cases, the imported dairy bulls are used haphazardly in cross-breeding. The major limitations of such cross-breeding schemes are that they are reliant on the continual importation of foreign genetics which represents a huge drain of financial resources and therefore are not sustainable in the long term; they lack any definite and well-designed breeding objectives underpinning the importation and genetic diversity is lost through the erosion of existing indigenous dairy breeds.

The question that arises is whether breeding programmes can be designed to utilize the crossbred animals that have resulted from the haphazard cross-breeding over the years and thus lay a foundation of a sustainable breeding programme that evaluates and disseminates superior and adapted cross-bred sires for use in Africa? The ADGG programme was designed to begin to address this challenge in East Africa by employing the rapidly expanding ICT infrastructure and developments in genomics. Usually, in the smallholder farming systems, parentage and breed composition are unknown due to a lack of adequate pedigree recording. Accurate estimation of the breed composition is important in determining the animals with the appropriate breed composition most suitable for the different agro-ecological zones. Using developments in genomic technologies, ADGG developed an assay of 200–400 SNPs for breed composition determination and parentage verification (Strucken et al. 2017). In addition, the lack of pedigree recording schemes in smallholder dairy systems constitutes a major bottleneck to genetic evaluation and the implementation of breeding programmes in developing countries (Kosgey and Okeyo 2007). The availability of genotypic information implies that the genomic relationship matrix can be computed thereby enabling the prediction of the genetic merit of animals, indicating less reliance on pedigree information. A simulation study by Kariuki et al. (2014) showed that genomic selection schemes for smallholder dairy production systems could yield higher responses to selection compared to strategies that used only pedigree data. Mrode et al. (2021), reported genomic prediction for daily milk yield and body weight for cross-bred cattle in Tanzania using genomic relationships constructed from genotypes. While the primary breeding goal in these cross-bred populations under ADGG is milk yield and body weight, attempts are being made to include fertility traits, milk fat, and protein percentages and somatic cell counts in the goals.

In contrast to dairy cattle, beef cattle have performance traits measurable in both sexes and growth rate, the major trait of interest is easy to capture. However, similar to dairy cattle, the establishment of systems to record performance routinely is also lacking. In addition, in many communities of Africa, several traits in beef cattle have societal and cultural values which are rather subjective. A classic example is the nature of horns in Ankole cattle found across several East African countries (Ndumu et al. 2008). Capturing these traits precisely and the value associated with them is a challenge. Some studies have employed discrete choice experiment methodology (Ndumu et al. 2008; Kugonza et al. 2012; Chawala et al. 2019) or focus group discussion to gauge the value of traits preferred by farmers.

16.1.1 Trends in Breeding Goals for Evolving Production Systems and Health and Welfare Considerations

Given strong trends in climate change and a society that is more aware of animal health and welfare issues, it is inevitable that the production systems in Africa will have to evolve to accommodate these changes. The contribution of African livestock production systems to GHG emissions and mitigation strategies are fully discussed in Sects. 14.5 and 14.8 of Chap. 14. Global surface temperatures are reported to be 1.09 °C higher between 2001 and 2020 compared with the period 1850–1900 (IPCC 2021). This has a huge potential impact on livestock in terms of feed resources, the emergence of new diseases and reduced productivity. Globally, methane gas (CH_4), generated through enteric fermentation in ruminants is reported to represent 17% of greenhouse gas (GHG) emissions (Knapp et al. 2014). The impact that the smallholder farming production systems for cattle have on climate is not clearly understood as data quantifying GHG emissions in these systems are limited. About 5% higher CH_4 emission per unit product has been attributed to smallholder systems compared to dairy and beef systems in Europe and North America (Gerber et al. 2013), mainly due to the low productive efficiency of the smallholder systems. Therefore, giving higher weights to traits associated with superior productivity and resilience in the breeding goal under conditions of climate change, such as heat and drought tolerance and resistance to certain diseases, will be important. In addition, the definition of breeding goals should directly include traits to mitigate GHG emissions. In developed countries, direct selection for CH_4 emission and the efficiency of feed intake have been used as approaches to mitigate CH_4 emission (Manzanilla-Pech et al. 2021). de Haas et al. (2021) reported that selection for feed efficiency was a good proxy for mitigating CH_4 emission resulting in about 26% reduction in methane over a 10-year period. Consequently, several countries have developed indices to improve feed efficiency (usually termed Feed saved or Feed advantage) in their dairy breeding programmes (Lidauer et al. 2019). There is currently no corresponding breeding goal being implemented in Africa; however, the ADGG programme has included body weight in a selection index for dairy cattle for Tanzania as an indirect approach to improve feed utilization (Mrode et al. 2021 https://cgspace.cgiar.org/handle/10568/113071).

There is ongoing research to improve the adaptive ability of beef and dairy cattle, mostly aimed at identifying the genomic regions that confer adaptiveness in indigenous breeds and for later incorporation of these regions in breeding programmes. Kim et al. (2017) reported selection signatures and genes or pathways controlling anaemia and feeding behaviour in the trypanotolerant N'Dama, coat colour and horn development in Ankole, and heat tolerance and tick resistance across African zebu cattle breeds such as Boran, Kenana and Ogaden. In a further analysis of 16 indigenous African cattle breeds, Kim et al. (2020) identified genomic regions composed of

16 loci associated with immune, heat-tolerance, trypanotolerance and reproduction-related adaptive characteristics of African cattle. This research is still at an early stage in terms of identifying with adequate precision the genomic regions which confer adaptiveness to the indigenous cattle breeds to enable their incorporation in any breeding programme. However, possible ways to incorporate such regions in future breeding programmes include gene editing to create surrogate sires or the inclusion of Bayesian or weighted GBLUP in the prediction of genomic breeding values (Cheruiyot et al. 2022).

16.2 Defining Breeding Goals for Cattle and Buffaloes in Various African Production Systems

This session examines the fundamental role of production systems and market situations in the design of breeding goals and the influence of some of the regional institutions and policies on breeding programmes in Africa. A detailed description of various production systems in Africa and the challenges and interventions for livestock development are presented in Sects. 2.3 and 2.4 of Chap. 2. It then outlines the breeding goals for beef and dairy cattle production and some lessons from the design of breeding goals from previous breeding programmes.

16.2.1 Outline of Market and Aesthetic Products from Dairy Cattle Systems

Prevailing and emerging market situations are fundamental to the design of efficient breeding goals for any livestock species. The drivers for growth in dairy production and markets include optimization of milk production systems, improved animal health and welfare, improved efficiencies in feeding and improved genetics. FAOSTAT (2021) reported that although dairy production improved in most continents globally, Africa's dairy output (excluding camel's milk) has adversely been affected by prevalent market conditions, stemming from economic meltdowns, arable land shortage, conflicts and displacements, and climate issues such as drought and floods in some regions (GDP F 2018).

Dairy and dairy products contribute 27% to global agricultural gross domestic product (AgGDP) and 5% to AgGDP in Africa. Dairy systems form part of the demand for dairy and dairy aesthetic products. The markets for dairy products in Africa can be organized into two main types; informal and formal markets. Alonso et al. (2018) defined the formal dairy sector as a dairy value chain that commercializes industrially processed and packed dairy products by legally registered value chain actors, while the informal dairy sector is described as value chains commercializing dairy products that are not industrially processed by either the legally licensed or unlicensed enterprises. Informally marketed products include raw, traditionally pasteurized milk, boiled milk and other traditionally derived milk products. The informal sector is mainly operationalized by small-scale dairy farmers, milk transport traders, milk bar operators and mobile milk traders with limited infrastructure such as electricity, water, sanitation, refrigeration and other infrastructure and amenities. The informal sector receives little or no support from the government and other development partners (Grace 2015; McDermott et al. 2010). The informal markets have diverse prices and are unstructured while for the formal market, prices for livestock products are streamlined and defined in a structured manner. Studies by Otieno and Owuor (2019) have demonstrated the relevance of policies in informal markets for the dairy sector so that monopolistic tendencies are restricted, thereby creating a level playing field for all milk traders in the industry.

However, despite changing market dynamics and dairy prices in Africa, there is there is still huge potential to develop systems to support the demand and supply of dairy and dairy products such as meat, milk, leather and other related-derived products as well as manure (bio-gas), draught power, art, musical and historic artefact products. The demand for livestock and livestock products in the future could be heavily moderated by socio-economic factors such as human health concerns, changing socio-cultural values (Steinfeld et al. 2006) and increased purchasing power (OECD, F 2016; OECD/FAO 2020). In addition, there is a growing emphasis on the sustainability of dairy and dairy products as customers are now well informed about dairy and dairy products available to them. People are concerned about the environment, animal welfare and the quality of their food in a bid to enhance climate-friendly environment for the dairy systems in Africa. However, all the aforementioned issues could be mitigated in Africa if there are adequate and functional systems for improvement in genetics, animal welfare, a streamlined market for dairy productions, and collaborations among stakeholders involved in dairy production system and national governments. Also, accelerating the adoption of existing best practices and technologies is essential to further improve production efficiency and overall dairy productivity in Africa.

16.2.2 Essential Elements of a Sustainable Breeding Goal: Lessons from Past and Ongoing Breeding Programmes

Breeding Programmes at Continental and Regional Levels

Breeding programmes in the continent of Africa have mainly been driven by governments and bilateral organizations seeking to change the productivity within the existing farming systems. Through these programmes the basic structures that have been put in place include; (i) policies to guide the adoption and use of different breeds in the target environments (AU-IBAR); (ii) the establishment of departments in key ministries to support the implementation of breeding activities and (iii) simulation studies and pilot projects to test select breeding goals in more controlled environments. Additionally, within the different regions of the continent, supported through the regional centres for agricultural research including; Centre for Coordination of Agricultural Research and Development for Southern Africa (CCARDESA), Association for Strengthening Agricultural Research in Eastern and Central Africa (ASARECA) and West and Central African Council for Agricultural Research and Development (CORAF), structures, policies and capacities have been developed to support the establishment of centres of excellence for livestock production (AUIBAR 2019). Details on these different organizations are outlined in Sects. 30.3 and 30.4 of Chap. 30.

The dearth of the documentation of indigenous breeding practices across the continent resulted in a concerted effort to characterize existing indigenous animal genetic resources, initially driven by the inability of many countries to provide information on their livestock populations for the 'State of the World Animal Genetic Resources' report prepared by the FAO in 2007. Changing agro-economic landscapes, climatic conditions and population pressures are however resulting in the transformation of cattle production in Africa with a greater need to adapt technologies and practices to derive increased products from existing animals. The increased demand for quality animal products against a low level of supply in the different markets requires critical evaluation of bottlenecks in adopting targeted breeding goals in cattle populations.

Most of the livestock in Africa are reared in rural communities and provide social, cultural and economic benefits for households. Any practical implementation of breed improvement towards a targeted breeding goal would require documented records on the animal performance. Efforts at large-scale recording and monitoring of performance and productivity in most African Countries are limited. By 2015, some countries in Southern Africa (Botswana, Lesotho, Namibia, South Africa, Zimbabwe), East Africa (Kenya) and West Africa (Cameroon) had operational national livestock recording systems in place (AUIBAR 2019). The proportion of the livestock populations documented through the systems is however very low (<1%) in most of the countries except South Africa.

Table 16.1 Example of breeding goals and breeds developed for meat production using indigenous breeds[a] in different regions of Africa

Breeding goals	Breeds developed	Regions	References
Adaptability to tropical conditions and meat quality	Africaner Renitelo Wakwa	Southern Africa Western Africa	Raeliarijaona (2017), Casey (2021), and Nsangou et al. (2021)
Adaptability to tropical conditions, growth rate and meat quality	Bonsmara, DrakensbergerBoran	Southern Africa Eastern Africa	Casey (2021) and Rewe (2009)
Disease tolerance and meat quality	Tswana	Southern Africa	Saki (2020)
Fertility, adaptability and growth rate	Nguni	Southern Africa	Saki (2020)

[a]Characteristics of different breeds and their development in the continent are presented in Chaps. 3 and 4, and also available through DAGRIS (http://dagris.ilri.cgiar.org/) and the Animal Genetics Training Resource (http://agtr.ilri.org/)

Across Africa, breeding goals for specific products from cattle (Table 16.1) have been adopted at different rates. The high focus on economic returns from animals while overlooking socio-cultural significance of the animal species in the diverse communities across the continent has resulted in the slow adoption of the concept of using breeding goals to drive cattle production in Africa.

Breeding Goals for Beef Production

The challenging climates with vector-transmitted diseases that tend to be better tolerated by existing indigenous breed types enabled interest in the indigenous breeds in different regions for beef production, rather than the use of purebred introduced temperate breeds as was the practice in the dairy sector. The development and use of breeding goals in the beef sector are most evident in the Southern Africa region where the establishment of breed societies was extensively used to enhance the image and use of different breeds using performance data (Casey 2021). In South Africa, state-initiated performance recording schemes were launched to improve production efficiency using targeted breeding goals.

In Eastern Africa, the improved Boran cattle have been selected for beef (Rewe 2009). Strategies adopted to develop the different breeds are outlined in Sect. 16.4 of this chapter.

Breeding Goals for Dairy Production

The lack of clear breeding objectives, recording of animal performance and large-scale uncontrolled cross-breeding of temperate *Bos taurus* breeds with local *Bos indicus* breeds has made it difficult to identify better-performing genotypes in the different farming systems of sub-Saharan Africa. Key issues noted to impact dairy production in 15 sub-Saharan African countries include (i) poor productivity of imported exotic breeds and their crosses in Africa (62.3%), (ii) fluctuations in milk prices in both formal and informal markets (50.9%), (iii) inadequate genetic evaluations of individual animals and sires (39.6%), (iv) poor management systems in terms of herd health, feeding and housing (32.1%) and (v) poor infrastructural facilities (30.3%) (Opoola et al. 2019).

The need to adopt alternative strategies to transform dairy production in the continent was magnified in the second report on the State of the World Animal Genetic Resources (FAO 2015). In 2013, the Centre for Development Innovation in Wageningen, The Netherlands, was tasked by the Inter-Agency Donor Group on pro-poor livestock research and development to study and document the state of dairy development in different countries of East Africa (Makoni et al. 2014) in order to identify opportunities to improve coordination in dairy development among development agencies and create synergies across the countries. Lack of data, information services, infrastructure and institutional capacity-building programmes for the dairy industry were identified as some of the critical challenges to be addressed. In the same year, different pilot projects were designed to test options for adapting genomic technologies to catalyse change among the large number smallholder dairy producers on the continent. The International Livestock Research Institute (ILRI) in partnership with different national partners implemented pilot projects in Eastern Africa (Box 16.1) and West Africa (Box 16.2).

Box 16.1: Genomics to Identify Appropriate Cross-Bred Dairy Cattle for Smallholder Farming Systems in East Africa—The Dairy Genetics East Africa Project

Challenge: Smallholder farmers mostly keep crosses between indigenous cattle and exotic dairy breeds such as Holstein, Friesian, Ayrshire and Jersey. There is however no systematic breeding of the cross-bred cattle, and farmers rarely keep pedigree or performance records. Most mating events involve local cross-bred or indigenous bulls, where the cross-bred bulls are of unknown breed composition.

Goal: To determine the most appropriate genotypes for the range of dairy production systems and levels of production operated by small-holder farmers in East Africa, and how the genotypes could be delivered to smallholders.

Methodology: Apply high-density single nucleotide polymorphism (SNP) technology to determine breed composition of cows owned by smallholders in combination with traditional and participatory appraisal of animal and farm performance to determine which genotypes are most profitable at different levels of production.

Working with smallholder farmers in targeted areas, performance data were collated on milk yield, reproduction events and disease incidences over 2 years. The recorded animals were genotyped using the Illumina bovine high-density SNP (HD,

780 k) assay with the HD SNP data used to perform admixture analyses, using the ADMIXTURE software (Alexander et al. 2009), to generate an estimate of the ancestral breed composition of each animal. This was the first accurate information on breed composition to be combined with in situ performance data from smallholder farming environments of East Africa.

Key results

A comparison of the farmer and extension agent assessment of breed composition based on phenotypic appearance and farmer recollection on cow origin, with the admixture determinations of actual breed composition revealed that phenotype-based assessments were poor predictors of actual breed composition ($R^2 = 0.16$).

Intermediate to low grade (<50% exotic breed ancestry) cows performed better in the majority of the smallholder farms where inputs were low, while animals with higher grades (>50%) performed better in the best environments where farmers invested in providing inputs for dairy production (Ojango et al. 2014).

The use of genomic testing demonstrated the value and feasibility of establishing long-term genetic improvement programmes, which led to the establishment of the Africa Dairy Genetic Gains (ADGG) programme in 2016.

Box 16.2: West Africa

A study similar to the DGEA (Box 16.1) was implemented for cattle reared in smallholder farming systems in Senegal (Marshall et al. 2019). The breeds evaluated were cross-breeds between indigenous Zebu and Guzerat, and indigenous Zebu and *Bos taurus* breeds (Montbéliard and

Jersey), pure indigenous Zebu breeds and pure *Bos taurus* breeds (Tebug et al. 2018). Genotyping in this case was performed using the Bovine 50 K SNP assay and admixture analysis was performed using the Bayesian Analysis of Population Structure software (Corrander et al. 2008).

In this study, cross-bred indigenous zebu by *Bos taurus* dairy cattle kept under better management produced up to 7.5 times higher milk yields, eightfold higher household profit, and threefold lower greenhouse gas emission intensity, per cow per annum, in comparison to indigenous Zebu kept under poorer management, for a typical herd size of eight animals (Marshall et al. 2016; Salmon et al. 2018).

Following successful pilot programmes on data capture, monitoring and feedback to inform and develop breeding programmes for dairy production in smallholder systems of Africa, from 2016 The African Dairy Genetic Gains (ADGG) programme was designed under the leadership of ILRI and different partners, supported by the Bill and Melinda Gates Foundation, to address limitations in data from smallholder dairy systems and to identify and promote the use of more productive dairy genetics suited to the local production environments (Box 16.3).

Box 16.3: The ADGG Programme (https://portal.adgg.ilri.org/)

The African Dairy Genetic Gains (ADGG) project was developed and from 2016 to identify and promote the use of more productive dairy genetics suited for local production conditions/systems through collaboration with local and international genetics and artificial insemination companies. The programme is developing and testing a multi-country genetic gains platform that uses on-farm performance information and basic genomic data to identify and provide superior cross-bred bulls for artificial insemination (AI) delivery and planned natural mating to smallholder farmers in Africa.

Western Africa

In Burkina Faso, the 'Vache du Faso' project funded by CEVA-SANTE ANIMALE was implemented between 2016 and 2019. The aim of the project was to improve the milk production capacity of indigenous zebu cattle breeds of Burkina Faso through cross-breeding with exotic French cattle breeds (Tarentaise and Montbeliarde). During the 3-year period of the project, the local project team conducted 6 separate reproduction (heat synchronization and reproduction) campaigns targeting 5 541 cows and 1500 F1 cross-bred calves were born between 2017 and 2019. However, there was no plan for performance recording and genetic evaluation of the F1 cross-breds.

Lessons in Developing Breeding Goals for Production Systems in Africa

1. Breeding goals should cater to both economic and socio-economic aspects related to the target products as the social role of cattle in different communities greatly influences the choice of breeds and the structure of the herd.
2. The emphasis on breeding goal traits should incorporate gendered preferences as the successful management of select animals is determined by the labour requirements from different gender groups. For instance, in smallholder zero grazing systems of Eastern Africa, the adoption of high-yielding dairy animals requires greater labour inputs to meet the animal's requirements for feed, water and milking. The stall-fed animals are often left in the care of women in households who have diverse demands on their time. This often results in periods when animals have no feed or water, leading to fluctuations in productivity. Lactation curves of animals within small-

holder farming systems reflect low levels of production following an initial high yield in the first few weeks following calving (Ojango et al. 2017; Migose et al. 2020).
3. Adoption of technologies to boost productivity must be preceded by capacity building within community groups for correct adoption.
4. Special interest groups are important for safeguarding and marketing cattle breeds. Without an interest group, selecting and enhancing breeding goal traits will not be implemented resulting in haphazard cross-breeding of animals in populations.
5. Supportive infrastructure and markets for products need to be in place to help drive the adoption of breeding goals.
6. A consistent level of public investment is required to enable the recording and monitoring of animal performance to support regular monitoring and evaluation of breeding goals. Support should include policies to govern custody and use of information collated, standards for data capture and availability of services to enhance productivity in targeted livestock enterprises.

16.3 Breeding Strategies for Cattle Improvement in Various Production Systems

In Chap. 15 and Sect. 15.2, the basic principles of defining breeding strategies were examined. In this section, we look at some peculiar features of such designs for cattle and buffaloes. As indicated in Sect. 16.1, most of the genetic improvement in dairy cattle and buffaloes (about 66%) comes from the selection of bulls and this has a huge influence in designing a breeding strategy for the dairy system. Strategies that will permit the estimation of the genetic merit of bulls and their selection for use early in life will reduce generation interval, for example, genomic selection and hence a higher rate of genetic progress. Similarly, strategies that will permit obtaining phenotypes on daughters of bulls early in life such as indirect prediction of fertility or will permit rapid multiplication of animals of superior genetic merit such as MOET are very important in improving the rate of genetic improvement in the dairy cattle or buffaloes. Initially, we examine past breeding strategies, lessons learnt and present some future perspectives.

16.3.1 Different Types of Breeding Strategies; Historical Lessons in Various Input Cattle Systems and Future Perspectives

This section considers breeding strategies aimed at promoting sustainable animal production systems in Africa. The African environment is quite diverse and characterized by nutritional diversity, geographical diversity, diversity of animal and human resources, severe poverty and hunger and a high level of indebtedness among others. Therefore, any breed improvements should be related to the environments where performance and production are expected to occur.

16.3.1.1 Breed Substitution and Cross-Breeding

Breed substitution has been defined as the import of males and females of a breed to replace local breed(s) in the pursuit of a production goal (Vaccaro and Steane 1990). At this level of breed substitution are the cases of South Africa in the 1800s where European settlers introduced dairy cattle (Holstein, Jersey, Guernsey, Ayrshire, Milking Shorthorn) (Gertenbach 2021). Today, these breeds are the only ones recognized as dairy cattle in South Africa. In Kenya, Holstein, Jersey, Guernsey and Ayreshire dairy cattle were also introduced from Europe in the nineteenth century (Initiative 2020). Today, they are the only recognized dairy cattle breeds in Kenya. A similar strategy has been followed by Morocco where massive imports of dairy cattle (350,000 since the 1970s) have been made in pursuit of a national dairy production goal (Taher 2011). Forty per cent (40%) of these imports died within 3 years of importation. This indicates that this level of breed substitution may be the least satisfactory. Otherwise, careful evaluation of environmental factors (biotic and abiotic) and necessary modifications need to be made before the introduction of exotic breeds.

Adaptive Breed Substitution

Adaptive breed substitution is the case of breed substitution where ethnically owned local cattle are substituted by another local breed when pastoralists, for one reason or another, move into a new environment with their cattle. This is the case of the Fulani of West Africa in Burkina Faso who moved with their cattle to an ecological zone infested by the tsetse fly, a vector of trypanosomiasis-causing parasite (Boutrais 2007). The cattle native to the new environment are trypanotolerant. The immigrant pastoralist cross-bred their Zebu (Bos indicus) cattle with the local trypanotolerant *Bos taurus* cattle. The crosses are relatively more tolerant to trypanosomiasis than the exotic local Zebu cattle. This is done purely by pastoralists without input from public research and policy sectors.

In another breed substitution strategy, the Fulanis of the Cameroon grasslands replaced their Red Mbororo Zebu breed with another Zebu breed (preferred), the Gudali of Adamawa. The decision was based mainly on the physical features of the Gudali (size, hump, coat colour, size and form of horns). This is implemented by importing the Gudali into their region or by cross-breeding (Gudali bulls with Red Mbororo females), a move encouraged by the colonial policy (Boutrais 2007). The Fulanis are generally regarded as people known for their ability to adapt breeds, refashion and construct them. Another example is the 'Bakalandji' cattle in the Adamawa region of Cameroon which have resulted from cross-breeding Gudali with the Red Mbororo cattle in order to adopt the breed qualities of the former without necessarily replacing the Red Fulani type.

Substitution by Upgrading or Cross-Breeding

Substitution by upgrading (cross-breeding) local breed(s) with the exotic breed(s) to combine or benefit from the productivity advantage(s) of two or more breeds involved. Usually, the levels of breed/genotype mix most suited to the desired goal are accepted after genetic, environmental and economic analyses have been made. These may be considered as synthetics/composites in which local breeds are only partially genetically substituted, or upgrades in which the local breeds are substantially genetically modified to look and perform like the exotic breed(s).

Briefly defined, cross-breeding is the mating of two or more breeds to exploit the merit(s) of each breed in the resulting product (composite). Attention is given to the choice of dam and sire breeds in an attempt to maximize the productive and reproductive efficiency of the cross-bred. The gain could be higher productivity of the cross-bred due to quantitative genetic inheritance and hybrid vigour/heterosis; therefore, cross-breeding combines the genetic advantages of both breeds and heterosis. Maximum heterosis is generated in the first filial generation (F1); hence, different systems of cross-breeding are usually practiced in an attempt to maximize the heterosis in other cross-breds. Cross-breeding cattle for improved beef and dairy production has been practiced in Africa with diverse results and outcomes.

Two to four case studies from each of the five sub-regions of Africa are handled below with two of the cases summarized in Boxes 16.4 and 16.5. Each case is considered from the point of view of goal(s), specific objectives, design, funding, implementation, outcomes, status and the way forward.

Cross-Breeding for Beef Production: South Africa Case Studies

A very good example of this include the Bonsmara, a synthetic breed in South Africa (AGTR: agtr.ilri-cgiar.org/bonsmara). At the policy level, the development of the breed was initiated at the Mara Research Station in Transvaal in 1936 by Bonsma. The goal was to improve beef production. The strategy was to cross-breed local Africander cows with European Shorthorn and Hereford bulls in a three-breed mating design. The details of this breed in terms of characteristics and production statistics are described in Sect. 4.4 of Chap. 4.

Another example is the Renitelo, a Synthetic Beef Breed in Madagascar (Gilibert 1974; Raeliarijaona 2017): The aim was to develop a cattle breed with improved beef production and it started at the Centre de Recherches zootechniques de Kianjasoa, Madagascar, in 1930. The strategy was a 3-breed 3-step mating plan involving the Madagascar Zebu, Limousin from France and Africander from South Africa. The purpose was to create a breed that would satisfy both the meat and draft power needs of the people of Madagascar. It was recognized as Renitelo breed in 1962. More information on this breed has been presented in Chap. 4 (Sect. 4.4).

Cross-Breeding for Beef Production: West and Central Africa Case Studies

> **Box 16.4: Wakwa, a Synthetic Beef Breed in Cameroon**
>
> According to Nsangou et al. (2021) at the policy level, the project started in 1952 as a successor to the Montbeliard project of the 1930s that failed to consider the local production system and the inputs of the farmers into the policy formulation process. The goal was to improve the local Zebu (Gudali) for beef production. The strategy/specific objective was to cross-breed the exotic Zebu (Brahman) with local Gudali to achieve a synthetic or composite of 50% Brahman and 50% Gudali.
>
> The design involved three steps: (1) production of F1; (2) inter se mating of F1 individuals to produce F2 and (3) selection and inter se mating of F2 to produce F3. These F3 were considered stabilized and were designated as the Wakwa breed (Fig. 16.1).
>
> The Wakwa cattle were disseminated among farmers who had a training period at the Fory Station of Livestock Production managed by the Ministry of Livestock, Fisheries and Animal Industries. Funding was by the colonial administration until the 1970s when independent Cameroon created the Institute of Animal Research (IRZ) and assumed financial responsibility in partnership with multiple partners (France, Germany, United Kingdom and the United States). The research was greatly reinforced, and economics and farmer inputs became a component of its activities.
>
> Analyses have shown that the Wakwa is more productive than the local Gudali despite greater susceptibility to dermatophilosis. Analyses of economic traits show that they are adequately heritable and genetic trends are positive (Tawah et al. 1993, 1994; Ebangi et al. 2001, 2002). This genetic potential was transmitted to the farmers as already indicated above.
>
> Sustainability of Wakwa: There was no plan to maintain the Wakwa blood levels in farmers's herds. Hence, it can be said that Wakwa blood in farmers' herds varied from 50% Brahman to 25% or less in subsequent generations as these animals were gradually backcrossed to the Gudali Zebu on-farm. As such, no sustainability strategy (e.g. formation of a breed association, cooperative society or common interest group) was made for the breed to survive the risk of extinction or dilution. Today, only the Tadu Dairy Cooperative Society (TDCS) produces F1 (Fig. 16.1) given the demand by farmers in the Northwest region of Cameroon.
>
> For the way forward, given the demand by producers, the Ministry in Charge of Livestock Production should strengthen its policy and partnership between research (institute) and management (livestock ministry). It should promote the creation of breed associations as well.

Fig. 16.1 Wakwa (Brahman (50%)–Gudali (50%)) bull. (Courtesy of Dr. David Mbah)

Box 16.5: Genetic Improvement of Cattle in Ghana

At the policy/decision level, the first efforts at cattle breed improvement in Ghana started with imports from Europe in 1909 followed by another import in 1919. The objective was to upgrade the local *Bos taurus* to the exotic breed(s). Both efforts failed (Okantah 2009) for lack of adaptability to the local pathological environment. Three institutions are important actors in the genetic improvement of cattle in Ghana. These include the University—Faculty of Agriculture, the Animal Research Institute and the Ministry of Food and Agriculture. Each of these institutions has its own objectives.

The objectives of the Ministry of Food and Agriculture include the extension of improvement to farmers, production of breeding stock for farmers and multiplication and supply of cross-breds to farmers. To achieve these objectives, the Ministry imported exotics from Europe and Australia and N'Dama from other West African countries (Okantah 2009).

The objectives of the Animal Research Institute include the development of the local production system into a dual-purpose (meat and milk) production system, cross-breeding local breeds with exotic breeds leading to a composite genotype of 50% local and 50% exotic, and selection of the composite.

The University of Ghana's objectives included cross-breeding exotic breeds with indigenous breeds leading to a composite of four breeds: 37.5% local breed and 62.5% exotic composed of three breeds. The goal was the improvement of beef production.

The efforts of the three actors were not coordinated. However, a unified breeding policy and animal breeding plans were developed in 1995 (Okantah 2009). The objectives were as follows: (1) The selection of the West African Shorthorn for meat and (2) The development of Sanga as a dual purpose (meat and milk) by cross-breeding the local Sanga with an exotic dairy cattle breed (Friesian).

Of the two breed improvement approaches (University of Ghana and Animal Research Institute), the most extensive was that of the University of Ghana (Okantah 2009). The design of the University of Ghana project (1967–1982) was as follows: (1) cross-breeding Jersey (J) with the West African Shorthorn or Ghana Shorthorn (GSH), (2) cross-breeding Jersey with Sokoto Gudali (G), (3) backcrossing the F1s to Jersey and (4) pure breeding local breeds (Rege et al. 1994). The resulting genotypes were GSH1/2J1/2, GSH1/4J3/4, G1/2J1/2 and G1/4J3/4.

Results (Rege et al. 1994): The Gudali was better than the Ghana Shorthorn in production traits for milk. The F1s were better than the purebred locals. The G1/4J3/4 was better than the corresponding F1s. For milk yield, G1/4J3/4 had no advantage over corresponding F1s. Heterosis was important for Shorthorn crosses but not for Gudali crosses for milk production traits. Heterosis effects were positive/larger for Gudali crosses than for Shorthorn crosses.

Impact: There was no impact of the efforts of the University—Faculty of Agriculture and Animal Research Institute on the local production system since the efforts were limited to research stations only (Okantah 2009). The efforts of the Ministry of Food and Agriculture impacted the local production system (results of Zebu—taurine cross-breeding adopted by farmers) (Okantah 2009).

Sustainability/way forward: the review recommends that genetic improvement in Ghana should be based on local/indigenous breeds—West African Shorthorn and Sanga—and a clear breeding policy. Sustainability can be assured by the existence of breed associations/cooperatives. More information has also been presented on the Ghana Sanga breeds in Chap. 4, Sect. 4.2.3.

Cross-Breeding for Dairy Production

By their nature, dairy cattle require more optimum environmental and management conditions to perform. These conditions include good quality nutrition, good health delivery and tolerable ambient temperatures among others (Osei-Amponsah et al. 2019). In the subtropical environment with shade and no shade temperatures of 28.4 and 36.7, respectively, respiration (54 vs. 82) and rectal temperatures (38.9 °C vs. 39.4 °C) of dairy cattle increased. Conception rates (44.4% vs. 25.3%) and milk yield (16.6 kg/day vs. 15.0 kg/day) were reduced (Roman-Ponce et al., 1976). The impact of the environment varies with the stage of the life cycle and the level of adaptation of the given breeds/genotypes (Collier et al. 1981).

Rectal temperature is moderately heritable (0.17 ± 0.13). It is genetically positively correlated with production traits (yields of milk, fat, protein, production life, net merit). It is, however, negatively genetically correlated with somatic cell score and daughter pregnancy rate. Hence, attention must be paid to these correlations when selection is contemplated (Dikmen et al. 2012).

Heat tolerance is due to low metabolic heat production and low milk yield (Collier et al. 1981). Hence, high metabolic production and high milk yield would suggest heat intolerance. Critical temperatures for dairy cattle vary with physiological state (Collier et al. 1982). For peak lactation, these are −25 °C (lower) and 25 °C (upper) (Collier et al. 1982). Heat stress leads to a reduction in calf birth weight and lower milk yield for cows not in shade. When lactating cows are exposed to environmental temperatures outside their thermoneutrality zone, they tend to adjust by reducing their metabolic activities (Collier et al. 1981). These adjustments include reduced feed intake, faecal water elimination, high increases of water intake and urine excretion from the body surface as well as respiration (Collier et al. 1981).

In the tropical highlands of Cameroon (Adamawa plateau), distribution of cattle by breed/genotype and heat level (shade and no shade) at 22 °C and 20.5% relative humidity under grazing conditions, Holsteins tended to be in the shade while Holstein × Gudali and Montbeliard × Gudali crosses tended to be in the sun (Mbah 1982). The Gudali crosses were significantly in the sun more than in the shade. They showed no evidence of heat stress. Under the same environmental conditions but with rain falling, all genetic groups were in the field and grazing. The rain apparently helped to reduce the effect of heat on the Holsteins.

The potential susceptibility to heat stress suggested here could be more severe in the hotter and drier environments to the north (Sudano-Sahelian region) and could last for about 7 months of the dry season. Hence, dairy cattle improvement strategy would need to pay attention to environmental modifications to handle potential heat stress, as well as nutrition and health stressors. Some case studies from some regions are treated hereafter (Boxes 16.6, 16.7, 16.8, and 16.9).

Cross-Breeding for Dairy Production: West and Central Africa Case Studies

Cross-Breeding for Dairy Production: East Africa Case Studies

---

Box 16.6: Genetic Improvement of Cattle for Dairy Production in Cameroon

Genetic improvement efforts in Cameroon started with imports of Brown Swiss cattle by the National Advanced School of Agriculture in Yaounde in 1986. This was followed by efforts with local breeds (While Fulani, Red Fulani and Gudali) at the Institute of Agricultural Research for Development (IRAD, https://irad.cm/) centres in Bambui on the Western Highlands (Munji 1973) (facility started in the 1940s as an Agricultural Experiment Station) and Wakwa on the Adamawa Highlands (Lhoste and Pierson 1975) (facility started in the 1930s as Cattle Production Improvement Station). These efforts were found inadequate for meeting national needs. Attention was quickly shifted to cross-breeding local breeds with exotic dairy breeds (Holstein, Jersey, Montbeliard). These involved importations of live animals and semen from the United States through Heifer Project International (a non-governmental organization, https://www.heifer.org/) and from France. The Dairy Research Program of the Institute of Animal and Veterinary Research is coordinated from Wakwa and is now within the Department of Animal and Fisheries Research of the Institute of Agricultural Research for Development. With funding from the World Bank, the country imported 165 gestating Montbeliard heifers from France in 2020 (MINEPIA 2020), https:www.prodel.cm/wp-content/uploads/2020/05/MAGAZINE-Fr-2020.pdf). It has as partners the Ministry in Charge of Livestock and Animal Industries, the ministry in charge of scientific research and funding partners.

Objectives: The objectives were the improvement of milk production by cross-breeding local Zebu cattle with exotic dairy breeds: the identification of adapted and productive genotypes and the evaluation of the economic impact of the research results.

The design was as follows:
(i) IRAD, Bambui Center: Holstein × Red Fulani, Holstein × Gudali, Jersey × White Fulani
(ii) IRAD, Wakwa Center: Holstein × Gudali, Montbeliard × Gudali

A dairy technology laboratory and a nutrition laboratory were developed at IRAD centres at Bambui and Mankon, respectively.

At both Research centres, breeding (artificial insemination at Bambui) was aimed at producing cross-breds with 50–75% exotic genes. For Montbeliard cross-breeding, some F2s were generated as well. The genotypes were evaluated on economic traits and adaptability. Farmers were trained in all aspects of dairy production. Developed technologies were disseminated through extension bulletins and field days.

Funding: Funding from the 1970s through the mid-1980s was by many partners including the Government of Cameroon, USA (Heifer Project International, Land O'Lakes), German Technical Cooperation (GTZ), UK (DFID) and France (CIRAD-IEMVT). Each partner was focused with a particular component of the National Agricultural Research Project, which implies that there was a lack of coordination.

Results: The cross-bred animals (Figs. 16.2 and 16.3) were significantly more productive (3–4 times) than the local breeds and more adapted than the pure exotic breeds. Estimates of heterosis were strongly positive for lactation length but strongly negative for milk yield. The heterosis effect was, as expected, reduced in the F2. These results were economically good (Goldman et al. 1985; Pingpoh et al. 2019). Tawah et al. (1996) reported on the performance of the crosses produced in Wakwa centre.

Sustainability: Farmers involved in dairy production founded the Bamenda Dairy Cooperative (Peri-urban dairy, under a zero-grazing production system), Ngaoundere Projet Laitier (later SOGELAIT) as well as Common Initiative Groups (GIC), Tadu Dairy Cooperative Society (rural, semi-intensive (low input, animal recording) production system and mastery of artificial insemination for production of seed stock as well as milk transformation.

Subsequently, the Tadu Dairy Cooperative Society target was a synthetic dairy breed that is Holstein 5/8 Gudali 3/8 (i.e. 62.5% Holstein and 37.5% Gudali) (Figs. 16.4a and 16.4b (MINEPIA 2008)). At the policy level, the ministry in charge of Livestock Production and Animal Industries is extending dairy production improvement to the northern regions of the country (165 gestating Montbeliard heifers recently imported were destined to the hotter and drier Sudano-Sahelian region) (more challenging environment for dairy cattle) and hot and humid south. Research to determine suitable genotypes and the type of environmental modifications required for the zones is needed.

Box 16.7: Cross-Breeding Azawak Zebu Cattle with Local Sudanese Fulani Cattle for Dairy Production in Burkina Faso
At the policy level, 'Project to support the dissemination of Azawak Zebu cattle implemented from 2000 to 2015', within the national policy for local dairy production (Ouédraogo et al. 2020) had three objectives: to support local dairy production, to use the genetic potential of local breeds to increase consumption of animal protein and to increase farmers' income.

The strategy was cross-breeding Azawak Zebu (*Bos indicus*) (sire breed) and local Sudanese Fulani Zebu (*Bos indicus*) (dam breed). The design was a 3-phase plan centred on an open nucleus. There was (1) The selection of Azawak Zebu (AZ) and (2) cross-breeding Azawak Zebu with the local Sudanese Fulani Zebu (SF) as follows:
1. AZ × SF = F1: SF1/2AZ1/2
2. SF1/2AZ1/2 × AZ2/2 = F2: AZ3/4SF1/4
3. AZ3/4SF1/4 × AZ4/4 = AZ7/8SF1/8 (i. e. Z87.5%SF12.5%)

Hence, the local Sudanese Fulani Zebu was upgraded to Azawak Zebu.

Funding was from the Belgian Technical Cooperation and Luxembourg Development Cooperation.

The results were as follows:

- Birth weight: 18.9 plus/minus 3.4 kg and 20.5 plus/minus 3.4 kg for Sudanese Fulani Zebu and pure Azawak Zebu, respectively. It was 20.6 plus/minus 3.7 kg and 20.9 plus/minus 2.4 kg for F1 and F2, respectively.
- Milk yield (during 186 days) was 625 plus/minus 198 kg for pure Azawak Zebu and 516 plus/minus 218 kg for F1 and 560 plus/minus 220 kg for Sudanese Fulani Zebu.

Comment: The F1 is less productive than the parent that was being upgraded!

Box 16.8: Genetic Improvement of Cattle for Dairy Production in Ethiopia

At the policy level, the dairy development effort in Ethiopia started with imports of 300 Holstein and Brown Swiss cattle in 1947 and was followed later by small-scale imports by missionaries and private organizations (Staal et al. 2008). Between 1959 and 1969, measures were taken to improve dairy production around Addis Ababa. From the 1960s to 1991, political/policy regimes changed from imperial through socialist to democratic. However, the dairy development objective remained essentially unchanged.

Improvement has since included more exotic cattle and cross-breeding with varying policy and operational measures. This saw the creation of the Dairy Development Agency (DDA) (Staal et al. 2008) which founded 30 dairy farms, each with 40 members. Subsequently, cooperatives and agricultural development projects with strong dairy components were funded by International Development Agency (IDA) and the African Development Bank. By 1972, the dairy industry was well established around Addis Ababa (Staal et al. 2008).

In 1990, another policy change led to the democratic re-organization of the socialist model cooperatives. This revived the 1991–1992 collapse of the industry as farms were individually owned coupled with the private sector involvement. This arrangement was handled by the Dairy Development Board.

The design of the Ethiopian Dairy Development is based essentially within the Ethiopian Agricultural Research Institute (EARI). Based on four different environments (Holeta, Bako, Adami Tulu and Melka), it involves cross-breeding three indigenous breeds (Ethiopian Boran, Horro, Barka) and three exotic breeds (Holstein, Jersey, Simmental). Coordinated from Holeta, the strategy involves producing various levels of cross-bred composition adapted to different environments/ecologies of Ethiopia. At Holeta, for example, cross-breeding Boran with Holstein along with coordination of pure Jersey improvement and selection of pure Boran, while at Debre Zeit, it is cross-breeding Holstein with Barka and Boran. The national partners include the Federal and Regional research institutions and universities, particularly Alemaya College of Agriculture (Debre Zeit Agriculture Research Centre) (Tadesse and Dessie 2003).

The results indicate that all Holstein crosses are best in milk production traits (milk yield, lactation length, age at first calving, calving interval): crosses carrying 50% local and about 50% exotic blood are good for use under favourable environmental/ecological and management conditions; Jersey crosses, given their smaller size and heat tolerance potential are better than Holstein crosses where feeding is a challenge (EARI 2021).

Dissemination of the various technologies developed is effected through farmer training, multiplication and distribution/supply of seed stock.

Sustainability is assured by the involvement of dairy cooperatives, and the way forward includes continuous research policy support. It has been indicated that the Ethiopian Boran can be improved for milk or meat (and that it is practically a dual-purpose breed (Haile 2011)).

Box 16.9: Genetic Improvement of Cattle for Dairy Production in Kenya
Dairy development in Kenya started with the introduction of dairy cattle breeds—Holstein, Ayrshire—essentially in the 1920s by European settlers in the Highlands (Staal et al. 2008; Thorpe et al. 2000).

In the 1950s, indigenous Kenyans were allowed to carry out commercial agriculture which was still largely handled by settlers (Staal et al. 2008). A 1954 plan allowed Kenyan indigenous farmers to keep improved dairy cattle. Today, the sector is dominated by smallholders due to several complementary strategies, among which are research, breeding, artificial insemination and extension.

In 1958, a Dairy Industry Act established the Kenya Dairy Board (revised several times since then) aimed at improving the performance of the sector, and, in 1997, the Cooperative Development Act was revised to reinforce farmer control of the cooperatives. The policy environment is dynamic and improving. Today, there is a whole institute, the Dairy Research Institute (within the Kenya Agriculture and Livestock Development Board (Staal et al. 2008). Various programmes of the institute aim at improving the dairy productivity of cattle.

At Naivasha, where the Dairy Research Institute is based, genetic improvement strategies (Sect. 16.10) include the following:

The objectives of the institute include genetic improvement of productivity through cross-breeding, artificial insemination, multiple ovulation and embryo transfer (MOET) and genotype × environment interaction, among others. The Livestock Development Policy of 1980 expanded breeding and selection (progeny testing) and the use of state farms for breeding and multiplication of improved stock (Staal et al. 2008). Resources and support were inadequate in the 1980s and 1990s. Privatization of artificial insemination in 1991 led to imports of genetic material by the private sector and 1998–2000 saw a shift from artificial insemination to natural service.

The status of the dairy improvement strategy, now handled by the Dairy Research Institute, suggests that technically, research will continue. The policy environment appears active and is constantly evolving. The presence of dairy cooperatives with reinforced farmer control as well as commercialization outlets promises to enhance sustainability.

Dual Purpose
The goals for dual-purpose cattle may include meat and milk as well as meat and work/traction. In all local African cattle production systems, meat is usually the primary goal while milk is usually a secondary goal. Efforts of combining meat and milk production in a breed/genotype are available from Kenya (Box 16.10). Many *Bos taurus* (dairy)–*Bos indicus* genetic improvement (cross-breeding) efforts for increased milk yield in Africa end up really as dual-purpose (meat and milk) efforts.

The Mpwapwa cattle breed illustrates another cross-breeding initiative in Eastern Africa. It was developed at the Livestock Production Research Institute in Mpwapwa, Tanzania, in the 1920s with the indigenous Shorthorn Zebu being crossed with other African Zebus (Boran and Ankole) and the Bos taurus bulls (particularly

Fig. 16.2 Holstein (67.5%) Gudali (32.5%) bull. (Courtesy Dr. David Mbah)

Fig. 16.3 A 3-year-old Holstein 5/8 Gudali 3/8 Milker. (Courtesy Dr. David Mbah)

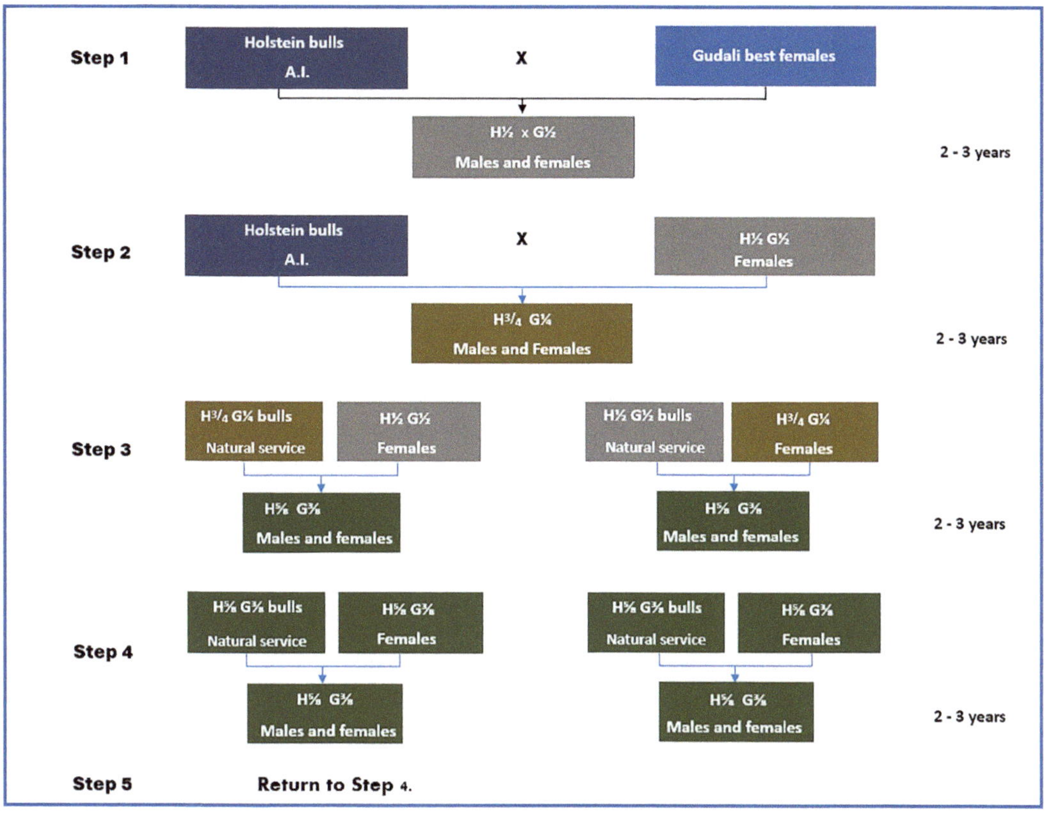

Fig. 16.4a Holstein-Gudali programme to produce H 5/8 G3/8 cross-breds [8–12 years]

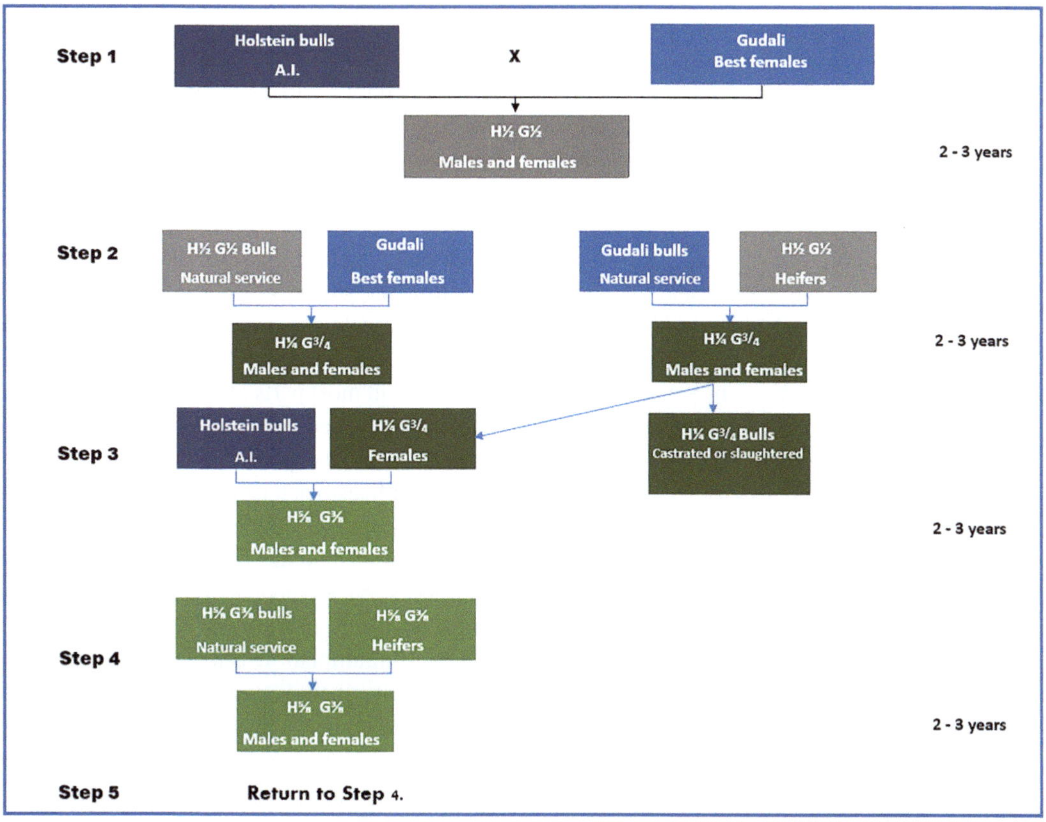

Fig. 16.4b Holstein-Gudali programme H ¼ G ¾ to produce H 5/8 G 3/8 cross-breds for farmers [8–12 years]

> **Box 16.10: Kenya Sahiwal Breed (Muhuyi et al. 1999)**
>
> The Sahiwal breed development started in 1939 with imports of pure Sahiwal from India. The main objective was to upgrade local indigenous Zebu cattle to the Sahiwal in marginal lands, while secondary objectives included the production of semen from selected bulls, production of breeding stock for farmers and conservation/improvement of the Sahiwal genetic resource at 13 Livestock Improvement Centres. A second set of imports included 60 bulls and 10 cows from Pakistan in 1945. A third round of imports included 15 bulls from India.
>
> The foundation stock included 10 Sahiwal cows and selected local Zebu cows. This was used in grading up local cattle breeds to the Sahiwal and their subsequent multiplication. By 1962, there were 2,500 Sahiwals located at 13 Livestock Improvement Centres. In 1991, there were further imports (1000 doses of semen) of Sahiwal bulls for use at the National Sahiwal Stud (1963) located at the National Animal Husbandry Research Centre at Naivasha in an ecologically favourable environment. The mating plan for genetic improvement was based on the closed nucleus approach to facilitate artificial insemination and animal recording. The target traits were milk and growth. Within the nucleus, selection was based on progeny testing for bulls. Young bulls were selected for growth rate while cows were selected based on milk yield. The design included animal identification and characterization. The preservation methods included purebred herds (how much Indian-Pakistani Sahiwal and how much local Kenya Zebu in the Kenya Sahiwal) on state farms and individual ranches and the creation of Sahiwal semen bank as well as use in smallholder mixed farming systems. Importantly, the stud at the Research Centre is used by the Sahiwal Breeders' Society of Kenya. Breed registration standards require detailed pedigree and performance data.
>
> The performance of the Kenya Sahiwal is indicated by the following average performance levels: milk yield: 1574 plus/minus 575 kg (heritability = 0.127 plus/minus 0.04); lactation length: 293 plus/minus 37.5 days; age at first calving: 40.1 plus/minus 3.5 months (heritability = 0.20 plus/minus 0.097) and gestation length: 287 plus/minus 5.1 days (heritability = 0.10 plus/minus 0.03).

Ayrshires) being introduced into the programme later (Syrstad, 1990). The objective was to develop cross-breed animals with improved meat and milk production. The details of the development of this breed have been outlined in Chap. 4, Sect. 4.4 on communal composite breeds.

16.3.1.2 Selection Approaches for Cattle Improvement

In developed economies, cattle genetic improvement programmes have considerably advanced because of the available infrastructure for planned breeding activities such as intensive performance recording, artificial insemination and embryo transfer coupled with advances in quantitative genetics, population genetics and statistics, which have made it possible to develop genetic improvement schemes applicable on a national and global scale. In contrast, such facilities are either lacking, have not been sustained or are inadequate in most parts of Africa. Moreover, the need for advanced breeding schemes has been constrained by the predominance of traditional methods of livestock husbandry practised, lack of livestock policies, lack of skilled manpower, weak/poor stakeholder institutions, inefficient extension services, poorly organized and poorly funded research institutions and projects. The dominant traditional livestock husbandry systems in Africa include pastoralism, agropastoralism, nomadism or backyard farming. Because of this backdrop, most African Governments have instituted state-sponsored livestock farms and animal research stations to effect genetic improvements in herds of limited cattle breeds with the view to passing on the improved stock, mostly males, to interested farmers.

It is also worth noting that in Africa animals are kept for multiple purposes: food, labour (traction power), wealth/income, hides and skins/or wool, manure for use as fertilizer for land or fuel for a fire or as construction materials, savings account (sell an animal when needed) and social status (more is better). Due to a lack of government policy and support, selective breeding is usually not well organized and structured in the smallholder systems as the required infrastructure is absent. However, an increasing number of selective breeding programmes have been developed in some African countries, some of which have been quite successful. The increasing level of education in these countries is an important factor in that success. Many African Governments are also investing in livestock ministries and laying down animal policies that are helping to improve enabling environment for the growth of the sector.

Indigenous cattle breeds are adapted to their local environment and so constitute a unique reservoir of genetic resources for the continuous improvement of livestock productivity in Africa (Osei-Amponsah et al. 2019; Hanotte et al. 2000). However, most of these breeds show low productivity primarily because of a lack of exploitation of their genetic potential, inadequate nutrition, weak health services and poor management. Breeding strategies in Africa encounter such challenges as (i) definition of unrealistic breeding objectives, (ii) limited involvement of stakeholders, (iii) poor infrastructure and (iv) lack of sustainable funding (in most cases funding is external and generally for short periods of time coupled with uncertainty of continuity; thus, they are unlikely to yield tangible expected genetic improvement).

Among the different ways of increasing the numbers of genetically improved animals are (i) within indigenous breed selection for desired traits, from which pool, animals can be sourced for various upgrading, cross-breeding or breed constitution programmes and (ii) crossing local females with superior imported sires has gained momentum with the advent of artificial insemination. Attempts at importing live exotic bulls failed because of inadaptability to the environmental constraints of tropical and subtropical diseases and poor infrastructure. It is important to note that because of the multiple-purpose nature of cattle production in Africa, selection strategies designed for western cattle production systems may not directly apply to Africa, especially as most of these traits are not directly measurable. Besides, performance recording is not practiced in most traditional farms. Also, the production conditions tend to vary considerably, making it quite costly to establish breed improvement facilities within each agro-ecological zone. However, few cattle breeding programmes in sub-Saharan Africa have been successful (Rewe et al. 2009). Diop (1997) and Rewe (2009) have carried out simulations that have looked at integrating the interests of all stakeholders in the formulation of breeding objectives and their effect on selection response. Rewe (2009) showed that restricting growth for milk production would be beneficial in meeting the multipurpose objectives of farmers in Kenya. Dempfle (1993) reported that it is possible to improve milk yield in N'Dama cattle by including growth and trypanotolerance in the breeding objectives.

Selection is the identification of desirable traits in plants and animals coupled with the steps taken to enhance and perpetuate those traits in future generations. Selection of cattle populations involves choosing and mating the best to produce the best in the next generation. Choosing the best requires animal identification, pedigree information, performance recording and analysis of performance records using advanced statistical and quantitative genetics techniques. The selected animals (best animals) are mated to produce the next generation, which is evaluated to determine the selection progress. In cattle populations, the rate of genetic change (selection progress) is a function of the accuracy of selection, selection intensity, genetic variation and generation interval (Bourdon and Bourbon 2000). The rate of genetic change can be increased by reducing the generation interval, or by increasing the selection intensity, accuracy of selection or genetic variation.

Genetic improvement of cattle populations requires identifying and selecting animals with the best breeding values to become parents for the next generation (Bichard 2002). On an individual basis, genetic improvement requires that the net benefit per animal be greater than the total cost of achieving the improvement (Garrick 2006). Therefore, the goal of the breeder is to produce future generations that are more profitable than the previous ones (Bichard 2002).

Accuracy of selection is defined as the correlation between the true breeding values and their predictions for a trait under selection (Bourdon and Bourbon 2000). Increasing accuracy can enhance the rate of genetic change by more appropriate estimates of true breeding values. Therefore, accurate estimation of genetic parameters can maximize response to selection as more is known about the animals used for breeding. Accuracy of breeding values for cattle can be improved by the use of statistical tools like Best Linear Unbiased Prediction (BLUP), and progeny testing (individuals with large numbers of progeny having more accurate records than animals where only the individuals are tested).

Selection is a function of reproduction, replacement rate and recognizing genetically superior animals early in life for breeding (Bichard 2002). The selection that drives genetic improvement is usually carried out in nucleus herds and the genetic improvement is transmitted to other herds via the multiplier herds and commercial herds. Selection is most effective and can be most intense when the traits are highly heritable (Wright 1923). Historically, selection has been more intense in males because they are able to produce more offspring during the same breeding period than females (Lush 1946). However, a negative effect of intense selection is the loss of genetic diversity (Cleveland et al. 2005).

Generation interval also affects the rate of genetic change, and it is desirable to lower the generation interval for a greater rate of genetic change. Generation interval is defined as the time required to replace one generation with the next. It is the average age of the parents at the birth of their offspring (Bourdon and Bourbon 2000). It is calculated by taking the age of each of the parents at the birth of their offspring and averaging over all the parents (Barker and Davey 1960).

Importance of Selecting for Indigenous Breeds of Cattle in Africa

As stated above, traditional systems of cattle husbandry tend to dominate the livestock industry in Africa. Therefore, the factors that advocate for the selection of indigenous cattle breeds in these conditions in Africa include the following:

(a) Adaptation to tropical/subtropical environments: Most indigenous cattle breeds have important adaptational qualities for disease and parasite resistance like trypanosomes and ticks; they can survive and reproduce in hot, humid environments and are able to exist on low-quality feed and limited water supply. Selection for increased productivity should, therefore, raise their potential for productive traits without seriously impacting their adaptational qualities.

(b) Likelihood of extinction of indigenous breeds: The heavy importation of germplasm of exotic breeds, either as live animals or frozen semen, into Africa, coupled with their systematic and random cross-breeding with indigenous stock poses a major threat to the survival of the latter. Consequently, the survival of indigenous cattle breeds can only be guaranteed by raising their genetic merit through selection (Cunningham 1979).

(c) Selection among indigenous cattle breeds: This should be an integral part of any cross-breeding programme.

Selection programmes in institutional herds in Africa should be simple in design and based on traits that are of economic value to farmers. Moreover, because of the harsh tropical/subtropical environments, adaptability traits should always be emphasized. Ensure that the population under selection is exposed to environmental conditions like those in which the offspring are expected to perform. For example, if the objective of selection in an experimental herd of beef cattle is to supply genetically superior bulls to farmers who keep their animals on range without supplementary feeding, it would be inappropriate to keep the experimental animals on high levels of concentrate. Every effort should be made to minimize the effect of genotype × environment interactions. The same argument also applies to experimental herds that observe stringent parasite control measures such as frequent treatment with acaricides or anthelmintics. Such considerations should, however, not lead to the other extreme of imposing excessively harsh conditions that would mask the expression of genetic variation and penalize the animals for their normal performance.

Selection programmes in most African countries suffer from inefficiencies at the institutional farms because of their small- or medium-sized herds and the lack of continuity in the pursuit of selection objectives due to unpredictable funding sources. Their effectiveness can be augmented by integrating different public and private farms of the same cattle breed through links with the use of commonly selected bulls (or sires) under similar standards of management or environment. Establishing links with private farmers is important not only to extend the genetic improvement to their herds but also to increase the effective size of the population under selection and to ensure that testing of animals is also done under similar on-farm conditions.

Potential Selection Strategies

The two major selection procedures used in ruminant livestock are performance testing and progeny testing. While the former is used for characters of high heritability and easily measurable in both sexes like growth rate, the latter is used for characters with low heritability and may be measurable in only one sex, like milk yield in female animals or after slaughter, like carcass quality. These principles of genetic improvement form the basis of different types of selection schemes. These may operate within individual herds, among a group of cooperating farmers or nationally. Some of the important ones are described below.

Testing Stations/Experimental Farms: Animals from different herds are assembled in testing stations and measured under a uniform environment. The influence of the herd environment is, therefore, minimized and the population under comparison is vastly increased. It is particularly suitable for situations where a large number of small herds are involved, each of which does not have the capacity to carry out proper within-herd comparison. Testing stations can be used either for performance testing, for example, growth rate or for progeny testing, for example, milk yield.

Progeny Testing Schemes: The progeny test depends on its accuracy in having a large progeny group per sire. Where a number of herds are involved in testing, the maximum efficiency is achieved when each sire has an equal number of progeny in each of the herds. The scheme usually operates by having a central station from where the semen of the different sires on test is distributed to a network of AI centres that carry out the inseminations.

Reference Sire Scheme: A modification of this scheme called the reference sire scheme enables comparisons to be made also among bulls that are used only within individual herds. The reference sires, which are usually progeny-tested animals and available only by AI, are used alongside the individual herd's own bulls in all herds. The breeding values of home-bred bulls are then estimated in relation to the reference sires. Comparisons among bulls in different herds are made through the linkages established by the reference sires. Thus, sire replacements can be made from among all herds.

Index Selection: Hazel (1943) and Hazel and Lush (1942) provided the basis for index selection, which is defined as the method of artificial selection in which several useful traits are selected simultaneously. Index selection may be termed an approach that is usually implemented within the various selection strategies described above. Each trait that is going to be selected is assigned a relative weight, which reflects the importance of the trait. In a profitability system, the weight is the relative economic value of the trait as the desire of the producer is to maximize profit (MacNeil et al. 1997). More details on economic selection indices are presented in Chap. 15. The index can also combine production and functionality traits (Missanjo et al. 2013). In Africa, where cattle breeding is based on multiple production, functionality and aesthetic traits, the use of the selection index would be ideal if there are possibilities of measuring all these traits. Rewe et al. (2006a, b, 2010) have demonstrated that index selection can be used in Africa for the selection of beef cattle in Kenya. The selection index combines both the production and functionality traits weighted with relative economic values. Diop et al. (1999) also showed through simulation studies that the selection index can be used to improve Gobra cattle in a nucleus breeding scheme in Senegal.

A few cases of selection for improved milk production in several regions of Africa are outlined below.

Selection of Gudali and Wakwa Cattle in Cameroon

Context and Breeding Objectives: The Gudali cattle are shorthorn Zebu herded primarily by Fulani herders in the Adamawa region of Cameroon. Its productivity has predominantly been low because of traditional husbandry practices. Efforts to improve its productivity started in the 1940s with the introduction of the semen of some exotic beef cattle breeds from France and inseminating selected indigenous Gudali and Bororo females at the Animal Production Station of the Ministry of Animal Production, Fisheries and Animal Industries in Wakwa, Cameroon (Mandon 1948). The resulting crosses turned out not to be adaptable to the disease situation in the Adamawa Highlands, and so cross-breeding with Gudali for beef production was temporarily shelved. This was resumed with the introduction of Montbeliard cattle in Wakwa as part of the research programme (Tawah et al. 1996).

Attempts to improve the beef performance of the Gudali continued with the importation between 1952 and 1958 of about 45 young Brahman bulls from the United States of America (Tawah 1992). These bulls were raised and crossed with selected local Gudali females (Le Mandon 1957). The resulting first filial generation (F1) was called Pre-Wakwa, which unfortunately suffered from dermatophylosis (a tropic skin disease) and tick-borne diseases and so many of them died (Lhoste and Pierson 1975; Le Mandon 1957). As a result, the Pre-Wakwa cattle were inter se mated to produce the second filial generation (F2) known as the Wakwa cattle, which tended to be more tolerant to the skin and tick-borne diseases. This composite breed, known as Wakwa, was eventually fixed at half Brahman and half Gudali through several generations of breeding and selection (Tawah and Mbah 1989).

Still, given the limited success in trying to improve the beef performance of the indigenous Gudali through cross-breeding, in 1969 the breeding strategy shifted to the selection of the Gudali and Wakwa cattle (LHoste 1977). The breeding objective was to improve the beef performance of the indigenous Gudali and the composite Wakwa cattle populations. The conditions at the Station in Wakwa have been well documented by Lhoste (1977, 1968, 1969). Overall, the management of the animals was like what prevails under semi-intensive conditions in the cattle-rearing regions of Cameroon. It is important to note that some of the selected bulls were sold to the two government ranches in the two major cattle-rearing regions of Cameroon (Adamawa and Northwest), where improved breeding and management were taking place. These ranches were designed to multiply and disseminate improved breeding materials to interested farmers at cost.

Breeding Scheme: The breeding scheme based on mass selection carried out at the Animal Production and Veterinary Station in Wakwa started in 1968. The Gudali and Wakwa breeding females were separately herded in fenced natural grazing areas at the station. The bulls were grazed away from the female herds. Breeding herds comprised gestating, non-gestating and lactating females. They were raised in assigned paddocks on natural pastures and routinely supplemented with concentrates and mineral licks, especially during the severe dry seasons in Wakwa. Calves were weaned at 8 months of age. Weaners were separated into male and female herds and raised on natural pastures with supplementary feeding depending on the severity of the dry season. Each breeding herd with at least 30 breeding females and a bull was reconstituted every year following pre-selection based on age, mortalities, conformation and reproduction and selection based on growth, disease tolerance and fertility. At the herd level, information was traditionally collected for the purpose of evaluating young bulls with a view to selecting for sire replacements and sale of breeding stock; young heifers with a view to selecting them for herd replacements and cows to decide whether to retain or cull them. The data collected were stored in herdbooks at the Wakwa Animal and Veterinary Research Centre and included pedigree information, month and year of calving, birth weight, weaning weight, yearling weight, 18-month weight and treatment records. Young bulls of between 3 and 5 years were randomly mated annually to a herd of about 30–40 selected breeding females for progeny testing. Heifers were usually mated at 3–4 years old or 250 kg minimum body weight. Breeding cows were annually culled with their offspring for poor calf weaning weight, individual performance (age, conformation, agalactia, hardiness, disease tolerance, maternal instinct, temperament, etc.) and failure to conceive after two consecutive matings (Tawah et al. 1994).

Selection Process: Recurrent mass selection was aimed at increasing growth rate and was based on a weight ratio (index) at weaning, 12, 24 and 36 months of age. The index was determined as a ratio of an individual weight to the age-sex-breed contemporary group average weight at weaning, yearling, two-yearling and three-yearling ages (Ebangi et al. 2002). Selection intensity was affected in males and females and breeds by reproductive rate (fertility), deaths, culls, age, sales, emergency slaughters, finances, replacement needs, disease tolerance or susceptibility. Young males and females were also pre-selected for conformation and physical or structural soundness.

Selection Performance: Tawah et al. (1993) and Ebangi et al. (2002) used BLUP and Animal Models, respectively, to estimate genetic parameters for Gudali and Wakwa cattle from the breeding data from the Wakwa Animal Research Station. Tawah et al. (1993) reported estimates of direct heritability for weaning weight of 0.27 and 0.65 and for maternal heritability of 0.20 and 0.27 for Gudali and Wakwa, respectively. Correlations between direct and maternal effects for weaning weight were −0.68 and −0.39 for Gudali and Wakwa, respectively. Ebangi et al. (2001) reported direct and maternal heritability estimates of 0.37 and 0.05, 0.24 and 0.17, 0.27 and 0.19, 0.51 and 0.20, and 0.18 and 0.02, respectively, for birth weight, pre-weaning average daily gain (ADG), weaning weight, yearling weight and 18-month weight of Gudali. The corresponding total heritability estimates for these traits were 0.21, 0.07, 0.11, 0.22 and 0.18 for Gudali. They also reported corresponding estimates of 0.55 and 0.23, 0.26 and 0.07, 0.28 and 0.09, 0.18 and 0.00 and 0.14 and 0.06 for birth weight, ADG, weaning weight, yearling weight and 18-month weight of Wakwa. The corresponding total heritability estimates for these traits were 0.18, 0.12, 0.15, 0.17 and 0.17 for Wakwa.

Selection Responses: Tawah et al. (1994) estimated selection responses for weaning weight using regression of average sire estimated transmitting abilities (ETAs) and average dam estimated breeding values (EBVs) on year of calving. The ETAs and EBVs were calculated using BLUP procedures. Also, Ebangi et al. (2002) estimated EBVs for birth, weaning, yearling and 18-month weights and pre-weaning daily gain (growth rate) for Gudali and Wakwa cattle involved in this selection programme using animal model BLUP procedures. Direct and maternal and total genetic trends were estimated for each of the Gudali and Wakwa body weights and growth rate.

Tawah et al. (1994) obtained genetic trends of 0.67 kg and 1.69 kg per year for sire ETAs and −0.03 kg and −0.24 kg per year for dam EBVs for the weaning weight of Gudali and Wakwa, respectively. Ebangi et al. (2001) reported significantly positive annual mean direct genetic trends for average pre-weaning daily gain (ADG), birth weight, weaning weight, yearling weight and 18-month weight in both breeds. Corresponding annual maternal genetic trends were significantly negative except for ADG in Gudali and yearling weight in Wakwa. Ebangi et al. (2001) reported total annual genetic trends for birth, weaning, yearly and 18-month weights of 0.29 kg, 2.39 kg, 3.08 kg and 3.94 kg, respectively, for Gudali. They also obtained corresponding total annual genetic trends for birth, weaning, yearling and 18-month weights of 0.20 kg, 2.90 kg, 4.03 kg and 4.88 kg for Wakwa. As for ADG, the total annual genetic gains were 0.01 kg/day and 0.01 kg/day, respectively, for Gudali and Wakwa. Despite the infrastructural, human resources and financial constraints for this breeding programme in Wakwa, the overall selection for growth yielded moderate selection responses in both Gudali and Wakwa. Phenotypic trends were positive except for the birth weight of the Wakwa. The above results show that selection for individual growth performance can produce substantial genetic gains in these indigenous cattle populations in a tropical environment.

Selection of Gobra Cattle in Senegal
Context and Breeding Objectives: The Gobra is a lyre-horned Zebu cattle raised in the north and centre regions of Senegal by Fulani herders. Genetic improvement of the local Gobra has involved upgrading with Indo-Pakistani breeds (Sahiwal and Guzerat) and within-breed selection. However, an evaluation of the upgrading programme resulted in its abandonment in favour of selection. The selection programme started in 1963 at the Dahra Centre for Animal Research in Senegal with a foundation herd established in 1952 with animals bought from local herds. The production environment was described by Sow et al. (1988) as semi-extensive. The climate is tropical and the natural pasture is the main source of feed. Because the nutritional value of the forage decreases in the dry season, supplementary feeding is provided, especially to suckling animals and weaned calves. Management of the herd was described by Sow et al. (1988) and is like the management in any semi-intensive system.

Breeding Scheme: Breeding females were randomly assigned to sires (30–40 cows per sire) for a breeding season that ran from December to March each year. Cows not pregnant 3 months after the launch of the breeding season were re-assigned to a different bull. The station conditions were semi-extensive such that the improved breeding animals could be used in similar conditions on-farm. This was aimed at minimizing the effects of genotype × environment interactions at the on-farm level.

According to Denis and Thiongane (1974), the selection of the Gobra cattle at Dahra (Senegal) comprised several phases. The first phase was phenotypic selection to constitute the foundation herd. It was based uniquely on conformation and other empirical criteria like sex, coat colour, and shape and length of the horn. The second phase was mass selection based on growth rate and weight from birth to 24 months of age. The selection decision points were set at 10% for males destined for breeding on the station and for progeny testing, 20% for males for distribution to the pastoralists, 70% for males for culling and 80% for females retained for breeding purposes. The males were also evaluated based on the growth performance of their progeny (progeny testing) as the selection was for meat production. This was unlike the normal progeny testing with the objective of selecting bulls for their dairy potential based on the dairy performance of their daughters. Females were pre-selected based on body weight at 24 months and conformation; about 20% of the females were eliminated at each selection phase. After weaning of three calves, the females were selected based on the growth performance of their calves from birth to weaning, sexual maturity, age at first calving and calving interval. This was considered as a proxy measure of the maternal performance of the females.

Selection Process: Selection as described by Denis and Thiongane (1974) was basically mass selection based on growth performance. Selection for both males and females was based on weight at 6 months (weaning). Males were then re-selected on their 18-month weight. Then, the best 10 males were subjected to a growth performance test before the final selection of 2–3 bulls to be used as replacement bulls. Females were selected as replacements based on 24-month weight. About 5% and 80% of males and females, respectively, selected after weaning, were used as replacements. Culling of cows was based on poor reproductive performance such as long calving interval or failure to calve after two breeding seasons or poor growth performance of offspring (based on the weaning weight of offspring).

Selection Performance: Diop and Van Vleck (1998) used animal models to estimate genetic parameters for growth traits of the Gobra cattle at the Dahra Station in Senegal. Direct and maternal heritability estimates were 0.07 and 0.04, 0.20 and 0.21, 0.24 and 0.21, and 0.14 and 0.16, respectively, for birth, weaning (6 months), yearling and final (18 months) weights. Diop et al. (1999) reported similar heritability estimates for these traits when grandmaternal effects were included in the model. Correlations between direct and maternal genetic effects were negative for all traits and large for weaning and yearling weights with estimates of −0.61 and −0.50, respectively. Diop and Van Vleck (1998) reported significant positive linear phenotypic trends for weaning and yearling weights. They reported that linear trends for additive direct and maternal breeding values were not significant for all traits except the maternal breeding value for yearling weight. A review of the over two decades of the selection of the Gobra in Dahra showed that its impact on the local herds has been limited, especially as the small size of the breeding herd (200–300 cows) has limited the number of selected bulls to be distributed to farmers (Diop 1997). Similarly, an evaluation of about three decades of genetic improvement of cattle in Senegal has raised the need for (i) improving cattle productivity, (ii) thorough genetic evaluation of the selection programme, (iii) concentrating on selection within local breeds and (iv) a change in the current breeding plan to make it more efficient by considering the production objectives of the producers and their participation in defining the breeding objectives (Diop 1997). The open nucleus breeding system is considered an option to allow for the participation of producers. It is with this backdrop that Diop (1997) carried out a simulation of new breeding plans for the selection of Gobra cattle for meat and milk production under an open nucleus breeding system (a three-tier system). The results showed greater rates of genetic gain and less variation in selection response than the closed nucleus system. When the migration of pre-nucleus (multiplier) females to the nucleus was below 40%, the open nucleus showed greater rates of genetic gain.

16.3.1.3 Nucleus or Group Breeding Schemes

This scheme is based on the principle that in each herd there is a small number of genetically very superior animals which if brought together will form a nucleus whose average genetic merit is far greater than that in any of the contributing herds. The important element in this scheme is, therefore, for a group of farmers to agree to pool their high-performing animals. Once the nucleus herd is assembled, an efficient system of recording and selection is implemented. The best males are kept for breeding in the nucleus while other selected males are given to the base herds for breeding. By this means, improvements are quickly spread throughout the group.

The nucleus may remain open to animals from the base herds, the best females from the latter being admitted periodically and compared with those in the nucleus. Usually, only females are transferred from the base to the nucleus since sire selection will not be practicable in base herds due to low management conditions. The main advantage of the nucleus scheme is that the genetic superiority of sire replacements coming into the base herds from the nucleus is far greater than what is achievable in each of the base herds. It is particularly attractive in situations where within-herd selection programmes are ineffective due to small population sizes or inadequate technical skills.

Since the nucleus breeding scheme shifts the onus of operating the breeding programme from the farmer to the nucleus herd, it seems an attractive method for the tropics and subtropics because of the limitations discussed earlier. However, the organization of the scheme may have to be under government control because cooperative ventures among farmers may not always be practicable. Practical breed improved constraints in West Africa have led to the evolution of the breeding systems of the N'Dama cattle in Mali and Senegal from a closed to an open nucleus breeding system. The open nucleus system allows animals to flow between the nucleus and the local populations in both directions, while the closed nucleus only allows animals to flow from the nucleus into the local populations. Although nucleus programmes can allow accurate recording of performance compared to the on-farm programmes, where such recording does not normally occur, they require adequate infrastructure and technical inputs to operate sustainably (Kosgey et al. 2006). Many of such programmes have failed because of the lack of sustainable support and active involvement of the communities (Sölkner et al. 1998, 2008; Wurzinger et al. 2011). A good example of the open nucleus breeding programme in West Africa that has succeeded is the Djallonke sheep breeding programme which had the support of the government and involved the active participation of the community (Yapi-Gnoare 2000). It is important to note that the failure of the closed nucleus programmes in Mali and Senegal was because of the high cost of maintaining and feeding animals in the breeding unit and the unwillingness of the participating farmers to bring their best animals to the breeding unit at central station (Kahi et al. 2005). A similar case occurred with the farmers involved with the Sudanese Fulani Zebu programme in Burkina Faso, who were eventually allowed to retain the bull candidates in their respective herds. A disadvantage of the nucleus programme is the fact that the improvement of local breeds in research stations (breeding units) may not be fully transferrable to farmers' herds.

As a result, many breeding programmes now tend to involve farmers and producers in defining breeding goals and selection criteria (Madalena et al. 2002). This is the community-based approach that has been recommended for traditional, low-input smallholder farming systems (Sölkner et al. 2008; Madalena et al. 2002; Haile et al. 2011, 2019; Mueller et al. 2015; Gizaw et al. 2018). This approach was adopted for the Baoule cattle and the Baoule × Zebu cross-breds in Burkina Faso. Contrary to the conventional top-down strategy, the community-based approach involves farmers in all stages of the process from design to implementation and considers the indigenous knowledge of breeding practices and multiple objectives. The programmes typically feature a single tier with no distinction between breeding and production units because the farmers involved are both breeders and producers (Gizaw et al. 2018). According to Ouedraogo et al. (2021a), the dispersed nucleus programmes may be less expensive because the animals are managed by the farmers themselves. Moreover, the practice of distributing genetic gain to village populations by interested farmers within the dispersed nucleus scheme tends to make the breeding programme more visible to the potential beneficiaries (Mueller et al. 2015).

Consequently, to improve indigenous cattle productivity in West Africa, diverse breeding strategies and policies have been introduced (Ouédraogo et al. 2021a), including centralized breeding schemes that are entirely managed by governments with minimal, if any, participation of farmers (Bosso et al. 2009a). The breeding objective was aimed at optimizing (i) meat and milk performance and (ii) trypanotolerance of the taurine cattle, and to achieve these objectives, the following breeding strategies were applied: (i) the closed nucleus scheme which proved to be limited, (ii) open nucleus scheme and (iii) community-based breeding programme (Ouédraogo et al. 2021a). A few of the Open Nucleus cattle breeding improvement programmes reported in West Africa include the breed improvement of N'Dama in Mali, Senegal and The Gambia. Some case studies in three African countries are outlined.

Open Nucleus D'Dama Breeding in The Gambia

Context and Breeding Objectives: The N'Dama cattle (longhorn taurine cattle) are widely distributed in West and Central Africa and reared in low-input systems under periodic extreme feed scarcity and the presence of trypanosomiasis and tick-borne diseases (Dempfle and Jaitner 2000). In 1995, the International Trypanotolerance Centre (ITC) based in The Gambia initiated the N'Dama purebreeding programme aimed at disseminating the genetics of adapted N'Dama cattle across Western Africa in order to improve the welfare of livestock owners and their families through better performance and increased livestock productivity and to conserve the endemic ruminant cattle (Bosso et al. 2009a; Dempfle and Jaitner 2000; Jaitner et al. 2003a; Bosso 2006; Bosso et al. (2007); Olaniyan 2015). The breeding objectives of increasing meat and milk production without compromising its adaptation and resistance to disease were defined in a participatory manner with the farmers (Dempfle and Jaitner 2000; Jaitner et al. 2003a).

Breeding Scheme: According to Agyemang (1997), an open nucleus (ONBS) with three-tier breeding programme involving nucleus, multiplier and commercial herds with a multiple breeding objective (meat, milk and disease resistance) was initiated in The Gambia in 1994. As stated above, while selection increased meat and milk production, an effort was made to ensure that the unique characteristics of the N'Dama, which are trypanotolerance and adaptive traits, were maintained. In this breeding scheme, priorities were given to phenotypic selection, screening, breeding and genetic improvement of the N'Dama cattle. To minimize genotype × environment interactions, feeding and other management systems in the nucleus herd were like those in the multiplier and commercial herds (on-farm situations). An important benefit of this scheme was that the genetic gain of the animals in the nucleus herd is permanent and cumulative and can easily be disseminated to the other farmers' herds.

Sustainability and within-breed diversity in the nucleus were guaranteed through screening and the introduction of outstanding offspring from farmers' herds into the nucleus herd (Olaniyan 2015). To ensure a low operating cost and shortened generation interval, a simple young sire system was preferred over a progeny testing scheme in the Open nucleus (Ouédraogo et al. 2021a). Moreover, building the capacity of local staff, farmers and multipliers and selling breeding stock as a source of income also ensured the sustainability of the programme (Dempfle and Jaitner 2000).

A close working relationship between researchers in the nucleus and participants in the multiplier and commercial herds resulted in the formation of indigenous Livestock Breeders Associations in 2002 (Bosso 2006). This led to the adequate involvement of multipliers under their association known as the Gambian Indigenous Livestock Multipliers' Association in the dissemination of improved elite bulls to participating multiplier and commercial herds (Olaniyan 2015).

Unfortunately, the trend of these activities has been drastically affected by limited funding, reduced number of nucleus animals and the collapse of the multipliers' associations. Moreover, inadequate funding by ITC, coupled with insufficient human capacity and extension services in the nucleus herd, has led to the suspension of the introduction of new animals from outside herds into the nucleus. The consequence is the limitation in the number of improved elite bulls disseminated to the multiplier and commercial herds and limited farmers' involvement in the whole breeding programme (Olaniyan 2015).

Open Nucleus N'Dama Breeding in Senegal
Context and Breeding Objective: The N'Dama cattle breeding programme started in 1972 in Casamance and Kolda in the southern sub-humid areas of Senegal. Its breeding objective was to improve beef performance of the N'Dama cattle (Camara et al. 2020a). To involve farmers in the breeding programme, an open nucleus genetic improvement system was adopted in 1991 (Ouédraogo et al. 2021a). Milk performance and trypanotolerance were included as breeding objectives (Camara et al. 2020a), although the latter trait was not directly considered in the selection process. Since 2008 the programme has received funding from the African Development Bank and the Food and Agriculture Organization (FAO).

Breeding Scheme: Several breeding schemes have been implemented since 1972. The new programme started in 2008 as a three-tier open nucleus scheme consisting of a selection unit and a reproduction unit comprising herds of farmers in a cooperative of N'Dama Cattle Breeders and a dissemination unit (village herds). The breeding unit was made of 200 females and 4 males with a change in inbreeding rate per generation of 0.039. It allowed the introduction and performance testing of 12–24-month-old bulls from village herds. All farmers could participate through their cooperatives by providing candidates for the breeding unit, managing the reproduction units and disseminating improved animals.

Selection Process: In the open nucleus, two pre-selection steps were performed before the final selection when bulls are 36 months old. Bulls between 6 and 18 months of age were eligible for pre-selection based on body weight and those weighing more than 150 kg at 18 months were pre-selected. The performance of pre-selected bulls was tested from 18 to 36 months of age based on daily gain. Selection criteria were expected to include haematocrit values as an indicator of trypanotolerance but this was never implemented, although the breeding programme claims indirect selection due to the positive correlation (0.40–0.70) between haematocrit value and growth. BLUP analysis reported a genetic gain of 0.43 kg per year for weight at 36 months, while heritability estimates for birth and 36-month weights were 0.07 and 0.12, respectively.

Open Nucleus N'Dama Breeding in Mali

Context and Breeding Objectives: Genetic improvement started in 1975 with the objective of improving and conserving trypanotolerant N'Dama cattle in their own niche. The selection was for beef performance. It was initially restricted to a government ranch but has since been replicated in other locations. The open nucleus scheme started in 2008 with funding from the AfDB and FAO (Camara et al. 2020a).

Breeding Scheme: A closed nucleus breeding scheme based on mass selection was carried out on the ranch from 1981 to 1986. Feed and budgetary limitations in the ranching system led to a shift towards participatory management involving livestock technical services and farmers. Consequently, selected animals on the ranch were transferred to village farms between 1991 and 1993 to test their adaptability in these village conditions and strengthen farmers' participation in the selection process by including their objectives and practices in the overall breeding objectives of the scheme. Dissemination was based on a contractual system in which farmers were to return the same number and sex of animals to the ranch after 10 years. A selection scheme based on an open village nucleus was eventually established, but it failed because of financial constraints. However, in 2008 through funding from the AfDB and FAO (PROGEBE), a new center based on an open nucleus selection scheme was established (Olaniyan 2015). The animals previously loaned to farmers were eventually used to reconstitute the selection unit at the 'Centre de Conservation et de multiplication du Betail Ruminant Endemique de Medina Diassa or Endemic Ruminant Livestock Reproduction Centre'.

Selection Process: Animals were selected based on coat colour and conformation. Unblemished fawn animals with massive stock conformation were preferred and animals meeting these conditions were selected based on their daily gain between 8 and 18 months and weight at 18 months exceeding 150kg. Trypanotolerance was included in the selection process and low drug absorption was screened as an auxiliary trait.

16.3.1.4 Community-Based Livestock Breeding Schemes (CBBP)

Community-Based Breeding Programmes (CBBP) are being promoted as a viable or sustainable alternative for improving livestock production under smallholder conditions and in low-input systems in the tropics. Mueller et al (2015) define CBBP as a low-input type livestock breed improvement system with farmers within delineated geographical boundaries having a common interest in working together to improve their genetic resources. The participating communities usually define their breeding objectives in a participatory process which are then applied in a one to two-tier small-scale breeding structures. Since the genetic resources are generally local, CBBP can also contribute to in situ conservation. The aim of this breeding strategy is to initiate systematic breed improvement at the community level, which may include organized animal identification and recording of performance and pedigree data. To guarantee its sustainability, building capacities and ownership participation are integral parts of the approach. Also, enablers like breeding policies, market integration, legal and institutional frameworks and sustained funding are pre-requisites to ensure their continuity (Haile et al. 2019; Mueller et al. 2015; Wurzinger et al. 2021). Case studies from Burkina Faso are briefly presented:

Community-Based Breeding of Azawak Zebu Cattle in Burkina Faso

Context and Breeding Objectives: According to Ouedraogo et al. (2021a) genetic improvement programmes started in Burkina Faso in the 1990s with the objectives of enhancing local dairy potential of indigenous breeds, exploit the genetic potential of local breeds, increase the consumption of animal proteins by rural populations and increase farmers' disposable income. Between 2000 and 2015 with funding from the Belgian Technical Cooperation, a project to support the dissemination of Azawak Zebu was implemented to improve their local dairy production. To this end, pure Azawak Zebu bulls were introduced and continuously backcrossed with the local Sudanese Fulani Zebu females. This was a three-phase project with the final phase (2011–2015) funded by the BKF programme of the Luxembourg Development Cooperation and focused on the restoration of degraded pastures. The project covered 11 sites in the Sahel region and involved 329 farmers and about 2400 animals. Farmers at each site were organized into 11 Azawak Zebu breed associations, which together constituted the National Union of Azawak Zebu Breeders. The goals of the programme were to (i) increase the population of the Azawak Zebu, (ii) establish a participatory genetic improvement scheme, (iii) optimize animal management to increase milk and meat production and (iv) increase the skills of farmers.

Breeding Scheme: According to Kahi et al. (2005) and Mueller et al. (2015), a dispersed nucleus scheme was implemented with pure Azawak Zebu subjected to within-breed selection, while the Sudanese Fulani Zebu females were backcrossed to selected Azawak bulls.

Selection Process: Pure 9–12 months old Azawak bulls were selected for performance testing at a central station equipped for performing artificial insemination. Testicle size, chest girth, body weight and libido were recorded on each bull. The bulls were trained for sperm collection and sperm quality was evaluated. Bulls 30 months old were ranked according to a selection index based on daily gain and libido and then selected by a committee of technicians and farmers. Best bulls were selected and re-introduced into the breeding herds as sires, while non-selected and cross-bred bulls were destocked.

Community-Based Breeding of Sudanese Fulani Zebu Cattle in Burkina Faso

Context and Breeding Objectives: According to Ouedraogo et al. (2021b), the Sudanese Fulani Zebu cattle are the most important cattle breed in Burkina Faso, which are traditionally raised in the Sahel region because of adaptation to the conditions there. However, traditional husbandry practices are such that bulls are used for 7–8 years and the selection of replacement bulls is among their offspring, resulting in increased inbreeding in the population and reduced productivity. Also, cows with low milk yield are not milked and so they return to reproduction early, leading to 12–18 month calving intervals, while those that produce more milk and are milked tend to have calving intervals of 18–24 months, a reflection of lactation anestrus. Therefore, to improve productivity, a genetic improvement strategy was introduced through the development project to Support the Development of the Sudanese Fulani Zebu in the Sahel funded by BTC and implemented by Veterinarians Without Borders—Belgium in collaboration with a local Non-Government Organization known as 'Association Nodde Noto' from 2005 to 2018.

The objective of the project was to combine farmers' traditional knowledge with technical selection criteria to set up a breeding programme that will minimize inbreeding and improve the productivity of the Sudanese Fulani Zebu. So, the project set out to increase milk production per cow per lactation, fertility and the number of weaned calves per cow per year. The above selection criteria were determined in a participatory approach based on a long list of possible traits from the farmers. The project team, representatives of the Ministry of Livestock Resources and the farmers ranked the animals and kept the best ones for breeding. The preferred traits were milk production of the dam of the sire, body size, head and neck profile, large ears, long tail and good conformation for the males and milk production, docility, large pelvis, large and well-fixed udder, long and soft teats, belly size and fertility for the females.

Breeding Scheme: According to Kahi et al. (2005) and Mueller et al. (2015) this project was implemented in a dispersed nucleus breeding system, where the selection unit was made up of elite farmers selected by the project team because their herds contained top-breeding females with good maternal lines and the owners employed good husbandry practices and were willing to adopt the project strategy and abide by its rules. Farmers in the reproduction unit were selected by the farmers' associations. The project started with 28 farmers in the selection unit and 233 breeding cows and only 6 bulls that complied with the project's criteria for good breeding. Additional bulls were procured by the project for all 28 herds. However, a performance testing unit planned for rearing young bull candidates for selection was aborted as it turned out too technical and unappealing to the farmers (Ouédraogo et al. 2021b). Interestingly, farmers preferred to keep the young candidate bulls in their own herds. Eventually, the farmers were authorized to keep their young bulls and local barns were erected for periodic testing of the performance of the young candidate bulls.

Selection process: A controlled mating system was introduced in the selection unit involving top females from good maternal lines and selected bulls. Offspring performance was recorded, and young bulls were ranked into four groups based on an index that factored in growth, size and sexual activity (libido). The top bulls (group 1) were distributed to the farmers in the selection unit, group 2 bulls to farmers in the reproduction unit and group 3 bulls to farmers who were not part of the breeding programme. Group 4 bulls were fattened and sold on the market for meat. A selection committee of three farmers and four representatives from the Ministry of Livestock Resources and other professional organizations was responsible for overseeing the bull selection.

Community-Based Breeding of Baoule Cattle and Baoule × Zebu Cross-Breds in Burkina Faso

Context and Breeding Objectives: According to Ouedraogo et al. (2021), the Baoule cattle, also known as Lobi cattle, are the most important taurine breed in Burkina Faso. Like most Central and West African taurine breeds, these small trypanotolerant breeds are found in the hot humid agroclimatic areas and are under threat of extinction because of indiscriminate cross-breeding with Zebu cattle. Ouedraogo et al. (Ouédraogo et al. 2020a, b), reported that community-based breeding programmes were implemented in 2016 to improve the local Boaule and cross-bred cattle in southwestern Burkina Faso with funding from the Austrian Development Agency. Their goal was to implement sustainable breeding programmes to preserve and improve the local cattle breeds. Three breeding programmes, involving altogether 100 farmers and 200 animals, were introduced in three communities each with its own production system. There was one breeding programme for pure Baoule cattle in a sedentary system in Bouroum-Bouroum, another one for cross-breds in a sedentary system in Loropeni and a third one for cross-breds in a transhumant system in Kampti. A participatory approach was used to define the breeding objectives: farmers were surveyed about their trait preferences and how they ranked their own herd animals (Ouédraogo et al. 2020a). Body size was the most important selection criterion, implying that the farmers were all interested in productive traits. Consequently, the objective of the programme was to improve body size by selecting for weight at a given age and daily gain (growth).

Breeding Scheme: According to Mueller et al. (2015), village breeding schemes were implemented following the community-based breeding approach, which aims to promote the strong participation of farmers at all stages from design to implementation. The village herd was taken as the selection unit and young males were selected.

Selection process: The body weight of young candidate bulls was recorded once and again 6 months later. An index was calculated that equally weighted current age corrected body weight and growth during the previous 6 months. Animals were classified into three groups based on the index: animals that were heavier and/or growing faster for their age were ranked higher. Trypanotolerance was not included in the index but faster-growing young bulls were assumed to be more trypanotolerant (Ouedraogo et al. 2020). A selection committee in each locality comprising 3 local men, 2 women and a youth (<25 years) was provided with weight and growth data for each of the three groups of bulls and asked to choose the best bull from each group and to justify their choices. By 2020, 3 rounds of selection involving 200 candidates were performed, resulting in the selection of 70 bulls that were distributed for breeding.

16.3.1.5 Unplanned Selection Programmes

In this section, a few cases of un-designed selection attempts for improved performance in cattle are presented.

Selection of Boran Cattle in Kenya

Beef cattle breeding started in the 1920s in Kenya with European ranchers crossing indigenous African cattle with European types (Homann et al. 2005). Subsequently, most of the research efforts focused on characterization and crossbreeding of indigenous African cattle (Trial and Gregory 1981). These studies showed that the contribution of exotic beef cattle and their crosses to meat production was either similar or worse than that of indigenous cattle. Consequently, attention was shifted to the selection of indigenous cattle with a particular focus on the indigenous Boran cattle. Rege (2001) has reported the pursuit of indigenous cattle improvement programmes in Kenya involving Boran and other Zebu breeds.

The Boran, which is a large East African Zebu cattle, is classified into improved and unimproved types. The unimproved Boran is common in subsistence and semi-commercial production systems in Ethiopia, Kenya and Somalia, where they are referred to as Borana, Boran or Awai, respectively. Further information on the Boran breed in terms of distribution and population statistics has been presented in Chap. 4, Sect. 4.1.1.1

The Kenyan Boran originated from the Borana, Somali and Orma Boran brought by European ranchers from central Kenya in the early twentieth century (Rege 2001). They are primarily reared by commercial beef producers in large-scale ranches in the semi-arid areas of Kenya. Their potential for beef production brought together cattle breeders in 1951 to form the Boran Cattle Breeders' Society (Rewe et al. 2007). Individual efforts of the Boran cattle farmers have contributed to the development of the improved Kenyan Boran. However, these independent efforts with respect to selection and genetic improvement have become a major challenge for the future of the Boran cattle (Rewe et al. 2009). Besides, the absence of an organized breeding programme has further compounded this problem and has resulted in the lack of quality breeding stock and the non-sustainability of the breeding programmes. This may explain why the improved Boran has an admixture of *Bos indicus* (zebu) and *Bos taurus* (taurine) types (Hanotte et al. 2003).

Since 1951, BCBS has developed guidelines for improving indigenous cattle and has worked hard to register the improved Boran as a distinct breed. They have also maintained the breeding objective of improving beef production while at the same time selecting for efficiency and adaptation to the harsh conditions in Kenya. However, years of selecting the Boran on growth have resulted in the breed becoming more susceptible to tropical diseases compared to its unimproved type (Hanotte et al. 2003). The improved and unimproved Boran differ in performance for growth, reproduction and fitness. The improved Boran is heavier at birth than the unimproved type and so tends to have heavier sale weights. It also has shorter calving intervals and calves first at younger ages than the unimproved Boran type (Okeyo et al. 1998).

To better organize the breeding improvement of the Boran under the ranching scheme, a beef industry development programme and a National Beef Research Centre (NBRC) were established in 1968. A bull performance testing scheme was carried out at the NBRC, where three zebu breeds (Boran, Sahiwal and the East African Zebu) were evaluated. However, the lack of well-defined breeding objectives, over-dependence on donor funding, and diversified and decentralized breeding, production and marketing systems led to the stagnation of the programme (Rewe 2004). Also, in the 1970s a recording scheme was initiated, and producers were routinely sending animal performance records to the Livestock Recording Centre for genetic evaluation. However, inconsistencies and delays in releasing evaluation results coupled with a high cost of recording forced many producers to opt out of the scheme (Rewe 2009). Similarly, the beef bull performance evaluation also failed because of the lack of a long-term strategy for human resources and financing. The above failures notwithstanding, individual efforts of Boran cattle farmers under the BCBS have resulted in the selective breeding of the improved Boran. Consequently, the experiences of the Kenyan Boran cattle farmers coupled with the unique herd characteristics of the Boran cattle have attracted keen interest in the Boran cattle from South Africa and Australia, resulting in the creation of the Boran Cattle Breeders' Society of South Africa in 2003 (Rewe 2009).

Rewe (2009) carried out a simulation of alternate closed and open nucleus breeding programmes involving large-scale commercial ranchers and low-input beef producers involving the Boran cattle reared in the semi-arid regions of Kenya to evaluate the genetic and economic benefits of these programmes. The study showed that open nucleus schemes resulted in greater genetic gains than the closed nucleus system but with a lower profit. The application of a combined growth and adaptability breeding objective resulted in a gain in adaptability (trypanotolerance) but a decline in sales weight gain compared to a growth breeding objective scheme. The most profitable options for the selection schemes were the 25% nucleus size, 70% genetic contribution to commercial herds and 10% nucleus opening.

Ankole Cattle of Eastern Africa

The Ankole cattle are indigenous long-horned (Sanga) cattle in Uganda, which are reputed for their adaptability to the local environment and their cultural significance to the local keepers. There has been severe indiscriminate crossbreeding of the Ankole with the more productive exotic breeds, which has constituted a threat to its existence. Therefore, conservation of the Ankole cattle through selective breeding to improve on its production traits through a nucleus breeding programme has been set up at the Nschaara Stock Farm in the traditional Ankole keeping area in Southwestern Uganda. Uganda through its Poverty Eradication Plan has mandated the National Animal Genetic Resources Centre and Databank (NAGRC&DB), a corporate body under the Ministry of Agriculture, Animal Industry and Fisheries (MAAIF) to oversee animal breeding activities in the country. So, NAGRC&DB intends to operate several large cattle breeding stock farms with different types of breeds including the Ankole (Nakimbugwe et al. 2005). Chapter 4, Sect. 4.2 presents further information on the Ankole breed in terms of production characteristics, distribution and population statistics.

Ethiopian Boran Cattle of Eastern Africa

Chapter 4, Sect. 4.1.1.1 presents a description of the physical characteristics of the Ethiopian Boran cattle. According to Haile et al. (2011), in 1960 a herd of 351 Boran cows and 12 bulls was purchased from the community and established in the Adami Tulu ranch in Southern Ethiopia. The objective was the genetic improvement of Boran under ranch management conditions. In

1971, the breeding programme was disrupted and some of the animals were distributed to farmers and some were transferred to Abernossa ranch, which was established in 1962 with the aim of improving the Boran cattle. Both ranches, which are not far from each other, were combined in 1975 and named Adami Tulu and Abernossa Cattle Improvement and Multiplication Center. The objective of the Center was to conserve and improve the Boran cattle through selection and controlled breeding. Since its establishment, the Boran herd has remained closed and the ranch operated two breeding units—one unit for selection of pure Boran cattle and the other unit for cross-breeding improved Boran cows with exotic *Bos taurus* bulls. The Boran cows were inseminated with frozen semen from Friesian bulls.

The Boran breeding unit used the natural mating of Boran cows with superior Boran bulls on all-year breeding, which in 1983 shifted to 3-4 months (September to December) controlled breeding season until the programme was transferred to the Dida Tuyera ranch. The breeding unit operated a single sire mating scheme with herds composed of 50 cows and raised in separate paddocks. In addition, there was a herd of breeding bulls, a herd of pre-selection young bulls and a herd of pre-selection young heifers. Heifers from the pre-selection herd were used either as replacements for the breeding unit or as replacement stock to produce F1 cross-breds in the cross-breeding unit. Boran heifers were bred at 24 months of age and 250 kg body weight, while young bulls were selected at about 12 months of age for breeding based on pedigree, birth and weaning weight, breed characteristics and general body conformation.

The Dida Tuyera ranch was established in 1987 with the objective of conserving and improving the Boran cattle using selection and controlled breeding. The ranch produced and distributed to farmers pure Boran bulls and supplied replacement heifers to Abernossa ranch for cross-breeding. However, the improvement programme was disrupted in 1991 by the change in government and the looting of the animals on the ranch. The latter was re-established in 1993.

16.3.2 Emerging Molecular and Reproductive Technologies in Breeding Strategies Relevant for Genetic Improvement in Various Production Systems for Cattle and Buffaloes

Recent advancements in molecular and reproductive technologies have contributed to enhancing animal productivity and reproductive efficiency in livestock production systems in developed countries. Their implementation of breeding objectives for African livestock production systems will contribute to enhancing livestock production efficiency and in achieving faster genetic gain in breeding programmes for the various livestock input systems. Molecular technologies like genotyping, next-generation sequencing, gene editing and genetic engineering, and reproductive technologies like artificial insemination, cloning, oestrus synchronization, embryo transfer, transgenesis and in vitro fertilization have been described in more detail in Chap. 21. Biotechnological advances enable individuals to improve their genetic merit and their potential to pass on their genes to future generations while limiting the influence of environmental conditions; and are better positioned to advance African animal agriculture. Initially, we will examine the application of molecular and genomic technologies and their role in genetic improvement in various production systems for cattle and buffaloes.

16.3.2.1 Molecular and Genomic Technologies for Breeding Strategies

Advances in molecular and genomics technologies have made it feasible to identify genomic variations and elements that influence livestock traits of interest, and their implementation in breeding programmes. As a result, new options for achieving balanced breeding goals that include highly heritable (e.g. milk yield, fat yield and milk protein yield), lowly heritable (e.g. health, longevity, calving, fertility, conformation and workability traits) and emerging (e.g. fitness, animal welfare, environmental adaptability, heat tolerance and feed efficiency) traits and in improving selection responsiveness have emerged. These technologies are particularly of great importance in the improvement of breeding of traits with low heritability, traits for which phenotype measurement is difficult, expensive, and only possible late in life, or traits that are not possible on selection candidates (Dekkers 2004; Miglior et al. 2017).

Advanced technologies like next-generation sequencing and genotyping technologies support the identification of genetic markers, factors and genes that influence production, health, welfare and adaptation traits, and implementation of marker-assisted selection (MAS), genomic breeding, gene editing and transgenes in breeding objectives for improving livestock production. The development of molecular genetic techniques and gene manipulation started with first-generation sequencing technologies like the Sanger sequencing method (Sanger et al. 1977), the breakthrough technology of polymerase chain reaction (PCR) amplification (Mullis et al. 1992) and several developments in recombinant deoxyribonucleic acid (DNA) technologies. PCR gives the ability to amplify specific regions of the genome in large quantities for use in a myriad of applications, and is considered the primary technology in genome manipulation. While the Sanger sequencing technology supports the sequencing of short genomic sequences after PCR amplification at single base resolution and identification of genetic variations, other breakthrough technologies have emerged. For instance, next-generation sequencing (second and third generation) have transformed the genomic landscape and made it possible to sequence target genome regions and whole genomes at unprecedented depths. Next generation sequencing led to detection of millions of sequence variations; whole genome gene expression (transcriptomics) (identifies genes associated with specific conditions); whole epigenome sequencing (identify epigenetic alterations) and to genome editing (ability to insert, delete, replace or modify a portion of an organism's genome). These developments (summarized in Table 16.2) have supported

Table 16.2 Breakthrough biotechnological innovations in cattle and buffalo genetics and breeding

Event	Year	References
Artificial insemination applied to cattle and other farm animals	Early 1900s	Perry (1945) and Ivanoff (1922)
Development of frozen semen technology (extenders, the addition of antibiotics to semen, etc.)	1930s 1940s	Moore and Hasler (2017a)
Development in sperm cryopreservation with glycerol	1949	Moore and Hasler (2017a)
Watson and Crick published the structure of DNA in 1953	1953	
Superovulation in cattle utilizing gonadotropins prior to oestrus is reported for the first time	1975	Bó and Mapletoft (2014) and Gordon (1975)
Fertilization of cow oocytes that have been developed in vitro	1978	Betteridge (2003a, b)
Superovulation in cattle has advanced to the point where 10 live calves can be produced	1981	Betteridge 2003a and Seidel Jr. (1981)
Development of the polymerase chain reaction	1983	Mullis et al. (1986)
Transgenic cow (Herman the Bull)	1990	Krimpenfort et al. (1991)
Development of DNA level markers: single nucleotide polymorphisms (SNPs), copy number variants, restriction fragment length polymorphisms, microsatellites, etc.	1990 onwards	Weller (2016)
Human lactoferrin produced in cow's milk	1994	Vàzquez-Salat and Houdebine (2013)
First maps of quantitative trait loci (QTL) for milk production traits become available	1995	Georges et al. (1995) and Meuwissen et al. (2001)
Approximately 500,000 cattle embryos have been transferred over the world	2001	Betteridge (2003a) and Thibier (2002)

(continued)

Table 16.2 (continued)

Event	Year	References
Start of development of SNP genotyping chips	2006	Weller (2016)
Introduction of genomic selection in dairy cow improvement	2008	García-Ruiz et al. (2016) and Wiggans et al. (2011)
Bos taurus (cow) genome sequence completed	2009	Elsik et al. (2009)
Developments in gene editing technologies, gene editing in cattle, buffalo and other livestock species	2010 onwards	Nakamura et al. (2021), Randhawa and Sengar (2021), Karavolias et al. (2021a), and Van Eenennaam et al. (2021)
Complete sequence of Zebu (Bos primigenius indicus) genome	2012	Canavez et al. (2012)
Start of mapping of epigenetic alterations or regulatory elements in livestock species	2015	Andersson et al. (2015)
Reports of epigenome editing	2016	Randhawa and Sengar (2021) and Kungulovski and Jeltsch (2016)
Water buffalo genome sequence completed	2019	Mintoo et al. (2019)
Egyptian buffalo genome sequence completed	2020	(El-Khishin et al. (2020)

many advances in cattle and buffalo breeding and genetic improvement. Moreover, these developments made it possible to detect all kinds of genomic variations (e.g. single nucleotide polymorphisms (SNPs), insertions, deletions, translocation and copy number variants), epigenomic alterations (DNA methylation, histone modifications, etc.) and genes influencing livestock traits of economic importance. To identify these variations in large populations and implementation in cattle and buffalo breeding programmes, genotyping assays of different sizes have been developed. Most available genotyping assays are for DNA variations while none is currently available for epigenome alterations.

Thanks to the breakthrough technologies of molecular biology, genomics, statistical methodologies for quantitative trait loci (QTL) mapping, bioinformatics tools for analysing big sequence data, genetic factors influencing economically important features may now be found. These discoveries have the potential to greatly accelerate the speed of genetic improvement in African livestock populations by MAS of specific loci, genome-wide selection, genomic breeding, gene introgression, positional cloning and gene editing. Specific cases of the application of genetic markers and breeding technologies in African cattle and buffalo populations for implementation in selection programmes are presented in Chap. 22.

16.3.2.2 Reproductive Technologies for Breeding Strategies for African Cattle and Buffalo

Reproductive technologies are among the modern technologies that have revolutionized livestock production and in particular, dairy production starting from the early 1900s. Among the reproductive technologies, artificial insemination (AI) is best mastered and the most widely used assisted reproduction technology (ART) in developed and developing countries. Major advances and implementation of reproductive technologies in livestock breeding have been summarized in Table 16.2 and discussed in more detail in Chap. 21. The consideration of ART in breeding objectives for African livestock improvement will no doubt lead to rapid improvements in desired traits.

Artificial Insemination (AI)

Artificial insemination stands as one of the first reproductive technologies to be mastered and widely used and also the most widely adopted assisted reproduction technology in Africa. Developments in AI technology have resulted in the low cost of semen production, insemination and insemination success. For example, a single bull sire can produce between '100,000 and 150,000' doses of semen over the course of his life (Vishwanath 2003) and which can be used to inseminate thousands of cows. Therefore, AI is widely regarded as a successful, inexpensive and accessible reproductive technology, which has been available to cattle producers for over a century. Artificial insemination facilitates genetic improvement through the introduction of high-quality genetic materials into thousands of females without the need to keep genetically superior males. The AI technique involves introducing the sperm of a superior male to the reproductive tract of many female animals without the process of natural mating. The practice of AI has extensive benefits, including reduction of disease transmission, reduced need for maintenance of breeding animals, increased accuracy of pedigree recording, decreased cost of the use of improved genetics, efficient use of high-quality bulls and decreased running cost associated with keeping fewer breeding bulls (Beilby et al. 2009; Lamb and Mercadante 2016; Moore and Hasler 2017b). The AI technology is used most efficiently in the development of desired livestock, and its use has led to improvements in semen freezing, cryopreservation, sperm sexing and the rapid dissemination of high-quality genetics. The adoption of the AI technique has enabled the breeding of 100s of cows from a single ejaculation and has allowed a single bull to father over 500,000 offspring (Norman et al. 2011). Details of steps in the application of AI are covered in greater depth in Sect. 20.4.1 of Chap. 20.

Artificial insemination is used extensively in dairy cattle breeding and has popularized the practice of using progeny tests to identify genetically superior bulls followed by extensive dissemination of those genetics through commercial dairy herds (Vishwanath 2003). In beef cattle, however, this practice is not as commonly used due to the temperament and large pastures often associated with this industry (Foote 2010). AI does prove useful in smaller beef herds kept in close confinement and offers a substantial advantage in cross-breeding programmes as an inexpensive alternative for maintaining bulls of many different breeds (Foote 2010).

Case studies: Increasing demand for milk and dairy products in sub-Saharan Africa has seen the rise in the adoption of the AI technique, although this demand is still low among small-scale farmers (Mugisha et al. 2014; Tefera 2013; Omondi et al. 2017; Mwanga et al. 2019). Several factors however limit the adoption of AI in breeding goals by small-scale African livestock producers including limited experience in dairy farming, limited ability to keep records, farm size, cost of semen and AI service, and lack of necessary institutional settings and support, among others.

Animal Cloning

Animal cloning has helped to rapidly incorporate improvements into livestock herds for more than two decades and has been an important tool for scientific research since the 1950s. Although the 1997 debut of Dolly (the cloned sheep) brought animal cloning to the public consciousness, Dolly was considered a scientific breakthrough not because she was a clone, but because the source of the genetic material used to produce Dolly was from an adult cell, not embryonic (Polejaeva and Campbell 2000). Recombinant DNA technologies in conjunction with animal cloning have provided researchers with excellent animal models for studying genetic diseases, ageing, cancer and drug discovery, and in the future, they will help in the evaluation of other forms of therapy, such as gene and cell therapy. Animal cloning also provides researchers with a tool to save endangered species. In Africa, cloning may be useful for the commercial production of cattle and buffaloes.

Generally, animal cloning is the production of a genetically identical individual, through the transfer of the nucleus of differentiated adult cells into an oocyte from which the nucleus has been removed (Paterson et al. 2003). This is known as 'nuclear transfer' and it is how Dolly the sheep was produced. Following the successful cloning of Dolly, somatic cell nuclear transfer (SCNT) has been used in conjunction with modern genetic engineering technologies to produce many genetically modified farm animals (Table 16.2). This possibility has unlocked new possibilities to study the function of genes and also the establishment of animal models for an assortment of human disease conditions, and avenues to improve livestock production, health and environmental adaptation traits (Karavolias et al. 2021b; Klinger and Schnieke 2021).

Case Studies: Africa's first successful cloned cattle was named 'Futhi' (meaning 'replicate' in Zulu language; Bartels et al. 2004) and denoted a major breakthrough in the development of animal biotechnology in Africa. DNA from the ear of a 9-year-old high milk-producing Holstein cow was introduced by nuclear transfer into the eggs from a second cow and the embryo was grown in a test tube for 8 days, and then transferred into a surrogate cow resulting in the birth of Futhi 9 months later (Vajta et al. 2004). The cloning of a Kenyan Boran bull by SCNT using primary embryonic fibroblasts (Yu et al. 2016) was Africa's second successful cloned cattle. The aim of the cloning was a proof-to-concept step towards the establishment of genetically modified Kenyan Boran with genome-modified fibroblasts for disease resistance and other traits. Embryonic fibroblasts used were from a 3-month-old male Kenyan Boran foetus and introduced by SCNT into oocytes from adult *B. indicus* cattle. Following the introduction of the cloned bull to a female Boran herd, two calves were born, both showing very low glucose (blood) levels (less than 30 mg/dl) and difficulty standing alone (Fig. 16.5). One calf died a few hours after birth due to low blood glucose while a second calf survived and remained healthy after supplemental glucose was administered (Fig. 16.5) (Yu et al. 2016). This example (successful cloning of a Kenyan Boran bull) opens up the possibilities of establishing genetically modified African livestock with desired genes or traits through genome-edited fibroblast cells followed by SCNT and cloning.

Embryo Transfer

Embryo transfer (ET) refers to the process of transferring a viable embryo from a donor female to a recipient female enabling the production of multiple offspring from one superior female donor (Mishra et al. 2017). The process involves the superovulation and insemination of donor females within the proper timeframe regarding ovulation. The induction of this superovulation in the donor female is monitored via ultrasound, to determine the number of oocytes that will be available, and through daily checking of the female's oestradiol levels as an indirect measure of the quality of available oocytes (DeCherney 1986). After insemination of the female, the embryos are flushed from the donor at 1 week of age, isolated and examined using microscopy for quantity and quality; the viable embryos are then inserted into the lining of the surrogate, or recipient, female's uterus (Mishra et al. 2017). Embryo transfer is widely used globally. In 2019, a total of 1,117,151 bovine embryos were transferred, including 3645 embryos transferred in Africa (Viana 2019). The ET technology provides many benefits, including decreasing the transfer of disease associated with natural mating, improving the reproductive success of superior females, facilitating the development of rare and economically important genetic stocks and contributing to livestock breeding research by the creation of numerous closely related and genetically similar animals (Moore and Hasler 2017b; Mishra et al. 2017). When compared to AI, ET is viewed as more genetically efficient due to the ability to introduce 100% of a new genome into a herd, compared to only 50% when using AI (Hazeleger and Kemp 2001). However, ET is not a widely adopted ART technique in Africa due to a wide range of challenges associated with the technology. For example, every single step in the process must be performed with success to yield offspring. Other challenges include the success rate, which tends to decrease as the donor females age due to a decrease in success of fertilization or increase in loss of the embryo in the oviduct, variability in response to superovulation, high cost associated with the technology, labour-intensive nature of superovulation procedures, time required between collections and technical knowhow (Ashry and Smith 2015; Hansen 2020). More details on ET are presented in Sect. 20.4.2 of Chap. 20.

16.3.2.3 Nucleus Breeding Schemes for Assisted Reproduction in Cattle and Buffalo

The practice of Nucleus Breeding Schemes (NBSs) allows the dissemination of superior genes in countries and herds that do not possess either the infrastructure or the financial means to

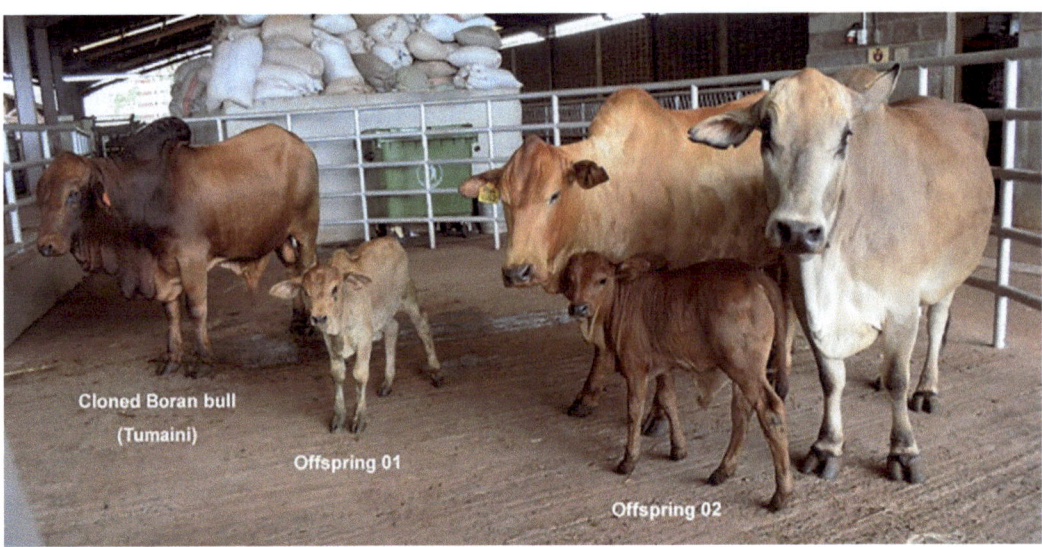

Fig. 16.5 A cloned Boran bull and its offspring (calves). Calf 1 died due to low blood glucose while calf 2 survived following supplemental glucose administration. (From Yu et al. (2016), licensed under CC-BY 4.0)

support other reproductive technologies such as AI and ET. The NBSs initially implemented within the structure of research and development projects include open nucleus, closed nucleus, dispersed nucleus and community-based breeding programmes (Ouédraogo et al. 2020). For dispersed NBS, genetic improvement is performed in the herds of elite farmers, from which improved breeding animals are distributed to other farms or herds (Kahi et al. 2005). For community-based NBS, the breeding animals are kept by elite farmers from which the best young animals are selected for breeding in a communal way (Mueller et al. 2015). The closed NBS is exclusively conducted at a central location or station from where improved animals are distributed to farmers. The open NBS on the other hand performs improved breeding of animals at a central location but also brings in select animals from farmers' herds (Kahi et al. 2005; Gicheha et al. 2006). In an open NBS, there are two types of herds: the nucleus herd, or the genetically superior herd, and the lower herd, also called the multiplier or commercial herd (Mishra et al. 2017). In this system, the genetics flow both ways between each herd, as superior animals from the nucleus herd are introduced into the lower herd to improve offspring, and members of the lower herd are introduced into the superior herd to reduce inbreeding. The open NBS can be carried out in a pedigreed herd at governmental or institutional farms, or through farming breed societies that form a cooperative breeding scheme (Mishra et al. 2017). When utilizing this system, a generation of offspring from the nucleus herd is raised, documented and the males of that generation are assessed based on the performance of their sibs, paternal half sibs or their own performance. The genetics of these superior males can then be disseminated through AI, natural service or ET based on the needs and capabilities of the producer (Mishra et al. 2017).

The open NBS system works very well for Africa and other developing countries lacking the infrastructure to support the extensive record-keeping required by AI and ET because record-keeping is only done in the nucleus herd (Kahi et al. 2005). Furthermore, an open system is more effective than a closed system due to the higher expected mean genetic value of nucleus replacements, and because the system encourages farmer participation, integrates farmer resources and reduces overhead costs (Bondoc and Smith 1993).

Case Studies: Dispersed NBS in Sudanese Fulani Zebu cattle in Burkina Faso—this was a project undertaken to multiply improved breeding stock for distribution to farmers (Kahi et al. 2005; Mueller et al. 2015). The selection unit consisted of elite farmers chosen because: their herds contained top-breeding females with good maternal lines, they practiced good husbandry, and they were willing to adopt the project strategy and follow its rules. Farms were chosen for the reproductive unit by farmers' groups. In the beginning, the project had 28 farms in the selection unit, 233 breeding cows and only 6 bulls who met the project's breeding criteria. In order to serve all 28 herds, the project purchased more bulls. A performance testing unit was developed for rearing young bull candidates for selection, but farmers found it too technical and uninteresting, preferring to keep the young bull candidates in their own herds. Farmers were eventually allowed to keep their young bulls, and local barns were established to examine the performance of young candidates on a regular basis. In the selection unit, a regulated mating system was created, comprising top females from good maternal lines and selected bulls. Young bulls were divided into four groups based on an index that took into account growth, size and sexual activity. The top bulls (group 1) were distributed to breeding unit farms; group two animals were distributed to farmers for reproduction and group three animals were distributed to farmers who were not part of the breeding programme. Animals from group four were fattened and sold for meat (Ouédraogo et al. 2021b). Bull selection was overseen by a commission of three farmers and four officials from the Ministry of Livestock Resources and other professional groups. Further examples of NBSs in cattle and other ruminants in have been presented by Mueller et al. (2015) and Ouédraogo et al. (2021b), and some examples for cattle are summarized in Table 16.3.

16.3.2.4 In Vitro Fertilization

In vitro fertilization (IVF) reproductive technology is used when other technologies fail due to anatomical issues, the presence of disease or issues with semen collection. In this technique, fertilization occurs outside the body, or in vivo, under controlled conditions (Mishra et al. 2017). Due to recent advances in 'embryo production and cryopreservation of reproductive cells', IVF has been performed with success in many species including rabbits (the first successful IVF procedure in 1959), 'hamsters, mice, cats, squirrels, rats, guinea pigs, cow, monkeys, pigs and humans' (Mishra et al. 2017). IVF allows producers to select superior genetics from both parents via the selection of semen and egg, affording this technique the potential for a large measure of genetic improvement (Mishra et al. 2017). The success of IVF culture is dependent upon the medium, maturity of the embryo at the beginning of the culture protocol and the length of the culture protocol (Mishra et al. 2017). The use of IVF offsets the lack of technological capabilities in many regions, as it is possible to fertilize a mature oocyte with just one spermatozoon, so it is possible to yield multiple offspring with relatively few sperm through IVF (Garner and Seidel 2003). The details of the methodology and the various stages involved are outlined in greater depth in Sects. 20.4.2 and 20.4.3 of Chap. 20.

Case Study: Successful implementation of IVF leading to the birth of live calves was reported in 2001 by scientists in South Africa (Arlotto et al. 2001). For the IVF procedure, oocytes (36.2 oocytes/donor) were obtained from the ovaries of six slaughtered Bovelder cows. Out of the collected oocytes, IVF with frozen semen from a Bovelder bull resulted in 43 blastocysts from five of the six donors. The top 11 blastocysts were implanted in oestrous synchronized recipient cows. Seven calves were born without any calving difficulty thus establishing the feasibility of IVF in commercial beef herds in South Africa (Arlotto et al. 2001).

16.3.2.5 Somatic Cell Nuclear Transfer

Somatic nuclear cell transfer (SCNT) is an ART that offers another technique to reproduce animals in an extraordinary way. Somatic nuclear cell transfer is a procedure whereby the nucleus or DNA of a somatic cell is placed into an oocyte that has had its own nucleus removed to create an individual that is genetically identical to the donor of the somatic cell (Tian et al. 2003). The well-known example of Dolly the Sheep was the

Table 16.3 Examples of nucleus/community breeding scheme for cattle improvement in West Africa

Parameter	N'Dama	N'Dama	N'Dama	Fulani Zebu	Baoulé and Baoulé × Zebu Crosses
Country	Senegal	Mali	Gambia	Burkina Faso	Burkina Faso
Period	1972–ongoing	1975–ongoing	1994–ongoing	2005–2018	2016–ongoing
Breeding objectives	Milk, meat and trypanotolerance	Meat and trypanotolerance	Milk, meat and trypanotolerance	Milk and control of inbreeding	Meat and trypanotolerance
Animals selected	Females and males	Males	Females and males	Females and males	Males
Selection criteria	Milk production: Females Meat and performance: Males (body weight at 18 months and daily gain from 18 to 36 months	Coat colour, Body weight at 18 months, daily gain from 18 to 36 months	Females: milk performance in the first 100 days of lactation. Males: daily gain from 0 to 10 months and 15 to 35 months	Females: own maternal line milk performance. Males: daily gain and libido at 30 months	Body weight at 3–5 years, Daily gain until 6 months
Breeding scheme	Open nucleus	Open nucleus	Open nucleus	Dispersed nucleus	Community or village breeding
Number of tiers	3	3	3	2	1
Location of nucleus	Research centre	Genetic improvement centre	Research centre	Elite farms in villages	No nucleus
Size of nucleus	200 females	Not indicated	Not indicated	233 females	not reported
	4 males			28 males	
Recording site	Station	Station	Station	Farm	Farm
Genetic evaluation	BLUP in 2012	In 1984	BLUP in 2007	No	Phenotype deviation
Participation of farmers	Breeding, reproduction and dissemination units	Reproduction and dissemination units	Reproduction and dissemination units	Selection criteria, Selection based on recorded performance (selection committee)	Selection criteria, Selection based on recorded performance (selection committee)
Stakeholders involved	NGO, National research centre	NGO, National genetic improvement centre, Government extension service	NGO, National research centre	NGO, Farmers' organization, Ministry of Livestock Resources and extension services, Genetic improvement center	National and foreign partner universities, National research institute, Government extension services
Breeders' association	Yes	No	Yes	Yes	Being set up
References	Camara et al. (2019a, b, 2020b)	Camara et al. (2019a, b, 2020b)	Bosso et al. (2007, 2009b) and Jaitner et al. (2003b)	Project reports	Ouédraogo et al. (2020b, b)

Adapted from Ouédraogo et al. (2021b)

first adult animal to be cloned using this technique, and her case used a 'differentiated adult mammary epithelial cell', which shows that this technique allows the reactivation of genes that were already inactivated on different tissues. This technology invites the potential to create clones of genetically superior animals and could be used in research to study the relationship between genotype and environment, as well as the dissemination of transgenics (Wheeler and Walters 2001). Somatic nuclear cell transfer finds application in genome editing by which genome-edited cells are turned into animals and it is also a route to reprogram somatic cells to pluripotency.

The SCNT technique may not be the most efficient on account of several factors. These include invasive micromanipulation nature of the technique; oocyte incompetence and variation in developmental efficiency; in vitro culture inconsistencies; early protocol deficiencies; high incidences of abortion and foetal mortality; the incidence of abnormal development due to incorrect reprogramming of nuclear DNA (epigenetic interference) and unusual conditions during in vitro processes (Tian et al. 2003; Gouveia et al. 2020). It is however worth noting that Africa's second successful cloned cattle, a Kenyan Boran bull, was by SCNT (Yu et al. 2016).

16.3.2.6 Multiple Ovulation and Embryo Transfer

Multiple ovulation and embryo transfer (*MOET*) is a reproductive technique that allows for the use of oocytes that would normally be wasted. This waste of oocytes can occur, especially in sheep, in the form of substantial loss of potentially valuable oocytes that normally undergo atresia (Naqvi et al. 2004). While an exciting and relatively newer reproductive technique, the procedure is laborious to implement by veterinarians and producers due to the high cost involved, inconsistent results and a general lack of knowledge of the procedure by producers (Mishra et al. 2017). Implementation of this procedure also requires proper facilities that are sanitary with good lighting, superovulation of the donor, early embryo collection and evaluation and transfer of embryos to synchronized recipients, which can induce further hurdles in the necessity of maintenance of both donor and recipient females (Betteridge 2003b). See Sect. 20.4.2 for more details.

Multiple ovulation and embryo transfer are not widely used for many reasons, one of which is the effects of inconsistent ovulatory responses to the hormonal treatment used to super-ovulate the female donors (González-Bulnes et al. 2004). This variability has both extrinsic and intrinsic factors; the extrinsic factors are the source, purity of gonadotropin and the protocol of administration, while the intrinsic factors are reproductive status, nutrition, age and breed (González-Bulnes et al. 2004).

Case Study: Multiple ovulation and embryo transfer technology were introduced in Kenya in the 1980s as a means of speeding up progress in milk production. An evaluation of the success of the MOET since its introduction indicated that super ovulations resulted in zero viable embryos to 13 transferable embryos and successful transfer rates of up to 67% (Kios et al. 2019). Problems encountered by the programme included the lack of suitable MOET protocols, poor technique implementation, lack of technical know-how, poor choice of donors/recipients and high cost associated with the technology (Kios et al. 2019).

16.3.3 Some Final Remarks and Future Perspective

An evaluation of past breeding programmes in Africa (Rewe et al. 2009; Ouédraogo et al. 2021a) has reported that many cattle genetic improvement programmes have failed for several reasons, including non-participation of producers/farmers in the setting of breeding objectives (Sölkner et al. 1998, 2008; Wurzinger et al. 2011), the central and sometimes unsustainable roles played by governments and international development agencies in setting up breeding herds (Rewe 2009; Ouédraogo et al. 2021a), and the limited role or absence of cattle breeders' societies or associations coupled with the absence of a clear demarcation between breeders and commercial producers resulting in producers playing the dual role of breeding and trading in beef (Rewe 2004). The organization of cattle genetic improvement programmes without the involvement of the livestock keepers is an impediment to their sustainability (Sölkner et al. 1998).

The selection of animals for breeding purposes tends to reflect the preferences of the products' market or consumers (commercial or market-oriented production systems) or farmers (subsistence production systems). Rewe (2009) reported that optimization of animal breeding programmes can take care of these preferences and ensure efficiency in the use of limited resources. Olesen et al. (2000) demonstrated that genetic improvement of animal performance for economic benefits is associated with the maximization of both production and profitability while the inclusion of farmers' preferences (non-market traits) limits this objective to the concept of optimization.

Farmers' knowledge and preferences about breeding goals are essential for breeding programmes that target extensive smallholder production, where animals are husbanded for multiple purposes (Mueller et al. 2015). Therefore, unlike conventional market-oriented breeding programmes, economic considerations are not the only criterion for such systems (Goddard 1998). In traditional animal husbandry systems, it is a challenge to define realistic, feasible and measurable breeding objectives that reflect the beneficiaries' needs. The breeding objectives for the above selection programmes have been meat and milk production as well as disease tolerance like trypanotolerance.

Unlike the programme involving the N'Dama cattle in Mali and Senegal where the objective of improving beef performance was decided without involving farmers, the objectives of breed improvement for the Sudanese Fulani Zebu, Baoule and Crossbreds in Burkina Faso and the N'Dama cattle in The Gambia were set after taking into consideration trait preferences of farmers in a participatory process at the start of the breeding programmes (Jaitner et al. 2003a; Ouedraogo et al. 2020). Wurzinger et al. (2021) identified limited funding and weak institutionalization as the major constraints for community-based livestock breeding programmes.

The Senegal programme eventually included milk production as a breeding objective based on farmers' interest (Camara et al. 2020a), although the goal was unrealistic as the breed's milk performance was low. Besides the productive traits, disease resistance was factored into the breeding programmes involving N'Dama, Baoule and Crossbreds in Burkina Faso. Note that disease resistance was measured in different ways across the different selection programmes. In the case of Mali and Senegal, the emphasis was on coat colour which may be attractive to tsetse flies and so influence trypanotolerance, while that of Burkina Faso considered the positive correlation between growth and disease resistance, and in The Gambia young bull candidates were reared in tsetse-infested areas. Also note that among the numerous trait preferences of farmers, it is advisable to focus on a few that represent breeding goals and are heritable and easy to measure. It is important to note that the Government of Senegal in March 2016 decided to relaunch the selection programme for both the Gobra (Dahra station in the north) and N'Dama (Kolda station in the south) cattle for the purpose of producing improved breeding stock for the cattle farmers.

The above re-emphasize the essential elements of a successful breeding strategy of farmers' participation at all stages, the free flow of information among four key partners (research (ministry/institute), development (livestock ministry) and producers (livestock farmers, and consumers, continuous evaluation and feedback to farmers researchers and development partners and increased productivity leading to improved revenue for farmers and fulfilment of several socio-cultural roles.

The breeding programmes reviewed above indicate that major emphasis has been placed on improving productive traits that drive profitability. However, changes in consumers' attitude, emerging markets, climate change and advances in technology will play a key role in breeding strategies in the future; similar to the trend in developed countries. Such breeding strategies, however, must also ensure food security and poverty alleviation in the smallholder systems, which are still dominant in the African production systems. Consumers' desire for more healthy diets and preferences for livestock raised in management systems with more welfare considerations are now apparent. Changes in the climate which could have huge impact on livestock productivity can no longer be ignored. Therefore, breeding strategies for the mitigation of climate change and the adaptation of livestock systems to accommodate climate change will play central roles in future breeding programmes. In addition, the incorporation of emerging reproductive, molecular and digital technologies must be essential elements in the design of future breeding goals and strategies in Africa to achieve efficiency in data capture, genomic predictions and rapid dissemination of superior genetics.

References

Agyemang K (1997) Village N'Dama cattle production in West Africa: six years of research in The Gambia. ILRI (aka ILCA and ILRAD)

Alexander DH, Novembre J, Lange K (2009) Fast model-based estimation of ancestry in unrelated individuals. Genome Res 19(9):1655–1664

Alonso S, Muunda E, Ahlberg S, Blackmore E, Grace D (2018) Beyond food safety: socio-economic effects of training informal dairy vendors in Kenya. Glob Food Sec 18:86–92

Andersson L, Archibald AL, Bottema CD, Brauning R, Burgess SC, Burt DW, Casas E, Cheng HH, Clarke L, Couldrey C et al (2015) Coordinated international action to accelerate genome-to-phenome with FAANG, the Functional Annotation of Animal Genomes project. Genome Biology 16(1):1–6, journal article. https://doi.org/10.1186/s13059-015-0622-4

Arlotto T, Gerber D, Terblanche S, Larsen J (2001) Birth of live calves by in vitro embryo production of slaughtered cows in a commercial herd in South Africa. J S Afr Vet Assoc 72(2):72–75

Ashry M, Smith G (2015) Application of embryo transfer using in vitro produced embryos: intrinsic factors affecting efficiency. Cattle Pract J Br Cattle Vet Assoc 23(Pt 1):1

AUIBAR (2019) The state of farm animal genetic resources in Africa. African Union Interafrican Bureau for Animal Resources (AUIBAR), Nairobi

Barker J, Davey GP (1960) The breed structure and genetic analysis of the pedigree cattle breeds in Australia. II. The Poll Hereford. Aust J Agric Res 11(6):1072–1100

Beilby K, Grupen C, Thomson P, Maxwell W, Evans G (2009) The effect of insemination time and sperm dose on pregnancy rate using sex-sorted ram sperm. Theriogenology 71(5):829–835

Betteridge KJ (2003a) A history of farm animal embryo transfer and some associated techniques. Anim Reprod Sci 79(3–4):203–244. https://doi.org/10.1016/s0378-4320(03)00166-0 From NLM

Betteridge KJ (2003b) A history of farm animal embryo transfer and some associated techniques. Anim Reprod Sci 79(3–4):203–244

Bichard M (2002) Genetic improvement in dairy cattle—an outsider's perspective. Livest Prod Sci 75(1):1–10

Bó GA, Mapletoft RJ (2014) Historical perspectives and recent research on superovulation in cattle. Theriogenology 81(1):38–48. https://doi.org/10.1016/j.theriogenology.2013.09.020 From NLM

Bondoc O, Smith C (1993) Deterministic genetic analysis of open nucleus breeding schemes for dairy cattle in developing countries. J Anim Breed Genet 110(1–6):194–208

Bosso N (2006) Genetic improvement of livestock in tsetse infested areas in West Africa. Wageningen University and Research

Bosso NA, Corr N, Njie M, Fall A, van der Waaij EH, van Arendonk JAM, Jaitner J, Dempfle L, Agyemang K (2007) The N'Dama cattle genetic improvement programme: a review. Anim Genetic Resour Infor 40:65–69. https://doi.org/10.1017/S1014233900002200 From Cambridge University Press Cambridge Core

Bosso N, Van der Waaij E, Kahi A, Van Arendonk J (2009a) Genetic analyses of N'Dama cattle breed selection schemes. Livest Res Rural Dev 21(8).:np-np

Bosso NA, Waaij EHvd, Kahi AK, Arendonk JAMvJLrfrd (2009b). Genetic analyses of N'Dama cattle breed selection schemes. 21

Bourdon RM, Bourbon RM (2000) Understanding animal breeding. Prentice Hall Upper Saddle River, NJ

Boutrais J (2007) The Fulani and cattle breeds: cross-breeding and heritage strategies. Africa 77(1):18–36

Camara Y, Moula N, Sow F, Sissokho MM, Antoine-Moussiaux N (**2019a**) Analysing innovations among cattle smallholders to evaluate the adequacy of breeding programs. Animal 13(2):417–426. https://doi.org/10.1017/S1751731118001544

Camara Y, Sow F, Govoeyi B, Moula N, Sissokho MM, Antoine-Moussiaux N (2019b) Stakeholder involvement in cattle-breeding program in developing countries: a Delphi survey. Livest Sci 228:127–135. https://doi.org/10.1016/j.livsci.2019.08.014

Camara Y, Sissokho MM, Sall M, Farnir F, Antoine-Moussiaux N (2020a) N'Dama cattle breeding programs in West Africa: case of Senegal, Mali and The Gambia. Cahiers Agric:29

Camara Y, Sissokho MM, Sall M, Farnir F, Antoine-Moussiaux NJCA (2020b) Programmes de sélection du bovin N'Dama en Afrique de l'Ouest : cas du Sénégal, du Mali et de la Gambie, 29, 11

Canavez FC, Luche DD, Stothard P, Leite KR, Sousa-Canavez JM, Plastow G, Meidanis J, Souza MA, Feijao P, Moore SS et al (2012) Genome sequence and assembly of Bos indicus. J Hered 103(3):342–348. https://doi.org/10.1093/jhered/esr153 From NLM

Casey NH (2021) A profile of South African sustainable animal production and greenhouse gas emissions. Anim Front 11(4):7–16

Chagunda MG, Mwangwela A, Mumba C, Dos Anjos F, Kawonga BS, Hopkins R, Chiwona-Kartun L (2016) Assessing and managing intensification in smallholder dairy systems for food and nutrition security in Sub-Saharan Africa. Reg Environ Chang 16:2257–2267

Chawala A, Banos G, Peters A, Chagunda M (2019) Farmer-preferred traits in smallholder dairy farming systems in Tanzania. Trop Anim Health Prod 51:1337–1344

Chawala A, Mwai A, Peters A, Banos G, Chagunda M (2020) Towards a better understanding of breeding objectives and production performance of dairy cattle in sub-Saharan Africa: a systematic review and meta-analysis. CABI Rev 2020

Cheruiyot EK, Haile-Mariam M, Cocks BG, MacLeod IM, Mrode R, Pryce JE (2022) Functionally prioritised whole-genome sequence variants improve the accuracy of genomic prediction for heat tolerance. Genet Sel Evol 54(1):1–18

Cleveland M, Blackburn H, Enns R, Garrick D (2005) Changes in inbreeding of US Herefords during the twentieth century. J Anim Sci 83(5):992–1001

Collier R, Eley R, Sharma A, Pereira R, Buffington D (1981) Shade management in subtropical environment for milk yield and composition in Holstein and Jersey cows. J Dairy Sci 64(5):844–849

Collier R, Beede D, Thatcher W, Israel L, Wilcox C (1982) Influences of environment and its modification on dairy animal health and production. J Dairy Sci 65(11):2213–2227

Corrander J, Marttinen P, Siren J, Tang J (2008) Enhanced Bayesian modelling in BAPS software for learning genetic structure of populations. BMC Bioinfor 9:539

Cunningham E (1979) The importance of continuous genetic progress in adapted breeds. In: Report of the FAO expert consultation on dairy cattle breeding in the humid tropics. FAO, Rome, pp 35–41

De Haas Y, Veerkamp R, de Jong G, Aldridge M (2021) Selective breeding as a mitigation tool for methane emissions from dairy cattle. Animal 15:100294

DeCherney AH (1986) In vitro fertilization and embryo transfer: a brief overview. Yale J Biol Med 59(4):409

Dekkers JC (2004) Commercial application of marker- and gene-assisted selection in livestock: strategies and lessons. J Anim Sci 82(suppl_13):E313–E328

Dempfle L (1993) Open nucleus breeding schemes principles, limits and practical aspects. Etude FAO: Production et Sante Animales (FAO)

Dempfle L, Jaitner J (2000) Case study about the N'Dama breeding programme at the international Trypanotolerance Centre (ITC) in The Gambia. ICAR Technical Series

Denis J, Thiongane A (1974) L'aptitude a la production de viande chez le zebu Gobra du Senegal. In: Proceedings of the first world congress on genetics applied animal production, Madrid, vol. 3, pp 889–897

Dikmen S, Cole J, Null D, Hansen P (2012) Heritability of rectal temperature and genetic correlations with production and reproduction traits in dairy cattle. J Dairy Sci 95(6):3401–3405

Diop M (1997) Design and analysis of open nucleus breeding systems for cattle in Senegal. The University of Nebraska-Lincoln

Diop M, Van Vleck L (1998) Estimates of genetic parameters for growth traits of Gobra cattle. Anim Sci 66(2):349–355

Diop M, Dodenhoff J, Van Vleck LD (1999) Estimates of direct, maternal and grandmaternal genetic effects for growth traits in Gobra cattle. Genet Mol Biol 22:363–367

EARI (2021) Ruminant Livestock Research. Ethiopian Agricultural Research Intitute (EARI). http://www.eiar.gov.et/holeta/index.php/ruminant-livestock-research. Accessed 9/3/2021

Ebangi A, Erasmus G, Mbah D, Tawah C, Messine O (2001) Genetic parameter estimates for growth traits in purebred Gudali and two-breed synthetic Wakwa beef cattle in a tropical environment. J Cameroon Acad Sci 1(1):86–95

Ebangi A, Erasmus G, Tawah C, Mbah D (2002) Genetic trends for growth in a selection experiment involving purebred and two-breed synthetic beef breed in a tropical environment= Progrès génétiques pour la croissance dans une expérience de sélection portant sur deux races bovines bouchères, pure et synthétique, dans un environnement tropical= Tendencias geneticas para el crecimiento en un experimento de seleccion, involucrando una raza de carne pura y una raza de carne sintética a partir de los razas en un medio tropical. Revue d'élevage et de médecine vétérinaire des pays tropicaux 55(4)

El-Khishin DA, Ageez A, Saad ME, Ibrahim A, Shokrof M, Hassan LR, Abouelhoda MI (2020) Sequencing and assembly of the Egyptian buffalo genome. PLoS One 15(8):e0237087. https://doi.org/10.1371/journal.pone.0237087

Elsik CG, Tellam RL, Worley KC, Gibbs RA, Muzny DM, Weinstock GM, Adelson DL, Eichler EE, Elnitski L, Guigo R et al (2009) The genome sequence of taurine cattle: a window to ruminant biology and evolution. Science 324(5926):522–528. https://doi.org/10.1126/science.1169588 From NLM

FAO (2015) The second report on the state of the world's animal genetic resources for food and agriculture. Food and Agriculture Organisation of the United Nations (FAO), Rome. http://www.fao.org/3/a-i4787e/index.html

FAOSTAT (2021) Food and Agriculture Organization of the United Nations FAOSTAT database. FAOSTAT, (FAO), F. a. A. O. o. t. U. N., Ed, Rome

Foote R (2010) The history of artificial insemination: selected notes and notables. J Anim Sci 80:1–10

García-Ruiz A, Cole JB, VanRaden PM, Wiggans GR, Ruiz-López FJ, Van Tassell CP (2016) Changes in genetic selection differentials and generation intervals in US Holstein dairy cattle as a result of genomic selection. Proc Natl Acad Sci 113(28):E3995–E4004. https://doi.org/10.1073/pnas.1519061113

Garner D, Seidel G Jr (2003) Past, present and future perspectives on sexing sperm. Can J Anim Sci 83(3):375–384

GDP, F (2018) Climate change and the global dairy cattle sector—the role of the dairy sector in a low-carbon future. The Food and Agriculture Organization, Rome

Gebreyohanes G, Okeyo Mwai A, Gebreyohanes G, Meseret S, Mrode RA, Ojango JM, Ekine C, Tessema E, Jufare B, Negussie E (2021) African dairy genetic gains (ADGG) project overview. ILRI, Nairobi

Georges M, Nielsen D, Mackinnon M, Mishra A, Okimoto R, Pasquino AT, Sargeant LS, Sorensen A, Steele MR, Zhao X et al (1995) Mapping quantitative trait loci controlling milk production in dairy cattle by exploiting progeny testing. Genetics 139(2):907–920. https://doi.org/10.1093/genetics/139.2.907 From NLM

Gerber PJ, Steinfeld H, Henderson B, Mottet A, Opio C, Dijkman J, Falcucci A, Tempio G (2013) Tackling climate change through livestock – a global assessment of emissions and mitigation opportunities. Food and Agriculture Organization of the United Nations (FAO), Rome. https://www.fao.org/3/i3437e/i3437e.pdf. Accessed 07-05-2022

Gertenbach, W. D. Dairying in Kwazulu-Natal, Breeds of dairy cattle; Department of Agriculture and Rural Development, Province of Kwazulu Natal, South Africa., 2021. https://www.kzndard.gov.za/images/Documents/RESOURCE_CENTRE/GUIDELINE.DOCUMENTS/. Accessed 21 Dec 2021

Gicheha M, Kosgey I, Bebe B, Kahi A (2006) Evaluation of the efficiency of alternative two-tier nucleus breeding systems designed to improve meat sheep in Kenya. J Anim Breed Genet 123(4):247–257

Gilibert, J. Une nouvelle race bovine: le Renitelo. 1974

Gizaw S, Abebe A, Bisrat A, Zewdie T, Tegegne A (2018) Defining smallholders' sheep breeding objectives using farmers trait preferences versus bio-economic modelling. Livest Sci 214:120–128

Goddard M (1998) Consensus and debate in the definition of breeding objectives. J Dairy Sci 81:6–18

Goldman M, Vabi B, Mbah A (1985) Semi-intensive commercial dairy farming in the Adamawa province, Republic of Cameroon: a case. Revue Science et Technique. Serie Sciences Agronomiques et Zootechniques (Cameroon)

González-Bulnes A, Baird DT, Campbell BK, Cocero MJ, García-García RM, Inskeep EK, López-Sebastián A, McNeilly AS, Santiago-Moreno J, Souza CJ (2004) Multiple factors affecting the efficiency of multiple ovulation and embryo transfer in sheep and goats. Reprod Fertil Dev 16(4):421–435

Gordon I (1975) Problems and prospects in cattle egg transfer. Ir Vet J (Ireland) 29:21–62

Gouveia C, Huyser C, Egli D, Pepper MS (2020) Lessons learned from somatic cell nuclear transfer. Int J Mol Sci 21(7):2314

Grace D (2015) Food safety in developing countries: an overview

Haile A (2011) Breeding strategy to improve Ethiopian Boran cattle for meat and milk production. ILRI (aka ILCA and ILRAD)

Haile A, Wurzinger M, Mueller J, Mirkena T, Duguma G, Okeyo Mwai A, Sölkner J, Rischkowsky BA (2011) Guidelines for setting up community-based sheep breeding programs in Ethiopia: lessons and experiences for sheep breeding in low-input systems. ICARDA tools and guidelines

Haile A, Wurzinger M, Mueller J, Mirkena T, Duguma G, Rekik M, Mwacharo JM, Okeyo Mwai A, Sölkner J, Rischkowsky BA (2019) Guidelines for setting up community-based small ruminants breeding programs

Hanotte O, Tawah CL, Bradley D, Okomo M, Verjee Y, Ochieng J, Rege J (2000) Geographic distribution and frequency of a taurine Bos taurus and an indicine Bos indicus Y specific allele amongst sub-Saharan African cattle breeds. Mol Ecol 9(4):387–396

Hanotte O, Ronin Y, Agaba M, Nilsson P, Gelhaus A, Horstmann R, Sugimoto Y, Kemp S, Gibson J, Korol A (2003) Mapping of quantitative trait loci controlling trypanotolerance in a cross of tolerant West African N'Dama and susceptible East African Boran cattle. Proc Natl Acad Sci 100(13):7443–7448

Hansen PJ (2020) The incompletely fulfilled promise of embryo transfer in cattle—why aren't pregnancy rates greater and what can we do about it? J Anim Sci 98(11):skaa288

Hazel LN (1943) The genetic basis for constructing selection indexes. Genetics 28(6):476–490

Hazel L, Lush JL (1942) The efficiency of three methods of selection. J Hered 33(11):393–399

Hazeleger W, Kemp B (2001) Recent developments in pig embryo transfer. Theriogenology 56(8):1321–1331

Homann S, Maritz J, Hülsebusch C, Meyn K, Zarate AV (2005) Boran and Tuli cattle breeds-origin, worldwide transfer, utilisation and the issue of access and benefit sharing. Verlag Grauer, Stuttgart

Initiative F (2020) Dairy cattle breeds reared in Kenya. In: The Standard. Paul Kangethe, Nairobi

IPCC (2021) Summary for policymakers. In: Climate change 2021. The Physical Science Basis

Ivanoff EI (1922) On the use of artificial insemination for zootechnical purposes in Russia. J Agric Sci 12(3):244–256. https://doi.org/10.1017/S002185960000530X From Cambridge University Press Cambridge Core

Jaitner J, Corr N, Dempfle L (2003a) Ownership pattern and management practices of cattle herds in The Gambia: implications for a breeding programme. Trop Anim Health Prod 35:179–187

Jaitner J, Corr N, Dempfle L (2003b) Ownership pattern and management practices of cattle herds in The

Gambia: implications for a breeding Programme. Trop Anim Health Prod 35(2):179–187. https://doi.org/10.1023/A:1022881703918

Kahi A, Rewe T, Kosgey I (2005) Sustainable community-based organizations for the genetic improvement of livestock in developing countries. Outlook Agric 34(4):261–270

Karavolias NG, Horner W, Abugu MN, Evanega SN (2021a) Application of gene editing for climate change in agriculture. Front Sustain Food Syst 5:296, Systematic Review. https://doi.org/10.3389/fsufs.2021.685801

Karavolias NG, Horner W, Abugu MN, Evanega SN (2021b) Application of gene editing for climate change in agriculture. Front Sustain Food Syst 5:685801

Kariuki C, Komen H, Kahi A, Van Arendonk J (2014) Optimizing the design of small-sized nucleus breeding programs for dairy cattle with minimal performance recording. J Dairy Sci 97(12):7963–7974

Kim J, Hanotte O, Mwai OA, Dessie T, Bashir S, Diallo B, Agaba M, Kim K, Kwak W, Sung S (2017) The genome landscape of indigenous African cattle. Genome Biol 18:1–14

Kim K, Kwon T, Dessie T, Yoo D, Mwai OA, Jang J, Sung S, Lee S, Salim B, Jung J (2020) The mosaic genome of indigenous African cattle as a unique genetic resource for African pastoralism. Nat Genet 52(10):1099–1110

Kios D, Tsuma V, Mutembei H (2019) Alternative follicle stimulating hormone dose rate for embryo production in dairy cattle. Dairy Vet Sci J 10(3):555787

Klinger B, Schnieke A (2021) 25th anniversary of cloning by somatic-cell nuclear transfer twenty-five years after Dolly: how far have we come? Reproduction 162(1):F1–F10

Knapp JR, Laur G, Vadas PA, Weiss WP, Tricarico JM (2014) Invited review: enteric methane in dairy cattle production: quantifying the opportunities and impact of reducing emissions. J Dairy Sci 97(6):3231–3261

Kosgey I, Okeyo A (2007) Genetic improvement of small ruminants in low-input, smallholder production systems: technical and infrastructural issues. Small Rumin Res 70(1):76–88

Kosgey I, Baker R, Udo H, Van Arendonk JA (2006) Successes and failures of small ruminant breeding programmes in the tropics: a review. Small Rumin Res 61(1):13–28

Krimpenfort P, Rademakers A, Eyestone W, van der Schans A, van den Broek S, Kooiman P, Kootwijk E, Platenburg G, Pieper F, Strijker R et al (1991) Generation of transgenic dairy cattle using 'in vitro' embryo production. Bio-technology (Nature Publishing Company) 9(9):844–847. https://doi.org/10.1038/nbt0991-844 From NLM

Kugonza DR, Nabasirye M, Hanotte O, Mpairwe D, Okeyo AM (2012) Pastoralists' indigenous selection criteria and other breeding practices of the long-horned Ankole cattle in Uganda. Trop Anim Health Prod 44:557–565

Kungulovski G, Jeltsch A (2016) Epigenome editing: state of the art, concepts, and perspectives. Trends Genet 32(2):101–113. https://doi.org/10.1016/j.tig.2015.12.001 From NLM

Lamb GC, Mercadante VR (2016) Synchronization and artificial insemination strategies in beef cattle. Vet Clinics Food Anim Pract 32(2):335–347

Le Mandon A (1957) zébu brahma au Cameroun: premiers résultats de son introduction en Adamawa. Rev Elev Med Vet Pays Trop 10(2):129–145

Lhoste P (1968) Seasonal behaviour of Zebu cattle in Adamawa, Cameroon. II. Preweaning growth of indigenous and half-bred Brahman calves. Revue Elev Méd Vét Pays trop 41:499–517

Lhoste P (1969) Cattle breeds of Adamawa (Cameroon). In: Colloque sur l'élevage, Fort-Larny, Tchad, pp 519–533

LHoste P (1977) Genetic improvement of Adamawa zebu (Cameroon) for beef production. In: First international workshop on livestock research in a humid tropical environment, Bouake, pp 761–769

Lhoste P, Pierson J (1975) Rapports Annuels CRZ de Wakwa. Ngaoundéré, Cameroun 1973(1974):1976

Lidauer MH, Leino A-M, Stephansen RS, Pösö J, Nielsen US, Fikse WF, Aamand GP (2019) Genetic evaluation for maintenance–towards genomic breeding values for saved feed in Nordic dairy cattle. Interbull Bull 55:21–25

Lowder SK, Sánchez MV, Bertini R (2021) Which farms feed the world and has farmland become more concentrated? World Dev 142:105455

Lush JL (1946) Chance as a cause of changes in gene frequency within pure breeds of livestock. Am Nat 80(792):318–342

MacNeil, M.; Nugent, R.; Snelling, W. Breeding for profit: an introduction to selection index concepts. 1997

Madalena F, Agyemang K, Cardellino R, Jain G (2002) Genetic improvement in medium-to low-input systems of animal production. Experiences to date. In: Proceedings of the 7th world congress on genetics applied to livestock production, Montpellier, France, August, 2002, pp 19–23

Makoni N, Mwai R, Redda T, van der Zijpp A, Van der Lee J (2014) White gold: opportunities for dairy sector development collaboration in East Africa. Centre for Development Innovation, Wageningen

Mandon A (1948) L'élevage des bovins et l'insémination artificielle en Adamaoua (Cameroun français). Rev Elev Med Vet Pays Trop 2(3):129–149

Manzanilla-Pech C, Gordo DM, Difford G, Pryce J, Schenkel F, Wegmann S, Miglior F, Chud T, Moate P, Williams S (2021) Breeding for reduced methane emission and feed-efficient Holstein cows: an international response. J Dairy Sci 104(8):8983–9001

Marshall K, Tebug S, Juga J, Tapio M, Missohou A (2016) Better dairy cattle breeds and better management can improve the livelihoods of the rural poor in Senegal. ILRI (aka ILCA and ILRAD)

Marshall K, Gibson JP, Mwai O, Mwacharo JM, Haile A, Getachew T, Mrode R, Kemp SJ (2019) Livestock genomics for developing countries – African examples in practice. Front Genet 10:297. https://

doi.org/10.3389/fgene.2019.00297 From NLM PubMed-not-MEDLINE

Mbah D (1982) Adaptation of dairy cattle to Wakwa (Adamawa) environment, I: resistance to cattle ticks. Revue Science et Technique 2(2/3):101–106

McDermott JJ, Staal SJ, Freeman HA, Herrero M, Van de Steeg J (2010) Sustaining intensification of smallholder livestock systems in the tropics. Livest Sci 130(1–3):95–109

Meuwissen TH, Hayes BJ, Goddard ME (2001) Prediction of total genetic value using genome-wide dense marker maps. Genetics 157(4):1819–1829. https://doi.org/10.1093/genetics/157.4.1819 From NLM

Miglior F, Fleming A, Malchiodi F, Brito LF, Martin P, Baes CF (2017) A 100-year review: identification and genetic selection of economically important traits in dairy cattle. J Dairy Sci 100(12):10251–10271

Migose S, van der Linden A, Bebe B, de Boer I, Oosting S (2020) Accuracy of estimates of milk production per lactation from limited test-day and recall data collected at smallholder dairy farms. Livest Sci 232:103911

MINEPIA (2008) Smallholder dairy development project: genetic, breeding and management programmes. Ministry of Livestock, Fisheries, and Animal Industries (MINEPIA), Yaounde

MINEPIA (2020) Dairy Sector: 165 cows to boost production. Ministry of Livestock, Fisheries, and Animal Industries (MINEPIA)

Mintoo AA, Zhang H, Chen C, Moniruzzaman M, Deng T, Anam M, Emdadul Huque QM, Guang X, Wang P, Zhong Z et al (2019) Draft genome of the river water buffalo. Ecol Evol 9(6):3378–3388. https://doi.org/10.1002/ece3.4965 PubMed

Mishra A, Reddy IJ, Gupta PSP, Mondal S (2017) Expression of apoptotic and antioxidant enzyme genes in sheep oocytes and in vitro produced embryos. Anim Biotechnol 28(1):18–25

Missanjo E, Imbayarwo-Chikosi V, Halimani T (2013) A proposed selection index for Jersey cattle in Zimbabwe. Int Sch Res Not 2013

Moore SG, Hasler JF (2017a) A 100-year review: reproductive technologies in dairy science. J Dairy Sci 100(12):10314–10331. https://doi.org/10.3168/jds.2017-13138 From NLM

Moore S, Hasler J (2017b) A 100-year review: reproductive technologies in dairy science. J Dairy Sci 100(12):10314–10331

Mrode R, Ojango J, Ekine-Dzivenu C, Aliloo H, Gibson J, Okeyo M (2021) Genomic prediction of crossbred dairy cattle in Tanzania: a route to productivity gains in smallholder dairy systems. J Dairy Sci 104(11):11779–11789

Mueller JP, Rischkowsky B, Haile A, Philipsson J, Mwai O, Besbes B, Valle Zarate A, Tibbo M, Mirkena T, Duguma G et al (2015) Community-based livestock breeding programmes: essentials and examples. J Anim Breed Genet 132(2):155–168. https://doi.org/10.1111/jbg.12136 From NLM Medline

Mugisha A, Kayiizi V, Owiny D, Mburu J (2014) Breeding services and the factors influencing their use on smallholder dairy farms in Central Uganda. Vet Med Int 2014

Muhuyi W, Lokwaleput I, Sinkeet SO (1999) Conservation and utilisation of the Sahiwal cattle in Kenya. Anim Genetic Resour/Resources génétiques animales/Recursos genéticos animales 26:35–44

Mullis K, Faloona F, Scharf S, Saiki R, Horn G, Erlich H (1986) Specific enzymatic amplification of DNA in vitro: the polymerase chain reaction. Cold Spring Harb Symp Quant Biol 51(Pt 1):263–273. https://doi.org/10.1101/sqb.1986.051.01.032 From NLM

Mullis K, Faloona F, Scharf S, Saiki R, Horn G, Erlich H (1992) Specific enzymatic amplification of DNA in vitro: the polymerase chain reaction. Biotechnol Ser:17–17

Munji MT (1973) Dairy production. In Annual report 1972/73. Animal Research Centre, Bambui

Mwanga G, Mujibi FDN, Yonah ZO, Chagunda MGG (2019) Multi-country investigation of factors influencing breeding decisions by smallholder dairy farmers in sub-Saharan Africa. Trop Anim Health Prod 51(2):395–409. https://doi.org/10.1007/s11250-018-1703-7 From NLM Medline

Nakamura M, Gao Y, Dominguez AA, Qi LS (2021) CRISPR technologies for precise epigenome editing. Nat Cell Biol 23(1):11–22. https://doi.org/10.1038/s41556-020-00620-7 From NLM

Nakimbugwe H, Sölkner J, Willam A (2005) Open nucleus cattle breeding programme in the Lake Victoria Crescent region of Uganda. Proceedings of the International Agricultural Research for Development:5–7

Naqvi S, Maurya V, Gulyani R, Joshi A, Mittal J (2004) The effect of thermal stress on superovulatory response and embryo production in Bharat merino ewes. Small Rumin Res 55(1–3):57–63

Ndumu D, Baumung R, Wurzinger M, Drucker AG, Okeyo A, Semambo D, Sölkner J (2008) Performance and fitness traits versus phenotypic appearance in the African Ankole Longhorn cattle: a novel approach to identify selection criteria for indigenous breeds. Livest Sci 113(2–3):234–242

Norman H, Hutchison J, VanRaden P (2011) Evaluations for service-sire conception rate for heifer and cow inseminations with conventional and sexed semen. J Dairy Sci 94(12):6135–6142

Nsangou A, Mbah D, Tawah C, Manchang T, Bah G, Manjeli Y, Njehoya C, Mfopit Y, Nguetoum C (2021) Amélioration génétique bovine par voie de croisement et de sélection en Afrique Tropicale: Expériences du Cameroun. J Cameroon Acad Sci 17(1):19–41

OECD, F (2016) Commodity snapshots. OECD Publishing, Paris, pp 110–129

OECD/FAO (2020) Dairy and dairy products. OECD-FAO agricultural outlook 2020–2029

Ojango JM, Marete A, Mujibi F, Rao E, Poole EJ, Rege J, Gondro C, Weerasinghe W, Gibson JP, Okeyo Mwai A (2014) A novel use of high density SNP assays to optimize choice of different crossbred dairy cattle genotypes in smallholder systems in East Africa. American Society of Animal Science

Ojango JM, Mrode R, Okeyo A, Rege J, Emerge-Africa K, Chagunda M (2017) Improving smallholder dairy farming in Africa. In: Achieving sustainable production of milk, volume 2. Burleigh Dodds Science Publishing, pp 371–396

Okantah S (2009) A review of studies on breed evaluation and genetic improvement of cattle in Ghana. Ghana J Agric Sci:42 (1-2)

Okeyo A, Mosi R, Langat L (1998) Effects of parity and previous parous status on reproductive and productive performance of Kenya Boran cows. (384) Tropical agriculture

Olaniyan OF (2015) Sustaining N'Dama cattle for the resource-poor farmers in The Gambia. Bull Anim Prod Health Afr 63(1):83–92

Olesen I, Groen AF, Gjerde B (2000) Definition of animal breeding goals for sustainable production systems. J Anim Sci 78(3):570–582

Omondi IA, Zander K, Bauer S, Baltenweck I (2017) Understanding farmers' preferences for artificial insemination services provided through dairy hubs. Animal 11(4):677–686

Opoola O, Mrode R, Banos G, Ojango J, Banga C, Simm G, Chagunda M (2019) Current situations of animal data recording, dairy improvement infrastructure, human capacity and strategic issues affecting dairy production in sub-Saharan Africa. Trop Anim Health Prod 51:1699–1705

Osei-Amponsah R, Chauhan SS, Leury BJ, Cheng L, Cullen B, Clarke IJ, Dunshea FR (2019) Genetic selection for thermotolerance in Ruminants. Animals (Basel) 9(11) https://doi.org/10.3390/ani9110948 From NLM PubMed-not-MEDLINE

Otieno D, Owuor PMG (2019) Structure of the small and medium informal dairy enterprises in Olenguruone and Bahati Sub-Counties. Structure 10(22)

Ouédraogo D, Soudré A, Ouédraogo-Koné S, Zoma BL, Yougbaré B, Khayatzadeh N, Burger PA, Mészáros G, Traoré A, Mwai OA et al (2020a) Breeding objectives and practices in three local cattle breed production systems in Burkina Faso with implication for the design of breeding programs. Livest Sci:232. https://doi.org/10.1016/j.livsci.2019.103910

Ouedraogo D, Soudre A, Ouedraogo-Kone S, Yougbaré B, Zoma BL, Tapsoba ASR, Mészáros G, Burger P, Khayatzadeh N, Wurzinger N et al (2020) Selection of breeding bulls in community-based cattle breeding programs in Burkina Faso. Bull Anim Health Prod Afr 68:103–111

Ouédraogo D, Soudré A, Ouédraogo-Koné S, Zoma BL, Yougbaré B, Khayatzadeh N, Burger PA, Mészáros G, Traoré A, Mwai OA et al (2020b) Breeding objectives and practices in three local cattle breed production systems in Burkina Faso with implication for the design of breeding programs. Livest Sci 232:103910. https://doi.org/10.1016/j.livsci.2019.103910

Ouédraogo D, Soudré A, Ouéderaogo-Koné S, Yougbaré B, Zoma BL, Tapsoba SAR, Mészáros G, Burger P, Khayatzadeh N, Wurzinger M et al (2020c) Selection of breeding bulls in community-based cattle breeding programs in Burkina Faso. Bull Anim Health Prod Afr 68:103–111

Ouédraogo D, Soudré A, Yougbaré B, Ouédraogo-Koné S, Zoma-Traoré B, Khayatzadeh N, Traoré A, Sanou M, Mészáros G, Burger PA (2021a) Genetic improvement of local cattle breeds in West Africa: a review of breeding programs. Sustain For 13(4):2125

Ouédraogo D, Ouédraogo-Koné S, Yougbaré B, Soudré A, Zoma-Traoré B, Mészáros G, Khayatzadeh N, Traoré A, Sanou M, Mwai OA (2021b) Population structure, inbreeding and admixture in local cattle populations managed by community-based breeding programs in Burkina Faso. J Anim Breed Genet 138(3):379–388

Paterson L, DeSousa P, Ritchie W, King T, Wilmut I (2003) Application of reproductive biotechnology in animals: implications and potentials: applications of reproductive cloning. Anim Reprod Sci 79(3):137–143. https://doi.org/10.1016/S0378-4320(03)00161-1

Perry EJ (1945) The artificial insemination of farm animals. Rutgers University Press

Pingpoh D, Mbah D, Tawah L (2019) Profitability of agricultural research: the case of genetic improvement of cattle for milk production in Cameroon. J Cameroon Acad Sci 15(1):3–8

Polejaeva I, Campbell K (2000) New advances in somatic cell nuclear transfer: application in transgenesis. Theriogenology 53(1):117–126

Raeliarijaona GM (2017) Etat des lieua et perspectives pour la conservation des troupeaux de bovins Renitelo a Kianjasoa. Universite D'Antananarivo, Antananarivo

Randhawa S, Sengar S (2021) Chapter One – The evolution and history of gene editing technologies. In: Ghosh D (ed) Progress in molecular biology and translational science, vol 178. Academic, pp 1–62

Rege J (2001) Zebu cattle of Kenya: uses, performance, farmer preferences, measures of genetic diversity and options for improved use. ILRI (aka ILCA and ILRAD)

Rege J, Aboagye G, Akah S, Ahunu B (1994) Crossbreeding Jersey with Ghana Shorthorn and Sokoto Gudali cattle in a tropical environment: additive and heterotic effects for milk production, reproduction and calf growth traits. Anim Sci 59(1):21–29

Rewe T (2004) Development of breeding objectives for production systems utilising the Boran Breed in Kenya. Egerton University Nakuru Kenya

Rewe T (2009) Breeding objectives and selection schemes for Boran cattle in Kenya

Rewe TO, Indetie D, Ojango JM, Kahi AK (2006a) Breeding objectives for the Boran breed in Kenya: model development and application to pasture-based production systems. Anim Sci J 77(2):163–177

Rewe T, Indetie D, Ojango JM, Kahi A (2006b) Economic values for production and functional traits and assessment of their influence on genetic improvement in the Boran cattle in Kenya. J Anim Breed Genet 123(1):23–36

Rewe T, Herold P, Kahi A, Valle Zaráte A (2007) Breeding programmes for sub-Saharan Africa. In: Book of abstracts of the 58th annual meeting of the European Association for Animal Production, Dublin, Ireland, vol. 2628, p 496

Rewe T, Herold P, Kahi A, Zárate AV (2009) Breeding indigenous cattle genetic resources for beef production in Sub-Saharan Africa. Outlook Agric 38(4):317–326

Rewe T, Herold P, Piepho H-P, Kahi A, Valle Zárate A (2010) Genetic and economic evaluation of a basic breeding programme for Kenya Boran cattle. Trop Anim Health Prod 42:327–340

Saki A (2020) Economic sustainability of extensive beef production in South Africa. Stellenbosch University, Stellenbosch

Salmon GR, Marshall K, Tebug S, Missohou A, Robinson TP, Macleod M (2018) The greenhouse gas abatement potential of productivity improving measures applied to cattle systems in a developing region. Animal 12(4):844–852

Sanger F, Nicklen S, Coulson AR (1977) DNA sequencing with chain-terminating inhibitors. Proc Natl Acad Sci 74(12):5463–5467

Schmidt GH, Van Vleck LD (1974) Principles of dairy science. WH Freeman and Company

Seidel GE Jr (1981) Superovulation and embryo transfer in cattle. Science 211(4480):351–358. https://doi.org/10.1126/science.7194504 From NLM

Sölkner J, Nakimbugwe H, Valle Zarate A (1998) Analysis of determinants for success and failure of village breeding programmes. In: Proceedings of the 6th world congress on genetics applied to livestock production, vol. 25, pp. 273–281

Sölkner J, Grausgruber H, Okeyo AM, Ruckenbauer P, Wurzinger M (2008) Breeding objectives and the relative importance of traits in plant and animal breeding: a comparative review. Euphytica 161:273–282

Sow R, Denis J, Trail J, Thiongane P, Mbaye M, Diallo I (1988) Productivité du zébu Gobra au centre de recherches zootechniques de Dahra (Sénégal). Etude et document 1(2):45

Staal SJ, Nin Pratt A, Jabbar M (2008) Dairy development for the resource poor. Part 2: Kenya and Ethiopia. Dairy development case studies. FAO PPLPI Working Paper 2008

Steinfeld H, Gerber P, Wassenaar TD, Castel V, Rosales M, Rosales M, de Haan C (2006) Livestock's long shadow: environmental issues and options. Food & Agriculture Org

Strucken EM, Al-Mamun HA, Esquivelzeta-Rabell C, Gondro C, Mwai OA, Gibson JP (2017) Genetic tests for estimating dairy breed proportion and parentage assignment in East African crossbred cattle. Genet Sel Evol 49:1–18

Syrstad O (1990) Mpwapwa cattle: an Indo-Euro-African synthesis. Trop Anim Health Prod 22(1):17–22

Tadesse M, Dessie T (2003) Milk production performance of Zebu, Holstein Friesian and their crosses in Ethiopia. Livest Res Rural Dev 15(3):1–9

Taher SM (2011) Dairy development in Morocco, Dairy Reports. Food and Agriculture Organization of the United Nations, Rome, pp 20–21

Tawah CL (1992) Genetic evaluation of growth performance of Gudali and two-breed synthetic Wakwa beef cattle populations under selection in Cameroon. Anim Sci 58:25

Tawah, C.; Mbah, D. Cattle breed evaluation and improvement in Cameroon: a review of the situation. 1989

Tawah C, Mbah D, Rege J, Oumate H (1993) Genetic evaluation of birth and weaning weight of Gudali and two-breed synthetic Wakwa beef cattle populations under selection in Cameroon: genetic and phenotypic parameters. Anim Sci 57(1):73–79

Tawah C, Rege J, Mbah D, Oumate H (1994) Genetic evaluation of birth and weaning weight of Gudali and two-breed synthetic Wakwa beef cattle populations under selection in Cameroon: genetic and phenotypic trends. Anim Sci 58(1):25–34

Tawah C, Mbah D, Lhoste P (1996) Effects of Bos taurus genes on pre-weaning growth of zebu cattle on the Adamawa highlands, Cameroon. Tropical agriculture

Tebug SF, Missohou A, Sourokou Sabi S, Juga J, Poole EJ, Tapio M, Marshall K (2018) Using body measurements to estimate live weight of dairy cattle in low-input systems in Senegal. J Appl Anim Res 46(1):87–93

Tefera SS (2013) Determinants of artificial insemination use by smallholder dairy farmers in Lemu-Bilbilo District, Ethiopia. Egerton University

Thibier M (2002) A contrasted year for the world activity of the animal embryo transfer industry – a report from the IETS Data Retrieval Committee. https://www.semanticscholar.org/paper/A-CONTRASTED-YEAR-FOR-THE-WORLD-ACTIVITY-OF-THE-A/887e61272188fd61272118e61277732afb2442748683fe2442745745

Thorpe W, Muriuki H, Omore AO, Owango M, Staal SJ (2000) Dairy development in Kenya: the past, the present and the future

Tian XC, Kubota C, Enright B, Yang X (2003) Cloning animals by somatic cell nuclear transfer–biological factors. Reprod Biol Endocrinol 1(1):1–7

Trial J, Gregory K (1981) Characterization of the Boran and Sahiwal breeds of cattle for economic characters. J Anim Sci 52(6):1286–1293

Vaccaro L, Steane DE (1990) Practical technologies and options for genetic improvement of livestock in developing countries. In: Proceedings of the FAO Expert Consultation on Strate gies for sustainable animal agriculture in developing countries. Rome, 10–14 December 1990; Mack, S., Ed.; Food and Agriculture Organization of the United Nations Rome (FAO): Rome Italy, 1993

Vajta G, Bartels P, Joubert J, de la Rey M, Treadwell R, Callesen H (2004) Production of a healthy calf by somatic cell nuclear transfer without micromanipulators and carbon dioxide incubators using the handmade cloning (HMC) and the submarine incubation system (SIS). Theriogenology 62(8):1465–1472

Van Eenennaam AL, De Figueiredo Silva F, Trott JF, Zilberman D (2021) Genetic engineering of livestock: the opportunity cost of regulatory delay. Annu Rev Anim Biosci 9:453–478. https://doi.org/10.1146/annurev-animal-061220-023052 From NLM

Vàzquez-Salat N, Houdebine LM (2013) Will GM animals follow the GM plant fate? Transgenic Res 22(1):5–13. https://doi.org/10.1007/s11248-012-9648-5 From NLM

Viana J (2019) 2018 Statistics of embryo production and transfer in domestic farm animals. Embryo Technology Newsletter 36(4):17

Vishwanath R (2003) Artificial insemination: the state of the art. Theriogenology 59(2):571–584

Weller JI (2016) Genomic selection in animals. Wiley, p 171

Wheeler M, Walters E (2001) Transgenic technology and applications in swine. Theriogenology 56(8):1345–1369

Wiggans GR, VanRaden PM, Cooper TA (2011) The genomic evaluation system in the United States: past, present, future. J Dairy Sci 94(6):3202–3211. https://doi.org/10.3168/jds.2010-3866

Wright S (1923) Mendelian analysis of the pure breeds of livestock: I. The measurement of inbreeding and relationship. J Hered 14(8):339–348

Wurzinger M, Sölkner J, Iñiguez L (2011) Important aspects and limitations in considering community-based breeding programs for low-input smallholder livestock systems. Small Rumin Res 98(1–3):170–175

Wurzinger M, Gutiérrez GA, Sölkner J, Probst L (2021) Community-based livestock breeding: coordinated action or relational process? Front Vet Sci 8:613505

Yapi-Gnoare C (2000) The open nucleus breeding programme of the Djallonké sheep in Côte D'Ivôire. ICAR Technical Series

Yu M, Muteti C, Ogugo M, Ritchie WA, Raper J, Kemp S (2016) Cloning of the African indigenous cattle breed Kenyan Boran. Anim Genet 47(4):510

Open Access This chapter is licensed under the terms of the Creative Commons Attribution 4.0 International License (http://creativecommons.org/licenses/by/4.0/), which permits use, sharing, adaptation, distribution and reproduction in any medium or format, as long as you give appropriate credit to the original author(s) and the source, provide a link to the Creative Commons license and indicate if changes were made.

The images or other third party material in this chapter are included in the chapter's Creative Commons license, unless indicated otherwise in a credit line to the material. If material is not included in the chapter's Creative Commons license and your intended use is not permitted by statutory regulation or exceeds the permitted use, you will need to obtain permission directly from the copyright holder.

17. Breeding Goals and Strategies for Improving Small Ruminant Productivity in African Input Systems

Julie M. K. Ojango, Richard Osei-Amponsah, Isidore Houaga, Timothy Gondwe, Moses Okpeku, and Donald R. Kugonza

Abstract

Africa's sheep and goat genetic resources are well adapted to their production environments, are important components of the food basket, a source of employment and wealth and fulfil socio-cultural and religious roles. This chapter provides information on the key components of sheep and goat breeding programmes, including various technologies adapted in breeding programmes in Africa. Section 17.1 presents a general overview of sheep and goat production in Africa to contextualize the importance of sheep and goat breeding programmes for the continent. This is followed by an outline of the key components required to define breeding goals for sheep and goats with examples of parameters developed and breeding practices adopted in different sheep and goat populations in Africa in Sect. 17.2. Breed-specific examples of successful breeding programmes and the lessons learned in relation to genotype–environment interaction, socio-economic aspects, and various strategies to ensure sustainable use and conservation of sheep and goat in Africa are discussed in Sects. 17.3 and 17.4. The final sections of this chapter (Sects. 17.5 and 17.6) present an overview of external factors that influence breeding structures and opportunities for adopting new technologies in sheep and goat improvement in Africa. An important recommendation is the need for governments in Africa to invest in human and institutional capacity development, encourage public–private partnerships, and embrace technological innovations in small ruminant production.

Keywords

Breeding technologies · Breeding products · selection · Cross-breeding · Genetic parameters

J. M. K. Ojango (✉)
International Livestock Research Institute ILRI, Nairobi, Kenya
e-mail: j.ojango@cgiar.org

R. Osei-Amponsah
Department of Animal Science, School of Agriculture, University of Ghana, Accra, Ghana

I. Houaga
Roslin Institute and Royal Dick School of Veterinary Studies, University of Edinburgh, Edinburgh, UK

T. Gondwe
Animal Science Department, Lilongwe University of Agriculture and Natural Resources, Lilongwe, Malawi

M. Okpeku
Discipline of Genetics, School of Life Sciences, University of Kwazulu-Natal, Durban, South Africa

D. R. Kugonza
College of Agricultural & Environmental Sciences CAES, Makerere University MAK, Kampala, Uganda

Acronyms

AGIN	African Goat Improvement Network
ARC	Agricultural Research Center
AU	African Union
AU-IBAR	African Union—Inter-African Bureau for Animal Resources
CAADP	Comprehensive Africa Agriculture Development Programme
CBBP	Community Based Breeding Programs
FAOSTAT	Food and Agricultural Organization Statistics
GE	Genotype by Environment
GPA	Global Plan of Action
ICARDA	International Center for Agricultural Research in the Dry Areas
ILRI	International Livestock Research Institute
INRA	Institut National de la recherche Agronomique
LiDESA	Livestock Development Strategy for Africa
SNP	Single Nucleotide Polymorphism
USDA	United States Department of Agriculture
USAID	United States Agency for International Development
WAD	West African Dwarf
YAN	Yankasa sheep

17.1 Background and Context

Across the continent of Africa, breeding programmes are dependent on the environment, the culture of the people raising the animals, available breeds, market demands for the products, and the prices offered for different goods. African countries are highly vulnerable to changes in climatic conditions because their economies and agricultural production systems heavily depend on rain. As a result, there is huge diversity in the adoption and implementation of livestock breed improvement programmes across the production systems. The continent hosts 33% of the world's sheep and 48% of the world's goat (FAOSTAT 2023). The regional distribution and trend in populations of sheep and goat—available through the FAOSTAT database from 2000 to 2021—are presented in Tables 17.1 and 17.2, respectively. Details on the characteristics of the sheep and goat populations and the use of the animals in different countries are outlined under Sect. 6.1 of Chap. 6 (African sheep genetic resources, diversity and unique features) and Sect. 5.1 of Chap. 5 (African goat genetic resources, diversity and unique features).

The sheep and goat populations in the different regions continue to increase over the years (Tables 17.1 and 17.2), except in Southern Africa, where the number of animals shows a decline over the years. West Africa hosts the largest numbers of both sheep and goat, followed by East Africa. Notably, the number of goats in North and Southern Africa is 50% less than the number of sheep in the same regions (Tables 17.1 and 17.2).

The large number of animals comprises an estimated **363** different breed types of sheep and **289** breed types of goat distributed across the regions, as illustrated in Table 17.3 with details on specific sheep and goat breed types available under Sect. 6.4 of Chap. 6 and Sect. 5.5 of Chap. 5, respectively. Most breed types are indigenous or local to the regions.

Table 17.1 Number of sheep in the different regions of Africa from 2000 to 2021[a]

Region	2000	2005	2010	2015	2021
Southern Africa	32,470,491	29,231,108	28,843,339	27,546,308	24,384,880
East Africa	42,107,173	55,751,971	66,076,032	87,928,460	103,647,389
Central Africa	14,006,924	18,484,823	24,429,815	33,257,740	48,386,131
West Africa	67,921,341	79,494,024	90,852,545	106,645,039	129,822,496
North Africa	96,530,009	103,523,500	112,734,569	105,924,641	110,724,177
Total Numbers	**253,035,938**	**286,485,426**	**322,936,300**	**361,302,188**	**416,965,073**
% of Global population	**23.1%**	**25.9%**	**29.4%**	**30.7%**	**32.9%**

[a]Data is from FAOSTAT (2023)

Table 17.2 Numbers of goats in the different regions of Africa from 2000 to 2021[a]

Region	2000	2005	2010	2015	2021
Southern Africa	11,707,731	11,444,650	11,463,001	10,027,826	9,254,419
East Africa	65,470,322	83,712,502	112,603,749	138,614,142	177,229,471
Central Africa	26,103,931	32,329,142	41,206,701	50,372,907	65,062,755
West Africa	87,694,541	104,645,932	121,837,350	153,350,262	181,263,644
North Africa	52,640,777	58,977,120	61,584,926	50,273,533	48,253,665
Total Number	**243,617,302**	**291,109,346**	**348,695,727**	**402,638,671**	**481,063,954**
% of Global population	**32.1%**	**34.2%**	**38%**	**40%**	**48%**

[a]Data is from FAOSTAT (2023)

Table 17.3 Number of different breeds of sheep and goats in the different regions of Africa

	Number of breed types[a]	
Region	Sheep	Goats
Southern Africa	109[b]	79
East Africa	99	101[b]
Central Africa	28	29
West Africa	62	47
North Africa	65	33
Total Number	**363**	**289**

[b]Indicates region with the largest number of different breeds in each species
[a]Data is adapted from AU-IBAR (2019)

> There is no better way to conserve a breed for future generations than to consistently keep the breed viable.

The sheep and goat are reared in production systems that vary from very low-input large-scale extensive pastoral production systems to high-input small-scale intensive production, as outlined in Sect. 2.3 of Chap. 2 on African Livestock production systems. In most systems, animal offtake rates are low <20% due to the slow growth rates of animals. Notably, 76% of the countries in Africa have no protocols established for monitoring their livestock genetic resources (James 1998; Haile et al. 2019).

In 2015, as part of the Agenda 2063 of the African Union (AU), the first 20-year Livestock Development Strategy for Africa (LiDESA) was endorsed under the Comprehensive Africa Agricultural Development Program (CAADP) (AU-IBAR 2015). Details on LiDESA are outlined in Sect. 30.3.5 of Chap. 30 (Policies, frameworks, strategies, and action plans for conservation and sustainable use of African animal genetic resources).

Economic and structural reforms have impacted market demands for different livestock products across the continent. A shift towards urban dwelling centres rather than country living means that the demand for livestock products is changing. In urban settings, the quality of products and the presentation of products affect their uptake and will have a long-term impact on how the products are produced at the farm level. Understanding the existing livestock populations, interventions undertaken to improve their productivity and learnings through past experiences is critical to devising options for harnessing desired productivity gains through breeding programmes.

> **Definitions** See Box 15.1, Chap. 15, Defining Breeding Goals and Breeding Strategies
> ***Breeding goals*** must be futuristic. Breeding goals generally include a combination of traits specified according to their relative importance. The outcome from breeding using defined breeding goals is realized many years after selection decisions are made. Hence, in developing goals, one must anticipate the demands and needs of future generations.
> A ***Breeding programme*** is aimed at generating the next-generation animals aligned to the overall breeding goal.
> ***Breeding strategies*** integrate the components of a breeding programme into a structured system for genetic improvement in order to maximize an overall objective.

17.2 Defining Breeding Goals for Small Ruminants in Africa

17.2.1 Components for Breeding Goals for Small Ruminants

For any livestock species, breeding goals should have a long-term perspective in line with national development plans (see Sect. 15.1.1 of Chap. 15). Such goals are key to addressing challenges in food production and determining incomes accruable from different product lines obtained from the targeted resource. Breeding goals should thus integrate characteristics aimed at improving productivity and efficiency of production, reproduction, health, and survival, and increasing resilience while reducing negative impacts on the environment. Before defining the breeding goals, it is important to understand the production environment, the purpose of keeping the animals, as well as the course for different products from the animals. Through mapping production systems to develop an outline of the activities and distribution channels involved in producing specific products, a better understanding of the opportunities for the breeding goal is obtained.

Breeding goals can be defined at the individual animal level or at the system level (see Sect. 15.1.3 of Chap. 15). Generally, breeding goals for small ruminant production in Africa are defined for production systems both in quantitative and qualitative terms. Economic values for different traits, though deemed important, are not routinely used in indexes for breeding sheep and goat in Africa (Bett et al. 2009a; Kosgey et al. 2003; Gebre et al. 2012). Due to limitations in data on animal performance in different production systems found in Africa, breeding goals are outlined in terms of the anticipated products or outcomes as follows:

> Approaches to improve sheep and goat productivity in Africa must be contextualized to consider the potential of indigenous breeds, the environmental constraints and the socio-economic demands (Philipsson et al. 2011).

- Increase meat production
- Increase milk production
- Increase both meat and milk production
- Improve resilience to diseases
- Enhance adaptability to different environmental conditions
- Preserve unique morphological attributes

Breeding goals usually involve a change in more than one trait simultaneously; see Sect. 15.1.2 of Chap. 15. To ensure progress is achieved in the desired direction, the relationships among traits need to be well understood and quantified. The magnitude and direction of correlations between breeding goal traits determine the rate of progress towards the overall objective. A breeding programme is normally designed and implemented to achieve the defined breeding goals. Outcomes from the breeding programme are realized several years after selection decisions are made in line with the breeding goal.

Sustainable breeding programmes in developing countries comprise several different components, as outlined below (Philipsson et al. 2011; Hutu et al. 2020):

- Characterize the production environment: existing policies, markets, socio-cultural values of communities, and the different farming systems
- Define the breeding goal
- Collate phenotypic information on the populations: the sheep and goat breeds, population characteristics, and productivity levels
- Develop infrastructure and institutional arrangements to support animal performance
- Adopt relevant breeding tools and appropriate technologies: selection criteria, reproduction methods, and evaluation methodologies
- Determine strategies for breeding: structure of the breeding programme and dissemination of genetic gains

Details on the production environments and systems under which sheep and goat are produced in Africa are outlined in Chap. 2 (Production Systems), while the breed types available and characteristics of the different populations are presented in Chap. 5 (African goat genetic resources) and Chap. 6 (African sheep genetic resources).

17.2.2 Local and Conventional Products from Sheep and Goat

The large populations of local indigenous sheep and goat breeds are used for several different products, generally produced in low quantities per individual animal when compared with exotic product-focused breeds found in temperate environments. The main motivation for sheep and goat production in Africa is to obtain meat and milk for human consumption and blood in some pastoral communities (Kahi and Wasike 2019; Muigai et al. 2018). Milk from sheep and goat is more readily digestible by infants and immune-compromised elderly people. The skins of sheep and goat are also important by-products in several countries (Wilson 1992). These skins are used to produce various leather products and are exported to specialized niche markets in different countries. Sheep and goat skins from Africa contribute up to 17.7% and 21.1% of the global production, respectively (FAOSTAT 2021). Sheep and goat also provide a source of employment and income for both men and women in addition to fulfilling different socio-economic roles within communities (Verbeek et al. 2007; Kosgey et al. 2008; Solomon et al. 2014).

The most desirable traits associated with the different products derived from sheep and goat are presented in Table 17.4.

17.2.3 Infrastructure and Institutional Arrangements

In most African countries, breeding programmes have been initiated by national governments in collaboration with bilateral donor organizations. However, the involvement of relevant stakeholders and adequate investments in appropriate infrastructure tend to be limiting (Philipsson et al. 2011). Critical infrastructure includes finance, trained personnel, breeding facilities, market channels, credit facilities, resources for data systems to enable animal data collection, monitoring and evaluation, and management structures that enable integration and inputs from different stakeholders, including governments, private sector service providers, investors, and farmers (Ayalew et al. 2003; Zonabend et al. 2013). The FAO reports on The State of the World's Animal Genetic Resources for Food and Agriculture of 2007 and 2015 (FAO 2007a, 2015) and the report on the state of animal genetic resources in Africa (AU-IBAR 2019) highlights the need for enhancing infrastructure for animal genetic resources at different levels across the continent of Africa.

Over the years, there have been several efforts to develop capacity in animal genetics and breed-

Table 17.4 Traits and performance indicators associated with different products from sheep and goat

Category	Product	Traits and performance indicators
Production	Milk	Milk yield and quality
	Meat	Weight at different ages: birth, weaning, yearling
		Growth rates, e.g. average daily gain
		Carcass quality and yield
		Fat deposition/marbling
	Hides/skins	Coat colour
		Skin quality
		Wool quality Sheep
	Manure	Quantity of manure
	Lambs/Kids	Mothering ability
Reproduction	Lambs/Kids Yearlings	Age at first lambing/kidding
		Lambing interval
		Lambing/kidding rates
		Litter size
		Weaning rate
	Yearlings Mature animals, Yearlings	Survival to yearling
		Survival longevity
Resilience	Mature animals, Yearlings	Disease resistance
		Adaptability to environments

ing in Africa (AU-IBAR 2019; Ojango et al. 2011),resulting in an increased awareness of the critical role of animal breeding in the livestock sector and increased efforts in the development of relevant policies (see Chaps. 27 and 28 on Capacity development in Africa). A greater impact of the capacity developed is evident in dairy cattle production and is expanding to small ruminant production systems at individual country levels as demonstrated in various community breeding programmes instituted recently (Mueller et al. 2015a; Haile et al. 2019; Kaumbata et al. 2020).

Additionally, interest groups are an important organizational component in developing breeds with specific traits. These comprise breeding organizations, breed societies, or different types of community innovation groups. The success of such stakeholder groups in developing and adopting breeding goals is evident in the development of the Dorper sheep and Boer goat in South Africa (Ramsay et al. 2000; Milne 2000),and of the improved strain of Djallonke sheep in Côte d'Ivoire (Yapi-Gnaore 2000).

17.2.4 Technologies Adopted in Sheep and Goat Production

Different technologies have been introduced to enhance the success of targeted small ruminant breeding programmes. These are generally implemented through pilot projects in institutionally owned flocks with limited prospects for expansion to the larger sheep and goat populations. Reviews by different authors on small ruminant breeding projects implemented in different countries of Africa illustrate the adoption of different breeding methodologies and tools. In Eastern Africa, overviews on successes and challenges in small ruminant breeding programmes are documented by Kosgey et al. (2006) and Nwogwugwu et al. (2018), while Bett et al. (2009a) and Solomon et al. (2014) reviewed different small ruminant improvement projects implemented. A review of the indigenous meat goat of Southern Africa and its genetic improvement was carried out by Visser (2019). The various reviews also highlight the need for focused and targeted genetic improvement programmes for indigenous sheep and goat populations in Africa.

The main technologies adopted to support the breeding interventions are (i) recording and monitoring of animal performance, (ii) genetic evaluation of performance in different traits with the selection of better performing animals based on the resultant breeding values, and (iii) crossbreeding. When adopting technologies, crossbreeding is generally used to introduce new breeds into the environment. In some regions, automated systems for monitoring feeding, optimizing the temperatures within housing environments, and monitoring animal health have been introduced on a small scale (see Sects. 21.3.2 and 21.4 of Chap. 21).

Technologies to enhance reproduction in small ruminant populations include artificial insemination AI, oestrus synchronization, and embryo transfer (Menchaca and Ungerfeld 2018). Although AI is the oldest reproductive technique available (see Sect. 20.4 on reproductive technologies to enhance meat and milk production in Chap. 20), its use in small ruminant populations of Africa is not widespread. In Eastern Africa, AI in goat is commercially available in different areas of Kenya to propagate improved dairy goat, while in other countries, AI has been used to introduce new temperate breeds into populations through specific donor-funded projects (Kifaro et al. 2007). Embryo transfer ET is not widely used in small ruminant production in Africa (AU-IBAR 2019).

The heritability of different traits is key to the possible response achievable in a selection or mating programme. Heritability provides information on the 'transmitablility' of a trait from one generation to another and is necessary for predicting the breeding values of animals. Estimates of heritability for different traits vary with populations, breeds, and environmental parameters as the additive genetic variance depends on gene frequencies, levels of inbreeding, and selection (see Chap. 15). The estimation of genetic parameters on a large scale in different small ruminant populations in Africa has been limited by the absence of data on the performance of animals across diverse environments and human and institutional resources. Published genetic parameter estimates for sheep and goat from select indigenous breeds are presented in Table 17.5. Parameter estimates obtained from the documented populations within a country are generally extrapolated to reflect the expected parameter estimates for animals with no performance information available that are reared under low-input systems. A review of genetic parameter estimates from sheep and goat populations in different environments across the world was presented by Shrestha and Fahmy (2005) while Jembere et al. (2017) presented a meta-analysis of genetic parameter estimates for goat populations around the world. The data highlight a paucity of parameter estimates for goat breeds of Africa. More recent estimates of genetic parameters for growth traits in indigenous small ruminant breeds in Africa are presented in Table 17.5.

Genomic tools have been used to characterize genetic diversity within small population clusters of sheep and goat in Africa. In a review of studies implemented, Visser (2019) noted that indigenous goat show moderate to high levels of genetic variation and relatively low inbreeding. The African Goat Improvement Network (AGIN) developed protocols for sampling goat in characterization studies and demonstrated the use of the Illumina goat 50k SNP panel in distinguishing indigenous African goat populations (Huson et al. 2014). Genetic diversity directly influences possible selection strategies and genetic progress achievable through breeding programmes. Using molecular markers, Yaro (2017) demonstrated substantial genetic similarities between Djallonke and Sahelian sheep breeds in Ghana. Using SNP

Table 17.5 Estimates of genetic parameters for growth traits in some sheep and goat breeds in Africa

	Heritability estimates (Standard deviation)						
Breed	Birth	Birth maternal	Weaning direct 120 days	Weaning maternal	Average daily gain to weaning	Yearling	References
Goat							
Boer	0.05–0.14		0.18–0.15	0.05–0.45	0.170	0.30 (0.07)	Visser and van Marle-Köster (2018)
West African Dwarf	0.50(0.05)		0.43(0.07)		0.32(0.08)		Bosso et al. (2007)
	0.45(0.15)		0.57(0.29)		0.55 (0.39)		Ofori and Hagan (2020)
Sheep							
Red Maasai Kenya	0.40 (0.08)	0.04 (0.05)	0.39 (0.08)	0.02			Oyieng et al. (2022)
Djallonke	0.39 (0.06)		0.54 (0.08)	–	0.54 (0.09)	0.21 (0.11)	Bosso et al. (2007)
Dorper Kenya	0.18 (0.01)	0.16 (0.01)	0.28 (0.05)	0.19 (0.04)	0.12 (0.05)	0.29 (0.09)	Kariuki et al. (2010)
	0.24 (0.09)	0.08 (0.06)	0.10 (0.07)	0.05 (0.03)			Oyieng et al. (2022)
Dorper South Africa			0.21 (0.01)	0.05 (0.01)			Zishiri et al. (2013)
D'man Sheep Morocco	0.05(0.02)	0.10 (0.02)	0.08 (0.03)	0.07 (0.03)			Boujenane et al. (2015)

markers, Huson et al. (2014) showed clear differentiation between populations from South Africa and North Africa, resulting from geographic isolation. The sequencing and analyses of diversity in indigenous sheep and goat breeds in Africa are however not complete (AU-IBAR 2019).

17.3 Breeding Strategies and Key Considerations in Small Ruminant Breeding Programmes

17.3.1 Breeding Strategies Adopted for Small Ruminants in Africa

Animal breeding involves selection within and across breeds and populations, and the mating of the selected individuals in order to obtain specific products. Breeding strategies generally involve using these two principles in various combinations. Differential breeding strategies tend to be adopted in line with the existing animal breeds, the resource endowment of the livestock keeper, and the type of production system. Concerted efforts have been made to document information on the characteristics of the diverse sheep and goat genetic resources of Africa, see Chaps. 3 and 4 in anticipation that the information generated will be used in defining breeding strategies for their sustainable use.

Selection (see Chap. 15) has been implemented differentially across sheep and goat populations in Africa. Ideally, parents for a breeding line are identified based on the assessment of the breeding values of individual animals with the aim of maximizing the response to selection in line with the breeding goal (Van der Werf 2000). Cognisance of the ancestry patterns of the parents involved is required in order to prevent the build-up of defective genes and inbreeding through the mating of closely related individuals. Decisions on which animals to be mated are often made in relation to dominance effects, which are significant particularly when the selected breeds are of different genetic lines.

Selection within specific breeds enables the maintenance of inherent traits in local populations, notably those related to adaptation and resilience to the prevailing environment. A major challenge in implementing selection programmes in Africa is the lack of appropriate data to make evidence-based decisions (Kosgey et al. 2006). Examples of breeds developed through selection within local indigenous sheep and goat populations in the different regions of the continent are presented in Table 17.6, and average performance in growth traits for the different breeds is presented in Table 17.7. Select case studies on the development of indigenous breeds through targeted selection that have achieved broad use across the continent are presented in Sect. 17.4.

> The best way to conserve a genetic resource is to ensure it is useful

> In addition to striving to obtain marketable products from animals, breeding strategies for sheep and goat production in Africa seek to address both environmental and socio-economic challenges in different regions.

Cross-breeding (see Chap. 15) has been utilized in many populations as a quick means of transferring genes and traits between breeds. Cross-breeding is generally practised between tropically adapted indigenous breeds and high-yielding breeds from more temperate zones in order to utilize heterosis in the offspring. Cross-breeding among sheep and goat populations in Africa has yielded mixed results. Technical aspects of cross-breeding as outlined in Chap. 15 are well understood, however, results from long-term cross-breeding in different sheep and goat populations tend to be highly variable. The choice of parental breeds or populations for cross-breeding depends largely on the availability of

Table 17.6 Indigenous breeds characterized and developed through planned breeding programmes

Strategy adopted	Breed name	Parent breed	Region/country	References
Sheep				
Within breed selection	Damara		Namibia, South Africa	Du Toit (n.d.), Sponenberg (2007), FAO (n.d.), Almeida (2011)
	Red Maasai		Kenya; Eastern Africa	Oyieng et al. (2022)
	Djallonke		Central and Western Africa; Cote d'Ivoire, Ghana, Togo	Yapi-Gnaore (2000)
	Bali Bali			
	Menz		Ethiopia, Eastern Africa	
Cross-breeding	Dorper	Black Head Persian x Dorset Horn	South Africa; Southern Africa	Milne (2000)
Goats				
Within breed selection	Kalahari Red		South Africa; Southern Africa	Visser and van Marle-Köster (2018)
	Galla		Kenya; Eastern Africa	
	Small East African		Kenya, Tanzania, Uganda; Eastern Africa	
	West African Dwarf		Ghana, Western Africa	Ayizanga et al. (2013)
	Mubende		Uganda, Eastern Africa	Kugonza et al. (2014)
Cross-breeding	Boer	Several indigenous breeds of S. Africa	South Africa; Southern Africa	Malan (2000), Visser and van Marle-Köster (2018)

healthy animals of appropriate breeding age and fiscal constraints, which usually result in the purchase of a limited number of unrelated animals for crossing, mostly sires.

The most successful sheep breed on the continent developed through cross-breeding is the Dorper Sheep in South Africa (Ramsay et al. 2000; Milne 2000) (Box 17.1). Dorper sheep are hardy, single-purpose hair sheep that are tolerant of harsh environments. Their adoption by livestock producers in many environments across the world emanates from their hardy nature, lower labour requirements, and high production efficiency (Knights 2010).

> **Box 17.1: Cross Breeding: Dorper Sheep in South Africa (Ramsay et al. 2000; Milne 2000)**
>
> The Dorper sheep was developed through crossing the indigenous fat-tailed Blackhead Persian with the British Dorset Horn following a series of cross-breeding trials by the South African Department of Agriculture from 1933 to 1946. The composite breed development was led by the Grootfontein Research Institute of the department, with the breeding goal to develop a breed able to survive in unfavourable conditions, reproduce regularly under local conditions, and produce lambs able to grow rapidly to a marketable size.
>
> Critical to the successful development of the Dorper was the formation of a breed society in 1950 that defined and bred for the desired characteristics and popularized the breed in markets for mutton production.
>
> Over the years, the Dorper has spread to different countries of Africa, with different strains popularized in different environments. A shorter legged more compact breed is popular in areas of higher rainfall, while a longer legged more hardy strain is found in more arid environments (Fig. 17.1).

Table 17.7 Mean performance of various indigenous sheep and goat breeds and their crosses in Africa

Goat breed local	Countries	Body weights in kg				Age at first kidding days	Fertility traits		Reference
		Birth	Weaning wt 3 months	Yearling Wt	Adult Wt >12 months		Kidding/lambing interval days	Litter size at birth	
Goats									
West African Dwarf Goat	The Gambia,	1.57 0.36	5.75 1.65	8.04 2.46		18-24			Bosso et al. (2007)
	Ghana	1.15 0.03	5.44 0.14	12.24 0.31		417.6 6.7	246.9 12.9	1.86	Ayizanga et al. (2018b)
Egyptian Nubian Goat	Egypt	1.94	15.4					2.1	Aboul-Naga et al 2012
Boer	South Africa		23-29 kg	70-120 kg					60,61
Red Sokoto Goat RSG	Nigeria	2.02	8.87	15.18			3278		Yusuf et al., 2015
WAD x RSG	Nigeria	1.81	7.41	14.87					Yusuf et al., 2015
Galla	Kenya	2.6	7.5	20.4	45				
Sheep									
Dorper D	Kenya South Africa	3.7 0.02	19.4* 30.9 4.5	36.7 45.14.5			298 44.4	1.23 0.38	Ayizanga et al. (2018b)
Djallonke sheep	Benin	1.74; 1.94	10.6 2.1	27.5 2.4		622.4 55.6	242.6 20.8	1.40.5	Gbangboche et al. (2006)
Red Maasai RM	Gambia	2.01 0.48	8.51 2.44	17.79 3.22					Bosso et al. (2007)
	Kenya	3.15 0.02	14.7 0.13						Oyieng et al. (2022)
RM x D	Kenya	3.43 0.02	16.6 0.11						Oyieng et al. (2022)
D'man	Morocco	2.92 0.03	20.3 0.14						Boujenane et al. (2015)

[a] Weaning weight was at 6 months of age

Fig. 17.1 Strains of Dorper sheep found in Kenya and South Africa

Cross-breeding practices using indigenous breeds have also led to improved productivity in the resultant cross-breds. In the hot and humid climatic conditions of southwest Nigeria, significantly higher milk production was recorded in cross-bred West African Dwarf WAD X Yankasa YAN sheep 273–385 ml day^{-1} than in either of the pure breeds (Adewumi and Olorunnisomo 2009). The daily milk production of WAD sheep ranged from 146 to 217 ml day^{-1}, while that from YAN sheep ranged between 210 and 327 ml day^{-1}. The higher milk yield in WAD x YAN reflects the ability of the crosses to combine the advantage of large body size in YAN with the adaptability of the WAD sheep to the environment (Adewumi and Olorunnisomo 2009).

Different cross-breeding programmes have been implemented for sheep and goat, targeting multiple traits in resultant populations through donor-funded projects. Through such projects, more productive animals have been developed at different scales for smallholder farming systems with mixed results. The smallholder goat improvement programme implemented across different countries in Eastern Africa by FARM Africa (2003) is illustrated in Box 17.2. Though the cross-breeding programme was successful, its small scale of implementation meant that there were too few improved cross-bred dairy goat to meet the demand. The farmers who managed improved animals sold the parent stock as the prices of improved animals were greatly inflated (Peacock et al. 2011; Ojango et al. 2010).

The degree to which the indigenous breeds within regions are interrelated is demonstrated in studies that seek to characterize populations using genomic tools (Huson et al. 2014; Yaro 2017).

Box 17.2: Cross-breeding for smallholder farming systems; *The FARM Africa Goat improvement programme*

A community-based goat improvement project that aimed to breed a dairy goat for smallholder farms was implemented in five countries through FARM Africa: Ethiopia, Tanzania, Kenya, Uganda, and South Africa. The genetic improvement model adopted to improve goat productivity evolved over 20 years of implementation across the countries (Peacock et al. 2011). The breeding strategy adopted involved cross-breeding local goats with Toggenburg bucks in community-managed buck stations, as illustrated in Fig. 17.2.

Central to implementing the breeding was community development to improve the livelihoods of smallholder farmers, improve family nutrition and incomes for poor farmers, create employment within target communities, and enhance market access by the poor. All key inputs were in the hands of the farmers, supported by private animal healthcare providers, who

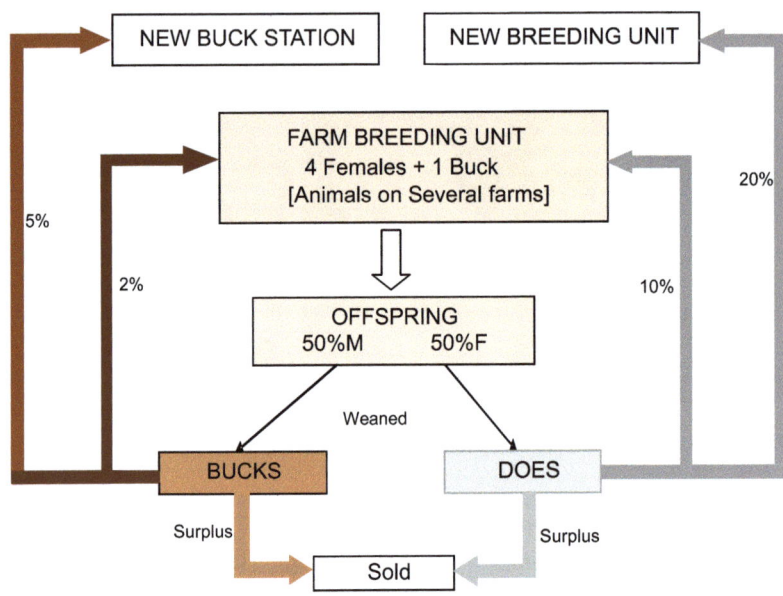

Fig. 17.2 Illustration of the FARM Africa farmer-based breeding strategy for goat improvement. Adapted from Ojango et al. (2010)

formed a breeders' association to manage the breeding of the improved goat genotypes identified. Successful development of a 'dairy goat' with requisite marketing of goat milk through project activities greatly contributed to transformation in dairy goat milk production in Eastern Africa, resulting in an increased demand for 'dairy goats'.

A high demand for breeding stock led to high off-take rates through sale of young goats at high prices. The smallholder farmers had to make choices of whether to maximize benefits through the lucrative sale of breeding animals or engage in a longer-term genetic improvement programme of their flocks. Their need for financial resources resulted in the first option, and animals were sold. This hindered continued genetic improvement of animals in the communities. Subsequent performance records of the animals that were sold were lost as the new owners generally did not follow up on keeping pedigree and performance records.

Composite Breeds Sheep and goat breed types that have been developed and stabilized as a result of crossing two or more breeds are referred to as 'composite breeds'. These are generally new breeds developed to meet the need to increase productivity while retaining the desirable characteristics of local breeds. An example mating scheme to develop a composite breed is presented in Fig. 17.3. Initially, producing the F1 cross-breds is undertaken in a base population while select female crosses with above-average performance are mated to produce the next-generation F2 of breeding females (Leymaster 2002). The composite breeds exhibit heterosis in different traits both at the maternal level and at the individual animal level. Animals selected to be parents should be screened to ensure their performance is above average for target traits in line with the breeding objective. Measures of performance should be routinely recorded and evaluated based on data collated within the flocks participating in the breeding programme.

17.3.2 Genotype x Environment Interactions in Breeding Interventions

African sheep and goat genetic resources have over time adapted to their production environments and acquired unique traits that enable them

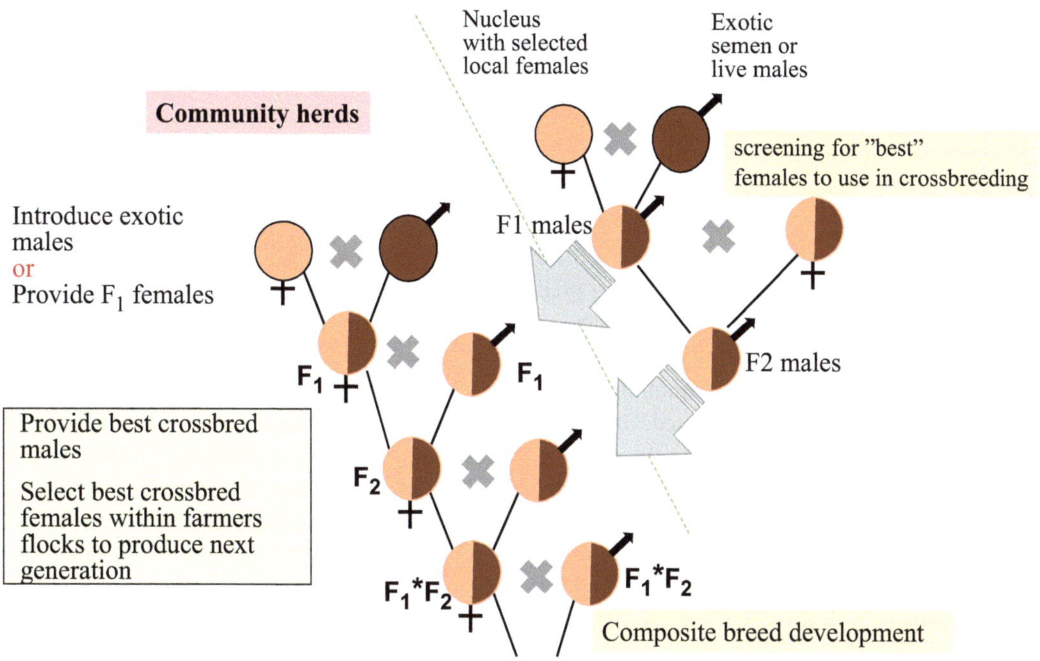

Fig. 17.3 Example of cross-breeding to develop a composite breed

to produce outputs with little inputs or modification of the environment. When importing external small ruminant genetic resources to the different production environments, genotype by environment interactions must be considered. Breeding strategies that involve the introduction of animals from more temperate environments have had varying levels of success under better managed more intensive production systems operational on national research stations and institutional farms.

The failure to incorporate Genotype by Environment GE interactions alongside long-term support for the establishment of breeding programmes has resulted in a variety of crossbred animals in different production systems of Africa. Bett et al. (2009a) reviewed nine dairy goat breeding interventions implemented in East and Central Africa that were initially aimed at reducing poverty and improving nutrition and health for women and children in households. Both two-way and four-way cross-breeding strategies were adopted for upgrading local goat breeds to performance levels of European breeds. Offspring from the two-breed crossing of indigenous and exotic breeds that demonstrate considerable potential for improved productivity are perceived by livestock keepers to be more productive under local conditions. It is notable that none of the projects included economic evidence on the performance of the different breed lines in the target environments prior to introducing cross-breeding. With the limited scale and the short term of interventions implemented, follow-up on performance, and breed development at the end of donor funding has not been available—the bane of many livestock improvement projects in Africa. This has resulted in a mosaic of crossbred animals with differential performance levels being reared across the smallholder farming systems from unplanned cross-breeding. In some populations, important characteristics of indigenous breeds such as adaptability, fecundity, and disease resistance have been lost or diluted by exotic germplasm.

17.3.3 Socio-Economic Aspects in Breeding Sheep and Goat

The sustainability of breeding programmes in different communities is highly influenced by socio-economic factors. These greatly affect access to resources and the use of the resultant products by men and women in communities. This is demonstrated through the level of adoption of breeding interventions. The impact of breeding interventions is dependent not only on their suitability to the bio-physical environment but also on the pathway for adoption which is greatly influenced by the way the breeding goal impacts gender norms and dynamics in a household. When the breeding goal does not suit the socio-economic context of the communities, its successful adoption will be limited. In many instances, little attention is given to the way the system of gender norms and dynamics at the household and community level affect the livestock enterprise in a household (Njuki and Sanginga 2013). For example, gender norms that reduce women's benefits from live animals that are controlled by their husbands or their male counterparts may reduce women's willingness to adopt a technological intervention because it would result in more work for the women who look after the animals with no commensurate benefits. The failure to align emerging technologies in livestock production, including breeding goals to gender constraints and opportunities, limits their adoption leading to low impacts on breeding in livestock systems. A study on gender dynamics in sheep and goat production (Kariuki et al. 2022) illustrated the influence of gender dynamics on flock management practices in pastoral production systems of Kenya. Results from the study demonstrated that despite women's contribution to sheep and goat management in the pastoral systems, men were perceived to exercise a greater variety of ownership rights, especially regarding decision-making over sheep and goat breeding. Prioritization of traits in animals differed between men and women depending on the division of labour and decision-making regarding the flocks. This would greatly influence the adoption of a breeding programme if implemented without considering gender dynamics. Strategies to ensure that sheep and goat improvement practices are compatible with the preferences of men and women are more likely to improve technology adoption rates (Kariuki et al. 2022; Marshall et al. 2014).

Involvement of women in setting the agenda for small ruminant breeding programmes is critical, as women are estimated to comprise 66% of the 752 million poor livestock farmers (Kristjanson et al. 2010). Women engage in livestock rearing so as to produce food, generate income, manage risks, and build assets; hence, their interests and concerns should be considered when developing livestock breeds. The idea of gender-responsive breeding of agricultural commodities in Africa has been of interest in recent years, and significant effort has been put into tooling and re-tooling of African researchers in approaching agricultural research from a gendered perspective. This involves conceptualizing and identifying relevant gender research questions, and the implementation of breeding programmes that are more appropriate to the needs of both women and men farmers. Gender-responsive breeding research must involve commitments of time, financial resources, and personnel (Mangheni et al. 2021) so as to cause long-term impacts for all actors.

Culture and beliefs across different communities tend to influence the type of livestock species raised and the breed preferences. Communities living side by side have been reported to either prefer sheep or goat in line with the use of the animals for cultural and religious ceremonies (Zonabend König et al. 2016; Ojango et al. 2016). During funerals and religious celebrations, such as *Eid-il Adha*, Christmas, and Easter festivities, either sheep or goats are slaughtered depending on the household's socio-cultural practices and preferences.

In pastoral communities, responsibilities for animals are assigned by age groups. For instance, young children are responsible for herding and grazing small ruminants, while more mature men herd the larger animals. The need for young children to attend school leaves the tending of small ruminants to women in households with less labour available to tend separated flocks. This has resulted in higher rates of inbreeding among sheep and goat in arid areas as mating is random in the mixed flocks (Oyieng et al. 2021).

17.3.4 Strategies That Enable Conservation of Indigenous Genotypes

Maintaining the diversity in African small ruminant populations through their sustainable use is an important aspect of the management of Africa's small ruminant genetic resources in line with the Global Plan of Action GPA (FAO 2007b). Indiscriminate cross-breeding and neglect of local small ruminant genetic resources lead to both the loss of important adaptive traits and of breeds unique to the diverse agroecologies in Africa. A conservation model for small ruminants that has shown the greatest traction in African production systems is the Community-Based Breeding Program CBBP approach (see Sect. 25.3.2 in Chap. 25).

Community-based breeding programmes typically target low-input systems, working with farmers within limited geographical boundaries who have a common interest in terms of improving their livestock genetic resources (Mueller et al. 2015a; Haile et al. 2011). In community-based breeding programmes, the livestock keepers in target areas are encouraged to engage in community meetings through which training and support are provided in relation to determining and adopting breeding objectives and strategies in raising sheep and goat genetic resources. Engagements with communities are aligned to a targeted impact pathway built around the African traditional concept of 'Ubuntu' understanding one another and working together for the common good. Various impact pathways adopted in communities are illustrated in a number of publications (AU-IBAR 2019; Mueller et al. 2015a; Ojango et al. 2018).

The breeding of improved animals through CBBP is implemented using a dispersed nucleus design such that better performing animals in different flocks are selected as potential 'sire-seed' animals for the larger community. When implementing CBBP, the technical expertise of scientists, extension workers, and other service providers is important for facilitation, capacity development, and adoption of critical breeding management practices such as performance recording, evaluation, and ranking of animals (Lamuno et al. 2018).

An overview of CBBP and their implementation processes is presented in the State of Farm Animal Genetic Resources in Africa by AU-IBAR (2019).

Through the African Goat Improvement Network AGIN under Feed the Future supported by USDA/ARC with USAID funding https://portal.nifa.usda.gov/web/crisprojectpages/0427456, two CBBP were implemented in goat in Uganda and Malawi from 2015. These CBBP followed the successful implementation of CBBP in sheep in Ethiopia supported by ICARDA (Mueller et al. 2015b; Gutu et al. 2015). Critical protocols in implementing CBBP as outlined by Haile et al. (2011) are as follows:

- Identify the breeding objectives at the community level
- Establish a system of selection over time and adopt objective measurement and recording of critical traits and performance measures
- Establish structures to support the implementation of the selection system determined
- Characterize the sheep and goat in the target communities
- Select the best animals within the flocks to provide 'seed sires'
- Manage the breeding and regularly evaluate the population in line with the breeding objectives
- Document progress
- Engage critical partners and out scale the breeding to other communities

CBBP have been shown to improve the productivity of goat within local populations in Malawi by 10–15% for growth and meat traits (Kaumbata et al. 2020). In Ethiopia, CBBP have been used to improve the productivity of various indigenous breeds of sheep under smallholder production systems (Weldemariam and Mezgebe 2021).

17.4 Examples of Indigenous Small Ruminant Breeds Developed or Improved Through Targeted Selection

Except for a very few cases, the implementation of breeding programmes for small ruminants in Africa over the years has mainly been driven by external funding, implemented at a small scale mainly at the institutional level with some community integration, promoting strategies geared towards traits that are perceived to be economically beneficial for the producers with little inputs from producers at the outset. In the absence of data, a large number of studies are based on questionnaires, obtaining information on trait preferences and characteristics valued by livestock keepers in the different regions.

There have been several programmes implemented to improve sheep and goat production in the different regions of Africa over time, with variable levels of success. For many small ruminant populations in Africa, records on relevant phenotypes are limited. The most common breeding system adopted to develop and disseminate improved genetics in small ruminant populations in Africa is an Open Nucleus Breeding system. Conventionally, animals in the open nucleus system belong to one of three tiers: nucleus-based animals that are used for generating the best animals of the selected breed, multiplier flocks that are used to multiply numbers of the improved animals derived in the nucleus for use in the larger community, or village flocks dispersed in different regions. Movement of improved genetic material across the tiers is generally through improved male animals. Animals in multiplier and community flocks with above-average performance in pre-defined traits can qualify as nucleus flock animals based on stringent selection measures. Examples of popular indigenous breeds of sheep and goat developed in Africa using variations in nucleus breeding schemes are presented in Boxes 17.3, 17.4, 17.5, 17.6, and 17.7.

> The ability to select and improve traits in livestock populations generally relies on recording and access to clearly defined, accurate phenotypes

17.4.1 Select Indigenous Sheep Breeds

Box 17.3: Improved Djallonke Sheep: Côte d'Ivoire Adapted from Yapi-Gnaore (2000)
Djallonke sheep also known as West African Dwarf sheep are thin-tailed hair sheep popular across several countries in West Africa for their prolificacy and disease tolerance http://agtr.ilri.cgiar.org/djallonke. Following a drought in the Sahel from 1972 to 1973, the government of Côte d'Ivoire prioritized livestock development and promoted a national sheep programme for the country. In 1977, a within-breed selection programme was instituted on two state-owned farms and select private farms across the country for the Djallonke breed. The objective of the programme was to improve the growth and liveweight of the animals. The programme involved centralized performance recording of rams in nucleus flocks, with breeding ewes retained in the select farmers' flocks. Ram lambs for the nucleus flocks were initially selected from farmers' flocks based on their individual performance. Ram lambs not earmarked as breeding animals were castrated and reared for mutton production.

Through a well-resourced and trained extension service, the capacity of the livestock keepers was enhanced, and the smallholder farmers adopted new and improved techniques for managing their sheep populations. Over time, the top select sire rams from smallholder farms were purchased for the nucleus flocks, while the second category of sire rams was sold to other commercial farmers for the production of animals for mutton. The base population of the improved sheep grew from 12,000 breeding ewes on 71 farms in 1992 to 17,000 breeding ewes on 143 farms in 1999. Key traits selected in the improved Djallonke sheep are fertility, growth rate, and trypanotolerance.

Box 17.4: D'Man Sheep in Morocco
Selection in D'Man sheep was implemented following a breed conservation effort by Institut National de la Recherche Agronomique INRA in Morocco in 1970. Prolific D'Man ewes with good growth potential were selected from populations retained in three research stations. Later in 1986, a national sheep development plan was instituted to enable planned use of the improved D'Man both as a pure-bred animal and in cross-breeding programmes with other indigenous sheep in farmers flocks. The INRA research station in Errachidia maintains a flock of 80 breeding D'man ewes under an accelerated reproduction system of three lambings in 2 years (Boujenane et al. 2015).

Key traits in selecting D'Mann sheep are ewe fertility prolificacy and growth performance.

17.4.2 Select Indigenous Goat Breeds

Box 17.5: The Boer Goat: South Africa (Malan 2000; Visser and van Marle-Köster 2018)
The best example of a new breed of meat goat that has demonstrated considerable promise for increased productivity is the 'Boer breed' developed in South Africa. In the early 1920s, the South African farmers from the Eastern Cape developed this breed by selecting and cross-breeding indigenous goat kept by the Hottentot and Bantu tribes in South Africa with imported Nubian and Indian goat. Successive years of recurrent selection for size and conformation resulted in a well-documented new breed of goat selected for commercial meat production supported by the South African Boer goat association that was founded in 1959. Key traits of Boer goat: High fertility and fecundity, adaptable and disease resistant, and good meat production.

Box 17.6: West African Dwarf WAD Goat
The WAD goat is known for its hardiness, prolificacy, and good mothering ability, making it the breed of choice among smallholder livestock farmers in West and Central Africa. Ghana's Ministry of Food and Agriculture has a breeding station at Kintampo in the Bono East Region aimed at improving the growth and reproductive performance of WAD goat. Ayizanga et al. (2018a) reported an average pre-weaning growth rate of 35.78 g/day and post-weaning daily gain of 26.49 g for the WAD goat in Ghana. Liberia Karnuah et al. (2018) reported two major ecotypes of WAD goat, corresponding to the humid zone and the savanna zone, with the latter being generally heavier with larger body size. The main coat colour types of the WAD goat are black and white 40%, white 14%, black 9%, black and brown 5%, and several other mixtures that gives a strong indication of admixture as a result of the uncontrolled random mating in the free-range extensive husbandry systems (Fig. 17.4).

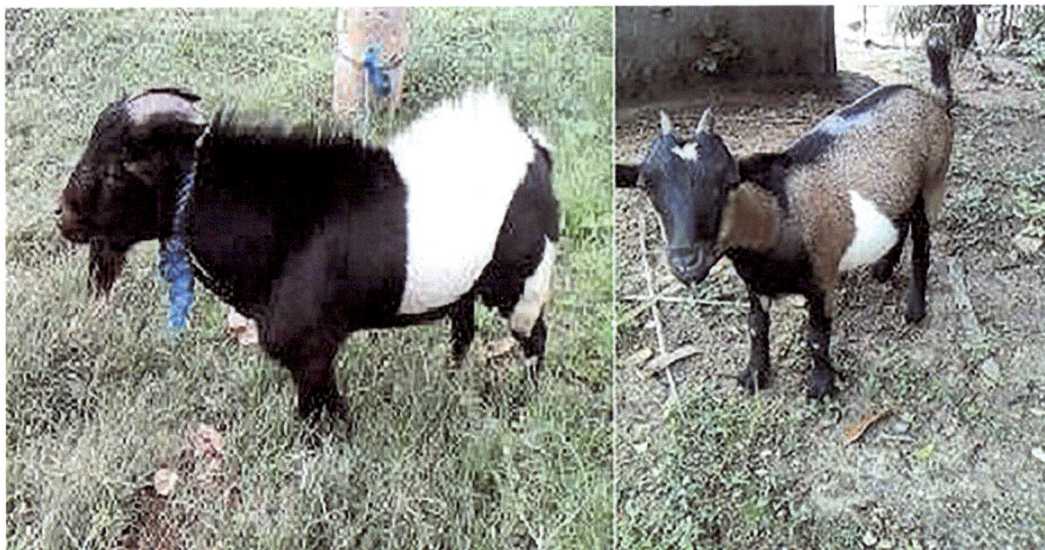

Fig. 17.4 West African Dwarf Goat in Ghana

Box 17.7: The Galla Goat: Kenya
The Galla goat occupies an important economic and ecological niche in the Arid and Semi-Arid Lands ASAL of Kenya. Female Galla goat has a long productive life and continue to breed and rear kids up to 10 years. The Galla are rarely culled due to loss of teeth and have a very strong compensatory growth after long dry seasons (Skea 1989). Since 1972, several projects have been implemented in Kenya to improve indigenous goat in different production systems (Fig. 17.3). The Galla goat has been a breed of focus as it is the most popular indigenous breed used in cross-breeding and upgrading initiatives. The cross-breeding programmes implemented have resulted in a mosaic of cross-bred Galla goat with varying levels of imported genes (Muigai et al. 2018; Bett et al. 2009b). Long-term strategies sustaining Galla goat in Kenya are implemented through the Dairy Goat Association of Kenya, which follows up on the registration and recording of pedigree and performance of the animals in the Kenya Stud book in line with breed standards defined in 1982. Key traits of the Galla goat: milk production, growth rate, drought resilience (Figs. 17.5 and 17.6).

17.5 Key External Factors That Influence Small Ruminant Breeding Structures in Africa

The improvement of animal genetic resources with genetic gains in production and productivity requires available genetic resources, technical know-how, collection and management of animal information, selection methods, estimation of breeding values, and organized mating systems. These should have an organizational structure with a financial resource base for long-term sustainability. Additionally, livestock breeding structures require individual, community, and public institutional buy-in.

The importance of public sector support through national government systems in providing functional and durable infrastructure, sup-

Fig. 17.5 Map of Kenya indicating the location of main goat projects from 1972 to 2010

Key

△ German Technical Corporation (GTZ)
△ Small Ruminant Collaborative Research Program (SRCRP)
△ Integrated Project in Arid Lands (IPAL)
△ Overseas Development Administration (ODA)
△ Integrated Small Livestock Improvement Program (ISLP)
△ FARM-Africa project
△ Heifer Projects International- Kenya
△ United Nations Development Program & Food and Agriculture Organization
△ Small Ruminants Trust project

portive policy, financial resources, and investment in training and capacity development of local expertise in animal breeding and management cannot be over-emphasized (Okpeku et al. 2019). Breeding structures should be set up to make good operational decisions and be able to model different options for strategic and informed decisions (van der Werf and Kinghorn 2017). The process of optimizing breeding structure across systems should strive to produce the best marketable product without adversely affecting genetic diversity within and between breeds and populations.

Optimizing breeding structures has to do with the choices made regarding a given species in a specified production system. Small ruminant breeding structures in Africa are, to a large extent, operationalized by livestock keepers with few

Fig. 17.6 Galla Goats in Kenya

animals under variable management conditions. Recommendations and simulations of optimizing breeding in these systems in line with pre-defined breeding objectives are documented in the literature; however, relatively little progress has been achieved in the larger small ruminant flocks found in rural agricultural systems of Africa (Okpeku et al. 2019).

With strengthening trade links across the African continent, it is important to identify livestock product demand drivers in different regions, taking cognizance of the differential production environments. Through inter-regional engagements, product exchange based on the relative advantages of production in different regions should be adopted. Relevant breeding structures could then be adopted based on demand and marketing structures. Livestock markets are vital for cost recovery and distribution of finished products.

17.6 Opportunities in Adopting New Technologies in Sheep and Goat Breeding Programmes of Africa

New genomic technologies such as high-density SNP microarrays (see Chap. 20 and 21 on the role of modern technologies) can be used to determine the breed composition in individual animals as well as, to some extent, the functional variation between breeds (Rupp et al. 2016; Kim et al. 2015). Genome-wide studies in small ruminant populations of Africa could provide valuable information on the diversity across the indigenous breeds, and possible populations in which genes associated with resilience and adaptability play an important role. Studies on possible identification of major genes associated traits in populations of Africa are limited, yet in different populations across the world major genes are being identified in populations and used in breeding programmes (Rupp et al. 2016). Incorporating genomic selection in small ruminant breeding programmes could allow countries in Africa to benefit from the on-going SNP discovery projects around the world (Kalds et al. 2019),and would contribute good-quality phenotype data on existing populations. Additionally, the use of Artificial Insemination (AI) for breeding across small ruminant populations is very limited, resulting in animals with unknown pedigree information in naturally mating populations. SNP technology could help to infer pedigree information that would guide the selection and use of unrelated males in breeding across populations. The adoption of SNP technology alongside the recording of animal performance data in small ruminant populations using hand-held information and communication technologies ICT could greatly ease the implementation of breeding programmes for improved productivity and resilience under low-input systems.

17.7 Conclusion and Future Perspectives

Breeding structures to sustain genetic improvement in different sheep and goat populations of Africa are either very weak or absent. The availability of improved breeding stock and infrastructure to support and guide productivity monitoring and improvement in most production systems is also limited, and in a few instances, where available, is inadequate to support upscaling of the monitoring. The adoption of ICT-based platforms with mobile phone technologies for learning and information exchange across the continent, however, offers a great opportunity to generate information from widespread sheep and goat populations in Africa. Such information, once harnessed, should be innovatively used to design and influence targeted breeding strategies for animals while providing relevant and strategic feedback to the producers based on the evaluation of the collated data.

There are also new opportunities emerging with the expanded use of more cost-effective genomic technologies for sheep and goat breeding programmes in Africa. Genomic tools that enable rapid characterization of populations enhance the ability for breeders to source relevant adaptive and productive animals from different environments that can better boost the performance of indigenous breeds without eroding their inherent adaptive capabilities.

It is notable that the use of reproductive technologies to disseminate superior genetics across small ruminant populations is rare, and most genetic improvement is based on natural mating programmes. Evolving reproductive technologies that enable the production of surrogate sires for different environments could bridge the gap in mating practices that promote desired genotypes for different environments.

With changing land ownership and use patterns across production systems in Africa, the formation of interest groups to promote and retain desired characteristics within breeds, either as communities or as breed societies governed through supportive policies, with adequate resourcing for breed development and capacity development of various stakeholders, is critical to successful breed improvement programmes. Enhanced capacity within African countries is required to adapt and contextualize evolving technologies for improved sheep and goat productivity.

Acknowledgements The authors acknowledge the support provided through the CGIAR Research Initiatives on Livestock, Climate and Systems Resilience (LCSR) and Sustainable Animal Productivity for Livelihoods, Nutrition and Gender Inclusion (SAPLING) in drafting this chapter. CGIAR research is supported by contributions to the CGIAR Trust Fund. CGIAR is a global research partnership for a food-secure future dedicated to transforming food, land, and water systems in a climate crisis.

References

Adewumi O, Olorunnisomo O (2009) Milk yield and milk composition of West African Dwarf, Yankasa and crossbred sheep in southwest of Nigeria. Livest Res Rural Dev 21:Article #42. Retrieved December 4, 2021

Almeida AM (2011) The Damara in the context of Southern Africa fat-tailed sheep breeds. Trop Anim Health Prod 43(7):1427–1441. https://doi.org/10.1007/s11250-011-9868-3

AU-IBAR (2015) The livestock development strategy for Africa 2015–2035, Nairobi, Kenya

AU-IBAR (2019) The state of farm animal genetic resources in Africa. AU-IBAR, Nairobi, Kenya

Ayalew W, King JM, Bruns EA, Rischkowsky B (2003) Economic evaluation of smallholder subsistence livestock production: lessons from an Ethiopian goat development program. Ecol Econ 45:473–485

Ayizanga R, Osei-amponsah R, Darfour-Oduro K, Aboagye GS, Ben A (2013) Growth performance and genetic parameter estimates of the West African Dwarf goat at Kintampo-Ghana. Ghanaian J Anim Sci 7:105–112

Ayizanga R, Tecku P, Obese F (2018a) Growth and reproductive performance of West African Dwarf goats at the Animal Research Institute. Katamanso Station, pp 43–53

Ayizanga RA, Tecku PKM, Obese FY (2018b) Growth and reproductive performance of West African Dwarf goats at the Animal Research Institute, Katamanso Station. Ghana J Agric Sci 52(1):43–53. https://doi.org/10.4314/gjas.v52i1

Bett RC, Kosgey IS, Kahi AK, Peters KJ (2009a) Realities in breed improvement programmes for dairy goats in east and Central Africa. Small Rumin Res 85:157–160

Bett RC, Kosgey IS, Kahi AK, Peters KJ (2009b) Analysis of production objectives and breeding practices of dairy goats in Kenya. Trop Anim Health Prod 41(3):307–320. https://doi.org/10.1007/s11250-008-9191-9

Bosso NA, Cissé MF, van der Waaij EH, Fall A, van Arendonk JAM (2007) Genetic and phenotypic parameters of body weight in West African Dwarf goat and Djallonké sheep. Small Rumin Res 67(2–3):271–278. https://doi.org/10.1016/j.smallrumres.2005.11.001

Boujenane I, Chikhi A, Ibnelbachyr M, Mouh FZ (2015) Estimation of genetic parameters and maternal effects for body weight at different ages in D'man sheep. Small Rumin Res 130:27–35. https://doi.org/10.1016/j.smallrumres.2015.07.025

Du Toit D (n.d.) The Damara of Southern Africa -an essay on the genetic variation within the breed. New Zealand rare breeds. https://www.rarebreeds.co.nz/damara2.html. Accessed 2023-03-13

FAO (2007a) In: Pilling BRD (ed) The state of the world's animal genetic resources for food and agriculture. Food and Agriculture Organization of the United Nations, Rome, Italy

FAO (2007b) Global plan of action for animal genetic resources and the Interlaken declaration. Food and Agriculture Organization of the United Nations, Rome, Italy

FAO (2015) In: Scherf BD, Pilling D (eds) The second report on the state of the world's animal genetic resources for food and agriculture. FAO Commissnion on Genetic Resources for Food and Agriculture Assessments, Rome, Italy

FAO Breed data sheet: Damara. Domestic Animal Diversity Information System of the Food and Agriculture Organization of the United Nations. https://dadis-breed-datasheet-ext-ws.firebaseapp.com/?country=ZAF&specie=Sheep&breed=Damara&lang=en

FAOSTAT (2021) FAO statistical database. FAO, Rome, Italy. https://faostat.fao.org. Accessed 25 Oct 2021

FAOSTAT (2023) Population of sheep and goats on sub-Saharan Africa. FAO statistical database. FAO, Rome, Italy. https://www.fao.org/faostat/en/#data/QCL. Accessed 2023-09-21

FARM-Africa (2003) Delivering affordable and quality animal health services to Kenya's rural poor. FARM-Africa's experiences. FARM-Africa, P. by, Ed

Gbangboche AB, Adamou-Ndiaye M, Youssao AKI, Farnir F, Detilleux J, Abiola FA, Leroy PL (2006) Non-genetic factors affecting the reproduction performance, lamb growth and productivity indices of Djallonke sheep. Small Rumin Res 64:133–142

Gebre KT, Fuerst-Waltl B, Wurzinger M, Philipsson J, Duguma G, Mirkena T, Haile A, Sölkner J (2012) Estimates of economic values for important traits of two indigenous Ethiopian sheep breeds. Small Rumin Res 105(1–3):154–160. https://doi.org/10.1016/j.smallrumres.2012.01.009

Gutu Z, Haile A, Rischkowsky B, Mulema AA, Kinati W, Kassie GT (2015) Evaluation of community-based sheep breeding programs in Ethiopia. Addis Ababa. https://hdl.handle.net/10568/76233. Accessed 2023-09-25

Haile A, Wurzinger M, Mueller J, Mirkena T, Duguma G, Mwai O, Sölkner J, Rischkowsky B (2011) Guidelines for setting up community-based sheep breeding programs in Ethiopia. ICARDA-t.; Azage Tegegne (IPMS/ILRI), Ed, Aleppo, Syria

Haile A, Gizaw S, Getachew T, Mueller JP, Amer P, Rekik M, Rischkowsky B (2019) Community - based breeding programmes are a viable solution for Ethiopian small ruminant genetic improvement but require public and private investments. Anim Breed Genet 136:319–328. https://doi.org/10.1111/jbg.12401

Huson HJ, Sonstegard TS, Silverstein J, Woodward-Greene MJ, Masiga C, Muchadeyi FC, Rees J, Sayre B, Elbetagy A, Rothschild MF (2014) Genetic and phenotypic characterization of African goat populations to prioritize conservation and production efforts for small-holder farmers in sub-Saharan Africa. American Society of Animal Science

Hutu I, Oldenbroek K, der Van WL (2020) Animal breeding and husbandry chapter 3. Breeding goals. In: Agroprint, Timişoara, Wageningen, pp 170–191

James JW (1998) Economic evaluations of breeding objectives in sheep and goats. General considerations. In: Proceedings of the 6TH world congress on genetics applied to animal production, 11–16th January, 1998, Armidale, Australia

Jembere T, Dessie T, Rischkowsky B, Kebede K, Okeyo AM, Haile A (2017) Meta-analysis of average estimates of genetic parameters for growth, reproduction and Milk production traits in goats. Small Rumin Res 153:71–80. https://doi.org/10.1016/j.smallrumres.2017.04.024

Kahi AK, Wasike CB (2019) Special issue — dairy goat production in sub-Saharan Africa: current status, constraints and prospects for research and development. Asian Australas J Anim Sci 32(8):1266–1274. https://doi.org/10.5713/ajas.19.0377

Kalds P, Zhou S, Cai B, Liu J, Wang Y, Petersen B, Sonstegard T, Wang X, Chen Y (2019) Sheep and goat genome engineering: from random transgenesis to the CRISPR era. Front Genet 10(JUL). https://doi.org/10.3389/fgene.2019.00750

Kariuki CM, Ilatsia ED, Kosgey IS, Kahi AK (2010) Direct and maternal (co)variance components, genetic parameters and annual trends for growth traits of Dorper sheep in semi-arid Kenya. Trop Anim Health Prod 42(3):473–481. https://doi.org/10.1007/s11250-009-9446-0

Kariuki J, Galie A, Birner R, Oyieng E, Chagunda MGG, Jakinda S, Milia D, Ojango JMK (2022) Does the gender of farmers matter for improving Small Ruminant productivity? A Kenyan case study. Small Rumin Res 206:106574. https://doi.org/10.1016/j.smallrumres.2021.106574

Karnuah AB, Osei-Amponsah R, Gregory D, Arthur W, Walter T, Boettcher P (2018) Phenotypic characterization of the West Africa Dwarf goats and the production system in Liberia. Int J Livestock Prod 9:221–231. https://doi.org/10.5897/IJLP2018.0496

Kaumbata W, Banda L, Mészáros G, Gondwe T, Woodward-Greene MJ, Rosen BD, Van Tassell CP, Sölkner J, Wurzinger M (2020) Tangible and intangible benefits of local goats rearing in smallholder farms in Malawi. Small Rumin Res 187(April):106095. https://doi.org/10.1016/j.smallrumres.2020.106095

Kifaro G, Eik L, Mtenga L, Mushi D, Safaril J, Kassuku A, Kimbiti E, Maeda-Machang'u A, Kanuya N, Muhikambele V, Ndemanisho E, Ulvund M (2007) The potential use of artificial insemination in sustainable breeding of dairy goats in developing countries: a case study of Norwegian dairy goats' in Tanzania. Tanz J Agric Sci 8(1):19–24

Kim E-S, Elbeltagy AR, Aboul-Naga AM, Rischkowsky B, Sayre B, Mwacharo JM, Rothschild MF (2015) Multiple genomic signatures of selection in goats and sheep indigenous to a hot arid environment. Heredity (Edinb) 116(February):1–10. https://doi.org/10.1038/hdy.2015.94

Knights R (2010) Dorper Sheep and the production of Lean Lamb in Arid Australia

Kosgey IS, Van Arendonk JAM, Baker RL (2003) Economic values of traits of meat sheep in medium to high production potential areas of the tropics. Small Rumin Res 50:187–202

Kosgey IS, Baker RL, Udo HMJ, Van Arendonk JAM (2006) Successes and failures of small ruminant breeding programmes in the tropics: a review. Small Rumin Res 61(1):13–28. https://doi.org/10.1016/j.smallrumres.2005.01.003

Kosgey IS, Rowlands GJ, van Arendonk JAM, Baker RL (2008) Small ruminant production in smallholder and pastoral/extensive farming systems in Kenya. Small Rumin Res 77(1):11–24. https://doi.org/10.1016/j.smallrumres.2008.02.005

Kristjanson P, Waters-bayer A, Johnson N, Tipilda A, Njuki J, Baltenweck I, Grace D, Macmillan S (2010) Livestock and women's livelihoods: a review of the recent evidence livestock and women's livelihoods: a review of the recent evidence, Nairobi, Kenya

Kugonza D, Stalder K, Rothschild M (2014) Effects of Buck and doe size on the growth performance and survival of their progeny. Livest Res Rural Dev 26:Article #47. Retrieved March 13, 2023

Lamuno D, Sölkner J, Mészáros G, Nakimbugwe H, Mulindwa H, Nandolo W, Gondwe T, Van Tassell CP, Gutiérrez G, Mueller J, Wurzinger M (2018) Evaluation framework of community-based livestock breeding programs. Livest Res Rural Dev 30:Article #47

Leymaster KA (2002) Fundamental aspects of crossbreeding of sheep: use of breed diversity to improve efficiency of meat production. Sheep Goat Res 17(3):50–59

Malan SW (2000) The improved Boer goat. Small Rumin Res 36(2):165–170

Mangheni MN, Boonabaana B, Asiimwe E, Tufan HA, Jenkins D, Garner E (2021) Developing a competency framework for trainers of gender-responsive agricultural research training programs. AgriGender, J Gender Agric Food Secur 6(2):41–57

Marshall K, Mtimet N, Wanyoike F, Ndiwa N (2014) The complex and gender differentiated objectives of livestock keeping for Somali pastoralists. In: Proceedings, 10th world congress of genetics applied to livestock production, 17th -24th august, Vancouver, Canada

Menchaca MA, Ungerfeld R (2018) Reproductive strategies for goat production in adverse environments. In: Simões J, Gutiérrez C (eds) Sustainable goat production in adverse environments: volume I. Springer, Cham, pp 71–88. https://doi.org/10.1007/978-3-319-71855-2

Milne C (2000) The history of the Dorper sheep. Small Rumin Res 36(2):99–102. https://doi.org/10.1016/S0921-4488(99)00154-6

Mueller JP, Rischkowsky B, Haile A, Philipsson J, Mwai O, Besbes B, Valle Zárate A, Tibbo M, Mirkena T, Duguma G, Sölkner J, Wurzinger M (2015a) Community-based livestock breeding programmes: essentials and examples. J Anim Breed Genet 132(2):155–168. https://doi.org/10.1111/jbg.12136

Mueller JP, Rischkowsky B, Haile A, Philipsson J, Mwai O, Besbes B, Valle Zarate A, Tibbo M, Mirkena T, Duguma G, Sölkner J, Wurzinger M (2015b) Community-based livestock breeding programmes: essentials and examples. J Anim Breed Genet 132:155–168

Muigai AWT, Okeyo AM, Ojango JMK (2018) Goat production in Eastern Africa: practices, breed characteristics, and opportunities for their sustainability. In: Simões J, Gutiérrez C (eds) Sustainable goat production in adverse environments, vol 1. Springer, Cham. https://doi.org/10.1007/978-3-319-71855-2_3

Njuki J, Sanginga PC (2013) Gender and livestock: key issues, challenges and opportunities. In: Women, livestock ownership and markets. Bridging the gender gap in Eastern and Southern Africa. Routledge & International Devlopment Research Centte, Ottawa, Canada

Nwogwugwu CP, Lee S, Freedom EC, Manjula P, Lee JH (2018) Review on challenges, opportunities and genetic improvement of sheep and goat productivity in Ethiopia. J Anim Breed Genom 2(1):1–8

Ofori SA, Hagan JK (2020) Genetic and non-genetic factors influencing the performance of the West African Dwarf (WAD) goat kept at the Kintampo goat Breeding Station of Ghana. Trop Anim Health

Prod 52(5):2577–2584. https://doi.org/10.1007/s11250-020-02276-9

Ojango JMK, Ahuya C, Okeyo AM, Rege JEO (2010) The FARM-Africa dairy goat improvement project in Kenya: a case study. In: Animal genetics training resource version 3. International Livestock Research Institute, Nairobi, Kenya and Swedish University of Agricultural Science (SLU), Uppsala, Sweden/Nairobi, Kenya, pp 3–8

Ojango JMK, Malmfors B, Mwai O, Philipsson J (2011) Training the trainers-an innovative and successful model for capacity building in animal genetic resource utilization in sub-Saharan Africa and Asia, Nairobi, Kenya

Ojango JMK, Audho J, Oyieng E, Recha J, Okeyo AM, Kinyangi J, Muigai AWT (2016) System characteristics and management practices for small ruminant production in "Climate Smart Villages" of Kenya. Anim Genet Resour/Ressources génétiques animales/Recursos genéticos animales 58(May):101–110. https://doi.org/10.1017/S2078633615000417

Ojango J, Oyieng E, Milia D, Audho J, Kariuki J, Jakinda S (2018) Best practices for selective breeding for improved livestock productivity: Module 2-Engage. ILRI, Nairobi, Kenya, p 25

Okpeku M, Ogah DM, Adeleke MA (2019) A review of challenges to genetic improvement of indigenous livestock for improved food production in Nigeria. Afr J Food Agric Nutr Dev 19(01):13959–13978. https://doi.org/10.18697/ajfand.84.BLFB1021

Oyieng E, Ojango J, Audho J, Gitau J, Gachora J (2021) Improving small ruminant productivity in pastoral systems of Kenya: baseline household survey report, Nairobi, Kenya

Oyieng E, Mrode R, Ojango JMK, Ekine-Dzivenu CC, Audho J, Okeyo AM (2022) Genetic parameters and genetic trends for growth traits of the Red Maasai sheep and its crosses to Dorper sheep under extensive production system in Kenya. Small Rumin Res 206:106588. https://doi.org/10.1016/j.smallrumres.2021.106588

Peacock C, Ahuya CO, Ojango JMK, Okeyo AM (2011) Practical crossbreeding for improved livelihoods in developing countries: the FARM Africa goat project. Livest Sci 136(1):38–44. https://doi.org/10.1016/j.livsci.2010.09.005

Philipsson J, Rege J, Zonabend E, Okeyo AM (2011) Sustainable breeding programmes for tropical low- and medium input farming systems. In: Ojango JM, Malmfors B, Okeyo AM (eds) Animal genetics training resource, version 3, 2011. International Livestock Research Institute, Nairobi, Kenya, and Swedish University of Agricultural Sciences, Uppsala, Sweden, pp 1–35. http://agtr.ilri.cgiar.org/index.php?option=com_content&view=article&id=27&Itemid=267

Ramsay K, Swart D, Oliver B, Hallowell G (2000) An evaluation of the breeding strategies used in the development of the Dorper sheep and improved Boar goat of South Africa. In: Proceedings of workshop on developing breeding strategies for lower input animal production environments, 22nd to 25th September 1999; Bella, Italy

Rupp R, Mucha S, Larroque H, McEwan J, Conington J (2016) Genomic application in sheep and goat breeding. Anim Front 6(1):39–44. https://doi.org/10.2527/af.2016-0006

Shrestha JNB, Fahmy MH (2005) Breeding goats for meat production: a review - 1. Genetic resources, management and breed evaluation. Small Rumin Res 58(2):93–106. https://doi.org/10.1016/S0921-4488(03)00183-4

Skea IN (1989) Keeping goats in Kenya, Nairobi

Solomon AK, Mwai O, Grum G, Haile A, Rischkowsky BA, Solomon G, Dessie T (2014) Review of goat research and development projects in Ethiopia review of goat research and development projects in Ethiopia

Sponenberg P (2007) Rainbow Livestock: Nguni Cattle, Damara Sheep and Indigenous Goats. South African Studbreeder Magazine, 2nd quarter edition, South Africa Studbook Authority

Van der Werf J (2000) Livestock straight breeding system structures for the sustainable intensification of extensive grazing systems. In: Proceedings of a workshop on developing breeding strategies for lower input animal production environments, 22nd - 25th September 1999, Bella, Italy, pp 105–178

van der Werf J, Kinghorn B (2017) Breeding program design principles. Genetic evaluation and breeding. University of New England. https://www.woolwise.com/wp-content/uploads/2017/07/GENE-422-522-11-T-15.pdf. Accessed 2023-09-25

Verbeek E, Kanis E, Bett RC, Kosgey IS (2007) Socio-economic factors influencing small ruminant breeding in Kenya. Livest Res Rural Dev 19(6):Article #77

Visser C (2019) A review on goats in Southern Africa: an untapped genetic resource. Small Ruminant Res Elsevier BV:11–16. https://doi.org/10.1016/j.smallrumres.2019.05.009

Visser C, van Marle-Köster E (2018) The development and genetic improvement of South African goats. Goat Sci. https://doi.org/10.5772/intechopen.70065

Weldemariam B, Mezgebe G (2021) Community based small ruminant breeding programs in Ethiopia: progress and challenges. Small Rumin Res 196:106264. https://doi.org/10.1016/j.smallrumres.2020.106264

Wilson RT (1992) Goat and sheep skin and fibre production in selected sub-Saharan African countries. Small Rumin Res 8(1–2):13–29. https://doi.org/10.1016/0921-4488(92)90003-M

Yapi-Gnaore CV (2000) The open nucleus breeding programme of the Djallonke sheep in Cote d'Ivoire. In: Galal S, Boyazoglu J, Hammond K (eds) Workshop on developing breeding strategies for lower input animal production environments; ICAR technical series 3, 22–25 September 1999. International Committee for Animal Recording: Bella, Italy, Rome, Italy, pp 283–292

Yaro M (2017) Characterisation of the Djallonke sheep breed in Ghana using molecular markers. Curtin University

Zishiri OT, Cloete SWP, Olivier JJ, Dzama K (2013) Genetic parameters for growth, reproduction and fitness traits in the South African Dorper sheep breed. Small Rumin Res 112(1):39–48. https://doi.org/10.1016/j.smallrumres.2013.01.004

Zonabend König E, Mirkena T, Strandberg E, Audho J, Ojango J, Malmfors B, Okeyo AM, Philipsson J (2016) Participatory definition of breeding objectives for sheep breeds under pastoral systems—the case of Red Maasai and Dorper sheep in Kenya. Trop Anim Health Prod 48(1). https://doi.org/10.1007/s11250-015-0911-7

Zonabend E, Okeyo AM, Ojango JMK, Hoffmann I, Moyo S, Philipsson J (2013) Infrastructure for sustainable use of animal genetic resources in Southern and Eastern Africa. Anim Genet Resour/Ressources génétiques animales/Recursos genéticos animales 53:79–93. https://doi.org/10.1017/S2078633613000295

Open Access This chapter is licensed under the terms of the Creative Commons Attribution 4.0 International License (http://creativecommons.org/licenses/by/4.0/), which permits use, sharing, adaptation, distribution and reproduction in any medium or format, as long as you give appropriate credit to the original author(s) and the source, provide a link to the Creative Commons license and indicate if changes were made.

The images or other third party material in this chapter are included in the chapter's Creative Commons license, unless indicated otherwise in a credit line to the material. If material is not included in the chapter's Creative Commons license and your intended use is not permitted by statutory regulation or exceeds the permitted use, you will need to obtain permission directly from the copyright holder.

Defining Breeding Goals and Breeding Strategies for Chicken Production Systems in Africa

Sunday O. Peters, Michael O. Ozoje,
Adeyemi S. Adenaike, Blaise A. Hako Touko,
Christian K. Tiambo, Matthew A. Adeleke,
Oluyinka Opoola, and Tadelle Dessie

Abstract

This chapter defines breeding goals and strategies for chicken production systems in Africa. Sections 18.1 and 18.2 introduces this chapter and highlight types of breeding goals at different trait levels, and critical components for breeding goals in different input systems of production. Sections 18.3 and 18.4 cover traits of economic importance with examples of various breeding programs for low-input systems (e.g., in Ethiopia as an example). Breeding objectives and strategies for adaptability and resilience for optimal productivity in low-input systems are covered in Sects. 18.5 and 18.6. Furthermore, the current practices for chicken breeding systems and the opportunities for the incorporation of omics technologies are discussed in Sects. 18.7 and 18.8. Also, strategic approaches toward integrating environmental, structural, and socioeconomic factors for chicken production to ensure effective multiplication, distribution, and delivery of improved chicken strains to end-users are extensively covered in Sect. 18.9. Lastly, a conclusion and future perspectives are presented in the last part of Sect. 18.9. Several attempts made to develop poultry production and productivity in Africa have concentrated on the introduction of exotic birds, improving management systems, and disease man-

---

S. O. Peters (✉)
Department of Animal Science, Berry College, Mount Berry, GA, USA
e-mail: speters@berry.edu

M. O. Ozoje · A. S. Adenaike
Department of Animal Breeding and Genetics, Federal University of Agriculture, Abeokuta, Ogun State, Nigeria

B. A. Hako Touko
Department of Animal Production, Faculty of Agriculture and Agricultural Science, University of Dschang, Dschang, Cameroon

C. K. Tiambo
Centre for Tropical Livestock Genetics and Health (CTLGH-ILRI), Nairobi, Kenya

M. A. Adeleke
Discipline of Genetics, School of Life Sciences, University of KwaZulu-Natal, Durban, South Africa

O. Opoola
The Global Academy for Agriculture and Food systems (GAAFS), Easter Bush Campus, University of Edinburgh, Edinburgh, UK

The Roslin Institute and Royal (Dick) School of Veterinary Studies, Easter Bush Campus, University of Edinburgh, Midlothian, UK

T. Dessie
International Livestock Research Institute (ILRI), Addis Ababa, Ethiopia

agement through vaccination among others. These strategies only contributed marginally to the increase in overall productivity among the exotic-indigenous chicken crossbreds which could not be sustained because of several environmental stressors. Indigenous chickens are crucial components of the sustainability of the African poultry industry. Selection pressure in African indigenous chickens is different compared to introduced exotic commercial breeds because selection footprint is associated with adaptation to local prevailing environmental conditions. Balancing characteristics related to growth with that of reproduction is a vital consideration in the development of African indigenous chicken breeds. In some traits, unfavorable genetic correlations, also called genetic antagonisms, cause a decrease in genetic merit when single-trait selection is practiced. In the same way, failure to consider selection response in correlated traits not directly under selection when selection is practiced also decreases genetic merit. The ability of improved chicken strains to thrive and benefit farmers in harsh environments serves as a success indicator to breeders in Africa. Indigenous and tropically adapted breeds were not given due attention until recently when the African Chicken Genetic Gains (ACGG) group (also recently known as Tropical Poultry Genetic Solutions; TPGS) encouraged the selection and use of indigenous chicken strains with high productivity in semi-scavenging systems in the tropics. Genomic selection is considered the most appropriate standard breeding method in livestock. It is particularly advantageous as it can be used for the selection of breeding animals at an early stage of life without reference to their own breeding or production records. The developments in omics technologies also made the estimations of breeding values (BVs) more accurate, increasing genetic gain by aiding in the selection of genetically superior, disease-free animals at an early stage of life for enhanced productivity and profitability.

Keywords

Poultry breeding · Breeding goals · Crossbreeding system · Breeding program

Abbreviations

ACCG	African Chicken Genetic Gains
TPGS	Tropical Poultry Genetic Solutions
BMGF	Bill and Melinda Gates Foundation
ILRI	International Livestock Research Institute
BVs	Breeding values
EBV/EBVs	Estimated breeding value/values
CCPS	Combined crossbred purebred selection
h^2_c	Heritability of crossbred
BLUP	Best linear unbiased prediction
DNA	Deoxyribonucleic acid
RRS	Reciprocal recurrent selection
PLS	Pure-line selection
FCR	Feed conversion ratio
RFC	Residual feed consumption
FE	Feed efficiency
FI	Feed intake
RFI	Residual feed intake
QTL	Quantitative trait loci
MAS	Marker-assisted selection
SNPs	Single nucleotide polymorphisms (SNPs)
GEBV	Genomic estimates of breeding value
PLBS	Producer Level Baseline Survey
FUNAAB	Federal University of Agriculture, Abeokuta
RFLPs	Restricted fragment length polymorphisms
ESTs	Expressed sequence tags
RAPDs	Random-amplified polymorphic DNAs
AFLP	Amplified fragment length polymorphism
FAO	Food and Agriculture Organization
GGP	Great grand parents
RIR	Rhode Island Red
G x E	Genetic by Environment
F1	First filial generation

F2	Second filial generation
RNA	Ribonuclease acid
mRNA	Messenger RNA
GWAS	Genome-wide association
WGS	Whole-genome sequencing
GEBVs	Genomic estimated breeding values
ssGBLUP	Single-step genomic best linear unbiased prediction
qRT-PCR	Quantitative real-time reverse transcription
PCA	Principal component analysis
HCA	Hierarchical cluster analysis
PLS-DA	Partial least-squares discriminant analysis
APPs	Acute phase proteins
IgA	Immunoglobulin A
LMIC	Low- and middle-income countries

18.1 Introduction

A wide range of actors influences poultry breeding, as the needs and priorities of several interest groups must be balanced. Poultry breeding is a continuous process where future developments must be anticipated, and planning must be done carefully to establish effective breeding programs. Breeding programs are designed to generate genetic improvement in a population, which aims to use genetic variation available within and between breeds effectively. Its goal is to improve production, quality and output using breeding principles and technological means to optimize the input of resources per unit of the output. A key factor in any breeding program is the estimation of breeding values (the predicted genetic values of an individual determined by the genes of its parents) and a well-structured breeding program is required to improve productivity without exhausting genetic variability. Modern poultry breeding programs constantly evolve with changes and advances in breeding theory, biotechnology, and genetics. A modern breeding program combines the recording of selected traits, estimation of breeding values, selection of potential parents, design, and implementation of the mating program for the selected parents with the aim of generating genetic gain that will lead to more economic value. A comprehensive review of specific goals and strategies targeted for poultry breeding has been covered in Chap. 15.

Directional progress in the genetics of animals in subsequent generations is the focus of any genetic improvement that aims to produce desired products more efficiently in a future social, economic, and ecological production environment. The direction of improvement is formalized in a "breeding goal" which is the first step in the design process of genetic improvement strategies. The definition of breeding goal is therefore an important area of decision-making, although it shifts over time in response to the changing requirements of poultry producers and consumer demands. Breeding goals should both be market and demand-driven, in direct connection with the preferred chicken types (meat, eggs, or dual purpose).

Breed improvement strategies are important in realizing developmental objectives because the state of the future generation depends partly on the state of the current one. The development of effective improvement strategies is therefore of great importance for the management of animal genetic resources (see Chap. 7 on poultry genetic resources). Furthermore, Chap. 7 covers extensively the various domestic poultry breeds, genetic diversity, and the unique characteristics that enable their adaptation and resilience in their diverse environments. The increase in the rate of genetic progress achieved in production traits in poultry today is attributable to the methodological revolution in poultry breeding. This has modified the physiology of birds and the genetic architecture of the traits in chicken population at an unprecedented rate.

From the mid-twentieth century, meat-type (broiler) chickens have been subjected to intense selection for juvenile growth rate, feed conversion ratio, and body composition. These entries describe how the production and functional traits of the individual bird and accompanying management practices have responded to this unremitting selection and this suggest what the future might hold for other agricultural animal species (Hutu et al. 2020; Gowane et al. 2019).

This chapter discusses breeding goals, programs, and strategies in chicken production systems. The main components of structured breeding programs and strategies with relevant traits using performance records and data analysis to identify superior birds and the use of superior birds to produce the next generation of chickens.

18.2 Breeding Goals

The poultry breeding process involves the development of breeding objectives described by a profit function that takes genetic values as inputs and profit as outcomes. It starts with setting goals to identify traits that deliver necessary improvements to significantly impact poultry performance. The development of a new generation of birds that will produce desired products more efficiently under a future farm's economic and social circumstances is the main goal of poultry breeding (Hutu et al. 2020; Rivero et al. 2013).

A breeding goal defines the (combination of) specific trait(s) with their desired direction of change over generations. Each new generation is a small step toward the ultimate goal. Breeding goal is a list of traits and their relative importance targeted by a breeding program and it includes a description of how they should be changed genetically (increased, decreased, or maintained) from time to time. It is a linear combination of traits that influences the profitability, weighed with their respective economic values of a given species (Ponzoni 1988). The list of traits in a breeding goal is normally based on the objective of the agricultural production system and the specific characterization of the animal production system (see Chap. 15). They could be extended to include consumer behavior regarding preferences of animal origin or societal wishes (Amer et al. 2015). However, a breeding goal should involve traits that can be cheaply collected. Their characteristics should be amenable to exact measurement or evaluation, such as egg (size, quantity, and color), body weight, shank (length and color), and comb type. To maximize egg and meat production, commercial poultry breeders create unique lines with varying breeding goals and then cross the different lines to produce the best possible offspring. Significant gains can be realized in each generation since just a handful of breeding goal features are in a targeted line. The breeding goals achieved to a high level in each line are integrated through line breeding. This approach is more fruitful than selecting all desirable phenotypes in a single breeding population. In the chicken production system, a four-way cross is commonly used, with the first two generations consisting of hens from a line selected for egg production and a cock from a line selected for size. Secondly, hens from a crossbreed are bred with cocks from a line bred for their high-quality meat production. Many healthy, strong offspring are produced as a result.

Typically, breeding goals include more than one trait. In addition, traits of economic importance in poultry species have been previously covered (see Chap. 7). However, balancing the emphasis on each trait in a breeding program is also part of the goal. It focuses on saving the inputs of resources, achieving sustainability, dynamic genetic improvements, and accommodating local production environments. Although breeding goals shift over time in response to the changing requirements of livestock producers and consumers' demand, genetic change is likely to be in the wrong direction, when the focus is put on the wrong traits or when important traits are left out of the breeding goal (Amer et al. 2015; Ponzoni 1984).

18.2.1 Types of Breeding Goals

Regardless of the species, the breeding goal focuses on the improvement of gross efficiency by improving productivity (higher yield), feed conversion, reproduction, health, and survivability. The traits in a breeding goal can either be quantitative or qualitative. Meat/egg production traits, body measurement traits, and performance expression traits are some examples of quantitative traits measured in units of gm or kg. Quantitative methods are used when breeding for desired characteristics. Product traits such as scores for body conformation, disease incidences and/or performance traits, meat quality, plumage color, and comb type are all examples of qualitative goals that might guide chicken breeding and be measured in classes. Some breeding goal traits cannot be measured when needed; for example, meat quality is an essential trait in meat production, but carcass composition cannot be measured when it is considered helpful. It can only be measured after the bird is slaughtered, breeding with that bird is impossible, but indicator traits obtained by scanning live birds for body composition before breeding decisions are made can aid in the prediction of carcass composition. Many underlying traits also complicate breeding goals. For example, overall reproduction capacity results from both male and female reproduction traits. In males, sperm quality and insemination results are part of their reproductive capacity, while age at first egg, egg size, and clutch size are underlying reproductive traits in females. The inclusion of reproductive capacity will therefore complicate the breeding goal (Okeno et al. 2013).

Traits in a breeding goal can be evaluated at four different levels:

1. Individual level: At the individual level, the focus is on how the improvement of a trait would affect the profitability of a bird.
2. Crossbreeding system level: At this stage, the focus is on the effect of selection in the grandparents on the profitability of the crossbred grand progeny in producing the final product.
3. Farm level: At the farm level, the interest is the effect on income per farm.
4. Production chain level: The effect on production and processing levels is the focus.

Each level generates different outcomes. When a producer of broilers obtains a contract to supply a fixed kilogram of broiler meat annually, selection for daily gain will result in heavier carcasses. When a producer subsequently keeps fewer animals than needed, there will likely be a surplus of feed that cannot be converted into valuable carcasses. As such, the profit from a higher daily gain will be lower at the farm level (Amer et al. 2015).

18.2.2 Key Elements of Appropriate Breeding Goals for Various Input Systems

A breeding system specifies how genetic resources are combined to attain the breeding goal by outlining the mating strategies and breeds or strains that will be used. Other higher levels include the farm, but a production system is an association of a finite number of parts analyzed and their interaction with the environment. The production system is defined by husbandry, feeding, and management practices. Two types of farmers work within the chicken industry: those raising chickens for sustenance and those raising chickens for profit. Due to the low costs and low risk involved in the subsistence system, it has been determined that aversion to risk is a major factor in the decisions made by farmers. Regarding input costs and final product pricing, marketing systems detail the channels via which items are sold to customers (Rivero et al. 2013; Ponzoni 1988).

Breeding objective denotes a profit function that takes genetic values as input and produces profit as outcome. It requires consideration of all potential income and expenditure. This is because an overestimation of genetic improvement's monetary value could occur from disregarding expenses that impair selection efficiency. Costs in commercial poultry operations are primarily attributable to inputs, while earnings directly result from the sale of outputs or products. A flock's fixed costs are unlikely to vary even if output levels increase, while the variable costs do alter. Growing a business to a larger size may increase fixed costs. Cost and revenue models from various agricultural businesses show that different systems typically rely on different sets of inputs. Each manufacturing system has its own set of costs from various places. Therefore, it is essential to analyze them individually. It may be challenging to put a price on the various economic benefits that the subsistence system provides. A chicken's biological features significantly impact the system's profitability (Okeno et al. 2013; Hietala et al. 2014).

18.2.3 Unintended Consequences of Breeding Goals

In recent times, modern farming technologies, including quantitative selection and breeding methods, have significantly increased the efficiency of production in many poultry species. However, selection for an increase in production output has trade-offs and negative outcomes often more pronounced in intensive systems of production (Berry and Evans 2014).

In the last century, the formulation of breeding objectives and their applications have dominated quantitative genetic selection practices. Adoption of a data-driven approach to the formulation of breeding goals resulted in the development of animal recording systems. However, Garrick and Golden (2009) emphasized the potential dangers associated with the use of only traits that are easily recorded, such as growth with moderate to high heritability estimates in selection programs because they result in relatively fast genetic progress.

The aim of the poultry industries is to maximize profitability with genetic progress in production traits (Miglior et al. 2017). This goal is, however, associated with a decline in other traits correlated mainly with reproduction and welfare. Historically, the driving force behind the setting of breeding objectives, and thus selection progress, is the profitability of the operation. The focus of the breeding goal on productivity in combination with a data-driven selection approach has resulted in unintended outcomes in several traits (Dawkins and Layton 2012). In all production systems in the poultry industry, fertility is considered the most important economic trait (Cammack et al. 2009), although, intensive selection for production traits in the past has unintentionally led to selection for impaired fertility as a result of the unfavorable genetic relationships between these components (Dawkins and Layton 2012; Berry et al. 2014). The emphasis of selection on increased muscularity and meat yield in the past had also inadvertently led to conditions of double muscling which has adverse effects on fertility (Dawkins and Layton 2012). As such, most breeding companies now increasingly incorporate health and welfare goals alongside economic ones in their breeding programs because animal health and welfare is now becoming a major concern in the different production systems.

18.3 Breeding Programs for Chicken Production

A breeding program is a set of actions outlined to achieve specific breeding goals and create a new animal generation. It is a set of systematic actions designed and implemented to enhance desired phenotypic and/or genotypic characteristics in the target breeding population (Ponzoni 1988; Declaration 2007).

To begin a breeding program, one must first establish a target for the program's offspring and then create a plan to achieve that target through genetic improvement. Structured breeding programs are comprised of a breeding goal with the required traits, collection of performance data, analysis of the data for the identification of superior chickens, and the use of superior chickens to produce the next generation (Fig. 18.1).

A breeding program is designed to create genetic progress that will lead to increased productivity and improvements in other areas with a strong environmental impact. All individual groups and activities contributing to the transformation process are regarded as elements of the breeding program (Ponzoni 1988; Simianer et al. 2021). The economic benefits of the genetic progress generated are generally difficult to quantify since the potential for monetary returns depends mainly on factors outside the breeding program (Dekkers and Shook 1990). The economic benefits of the genetic gains in a breeding program can be calculated for a nation's economy, a sector of a national economy, or different institutional actors in multi-action breeding programs (Hayes et al. 2009).

Generational shifts in breeding objectives do not lead to noticeable gains in overall success. Economic benefit and heritability are two factors that should be considered when deciding whether a trait should or should not be included in the

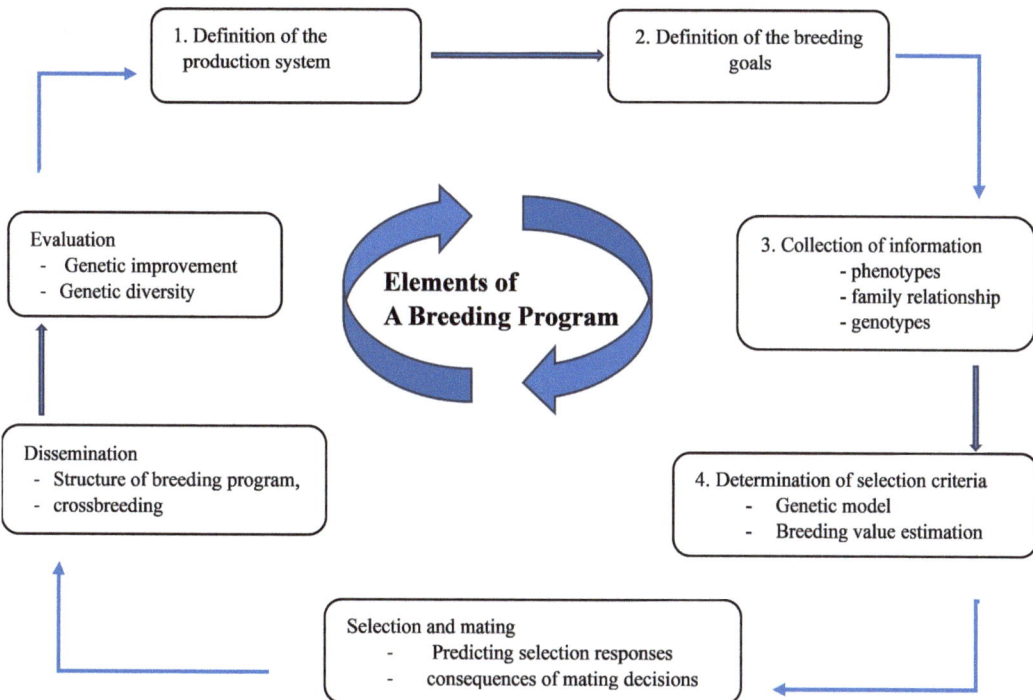

Fig. 18.1 The cycle of a breeding program

breeding goal. Both can be used to express breeding goals. The traits selected for breeding could be limited to the breeder's desires, the needs of the producers and processors, or the preferences of the general public. However, including more traits in the breeding goal will result in slower improvement of each trait with each successive generation. A breeding program's selection decisions usually take a long time (multiple generations of selective breeding) to bear fruit, therefore, emphasizing the importance of considering long-term financial benefits during the breeding goals' establishment. The success of a breeding program resides in solid plans, without frequent to no revisions.

With advances in animal breeding theory, biotechnology techniques, and genetics, modern poultry breeding programs consistently evolve. They are set up to sustainably exploit the existing genetic variation between and within traits (Harris et al. 1984). Both genetics and reproduction techniques tend to influence the optimal design of the breed improvement and dissemination program. Breed improvement programs are designed within and between breeds, enhancing the sustainability and competitiveness of animal production. A key factor in any breeding program is the estimation of breeding values, which is the genetic merit of an individual (Simianer et al. 2021; Kessner and Novembre 2014). Relevant tools and methodologies utilized in bringing about genetic improvement in poultry breeding have been extensively discussed in Chap. 23.

Breeding programs are natural stochastic processes with implicit randomness. A systematic and comprehensive approach to the design of breeding programs, described by Harris et al. (1984), consists of nine steps that begins with the analysis of the production system. Other steps include the definition of breeding goals, the choice of the breeding system and breeds to be used, data recording (both phenotypic and genomic data), estimations of heritability and genetic associations, breeding value estimations including large data sets using genomic information, selection of reproducing animals, mating of reproducing animals, and evaluation of the effects selection practice and its consequences. The design also addressed the need to compare alternative forms of breeding programs with respect to genetic progress and costs. Kinghorn et al. (2000) differentiated between two variant procedures of designing breeding programs: the bottom-up and the top-down approaches. In the top-down approach, the different steps are defined and implemented by rules, but in the bottom-up approach, available animal material and technologies are made use of in a beneficial manner. It is also called "tactical implementation of breeding programs," in that all available technologies are used and combined optimally. However, in many real-life cases, implemented breeding programs are a mixture of both approaches (Simianer et al. 2021).

To fully design a breeding program, according to Simianer et al. (2021), both the breeding environment and the structure of the breeding program are needed. The founder population(s) must be defined along with the relevant traits in the context of performance testing and the estimation of breeding values included in the breeding goal. For a joint set of traits, phenotypic mean and variance in the founder population, heritability, repeatability, and genetic and phenotypic correlations, which are all the required parameters must also be defined. If the objective of the breeding program is to make genetic progress, the breeding goal, which can be implicitly formulated in an economical manner by providing marginal economic values per natural unit or genetic standard deviation of each trait, must be defined also (Simianer et al. 2021; Faux et al. 2016; Greenbaum and Fefferman 2017; Schlather 2020).

There are typical phenotyping and selection scenarios in many breeding programs at certain stages. For example, a specific set of traits may be recorded for a particular group of individuals, which may depend on combinations of data that can be own or relatives' phenotypes, genotypes, etc. This must be well defined together with the type of breeding value estimation method applied at a given stage. Once the breeding environment has been properly defined, the breeding program structure can then be represented as a graph that is composed of nodes, made up of elementary objects and edges linking the nodes in a directional way (Schlather 2020; Pook et al. 2020).

An elementary object in a breeding program is the smallest unit to be addressed. It can exist in different types, either as "individual" or "gamete" with a certain set of attributes (e.g., "sex," for individual or "age," for "phenotype"). A group of elementary objects of the same type is called a node which can also have two attributes; size, a positive integer, and duration, which reflects the time allocated to its "lifespan." A node can either be a founder node or a derived node, resulting from one or several other nodes.

Edges exist in different classes. For example, it can be "aging," "selection," or "reproduction." It is a directed link from an origin node to a target node. A given edge class has a specific set of attributes. For example, selection as an edge can have the attribute of selected proportion and selection type that can either be random or phenotypic. An edge of a specific class defines the transformation rule of the set of elementary objects in the node of origin. The transformation rule is applied uniformly and in a probabilistic way to all elementary objects of the node of origin to prevent systematic discrimination between the elementary objects in that node (Simianer et al. 2021; Faux et al. 2016; Pook et al. 2020).

The genetics of the founder populations in addition to the correlations (genetic and phenotypic covariances) among traits critically determines the success of any breeding program since the founder population genetics provides the range of genotypes that can be selected. This is a particularly important consideration in breeding programs involving broilers because many generations of selection for a narrow range of traits may have eliminated the variation necessary for breeding for a broader range (Dawkins and Layton 2012).

In a formal assessment of a breeding program, an essential requirement is that one can predict genetic progress achieved through the program by regressing average phenotypic and breeding values on the year of birth. In addition, regular internal and external performance testing can be performed by the breeder which must cover various production environments to ensure that selected animals perform well under various conditions. Other sources of information for regular assessments are field results and customer feedback, which are probably the most important. The impact of a breeding program depends on the dissemination of genetic progress to customers (Simianer et al. 2021; Pook et al. 2020).

18.3.1 Breeding Programs for Local/Native Chicken in Low-Input System

To improve the productivity of native chickens with respect to egg production, growth rate, and feed efficiency without compromising on crucial genetic features like product quality, disease resistance, or adaptation, well-structured breeding programs are required. Determining specific breeding objectives that meet the needs of the production systems that will use indigenous chicken is essential when establishing breeding programs. According to Hazel (1943), breeding objectives are a linear collection of qualities that impact profitability weighed with individual respective economic worth. Therefore, genetic change is likely to be in the wrong direction or worse than none when the focus is placed on false attributes or when crucial traits are left out of the breeding aim. Most African countries lack indigenous chicken breeding schemes due to unclear breeding goals (Manyelo et al. 2020). Selecting native chicken requires the development of sustainable breeding goals.

The chickens' end-use and users' preferences heavily influence the breeding objective and plans. Many questions about the production system must be answered before defining the breeding aim. Why do you have chickens? How do you advertise your chickens and other products? What factors are most critical for husbandry and management practices? How regulated are the breeders? Is there a plan in place to start a breeding program? How observable are the characteristics? Because of this, the potential for breeding programs and the selection of breeding goal traits is determined by these aspects of production systems.

The first stage in starting a breeding program is selecting the best breed to use in the target environment or production system, focusing on the breed's adaptive performance. There have been failed attempts to transplant high-yielding chicken breeds (such as Rhode Island Red) to tropical regions. Lack of adaptation to the heat means the chickens have trouble reproducing, and that heat stress makes it difficult to raise high production rates. There is also a significant death rate due to various tropical diseases.

18.3.2 Examples of Breeding Programs in Low-Input System

Over 90% of Ethiopia's chicken meat and egg production comes from local poultry systems using indigenous breeds (Dana et al. 2011). Characteristics of this system include keeping a small flock size per household, maintaining birds in scavenging conditions, not providing any supplementary feeding, not providing any separate shelters other than night enclosures in the family house, and not providing any medical attention to the flock. A survey examined the socioeconomic aspects of the production system to determine and rank the breeding goals and trait preferences of village poultry producers. The breeders' primary traits are disease and stress resistance, the ability to evade predators, scavenging vigor, and live weight development and egg production. Hereafter, they will consider reproduction potential, hatchability of eggs, or size and color. In addition, the study found that villager poultry breeders preferred local poultry breeds over a reference modern breed for various traits, including resistance to disease and stress, ability to avoid predators, minimal care requirements, scavenging tendencies, egg hatchability, and meat flavor. These results resulted in adopting a breeding program utilizing mass selection live weight at 16 weeks for males, age at first egg, and egg production up to 45 weeks for females. The productivity considerably increased in just five generations, demonstrating that a "complex" breeding program may be unnecessary. It must be adaptable to regional conditions (small holdings) and contribute to the existing value structure.

18.4 Economic Traits in Chicken Breeding Programs

The principles of population genetics through quantitative selection practices have been in use since the 1940s for improving the productivity and efficiency of production in poultry breeding programs (Hunton 2006). Developing genetically superior stocks capable of higher production has transformed poultry production, and the poultry industry has grown exponentially over the years. Increased genetic selection pressure for economically important traits and traits associated with carcass-processing characteristics in broilers and egg-laying characteristics in layers have contributed significantly to the productivity and efficiency of the poultry industry. The development of high-yielding layers (310–340 eggs) and broilers (2.4–2.6 kg at 6 weeks) largely attributed to a combination of crossbred and purebred selection practices have significantly contributed to spectacular growth in poultry production (Rajkumar et al. 2021; Saxena and Kolluri 2018). Nevertheless, traits' response to selection depends on population size, gene frequency, mutation, allelic fixation, random drift, and physiological limits. The development of effective breeding strategies for improved productivity depends heavily on the magnitude and direction of correlated responses (Rajkumar et al. 2021; Chatterjee and Rajkumar 2015).

Through a unique selection program, specialized sire and dam lines for different traits have been developed over the years. The dam lines were developed through intense selection practice for reproductive traits, for example, egg production, egg size, egg weight, shell quality, age at sexual maturity, and hatchability besides juvenile growth. In contrast, the sire lines were developed and selected for improved growth rate, body confirmation, feed conversion ratio, carcass quality, and fertility. Mating of these genetically diverse lines resulted in several kinds of gene recombination, producing heterotic effects in the progenies for different economic traits. Increased selection pressure within pure lines and crossing those that are genetically diverse has been the most characteristic feature of the broiler breeding program (Saxena and Kolluri 2018).

Layer breeders have improved over 30 traits of commercial importance for egg production through the application of selection pressure over time. This intense selection for egg production has significantly reduced genetic and phenotypic variations in egg production traits. Peak production is now very close to the biological limit of one egg a day per bird. However, genetic variation at early production (at sexual maturity) and late production (persistency) is still high (Thiruvenkadan et al. 2010; Schmidt and Figueiredo 2005). Nowadays breeders select for age at sexual maturity, laying rate, livability, egg weight, body weight, feed conversion, shell color, shell strength, albumen height, egg inclusions (blood and meat spots), and temperament (Saxena and Kolluri 2018). Residual feed consumption has been used as the major selection criteria for improving feed efficiency because of high heritability and the absence of any significant negative effects on production parameters (see Chap. 23 for modern technologies used in the improvement of feeds, targeted feeding, and poultry nutrition).

The breed improvement program has concentrated more on rapid growth and carcass traits in broiler chickens. Selection at a commercial weight in broiler birds applies selection pressure at a weight that matches the market weight. As such, the age at selection becomes progressively earlier as growth potential increases (Tavárez and Solis de los Santos 2016). As research in genetics and breeding continues, productive performance in the broiler industry also continues to evolve. The industry has, therefore, been characterized as an age-for-weight market industry with the slaughter at a fixed target weight; thereby, emphasizing rapid growth and early carcass development (Zuidhof et al. 2014; Makanjuola et al. 2021a). According to Tavárez and Solis de los Santos (2016), it is difficult to accept that genetic improvement in broiler chicken has not reached a biological threshold. By the recent projections from Agri-Stats, Inc., broiler body weight of 2.34-kg (5-lb) will be attained at the end of a 28.8-day growing period in the year 2034. This will result from genetic improvement programs. The growth period is therefore projected to reduce at a rate of 0.56 days/year and the main benefit of selecting birds for faster growth will be improved feed efficiency.

Maternal effects also play an important role in the development and expression of economic traits as a result of genetic or environmental differences between dams or by the combination of both genetic and environmental differences (Mohammadi et al. 2018). Maternal effect is a situation wherein the genotype and environment of the mother influence traits. Egg weight is a major factor resulting from maternal effects on body weight. Other traits, such as hatch weight, incubation conditions, and nutrition, may also play a role. Correlations of egg weight on hatching weight and the subsequent body weight of chicks also reflect maternal effects. Therefore, including maternal effects in breeding programs will reduce bias and increase the precision of genetic parameters (Rajkumar et al. 2021; Ghorbani et al. 2012).

18.4.1 Trait Antagonism

In livestock production, selection decisions often impact traits that may not necessarily be under selection. These unintended consequences result from genetic correlation between traits. Genetic correlation provides information on whether two heritable traits share genes, and this means that selection for one will cause changes in the other. When selection is carried out for a trait that is highly correlated with another, substantial changes can be expected in the other trait, simply because of the strong genetic correlation between them. A negative genetic correlation indicates that the other trait tends to decrease as one trait increases. When selection for one trait produces a desirable outcome in another trait, then the genetic correlation is said to be favorable. As such, a trait can either have a favorable positive correlation or a favorable negative correlation. Unfavorable genetic correlations are sometimes referred to as genetic antagonisms. In some traits, when single-trait selection is practiced, genetic antagonisms can cause a decrease in genetic merit. Similarly, when one fails to consider selection responses in correlated traits that are not directly under the selection, genetic antagonisms can also result in a decrease of genetic merit (Tavárez and Solis de los Santos 2016). In poultry, two of the most important reproduction traits are clutch length and broodiness. Clutch length refers to the number of eggs laid in a single brood, while broodiness is the behavioral tendency to sit on a clutch of eggs to incubate them (Ohkubo 2017) Broodiness is negatively correlated with egg number, and clutch length is also negatively correlated with the body weight of females. This is an undesirable genetic correlation between production traits in poultry according to Jambui et al. (2017) and Makanjuola et al. (2021a). Similarly, some low-fitness traits have emerged over generations of breeding programs in broiler birds where early growth is considered a priority. Slower skeletal development has put immense pressure on broiler birds' gait quality and leg integrity as compared to the rapid growth of breast muscles and this has led to well-documented disorders such as tibial dyschondroplasia, bone deformities, and valgus-varus deformities (Avendaño et al. 2017). In these birds, cardiovascular efficiency has also been compromised, leading to the rise in myopathies causing defects in meat quality. There is also reported antagonism between early growth and reproduction traits in broilers (Tavárez and Solis de los Santos 2016). With osteoporosis in commercial layers, several "unfavorable trends" have also emerged because of "major pressure put on calcium metabolism and bone remodeling by eggshell synthesis." Bird selection has also been linked with injurious feather pecking, with one of the hypotheses stating that the trait for total bird egg numbers is linked to social dominance (Makanjuola et al. 2021a). According to Jambui et al. (2017), the antagonistic relationship between productive and reproductive traits has consequently changed the breeding program goals over time to a more balanced approach that considers production and reproductive traits without trading off on animal health and welfare

(Avendaño et al. 2017; Makanjuola et al. 2021b). However, several individual traits, including egg production, hatchability, and fertility, affect overall reproductive performance.

Increasingly, breeders had to deal with antagonisms between different traits as breeding objectives expanded. Selecting one trait can bring a correlated beneficial response in the other trait when the genetic correlation between two traits is favorable. Nevertheless, selecting one trait will lead to an undesirable response in the other when traits are antagonistically correlated. It is common practice in such a case to include both traits in the selection objective and select animals with desirable attributes for both traits. This is a strategy that allows all traits to be improved over time (Neeteson-van Nieuwenhoven et al. 2013). Typically, combining traits in a "selection index" is the most efficient way to select multiple traits (Phocas et al. 2013). In a selection index, traits are weighted according to index coefficients that consider the economic importance of such traits and their genetic relationships and in addition, maximize the correlation between the selection index and the breeding goal (see Chap. 15). However, the outcomes of breeding programs are realized many years after selection decisions are made. A genetic change implemented in a breeding nucleus in poultry will take at least 3 years to have a noticeable effect at the commercial level, which underlines the need to anticipate future demands when defining breeding goals. The development of specialized male and female lines in layer and broiler stocks was a result of negative correlations realized in production and reproduction traits. These specialized lines developed in both meat-type and egg-type stocks usually have very different foundation genetic sources (Chambers 1990; Crawford 1990; O'Sullivan et al. 2010).

18.5 Breeding Strategies in Chicken Production Systems

Breeding objectives address the question of the direction of production, but breeding strategies describe "how to get there." (Kinghorn et al. 2000) The breeding strategy is made up of five building blocks, namely, trait measurement, estimation of breeding values, reproductive technologies, selection, culling, and mating. These building blocks are divided into three major categories: selection between breeds, selection within breeds or lines, and crossbreeding. The production system characteristics and the available animal type determine the choice of strategy to pursue (Declaration 2007).

The domestication of the jungle fowl and identification of superior birds based on performance or phenotype marked the beginning of selective breeding in poultry. This practice was, however, carried out without knowing of the underlying principles of genetics until 1866 when Mendel reported that the inheritance of traits from parents is governed by particles of inheritance later called genes and the "law of heredity" gave the scientific basis for their inheritance (Saxena and Kolluri 2018; Singh n.d.).

As knowledge and needs increased, selection and breeding programs have evolved with time in the poultry business. The concept of two- or multi-way crosses has transformed the poultry breeding output, resulting in high-yielding modern layer and broiler strains. This concept was adopted in the 1980s, from a maize improvement program. In addition, commercial hybrids have replaced pure breeds as terminal crosses, while specialized egg and meat-type birds replaced dual-type birds. Specialized male and female lines in layer and broiler stock with different foundation genetic sources were also developed because of the negative relationships in production and reproductive traits (Chambers 1990; Crawford 1990; O'Sullivan et al. 2010).

The inheritance pattern of quantitative traits is multigenic with minor allelic and additive effects in addition to substantial environmental influence. Identifying or selecting superior individuals for these traits required different procedures based on the theories and principles of quantitative genetics. Chicken breeds show a range of traits that encompasses body size, egg production, and egg color. Large-scale industrial poultry breeding started with the hybridization of selected pure breeding lines. Different breeding and selection technologies for genetic improvement were introduced at different times. Techniques, such as mass selection, hybridization, pedigree selection, artificial insemination, Osborne index, family feed conversion testing, selection index, individual feed conversion testing, BLUP breeding value estimation, and DNA markers have been used over decades (Saxena and Kolluri 2018).

The development and utilization of synthetic lines from specialized selection programs and crossbreeding processes were vital for the progress made in poultry production over the years. The inclusion of performance record information on pure and crossbreds in selection criteria helped improve responses to selection (Baumung et al. 1997; Uimari and Gibson 1998). Therefore, estimated breeding value (EBV) for crossbreed performance can be determined using the knowledge obtained from both pure and crossbred animals, which may also be used as the basis for selection in a strategy known as "combined crossbred purebred selection (CCPS) strategy" (Uimari and Gibson 1998). Genetic correlation between the heritability of crossbreds (h^2_C) and the performances of pure and crossbreds, especially when combined crossbred and purebred selection method is applied to achieve genetic progress in crossbreds, are important parameters for optimizing and evaluating crossbreeding systems (Bijma and Van Arendonk 1998; Wei and van der Werf 1994). When crossbred performance is the breeding goal and the genetic correlation between purebred and crossbred performance is low, the EBVs of selected candidates will be dominated by the information obtained from the crossbred half-sibs (Bell 1982). Low or negative estimates of crossbred performance indicate a nonadditive genetic effect that suggests a more effective reciprocal recurrent selection (RRS).

Both pure-line selection (PLS) and crossbreeding in a combined crossbred and purebred selection (CCPS) program are used in modern poultry breeding. Purebred and crossbred performance are treated as genetically correlated traits assuming the infinitesimal model (Bell 1982). Depending on estimated genetic parameters (e.g., heritability and correlations) involved, the phenotypic selection method is primarily adopted for improving body weights. The selection index (Osborne index) method is adopted in PLS to improve egg production. Several traits are now being included in the selection program for poultry birds. However, modern breeding programs rely on breeding value estimation with an animal model of the best linear unbiased predictor (BLUP), whereas genetic evaluations based on traditional BLUP utilize average relationships based on animals' pedigrees. Genome-enhanced BLUP utilizes the actual genomic relationship between the animals, which allows for a more précised estimation (Meuwissen et al. 2001).

The maximum number of saleable eggs per hen housed at a low feed cost per egg is the main breeding goal for laying birds. In addition, the eggs should have optimal internal and external qualities, and the stock should have a low mortality rate and high adaptability to different environments. Through the application of selection pressure, layer breeders have improved over 30 traits of commercial importance egg-producing birds. Age at sexual maturity, rate of lay, livability, egg weight, body weight, feed conversion, shell color, shell strength, albumen height, egg inclusions (blood and meat spots), and temperament have been included in selection programs by modern poultry breeders. The strategy for the improvement of egg production also included part-time egg production records, persistency of lay, clutch length, FCR/Residual feed consumption (RFC), and skeletal problems (Saxena and Kolluri 2018).

Selection strategies for broilers concentrated on rapid growth rate and carcass traits. "Selection at commercial weight," is the most practiced strategy in pure line selection. This method employs selection principles at a weight that matches the market weight and the age at selection becomes progressively earlier as growth potential increases. The other strategies practiced are the multi-stage selection strategies and selection at the commercial age (Saxena and Kolluri 2018; Thiruvenkadan et al. 2010). Breast muscle weight, meat quality, and FCR are traits considered. Selection based on breast area determined by measuring the length and width of the breast using a pachymeter and body weight resulted in a genetic gain of about 277% per generation while keeping feed conversion and fertility in their actual levels (Glasbey and Robinson 2002). Recently attention has also been given to skeletal abnormalities, metabolic disorders, and welfare.

Within 30–35 years, the development of genetically superior poultry birds capable of higher production, even under adverse climatic conditions, revolutionized poultry production and transformed it from a rural farming system to a full-fledged industry. The increase in production volume and productivity per bird were largely attributed to the method of "combined crossbred and purebred selection (CCPS)." Constant evaluation of the superior purebred lines for nicking ability and best nicking male and female lines by a specialized crossbreeding program was used for the development of four-way commercial crosses. DNA marker technology emerged as a finer tool for assessing genetic variability with advances in molecular biology techniques. Genome-wide scans identified quantitative trait loci (QTL) for use in marker-assisted selection (MAS). The discovery of single nucleotide polymorphisms (SNPs) in addition to "next-generation sequencing" technique led to the development of high-density SNP arrays as a powerful tool for genetic analysis. The prediction of genomic estimates of breeding value (GEBV) of individuals using SNPs across the whole genome paved the way for conceptualizing "genomic selection," which has emerged as the most advanced technology that revolutionizes animal production (Saxena and Kolluri 2018; Meuwissen et al. 2001).

18.5.1 Breeding Strategies for Low-Input Systems

Though breeding and genetic improvement of local chickens are complex processes, they are crucial components of sustainability of the African poultry industry (Manyelo et al. 2020). Breeding strategies are based on the careful selection of breeds or genes to be used and require a good understanding of the specific attributes or phenotype of each contributing breed, as well as its input to the general objective of the local poultry industry development.

The general aim of poultry breeding strategies should be to obtain generations of animals in the future that will produce more efficiently under future production, market, and consumption circumstances. For a breeder, this will mean firstly designing tactics to adopt and integrate new technologies and/or to improve old local proven ones to maximize the performance of the existing stock, and secondly integrating the components of a breeding program into a structured system for genetic improvement, mating the best to the best, to optimize the overall objective.

The basic components of a successful poultry breeding strategy for Africa can be structured and integrated, as illustrated in Fig. 18.2.

18.5.1.1 Defining Chicken Breeding Objectives in African Input Systems

In Africa, poultry production is carried out in a wide range of production systems which vary in both scale and resource utilization. Each system performs a valuable role in people's lives so that they will co-exist rather than transit from one to another. Africa, therefore, requires a breed or breed of chickens that will perform in these different production systems (Crawford 1990). Breeding objectives need to be clearly defined with specific and measurable selection criteria to ensure genetic improvement (Garrick and Golden 2009; Crawford 1990). The definition of breeding objective should be market-led and demand-led, directly connected with the preferred types of chicken (meat, eggs, or dual purpose). Still, it should also consider economic productivity, conservation needs, or other sociocultural considerations that could ensure the sustainability of the systems. All these parameters are interdependent with the local policy and other support mechanisms from the local governments. According to Olori (2012), the success of a global leader like Aviagen has resulted in a thorough understanding of the requirements in each of the very diverse production systems and environments that a global market entails and the capacity to meet the

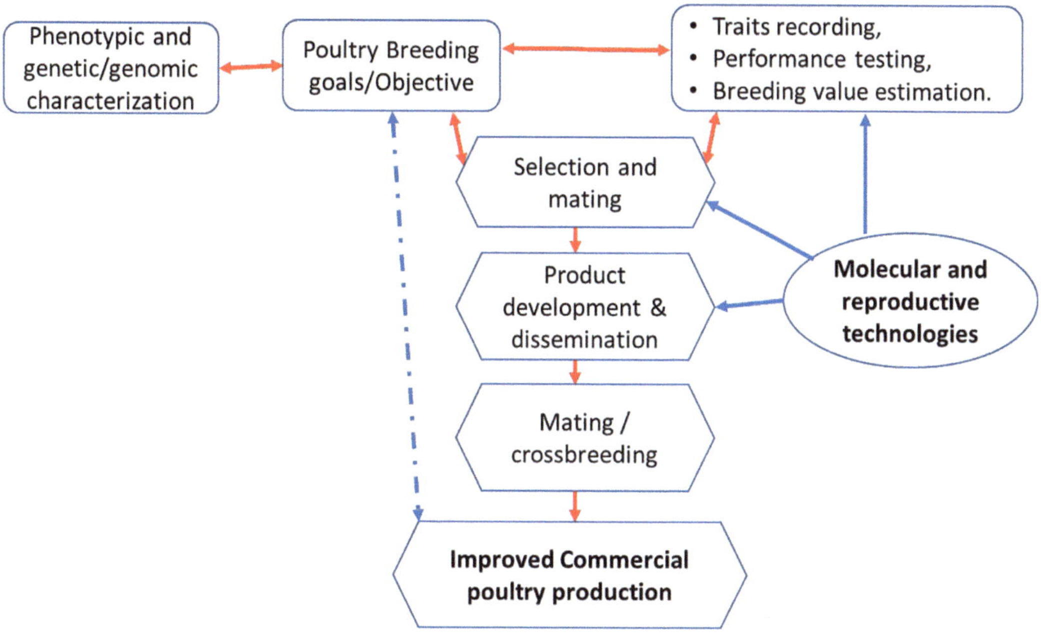

Indicates the continues developing technologies for improving breeding strategies.

Indicates that the breeding objectives/goals can be adjusted from time to time based on market preference/ consumer interest

Fig. 18.2 Basic components for a successful poultry breeding strategy for Africa. (Adapted from Jack Dekkers 1990)

challenges of producing a suitable product for each market. This also applies to the future African poultry industry.

Trait Recording, Performance Testing, and Breeding Value Estimations

Trait recording is the ground for identifying animals with the "best" genetics in relation to the breeding objectives. Figure 18.3 shows some breeding objective traits normally included in a typical broiler breeding program of a breeding company (Olori 2012; Laughlin and MIBiol 2007).

Data recording is the primary objective for breeding and genetic improvement within breeding populations with pedigreed records in dispersed facilities that comprise the breeding nucleus (Olori 2012). Precision breeding robustness that will meet the needs of the future African poultry industry has been facilitated by the impact of technological advancement, especially on data recording and genomic selection which has provided poultry breeders with additional tools for breeding program efficiency and accurate selection decisions on a broad range of traits. Animal identification and traceability remain major hindrance to livestock and poultry development in Africa. This is the most basic and important information in genetic evaluation because of the need for accurate pedigree registration.

Population-specific parameter estimates require pedigree and performance recording, which is one of the biggest challenges in developing countries owing to small flock sizes and lack of commitment by smallholder farmers (Wasike et al. 2011). Performance testing will provide the essential precondition for breeders to improve economically important traits in African poultry. An on-farm performance testing was recently conducted by Abegaz et al. (2019) on tropically adaptable chicken strains under smallholders management in three countries of sub-Saharan Africa under ILRI's African Chicken Genetic

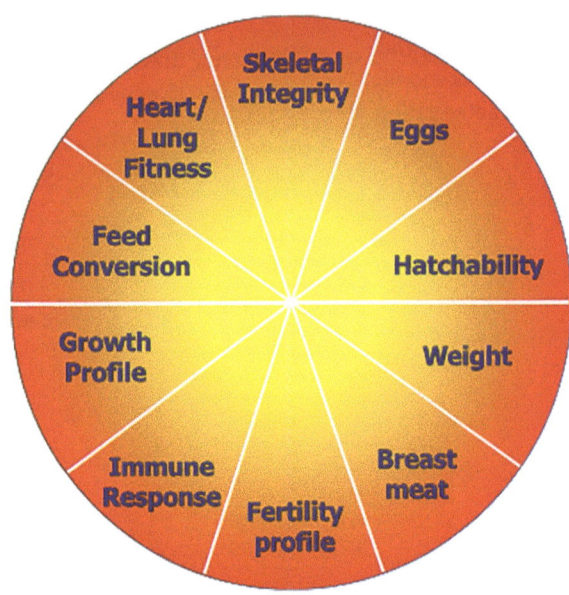

Fig. 18.3 Breeding objective traits in a typical broiler breeding program. (*Source*: Olori (2012))

Gains (ACGG) Project. This can potentially define the chicken breeds, phenotypes, and genotypes preferred by African smallholder farmers in terms of bird color, body conformation and temperament, egg and meat productivity, and overall tropical adaptability under low-input production systems. The design of this chicken performance test was informed by the ACGG Producer Level Baseline Survey (PLBS). The information generated from the performance tests prepared a roadmap for long-term chicken genetic gains programs in the respective environment or countries (Dessie 2018). A similar study in Nigeria by Bamidele et al. (2020) identified FUNAAB Alpha and the Noiler chicken breed as more suitable for dual-purpose functions (egg and meat). In contrast, Sasso and Kuroiler (meat) and Shika-Brown (egg) were reported to be better suited for a single-purpose function. These findings could guide the introduction of smallholder poultry-specific hybrid germplasms for developing a smallholder poultry production system in Nigeria.

Accessing farmers' and breeders' data for recording and performance testing should generally lead to the development of estimated breeding values (EBVs), resulting in genetic improvement in most of the preferred production traits with high heritability. The lower the heritability of a trait under selection, the slower the herd will improve (Wolc 2014).

Selection Approaches

Breeding of indigenous chickens in Africa, in general, has mainly been done by smallholder farmers who raise the birds in extensive and scavenging systems characterized by low inputs, high disease pressure, and adverse climatic conditions (Ngeno 2015). This has resulted in natural selection with the development of specific genetic adaptations for challenging environments. Natural selection is most likely a major factor that shaped the genomic variation of the rural African indigenous chicken, driving the development of their genetic footprints (Elbeltagy et al. 2019). Expectedly, selection footprints are associated with adaptation to locally prevailing environmental stressors, including high altitude, disease resistance, poor nutrition, and oxidative and heat stresses.

Although artificial selection driven by human needs has resulted in phenotypic, physiological, and behavioral changes in chickens (Rubin et al. 2010), the selection pressure has been different in African local chicken compared to the commercial breeds.

Identification of superior animals on the basis of performance for breeding has been in practice for ages (Saxena and Kolluri 2018). Phenotypic selection allowed for the creation of native breeds from wild breeds. Selection strategies for improving egg production traits in African chicken should include part-time egg production records, persistency of lay, clutch length, FCR/Residual feed consumption (RFC), and skeletal problems (Thiruvenkadan et al. 2010). For meat-producing chicken, selection strategies should concentrate on rapid growth and carcass traits. In developing African poultry breeds intended to be dual-purpose and tropical adaptation, geneticists must consider a balance of characteristics related to growth and reproduction (Table 18.1).

At different periods of time, contrasting selection and breeding methods that gradually led to the exploitation of quantitative and molecular selection methods were employed for the genetic improvement of poultry. Some of these include mass selection, trap nesting, hybridization, artificial insemination, Osborne index in layers, family feed conversion testing, selection index, individual feed conversion testing, BLUP breeding value estimation, and DNA markers techniques.

Table 18.1 Characteristics to be balanced when selecting for dual purpose and tropically adapted chicken breeds

Growth and meat-related traits	Reproduction and egg-related traits
Growth rate	Acceptance laying in nests
Weight-for-age	Adaptation to heat distress
Feed efficiency	Aggressiveness
Meat (breast) and carcass yield and body conformation	Feed conversion per kg of egg
Livability	Disease resistance
Skeletal integrity	Egg number
Feathering-cover rate and color	Marketable egg produced per bird
Adaptation to heat distress	Egg size
Daily gain weight	Egg weight
Growth (body weight at different ages)	Fertility
Feed efficiency	Hatchability
Viability	Internal egg quality (Haugh units)
Bone density	Absence of itching
Feather: coverage	Libido
Heat adaptation (heat tolerance)	Mature weight and age
Disease resistance	Persistence of laying without broodiness
	Sexual precocity
	Shell quality (strength and thickness)
	Visual sexing

Utilization of Molecular Information and Tools Relevant to the Various Input Systems

Molecular information and techniques has been used in the selection and utilization of specific genes that affect the anatomical development of breast and/or calcium metabolism in hybrids, to the point of reaching the limit of pathological situations (Castellini et al. 2008). There are, however, limited studies involving molecular tools with high genomic coverage that will perform genome-wide scans of genomic variations and also reveal how these genomic variations can be associated with variability in resistance and susceptibility of African local chickens to rusticity and adaptability. Several first, second, and third-generation genomic markers (RFLPs, ESTs, SNP, RAPDs, micro- and minisatellites, AFLP, etc.) have considerably contributed to the development of poultry breeding (Mpenda et al. 2019; Vaarst et al. 2015; Oldenbroek and van der Waaij 2014; Oluyemi et al. 1979). Today, SNPs have become very popular with various genetic applications, including quantitative trait loci (QTL) identification and genome-wide scans. Candidate gene and high-density SNP genotyping for whole-genome selection are also in use and according to Mpenda et al. (2019), the genetic/genomic selection of resistant chicken to viral infections is a promising strategy currently in use. Reports have revealed that African local chickens are highly genetically diverse and adapted to harsh tropical environmental conditions. In the context of climate change, therefore, the African local chicken can be a potential candidate chicken population for selection of resistance to disease infections (Vaarst et al. 2015).

18.5.1.2 Chicken Breeds Development and Dissemination

The choice of breeding goals, selection criteria, the design of the breeding scheme, identification of objective traits, derivation of appropriate selection criteria, and accurate genetic evaluation are some of the core activities of the primary breeder, which determines its effectiveness and the program's success. For its sustainable and independent development, it is critically important for African livestock, in general, and the poultry sector, in particular, to be structured as recommended by FAO (Declaration 2007) and create specialization along the value chain (Fig. 18.4). However, the sustainability of such a system will require an understanding of the intrinsic and extrinsic factors that affect poultry production and productivity in Africa, such as the climate, production systems, and available feed resources.

Considering the proportion of farmers in the low-input production system, Africa will have hardy, robust poultry breeds that can resist or survive disease outbreaks and cope with harsh weather conditions and should be resilient to climate change.

The dissemination aspect following product development entails programs for marketing and distributing superior genes into the commercial sector as parental stock. Prerequisites to successful dissemination include an efficient progeny test, applying practical breed multiplication tools, and installing trusted multipliers. The selection response dissemination depends on the breeding program's structure (Oldenbroek and van der Waaij 2014). Various intervention programs including crossbreeding/upgrading, breed substitution, and selection within the population intended to enhance the productivity of African chickens have been attempted in the past.

Mating/Crossbreeding

Crossbreeding/upgrading of indigenous chickens with exotic chickens through cockerels or pullets exchange was another genetic intervention method implemented in several African countries at a time. These interventions started in 1937 in Tanzania, in the 1950s in Nigeria and Malawi, and 1956 in Sudan, and were extended to many other African countries where local chickens were mated with such breeds as Rhode Island Red (RIR), Light Sussex, and Black Australorp chicken (Ponzoni 1984; Oluyemi et al. 1979). According to these authors, crossbreds demonstrated superior performance, but their survivability was poor. The intervention was therefore categorized as unsuccessful.

Breed Substitution

Intensification of African chicken replacement by exotic breeds started around the 1960s and has culminated with the most recent introduction of the Kuroiler chicken. In principle, the main objective of the substitution was to have chickens with better growth and superior egg production

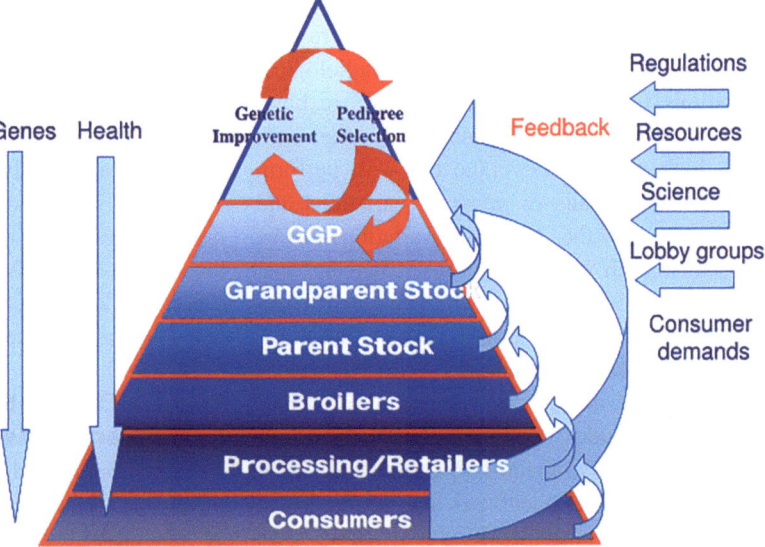

Fig. 18.4 Pyramidal structure of poultry breeding. (*Source*: Olori (2008)

performances. However, the adaptability of the exotic chickens was poor and could have been improved under the prevailing tropical conditions of production systems before distribution. Moreover, substituting locally adapted breeds with exotic breeds that are more susceptible to the different environmental stressors, though more productive was contrary to the global trend on the conservation of indigenous genetic resources because it was threatening the existence of the indigenous breeds (Kosgey et al. 2006; Hanotte et al. 2010). The only way to prevent local poultry breeds' substitution from occurring would be to make them more valuable to farmers. This can be realized by genetic improvement through careful within-breed selection.

Within-Breed Selection

Within-breed selection has been one of the major strategies used in some African countries, like Egypt, Nigeria, and recently Ethiopia, for the genetic improvement of indigenous chicken (Hossaryl and Galal 1994; Wondmeneh et al. 2014). In Egypt, this strategy was successful as it created two pure lines strains of the Fayoumi breed of chicken with a 60% higher egg production rate when compared with the original indigenous breed (Hossaryl and Galal 1994). In Ethiopia, the selection of Horro local chicken started in 2000 at the Debre Zeit Research station. This program has successfully increased the egg production rate by 123.5% to 75 eggs at week 45 and reduced the age of the first egg from 203 to 148 days by the fifth generation. It also led to an increase in the body weight of the birds (Wondmeneh et al. 2014). The within-breed selection is a promising strategy for improving the productivity of African chickens compared to crossbreeding or upgrading systems. As a baseline consideration, the lessons from the failed crossbreeding/upgrading programs should be put into consideration when initiating breeding programs for within-breed selection. These failures demonstrate that success in any crossbreeding/upgrading program requires proper planning, infrastructural development, sustainable funding,

and an understanding of the production environment.

Under the smallholder farming system, functional breeding programs remain a big challenge. Earlier native chicken genetic improvement efforts mostly failed for the following reasons among other things: (i) dependence on breed replacement and/or indiscriminate crossbreeding with exotic breeds that had no sustainable replacement plan and no intentional selection on the dam line; (ii) incompatibility between the exotic genotypes and the producers' breeding objectives, management practice, and environmental conditions; (iii) adoption of complex breeding programs that require and technological development and many logistics; and (iv) insufficient/ no systematic breed evaluation studies that ensure a fair comparison of the relative merit of the indigenous and the exotic breeds that considers the role of genetics × environment interactions (Wurzinger et al. 2014). Smallholder production systems are normally characterized by small population size, single-sire herds/flocks, lack of animal identification, absence of performance and pedigree recording, illiteracy, poor infrastructure, and ill-functioning institutions. These in addition to the mobility of pastoral flocks pose additional difficulties in data recording and selection practice.

18.6 Increasing Productivity Without Sacrificing the Adaptability of Traits to the Production Environment

The success or failure of a breeding program is profoundly affected by obstacles encountered during the breeding process. Chicken farmers, chicken consumers, the food business, and, increasingly, the general public all have a hand in setting these standards. Finding that sweet spot between competing needs is an ongoing effort that calls for forethought about the future and systematic planning of breeding initiatives. In chicken breeding programs, beyond the apparent breeding objective features, other traits, such as the chickens' health and welfare and their ability to adjust to the feed of lesser quality and harsh climates, play a crucial influence in the birds' ability to produce and reproduce. Robustness in performance, according to Niyas et al. (2015), is "the ability to combine a high production potential with resilience to stressors, allowing for the unproblematic expression of a high production potential in a wide variety of environmental conditions." Robustness and disease resistance have improved significantly through selective breeding programs that use both genomics and new quantitative genetic theory to improve the response of birds to environmental changes as well as predicting adaptation responses (Burrow 2012).

The major environmental stressors affecting productive and reproductive traits are heat and nutrition related. Heat stress is the result of high ambient temperature which adversely affects both production and reproduction. The effect of heat stress on livestock is reflected by reduced feed intake, poor growth performance, panting, increased rectal temperature, respiratory rate, and water intake (Augère-Granier 2019). Changes in hematological parameters, electrolytes, metabolites, increased mortality and morbidity, and reduced immune function also result from the effect of heat stress (Sejian et al. 2010). Fitness and adaptation, which determine an animal's tolerance to adverse conditions, are influenced by individual genetic makeup. Adaptability measures the potential or actual capacity to adapt (Hoffmann 2010). Many breeds have developed adaptive traits that increase their survivability in harsh environmental conditions (Mirzaie et al. 2018; Barker 2009). The range of unique adaptive characteristics of poultry birds which evolves in stressful environments enables them to survive and be productive in adverse environmental conditions.

Initial steps in a breeding program should focus on selecting the most adaptable breed for the target environment or production system. If this information is disregarded, the animals' fitness will decline. An increase in the susceptibility of birds to environmental challenges resulted from the selection for high production, according to (Gaughan and Cawsell-Smith 2015). Although optimum production is easier to achieve under controlled environmental conditions, which is common in poultry production, it is not necessarily the most economical. Adaptation involves trade-offs that must be considered when selecting animals for use in breeding programs. Broadly, adaptation is a short-term, long-term, or generational (genetic) response to challenges (stressors). Short-term responses to a stressor encompass reduced feed intake and increased respiration rates when birds are exposed to high ambient temperatures. However, short-term responses also have a genetic basis, with some animals able to cope better than others when exposed to the same stressors. Inherent genetic variation is the primary factor that determines the ability to adapt as measured by survival and reproductive rates which interact with the constraints of the environment creating phenotypic variation (Mirzaie et al. 2018). Adaptation is often at the expense of performance, and survival rate is often better in animals with "low" performance because their input needs (mainly feed) and internal heat production are lower (Gaughan and Cawdell-Smith 2015). However, the issue of whether to directly select for adaptive traits, and production, reproduction, and growth traits in harsh environments is debatable. Generally, favorable correlations suggest that adaptive traits would not be compromised when selection is emphasized for performance traits (Burrow et al. 1991). Indications exist to suggest that selection for better performance traits (e.g., reproduction, survival, and growth) in stressful environments will lead to selection for the most suitable animals (Turner 1984).

One of the most cost-effective tools for achieving a permanent change in bird tolerance to heat stress is the application of genetic principles of selection. Although its implementation strategies are challenging because of the complexity of responses to heat stress and the antagonism between heat tolerance and productivity. Variability in the population is the basis for any genetic improvement which is primarily determined by response to selection in the trait under consideration and other correlated traits of economic importance. The magnitude and the direction of correlated responses play a major role in the development of effective breeding strategies for improved productivity (Hassan et al. 2009; Boki 2000; Hill and Mackay 2004). However, as the number of traits in a selection program increases, the amount of genetic progress that can be made in the improvement of any one trait slows down unless the traits are highly genetically correlated (Bamidele et al. 2020). The use of genetic tests based on single nucleotide polymorphisms to identify genetic markers that predict thermotolerance (Dikmen et al. 2008) in addition to genomic selection offers new opportunities for commercial breeders to produce better breeds and manage their birds in the future (Niyas et al. 2015; Burrow 2012).

There is, however, the need to duly emphasize environmental effects when selecting breeding stocks and planning breeding strategy since the phenotype depends on genotype and environment. Most breeding companies need to pay more attention to the impact of the genotype x environment (G × E) interaction because having a selection environment that accounts for the traits of importance is inconvenient (Sørensen 2005). A breeding scheme should evolve animals that perform optimally under specific climatic conditions. The chance for a G × E interaction increases as the difference between the environments grows, and the genetic distance between the breeds is large (Sørensen 2005). Therefore, every breeding animal must be evaluated under specific climatic conditions before being released for commercial use. The success of a breed or strain in a particular environment depends on its ability to adapt and perform in that environment/climatic zone. The available evidence for genetics × environment interactions in performance analysis of poultry birds in various suboptimal conditions emphasizes the need for breeding programs to improve performance in a particular environment.

18.7 Breeding Systems

Regardless of the character involved, the kind of progeny (young ones) produced determines whether a male or female will be kept for breeding purposes. When a male mated to a female with a high egg production record is a progeny of a high egg-producing dam, the daughters frequently produce a high number of eggs. The results obtained from a given mating are largely determined by the genetic contribution of the birds that mated. Breeding systems can, therefore, be classified depending on the aim, which can either be to increase homozygosity or heterozygosity in random mating, inbreeding, or outbreeding (Saxena and Kolluri 2018).

18.7.1 Random Mating

This involves individual mating that does not involve any form of selection. It is used to develop a control population for comparison and measuring the effects of other breeding systems. Controlling population also helps in the estimation of the effects of the environment, which in turn, aids in the estimation of true genetic gain through any breeding method.

18.7.2 Inbreeding

Inbreeding is the mating between individuals more closely related than the average relationship between all individuals in a population. Inbreeding has its own benefits and can be consistently carried out for several generations with benefits, although the consequences can be damaging.

There are three distinct ways inbreeding and be practiced. These are as follows:
1. Close inbreeding: This is mating between siblings, parents, and progenies.
2. Strain formation: This involves developing small groups within a breed for a special desired character. It is usually regarded as a mild form of inbreeding.

3. Line breeding is a form of inbreeding with an ancestral line where an attempt is made to concentrate the inheritance potential of a common ancestor (usually a male) on an individual. It is backcrossing to the same parent for several generations in succession, an intensive form of back-crossing.

18.7.3 Outbreeding

Outbreeding is the opposite of inbreeding, and it is the interbreeding of individuals that are relatively unrelated as compared to the average relationship of individuals within the population. Mating between strains/inbred lines is a form of outbreeding (Hutu et al. 2020).

The methods of outbreeding (crossbreeding) include the following:

(a) Single or two-way cross: When two populations of the same or different breeds are crossed to produce an F1 generation not intended for breeding purposes, the progenies usually exhibit hybrid vigor, especially when inbred lines are involved. When two inbred lines of the same breed are crossed, the progeny is said to be crossbreds.
(b) Three-way cross: In a three-way cross, F1 crossbred females or males are mated to males/females of a third line to obtain an F2 progeny.
(c) Four-way cross: In a four-way cross, two different single crosses (AB and CD) are made to obtain ABCD. Four-way cross is usually practiced in poultry breeding program between inbred lines of low viabilities since only a relatively small number of animals of lines need to be maintained.
(d) Crossing for a new breed: To produce modern-day breeds of poultry birds, different breed types have been crossed to combine desirable traits from many sources. The foundation crosses were subjected to inbreeding combined with selection practice to consolidate them into true breeding populations to meet desired needs (breeds) (Singh n.d.).

18.7.4 Mating Methods

Mating is the act of pairing a male and female for the purpose of reproduction. Different methods of mating are commonly practiced in poultry.

1. *Pen mating*: It is a mating system where a male is mated to a group of females in a pen during the breeding season. It is sometimes described as a mass mating system where two or more males are mated with several females in a single pen. Male to female ratio is generally high in pen mating. It ranges from one male to 12 to 15 females in egg-type chickens and 10 to 12 for meat-type chickens.
2. *Stud mating*: In this case, the males are usually confined to individual pens (studs) within a large laying pen and the females are held in the stud for a known period of time, till they are mated. They are then replaced by other ones. In this method, mating at least once every 5 days or twice a week is desirable for optimum fertility.
3. *Artificial insemination*: Artificial insemination (AI) is the technique where semen collected from a male is introduced or deposited in the female reproductive tract using a pipette. It is usually a three-step procedure that involves semen collection, semen dilution, and insemination. However, undiluted semen can be used directly within 30 min of collection. On average, one cock yields about 0.5–1.0 ml of semen, depending on the body weight. Usually, about 0.05–0.10 ml of semen is enough to inseminate one hen.
4. *Flock mating*: In this mating type, large numbers of hens are kept with all the breeding males in a single flock at a ratio of approximately 10 hens per cock. Record keeping is minimal in this mating system as the lineage of individual chick is relatively unknown. Because of the large number of birds per unit, flock mating reduces the costs of operation and it is preferred where pedigree records are not maintained (Singh n.d.).

18.8 Omics Technologies in Selective Breeding for Economic Traits in Chicken Production

Genetic selection and breeding are crucial breed development tools that have resulted in genetically superior and disease-free animals with improved productivity and efficiency in various livestock species (Rexroad et al. 2019; Erasmus and van Marle-Köster 2021). Conventional animal selection and breeding methods were based on observed production trait characteristics and the estimations of the breeding value (BV) of economic traits included in genetic improvement programs (Brito et al. 2021). However, there are limitations in these methods, particularly for traits that are sex-limited, traits with low heritability, and traits expressed later in life (e.g., carcass traits). The consequence is that their genetic gain is slow because of high generation intervals. However, with the onset of high-throughput omics techniques, discoveries and availability of multi-omics technologies, and development of sophisticated analytic packages, several promising tools and methods for the estimation of the actual genetic potential of animals have been developed. In the last two decades, landmark developments in omics technologies have revolutionized genetic selection for animal breeding. It has now become possible to collect and access large and complex datasets which provides new opportunities for a better understanding of the mechanisms regulating actual animal performance. Large and complex datasets comprising different omics (genomics, transcriptomics, proteomics, metabolomics, and phenomics) data as well as animal-level data (such as longevity, behavior, and adaptation) can now be easily accessed and studied to provide a better understanding of the regulatory mechanisms of animal performance.

Better and more accurate breeding values (BVs) can now be estimated with data generated through omics technologies to increase genetic gain by assisting the selection of genetically superior, disease-free birds at an early stage of life for enhancing productivity and profitability. Selection and breeding programs utilize omics-generated data to estimate more accurately breeding values for early selection and to achieve reduced generation interval and increased rate of genetic gain (Krassowski et al. 2020; Chakraborty et al. 2022). Several omics tools have been developed in the last two decades for the collection and analysis of high-throughput data in proteins (proteomics), mRNA transcripts (transcriptomics), gene sequences (genomics), microbial diversity (metagenomics), epigenetic regulation of gene expression (epigenomics), metabolic profile (metabolomics), etc. (Chakraborty et al. 2022). Omics technology applications allowed for extraordinary progress in the study and understanding of many quantitative traits in broilers (see Chap. 23). The approaches used in transcriptomics, proteomics, and metabolomics have been successfully applied for investigating the molecular basis of complex traits, such as feed efficiency and muscle myopathies, and for the assessment of the molecular responses to nutritional treatments and evaluation of important aspects relating to immunity and disease resistance (Zampiga et al. 2018).

18.8.1 DNA Markers and Genomics Selection

Tremendous advancements in molecular genetics have provided a better understanding of quantitative economic traits in animal production, ever since Chang et al. (2003) reported that DNA markers is advantageous in animal breeding to foretell genetic makeup and animal performance. Several genes and gene combinations have been found to be directly associated with animal performance and production efficiency (Rexroad et al. 2019; Ruan et al. 2021). In addition, many quantitative trait loci (QTL) responsible for trait diversity have been identified for various production and reproductive traits and are in use for selection and breeding decisions (Zhang et al. 2021). Several genetic markers have also been discovered and used in marker-assisted selection (MAS) of breeding stock (Ma et al. 2021). Marker-assisted selection (MAS) uses genomic and phenotypic information to increase reliability in the selection process for the genetic improvement program. Identified quantitative trait loci (QTLs) and candidate genes were crucial for producing chickens with more desirable meat characteristics.

The decoding of the genomic sequence soon after releasing the draft chicken genome sequence (Hillier et al. 2005) showed that there are millions of places within the chicken genome where poultry birds differ from each other at a single nucleotide (Wolc 2015). The search for differences between individuals in their DNA sequence and differences in observed phenotypic traits resulted in the identification of single nucleotide polymorphisms (SNPs). In the chicken, 2.8 million SNP markers have been mapped and are available for evaluation. The first "high-density" panel of genetic markers for chicken 6000 single nucleotide polymorphisms (SNPs) was released in 2007 (Andreescu et al. 2007), and a 600,000 SNPs panel (Kranis et al. 2013) was released in 2013. Genome-wide association studies (GWAS) demonstrated the link between SNPs and QTLs Field 105 using high-density SNP chips. Whole-genome sequencing (WGS) was a breakthrough in detecting molecular signatures for selecting and breeding animals (Bovo et al. 2020). The genome is a complete set of genetic material present in an organism.

The paradigm shift in QTL- and candidate gene-based approaches for genetic selection leveraged on the landscape of genome sequence. In genomic selection technology, a sample DNA can be used to calculate the breeding values of animals. Meuwissen et al. (2001) initially proposed the method, which involves scanning entire genome-wide genetic markers to generate the breeding value of the selected candidates. This method is based on certain assumptions such that linkage disequilibrium between the markers and the polymorphisms causes variation in important traits (Meuwissen et al. 2001). In genomic selection, the underlying principle is that information generated from many markers can be used to predict genomic estimated breeding values (GEBVs) without understanding the specific location of the gene on the genome length. In the selection of traits of economic importance relatively governed by many genes, sex-limited traits (such as egg production and egg quality), and traits with low heritability (such as disease resistance), this method has shown higher reliability and efficiency when compared to the traditional method (Khare and Khare 2017). This technique also increases the accuracy of selection for new breeding animals by selecting beneficial genes. The availability of genomic information has recently permitted the simultaneous combination of pedigree and genomic relationships in a single-step genomic evaluation method (ssGBLUP). Using a random regression model, an increase of about 23% in heritability estimates for fertility and hatchability was obtained using this method. In addition, the ssGBLUP method resulted in higher prediction accuracy when compared with the traditional pedigree method (Makanjuola et al. 2021b).

Genomic selection is considered the most appropriate standard breeding method in livestock. Poultry breeders have taken a huge interest in its application and execution after numerous successful validation studies (Wolc 2015; Wang et al. 2013) with the main aim of yielding rapid genetic progress in its selection program. Increased accuracy in genomic prediction is the main advantage exploited by implementing the genomic selection program in the poultry industry (Kaisa et al. 2020). The revolution of genomic selection has already been described as an efficient technology in poultry improvement programs, justified by its increased rate of genetic gain.

Genomic selection is considered particularly advantageous as it can be used to select animals for breeding at an early stage of life without reference to their own breeding or production records (Fig. 18.5).

Genomic selection has become widely accepted for selecting animals for breeding with advances in high-throughput omics technologies which help in the identification of functional SNPs and their priority to increase the accuracy of the genetic selection (Chakraborty et al. 2022; Tan et al. 2017; Chang et al. 2019). Further, population-level omics (e.g., population genom-

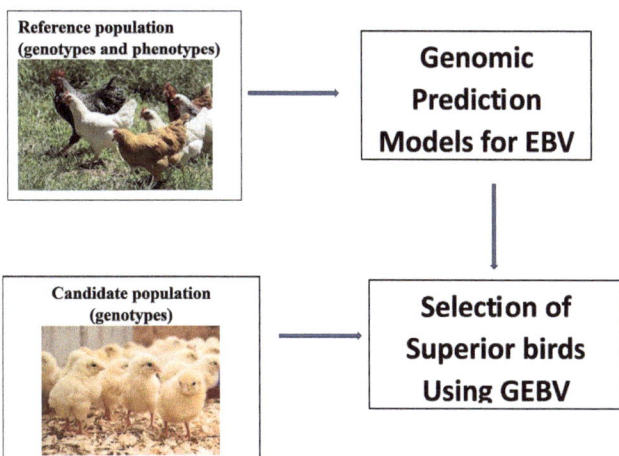

Fig. 18.5 Genomic selection in poultry

ics) hold tremendous potential for classifying individuals based on allelic diversity and the identification of genetically related individuals (Lippert et al. 2017).

18.8.2 Biomarkers

Biomarkers are quantifiable biological molecules representing specific physiological states' characteristics that reflect the molecular changes caused by the interactions between organisms and environmental factors. The identification and validation of biomarkers (genes, transcripts, proteins, and metabolites) associated with fertility have shown great potential for the improvement of reproductive efficiency in poultry. Biomarkers play key roles in the definition of animal diseases, although, some have poor diagnostic specificity and are not pathognomonic for diseases (Ali et al. 2020). The biomarker technique is a post-genome technology induction. It acts precisely on a specific biomarker in a pool of molecules inside a tissue or fluid. In the biomarker field, the development of omics is a remarkable discovery in the assessment of animal health (Ali et al. 2020; Moore et al. 2007; van Ginneken 2017). The rapid advancement of post-genomic technologies in transcriptomics, proteomics, and metabolomics has allowed for the development of strategies that help to identify biomarkers from the thousands of molecules in a tissue or biological fluid (Moore et al. 2007). Based on the analysis of RNA-level gene expressions, the occurrence and change of rules of the transcriptome in biological cells is what is generally called transcriptomics (Oskoueian et al. 2016; Wang et al. 2009; Lockhart and Winzeler 2000). Large-scale gene expression profile techniques, qRT-PCR, and biostatistical methods such as principal component analysis (PCA), hierarchical cluster analysis (HCA), and partial least squares discriminant analysis (PLS-DA) (Oladokun and Adewole 2022) are used for the determination of transcriptional biomarkers on the transcriptomic platform. The use of biomarkers will increase the rate of genetic improvement due to more accurate selection. Biomarker techniques have been successfully applied in molecular medicine, disease diagnosis, and prediction. Current biological processes can be diagnosed and monitored promptly by identifying and monitoring biomarkers (Oskoueian et al. 2016; Su et al. 2022). Blood biomarkers enable an early-warning system in poultry, detecting health challenges like coccidiosis and heat stress. Biomarkers are also currently being used in precision nutrition studies in broilers to identify amino acid imbalance and monitor intestinal health in poultry birds (Hassan et al. 2009; van Marle-Koster and Nel 2000; van Marle-Koster 2008; Ramadam et al. 2012)

Along with platform sensitivity and data analysis, technological advances in extraction and fractionation techniques have enabled the discovery of next-generation biomarkers with high sensitivity, specificity, and precision that are useful for monitoring the health and well-being of poultry birds (Al-Dawood 2016). These biomarkers are also used for surveillance against pathogens, assessing pharmacologic responses to therapeutic and directing genetic selection and breeding. A number of proteomic biomarkers, for example, have been identified and used for the diagnoses of gastrointestinal, respiratory, diseases, and heat stress in poultry (van Marle-Koster and Nel 2000; Ramadam et al. 2012; Al-Dawood 2016). The acute phase proteins (APPs) are blood biomarkers associated with innate immune response and they are responsible for systemic reactions to inflammations. Increases in the concentrations of APPs are valuable biomarkers of inflammation in the body (van Marle-Koster 2008). When physiological parameters deviate from normal, biomarkers are useful in determining the degree of physiological imbalance and early detection reduces disease risk and improves production and reproduction performance (van Marle-Koster 2008; Ramadam et al. 2012).

18.9 Strategic Approach to Sustainable Use and Genetic Improvement

18.9.1 Accounting for Major Environmental, Structural, and Socioeconomic Differences Among Production Systems

Although the FAO recognizes four poultry production systems, clear differentiation can be made between the commercial/intensive systems and the extensive system. The poultry production system in Africa and elsewhere in the developing world includes both commercial (broilers and layers) and traditional birds relying mainly on indigenous and improved dual-purpose breeds kept primarily in a free-range system. Commercial poultry production is mostly practiced in urban and peri-urban areas, while the traditional poultry system is confined to rural areas. The traditional production system, however, contributes most of the flock and supplies the bulk of poultry meat and eggs consumed in both rural and urban areas. Poultry production's impact on the environment is less problematic compared to other livestock species. Structurally, most commercial setups are more vertically integrated than the extensive production system, including breeder farms, hatcheries, layer and broiler farms, traders, and processors. Intensification demands keeping big flocks of chickens, whose feed and water consumption remains high. Commercial or intensive poultry production systems are mainly profit-oriented and have a less direct socioeconomic contribution to individual households like the smallholder or the extensive poultry production system.

18.9.2 Ensuring Participation of End Users (the Chicken Keepers)

End users (both men and women) should have access to relevant information and be involved in policies formulating policies and planning with ample opportunities to give personal opinions. Women comprise most of the world's poor and poor livestock keepers (Bagnol 1998), who generally own and manage village chickens. Village chickens are normally the most essential assets for female-headed households, who have the power to control the income generated from them (Guèye 2000; Donadeu et al. 2019; MacVicar 2020). They are easier to manage and place little or no additional burdens on these women. Apart from the provision of household food scraps, these birds find their feed and require little supervision (Donadeu et al. 2019; Alders et al. 2010). Rural poor women livestock keepers in low- and middle-income countries (LMIC) are more likely to own chickens and goats as assets that generate income, provide high-protein food, accumulate wealth, confer social status, protect against economic shocks, and act as financial reserves for the family (McKune et al. 2021). The poorest smallholder farmers in developing countries tend to own only chickens, and as they progress out of poverty, pigs and small ruminants (goats or sheep) are added, with cattle, camels, or buffalo being the most desirable animals (Donadeu et al. 2019). Most of the household men, head the households. Although decisions are expected to be made in consultation, women often accept the decisions of their spouses. In villages, men get interviewed about the poultry intervention, while women responsible for poultry are not given a chance. Women might prefer technologies that do not require much hard work, which would not burden their already heavy workload.

18.9.3 Multiplication and Delivery of Improved Strains

One of the main tasks of breeding programs is the distribution of the improved chicken strains to users across broader geographies. The ability of the improved chicken strains to strive and benefit farmers in harsh environments serves as a success indicator for breeders and farmers alike. Commercial chicken strains or exotic breeds introduced in Africa failed to bring the expected result due to high feed, veterinary, and energy costs (Sonaiya and Swan 2007). Exotic birds were not adapted to the local conditions and demanded high feed investments, veterinary support, and housing conditions. Local and tropically adapted breeds were not given due attention until recently. A good example is the BMGF-funded, ILRI-led African Chicken Genetic Gains project. ACGG blends new chicken strains, enhancing delivery systems to support the adoption of highly productive birds in semi-scavenging systems in the tropics. The strains were selected based on their previous history of being used in harsh environments. Brooded and vaccinated birds 45 days old were widely tested in Ethiopia, Nigeria, and Tanzania (Marshall et al. 2019). The key to the success of such projects or sustainable distribution of improved genetics will be its ability to deliver more productive chickens on a large scale and reach the users whenever demanded. The poultry industry is a private domain whose motivation is essentially profit. Distribution models differ from location to location depending on the existence of the performing value chain. However, sustainable distribution should be identified after several models have been tested. Prior identification of the right and preferred chicken strain traits and strategically supporting actors along the value chain will help reach the poultry producers effectively. Transfer and adoption of livestock technologies are particularly limited because most entail very high costs and require institutional linkages and partnerships to bridge this gap and transform traditional subsistence production into new production objectives targeting the market (Dana et al. 2006). Poultry development programs were possible through the support of donors. The programs reached out to several households in different geographies. The sustainability of such programs can be possible if there is a buy-in from the governments.

18.9.4 Promote Step-By-Step Development and the Sustainability of the Actions Implemented

Several attempts were made in Africa to develop poultry production and productivity. The intervention involves improving the management of chickens, introducing exotic blood, vaccination, and so on. Most of the interventions have contributed to the overall productivity level but could not sustain (Kitalyi 1998; Boki 2000). The results observed in limited locations or research stations may indicate the potential of the interventions but can only bring about a change if scaled out in huge numbers. Different interventions need to be brought together into a package. Most technological interventions should be supported by coordination, planning, marketing, etc. Scaling out of a particular technology can best be promoted in a public–private partnership model involving local agriculture businesses as paid technology and input agents. An institutional mechanism needs to be established to arrange accessible credit facilities for the farmers and local entrepreneurs to scale up proven technologies.

18.9.5 Conclusion and Future Perspectives

Africa is home to a diverse poultry population. Indigenous poultry resources have never been fully utilized, although they remain the primary source of chicken meat and eggs. Farmers have never been given a chance in the formulation of development policies that led to the introduction of commercial chicken breeds from overseas. Commercial chicken production has not benefited smallholder farmers due to their need for management inputs that remain challenging in low-input systems. Breeding companies can design products for specific ago-ecologies in Africa. The approach is expensive as agro-ecologies are quite diverse, and the demand shifts every time. The logical path is to consider breeding programs by countries involving indigenous chickens. Breeding programs require proper planning, understanding, and execution capabilities. It will be essential to understand the production environments. The necessary steps need to be followed, and modern technologies should be considered at the right time and stage.

Several attempts were made in Africa to develop poultry production and productivity. The interventions involved improving the management of chickens, introducing exotic chicken genetics, vaccination, and so on. Most of these interventions contributed marginally to overall productivity, especially among the crossbreds which could not even be sustained. An institutional mechanism must also arrange accessible credit facilities for farmers and local entrepreneurs to scale up proven technologies. Locally developed indigenous chicken can be important in poultry development under low-input production environments.

Poultry production systems in Africa and other developing parts of the world include both commercial (broilers and layers) and traditional birds relying mainly on the indigenous and improved dual-purpose breeds primarily kept on free-range. End users (farmers) should have access to relevant information, be involved in policy formulation and planning, and be given ample opportunities to air their opinions. The ability of the improved chicken strains to strive and benefit farmers in low-input environments serves as a success indicator. Breeders/farmers will be able to design products specific to each low-input agro-ecology. This provides room for countries to develop their breeds involving indigenous chicken. Commercial chicken strains or exotic breeds introduced in Africa failed to bring the expected result because of high feed, veterinary, energy costs, and the need for specific infrastructure. These birds were not adapted to local environmental conditions and demanded high feed investments, veterinary support, and housing conditions. Transfer and adoption of new livestock technologies are limited because most entail very high costs and require institutional linkages and partnerships to bridge the gap and transform traditional subsistence production into a market-oriented production system.

This chapter extensively covers breeding goals, strategies, and breeding programs relevant for chicken production systems in Africa. This approach therefore presents the opportunity to harness the potential of poultry genetics resources that could adapt efficiently to the tropical environment while enhancing food security, farmer livelihoods, and income opportunities to those that need it the most.

References

Abegaz S, Esatu W, Assefa G, Goromela E, Sonaiya EB, Mbaga S, Adebambo O, Bamidele O, Teressa A, Bruno J (2019) On-farm performance testing of tropically adaptable chicken strains under smallholder management in three countries of sub-Saharan Africa

Al-Dawood A (2016) Application of acute phase proteins as biomarkers in chicken: a review. Egypt J Agric Sci 67:193

Alders R, Bagnol B, Young M (2010) Technically sound and sustainable Newcastle disease control in village chickens: lessons learnt over fifteen years. Worlds Poult Sci J 66(3):433–440

Ali W, Ali M, Ahmad M, Dilawar S, Firdous A, Afzal A (2020) Application of modern techniques in animal production sector for human and animal welfare. Turk J Agric Food Sci Technol 8(2):457–463

Amer P, Allain DD, Avendaño S, Baselga M, Boettcher P, Dürr JW, Garreau HH, Gootwine E, Reynoso GAG, Knap PW (2015) Breeding strategies and programmes.

The Second Report on the State of the World's Animal Genetic Resources for Food and Agriculture

Andreescu C, Avendano S, Brown SR, Hassen A, Lamont SJ, Dekkers JC (2007) Linkage disequilibrium in related breeding lines of chickens. Genetics 177(4):2161–2169

Augère-Granier, M.-L. The EU poultry meat and egg sector: Main features, challenges and prospects. 2019

Avendaño S, Neeteson AM, Fancher B (2017) Broiler breeding for sustainability and welfare–are there trade-offs? In Proceedings of the New Zealand Poultry Beyond 2023 Conference, vol 7

Bagnol B (1998) The social impact of Newcastle disease control. In: ACIAR proceedings, 2001; ACIAR, pp 69–75

Bamidele O, Sonaiya E, Adebambo O, Dessie T (2020) On-station performance evaluation of improved tropically adapted chicken breeds for smallholder poultry production systems in Nigeria. Trop Anim Health Prod 52:1541–1548

Barker JS (2009) Defining fitness in natural and domesticated populations. In: Adaptation and fitness in animal populations: evolutionary and breeding perspectives on genetic resource management. Springer, Dordrecht, pp 3–14

Baumung R, Sölkner J, Essl A (1997) Selection response according to the use of pure-and/or crossbred information in crossbred animals with varying levels of dominance and additive by additive effects. In: Book of abstracts of the 48th annual meeting of the European Association for Animal Production, August 25–28, Vienna, Austria

Bell A (1982) Selection for heterosis-results with laboratory and domestic animals. In: Proceeding of 2nd world congress on genetics applied to livestock production, vol. 6, pp 206–227

Berry D, Evans R (2014) Genetics of reproductive performance in seasonal calving beef cows and its association with performance traits. J Anim Sci 92(4):1412–1422

Berry D, Wall E, Pryce J (2014) Genetics and genomics of reproductive performance in dairy and beef cattle. Animal 8(s1):105–121

Bijma P, Van Arendonk J (1998) Maximizing genetic gain for the sire line of a crossbreeding scheme utilizing both purebred and crossbred information. Anim Sci 66(2):529–542

Boki KJ (2000) Poultry industry in Tanzania with emphasis on small-scale rural poultry. In Possibilities for smallholder poultry projects in eastern and Southern Africa, Morogoro (Tanzania), 22–25 May 2000

Bovo S, Ribani A, Muñoz M, Alves E, Araujo JP, Bozzi R, Čandek-Potokar M, Charneca R, Di Palma F, Etherington G (2020) Whole-genome sequencing of European autochthonous and commercial pig breeds allows the detection of signatures of selection for adaptation of genetic resources to different breeding and production systems. Genet Sel Evol 52(1):1–19

Brito L, Bédère N, Douhard F, Oliveira H, Arnal M, Peñagaricano F, Schinckel A, Baes CF, Miglior F (2021) Genetic selection of high-yielding dairy cattle toward sustainable farming systems in a rapidly changing world. Animal 15:100292

Burrow H (2012) Importance of adaptation and genotype× environment interactions in tropical beef breeding systems. Animal 6(5):729–740

Burrow H, Seifert G, Hetzel D (1991) Consequences of selection for weaning weight in zebu, Bos taurus and zebu× Bos taurus cattle in the tropics. Aust J Agric Res 42(2):295–307

Cammack K, Thomas M, Enns R (2009) Reproductive traits and their heritabilities in beef cattle. Prof Anim Sci 25(5):517–528. Walsh, S.; Williams, E.; Evans, A. A review of the causes of poor fertility in high milk producing dairy cows. Animal Reproduction Science 2011, 123 (3–4), 127–138

Castellini C, Berri C, Le Bihan-Duval E, Martino G (2008) Qualitative attributes and consumer perception of organic and free-range poultry meat. Worlds Poult Sci J 64(4):500–512

Chakraborty D, Sharma N, Kour S, Sodhi SS, Gupta MK, Lee SJ, Son YO (2022) Applications of omics technology for livestock selection and improvement. Front Genet 13

Chambers J (1990) Genetics of growth and meat production in chickens. In: Poultry breeding and genetics. Elsevier, Amsterdam

Chang K, Beuzen N, Hall A (2003) Identification of microsatellites in expressed muscle genes: assessment of a desmin (CT) dinucleotide repeat as a marker for meat quality. Vet J 165(2):157–163

Chang L-Y, Toghiani S, Aggrey SE, Rekaya R (2019) Increasing accuracy of genomic selection in presence of high density marker panels through the prioritization of relevant polymorphisms. BMC Genet 20(1):1–10

Chatterjee R, Rajkumar U (2015) An overview of poultry production in India. Indian J Anim Health 54(2):89–108

Crawford R (1990) Chapter 1. Origin and history of poultry species. In: Crawford RD (ed) Poultry breeding and genetics. Elsevier Science Publishers, Amsterdam

Dana N, Duguma R, Aliye S, YVold HT, Kumsa T, Assefa S, Yami A, Hassen F, Work KK, Asfaw A (2006). Promotion of Improved Poultry Technologies in East Shewa Central Ethiopia. What the Farmer Said, p 193

Dana N, Vander Waaij E, Van Arendonk JA (2011) Genetic and phenotypic parameter estimates for body weights and egg production in Horro chicken of Ethiopia. Trop Anim Health Prod 43:21–28

Dawkins M, Layton R (2012) Breeding for better welfare: genetic goals for broiler chickens and their parents. Anim Welf 21(2):147–155

Declaration I (2007) Global plan of action for animal genetic resources and the Interlaken Declaration. FAO Commission on Genetic Resources for Food and Agriculture, FAO, Rome

Dekkers J, Shook G (1990) Economic evaluation of alternative breeding programs for commercial artificial insemination firms. J Dairy Sci 73(7):1902–1919

Dessie T (2018) Progress narrative: African chicken genetic gains program

Dikmen S, Alava E, Pontes E, Fear J, Dikmen B, Olson T, Hansen P (2008) Differences in thermoregulatory ability between slick-haired and wild-type lactating Holstein cows in response to acute heat stress. J Dairy Sci 91(9):3395–3402

Donadeu M, Nwankpa N, Abela-Ridder B, Dungu B (2019) Strategies to increase adoption of animal vaccines by smallholder farmers with focus on neglected diseases and marginalized populations. PLoS Negl Trop Dis 13(2):e0006989

Elbeltagy AR, Bertolini F, Fleming DS, Van Goor A, Ashwell CM, Schmidt CJ, Kugonza DR, Lamont SJ, Rothschild MF (2019) Natural selection footprints among African chicken breeds and village ecotypes. Front Genet 10:376

Erasmus L-M, van Marle-Köster E (2021) Moving towards sustainable breeding objectives and cow welfare in dairy production: a South African perspective. Trop Anim Health Prod 53(5):470

Faux AM, Gorjanc G, Gaynor RC, Battagin M, Edwards SM, Wilson DL, Hearne SJ, Gonen S, Hickey JM (2016) AlphaSim: software for breeding program simulation. Plant Genome 9(3)., plantgenome2016.2002.0013

Garrick D, Golden B (2009) Producing and using genetic evaluations in the United States beef industry of today. J Anim Sci 87(suppl_14):E11–E18

Gaughan, J.; Cawdell-Smith, A. Impact of climate change on livestock production and reproduction. Climate change impact on livestock: adaptation and mitigation 2015, 51–60 Springer, New Delhi

Ghorbani S, Kamali M, Abbasi M, Ghafouri KF (2012) Estimation of maternal effects on some economic traits of north Iranian native fowls using different models

Glasbey C, Robinson C (2002) Estimators of tissue proportions from X-ray CT images. Biometrics 58(4):928–936

Gowane GR, Kumar A, Nimbkar C (2019) Challenges and opportunities to livestock breeding programmes in India. J Anim Breed Genet 136(5):329–338

Greenbaum G, Fefferman NH (2017) Application of network methods for understanding evolutionary dynamics in discrete habitats. Mol Ecol 26(11):2850–2863

Guèye E (2000) The role of family poultry in poverty alleviation, food security and the promotion of gender equality in rural Africa. Outlook Agric 29(2):129–136

Hanotte O, Dessie T, Kemp S (2010) Time to tap Africa's livestock genomes. Science 328(5986):1640–1641

Harris D, Stewart T, Arboleda C (1984) Animal breeding programs: a systematic approach to their design [Mathematical models]. Advances in Agricultural Technology (USA). no. 8

Hassan H, Neger FWC, de Kock A, van Maele-Koster E (2009) Study of the genetic diversity of native chicken in Northwest Ethiopia using microsatellite markers, Africa. J Biotechnol 8(7):1347–1353

Hayes BJ, Bowman PJ, Chamberlain AJ, Goddard ME (2009) Invited review: genomic selection in dairy cattle: progress and challenges. J Dairy Sci 92(2):433–443

Hazel LN (1943) The genetic basis for constructing selection indexes. Genetics 28(6):476–490

Hietala P, Wolfová M, Wolf J, Kantanen J, Juga J (2014) Economic values of production and functional traits, including residual feed intake, in Finnish milk production. J Dairy Sci 97(2):1092–1106

Hill WG, Mackay TF (2004) DS falconer and introduction to quantitative genetics. Genetics 167(4):1529–1536

Hillier LW, Miller W, Birney E, Warren W, Hardison RC, Ponting CP, Bork P, Burt DW, Groenen MA, Delany ME (2005) Erratum: sequence and comparative analysis of the chicken genome provide unique perspective on vertebrate evolution (Nature (2004) 432 (695–716)). Nature 433(7027):777–777

Hoffmann I (2010) Climate change and the characterization, breeding and conservation of animal genetic resources. Anim Genet 41:32–46

Hossaryl M, Galal E (1994) Improvement and adaptation of the Fayoumi chicken. Anim Genetic Resour/ Resources génétiques animales/Recursos genéticos animales 14:33–39

Hunton P (2006) 100 years of poultry genetics. Worlds Poult Sci J 62(3):417–428

Hutu I, Oldenbroek K, van der Waaji L (2020) Chapter II.3 breeding goals. In: Animal breeding and husbandry textbook. Agroprint Publishing House, Timisoara

Jambui M, Honaker C, Siegel P (2017) Correlated responses to long-term divergent selection for 8-week body weight in female white Plymouth rock chickens: sexual maturity. Poult Sci 96(11):3844–3851

Kaisa K, Kumar H, Saravanan K, Rajawat D, Bhushan B, Kumar P, Panigrahi M (2020) Concepts of genomic selection in poultry and its applications. Int J Livest Res 10(10):32–42

Kessner D, Novembre J (2014) Forqs: forward-in-time simulation of recombination, quantitative traits and selection. Bioinformatics 30(4):576–577

Khare V, Khare A (2017) Modern approach in animal breeding by use of advanced molecular genetic techniques. Inter J Livestig Res 7:1–22

Kinghorn B, van der Werf J, Ryan M (2000) Animal breeding: use of new technologies: a textbook for consultants, farmers, teachers and for students of animal breeding. Purdue University Press

Kitalyi AJ (1998) Village chicken production systems in rural Africa: household food security and gender issues. Food & Agriculture Org

Kosgey I, Baker R, Udo H, Van Arendonk JA (2006) Successes and failures of small ruminant breeding programmes in the tropics: a review. Small Rumin Res 61(1):13–28

Kranis A, Gheyas AA, Boschiero C, Turner F, Yu L, Smith S, Talbot R, Pirani A, Brew F, Kaiser P (2013) Development of a high density 600K SNP genotyping array for chicken. BMC Genomics 14(1):1–13

Krassowski M, Das V, Sahu SK, Misra BB (2020) State of the field in multi-omics research: from computational needs to data mining and sharing. Front Genet 11:610798

Laughlin K, MIBiol C (2007) The evolution of genetics, breeding and production. Temperton Fellowship Report, 15

Lippert C, Sabatini R, Maher MC, Kang EY, Lee S, Arikan O, Harley A, Bernal A, Garst P, Lavrenko V (2017) Identification of individuals by trait prediction using whole-genome sequencing data. Proc Natl Acad Sci 114(38):10166–10171

Lockhart DJ, Winzeler EA (2000) Genomics, gene expression and DNA arrays. Nature 405(6788):827–836

Ma B, Khan R, Raza SHA, Gao Z, Hou S, Ullah F, Hassan MM, Hassan MM, AlGabbani Q, Alotaibi MA (2021) Determination of the relationship between class IV sirtuin genes and growth traits in Chinese black Tibetan sheep. Anim Biotechnol 34:1–7

MacVicar I (2020) Fact check 9: Women Livestock Keepers

Makanjuola BO, Abdalla EA, Baes CF, Wood BJ (2021a) Selection for reproductive efficiency in turkeys and broiler chickens: egg production, hatchability and fertility

Makanjuola BO, Olori VE, Mrode RA (2021b) Modeling genetic components of hatch of fertile in broiler breeders. Poult Sci 100(5):101062

Manyelo TG, Selaledi L, Hassan ZM, Mabelebele M (2020) Local chicken breeds of Africa: their description, uses and conservation methods. Animals 10(12):2257

Marshall K, Gibson JP, Mwai O, Mwacharo JM, Haile A, Getachew T, Mrode R, Kemp SJ (2019) Livestock genomics for developing countries–African examples in practice. Front Genet 10:297

McKune S, Serra R, Touré A (2021) Gender and intersectional analysis of livestock vaccine value chains in Kaffrine, Senegal. PLoS One 16(7):e0252045

Meuwissen TH, Hayes BJ, Goddard M (2001) Prediction of total genetic value using genome-wide dense marker maps. Genetics 157(4):1819–1829

Miglior F, Fleming A, Malchiodi F, Brito LF, Martin P, Baes CF (2017) A 100-year review: identification and genetic selection of economically important traits in dairy cattle. J Dairy Sci 100(12):10251–10271

Mirzaie S, Zirak-Khattab F, Hosseini SA, Donyaei-Darian H (2018) Effects of dietary spirulina on antioxidant status, lipid profile, immune response and performance characteristics of broiler chickens reared under high ambient temperature. Asian Australas J Anim Sci 31(4):556

Mohammadi A, Naderi Y, Nabavi R, Jafari F (2018) Determination of the best model for estimation of genetic parameters on the Fars native chicken traits using Bayesian and REML methods. Genetika 50(2):431–447

Moore RE, Kirwan J, Doherty MK, Whitfield PD (2007) Biomarker discovery in animal health and disease: the application of post-genomic technologies. Biomark Insights 2:117727190700200040

Mpenda F, Schilling M, Campbell Z, Mngumi E, Buza J (2019) The genetic diversity of local african chickens: a potential for selection of chickens resistant to viral infections. J Appl Poult Res 28(1):1–12

Neeteson-van Nieuwenhoven A-M, Knap P, Avendaño S (2013) The role of sustainable commercial pig and poultry breeding for food security. Anim Front 3(1):52–57

Ngeno K (2015) Breeding program for indigenous chicken in Kenya. Wageningen University

Niyas P, Chaidanya K, Shaji S, Sejian V, Bhatta R (2015) Adaptation of livestock to environmental challenges. J Vet Sci Med Diagn 4(3):2

O'Sullivan N, Preisinger R, Koerhuis A (2010) Combining pure-line and cross-bred information in poultry breeding. In: Proceedings of the world congress on genetics applied to livestock production, volume genetic improvement programmes: design of selection schemes exploiting additive and/or non-additive effects–lecture sessions, p 0984

Ohkubo T (2017) Neuroendocrine control of broodiness. In: Avian reproduction: from behavior to molecules. Springer, Singapore, pp 151–171

Okeno T, Kahi A, Peters K (2013) Evaluation of breeding objectives for purebred and crossbred selection schemes for adoption in indigenous chicken breeding programmes. Br Poult Sci 54(1):62–75

Oladokun S, Adewole DJ (2022) Biomarkers of heat stress and mechanism of heat stress response in avian species: current insights and futre perspectives from poultry science. J Therm Biol 110:103332

Oldenbroek K, van der Waaij L (2014) Textbook animal breeding: animal breeding and genetics for BSc students

Olori V (2008) Breeding broilers for production Systems in Africa. Niger Poult Sci J 5(4):173–180

Olori, V. Data recording and profitability in livestock breeding: lessons from poultry breeding. 2012

Oluyemi J, Adene D, Ladoye G (1979) A comparison of the Nigerian indigenous fowl with white rock under conditions of disease and nutritional stress. Trop Anim Health Prod 11(4):199–202

Oskoueian E, Eckersall PD, Bencurova E, Dandekar T (2016) Application of proteomic biomarkers in livestock disease management. In: Agricultural proteomics volume 2: environmental stresses. Springer, Cham, pp 299–310

Phocas F, Brochard M, Larroque H, Lagriffoul G, Labatut J, Guerrier J (2013) Etat actuel et perspectives d'évolution des objectifs de sélection chez les ruminants. Rencontres autour des Recherches sur les Ruminants, Paris, pp 129–132

Ponzoni R (1984) Breeding objectives and selection procedures with special reference to Merino sheep. In Proceedings of the 4th conference of the association for the advancement of animal breeding and genetics, pp 229–237

Ponzoni RW (1988) Accounting for both income and expense in the development of breeding objectives. In: Conference Australian of assessment in animal breeding and genetic, vol 7, pp 55–66. Menge, E. O. Breeding goals for production systems utilizing indigenous chicken in Kenya. Egerton University, 2008

Pook T, Schlather M, Simianer H (2020) MoBPS-modular breeding program simulator. G3 Genes Genomes Genetics 10(6):1915–1918

Rajkumar U, Prince LLL, Rajaravindra K, Haunshi S, Niranjan M, Chatterjee R (2021) Analysis of (co) variance components and estimation of breeding value of growth and production traits in Dahlem red chicken using pedigree relationship in an animal model. PLoS One 16(3):e0247779

Ramadam S, Kayang B, Inoue E, Nirasawa K, Hayakawa H (2012) Evaluation of genetic diversity and conservation priorities for Egyptian chickens. Open J Anim Sci 2:183–190

Rexroad C, Vallet J, Matukumalli LK, Reecy J, Bickhart D, Blackburn H, Boggess M, Cheng H, Clutter A, Cockett N (2019) Genome to phenome: improving animal health, production, and well-being–a new USDA blueprint for animal genome research 2018–2027. Front Genet 10:327

Rivero J, Hodgkinson SM, López-Villalobos N (2013) Definition of the breeding goal and determination of breeding objectives for European wild boar (Sus scrofa L.) in a semi-extensive production system. Livest Sci 157(1):38–47

Ruan D, Zhuang Z, Ding R, Qiu Y, Zhou S, Wu J, Xu C, Hong L, Huang S, Zheng E (2021) Weighted single-step GWAS identified candidate genes associated with growth traits in a Duroc pig population. Genes 12(1):117

Rubin C-J, Zody MC, Eriksson J, Meadows JR, Sherwood E, Webster MT, Jiang L, Ingman M, Sharpe T, Ka S (2010) Whole-genome resequencing reveals loci under selection during chicken domestication. Nature 464(7288):587–591

Saxena VK, Kolluri G (2018) Selection methods in poultry breeding: from genetics to genomics. In: Application of genetics and genomics in poultry science. InTech, pp 19–32

Schlather M (2020) Efficient calculation of the genomic relationship matrix. bioRxiv 2020, 2020.2001.2012.903146

Schmidt GS, Figueiredo E (2005) Selection for reproductive traits in white egg stock using independent culling levels. Braz J Poult Sci 7:231–235

Sejian V, Maurya VP, Naqvi SM (2010) Adaptive capability as indicated by endocrine and biochemical responses of Malpura ewes subjected to combined stresses (thermal and nutritional) in a semi-arid tropical environment. Int J Biometeorol 54:653–661

Simianer H, Büttgen L, Ganesan A, Ha NT, Pook T (2021) A unifying concept of animal breeding programmes. J Anim Breed Genet 138(2):137–150

Singh R (n.d.) Homeopathy–paving ways into the animal world

Sonaiya EB, Swan S (2007) Small scale poultry production: technical guide

Sørensen P (2005) Breeding strategies and genetic adaptation of meat type and layers type birds used for organic production. In: Proceedings of the XVII symposium on the quality of poultry meat and XI European symposium on the quality of eggs and egg production, pp 23–26

Su Z, Bai X, Wang H, Wang S, Chen C, Xiao F, Guo H, Gao H, Leng L, Li H (2022) Identification of biomarkers associated with the feed efficiency by metabolomics profiling: results from the broiler lines divergent for high or low abdominal fat content. J Anim Sci Biotechnol 13(1):1–14

Tan C, Bian C, Yang D, Li N, Wu Z-F, Hu X-X (2017) Application of genomic selection in farm animal breeding. Yi Chuan= Hereditas 39(11):1033–1045. Yang, A.-Q.; Chen, B.; Ran, M.-L.; Yang, G.-M.; Zeng, C. The application of genomic selection in pig cross breeding. Yi Chuan= Hereditas 2020, 42(2), 145–152

Tavárez MA, Solis de los Santos F (2016) Impact of genetics and breeding on broiler production performance: a look into the past, present, and future of the industry. Anim Front 6(4):37–41

Thiruvenkadan A, Panneerselvam S, Prabakaran R (2010) Layer breeding strategies: an overview. Worlds Poult Sci J 66(3):477–502

Turner H (1984) Variation in rectal temperature of cattle in a tropical environment and its relation to growth rate. Anim Sci 38(3):417–427

Uimari P, Gibson J (1998) The value of crossbreeding information in selection of poultry under a dominance model. Anim Sci 66(2):519–528

Vaarst M, Steenfeldt S, Horsted K (2015) Sustainable development perspectives of poultry production. Worlds Poult Sci J 71(4):609–620

van Ginneken V (2017) Are there any biomarkers of aging? Biomarkers of the brain. Biomed J Sci Tech Res 1(1):2017

van Marle-Koster E (2008) Chicken population: evaluation and selection of polymorphic microsatellite markers. South Afr J Anim Sci 3894:1–11

van Marle-Koster E, Nel LH (2000) Genetic divisity and population structure of locally adapted South African chicken line: implication for conservation. South Afr J Anim Sci 30(1)

Wang Z, Gerstein M, Snyder M (2009) RNA-Seq: a revolutionary tool for transcriptomics. Nat Rev Genet 10(1):57–63

Wang C, Habier D, Peiris B, Wolc A, Kranis A, Watson K, Avendano S, Garrick D, Fernando R, Lamont S (2013) Accuracy of genomic prediction using an evenly spaced, low-density single nucleotide polymorphism panel in broiler chickens. Poult Sci 92(7):1712–1723

Wasike CB, Magothe TM, Kahi AK, Peters KJ (2011) Factors that influence the efficiency of beef and dairy cattle recording system in Kenya: a SWOT–AHP analysis. Trop Anim Health Prod 43:141–152

Wei M, van der Werf JH (1994) Maximizing genetic response in crossbreds using both purebred and crossbred information. Anim Sci 59(3):401–413

Wolc A (2014) Understanding genomic selection in poultry breeding. Worlds Poult Sci J 70(2):309–314

Wolc A (2015) Genomic selection in layer and broiler breeding. Lohmann Information 49(1):1–11

Wondmeneh E, Van der Waaij E, Dessie T, Okeyo Mwai A, van Arendonk JA (2014) A running breeding program for indigenous chickens in Ethiopia: evaluation of success. Am Soc Anim Sci

Wurzinger M, Mirkena T, Sölkner J (2014) Animal breeding strategies in Africa: current issues and the way forward. Wiley Online Library 131:327–328

Zampiga M, Flees J, Meluzzi A, Dridi S, Sirri F (2018) Application of omics technologies for a deeper insight into quali-quantitative production traits in broiler chickens: a review. J Anim Sci Biotechnol 9(1):1–18

Zhang S, Gao X, Jiang Y, Shen Y, Xie H, Pan P, Huang Y, Wei Y, Jiang Q (2021) Population validation of reproductive gene mutation loci and association with the litter size in Nubian goat. Arch Anim Breed 64(2):375–386

Zuidhof M, Schneider B, Carney V, Korver D, Robinson F (2014) Growth, efficiency, and yield of commercial broilers from 1957, 1978, and 2005. Poult Sci 93(12):2970–2982

Open Access This chapter is licensed under the terms of the Creative Commons Attribution 4.0 International License (http://creativecommons.org/licenses/by/4.0/), which permits use, sharing, adaptation, distribution and reproduction in any medium or format, as long as you give appropriate credit to the original author(s) and the source, provide a link to the Creative Commons license and indicate if changes were made.

The images or other third party material in this chapter are included in the chapter's Creative Commons license, unless indicated otherwise in a credit line to the material. If material is not included in the chapter's Creative Commons license and your intended use is not permitted by statutory regulation or exceeds the permitted use, you will need to obtain permission directly from the copyright holder.

19

Defining Breeding Goals and Breeding Strategies for Pigs in Various African Production Systems

Donald R. Kugonza and Richard Osei-Amponsah

Abstract

Pigs (Sus scrofa) are an important animal genetic resource in Africa with many different genotypes adapted to various environments and production systems. Sustainable animal breeding programmes seek to correctly evaluate breeding candidates and select the ones most appropriate to be parents of the next generation in order to achieve desired breeding goals and objectives. This chapter presents an overview and significance of pig production in Africa as well as components of sustainable breeding programmes in the context of farmer-preferred traits. Cognisant of genotype–environment interactions, there is a need to select appropriate genotypes for specific breeding objectives and production systems. Additionally, we discuss the changing environmental conditions and market demands and make a case for the need to breed more productive, resilient pigs for both current and future generations. However, this must be done in consonance with available genetics, feed, housing, healthcare and the targeted markets. The importance of indigenous and locally adapted pig breeds and their sustainable utilisation and conservation as important players in protein-source food security, livelihoods and wealth creation are presented. The need for all stakeholders to be actively involved in defining objectives for pig breeding, human and institutional capacity building as well as sustainable government policies cannot be overemphasised. Finally, we posit that any pig breeding programme aimed at improving the productive and adaptive qualities of pigs in Africa should aim at conserving locally adapted genotypes while making use of emerging innovative technologies for greater efficiency.

Keywords

Adaptive traits · Breeding goal · Indigenous pig breeds · Genotype-by-environment interaction

D. R. Kugonza (✉)
Department of Animal and Range Sciences (DARS), School of Agricultural Sciences (SAS), College of Agricultural & Environmental Sciences (CAES), Makerere University (MAK), Kampala, Uganda
e-mail: rugira.kugonza@mak.ac.ug

R. Osei-Amponsah
Department of Animal Science, School of Agriculture, College of Basic and Applied Sciences, University of Ghana, Legon, Accra, Ghana

19.1 Introduction

Pork production is an important animal protein food worldwide, accounting for more than two-fifths of all red meat consumed worldwide and in most African countries, a source of income, livelihoods and socio-cultural values. Pig husbandry has been an economic activity over the past seven millennia but has undergone transformation mainly driven by shifts in pig product valuation and processes to develop preferred pig genotypes. A host of pig breeds have been developed or evolved in and beyond Africa and each breed development process has focussed on a set of breeding objectives. Pigs are reared alongside a wide array of other livestock species in varying production systems and with different production goals and priorities, as well as management strategies. In all the production systems, pigs provide cash income from animal and pork sales, meat for home consumption and manure to enhance soil fertility, while in a few systems particularly in smallholder semi-intensive management, pigs are also used as a means to save the farmers' wealth and secure their livelihoods. A breeding objective specifies the ideal pig that a producer aims to breed and such a pig is often identified through selection based on the traits of interest. It is pertinent that all pig breeding enterprises establish breeding objectives and goals, and then implement selection processes that aim to meet the identified objectives as part of the general planning process of the business. The determination of the breeding goal is one of the most critical elements and foundation of breeding as it determines the direction, process and success of genetic improvement. Breeding goals have over time shifted from principally production-driven targets to more balanced breeding goals that target simultaneous improvement of production, efficiency, health, functional (Miglior et al. 2017) and adaptive traits.

The primary goal of genetic improvement is the selection of the best animals of each generation to be parents of the next generation. Selection is a major component of the breeding programme which describes the process of choosing individuals that meet the requirements of the breeding objective to enable them to pass traits of interest to their offspring within a breeding enterprise. The process considers both traits of a subjective nature such as visual assessment and objectively measured traits such as growth rate and genetic attributes including heritability.

19.2 What Are the Key Considerations/Drivers for Implementing Sustainable Pig Breeding Programmes in Africa?

19.2.1 Farmer-Preferred Breed and Trait Preferences

Breeding programmes should consider the breeding goals and objectives of the farmer. Commercial pig farmers normally focus on a few traits of economic importance, such as fast growth rates, larger carcasses and disease tolerance. On the other hand, goals for smallholders include aesthetic (colour and patterning), behavioural aspects (temperament, mothering ability, foraging behaviour, herd-ability and other aspects that minimise labour inputs), adaptability, in particular the ability to survive under low-cost management levels (e.g. inadequate feeding and housing) (Halimani et al. 2020). Traits of religious and socio-cultural importance are also of importance to local farmers. Therefore, participatory approaches to designing breeding programmes that involve livestock breeders and farmers at every stage of the planning and implementation of the breeding programme, including the definition of the breeding objective, are necessary (Roessler 2019). A programme that meets the need to develop a sustainable industry must consider the prevailing conditions of the piggery sector—the strengths and opportunities—as well as the challenges reflected in the weaknesses and threats (Table 19.1).

19.2.2 Genotype by Environment Interactions

These arise when the best genotype in one environment is not the best genotype in another environment or when genotypes differ in their

Table 19.1 A SWOT analysis of a typical piggery value chain

Strength	Weakness
A growing market for animal proteins due to the increasing human population and higher incomes	Relatively low knowledge level, especially on disease and farm management
Pig farmers getting more organised	Lack of practical education in livestock management
Government policies towards piggery are favourable in several countries	Few farmers trained in livestock management
Changing consumption patterns towards quality products for a growing middle class	Limited of rural veterinary and artificial insemination extension services
High domestic production of maize	Lack of independent testing facilities for feed and veterinary inputs
Growing awareness and interest in improved technologies by farmers	Limited of applied biosecurity on farms
Growing investments into piggery business	Lack of quality standards and enforced slaughter regulations
Growing consumption and demand for pork and pork products	Lack of specialised pig feed supply market
Increasing interest into piggery by youth farmers	Underdeveloped market for quality pork meat
Improving the quality of pig carcasses	Mostly practiced as non-commercial farming
Availability of prophylactic treatment options for pigs increasing	Lack of abattoirs Transport of pigs is not allowed by night according to laws, leading to informal fines
Opportunity	**Threat**
Great potential to become a strong export market due to relatively low cost of production	Seasonality in input prices
Increasing demand for pig products	High reliance on the domestic market
Increasing demand for value-added goods	High animal disease prevalence, especially African swine fever
Improvement in waste management practices can greatly benefit the sector	Pig farming is not a priority for some national governments
Sector growth allows for local feed production	Traders have excessive market power, leading to low farmer profitability
Genetics/breeding certification and registration agency exists	Market flooded with under-size piglets of non-descript genetics

Source: Modified from Larive International (2020)

response to environmental factors (Mulder and Rashidi 2017a). Given the wide variety of Sus genotypes and production systems (Chaps. 2 and 8), Africa needs to breed pigs that perform optimally in a wide range of environmental circumstances and meet the needs of stakeholders. The genetic correlation between the same trait across different environments (genotype by environment interaction) determines the robustness or sensitivity of the genetic potential of a specific trait to the environment to which it is exposed.

According to de Leon et al. (2016), genotype-by-environment interactions include the ability of genotypes to express different production potentials when subjected to different environments and production systems. Some pig breeds are most suited to perform in particular environments and suffer excessive stress when transferred to other environments. A common example is the improved performance of exotic and crossbred pigs raised at research centres or on government farms compared to their relatively inferior performance under local farm conditions. Most exotic pig breeds have been farmed in Africa but their full genetic potential is only achieved when they are kept under intensive husbandry conditions with adequate nutrition, healthcare and appropriate housing. Thus, it is important to ensure that the breeds or genotypes used in the breeding programme will be available for the foreseeable future and farmers can provide the minimum inputs for maintaining them. One should therefore carefully consider the production environment in terms of breeding infrastructure, feed and health services, and ease of accessing the targeted market as part of any pig improvement programme.

19.2.3 Conservation of Indigenous Pig Genetic Resources

In most African countries, indigenous pig breeds are generally kept under smallholder husbandry systems (see Chap. 2) and have slower growth rates than exotics. However, they have adaptive traits (such as reduced deposition of adipose tissue, adaptation to extreme weather conditions—hot and very cold, and ability to withstand heavy parasite load) that give them an advantage in low-intensively management smallholder systems. Invariably, these adaptive traits are constantly threatened by genetic erosion caused by indiscriminate crossbreeding with exotic breeds (Halimani et al. 2020; Osei-Amponsah et al. 2017). Sustainable pig breeding programmes should therefore aim at improving indigenous pigs through selection and crossbreeding programmes that combine the strengths of indigenous and imported pig breeds. For selective breeding to happen, there should be concerted efforts to conserve the indigenous and the naturalised breeds and genotypes so that these are reservoirs for those special genes, particularly for given locations.

The Large black and Saddleback pigs for instance were introduced in Uganda over 70 years ago. Due to several management bottlenecks, the prevailing environmental conditions and genetic erosion, individual animals of these breeds are currently generally small in size and are grouped by farmers among native and or local pig breeds. The many small-scale pig farmers in Africa depend on the indigenous pig on account of its adaptive, disease resistance, ease of management, and various roles in sustaining livelihoods and socio-cultural practices. Therefore, all breed improvement programmes should ensure that indigenous pig breeds are conserved and protected from excessive genetic dilution and indiscriminate crossbreeding. National governments need to ensure that indigenous pigs are conserved in situ and ex situ in vivo on government and institutional farms. Additionally, pig breeding improvement programmes should target traits that will help conserve the unique adaptive traits of indigenous pigs and give them an upper hand on niche markets.

19.2.4 Feed Resources

For a breeding programme to be sustainable, there has to be a deep appreciation of ensuring that there is feed for the breeds being promoted/developed and this should be in synch with the circumstances of the farming communities and the environment. It is defeating to develop high-performing (fast-growing, very prolific) pigs to be distributed to smallholder farmers who have neither feed production capacity nor capital to invest in the procurement of processed feeds. The likely result would be high-performing animals that farmers cannot maintain, and such breeding programme(s) would ultimately become unsustainable. Purchased feeds can be a complete substitute for land and this can spur small and medium size intensively managed pig operations. In many ways, the production of feed resources and the availability of land on which to grow/produce the feed resources are almost inseparable. The exception is the intensive system that is designed to rely on purchased feeds that may be imported or locally sourced (Mwesigwa et al. 2013a, b). Systems that rely on this option are generally intensively managed and are almost exclusively driven by demand for pork. In many African countries, pigs are reared for income from the sale of weaned pigs for fattening or mature pigs for slaughter; significant income also comes from the sale of pigs for use in breeding. Despite this, most female and male pig farmers' rank the keeping of pigs as a saving/insurance asset (Babigumira et al. 2021), and such farmers may not invest in the best feeding options available. Research into alternative feed sources to replace important but expensive components in particular protein and carbohydrate sources should be encouraged in Africa.

19.2.5 Socio-economic Factors

Socio-economic factors like gender, religion, urbanisation, market, consumer preferences and labour availability are valuable elements in the pig production value chain. Female and male pig farmers have the same level of knowledge regarding available breeds and lines. A Ugandan study reported that 69% of female and 70% of male pig farmers are very familiar with local pigs, whereas their knowledge of exotic and crossbred pigs is very low. Women and men do not differ much in their pig breed preferences (Babigumira et al. 2021). Both women and men eat pork although the proportion of men is slightly higher; and pork ranks second to chicken as preferred meat, mainly due to its special taste. The choice of pig breed is not necessarily influenced by the religion/faith of the farmer, notwithstanding that some faiths such as Islam, Judaism, and Seventh Day Adventist, Ethiopian and Eritrean Orthodox churches consider pigs and pork a taboo—not rearing them, associating with them or eating them. Urbanisation to a very large extent influences the choice of breed and invariably the type of pig breeding programme. In Uganda where a pig breeding programme supported by artificial insemination initiatives thrives, there is a clear distinction in breed choices. Urban pork consumers demand lean meat, whereas rural pork consumers prefer fatty pork and as result the trend is to rear more indigenous pigs in rural areas as these tend to have much fat. In urban and peri-urban areas, the market forces demand that farmers rear exotic breeds or crossbreds to produce lean carcasses that the market prefers.

19.2.6 Smallholder Versus Industrial Pig Production Systems

Smallholder farming of pigs is compatible with deliberate low-input production systems, whereas industrial-scale piggeries demand that the most productive systems are deployed to maximise growth, reproductive ability and overall performance. Farms that focus on economies of scale, maximum production efficiency per animal or per unit of land would choose pig breeds that can pay back the investment. Such breeds of high-payback potential would possibly not be well adapted to a low-input farming system. On the other hand, farms that focus on product quality, self-sufficiency and niche marketing are more likely to target breeds that are resilient to feeding systems and environmental conditions that are not necessarily top-of-the-range. Selecting pig breeds that are appropriate to the local environment safeguards animal health and welfare and leads to long-term and sustained production.

19.2.7 Partnerships that Enhance Efforts of Actors

Implementing sustainable pig breeding programmes in and for Africa will require partners-researchers, breeders, funders, promoters and other enhancers in all African nations to collaborate, and above others, share resources, expertise and lessons learned during their professional careers and demonstrated best practices (see Chap. 31). Collaboration with developing partners in human and institutional capacity building should help Africa take advantage of modern innovative breeding technologies (see Chaps. 20 and 21) for improved sustainable pig production.

Pigs, like other animal genetic resources in Africa, continue to suffer from both direct and indirect effects of climate change particularly global warming and heat stress. This is one of the several areas of pig production that should be targeted by partnership collaborations. Generally and as elaborated in Chap. 8, indigenous pigs such as the Ashanti Dwarf of Ghana, the Ugandese black pig and the Bakosi of Gabon and Cameroon are well adapted to their production environments compared to exotic breeds such as Large White and the Duroc. Other efforts that will deliver sustainable pig breeding programmes include molecular characterisation of indigenous pigs of Africa in order to identify the genes responsible for various traits of economic importance to help match them to appropriate genotypes in crossbreeding programmes.

Additionally, partnerships under the auspices of continental agencies such as the African Forum for Agricultural Advisory Services (AFAAS) and the Forum for Agricultural Research in Africa (FARA) should target pig farmers for training and support to stock and keep appropriate pig genotypes while ensuring the provision of adequate feed, housing and healthcare in order to increase their production levels. Pig breeding programmes in Africa should also include resilience as a trait of economic importance in order to breed more robust productive pigs for sustainable protein food security.

19.3 Definitions and Considerations for Pig Production

19.3.1 Definition of Breeding Goals, Objectives and Strategies

A breeding goal specifies the traits to be improved including the emphasis given to each trait as well as the desired direction of the improvement of the population. The breeding goal as defined by the breeder indicates what is considered important in the future including expected customer demands. The breeding goal should always consider the unique production environments, namely high-, medium- and low-input production systems in order to be sustainable and useful to stakeholders. In this regard, breeding goals set for some pig genotypes in intensive production systems in other parts of the world are bound to fail if implemented in extensive production systems in parts of sub-Saharan Africa.

The breeding goal and the corresponding selection index determine the direction of a genetic improvement programme. In all species, breeding goals have moved from primarily production-driven breeding goals to balanced breeding goals that aim for simultaneous improvement of production, efficiency, and health and functional traits (Miglior et al. 2017; Berghof et al. 2019). A breeding objective specifies those characteristics that mostly affect profit in a particular pig business, in addition to how each characteristic is important to the realised profit. Every pig breeding objective should be specific, measurable, realistic and must be time bound. Pig breeding objectives focus on traits of interest and their relative importance with the strategy of achieving them set out in the breeding programme (Merks et al. 2012). The breeding objective should thus be based on projected profits under future conditions of production, not merely on the potential to change traits genetically. It requires an idea about the expected developments in production systems and regulations; new developments related to housing systems, nutrition and how are they expected to influence the performance of selected animals; (inter)national government regulations that may limit the current production system.

Some critical considerations when developing pig breeding objectives include the following:

(a) Understanding the customer and market requirements, as each breeding objective is generally specific to a particular market.
(b) The economic importance of traits depending on the target market.
(c) Production performance of the breeding stock needs to be evaluated continuously while monitoring the prevailing market demands to adjust the breeding programme accordingly. More health-conscious consumers may, for example, lead to adjusting the breeding goal to include pig breeds which will produce leaner carcasses.
(d) Some of the target traits in pigs are highly heritable and these should be targeted to attain the greatest progress towards breeding objectives.
(e) The need to focus on traits that have economic value instead of traits that have more to do with personal preferences and cultural or traditional roles.

Whenever pig producers determine the requirements of the target pig and pork market, and have refined breeding objectives that are well aligned with market requirements, they can start the selection of pigs that meet their preferred breeding objectives, considering both subjectively and objectively assessed traits. This selection process is most effectively done when some records are available on the traits of interest. Farmer and extension service personnel's capacity for record keeping should thus be encouraged by national governments. At all times, one should embrace innovative data recording technology such as the use of weigh, or calibrated chest girth tapes (Marshall et al. 2024) to determine body weight, ultrasound to measure fat and loin eye area, computerised tomography scans for individualised primal yield of carcass components and sophisticated scoring methods for anatomical traits. Additionally, proper pig pedigree records help to avoid mating of relatives which will lead to inbreeding with its undesirable effects.

Subjective visual assessment involves assessing pigs on the basis of what can be physically seen, with a main focus on the conformation of the pig for instance the body condition scoring or body muscling; and on the pig structure for instance ear size, ear shape and sometimes hair coat colour. Objective genetic assessment makes use of real measurements to assess the worth of selection candidates relative to other pigs on the farm. Genetic evaluation is one of the available forms of objective assessment and it provides an understanding of the genetic merit of pigs being farmed. When boar acquisition for the improvement of a pig herd is based on the breeding objectives of the pig farm, objective assessments are necessary.

The definition of the best animal/breed is subjective, based on the function of the animal/breed, culture, market structure, production environment, legal framework, population structure and environment limitations. In this regard, the best animal may not necessarily be a high-performance animal for a particular animal product (meat or litter size) but could be an average performance animal with reasonable resistance to an endemic disease. Defining the best animal thus requires inputs from all stakeholders, including farmers, consumers, breeders and researchers. Additionally, there is a need for adequate data/records on candidates for selection to make objective assessments and then select the 'best' animals/genotypes. It is for this reason that any serious genetic improvement programme needs infrastructure for collecting data. Without data collection, it is almost impossible to undertake any form of tractable genetic improvement.

Based on the breeding goal, phenotypic data that can help to establish the value of selection candidates should be collected. For instance, when the fertility of pigs is in the breeding goal, litter traits should be recorded. Pig genetic improvement programmes should have the infrastructure for collecting data to trace genetic improvements. Involving animal keepers in a genetic improvement programme offers an opportunity to collect data on their animals. Additionally, a data repository centre with high storage and computing ability is essential in developing any improvement programme. Based on an appropriate genetic model and a statistical model including pedigree information, a breeding value for a trait is estimated. Estimated breeding value indicates the value of the animal with respect to the breeding goal: the lowest ones will have a negative effect on the breeding goal traits and the highest ones will improve breeding goal traits.

19.3.2 Pig Production Systems in Africa in Relation to Heat Stress and Environmental Impacts

The predominant pig production system in Africa is the smallholder system characterised by minimum housing made from locally available materials, scavenging for food with some supplementation from household wastes from time to time, and little or no veterinary care. Often the environment is harsh with little or no veterinary care exposing the animals to disease pathogens, bad weather and predation. The little veterinary care sometimes comes in the form of herbal concoctions. African indigenous pig breeds such as the Ashanti Dwarf pig of Ghana are generally adapted to the production environment, survive under poor management, are more resistant to heat stroke and remain more active than introduced exotic breeds across a range of environments (Osei-Amponsah et al. 2017). The increasing use of exotic genotypes in low-input pig production systems in Africa calls for strategic breeding to reduce the negative effects of heat stress on animal welfare, production and reproduction. Exotic pig breeds are often kept in medium- to large-scale production systems where they are provided with adequate housing, feeding and veterinary care.

A recent spatial analysis has identified heat stress hotspots in Uganda where pig production could be affected significantly (Mutua et al. 2020a), with some regions reporting severe heat stress conditions over 183 days per year (Fig. 19.1).

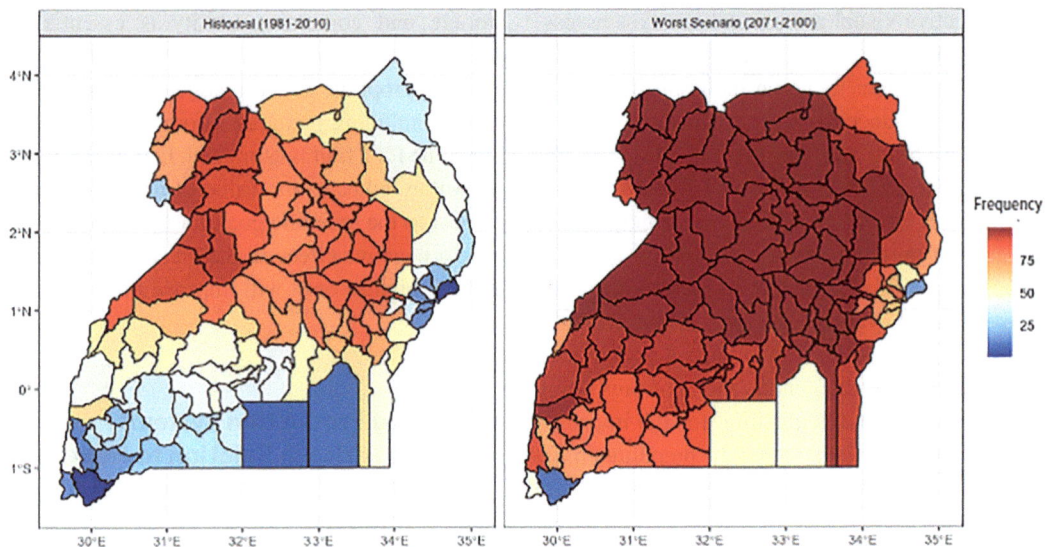

Fig. 19.1 Frequency of severe heat stress for past and future periods for pigs in Uganda. (From Mutua et al. (2020b), licensed under a CC BY 4:0)

Heat stress risk factors vary in impact and effect and include climate (air temperature, humidity), management system (specifically free-range pig management) and pig characteristics (age, lactation, stage, dark hair colour, overweight, breed, and pregnancy) (Fig. 19.2).

Resilience is the capacity of an animal to be minimally affected by disturbances or to rapidly return to the state pertained before exposure to a disturbance (Berghof et al. 2018; Colditz and Hine 2016) In the swine industry, economic losses associated with heat stress are mainly explained by reduced and inconsistent growth, decreased feed efficiency, decreased carcass quality, poor sow performance, increased mortality and morbidity, and decreased facility efficiency (Zaake et al. 2020; Baumgard and Rhoads Jr 2013). Reduced reproductive performance is characterised by anoestrus, increased wean-to-oestrus interval, decreased farrowing rate and reduced litter size (Ross et al. 2015). Similarly, poor semen production and quality occur in boars exposed to heat stress. Thus, heat stress compromises almost every economically important phenotype within the industry (Mayorga et al. 2019).

Selection for improved heat tolerance in pigs based on reproductive performance and temperature-humidity indices based on public weather station data is possible and should be encouraged in porcine breeding programmes. Heat susceptibility appears to be a heritable trait in finishing pigs, and therefore, genetics may offer a viable strategy to improve production during periods of high temperature–humidity indices. Heat tolerance based on reproductive performance indicators is a heritable trait and genetic progress for heat tolerance can be achieved through genetic or genomic selection and genomic regions and candidate genes with important biological functions have been identified (Tiezzi et al. 2020). Other heat stress management options that can be up-scaled are summarised in Table 19.2. One of these, the indigenous micro-organism (IMO) technology involves using a diverse mixture of beneficial native micro-organisms to enhance the natural processes that occur in the environment (Kumar and Gopal 2015; Yadav et al. 2020). The IMO is added to a natural bedding material such as sawdust, rice hulls and crushed corn cobs, and these are put on the pig sty floor. The set-up controls odours, reduces the spread of pathogens and provides a comfortable and safe environment for the pigs.

Current developments in big data collection give opportunities to determine new resilience

Fig. 19.2 Risk factors that cause heat stress in pigs. (From Mutua et al. (2020b), licensed under a CC BY 4:0)

Table 19.2 Current and futuristic adaptations to heat stress management for pigs (Mutua et al. 2020b)

Pig value chain node	Current adaptations	Futuristic adaptations
Input supply	Community water management committees for sustainable use of water resources	Farm-based water points
	Increased investment in water for production, solar use	Private sector-led investments in water for production
	Adoption of intensive management practices	Further research in pig house construction
	Provision of general extension services	Recruitment of specialised extension staff
On-farm pig production	Capacity development of farmers	Strengthen farmer-extension-research linkage
	Promotion of tolerant feed resource plant varieties and feed conservation	Emphasise capacity building among smallholder farmers on feed conservation
	Indigenous micro-organism (IMO) technologies	More research and capacity development in IMO
Transport	Day time transportation of pigs	Capacity building for transporters and policymakers to appreciate the benefit of transporting pigs at night
	Holding pigs under trees and other shades	Research in appropriate pig transportation and holding facilities
Output market	Increased establishment of pig market associations and farmer trade relations	More awareness of climate change information knowledge exchange with pig market actors
	Slaughtering of pigs in basic slaughter facilities	Increased innovation by the private sector on appropriate structures and abattoirs
	Implementation of animal movement and certification	Review of regulations to allow for animal movements at any time of day based on circumstances

indicators based on longitudinal data, which can aid the incorporation of resilience in animal breeding goals (Berghof et al. 2019). Many significant genomic regions in relation to heat tolerance have been identified in pigs (Riquet et al. 2017). This new genomic information could be used in the future to identify pigs capable of maintaining high levels of productivity during heat stress (Mayorga et al. 2019). Improving swine climatic resilience through genomic selection has the potential to minimise welfare issues and increase industry profitability (Tiezzi et al. 2020).

19.3.3 Pig Genetic Improvement and Conservation

In general, small-scale farmers dominate the pig production landscape in Africa. These farmers often keep indigenous pigs, and provide little or no housing, feed and veterinary care. Increasingly, commercial pig producers operating in the semi-intensive and intensive production systems utilise mostly exotic pig breeds and, in some cases, crosses between indigenous and exotic pig breeds. These commercial farmers have often been the targets of various pig genetic improvement programmes in Africa. The open nucleus breeding strategy is often employed to disseminate superior genetics to pig farmers and other stakeholders. Generally, the breeding of female pigs is organised into a three-step system consisting of breeding herds, multiplier herds and production herds. In the breeding herds, purebred sows and boars are selected and bred according to a specific selection criterion defined by the company's or producer's strategy (Pedersen 2018).

The government and other stakeholders often import exotic breeds and raise them in a nucleus herd on institutional or research station farms. The best-performing offspring from these pigs are distributed to grower farmers, who are carefully selected based on the availability of housing, ability to feed the pigs, keep records and provide significant veterinary care. The outgrower units multiply the stock supplied to them and the performance of the offspring is monitored regularly. Superior offspring are later given to other outgrower farmers or sent back to the nucleus station and the rest are sold for slaughter. Apart from such organised breeding programmes, there are many instances of indiscriminate crossbreeding between exotic and indigenous pig breeds.

Pig conservation (see Chap. 25) efforts have been very limited hitherto the establishment of the five African Union Animal Resources Seed Centres of Excellence (AU-ARSCoE). In most of Africa, pig conservation has been a preserve of the farmers who rear them. Whereas this in situ conservation strategy is practically the best, the lack of institutional support mechanisms means that farmers may change their production objectives and with that, the entire conservation effort. Recent efforts have seen the inclusion of pigs in the national strategies and action plans on the conservation and sustainable utilisation of indigenous animal genetic resources. For instance, in Uganda, the National Agricultural Research Organisation (NARO) is establishing a conservation scheme for the Ugandan local black pigs. This is to include a managed station herd and a registration of active farmers. The AU-ARSCoE gene banks will receive genetic material for all livestock species from the countries in the five geographical sub-regions of Africa and manage these collections on behalf of the countries in the various sub-regions. This ex situ in vitro conservation plan is expected to stave the real threat to African indigenous animal genetic resources that are continually threatened by a consistently changing animal agriculture landscape.

19.4 Defining Breeding Goals for Pigs for African Low-, Mid- and High-Input Systems

19.4.1 General Overview of the Pig Production System

The low input system, generally referred to as the smallholder pig production system (Table 19.3), is still dominant on the African continent (see

Table 19.3 Pig production system considering the scale of operation—the Uganda case

Parameter	Small-scale farming	Medium scale breeders and farmers	Large scale integrators
Source genetics	Free mating, NAADS[a], local breeder	Commercial farm or imported	Imported
Production system	Free range/wooden pens	Concrete pens or IMO system[b]	Indoor pens
Knowledge level	Very low	Low but increasing	High
Source of information	Other farmers, NGO[c] training	NGO training, other farmers	Foreign suppliers
Biosecurity	Non-existent	Low to medium	High
Disease prevalence	High	High	Medium
Veterinary service	None	Local veterinarians	Local veterinary/on-farm veterinarian assistants
Source of medicine	City market	City market	City market or imported
Feed type	Household waste or other waste products	Compounded or concentrate-based	Concentrate/ premix own feed production
Source of feed	Household or farm waste	On-site feed mixing or Private Feed Companies e.g. Devenish	On-site feed mixing
Harvest size of pig	30–60 kg	50–90 kg	80–125 kg

Adapted from: Larive International (2020)
[a]*NAADS* National Agricultural Advisory Services, a central government agency under the Ministry of Agriculture, Animal Industry and Fisheries
[b]*IMO* Indigenous Microorganisms Deep Litter System—involves the use of litter such as wood shavings impregnated with bacteria to breakdown wastes
[c]*NGO* Non-Governmental Organisation

Chap. 2). A significant proportion of the pig herd on the continent is managed within this system. The small-scale pig production system is generally characterised by pigs reared outdoors, free range on pasture all the time or on a tether under a tree shade. This system also extends to the free-range scavenging type of pig rearing where pigs are left to freely move all over the home, the gardens, waste bins, wastelands and the entire neighbourhood searching for food. Pigs that are tethered may be kept on a leash during the day but very little is provided in the form of feed (such as kitchen and table leftover food, swill, crop residues and water) (Pezo and Waiswa 2012). The free-range system is practiced by extremely resource poor dwellers in peri-urban and rural settings. The pigs reared under this system are almost entirely indigenous breeds or non-descript types, and no formal breeding programmes exist for these producers. The sale of pigs is not systematic, being done when there is a need for cash to solve a problem; and not when the pigs reach market weight/size. In general, smallholder pig farmers typically keep 1–3 sows or less than 10 pigs of all ages at any given time. Many smallholder pig farmers operate under the low-input mode, but a significant number, particularly in peri-urban areas, do invest in some form of housing and feeding and in return, reap good levels of income by selling piglets, porkers and spent sows/boars to butcheries and pork-eating joints that are frequented by the low- and middle-class residents of these locations. Piglets are sold mostly to other smallholder farmers who specialise in growing and fattening pigs. Smallholder pig farming is also characterised by inadequate feeding regimes, below optimal reproductive performance, slow growth rates, relatively low-grade carcasses, and poor disease and parasite management, which is typically a precursor for rampant disease and parasite outbreaks and spread. African swine fever, porcine cysticercosis and other major diseases tend to be rampant under this system since biosecurity measures are regularly below acceptable minimums.

In the medium/semi-intensive input system (Table 19.3), pigs are housed for part of the year and when they are dry or in gestation, they are put on pasture. The system is desirable in areas where the weather is not too hot for open air/paddocking of pigs and where there is land for putting the pigs on range. It is a system desired by some producers as it allows for keeping costs low for part of the year (savings on processed feed and labour); but it is also associated with the production of organic pork that relies much less on industrial processes inputs like feed, drugs and equipment. There is improved management and more investment in this system than in the low-input system—characterised by better feeding, better disease and parasite prevention and control and heat stress management leading to better growth rates, animal welfare and production. Farm income can be enhanced if marketing is strategically planned and sequenced to match high production levels with peak market demands. In this system, gilts and sows in late gestation and sows in lactation, as well as barrows and castrates under finishing are kept under intensive management, while boars, dry sows that have recently weaned piglets, and pigs in gestation are put on the semi-intensive component of the system so as to take advantage of the afore-mentioned benefits of keeping costs low and maximising on land utility.

The key feature of the high-input system is adequate and continual housing where pigs are provided with feed, water and protection from extreme weather (hot sun and rain showers) or from getting into contact with diseased/stray animals. The labour needs and other inputs, such as feed and water for both pig consumption and cleaning purposes, are high. Notably, the system provides the highest level of farm output compared to other systems and it is the most commercially viable. In much of Africa, this system accounts for a very small portion of pig production (about a tenth), partly because of the high investment capital needed, superior management skills and a well-connected marketing system. High-input systems invest in high-grade breeding stock (fast-growing and high carcass quality pigs) and also set up on-farm selection programmes. They are also characterised by effective disease and parasite control practices, good hygiene and low piglet mortality that mainly arises from poorly constructed farrowing pens in the other systems, and or starvation and cannibalism. Though of very low incidence, farms that use these systems occasionally experience devastating incidence of African Swine Fever (ASF), particularly when there is a gap in the year-round biosecurity system of the farms.

Industrial pig farms/large integrators are a type of high-input system (Table 19.3) that typically rear 500 or more pigs, mainly for the commercial sale of pork or breeding stock (Mutetikka and Kugonza 2009). These farms are near large cities and supply most of the pork produced to the formal sector, such as high-end butcheries, restaurant chains, supermarket chains and for processing into other pork products.

19.4.2 Production Characteristics, Breeds, Markets and Farmer Organisations for Pig Genetic Improvement

19.4.2.1 Production Characteristics for Pig Improvement

Identification of pig breeds, characterisation of the production environment, the definition of production objectives and a database on the production performance of animals is a foundational step in any pig improvement programme. In most African pig production systems, increasing the quantity and quality of meat is a primary breeding goal. This has been largely achieved through improved litter size at birth and weaning as well as litter weight at birth and weaning. Additionally, the emergence of educated and health-conscious consumers has created a market for lean carcasses, and thus, the breeding for lean meat. The diversity of pig genetic resources in Africa is enormous (see Chap. 8). There is thus a need for phenotypic and genetic characterisation of pigs in Africa using appropriate tools so that metadata analysis across countries and regions can be carried out and data used as a base for informed decisions on breeding strategies. Advances in molecular biology and the availability of various genomic tools and technology provide greater opportunity for the characterisation, sustainable use and conservation of Africa's pig genetic resources (see Chap. 21).

19.4.2.2 Breeds Targeted for Pig Improvement

Although most indigenous pig breeds are valued on account of their adaptation to the production environments and their tastier carcasses, they do not produce as much meat as exotic genotypes (particularly under intensive production systems). Indigenous breeds are the most efficient under the prevalent traditional extensive systems of production on account of their better adaptation, ability to survive under harsh conditions and scarce feed resources, and resistance to various endemic diseases. Indigenous pig genetic resources are also superior in terms of some carcass and organoleptic attributes including taste, which makes them popular delicacies for consumers. These are important traits and attributes which call for an in-depth characterisation and utilisation of indigenous pig genotypes in breeding programmes to develop pig genotypes for specific purposes and for various production systems. A locally adapted exotic genotype such as the Large White is already useful under intensive production systems and may be considered in planned crossbreeding programmes with indigenous pigs for the development of pig genotypes with improved production and adaptability characteristics.

19.4.2.3 Markets for Improved Pig Breeds and Pig Products

Improved access to livestock genetics to international markets and the growth of world trade have enabled livestock breeding firms to sell their products worldwide. The increase in the size of the potential market strengthens incentives to invest in research to develop new animal production inputs. For example, several multinational companies have emerged over the past several years to become major developers and suppliers of improved pig breeds for producers including Babcock Swine (US), Cotswold USA (Heartland Pork Enterprises, UK), DeKalb Swine Breeders (Monsanto, US), DanBred (Denmark), Farmers Hybrid (US), Genetic Improvement Service (US), Genetic Porc (Canada), Nesham Hybrids (US), PIC-USA (London-Dalgery PLC, UK) and Seghers Hybrid (Belgium) (Narrod and Fuglie 2000).

Breeding schemes aim at utilisation of the between- and within-breed genetic diversity. Breeding schemes do not aim at a fixed target; breeding organisations are dynamically searching for improvements. Differences in economic, social and ecological production environments give rise to different desired directions of change. The desired direction of change of a particular breed might differ between regions and change over time. Changes over time are strongly driven by the needs of consumers and society. Breeding organisations are increasingly aware of this and are changing their breeding objectives by including traits related to animal welfare and the quality of the product. The reproductive capacities of animals put a major constraint on any animal breeding operation. Reproductive techniques like artificial insemination and embryo transfer can be used to overcome these constraints. These techniques play an important role in the activities of breeding organisations. In essence, the most basic effect of reproductive technologies is to increase fecundity. This means that fewer parents are needed to produce a given number of offspring. The application of reproductive techniques has had a major impact on the structure of breeding programmes, the rate of genetic gain and the dissemination of genetic gain in livestock production (Nimbkar and van Arendonk 2010).

Globalisation represents both an opportunity and a threat for developing countries' agriculture and in particular sustainable use of animal genetic resources. It offers the potential to serve markets beyond local and domestic ones, but at the same time exposes producers to the full force of international competition, including highly efficient low-cost producers from other developing regions. There is a real risk that poor, smallholder livestock producers including the many pig farmers will lose out as they fail to compete in the 'free market' and will be replaced by either large-scale local operations or imported breeds (Rege et al. 2011). Therefore, there is a need for pro-poor animal breeding strategies to ensure that the many local small-scale farmers and their breeds and not disadvantaged as a result of global competition. In targeting markets for genetic improvement in Africa, therefore, sustainable intervention should be included to sustain use and conserve local pig genetic resources and their unique production systems.

Concerns about boar taint and aggressive behaviour are being addressed through breeding. To support the transition to the abandonment of castration and the rearing of entire males in pig production, genetic selection brings the medium- to long-term solutions to control the problems of boar taint and aggressive behaviour. Some breeds have been identified as a low-taint risk and can be used for crossbred pigs. Because of the knowledge developed over many years on the genetic determinism of skatole and especially androstenone, breeding organisations have recently developed selection strategies to propose improved lines for the breeding of entire males (Larzul 2021).

19.4.2.4 Farmer Organisations for Pig Improvement

Organised pig farmer organisations exist in many African countries such as Ghana, Kenya, Nigeria, South Africa, Tanzania and Uganda (see Chap. 31). Through these associations, farmers build capacity, get access to improved genetics, bulk purchase of inputs, and have a collective voice on issues germane to improved pig production, healthcare and marketing. Under the umbrella of an organisation, farmers can be easily consulted and their inputs taken into account on varied issues, including setting breeding goals, selecting breeding stock and determining appropriate breeding programmes based on their preferred traits. Breeding goals are also dependent on the targeted market and in some cases when value addition is to be considered. Furthermore, sensitisation of local farmers on the important adaptive attributes of indigenous pig breeds and the need to ensure that they remain an important/useful resource in their communities should remain on the agenda.

19.4.2.5 Future Perspectives

Officers will provide information on the most suitable pig genetic resource and important part of the production system will maintain/conserve pig biodiversity for future needs. Moreover, utilisation of exotic genotypes in crossbreeding with indigenous breeds must be well designed/planned to ensure that the unique and important attributes of indigenous breeds are not lost. The benefits of organised pig farmer organisations should be extended to countries where they do not exist. To start pig farming, and particularly a pig breeding company, it is important to formalise operations from the outset. The starting point is to contact the local government authority in charge of Animal Production. Often, Animal Production is a Directorate or a Unit under the Ministry of Agriculture in the different African countries. Animal Production information on service providers (feed millers, and veterinary practitioners and markets among others). Supply or access to the appropriate swine genetic resources in the intended environment as well as the market demands are vital aspects that must be carefully considered by potential producers.

19.4.3 Products from Pig Production Systems

19.4.3.1 Pork and Pork-Derived Products

Pork is the meat harvested from pigs and it is the most commonly consumed meat globally. Pork is consumed either as freshly cooked, or as preserved or cured meat. A 100 kg pig at a market age of about seven months yields about 70 kg of pork (70% dressing percentage). Fresh pork may be cooked by roasting, frying, stewing or steaming. Preserved pork-derived products on the African market include bacon, gammon, ham, hot dogs, salami, sausages, smokies, smoked pork, pork chops and others. Additionally, pork and pork products are important components of many local and continental dishes creating a vibrant market for the industry.

19.4.3.2 By-Products from the Pig Industry

By-products from the pig industry include gelatin, wax, adhesive, bio-fertiliser, lard, soap, leather, hair, industrial lubricant and pet food. Lard, the fat harvested from the abdomen of pig, is particularly useful as a substitute for vegetable oils but its high price puts it beyond the reach of many households on the African continent. Lard is also a major ingredient in the manufacture of soaps, creams, beauty products, baked goods and other foods. Whereas most pigs reared in Africa produce fatty pork, consumer preference is shifting towards lean pork and, hence, the production of lard will eventually drop. Pig by-products are also used in the manufacture of water filters, insulation materials, rubber, antifreeze (used in various equipment), plastic, wax for flooring, crayon and chalk (USDA 2016). A growing sector that offers much promise for the pig industry is the human medicine organ and parts industry. In January 2022, the first whole heart transplant from a pig to a human was performed (www.floridatoday.com/story/life/wellness/2022/02/15). Although the recipient lived only two months post-surgery, history was made as the future of the pig industry now extends beyond food and general-purpose functions. Additional valuable products of direct human health importance harvested from the pigs include insulin for regulating diabetes and valves for human heart function. Pig genetic resources and their diversity in Africa therefore need to be sustainably used and appropriately conserved not only for protein food security and wealth creation but also for the future survival of humanity.

19.4.3.3 Manure and Biogas

It is estimated that a herd of 500 finishing pigs produces almost 84,950 litres of manure per month. This is a high-grade fertiliser for use in crop production. With the estimated 40 million pigs on the African continent, 6.8 million M³ of manure are available to substitute for some of the millions of tons of imported inorganic fertilisers. To ensure efficient use of pig manure, there is a need to have innovative and cost-effective means of collecting, storing and transporting it to the point of use. This requires the education and training of pig farmers and the provision of appropriate technology. Additionally, end users or crop farmers also need to be sensitised on the value and importance of using pig manure on their farms instead of inorganic fertilisers.

19.5 Traits of Economic Importance in Pig Production

Traits of economic importance in pig breeding include litter size, growth rate, feed conversion efficiency and back fat thickness. The main breeding objective of pig production is producing large numbers of fast and efficiently growing lean progeny. Although this strategy has been highly successful and greatly increased the weight of lean meat produced per sow per year, negative correlated consequences with welfare traits have been recorded (Turner et al. 2018). The impacts of pig production on the environment call for attention to traits that ensure environmental sustainability in pig breeding and improvement.

19.5.1 Litter Size

Litter size is a major component of sow productivity, and it is the main driver of the profitability associated with pig farming. As litter size increases, the fixed on-farm production cost per pig is reduced to the point at which the entire enterprise first becomes economically neutral and then realises a profit (Zak et al. 2017). Reproduction traits such as litter size is heritable (h^2 = 10–15% across breeds) and can enhance the genetic gain within a population by improving selection intensity or reducing generation interval (Zak et al. 2017). Targeting litter size for improvement in a breeding programme requires the availability of the necessary infrastructure—

such as adequate housing and nutrition, improved management and veterinary care—to support larger litter sizes.

19.5.2 Number of Pigs per Year

This is directly related to litter size and also to the number of piglets born alive. A large litter size and more pigs in an operation per year enable farmers to make more economic use of labour and other production inputs, including veterinary care and high returns. In terms of breeding, sows bearing large litter sizes are preferred to enable farmers to increase the number of pigs in an operation per year. Moreover, the most appropriate genotypes or high-producing genotypes that are supported by the prevailing environmental conditions, availability of inputs and technical know-how are important considerations for having a high number of pigs per year.

19.5.3 Carcass Traits

Superior carcass traits like carcass yield will increase farm productivity. An increase in the average weight of carcasses will ultimately lead to the reduction of the herd size since fewer animals have to be slaughtered to attain the amount of meat demanded. However, on the contrary, with increasing urbanisation and population growth, demand for pork and pork products will ultimately increase and farmers will respond by increasing the pig population on the African continent. With better nutrition and disease management, carcass weights will keep improving as long as the quality of the carcasses is sustained. Superiority in carcass yield should translate into profit and create more wealth for the pig farmer. An important aspect of the carcass in terms of quality is the leanness of the carcass. Pig improvement programmes should target lean carcasses meaning that breeds with a high meat-to-fat ratio should be prioritised. Another important trait related to carcass yield is the palatability and organoleptic attributes. The breeding objective should therefore prioritise breeds which produce more tender and marbling carcasses to improve carcass quality.

19.5.4 Welfare and Resilience

The welfare of pigs is related to their ability to cope with their environment while resilience is the ability to cope with and quickly recover from stressors (Broom 1994). It is also regarded as the capacity of a pig to be minimally affected by disturbances or their ability to quickly return to normalcy following exposure to a disturbance (Berghof et al. 2019). Typical stressors for pigs include respiratory diseases, bacteria diseases, mycotoxins, heat stress, poor human handling, environmental pollutants (e.g. dust and ammonia) inadequate feeding and lack of water or poor water quality. The growing concern about pig welfare and hence the need to consider welfare traits in breeding programmes can be attributed to the increased cost of pig production that often leads to poor husbandry conditions resulting in extra stress to animals; decreasing the capability of pigs to cope with changing environmental conditions; increasing ethical consciousness of the human populace; and the increased availability of pig welfare information (Kanis et al. 2004). Breeding strategies that produce more resilient pigs increase the capacity of pigs to attain their full potential under commercial conditions and ensure efficient and sustainable production. In pig improvement programmes, it is important to incorporate indigenous genetics to ensure the adaptability of piglets, particularly in the face of the negative direct and indirect impacts of climate change on exotic pigs under African environmental and management conditions.

Growing attention on aspects of pig welfare and resilience is born from evidence showing that pig welfare aspects have a genetic basis and the selection of pigs for increased production can have negative side effects on animal welfare. Therefore, like other negatively correlated traits, considering them alongside production and health traits can lead to faster gains. Significant negative genetic correlations between leg quality and lean meat percentage, and between leg quality and daily pig growth rate are typical examples of how selection for one trait can affect another; and in particular, unfavourable genetic correlations between lean meat percentage and stress susceptibility (Rauw et al. 1998). Genetic selection for production and reproduction traits possibly results in welfare problems in pigs. For example, selection against nest-building behaviour before birth could lead to pigs with little or no oestrus behaviour (Wischner et al. 2009) and hence reduced fitness.

Several adaptation and coping strategies to welfare challenges in pigs have been documented and include a reduction in feed intake and an increase in maintenance energy among others (Mayorga et al. 2019).

Breeding for improved pig welfare requirements should be matched with improvements in productivity as well as address general society and ethical concerns. Other resilience-associated concepts for consideration include robustness, tolerance, resistance, genotype by environment interaction, genetic heterogeneity of environmental variance, plasticity, environmental sensitivity, canalisation and stability (Berghof et al. 2019).

19.5.5 Piglet Survival and Morbidity

Piglet survival is important to ensure increased profitability of the herd. Although breeding programmes have focused on improving the total number of piglets born, this has led to a concomitant increase in piglet mortality (Rutherford et al. 2013). Consequently, adjusting selection criteria to include neonatal survival, in addition to the number born, is a more sustainable strategy for improving piglet survival rates (Roehe et al. 2010). Closely linked to this is the need to breed for improved maternal behaviour to ensure piglet survival (Baxter et al. 2011).

Resistance is the ability of a host to restrict an invading pathogen's life cycle. An animal with greater resistance will have a lower pathogen burden (Mulder et al. 2013). Pathogen burden is the amount of pathogen in the animal's body. Tolerance is the animal's ability to minimise the detrimental impact of infection on performance. A tolerant animal maintains performance despite the pathogen burden (Mulder and Rashidi 2017b). Resistance and tolerance are both important components of resilience which in recent times have become an important trait on account of climate change and its adverse effects on production environments. Selection of the Expected Breeding Value (EBV) for resilience will target general disease tolerance and resistance rather than specific disease resistance or tolerance. Indigenous pig genetic resources are known for their resistance to most endemic diseases and their ability to survive under harsh environments including feed scarcity and heat stress (Osei-Amponsah et al. 2017). Therefore, in considering pig breeds for various improvement programmes, there is a need to consider indigenous pig genetic resources to ensure resilience and sustainability.

19.5.6 Feed Conversion Efficiency

Increasing feed efficiency is a major goal of breeders as it helps reduce production costs and energy consumption (Ding et al. 2018). Increasing feed efficiency not only reduces feed consumption while decreasing farming cost and energy use but also lowers manure production and the total amount of potential greenhouse gas emission. Feed conversion ratio (FCR) and residual feed intake (RFI) are two traits that have been used to evaluate feed efficiency (Lu et al. 2017). FCR (the ratio of feed intake to output) is widely used to estimate feed efficiency in pig breeding because of its simplicity in calculation and its correlation with growth rate and body weight (BW). Feeding behaviour is one of the most important factors affecting feed efficiency (Ding et al. 2018).

Selection for feed efficiency (*FE*) is a strategy to reduce the production costs per unit of animal product, which is one of the major objectives of current animal breeding programmes. In pig breeding, selection for feed efficiency and other traits traditionally takes place based on purebred pig performance at the nucleus level, while pork production typically makes use of crossbred animals. The success of such selection, therefore, depends on the genetic correlation between the performance of purebred pig and crossbred animals ($r$pc) and on the genetic correlation (rg) between feed efficiency and the other traits that are under selection (Godinho et al. 2018). Pig farmers in Africa and other stakeholders should be trained and given the right tools to enable them to monitor and take appropriate data on feed consumption and costs regularly. This will help in both breeding and management decisions.

Breeding schemes so far have focused on improving the performance of pig genotypes with little or no attention given to the feeding regimes in the selection environment and production environment (Nirea and Meuwissen 2017). Therefore, pig breeding programmes should target selection for robust animals that are genetically superior and perform best under lower-quality feed resources. The costs of feed can be reduced; therefore, any improvement in feed efficiency made will have a significant impact on economic return (Nirea and Meuwissen 2017).

19.6 Breeding Strategies for Pigs

19.6.1 Past and Ongoing Breeding Programmes in Africa or for Africa

19.6.1.1 Selection Programmes/ Initiatives

Indigenous pig breeds possess unique genetic characteristics that are crucial for the production of future breeds with production traits for survival in different geographical locations (Buckner et al. 2016). However, this genetic diversity is not well utilised since the majority of the smallholders' pig farmers in Africa are not engaged in well-defined pig selection programmes neither are they following concrete breeding programmes (Halimani et al. 2012). Lack of access to production information, poverty and/or resource constraints, ill-defined government policies and programmes coupled with low-input production systems have hindered the development of adequate breeding programmes in indigenous pig communities (Buckner et al. 2016; Halimani et al. 2012). In developing a regional strategy for the control of African swine fever in Africa, an FAO-ILRI-AU-IBAR consortium initiative has recommended selective breeding programmes targeting well-adapted animals and tailored for improved productivity, with adaptation to diseases being one of the lead targets of such an initiative (FAO/ILRI/AU-IBAR 2017).

19.6.1.2 Crossbreeding Programmes/ Initiatives

The ever-increasing market demands for pork and pig products have led to the commercialisation of the pig industry in the majority of African countries. The indigenous pig breeds, which are regarded as unsuited for intensive commercial production (Swart et al. 2010), are being replaced by exotic species that are fast growing and have high carcass weights (Buckner et al. 2016). Breeds such as the Landrace, Large White, Duroc, Comborough, and Pietrain (AU-IBAR 2019) have been introduced through various routes for crossbreeding purposes. However, these initiatives have also facilitated the loss of indigenous genetic diversity. A typical three-way breed crossing programme has been initiated by the National Animal Genetic Resource Centre and Data Bank (NAGRC&DB) in Uganda. The programme is producing a hybrid from Large White sows mated to Landrace sires, with the resultant crossbred sows mated to a Duroc terminal sire to produce the NAGR-Pig that holds quite a significant promise for the Ugandan pig industry (NAGRC&DB 2022).

19.6.1.3 Community-Based Breeding Programmes

Community-based breeding programmes (CBBP) are promoted mainly as a strategy for smallholder farmers to improve their livestock breeds. These programmes are defined as being typically related to low-input systems with farmers within geographical boundaries having a common interest in working together for the improvement of their genetic resources

(Wurzinger et al. 2021). Community-based breeding programmes can also offer the opportunity for in situ conservation of indigenous pig genetics. However, constraints such as a lack of marketing strategies have greatly hindered its development as a breeding strategy when compared to community-based breeding schemes for other livestock species. The use of communal boars is widespread among the smallholder pig farmers of Africa. In several countries, for instance, Uganda, a household may own a boar that is used by home studs for natural servicing, being that over 52% of all smallholder pig farmers do not keep their own boar. However, the main challenge with these typical communal boars is that they lack the improved genetic qualities and they do aid the spread of African Swine Fever (ASF). Furthermore, farmers tend to select replacement stock from within their herd; thus resulting in the maintenance of a specific breed. This may lead to a very high risk of inbreeding that might negate the efforts to conserve and utilise the breed (Halimani et al. 2012). However, there is limited literature on the aspect of community-based pig breeding programmes in Africa.

19.6.1.4 Efforts to Breed for Swine Fever Tolerance

There is wide genetic variation in the extent of susceptibility to diseases and parasites among different pig breeds, particularly on the African continent. The indigenous pig breeds have shown better chances of survival to various disease outbreaks and have a higher capacity to utilise fibrous and poor-quality feed resources compared to the exotic breeds (FAO/ILRI/AU-IBAR 2017). Furthermore, indigenous pig breeds have been reported to survive outbreaks of African swine fever in Mchinje District of Malawi and Angonia District in Mozambique (Halimani et al. 2010). These present opportunities for selection within the tolerant breeds for resistant animals, introgression of the resistance genes into their commercial counterparts and/or identification of the genes for inclusion in other strains (Wurzinger et al. 2021). However, these qualities and breeds that exhibit these qualities have been understudied while the genetic progress in resistance to disease is low. A study conducted in Northern Mozambique to evaluate the heritability of ASF resistance genes by expositing offspring of outbreak survivors to African swine fever virus indicated that ASF resistance was not inherited and hence not genetic but rather a result of enzootic stability between the indigenous pigs and the virus (Penrith et al. 2004). Efforts elsewhere particularly in mainland Europe and Asia indicate the possibility of disease control through selective breeding for African swine fever tolerance, however, the veracity of results so far obtained is still suspect (Guinat et al. 2017).

19.6.1.5 Breeding Programmes for Breeder Companies

Breeding programmes of private pig breeding companies are considered trade secrets and hence these are not publicly provided. Nevertheless, information on the breeds and commercial pig lines of focus and the locality of the breeder companies is provided in Table 19.4.

The Pig Improvement Company (PIC), at its establishment towards the end of the 1950s, aimed at exploring the benefits of a high health status breeding environment and genetic improvement. Large White (LW) females and Landrace (LR) males were used to start the initial herd. To ensure that the progeny was free of pathogens, for example, the hysterectomy technique was used. The womb was surgically removed just prior to giving birth naturally avoiding the contamination of the offspring. With this technology, 80 disease-free piglets were produced to start the breeding programme. The mating of LW females with LR males resulted in a 10% increase in the pigs' size, feed conversion, and live weight gain as a result of 'hybrid vigour'. It was necessary to include animals with good fertility characteristics because litter size is only inheritable by 10%, whereas feed conversion and live weight gain can be up to 30% inherited (Zylbersztajn et al. 2000).

The internationalisation strategy for PIC was developed in the 1970s. It necessitated the development of a range of selection standards directed at specific countries with their own particular

Table 19.4 Pig breeding companies operating in Africa

Nr	Company name	Head office (origin)	African office location	Breeds/lines	Website
1	Pig Improvement Company (PIC)	Europe: PIC UK	South Africa:	Gilts; Camborough. Boars; PIC337, PIC 410, PIC800	https://www.picrsa.co.za/
2	DanBred	Europe, Denmark	South Africa: DANBRED AFRICA (PTY) LTD.	DanBred landrace (LL), DanBred Yorkshire (YY), and DanBred Duroc (DD), DanBred hybrid	https://danbred.com/
3	Topigs Norsvin	Norway and Netherlands	South Africa	Terminal boars: TN select, TNTraxx, TN talent, TN tempo, Norsvin Duroc, TN Duroc. Parent sows TN70, TN60	https://topigsnorsvin.co.za/
4	Choice Genetics	USA:	No office	Parent gilt: CG36, grand parent females: M3 Largewhite, M6 landrace terminal boar: P26, P81, P90	https://choice-genetics.com/en/
5	Alliance Genetics South Africa	South Africa	South Africa Pretoria	Landrace, Largewhite, Duroc, F1 crossbred gilts	https://www.alliancegenetics.co.za/index.htm

production system and distinct market requirements. As a result of this, races of pigs such as the Pietrain were introduced that had desirable production traits despite being susceptible to the Pale Soft Exudative (PSE) muscle. The Duroc race proved to be particularly promising when introduced in systems that require robustness in a harsher environment and improved meat quality. International PIC believed that the main role of the associated companies was to guarantee that the clients adopted the standards considered necessary for business success (Zylbersztajn et al. 2000).

Improvements in pigs at the international level have been constant over the past years, but new technology provided a powerful tool for accelerating the process. PIC patented an identification method for isolating the ESR gene whose occurrence is associated with an increase in reproductive capacity. PIC-sponsored researchers had identified an association of the RYR gene to high-quantity non-fat meat and also high susceptibility to stress; the RN gene with acid meat, the K88 gene with resistance to *Escherichia coli* and the MHC gene with offspring size, bacon thickness and piglet mortality. Improvement programmes could now be accelerated and targeted more accurately by 'gene tagging'.

PIC also collaborates with its international partners, stakeholders and clients to ensure that the environment and human resources needed to achieve the best out of its products are in place. This is because all the investment in pork improvement could be wasted without suitable management of the herd with appropriate facilities run by skilled personnel using good-quality inputs (Zylbersztajn et al. 2000). This is a lesson Africa should be mindful of before importing any exotic animal genetic resource.

The pig sector has a pyramidal structure and some cooperative companies of pig breeders such as Topigs are involved in breeding. There are more breeding companies in the pig sector than in poultry although a few of them have large market shares such as PIC. As in poultry, the organisations involved in genetic improvement have worldwide distribution networks. Compared to poultry breeding, the number of breeders, that is, people owning male and or female animals, is much larger in pig breeding (Nimbkar and van Arendonk 2010). Vertical integration from genetics to pork products is high in North America, and fast growing in many European countries (Gura 2007). Commercial producers buy both sows and boars of specialised lines or crosses from breeding companies and crossbreed them to produce pigs for slaughter. Artificial insemination (AI) is used in the pig industry. In contrast to poultry, there are still breeding associations for pigs in many countries (Nimbkar and van Arendonk 2010).

Leading pig breeding companies have implemented single-step evaluation for genomic selection. Overall, this increases EBV accuracies by half. Further improvement requires focus on trait- and line-specific QTLs, exploitation of crossbred performance for non-additive genetic effects, and training for hard-to-measure traits. The multi-breeding-company multi-line crossbred pig production structure limits very high accuracies. Increased genotyping and phenotyping will lead to improved training data and may increase accuracies by two-thirds rather than half. Novel technologies will allow for genotyping of all selection candidates, reducing the generation interval and emphasising the need for inbreeding control, more efficient breeding structures, and higher selection intensities (Knol et al. 2016).

PIC markets approximately 2 million breeding animals with a volume of sales approaching US $400 million a year. PIC has 30 to 40% of the market in North America and 11% of the market in Europe, and is represented in around 30 countries with more than 1500 employees. 1,6 million breeding sows are sold each year, raised on some 40 farms. Gross margin is at 35%. PIC belonged to Sygen (turnover 129 million USD) until in 2005 the UK-based Genus plc, owner of the world's largest cattle breeding company, ABS, bought Sygen, a specialist in quantitative genetics of pig and shrimps, with its daughter, PIC, the world's largest pig breeding company (Gura 2007).

Hypor, the world's second-largest pig breeding company, still belongs to Nutreco, based in

NL, which is Europe's largest animal compound feed and fish feed producer. Its breeding division, Euribrid, also comprises the world's second largest turkey breeding company, Hybrid, and the fourth largest broiler breeder, Hybro. The total turnover of Hypor is approximately 35 million €. Hypor has around 250 employees and is represented in Canada, Spain and Belgium, with a market share between 20 and 24%. It also holds substantial market shares in the Netherlands, Italy, Germany, Poland, Japan, Mexico and the Philippines (Gura 2007).

The Dutch cooperative *Topigs* is globally the third largest pig breeding organisation, producing almost 850,000 gilts per year. Topigs is a subsidiary of the Pigture Group Pig Breeders Co-operative which is owned by 3000-member pig farmers in the Netherlands. The Pigture Group Pig Breeders Co-operative owns 77.5% of Topigs; 22.5% are owned by Europe's largest fresh-meat processor Vion Food Group. The Pigture Group has around 400 employees and a turnover of 103 million €. In the Netherlands, Topigs has a market share of over 80%, and with a line well suited for Parma ham, it leads the Italian market. In 2006, it opened nucleus farms in Russia and Croatia. Production and distribution of the breeding material are based on a franchise system. Topigs highly values its independence and, therefore, makes its genetics freely available. Topigs is the first European pig breeding organisation that has been certified for the Code of Good Practice for Farm Animal Breeding and Reproduction Organisations, Code-EFABAR (Gura 2007).

Monsanto's share in the US pig genetics market currently is about 10%. In 1998, Monsanto acquired DeKalb with, among others, their pig breeding sector, and in 2001, Monsanto purchased the Canadian pig breeding enterprise Unipork. Monsanto also is the exclusive distributor of the 'Genepacker' boar of JSR Genetics, UK. Monsanto has license contracts with Metamorphix, which in turn has near to exclusive access to the pig genome (Gura 2007).

About 42% of global pig production is industrial, with five dominating breeds (Large White, Duroc, Landrace, Hampshire and Pietrain). According to FAO, 66% of the mothers of European fattening pigs are hybrid crosses of the 'Large White' and 'Landrace' breeds. Effective population size, a parameter used in breed conservation to calculate genetic diversity, in pigs was found to be 71, 74 and 61 animals for the Yorkshire, Hampshire and Duroc, respectively. While these effective population sizes are somewhat larger than those reported for Holstein, Brown Swiss and Jersey cattle, they are still under the 100 head which is considered a critical level for maintaining genetic diversity (Gura 2007).

19.6.2 Molecular and Assisted Reproductive Innovations for Pig Improvement

19.6.2.1 Genomic Selection

The molecular and assisted reproduction technologies relevant to livestock improvement have been described in more detail in Chap. 21. Although the capacity for implementation of genomic selection (Lillehammer et al. 2011) is still in development in Africa, it has been shown to increase rates of genetic gain in pig nucleus breeding schemes (Halimani et al. 2012). For instance, GS has the potential to improve feed efficiency traits, minimise the cost of recording and account for environmental sensitivity that occurs due to differences in feeding systems between environments (Lillehammer et al. 2011). To utilise the potential of GS, it is essential to include animals in the reference population from both the nucleus and production environments. For example, when implementing genomic selection for feed efficiency traits, the reference population should be based on the feeding and management structure of the population and the reference population should involve animals from the environment where the breeding goal is targeted (Swart et al. 2010). In future years, there is a need to broaden pig breeding goals to include welfare-relevant traits in order to counter the effects of selection on productivity and to more directly improve welfare traits in their own right, environmental adaptation traits, disease-relevant traits and other lowly heritable traits.

19.6.2.2 Nanotechnology

Evidence exists that shows that nanoparticles increase the life of spermatozoa and improve the general fertility of sires. An extensive review of the applications of nanotechnology to animal reproduction specifically on the management of mammalian spermatozoa has been provided by Feugang and colleagues (Feugang et al. 2019). They did show that treatment of boar sperm with nanoparticles enhances fertility and provides a practical biomedical application to improve economic productivity for farmers. More stakeholder collaborative efforts and research in this direction within the African context are therefore recommended.

19.6.2.3 Clustered Regularly Interspaced Short Palindromic Repeats/Cas9 (CRISPR-Cas9) System

This technique is being used for livestock to enable the production of resilience against diseases, enhance reproductive traits, and act as a model for biomedical research (Penrith et al. 2004). One application of Cas-9-based gene modification has been in the production of pigs with the activity of porcine endogenous retroviruses made non-functional, hence a pathway for transplantation of organs from pigs to humans. Alterations that have been done include knocking out and in of short palindromic repeats associated proteins into pig genetic material to insert pre-confirmed alterations (Jabbar et al. 2021) to enhance production parameters such as the formation of standard heat, improve food digestion and increase resistance against pig ailments.

The Cas-9 technique has enabled knocking out the sexually determined region of the Y gene of pigs by micro inoculating Cas plasmid cytoplasm. Sex change is so valuable as it influences economic gain due to varied economic benefits harvested from wither female or male pigs.

19.6.2.4 Transgenesis

Genetic engineering has been used for the introduction of foreign genes such as genes controlling growth to increase growth and improve carcass quality (Madan 2005). Transgenesis refers to the process of transferring an exogenous DNA segment or gene (transgene) to an animal so that it is able to pass the transgene on to all of its offspring. In this way, a transgenic pig is one that has stably incorporated engineered DNA into its germline and is thus a genetically modified organism. Where the appropriate institutional and human capacities exist, transgenesis could be used to introduce useful genes such as those for improved growth, production, disease resistance and thermotolerance into livestock species to enable them to adapt and produce in various environments.

19.6.2.5 Assisted Reproductive Technologies

Reproductive biotechnologies are of critical value to African pig genetic improvement and have had varying levels of success in different African countries. The primary goals of the assortment of technologies/innovations are increasing production, reproductive efficiency and enhancing the rates of genetic improvement (Madan 2005). Additional techniques that further enhance reproductive efficiency and pregnancy rates of pigs are the evaluation of sperm fertilisation capacity, sperm sexing, oestrus synchronisation, fixed-time insemination, embryo splitting and in vitro embryo production.

The uptake of pig artificial insemination (AI) in Africa is still very low compared to progress with bovids, but successful conception rates of field AI programmes in a few countries are quite impressive (Niyiragira et al. 2018). Whereas bovine AI is mainly challenged by high technical expertise of the practice and the low spread of cryo-storage facilities, pig AI is limited by the limited availability of inseminators, insemination kits and a poor fresh semen distribution mechanism in most countries. The biggest boost of AI in swine is the lack of boars in most smallholder pig farms (Bamundaga et al. 2018). To reduce costs associated with multiple inseminations that are conventionally required, research has recently focussed on improving oestrus detection followed by carefully timed single-dose insemination. Results of farrowing rates and litter sex ratios have been quite comparable for single and double AI and adoption of this biotechnology in Eastern Africa is on the rise (Bamundaga et al. 2018).

Oestrus synchronisation adds value and success rate to swine artificial insemination. The major impetus of this technology is the limited skill to detect oestrus by most smallholder farmers in Africa. The oestrus cycle in domestic pigs runs for 18–24 days consisting of a follicular phase lasting 5–7 days and a luteal phase of 13–15 days (Soede et al. 2011). The follicular phase enables small antral follicles to develop into large, pre-ovulatory follicles. Pigs are able to produce between 15 and 30 follicles, the number is influenced by several factors including age and nutritional status among others (Blödow et al. 1990). Gilts are usually inseminated at their second or third oestrus cycle after the onset of puberty while sows should be served at the first oestrus following the weaning of its litter. Oestrus and ovulation occur 4–7 days after weaning, with the cessation of lactational anoestrus (Brüssow and Wähner 2011). This means if the pig in the luteal phase receives a dose of prostaglandin, it will come to oestrus within 72 hours and this is the window of action for insemination.

19.7 Prospects for Sustainable Utilisation and Conservation of Indigenous Pigs of Africa

African pig genetic resources constitute an important component of animal agriculture contributing to food security, wealth creation and sustaining livelihoods. African indigenous pig breeds remain vital functional components of animal agriculture and it is important to recognise this fact and put in place appropriate breeding strategies for their sustainable utilisation and conservation. National governments should ensure the implementation of the pig component of their strategic plans on animal genetic resources improvement and conservation. Programmes favouring in situ conservation of indigenous pig genetic resources within their production systems should be promoted. Additionally, depending on the location, production system and the availability of resources, ex situ conservation options should also be explored to maintain the pigs for future generations.

Ex situ in vitro conservation may be expensive for some African countries, especially those that do not have national gene banks in operation. It is, therefore, imperative that the five African sub-regional gene banks, now rebranded as African Union Animal Resources Seed Centres of Excellence (AU-ARSCoE), prioritise pig genetic resources from such countries. These centres are located in Entebbe, Uganda for Eastern Africa Region, Bambui, Cameroun for Central Africa, Gabarone, Botswana for Southern Africa, Bobo-Dioulasso, Burkina Faso for Western Africa and Tunis, Tunisia for Northern Africa. Pig genetic material that could be conserved cryogenically at these regional centres includes embryo, ova, germinal cells, tissues, DNA and semen.

19.8 Capacity Building for Pig Genetics and Breeding in Africa

In order to have vibrant pig breeding programmes, there is a need for higher degree training (certificate, diploma and bachelor and post-graduate degree level) in human capacity. Institutions of higher learning in Africa, particularly those focusing on Animal Science, should train more animal breeders and geneticists to lead the crusade for sustainable use and conservation of Africa's pig genetic resources. Africa's human and institutional capacity needs to be upscaled to enable the continent to make good use of innovative technologies in pig breeding and genetics for increased production. In consultation with the government, industry and the relevant stakeholders, universities should put up relevant and practical graduate programmes across Africa in modern technologies as applied to pig breeding in particular and animal breeding in general. Additionally, AU-IBAR (Wurzinger et al. 2021) indicates the need for strengthening the various stakeholder institutions including teaching, research, extension, trade and marketing as well as those dealing with policy and legal aspects. Furthermore, all stakeholders should play their unique roles in the management of farm animal genetic resources while exploring public–private partnerships in capacity building. There is also a need for people trained in modern technologies and operationalisation of established regional gene banks (the African Union Animal Resources Seed Centres of Excellence) and national AI centres across the continent. Existing partnerships between African scientific organisations with companies that specialise in pig reproduction technologies such as Pig Improvement Company (PIC), KUBUS Spain, Minitube GMBH and Hendrix Genetics should be strengthened and new partnerships explored.

In many African countries, there is a general lack of practical livestock-specific education, and many private education initiatives that provide post-primary and post-secondary school vocational education in Agriculture find difficulty in gaining accreditation. Very few institutions offer a livestock and animal science-specific programme and where they exist there is no specialisation in the field of piggery biotech and business. Where they exist, the programmes are highly theoretical and, hence, students find difficult to utilise the knowledge acquired. For example, Makerere University in Uganda, the University of Ghana and many other African universities are located in the capital cities and their teaching farms are located far from the campus and students' residences which makes it difficult to acquire practical skills in pig management and production on a sustainable basis. There is therefore a need to rethink the location and programming of agricultural training institutions. Integration of field attachments and internships into curricula could go a long way to ameliorate the situation. These could be offered at private agricultural establishments, research institutes, and within and across African states.

19.9 Conclusion and Future Perspectives

Africa boasts of a wide diversity of pig genetic resources including indigenous breeds, locally adapted exotic genotypes, and crosses between indigenous and introduced exotic breeds. Pigs contribute greatly to wealth creation, meat production, employment and various industrial products such as gelatin, wax, soap, biofuels and manure, but in many parts of Africa, the full benefits from the pig are not yet realised. Breeding programmes are often based on specific production objectives and the prevailing production systems. Pig production systems and farmers are unique and vary from one region of Africa to another, and also from country to country in a given sub-region. These systems have market-oriented breeding objectives that unfortunately often fail to match the available improved genotypes to favourable production environments. There are many examples of locally adapted exotic pig breeds not producing to their maximum because of the poor environment and management. Over the years, pig breeding in Africa has shifted from breeding for just high production to breeding for production efficiency, health, functional, resilient and adaptive traits. The African pig production value chain is bridled with various challenges from lack of planned breeding programmes, poor conservation of indigenous pig genetic resources, poor pig housing, high feed and veterinary costs as well as diseases, particularly the dreaded African Swine Fever. At the same time, emerging innovative technologies provided with the right human and institutional resources can help turn most of the challenges into opportunities and in the process help improve pig breeding in Africa.

Sustainable pig breeding in Africa will require the conservation of indigenous pig genetic resources through their sustainable use and ensuring that they remain vital components of the production system. In this regard, future pig breeding goals should focus not only on large litter sizes, feed conversion efficiency and growth rates but also on welfare traits like heat tolerance, adaptation as well as carcass quality and traits that will require indigenous pig breed contributions. Sustainable pig breeding programmes will also require infrastructure for data collection and their evaluation to guide future initiatives. We need to learn from best pig breeding practices and take advantage of available innovative reproductive technologies including artificial insemination and embryo transfer to speed up pork production for food security and wealth creation. National governments should conserve a wide variety of pig genetic resources to better respond to future environmental uncertainties and changing market demands. Individual farmers may invest in conservation efforts but government departments must be in the vanguard of this deliberate action. Pig farmer organisations should be capacitated and encouraged to be part of a national pig production platform to ensure that agreed-upon farmer-trait preferences are considered in future pig breeding programmes. At the same time, multinational pig breeding companies doing business in Africa should be mandated to contribute towards sustainable pig breeding programmes by contributing towards in situ and ex situ conservation of pigs in Africa. Finally, Africa needs to invest in its human and institutional resources to take advantage of emerging technologies such as genomic selection, transgenesis and assisted reproductive technologies including artificial insemination and embryo transfer. This will require comprehensive training programmes from the secondary to tertiary levels to ensure sustainable pig breeding as well as sustainable use and conservation of pig resources in Africa.

References

AU-IBAR (2019) The state of farm animal genetic resources in Africa. AU-IBAR Publication, p 304

Babigumira BM, Sölkner J, Mészáros G, Pfeiffer C, Lewis C, Ouma E, Wurzinger M, Marshall K (2021) A mix of old British and modern European breeds: genomic prediction of breed composition of smallholder pigs in Uganda. Front Genet 12:1056. https://doi.org/10.3389/fgene.2021.676047

Bamundaga GK, Natumanya R, Kugonza DR, Owiny DO (2018) Reproductive performance of single and double artificial insemination protocol in swine. Bull Anim Health Prod Afr 66(1):143–157. https://www.ajol.info/index.php/bahpa/article/view/167855

Baumgard LH, Rhoads RP Jr (2013) Effects of heat stress on post absorptive metabolism and energetics. Ann Rev Animal Biosci 1:311–337. https://doi.org/10.1146/annurev-animal-031412-103644

Baxter EM, Jarvis S, Sherwood L, Farish M, Roehe R, Lawrence AB, Edwards SA (2011) Genetic and environmental effects on piglet survival and maternal behaviour of the farrowing sow. Appl Anim Behav Sci 130:28–41

Berghof TVL, Poppe M, Mulder HA (2018) Opportunities to improve resilience in animal breeding programs. Front Genet 9:692. https://doi.org/10.3389/fgene.2018.00692

Berghof TVL, Poppe M, Mulder HA (2019) Opportunities to improve resilience in animal breeding programs. Front Genet 9:e692. 15 pp

Blödow G, Bergfeld J, Kitzig M, Brüssow KP (1990) Steroid hormone levels in follicular fluid of swine with spontaneous estrus and ovulation synchronization. Archiv fur Experimentelle Veterinarmedizin 44(4):611–620

Broom DM (1994) The effects of production efficiency on animal welfare. In: Huisman EA, Osse JWM, van der Heide D, Tamminga S, Tolkamp BJ, Schouten WGP, Hollingworth CE, van Winkel GL (eds) Biological Basis of Sustainable Animal Production. Proceedings of the 4th Zodiac Symposium. Wageningen Press, Wageningen, pp 201–211. EAAP publication no. 67

Brüssow KP, Wähner M (2011) Biological and technological background of estrus synchronization and fixed-time ovulation induction in the pig. Biotechnol Animal Husband 27(3):533–545. https://doi.org/10.2298/bah1103533B

Buckner CA, Lafrenie RM, Dénommée JA, Caswell JM, Want DA, Gan GG, Leong YC, Bee PC, Chin E, Teh AKH, Picco S, Villegas L, Tonelli F, Merlo M, Rigau J, Diaz D, Masuelli M, Korrapati S, Kurra P, Mathijssen RHJ (2016) Tracing the Domestic Pigs in Africa. IntechOpen. 11 (tourism), 13. https://www.intechopen.com/books/advanced-biometric-technologies/liveness-detection-in-biometrics.pdf

Colditz IG, Hine BC (2016) Resilience in farm animals: biology, management, breeding and implications for animal welfare. Animal Product Sci *56*(12). https://doi.org/10.1071/AN1529

de Leon N, Jannink J-L, Edwards JW, Kaeppler SM (2016) Introduction to a special issue on genotype by environment interaction. Crop Sci 56:2081–2089

Ding R, Yang M, Wang X, Quan J, Zhuang Z, Zhou S, Li S, Xu Z, Zheng E, Cai G, Liu D, Huang W, Yang J, Wu Z (2018) Genetic architecture of feeding behavior and feed efficiency in a Duroc pig population. Front Genet 9:220. https://doi.org/10.3389/fgene.2018.00220

FAO/ILRI/AU-IBAR (2017) Regional strategy for the control of African swine fever in Africa. FAO, p 45

Feugang JM, Rhoads CE, Mustapha PA, Tardif S, Parrish JJ, Willard ST, Ryan PL (2019) Treatment of boar sperm with nanoparticles for improved fertility.

Theriogenology 137:75–81. https://doi.org/10.1016/j.theriogenology.2019.05.040

Godinho RM, Bergsma R, Silva FF, Sevillano CA, Knol EF, Lopes MS, Lopes PS, Bastiaansen JWM, Guimarães SEF (2018) Genetic correlations between feed efficiency traits, and growth performance and carcass traits in purebred and crossbred pigs. J Anim Sci 96:817–829. https://doi.org/10.1093/jas/skx011

Guinat C, Vergne T, Jurado-Diaz C, Sánchez-Vizcaíno JM, Dixon L, Pfeiffer DU (2017) Effectiveness and practicality of control strategies for African swine fever: what do we really know? Vet Rec 180:97–97. https://doi.org/10.1136/vr.103992

Gura S (2007) Livestock genetics companies concentration and proprietary strategies of an emerging power in the global food econo. League for Pastoral Peoples and Endogenous Livestock Development, Ober-Ramstadt

Halimani TE, Muchadeyi FC, Chimonyo M, Dzama K (2010) Pig genetic resource conservation: the Southern African perspective. Ecol Econ 69(5):944–951. https://doi.org/10.1016/J.ECOLECON.2010.01.005

Halimani TE, Muchadeyi FC, Chimonyo M, Dzama K (2012) Opportunities for conservation and utilisation of local pig breeds in low-input production systems in Zimbabwe and South Africa. Trop Anim Health Prod 45(1):81–90. https://doi.org/10.1007/s11250-012-0177-2

Halimani TE, Mapiye O, Marandure T, Januarie D, Imbayarwo-Chikosi VE, Dzama K (2020) Domestic free-range pig genetic resources in Southern Africa: Progress and prospects. Diversity 12:68. https://doi.org/10.3390/d12020068

Jabbar A, Zulfiqar F, Mahnoor M, Mushtaq N, Zaman MH, Salah Ud din A, Khan MA, Ahmad HI (2021) Advances and perspectives in the application of CRISPR-Cas9 in livestock. Mol Biotechnol 63:757–767. https://doi.org/10.1007/s12033-021-00347-2

Kanis E, van den Belt H, Groen AF, Schakel J, de Greef KH (2004) Breeding for improved welfare in pigs: a conceptual framework and its use in practice. Anim Sci 78:315–329

Knol EF, Nielsen B, Knap PW (2016) Genomic selection in commercial pig breeding. Anim Front 6(1):15–22. https://doi.org/10.2527/af.2016-0003

Kumar BL, Gopal DVRS (2015) Effective role of indigenous microorganisms for sustainable environment. 3 Biotech 5:867–876. https://doi.org/10.1007/s13205-015-0293-6

Larive International, Piggery and poultry market roadmap for sustainable value chain development: identifying opportunities in the Uganda poultry and piggery sectors. Embassy of the Kingdom of the Netherlands in Uganda 2020, 201911124 / PSS19UG02

Larzul C (2021) How to improve meat quality and welfare in entire male pigs by genetics. Animals 11(3):699

Lillehammer M, Meuwissen THE, Sonesson AK (2011) Genomic selection for maternal traits in pigs. J Anim Sci 89:3908–3916

Lu D, Jiao S, Tiezzi F, Knauer M, Huang Y, Gray KA (2017) The relationship between different measures of feed efficiency and feeding behavior traits in Duroc pigs. J Anim Sci 95:3370–3380. https://doi.org/10.2527/jas.2017.1509

Marshall K, Poole J, Oyieng E, Ouma E, Kugonza DR (2023) A farmer-friendly tool for estimation of weights of pigs kept by women and men smallholders in Uganda. Tropical Animal Health & Production, 55:219, 1-12. https://doi.org/10.1007/s11250-023-03561-z

Madan ML (2005) Animal biotechnology: applications and economic implications in developing countries. Revue Scientifique et Technique de l'Office International des Epizooties (Scientific and Technical Review of the International Office of Epizootics) 24(1):127–139

Mayorga EJ, Renaudeau D, Ramirez B, Jason W, Ross JW, Lance H, Baumgard LH (2019) Heat stress adaptations in pigs. Anim Front 9(1):54–61. https://doi.org/10.1093/af/vfy035

Merks JWM, Mathur PK, Knol EF (2012) New phenotypes for new breeding goals in pigs. Animal 6(4):535–543. https://doi.org/10.1017/S1751731111002266

Miglior F, Fleming A, Malchiodi F, Brito LF, Martin P, Baes CF (2017) A 100-year review: identification and genetic selection of economically important traits in dairy cattle. J Dairy Sci 100:10251–10271. https://doi.org/10.3168/jds.2017-12968

Mulder HA, Rashidi H (2017a) Selection on resilience improves disease resistance and tolerance to infections. J Anim Sci 95:3346–3358. https://doi.org/10.2527/jas2017.1479

Mulder HA, Rashidi H (2017b) Selection on resilience improves disease resistance and tolerance to infections. J Anim Sci 95(8):3346–3358. https://doi.org/10.2527/jas.2017.1479

Mulder HA, Rönnegård L, Fikse WF, Veerkamp RF, Strandberg E (2013) Estimation of genetic variance for macro- and micro-environmental sensitivity using double hierarchical generalized linear models. Genet Sel Evol 45:23. https://doi.org/10.1186/1297-9686-45-23

Mutetikka D (2009) A guide to pig production at farm level. Fountain Publishers, Kampala

Mutetikka D, Kugonza DR (2009) Okuganyulwa mu Bulunzi bw'Embizzi (How to benefit from piggery business). Fountain Improved Farming Series. Farming Guide 11. Fountain Publishers, Kampala. 39 pages. ISBN: 978-9970-02-967-9.

Mutua JY, Marshall K, Paul BK, Notenbaert AMO (2020a) A methodology for mapping current and future heat stress risk in pigs. Animal 14(9):1952–1960. https://doi.org/10.1017/S1751731120000865

Mutua JY, Zaake P, Notenbaert AMO (2020b) Reducing climate-induced heat stress in pigs in Uganda: policy implications. CGIAR Research Program on Livestock, ILRI

Mwesigwa R, Mutetikka D, Kugonza DR (2013a) Performance of growing pigs fed diets based on by-products of maize and wheat processing. Tropical Animal Health and Production, 45(2):441–446. https://doi.org/10.1007/s11250-012-0237-7

Mwesigwa R, Mutetikka D, Kabugo S, Kugonza DR (2013b) Varying levels of wheat processing by-products in diets for growing pigs: effect on growth and carcass Traits. Tropical Animal Health and Production, 45(8):1745–1749. https://doi.org/10.1007/s11250-013-0425-0

NAGRC&DB (National Animal Genetic Resources & Data Bank) (2022) National Animal Breeding Strategy and action plan for Uganda. NAGRC&DB, Entebbe, p. 110

Narrod CA, Fuglie KO (2000) Private investment in livestock breeding with implications for public research policy. Agribusiness 16(4):457–470. https://doi.org/10.1002/1520-6297(200023)16:4<457::Aid-agr5>3.0.Co;2-7

Nimbkar C; van Arendonk JAM (2010) Recent trends in the global organization of animal breeding. In International Technical Expert Workshop on "Exploring the need for specific measures for access and benefit-sharing of animal genetic resources for food and agriculture", Wageningen, The Netherlands

Nirea KG, Meuwissen THE (2017) Improving production efficiency in the presence of genotype by environment interactions in pig genomic selection breeding programmes. J Anim Breed Genet 134:119–128

Niyiragira V, Kugonza DR, Hirwa CD (2018) Success drivers of pig artificial insemination based on imported fresh semen. Int J Livestock Product 9(6):102–107

Osei-Amponsah R, Skinner BM, Adjei OD, Bauer J, Larson G, Affara NA, Sargent CA (2017) Origin and phylogenetic status of the local Ashanti Dwarf pig (ADP) of Ghana based on genetic analysis. BMC Genomics 18:193. 12 pp. https://doi.org/10.1186/s12864-017-3536-6

Pedersen LJ (2018) Overview of commercial pig production systems and their main welfare challenges. In: Spinka M (ed) Advances in pig welfare. Woodhead Publishing, pp 3–26. https://doi.org/10.1016/B978-0-08-101012-9.00001-0

Penrith ML, Thomson GR, Bastos ADS, Phiri OC, Lubisi BA, Du Plessis EC, Macome F, Pinto F, Botha B, Esterhuysen J (2004) An investigation into natural resistance to African swine fever in domestic pigs from an endemic area in southern Africa. Revue Scientifique et Technique, (International Office of Epizootics) 23(3):965–977. https://doi.org/10.20506/RST.23.3.1533

Pezo D, Waiswa C (2012) Farming systems perspectives: Lessons for managing health risks in smallholder pig systems. In Paper presented in the South–South symposium managing risks in emerging pork markets, Hanoi, Viet Nam 23–25 April, 2012. Nairobi, Kenya: international livestock research institute (ILRI) and Kampala, Uganda: Makerere University

Rauw W, Kanis E, Noordhuizen-Stassen E, Grommers F (1998) Undesirable side effects of selection for high production efficiency in farm animals: a review. Livest Prod Sci 56:15–33

Rege JEO, Marshall K, Notenbaert A, Ojango JMK, Okeyo AM (2011) Pro-poor animal improvement and breeding — what can science do? Livest Sci 136(1):15–28. https://doi.org/10.1016/j.livsci.2010.09.003

Riquet J, Gilbert H, Feve K, Labrune Y, Rose R, Billon Y, Giorgi M, Loyau T, Gourdine JL, Renaudeau D (2017) Genetic dissection of mechanisms underlying heat adaptation in pigs. In: Proceedings of 36th conference of the International Society for Animal Genetics (ISAG), Illinois, USA, p 133

Roehe R, Shrestha NP, Mekkawy W, Baxter EM, Knap PW, Smurthwaite KM, Jarvis S, Edwards AB, Smurthwaite SA (2010) Genetic parameters of piglet survival and birth weight from a two-generation crossbreeding experiment under outdoor conditions designed to disentangle direct and maternal effects. J Anim Sci 88:1276–1285

Roessler R (2019) Selection decisions and trait preferences for local and imported cattle and sheep breeds in peri–/urban livestock production systems in Ouagadougou, Burkina Faso. Animals 9:207. https://doi.org/10.3390/ani9050207

Ross JW, Hale BJ, Gabler NK, Rhoads RP, Keating AF, Baumgard LH (2015) Physiological consequences of heat stress in pigs. Anim Prod Sci 55:1381–1390. https://doi.org/10.1071/an15267

Rutherford KMD, Baxter EM, D'Eath RB, Turner SP, Arnott G, Roehe R, Ask B, Sandøe P, Mousten VA, Thorup F, Edwards SA, Berg P, Lawrence AB (2013) The welfare implications of large litter size in the domestic pig I: biological factors. Anim Welf 22:199–218. https://doi.org/10.7120/09627286.22.2.199

Soede NM, Langendijk P, Kemp B (2011) Reproductive cycles in pigs. Anim Reprod Sci 124(3–4):251–258. https://doi.org/10.1016/j.anireprosci.2011.02.025

Swart H, Kotze A, Olivier PAS, Grobler JP (2010) Microsatellite-based characterization of Southern African domestic pigs (*Sus scrofa domestica*). South Afr J Animal Sci 40(2):121–132. https://doi.org/10.4314/sajas.v40i2.57280

Tiezzi F, Brito LF, Howard J, Huang YJ, Gray K, Schwab C, Fix J, Maltecca C (2020) Genomics of heat tolerance in reproductive performance investigated in four independent maternal lines of pigs. Front Genet 11:629. https://doi.org/10.3389/fgene.2020.00629

Turner SP, Camerlink I, Baxter EM, D'Eath RB, Desire S, Roehe R (2018) Breeding for pig welfare: opportunities and challenges. In: Advances in Pig Welfare. https://doi.org/10.1016/B978-0-08-101012-9.00012-5

USDA., US monthly Federally inspected average dressed weight hogs. 2016., Available at: https://www.nationalhogfarmer.com/

Wischner D, Kemper N, Krieter J (2009) Nest-building behaviour in sows and consequences for pig husbandry. Livest Sci 124(1–3):1–8. https://doi.org/10.1016/j.livsci.2009.01.015

Wurzinger M, Gutiérrez GA, Sölkner J, Probst L (2021) Community-based livestock breeding: coordinated action or relational process? Front Veterin Sci

8:443. https://doi.org/10.3389/FVETS.2021.613505/BIBTEX

Yadav S, Bharti PK, Gaur GK, Devi B, Abhishek, Sahoo NR, Somagond CA, Antil M (2020) Standardisation and categorization of indigenous microorganisms (IMOs) for inoculated deep litter piggery in India. Indian J Animal Sci 90(4):530–534

Zaake P, Mugumya R, Rubayiza I, Wairagala P, Paul BK (2020) Reducing climate-induced heat stress in pigs in Uganda: policy actions. Policy engagement workshop report, 10 December. Alliance of Bioversity International and CIAT, Nairobi, Kenya

Zak LJ, Gaustad AH, Bolarin A, Broekhuijse MLWJ, Walling GA, Knol EF (2017) Genetic control of complex traits, with a focus on reproduction in pigs. Mol Reprod Dev 84:1004–1011

Zylbersztajn D, Turner JC, Jones JVH (2000) Agroceres-PIC. J Bus Res 50(1):71–81. https://doi.org/10.1016/S0148-2963(98)00113-1

Open Access This chapter is licensed under the terms of the Creative Commons Attribution 4.0 International License (http://creativecommons.org/licenses/by/4.0/), which permits use, sharing, adaptation, distribution and reproduction in any medium or format, as long as you give appropriate credit to the original author(s) and the source, provide a link to the Creative Commons license and indicate if changes were made.

The images or other third party material in this chapter are included in the chapter's Creative Commons license, unless indicated otherwise in a credit line to the material. If material is not included in the chapter's Creative Commons license and your intended use is not permitted by statutory regulation or exceeds the permitted use, you will need to obtain permission directly from the copyright holder.

If you have any concerns about our products,
you can contact us on
ProductSafety@springernature.com

In case Publisher is established outside the EU,
the EU authorized representative is:
**Springer Nature Customer Service Center GmbH
Europaplatz 3, 69115 Heidelberg, Germany**

Printed by Libri Plureos GmbH
in Hamburg, Germany

Sustainable Development Goals Series

The **Sustainable Development Goals Series** is Springer Nature's inaugural cross-imprint book series that addresses and supports the United Nations' seventeen Sustainable Development Goals. The series fosters comprehensive research focused on these global targets and endeavours to address some of society's greatest grand challenges. The SDGs are inherently multidisciplinary, and they bring people working across different fields together and working towards a common goal. In this spirit, the Sustainable Development Goals series is the first at Springer Nature to publish books under both the Springer and Palgrave Macmillan imprints, bringing the strengths of our imprints together.

The Sustainable Development Goals Series is organized into eighteen subseries: one subseries based around each of the seventeen respective Sustainable Development Goals, and an eighteenth subseries, "Connecting the Goals", which serves as a home for volumes addressing multiple goals or studying the SDGs as a whole. Each subseries is guided by an expert Subseries Advisor with years or decades of experience studying and addressing core components of their respective Goal.

The SDG Series has a remit as broad as the SDGs themselves, and contributions are welcome from scientists, academics, policymakers, and researchers working in fields related to any of the seventeen goals. If you are interested in contributing a monograph or curated volume to the series, please contact the Publishers: Zachary Romano [Springer; zachary.romano@springer.com] and Rachael Ballard [Palgrave Macmillan; rachael.ballard@palgrave.com].

Eveline M. Ibeagha-Awemu
Sunday O. Peters • Appolinaire Djikeng
John E. O. Rege
Editors

African Livestock Genetic Resources and Sustainable Breeding Strategies

Unlocking a Treasure Trove and Guide for Improved Productivity

Volume 2

Editors
Eveline M. Ibeagha-Awemu
Sherbrooke Research and Development Centre
Agriculture and Agri-Food Canada
Sherbrooke, QC, Canada

Appolinaire Djikeng
International Livestock Research Institute (ILRI)
Nairobi, Kenya

Sunday O. Peters
Department of Animal Science
Berry College
Mount Berry, GA, USA

John E. O. Rege
Emerge Centre for Innovations–Africa
Nairobi, Kenya

ISSN 2523-3084   ISSN 2523-3092   (electronic)
Sustainable Development Goals Series
ISBN 978-3-031-92075-2   ISBN 978-3-031-92076-9   (eBook)
https://doi.org/10.1007/978-3-031-92076-9

International Livestock Research Institute
This work was supported by the International Livestock Research Institute (ILRI) and its funders.

This book is an open access publication.

Color wheel and icons: From https://www.un.org/sustainabledevelopment/
Copyright © 2020 United Nations. Used with the permission of the United Nations.

The content of this publication has not been approved by the United Nations and does not reflect the views of the United Nations or its officials or Member States.

© Crown 2026

Open Access This book is licensed under the terms of the Creative Commons Attribution 4.0 International License (http://creativecommons.org/licenses/by/4.0/), which permits use, sharing, adaptation, distribution and reproduction in any medium or format, as long as you give appropriate credit to the original author(s) and the source, provide a link to the Creative Commons license and indicate if changes were made.
The images or other third party material in this book are included in the book's Creative Commons license, unless indicated otherwise in a credit line to the material. If material is not included in the book's Creative Commons license and your intended use is not permitted by statutory regulation or exceeds the permitted use, you will need to obtain permission directly from the copyright holder.
The use of general descriptive names, registered names, trademarks, service marks, etc. in this publication does not imply, even in the absence of a specific statement, that such names are exempt from the relevant protective laws and regulations and therefore free for general use.
The publisher, the authors and the editors are safe to assume that the advice and information in this book are believed to be true and accurate at the date of publication. Neither the publisher nor the authors or the editors give a warranty, expressed or implied, with respect to the material contained herein or for any errors or omissions that may have been made. The publisher remains neutral with regard to jurisdictional claims in published maps and institutional affiliations.

Editorial Contact: Annette Klaus

This Springer imprint is published by the registered company Springer Nature Switzerland AG
The registered company address is: Gewerbestrasse 11, 6330 Cham, Switzerland

If disposing of this product, please recycle the paper.

Preface

For Africa, the journey to achieving the aspirations of the African Union Agenda 2063 (Goal 1 on quality of life and well-being, and Goal 2 on health and nutrition) and the corresponding ambitions of the United Nation's Sustainable Development Goals (SDGs: SDG1 on Poverty alleviation, SDG2 on ending Hunger, etc.) must consider livestock production and productivity. Understanding the key types, availability, and use options of the livestock genetic resources in Africa is a first step in taking measures toward sustainable increased productivity and resilience in the face of a myriad of current and future challenges. In this connection, members of the executive committee of the African Animal Breeding Network (AABNet, https://www.animalbreeding-africa.org/index.html) started discussions in mid-2019 to explore the landscape and review key drivers and opportunities for establishing breeding goals for adapted and resilient African livestock to meet increasing demand for animal source food, the challenge of climate change, and environmental concerns. As discussions progressed, it became clear that the substantial information gaps on African livestock genetic resources could not be adequately covered, in the depth and breadth we had envisaged, in a single review paper, and the idea to write this book on "African livestock genetic resources and sustainable breeding strategies" was born, and the writing started in earnest in late 2020 involving 83 contributing authors—professionals and experts from animal science, animal breeding, genetics, genomics, and other science disciplines—from 24 countries.

Authored by African livestock professionals representing a wide range of disciplines and focused on an examination of efforts toward understanding indigenous African livestock, their uses and systems of production, improvement challenges, and achievements made to date, this book presents a rich resource, across livestock species, for formulating technical, institutional, and policy interventions going forward and that leverage on rapidly changing technologies and responds to evolving human nutritional needs and other trends.

For thousands of years, natural selection and human activities have generated genetically diverse breeds of domesticated farm animals, which can significantly contribute to the livelihoods of millions of present-day Africans. Indeed, they are a treasure trove. Africa's indigenous livestock are particularly hardy and well-adapted to local production contexts, having developed adaptations to the continent's diverse climatic conditions and environmental pressures. Despite the wealth of desirable genetic traits, some of Africa's iconic and lesser-known livestock are disappearing at an alarming rate. Despite the increasing recognition of the benefits of this diversity, little has been done to understand and optimally harness the full potential of these genetic resources. This book catalogues the main farm animal genetic resources (cattle, sheep, goat, pig, chicken, dromedary, horse, donkey, buffalo, turkey, duck, and geese) and some non-conventional animal genetic resources (frog, rabbit, snail, and honeybee) in Africa and the opportunities that can be leveraged with available technologies and knowledge for achieving rapid genetic gain and improved productivity. While the structure of the chapters differ considerably, they cover five broad areas in the context of animal genetic resources in Africa: their characterization (to understand the unique genetic composition of what we have, where it might have originated from, and its present-day distribution); conservation; use (efforts and approaches to enhance sustainable utilization, focusing on improving production and productivity, and product quality to meet evolving human needs); technological support (available technological developments that support sustainable livestock improvement); and the institutions and institutional arrangements needed to support their characterization, conservation, and sustainable use.

This book is the first to present as one source, the diversity and uniqueness of African farm animal genetic resources, and the possibilities that new technologies present, which will give a better appreciation of the existing resources to a wide range of users. It is a valuable manual for students, professors, researchers, animal science professionals, livestock farmers and farmer organizations, civil society organizations, non-governmental organizations, business professionals, government agencies, including policymakers, the international development community, and anyone who seeks to understand the uniqueness of African livestock genetic resources, production systems, and strategies for sustainable improvement for the African environment.

Sherbrooke, QC, Canada	Eveline M. Ibeagha-Awemu
Mount Berry, GA, USA	Sunday O. Peters
Nairobi, Kenya	Appolinaire Djikeng
Nairobi, Kenya	John E. O. Rege

Acknowledgements

For this book to have its desired impact, it must be freely available to all interested readers. Open access availability of this book was made possible through financial support from the International Livestock Research Institute (ILRI, https://www.ilri.org/) and its funders.

We acknowledge with profound gratitude all the institutions of contributing authors for supporting them during the production of this book.

Eveline M. Ibeagha-Awemu
Sunday O. Peters
Appolinaire Djikeng
John E. O. Rege

Contents of Volume 2

Part III Leveraging Modern Technologies for Improved Utilization of African Livestock Genetic Resources

20 Role of Modern Technologies for Sustainable Genetic Improvement of African Livestock 851
Eveline M. Ibeagha-Awemu, Sunday O. Peters, Martha N. Bemji, Jean M. Feugang, Richard Osei-Amponsah, Daniel Trocmé, and Raphael Mrode

21 The Role of Modern Technologies for Improving the Production Environment of Livestock in Africa 909
Eveline M. Ibeagha-Awemu, Faith A. Omonijo, Martha N. Bemji, Obioha Duranna, Iliya D. Kwoji, Michael O. Ozoje, and Richard Osei-Amponsah

22 Prospects for Utilization of Modern Technologies for Cattle Improvement in Africa 991
Sunday O. Peters, Eveline M. Ibeagha-Awemu, Iliya D. Kwoji, Raphael Mrode, Michael O. Ozoje, David A. Mbah, Isidore Houaga, Moses Okpeku, Peter O. Fayemi, Daniel Trocmé, Oluyinka Opoola, and Matthew A. Adeleke

23 Prospect of Modern Technologies for Poultry Improvement in Africa 1021
Victor E. Olori, Sunday O. Peters, and Matthew A. Adeleke

24 Prospect for Utilization of Modern Technologies for Small Ruminant and Pig Improvement in African Input Systems: Case Studies 1049
Jean M. Feugang, Othman E. -M. Othman, Wilson Nandolo, Donald R. Kugonza, and Richard Osei-Amponsah

Part IV Conservation of African Livestock Genetic Resources

25 Conservation and Management of Animal Genetic Resources in the Context of African Livestock Production Systems: The Case for In Situ and Ex Situ Conservation 1071
Jean M. Feugang, Richard Osei-Amponsah, John E. O. Rege, Khaled Fantazi, Christian K. Tiambo, Felicien Shumbusho, Isidore Houaga, Derradji Harek, Notsile H. Dlamini, and Semir B. S. Gaouar

26 Economic Considerations and Framework of Conservation of African Animal Genetic Resources 1091
Wilson Kaumbata, Maria Wurzinger, and John E. O. Rege

Part V Institutional Arrangements and Enabling Environment for Enhanced Utilization of African Livestock Genetic Resources

27 Capacity Strengthening of Animal Genetic Improvement Education in Africa 1109
Samuel E. Aggrey, Richard Osei-Amponsah, Donald R. Kugonza, Raphael A. Mrode, Romdhane Rekaya, and John E. O. Rege

28 Capacity Building in Livestock Breeding and Genetic Improvement in Achieving UN Sustainable Development Goals for Africa 1123
Christian K. Tiambo, Oluyinka Opoola, Moses Okpeku, Isidore Houaga, Blaise A. Hako Touko, and Tadelle Dessie

29 Harnessing Multi-Country Cooperations, Initiatives, Facilities, and Technologies for Advancing Livestock Genetic Improvement in Africa 1149
Ntanganedzeni O. Mapholi, Cuthbert Banga, Raphael Mrode, Oluyinka Opoola, Isidore Houaga, Lucy T. Nesengani, and Eveline M. Ibeagha-Awemu

30 Policies, Frameworks, Strategies, and Action Plans for Conservation and Sustainable Use of African Animal Genetic Resources 1185
John E. O. Rege and Mure Agbonlahor

31 Resourcing and Institutional Arrangements to Deliver Sustainable Animal Genetic Improvement in Africa 1233
Eveline M. Ibeagha-Awemu, Victor Olori, Ismail Muritala, Olubunmi I. Duduyemi, Mizeck G. G. Chagunda, and John E. O. Rege

Index 1279

Contents of Volume 1

Part I African Livestock Production Systems and Genetic Resources

1. **An Overview of African Livestock Genetic Resources and Improvement Strategies** 3
   Eveline M. Ibeagha-Awemu, Sunday O. Peters, Appolinaire Djikeng, and John E. O. Rege

2. **African Livestock Production Systems: The Past, Present and the Projected Future** 13
   Eveline M. Ibeagha-Awemu, Richard Osei-Amponsah, and Martha N. Bemji

3. **The History, Geography, and Characteristics of Indigenous African Taurine Cattle** 65
   John E. O. Rege, Chi L. Tawah, Isidore Houaga, and Eveline M. Ibeagha-Awemu

4. **The History, Geography, and Characteristics of African Zebu, Zebu–Taurine Derivatives, and Well-Established Exotic Cattle Breeds** 117
   John E. O. Rege, Chi L. Tawah, Donald R. Kugonza, Mizeck G. G. Chagunda, Isidore Houaga, Oluyinka Opoola, and Eveline M. Ibeagha-Awemu

5. **African Goat Genetic Resources, Diversity and Unique Features** 185
   Moses Okpeku, Martha N. Bemji, Isidore Houaga, Khaled Fantazi, Liveness J. Banda, Timothy Gondwe, Sebastine Chenyambuga, Sahar A. Elnahta, Doctor M. N. Mthiyane, Shumuye Belay, Tadelle Dessie, Taiye S. Adewumi, and Oliver Hanotte

6. **African Sheep Genetic Resources, Diversity and Unique Features** 239
   Martha N. Bemji, Semir B. S. Gaouar, Abdelkader Ameur Ameur, Fatima Z. Belharfi, Isidore Houaga, and Anne W. T. Muigai

7   **African Domestic Poultry Genetic Resources, Diversity, and Unique Features**.................................... 327
    Tadelle Dessie, Christian K. Tiambo, Liveness J. Banda,
    Raman A. Lawal, Sheila C. Ommeh, Timothy Gondwe,
    Esatu Wondmeneh, Matthew A. Adeleke, and Olivier Hanotte

8   **Pig Genetic Resources of Africa**........................... 357
    Richard Osei-Amponsah and Wilson S. Kaumbata

9   **African Dromedary Genetic Resources, Diversity and Breeding Systems**.................................. 395
    Semir B. S. Gaouar, Imane Meghelli, Zoubeyda Kaouadji,
    Félix Meutchieye, and Djalel E. Gherissi

10 **African Donkey Genetic Resources, Diversity and Breeding Strategies**................................. 451
    Martha N. Bemji, Eveline M. Ibeagha-Awemu,
    Abdelkader Ameur Ameur, Albano Beja-Pereira,
    Madani Labbaci, Liveness J. Banda, and Semir B. S. Gaouar

11 **African Horse Genetic Resources and Breeding Strategies for African Input Systems** 497
    Djalel E. Gherissi, Ginette Aumassip-Kadri,
    Mohammed E. A. Benhamadi, Félix Meutchieye,
    Yassine H. Jamali, and Semir B. S. Gaouar

12 **African Water Buffalo Genetic Resources, Diversity, and Unique Features**.................................. 593
    Sahar S. E. Ahmed, Amal A. M. Hassan,
    Ibrahim A. H. Barakat, and Hassan M. Al Ashmaoui

13 **Nonconventional Animal Genetic Resources, Diversity, and Unique Features**.................................. 619
    Kingsley A. Etchu, Abdelkader Ameur Ameur, M. Chahbar,
    Félix Meutchieye, Annick N. Enangue Njembele,
    Gerald C. Tasse Taboue, and Semir B. S. Gaouar

14 **Contributions of African Livestock Production Systems to Greenhouse Gas Emissions and Global Warming in the Face of Climate Change**......................... 675
    Mizeck G. G. Chagunda, Kingsley A. Etchu,
    Kwamboka Tirimba, and Okeyo Mwai

Part II   Opportunities for Improved Utilization of African Livestock Genetic Resources

15 **Defining Breeding Goals and Breeding Strategies for Improving Livestock Under Various Production Systems in Africa: Concept and Brief Overview** 691
    Raphael Mrode, David A. Mbah, and Julie M. K. Ojango

16 **Defining Breeding Goals and Breeding Strategies for Improving the Productivity of Cattle Breeds and Buffaloes in African Production Systems** 705
Raphael Mrode, David A. Mbah, Chi L. Tawah,
Julie M. K. Ojango, Sunday O. Peters,
Eveline M. Ibeagha-Awemu, Oluyinka Opoola,
Isidore Houaga, Richard Osei-Amponsah, Moses Okpeku,
and John E. O. Rege

17 **Breeding Goals and Strategies for Improving Small Ruminant Productivity in African Input Systems** 759
Julie M. K. Ojango, Richard Osei-Amponsah, Isidore Houaga,
Timothy Gondwe, Moses Okpeku, and Donald R. Kugonza

18 **Defining Breeding Goals and Breeding Strategies for Chicken Production Systems in Africa** 785
Sunday O. Peters, Michael O. Ozoje, Adeyemi S. Adenaike,
Blaise A. Hako Touko, Christian K. Tiambo,
Matthew A. Adeleke, Oluyinka Opoola, and Tadelle Dessie

19 **Defining Breeding Goals and Breeding Strategies for Pigs in Various African Production Systems** 819
Donald R. Kugonza and Richard Osei-Amponsah

List of Figures

Fig. 1.1	Africa's total, rural, and urban population statistics for the years from 2010 to 2021.	6
Fig. 2.1	(**a**) Cattle, sheep and goat population trends in Africa and regions from 2005 to 2021. (**b**) Pig and chicken population trends in Africa and regions from 2005 to 2021	17
Fig. 2.2	(**a**) The main agro-ecological zones in Africa (Source: Sebastien (*2009*)). (**b**) Livestock production systems by agro-ecological zones in Africa. (Source: Robinson et al. (*2011*). Available freely through ILRI on Flickr https://www.flickr.com/photos/ilri/14509985147/)	20
Fig. 2.3	Progression of African main livestock production systems according to land availability and level of inputs	24
Fig. 2.4	Pastoralism in West Africa, specifically in the Bamenda Grasslands of Cameroon (**a**) and the Sahel region of Burkina Faso (**b**). (Photos by Dr. Eveline M. Ibeagha-Awemu).	28
Fig. 2.5	Positive synergies between crops, livestock and households in mixed crop–livestock systems	31
Fig. 2.6	Small intensively managed backyard pig farm (**a**) and poultry farm (**b**) in Bambili, Cameroon. (Photos by Azenui B. Abongban).	33
Fig. 2.7	A peri-urban medium-scale, commercial, intensively managed poultry farm in Nkwen, Cameroon. (Photo by Azenui B. Abongban)	35
Fig. 2.8	Large-scale, intensively managed poultry operations: (**a**) broiler, (**b**) layer, (**c**) meat processing and (**d**) chick units at commercial farms in Nigeria.	38
Fig. 2.9	Transformational drivers of current and future African high-producing and resilient livestock production systems.	53

Fig. 3.1	Origin and migration patterns of domestic cattle in Africa. (Source: Modified from Mwai et al. (*2015*), licensed under CC-BY 4.0)	69
Fig. 3.2	Distribution of cattle breed types in Africa. *Dots indicate the type of cattle in a region (cattle of North Africa and exotic cattle breeds are not shown). (Source: Mwai et al. (*2015*), licensed under CC-BY 4.0)	70
Fig. 3.3	A typical N'Dama bull and herd at Nsukka, Nigeria. (Photos by Dr E.M. Ibeagha-Awemu)	73
Fig. 3.4	A Kuri cattle herd in its Lake Chad environment and typical Kuri animals. (Courtesy: ILRI)	77
Fig. 3.5	Doayo (Namchi) cattle at Wakwa, Cameroon. (Photos by Dr E.M. Ibeagha-Awemu)	85
Fig. 3.6	A Muturu herd (**a**), bull (**b**), and cow with calf (**c**), showing typical black and white colour patterns. (Courtesy: Prof. O. A. Osinowo)	86
Fig. 3.7	A herd of Ghana Shorthorn cattle. (Source: ILRI)	90
Fig. 3.8	A Lagune bull. (Source: Ahozonlin et al. (*2022*), licensed under CC-BY 4.0)	90
Fig. 3.9	Typical Somba (**a**) and Lagune (**b**) cows. (Source: Vanvanhossou et al. (*2021a*), licensed under CC-BY 4.0)	91
Fig. 3.10	A Baoulé cow. (Courtesy: Albert Soudre', Open Source: https://www.eurekalert.org/multimedia/722329)	92
Fig. 3.11	Kapsiki (**a**), Namchi (**b**), Bamenda (**c**), and Bakosi (**d**) cows. (Source: Ojong et al. (*2021*), licensed under CC-BY 4.0)	94
Fig. 3.12	A Sheko cattle herd. (Source: ILRI)	109
Fig. 4.1	A Butana cow with a calf and a Kenana cow. (Source: https://twitter.com/ensembl/status/951859117488005120 - Open Access)	123
Fig. 4.2	A herd of Baggara cattle. (Courtesy: AGTR, ILRI)	123
Fig. 4.3	Typical Karamojong cows with suckling calves. (Courtesy: Dr. Donald Kugonza)	125
Fig. 4.4	A Boran bull (**a**) and herd (**b**) at the Sosian ranch in Kenya and typical Ethiopian Boran bulls in Yabello, Borana, southern Ethiopia. (Courtesy: ILRI)	127
Fig. 4.5	A herd of Small East African Zebu in Tanzania. (Courtesy: Dr. Gladness Mwanga)	130
Fig. 4.6	Angoni bull (**a**) and a cow with a calf (**b**). (Courtesy: Dr. Maria da Gloria Taela)	131
Fig. 4.7	(**a**) A herd of Madagascar Zebu. (Courtesy: Dr. Maria da Gloria Taela). (**b**) A herd of Malawi Zebu. (Courtesy: Dr. Patricia Mayuni)	133

Fig. 4.8	(**a**) Typical Banyo Gudali cow and calf and (**b**) mature bull at Sabga, Cameroon, portraying the typical physical characteristics of the breed. (Courtesy: Dr. Eveline M. Ibeagha-Awemu). (**c**) Sokoto Gudali bull (Nigeria). (Courtesy: Dr. Eveline M. Ibeagha-Awemu). (**d**) Typical Ngaoundere Gudali cow and calf and (**e**) bull (Cameroon). (Courtesy: Dr. Eveline M. Ibeagha-Awemu).	135
Fig. 4.9	Wadara or Shuwa cattle: a bull and a herd	139
Fig. 4.10	Azawak cattle in Niger. (Courtesy: Larry W. Harms, Virginia)	140
Fig. 4.11	Maure cattle. (Courtesy: Dr. Larry W. Harms, Virginia)	141
Fig. 4.12	Gobra cattle. (Courtesy: Dr. Larry W. Harms Virginia)	142
Fig. 4.13	White Fulani bull (**a**) and cow and calf (**b**). (Courtesy: Dr. E. M. Ibeagha-Awemu)	144
Fig. 4.14	Red Fulani herd (**a**) and bull (**b**). (Courtesy: Dr. E. M. Ibeagha-Awemu)	146
Fig. 4.15	Sudanese Fulani cattle. (Courtesy: ILRI)	147
Fig. 4.16	Djelli cattle. (Courtesy Dr. Larry W. Harms, Virginia)	148
Fig. 4.17	A Danakil cow and a Raya Azebo herd. (Courtesy: AGTR, ILRI)	151
Fig. 4.18	Ankole bull and herd showing their characteristic lyre horns. (Courtesy: Dr. Donald Kugonza)	153
Fig. 4.19	An Abigar bull and a herd of Abigar cattle. (Courtesy: AGTR, ILRI)	154
Fig. 4.20	A Landim cattle herd and a heifer in Mozambique. (Courtesy: Maria da Gloria Taela)	155
Fig. 4.21	A typical Nguni bull (**a**) and a cow with a calf (**b**). (AGTR, ILRI)	155
Fig. 4.22	A herd of Tonga cattle. (Courtesy: DAGRIS, ILRI)	157
Fig. 4.23	A herd of Barotse cattle. (Courtesy: Dr. Denis Lembani)	157
Fig. 4.24	A herd of Tswana cattle. (Courtesy: ILRI)	159
Fig. 4.25	A Tuli herd and a bull. (Courtesy: Tuli Cattle Society of Southern Africa)	160
Fig. 4.26	Afrikaner bull and herd. (Courtesy: Afrikaner Cattle Breeders' Society of South Africa)	161
Fig. 4.27	Typical Keteku (**a**) and Djakore (**b**) cattle herds. (Courtesy: AGTR, ILRI)	163
Fig. 4.28	Fogera cow (**a**), bull (**b**), and a cow with a calf (**c**). (Source: Tesfa et al. (*2022*)) (attributed to ILRI)	167
Fig. 4.29	The Ethiopian Horro cattle. (Courtesy: DAGRIS, ILRI)	168
Fig. 4.30	A typical Nganda cow. (Courtesy: Dr. Donald Kugonza)	168
Fig. 4.31	Herds of Tete cattle. (Courtesy: AGTR, ILRI, and Dr. Maria da Gloria Taela)	168
Fig. 4.32	A Bonsmara bull (**a**) and a cow with a calf (**b**). (Courtesy: Bonsmara Cattle Breeders' Society, South Africa)	169

Fig. 4.33	An Mpwapwa bull (**a**) and a cow (**b**). (Courtesy: Aluna Chawala)	169
Fig. 4.34	A herd of Renitelo cattle of Madagascar. (Courtesy: Dr. Maria da Gloria Taela)	171
Fig. 4.35	A herd of Wakwa cattle. (Courtesy: Mr. Lawrence Shang of the TADU Dairy Cooperative, Bui Division, North-West Region of Cameroon)	172
Fig. 4.36	A herd of Drakensberger cows (**a**) and a bull (**b**). (Courtesy: Drakensberger Cattle Breeders' Society)	172
Fig. 5.1	Nubian goat	190
Fig. 5.2	(**a**) Begait does with its kids. (Photo by Shumuye Belay). (**b**) Begait buck. (Photo by Shumuye Belay)	194
Fig. 5.3	(**a**) Kalahari Red bucks at Federal University of Agriculture, Abeokuta (FUNAAB). (Source: Bemji et al. *2014*). (**b**) Kalahari Red doe and kids at FUNAAB. (Photo by O. A. Osinowo)	195
Fig. 5.4	A herd of Abergelle goats. (Photo by Shumuye Belay)	200
Fig. 5.5	A herd of Afar goats. (Anne W.T Muigai)	204
Fig. 5.6	Sudanese hill goat. (Photo: Y. A. Hassan)	205
Fig. 5.7	A Western lowland or Shankela doe. (Photo: ILRI)	207
Fig. 5.8	(**a**) Keffa Central highland goat. (Photo: ILRI). (**b**) Keffa Central highland goat. (Photo: ILRI)	208
Fig. 5.9	Cameroon Grassland Dwarf. (Photo by Jules Fon)	209
Fig. 5.10	West African Dwarf doe at Southwest Nigeria. (Photo by M.N. Bemji)	209
Fig. 5.11	(**a**) A Mubende doe. (Photo by D. Kugonza). (**b**) A Mubende buck. (Photo by D. Kugonza)	210
Fig. 5.12	Small East African goat	211
Fig. 5.13	A herd of Red Sokoto does at the National Animal Production Research Institute (NAPRI), Zaria, Nigeria. (Photo: M.N. Bemji)	212
Fig. 5.14	West African Long-legged at the National Animal Production Research Institute (NAPRI), Zaria, Nigeria. (Photo by J.C. Amabo)	213
Fig. 5.15	Sudanese Desert goat. (Photo: Y.A. Hassan)	214
Fig. 5.16	Angora doe with its kid (Visser and Van Marle-Köster *2018*)	215
Fig. 5.17	(**a**) Damascus buck. (ESGPIP *2008*). (**b**) Damascus doe (ESGPIP *2008, 1971*)	216
Fig. 5.18	Boer goat	222
Fig. 5.19	Original *Bouhezza* cheese from Algeria	229
Fig. 6.1	West African Dwarf Ram at Federal University of Agriculture, Abeokuta, Nigeria. (Photo: Ismaila Muritala)	253
Fig. 6.2	Cameroon Blackbelly Ewe. (Photo: Julius Félix Meutchieye)	254

Fig. 6.3	Cameroon Blackbelly Ram. (Photo: Julius Félix Meutchieye)	255
Fig. 6.4	Balami Ewes at National Animal Production Research Institute, Ahmadu Bello University, Zaria, Nigeria. (Photo: Julius C. Amabo)	255
Fig. 6.5	Uda Ewes at National Animal Production Research Institute, Ahmadu Bello University, Zaria, Nigeria. (Photo: Julius C. Amabo)	256
Fig. 6.6	Yankasa Ewes at National Animal Production Research Institute, Ahmadu Bello University, Zaria, Nigeria. (Photo: Julius C. Amabo)	257
Fig. 6.7	D'man. (From Djaout et al *2017*, licensed under CC-BY 4.0)	258
Fig. 6.8	Bélier Ouled Djellal ram at Biskra. (From Djaout et al. *2017*; licensed under CC-BY 4.0)	262
Fig. 6.9	Peul-Peul sheep in the sylvo-pastoral zone of Senegal. (From Ndiaye et al. *2018*, licensed under CC-BY 4.0)	263
Fig. 6.10	Sudan Desert sheep reared in the semi-desert belt of Sudan. (From Ali et al. *2016*, licensed under CC-BY 4.0)	264
Fig. 6.11	Berbere sheep of the Bouhadjar mountains (El-Tarf), Algeria. (From Djaout et al. *2017*, licensed under CC-BY 4.0)	267
Fig. 6.12	Hamra Ewes in Mecheria City, Nâama Province, Algeria. (From (Djaout et al. *2017*, licensed under CC-BY 4.0)	268
Fig. 6.13	Beni Guil ram of Morocco. (From Bechchari *2009*, published in an open-source report published by the Ministry of Agriculture, Kingdom of Morocco)	270
Fig. 6.14	A Saidi ram. (From Elshazly and Youngs *2019*, licensed under CC-BY 4.0)	272
Fig. 6.15	A Saidi ewe. (From Elshazly and Youngs *2019*, licensed under CC-BY 4.0)	272
Fig. 6.16	Farafra ram (left) and ewe (right) found in the El-Farafra Oasis, Western Egypt. (From Elshazly and Youngs *2019*, licensed under CC-BY 4.0)	273
Fig. 6.17	Afar ram in the Afar State of Ethiopia. (From Getachew *2008*, through ILRI repository; CGSPACE, an open-source repository for Agricultural Research Output)	273
Fig. 6.18	Afar ewe in the Afar State of Ethiopia. (From Getachew *2008*, through ILRI repository; CGSPACE, an open-source repository for Agricultural Research Output)	274
Fig. 6.19	Washera ram (left) and ewe (right) in the Amhara state in Ethiopia. (From Ferede et al. *2014*, licensed under CC-BY 4.0)	275

Fig. 6.20	Horro ewe; Photo: ILRI/Apollo Habtamu. (Available on Flickr for free use under the Creative Commons Attribution Licence. (Image can be viewed here)).	275
Fig. 6.21	Adilo ewe. (Photo: ILRI\Zerihun Sewunet) (Available on Flickr for free use under the Creative Commons Licence (Image can be viewed here))	276
Fig. 6.22	A Bonga ram. (Photo: ILRI/Apollo Habtamu) (available on Flickr for free use under the Creative Commons Licence (Image can be viewed here)).	277
Fig. 6.23	Red Maasai sheep, Photo: ILRI. (Available on Flickr for free use under the Creative Commons Licence (Image can be viewed here))	278
Fig. 6.24	Tanzanian long fat-tailed sheep, Photo: ILRI/Cleaned VC. (Available on Flickr for free use under the Creative Commons Licence. (Image can be viewed here))	279
Fig. 6.25	Indigenous Menz sheep herd in Amhara Region, Ethiopia; Photo: ICARDA\Getachew Mengistu. (Available on Flickr for free use under the Creative Commons Licence. (Image can be viewed here))	286
Fig. 6.26	Menz ram in Menz Area, Ethiopia. (From Ferede et al. *2014*, licensed under CC-BY 4.0)	287
Fig. 6.27	Menz ewe in Menz Area, Ethiopia. (From Ferede et al. *2014*, licensed under CC-BY 4.0)	287
Fig. 6.28	Farta ram in Farta District, Ethiopia. (From Ferede et al. *2014*, licensed under CC-BY 4.0)	289
Fig. 6.29	Farta ewe in Farta District, Ethiopia. (From Ferede et al. *2014*, licensed under CC-BY 4.0)	289
Fig. 6.30	Barki ram. (From Elshazly and Youngs *2019*, licensed under CC-BY 4.0)	291
Fig. 6.31	Barki ewe. (From Elshazly and Youngs *2019*, licensed under CC-BY 4.0)	291
Fig. 6.32	Blackhead Somali sheep in a livestock market in Somaliland, Photo: ILRI. (Available on Flickr for free use under the Creative Commons Licence (Image can be viewed here))	292
Fig. 6.33	Bergui ewes. (Queue Fine de l'Ouest) (Tozeur). (From Khaldi et al. *2011*, licensed under CC-BY 4.0)	296
Fig. 6.34	Rahmani ram. (From Elshazly and Youngs *2019*, licensed under CC-BY 4.0)	296
Fig. 6.35	Rahmani ewe. (From Elshazly and Youngs *2019*, licensed under CC-BY 4.0)	296
Fig. 6.36	Taâdmit ram in Djelfa, Taâdmit. (From Djaout et al. *2017*, licensed under CC-BY 4.0)	300
Fig. 6.37	Noire de Thibar in Zaghouène in Tunisia. (From Djaout et al. *2017*, licensed under CC-BY 4.0)	301
Fig. 6.38	Sicilo Sarde. (From Aloulou et al. *2018*, licensed under CC-BY 4.0)	301

List of Figures

Fig. 6.39	Sheep milk production in Africa (in millions of tonnes), 2009–2019 (FAOSTATS *2021*)	311
Fig. 6.40	Sheep meat production in Africa (in millions of tonnes), 2009–2019 (FAOSTATS *2021*)	312
Fig. 6.41	Sheep greasy wool production in Africa (in millions of tonnes), 2009–2019 (FAOSTATS *2021*)	313
Fig. 7.1	Fayoumi chicken. (Courtesy Dr. Wondmeneh Esatu)	335
Fig. 7.2	Horro chicken. (Courtesy Dr. Wondmeneh Esatu)	338
Fig. 7.3	Konso chicken. (Source: Dana et al. *2010*), Food and Agricultural Organization, reproduced with permission)	339
Fig. 7.4	Tepi chicken. (Courtesy: Dr. Wondmeneh Esatu)	340
Fig. 7.5	Tilili chicken. (Courtesy: Dr. Wondmeneh Esatu)	340
Fig. 7.6	Indigenous Rwandan chickens. (Source: Hirwa et al. 2019, licensed under CC BY 4.0)	341
Fig. 7.7	Ugandan chicken. (Source: Yussif et al. 2023) (Licensed under C-C By 4:0)	341
Fig. 7.8	Ovambo. (Source: Grobbelaar et al. *2010*, Food and Agricultural Organization, reproduced with permission)	341
Fig. 7.9	Potchefstroom Koekoek chicken. (Courtesy: Dr. Wondmeneh Esatu)	342
Fig. 7.10	Venda chicken. (Source: Grobbelaar et al. *2010*, Food and Agricultural Organization, reproduced with permission)	342
Fig. 7.11	Boschveld chicken. (Source: http://boschveld.co.za/)	343
Fig. 8.1	Major centers of livestock domestication based on archaeological and molecular information. From FAO 2007a, licensed under Copyright © FAO 2007	360
Fig. 8.2	Suiforme diversity and phylogenetic relationship. (From Chen et al. *2007*, licensed under CC-BY 4.0). (Source: https://untamedscience.com/family/suidae/)	361
Fig. 8.3	Ashanti Dwarf pig of Ghana	367
Fig. 8.4	Burkina Faso pig. (From Kiendrebeogo et al. *2012*, licensed under CC-BY 4.0)	368
Fig. 8.5	Korhogo pig of Burkina Faso and Cote d'Ivoire. (From Kiendrebeogo et al. *2012*), licensed under CC-BY 4.0)	369
Fig. 8.6	Pigs in local production systems in Liberia. (**a**) Libera native pig; (**b**) crossbred; (**c**) Exotic breed	369
Fig. 8.7	Indigenous pig breeds of Cameroon. (**a**) Mankon long nose; (**b**) Bakosi; (**c**) Bamiléké. (From Motsa'a et al. *2021*, licensed under CC-BY 4.0)	370
Fig. 8.8	Indigenous pig of Uganda. (Ugandan pig). (From Marshall *2022*, licensed under CC-BY 4.0)	371

Fig. 8.9	The Bantu pig of South Africa. (Source: Lilongwe University of Agriculture and Natural Resources 2011, licensed under CC-BY 4.0)	372
Fig. 8.10	The Mukota pig breed of Zimbabwe. (From https://www.petmapz.com/breed/mukota-pig/#, licensed under CC-BY 4.0)	373
Fig. 8.11	The Kolbroek pig breed of South Africa. (From http://farmersweekly.co.za, licensed under CC-BY 4.0)	373
Fig. 8.12	The Duroc pig breed. (Credit: Prof. R. Osei-Amponsah)	375
Fig. 8.13	The Large White pig breed. (Credit: Prof. R. Osei-Amponsah)	376
Fig. 8.14	The Pietrain pig breed. (Credit: Prof. R. Osei-Amponsah)	377
Fig. 8.15	A Hampshire sow with piglets. (From Source: https://agro4africa.com/hampshire-pig-breed/, licensed under CC-BY 4.0)	377
Fig. 8.16	The Landrace pig breed. (Credit: Prof. R. Osei-Amponsah)	377
Fig. 8.17	Extensive pig production practices showing poor housing (**a** and **b**), feeding/watering (**c**), and feed resources	379
Fig. 8.18	A farmer feeding pigs in a small-scale semi-intensive production system. (Credit: Prof. R. Osei-Amponsah)	380
Fig. 8.19	Simplified marketing channel for indigenous pigs from the free-range/scavenging and smallholder semi-intensive systems	384
Fig. 8.20	Simplified marketing channel for pigs from medium and large-scale intensive farms	384
Fig. 9.1	Distribution of large camelids in the world. (From FAOSTAT *2019*, licensed under Open Government License OGLv3.0)	397
Fig. 9.2	Migration map of the historical camelid family. (From Burger et al. *2019*, licensed under CC-BY 4.0) The current distribution of dromedaries and Bactrian camels is presented in red and green colours. The last refugia of the wild two-humped camels in China and Mongolia are shown as dark-green patches. The map was adapted from Mesa Schumacher/Aramco World (https://www.aramcoworld.com/en-US/Articles/November-2018/The-Magnificent-Migration). Reprint permits were granted by AramcoWorld on March 6, 2019	399
Fig. 9.3	Major coat colours found in dromedary populations in Algeria. (Photos by Gaouar S.B.S)	403
Fig. 9.4	Dromedary populations in the southwest of Algeria. (Photos by Gaouar S.B.S)	404

Fig. 9.5	Coat colours as they are called by Chadian dromedary herders. (From Djomtchaigue et al. *2015*, licensed under CC-BY 4.0)	407
Fig. 9.6	Somali dromedary. (From Tilahun et al. *2020*, licensed under CC-BY 4.0)	409
Fig. 9.7	Rendille dromedary. (From Sun *2005*, licensed under CC-BY 4.0). (**a**) A warrior driving camel herds out of the camp for day's herding. (**b**) Transporting wedding goods	409
Fig. 9.8	Turkan young camels with female, in Prosopis thicket at lake, Northern Turkana. (From Carr *2017*, licensed under CC-BY 4.0)	410
Fig. 9.9	Afar dromedary in Afar Region, Northeast Ethiopia (Melkamu et al. *2022*). (From Melkamu et al. *2022*, licensed under CC-BY 4.0)	411
Fig. 9.10	Maghrebi dromedary ecotypes in southern Tunisia according to their tribal affiliation. (From Chniter et al. *2013*, licensed under CC-BY 4.0)	415
Fig. 9.11	Dromedary coat colours nomenclature used by Tunisian dromedary herders. (From Chniter et al. *2013*, licensed under CC-BY 4.0)	416
Fig. 9.12	Female dromedary of Chalfi and Kawa types. (From Chniter et al. *2013*, licensed under CC-BY 4.0)	417
Fig. 9.13	Moroccan dromedary breeds. (From FAOSTAT *2019*, licensed under CC-BY 4.0)	417
Fig. 9.14	Pakistani dromedary. (Photo by: Dr. Asim Faras)	418
Fig. 9.15	Traditional milk production. (From Cherifi et al. *2013*, licensed under CC-BY 4.0)	424
Fig. 9.16	Dromedary slaughter and carcass/meat handling. (From Cherifi et al. *2013*, licensed under CC-BY 4.0)	426
Fig. 10.1	Origin of domestic donkey and routes of dispersion in Africa, Asia and Europe	456
Fig. 10.2	Idabari jenny (**a**) and jack (**b**) at the Federal University of Agriculture, Abeokuta, Nigeria. (Photos by M.N. Bemji)	458
Fig. 10.3	Dunni donkeys at Zaria, Nigeria. (**a**) brown, and (**b**) black types. (Photo by DR. E. M. Ibeagha-Awemu)	460
Fig. 10.4	Fari donkeys at Zaria, Nigeria. (**a**) White with brown patches, and (**b**) pale cream. (Photo by Dr. E. M. Ibeagha-Awemu)	460
Fig. 10.5	Egyptian donkey. (Source Wikipedia: photo Jerome Bon by https://en.wikipedia.org/wiki/List_of_donkey_breeds#/media/File:Egyptian_donkey_(2428011421).jpg, CC-BY-2.0 license)	461

Fig. 10.6	Libyan donkey. (Photo Source: Wikipedia and available at https://en.wikipedia.org/wiki/List_of_donkey_breeds#/media/File:Donkey_in_Wasita_Libya.jpg, CC-BY-4.0 license)	462
Fig. 10.7	Tunisian. (Source: Wikipedia and available at https://en.wikipedia.org/wiki/List_of_donkey_breeds#/media/File:Barrage_bni_mtir_456789_13.JPG, CC-BY-3.0 license/)	463
Fig. 10.8	Moroccan. (Source: Wikipedia, photo by Dale Harvey and available at https://en.wikipedia.org/wiki/List_of_donkey_breeds#/media/File:Tamraght-daleharvey-10-donkey.jpg, CC-BY-2.0 license).	463
Fig. 10.9	Kassala donkey carrying grains from a market in Keren, Eritrea. (Source: Wikipedia- photo by David Standley and available at https://en.wikipedia.org/wiki/List_of_donkey_breeds#/media/File:Donkey_Transport_(8383370559).jpg, CC-BY-2.0 license)	464
Fig. 10.10	Sudanese Pack. (Source: Wikipedia- photo by David Haberlah and available at https://en.wikipedia.org/wiki/List_of_donkey_breeds#/media/File:Donkey_carrying_Rahal_basket.jpg, CC BY-SA 3.0 license)	465
Fig. 10.11	Abyssinian donkeys grazing (**a**), slate-grey Abyssinian adult donkey (**b**), and chestnut-brown Abyssinian jenny and foal (**c**). (Photos by Dr. E. Kefena, Agricultural Transformation Institute, Ethiopia)	467
Fig. 10.12	Sinnar donkeys displaying different coat colours: (**a**) white, (**b**) brown, and (**c**) fawn. (Photos by Dr. E. Kefena, Agricultural Transformation Institute, Ethiopia)	468
Fig. 10.13	A herd or drove of Afar donkeys (**a**) and Afar Jenny and foal (**b, c**). (Photos by Dr. E. Kefena, Agricultural Transformation Institute, Ethiopia)	469
Fig. 10.14	A herd or drove of Omo donkeys (**a**) and fawn-coloured Omo jack (**b**) and jenny (**c**). (Photos by Dr. E. Kefena, Agricultural Transformation Institute, Ethiopia)	471
Fig. 10.15	A Maasai donkey (Source: Wikipedia. Photo by Daryona, https://commons.wikimedia.org/w/index.php?curid=10558134, CC BY-SA 3.0)	472
Fig. 10.16	A Somali donkey. (Photo by Dr. E. Kefena, Agricultural Transformation Institute, Ethiopia)	472
Fig. 10.17	Two Somali Wild Asses (*Equus africanus somaliensis*) in Marwell Zoo, Hampshire, UK. (Source: Wikipedia- photo by Chris Keating, CC BY-SA 3.0, https://commons.wikimedia.org/w/index.php?curid=15354236)	473
Fig. 10.18	Muscat donkey in Tanzania. (Source: Wikipedia- photo by Nevit Dilmen (talk), CC BY-SA 3.0, https://commons.wikimedia.org/w/index.php?curid=11320901)	474

List of Figures

Fig. 10.19 Tswana adult donkey showing typical features. (Photo by Dr. E. M. Ibeagha-Awemu) 475

Fig. 10.20 Greyish-brown Tswana donkey ready to transport water at a collection point near Opuwo in Namibia. (© The Donkey Sanctuary 2024) 475

Fig. 10.21 Donkeys grazing in Zaria, Nigeria (**a**) (Photo by Dr. E. M. Ibeagha-Awemu) and in Ethiopia (**b**) (Photo by Dr. E. Kefena) 476

Fig. 10.22 Mule breeding in Ethiopia. Female horse (left) and male donkey (right) ready to mate. (Photo by Dr. E. Kefena, Agricultural Transformation Institute, Ethiopia) 477

Fig. 10.23 Donkeys at a livestock market in Niger. (Source: Wikipedia.com- photo by Vincent van Zeijst, CC BY-SA 4.0, https://commons.wikimedia.org/w/index.php?curid=82422085) 480

Fig. 10.24 Women fetching water to be transported by donkeys at a collection point near Opuwo in Namibia. (©The Donkey Sanctuary 2024) 481

Fig. 10.25 Donkey pulling cart with woman in Somali. (Source: Wikipedia and available at https://en.wikipedia.org/wiki/List_of_donkey_breeds#/media/File:AMISOM_Djiboutian_Contingent_in_Belet_Weyne_11_(8212396061).jpg, CC0 1.0 license) 482

Fig. 10.26 Donkeys ploughing a field at a Fulani settlement in Zaria, Nigeria. (Photo by Dr. E.M. Ibeagha-Awemu) 483

Fig. 10.27 Donkeys transporting humans, goods and animals at Zaria, Nigeria. (Photo by Dr. E.M. Ibeagha-Awemu) 484

Fig. 10.28 Donkey racing during the celebration of Lamu's annual Maulidi festival. (Photos by Zollo Nyambu, https://africaviewfacts.com/facts/lamu-donkey-race-celebrating-tradition-and-community-in-kenya/) 486

Fig. 10.29 The donkey as essential means of transportation of persons and goods in old town of Lamu. (Photo by Alex Malombo, https://africaviewfacts.com/facts/lamu-donkey-race-celebrating-tradition-and-community-in-kenya/) 487

Fig. 10.30 Pictures show the poor slaughter, handling and disposal of donkeys in Nigeria, Kenya and Ghana as deplorable activities associated with the donkey skin trade. (©The Donkey Sanctuary 2024) 489

Fig. 11.1	Barb horses. (**a**) Barb horse from the national stud farm "Chaou-Chaoua" in Tiaret, Algeria (Chikhaoui et al. *2018*) (Image credit: (Chikhaoui et al. *2018*)); (**b**) Moroccan Barb horse used for Fantasia (Talley *2015*) (Image credit: Gwyneth Ursula Jean Talley); (**c**) Barbe horse breed in Mauritania (Photo Courtesy: Ministry of Rural Development, Mauritania) (AU-IBAR *2019*), (**d**) Mounted Barb horse in Nigeria (Image credit: Eveline M. Ibeagha-Awemu) 504
Fig. 11.2	Dongola horse in Eritrea (Ministry of information of Eretria *2021*) 507
Fig. 11.3	Dongola horse (Koçkar *2012*). (Reproduced with permission CC-BY 4.0 license) 507
Fig. 11.4	(**a**) Algerian Arabian Stallion "SAKHI" with his owner Ahmed Feghouli, WAHO 2010 trophy winner (**b**) Diable Du Desert. Breeder: Amraoui Belkasem. Owner: Bouteldja Mohamed, WAHO 2014 trophy winner; (**c**) FOUSHA. Mare, 1999., Breeder: Haras Nationaux, Jumenterie de Tiaret with Mr. Ahmed Bouakkaz, General Director and Head of Stud Book at ONDEEC, and the owner Mr. Zeguaoui Abed, WAHO 2016 trophy winner 508
Fig. 11.5	Typical Selale riding horse (Kefena et al. *2012*) 509
Fig. 11.6	Kafa stallion (Kefena et al. *2012*) 511
Fig. 11.7	Director of Parks Mather insists on A. W.'s being properly mounted on an Abyssinian horse. (Harris *1952*) 512
Fig. 11.8	Pair of Abyssinian horses engaged in cropland cultivation. (Kefena et al. *2012*)...................... 513
Fig. 11.9	Bay Bale horse. (Image credit: Laika ac from UK 2014). Reproduced with permission—CC-BY 4.0 license 513
Fig. 11.10	Borana horse riding. (Sirikoi *2023*) 514
Fig. 11.11	Horro horse in Street Jimma, Ethiopia. (Image credit: Rod Waddington 2014) 516
Fig. 11.12	Kundudo mare with foal. (Paris *2014*)................. 517
Fig. 11.13	A thin white Kundudo mare standing on a rocky mountain slope with few trees. (Paris *2014*) 517
Fig. 11.14	Gharkawi horse. (Image credit: Sean Woo *2004*) 518
Fig. 11.15	Tawleed horse. (Image Credit—Omdurman-Tuti Island jetty, Sudan)................................. 519
Fig. 11.16	Poneys musey à Gobo (Arthus-Bertrand 2003). (Image credit: Seignobos C). (Arthus-Bertrand *2003*)...... 520
Fig. 11.17	Coat color diversity of the Kirdi pony in the Sudano-Sahelian region, Kabbia South-West of Chad. (Image credit: authors) 522
Fig. 11.18	Bélédougou horse used for carriage and pulling. (Image credit: Ferdinand Reus) (Claudot-Hawad *1988*) 523

Fig. 11.19 Arewa stallion. (Edward and Bukola *2023*).............. 525
Fig. 11.20 White Bobo horse harnessed for a wedding procession, in the enclosure of Ouagadougou, Burkina Faso. (Image credit: Amélie Tsaag Valren).................. 525
Fig. 11.21 M'bayar horse served for light draft in Pikine, Sénégal. (Source: Lejosne *2023*)............................ 527
Fig. 11.22 Fleuve horse breed from Sénégal in West Africa.......... 527
Fig. 11.23 A Foutanke or Narougor horse in Pikine, Sénégal. (Lejosne *2023*).................................... 529
Fig. 11.24 Mogods pony. (Chabchoub et al. *2006*)............... 531
Fig. 11.25 Cape Boerperd mares and foals at pasture (Van der Merwe and Martin *2002*). (Image credit: Van der Merwe and Martin *2002*).................... 534
Fig. 11.26 Basuto pony (Van der Merwe and Martin *2002*). (Image credit: Martin J. in Van der Merwe and Martin *2002*)................................... 535
Fig. 11.27 The South African Nooitgedacht stallion named Mac. (FAO *2002*)In 1958, the birth of the stallion Mac marked a pivotal moment in the development of the Nooitgedacht horse breed. Mac exemplified the ideal general-purpose horse sought by project managers, displaying key physical traits such as a strong shoulder, well-muscled back, and robust legs with hard hooves. Additionally, Mac embodied the distinctive characteristics of the Nooitgedacht breed, with his small pointed ears, pronounced brows, and straight to slightly concave profile. This iconic horse left a lasting legacy in the breed's evolution (Photo: Landbouweekblad).................. 536
Fig. 11.28 A South African Vlaamperd stallion. (SAVBS *2019*)...... 537
Fig. 11.29 Algerian native Arabian stallion "SAKHI", senior champion stallion of World Arabian Horse Organization (WAHO) 2014 (http://www.waho.org/algeria)............ 539
Fig. 11.30 Anglo-Arab horse. (Image credit: Personal image)........ 542
Fig. 11.31 Horse populations from southwestern Ethiopia (Mustefa et al. *2022*). (**a**) Telo stallion; (**b**) Masha stallion; (**c**) Gesha stallion; (**d**) Gesha mare (Mustefa et al. *2022*) ... 544
Fig. 11.32 Kundido feral horses from Ethiopia. (Kefena et al. *2012*)... 548
Fig. 11.33 Namibian wild horses. (Pütz and Schlottmann *2020*)...... 549
Fig. 11.34 Feeding of the horses with hay. (Pütz and Schlottmann *2020*)......................... 550
Fig. 11.35 Plowing with horses. (Image credit: Author's Own 2022) (Kebede *2023*).................................... 556
Fig. 11.36 (**a**) Barb or Arab-Barb horse-drawn cart for passenger transport; (**b**) Horse-drawn cart with pneumatic wheels and elevated platform used for transporting goods in Tozeur region of Tunisia. (Charles *2022*).............. 556
Fig. 11.37 Weeding festivity in Chechar city, Khenchla (Algeria)..... 558

Fig. 11.38 Rider and horse (caparisoned for a wedding ceremony) performing the pace walk. (Baynes-Rock and Teressa *2021*). 559

Fig. 11.39 Awi Agew horse riders during annual festival. (Kebede *2023*) 559

Fig. 11.40 Fulani horse dancing traditionally at the conclusion of the Ramadan. (Image credit: Flickr/Rosemary Lodge) (DLIFLC *2018*) 560

Fig. 11.41 Ndop loincloth during funerals, Cameroun. (Image credit: Foréké Dschang 2017) (Noel *2021*) 560

Fig. 11.42 Fantasia in Northern Cameroon. (Image credit: Félix Meutchieye) 562

Fig. 11.43 Horse show at Jida horse market in Ethiopia. (Kefena et al. *2012*) 562

Fig. 11.44 Horse parade to celebrate the 123rd anniversary of Adwa victory as led by prominent Ethiopian musician Hachalu Hundessa. (Kefena et al. *2021*) 563

Fig. 11.45 Durbar festival: an annual equestrian event in Northern Nigeria. (Yusuf et al. *2023*) 563

Fig. 11.46 Lambert Leclezio first Mauritian athlete (Mauritius is a country in eastern Africa) to compete at the FEI Open European Vaulting Championships for Juniors 2011. (FEI *2012*) 567

Fig. 11.47 Fantasia parade in Mekhatria, at 10 kms north of the seat of the wilaya of Ain Defla, Algeria. (Image credit: Bisker Mohamed) (Bisker *2017*). 569

Fig. 11.48 The Fantasia equestrian parade with warrior inspiration (Blog du cheval) (https://blog.cheval-daventure.com/fr/post/140/la-fantasia-une-tradition-equestre-dinspiration-guerriere) 569

Fig. 11.49 Saddle, breast strap, and bridle on a Barb horse Capital Festival of traditional horseback riding, Rabat, August 2014. (Image credit: Gwyneth Ursula Jean Talley) (Talley *2015*) 570

Fig. 11.50 Lesotho's mountain jockeys race. (Kuwait times *2022*) 571

Fig. 11.51 Tourism activity by riding Nooitgedacht horses in South Africa. (The figure illustrates a scene where a game warden and a tourist, riding Nooitgedacht horses, are approaching a female rhinoceros with a calf visible in the background. These South African horse breeds have demonstrated their exceptional suitability for guiding inexperienced riders during game safaris. (Photo by F.J. van der Merwe) (FAO *2002*)). 572

Fig. 11.52 Etalon Arabe-Barbe alezan attaché devant le domicile de son propriétairefromTozer, Tunisia. (Charles *2022*) 577

List of Figures   xxix

Fig. 12.1   Distribution of wild and domestic water buffalo in Africa. (From Furstenburg (*2007*), licensed under CC-BY 4.0, modified by adding distribution of water buffalo) 596

Fig. 12.2   Phylogeny of Bubalus based on complete mitochondrial genomes. The tree was reconstructed with MrBayes software using the 118 mitogenomic haplotypes of Bubalus identified in our alignment of 16,356 nucleotides. The tree was rooted with Syncerus (not shown). Haplogroups showing more than 0.5% of nucleotide divergence are highlighted by different colored rectangles, and those found in the swamp buffalo are named SBa to SBe (the name of the haplotype is followed by a dash and a number when the same haplotype was found in least two individuals). Among the 29 mitogenomes of Egyptian river buffalo, the 21 samples collected from North are indicated in red, whereas the eight samples collected from South are indicated in green. The four mitogenomes of river Buffalo assembled using SRA data available in GenBank are named with the SAMN accessions. For nodes supported by bootstrap percentage (BP) _ 90 in the RAxML analysis (see details in Material and Methods), the two values correspond to the posterior probability (PP calculated using MrBayes software, left of the slash) and BP (right of the slash). An asterisk is used when both MrBayes and RAxML analyses provided maximal support values, i.e., PP ¼ 1 and BP ¼ 100, respectively. No information was provided for nodes supported by BP <90. (From Youssef et al. (*2021*), licensed under C-C BY 4.0). 598

Fig. 12.3   Phylogeny of genus *Bubalus* based on complete mitochondrial genomes. The tree was rooted with African buffalo. (From Hassan et al. (*2022*), licensed under C-C BY 4.0). 600

Fig. 12.4   A phylogenetic tree assembled using the 5′ half of CO1 gene sequence (650-bp) in all the investigated species and the donkey sequence as an outgroup. Accession numbers are given to all samples. (From Oraby et al. (*2011*), licensed under C-C BY 4.0). 601

Fig. 12.5   COI (mitochondrial cytochrome C oxidase subunit 1) barcode-based phylogenetic tree constructed using sequences from four buffalo taxa groups and a sequence of goat (*Capra hircus*) as an outgroup. *Hap*. haplotype. (Hassan et al. *2018*, licensed under C-C BY 4.0). 602

Fig. 12.6   Egyptian buffalo types 603

Fig. 13.1   An adult Goliath Frog 622
Fig. 13.2   Nest of Goliath frog 623

Fig. 13.3	Median-joining network for the cyt b mtDNA haplotypes of *C. aspersum*. (Guiller and Madec *2010*)	625
Fig. 13.4	The most common snail varieties in northern Algeria. (Adapted from Bouchiba et al. *2021* and Mezouar et al. *2021*)	626
Fig. 13.5	Distribution of snails according to shell color in Northern Algeria	628
Fig. 13.6	Picture comparing two Achatinadae snails	629
Fig. 13.7	Some species of land snail in South Africa. (Herbert *2010*)	630
Fig. 13.8	Picture comparing two Achatinadae snails: a. *Achatina achatina* snails, b. *Acharchatina marginata* snails	636
Fig. 13.9	Snail nutrition	636
Fig. 13.10	Reproduction of *Archachatina marginata* snails	637
Fig. 13.11	The Kabyle rabbit phenotypes. (Carneiro et al. *2011*)	640
Fig. 13.12	Phenotypes of the white breed. (Gacem et al. *2008*)	640
Fig. 13.13	Phenotypes of the synthetic breed	641
Fig. 13.14	Tadla breed	641
Fig. 13.15	Zemmouri breed	641
Fig. 13.16	Bauscat breed. (El Raffa et al. *2002*)	642
Fig. 13.17	Giza White Breed. (El Raffa et al. *2002*)	642
Fig. 13.18	Chinchilla breed. (Bolet et al. *2012*)	642
Fig. 13.19	Baladi Breed. (El Raffa et al. *2002*)	643
Fig. 13.20	Gabali Breed	643
Fig. 13.21	Phendula	644
Fig. 13.22	The origin of the honeybee, *Apis mellifera,* according to three hypotheses. (Gupta et al. *2014*)	651
Fig. 13.23	Harvesting of wax by African beekeepers. (Mayazine *2019*)	654
Fig. 13.24	a. Traditional log hive for *A. mellifera* bees; b. modern hive; c. transitional Kenyan top bar hive; d. clay pot hive for stingless bees. (Leven et al. *2005*)	655
Fig. 13.25	The clinical signs of American foulbrood *P. larvae* observed in the field. (Onions *1912*)	661
Fig. 14.1	Various methane emission measuring techniques at Mazingira centres of the International Livestock Research Institute (ILRI). Cattle metabolic chamber (**a**), SF6 gas cylinder (**b**), sheep metabolic chamber (**c**) and gas analysis kit (**d**). (Photos by Mizeck Chagunda)	681
Fig. 14.2	Laser methane detector. On the left, a scientist taking methane measurements and on the right, two different models of the laser methane detector. (Photo by A. Ross)	682
Fig. 14.3	Examples of cattle under extensive production systems. Top, Kenya and bottom, Zambia. (Photos by Mizeck Chagunda)	685

Fig. 15.1	A schematic illustration of the main components of a breeding strategy	692
Fig. 15.2	Impact pathway for best practices for selective breeding for improved livestock (Ojango et al. *2018*)	701
Fig. 16.1	Wakwa (Brahman (50%)–Gudali (50%)) bull. (Courtesy of Dr. David Mbah)	718
Fig. 16.2	Holstein (67.5%) Gudali (32.5%) bull. (Courtesy Dr. David Mbah)	724
Fig. 16.3	A 3-year-old Holstein 5/8 Gudali 3/8 Milker. (Courtesy Dr. David Mbah)	724
Fig. 16.4a	Holstein-Gudali programme to produce H 5/8 G3/8 cross-breds [8–12 years]	725
Fig. 16.4b	Holstein-Gudali programme H ¼ G ¾ to produce H 5/8 G 3/8 cross-breds for farmers [8–12 years]	725
Fig. 16.5	A cloned Boran bull and its offspring (calves). Calf 1 died due to low blood glucose while calf 2 survived following supplemental glucose administration. (From Yu et al. (*2016*), licensed under CC-BY 4.0)	746
Fig. 17.1	Strains of Dorper sheep found in Kenya and South Africa	769
Fig. 17.2	Illustration of the FARM Africa farmer-based breeding strategy for goat improvement. Adapted from Ojango et al. (*2010*)	770
Fig. 17.3	Example of cross-breeding to develop a composite breed	771
Fig. 17.4	West African Dwarf Goat in Ghana	776
Fig. 17.5	Map of Kenya indicating the location of main goat projects from 1972 to 2010	777
Fig. 17.6	Galla Goats in Kenya	778
Fig. 18.1	The cycle of a breeding program	791
Fig. 18.2	Basic components for a successful poultry breeding strategy for Africa. (Adapted from Jack Dekkers *1990*)	799
Fig. 18.3	Breeding objective traits in a typical broiler breeding program. (*Source*: Olori (*2012*))	800
Fig. 18.4	Pyramidal structure of poultry breeding. (*Source*: Olori (*2008*))	802
Fig. 18.5	Genomic selection in poultry	809
Fig. 19.1	Frequency of severe heat stress for past and future periods for pigs in Uganda. (From Mutua et al. (*2020b*), licensed under a CC BY 4:0)	826
Fig. 19.2	Risk factors that cause heat stress in pigs. (From Mutua et al. (*2020b*), licensed under a CC BY 4:0)	827
Fig. 20.1	Benefits and beneficiaries of a livestock identification and management system	859

Fig. 20.2	Wearable sensors such as milking collar, leg band and ear tags, temperature and proximity sensors; image-based monitoring systems and internal sensors collect data (**a**) which is analysed (**b**) for informed management decisions. (From Siberski-Cooper and Koltes (Siberski-Cooper and Koltes *2021*), licensed under CC-BY 4.0)	860
Fig. 20.3	iCOW system on mobile phones used for farmer training and advisory applications	867
Fig. 20.4	Genome editing platforms and mechanisms for DSB repair with endogenous DNA. Genome editing nucleases (ZFNs, TALENs and CRISPR/Cas9) induce DSBs at targeted sites. DSBs can be repaired by NHEJ or, in the presence of donor template, by HDR. Gene disruption by targeting the locus with NHEJ leads to the formation of indels. When two DSBs target both sides of the amplification or insertion, a therapeutic deletion of the intervening sequences can be created, leading to NHEJ gene correction. In the presence of a donor-corrected HDR template, HDR gene correction or gene addition induces a DSB at the desired locus. DSB double-stranded break, ZFN zinc-finger nuclease, TALEN transcription activator-like effector nuclease, CRISPR/Cas9 clustered regularly interspaced short palindromic repeat associated 9 nuclease, NHEJ nonhomologous end-joining and HDR homology-directed repair. (From Li et al. (*2020*), licensed under CC-BY 4.0)	877
Fig. 20.5	Diagrammatic representation of CRISPR/Cas9 gene editing using either zygote micromanipulation by microinjection or electroporation, or somatic cell nuclear transfer (SCNT) for the generation of livestock animals for various applications. (From Parisse et al. (*1713*), licensed under CC-BY 4.0)	882
Fig. 20.6	Cellular agriculture uses cells from animals or precision fermentation technologies to produce products of animal origin (meat, eggs, milk) and fermentation products for use as ingredients for the creation of varied products including textiles.	883
Fig. 20.7	Main steps involved in the production of cell-cultured meat	884
Fig. 20.8	Schematic representation of improved farm management practices and technologies that have led to enhanced genetic gain in dairy cattle breeding and improvement in milk production. (From Ibeagha-Awemu and Yu (*2021*), licensed under CC-BY 4.0)	891

List of Figures

Fig. 21.1	A precision livestock farm uses state-of-the-art technologies to continuously collect data on animal (microphones, cameras, sensors) followed by continuous data analysis to provide real time information or data to support management and production decisions as well as resolve health and welfare issues.	916
Fig. 21.2	The one health concept.	919
Fig. 21.3	Agricultural waste management toward a circular bioeconomy. Digestate (10% dilution) resulting from anaerobic digestion of chicken manure was used as nutrients for microalgal (strain *Chlorella vulgaris* CPCC 90) growth resulting in production of biofuel and bioproducts that can be used as fertilizer (for soil treatment), bio-pesticide and as animal feed. (From Rajagopal et al. *2021*), licensed under CC-BY 4.0)	929
Fig. 21.4	Modern techniques for the improvement of livestock health	954
Fig. 21.5	Tie-stall barn (**a**), cubicle barn (**b**), and free walk barn (**c**) dairy cow housing systems	961
Fig. 21.6	A typical portable milking machine	963
Fig. 21.7	Different parlor milking systems	965
Fig. 21.8	Sample pasteurization system for large-scale use (Tubular Aseptic UHT Pasteurizer with Vacuum Deaerator).	968
Fig. 22.1	IBIS in Burundi: identification, insemination, and consanguinity management. IFAD, Ministry of Livestock, Burundi and Adventiel	1010
Fig. 22.2	Zebuscan, a mobile application for animal traceability in Madagascar 2019, World Bank, Ministry of Agriculture, Livestock and Fisheries, Adventiel	1010
Fig. 22.3	Example of the modular structure of a Livestock Identification and Management System	1011
Fig. 23.1	A representative SNP Chip	1026
Fig. 23.2	A Barcode Scanner.	1030
Fig. 23.3	A 4G mobile data sim wireless router useful for digital telephony and data transfer	1031
Fig. 23.4	A typical smartphone suitable for communication, data capture, and transfer	1031
Fig. 23.5	Programmable hand-held terminals that can be used for data recording	1032
Fig. 23.6	A miniature mobile printer that can be used to print instant labels during recording.	1032
Fig. 23.7	A large-capacity portable data storage device (1 TB)	1033

Fig. 28.1	Contribution of animal production to the different Sustainable Development Goals (SDGs). (Source: FAO (*2015*)).	1127
Fig. 28.2	Eight capacity building-related domains to address the causes of genetic erosion in African livestock genetic diversity. (Source: FAO (*2021*)).	1137
Fig. 28.3	The cyber-physical management cycle of smart farming enhanced by cloud-based event and Big Data management. (Source Wolfert et al. (*2014*)).	1141
Fig. 28.4	The data chain of Big Data applications in livestock breeding. APIs Application Programming Interface, PRA Data Collectors/Performance Recording Agents, DA Direct Agreement between firm/farm & the programme, FA Farmer Assistants, MTA Material Transfer Agreement, CRA Collaborative Research Agreement, DSA Data Sharing Agreements, NDA Non-Disclosure Agreement	1141
Fig. 28.5	The collection of technologies that we refer to as advanced technologies can help animal farmers create better outcomes	1142
Fig. 28.6	Key domains of capacity building on machine learning algorithms to create optimal performance in livestock breeding. (Adapted from Neethirajan *2020*).	1143
Fig. 29.1	Organization and core elements of a typical Genome Core facility	1158
Fig. 29.2	Overview of high-performance computing options showing common clusters (e.g. GPUs, FPGAs, and cloud solutions) and highlighting differences in flexibility, performance, and options for custom design. CPU: Central Processing Units; GPU: Graphics Processing Unit; FPGA: Field Programmable Gate Arrays; HDL: Hardware Description Language; RTL: Register-Transfer Level. (From Lightbody et al. (*2019*), licensed under a CC BY 4:0)	1165
Fig. 29.3	Overview of a high-performance computing core with service centres in other locations	1167
Fig. 29.4	Mapping the five African Regional Animal Genebanks. African regional Genebanks (large triangles-pink, yellow, red, blue, and green colours) showing interconnections between them (blue lines) and countries served by each Genebank (small circles in the colours of their respective regional Genebanks)	1174
Fig. 30.1	The Regional Economic Communities in Africa	1210

Fig. 31.1	Recipients of piglets and feed through The Pig Van Djouke POG (Pass on the Gift) Project. (Photo by NOWEPIFAC)	1240
Fig. 31.2	Imported Holstein (**a**) and Jersey (**b**) and Holstein (1995 batch) (**c**) at IRZV Bambui centre, as well as imported hens (**d**). (Photos by HPI, extracted from HPI Cameroon archives and provided by Emmanuel Bassam (HPI Director of Program))	1256
Fig. 31.3	Farmers receive Holstein-Friesian cattle at Nseh Mbabu village (**a**), Holstein-Friesian under semi-intensive system of management at Bamendakwe (**b**, **c**); and HPI staff at a pasture and forage development site at Fonta, Bambui (**d**). (Photos by HPI, culled from HPI Cameroon archives and provided by Emmanuel Bassam (HPI Director of Program))	1259
Fig. 31.4	A ram from HPI Project in the far North Region of Cameroon (**a**); Sheep placement in the Mbam Upper Sanaga Valley Integrated Sheep/Goat Project in the Center Region of Cameroon by HPI (**b**); and Sheep placement in the Bui Donga Small Holder Integrated Sheep & Goat Project in the Northwest Region of Cameroon by HPI (**c**). (Photos by HI, extracted from HPI Cameroon archives and provided by Emmanuel Bassam (HPI Director of Program))	1260
Fig. 31.5	Boran animals imported from Kenya (**a**, **b**) and offspring of Boran × Zebu (**c**) distributed to farmers under "*Lake Nyos Livestock Restocking Project*". (Photos by HI, extracted from HPI Cameroon archives and provided by Emmanuel Bassam (HPI Director of Program))	1261
Fig. 31.6	Livestock farmers showing off their milk products to Cameroon government officials and veterinarians during Heifer International Cameroon's 40th Anniversary Celebration in Bamenda in 2015. (Photo by HPI, extracted from HPI Cameroon archives and provided by Emmanuel Bassam (HPI Director of Program))	1262
Fig. 31.7	Collaboration web for successful delivery of sustainable livestock development in Africa	1274

List of Tables

Table 1.1	The number of livestock breeds and breed types by species in Africa relative to other regions of the world	4
Table 1.2	The number of livestock breeds and breed types in Africa and regions	5
Table 2.1	Livestock populations[a] in Africa and regions in the year 2021	16
Table 2.2	Main characteristics of the agro-ecological zones of Africa[a]	21
Table 2.3	Key features of African livestock production systems[a]	25
Table 2.4	Major mixed crop–livestock farming systems of sub-Saharan Africa described by Dixon et al. (*2001*)	32
Table 2.5	Major challenges limiting animal performance under current livestock production systems and possible solutions for the sustainable development of improved livestock productivity	45
Table 3.1	Humpless cattle breeds of Sub-Saharan Africa	71
Table 3.2	Adult linear body measurements by sex and breed of Shorthorn cattle	87
Table 3.3	Special genetic characteristics of some Shorthorn cattle breeds and stabilized crosses	89
Table 3.4	Mean live weight by sex of some Shorthorn cattle breeds and stabilized crosses	96
Table 3.5	Estimates of reproductive parameters of some Shorthorn cattle breeds and stabilized crosses	97
Table 3.6	Estimates of milk production by milking and management system for some Shorthorn cattle breeds	98
Table 3.7	Live weights, carcass weights, and dressing-out percentage of Baoulé, Muturu, and Keteku cattle by sex and age	99
Table 3.8	Birthweight and linear body measurements of Baoulé cattle in Côte d'Ivoire	104
Table 3.9	Best linear unbiased estimates of live weight and body measurements of Namchi and Kapsiki cattle under the natural environment in Cameroon	107

Table 3.10	Population estimates of Shorthorn cattle types of West and Central Africa	110
Table 4.1	Zebu cattle breeds of Eastern and Southern Africa	120
Table 4.2	West and Central African Zebu Cattle	134
Table 4.3	The Sanga cattle breeds of eastern and southern Africa	149
Table 4.4	The pseudo-Sanga of West Africa[a]	161
Table 4.5	Zenga (Zebu–Sanga) cattle, recent derivatives, and synthetic breeds	166
Table 4.6	Summary of exotic cattle breeds used in Africa	175
Table 5.1	Five sub-species of *Capra* (domesticated goat)	188
Table 5.2	African goat genetic resources: breed groups, subgroups and breeds	198
Table 6.1	Position of Africa in the world in terms of sheep production and processed products for the year 2019 (FAOSTATS *2019*)	241
Table 6.2	Sheep production by regions in Africa for the period 2018–2019 (FAOSTATS *2019*)	242
Table 6.3	Classes of sheep breeds in Africa	247
Table 6.4	Threatened conventional sheep breeds in Africa that need conservation (AU-IBAR *2019*)	303
Table 6.5	Trait preference, selection criteria and breeding strategies for genetic improvement of economically important traits in sheep	305
Table 6.6	Heritabilities, expected direct and correlated responses to selection for some reproductive traits in sheep	310
Table 6.7	Global sheep raw skin production (tonnes) 2000–2012 (FAOSTATS *2021*)	314
Table 7.1	Identified characteristics of some African indigenous chicken breeds	333
Table 7.2	Examples of genes in indigenous chicken populations that may be linked to the adaptation to tropical conditions	333
Table 7.3	FAO chicken production systems classification	349
Table 7.4	Examples of chicken genetic improvement programs in Africa	350
Table 8.1	Pig breeds of Africa by country and region	365
Table 8.2	Key outputs, activities, and actors for sustainable utilization and conservation of pig genetic resources in Africa	382
Table 9.1	Number of dromedaries in different African countries (FAOSTAT *2019*)	420
Table 9.2	Evolution of African livestock animal numbers between 1969 and 2019 (FAOSTAT *2019*)	421

Table 10.1	Donkey populations by regions in Africa for the period 2018–2022 (FAOSTAT *2024*)	453
Table 10.2	Distribution of donkey types in Nigeria based on coat colours	457
Table 11.1	Horse population by African countries (FAO DAD-IS *2021*)	510
Table 11.2	Risk statue of African horse breeds (FAO *2019*)	578
Table 12.1	Genetic diversity of milk traits genes and association with different milk production measures	605
Table 12.2	Genetic diversity of fertility traits genes associated with different fertility measures	608
Table 13.1	Descriptive analysis of body measurements (height, length, and width of shell and body weight) between different species in Algeria (Bouchiba et al. *2021*)	627
Table 15.1	Summary of main breeding goals by species	696
Table 16.1	Example of breeding goals and breeds developed for meat production using indigenous breeds[a] in different regions of Africa	712
Table 16.2	Breakthrough biotechnological innovations in cattle and buffalo genetics and breeding	742
Table 16.3	Examples of nucleus/community breeding scheme for cattle improvement in West Africa	748
Table 17.1	Number of sheep in the different regions of Africa from 2000 to 2021[a]	760
Table 17.2	Numbers of goats in the different regions of Africa from 2000 to 2021[a]	761
Table 17.3	Number of different breeds of sheep and goats in the different regions of Africa	761
Table 17.4	Traits and performance indicators associated with different products from sheep and goat	763
Table 17.5	Estimates of genetic parameters for growth traits in some sheep and goat breeds in Africa	765
Table 17.6	Indigenous breeds characterized and developed through planned breeding programmes	767
Table 17.7	Mean performance of various indigenous sheep and goat breeds and their crosses in Africa	768
Table 18.1	Characteristics to be balanced when selecting for dual purpose and tropically adapted chicken breeds	801

Table 19.1	A SWOT analysis of a typical piggery value chain.	821
Table 19.2	Current and futuristic adaptations to heat stress management for pigs (Mutua et al. *2020b*)	827
Table 19.3	Pig production system considering the scale of operation—the Uganda case	829
Table 19.4	Pig breeding companies operating in Africa	838
Table 20.1	Production share of live animals and some products by continent in 2019	854
Table 20.2	Technologies and sequencing systems for next-generation sequencing	869
Table 20.3	Publication of the whole-genome and pangenome sequences of some livestock species	871
Table 20.4	Genome-wide genotyping chips for various livestock species	872
Table 20.5	Quantitative trait loci mapped for various livestock species	873
Table 20.6	Technological developments in gene editing and application in animal production	875
Table 20.7	Summary of gene editing in livestock species	878
Table 20.8	Sample cellular agriculture companies producing a wide range of cultivated- and precision fermentation-based products	885
Table 21.1	Number of poultry and livestock species and the estimated amount of manure produced, and nitrogen and phosphorus excreted in the USA in year 2017	920
Table 21.2	Some chemical preservatives for meat and target organisms	942
Table 21.3	Practical feed improvement technologies for smallholder livestock production systems and their expected or observed impacts	947
Table 21.4	Categories of some of the novel feed ingredients or supplements in livestock feed. Adapted from (Chisoro et al. *2023*)	951
Table 22.1	Examples of diagnostic tests available in South Africa	997
Table 22.2	Examples of gene editing application and transgenes in cattle	1000
Table 24.1	Examples of precision livestock farming technologies	1052
Table 25.1	Constraints of animal selection and crossbreeding in Africa	1075

List of Tables

Table 26.1	Values of indigenous animal genetic resources	1092
Table 26.2	A summary of the main findings of economic valuation studies that have been undertaken in Africa	1096
Table 28.1	Representativity of top-50 leading agricultural universities and colleges in Africa in 2021	1132
Table 28.2	Objectives and expected impacts of capacity building for sustainable intensification of livestock in Africa	1139
Table 28.3	Examples of Big Data applications/aspects in livestock and fishery smart farming processes	1142
Table 29.1	Documents and standard operating procedures	1159
Table 29.2	Some centres that provide genomic and high-performance computing services in Africa	1168
Table 30.1	Status of National Biosafety Frameworks in SSA countries	1195
Table 30.2	Classification of SSA countries on the basis of enabling environment for applications of biotechnology in livestock	1196
Table 30.3	Strategic Priority Areas (SPA) and Strategic Priorities (SP) of the Global Plan of Action for AnGR	1202
Table 30.4	Summary of achievements of AU-IBAR's Animal Genetic Resources Project	1206
Table 30.5	Summary of status of policies and implementation instruments by African countries	1218
Table 30.6	Number of livestock breeds and breed societies in South Africa	1225
Table 30.7	Cartagena Protocol on Biosafety and Nagoya Protocol on Access and Benefit-sharing: status of ratification by African countries and entry into force	1229
Table 31.1	Sample farmer organizations in Africa	1239
Table 31.2	Sample livestock breed societies in Africa	1242
Table 31.3	Sample professional animal production-based societies in Africa and internationally	1247
Table 31.4	Sample FAO books, reports, and databases on animal genetic resources and their management and improvement strategies	1270

List of Boxes

Box 2.1	Forms of pastoralism	28
Box 2.2	Summary of case studies on the impact of livestock grazing system (LGS) management using a multifunctional approach (Ickowicz et al. 2022)	57
Box 4.1	Summary of the risk status of cattle breeds in Africa	176
Box 15.1	Some Useful Terms in Defining Breeding Goals and Breeding Strategies	692
Box 15.2	The Global Plan of Action (GPA, 2007)	697
Box 15.3	AU-IBAR Animal Genetics Project	698
Box 16.1:	Genomics to Identify Appropriate Cross-Bred Dairy Cattle for Smallholder Farming Systems in East Africa—The Dairy Genetics East Africa Project	713
Box 16.2:	West Africa	713
Box 16.3:	The ADGG Programme (https://portal.adgg.ilri.org/)	714
Box 16.4:	Wakwa, a Synthetic Beef Breed in Cameroon	717
Box 16.5:	Genetic Improvement of Cattle in Ghana	718
Box 16.6:	Genetic Improvement of Cattle for Dairy Production in Cameroon	720
Box 16.7:	Cross-Breeding Azawak Zebu Cattle with Local Sudanese Fulani Cattle for Dairy Production in Burkina Faso	721
Box 16.8:	Genetic Improvement of Cattle for Dairy Production in Ethiopia	722
Box 16.9:	Genetic Improvement of Cattle for Dairy Production in Kenya	723
Box 16.10:	Kenya Sahiwal Breed (Muhuyi et al. 1999) Definitions See Box 15.1, Chap. 15, Defining Breeding Goals and Breeding Strategies	726 … 761
Box 17.1:	Cross Breeding: Dorper Sheep in South Africa (Ramsay et al. 2000; Milne 2000)	767
Box 17.2:	Cross-breeding for smallholder farming systems; *The FARM Africa Goat improvement programme*	769
Box 17.3:	Improved Djallonke Sheep: Côte d'Ivoire Adapted from Yapi-Gnaore (2000)	774

xliii

Box 17.4:	D'Man Sheep in Morocco	775
Box 17.5:	The Boer Goat: South Africa (Malan 2000; Visser and van Marle-Köster 2018)	775
Box 17.6:	West African Dwarf WAD Goat	775
Box 17.7:	The Galla Goat: Kenya	776
Advantages of Genomic Evaluation		1025
Box 27.1:	Collaborative Masters in Agricultural and Applied Economics (MAAE)	1118
Box 30.1	Definition of Terms	1189
Box 30.2	Treaty State Descriptors—Ratification, Accession, Approval, and Acceptance	1191
Box 31.1	HPI-Specific Objectives in Cameroon	1257
Box 31.2		1263

Part III

Leveraging Modern Technologies for Improved Utilization of African Livestock Genetic Resources

Role of Modern Technologies for Sustainable Genetic Improvement of African Livestock

Eveline M. Ibeagha-Awemu, Sunday O. Peters, Martha N. Bemji, Jean M. Feugang, Richard Osei-Amponsah, Daniel Trocmé, and Raphael Mrode

Abstract

Evidence of the impact of advanced technologies in genetics, nutrition and health management in the last seven decades has led to a twofold increase in the milk produced by a Holstein cow and a fourfold increase in the size of a broiler chicken in Western countries. Therefore, harnessing breakthrough technologies of genomic breeding, reproduction, management practices (health and environment), etc., with supporting infrastructure, funding and technical know-how is key to solving current problems and achieving rapid developments in the African livestock sector. This chapter presents an overview of some modern technologies that have played and continue to play significant roles in advancing sustainable genetic improvement in the livestock sector. Firstly, Section 20.2 presents the key issues impeding sustainable livestock improvement within Africa's complex livestock production systems. Then the phenomics technologies used for data capture to support genetic selection are presented in Section 20.3 followed by the breakthrough genomics technologies responsible for sustained genetic improvements in livestock traits (Section 20.4). The emerging technologies of genome editing (Section 20.4.3) and cellular agriculture (Section 20.4.4) are also presented. Breakthrough reproductive technologies that have been responsible for the spread of high-

E. M. Ibeagha-Awemu (✉)
Sherbrooke Research and Development Centre, Agriculture and Agri-Food Canada, Sherbrooke, Quebec, Canada
e-mail: Eveline.ibeagha-awemu@agr.gc.ca

S. O. Peters
Department of Animal Science, Berry College, Mount Berry, GA, USA

M. N. Bemji
Department of Animal Breeding and Genetics, University of Agriculture Abeokuta, Abeokuta, Nigeria

J. M. Feugang
Department of Animal and Dairy Sciences, Mississippi State University, Mississippi State, MS, USA

R. Osei-Amponsah
Department of Animal Science, School of Agriculture, University of Ghana, Legon, Accra, Ghana

D. Trocmé
Strategy and Development Director. Adventiel, Pace, France

R. Mrode
International Livestock Research Institute, Nairobi, Kenya

Scotland Rural College, Edinburgh, UK

merit genetics are examined in Section 20.5. This chapter concludes with future perspectives (Section 20.6) on how the adoption and implementation of modern genomic technologies can accelerate desired improvements in the African livestock sector.

Keywords

Sustainable genetic improvement · Genomics · Gene editing · Reproduction · Phenomics

Abbreviations

ADGG	African dairy genetics gains
AI	Artificial insemination
ART	Assisted reproduction technologies
CA	Cellular agriculture
Cas9	CRISPR-associated protein 9
CH4	Methane
CNV	Copy number variations
CRISPR	Clustered regularly interspaced short palindromic repeats
CT	X-ray-computed tomography
DITI	Digital infrared thermal imaging
DNA	Deoxyribonucleic acid
DSB	Double-stranded break
EBV	Estimated breeding values
EC	Electrical conductivity
ECOWAS	Economic Community of West African States
eQTL	Expression quantitative trait loci
FAO	Food and Agriculture Organization
GBLUP	Genomically best linear unbiased prediction
GEBV	Genomic estimated breeding value
GHG	Greenhouse gas
HDR	Homology-directed repair
HMC	Handmade cloning
HR	Homologous recombination
ICSI	Intracytoplasmic sperm injection
ILRI	International livestock research institute
IVF	In vitro fertilization
IVM	In vitro maturation
IVP	In vitro produced
K^+	Potassium ion
MAS	Marker-assisted selection
MI	Microinjection
MOET	Multiple ovulation and embryo transfer
MUN	Milk urea nitrogen
NA^+	Sodium ions
NGS	Next generation sequencing
NHEJ	Nonhomologous end-joining
OPU	Ovum pick-up
PGF2α	Prostaglandin
PMSG	Pregnant mare's serum gonadotropin
QTL	Quantitative trait loci
RNA	Ribonucleic acid
SCNT	Somatic cell nuclear transfer
SF6	Sulphur hexafluoride
SMRT	Single molecule real time
SNP	Single nucleotide polymorphism
TALEN	Transcription activator-like effector nucleases
UK	United Kingdom
VIA	Video image analysis
WGS	Whole-genome sequence
ZFN	Zinc finger nucleases

20.1 Introduction

The rapid growth of Africa's population, currently 1.4 billion and estimated to reach 2.5 billion by 2050, is not matched by growth in agricultural productivity, including livestock production. Despite the high genetic diversity (species and numbers) of African livestock populations (see Chapters 2, 3, 4, 5, 6, 7, 8, 9, 10, 11, 12, and 13), productivity is low, plagued by a myriad of problems, and unable to sustain lives and economic development. Sustainable improvement of African livestock species is urgently needed to meet the food needs of the people as well as for sustained economic development of the various countries.

To advance the needed solutions to Africa's food and health needs, it is generally accepted that modern technologies will play a big role. In particular, developments in genomics, reproduction, growth, phenomics, health and management technologies, etc., will support rapid growth in the livestock sector. Effective research and development activities in solving even basic problems require multidisciplinary, problem-solving and technological approaches. For instance, major breakthroughs and scientific progress in deciphering whole-genome sequences to developing vaccines for the control of diseases resulted from harnessing new technologies, supporting infrastructure and the joint efforts of large multidisciplinary teams of professionals of varied backgrounds, such as biologist, clinicians, bioinformaticians, mathematicians, computer scientists and project administrators among others.

In particular, the implementation of genetic improvement has been the most efficient strategy used to increase the biological performance, efficiency and sustainability of animal production systems in recent decades. The genetic improvement science evolved through the application of statistical approaches, quantitative genetic theories, artificial insemination and organized breeding practices, which resulted in rapid gains in livestock traits in the past decades, and in recent times more genetic progress in livestock traits is achieved with the implementation of genomic selection and advanced reproductive technologies.

Technological developments of genomic tools such as next-generation and third-generation sequencing approaches, and dense single nucleotide polymorphism (SNP) genotyping panels have been crucial to both advance basic knowledge on the genomic bases of complex traits and increase the rate of genetic progress in livestock production and health. Aided by these and other technologies, genomic information is now routinely used in selective breeding to support livestock production, in species like cattle, chicken and pig (Georges et al. 2019). Genomic technologies have consequently transformed the livestock industry in terms of increased productivity, environmental adaptation and health management. Therefore, efforts to advance livestock improvement in Africa will require multiple technologies and approaches, including genomics, reproduction, nutrition management, phenomics, statistical/bioinformatics management of data, etc., and the involvement of multidisciplinary teams with the required supporting infrastructure, government policies and financial investments. This chapter will focus on phenomics, genomics, reproduction and other emerging (gene editing, cellular agriculture) technologies that can support sustainable genetic improvement in African livestock production systems, while technological developments supporting other aspects of livestock production that are necessary for optimal genotype x environment interaction for optimal phenotypic expression, like nutrition and health, have been presented in Chapter 21.

The adoption of technology in various aspects of livestock farming has been extensively studied (Feder et al. 1985; El-Osta and Morehart 2000). It is stipulated that larger farms tend to be at the forefront of adopting technologies which have helped to increase profitability and feed quality, and reduced production costs and incidence of disease outbreaks. In dairy production, the late adopters of technology are small-scale milk producers because most of the technologies are capital-intensive and not easily affordable (El-Osta and Morehart 2000), which is one of the issues facing livestock farming in Africa. However, with the current low levels of productivity of most African livestock, the adoption and use of modern technologies are necessary to drive and sustain improved management and productivity. This chapter will focus on different technologies for sustainable genetic improvement adopted by low-, medium- and large-scale livestock production systems.

20.2 Key Issues Related to Sustainable Livestock Improvement in African Livestock Production Systems

In Africa, 70 percent of the resource-poor farmers own animals which represent about half of their assets and account for 20% agricultural output (The World Bank 2018). Resource-poor small-holder farmers constitute the largest labour force that produces most of the milk and meat in Africa. Given that farming is subsistence in nature, sustainable livestock production and improvement are marred by several factors. These limitations (e.g. low productivity, genetic erosion through indiscriminate crossbreeding, lack of performance recordings, limited feed resources, among others) are examined below to provide recommendations for affordable technologies that could promote the improvement and sustainability of the livestock enterprise.

20.2.1 Low Productivity of Indigenous Animal Genetic Resources

Livestock productivity is generally low across the African's production systems. Current indicators show that agricultural output in terms of meat, milk, hide and skin, and draft power is generally low, irrespective of the type of production system (pastoral, agro-pastoral, mixed small holder farming, urban or peri-urban) (Ibeagha-Awemu et al. 2019). Considering the predominant meat and milk-producing livestock species (cattle, goat, pig, poultry and sheep), Africa has the highest number of cattle breeds (44.6%) in the world followed by Europe (25.1%) and Asia (16.2%), while North America has the least numbers (0.9%) (Table 20.1). However, Africa has only 19.9% of live cattle and ranks least in terms of milk and butter production and only better than Oceania by 3.7% in meat production in 2019 (Table 20.1). The productivity of chickens and pigs is considerably lower compared with other continents. Although Africa ranked second in the number of live sheep and goats, meat production from these species is proportionately lower than in the Asian continent (Table 20.1). The low productivity of the African indigenous livestock resources is generally due to poor nutrition (low quality and quantity of feed materials, limited grazing land, low-quality pasture), poor management (e.g. disease) and lack of sustained genetic selection of high-producing individuals. Production is characterized by extensive systems of management that are not designed to optimize animal growth and productivity, lack of data recording schemes as means to judge animal performance, unimproved breeding animals, indiscriminate mating and crossbreeding within and between breeds, high disease burden, limited water resources, desertification and increasingly less rainfall in many parts of Africa, limited support from national governments and lack of appropriate legislation promoting livestock improvement by most governments.

Implementation of innovative technologies, including genomic and management technologies to increase milk and meat production, technologies to improve nutrition and health, and phenomics technologies for data recording and reproduction technologies (discussed in this chapter and Chapter 21) will go a long way in advancing the livestock sector in Africa.

20.2.2 Erosion of Diversity Through Indiscriminate Crossbreeding

In most African livestock production systems, directional or intentional breeding of animals is not practised. Rather, animals breed indiscrimi-

Table 20.1 Production share of live animals and some products by continent in 2019

Continent	Live animals (%)					Meat production (%)					Milk/Butter (%)
	Cattle	Goat	Sheep	Pig	Chicken	Cattle	Goat	Sheep	Pig	Chicken	Cattle
Europe	9.9	2.1	12.7	21	10.6	19.5	2.7	15.9	26.4	16.8	53.3
Americas	35.6	4.2	8.0	17.1	26.6	48.1	2.9	5	17.4	44	20.3
Asia	31.9	56.9	40.6	58.4	54.5	19.4	70.5	47.2	54.7	33	12.5
Oceania	2.7	0.4	11.6	0.6	0.6	4.6	0.4	13.9	0.5	1.3	10.9
Africa	19.9	36.4	27.1	3.0	8.1	8.3	23.4	18.1	1.0	5.0	3.0

Source: Data used to generate this table was from FAOSTATS (2019)

nately within and between populations. As a result, frequencies of genes that favour increased genetic gains in many traits are constantly diluted, resulting in no genetic gain in traits over the years. Crossbreeding activities varied widely across species and countries (Leroy et al. 2016) due to indiscriminate use of breeding animals, thereby causing high gene flow between populations and erosion of animal genetic diversity in cattle (Traoré et al. 2017; Ibeagha-Awemu et al. 2004; Mwai et al. 2015), sheep (Brahi et al. 2015), chicken (Eltanany et al. 2011; Kejela 2020) and pig (Ramirez et al. 2009). Given the poor performance of indigenous livestock, some African countries adopted crossbreeding of locally adapted breeds with improved exotic breeds as the quickest approach to enhance productivity. According to the history of animal genetic improvement, mostly crossbreeding with exotic breeds started in West Africa and extended to other regions of sub-Saharan Africa by colonial settlers (Madalena et al. 2002). In Cameroon, Nigeria and Ghana, crossbreeding with exotic and indigenous breeds for improvement in meat and milk production took place in the early decades of the twentieth century. However, crossbreeding programmes were not sustainable due to the poor adaptation of the exotic and crossbreds to low-quality feed and the environmental conditions (Leroy et al. 2016; Madalena et al. 2002). Although attention shifted from crossbreeding to straight breeding in the 1980s following the recognition of adaptive features of indigenous breeds (Madalena et al. 2002), their genetic uniqueness and adaptation to local conditions have been compromised, giving way to genetically mixed populations with weak differentiation between sub-populations in chicken (Eltanany et al. 2011), pig (Ramirez et al. 2009) and sheep (Brahi et al. 2015; Agaviezor et al. 2012). Carefully designed breeding programmes (see Chapters 15, 16, 17, 18, and 19) and implementation will limit the effects of indiscriminate crossbreeding and promote sustainable genetic gain increases in desired traits and improvement in overall livestock productivity.

20.2.3 Scattered, Small Herd Sizes and Lack of Performance Recordings

Information on the phenotype is fundamental for understanding the genetic basis of performance and for making informed decisions for effective herd and flock management. Unfortunately, African subsistence agricultural systems are characterized by scattered, small herd sizes, and most smallholder farmers make use of indigenous and nondescript crossbreds with no animal recording (Ibeagha-Awemu et al. 2019; Zonabend et al. 2013a; Marshall et al. 2019). This is further compounded by a lack of coordination among the institutions that are set up to support animal breeding programmes and a lack of skilled manpower (Zonabend et al. 2013a).

While adequate and sustained animal performance recording is the backbone of livestock improvement schemes in industrialized nations, livestock data recording is not commonly practised in most African countries. This lack of sustained data recording schemes in Africa is caused by many factors, including poor animal identification systems and inadequate infrastructure for collecting and processing data (Ojango et al. 2017). Only a few countries in Africa (e.g. South Africa, Kenya, Ethiopia, Botswana and Zimbabwe) have organized and sustained or on-and-off or collapsed data recording schemes for performance monitoring. The adoption of phenomics technologies suitable for smallholder farmers and for all production systems will facilitate data collection, which is the basic requirement for performance evaluation and monitoring.

20.2.4 Limited Quantity and Quality of Feed Resources, and Low Efficiency of Feed Utilization

Many forage-based animal production systems in the tropics grow slowly and produce small amounts of milk and meat because their diets are inadequate in protein, energy and micronutrients (National Research Council 2009; Muyekho et al. 2014). Many ruminant animals usually receive or graze only plant residues (e.g. wheat straw and corn straw) in the dry season. About 72% to 93% of the feed consumed by ruminants in Africa is made up of low-quality natural pastures and crop residues (FAO 2014). Supplemental feeding with concentrates is common in semi-intensive and intensive systems of production, which caters to only ~10% of the ruminant livestock produced in Africa. Improving the nutritional quality of forage and crop residues is necessary for sustainable livestock production in Africa. Common methods for increasing digestibility and protein content of crop residues include physical (grinding, chaffing and chopping of roughage or fibrous feeds and pelleting), biological or chemical treatment (e.g. ammonization, alkali and urea) (Valli 2020). Forage can be improved by breeding and the creation of improved pastures for grazing. In particular, the implementation of genomic breeding technologies for improving the nutritional and yield qualities of forage crops holds great potential for improving livestock productivity in Africa.

20.2.5 Animal Health and Welfare

Livestock productivity is hindered by several constraints, including a high prevalence of diseases and parasites irrespective of production systems (Santoze and Gicheha 2019). Certain types of livestock systems are associated with zoonotic and emerging infectious diseases or food-borne illnesses (Kristjanson et al. 2010). This has implications for the high cost of disease control, as witnessed by improved dairy production in Kenya (Muyekho et al. 2014). According to the Food and Agricultural Organization (FAO) (FAO 2018), poor awareness of animal health in subsistence-oriented production systems, combined with poor communication and transport infrastructure, often translates into ill-functioning private and public animal health services, thereby calling for the strengthening of existing health services and the establishment of alternative forms of animal health services.

20.2.6 Environmental Pollution/ Degradation

Livestock and livestock systems can have negative effects locally which include land conversion and land degradation. Twenty percent of the world's pastures and rangelands have been degraded to some extent through overgrazing, compaction, and erosion caused by livestock activities. Poultry, pig, sheep, goat, cattle and other domesticated animals generate around 85 percent of the world's animal faecal waste (The World Bank 2018) which can contribute to environmental pollution if not properly disposed of or managed. Most cities in Africa are characterized by the constant presence of livestock which is a source of pollution and may be reservoirs of diseases including zoonotic diseases (Wilson 2018). It has been estimated that the livestock sector in Africa contributes to more than 70% of the continent's agricultural greenhouse gas (GHG) emissions, mainly attributed to methane and nitrous oxide from cattle and their manure (ILRI 2021). However, due to huge research gaps, understanding of Africa's contribution to global GHG emissions remains incomplete and highly uncertain (Kim et al. 2016).

20.2.7 Farmers' and Herders' Clashes

Insecurity has discouraged both farming and herding in many areas, with serious effects on livelihoods and food supply (Blench 2004). The presence of animals in many African cities is a major source of conflict (Wilson 2018). In Nigeria and Ghana, violent herdsmen and smallholder farmer conflicts are mainly due to the migration of herdsmen from the northern regions, which is aggravated by climate change and diminishing agricultural lands, to the southern regions in search of pasture for their animals and low adherence to the ECOWAS (Economic

Community of West African States) Protocol on Transhumance (Asueni and Godknows 2019; Majekodunmi et al. 2014). It has been noted that other important aspects of violence beyond 'farmer-herder clashes' include revenge for past attacks, economies of conflict (cattle rustling and armed robbery for gain), resource scarcity, decreasing interdependence of pastoral and agricultural economies, cultural/religious differences between herders and farmers and institutional failures to resolve conflicts as well as the broader political and historical contexts (Asueni and Godknows 2019; Higazi 2013). Addressing this issue is paramount to supporting livestock and crop production, as well as saving lives. The local, national and regional governments in affected areas must take concrete steps to addressing farmer-herder conflicts by setting supporting livestock policies, creating designated grazing lands and grazing corridors away from crop farm lands and taking adequate measures to protect the lives of crop farmers.

20.2.8 Marketing Access

Livestock marketing faces many challenges in Africa, including the lack of marketing facilities, inadequate marketing channels and infrastructures, organization and methods, inadequate government policies and limited access to across country and international markets (Ibeagha-Awemu et al. 2019). These challenges must be addressed by all African countries and livestock production must be recognized as a viable business to spur increased implementation of modern technologies for increased productivity.

20.2.9 Funding

The implementation of modern technologies described in the sections below is dependent on available/adequate and sustainable funding of livestock research, farming and associated aspects, including capacity building, health management, feeding management and market access among others. Livestock production is an important source of food for the population and merits government funding as does services like education, health and the military, among others.

20.2.10 Government Policies

Policies aimed at sustainable livestock production and mitigation of the impact of animal production are largely aspirational goals. While many countries in Africa recognize the need for livestock breeding and management policies for setting priorities and directing activities in livestock breeding (Zonabend et al. 2013b), the implementation of policies is lacking. Recently, the Africa Union InterAfrican Bureau for Animal Genetic Resources (AU-IBAR) undertook a study with the aim to facilitate or fast-track implementation of the Global Plan of Action for the sustainable use of animal genetic resources in Africa (AU-IBAR 2018). Despite such initiatives, livestock production in most of Africa is unregulated and saddled with low human and institutional capacity to take advantage of appropriate innovative technologies. Therefore, putting in place enabling national, regional and continent-wide policies aimed at supporting growth of the livestock sector are necessary for the growth and expansion of livestock production and marketing in Africa.

20.3 Information and Phenomic Technologies in Animal Production

Phenomics may be broadly defined as the application of technologies to enable the collection of phenotypes cheaply, easily, more efficiently and in large volumes. It is therefore very important as phenotypic data play a key role in the design of effective herd and flock management systems and in understanding the genetic basis of livestock performance. In addition, at the national level, capturing phenotypes is fundamental in national aggregates of produce that underlines government agricultural policies and projections (Mrode et al. 2020). Therefore, at the farm level, phenomics is important for profitability and in the design of breeding programmes, and at the national level for effective government agricultural policies. Labour accounts for about 70% of the production cost of most livestock products, therefore in an attempt to reduce cost, there have been significant developments in the last two decades in sensors, digital communications and other technologies that enable cheaper and more accurate recording of existing phenotypes and access to new ones. Also, the more accurately we can capture phenotypic records, the higher the rate of genetic progress that can be achieved. The elaborate efforts on technologies to capture data or performance are proving valuable for data collection, especially for difficult and expensive to measure traits such as reproductive and health events, feed intake and carcass traits.

In the following sections, we examine existing and emerging technologies for information and phenomics, and in particular their suitability and usage in African production systems, given that systematic data capture has been a missing foundational stone that has hindered sustainable livestock genetic improvement programmes in Africa. In general, the degree to which these technologies are used is influenced by cost, practicality and portability.

20.3.1 Information Systems in Livestock Management

What has information technologies to do with the development of livestock breeding and farming activities? Economically, the production and income of farmers can grow only with their ability to sell quality meat and dairy products on the national and international markets at competitive prices. But nowadays, having access to these markets requires providing quality proof. It thus requires animal registration and traceability as well as performance management. The interest of farmers in high-performance breeds will be strictly proportional to their awareness of this economic opportunity.

This is where integrated information systems come into play. Their ability to integrate traceability, health, reproduction, performance recording, genetics and marketing of products in a unique system, and to provide each type of user their rights, is the main key to success. Indeed, the development of the livestock sector needs all these aspects to be synchronized.

France was the first country to identify its cattle in the 1970s, creating a strong link between identification, health and genetic systems. The European Union has progressively adopted restrictive regulations due to the bovine spongiform encephalopathy crisis and the growing demand for traceability. The whole world has taken inspiration from these principles of regulation to increase their ability to export their products.

In 2015, Kyrgyzstan was assisted by FAO in implementing an integrated system of animal identification and traceability. The integrated identification and traceability system was developed by Adventiel (https://www.adventiel.com/) in collaboration with public organizations. The multi-partner system created the ability to collect continuously accurate information on livestock on the field (which gave a modular structure to the system): with a mobile application, a web portal and a common database. Specifically, it records:

- Identity of all farmers, farms, premises, marketplaces and slaughterhouses.
- Identity of animals, unique from birth to death (in line with the International Committee on Animal Recording guidelines).
- Traceability of the movement of live animals.
- Record of animal health events and status (prevention, diseases, treatments, etc.)

These modules are the core of the system. Additional components are as follows:

- Marketing modules such as traceability of animal products' batches.
- Reproduction and insemination modules.
- Performance recording field applications.
- Genetic, genotyping and pedigree management module. Genetic improvement is impossible at a large scale without extensive identification of the animals.

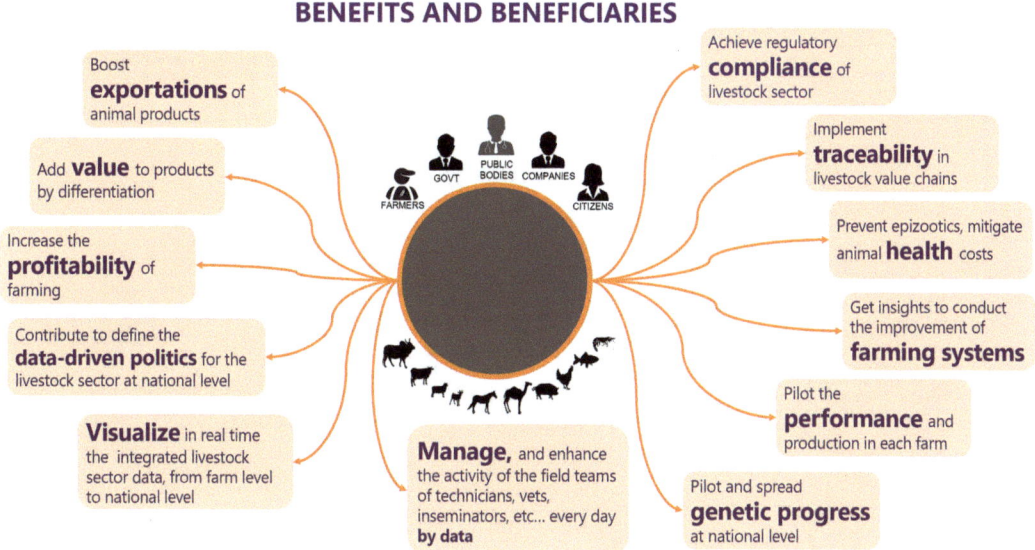

Fig. 20.1 Benefits and beneficiaries of a livestock identification and management system

Next is the ability to exchange data with other systems in a secure manner at the national and regional scales and to offer reliable guarantees regularly to users (buyers and/or foreign authorities). This possibility to provide guarantees from the suppliers to the users accords numerous benefits (Fig. 20.1):

- Foster long-term commercial relationships and alliances in the region and beyond.
- Dramatically reduce the restrictions and prohibitions on animals and products.
- Allow economic players to pay equitable prices to farmers for highly valuable products.

20.3.2 Sensors for Fertility, Health and Body Weight Traits

Sensors are defined as devices that measure a physiological or behavioural parameter of an individual animal that might be related to fertility or health or any other trait, and that enable automated, on-farm detection of changes that warrant a management intervention (Rutten et al. 2013). Sensors can be classified into two categories: attached and non-attached. The attached sensors are either on-animal sensors that are fitted on the outside of the animal's body, or in-animal sensors that are inside the body such as rumen implants or bolus. On-animal sensors such as wearable sensors can monitor the overall well-being of the animal, such as milk quality, pregnancy hormones, rumination, illness, and lameness efficiently and more accurately (Fig. 20.2) (Siberski-Cooper and Koltes 2021). For non-attached sensors, animals may pass by, over, or through these for the measurements to be taken. This class of sensors includes on-line sensors that automatically take a sample such as milk that is analysed by the sensor or in-line sensors that take measurements in a continuous flow of a product (e.g. milk from cows during milking) (Simm et al. 2021). In general, sensors monitor the daily activity and health-related issues of animals and at the same time generate data for the entire herd, which is analysed and used for herd/farm management decisions. The performance of wearable sensors commonly used to collect behavioural and physiological data for precision livestock farming was recently reviewed (Lee and Seo 2021).

Most sensors have been developed for livestock management with the aim of reducing labour

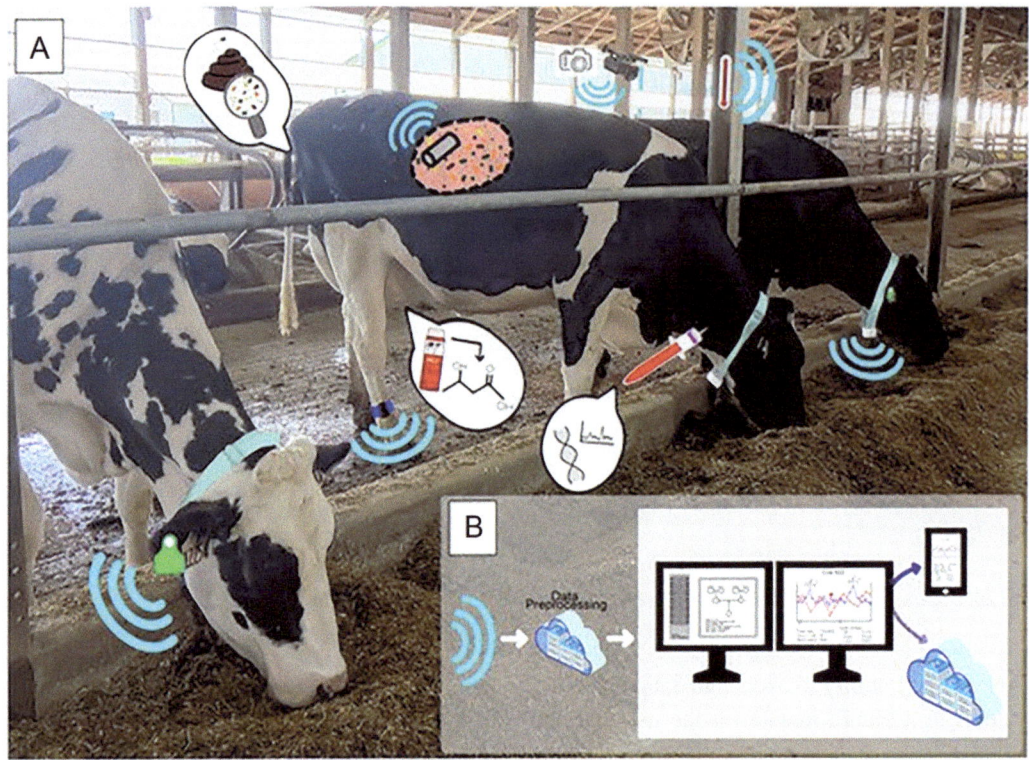

Fig. 20.2 Wearable sensors such as milking collar, leg band and ear tags, temperature and proximity sensors; image-based monitoring systems and internal sensors collect data (**a**) which is analysed (**b**) for informed management decisions. (From Siberski-Cooper and Koltes (Siberski-Cooper and Koltes 2021), licensed under CC-BY 4.0)

requirements and costs. There are several sensors for the automated detection of oestrus, based on either the activity of the cow or on progesterone level of milk. Also, electronic Alpha-Detectors (e.g. Alpha-D) detect the animal's electronic identification and transmit the data on the frequency of mating, true and false coverings and the number of animals mated to a centralized computer. This type of sensor therefore interacts with individual animal electronic identification and collects data to feedback systems. The sensors that monitor the activity of cows are pedometers, activity meters (sometimes referred to as activometers) and three-dimensional (3D)-accelerometers (Rutten et al. 2013). These are usually on-cow sensors with pedometers and 3D accelerometers usually attached to one of the cow's legs and the activity meters to a neck collar.

The automatic monitoring of progesterone levels in collected milk samples is achieved using on-line sensors such as biosensors and immunostrips. An oestrus alert is provided for the farmer with these sensor systems and in some cases with a probability measure of the alert. Some sensors monitor and report on the reproductive state of the cow such as postpartum anoestrus, oestrus cycling or potentially pregnant.

Several activity meters have also been used for the automatic detection of health traits such as locomotion problems in dairy cows. The walking behaviour/gait has been measured in cattle and poultry using sensors such as 3D accelerometers and/or video cameras. For example, accelerometers (tied around the foot, neck or head) measure the movement, speed and direction of animals (Roland et al. 2018). Three axial accelerometers for instance record movement patterns and behaviour, such as resting, grazing, moving and running/playing or lameness in animals in a three-dimensional pattern (Fogarty et al. 2018). A recent study demonstrated the use of axial accelerometers for reproduction control/management in dairy sheep (Mozo et al. 2019). In addition, the detection of locomotion problems has been automated by means of several off-animal sensors such as four balance-weighing floors, weighing platforms, parallel force plates and force distribution plates by studying the weight distribution among the cow's legs (Rutten et al. 2013). In these systems, animals are required to stand on or walk over these sensors and an alert for abnormal locomotive behaviour is provided to the farmer, with some measure of the degree of reliability or analyses of processed sensor data such as walking activity or weight over time.

Rumination activity collars and sensors assist in detecting behavioural and activity changes in cows, and early indication of sickness or infection. Devices like pedometers effectively track specific cow actions like its readiness for breeding or going into labour, facilitating the farmers' help for opportune assistance. The rumination collar uses a microphone to measure a dairy cow's rumination to ensure a smooth transition. Changes in rumination can be an early sign of calving diseases, negative impacts of ration changes, cow discomfort, etc. Knowing normal rumination patterns for a specific dairy cow helps mitigate the effects of manageable stressors (Groher et al. 2020; Silvi et al. 2021).

A rumen bolus measures rumen temperature and pH levels to help identify systemic infections that need to be treated. If a rumen bolus tells a dairy producer that a cow's pH is dropping, it is a sign that a total mix ration (TMR) is not thoroughly mixed, and a ration change has negative impacts. This can help producers make appropriate changes to their dairy cattle nutrition programme (Krpalkova et al. 2021; Simitzis et al. 2021).

Increasingly, automated weighing techniques are being used to monitor growth or to monitor changes in weight in several species that may provide an early indicator of health issues. Typically, animals walk over platforms mounted on weigh cells, with both weights and animal identities being captured using non-attached sensors (Simm et al. 2021).

The electrical conductivity (EC) of milk has been used for routine in-line tests for the diagnosis of sub-clinical mastitis in dairy cattle for some years (Milner et al. 1996). Typically, milk EC is determined by measuring the concentration of cations and anions. During infection, milk concentrations of Na^+ and Cl^- ions increase while concentrations of lactose and K^+ are decreased. The sensor system uses the statistical relationship between the EC measurements and a gold standard measure for mastitis, such as the California Mastitis Test, to provide the farmer with a mastitis alert and an associated degree of reliability.

20.3.3 Animal Position Monitoring

Monitoring animal position, foraging and other behaviours can benefit animal health and welfare by continuously monitoring each animal in the flock. Any small deviation from 'normal' behaviour (for that individual animal) can be quickly identified and flagged to the farmers. Some farmers use Radio Frequency Identification tags to track animals' origins, movements and whereabouts. Some farmers use pedometers (like Fitbits for cows) to track the daily activities of each cow and to identify cows that may need a little extra care or help (Silvi et al. 2021; Krpalkova et al. 2021). Other times, farmers track, monitor, and control nutrition, behaviour, pregnancy, milking frequency, milk output anomaly and activity level in real-time owing to wearable animal gadgets similar to human fitness trackers. These intelligent animal trackers may be placed on the ears, tails, legs, neck, or any other portion of the body (Silvi et al. 2021; Shalloo et al. 2018). Cattle monitoring drones track the cattle and herd them back from the fields to the barns. Some drones are outfitted with thermal sensing equipment, which allows them to track cattle based on their body heat. Drones may also scan pasture lands and communicate information about whether they are suitable for cattle grazing (Joshi et al. 2019).

A computer-controlled gate system opens and closes electronically in automated cow traffic management system. These gates can categorize livestock based on their readiness to milk. The animals ready to be milked are transferred to the milking area, while the others are placed in the waiting area or back to their barns. Global navigation satellite system technology also allows for characterizing grazing behaviour, including grazing patterns, paths and favoured areas. Grazing activities can also be differentiated based on the speed of movements. Tracking location on pasture, through global positioning system sensors, has been successfully used to detect static or dynamic unitary behaviours differentiated through changes in path speeds, such as foraging or grazing, resting and walking. Likewise, the use of global positioning system 'collars' for livestock has opened the possibility of recording detailed position data for extended periods, thus allowing for a complete understanding of the habits and causes of the spatial distribution of ruminants (Shalloo et al. 2018; Monteiro et al. 2021).

20.3.4 Automated Systems for Measuring Feed Intake

Feed intake constitutes about 70% of the production costs in most livestock systems, therefore improvement in feed efficiency will enhance farm profitability and reduce the environmental impact of livestock production. However, capturing individual animal feed intake is very expensive and difficult. The earlier systems for capturing feed intake involved the use of specialized equipment that consisted of an individual bin for an animal. Then the access of each animal is restricted to only one bin through the use of an electronic identification device mounted in a collar. Initially, a certain amount of weighed feed is added to the bin and after a specified time interval (12 hours, for instance) the feed remaining in the bin is weighed and discarded. Therefore, the feed consumed by the animal with access to the bin was calculated as the difference between the two feed weights. However, this system, although cheap, is labour intensive as feed has to be manually weighed in and out of the bin, and it might also influence the cows' feeding behaviour.

Advances in technology have resulted in the emergence of several automatic systems to measure individual animal intake, but most of these systems are rather expensive and tend to be used mostly in nucleus breeding populations or research farms. In general, these systems have feed bins mounted on weigh cells that are constantly monitored. The identity of animals accessing a bin is captured from an electronic identity device either on the neck collar or ear tag by an aerial device mounted on the entry to the bin. The weight of the feed bin is constantly monitored, therefore, the feed intake for a particular animal is recorded as simply the change in weight of the bin during the period the animal is registered as present at the feed bin. The system offers flexibility as animals can feed at any bin and individual animal feeding behaviour can be studied from additional records on animals such as the daily profiles of feed intake per animal, meal size, meal duration and the frequency of feeding.

Some examples of modern technologies for measuring individual feed intake and in some cases water intake, include GrowSafe (GrowSafe Systems 2021), Calan Broadbent Feeding System (American Calan Inc 2019) and Insentec Roughage Intake Control system (Hokofarm Group 2021).

20.3.5 In Vivo Techniques for the Prediction of Carcass Traits

Carcass and meat quality traits are very important traits in a society that is becoming more health conscious by considering which animal products to consume. However, these traits are expensive and difficult to measure and direct measurement can only take place after slaughter, which usually precludes such measurements on selection candidates in the breeding programmes. In the last decade, much research has been tailored in finding non-invasive techniques to predict various carcass traits, such as carcass fat and lean weight or proportion or distribution, on live animals (in vivo). A detailed review of non-invasive techniques for carcass and meat quality trait measurements was presented by Scholz and colleagues (Scholz et al. 2015). In general, these techniques involve the use of electromagnetic or mechanical energy, which is able to pass completely or partially through body or carcass tissues, like muscle, fat and bone. As indicated by Scholz and colleagues (Scholz et al. 2015), the electromagnetic signal produced by the instrument may originate from mechanical energy such as 'photon' radiation (X-ray-computed tomography (CT), dual-energy X-ray absorptiometry) or sound waves (ultrasound) or radio frequency waves (magnetic resonance imaging). The signal interacts with tissues in the body or carcass at the atomic or molecular level, resulting in secondary signals that are detected by the instruments and processed to measure tissue depth, area, volume or distribution of fat, muscle and bone or bone mineral. The measurements from these devices are usually validated through a regression equation; with these measures used to predict the measurements obtained directly from the carcass or with proportions or weights of tissues from carcass dissection. The measure of accuracy of prediction by the device is measured by the R^2 of the regression analysis, with high R^2 values indicating high precision of prediction.

Ultrasonic measures have been the most widely used, especially in pig, for in vivo prediction of meat quality, as a result of being relatively cheap, mobile and moderately accurate. More recently, especially for sheep meat, CT has been used more widely in breeding programmes for livestock and it is considered to be the most accurate of the methods discussed above. The details of the application of CT scanning for various carcass traits and utilization in sheep breeding programmes were presented recently (Simm et al. 2021). In summary, CT scanning has been used to estimate carcass composition and muscularity in UK terminal sire breeding programmes since 2000 (Bunger et al. 2014). More recently, intra-muscular fat in live lambs and carcass cuts has been measured by CT scanning as a measure of meat quality (Lambe et al. 2017). The results of Lambe and colleagues showed that utilizing CT measurements in a selection index can maintain levels of intra-muscular fat, hence improving eating quality, while also improving carcass quality through increased lean per cent and decrease in total fat in commercial slaughter lambs (Lambe et al. 2017).

20.3.6 Video Image Analysis/ Computer Vision Technologies for Weight, Conformation and Carcass Traits

The prediction of carcass weight as well as the weight of individual primal cuts and conformation and fat class through automated video image analysis (VIA) carcass grading technologies has been implemented in a number of countries. For instance, the VBS2000 VIA machines (E + V Technology 2021) are used in several abattoirs in the UK for this purpose. The automated VIA carcass grading technology involves the use of video cameras on the slaughter line and specialist software to classify the carcasses. In practice, one side of the carcass is suspended on a holding frame while a digital camera on the VIA machine takes a 2D (under normal lighting) and 3D (under striped lighting) image, using a previously calibrated lighting arrangement. These images are analysed to predict the weight of the carcass, the individual primal cuts, conformation and fat class. The prediction of the weights for the carcass cuts is based on a multiple-regression method using both the carcass weight and the VIA information. The precision of prediction (R^2 values) reported by Pabiou and colleagues (Pabiou et al. 2011) ranged from 0.65 for the predictions of low-value cuts (lean trimmings, ribs, flank, and brisket) in heifers to 0.93 for predictions of high-value cuts (silverside, topside, knuckle, salmon cut) in steers. The automation of this process means that carcass weights and primal cuts can be measured in a large number of animals thus enabling genetic evaluations for these traits. Medium estimates of heritability (0.40 to 0.46) were reported (Moore et al. 2017) for a range of cuts in crossbred beef cattle in the UK. Breeding values are being routinely calculated for various prime cuts in the UK and this implies that breeding animals can be selected for a higher proportion of high-priced cuts. Moreover, the use of these traits in genomic selection means selection for carcass weight and primal cuts was carried out early in life, thereby reducing the generation interval.

More recent application of VIA involves the use of 3D imaging technology and machine learning algorithms (artificial neural networks) to predict live weight and carcass characteristics of live beef animals. The benefits of these approaches include the ability to obtain multiple measurements of traits of interest with minimal animal stress and use the repeated measurements to monitor changes in weight and conformation and predict the optimal time of slaughter. Using three-dimensional images, liveweight on finishing steer and heifer beef cattle of various breeds and machine learning algorithms, Miller and colleagues (Miller et al. 2019) predicted liveweight and several carcass characteristics for the animals. The accuracy of the prediction models they developed based on R^2 was 0.7 for liveweight, 0.88 for cold carcass weights and 0.73 for saleable meat yield. The use of 3D imaging technology has also been used to predict liveweight and various body measurements such as body volume, area, and length in pigs (Wang et al. 2008) and broilers (Mortensen et al. 2016).

20.3.7 Milk Mid-Infrared Spectral Data

There has been an increasing use of milk spectral data as an innovative and low-cost phenotyping strategy in dairy cattle for health and resilience traits. Milk contains molecular signatures or chemical properties reflecting the physiological status of the cow, their performance, their behaviour and environmental impact which can and is being increasingly explored by mid-infrared (MIR). The MIR spectrum represents the absorptions of infrared radiation at frequencies correlated to the vibration of specific chemical bonds within a molecule. Therefore, the MIR milk spectrum represents the chemical composition of milk. The ability of mid-infrared to explore these properties in the milk has resulted in the usefulness of spectra data for various purposes. Major milk components such as fat, protein, urea and lactose contents have been routinely predicted by MIR spectrometry globally (Luinge et al. 1993). The use of MIR has been extended to obtain other phenotypes and has been used in equations to predict several traits of economic importance in dairy cattle. Studies have indicated the ability of MIR to predict individual and groups of milk fatty acids with moderate to high accuracy (Soyeurt et al. 2011; Maurice-Van Eijndhoven et al. 2013), milk technological traits such as milk coagulation properties (Visentin et al. 2015), milk titratable acidity (Visentin et al. 2015; Toffanin et al. 2015) and milk minerals (Toffanin et al. 2015).

Secondly, traits of importance in terms of the resilience of cows and environmental impact can also be predicted based on the longitudinal nature of the relationship between the trait of interest and the MIR spectra. Two studies have documented the ability of the MIR spectrum to predict methane emissions in dairy cows (Dehareng et al. 2012). This link materializes from ruminal fermentation resulting in the production of de novo milk fatty acids as well as the eructation of methane emissions (Chilliard et al. 2001). Nitrogen in the urine of grazing animals can be an important source of groundwater pollution. Based on MIR analyses, Beatson and colleagues (Beatson et al. 2019) reported a heritability of 0.22 for measures of milk urea nitrogen (MUN) concentration, indicating the possibility of selection for lower MUN.

Of much interest also is the use of MIR to predict health traits. Clinical and sub-clinical ketosis results in elevated concentrations of ketone bodies such as acetone and β-hydroxybutyrate in blood and milk. The usefulness of MIR-predicted contents of acetone and β-hydroxybutyrate in milk as a tool for screening for cows less susceptible to ketosis was demonstrated by de Roos and colleagues (de Roos et al. 2007).

In general, MIR has been applied in the measurement of individual milk fatty acids, milk coagulation properties and milk mineral content (Soyeurt et al. 2011; Toffanin et al. 2015); cow feed intake and feed efficiency (McParland et al. 2014); susceptibility to ketosis (de Roos et al. 2007) and methane emissions (Vanlierde et al. 2018). The estimates of correlation coefficient squared (R^2) between various milk fatty acids predicted from MIR and actual measurements in the validation data sets reported by Soyeurt et al. (Soyeurt et al. 2011) varied from 0.38 to 0.98. Using MIR to predict methane emissions, Vanlierde and colleagues (Vanlierde et al. 2018) reported a validation accuracy of 0.57 compared to methane measurements in a respiration chamber. These authors concluded that MIR spectra could be used as a potential proxy to estimate daily methane emissions from dairy cows, cheaply, reliably, rapidly and on a large scale.

20.3.8 Measuring Greenhouse Gas Emissions

In ruminants, methane is produced as a natural part of the digestion process, where microbes in the rumen break down feed and release methane as a by-product, primarily through eructation or belching. Climate change mitigation efforts have focused attention on the greenhouse gas produced by the livestock sector. Globally, agriculture is directly responsible for 14% of annual GHG emissions (Vermeulen et al. 2012), with methane emissions from ruminants accounting for around 80% of all livestock emissions and cattle is responsible for the majority of the emission (Gerber et al. 2013). These observations have resulted in attempts to develop techniques for direct and indirect measures of CH_4 emitted at the individual animal level to enable the development of various mitigation approaches including genetic improvement. The gold-standard method for measuring the amount of CH_4 is the use of respiration chambers and there are two types: the closed-circuit chamber, where the increasing concentration of methane is measured (Boadi et al. 2002) and an open-circuit chamber which involves analysing the composition of the in-flowing and out-flowing air, which are then compared. However, respiratory chambers are expensive and not very suitable for measuring a large number of animals; therefore, various proxy methods are being investigated. Some of these include the SF6 (sulphur hexafluoride) tracer technique, which involves inserting a calibrated source of SF6 into the rumen of the animal being measured. The ratio of CH_4 to SF6 in the breath of an animal is measured over a 24-hour period, at regular intervals, and corrected with reference to the background methane concentration. Since the concentration of the tracer is known, the rate of production of CH_4 can be inferred. In general, repeated 24-h samples collected over 5 successive days show good day-to-day consistency in daily emission for each animal. The disadvantage of the technique is that the marker SF6 is itself a potent greenhouse gas.

Another instrument that has been used is the hand-held laser methane detector, or laser gun, to measure methane outputs from an animal over short periods of time. The laser methane detector is positioned about one metre distance from the mouth and nostril area of the animal and methane concentrations are recorded at regular short intervals (Chagunda and Yan 2011). Various indirect predictions for the amount of methane an animal produces have also been studied by examining feed intake, milk production and body weight. The methods used to monitor greenhouse gas emissions and indirect predictors were recently reviewed (Hill et al. 2016).

20.3.9 Phenotyping of Livestock Welfare

Recent greater public awareness of animal welfare issues and demand for better raising conditions for food-producing animals have pushed livestock producers to place increasing emphasis on welfare traits. Consequently, management and breeding technologies and practices developed in recent years have been tailored to enhance animal welfare. Thus, an aggregation of multiple welfare indicators for facilitating the overall assessment of animal welfare is of great significance (Botreau et al. 2007). Such indicators will help to identify or develop the most resilient animals (e.g. disease resilient and heat resilient) and to develop mitigation strategies. Some indicators of animal welfare can be assessed prior to clinical signs (e.g. milk somatic cell count and mastitis). Animal welfare assessment depends on some key indicators developed in a Welfare Quality Project by Rushen et al. (Rushen et al. 2011) including good feeding, good health condition, adequate housing and appropriate behaviour. These conditions can be assessed using various parameters such as blood parameters (König and May 2019); body condition score (Roche et al. 2009); cannibalism (Lambton et al. 2015); feeding behaviour such as active chewing time, standing and lying time and rumination time (Ding et al. 2018); proportion of time active and posture (Vasseur et al. 2012); immune response (Kovács et al. 2014) and response to infection (Nyman et al. 2014). Various technologies used to measure these variables were recently reviewed (Brito et al. 2020) including technological devices such as sensors (cameras, microphones) to capture vocalizations, thermometers, automatic scales to measure body weight and lean-fat ratios, automated feeding and milking systems, milk spectral data, electrodes to detect heart rate and skin conductivity and accelerometers.

20.3.10 Digital Capture of Phenotypic Data

There has been an increasing trend in the application of digital tools and mobile apps to capture livestock performance data for farmers in order to increase accuracy, avoid translation errors and reduce costs. The approach circumvents the settling up of elaborate infrastructure to capture performance data and has therefore become very important in low- and middle-income countries where herd sizes are small and very dispersed.

The African Dairy Genetics Gains (ADGG) project has implemented the use of digital tools to collect milk yield and other performance data in small dairy cattle systems in East Africa since 2016. The system is based on the Open Data Kit which is formatted to collect performance data and other farm characteristics. The data are automatically relayed to a database for processing and analysis (Ojango et al. 2018) and this implies the availability of a reliable internet service. Usually, this is a challenge in most developing countries as the connection could be intermittent. The Community-Based Breeding Program in Ethiopia has adopted another approach using tools (Anicloud and AniCapoture) developed by AbiocusBio Limited (https://abacusbio.com/) that permit offline performance data capture for small ruminants.

In addition, ADGG collaborating with Green Dreams Tech Ltd. and operating under the name iCow in Kenya, Ethiopia and Tanzania (Green Dreams Tech 2019) has used the mobile phone for a range of farmer training and advisory applications by sending specific training materials and alerts to improve management (Fig. 20.3).

Some of the hand-held tools are also employed for data capture and data viewing in developed countries including the Pocket Companion used to capture milk performance data by National Milk Records, one of the milk recording agents in the UK (National Milk Records 2021). Similarly, The Cattle Information Service, another milk recording company in the UK, has developed a mobile app for iPhone or Android called Mobile Herd (Cattle Information Service 2021) which enables data to be collected and analysed while out and about on the farm.

20.4 Genomic and Emerging Technologies

The genome holds the blueprint information for characters and traits in all living organisms and genomics encompasses the study of the complete genetic information of living organisms embedded in their DNA and related molecules that result from reading the genome like RNA and proteins. Starting with the launch of the human genome sequencing project in 1990, then an ambitious project to sequence the entire human genome, rapid developments in genome sequenc-

Fig. 20.3 iCOW system on mobile phones used for farmer training and advisory applications

ing applications have made it possible to unravel the mystery of the genomes of living organisms. The sequencing of the human genome (1990 to 2003) provided unprecedented insight into the genetic blueprint of a human and the unravelling of the genetic bases of diseases, as well as understanding the impact of living conditions (or lifestyle) and the environment on the genome and health (Gates et al. 2021). Thus, insights garnered from genomics hold the keys to attending to rapid innovations across many sectors, including animals and plants, and present huge opportunities to advance human health, animal health, agricultural productivity and, in particular, animal production.

20.4.1 Genomic Technologies

To enable effective application of genomic information in human, plant, and animal health management, and livestock and crop improvement, sequencing and genotyping technologies have been developed to support genome analysis including sequencing of portions of the genome and whole organism sequencing, and genotyping of sequence variations in a few samples to large number of samples.

20.4.1.1 Genome Sequencing for Detection of Sequence Variations

Following the unravelling of the three-dimensional structure of the DNA molecule by Watson and Crick in 1953 (Watson and Crick 1953), methodology developments for the study of the nature of DNA and proteins started. However, strategies to sequence the DNA and RNA molecules started in the 1970s, culminating with the development of Sanger's chain-termination breakthrough technology in 1977 (Maxam and Gilbert 1977) and the birth of first-generation sequencing. Further developments of the Sanger technique led to the emergence of commercial and automated DNA sequencing systems (Hunkapiller et al. 1991) which facilitated genome sequencing of complex species at an increased speed. First-generation sequencers generated short reads necessitating techniques like 'shotgun sequencing' (Anderson 1981) to support the sequencing of long fragments (long fragments are fragmented into short sequences, cloned, sequenced and overlapping sequences assembled into contigs) of the genome. The study of the genome sequence was further aided by the development of the polymerase chain reaction, recombinant DNA technology (Saiki et al. 1988) and other techniques in the 1970s and 1980s, which made possible the generation of high concentrations of desired DNA fragments for sequencing. Further developments, especially the development of the pyrosequencing method which uses the same technique of sequencing by synthesis (SBS) as Sanger sequencing, heralded Next-Generation-Sequencing (NGS) technology. The first successful NGS system, the 454 Life Sciences/Roche system used the pyrosequencing technology (Margulies et al. 2005). These developments were fuelled by the need for large-scale collaborative science and genome projects of the 1980s and 1990s, in particular, the human genome project which brought together a large group of researchers from many institutions across the globe who contributed to the development of tools, methods and technologies for genome interrogation and sequencing. The high-throughput capability of NGS through performing high numbers of sequencing reactions in parallel on a micrometre scale made possible by improvements in microfabrication and high-resolution imaging heralded second-generation DNA sequencing (Shendure and Ji 2008) and the emerging of NGS systems using various sequencing technologies, and mostly dominated by Illumina (www.illumina.com). Further developments including the ability to sequence single molecules without the need for DNA amplification shared by previous technologies are considered the birth of third-generation sequencing (Heather and Chain 2016). The widely used third-generation technologies include the single molecule real-time (SMRT) technology developed by Pacific Biosciences (https://www.pacb.com/) and nanopore sequencing technology developed by Oxford Nanopore (https://nanoporetech.com/). While second-generation sequencers produce short reads of up to 1000 bp, the third-generation range of sequencers can sequence full-length transcripts or generate reads up to >4 MB. The evolution of available NGS systems, their characteristics and capabilities are shown in Table 20.2. Concurrent with the development of NGS technology and systems were progressive developments in sequencing chemistry and procedures. Notable applications of NGS systems include amplicon DNA and RNA sequencing, custom sequencing, targeted DNA sequencing and whole-genome DNA and RNA transcriptome sequencing. With these tools, the genomes of animals, plants and humans have been studied in greater detail and the information generated is being used to manage the health of humans, animals and plants, and in animal and plant improvement. Aided by these tools, the whole-genome and pangenome sequences of several livestock species have been published (Table 20.3). Moreover, these tools also facilitate the unravelling of the transcriptome, epigenome, metabolome, metagenome, proteome, etc., structures and variations under various conditions and influencing factors, and use in designing production and disease management strategies.

With these technologies, the genomes (DNA and expressed RNA molecules) and epigenomes of several African livestock species have been studied and all types of sequence variations detected (Paguem et al. 2020; Ayalew et al. 2024a, b; Gheyas et al. 2022; Mauki et al. 2022; Tijjani et al. 2024).

20.4.1.2 Genotyping Technologies

The developments of sequencing technologies enabled the sequencing of animal and plant genomes and the detection of all kinds of sequence variations including single nucleotide polymorphisms (SNPs), indels (insertion or deletion of bases in a sequence), copy number variations (CNVs) (where portions of the genome are repeated), translocations and transversions. Genome sequencing to detect sequence variations, irrespective of the method and scale is expensive and is mostly feasible on a limited number of samples. To apply knowledge of sequence variations for population use or on a large number of samples, genotyping methods were developed. Genotyping techniques enable the detection of an individual's (animal/plant) genotype at a locus or loci of interest, thus facili-

Table 20.2 Technologies and sequencing systems for next-generation sequencing

Provider	Instrument/label	Reads per run	Read length	Chemistry
Illumina	iSeq 100	4 M	2×150 bp	Sequencing by synthesis
	MiniSeq	25 M	2×150 bp	Sequencing by synthesis
	MiSeq Series	25 M	2×300 bp	Sequencing by synthesis
	HiSeq 2000/2500	4 B	100–125 bp	Sequencing by synthesis
	HiSeq 2500 RR	600 M	150–250 bp	Sequencing by synthesis
	HiSeq 4000	5 B	150 bp	Sequencing by synthesis
	HiSeq X	6 B	150 bp	Sequencing by synthesis
	NextSeq 550 series+	400 M	2×150 bp	Sequencing by synthesis
	NextSeq 1000 & 2000	100 M to 1.8 B	1×50 to 2×300 bp	Sequencing by synthesis
	NextSeq P1	100 M	2×300	Sequencing by synthesis
	NextSeq P2	200 M	2×300	Sequencing by synthesis
	NovaSeq 6000	20 B	2×150 bp	Sequencing by synthesis
	Genome Analyzer IIx	138–168 M	1×35 to 2×75 bp	Sequencing by synthesis
Applied Biosystems/ ThemoFisher Scientific	SOLiD 1	40 M	25 bp	Sequencing by ligation
	SOLiD 2	115 M	35 bp	Sequencing by ligation
	SOLiD 3	320 M	50 bp	Sequencing by ligation
	SOLiD 4	2 B	50 bp	Sequencing by ligation
	SOLiD 5500xl	3 B	60 bp	Sequencing by ligation
	SOLiD 5500xl W	3 B	75 bp	Sequencing by ligation
454 Life Sciences/ Roche	IonTorrent PGM 314 chip	100,000	100 bp	Ion-semiconductor (IS) Sequencing technology
	IonTorrent PGM 316 chip	1 M	100 bp	IS sequencing technology
	IonTorrent PGM 318 chip V2	5 M	400 bp	IS sequencing technology
	IonTorrent Proton PI	50 M	200 bp	IS sequencing technology
	IonTorrent S5/ S5XL 530 chip	20 M	400 bp	IS sequencing technology
	Ion GeneStudio S5 System	2–130 M	200–600 bp	IS sequencing technology
	Ion GeneStudio S5 Plus System	2–130 M	200–600 bp	IS sequencing technology
	Ion GeneStudio S5 prime system	2–130 M	200–600 bp	IS sequencing technology
Singular genomics	G4 and G4X Spatial Sequencers	250–450 M	Sample dependent	Four-colour sequencing by synthesis (SBS)
Element Biosciences	AVITI	75 to 300 Gb	2×75 to 2×300 bp	Avidite base chemistry (ABC)
Pacific Biosciences	PacBio Sequel System	500,000	Full-length transcripts (FLT)	Single molecule, real-time (SMRT) sequencing technology
	PacBio Sequel II System	8 M	FLT	SMRT sequencing technology
	PacBio Sequel IIe System	8 M	FLT	SMRT and HIFI sequencing technology
	PacBio Sequel revio system	8 M	FLT	SMRT and HIFI sequencing technology

(continued)

Table 20.2 (continued)

Provider	Instrument/label	Reads per run	Read length	Chemistry
	ONSO System	400–1000 M	2×100 to 1×200 bp	Sequencing by binding (SBB)
OxFord Nanopore	GriDION MK1	250 Gb	>4 Mb	Protein nanopores (PN) sequencing technology
	PromethION 2	580 Gb	>4 Mb	PN sequencing technology
	PromethION 48	7 Tb	>4 Mb	PN sequencing technology
	PromethION 48	14 Tb	>4 Mb	PN sequencing technology

NB: Information in this table only serves as a guide and it does not represent a complete list of sequencers nor providers. Consult providers of the technologies and instruments for more detailed, precise and specific information on the platforms: ThermoFisher Scientific (https://www.thermofisher.com/); Roche (https://www.roche.com/); Illumina (https://www.illumina.com/); Singular Genomics (https://singulargenomics.com/); Element Biosciences (https://www.elementbiosciences.com/); Pacific Biosciences (https://www.pacb.com/) and Oxford Nanopore Technologies (https://nanoporetech.com/)

tating its application in a large number of samples and at cheaper cost.

Methodological developments in the last 50 years saw progression from methods capable of genotyping single loci to whole genome sequence (WGS) variants. The methods include those that (1) detect single loci: random amplification of polymorphic DNA, amplified fragment length polymorphism, restriction fragment length polymorphism, single-strand conformation polymorphism; (2) detect simple sequence repeats like microsatellites: short tandem repeats, simple sequence repeats, variable number of tandem repeats, etc.; (3) detect multiple loci: low- and high-density DNA genotyping chips or DNA microarrays (see Table 20.4) and genotyping-by-sequencing, and (4) detect whole genome sequence variants: whole-genome deep sequencing by NGS. Notable examples of available whole-genome-wide genotyping arrays or microarrays for livestock species are listed in Table 20.4.

These genotyping tools are efficient for assessing the relationship of sequence variations with traits of interest, biodiversity, detection of signatures of selection and quantitative trait loci mapping, and for making decisions regarding breeding, breed conservation and development of new breeds or genetic lines. Consequently, information obtained from sequence variant genotyping provides data for breeding decisions, breed conservation and development of new breeds or genetic lines. A major pitfall of currently available high-throughput genotyping arrays (Table 20.4) is the bias towards Western breeds or populations used to establish such arrays. Therefore, most of the genotyping applications currently available in the market (Table 20.4) do not adequately represent the sequence variations in the genomes of African livestock species as few African livestock breeds were included in the development of the assays. Consequently, limited African livestock-specific variations are included in the majority of the available genotyping assays. Therefore, a combination of technologies like whole-genome sequencing, genotyping assays, genotyping-by-sequencing and genotype imputation could be a reliable approach to identifying population (e.g. *Bos indicus*) specific variations that are associated with traits of economic importance in African livestock. Alternatively, genotyping assays containing genetic variation information specific to African indigenous livestock populations should be developed

20.4.1.3 Quantitative Trait Loci Mapping

Since the genetic variation that underlies many livestock traits (e.g. milk yield, growth, wool production or quality and meat quality), disease susceptibility and morphology is complex and directed by loci that have effects on the phenotype, quantitative trait loci (QTL) mapping or analysis is undertaken for use in breeding programmes. Thus, all kinds of genomic variations (SNPs, CNVs, microsatellites, WGS variants,

Table 20.3 Publication of the whole-genome and pangenome sequences of some livestock species

Livestock species	Year	Reference
Whole-genome sequences		
Cattle (*Bos taurus*)	2009	Elsik et al. (2009)
Cattle (*Bos indicus*)	2012	Canavez et al. (2012)
Asian buffalo leech (*Hirudinaria manillensis*)	2019	Guan et al. (2019); Rehman et al. (2021)
Egyptian water buffalo (*Bubalus bubalis*)	2020	Rehman et al. (2021); El-Khishin et al. (2020)
Bactrian camel (*Camelus bactrianus*)	2014	Wu et al. (2014)
Dromedary (*Camelus dromedarius*)	2014	Wu et al. (2014)
Alpaca (*Vicugna pacos*)	2014	Wu et al. (2014)
Chicken (*Gallus Gallus*)	2004	ICGSC (2004)
Pig (*Sus scrofa*)	2008	Humphray et al. (2007)
Sheep (*Ovis aries*)	2008/2009	Jiang et al. (2014)
Goat (*Capra hircus*)	2012	Dong et al. (2013)
Rabbit (*Oryctolagus cuniculus*)	2005/2009	Bai et al. (2021)
Atlantic Salmon (*Salmo salar*)	2016	Lien et al. (2016)
Channel Catfish (*Ictalurus punctatus*)	2016	Liu et al. (2016)
Horse (*Equus caballus*)	2009	Wade et al. (2009)
Single hump dromedary (*Camelus dromedaries*)	2016	Fitak et al. (2016)
Western Clawed Frog (*Xenopus tropicalis*)	2010	Hellsten et al. (2010)
Pangenomes		
Sheep (*Ovis aries*)	2023	Li et al. (2023)
Cattle	2022	Zhou et al. (2022)
Goat	2019	Li et al. (2019)
Pig	2023	Jiang et al. (2023)

etc.) are used in QTL analysis to determine a link/association between a trait or phenotype and genotype (genomic variation) in order to explain or elucidate the genetic variation underlying different traits (Leal, 2001). Therefore,

QTL mapping attempts to identify regions of the genome or DNA closely linked to genes and underlying a specific trait through statistical analysis of genomic variations and traits in populations. QTL mapping dissects complex traits into its component alleles and also quantifies the relative effects of alleles on the traits as well as locates the genomic regions of marker–trait association. QTL mapping provided the basis for marker-assisted selection (MAS) to expedite the breeding process of specific traits following proper estimation of QTL position and effects.

QTL mapping in livestock started in the 1980s and today, thousands of QTLs have been mapped for various traits in livestock (Table 20.5) (Hu et al. 2022). It was predicted that the application of genomic markers linked to QTLs in dairy and other livestock in breeding programmes could increase the response to selection by up to 30% (Kashi et al. 1990). The potential benefits of MAS depend on the QTL effect and the strength of the marker–QTL linkage.

20.4.2 Breeding Technologies

The dairy cattle population has been undergoing selection for milk production and other traits for many decades. Prior to 2009, the methodologies used cantered on phenotypic measurements of traits on animals as the only way of measuring genetic differences between animals, but genomic technologies are helping to address new breeding objectives faster. Thus, knowledge of the association of sequence variations with livestock traits of economic importance opened up the possibility of including them in breeding programmes and disease management. Genetic markers associated with traits of interest are used to guide management or future selection decisions because inheritance of the marker automatically confers the desired positive effect. Genetic markers have over the years been used in different programmes aimed at genetic improvement of desired traits such as marker-assisted selection and genomic breeding. Recent advancements are making it possible to study

Table 20.4 Genome-wide genotyping chips for various livestock species

Species	Genotyping chip or array	Number of markers[a]	Manufacturer[b]
Cattle	BovineSNP50 V1 Genotyping BeadChip	54,001	Illumina
	BovineSNP50 v2 Genotyping BeadChip	54,609	Illumina
	BovineSNP50 v3 Genotyping BeadChip	53,218	Illumina
	BovineLD BeadChip	7000	Illumina
	Illumina Bovine HD (777 K) chip	777,962	Illumina
	Infinium iSelect custom target genotyping arrays (available for many species including bovine, ovine, porcine and chicken)	Up to 700,000	Illumina
	Infinium XT ((available for many species including bovine, ovine, porcine and chicken)	Up to 50,000	Illumina
	GeneSeek Genomic Profiler (GGP) LD v4 (40 K) SNP chip	40,660	Neogen GeneSeek
	GGP 50 K version 1 SNP chip	49,463	Neogen GeneSeek
	GGP HD (80 K) SNP chip	77,376	Neogen GeneSeek
	GGP 100 K	100,000	Neogen GeneSeek
	GGP UHD (150 K) SNP chip	150,000	Neogen GeneSeek
	Axiom Genome-Wide BOS 1 Bovine Array Kit	648,855	[c]Thermofisher Scientific
	Axiom Bovine Genotyping V2 Array	>67,000	[c]Thermofisher Scientific
	Axiom Bovine Genotyping V3 Array	>63,000	[c]Thermofisher Scientific
	Axiom Bovine-Ovine-Caprine Genotyping Array	54,560 (bovine), 60,000 (ovine) & 54,236 (caprine)	[c]Thermofisher Scientific
Buffalo	Axiom Buffalo Genotyping Array	90,000	[c]Thermofisher Scientific
Sheep	Ovine Infinium HD SNP BeadChip (600 K)	606,006	Illumina
	OvineSNP50 SNP BeadChip	54,241	Neogen GeneSeek
	Axiom Ovine Genotyping Array	51,000	[c]Thermofisher Scientific
Goat	Caprine 50 K SNP	537,145	Illumina
Pig	GGP Porcine v2	51,558	Neogen GeneSeek
	PorcineSNP60	63,480	Neogen GeneSeek
	GGP Porcine HD v1	70,231	Neogen GeneSeek
	Axiom Porcine HD Genotyping Array	658,692	[c]Thermofisher Scientific
	Axiom Porcine Breeders Genotyping Array	55,232	[c]Thermofisher Scientific
Chicken	Axiom Genome-Wide Chicken Array Kit (600 K Affymetrix® Axiom® HD genotyping array)	580,000	[c]Thermofisher Scientific
	Illumina 60 K SNP BeadChip	57,636	Illumina
	Chicken 55 K SNP genotyping array	52,184	Liu et al. (2019a)

(continued)

Table 20.4 (continued)

Species	Genotyping chip or array	Number of markers[a]	Manufacturer[b]
Turkey	Axiom Turkey Genotyping Array	643,845	[c]Thermofisher Scientific

[a]The marker types are SNPS (all or the majority of the markers in the arrays), insertions, deletions and copy number variants
[b]Visit web site of providers for more precise and up-to-date information: Illumina (https://www.illumina.com/); Neogen GeneSeek (Neogen GeneSeek (https://www.neogen.com/) and Thermofisher Scientific (https://www.thermofisher.com/)
[c]Initially Applied Biosystem and bought over by ThermoFisher Scientific

Table 20.5 Quantitative trait loci mapped for various livestock species

Species	Number of QTLs[a]	Number of base traits (trait variants)	eQTLs[b]	Number of publications
Cattle	193,453	558 (417)	346	1206
Sheep	5,417	178 (264)	0	289
Goat	2,713	90 (120)	0	47
Pig	57,041	406 (1088)	240	854
Chicken	29,116	246 (246)	40	416
Horse	2,486	71 (14)	0	129
Rainbow Trout	2,201	35 (6)	0	23

[a]QTL—Quantitative trait loci (data from https://www.animalgenome.org/cgi-bin/QTLdb/index; accessed on 7 July 2025)
[b]eQTL—Expression quantitative trait loci. The number of eQTLs is included in the total number of QTLs

epigenomic variations and to edit the genomes of animals and this information may soon find application in breeding.

20.4.2.1 Marker-Assisted Selection Technology

Marker-assisted selection is a concept that utilizes the information of polymorphic loci (biomarkers: DNA, RNA, proteins, etc.,) as an aid to selection. Thus, MAS also known as marker-aided selection is a process whereby a trait of interest is selected based on its association with a polymorphic biomarker. MAS is an indirect selection process because a trait of interest is selected based on a marker linked to it and not based on the trait itself. The usefulness of MAS is that the effect of genes or markers on health and production is directly measured based on the genome information or genetic makeup of the individual and not on the phenotype. MAS in addition to its use in breeding for specific traits is also used for ancestry determination and in conservation programmes by ensuring that particular traits or useful genes in a population are maintained for future exploitation.

The application of genetic markers in MAS for livestock genetic improvement was made possible by developments of molecular genetic techniques from the 1970s and relied on the genotypes of individuals at specific or limited loci (Hill 2014). MAS that relies on markers with major effects prompted linkage mapping studies, the development of microsatellite markers and the detection of QTL in livestock species from the 1980s onwards.

The use of MAS to increase the rate of genetic progress in breeding populations is reliant on the target trait as well as the information on the marker–QTL relationship (Wakchaure et al. 2015). Sexing embryos is a unique example of marker-assisted selection in the cattle industry. In selection programmes, MAS is extremely useful since it overcomes the constraints of traditional breeding methods. It increases the genetic gain through reducing generation interval but requires information on alleles or markers associated with the target traits jointly with quantitative estimates of these associations prior to application to a specific animal population and it increases the rate of introgression of favourable alleles into desired populations. Genetic markers that indicate desirable alleles of economically important attributes might be used in conjunction with estimated breeding values (EBV) to guide mating decisions, resulting in larger genetic gains over a wider range of traits (Calus 2010). Economic trait loci must be found, verified and described for efficacy in enhancing genetic gain before MAS may be employed in commercial production (e.g. dairying) (Sonstegard et al. 2001). The development of MAS programmes is divided into three stages: (1) Detection phase: Polymorphisms in DNA sequences are employed as linked or direct markers in the detection phase to find QTL segregating in certain populations with specified allele frequencies. One or more markers related to QTL are used to quantify the magnitude of QTL allele effects and the position of the QTL in the genome (Alvarez-Castro and Carlborg 2007) (2). Evaluation phase: The associated markers are tested in target populations or families during the assessment phase to see if the discovered QTLs are segregating in those groups (Lynch and Walsh 1998) (3). Implementation phase: To establish a genotyping database, linked markers that have been shown to be predictive in a population are utilized within families, and direct markers are used across families. These data are combined with phenotypic and pedigree information in genetic assessment to anticipate the genetic merit of individuals within a group (Davis and DeNise 1998).

Linked and direct markers are used in MAS. Linked markers are genetic markers that are close enough to the trait gene on the chromosome to be transmitted or inherited simultaneously. Linked markers are limited in that they cannot be used to predict phenotype until the inheritance of markers associated with genes of interest is researched within families, which can alter due to recombination events. Direct markers on the other hand are functional polymorphisms in the gene that determine phenotypic variation. Once the functional polymorphism is understood, it is easy to predict the effect of certain alleles in animal populations. As a result, 'direct' markers outperform 'linked' markers when it comes to forecasting phenotypic variance of target attributes within a population (Dekkers 2004).

Molecular markers can be used to track down and identify genetic diseases and morphological deformities in livestock animals. The cause and origin of these problems may be traced back to genetic changes and DNA mutations that affect protein structure and function. This knowledge can help breeders avoid such faults to keep their animals healthy.

20.4.2.2 Genomic Breeding

From the 1980s to the early 2000s, QTL data were incorporated into linear models for genetic evaluations. Estimation of the breeding value of a trait under selection was determined by summing the estimated effects of several markers and QTLs (Weigel et al. 2017) for use in MAS of young sires, in inter-breed introgression programmes for loci that alter performance attributes or disease resistance/susceptibility, etc. During this time, genetic markers used for QTL detection were mostly restriction length fragment polymorphisms and microsatellites. With the development of next-generation sequencing technologies and genotyping capabilities of all kinds of sequence variations in a large number of samples, it was possible to predict breeding values from dense markers and use them in breeding programmes in a process called genomic selection (GS) (Meuwissen et al. 2016). Therefore, genomically best linear unbiased prediction (GBLUP) was developed based on the understanding that quantitative traits are controlled by many loci that contribute equally to a trait meaning homogeneous variance among markers (Goddard 2009). In the same light, genomic estimated breeding value (GEBV), regarded as the sum of the effects of dense or thousands of genetic markers as well as their haplotypes across the whole genome, were calculated or predicted and applied in selection programmes (Meuwissen et al. 2013). Therefore, selection decision based on GEBV is known as genomic selection. Application of GS starting from 2009 has revolutionized dairy cattle (Wiggans et al. 2017a), pig (Zhang et al. 2018a) and poultry (Avendaño et al. 2010) breeding in several Western countries.

Implementation of genomic breeding, which relies on the use of genomic information in the form of SNPs resulted in rapid genetic improvement in lowly heritable traits (e.g. fertility, lifespan and health traits), shortened generation interval, selection of animals at an early age, higher rate of genetic gain, reduced cost of progeny testing as GEBVs are available early in life, increased reliability of predicting breeding value and higher intensity of selection (Wiggans et al. 2017b).

Genome variation currently exploited for genomic breeding explains only a portion of the phenotypic variance or heritability in traits (Ibeagha-Awemu and Khatib 2017). The missing portion of variation may be due to yet to be identified large number of variants with smaller effects; low frequency or rare variants with potentially large effects that are not represented in currently available genotyping arrays that focus on common variants with a population frequency of five percent or more; gene × gene interactions, copy number variants; inadequate documentation of shared environment among relatives; epigenetic variation, epigenetic effects and noncoding RNA regulation of phenotypic expression (Eichler et al. 2010). Moreover, it is generally accepted that epigenetic mechanisms form part of the uncovered players in the expression of complex animal health and production traits, and are likely responsible for a portion of the missing variation in livestock traits and also for the non-accounted portion of trait heritability in current genetic improvement schemes (Ibeagha-Awemu and Zhao 2015). Therefore, the detailed characterization of all kinds of livestock functional genomic variations, epigenomic variations/modifications, non-coding RNA expression and variation, and their interaction with the environment will ultimately provide a holistic view of information for improvement decisions and application in African livestock populations.

20.4.3 Genome and Epigenome Editing Technologies

Advances in DNA and RNA manipulation starting in the 1970s opened up new frontiers in genome editing enabling a better understanding of the contribution of individual gene products to disease development and other traits. The technologies for editing the genome (Table 20.6) developed rapidly in the last 30 years and have wide applications in basic to applied biotechnology, agriculture and biomedicine. Major milestones starting with the development of DNA strand break and repair to the development of the

Table 20.6 Technological developments in gene editing and application in animal production

Year	Technological development	Reference
1985	Transgenic rabbits, sheep and pigs produced by microinjection, Zing fingers discovered	Hammer et al. (1985) Miller et al. (1985)
1987	CRISPR discovered in *Escherichia coli*	Ishino et al. (1987)
1996	ZFNs are developed and the first successful cloning of Dolly (sheep)	Kim et al. (1996), Campbell et al. (1996)
2005	Cas9 discovered.	Bolotin et al. (2005)
2008	Ability of CRISPR to target DNA discovered	Marraffini and Sontheimer (2008)
2010	Ability of CRISPR/Cas9 to cleave DNA demonstrated	Garneau et al (2010).
2010	Use of zing fingers in genome manipulation	Klug (2010)
2011	CRISPR/Cas9 complex assembled and tracrRNA discovered	Deltcheva et al. (2011)
2011	TELENS used for gene editing	Li et al. (2011)
2012	TALEN-mediated gene knockout in livestock	Carlson et al. (2012)
2013,	CRISPR/Cas9 system adapted for genome editing	Cong et al. (2013)
2016	Developments in epigenome editing and application in livestock	Kungulovski and Jeltsch (2016)
2017	Method for CRISPR/Cas9-mediated knock-in to generate stable cell lines with quantifiable protein production developed	Lo et al. (2017)
2022	Role of CRISPR gene editing in tackling global food crisis	Straiton (2022), Gostmskaya (2022)

CRISPR clustered regularly interspaced short palindrome repeats, *ZFNs* zinc finger nucleases, *Cas9* CRISPR-associated protein 9, *TELENS* transcription activator-like effector nucleases

highly efficient CRISPR/Cas system and the production of transgenic livestock are summarized in Table 20.6.

Major advances in genetically engineered farm animals by DNA microinjection into the pronucleus of zygotes lead to the production of transgenic animals including livestock animals (cattle, goat, pig, rabbit and sheep) (Hammer et al. 1985). Several methods are used for transgenic animal production, including recombinant retroviruses, sperm-mediated DNA transfer, germ cell transplantation, nuclear transfer cloning, pronuclear injection and embryonic stem cells (Wheeler and Walters 2001; Wall 2002).

Gene editing or genome editing, or genome engineering is a procedure by which DNA is inserted, deleted, replaced or modified in an individual's genome. Genome editing can be performed in vivo or in vitro by delivering an editing apparatus in situ, which deletes, adds or modify genes as well as performs more targeted genomic modifications. Generally, systems like zinc finger nucleases (ZFN), transcription activator-like effector nucleases (TALEN) and Clustered Regularly Interspaced Short Palindromic Repeats (CRISPR)/Cas9 are commonly used to achieve efficient genome editing by introducing genes or alleles of interest into a recipient genome, by editing a section of the genome, deleting a portion of the genome or by switching genes on or off (Randhawa and Sengar 2021; Nakamura et al. 2021; Perisse et al. 1713).

In general, targeted DNA or genome editing or alteration starts with the generation of a nuclease-induced double-stranded break (DSB), which causes the stimulation of highly efficient recombination mechanisms of cellular DNA in mammalian cells, followed by repairs using either of the two mechanisms, nonhomologous end-joining (NHEJ) or homology-directed repair (HDR), which occur in most cell types and organisms (Fig. 20.4) (Randhawa and Sengar 2021; Li et al. 2020). Following the successful generation of a targeted DSB, the repair mechanisms can be used to restructure the cleaved DNA by introducing a desired change (delete or introduce) in the genetic code using an exogenous nucleic acid template. Therefore, genome editing can enable the alteration of an animal's genome by altering a section (a single base pair or a stretch of nucleic acids) of the genome or by introducing a stretch of nucleic acids from another source.

In livestock animals, the main method used to produce livestock harbouring nuclease-mediated, genetic changes in their genomes is cell nuclear transfer of edited somatic cells, particularly homology-directed repair donor-template alterations (Tan et al. 2016). Genome editing therefore offers the livestock industry efficient opportunities to introduce desired genetic variation into livestock breeding programmes through targeted inactivation of gene function and/or through allele introgression while avoiding the undesired linkage drag (Van Eenennaam et al. 2021).

Genome editing and its application in livestock have been discussed in several recent reviews (Perisse et al. 1713; Van Eenennaam et al. 2021; Karavolias et al. 2021). In summary, various gene editing techniques have been used to target certain genes in livestock species and to generate transgenic animals, like the MSTN or double muscling gene in cattle, sheep and pigs; the polled allele in cattle; the LDLR gene in a pig model of atherosclerosis; lysostaphin (LSS) gene for mastitis resistance in cattle, etc., to address disease resistance, productivity, product quality and environmental adaptation. Some genes that have been edited in livestock are summarized in Table 20.7.

Similar to genome editing, epigenome editing at epigenetic loci of interest represents an innovative method that might selectively and heritably alter gene expression. In contrast to genome editing, epigenome editing technology changes the chromatin state without changing the DNA sequence (Lau and Suh 2018). The feasibility of adding or removing epigenetic marks by different compounds has made it possible to introduce a wide range of epigenetic drugs for human disease treatment, among other applications (Montalvo-Casimiro et al. 2020). An example is the FDA-approved DNMTi 5-aza-2'-deoxycytidine, a DNA methylation inhibitor, known as a cancer drug since 1968 because of its cytotoxic effects on cancer cells (Heerboth et al. 2014). However, a proof-of-concept of this technology was dem-

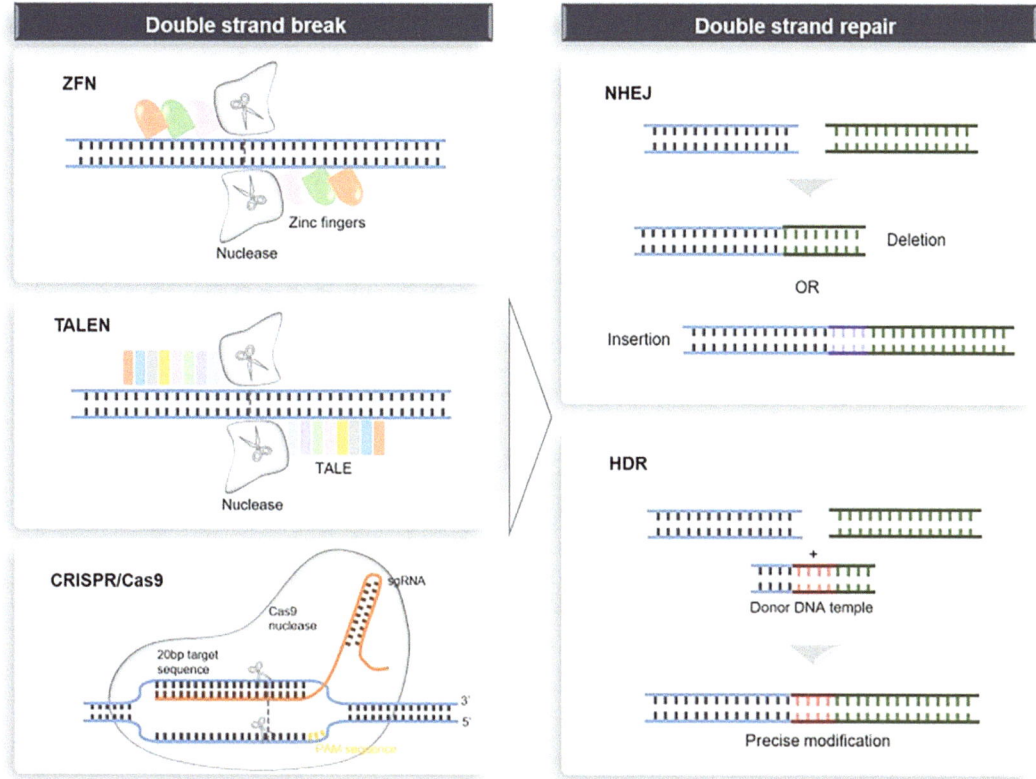

Fig. 20.4 Genome editing platforms and mechanisms for DSB repair with endogenous DNA. Genome editing nucleases (ZFNs, TALENs and CRISPR/Cas9) induce DSBs at targeted sites. DSBs can be repaired by NHEJ or, in the presence of donor template, by HDR. Gene disruption by targeting the locus with NHEJ leads to the formation of indels. When two DSBs target both sides of the amplification or insertion, a therapeutic deletion of the intervening sequences can be created, leading to NHEJ gene correction. In the presence of a donor-corrected HDR template, HDR gene correction or gene addition induces a DSB at the desired locus. DSB double-stranded break, ZFN zinc-finger nuclease, TALEN transcription activator-like effector nuclease, CRISPR/Cas9 clustered regularly interspaced short palindromic repeat associated 9 nuclease, NHEJ nonhomologous end-joining and HDR homology-directed repair. (From Li et al. (2020), licensed under CC-BY 4.0)

onstrated in human cell lines, where targeted acetylation of gene promoters and enhancers resulted in the successful transcriptional activation of several genes (Hilton et al. 2015). For livestock breeding purposes, these technologies can be used to introduce or fix favourable alleles and epigenetic marks for addressing disease resistance, increased productivity and environmental adaptation. Although epigenome editing is yet to be performed in livestock, a simulation study by Jenko et al. (Jenko et al. 2015) demonstrated that genome editing to promote alleles of polygenic traits could double genetic gains when compared to conventional genomic selection.

The technologies of genome and epigenome editing could be applied in livestock management to (1) move genetic elements between breeds and species; (2) alter existing genetic elements to create new ones, effectively expanding the available genetic diversity (σA); (3) modify genes to address traits ranging from production efficiency to animal welfare, environmental sustainability and disease resistance; and (4) introduce useful alleles in the absence of linkage drag or other unwanted traits. The various techniques that have been applied to generate gene-edited livestock are summarized in Figure 20.5 (Perisse et al. 1713). Therefore, the adoption of epigenome and

Table 20.7 Summary of gene editing in livestock species

Species	Genes edited	Purpose of manipulation	Method[a]	Editor	Reference
Buffalo	GDF8	Meat yield	ICSI	CRISPR/Cas9	Su et al. (2018a)
Cattle	NANOS2	Surrogate sires for genetic dissemination	MI	CRISPR/Cas9	Ciccarelli et al. (2020)
	SLICK	Thermotolerance	SCNT	TALE nickase	Bellini (2018)
	Lysostaphin, hLYZ	Mastitisresistance,milkcomposition-	SCNT	ZFN	Liu et al. (2014)
	SP110	Bovine tuberculosis resistance	SCNT	TALE nickase	Wu et al. (2015)
	CD18	Leukotoxin resistance	SCNT	ZFN	Shanthalingam et al. (2016)
	NRAMP1	Tuberculosis resistance	SCNT	CRISPR/Cas9	Gao et al. (2017)
	MSTN	Meat yield	SCNT, MI	ZFN, TALE nickase	Luo et al. (2014); Proudfoot et al. (2015)
	BLG	Reduction of milk allergen	SCNT	ZFN	Yu et al. (2011)
	LacS	Reduction of milk allergen	SCNT	TALE nickase	Su et al. (2018b)
	PRNP (knockout)	Prevent prion diseases	Not specified	Knockout vectors	Richt et al. (2007)
	CSN2, CSN3	Milk composition	NT	DNA construct	Brophy et al. (2003)
	Mfat-1	Milk composition	NT	Vector construct	Wu et al. (2012)
	BLG	Milk composition	NT	Tandem construct	Jabed et al. (2012)
	Myostatin (MSTN) shRNA	Increased muscle yield	SCNT	Recombinant lentiviral vector	Tessanne et al. (2012)
	Polled Celtic variant knock-in	Polled cow	SCNT	CRISPR/Cas12a	Schuster et al. (2020)
Sheep	ASIP	Coat colour pattern	MI	CRISPR/Cas9	Zhang et al. (2017)
	MSTN	Improve skeletal muscle satellite cell differentiation	SCNT	CRISPR/Cas9	Zhang et al. (2018b)
	FGF5	Wool growth	MI	CRISPR/Cas9	Hu et al. (2017)
	MSTN, ASIP, and BCO2	Economically important traits	MI	CRISPR/Cas9	Wang et al. (2016)
	MSTN	Meat production	MI or SCNT	CRISPR/Cas9, TALE nickase	Proudfoot et al. (2015); Wang et al. (2016); Crispo et al. (2015); Zhang et al. (2018b)
	Omega-3 fatty acid desaturase fat-1 (fat-1)	Meat composition	HMC	Vector construct	Zhang et al. (2013)
	Prion protein (PRNP)	Xenotransplantation, Prion disease resistance	SCNT	Promoterless vectors	Denning et al. (2001)
	Growth hormone (GH), growth hormone-releasing hormone (GHRH)	Growth rate	Ova injection	Ovine metallothionein promoter, cysteine biosynthetic pathway	Rexroad Jr. et al. (1989)

	Gene	Trait	Method	Vector/System	Reference
Goat	Insulin-like growth factor 1 (*IGF1*), keratin intermediate filament type II (*KRT2.10*)	Wool growth	Pronuclear MI	Ultra-high-sulfur keratin promoter	Damak et al. (1996)
	BLG	Milk quality	MI	CRISPR/Cas9	Zhou et al. (2017)
	FGF5 and/or MSTN	Meat and cashmere production	MI or SCNT	CRISPR/Cas9, TALE nickase	Zhang et al. (2018b); Guo et al. (2016); He et al. (2018)
	Myostatin (*MSTN*) shRNA	Increased muscle yield	SCNT	Myostatin-targeted vector	Zhou et al. (2013)
	NANOS2	Surrogate sires for genetic dissemination	SCNT	CRISPR/Cas9	Ciccarelli et al. (2020)
	EDAR	Cashmere yield	SCNT	CRISPR/Cas9	Hao et al. (2018)
Goat	GDF9	Litter size	MI	CRISPR/Cas9	Niu et al. (2018)
	Fat-1 into MSTN	Milk composition	SCNT	CRISPR/Cas9	Zhang et al. (2018c)
	Lysozyme (*LYZ*)	Milk safety and production	Pronuclear microinjection	DNA construct	Maga et al. (2006)
	SCD	Milk composition	Pronuclear microinjection	Bovine beta-lactoglobulin promoter-rat stearoyl-CoA desaturase cDNA construct	Reh et al. (2004)
	LF	Pathogen resistance	Pronuclear microinjection	pGEM-T easy vector or pBC1-hLf plasmid vector	Zhang et al. (2008)
	DEFB103A	Milk composition	SCNT	pEBB vector, electroporation	Liu et al. (2013)
	PRNP shRNA	Prion protein and caprine reduction	SCNT	RNA interference, Lentiviral vector	Golding et al. (2006)
	BLG	Milk quality	Microinjection	CRISPR/Cas9	Zhou et al. (2017)
Pig	NANOS2	Surrogate sires for genetic dissemination	SCNT or MI	CRISPR/Cas9	Ciccarelli et al. (2020); Park et al. (2017)
	IGF2 regulatory element	Meat production	MI (nCas9)	CRISPR/Cas9	Xiang et al. (2018)
	ZBED6	Improve lean meat production	SCNT	CRISPR/Cas9	Liu et al. (2019b)
	ANPEP	Transmissible gastroenteritis viral resistance	MI	CRISPR/Cas9	Whitworth et al. (2019)
	CD163, CD1D	Resistance to porcine respiratory syndrome virus (PPRV), innate immunity	MI, EP, or SCNT	CRISPR/Cas9	Whitworth et al. (2014)
	CD163	Resistance to PRRSV, innate immunity	MI/SCNT	CRISPR/Cas9	Whitworth et al. (2016)
	IRX3	Reduced fat content in Bama minipigs	SCNT	CRISPR/Cas9	Zhu et al. (2020)

(continued)

Table 20.7 (continued)

Species	Genes edited	Purpose of manipulation	Method[a]	Editor	Reference
	MSTN	Meat production	SCNT, EP	CRISPR/Cas9, TALENs, ZFN,	Wang et al. (2017)
	CD163 and pAPN	Viral resistance	SCNT	CRISPR/Cas9	Xu et al. (2020)
	FBXO40	Meat production	SCNT	CRISPR/Cas9	Zou et al. (2018)
	shRNA knock-in to Rosa26 locus	Classical swine fever virus	SCNT	CRISPR/Cas9/RNA-i	Xie et al. (2018)
	FAT-1 knock-in to ROSA26 locus	Classical swine fever, fat-1 (nutrition)	SCNT	CRISPR/Cas9	Li et al. (2018)
	FAT-1	Meat composition	Electroporation	Pgk-neo expression cassette	Lai et al. (2006)
	LYZ	Piglet survival	SCNT	pBC1-hLZ-GFP-Neo vector	Tong et al. (2011)
	GH1, GHRH, IGF1	Growth rate	Zygote, ova	Chimaeric constructs, promoter plasmid vector	Wieghart et al. (1990)
	SKI proto-oncogene (*SKI*)	Muscle development	Not specified	Not specified	Pursel (1992)
	LALBA	Milk composition, milk yield and growth rate	Pronuclear microinjection	a-lactalbumin construct.	Wheeler et al. (2001)
	Mx1	Influenza virus infection resistance	Microinjection	Interferon (IFN), virus-inducible mMx::Mx construct	Müller et al. (1992)
	Foot-and-mouth disease virus (FMDV) antiviral small hairpin RNAs (shRNAs)	FMDV resistance	Intramuscular injection	shRNA vector	Hu et al. (2015)
	Ucp1	Enhanced thermoregulation	SCNT	CRISPR/Cas9	Zheng et al. (2017)
	Classical swine fever virus (CSFV) antiviral small hairpin RNAs (shRNAs)	CSF resistance	SCNT	CRISPR/Cas9	Xie et al. (2018)
	FBXO40	Meat production	SCNT	CRISPR/Cas9	Zou et al. (2018)
	TGEV and PEDV	Coronavirus resistance	MI/ oral injection	CRISPR/Cas9	Whitworth et al. (2019)
Chicken	chNHE1	Avian leukosis	MI	CRISPR/Cas9	Koslová et al. (2020)
	G0S2	Less abdominal fat	Embryo injection	CRISPR/Cas9	Park et al. (2019)
	Fluorescent protein into sex chromosome	Sex determination	In vitro cell transfection by lipofectamine	CRISPR/Cas9 (Hu et al. 2015)	Lee et al. (2019)
	Alv6	Avian leukosis resistance	Not specified	Not specified	Salter and Crittenden (1989)
	Influenza A virus polymerase shRNA	Viral disease resistance	Microinjection	Retroviral vector	Lyall et al. (2011)
	β-Galactosidase (*lacZ*)	Beta-galactosidase expression	Microinjection	Lentiviral vector	Mozdziak et al. (2003)

Species	Gene	Trait	Method	Technology	Reference
Grass carp	gcJAM-A	Grass carp reovirus	Not specified	CRISPR/Cas9	Ma et al. (2018)
Carp	sp7, MSTN	Meat yield	MI	CRISPR/Cas9, TALE nickase	Zhong et al. (2016)
	Growth hormone (*GH1*)	Growth rate	Microinjection	Recombinant vector	Wu et al. (2005)
	Lactotransferrin (*LTF*)	Resistance to *Aeromonas hydrophila* infection and improve phagocytosis	Sperm electroporation	Recombinant vector	Weifeng et al. (2004)
Catfish	MSTN	Growth rate	MI	CRISPR/Cas9	Khalil et al. (2017)
	Cecropin-B (*CecB*)	*Flavobacterium columnare* resistance	Not specified	Cytomegalovirus promoter	Dunham et al. (2002)
Oyster	MSTN	Meat yield	MI	CRISPR/Cas9	Yu et al. (2019)
Rabbit	MSTN (Rahman et al. 1998)	Meat yield	MI	CRISPR/Cas9	Guo et al. (2016)
Salmon	Growth hormone 1 (*gh1*)	Growth rate	Micropyle MI	Ocean pout antifreeze protein gene (AFP) promoter	Du et al. (1992)
	Lysozyme (*lyz2*)	Boost immunity	Not specified	Ocean pout antifreeze protein gene promoter	Fletcher et al. (2011)
	Liver-type antifreeze protein (*wflAFP-6*)	Cold tolerance	Not specified	Genomic clone 2A-7 coding	Hew et al. (1999)
Tilapia	Growth hormone 1 (*gh1*)	Growth rate	MI	Ocean pout antifreeze promoter	Rahman et al. (1998)
Trout	Follistatin (*fst*)	Muscle development	Micropyle MI	5′ rat myosin light chain (rmylc) promoter	Medeiros et al. (2009)

[a]*SCNT* somatic cell nuclear transfer, *MI* zygote or tunica albuginea microinjection, *EP* zygote electroporation, *nCas9* Cas9 nickase, *ZFN* zinc finger nickase, *HMC* handmade cloning, *HR* homologous recombination, *N/A* not applicable, *ICSI* intracytoplasmic sperm injection

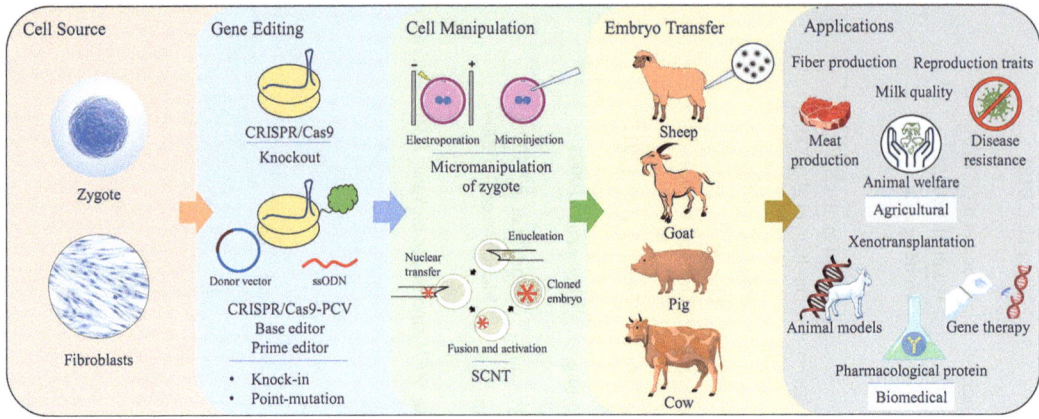

Fig. 20.5 Diagrammatic representation of CRISPR/Cas9 gene editing using either zygote micromanipulation by microinjection or electroporation, or somatic cell nuclear transfer (SCNT) for the generation of livestock animals for various applications. (From Parisse et al. (1713), licensed under CC-BY 4.0)

genome editing technologies could significantly impact animal health, production, reproduction and environmental adaptation using targeted manipulation of candidate genes affecting corresponding traits.

20.4.4 Cellular Agriculture

To satisfy the increasing demand for animal food products by an ever-increasing human population, more efficient ways of making available animal-based foods are being developed to sustain growing demand while tackling challenges associated with traditional livestock farming such as animal welfare and environmental issues (pollution/degradation). Among the solutions, the production of livestock products from animal cells or engineered microbes is presented as a viable alternative to increasing the production of animal products without the negative effects associated with traditional livestock farming and for consumers who advocate for responsible livestock farming but do not wish to change their dietary habits. Cellular agriculture (CA), a term coined by Isha Datar, a Canadian molecular biologist (Groen 2022) is generally defined as the production of animal products (e.g. eggs, meat, milk, and seafood) and ingredients from cell cultures in vitro, artificially or lab-grown rather than from farmed animals (Ong et al. 2021). According to the production method, CA is either cellular (cell cultivation or tissue engineering) or acellular (precision fermentation) based (Stephens et al. 2018). CA therefore comprises numerous innovative technologies and approaches that use cells, cell culture, tissue engineering or precision fermentation to make food products (e.g., meat and milk, etc.) and other materials (Fig. 20.6). Although CA is an emerging technology that is still in the development phase, it has attracted the attention of many industrialists, investors and governments who are scrambling to support the growth of the industry and to put in place necessary regulations. Although animal production is still developing in Africa, CA may offer opportunities to diversify and create products with high nutritional content without the many problems currently facing livestock production. The economic opportunity presented by CA was recently analysed and estimated as a $12.5 billion opportunity in food innovation in Canada creating about 30,000 to 142,000 jobs (OG 2021).

20.4.4.1 Cellular or Cultivation-Based Cellular Agriculture

Cultivation-based CA (e.g. cultured meat and milk, etc.) uses cells or cell lines taken from living animals, cultured on engineered tissue (such as scaffolds) in bioreactors (in vitro) to produce products with lesser amounts of life animal tissue input and other inputs compared to conventional

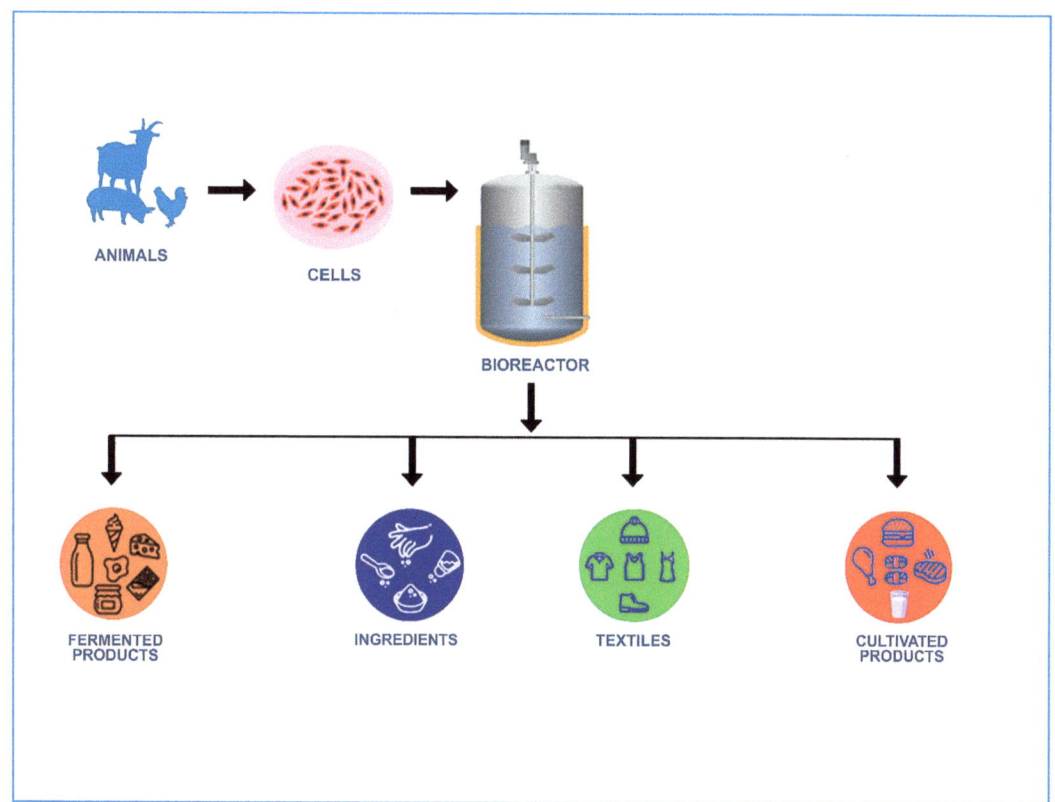

Fig. 20.6 Cellular agriculture uses cells from animals or precision fermentation technologies to produce products of animal origin (meat, eggs, milk) and fermentation products for use as ingredients for the creation of varied products including textiles

production methods. The cells used as starting material could be taken from the animal through the biopsy approach and grown in vitro or genetically modified (immortalized cell line) for perpetual growth in culture. The steps involved in cell-cultured meat and seafood technologies were described recently (Ong et al. 2021; Kwon et al, 2024; Reiss et al, 2021; Soleymani et al, 2024; Munteanu et al. 2021); and it involves (1) obtaining the cell from a live or slaughtered animal (e.g. embryonic stem cells, mesenchymal stem cells, induced pluripotent stem cells, myosatellite cells (adult stem cells) or primary cells lines); (2) cells are isolated and (3) culture conditions optimized. Moreover, genetic modifications of desired characters can be performed (4). Biomass production (includes proliferation, differentiation and/or maturation of cells in bioreactors); (5) on attaining the desired biomass, the product is harvested and processed by the addition of other ingredients for the manufacture of various products (Fig. 20.7).

Several companies are today using cultivation-based CA to manufacture a variety of products such as cell-cultured meat, egg white and milk (Table 20.8). Singapore was the first country to grant approval for a cell-cultured chicken used as an ingredient in a hybrid product made with plant protein (SFA 2020). Today, several countries including Isreal, the USA and Australia have granted regulatory approval for cell cultivated products such as cultivated chicken, quail and beef, and cultivated pork fat.

20.4.4.2 Acellular or Precision Fermentation-Based Cellular Agriculture

Precision Fermentation-based CA manufactures by precision fermentation using genetic engineering approaches or industrial biotechnology

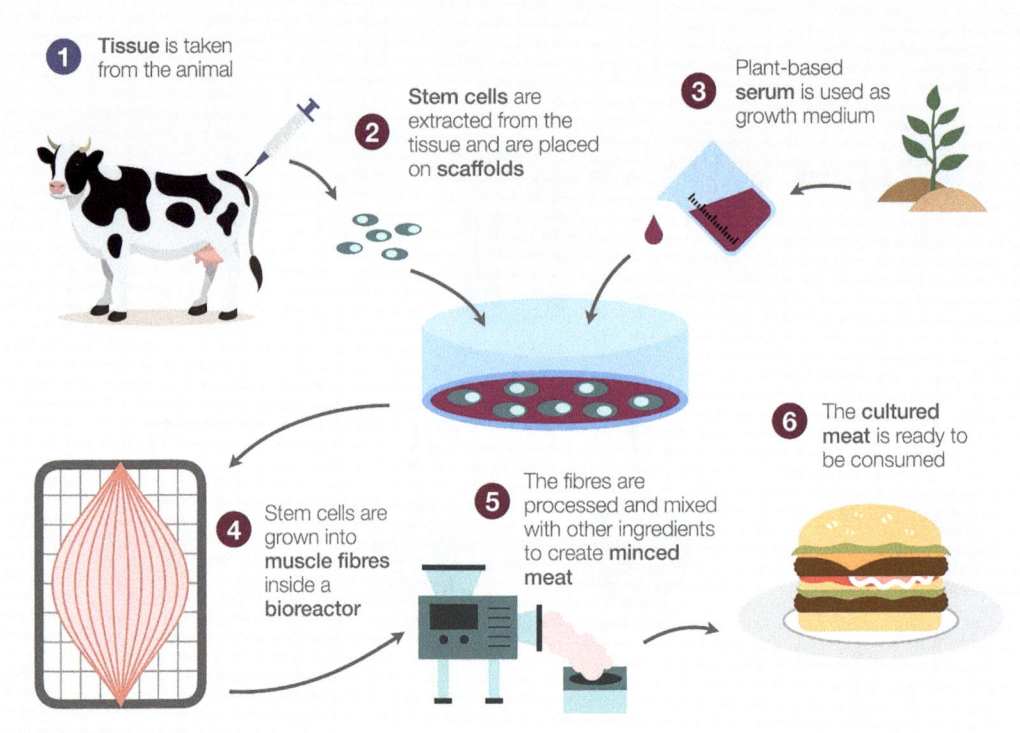

Fig. 20.7 Main steps involved in the production of cell-cultured meat

and microorganisms such as bacteria, yeast or algae. These microorganisms are typically genetically modified by adding recombinant DNA bearing a desired gene and mass produced in biorecators to produce specific proteins (caseins, ovalbumin, etc.), fats, collagen and other molecules that can be used to biofabricate animal products such as gelatine, enhanced milk products, chocolate, honey and leather, among others. The process of precision fermentation-based cellular agriculture has been discussed in recent reviews (Eastham and Leman 2024; Dupuis et al., 2023; Oliveira et al., 2024; Knychala et al., 2024) Some companies today use precision fermentation-based systems to produce a variety of products. For example, Meadow company uses a precision fermentation system harbouring genetically modified cells with instructions to produce a specific type of collagen for the manufacture of leather. Other examples include The EVERY Company (egg white), Perfect Day (milk), Pembient (Rhinoceros horn), etc. (Table 20.7).

20.4.4.3 Potential Benefits of Cellular Agriculture

The potential benefits of CA products have been summarized in many review articles (Stephens et al. 2018; El Wali et al. 2024; Munteanu et al. 2021; Kadim et al. 2015; Treich 2021) and include (1) potential to increase the availability of animal-based products and to meet the needs of consumers concerned about the environmental impact of conventional livestock production; (2) potential to support the production of specific livestock products in larger quantities (e.g. lactose free milk, casein proteins); (3) it offers wide possibilities to edit or modify genes to support the production of desired livestock phenotypes; (4) potential to reduce greenhouse gas emissions as well as solve other environmental issues like soil and water pollution compared with conven-

Table 20.8 Sample cellular agriculture companies producing a wide range of cultivated- and precision fermentation-based products

Product category	Product	Company	Website
Precision fermentation products	Dairy	Perfect Day	https://perfectday.com/
		TurtleTree	https://turtletree.com/
	Eggs	The EVERY Company (previously Clara Foods)	https://theeverycompany.com/
	Gelatin	Geltor	https://geltor.com/
	Myoglobin	Motif FoodWorks	https://madewithmotif.com/
	Honey	MeliBio, Inc	https://www.melibio.com/
Cultivated products	Beef	Mosa Meat	https://mosameat.com/
		Aleph Farms	https://www.aleph-farms.com/
		GOOD Meat	https://goodmeat.co
	Dairy (milk)	Opalia (previously Bettermilk)	https://www.opaliafoods.com/
		Perfect Day	https://perfectday.com/
	Chicken	JUST	http://www.ju.st/
		SuperMeat	https://supermeat.com/
		Upside Foods—previously Memphis Meats	https://upsidefoods.com/
	Seafood	Wildtype	https://www.wildtypefoods.com/
		Finless Foods	https://finlessfoods.com/
		BlueNalu Inc.	https://www.bluenalu.com/
		Shiok	https://shiokmeats.com/
Production applications	Leather	Modern meadow	https://www.modernmeadow.com/
	Silk	Bolt Threads	https://boltthreads.com/
		Spiber	https://spiber.inc/
	Cotton	GALY	https://recruit.galy.co/
	Horn	Pembient	https://www.pembient.com/

tional livestock production; (5) potential to reduce the risks of emerging infectious diseases associated with the storage, production and consumption of animal food resulting from conventional farming; (6) potential to support the production of products of desired nutritional quality; (7) potential to ensure food safety as lab-grown products are kept in controlled environments with close monitoring to easily detect microbial contamination and to remove such, compared with conventional farming which can introduce disease pathogens or antimicrobial resistant genes into the food chain; (8) less space requirement is needed for lab grown-food thus sparring the environment; (9) potential to lower global food insecurity since indoor production of CA products can still occur even during unfavourable conditions like natural disasters and pandemics; (10) CA products have strong moral implications as their production is largely not based on the slaughter of animals. It should be noted that many of these benefits are not yet supported by scientific or substantial evidence.

20.4.4.4 Challenges of Cellular Agriculture

Cellular agriculture is a developing technology and it is plagued by a myriad of challenges which have been discussed extensively by several recent review articles (Munteanu et al. 2021; Treich 2021; Chriki et al. 2022). The major concern relates to food safety considerations, contamination during the manufacturing processes, consumer acceptance/perception, ethical issues and many unknowns like the economic benefits and environmental issues associated with CA. Many governments such as Singapore, Canada, United States, Isreal and the European Union have or are working to put in place safety regulations to guide CA processes and ensure the safety of CA products.

20.5 Reproductive Technologies to Enhance Meat and Milk Production

Compared to developed and emerging countries, limited impacts have been observed in the application of reproductive biotechnologies in African livestock. However, various biotechnological techniques are increasingly used in African cattle breeds, being the most used livestock species for all assisted reproduction technologies (ART). These include oestrus synchronization, superovulation and ovum pick-up (OPU), semen collection and artificial insemination (AI), in vitro fertilization (IVF) and embryo culture, and in vivo embryo flushing and transfer. The sections below describe the use of these technologies in livestock under various production systems in Africa.

20.5.1 Artificial Insemination

Among the ART adopted to improve the livestock industry's viability, AI is one of the most powerful technologies; the first ART generation tool applied in livestock production systems to advance genetic improvement and selection of desired animals. Cattle, pig and sheep/goat account for most mammalian inseminations, the successes of which rely on the efficiency of semen collection, handling and preservation (Ugur et al. 2019). AI is the most widely used ART in African livestock systems.

20.5.1.1 Semen Collection
There are three methods of semen collection used in artificial insemination (Patel et al. 2017). First, using an artificial vagina is the most common method for semen collection from most domestic livestock. This method requires the male to mount a teaser female or dummy, and the penis is diverted into the artificial vagina, giving a thrust to semen release or ejaculation. Second, electroejaculation is a reliable method to stimulate the male by using a low-voltage current to nerves that control ejaculation and emission. It is commonly applied to rams and bulls; however, this technique is unaffordable to most smallholder farmers, and its use may be prohibited in several countries. Third, the transrectal massage method involves stroking the ampullae, seminal vesicles and prostate gland through the rectum wall in bulls (Rouge and Bowen 2002).

20.5.1.2 Oestrous Synchronization
Oestrous synchronization involves manipulating the reproductive processes using a combination of hormones that bring several females into oestrus (heat) at a similar time. Various oestrous synchronization protocols that utilize different hormones are used in small ruminants, beef heifers and replacement dairy heifers (Rodning et al. 2012). Prostaglandin (PGF2α), progestins and gonadotropin-releasing hormone are used in oestrus synchronization protocols. Prostaglandin is injected intramuscularly and subcutaneously to cause luteolysis of the corpus luteum and induce oestrus behaviour. The PGF2α is commercially available as Lutalyse® (Zoetis). Synthetic analogues of prostaglandins such as Fenprostenol®, Estrumate® and estroPlan® have been found to have a more rapid and dramatic effect in synthesizing progesterone (Omontese 2018). For better response, two injections of prostaglandins are administered 9–11 days apart.

Progestins include melengestrol acetate, ingested in feed and progesterone, administered vaginally through a controlled intravaginal drug release device. It inhibits heat; a cow or heifer usually comes into heat 1 to 3 days after removal. Moreover, a gonadotropin-releasing hormone injection administered 48 hours after a prostaglandin injection has provided a more concise synchrony of ovulation (Rodning et al. 2012).

20.5.1.3 Insemination Versus Natural Breeding

Compared to natural breeding, AI has revolutionized the livestock industry by promoting the utilization of genetically superior males (Paxton 2021). Using AI in farms eliminates the need for a breeding male whose semen is commercially available in various specialized AI centres and reduces the risks of transmission of genital diseases. Hence, removing the costs associated with the maintenance of breeding bulls leads to increased profitability (TNAU 2009). Furthermore, the possibility of using frozen and extended semen has made it possible to extend the shelf-life and maintain the fertilizing potential of harvested spermatozoa, therefore preserving the fertility of selected males used to sire many females (Ugur et al. 2019).

In many African countries, AI is limited due to several constraints (see Chapter 24). Consequently, AI is less affordable for most African smallholder farmers who rely upon natural mating as a standard and cheaper alternative breeding practice, requiring only the maintenance of genetically superior males on the farms during the breeding season. Most importantly, it eliminates the use of well-trained technicians and the knowledge of the structure and function of reproduction on the operator's part (TNAU 2009). Today, the AI technique in Africa is used especially in crossbreeding between exotic sires and indigenous African cattle to increase milk production and beef potential of local cattle breeds. AI has the potential to increase genetic gains using internationally proven exotic sires. Thus, the ADGG project has used AI to produce crossbred dairy cattle that combine the productivity of Western breeds and the tropically adapted East-African zebus in Tanzania and Ethiopia. In Tanzania, 236,335 doses of semen from Boran, Sahiwal, Mpwapwa, Ayrshire, Friesian, Jersey and Simmental were distributed to different organizations or farmers operating medium- and large-scale dairy and beef production farms to upgrade the indigenous cattle breeds (Msalya 2017). The combination of AI with oestrus synchronization allows the insemination of a larger number of cows on a predetermined date. The use of oestrus synchronization in breeding programmes has shown conception rates comparable to natural mating on the African continent (Van Marle-Köster and Webb 2014).

20.5.2 Embryo Transfer

20.5.2.1 Superovulation

Superovulation is a reproductive technology used in the dairy industry to increase the reproductive rate of superior females. It is an essential process of embryo transfer that increases the number of oocytes available at ovulation and for fertilization (Lonergan and Boland 2011).

The use of hormones such as equine chorionic gonadotrophin, follicle-stimulating hormone, pregnant mare's serum gonadotropin (PMSG) and horse anterior pituitary extracts permits the growth of subordinate follicles alongside their dominant counterpart. Nevertheless, the quality of gonadotropin preparations influences the effectiveness of superovulation. For example, the combination of follicle-stimulating hormone and luteinizing hormone plays an important role as they control the growth of ovarian follicles. Superovulation using human menopausal gonadotrophin has revealed effectiveness in cattle, sheep and goat (González et al. 2001). Its application in cattle has achieved an average of six transferable embryos per session. In goat, superovulation by PMSG resulted in lower fertilization and embryo recovery rates than the horse anterior pituitary (González et al. 2001).

20.5.2.2 Insemination and Embryo Flushing

Heat or oestrus detection is critical for efficient reproduction management. The difficulty in determining oestrus (onset and end) requires careful observation of the animal's behaviour using visual and/or sensing aids. Insemination should be done sometime between the onset of heat activity (oestrus) and ovulation, with the morning (AM)/evening (PM) rule being the easiest practical method of timing insemination. Insemination is performed by natural mating or AI using frozen, chilled or fresh semen.

Insemination is done approximately 4, 12, 24 or 36 hours after the onset of standing oestrus (Hasler 2004). In sows, inseminations occur 36 and 42 hours after human chorionic gonadotropin injection. In the ewe, however, the risk of fertilization failures in ewes due to higher ovarian stimulation and response to gonadotrophin oblige for surgical insemination directly into the uterus using laparotomy or laparoscopy (Ishwar and Memon 1996). For further improvement of reproductive success, many commercial farms are adopting fixed-timed insemination requiring a cocktail of hormonal treatment of females prior to insemination (Lamb et al. 2010; De Rensis and Kirkwood 2016).

In vivo-produced embryos may be collected through flushing (cows) or laparoscopic surgery in ewe, doe and sow (Besenfelder et al. 1997). In cows and heifers, a Foley rubber catheter is mainly used for insertion through the cervix of the donor cow (Callesen et al. 1996), and through which a buffer medium (e.g. Dulbecco's phosphate-buffered saline supplemented with 1% faecal calf serum) is flushed into and out of the uterus to collect the embryos seven days after oestrus. Each uterine horn is filled and emptied five to ten times with 30–200 mL of the buffer medium each time, according to the size of the uterus. Embryos are then flushed out, collected with the liquid and separated from the flush media for examination under a stereomicroscope to determine their quality and stage of development (Hasler 2004). In the ewe and doe, embryos are surgically collected 3 to 7 days after oestrus under general anaesthesia through a midventral laparotomy (Moore 1982).

20.5.2.3 Transfer to Surrogates

In cattle, each embryo is loaded into a 0.25 mL insemination straw. This procedure is done under microscopic viewing with the aid of a 1 mL syringe and requires considerable practice, patience and skill. Before embryo transfer, the recipient's ovaries are palpated rectally to determine which ovary has ovulated. Next, a transfer gun, also called an insemination rod, is passed through the cervix, with the tip of the rod allowed to slide into the horn of the ovary possessing the active corpus luteum. Finally, the embryo is gently expelled into the forward end of that uterine horn.

In small ruminants and sows/gilts, each embryo is loaded into a 0.25 mL sterile insemination straw. The straw is placed in an ordinary insemination pipette or a specially designed embryo transfer catheter, a thinner and more extended version. Laparoscopy has been reported effective in these species but requires more experience to obtain reasonable pregnancy rates (Besenfelder et al. 1997). In addition, the oestrous cycle of the recipients must be synchronized with that of the donor. Before transfer, it must be ascertained that oestrus has occurred within ±24 hours of the donor and that a functional corpus luteum is present (Hasler 1998).

Multiple ovulation and embryo transfer (MOET) in combination with cloning and embryo sexing can potentially increase reproduction efficiency and genetic gains. The MOET has been used in various livestock species and wildlife on the African continent and has increased the conception rates to above 50% when combined with better embryo sorting and transfer of more than one embryo per recipient female (Van Marle-Köster and Webb 2014).

20.5.3 In Vitro Fertilization and Embryo Culture

20.5.3.1 Oocyte Collections: Post-mortem Versus Alive Animals

Oocyte collection is either from the post-mortem (slaughterhouse)-derived ovaries or living animals after ovum-pick-up (OPU). In cows, follicular aspiration of collected ovaries may vary from eight to nine oocytes per ovary and around 15 developmentally competent cumulus–oocytes complexes per animal (Fair et al. 1995).

In contrast, the OPU technique, which consists of the transvaginal removal of the oocytes by aspiration of the ovarian follicles under ultrasonography guidance, is a harmless procedure performed on donor cows constrained in a chute. This procedure is a modified technique of human reproduction that has become the tool of choice for ovarian oocyte retrieval in veterinary-assisted reproduction (Pieterse et al. 1988). The OPU is performed on stimulated and non-stimulated cows with a variable frequency that depends on the utilized protocols (e.g. one or two sessions/week). Higher numbers of puncturable antral follicles of 5–10 mm (6 to 16), oocytes (6 to 12) and blastocysts (>2) are generated per OPU session (Stubbings and Walton 1995). Although OPU has a great advantage in producing many oocytes per donor, there is a need for special care such as animal welfare, repetitive epidural anaesthesia, ovary stroma integrity and adhesions must be considered (McEvoy et al. 2006).

On the other hand, laparoscopy-guided follicular puncture is also used in large and small ruminants (Baldassarre et al. 2001). This technique has the advantage of doubling the efficiency of oocyte collection, but adhesions may limit their use on donors. In vitro-produced embryos based on laparoscopic-aided ovum pick-up obtained oocytes may become an alternative to conventional embryo production in small ruminants, as it may significantly improve the problems in vivo technology carried over the years (Cognie 1999).

Regardless of their origins, oocytes enclosed with more than three layers of cumulus cells, forming the cumulus–oocyte complexes, are selected under a stereomicroscope for in vitro maturation (IVM).

20.5.3.2 In Vitro Maturation

At birth, oocytes are blocked at prophase I of meiosis. Their developmental competence is progressively acquired during intrafollicular growth, with the oocyte accumulating various compounds during its development (Fair et al. 1995; Lonergan et al. 1994). In most ruminants, the oocyte will undergo final maturation during the last 24 h following the luteinizing hormone surge and ovulation. In contrast, this maturation will resume spontaneously in vitro after removing the cumulus–oocytes complexes from its surrounding follicular fluid, known to contain a maturating inhibiting factor. The maturing oocyte undergoes both cytoplasmic and nuclear maturation.

Cytoplasmic maturation is characterized by changes in organelles, proteins and transcripts (Hyttel et al. 1997). Meanwhile, the progression of meiosis from prophase I to the metaphase of the second meiotic division corresponds to the nuclear maturation that makes a second arrest while waiting for fertilization. In livestock, around 85% to 90% of the cultured immature oocytes will reach the second meiotic division at the end of IVM under proper conditions (Lonergan and Fair 2016).

20.5.3.3 In Vitro Fertilization

Upon completion of IVM, oocytes are co-incubated with spermatozoa for up to 18 to 24 h while undergoing IVF. The IVF usually occurs in a culture dish where both male and female gametes are placed into a microenvironment of 50 to 100 μl, for low numbers of OPU donor oocytes or > 400 μl, for abattoir mass-collected oocytes (Gordon 2003). Spermatozoa are washed and selected (e.g. swim-up or density gradient centrifugation procedures) to remove freezing media, seminal plasma, debris and dead spermatozoa. The purified motile spermatozoa must be treated with capacitating agents (e.g. heparin) to gain the ability to penetrate the zona pellucida of the oocyte (Parrish et al. 1986). Large numbers of oocytes can be fertilized in groups using conventional frozen semen. Below are further adjustments to ensure high fertilization rates:

- Selecting high-fertile bulls.
- Adjusting capacitation agent concentration.
- Optimizing the composition and volume of the fertilization medium.
- Adjusting oocyte density (number per volume of medium and per spermatozoa).

A major advantage of IVF is that it requires a small number of spermatozoa to fertilize the collected oocytes (Van Soom et al. 1991).

Following interaction with the spermatozoon, meiosis of the oocyte will resume for completion, and the second polar body will be extruded within the perivitelline space. Thus, the oocyte is a proper haploid cell in which the remaining set of chromosomes will pair with their sperm-derived homologs. This process is known as amphimixis, which establishes fertilization and returns to the new zygote's diploid status (Sirard 2001).

Thereafter, the presumptive zygotes are denuded from their surrounding cumulus cells and transferred into the culture medium for additional culture in defined, semi-defined or complex media until the blastocyst stage, which is the last stage for in vitro embryo culture for most livestock animals (Feugang et al. 2009). Despite the efforts, the proportion of fertilized oocytes or zygotes that develop to the blastocyst stage seems to plateau around 30–40% (Rizos et al. 2002). Currently, the quality of oocytes remains the main limiting factor to the efficiency of in vitro embryo production. Indeed, studies show that in vivo matured oocytes yield more blastocysts than their IVM counterparts (Sirard and Blondin 1996). It is essential to adjust the duration of the IVM phase according to follicular maturity to the in vivo-maturation period for that species and to culture conditions (Hyttel et al. 1997).

The in vitro embryo production technique has several challenges. For instance, the complexity of culture conditions requiring a fully controlled environment (gas and culture compositions), the need for highly qualified personnel, and related reagents and equipment costs are significant limitations for implementing the technique in numerous African countries.

20.5.4 Pregnancy Detection and Monitoring

Pregnancy diagnosis in domestic livestock is a vital fertility management tool, and there is a need to check animals for pregnancy early (Bekele et al. 2016). Methods of pregnancy detection include transrectal palpation, ultrasonography and visual techniques such as non-return to oestrus and exposure to a bull. Oestrus detection after insemination is beneficial as it allows for detecting non-pregnant cows and re-synchronization. Non-return to oestrus assumes that the animal is pregnant. Transrectal palpation involves palpating the reproductive tract through the rectal wall 35 days post-breeding (Youngquist 2006).

On the other hand, pregnancy detection relies on A-mode, Doppler ultrasound, and B-mode or real-time ultrasonography. In addition, the most reliable indicator of pregnancy is blood progesterone, which can be quantified in a laboratory using an RIA or ELISA test (Fricke 2010). However, sheep and goats' plasma or milk progesterone assay is not accurate for pregnancy diagnosis (Goel et al. 2009).

Furthermore, digital infrared thermal imaging (DITI) has been introduced into livestock for the improved management of environmental changes, (re)production and health conditions (Hellebrand et al. 2003). The DITI method relies on detecting infrared energy emitted from an object and converting it into temperature displayed as a temperature gradient or thermogram. Despite the challenges of interpreting data due to factors such as coat colour and environmental variables, the DITI is a non-invasive, non-destructive and effortless method capable of detecting scrotal temperatures, health, oestrus and pregnancy statuses of various livestock (Stelletta et al. 2012). However, the DITI utilization in livestock farms is still limited and inexistent in African countries.

20.6 Future Perspectives and Conclusion

The factors impeding the implementation of modern technologies for livestock genetic improvement have been discussed in Section 20.2 and Chapter 2. Addressing these challenges will be an important step towards advancing African livestock genetic improvement. For example, the implementation of genomic selection in African livestock will require breed-specific tools; like genotyping arrays specific for African livestock. Characterizing the genomic variation in African livestock and association with traits of interest is an important step in prioritizing genetic markers for breeding and health management. Recent initiatives like the African BioGenome project (Ebenezer et al. 2022), which has as its goal to sequence the genomes of plants and animals in Africa, will facilitate the catalogu-

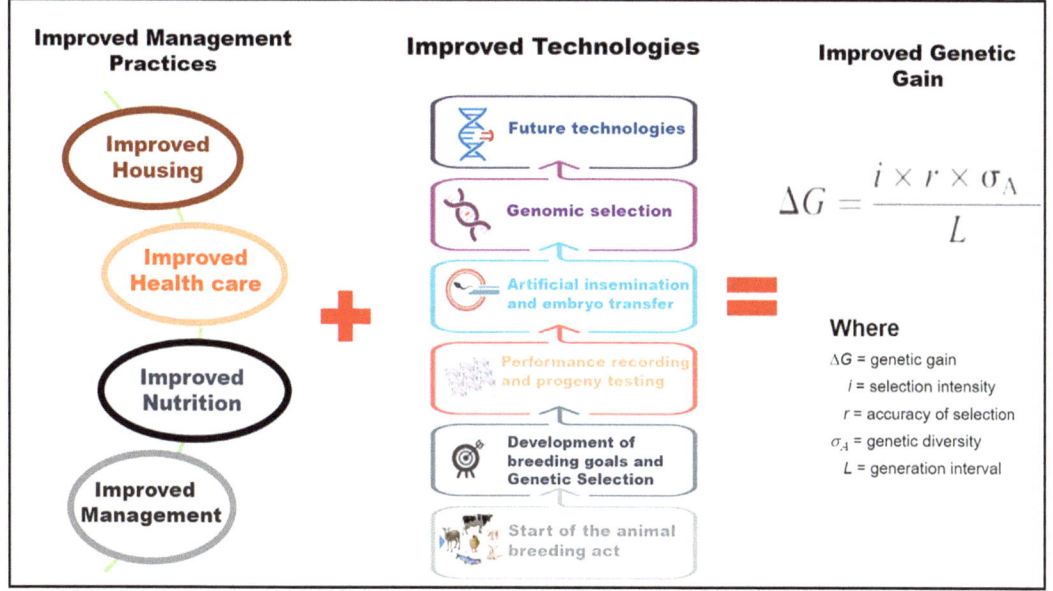

Fig. 20.8 Schematic representation of improved farm management practices and technologies that have led to enhanced genetic gain in dairy cattle breeding and improvement in milk production. (From Ibeagha-Awemu and Yu (2021), licensed under CC-BY 4.0)

ing of the genomic variations necessary for the implementation of genomic breeding in the continent.

The implementation of modern technologies has been the bedrock of improved productivity in the dairy, swine and poultry sectors in Western nations. The improvements were through advances in all aspects of livestock management such as improved management practices, improved nutrition, improved health care, improved genetic/genomic selection, reproduction technologies, etc. (Fig. 20.8) (Ibeagha-Awemu and Yu 2021).

The introduction of modern technologies in dairy production in some African countries has resulted in increased milk production and farm incomes. A typical example is the African Dairy Genetic Gains (ADGG) (https://www.ilri.org/research/projects/african-dairy-genetic-gains) project by the International Livestock Research Institute (ILRI) (Kenya). The ADGG project is operating in seven countries (Ethiopia, Tanzania Uganda, Kenya, Nigeria, Zambia and Nepal) and its main goal is to address some of the challenges faced by smallholder livestock producers through the application of ICT and genomic technologies, among others. The ADGG project has developed a robust, agile and scalable data platform and capture tools. It has also developed and promoted several digital tools (e.g. ADGG dairy tool: Helping your cow to calve- (Ojango et al. 2020a); ADGG dairy tool: Ensuring dairy herd hygiene (Ojango et al. 2020b)). The ADGG project is therefore providing dairy cattle genetic gains with the potential to transform the lives of millions of smallholder dairy families in the participating countries with the potential to extend to other African countries.

The ADGG is a practical example of how modern technologies and technical know-how with the participation of relevant expert institutions harnessed in a farmer-oriented manner delivered and achieved increased livestock productivity and farm incomes. The significant improvement in productivity is attributable to the practical application of the know-how of participating researchers and expert institutions/bodies instituted systems that applied modern tools of milk production technologies (genetics, ICT, data processing, AI, etc.) spanning individual farm levels to group levels to country/regional levels by collecting performance data and sharing information for the benefit of participants at all levels within the ADGG project. The ADGG instituted system acts at a national/regional scale to provide official standardized identification, data collection structure covering all participating farms, exchange of data with extension personnel, veterinarians and farmers, interpretation of performance recording data, provide information/advice on nutrition and insemination frequency, enhance follow up of health, production, insemination as well as connect farmers with key service providers among others.

The application of in vitro embryo production (IVEP) for cheaper production of a predictable supply of embryos from the ovaries of live or slaughtered cows in the Boran cattle was demonstrated (Mutembei et al. 2015). Using IVEP and OPU techniques, the authors reported that in vitro embryo production obtained an oocyte maturation rate of over 90%, over 70% cleavage rate and an embryo output of 30–50% blastocyst rate. The conception rate from embryo transfer they reported was 45–50%, with 17 calves born reaching puberty without any abnormalities.

The use of microsatellites for parentage verification was demonstrated in the Boran breed by Kios and colleagues (Kios et al. 2012). A microsatellite marker panel consisting of 14 markers has been developed and utilized for parentage verification in the Angora Breed (Visser 2011).

The use of high-density markers (SNPs) allowed accurate estimation of breed composition and parentage assignment in smallholder dairy farms in Africa which do not perform pedigree recording (Strucken et al. 2017). Genotyping 2940 crossbred dairy cattle in Kenya, Uganda, Ethiopia and Tanzania with 777 K Illumina BovineHD SNP chip, it was demonstrated that a panel of 200 SNPs gave the best prediction of breed composition while another set of 200 SNPs was the most accurate at parentage assignment (Strucken et al. 2017).

Further practical examples of the prospects of modern technologies in African livestock genetic improvement are presented in Chapters 22, 23 and 24.

In conclusion, modern technologies have been the main tools behind livestock genetic advances achieved in the last seven decades. Although these technologies have been applied to a lesser extent in African livestock, their adoption is feasible and will facilitate rapid progress as demonstrated with the ADGG project.

References

Agaviezor BO, Adefenwa MA, Peters SO, Yakubu A, Adebambo OA, Ozoje MO, Ikeobi CON, Ilori BM, Wheto M, Ajayi OO, Amusan SA, Okpeku M, De Donato M, Imumorin IG (2012) Genetic diversity analysis of the mitochondrial D-loop of Nigerian indigenous sheep. Animal Genetic Resources/Ressources génétiques animales/Recursos genéticos animales 50:13–20. https://doi.org/10.1017/s2078633612000070.

Alvarez-Castro JM, Carlborg O (2007) A unified model for functional and statistical epistasis and its application in quantitative trait Loci analysis. Genetics 176(2):1151–1167. https://doi.org/10.1534/genetics.106.067348. From NLM

American Calan Inc (2019) Calan Broadbent feeding system. American Calan Inc. http://americancalan.com/code/template.cfm?page=2. Accessed 14/10/2021

Anderson S (1981) Shotgun DNA sequencing using cloned DNase I-generated fragments. Nucleic Acids Res 9(13):3015–3027. https://doi.org/10.1093/nar/9.13.3015. %J Nucleic Acids Research (acccessed 4/12/2022)

Asueni O, Godknows N (2019) Climate change and social conflict: migration of FULANI herdsmen and the implications in Nigeria. Br J Educ 7(5):82–93

AU-IBAR (2018) Project to facilitate and fast track the implementation of the Global Plan of Action (GPA) for sustainable use of AnGR in Africa. https://www.au-ibar.org/au-ibar-projects/angr. Accessed 15 July 2022

Avendaño S, Watson KA, Kranis A (2010) Genomics in poultry breeding—from utopias to deliverables. In: Proceedings of the 9th World Congress on Genetics Applied to Livestock Production: 1–6 August 2010. Leipzig, Germany. Session 07-01. http://wcgalp.org/system/files/proceedings/2010/genomics-poultry-breeding-ndash-utopias-deliverables.pdf (accessed on 13 September, 2022).

Avendaño S, Watson K A, Kranis A (2012) Genomics in poultry breeding—Into consolidation phases. In: Proceedings of the 24th World Poultry Congress: 5–9 August 2012. Salvador, Brazil. P.36. https://lohmann-breeders.com/lohmanninfo/genomic-selection-layer-broiler-breeding/ (accessed on 13 September 2022).

Bai Y, Lin W, Xu J, Song J, Yang D, Chen YE, Li L, Li Y, Wang Z, Zhang J (2021) Improving the genome assembly of rabbits with long-read sequencing. *Genomics* 113(5):3216–3223. https://doi.org/10.1016/j.ygeno.2021.05.031.

Baldassarre H, Keefer C, Gauthier M, Bhatia B, Begin I, Pierson J, Laurin D, Trigg T, Downey B, Karatzas C (2001) Laparoscopic ovum pick-up and zygote recovery in goats treated with deslorelin implants before superovulation. Theriogenology 55(1):510

Bawden CS, Powell BC, Walker SK, Rogers GE (1998) Expression of a wool intermediate filament keratin transgene in sheep fiber alters structure. Transgenic Research 7(4):273–287. https://doi.org/10.1023/a:1008830314386.

Beatson P, Meier S, Cullen N, Eding HJA (2019) Genetic variation in milk urea nitrogen concentration of dairy cattle and its implications for reducing urinary nitrogen excretion. Animal 13(10):2164–2171. https://doi.org/10.1017/S1751731119000235

Bekele N, Addis M, Abdela N, Ahmed WM (2016) Pregnancy diagnosis in cattle for fertility management: a review. Global Veterinaria 16(4):355–364. https://doi.org/10.5829/idosi.gv.2016.16.04.103136

Bellini J (2018) This gene-edited calf could transform Brazil's beef industry. Wall Street Journal of October 1, 2018. Available at https://www.wsj.com/video/series/moving-upstream/this-gene-edited-calf-could-transform-brazil-beef-industry/D2D93B49-8251-405F-BC35-1E5C33FA08AF?page=1&pos=11. Accessed 14 June 2023.

Besenfelder U, Mödl J, Müller M, Brem G (1997) Endoscopic embryo collection and embryo transfer into the oviduct and the uterus of pigs. Theriogenology 47(5):1051–1060. https://doi.org/10.1016/s0093-691x(97)00062-9

Bi Y, Hua Z, Liu X, Hua W, Ren H, Xiao H, Zhang L, Li L, Wang Z, Laible G et al. (2016) Isozygous and selectable marker-free MSTN knockout cloned pigs generated by the combined use of CRISPR/Cas9 and Cre/LoxP. *Sci Rep* 6:31729. https://doi.org/10.1038/srep31729.

Bleck G, White B, Hunt E, Rund L, Barnes J, Bidner D, Bremel R, Wheeler MJT (1996) Production of transgenic swine containing the bovine α-lactalbumin gene. 1(45):347. https://doi.org/10.2527/1998.76123072x

Blench R (2004) Natural resource conflicts in north-Central Nigeria: a handbook and case studies. Mallam Dendo Ltd., Cambridge, UK

Boadi D, Wittenberg K, Kennedy AJCJ (2002) Validation of the sulphur hexafluoride (SF6) tracer gas technique for measurement of methane and carbon dioxide production by cattle. Canadian Veterinary Journal 82(2):125–131. https//doi.org/10.4141/A01-054.

Bolotin A, Quinquis B, Sorokin A, Ehrlich SD (2005) Clustered regularly interspaced short palindrome repeats (CRISPRs) have spacers of extrachromosomal origin. Microbiology 151(8):2551–2561. https://doi.org/10.1099/mic.0.28048-0

Botreau R, Bonde M, Butterworth A, Perny P, Bracke MB, Capdeville J, Veissier I (2007) Aggregation of measures to produce an overall assessment of animal welfare. Part 1: a review of existing methods. Animal 1(8):1179–1187. https://doi.org/10.1017/s1751731107000535.

Brahi OHD, Xiang H, Chen X, Farougou S, Zhao X (2015) Mitogenome revealed multiple postdomestication genetic mixtures of west African sheep. J Anim Breed Genet 132(5):399–405. https://doi.org/10.1111/jbg.12144

Brito LF, Oliveira HR, McConn BR, Schinckel AP, Arrazola A, Marchant-Forde JN, Johnson JS (2020) Large-scale phenotyping of livestock welfare in commercial production systems: a new frontier in animal breeding. Front Genet 11:793. https://doi.org/10.3389/fgene.2020.00793.

Brophy B, Smolenski G, Wheeler T, Wells D, L'Huillier P, Laible G (2003) Cloned transgenic cattle produce milk with higher levels of beta-casein and kappa-casein. Nat Biotechnol 21(2):157–162. https://doi.org/10.1038/nbt783.

Burkard C, Opriessnig T, Mileham AJ, Stadejek T, Ait-Ali T, Lillico SG, Whitelaw CBA, Archibald AL (2018) Pigs Lacking the Scavenger Receptor Cysteine-Rich Domain 5 of CD163 Are Resistant to Porcine Reproductive and Respiratory Syndrome Virus 1 Infection. *J Virol* 92(16). https://doi.org/10.1128/jvi.00415-18

Burkard C, Lillico SG, Reid E, Jackson B, Mileham AJ, Ait-Ali T, Whitelaw CB, Archibald AL (2017) Precision engineering for PRRSV resistance in pigs: Macrophages from genome edited pigs lacking CD163 SRCR5 domain are fully resistant to both PRRSV genotypes while maintaining biological function. *PLoS Pathog* 13(2), e1006206. https://doi.org/10.1371/journal.ppat.1006206.

Bünger L, Clelland N, Moore K, McLean K, Kongsro J, Lambe N (2014) Integrating computed tomography (CT) into commercial sheep breeding in the UK: cost and value. Farm Animal Imaging Copenhagen 2014:22-27.

Callesen H, Liboriussen T, Greve T (1996) Practical aspects of multiple ovulation-embryo transfer in cattle. Anim Reprod Sci 42(1–4):215–226. https://doi.org/10.1016/0378-4320(96)01513-8

Calus MP (2010) Genomic breeding value prediction: methods and procedures. Animal 4(2):157–164. https://doi.org/10.1017/s1751731109991352.

Campbell KH, McWhir J, Ritchie WA, Wilmut I (1996) Sheep cloned by nuclear transfer from a cultured cell line. Nature 380(6569):64–66. https://doi.org/10.1038/380064a0.

Canavez FC, Luche DD, Stothard P, Leite KR, Sousa-Canavez JM, Plastow G, Meidanis J, Souza MA, Feijao P, Moore SS et al (2012) Genome sequence and assembly of Bos indicus. J Heredity 103(3):342–348. https://doi.org/10.1093/jhered/esr153.

Carlson DF, Tan W, Lillico SG, Stverakova D, Proudfoot C, Christian M, Voytas DF, Long CR, Whitelaw CBA, Fahrenkrug SC (2012) Efficient TALEN-mediated gene knockout in livestock. Proc Natl Acad Sci USA 109(43):17382–17387. https://doi.org/10.1073/pnas.1211446109.

Carter S, Sutton A, Stenglein R. (2022) Diet and feed management to mitigate airborne emissions. eXtension Air Quality in Animal Agriculture, 10 (accessed July 10, 2022). https://lpelc.org/wp-content/uploads/2019/03/Dietand-Feed-FINAL.pdf.

Cattle Information Service (2021) Mobileherd app – your date on the hoof. Cattle Information Service. https://www.thecis.co.uk/services/MobileHerd. Accessed 14/10/2021

Chagunda M, Yan TJAFS (2011) Do methane measurements from a laser detector and an indirect open-circuit respiration calorimetric chamber agree sufficiently closely? Animal Feed Sci Technol 165(1–2):8–14

Chilliard Y, Ferlay A, Doreau MJL (2001) Effect of different types of forages, animal fat or marine oils in cow's diet on milk fat secretion and composition, especially conjugated linoleic acid (CLA) and polyunsaturated fatty acids. Livestock Product Sci 70(1–2):31–48

Chriki S, Ellies-Oury M-P, Hocquette J-F (2022) Is "cultured meat" a viable alternative to slaughtering animals and a good comprise between animal welfare and human expectations? Animal Front 12:35–42. Chriki, S.; Hocquette, J.-F. The Myth of Cultured Meat: A Review. Frontiers in Nutrition 2020, 7, 7. https://doi.org/10.3389/fnut.2020.00007.

Ciccarelli M, Giassetti MI, Miao D, Oatley MJ, Robbins C, Lopez-Biladeau B, Waqas MS, Tibary A, Whitelaw B, Lillico S et al (2020) Donor-derived spermatogenesis following stem cell transplantation in sterile NANOS2 knockout males. Proc Natl Acad Sci 117(39):24195. https://doi.org/10.1073/pnas.2010102117

Cognie Y (1999) State of the art in sheep-goat embryo transfer. Theriogenology 51(1):105–116. Cognie, Y.; Baril, G.; Poulin, N.; Mermillod, P. Current status of embryo technologies in sheep and goat. *Theriogenology* 2003, 59 (1), 171–188. https://doi.org/10.1016/s0093-691x(02)01270-0

Cong L, Ran FA, Cox D, Lin S, Barretto R, Habib N, Hsu PD, Wu X, Jiang W, Marraffini LA et al (2013) Multiplex genome engineering using CRISPR/Cas Systems. Science 339(6121):819–823. https://doi.org/10.1126/science.1231143

Crispo M, Mulet AP, Tesson L, Barrera N, Cuadro F, dos Santos-Neto PC, Nguyen TH, Crénéguy A, Brusselle L, Anegón I et al (2015) Efficient Generation of Myostatin Knock-Out Sheep Using CRISPR/Cas9 Technology and Microinjection into Zygotes. PLOS One 10(8):e0136690. https://doi.org/10.1371/journal.pone.0136690.

Damak S, Su H, Jay NP, Bullock DW (1996) Improved wool production in transgenic sheep expressing insulin-like growth factor 1. Nature Biotechnology 14(2):185–188. https://doi.org/10.1038/nbt0296-185.

Davis GP, DeNise SK (1998) The impact of genetic markers on selection. J Anim Sci 76(9):2331–2339. https://doi.org/10.2527/1998.7692331x.

De Rensis F, Kirkwood RN (2016) Control of estrus and ovulation: fertility to timed insemination of gilts and sows. Theriogenology 86(6):1460–1466. https://doi.org/10.1016/j.theriogenology.2016.04.089. From NLM

de Roos AP, van den Bijgaart HJ, Hørlyk J, de Jong G (2007) Screening for subclinical ketosis in dairy cattle by Fourier transform infrared spectrometry. J Dairy Sci 90(4):1761–1766. https://doi.org/10.3168/jds.2006-203. From NLM

De Roover R, Feugang J, Bols P, Genicot G, Hanzen C. (2008) Effects of ovum pick-up frequency and FSH stimulation: a retrospective study on seven years of beef cattle in vitro embryo production. *Reproduction in Domestic Animals* 43 (2):239–245. https://doi.org/10.1111/j.1439-0531.2007.00873.x

De Roover R, Genicot G, Leonard S, Bols P, Dessy F (2005) Ovum pick up and in vitro embryo production in cows superstimulated with an individually adapted superstimulation protocol. *Animal Reproduction Science* 86(1–2):13–25. https://doi.org/10.1016/j.anireprosci.2004.05.022

Dehareng F, Delfosse C, Froidmont E, Soyeurt H, Martin C, Gengler N, Vanlierde A, Dardenne PJA (2012) Potential use of milk mid-infrared spectra to predict individual methane emission of dairy cows. Animal 6(10):1694–1701. https://doi.org/10.1017/S1751731112000456.

Dekkers JC (2004) Commercial application of marker- and gene-assisted selection in livestock: strategies and lessons. J Anim Sci (82 E-Suppl):E313–E328. https://doi.org/10.2527/2004.8213_supplE313x.

Deltcheva E, Chylinski K, Sharma CM, Gonzales K, Chao Y, Pirzada ZA, Eckert MR, Vogel J, Charpentier E (2011) CRISPR RNA maturation by trans-encoded small RNA and host factor RNase III. Nature 471(7340):602–607. https://doi.org/10.1038/nature09886

Denning C, Burl S, Ainslie A, Bracken J, Dinnyes A, Fletcher J, King T, Ritchie M, Ritchie WA, Rollo M et al (2001) Deletion of the alpha(1,3)galactosyl transferase (GGTA1) gene and the prion protein (PrP) gene in sheep. Nat Biotechnol 19(6):559–562. https://doi.org/10.1038/89313.

Ding R, Yang M, Wang X, Quan J, Zhuang Z, Zhou S, Li S, Xu Z, Zheng E, Cai G et al (2018) Genetic architecture of feeding behavior and feed efficiency in a Duroc pig population. Front Genet 9. Original Research

Ding Y, Zhou S-w, Ding Q, Cai B, Zhao X-e, Zhong S, Jin M-h, Wang X-l, Ma B-h, Chen Y-l (2020) The CRISPR/Cas9 induces large genomic fragment deletions of MSTN and phenotypic changes in sheep. J. Integr. Agric. 19(4):1065–1073. https://doi.org/10.1016/S2095-3119(19)62853-4

Doormaal BV (2019) 10 Years of Genomic Selection: What's Next? *Lactanet's Information Articles (Canada)*, https://www.cdn.ca/document.php?id=530.

Dong Y, Xie M, Jiang Y, Xiao N, Du X, Zhang W, Tosser-Klopp G, Wang J, Yang S, Liang J et al

(2013) Sequencing and automated whole-genome optical mapping of the genome of a domestic goat (Capra hircus). Nat Biotech 31(2):135–141. Research. https://doi.org/10.1038/nbt.2478; http://www.nature.com/nbt/journal/v31/n2/abs/nbt.2478.html#supplementary-information

Du SJ, Gong ZY, Fletcher GL, Shears MA, King MJ, Idler DR, Hew CL (1992) Growth enhancement in transgenic Atlantic salmon by the use of an "all fish" chimeric growth hormone gene construct. Bio/Technology (Nature Publishing Company) 10(2):176–181. https://doi.org/10.1038/nbt0292-176.

Dunham RA, Warr GW, Nichols A, Duncan PL, Argue B, Middleton D, Kucuktas H (2002) Enhanced bacterial disease resistance of transgenic channel catfish Ictalurus punctatus possessing cecropin genes. Marine Biotechnol (New York, N.Y.) 4(3):338–344. https://doi.org/10.1007/s10126-002-0024-y.

E + V Technology (2021) A quantum leap in the automation of the food industry. E + V Technology. https://www.eplusv.com/en/. Accessed 14/10/2021

Ebenezer TE, Muigai AWT, Nouala S, Badaoui B, Blaxter M, Buddie AG, Jarvis ED, Korlach J, Kuja JO, Lewin HA et al (2022) Africa: sequence 100,000 species to safeguard biodiversity. Nature 603(7901):388–392. https://doi.org/10.1038/d41586-022-00712-4.

Edea Z, Dessie T, Dadi H, Do K-T, Kim K-S (2017) Genetic Diversity and Population Structure of Ethiopian Sheep Populations Revealed by High-Density SNP Markers. Frontiers in Genetics 8:218. https://doi.org/10.3389/fgene.2017.00218.

Eichler EE, Flint J, Gibson G, Kong A, Leal SM, Moore JH, Nadeau JH (2010) Missing heritability and strategies for finding the underlying causes of complex disease. Nat Rev Genet 11(6):446–450. https://doi.org/10.1038/nrg2809.

El-Khishin DA, Ageez A, Saad ME, Ibrahim A, Shokrof M, Hassan LR, Abouelhoda MI (2020) Sequencing and assembly of the Egyptian buffalo genome. PLoS One 15(8):e0237087. https://doi.org/10.1371/journal.pone.0237087.

El-Osta HS, Morehart MJ (2000) Technology adoption and its impact on production performance of dairy operations. Review of Agricultural Economics 22(2):477–498. https://doi.org/10.2307/1349806.

El-Osta HS, Johnson JD. (1998) Determinants of financial performance of commercial dairy farms. Resource Economics Division, Economic Research Service, U.S. Department of Agriculture. Technical Bulletin No. 1859. https://www.ers.usda.gov/publications/pub-details?pubid=47192

Elsik CG, Tellam RL, Worley KC, Gibbs RA, Muzny DM, Weinstock GM, Adelson DL, Eichler EE, Elnitski L, Guigo R et al (2009) The genome sequence of taurine cattle: a window to ruminant biology and evolution. Science 324(5926):522–528. https://doi.org/10.1126/science.1169588.

Eltanany M, Philipp U, Weigend S, Distl O (2011) Genetic diversity of ten Egyptian chicken strains using 29 microsatellite markers. Anim Genet 42(6):666–669. https://doi.org/10.1111/j.1365-2052.2011.02185.x

Fair T, Hyttel P, Greve T (1995) Bovine oocyte diameter in relation to maturational competence and transcriptional activity. Mol Reprod Dev 42(4):437–442. https://doi.org/10.1002/mrd.1080420410

FAO (2014) World mapping of animal feeding systems in the dairy sector. Food and Agricultural Organization of the United Nations [accessed July, 2022]. http://www.fao.org/3/a-i3913e.pdf

FAO (2018) World livestock: transforming the livestock sector through the sustainable development goals. Food and Agriculture Organization of the United Nations. Available at https://livestockdata.org/publications/transforming-livestock-sector-through-sustainable-development-goals.

FAOSTAT (2019) Food and Agricultural Organization of the United Nations Statistics. https://www.fao.org/faostat/ (Accessed August 11, 2019).

Feder G, Just RE, Zilberman D (1985) Adoption of agricultural innovations in developing countries: a survey. Econ Develop Cult Change 33(2):255–298. https://documents1.worldbank.org/curated/en/426691468325192369/pdf/multi0page.pdf

Fernandes AFA, Dórea JRR, Fitzgerald R, Herring W, Rosa GJMA (2019) novel automated system to acquire biometric and morphological measurements and predict body weight of pigs via 3D computer vision. J Anim Sci, 97 (1), 496–508. https://doi.org/10.1093/jas/sky418.

Feugang JM, Camargo-Rodríguez O, Memili E (2009) Culture systems for bovine embryos. Livest Sci 121(2–3):141–149

Fitak RR, Mohandesan E, Corander J, Burger PA (2016) The de novo genome assembly and annotation of a female domestic dromedary of North African origin. Mol Ecol Resour 16(1):314–324. https://doi.org/10.1111/1755-0998.12443.

Fletcher GL, Hobbs RS, Evans RP, Shears MA, Hahn AL, Hew CLJAR (2011) Lysozyme transgenic Atlantic salmon (Salmo salar L). Aquacult Res. 42(3):427–440

Fogarty ES, Swain DL, Cronin G, Trotter M (2018) Autonomous on-animal sensors in sheep research: a systematic review. Comput Electron Agric 150:245–256. https://doi.org/10.1016/j.compag.2018.04.017

Fricke P (2010) Methods for diagnosis and monitoring of pregnancy in dairy cattle and their implementation. University of Wisconsin, Madison. National extension. org/page/11200

Gao Y, Wu H, Wang Y, Liu X, Chen L, Li Q, Cui C, Liu X, Zhang J, Zhang Y (2017) Single Cas9 nickase induced generation of NRAMP1 knockin cattle with reduced off-target effects. Genome Biol 18(1):13. https://doi.org/10.1186/s13059-016-1144-4.

Garneau JE, Dupuis M-È, Villion M, Romero DA, Barrangou R, Boyaval P, Fremaux C, Horvath P, Magadán AH, Moineau S (2010) The CRISPR/Cas bacterial immune system cleaves bacteriophage and plasmid DNA. Nature 468(7320):67–71. https://doi.org/10.1038/nature09523

García-Ruiz A, Cole JB, VanRaden PM, Wiggans GR, Ruiz-López FJ, Van Tassell CP (2016) Changes in genetic selection differentials and generation intervals in US Holstein dairy cattle as a result of genomic selection. *Proceedings of the National Academy of Sciences* 113(28):E3995–E4004. https://doi.org/10.1073/pnas.1519061113.

Gates AJ, Gysi DM, Kellis M, Barabási AL (2021) A wealth of discovery built on the human genome project – by the numbers. Nature 590(7845):212–215. https://doi.org/10.1038/d41586-021-00314-6.

Gaouar SBS, Da Silva A, Ciani E, Kdidi S, Aouissat M, Dhimi L, Lafri M, Maftah A, Mehtar N (2015) Admixture and Local Breed Marginalization Threaten Algerian Sheep Diversity. PLOS ONE 10(4):e0122667. https://doi.org/10.1371/journal.pone.0122667.

Georges M, Charlier C, Hayes B (2019) Harnessing genomic information for livestock improvement. Nat Rev Genet 20(3):135–156. https://doi.org/10.1038/s41576-018-0082-2

Gerber PJ, Steinfeld H, Henderson B, Mottet A, Opio C, Dijkman J, Falcucci A, Tempio G (2013) Tackling climate change through livestock: a global assessment of emissions and mitigation opportunities. Food and Agriculture Organization of the United Nations (FAO)

Gertz EM, Schäffer AA, Agarwala R, Bonnet-Garnier A, Rogel-Gaillard C, Hayes H, Mage RG (2013) Accuracy and coverage assessment of Oryctolagus cuniculus (rabbit) genes encoding immunoglobulins in the whole genome sequence assembly (OryCun2.0) and localization of the IGH locus to chromosome 20. Immunogenetics 65(10):749–762. https://doi.org/10.1007/s00251-013-0722-9

Gilchrist RB, Thompson JG (2007) Oocyte maturation: emerging concepts and technologies to improve developmental potential in vitro. *Theriogenology* 67(1):6–15. https://doi.org/10.1016/j.theriogenology.2006.09.027.

Goddard M (2009) Genomic selection: prediction of accuracy and maximisation of long term response. Genetica 136(2):245–257. https://doi.org/10.1007/s10709-008-9308-0. From NLM

Goel A, Kharche S, Jindal S (2009) Determination of early pregnancy and embryonic development by transrectal ultrasonography in goats. Indian J Animal Sci 79(5):476–478

Golding MC, Long CR, Carmell MA, Hannon GJ, Westhusin ME (2006) Suppression of prion protein in livestock by RNA interference. Proc Natl Acad Sci USA 103(14):5285–5290. https://doi.org/10.1073/pnas.0600813103.

González F, Calero P, Beckers J-F (2001) Induction of superovulation in domestic ruminants. In: Biotechnology in animal husbandry. Springer, pp 209–223

Gordon I (2003) In vitro fertilization. Laboratory production of cattle embryos, (Ed. 2), 176–219

Gostimskaya I (2022) CRISPR–Cas9: A history of its discovery and ethical considerations of its use in genome editing. Biochem Mosc 87(8):777–788. https://doi.org/10.1134/S0006297922080090

Green Dreams Tech (2019) iCow. Engineering for change. https://www.engineeringforchange.org/solutions/product/icow/. Accessed 14/10/2021

Groen D (2022) Lab-grown meat can help save our planet—if we can get people to eat it. Canbadian Business Editorial Newsletter, May 3, 2022. https://www.canadianbusiness.com/ideas/meat-alternatives-lab-grown-protein/

Groenen MAM, Archibald, A. L, Uenishi, H, Tuggle, C. K, Takeuchi, Y, Rothschild, M. F, Rogel-Gaillard, C, Park, C, Milan, D, Megens, H.-J et al. (2012) Analyses of pig genomes provide insight into porcine demography and evolution. *Nature* 491 (7424):393–398, https://doi.org/10.1038/nature11622

Groher T, Heitkämper K, Umstätter C (2020) Digital technology adoption in livestock production with a special focus on ruminant farming. Animal 14(11):2404–2413. https://doi.org/10.1017/S1751731120001391.

GrowSafe Systems (2021) Vytelle sense platform. https://vytelle.com/vytelle-sense/. Accessed 14/10/2021

Guan DL, Yang J, Liu YK, Li Y, Mi D, Ma LB, Wang ZZ, Xu SQ, Qiu Q (2019) Draft genome of the Asian Buffalo leech Hirudinaria manillensis. Front Genet 10:1321. https://doi.org/10.3389/fgene.2019.01321.

Guo R, Wan Y, Xu D, Cui L, Deng M, Zhang G, Jia R, Zhou W, Wang Z, Deng K et al (2016) Generation and evaluation of Myostatin knock-out rabbits and goats using CRISPR/Cas9 system. Sci Rep 6:29855. https://doi.org/10.1038/srep29855.

Hammer RE, Pursel VG, Rexroad CE, Wall RJ, Bolt DJ, Ebert KM, Palmiter RD, Brinster RL (1985) Production of transgenic rabbits, sheep and pigs by microinjection. Nature 315(6021):680–683. https://doi.org/10.1038/315680a0

Hao F, Yan W, Li X, Wang H, Wang Y, Hu X, Liu X, Liang H, Liu D (2018) Generation of cashmere goats carrying an *EDAR* gene mutant using CRISPR-Cas9-mediated genome editing. *Int J Biol Sci* 14(4):427–436. https://doi.org/10.7150/ijbs.23890.

Hasler JF (1998) The current status of oocyte recovery, in vitro embryo production, and embryo transfer in domestic animals, with an emphasis on the bovine. J Anim Sci 76(suppl_3):52–74. https://doi.org/10.2527/1998.76suppl_352x.

Hasler J (2004) Factors influencing the success of embryo transfer in cattle. Medecin Veterinaire Du Quebec 34:66–66

He Z, Zhang T, Jiang L, Zhou M, Wu D, Mei J, Cheng Y (2018) Use of CRISPR/Cas9 technology efficiently targetted goat myostatin through zygotes microinjection resulting in double-muscled phenotype in goats. Biosci Rep 38(6). https://doi.org/10.1042/bsr20180742.

Heather JM, Chain B (2016) The sequence of sequencers: the history of sequencing DNA. Genomics 107(1):1–8. https://doi.org/10.1016/j.ygeno.2015.11.003.

Heerboth S, Lapinska K, Snyder N, Leary M, Rollinson S, Sarkar S (2014) Use of epigenetic drugs in disease: an overview. Genet Epigenet 6:9–19. https://doi.org/10.4137/GEG.S12270.

Hellebrand H, Brehme U, Beuche H, Stollberg U, Jacobs H (2003) Application of thermal imaging for cattle management. In: Proc., 1st European conference on precision livestock farming, Berlin, Germany, pp 761–763

Hellsten U, Harland RM, Gilchrist MJ, Hendrix D, Jurka J, Kapitonov V, Ovcharenko I, Putnam NH, Shu S, Taher L et al (2010) The genome of the Western clawed frog Xenopus tropicalis. Science (New York, N.Y.) 328(5978):633–636. https://doi.org/10.1126/science.1183670.

Hew C, Poon R, Xiong F, Gauthier S, Shears M, King M, Davies P, Fletcher G (1999) Liver-specific and seasonal expression of transgenic Atlantic salmon harboring the winter flounder antifreeze protein gene. Transgenic Res 8(6):405–414. https://doi.org/10.1023/a:1008900812864.

Higazi A (2013) Rural insecurity on the Jos plateau, Nigeria: livelihoods, land, and religious reform among the Berom, Fulani, and Hausa. Interfaith Relations in northern Nigeria. Policy Paper 2:5–7

Hill WG (2014) Applications of population genetics to animal breeding, from Wright, Fisher and Lush to genomic prediction. Genetics 196(1):1–16. https://doi.org/10.1534/genetics.112.147850.

Hill J, McSweeney C, Wright AG, Bishop-Hurley G, Kalantar-Zadeh K (2016) Measuring methane production from ruminants. Trends Biotechnol 34 (1), 26–35. https://doi.org/10.1016/j.tibtech.2015.10.004.

Hilton IB, D'Ippolito AM, Vockley CM, Thakore PI, Crawford GE, Reddy TE, Gersbach CA (2015) Epigenome editing by a CRISPR-Cas9-based acetyltransferase activates genes from promoters and enhancers. *Nat Biotech* 33(5):510–517. Research. https://doi.org/10.1038/nbt.3199; http://www.nature.com/nbt/journal/v33/n5/abs/nbt.3199.html#supplementary-information

Hokofarm Group (2021) IVOG. IVOG. https://hokofarm-group.com/products/ivog/. Accessed 14/10/2021

Hong bing HAN YM, Tao WANG, Ling LIAN, Xiuzhi TIAN, Rui HU, Shoulong DENG, Kongpan LI, Feng WANG, Ning LI, Guoshi LIU, Yaofeng ZHAO, Zhengxing Lian (2014) One-step generation of myostatin gene knockout sheep via the CRISPR/Cas9 system *1*(1):2–5. https://doi.org/10.15302/j-fase-2014007

Hong TK, Shin D-M, Choi J, Do JT, Han SG (2021) Current Issues and Technical Advances in Cultured Meat Production: A Review. *Food science of animal resources*, 41 (3), 355–372. https://doi.org/10.5851/kosfa.2021.e14.

Hu S, Qiao J, Fu Q, Chen C, Ni W, Wujiafu S, Ma S, Zhang H, Sheng J, Wang P et al (2015) Transgenic shRNA pigs reduce susceptibility to foot and mouth disease virus infection. eLife *4*:e06951. https://doi.org/10.7554/eLife.06951.

Hu R, Fan ZY, Wang BY, Deng SL, Zhang XS, Zhang JL, Han HB, Lian ZX (2017) RAPID COMMUNICATION: generation of FGF5 knockout sheep via the CRISPR/Cas9 system12. J Animal Sci 95(5):2019–2024. https://doi.org/10.2527/jas.2017.1503.

Hu Z-L, Park CA, Reecy JM (2022) Bringing the animal QTLdb and CorrDB into the future: meeting new challenges and providing updated services. Nucleic Acids Res 50(D1):D956–D961. https://doi.org/10.1093/nar/gkab1116. (acccessed 8/22/2022)

Humphray SJ, Scott CE, Clark R, Marron B, Bender C, Camm N, Davis J, Jenks A, Noon A, Patel M et al (2007) A high utility integrated map of the pig genome. Genome Biol 8(7):R139. https://doi.org/10.1186/gb-2007-8-7-r139.

Hunkapiller T, Kaiser RJ, Koop BF, Hood L (1991) Large-scale and automated DNA sequence determination. Science 254(5028):59–67. https://doi.org/10.1126/science.1925562

Hyttel P, Fair T, Callesen H, Greve T (1997) Oocyte growth, capacitation and final maturation in cattle. Theriogenology 47(1):23–32

Ibeagha-Awemu EM, Khatib, H (2017) Epigenetics of Livestock Breeding In Handbook of Epigenetics. The New Molecular and Medical Genetics Second ed., Tollefsbol, T. O. Ed., Academic Press, pp 441–463. https://doi.org/10.1016/B978-0-12-805388-1.00029-8.

Ibeagha-Awemu EM, Yu Y (2021) Consequence of epigenetic processes on animal health and productivity: is additional level of regulation of relevance? Anim Front 11(6):7–18. https://doi.org/10.1093/af/vfab057. (acccessed 12/21/2021)

Ibeagha-Awemu EM, Zhao X (2015) Epigenetic marks: regulators of livestock phenotypes and conceivable sources of missing variation in livestock improvement programs. Front Genet 6:302. https://doi.org/10.3389/fgene.2015.00302.

Ibeagha-Awemu EM, Jann OC, Weimann C, Erhardt G (2004) Genetic diversity, introgression and relationships among west/central African cattle breeds. Genet Sel Evol 36(6):673. https://doi.org/10.1186/1297-9686-36-6-673

Ibeagha-Awemu EM, Peters SO, Bemji MN, Adeleke MA, Do DN (2019) Leveraging available resources and stakeholder involvement for improved productivity of African livestock in the era of genomic breeding. Front Genet 10:357. https://doi.org/10.3389/fgene.2019.00357

ICGSC (2004) Sequence and comparative analysis of the chicken genome provide unique perspectives on vertebrate evolution. Nature 432(7018):695–716. https://doi.org/10.1038/nature03154.

ILRI (2021) Science helps tailor livestock-related climate change mitigation strategies in Africa. International Livestock Research Institute, Nairobi

Ishino Y, Shinagawa H, Makino K, Amemura M, Nakata A (1987) Nucleotide sequence of the iap gene, responsible for alkaline phosphatase isozyme conversion in Escherichia coli, and identification of the gene product. J Bacteriol 169(12):5429–5433. https://doi.org/10.1128/jb.169.12.5429-5433.1987.

Ishwar A, Memon M (1996) Embryo transfer in sheep and goats: a review. Small Rumin Res 19(1):35–43

Jabed A, Wagner S, McCracken J, Wells DN, Laible G (2012) Targeted microRNA expression in dairy cattle directs production of β-lactoglobulin-free, high-casein milk. Proc Natl Acad Sci USA 109(42):16811–16816. https://doi.org/10.1073/pnas.1210057109.

Jabbar A, Zulfiqar F, Mahnoor M, Mushtaq N, Zaman MH, Din ASU, Khan MA, Ahmad HI (2021) Advances and Perspectives in the Application of CRISPR-Cas9 in Livestock. Mol Biotechnol 63(9):757–767. https://doi.org/10.1007/s12033-021-00347-2.

Jenko J, Gorjanc G, Cleveland MA, Varshney RK, Whitelaw CBA, Woolliams JA, Hickey JM (2015) Potential of promotion of alleles by genome editing to improve quantitative traits in livestock breeding programs. Genet Select Evolut 47(1):55. https://doi.org/10.1186/s12711-015-0135-3.

Jiang Y, Xie M, Chen W, Talbot R, Maddox JF, Faraut T, Wu C, Muzny DM, Li Y, Zhang W et al (2014) The sheep genome illuminates biology of the rumen and lipid metabolism. Science 344(6188):1168–1173. https://doi.org/10.1126/science.1252806

Jiang Y-F, Wang S, Wang C-L, Xu R-H, Wang W-W, Jiang Y, Wang M-S, Jiang L, Dai L-H, Wang J-R et al (2023) Pangenome obtained by long-read sequencing of 11 genomes reveal hidden functional structural variants in pigs. iScience 26(3):106119. https://doi.org/10.1016/j.isci.2023.106119

Joshi S, Mobeen A, Jan K, Bashir K, Azad ZAA (2019) Emerging technologies in dairy processing: present status and future potential. In: Health and safety aspects of food processing technologies. Springer, pp 105–120

Kadim IT, Mahgoub O, Baqir S, Faye B, Purchas R (2015) Cultured meat from muscle stem cells: a review of challenges and prospects. J Integr Agric 14(2):222–233. https://doi.org/10.1016/S2095-3119(14)60881-9

Kalds P, Gao Y, Zhou S, Cai B, Huang X, Wang X, Chen Y (2020) Redesigning small ruminant genomes with CRISPR toolkit: Overview and perspectives. *Theriogenology 147:*25–33. https://doi.org/10.1016/j.theriogenology.2020.02.015.

Karavolias NG, Horner W, Abugu MN, Evanega SN (2021) Application of Gene Editing for Climate Change in Agriculture. Front Sustain Food Syst 5:296. Systematic Review. https://doi.org/10.3389/fsufs.2021.685801.

Kashi Y, Hallerman E, Soller M (1990) Marker-assisted selection of candidate bulls for progeny testing programmes. Anim Sci 51(1):63–74. https://doi.org/10.1017/S0003356100005158. From Cambridge University Press Cambridge Core

Kejela Y (2020) Introduction of the exotic breeds and cross breeding of local chicken in Ethiopia and solution to genetic erosion: A review. Afr J Biotechnol 19(2):92–98. https://doi.org/10.5897/ajb2019.16748

Khalil K, Elayat M, Khalifa E, Daghash S, Elaswad A, Miller M, Abdelrahman H, Ye Z, Odin R, Drescher D et al (2017) Generation of myostatin gene-edited channel catfish (Ictalurus punctatus) via zygote injection of CRISPR/Cas9 system. Sci Rep 7(1):7301. https://doi.org/10.1038/s41598-017-07223-7.

Khanal AR, Gillespie J, MacDonald J (2010) Adoption of technology, management practices, and production systems in US milk production. J Dairy Sci 93(12):6012–6022. https://doi.org/10.3168/jds.2010-3425.

Kim Y-G, Cha J, Chandrasegaran S (1996) Hybrid restriction enzymes: zinc finger fusions to Fok I cleavage domain. Proc Natl Acad Sci 93(3):1156–1160

Kim D-G, Thomas AD, Pelster D, Rosenstock TS, Sanz-Cobena A (2016) Greenhouse gas emissions from natural ecosystems and agricultural lands in sub-Saharan Africa: synthesis of available data and suggestions for further research. Biogeosciences 13(16):4789–4809. https://doi.org/10.5194/bg-13-4789-2016

Kim J, Hanotte O, Mwai OA, Dessie T, Bashir S, Diallo B, Agaba M, Kim K, Kwak W, Sung S, et al. (2017) The genome landscape of indigenous African cattle. *Genome Biology 18*(1):34. https://doi.org/10.1186/s13059-017-1153-y

Kios D, van Marle-Köster E, Visser C (2012) Application of DNA markers in parentage verification of Boran cattle in Kenya. Trop Anim Health Prod 44(3):471–476. https://doi.org/10.1007/s11250-011-9921-2.

Kilpinen H, Dermitzakis, ET (2012) Genetic and epigenetic contribution to complex traits. Human Molecular Genetics *21*(R1), R24-R288. https://doi.org/10.1093/hmg/dds383.

Klug A (2010) The discovery of zinc fingers and their development for practical applications in gene regulation and genome manipulation. Q Rev Biophys 43(1):1–21. https://doi.org/10.1017/S0033583510000089. From Cambridge University Press Cambridge Core

König S, May K (2019) Invited review: phenotyping strategies and quantitative-genetic background of resistance, tolerance and resilience associated traits in dairy cattle. Animal 13(5):897–908. https://doi.org/10.1017/S1751731118003208

Koufariotis L, Hayes B J, Kelly M, Burns BM, Lyons R, Stothard P, Chamberlain AJ (2018) Moore, S. Sequencing the mosaic genome of Brahman cattle identifies historic and recent introgression including polled. Sci Rep 8(1):17761. https://doi.org/10.1038/s41598-018-35698-5.

Koslová A, Trefil P, Mucksová J, Reinišová M, Plachý J, Kalina J, Kučerová D, Geryk J, Krchlíková V, Lejčková B et al (2020) Precise CRISPR/Cas9 editing of the NHE1 gene renders chickens resistant to the J subgroup of avian leukosis virus. Proc Natl Acad Sci USA 117(4):2108–2112. https://doi.org/10.1073/pnas.1913827117.

Kovács L, Jurkovich V, Bakony M, Szenci O, Póti P, Tőzsér J (2014) Welfare implication of measuring heart rate and heart rate variability in dairy cattle: literature review and conclusions for future research. Animal 8(2):316–330. https://doi.org/10.1017/S1751731113002140

Kristjanson P, Waters-Bayer A, Johnson N, Tipilda A, Njuki J, Baltenweck I, MacMillan S (2010) Livestock and women's livelihoods: a review of the recent evidence. International Livestock Research Institute, Nairobi

Krpalkova L, O'Mahony N, Carvalho A, Campbell S, Corkery G, Broderick E, Walsh J (2021) Decision-making strategies on smart dairy farms: a review. Int J Agric Biosyst Eng 15(11):138–145

Kruip TA, Boni R, Wurth,Y, Roelofsen M, Pieterse M (1994) Potential use of ovum pick-up for embryo production and breeding in cattle. Theriogenology 42(4):675–684. https://doi.org/10.1016/0093-691X(94)90384-U.

Kungulovski G, Jeltsch A (2016) Epigenome editing: state of the art, concepts, and perspectives. Trends Genet 32(2):101–113. https://doi.org/10.1016/j.tig.2015.12.001.

Lai L, Kang JX, Li R, Wang J, Witt WT, Yong HY, Hao Y, Wax DM, Murphy CN, Rieke A et al (2006) Generation of cloned transgenic pigs rich in omega-3 fatty acids. Nat Biotechnol 24(4):435–436. https://doi.org/10.1038/nbt1198.

Langevin SM, Kelsey KT (2013) The fate is not always written in the genes: Epigenomics in epidemiologic studies. Environmental and Molecular Mutagenesis 54(7):533–541. https://doi.org/10.1002/em.21762.

Lamb GC, Dahlen CR, Larson JE, Marquezini G, Stevenson JS (2010) Control of the estrous cycle to improve fertility for fixed-time artificial insemination in beef cattle: a review. J Anim Sci 88(13 Suppl):E181–E192. https://doi.org/10.2527/jas.2009-2349.

Lambe NR, McLean KA, Gordon J, Evans D, Clelland N, Bunger L (2017) Prediction of intramuscular fat content using CT scanning of packaged lamb cuts and relationships with meat eating quality. Meat Sci 123:112–119. https://doi.org/10.1016/j.meatsci.2016.09.008.

Lambe N, McLaren A, Clelland N, Kaseja K, Boon S, Bunger L (2018) Using CT scanning to simultaneously breed UK slaughter lambs for improved carcass and meat quality. In Proceedings of the World Congress on Genetics Applied to Livestock Production 11:295. https://www.wcgalp.org/proceedings/2018/using-ct-scanning-simultaneously-breed-uk-slaughter-lambs-improved-carcass-and-meat.

Lambton SL, Knowles TG, Yorke C, Nicol CJ (2015) The risk factors affecting the development of vent pecking and cannibalism in free-range and organic laying hens. Anim Welf 24(1):101–111. https://doi.org/10.7120/09627286.24.1.101

Lau CH, Suh Y (2018) In vivo epigenome editing and transcriptional modulation using CRISPR technology. Transgenic Res 27(6):489–509. https://doi.org/10.1007/s11248-018-0096-8.

Lee K, Uh K, Farrell K (2020) Current progress of genome editing in livestock. Theriogenology 150:229–235. https://doi.org/10.1016/j.theriogenology.2020.01.036.

Lee M, Seo S (2021) Wearable wireless biosensor technology for monitoring cattle: a review. Animals 11(10). https://doi.org/10.3390/ani11102779.

Lee HJ, Yoon JW, Jung KM, Kim YM, Park JS, Lee KY, Park KJ, Hwang YS, Park YH, Rengaraj D et al (2019) Targeted gene insertion into Z chromosome of chicken primordial germ cells for avian sexing model development. FASEB J 33(7):8519–8529. https://doi.org/10.1096/fj.201802671R.

Leroy G, Baumung R, Boettcher P, Scherf B, Hoffmann I (2016) Review: sustainability of crossbreeding in developing countries; definitely not like crossing a meadow.... Animal 10(2):262–273. https://doi.org/10.1017/s175173111500213x

Li R, Zeng W, Ma M, Wei Z, Liu H, Liu X, Wang M, Shi X, Zeng J, Yang L, et al. (2020) Precise editing of myostatin signal peptide by CRISPR/Cas9 increases the muscle mass of Liang Guang Small Spotted pigs. Transgenic Research 29(1):149–163. https://doi.org/10.1007/s11248-020-00188-w.

Li T, Huang S, Zhao X, Wright DA, Carpenter S, Spalding MH, Weeks DP, Yang B (2011) Modularly assembled designer TAL effector nucleases for targeted gene knockout and gene replacement in eukaryotes. Nucleic Acids Res 39(14):6315–6325. https://doi.org/10.1093/nar/gkr188.

Li M, Ouyang H, Yuan H, Li J, Xie Z, Wang K, Yu T, Liu M, Chen X, Tang X et al (2018) Site-specific Fat-1 knock-in enables significant decrease of n-6PUFAs/n--3PUFAs ratio in pigs. G3 (Bethesda) 8(5):1747–1754. https://doi.org/10.1534/g3.118.200114.

Li, W.-R, Liu, C.-X, Zhang, X.-M, Chen, L, Peng, X.-R, He, S.-G, Lin, J.-P, Han, B, Wang, L.-Q, Huang, J.-C et al. (2017) CRISPR/Cas9-mediated loss of FGF5 function increases wool staple length in sheep. 284(17):2764–2773. https://doi.org/10.1111/febs.14144.

Li R, Fu W, Su R, Tian X, Du D, Zhao Y, Zheng Z, Chen Q, Gao S, Cai Y et al (2019) Towards the complete goat Pan-genome by recovering missing genomic segments from the reference genome. Front Genet 10:1169. https://doi.org/10.3389/fgene.2019.01169.

Li H, Yang Y, Hong W, Huang M, Wu M, Zhao X (2020) Applications of genome editing technology in the targeted therapy of human diseases: mechanisms, advances and prospects. Signal Transduct Target Ther 5(1):1. https://doi.org/10.1038/s41392-019-0089-y

Li R, Gong M, Zhang X, Wang F, Liu Z, Zhang L, Yang Q, Xu Y, Xu M, Zhang H et al (2023) A sheep pangenome reveals the spectrum of structural variations and their effects on tail phenotypes Genome Res. 33:463. https://doi.org/10.1101/gr.277372.122

Lien S, Koop BF, Sandve SR, Miller JR, Kent MP, Nome T, Hvidsten TR, Leong JS, Minkley DR, Zimin A et al (2016) The Atlantic salmon genome provides insights into rediploidization. Nature 533(7602):200–205. https://doi.org/10.1038/nature17164.

Liu J, Luo Y, Ge H, Han C, Zhang H, Wang Y, Su J, Quan F, Gao M, Zhang Y (2013) Anti-bacterial activity of recombinant human β-defensin-3 secreted in the milk of transgenic goats produced by somatic cell nuclear transfer. PLoS One 8(6):e65379. https://doi.org/10.1371/journal.pone.0065379. From NLM

Liu X, Wang Y, Tian Y, Yu Y, Gao M, Hu G, Su F, Pan S, Luo Y, Guo Z et al (2014) Generation of mastitis resistance in cows by targeting human lysozyme gene to β-casein locus using zinc-finger nucleases. Proc R Soc B Biol Sci 281(1780). Article. https://doi.org/10.1098/rspb.2013.3368

Liu Z, Liu S, Yao J, Bao L, Zhang J, Li Y, Jiang C, Sun L, Wang R, Zhang Y et al (2016) The channel catfish genome sequence provides insights into the evolution of scale formation in teleosts. Nat Commun 7(1):11757. https://doi.org/10.1038/ncomms11757

Liu R, Xing S, Wang J, Zheng M, Cui H, Crooijmans RPMA, Li Q, Zhao G, Wen J (2019a) A new chicken 55K SNP genotyping array. BMC Genomics 20(1):410. https://doi.org/10.1186/s12864-019-5736-8

Liu X, Liu H, Wang M, Li R, Zeng J, Mo D, Cong P, Liu X, Chen Y, He Z (2019b) Disruption of the ZBED6 binding site in intron 3 of IGF2 by CRISPR/Cas9 leads to enhanced muscle development in Liang Guang Small Spotted pigs. Transgenic Res 28(1):141–150. https://doi.org/10.1007/s11248-018-0107-9. From NLM

Lo C-A, Greben AW, Chen BE (2017) Generating stable cell lines with quantifiable protein production using CRISPR/Cas9-mediated knock-in. BioTechniques 62(4):165–174. https://doi.org/10.2144/000114534. (acccessed 2024/01/04)

Lonergan P, Boland M (2011) Gamete and embryo technologyl multiple ovulation and embryo transfer

Lonergan P, FairT (2016) Maturation of Oocytes in Vitro. Ann Rev Animal Biosci 4(1):255–268. https://doi.org/10.1146/annurev-animal-022114-110822.

Lonergan P, Monaghan P, Rizos D, Boland M, Gordon I (1994) Effect of follicle size on bovine oocyte quality and developmental competence following maturation, fertilization, and culture in vitro. Mol Reprod Dev 37(1):48–53. https://doi.org/10.1002/mrd.1080370107

Luinge H, Hop E, Lutz E, Van Hemert J, De Jong EJACA (1993) Determination of the fat, protein and lactose content of milk using Fourier transform infrared spectrometry. Anal Chim Acta 284(2):419–433. https://doi.org/10.1016/0003-2670(93)85328-H.

Luo J, Song Z, Yu S, Cui D, Wang B, Ding F, Li S, Dai Y, Li N (2014) Efficient generation of myostatin (MSTN) biallelic mutations in cattle using zinc finger nucleases. PLoS One 9(4):e95225. https://doi.org/10.1371/journal.pone.0095225. From NLM

Lyall J, Irvine RM, Sherman A, McKinley TJ, Núñez A, Purdie A, Outtrim L, Brown IH, Rolleston-Smith G, Sang H et al (2011) Suppression of avian influenza transmission in genetically modified chickens. Science (New York, N.Y.) 331(6014):223–226. https://doi.org/10.1126/science.1198020. From NLM

Lynch M, Walsh B (1998) Genetics and analysis of quantitative traits. Sinauer Associates, Inc., Sunderland, p 980

Ma J, Fan Y, Zhou Y, Liu W, Jiang N, Zhang J, Zeng L (2018) Efficient resistance to grass carp reovirus infection in JAM-A knockout cells using CRISPR/Cas9. Fish Shellfish Immunol 76:206–215. https://doi.org/10.1016/j.fsi.2018.02.039. From NLM

Madalena F, Agyemang K, Cardellino R, Jain G (2002) Genetic improvement in medium-to low-input systems of animal production. Experiences to date. In Proceedings of the 7th World Congress on Genetics Applied to Livestock Production, Montpellier, France, August, pp 19–23

Maga EA, Shoemaker CF, Rowe JD, Bondurant RH, Anderson GB, Murray JD (2006) Production and processing of milk from transgenic goats expressing human lysozyme in the mammary gland. J Dairy Sci 89(2):518–524. https://doi.org/10.3168/jds.S0022--0302(06)72114-2. From NLM

Majekodunmi AO, Fajinmi A, Dongkum C, Shaw APM, Welburn SC (2014) Pastoral livelihoods of the Fulani on the Jos plateau of Nigeria. Pastoralism 4(1). https://doi.org/10.1186/s13570-014-0020-7

Manolio TA, Collins FS, Cox NJ, Goldstein DB, Hindorff LA, Hunter DJ, McCarthy MI, Ramos EM, Cardon LR, Chakravarti A, Cho JH, Guttmacher AE, Kong A, Kruglyak L, Mardis E, Rotimi CN, Slatkin M, Valle D, Whittemore AS, Boehnke M, Clark AG, Eichler EE, Gibson G, Haines JL, Mackay TF, McCarroll SA, Visscher PM (2009) Finding the missing heritability of complex diseases. Nature 461(7265):747–753. https://doi.org/10.1038/nature08494.

Margulies M, Egholm M, Altman WE, Attiya S, Bader JS, Bemben LA, Berka J, Braverman MS, Chen Y-J, Chen Z et al (2005) Genome sequencing in microfabricated high-density picolitre reactors. Nature 437(7057):376–380. https://doi.org/10.1038/nature03959.

Marraffini LA, Sontheimer EJ (2008) CRISPR interference limits horizontal gene transfer in staphylococci by targeting DNA. Science 322(5909):1843–1845. https://doi.org/10.1126/science.1165771

Marshall K, Gibson JP, Mwai O, Mwacharo JM, Haile A, Getachew T, Mrode R, Kemp SJ (2019) Livestock genomics for developing countries—African examples in practice. Front Genet 10(297). https://doi.org/10.3389/fgene.2019.00297.

Maurice-Van Eijndhoven MH, Soyeurt H, Dehareng F, Calus MP (2013) Validation of fatty acid predictions in milk using mid-infrared spectrometry across cattle breeds. Animal 7(2):348–354. https://doi.org/10.1017/s1751731112001218. From NLM

Maxam AM, Gilbert W (1977) A new method for sequencing DNA. Proc Natl Acad Sci 74(2):560–564. https://doi.org/10.1073/pnas.74.2.560. Sanger, F.; Coulson, A. R. (1975) A rapid method for determining sequences in DNA by primed synthesis with DNA polymerase. *Journal of Molecular Biology* 94(3):441–448. https://doi.org/10.1016/0022-2836(75)90213-2.

McEvoy TG, Alink FM, Moreira VC, Watt RG, Powell KA (2006) Embryo technologies and animal health—consequences for the animal following ovum pick-up, in vitro embryo production and somatic cell nuclear transfer. Theriogenology 65(5):926–942. https://doi.org/10.1016/j.theriogenology.2005.09.008

McFarlane GR, Salvesen HA, Sternberg A, Lillico SG (2019) On-Farm Livestock Genome Editing Using Cutting Edge Reproductive Technologies. Front Sustain Food Syst 3:(106). https://doi.org/10.3389/fsufs.2019.00106.

McParland S, Banos G, McCarthy B, Lewis E, Coffey MPO, Neill B, O'Donovan M, Wall E, Berry DP (2012) Validation of mid-infrared spectrometry in milk for predicting body energy status in Holstein-Friesian cows. *J Dairy Sci* 95(12):7225–7235. https://doi.org/10.3168/jds.2012-5406.

McParland S, Lewis E, Kennedy E, Moore SG, McCarthy B, O'Donovan M, Butler ST, Pryce JE, Berry DP (2014) Mid-infrared spectrometry of milk as a predictor of energy intake and efficiency in lactating dairy cows. J Dairy Sci 97(9):5863–5871. https://doi.org/10.3168/jds.2014-8214.

Medeiros EF, Phelps MP, Fuentes FD, Bradley TM (2009) Overexpression of follistatin in trout stimulates increased muscling. Am J Physiol Regul Integr Comp Physiol 297(1):R235–R242. https://doi.org/10.1152/ajpregu.91020.2008. From NLM

Menchaca A, Dos Santos-Neto PC, Mulet AP, Crispo M (2020) CRISPR in livestock: From editing to printing. *Theriogenology* 150:247–254. https://doi.org/10.1016/j.theriogenology.2020.01.063.

Meuwissen T, Hayes B, Goddard M (2013) Accelerating improvement of livestock with genomic selection. Annu Rev Anim Biosci 1:221–237. https://doi.org/10.1146/annurev-animal-031412-103705. From NLM

Meuwissen TH, Hayes BJ, Goddard ME (2001) Prediction of total genetic value using genome-wide dense marker maps. *Genetics*, 157 (4), 1819–1829. https://doi.org/10.1093/genetics/157.4.1819

Meuwissen T, Hayes B, Goddard M (2016) Genomic selection: A paradigm shift in animal breeding. Anim Front 6(1):6–14. https://doi.org/10.2527/af.2016-0002. (acccessed 8/30/2021)

Miglior F, Fleming A, Malchiodi F, Brito LF, Martin P, Baes CFA (2017) 100-Year Review: Identification and genetic selection of economically important traits in dairy cattle. *Journal of Dairy Science* 100(12):10251–10271. https://doi.org/10.3168/jds.2017-12968

Miller J, McLachlan A, Klug A (1985) Repetitive zinc-binding domains in the protein transcription factor IIIA from Xenopus oocytes. EMBO J 4(6):1609–1614. https://doi.org/10.2144/000114534

Miller GA, Hyslop JJ, Barclay D, Edwards A, Thomson W, Duthie C-A (2019) Using 3D imaging and machine learning to predict liveweight and carcass characteristics of live finishing beef cattle. Front Sustain Food Syst 3:30

Milner P, Page KL, Walton AW, Hillerton JE (1996) Detection of clinical mastitis by changes in electrical conductivity of foremilk before visible changes in milk. J Dairy Sci 79(1):83–86. https://doi.org/10.3168/jds.S0022-0302(96)76337-3. From NLM

Montalvo-Casimiro M, González-Barrios R, Meraz-Rodriguez MA, Juárez-González VT, Arriaga-Canon C, Herrera LA (2020) Epidrug repurposing: discovering new faces of old acquaintances in cancer therapy. Front Oncol 10(2461) Review:2461. https://doi.org/10.3389/fonc.2020.605386

Monteiro A, Santos S, Gonçalves P (2021) Precision agriculture for crop and livestock farming—brief review. Animals 11(8):2345

Moore NW (1982) Egg transfer in the sheep and goat. In: Adams CE (ed) Mammalian egg transfer. CRC Press, Boca Raton, pp 119–133

Moore KL, Mrode R, Coffey MP (2017) Genetic parameters of visual image analysis primal cut carcass traits of commercial prime beef slaughter animals. Animal 11(10):1653–1659. https://doi.org/10.1017/s1751731117000489. From NLM

Mortensen AK, Lisouski P, Ahrendt PJC, Agriculture EI (2016) Weight prediction of broiler chickens using 3D computer vision. Comput Electron Agric 123:319–326

Mozdziak PE, Pophal S, Borwornpinyo S, Petitte JN (2003) Transgenic chickens expressing beta-galactosidase hydrolyze lactose in the intestine. J Nutr 133(10):3076–3079. https://doi.org/10.1093/jn/133.10.3076.

Mozdziak PE, Borwornpinyo S, McCoy DW, Petitte JN (2003) Development of transgenic chickens expressing bacterial beta-galactosidase. *Developmental dynamics : an official publication of the American Association of Anatomists* 226(3):439–445. https://doi.org/10.1002/dvdy.10234.

Mozo R, Alabart JL, Rivas E, Folch J (2019) New method to automatically evaluate the sexual activity of the ram based on accelerometer records. Small Rumin Res 172:16–22. https://doi.org/10.1016/j.smallrumres.2019.01.009

Mrode R, Ekine Dzivenu C, Marshall K, Chagunda MGG, Muasa BS, Ojango J, Okeyo AM (2020) Phenomics and its potential impact on livestock development in low-income countries: innovative applications of emerging related digital technology. Animal Front 10(2):6–11. https://doi.org/10.1093/af/vfaa002. From NLM

Msalya G (2017) Possibilities of utilizing biotechnology to improve animal and animal feeds productivity in Tanzania—review of past efforts and available opportunities. J Dairy Veterin Animal Res 5(5). https://doi.org/10.15406/jdvar.2017.05.00155

Müller M, Brenig B, Winnacker EL, Brem G (1992) Transgenic pigs carrying cDNA copies encoding the murine Mx1 protein which confers resistance to influenza virus infection. Gene 121(2):263–270. https://doi.org/10.1016/0378-1119(92)90130-h. From NLM

Munteanu C, Mireşan V, Răducu C, Ihuţ A, Uiuiu P, Pop D, Neacşu A, Cenariu M, Groza I (2021) Can cultured meat be an alternative to farm animal production for a sustainable and healthier lifestyle? Front Nutr 8:749298–749298. https://doi.org/10.3389/fnut.2021.749298. PubMed

Mutembei HM, Tsuma VT, Muasa BT, Muraya J, Erastus RM (2015) Bovine in-vitro embryo production and its contribution towards improved food security in Kenya. African J Food Agric Nutrit Develop 15:9722. https://doi.org/10.18697/ajfand.68.14040.

Muyekho N, Siamba DN, Humphrey H, Nyagol D, Karugara D (2014) Characterization of the livestock production system and potential to enhance productivity through improved feeding in Kisumu West sub-County of Kisumu County, Kenya. Nairobi: ILRI. Available a https://hdl.handle.net/10568/112970.

Mwai O, Hanotte O, Kwon Y-J, Cho S (2015) African indigenous cattle: unique genetic resources in a rapidly changing world. Asian Australas J Anim Sci 28(7):911–921. https://doi.org/10.5713/ajas.15.0002R. PMC

Nakamura M, Gao Y, Dominguez AA, Qi LS (2021) CRISPR technologies for precise epigenome editing. Nat Cell Biol 23(1):11–22. https://doi.org/10.1038/s41556-020-00620-7.

National Milk Records (2021) Pocket companion. https://www.nmr.co.uk/software/pocket-companion. Accessed 14/10/2021

National Research Council (2009) Emerging technologies to benefit farmers in sub-Saharan Africa and South Asia. National Academies Press

Navarro-Serna S, Vilarino M, Park I, Gadea J, Ross PJ (2020) Livestock Gene Editing by One-step Embryo Manipulation. *Journal of Equine Veterinary Science* 89:03025. https://doi.org/10.1016/j.jevs.2020.103025.

Negussie E, Lehtinen J, Mäntysaari P, Bayat AR, Liinamo AE, Mäntysaari EA, Lidauer MH (2017) Non-invasive individual methane measurement in dairy cows. *Animal* 11(5):890–899. https://doi.org/10.1017/s1751731116002718

Ni W, Qiao J, Hu S, Zhao X, Regouski M, Yang M, Polejaeva IA, Chen C (2014) Efficient Gene Knockout in Goats Using CRISPR/Cas9 System. PLOS ONE 9(9): e106718 https://doi.org/10.1371/journal.pone.0106718

Niu Y, Zhao X, Zhou J, Li Y, Huang Y, Cai B, Liu Y, Ding Q, Zhou S, Zhao J et al (2018) Efficient generation of goats with defined point mutation (I397V) in GDF9 through CRISPR/Cas9. Reprod Fertil Dev 30(2):307–312. https://doi.org/10.1071/rd17068.

Nyman AK, Persson Waller K, Bennedsgaard TW, Larsen T, Emanuelson U (2014) Associations of udder-health indicators with cow factors and with intramammary infection in dairy cows. J Dairy Sci 97(9):5459–5473. https://doi.org/10.3168/jds.2013-7885. (acccessed 2022/07/19)

OG (2021) Cellular Agriculture: Canada's $12.5 Billion Opportunity in Food Innovation. Ontario Genomics Report, 22 November 2021, https://www.ontariogenomics.ca/news-events/cellular--agriculture-report-identifies-billions-in-annual-economic-opportunity-for-canada-over-the-next-decade/#:~:text=AcCELLerate%2DON-,Cellular%20Agriculture%20Report%20Identifies%20Billions%20in%20Annual%20Economic,Canada%20Over%20the%20Next%20Decade&text=Momentum%20in%20cellular%20agriculture%20is,food%20and%20create%20142%20142C20000%20120jobs

Ojango JM, Mrode R, Okeyo A, Rege J, Emerge-Africa K, Chagunda M (2017) Improving smallholder dairy farming in Africa. In: Achieving sustainable production of milk, vol 2. Burleigh Dodds Science Publishing, pp 371–396

Ojango JMK, Audho JO, Mogaka D, Oyieng EP (2018) ADGG-ODK-1 installing ODK and update tools

Ojango JMK, Chinyere E, Rao J, Kangethe E, Okeyo AM (2020a) ADGG dairy tool: helping your cow to calve. ILRI. https://cgspace.cgiar.org/handle/10568/110179, Nairobi, Kenya

Ojango JMK, Rao J, Bett B, Omore A, Kangethe E, Okeyo AM (2020b) ADGG dairy tool: ensuring dairy herd hygiene. ILRI. https://cgspace.cgiar.org/handle/10568/110180, Nairobi, Kenya

Omontese BO (2018) Estrus Synchronization and Artificial Insemination in Goats. in Goat Science. S. Kukovics, ed. IntechOpen, Rijeka. https://doi.org/10.5772/intechopen.74236

Ong KJ, Johnston J, Datar I, Sewalt V, Holmes D, Shatkin JA (2021) Food safety considerations and research priorities for the cultured meat and seafood industry. Compr Rev Food Sci Food Saf 20(6):5421–5448. https://doi.org/10.1111/1541-4337.12853. From NLM

Pabiou T, Fikse W, Cromie A, Keane M, Näsholm A, Berry DJLS (2011) Use of digital images to predict carcass cut yields in cattle. Livest Sci 137(1–3):130–140

Paguem A, Abanda B, Achukwi MD, Baskaran P, Czemmel S, Renz A, Eisenbarth A (2020) Whole genome characterization of autochthonous Bos taurus brachyceros and introduced Bos indicus indicus cattle breeds in Cameroon regarding their adaptive phenotypic traits and pathogen resistance. BMC Genet 21(1):64. https://doi.org/10.1186/s12863-020-00869-9.

Park KE, Kaucher AV, Powell A, Waqas MS, Sandmaier SE, Oatley MJ, Park CH, Tibary A, Donovan DM, Blomberg LA et al (2017) Generation of germline ablated male pigs by CRISPR/Cas9 editing of the NANOS2 gene. Sci Rep 7:40176. https://doi.org/10.1038/srep40176. From NLM

Park TS, Park J, Lee JH, Park JW, Park BC (2019) Disruption of G(0)/G(1) switch gene 2 (G0S2) reduced abdominal fat deposition and altered fatty acid composition in chicken. FASEB J 33(1):1188–1198. https://doi.org/10.1096/fj.201800784R. From NLM

Parrish JJ, Susko-Parrish JL, Leibfried-Rutledge ML, Critser ES, Eyestone WH, First NL (1986) Bovine in vitro fertilization with frozen-thawed semen. Theriogenology 25(4):591–600. https://doi.org/10.1016/0093-691x(86)90143-3

Patel GK, Haque N, Madhavatar M, Kumar Chaudhari A, Kumar Patel D, Bhalakiya N, Jamnesha N, Rajesh

Kumar PP (2017) Artificial insemination:a tool to improve livestock productivity. J Pharmacogn Phytochem SP1:307–313

Paxton M (2021) Artificial insemination and breeding animals. Career with animals

Perisse IV, Fan Z, Singina GN, White KL, Polejaeva IA (1713), Review) Improvements in gene editing technology boost its applications in livestock. Front Genet 2021:11. https://doi.org/10.3389/fgene.2020.614688

Pieterse MC, Kappen KA, Kruip TAM, Taverne MAM (1988) Aspiration of bovine oocytes during transvaginal ultrasound scanning of the ovaries. Theriogenology 30(4):751-762. https://doi.org/10.1016/0093-691x(88)90310-x.

Pressman E, Schaefer JM, Kebreab E. (2018) Mitigation of enteric methane emissions from dairy cattle in East Africa through urea treatment of crop residue feeds. American Geophysical Union Fall 2018 Meeting Abstract GC51K-0929. https://agu.confex.com/agu/fm18/meetingapp.cgi/Paper/412733.

Proudfoot C, Carlson DF, Huddart R, Long CR, Pryor JH, King TJ, Lillico SG, Mileham AJ, McLaren DG, Whitelaw CB et al (2015) Genome edited sheep and cattle. Transgenic Res 24(1):147–153. https://doi.org/10.1007/s11248-014-9832-x. From NLM

Pursel VJT (1992) Transfer of c-ski gene into swine to enhance muscle development. Theriogenology 37:278–282

Pursel VG, Hammer RE, Bolt DJ, Palmiter RD, Brinster RL (1990) Integration, expression and germ-line transmission of growth-related genes in pigs. Journal of reproduction and fertility. Supplement 41:77–87. PMID: 2213718.

Pursel VG, Pinkert CA, Miller KF, Bolt DJ, Campbell RG, Palmiter RD, Brinster RL, Hammer RE (1989) Genetic engineering of livestock. Science (New York, N.Y.) 244(4910):1281–1288. https://doi.org/10.1126/science.2499927.

Pryce JE, Daetwyler HD (2012) Designing dairy cattle breeding schemes under genomic selection: a review of international research. Ani. Prod. Sci. 52(3):107–114. https://doi.org/10.1071/AN11098

Qian L, Tang M, Yang J, Wang Q, Cai C, Jiang S, Li H, Jiang K, Gao P, Ma D, et al. (2015) Targeted mutations in myostatin by zinc-finger nucleases result in double-muscled phenotype in Meishan pigs. Sci Rep 5:14435. https://doi.org/10.1038/srep14435

Rahman MA, Mak R, Ayad H, Smith A, Maclean N (1998) Expression of a novel piscine growth hormone gene results in growth enhancement in transgenic tilapia (Oreochromis niloticus). Transgenic Res 7(5):357–369. https://doi.org/10.1023/a:1008837105299.

Ramirez O, Ojeda A, Tomas A, Gallardo D, Huang LS, Folch JM, Clop A, Sánchez A, Badaoui B, Hanotte O et al (2009) Integrating Y-chromosome, mitochondrial, and autosomal data to analyze the origin of pig breeds. Mol Biol Evolut 26(9):2061. https://doi.org/10.1093/molbev/msp118.

Randhawa S, Sengar S (2021) Chapter One – The evolution and history of gene editing technologies. In: Ghosh D (ed) Progress in molecular biology and translational science, vol 178. Academic Press, pp 1–62

Reh WA, Maga EA, Collette NM, Moyer A, Conrad-Brink JS, Taylor SJ, DePeters EJ, Oppenheim S, Rowe JD, BonDurant RH et al (2004) Hot topic: using a stearoyl-CoA desaturase transgene to alter milk fatty acid composition. J Dairy Sci 87(10):3510–3514. https://doi.org/10.3168/jds.S0022-0302(04)73486-4.

Rehman SU, Hassan FU, Luo X, Li Z, Liu Q (2021) Whole-genome sequencing and characterization of Buffalo genetic resources: recent advances and future challenges. Animals 11(3). https://doi.org/10.3390/ani11030904.

Rexroad CE Jr, Hammer RE, Bolt DJ, Mayo KE, Frohman LA, Palmiter RD, Brinster RL (1989) Production of transgenic sheep with growth-regulating genes. Mol Reprod Dev 1(3):164–169. https://doi.org/10.1002/mrd.1080010304.

Richt JA, Kasinathan P, Hamir AN, Castilla J, Sathiyaseelan T, Vargas F, Sathiyaseelan J, Wu H, Matsushita H, Koster J et al (2007) Production of cattle lacking prion protein. Nat Biotechnol 25(1):132–138. https://doi.org/10.1038/nbt1271.

Rizos D, Ward F, Duffy P, Boland MP, Lonergan P (2002) Consequences of bovine oocyte maturation, fertilization or early embryo development in vitro versus in vivo: implications for blastocyst yield and blastocyst quality. Mol Reprod Dev 61(2):234–248. https://doi.org/10.1002/mrd.1153

Roche JR, Friggens NC, Kay JK, Fisher MW, Stafford KJ, Berry DP (2009) Invited review: body condition score and its association with dairy cow productivity, health, and welfare. J Dairy Sci 92(12):5769–5801. https://doi.org/10.3168/jds.2009-2431. (acccessed 2022/07/19)

Rodning S, Michelle F, Misty A, Elmore J, Gard J, Lovelady A (2012) Estrus Synchronization and Artificial Insemination Programs for Beef Cattle. Livestock & Poultry – The Alabama Cooperative Extension System. Available at https://www.aces.edu/wp-content/uploads/2018/09/ANR-1027.REV_.4.pdf. Accessed 14 June 2023.

Roland L, Lidauer L, Sattlecker G, Kickinger F, Auer W, Sturm V, Efrosinin D, Drillich M, Iwersen M (2018) Monitoring drinking behavior in bucket-fed dairy calves using an ear-attached tri-axial accelerometer: a pilot study. Comput Electron Agric 145:298–301. https://doi.org/10.1016/j.compag.2018.01.008

Ronaghi M, Uhlén M, Nyrén, P. (1998) A sequencing method based on real-time pyrophosphate. Science 281(5375):363–365. https://doi.org/10.1126/science.281.5375.363.

Rouge M, Bowen R (2002) Collection and evaluation of semen: introduction and index. Clin Tech Small Anim Pract 17(3):104-107. https://doi.org/10.1053/svms.2002.34326

Rushen J, Butterworth A, Swanson JC (2011) Animal behavior and well-being symposium: farm animal welfare assurance: science and application1. J Anim Sci 89(4):1219–1228. https://doi.org/10.2527/jas.2010-3589.

Rutten CJ, Velthuis AGJ, Steeneveld W, Hogeveen H (2013) Invited review: sensors to support health management on dairy farms. J Dairy Sci 96(4):1928–1952. https://doi.org/10.3168/jds.2012-6107.

Saiki RK, Gelfand DH, Stoffel S, Scharf SJ, Higuchi R, Horn GT, Mullis KB, Erlich HAJS (1988) Primer-directed enzymatic amplification of DNA with a thermostable DNA polymerase. Science 239(4839):487–491. https://doi.org/10.1126/science.2448875.Cohen, S. N.; Chang, A. C. Y.; Boyer, H. W.; Helling, R. B. (1973) Construction of Biologically Functional Bacterial Plasmids *In Vitro*. *70*(11):3240–3244. https://doi.org/10.1073/pnas.70.11.3240

Salter DW, Crittenden LB (1989) Artificial insertion of a dominant gene for resistance to avian leukosis virus into the germ line of the chicken. Theor Appl Genet 77(4):457–461. https://doi.org/10.1007/bf00274263.

Sanger F, Nicklen S, Coulson AR (1977) DNA sequencing with chain-terminating inhibitors. *Proc Natl Acad Sci US 74*(12):5463–5467. https://doi.org/10.1073/pnas.74.12.5463.

Santoze A, Gicheha M (2019) The status of cattle genetic resources in West Africa: a review. Adv Anim Vet Sci 7(2):112–121. https://doi.org/10.17582/journal.aavs/2019/7.2.112.121.

Scholz AM, Bünger L, Kongsro J, Baulain U, Mitchell AD (2015) Non-invasive methods for the determination of body and carcass composition in livestock: dual-energy X-ray absorptiometry, computed tomography, magnetic resonance imaging and ultrasound: invited review. Animal 9(7):1250–1264. https://doi.org/10.1017/s1751731115000336.

Schuster F, Aldag P, Frenzel A, Hadeler KG, Lucas-Hahn A, Niemann H, Petersen B (2020) CRISPR/Cas12a mediated knock-in of the Polled Celtic variant to produce a polled genotype in dairy cattle. Sci Rep 10(1):13570. https://doi.org/10.1038/s41598-020-70531-y.

SFA (2020) Singapore Food Authority: Requirements for the Safety Assessment of Novel Foods and Novel Food Ingredients. https://www.sfa.gov.sg/docs/default-source/food-import-and-export/Requirements-on-safety-assessment-of-novel-foods_22Apr.pdf. Accessed on 10 May 2022

Shalloo L, O'Donovan M, Leso L, Werner J, Ruelle E, Geoghegan A, Delaby L, O'leary, N. (2018) Grass-based dairy systems, data and precision technologies. Animal 12(s2):s262–s271. https://doi.org/10.1017/S175173111800246X.

Shanthalingam S, Tibary A, Beever JE, Kasinathan P, Brown WC, Srikumaran S (2016) Precise gene editing paves the way for derivation of Mannheimia haemolytica leukotoxin-resistant cattle. Proc Natl Acad Sci USA 113(46):13186–13190. https://doi.org/10.1073/pnas.1613428113.

Shendure J, Ji H (2008) Next-generation DNA sequencing. Nat Biotechnol 26(10):1135–1145. https://doi.org/10.1038/nbt1486.

Siberski-Cooper CJ, Koltes JE (2021) Opportunities to harness high-throughput and novel sensing phenotypes to improve feed efficiency in dairy cattle. Animals *12*(1). https://doi.org/10.3390/ani12010015.

Silvi R, Pereira LGR, Paiva CAV, Tomich TR, Teixeira VA, Sacramento JP, Ferreira RE, Coelho SG, Machado FS, Campos MM (2021) Adoption of precision technologies by Brazilian dairy farms: the Farmer's perception. Animals 11(12):3488. https://doi.org/10.3390/ani11123488.

Simitzis P, Tzanidakis C, Tzamaloukas O, Sossidou E (2021) Contribution of precision livestock farming Systems to the improvement of welfare status and productivity of dairy animals. Dairy 3(1):12–28. https://doi.org/10.3390/dairy3010002

Simm G, Pollott G, Mrode R, Houston R, Marshall K (2021) Genetic Improvement of Farmed Animals. Wallingford, U.K: CABI

Sirard MA (2001) Resumption of meiosis: mechanism involved in meiotic progression and its relation with developmental competence. Theriogenology 55(6):1241–1254. https://doi.org/10.1016/s0093-691x(01)00480-0

Sirard MA, Blondin P (1996) Oocyte maturation and IVF in cattle. Anim Reprod Sci 42(1–4):417–426. https://doi.org/10.1016/0378-4320(96)01518-7

Sun Z, Wang M, Han S, Ma S, Zou Z, Ding F, Li X, Li L, Tang B, Wang H et al. (2018) Production of hypoallergenic milk from DNA-free beta-lactoglobulin (BLG) gene knockout cow using zinc-finger nucleases mRNA. *Sci Rep 8*(1):15430. https://doi.org/10.1038/s41598-018-32024-x.

Sykes DJ, Couvillion JS, Cromiak A, Bowers S, Schenck E, Crenshaw M, Ryan PL (2012) The use of digital infrared thermal imaging to detect estrus in gilts. *Theriogenology 78* (1):147–152. https://doi.org/10.1016/j.theriogenology.2012.01.030

Sonstegard TS, Van Tassell CP, Ashwell MS (2001) Dairy cattle genomics: tools to accelerate genetic improvement? J Anim Sci 79(suppl_E):E307–E315. https://doi.org/10.2527/jas2001.79E-SupplE307x.

Soyeurt H, Dehareng F, Gengler N, McParland S, Wall E, Berry DP, Coffey M, Dardenne P (2011) Mid-infrared prediction of bovine milk fatty acids across multiple breeds, production systems, and countries. J Dairy Sci 94(4):1657–1667. https://doi.org/10.3168/jds.2010-3408.

Stelletta, C., M. Morgante, M. Gianesella, J. Vencato, and E. Fiore. (2012) Thermographic Applications in Veterinary Medicine. In R. V. Prakash, ed. Infrared Thermography, IntechOpen, Rijeka. https://doi.org/10.5772/29135.Gabor, G.; Sasser, R.; Kastelic, J.; Coulter, G.; Falkay, G.; Mézes, M.; Bozo, S.; Völgyi-Csík, J.; Bárány, I.; Szász Jr, F. (1998) Morphologic, endocrine and thermographic measurements of testicles in comparison with semen characteristics in mature Holstein–Friesian breeding bulls. *Animal Reproduction Science* 51(3):215–224. https://doi.org/10.1016/s0378-4320(98)00077-3

Stephens N, Di Silvio L, Dunsford I, Ellis M, Glencross A, Sexton A (2018) Bringing cultured meat to market: technical, socio-political, and regulatory challenges in cellular agriculture. Trends Food Sci Technol 78:155–166. https://doi.org/10.1016/j.tifs.2018.04.010

Straiton J (2022) Biohacking the food chain: using CRISPR to combat the global food crisis. BioTechniques 73(4):159–161. https://doi.org/10.2144/btn-2022-0102

Strucken EM, Al-Mamun HA, Esquivelzeta-Rabell C, Gondro C, Mwai OA, Gibson JP (2017) Genetic tests for estimating dairy breed proportion and parentage assignment in East African crossbred cattle. Genet Sel Evol 49(1):67. https://doi.org/10.1186/s12711-017-0342-1.

Stubbings R, Walton J (1995) Effect of ultrasonically-guided follicle aspiration on estrous cycle and follicular dynamics in Holstein cows. Theriogenology 43(4):705–712. https://doi.org/10.1016/0093-691x(95)00013-x

Su X, Cui K, Du S, Li H, Lu F, Shi D, Liu Q (2018a) Efficient genome editing in cultured cells and embryos of Debao pig and swamp buffalo using the CRISPR/Cas9 system. In Vitro Cell Dev Biol Anim 54(5):375–383. https://doi.org/10.1007/s11626-018-0236-8.

Su X, Wang S, Su G, Zheng Z, Zhang J, Ma Y, Liu Z, Zhou H, Zhang Y, Zhang L (2018b) Production of microhomologous-mediated site-specific integrated LacS gene cow using TALENs. Theriogenology 119:282–288. https://doi.org/10.1016/j.theriogenology.2018.07.011.

Tan W, Proudfoot C, Lillico SG, Whitelaw CB (2016) Gene targeting, genome editing: from Dolly to editors. Transgenic Res 25(3):273–287. https://doi.org/10.1007/s11248-016-9932-x.

Tanihara F, Takemoto T, Kitagawa E, Rao S, Do LT, Onishi A, Yamashita Y, Kosugi C, Suzuki H, Sembon S et al. (2016) Somatic cell reprogramming-free generation of genetically modified pigs. *Science Advances* 2(9):e1600803. https://doi.org/10.1126/sciadv.1600803.

Tanihara F, Hirata M, Nguyen NT, Le QA, Wittayarat M, Fahrudin M, Hirano T, Otoi T (2021) Generation of CD163-edited pig via electroporation of the CRISPR/Cas9 system into porcine in vitro-fertilized zygotes. *Animal Biotechnology* 32(2):147–154. https://doi.org/10.1080/10495398.2019.1668801

Tait-Burkard C, Doeschl-Wilson A, McGrew MJ, Archibald AL, Sang HM, Houston RD, Whitelaw CB, Watson M (2018) Livestock 2.0 - genome editing for fitter, healthier, and more productive farmed animals. *Genome Biol* 9(1):204. https://doi.org/10.1186/s13059-018-1583-1.

Tawfik DS, Griffiths AD (1998) Man-made cell-like compartments for molecular evolution. Nat. Biotechnol 16(7):652–656. https://doi.org/10.1038/nbt0798-652.

Tessanne K, Golding MC, Long CR, Peoples MD, Hannon G, Westhusin ME (2012) Production of transgenic calves expressing an shRNA targeting myostatin. Mol Reprod Dev 79(3):176–185. https://doi.org/10.1002/mrd.22007.

The World Bank (2018) Moving towards sustainability: the livestock sector and World Bank. Brief, October 18, 2021. Available at https://www.worldbank.org/en/topic/agriculture/brief/moving-towards-sustainability-the-livestock-sector-and-the-world-bank. Accessed 18 June 2023

Tijjani A, Utsunomiya YT, Ezekwe AG, Nashiru O, Hanotte O (2019) Genome Sequence Analysis Reveals Selection Signatures in Endangered Trypanotolerant West African Muturu Cattle. *Front Genet* 10:442. https://doi.org/10.3389/fgene.2019.00442.

TNAU, T. N. A. U. (2009) Artificial insemination. In TNAU Agric Portal. Available at https://agritech.tnau.ac.in/ta/animal_husbandry/majoractivities_artificialinsemination.html. Accessed 18 June 2023

Toffanin V, De Marchi M, Lopez-Villalobos N, Cassandro MJIDJ (2015) Effectiveness of mid-infrared spectroscopy for prediction of the contents of calcium and phosphorus, and titratable acidity of milk and their relationship with milk quality and coagulation properties. Int Dairy J 41:68–73. https://doi.org/10.1016/j.idairyj.2014.10.002.

Tong J, Wei H, Liu X, Hu W, Bi M, Wang Y, Li Q, Li N (2011) Production of recombinant human lysozyme in the milk of transgenic pigs. Transgenic Res 20(2):417–419. https://doi.org/10.1007/s11248-010-9409-2.

Traoré SA, Markemann A, Reiber C, Piepho HP, Valle Zárate A (2017) Production objectives, trait and breed preferences of farmers keeping NDama, Fulani Zebu and crossbred cattle and implications for breeding programs. Animal 11(4):687. https://doi.org/10.1017/S1751731116002196.

Treich N (2021) Cultured meat: promises and challenges. Environ Resour Econ (Dordr) 79(1):33–61. https://doi.org/10.1007/s10640-021-00551-3. PubMed.

Ugur MR, Saber Abdelrahman A, Evans HC, Gilmore AA, Hitit M, Arifiantini RI, Purwantara B, Kaya A, Memili E (2019) Advances in cryopreservation of bull sperm. Front Veterin Sci 6(268). https://doi.org/10.3389/fvets.2019.00268.

Valli C (2020) Mitigating enteric methane emission from livestock through farmer-friendly practices. In: Venkatramanan V, Shah S, Prasad R (eds) Global climate change and environmental policy: agriculture perspectives Springer Singapore, pp 257–273. https://doi.org/10.1007/978-981-13-9570-3_8

Van Eenennaam AL, De Figueiredo Silva F, Trott JF, Zilberman D (2021) Genetic engineering of livestock: the opportunity cost of regulatory delay. Annu Rev Anim Biosci 9:453–478. https://doi.org/10.1146/annurev-animal-061220-023052.

Van Marle-Köster E, Webb EC (2014) A perspective on the impact of reproductive technologies on food production in Africa. Springer, New York, pp 199–211

Van Soom A, Van Vlaenderen IM, de Kruif A (1991) Bull specific effect on in vitro penetration and fertilization of bovine oocytes using various levels of heparin. Assisted Reproduct Technol/Androl 2(2):105–106

Vanlierde A, Soyeurt H, Gengler N, Colinet FG, Froidmont E, Kreuzer M, Grandl F, Bell M, Lund P, Olijhoek DW et al (2018) Short communication: development of an equation for estimating methane emissions of dairy cows from milk Fourier transform mid-infrared spectra by using reference data obtained exclusively from respiration chambers. J Dairy Sci 101(8):7618–7624. https://doi.org/10.3168/jds.2018-14472.

Vanlierde A, Vanrobays M. L, Dehareng F, Froidmont E, Soyeurt H, McParland S, Lewis E, Deighton MH, Grandl F, Kreuzer, M et al. (2015) Hot topic: Innovative lactation-stage-dependent prediction of methane emissions from milk mid-infrared spectra. J Dairy Sci 98(8): 5740–5747. https://doi.org/10.3168/jds.2014-8436.

Vasseur E, Rushen J, Haley DB, de Passillé AM (2012) Sampling cows to assess lying time for on-farm animal welfare assessment. J Dairy Sci 95(9):4968–4977. https://doi.org/10.3168/jds.2011-5176.

Vermeulen SJ, Campbell BM, Ingram JSI (2012) Climate change and food systems. Annu Rev Environ Resour 37:195–222

Visentin G, McDermott A, McParland S, Berry DP, Kenny OA, Brodkorb A, Fenelon MA, De Marchi M (2015) Prediction of bovine milk technological traits from mid-infrared spectroscopy analysis in dairy cows. J Dairy Sci 98(9):6620–6629. https://doi.org/10.3168/jds.2015-9323.

Visser C (2011) Parentage verification of South African angora goats, using microsatellite markers. S Afr J Animal Sci 41:250–255

Wade CM, Giulotto E, Sigurdsson S, Zoli M, Gnerre S, Imsland F, Lear TL, Adelson DL, Bailey E, Bellone RR et al (2009) Genome sequence, comparative analysis, and population genetics of the domestic horse. Science 326(5954):865–867. https://doi.org/10.1126/science.1178158.

Wakchaure R, Ganguly S, Praveen PK, Kumar A, Sharma S, Mahajan T (2015) Marker assisted selection (MAS) in animal breeding: a review. J Drug Metab Toxicol 6:5. https://doi.org/10.4172/2157-7609.1000e127

Wall RJ (2002) New gene transfer methods. Theriogenology 57(1):189–201. https://doi.org/10.1016/s0093-691x(01)00666-5.

Wang K, Ouyang H, Xie Z, Yao C, Guo N, Li M, Jiao H, Pang D (2015) Efficient Generation of Myostatin Mutations in Pigs Using the CRISPR/Cas9 System. Scientific Reports (1): 6623. https://doi.org/10.1038/srep16623.

Wang M, Ibeagha-Awemu EM (2021) Impacts of Epigenetic Processes on the Health and Productivity of Livestock. Frontiers in Genetics 11:1812. https://doi.org/10.3389/fgene.2020.613636

Wang Y, Yang W, Winter P, Walker LJBE (2008) Walk-through weighing of pigs using machine vision and an artificial neural network. Biosyst Eng 100(1):117–125.

Wang X, Niu Y, Zhou J, Yu H, Kou Q, Lei A, Zhao X, Yan H, Cai B, Shen Q et al (2016) Multiplex gene editing via CRISPR/Cas9 exhibits desirable muscle hypertrophy without detectable off-target effects in sheep. Sci Rep 6:32271. https://doi.org/10.1038/srep32271.

Wang K, Tang X, Xie Z, Zou X, Li M, Yuan H, Guo N, Ouyang H, Jiao H, Pang D (2017) CRISPR/Cas9-mediated knockout of myostatin in Chinese indigenous Erhualian pigs. Transgenic Res 26(6):799–805. https://doi.org/10.1007/s11248-017-0044-z.

Wang X, Yu H, Lei A, Zhou J, Zeng W, Zhu H, Dong Z, Niu Y, Shi B, Cai B, et al. (2015) Generation of gene-modified goats targeting MSTN and FGF5 via zygote injection of CRISPR/Cas9 system. Sci Rep 5:13878. https://doi.org/10.1038/srep13878.

Ward KA, Brown BW (1998) The production of transgenic domestic livestock: successes, failures and the need for nuclear transfer. *Reproduction, fertility, and development*, 10 (7–8), 659–665. https://doi.org/10.1071/rd98074.

Watson JD, Crick FHC (1953) Molecular structure of nucleic acids: a structure for deoxyribose nucleic acid. Nature 171(4356):737–738. https://doi.org/10.1038/171737a0

Weersink A, Tauer LW (1991) Causality between dairy farm size and productivity. *Am J Agric Econ* 73(4), 1138–1145.

Weifeng M, Yaping W, Wenbo W, Bo W, Jianxin F, Zuoyan ZJA (2004) Enhanced resistance to Aeromonas hydrophila infection and enhanced phagocytic activities in human lactoferrin-transgenic grass carp (Ctenopharyngodon idellus). Aquaculture 242(1–4):93–103. https://doi.org/10.1016/j.aquaculture.2004.07.020.

Weigel KA, VanRaden PM, Norman HD, Grosu H (2017) A 100-year review: methods and impact of genetic selection in dairy cattle-from daughter-dam comparisons to deep learning algorithms. J Dairy Sci 100(12):10234–10250. https://doi.org/10.3168/jds.2017-12954.

WheelerM, Walters E (2001) Transgenic technology and applications in swine. Theriogenology 56(8):1345–1369. Wall, R. New gene transfer methods. *Theriogenology* 2002, 57 (1), 189–201. https://doi.org/10.1016/s0093-691x(01)00635-5

Wheeler MB, Bleck GT, Donovan SM (2001) Transgenic alteration of sow milk to improve piglet growth and health. Reproduction (Cambridge, England) Suppl 58:313–324. PMID: 11980200.

Whitworth KM, Lee K, Benne JA, Beaton BP, Spate LD, Murphy SL, Samuel MS, Mao J, O'Gorman C, Walters EM et al (2014) Use of the CRISPR/Cas9 system to produce genetically engineered pigs from in vitro-derived oocytes and embryos. *Biol Reprod* 91(3):78. https://doi.org/10.1095/biolreprod.114.121723.

Whitworth KM, Rowland RR, Ewen CL, Trible BR, Kerrigan MA, Cino-Ozuna AG, Samuel MS, Lightner JE, McLaren DG, Mileham AJ et al (2016) Gene-edited pigs are protected from porcine reproductive and respiratory syndrome virus. *Nat Biotechnol* 34(1):20–22. https://doi.org/10.1038/nbt.3434.

Whitworth KM, Rowland RRR, Petrovan V, Sheahan M, Cino-Ozuna AG, Fang Y, Hesse R, Mileham A, Samuel MS, Wells KD et al (2019) Resistance to coronavirus infection in amino peptidase N-deficient pigs. Transgenic Res 28(1):21–32. https://doi.org/10.1007/s11248-018-0100-3

Wieghart M, Hoover JL, McGrane MM, Hanson RW, Rottman FM, Holtzman SH, Wagner TE, Pinkert CA (1990) Production of transgenic pigs harbouring a rat phosphoenolpyruvate carboxykinase-bovine growth hormone fusion gene. J Reprod Fertility Suppl 41:89–96. PMID: 2213719

Wiggans GR, VanRaden PM, Cooper TA (2011) The genomic evaluation system in the United States: Past, present, future. *Journal of Dairy Science* 94 (6), 3202–3211. https://doi.org/10.3168/jds.2010-3866

Wiggans GR, Cole JB, Hubbard SM, Sonstegard TS (2017a) Genomic selection in Dairy Cattle: the USDA experience. Annu Rev Anim Biosci 5:309–327. https://doi.org/10.1146/annurev-animal-021815-111422.

Wiggans GR, Cole JB, Hubbard SM, Sonstegard TS (2017b) Genomic selection in dairy cattle: the USDA experience. Annu Rev Anim Biosci 5(1):309–327. https://doi.org/10.1146/annurev-animal-021815-111422.

Wilson RT (2018) Domestic livestock in African cities: production, problems and prospects. Open Urban Stud Demograp J 4(1):1. From http://worldcat.org/z-wcorg/

Wolc A, Kranis A, Arango J, Settar P, Fulton JE, O'Sullivan NP, Avendano A, Watson KA, Hickey JM, de los Campos G et al. (2016) Implementation of genomic selection in the poultry industry. Animal Frontiers 6(1):23–31. https://doi.org/10.2527/af.2016-0004 (accessed 9/13/2022).

Woyengo TA, Gachuiri CK, Wahome RG, Mbugua PN (2004) Effect of protein supplementation and urea treatment on utilization of maize stover by Red Maasai sheep. *South African Journal of Animal Science* 34(1):23–30. https://doi.org/10.4314/sajas.v34i1.3806

Wu B, Sun YH, Wang YW, Wang YP, Zhu ZY (2005) Characterization of transgene integration pattern in F4 hGH-transgenic common carp (Cyprinus carpio L.). Cell Res 15(6):447–454. https://doi.org/10.1038/sj.cr.7290313.

Wu X, Ouyang H, Duan B, Pang D, Zhang L, Yuan T, Xue L, Ni D, Cheng L, Dong S et al (2012) Production of cloned transgenic cow expressing omega-3 fatty acids. Transgenic Res 21(3):537–543. https://doi.org/10.1007/s11248-011-9554-2.

Wu H, Guang X, Al-Fageeh MB, Cao J, Pan S, Zhou H, Zhang L, Abutarboush MH, Xing Y, Xie Z et al (2014) Camelid genomes reveal evolution and adaptation to desert environments. Nat Commun 5(1):5188. https://doi.org/10.1038/ncomms6188

Wu H, Wang Y, Zhang Y, Yang M, Lv J, Liu J, Zhang Y (2015) TALE nickase-mediated SP110 knockin endows cattle with increased resistance to tuberculosis. Proc Natl Acad Sci USA 112(13):E1530–E1539. https://doi.org/10.1073/pnas.1421587112.

Xiang G, Ren J, Hai T, Fu R, Yu D, Wang J, Li W, Wang H, Zhou Q (2018) Editing porcine IGF2 regulatory element improved meat production in Chinese Bama pigs. Cell Mol Life Sci 75(24):4619–4628. https://doi.org/10.1007/s00018-018-2917-6.

Xie Z, Pang D, Yuan H, Jiao H, Lu C, Wang K, Yang Q, Li M, Chen X, Yu T et al (2018) Genetically modified pigs are protected from classical swine fever virus. PLoS Pathog 14(12):e1007193. https://doi.org/10.1371/journal.ppat.1007193.

Xu K, Zhou Y, Mu Y, Liu Z, Hou S, Xiong Y, Fang L, Ge C, Wei Y, Zhang X et al (2020) CD163 and pAPN double-knockout pigs are resistant to PRRSV and TGEV and exhibit decreased susceptibility to PDCoV while maintaining normal production performance. *eLife* 9. https://doi.org/10.7554/eLife.57132.

Yang H, Zhang J, Zhang X, Shi J, Pan Y, Zhou R, Li G, Li Z, Cai G, Wu Z. (2018) CD163 knockout pigs are fully resistant to highly pathogenic porcine reproductive and respiratory syndrome virus. *Antiviral Research* 151:63–70. https://doi.org/10.1016/j.antiviral.2018.01.004

Youngquist RS (2006) Pregnancy Diagnosis. Applied reproductive strategies in beef cattle. Department of Veterinary Medicine and Surgery, University of Missouri, Columbia, pp 329–338

Yu S, Luo J, Song Z, Ding F, Dai Y, Li N (2011) Highly efficient modification of beta-lactoglobulin (BLG) gene via zinc-finger nucleases in cattle. Cell Res 21 (11), 1638–1640. https://doi.org/10.1038/cr.2011.153.

Yu H, Li H, Li Q, Xu R, Yue C, Du S (2019) Targeted gene disruption in Pacific oyster based on CRISPR/Cas9 ribonucleoprotein complexes. Marine Biotechnol (New York, N.Y.) 21(3):301–309. https://doi.org/10.1007/s10126-019-09885-y.

Yu B, Lu R, Yuan Y, Zhang T, Song S, Qi Z, Shao B, Zhu M, Mi F, Cheng Y (2016) Efficient TALEN-mediated myostatin gene editing in goats. BMC Dev Biol 16(1):26. https://doi.org/10.1186/s12861-016-0126-9.

Zhang J, Li L, Cai Y, Xu X, Chen J, Wu Y, Yu H, Yu G, Liu S, Zhang A et al (2008) Expression of active recombinant human lactoferrin in the milk of transgenic goats. Protein Expr Purif 57(2):127–135. https://doi.org/10.1016/j.pep.2007.10.015.

Zhang P, Liu P, Dou H, Chen L, Chen L, Lin L, Tan P, Vajta G, Gao J, Du Y et al (2013) Handmade cloned transgenic sheep rich in omega-3 fatty acids. PLoS One 8(2):e55941. https://doi.org/10.1371/journal.pone.0055941.

Zhang, R, Li, Y, Jia, K, Xu, X, Li, Y, Zhao, Y, Zhang, X, Zhang, J, Liu, G, Deng, S, et al. (2020) Crosstalk between androgen and Wnt/β-catenin leads to changes of wool density in FGF5-knockout sheep. Cell Death & Disease *11*(5):407. https://doi.org/10.1038/s41419-020-2622-x.

Zhang X, Li W, Liu C, Peng X, Lin J, He S, Li X, Han B, Zhang N, Wu Y et al (2017) Alteration of sheep coat color pattern by disruption of ASIP gene via CRISPR Cas9. Sci Rep 7(1):8149. https://doi.org/10.1038/s41598-017-08636-0.

Zhang C, Kemp RA, Stothard P, Wang Z, Boddicker N, Krivushin K, Dekkers J, Plastow G (2018a) Genomic evaluation of feed efficiency component traits in Duroc pigs using 80K, 650K and whole-genome sequence variants. Genet Sel Evol 50(1):14. https://doi.org/10.1186/s12711-018-0387-9

Zhang Y, Wang Y, Yulin B, Tang B, Wang M, Zhang C, Zhang W, Jin J, Li T, Zhao R et al (2018b) CRISPR/Cas9-mediated sheep MSTN gene knockout and promote sSMSCs differentiation. J Cell Biochem 120:1794. https://doi.org/10.1002/jcb.27474.

Zhang J, Cui ML, Nie YW, Dai B, Li FR, Liu DJ, Liang H, Cang M (2018c) CRISPR/Cas9-mediated specific integration of fat-1 at the goat MSTN locus. FEBS J 285(15):2828–2839. https://doi.org/10.1111/febs.14520.

Zheng Q, Lin J, Huang J, Zhang H, Zhang R, Zhang X, Cao C, Hambly C, Qin G, Yao J et al (2017) Reconstitution of UCP1 using CRISPR/Cas9 in the white adipose tissue of pigs decreases fat deposition and improves thermogenic capacity. Proc Natl Acad Sci USA 114(45):E9474–e9482. https://doi.org/10.1073/pnas.1707853114.

Zhong Z, Niu P, Wang M, Huang G, Xu S, Sun Y, Xu X, Hou Y, Sun X, Yan Y et al (2016) Targeted disruption of sp7 and myostatin with CRISPR-Cas9 results in severe bone defects and more muscular cells in common carp. Sci Rep 6:22953. https://doi.org/10.1038/srep22953.

Zhao J, Lai L, Ji W, Zhou Q (2019) Genome editing in large animals: current status and future prospects. *National Science Review* 6(3):402–420. https://doi.org/10.1093/nsr/nwz013.

Zhou ZR, Zhong BS, Jia RX, Wan YJ, Zhang YL, Fan YX, Wang LZ, You JH, Wang ZY, Wang F (2013) Production of myostatin-targeted goat by nuclear transfer from cultured adult somatic cells. Theriogenology 79(2):225–233. https://doi.org/10.1016/j.theriogenology.2012.08.006.

Zhou W, Wan Y, Guo R, Deng M, Deng K, Wang Z, Zhang Y, Wang F (2017) Generation of beta-lactoglobulin knock-out goats using CRISPR/Cas9. PLoS One 12(10):e0186056. https://doi.org/10.1371/journal.pone.0186056

Zhou Y, Yang L, Han X, Han J, Hu Y, Li F, Xia H, Peng L, Boschiero C, Rosen BD et al (2022) Assembly of a pangenome for global cattle reveals missing sequences and novel structural variations, providing new insights into their diversity and evolutionary history. Genome Res 32:1585. https://doi.org/10.1101/gr.276550.122

Zhu X, Wei Y, Zhan Q, Yan A, Feng J, Liu L, Tang D (2020) CRISPR/Cas9-mediated biallelic knock-out of IRX3 reduces the production and survival of somatic cell-cloned bama minipigs. Animals (Basel) 10(3):501. https://doi.org/10.3390/ani10030501.

Zonabend E, Okeyo A, Ojango JM, Hoffmann I, Moyo S, Philipsson J (2013a) Infrastructure for sustainable use of animal genetic resources in southern and eastern Africa. Animal Genetic Resources/Resources génétiques animales/Recursos genéticos animales 53:79–93. https://doi.org/10.1017/S2078633613000295.

Zonabend E, Okeyo AM, Ojango JMK, Hoffmann I, Moyo S, Philipsson J (2013b) Infrastructure for sustainable use of animal genetic resources in Southern and Eastern Africa. Animal Genetic Resources/Ressources Génétiques Animales/Recursos Genéticos Animales 53:79–93. https://doi.org/10.1017/S2078633613000295.

Zou Y, Li Z, Zou Y, Hao H, Li N, Li Q (2018) An FBXO40 knockout generated by CRISPR/Cas9 causes muscle hypertrophy in pigs without detectable pathological effects. Biochem Biophys Res Commun 498(4):940–945. https://doi.org/10.1016/j.bbrc.2018.03.085

Wondossen, Ayalew Wu, Xiaoyun Getinet Mekuriaw, Tarekegn Rakan, Naboulsi Tesfaye, Sisay Tessema Renaud, Van Damme Erik, Bongcam-Rudloff Min, Chu Chunnian, Liang Zewdu, Edea Solomon, Enquahone Yan, Ping (2024a) Whole genome sequences of 70 indigenous Ethiopian cattle Abstract Scientific Data 11(1) 10.1038/s41597-024-03342-9

Wondossen, Ayalew Wu, Xiaoyun Getinet Mekuriaw, Tarekegn Tesfaye Sisay, Tessema Min, Chu Chunnian, Liang Rakan, Naboulsi Renaud, Van Damme Erik, Bongcam-Rudloff Yan, Ping (2024b) Whole-genome sequencing of copy number variation analysis in Ethiopian cattle reveals adaptations to diverse environments Abstract BMC Genomics 25(1) 10.1186/s12864-024-10936-5

Open Access This chapter is licensed under the terms of the Creative Commons Attribution 4.0 International License (http://creativecommons.org/licenses/by/4.0/), which permits use, sharing, adaptation, distribution and reproduction in any medium or format, as long as you give appropriate credit to the original author(s) and the source, provide a link to the Creative Commons license and indicate if changes were made.

The images or other third party material in this chapter are included in the chapter's Creative Commons license, unless indicated otherwise in a credit line to the material. If material is not included in the chapter's Creative Commons license and your intended use is not permitted by statutory regulation or exceeds the permitted use, you will need to obtain permission directly from the copyright holder.

The Role of Modern Technologies for Improving the Production Environment of Livestock in Africa

21

Eveline M. Ibeagha-Awemu, Faith A. Omonijo, Martha N. Bemji, Obioha Duranna, Iliya D. Kwoji, Michael O. Ozoje, and Richard Osei-Amponsah

Abstract

Technology was instrumental in advancing human civilization, and present-day modern technologies are enabling breakthrough innovations in medicine, agriculture, and other science sectors in Western countries. Current levels of livestock productivity in Africa are suboptimal and marred by a myriad of problems requiring urgent intervention. Modern technologies, known to be instrumental in advancing the livestock sector in western countries, will play a big role in supporting rapid growth in the African livestock sector. This chapter presents modern technologies that have advanced many aspects of livestock management in recent decades. First, Section 21.2 presents the technologies used for assessment and improvement of production efficiencies followed by examination of the technologies in animal welfare management (Section 21.3), while some of the technologies enhancing meat production, processing, and preservation are presented in Section 21.4. This is followed by an overview of the technologies for livestock nutrition and feeding (Section 21.5), livestock health management (Section 21.6), housing and milking management (Section 21.7), and technologies in pastoral livestock production systems (Section 21.8). To fully benefit from these technologies, current challenges (Section 21.9) must be addressed to pave the way for accelerated improvements in the African livestock sector following their adoption and implementation.

Keywords

Sustainable livestock production · Modern technology · Production efficiency · Welfare and management · Nutrition, health, and growth technologies

E. M. Ibeagha-Awemu (✉) · F. A. Omonijo
Sherbrooke Research and Development Centre,
Agriculture and Agri-Food Canada,
Sherbrooke, Quebec, Canada
e-mail: Eveline.ibeagha-awemu@agr.gc.ca

M. N. Bemji
Department of Animal Breeding and Genetics,
University of Agriculture Abeokuta,
Abeokuta, Nigeria

O. Duranna
Lakeland College, Vermilion, Alberta, Canada

I. D. Kwoji
Discipline of Genetics, School of Life Sciences,
University of Kwa-Zulu Natal Westville Campus,
Durban, South Africa

M. O. Ozoje
Department of Animal Breeding and Genetics,
Federal University of Agriculture, Abeokuta, Nigeria

R. Osei-Amponsah
Department of Animal Science, School of
Agriculture, University of Ghana,
Legon, Accra, Ghana

Abbreviations

ACGG	African chicken genetic gains
AD	Anaerobic digester
AAT	African animal trypanosomiasis
$AlCl_3$	Aluminum chloride
Al_2SO_4	Aluminum sulfate or alum
APP	Air-permeable packaging
AMS	Automated robotic milking systems
BVD	Bovine viral diarrhea
ATOs	Automatic take-offs
AU-IBAR	Africa union—inter African bureau for animal genetic resources
C	Carbon
CA	Cellular agriculture
$Ca(OH)_2$	Calcium hydroxide
Ca-P	Calcium phosphorus
$Ca(OH)_2$	Calcium hydroxide
$CaSO_4$	Calcium sulfate
CB	Cubicle barns
CBPP	Contagious bovine pleuropneumonia
CD8+	Cluster of differentiation 8
CH4	Methane
Cl^+	Chlorine ion
CIS	Cattle information system
CO_2	Carbon dioxide
CNV	Copy number variation
CP	Crude protein
CRFI	Controlling and recording feed intake
DDGS	Distillers dried grains with soluble
DM	Dry matter
DNA	Deoxyribonucleic acid
EC	Electrical conductivity
ECOWAS	Economic Community of West African States
eDHI	Electronic dairy Herd improvement
EID	Electronic animal identification
EVPs	Ethnoveterinary preparations
FAO	Food and Agricultural Organization
FCR	Feed conversion ratio
$FeCl_3$	Ferric chloride
$FeSO_4$	Ferric sulfate
FEAST	Feed assessment tool
FL	Fidelity level
FW	Free walk housing systems
GH	Grasshopper
GHG	Greenhouse gas
GFR	Gain to feed ratio
GIT	Gastro intestinal tract
GM-CSF	Granulocyte-macrophage colony-stimulating factor
H_2O	Water
H_2O-P	Water-phosphorus
H_2S	Hydrogen sulfide
HTST	High temperature short time
HAP	High available phosphorus
HF	High frequency
HFCs	Hydrofluorocarbon
IBR	Infectious bovine rhinotracheitis
IL-2	Interleukin-2
ICF	Informant consensus factor
ICT	Information communication technology
ILRI	International livestock research institute
K^+	Potassium ion
LF	Low frequency
LMIC	Low- and mid-income countries
MAP	Modified atmosphere packaging
MAPHEX	Manure phosphorus extraction
$MgHPO_4 \cdot 3H_2O$	Magnesium hydrogen phosphate trihydrate
MHC-1	Major histocompatibility complex 1
MJ/kg	Megajoules per kilogram
mRNA	Messenger ribonucleic acid
MIR	Mid-infrared
N	Nitrogen
Na^+	Sodium ions
N_2O	Nitrous oxide
$NaHCO_3$–P	Sodium hydrogen carbonate-phosphorus
NGO	Non-governmental organization
NH3	Ammonia
NH3-N	Ammoniacal nitrogen
NH4	Ammonium
NH_4^+	Ammonium-ion
NIRS	Near infrared reflectance spectroscopy
NO_3-N	Nitrate nitrogen
NRC	National Research Council
NUE	Nitrogen use efficiency
O_3	Ozone
OHA	One health approach

P	Phosphorus
PLF	Precision livestock feeding
PPPs	Public private partnerships
P_2O_5	Phosphorus pentoxide
QR	Quick response
PCR	Polymerase chain reaction
PGF2α	Prostaglandin F2 alpha
RBG	Residual body gain
RFI	Residual feed intake
RFID	Radio frequency identification
RIG	Residual feed intake and gain
SARS	Severe acute respiratory disease
SSA	Sub-Saharan Africa
RNA	Ribonucleic acid
SF6	Sulfur hexafluoride
SMRT	Single molecule real time
TPGs	Tropical poultry genetic solutions
TS	Tie stall
TKN	Total Kjeldahl nitrogen
TP	Total phosphorus
TP	Tropical poultry
TSS	Total suspended solids
UHF	Ultra-high frequency
UHT	Ultra-high temperature
UN	United Nations
USA	United States of America
USDA	United States department of agriculture
UV	Ultraviolet
VIA	Video image analysis
WEP	Water extractable phosphorus
WHO	World Health Organization

21.1 Introduction

As pointed out in Chapter 20, genetic improvement is only one part of overall livestock improvement. While productivity and production require appropriate genetics, the expression of the genetics of the animal (to its fullest potential) is dependent on provision of a suitable production environment, including quality feeds and feeding regimen, health, housing, and overall management, including such operations as milking, and specific considerations for management under pastoral conditions. Over the years, demand for food and products of animal origin are increasing drastically worldwide owing to an increasing human population. For instance, the global meat consumption was reported to increase from 23.1 kg to 42.2 kg per person per year from 1961 to 2011 (Sans and Combris 2015). This growth in demand for livestock-derived foods will likely remain strong in low- and middle-income countries, increasing from 29% to 35% by 2030 and to 37% by 2050 in sub-Saharan Africa (Enahoro et al. 2019). Current meat production in Africa remains insufficient to satisfy demand. At the same time, the direct and indirect negative impacts of climate change on livestock production, including desertification and dwindling water resources, will remain perennial challenges in most production systems. Under these conditions, intensification of livestock production via the use of available technologies and innovations will be a viable strategy to facilitate the rapid production of livestock products (meat, milk, etc.) to meet the ever-increasing demand in Africa. Therefore, to address the animal protein needs of Africans in response to UN sustainable development goals and objectives of making food available to all by 2063, it is generally acknowledged that modern technologies will play a major role.

The adoption of technology in various aspects of livestock farming has been extensively studied (El-Osta and Morehart 2000; Khanal et al. 2010; Groher et al. 2020). It is stipulated that larger farms tend to be at the forefront of adopting technologies which have helped to increase profitability and feed quality, and reduce production costs and incidence of disease outbreaks. In dairy production, the late adopters of technology are small-scale milk producers, because most of the technologies are capital-intensive and not easily affordable (El-Osta and Morehart 2000), which is one of the issues faced by livestock farming in Africa. However, with the current low levels of productivity of most African livestock, the adoption and use of modern technologies are necessary to drive and sustain improved management and productivity. The implementation of modern technologies has been the driving force behind rapid improvements in the health and productivity in the beef, dairy, swine, and poultry industries in many western nations. These improvements were achieved through advances in all aspects of livestock management such as advances in management practices, improved nutrition, improved health care, improved housing, improved genetic/genomic selection strategies, and improvements in reproduction management (Ibeagha-Awemu and Yu 2021). While Chapter 20 addressed the key issues related to sustainable livestock improvement and the role of modern technologies in driving sustained genetic improvement in livestock, this chapter will focus on the different technologies facilitating progress in livestock nutrition, health, housing, and overall management, including such operations as milking and specific considerations for management under pastoral conditions.

21.2 Technologies for Assessment and Improvement of Production Efficiency

The current low productivity levels of most African livestock can be oriented toward more sustainable systems and increased efficiency outcomes by adopting technologies that improve livestock farming practices, feed efficiency, precision feeding, welfare management, breeding and improvement management, disease management, and health and environment management. These provide opportunities to move the African livestock sector toward more sustainable development and improved contribution to livelihoods.

21.2.1 Feed and Growth Efficiency in Livestock Systems

Efficiency in broad terms refers to the ratio of output to input, intending to minimize the inputs while maximizing outputs. Production efficiency involves better feed utilization, accompanied by optimum growth potentials and high reproductive abilities. Therefore, production efficiency describes the measures that showcase the reproductive, feed, and growth outcomes in livestock operations. These production efficiency measures have important implications for individual livestock operations and the general public, such as providing food security for a growing world population (David et al. 2021), projected to reach 9.8 billion in 2050 by the United Nations, and would require about 70% more food.

Feed constitutes a significant expense in any livestock operation and is a primary driver of growth and production of products such as meat, eggs, and wool. Other items that constitute the total variable costs in livestock operations include labor, fuel, yardage, bedding, and so on. Feed accounts for a majority of these variable costs in livestock operations; at least 50% in cattle operations (Kennedy et al. 1993), up to 85% in pig operations (McGlone and Pond 2003), and about 65% of egg production costs (Aggrey et al. 2010). For cattle operations, about 75–85% of the total energy intake is channeled toward the maintenance requirements of beef cattle females (Durunna et al. 2011b). The significance of the cowherd expenses on the entire enterprise is due to their longevity, size, and varying feed requirements to meet nutrient needs for different stages of production (such as pregnant, lactating, or open cows). From Western Canadian data (WCCCS 2014), the cow needs to produce about five calves to recover all her costs or to break even on the heifer-development costs, which may translate to staying on the herd for at least 6 years. In addition, reproductive soundness is a significant determinant of the culling decisions in most herds while some productive cows (calving annually) have been retained for over 12 years in some operations (WCCCS 2014).

Feed intake is usually estimated as a percentage of the body size, implying that as cows approach their mature body sizes, their expected feed intake will be maximum. Further, as the cows move through different production phases (from open cow to calving to lactating cow), the feed quality and quantity also determine their abilities to rebreed in subsequent years, hence their longevity in the operations may be affected. The availability of sufficient quantity and quality of feed will support better growth. However, the ability of the animal to convert the feed materials into products or "farm currencies" (meat, egg, milk, etc.) can also depend on the genetic potential of the animals to utilize feed effectively. Significant variations exist in individual animals' ability to use feed (Durunna et al. 2011a; Nkrumah et al. 2007), which means that individual differences in feed intake can be exploited to increase farm profitability by selecting animals that are more efficient at feed utilization. The efficiency of converting consumed feed into products, especially when feed resources are limited, enhances the profitability and sustainability of livestock enterprises. The issues may be exacerbated if stored forages are insufficient to meet the animal's growth requirements due to quality or quantity issues. The producers may reprioritize their objectives toward low-cost feeds that support maintenance requirements rather than growth. Unplanned disposal or sale of livestock (at below-market prices) may occur if feed becomes unaffordable or unavailable. Using efficient replacement candidates in breeding plans will reduce the long-term production and product costs of the livestock enterprises, thus animals with better production efficiency potentials may offset the potential excesses from high feed costs.

21.2.2 Strategies to Improve the Efficiency of Livestock Operations

Strategies that will facilitate improved livestock production (feed, growth, and reproductive) efficiencies in developing countries include adoption of improved production management, improved plant and animal breeding management, improved feed processing and forage utilization, effective utilization of human food waste or by-products, and improved health and immunity of animals (Terry et al. 2020).

Some production management strategies that maximize carcass weight within the shortest time enhance the production efficiency of livestock. These practices generally have positive impacts on the operation's profitability and the environment. Such management strategies include the incorporation of growth implants, ionophores, and beta-agonists to improve growth rate and feed efficiency or deploying precision feeding technologies in intensive production systems to ensure that nutrient/feed demands are met without overfeeding. Even though small-holder operations would benefit from the impact of these strategies, the costs associated with implementing them may exceed the benefits, thereby limiting the adoption of these technologies by the small operations.

Growth implants promote protein deposition rather than fat, thereby improving weight gain, total meat yield, and ribeye area. The growth implants could contain natural or synthetic hormones that influence the hormone content of the animal. Two major classes of growth implants available in Canada include estrogenic (e.g., estradiol, estradiol benzoate, or zeranol) and androgenic (e.g., trenbolone acetate) compounds. Results from a feedlot phase study on steers that received growth promotants (implants with trenbolone acetate + estradiol, and monensin in the feed) showed that body gain increased by about 21% while feed efficiency improved by about 23% compared to steers that received grass silage alone (Berthiaume et al. 2006).

Ionophores (such as monensin, lasalocid, or laidlomycin propionate) belong to a class of antimicrobials that inhibit the ability of bacteria to reproduce. They affect the ion gradient in microorganisms, making them expend more energy by preventing them from maintaining normal metabolism. The ionophores' inhibition of acetogenic bacteria in the rumen increases the amount of propionate generated. Beta-agonists are compounds added into feed, but they do not influence the hormone levels in the animal but rather redirect nutrients at the cellular level such that more growth occurs at the muscle tissues, improving feed efficiency and carcass leanness. While implants have been available since 1975, beta-agonists have only been available since 2004 (Radunz 2011).

In-feed antibiotics in poultry increase growth and improve feed efficiency; however, concerns over the development of antimicrobial resistance have led to the search for alternatives in the poultry industry. Alternatives to in-feed antibiotics use in the poultry industry include prebiotics, probiotics, enzymes, phytogenics, and organic acids. Prebiotics are non-digestible feed ingredients that enhance the growth or activities of some gut bacteria. Probiotics are live microbial feed supplements that create intestinal microbial balance. Enzymes such as phytase, carbohydrases, and proteases are biologically active proteins derived from microbes that facilitate digestion and absorption of nutrients. Phytogenics are bioactive compounds of plant descent added into livestock feed (especially pigs and poultry) to facilitate productivity. Organic acids such as simple monocarboxylic (e.g., acetic, propionic, or butyric acid) or carboxylic (e.g., lactic, malic, or citric acids) acids improves growth performance and efficiency in broilers. Animal breeding programs that select animals with genetic potential for faster growth while receiving the most abundant or economic feed options have positive impacts on production efficiency and farm profitability. The genetic approach exploits the genetic variations within the replacement candidates and can pass on those beneficial traits to their offspring. The efficiency traits, such as residual feed intake and residual body gain (RBG) are moderately heritable (Durunna et al. 2012) and can be included in breeding programs. The strategy has been applied in ruminants and non-ruminants (including pig and poultry) and it is more sustainable (especially for developing countries in Africa) because the improvements are additive, cumulative, and permanent since the traits are passed on from one generation to another. On the other hand, plant breeding programs that develop plant varieties with higher yield and greater digestibility potential increase the production efficiency of livestock (Terry et al. 2020).

21.2.2.1 Technologies to Assess and Improve Production Efficiency

Traditional Tools to Assess Production Efficiency

There are several measures to assess feed and growth efficiency. As mentioned earlier, the more common assessment expresses the outcomes as ratio traits where inputs such as feeds are expressed relative to the outputs such as meat, milk, wool, or other livestock products. Animals that produce greater output per input (e.g., feed consumed) are more desirable. Common traits used to evaluate performance metrics in developing and developed countries include body weight, milk yield, and body condition score. The most common method to assess feed efficiency is feed conversion ratio (FCR) or its inverse, gain to feed ratio (GFR). The FCR refers to the quantity of feed consumed per body gain, while the gain to feed ratio is the quantity of body weight gain per feed intake. These two interpretations are well understood and adopted by producers in developing and developed countries. They are easy to measure even with limited technologies. For example, a cattle or small-ruminant producer may measure the quantity of feed consumed by an animal by summing the number of bags of commercial feed (with known weight), while the body weight(s) could be estimated using a weigh tape by measuring the heart girth. Alternatively, the actual weight could be captured with a weigh scale.

However, incorporating ratio traits such as FCR in breeding or selection programs may present some issues (Gunsett 1986) that may have long-term implications on the livestock enterprise from a feed efficiency standpoint. One of the major concerns is that two animals with identical FCR might differ in their average feed intake or daily gain. In such scenarios, unanticipated outcomes could arise when different selection pressures are applied to the numerator or denominator (i.e., the feed or gain) components (Durunna et al. 2011a; Nkrumah et al. 2007). Applying more selection pressure on the trait with higher variance may result in the indirect selection of animals with higher values. For example, broader variance in body gain may result in the indirect selection of animals with larger frame sizes (because they put on more weight than the others), and not necessarily that they have better feed efficiency potentials. The long-term implication of animals with larger body sizes may result in higher feed intake requirements by such animals.

Some traditional measures of growth efficiency include the Kleiber ratio and relative growth rate. The Kleiber ratio measures growth efficiency independent of feed intake (Hoque et al. 2009). It is calculated as the ratio of the average daily gain to the animal's metabolic weight. On the other hand, the relative growth rate measures the rate of growth of each animal relative to its final weight in the measuring period and may be expressed as the percentage gain (Winder et al. 1990).

Modern Tools and Technologies to Assess Production Efficiency

The recent and more popular feed and growth efficiency measures are residual feed intake (RFI) and residual body gain (RBG). RFI is the difference between the actual feed intake and expected intake based on estimated maintenance requirements and production potentials (Durunna et al. 2011a). It provides the best opportunity to select animals with efficiency measures independent of other traits considered. RFI models can accommodate additional adjustments for body composition measures such as backfat thickness (Durunna et al. 2012; Basarab et al. 2011) and activity levels (Durunna et al. 2012; Basarab et al. 2011). These adjustments ensure that the efficiency measures are phenotypically independent of the traits included in the models. For example, positive relationships exist between RFI and backfat, indicating that inefficient animals, i.e., animals with positive RFI will tend to have higher backfat thickness than their efficient contemporaries. Adjusting the models for the backfat level also ensures that the selection efforts do not discriminate against backfat thickness.

The RBG is the difference between the actual gain and the expected gain based on feed intake and other energy sinks adjusted in the model. Residual feed intake and gain (RFIG) is another recent measure that combines the desirable features of residual feed intake (RFI) and RBG in one trait.

Using these recent efficiency measures in Africa requires individual animal data, which is cumbersome in a large cohort if manually done. For example, manual measurement of delivered feed and the refusals for 50 heads housed in individual pens will require significant hours of labor for an individual. Technician bias may also be introduced if several people collect the data from different animals. Another implication of using a trait like RFI or RBG is the additional need to measure body composition, such as ultrasound backfat thickness. Measuring the backfat thickness, especially in cattle, will require handling facilities to restrain the animals and trained technicians to scan and interpret the results.

Deploying automatic feeding systems that can measure individual animal feed intake in a group setting will significantly reduce the errors that may arise. Some examples of such systems for sheep, goat, and cattle include Vytelle Sense ® (previously GrowSafe® Systems) (Vytelle, Calgary, Alberta, Canada), Insentec® (Hokofarm Group, Emmeloord, The Netherlands), SmartFeed® (C-Lock Inc. Rapid City, South Dakota, USA), Calan Broadbent (American Calan, Northwood, New Hampshire, USA), or BioControl CRFI systems (BioControl, Rakkestad, Norway). Few automatic feeding systems exist for poultry but the Bird Individual Ration Dispenser-electronic (Pampouille et al. 2021) may have commercial utility. The disadvantage is that these automatic units are expensive to purchase and maintain.

Furthermore, Mrode and colleagues (2020) identified insufficient funding as a critical bottleneck hindering livestock improvement programs in some African countries. Sustaining data collection efforts toward a livestock improvement program requires a good understanding of the values of the phenotypes collected toward future infrastructure or phenotyping activities. The high cost of measuring individual animal performance has often left producers in these countries to use subjective means, such as guess-estimates of bodyweight or milk yield. Other products such as eggs from laying hens are easier to count and record; however, tracking the individual feed consumption of livestock in small-holder farms in Africa is not common.

Poultry remains the most popular source of proteins and revenue in developing countries but the rising cost of poultry feed echoes the need for systems or tools that will identify and develop more efficient (feed and growth) birds. The birds are usually selected for traits such as faster growth, increased yield and feed efficiency. Selecting efficient poultry has moved from labor-intensive traditional phenotyping of large number of birds raised in cages to the use of individual identification through radio frequency identification (RFID) tags.

21.2.2.2 Precision Livestock Farming, Machine Learning, and Artificial Intelligence

Precision livestock farming, which uses state-of-the art technologies for monitoring and data capture, provide the farmer with more objective data on animal and its environment for informed decisions and choices about the sustainability of farming activities and production system (Fig. 21.1). Precision livestock farming is developing rapidly and appears to be the main driving force behind the industrialization of animal farming (Werkheiser 2018). Precision livestock farming means using automated remote detection and monitoring of individual animals for health and welfare factors through capture and analysis of real-time images, sounds, tracking data, body weight, and body condition, as well as biological metrics, etc., which helps in the early detection of diseases, physiological status, welfare issues, etc., at the farm level (Berckmans 2014; Benjamin and Yik 2019). Generally, precision livestock feeding (PLF) technology is regarded as a means of narrowing the gap between animals and producers in the face of expanding systems (Norton et al. 2019). Thus, the main goal of PLF is to develop livestock monitoring and management systems with technologies to support the farmer (Berckmans 2014). Examples are sensor technologies for observing animals, the use of modern control theory to improve automation of the production process in welfare, health and production monitoring, and meat processing, and the application of artificial intelligence or advanced data processing methods to synthesize and combine different types of production and processing data (Barbar et al. 2022; Park et al. 2022; Fuentes

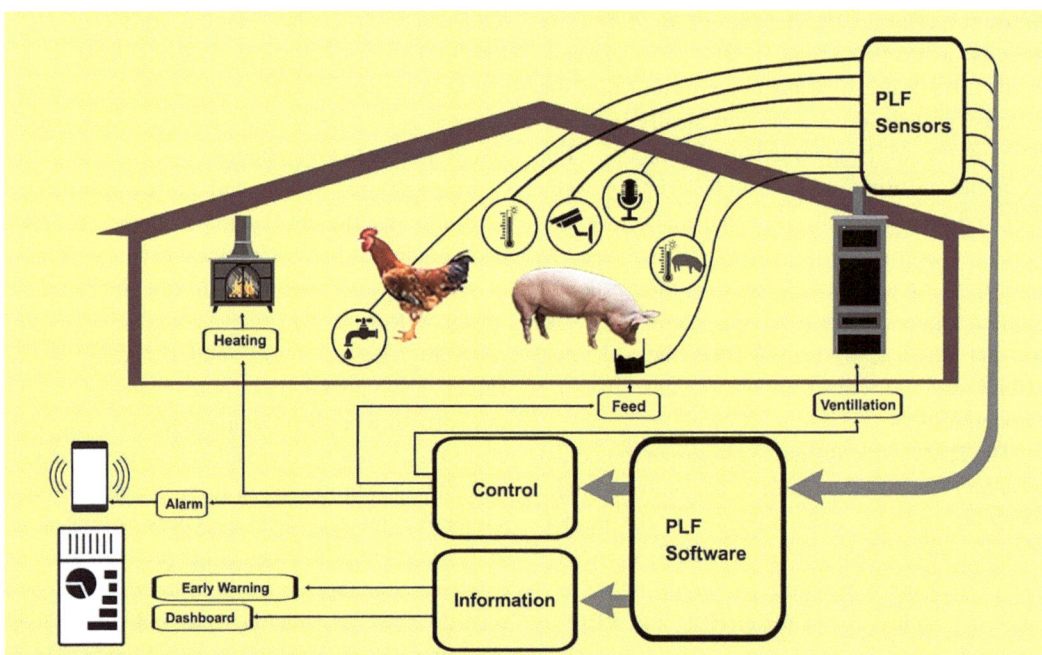

Fig. 21.1 A precision livestock farm uses state-of-the-art technologies to continuously collect data on animal (microphones, cameras, sensors) followed by continuous data analysis to provide real time information or data to support management and production decisions as well as resolve health and welfare issues

et al. 2022). Precision livestock farming is therefore an interface between different scientific disciplines and stakeholders in the livestock industry and requires the fulfilment of certain conditions to achieve sufficient levels of monitoring and management (Berckmans 2009), which include (1) animal variables, such as parameters related to behavior, physiological state, and welfare of animals, which requires continuous measurement with accurate and cost-effective sensor technologies; (2) reliable prediction models on how animal variables vary and how animals will respond at every moment, and (3) integration of predictions and on-line measurements for automatic monitoring and/or management. A recent review discussed livestock farming digital transformation under three themes with several examples: (i) biometric techniques for assessing health and welfare, (ii) animal identification and traceability, and (iii) application of machine and deep learning application to address complex problems (Fuentes et al. 2022). Therefore, machine learning and artificial intelligence are crucial aspects of PLF. Several examples of the incorporation of artificial intelligence-driven automation for achieving efficiency in meat/poultry processing, poultry breeder and broiler management, etc., have been described (Barbar et al. 2022; Park et al. 2022). Furthermore, case studies of the use of digital livestock tools in Kenya and India were recently presented (Daum et al. 2022).

Machine learning is an aspect of artificial intelligence that makes use of algorithms for predicting statistics and inference (Morota et al. 2018). Machine learning is gaining interest in the area of PLF because it enables computer algorithms to progressively learn from big data sets generated from sensor and improve themselves accordingly, thereby eliminating the use of human data analysts. Machine learning helps animal geneticists to predict phenotype based on prior genetic information, identification of outliers in a population and imputation of genotype. Machine learning can help in detecting disease (e.g., mastitis) from automated milking machines in dairy herds by estimating animal body weight and monitoring of gut health microbiome. Furthermore, machine learning has been recognized as a tool for monitoring animal welfare by predicting the early chance of dairy cattle lameness and mastitis, which can have an adverse effect on milk quantity and quality (Warner et al. 2020).

21.2.2.3 Blockchain Technology

Blockchain technology is an advanced tool that can be used to promote transparency on the history of an animal (Picchi et al. 2019). This is because consumers are keen on knowing the prior health condition of an animal before it is slaughtered for human consumption. The increase in foodborne illnesses has increased consumers' distrust of food labels. With the use of blockchain, there is an increased likelihood of regaining consumer's trust by offering product traceability from producer to retailer (Connolly 2018). Blockchain technology is an electronic ledger comprising unique QR codes attached to livestock products' labels. These codes when scanned with mobile devices can provide details regarding the cattle feed, any plausible administration of a treatment, rearing process, an incident of slaughter, time and location of milk production, processing techniques, added ingredients, the product's market release date, the origin of the milk, how old it is, what type of transportation and cold milk chain facilities were used, etc. The consumers can scan the codes using their smartphones, tracking the entire milk production process thread leading to the final product. Customers nowadays want to know where their dairy products come from. This necessitates end-to-end supply chain transparency to increase client confidence (Simitzis et al. 2021). Many dairy producers, suppliers, and stakeholders utilize blockchain technology to provide customers with real-time data about their products. Blockchain technology offers numerous benefits to livestock agriculture, such as automated transactions, decentralized systems that could contribute to more efficient systems of auditing for certification and regulatory organizations, system integration and proper record keeping of chain transactions throughout the lifecycle of an animal from farm to fork. This will, in turn yield greater traceability and transparency within livestock agriculture. Blockchain technology could as well proffer solutions to sustainability and ethical concerns raised by consumers on how food animals are raised. Furthermore, blockchain technology could be used in tracking livestock disease breakouts, including salmonella outbreak, H1N1 swine flu, foot-and-mouth and mad cow diseases, and Avian influenza (Lin et al. 2018). In the event of a livestock disease outbreak, farmers all over the world can access data on the disease (the cause of the disease outbreak, how animals were treated, and precautionary measures to be considered). All these will help livestock farmers make informed decisions on what to do during a disease outbreak or an outbreak that is likely to reach their farms.

21.3 Technologies in Animal Welfare and Management

The safety and health of animals and farm workers are paramount to addressing animal welfare issues and public acceptance. In recent years, great strides have been achieved in developing technologies to address improved productivity, efficiency, safety, and animal welfare, including remote monitoring technologies, automated dairy installations, feeding technologies, pasture-based technologies, automated cleaning systems and herd management systems. Good animal welfare practices support livestock operations to improve productivity, efficiency, profitability, sustainability, and public perception or social license (Grandin 1995). Animal welfare measures can be classified into animal-based, management-based, and resource-based measures (Forkman and Lund 2017). These measures address the overall status-assessment of the animals, management routines (such as appropriate use of antibiotics or analgesics) that maintain animal health, and provide enhanced habitation resources that do not expose animals to risk factors such as injury or disease (Madzingira 2018). Understanding the impact of sound welfare practices on production efficiency will reduce associated morbidity or mortality of the animals and minimize safety concerns for the handlers. Enhanced welfare practices also positively impact reproduction, growth, productivity, and meat quality, which influence farm profitability (Rushen 2001). Addressing welfare concerns related to environmental conditions will improve feed and growth efficiencies.

Heat stress issues resulting from elevated ambient temperatures, relative humidity, or stocking rates will reduce the performance of animals. Livestock operations with poor air quality, ventilation systems, and inadequate provision of shades will exacerbate the negative impacts of inclement environmental conditions. Providing suitable bedding materials promotes health and improves production efficiency through comfortable rests for the animals. Hard, slippery, uneven, or muddy surfaces may expose the animals to injuries and diseases such as sores, lameness, foot rot, or mastitis. It is also essential to ensure that facilities and equipment used in managing animals do not predispose the animals to stress, injuries, or death; instead, such tools should reduce excitement and agitation in animals. Technologies and methods that reduce stress in livestock will support improved productivity and efficiency. These include incorporating excellent stockmanship procedures that improve human-livestock interactions during herding and handling (Petherick 2005). Low-stress techniques will minimize unnecessary anxiety tendencies that may disrupt the normal state of the animals. In the absence of less-painful husbandry practices, pain mitigation methods should be adopted to reduce the pain duration and facilitate the animals' recovery.

21.3.1 Technologies to Reduce the Impact of Animal Production on Animal, Human, and Environment Health (the One Health Concept)

In recent times, the One Health Approach (OHA) has been adopted to support global health security through collaborative and multi-disciplinary approaches cutting across boundaries of the

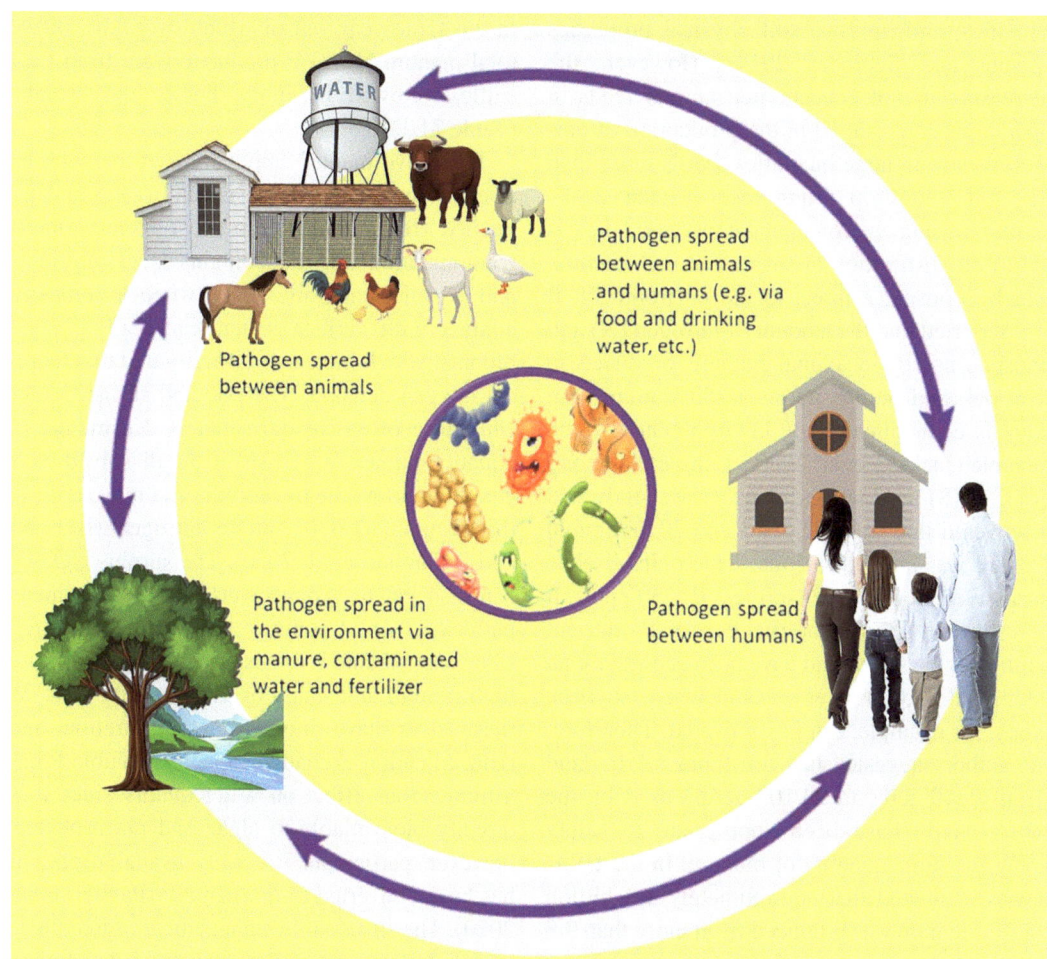

Fig. 21.2 The one health concept

human-animal-environment interface to tackle shared health threats such as zoonotic diseases, antimicrobial resistance, food safety, and so on (Sinclair 2019). The OHA is depicted in Fig. 21.2. Since the first use of the term "One-Health" following the outbreak of the SARs (severe acute respiratory disease) in early 2003, many countries over the last decade have implemented the OHA with demonstrated benefits (Sinclair 2019). The principles of the OHA will benefit efforts to develop plans for response and control of zoonotic diseases for the health benefit of animals, the environment, and the society. The impact of animal production on the health of animals, the environment, and humans is through the production of varied products, manure (which contains pathogens and excess nutrients), greenhouse gases, etc., and their proper management is necessary to safeguard animal, environment, and human health. Various technologies that ensure the safety of by-products of animal farming, such as pathogens, manure, and greenhouse gases (GHG), will be examined in this section.

21.3.1.1 Technologies to Handle Farm Animal Manure and Storage

Animal manure is a mix of animal feces and/or urine and bedding materials, and it is an integral part of sustainable crop production. Animal manure is mostly applied to soils as a form of fertilizer to enhance crop yield, build organic soil matter content, buffer soil pH, increase biological activities, and improve soil physical properties (Zhang and Schroder 2014). However, the increasing use of concentrated animal feeds in operations has resulted in the production of animal manure in large quantities that is more than the soil and crop requirement (Table 21.1), thereby raising environmental concerns. This has led to the disposition of surplus animal manure, which is posing a threat to the environment. It was reported that the amount of nitrogen (N) and phosphorus (P) in animal manure produced in livestock and poultry farms is 50% more than what is economically and environmentally sustainable (Jackson et al. 2000). Long et al. (Long et al. 2018) used ten farms as a case study and discovered that between 2013 and 2015, animal manure was applied to soils, which already had excessive amounts of P (soil test $P > 50$ ppm). It was further stated that the excess P nutrient applied to the soil would have been enough for an additional 10,000 acres of land where soybean crops, for example, can be cultivated. Therefore, the author suggested that ten times of the land (from 2440 acres to 23,109 acres) used by the farms would have been appropriate to safely apply the same quantity of manure. In the USA, it was estimated that approximately 1.4 billion tons of manure was produced from more than 9.8 billion heads of livestock (beef cattle, dairy cows, swine, poultry, goat, sheep, horse, etc.) in the year 2017 (Pagliari et al. 2020). Likewise, 8.1 million tons of manure N and 2.6 million tons of manure P were excreted by poultry and livestock in the same year, out of which over 75% of the total manure N and P produced were from beef cattle, followed by dairy, horse, and broilers (Table 21.1).

Excessive manure application to the soil can affect water (surface and groundwater) and air quality. Heavy rainfall can cause surface runoff from manured land that contains excess nutrients and organic materials, which when it comes in contact with surface water, can lead to algal blooms, which increases turbidity and biochemical oxygen demand. The polluted surface water can cause offensive odor that is detrimental to aquatic animals (e.g., fish) if the dissolved oxygen goes below the thresholds; and may result to the accumulation of nitrate–nitrogen (NO_3–N) and P (Pagliari et al. 2020). The surplus NO_3–N can also get to ground water through drainage pathways or leaching into ground water. Likewise, soluble P losses are more in areas with high rainfall and sandy or loamy sandy soils, as these soils have more water infiltration and reduced cation exchange capacity. Soluble P has a tremendous effect on water quality since it is directly bioavailable to algae and macrophytes; however, particulate P is only available once it has been converted to inorganic P (Pagliari et al. 2020). The sustainable management of manure is a global issue, particularly in most developing countries, where conventional management practices are mainly limited to land application and stockpiling (Fernandez-Lopez et al. 2015; Adghim et al. 2020). Improper manure manage-

Table 21.1 Number of poultry and livestock species and the estimated amount of manure produced, and nitrogen and phosphorus excreted in the USA in year 2017

Livestock[a]	Number of animals (1000 head)	Manure produced (tons/y)	Nitrogen produced (tons/y)	Phosphorus produced (tons/y)
Beef cattle	108,405	1,200,000,000	6,294,416	2,233,503
Dairy	14,181	145,000,000	1,663,735	258,803
Swine	129,388	596,000	3967	1511
Chicken (layer)	375,845	49,000	672	251
Chicken (broiler)	8,913,000	1,050,000	14,438	4463
Turkey	242,500	35,000	594	225
Horse	3914	36,000,000	197,647	141,176
Total	9,822,460	1,382,730,000	8,175,469	2,639,932

[a]Adapted from Pagliari et al. (2020)

ment causes issues related to environmental degradation, climate change, resource depletion, acidification, eutrophication potential, and water scarcity (Adghim et al. 2020). The importance of effective manure management and utilization has gained the attention of most national governments, with many having national policies related to manure management. To reduce the impact of animal manure application to land on the quality of air and water, several management practices have been instituted including application of the exact amount of nutrients needed by the soil. This can be achieved by performing soil testing and animal manure nutrient testing to know the exact nutrient content of the soil and manure before animal manure application. In the absence of manure testing facilities, book values of manure nutrient contents can be used (Christensen and Sommer 2013). However, it is important to keep in mind that manure nutrients might change due to several factors like efficient use of nutrients by animals due to breeding efforts, manure management systems, and animal feeding practices. It is expedient to follow best management practices to maximize the use of manure for food production without jeopardizing the environment. Therefore, this section focuses on several technologies for the best management practices that can be adopted to reduce the impact of animal manure produced in the environment.

Manure storage (includes collection, handling, and storage) has been very helpful in managing manure nutrients. Manure storage helps farmers to apply manure when the soil and plants can best make use of it, which could help prevent excessive application of manure to soils and in turn enhance crop production. The manure storage facilities must be well designed and structured in a way that will prevent environmental pollution (e.g., odor, flies, and pollution of ground and surface waters). Manure can be solid (greater than 15% dry matter), slurry (5–10% dry matter), or liquid (less than 5% dry matter) depending on the content of manure solids and this will impact the options to consider during manure handling and storage (Janni and Cortus 2020). It could also affect the type of equipment that will be used for handling, collection, and land application of the manure. Approximately 85% of swine production systems in the USA produce slurry or liquid manure. In such systems, manure (feces and urine) can be collected in a storage area below the slotted floor and the manure can be stored either for a short (one day) or long term (greater than 6 months) period (Pagliari et al. 2020). Manure can be stored at short term in gutters or shallow pits or at long term in deep pits, above ground tanks and lined earthen basins. In long-term storage facilities, the manure is agitated to create a more uniform slurry before land application (Pagliari et al. 2020).

In poultry production, litters are stored under a stacking shed with concrete floor and walls (the walls can be either short or full and covered with a roof). The stacking shed helps to prevent nutrient seepage if the litter is wet and to keep the litter dry. Some poultry barns have manure belts below each level of cages that catches the manure and the belts transports the manure away for storage located outside of the barn. Manure from layer houses with manure belts is commonly removed from the layer house every day to every third day and then stored in a stacking shed. Some layer houses have flush systems that handle the manure as a liquid and the liquid manure can be stored in lined earthen basins, or steel or concrete structures for few months to a year before land application (Janni and Cortus 2020). Dairy operations generate manure in the form of solid, slurry, or liquid. Concrete floors and walls covered with roof are mostly used to store solid manure. Liquid and slurry manure from dairy operations are stored in similar structures, such as below-ground tanks, lined-earthen basins, and above ground tanks (Janni and Cortus 2020). These systems of storage may include multiple basins that provides separation of manure solids and also allow wastewater to be used for flushing manure alleys or manure flumes. Most liquid and slurry manure systems of storage are designed to store manure for at least a year before it can be land applied.

In beef production, feedlot runoff is often handled using a two-stage system. The first stage is the use of a smaller settling basin designed to separate solids from the liquid. Afterward, the liquid portion is transferred to a larger, longer-term storage in a larger basin. Solids need to be periodically emptied from the settling basin. The basin needs enough storage to accommodate normal and storm-event weather conditions in all the entire feedlot area between the period of time when the liquid can be land applied. The basin may be referred to as a detention basin, holding pond, or evaporation pond in different areas and for different modes of management. The low solids and nutrient content of the basin liquid is liable to be distributed by irrigation. However, evaporation from the basin may be significant in dry climates (Janni and Cortus 2020). Furthermore, solid manure removed from open feedlots or barns with solid floors can be stored in uncovered stacks on impervious surfaces that prevent nutrient from leaching into the soil. Uncovered stacks are mostly required to have runoff collection systems and storage to prevent pollution of surface water. Also, solid manure from beef operations can be stored in covered stacks under sheds that limit runoff. Manure collected in deep pits needs to be agitated before removal. Most times, additional water may be added to reduce the solids content of the manure for easy removal by pumps. The manure is directly applied to cropland or soil according to nutrient management plans.

21.3.1.2 Technologies to Reduce Nutrient Excretion and Manure Treatment

Feeding Animals Based on Dietary Requirement

Nitrogen is an important element for proteins, amino acids, nucleic acids, and energy transfer in vertebrate animals; thus, large amounts of dietary N are required to produce profitable quantities of meat and milk and ensure successful stock procreation. When dairy rations are adequately balanced for energy, crude protein, and minerals, dietary N is metabolized and transferred to milk, urine, and feces in approximately equal proportions (Powell et al. 2011). However, feeding excess dietary nitrogen to cattle can lead to increased concentrations of urine-nitrogen, and to a lesser extent fecal-nitrogen and as much as 75–95% of nitrogen fed can be excreted (Cole et al. 2006; Selbie et al. 2015). For beef cattle, the nitrogen use efficiency (NUE) is generally low, of which 80–90% of the dietary feed N can be excreted in manure (Cole et al. 2006). When high dietary CP concentrations are fed to cattle, most excess N is excreted in urine (Waldrip et al. 2013). For example, Cole and colleagues (Cole et al. 2006) observed that urinary nitrogen content increased from 84 to 94 g/d, while fecal-nitrogen increased from 39 to 65 g/d, when dietary CP was increased from 11.5 to 13.0% (this is within the National Research Council (NRC) recommended CP levels for finishing cattle). Meanwhile, Cole et al. (Cole et al. 2005) reported 140% increased NH_3 emissions when beef cattle were fed 13.0% CP, compared to 11.5% CP. As a result, cattle NUE, and impact of diet composition on N use have gained a lot of interest for livestock managers. Interestingly, it was estimated that a diet containing 23% CP could facilitate optimum milk production (NRC 2001). Whereas, overfeeding protein nitrogen can have negative effects on milk nitrogen content. For instance, it was reported that increasing the intake of CP from 15 to 19% led to 18% reduction in milk nitrogen from 0.27 to 0.22 g/d (Groff and Wu 2005).

In contrast to nitrogen excretion, most excess phosphorus is excreted in cattle feces, rather than urine, which can be partitioned differently depending on diet and animal characteristics. The dietary phosphorus requirement for a feedlot steer is 0.30% of the diet dry matter (DM) (NRC 2000), but could be as low as 0.16% of dietary DM (Erickson et al. 2005). The phosphorus use efficiency of lactating cows is relatively low, and ~67% of total consumed phosphorus is excreted (Wu et al. 2003). The level of manure phosphorus excreted in feces can be reduced by decreasing dietary phosphorus levels (Feng et al. 2015), either by manipulating the amount of supplemental P or changing the ratio of silage to forage being fed. Cereal grains, silage, and other feedstuffs contain higher phosphorus levels than forages (Singh et al. 2018). For instance, steam-flaked corn contains approximately 0.25% phosphorus, while meadow hay contains approximately 0.18% phosphorus (Wilson et al. 2020). Generally, grains consist of 3.5–4.5 g of phosphorus/kg DM, while straw contains about less than 1 g/kg DM (Singh et al. 2018). Knowlton and Herbin (Knowlton and Herbein 2002) reported that decreasing dietary P content from 0.67 to 0.34% DM decreased daily phosphorus excretion from 113 g/d/cow to 43 g/d/cow. Similarly, Wu and colleagues (Wu et al. 2003) observed that decreasing dietary phosphorus from 0.42 to 0.33% DM caused a 25% decrease in phosphorus excreted in the feces of lactating cows. The authors further reported that there was a positive linear relationship between the amount of phosphorus in the diet and feces-phosphorus, where increase in diet phosphorus content by 0.01% elevated feces-phosphorus by an average of 0.02% (Wu et al. 2000).

Inclusion of Feed Additives to Improve Digestibility

Feed additives may be beneficial for decreasing phosphorus and nitrogen contents of litter, as well as substantially reducing the odor of manure. One of the management practices to reduce phosphorus runoff from poultry litter is to decrease the amount of phosphorus in feed. In the past, more phosphorus was supplemented in poultry rations because most grains store around 80–90% of their phosphorus as phytate (Turner et al. 2002). Poultry birds cannot easily digest phytate-phosphorus because they are relatively stable. However, when phytase enzymes are added to poultry diet, it cleaves the phosphorus from the phytate molecule, making it readily available. Phytase can degrade digestible fibers from indigestible grain fractions and decrease the need for phosphorus supplementation, which in turn minimizes the phosphorus level in poultry manure. Furthermore, the addition of amino acids to poultry ration has been shown to reduce phosphorus excretion by nearly 50%, although this increases the production costs (Keshavarz and Austic 2004). Another diet modification technology that can be potentially used to decrease the amount of dicalcium phosphate added in the feed is the use of high available phosphorus (HAP) corn, which are varieties that have been selected or bred for their ability to store P in forms that is more bioavailable than phytate (Raboy 2002). Miles and colleagues (Miles et al. 2003) showed that poultry diets containing phytase and HAP corn produced manure with decreased phosphorus content resulting in lower phosphorus runoff (Smith et al. 2004). Distillers dried grains with soluble (DDGS) is a fermented product with low concentrations of phytate-bound phosphorus, that is mostly available as a by-product of ethanol production (Stein and Shurson 2009). It was reported that phosphorus concentrations in manure are mostly reduced when DDGS is added to the diet as long as the total dietary phosphorus is adjusted to account for the higher digestibility of phosphorus in DDGS (Stein and Shurson 2009).

There are other feed additives that have been used to mitigate methane (CH_4) emissions and improve feed efficiency, including seaweed (Vijn et al. 2020), secondary plant metabolites, such as tannins, saponins, essential oils, and flavonoids, as well as lipids, ionophores, nitrogen sources, and biocarbon (Ku-Vera et al. 2020; Villar et al. 2020). The most common lipids used for CH_4 emission mitigation are oil-rich seeds, such as cottonseed (Beck et al. 2018), linseed (Doreau et al. 2018), and canola oil (Villar et al. 2020). Ionophores are antimicrobials mixed with cattle feed to improve nutrient availability and prevent diseases like coccidiosis. Interestingly, when ionophores are mixed with cattle feed, it enhances feed efficiency and weight gain by selectively inhibiting methanogenic bacteria, thereby allowing the beneficial rumen bacteria to make more feed energy available to the animal. Monensin is the major ionophore used in ruminant diets to increase feed efficiency by limiting gram-positive bacteria and protozoa (Witzig et al. 2018).

Steroid hormone implants (estrogens, androgens, or their combination) have been commonly used in U.S. cattle production systems for a very long time, and are known to increase the synthesis of muscle protein, reduce protein degradation, and improve average daily gain (Beck et al. 2014; Webb et al. 2017; Cleale et al. 2018); as well as reduce the GHG emissions and resource use per kg of beef produced (Webb et al. 2017). Implants have also been found to improve productivity in both feedlot and pasture-based steers, heifers, and bulls, which in turn improve their economic returns (Beck et al. 2014; Al-Husseini et al. 2014; Smith and Johnson 2020). Combining growth implants with ionophore feed additives has been shown to be effective in feedlot programs. It was shown that growth implants improved daily gains and feed efficiency while ionophore feed additives in combination with growth implants reduced the amount of feed needed for a given amount of gain by an additional 7–8% (Al-Husseini et al. 2014). Recombinant bovine somatotropin was among the first agricultural biotechnology products to be approved by the United States Food and Drug Administration (FDA). Somatotropin is a key homeorhetic control in regulating nutrient partitioning because when administered to dairy cows, it enhances milk production by 15–20% (Etherton et al. 1993), improves milk synthesis efficiency by 10% (Bauman 1999) and at the same time maintains the health and overall well-being of dairy cows (Bauman 1999; NRC. National Research Council 1994). When recombinant bovine somatotropin was administered to growing pigs, carcass fat content was reduced by 70–80% and productive efficiency was improved by 15–35% (Buzby et al. 2006). However, hormone implants are not widely accepted in other countries and generally, there is negative public perception regarding the use of hormone implants in animal production.

Beta-agonists are a class of non-hormonal compounds fed to cattle. They work by binding to receptors on fat cells in the animals' body and redirect and reduce fat metabolism so that less fat is produced and/or stored in the carcass. Also, beta-agonists bind to receptors on muscle cells and redirect and increase the size of muscle fibers, which replaces some of the fat, thus increasing the percentage of lean muscle. In doing so, the energy supplied by the feed to produce weight gain is reduced. Therefore, more weight is produced by the same level of feed intake and feed efficiency is increased (Felix 2017). There are two compounds available to cattle feeders: Zilmax© (zilpaterol) made by Merck and Co. (Merck and Co Inc., New Jersey, USA), and Optaflexx© (ractopamine) made by Eli Lilly Co. (Eli Lilly and Company, Indiana, USA). Both products are fed at very low levels (200 mg/head/day for Optaflexx© and 60–90 mg/head/day of Zilmax©) for a short period of time just before slaughter (25–35 days) (Felix 2017). A recent report showed that 60–80% of feedlot cattle in the USA are fed a beta-agonist (Felix 2017).

Phase Feeding

Phase feeding involves formulating animal diet closer to their nutritional needs and this could reduce N and P excretions by 10–15%. It involves the adjustment of dietary CP to meet, but not exceed, animal CP requirements at different stages of growth and production. For instance, the nutritional needs of swine varies through different life stages (NRC 2012); thus, manure characteristics will change over time. To meet these nutritional changes, phase feeding is often employed. Swine diets are formulated based on amino acid balance needs of the pig. Although some feedstuffs like soybean meal provide the required distribution of amino acids, the addition of synthetic amino acids into swine diet can more closely meet the pig requirements, thereby reducing excretion of NH_4–N (Liu et al. 2017) and K (Li et al. 2015). In general, phase feeding can reduce N and P by approximately 5% (Sutton and Lander 2003). Knowlton and colleagues found that moving from a one-phase feeding to a three-phase feeding system, helped to better meet the exact nutritional needs of growing pigs, thus P excretion was reduced by 12.5% (Knowlton et al. 2004).

Manure Management to Reduce Nutrient Losses and Improve Production

Atmospheric emissions of compounds like NH_3 from animal manure cause production problems and environmental issues such as soil acidification, formation of fine particulate matter, and excessive N deposition into aquatic and terrestrial ecosystems (He et al. 2016). During the winter months, gaseous NH_3 concentrations can reach very high levels in poultry houses, which can sometimes exceed 100 ppm. When atmospheric ammonia is deposited onto soil, it is converted to nitrate via nitrification, which is an acid-generating process (Ashworth et al. 2020). Presently, the only technology adopted by the poultry industry to decrease NH_3 volatilization from manure is acidification with acid salts, such as aluminum sulfate (alum), sold under the tradename of Al$^+$Clear or sodium bisulfate sold under the tradename PLT® (Ashworth et al. 2020). When litter is acidified, it pushes the ammonia/ammonium equilibrium toward ammonium, which is not volatile.

The two phosphorus fractions that are mostly abundant in cattle manure are water-phosphorus (H_2O-P) and sodium hydrogencarbonate-phosphorus ($NaHCO_3$–P). These two phosphorus fractions are generally considered labile phosphorus fractions and should be readily available for plant growth (Pagliari 2014). H_2O-P content is a critical indicator of the potential of phosphorus runoff (Kleinman et al. 2020). The higher amount of H_2O-P and $NaHCO_3$–P in cattle manure indicates that the management of P should focus on their immediate impacts on water quality and plant response. Cattle production systems generally capture and store manure. It has been shown that manure storage systems are effective at decreasing runoff potential (Janni and Cortus 2020). Hong and colleagues (Hong et al. 2018) characterized P in the sediment sludge and crust formed on retention ponds of a dairy manure storage by sequential fractionation method. It was reported that pond sludge and crust contained higher amounts of labile phosphorus (H_2O-P and $NaHCO_3$–P). However, there were lower levels of H_2O-P in sludge and crust than what was found in raw manure, which indicates that the use of sludge and crust rather than raw manure as a soil amendment could decrease phosphorus loss by surface runoff and leaching, as well as providing soluble P for immediate plant use. In addition, most of the organic compounds containing phosphorus in manure can highly interact with soil clay particles and they will bind to soil clay particles when the conditions are appropriate. When P binds to soil particles, it becomes more resistant to hydrolysis and therefore not readily available (Karathanasis and Shumaker 2009). Therefore, understanding the forms of phosphorus in manure can aid to develop best management practices, which could enhance the beneficial reuse of manure as a nutrient source in crop production.

Use of Water Extractable Phosphorus

Water extractable phosphorus (WEP) is used to predict the potential for manure applied on land, particularly manure applied to the surface of soil, to directly transfer dissolved forms of P to runoff water (Ashworth et al. 2020). Water extractable P can be used to adjust environmental recommendations for land application of manure (Elliott et al. 2006), and WEP is an essential factor in the most sophisticated computational models when considering (WEP) runoff (Vadas et al. 2009). Just as total P differs from one animal species to the other, so too does WEP. In a recent survey, Liu et al. (2018) reported that in average, WEP was highly concentrated in turkey manure and swine manure, followed by chickens (layers and broilers), dairy, beef, and horse manures. Generally, composted manures had lesser WEP concentrations. It was estimated that WEP comprised of 11–58% of total P in manure and compost.

Metal salts can be potentially used to reduce manure WEP. The most accepted metal salts is alum ($Al_2(SO_4)_3$), which is used in the production of over 1 billion broiler chickens in the United States alone. The use of alum in the broiler industry reflects its contribution to ammonia conservation, and improvement on the growth and health of housed birds (Choi and Moore 2008). Other examples of metal salts that have been experimented include ferric chloride ($FeCl_3$), ferric sulfate ($FeSO_4$), aluminum chloride ($AlCl_3$), and calcium sulphate (gypsum) ($CaSO_4$), all of which have shown great potential to lower WEP in manure but none of them gained widespread acceptance (Kleinman et al. 2020). In general, a decrease in the solubility of P in manure when salt is added does not significantly affect P availability to crops, although the added P adsorption capacity from salt amendments in manure can also decrease P solubility in soils, as shown by long-term studies with alum-treated poultry litter (Huang et al. 2016).

Housing Management

The cleanliness of housing could affect the manure properties during storage. Li and colleagues (2017) compared dirty pig housing (sprayed with manure) versus clean housing (washed daily), and found that when stored in simulated deep pits, there was an increase in manure NH_3, nitrous oxide (N_2O), and hydrogen sulfide (H_2S) gas emissions from the dirty housing system. Manure is generally considered to be made up of feces and urine, but it also includes waste food and water. Food spillage can occur during pig feeding, and it is estimated that nitrogen and phosphorus in manure can increase by 1.5% for every 1% of food spilled (Ferket et al. 2002). Also, nipple drinkers have more wastage compared to cup drinkers or wet and/or dry feeders (Matlock et al. 2014) and it is estimated that adequate management and selection of appropriate drinkers could reduce water wastage by 30% (Muhlbauer et al. 2011).

Management of environmental factors within housing, like temperature and humidity, is also very important as this could impact manure properties. Temperature and humidity not only affect the food intake and manure production of pigs (NRC 2012) but can also influence emissions of gases and nutrient transformations of manure stored in pits. In a meta-analysis conducted on different types of swine production facilities in the United States, Liu et al. (2014) stated that NH_3 emissions from all types of swine facilities, such as deep pits and recharge pits increased with increasing temperature, but this does not affect H_2S emissions. However, Thorne et al. (2009) reported that H_2S emissions in barn were affected by number of pigs housed, wind, and relative humidity.

Handling of Manure and Treatment

Handling and moving manure into storage can affect manure characteristics, particularly nutrients and DM content. For instance, in flush systems, water is used for moving swine manure from temporary storage under the confinement building to storage outside the barn. When manure is flushed, lots of gaseous H_2S is emitted, but regular (daily) flushing compared to once weekly or bi-weekly flushing was found to reduce overall H_2S and NH_3 gas losses (Lim et al. 2004). In flush systems with pit recharge (the pit is partially filled with water for the manure to fall into), the rate of pit recharge (or how often manure is removed) can also impact H_2S emissions (Liu et al. 2014). For instance, if the manure stays for a long time in the pit before being flushed, the more H_2S is released. Agitation is often used to mix manure before crop or land application so that nutrients are redistributed uniformly. In a research carried out on a deep pit system, it was discovered that the average H_2S concentration in the barn increased by approximately 62 times while NH_3 increased by ten times before the manure was agitated (Hoff et al. 2006).

Manure disposal and management are among the most important issues facing producers, as improper manure treatment and management can lead to excessive odor, fly breeding, and environmental pollution. Anaerobic lagoons are system used only in warm climates for treating manure because the microorganisms are more active under warm climatic conditions to break down the manure accumulated in an appropriately sized lagoon. This is because, a small-sized lagoon with excess daily manure loading in cold temperature can overwhelm the microbial activity of the lagoon, which can lead to odor problems and poor treatment of manure anaerobically (Vanotti et al. 2020). Other manure treatment systems have been proposed that recover calcium-phosphorus (Ca-P) compounds. For example, in South Carolina, Vanotti and colleagues (2010) developed a full-scale, two-stage system for a swine production facility that can treat raw swine manure. This system uses a nitrification bioreactor to decrease carbonate and ammonium buffers, followed by the addition of calcium hydroxide ($Ca(OH)_2$) to precipitate Ca–P (Vanotti et al. 2005). Manure solids containing about 95% of the phosphorus are collected once the treatment process is completed and are exported to end users as compost or low-solubility fertilizer. This system can also be used to remove odors and pathogens from manure. It is noteworthy that many calcium-based phosphorus recovery systems and anaerobic digesters generally volatilize much of the manure nitrogen content as they adjust pH to favor Ca-P precipitation unless recovery systems are included to capture NH_3 off-gassed (Karunanithi et al. 2015). Wei and colleagues (Wei et al. 2022) showed that nitrogen losses from land application were significantly reduced (52.6%) when compared to the level of discharge consent (75.9%). It was further stated that ammoniacal nitrogen (was the major source of nitrogen loss accounting for 72.9% and 50.2% in land applied manure and in discharge consent, respectively. It is imperative to control ammonia emission for sustainable manure management (Wei et al. 2022).

In-Field Stacking of Manure

Stacking of dry manure is a common short-term management practice in which dry manure is stacked in the field before land-application. Even though it is recommended to store manure under a permanently roofed structure, which could prevent litter from direct exposure to rain; however, it is still common to stack poultry litter in crop fields when the manure will be used within a short time frame (Kleinman et al. 2020). To minimize nutrient runoff from the manure stacked in the fields, most developed countries like USA states and Canadian provinces have established relevant management guidelines, which involves covering manure stacked on the field with plastic sheeting or with reinforced, ultraviolet-resistant cover (resistant to environmentally sensitive areas like streams, water wells, drainage ditches, and sinkholes), and with the manure stacked on lands with gentle slopes to prevent runoff (Kleinman et al. 2020). It is economical to use plastic sheet for covering manure stacks to reduce P and N runoff while better preserving the N value in the manure. Although some studies have several concerns associated with covering a manure stack under certain circumstances, including elevated nutrient leaching by water generated by microorganism respiration under aerobic conditions (Dewes 1995), most of the studies have also pointed out the benefits of covering the manure stacks (Doody et al. 2012; Liu et al. 2015). For example, in a field study conducted on poultry litter stacks, it was reported that although P losses from both covered and uncovered stacks were very low due to their high capacity to hold precipitation water, covering stacks reduced the total P leaching by 25–100 times such that leachate P losses from covered stacks was similar to the controls with no manure stacking (Liu et al. 2015). Moreover, it is worthy to note that manure stacking could result in the development of P "hotspots" in fields (Liu et al. 2015). Liu et al. (2015) observed that after 2 years of poultry litter stacking, WEP concentrations in upper 5-cm soils were increased (120–240 mg/kg) under the covered stack and under the uncovered stack (140–250 mg/kg). Therefore, it was suggested that relocating stack sites is necessary when field stacking of manure is mostly practiced.

Aerobic and Anaerobic Digestion of Manure

Aerobic digestion is described as a process in which animal manure is agitated with air or oxygen for a specific solids retention time at a set temperature, which could work as a batch or continuous system. The specifications for the needed mean cell residence time include 40 days at a temperature of 20 °C and 60 days at 15°C (Bekoe et al. 2018). Bekoe et al. (2018) reported that a 14-day aerobic treatment reduced the total solid content of swine manure by >15%, and the aerobic treatment of swine manure before anaerobic digestion showed a higher CH_4 yield. Anaerobic digestion is the microbial decomposition of organic materials void of oxygen, producing mainly CH_4 (55–65%) and carbon dioxide (CO_2) (30–35%) gases, commonly known as biogas (Gerber et al. 2013). Anaerobic digesters can be classified as passive systems, low-rate systems, or high-rate systems (Harrison and Ndegwa 2020). A passive system is when biogas recovery is achieved with an existing treatment system, that could be accomplished similar to a covered manure lagoon. A low-rate system consists primarily of manure as the major source of CH4-

forming organisms. Designs of low-rate system include complete mix and plug flow. A high-rate system traps CH4-forming microorganisms in the digester and includes recycling of solids, fixed film, suspended media, and sequencing batch digesters. Anaerobic digesters (AD) provide many benefits including: odor management, reduced emission of greenhouse gases, pathogen control, and production of biogas (CH_4) for use as fuel (Harrison and Ndegwa 2020).

Anaerobic digestion of livestock manure coupled with addition of biogas production is thus effective in decreasing CH_4 emission because this decreases readily available carbon for microbial processes to produce CH_4, either during subsequent storage or following land application of the digester effluent (Harrison and Ndegwa 2020). Furthermore, when CH_4 is captured and used as a renewable energy source, it can serve as an alternative to non-renewable fossil fuels, further decreasing emissions of GHG, nitrogen oxides, hydrocarbons, and particulate matter (Harrison and Ndegwa 2020). The economic benefit of converting fibrous materials, such as manure, food waste, and by-product feeds into CH4 is mostly linked to fuel production for generating electricity, transporting vehicles, and providing energy for home heating and cooking. In addition, the liquid after anaerobic digestion can be utilized as a nutrient source for crop production, while solids can be used either as bedding for animals or as a soil amendment. The environmental and societal benefit of AD is the diversion of wastes away from landfills (Gould 2012). Anaerobic digesters are used in the management of manure associated with dairy farming and waste water at swine and poultry operations (Harrison and Ndegwa 2020). In a recent study, Rajagopal and colleagues (2021) demonstrated that the liquid digestate (leachate) resulting from a two-stage liquid-solid anaerobic digestion (at 20 ± 1 °C) of chicken manure can be used to grow microalgae (Fig. 21.3). Microalgae are photosynthethic microorganisms with capability to grow and multiply rapidly by utilizing nutrients (e.g., nitrogen, phosphorus, and potassium) from water and producing organic exploitable biomass (e.g., biofuel) (Cai et al. 2013). Considering the high nutrient requirement by microalgae, Rajagopal and colleagues (2021) proposed that enhanced growth of microalgae at higher concentrations of manure digestate and better adaptation of microalgae/microalgae-bacteria consortia to high-N content wastes will positively influence the circulation bioeconomy framework.

Small-Scale Anaerobic Digestion in Developing Regions

Although developed countries have the concentrated animal density and infrastructure to promote the utilization of highly engineered AD systems, developing countries must make use of less sophisticated models (McCord et al. 2020). Common feedstocks for use in ADs in Uganda, Rwanda, and Bolivia include manure, food waste, slaughterhouse waste, and human waste (Vogeli et al. 2014). The designs often times consist of underground domes made of brick and cement, above ground poly-tanks, and bio digesters (Vogeli et al. 2014). Most of these digesters are simple pass-flow reactors that do not have mixing components, do not need elaborate continuous monitoring, and can be used in tropical weather conditions (Harrison and Ndegwa 2020). The most common use of the biogas is for cooking, which provides for improved indoor air quality since it is used as wood replacement (McCord et al. 2017). The replacement of wood with biogas reduces deforestation, which serves as a benefit for the environment. In addition, a further technology that has been adopted is the use of absorption coolers for refrigeration that are operated on biogas (Harrison and Ndegwa 2020).

Pasteurization

Pasteurization is the process by which manure is heated above a certain predetermined temperature for a short period of time. The pasteurization process is accomplished mostly with heated exchangers or by steam injection. Moreover, it eradicates zoonotic pathogens without spores, but the bacteria harboring spores are only decreased. The pasteurization method can be used as a pre-treatment method before the application of other known methods of manure treatment. However, pasteurization can be conducted

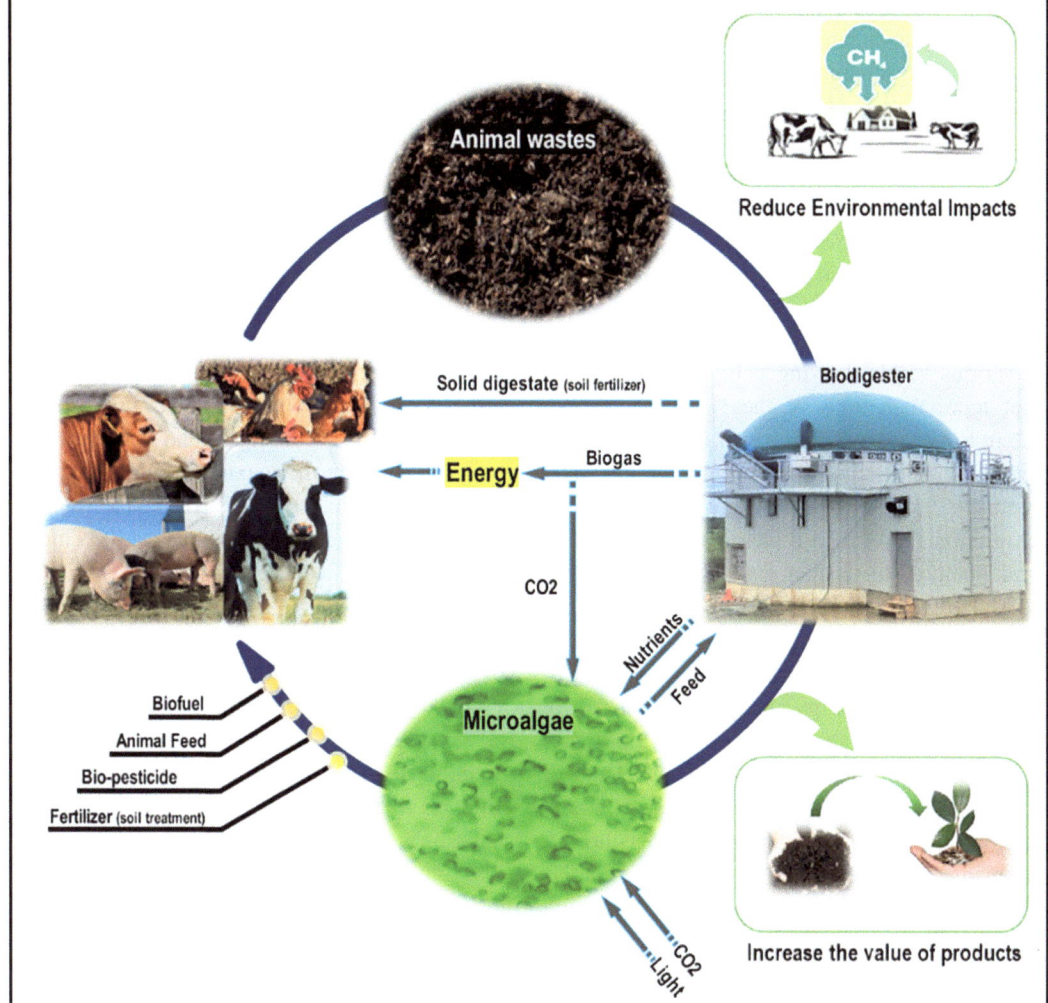

Fig. 21.3 Agricultural waste management toward a circular bioeconomy. Digestate (10% dilution) resulting from anaerobic digestion of chicken manure was used as nutrients for microalgal (strain *Chlorella vulgaris* CPCC 90) growth resulting in production of biofuel and bioproducts that can be used as fertilizer (for soil treatment), biopesticide and as animal feed. (From Rajagopal et al. 2021), licensed under CC-BY 4.0)

before or after anaerobic digestion in a batch or continuous mode. Mostly, the batch operation is usually recommended because the temperature and time can be controlled easily (Marañón et al. 2006). The pre-treatment of animal waste is done at 70 °C for 60 min or 30 min before mesophilic and thermophilic digestion, respectively (Colleran 2016). For example, Marañón et al. (Marañón et al. 2006) pre-pasteurized cattle manure for 2 h, followed by mesophilic anaerobic digestion in an up flow anaerobic sludge blanket and the result showed that some pathogenic bacteria (*Yersinia, Pseudomonas, Enterococcus,* and *Coliforms*) were removed without spores.

Liquid–Solid Separation

There are different methods used for liquid–solid separations and these include incline screens, rotary screens, screw presses, filter presses, rotary presses, and dissolved air flotation devices. Two liquid–solid separators were evaluated, a screw press separator (Eys, Daritech Inc., Lynden, WA), and rotating drum screen separator (DT-360, Daritech Inc., Lynden, WA) and it was reported that nutrient partitioning after liquid–solid separation indicate a range in solids separation of 13–25%, in N separation of 4.3–12.9%, and in P separation of 9.2–21.5% (Vanotti et al. 2020). Church and colleagues (2016) demonstrated that up to 60% of P could be recovered from dairy slurry when liquid–solid separation technologies are used (screw press followed by centrifugation). Interestingly, it was reported that when these technologies were combined with chemical treatment, up to 99% of manure P was recovered in various solids, leaving a liquid effluent that kept more than 90% of the N in the manure (Church et al. 2016; Church et al. 2018). A mobile version of the solid separation and chemical treatment system, MAnure PHosphorus EXtraction (MAPHEX), was also developed to treat manure slurries on small dairies (Church et al. 2016; Church et al. 2018).

Natural flocculants (e.g., chitosan) may also have important roles in waste management due to increased cost of energy and renewed interest in organic farming systems. Chitosan is a polymer found in certain fungi and in the exoskeleton of arthropods such as shrimp and crab shell waste. For instance, it was demonstrated that naturally-occurring flocculants like chitosan can be as effective as synthetic polymers for separating solids and nutrients from concentrated dairy manure effluents (Garcia et al. 2009). The author used various rates of chitosan to flocculate mixtures of dairy manure and the lagoon supernatant used to flush free stall alleys. The flocculated manure was dewatered using 1-mm and 0.25-mm screens. The results showed that separation by screening only (1-mm screen) was not effective (average efficiencies were 60% for total suspended solids (TSS), 22% for total Kjeldahl nitrogen (TKN), and 25% for total phosphorus (TP)), as compared to the mixture of chitosan before screening, which significantly enhanced efficiency of separation (97% for TSS, 79% for TKN, and 58% for TP).

Use of Polyacrylamide

Organic polymers such as polyacrylamide are mostly used to increase separation of suspended solids and carbon compounds from liquid swine and dairy manure. Polyacrylamides are characterized as high molecular weight, long-chained, water-soluble organic polymers. The long polymer molecules affect the stability of suspended, charged particles by binding to them and building bridges between several suspended particles. Polyacrylamides have mostly been used to increase the performance of several mechanical and physical manure separators including incline screens, rotary screens, screw presses, filter presses, rotary presses, and dissolved air flotation devices (Vanotti et al. 2020). The maximum rate of metal coagulants (aluminum sulfate and ferric chloride) is 200–4000 mg/L, and it is comparatively higher than the maximum rate of application for organic polymers in swine manure (150 mg/L) and dairy manure (300 mg/L) (Vanotti et al. 2020).

21.3.2 Technologies to Reduce Greenhouse Gas Emissions

Greenhouse gases are referred to as gases in the atmosphere that absorb and emit heat (infrared radiation). As the sun's energy (i.e., solar radiation) travels to earth, some of it is reflected at both the earth's surface and atmosphere, while most is absorbed by the earth's surface, thus warming it. The warmed earth's surface then emits infrared radiation back into the atmosphere. Some of the infrared radiation escapes back to space, but the greenhouse gases, in the atmosphere, absorb some and re-emit it in all directions, which warms the earth's surface and the lower atmosphere (Harrison and Ndegwa 2020). The common GHG in the earth's atmosphere are CO_2, CH_4, N_2O, hydrofluorocarbons (HFCs), water vapor (H_2O), and ozone (O_3) (Harrison and Ndegwa 2020). The amount of heat trapped in the earth's surface depends on the concentrations of GHG in the atmosphere and recent increases in the global temperature has been associated with an increase in the release of these gases into the atmosphere. Particularly, agriculture contributes approximately 9% of GHG emissions, and this is majorly from livestock, soils, and rice production (USEPA 2017). Whether or not the soil acts as a net source of CO_2 or a net sink for CO_2, can be influenced depending on soil management. When the level of soil organic matter is increased (process known as carbon sequestration), CO_2 can be decreased while increasing the soil carbon sink (Harrison and Ndegwa 2020). Soil organic matter levels can be increased by producing healthier crops and decrease soil tillage operations. It has been shown that healthy crops not only increase yield but they also decrease GHGs by trapping more carbon in their roots, some of which is converted to more stable soil organic matter. Reducing soil tillage operations increases soil organic matter levels by decreasing the amount of organic matter that is oxidized and released to the atmosphere as CO_2 (GM 2022).

The two major potential GHGs from dairy and swine facilities are CH_4 and N_2O. These two GHGs (CH_4 and N_2O) are extremely potent GHGs with global warming potentials of 25 and 298 times that of CO_2, respectively (Montes et al. 2013). The process of managing dairy and swine manures largely dictates the amount of CH_4 and N_2O emissions; hence, their respective contributions to GHG in the atmosphere. Nitrous oxide can be produced enterically in minute quantities; however, most N_2O is emitted from excreted manure. The manure or slurry accumulated in lagoons and retention ponds are often anaerobic and serve as the main source of N_2O produced from feedlots and dairies. Different methods of manure treatments and storage determines the amount of these GHG produced. The management of manure account for about 15% of the total GHG emissions from the agricultural sector in the United States (Harrison and Ndegwa 2020). The main methods for controlling GHG emissions from livestock operations are: (i) controlling decomposition of manure to reduce emissions of N_2O and CH_4; (ii) capturing CH_4 from manure decomposition and using it as renewable energy; (iii) nutritional changes that reduce GHG (e.g., increase forage quality, reduce age to harvest forage, use high edible oils and high energy grains, and rotational grazing); (iv) improve livestock fertility and survival rate; and (v) genetic selection for feed efficient and/or longer lived animals. The major impacts of anaerobic digestion in GHG emissions mitigations are from: (i) replaced fossil fuel consumption, and (ii) reduced emissions due to reduced fertilizer use and production (Kaparaju and Rintala 2011). Therefore, the technologies that maximize these processes, will ultimately lead to reductions in GHG emissions from livestock operations.

21.3.2.1 Composting

Composting manure is a process by which animal feces, urine, and bedding are stacked, turned, and managed in a way that enhances manure decomposition. The finished product has a soil-like texture and odor, and may be applied on crop land or as a pasture nutrient source. Composting is highly beneficial including destruction of harmful pathogens and weed seeds, and decrease in overall volume and odor emitted. The heat generated by composting manure can kill human and animal pathogens that are present in raw manure. This can help limit the risk of infection for animals and farmers that come in close contact with the manure; and also for members of the society that might encounter manure runoff that has escaped into a waterway or food source (Modderman 2020). Furthermore, composted manure has a less-offensive odor than raw manure. This makes composted manure more desirable to farmers with neighbors that might dislike the smell of raw manure. Composted manure helps prevent direct and indirect environmental nutrient pollution. The direct prevention of environmental pollution involves increasing the stability of nitrogen in the manure, thus reducing nitrogen emission to the environment (Eghball 2000). Indirectly, composted manure decreases the likelihood of pollution as a result of over-application on crop or pasture land. Manure can be over-applied to land when the actual nutrient content of the manure varies, and therefore, unknown; and raw manure has a high intensity of inherent differences both spatially and over time. However, composted manure's nutrients are more uniform and stable than raw manure (Lory 2008). Thus, the farmer is more informed on the actual nutrients being applied when using composted manure than raw manure, which leads to more-accurate calculations and rates of application;

resulting in decreased risk of over-application of manure (Rynk 1992).

There are biological, organic, and inorganic additives that alter the state of compost and aid in fast decomposition of manure, as well as increase the overall quality of the finished product. When compost is amended with these additives, components such as temperature, moisture content, pH, and nutrient availability may be altered. Inorganic additives are typically mineral and include chemicals, lime, clay, and industrial waste. Often, compost is amended with an industrial by-product, making it easily accessible and at low cost. Organic additives are mostly plant materials including residual straw from crops, grass clippings, bark, cornstalks, and biochar. These organic additives contain their own nutrients, so when using them, adequate care must be taken to prevent unbalancing of the overall calcium to nitrogen ratio of a compost pile (Doublet et al. 2011). At the same time, organic additives can be used to rectify an unbalanced C to N ratio. For instance, when the calcium/nitrogen of a manure pile is too low, carbon sources, such as straw or wood chips, can be added to the pile. On the other hand, if the calcium to calcium ratio is too high, nitrogen sources, such as grass clippings or hay, can be added (Rynk 1992). Biological additives are microorganisms that are added to a compost pile. These are usually commercial products that are marketed for aiding decomposition or removal of odor and gases. Using microbiological additives comes with lots of uncertainty since private companies are not obliged to disclose the active ingredients, and most of the times they describe their product with vague terms such as "over 60 strains of active bacteria" (Wakase et al. 2008). However, DNA assays have shown that most of the biological additives contain some bacterial genera like *Alcaligenes, Bacillus, Clostridium, Enterococcus*, and *Lactobacillus* (Sasaki et al. 2006).

21.3.2.2 Biochar and Biogas Production

This involves the thermochemical processing of animal manure (poultry litter, cattle manure, swine manure), and sewage sludge for the production of bioenergy and biochar. Animal manure consisting of high organic carbon and mineral nutrients is an available feedstock for bioenergy and biochar. Pyrolysis, gasification, hydrothermal liquefaction, and combustion are the thermochemical techniques used for converting animal manure into valuable bioenergy and biochar products (Guo et al. 2020). Dry manure solids range in energy density from 13 to 19 MJ/kg and can be combusted directly as an alternative to firewood. The major products produced from pyrolysis of manure are biochar and bio-oil; products from gasification are syngas and biochar; and products from hydrothermal liquefaction are biocrude oil and hydrochar. Biochar and hydrochar derived from manure carry significant nitrogen, phosphorus, and potassium nutrients, and both are promising soil amendments. Bio-oil, biocrude oil, and syngas derived from manure contain lots of impurities and this may pose challenges for subsequent upgrading processes (Guo et al. 2020). Thermochemical processing of manure may cause 100% of feed nitrogen loss, which is the main concern for such treatment. Pyrolysis at the temperature range from 400 to 450 °C is recommended for use in producing quality biochar from manure. Presently, thermochemical conversion of animal manure to bioenergy and biochar is technically feasible, but this process needs to be enhanced in order to be economically viable (Guo et al. 2020).

21.3.2.3 Mitigation of Odor and Gas Emissions

During the process of composting animal manure, unpleasant odors may be generated, which might not be palatable for neighbors. Ammonia and sulfur-containing compounds are the primary sources of foul odor in composting, and there are additives to mitigate these odors. Biochar added to compost can decrease ammonia emission and odor in manure containing high amount of nitrogen, such as poultry litter (Steiner et al. 2010). Organic additives that elevate aeration will catalyze microbial activity, thereby increasing ammonia loss and odor. It was found that biochar might potentially enhance ammonia production and odor because of increased microbial activity

(Waqas et al. 2018). Similarly, Jiang and colleagues (Jiang et al. 2015) demonstrated the same effect with cornstalk. Addition of inorganic additives to compost may produce better results since they typically are not bulking agents. For example, adding zeolite to compost has been shown to decrease both ammonia and sulfurous odors, with the concentration of added zeolite being directly proportional to the decrease in odor (Lefcourt and Meisinger 2001). Addition of sodium nitrate and sodium nitrite reduced sulfurous odors in swine manure composting significantly by lowering emissions of dimethyl sulfide and dimethyl disulfide (Zang et al. 2017). Furthermore, addition of magnesium hydroxide and phosphoric acid, along with other magnesium and phosphate salts to compost manure, have also been shown to reduce loss of ammonia and odor (Zhang et al. 2017).

21.3.2.4 Pelletizing Manure

Pelletization or granulation of manure provides a greater opportunity to fight against challenges faced in both on- and off-farm manure utilization by creating a product that is easier to handle, transport and apply to land or soil. Pelletizing animal manure removes its moisture, decreases its weight, volume, and odor, thereby creating a product that is more preferable than raw manure because of its reduced cost of storage, easy transportation, handling, and application. Blair et al. (Blair et al. 2014) compared the effects of pelletized poultry litter with inorganic Osmocote fertilizer on the availability of phosphorus nutrient and activities of enzyme over a period of growing season and reported that at high soil phosphorus conditions, pelletized poultry litter had higher activity of alkaline phosphatase and released more phosphate than the inorganic Osmocote fertilizer. Application of pelletized poultry compost causes greater availability of liable phosphorus in soil than the unpelletized form based on a 56-days lab incubation study (Takahashi et al. 2016). Poultry litter is the most common animal manure to be pelletized; however, cattle, swine, and horse manure have also been pelletized with the addition of bulking agents or other supplements.

21.3.2.5 Technologies to Recover Phosphorus

Use of Iron and Aluminum Phosphates

Metal salt is the main chemical used to precipitate phosphorus so as to remove phosphorus from municipal wastewater. The process has been confirmed for the removal of phosphorus from manure making use of different doses of metal salts. Other chemicals that have been considered for treating manure are aluminum sulfate or alum (Al_2SO_4), $AlCl_3$, $Fe_2(SO_4)_3$, and $FeCl_3$, which can be used singly or in combination with polymers; and can be applied both in wet and dry manure systems. The most common types of chemicals used in recovering phosphorus are alum due to its lower cost, and $FeCl_3$ because it is effective over a wide range of pH (4.0-12) (Chastain 2013). These two metal salts (alum and ferric chloride) can reduce phosphorus solubility but also destabilize colloidal particles through coagulation chemical process. Different methods can be used to apply these metal salts to precipitate phosphate from manure; these include injection, mixing, and separation by sedimentation (Vanotti et al. 2020).

Use of Calcium Phosphates

Lime is used to significantly remove carbon and ammonium interferences before precipitating phosphate as calcium phosphate. This results in distinct chemical equilibrium between phosphorus and calcium ions (Vanotti et al. 2020). The presence of bicarbonate in wastewater interferes with the production of high-grade phosphates. In wastewater containing high NH_4^+ concentration, huge amounts of hydrated lime are needed to increase the pH to required values because the NH_4^+ reaction could potentially neutralize the hydroxyl ions (Vanotti et al. 2020).

Other Phosphorus Recovery Methods

Other methods used to remove the inhibitory effect of inorganic carbon and to increase the dissolution of P conditions include: (i) chemical acidification, which is the process of adding mineral or organic acids to acidify the manure before the P precipitation step with lime; (ii) biological acidification, which involves precipitating the P immediately after the acidification phase of anaerobic digestion during a multiphase anaerobic process of manure digestion (Barack 2013); (iii) physical acidification, which entails the addition of P precipitating compounds (calcium or magnesium) to the wastewater after the carbonate alkalinity and ammonia have been significantly decreased using gas-permeable membranes (Vanotti et al. 2018). It was reported that, when magnesium is added to wastewater after the removal of alkalinity and ammonia, the phosphates generated are very high grade (46% P_2O_5, >98% available), which is similar to the composition of the biomineral newberyite ($MgHPO_4 \cdot 3H_2O$) found in guano deposits (Vanotti et al. 2017).

21.3.2.6 Technologies to Recover Nitrogen

Use of Air Scrubbers in Barns

One of the methods for decreasing or limiting NH_3 emissions from livestock production is the use of chemical and/or biological air scrubbers and biofilters, in which exhausted air from the animal houses is allowed to pass through a wet packed bed so that NH_3 and other water-soluble components can be removed (Van der Heyden et al. 2015). In a chemical air scrubber, an acid, e.g., sulfuric acid is added to the washing water to maintain pH < 4, shifting the equilibrium NH_3/NH_4^+ toward NH_4^+ as the dissolved NH_3 is captured by the acid, thus forming an ammonium salt (Vanotti et al. 2020). The suitability of acid salts, such as aluminum sulfate (alum), sodium bisulfate, potassium bisulfate, $FeCl_3$, and ferric sulfate have been confirmed, as well as strong acids (hydrochloric, phosphoric, and sulfuric) for capturing NH_3 (Moore Jr et al. 2018). In a biological air scrubber, NH_3 captured in the washing water is oxidized by bacteria using nitrification process according to nitratation by ammonium oxidizing bacteria and nitratation by nitrite oxidizing bacteria. In a biotrickling filter, these bacteria are immobilized in a biofilm on the packing material, while in a bioscrubber, bacteria are contained in a separated bioreactor where nitrification is performed. In the case of biofilters, they contain a humid filter bed of organic material, but not completely wet to support bacterial growth. Full-scale spray scrubbers have been produced for the recovery of ammonia at poultry facilities having a removal efficiency of 71–81% (Hadlocon et al. 2015), and also for deep-pit swine finishing facilities with ammonia removal efficiency of 88% (Hadlocon et al. 2014). It was found that ammonia removal efficiencies ranged from 91% to 99% when acid scrubbers were used, and from 35% to >90% when biotrickling biofilters were used (Melse and Ogink 2005).

Ammonia stripping by air or steam is a method that can efficiently recover ammonia from liquid manure. Ammonia striping by air or steam are similar gas-liquid mass transfer processes. Steam stripping is mostly a distillation process that uses higher temperatures than air stripping (Vanotti et al. 2020). The air stripping process involves the transfer of ammonia from the waste steam into the air, then absorbed from the air into a strong acid solution producing an ammonium salt. Nitrogen in liquid manure is mostly present as ammonium ions (NH_4^+) and aqueous ammonia (NH_3), which are highly volatile. The equilibrium between both forms of N strongly depends on pH and temperature. Ammonium ions are more dominant when pH is less than 7, regardless of temperature; while NH_3 increases as the pH increases, thus shifting the equilibrium. Similarly, the concentration of NH_4^+ reduces when the temperature increases, thereby favoring the production of NH_3. Therefore, high pH and temperature will favor the process of ammonia stripping. The quantity of NH_3 that can be gotten from liquid manure, or absorbed in the acidic solution depends on two equilibria: NH_3 gas/liquid equilibrium and NH_3 dissociation equilibrium in the liquid (Vanotti et al. 2020). It is stipulated that the efficiency of air stripping majorly depends on the following: pH, temperature, ratio of air to liquid volume, and characteristics of liquid. This process has been produced at industrial scale, involving the utilization of stripping towers, compressors, or pumps that are used to introduce air or steam into the liquid phase. Condensation or adsorption equipment are needed to recover

NH_3 from the gas phase when air stripping is used. For steam stripping, condensation or absorption equipment is used to recover NH_3, without requiring any further post-treatment of exhaust gases (Zeng et al. 2006).

Use of Gas-Permeable Membranes

The gas-permeable membrane process involves the transportation of gaseous NH_3 via a microporous hydrophobic membrane followed by NH_3 capture and concentration in an acidic stripping solution located on the other side of the membrane. The process can be used for the removal and recovery of nitrogen from liquid manures in storage tanks (Karunanithi et al. 2015), and from the air emitted from poultry and animal barns (Szögi et al. 2014).

Use of Ion Exchange–Zeolites

Zeolites are crystalline, hydrated aluminosilicates of alkali and alkaline earth cations, with infinite, 3D structures. They consist of varied properties, such as adsorption, cation exchange, dehydration–rehydration, and catalysis properties, contributing to a high variety of applications (Vanotti et al. 2020). For example, zeolite has been used in water and wastewater treatment, and also in animal waste treatment because of its high interest and selectivity for NH_4^+ ions. Addition of 6.25% zeolite to dairy slurry led to a 50% decrease in NH_4^+ volatility (Lefcourt and Meisinger 2001). Likewise, the addition of zeolite to swine manure before applying it to soil decreased NH_4^+ volatility between 65% and 71% (Portejoie et al. 2003). Similarly, when a layer of zeolite was added to the surface of composting poultry litter, NH_4^+ emissions was reduced by 44% (Kithome et al. 1999), and in a laboratory experiment, it was reported that application of a 5% (w/w) rate of zeolite to stored poultry manure caused 81% decrease in NH_4^+ emission (Li et al. 2006). Other authors have also confirmed that the addition of zeolite to cattle manure (2% rates) was able to counteract the inhibitory effect of NH_3 during anaerobic digestion (Borja et al. 1996), and when zeolite was used as air scrubber packing material and as a filtration agent in deep-bedded cattle housing (Milán et al. 1999).

Manure Pathogens Treatment

Indigenous microorganisms (a group of innate microbial consortium that live in the soil and the surfaces of all living things, having the capacity to biodegrade, bioleach, biocompost, fix nitrogen, improve soil fertility and produce hormones for plant growth) can prevent the survival of manure-borne pathogens because of predation, substrate competition, and antagonism (Chen et al. 2020). Therefore, enforcing a waiting period for raw manure application before harvesting crops (90 and 120 days for crops not in contact or in contact with soils, respectively) may be helpful in reducing the risk of exposure to manure-borne pathogens. Manure-borne pathogens survival in soils can last for few days to several months, indicating that a waiting period of 120 days may not be long enough to completely eliminate manure pathogens contamination in fresh produce. Thus, to further reduce the levels of pathogens in manure, pre-treatment of raw animal manure is required before manure is land-applied (Chen et al. 2020). The US Food and Drug Administration states that management of manure is acceptable before manure is land-applied if the amount of indicator pathogens in manure are decreased below the recommended levels (Chen et al. 2020). The threshold levels of manure-borne pathogens differ in many countries. For example, in France and the United Kingdom, the threshold level of *Salmonella* in manure is absence (not present) per 25 g fresh mass (wet weight basis) (Manyi-Loh et al. 2016). Whereas the threshold level for *Escherichia coli* is 100 and <1000 per g fresh mass (wet weight basis) in France and United Kingdom, respectively. Common methods for pre-treatment of animal manure include chemical, physical, and biological methods or the combination of any of the methods, which have all been found to be effective for reducing the levels of pathogens (Chen et al. 2020).

Exposure of Manure to Ultraviolet Irradiation

The sun is the main source of ultraviolet radiation. Ultraviolet radiation is a promising technology involved in the disinfection of cattle manure for the control of pathogens. Ultraviolet radiation could be characterized into UV C, UV B, and UV A depending on the wavelengths and energy intensity. However, because of the high energy level of UV C, it is absorbed by the earth stratosphere ozone layer (Manyi-Loh et al. 2016). UV A and UV B (280–315 nm) radiations are used in the process of disinfection, but Willey and colleagues (2011) observed that UV radiation around 265 nm is somewhat lethal. Oni and colleagues (2013) reported that a 5 log decrease in *Salmonella* cells was noted when exposed to UV A (365 nm) in a control medium compared to a 1.5 log reduction in a manure dust matrix. The main significance of UV irradiation is that it is capable of disinfecting the manure without using any chemicals, consequently, no disinfection by-products are produced which could alter the physicochemical and nutritional properties of the manure after treatment (Spiehs and Goyal 2007). Interestingly, only the microbial cell that has absorbed the energy at the set wavelength is destroyed.

21.4 Growth-Enhancing Technologies in Meat Production, Processing, and Preservation

One of the major products of any livestock system is meat, which comes from an increase in size or multiplication of muscle fibers. Farm animal production has long had the objective of producing high-quality animal protein for human consumption and raw materials for the food industry, providing employment for millions on the meat supply value chain and sustaining livelihoods. In general, meat production in Africa has largely relied on extensive, subsistent, and unconventional traditional systems or tools using handed-down practices with little or no incorporation of state-of-the-art technologies. Current realities in terms of increasing human population, average per capita income, and urbanization changes have led to projected increased demand for livestock and livestock products, including meat, warranting the use of modern technologies to increase productivity. In this section, we review some of the technological innovations that can be targeted to achieve increased meat production in Africa.

Stakeholders led by national governments and non-governmental organizations have collaborated to find alternative means of increasing the growth performance of farmed animals. In the process, appropriate and relevant technologies have been encouraged worldwide, including crossbreeding, improved production systems and management practices, inclusion of approved growth promoters (probiotics and prebiotics), housing and keeping of improved breeds (disease resistant breeds), among others to increase growth rates and improve overall efficiency and product quality. The most extensively utilized growth promoters are feed additives, anabolic implants (both estrogenic and androgenic), bovine somatotropin, repartitioning agents (beta-agonists), and probiotics. The use of non-nutrient feed additives such as antibiotics and exogenous enzymes that improve animal growth (growth promoters) is fast gaining ground in Africa.

21.4.1 Use of Feed Additives: Probiotics and Prebiotics

According to the World Gastroenterology Organization Global Guidelines on Probiotics and Prebiotics of February 2017 (https://www.worldgastroenterology.org/guidelines/probiotics-and-prebiotics), probiotics are live microorganisms that confer health benefits on the host when administered adequately. Probiotics consist of various types of microorganisms that improve gut microflora and affect both the local and systemic immune systems by secreting beneficial enzymes, organic acids, vitamins, and nontoxic antibacterial substances upon ingestion (Popova 2017). Probiotics affect the host upon ingestion positively by improving the balance of the intestinal microflora. Not only that, probiotics stabilize rumen pH, reduce gastrointestinal tract-invading pathogens, increase volatile fatty acids production, and stimulate lactic acid utilizing protozoa, resulting in a highly efficient rumen function. Probiotics release anti-inflammatory factors, downregulating pro-inflammatory cytokines and thwarting intestinal inflammation (Alfonsetti et al. 2022). Particularly, probiotics can improve phagocytosis and enhance the secretion of antibodies, generating increased immunological defenses against pathogens (Maldonado Galdeano et al. 2019). Probiotics interact with bile acids in the gut lumen, modifying bile acid metabolism and inducing cholesterol absorption (Plaza-Diaz et al. 2019). Probiotics positively affect digestive processes such as cellulolysis and the synthesis of microbial proteins (Uyeno et al. 2015). It serves as prophylaxes and therapeutic purposes in clinical and veterinary practices (Srinivas et al. 2017). They have shown the ability to secrete hydrolytic enzymes against bacterial toxins to inactivate toxin receptors that can cause toxin-mediated infections in livestock animals (Hossain et al. 2017).

Probiotics are widely used in animals to improve growth rate, increase milk yield, meat, and egg production. Yeast culture supplements, containing *Saccharomyces cerevisiae*, are known to be a rich source of enzymes, vitamins, other nutrients, and important co-factors, and have been reported to produce a variety of beneficial production responses (Markowiak and Śliżewska 2018). Beneficial responses include growth rate, feed intake, feed efficiency, milk composition, egg production, and reproduction in ruminants, poultry, pigs, and horses. The most commonly used probiotics are *Lactobacillus acidophilus, Lactobacillus lactis, Lactobacillus plantarum, Lactobacillus bulgaricus, Lactobacillus casei, Lactobacillus helveticus, Lactobacillus salivarius, Bifido bacterium* spp., *Bacillus* spp., *Enterococcus faecium, Enterococcus faecalis, Streptococcus thermophilus, Escherichia coli* bacteria, and other probiotic fungi such as *Saccharomyces cerevisiae* and *Saccharomyces boulardii* (Al-Shawi et al. 2020).

Non-digestible but fermentable food ingredients that beneficially affect the host by selectively stimulating the growth and/or activity of one or a limited number of bacteria in the colon are known as prebiotics (Davani-Davari et al. 2019). Prebiotics are dietary substances consisting mostly of non-starch polysaccharides and oligosaccharides. Commonly used prebiotics are oligosaccharides, i.e., Mannan-oligosaccharides, fruto-oligosaccharides, galacto-oligosaccharides, chito-oligosaccharide, isomalto-oligosaccharides, pectic-oligosaccharides, xylo-oligosaccharides, lactulose, inulin, and breast milk oligosaccharides. Probiotics and prebiotics both act by modulating the balance of microbiota within the gastrointestinal tract (GIT). While probiotics add beneficial gut bacteria, prebiotics enable existing microbes to multiply. Butyric acid stimulates the proliferation and differentiation of epithelial cells in the GIT to increase surface area for nutrient absorption (Pierce et al. 2018). Probiotics are live microorganisms that can replace antibiotics in animal husbandry as a safe and viable alternative to antibiotics, which can confer health benefits on the host when administered in adequate dosage (Alayande et al. 2020).

Probiotics stabilize rumen pH, reduce gastrointestinal tract-invading pathogens, increase volatile fatty acids production, and stimulate lactic acid utilizing protozoa, resulting in a highly efficient rumen function. Probiotics release anti-inflammatory factors, downregulating pro-inflammatory cytokines and thwarting intestinal inflammation (Alfonsetti et al. 2022). Particularly, probiotics can improve phagocytosis and enhance the secretion of antibodies, generating increased immunological defenses against pathogens (Maldonado Galdeano et al. 2019). Probiotics interact with bile acids in the gut lumen, modifying bile acid metabolism and inducing cholesterol absorption (Plaza-Diaz et al. 2019). Probiotics positively affect digestive processes such as cellulolysis and the synthesis of microbial proteins (Uyeno et al. 2015). It serves as prophylaxes and therapeutic purposes in clinical and veterinary practices (Srinivas et al. 2017). They have shown the ability to secrete hydrolytic enzymes against bacterial toxins to inactivate toxin receptors that can cause toxin-mediated infections in livestock animals (Hossain et al. 2017).

The inclusion of probiotics and prebiotics has played a pivotal role in animal health thereby contributing tremendously to the meat industry. Directly or indirectly, this technology has had some positive impacts on some key production elements such as growth performance and feed efficiency in farmed animals. Furthermore, the increase in meat demand has resulted in increased use of antibiotics as growth promoters and as prophylaxis and metaphylaxis in food animals, a situation worsened by inadequate biosecurity and good animal husbandry practices (Abdalla et al. 2021). Over the years, concerns have been raised on the risk of developing cross-resistance and multiple antibiotic resistance in pathogenic bacteria in both human and livestock, increase in foodborne allergies, negative impacts on the environment and their potential effects on human health owing to persistent use of antibiotics and chemotherapeutics in prophylactic doses used in animals (Popova 2017; Al-Shawi et al. 2020; Rai et al. 2013). An alternative to these problems is the use of growth promoters like probiotics and prebiotics which have positive effects on animal health and growth performance.

21.4.2 Impact of Growth Promoting Technologies on Meat Production

Growth performance is a combined result of several factors including but not limited to the genetic potential for growth, optimal environmental conditions, nutritionally complete feeds and good farm management practices. There have been several reports on improved growth performance, carcass yield, and meat quality owing to the use and application of growth promoting technologies in meat production, particularly with the use of probiotics and prebiotics (Valenzuela-Grijalva et al. 2017). Over the years, the effects of dietary supplementation of probiotics on growth performance of poultry have been extensively investigated. Most of the studies found that probiotics displayed great efficacy in promoting animal growth. For instance, Jin and colleagues (Jin et al. 1998) reported that dietary inclusion of probiotics *Lactobacillus* increased body weights and feed to gain ratio when compared to control broilers. In addition, intestinal immunity was increased in chickens fed with diets supplemented with yeast product and *Lactobacillus*-based probiotic culture (Gao et al. 2008; Higgins et al. 2008). Another study found that *Lactobacillus* inclusion in broiler nutrition resulted in higher broiler productivity index which was measured based on daily weight gain, feed efficiency, and mortality (Timmerman et al. 2006). Again, probiotics supplementation improved feed intake, feed efficiency, and carcass yield of broilers (Denli et al. 2003). Therefore, feed additive technology, in particular, the use of probiotics in animal feed should be encouraged by national governments and other key stakeholders to boost meat production in Africa.

Meat quality comprises of several factors including pH, tenderness, odor, tastes, flavor, characteristics, and intramuscular lipid content. The use of probiotic (*Lactobacillus fermentum*) in the water of broilers significantly increased the redness in breast, while there was no effect on the yellowness and lightness in breast and thighs (Popova 2017). Also, *Bacillus licheniformis* treatment increased significantly the protein content and respective contents of essential and flavor amino acids, while on the other hand, the fat content was decreased (Popova 2017). The use of probiotics in fresh and fermented meat products has been shown to reduce pathogenic and spoilage microorganisms and improve sensory characteristics (Al-Shawi et al. 2020). Additionally, probiotics improves feed efficiency, weight gain, and immune response with its overall effectiveness dictated by factors such as optimal selection of microbial strains, the use of a suitable dose, and the species and age of the host. Use of probiotics has led to improvement in product quality and safety, extending shelf-life, imparting unique sensory qualities, and providing health benefits. The use of probiotics has been shown to improve water holding capacity and tenderness in meat. Furthermore, the application of probiotics in animal nutrition aims to promote production performance and prevent diseases via the maintenance of a healthy gastrointestinal environment and improvement of intestinal function. The mechanism of probiotic effect on animal health is primarily based on the competition between beneficial bacteria and pathogens, and the replacement of pathogens by probiotic bacteria.

Prebiotics have been shown to prevent the colonization of the digestive system with pathogens by creating unfavorable condition like altering the pH of intestinal content. Likewise, *Bifidobacterium* and *Lactobacillus* found in the digestive system has Manase enzyme used in breaking down polysaccharides. Plant extracts also known as phytobiotics are another potential alternatives for antibiotics because of their antimicrobial, anti-inflammatory, antioxidant, and antiparasitic activities, and they have been used successfully in poultry production for many years (Andrew Selaledi et al. 2020). In African countries, plant extracts from aromatic spices (cinnamon, clove, etc.), pungent spices (pepper, garlic, and ginger), and herb spices (rosemary, thyme, mint, etc.) have received increased attention over antibiotics because they are cheaper and naturally available, and they have been shown to improve poultry production and health status. Current research on alternatives to antibiotic use in food animals is slow, especially in the African continent. The available alternatives to antibiotics have a potential role in lowering dependency on antimicrobial substances. Failure to implement the WHO antimicrobial usage recommendations could deteriorate the situation and intensify the burden of diseases or increase the mortality rate in Africa and other continents (Andrew Selaledi et al. 2020).

Antibiotics function by eliminating all bacteria and depending on the spectrum of its activity, both beneficial and pathogenic microorganisms, and may lead to bacterial resistance (Michalak et al. 2021). Over the years the growth and health of food-producing animals have been enhanced by the use of antibiotics. These have helped reduce on-farm mortalities, lower incidences of diseases and more importantly improve productivity. Generally, the utilization of antibiotics in feed has been re-evaluated since bacterial pathogens have established and shared a variety of antibiotic resistance mechanisms that can easily be spread within microbial communities. Apart from developing antibiotic resistance, people can also develop allergic reaction or liver damage on consuming antibiotic residues in animal products (Andrew Selaledi et al. 2020). Therefore, antibiotics that are important for treating humans must be prohibited from being used in the feed as growth-promotants. Africa produces fewer antibiotics as compared to other continents. Nonetheless, many antibiotics can be bought over the counter in many African countries and this practice could play a key role in worsening antibiotic resistance. Most farmers in Africa purchase antimicrobial agents without consulting animal health professionals because they are not accessible/available, or farmers do not have the means to reach them. Antimicrobial resistance is prevalent in Africa and pose a threat to food safety and security. The one health approach concept simply emphasizes that the health of the environment, animals and people are connected and could be used to manage antimicrobial-resistant and food safety concerns (Andrew Selaledi et al. 2020). Alternatives to antibiotics which include probiotics, prebiotics, enzymes, and organic acids, among others are found to have the ability to replace antibiotics (Andrew Selaledi et al. 2020; Alayande et al. 2020).

21.4.3 Meat Processing and Preservation Technologies

The purpose of meat processing and preservation is to increase meat shelf life, ensure that the nutrient in meat is preserved and readily available for human consumption (Heinz and Hautzinger 2007). Meat processing involves a broad range of methods, including chemical and physical treatment. Thus, meat processing technologies include technical processes like cutting, chopping, or comminuting (i.e., reduction in size), mixing or tumbling, stuffing or filling of semi-fabricated mixtures of meat into cans, synthetic films, casings etc., and treatment with heat. Meat processing also includes biochemical or chemical processes, which are mostly used in conjunction with technical processes, such as salting, curing, the addition of spices and additives, fermentation, drying, and smoking (Heinz and Hautzinger 2007; Balny 2002).

21.4.3.1 Technical Processes Involved in Meat Processing

In meat processing, especially red meat, most of the steps involved can be mechanized. Different methods of meat processing with various equipment are summarized below. Recently, a robotic workbench for poultry meat processing was described (Ahlin 2022). This system can accomplish disparate tasks at multiple stages of meat processing. Importantly, it allows the integration of flexible manufacturing into assembly line procedure, thereby promoting greater agility, scalable manufacturing, and hazard mitigation (Ahlin 2022).

Meat cutting/chopping/comminuting (reduction in size) are achieved by different methods: (1) mincing or grinding of lean and fatty meat, (2) chopping of meat in bowl cutter, (3) use of emulsifying machines for chopping meat, and (4) cutting of frozen meat and fatty tissues. Meat grinders are used for reducing meat sizes and interestingly, some meat grinders are also specially designed for cutting frozen meat while some grinding machines can automatically separate bones and tendons from soft tissues. Meat grinders are either manually or automatically operated and they work to forcefully push meat or meat trimmings through a barrel that is mounted horizontally (Sastry 2008).

Bowl cutters are the most commonly used equipment for chopping lean meat and fat tissues into small sizes (Zheng and Sun 2006). They are also used for mixing fresh or frozen lean meat and fatty meat with water (ice water), extenders, and functional ingredients (additives, curing agents, and salt) (Rivas-Cañedo et al. 2009). Chopping of meat by emulsifying machines is done by first premixing meat with functional ingredients, seasoning, and other raw materials and second, the premixed meat are pre-cut using grinders or bowl cutter. Afterward, the pre-cut meat are passed through emulsifiers, which is also known as colloid mills so as to attain the finely chopped or emulsified meat mix (Gola et al. 2000). Frozen meat tissues can be cut into slices, cubes, or flakes by frozen meat cutters. The frozen meat can be chopped directly in bowl cutters without thawing to prevent drip loss, meat discoloration, and bacteria growth (Gentry and Roberts 2005). Meanwhile, meat mixing involves the blending of coarse and finely chopped meat and spices together to achieve a desired meat texture and color. Another process involves stuffing or filling of meat into containers or casings followed by chopping to achieve finely comminuted fat or lean meat pieces (Knorr et al. 2004).

21.4.3.2 Chemical or Biochemical Processes Involved in Meat Processing

These involve the use of functional ingredients, smoking, fermentation, and drying (Heinz and Hautzinger 2007).

Salt is a very important ingredient in meat preservation. For example, salt (1.5–3%) is added to meat to increase the water holding capacity up to 5% (Beaufort et al. 2009). Spices used in meat preservation include pepper, paprika, nutmeg, mace, cloves, ginger, cinnamon, cardamom, chili, coriander, cumin, and pimento, which are added to meat to achieve a desired flavor. Pepper is the most used spice during processing of meat into sausage. Most spices used during meat processing are always in the ground form with particles size ranging from 0.1 to 1 mm (Diez et al. 2009). Smoking of meat using natural or liquid smoke is done to achieve meat and meat products of desired color and flavor. Smoking is a good source of antioxidant (phenols) and antimicrobial (phenols + acids) for meat (Bertrand 2005). Likewise, fermentation and drying are other non-technical methods mostly used during meat processing and for meat preservation.

21.4.3.3 Technologies for Meat Preservation

Meat preservation is another method of meat processing that is mostly used for the control or elimination of microbial agents that causes spoilage. The principles of meat preservation include (i) using aseptic techniques for the prevention or delay of microbial growth in meat, removing microorganisms from meat, preventing microorganisms growth and activity by storing meat in the refrigerator, freezer, drying of meat, use of chemicals, storing of meat in an anaerobic condition, and killing of microorganisms by heat or irradiation (Andersen et al. 2014); (ii) destruction or inactivation of meat enzymes that cause meat decomposition using blanching methods or chemicals (e.g., antioxidant to prevent or delay oxidation) (Tassou et al. 2010); (iii) using chemicals or storing of meat in airtight containers to prevent insect or animal damage (Gupta and Gould 1997). The preservation technologies, such as freezing, heating, dehydration, chemical, cooking, fermentation, and hurdle technology of meat, are further discussed below.

Meat storage at low freezing temperatures (0 °F and below) can be used to stop or inhibit enzyme activity that causes meat deterioration or spoilage. Freezing meat also helps to inhibit the growth of microbes, yeasts, and mold that causes meat spoilage. It has been shown that freezing extends the shelf life of beef to 12 months, pork to 6 months, poultry from 3 to 6 months, and lamb from 6 to 9 months. There might be noticeable changes in red meat during freezing and these include lipid oxidation, recrystallization, meat dehydration (loss of water from meat to the environment), and enzymatic reaction (proteolytic and lipolytic enzymes) (De Mey et al. 2017).

Heat treatment is applied to kill microorganisms responsible for meat spoilage. Heat treatment methods include cooking and canning. The process of cooking uses very high temperature to heat meat and meat products in order to eradicate spoilage of microorganisms. Two main cooking methods are pasteurization and sterilization. Pasteurization involves the use of very high temperatures from 66 °C to 77 °C (150–170 °F) to kill most of the microorganisms found in meat (Sikes et al. 2010) while sterilization is the process of cooking meat under pressure to 121 °C (250 °F). In sterilization method, all the microorganisms are killed in meat and meat products, thus prolonging meat shelf life. The canning processes involves putting meat in cans or jars and heating at very high temperature to destroy all the microorganisms causing meat spoilage. The heating process ensures that air is completely removed from the container and vacuum sealed to prevent air from penetrating back into the container. This is because air is favorable for microbial growth. Two methods used for meat canning are boiling water bath method and pressure canner method. In the former, containers containing red meat are heated and totally covered with boiling water at 100 °C (212 °F) and cooked for some time while in the latter method, the containers containing meat are placed in about 2–3 inches of water in a specialized pressure cooker. The pressure cooker is then heated to a very high temperature of 116 °C (240 °F) and above (Hugas et al. 2002).

Chemical treatment is used to inhibit microbial growth, enhance meat flavor, color, and shelf-life. The chemicals used for meat treatment are salt, sodium nitrite, and sodium lactate. Salt has very good water holding capacity because it binds with water molecules, therefore acting as a dehydrating agent and at the same time preventing the growth of microbes. Meat can be dipped into water containing salt, or by hand rubbing of salt on meat or by injection into meat (Ahmed et al. 2003). Some chemical preservatives used in meat preservation are listed in Table 21.2. Microorganisms find it difficult to survive at low pH; thus, the use of acids like vinegar or citric acid helps to lower the pH of meat making it uncomfortable for microbial growth on meat.

The fermentation process uses bacteria or yeast to convert carbohydrate to organic acids in an anaerobic condition (i.e., very low or no oxygen). Fermentation inhibits the growth of bacteria in containers containing red meat. The container is void of air and or filled with nitrogen or CO_2 (Jeong et al. 2010).

Sugar is also used to preserve meat. When sugar is added to meat up to 65% and above, it can prevent the growth of microorganisms (Pawar et al. 2000).

Meat preservation by irradiation is a process of heating up meat by radiant energy. Irradiation, such as Gamma rays, electromagnetic X-rays, and ultraviolet radiations are commonly utilized in preserving food by killing microorganisms.

High-pressure processing or high hydrostatic pressure processing or ultra-high-pressure

Table 21.2 Some chemical preservatives for meat and target organisms

Chemical preservative	Target organism	Effect
Sulfites	Yeast and bacteria	Antioxidant
Propionic acid	Mold	Antimicrobial
Sorbic acid	Mold	Antimicrobial
Benzoic acid	Yeast and mold	Antimicrobial
Sodium nitrate	Bacteria	Antimicrobial

Adapted from Ahmed et al. (2003)

processing uses increased pressures of up to 600 MPa either with the use of external heat or without the use of external heat. High pressure is used purposely for inactivating microorganisms in meat and meat products (Carlez et al. 1995).

Microwave heating makes use of energy (300–300,000 MHz) that produces heat in dielectric materials, such as foods, through dipole rotation and ionic polarization. Microwave uses rapid volumetric heating which helps to decrease the amount of time needed to attain a temperature of interest, thereby decreasing the time needed for cumulative thermal treatment and at the same time preserving meat nutrients (Ramaswamy and Tang 2008).

Pulsed electric field processing is the use of a very high voltage electric field (20–70 kV/cm) for a few microseconds on meat. The processing parameters used in governing the microbial safety of the processed meat, e.g., red meat, include electric field strength, treatment temperature, flow rate or treatment time, pulse shape, pulse width, frequency, and pulse polarity (Yousef and Zhang 2006).

High power ultrasound processing or sonication is a technology used to inactivate microorganisms that promote meat spoilage. Ultrasound generates energy by sound wave if the frequency is more than 16 kHz. The wave penetrates via a medium causing compression and reduction in medium particles to create micro-bubbles. Acoustic cavitation is what ultrasound uses to exhibit their antimicrobial properties (Piyasena et al. 2003).

Hurdle technology is the process whereby known and novel techniques are combined together to improve stability of microbes, sensory quality, nutritional quality, and economic properties (Leistner 2000). Microwave and radio-frequency heating, treatment with pulse electric field, processing with high pressure, irradiation and oscillating magnetic fields, and ultrasonic applications are now mostly used to improve, replace, or use in conjunction with conventional meat processing techniques (Heinz and Hautzinger 2007; Brewer 2009).

21.4.3.4 Meat Packaging and Labeling

Packaging of meat involves the use of protective materials, such as plastic films or foils to wrap meat products. Packaging helps to conserve meat color, flavor, odor, and texture as well as prevent meat contamination with biological, chemical, or physical contaminants, maintain oxygen and water level in meat, prevent meat shrinkage, and has labels on which information like ingredients and nutrition facts are written (Day 1990).

Labeling is a way of displaying pertinent information about the meat product. The nutritional information on labels, for example, informs consumers about the nutritional contents of the meat product thereby helping them to make healthy food choices (Guthrie et al. 1995). Meat product labeling schemes are in place in most countries to help consumers make good food choices, and to prevent fake or misleading nutrition labels. Labeling encourages meat producers and processors to apply adequate principles of nutrition when formulating meat products.

21.4.3.5 Meat Preservation in Africa: Ethiopia as a Case Study

Ethiopia is the largest producer of livestock animals in Africa and the tenth largest world producer. Ethiopia has 59.5 million heads of cattle, 30.70 million heads of sheep, 30.20 million heads of goats, 56.53 million heads of poultry, and 1.21 million heads of camel (CSA 2017). Ethiopians make use of conventional methods of meat processing such as drying to make "quanta," application of salt on the meat surface, and use of honey and spices. In pastoral and nomadic systems, the main meat preservation methods are sun-drying and deep frying. Ethiopia however contributes only 0.2% of total world red meat production despite being the tenth largest world producer of livestock (mostly sheep and goat). This is because the meat processing and preservation techniques are not optimal or do not support longer shelf life. As a result, they are ranked the 55th largest meat producing country in the world. Ethiopia like most African countries lacks novel techniques of meat processing and preservation thus reducing their capacity to meet international standards and the demand for meat and meat products (Filip 2006). However, in recent times, the Ethiopian government is using agriculture development led industrialization to improve their economic sector and the meat processing industry is growing gradually.

In conclusion, meat processing and preservation is a very vital part of meat production because it improves meat shelf life during storage, promotes meat quality, texture, flavor, and color. Therefore, it is expedient for developing countries, mostly African countries to establish proper meat processing and packaging facilities in order to achieve meat processing standard which can in turn improve public health.

21.4.4 Technology to Improve Product Shelf-Life Extension

The shelf life of the perishable animal and dairy products is limited by factors that generally bring changes in odor, flavor, color, and texture resulting in their complete unacceptability (Sridhar et al. 2021; Lee 2018). The most important factors are oxygen, light, spoilage bacteria/microorganism, enzymes, and inadequate cooling (Eie et al. 2007). Over the centuries, different processing techniques have been developed to secure a constant availability of food products through increase in shelf life. However, both intrinsic, as well as extrinsic factors influence the quality of food and thus its shelf-life (Lee 2018), which can be defined as the period a food maintains its safety and/or quality under reasonably foreseeable conditions of distribution storage and use (Singh et al. 2017). Intrinsic factors, among others, include pH, water activity, initial microbial population, redox potential value, and nutrient content. These determine the nature of the decay mechanisms of a food product.

On the other hand, extrinsic factors determine how fast decay mechanisms proceed. Typical examples include the atmosphere, climatic conditions, and illumination (Sridhar et al. 2021; Bauer et al. 2022). Packaging is the main tool that prevents product deterioration and prolongs shelf life (Bauer et al. 2022). The package protects these products against physical, chemical, and biological damage. It also acts as a physical barrier to oxygen, moisture, volatile chemical compounds, and microorganisms that are detrimental to these products. Packaging also acts as a mediator or separator between intrinsic and extrinsic systems. The traditional role of product packaging is restricted to the preservation and protection of quality from chemical, physical, and biological deterioration. According to this concept, spoilage retardation, shelf-life extension, and quality preservation of packaged food, meat, and dairy products are a priority. Choosing the right packaging material can positively affect quality maintenance and shelf-life (Eie et al. 2007).

Traditional packaging methods include air-permeable packaging (APP), vacuum packaging, and modified atmosphere packaging (MAP). These packaging technologies have been successfully devised for fresh and processed meat products. Vacuum packaging systems are primarily used for wholesale products, while APP is the most popular packaging method for retail products. MAP is also used for these purposes, though less frequently used (Sridhar et al. 2021; Lee 2018).

The packaging films used for vacuum packaging systems have low gas permeability. They are usually multi-layered, with a polyamide layer as a gas barrier and a polyethene layer as a heat sealer. Less frequently, however, polyethene terephthalate or polyvinyl chloride are used as layers of barrier, while polypropylene is used as the sealing layer. Ethylene-vinyl acetate and ionomer films are sometimes used for better seal ability. Ethylene vinyl alcohol and polyvinylidene chloride layers can also be incorporated to enhance gas barrier properties (Sridhar et al. 2021; Bauer et al. 2022). Extraction, lamination, and/or coating are combined to create the desired properties. Recently, composite films incorporated with inorganic fillers (clay, glass flakes, and nanoparticles) are becoming popular in the market of fresh and processed meat products due to their advantages in microwave ability (Lee 2010). The advantage of applying a vacuum in fresh meat packaging is a longer shelf-life and improved tenderness. The increasing freshness of the retail market vacuum-packaged meat is due to its extended shelf-life (Sridhar et al. 2021; Lee 2018). In the air-permeable packaging system, wrapping films are typically made of plasticized polyvinyl chloride. The trays are mostly made of polystyrene paper, pulp mold, and rigid or foamed polyethene or polypropylene. These packages can also be produced from trays made up of relatively thick gauges of polypropylene and/or polyethene sheets that are hermetically sealed with a top film made of PP and high-density polyethylene. However, polyvinyl chloride films are still the most prevalent in the retail meat markets as wrapping films because of their superior mechanical properties and cheap prices compared to other alternative films (Eilert 2005).

MAP is a process where O_2 is often substituted by N_2. In MAP, CO_2 is the most important component in the gas mixture used because of its antimicrobial activity. Increased concentration of CO_2 leads to increased inhibition of bacterial growth. Three major gases are used in MAP, either individually or in combination. Carbon dioxide is used to provide an antimicrobial effect, that is, to suppress aerobic putrefactive spoilage bacteria, while N_2 act as an inert gas and is used in MAP as a filler either to substitute other gases or to prevent the package from deformation and O_2 is to convert purplish red myoglobin to bright red oxymyoglobin (Sridhar et al. 2021; Lee 2018). According to Bauer et al. (2022), argon was recently recommended for use in MAP because of its better effectiveness in retarding enzymatic activities, microbial growth, and chemical spoilage than N_2. Gas compositions used in MAP vary depending on the product species. Compared to vacuum packaging, MAP is less efficient because it needs more time to package the product and more investment in its operation. However, the application of MAP is expanding due to its value-added retail format, particularly regarding shelf-life, satisfying the requirements of both the consumer and the retailer (Sridhar et al. 2021; Lee 2018). The shelf-life of the vacuum-packaged product is preferentially influenced by factors such as storage temperature, size of meat cut, initial levels of contaminating microorganisms, and the O_2 permeability of packaging material (Koch et al. 2009). Longer shelf-life of fresh meat, processed meat, and dairy products can be achieved with a lower O_2 permeability of packaging film, storage temperature approaching a freezing point, and a lower initial bacterial load before the packaging (Lee 2018; Bauer et al. 2022).

However, these packaging technologies are continuously being developed to improve equipment, packaging material, and methodology. These packaging systems are typically characterized by the concentration and composition of gas inside the package and the packaging materials used. Recently though, various other innovative packaging methods have been introduced that utilize technologies such as barrier films, active packaging, nanotechnology, microperforated films, far-infrared radiations, and plasma treatment to extend shelf-life and preserve the quality of fresh and processed meat and dairy products (Sridhar et al. 2021; Lee 2018).

The positive effects of these various innovative packaging technologies for quality improvement and shelf-life extension have been verified (Lee 2010). The concepts of these technologies are characterized by how they regulate gas permeability or water vapor transmission rate, otherwise known as passive packaging, and how they incorporate bioactive ingredients into or onto the packaging materials, a technique known as active packaging. Some innovative approaches have also been developed by improving the control of gas permeability or water vapor transmission rate (micro-perforation of film, high gas-barrier film, nanotechnology etc.) by functional improvement in the packaging material itself (nanotechnology, plasma treatment, and irradiation) and by the application of active packaging systems (Kanter et al. 2018). Active packaging systems can be classified in terms of the mode of application, which includes the direct incorporation of active agents into the packaging materials and the use of edible films and coatings with active agents (Lee 2018).

In milk and dairy products, it is generally accepted that riboflavin plays a major role as a photosensitizer. Riboflavin absorbs light in the UV region of the spectrum up to about 500 nm (violet and blue light). Studies have shown that the 400–500 nm region is the most harmful part of the visible spectral region regarding photooxidation in dairy products (Bosset et al. 1994). In addition to riboflavin, Wold et al. found that chlorophylls and porphyrins probably also act as photosensitizers in cheese (Wold et al. 2005). It was shown that the sensory-measured photooxidation in a cheese induced by red light did not differ significantly from that induced by blue light. Its most degraded quality was caused by violet light, while green and yellow light (500–600 nm) gave the least adverse effects. Similar results were reported for milk by Hansen, Turner and Aurand (Hansen et al. 1975). These results support the hypothesis that porphyrins and chlorophylls are active photosensitizers in cheese. Similar findings reported by Borle, Siebert and Bosset (Borle et al. 2001) indicated that green and yellow package material could protect milk products to a certain extent because none of these photosensitizers has pronounced light absorption peaks in the green and yellow part of the spectrum (520–610 nm).

Today's efforts to improve packaging performance with clear effects on raw and processed product quality are directed toward green polymers and active packaging. The performance expected from bio-plastic materials is to protect the product from the environment while maintaining its quality. It is obvious that to perform these functions. It is important to control and modify their mechanical and barrier properties, which consequently depend on the structure of the polymeric packaging material (Bauer et al. 2022). Active packaging is the most relevant innovative idea applied to consumer satisfaction recently. It has been defined as a system in which the product, the package and the environment interact in a positive way to extend the shelf life of the product or to achieve some characteristics that cannot be obtained otherwise. In many present-day active packaging technologies, the active agent is placed in the package with the food, in a small sachet, pad, or device manufactured from a permeable material, which allows the active compound to achieve its purpose but prevents direct contact with the food product, protecting it from contamination or degradation (Sridhar et al. 2021; Bauer et al. 2022).

21.5 Innovative Technologies for Livestock Nutrition and Feeding

Feed plays a vital role in various aspects of livestock production, including productivity, profitability, environmental impact, human food and nutrition security, animal welfare, and health. It constitutes a significant portion of variable costs in livestock production, up to 70% and even 90% in intensive systems (Makkar 2016). The global value of compound feed relative to total animal output averages around 30%, varying by species (FEFAC 2016). High-quality feed improves livestock productivity, leading to lower age at first calving and shorter inter-calving intervals, thus enhancing profitability. Proper feeding also boosts animal immunity, health, welfare, and reproductive performance. It contributes to environmental sustainability by converting nutrients from non-arable land and inedible by-products into nutritious food (Balehegn et al. 2020). Forage-based systems, including silvopastoral systems, can mitigate emissions from livestock, particularly CH_4 (Garg et al. 2013), which is crucial as feed production and enteric fermentation account for a significant portion of total emissions from livestock production (Garg et al. 2013; Rao et al. 2015). Despite its importance, challenges such as high feed costs, limited availability, and low adoption of feed improvement technologies persist in low and mid-income countries (LMICs) (Balehegn et al. 2020). Feed development interventions must consider the complex socioeconomic contexts and systemic constraints faced by smallholders.

Successful feed enhancement interventions in LMICs have been achieved through technologies that align with local conditions and needs, often involving participatory approaches and collaboration between stakeholders (Balehegn et al. 2020). Commonly applied technologies include introducing improved forage varieties, enhancing the quality of crop residues and roughages, improving the production and utilization of processed concentrates and agro-industrial by-products, and involving the private sector in supplying inputs or processing, preserving, and marketing feeds (Table 21.3). The exploration and utilization of locally available, novel feed options that are economical, sustainable, and accessible are more sustainable options. These resources include agriculture remnants, invasive plant species, and food processing by-products (Table 21.4). Numerous studies have explored the utilization of unconventional feed sources in pig and poultry production (Baruah et al. 2022; Mbukwane et al. 2022) . Research has shown that incorporating ingredients such as rice bran, palm kernel meal, common vetch, duckweed, Azolla, baobab leaves and seeds, sweet potato vines, cassava leaves, marula seeds, and moringa leaves can enhance the growth and productivity of pigs and broiler chickens. Additionally, banana stems and oil palm fronds have been identified as beneficial fiber sources for both ruminants and non-ruminants (Liao et al. 2022). However, incorporating these novel feed sources requires careful consideration of factors like nutrient composition, antinutritional elements, digestibility, and palatability to optimize animal health and productivity (Chisoro et al. 2023). Also, attention must be paid to proper processing and feeding methods.

Some companies like Kenya's InsectiPro, Ecodudu, and Rwanda's Magofarm are leading the charge in using insects as alternative animal feed, converting food waste into nutritious blackfly larvae (TFF 2024). These larvae, rich in protein and nutrients like zinc, offer a sustainable and eco-friendly solution for feeding farmed fish and livestock. Although insect feed may currently be more expensive than traditional soy meal, it is considered an investment in the future of sustainable animal feed production (TFF 2024). Beyond insects, fungi-based meat alternatives are gaining traction, with Zambian research exploring the use of gourmet oyster mushrooms as a supplement for goats (TFF 2024). Additionally, synthetic biology companies are developing bacterial and fungus-derived proteins as alternatives to fishmeal, utilizing industrial waste and renewable energy sources for production (TFF 2024). Seaweed and algae are also emerging as valuable feed supplements, improving gut health in pigs and poultry while reducing CH_4 emissions in cows (TFF 2024). The cultivation of seaweed and microalgae presents significant opportunities for the aquaculture sector. Further innovative solutions aimed at improving feed availability includes straw ammoniation, silage production, veld reinforcement, rehabilitation, and strategic destocking. In West African Sahel, the Common feed resources in West Africa Sahelian region are crop residues (millet, sorghum, etc.), rangeland herbaceous legumes, rangeland grasses, browse species, fodder banks, agro-industrial by-products (cotton seed cake, bran of millet, sorghum, maize, wheat, and concentrated feeds) (Amole et al. 2022).

Innovations in forage crops are enhancing the nutritional value of feed for grazing livestock, with initiatives like irrigated Napier grass proving beneficial for cattle productivity (TFF 2024). Technologies such as computational DNA analysis are driving advancements in breeding forage crops that can withstand the impacts of climate change (TFF 2024). The determination of nutrient requirements, development of the capacity for feed analysis using near-infrared reflectance spectroscopy, deployment of ration formulation apps/software, and examination of the effects of strategic supplementation on livestock performance are necessary elements to adopt (Balehegn et al. 2020). Feed Assessment Tool (FEAST) developed to involve farmers more directly in decision-making regarding livestock feed improvement is another innovative technology used for enhancing livestock feeding (Duncan

Table 21.3 Practical feed improvement technologies for smallholder livestock production systems and their expected or observed impacts

Category	Technology	Production system[a]	Description of technologies	Observed or expected impact	Ref
Improving feed availability/ productivity	Improved forage plants	Small holder mixed crop-livestock, semi-intensive and intensive	Introducing higher yielding and higher-quality forage species including legumes	Increase forage availability and or nutritive value, reduce seasonal fluctuation in availability	Foster et al. (2009); Alemayehu et al. (2017)
	Conservation-based forage development	Small holder mixed crop-livestock, pastoral	Introduction of forage plants in natural resource conservation structures such as gullies and terraces, which serve as a source of feed, while reinforcing soil and water conservation	Protect soil loss and land degradation while improving feed availability	Alemayehu et al. (2017)
	Silvopastures / agro-forestry	Small holder mixed crop-livestock, pastoral, semi-intensive peri-urban	Use pasture, farmlands, and degraded areas for growing trees that synergistically impact pasture productivity	Improved fodder biomass productivity by up to 500% compared to conventional fodder tree growing strategies	Balehegn et al. (2014)
	Food-feed crop integration	Small holder mixed crop-livestock	Intercropping or alley farming to exploit synergies in pest protection and soil and water conservation, while improving availability of forage	Improved soil fertility, reduced pest load to food crops, while improving feed availability	Lenné et al. (2003)
	Protected grazing (exclosures, zero-grazing, cut and carry, rotational grazing, deferred grazing	Small holder mixed crop-livestock, pastoral, semi-intensive peri-urban	Protection or prescribed grazing on range and grazing lands to protect degraded areas and allow for natural regeneration of forage and improvement of forage production	Improved grazing land productivity, forage biomass, quality of forage produced, and reduce grazing land degradation	Yayneshet et al. (2009)
	Protected agriculture, e.g., hydroponics, greenhouse forage production.	Semi-intensive urban and peri-urban	Producing forage under protected conditions in areas and localities where conventional way of production is not possible or ineffective	Enable forage production in small areas of land or in soilless agriculture. Improved availability of green fodder	Masud and Bhowmik (2018)
	Use of underutilized locally available feed resources	All systems	The use of underutilized locally available feed resources including indigenous fodder species, local brewery residues, etc.	Improved feed availability, reduced need for commercial concentrates, improved farm profitability	Balehegn et al. (2014)
	Improved human-food-waste processing	Intensive urban and peri-urban	The use of affordable drying, cleaning, sorting, and processing technologies that enable safe use of human food-waste for livestock feeding	Increased supply of feed in areas where there is resource limitation for growing forage	Makkar (2016)
	Agronomic interventions on cultivated pastures	All	These include the use of recommendations on sowing rates, spacing, species mixing and association, seed treatment, weed and pest control, irrigation land preparation, shading, control of water logging	Improved forage yields	Falvey (1999)

(continued)

Table 21.3 (continued)

Category	Technology	Production system[a]	Description of technologies	Observed or expected impact	Ref
Enhancing feed quality	Chemical treatment of crop residues	Intensive commercial, semi-intensive urban and peri-urban, small holder mixed crop-livestock	Involves treating crop residues with urea and spraying or soaking in dilute acid and alkaline solutions, etc.	Improved crude protein content (with urea treatment); improved intake and digestibility of crop residues	Sarnklong et al. (2010)
	Biological treatment of crop residues	Intensive commercial, semi-intensive urban and peri-urban, small holder mixed crop-livestock	Involves treating crop residues with enzymes, bacterial inoculants or white /brown rot fungi	Improved digestibility and intake of crop residues	Li et al. (2010)
	Reducing particle size of crop residues	Intensive commercial, semi-intensive urban and peri-urban, small holder mixed crop-livestock	Chopping and grinding crop residues	Improves intake by animals, reduces bulkiness	Hamed and Elimam (2009)
	Fertilization of crops	Intensive commercial, small holder mixed crop-livestock	Applying fertilizers to improve the nutrient content (mainly CP) of crop residues	Fertilization of crops improves quality (improve CP and digestibility and reduce crude fiber) of crop residues as livestock feed resulting in up to 40% greater milk production	Reddy et al. (2003); Haileslassie et al. (2013)
	Forage crop breeding	Intensive commercial, small holder mixed crop-livestock	Selective breeding of forages for developing high yielding and better-quality forage accessions	Improved biomass productivity, feed quality and thus improved livestock productivity	Den Hartog and Sijtsma (2013)

Maintain or conserve feed quality	Correct timing of forage harvesting	Intensive commercial, small holder mixed crop-livestock	Harvesting forages when the nutritional content of the forage is at optimal or when both nutritional value and biomass yield are optimal	Improved intake, digestibility, and livestock productivity	Makkar (2016)
	Silage making	Intensive commercial, semi-intensive peri-urban	Storing fresh fodder under anaerobic conditions to preserve the quality	Conserved fodder with minimal energy and nutrient loss and spoilage	Titterton and Bareeba (2000)
	Hay making	Intensive commercial, semi-intensive urban and peri-urban, small holder mixed crop-livestock	Reducing loss of nutrients from green fodder by drying	Conserved fodder with minimal energy and nutrient loss and spoilage	Klinner and Shepperson (1975)
	Using preservatives	Intensive commercial, semi-intensive urban and peri-urban,	Using microbes or chemicals that inhibit spoilage organisms and preserve the quality of fresh fodder	Conserved fodder with minimal energy and nutrient loss and spoilage	Wang et al. (2018)
Improve the nutritional status of animals	Balanced and or phased rationing or ration formulation	Intensive commercial, semi-intensive urban and peri-urban, small holder mixed crop-livestock	Feeding a balanced ration formulated to meet the nutrient requirements of the animal or targeting rations to animals at specific levels of performance	Improved milk yield by 2–14%, improve net daily income 10–15%; reduced emission of greenhouse gases by 15–20%	Garg et al. (2013)
	Supplementation with concentrates	All systems	Supplementing low quality basal diets of animals with nutritious concentrates	Improved intake, digestibility, body weight gain, and milk yield	Selemani and Eik (2016)
	Supplementation with multi-nutrient blocks	Semi-intensive urban and peri-urban, small holder mixed crop-livestock	Providing animals on low quality basal diets multi-nutrient blocks that provide needed supplementary nutrients	Enabled production of the same amount of milk when 50% less green fodder and 30% less protein supplement was fed; improved feed intake and protein supply, increased milk yield by 1–1.5 L per day and enhanced reproductive performance in cattle	Makkar et al. (2007)
	Supplementation with feed additives	Intensive commercial, semi-intensive urban and peri-urban	Enzymes, probiotics, yeast, and other products that are added to feeds to help improve the ability of an animal to digest and assimilate feeds	Improved feed intake, digestion and performance	Ramirez (2014)

(continued)

Table 21.3 (continued)

Category	Technology	Production system[a]	Description of technologies	Observed or expected impact	Ref
Analytical and operational technologies	Near infrared reflectance spectroscopy (NIRS)	Intensive commercial, semi-intensive urban, and peri-urban	The use of NIRS technology that enables a quick and affordable assessment of the nutritional quality of various types of feeds without reagents	Improved efficiency and cost-effectiveness of feed analysis; may be used to enhance feed marketing	Modroño et al. (2017)
	User friendly ration formulation tools	Intensive commercial, semi-intensive urban and peri-urban, small holder mixed crop-livestock	The development and use of simple ration formulation tools such as excel or mobile phone applications that enable farmers to formulate effective rations based on locally available feed resources	Make ration formulation easier and facilitate its adoption	Chakeredza et al. (2008)
	Livestock/feed management applications	Intensive commercial, semi-intensive urban and peri-urban	These are user friendly mobile phone-based applications to monitor, track, and analyze feed consumed, produced, wasted, etc.	Help improve livestock farm profitability	Hwang et al. (2011)

Adapted from Balehegn et al. (2020)

Table 21.4 Categories of some of the novel feed ingredients or supplements in livestock feed. Adapted from (Chisoro et al. 2023)

Category	Examples	Justification	Ref
Oilseeds and leaf meals of indigenous fruit-bearing trees	Neem seeds, baobab leaves and seeds, red sour plum seeds, Moringa leaves and seeds, etc.	Indigenous trees have many properties and most bear fruit that have been eaten and used since ancient times. Some have been used as food and medicine in African rural communities. They have great potential as feed resources in particular their seeds, leaf, and oilcake meals	Chisoro and Nkukwana (2020)
Grain and pulses crop leftovers	Rice, sorghum, millet, groundnut Stover, wheat hay, etc.	Crop leftovers are by-products obtained after crop harvest, especially for grain crops. These can be used as partial replacers for more expensive feeds like maize or wheat because they are excellent sources of fiber and carbs in livestock diets	Sebola et al. (2021)
Pasture and forage crops	Alfalfa, lucerne, rye grass, black clover, etc.	These are plants, the likes of legumes, grasses, and browse species, that are planted to be fed to livestock. Mostly they are utilized as significant sources of protein and other nutrients to enhance the nutritional value of livestock diets	Chioma et al. (2021)
Agro-industrial by-products	Dried distillers' grains, peanut shells, cottonseed cake, palm kernel cake, etc.	These are substances produced during the processing of agricultural products like, peanut butter production, cooking oil, and beer manufacture. They can be used to enhance livestock performance and in diets as valuable sources of fiber, energy, and protein	Girma (2019)
Insects	Black solider fly, silkworm, locust, crickets, etc.	These have taken the feed sector by storm, due to their high protein value, ease, and sustainable propagation from organic waste, insects have gained popularity as a source of protein in livestock feed especially for pigs and poultry	Khan (2018)

(continued)

Table 21.4 (continued)

Category	Examples	Justification	Ref
Aquatic plants and animals	Duckweed, azola, water hyacinth, water lettuce, etc.	These include plants, fish, and other aquatic creatures. In aquatic systems, they are often regarded as "waste or weeds"; however, they can be used as sources of nutrients for livestock	Mahgoub and Al-Mutairi (2015)
Leftovers of food processing and manufacturing factories	Biscuit meal, candy and potato chips by-products, etc.	In the making of candy, biscuits, potato chips, and other foods they produce a lot of by-products, and all the broken and deformed products that do not meet market and production standards are left behind (Mbele et al. 2019). They still contain valuable nutrients that are still useful for use in livestock feed	Mbele et al. (2019); Ominski et al. (2021)

et al. 2023). FEAST is a participatory tool that utilizes focus group discussions and individual farmer interviews to develop a comprehensive overview of the livestock farming system (Duncan et al. 2023). Over the past decade, FEAST has been extensively applied in various countries, including Kenya and Rwanda, notably within initiatives like the Accelerated Value Chain Development Project and the Rwanda Dairy Development Project. In these cases, FEAST informed feed options with substantial input from farmers. While FEAST primarily supports improved feeding strategies at the farm level, the data collected through the FEAST app and published in reports serve as valuable resources for understanding broader livestock feed issues at the system level (Duncan et al. 2023). FEAST exemplifies a participatory tool that empowers farmers in decision-making while providing insights to researchers across farming systems. Its widespread adoption highlights its significance in addressing gaps in livestock feed development. Its innovation lies in bridging the knowledge gap between livestock researchers and small-scale livestock keepers, contributing to more informed and sustainable livestock farming practices (Duncan et al. 2023).

Production efficiency can be improved by deploying automated systems or smart machines in livestock operations that can deliver a precise amount of feed to each animal to meet its nutrient requirement for that particular production stage. Precision livestock feeding or nutrition is part of PLF approach involving techniques that allow feeding of the proper amount of feed of suitable composition and in a timely manner to individual animals or groups of animals (Andretta et al. 2014; Niemi et al. 2010). Precision livestock feeding matches nutrient requirements with nutrient supply based on real-time factual data determined by sensors (Zuidhof 2020). Such systems combine specific management approaches to each animal such that the feed it receives (including appropriate use of any feed additive) contains the appropriate nutrients required at the correct dosage to the appropriate genotype and physiological stage while employing effective processing methods and minimal antinutritive factors (Banhazi et al. 2012). Apart from improved efficiency of feed utilization, other important benefits from these systems are being driven by increased animal productivity, profit maximization, and reduced negative impact on the environment via lowering nutrient content of excreta. Using precision systems will reduce the oversupply of feed, which is a precursor to overfeeding and the over-excretion of wastes that exceed the animal's requirement. The lower cost of advanced technology has encouraged the exploration and applications of these innovative technologies in different livestock species. The application of precision feeding on-farm requires the design and development of measuring devices that determine animal feed intake and weight, computation methods that estimate the nutrient requirement of animals (based on actual growth of animal) in a timely manner and systems of feeding with capability to make available the required composition and amount of feed needed to achieve the desired production goal (Pomar et al. 2019).

There are factors affecting the adoption of innovative technologies, these includes (Baltenweck et al. 2020): (1) management of technology: Implementing feed interventions often requires specific skills and access to technology, extension services, and training. Farmers need to acquire and apply these skills effectively to adopt new feeding practices successfully; (2) perceived benefits and alignment with objectives: Farmers must perceive the benefits of adopting new feed technologies and understand how they align with their objectives. However, the link between improved feeding practices and financial gains may not always be evident to farmers, especially considering that livestock serve various purposes beyond meat and milk production. Development agents must recognize and address farmers' diverse objectives to promote successful adoption; (3) resource availability and trade-offs. The availability of key resources like land and labor, and the trade-offs imposed by feed technologies, significantly impact adoption. Adequate availability of resources, particularly labor, tends to facilitate adoption. However, in some contexts, farmers may prioritize allocating resources to non-livestock activities, requiring careful consideration of trade-offs when promoting feed interventions. Therefore, interventions for addressing sustainable feed resources should focus on understanding and addressing sociocultural and institutional factors, as well as ensuring private sector engagement and policy reforms. By prioritizing these aspects and developing innovative solutions, the livestock sector in LMICs can improve productivity, contribute to food security, and alleviate poverty (Balehegn et al. 2020).

21.6 Innovative Technologies for Livestock Health Management

Health in livestock is defined based on multifactorial indices to denote the different states of the animal. Hence, a more comprehensive definition is the "state of normal biological functions and homeostasis, the absence of physical and psychological stress with optimum productivity, including reproduction.". Health in animals is more than a mere absence of disease but encompasses an improvement of animal welfare through health and stress monitoring for sustainable output (Giannuzzi et al. 2022). One of the descriptors of animal performance includes the animal's live weight. This is influenced by nutrition, welfare, and disease (Key 2018). Other factors taking a toll on cattle production include environmental changes due to climate changes (El-Sayed and Kamel 2020). The rise in the global population has necessitated an increased demand for animal-based proteins, resulting in pressure on animal production globally, including in tropical and sub-tropical developing countries (Renaudeau et al. 2012). Another confronting factor in cattle husbandry with a somewhat negative impact is using antimicrobials as growth promoters (Ma et al. 2021). Adopting new and sustainable technologies to boost livestock production with animal welfare is imperative to satisfy the increasing demand for animal-sourced foods.

21.6.1 Surveillance and Disease Forecasting

According to the World Health Organization (WHO), health surveillance is "the continuous, systematic collection, analysis, and interpretation of health-related data needed for the planning,

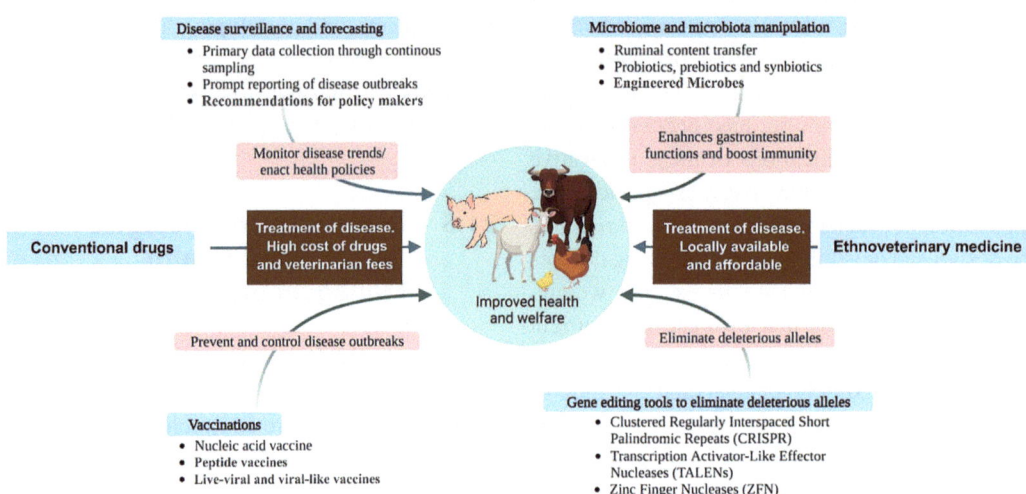

Fig. 21.4 Modern techniques for the improvement of livestock health

implemention, and evaluation of public health practice" (http://www.who.int/topics/public_health_surveillance/en/). In livestock, disease surveillance is planned to detect the early onset of diseases and evaluate the effectiveness of control measures. The data generated from disease surveillance is used to mount national control and eradication programs (Wakhusama et al. 2019). Disease surveillance is important in safeguarding animal and human health and welfare. Disease surveillance depends on notifiable disease reporting and secondary data analysis in humans. In livestock, primary data are obtained through active monitoring and sampling for the early detection of diseases (Wakhusama et al. 2019). Active disease surveillance in livestock aids in fulfilling trade requirements and certifications of animals and animal products to ensure safety (Comin et al. 2019). The current animal disease surveillance system in sub-Saharan Africa is inefficient because of its weakness, poor funding, and inability to provide quality and sensitive information on the status and trends of diseases. Hence, the need for urgent improvement in the specificity, sensitivity, and diagnostic flexibility for detecting silent epidemics and outbreaks (Abubakar et al. 2022). Animal disease surveillance requires improving the sensitivity of the reporting system, diagnostic accuracy, and disease containment activities using an integrated approach. It should be early alert and able to provide an understanding of disease dynamics with priority for intervention plans to prevent epidemics or pandemics and transmission of diseases from animals to humans (Abubakar et al. 2022). The ecology of Africa is complex and needs a specialized form of surveillance to document intervention impacts, track elimination progress, and monitor the epidemiological dynamics to determine health policy and practice. Disease surveillance is used to track and monitor the zoonotic transmission and progression of several diseases, such as bovine tuberculosis (Menghistu et al. 2018), schistosomiasis, leishmaniasis, helminthiasis (Fesseha et al. 2022), and brucellosis (Tadesse 2016) in humans. Therefore, mounting effective animal disease surveillance systems will help prevent and control disease transmission from animals to humans, boost production through improved health and welfare of the animals, and an overall increase in the gross domestic product through increased export of animals and animal products (Fig. 21.4). With the recent technological development and increased access to handheld computing devices such as smartphones, tablets, and portable computers, several software are being developed to aid in effective disease surveillance and forecasting (Adepetun 2015).

21.6.2 Microbiome and Microbiota Manipulation

The rumen is a complex ecosystem with an intricate microbial community consisting of bacteria, viruses, archaea, protozoans, and fungi coexisting in a symbiotic manner (Ribeiro et al. 2017). These microbes interact symbiotically to break down complex fibrous substrates into volatile fatty acids and microbial proteins the host utilizes for growth, lactation, and general maintenance (Morgavi et al. 2013). The rumen microbiome digests dietary polysaccharides, which the rumen cannot break down. This process largely depends on the structure and diversity of the microbiota present in the rumen (Uyeno et al. 2015). The cattle rumen is an efficient system for degrading plant-cell walls to generate energy from forages. However, less than 50% of the carbohydrate from the cell wall are digested in low-quality forage (Ribeiro et al. 2017). Hence, improving carbohydrate metabolism in the rumen will produce adequate energy for health maintenance and enhanced productivity in the beef and dairy industries (Ribeiro et al. 2017). One way to improve the microbiome function is to manipulate the ruminal microbiota through probiotics administration or ruminal contents transfer.

Direct-fed microbes to ruminants manipulate the ruminal microbial community and alleviate dysbiosis. Probiotics enhance gastrointestinal health by stimulating and developing a healthy microbiome, inhibiting enteric pathogens, enhancing feed digestibility, lowering gastrointestinal pH, and modulation of mucosal immunity (Uyeno et al. 2015). The administration of probiotics in livestock has gained acceptance to replace low-dose antibiotics to improve animal health with benefits such as enhanced feed efficiency, body weight gain, and performance (Buntyn et al. 2016). Probiotics in adult ruminants are selected based on their ability to aid the breakdown of fibers, particularly cellulose (Uyeno et al. 2015). The actions of probiotics are associated with the microbial genera, species, or strains with multi-strain preparations recommended for a broader spectrum of action (Kwoji et al. 2021). Colonizing young animals with probiotics early in life is important as they modulate the expression of genes that support the establishment of beneficial commensals microbes (Siggers et al. 2007). The activities of probiotics are enhanced through prebiotics ingestion (Gibson et al. 2017). Common prebiotic substrates include carbohydrates, such as oligosaccharides or dietary fibers, with low digestibility (Uyeno et al. 2015). A combination of probiotics and prebiotics formulations are administered as symbiotics to animals to improve the survivability of the probiotics microbes (Gaggìa et al. 2010). The administration of probiotics in cattle, especially at the pre-ruminant stage, alleviates the incidence and severity of diarrhea and the carriage of pathogenic microbes (Gaggìa et al. 2010). Probiotics are used for prophylaxis and treatment of neonatal diarrhea in calves (Galvão et al. 2005). To protect probiotics during transit along the gastrointestinal tract, a single preparation of prebiotics and probiotics are formulated as synbiotics. A synbiotic is "a mixture of probiotics and prebiotics that beneficially affects the host by improving the survival and implantation of live microbial dietary supplements in the gastrointestinal tract, by selectively stimulating the growth and activating the metabolism of one or a limited number of health-promoting bacteria, and thus improving host welfare" (Markowiak and Śliżewska 2018).

Gastrointestinal microbial colonization is important during the rumen development of livestock (Sommer and Bäckhed 2013). This is because the rumen microbes supply approximately 70% of the daily energy needs of ruminants (Yeoman and White 2014). Different feeding programs are associated with variations in the microbial communities established in the rumen of young animals (Abecia et al. 2014). However, the composition of the ruminal microbiome may differ among cattle, even those fed with the same feed (Weimer et al. 2010a). Thus, indicating the possibility of the microbiome host-specificity and likelihood of resistance to colonization by strains not adapted to the recipient host (Weimer et al. 2010b). Moreover the microbiome host-specificity, feeding method, and pre-weaning diets greatly influence the ruminal microbiome (Abecia et al. 2013). For example, introducing calves to solid feed early in life resulted in a diverse microbial community compared to those conventionally weaned at 6 weeks (Li et al. 2019).

21.6.3 Vaccination

Animal health is important for an efficient transition toward large, healthy, and sustainable food systems with animal welfare standards. Animal health is necessary for global food security, quality, and poverty alleviation (Pradere 2014). Therefore, preventing and controlling animal diseases, including zoonoses, and appropriately managing emerging diseases, pandemic threats, and antimicrobial and antiparasitic resistances in livestock are pertinent (FAO 2019). Vaccines in veterinary medicine are vital to preventing, controlling, and protecting animal and human health against diseases (Meeusen et al. 2007). Vaccines in animal disease control are sustainable and provide lasting protection without leaving pharmaceutical residues in the environment while reducing the need for antimicrobials (Hoelzer et al. 2018). The need for vaccines to prevent existing and emerging diseases in livestock has necessitated research in identifying protective antigens and virulence mechanisms using genomic, proteomic, immunological, biological, and bioinformatic techniques and suitable delivery methods (Charlier et al. 2022). Vaccines have reduced the burden of numerous diseases, such as smallpox, tetanus, polio, diphtheria, and measles (Younger et al. 2016). In cattle, several vaccines have been developed to prevent infections, including bovine tuberculosis (Vordermeier et al. 2016), contagious bovine pleuropneumonia (CBPP), bovine viral diarrhea (BVD) (Van Oirschot et al. 1999), and infectious bovine rhinotracheitis (IBR) (Nardelli et al. 2008) among others.

There is also the need for new vaccines with effective, long-lasting single shots to be developed due to the short lifespan of livestock (Charlier et al. 2022). For effective vaccine development, the host-immune mechanisms and means of immune evasion by pathogens must be understood (Entrican et al. 2020). Vaccinology is a dynamic research area whereby new technologies, including nucleic acid, peptide, live-viral, and virus-like particle vaccines, have been developed to induce high protection, and they constitute affordable vaccines (Francis 2018). Vaccines used for prophylaxis or therapeutics are grossly grouped as live attenuated (weakened microorganisms), inactivated (killed microbes), toxoid (inactivated bacterial toxins), and sub-unit (purified antigens) vaccines (Wadhwa et al. 2020).

21.6.3.1 Nucleic Acid Vaccine

Nucleic acid vaccines employ genetic material (DNA or mRNA) from a pathogen to elicit a specific immune response. Nucleic acid vaccines are becoming versatile and robust in their development and are potentially safe, cost-effective, and efficient in combating diseases (Qin et al. 2021). DNA vaccines are made from bacterial plasmids that encode immunostimulatory antigens such as Interleukin-2 (IL-2) and Granulocyte-macrophage colony-stimulating factor (GM-CSF) (Yankauckas et al. 1993). DNA vaccines are administered through various routes to the host, including intramuscular, transdermal, and intramucosal. These vaccines are efficient and elicit both humoral and cellular immunity in the host. Following administration, DNA vaccines are transcribed and translated into the nucleus and cytoplasm, respectively (Bai et al. 2017).

21.6.3.2 Peptide Vaccines

Peptide vaccines are subunit vaccines made up of mainly specific epitopes capable of inducing desirable B- and T-cells mediated immune responses (Li et al. 2014). The peptides are usually 20–30 amino acid sequences representing the immunogenic molecule of the specific antigen-binding site (epitope) (Li et al. 2014). Peptide vaccines can stimulate humoral and cellular responses because the epitopes are the antigenic determinants of a larger protein (Bijker et al. 2007). The development of peptide vaccines was birthed out of the discovery of the ability of major histocompatibility complex-I (MHC-1) to restrict the recognition of $CD8^+$ T-Cells by target cells (Zinkernagel and Doherty 1974) and the ability of the $CD8^+$ T-Cells to detect short peptide sequences of 11–16 in length generated from the proteolytic degradation of target antigen presented to MHC-I (Townsend et al. 1984). It was further discovered that the MHC-I recognizes even smaller peptides of the $CD8^+$ T-cells epitopes of eight amino acids (Rötzschke et al. 1990). The use of peptide vaccines reduces the cumbersomeness of multiple vaccinations because a conserved peptide sequence region of the epitopes from different strains, serovars, or species of a pathogen could be used in designing the vaccine (Li et al. 2014).

21.6.3.3 Live-Attenuated Vaccines

Live-attenuated vaccines involve the use of weakened microbes. The pathogen is weakened to reduce its disease-causing ability but retains its immunogenic ability (Kollaritsch and Rendi-Wagner 2012). During the formulation of live-attenuated vaccines, the microbe is passaged by serially culturing in cell cultures or animal embryos (mostly chicken embryos). Attenuating viruses is carried out by culturing on cells that have not been grown for so many generations (Dai et al. 2019). Live-attenuated vaccines can elicit higher immunogenicity than inactivated vaccines since the process mimics natural infection to stimulate cellular and humoral responses (Kollaritsch and Rendi-Wagner 2012). Also, the immunity conferred by live-attenuated vaccines is long-lasting and sometimes lifelong (Fuertes Marraco et al. 2015). However, safety is the major disadvantage of live-attenuated vaccines as the pathogens may revert to virulence through back mutations and cause classical diseases like the wild pathogen, especially in stressed or immunocompromised hosts (Kollaritsch and Rendi-Wagner 2012).

21.6.4 Conventional Drugs

Addressing livestock health issues requires a holistic approach considering the diverse roles and perspectives within livestock production systems and the challenge of access to drugs and cost of treatment. As an example, African animal trypanosomosis (AAT), caused by the protozoan parasite *Trypanosoma congolense*, poses a significant challenge to improving animal productivity and agricultural production in sub-humid zones, as evidenced by multiple studies conducted over the past two decades (Swallow 2000; Chanie et al. 2013). The FAO estimates the annual economic losses due to AAT in Africa at $1–1.5 billion (Swallow 2000). AAT poses a significant challenge to agricultural productivity and livestock welfare, especially given that 70% of agricultural land is infested with tsetse flies (vector of Trypanosome spp), putting over 60 million animals at risk (Ouédraogo et al. 2006). In Africa, the tsetse fly-infested zone spans approximately 8.7 million km (Enahoro et al. 2019), with nearly 46 million livestock animals affected (Swallow 2000). The use of trypanocidal products to control AAT is widespread, with over 35 million doses administered annually in Africa. A study in Burkina Faso showed promising results, with a 56% reduction in animal mortality rates achieved through combined trypanocide treatments and trapping (Kamuanga et al. 2001). In the Province of Kénédougou, it was revealed that the prevalence of sick animals plays a crucial role in treatment decisions, with farmers adopting a recursive strategy for treating trypanosomosis (Ouédraogo et al. 2006). A higher prevalence of affected animals prompt farmers toward seeking professional care. Likewise, resistance to trypanocides also influences this decision, with repeated ineffective treatments leading to professional consultation (Ouédraogo et al. 2006). Agropastoralism has a significant negative correlation with consulting animal health professionals for trypanocide treatment, with herders more likely to seek professional care than agropastoral farmers. The size of agricultural households also plays a role, as larger farms tend to consult professionals due to lower labor opportunity costs. Socioeconomic and institutional factors such as the number of agricultural workers, transaction costs, and transport effectiveness interact with biological variables like disease prevalence and resistance to shape treatment choices (Ouédraogo et al. 2006). Therefore, it is important to consider these factors in rural development policies to enhance livestock health coverage and increase agropastoral farming productivity in the region.

In South Africa, granting farmers access to veterinary medicines without oversight addresses the scarcity of veterinary services and fostering sustainable farming. However, this practice assumes farmers can accurately diagnose and treat animals independently. Yet, like any medication, veterinary drugs can be used rationally or irrationally (Gulwako et al. 2023). Rational usage, endorsed by the WHO, involves appropriate drug selection, dosage, and cost. Conversely, irrational use risks ineffective treatment, harming animals, and wasting resources (Gulwako et al. 2023). Kirui et al. reported that many farmers in sub-Saharan African (SSA; Ethiopia, Malawi, Nigeria, and Tanzania) lack proper knowledge and training in administering veterinary medicines, leading to incorrect usage (Kirui 2019). Issues such as incorrect route of administration, drug storage, dose calculation, withdrawal periods, and disposal are prevalent, especially among those without formal education (Kirui 2019). Similar findings were noted in Bangladesh and Ethiopia, highlighting a global concern (Khatun et al. 2016). Misuse of veterinary drugs, particularly in food-producing animals, raises concerns due to potential residues in animal products, jeopardizing human health. This misuse exacerbates antimicrobial resistance, limiting treatment options and posing direct health risks. Understanding farmers' knowledge of safe drug use is crucial (Gulwako et al. 2023). Training programs, though beneficial, often fail to adequately cover food safety aspects like withdrawal periods. Therefore, farmers need to be properly trained for proper management of production animals to ensure the proper utilization of available medications. Targeted training on safety and storage aspects of drug use, along with strategic allocation of farm subsidies, could help address these challenges effectively (Gulwako et al. 2023).

In Uganda, Tanzania, Nigeria, and Kenya, veterinary drugs market remains peripheral in the global market for modern veterinary drugs (Gulwako et al. 2023). Unlike emerging Asian countries, SSA lacks appeal for international pharmaceutical companies. The market is characterized by weak regulation, fragmented supply chains, and a dual system for veterinary medicines. In SSA, the public sector supplies small-scale farmers in rural areas with limited resources, focusing mainly on vaccines and parasiticides through large-scale campaigns. In contrast, the largely unregulated private sector serves commercial and industrial livestock farming, relying on private veterinarians and various retailers predominantly located in urban and peri-urban areas. International efforts have been made to improve drug access and supply chain efficiency in SSA. These efforts include support for national legislation on veterinary drugs, harmonization of registration procedures, and arrangements to enhance availability through public-private partnerships (PPPs) and involvement of pharmaceutical companies (Gulwako et al. 2023). Jaime et al. emphasizes the need for policies informed by a better understanding of the drivers behind access to veterinary drugs (Jaime et al. 2022). The authors suggest analyzing factors affecting drug accessibility, including availability, quality, and economic affordability. Economic studies on affordability are crucial to understand pricing dynamics and stakeholder decisions. Additionally, assessments should consider the supply chain's capacity to handle epidemics and emergencies. The evolving international policy landscape highlights the increasing influence of commercial actors in drug availability decisions, necessitating a balance of stakeholders' interests. Furthermore, policies should avoid bias toward certain diseases at the expense of others important for veterinary public health, thus ensuring a holistic approach to animal health management in SSA (Organization, W. H 2004).

21.6.5 Ethnoveterinary Medicine

Ethnoveterinary preparations (EVPs) play crucial roles in maintaining the health and productivity of livestock in many developing countries, where access to conventional veterinary medicine may be limited (Wendimu et al. 2023). EVPs have been utilized for animal health management since ancient times. Historically, EVPs such as medicinal plants and herbs were prevalent before the industrial revolution but declined with the rise of conventional medicine. However, there is now a renewed interest in EVPs due to their sustainability, cost-effectiveness, availability, and environmental friendliness, particularly in addressing issues arising from conventional medicine use. This shift toward EVPs is driven by the need for organic solutions and a comprehensive approach to health problems (Tumasang et al. 2023). Plant materials and derivatives have a long history of use in human and animal health interventions, with extracts from plants being utilized to create remedies. Moreover, there is high demand for healthier organic alternatives in livestock products, driven by consumer preferences for pasture-fed livestock products free from synthetic residues. Additionally, there is a growing focus on ethically raised farm animals, prompting consumers to pay premium prices for products from certified farm sources, along with increased awareness from animal rights activism (Nwafor and Nwafor 2022). The rising financial costs associated with orthodox practices for treating livestock infections and diseases, including high drug and medication costs and fees payable to veterinarians and technicians, have further propelled the use of plant-derived remedies. With estimates from the FAO indicating significant losses due to livestock diseases among smallholder farmers in developing countries, the adoption of alternative practices involving plant-derived remedies presents a way to mitigate these losses (FAO 2018). This trend aligns with observations that animals in their natural habitat consume certain plants to alleviate ailments, and many farmers have witnessed improvements in livestock health after using medicinal plants or extracts (Nwafor and Nwafor 2022). Ethnoveterinary practices and products are widely utilized among rural smallholder livestock farmers in many SAA countries. These farmers often resort to herbal remedies when conventional knowledge is lacking, veterinary services are inaccessible or unaffordable, or when alternative remedies are more available and affordable. The availability and affordability of such alternative remedies are key factors driving the use of ethnoveterinary medicine among farmers. However, despite the availability of herbal remedies, some livestock owners still opt for orthodox treatments due to unsatisfactory experiences with herbal remedies or a perception that traditional practices are outdated and ineffective. Additionally, some farmers choose to use a combination of orthodox treatments and ethnoveterinary medicine for treating their livestock. This underscores the complex decision-making process involved in animal healthcare among rural smallholder farmers, who weigh factors such as effectiveness, availability, affordability, and personal experience when choosing treatment options for their livestock (Nwafor and Nwafor 2022).

The profile of users and adopters of ethnoveterinary products and practices varies based on socioeconomic factors and cultural perspectives. In developed economies, farmers may utilize phyto-therapeutic products to comply with organic livestock treatment directives. In contrast, in developing countries, ethnoveterinary practices are popular among rural livestock owners who practice extensive farming, rely on communal pastures, and face resource constraints. These farmers often gather herbs and plant materials from the wild due to their inability to afford conventional veterinary services and products, which may not be readily available in rural communities. Not only that, there are some smallholder or emerging livestock farmers who utilize a combination of bio-medicine and ethnoveterinary practices, known as medical pluralism. Among this group, ethnoveterinary practices are often used in conjunction with pharmaceuticals rather than exclusively (Nwafor and Nwafor 2022).

Ethnoveterinary medicine is rooted in traditional beliefs and indigenous knowledge and it is widely utilized globally to treat livestock diseases (Rahman et al. 2022). In a study conducted in Namibia's Omusati and Kunene regions, a total of 100 people were interviewed using a semi-structured questionnaire to identify ethnoveterinary medications used for treating livestock ailments (Eiki et al. 2022). The study identified 15 veterinary medicinal plant species from 10 families. Leaves were the most widely used plant parts for ethnoveterinary medicine (71%), followed by bark (14%), stem (8%), and root (7%). Fresh components were preferred in medical compositions. Oral administration was the most common method (42.76%), followed by cutaneous (topical) administration (36.18%) (Eiki et al. 2022). The majority of ethnoveterinary medications were administered orally or topically after being crushed and soaked in water. Specific plants like *Ziziphus mucronate*, *Combretum collinum*, and *Colophospermum mopane* are used to treat diarrhea, *Ziziphus mucronate* is used to treat mastitis, *Aloe esculenta* and *Salvadora persica* are used to treat skin infections and *Ximenia americana* and *C. imberbe* are used to treat eye infections in cattle, sheep, and goat (Eiki et al. 2022). The authors stated that further research is needed to determine the chemical components, minimum inhibitory concentrations, biological activities, and toxicities of these plants, as well as to explore plausible mechanisms for achieving the reported therapeutic effects (Eiki et al. 2022). Another field survey in the Wolaita Zone of south Ethiopia involving 90 proficient healers (Wendimu et al. 2023) revealed that 54 plants from 28 different families were used to treat 39 livestock illnesses. Leaves constituted the majority of plant parts used (49%), followed by herbs (9%). Remedies were administered through various routes including nasal, oral, topical/dermal, and ocular. Blackleg, bloat, and endoparasites showed the highest Informant Consensus Factor (ICF) values. *Withania somnifera* was identified as the most potent remedy for blackleg, while *Zingiber officinale* had the highest fidelity level (FL = 94%) for treating bloat (Wendimu et al. 2023). *Croton macrostachyus* was widely known and utilized in the community for various purposes. *Stephania abyssinica* had the highest mean cultural importance, followed by *Pentas schimperina L.* The main threats to medicinal plants in the study area were agricultural expansion, drought, and construction. Thus, conservation efforts for therapeutic plants are deemed essential, with responsibilities lying with local communities and other relevant organizations (Wendimu et al. 2023). In northern Nigeria, where modern veterinary services are scarce and costly, a study involving 50 livestock farmers, reported that traditional remedies, comprising plant extracts, seeds, leaves, tree barks, tubers, and roots, are widely used due to their local availability and affordability (Alawa et al. 2002). The combination of these ingredients suggested potential additive, synergistic, and nutritional effects in treating various animal diseases, highlighting the importance of indigenous knowledge alongside conventional methods (Alawa et al. 2002). In Cameroon, the use of EVPs among smallholder farmers declined following the introduction of orthodox veterinary medicine during colonization (Tumasang et al. 2023). However, there has been a resurgence of interest in ethnoveterinary practices in recent decades, with efforts to document and validate traditional knowledge (Tumasang et al. 2023). A plethora of studies from other regions of the world (Luo et al. 2022; Rehman et al. 2022) underscore the importance of ethnoveterinary knowledge in sustainable livestock production and highlight the potential of traditional herbal treatments in addressing livestock health issues globally.

21.7 Technologies for Housing and Milking Management in Various Input Systems

Housing is required to protect animals from inclement weather and provide clean, comfortable stay for good health and efficient management. Different housing systems exist for cattle, poultry, and pig. However, the evolution toward more sustainable farming systems is occurring worldwide, and buildings are becoming an integral part of modern farming systems, which contribute greatly to the efficiency of operation, the quality of the products, and the health and comfort of livestock and workers. The sustainability of animal production depends on using modern housing systems, which ensure high productivity and quality of products (meat, milk and eggs) with the possibility of robotization of work, simultaneously minimizing the negative impact on the environment, including energy inputs. New technologies are rapidly integrating into agricultural systems and have altered the standard ways livestock, especially dairy cattle and chicken, are housed (Galama et al. 2020; Klopčič et al. 2019; Beaver et al. 2019).

21.7.1 Housing Innovations

Recent changes and expected developments in housing systems for dairy cows create an appropriate production environment for modern high-producing dairy cows and stimulate dairy farming-related developments (Klopčič et al. 2019). Before the 1970s, tie-stall (TS) housing systems were common (Fig. 21.5a). The TS system fits bucket milking systems and hay feeding and makes it relatively easy to keep the animals clean; however, it is not ideal for animal welfare because of the lack of freedom of movement, limited space per cow, and the potential for teat and leg injuries (Galama et al. 2015). Moreover, TS are not optimal for labor efficiency (Beaver et al. 2019). These factors explain the gradual change from TS to Cubicle barns (CB) since the 1970s (Bewley et al. 2017). Increased labor efficiency has been the most important driving force of the change from TS barns to cubicle barns (also known as free-stall barns). The transition to CB (Fig. 21.5b) was stimulated by developing parlor milking systems, milk cooling in a tank, availability of grass silages, and other innovations. Bewley and colleagues (Bewley et al. 2017) also described the ease of separating lactating cow groups and feeding in CB. The higher labor efficiency in CB has allowed for farms to be scaled up. Nevertheless, cows walk on a concrete floor, which can cause hoof problems (Kester et al. 2014). In addition, the manure product from these cows changes from feces mixed with straw and urine in TS to slurry in CB. Since the mid-1980s, after approximately 15 years of experience with CB, milk production has increased through improved technology, genetics, milking, and feeding management (Khanal et al. 2010). Still, animal welfare issues and minimizing ammonia emissions during field manure application increased in importance. The need for cow comfort was addressed by innovations such as mattresses in stalls; the use of deep sand, straw, and manure solids as bedding; innovative cubicle partitions that decrease cow injuries; open sides of barns; curtains to regulate ventilation; higher and insulated roofs; and wider walking areas (Galama et al. 2015). Sand is considered the gold standard for deeply bedded cubicles, but handling sand-loaded manure poses some challenges (Palmer and Holmes 2005). Environmental and cattle welfare requirements are becoming even more stringent in any production system. In addition to ammonia emissions, decreasing the emission of GHG (related to climate change) has gained importance (Steinfeld et al. 2006). New techniques, such as capturing emissions in the barn and separating feces and urine, are being tried. However, as the aesthetics of buildings in the landscape became an issue due to the con-

Fig. 21.5 Tie-stall barn (**a**), cubicle barn (**b**), and free walk barn (**c**) dairy cow housing systems

struction of larger barns and the need for more space per cow and increasing herd sizes, farmers started searching for alternatives to CB. As a result, the option of free-walk barns, such as bedded-pack barns, became a research focus. Proposed future housing systems with indicated benefits of animal welfare and manure quality as well as trade-offs (Galama et al. 2015) consider the following: (1) more space per cow versus more ammonia and nitrous oxide emission, (2) using waste materials as bedding versus increased food safety risk, (3) limited availability of bedding material (mostly wood chips or sawdust) at an affordable price, and (4) larger buildings versus landscape quality (Galama et al. 2020).

The natural behavior of cows, climate control, emissions of ammonia and GHG, reuse of waste, manure quality, the aesthetics of buildings in the landscape, and capital efficiency are now very important elements in modern cattle barns. Although technical innovations in feeding and milking made the transition to CB possible, the prevalence of lameness and hock lesions can be high in CB, especially on concrete flooring, including both slatted and solid floors (Leso et al. 2020). However, free walk (FW) housing systems, especially bedded-pack barns, have become of interest globally (Bewley et al. 2017) (Fig. 21.5c). Nevertheless, the implementation of FW systems is still in development. Even so, FW systems are spreading in several areas, where the evolution toward more sustainable farming systems is occurring worldwide.

Writing about future housing systems for cattle, Galama, et al. (2020) reported that approximately 55 Dutch farms as of 2019 have built an FW bedded-pack barn system, mostly with wood chips as bedding material. Half of these farms have adopted a greenhouse-type building, resulting in low roof construction investment to compensate for the greater area (m^2) per cow. A European Union Free Walk program involving eight European countries recently evolved due to knowledge exchange intensification between farmers. The project aims to develop further FW cattle housing systems, which are expected to improve animal welfare and soil structure, utilize waste products, and have greater public support than CB systems. The smart technology objectives of the FW system are climate-neutral development (20% reduction in GHG compared with 1990 data; 16% sustainable energy; energy-efficiency +2%/year); maintain grazing (81.2% of farms should use grazing); continuous improvement of animal welfare and health (70% reduction in antibiotic use; longevity +6 month compared with 2011); improve welfare scores (in operation by 2017); maintenance of the environment (100% use of responsible soy; national phosphate usage at 2002 level; and a 5-kt reduction in ammonia emissions compared with 2011); and improve biodiversity (biodiversity tool available in 2017) (Galama et al. 2015).

Other practices to increase milk production on dairy farms include nutrition, disease, and milking process management.

21.7.2 Milking Preparation

Since dairy cows are kept with the primary objective of producing milk, optimal milk removal from the udder is necessary. Under low input systems with few cows, milk removal is by hand or with portable milking systems. With increased herd size for medium and large operations or systems, more productive and efficient systems are used. An important step in the milking process, whether by hand or with the use of milkers, is udder premilking preparation, which is key to producing milk of high quality while minimizing the risk of contamination and infections. Therefore, methods for udder premilking preparation should ensure that teats are properly cleaned and dried beforehand milking or machine attachment to minimize sediments and high bacterial counts in milk. Procedures for effective udder cleaning with approved udder wash sanitizers are in effect in many high milk producing countries. For example, the majority of approved udder sanitizers by the Dairy Farmers of Canada have chlorhexidine acetate and/or iodine and/or glycerin and/or hydrogen peroxide, as the main components. Steps for pre-milking and post milking teat disinfection by the Canadian Mastitis Research Network (http://www.mastitisnetwork.org/) include: (1) clean teat thoroughly with clean water using wet wipes/paper towel (disposable)

Fig. 21.6 A typical portable milking machine

to remove surface dirt, which can reduce the efficiency of disinfectant; (2) submerge teats entirely in to an approved dip solution for at least 30 s; (3) wipe teats thoroughly with a disposable paper towel to remove the dip solution, followed by milking (attach milking unit or by hand); and (4) post-milking, coat the milk film on the teat by dipping in to an antibacterial solution (this prevents the transmission of pathogenic contagious bacteria (e.g., *Staphylococcus aureus*) and improves teat condition.

Before 1993, cows' udders were mostly hand cleaned in the USA, but in 1993, the udder washer system was developed which facilitated udder cleaning in the holding pen before entering the milking parlor (Khanal et al. 2010). Different types of udder washers include water heater with automatic teat spraying systems, spray guns, or teat dipping operations. In 1993, approximately 3% of US farms used udder washers and 71% of those farms had herd sizes of more than 300 cows (Khanal et al. 2010; Short 2000).

21.7.3 Milking Management Technologies for Small-Scale Farms

Depending on the size of the herd or scale of operation, various systems have been used to accomplish the milking process. For small-scale producers, the milking process can be accomplished using portable milking machines, barn milking systems or milking with automatic take-offs (ATOs).

21.7.3.1 Portable Milking System

Portable milking machine or portable milker (e.g., bucket milker) is an easy system of milking for small cow herds. It is an efficient way of milking cows that are reared in extensive production systems. It is also suitable for nomads or pastoral systems because the portable milking machine can be moved from one place to another. Also, this system can be used to milk cows in the barn, field or to milk sick animals in isolated pens (Patil 2013a). A typical portable milking machine (Fig. 21.6) has oil free vacuum pump that is controlled by an electric motor; vacuum regulator and a gauge that helps to monitor and maintain the vacuum level during the process of milking; and pneumatic pulsator that can be easily adjusted with an adjustment screw located on the pulsator. The pulsator is very reliable, precise, studded, and simple to use. It is recommended that the vacuum level for milking a cow with a portable milker should be 48–50 kPa and the pulsation rate should be set to 60 ppm. In places where there is no electricity, a diesel engine or batteries can be used to power the portable milker. The milk can be collected in a sturdy stainless steel bucket mounted on a lightweight but very strong trolley for easy movement (Patil 2013a).

21.7.3.2 Barn Milking Systems

Barn milking systems are mostly used in tied stall barns having herd sizes of 100 or more cows. The milking is done by a portable milking unit, which provides pulsation and allows the vacuum to be delivered to the end of the teat. The portable milking unit is connected to the milking station and fixed on the milk line. In barn milking systems, a milking station can serve two cows. The milk moves to the milk receiving set or receiving vessel having a liquid control level, sanitary tap, milk pump, and milk filter. Afterward, the milk flows from the milk receiving vessel into the milking tank (Patil 2013b). The portable machine has vacuum pumps that is regulated by vacuum regulator in order to maintain a desired level. It is recommended that one vacuum line loop should be used for every 50 cows. This makes it easy for the milk to be pumped from the receiving set installed on the vacuum loop directly to the milk cooling tank. Automated washing system can be installed in the barn for adequate cleaning after milking to wash milking equipment that was used in milking including the portable milking machine. The water slug can be fixed in the milk line using air injector valve for proper washing (Patil 2013b).

21.7.3.3 Milking with Automatic Take-Offs

Mastitis, an inflammation of the mammary gland, can occur in cases where cow's udder is milked beyond the point where they have stopped releasing milk. Also, under-milking cows might cause reduction in milk yield and discomfort to cows. Therefore, ATOs are very essential because they have sensors that sense when milk flow has stopped. Once the milk flow stops, the ATOs switches off the vacuum and automatically removes the milking unit from the udder. Mechanical or electronic ATOs are used in milking parlors to ease human labor by reducing the number of times the operator needs to bend on the knees to examine if milk flow has stopped and it increases the number of cows to be milked (Tranel 2008).

21.7.4 Milking Management Technologies in Medium to Large Production Systems

21.7.4.1 Milking Parlors

Milking parlors are mostly used on medium-scale, large-scale, and sometimes on small-scale dairy milk production farms. In milk parlor system, cows are milked in stalls mostly raised above the ground for easy milking. Milking parlors are arranged in herringbone, swingover, rapid exit parallel, flat, or rotary manner. Milking parlors in general are cost-effective for large farms with 100 or more cows while automated flat parlors are used mostly by small dairy farmers wishing to change from stall to free stall housing (Tauer 1998). Milking by majority of farmers is done twice a day, in the morning and evening. In order to use milking parlors more efficiently and increase milk production per cow, milking can be performed three times daily. Research has shown a 6–19% increase in milk production when cows were milked three times per day (Gisi et al. 1986). Milking cows three times daily tends to maintain milk quality in late lactation stage (Sorensen et al. 2001). Another alternative to the milking parlor system is the bucket milkers or pipeline systems common in stanchion or tie-stall barns. Sample milking parlors are shown in Fig. 21.7.

Parallel Milking Parlor In this system, more cows can be loaded in the stall and within a short time. It consists of a vacuum operated gate lift and the building is very wide in length. This system can easily locate the exit gate in front of the cow stall making it a fast and efficient way of exiting and reloading cows (Patil 2013a). A recent parallel milking design by DeLaval company (DeLaval's P500 parallel parlor, www.delaval.com) has independently controlled, suspended, non-overlapping sequence gates, with ability to lift up during milking, allows easy and calm movement of cows (minimize stress on

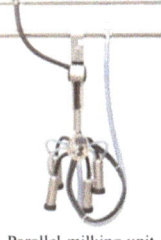

Fig. 21.7 Different parlor milking systems

cows) as well as reduce the size of parlor footprints.

Herringbone Milking Parlors This is the most used milking parlor in medium-scale dairy farms housing about 600 cows. With this system, the milkers or milking machines are positioned in the center or the spine (like spine of a fish) or center aisle, between two aisles with enough space or room for the cows. During the milking hour, the animals enter the parlor and line up between each spine, creating two rows of cows followed by milking by the milker. The milker washes the cow's udder and teats and then fix the cups to the udder from the rear of the hind legs. The size of the herringbone milking parlor ranges from a double 8/8 to 20/20, which means that both sides of the milking parlor have the same number of stalls, with each stall angle ranging from 30 to 90 degrees. Only one milking control is required for each stall in a double parlor (Patil 2013a).

Swingover Milking Parlor The Swingover parlor is similar to the herringbone parlor, except that two cows on the same row share a set of milking cups. This means that two cows can be milked at the same time and if one cow is done milking, it can be moved out for another to enter. In the swingover style, two sides of the parlor have the same control and therefore, it is mounted at the center of the pit. With the ram and the cluster fixed to it, it is able to move from one side of the parlor to the other side; hence, making it cost-effective for a double parlor. The swingover parlor is cheaper to install.

Rotary Milking Parlors The rotary milking parlor is a circular raised platform on which the cow stands, allowing attachment of the milking machine by the farmer or operator from below. During use, the system rotates slowly allowing cows to enter and exit the platform easily when a stall is free or at regular intervals. The rotary milking parlor is the least labor-intensive form of milking, and it is suitable for medium to large farms. It is bigger (can milk more than 1000 cows) and milking is also quicker than in the herringbone parlor. Herringbone parlor mostly accommodates 20–40 points per side whereas a typical rotary parlor is made up of a turntable having 12–100 individual stalls for cows around the outer edge (Patil 2013a). The turntable is controlled by an electric motor drive and it is designed in a way that one turn represents the time in which a cow is totally milked. Once an empty stall passes through the entrance, a cow steps on it facing the center and then the turntable rotates with the cow. This is repeated until all the cows are milked. The milker washes the cow udder before fixing the milking machine cups. Afterward, the cows are milked as the platform rotates and once the udder stops releasing milk, the ATOs or milker removes the cups from the cow's udder, and the cow is allowed to exit the stall just before the entrance. Two operators can milk 250 cows per hour on a 60 point platform (Patil 2013a).

Pipeline or Milkline Milking System The pipeline milking system is commonly used in tie-stall barns and milking parlors. It is a portable milking unit that provides pulsation, delivers vacuum to the teat end and at the end of milking, delivers the milk directly to the cooling tank through a pipeline or milkline system. In organized milk recording systems, a milk meter flask-based is installed at every milking stall for the measurement of milk produced by each cow.

The pipeline milking systems are the most predominant type of milkline systems on dairy farms while some producers make use of a less common milking system known as weigh jar systems. The pipeline milking systems can perform two functions at a time including vacuum supply and the movement of milk from the milk unit to the receiver in milklines. While in weigh jar milking systems, only milk can be transported from weigh jars to the receiver while vacuum lines move only air from weigh jars to the sanitary trap. Furthermore, the milk is transported from weigh jars via the sanitary trap by the air pressure in the weigh jar and vacuum pressure in the milk receiver. Another sanitary airline provides milking vacuum to the weigh jar during the process of milking cows and also supply water to the weigh jar during cleaning (Reinemann 2013). The weigh jars are placed below or almost at the same position of the cow udder for easy milking. Even though the size of the milk transfer lines does not affect the milking performance, because they are only used to empty weigh jars after milking and also looping (enhances cleaning process and emptying of weigh jars), it is expedient that the capacity of the milk transfer line is as large as the line supplying water used for cleaning the weigh jars and milk meters (Reinemann 2013).

In dual-purpose milk pipeline systems, the pipeline uses gravity to transport milk from the milkline to the receiver. Low-level, high-level, and mid-level pipeline systems are used in milking parlors and they are located under the cow udder, a little above the cow udder and high above the cow udder, respectively. The dual milk pipeline systems are also used in round-the-barn milking system, which is a type of facility that milks cow in the same place (tie-stall or stanchion barns) where the cows are kept (Reinemann 2013).

21.7.4.2 Computerized Milking Systems or Robotic Milking and Feeding Systems

Computerized milking systems also known as automated robotic milking systems is a system that automatically attaches cups to the udder teats of cows. AMS allows voluntary milking of motivated cows three to four times daily as it pleases them, which has been shown to increase milk production by 5–10% (De Koning 2010). AMS help cows to relieve the pressure of milk in their udders. However, if there is no milk in the cow udder and the cow enters the AMS, the cow will be automatically turned away. The AMS are also connected to automatic feeding and cleaning systems. Lactanet (www.lactanet.ca), a Canadian Dairy Production Center of excellence and established by the Dairy Farmers of Canada, has introduced a new electronic milk recording service called electronic Dairy Herd Improvement, which allows farmers to record herd data remotely without the help of a technician, but this service is only available to farmers using AMS and milk parlor systems on their farms (Lactanet 2021). In New Zealand, the AMS was used on cows reared in paddocks or on pasture-based systems following the Greenfield project in 2001 at DairyNZ's Greenfield Project farm. In this scenario, the cows on the Greenfield farm move from the paddock to the AMS and then back to the paddock (Science Learning Hub 2017). The movement of the cows is directed by installation of temporary fences and computer-controlled gates. Also, the cows are moved to the robotic milking stall paddock by giving them incentives, such as water or fresh grass. The AMS is used in identifying individual cows, milking cows, recording individual cow behavior, milk quality and quantity, and cow's health status. After the 2001 Greenfield project, several farms (organic pastoral farms, fully housed systems, and herds ranging from 180 to 800 cows) in New Zealand are now making use of AMS for milking their cows (Science Learning Hub 2017).

In Sweden, a large Swedish producer of dairy and farming equipment, DeLaval International installed 24 AMS (Robotic Voluntary Milking System) that are capable of milking 1500 cows at TDI Farms LLC (Westphalia, Germany) (Amelinckx 2016). The producer installed the AMS to increase the efficiency of his farm through low cull rates and high cow longevity, since the AMS has been programmed to monitor udder health and identify mammary gland infections at an early stage. This firm also installed another 64 Robotic Voluntary Milking Systems in Los Angeles that can milk 4500 cows (Amelinckx 2016). In Michigan, there are presently 243 AMS installed on 55 farms (Michigan Department of Agriculture and Rural Development data (Malacco 2022)). The reasons for adopting AMS are different from one farm to the other. Small dairy farms adopt AMS to attain more flexibility in work schedules and better well-being of their workers while large dairy farms adopt AMS because of shortage of labor as well as to reduce the number of hours spent on cows per day (Malacco 2022). In general, the use of AMS on farms has resulted to reduction in the cost of labor (84.6%), improves welfare of cows (76.9%), reduce human labor (69.2%), improves overall cow well-being (50%), improves cow reproductive performance (36%), and improves herd performance (73.1%) (Malacco 2022). Although the use of AMS is very efficient, the decision to transit from conventional methods to AMS depends on the size, needs, and objectives of the farm operation (Malacco 2022).

The production of AMS is undertaken by many companies including Lely (Lely 2022), DeLaval (Amelinckx 2016), and GEA (GEA 2022a). For instance, GEA Farm technologies, a German company has built DairyProQ, a robotic rotary system that is in use on two dairy farms in Germany and two in Canada (GEA 2022b), and also GEA Dairyrobot 9500 robotic milking system that has been engineered to improve the milking process (GEA 2022c). Due to the flexibility in the machine design, it supports voluntary milking of cows and the milking of cows in group at specific times of the day. The system can also be used in milk parlors. The GEA's in-liner performs all the steps involved in milking cows, such as stimulation, teat cleaning, fore-stripping, milk harvest, and post-dipping just in a single attachment. The GEA DairyMilk M6850 sensor technology monitors individual cow udder quarter during milking without using chemical consumables or reagents. This machine facilitates early detection of infected quarter(s) (infected with mastitis pathogens) due to their pioneering sensor-based technology (GEA 2022c).

At the University of Melbourne Dookie Dairy Farm, daily milk production, milk temperature, milk quality (somatic cell count, milk fat, and protein%), cow weight, and concentrate intake are collected automatically using a Lely Automatic Milking System, which also identifies individual cows via RFID ear tags (Osei-Amponsah et al. 2020). Additionally, each cow is fitted with a transponder (Qwes-HR, Lely) that contains a rumination monitor. The rumination monitor uses a microphone to detect chewing sounds and differentiates between eating and rumination time (Osei-Amponsah et al. 2020).

21.7.4.3 Using a Feed Nutrition Expert

In large-scale milk production, it is very essential to understand and provide adequate nutrition to improving cow health and milk yield. Proper feed management practices can help to reduce the excretion of specific nutrients in manure. The USDA-Natural Resources Conservation Service defined feed management practice standard as managing "the quality of available nutrients fed to livestock for their intended purpose." As a result, it is important to have an animal nutritionist with experience in dairy feed formulation and

Fig. 21.8 Sample pasteurization system for large-scale use (Tubular Aseptic UHT Pasteurizer with Vacuum Deaerator)

nutrient management to improve animal health and promote milk yield.

21.7.4.4 Technologies for Pasteurization, Cooling, and Storage in Low to High Input Systems

Milk pasteurization is a process by which milk is subjected to heat treatment to kill or inactivate organisms that cause spoilage, which in turn extend milk shelf life and promote food safety. Different thermal processing methods are commonly used, such as thermization (57–68 °C for 15 min), batch pasteurization (63 °C for 30 min), flash pasteurization (72–74 °C for 15–20 s), ultra-high temperature (UHT) pasteurization (135–140 °C for 2–4 s) (Fig. 21.8), canned pasteurization (115–121 °C for 10–20 min), and high temperature short time (HTST) pasteurization (72 °C or 161 °F for 15 s or above) systems (Aryal 2019). Generally, the steps in milk pasteurization include chilling of milk (2–5 °C) (to arrest bacterial growth), pre-heating (40 °C), and standardization to separate butter fat, clarification (to remove milk sediments), homogenization (to break down the milk fat globules into tiny droplets to prevent milk separation), heating (from 60–72 °C in order to kill microbial spores, such as *Clostridium botulinum* spores), milk holding (where milk is held at a constant temperature for at least 16 s in the holding tubes) and milk cooling/chilling of pasteurized milk to 4 °C before pumping to the packaging machines for aseptic packaging and storage in the cold room (Aryal 2019). Pasteurization systems for small scale use are also available, such as P-3000 SafGard 2 gallon and Milky FJ50PF multipurpose 12-gallon milk pasteurizers (Farmandranchdepot.com). There are different types of commonly used cooling systems including direct expansion cooling system with pre-cooling of milk, ice bank cooling system without pre-cooling, ice bank with pre-cooling of milk, and solar thermal solutions (Upton et al. 2015).

21.8 Technologies to Improve Competitiveness in Pastoral Production Systems

Pastoral farming, also known as nomadic farming, refers to a farming system whereby farmers rear animals (e.g., cattle, sheep, goat, camel and donkey) on a large scale and move them from one place to another in search of water and food (see Chapter 2 for more information on pastoralism in Africa). Pastoralism is a traditional system of livestock farming mostly practiced in the arid and semi-arid areas of East Africa, Southern Africa, North Africa, West Africa, and Central Africa. Pastoralism is mostly practised by persons belonging to some ethnic groups including the Tuareg, Fulani, Mbororo, Peul, Maure, Ababdeh, Afars, Beja, Berbers, Borona Oromo, Dinka, Gabra, Karamojong, Maasai, Mrazig of Tunisia, Nuer, Pokot, Rendille, Sahrawis, Samburu, Somali, Toubou, Trekboers, and Turkana. More than 90% of meat and over 50% of milk produced in East Africa comes from pastoralists (FAO 2022). Pastoral livestock (e.g., cattle) is not only meant for food production, but also for social status satisfaction, domestic fuel, and farm power (Suleiman et al. 2015). Countries such as Sudan, Kenya, Somalia, and Ethiopia generate revenue from exporting livestock from pastoralists to Gulf countries (FAO 2022). However, pastoral farmers are faced with a lot of challenges because pastoral farming is mostly practiced in the dry areas where crop production is less favorable, thereby causing an increase in livestock competition for land use. The growth in human population has resulted in an increase in the use of land meant for livestock grazing for other purposes, such as infrastructure, intensive crop agriculture, and housing. Clashes between pastoralists and crop farmers are becoming more rampant mostly when cattle feed on crops cultivated for human consumption or when a grazing land has been converted to cropland. Likewise, irrigation farming can reduce access to communal water and prevent dry season grazing for pastoral farmers. This has triggered the migration of some pastoralists to other parts of the country where they can have access to green pasture and water for their animals. For example, in Nigeria, the movement of cattle from the northern parts to the southern tsetse fly-infested zones may increase exposure of the animals to tsetse fly bites and eventually trypanosomiasis. The animals can also be exposed to other diseases like foot-and-mouth disease, mastitis, pneumonia; therefore, reducing animal productivity, and numbers through death of animals (Suleiman et al. 2015). The improvement of pastoral farming is therefore necessary for optimum animal performance and income generation.

21.8.1 Best Management Practices for Pastoral Farming

Rotational grazing is one of the best management practices that pastoral farmers can utilize for better animal management (MacPhail and Kyle 2012). It is a situation whereby cows are moved from one area to another with the intent of allowing only a portion of the grazing area to be grazed at a time while the other portions rest (Undersander et al. 2002). This is to give the pasture and soil enough time to recover. Rotational grazing helps to increase pasture quality, promote cattle health, and lead to sustainable environment. For good pasture management, the number of animals must be balanced with available forage supply, livestock must be evenly distributed but should be excluded from sensitive areas when conditions are not suitable for grazing (pastures too wet or dry), and there must be balance between the period of grazing and rest to maintain the pasture. Paddock (an enclosed field of grassland that is meant purposely for pasturing or grazing livestock) is increasingly finding being adopted for pastoral livestock management. Fences, such as permanent page-wire, barbwire, or high tensile strength electric fences can be used to surround the grazing field while temporary poly-wire fences can be used to subdivide fields into smaller paddocks (MacPhail and Kyle 2012). There must be access to drinking water situated in close proximity (about 800 feet is recommended) in each paddock. Windbreaks can be planted around the paddocks. Also, mixture of forages should be grown in each paddock because every forage has their own susceptibility and resistance to insect, disease, or environmental conditions (mixture of legume and grasses) (MacPhail and Kyle 2012; Undersander et al. 2002).

In Africa, pastoral cattle feed mostly on native pastures; therefore, there is a need for adaptation of the technology that can improve pasture quality, preserve natural resources, and boost soil health. Farmers in New Zealand are the first adopters of technologies to improve forage quality (Caradus et al. 2020). This is because farmers in New Zealand practice more of pastoral farming when compared to other developing countries. For instance, the cultivation of the herb, chicory, is gaining importance in New Zealand because it can be used to improve soil fertility. Chicory has high mineral content, and it is deeply rooted and can take up minerals and nutrients in the soil for proper growth and development. The perennial ryegrass is also another forage crop used by pastoral farmers for cattle feeding (Caradus et al. 2020). It lives in a symbiotic relationship with a fungus known as "endophyte" to protect itself from insect pests (such as argentine stem weevil, black beetle, root aphid, pasture mealy bug, and porina). AR37 is the most recent endophyte of ryegrass commercialized in New Zealand. It does not produce alkaloid compounds lolitrem, peramine, or ergovaline, but it production of indole diterpene like compound, epoxy-janthitrem, makes it unique when compared to another endophyte, AR1 endophyte which produces more of ergovaline. Re-grassing of ryegrass with AR37 only requires 8 years interval. Research has shown that there was a 12% increase in the DM of ryegrass containing AR37 endophyte. Likewise, cattle fed with AR37 ryegrass had increased production of milk solids with no adverse effect on animal health (Caradus et al. 2020).

21.8.2 Precision Livestock Technologies for Better Productivity in Pastoral Farming

Precision livestock farming is the proper use of data analysis and modeling for the purpose of informing management about animal reproduction status, feed regimen, and productivity status. All the information gathered can be used to understand and predict animal health and welfare (Neethirajan and Kemp 2021). Precision livestock technology is a powerful tool that can improve pastoral farming in terms of animal productivity capacity, welfare, or management (Odintsov Vaintrub et al. 2021). Recently, technologies, such as biometric sensors, have been developed to monitor behavioral and physiological parameters of livestock. These enable farmers to monitor animal's health and welfare remotely. Biometric sensors can either be invasive or non-invasive. Invasive sensors are used for monitoring the internal behavior of the animal, such as the rumen health, body temperature, and vaginal pressure. However, they are not mostly used in animal production, because they are typically swallowed or implanted in the animal. The predominant sensors used in the livestock industry are non-invasive sensors which are also suitable in pastoral livestock farming. These include surveillance cameras, sensors to measure the feeding system of the animal, and to monitor animal weight gain and feed intake (Neethirajan and Kemp 2021). Other technologies that can be adopted by pastoral farmers include drones for checking fence lines, water-troughs, and for monitoring pasture quality more effectively through aerial images and video; RFID for identifying or tracking tags on animals, to monitor human-animal or cow-calf interactions, image analysis for detecting excessive mounting, accelerometer for estimating locomotion, and so on. However, all the information gathered by this technology needs proper interpretation for the purpose of making informed decisions (Odintsov Vaintrub et al. 2021).

21.8.2.1 Electronic Animal Identification, Sensors, and Automatic Drafter and Weighing Systems in Pastoral Farming

Electronic identification, like RFID system, is used in pastoral farming for independent identification of individual animals for the purpose of collecting data and storing them for decision making (Odintsov Vaintrub et al. 2021). The RFID operates on divergent level of radio frequency (noise to signal ratio): low frequency (LF, 125–134.2 kHz), high frequency (HF, 13.56 MHz), and ultra-high frequency (UHF, 860–915 MHz). The HF and UHF allow anti-collision, longer distance, less noise, and stable connection. However, LF is less stable, but it can pass through obstacles when compared with HF and UHF if a larger antenna is used (Odintsov Vaintrub et al. 2021). Passive EIDs tags can be used for identifying animals and it involves the use of copper coils that charge transmitters through the energy that passes through an active reader to store information. Passive EID does not require a battery, and this makes it ideal for pastoral farmers (Holje 2012). However, ear tagging is the most common method used by farmers for animal identification because it is easy to use, less costly, and does not require a high level of expertise. The EID can also be enclosed in a ceramic bolus and inserted into the rumen. This bolus EID is permanent and cannot easily get lost or malfunction. However, bolus EID is only suitable for matured animals, while smaller boluses can be inserted in smaller animals for greater efficiency (Hentz et al. 2014).

Sensors (discussed in Chapter 20), are increasingly used in pastoral farming. For example, sensors can be used to create virtual fencing for pastoral livestock management. Virtual fencing serves as electronic barrier in place of physical barrier. Animal movements are restricted by a visible and/or audible system attached to an electric stimulus. Virtual fences are used in moving animals wearing a sensor remotely from one area of pasture to another (Campbell et al. 2020). However, the use of virtual fencing may pose risk to animal due to their long-term exposure to the electric stimulus; thereby affecting their welfare and also arousing public concerns (Lee et al. 2018). With the use of global positioning system, collar sensors connected to the global system for mobile communication or connected to a Sigfox network, and with low-cost Bluetooth tags, animals' movement in the grazing area is tracked remotely using a software, thereby keeping farmer informed of animals' behavior and location at all times.

Automatic drafter, weighing crates, or walk overweight are examples of stationary tools that are placed in the gate of a paddock for the purpose of distinguishing and directing the movement of animals. Automatic drafter recognizes the EID of individual animals which facilitates data collection, feeding control and labor efficiency during animal selection for sale, treatment of diseases, and sheering. Other systems like walk overweight, which only allows animal to pass through one way for the purpose of weighing, is placed at a specific angle of a paddock for communication with individual animal EID for data collection.

21.8.2.2 Machine Learning and Artificial Intelligence in Pastoral Farming

Machine learning (discussed in Chapter 20) is valuable in pastoral farming in the areas of animal monitoring, data capture, and prediction of disease and soil properties for pasture management. Regression algorithms are the backbone for predicting soil properties, weather, and crop yield. Deep learning algorithms (convolutional neutral network) and machine learning (support vector machines and random forest) can be used in identifying disease and weeds in forages planted (Sharma et al. 2021). Machine learning algorithms can be used to determine the maximum amount of fertilizer to be applied to soils before planting forages for animal grazing. Cameras attached to drones are used for monitoring pasture growth, spray pesticides, and for drip irrigation. Deep learning can therefore be used to examine all the images captured by drones for the forages grown, for disease and weed identification. The drones can then be used to spray pesticides on infected grasses and weeds. Drip irrigation attached to drones is an effective system of irrigation because they can denote weather pattern and can reduce water problems for farmers drastically. Likewise, robots can be used alongside with drones for field monitoring (Sharma et al. 2021).

There are different technologies that can also be used to monitor soil and pasture quality. Internet of Things (IoT) is a smart farming system that can be used to monitor soil nutrients and soil moisture content using sensors (Sharma et al. 2021; Tekin et al. 2021). The use of remote sensing, such as unmanned aerial vehicle, manned aircraft, and satellite imagery can be used to predict grass yield and quality. Automated grass measurement system, Grasshopper (GH), is another tool for pasture management (Odintsov Vaintrub et al. 2021). It evaluates herbage mass by means of a microsonic measurement device and transmission of real-time data. The GH is used in measuring the time of return of a sonic transmission from its reflective circulate plate that is placed on a grass emitter (Odintsov Vaintrub et al. 2021). Global positioning system is used for the integration of data that provide information needed for the planning of animal movement and pasture availability for the year (di Virgilio et al. 2018).

21.9 Challenges, Future Perspectives, and Conclusion

Livestock farming is the cornerstone of agriculture in Africa, engaging about 90% of Africa's rural population. However, sustainable smallholder livestock farming is hindered by many factors hitherto discussed in Chapter 20 (Section 20.2) including (1) limited access to modern technologies and information; (2) limited technical and technological capacity; (3) poor/lack of supporting infrastructure; (4) limited access to markets and/or lack of market orientation and/or lack of market infrastructure and/or weak market linkages; (5) inefficient/insufficient input and service delivery; (6) limited feed resources and grazing lands; (7) limited and/or lack of private sector investment; (8) limited financial capacity; (9) limited government support; (10) weak regulatory institutions; (11) lack of supporting government policies; (12) limited or lack of robust collaboration structures among relevant stakeholders (farmers/producers, research institutions, agro-industries, and Government) in the livestock sector; (13) climate change, other environmental factors, etc. (Gebreyohanes and Mwai 2021). The adoption of modern technologies and in particular, genetic technologies have been responsible for majority of improved gains realized in several livestock traits. For example, genetics selection alone accounts for 85–90% of the changes in growth rate of broiler chickens irrespective of the significant improvements in poultry nutrition and management (Havenstein et al. 2003). Moreover in year 2001, a typical broiler chicken was ~5 times as large as a 1957 broiler at 42 days of age (Havenstein et al. 2003). In general, the adoption of modern technologies in all areas of livestock management and breeding will positively impact animal agriculture in Africa. For example, the results of a recent study of 291 smallholder livestock producers in four communities in Zimbabwe, indicated that emerging technologies have significant and positive impacts on risk management activities since the use of digital technologies enhanced risk mitigation and value chains in smallholder livestock farming, thus indicative of a strong need for the adoption of emerging technologies in smallholder livestock farming (Gwaka and Dubihlela 2020). The outcome of this study is not surprising since Delgado and colleagues suggested in 1999 that progress in technological innovations in the production, processing, and distribution of livestock products will be central to achieving a positive outcome of the "Livestock to 2020 : the next food revolution" (Delgado et al. 1999). Within the Africa Union Agenda 2063, increasing livestock productivity is a key component and the adoption of innovative technologies will be crucial in attaining objectives of the livestock sector.

The efficient application of modern technologies in enhancing animal productivity requires a commercial setting. The dominant smallholder systems in the livestock sector in African possess a challenge in terms of the cost-efficient application of these technologies due to high cost and inadequate technological skills. However, several pilots of these technologies or application of these technologies have been attempted for the dairy, beef, and poultry sectors in Africa.

Over the past decades, the adoption of innovative modern technologies has seen transformation in various livestock sub-sectors including a four-fold increase in broiler chicken body size with the same amount of feed input and doubling in the amount of milk produced by a Holstein cow due to implementation of technologies, discussed in sections above, to improved feed efficiency and milk production (Ibeagha-Awemu and Yu 2021). It was indicated recently that improved livestock productivity relies on enhancing the rate of genetic gain for target phenotypes or outcomes and breeding programs, for example, will embrace enhanced measures that increase selection accuracy, intensity of selection, and genetic diversity while reducing the generation interval, such as improved management practices, improved nutrition, improved health care, improved genetic/genomic selection, and reproduction technologies (Ibeagha-Awemu and Yu 2021)..

In some African countries, the introduction of modern technologies has resulted in increased livestock productivity and farm incomes. In Senegal, the implementation of a dairy processor and a milk collection center, representing a new format of smallholder livestock development, created a new dynamism in production activities such as extension services and a consistent market opportunity through increased local productivity and increased incomes (Habanabakize et al. 2022). To harness the growth and opportunities, however, it is imperative to establish robust and continuous collaboration structures between the Government, agro-industries, and producers or farmer clusters in a public-private partnership that will help producers lower the cost of inputs and services and increase productivity while boosting local dairy production (Habanabakize et al. 2022).

The African Chicken Genetic Gains (ACGG) program, now Tropical Poultry Genetic Solutions (TPGS) (https://africacgg.net/) is a practical example of how modern technologies and technical know-how with the participation of relevant expert institutions harnessed in a farmer-oriented manner delivered and achieved increased livestock productivity and farm incomes. The ACGG was initiated in 2014 in Ethiopia, Nigeria, and Tanzania, and recently extended to include Kenya, Ghana, Zimbabwe, Cambodia, Myanmar, and Vietnam, leverages numerous partnerships and innovation platforms to engage farmers, private sector actors, policymakers, and professionals (geneticists, human and animal nutritionists, and social economists) to deliver better and more contextually appropriate chicken genetics to smallholder farmers. The TPGS is working to characterize chicken ecotypes based on farmer preferences in order to develop chicken lines that are better-producing birds with low inputs and environmentally resilient (Dessie 2021).

Given the limitations of the smallholder system, large scale and cost-efficient application of modern technologies will require public-private partnerships between farmers, private industry, and embryo transfer practitioners as the case in Brazil (Viana et al. 2017) to drive productivity in the livestock sector. In addition, the formation of farmers' co-operatives will facilitate large scale and efficient applications of modern technologies among members and provide market outlets for their products. More in-depth practical examples of the application of modern technologies in African livestock production systems are discussed in Chapters 22, 23, and 24.

These practical examples in Africa and described technologies in sections above demonstrate that the modern technologies necessary to advance African livestock productivity are available and applicable in African livestock production systems.

In conclusion, the modern technologies presented above and technical know-how have been key drivers to increased livestock productivity in Western countries, and practical examples in

Africa demonstrates that their adoption is feasible and can facilitate rapid progress in efforts to increase livestock productivity in various African livestock production systems. The presented technologies are not an exhaustive list of existing and future technologies but presents opportunities for their implementation for sustainable livestock production within the various livestock production systems in Africa.

References

Abdalla SE, Abia ALK, Amoako DG, Perrett K, Bester LA, Essack SY (2021) From farm-to-fork: E. coli from an intensive pig production system in South Africa shows high resistance to critically important antibiotics for human and animal use. Antibiotics 10(2):178. https://doi.org/10.3390/antibiotics10020178.

Abecia L, Martín-García A, Martínez G, Newbold C, Yáñez-Ruiz DR (2013) Nutritional intervention in early life to manipulate rumen microbial colonization and methane output by kid goats postweaning. J Anim Sci 91(10):4832–4840. https://doi.org/10.2527/jas.2012-6142

Abecia L, Waddams KE, Martínez-Fernandez G, Martín-García AI, Ramos-Morales E, Newbold CJ, Yáñez-Ruiz DR (2014) An antimethanogenic nutritional intervention in early life of ruminants modifies ruminal colonization by Archaea. Archaea 2014:841463. https://doi.org/10.1155/2014/841463

Abubakar AT, Ogundijo OA, Al-Mustapha AI (2022) Animal disease surveillance system key to understanding disease dynamics. Nature Africa, Comment 14 October 2022. https://doi.org/10.1038/d44148-022-00142-4

Adepetun A (2015) Africa's mobile phone penetration now 67%. The Guardian, 17 June 2015, Available at https://guardian.ng/technology/africas-mobile-phone-penetration-now-67/ (accessed 14 June 2022)

Adghim M, Abdallah M, Saad S, Shanableh A, Sartaj M, El Mansouri AE (2020) Comparative life cycle assessment of anaerobic co-digestion for dairy waste management in large-scale farms. J Clean Prod 256:120320. https://doi.org/10.1016/j.jclepro.2020.120320

Aggrey SE, Karnuah AB, Sebastian B, Anthony NB (2010) Genetic properties of feed efficiency parameters in meat-type chickens. Genet Sel Evol 42(1):1–5. https://doi.org/10.1186/1297-9686-42-25.

Ahlin K (2022) The robotic workbench and poultry processing 2.0. Animal Front 12(2):49–55. https://doi.org/10.1093/af/vfab079.

Ahmed SN, Chattopadhyay UK, Sherikar AT, Waskar VS, Paturkar AM, Latha C, Munde KD, Pathare NS (2003) Chemical sprays as a method for improvement in microbiological quality and shelf-life of fresh sheep and goat meats during refrigeration storage (5–7 C). Meat Sci 63(3):339–344. https://doi.org/10.1016/s0309-1740(02)00091-8

Alawa JP, Jokthan GE, Akut K (2002) Ethnoveterinary medical practice for ruminants in the subhumid zone of northern Nigeria. Prev Vet Med 54(1):79–90. https://doi.org/10.1016/s0167-5877(01)00273-2.

Alayande KA, Aiyegoro OA, Ateba CN (2020) Probiotics in animal husbandry: applicability and associated risk factors. Sustain For 12(3):1087. https://doi.org/10.3390/su12031087. WorldCat.org

Alemayehu M, Gezahagn K, Fekede F, Getnet A (2017) Overview of improved forage and forage seed production in Ethiopia: lessons from fourth livestock development project. Int J Agric Biosci 6(4):217–226

Alfonsetti M, Castelli V, d'Angelo M (2022) Are we what we eat? Impact of diet on the gut–brain axis in Parkinson's disease. Nutrients 14(2):380. https://doi.org/10.3390/nu14020380

Al-Husseini W, Gondro C, Quinn K, Cafe LM, Herd RM, Gibson JP, Greenwood PL, Chen Y (2014) Hormonal growth implants affect feed efficiency and expression of residual feed intake-associated genes in beef cattle. Anim Prod Sci 54(5):550. https://doi.org/10.1071/an12398

Al-Shawi SG, Dang DS, Yousif AY, Al-Younis ZK, Najm TA, Matarneh SK (2020) The potential use of probiotics to improve animal health, efficiency, and meat quality: a review. Agriculture 10(10):452. https://doi.org/10.3390/agriculture10100452

Amelinckx A (2016) Rise of the (Cow Milking) Robots. https://modernfarmer.com/2016/12/rise-cow-milking-robots/. Accessed 18 June 2022

Amole T, Augustine A, Balehegn M, Adesogoan AT (2022) Livestock feed resources in the West African Sahel. Agron J 114(1):26–45. https://doi.org/10.1002/agj2.20955.

Andersen M-BS, Rinnan Å, Manach C, Poulsen SK, Pujos-Guillot E, Larsen TM, Astrup A, Dragsted LO (2014) Untargeted metabolomics as a screening tool for estimating compliance to a dietary pattern. J Proteome Res 13(3):1405–1418. https://doi.org/10.1021/pr400964s

Andretta I, Pomar C, Rivest J, Pomar J, Lovatto PA, Radünz Neto J (2014) The impact of feeding growing-finishing pigs with daily tailored diets using precision feeding techniques on animal performance, nutrient utilization, and body and carcass composition. J Anim Sci 92(9):3925–3936. https://doi.org/10.2527/jas.2014-7643.

Andrew Selaledi L, Mohammed Hassan Z, Manyelo TG, Mabelebele M (2020) The current status of the alternative use to antibiotics in poultry production: an african perspective. Antibiotics (Basel, Switzerland) 9(9). https://doi.org/10.3390/antibiotics9090594.

Aryal S (2019) Milk pasteurization- definition, methods, steps, significance. https://microbenotes.com/milk-pasteurization-methods-steps-significance/. Accessed 30-05-2022

Ashworth AJ, Chastain JP, Moore PA Jr (2020) Nutrient characteristics of poultry manure and litter. In: Animal manure: production, characteristics, environmental concerns, and management, in Animal Manure, vol 67. American Society of Agronomy and Soil Science Society of America, pp 63–87. https://doi.org/10.2134/asaspecpub67.c5

Bai H, Lester GMS, Petishnok LC, Dean DA (2017) Cytoplasmic transport and nuclear import of plasmid DNA. Biosci Rep 37(6):BSR20160616. https://doi.org/10.1042/BSR20160616

Balehegn M, Eik LO, Tesfay Y (2014) Replacing commercial concentrate by Ficus thonningii improved productivity of goats in Ethiopia. Trop Anim Health Prod 46:889–894. https://doi.org/10.1007/s11250-014-0582-9

Balehegn M, Duncan A, Tolera A, Ayantunde AA, Issa S, Karimou M, Zampaligré N, André K, Gnanda I, Varijakshapanicker P et al (2020) Improving adoption of technologies and interventions for increasing supply of quality livestock feed in low- and middle-income countries. Glob Food Sec 26:100372. https://doi.org/10.1016/j.gfs.2020.100372

Balny C (2002) High pressure and protein oligomeric dissociation. Int J High Press Res 22(3–4):737–741. https://doi.org/10.1080/08957950212447

Baltenweck I, Cherney D, Duncan A, Eldermire E, Lwoga ET, Labarta R, Rao EJO, Staal S, Teufel N (2020) A scoping review of feed interventions and livelihoods of small-scale livestock keepers. Nature Plants 6(10):1242–1249. https://doi.org/10.1038/s41477-020-00786-w

Banhazi TM, Tscharke M, Babinszky L, Halas V (2012) Precision livestock farming: precision feeding technologies and sustainable livestock production. Int J Agric Biol Eng 5(4):54–61. https://doi.org/10.3965/j.ijabe.20120504.006.

Barack P (2013) Phosphate recovery from acid phase anaerobic digesters. 8568,590B2

Barbar C, Bass PD, Barbar R, Bader J, Wondercheck B (2022) Artificial intelligence-driven automation is how we achieve the next level of efficiency in meat processing. Anim Front 12(2):56–63. https://doi.org/10.1093/af/vfac017.

Baruah KK, Khargharia G, Deori S, Kadirvel G, Baruah A, Abedin SN (2022) Effect of feeding Sun dried Banana Pseudostem as a partial replacement of dietary maize on performance and production economics of crossbred grower pigs. Indian J Anim Res. https://doi.org/10.18805/IJAR.B-4975

Basarab J, Colazo M, Ambrose D, Novak S, McCartney D, Baron V (2011) Residual feed intake adjusted for backfat thickness and feeding frequency is independent of fertility in beef heifers. Can J Anim Sci 91(4):573–584. https://doi.org/10.4141/cjas2011-010.

Bauer A-S, Leppik K, Galić K, Anestopoulos I, Panayiotidis MI, Agriopoulou S, Milousi M, Uysal-Unalan I, Varzakas T, Krauter V (2022) Cereal and confectionary packaging: background, application and shelf-life extension. Food Secur 11(5):697. https://doi.org/10.3390/foods11050697

Bauman DE (1999) Bovine somatotropin and lactation: from basic science to commercial application. Domest Anim Endocrinol 17(2–3):101–116. https://doi.org/10.1016/s0739-7240(99)00028-4.

Beaufort A, Cardinal M, Le-Bail A, Midelet-Bourdin G (2009) The effects of superchilled storage at− 2 C on the microbiological and organoleptic properties of cold-smoked salmon before retail display. Int J Refrig 32(7):1850–1857. https://doi.org/10.1016/j.ijrefrig.2009.07.001.

Beaver A, Ritter C, von Keyserlingk MA (2019) The dairy cattle housing dilemma: natural behavior versus animal care. Vet Clin N Am Food Anim Pract 35(1):11–27. https://doi.org/10.1016/j.cvfa.2018.11.001.

Beck P, Hess T, Hubbell D, Hufstedler GD, Fieser B, Caldwell J (2014) Additive effects of growth promoting technologies on performance of grazing steers and economics of the wheat pasture enterprise. J Anim Sci 92(3):1219–1227. https://doi.org/10.2527/jas.2013-7203.

Beck MR, Thompson LR, White JE, Williams GD, Place SE, Moffet CA, Gunter SA, Reuter RR (2018) Whole cottonseed supplementation improves performance and reduces methane emission intensity of grazing beef steers. Profess Animal Sci 34(4):339–345. https://doi.org/10.15232/pas.2018-01722

Bekoe D, Wang L, Zhang B, Scott Todd M, Shahbazi A (2018) Aerobic treatment of swine manure to enhance anaerobic digestion and microalgal cultivation. J Environ Sci Health B 53(2):145–151. https://doi.org/10.1080/03601234.2017.1397454. From NLM

Benjamin M, Yik S (2019) Precision livestock farming in swine welfare: a review for swine practitioners. Animals 9(4):133. https://doi.org/10.3390/ani9040133

Berckmans D (2009) Automatic on-line monitoring of animals by Precision Livestock Farming. Pages 287-294 in R. Geers and F. Madec (eds) Livestock Production and Society. MyBook. https://doi.org/10.3920/9789086865673_023

Berckmans D (2014) Precision livestock farming technologies for welfare management in intensive livestock systems. Revue scientifique et technique (International Office of Epizootics) 33(1):189–196. https://doi.org/10.20506/rst.33.1.2273. From NLM

Berthiaume R, Mandell I, Faucitano L, Lafrenière C (2006) Comparison of alternative beef production systems based on forage finishing or grain-forage diets with or without growth promotants: 1. Feedlot performance, carcass quality, and production costs. J Anim Sci 84(8):2168–2177. https://doi.org/10.2527/jas.2005-328.

Bertrand K (2005) Microwavable foods satisfy need for speed and palatability. Food Technol 59:30–34

Bewley J, Robertson L, Eckelkamp E (2017) A 100-year review: lactating dairy cattle housing management. J Dairy Sci 100(12):10418–10431

Bijker MS, Melief CJ, Offringa R, Van Der Burg SH (2007) Design and development of synthetic peptide vaccines: past, present and future. Expert Rev Vaccines 6(4):591–603

Blair RM, Savin MC, Chen P (2014) Phosphatase activities and available nutrients in soil receiving pelletized poultry litter. Soil Sci 179(4):182–189

Borja R, Sánchez E, Duran MM (1996) Effect of the clay mineral zeolite on ammonia inhibition of anaerobic thermophilic reactors treating cattle manure. J Environ Sci Health Part A 31(2):479–500

Borle F, Siebert R, Bosset J-O (2001) Photo-oxidation and photoprotection of foods, with particular reference to dairy products. An update of a review article (1993–2000). Sciences des Aliments (France)

Bosset J, Gallmann P, Sieber R (1994) Influence of light transmittance of packaging materials on the shelf-life of milk and dairy products—a review. In: Food packaging and preservation. Springer, pp 222–268

Brewer MJMS (2009) Irradiation effects on meat flavor: a review. Meat Sci 81(1):1–14

Buntyn JO, Schmidt T, Nisbet DJ, Callaway TR (2016) The role of direct-fed microbials in conventional livestock production. Ann Rev Animal Biosci 4:335–355

Buzby JC, Wells HF, Vocke G (2006) Possible Implications for U.S. Agriculture From Adoption of Select Dietary Guidelines

Cai T, Park SY, Li Y (2013) Nutrient recovery from wastewater streams by microalgae: status and prospects. Renew Sust Energ Rev 19:360–369. https://doi.org/10.1016/j.rser.2012.11.030

Campbell DL, Ouzman J, Mowat D, Lea JM, Lee C, Llewellyn RS (2020) Virtual fencing technology excludes beef cattle from an environmentally sensitive area. Animals 10(6):1069

Caradus J, Lovatt S, Belgrave B (2020) Adoption of forage technologies by New Zealand farmers–case studies

Carlez A, Veciana-Nogues T, Cheftel J-C (1995) Changes in colour and myoglobin of minced beef meat due to high pressure processing. LWT-Food Sci Technol 28(5):528–538

Chakeredza S, Akinnifesi FK, Ajayi OC, Sileshi G, Mngomba S, Gondwe FM (2008) A simple method of formulating least-cost diets for smallholder dairy production in sub-Saharan Africa. Afr J Biotechnol 7(16):2925–2933

Chanie M, Adula D, Bogale B (2013) Socio-economic assessment of the impacts of trypanosomiasis on cattle in Girja District, Southern Oromia Region, Southern Ethiopia. Acta Parasitologica Globalis 4(3):80–85

Charlier J, Barkema HW, Becher P, De Benedictis P, Hansson I, Hennig-Pauka I, La Ragione R, Larsen LE, Madoroba E, Maes D (2022) Disease control tools to secure animal and public health in a densely populated world. Lancet Planet Health 6(10):e812–e824

Chastain JP (2013) Solid-liquid separation alternatives for manure handling and treatment. USDA Natural Resources Conservation Service, Washington, D.C.

Chen C, Hilaire S, Xia K (2020) Veterinary pharmaceuticals, pathogens and antibiotic resistance. In: Animal manure: production, characteristics, environmental concerns, and management, vol 67. American Society of Agronomy and Soil Science Society of America, pp 385–407

Chioma IC, Nwankwo CN, Ezeaku IE (2021) Novel feed resources: prospects for sustainable livestock production in Africa. Sustain For 13:2635–2642

Chisoro P, Nkukwana TT (2020) Oilseeds of indigenous fruit bearing trees as alternative feedstuffs or supplements in livestock diets. J Dairy Vet Sci 1:1014

Chisoro P, Jaja IF, Assan N (2023) Incorporation of local novel feed resources in livestock feed for sustainable food security and circular economy in Africa. Front Sustain 4. https://doi.org/10.3389/frsus.2023.1251179. Review

Choi IH, Moore PA (2008) Effect of various litter amendments on ammonia volatilization and nitrogen content of poultry Litter1. J Appl Poult Res 17(4):454–462. https://doi.org/10.3382/japr.2008-00012

Christensen ML, Sommer SG (2013) Manure characterisation and inorganic chemistry. In: Jensen LS, Christensen ML, Sommer SG, Schmidt T (eds) Animal Manure: recycling, treatment, and management. Wiley, pp 41–65

Church CD, Hristov AN, Bryant RB, Kleinman P, Fishel SK (2016) A novel treatment system to remove phosphorus from liquid manure. Appl Eng Agric 32:103–112

Church CD, Hristov AN, Kleinman PJ, Fishel SK, Reiner MRA, Bryant RB (2018) Versatility of the MAnure PHosphorus EXtraction (MAPHEX) system in removing phosphorus, odor, microbes, and alkalinity from dairy manures: a four-farm case study. Appl Eng Agric 34(3):567–572

Cleale RM, Hilbig DR, Short TH, Sweiger SH, Gallery T (2018) Effects of Synovex one grass, Revalor-G, or encore implants on performance of steers grazing for up to 200 days. Profess Animal Sci 34(2):192–201. https://doi.org/10.15232/pas.2017-01685

Cole NA, Clark RN, Todd RW, Richardson CR, Gueye A, Greene LW, McBride K (2005) Influence of dietary crude protein concentration and source on potential ammonia emissions from beef cattle manure. J Anim Sci 83(3):722–731

Cole NA, Defoor PJ, Galyean ML, Duff GC, Gleghorn JF (2006) Effects of phase-feeding of crude protein on performance, carcass characteristics, serum urea nitrogen concentrations, and manure nitrogen of finishing beef steers. J Anim Sci 84(12):3421–3432

Colleran E (2016) Hygienic and sanitation requirements in biogas plant treating animal manure or mixtures of manure and other organic wastes. http://www.ava1.de/botulinum/DS4_Colleran-1.pdf. Accessed

Comin A, Grewar J, Schaik GV, Schwermer H, Paré J, El Allaki F, Drewe JA, Lopes Antunes AC, Estberg L, Horan M (2019) Development of reporting guidelines for animal health surveillance—AHSURED. Front Veterin Sci 6:426

Connolly A (2018) 8 digital technologies for a new era of beef production. Progressive Cattle

CSA (2017) Agricultural sample survey 2016/2017. Report on livestock and livestock characteristics; Central Statistical Agency, Statistical bulletin

Dai X, Xiong Y, Li N, Jian C (2019) Vaccine types. In: Vaccines-the history and future. IntechOpen

Daum T, Ravichandran T, Kariuki J, Chagunda M, Birner R (2022) Connected cows and cyber chickens? Stocktaking and case studies of digital livestock tools in Kenya and India. Agric Syst 196:103353. https://doi.org/10.1016/j.agsy.2021.103353

Davani-Davari D, Negahdaripour M, Karimzadeh I, Seifan M, Mohkam M, Masoumi S, Berenjian A, Ghasemi Y (2019) Prebiotics: definition, types, sources, mechanisms, and clinical applications. Food Secur 8(3):92. https://doi.org/10.3390/foods8030092

David I, Huynh Tran V-H, Gilbert H (2021) New residual feed intake criterion for longitudinal data. Genet Sel Evol 53(1):53. https://doi.org/10.1186/s12711-021-00641-2

Day BPF (1990) Perspective of modified atmosphere packaging of fresh produce in Western Europe. Food Sci Technol Today 4:215–221

De Koning C (2010) Automatic milking–common practice on dairy farms

De Mey E, De Maere H, Paelinck H, Fraeye I (2017) Volatile N-nitrosamines in meat products: potential precursors, influence of processing, and mitigation strategies. Crit Rev Food Sci Nutr 57(13):2909–2923

Delgado C, Rosegrant M, Steinfeld H, Ehui S, Courbois C (1999) Livestock to 2020: the next food revolution. IFPRI food, agriculture, and the environment discussion paper 28. Washington, D.C. (USA): IFPRI. 83 pp. Acccessed 2022/06/28

Den Hartog L, Sijtsma R (2013) Challenges and opportunities in animal feed and nutrition

Denli M, Okan F, Celik K (2003) Effect of dietary probiotic, organic acid and antibiotic supplementation to diets on broiler performance and carcass yield. Pak J Nutr. https://doi.org/10.3923/pjn.2003.89.91

Dessie T (2021) Tropical Poultry Genetic Solutions (TPGS): Delivering farmer preferred, productive and ecologically adapted poultry to smallholders. Presented at the CTLGH Virtual Development Meeting, 2–3 December 2021. Nairobi, Kenya: ILRI. https://hdl.handle.net/10568/117265

Dewes T (1995) Nitrogen losses from manure heaps. Biol Agric Horticul 11(1–4):309–317. https://doi.org/10.1080/01448765.1995.9754715

di Virgilio A, Morales JM, Lambertucci SA, Shepard EL, Wilson RP (2018) Multi-dimensional precision livestock farming: a potential toolbox for sustainable rangeland management. PeerJ 6:e4867

Diez AM, Santos EM, Jaime I, Rovira J (2009) Effectiveness of combined preservation methods to extend the shelf life of Morcilla de Burgos. Meat Sci 81(1):171–177

Doody DG, Foy RH, Bailey JS, Matthews D (2012) Minimising nutrient transfers from poultry litter field heaps. Nutr Cycl Agroecosyst 92(1):79–90

Doreau M, Arbre M, Popova M, Rochette Y, Martin C (2018) Linseed plus nitrate in the diet for fattening bulls: effects on methane emission, animal health and residues in offal. Animal 12(3):501–507. https://doi.org/10.1017/s1751731117002014. From NLM

Doublet J, Francou C, Poitrenaud M, Houot S (2011) Influence of bulking agents on organic matter evolution during sewage sludge composting; consequences on compost organic matter stability and N availability. Bioresour Technol 102(2):1298–1307

Duncan AJ, Lukuyu B, Mutoni G, Lema Z, Fraval S (2023) Supporting participatory livestock feed improvement using the feed assessment tool (FEAST). Agron Sustain Dev 43(2):34. https://doi.org/10.1007/s13593-023-00886-9

Durunna ON, Plastow G, Mujibi FD, Grant J, Mah J, Basarab JA, Okine EK, Moore SS, Wang Z (2011a) Genetic parameters and genotype x environment interaction for feed efficiency traits in steers fed grower and finisher diets. J Anim Sci 89(11):3394–3400. https://doi.org/10.2527/jas.2010-3516. WorldCat.org

Durunna, O. N.; University of Alberta. Department of Agricultural, F.; Nutritional, S (2011b) Genetics of feed efficiency and feeding behavior in crossbred beef steers with emphasis on genotype-by-environment interactions. University of Alberta, Edmonton, Alta. https://central.bac-lac.gc.ca/.item?id=TC-AEU-29450&op=pdf&app=Library; http://hdl.handle.net/10048/2052; http://hdl.handle.net/10402/era.27450; https://era.library.ualberta.ca/public/datastream/get/uuid:35b625a2-9633-41af-bb06-79a37369fa56/DS1

Durunna ON, Colazo MG, Ambrose DJ, McCartney D, Baron VS, Basarab JA (2012) Evidence of residual feed intake reranking in crossbred replacement heifers. J Anim Sci 90(3):734–741. https://doi.org/10.2527/jas.2011-4264. WorldCat.org

Eghball B (2000) Nitrogen mineralization from field-applied beef cattle feedlot manure or compost. Soil Sci Soc Am J 64(6):2024–2030

Eie T, Larsen H, Sørheim O, Pettersen MK, Hansen A, Wold J, Naterstad K, Mielnik M (2007) New technologies for extending shelf life. Italian J Food Sci 19(2):127–152

Eiki N, Maake M, Lebelo S, Sakong B, Sebola N, Mabelebele M (2022) Survey of ethnoveterinary medicines used to treat livestock diseases in Omusati and Kunene Regions of Namibia. Front Veterin Sci:9. https://doi.org/10.3389/fvets.2022.762771. Original Research

Eilert S (2005) New packaging technologies for the 21st century. Meat Sci 71(1):122–127

Elliott HA, Brandt RC, Kleinman PJA, Sharpley AN, Beegle DB (2006) Estimating source coefficients for phosphorus site indices. J Environ Qual 35(6):2195–2201

El-Osta HS, Morehart MJ (2000) Technology adoption and its impact on production performance of dairy operations. Appl Econ Perspect Policy 22(2):477–498

El-Sayed A, Kamel M (2020) Climatic changes and their role in emergence and re-emergence of diseases. Environ Sci Pollut Res 27(18):22336–22352

Enahoro D, Mason-D'Croz D, Mul M, Rich KM, Robinson TP, Thornton P, Staal SS (2019) Supporting sustainable expansion of livestock production in South Asia and Sub-Saharan Africa: scenario analysis of investment options. Global Food Secur 20:114–121. https://doi.org/10.1016/j.gfs.2019.01.001. WorldCat. org

Entrican G, Lunney JK, Wattegedera SR, Mwangi W, Hope JC, Hammond JA (2020) The veterinary immunological toolbox: past, present, and future. Front Immunol 11:1651

Erickson GE, Klopfenstein TJ, Adams DC, Rasby RJ (2005) Utilization of corn co-products in the beef industry

Etherton TD, Kris-Etherton PM, Mills EW (1993) Recombinant bovine and porcine somatotropin: safety and benefits of these biotechnologies. J Am Diet Assoc 93(2):177–180. https://doi.org/10.1016/0002-8223(93)90835-9. From NLM

Falvey L (1999) Smallholder dairying in the tropics; ILRI (aka ILCA and ILRAD)

FAO (2018) World Livestock: Transforming the livestock sector through the Sustainable Development Goals, 222. https://www.fao.org/documents/card/en?details=ca1201en. Accessed 23 Apr 2024

FAO (2022) Pastoralist knowledge hub. Food and Agricultural Organization

FAO, U (2019) Protecting people and animals from disease threats

FEFAC, E (2016) Feed, Food statistical yearbook 2016. European Feed Manufactures Federation Bruxelles[Google Scholar]

Felix TL (2017) Use of Beta-agonists in cattle feed. PennState Extension. https://extension.psu.edu/use-of-beta-agonists-in-cattle-feed. Accessed 2022/07/10

Feng X, Ronk E, Hanigan MD, Knowlton KF, Schramm H, McCann M (2015) Effect of dietary phosphorus on intestinal phosphorus absorption in growing Holstein steers. J Dairy Sci 98(5):3410–3416

Ferket PR, van Heugten E, van Kempen TATG, Angel R (2002) Nutritional strategies to reduce environmental emissions from nonruminants. J Anim Sci 80:E168–E182. https://doi.org/10.2527/animalsci2002.80E-Suppl_2E168x

Fernandez-Lopez M, Puig-Gamero M, Lopez-Gonzalez D, Avalos-Ramirez A, Valverde J, Sanchez-Silva L (2015) Life cycle assessment of swine and dairy manure: pyrolysis and combustion processes. Bioresour Technol 182:184–192. https://doi.org/10.1016/j.biortech.2015.01.140

Fesseha H, Kefelegn T, Mathewos M (2022) Animal care professionals' practice towards zoonotic disease management and infection control practice in selected districts of Wolaita zone, Southern Ethiopia. Heliyon 8:e09485

Filip C (2006) Ethiopian Borena and southern Somali areas livestock value chain analysis report. ACDI/VOCA Pastoralist Livelihood Initiative Livestock Marketing Project

Forkman B, Lund VP (2017) A short introduction to animal welfare assessment. In: DCAW and NordCAW animal welfare conference. DCAW—Danish Centre for Animal Welfare & NordCAW—Nordic Network for Communicating Animal Welfare; 3–4 October. University of Copenhagen, Copenhagen, pp 5–6

Foster J, Adesogan A, Carter J, Blount A, Myer R, Phatak S (2009) Intake, digestibility, and nitrogen retention by sheep supplemented with warm-season legume hays or soybean meal. J Anim Sci 87(9):2891–2898

Francis MJ (2018) Recent advances in vaccine technologies. Veterin Clin Small Animal Pract 48(2):231–241

Fuentes S, Gonzalez Viejo C, Tongson E, Dunshea FR (2022) The livestock farming digital transformation: implementation of new and emerging technologies using artificial intelligence. Anim Health Res Rev 23:1–13. https://doi.org/10.1017/S1466252321000177. From Cambridge University Press Cambridge Core

Fuertes Marraco SA, Soneson C, Cagnon L, Gannon PO, Allard M, Maillard SA, Montandon N, Rufer N, Waldvogel S, Delorenzi M (2015) Long-lasting stem cell–like memory CD8+ T cells with a naïve-like profile upon yellow fever vaccination. Sci Translat Med 7(282):282ra248

Gaggìa F, Mattarelli P, Biavati B (2010) Probiotics and prebiotics in animal feeding for safe food production. Int J Food Microbiol 141:S15–S28

Galama P, De Boer H, Van Dooren H, Ouweltjes W, Driehuis K (2015) Sustainability aspects of ten bedded pack dairy barns in The Netherlands. Wageningen UR Livestock Research

Galama P, Ouweltjes W, Endres M, Sprecher J, Leso L, Kuipers A, Klopčič M (2020) Symposium review: Future of housing for dairy cattle. J Dairy Sci 103(6):5759–5772

Galvão KN, Santos JE, Coscioni A, Villaseñor M, Sischo WM, Berge ACB (2005) Effect of feeding live yeast products to calves with failure of passive transfer on performance and patterns of antibiotic resistance in fecal Escherichia coli. Reprod Nutr Dev 45(4):427–440

Gao J, Zhang HJ, Yu SH, Wu SG, Yoon I, Quigley J, Gao YP, Qi GH (2008) Effects of yeast culture in broiler diets on performance and immunomodulatory functions. Poult Sci 87(7):1377–1384. https://doi.org/10.3382/ps.2007-00418. WorldCat.org

Garcia MC, Szogi AA, Vanotti MB, Chastain JP, Millner PD (2009) Enhanced solid–liquid separation of dairy manure with natural flocculants. Bioresour Technol 100(22):5417–5423. https://doi.org/10.1016/j.biortech.2008.11.012

Garg M, Sherasia P, Bhanderi B, Phondba B, Shelke S, Makkar H (2013) Effects of feeding nutritionally balanced rations on animal productivity, feed conversion efficiency, feed nitrogen use efficiency, rumen microbial protein supply, parasitic load, immunity and enteric methane emissions of milking animals under field conditions. Anim Feed Sci Technol 179(1–4):24–35

GEA (2022a) EDITION 2021 for our DairyRobot product line—The next step in automated milking. https://www.gea.com/en/articles/dairyrobot-r9500-edition-2021/index.jsp. Accessed

GEA (2022b) GEA DAIRYPROQ ROBOTIC rotary milking parlor. https://www.gea.com/en/products/milking-farming-barn/dairyrobot-automated-milking/rotary-parlor-dairyproq.jsp?utm_source=google&utm_medium=cpc&utm_campaign=R9900_en. Accessed 30-05-2022

GEA (2022c) GEA DAIRYROBOT R9500 robotic milking system. https://www.gea.com/en/products/milking-farming-barn/dairyrobot-automated-milking/dairyrobot-r9500-robotic-milking-system.jsp. Accessed

Gebreyohanes G, Mwai O (2021) African dairy genetic gains project overview. Presented at the CTLGH Virtual Development Meeting, 2–3 December 2021. Nairobi, Kenya: ILRI. https://hdl.handle.net/10568/117266

Gentry TS, Roberts JS (2005) Design and evaluation of a continuous flow microwave pasteurization system for apple cider. LWT-Food Sci Technol 38(3):227–238

Gerber PJ, Henderson B, Makkar HPS (2013) Mitigation of greenhouse gas emissions in livestock production: A review of technical options for non-CO2 emissions. FAO Anim Prod Health Pap 184:226

Giannuzzi D, Mota LFM, Pegolo S, Gallo L, Schiavon S, Tagliapietra F, Katz G, Fainboym D, Minuti A, Trevisi E (2022) In-line near-infrared analysis of milk coupled with machine learning methods for the daily prediction of blood metabolic profile in dairy cattle. Sci Rep 12(1):1–13

Gibson GR, Hutkins R, Sanders ME, Prescott SL, Reimer RA, Salminen SJ, Scott K, Stanton C, Swanson KS, Cani PD (2017) Expert consensus document: the international scientific Association for Probiotics and Prebiotics (ISAPP) consensus statement on the definition and scope of prebiotics. Nat Rev Gastroenterol Hepatol 14(8):491–502

Girma G (2019) An overview of locally available feed resources for ruminant livestock in Ethiopia. Int J Livest Prod 10:20–30

Gisi DD, DePeters EJ, Pelissier CL (1986) Three times daily milking of cows in California dairy herds. J Dairy Sci 69(3):863–868

GM (2022) Agriculture soil management guide: greenhouse gases in agriculture. Government of Manitoba. https://www.gov.mb.ca/agriculture/environment/soil-management/soil-management-guide/print,greenhouse-gases-in-agriculture.html. Accessed 2022/07/15

Gola S, Mutti P, Manganelli E, Dazzi M, Squarcina N, Ghidini M, Rovere P (2000) Behaviour of pathogenic E. coli in a model system and in raw minced meat treated by HP. Microbiological and Technical Aspects Industria Conserve (Italy)

Gould C (2012) Introduction to anaerobic digestion; Bioenergy Training. University of Wisconsin Extension

Grandin T (1995) The economic benefits of proper animal welfare. In: Reciprocal meat conference proceedings. American Meat Science Association, pp 122–127

Groff EB, Wu Z (2005) Milk production and nitrogen excretion of dairy cows fed different amounts of protein and varying proportions of alfalfa and corn silage. J Dairy Sci 88(10):3619–3632

Groher T, Heitkämper K, Umstätter C (2020) Digital technology adoption in livestock production with a special focus on ruminant farming. Animal 14(11):2404–2413. https://doi.org/10.1017/S1751731120001391. PubMed

Gulwako MS, Mokoele JM, Ngoshe YB, Naidoo V (2023) Evaluation of the proper use of medication available over the counter by subsistence and emerging farmers in Mbombela Municipality, South Africa. BMC Veterin Res 19(1):83. https://doi.org/10.1186/s12917-023-03634-z

Gunsett F (1986) Problems associated with selection for traits defined as a ratio of two component traits. In: Proceedings of the Third World Congress on Genetics Applied to Livestock Production, pp 437–442

Guo M, Li H, Baldwin B, Morrison J (2020) Thermochemical processing of animal manure for bioenergy and biochar. In: Animal Manure, pp 255–274. https://doi.org/10.2134/asaspecpub67.c21

Gupta PB, Gould SJ (1997) Consumers' perceptions of the ethics and acceptability of product placements in movies: product category and individual differences. J Curr Issues Res Advertis 19(1):37–50

Guthrie JF, Fox JJ, Cleveland LE, Welsh S (1995) Who uses nutrition labeling, and what effects does label use have on diet quality? J Nutr Educ 27(4):163–172

Gwaka L, Dubihlela J (2020) The resilience of smallholder livestock farmers in Sub-Saharan Africa and the risks imbedded in rural livestock systems. Agriculture 10(7). https://doi.org/10.3390/agriculture10070270

Habanabakize E, Ba K, Corniaux C, Cortbaoui P, Vasseur E (2022) A typology of smallholder livestock production systems reflecting the impact of the development of a local milk collection industry: case study of Fatick region, Senegal. Pastoralism 12(1):22. https://doi.org/10.1186/s13570-022-00234-8

Hadlocon LJS, Zhao L, Manuzon RB, Elbatawi IE (2014) An acid spray scrubber for recovery of ammonia emissions from a deep-pit swine facility. Trans ASABE 57(3):949–960

Hadlocon LJS, Manuzon RB, Zhao L (2015) Development and evaluation of a full-scale spray scrubber for ammonia recovery and production of nitrogen fertilizer at poultry facilities. Environ Technol 36(4):405–416. https://doi.org/10.1080/09593330.2014.950346

Haileslassie A, Blümmel M, Wani S, Sahrawat KL, Pardhasaradhi G, Samireddypalle A (2013) Extractable soil nutrient effects on feed quality traits of crop residues in the semiarid rainfed mixed crop–

livestock farming systems of Southern India. Environ Dev Sustain 15:723–741. https://doi.org/10.1007/s10668-012-9403-3

Hamed AHM, Elimam ME (2009) Effects of chopping on utilization of sorghum Stover by Nubian goats. Pak J Nutr 8(10):1567–1569

Hansen A, Turner L, Aurand L (1975) Fluorescent light-activated flavor in milk. J Milk Food Technol 38(7):388–392

Harrison JH, Ndegwa PM (2020) Anaerobic digestion of dairy and swine waste. In: H.M. Waldrip, P.H. Pagliari, Z. He (Eds.) Animal Manure: production, characteristics, environmental concerns, and management, vol 67. American Society of Agronomy and Soil Science Society of America, pp 115–127. https://doi.org/10.2134/asaspecpub67.c13

Havenstein GB, Ferket PR, Qureshi MA (2003) Growth, livability, and feed conversion of 1957 versus 2001 broilers when fed representative 1957 and 2001 broiler diets. Poult Sci 82(10):1500–1508. https://doi.org/10.1093/ps/82.10.1500. From NLM

He Z, Pagliari PH, Waldrip HM (2016) Applied and environmental chemistry of animal manure: a review. Pedosphere 26(6):779–816. https://doi.org/10.1016/S1002-0160(15)60087-X

Heinz G, Hautzinger P (2007) Meat processing technology for small to medium scale producers. Food and Agriculture Organization (FAO) of the United Nations. Available at https://openknowledge.fao.org/server/api/core/bitstreams/4cfabbd3-16aa-47f8-ac6f-b54a48cb8abd/content. Accessed 14 June 2023.

Hentz F, Umstätter C, Gilaverte S, Prado OR, Silva CJA, Monteiro ALG (2014) Electronic bolus design impacts on administration1. J Anim Sci 92(6):2686–2692. https://doi.org/10.2527/jas.2013-7183

Higgins SE, Higgins JP, Wolfenden AD, Henderson SN, Torres-Rodriguez A, Tellez G, Hargis B (2008) Evaluation of a Lactobacillus-based probiotic culture for the reduction of Salmonella enteritidis in neonatal broiler chicks. Poult Sci 87(1):27–31. https://doi.org/10.3382/ps.2007-00210

Hoelzer K, Bielke L, Blake DP, Cox E, Cutting SM, Devriendt B, Erlacher-Vindel E, Goossens E, Karaca K, Lemiere S (2018) Vaccines as alternatives to antibiotics for food producing animals. Part 2: new approaches and potential solutions. Vet Res 49(1):1–15

Hoff SJ, Bundy DS, Nelson MA, Zelle BC, Jacobson LD, Heber AJ, Ni J, Zhang Y, Koziel JA, Beasley DB (2006) Emissions of ammonia, hydrogen sulfide, and odor before, during, and after slurry removal from a deep-pit swine finisher. J Air Waste Manage Assoc (1995) 56(5):581–590. https://doi.org/10.1080/10473289.2006.10464472. From NLM

Holje L (2012) Bio-thermo microchip. Swedish University of Agricultural Sciences, Uppsala, Sweden

Hong WT, Hagare D, Khan M, Fyfe JJW, Air, Pollution, S (2018) Phosphorus characterisation of sludge and crust produced by stabilisation ponds in a dairy manure management system. Water Air Soil Pollut 229:1–8

Hoque M, Hosono M, Oikawa T, Suzuki K (2009) Genetic parameters for measures of energetic efficiency of bulls and their relationships with carcass traits of field progeny in Japanese Black cattle. J Anim Sci 87(1):99–106

Hossain MI, Sadekuzzaman M, Ha S-D (2017) Probiotics as potential alternative biocontrol agents in the agriculture and food industries: a review. Food Res Int 100:63–73

Huang L, Moore PA, Kleinman PJ, Elkin KR, Savin MC, Pote DH, Edwards DR (2016) Reducing phosphorus runoff and leaching from poultry litter with alum: twenty-year small plot and paired-watershed studies. J Environ Qual 45(4):1413–1420. https://doi.org/10.2134/jeq2015.09.0482. From NLM

Hugas M, Garriga M, Monfort JM (2002) New mild technologies in meat processing: high pressure as a model technology. Meat Sci 62(3):359–371

Hwang J, Jeong H, Yoe H (2011) Design and implementation of smart phone application for effective livestock farm management. In: International conference on technology systems and management. Springer, pp 285–290

Ibeagha-Awemu EM, Yu Y (2021) Consequence of epigenetic processes on animal health and productivity: is additional level of regulation of relevance? Anim Front 11(6):7–18. https://doi.org/10.1093/af/vfab057. (acccessed 12/21/2021)

Jackson LL, Keeney DR, Gilbert EM (2000) Swine manure management plans in north-central Iowa: nutrient loading and policy implications. J Soil Water Conserv 55(2):205–212

Jaime G, Hobeika A, Figuié M (2022) Access to veterinary drugs in sub-Saharan Africa: roadblocks and current solutions. Front Veterin Sci 8. https://doi.org/10.3389/fvets.2021.558973

Janni K, Cortus E (2020) Common animal production systems and manure storage methods. In: Animal Manure. Wiley, pp 27–43

Jeong S-H, Kang D, Lim M-W, Kang CS, Sung HJ (2010) Risk assessment of growth hormones and antimicrobial residues in meat. Toxicol Res 26(4):301–313

Jiang T, Li G, Tang Q, Ma X, Wang G, Schuchardt F (2015) Effects of aeration method and aeration rate on greenhouse gas emissions during composting of pig feces in pilot scale. J Environ Sci 31:124–132

Jin LZ, Ho YW, Abdullah N, Jalaludin S (1998) Growth performance, intestinal microbial populations, and serum cholesterol of broilers fed diets containing Lactobacillus cultures. Poult Sci 77(9):1259–1265. WorldCat.org

Kamuanga M, Sigué H, Swallow B, Bauer B, d'Ieteren G (2001) Farmers' perceptions of the impacts of tsetse and trypanosomosis control on livestock production: evidence from southern Burkina Faso. Trop Anim Health Prod 33:141–153

Kanter R, Vanderlee L, Vandevijvere S (2018) Front-of-package nutrition labelling policy: global progress and future directions. Public Health Nutr 21(8):1399–1408

Kaparaju P, Rintala J (2011) Mitigation of greenhouse gas emissions by adopting anaerobic digestion technology on dairy, sow and pig farms in Finland. Renew Energy 36(1):31–41

Karathanasis AD, Shumaker PD (2009) Preferential sorption and desorption of organic and inorganic Phospates by soil Hydroxyinterlayered minerals. Soil Sci 174(8):417–423

Karunanithi R, Szogi AA, Bolan N, Naidu R, Loganathan P, Hunt PG, Vanotti MB, Saint CP, Ok YS, Krishnamoorthy S (2015) Chapter Three – Phosphorus recovery and reuse from waste streams. In: Sparks DL (ed) Advances in agronomy, vol 131. Academic Press, pp 173–250

Kennedy B, Van der Werf J, Meuwissen T (1993) Genetic and statistical properties of residual feed intake. J Anim Sci 71(12):3239–3250

Keshavarz K, Austic RE (2004) The use of low-protein, low-phosphorus, amino acid- and Phytase-supplemented diets on laying hen performance and nitrogen and phosphorus excretion. Poult Sci 83(1):75–83. https://doi.org/10.1093/ps/83.1.75

Kester E, Holzhauer M, Frankena K (2014) A descriptive review of the prevalence and risk factors of hock lesions in dairy cows. Vet J 202(2):222–228

Key VI (2018) Selected highlights from other journals. Sci Rep 8:9756

Khan SH (2018) Recent advances in role of insects as alternative protein source in poultry nutrition. J Appl Anim Res 46(1):1144–1157

Khanal AR, Gillespie J, MacDonald J (2010) Adoption of technology, management practices, and production systems in US milk production. J Dairy Sci 93(12):6012–6022

Khatun R, Howlader MAJ, Ahmed S, Islam MN, Hasan MA, Haider MS, Mahmud MS (2016) Impact of training and monitoring of drug used by small scale poultry farmers at different location of Bangladesh. Am J Food Sci Health 2(6):134–140

Kirui O (2019) The complementarity of education and use of productive inputs among smallholder farmers in Africa. SSRN Electron J. https://doi.org/10.2139/ssrn.3372123

Kithome M, Paul JW, Bomke AA (1999) Reducing nitrogen losses during simulated composting of poultry manure using adsorbents or chemical amendments. Wiley Online Library

Kleinman PJA, Spiegal S, Liu J, Holly M, Church C, Ramirez-Avila J (2020) Managing animal manure to minimize phosphorus losses from land to water. In: Animal Manure, pp 201–228

Klinner W, Shepperson G (1975) The state of haymaking technology—a review. Grass Forage Sci 30(3):259–266

Klopčič M, Galama P, Kuipers A (2019) Innovation in housing and management systems for dairy cows. In Proc. 28th Int. Sci. Symp. Nutr. Farm Animals. T. Čeh, and S. Kapun ed. Zadravec-Erjavec Days, Radenci, Slovenia. [Summary in English], pp 95–100

Knorr D, Zenker M, Heinz V, Lee D-U (2004) Applications and potential of ultrasonics in food processing. Trends Food Sci Technol 15(5):261–266

Knowlton KF, Herbein JH (2002) Phosphorus partitioning during early lactation in dairy cows fed diets varying in phosphorus content. J Dairy Sci 85(5):1227–1236

Knowlton KF, Radcliffe JS, Novak CL, Emmerson DA (2004) Animal management to reduce phosphorus losses to the environment. J Anim Sci 82 E-Suppl:E173–E195. https://doi.org/10.2527/2004.8213_supplE173x. From NLM

Koch G, Christensen H, Eides S, Meinert L (2009) Requirements to shelf-life of fresh meat and meat products. In Proceedings of the 55th International Congress of Meat Science and Technology (ICoMST), Copenhagen, Denmark, pp 16–21

Kollaritsch H, Rendi-Wagner P (2012) Principles of immunization. Travel Med Expert Consult-Online Print 67

Ku-Vera JC, Jiménez-Ocampo R, Valencia-Salazar SS, Montoya-Flores MD, Molina-Botero IC, Arango J, Gómez-Bravo CA, Aguilar-Pérez CF, Solorio-Sánchez FJ (2020) Role of secondary plant metabolites on enteric methane mitigation in ruminants. Front Vet Sci 7:584. https://doi.org/10.3389/fvets.2020.00584. From NLM

Kwoji ID, Aiyegoro OA, Okpeku M, Adeleke MA (2021) Multi-strain probiotics: synergy among isolates enhances biological activities. Biology 10(4):322

Lactanet (2021) Milk recording and analysis. https://lactanet.ca/en/milk-recording/edhi/. Accessed 26/05/2022

Lee KT (2010) Quality and safety aspects of meat products as affected by various physical manipulations of packaging materials. Meat Sci 86(1):138–150. https://doi.org/10.1016/j.meatsci.2010.04.035

Lee KT (2018) Shelf-life extension of fresh and processed meat products by various packaging applications. Korean J Packag Sci Technol 24:57–64

Lee C, Colditz IG, Campbell DL (2018) A framework to assess the impact of new animal management technologies on welfare: a case study of virtual fencing. Front Veterin Sci 5:187

Lefcourt AM, Meisinger JJ (2001) Effect of adding alum or zeolite to dairy slurry on ammonia volatilization and chemical composition. J Dairy Sci 84(8):1814–1821

Leistner L (2000) Basic aspects of food preservation by hurdle technology. Int J Food Microbiol 55(1–3):181–186

Lely (2022) Mlking. https://www.lely.com/ca/en/solutions/milking/?ds_rl=1288652&gclid=Cj0KCQjw1t-GUBhDXARIsAIJx01l2S7FiLRSWrxlzwqT7kEFJZlTbvPoMvIM3DXsiwRmq5CMJhE_mbUAaAoOrE-ALw_wcB&gclsrc=aw.ds. Accessed 30-05-2022

Lenné JM, Fernandez-Rivera S, Blümmel M (2003) Approaches to improve the utilization of food-feed crops—synthesis. Field Crop Res 84(1–2):213–222

Leso L, Barbari M, Lopes M, Damasceno F, Galama P, Taraba J, Kuipers A (2020) Invited review: compost-bedded pack barns for dairy cows. J Dairy Sci 103(2):1072–1099

Li H, Xin H, Burns RT (2006) Reduction of ammonia emission from stored poultry manure using additives: Zeolite, Al+ clear, Ferix-3 and PLT. American Society of Agricultural and Biological Engineers, p 1

Li J, Shen Y, Cai Y (2010) Improvement of fermentation quality of rice straw silage by application of a bacterial inoculant and glucose. Asian Australas J Anim Sci 23(7):901–906

Li W, Joshi MD, Singhania S, Ramsey KH, Murthy AK (2014) Peptide vaccine: progress and challenges. Vaccine 2(3):515–536

Li QF, Trottier N, Powers W (2015) Feeding reduced crude protein diets with crystalline amino acids supplementation reduce air gas emissions from housing. J Anim Sci 93(2):721–730. https://doi.org/10.2527/jas.2014-7746. From NLM

Li MM, Seelenbinder KM, Ponder MA, Deng L, Rhoads RP, Pelzer KD, Radcliffe JS, Maxwell CV, Ogejo JA, White RR et al (2017) Effects of dirty housing and a Typhimurium DT104 challenge on pig growth performance, diet utilization efficiency, and gas emissions from stored manure. J Anim Sci 95(3):1264–1276. https://doi.org/10.2527/jas.2016.0863. From NLM

Li W, Edwards A, Riehle C, Cox MS, Raabis S, Skarlupka JH, Steinberger AJ, Walling J, Bickhart D, Suen G (2019) Transcriptomics analysis of host liver and meta-transcriptome analysis of rumen epimural microbial community in young calves treated with artificial dosing of rumen content from adult donor cow. Sci Rep 9(1):1–11

Liao S, Liao L, Huang P, Wang Y, Zhu S, Wang X, Lv T, Li Y, Fan Z, Liu T (2022) Effects of different levels of garlic straw powder on growth performance, meat quality, antioxidant and intestinal mucosal morphology of yellow-feathered broilers. Front Physiol 13:902995

Lim TT, Heber AJ, Ni J-Q, Kendall DC, Richert BT (2004) Effects of manure removal strategies on odor and gas emissions from swine finishing. Trans ASABE 47:2041–2050

Lin J, Shen Z, Zhang A, Chai Y (2018) Blockchain and IoT based food traceability for smart agriculture. In Proceedings of the 3rd International Conference on Crowd Science and Engineering, pp 1–6

Liu Z, Powers W, Murphy J, Maghirang R (2014) Ammonia and hydrogen sulfide emissions from swine production facilities in North America: a meta-analysis. J Anim Sci 92(4):1656–1665. https://doi.org/10.2527/jas.2013-7160. From NLM

Liu J, Kleinman PJ, Beegle DB, Weld JL, Sharpley AN, Saporito LS, Schmidt JP (2015) Phosphorus and nitrogen losses from poultry litter stacks and leaching through soils. Nutr Cycl Agroecosyst 103(1):101–114

Liu S, Ni JQ, Radcliffe JS, Vonderohe CE (2017) Mitigation of ammonia emissions from pig production using reduced dietary crude protein with amino acid supplementation. Bioresour Technol 233:200–208. https://doi.org/10.1016/j.biortech.2017.02.082. From NLM

Liu J, Spargo JT, Kleinman PJA, Meinen R, Moore PA Jr, Beegle DB (2018) Water-extractable phosphorus in animal manure and manure compost: quantities, characteristics, and temporal changes. J Environ Qual 47(3):471–479

Long CM, Muenich RL, Kalcic MM, Scavia D (2018) Use of manure nutrients from concentrated animal feeding operations. J Great Lakes Res 44(2):245–252

Lory JA (2008) 8. Using manure as a fertilizer for crop production

Luo B, Hu Q, Lai K, Bhatt A, Hu R (2022) Ethnoveterinary survey conducted in Baiku Yao Communities in Southwest China. Front Veterin Sci:8. https://doi.org/10.3389/fvets.2021.813737. Original Research

Ma F, Xu S, Tang Z, Li Z, Zhang L (2021) Use of antimicrobials in food animals and impact of transmission of antimicrobial resistance on humans. Biosafety Health 3(01):32–38

MacPhail VA, Kyle J (2012) Rotational grazing in extensive pasture. Accessed 1-21-2022

Madzingira O (2018) Animal welfare considerations in food-producing animals. Animal Welfare; IntechOpen, London, p 99

Mahgoub SA, Al-Mutairi SE (2015) Use of alternative feeds for sustainable aquaculture production in Saudi Arabia. World Aquacult 46:16–19

Makkar HP (2016) Smart livestock feeding strategies for harvesting triple gain–the desired outcomes in planet, people and profit dimensions: a developing country perspective. Anim Prod Sci 56(3):519–534

Makkar HP, Sánchez M, Speedy AW (2007) Feed supplementation blocks: urea-molasses multinutrient blocks: simple and effective feed supplement technology for ruminant agriculture. Food & Agriculture Org

Malacco V (2022) Promises and potential of automated milking systems. https://www.canr.msu.edu/news/promises-and-potential-of-automated-milking-systems. Accessed

Maldonado Galdeano C, Cazorla SI, Lemme Dumit JM, Vélez E, Perdigón G (2019) Beneficial effects of probiotic consumption on the immune system. Ann Nutr Metab 74(2):115–124

Manyi-Loh CE, Mamphweli SN, Meyer EL, Makaka G, Simon M, Okoh AI (2016) An overview of the control of bacterial pathogens in cattle manure. Int J Environ Res Public Health 13(9):843. https://doi.org/10.3390/ijerph13090843.

Marañón E, Castrillón L, Fernández JJ, Fernández Y, Peláez AI, Sánchez J (2006) Anaerobic mesophilic treatment of cattle manure in an upflow anaerobic sludge blanket reactor with prior pasteurization. J Air Waste Manage Assoc (1995) 56(2):137–143. https://doi.org/10.1080/10473289.2006.10464448. From NLM

Markowiak P, Śliżewska K (2018) The role of probiotics, prebiotics and synbiotics in animal nutrition. Gut Pathogens 10:1–20

Masud MT, Bhowmik S (2018) Feasibility study of solar-powered hydroponic fodder machine in Bangladesh. In: Renewable energy in developing countries: local development and techno-economic aspects. Springer, pp 85–94

Matlock M, Greg Thoma B, Eric Boles Mansoor Leh, P, Sandefur Rusty Bautista, H, Rick Ulrich P (2014) A life cycle analysis of water use in U.S. In: Pork production: comprehensive report. Pork Checkoff, Fayetteville, AR

Mbele GO, Mellau LSB, Mwanri AW, Kifaro GC (2019) Agro-industrial by-products: potential feed resources for improving livestock productivity and environmental conservation in Tanzania. J Agricultural Sci 11:10–20

Mbukwane MJ, Nkukwana TT, Plumstead PW, Snyman N (2022) Sunflower meal inclusion rate and the effect of exogenous enzymes on growth performance of broiler chickens. Animals (Basel) 12(3). https://doi.org/10.3390/ani12030253. From NLM

McCord AI, Stefanos SA, Tumwesige V, Lsoto D, Meding AH, Adong A, Schauer JJA, Larson RA (2017) The impact of biogas and fuelwood use on institutional kitchen air quality in Kampala, Uganda. Indoor Air 27(6):1067–1081

McCord AI, Stefanos SA, Tumwesige V, Lsoto D, Kawala M, Mutebi J, Nansubuga I, Larson RA (2020) Anaerobic digestion in Uganda: risks and opportunities for integration of waste management and agricultural systems. Renew Agric Food Syst 35(6):678–687. https://doi.org/10.1017/S1742170519000346. From Cambridge University Press Cambridge Core

McGlone J, Pond W (2003) Pig production: biological principles and applications. Delmar Learning. Inc., Clifton Park, pp 248–250

Meeusen EN, Walker J, Peters A, Pastoret P-P, Jungersen G (2007) Current status of veterinary vaccines. Clin Microbiol Rev 20(3):489–510

Melse RW, Ogink NWM (2005) Air scrubbing techniques for ammonia and odor reduction at livestock operations: review of on-farm research in The Netherlands. Trans ASAE 48(6):2303–2313

Menghistu HT, Hailu KT, Shumye NA, Redda YT (2018) Mapping the epidemiological distribution and incidence of major zoonotic diseases in South Tigray, North Wollo and Ab'ala (Afar), Ethiopia. PLoS One 13(12):e0209974

Michalak M, Wojnarowski K, Cholewińska P, Szeligowska N, Bawej M, Pacoń J (2021) Selected alternative feed additives used to manipulate the rumen microbiome. Animals 11(6). https://doi.org/10.3390/ani11061542. WorldCat.org

Milán Z, Sanchez E, Borja R, Ilangovan K, Pellon A, Rovirosa N, Weiland P, Escobedo R (1999) Deep bed filtration of anaerobic cattle manure effluents with natural zeolite. J Environ Sci Health Part B 34(2):305–332

Miles DM, Moore PA Jr, Smith DR, Rice DW, Stilborn HL, Rowe DR, Lott BD, Branton SL, Simmons JD (2003) Total and water-soluble phosphorus in broiler litter over three flocks with alum litter treatment and dietary inclusion of high available phosphorus corn and phytase supplementation. Poult Sci 82(10):1544–1549. https://doi.org/10.1093/ps/82.10.1544. From NLM

Modderman C (2020) Composting with or without additives. In: Animal Manure. Wiley, pp 245–254

Modroño S, Soldado A, Martínez-Fernández A, de la Roza-Delgado B (2017) Handheld NIRS sensors for routine compound feed quality control: real time analysis and field monitoring. Talanta 162:597–603

Montes F, Meinen R, Dell C, Rotz A, Hristov AN, Oh J, Waghorn G, Gerber PJ, Henderson B, Makkar HP et al (2013) Special topics–mitigation of methane and nitrous oxide emissions from animal operations: II. A review of manure management mitigation options. J Anim Sci 91(11):5070–5094. https://doi.org/10.2527/jas.2013-6584. From NLM

Moore PA Jr, Li H, Burns R, Miles D, Maguire R, Ogejo J, Reiter MS, Buser MD, Trabue S (2018) Development and testing of the ARS air scrubber: a device for reducing ammonia emissions from animal rearing facilities. Front Sustain Food Syst 2:23

Morgavi D, Kelly W, Janssen P, Attwood G (2013) Rumen microbial (meta) genomics and its application to ruminant production. Animal 7:184–201. https://doi.org/10.1017/S1751731112000419

Morota G, Ventura RV, Silva FF, Koyama M, Fernando SC (2018) Big data analytics and precision animal agriculture symposium: machine learning and data mining advance predictive big data analysis in precision animal agriculture. J Anim Sci 96(4):1540–1550

Mrode R, Ekine Dzivenu C, Marshall K, Chagunda MGG, Muasa BS, Ojango J, Okeyo AM (2020) Phenomics and its potential impact on livestock development in low-income countries: innovative applications of emerging related digital technology. Anim Front 10(2):6–11. https://doi.org/10.1093/af/vfaa002. %J Animal Frontiers (acccessed 8/10/2020)

Muhlbauer, R. V.; Moody, L. B.; Burns, R. T.; Harmon, J., and; Stalder, K. Water consumption and conservation techniques currently available for swine production National Pork Board; Des Moines, IA., 2011

Nardelli S, Farina G, Lucchini R, Valorz C, Moresco A, Dal Zotto R, Costanzi C (2008) Dynamics of infection and immunity in a dairy cattle population undergoing an eradication programme for infectious bovine Rhinotracheitis (IBR). Prev Vet Med 85(1–2):68–80

Neethirajan S, Kemp B (2021) Digital livestock farming. Sens Bio-Sens Res 32:100408

Niemi JK, Sevón-Aimonen M-L, Pietola K, Stalder KJ (2010) The value of precision feeding technologies for grow–finish swine. Livest Sci 129(1):13–23. https://doi.org/10.1016/j.livsci.2009.12.006

Nkrumah JD, Basarab JA, Wang Z, Li C, Price MA, Okine EK, Crews DH, Moore SS (2007) Genetic and phenotypic relationships of feed intake and measures of efficiency with growth and carcass merit of beef cattle1. J Anim Sci 85(10):2711–2720. https://doi.org/10.2527/jas.2006-767

Norton T, Chen C, Larsen MLV, Berckmans D (2019) Review: precision livestock farming: building 'digital representations' to bring the animals closer to the farmer. Animal 13(12):3009–3017. https://doi.org/10.1017/s175173111900199x. From NLM

NRC (2000) Nutrient requirements of beef cattle, 7th rev. edn. National Academy Press, Washington, DC

NRC (2001) Nutrient requirements of dairy cattle. National Research Council, p 519

NRC (2012) Nutrient requirements of swine. National Academies Press, Washington, D.C.

NRC. National Research Council (1994) Metabolic modifiers: effects on the nutrient requirements of food-producing animals. National Academies

Nwafor IC, Nwafor CU (2022) African smallholder farmers and the treatment of livestock diseases using ethnoveterinary medicine: a commentary. Pastoralism 12(1):29. https://doi.org/10.1186/s13570-022-00244-6. From NLM

Odintsov Vaintrub M, Levit H, Chincarini M, Fusaro I, Giammarco M, Vignola G (2021) Review: precision livestock farming, automats and new technologies: possible applications in extensive dairy sheep farming. Animal 15(3):100143. https://doi.org/10.1016/j.animal.2020.100143. WorldCat.org

Ominski K, McAllister T, Stanford K, Mengistu G, Kebebe EG, Omonijo F, Cordeiro M, Legesse G, Wittenberg K (2021) Utilization of by-products and food waste in livestock production systems: a Canadian perspective. Anim Front 11(2):55–63. https://doi.org/10.1093/af/vfab004. (acccessed 4/29/2024)

Oni RA, Sharma M, Micallef SA, Buchanan RL (2013) The effect of UV radiation on survival of salmonella enterica in dried manure dust. http://iafp.confex.com/iafp/2013/.../Paper4090.html.ExhbitHall/CharlotteConventionCentre. Accessed 2022/11/07

Organization, W. H (2004) Annual report 2003: essential drugs and medicines policy. World Health Organization

Osei-Amponsah R, Dunshea FR, Leury BJ, Cheng L, Cullen B, Joy A, Abhijith A, Zhang MH, Chauhan SS (2020) Heat stress impacts on lactating cows grazing Australian summer pastures on an automatic robotic dairy. Animals (Basel) 10(5):869

Ouédraogo D, Kamuanga M, Savadogo K, McDermott J, Woitag T (2006) Determinants of therapeutic choices and cattle health management in Western Burkina Faso. Revue d'économie du développement 14(5):93–111. https://doi.org/10.3917/edd.205.0093. From Cairn.info Cairn.info

Pagliari PH (2014) Variety and solubility of phosphorus forms in animal manure and their effects on soil test phosphorus. In: He Z, Zhang H (eds) Applied manure and nutrient chemistry for sustainable agriculture and environment. https://doi.org/10.1007/978-94-017-8807-6_8

Pagliari P, Wilson M, He Z (2020) Animal Manure production and utilization: impact of modern concentrated animal feeding operations. In: Waldrip, H.M., Pagliari, P.H., He, Z., editors. Animal Manure: Production, Characteristics, Environmental Concerns and Management. ASA Special Publication 67. Madison, WI: ASA and SSSA. p. 1–14. https://doi.org/10.2134/asaspecpub67.c1.

Palmer R, Holmes B (2005) Cow comfort issues in free stall barns. In Proceedings of the 7th Western Dairy Management Conference, pp 9–11

Pampouille E, Mika A, Bernard J, Guettier E, Berger Q, Bouvarel I, Mignon-Grasteau S (2021) BIRD-e -The Poultry E-Feeding System: Basis and Applications BIRD-e -Le système d'alimentation électronique pour volaille : fondements et applications. In: Animal nutrition conference of Canada, En ligne, France

Park M, Britton D, Daley W, McMurray G, Navaei M, Samoylov A, Usher C, Xu J (2022) Artificial intelligence, sensors, robots, and transportation systems drive an innovative future for poultry broiler and breeder management. Anim Front 12(2):40–48. https://doi.org/10.1093/af/vfac001. (acccessed 7/18/2022)

Patil VM (2013a) Milking Parlours. Karnataka Veterinary, Animal & Fisheries Sciences University, Post Box No. 6, Nandinagar, Bidar, India – 585401. https://sites.google.com/site/viveklpm/cattle-and-buffalo--production-management/milking-machine/types-of-milking-machinessystems/milking-parlours. Accessed 26/05/2022

Patil VM (2013b) Barn milking systems. https://sites.google.com/site/viveklpm/cattle-and-buffalo--production-management/milking-machine/types-of-milking-machinessystems/barn-milking-systems. Accessed 26/05/2022

Pawar DD, Malik SVS, Bhilegaonkar KN, Barbuddhe SB (2000) Effect of nisin and its combination with sodium chloride on the survival of listeria monocytogenes added to raw buffalo meat mince. Meat Sci 56(3):215–219

Petherick JC (2005) Animal welfare issues associated with extensive livestock production: the northern Australian beef cattle industry. Appl Anim Behav Sci 92(3):211–234

Picchi VV, de Castro EF, Marino FC, Ribeiro SL (2019) Increasing the Confidence of the Brazilian Livestock Production Chain using Blockchain. In Proceedings of the 2019 2nd International Conference on Blockchain Technology and Applications, pp 93–98

Pierce MD, Dzama K, Muchadeyi FC (2018) Genetic diversity of seven cattle breeds inferred using copy number variations. Front Genet 9:163. https://doi.org/10.3389/fgene.2018.00163

Piyasena P, Mohareb E, McKellar RC (2003) Inactivation of microbes using ultrasound: a review. Int J Food Microbiol 87(3):207–216

Plaza-Diaz J, Ruiz-Ojeda FJ, Gil-Campos M, Gil A (2019) Mechanisms of action of probiotics. Adv Nutr 10(suppl_1):S49–S66

Pomar C, van Milgen J, Remus A (2019) Precision livestock feeding, principle and practice. In: Poultry and pig nutrition. Wageningen Academic Publishers

Popova T (2017) Effect of probiotics in poultry for improving meat quality. Curr Opin Food Sci 14:72–77. https://doi.org/10.1016/j.cofs.2017.01.008. WorldCat.org

Portejoie S, Martinez J, Guiziou F, Coste CM (2003) Effect of covering pig slurry stores on the ammonia emission processes. Bioresour Technol 87(3):199–207

Powell JM, Aguerre MJ, Wattiaux MA (2011) Dietary crude protein and tannin impact dairy manure chemistry and ammonia emissions from incubated soils. J Environ Qual 40(6):1767–1774

Pradere J (2014) Improving animal health and livestock productivity to reduce poverty. Rev Sci Tech 33(3):735–744

Qin F, Xia F, Chen H, Cui B, Feng Y, Zhang P, Chen J, Luo M (2021) A guide to nucleic acid vaccines in the prevention and treatment of infectious diseases and cancers: from basic principles to current applications. Front Cell Dev Biol 830

Raboy V (2002) Progress in breeding low phytate crops. J Nutr 132(3):503s–505s. https://doi.org/10.1093/jn/132.3.503S. From NLM

Radunz AE (2011) Use of beta agonists as a growth promoting feed additive for finishing beef cattle. J Anim Sci 88:2476–2485

Rahman IU, Ijaz F, Bussmann RW (2022) Editorial: Ethnoveterinary practices in livestock: animal production, healthcare, and livelihood development. Front Vet Sci 9:1086311. https://doi.org/10.3389/fvets.2022.1086311. From NLM

Rai V, Yadav B, Lakhani GP (2013) Applications of probiotic and prebiotic in animal production: a review. Environ Ecol 31:873–876

Rajagopal R, Mousavi SE, Goyette B, Adhikary S (2021) Coupling of microalgae cultivation with anaerobic digestion of poultry wastes: toward sustainable value added bioproducts. Bioengineering 8(5). https://doi.org/10.3390/bioengineering8050057

Ramaswamy H, Tang J (2008) Microwave and radio frequency heating. Food Sci Technol Int 14(5):423–427

Ramirez S (2014) Feed additives: do they work and how do they work? – a practical Review

Rao IM, Peters M, Castro A, Schultze-Kraft R, White D, Fisher M, Miles JW, Lascano Aguilar CE, Blümmel M, Bungenstab D (2015) LivestockPlus: The sustainable intensification of forage-based agricultural systems to improve livelihoods and ecosystem services in the tropics

Reddy B, Reddy PS, Bidinger F, Blümmel M (2003) Crop management factors influencing yield and quality of crop residues. Field Crop Res 84(1–2):57–77

Rehman S, Iqbal Z, Qureshi R, Rahman IU, Sakhi S, Khan I, Hashem A, Al-Arjani A-BF, Almutairi KF, Abd-Allah EF et al (2022) Ethnoveterinary practices of medicinal plants among tribes of Tribal District of North Waziristan, Khyber Pakhtunkhwa, Pakistan. Front Veterin Sci 9. https://doi.org/10.3389/fvets.2022.815294. Original Research

Reinemann DJ (2013) Chapter 8 – Milking machines and milking parlors. Academic Press. https://doi.org/10.1016/B978-0-12-385881-8.00008-2

Renaudeau D, Collin A, Yahav S, De Basilio V, Gourdine J-L, Collier R (2012) Adaptation to hot climate and strategies to alleviate heat stress in livestock production. Animal 6(5):707–728

Ribeiro GO, Oss DB, He Z, Gruninger RJ, Elekwachi C, Forster RJ, Yang W, Beauchemin KA, McAllister TA (2017) Repeated inoculation of cattle rumen with bison rumen contents alters the rumen microbiome and improves nitrogen digestibility in cattle. Sci Rep 7(1):1–16

Rivas-Cañedo A, Fernández-García E, Nuñez M (2009) Volatile compounds in dry-cured serrano ham subjected to high pressure processing. Effect of the packaging material. Meat Sci 82(2):162–169

Rötzschke O, Falk K, Deres K, Schild H, Norda M, Metzger J, Jung G, Rammensee H-G (1990) Isolation and analysis of naturally processed viral peptides as recognized by cytotoxic T cells. Nature 348(6298):252–254

Rushen J (2001) Assessing the welfare of dairy cattle. J Appl Anim Welf Sci 4(3):223–234

Rynk R (1992) On-farm composting handbook. Northeast Regional Agricultural Engineering Service Pub. No. 54. Cooperative Extension Service. Ithaca, NY, 186 pp. A classic in on-farm composting. Website: www. nraes. org. diakses tanggal

Sans P, Combris P (2015) World meat consumption patterns: an overview of the last fifty years (1961–2011). Meat Sci 109:106–111. https://doi.org/10.1016/j.meatsci.2015.05.012. WorldCat.org

Sarnklong C, Cone J, Pellikaan W, Hendriks W (2010) Utilization of rice straw and different treatments to improve its feed value for ruminants: a review. Asian Australas J Anim Sci 23(5):680–692

Sasaki H, Kitazume O, Nonaka J, Hikosaka K, Otawa K, Itoh K, Nakai Y (2006) Effect of a commercial microbiological additive on beef manure compost in the composting process. Anim Sci J 77(5):545–548

Sastry S (2008) Ohmic heating and moderate electric field processing. Food Sci Technol Int 14(5):419–422

Science Learning Hub (2017) Robotic milking. https://www.sciencelearn.org.nz/resources/2089-robotic-milking. Accessed 30-05-2022

Sebola NT, Moroke T, Liphadzi SE (2021) Prospects of sub-Saharan African unconventional feed resources as sources of livestock feed: a review. Livestock Res Rural Dev 33:1–14

Selbie DR, Buckthought LE, Shepherd MA (2015) The challenge of the urine patch for managing nitrogen in grazed pasture systems. Adv Agron 129:229–292

Selemani IS, Eik LO (2016) The effects of concentrate supplementation on growth performance and behavioral activities of cattle grazed on natural pasture. Trop Anim Health Prod 48:229–232

Sharma A, Jain A, Gupta P, Chowdary V (2021) Machine learning applications for precision agriculture: a comprehensive review. IEEE Access 9:4843–4873. https://doi.org/10.1109/access.2020.3048415

Short SD (2000) Structure, management, and performance characteristics of specialized dairy farm businesses in the United States. US Department of Agriculture, Economic Research Service

Siggers R, Thymann T, Siggers J, Schmidt M, Hansen A, Sangild P (2007) Bacterial colonization affects early organ and gastrointestinal growth in the neonate. Livest Sci 109(1–3):14–18

Sikes A, Tornberg E, Tume R (2010) A proposed mechanism of tenderising post-rigor beef using high pressure–heat treatment. Meat Sci 84(3):390–399

Simitzis P, Tzanidakis C, Tzamaloukas O, Sossidou E (2021) Contribution of precision livestock farming systems to the improvement of welfare status and productivity of dairy animals. Dairy 3(1):12–28

Sinclair JR (2019) Importance of a one health approach in advancing global health security and the sustainable development goals. Revue scientifique et technique (International Office of Epizootics) 38(1):145–154. https://doi.org/10.20506/rst.38.1.2949. From NLM

Singh P, Wani AA, Langowski H-C (2017) Food packaging materials: testing & quality assurance. CRC Press

Singh J, Hundal J, Sharma A, Singh U, Sethi A, Singh PJIJCMAS (2018) Phosphorus nutrition in dairy animals: a review. Int J Curr Microbiol App Sci 7(4):3518–3530

Smith ZK, Johnson BJ (2020) Mechanisms of steroidal implants to improve beef cattle growth: a review. J Appl Anim Res 48(1):133–141. https://doi.org/10.1080/09712119.2020.1751642

Smith DR, Moore PA Jr, Miles DM, Haggard BE, Daniel TC (2004) Decreasing phosphorus runoff losses from land-applied poultry litter with dietary modifications and alum addition. J Environ Qual 33(6):2210–2216. https://doi.org/10.2134/jeq2004.2210. From NLM

Sommer F, Bäckhed F (2013) The gut microbiota—masters of host development and physiology. Nat Rev Microbiol 11(4):227–238

Sorensen A, Muir DD, Knight CH (2001) Thrice-daily milking throughout lactation maintains epithelial integrity and thereby improves milk protein quality. J Dairy Res 68(1):15–25

Spiehs MJ, Goyal SM (2007) Best management practices for pathogen control in manure management systems. Minnesota Extension

Sridhar A, Ponnuchamy M, Kumar PS, Kapoor A (2021) Food preservation techniques and nanotechnology for increased shelf life of fruits, vegetables, beverages and spices: a review. Environ Chem Lett 19:1715–1735

Srinivas B, Rani GS, Kumar BK, Chandrasekhar B, Krishna KV, Devi TA, Bhima B (2017) Evaluating the probiotic and therapeutic potentials of Saccharomyces cerevisiae strain (OBS2) isolated from fermented nectar of toddy palm. AMB Express 7(1):1–14

Stein HH, Shurson GC (2009) Board-invited review: the use and application of distillers dried grains with solubles in swine diets. J Anim Sci 87(4):1292–1303. https://doi.org/10.2527/jas.2008-1290. From NLM

Steiner C, Das KC, Melear N, Lakly D (2010) Reducing nitrogen loss during poultry litter composting using biochar. J Environ Qual 39(4):1236–1242

Steinfeld H, Gerber P, Wassenaar TD, Castel V, Rosales M, Rosales M, de Haan C (2006) Livestock's long shadow: environmental issues and options. Food & Agriculture Org

Suleiman A, Jackson EL, Rushton J (2015) Challenges of pastoral cattle production in a sub-humid zone of Nigeria. Trop Anim Health Prod 47(6):1177–1185. https://doi.org/10.1007/s11250-015-0845-0. WorldCat.org

Sutton, A., and; Lander, C. Effects of diet and feeding management on nutrient content; Washington, D.C., 2003

Swallow BM (2000) Impacts of trypanosomiasis on African agriculture. Food and Agriculture Organization of the United Nations Rome, Italy

Szögi AA, Vanotti MBA, Rothrock MJ (2014) Gaseous ammonia removal system, USA. 8906,332 B2

Tadesse G (2016) Brucellosis seropositivity in animals and humans in Ethiopia: a meta-analysis. PLoS Negl Trop Dis 10(10):e0005006

Takahashi S, Ihara H, Karasawa T (2016) Compost in pellet form and compost moisture content affect phosphorus fractions of soil and compost. Soil Sci Plant Nutr 62(4):399–404

Tassou SA, Lewis JS, Ge YT, Hadawey A, Chaer I (2010) A review of emerging technologies for food refrigeration applications. Appl Therm Eng 30(4):263–276

Tauer LW (1998) Cost of production for stanchion versus parlor milking in New York. J Dairy Sci 81(2):567–569. https://doi.org/10.3168/jds.S0022-0302(98)75609-7

Tekin K, Dikmen BY, Kanca H, Guatteo R (2021) Precision livestock farming technologies: novel direction of information flow. Ankara Üniversitesi Veteriner Fakültesi Dergisi 68(2):193–212. https://doi.org/10.33988/auvfd.837485

Terry SA, Basarab JA, Guan LL, McAllister TA (2020) Strategies to improve the efficiency of beef cattle production. Can J Anim Sci 101(1):1–19. https://doi.org/10.1139/cjas-2020-0022.

TFF (2024) Thought for Food (TFF). What's the future of animal feed in Africa? 2022; Vol. April 28. https://thoughtforfood.org/content-hub/whats-the-future-of-animal-feed-in-africa/

Thorne PS, Ansley AC, Perry SS (2009) Concentrations of bioaerosols, odors, and hydrogen sulfide inside and downwind from two types of swine livestock operations. J Occup Environ Hyg 6(4):211–220. https://doi.org/10.1080/15459620902729184.

Timmerman H et al (2006) Mortality and growth performance of broilers given drinking water supplemented with chicken-specific probiotics. Poult Sci 85(8):1383. From http://worldcat.org/z-wcorg/

Titterton M, Bareeba F (2000) Grass and legume silages in the tropics. FAO Plant Production and Protection Papers, pp 43–50. https://www.fao.org/4/x8486e/x8486e0c.htm. Accessed 27 June 2023

Townsend A, McMichael A, Carter N, Huddleston J, Brownlee G (1984) Cytotoxic T cell recognition of the influenza nucleoprotein and hemagglutinin expressed in transfected mouse L cells. Cell 39(1):13–25. https://doi.org/10.1016/0092-8674(84)90187-9

Tranel L (2008) Automatic take-offs—are they for you. Iowa State Univ. Fact Sheet LT-06. Iowa State University, Ames

Tumasang T, Awah Ndukum J, Niba A, Kameni S, Jemima G (2023) Ethnoveterinary practices: a review of phytotherapeutical approaches in the treatment of livestock in Africa: case of Cameroon. Int J Res GRANTHAALAYAH:*10*. https://doi.org/10.29121/granthaalayah.v10.i12.2022.4679

Turner BL, Papházy MJ, Haygarth PM, McKelvie ID (2002) Inositol phosphates in the environment. Philos Trans R Soc Lond B Biol Sci 357(1420):449–469. https://doi.org/10.1098/rstb.2001.0837.

Undersander DJ, Albert B, Cosgrove D, Johnson D, Peterson P (2002) Pastures for profit: A guide to rotational grazing. University of Wisconsin-Extension. https://ucanr.edu/sites/default/files/2020-12/341086.pdf

Upton J, Murphy M, De Boer IJM, Koerkamp PWGG, Berentsen PBM, Shalloo L (2015) Investment appraisal of technology innovations on dairy farm electricity consumption. J Dairy Sci 98(2):898–909. https://doi.org/10.3168/jds.2014-8383

USEPA (2017) Inventory of U.S. greenhouse gas emissions and sinks. U.S. Environmenal Protecion Agency, Washington, D.C. https://www.epa.gov/ghgemissions/inventory-us-greenhouse-gasemissions-and-sinks. Accessed 2022/08/27

Uyeno Y, Shigemori S, Shimosato T.(2015) Effect of Probiotics/Prebiotics on Cattle Health and Productivity. Microbes Environ *30*(2):126-32. https://doi.org/10.1264/jsme2.ME14176.

Vadas PA, Good LW, Moore PA Jr, Widman N (2009) Estimating phosphorus loss in runoff from manure and fertilizer for a phosphorus loss quantification tool. J Environ Qual 38(4):1645–1653. https://doi.org/10.2134/jeq2008.0337.

Valenzuela-Grijalva NV, Pinelli-Saavedra A, Muhlia-Almazan A, Domínguez-Díaz D, González-Ríos H (2017) Dietary inclusion effects of phytochemicals as growth promoters in animal production. J Animal Sci Technol 59(1). https://doi.org/10.1186/s40781-017-0133-9.

Van der Heyden C, Demeyer P, Volcke EIP (2015) Mitigating emissions from pig and poultry housing facilities through air scrubbers and biofilters: state-of-the-art and perspectives. Biosyst Eng 134:74–93. https://doi.org/10.1016/j.biosystemseng.2015.04.002

Van Oirschot J, Bruschke C, Van Rijn P (1999) Vaccination of cattle against bovine viral diarrhoea. Vet Microbiol 64(2–3):169–183. https://doi.org/10.1016/s0378-1135(98)00268-5

Vanotti MB, Szogi AA, Hunt PG (2005) Wastewater treatment system. US Patent No. 6,893,567

Vanotti MB, Szogi AA, Fetterman LM (2010) Wastewater treatment system with simultaneous separation of phosphorus and manure solids. USDA Patent No: US7674379B2. https://patents.google.com/patent/US20080314837A1/en

Vanotti MB, Dube PJ, Szogi AA, García-González MC (2017) Recovery of ammonia and phosphate minerals from swine wastewater using gas-permeable membranes. Water Res 112:137–146. https://doi.org/10.1016/j.watres.2017.01.045

Vanotti MB, Szögi AA, Dube PJ (2018) Systems and methods for recovering ammonium and phosphorus from liquid effluents, Washington, D.C.. 9926,213 B2

Vanotti MB, García-González MC, Szögi AA, Harrison JH, Smith WB, Moral R (2020) Removing and recovering nitrogen and phosphorus from animal manure. In: Animal manure: production, characteristics, environmental concerns, and management, vol 67. American Society of Agronomy and Soil Science Society of America, pp 275–321. https://doi.org/10.2134/asaspecpub67.c22

Viana JHM, Figueiredo ACS, Siqueira LGB (2017) Brazilian embryo industry in context: pitfalls, lessons, and expectations for the future. Anim Reprod 14(3):476–781. https://doi.org/10.21451/1984-3143-AR989

Vijn S, Compart DP, Dutta N, Foukis A, Hess M, Hristov AN, Kalscheur KF, Kebreab E, Nuzhdin SV, Price NN et al (2020) Key considerations for the use of seaweed to reduce enteric methane emissions from cattle. Front Vet Sci 7:597430. https://doi.org/10.3389/fvets.2020.597430.

Villar ML, Hegarty RS, Nolan JV, Godwin IR, McPhee M (2020) The effect of dietary nitrate and canola oil alone or in combination on fermentation, digesta kinetics and methane emissions from cattle. Anim Feed Sci Technol 259:114294. https://doi.org/10.1016/j.anifeedsci.2019.114294

Vogeli, Y.; Lohri, C. R.; Gallardo, A.; Diener, S., and ; Zurbugg, C. Anaerobic digestion of biowaste in developing countries. Practical information and case studies; Eawag– Swiss Federal Institute of Aquatic Science and Technology, Department of Water and Sanitation in Developing Countries (Sandec), Dubendorf, Switzerland, 2014

Vordermeier HM, Jones GJ, Buddle BM, Hewinson RG, Villarreal-Ramos B (2016) Bovine tuberculosis in cattle: vaccines, DIVA tests, and host biomarker discovery. Ann Rev Animal Biosci 4:87–109. https://doi.org/10.1146/annurev-animal-021815-111311

Wadhwa A, Aljabbari A, Lokras A, Foged C, Thakur A (2020) Opportunities and challenges in the delivery of mRNA-based vaccines. Pharmaceutics 12(2):102. https://doi.org/10.3390/pharmaceutics12020102

Wakase S, Sasaki H, Itoh K, Otawa K, Kitazume O, Nonaka J, Satoh M, Sasaki T, Nakai Y (2008) Investigation of the microbial community in a microbiological additive used in a manure composting process. Bioresour Technol 99(7):2687–2693. https://doi.org/10.1016/j.biortech.2007.04.040

Wakhusama SW, Bastiaensen P, Bouguedour R, Letshwenyo M, Tounkara K, Mérot J, Ripani A, Gregorio T, Renaudin S, Bertrand-Ferrandis C (2019) The world Organization for Animal Health (OIE)'s engagement towards rabies elimination in Africa. J Vet Med One Health 1(1):3–14. https://doi.org/10.36856/jvmoh.v1i1.11

Waldrip HM, Todd RW, Cole NA (2013) Prediction of nitrogen excretion by beef cattle: a meta-analysis. J Anim Sci 91(9):4290–4302. https://doi.org/10.2527/jas.2012-5818

Wang M, Yu Z, Wu Z, Hannaway DB (2018) Effect of Lactobacillus plantarum 'KR107070'and a propionic acid-based preservative on the fermentation characteristics, nutritive value and aerobic stability of alfalfa-corn mixed silage ensiled with four ratios. Grassl Sci 64(1):51–60. https://www.researchgate.net/publication/333447239_Effect_of_Lactobacillus_plantarum_'KR107070'_and_a_propionic_acid-based_preservative_on_the_chemical_and_mycotoxin_composition_in_alfalfa-corn_mixed_silage_exposed_to_air

Waqas M, Nizami AS, Aburiazaiza AS, Barakat MA, Ismail IMI, Rashid MI (2018) Optimization of food waste compost with the use of biochar. J Environ Manag 216:70–81. https://doi.org/10.1016/j.jenvman.2017.06.015

Warner D, Vasseur E, Lefebvre DM, Lacroix R (2020) A machine learning based decision aid for lameness in dairy herds using farm-based records. Comput Electron Agric 169:105193. https://doi.org/10.1016/j.compag.2019.105193

WCCCS (2014) Western Canadian Cow-Calf Survey http://westernbeef.org/pdfs/economics/WCCCS_Summary_Overall_Jun2015.pdf

Webb MJ, Pendell DL, Harty AA, Salverson RR, Rotz CA, Underwood KR, Olson KC, Blair AD (2017) Influence of growth promoting technologies on animal performance, production economics, environmental impacts and carcass characteristics of beef. Meat and Muscle Biology 1(3):23–24. https://doi.org/10.22175/rmc2017.022

Wei S, Chadwick DR, Amon B, Dong H (2022) Comparison of nitrogen losses from different manure treatment and application management systems in China. J Environ Manag 306:114430. https://doi.org/10.1016/j.jenvman.2022.114430. From NLM

Weimer P, Stevenson D, Mertens D (2010a) Shifts in bacterial community composition in the rumen of lactating dairy cows under milk fat-depressing conditions. J Dairy Sci 93(1):265–278. https://doi.org/10.3168/jds.2009-2206

Weimer P, Stevenson D, Mantovani H, Man S (2010b) Host specificity of the ruminal bacterial community in the dairy cow following near-total exchange of ruminal contents. J Dairy Sci 93(12):5902–5912. https://doi.org/10.3168/jds.2010-3500

Wendimu A, Bojago E, Abrham Y, Tekalign W (2023) Practices of ethnoveterinary medicine and ethnobotanical knowledge of plants used to treat livestock diseases, Wolaita zone, Southern Ethiopia. Cogent Food Agric 9(1):2248691. https://doi.org/10.1080/23311932.2023.2248691

Werkheiser I (2018) Precision livestock farming and farmers' duties to livestock. J Agric Environ Ethics 31(2):181–195. https://doi.org/10.1007/s10806-018-9720-0

Willey J, Sherwood LM, Woolverton CJ, Prescott LM (2011) Control of microorganisms in the environment. McGraw-Hill Companies Inc, New York, pp 190–207

Wilson ML, Niraula S, Cortus EL (2020) Nutrient characteristics of swine manure and wastewater. In: Animal Manure: production, characteristics, environmental concerns, and management, vol 67. American Society of Agronomy and Soil Science Society of America, pp 89–113. https://doi.org/10.2134/asaspecpub67.c6

Winder JA, Brinks JS, Bourdon RM, Golden BL (1990) Genetic analysis of absolute growth measurements, relative growth rate and restricted selection indices in red Angus cattle. J Animal Sci 68(2):330–336. https://doi.org/10.2527/1990.682330x

Witzig M, Zeder M, Rodehutscord M (2018) Effect of the ionophore monensin and tannin extracts supplemented to grass silage on populations of ruminal cellulolytics and methanogens in vitro. Anaerobe 50:44–54. https://doi.org/10.1016/j.anaerobe.2018.01.012

Wold JP, Veberg A, Nilsen A, Iani V, Juzenas P, Moan J (2005) The role of naturally occurring chlorophyll and porphyrins in light-induced oxidation of dairy products. A study based on fluorescence spectroscopy and sensory analysis. Int Dairy J 15(4):343–353. https://doi.org/10.1016/j.idairyj.2004.08.009

Wu Z, Satter LD, Sojo R (2000) Milk production, reproductive performance, and fecal excretion of phosphorus by dairy cows fed three amounts of phosphorus. J Dairy Sci 83(5):1028–1041. https://doi.org/10.3168/jds.S0022-0302(00)74967-8

Wu Z, Tallam SK, Ishler VA, Archibald DD (2003) Utilization of phosphorus in lactating cows fed varying amounts of phosphorus and forage. J Dairy Sci 86(10):3300–3308. https://doi.org/10.3168/jds.S0022-0302(03)73931-9

Yankauckas MA, Morrow JE, Parker SE, Abai A, Rhodes GH, Dwarki VJ, Gromkowski SH (1993) Long-term anti-nucleoprotein cellular and humoral immunity is induced by intramuscular injection of plasmid DNA containing NP gene. DNA Cell Biol 12(9):771–776. https://doi.org/10.1089/dna.1993.12.771

Yayneshet T, Eik L, Moe S (2009) The effects of exclosures in restoring degraded semi-arid vegetation in communal grazing lands in northern Ethiopia. J Arid Environ 73(4–5):542–549. https://doi.org/10.1016/j.jaridenv.2008.12.002

Yeoman CJ, White BA (2014) Gastrointestinal tract microbiota and probiotics in production animals. Annu Rev Anim Biosci 2(1):469–486. https://doi.org/10.1146/annurev-animal-022513-114149

Younger DS, Younger AP, Guttmacher S (2016) Childhood vaccination: implications for global and domestic public health. Neurol Clin 34(4):1035–1047. https://doi.org/10.1016/j.ncl.2016.05.004

Yousef AE, Zhang HQ (2006) Microbiological and safety aspects of pulsed electric field technology. ACS Publications

Zang B, Li S, Michel FC, Li G, Zhang D, Li Y (2017) Control of dimethyl sulfide and dimethyl disulfide odors during pig manure composting using nitrogen amendment. Bioresour Technol 224:419–427. https://doi.org/10.1016/j.biortech.2016.11.023

Zeng L, Mangan C, Li X (2006) Ammonia recovery from anaerobically digested cattle manure by steam stripping. Water Sci Technol 54(8):137–145. https://doi.org/10.2166/wst.2006.852

Zhang H, Schroder J (2014) Animal manure production and utilization in the US. In: Applied manure and nutrient chemistry for sustainable agriculture and environment. Springer, pp 1–21. https://doi.org/10.1007/978-94-017-8807-6_1

Zhang D, Luo W, Yuan J, Li G, Luo Y (2017) Effects of woody peat and superphosphate on compost maturity and gaseous emissions during pig manure composting. Waste Manag 68:56–63. https://doi.org/10.1016/j.wasman.2017.05.042

Zheng L, Sun D-W (2006) Innovative applications of power ultrasound during food freezing processes—a review. Trends Food Sci Technol 17(1):16–23. https://doi.org/10.1016/j.tifs.2005.08.010

Zinkernagel RM, Doherty PC (1974) Restriction of in vitro T cell-mediated cytotoxicity in lymphocytic choriomeningitis within a syngeneic or semiallogeneic system. Nature 248(5450):701–702. https://doi.org/10.1038/248701a0

Zuidhof MJ (2020) Precision livestock feeding: matching nutrient supply with nutrient requirements of individual animals. J Appl Poult Res 29:11–14. https://doi.org/10.1016/j.japr.2019.12.009

Open Access This chapter is licensed under the terms of the Creative Commons Attribution 4.0 International License (http://creativecommons.org/licenses/by/4.0/), which permits use, sharing, adaptation, distribution and reproduction in any medium or format, as long as you give appropriate credit to the original author(s) and the source, provide a link to the Creative Commons license and indicate if changes were made.

The images or other third party material in this chapter are included in the chapter's Creative Commons license, unless indicated otherwise in a credit line to the material. If material is not included in the chapter's Creative Commons license and your intended use is not permitted by statutory regulation or exceeds the permitted use, you will need to obtain permission directly from the copyright holder.

Prospects for Utilization of Modern Technologies for Cattle Improvement in Africa

Sunday O. Peters, Eveline M. Ibeagha-Awemu, Iliya D. Kwoji, Raphael Mrode, Michael O. Ozoje, David A. Mbah, Isidore Houaga, Moses Okpeku, Peter O. Fayemi, Daniel Trocmé, Oluyinka Opoola, and Matthew A. Adeleke

Abstract

This chapter discusses the prospects of the utilization of modern technologies for cattle improvement in African livestock production systems. The first section gives a general overview of modern technologies and their application for improving cattle in the African continent. Section 22.1 discusses the general overview of how different molecular information improves the development of large ruminants in Africa. The section outlines some practical case studies where molecular information has been used for the detection of genomic variations in African cattle and buffalo, the use of genetic markers to assign individuals to breeds, and parentage identification. It concludes with the use of molecular information in gene editing and transgenic animals in animal production. In Sect. 22.3, the prospect for the application of modern technology in developing livestock feeds and feeding

---

S. O. Peters (✉)
Department of Animal Science, Berry College, Mount Berry, GA, USA
e-mail: speters@berry.edu

E. M. Ibeagha-Awemu
Sherbrooke Research and Development Centre, Agriculture and Agri-Food Canada, Sherbrooke, QC, Canada

I. D. Kwoji · M. Okpeku · M. A. Adeleke
Discipline of Genetics, School of Life Sciences, University of Kwa-Zulu Natal Westville Campus, Durban, South Africa

R. Mrode
International Livestock Research Institute, Nairobi, Kenya

M. O. Ozoje
Department of Animal Breeding and Genetics, Federal University of Agriculture, Abeokuta, Nigeria

D. A. Mbah
Cameron Academy of Sciences, Yaounde, Cameroon

Scotland Rural College, Edinburgh, UK

I. Houaga
The Roslin Institute and Royal (Dick) School of Veterinary Studies, Easter Bush Campus, University of Edinburgh, Midlothian, UK

Centre for Tropical Livestock Genetics and Health (CTLGH), Roslin Institute, Easter Bush Campus, University of Edinburgh, Edinburgh, UK

P. O. Fayemi
Beta-Letters Agrinextiomics Ltd MoA, Durban, South Africa

D. Trocmé
Adventiel, Pacé, France

O. Opoola
Scotland Rural College, Edinburgh, UK

The Global Academy for Agriculture and Food Systems (GAAFS), Easter Bush Campus, University of Edinburgh, Edinburgh, UK

© The Author(s) 2026
E. M. Ibeagha-Awemu et al. (eds.), *African Livestock Genetic Resources and Sustainable Breeding Strategies*, Sustainable Development Goals Series, https://doi.org/10.1007/978-3-031-92076-9_22

is discussed with practical examples. The subsequent sections discuss the application of modern technologies in improving livestock housing and husbandry systems (Sect. 22.4). Other modern technological advancements, including the development of diagnostic tests for detecting genetic defects using single nucleotide polymorphisms (SNPs), using genomic markers for selective breeding, and genome editing to improve productivity and control of deleterious alleles in livestock, are also discussed. The chapter also reviews the application of modern technologies to improve livestock health. In each case, practical examples are outlined as case studies where the technologies are applied in the African setting. The chapter concludes by looking at challenges and constraints in the implementation of modern technologies in cattle and buffalo production in Africa. Potential solutions are suggested.

Keywords

Database · Genetic improvement · Microbiome · Probiotics · Disease surveillance · Information system · Modular system · Mobile application · Animal identification

22.1 Introduction

The development and application of modern technologies in science have been instrumental in driving every sector of human endeavours for sustainable development and understanding complex processes. The application of technologies has made everyday work easier in the agricultural sector 1. An extensive review of modern technologies and how they are being applied to improve productivity in the livestock sector in Africa and worldwide is presented in Chap. 21. Advancements in molecular biology have aided in understanding the genome and its influence on phenotypes. Further developments in the milking process have led to the introduction of milking robots, which have brought new advantages to the farmers, such as labour efficiency and automatic recording of several animal-related parameters (Groher et al. 2020). Today, farmers are faced with a wide range of new technologies and the uncertainty of their effects on the sustainability of farming systems (Anderson 2010; Frezal et al. 2021). Technological innovations and developments are rapidly evolving. Sensors of various kinds are being developed to monitor real-time milk quality, health, and pregnancy hormones. In addition, virtual fences can remotely move animals wearing these sensors from one pasture area to another. Even robotics is advancing fast in the livestock industry, addressing the challenges of labour shortages in livestock farms. About 12% of dairy farms are using robots, which is expected to grow to 20% in the next 5 years, according to Collado et al. (2018). However, several factors facilitate the adoption of new technologies in sustainable farming systems.

There is a significant interest in applying advanced technologies for the improvement of livestock production in Africa. Demand for animal-based source foods (ASFs) such as meat, milk, and eggs in Africa is increasing (Opoola et al. 2022) and therefore the current ASFs produced do not meet the supply for the growing human population. However, the level of productivity of most African livestock is low, and the adoption of modern technologies will help to drive sustained growth in cattle under different input systems. This chapter reviews the practical application of modern technologies like genomic information, reproduction technologies, growth technologies, and phenomics technologies in cattle production in Africa and other developing countries. References are made to buffalo where appropriate.

22.2 Application of Genomic Technologies in Large Ruminant Production

Traits of economic importance in livestock in terms of productivity, disease resistance, behaviour, and adaptability are the product of the actions and interactions of many genes combined with environmental factors. Understanding and identifying the genetic basis that controls genetic variation at the DNA level and incorporating such molecular information in animal breeding programmes represent the optimum path for the rapid implementation of livestock improvement. However, unravelling the genetic control of such complex traits at the genome level remains challenging. Still, tremendous progress has been made in the last few decades, especially from the attempts to sequence the genomes of many livestock species. Recent advances in genomic technologies have been presented in Chap. 20. This section examines their various application levels and potential roles in driving genetic improvement and the genetic basis of the adaptation of African indigenous breeds. African cattle breeds have been extensively covered in Chaps. 3 and 4.

22.2.1 Detection of Genomic Variations in African Cattle and Buffalo Populations

The discovery of SNPs during the sequencing of the human genome shifted attention to their use as markers due to their abundance in the genome. With the completion of the first human genome draft in the early 2000s (Wakeley et al. 2001), much effort was exerted to sequence the genomes of commonly farmed animal species. Starting in December 2003, the efforts of Richard Gibbs and George Weinstock (Baylor College of Medicine's genome sequencing centre, Houston, Texas, USA) culminated with the sequencing and publication of the first draft of the Herford genome in 2009 (Sequencing et al. 2009). In the same year, another group at the University of Maryland published the genome sequence of the domestic cow (Zimin et al. 2009). Following were the publication of the Bos indicus genome in 2012 (Canavez et al. 2012) and the genomes of the Asian buffalo leech and Egyptian water buffalo in 2019 and 2020, respectively (Canavez et al. 2012; Tijjani et al. 2022). Subsequently, cattle genomes of different breeds have been sequenced including initiatives like the 1000 Bull Genomes project (Hayes and Daetwyler 2019) which resulted in the genome sequencing of animals from different cattle breeds in an attempt to better understand cattle genomes and complex traits. The incorporation of this information into breeding strategies to improve the rate of genetic improvement. More recent efforts include building the pangenomes of cattle breeds to better catalogue genomic variations in cattle and buffalo populations worldwide (Zhou et al. 2022; Smith et al. 2023).

Depending on the species and population, SNPs, the most abundant form of genomic variation, occur in every 1000 nucleotides on average, implying that there are approximately four (4) to five (5) million SNPs in a typical bovine genome. The abundance of SNPs and the plausibility of genotyping SNPs in large numbers makes them very useful as biological markers in mapping quantitative trait loci (QTL) that are associated with traits of economic importance in livestock and in improving the accuracy of breeding value prediction.

Consideration for the inclusion of genetic marker information in breeding programmes requires knowledge of the available genomic variation and their relationship with traits of interest (Ibeagha-Awemu et al. 2004). Therefore, the first step in deciding to include genetic information in breeding objectives and plans is detecting sequence variations and association with traits. Genomic variations in the form of SNPs, microsatellites, and copy number variants have been detected in some African cattle and buffalo populations and, in some cases, are associated with production traits.

22.2.1.1 Case Studies
Characterizing 28 autosomal markers (microsatellites, SNPs) in *Bos indicus* and *Bos taurus* cattle breeds of Nigeria and Cameroon Ibeagha-Awemu et al. (2004) reported high genetic polymorphisms amongst the breeds, as well as close relationships due to high gene flow between breeds. Rushdi et al. (2017) evaluated polymorphism at eight polymorphic microsatellite loci (BM143, ETH2, BM1706, ETH131, BMS711, BM415, BM6438, and BM1443) and showed their association with milk production

traits in the Egyptian buffaloes (Rushdi et al. 2017). Whole genome SNP panel (Illumina BovineSNP50 BeadChip) genotyping in six South African cattle breeds (Afrikaner, Nguni, Drakensberger, Bonsmara, Angus, and Holstein) indicated that ~56% of SNPs (21,290 SNPs) were polymorphic in the breeds. In contrast, 41% exhibited a high degree of polymorphism (minor allele frequency ≥ 0.05) (Makina et al. 2014). Furthermore, 33 genomic regions harbouring putative signatures of selection and genes associated with adaptation to tropical climates (KRT25, KRT24, KRT26, KRT27, KRT222, and HSPB9), immune response (CDC6, CYM, and CDK10), and production (MTPN, AJAP1, TGFB1, and IGFBP4) were identified in the same South African cattle populations (Makina et al. 2015).

Further genome characterization of cattle breeds from North Africa, West Africa, East Africa, and South Africa with Illumina BovineHD or BovineSNP50 Genotyping BeadChips revealed high levels of genetic polymorphisms, candidate genes for various traits and positional positive selection regions harbouring genes and QTL for reproduction traits, milk traits, immunity, heat stress, skin and hair properties, genes associated with adaptation to marginal environments and copy number variations (Gautier et al. 2009; Bahbahani et al. 2018). Using the method of whole genome sequencing to characterize the genomes of five indigenous African cattle breeds: Ankole (African Sanga cattle), Boran (East African zebu), N'Dama (West African taurine), Kenana (East African zebu), and Ogaden (East African zebu), Kim and colleagues Kim et al. (2017) reported a high number of genomic variations (mostly SNPs), as well as SNPs specific for the breeds. Compared with three European breeds (Holstein, Angus, and Jersey) subjected to intensive selection for various traits over generations, the African breeds presented higher nucleotide diversity (Kim et al. 2017). From these reports, high levels of genome variation are evident within African cattle and buffalo populations attributable to domestication, migrations, selection, and environmental adaptation to diverse environmental exposures (cold, dry, hot, or humid tropical climatic conditions) and indicative of their potential application in breed improvement for dairy, breed and environmental adaptation traits.

Evaluation of genotyping-by-sequencing (GBS) data of 193 Nigerian cattle from different breeds against whole-genome sequencing data, Mauki (Mauki et al. 2022) and colleagues recorded a high rate of sequence variations and genomic regions harbouring imprints of adaptation and genes associated with growth, reproduction, the immune response, efficiency of feeds utilization, and thermotolerance.

22.2.2 Application of Genetic Markers for Breed Group Assignation and Parentage Identification

African cattle and buffalo populations are mostly under traditional management systems where breeding is not controlled, and gene flow between breeds is high. Recent studies have reported high levels of gene flow and admixtures between African cattle populations (Ibeagha-Awemu et al. 2004; Gebrehiwot et al. 2020). Moreover, the introduction of exotic cattle breeds (e.g. Holstein, Ayrshire, and Montbeliarde) for increasing the productivity of indigenous African populations in the last seven decades has resulted in high levels of exotic genes in local populations. As such, many individual cattle in the continent are of uncertain genetic composition, even when main physical characteristics qualify assignation to specific breed types. With a lack of structured genetic evaluations of the cattle populations (indigenous breeds and the crossbreds) and pedigree records, as well as the inability to assign individuals to specific breed types based on visual traits, genomic variation has been used to assign individual cattle to their respective breed groups or compositions, parentage verification and to make informed decisions on breeding programmes for specific traits.

22.2.2.1 Case Studies

Using SNP genotype data (Illumina BovineSNP50 and BovineHD Beadchips) on 4231 individual animals from various cattle populations in Africa (African *Bos taurus* and *Bos indicus*, European *Bos taurus* and crossbreds), Gebrehiwot et al. (2020) identified 200 SNPs with high prediction accuracy for breed composition. Applying SNP data generated with Illumina BovineHD Beadchip followed by admixture analysis provided accurate information on the genetic composition of breeds and the breed composition best suited for different smallholder production environments (Alexander et al. 2009). A study on the admixture, population structure, and levels of inbreeding in Baoulé and Baoulé x Zebu crossbreds under traditional management systems in Burkina Faso using data on 38,207 SNP genotypes, correctly assigned analysed individuals to their main genetic backgrounds (*Bos indicus* and *Bos taurus*) as well as defined their levels of admixture (Ouédraogo et al. 2021).

In pure breeds of beef cattle, especially those with herd books, accurate pedigree information is essential for the success of their seed stock market and the accurate ranking of breeding stock. For example, using microsatellite markers for parentage identification, incorrect and incomplete parentage recording of up to 14% of individuals was found, resulting in significant re-ranking of the Angora goat sires (Visser et al. 2011; Garritsen et al. 2015). Similarly, Kios et al. (2012) 55.2% of the misidentification of sires and 2.3% of dams in the Boran seed stock in Kenya were reported using DNA-based testing. Visser et al. (2011) reported that about 35% of cattle breeders use DNA parentage testing on a routine basis, especially in larger herds where multi-sire mating is performed.

Working with crossbred dairy cattle in Senegal, genotyped with the Bovine 50 K SNP assay and using population Structure analysis, Tebug et al. (2018) assigned animals to different proportions of ancient Zebu, recent Zebu, ancient Taurine, and recent Taurine, and subsequently defined different breed groups. Based on the breed-composition assignment, Marshal et al. (2017) undertook a trade-off analysis for the various household dairy systems and determined the most productive bred type for each management or production system.

Similarly, for crossbred dairy cattle in Eastern Africa, Strucken et al. (2017) developed an assay of 200 SNPs to determine the breed composition of animals. However, the number of SNPs would increase to 400 if parentage verification was to be incorporated. Also, using SNP data, Al Kalaldeh et al. (2021a) estimated breed proportions for genomic prediction for smallholder crossbred dairy production in India and in determining the performance of these crossbred animals under different environments.

22.2.3 Application of Genetic Markers for QTL Mapping, Marker-Trait Associations, and Detection of the Signature of Selection

Linkage mapping of QTLs using molecular markers is commonly used in conventional breeding. With developments in MAS from the 1980s, information on QTL was incorporated into linear models for genetic evaluations (Weigel et al. 2017). Typically, the estimated breeding value (EBV) of a candidate under selection is determined by summing the estimated effects of several QTLs. Thus, QTL information was employed in MAS of young sires or breed introgression programmes, especially for loci that alter performance attributes or disease resistance/susceptibility (Mitra et al. 1999). Moreover, by determining the coupling linkage relationship between specific alleles at marker loci, breeders can identify genetic markers and QTL that confer favourable traits of interest, increasing the accuracy of selection and the rate of introgression of favourable alleles into desired populations (Soller 1994).

From the 1980s to the early 2000s, genetic markers used for QTL detection were mostly microsatellites and restriction-length fragment polymorphisms. With the developments of next-generation sequencing technologies and the availability of whole genome genotyping assays, more sequence variations, mostly SNPs and copy number variants, are now applied to map QTLs, detect-market trait associations, and signatures of selection at unprecedented depth.

22.2.3.1 Case Studies

The application of QTL mapping to guide the selection and genetic improvement of African cattle is exemplified by the study of Hanotte et al. (2003) involving the crossing of tolerant West African N'Dama (*Bos taurus*) and trypan-susceptible East African Boran cattle (*Bos indicus*) to identify QTL controlling resistance to trypanosomiasis. To test their hypothesis, 177 F2 animals and their parents and grandparents were genotyped at 477 microsatellite loci covering 29 autosomes. Several identified QTLs were mapped to 18 autosomes, and individual QTL effects ranged from 6 to 20% of phenotypic variants of the trait (trypanotolerance). Results of the study suggest that selection for trypanotolerance within F2 populations could produce synthetic breeds with higher trypanotolerance levels than their parents.

Vohra et al. (2021) Vohra and colleagues conducted a genome-wide association study (GWAS) in Murrah Indian buffalo using 38,000 SNPs obtained by double digestion restriction associated DNA (RAD) tag genotyping-by-sequencing and identified genomic regions associated with lactation (305-day milk yield, lactation persistency, breeding efficiency) and fertility (postpartum breeding interval and age at sexual maturity) traits (Vohra et al. 2021). The same study identified many other potential genomic regions associated with milk yield and fertility in Murrah buffalo.

Following Illumina BovineSNP50 genotyping in East African Shorthorn Zebu, Butana, and Kenana zebu breeds, several regions under selection and intersecting with genes and QTL for reproduction performance, production and environmental adaptation to stress (heat stress and immunity) and dairy traits were found (Bahbahani et al. 2015, 2018). Analysing the whole genome sequences of indigenous breeds of cattle from East, West, and Southern Africa, Kim and colleagues (Kim et al. 2017) identified signatures of selection as well as genes and pathways controlling anaemia, drinking behaviour, feeding behaviour, and circadian rhythm in the N'Dama, heat tolerance or thermoregulation and tick resistance in Ogaden, Boran and Kenana cattle and horn development and coat colour in Ankole (Bahbahani et al. 2018). Investigated the trypanotolerant basis of the Following genotyping of the Sheko breed of Ethiopia (the breed shows better trypanotolerance than other breeds) with Illumina BovineHD Genotyping Beadchip, Mekonnen et al. (2019) reported that the Sheko genome is an admixture of taurine and indicine genomes or ancestries (Mekonnen et al. 2019). Moreover, 99 genomic regions harbouring 364 signature genes and several pathways related to oxidative stress response, which possibly modulates oxidative stress following trypanosome infection, were identified (Mekonnen et al. 2019).

22.2.4 Application of Molecular Markers for Marker-Assisted Selection

By discovering genomic diversity and identifying important genes and genomic variations controlling traits, it is now feasible to implement more informed selection programmes for cattle and buffalo improvement. Therefore, the availability of molecular information, environmental factors, generation intervals, and the cost of rearing cattle or buffalo to a specific age before phenotypes are manifested have all been removed from the selection challenge. Thus, MAS is based on detecting connections between genetic markers, linked QTL, and traits. The distance between the marker and the target trait impacts the marker-QTL relationship. Once QTL markers have been identified, they may be used in selection procedures. Generally, molecular markers find applications in MAS, marker-assisted introgression (MAI) breeding, screening undesirable traits, crossbreeding, and genomic breeding.

Marker-assisted introgression is based on the tandem selection in a multigenerational backcrossing programme, in which a marker based on the presence of donor breed alleles at or near the target gene is used in the first selection step (foreground selection), followed by background selection of markers based on the presence or absence of recipient alleles at markers spread across the genome, on phenotype, or a combination of the two (Montaldo and Meza-Herrera 1998). Backcrosses with molecular markers may also introduce a significant gene into a population through introgression (Hayes and Daetwyler 2019). When used with targeted selection for specific advantageous alleles known to exist in the donor line or with closely related markers, traditional introgression strategies (introgressing specific QTL alleles) are most likely to succeed (Barton and Keightley 2002). Genomic selection may be used to select all marker alleles in linkage disequilibrium (LD) with beneficial QTL; this method may be especially effective when several QTLs underlying the genetic variation of the trait are of interest (Hayes and Goddard 2001). Marker-assisted selection deployment necessitates developing and integrating procedures and logistics for DNA collection and storage, genotyping, and data processing. Response to MAS is usually at its greatest in the first generation (Goddard 1996). Still, there is a steady drop in response to MAS in succeeding generations, possibly due to increased recombination events leading to linkage equilibrium.

22.2.4.1 Case Study

A case study of MAI was reported by Rothschild et al. (1994) a researcher who found evidence for a QTL affecting litter size in a cross between Large White and Chinese Meishan pigs, which increased litter size by one piglet per litter. This QTL for litter size was found to have originated from the non-commercial Meishan breed. It would benefit commercial populations if introduced into commercial lines, which are superior for meat traits. Other studies involving non-commercial Meishan pigs have found a QTL on chromosome 7 that decreases the backfat depth (Walling et al. 1998). This finding is interesting because the Meishan is significantly fatter than leaner commercial Western breeds; hence, the allele is 'cryptic' (i.e. the allele improved a trait despite being at a high frequency in a breed with poorer commercial performance for the trait).

22.2.5 Application of Molecular Markers for the Detection of Genetic Defects

In addition to genetic selection, genomic variants have rapidly facilitated the detection and prediction of carriers of genetic defects (Biscarini et al. 2016). This has led to the development of various diagnostic tests that offer stud breeders and livestock producers the tool to identify carriers of undesired mutations and select them out of the population. Some methods for detecting genetic defects include haplotype-based predictions (Pirola et al. 2013) and linear discriminant analysis (Biffani et al. 2015). The utilization of lower-density SNP chips for detecting genetic defects has yielded accuracies comparable to those from the 54K Bovine SNP array (Biscarini et al. 2016). Based on SNP genotypes, several diagnostics tests have been developed for cattle. Examples of such tests (Table 22.1) available in South Africa for cattle were summarized by Visser et al. (2011). The tests are relatively cost-effective and find application in both the commercial and emerging farmer sectors in South Africa.

22.2.6 Application of Molecular Markers for Genomic Breeding

While MAS strategies consider one or a few markers with mostly large effects on a trait, whole genomic selection considers thousands of markers and even whole genome data. Thus, genomic selection (GS) involves estimating breeding values using markers spanning the entire genome. Genomic selection implementa-

Table 22.1 Examples of diagnostic tests available in South Africa

Diagnostic test	Company
Double muscling/Myostatin	Unistel, Clinomics
Curly calf syndrome	Unistel
Polled, scurred, horned	Unistel
Bulldog mutation screening	Clinomics, Unistel
FreeMartin	Unistel
Cytogenetics: 1/29 Translocation	Unistel

tion in developed countries has consequently resulted in rapid genetic improvement in all traits, particularly lowly heritable traits (e.g. health traits, fertility, and lifespan), selection of animals at an early age, higher rate of genetic gain, shortened generation interval, higher intensity of selection and increased reliability of predicting breeding value (Miglior et al. 2017). Genomic selection also resulted in more artificial insemination (AI) active sires in the United States (Hutchison et al. 2014). For example, in 2007, adolescent bulls accounted for 28% and 25% of Holstein and Jersey inseminations, respectively. Still, these percentages jumped to 51 and 52% in 2012 due to the use of genomic-verified young bulls (Hutchison et al. 2014). In many countries, well-established conventional genetic evaluation procedures have given a solid foundation for the success of GS.

22.2.6.1 Case Studies

Several case studies of GS on African cattle have been reported. Brown et al. (2016) used milk test day information on 1034 cows and genotypes (Illumina BovineHD BeadChip) within the Dairy Genetics East Africa (DGEA) project to investigate the accuracy of genomic predictions for cows of various breed compositions (crosses between indigenous African cattle—N'Dama (*Bos taurus*), and Nellore (*Bos indicus*) and exotic dairy breeds (*Bos taurus*)—Holstein, Friesian, Ayrshire, Jersey, and Guernsey) using genomic best linear unbiased prediction (GBLUP) and Bayes C models (Brown et al. 2016). The study found that genetic prediction accuracy ranged from 0.30 to 0.40 (Brown et al. 2016). Mrode et al. (2018) looked at models with dominance impact and a multi-trait approach that fit breed proportion as independent traits using the same dataset. Although the dominance effects were zero, presumably due to the small size of the dataset, the multi-trait strategy resulted in a tiny improvement in the model's predictive ability, though not in prediction accuracy, when compared to the results of Brown et al. (2016). Another study on genomic prediction of crossbred cattle in Tanzania and within DGEA, consisting of 134,987 test day milk records of 14,741 cows (from 8735 herds) and genotypes obtained with Illumina BovineSNP50 Beadchip reported moderate to high levels of genomic prediction accuracy (0.53–0.83) for body weight and milk yield (Mrode et al. 2021). Similar genomic predictions in crossbred dairy cattle in India have been reported for daily milk yield using only genotypic data (Al Kalaldeh et al. 2021b).

22.2.7 Application of Molecular Information for Gene Editing and Transgenic Animal Production for Livestock Improvement

Developments in gene editing or genetic engineering techniques have increased their application in livestock animals (Perisse et al. 2021; Rieblinger et al. 2021; McFarlane et al. 2019). Gene editing (see Chap. 21) involves the introduction of single or multiple edits (insertions or deletions) or long DNA sequences into specific genome locations simultaneously and precise nucleotide transitions and transversions. Meanwhile, genetically modified or transgenic animal/livestock production is a method by which 'new' or modified genes and DNA sequences can be rapidly introduced into animals without crossbreeding or hybridizing (Wheeler et al. 2003). Transgenic livestock is used to (1) produce new animal products, (2) study the genetic basis of physiological systems, (3) improve animal production, (4) health, reproduction, and adaptation traits, and (5) build human genetic disease models. Thus, in addition to the obvious scientific interest in the study of genes and their regulation, transgenic animal technologies have been proposed as a way to accelerate livestock improvement by introducing new genes or modifying the expression of endogenous genes that regulate economically important traits like milk production traits (Wheeler et al. 2003). Therefore, livestock genome editing and transgene technologies will allow breeders to improve animal health, welfare, performance, environmental adaptation, and efficiency at unprecedented speed, opening the path for more sustainable livestock farming in Africa.

Many livestock traits have been targeted for genetic modification, including traits for disease resistance, nutrient enhancement, milk composition, growth rates, carcass composition, environmental impact, reproductive performance, hair and fibre quality, animal welfare, and climate change (Perisse et al. 2021; Karavolias et al. 2021; Van Eenennaam et al. 2021). While gene editing has been successfully implemented for crop improvement in Africa and proof of concept has been established for resistance to trypanosomiasis in goats, no specific example exists for African cattle (Karavolias et al. 2021).

22.2.7.1 Case Studies

Examples of gene editing applications and transgenic production in cattle are summarized in Table 22.2. For instance, using the CRISPR/Cas12a system, a polled Celtic variant from the genome of an Angus cow was extracted and integrated into the genome of fibroblasts derived from a horned bull, followed by somatic cell nuclear transfer (SCNT) (Schuster et al. 2020). The work effectively established CRISPR/Cas12a's practical application in dairy farming. Furthermore, Sun et al. (2018), generated DNA-free β-lactoglobulin (BLG) bi-allelic knockout cow by ZFNs mRNA, which produced hypoallergenic milk (Sun et al. 2018). ZFNs and other genome-editing methods can establish alternatives to transgenic (genetic modification)-based methods for the genetic improvement of livestock and the production of 'designer' milk for human health benefits.

22.3 Application of Modern Technologies in Feeds and Feeding

The humped (*Bos indicus*) and non-humped (*Bos taurus*) cattle breeds are raised in many agroecological zones for milk, meat, manure, hide, and draught power generation. Details of breeds and classification of cattle have been provided in Chaps. 3 and 4. Bovine meat is the third muscle food consumed worldwide, following pork and chicken. As global meat production is projected to reach 44 M tonnes by 2030 (Frezal et al. 2021), the aggregate consumption of beef veal remains on an upward trajectory. Global milk production is estimated to grow at 1.7% annually and reach 182 M. tonnes by 2030 (Frezal et al. 2021). These estimates considerably depend on the nutritional worth of feedstuffs that constitute the bovine ration at different stages of growth across breeding seasons. Invariably, bioactive compounds in the diet do not only affect feed intake but also the welfare of the animal and the biological value of its by-products. Producers that would spend 55–75% of the total cost of production on feeding would opt for an ideal feeding regime and metabolic modifiers that can inhibit pathogenic invasion, enhance feed efficiency, boost average body gain, carcass weight, milk yield, and profitability of the enterprise (Lamb and Maddock 2009).

Aside from the estimated breeding values, feed efficiency (expressed per unit of dry matter consumed, per hectare of pastureland, per unit of nitrogen fed, and per methane produced) is a key performance indicator to determine the genetic merit of each trait and profitability of beef-dairy cattle husbandry. To optimize beef-dairy productivity indices, metabolic modifiers that can aid ruminal microflora, the bioactivity of beta-adrenergic agonists, and hormone-like growth enhancers have been prioritized in beef-dairy cattle feeding regimes.

22.3.1 Probiotics and Prebiotics Supplements

Maintaining microbial balance in the gastrointestinal tract (GIT) is vital in nutrient utilization. On a broad scale, the dynamics of the GIT network interact with the circulatory, endocrine, immune, and nervous systems to modulate the immunologic and metabolic markers of the host (Alfonsetti et al. 2022). The significance of gut-intestinal microbiota in health promotion makes the utilization of probiotics, prebiotics, and symbiotics, a better alternative for realizing feed effi-

Table 22.2 Examples of gene editing application and transgenes in cattle

Transgenic animal and gene manipulation outcome	Technique	Trait	References
Human recombinant lactoferrin expressed in bovine milk	Somatic cell nuclear transfer (SCNT)mmm	Innate host defence, milk composition	Van Berkel et al. (2002)
Milk with higher levels of CSN2 (β-casein) and CSN3 (κ-casein) proteins	Gene knock-in	Milk composition	Brown et al. (2016)
Cows secreting lysostaphin in milk	Somatic cell nuclear transfer (SCNT)	Enhance mastitis resistance	Wall et al. (2005)
Expression of human recombinant lysozyme in bovine milk	Somatic cell nuclear transfer	Animal health, disease resistance, milk composition	Yang et al. (2011)
Dairy cows expressing omega-3 fatty acid desaturase gene	Nuclear transfer	Tissue and milk fatty acid composition	Wu et al. (2012)
MiRNA-directed depletion of β-lactoglobulin (BLG) (an allergen) results in the production of BLG-free milk with high-casein content	Gene knockdown	BLG-free milk, milk composition	Jabed et al. (2012)
Animals expressing shRNAs targeting the myostatin (MSTN) gene	Suppress MSTN expression by RNA interference and somatic cell nuclear transfer	Enhance muscle development	Tessanne et al. (2012)
Beef cattle expressing omega-3 fatty acid desaturase (FAD3)	Somatic cell nuclear transfer	Muscle tissue fatty acid composition	Cheng et al. (2015)
Cattle with site-specific knockin of transcription activator-like effector (TALE) nickase-mediated nuclear body protein gene (SP110)	Transcription activator-like effector nuclease (TALEN)-mediated editing (gene knockin)	Enhance resistance to tuberculosis (tuberculosis-resistant cattle)	Wu et al. (2012)
Cattle with Cas9 nickase (Cas9n) induced generation of the natural resistance-associated macrophage protein (NRAMP1) gene	CRISPR/Cas9n mediated gene knocking and SCNT	Enhance resistance to tuberculosis	Gao et al. (2017)
Cow-producing BLG free-milk	zinc-finger nuclease (ZFNs) knockout of BLG and nuclear transfer	Hypoallergenic milk, milk composition	Sun et al. (2018)
Cattle harbouring knockout alleles of the Nanos C2HC-Type Zinc Finger 2 (NANOS2) gene	CRISPR-Cas9 editing and spermatogonial stem cell transplantation	Male germline ablation, surrogate sires for genetic dissemination	Nigam and Bhoomika (2022)
Polled offspring with Polled Celtic variant (202 base pair insertion-deletion (indel) on bovine chromosome 1)	CRISPR/Cas12a knock-in and SCNT	Polled dairy cattle	Schuster et al. (2020)
Deletion of three bp in pre-melanosomal protein 17 genes (PMEL), the causative variant of semi-dominant colour dilution phenotype	gRNA/Cas9-mediated editing	Adaptation to climate change	Laible et al. (2021)

ciency in cattle production (Atılgan Türkmen 2020). Bifidobacteria, Enterococci, Lactobacilli, yeasts, and combinations of different beneficial strains are viable probiotic bacteria and gut modifiers commonly used to stimulate efficient feed conversion and average daily gain in feedlots (Varankovich et al. 2015). *Aspergillus oryzae, Saccharomyces cerevisiae, Lactobacillus acidophilus*, and other multistrain probiotics are known for stabilizing gut microbiota, improving ruminal fermentation, limiting the risk of pathogen colonization, enhancing growth and the production of milk during early-lactation in cows (Singh et al. 2015). For instance, *Lactobacillus acidophilus* BT-1386 has been used in finishing feedlot cattle to increase feed efficiency, average daily weight gain, and reduce diarrhoea in ruminants and milk yield (Lambo et al. 2021).

A plethora of investigations have reported on the benefits of supplementing the diets of livestock with various feed supplements (Singh et al. 2015). Despite the derivable benefits from the incorporation of probiotics in the diets of ruminants, very little exists in the literature on the implementation of this technology in ruminant production on the African continent apart from a few reviews (Ibeagha-Awemu et al. 2024). However, the use of probiotics in poultry (Reid et al. 2014) and swine (Ohimain and Ofongo 2012) production is more popular in Africa. Research efforts to implement the use of these supplements in ruminant production in Africa are much needed.

With proper compliance to standard practices, conformity to ethical practices, and effective consumer cum stakeholder orientation, a beef-diary enterprise (Kwoji et al. 2021) can meet the millennial goal of meeting dietary proteins by all categories of consumers in 2030. This is also poorly explored in cattle production systems in Africa, hence should be encouraged to improve productivity.

22.3.2 Technologies for Dry Season Feed Production

The demand for improved pasture, especially for milk production and sustenance of the African beef industry, has grown considerably. A thriving but ancient traditional method of combatting dry season feeding challenge for ruminants is planting deciduous trees, which produce lots of green leaves all year round from where leaves are harvested to supplement feeding for cattle. Although this is very productive in keeping animals alive, it is more suited for raising small ruminants like goats, which are more sure-footed in climbing trees to reach even the farthest leaves. Nevertheless, this is less suitable for dairy products as the nutritive value of such tree plants begins to decline with prolonged periods of the dry season.

A large portion of the grazing areas in Africa are subjected to numerous climatic challenges that limit the year-round quality and quantity of available pasture and forages, especially in the dry season, which is more prolonged in some African countries than elsewhere in the world. Adopting local preservation technology, such as silage-making (Ntakyo et al. 2020), from crop residues and farm waste in a time of plenty and kept for future use is common and very helpful. However, making well-preserved silage requires skills and experience; the acidity level in the final product must be right to avoid acidosis. Bacteria treatment and incorporation of nutritive additives provide quality material for storage and extended shelf life. Quality and shelf life depend on the feeding value of ensiled materials. One big challenge would include storage space and the availability of quality storage materials, especially in rural agricultural systems. The technology is common in all ruminant-producing regions of Africa and is the oldest and most explored technology in ruminant production (Ntakyo et al. 2020).

22.4 Application of Digital and Milking Technologies in African Large Ruminant Production Systems

Advances in digital technologies are helping modern farms optimize economic contribution per animal, reduce the drudgery of repetitive farming tasks, and overcome less effective isolated solutions. Farm management and operations will drastically change due to access to real-time data, real-time forecasting, and tracking of physical items in combination with Internet of Things (IoT) developments to further automate farm operations (Akbar et al. 2020; Nasirahmadi and Hensel 2022).

Many new technologies intended to increase the effectiveness of milk generation, improve livestock administration and management, and increase the productivity and profitability of farm animals and animal products (Groher et al. 2020; Petersburg-Pushkin 2015), Joshi et al. (2019) such as Robotic/Automatic milking systems (AMS), cow activity sensors, rumination sensors, programmed calf feeders, Blockchain, Artificial Intelligence (AI), and machine learning (ML) are among other precision technologies becoming progressively popular in the operations of dairy farming in developed countries and in some parts of Africa. For instance, the Africa Asia Dairy Genetic Gains (AADGG) programme utilizes digital technologies to ensure cutting-edge genetic (and genomic) technologies, farmer training, digital monitoring, and mobile phone technology that supports thousands of farmers across Tanzania, Kenya, Ethiopia, Uganda, and Nepal to rear productive, adaptive and resilient dairy cattle (Mwai et al. 2023). The use of sensor (wearable) technologies for digital monitoring of dairy cattle can greatly impact Africa's extensive cattle production systems by decreasing operational costs due to timely upkeep and monitoring of cattle's welfare and health such as illness, lameness, feed efficiency, and rumination effectively and accurately.

The Digital Twins (DT) in dairy farming hold significant promise to greatly shape the future of animal welfare, production efficiency, sustainability, and management practices (Schillings et al. 2021) if deployed for cattle production systems in Africa. The DT can address key constraints and challenges related to the intensive nature of livestock production systems, which are typically characterized by high levels of input and output (Zhang et al. 2023).

Automated milking systems (See Chap. 21) drastically reduce the labour needed for milking cows. It also allows farmers to maintain a sanitary milking procedure, milk cows at any time of day, and increase milk output. Automatic milking systems are popular in the intensive milk production industry in South Africa and some southern and eastern African countries.

22.4.1 Feasibility of Application

One of the biggest challenges our society is facing and will continue to face is the ability to feed its growing population while minimizing environmental impacts and ensuring human health because, by the year 2050, it has been estimated that the world population will be about 9.7 billion and food production will need to increase up to 60% more to meet the demand (Symeonaki et al. 2024). Similarly, global meat production is expected to double by 2050. This increase in production will only be achieved by a combination of expansion in animal numbers and increased productivity (Food; Organization., A 2018). Livestock production would likely intensify, increasing animal density and lowering the stock person per animal ratio. This will result in less time to monitor and properly manage individual animals. An increase in animal numbers could make their management more challenging, especially if, as was observed in the EU recently, the number of farmers continues to decrease by (Cook 2020). In the UK, while livestock numbers remained stable between 2018 and 2019, the labour force on commercial holdings decreased by 0.3% (DEFRA, U 2020). Having fewer farmers to look after larger numbers of animals likely makes it more difficult to address animal health and welfare challenges. Therefore, farmers need reliable and affordable technologies to assist them in daily management, guaranteeing accurate and continuous individual animal monitoring.

Farmers have always looked to new technologies to increase production efficiency and reduce costs.

There is a growing interest in automating productivity, animal health, and welfare using precision livestock farming (PLF), which increases the farmer's ability to keep contact with individual animals in the growing livestock production intensification (Vranken and Berckmans 2017). PLF technologies are designed to support farmers in livestock management by monitoring and controlling animal productivity, environmental impacts, and health and welfare parameters in a continuous, real-time, and automated manner, offering the opportunity to improve productivity and detect health issues early. The premise of PLF systems is that fully automated continuous monitoring of individual animals will enhance the ability of the producers to detect and manage animal health, productivity/reproduction, and environmental aspects of their livestock operations (Berckmans 2017). Various systems using technologies such as sensors, cameras, or microphones can directly alert farmers via connected devices (e.g. phones, computers, or tablets) about detected anomalies, thus allowing farmers to intervene early. Research is pointing toward the great potential for these 'smart technologies' to help livestock farmers monitor the welfare of their animals, and several countries are already investing in their development, reflecting their potential to be part of strategies to move toward sustainable agriculture (Rose and Chilvers 2018). Silvi et al. (2021) recently reported on the various levels of adoption of precision livestock technology in Brazil and concluded that understanding the perception of dairy farmers regarding precision livestock technologies is crucial for creating strategic actions that will increase the adoption and usage rate of such technologies.

Understandably, adopting technologies for sustainable farming systems is a challenging and dynamic issue for farmers. The agricultural sector needs to employ a wide range of evolving technologies and farm practices across many different farming systems and structures to meet various changing and heterogeneous demands for food products. Farmers need reliable and affordable technologies in modern livestock production systems to assist them in daily management tasks, guaranteeing accurate and continuous individual animal monitoring (Groher et al. 2020). Undoubtedly, the scientific way of farming and the adoption of viable technologies will enable the farmers to match the pace with the changing agricultural scenario. Livestock farmers want each animal to receive the right care at the right time, place, and amount. These new technologies will help make this happen. When precision livestock farming technology is implemented and understood, farmers can create their data to manage their herds better (Nasirahmadi and Hensel 2022; Krpalkova et al. 2021).

Technology adoption, however, does not guarantee that the technologies will be used optimally. However, the potential of PLF to help reduce the duration and/or severity of diseases and injuries in livestock farming systems is a promising technology that can detect health issues at an early stage and help ensure optimal environmental conditions (Nasirahmadi and Hensel 2022; Buller et al. 2020).

While genetic engineering/transgenesis in cattle improvement is at the commercialization level, gene editing is not yet at that level. The regulatory status still needs to be completed. Genetic engineering/transgenesis may have concerns/risks, but technical and regulatory mechanisms exist in many African countries (CBD 1992) (https://www.cbd.int/doc/legal/cbd-en.pdf). The regulations consider the stipulations of the related Cartagena Protocol and its supplementary Nagoya-Kuala Lumpur protocol to the CBD.

22.4.1.1 Case Study: Genetically Modified Dairy Cattle

Wall et al. (1997) reviewed transgenic methods for improving livestock and dairy cattle. The goals included the search for new products and production efficiency and cost. The strategy was the production of pharmaceuticals for human health and engineering of the mammary gland of cattle to change milk composition for human consumption: alteration of protein and fat levels as well as alteration of milk composition to be similar to that of humans. This could reshape the compensation given to dairy farmers. Since then, several laboratories and companies have pursued these goals and objectives.

The main problems involved the long generation interval in producing transgenic cattle, the very high cost of producing a transgenic cow, inefficiencies (low gene integration rates, low survival of embryos, and unpredictability of the behaviour of the transgene involved.

The bovine somatotropin gene was isolated and inserted in bacterial cells for massive production of recombinant BST, and cows injected with 30 mg of it increased milk yield by 10–30% (Wall et al. 1997). This increase was more dramatic in the zebu cattle (Mbah 2006).

Brophy et al. (2003) showed the potential to enhance milk composition and processing through increased casein proportion. The introduction of additional copies of genes encoding cattle beta-casein (*CSN2*) and cattle kappa–casein (*CSN3*) into cow fibroblasts resulted in 11 transgenic calves. Analysis of hormonally induced milk from these transgenics revealed high levels of transgene–derived caseins in the milk. Nine cows of the high expressing lines gave milk containing an 8–20% increase in beta-casein and a two-fold increase in kappa–casein. These results indicated the potential to alter major milk components by transgenic methods and improve milk composition.

Assessment of differences in the composition of milk and cheese from transgenic and cloned cattle has been made by Laible et al. (2021). The components involved were amino acids, fatty acids, minerals and vitamins, and colostrum. Results indicated that levels of these components in milk were similar between transgenic and non-transgenic cows. Colostrum IgG levels were also similar. In cheese, fatty acid levels were lower than in non–transgenic cheese. Salt levels were higher in transgenic cheese than in non-transgenic cheese. Amino acid levels showed small differences between transgenic cheese and non-transgenic cheese.

Yang et al. (2008) reported using cattle mammary bioreactor to produce recombinant human lactoferrin (rhLF) through transgenic cloning. Using 2 transgenic cows secreting rhLF at good levels, they found that glycosylation was similar to that of the lactoferrin of man (hLF). Its proteolytic susceptibility was also similar to that of man. Further analysis revealed that the properties of iron-binding/releasing were similar to those of non-recombinant hLF. Thus, the resulting bioreactor led to the production of rhLF at an industrial level.

Furthermore, Yang et al. (2008) characterized human lysozyme expressed in cloned transgenic cattle milk. Lysozyme, a bactericidal protein highly produced in human milk but very little in cow milk, protects infants from microbial infections. Focusing on 4 transgenic cattle expressing recombinant human lysozyme, they reported an expression level of 25,96 mg/l. The physicochemical properties (molecular mass, bacterial lysis) were similar to those of the natural (human) lysozyme, and there were no significant differences between transgenic and non–transgenic cattle milk in composition (lactose level, total protein, total fat, total solids). They concluded that milk from transgenic cattle has similar nutritional value as human milk.

Jabed et al. (2012) experimented with decreasing the level of beta-lactoglobulin, an allegen, in cow milk. Using RNAi technology, they produced a transgenic calf expressing miRNAs. Analysis of hormonally produced milk from the transgenic calves showed the absence of beta-lactoglobulin and increased milk casein proteins. It was concluded that miRNA mediated the elimination of the allergen effectively and that miRNA expression is an effective strategy to change/improve milk composition (and perhaps other characteristics).

The transgene's effect on milk components was further assessed by Wu et al. (2012). They compared proteins and colostrum from transgenic cloned cattle (TC) expressing α–lactalbumin (TC-LA), lactoferrin (TC-LF), lysozyme (TC-LZ) and cloned non-transgenic cattle (C), and conventionally bred cattle(N). The results indicated an altered expression level in the profiles of 37 proteins in the TC and C groups relative to the N control group. Evaluation of the transgene effect and variability in milk protein profiles indicated that variability in the TC was similar to that of control groups. At least 50 parameters of colostrum and milk were used to compare TC and control cows. Results indicated that the average values of the parameters in RC cows were similar to those of N cows. It was concluded that milk components were not affected/modified by the transgene.

On the prospect of using transgenesis/genetic engineering/genomic technology to improve cattle productivity and production, African Science Academies (NASAC) has taken a position on modern agricultural biotechnology (NASAC 2015; www.nasaconline.org). This could have a positive effect on animal/livestock biotechnology. African countries are parties to the Convention on Biological Diversity and related protocols on biosafety and access to genetic resources. As parties to these, they have developed regulatory mechanisms to enable sustainable use of the resources and use of genetic modification (genetic engineering) on a case-by-case basis based on risk analyses. However, while easily accepting this in the health sector, public acceptance has concerns with the food sector's technology. Hence, the utilization of genetic engineering/genomic technology in African cattle production systems may not be immediate but may change with problem-free use of it in the health sector.

Capacity (human resources skilled in genetics, reproductive physiology, nutrition, health, and economics and infrastructural developments (such as laboratories) must be improved for effective implementation of genomic technologies in African livestock improvement. Public policy must be in place, guided by stipulations of the Cartagena Protocol to the Convention on Biological Diversity (CBD 2000; https://www.cbd.int/doc/legal/cbd-en-pdf).

22.5 Application of Modern Technologies in Reproductive Efficiency in African Cattle

The inefficiency of reproduction is one of the most significant causes of economic losses in animal husbandry globally (Bianchi et al. 1986). Although there have been significant developments in reproductive physiology in recent times, the issues of infertility caused by low conception rates and high rates of embryonic mortality continue to be a major concern (Hafez 2015). The need to increase animal productivity per animal unit without increasing the herd population size necessitates using modern technology applications to ensure sustainable livestock production. Introducing three cutting-edge reproductive technologies—zygote electroporation, recombinant adeno-associated virus (rAAV) zygote transduction, and surrogate sire technology give cattle producers a new toolkit of genome editing delivery options (Wheeler et al. 2003). The ease with which these technologies may be integrated into present breeding methods will enable extensive on-farm adoption in African livestock species. Therefore, modern reproductive technologies, including artificial insemination (AI), multiple ovulation and embryo transfer, and embryonic micromanipulation, described in Chap. 21, are being advocated.

22.5.1 Artificial Insemination

The most widely applied reproductive technology in African cattle and buffalo populations is AI. The application of AI was made possible through advances in semen technology including collection and facilities, evaluation, processing and packaging, and preservation/storage. Practical AI service in the field involves heat detection, semen handling, insemination, and advice to farmers. The delivery chain includes AI services/centres, cooperatives, farmer organizations, and products (milk, meat, semen). Services at the management level include records and economics, heat detection, insemination and motivation, pregnancy diagnosis, and recording (Verma et al. 2012).

22.5.1.1 *Case Study 1*: Kenya National Artificial Insemination Service

The historical perspectives of AI of cattle in Kenya are well documented in these reports by (Joint 2005; KDDP/SDP 2004; Competition Authority of Kenya (CAK) 2014). Artificial Insemination (AI) was first introduced in Kenya in the 1940s and has risen to become the most relevant approach for breed improvement. According to Duncanson (1977), the AI of cattle in Kenya started with the collection of semen from sires on private farms for use in the same farms. In 1941, an AI scheme was set up based on a community bull scheme, followed by a farmer association (Limuru Cattle Breeders

Association) in 1942. With AI for breed improvement, semen is imported mainly from South Africa, Europe, and North America or locally produced semen collected from locally bred bulls. The dominant breeding method applied by farmers is the natural service whereby cows are bred by local bulls. Currently, AI services are provided in partnership between County Governments and the Kenya Animal Genetic Resource Centre (KAGRIC). The utilization of AI in Kenya is to optimize productivity, which will translate to income and revenue for farmers. The Kenya AI service is funded by the government of Kenya and various Swedish grants and expertise organized as follows:

The main goals of the AI service include increasing milk and meat production, improving cattle productivity nationally, upgrading local zebu cattle, and controlling diseases transmitted through natural breeding. The service has two sections—the field services section and the planning and development section. Each section is well equipped (human and infrastructural capacity and logistics). Located at Kabete, the National AI Station has bulls composed of exotic dairy breeds (Holstein, Jersey, Ayrshire, and Guernsey) and beef breeds (*Bos indicus*: Boran, Sahiwal; *Bos taurus*: Simmental, Hereford, and Charolais). Semen from top bulls are also imported for use on top-producing cows. Semen produced at the station is mainly for national use while only about 1% is exported. From 1993, about 20% of collected semen was frozen in liquid nitrogen, while 80% was kept at room temperature and delivered to schemes twice a week.

The impact of the AI service is significant given that its annual production of improved dairy heifer calves by farmers is reported to be 100,000 crossbreds and upgrades. Furthermore, this is at a low cost to the farmers, who pay only 4.5% of the total cost. Hence, adequate funding will enable the sustainability of the service.

Suggestions for the way forward include the need to use cars for daily activities (i.e. the need to increase efficiency), use deep frozen semen only, and improve recording (AI and milk).

22.5.1.2 *Case Study 2:* Tadu Dairy Cooperative Society Artificial Insemination Service, Cameroon

According to Shange (2021), the Tadu Dairy Cooperative Society (TDCS), located at Kumbo in the Northwest Region of Cameroon is composed of about 200 male and female members and was founded in 1990 by pastoralists interested in AI as a tool to improve cattle. It is a multipurpose cooperative through which farmers can access new breeds through AI for greater and better-quality milk production, transformation, and commercialization of dairy products.

Structurally, men handle cattle while women handle milk and related products. It comprises 2 clusters: (i) five groups of grassroots cattle farmers interested in AI services, and (ii) five (5) dairy women groups affiliated with the cattle farmers groups.

The 5 cattle breeders groups live in 5 different TDCS Districts. Each district is provided with a crossbreeding station. AI services are delivered at the station managed by TDCS. Each station has a multipurpose corral and grazing land and is equipped with a 20/20 cryogenic container and a loaded AI breeding kit. The resident inseminator/herdsman resides at the station.

Cooperative members provide cows for insemination. The cows are held for a month, during which time they are observed. After insemination, they are held for another month. Those indicating heat are given a second insemination and returned to their owners. Members presenting cows for insemination pay 50% of the cost of AI, and the rest is paid when the calf is born.

The management of each station is by the inseminator/herdsman trained and paid by the cooperative, which also furnishes them with AI equipment, supplies, and logistics. The animal recording is at 3 levels: on arrival at the AI station, after successful insemination, and when a calf is born. According to advice from research personnel at the Institute of Agricultural Research for Development (IRAD) Bambui and Wakwa

Centres, the TDCS Dairy Performance Enhancement Strategy involves the insemination of local Gudali and Red Fulani females with Holstein semen. The TDCS beef improvement strategy involves the insemination of Gudali and Red Fulani with Brahman semen. All semen is imported from the USA.

The funding partners of TDCS activities include USAID, IFC, World Bank and Land O'Lakes, IRAD, and the Cameroon Ministry of Livestock, Fisheries and Animal Industries. USAID and Land O'Lakes, in particular, supported the training of inseminators in semen technology in the USA.

The success of the service is reflected by the results obtained within and outside the cooperative. Within the TDCS, the AI performance is 85.05% AI success, 2.03 straws per insemination, 90.78% calving rate, 89.84% calf weaning rate and 2.49 straws per calf weaned. Outside the cooperative (contract to produce crossbred calves at the state ranch, SODEPA), the performance is 89.3% AI success (significantly less than 2 inseminations), 94.9% calving rate of cows confirmed pregnant, 91.5% weaning rate of calves born and 8.5% pre-weaning calf mortality. The crossbred animals produce 3 to 4 times more milk than the local zebus and are more adapted than the purebred exotics.

The major challenges include an absence of a coherent cattle crossbreeding strategy. This leads to potential genetic erosion, dysfunctional crossbreeding regulatory mechanisms, seasonal fluctuation of feeding resources, high cost of AI inputs, sporadic access to electrical power, non-replacement of the equipment pool, absence of a functional cold chain between the milk factory and market outlets, high taxes, inadequate access to key inputs such as starter cultures and prolonged regional socio-political crisis. The way forward includes establishing a coherent national cattle breeding policy, improving access to electrical power, strengthening links with the Livestock Research Department of IRAD, and introducing a cold chain between the dairy processing factory and marketing outlets.

From these examples, it is feasible to apply AI upon observation of natural or synchronized heat in private and state farms to facilitate crossbreeding local cattle with exotic semen or semen of improved local Zebu males on improved local females. This requires mastery of semen processing technologies like semen collection and preservation, multiple ovulation, and embryo transfer among others. These technologies are also essential in conservation/regeneration/repopulation efforts for breeds at risk of extinction or synthetics/upgrades.

22.5.2 Multiple Ovulation and Embryo Transfer

Multiple ovulation and embryo transfer (MOET) technologies are regarded as second-generation reproductive biotechnologies. The MOET involves the super-ovulation of improved females (donors, which can yield 10 eggs each per oestrus with an average of 5–7 eggs for dairy cows and 6-6 for beef cows), followed by AI and transfer of fertilized eggs into surrogate mothers (recipients) (Mbah 2006; Kidie 2019). There are two methods of producing embryos from the donors: in vivo and in vitro techniques (Kidie 2019). MOET programmes require two or more nucleus herds (donor herd, and receiver herd). The genetically improved embryos are disseminated through transfer or IA or young bulls for natural service.

22.5.2.1 *Case Study 1:* Multiple Ovulation and Embryo Transfer in Kenya

In a simulation model, Sagwa et al. (2019) concluded that an increase in the rates of reproduction of both male and female dairy cattle increases selection response. They used conventional semen (CS), and X-chromosome sorted semen (XS) and compared them with the most used strategy in Kenya, AI as follows: AI-CS, AI-XS, MOET-CS, and MOET-XS. About 5% of cows were in the nucleus herd and 95% were in farmers' herds. The improved genetic material in the nucleus herd was disseminated to all the strategies while dissemination in farmers' herds was based on AI–CS only. Results indicated that

MOET-CS and MOET-XS increased the reproduction rate of both males and females. This led to an increased response to selection. This was evident in annual genetic gain and economic benefits/cow/year.

While MOET has grown fast since the first successful case in 1973, it is still at an embryonic stage in Africa relative to the number of countries engaged in embryo production and transfer (Brophy et al. 2003). In 2015, two out of 57 African countries submitted data on MOET, while Europe (71.8%) and North America (100%) were at the top. Only South Africa produced in vivo derived (IVD) from in vitro fertilization (IVF) (0.84%) of world embryo production, while North America produced 54.5% and Europe (19.4%). In the same year (2015), Africa transferred 183 embryos (fresh and frozen) while North America, South America, and Europe produced 97,871, 22,887, and 14,502, respectively. Countries exporting IVD cattle embryos include South Africa (659), Canada (12,758), the USA (15,896), Australia (2426), New Zealand (28), Argentina (2946), and Canada (1228, IVF) and Dominican Republic (590, IVF) (Brophy et al. 2003).

It is evident that embryo transfer in Africa is below demand. This is due to challenges such as capacity (human, infrastructural, funding, policy) insufficiency. Apart from South Africa, Ethiopia has affected ET at the Animal Research Centre (Adanu Tulu) (Brophy et al. 2003). Despite the potential for multiplication and dissemination of genetic material, regeneration of breeds at risk, etc., the prospect for utilization for cattle improvement in Africa is not yet realized.

22.5.2.2 *Case Study 2:* Multiple Ovulation and Embryo Transfer in Burkina Faso

Through a Ceva (Ceva Santé Animale) and CMAP (Centre National de Multiplication des Animaux Performants) partnership, F1 calves were produced using MOET technology under the auspices of the Vache du Faso project (Bianchi et al. 1986). The CMAP is a public centre under the Ministry of Livestock with the mandate of genetic improvement of the national cattle herd in Burkina Faso, while Ceva is a multinational animal health company established in 1999 in France that undertakes research, development, production, and marketing of pharmaceutical products, vaccines and equipment for ruminants, poultry, swine, and companion animals. The transfer of pure Montbéliard and Tarentaise embryos was performed on local cows at the CMAP station in Burkina Faso. Over 3 years, 31 embryo transfers were undertaken leading to the birth of 7 live calves (23% calving rate), of which only three survived (one male Tarentaise, one female Tarentaise, and one male Montbéliard). Despite the low survival rate, the experiment paved the way for the feasibility of MOET in local cattle breeds in Burkina Faso.

22.5.2.3 *Case Study 3:* Sperm Sorting at First African Sorting Lab

The first African sorting lab has been recently established in South Africa through a partnership between Sexing Technologies® (ST) and RAMSEM (STgenetics, 2021). RAMSEM is a pioneering livestock artificial insemination company in South Africa, while ST is a global livestock semen sorting leader and innovator. The new sexing lab is at RAMSEM's facility near Bloemfontein, South Africa, and will produce fresh and frozen sex-sorted semen from sheep, goats, and cattle. This will provide customers with semen that is more than 90 percent accurate for the desired gender, with conception rates comparable to conventional (unsorted) semen used for AI. This introduction of semen sorting into the African continent is set to revolutionize the breeding industry for African cattle, sheep, and goats.

22.5.2.4 *Case Study 4:* Birth of Live Calves by in Vitro Embryo Production in a Commercial Herd in South Africa

A successful story of bovine in vitro fertilization and the birth of live calves under typical South African conditions has been reported (Arlotto et al. 2001). Oocytes for in vitro fertilization (IVF) were collected from six slaughtered Bovelder beef cows. The oocytes produced 43 blastocytes from five donors using frozen Bovelder semen. The best 11 produced embryos were transferred into oestrous synchronized Bovelder recipients in the same herd. Results showed that seven calves were born without calving difficulties from four of the original donors resulting in a 64% calving rate. The calves had a normal birth weight apart from the mean of the gestation length, which was significantly longer in males than the herd average (291.6 vs. 285.2 days, respectively).

22.6 Application of Information Systems and Phenomics Technologies in Large Ruminant Improvement in African Input Systems

For many livestock stakeholders, information technologies remain largely unknown or blurry. Nevertheless, examples in several countries show interesting paths that can used to inspire future progress in Africa in the context of the steep growth of animal protein demand. Considering the advantages offered by information systems, the development of animal farming in Africa will increasingly depend on the ability of public bodies to implement multi-partner information systems.

22.6.1 Case Study 1: Information System for Livestock Management in Burundi and Madagascar

In 2012, in the context of the re-population of cattle in Burundi, the International Fund for Agricultural Development (IFAD) supported the creation of a digital system to register the identity of imported animals and the insemination and production recording in Burundi. Adventiel (https://www.adventiel.com/) developed a web application called IBIS (Fig. 22.1) and later a tablet application connected to the same database.

This IBIS system (still in operation today), allows agricultural technicians to record field data and to manage the efficiency of their programmes. It allows the central management and distribution of semen in participating districts, and to plan individual inseminations according to the pedigree of each animal, and thus prevent consanguinity.

The planned extension of the system to all livestock species is yet to be realized due to difficulties accessing farms in remote locations and organizational and regulatory impediments. Nevertheless, it is the first experience for the Ministry of Livestock and Fisheries of Burundi to manage livestock herds in a digital system.

In 2019, the Ministry of Agriculture, Livestock and Fisheries of Madagascar decided to improve the traceability of cattle, fight against the theft of animals, and foster exportations of animals to high-value markets such as Saudi Arabia. For this reason, Adventiel developed ZebuScan (Fig. 22.2), a mobile traceability application designed for this purpose. Today, this system is operational in the field, and the Ministry is able to build on this first system, to add other modules, such as performance recording and genetic improvement, which are so important to foster food security in the country. Indeed, building livestock identification and management systems is a progressive effort.

The chart in Fig. 22.3 gives an example of the structure of a complete system for a developing country. According to the objectives of the genetic organizations and the public bodies in charge, the system can take different physiognomies and offer various functionalities. The most important is to start simple, with animal traceability on a solid regulatory base, and then, with experience, to build on, and offer a connection progressively to the public bodies and various stakeholders of the livestock value chains.

The benefits are huge, as one can see in developed countries. For example, in France today, the Ministry of Agriculture and Food is organizing the renovation of the whole information system for all species. The French system is much more sophisticated than the chart shown in Fig. 22.3 because it comprises several sub-systems. It automates many tasks and data flows and considers the numerous structures involved with its

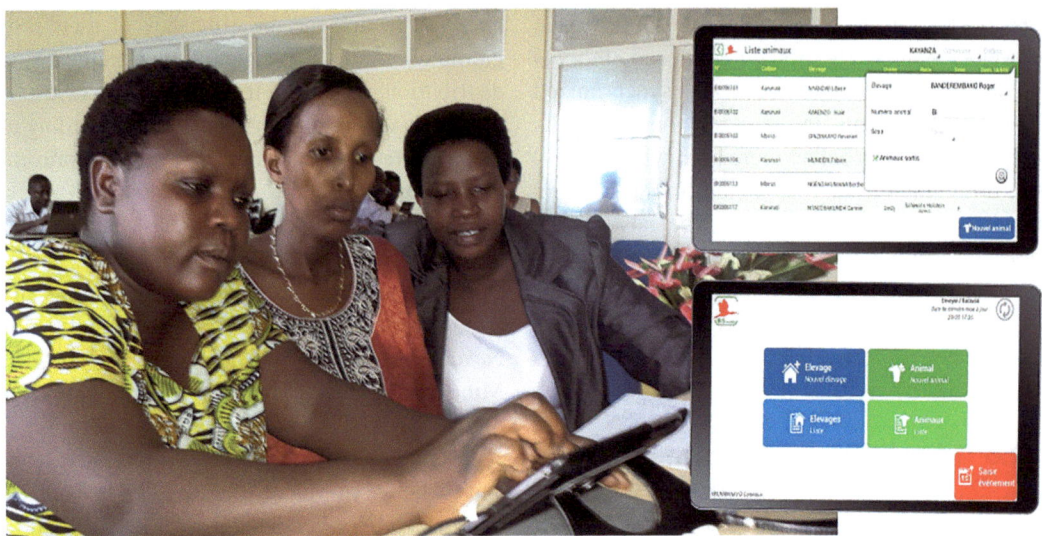

Fig. 22.1 IBIS in Burundi: identification, insemination, and consanguinity management. IFAD, Ministry of Livestock, Burundi and Adventiel

Fig. 22.2 Zebuscan, a mobile application for animal traceability in Madagascar 2019, World Bank, Ministry of Agriculture, Livestock and Fisheries, Adventiel

own information systems. The goal of the Ministry is to reduce costs, foster traceability, animal health, genetic improvement, and market access, and allow the private sector to connect to the system daily to improve the efficiency of the whole value chain.

Today, the digital management of the meat and dairy value chains is a fundamental criterion of economic performance. It has become indispensable for genetic companies and farmers, dairies, slaughterhouses, and all economic stakeholders.

Today, most African countries do not benefit from such systems. However, this is probably one of the most valuable investments they can perform for the genetic improvement of their herds, the profitability of their livestock value chains, and food sovereignty.

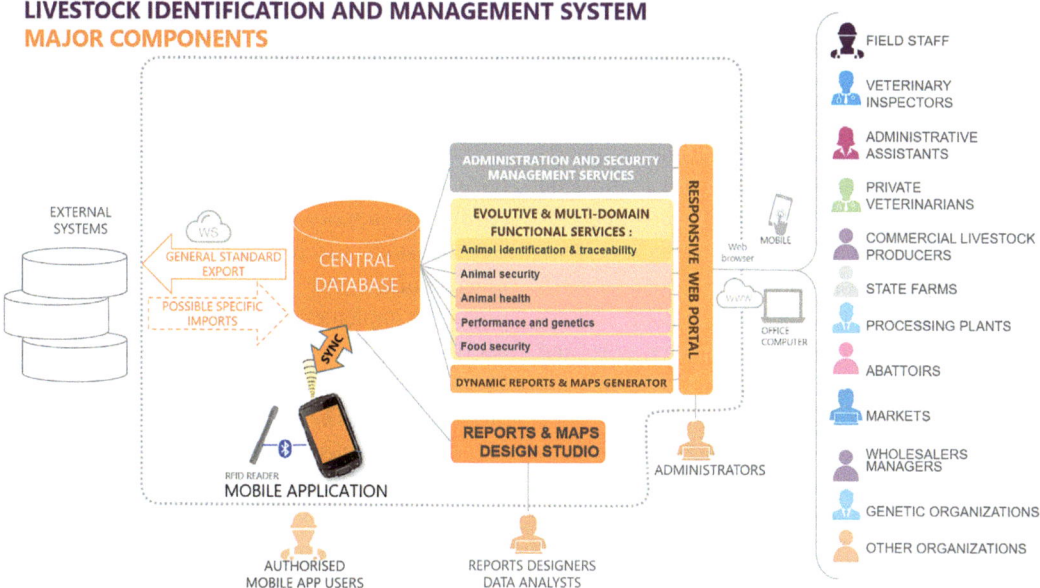

Fig. 22.3 Example of the modular structure of a Livestock Identification and Management System

22.7 Application of Modern Technologies to Improve the Health of Cattle

Local farmers and African governments recognize that the health of cattle is an important aspect of livestock growth, development and productivity, and farm profitability. In an effort to improve the health of livestock, many African governments have dedicated departments of livestock health to support livestock management through access to veterinarians and livestock vaccination programmes, training, and in some cases drugs. However, poor health care management is one of the factors affecting livestock improvement in Africa. In the past and recent times, African governments with funding support from international partners have implemented programmes to enhance livestock health.

22.7.1 Case Study: The Use of a Smartphone App (AfyaData) in Disease Surveillance (One Health Approach) in Africa

An information technology-based system was developed to enhance the detection of human and animal community-based disease outbreaks in Eastern and Southern Africa. The system aims to aid in the training and dissemination of information by health reporters to improve disease surveillance at the national, regional, and global scales.

The system was developed by an EpiHack team comprising human and animal health experts and ICT programmers following a meeting in Tanzania (2014) that identified the major challenges confronting early detection, timely reporting, and prompt response to disease outbreaks or events. The software and ICT tools were developed and packaged as the *AfyaData* app to support One Health disease surveillance (Karimuribo et al. 2017). The software supports close-to-real-time data collection and submission at the community and healthcare facility and provides reporters with feedback. The software proved effective following the reports of 1915 and 1816 cases in livestock and humans following deployment in and usage in Morogoro and Ngorongoro communities of Tanzania over 5 months (August to December 2016). Hence indicating an opportunity for one health engagement of people in their local communities to detect infectious human and animal diseases (Karimuribo et al. 2017).

22.7.2 Case Study: Livestock Vaccination Programmes in Africa

In September 2015, the FAO and the government of Uganda launched a livestock vaccination campaign (https://www.fao.org/africa/news/detail-news/en/c/327820/) with the goal of controlling transboundary animal diseases (TBAD). The 3-month programme funded by the United Kingdom's Department of International Development (DFID) vaccinated about 500,000 cattle against Contagious Bovine Pleuropneumonia (CBPP), 240,000 cattle against Foot and Mouth Disease (FMD), 1000,000 goats and sheep against Sheep/Goat Pox and Peste des Petits Ruminants (PPR) and 400,000 goats against Contagious Caprine Pleuropneumonia (CCPP). The livestock production system of the Karamoja Region of Uganda is predominantly pastoral characterized by frequent movement of livestock and frequent outbreaks of TBADs such as FMD and PPR. The programme's main goal was to control and prevent future disease outbreaks in the area and disease spread to other regions of the country.

Through a recent programme termed 'The SheVax Story in East Africa' (https://afrohun.org/shevax/), livestock vaccines are being made available to marginalized and women livestock farmers in East Africa.

A Livestock Vaccine Innovation Fund (LVIF) funded by the Canada International Development Research Centre (IDRC), Global Affairs Canada (GAC), and Bill and Melinda Gates Foundation (BMGF) is supporting the development of affordable livestock vaccines as well as diseases that affect women and men livestock keepers. The project is implemented through a set of research action-based accomplishments including the SHeVax+ research (AFROHUN 2023). *SHeVax+ means hearing Their Voices and action Research to Support Women's Agency and Empowerment in Livestock Vaccine Distribution, delivery, and Use.* The SHeVax+ research is a partnership between the Cumming School of Veterinary Medicine at Tufts University, the Africa One Health University Network (AFROHUN), Makerere University, the University of Nairobi and the University of Rwanda. In a bid to improve livestock production and livelihoods in Uganda, Kenya, and Rwanda, the project aimed to understand the state of women's empowerment and gender parity on the vaccine value chain and identify key entry points among women smallholder farmers and entrepreneurs to contribute to and benefit from livestock vaccines.

The project's successes since its implementation include (1) training of 24 animal health service providers and well equipped to serve 140,000 households; (2) animal health training provided to over 1500 persons; and (3) dissemination of over 1200 Vaxxer calendars (a calendar with information on vaccination, animal health, and gender issues). Thus, the SheVax+ research is looking for new ways to improve women's participation, and how women can better benefit from livestock vaccine value chains in Kenya, Rwanda, and Uganda.

In order to transform livestock health services and make veterinary extension services more gender aware, responsive and inclusive, and accessible, the SheVAx programme recommends as follows (AFROHUN 2023): (1) put in place gender-inclusive livestock policies into practice through actual procedures or processes to improve vaccination programmes; (2) reach various stakeholders by fostering collaborations between animal health technical providers and community development service providers; (3) make gender responsiveness of service provision an integral part of animal health technical training; (4) improve access by smallholder farmers to efficient vaccination programmes by including women in animal and poultry vaccine value chain; (5) private animal health service providers and governments should recognize and solve the socio-cultural issues at household and community levels that prevent women and other marginalized small holders from using and engaging as providers of livestock vaccine services; (6) committed leaders from government agencies and relevant departments are necessary for public-private collaborations within livestock vaccine value chains; (7) invest in the integration of gender knowledge and communication skills into curriculum and training of animal service providers and veterinarians; (8) encourage women and marginalized groups to become animal health service providers; (9) enrich training of community animal health workers with competence-based learning frameworks and gender and soft skills modules; (10) make animal health welfare services (e.g. veterinary care, drugs and information and technologies) accessible to women, the elderly, the poor and disable livestock keepers; (11) invest in gender transformative approaches in veterinary medicine and extension training; and (12) sensitize community animal health workers, community leaders, and other key community stakeholders on the importance of adequate livestock health management.

22.8 Challenges and Constraints in the Implementation of Modern Technologies in Cattle and Buffalo Production in Africa

Modern technologies have the potential to revolutionize cattle and buffalo production in Africa by increasing productivity, improving animal health, and reducing labour costs. However, there are several challenges and constraints that need to be addressed to effectively implement these technologies in the region. Some of these obvious challenges and constraints are:

1. Limited access to technology: One of the main challenges in implementing modern technologies in cattle and buffalo production in Africa is limited access to technology. Many small-scale farmers in Africa do not have the financial resources to invest in modern equipment and tools, such as automated feeding systems, digital health monitoring devices, and genetic testing technologies. This limits their ability to improve the efficiency and productivity of their operations.
2. Lack of infrastructure: Another major constraint in implementing modern technologies for cattle and buffalo production in Africa is the lack of adequate infrastructure. Many rural areas in the region lack reliable electricity, internet connectivity, and transportation networks, which are essential for the adoption of modern technologies. Without these basic infrastructure elements, farmers are unable to access and utilize the latest innovations in the industry.
3. Limited technical knowledge: In addition to limited access to technology and infrastructure, many cattle farmers and producers in Africa also lack the technical knowledge and skills needed to effectively implement and maintain modern technologies in their operations. Training programmes and extension services are often not adequate in the region, making it difficult for farmers to learn how to use new technologies and troubleshoot problems that may arise.
4. High cost of technology: The high cost of modern technologies is another significant constraint in the implementation of these technologies for cattle and buffalo production in Africa. Many farmers simply cannot afford to purchase expensive equipment and software, which limits their ability to improve the efficiency and productivity of their operations. In addition, the ongoing costs of main-

taining and upgrading technology can be prohibitive for small-scale farmers with limited financial resources.
5. Regulatory challenges: Regulatory challenges, such as import restrictions and licensing requirements, can also hinder the implementation of modern technologies in cattle and buffalo production in Africa. In some cases, government policies and regulations may be outdated or overly restrictive, making it difficult for farmers to access and utilize new technologies that could benefit their operations.

22.9 Conclusion and Future Perspectives

Advances in technology introduced into the cattle industry are more pronounced in countries with advanced economies. Africa has the potential to adopt and tweak these technologies to suit local production. Yet, the absence of such technology in most African farms is a testimony to how much there is to be accomplished on the African soil if African farmers will ever run in the production league with the more technologically advanced countries.

While modern technologies have the potential to transform cattle and buffalo production in Africa, there are several challenges and constraints that need to be addressed in order to effectively implement innovative tools in the region. Limited access to technology, lack of infrastructure, limited technical knowledge, high cost of technology, and regulatory challenges are some of the key issues that must be overcome to realize the full potential of modern technologies in the large ruminant industry.

The adoption and implementation of finished/imported technology are expensive. Investment in animal agriculture in Africa still needs improvement, and government presence and support from financial houses are equally poor. In addition, adopting and modifying advanced technology to suit local production requires capacity development in technological know-how, funding, and sound ethics backed by policy. For this to happen, a united front and government presence are undeniably very important, and a strong public-private partnership (Okpeku et al. 2019) investment into livestock production cannot be compromised. By addressing these challenges and constraints, policymakers, researchers, and industry stakeholders can help to ensure that African cattle and buffalo farmers have the tools, resources, and range of technologies they need to improve the efficiency, productivity, and sustainability of their cattle and buffalo operations.

22.9.1 Potential Solutions and Way Forward

1. Promoting access to technology: To address the limited access to technology, governments, non-governmental organizations, and private sector stakeholders can work together to provide financial assistance, subsidies, or low-interest loans to small-scale cattle and buffalo farmers to help them invest in modern equipment and tools. Additionally, technology companies can develop more affordable and scalable solutions tailored to the needs and budgets of African farmers.
2. Improving infrastructure: Improving infrastructure, such as electricity, internet connectivity, and transportation networks, is essential for the successful implementation of modern technologies in cattle and buffalo production. Governments and international organizations can invest in infrastructure development projects in rural areas to ensure that farmers have the necessary resources to access and utilize new technologies.
3. Enhancing technical knowledge: Training programmes, workshops, and extension services can help farmers improve their technical knowledge and skills in using modern technologies in cattle and buffalo production. Collaboration between research institutions, universities, and agricultural extension services can provide cattle and buffalo farmers with the necessary information and support to effectively implement and maintain new technologies on their farms.

4. Reducing the cost of technology: Governments and industry stakeholders can work together to reduce the cost of modern technologies for cattle and buffalo through subsidies, incentives, and partnerships with technology providers. By making technology more affordable and accessible to small-scale farmers, the adoption rate of modern tools in cattle and buffalo production can be increased, leading to improved efficiency and productivity in the industry.
5. Addressing regulatory challenges: Governments in African Countries can review and update regulatory policies to facilitate the importation and licensing of modern technologies in cattle and buffalo production. By streamlining regulatory processes and reducing bureaucratic hurdles, cattle farmers can more easily access and utilize new technologies that can benefit their operations.

On a final note, while there are significant challenges and constraints in implementing modern technologies in cattle and buffalo production in Africa, there are also potential solutions and a clear way forward to overcome these obstacles as discussed above. By promoting access to technology, improving infrastructure, enhancing technical knowledge, reducing the cost of technology, and addressing regulatory challenges, stakeholders can help African cattle and buffalo farmers harness the full potential of modern technologies that were extensively discussed in Chap. 21. By working together and taking proactive measures, policymakers, researchers, and industry stakeholders can pave the way for a more sustainable, efficient, and productive cattle and buffalo production sector in Africa.

References

AFROHUN (2023) Making vaccines available and accessible to women and other marginalized livestock farmers: The SheVax Story in East Africa. Available at https://afrohun.org/making-vaccines-available-and-accessible-to-women-and-other-marginalized-livestock-farmers-the-shevax-story-in-east-africa/. Accessed on 6 March 2024

Akbar MO, Saad M, Khan S, Ali MJ, Hussain A, Qaiser G, Pasha M, Pasha U, Missen MS, Akhtar N (2020) IoT for development of smart dairy farming. J Food Qual 2020:8

Al Kalaldeh M, Swaminathan M, Gaundare Y, Joshi S, Aliloo H, Strucken EM, Ducrocq V, Gibson JP (2021a) Genomic evaluation of milk yield in a smallholder crossbred dairy production system in India. Genet Sel Evol 53(1):1–14

Al Kalaldeh M, Swaminathan M, Gaundare Y, Joshi S, Aliloo H, Strucken EM, Ducrocq V, Gibson JP (2021b) Genomic evaluation of milk yield in a smallholder crossbred dairy production system in India. Genet Sel Evol 53:1–14

Alexander DH, Novembre J, Lange K (2009) Fast model-based estimation of ancestry in unrelated individuals. Genome Res 19(9):1655–1664

Alfonsetti M, Castelli V, d'Angelo M (2022) Are we what we eat? Impact of diet on the Gut–Brain Axis in Parkinson's disease. Nutrients 14(2):380

Anderson K (2010) Globalization's effects on world agricultural trade, 1960–2050. Philos Trans R Soc B Biol Sci 365(1554):3007–3021

Arlotto T, Gerber D, Terblanche S, Larsen J (2001) Birth of live calves by in vitro embryo production of slaughtered cows in a commercial herd in South Africa. J S Afr Vet Assoc 72(2):72–75

Atılgan Türkmen B (2020) Renewable energy applications for sustainable agricultural systems. Int J Innov Approach Agric Res 4:497

Bahbahani H, Clifford H, Wragg D, Mbole-Kariuki MN, Van Tassell C, Sonstegard T, Woolhouse M, Hanotte O (2015) Signatures of positive selection in East African Shorthorn Zebu: a genome-wide single nucleotide polymorphism analysis. Sci Rep 5(1):11729

Bahbahani H, Salim B, Almathen F, Al Enezi F, Mwacharo JM, Hanotte O (2018) Signatures of positive selection in African Butana and Kenana dairy zebu cattle. PLoS One 13(1):e0190446

Barton NH, Keightley PD (2002) Understanding quantitative genetic variation. Nat Rev Genet 3(1):11–21

Berckmans D (2017) General introduction to precision livestock farming. Anim Front 7(1):6–11

Bianchi M, Chicoteau P, Cloé C, Bassinga A (1986) Premiers essais de transferts d'embryons sur bovins de race Baoulé au Burkina Faso [Preliminary trials of embryo transfers in Baoulé cattle in Burkina Faso]. Rev Elev Med Vet Pays Trop 39(1):139–144. French. PMID: 3562990

Biffani S, Dimauro C, Macciotta N, Rossoni A, Stella A, Biscarini F (2015) Predicting haplotype carriers from SNP genotypes in Bos taurus through linear discriminant analysis. Genet Sel Evol 47:1–11

Biscarini F, Schwarzenbacher H, Pausch H, Nicolazzi EL, Pirola Y, Biffani S (2016) Use of SNP genotypes to identify carriers of harmful recessive mutations in cattle populations. BMC Genomics 17(1):1–17

Brophy B, Smolenski G, Wheeler T, Wells D, L'Huillier P, Laible (2003) Cloned transgenic cattle produce milk with higher levels of beta-casein and kappa-casein. Nat Biotech 21:157–142

Brown A, Ojango J, Gibson J, Coffey M, Okeyo M, Mrode R (2016) Genomic selection in a crossbred cattle population using data from the dairy genetics East Africa project. J Dairy Sci 99(9):7308–7312

Buller H, Blokhuis H, Lokhorst K, Silberberg M, Veissier I (2020) Animal welfare management in a digital world. Animals 10(10):1779

Canavez FC, Luche DD, Stothard P, Leite KR, Sousa-Canavez JM, Plastow G, Meidanis J, Souza MA, Feijao P, Moore SS (2012) Genome sequence and assembly of Bos indicus. J Hered 103(3):342–348

CBD (1992) Convention on Biological Diversity. Text and Annexes, United Nations

CBD (2000) Convention on Biological Diversity. Text and Annexes. United Nations

NASAC (2015) Biodiversity and Biotechnology: Socioeconomic and Environmental Impacts of Biotechnology on the Agricultural Sector in African Countries, Network of African Science Academies

Cheng G, Fu C, Wang H, Adoligbe C, Wei S, Li S, Jiang B, Wang H, Zan L (2015) Production of transgenic beef cattle rich in n-3 PUFAs by somatic cell nuclear transfer. Biotechnol Lett 37:1565–1571

Collado E, Fossatti A, Saez Y (2018) Smart farming: a potential solution towards a modern and sustainable agriculture in Panama. AIMS Agric Food 4(2):266–284

Competition Authority of Kenya (CAK) (2014) Benchmarking the Kenyan artificial insemination service sub-industry. Submitted to the competition authority of Kenya for Kenya Markets Trust

Cook E (2020) Agriculture, forestry and fishery statistics: 2020 edn. Publications Office of the European Union

DEFRA, U (2020) The path to sustainable farming: an agricultural transition plan 2021 to 2024. Department for Environment, Food and Rural Affairs, London

Duncanson GR (1977) The Kenya National Artificial Insemination Service. Published in animal breeding: selected articles from the World Animal Review. FAO

Food; Organization., A (2018) Future of food and agriculture 2018: alternative pathways to 2050. Food & Agriculture Org

Frezal C, Gay SH, Nenert C (2021) The Impact of the African Swine Fever outbreak in China on global agricultural markets

Gao Y, Wu H, Wang Y, Liu X, Chen L, Li Q, Cui C, Liu X, Zhang J, Zhang Y (2017) Single Cas9 nickase induced generation of NRAMP1 knockin cattle with reduced off-target effects. Genome Biol 18(1):1–15

Garritsen C, Van Marle-Köster E, Snyman M, Visser C (2015) The impact of DNA parentage verification on breeding value estimation and sire ranking in South African Angora goats. Small Rumin Res 124:30–37

Gautier M, Flori L, Riebler A, Jaffrézic F, Laloé D, Gut I, Moazami-Goudarzi K, Foulley J-L (2009) A whole genome Bayesian scan for adaptive genetic divergence in West African cattle. BMC Genomics 10(1):1–18

Gebrehiwot NZ, Strucken E, Aliloo H, Marshall K, Gibson JP (2020) The patterns of admixture, divergence, and ancestry of African cattle populations determined from genome-wide SNP data. BMC Genomics 21(1):1–16

Goddard M (1996) The use of marker haplotypes in animal breeding schemes. Genet Sel Evol 28(2):161–176

Groher T, Heitkämper K, Umstätter C (2020) Digital technology adoption in livestock production with a special focus on ruminant farming. Animal 14(11):2404–2413

Hafez YM (2015) Assisted reproductive technologies in farm animals. ICMALPS, Alexandria University, Egypt

Hanotte O, Ronin Y, Agaba M, Nilsson P, Gelhaus A, Horstmann R, Sugimoto Y, Kemp S, Gibson J, Korol A (2003) Mapping of quantitative trait loci controlling trypanotolerance in a cross of tolerant West African N'Dama and susceptible East African Boran cattle. Proc Natl Acad Sci 100(13):7443–7448

Hayes BJ, Daetwyler HD (2019) 1000 bull genomes project to map simple and complex genetic traits in cattle: applications and outcomes. Ann Rev Anim Biosci 7:89–102

Hayes B, Goddard ME (2001) The distribution of the effects of genes affecting quantitative traits in livestock. Genet Sel Evol 33:1–21

Hutchison J, Cole J, Bickhart D (2014) Use of young bulls in the United States. J Dairy Sci 97(5):3213–3220

Ibeagha-Awemu EM, Jann OC, Weimann C, Erhardt G (2004) Genetic diversity, introgression and relationships among West/Central African cattle breeds. Genet Sel Evol 36:1–18

Ibeagha-Awemu EM, Omonijo F, Piché L, Vincent A (2024) Alternatives to antibiotics for sustainable livestock production in the context of the One-Health-Approach: tackling common foes. Animal (submitted)

Jabed A, Wagner S, McCracken J, Wells DN, Laible G (2012) Targeted microRNA expression in dairy cattle directs production of β-lactoglobulin-free, high-casein milk. Proc Natl Acad Sci 109(42):16811–16816

Joint F (2005) Improving artificial breeding of cattle in Africa. Guidelines and recommendations. A manual prepared under the framework of an IAEA technical cooperation regional AFRA project on increasing and improving milk and meat production. International Atomic Energy Agency

Joshi S, Mobeen A, Jan K, Bashir K, Azad ZAA (2019) Emerging technologies in dairy processing: present status and future potential. In: Health and safety aspects of food processing technologies. Springer, Cham, pp 105–120

Karavolias NG, Horner W, Abugu MN, Evanega SN (2021) Application of gene editing for climate change in agriculture. Front Sustain Food Syst 5:685801

Karimuribo ED, Mutagahywa E, Sindato C, Mboera L, Mwabukusi M, Njenga MK, Teesdale S, Olsen J, Rweyemamu M (2017) A smartphone app (AfyaData) for innovative one health disease surveillance from community to national levels in Africa: intervention in disease surveillance. JMIR Public Health Surveill 3(4):e7373

KDDP/SDP (2004) Data and Statistics from the KDDP breeding assessment study

Kidie HA (2019) Review on growth and development of multiple ovulation and embryo transfer technology in cattle. World Sci News 127(3):199

Kim J, Hanotte O, Mwai OA, Dessie T, Bashir S, Diallo B, Agaba M, Kim K, Kwak W, Sung S (2017) The genome landscape of indigenous African cattle. Genome Biol 18:1–14

Kios D, van Marle-Köster E, Visser C (2012) Application of DNA markers in parentage verification of Boran cattle in Kenya. Trop Anim Health Prod 44:471–476

Krpalkova L, O'Mahony N, Carvalho A, Campbell S, Corkery G, Broderick E, Walsh J (2021) Decision-making strategies on smart dairy farms: a review. Int J Agric Biosyst Eng 15(11):138–145

Kwoji ID, Aiyegoro OA, Okpeku M, Adeleke MA (2021) Multi-strain probiotics: synergy among isolates enhances biological activities. Biology 10(4):322

Laible G, Cole SA, Brophy B, Wei J, Leath S, Jivanji S, Littlejohn M, Wells D (2021) Holstein Friesian dairy cattle edited for diluted coat color as a potential adaptation to climate change. BMC Genomics 22(1):1–12

Lamb GC, Maddock T (2009) Feed efficiency in cows. In: Florida beef cattle short course, pp 35–42

Lambo MT, Chang X, Liu D (2021) The recent trend in the use of multistrain probiotics in livestock production: an overview. Animals 11(10):2805

Makina SO, Muchadeyi FC, van Marle-Köster E, MacNeil MD, Maiwashe A (2014) Genetic diversity and population structure among six cattle breeds in South Africa using a whole genome SNP panel. Front Genet 5:333

Makina SO, Muchadeyi FC, van Marle-Köster E, Taylor JF, Makgahlela ML, Maiwashe A (2015) Genome-wide scan for selection signatures in six cattle breeds in South Africa. Genet Sel Evol 47(1):1–14

Marshall K, Tebug S, Salmon GR, Tapio M, Juga J, Missohou A (2017) Improving dairy cattle productivity in Senegal. ILRI Policy Brief, p 22

Mauki DH, Tijjani A, Ma C, Ng'ang'a SI, Mark AI, Sanke OJ, Abdussamad AM, Olaogun SC, Ibrahim J, Dawuda PM (2022) Genome-wide investigations reveal the population structure and selection signatures of Nigerian cattle adaptation in the sub-Saharan tropics. BMC Genomics 23(1):306

Mbah D (2006) Biotechnology and animal production. J Cameroon Acad Sci 6(1):29–40

McFarlane GR, Salvesen HA, Sternberg A, Lillico SG (2019) On-farm livestock genome editing using cutting edge reproductive technologies. Front Sustain Food Syst 3:106

Mekonnen YA, Gültas M, Effa K, Hanotte O, Schmitt AO (2019) Identification of candidate signature genes and key regulators associated with Trypanotolerance in the Sheko Breed. Front Genet 10:1095

Miglior F, Fleming A, Malchiodi F, Brito LF, Martin P, Baes CF (2017) A 100-year review: identification and genetic selection of economically important traits in dairy cattle. J Dairy Sci 100(12):10251–10271

Mitra A, Yadav BR, Ganai NA, Balakrishnan C (1999) Molecular markers and their applications in livestock improvement. Curr Sci 77(8):1045–1053

Montaldo HH, Meza-Herrera CA (1998) Use of molecular markers and major genes in the genetic improvement of livestock. Electron J Biotechnol 1(2):15–16

Mrode R, Coffey M, Ojango J, Mujibi D, Okeyo M, Strucken E, Gibson JP, Aliloo H (2018) The impact of modelling and pooled data on the accuracy of genomic prediction in small holder dairy data. In: Proceedings of the world congress on genetics applied to livestock production. Massey University

Mrode R, Ojango J, Ekine-Dzivenu C, Aliloo H, Gibson J, Okeyo M (2021) Genomic prediction of crossbred dairy cattle in Tanzania: a route to productivity gains in smallholder dairy systems. J Dairy Sci 104(11):11779–11789

Mwai A, Gebreyohanes G, Ojango JMK, Mrode R, Chinyere E, Jabes Y, Kipkosgei G, Mogaka D, Abdulkadir KN, Agasi H, Meseret S, Eliamoni L, Kemp S (2023) The Africa Asia Dairy Genetics Gains (AADGG) Platform, ILRI project profile. ILRI, Nairobi, Kenya

Nasirahmadi A, Hensel O (2022) Toward the next generation of digitalization in agriculture based on digital twin paradigm. Sensors 22(2):498

Nigam D, Bhoomika K (2022) Targeted genome editing by CRISPR/Cas9 for livestock improvement. In: Emerging issues in climate smart livestock production. Elsevier, pp 415–447

Ntakyo PR, Kirunda H, Tugume G, Natuha S (2020) Dry season feeding technologies: assessing the nutritional and economic benefits of feeding hay and silage to dairy cattle in South-Western Uganda. Open J Anim Sci 10(3):627–648

Ohimain EI, Ofongo RT (2012) The effect of probiotic and prebiotic feed supplementation on chicken health and gut microflora: a review. Int J Anim Vet Adv 4(2):135–143

Okpeku M, Ogah DM, Adeleke MA (2019) A review of challenges to genetic improvement of indigenous livestock for improved food production in Nigeria. Afr J Food Agric Nutr Dev 19(1):13959–13978

Opoola O, Shumbusho F, Hambrook D, Thomson S, Dai H, Chagunda MGG, Capper JL, Moran D, Mrode R, Djikeng A (2022) From a documented past of the Jersey breed in Africa to a profit index linked future. Front Genet 13:881445

Ouédraogo D, Ouédraogo-Koné S, Yougbaré B, Soudré A, Zoma-Traoré B, Mészáros G, Khayatzadeh N, Traoré A, Sanou M, Mwai OA (2021) Population structure, inbreeding and admixture in local cattle populations managed by community-based breeding programs in Burkina Faso. J Anim Breed Genet 138(3):379–388

Perisse IV, Fan Z, Singina GN, White KL, Polejaeva IA (2021) Improvements in gene editing technology boost its applications in livestock. Front Genet 11:614688

Petersburg-Pushkin S (2015) Environmentally friendly agriculture and forestry for future generations

Pirola Y, Della Vedova G, Bonizzoni P, Stella A, Biscarini F (2013) Haplotype-based prediction of gene alleles using pedigrees and SNP genotypes. In: Proceedings of the international conference on bioinformatics, computational biology and biomedical informatics, pp 33–41

Reid G, Nduti N, Sybesma W, Kort R, Kollmann TR, Adam R, Boga H, Brown EM, Einerhand A, El-Nezami H (2014) Harnessing microbiome and probiotic research in sub-Saharan Africa: recommendations from an African workshop. BioMed Central

Rieblinger B, Sid H, Duda D, Bozoglu T, Klinger R, Schlickenrieder A, Lengyel K, Flisikowski K, Flisikowska T, Simm N (2021) Resources for genome editing in livestock: Cas9-expressing chickens and pigs. Proc Natl Acad Sci 118(10):1–9

Rose DC, Chilvers J (2018) Agriculture 4.0: broadening responsible innovation in an era of smart farming. Front Sustain Food Syst 2:87

Rothschild M, Jacobson C, Vaske D, Tuggle C, Short T, Sasaki S, Eckardt G, McLaren D (1994) A major gene for litter size in pigs. In: Proceedings of the 5th world congress on genetics applied to livestock production, vol 21. University of Guelph, Canada, Guelph, pp 225–228

Rushdi H, Moghaieb REA, Abdel-Shafy H, Ibrahim M (2017) Association between microsatellite markers and milk production traits in Egyptian buffaloes. Czeh J Anim Sci 62(9):384–391

Sagwa C, Okeno T, Kahi A (2019) Increasing reproductive rates of both sexes in dairy cattle breeding optimizes response to selection. S Afr J Anim Sci 49(4):654–663

Schillings J, Bennett R, Rose DC (2021) Exploring the potential of precision livestock farming technologies to help address farm animal welfare. Front Anim Sci 2:639678

Schuster F, Aldag P, Frenzel A, Hadeler K-G, Lucas-Hahn A, Niemann H, Petersen B (2020) CRISPR/Cas12a mediated knock-in of the Polled Celtic variant to produce a polled genotype in dairy cattle. Sci Rep 10(1):13570

Sequencing BG, Consortium A, Elsik CG, Tellam RL, Worley KC, Gibbs RA, Muzny DM, Weinstock GM, Adelson DL, Eichler EE et al (2009) The genome sequence of taurine cattle: a window to ruminant biology and evolution. Science 324(5926):522–528

Shange LB (2021) Tadu Dairy Cooperative Society Model, Jakiri, Cameroon, p 15

Silvi R, Pereira LGR, Paiva CAV, Tomich TR, Teixeira VA, Sacramento JP, Ferreira RE, Coelho SG, Machado FS, Campos MM (2021) Adoption of precision technologies by Brazilian dairy farms: the farmer's perception. Animals 11(12):3488

Singh P, Wani AA, Langowski H-C (2015) Food packaging materials: testing & quality assurance; CRC Press, 2017. Uyeno, Y.; Shigemori, S.; Shimosato, T. Effect of probiotics/prebiotics on cattle health and productivity. Microbes Environ 30(2):126–132

Smith TP, Bickhart DM, Boichard D, Chamberlain AJ, Djikeng A, Jiang Y, Low WY, Pausch H, Demyda-Peyrás S, Prendergast J (2023) The Bovine Pangenome Consortium: democratizing production and accessibility of genome assemblies for global cattle breeds and other bovine species. Genome Biol 24(1):139

Soller M (1994) Marker assisted selection-an overview. Anim Biotechnol 5(2):193–207

Strucken EM, Al-Mamun HA, Esquivelzeta-Rabell C, Gondro C, Mwai OA, Gibson JP (2017) Genetic tests for estimating dairy breed proportion and parentage assignment in East African crossbred cattle. Genet Sel Evol 49:1–18

Sun Z, Wang M, Han S, Ma S, Zou Z, Ding F, Li X, Li L, Tang B, Wang H (2018) Production of hypoallergenic milk from DNA-free beta-lactoglobulin (BLG) gene knockout cow using zinc-finger nucleases mRNA. Sci Rep 8(1):15430

Symeonaki E, Maraveas C, Arvanitis KG (2024) Recent advances in digital twins for agriculture 5.0: applications and open issues in livestock production systems. Appl Sci 14:686

Tebug SF, Missohou A, Sourokou Sabi S, Juga J, Poole EJ, Tapio M, Marshall K (2018) Using body measurements to estimate live weight of dairy cattle in low-input systems in Senegal. J Appl Anim Res 46(1):87–93

Tessanne K, Golding M, Long C, Peoples M, Hannon G, Westhusin M (2012) Production of transgenic calves expressing an shRNA targeting myostatin. Mol Reprod Dev 79(3):176–185

Tijjani A, Salim B, da Silva MVB, Eltahir HA, Musa TH, Marshall K, Hanotte O, Musa HH (2022) Genomic signatures for drylands adaptation at gene-rich regions in African zebu cattle. Genomics 114(4):110423

Van Berkel PH, Welling MM, Geerts M, van Veen HA, Ravensbergen B, Salaheddine M, Pauwels EK, Pieper F, Nuijens JH, Nibbering PH (2002) Large scale production of recombinant human lactoferrin in the milk of transgenic cows. Nat Biotechnol 20(5):484–487

Van Eenennaam AL, De Figueiredo Silva F, Trott JF, Zilberman D (2021) Genetic engineering of livestock: the opportunity cost of regulatory delay. Ann Rev Anim Biosci 9:453–478

Varankovich NV, Nickerson MT, Korber DR (2015) Probiotic-based strategies for therapeutic and prophylactic use against multiple gastrointestinal diseases. Front Microbiol 6:685

Verma O, Kumar R, Kumar A, Chand S (2012) Assisted reproductive techniques in farm animal-from artificial insemination to nanobiotechnology. Veterin World 5(5):301

Visser C, van Marle-Koster E, Friedrich H (2011) Parentage verification of South African Angora goats, using microsatellite markers. S Afr J Anim Sci 41(3):250–255

Vohra V, Chhotaray S, Gowane G, Alex R, Mukherjee A, Verma A, Deb SM (2021) Genome-wide association studies in Indian Buffalo revealed genomic regions for lactation and fertility. Front Genet 12:696109

Vranken E, Berckmans D (2017) Precision livestock farming for pigs. Anim Front 7:32–37

Wakeley J, Nielsen R, Liu-Cordero SN, Ardlie K (2001) The discovery of single-nucleotide polymorphisms—and inferences about human demographic history. Am J Hum Genet 69(6):1332–1347

Wall R, Kerr D, Bondioli K (1997) Transgenic dairy cattle: genetic engineering on a large scale. J Dairy Sci 80(9):2213–2224

Wall RJ, Powell AM, Paape MJ, Kerr DE, Bannerman DD, Pursel VG, Wells KD, Talbot N, Hawk HW (2005) Genetically enhanced cows resist intramammary Staphylococcus aureus infection. Nat Biotechnol 23(4):445–451

Walling G, Archibald A, Visscher P, Haley C (1998) Mapping of quantitative trait loci on chromosome 4 in a large White x Meishan pig F2 population. In: 6th world congress on genetics applied to livestock production

Weigel K, VanRaden P, Norman H, Grosu H (2017) A 100-year review: methods and impact of genetic selection in dairy cattle—from daughter–dam comparisons to deep learning algorithms. J Dairy Sci 100(12):10234–10250

Wheeler M, Walters E, Clark S (2003) Transgenic animals in biomedicine and agriculture: outlook for the future. Anim Reprod Sci 79(3–4):265–289

Wu X, Ouyang H, Duan B, Pang D, Zhang L, Yuan T, Xue L, Ni D, Cheng L, Dong S (2012) Production of cloned transgenic cow expressing omega-3 fatty acids. Transgenic Res 21:537–543

Yang P, Wang J, Gong G, Sun X, Zhang R, Du Z, Liu Y, Li R, Ding F, Tang B (2008) Cattle mammary bioreactor generated by a novel procedure of transgenic cloning for large-scale production of functional human lactoferrin. PLoS One 3(10):e3453

Yang B, Wang J, Tang B, Liu Y, Guo C, Yang P, Yu T, Li R, Zhao J, Zhang L (2011) Characterization of bioactive recombinant human lysozyme expressed in milk of cloned transgenic cattle. PLoS One 6(3):e17593

Zhang Y, Zhang Y, Gao M, Dai B, Kou S, Wang X, Fu X, Shen W (2023) Digital twin perception and modeling method for feeding behavior of dairy cows. Comput Electron Agric 214:108181

Zhou Y, Yang L, Han X, Han J, Hu Y, Li F, Xia H, Peng L, Boschiero C, Rosen BD (2022) Assembly of a pangenome for global cattle reveals missing sequences and novel structural variations, providing new insights into their diversity and evolutionary history. Genome Res 32(8):1585–1601

Zimin AV, Delcher AL, Florea L, Kelley DR, Schatz MC, Puiu D, Hanrahan F, Pertea G, Van Tassell CP, Sonstegard TS (2009) A whole-genome assembly of the domestic cow, Bos taurus. Genome Biol 10:1–10

Open Access   This chapter is licensed under the terms of the Creative Commons Attribution 4.0 International License (http://creativecommons.org/licenses/by/4.0/), which permits use, sharing, adaptation, distribution and reproduction in any medium or format, as long as you give appropriate credit to the original author(s) and the source, provide a link to the Creative Commons license and indicate if changes were made.

The images or other third party material in this chapter are included in the chapter's Creative Commons license, unless indicated otherwise in a credit line to the material. If material is not included in the chapter's Creative Commons license and your intended use is not permitted by statutory regulation or exceeds the permitted use, you will need to obtain permission directly from the copyright holder.

Prospect of Modern Technologies for Poultry Improvement in Africa

23

Victor E. Olori, Sunday O. Peters, and Matthew A. Adeleke

Abstract

Poultry production is the most extensive commercial livestock production sector in Africa. It has the highest potential to meet the projected increase in demand for animal protein in the continent in the next decade. This chapter focuses on technologies used in different areas of poultry production. The emphasis is on the prospect of using these technologies to improve poultry production. Section 23.1 gives a brief introduction of the topic indicating that technology has a key role to play in all aspects of poultry production. Its deployment holds the key to the ability of the poultry sector to fulfil the expectation of being the main source of animal protein in Africa. Genetic improvement and the supply of high-quality day-old chicks, which is the bedrock of commercial poultry production, rely on the deployment of various technologies. Section 23.2 gives a detailed insight into this. Other technologies are essential in maintaining the good health and wellbeing of birds which is critical in the enhancement of productivity. This is the focus of Sect. 23.3. It includes a subsection on the success of vaccination programmes and other animal health initiatives in Africa. Section 23.4 describes technologies useful in the improvement of nutrition, feed, and feeding of poultry, including a subsection on technology used in feed manufacturing and preservation as well as the biological interaction between feed and bird performance. Section 23.5 describes various technologies that can be deployed to create a good and safe production environment. These include the housing of birds, their nutrition, and health monitoring. These technologies support human skill and knowledge to create and maintain a good production environment. Concluding remarks and perspectives for the future are presented in Sect. 23.6, pointing out that the utilization of technology has the potential to facilitate a sustainable increase in poultry production capacity, and the productivity of farms and individual birds.

V. E. Olori (✉)
Research and Development Department, Aviagen Ltd., Newbridge, Scotland, UK

S. O. Peters
Department of Animal Science, Berry College, Mount Berry, GA, USA

M. A. Adeleke
Discipline of Genetics, School of Life Sciences, University of KwaZulu-Natal, Durban, South Africa

Keywords

Animal relationship Matrix · Newcastle Disease · Genomic BLUP · Best Linear Unbiased Prediction · Poultry production

Abbreviations

A-Matrix	Animal relationship Matrix
AMR	Anti-Microbial Resistance
AOAC	Association Of Analytical Chemists
BLUP	Best Linear Unbiased Prediction
BV	Breeding Value
CT	Computerized Tomography
DEXA	Dual Energy X-ray Absorptiometry
DF-REML	Derivative Free REstricted Maximum Likelihood
DGV	Direct Genomic Value
DNA	Deoxyribonucleic acid
ELISA	Enzyme-Linked Immunosorbent Assay
FAO	Food and Agriculture Organisation
FIPS-Africa	Farm Inputs Promotions Africa
G-Matrix	Genomic relationship Matrix
GBLUP	Genomic BLUP
GEBV	Genomic-enhanced Estimated Breeding Values
GIS	Geographic Information Systems
GWAS	Genome-Wide Association
HHT	Hand Held Terminal
HPAI	High Pathogenic Avian Influenza
HPLC	High-Performance Liquid Chromatography
ICT	Information Communication Technology
IPRAV	Ivory Coast Poultry Inter-Professional Association
IRTI	Infrared Thermal Imaging
LD	Linkage Disequilibrium
MDR	Multi-Drug Resistance
mRNA	messenger Ribonucleic Acid
ND	Newcastle Disease
NDAFC	Newcastle Disease and Avian Flu Control
NIR	Near Infra-Red
NMR	Nuclear Magnetic Resonance
NTS	Non-Typhoidal Salmonella
OECD	Organisation for Economic Cooperation and Development
OIE	'Office International des epizooties' (World organization for Animals health)
PLF	Precision Livestock Farming
QTL	Quantitative Trait Loci
RAM	Ready Access Memory
RFID	Radio Frequency Identification
RRM	Random Regression Method
SANDCP	Southern African Newcastle Disease Control Project
SNP	Single Nucleotide Polymorphism
SNP-BLUP	See SNP and BLUP
SPC	Statistical Process Control
SSR	Simple Sequence Repeats
TB	Terra Byte
USAID	United States Agency for International Development
VFD	Variable Frequency Drive
WGS	Whole Genome Sequence
WiFi	Wireless Fidelity

23.1 Introduction

There is increasing interest in the contribution of the livestock sector in various national economies of African countries. This has come with the need to diversify these economies in order to generate revenue. The interest in the sector is gradually shifting focus from improving livestock performance and their production environment to achieving a sustainable increase in productivity as well as the development of the overall value chain. Sustainable increase in productivity requires improvement in the biological efficiency of individual animals, improvement in productive efficiency of individual livestock farms, and increase in capacity, and diversity of the livestock sector. To this end, the role of proven and emerging technologies cannot be over-emphasized.

The objective of this chapter is to illustrate the role technology can play and the prospect for its utilization in improving the capacity of poultry production in Africa. Although technology has played and will continue to play a key role in the improvement of poultry production, its impact has not been felt in Africa as it has in Western and some Middle Eastern countries. The use of technology in different aspects of poultry production in the aforementioned countries will therefore serve as a case study in the discussion of its prospect in African countries. A more detailed overview of modern technologies relevant to livestock production in general is presented in Chap. 21.

The most common poultry species farmed in Africa are chickens, turkeys, ducks, and guinea fowls. Both indigenous and imported breeds of these species are reared in different production systems ranging from small-scale extensive family production systems to large-scale intensive commercial systems. Chapter 2 gives a more detailed discussion of livestock production systems in Africa. The most common factor distinguishing these systems is the level of input use, particularly the quality and availability of day-old chicks that are the primary seed stock in the case of poultry farms. These inputs include the breeds, feed resources, and provision of housing, health care, and biosecurity. To a great extent, it also includes human capacity in terms of knowledge and skill level of the management staff. All of these aspects involve the use of some aspects of technology.

The biggest challenge to poultry production in Africa is the high cost of inputs such as the seed-stock which are generally imported and feed. Increased efficiency with technology can create a suitable production environment while the use of genetically improved feed-efficient birds will minimize feed requirement and hence, cost. Another major challenge is disease burden and associated health challenges on the birds. The tropical environment in most of Africa supports the prevalence and persistence of disease-causing pathogens. Genetic improvement of bird resilience and the use of innovative health care and biosecurity approaches are tools that have helped poultry production in other climes. These will be needed in Africa to bring about sustainable improvement. Finally, production capacity in Africa is low mostly due to poor investment in technology that can help in the management of farms with large numbers of birds. To improve production capacity in Africa, technology uptake must increase. Prospective technologies for poultry farm management, health, nutrition, and genetic improvement are discussed in the following sections of this chapter.

23.2 Improvement of the Birds' Biological Efficiency

23.2.1 The Use of Technology in the Genetic Improvement of Birds

Chapter 21 above gives a vivid picture of the low productivity of indigenous livestock resources used in African production systems described in Chap. 2. Chapter 7 highlights the abundant indigenous poultry resources with potential for improvement. Improved productivity however required purposeful genetic improvement efforts. The process of genetic improvement, which brings about a cumulative permanent change in performance or other features of livestock, relies on the application of several technologies. The adoption of a closed nucleus breeding programme is very common in poultry breeding. This requires that birds in the nucleus are housed, fed, and managed with all the skills and tools deployed in commercial poultry production. In addition to this, data required for genetic evaluation in the breeding programme has to be collected from various stages of the bird's life; from hatching through breeding stages. Data is collected on individual animal's performance, pedigree, and genomic information. Available data on the phenotype (observed performance) and the genotype (derived genomic composition) needs to be stored, maintained, and analysed to derive the required information for selection and breeding. In this regard, the process of data collection, transfer, storage, analysis, and dissemination of information all require technologies. In order to ensure a high rate of genetic gain, selection intensity is high and necessarily unequal between males and females. A few numbers of males are usually bred to a large number of females. This requires the use of reproductive technologies that, not only aid successful breeding but also sex and pedigree determination in chicks soon after hatch.

Because technology applies to innovative methodologies as well as tools and equipment, it is pertinent that the methodologies used to bring about genetic improvement in poultry breeding should be discussed herein. This includes methods and tools used for data collection, transfer, storage, and maintenance as well as the quantitative, biometric, and statistical tools used to analyse both phenotypic and molecular data in poultry breeding. While it is well understood from basic genetic principles that genes control traits, the identification of the specific genes or chromosome sections (haplotypes) responsible for many traits was largely unknown until recently. Despite this, many traits or features are known to be controlled by numerous genes all having small effects, which together determine the quantitative variation observed in the population. It was and still is impossible to precisely identify all the gene variants that affect a given trait and the physiological pathways underlying their effect. Over the years, available technology at the time has facilitated the development of methodologies, quantitative and molecular, to facilitate the elucidation of the underlying genetic composition or genotype of individual birds.

23.2.2 Quantitative Versus Molecular Evaluation

In Chap. 15, an overview of the strategies used in bringing about genetic improvement including the definition of breeding goals was discussed. Subsequent chapters discussed these in relation to cattle (Chap. 16), small ruminants (Chap. 17), and poultry (Chap. 18). In this chapter we give detailed emphasis on genetic evaluation, the process which allows us to bring about improvement through selection in line with the breeding objective. Genetic improvement is achieved through the concentration of the alleles that impact superiority with regard to certain traits, in the next generation. This can be done in two or more ways each adopting various technologies. These include the following:

- Quantitative analysis of variation in phenotypic data based on pedigree relationships and derived population parameters. This is referred to as polygenic evaluation.
- Quantitative analysis of molecular variation based on identified gene markers, reference population, marker associations, and genomic relationship parameters. This is referred to as genomic evaluation.

Genetic enhancement of animal features, traits, or weaknesses, such as susceptibility to a particular disease is also possible through the modification of the genotype and hence the gene variants controlling specific features of the animal. This is generally referred to as gene editing where the individual's genotype is modified through specific gene deletion, silencing, or activation. It is generally referred to as genetic modification where a gene or genetic material from a different breed or species is introduced to the individual's genome through gene insertion by different methods. Individuals whose features have been significantly modified through the introduction of foreign genes from other species or organisms are referred to as transgenic animals.

23.2.2.1 Polygenic Evaluation
The Best Linear Unbiased Prediction Method
The selection of animals to produce the next generation is at the core of the genetic improvement of livestock. The ranking of selection candidates is based on the use of quantitative statistical and mathematical methodologies to decipher the genetic merit of individual animals from observed performance (phenotype). The underlying genotype or genetic merit of an individual with respect to a given trait is measured in terms of the breeding value. It is estimated using information on the observed performance of the individual and its relatives, information on the relationship between animals in the population captured in the pedigree, and population parameters which provide an indication of the total variation of the trait in the given population and components of this variation attributable to genetic and non-genetic factors. The key technological breakthroughs and the evolution of the methodologies for genetic

evaluation of farm animals have been extensively reviewed (Grosu et al. 2013).

Of note are the breakthroughs and people that led to the development of the Best Linear Unbiased Prediction (BLUP) methodology which has become the standard for poultry genetic evaluation. These include J. Lush and C.R. Henderson who developed and with others expanded methodologies for predicting breeding values using the BLUP procedure (Grosu et al. 2013). Although the principles of BLUP and the derivation of its equations were known by 1949, it required improvement in computing capacity brought about by technological breakthroughs in the early 1970s before it could be implemented. Since then, BLUP has been implemented in several models (Grosu et al. 2013; Mrode 2014), whenever phenotypic data from a large population of birds are available to estimate population parameters, and together with relationship information, used to predict breeding values (Simm et al. 2020). The steps in the computation of breeding values with different models and for different sets of data have been well described (Mrode 2014). These include single trait models, multivariate animal models, which allow simultaneous estimation of breeding values for group traits, repeatability, and random regression models which are particularly suitable for traits measured repeatedly over time or space, also called longitudinal traits. Examples of these include body weight at different ages, repeated measures of egg weight, or monthly fertility values as the flock ages.

23.2.2.2 Genomic Evaluation

Genomic evaluation refers to methodologies that allow us to predict the genetic merit of individual animals, quantitatively, with information on the phenotype combined with information on genetic markers. A genetic marker refers to a heritable piece of DNA that can be mapped to a specific location on the chromosome of an organism. Advances in molecular technology have allowed repeated sequences of DNA bases, so-called microsatellites, or simple sequence repeats (SSR), which can be found all over the chromosome, to be mapped. Advances in technology have permitted full sequencing of the genome of many organisms and this has led to the identification of variation arising from changes at the single nucleotide level. These single nucleotide polymorphisms (SNPs) have become the standard genetic markers used in the genomic evaluation of animals.

The derivation of genomic breeding values using genetic markers is based on the principle that these markers are non-randomly associated with, or as we say, in linkage disequilibrium (LD), with loci that affect variation in quantitative traits. SNPs are more numerous than other markers known to date and distributed more widely over the entire genome. They are thus more likely to be in LD with all quantitative trait loci (QTLs). This association between SNP and QTL in the same chromosome segments allows us to track all QTLs and make the evaluation of different traits of the animal possible with the same set of SNPs captured in a 'SNP chip'. Such chip should be of reasonable density, i.e. have a large number of SNPs, to capture QTLs that may be spread across the entire genome. Figure 23.1 shows a picture of a typical commercial SNP chip. Similar chips exist for humans and different livestock species (see Chap. 21).

Advantages of Genomic Evaluation
- The same tissue sample and derived genotype can be used to evaluate several traits or features of the animal, i.e. one sample is sufficient to evaluate several phenotypes because the SNPs captured in a dense chip include some that can explain variation between individuals in a range of traits.
- It can be used to evaluate animals much earlier in life, i.e. as soon as a tissue sample can be obtained and used to genotype the animal. This saves cost of rearing animals to later ages. This is particularly important in traits that can only be measure late in an animal's life such as reproductive traits observed only after sexual maturity. These include egg

Fig. 23.1 A representative SNP Chip

The above indicates that once an animal is sampled (i.e. a tissue sample is taken) and genotyped using a reasonably dense SNP chip, the genotype, described in terms of the total number of SNPs on the chip, can be used to evaluate several traits or features of the animals. These include sex-limited traits, such as egg production and milk yield that are recorded only in females and in males based on the performance of their sisters and daughters in future generations. With genomic evaluation, the genetic merit of males for such traits can be determined much earlier in life with significant savings in cost. By evaluating animals much earlier in life, genomics offers us the opportunity to reduce the generation interval resulting in a faster rate of gain. In broiler and beef breeding, meat yield and other processing traits are very important traits of economic importance and can be measured directly only on individuals at slaughter. Evaluation of such traits amongst live selection candidates relies on SIB information or the availability of predictors. Genomics offers a chance to evaluate the direct genetic potential of all selection candidates for such traits with a higher level of accuracy. Similarly, genomic evaluation can be applied to complex traits such as disease resistance, robustness, animal welfare, and behavioural traits which are difficult or impossible to phenotype.

For genomic evaluation, once phenotypic information on relatives is available to form a reference population, individuals to be evaluated or selected do not need to have the phenotype before a genomic breeding value can be predicted on them. All that is required is that selection candidates have a genotype derived from a tissue sample, which can be taken very early in life and from all individuals.

The accuracy of genetic evaluation, based on a combination of genetic markers and phenotypes, tends to be higher than the accuracy obtained from polygenic evaluation without the genetic markers' information. This may, however, depend on the genetic architecture of the trait and the model/prediction method (Daetwyler et al. 2010; Wimmer et al. 2013). In addition, genetic markers can be used in genome-wide association studies (GWAS). This has facilitated the mapping of

production, fertility and hatchability in poultry, carcass traits in beef and poultry.
- It allows all animals to be evaluated for traits that can only be measure on some animals, e.g. sex limited traits, such as egg production and milk yield.
- Genomic breeding values can distinguish between full sibs with no data where in traditional evaluation, they will both get the same breeding value based on the parent average (PA)
- It allows faster genetic progress in age limited traits due to the reduction of generation interval, and increase in accuracy of predicted BVs.
- It is an alternative selection tool that can be used, in combination with the traditional method, to increase the accuracy of selection.

genetic pathways underlying many biological processes thus increasing our understanding of the genetic control of traits and especially, complex diseases (Wang et al. 2010a; Tam et al. 2019). GWAS can also be used to detect genes with major effects.

23.2.3 Genetic Evaluation Software and Models

The Best Linear Unbiased Prediction (BLUP) methodology has been deployed with a wide range of models in custom software that are widely available and hence transferable for genetic evaluation. These use different mathematical algorithms to solve the linear equations that characterize the BLUP methodology (Mrode 2014). The models to fit depend on the traits, source and structure of available data. Examples of such custom software used in poultry breeding include VCE® for variance component estimation (Groeneveld et al. 2008), PEST® for breeding value prediction (Groeneveld 2006), DMU® (Madsen and Jensen 2013), ASReml® available from VSN® International (Gilmour et al. 2021) which can estimate variance components and also predict BVs, Mix99 (Tam et al. 2019) and MixBLUP (ten Napel et al. 2020). Others include WOMBAT® (http://didgeridoo.une.edu.au/km/wombat.php), which has replaced the old DF-REML (Meyer 2018). Other software such as R® (Team, R. D. C. R 2010), which is freely available, Python® (Van Rossum and Drake 2009), SAS® (Delwiche and Slaughter 2019), and Fortran® (Lahey and Ellis 1994) are used in programming. They are thus useful in developing genetic evaluation pipelines as well as data manipulation and set-up for the various evaluation models. All these software vary in the models they can fit and the volume of data they can handle at a given time. While some can be deployed for genomic evaluation, others cannot. There are yet other software used specifically for genomic evaluation especially those required to fit the available array of Bayesian models for genomic evaluation using various regression models.

The array of models used in farm animal evaluation has been described in detail by Mrode (2014) and Simm et al. (2020). The ability of the custom software to fit these models has evolved with time and in particular with improvements in computing capacity, information communication technology, and data recording. As computing capacity increases in terms of processor speed, ready access memory (RAM), and storage capacity, more complex models can be fitted. For example, more computing capacity is required to fit a multivariate animal model compared to a single trait model. As data and pedigree size increases, more memory is required and more processing speed will facilitate a quick turnaround time in solving the equations for the same number of traits. While multivariate evaluations with the animal model are generally suitable for evaluating most traits in poultry and pig production, the availability of longitudinal data on traits recorded repeatedly over time or space has led to the introduction of random regression models (RRM). Threshold models are particularly useful in evaluating binomial traits which has been included in the F90 Suite of programmes (Misztal 1999). The deployment of different software for routine genetic evaluation depends on human capacity and knowledge, the availability, and the cost of licensing. However, these technologies are highly portable.

Genomic Evaluation Models

Prediction of BLUP breeding values in a polygenic evaluation is based on data on the observed performance of the animals and their relatives. Relationships are based on a detailed pedigree which enables us to derive a numeral value of the predicted relationship between animals based on the assumption that a parent transmits half of his/her alleles to each of its progeny and that each progeny may not inherit exactly the same set of alleles from its parents. The deviation of the average allelic effects a progeny receives from both parents from the average allelic effects from the parents common to all progeny is referred to as the Mendelian sampling. This pedigree relationship matrix is generally referred to as the '**A**' matrix (Mrode 2014). When animals are genotyped, the true observed relationship between genotyped animals can be derived based on the common SNPs they carry. This relationship matrix, based on genetic markers, is referred to as the 'G' matrix. Increased reliabilities from genomic selection could in part, be attributed to better estimates of the Mendelian sampling (Hayes et al. 2009).

A common genomic evaluation model or method, which simply uses the G matrix to replace the A matrix in a linear model is referred to as GBLUP. Equations are set up for each animal resulting in the estimation of a direct genomic value (DGV) calculated for each animal which can be used to rank animals for selection. The drawback however is that only genotyped animals can be included in a GBLUP run and hence non-genotyped animals, which may include older relatives of the selection candidates cannot contribute information and cannot get a DGV. This limits the use of GBLUP in a situation where cost and other factors prevent the genotyping of all available animals. In an animal model context, the mixed linear model equation for GBLUP can be written as;

$$y = Xb + Wa + e \quad (23.1)$$

Where y is a vector of observed data, X and W are design matrices relating data to fixed effects and random animals respectively. e represents the random error term and a is a vector of DGV with mean zero and variance; $\text{Var}(a) = G\sigma_a^2$

The mixed model equations for obtaining the solution for Eq. 23.1 are presented in Mrode (2014). A method of genomic evaluation whereby the linear mixed model equations are set up for each SNP rather than each animal is referred to as 'SNP-BLUP'. In this case, solutions are computed for each SNP. To get the DGV for each animal, The SNP genotype at each marker locus, (coded as 2, 1, or 0 depending on the number of the dominant alleles in that locus) is multiplied by the SNP solution for that SNP and summed over all SNPs for the animals. Again, DGVs derived this way are only available for genotyped animals (Mrode 2014).

A two-step genomic evaluation approach can be used to derive Genomic Breeding values (GEBV) for genotyped animals by combining the DGV for each genotyped animal with the estimated (BLUP) breeding values (EBV) based on the data and numerator relations ship matrix or a traditional BLUP run which combines all animals with and without data. The GEBVs for the genotyped animals benefit from the data and pedigree relationship information from related non-genotyped animals as well as the genomic relationship (if derived from GBLUP) or both genomic relationship and SNP solutions if derived from SNP-BLUP.

Misztal et al. (2009) developed the single-step G-BLUP by combining the A matrix of both genotyped and non-genotyped animals and the G-matrix of the genotyped animals into a hybrid matrix, H. This approach allows for the estimation of GEBV for both genotyped and non-genotyped animals. For a detailed understanding of genomic breeding value predictions, see Simm et al. (2020).

23.2.4 Selection

For many generations, the superiority of the next generation was achieved by ensuring that only a few individuals who perform above the population average, are selected and allowed to mate on a pre-determined scheme, i.e. none randomly, to breed the next generation. Selection of the breeding population is based on the ranking of all individuals (males and females) on a genetic merit scale determined by genetic and genomic evaluation described above. Selective breeding is the vehicle through which genetic progress is achieved in traits of economic importance. The efficiency of this process depends on the accuracy of the breeding values in identifying the true genetic merit and hence the ranking of animals for selection. A key aspect of selective breeding is the ability to restrict breeding to only the selected individuals by preventing all non-selected individuals from contributing offspring to the next generation. This requires technologies supporting efficient culling, restriction of movement, and separate housing of different categories of animals. In this regard, building materials which allow fencing, and flexible partitioning of pens are essential. These include technologies for feeding and supplying water to birds in different sections of a building shed.

23.2.5 Data Recording

The traits of interest in farm animal breeding are mostly quantitative in nature. This means they vary minutely between individuals on a continuum from a physiological minimum to a physiological maximum. For example, in meat-type animals (broilers, beef, pigs, etc.), live weight at a specified age; for dairy, milk yield, milk components, etc. Other traits are binomial in nature, e.g. mortality. (An animal is either alive or dead at a certain age). There are also traits that are categorical in nature, i.e. can be split into a few classes that may be categorical (discrete) or continuous in nature, e.g. carcass grade, feather coverage, walking ability, and lesion score. The description of the observed state, feature, or performance of an individual relative to one or more traits is generally described as phenotyping. The measurement and recording of data relative to the observed, allows these phenotypes and endophenotypes to be captured in a form that can be stored, reviewed, and analysed to provide further information about an individual or group of animals.

In farm animal breeding, data recording is critical as it facilitates the derivation of genetic information useful for bird selection and breeding. Recorded data includes phenotype as well as the genetic composition of the animal (genotype) described in terms of a set of genetic markers. Some phenotypes can be recorded directly by measuring the observed with recording equipment, e.g. recording body weight on a scale. Other phenotypes are derived from one or more recorded traits, e.g. percentage carcass yield can be derived from the eviscerated carcass weight and the body weight. Also, egg weight loss during incubation is derived from the weights of the egg at a set and transfer to the hatcher after 18 days. Male and or female fertility is derived from the total number of eggs set and the total found to be fertile following candling after 7–8 days of incubation. Thus sometimes, useful traits including survivability can be derived from simple counts.

Obtaining a record of the genetic makeup of an individual, or its genotype, is a little more involved than a simple measurement as is possible for many phenotypes. The true genotype of an individual animal is based on its genetic constitution which involves potentially thousands of genes located in a varying number of chromosomes depending on the species. In chickens, the genetic makeup comprises genes located across 39 chromosome pairs. Cattle on the other hand has 30 pairs of chromosomes. The process of determining the genetic makeup of an individual is referred to as genotyping. It is currently impossible to identify the full set of genes a particular farm animal carries. However, the genotype can be expressed in terms of hundreds of thousands of known genetic markers that have been identified and distributed across all chromosomes. The most common of these markers now used to genotype animals are SNPs described in Chap. 21. Hence genotyping is presently done relative to a fixed number of SNP genetic markers previously identified and constituted or described in a SNP Chip. The available chips vary by species and in terms of density, i.e. the number of different markers known across all chromosomes and by the proprietor or group that developed the chipset. Examples include the Affimetrix® medium density 55K SNP Chip (Liu et al. 2019), the Affymetrix High density 600K SNP Chip (Kranis et al. 2013), and the Illumina medium density 60K SNP Chip (Groenen et al. 2011). Further information on DNA genotyping chips can be seen in other chapters. One key feature of genomic data is that it is quite large. Technically speaking, you have as many data points as the number of SNPs used for genotyping per animal hence storage of this data requires a large computing/database infrastructure. Also deriving the data per animal is expensive because it involves a lot of stages with an array of technologies from tissue sampling, and DNA extraction to genotyping.

23.2.5.1 Evolution of Recording Equipment and Breeding

Technological advancement in the development of recording equipment has made data capture in farm animal breeding easier, less tasking, and faster. These attributes are important because breeding involves the recording of data in a very large population over a period of time. For example, measuring the weight of an animal requires a scale. However, scales that can accurately capture the weight of a 7-day-old chick need to be more sensitive than one required to weigh a 7-day-old calf. The sensitivity of scales has improved with technology from balanced weight-based scales to digital scales of different ranges. This has made it possible for the low weight of an egg to be captured as accurately as the high weight of a mature beef cattle.

Another technological development relevant to data recording is the possibility of integrating measuring instruments with other instruments or equipment, which allows the capture of animal identity or the transfer of captured data to a storage system. Such integration saves time and stress by allowing the automatic capture of multiple pieces of information seamlessly. For example, the integration of a barcode scanner (Fig. 23.2) with a scale and a laptop allows bird identification and body weight to be recorded efficiently within a short time and the obtained data safely transferred to a database. Such integration eliminates the need for writing down or typing in records hence avoiding human error or mistakes. It facilitates the recording of phenotypes on hundreds of birds by one individual in the shortest possible time, thus maximizing the efficiency of available labour. This is critical in pig and poultry breeding which requires the recording of numerous individuals within a short time.

Developments in information communication technology (ICT) have enhanced the capture, but more importantly the seamless transfer of data between recording equipment or location and storage database. Compared to wired connectivity, the ability for equipment to communicate wirelessly with each other through a router (Fig. 23.3) via wireless radio signal (WiFi) or

Fig. 23.2 A Barcode Scanner

directly through Bluetooth connectivity has revolutionized data capture and transfer. It provides data recorders more freedom and mobility in data capture while still allowing the safe transfer of recorded data to a centralized storage device. This is particularly useful in extensive production systems or in obtaining data from highly dispersed small-scale producers. Wireless communication technology has allowed an increase in the variation of equipment that can be used to capture data. For example, the increase in processing and storage capability of mobile phones (Fig. 23.4) makes them now suitable for a whole range of data capture and together with their WiFi and Bluetooth capability, makes them particularly powerful tools for capturing data in remote locations.

23.2.5.2 Equipment Miniaturization

Over the years, advances in various technologies especially in metallurgy, mechanical and electrical engineering as well as computer technology have facilitated the miniaturization of animal recording equipment. The size of the recording

Fig. 23.3 A 4G mobile data sim wireless router useful for digital telephony and data transfer

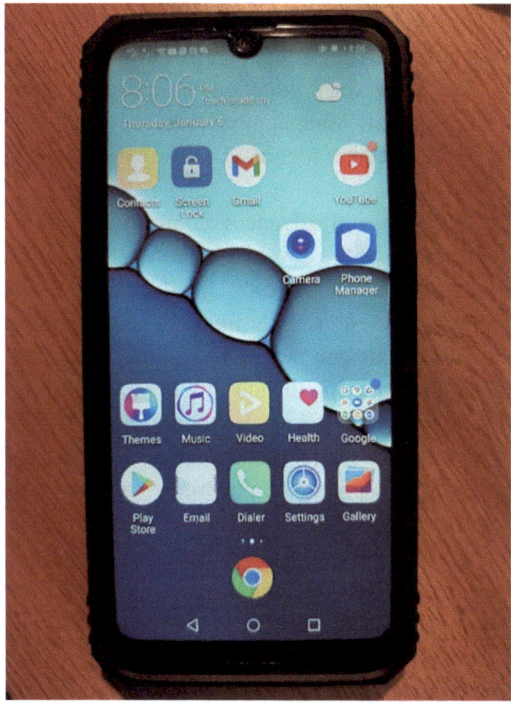

Fig. 23.4 A typical smartphone suitable for communication, data capture, and transfer

equipment affects where and how data is recorded. For example, measuring the body weight of a full-grown cow requires a big and heavy scale hence, this is often placed in one location on the farm and the animals are driven to the scale. Similarly, recording individual animals' feed intake or Methane emission may require the animals to be taken to a dedicated performance-testing pen. Miniaturization of the recording equipment allows the equipment to be taken to the field where the animals are, to record performance rather than moving the animals to the equipment. It facilitates the attachment of recording equipment to individual animals for automatic monitoring, data capture, and direct transfer of information to a central receiving database.

In poultry breeding, a large number of birds are recorded in the shed where they are housed or in the field in a free-range production system. Taking the recording equipment into the shed or field is thus the norm. So, the development of hand-held terminals (HHTs) which are miniature programmable computers with sufficient processing and storage capacity to hold recorded data temporarily makes data recording feasible. Such dedicated recording equipment is often made to be robust to the environment and sturdy enough to cope with knocks should they be dropped. Figure 23.5 shows some HHTs that can be programmed to record data. Figure 23.6 shows a mini printer that can be carried on the person and used to print information captured instantly while Fig. 23.7 shows a USB storage device with the capacity to hold one terabyte (TB) of data and allow such huge data to be transferred easily or even by post.

Fig. 23.5 Programmable hand-held terminals that can be used for data recording

Fig. 23.6 A miniature mobile printer that can be used to print instant labels during recording

Animal Identification

Individual animal identification is critical in the capture and storage of individual animal data. Hence innovations in animal identification systems make data recording on individual animals more feasible. Unique animal identification can be assigned to each animal on paper but for it to be effective, it needs to be associated with or attached to the live animal. For meat-type birds, this means from the day of hatch or birth until slaughter or death. So the form or means by which each bird is identified needs to meet a few conditions.

- It must be light enough to be carried around by the animal without impeding its natural behaviour and welfare.
- It must be rugged enough to survive the wear and tear of the production environment.
- The identification number must remain legible to the data recorder at all times.

Technological advances contributed significantly towards the development of tags that meet these conditions thereby facilitating animal identification and recording. In the simplest cases, animals can be marked simply by colour code (e.g. spraying the tail or wing feathers with different colour paint). Further miniaturization plus advances in electronic engineering have led to the development of electronic tags such as radio-frequency identification (RFID) tags. These are programmable tags that contain a radio transponder, receiver, and transmitter hence capable of transmitting data wirelessly to a designated RFID reader or antennae when triggered. Passive RFIDs do not need batteries but get their power from radio waves from the reader. Active RFIDs on the other hand have embedded batteries to power them. Electronic tags like these allow data to be captured even when the recorder cannot visualize the tag and unlike wing bands, do not need the physical immobilization of the animal either to read it manually or scan the barcode.

Fig. 23.7 A large-capacity portable data storage device (1 TB)

23.3 Improvement of Poultry Health and Welfare

Poultry health and welfare are the main pillars of poultry production where a lot of advancement has been made and they are essential to the sustainability of commercial poultry production (Ben Sassi et al. 2016). The World Organization for Animal Health states that 'An animal is in a good state of welfare if it is healthy, comfortable, well nourished, safe, able to express innate behavior and if it is not suffering from unpleasant states such as pain, fear and distress' (Terrestrial Animal Health Code). This applies also to all farm animals. Health is interconnected with welfare and it is critical to the strategy of sustainability because it directly affects human and animal welfare, economic viability, and environmental impact (Buller et al. 2018). A disease can be defined as any condition that interferes with the normal functioning of the cells, tissues, organs, and whole body systems of the animal. Prevention of disease is therefore one of the most effective ways to keep the birds healthy.

23.3.1 The Use of Technology in Various Aspects of Farm Animal Health and Welfare

There are many factors that affect animal health and welfare. These include stocking, density, carrying capacity, thermal stress, difficulties in accessing essential resources, environmental deterioration, and unsuitable social environments (Applegate et al. 2003; Muiruri and Harrison 1991; Tactacan et al. 2009). While many of these factors can be controlled by good management practices, recent advancements in technology have made the reduction in health and welfare problems in animal production achievable through the adoption of technologies in Precision Livestock Farming (PLF). This involves the use of process engineering that is based on automatic data acquisition, access, and processing (Mollo et al. 2009; Wathes et al. 2008). Data from diverse sources can be collected through smart sensors and compiled into a database where they will be later analysed to create an automatic management system based on real-time monitoring to control animal performance, health, and welfare (Aydin and Berckmans 2016).

23.3.1.1 Monitoring Sensors

Sensors are a good example of new technologies used in precision livestock farming. Great advances have been made in sensing technology in terms of diversity, accuracy, and affordability (Ben Sassi et al. 2016). Wireless sensors have a wide range of applications in engineering and agriculture. They were initially adopted to reduce livestock production costs while improving the health of the animals (Ruiz-Garcia et al. 2009). Environment monitoring sensors alert managers to deterioration in pen condition, such as temperature changes or air quality, allowing timely corrective measures to be taken before they cause harm to animals. Movement and health sensors indicate the condition of individual animals so that sick or incapacitated individuals can be promptly identify and isolated for treatment or special care. Such prompt action help improve overall flock performance, health, and welfare. Some sensors can be used to predict broiler body weight while others can help in monitoring hatching windows during egg incubation to maximize chick quality at the hatch. Sound sensors can be used to measure feed intake, predict growth, and estimate thermal comfort within farms in broiler production (Aydin et al. 2014; Bright 2008; Moura et al. 2008). Sound sensors have been used in laying pens to detect stress induced by environmental temperature variation and fear. They also help monitor conditions that encourage or induce feather pecking (Zimmerman et al. 2000; Pereira et al. 2014).

Locomotion sensors help assess movement deficiency in broilers while Geographic Information Systems (GIS) allow us to evaluate space use as well as different behaviours in laying hens (Nääs et al. 2010; Daigle et al. 2014). Locomotion sensors have also been used to study the jumping of hens between perches and its impact on the occurrence of bone breakage. They are useful in studying hens' use of pop holes and their effects on keel fracture incidence in hens tagged with radio frequency identification (RFID) transponders (Richards et al. 2012; Banerjee et al. 2014).

The environmental conditions, with specific reference to temperature, relative humidity, and length of exposure to these factors, influence the welfare, performance, and health of broiler chickens (Stamp Dawkins et al. 2004; Jones et al. 2005). For example, growth, feed conversion, and immunological response of birds exposed to elevated levels of carbon dioxide (CO_2) and Ammonia (NH_3) is impaired (Wang et al. 2010b; Ben Sassi et al. 2016). The adverse effect of these factors can be ameliorated with the use of sensors which can detect these factors early before their negative outcome manifests. Wireless technology armed with temperature sensors and accelerators has been used to detect highly pathogenic avian influenza (HPAI) infected chickens (Okada et al. 2009). A more sensitive technology based on signals from wireless 3-axis accelerometers and a radial lead thermistor has been used (Okada et al. 2014), to detect unusual responses caused by HPAI using data on the temperature and activities of birds transmitted to the wireless nodes. Sensors can therefore assist in detecting diseases prior to the manifestation of an outbreak which will subsequently prevent or limit economic losses. However, the cost of using sensors makes it difficult for general adoption (Ben Sassi et al. 2016), but recent developments in sensor techniques may help overcome this barrier.

23.3.1.2 In Ovo Technologies

Lack of access to feed for several hours (up to 72 hours) during transportation from the hatchery to the production farm can affect the health of poultry birds. This has adverse effects on the microbiome, immune system, nutrient intake, and utilization. To combat this, *in ovo* technology has been deployed which supplies the developing embryo with nutrients early to stimulate functional gut microflora. This provides necessary nutrients for the growth and development of the gastrointestinal tracts in poultry, thereby enhancing nutrient utilization and growth of the animal (Das et al. 2021). *In ovo* technology generally involves the administration of bioactive compounds, such as vaccines, nutrients, prebiotics, probiotics, creatine, follistatin, growth hormone, polyclonal antimyostatin antibody, and insulin-like growth factor-1, into the embryo. They can be applied during the late stages of incubation to enhance hatchability, performance, immune responses, gut morphology and microbiome, and overall health of birds (Das et al. 2021). The route, dosage, method, time of *in ovo* injection, and the host affect the outcome of this technology. Using this technology in vaccination is time and labour-efficient. It also increases the accuracy of vaccination and the development of immunity, without adversely affecting the hatchability of eggs in large-scale hatcheries.

23.3.1.3 Thermal Imaging

Preventing heat stress is crucial to farm animal welfare as it impacts behaviour, immunity, and physiological processes that can increase the mortality rate. Infrared thermal imaging (IRTI) technology has been used in this area for optimum performance. IRTI creates infrared images that reveal the distribution of the body's superficial temperature from the infrared radiation emitted by objects, which is then converted into electrical signals (Ben Sassi et al. 2016). In a study where optimum temperature and relative humidity were maintained, the optimal air velocity for the thermoregulation of broiler chickens was determined to be 2 m/s using the IRTI technique (Yahav et al. 2004). The body heat loss was calculated by radiation and convection using the IRTI methodology. It was reported (Giloh et al. 2012) that surface temperature detected by IRTI was more useful than ambient temperature in assessing the bird's status under experimental conditions. The IRTI technology has been utilized (Shinder et al. 2009) to influence the body temperature during the final phase of incubation and showed that chicks exposed to short periods of cold stress during incubation (days 18 and 19) had 13% to 18% lower incidence of ascites when compared to the control birds.

23.3.2 The Role of Technology in Handling Drug Resistance in Poultry

Poultry production has been used as a means to alleviate poverty and enhance economic development in Africa. With a rising population and increasing demand for animal protein, there is likely going to be a gradual shift from subsidence to intensive poultry production in Africa with a consequent increase in the usage of drugs and antibiotics (Hedman et al. 2020). Antimicrobial use increased between 2000 and 2015 by 65% (Klein et al. 2018). It is estimated to increase by another 67% by 2030 (Van Boeckel et al. 2019) due to an increase in intensive agriculture with low to medium-income countries leading the charge. This could increase antimicrobial resistance, especially in areas without effective biosafety and biosecurity measures (Hedman et al. 2020; Van Boeckel et al. 2019). It is therefore expedient to study the emergence of drug/antimicrobial resistance in poultry production (Ben Sassi et al. 2016; Van Boeckel et al. 2019; Food and Agriculture Organization (FAO) 2014) and evaluate the role of technology associated with it.

Drug/antimicrobial resistance is a growing global concern for human and animal health. It makes it difficult to treat bacterial infections while increasing morbidity and mortality risk caused by resistant bacteria (Hedman et al. 2020; Tang et al. 2017). The emergence of agricultural antimicrobial resistance in clinical settings has been reported by several authors (Van Boeckel et al. 2019; Allen 2014; Bengtsson-Palme et al. 2018; Robinson et al. 2016). More than 70% of antimicrobials produced globally are used in food-animal production (Van Boeckel et al. 2019; Food and Agriculture Organization (FAO) 2014) and poultry production represents a significant part of that. Although the use of antibiotics as a growth promoter has been banned by some organizations like the European Union, regulating this globally remains an issue (Allen 2014), because antibiotics are used to aid growth, prevent disease and treat infections (Laxminarayan et al. 2013). This makes the emergence of drug resistance a growing challenge. Several methods used in combating this include the use of probiotics, prebiotics, synbiotics, and phytobiotics. Advances in biotechnology have helped to improve the efficacy of these alternatives to antibiotic growth promoters.

23.3.2.1 Vaccines and Vaccination

The use of vaccines can offer protection against certain infections in farm animals. Because of this, the use of antibiotics in farm animals can be reduced without adversely affecting productivity. Vaccines cause a reduction in the symptoms that normally make producers use antibiotics. This results in an indirect reduction of antibiotic usage. When properly used, vaccines inhibit bacteria proliferation, keeping their population to the barest minimum and preventing them from causing havoc in farm animals. (Rappuoli et al. 2017).

Existing vaccines can prevent infections caused by antimicrobial pathogens (*Streptococcus pneumoniae* and *Haemophilus influenza*) (Rappuoli et al. 2017). Just as in humans, farm animal producers should be encouraged to adopt mass vaccination to ensure animal health. The risk of infection reduces as the use of vaccines increases (Woolhouse et al. 2015).

23.3.2.2 Bacteriophages

The use of bacteriophages to control pathogens is a promising aspect of handling drug resistance. Bacteriophages are bacteria-specific and possess the ability to infect only their target species or strain. Because of this specificity, bacteriophages do not destroy the intestinal microflora, hence they are safe and highly efficient. Bacteriophages cannot replicate in eukaryotic cells thereby resulting in a reduction of pathogenic bacteria (Wernicki et al. 2017). The population of pathogenic bacteria in the digestive tract of one-day-old broiler chickens was reduced when a high titre single dose of bacteriophage suspension was used (Fiorentin et al. 2005) while Salmonella concentration and mortality rate decreased (Lim et al. 2012) when 6-week-old broiler chickens received bacteriophages as a feed additive after having contact with salmonella. S. *enteritidis* strains were totally eliminated from the tonsils, 6 hours after application of bacteriophage suspension (Ahmadi et al. 2016). All these show that the use of drugs or antimicrobials can effectively be reduced with the use of phage therapy.

23.3.2.3 Biosecurity

Activities occurring on farms have a direct impact on the level of medications required. Observing adequate biosecurity measures on farms will significantly reduce the risk of infectious diseases. It is a way to minimize the usage of antimicrobials in animals and to prevent infections. Good husbandry and hygiene practices in conjunction with immunization can prevent the emergence and transmission of infectious diseases (Founou et al. 2021). Prevention of these diseases will have a direct impact on medication and the use of antimicrobials on farm animals.

23.3.2.4 Genomics

The use of genomics to solve the problem of drug resistance in poultry and other livestock species has been reported (Hu et al. 2020; Duggett et al. 2020; Delgado-Suárez et al. 2021). This includes the deployment of a genomic surveillance programme and One Health Initiative to tackle the problem of antimicrobial resistance (AMR), in humans, animals, and the environment (Duggett et al. 2020). The utility of whole genome data from both short and long-read sequences in bacteria collected from national surveillance activities has been demonstrated (Duggett et al. 2020). Such surveillance activities provide a wealth of data to assist in the monitoring of long-term national AMR trends more accurately including rapid retrospective analysis of a particular gene or AMR profile.

Whole genome sequencing (WGS) has been used to identify Multi-Drug Resistance (MDR) Non-Typhoidal Salmonella (NTS) from isolates in Mexico (Delgado-Suárez et al. 2021). They further reported that those from poultry and cattle had the highest population of MDR genotypes. The authors suggested that attaining significant improvement in AMR meat safety requires the identification and removal (or treatment) of products harbouring MDR NTS instead of screening for *Salmonella spp* or for isolates showing resistance to individual antibiotics. Further more, they (Delgado-Suárez et al. 2021) concluded that massive interrogation of WGS technologies in AMR surveillance provides the shortest path to accomplish these goals. Having genomic information provides enormous value for rapid retrospective molecular characterization to determine prevalence and transmission and facilitate comparison with new or current datasets in the event of an outbreak (Duggett et al. 2020).

23.3.3 Examples and Success Stories of Poultry Vaccination Programmes, Biosecurity, and Other Health Initiatives in Africa

A myriad of factors affect the health and welfare of animals in livestock production including intensive poultry production. While environmental factors and the effect of stocking density can be controlled through well-established management practices, temperature, and relative humidity control require improved technology for optimum performance. Technology has aided the evolution of intensive poultry production systems and the resultant improvement in the performance of poultry. It is equally vital in transforming subsistence and commercial poultry production systems, especially in developing countries (Terfa et al. 2018). However, disease outbreak is a big obstacle that could alter the gains of improved management. Vaccination programmes, tight biosecurity, and several health initiatives are deployed to prevent disease outbreaks. Prominent among the diseases is the Newcastle disease (ND). It is endemic in subsistence production systems relative to the intensive system (Terfa et al. 2018). While the use of vaccines has proven effective in intensive production systems, its application has not been equally successful at the subsistence level due to high costs and poor development of the required cold chain storage facility for vaccines (Terfa et al. 2018; Ideris et al. 1990).

Attempts to improve the control of Newcastle disease include the development of vaccines with thermo-tolerant strains of ND virus, artificial selection for enhanced thermo-tolerance, delivery of vaccines in food, and the use of additives to suppress the effect of heat on vaccines. Low-resource areas have recorded success in vaccination against Newcastle disease via the use of avirulent, thermo-tolerant Newcastle disease vaccines, administered to chickens of all ages via eye-drops every 4 months. If freeze-dried, some thermo-tolerant variants (NDV4-HR and I-2 vaccines) can remain potent between 9 and 29 °C for about 2 months, or for about 2 weeks between 30 and 37 °C (Alders et al. 2002). A low-dose, fast-dissolving tablet vaccine that could retain its activity for over 6 months at 4 °C was reported by Lal et al. (2014). This makes the utilization of this vaccine less effective in developing countries where storage might be a problem.

Africa has experienced some progress in the area of vaccine developments in poultry. This has aided the improvement of the performance and health of farm animals. Some examples are as follows;

The Newcastle Disease and Avian Flu Control Project implemented Newcastle disease virus vaccination and general poultry health and management programme in three districts of Tanzania. The scheme utilized the I_2 variant of the Newcastle disease virus using a multi-level community approach. Vaccinators, household leaders, and data recorders were trained and the vaccination regime was followed. The project was successful as the survey on the outbreak of Newcastle disease from household respondents decreased from 53.1% in 2006 (prior to vaccine training and administration) to 15.7% in 2007 (after training and the first round of vaccine administration). By December 2007, all farmers contacted had no reported cases of Newcastle disease outbreak in their flock (Msoffe et al. 2010).

1. In September 2011, Farm Input Promotions Africa (FIPS-Africa), a Nairobi-based non-profit organization, employed a model of ND vaccine delivery that relied on district coordinators and village-based advisors (VBAs) purchasing the 'La Sota' vaccine from local Agro-vets to vaccinate the birds in the villages.
2. Between 2003 and 2005, the Southern African ND Control Project (SANDCP) was active in four villages in Tanzania's *Chamwino* District, promoting ND vaccination in conjunction with the Tanzanian government.

Based on the positive outcome of intervention provided by these bodies, it was concluded (Lindahl et al. 2019) that interventions in which households are given assistance for vaccination of their birds, even for a short time, could drastically change their perspective, knowledge, and attitudes toward vaccines, resulting in a large decrease in bird mortality.

The absence of adequate ventilation (windows/openings) in the farm, the number of birds per production cycle, and the existence of duck farms in the village were used as measures for the biosecurity levels in an investigation of biosecurity and High Pathogenic Avian Influenza (HPAI) vaccination programme in Egypt (Artois et al. 2018). They concluded that the presence of the HPAI A (*H5N1*) virus on poultry farms depended on the type of production system, vaccination status, and the biosecurity level on such farms. The HPAI vaccination study revealed a decline in the probability of detecting the HPAI A (*H5N1*) virus in ducks with vaccination. The virus was only detected in about 22.08% of the duck farms that were vaccinated while 91.71% of unvaccinated farms reported an outbreak of the virus. The authors reported a few cases in which vaccines appeared not to prevent the virus from infecting the ducks. Statistical analysis found a significant effect of poor ventilation (window opening) in increasing the risk of HPAI A (*H5N1*) infection.

In a study conducted among local hatcheries in Ghana (Agbehadzi et al. 2019) following proper vaccination and biosecurity measures, maternal immunity against a virus was found to enhance the efficacy of vaccines against such virus in the offspring. Birds were vaccinated against Newcastle and Infectious Bursal Disease virus and the authors reported that vaccines, in addition to proper management practices, are efficient in preventing disease outbreaks and ensuring profitability in poultry farming.

In another study (Aliyu et al. 2016), *La Sota* and *Komarov* strains were given as secondary vaccinations at 3 and 6 weeks of age respectively to ISA brown pullets that had been given Newcastle disease vaccine at day-old. They observed a better immune response in the birds vaccinated with the *Hitchner B1* strain of Newcastle disease virus in Nigeria. All these indicate that vaccine development is advancing and Africa is gradually adopting the use of vaccines on a large scale.

Samples collected during an investigation into the high poultry mortalities recorded on the 10th of August 2021 in Mondoukou, Ivory Coast (Lo et al. 2022), were positive for *H5N1* HPAI as confirmed by the national Central Veterinary Laboratory of Bingerville and the reference laboratory in Padova, Italy. Additional samples taken from live poultry markets in various locations, and from migratory birds in water flyways in the initial outbreak areas following preliminary examination of the H5N1 virus's incomplete HA gene sequence, reveal that the virus was highly pathogenic and belonged to the clade 2.3.4.4.b. More than 300,000 fowl were slaughtered on impacted poultry farms by veterinary services as part of ongoing sanitary measures to restrict the virus's spread. Veterinary services further undertook cleaning and disinfection operations in live poultry markets in the Abidjan District. They also conducted countrywide sensitization of Poultry Sanitary Defence Groups and poultry breeders to strengthen biosecurity measures, disease surveillance, and real-time reporting in poultry farms and markets from November 29 to December 18, 2021 (Food and Agriculture Organization (FAO) 2021),with the support of FAO, USAID, and active participation of the Ivorian Poultry Inter-Professional Association (IPRAVI).

23.4 Improvement of Poultry Nutrition, Feed and Feeding

The anatomy and physiology of the poultry bird digestive system is such that formulation of feed for captive birds is complex requiring the supply and balancing of all essential nutrients. This is further complicated by the need for high genetic merit birds which are high in order for them to produce to their genetic potential. While free-roaming birds traverse far to feed on various feed sources to balance their needs, birds housed in poultry sheds require the nutrients to be supplied in their feed. Various technologies have been deployed to facilitate the understanding of nutritional requirements as well as the processing and compounding of feed ingredients, feed storage, and delivery to the birds.

23.4.1 Technology and Poultry Nutrition

Nutritional research revolves around detailed knowledge of ingredients to promote performance, and nutritional requirements of various poultry types at different stages of development. In general, poultry nutrition aims at feed utilization efficiency, protein efficiency, gut health management, feed enrichment, enzyme and mineral management, and mycotoxins under both conventional and antibiotic-free production systems. Nutritional science is an integration of biology, chemistry, physics, engineering, management, disease control, microbiology, toxicology, and in recent times, molecular biology. Determination of components of ingredients relies on analytical chemistry established in the Official Methods of Analysis published by the Association of Analytical Chemists (AOAC) (2019). Various types of ovens are employed to determine the moisture content of feed ingredients and feed. The traditional chemistry of the Kjeldahl analyser for determining protein content is expensive, laborious, and time-consuming. Technological development using the Dumas technology (LECO®) makes it easier to determine the amino acid and protein content of feed ingredients and feed. Traditional methods for determining ingredient composition have been replaced with near-infra-red (NIR) spectrophotometry, high-performance liquid chromatography (HPLC), plasma optical emission spectrometry, and ELISA methods. Innovations in chemical methods and instrumentations can directly and indirectly affect the diet. Accurate determination of feed utilization and performance are needed parameters for efficient poultry breeding. Therefore, to understand the nutritional needs of the various chicken types at different stages of development, multidimensional modelling and statistical tools are employed to model both biological and economic outcomes especially as nutrition can affect fat-free body composition (Eits et al. 2002). The use of computerized tomography (CT), dual-energy X-ray absorptiometry (DEXA), and nuclear magnetic resonance (NMR) technologies are employed to measure body composition. The use of these technologies and mathematical models can aid in precision nutrition.

23.4.2 Technology Used in Feed Manufacturing, Processing and Preservation

Feed manufacturing is an integral part of the farm value chain which traditionally has comprised grinding, batching, mixing, and pelleting. With innovations in information technology, emphasis is being placed on efficiency and precision in the manufacturing process. A major component of feed manufacturing is centred around ingredients. Ingredient composition has been the basis for diet formulation for years, however, advanced technologies for determining digestible components of nutrients have created a shift from the traditional ingredient compositionally-based formulations. By-product and rendered ingredients are usually not consistent in their components and quality thereby creating challenges in diet formulation. The NIR technology is already readily used in feed mills to analyse ingredient composition (moisture, protein, fat, fibre, starch, and amino acids) in real time. Additional biological information can be extracted by using in-vitro digestion modelling systems that can further improve the nutritional value of feed mill raw materials. Precision feed manufacturing is a quintessential part of feeding to meet the actual requirements of animals to optimize performance. Significant advancements in roller mills and hammer mills have improved the efficiency of grinding ingredients. Automation and adjustments in feed mill machinery have allowed for consistent particle size. Traceability of lot ingredients and usage of barcode readers for incorporating ingredients have benefited from programmable logic computers and variable frequency drive (VFD) controllers (Stark 2014). Enzymes, synthetic amino acids, pre-and probiotics, and other feed additives have significantly increased the number of ingredients used in feed manufacturing. The wide variability of inclusion rate among ingredients from macro ingredients like corn and soybean to micro-ingredients like enzymes requires more bins for micro-ingredients and VFD controllers to increase the accuracy of ingredient incorporation. Automatic systems allow feed mill operators to perform statistical process control (SPC) analysis to ensure consistency in feed quality from batch to batch. The data from SPC also assists in the decisions on equipment updates to improve precision and accuracy of operation.

The demand for high-quality animal feed requires the conversion of ground mash feed into dense pellets. Pelleting allows for the agglomeration of ingredients of different sizes, densities, and flow characteristics. The process involves steam injection and mechanical pressure. Pellet quality is expressed as the pellet durability index which measures the ratio of intact pellet before and after tumbling. Pellet quality is affected by the type of raw ingredients and additives, diet formulation, ingredient particle size and grinding, pressure steam and duration of the conditioning, roll pressing of the hot mash against metal die, and cooling. Changes in ingredients and formulation can complicate the pelleting process. Technological innovations in the conditioning, roll/die, and cooling process have contributed to the adjustments needed to accommodate changes in ingredients, particle size, and formulation. There are several studies that have demonstrated that feeding broiler chickens pelleted rather than mashed feed improved feed intake, weight gain, and feed efficiency (Jafarnejad et al. 2010; Chewning et al. 2012; Abdollahi et al. 2013; Massuquetto et al. 2019). Technological improvements in the pelleting process can directly affect the determination of feed efficiency values used in genetic improvement programmes.

Differences in feed quality between the time of manufacture and feeding can affect productivity. It is therefore imperative that the quality of feed is preserved in order to attain the expected performance of animals. The spoilage process can negatively affect the nutrient value of feed. Microbes can degrade feed and the degradation products are often unpalatable (Frey-Klett et al. 2011; Amit et al. 2017). Improper preservation of feed can affect feed intake and microbial by-products can be toxic to the animal. Mycotoxins (aflatoxins) produced by moulds can react with lipids and cause peroxidation. Prevention of nutrient degradation also means a more efficient use of resources. Feed contaminated with *Salmonella* or *E. coli* causes significant risk to both animals and humans (Crump et al. 2002; Heredia and García 2018). Feed preservation requires (1) the proper storage of feed ingredients to prevent spoilage, (2) protection of feed ingredients during the processing steps, (3) proper storage of feed to prevent spoilage, and (4) appropriate use of mould inhibitors and antioxidants to extend the shelf life of feed. Feed ingredients should be stored at the required temperature and moisture, under hygienic conditions, and applying preservatives at the required levels. Technological innovations in feed ingredient and feed storage equipment and computer-based sensor monitoring can maintain feed quality.

23.4.3 Interaction Between Feed and Bird Performance/Genetics Targeted Feeding

Even with all the advances in high throughput omics technologies, estimation of genetic parameters and genetic evaluation of breeding programmes cannot occur without phenomics. Precise and accurate measurement of phenotypes is essential for genetic improvement. Within a defined population, minimizing both environmental and genotype by nutrition interactions is essential in determining the genetic potential of an animal without these confounding effects. Computer-based controls can minimize fluctuations in the ambient environment under which poultry is raised, however, to minimize genotype by nutrition interactions within a population, precision feeding is required. Precision feeding aligns nutrient supply with the nutrient requirement of individual animals without unnecessary ingredient waste, based on a real-time sensor-based feedback system (Zuidhof 2020). Nutrient requirements are complex and depend on the genotype (genome) and the stage of development, assuming uniformity in the production environment (temperature, humidity, air quality, litter quality, disease control, etc.). Thus, in poultry, for a breeding population of the same age, nutrient requirements have to be tailored to the sex of the bird, growth, composition of gain, carcass composition, plumage coverage, and growth potential and efficiency of growth for meat-type birds (broilers), and the level and stage of egg production and egg quality for egg-type birds (layers). Emerging technologies (artificial intelligence, sensors, etc.) will have to bridge the gap between assessing the individual chicken's nutrient requirement and delivering the required diet in real time.

Precision feeding should include evaluating the nutri-physiological potential of feed ingredients, precise determination of nutrient requirement, and formulating balanced diets that limit the amount of excess nutrients with concomitant adjustment of nutrient supply to match the requirement of the animal fed (Pomar et al. 2011). Precision feed, therefore, requires extensive data collection, and large data analytics including artificial intelligence and mathematical modelling, which can be empirical, mechanistic, deterministic, or stochastic, and should be dynamic in real-time (Thornley and France 2007).

Nitrogen (protein) retention averaged 60.2% in broilers, 56.8% in turkeys, and 45.6% in laying hens (Applegate et al. 2003). Advances in nutrition by the usage of synthetic amino acids and dietary additives (enzymes, antioxidants, prebiotics, and probiotics) together with precision feeding can improve nutrient retention and reduce nitrogen and phosphorus in farm animal waste. This also requires concurrent technological innovations in the feed formulation system. Feed formulation has to be dynamic to respond to an individual animal's nutritional requirement in real time. The next-generation feed-formulation software should allow for the formulation of multi-blends, allocating ingredients according to cost and nutrient requirements and providing precision ingredient blends.

It has been shown that feeding pigs to meet their daily requirements significantly reduced their protein and phosphorus excretion by 25% and 29% respectively, and nutrient excretion was decreased by more than 38%. Furthermore, feed cost was reduced by about 11% (Pomar et al. 2011) and precision feeding on broiler breeders improved uniformity (Kutlu and Şahin 2017). This clearly demonstrates that precision feeding can lead to improved production efficiency and sustainability.

23.4.4 Advances in Farm Animal Nutrition Science and the Role of Technology

In order to support life in the form of maintenance and performance, farm animals utilize feed. The utilization of feed is a complex process including ingestion, mastication, digestion, absorption, assimilation, and biosynthesis. The dietary need of farm animals unlike other life forms is optimized according to the expected performance above maintenance requirements. Unlike classical nutrition which focuses on the diet or components of the diet and performance, molecular nutrition seeks to provide linkages between the diet through a stagnant genome and its dynamic genome expressions and performance. Thus, the conversion of the diet to animal protein is rather a complex process involving multi-organ interactions which are regulated by the genome and genome expressions. The study of the interaction between variations in the host genome (SNPs, gene variants, copy number variation, etc.) and the entire diet, components of the diet, bioactives, or nutriceuticals is often referred to as nutrigenetics. On the other hand, the effect of diet, components of the diet, bioactives, or nutriceuticals on genome expression (regulation of mRNA, protein, and blood or organ metabolites) is referred to as nutrigenomics. The use of omics to interrogate nutrition offers a huge potential to feed animals based on their individualized genomic signature to optimize sustainable (performance, health, and welfare) outcomes.

Nutrients regulate metabolic and other biological processes by altering genome expressions (Agyare et al. 2018). The technological platforms for determining genotype or sequencing the entire genome are similar regardless of the biological discipline or species. Next-generation sequencing of either short or long reads has become the standard technological platform for sequencing (Blankenburg et al. 2009; Judes et al. 2016; Karahalil 2016). Similarly, the platforms for metabolomics, targeted or global gene expression, and their associated databases and bioinformatics tools are the same for the different disciplines or species.

The use of the Internet of Things (IoT), which are physical objects that are embedded with sensors, processors, software, artificial intelligence, and data exchange in real-time over the internet, cloud, or other localized communication systems, together with the use of nutri-physical properties of a feed ingredient, smart feed formulation, and precision feeding will allow for accurate measure of phenotypes for genetic improvement.

23.5 The Technology Used to Support Human Capacity in Farm Management

In Chaps. 27 and 28, the human capacity needs for livestock improvement and production in Africa, the state of play and efforts, and human capacity development were discussed. This section focuses on the technologies that can be deployed to support and enhance the efforts, skills, and knowledge of those working in the poultry sector to improve production efficiency. Intensively managed farm animals require a lot of inputs to be efficiently utilized and managed for optimal production and profitability of the animal industry. Besides, human activities are required at different production and processing stages. These production and processing stages are labour-intensive and quite repetitive. For efficient management of staff and these processes, precision animal production becomes imperative. Technology is currently being employed to reduce the amount of labour in relation to output, improve the management of birds, and increase product uniformity and quality. Technology is also being used to capture, transmit, store, and analyse data as well as derive useful information for human and farm management. Most production chains are automated. For example, feeds and water are supplied to birds through automatic feeders and drinkers. As technology advances, sound technology, sensors, and cameras are employed to automatically collect and process data for efficient farm management and optimal production.

23.5.1 The Use of Information and Communication Technology in Poultry Production

The deployment of new technologies in poultry housing and equipment is critical in curbing increasing utility bills and labour costs. It also helps with animal welfare as well as facilitates automation of operations which helps to mitigate the effects of climate change. Information and communication, ventilation, heating, and cooling systems are very important factors for consideration in a modern poultry house. The extent to which a poultry house is information-based will determine the efficiency of production and management (Zheng et al. 2021). The information and communication system should connect all sections of the farm to allow easy management of staff, equipment, environment, inputs, and animals. For example, ventilation and air inlet systems including roofing technologies had the greatest influence on the mechanization coefficient (Samadpour et al. 2018) and thereby improved production efficiency and economic performance of poultry farms. An efficient ventilation system alters air composition by letting in fresh air and getting rid of carbon dioxide, ammonia, particulate matter, and moisture. These would aid the survival and productivity, especially in pig and poultry houses as well as keeping airborne diseases to a level that can be tolerated by the animals (Samadpour et al. 2018). The ventilation system is critical in densely populated and in intensively managed poultry and pig units. The prevailing environmental conditions in the geographical location of farms and the intensity of production will determine the type of ventilation systems to install. In modern commercial pig and poultry houses, a mechanical ventilation system is used because of its efficiency.

Given the increasing cost of energy and fuel globally, an efficient and cost-effective heating system needs to be in place in pig and poultry houses. Transmission and ventilation are two means of heat loss in pig and poultry houses that should be balanced to ensure that there is moderate heat for cost-effectiveness and the animal's optimal performance (Hulzebosch 2006). In a review of renewable and sustainable heating systems for poultry production towards energy saving and reduction in carbon dioxide emission (Cui et al. 2020), It was concluded that about 85% energy saving is achievable by applying technologically advanced ventilation and heating systems. As already indicated, ventilation helps with the circulation of fresh air, it is also a means of heat loss. Different technologies are being used to ensure a temperature that is tolerable by birds (Cui et al. 2020).

Automatic feeders and drinkers are used to supply feeds and water to birds in an efficient manner. This automation system reduces labour intensity, minimizes feeds and water wastage, reduces litter wetness, and thereby makes the litter less comfortable for microorganisms to thrive and cause diseases in poultry houses, especially from the floor of the pens.

23.5.2 Technology Used in Data Recording, Storage, Management, Analysis and Derivation of Information

Due to the large amount of data needed to be collected for efficient farm management; sound, sensors, and camera technologies are now playing critical roles in recording, storage, management, and analysis to derive useful information. A sound-based monitoring system can allow accurate detection of short-term feeding behaviours (Aydin and Berckmans 2016) at the group level and feed intake in broilers (Aydin et al. 2014, 2015) Individual meal size, meal duration, number of meals per day and feeding rate in broiler chickens can thus be determined in real-time with such sound processing technology. There has also been an attempt to use sound analysis to model the weight of broiler chickens (Fontana et al. 2017).

Sensor technology is important in data collection. These are devices that are positioned to collect data of different sorts that define animal or house stage (Jackman et al. 2020). There exist chemical and physical-based sensors that could be used in data collection from poultry houses and chickens for transmission and analysis. Relatively, camera-based weighing of broilers is potentially able to be used to determine the weight of broilers where contact is to be avoided (such as overweight and sick birds) and traditional automatic weighing machines are not suitable (Mortensen et al. 2016).

Computer vision systems are being deployed in poultry production for data capturing, transmission, and storage. Data can be collected from poultry houses on environmental variables as well as on birds (reproduction, production, health, and animal behaviour parameters). Components of computer vision systems include camera sensor, lighting, image processing board, software, and hardware (Nyalala et al. 2021)

In the fourth industrial revolution, there was huge interest in the application of robotics with the example of Gohbot®, ChickenBoy®, and ALIS chirpy sensor (Poultry Farming 2023). These technologies use machine learning, artificial intelligence, and sensors. With these, huge amounts of data can be generated in a very efficient and automated manner using powerful algorithms. Therefore, artificial intelligence is now available to the poultry industry to process and analyse large datasets being generated with these technologies.

The introduction of these technologies particularly in data analysis and analytics is not without challenges, particularly human capacity development. Hence there is a need for training and retraining for personnel in the poultry industry to be able to acquire and apply these technologies, especially in the area of data analysis involving programming and cloud computing since the majority of these large datasets are stored in the cloud. This can be in the form of workshops. Besides, curriculum review of Animal Science departments at Universities will be critical to training students with current skills that will be of immediate relevance to the poultry industry.

23.6 Conclusion and Future Perspectives

23.6.1 Conclusion

Africa as a continent is well endowed with a big variation in poultry genetic resources. This includes a wide variety of Indigenous, localized, and exotic poultry species like Chickens, Ducks, Geese, Turkeys, Guinea fowls, Ostriches Emus, and breeds/strains within species. There is thus a high prospect for developing breeds and strains of these species adapted to the widely different ecological and geographical regions of Africa given the will and the right technology. Africa also has a very large pool of human capacity, at home and in the diaspora, with the knowledge and skill to utilize the most appropriate technologies applied to poultry improvements and livestock development in general. Where such knowledge and skill are unavailable, it can be acquired because technology can be bought and skill is transferable given the means.

The prospect for adopting and adapting technology and technical innovation to develop poultry production in Africa therefore depends on having access to the technologies that have been used worldwide, to successfully develop the poultry industry, including the genetic improvement of productivity, housing, health, and welfare of its poultry population. There is also a requirement for development and or access to the human resources needed to deploy these technologies. Success will depend on the creation of the appropriate enabling environment (Olori 2022) to facilitate the application of technological innovations in poultry improvement and production. Deployment of these technologies will aid sustainability, encourage youth involvement in poultry production and minimize the environmental footprint of poultry production. This is critical if poultry production in Africa is to meet its share of the projected increase in poultry meat demand (Food and Agriculture Organization (FAO) 2021) in Africa.

23.6.2 Future Perspectives

Technologies change rapidly. It is envisaged that new methods and tools will become available for livestock production in the future. These may or may not totally replace what is available today. So skills in adapting to changes in technology will be essential. Sometimes new technologies still depend on old and basic things. For example, the application of genomic selection still requires traditional data collection and it can be enhanced with full pedigree knowledge. So whatever technologies are adopted today will not necessarily become obsolete in the future. New discoveries may lead to new ways of feeding animals just as the introduction of nipple drinkers revolutionized the way birds are given water to drink.

With climate change and the stress on agricultural resources, it is envisaged that new housing technologies may become available to allow more intensive control of the livestock housing environment. For example, the menace of Avian Influenza and other diseases has led to the development and deployment of air filtration systems in poultry houses to improve biosecurity. Rising temperatures and increasing use of solar power technologies may see the deployment of cooling systems in managing poultry stock in African countries where open-sided houses currently suffice.

Automation in many operations can be envisaged as labour shortages increase in the future. In big operations, the increase in the use of robotics cannot be ruled out. These have been deployed in milking cows so it is possible that as costs go down in the future, similar robotics could be deployed in egg collection grading and packaging, feeding, water dispensing, and waste management in livestock farms. The use of modern technologies may require new or different skill sets, so further training and re-training of farm managers and staff can be envisaged.

This chapter has described technologies that are currently available and being deployed for efficient poultry production in many Western countries. These technologies if adopted will enhance the efficiency of poultry production in Africa. Globally, genetic improvement of poultry birds has been mastered by commercial breeding companies and they have a good distribution network that makes highly improved poultry day-old chicks available in every locality. Deploying technologies described above for poultry feed and nutrition, health and biosecurity as well as housing and management will lead to massive improvement in poultry production capacity in Africa.

References

Abdollahi M, Ravindran V, Svihus B (2013) Pelleting of broiler diets: an overview with emphasis on pellet quality and nutritional value. Anim Feed Sci Technol 179(1–4):1–23

Agbehadzi R, Hamidu J, Adomako K, Enu R (2019) Economic contribution of local hatchery performance in the poultry value chain in Ghana. Poult Sci 98(6):2399–2404

Agyare C, Boamah VE, Zumbi CN, Osei FB (2018) Antibiotic use in poultry production and its effects on bacterial resistance. In: Antimicrobial resistance—a global threat, pp 33–51

Ahmadi M, Karimi Torshizi MA, Rahimi S, Dennehy JJ (2016) Prophylactic bacteriophage administration more effective than post-infection administration in reducing Salmonella enterica serovar Enteritidis shedding in quail. Front Microbiol 7:1253

Alders R, Anjos F, Bagnol B, Fumo A, Mata B, Young MB (2002) Controlling Newcastle disease in village chickens-a training manual

Aliyu H, Sa'idu L, Abdu P, Oladele S (2016) Humoral immune response of chickens following vaccination with different Newcastle disease vaccines. Niger Vet J 37(2):102–108

Allen HK (2014) Antibiotic resistance gene discovery in food-producing animals. Curr Opin Microbiol 19:25–29

Amit SK, Uddin MM, Rahman R, Islam SR, Khan MS (2017) A review on mechanisms and commercial aspects of food preservation and processing. Agric Food Secur 6:1–22

Applegate T, Potturi L, Angel R (2003) Model for estimating poultry manure nutrient excretion: a mass balance approach. In: Animal, agricultural and food processing wastes-IX. American Society of Agricultural and Biological Engineers, p 1

Artois J, Ippoliti C, Conte A, Dhingra MS, Alfonso P, Tahawy AE, Elbestawy A, Ellakany HF, Gilbert M (2018) Avian influenza A (H5N1) outbreaks in different poultry farm types in Egypt: the effect of vaccination, closing status and farm size. BMC Vet Res 14(1):1–10

Association of Analytical Chemists (AOAC), L., D. G. Ed (2019) Official methods of analysis

Aydin A, Berckmans D (2016) Using sound technology to automatically detect the short-term feeding behaviours of broiler chickens. Comput Electron Agric 121:25–31

Aydin A, Bahr C, Viazzi S, Exadaktylos V, Buyse J, Berckmans D (2014) A novel method to automatically measure the feed intake of broiler chickens by sound technology. Comput Electron Agric 101:17–23

Aydin A, Bahr C, Berckmans D (2015) A real-time monitoring tool to automatically measure the feed intakes of multiple broiler chickens by sound analysis. Comput Electron Agric 114:1–6

Banerjee D, Daigle CL, Dong B, Wurtz K, Newberry RC, Siegford JM, Biswas S (2014) Detection of jumping and landing force in laying hens using wireless wearable sensors. Poult Sci 93(11):2724–2733

Ben Sassi N, Averós X, Estevez I (2016) Technology and poultry welfare. Animals 6(10):62

Bengtsson-Palme J, Kristiansson E, Larsson DJ (2018) Environmental factors influencing the development and spread of antibiotic resistance. FEMS Microbiol Rev 42(1):fux053

Blankenburg M, Haberland L, Elvers HD, Tannert C, Jandrig B (2009) High-throughput omics technologies: potential tools for the investigation of influences of EMF on biological systems. Curr Genom 10(2):86–92

Bright A (2008) Vocalisations and acoustic parameters of flock noise from feather pecking and non-feather pecking laying flocks. Br Poult Sci 49(3):241–249

Buller H, Blokhuis H, Jensen P, Keeling L (2018) Towards farm animal welfare and sustainability. Animals 8(6):81

Chewning C, Stark C, Brake J (2012) Effects of particle size and feed form on broiler performance. J Appl Poult Res 21(4):830–837

Crump JA, Griffin PM, Angulo FJ (2002) Bacterial contamination of animal feed and its relationship to human foodborne illness. Clin Infect Dis 35(7):859–865

Cui Y, Theo E, Gurler T, Su Y, Saffa R (2020) A comprehensive review on renewable and sustainable heating systems for poultry farming. Int J Low-Carbon Technol 15(1):121–142

Daetwyler HD, Pong-Wong R, Villanueva B, Woolliams JA (2010) The impact of genetic architecture on genome-wide evaluation methods. Genetics 185(3):1021–1031

Daigle CL, Banerjee D, Montgomery RA, Biswas S, Siegford JM (2014) Moving GIS research indoors: spatiotemporal analysis of agricultural animals. PLoS One 9(8):e104002

Das R, Mishra P, Jha R (2021) In ovo feeding as a tool for improving performance and gut health of poultry: a review. Front Veterin Sci 8:754246

Delgado-Suárez EJ, Palós-Guitérrez T, Ruíz-López FA, Hernández Pérez CF, Ballesteros-Nova NE, Soberanis-Ramos O, Méndez-Medina RD, Allard

MW, Rubio-Lozano MS (2021) Genomic surveillance of antimicrobial resistance shows cattle and poultry are a moderate source of multi-drug resistant non-typhoidal Salmonella in Mexico. PLoS One 16(5):e0243681

Delwiche LD, Slaughter SJ (2019) The little SAS book: a primer. SAS Institute

Duggett N, AbuOun M, Randall L, Horton R, Lemma F, Rogers J, Crook D, Teale C, Anjum MF (2020) The importance of using whole genome sequencing and extended spectrum beta-lactamase selective media when monitoring antimicrobial resistance. Sci Rep 10(1):19880

Eits RM, Kwakkel RP, Verstegen MW (2002) Nutrition affects fat-free body composition in broiler chickens. J Nutr 132(8):2222–2228

Fiorentin L, Vieira ND, Barioni W Jr (2005) Oral treatment with bacteriophages reduces the concentration of Salmonella Enteritidis PT4 in caecal contents of broilers. Avian Pathol 34(3):258–263

Fontana I, Tullo E, Carpentier L, Berckmans D, Butterworth A, Vranken E, Norton T, Berckmans D, Guarino M (2017) Sound analysis to model weight of broiler chickens. Poult Sci 96(11):3938–3943

Food and Agriculture Organization (FAO) (2014) Animal production and health guidelines. FAO, Italy

Food and Agriculture Organization (FAO) (2021) Emergency prevention systems for animal health: Sub-Saharan Africa HPAI situation update. https://www.fao.org/ag/againfo/programmes/en/empres/HPAI_Africa/situation_update.html

Founou LL, Founou RC, Essack SY (2021) Antimicrobial resistance in the farm-to-plate continuum: more than a food safety issue. Future Sci OA 7(5):FSO692

Frey-Klett P, Burlinson P, Deveau A, Barret M, Tarkka M, Sarniguet A (2011) Bacterial-fungal interactions: hyphens between agricultural, clinical, environmental, and food microbiologists. Microbiol Mol Biol Rev 75(4):583–609

Gilmour A, Gogel B, Cullis B, Welham S, Thompson R, Butler D, Cherry M, Collins D, Dutkowski G, Harding SA-S (2021) User guide release 4.2 functional specification. VSN International Ltd Amberside House, Wood Lane, Paradise Industrial Estate, Hemel Hempstead, UK

Giloh M, Shinder D, Yahav S (2012) Skin surface temperature of broiler chickens is correlated to body core temperature and is indicative of their thermoregulatory status. Poult Sci 91(1):175–188

Groenen MA, Megens H-J, Zare Y, Warren WC, Hillier LW, Crooijmans RP, Vereijken A, Okimoto R, Muir WM, Cheng HH (2011) The development and characterization of a 60K SNP chip for chicken. BMC Genomics 12:1–9

Groeneveld E (2006) PEST user's manual. Institute of Animal Science, Neustadt, Germany

Groeneveld E, Kovac M, Mielenz N (2008) VCE user's guide and reference manual: institute of farm animal genetics. Friedrich Loeffler Institute (FLI) Germany, *Mariensee*

Grosu H, Lungu S, Oltenacu PA (2013) History of genetic evaluation methods in dairy cattle i. daughter-dam comparisons. Scientific Papers, p 25

Hayes BJ, Bowman PJ, Chamberlain AJ, Goddard ME (2009) Invited review: genomic selection in dairy cattle: Progress and challenges. J Dairy Sci 92(2):433–443

Hedman HD, Vasco KA, Zhang L (2020) A review of antimicrobial resistance in poultry farming within low-resource settings. Animals 10(8):1264

Heredia N, García S (2018) Animals as sources of food-borne pathogens: a review. Animal Nutr 4(3):250–255

Hu L, Cao G, Brown EW, Allard MW, Ma LM, Khan AA, Zhang G (2020) Antimicrobial resistance and related gene analysis of Salmonella from egg and chicken sources by whole-genome sequencing. Poult Sci 99(12):7076–7083

Hulzebosch J (2006) Effective heating systems for poultry houses. World Poultry 22(2):19

Ideris A, Ibrahim AL, Spradbrow P (1990) Vaccination of chickens against Newcastle disease with a food pellet vaccine. Avian Pathol 19(2):371–384

Jackman P, Penya H, Ross R (2020) The role of information and communication technology in poultry broiler production process control. Agric Eng Int CIGR J 22(3):284–299

Jafarnejad S, Farkhoy M, Sadegh M, Bahonar A (2010) Effect of crumble-pellet and mash diets with different levels of dietary protein and energy on the performance of broilers at the end of the third week. Veterin Med Int 2010:328123

Jones T, Donnelly C, Dawkins MS (2005) Environmental and management factors affecting the welfare of chickens on commercial farms in the United Kingdom and Denmark stocked at five densities. Poult Sci 84(8):1155–1165

Judes G, Rifaï K, Daures M, Dubois L, Bignon YJ, Penault-Llorca F, Bernard-Gallon D (2016) High-throughput «Omics» technologies: new tools for the study of triple-negative breast cancer. Cancer Lett 382(1):77–85

Karahalil B (2016) Overview of systems biology and omics technologies. Curr Med Chem 23(37):4221–4230

Klein EY, Van Boeckel TP, Martinez EM, Pant S, Gandra S, Levin SA, Goossens H, Laxminarayan R (2018) Global increase and geographic convergence in antibiotic consumption between 2000 and 2015. Proc Natl Acad Sci 115(15):E3463–E3470

Kranis A, Gheyas AA, Boschiero C, Turner F, Yu L, Smith S, Talbot R, Pirani A, Brew F, Kaiser P (2013) Development of a high density 600K SNP genotyping array for chicken. BMC Genomics 14(1):1–13

Kutlu HR, Şahin A (2017) Kanatlı beslemede güncel çalışmalar ve gelecek için öneriler. Hayvansal Üretim 58(2):66–79

Lahey TM, Ellis T (1994) Fortran 90 programming. Addison-Wesley Longman Publishing Co., Inc

Lal M, Zhu C, McClurkan C, Koelle D, Miller P, Afonso C, Donadeu M, Dungu B, Chen D (2014) Development of a low-dose fast-dissolving tablet formulation of

Newcastle disease vaccine for low-cost backyard poultry immunisation. Vet Rec 174(20):504–504

Laxminarayan R, Duse A, Wattal C, Zaidi AK, Wertheim HF, Sumpradit N, Vlieghe E, Hara GL, Gould IM, Goossens H (2013) Antibiotic resistance—the need for global solutions. Lancet Infect Dis 13(12):1057–1098

Lim T-H, Kim M-S, Lee D-H, Lee Y-N, Park J-K, Youn H-N, Lee H-J, Yang S-Y, Cho Y-W, Lee J-B (2012) Use of bacteriophage for biological control of Salmonella Enteritidis infection in chicken. Res Vet Sci 93(3):1173–1178

Lindahl JF, Young J, Wyatt A, Young M, Alders R, Bagnol B, Kibaya A, Grace D (2019) Do vaccination interventions have effects? A study on how poultry vaccination interventions change smallholder farmer knowledge, attitudes, and practice in villages in Kenya and Tanzania. Trop Anim Health Prod 51:213–220

Liu R, Xing S, Wang J, Zheng M, Cui H, Crooijmans RP, Li Q, Zhao G, Wen J (2019) A new chicken 55K SNP genotyping array. BMC Genomics 20(1):1–12

Lo FT, Zecchin B, Diallo AA, Racky O, Tassoni L, Diop A, Diouf M, Diouf M, Samb YN, Pastori A (2022) Intercontinental spread of Eurasian highly pathogenic avian influenza A (H5N1) to Senegal. Emerg Infect Dis 28(1):234

Madsen P, Jensen J (2013) A user's guide to DMU. A package for analysing multivariate mixed models. Version 6, release 5.2. Center for Quantitative Genetics and Genomics, University of Aarhus, Tjele, Denmark

Massuquetto A, Panisson JC, Marx FO, Surek D, Krabbe EL, Maiorka A (2019) Effect of pelleting and different feeding programs on growth performance, carcass yield, and nutrient digestibility in broiler chickens. Poult Sci 98(11):5497–5503

Meyer K (2018) Wrestling with a WOMBAT: selected new features for linear mixed model analyses in the genomic age. In: Proceedings of the 11th world congress on genetics applied to livestock production, pp 11–16

Misztal I (1999) Complex models, more data: simpler programming? Interbull Bullet 20:33–33

Misztal I, Aguilar I, Johnson D, Legarra A, Tsuruta S, Lawlor T (2009) A unified approach to utilize phenotypic, full pedigree and genomic information for a genetic evaluation of Holstein final score. Interbull Bullet 40:240–240

Mollo M, Vendrametto O, Okano M (2009) Precision livestock tools to improve products and processes in broiler production: a review. Braz J Poult Sci 11:211–218

Mortensen AK, Lisouski P, Ahrendt P (2016) Weight prediction of broiler chickens using 3D computer vision. Comput Electron Agric 123:319–326

Moura DJD, Nääs IDA, Alves ECDS, Carvalho TMRD, Vale MMD, Lima KAOD (2008) Noise analysis to evaluate chick thermal comfort. Sci Agric 65:438–443

Mrode RA (2014) Linear models for the prediction of animal breeding values. CabinetMaker

Msoffe PL, Bunn D, Muhairwa A, Mtambo M, Mwamhehe H, Msago A, Mlozi M, Cardona CJ (2010) Implementing poultry vaccination and biosecurity at the village level in Tanzania: a social strategy to promote health in free-range poultry populations. Trop Anim Health Prod 42:253–263

Muiruri HK, Harrison P (1991) Effect of peripheral foot cooling on metabolic rate and thermoregulation of fed and fasted chicken hens in a hot environment. Poult Sci 70(1):74–79

Nääs IDA, Paz ICDLA, Baracho MDS, Menezes AGD, Lima KAOD, Bueno LGDF, Mollo Neto M, Carvalho VCD, Almeida ICDL, Souza ALD (2010) Assessing locomotion deficiency in broiler chicken. Sci Agric 67:129–135

Nyalala I, Okinda C, Kunjie C, Korohou T, Nyalala L, Chao Q (2021) Weight and volume estimation of poultry and products based on computer vision systems: a review. Poult Sci 100(5):101072

Okada H, Itoh T, Suzuki K, Tsukamoto K (2009) Wireless sensor system for detection of avian influenza outbreak farms at an early stage. In: Sensors, 2009 IEEE. IEEE, pp 1374–1377

Okada H, Suzuki K, Kenji T, Itoh T (2014) Applicability of wireless activity sensor network to avian influenza monitoring system in poultry farms. J Sensor Technol 4:18

Olori VE (2022) Enabling environment for uptake of biotechnological innovations for livestock production in Africa. In: Agricultural biotechnology, biodiversity and bioresources conservation and utilization. CRC Press, pp 299–313

Pereira EM, Nääs IDA, Garcia RG (2014) Identification of acoustic parameters for broiler welfare estimate. Engenharia Agrícola 34:413–421

Pomar C, Hauschild L, Zhang G, Pomar J, Lovatto P (2011) Precision feeding can significantly reduce feeding cost and nutrient excretion in growing animals. In: Modelling nutrient digestion and utilisation in farm animals, pp 327–334

Poultry Farming: The robots are coming. https://www.wattagnet.com/articles/35589-poultry-farming-the-robots-are-coming Accessed Accessed May 2023

Rappuoli R, Bloom DE, Black S (2017) Deploy vaccines to fight superbugs. Nature 552(7684):165–167

Richards G, Wilkins L, Knowles T, Booth F, Toscano M, Nicol C, Brown S (2012) Pop hole use by hens with different keel fracture status monitored throughout the laying period. Vet Rec 170(19):494–494

Robinson TP, Bu D, Carrique-Mas J, Fèvre EM, Gilbert M, Grace D, Hay SI, Jiwakanon J, Kakkar M, Kariuki S (2016) Antibiotic resistance is the quintessential One Health issue. Trans R Soc Tro Med Hygiene 110(7):377–380

Ruiz-Garcia L, Lunadei L, Barreiro P, Robla JI (2009) A review of wireless sensor technologies and applications in agriculture and food industry: state of the art and current trends. Sensors 9(6):4728–4750

Samadpour E, Zahmatkesh D, Nemati M, Shahir M (2018) Determining the contribution of ventilation and insulation of broiler breeding houses in production performance using analytic hierarchy process (AHP). Braz J Poultry Sci 20:211–218

Shinder D, Rusal M, Giloh M, Yahav S (2009) Effect of repetitive acute cold exposures during the last phase of broiler embryogenesis on cold resistance through the life span. Poult Sci 88(3):636–646

Simm G, Pollott G, Mrode R, Houston R, Marshall K (2020) Genetic improvement of farmed animals. CABI

Stamp Dawkins M, Donnelly CA, Jones TA (2004) Chicken welfare is influenced more by housing conditions than by stocking density. Nature 427(6972):342-344

Stark C (2014) Statistical process control in feed mill. Feedstuffs 86(35):16–17

Tactacan GB, Guenter W, Lewis N, Rodriguez-Lecompte J, House J (2009) Performance and welfare of laying hens in conventional and enriched cages. Poult Sci 88(4):698–707

Tam V, Patel N, Turcotte M, Bossé Y, Paré G, Meyre D (2019) Benefits and limitations of genome-wide association studies. Nat Rev Genet 20(8):467–484

Tang KL, Caffrey NP, Nóbrega DB, Cork SC, Ronksley PE, Barkema HW, Polachek AJ, Ganshorn H, Sharma N, Kellner JD (2017) Restricting the use of antibiotics in food-producing animals and its associations with antibiotic resistance in food-producing animals and human beings: a systematic review and meta-analysis. Lancet Planet Health 1(8):e316–e327

Team, R. D. C. R (2010) A language and environment for statistical computing. No Title

ten Napel JVJ, Lidauer M, Stranden I, Taskinen M, Mantysaari E, Calus MPL, Veerkamp RF (2020) Mixblup 2.2.0 Manual

Terfa Z, Garikipati S, Kassie G, Bettridge J, Christley R (2018) Eliciting preferences for attributes of Newcastle disease vaccination programmes for village poultry in Ethiopia. Prev Vet Med 158:146–151

Terrestrial Animal Health Code. Accessed

Thornley JH, France J (2007) Mathematical models in agriculture: quantitative methods for the plant, animal and ecological sciences. Cabi

Van Boeckel TP, Pires J, Silvester R, Zhao C, Song J, Criscuolo NG, Gilbert M, Bonhoeffer S, Laxminarayan R (2019) Global trends in antimicrobial resistance in animals in low-and middle-income countries. Science 365(6459):eaaw1944

Van Rossum G, Drake F (2009) Python 3 reference manual. CreateSpace, Scotts Valley. Google Scholar

Wang K, Li M, Hakonarson H (2010a) Analysing biological pathways in genome-wide association studies. Nat Rev Genet 11(12):843–854

Wang Y, Meng Q, Guo Y, Wang Y, Wang Z, Yao Z, Shan T (2010b) Effect of atmospheric ammonia on growth performance and immunological response of broiler chickens. J Anim Vet Adv 9(22):2802–2806

Wathes CM, Kristensen HH, Aerts JM, Berckmans D (2008) Is precision livestock farming an engineer's daydream or nightmare, an animal's friend or foe, and a farmer's panacea or pitfall? Comput Electron Agric 64(1):2–10

Wernicki A, Nowaczek A, Urban-Chmiel R (2017) Bacteriophage therapy to combat bacterial infections in poultry. Virol J 14(1):1–13

Wimmer V, Lehermeier C, Albrecht T, Auinger HJ, Wang Y, Schön CC (2013) Genome-wide prediction of traits with different genetic architecture through efficient variable selection. Genetics 195(2):573–587

Woolhouse M, Ward M, Van Bunnik B, Farrar J (2015) Antimicrobial resistance in humans, livestock and the wider environment. Philos Trans R Soc B Biol Sci 370(1670):20140083

Yahav S, Straschnow A, Luger D, Shinder D, Tanny J, Cohen S (2004) Ventilation, sensible heat loss, broiler energy, and water balance under harsh environmental conditions. Poult Sci 83(2):253–258

Zheng H, Zhang T, Fang C, Zeng J, Yang X (2021) Design and implementation of poultry farming information management system based on cloud database. Animals 11(3):900

Zimmerman PH, Koene P, van Hooff JA (2000) The vocal expression of feeding motivation and frustration in the domestic laying hen, Gallus gallus domesticus. Appl Anim Behav Sci 69(4):265–273

Zuidhof M (2020) Precision livestock feeding: matching nutrient supply with nutrient requirements of individual animals. J Appl Poult Res 29(1):11–14

Open Access This chapter is licensed under the terms of the Creative Commons Attribution 4.0 International License (http://creativecommons.org/licenses/by/4.0/), which permits use, sharing, adaptation, distribution and reproduction in any medium or format, as long as you give appropriate credit to the original author(s) and the source, provide a link to the Creative Commons license and indicate if changes were made.

The images or other third party material in this chapter are included in the chapter's Creative Commons license, unless indicated otherwise in a credit line to the material. If material is not included in the chapter's Creative Commons license and your intended use is not permitted by statutory regulation or exceeds the permitted use, you will need to obtain permission directly from the copyright holder.

Prospect for Utilization of Modern Technologies for Small Ruminant and Pig Improvement in African Input Systems: Case Studies

24

Jean M. Feugang, Othman E. -M. Othman, Wilson Nandolo, Donald R. Kugonza, and Richard Osei-Amponsah

Abstract

Small ruminants and pig production are crucial elements of the agricultural value chain. Several innovative technologies have been developed globally since the 1980s to improve the animal resources sector's efficiency, productivity, and profitability. However, the diverse mixed farming systems and their limitations in Africa are very poignant, and the continent has yet to tap into its animal agriculture technology use, including digitization to reach optimal production and satisfy its markets. Modern biotechnologies are used to very limited extent, especially in sheep and goat systems that are dominated by local breeds reared in extensive or semi-intensive production systems. On the other hand, the high prolificacy of pigs and the relatively higher profitability of pig farming create more interest in intensive commercial production systems. Furthermore, the current reliance on pig production, mostly on imported or exotic breeds, motivates the introduction of various innovative technologies to enhance animal welfare and sustainable productivity. There is significant oppotunity for African smallholder farmers to harness available modern technologies to improve reproduction, production, farm management, or environmental monitoring. These technologies have great potential to transform livestock operations to become more profitable and improve livelihood and social conditions of the people. This chapter explores the prospects of utilizing these technologies in African small ruminants and pig farms for sustainable production, including genetic resource conservation.

Keywords

Animal health · Animal production · Animal reproduction · Biotechnologies · Genetics · Genomics · Livestock · Machine learning · Precision farming · Precision feeding · Tropical climate

J. M. Feugang (✉)
Department of Animal & Dairy Sciences, Mississippi State University, Mississippi State, MS, USA
e-mail: j.feugang@msstate.edu

O. E. undefined.-M. Othman
Department of Cell Biology, National Research Centre, Dokki, Egypt

W. Nandolo
Department of Animal Science, Lilongwe University, Lilongwe, Malawi

D. R. Kugonza
Department of Animal Biosciences, University of Makerere, Kampala, Uganda

R. Osei-Amponsah
Department of Animal Sciences, University of Ghana, Legon, Accra, Ghana

Abbreviations

AI	Artificial Insemination
AnGR	Animal Genomic Resources
ART	Assisted Reproductive Technologies
AU-IBAR	African Union-InterAfrican Bureau of Animal Resources
CIDR	Controlled Intravaginal Drug Release device
CTLGH	Centre for Tropical Livestock Genetics and Health
DITI	Digital Infrared Thermal Imaging
DNA	Deoxyribonucleic Acid
e/hCG	Equine/Human Chorionic gonadotropin
ES	Estrus Synchronization
ET	Embryo Transfer
FAO	Food and Agricultural Organization
FCR	Feed Conversion Rate
FE	Feed Efficiency
GNSS	Global Navigation Satellite Systems reflectometry
GnRH	Gonadotropin-releasing hormone
GPS	Global Positioning System
ILRI	International Livestock Research Institute
IVP	In Vitro Embryo Production
LH	Lutein Hormone
MOET	Multiple Ovulation-Embryo Transfer
PGF2α	Prostaglandin 2 alpha
PLF	Precision Livestock Farming
RFLP	Restriction Fragment Length Polymorphism
RFID	Radio Frequency Identification Technology
RNA	Ribonucleic Acid
SNP	Single Nucleotide Polymorphism
SSTR	Simple Sequence Tandem Repeats
STR	Short Tandem Repeats
UN	United Nations
US	Ultrasonography
GWAS	Genome-Wide Associated Studies
WGS	Whole-Genome Sequencing

24.1 Introduction

Small ruminants (sheep and goats) and pigs are essential animal genetic resources (AnGR) in Africa and significant contributors to rural livelihoods (Nantoumé 2020). With their hardiness and adaptability, goats and pigs, efficient feed converters with low initial investment requirements, are ideal candidates for helping farmers graduate from poverty. Both are profitable for small-scale rural farmers and well suited to Africa's harsh environmental conditions (e.g., climate, food availability, extensive farming systems, infectious diseases). Due to their higher prolificacy and fertility rates and little related capital investment, goats and pigs are likely to be dominant players in the livestock sector in Africa, at least until 2050 (Herrero et al. 2014), providing clear advantages for income generation to smallholder rural farmers. The pork business, which provides food and income to about 70% of rural farmers, could see further improvements in production efficiency with the introduction of modern biotechnologies.

Biotechnological advances have improved production efficiencies in livestock, including sheep, goats, and pigs. However, their adoption in African countries needs improvement despite their potential to address specific challenges (Simpkin et al. 2020). Livestock biotechnologies can be categorized as (1) monitoring production conditions for effective use of resources; (2) monitoring environments for effective animal health management; (3) identifying superior animals efficiently; and (4) quickly multiplying animals to increase the pool for selection and/or reproduction of superior animals. The first two categories fall under precision livestock farming (PLF) (Berckmans 2017), which is still limited in Africa (Banhazi et al. 2012; Andonovic et al. 2018). However, their combination with other technologies, such as genomics tools and assisted reproductive technologies (ART), can enhance livestock production in Africa, especially small ruminants and pigs having higher prolificacy. The combined use of these technologies is gaining popularity in countries like Ethiopia, Kenya, and Senegal, for improved breeding programs, breed selection in East African dairy cattle farms, and the creation of new breeds like Trypanosome-resistant cattle (Marshall et al. 2019).

24.2 Challenges Related to Small Ruminant and Pig Farming in Africa

Challenges such as the lack of sustained conservation programs (e.g., infrastructures, human capacity, regulations), genetic resource characterization, cultural issues, persisting climate change, and regulatory frameworks hinder the productivity of small ruminants and pigs in Africa (Visser 2019; Van Eenennaam et al. 2021; Havlík et al. 2015; Batlang et al. 2014; Adeyemo and Silas 2020). Continuous crossbreeding with imported breeds (e.g., Large White and Landrace pigs) increases productivity but leads to offspring less adapted to local conditions (Oluwole and Omitogun 2015). These offspring face issues like low-quality feed (e.g., reduced crop yields and access to water) and susceptibility to diseases such as African Swine Fever, which impacts global food security (Costard et al. 2009).

In Africa, the genetic diversity of small ruminants and pigs comes from controlled and unintended breeding with European and local breeds, allowing adaptability to the harsh environment and improved productivity (Amills et al. 2013; Mrode et al. 2018). Most African smallholder farmers practice in low-input settings (e.g., insufficient resources and poor feed and quality breeds) without any straightforward breeding program favoring genetic improvement (e.g., disease resistance, efficient growth rate, high meat quality, and reproductive performance), underscoring the urgent need for structured breeding programs. Some African governments (e.g., Algeria, Ethiopia, Morocco, and South Africa) and research institutes are working on genetic resistance to tropical diseases. These programs have been undertaken mostly through *in situ* or *ex situ* conservations (See Chap. 25). Sheep and goats are the preferred choice for smallholder farmers in Africa due to their economic benefits and resilience to environmental challenges. Community-based livestock breeding programs (CBBPs) have successfully preserved indigenous small ruminants and pigs in Malawi and Uganda, potentially revolutionizing future livestock breeding practices (Kaumbata et al. 2021; Demir et al. 2022). Initiatives to enhance the productivity of Indigenous goats through selective breeding and improved animal care have been introduced in Malawi and Uganda under the Feed the Future Initiative led by the United States Agency for International Development (USAID) and the African Goat Improvement Network (AGIN). These collaborative efforts among government agencies, stakeholders, smallholder farmers, and research institutions aim to leverage indigenous breeds and modern technologies to improve livestock productivity and uplift the livelihoods of farmers and stakeholders in Africa (Herrero et al. 2014; Latino et al. 2020). Therefore, smallholder pig, goat, and sheep producers in Africa must consider technological changes to leave the vicious cycle of poverty (Batlang et al. 2014). Despite the availability of various technologies (see Chaps. 20 and 21) to improve animal breeding, feeding, health care, traceability, record-keeping, and environmental monitoring in Africa, their adoption and use by stakeholders in the region are relatively low compared to other parts of the world. This chapter highlights successful cases of technology implementation and explores strategies to ensure their sustainable use, paving the way for a brighter future in African livestock farming.

24.3 Modern Technologies for Small Ruminants and Pigs Farming in Africa

Numerous technological approaches are available, and several have been applied to African livestock. For example, precision livestock farming (PLF) utilizes artificial intelligence and machine vision to automatically monitor individ-

Table 24.1 Examples of precision livestock farming technologies

Application	Technologies	Main objective/Function
Animal behavior	GPS sensors	Tracking location
	A neck collar with a series of sensors	Detection of estrus events through analysis of rumination rate, and the feeding and resting behavior
	Cameras and microphones Sound tool based on an algorithm	Find a correlation between vocalization and behavior
	A noninvasive imaging system such as VGG-face model, Fisherfaces, and convolutional neural networks	Pig-face recognition
Animal Health and welfare	A machine learning method	Pig cough detection-processing all incoming sounds and automatically identifying the number of coughs
	Air sensors	Prediction of the onset of Coccidiosis by monitoring the concentration of volatile organic compounds in the air
Feed management	An automated feeding system	Control the amount of feed provided, and the ambient temperature to optimize animal growth and reduce ammonia emission.
	A feed sensor	Measure and control the amount of feed delivered to individual feeders
	A next-generation feeding system	Provide feed with a variety of nutrient specifications to tailor both the amount and composition of the feed.
	A computer vision-based system CNN models using a low-cost RGB-D camera	Measures individual feed intake
	NIRS technology	Evaluation of physio-chemical composition of TMR and manure in dairy farms
Weight management	Weighing system based on image analysis	Determine the weight of an individual or group of animals (specifically pigs)

Adapted from Monteiro et al. (2021)

ual animals, boost production, and improve well-being (Monteiro et al. 2021; Berckmans 2017; Vranken and Berckmans 2017). The PLF relies on digital technologies (e.g., chips, wearable sensors, microphones, internet connections, smartphones, and video imaging) to track behavior, optimize feed, and evaluate animal well-being (Table 24.1). PLF's potential to significantly enhance livestock production (Monteiro et al. 2021; Racewicz et al. 2021), is particularly relevant in regions facing rapid population growth, including African countries such as Ethiopia, Kenya, Nigeria, and South Africa. Additionally, integrating genomic regions related to environmental adaptation into genomic selection practices holds promise for enhancing the sustainability of animal husbandry, mainly sheep and goats. Combining assisted reproductive technologies with genomic findings ensures further dissemination of highly genetic-merit animals with better commercial indexes and environmental adaptability.

24.3.1 Technologies for Environmental Mitigation and Precision Feeding

24.3.1.1 Animal Environment and Housing Monitoring

In Africa, persistent hot ambient temperatures require critical environmental modifications to animal housing to improve convection, conduction, radiation, and evaporative heat loss in animals. Additional cooling systems such as fogging, misting, showers, and evaporative pads have been developed to mitigate the negative effects of high temperatures on Africa's already insufficient livestock production systems. The use of these engineered systems would be beneficial for both small ruminants and pigs.

Small ruminants have the advantage of being highly versatile and tolerant to rising temperatures. Approximately 50% of the total populations of both goats and sheep are found in the world's arid regions (Gowane et al. 2017). About 80% of the total dairy goat population is distributed in the tropical areas of Africa and Asia, meeting the demands of low-income communities (Silanikove et al. 2010) and supporting their livelihoods. In most African countries, small ruminants are reared in rangelands, requiring less human input. However, Precision Livestock Farming (PLF) technologies such as Global Navigation Satellite Systems (GNSS) reflectometry may help characterize grazing sheep behavior for data-based decisions in pasture management.

Additionally, the Global Positioning System (GPS) collar sensors allow the location of animals on pasture to help better understand their habits and spatial distribution. These tools, along with the widespread use of cell or mobile phones, are bridging the rural-urban and rich-poor divides, creating a sense of community among African smallholder farmers. Moreover, mobile phones have greatly benefited African smallholder farmers, allowing them to communicate on flocks' movements and theft protection or livestock market activities (Aker and Mbiti 2010).

In the hot tropical climate of Africa, the potential of PLF technologies to revolutionize pig farming is immense. Most animal buildings have undergone structural changes and flooring types to help decrease indoor air temperatures, such as tunnel ventilation and evaporative pads, reducing heat stress impact on animals. During periods of heat, some animal houses are equipped with sprinkling nozzles and high-velocity air streams or conductive cooling pads to alleviate the effects of indoor heat (Godyń et al. 2020). However, most African small ruminants and hog operations are open barns often built with cheaper local materials (e.g., wood, straws, and bricks), allowing full ventilation and preventing direct exposure to sunlight. Upgrading these facilities to modern environmentally regulated buildings with advanced technologies such as PLF is unaffordable to most African smallholder farmers despite the foreseeable positive impacts on animal growth and well-being. Indeed, pigs in such improved housing conditions show higher feed efficiency than their counterparts in older (Godyń et al. 2020). Renaudeau et al. reported that feed intake and average daily weight gain of growing-finishing pigs decrease with increasing temperature starting from 20 °C (Renaudeau et al. 2011). A later report indicated the beneficial influence of precision feeding in improving feed intake. However, in Africa, the widespread small and medium-scale extensive farming systems and the more conservative behavior of farmholders and consumers do not favor adopting new technologies.

24.3.1.2 Animal Feeding Performance

The demand for livestock feed has led to the development of technologies to increase the production of high-quality feed with limited resources. These technologies fall into five main groups: feed productivity improvement, feed quality enhancement, feed quality maintenance or preservation, enhancing the nutritional status of animals, and analytical and operational technologies (Balehegn et al. 2020). Small ruminants have shown higher efficiency in converting feed into weight gain (FCR or Feed Conversion Rate) and high-capacity conversion of low-quality feed into quality products, which is critical for low-income smallholder farmers in Africa. In these countries, sheep and goats are farmed on grazing land in variable numbers, without any or no feed supplements, and relying on low feed, water, and labor inputs. Many smallholder farmers rely on naturally selected breeds (e.g., Anglo-Nubian goat breed and Malabari, Malpura, Salem Black, and Damara sheep breeds), showing more tolerance to arid conditions while maintaining good productivity (e.g., meat and milk) and greater adaptability and survival to extreme environmental conditions (e.g., high temperature, feed, and water scarcity). These conditions make introducing modern technologies in extensively reared flocks potentially less challenging for optimal nutrition and production achievements. Indeed, the existing technologies to monitor farm animals have the potential to ensure improved production and protection from thieves in conflict zones. In addition, these technologies may enable the characterization of grazing sheep behavior (i.e., GNSS) for database decisions in pasture management and animal location on pasture (i.e., GPS collar sensors) better to understand the habits and causes of the spatial distribution. Collar sensors are more accurate than human visual observations for tracking animal behavior to feeders, resting behavior, and frequency (Monteiro et al. 2021; Gaillard et al. 2021). In addition, noninvasive images and sound recording systems have a promising future, given the perfect correlation between animal sound and behavior (Berckmans 2017; Banhazi et al. 2012; Gaillard et al. 2021). Furthermore, technologies based on automatic identification have used approaches such as ear tags containing an RFID chip allowing recognition of individual animals (e.g., sheep, goats, and pigs) at electronic feeding stations (Banhazi et al. 2012; Andonovic et al. 2018; Vranken and Berckmans 2017).

Pigs lack natural adaptation mechanisms to environmental changes, requiring more human intervention in intensive production systems. The PLF in pig farming improves sustainability by minimizing losses, enhancing genetic identification, and reducing waste. Precision feeding improves nutrient utilization and reduces environmental impacts such as high altitude, climate change, environmental adaptation, and disease resistance (Rauw et al. 2020).

In Africa, the underutilization of feed improvement technologies is due to systemic constraints. Successful technology adoption is driven by financial incentives and the ability to address multiple problems. Engaging stakeholders in technology development is crucial. Diversifying suitable technologies for specific agroecologies and production systems is essential (Balehegn et al. 2020).

24.3.2 Machine Learning Systems and Growth Technologies

Feed efficiency (FE) is the ratio of daily weight gain to daily feed consumption over a specific period. Improving FE is a primary goal in livestock production to reduce costs and environmental waste. Precision feeding considers individual nutrient requirements within a group to adjust the quantity, quality, and timing of feed supplied, aiming to improve efficiency and reduce farm costs (Hendriks et al. 2019).

24.3.2.1 Feed Quality Improvement and Higher Feed Efficiency

The current meat production in Africa has relied on extensive traditional systems in challenging socio-economical environments that lead to insufficient supplies to meet market demands in Africa (Enahoro et al. 2019; Osei-Amponsah et al. 2019; Abdalla et al. 2021). Embracing technological advancements and innovative solutions to intensify livestock production offers a promising strategy to boost meat production rapidly. Critical factors such as genetic potential, environmental conditions, complete nutrition, and practical farm management practices influence the growth performance of livestock. In the Sahelian region of West Africa, 135 million people raise 173 million ruminant livestock, primarily sheep and goats. These animals rely on pastures and crop residues for food, while commercially formulated feeds are used for poultry and pig production. The availability and quality of feed resources vary across different agricultural areas and seasons. To meet the increasing demand for animal-source food, enhancing the nutritional value of crop residues, preserving fodder, and promoting feed markets is crucial (Amole et al. 2022). Smallholder farmers worldwide are encouraged to consider affordable alternatives such as growth promoters, improved breeds, housing, feed additives, hormones, and antibiotics to enhance the growth performance of farm animals (Gadde et al. 2017).

Various feed additives, such as nutrients (like hormones, pro/pre-biotics) and non-nutrients (such as antibiotics and exogenous enzymes), have been found to have positive effects on feed efficiency, growth performance, and meat quality improvement (Valenzuela-Grijalva et al. 2017; Suryadi et al. 2019). However, the continuous use of hormones and antibiotics in feed has raised serious concerns. This has led to the One Health concept, which emphasizes the connection between environmental, animal, and human health in managing antimicrobial-resistant and food safety concerns (Andrew Selaledi et al. 2020). The increasing use of antibiotics as growth promoters for prophylaxis and metaphylaxis has been blamed for cross-resistance, multiple antibiotic resistance of pathogenic bacteria in humans and livestock, increased foodborne allergies, and negative environmental impacts (Rai 2013; Popova et al. 2017; Al-Shawi et al. 2020; Selaledi et al. 2020). As a result, the use of antibiotics in feed is being re-evaluated worldwide. Meanwhile, their increased "unsupervised" uses in Africa pose a threat to food safety and security, prompting stakeholders to consider alternative solutions for healthy meat production. Among various alternatives (such as enzymes and organic acids), probiotics and prebiotics have emerged as safe and viable options for increasing performance in livestock (Alayande et al. 2020).

24.3.2.2 Case of Pro- and Prebiotics, and Plant Extracts

Several reports indicated growth performance, carcass yield, and meat quality improvements following the application of growth technologies in meat production, particularly with probiotics and prebiotics (Valenzuela-Grijalva et al. 2017), which are amongst livestock's most used growth promoters (i.e., feed additives and hormones). Probiotics are live microorganisms conferring health benefits when consumed in adequate amounts. Upon ingestion, probiotics help to balance the intestinal microflora and impact the local and systemic immune systems by producing beneficial enzymes, organic acids, vitamins, and nontoxic antibacterial substances (Popova et al. 2017). Common probiotics include Lactobacillus, Bifidobacterium, *Saccharomyces boulardii* yeast, *Clostridium butyricum*, and some *E. coli* and Bacillus species. The most commonly used are *Lactobacillus acidophilus, Lactobacillus lactis, Lactobacillus plantarum, Lactobacillus bulgaricus, Lactobacillus casei, Lactobacillus helveticus, Lactobacillus salivarius, Bifido bacterium spp., Enterococcus faecium, Enterococcus faecalis, Streptococcus thermophilus, Escherichia coli* bacteria, and other probiotic fungi such as *Saccharomyces cerevisiae* and *Saccharomyces boulardii* (Al-Shawi et al. 2020). Probiotics are widely used in livestock animals (e.g., ruminants, poultry, pigs, and horses) to enhance growth rate, increase milk yield, meat, and egg production, and improve reproduction (Jin and Speers 1998; Denli et al. 2003; Higgins et al. 2008; Timmerman et al. 2006; Markowiak and Śliżewska 2018).

Prebiotics are non-digestible but fermentable food ingredients. They affect the host by selectively stimulating the growth and/or activity of specific bacteria in the colon (Davani-Davari et al. 2019). Prebiotics are dietary substances consisting mainly of non-starch polysaccharides and oligosaccharides, commonly used as oligosaccharides (e.g., fructans, oligofructose, fructooligosaccharides, galactan, galactooligosaccharides), resistant starch, pectin, fiber components, insulin, and breast milk oligosaccharides. Additionally, prebiotics help prevent the colonization of the digestive system by pathogens by creating unfavorable conditions, such as altering the pH of intestinal content.

Plant extracts are alternatives to antibiotics, known as phytobiotics, which could be affordable for African smallholder farmers. These plants possess antimicrobial, anti-inflammatory, antioxidant, and antiparasitic activities and have been used in poultry production for many years (Andrew Selaledi et al. 2020). Other plant extracts found in Africa include aromatic spices (cinnamon, clove, etc.), pungent spices (pepper, garlic, and ginger), and herbs spices (rosemary, thyme, mint, etc.). They have received increased attention due to their low costs and availability. Therefore, they are viable alternatives to meet the WHO antimicrobial usage recommendations in Africa. This is necessary to avoid the growing deterioration of livestock health and mortality rates, jeopardizing food security attainment.

Probiotics, prebiotics, and plant extracts are essential in animal production and health, contributing tremendously to the meat industry. Feed additive technologies have had positive impacts on feeding efficiency and growth performance. More studies are needed to evaluate their efficacy in small ruminants and pigs in Africa's tropical climate, and stakeholders should support such initiatives. National governments and other key stakeholders should encourage smallholder farmers to implement their plans to boost meat production in Africa.

24.3.3 Technologies for Animal Health, Welfare, and Disease Prevention Monitoring

Animal health is crucial to the One Health concept, impacting human health and livestock productivity through the misuse of antibiotics. Therefore, advanced biotechniques are being developed to monitor animal physiological and genetic changes for proactive health management. Modern livestock systems use noninvasive monitoring devices (e.g., cameras, microphones, and detectors) and information technology to support farmers and improve animal welfare. Precision farming technologies also help address production challenges and meet consumer demands (Racewicz et al. 2021).

24.3.3.1 Wireless and Wearable Sensors

Precision livestock farming uses advanced technology to monitor individual animals' health and welfare in real time. It involves analyzing images, sounds, tracking data, weight, and biological metrics for early illness detection and continual monitoring of physiological status (Benjamin and Yik 2019). This approach is particularly beneficial in pig production, where imaging technology helps with weight estimation, behavior monitoring (e.g., aggressivity, walking patterns, and posture), and lactation observation. Simple microphones and infrared cameras further enable the detection of specific acoustic events and physiological changes (e.g., stressful situations, coughs, and respiratory diseases), ultimately enhancing animal welfare and agricultural efficiency (Benjamin and Yik 2019).

Available devices can be classified into three categories based on the methodology and range of technology applications. *The first category of devices* detects specific animal behavior using special sensors like automatically measuring the frequency of pig visits to the feeder and the time taken to feed through radio frequency identification (RFID) technology. This category uses real-time video visualization using conventional monochromatic, color, or 3D cameras to depict activity level, area occupancy, aggression, gait scores, resource use, and posture. *The second category of devices* detects and records specific behaviors, such as drinking, feeding, and spatial distribution. The data collected is then processed into numeric form and presented graphically. This allows for the identification of changes in animal behavior, with the interpretation of data being a crucial step in understanding the implications of these changes. *The third category of devices* involves intelligent production systems that automatically analyze recorded physiological and behavioral changes. Based on the optimal settings of the farm environment, these systems can extract deviations from these settings and automatically make decisions to adapt the production environment to optimal production conditions (Racewicz et al. 2021). Using audio recordings, voice analysis, and machine learning algorithms, this category can detect animal heat stress, illness, or discomfort. Such a combination could benefit the detection of respiratory diseases or discomfort due to poor air quality, signaled by changes in vocal characteristics and acoustic signs such as coughing and sneezing. Specialized microphones' capability to differentiate between infectious coughs and those caused by environmental factors, such as accumulated ammonia or dust, underscores their incredible potential for health assessment (Racewicz et al. 2021).

24.3.3.2 Imaging Technologies

Extensive studies have investigated the behaviors and welfare of small ruminants and pigs, focusing on early disease detection using 2D/3D cameras. Video monitoring and imaging are the most suitable technologies to enhance livestock production efficiency within PLF methods. These technologies offer noninvasive and effective tools for recording both individual and group animal behaviors. For instance, imaging can successfully assess body weight, detect lameness, and aggressive behavior, and identify heat in pigs (Racewicz et al. 2021). Digital infra-red imaging (DITI) is one of the latest imaging technologies adopted in veterinary medicine (Stelletta et al. 2012). It is an effortless, pain-free, and remote tool that can detect symmetric and asymmetric temperature gradients on the surface of the skin. This noninvasive nature of DITI reassures the audience about its animal-friendly application. This enables an accurate measurement of temperature gradients through thermal imaging. DITI is used in livestock production and reproduction systems to evaluate animal health and welfare (Soerensen and Pedersen 2015; Rekant et al. 2016; Nääs et al. 2020), with the ability to identify thermal abnormalities of the body and diagnose injuries and diseases in pigs and small ruminants. Recent studies have demonstrated the relevance of DITI in detecting temperature changes of animal bodies in tropical regions (Popoola et al. 2022; Leles et al. 2017), indicating its potential use in veterinary medicine for conditions such as heat stress, inflammations, and sickness.

Ultrasonography, an integral part of contemporary veterinary medicine, plays a crucial role in disease diagnosis. It involves using different transducers to examine structures close to the body surface (e.g., 5.0, 7.5, or 10.0 MHz transducers) or over 20 cm depth (e.g., 2.5 and 3.5 MHz). This inverse relationship between the depth of sound waves' penetration (frequency) and the image resolution means that low frequencies, permitting more significant penetrations, are associated with lower image resolution. Ultrasonography is widely applied in veterinary clinics to examine various areas and abdominal organs such as the thorax, spleen, digestive and urinary tracts, and reproductive organs for pathology detection (Cartee et al. 1993). Most organs can be imaged using the 3.5 MHz transducer (Braun 2004). Compared to routine histopathological analyses, it is an efficient and fast method for diagnosing diseases (e.g., endometritis) (Meira et al. 2012).

Digital Infrared Thermal Imaging and ultrasonography have revolutionized healthcare since their introduction over 40 years ago. Their implementation in Africa can enhance epidemic surveillance, improve animal welfare, and elevate reproductive management. For example, Precision Livestock Farming methods like electronic identification devices (i.e., RFID) are gaining popularity for identifying pigs and detecting diseases early (Arulmozhi et al. 2021).

24.3.4 Genomics Tools for Genetic Improvement

Molecular biology is driven by the transfer of genetic information from DNA to RNA and proteins, influenced by environmental factors. Gene transcription into RNA molecules is crucial for decoding the genome and regulating cell activities.

24.3.4.1 Natural Breeding and Genetics

Insemination by sexual intercourse, natural insemination, or mating is the common practice of animal breeding in most African countries, which lack sufficient infrastructure for assisted reproduction, such as artificial insemination (AI). In small ruminant and pig production systems, natural mating is cheaper and noninvasive. Still, it has less genetic input in livestock production. Furthermore, the lack of knowledge and absence of proactive government policies and legal frameworks to favor selective breeding hamper the generation of improved genetic animals for better adaptation to local conditions.

Nonetheless, genetics to improve livestock breeds and productivity is the focal point for scientific research and innovation. Most of the world's livestock breeding improvements rely on artificial insemination (AI), a powerful breeding tool in Africa, especially in small ruminants and pigs. Unlike natural mating, AI of synchronized females has higher pregnancy rates (Sh et al.

2020). Still, the existing constraints limit the widespread of AI in Africa. For example, a crucial need is for well-trained operators, accurate estrus detection, high-quality and good semen, an unbroken chain of liquid nitrogen storage, and appropriate equipment for enhanced precision breeding. Therefore, the recent progress in genetics and genomics opens exciting possibilities for creating new breeds that exhibit potential resistance to local factors (e.g., environment and disease). These innovations hold the promise of providing new approaches for African smallholder farmers to better manage animals in harsh environmental conditions. Genetic advances, stemming from increased knowledge in epigenetics (Zsidó and Hetényi 2020), DNA Methylation (Jang et al. 2017), histone modification (Alaskhar Alhamwe et al. 2018), RNA sequencing (Thorstensen et al. 2021), noncoding RNA (Mercer and Mattick 2013), and gene editing (Li et al. 2020a), are paving the way for a brighter future in livestock breeding in Africa.

24.3.4.2 Genetic Tools for Precision Breeding

Genomics, a crucial component of the African Strategy for biodiversity preservation, is a field that offers numerous biological techniques, particularly in genetics. The use of DNA, which serves as the stable fingerprint of an individual among cellular components, is of particular interest in the context of genomics conservation.

Restriction Fragment Length Polymorphism (RFLP), also known as DNA Fingerprinting, is a technique that involves the digestion of genomic DNA with restriction enzymes, followed by gel electrophoresis to visualize different DNA sequence lengths (Powell et al. 1996). This method, which exploits homologous DNA sequences, is widely used in forensic investigations and the mapping of hereditary diseases (Allen et al. 2000; Van Marle-Koster and Nel 2003).

Microsatellites are simple sequence tandem repeats (SSTRs) or short tandem repeats (STRs) corresponding to short (1–9 bp) and repetitive (5–50 times) segments of DNA scattered throughout the genome in noncoding regions between or within genes (introns) (Li et al. 2004). This technique locates mutations and identifies gene(s) responsible for a given trait or disease. It can be helpful for phylogeographic studies (Wright and Bentzen 1995). The literature lecture indicates that the genetic characterization of local tropical breeds is mainly achieved through microsatellites, especially with small ruminants such as the indigenous Moroccan sheep (Gaouar et al. 2016) and the goats in Southern Nigeria (Okpeku et al. 2011). Microsatellites and SNPs have been crucial in using and conserving animal genetic resources. Now, with the affordability of these new sequencing technologies, whole-genome sequencing (WGS) and genome editing have the potential to revolutionize livestock population studies, making advanced genomic techniques more accessible to African research institutions (Cortes et al. 2022).

Mitochondrial DNA (mtDNA) is an extranuclear double-stranded and small DNA with 2–10 copies in each mitochondrion. This technique detects population changes along with their maternal origin and migration routes (Moritz 1994; Wong and Boles 2005). It is less common in tropical breeds than the microsatellite technique.

Whole-genome sequencing (WGS) allows the determination of the order of the bases in an organism's genome in one process. As a result, data can potentially discover all genetic variants, showing the power of genome-wide associated studies (GWAS). Despite its costs for a large number of sample analyses, GWAS permits the identification of trait loci and nucleotides related to a particular trait or disease (Molotsi et al. 2017; Ramos-Onsins et al. 2014).

Microarrays are microchips containing short-sequence DNA (cDNA, oligo DNA, BAC, and SNP) or RNA to simultaneously detect polymorphisms or expression profiles of thousands of genes within a population (Heller 2002). The single nucleotide polymorphism (SNP) arrays are variants of DNA microarrays used in WGS for DNA genotyping to help predict individual responses to stimuli in GWAS (e.g., disease, drugs, and environmental factors) (Gunderson et al. 2006). Millions of equally dispersed biallelic SNPs within the animal genome (i.e., 335 million SNPs in humans) make this technical approach far better than microsatellites for gene expression studies (Cánovas et al. 2013). Still, many SNP markers are required to achieve valuable and comparable information to microsatellites.

RNA-Sequencing is based on labeling fragmented RNA (50–500 bp) with adapter sequences, sequence reads counts, and full-length transcripts assembly (Mantione et al. 2014). It offers improved specificity for detecting transcripts and isoforms better than microarrays while offering high sensibility detection of differential expression. As a result, RNA-Seq is more likely to be used routinely than microarray for gene expression profiling and has a bright future in bioinformatics data collection (Griffith et al. 2010; Cui et al. 2014; Toms et al. 2021).

Gene editing, the ability to make specific changes in the DNA sequence of a living organism, is a complex process that uses engineered nuclease enzymes to target a particular DNA sequence by removing, adding, or replacing a specific DNA segment spot. CRISPR (clustered regularly interspaced short palindromic repeats) technology, known for simplicity, is the top choice for genome editing. The gene editing technology is CRISPR-Cas9, a sophisticated genetic tool that plays a crucial role in gene inactivation (Li et al. 2020b). Inactivating genes (e.g., slick, NANOS2) may create novel farm animals with a modified genetic pool. These newly engineered animals may better adapt to tropical conditions, such as heat stress and high resistance exposure (Park et al. 2017; Ciccarelli et al. 2020; Karavolias et al. 2021). Unfortunately, few African countries (e.g., Nigeria and Kenya) have legislative interpretations of new genome editing techniques.

Bioinformatics or computational biology is an interdisciplinary science that models biological systems combining mathematics/physics, computer science, medicine, and biology85. According to animal species, the availability of various software, algorithms, and bioinformatics permits the interpretation of generated biological datasets (e.g., genomics, transcriptomics, proteomics, and metabolomics).

The abovementioned technical tools have been utilized for genomic selection in various African livestock animals. Unfortunately, only a few African countries have functional research laboratories or centers capable of completing genomics analyses to improve livestock. However, these techniques are gaining popularity across the continent. For instance, in Ghana (Osei-Amponsah et al. 2017), livestock genomic technologies have revealed the origin and phylogenetic status of the Ashanti black forest Dwarf Pig, which has several advantages over exotic European breeds, such as better survival under poor input environments, less susceptibility to local diseases and parasites, and better meat quality. In Ethiopia, these technologies were used to estimate breed proportions in the local sheep population (Marshall et al. 2019). Various research centers collaborate to share expertise and reduce costs. Genomic tools assess genetic diversity, improve animal genetics, trace phylogenetic relationships, and study geographical distribution and origins (Ibeagha-Awemu et al. 2019). The significant role of national governments (e.g., Algeria, Ethiopia, Kenya, Morocco, Rwanda, Senegal, and South Africa), international organizations or foundations (e.g., USAID, FAO, AU-IBAR, and Bill & Melinda Gates), and research organizations (e.g., ILRI and CTLGH) cannot be overstated. Their efforts have been instrumental in genetically improving small ruminants and pigs for increased milk or meat production through specific genetic improvement programs, such as Nucleus-based breeding schemes (Kosgey et al. 2006). As a result, more animals on the continent are genotyped and sequenced to enable precision farming (e.g., breeds and geographic matches, enhanced breeding programs) or to create new breeds (e.g., heat resilient or trypanotolerant breeds).

Genomic selection has transformed breeding in advanced economies and holds promise for African countries. Despite challenges, collaborative efforts are overcoming barriers for smallholder farmers. These hurdles, which make traditional genetic programs impractical for smallholder farmers, are being overcome by the collective efforts of stakeholder-led initiatives (e.g., ILRI) and projects like ADGG and iCow (http://www.icow.co.ke/). The lack of infrastructure and human capacity are significant limitations, but the establishment of the African Animal Breeders' Network (AABN—http://animalbreeding-africa.org/) is a crucial step toward building the necessary human capacity to drive genetic improvement programs across the continent (Burrow et al. 2021). Various chapters of this book discuss further applications of genomic tools with benefits in Africa and reviewed elsewhere (Marshall et al. 2019; Amills et al. 2013; Mrode et al. 2018; Muchadeyi et al. 2020).

24.3.5 For Reproductive Performance Improvement

The increasing worldwide demand for animal-source food for the human population drives the widespread use of ARTs for massive livestock production. These technologies, such as artificial insemination (AI), estrus synchronization (ES), pregnancy detection, *in vitro* embryo production (IVP), multiple ovulations coupled with AI, embryo collection and transfer to recipients (MOET), and IVP—embryo transfer (IVP–ET) hold the key to revolutionizing genetic improvement and reproductive processes in various livestock species90. In addition, imaging technologies such as Digital Infrared Thermal Imaging (DITI) and ultrasonography have been adapted for further reproductive improvement in various livestock species, including small ruminants and pigs. While developed nations have embraced ARTs to drive productivity and reduce production costs, the potential of these technologies in developing countries, particularly in Africa, remains largely untapped. Countries like Ethiopia, Kenya, Nigeria, and South Africa have seen varying degrees of success, primarily focusing on cattle. However, these technologies have immense potential to drive significant advancements in livestock production on the continent (Balehegn et al. 2020; Yadeta 2020; Niyiragira and Rugira 2018; Celestin et al. 2019).

24.3.5.1 Assisted Reproductive Technologies.

Artificial Insemination (AI) and Estrus Synchronization (ES) These are the popular technologies to disseminate improved genetics in livestock species in a broader reach to farmers in communities, facilitating access to genetic gain while reducing infectious reproductive diseases' transmission. Intensive pig production systems apply AI to introduce superior genes into sow herds. The AI is primarily used in various African countries in crossbreeding programs with freshly collected (indigenous breeds) or imported (exotic breeds such as Landrace and Large White). Boar semen is relatively easy to harvest by hand-gloved technique, initiating semen release (King and Macpherson 1973; Ko et al. 1989). Single ejaculates contain enough sperm to inseminate 15–25 sows through conventional cervical AI (Maes et al. 2011). Semen is used to inseminate females (gilts or sows) following naturally occurring estrus or hormonal synchronization. The combination of equine (eCG) and human (hCG) chorionic gonadotropin for synchronization induces better results than the boar exposure (e.g., sound, smell, and physical presence), inducing GnRH secretion, estrus, and ovulation. The success rate of pig AI (~90%) with fresh extended semen is comparable to natural mating (Roca et al. 2006; Knox 2016). The AI technology in pigs is adaptable and has been successful in many countries (Waberski et al. 2019; Vishwanath 2003). Estrus synchronization is the second most important technology for improving animal productivity. It brings females together for parturition, reducing labor and financial resources, shortening the calving period, and ensuring more consistent weaning weights. Additionally, it eliminates the need for estrus detection in AI, making the process more efficient and cost-effective. Both AI and ES techniques are sustainably adopted in many African countries for intensive and extensive (smallholder backyard) pig production systems, with a practice in cattle that goes back to 1938 in Ethiopia

(Bamundaga et al. 2018; Oke-Egbodo et al. 2018; Niyiragira et al. 2018).

For sheep and goats, AI is mainly practiced on a commercial level. Fresh or frozen semen is collected using an artificial vagina or electrical stimulation that gives a 10 to 15-volt output (Steyn 2003). Inseminations using estrus-induced ewes are done through cervical and intrauterine (or laparoscopic) routes. Sheep farming is of great economic, social, and environmental interest in the southern Mediterranean countries, such as Morocco, Algeria, and Tunisia. Artificial insemination (AI) of sheep has been studied since the 1980s, with AI centers established by public authorities in Morocco. Nonetheless, the main challenge in using AI in sheep selection programs is preserving ram semen, and the successful cryopreservation of semen for non-surgical AI is seen as the golden way to enhance fertility outcomes for rapid genetic progress (Atigui and Chniter 2022). In Ethiopia, the development of low-cost, low-infrastructure AI protocols supported sheep and goat breeding programs involving various stakeholders (i.e., research centers, universities, national AI centers, and veterinarians). Due to better management, this innovation improved conception rates (50–60%), overall fertility, and the number of offspring produced in the Community-Based Breeding Programs' flocks. The program has low infrastructure needs and applies to both sheep and goats. Stakeholders should ensure that any AI system put in place is reliable and easily accessible by farmers. Meanwhile, in Rwanda, the use of drone technology to deliver semen straws should be encouraged.

Several ES protocols are available to stimulate follicular development and regulate ovulation. Protocols use intramuscular prostaglandins F2α (PGF2α) administrations alone or combined with eGC, progesterone, or analogs through controlled intravaginal drug release device (CIDR and sponge), ensuring over 72% estrus induction (Ramukhithi et al. 2012; Gore et al. 2020a; Fierro et al. 2013; Bukar et al. 2012). Nonetheless, progestogen devices induce vaginitis in females, reducing fertilization rates and making the PGF2α administration a safe alternative to progestogens. Estrus synchronized ewes are frequently inseminated with fresh semen using the intrauterine route (Gore et al. 2020a), reaching the highest pregnancy rates (60–80%). Both ES and AI have been practiced in African breeds of sheep and goats but remain experimental or low-scale in various countries (Omontese et al. 2016; Mekuriaw et al. 2016; Abu et al. 2008; Gore et al. 2020b; Godfrey et al. 1999; Habeeb and Kutzler 2021). The most common estrus synchronization procedure in the Maghreb region involves using vaginal devices (sponges) impregnated with 30–40 mg of fluorogestone acetate progesterone implants for 14 days, along with an equine Chorionic Gonadotropin (eCG) intra-muscular injection on the removal day. Two synchronization treatments were tested for Moroccan sheep breeds using progesterone implants combined with either eCG or prostaglandin analog injection. The lambing rate of D'man ewes was 34.9% and 21.7% for ewes treated with PGF2α and progesterone, respectively. For the Timahdite breed, the rate was 39.1% and 13% for ewes treated with PGF2α and progesterone, respectively (Atigui and Chniter 2022).

Pregnancy Detection Pregnancy detection and monitoring are essential diagnostic procedures undertaken in livestock as part of fertility management. However, routine pregnancy diagnostic methods in small ruminants and pigs are more indirect. These methods include endocrine hormone detection (e.g., progesterone, estrogens, PGF2α, and estrone sulfate concentrations) (Goel et al. 2009; Williams et al. 2008) and visual systems, the most used approach in farms (Goel et al. 2009; Bharti and Jacob 2019; Kauffold et al. 2019). Hence, the non-return rates to estrus and teaser males are routine visual methods applied in smallholder farms to detect non-pregnant females within the herds. The transabdominal visualization of the reproductive tract through ultrasonography is another visual method (A-mode, Doppler, and B-mode ultrasound). Furthermore, the direct method of pregnancy diagnosis consists of rectal palpation of sheep and goats by placing fingers in the rectum and lifting the abdomen (Kutty 1999). Although pregnancies can be detected as early as

25–30 days in small ruminants and pigs, the most accurate detection appears to be around and above 45 days of gestation (Williams et al. 2008; Gonzalez-Bulnes et al. 2010).

Multiple Ovulation (MO), In Vitro Embryo Production (IVP), and Embryo Transfer (ET) Hormonal induction of multi-ovulation (e.g., PMS, FSH) may follow synchronization for in vivo embryo production. Pigs (gilts/sows) and small and small ruminants (does and ewe) are naturally multi-ovulating animals. Hormonal treatments (e.g., hCG and LH) allow more eggs to ovulate from the donor females during natural heat. The insemination of donors will permit egg fertilization and embryo development until collection through oviductal flushing. Contrarily to large animals (e.g., cows and mares), embryo flushing is a surgical procedure in small ruminants and pigs. This procedure limits the MOET application in most African countries, needing more resources and trained personnel. Despite the recent progress in non-surgical procedures such as intracervical catheter insertion in pigs (Martinez et al. 2001; Yoshioka et al. 2012), embryo recovery rates remain low.

Hence, the MOET using fresh or frozen-thawed embryos can help increase genetically superior females or breed restocking in Africa. Furthermore, IVP's potential for additional embryo production is generally low and more costly than AI, limiting its application in the continent (Graff et al. 1999; Cognie et al. 2003). Indeed, various limitations affect the successful application of IVP in Africa due to its reliance on immature oocytes harvested from living (e.g., laparoscopy and ultrasound-guided ovum pick-up) or post-mortem (e.g., slaughtered females at abattoirs) females (Fowler et al. 2018; Cognié et al. 2004; Ikoma et al. 2014).

24.3.5.2 Imaging Detection Techniques

Researchers have used DITI and ultrasonography to diagnose the reproductive status of livestock animals (Rekant et al. 2016). For example, the DITI has been used to detect estrus in gilts (Sykes et al. 2012), ovulation, and decreased vulvar temperature of sows during the periovulatory period (Clark 2011). In a recent study, we found the effectiveness of DITI in examining scrotal and udder temperatures of Red Sokoto goats and Uda sheep under semi-intensive systems during the hottest climate in Nigeria (Popoola et al. 2022). In this later study, bucks and rams exposed to high ambient temperature showed higher scrotal temperatures, which may negatively influence male fertility due to reduced semen outputs and compromised sperm quality.

Since 1983, transcutaneous B-mode ultrasonography has successfully diagnosed pregnancy in pigs (Inaba et al. 1983) and small ruminants (Tainturier et al. 1983; Baronet and Vaillancourt 1989). Its introduction in pig facilities has shown to be helpful in various applications, such as monitoring ovarian activity and estimation time of ovulation, detection of puberty onset, accurate pregnancy diagnosis around 30 days post-insemination, or end of estrus (Kauffold and Althouse 2007). Although transrectal examinations provided the most accurate diagnoses, the transcutaneous B-mode US at 5.0 MHz transducer also provides correct diagnoses. Studies show that successful pregnancy diagnosis in early gestation depends upon transducer frequency, application route, technical expertise, experience, and other factors inherent to the animal. Therefore, B-mode ultrasonography is the most valuable tool for scanning small ruminants and pigs. It is a relatively simple, fast, and safe procedure that does not cause additional distress to animals.

24.4 Capacity Development for Technology Adoption

The training programs for small ruminants and pigs in Africa can bring economic, environmental, and social benefits. Educating farmers about sustainable practices like biosecurity and water consumption is essential. It is crucial to increase awareness, training, and skills programs for farmers and stakeholders to promote the adoption of modern technology. Governments should focus on building local capacity in innovative water and agricultural solutions. Indigenous breeds can be crucial in adapting to climate change, and capacity development is essential to improve husbandry. Engaging pastoral communities through gender-inclusive groups is crucial for accelerating the adoption of new practices. Uncontrolled animal mating can lead to inbreeding, so improving animal husbandry practices and implementing biosecurity measures is essential. Omics technologies offer the potential for understanding how resilient breeds adapt to environmental changes, but their use in tropical animal production is limited. Advanced technologies such as PLF, omics, remote temperature monitoring, and GPS animal tracking have revolutionized sheep and goat farming. Further research is needed to gauge African farmers' awareness and willingness to adopt these technologies.

24.5 Conclusion

Small ruminant and pig genetic resources are crucial for African agriculture, but traditional methods must be revised to meet rising demands. Biotechnology and other innovations, including digital technologies, offer hope for achieving the "UN Sustainable Development Goals related to food security, poverty reduction, improved livelihoods, and the One Health Concept." These goals are achievable, and action is required. Governments must invest in capacity building, enact supportive policies, and leverage indigenous livestock diversity in production systems.

Acknowledgment This work was partly supported by the U.S. Department of Agriculture, Agricultural Research Service, project 6066-31320-017-000D, and the International Livestock Research Institute, Nairobi, Kenya.

References

Abdalla SE, Abia ALK, Amoako DG, Perrett K, Bester LA, Essack SY (2021) From Farm-to-Fork: E. coli from an intensive pig production system in South Africa shows high resistance to critically important antibiotics for human and animal use. Antibiotics 10(2):178

Abu A, Iheukwumere F, Onyekwere M (2008) Effect of exogenous progesterone on oestrus response of West African Dwarf (WAD) goats. Afr J Biotechnol 7(1):59

Adeyemo AA, Silas E (2020) The role of culture in achieving sustainable agriculture in South Africa: examining Zulu cultural views and management practices of livestock and its productivity. In: Regional development in Africa. IntechOpen, p 183

Aker JC, Mbiti IM (2010) Mobile phones and economic development in Africa. J Econ Perspect 24(3):207–232

Alaskhar Alhamwe B, Khalaila R, Wolf J, von Bülow V, Harb H, Alhamdan F, Hii CS, Prescott SL, Ferrante A, Renz H (2018) Histone modifications and their role in epigenetics of atopy and allergic diseases. Allergy, Asthma Clin Immunol 14(1):1–16

Alayande KA, Aiyegoro OA, Ateba CN (2020) Probiotics in animal husbandry: applicability and associated risk factors. Sustain For 12(3):1087

Allen RW, Traver M, Pritchard J (2000) DNA analysis in a paternity case involving a triploid fetus. Transfusion 40(2):240–244

Al-Shawi SG, Dang DS, Yousif AY, Al-Younis ZK, Najm TA, Matarneh SK (2020) The potential use of probiotics to improve animal health, efficiency, and meat quality: a review. Agriculture 10(10):452

Amills M, Ramírez O, Galman-Omitogun O, Clop A (2013) Domestic pigs in Africa. Afr Archaeol Rev 30(1):73–82

Amole T, Augustine A, Balehegn M, Adesogoan AT (2022) Livestock feed resources in the West African Sahel. Agron J 114(1):26–45

Andonovic I, Michie C, Cousin P, Janati A, Pham C, Diop M (2018) Precision livestock farming technologies. Global Internet of Things Summit (GIoTS), IEEE, pp 1–6

Andrew Selaledi L, Mohammed Hassan Z, Manyelo TG, Mabelebele M (2020) The current status of the alternative use to antibiotics in poultry production: an African perspective. Antibiotics 9(9):594

Arulmozhi E, Bhujel A, Moon B-E, Kim H-T (2021) The application of cameras in precision pig farming: an overview for swine-keeping professionals. Animals 11(8):2343

Atigui M, Chniter M (2022) Ovine artificial insemination in the Maghreb region: present status and future prospects. In: Sheep farming: herds husbandry, management system, reproduction and improvement of animal health, p 113

Balehegn M, Duncan A, Tolera A, Ayantunde AA, Issa S, Karimou M, Zampaligré N, André K, Gnanda I, Varijakshapanicker P (2020) Improving adoption of technologies and interventions for increasing supply of quality livestock feed in low-and middle-income countries. Glob Food Sec 26:100372

Bamundaga G, Natumanya R, Kugonza D, Owiny D (2018) Reproductive performance of single and double artificial insemination protocol in swine. Bull Anim Health Prod Afr 66(1):143–157

Banhazi TM, Lehr H, Black J, Crabtree H, Schofield P, Tscharke M, Berckmans D (2012) Precision livestock farming: an international review of scientific and commercial aspects. Int J Agric Biol Eng 5(3):1–9

Baronet D, Vaillancourt D (1989) Diagnostic de gestation par échotomographie chez la chèvre. Méd Vét Québec 19:67–72

Batlang U, Tsurupe G, Segwagwe A, Obopile M (2014) Development and application of modern agricultural biotechnology in Botswana: the potentials, opportunities and challenges. GM Crops Food 5(3):183–194

Benjamin M, Yik S (2019) Precision livestock farming in swine welfare: a review for swine practitioners. Animals 9(4):133

Berckmans D (2017) General introduction to precision livestock farming. Anim Front 7(1):6–11

Bharti MK, Jacob N (2019) Laboratory and imaging techniques for pregnancy diagnosis in animals. J Entomol Zool Stud 7(5):8

Braun U (2004) Diagnostic ultrasonography in bovine internal diseases. MEDECIN VETERINAIRE DU QUEBEC 34:13–14

Bukar MM, Yusoff R, Dhaliwal GK, Khan MAG, Omar MA (2012) Estrus response and follicular development in Boer does synchronized with flugestone acetate and PGF 2α or their combination with eCG or FSH. Trop Anim Health Prod 44(7):1505–1511

Burrow HM, Mrode R, Mwai AO, Coffey MP, Hayes BJ (2021) Challenges and opportunities in applying genomic selection to ruminants owned by smallholder farmers. Agriculture 11(11):1172

Cánovas A, Rincón G, Islas-Trejo A, Jimenez-Flores R, Laubscher A, Medrano JF (2013) RNA sequencing to study gene expression and single nucleotide polymorphism variation associated with citrate content in cow milk. J Dairy Sci 96(4):2637–2648

Cartee RE, Hudson JA, Finn-Bodner S (1993) Ultrasonography. Vet Clin N Am Small Anim Pract 23(2):345–377

Celestin M, Valentine N, Isaac M, Fabrice M, Oscar N, FranÃ B, Vianey MJM (2019) Factors influencing success of artificial insemination of pigs using extended fresh semen in rural smallholder pig farms of Rwanda. Int J Livestock Prod 10(4):101–109

Ciccarelli M, Giassetti MI, Miao D, Oatley MJ, Robbins C, Lopez-Biladeau B, Waqas MS, Tibary A, Whitelaw B, Lillico S, Park C-H, Park K-E, Telugu B, Fan Z, Liu Y, Regouski M, Polejaeva IA, Oatley JM (2020) Donor-derived spermatogenesis following stem cell transplantation in sterile NANOS2 knockout males. Proc Natl Acad Sci 117(39):24195

Clark S (2011) Vulvar skin temperature changes significantly during estrus in swine as determined by digital infrared thermography. J Swine Health Prod 19(3):151

Cognie Y, Baril G, Poulin N, Mermillod P (2003) Current status of embryo technologies in sheep and goat. Theriogenology 59(1):171–188

Cognié Y, Poulin N, Locatelli Y, Mermillod P (2004) State-of-the-art production, conservation and transfer of in-vitro-produced embryos in small ruminants. Reprod Fertil Dev 16(4):437–445

Cortes O, Cañon J, Gama LT (2022) Applications of microsatellites and single nucleotide polymorphisms for the genetic characterization of cattle and small ruminants: an overview. Ruminants 2(4):456–470

Costard S, Wieland B, De Glanville W, Jori F, Rowlands R, Vosloo W, Roger F, Pfeiffer DU, Dixon LK (2009) African swine fever: how can global spread be prevented? Philos Trans R Soc B Biol Sci 364(1530):2683–2696

Cui X, Hou Y, Yang S, Xie Y, Zhang S, Zhang Y, Zhang Q, Lu X, Liu GE, Sun D (2014) Transcriptional profiling of mammary gland in Holstein cows with extremely different milk protein and fat percentage using RNA sequencing. BMC Genomics 15(1):1–15

Davani-Davari D, Negahdaripour M, Karimzadeh I, Seifan M, Mohkam M, Masoumi SJ, Berenjian A, Ghasemi Y (2019) Prebiotics: definition, types, sources, mechanisms, and clinical applications. Food Secur 8(3):92

Demir E, Ceccobelli S, Bilginer U, Pasquini M, Attard G, Karsli T (2022) Conservation and selection of genes related to environmental adaptation in native small ruminant breeds: a review. Ruminants 2(2):255–270

Denli M, Okan F, Celik K (2003) Effect of dietary probiotic, organic acid and antibiotic supplementation to diets on broiler performance and carcass yield. Pak J Nutr 2:89

Enahoro D, Mason-D'Croz D, Mul M, Rich KM, Robinson TP, Thornton P, Staal SS (2019) Supporting sustainable expansion of livestock production in South Asia and Sub-Saharan Africa: scenario analysis of investment options. Glob Food Sec 20:114–121

Fierro S, Gil J, Viñoles C, Olivera-Muzante J (2013) The use of prostaglandins in controlling estrous cycle of the ewe: a review. Theriogenology 79(3):399–408

Fowler KE, Mandawala AA, Griffin DK, Walling GA, Harvey SC (2018) The production of pig preimplantation embryos in vitro: current progress and future prospects. Reprod Biol 18(3):203–211

Gadde U, Kim W, Oh S, Lillehoj HS (2017) Alternatives to antibiotics for maximizing growth performance and feed efficiency in poultry: a review. Anim Health Res Rev 18(1):26–45

Gaillard C, Durand M, Largouët C, Dourmad J-Y, Tallet C (2021) Effects of the environment and animal behavior on nutrient requirements for gestating sows: future improvements in precision feeding. Anim Feed Sci Technol 279:115034

Gaouar SBS, Kdidi S, Ouragh L (2016) Estimating population structure and genetic diversity of five Moroccan sheep breeds by microsatellite markers. Small Rumin Res 144:23–27

Godfrey R, Collins J, Hensley E, Wheaton J (1999) Estrus synchronization and artificial insemination of hair sheep ewes in the tropics. Theriogenology 51(5):985–997

Godyń D, Herbut P, Angrecka S, Corrêa Vieira FM (2020) Use of different cooling methods in pig facilities to alleviate the effects of heat stress—a review. Animals 10(9):1459

Goel A, Kharche S, Jindal S (2009) Determination of early pregnancy and embryonic development by transrectal ultrasonography in goats. Indian J Anim Sci 79(5):476–478

Gonzalez-Bulnes A, Pallares P, Vazquez M (2010) Ultrasonographic imaging in small ruminant reproduction. Reprod Domest Anim 45:9–20

Gore DLM, Mburu JN, Okeno TO, Muasya TK (2020a) Short-term oestrous synchronisation protocol following single fixed-time artificial insemination and natural mating as alternative to long-term protocol in dairy goats. Small Rumin Res 192:106207

Gore D, Muasya T, Okeno T, Mburu J (2020b) Comparative reproductive performance of Saanen and Toggenburg bucks raised under tropical environment. Trop Anim Health Prod 52:2653–2658

Gowane GR, Gadekar YP, Prakash V, Kadam V, Chopra A, Prince LLL (2017) Climate change impact on sheep production: growth, milk, wool, and meat. In: Sejian V, Bhatta R, Gaughan J, Malik PK, Naqvi SMK, Lal R (eds) Sheep production adapting to climate change. Springer Singapore, Singapore, pp 31–69

Graff K, Meintjes M, Dyer V, Paul J, Denniston R, Ziomek C, Godke R (1999) Transvaginal ultrasound-guided oocyte retrieval following FSH stimulation of domestic goats. Theriogenology 51(6):1099–1119

Griffith M, Griffith OL, Mwenifumbo J, Goya R, Morrissy AS, Morin RD, Corbett R, Tang MJ, Hou Y-C, Pugh TJ (2010) Alternative expression analysis by RNA sequencing. Nat Methods 7(10):843–847

Gunderson KL, Steemers FJ, Ren H, Ng P, Zhou L, Tsan C, Chang W, Bullis D, Musmacker J, King C (2006) Whole-genome genotyping. Methods Enzymol 410:359–376

Habeeb HMH, Kutzler MA (2021) Estrus synchronization in the sheep and goat. Veterina Clin Food Anim Pract 37(1):125–137

Havlík P, Leclère D, Valin H, Herrero M, Schmid E, Soussana J, Müller C, Obersteiner M (2015) Global climate change, food supply and livestock production systems: a bioeconomic analysis. In: Climate change and food systems: global assessments and implications for food security and trade. Food Agriculture Organization of the United Nations (FAO), Rome

Heller MJ (2002) DNA microarray technology: devices, systems, and applications. Annu Rev Biomed Eng 4(1):129–153

Hendriks WH, Verstegen MW, Babinszky L (2019) Poultry and pig nutrition: challenges of the 21st century. Wageningen Academic Publishers

Herrero M, Havlik P, McIntire J, Palazzo A, Valin H (2014) African Livestock Futures: realizing the potential of livestock for food security, poverty reduction and the environment in Sub-Saharan Africa

Higgins V, Dibden J, Cocklin C (2008) Building alternative agri-food networks: certification, embeddedness and agri-environmental governance. J Rural Stud 24(1):15–27

Ibeagha-Awemu EM, Peters SO, Bemji MN, Adeleke MA, Do DN (2019) Leveraging available resources and stakeholder involvement for improved productivity of African livestock in the era of genomic breeding. Front Genet 10:357

Ikoma E, Suzuki C, Ishihara Y, Komura K, Ookoda T, Maruno H (2014) Transvaginal ultrasound-guided ovum pick up (OPU) in Berkshire breed during natural estrous cycle. Japan J Swine Sci 51(2):45–53

Inaba T, Nakazima Y, Matsui N, Imori T (1983) Early pregnancy diagnosis in sows by ultrasonic linear electronic scanning. Theriogenology 20(1):97–101

Jang HS, Shin WJ, Lee JE, Do JT (2017) CpG and non-CpG methylation in epigenetic gene regulation and brain function. Genes 8(6):148

Jin Y-L, Speers RA (1998) Flocculation of Saccharomyces cerevisiae. Food Res Int 31(6–7):421–440

Karavolias NG, Horner W, Abugu MN, Evanega SN (2021) Application of gene editing for climate change in agriculture. Front Sustain Food Syst 5:296

Kauffold J, Althouse GC (2007) An update on the use of B-mode ultrasonography in female pig reproduction. Theriogenology 67(5):901–911

Kauffold J, Peltoniemi O, Wehrend A, Althouse GC (2019) Principles and clinical uses of real-time ultrasonography in female swine reproduction. Animals 9(11):950

Kaumbata W, Nakimbugwe H, Nandolo W, Banda LJ, Mészáros G, Gondwe T, Woodward-Greene MJ, Rosen BD, Van Tassell CP, Sölkner J (2021) Experiences from the implementation of community-based goat breeding programs in Malawi and Uganda: a potential approach for conservation and improvement of indigenous small ruminants in smallholder farms. Sustain For 13(3):1494

King GJ, Macpherson JW (1973) A comparison of two methods for boar semen collection. J Anim Sci 36(3):563–565

Knox RV (2016) Artificial insemination in pigs today. Theriogenology 85(1):83–93

Ko J, Evans L, Althouse G (1989) Toxicity effects of latex gloves on boar spermatozoa. Theriogenology 31(6):1159–1164

Kosgey IS, Baker RL, Udo HMJ, Van Arendonk JAM (2006) Successes and failures of small ruminant breeding programmes in the tropics: a review. Small Rumin Res 61(1):13–28

Kutty CI (1999) Gynecological examination and pregnancy diagnosis in small ruminants using bimanual palpation technique: a review. Theriogenology 51(8):1555–1564

Latino LR, Pica-Ciamarra U, Wisser D (2020) Africa: the livestock revolution urbanizes. Glob Food Sec 26:100399

Leles JS, Rodrigues ICS, Neto MFV, Neto AMV, da Rocha DR, da Costa ANL, Salles MGF, de Araújo AA (2017) Heat stress and body temperature in Brown Swiss cows raised in semi-arid climate of Ceará State, Brazil. Acta Sci Vet 45:1–8

Li Y-C, Korol AB, Fahima T, Nevo E (2004) Microsatellites within genes: structure, function, and evolution. Mol Biol Evol 21(6):991–1007

Li H, Yang Y, Hong W, Huang M, Wu M, Zhao X (2020a) Applications of genome editing technology in the targeted therapy of human diseases: mechanisms, advances and prospects. Signal Transduct Target Ther 5(1):1–23

Li A, Tanner MR, Lee CM, Hurley AE, De Giorgi M, Jarrett KE, Davis TH, Doerfler AM, Bao G, Beeton C (2020b) AAV-CRISPR gene editing is negated by pre-existing immunity to Cas9. Mol Ther 28(6):1432–1441

Maes D, Lopez Rodriguez A, Rijsselaere T, Vyt P, Van Soom A (2011) Artificial insemination in pigs. In: Artificial insemination in farm animals. In-Tech, pp 79–94

Mantione KJ, Kream RM, Kuzelova H, Ptacek R, Raboch J, Samuel JM, Stefano GB (2014) Comparing bioinformatic gene expression profiling methods: microarray and RNA-Seq. Med Sci Monit Basic Res 20:138

Markowiak P, Śliżewska K (2018) The role of probiotics, prebiotics and synbiotics in animal nutrition. Gut Pathogen 10(1):1–20

Marshall K, Gibson JP, Mwai O, Mwacharo JM, Haile A, Getachew T, Mrode R, Kemp SJ (2019) Livestock genomics for developing countries—African examples in practice. Front Genet 10:297

Martinez E, Vazquez JM, Roca J, Lucas X, Gil MA, Vazquez J (2001) Deep intrauterine insemination and embryo transfer in pigs. Reprod Cambridge-Suppl 58:301–311

Meira EBS, Henriques LCS, Sá LRM, Gregory L (2012) Comparison of ultrasonography and histopathology for the diagnosis of endometritis in Holstein-Friesian cows. J Dairy Sci 95(12):6969–6973

Mekuriaw Z, Assefa H, Tegegne A, Muluneh D (2016) Estrus response and fertility of Menz and crossbred ewes to single prostaglandin injection protocol. Trop Anim Health Prod 48(1):53–57

Mercer TR, Mattick JS (2013) Structure and function of long noncoding RNAs in epigenetic regulation. Nat Struct Mol Biol 20(3):300–307

Molotsi A, Dube B, Oosting S, Marandure T, Mapiye C, Cloete S, Dzama K (2017) Genetic traits of relevance to sustainability of smallholder sheep farming systems in South Africa. Sustain For 9(8):1225

Monteiro A, Santos S, Gonçalves P (2021) Precision agriculture for crop and livestock farming—brief review. Animals 11(8):2345

Moritz C (1994) Applications of mitochondrial DNA analysis in conservation: a critical review. Mol Ecol 3(4):401–411

Mrode R, Tarekegn GM, Mwacharo JM, Djikeng A (2018) Invited review: genomic selection for small ruminants in developed countries: how applicable for the rest of the world? Animal 12(7):1333–1340

Muchadeyi FC, Ibeagha-Awemu EM, Javaremi AN, Gutierrez Reynoso GA, Mwacharo JM, Rothschild MF, Sölkner J (2020) Editorial: why livestock genomics for developing countries offers opportunities for success. Front Genet 11:626

Nääs IA, Garcia RG, Caldara FR (2020) Infrared thermal image for assessing animal health and welfare. J Anim Behav Biometeorol 2(3):66–72

Nantoumé H (2020) Sheep feeding in the Sahel countries of Africa. In: Sheep farming-an approach to feed, growth and health

Niyiragira V, Rugira KD (2018) Success drivers of pig artificial insemination based on imported fresh semen. Int J Livestock Prod 9(6):102–107

Niyiragira V, Rugira KD, D'Andre HC (2018) Success drivers of pig artificial insemination based on imported fresh semen. Int J Livestock Prod 9(6):102–107

Oke-Egbodo B, Nwannenna A, Bawa E, Hassan R, Bello T, Bugau J, Oke P, Rekwot G (2018) Refractoriness of Porcine Corpora Lutea to Cloprostenol Sodium and Dinoprost Tromethamine treatments at day 7 of Oestrous cycle. Int J Agric Forest 8(3):119–123

Okpeku M, Peters S, Ozoje M, Adebambo O, Agaviezor B, O'Neill M, Imumorin I (2011) Preliminary analysis of microsatellite-based genetic diversity of goats in southern Nigeria. Anim Genet Resour/Resources génétiques animales/Recursos genéticos animales 49:33–41

Oluwole O, Omitogun GO (2015) Phenotypic evaluation of Nigerian Indigenous Pigs (NIP), its hybrid and backcross for Litter and reproductive traits during dry and wet season

Omontese B, Rekwot P, Ate I, Ayo J, Kawu M, Rwuaan J, Nwannenna A, Mustapha R, Bello A (2016) An update on oestrus synchronisation of goats in Nigeria. Asian Pacific J Reprod 5(2):96–101

Osei-Amponsah R, Skinner BM, Adjei DO, Bauer J, Larson G, Affara NA, Sargent CA (2017) Origin and phylogenetic status of the local Ashanti Dwarf pig (ADP) of Ghana based on genetic analysis. BMC Genomics 18(1):193

Osei-Amponsah R, Chauhan SS, Leury BJ, Cheng L, Cullen B, Clarke IJ, Dunshea FR (2019) Genetic selection for thermotolerance in ruminants. Animals 9(11):948

Park K-E, Kaucher AV, Powell A, Waqas MS, Sandmaier SES, Oatley MJ, Park C-H, Tibary A, Donovan DM, Blomberg LA, Lillico SG, Whitelaw CBA, Mileham A, Telugu BP, Oatley JM (2017) Generation of germline ablated male pigs by CRISPR/Cas9 editing of the NANOS2 gene. Sci Rep 7(1):40176

Popoola MA, Bogoro SE, Feugang JM (2022) Biothermoimaging tools for management of climate smart and precision livestock-assisted reproduction.

In: Agricultural biotechnology, biodiversity and bioresources conservation and utilization. CRC Press, pp 315–334

Popova LV, Dugina TA, Skiter NN, Panova NS, Dosova AG (2017) New forms of state support for the agro-industrial complex in the conditions of digital economy as a basis of food security provision. In: International conference on humans as an object of study by modern science. Springer, pp 681–687

Powell W, Morgante M, Andre C, Hanafey M, Vogel J, Tingey S, Rafalski A (1996) The comparison of RFLP, RAPD, AFLP and SSR (microsatellite) markers for germplasm analysis. Mol Breed 2(3):225–238

Racewicz P, Ludwiczak A, Skrzypczak E, Składanowska-Baryza J, Biesiada H, Nowak T, Nowaczewski S, Zaborowicz M, Stanisz M, Ślósarz P (2021) Welfare health and productivity in commercial pig herds. Animals 11(4):1176

Rai N (2013) Impact of advertising on consumer behaviour and attitude with reference to consumer durables. Int J Manag Res Busin Strat 2(2):74–79

Ramos-Onsins SE, Burgos-Paz W, Manunza A, Amills M (2014) Mining the pig genome to investigate the domestication process. Heredity 113(6):471–484

Ramukhithi FV, Nedambale TL, Sutherland B, Greyling JPC, Lehloenya KC (2012) Oestrous synchronisation and pregnancy rate following artificial insemination (AI) in South African indigenous goats. J Appl Anim Res 40(4):292–296

Rauw WM, Rydhmer L, Kyriazakis I, Øverland M, Gilbert H, Dekkers JC, Hermesch S, Bouquet A, Gómez Izquierdo E, Louveau I (2020) Prospects for sustainability of pig production in relation to climate change and novel feed resources. J Sci Food Agric 100(9):3575–3586

Rekant SI, Lyons MA, Pacheco JM, Arzt J, Rodriguez LL (2016) Veterinary applications of infrared thermography. Am J Vet Res 77(1):98–107

Renaudeau D, Gourdine JL, St-Pierre NR (2011) A meta-analysis of the effects of high ambient temperature on growth performance of growing-finishing pigs. J Anim Sci 89(7):2220–2230

Roca J, Hernandez M, Carvajal G, Vazquez JM, Martinez EA (2006) Factors influencing boar sperm cryosurvival. J Anim Sci 84(10):2692–2699

Selaledi L, Mbajiorgu C, Mabelebele M (2020) The use of yellow mealworm (T. molitor) as alternative source of protein in poultry diets: a review. Trop Anim Health Prod 52(1):7–16

Sh B, Bisrat A, Ch D, Abebe A, Abebe A, Sh G, Zewude T (2020) Comparative advantages of cervical insemination over natural mating on. J World Poult Res 10(2):66

Silanikove N, Leitner G, Merin U, Prosser CG (2010) Recent advances in exploiting goat's milk: quality, safety and production aspects. Small Rumin Res 89(2–3):110–124

Simpkin P, Cramer L, Ericksen P, Thornton PK (2020) Current situation and plausible future scenarios for livestock management systems under climate change in Africa. CCAFS Working Paper

Soerensen DD, Pedersen LJ (2015) Infrared skin temperature measurements for monitoring health in pigs: a review. Acta Vet Scand 57(1):5

Stelletta C, Gianesella M, Vencato J, Fiore E, Morgante M (2012) Thermographic applications in veterinary medicine. In: Infrared thermography. In Tech, China, pp 117–140

Steyn J (2003) Application of Artificial Insemination (AI) on commercial sheep and goat production. Proceeding Simpósio Internacional sobre Caprinos e Ovinos de Corte 2:367–379

Suryadi U, Nugraheni YR, Prasetyo AF, Awaludin A (2019) Evaluation of effects of a novel probiotic feed supplement on the quality of broiler meat. Veterin world 12(11):1775

Sykes DJ, Couvillion JS, Cromiak A, Bowers S, Schenck E, Crenshaw M, Ryan PL (2012) The use of digital infrared thermal imaging to detect estrus in gilts. Theriogenology 78(1):147–152

Tainturier D, Lijour L, Chaari M, Sardjana K, Denis B (1983) Diagnostic de la gestation chez la brebis par échotomographie. Rev Med Vet 134:523–526

Thorstensen MJ, Baerwald MR, Jeffries KM (2021) RNA sequencing describes both population structure and plasticity-selection dynamics in a non-model fish. BMC Genomics 22(1):1–12

Timmerman H, Veldman A, Van den Elsen E, Rombouts F, Beynen A (2006) Mortality and growth performance of broilers given drinking water supplemented with chicken-specific probiotics. Poult Sci 85(8):1383–1388

Toms D, Pan B, Bai Y, Li J (2021) Small RNA sequencing reveals distinct nuclear microRNAs in pig granulosa cells during ovarian follicle growth. J Ovarian Res 14(1):1–12

Valenzuela-Grijalva NV, Pinelli-Saavedra A, Muhlia-Almazan A, Domínguez-Díaz D, González-Ríos H (2017) Dietary inclusion effects of phytochemicals as growth promoters in animal production. J Anim Sci Technol 59(1):1–17

Van Eenennaam AL, De Figueiredo Silva F, Trott JF, Zilberman D (2021) Genetic engineering of livestock: the opportunity cost of regulatory delay. Ann Rev Anim Biosci 9:453–478

Van Marle-Koster E, Nel L (2003) Genetic markers and their application in livestock breeding in South Africa: a review. S Afr J Anim Sci 33(1):1–10

Vishwanath R (2003) Artificial insemination: the state of the art. Theriogenology 59(2):571–584

Visser C (2019) A review on goats in southern Africa: an untapped genetic resource. Small Rumin Res 176:11–16

Vranken E, Berckmans D (2017) Precision livestock farming for pigs. Anim Front 7(1):32–37

Waberski D, Riesenbeck A, Schulze M, Weitze KF, Johnson L (2019) Application of preserved boar semen for artificial insemination: past, present and future challenges. Theriogenology 137:2–7

Williams SI, Piñeyro P, de la Sota RL (2008) Accuracy of pregnancy diagnosis in swine by ultrasonography. Can Vet J 49(3):269

Wong L-JC, Boles RG (2005) Mitochondrial DNA analysis in clinical laboratory diagnostics. Clin Chim Acta 354(1–2):1–20

Wright JM, Bentzen P (1995) Microsatellites: genetic markers for the future. In: Molecular genetics in fisheries. Springer, pp 117–121

Yadeta JF (2020) Review on potential of reproductive technology to improved ruminant production in Ethiopia. J Biol Agric Healthc 10(8):10–24

Yoshioka K, Noguchi M, Suzuki C (2012) Production of piglets from in vitro-produced embryos following non-surgical transfer. Anim Reprod Sci 131(1):23–29

Zsidó BZ, Hetényi C (2020) Molecular structure, binding affinity, and biological activity in the Epigenome. Int J Mol Sci 21(11):4134

Open Access This chapter is licensed under the terms of the Creative Commons Attribution 4.0 International License (http://creativecommons.org/licenses/by/4.0/), which permits use, sharing, adaptation, distribution and reproduction in any medium or format, as long as you give appropriate credit to the original author(s) and the source, provide a link to the Creative Commons license and indicate if changes were made.

The images or other third party material in this chapter are included in the chapter's Creative Commons license, unless indicated otherwise in a credit line to the material. If material is not included in the chapter's Creative Commons license and your intended use is not permitted by statutory regulation or exceeds the permitted use, you will need to obtain permission directly from the copyright holder.

Part IV

Conservation of African Livestock Genetic Resources

Conservation and Management of Animal Genetic Resources in the Context of African Livestock Production Systems: The Case for In Situ and Ex Situ Conservation

Jean M. Feugang, Richard Osei-Amponsah, John E. O. Rege, Khaled Fantazi, Christian K. Tiambo, Felicien Shumbusho, Isidore Houaga, Derradji Harek, Notsile H. Dlamini, and Semir B. S. Gaouar

Abstract

Conserving and managing African animal genetic resources requires understanding how these resources evolve (political, environmental, cultural, and social). Factors associated with this context are unfavorable for sound conservation and management in most African countries. This chapter presents the major threats to indigenous animal genetic resources (Sect. 25.2) in Africa; the in situ and ex situ conservation methods practiced (Sects. 25.3 and 25.4), operational guidelines (Sect. 25.5), and the opportunities presented by modern technologies (Sect. 25.6). More research structures and skills in reproductive biotechnology and genetics are needed in the continent to ensure the sustainability of animal resources, in particular those with low numbers or less economic interest. Some African countries with constituted herds for conservation (ex situ in vivo) have also taken the initiative to generate gametes, embryos, tissue, and DNA banks.

J. M. Feugang (✉) · N. H. Dlamini
Department of Animal and Dairy Sciences, Mississippi State University, Mississippi State, MS, USA
e-mail: j.feugang@msstate.edu

R. Osei-Amponsah
Department of Animal Science, University of Ghana, Accra, Ghana

J. E. O. Rege
Emerge Centre for Innovations-Africa, Nairobi, Kenya

K. Fantazi · D. Harek
Animal Production, INRAA, Algiers, Algeria

C. K. Tiambo
Livestock Genetics, Centre for Tropical Livestock Genetics and Health/International Livestock Research Institute, Nairobi, Kenya

F. Shumbusho
Animal Resources, Rwanda Agriculture and Animal Resources Development Board (RAB), Kigali, Rwanda

I. Houaga
Centre for Tropical Livestock Genetics and Health, The Roslin Institute, University of Edinburgh, Edinburgh, UK

S. B. S. Gaouar
Applied Genetic in Agriculture, Ecology and Public Health Laboratory, Faculty SN/STU, University of Tlemcen, Tlemcen, Algeria

The multi-stakeholder breeders-researchers-decision-makers approach remains the most robust solution for sound management and preservation of biological units.

Keywords

Animal genetic resources (AnGR) · Community-based breeding programs · Gene-bank · Genomics · Indigenous livestock · In situ conservation · Ex situ conservation

Abbreviations

AI	Artificial Insemination
AnGR	Animal Genetic Resources
ANOC	Association Nationale des Ovins et Caprins (National Association of Sheep and Goat)
APA	Accès et Partage or Access and Benefit-Sharing
APRODJALCI	Association des Producteurs de chèvres Djallonké de Côte d'Ivoire
ART/ARTs	Assisted Reproductive Technologies
ASF	Animal Source Food
AU	African Union
AUC	African Union Commission
AU-IBAR	African Union-InterAfrican Bureau for Animal Resources
BNG	Banque Nationale de Gènes or National Genebank
BRG	Bank of Genetic Resources
CBBP	Community-Based livestock Breeding Programs
CBD	Convention on Biological Diversity
CGIAR	Consultative Group International Agricultural Research
CIRDES	Centre International de Recherche-Développement sur l'Elevage en zone Subhumide
COVID	Coronavirus Disease
CTLGH	Centre for Tropical Livestock Genetics and Health
DNA	Deoxyribonucleic Acid
EBV	Estimated Breeding Value
ECOWAS	Economic Community of West African States
ET	Embryo Transfer
EU	European Union
FAO	Food and Agricultural Organization
GWA	Genome-Wide Association
ILRI	International Livestock Research Institute
mtDNA	Mitochondrial DNA
NSAP	National Strategic Action Plans
OPU	Ovum-Pick Up
PNSO	Programme National de Selection Ovine
RFLP	Restricted Fragment Length Polymorphism
RICBO	Rwanda Indigenous Cattle Breeds Organization
RNA	Ribonucleic Acid
SDDP	Samburu District Development Program
SNP	Single nucleotide polymorphism
UNCED	United Nations Conference on Environment and Development
USAID	United States Agency for International Development

25.1 Introduction

Livestock production faces significant challenges associated with critical drivers of change in the global livestock sector today. These trends include changing market demands, more specifically the *livestock revolution*—the unprecedented global shift in demand for livestock products such as meat, milk, and eggs; economic development and globalization; significant changes in the demographic as well as the regional distribution of these demands; the need to reduce poverty in rural communities by supporting sustainable livelihoods; a growing awareness of the need to reduce the environmental impact of livestock production; the uncertainties of climate change;

and science and technology innovations. The need for legislation on livestock breeding and the equitable distribution of grazing lands and water resources is a major challenge for development in Africa. The continent remains home to significant genetic diversity, which trends affect livestock's ability to improve livelihoods and reduce poverty and the use of natural resources. In the developed world, the narrowing animal genetic resources (AnGR) base in industrial livestock production systems also raises the need to maintain a broader range of AnGR to deal with future uncertainties, such as climate change and zoonotic diseases. The implications are that breeds cannot adapt in time to meet new circumstances. Therefore, strategic interventions are necessary to improve the management of animal genetic resources in situations where these genetic resources are most at risk (Seré et al. 2008).

The worldwide growing industrial livestock production systems have a limited demand for biodiversity. In contrast, crop-livestock and pastoral systems, the primary livestock systems in Africa, will continue to rely on biodiversity in diverse genotypes of improved productivity under different and constantly changing environmental and socio-economic conditions. However, all systems will require genetic diversity, albeit to varying degrees, to cope with expected climate change and other insecurities (Osei-Amponsah et al. 2019). Thus, threats to livestock genetic diversity, wherever it is occurring, should be a global concern. Furthermore, the multiple values, functions, and consequences of the different livestock production systems and their rapid rate of change imply diverse interests within and among countries.

Conversely, uncertainties about the implications of rapid, multifaceted global change for each livestock production system, the livestock species and breeds within it, and the resulting future changes in the required genetic make-up of AnGR make collective action to tackle their conservation a long-term, global public good. In this context and a spirit of complementarity, North-South and South-South win-win relations are to be encouraged and initiated quickly by international organizations (e.g., Food Agricultural Organization or FAO) or research institutes (e.g., International Livestock Research Institute or ILRI). Of course, conserving AnGR will not, by itself, solve these problems. Still, it is an essential first step toward maintaining future options (Seré et al. 2008). The conservation of AnGR refers to all human activities (i.e., strategies, plans, policies, and actions) undertaken to maintain the diversity of AnGR for food and agricultural production and productivity or other present and future values of these resources (e.g., ecological, cultural) (Rischkowsky and Pilling 2007). It consists of both in situ and ex situ conservation methods. The latter is subdivided into ex situ in vivo and ex situ in vitro.

This chapter examines significant threats to African indigenous AnGR. Then, it presents options for conservation interventions, focusing on community-based approaches designed to be owned and driven by local farming communities.

25.2 Major Causes of Threat to AnGR

Considerable literature describes numerous threats to AnGR. For example, in an early study on African cattle genetic resources, 32% (47/145) of total cattle breeds in the continent were at risk of extinction, while about 15% (22/145) were extinct in the last century (Rege 1999). In 2007, the FAO provided the first global assessment and analysis of the status and trends in AnGR that raised worldwide awareness and understanding of the ongoing erosion (Food Organization, A 2007), with existent threats for indigenous livestock breeds in Africa. Several factors, such as the replacement of indigenous breeds by others, crossbreeding practices, local conflicts, animal diseases, neglected or lack of sustained breeding programs, and loss or restriction of habitats, gradually influence the distribution and characteristics of production systems (Alexandratos and Bruinsma 2012). Such is the case of various species, including the Bakosi cattle in Cameroon, the Ouled-Djellal sheep in Algeria, the Kuri cattle of Lake Chad, and the Walia Ibex goat of Ethiopia (Rege 1999; Gaouar 2002; Carson et al. 2009;

Iniguez 2005, 2011; Tawah et al. 1997). In the 2009 FAO global survey report, the economic and market drivers (28.5%), poor livestock sector policies (20.2%), and poor conservation strategies (14.5%) were the three most frequently mentioned threats (Hoffmann 2010a). Later, the FAO alarmed the world about the increased proportion of breeds classified in danger of extinction, from 15 to 17% (249/1,458 total breeds) between 2005 and 2014 (Scherf and Pilling 2015). This increase persisted in 2019, with 26% of the world's livestock breeds being at risk of extinction (Pilling et al. 2020; Bélanger and Pilling 2019).

The causes of threats to AnGR are multifactorial and global due to the following: (1) human and bioclimatic activities of the surrounding environments; (2) visual and highly unpredictable climate change impacts (i.e., high frequency, severity and duration of drought, high temperatures, and soil erosion) in various regions of Africa (Köhler-Rollefson 2004; Famien 2020); (3) resurgence/persistence of diseases affecting animals, while pandemics such as COVID-19 are putting the world in a public health crisis requiring a coordinated and multisectoral response (Baptista et al. 2021; Obese et al. 2021); (4) worldwide fast-growing human populations creates unprecedented demands for animal-source food (ASF); (5) indiscriminate crossbreeding between local breeds, and with exotic breeds; (6) replacement of local breeds by exotic breeds; and (7) the lack of continued efforts for management and conservation of local AnGR.

Since the Convention on Biological Diversity (CBD) in 2002, numerous global, regional, and national efforts have been undertaken to reverse the current trend of biodiversity loss (Hoffmann 2010a). However, the results of these efforts remain challenging to measure, and there is still a belief of an unimaginable loss of the diversity of endemic livestock affecting AnGR. For example, in Africa, the existing animal populations are increasingly losing their purity, and the notion of race no longer has any meaning. Nonetheless, FAO has made significant progress in assessing biodiversity. First, by publishing "The State of the World for Animal Genetic Resources" and second, by maintaining the Domestic Animal Diversity Information System (DAD-IS) as the clearinghouse mechanism for AnGR recognized by the CBD (Rischkowsky and Pilling 2007; Food Organization, A 2007; Hoffmann 2010b).

In a joint study, FAO, ILRI, and AU-IBAR (African Union-InterAfrican Bureau of Animal Resources) have estimated that the meat and milk markets will reach 34.8 and 82.6 million tons by 2050, corresponding to a respective increase of 145% and 155% compared to the levels of 2005/2007 (Ugo et al. 2013). As a result, consumption will surpass production. Africa will become a net importer of ASF if national governments do not make solid decisions for livestock development. For example, many countries use exotic breeds to improve productivity through crossbreeding programs with local breeds. This breeding approach is the preferred continent's tool for carrying out genetic improvement within the framework of controlled programs to preserve local diversity. Unfortunately, progeny from F2 crossbreeding perform poorly in most cases due to the loss of adaptive traits in African production systems, which undermines the sustainability of breeding programs. Nonetheless, various genetic conservation programs are established in response to the existing constraints related to crossbreeding (Table 25.1). Several priorities and complementary interventions are identified to improve the management of AnGR and their maintenance for future uses. The "keep it on the roof," encouraging the continuing sustainable use of traditional breeds (in situ conservation), and the "Put some in the bank," using new technologies for ex situ in vitro conservation of AnGR, are among those strategies.

25.3 In Situ Conservation Methods

This conservation method refers to preserving genetic resources through living animals evolving or rearing in cost-effective and safe natural environments (Rischkowsky and Pilling 2007; Hoffmann 2011). *Most African livestock* genetic resources *are maintained* in situ *in living populations, as this* conservation method provides sev-

Table 25.1 Constraints of animal selection and crossbreeding in Africa

Crossing constraints	Selection constraints
Knowledge of local genetic resources by population size, genetic characteristics, and average performance	Marginalization of local breeds in favor of exotic breeds
Low involvement of breeders	Lack of information and exhaustive knowledge of local resources and their qualities, characteristics, and precise data on numbers by breed
Supply of genetic resources (exotic and indigenous), food resources, and healthcare	Lack of a selection program for the conservation of biodiversity
Uncontrolled technicality for a more demanding product	Insufficient technical expertise
Satisfaction of the breeder with the product (cost/product ratio)	Lack of technological equipment (IA, MOET, and molecular tools)
Difficult flow of products	Weak monitoring and control
Lack of clear and guaranteed programs for elements that may affect the sustainability of the program	Poor collection and dissemination of information

eral advantages (Rege and Gibson 2003; Jiri et al. 2017). A few examples are as follows: (1) management of a wide range of genetic and species-diverse populations in their natural environment, thus a substantial genetic base (e.g., indigenous breeds and wild species co-conservation); (2) maintenance of an integrated system supporting continuous environmental adaptation; (3) preservation of animals selected based upon their adaptive traits (e.g., low feed requirement, ability to utilize poor quality feeds, thermo- and hydric-tolerance, and disease and parasite resistance); (4) introduction of high-performance breeds for genetic improvement of local breeds; and (5) low cost and effective gene conservation method.

In parallel, some drawbacks to the in situ conservation method can severely hamper farm success. Indeed, the same factors that allow for dynamic, holistic agroecosystem conservation may threaten the security of breeds (e.g., natural disasters, pathogen outbreaks, genetic problems, breeding cessation), altering genetic diversity (Gaouar 2002; Tiambo et al. 2019). Also, social and economic changes due to government policy changes (e.g., subsidies or direct interventions) may either foster or hinder on-farm AnGR conservation over time. Numerous examples indicate the reliance of this conservation method on smallholders' encouragement through market-driven incentives supported by appropriate public policies.

25.3.1 Challenges of Applying In Situ Conservations in African Countries

Many challenges are associated with in situ conservation methods in Africa (Gaouar 2002; Rege and Gibson 2003; Tiambo et al. 2019; Rege 2003; Gaouar 2009). The following examples can engender secondary causes of breed extinction, production, and productivity threats. (1) Economic and market drivers are significant threats that act through other more direct causes, such as crossbreeding and breed replacement, considering the spatial and temporal dynamics. (2) Significant gaps in regulatory policies and institutional arrangements directly relate to smallholders, generally rearing indigenous livestock breeds. (3) Insufficient knowledge of local breeds can lead to inappropriate crossbreeding, negatively impacting the hybrid vigor of the offspring. Indeed, most African farmers do not keep detailed records of each animal (e.g., pedigree, performance, and tags) to enable better breeding results. These farmers mostly use a complex set of traits (e.g., egg production, size, shape, color, milk yield and quality, aspects of adaptation, and the source of the breeding material) to name their animals as the basis of selection decisions and management. (4) Policies must address disease management. (5) Lack of risk awareness of climate change impacts and rigorous interventions for long-term mitigations. (6) Social and cultural contexts of breeds are critical for selection, often associated with traditional practices and beliefs having important implications [e.g., *Muturu* cattle considerations in various Nigerian communities (Adebambo 2001)]. (7) Lack of social organizations and institutions that can influence farmers' access to, and management of, household and community-level resources (e.g., equitability of land tenure and ownership distribution in the community, size, and the number of parcels of household land, intrahousehold access to land), and actions regarding sustainable use and maintenance of farm animal genetic diversity. (8) Economic reasons do not always drive breeding objectives and may take many forms [e.g., the coat color patterns in the Nguni cattle in South Africa (Tada et al. 2013)]. (9) The inconsistency in naming breeds or strains. For example, geographical connotations, tribes, or ethnic communities are associated with breed names (e.g., *Boran* cattle, *Mashona* cattle, *Nguni* cattle, and *Somali* sheep), while some generic names such as the *Djallonke* breed of sheep distributed in the whole coastal West and Central Africa may be overshadowing several sub-populations due to local environment and breeding practices (Wafula et al. 2005; Hoffmann and Scherf 2005).

In addition, western countries' typical industrial livestock systems focus on enhanced control of the production environment, favoring a narrower range of high-producing breeds and increasingly limited within-population diversity. In Africa, farming with imported breeds has largely been a failure in smallholder production systems. Instead, there is more attention to using indigenous breeds and crossbreds while increasing related scientific knowledge. Therefore, livestock development programs in Africa remain driven mainly by imported technological packages (e.g., artificial insemination, exotic germplasm) with minimal involvement of communities in their implementation.

25.3.2 Community-Based Management of Animal Genetic Resources

The "community-based management" emphasizes the involvement of local communities in AnGR conservation. Its contribution to "in situ conservation" can be valuable in an integrated system for maintaining genetic resources. In addition, community-based management is critical for conserving breeds or strains in areas with rich diversity, allowing continued adaptation to changing climate and evolution (Jiri et al. 2017).

In recent decades, the FAO has been working with national and intergovernmental levels to promote international collaborations on AnGR. This effort focuses on institutional and policy frameworks required to improve management worldwide. Strategies are based on currently valuable livestock species in their ecosystems (crops, forages, agroforestry species, and other animal species) to underpin an on-farm conservation program. This strategy helps (1) conserve processes of adaptation, evolution, and diversity, (2) integrate farmers into a national AnGR system, and (3) improve the livelihood of resource-poor farmers through economic and social development.

The "in situ conservation" and "community-based management" of AnGR are conceptually similar. Community-based livestock breeding programs (CBBP) are good examples of community-based management. These programs rely on defined geographical boundaries with farmers of common interests working together to design breeding objectives and execute genetic improvement. Livestock species are generally selected based on community needs and wants, as listed in Sect. 25.3.1. Several African countries have implemented the CBBP as an Open Nucleus Breeding Scheme (ONBS). These strategies comprise nucleus herds established from the best animals selected through screening from the base population of farmers' flocks. Programs started in various African countries in the early 1980s as alternatives to centralized government-controlled breeding schemes with governmental (e.g., research centers) and non-governmental (e.g., farmers) contributions.

The Ivory Coast government established the PNSO (Programme National de Selection Ovine) to improve the growth and live weight of indigenous *Djallonkè* sheep (Yapi-Gnaoré et al. 1997). This program involved approximately 143 farms, of which 88% and 12% were from farmers' flocks and state-owned, respectively. It resulted in the association of the *Djallonke* sheep producers of Ivory Coast (APRODJALCI) and a slightly increased genetic value during the selection period despite the deteriorating environment.

In Morocco, a similar development program termed "Moutonnier" is an excellent example of promoting AnGR conservation and improving the national genetic heritage (Boujenane 2005). This program involved the Moroccan government and the National Association of Sheep and Goat Breeders (ANOC). Local breeds (e.g., the *Barcha*, *Noire*, and *Chefchaouen* goats) of various stakeholders (e.g., community officials, industries, and farmers) were maintained in their cradle zones of origin to ensure the effectiveness of selection and crossbreeding with parents of high genetic value and replication in other geographical sites (e.g., oases of the southern, middle atlas, and the high plateau of Morocco, and the Wadi Saorua Valley of South Algeria) throughout the country (Boujenane 2005; Boukallouch 2006). This project has enabled the production of high-quality goat's milk, butter, buttermilk, and cheese marketed with the label "Ajbane Chefchaouen," the name of the city of production "Ajbane Chefchaouen," the name of the city of production (Boukallouch 2006).

In Ethiopia, established CBBP with small ruminants achieved a substantial genetic improvement of Bonga, Horro, and Menz breeds of sheep (Haile et al. 2020), with a net gain of one to three sheep per year at slaughter, having significant impacts on the livelihood of farmers. As a result, there has been a demand for improved rams and bucks within (e.g., NGOs and other local communities) and across the neighboring countries (e.g., Iran, Malawi, South Africa, Sudan, Tanzania, and Uganda) http://www.icarda.org/update/ethiopia-community-based-breeding-genetic-improvement-sheep-and-goats.

In Rwanda, two cattle breeds, Inyambo and Inkungu, are under in situ conservation and utilization by the Rwanda Indigenous Cattle Breeds Organization (RICBO) and in partnership with the Ministry of Agriculture and Animal Resources.

Despite the real-life successes, in situ conservation through CBBP requires substantial monitoring to ensure that farmers keep or continue keeping indigenous breeds in favor of new ones. Also, incentives may be necessary to encourage farmers to retain and continue traditional practices and know-how. There is a clear need for integrated conservation approaches that combine a range of available in situ and ex situ options. The current technological advances are making the ex situ conservation methods the viable alternatives for AnGR success through live-animal conservation (in situ option) and in vitro manipulations of various animal samples (ex situ option) in designated localities.

25.4 Ex Situ Conservation Methods

Ex situ conservation involves preserving genetic resources off-site or outside their natural habitats. Ex situ collections of AnGR offer the potential to reduce extinction risks, affording the option to maintain future breeding opportunities for productivity and heritage traits (De Oliveira Silva et al. 2019). Ex situ conservation could be in vivo (i.e., live animals in the form of organized herd/flock maintained in zoos, on-farm, or on station) or in vitro (i.e., cryo–conservation of a wide variety of living cells or tissues) for long periods.

25.4.1 Sample Candidates for Conservation

Nowadays, there is increasing recognition by national governments that livestock diversity has political, social, and economic utility. Existing works throughout many African countries indicate the reliance on AnGR conservation sustainability in manipulating various sample types. The harvest of such samples from living or dead animals facilitates their transportation, long-term diversity conservation, and marketability of genetic materials.

Sample candidates for conservations are gonadal, gametic, embryonic, and somatic tissues, and DNA (Bolaji et al. 2021; Morrell and Mayer 2017). Gonadal samples are both testicular and ovarian tissues that can be frozen for future uses through re-implantation or in vitro production of mature gametes. Both female (oocytes) and male (spermatozoa) gamete collections permit further in vitro quality improvement for viable embryo production and generation of somatic cells exhibiting adult embryonic stem cells. Samples are harvested from living or postmortem animals to improve understanding of their biological characteristics and quality. Collected samples should possess a detailed case history to ensure clear identification [e.g., owner's name, species, geographic location, sex, age, and sample type] and management [e.g., handling and breeding techniques (Alhajeri et al. 2021)] using software such as the *Epicollect5* (https://five.epicollect.net/). These cells are kept for long-term storage in cryogenic conditions (freezing at −80 °C or deep-freeze at −196 °C in liquid nitrogen), with relative success across species and sample types. This capacity for conserving samples is often called Genetic Resource Banking (GRB). Numerous studies have demonstrated the ability of fresh and frozen-thawed cells to generate new individuals via assisted reproductive technologies (ARTs), having crucial roles during livestock production or the repopulation of endangered animal species. ARTs are helpful in both in vivo and in vitro options of ex situ conservations.

Freezing various sample types offers many possible strategies for countries to address their AnGR conservation needs (Leroy et al. 2019). Each type of sample has its advantages (e.g., long-term freezing storage—from −20 °C, −80 °C, or −196 °C), disadvantages (e.g., different freezing tolerance, different freezing protocols, and ingredients), and outright (e.g., equipped facilities) limitations, depending upon the species, the cryopreservation, and the technical capacity of the country. For many livestock species, semen collection and cryopreservation are common, inexpensive, and represent simple options for preserving animal genetic resources. However, if semen cryopreservation makes it possible to recreate the descent of the breed, the reconstitution from exclusive semen requires many generations of backcrossing. In parallel, oocyte cryopreservation is now commonly performed with many species. However, cryopreservation of exclusive oocytes results in the loss of the breed's Y-chromosome in mammalian species. When cryopreserved embryos are considered, their preservation captures the diploid genome of an animal. It can therefore be immediately utilized to meet a particular need. As with most germplasm samples, the success of using frozen-thawed or vitrified embryos for regenerating live animals depends on many factors, such as the quality of post-thawed samples, qualified personnel for appropriate manipulation of post-thawed samples, and high expenses. Therefore, using germplasms to generate viable offspring requires surgical expertise and high costs associated with producing and preserving samples (e.g., embryos), limiting the widespread use of this option in many African countries.

The use of harvested samples for artificial insemination (AI) and embryo transfer (ET), the most popular breeding tools of ARTs in Africa, have been effective for genetic improvements worldwide. However, these technologies remain luxuries in most African countries due to low human and institutional capacities, lack of infrastructure, high cost, and poorly designed breeding programs (AU-IBAR 2019). Nevertheless, South Africa and Namibia appear as the only African countries reporting cases of in vivo- and in vitro-produced transferrable bovine embryos for breeding in both dairy and beef cattle farms (Viana 2019).

25.4.2 Animal Management and Sample Collections

The conservation of threatened and underutilized AnGR is a top priority. The loss of a typical breed may be crucial for some African communities as it may reflect the local cultural and historical identity or the genetic heritage of farm animals. Therefore, it is strategic to safeguard animal breeds for subsequent needs through ex situ conservation methods, either in vivo (populations of live animals not reared under typical management conditions) or in vitro (artificial manipulations, under cryogenic conditions, providing long-term insurance against future shocks)

(Rischkowsky and Pilling 2007). Regardless of the chosen approach, animal resource managers face many responsibilities and challenges when overseeing animal management programs. These challenges dependent on the type of facilities and species managed, and the kind of research conducted (Stark et al. 2010). All stakeholders should receive proper gene bank management training regarding human and institutional capacity building. Genetics researchers can help and train in developing and utilizing gene bank collections. Animal resource management programs using software appear more effective when they integrate researchers, decision-makers, and smallholder farmers with an active effort to sensibilize the breeders on the extinction risks of the breeds (Gaouar 2002, 2009; Alhajeri et al. 2021). Overall, the need for a statewide identification policy in most African countries is a critical problem in managing and developing domestic animals.

25.4.3 Extended Preservation for Immediate Utilization

Various germplasms (sperm, oocytes, and embryos) of cattle, small ruminants, pigs, and chickens are successfully conserved for AnGR and commercial utilization for ARTs. Since the 1960s, the practice of AI in African countries has boosted the use of livestock semen and the easiness of reproductive biotechnology, such as estrus synchronization, cryopreservation, and embryo production. The dilution of freshly collected semen of small ruminants (sheep and goats) and pigs in commercial extenders containing specific additives (e.g., nutrients, antibiotics, antioxidants, and chelating agents) provides prolonged storage in the cold (4–8 °C, for goats and stallions) and chilled temperatures (16–20 °C for boars) during one to five days post-collection (Johnson et al. 2000; Roca et al. 2016; Falchi et al. 2018). Such semen conditioning is known as liquid semen, which utilization right after collection or refrigerated storage results in satisfactory fertility rates (>65% births) after intra-cervical inseminations. However, these commercial extenders are only affordable to a few African smallholder farmers. Therefore, the large majority of smallholder farmers interested in the AI technique may rely on basic extenders supplemented with cheap compounds, such as indigenous plant extracts (e.g., cactus extracts, plant oils, rooibos, and honeybush) to counteract the progressive decrease of semen quality during storage (Hiemstra et al. 2006; Kameni et al. 2021; Ros-Santaella et al. 2020; Ros-Santaella and Pintus 2017). In addition, the poor conditions of post-collection semen manipulations (e.g., packaging, storage, and handling) contribute to the rapid loss of semen quality, drastically limiting the widespread application of liquid semen in African countries.

In contrast, the use of livestock embryos for transfer to take advantage of the genetic merit of females faces several technological and institutional barriers. Consequently, the application of embryo transfer remains limited in African countries.

25.4.4 Cryopreservation for Biobanking

Cryopreservation for ex situ in vitro conservation is a critical tool for genetic improvement and sustainable protection. Both gametes and embryos are the most vital genetic materials, but their cryopreservation requires the installation of processing, storage, and packaging facilities (Hiemstra et al. 2006; Leibo and Songsasen 2002). Unfortunately, few African countries abide by these rules due to a lack of resources (e.g., material, equipment, and local organization). Increasing government and FAO initiatives are creating more gene banks worldwide (Rischkowsky and Pilling 2007). As a result, more African countries now possess institutional facilities for gene banking (e.g., Botswana, Burkina Faso, Egypt, Rwanda, South Africa, and Tunisia). Meanwhile, the AU-IBAR facilitates regional gene banks storing genetic materials, resulting in universities and governments establishing cooperations and collaborations to develop joint training programs (AU-IBAR

2019). However, these supra-national or regional implementations require a fair and equitable sharing of the benefits of using these resources for research, commercial, or other purposes.

Semen of most livestock species can be frozen adequately using pre-existing or adapted protocols with commercial/homemade extenders and deep-freezing storage in liquid nitrogen (−196 °C). Frozen-thawed semen and embryos have served for AI or ET with relative success in cows, goats, and sheep (Viana 2019). Compared to other livestock species, ruminant spermatozoa are the most resistant to cryopreservation. In most African countries, efforts are made toward freezing local bull semen (e.g., Nguni cattle) for appreciable post-thaw motility (>70%) and fertility (>37%) rates following AI (Nengovhela et al. 2021; Magopa et al. 2021). In contrast, the ET is still limited to a few African countries. Only South Africa has reported cases of successful transfers of frozen in vivo-produced embryos from domestic dairy and beef cows (1107) and in vitro-produced by ovum-pick-up (OPU) from domestic beef cows (1976) (Viana 2019).

In summary, there is still a need to find solutions to overcome the current institutional and personnel challenges to establish successful cryobanking of farm animals' germplasms in Africa. In addition, researchers should consider the cost of specimen cryopreservation and the likelihood of successful outcomes following utilization. Despite the structural conditions in most African countries, other advanced biotechnologies of reproduction, such as cloning and DNA microinjection, having several advantages for ex situ in vitro conservation, are progressively penetrating the continent.

25.4.5 Current Efforts and Challenges in the Conservation of AnGR in Africa

The Convention on Biological Diversity (CBD) recommends the implementation of any ex situ conservation initiatives within the geographical location of the targeted species. In early 2010, ILRI launched the World's first livestock gene bank (Scidev.net 2013a, b), known as Azizi biorepository (http://azizi.ilri.cgiar.org/), ensuring the safety, security, and efficient collection and storage of biological materials and related data. Samples currently held include semen, embryos, blood, serum, pathogens, tissues (lung, brain), arthropod vectors, cultured cells, primers, plasma, milk, DNA, RNA, and swabs. These materials come from a wide range of species and environments. The Azizi biorepository offers African livestock the following services: (1) a platform for long-term storage of materials, (2) a platform for proper labeling of samples from the field, (3) support to livestock projects for proper sample and data collection (4) retrieving of samples and data in the repository, and (5) source of liquid nitrogen to various projects.

The technical office of the African Union Commission (AUC), the African Union InterAfrican Bureau for Animal Resources (AU-IBAR), has established five (5) Regional Multi-Purpose Animal Gene Banks under the project known as "Strengthening the Capacity of African Countries to Conservation and Sustainable Utilization of African Animal Genetic Resources." Infrastructures are located in Eastern Africa (the National Animal Genetic Resource Centre and Data Bank, Entebbe, Uganda); in Northern Africa (National Gene Bank of Tunis or BNG, Tunisia); in Western Africa (Centre International de Recherche-Développement sur l'Elevage en zone Subhumide or CIRDES, Bobo-Dioulasso, Burkina Faso); in Southern Africa (Department of Agricultural Research, Gaborone, Botswana); and in Central Africa (University of Dschang, Cameroon).

Several challenges contribute to the need for multi-country animal gene banks in Africa. Amongst them are the need for more funding, regulations on regional and continental exchange of animal genetic material, and a lack of consensus on procedures for operating gene banks. Coordination between countries is highly required for transboundary animal breed conservation. Although all countries agreed that regional African animal gene banks are a practical approach to conserving animal genetic resources, negotiations for their operationalization are lengthy and complex, mainly because gene banks should operate under internationally agreed legal protocols (Nagoya Protocol). In contrast, though signatory countries of the Convention on Biological Diversity and Party of the protocol, the various countries have different implementation frameworks. Legal protocols require that the samples in a regional animal gene bank remain in the ownership of the country of origin with provision made for appropriate access by interested parties presenting valid claims for its use. Users will also be expected to replenish the gene bank materials when possible.

For example, at the regional gene bank set-up at CIRDES, the legal and regulatory frameworks to govern the collection, use, and exchange of genetic materials between the Economic Community of West African States (ECOWAS) countries are required but have yet to be available. Hence, the gene bank still needs to be operational to receive genetic materials. At present, semen has been collected from some West African transboundary cattle breeds (N'Dama, Baoulé, Borgou, Somba, and Zébu Peulh Soudanais) available on the experimental farm of CIRDES.

25.5 Operation Procedures and Guidelines—Past and Current Efforts in Africa

The AU-IBAR has created five regional gene banks in Africa. For example, the AU-IBAR inaugurated the North Africa gene bank in Tunis (Tunisia), regrouping six North African countries in 2016, following the 6th General Assembly of the sub-regional focal point for genetic resources. This action is part of the action plan drawn up to establish a network of regional gene banks, issuing smooth running and sustainability, namely:

1. The NAGOYA and APA (Access and Benefit-Sharing) protocols are signed by all countries on a working basis.
2. Any samples and genetic materials transmissions to the BRG (Bank of Genetic Resources) must be done after declaration and authorization by the concerned departments.
3. Workshops on the development of the legal and contractual framework for the gene bank will be scheduled with technical experts and lawyers' participation to develop a draft cooperation convention to be submitted to member countries.
4. Regional value chains should be discussed and amended within the 'Live2Africa' project, established by the European Union and AU-IBAR, to "support the transformation of the African livestock sector for enhanced contribution to environmentally sustainable, climate resilient, socio-economic development and equitable growth." The ANOC (National Association of sheep and goats) of Morocco was cited as an example to promote these value chains.
5. Consider the contexts and realities of livestock farming in each country, the case of Mauritania, where livestock systems are mainly based on livestock mobility, especially transhumance.
6. Transboundary breeds are a priority in the action plans of the regional gene bank and breeds, or populations threatened with extinction, considering aquaculture and beekeeping.
7. Sample characterization and inventory must be done in the countries of origin with two regulatory and technical aspects.
8. The health aspect is important for exchanges between countries and the necessary health certification mechanisms, for example, during the COVID-19 pandemic.

25.6 Further Opportunities via Genomics

25.6.1 Existing and Explorative Genomic Tools

Advances in genetics and genomics tools have permitted remarkable improvements and discoveries in biological sciences. These techniques are based on the three pillars of the dogma of biology, namely DNA, RNA, and proteins.

The corresponding molecular methods could be listed as Restriction Fragment Length Polymorphism or DNA Fingerprinting (RFLP), microsatellites, mitochondrial DNA, whole-genome and RNA sequencing, microarrays, and proteomics, which are described in other chapters of the book. The high heterogeneity of African livestock breeds permits practical adaptations to the extremely changing environmental conditions of the continent. Yet, the benefits of genomics advances applied to livestock species are not widespread and harmonized. Most existing research is performed through foreign collaborations due to various limitations (e.g., nationally equipped facilities, qualified personnel, and high-related costs) within the continent (Kosgey et al. 2005; Rege et al. 2011; Marshall 2014). Only a few countries possess functional research laboratories or centers capable of complete livestock genomics analyses (e.g., Biotechnology Research Center or CRBt in Algeria, BNG in Tunisia; Veterinary Genomics Laboratory in Morocco, several biotechnology laboratories in South Africa—Inqaba Biotech; Agricultural Research Center & Biotechnology Platform or ARC-BTP; Africa Centre for Gene technologies or ACGT—African Biosciences Limited, Nigeria; and ILRI in Kenya). Nevertheless, the usefulness of these genomic tools in livestock has been explored in various programs to characterize and conserve animal genetic resources in Africa (Marshall et al. 2019).

25.6.2 Monitoring and Identification of Resilient Animals and Sampling

Explorative studies using genomics tools have found benefits in various economic and ecological traits by identifying the most adaptable and profitable breeds within several African countries' livestock production systems.

In small ruminants, researchers used various genomic tools to assess the genetic diversity of Nigerian Balami, Yankasa, Ouda, and West African Dwarf breeds of sheep (RFLP (Geka Rukayat et al. 2020)); genetic variability and phylogenetic relationships within and between various Algerian and Moroccan sheep populations (Ouled-Djellal, Hamra, Sidaoun, Taâdmit, D'men, Beni-Ighil, Foro-Foro, Rembi, Sardi, Boujaad, Timadhite, Barbarine, Ifilene, Srandi, Darâa, Berbere, BembiRembi, and Tazegzawt) sheep populations [Microsatellite markers (Ameur et al. 2020; Abdelkader et al. 2018; Gaouar et al. 2005, 2014, 2015, 2016)—mtDNA (Ghernouti et al. 2017)—Sequencing and Genome Wide Association (GWA) analyses (Djaout et al. 2018)—SNP (Belabdi et al. 2019; Gaouar et al. 2017)]. The studies revealed (1) high and alarming genetic dilution of traditional breeds due to uncontrolled crossbreeding (e.g., Barbarine, Berber, Rembi, Ouled-Djellal), (2) genetic origins that would be given priority for conservation programs (e.g., Tazegzawt, Hamra, D'men, Sidaoun), (3) geographical distributions (e.g., Ouled-Djellal in the entire Maghreb region), and (4) robust genetic proximity between Algerian sheep with the Caribbean and Brazilian breeds, indicating the immense influence of the Atlantic slave trade on their genetic make-up. More interestingly, the use of SNP assay in an Ethiopian project, aiming to upgrade local (Menz and Wollo) breeds through repeated backcrossing with the exotic Awassi breed, identified animals of 37.5–50% Awassi genes as the most productive (lamb weight) at 8 months of age (Getachew et al. 2017). Genomic tools have been used on local goat breeds (e.g., Naine de Kabylie, Arbia, Mekatia, M'zabite, West African Dwarf, Red Sokoto, South Africa Kalahari breed) for various purposes, such as the (1) genetic characterization and diversity using microsatellites (Tefiel et al. 2018, 2020) and SNP (Fantazi et al. 2018); (2) polymorphism associations of various genes, i.e., BMP15, BMPR1B, GDF9, and POU1F1, with prolificacy of Cameroon and local Nigerian goats using microsatellites (Wouobeng et al. 2020) and SNP (Olasege et al. 2021); and polymorphism association of the PRNP gene with scrapie resistance in Algerian sheep breeds using SNP (Djaout et al. 2018; Fantazi et al. 2018).

In cattle, genetic studies of genetic diversity and structure were achieved using four indigenous Algerian populations [Microsatellites (Rahal et al. 2020)]. At the same time, an analysis of mtDNA markers focusing on the Haplogroup T1 attributed the origin of two taurine breeds of Cameroon (Namchi Poli and Ngaoundere) to South-West Asia. However, the White and Red Fulani zebus gene flows may have contributed to their genetic diversities (Patrick Jolly et al. 2018). Another study used the SNP approach to investigate the population structure and genetic diversity of the Algerian Guelmoise cattle breed (Ourida et al. 2017). Findings indicated that the Guelmoise breed is an admixed population with strong genetic similarity to the Tunisian cattle breed but less permissivity to foreign breed introgression. Finally, regarding dairy cattle production, the use of bovine SNP arrays in programs such as the Kenya "Dairy Genetics East Africa project" and "Senegal Dairy Genetics" revealed the benefits of genomic tools in multi-objective studies (Marshall et al. 2019). The study used African indigenous Zebus, exotic *Bos taurus* breeds (e.g., Montbeliard and Jersey), and crossbreeds with various proportions of indigenous breeds to reveal SNP markers as predictors of high-level milk-producing cows.

Additionally, SNP assays enhance estimated breeding value (EBV) accuracy by assigning parentage that accelerates genetic improvements in African countries lacking reliable pedigree records of livestock animals. Two ongoing CBBP programs in four African countries (Kenya, Uganda, Ethiopia, and Tanzania) are conducted to validate SNP candidates for enhancing breeding programs: the "Dairy Genetics East Africa" and the "African Dairy Genetic Gains" (Strucken et al. 2017). In addition, there are great expectations to use the SNP as EBV predictors to enable the selection and recruitment of young bulls for national artificial insemination programs (Strucken et al. 2017; Mrode et al. 2018; Aliloo et al. 2018).

Various genomic tools have been applied in other domesticated animal breeds such as dromedary/camels (Harek et al. 2015; Cherifi et al. 2017), chickens (Mahammi et al. 2016; Al-Jumaili et al. 2020; Boudali et al. 2020), pigs (Osei-Amponsah et al. 2017), and horses (Benhamadi et al. 2020) for different purposes. In addition, new biotechnologies like gene editing can create new "breed types" of desired phenotypes.

Overall, informative utilization of the above-mentioned genomic tools relies on optimal farm management. At the same time, their incorporation into established breeding programs is still limited in most African countries.

25.6.3 Restocking and Improving Farm Populations from Conserved Materials

The sustainable management of local livestock species is a central issue in livestock animal conservation. Furthermore, because of many challenges related to climate change (e.g., poverty, conflict, drought, and disease), the demand for programs for genetic animal conservation is increasing (Olney et al. 1993). The restocking of genetically important (e.g., cultural, economic) livestock animals is achieved through in situ and/or ex situ conservations.

Many public and private initiatives have applied community-based approaches and restocking to rebuild herds and livelihoods as emergency programs to fight poverty in Africa. Initiatives included beneficiary governments and international public partners (e.g., USAID, AU, and EU), research organizations (e.g., ILRI and CTLGH), and non-governmental organizations (e.g., Heifer International). As a result, programs such as Girinka "One-Cow-per-poor family" and "Send-A-Cow" have been adopted in many African countries, such as Rwanda (Argent et al. 2014; Klapwijk et al. 2014), Ethiopia (Flax et al. 2021), and Tanzania (De Vries 2008), while many others received veterinary and extension services. Nevertheless, the distribution of animals maintained in community-based structures remains the most popular method for restocking livestock farms. Restocking has been primarily used as a relief program to assist victims of natural disasters or diseases while considering the beliefs and culture of farmers. For example, the Kenyan government first applied this program to provide livestock to victims of the devastating droughts of 1983–1984 and 1990–1993 through the Samburu District Development Program (SDDP) between the Kenyan and German governments from 1993 to 2000 (Lesorogol 2009; Hogg 1985). As results (Lesorogol 2004), beneficiary farmer families (~326 from four different communities) received 15–40 small ruminants (sheep or goats) per family per year with little assistance (e.g., food, animal health worker), with significant measures to ensure accountability (e.g., retaking animals from restocked families in case of no appropriate care taken, training restocked families, provide proofs of dead animals), translation into more milk production (>10% increase), poverty alleviation (e.g., substantial rose of animal heads, >10% herd growth rate), restoring social standing in the community (about 80% of restocked families), and empowering beneficiary women (~27% of total beneficiaries, mainly widows).

Nonetheless, the enhancement of livestock populations by restocking, however, remains both controversial and challenging for livestock repopulation. *Firstly*, stocking well-identified animal breeds is crucial to meet and exceed the capital value of the selected livestock chain of value, requiring an accurate genetic characterization and selection of animals with added genetic merit. *Secondly*, using ex situ/in vitro conservation approaches with cryopreserved genetic materials poses several institutional and personnel capacity-building and development challenges. Indeed, the lack of qualified personnel and shortages in materials for cryopreservation are other barriers to successful restocking. As mentioned above, there are continued efforts throughout the continent for intensive cooperations and collaborations to improve the cryopreservation practice. *Thirdly*, cryopreserved germplasm requires a solid grasp of assisted reproductive technologies or ART to ensure successful farm repopulations with improved animal breeds. While reintroducing animals, avoiding the potential risk of density-dependent mortality in some small ruminant breeds is crucial. This phenomenon is commonly observed in fish (Nickelson 2003; Levin et al. 2001). *Lastly*, breeders should consider the characteristics of newly crossbred animals during the stocking and restocking of farm populations to avoid ecological competition between local and crossbred animals (Goodman 2005).

25.7 Policies and Regulations Governing Conservation and Improved Management of AnGR

Most African countries lack appropriate livestock policies, and where they exist, most are embedded in the general agricultural policy. At the same time, regional policies on the management and sustainable use of AnGR, the development of which is being spearheaded by AU-IBAR remain largely unimplemented (see Chap. 30). For example, the lack of policies and legislation directly formulated and designed for farm AnGR is due to the policymakers being unaware of the economic contribution of AnGR in national incomes. Additionally, governments need to have complete control of the benefits of regional gene banks, which are controlled by international agreements on biodiversity following the Nagoya Protocol on Access to Genetic Resources and Fair and Equitable Sharing of Benefits.

Most countries in Africa still need formal policies and legislation on crossbreeding (AU-IBAR 2019). The few with stand-alone AnGR policies have legislation prohibiting imports and exports of farm animal genetic resources for the control of zoonotic diseases. Joint efforts between the EU and AU-IBAR assist several countries in developing National Strategic Action Plans (NSAP) to manage their AnGR. At the same time, creating platforms for governments and institutions with already set policy, legislative, and regulatory frameworks will permit the sharing of lessons learned and best practices with partners. Rwanda is one of the few countries with policies that give direction on crossbreeding and a strategy for genetic improvement (MAAR 2017). In addition, Ghana's National Strategic Action Plan (NSAP) seeks to promote the conservation and sustainable use of its local AnGR and the development of stabilized crosses, especially for peri-urban (Ahiagbe et al. 2021). Kenya defined its breeding policy in 1980, and the major national objectives targeted (fight against poverty, production of surpluses for export, development of conservation plans, etc.) (Anyango et al. 1989; Kenya, R. o 2013). Tanzania's priority investments through the Tanzanian Livestock Modernization Initiative include various combinations of nutrition, animal health, and animal genetics (e.g., AI and multi-breeding approaches) to improve the quality and performance of Tanzanian livestock populations (MoLFD 2015; Michael et al. 2018).

25.8 Conclusions and Future Opportunities

African smallholder farmers have multifactorial relationships with livestock animals, which, together with the environmental conditions, may jeopardize animal diversity maintenance. Hence, most endangered African AnGR are maintained in situ in living populations, which carries risks of loss due to various factors. The availability of the most advanced technologies of assisted reproduction and genetics tools makes the ex situ conservation method a viable complementary to in situ actions for successfully keeping animal biodiversity in Africa. Nonetheless, the deterioration and loss of valuable genetic material is a continuous and fast process. Therefore, it is urgent to rapidly overcome all existing political, structural, and infrastructural barriers and to develop the needed human and institutional capacity for increased productivity and successful conservation and management of AnGR in Africa. Finally, the best form of conservation of Africa's unique AnGR is to ensure their continued sustainable use in the various agro-ecological zones. Therefore, there will be a need to select breed(s) for development, improve husbandry practices based on innovative technological strategies and value-addition to their products, and consider government-inspired private-sector-led stakeholder collaborations.

Acknowledgment This work was partially supported by the U.S. Department of Agriculture, Agricultural Research Service, project 6066-31000-015-00D, and the International Livestock Research Institute, Nairobi, Kenya.

References

Abdelkader AAA, Benyoucef MTDA, Azzi N, Yilmaz O, Cemal I, Gaouar SBS (2018) New genetic identification and characterisation of 12 Algerian sheep breeds by microsatellite markers. Ital J Anim Sci 17(1):10

Adebambo OA (2001) The Muturu: a rare sacred breed of cattle in Nigeria. Anim Genet Resour Inf 31:27–36

Ahiagbe M, Shaibu M, Avornyo F, Ayantunde AA, Panyan E (2021) A guide to developing the small ruminant value chain in northern Ghana: a value chain approach. IITA, Ibadan, Nigeria

Alexandratos N, Bruinsma J (2012) World agriculture towards 2030/2050: the 2012 revision

Alhajeri BH, Alhaddad H, Alaqeely R, Alaskar H, Dashti Z, Maraqa T (2021) Camel breed morphometrics: current methods and possibilities. Trans R Soc S Aust 145:1–22

Aliloo H, Mrode R, Okeyo A, Ni G, Goddard M, Gibson JP (2018) The feasibility of using low-density marker panels for genotype imputation and genomic prediction of crossbred dairy cattle of East Africa. J Dairy Sci 101(10):9108–9127

Al-Jumaili AS, Boudali SF, Kebede A, Al-Bayatti SA, Essa AA, Ahbara A, Aljumaah RS, Alatiyat RM, Mwacharo JM, Bjørnstad G (2020) The maternal origin of indigenous domestic chicken from the Middle East, the north and the horn of Africa. BMC Genet 21(1):1–16

Ameur AA, Onur Y, A. T. A. N, Cemal I, Bechir S, Suheil G (2020) Assessment of genetic diversity of Turkish and Algerian native sheep breeds. Acta Agric Slovenica 115(1):9

Anyango G, Downing T, Getao C, Gitahi M, Kabutha C, Kamau C, Karanja M, Mbarire S, Muente S, Mutero W (1989) Drought vulnerability in central and eastern Kenya. Coping with drought in Kenya: national and local strategies. Lynne Rienner, Boulder, CO, pp 169–210

Argent J, Augsburg B, Rasul I (2014) Livestock asset transfers with and without training: evidence from Rwanda. J Econ Behav Organ 108:19–39

AU-IBAR (2019) The state of farm animal genetic resources in Africa. (AU-IBAR), A. U. I. B. f. A. R., Ed, Nairobi, Kenya

Baptista J, Blache D, Cox-Witton K, Craddock N, Dalziel T, de Graaff N, Fernandes J, Green R, Jenkins H, Kahn S (2021) Impact of the COVID-19 pandemic on the welfare of animals in Australia. Front Veterin Sci 7:1219

Belabdi I, Ouhrouch A, Lafri M, Gaouar SBS, Ciani E, Benali AR, Ould Ouelhadj H, Haddioui A, Pompanon F, Blanquet V, Taurisson-Mouret D, Harkat S, Lenstra

JA, Benjelloun B, Da Silva A (2019) Genetic homogenization of indigenous sheep breeds in Northwest Africa. Sci Rep 9(1):7920

Bélanger J, Pilling D (2019) The state of the world's biodiversity for food and agriculture. Food and Agriculture Organization of the United Nations (FAO)

Benhamadi MEA, Berber N, Benyarou M, Ameur AA, Haddam HY, Piro M, Gaouar SBS (2020) Molecular characterization of eight horse breeds in Algeria using microsatellite markers. Biodiversitas J Biol Divers 21(9):4107

Bolaji U-FO, Ajasa AA, Ridwan OA, Bello SF, Ositanwosu OE (2021) Cattle conservation in the 21st century: a mini review. Open J Anim Sci 11(02):304

Boudali SF, Al-Jumaili AS, Bouandas A, Mahammi FZ, Tabet Aoul N, Hanotte O, Gaouar SBS (2020) Maternal origin and genetic diversity of Algerian domestic chicken (Gallus gallus domesticus) from North-Western Africa based on mitochondrial DNA analysis. Anim Biotechnol 33:1–11

Boujenane I (2005) Small ruminant breeds of Morocco. In: Characterization of small ruminant breeds in West Asia and North Africa, vol 2. ICARDA, Aleppo, Syria, p 48

Boukallouch A (2006) Goat milk cheese: a mean of development of the Northern Moroccan provinces. In: Animal products from the Mediterranean area, vol 119. Publication-European Association For Animal Production, p 83

Carson A, Elliott M, Groom J, Winter A, Bowles D (2009) Geographical isolation of native sheep breeds in the UK—evidence of endemism as a risk factor to genetic resources. Livest Sci 123(2–3):288–299

Cherifi YA, Gaouar SBS, Guastamacchia R, El-Bahrawy KA, Abushady AMA, Sharaf AA, Harek D, Lacalandra GM, Saïdi-Mehtar N, Ciani E (2017) Weak genetic structure in northern African dromedary camels reflects their unique evolutionary history. PLoS One 12(1):e0168672

De Oliveira Silva R, Ahmadi BV, Hiemstra SJ, Moran D (2019) Optimizing ex situ genetic resource collections for European livestock conservation. J Anim Breed Genet 136(1):63–73

De Vries J (2008) Goats for the poor: some keys to successful promotion of goat production among the poor. Small Rumin Res 77(2):221–224

Djaout A, Chiappini B, Gaouar S-B-S, Afri-Bouzebda F, Conte M, Chekkal F, El-Bouyahiaoui R, Boukhari R, Agrimi U, Vaccari G (2018) Biodiversity and selection for scrapie resistance in sheep: genetic polymorphism in eight breeds of Algeria. J Genet 97(2):453–461

Falchi L, Khalil WA, Hassan M, Marei WFA (2018) Perspectives of nanotechnology in male fertility and sperm function. Int J Veterin Sci Med 6(2):265–269

Famien AM (2020) Analyse de la variabilité décennale et du changement climatique en Afrique de l'ouest à l'aide des produits CMIP5-Application à l'estimation des rendements agricoles à la fin du siècle. Université Félix Houphouët Boigny Abidjan (Côte d'Ivoire)

Fantazi K, Migliore S, Kdidi S, Racinaro L, Tefiel H, Boukhari R, Federico G, Di Marco Lo Presti V, Gaouar SBS, Vitale M (2018) Analysis of differences in prion protein gene (PRNP) polymorphisms between Algerian and Southern Italy's goats. Ital J Anim Sci 17(3):578–585

Flax VL, Ouma E, Izerimana L, Schreiner M-A, Brower AO, Niyonzima E, Nyilimana C, Ufitinema A, Uwineza A (2021) Animal source food social and behavior change communication intervention among Girinka livestock transfer beneficiaries in Rwanda: a cluster randomized evaluation. Glob Health Sci Pract 9(3):640–653

Food; Organization., A (2007) Global plan of action for animal genetic resources and the Interlaken declaration. FAO, Commission on Genetic Resources for Food and Agriculture, Rome

Gaouar SBS (2002) Contribution à l'étude de la variabilité génétique des races ovines par l'utilisation des microsatellites : Caractérisation de deux races ovine algériennes Hamra et Ouled-Djellal. Université d'Es-Sénia, Oran

Gaouar SBS (2009) Etude de la biodiversité : Analyse de la variabilité génétique des races ovines algériennes et de leurs relations phylogénétiques par l'utilisation de microsatellites. Université d'Es-Sénia, Oran

Gaouar S, Tabet-Aoul N, Derrar A, Goudarzy-Moazami K, Saïdi-Mehtar N (2005) Genetic diversity in Algerian sheep breeds, using microsatellite markers. Springer Netherlands, Dordrecht, pp 641–644

Gaouar SBS, Kdidi S, Tabet AN, Aït-Yahia RB, Aouissat N, Dhimi L, Yahyaoui MH, Saidi-Mehtar N (2014) Genetic admixture of North-African ovine breeds as revealed by microsatellite loci. Livest Res Rural Dev 26(7):1

Gaouar SBS, Da Silva A, Ciani E, Kdidi S, Aouissat M, Dhimi L, Lafri M, Maftah A, Mehtar N (2015) Admixture and local breed marginalization threaten Algerian sheep diversity. PLoS One 10(4):e0122667

Gaouar SBS, Kdidi S, Ouragh L (2016) Estimating population structure and genetic diversity of five Moroccan sheep breeds by microsatellite markers. Small Rumin Res 144:23–27

Gaouar S, Lafri M, Djaout A, El-Bouyahiaoui R, Bouri A, Bouchatal A, Maftah A, Ciani E, Da Silva A (2017) Genome-wide analysis highlights genetic dilution in Algerian sheep. Heredity 118(3):293–301

Geka Rukayat A, Anthony E, Stephen Sunday Acheneje E, Abdulkadir U (2020) Genetic diversity of Nigerian indigenous sheep breeds at the β-Lactoglobulin gene locus. Genet Biodivers J 4(2):40–49

Getachew T, Huson HJ, Wurzinger M, Burgstaller J, Gizaw S, Haile A, Rischkowsky B, Brem G, Boison SA, Mészáros G, Mwai AO, Sölkner J (2017) Identifying highly informative genetic markers for quantification of ancestry proportions in crossbred sheep populations: implications for choosing optimum levels of admixture. BMC Genet 18(1):80

Ghernouti N, Bodinier M, Ranebi D, Maftah A, Petit D, Gaouar S (2017) Control region of mtDNA identifies three migration events of sheep breeds in Algeria. Small Rumin Res 155:66–71

Goodman D (2005) Selection equilibrium for hatchery and wild spawning fitness in integrated breeding programs. Can J Fish Aquat Sci 62(2):374–389

Haile A, Getachew T, Mirkena T, Duguma G, Gizaw S, Wurzinger M, Soelkner J, Mwai O, Dessie T, Abebe A (2020) Community-based sheep breeding programs generated substantial genetic gains and socioeconomic benefits. Animal 14(7):1362–1370

Harek D, Berber N, Cherifi YA, Yakhlef H, Bouhadad R, Arbouche F, Sahel H, Djellout NE, Saidi-Mehtar N, Gaouar SBS (2015) Genetic diversity and relationships in saharan local breeds of Camelus dromedarius as inferred by microsatellite markers. J Camel Pract Res 22(1):1–9

Hiemstra SJ, van der Lende T, Woelders H (2006) The potential of cryopreservation and reproductive technologies for animal genetic resources conservation strategies. In: The role of biotechnology in exploring and protecting agricultural genetic resources. FAO, pp 45–60

Hoffmann I (2010a) Climate change and the characterization, breeding and conservation of animal genetic resources. Anim Genet 41:32–46

Hoffmann I (2010b) Livestock biodiversity. Rev Sci Tech 29(1):73

Hoffmann I (2011) Livestock biodiversity and sustainability. Livest Sci 139(1–2):69–79

Hoffmann I, Scherf B (2005) Management of farm animal genetic diversity: opportunities and challenges. In: WAAP book of the 2005. Wageningen Academic Press, pp 221–245

Hogg R (1985) Restocking pastoralists in Kenya: a strategy for relief and rehabilitation

Iniguez L (2005) Characterization of small ruminant breeds in West Asia and North Africa, vol 1. ICARDA Aleppo

Iñiguez L (2011) The challenges of research and development of small ruminant production in dry areas. Small Rumin Res 98(1–3):12–20

Jiri O, Mafongoya P, Musundire R (2017) The use of underutilised crops and animal species in managing climate change risks. In: Indigenous knowledge systems and climate change management in Africa, p 115

Johnson LA, Weitze KF, Fiser P, Maxwell WMC (2000) Storage of boar semen. Anim Reprod Sci 62(1–3):143–172

Kameni SL, Meutchieye F, Ngoula F (2021) Liquid storage of Ram Semen: associated damages and improvement. Open J Anim Sci 11(3):473–500

Kenya, R. o (2013) Sector plan for Science Technology and Innovation 2013-2017: Revitalizing and Harnessing Science, Technology and Innovation for Kenya's Prosperity and Global Competitiveness. https://vision2030.go.ke/publication/science-technology-and-innovation-2013-2017/

Klapwijk CJ, Bucagu C, van Wijk MT, Udo HMJ, Vanlauwe B, Munyanziza E, Giller KE (2014) The 'One cow per poor family' programme: current and potential fodder availability within smallholder farming systems in southwest Rwanda. Agric Syst 131:11–22

Köhler-Rollefson I (2004) Farm animal genetic resources: Safeguarding national assets for food security and trade

Kosgey I, Kahi A, Van Arendonk J (2005) Evaluation of closed adult nucleus multiple ovulation and embryo transfer and conventional progeny testing breeding schemes for milk production in tropical crossbred cattle. J Dairy Sci 88(4):1582–1594

Leibo S, Songsasen N (2002) Cryopreservation of gametes and embryos of non-domestic species. Theriogenology 57(1):303–326

Leroy G, Boettcher P, Besbes B, Danchin-Burge C, Baumung R, Hiemstra SJ (2019) Cryoconservation of animal genetic resources in Europe and two African countries: a gap analysis. Diversity 11(12):240

Lesorogol C (2004) Asset building through community participation: re-stocking pastoralists following drought in Northern Kenya

Lesorogol CK (2009) Asset building through community participation: restocking pastoralists following drought in northern Kenya. Soc Work Public Health 24(1–2):178–186

Levin PS, Zabel RW, Williams JG (2001) The road to extinction is paved with good intentions: negative association of fish hatcheries with threatened salmon. Proc R Soc Lond Ser B Biol Sci 268(1472):1153–1158

MAAR (2017) National Agriculture Policy. Ministry of Agriculture and Animal Resources (MAAR), Republic of Rwanda

Magopa TL, Mphaphathi ML, Mulaudzi T, Ramukhithi FV, Tshabalala MM, Raphalalani ZC, Sebopela MD, Nkadimeng N, Sithole SM, Nedambale TL (2021) 108 synchronization and artificial insemination of South African communal cattle. Reprod Fertil Dev 33(2):161–161

Mahammi F, Gaouar S, Laloë D, Faugeras R, Tabet-Aoul N, Rognon X, Tixier-Boichard M, Saidi-Mehtar N (2016) A molecular analysis of the patterns of genetic diversity in local chickens from western Algeria in comparison with commercial lines and wild jungle fowls. J Anim Breed Genet 133(1):59–70

Marshall K (2014) Optimizing the use of breed types in developing country livestock production systems: a neglected research area. J Anim Breed Genet 131(5):11

Marshall K, Gibson JP, Mwai O, Mwacharo JM, Haile A, Getachew T, Mrode R, Kemp SJ (2019) Livestock genomics for developing countries—African examples in practice. Front Genet 10:297

Michael S, Mbwambo N, Mruttu H, Dotto M, Ndomba C, Silva MD, Makusaro F, Nandonde S, Crispin J, Shapiro BI (2018) Tanzania livestock master plan. ILRI, Nairobi, Kenya

MoLFD (2015) Tanzania livestock modernization initiative. Ministry of Livestock and Fisheries Development Dar es Salaam

Morrell J, Mayer I (2017) Reproduction biotechnologies in germplasm banking of livestock species: a review. Zygote 25(5):545–557

Mrode R, Coffey M, Ojango J, Mujibi D, Okeyo M, Strucken E, Gibson JP, Aliloo H (2018) The impact of modelling and pooled data on the accuracy of genomic prediction in small holder dairy data. Proceedings of the World Congress on Genetics Applied to Livestock Production

Nengovhela NB, Mugwabana TJ, Nephawe KA, Nedambale TL (2021) Accessibility to reproductive technologies by low-income beef farmers in South Africa. Front Veterin Sci 8:611182

Nickelson T (2003) The influence of hatchery coho salmon (Oncorhynchus kisutch) on the productivity of wild coho salmon populations in Oregon coastal basins. Can J Fish Aquat Sci 60(9):1050–1056

Obese FY, Osei-Amponsah R, Timpong-Jones E, Bekoe E (2021) Impact of COVID-19 on animal production in Ghana. Anim Front 11(1):43–46

Olasege BS, Bemji MN, Ibeagha-Awemu EM, Isa AM, Wheto M, Sulaimon GD, Ayinde MO, Sodimu BO, Ogunniyi DA, Okwelum N (2021) Genetic polymorphism in the POU1F1 gene in Kalahari Red and two Nigerian goat breeds and their relationship with litter size. Genet Biodivers J 5(1):42–59

Olney PJ, Mace G, Feistner A (1993) Creative conservation: interactive management of wild and captive animals. Springer Science & Business Media

Osei-Amponsah R, Skinner BM, Adjei DO, Bauer J, Larson G, Affara NA, Sargent CA (2017) Origin and phylogenetic status of the local Ashanti Dwarf pig (ADP) of Ghana based on genetic analysis. BMC Genomics 18(1):1–12

Osei-Amponsah R, Chauhan SS, Leury BJ, Cheng L, Cullen B, Clarke IJ, Dunshea FR (2019) Genetic selection for thermotolerance in ruminants. Animals 9(11):948

Ourida R, Chadli A, M'hamed E, Hossem S, Elena C, Semir Bechir Suheil G (2017) A comprehensive characterization of Guelmoise, a native cattle breed from Eastern Algeria. Genet Biodivers J 1(1):31–42

Patrick Jolly NE, Félix M, Christian KT, Yacouba M (2018) Genetic diversity and origin of Namchi cattle breed inferred by matrilineage analyses. Genet Biodivers J 2(2):17–25

Pilling D, Bélanger J, Hoffmann I (2020) Declining biodiversity for food and agriculture needs urgent global action. Nat Food 1(3):144–147

Rahal O, Aissaoui C, Ata N, Yilmaz O, Cemal I, Ameur Ameur A, Gaouar S (2020) Genetic characterization of four Algerian cattle breeds using microsatellite markers. Anim Biotechnol 32:1–9

Rege JEO (1999) The state of African cattle genetic resources I. Classification framework and identification of threatened and extinct breeds. Anim Genet Resour/Resources génétiques animales/Recursos genéticos animales 25:1–25

Rege JEO (2003) Defining livestock breeds in the context of community-based management of farm animal genetic resources. In: Community-based management of animal genetic resources, pp 27–36

Rege JEO, Gibson JP (2003) Animal genetic resources and economic development: issues in relation to economic valuation. Ecol Econ 45(3):319–330

Rege JEO, Marshall K, Notenbaert A, Ojango JMK, Okeyo AM (2011) Pro-poor animal improvement and breeding — what can science do? Livest Sci 136(1):15–28

Rischkowsky B, Pilling D (2007) The state of the world's animal genetic resources for food and agriculture. Food & Agriculture Org

Roca J, Parrilla I, Bolarin A, Martinez EA, Rodriguez-Martinez H (2016) Will AI in pigs become more efficient? Theriogenology 86(1):187–193

Ros-Santaella JL, Pintus E (2017) Rooibos (Aspalathus linearis) extract enhances boar sperm velocity up to 96 hours of semen storage. PLoS One 12(8):e0183682

Ros-Santaella JL, Kadlec M, Pintus E (2020) Pharmacological activity of honeybush (Cyclopia intermedia) in boar spermatozoa during semen storage and under oxidative stress. Animals 10(3):463

Scherf BD, Pilling D (2015) The second report on the state of the world's animal genetic resources for food and agriculture

Scidev.net (2013a) World's first livestock gene bank https://www.scidev.net/global/multimedia/first-livestock-gene-bank/ [Online]

Scidev.net (2013b) Kenyan team leads plans for livestock genebank. https://www.scidev.net/global/news/kenyan-team-livestock-genebank/ [Online]

Seré C, van der Zijpp A, Persley G, Rege E (2008) Dynamics of livestock production systems, drivers of change and prospects for animal genetic resources. Anim Genet Resour Inf 42:3–24

Stark S, Petitto J, Darr S (2010) Animal research facility. Whole Building Design Guide, a program of the National Institute of Building Sciences. https://www.wbdg.org/building-types/research-facilities/animal-research-facility [Online]

Strucken EM, Al-Mamun HA, Esquivelzeta-Rabell C, Gondro C, Mwai OA, Gibson JP (2017) Genetic tests for estimating dairy breed proportion and parentage assignment in East African crossbred cattle. Genet Sel Evol 49(1):1–18

Tada O, Muchenje V, Dzama K (2013) Preferential traits for breeding Nguni cattle in low-input in-situ conservation production systems. Springerplus 2(1):195

Tawah C, Rege J, Gertrude S, Aboagye S (1997) A close lood at a rare African breed-the Kuri cattle of Lake Cha Basin: origin, distribution, production and adaptive characteristics. S Afr J Anim Sci 27(2):31

Tefiel H, Ata N, Chahbar M, Benyarou M, Fantazi K, Yilmaz O, Cemal I, Karaca O, Boudouma D, Gaouar SBS (2018) Genetic characterization of four Algerian goat breeds assessed by microsatellite markers. Small Rumin Res 160:65–71

Tefiel H, Ata N, Fantazi K, Yilmaz O, Cemal I, Karaca O, Chahbar M, Ameur AA, Gaouar S (2020) Microsatellite based genetic diversity in indigenous goat breeds reared in Algeria and Turkey. J Anim Plant Sci 30(5):1115–1122

Tiambo CK, Yaojing Y, Egesa P, Kemp SJ (2019) Training on reproduction technologies for cryo-conservation of African animal genetic resources, 16–20 July, 2019, Nairobi, Kenya

Ugo PC, Baker D, Morgan N, Ly C, Nouala S (2013) Investing in African livestock: business opportunities in 2030–2050

Viana J (2019) Statistics of embryo production and transfer in domestic farm animals: divergent trends for IVD and IVP embryos. International Embryo Technology Society (IETS)

Wafula P, Jianlin H, Sangare N, Sowe J, Coly R, Diallo B, Hanotte O (2005) Genetic characterization of West African Djallonke sheep using microsatellite markers. In: International workshop "The role of biotechnology for the characterization and conservation of crop, forestry, animal and fieshery genetic resources", Villa Gualino, Turin, pp 177–178

Wouobeng P, Kouam Simo J, Meutchieye F, Yacouba M, Achieng G, Tutah J, Mutai C, Agaba M (2020) Polymorphism of prolificacy genes (BMP15, BMPR 1B and GDF9), in the native goat (Capra hircus) of Cameroon. Genet Biodivers J 4(1):28–43

Yapi-Gnaoré C, Rege J, Oya A, Alemayehu N (1997) Analysis of an open nucleus breeding programme for Djallonké sheep in the Ivory Coast. 2. Response to selection on body weights. Anim Sci 64(2):301–307

Open Access This chapter is licensed under the terms of the Creative Commons Attribution 4.0 International License (http://creativecommons.org/licenses/by/4.0/), which permits use, sharing, adaptation, distribution and reproduction in any medium or format, as long as you give appropriate credit to the original author(s) and the source, provide a link to the Creative Commons license and indicate if changes were made.

The images or other third party material in this chapter are included in the chapter's Creative Commons license, unless indicated otherwise in a credit line to the material. If material is not included in the chapter's Creative Commons license and your intended use is not permitted by statutory regulation or exceeds the permitted use, you will need to obtain permission directly from the copyright holder.

Economic Considerations and Framework of Conservation of African Animal Genetic Resources

26

Wilson Kaumbata, Maria Wurzinger, and John E. O. Rege

Abstract

This chapter provides an overview of the economic aspects of animal genetic resources (AnGRs) and an economic framework for the sustainable use and conservation of AnGRs in Africa. It begins with a look at the role of economics in the sustainable use and conservation of AnGRs. The methods and tools used to quantify the economic value of AnGRs are discussed both from a theoretical and practical points of view. Drawing from literature and examples of experiences from practical applications, this chapter presents decision-making information on the economic and social benefits and costs of AnGRs conservation. The chapter concludes by discussing the framework that theoretically underpins not only the economic valuation of AnGRs but also other economic aspects of AnGRs' use and conservation.

Keywords

Economic valuation · Conservation and utilization · Animal genetic resources · Total economic value · Indigenous breeds · Social and economic benefits

Acronyms

AnGRs	Animal Genetic Resources
ARIS	Animal Resource Information System
AU-IBAR	African Union—Inter-African Bureau for Animal Resources
CBD	Convention on Biological Diversity
CVM	Contingent Valuation Method
DAD-IS	Domestic Animal Diversity Information System
FAO	Food and Agricultural Organization
ILRI	International Livestock Research Institute
LIMS	Livestock Information Management System
REC	Regional Economic Community
SADC	Southern Africa Development Community
TEV	Total Economic Value

W. Kaumbata
Department of Animal Science, Bunda College, Lilongwe University of Agriculture and Natural Resources, Lilongwe, Malawi

M. Wurzinger
Division of Livestock Sciences Department of Sustainable Agricultural Systems, BOKU-University of Natural Resources and Life Sciences, Vienna, Vienna, Austria

J. E. O. Rege (✉)
Emerge Centre for Innovations-Africa, Nairobi, Kenya
e-mail: ed.rege@emerge-africa.org

WTA   Willingness to Accept
WTP   Willingness to Pay

26.1 Introduction

Africa's indigenous livestock breeds have many unique characteristics and represent the heritage of Africa's diverse rural communities (AU-IBAR 2019). The current manifold functions, genetic attributes, and potential future uses of animal genetic resources (AnGR) described in Chaps. 2 to 14 of this book underpin the significant economic values of Africa's indigenous AnGRs, most of which are not captured in the marketplace (Drucker et al. 2001). These values are usually categorized into use and non-use values. Economic valuation of AnGRs usually considers the total economic value (TEV) of the genetic resource. The TEV is equal to the sum of all direct and indirect use values plus non-use and option values as described in Table 26.1.

A review of past economic assessments (Drucker 2010; FAO 2007; Roosen et al. 2005) showed that conservation decisions regarding native animal genetic resources have been largely based on the first category, that is, direct use values, although the other categories may be of equal or greater importance. For example, it has been estimated that approximately 80 percent of the value of livestock in low-input systems in developing countries can be attributed to non-market roles, while only 20 percent is attributable to direct production outputs (FAO 2007). By focusing exclusively on direct use values, native animal genetic resources have been consistently undervalued, resulting in a bias towards activities that promote the erosion of such resources.

The main economic reason for the erosion of AnGRs is that there is an underlying disparity between private (financial) and social (economic) costs and benefits of genetic resource use and conservation due to the existence of externalities (Drucker et al. 2001). Externalities are the losses to society (social economic costs) due to activities of individuals aimed at maximizing personal profits (private financial benefits). The divergence between private and social costs and benefits exists because the costs to society are not usually taken into account by the individuals as they do not represent a cost of production to the personal enterprises. Hence the existence of externalities, the divergence between private and social costs, and a focus on direct-use values only, contribute to the apparent undervaluation of local AnGRs and thus favour activities that promote the erosion of such resources. The economic valuation of AnGRs is thus important from a policy and management decisions perspective because it reveals the true value of the genetic

Table 26.1 Values of indigenous animal genetic resources

Use value	Direct use value	Refers to the benefits resulting from actual uses, such as for food, direct income, draught power, socio-economic (finance and insurance functions) and socio-cultural, religious, manure, hides, skins, fibre, meat, milk, and eggs
	Indirect use value	Are the benefits deriving from ecosystem functions, such as the maintenance of genetic stock and other important interactions between these breeds and the ecosystem. For example, some animals play a key role in the dispersion of certain plant species
	Option value	Are derived from the value given to safeguarding an asset for the option of using it at a future date. It is a kind of insurance value against unfavourable future events, for example, the occurrence of new animal disease or drought/climate change
Non-use values	Bequest values	Are non-use values that measure the benefit accruing to any individual from the knowledge that others might benefit from a resource in the future
	Existence values	Are non-use values derived simply from the satisfaction of knowing that a particular asset exists (e.g. blue whales, elephants, or N'Dama cattle)

Note that some asset values may overlap between these categories and double counting must be avoided (FAO 2007)

resources and can play a key role in translating such social values into efficient incentives and institutional arrangements (Drucker et al. 2001).

Establishing the total economic value of AnGRs is important (Drucker et al. 2001) because this can:

(i) Guide resource allocation between conservation of various types of AnGRs, research and development, and other socially valuable endeavours. Given that the benefits of genetic resource conservation and utilization are currently undervalued, taking their 'true' value into account in investment decisions would lead to a relative shift in the allocation of resources towards genetic diversity conservation because the marginal benefits to investment in this sector would now be more attractive.
(ii) Assist in the design of economic incentives and institutional arrangements for farmers/genetic resource managers and breeders. For example, landscape maintenance, cultural, existence, and future option values of local cattle breeds have been shown to account for 80% of their total economic value (Zander et al. 2013; Martin-Collado et al. 2014), and this justifies the need for incentive mechanisms that motivate farmers to capture some of these public good values and hence motivate them to undertake conservation-related activities. However, Signorello and Pappalardo (2003) have argued that such payment schemes are often insufficient to cover the true financial opportunity costs faced by the keepers of the local cattle breeds.

Moreover, in the face of strengthening laws for intellectual property protection of germplasm under the Convention on Biological Diversity (CBD), it became apparent that the past collegial system of free exchange of germplasm among stakeholders was going to break down fast, requiring that some mechanism (e.g. through mutually agreed terms) for genetic resources movement across borders would be needed—both for research and for commercial purposes

(Roosen et al. 2003). It was recognized that, because markets exist only for finished or nearly finished commercial genetic resources, the value of unimproved materials and the value added that were not 'valued by commercial interests' in the long process of breeding (including through efforts of indigenous communities) could not be directly measured. Yet this was going to be critical under the CBD regime. Hence, it is important that 'parties' (e.g. communities and countries) have a way of determining the values of the animal genetic resources involved in the 'exchange', whether within countries or across borders. Of particular interest is the situation in developing countries where on the one hand, livestock make important contributions to human livelihoods and food security while on the other, genetic erosion continues to place many breeds that are adapted to low-input agriculture and extreme environments, typical of these countries, at risk of loss. Mendelsohn (1999) has provided at least three arguments for the economic valuation of AnGR: (i) to inform the development of effective breeding programmes; (ii) to guide the choice of breeds in conservation programmes; and (iii) to facilitate benefit sharing in the context of the CBD.

26.2 Methods and Tools for Economic Valuation of AnGRs and Examples of Practical Application

How can we measure the value of AnGRs and which valuation methodologies are the most appropriate? A range of valuation methodologies exists. Most of these methods were initially developed for applications in plant genetic resources. Following the project on 'Economic Valuation of Farm Animal Genetic Resources' which was implemented by the International Livestock Research Institute (ILRI) (Mendelsohn 1999), a wide range of studies have been undertaken in the African continent during recent years to establish the true values of local AnGRs to guide conservation and sustainable use. Reviews of available valuation methodologies (Drucker et al. 2001) have concluded that they do provide good estimates of values which are useful for answering policy-relevant questions and as decision-supporting tools regarding conservation options. Available methods can broadly be categorized into three categories on the basis of the practical purpose for which they may be applied: (i) determining the appropriateness of AnGR conservation programme costs (i.e. consideration of environmental values); (ii) determining the actual economic importance of the breed at risk (i.e. a focus on breed values); and/or (iii) priority setting in AnGR breeding programmes (i.e. a focus on trait values).

The Contingent Valuation Method (CVM) is one of the most commonly cited approaches with the potential for determining the appropriateness of conservation programme costs—purpose category (i) above—although there are hardly any examples of cases where it has been applied in practice. CVM relies on questionnaires about willingness to pay (WTP) or willingness to accept (WTA) payment for conservation. It is considered a promising option for biodiversity valuation in general because: it is the only way to elicit non-use values directly; the potential for information provision and exchange during the survey process offers scope to experiment with respondent knowledge and understanding of biodiversity; and it can be used as a surrogate referendum on determining conservation priorities based on public preferences. Hypothetically, farmers might be asked about their willingness to accept payment for on-farm maintenance of AnGR, and the general public might be queried on WTP for maintenance on-farm or in gene banks. In this way, an upper bound to the costs that society is willing to confront for AnGR conservation could be determined. An alternative approach to defining an upper bound for economically justifiable conservation costs is to identify the minimum that society could economically justify based on a measure of production loss averted (Drucker et al. 2001). This approach attempts to identify the magnitude of potential production losses in the absence of AnGR conservation.

Examples of methods for determining the economic values of breeds at risk—purpose category (ii)—include the following:

- Econometric estimation of aggregate demand and supply curves can provide a measure of consumer and producer surplus based on the fact that changes in the traits or the composition of breeds will produce shifts in the estimated functions, which in turn will bring about a change in consumer and producer surplus (Mendelsohn 1999). Where multiple demand equations (one for each breed) can be estimated, the substitution effects across breeds can be explicitly modelled providing the most comprehensive evaluation of breeds while capturing substitution effects as well.
- Cross-sectional household and farm studies can also be applied to construct demand and supply functions.
- Market share analysis is a simpler but conceptually inferior approach that involves identifying the total share of market value that can be attributed to a given breed as a measure of the value to society of the bundle of traits embedded in the breed. The approach does not provide a consumer/producer surplus measure of value.
- The existing or potential value of Intellectual Property Rights and/or Contracts for AnGR use and conservation could also be used as an indication of the economic importance of given breeds.

Methods for priority setting in AnGR breeding programmes—purpose category (iii)—a focus on trait values are crucial because they speak to conservation through use. Specifically, it is the means by which indigenous breeds are maintained by ensuring that they remain a functioning part of the production system. Examples of approaches include the following:

- *Breeding programme evaluation approaches.* These are used to assess the costs versus benefits of breeding programmes and/or alternative genotypes or breeds. For example, the benefits of genetic material could be valued assuming that the yield effects of successive breeding stages and the necessary input cost information can be identified using the difference between the benefits of an improved breed (based on yield differential and associated prices) and the costs of all other factors employed in breeding operations such as capital and labour. The value of using alternative inputs/traits could then be compared to see how they affected economic returns.

- *Genetic production function models* are similar to breeding programme evaluation approaches, but their focus is on predicting potential future values rather than using the actual results of breeding programmes.
- *Hedonic valuation of animal characteristics* attempts to decompose the total value (price) of the single animal transacted into its relevant traits. In principle, this technique could also be used to value breeds (Mendelsohn 1999). It relies on having enough variability in the relevant vector of phenotypic (or genetic) traits of the animals involved.
- *Farm-level simulation models* of animal production can also be used by breeding programmes in order to ensure that breed benefits are being maximized by directly modelling the effects of improved animal characteristics on the economics of farms.

Table 26.2 presents a more detailed summary of the economic valuation methods and findings from the application of the methods in some African countries. Major findings include the following:

- AnGRs have economic values beyond those related to direct use (i.e. benefits resulting from actual uses, such as for food, direct income, drought power, socio-economic, finance and insurance functions and socio-cultural, religious, manure, hides and skins, fibre, etc.), The non-use values include those related to ecosystem functions, the option and knowledge that the resource is available for future use, and the satisfaction of knowing that a particular resource exists.
- The economic valuation of AnGR is important from a policy perspective because it can play a key role in translating such social values into efficient incentives and institutional arrangements for farmers/genetic resource managers and breeders.
- AnGRs conservation programmes have costs, and it is most cost-effective to conserve in situ.
- In a low-input farming system, attributes related to the subsistence functions of local AnGRs (reproductive, adaptive, and fitness traits) are more valued than attributes that directly influence marketable products of indigenous animals. These findings suggest the need to invest in the improvement of attributes that enhance the subsistence functions of AnGRs that the keepers accord higher priority to support their livelihoods than they do to tradable products.
- Phenotypic traits of traded indigenous animals (e.g. age, colour, body size, and tail condition in sheep) are major determinants of price and point to the importance of trait preferences in determining the price of AnGRs in local markets. Season and market locations are also very important price determinants.
- Production traits such as milk yield, fertility, and body size rank highly.
- A range of valuation methodologies drawn from the literature on plant genetic resources are available for consideration of potential application to AnGRs. It is concluded that a broad range of these tools needs to be field tested in order to determine which is best or most suitable for differing circumstances.

Data on available valuation methodologies suggests that they have both strengths and weaknesses. The decision on which technique to use for a particular application requires experience and judgement on the part of the analyst (Drucker et al. 2001). Data availability and the potential for acquiring relevant data will clearly be an impor-

Table 26.2 A summary of the main findings of economic valuation studies that have been undertaken in Africa

Method	Country (specie/breed)	Summary of main findings	References
Choice experiment	Benin (Chickens)	Chickens have values beyond growth and production performance, namely those related to religious beliefs and cultural ceremonies (peace) and the health status of animals.	Faustin et al. (2010)
	Ethiopia, Kenya (Cattle/Boran)	The costs of a community-based conservation programme split between Ethiopia and Kenya, based on safe minimum herd size, would require €25,400 per year in terms of direct support payments and management and monitoring costs. It is most cost-effective to conserve in situ the Ethiopian Borana subtype in Ethiopia and the Somali Borana subtype in Kenya.	Zander and Drucker (2008), Zander et al. (2009)
	Ethiopia (Cattle)	Attributes related to the subsistence functions of cattle (reproductive and adaptive traits) are more valued than attributes that directly influence the marketable products of the animals. The findings imply the strong need to invest in the improvement of attributes of cattle in the study area that enhance the subsistence functions of cattle that their owners accord higher priority to support their livelihoods than they do to tradable products.	Kassie et al. (2009, 2010)
	Kenya (Cattle/Zebu)	There are at least three classes of buyers with distinct preferences for cattle traits and most buyers favour exotic rather than indigenous breeds. Such preferences have implications for the conservation of indigenous cattle in Kenya and in other developing countries and suggest that some form of intervention may be required to ensure the preservation of this important animal genetic resource.	Ruto et al. (2008)
	Kenya (Cattle)	There is preference heterogeneity based on cattle production systems. Highly valued cattle traits for the cropping systems include traction fitness and trypanotolerance, while traits associated with herd increase are considered important in pastoral systems.	Ouma et al. (2007)
	Kenya (Goats)	Disease resistance is the most highly valued trait whose resultant augmentation results in a welfare improvement of up to KShs.2899. Drought tolerance and milk traits were found to be implicitly valued at KShs.2620 and 1179, respectively.	Omondi et al. (2008a)
	Kenya (Sheep	Disease resistance is the most highly valued trait whose resultant increment results in a welfare improvement of up to KShs.1537. Drought tolerance and fat deposition traits were found to be implicitly valued at KShs.694 and 738, respectively.	Omondi et al. (2008b)
	South Africa Pigs	Farmers in market-oriented production systems derived more benefits from productive traits such as heavier slaughter weights and large litter sizes than subsistence-oriented farmers. Under the subsistence-oriented production system, farmers in CSF-affected areas placed higher prices on adaptive traits than in the unaffected areas. Subsistence-oriented farmers who were affected by CSF wanted a total compensation price of US$1563.43 for keeping a pig genotype with unfavourable traits when compared to US$605.00 for their CSF-unaffected counterparts.	Madzimure (2011)
	Ethiopia (Sheep/Afar, Bonga, Horro, and Menz)	Producers' trait preferences were heterogeneous except for body size in rams and mothering ability in ewes where nearly homogeneous preferences were investigated. In the pastoral production system, attention was given to the coat colour of both breeding rams and ewes, favouring brown and white colours over black. Breeders in all areas attempt to combine production and reproduction traits as much as they can in order to maximize benefits from their sheep.	Duguma et al. (2011)

(continued)

Table 26.2 (continued)

Method	Country (specie/breed)	Summary of main findings	References
Conjoint analysis	Kenya (Chickens)	The study identified traits preferred by the farmers based on their current low-input production circumstances. Using these traits, indigenous chickens can be selected for higher productivity and performance while retaining their diversity and adaptability.	Bett et al. (2011)
	United Republic of Tanzania (Cattle/Tarime Zebu)	Livestock keepers in the Tarime district prefer Tarime cattle to exotic dairy cattle. The preference for indigenous cattle by most farmers should be viewed as the most favourable starting point for the conservation of Tarime cattle through sustainable utilization.	Ngowi et al. (2008)
Contingent valuation	Ethiopia (cattle)	Season, market location, class of cattle, body size, and age are very important determinants of the cattle price. The relative weight of the phenotypic characteristics of the animals is among the highest of all the factors considered. These preferences at the farmers' and farmer traders' levels are the ones that matter most in shaping the diversity of animals kept at the farm level, and the diversity of cattle genetic resources is quite essential for generating or identifying the best-suited breeds of cattle, given the livelihood objectives of the target community.	Kassie et al. (2011)
Hedonic pricing	Ethiopia (sheep)	Phenotypic traits of traded indigenous sheep (age, colour, body size, and tail condition) are major determinants of price implying the importance of trait preferences in determining the price of sheep in local markets. Season and market locations are also very important price determinants.	Terfa et al. (2013)
	Burundi, Rwanda, Uganda, United Republic of Tanzania (cattle/Ankole)	In Burundi, Rwanda, and parts of Uganda, livestock keepers are sedentary and herds are small, whereas in the other areas, Ankole cattle are kept in large herds. Productive traits such as milk yield, fertility, and body size were ranked highly. For bulls, the trait 'growth' was ranked highly in all study areas. Phenotypic features (coat colour, horn shape, and size) and ancestral information are more important in bulls than in cows.	Wurzinger et al. (2006)

(continued)

Table 26.2 (continued)

Method	Country (specie/breed)	Summary of main findings	References
Preference ranking	Ethiopia (poultry)	Farmers' ratings of indigenous chickens with respect to modern breeds showed the highest significance of the adaptive traits in general and the superior merits of indigenous chickens to high-yielding exotic breeds in particular. Adaptation to the production environment was the most important attribute of chickens in all the study areas. The high significance attributed to reproduction traits indicates the need for maintaining broody behaviour and a high level of hatchability while breeding for improved productivity of indigenous chickens for village conditions.	Dana et al. (2010)
	Ethiopia (cattle/Zebu/Sheko)	Overall, farmers showed slightly more preference for local Zebus over the Sheko breed. This is due to the voracious feeding behaviour of Sheko cattle, which makes them less preferable in the face of worsening feed shortage, and due to the aggressive temperament of Sheko cattle. This is despite Sheko's outperforming potential over local Zebus in their milk production, draft power, and hardiness.	Desta et al. (2011)
	South Africa (cattle/Nguni)	Trait preference in breeding bulls and cows is significantly influenced by socio-economic and demographic factors. It is recommended to consider farmer preferences in trait selection and designing communal breeding programmes.	Tada et al. (2013)
	Uganda (cattle/Ankole)	In the selection of cows, performance and fitness traits are emphasized by the cattle keepers while in the selection of bulls, the phenotypic appearance of the animal plays an important role. Individual fertility followed by milk performance are the main criteria for selecting cows, resistance to East Coast Fever was of the highest importance in bulls. In both sexes, a dark red coat colour was highly rated.	Ndumu et al. (2008)
	Zimbabwe (chickens)	Chicken body size was ranked the major determinant in choosing breeding animals followed by mothering ability, and fertility. More households culled chickens associated with poor reproductive performance, poor growth rates, and those intolerant to disease pathogens.	Muchadeyi et al. (2009)
	Ethiopia (cattle)	Dairy cow owners should be advised to use the optimum levels of inputs and replace their indigenous cows with cross-breed cows. Moreover, the herds should be medium size and feeding mainly depends on concentration.	Dayanandan (2011)
Production function/ gross margin analysis	Kenya (cattle/Orma and Sahiwal Zebu)	The Orma/zebu and Sahiwal/zebu breeds had comparable economic benefits, hence a pastoralist in the Magadi division is likely to get similar returns from both breeds. This study therefore recommends the adoption of not only the Sahiwal/zebu but also the Orma/zebu breed for cattle improvement in trypanosomosis endemic areas and conservation of indigenous genetic resources.	Maichomo et al. (2009)
	Malawi (Goats/Small East African)	Local goat enterprises in smallholder farms are profitable and economically viable. The mean annual net profit per flock and per goat was MK54,406 and MK11, 140 (€1 = MK830.00), respectively. The average return on capital invested was 24.6%, exceeding the prevailing average commercial deposit rate (8%) by several folds. The inclusion of intangible benefits of goats significantly increased the mean annual net profit and the return on capital by 60.3%, reflecting the importance of socioeconomic roles goats play in providing current and future economic stability to rural households' economy.	Kaumbata et al. (2020)

Adapted from the second report of the State of the World's Animal Genetic Resources (FAO 2015)

tant determinant, especially given the problems of missing markets for non-marketable values of animal genetic resources and market imperfections commonly encountered in developing country situations. Furthermore, despite the difficulties associated with the complexity of the economic valuation of farm AnGRs, the results of the studies reviewed suggest that the integration of methods based on expressed preferences, such as contingent valuation and choice experiments, can produce valuable information to assess breeding schemes and/or the cost-benefit analysis of conservation programmes for breeds at risk of extinction (Drucker et al. 2001).

26.3 Estimating the Economic and Social Benefits and Costs of AnGR Conservation

The economic, social, cultural, and environmental benefits of AnGRs are their capacity to produce food and other goods and services that can be sold or used at home and are generally the main reason why people choose to raise them. Estimating the value of these benefits and the associated costs of maintaining the diversity of AnGRs is fundamental for supporting decision-making regarding conservation, improvement, and sustainable use. The benefits constitute different components of TEV of AnGRs. As described in Sect. 26.2, several valuation methodologies have been suggested to estimate the different components of the TEV of AnGRs. Unfortunately, there is no single methodology that can be used to estimate all four 'values' (use, non-use, option, and quasi-option value). Moreover, a number of these methodologies have significant conceptual shortcomings and intensive data requirements (Drucker et al. 2001). However, they have been shown to produce useful estimates of the values that are placed on the market, non-market, and potential breed attributes useful for designing breeding and conservation strategies. The integration of methods based on expressed preferences, such as the contingent valuation methods (CVM), with bio-economic models (Cicia et al. 2003) can produce valuable information to assess the cost-benefit analysis of conservation programmes for breeds at risk of extinction. Even though the CVM cannot account for some types of social values, such as quasi-option value, which in the valuation of AnGRs can play an important role, components such as use, passive-use, and option values, have a relevance that alone may justify a conservation policy. Drawing from a study conducted by Drucker and Anderson (2004) to assess the benefits and costs of conservation of the endangered Box Keken pig breed using three valuation methodologies, FAO (2007) opines that despite the shortcomings of the employed methodologies and the fact that values can only be approximated, benefits of conservation clearly outweigh the costs. In a similar study (Zander et al. 2009), the cost of conserving the Ethiopian Borana cattle breed in Ethiopia and Kenya, based on opportunity cost, was estimated at €25,400 per annum. The benefits of conservation (e.g. maintaining option values) and sustainable use (e.g. poverty alleviation, improved food security, maintenance of Borana cattle for traditional and cultural uses) would need to be weighed up against these costs. Generally, conservation costs have been shown, in a number of case studies (Drucker and Anderson 2004), to be relatively small, both when compared to the size of subsidies currently being provided to the commercial livestock sector, and with regard to the benefits of conservation. These costs are lowest in developing countries which is encouraging given that an estimated 70 percent of the indigenous livestock breeds existing today are in developing countries, and that this is where the risk of loss is highest (Rege and Gibson 2003). Despite this, studies specifically tailored to comparing the benefits and costs of AnGRs conservation in Africa to inform decision-making are lacking. However, several African countries, have initiated actions aimed at conserving some endangered livestock breeds, using either in situ or ex situ methods. The major agencies responsible for AnGRs conservation programmes are governments, breeders' associations, private farms, and research institutes or universities. Major species of focus for in situ conservations include cattle, sheep, goats, pigs, poultry, and camels as well as emerging or non-conventional species such as duck, rabbit, pigeon, and guinea fowl. These include both locally adapted indigenous and exotic breeds. Traditionally, smallholder farmers have, for millennia, played key roles in conserving and using native breeds to support their livelihoods. They have maintained the rich biodiversity and the associated local knowledge in addition to the crucial role in sustaining livelihoods and other environmental benefits (AU-IBAR 2019). Conservation interventions should, therefore, aim to support these efforts in order to maintain the existing genetic wealth. Currently, ex situ conservation of AnGRs is quite low due to a lack of appropriate technology and expertise. Only 12 African countries reported that they undertake in vitro activities (AU-IBAR 2019), whereas 28 countries are planning to establish similar centres. The lack of prioritization of ex situ conservation programmes results in limited budget allocations.

Virtually, in all studies where breed comparisons have been conducted, indigenous breeds have shown superiority over exotic breeds in adaptive characteristics, such as fertility, survivability, disease resistance, heat and drought tolerance, and the ability to produce under conditions of nutritional inadequacy. Current models used for predicting climate change show that for the arid and semi-arid areas in which a large part of Africa falls, there is likely to be an increase in the frequency, severity, and length of droughts as well as significant increases in ambient temperatures (Drucker 2010). Under these aggravating conditions, the maladapted exotic breeds will be even more challenged. Indigenous AnGRs, therefore, represent a distinct economic resource to the continent, which must be conserved (see Chap. 24 on conservation and management of animal genetic resources). They also represent a valuable contribution to biological diversity at community, national, regional, and global levels. The arguments for the conservation and sustainable use of indigenous and the design of policies to counteract trends towards marginalization of adapted breeds are therefore compelling.

26.4 Economic Framework for Sustainable Use and Conservation of AnGRs

Economic arguments for the conservation and sustainable use of AnGRs can be an effective means of garnering the necessary public and political support, including the development of appropriate policies (Rege and Gibson 2003). As alluded to in Sect. 26.3, a range of methods and analytical approaches are available that could be used to value livestock breeds or traits. However, the practical applications of these tools and methods in contexts that could influence policymaking and livestock keepers' livelihoods remain limited (Drucker 2010). To translate the existing knowledge and recognition of the importance of economics within the Global Plan of Action on AnGRs into mainstream activities, the following should be ensured: (i) awareness creation; (ii) capacity development; (iii) availability and access to appropriate data; (iv) resources and incentives for conservation; and (v) multidisciplinary approach to conservation.

26.4.1 Awareness Creation

In most African countries, one of the main reasons for the lack of policies and legislations directly formulated and designed for the management of AnGRs is that policymakers are usually not fully aware of the economic value and contribution of AnGRs to national economies (AU-IBAR 2019) and the commensurate need to provide policy and budgetary support. Hence there is a need to raise awareness (through effective communication channels) of researchers, breeders, development practitioners, policymakers, and farmers regarding the important role of economic analysis in improving the sustainable use and conservation of AnGRs. Correct and accurate information about the value and economic contribution of AnGRs at household and national levels in Africa can help to design appropriate conservation strategies and action plans. Effective communication and coherence between ministries responsible for AnGRs and other stakeholders in the sector is essential. The methodologies and tools to undertake such interventions should be readily available to all relevant stakeholders.

26.4.2 Capacity Development

Beyond awareness, and the need for attention to technical capacities covered in Chaps. 26 and 27, there is a need to strengthen national capacities in the area of economic valuation so that relevant methods and decision-support tools can be applied. There is also a need to integrate such tools and methods into wider national livestock development programmes, including designing appropriate incentive mechanisms (FAO 2015). Institutions (teaching and research, extension, trade and marketing, policy and legal entities) that deal with AnGRs and allied sectors need to be strengthened. The idea of developing some selected African institutions at a national and regional level to become effective centres of excellence could be supported by regional economic communities (RECs). Farmers' and breeders' associations and/or organizations should be strengthened as they can potentially play important roles, including representation of important actors involved in the management of AnGRs and various commodity value chains. Capacity of field technical staff responsible for data collection, analysis, and reporting at the breed and household levels should also be strengthened (AU-IBAR 2019).

26.4.3 Availability and Access to Appropriate Data

Closely connected to capacity development is data availability. Data for economic analysis are either not available or perceived as expensive to obtain, perhaps also contributing to a misperception that the tools are too technically complex to apply (Drucker 2010). The absence of data/information regarding local AnGRs shakes the very foundation of the understanding of what the genetic resources are and how best to conserve and sustainably improve and use them (AU-IBAR 2019). Until breed-level data are routinely collected as part of the national statistics, there will continue to be a need to undertake intensive primary data collection for specific purposes. This is likely to require the application of participatory rural appraisal techniques for which, once again, capacity is limited in many countries (Drucker 2010). Research and development in AnGRs are critical for up-to-date evidence-based information and data on the usefulness and importance of African AnGRs. There should be national and regional monitoring mechanisms that will function as early warning systems particularly on breed populations to flag and respond to emerging risk of breed loss. These should be put in place through the collaboration of national governments and RECs to ensure breeds at risk are monitored and populations restored when necessary (AU-IBAR 2019).

26.4.4 Resources and Incentives for Conservation of AnGRs

The need and methods for the conservation of animal genetic resources are covered in Chap. 24. Major funding for AnGRs conservation programmes in Africa is usually provided by the government and international donor agencies. However, funding by governments has generally been both inconsistent and inadequate (AU-IBAR 2019). In most African countries, few nucleus herds of local animals exist in government or research institutional farms. Conservation activities are usually restricted to a few breeds of major livestock species such as cattle, goats, sheep, chickens, and pigs. Currently, ex situ conservation is quite low due to a lack of appropriate technology and expertise. A few countries have national gene banks, others have specialized agricultural research institutes responsible for maintaining genetic resources, and many do not have any specific programmes on the characterization and conservation of AnGRs. Incentive payments aimed at compensating livestock keepers for the financial opportunity costs of continuing to maintain locally adapted AnGRs are hardly found in

African settings. The Global Plan of Action on AnGR however expects governments to demonstrate sustained political will and mobilize adequate resources needed for successful implementation. This will require regional and international cooperation. FAO and other relevant international organizations, countries, the scientific community, donors, civil society organizations, and the private sector, all have important roles to play. Indeed, there are both moral and practical imperatives to provide support to livestock keepers and breeders, who are the custodians of Africa's diverse animal genetic resources. Failure to take these actions will mean continued erosion as Africa's livestock biodiversity remains underutilized and underdeveloped (Kaumbata et al. 2020), and at-risk and endangered breeds will disappear.

26.4.5 Multi-Disciplinary Approach to Conservation

There are multiple deficiencies impeding the uptake of economic tools and methods regarding the valuation of AnGRs and the design and implementation of conservation plans. Biodiversity conservation programmes tend to be deficient in their design as a result of a lack of consideration of a broad range of technical issues, of which economics is just one (Drucker 2010). The national reports and Global Plan of Action tend to confirm this view as characterization and improvement use strategies identify priorities that go well beyond economic valuation, and include molecular analysis, performance evaluation, and monitoring. Given that the economics of AnGRs work is best carried out within a production systems' context and requires informational inputs from a range of disciplines, overcoming other deficiencies influencing the conservation and use of AnGRs will play a contributory role. Such multi-disciplinary approaches may include the following:

(i) *Development of national and regional livestock information systems.* Livestock information systems are vital for data/information collection storage and retrieval. At the global level, the Domestic Animal Diversity Information System (DAD-IS), developed and maintained by FAO, provides access to searchable databases of breed-related information and photos and links to other online resources on livestock diversity. It facilitates analyses of the diversity of livestock breeds at national, regional, and global levels, including the status of breeds regarding their risk of extinction. ILRI has also developed the Domestic Animal Genetic Resources Information System which focuses on technical (characterization) information and focuses on Africa and Asia. The AU-IBAR also hosts the Animal Resources Information System (ARIS) as the main information system specifically on Africa's AnGRs (AU-IBAR 2019). Similarly, some countries and regions have established information networks. In the SADC region, for example, the Livestock Information Management Systems (LIMS) is hosted by the SADC Secretariat. Despite the availability of these information systems, countries generally do not routinely provide data and hence there are significant data gaps in these databases. This is in significant part because most of the countries do not have dedicated in-country organizations, networks, and initiatives for sustainable use, breeding, and conservation of AnGRs (AU-IBAR 2019). Where they exist, they require strengthening.

(ii) *Participation of stakeholders, including the private sector.* Stakeholders play important roles in the implementation of the various activities in the management of AnGRs. In most African countries steps have been taken to engage or empower relevant stakeholders in the management of AnGRs (AU-IBAR 2019; Drucker 2010). The establishment and support of permanent and dedicated supporting organizations and other stakeholders provide sustainable institutional support (Wurzinger et al. 2021) instrumental for the establishment and growth of breeding (and conservation) programmes. The private sector participation in the development of

AnGRs institutions and networks is happening in some countries, but there is still much to be done in this regard. A value chain approach to the development of the AnGRs subsector requires that all value chain actors, including those in the private sector, play an active role in the governance as well as identifying opportunities for upgrading the AnGRs value chain, including through the introduction of higher-value products (AU-IBAR 2019). Product processing and trade in livestock products are key areas of business opportunities for the private sector and may have a lasting impact on conservation and breeding programmes, especially through investments in traits unique to indigenous breeds. For example, participation of the private sector through the establishment of small ruminant abattoirs, and engagement of processors and retailers of indigenous small ruminants' products boosted the incomes of smallholders (Kaumbata et al. 2020) contributing to the sustainability of small ruminant breeding programmes in smallholder farms. Public-private partnership arrangements are also possible and viable in some situations and must be pursued in the development of programmes and projects. The private sector stakeholders in the AnGRs sub-sector also have a role in the policy development process by bringing on board their unique perspectives, for example, experiences on the input supply and marketing of products as well as financing.

mote the erosion of such resources. Establishing the true total economic value of AnGRs is important because it guides resource allocation between the conservation of various types of AnGRs and assists in the design of economic incentives for various stakeholders.

A range of valuation methodologies exist most of which provide good estimates of values that are useful for answering policy-relevant questions and as decision-supporting tools regarding conservation options. Data on the application of the available valuation methodologies in Africa suggests that they have both strengths and weaknesses. The decision on which technique to use for a particular application requires experience and judgement on the part of the analyst. The integration of various valuation methodologies is also recommended since no single methodology has the capacity to estimate all four components of the TEV.

Although a range of methods and analytical approaches that could be used to value livestock breeds are available, the practical applications of these tools and methods in the contexts that could influence policymaking and livestock keepers' livelihoods remain limited. To translate the existing knowledge and recognition of the importance of economics within the Global Plan of Action on AnGRs into mainstream activities, the following should be ensured: awareness creation; capacity development; availability and access to appropriate data; resources and incentives for conservation; and multi-disciplinary approach to conservation.

26.5 Conclusions

The current manifold functions and potential future uses of animal genetic resources, underpin the significant economic values of Africa's indigenous AnGRs. The economic valuation of AnGRs takes into account the total economic value of the genetic resources. A review of past economic assessments showed that conservation decisions have been made based on direct use values alone leading to the undervaluation of Africa's AnGRs, which results in a bias towards activities that pro-

References

AU-IBAR (2019) The state of farm animal genetic resources in Africa: towards accelerated agricultural growth and transformation by the year 2025. AU-IBAR, Nairobi

Bett HK, Bett RC, Peters KJ, Kahi AK, Bokelmann W (2011) Estimating farmers' preferences in selection of indigenous chicken genetic resources using non-market attributes. Anim Genet Resour 49:51–63. Available at http://www.fao.org/docrep/014/ba0128t/ba0128t00.pdf

Cicia G, D'Ercole E, Marino D (2003) Costs and benefits of preserving farm animal genetic resources from extinction: CVM and bio-economic model for valuing a conservation program for the Italian Pentro Horse. Ecol Econ 45(3):445–459. https://doi.org/10.1016/S0921-8009(03)00096-X

Dana N, van der Waaij L, Dessie T, van Arendonk J (2010) Production objectives and trait preferences of village poultry producers of Ethiopia: implications for designing breeding schemes utilizing indigenous chicken genetic resources. Trop Anim Health Prod 42:1519–1529

Dayanandan R (2011) Production and marketing efficiency of dairy farms in highland of Ethiopia—an economic analysis. Int J Enterp Comput Busin Syst. [online]. Available at http://www.ijecbs.com/July2011/16.pdf

Desta T, Ayalew W, Hegde BP (2011) Breed and trait preferences of Sheko cattle keepers in southwestern Ethiopia. Trop Anim Health Prod 43:851–856

Drucker AG (2010) Where's the beef? The economics of AnGR conservation and its influence on policy design and implementation. Anim Genet Resour/Ressources génétiques animales/Recursos genéticos animales 47:85–90. https://doi.org/10.1017/s2078633610000913

Drucker AG, Anderson S (2004) Economic analysis of animal genetic resources and the use of rural appraisal methods: lessons from Southeast Mexico. Int J Agric Sustain 2(2):77–97. https://doi.org/10.1080/14735903.2004.9684569

Drucker AG, Gomez V, Anderson S (2001) The economic valuation of farm animal genetic resources: a survey of available methods, vol 36. www.elsevier.com/locate/ecolecon

Duguma G, Mirkena T, Haile A, Okeyo AM, Tibbo M, Rischkowsky B, Sölkner J, Wurzinger M (2011) Identification of smallholder farmers and pastoralists' preferences for sheep breeding traits: choice model approach. Animal 5:1984–1992

FAO (2007) The state of the world's animal genetic resources for food and agriculture. Commission on Genetic Resources for Food and Agriculture, Food and Agriculture Organization of the United Nations

FAO (2015) The second report on the state of the world's animal genetic resources for food and agriculture

Faustin V, Adégbidi AA, Garnett ST, Koudandé DO, Agbo V, Zander KK (2010) Peace, health or fortune? Preferences for chicken traits in rural Benin. Ecol Econ 69:1848–1857. Available at http://www.sciencedirect.com/science/article/pii/S0921800910001862

Kassie GT, Abdulai A, Wollny C (2009) Valuing traits of indigenous cows in central Ethiopia. J Agric Econ 60:386–401

Kassie GT, Abdulai A, Wollny C (2010) Implicit prices of indigenous bull traits in crop-livestock mixed production systems of Ethiopia. Afr Dev Rev 22:482–494

Kassie G, Abdulai A, Wollny C (2011) Heteroscedastic hedonic price model for cattle in the rural markets of central Ethiopia. Appl Econ 43:3459–3464

Kaumbata W, Nakimbugwe H, Haile A, Banda L, Mészáros G, Gondwe T, Woodward-Greene MJ, Rosen BD, van Tassell CP, Sölkner J, Wurzinger M (2020) Scaling up community-based goat breeding programmes via multi-stakeholder collaboration. J Agric Rural Dev Trop Subtrop 121(1):99–112. https://doi.org/10.17170/kobra-202005281298

Madzimure J (2011) Climate change adaptation and economic valuation of local pig genetic resources in communal production systems of South Africa. Alice, South Africa, University of Fort hare, Phd thesis. Available at http://tinyurl.com/ojg6jpk

Maichomo MW, Kosura WO, Gathuma JM, Gitau GK, Ndung'u JM, Nyamwaro SO (2009) Economic assessment of the performance of trypanotolerant cattle breeds in a pastoral production system in Kenya. J S Afr Vet Assoc 80:157–162

Martin-Collado D, Diaz C, Drucker AG, Carabaño MJ, Zander KK (2014) Determination of non-market values to inform conservation strategies for the threatened Alistana-Sanabresa cattle breed. Animal 8(8):1373–1381. https://doi.org/10.1017/S1751731114000676

Mendelsohn R (1999) Economic valuation of animal genetic resources, a synthesis report of the planning workshop. In: Rege JEO (ed) Economic valuation of animal genetic resources: proceedings of an FAO/ILRI workshops; held in FAO Headquarters, Rome Italy, 15–17 March 1999. Food and Agriculture Organization of the United Nations

Muchadeyi FC, Wollny CBA, Eding H, Weigend S, Simianer H (2009) Choice of breeding stock, preference of production traits and culling criteria of village chickens among Zimbabwe agro-ecological zones. Trop Anim Health Prod 41:403–412

Ndumu D, Wurzinger M, Baumung R, Drucker A, Mwai O, Semambo D, Sölkner J (2008) Performance and fitness traits versus phenotypic appearance in the African Ankole Longhorn cattle: a novel approach to identify selection criteria for indigenous breeds. Livest Sci 113:234–242

Ngowi EE, Chenyambuga SW, Gwakisa PS (2008) Socio-economic values and traditional management practices of Tarime zebu cattle in Tanzania. Livest Res Rural Dev 20:Article # 94. Available at http://www.lrrd.org/lrrd20/6/ngow20094.htm

Omondi I, Baltenweck I, Drucker A, Obare G, Zander K (2008a) Valuing goat genetic resources: a pro-poor growth strategy in the Kenyan semi-arid tropics. Trop Anim Health Prod 40:583–596

Omondi I, Baltenweck I, Drucker A, Obare G, Zander K (2008b) Economic valuation of sheep genetic resources: implications for sustainable utilization in the Kenyan semi-arid tropics. Trop Anim Health Prod 40:615–626

Ouma E, Abdulai A, Drucker A (2007) Measuring heterogeneous preferences for cattle traits among cattle-keeping households in east Africa. Am J Agric Econ 89:1005–1019

Rege JEO, Gibson JP (2003) Animal genetic resources and economic development: issues in relation to economic valuation. Ecol Econ 45(3):319–330. https://doi.org/10.1016/S0921-8009(03)00087-9

Roosen J, Fadlaoui A, Bertaglia M (2003) Economic evaluation and biodiversity conservation of animal genetic resources; www.econstor.eu

Roosen J, Fadlaoui A, Bertaglia M (2005) Economic evaluation for conservation of farm animal genetic resources

Ruto E, Garrod G, Scarpa R (2008) Valuing animal genetic resources: a choice modeling application to indigenous cattle in Kenya. Agric Econ 38:89–98

Signorello G, Pappalardo G (2003) Domestic animal biodiversity conservation: a case study of rural development plans in the European Union. Ecol Econ 45(3):487–499. https://doi.org/10.1016/S0921-8009(03)00099-5

Tada O, Muchenje V, Dzama K (2013) Preferential traits for breeding nguni cattle in low-input *in-situ* conservation production systems. Springerplus 2:195

Terfa Z, Haile A, Baker D, Kassie G (2013) Valuation of traits of indigenous sheep using hedonic pricing in central Ethiopia. Agric Food Econ 1:6

Wurzinger M, Ndumu D, Baumung R, Drucker A, Mwai O, Semambo D, Byamungu N, Sölkner J (2006) Comparison of production systems and selection criteria of Ankole cattle by breeders in Burundi, Rwanda, Tanzania and Uganda. Trop Anim Health Prod 38:571–581

Wurzinger M, Gutiérrez GA, Sölkner J, Probst L (2021) Community-based livestock breeding: coordinated action or relational process? Front Vet Sci:8. https://doi.org/10.3389/fvets.2021.613505

Zander KK, Drucker AG (2008) Conserving what's important: using choice model scenarios to value local cattle breeds in east Africa. Ecol Econ 68:34–45

Zander KK, Drucker AG, Holm-Müller K (2009) Costing the conservation of animal genetic resources: the case of Borana Cattle in Ethiopia and Kenya. J Arid Environ 73(4–5):550–556. https://doi.org/10.1016/j.jaridenv.2008.11.003

Zander KK, Signorello G, de Salvo M, Gandini G, Drucker AG (2013) Assessing the total economic value of threatened livestock breeds in Italy: implications for conservation policy. Ecol Econ:219–229. https://doi.org/10.1016/j.ecolecon.2013.06.002

Open Access This chapter is licensed under the terms of the Creative Commons Attribution 4.0 International License (http://creativecommons.org/licenses/by/4.0/), which permits use, sharing, adaptation, distribution and reproduction in any medium or format, as long as you give appropriate credit to the original author(s) and the source, provide a link to the Creative Commons license and indicate if changes were made.

The images or other third party material in this chapter are included in the chapter's Creative Commons license, unless indicated otherwise in a credit line to the material. If material is not included in the chapter's Creative Commons license and your intended use is not permitted by statutory regulation or exceeds the permitted use, you will need to obtain permission directly from the copyright holder.

Part V

Institutional Arrangements and Enabling Environment for Enhanced Utilization of African Livestock Genetic Resources

Capacity Strengthening of Animal Genetic Improvement Education in Africa

27

Samuel E. Aggrey, Richard Osei-Amponsah, Donald R. Kugonza, Raphael A. Mrode, Romdhane Rekaya, and John E. O. Rege

Abstract

Africa is home to a rich diversity of farm animal genetic resources which play major roles in poverty alleviation, food, and nutritional security as well as helping meet various socio-cultural needs. These valuable animal resources range from the major species such as pigs, cattle, sheep, goats, donkeys, camelids, poultry, aquatic resources, and bees to micro-livestock such as guinea fowl, grasscutters, and cavies. In line with the Global Plan of Action on Animal Genetic Resources, there is a need for Africa to sustainably manage these resources and help ensure their conservation. Genetic improvement of Africa's animal resources will be key to breeding resilient, productive, and climate-adaptive animals for the current and future generations. Unfortunately, Africa continues to lag behind in both the institutional and human resource capacity for animal genetic resources management, including improvement. This chapter presents a brief overview of human resources capacity challenges, limitations in animal breeding and genetics and subsequently suggests how to turn some of the challenges into opportunities. The basic concepts and misconceptions in animal breeding, lessons learned from past capacity-building initiatives, and the need to improve agricultural education in primary and secondary schools are covered. Additionally, we review post-secondary technical education and graduate education in animal breeding. Finally, we highlight the need to embrace good collaborative partnerships to enable Africa to boost its human and institu-

S. E. Aggrey
Department of Poultry Science, University of Georgia, Athens, GA, USA

Institute of Bioinformatics, University of Georgia, Athens, GA, USA

R. Osei-Amponsah (✉)
Department of Animal Science, School of Agriculture, University of Ghana, Legon, Ghana
e-mail: ROsei-Amponsah@ug.edu.gh

D. R. Kugonza
Department of Animal Biosciences, University of Makerere, Kampala, Uganda

R. A. Mrode
Department of Animal and Veterinary Science, Scotland Rural College, Scotland, UK

International Livestock Research Institute, Nairobi, Kenya

R. Rekaya
Institute of Bioinformatics, University of Georgia, Athens, GA, USA

Department of Animal and Dairy Science, University of Georgia, Athens, GA, USA

J. E. O. Rege
Emerge Centre for Innovation-Africa, Nairobi, Kenya

Centre of Excellence for Livestock Innovation and Business, Egerton University, Egerton-Njoro, Kenya

tional capacity in order to take advantage of current and emerging innovative technologies in animal breeding education.

Keywords

Animal agriculture · Breeding · Genetics · Human resource development

27.1 Introduction

Animal genetic improvement can be viewed in general terms as the science of increasing the genetic merit of the progeny population through the selection and the appropriate mating of genetically superior parents. Increasing the genetic merit of animals within a population can lead to significant improvements in traits that are directly or indirectly under selection pressure. Mastering this know-how requires the acquisition of a diverse set of skills and knowledge. This is a multi-level education and training process that integrates knowledge from quantitative genetics, population genetics, statistics, and computer science. This chapter reviews the current status of animal genetic improvement education in Africa, identifies gaps and deficiencies, and makes recommendations to strengthen the existing capacity.

As clearly articulated in the next chapter (Chap. 28, Sect. 28.2), the importance of animal genetic resources (AnGR) to the attainment of the United Nation's Sustainable Development Goals (SDGs) and the Livestock Development Strategy of Africa cannot be overemphasized. This calls for concerted stakeholder actions to ensure sustainable use, improvement, and conservation of Africa's AnGR. National animal genetic improvement programmes that have a significant and continuous impact on animal production systems in Africa are, at best, scarce, but generally non-existent. While there are several reasons that explain the lack of successful genetic selection programmes in Africa, this chapter focuses on issues pertaining to animal breeding education. Among other things, the scientific and technical knowledge of locally trained personnel in charge of planning, implementing, and managing farm animal improvement and conservation programmes is abysmal in the whole of Africa. Without adequately trained personnel to develop and implement efficient and sustainable animal improvement programmes, Africa will continue to be a net importer of farm animals and animal products. Continuous importation is not only unsustainable but also provides no path towards long-term animal protein food security and self-sufficiency.

27.2 Concepts and Misconceptions of Animal Genetic Improvement

The definition of 'animal genetic improvement' is implicit within the phraseology. It means positive changes in the productivity of animals within a population resulting from genetic selection. Genetic improvement usually applies to a population; therefore, the change must increase the genetic merit of that particular population. Selection pressure will lead to changes in the phenotype of traits defined within the population, as well as the population's genetic composition and parameters. To effect genetic improvement, a breeding scheme needs to be in place. The breeding scheme may include the following: (i) the breeding goals, (ii) trait measurement and data analysis, (iii) selection and mating, and (iv) genetic evaluation. The purpose of a breeding scheme is to efficiently integrate all components to achieve the optimum genetic gain in the progeny generation of a population. The basic understanding of what genetic improvement is should guide how academic and non-academic education and training are conducted, and how animal genetic improvement programmes are designed, executed, and evaluated. Classically, genetic improvement relied heavily on quantitative and population genetics. More recently, new fields of knowledge, including molecular genetics, microbiology, physiology, statistics, bioinformatics, phenomics, and artificial intelligence are playing increasingly important roles in the implementation of genetic improvement programmes. To improve the accuracy of genetic predictions, geneticists have sought to include genetic markers and sequence data, microbiome, and metabolites information, and they are on the verge of adopting sophisticated analytics tools, including artificial intelligence and machine learning for data analysis. Thus, animal improvement professionals have welcomed inputs from other disciplines to advance their cause.

Why then are there misconceptions about animal genetic improvement and why are they important? At the very minimum, one progeny generation is required to evaluate the success or failure of a genetic improvement programme. Reproducing an offspring generation from the parental generation can be synonymous with the term 'breeding'. It is common that animal geneticists in pursuit of livestock improvement are referred to as breeders; the same label (title) is assigned to personnel performing animal mating and reproduction who are also called breeders. The layman does not know the difference. By definition, a breed is a group of organisms with distinctive appearance or characteristics. Whereas beef and dairy producers use predominantly distinct animal breeds, poultry, and pig producers often use breeds, strains, lines, or ecotypes. Comparisons among breeds for different goals are important for decision-making on different levels. However, mere breed, strain, or ecotype comparison for performance or other non-genetic criteria is not breeding and the practitioner of such activity does not meet the general expertise and qualification of a geneticist. The inability to distinguish between the different definitions of 'breeders' can be consequential in education and training as well as in the implementation of national animal genetic improvement programmes. According to the Career Explorer website, 'There is no educational requirement for animal breeders—anyone can become a breeder with extensive knowledge of animal breed' (https://www.careerexplorer.com/careers/animal-breeder/how-to-become/). Such misconception can go a long way to affect the development of personnel for animal genetic improvement. Under these circumstances, non-geneticists who are erroneously labelled as breeders may be chosen to lead an important genetic improvement programme.

Genetic improvement can be simplified to mean the increase in the frequency of favourable alleles and a simultaneous decrease in the frequency of unfavourable alleles in the progeny population based on the traits defined in the breeding goal. The education of animal genetic improvement geneticists should encompass the fundamental principles of population genetics, quantitative genetics, computational science, and an understanding of the other critical complementary disciplines such as molecular genetics, statistics, microbiome, metabolomics, artificial intelligence, and machine learning. It is imperative to understand the core and supportive drivers of modern breeding for the successful implementation of an animal genetic improvement programme. The study of genetics encompasses several disciplines or specializations. Specializations could include cytogenetics, molecular genetics, molecular ecology, veterinary genetics, microbial genetics, population genetics, quantitative genetics, and evolutionary genetics, to name a few. Thus, scientists educated and trained in the relevant expertise areas within a well-defined structure and by competent personnel should lead genetic improvement programmes on the continent. Genetic improvement is a component of the overall animal production system, and when executed properly should contribute positively to the overall performance of farm animals. The genetic gains together with the appropriate management of genetic resources should translate into positive economic, nutritional, and environmental impacts for the citizenry.

Genetic improvement education should be viewed in the context of the projected future demand and supply of animal protein, the nutritional requirements of the population, input resources for animal production, technological changes, genetic diversity (including conservation), climate change, and the appropriate strategy for the broader agro-ecological and socio-cultural environments in the continent more generally, and the productions systems of the specific countries. The graduates of animal genetic improvement education must be able to balance the aforementioned factors in an economic, socio-cultural, and environmentally sustainable manner.

27.3 Lessons Learned from Past Capacity Strengthening Initiatives

Most institutions in Africa with the mandate for human capacity development in genetic improvement of animal resources lack the expertise in all disciplines in breeding and genetics. This has led to limited trained personnel, both in terms of numbers and skills. Even for those trained, the rapid increase in new knowledge and technological advances in tools for executing animal genetic improvement requires continuous education at all levels of the human capital involved in animal breeding and genetics. Capacity gaps in animal breeding training were recognized on the African continent about 40–50 years ago. As far back as the mid-1980s, the International Livestock Centre for Africa (ILCA) would organize training for the national agricultural research scientists (NARS) by hosting about 25–30 scientists at a time training them on the analysis and interpretation of data from on-station and on-farm breeding programmes. ILCA (which was headquartered in Addis Ababa, Ethiopia) and the International Laboratory for Research on Animal Diseases (ILRAD) which was headquartered in Nairobi, Kenya, merged to form the International Livestock Research Institute (ILRI) in 1995. Since its formation, ILRI has been the primary institution operating at the continental level to support significant programmes producing new MSc and PhD cadre personnel as well as a retooling of practising animal breeding/genetics professionals in both research institutes and universities. ILRI has also worked extensively with both national and international partners in areas of biotechnology for genetic characterization and improvement in farm animals, and in the process has trained/mentored a number of professionals in these disciplines.

Capacity strengthening in animal genetics and breeding was implemented under two main categories: (i) instruction only, and (ii) research and instruction. ILRI alone or in collaboration with African National Agricultural Research Systems (NARS) provided training for graduate students, group training courses, and individual short-term training. ILRI also offered fellowships and hosted fellows for part of their graduate research or postdoctoral research. These fellows were mentored with assistance from national and international networks of experts. In addition to pursuing training to meet their specific research goals, these fellows also participated in diverse programmes related to the work of ILRI and hence had significant international exposure. In addition to ILRI research programmes, the Biosciences Eastern and Central Africa (BecA) hosted by ILRI has also contributed significantly to the training of students and scientists at all levels in modern bioscience applications in agricultural sciences, including animal genetics and breeding.

ILRI has trained many MSc and PhD students in genetics and breeding, either through fellowships or in collaboration with NARS where ILRI hosted the fellows. Since 2004, the BecA-ILRI facility, especially the genomics and bioinformatics platforms, has been a major means of delivering high-quality training, including exposure to cutting-edge research facilities and methodologies (Rege et al. 2021). Many of the BecA fellows subsequently became research leaders in their own institutions, and some have retained intensive collaboration with ILRI and other international institutions. Consequently, many among the current generation of leading livestock geneticists in ILRI have trained an estimated 200 BSc, over 69 MSc, and more than 66 PhD graduates, as well as over 35 postdoctoral fellows in livestock genetics and breeding. This training model is one that can be built upon by all stakeholders.

In 1999, ILRI and the Swedish University of Agricultural Sciences (SLU) with funding from the Swedish International Development Agency partnered to provide capacity building on the sustainable use of Animal Genetic Resources (AnGR). The focus of this partnership was on 'training the trainers' to translate concepts and knowledge to students, researchers, and policymakers (Ojango et al. 2011). The objectives for the ILRI-SLU project were as follows: (i) strengthen subject knowledge and skills of NARS scientists in teaching, research, and supervision of animal breeding and genetics, (ii) strengthen communication skills of these teachers and researchers, (iii) catalyse curriculum development, review of course content, and use of new and expanded teaching methods in university education, (iv) develop computer-based training resources relevant for use by NARS teachers and researchers, (v) stimulate contacts and exchange of experiences and ideas among teachers/researchers from developing countries on research and training of students in animal breeding and genetics, and (vi) strengthen the human capacity base for work on animal genetic resources in developing countries. Between 2000 and 2010, the ILRI-SLU project trained 138 university lecturers and researchers across 46 countries in Africa and Asia in animal breeding and genetics (Ojango et al. 2011). After the training courses, 13 and 5 participants from Sub-Saharan Africa and Asia, respectively, enrolled in PhD education in animal breeding and genetics.

Despite these efforts by ILRI and those by national governments and other development partners, capacity in animal genetics and breeding remains well below what is expected across Africa. While rapid advances in the last decade in molecular and high throughput technologies have made it possible to accelerate farm animal genetic improvement, not much progress can be made in the absence of skilled professionals, especially in the areas of quantitative genetics, statistical genomics, and bioinformatics. Geneticists with the competency to combine classical and modern molecular tools to rapidly improve farm animal productivity and to develop and evaluate improvement programmes are a scarce commodity even in better-endowed African countries. Between 2015 and 2017, the BecA-ILRI facility, SRUC, and the University of Georgia organized training in quantitative and molecular genetics for 45 field personnel from 14 countries in Africa. The course was highly received, and the post-course reviews clearly pointed to a significant need for broader and long-term institutionalized training in Africa to build capacity in farm animal genetic genomics. In March 2022, the African Animal Breeding Network (AABNet) organized capacity training in Animal Breeding for 49 personnel from 25 countries representing all regions in Africa (West Africa, North Africa, East Africa, Central Africa, and Southern Africa). There are other short-term projects with animal breeding training content in Africa. One example was the University of Natural Resources and Life Sciences in Vienna, Austria, which led the training in the design of community-based breeding strategies for indigenous sheep breeds of smallholders in Ethiopia between 2007 and 2011.

Farm animal genetic improvement and conservation are not the responsibility of only technical personnel. Field technical staff conducting inseminations and collecting phenotypes, as well as laboratory technicians processing biological samples are equally important. Computational specialists and bioinformaticians are equally crucial for the successful implementation of animal breeding and conservation programmes. To that end, the ILRI livestock genetics team has trained a large number of support personnel from various Sub-Saharan African institutions. The BecA-ILRI facility has hosted large numbers of NARI scientists and technical associates undertaking non-degree related training in research methods, molecular laboratory skills, and bioinformatics.

27.4 Education in Animal Breeding—A Case for the Introduction of Agriculture Education in Primary School Curricula

Education in Animal breeding has to be viewed in the broader context of Agriculture Education in Africa. Whereas most disciplines including languages that are even foreign to Africa are taught in primary schools, agriculture is offered as part of electives in some High School programmes. Agriculture education has become either an after-thought, an add-on, or the quintessential orphan child of STEM (Science,

Technology, Engineering and Mathematics). Yet, it is an established fact that agricultural education contributes to poverty reduction (Wallace 2007). This therefore begs the question, why is agriculture not taught in primary schools in Africa? The knowledge needed for agricultural productivity within the socio-cultural environment and agro-ecological regions of Africa should start at the Primary School. Concurrent progression in advanced knowledge in STEM along the educational ladder can only create innovation in agriculture and shift the focus from agriculture as a means of food production to the broader goal of rural or suburban development. On the contrary, agriculture education is common in elementary, middle, and high school in countries in the global North, including the United States (Phipps et al. 2008). Incorporation of agriculture in elementary schools in rural America significantly spurred achievement and improved other STEM skills (Education 2011). There are relatively few agricultural high schools and agricultural colleges in most African countries, however, most universities offer agricultural science degrees with different specializations, such as Animal Science, Crop Science, Soil Science, Agricultural Engineering, Agricultural Economics, Agribusiness, and Agricultural Extension. The mere expansion of agricultural education at all educational levels may not improve productivity or alleviate poverty. Rather, agricultural education should be backed with a national strategy, infrastructure, both public and private sector investments, and research. Even though Malaysia and Ghana both gained their independence in 1957, per capita income in Malaysia is significantly higher than that of Ghana because of heavy investments in agricultural research and education among other areas (Diao et al. 2019). The development of incentives for retaining scientific and support personnel will help reduce brain drain. Such a strategy was pursued in Brazil. Public sector investments in the Brazilian Agriculture Research and Development Organization (EMBRAPA) have resulted in significant development of livestock and poultry in the country.

27.5 Post-Secondary Technical Education in Animal Breeding

Educational training at the post-secondary level is generally very important for a successful and progressive agriculture that is capable of identifying bottlenecks and applying the appropriate technological solutions in terms of advances in farming techniques, management, and technologies. The curriculum at this level is therefore aimed at producing individuals who have the practical knowledge of farm management and technical skills to be proficient in addressing practical problems.

In terms of animal breeding, some of the essential post-secondary skills needed to support breeding activities include heat detection, designing mating plans to avoid inbreeding, understanding artificial insemination, understanding the concept of estimated breeding values, and therefore the skill to select superior animals as parents of the next generation. Data is primary to every farm operation including breeding activities. Therefore, additional skills in the use of ICT to identify individual animals, record the performance of livestock on the farm, use summary statistics to undertake projections, and advances in reproductive physiology technologies such as the use of oestrus synchronization to produce calves targeted for a particular market or season should be part of such post-secondary training.

In general, these courses could be of short duration; for example, one-year duration leading to national certificates or diplomas in specific fields or part-time training for two years and should be very hands-on. These courses could be work-based learning (such as Modern Apprenticeships) and short skills-based training courses in an institute with the appropriate infrastructure. Throughout the course, students should be given the opportunity to develop independent learning skills and teamwork skills using a mixture of classroom-based activities and practical hands-on training on modern and progressive farms in local contexts. Several Agricultural colleges and Institutes offer a number of these short certificate or diploma courses with some content on activities related to animal breeding. For example, the Technical and Vocational and Training Authority recognizes the Kenyan YMCA College of Agriculture and Technology middle-level agricultural and technology courses that are needed to improve the overall workforce in Agriculture. The Zambia College of Agriculture (https://www.zcamonze.edu.zm/) is another example of colleges in Africa offering certificate and diploma courses in General Agriculture, both on campus and through Distant Learning. The Kita College (https://kitaghana.org/) is also a private agricultural college in Ghana that offers a wide range of technical and vocational training in Agriculture.

27.6 Graduate Education in Animal Breeding

According to the African Union Inter-African Bureau for Animal Resources (AUIBAR) (2019), the perceived complexity of courses in animal breeding and genetics, including biometrics, has been a contributing factor to low interest among potential students. Furthermore, some universities train students in animal breeding and genetics only at the PhD Level. These bring to the fore the need to introduce the subject in the early levels of university education. The pedagogic approach in training the next generation of animal breeders and geneticists also needs to be fine-tuned by revising the curricula and mode of teaching while making available technological tools including appropriate software to help students enjoy the uniqueness of animal breeding and genetics. The theory of both population and quantitative genetics hinges on a good understanding of statistics. Therefore, statistical genetics should be introduced to students of agriculture at the undergraduate level.

Most graduate students specializing in animal breeding first obtain a bachelor's degree in Animal Sciences or related degrees. Historically, the overview of genetic improvement is taught at the undergraduate level as Animal Breeding. However, at the graduate level, the specialization is often referred to as Quantitative or Statistical Genetics. Advances in molecular and high throughput technologies in the omics (e.g. sequences, transcriptomes, metabolome, and microbiome, bioinformatics, and artificial intelligence have made it possible to accelerate animal genetic improvement. However, in order to harness these technological advances for animal genetic improvement applications, highly skilled professionals and appropriate infrastructure are needed. Geneticists with the competency to combine classical and modern molecular and artificial intelligence tools to rapidly improve farm animal productivity and develop and evaluate improvement programmes are a scarce commodity even in the most developed countries.

Based on current and emerging opportunities and needs, a well-grounded animal breeder should acquire competencies in the following:

1. Population genetics
2. Quantitative genetics
3. Principles of selection and breeding strategies
4. Computational and programming skills, including artificial intelligence and machine learning
5. Setting up a breeding programme
6. Management and analysis of sparse and complete data
7. Understanding family relationships and estimation of genetic relationships, inbreeding coefficients in pedigrees
8. Evaluation of breeding programmes for genetic gain, diversity, and environmental impact
9. Understanding of traits and estimation of genetic parameters
10. Dissemination of improved genetics to end-users

As alluded to above, African research institutions generally lack opportunities for developing skills of both research staff (scientists) and support staff, including technicians and administrators. Inadequate numbers of staff with appropriate levels of research skills, science communication, project and team leadership, and management, coupled with poor staff retention continue to bedevil the continent. Efforts for generation-long skilling and capacity development are generally lacking and instead, there are piecemeal ad-hoc training interventions not institutionalized and connected to clear long-term strategies. Higher degree training overseas that does not provide a conducive environment for return home is not of much help either.

It is highly unlikely that an Animal Science department in an African University can offer courses or programmes to cover all the required competency areas given some of the aforementioned limitations. As such, graduate-level animal breeding education in Africa should be re-imagined. Animal breeding education can be achieved through collaboration among institutions offering complimentary programmes or through well-designed programmes at regional hubs. Graduate-level education in animal breeding in Africa should also consider the heterogeneous production systems, socio-cultural value of animals, existing and future infrastructure, climate change, and the multipurpose goals for animal production in different countries, production systems, and economic situations.

Animal breeding education should explicitly cover the conservation of animal genetic diversity. The Food and Agricultural Organization of the United Nations (FAO 2007) recommends strengthening training and technology transfer programmes, and information systems for the inventory, characterization, and monitoring of trends and associated risks; sustainable use and development; and conservation, particularly in developing countries and countries with economies in transition. Furthermore, Africa needs to strengthen collaborative networks of researchers, breeders and conservation organizations, and other public, civil, and private actors, within and between countries, for information and knowledge exchange for sustainable use, breeding, and conservation.

Internships achieve several objectives including enabling students to obtain hands-on experience. They provide an opportunity for students to implement some of the theoretical concepts they have studied. Internships also provide an opportunity for students and their academic staff supervisors to interact with the stakeholders and potential employers to appreciate practical field situations. A typical internship programme that combines formal instruction and an opportunity to learn less formally, or more informally while serving in a particular community in order to provide a pragmatic and progressive learning experience is termed as 'service learning' (Nonnecke et al. 2015). This kind of internship ought to properly connect traditional classroom experiences with out-of-class or 'real-life' learning that results from the service. The need for a continental broad-based stakeholder strategy leveraging on higher learning and research institutes (HLRI) as well as networking in livestock breeding and genetic improvement in Africa has also been recommended (Refer to Chap. 28, Sect. 28.3) and must be pursued as a matter of urgency.

27.7 Retention of Animal Breeding Labour Force

Among the many constraints for efficient human and institutional capacity building for animal breeding development (Chap. 28, Sect. 28.3.2) are the lack of skilled personnel, facilities, inappropriate curricula, and teaching methods, as well as inadequate targeted research and collaboration among stakeholders. A recent assessment of African research organizations (universities, public and private research institutions, not-for-profit organizations) by the African Development Bank posited that the major hindrances to retention of animal breeding personnel are insufficient and irregular funding, high staff turnover due to low remuneration, and financial instability. The assessment presents a gloomy picture. Despite numerous efforts to strengthen capacity, knowledge-based institutions in sub-Saharan Africa remain small in size, scope, and influence. If Africa's capacity is to be transformed, these and other stumbling blocks need to be lifted or else, the status quo will remain for the foreseeable future. There should be an increase in fellowships for the labour force to upgrade their skills and obtain new tools. Fellowships take different forms, however, those that are designed to enable selected graduate students to undertake advanced studies at a university or research institution could be more impactful. Well-designed fellowships enable the strengthening of linkages and collaboration between research institutions and or universities.

There are only a few countries with some form of grant system or a national strategy for animal genetics resource improvement. Scientists should be encouraged to take advantage of the current existing and emerging opportunities. For example, the current seed grant for new African principal investigators under the World Academy of Sciences supports early career returnees with grant support for equipment start-up, consumables, degree training, international conferences, industrial link, collaborative mobility, open-access international publication, skills-building workshop, and female scientist-after-child grants.

27.8 Proposing a Collaborative Animal Genetics and Breeding Graduate Training Model for Africa

More often than not, the agricultural education system does not align with the agricultural industry. In some cases, agricultural science and animal breeding graduates in particular do not have the skills required by industry. At the same time when skilled graduates are produced, the industry does not have the capacity to hire new graduates. This among many other factors has contributed in part towards graduate unemployment in Africa (Ogege 2011). For animal breeding education to contribute towards development, linkages between the education and local animal industry should be developed. The curricula need to be current and connected to the economy. In addition to the scientific and technical knowledge, the animal breeding entrepreneur and business value chain (opportunities) should be part of the overall curricula. The Institute of Agribusiness Management Nigeria (IAMN) (https://agribusinessnigeria.org/) has taken the lead in training organizations, business executives, farmers, agriculturalists, researchers, consultants, practitioners, etc., on the advancement of trade, capacity, and enterprise activities with the agribusiness industry in Nigeria. Animal breeding education should interface with a similar type of agribusiness education to spur growth and productivity in the animal industry. Vice Chancellors (VCs), Deans, and Head of Animal Science Departments in Africa should form strong collaborative networks with all stakeholders in order to obtain support to train relevant Animal Breeders for the African context. National Governments, international research organizations, and their donor partners should strategically support human and institutional capacity development for improved animal breeding that is able to take advantage of modern innovative approaches to enhance animal genetic resource development and improvement in Africa.

In pursuit of context relevance, effectiveness, efficiency, and sustainability, several collaborative models have been tried for MSc and/or PhD training in various fields of agriculture in Africa. Box 27.1 presents a case example—the Collaborative Masters in Agricultural and Applied Economics (MAAE). Given the capacity challenges that continue to constrain animal genetics and breeding and considering the fast pace of developments in disciplines critical for shaping genetic improvement of livestock, Africa needs a well-articulated set of collaboratives involving key institutions in each sub-region of the continent—Eastern, Southern, Central, West and North—building on experiences from these case examples.

The trend towards joint degree programmes offers an opportunity for quality training and enhancing collaboration instead of competition (Adipala et al. 2010). Through networking opportunities, universities in countries where political or economic conditions are unfavourable can be kept operating at the highest possible standard. This strategy can also be refined and developed to help participating universities adapt to change and expand their role within the National Agricultural Research and Extension Systems (NARES).

Africa can also make good use of its external partnerships particularly to expose Animal Breeding personnel to the current state-of-the-art technologies in the subject. In this regard AU-IBAR and national governments should encourage scientists, particularly students and faculty of animal breeding, to take advantage of opportunities such as the Australia Awards Africa Graduate and Post-Doctoral Training Fellowships (https://www.dfat.gov.au/people-to-people/australia-awards/australia-awards-scholarships) among others, to train a critical mass of skilled personnel in Animal Breeding and Genetics at all times. In this regard, we need to encourage North-South and South-South collaborative Capacity Building Programmes which have some bias towards animal breeding training and development in Africa. Best Practices and Lessons Learnt in past and ongoing programmes should be brought to bear on any future capacity-building programmes in Animal Breeding and Genetics Training in Africa. Other such useful capacity-building opportunities for Animal Breeding and Genetics in Africa include:

European Master in Animal Breeding and Genetics (EMABG) is a two-year MSc programme provided by six EU groups/universities and recognized by the EU as an Erasmus Mundus MSc programme. More information on this programme is available at https://lohmann-breeders.com/de/lohmanninfo/european-master-in-animal-breeding-and-genetics-international-collaboration-to-face-future-needs/; EU-funded Masters (MSc) Scholarships in Animal Breeding and Genetics available for developing country nationals https://www.advance-africa.com/MSc-Scholarships-in-Animal-Breeding-and-Genetics.html.

A few Animal Science graduates have already benefited from these programmes but annually many are unsuccessful because of the intense competition. Therefore, we suggest that similar models should be put in place by the African Union for African students to help train many more Animal Breeders and Geneticists for Africa.

Finally, and as recommended in Chap. 28 (Sect. 28.3.3) training of stakeholders through people-centred livestock community-based initiatives, training of trainers, and continuous education should be pursued. In all these, the roles of the AUIBAR, Pan African University, and the African Animal Breeding Network (AABNet) (Chap. 28, Sects. 28.3.4 and 28.3.5) will be crucial to ensure sustainable human and institutional capacity development for the improvement of animal genetic resources (AnGR) in Africa.

> **Box 27.1: Collaborative Masters in Agricultural and Applied Economics (MAAE)**
> CMAAE builds the capacity to conduct policy research in agricultural and applied economics to address food security, agricultural productivity, and environmental management. The specific programme objectives are to:
>
> - Produce graduates in agricultural and applied economics with the knowledge and skills for transforming the agro-food sectors and the rural economies of the region in an environmentally sustainable fashion. By doing so, helping to fulfil the Sustainable Development Goals (SDGs) and addressing the persistent problems of food insecurity and poverty in Africa.
> - Upgrade the teaching and research capacity of departments currently in the CMAAE Programme while initiating planning for a system to scale out the programme to other regions of Sub-Saharan Africa.
> - Strengthen the research network to promote agricultural development.
>
> The typical degree programme takes 18–24 months and is divided into three components:
>
> (a) *Core Courses*
>
> In the first year of study, students are required to complete eight core courses and one common course in Institutional and Behavioural Economics all scheduled for semesters 1 and 2.

Semester 1 (13–15 weeks): Microeconomics, Statistics, Mathematics, and Issues in Agricultural and Applied Economics.

Semester 2 (13–15 weeks): Production Economics, Econometrics, Macroeconomics, Research Methods, Computer Applications, and Institutional and Behavioural Economics.

(b) *Shared Facility for Specialization and Electives (SFSE)*

During the third semester covering four months, students from all the accredited universities converge at an SFSE to teach foundation and elective courses by visiting professors drawn from all over the world. Each student is expected to take one foundation course in his/her chosen field of specialization from the following four areas: Agribusiness Management, Agricultural and Rural Development, Agricultural Policy Analysis, and Environmental and Natural Resource Management. In addition, students take one common course and at least one elective course in the specific cluster/area of specialization. Over the years, SFSE has been held at the University of Pretoria (UP) in South Africa. This is based on the fact that UP has a world-class facility with an environment that is conducive to learning.

(c) *Thesis Research*

The last part—thesis research—starts at the SFSE where students present their concept notes during weekly seminar sessions. The concepts are developed in line with the four areas of specialization. The students return to their respective accredited universities to write their thesis. All students are required to complete a thesis and undertake an oral examination in their fourth and fifth semesters. Thesis supervision is provided by a designated supervisor, with the assistance of an additional thesis committee member, from within or outside the home university, providing specialized guidance.

Upon completion, all CMAAE graduates are encouraged to belong to an alumni association. To facilitate the association, the Programme runs an interactive alumni page which provides alumni with an opportunity to network and exchange ideas and experiences in their career path. https://aercafrica.org/training/cmaae/

27.9 Conclusion and Future Opportunities

Genetic improvement of Africa's animal resources will be key to breeding resilient, productive, and climate-adaptive animals for the current and future generations. Unfortunately, Africa continues to lag behind in both the institutional and human resource capacity for animal genetic improvement. Scientific and technical knowledge of locally trained personnel in charge of planning, implementing, and managing farm animal improvement and conservation programmes is abysmal in the whole of Africa. Most institutions in Africa with the mandate for human capacity development in genetic improvement of animal resources lack the expertise in all disciplines in breeding and genetics. Livestock Genetic improvement requires skilled knowledge in quantitative and population genetics, molecular genetics, microbiology, physiology, statistics, bioinformatics, phenomics, and artificial intelligence. Thus, scientists educated and trained in the relevant expertise areas within a well-defined structure and by competent personnel should lead genetic improvement programmes on the continent. At the same time, Africa has relatively few agricultural high schools and agricultural colleges although most universities offer agricultural science degrees with different specializations, such as Animal Science, Crop Science, Soil Science, Agricultural Engineering, Agricultural Economics, Agribusiness, and Agricultural Extension. Unfortunately, the mere expansion of agricultural education at all educational levels may not improve productivity or alleviate poverty.

Implementation of the Global Plan of Action on Animal Genetic Resources for sustainable food security, wealth creation, and enhanced livelihoods will require the training of a critical mass of Animal Geneticists. We therefore recommend that agricultural education should be backed with a national strategy, infrastructure, both public and private sector investments, and research. To fully characterize, sustainably use, and conserve its animal genetic resources, Africa needs to prioritize human and institutional capacity building in line with emerging and innovative technologies. On its part, ILRI has trained an estimated 200 BSc, over 69 MSc, and more than 66 PhD graduates, as well as over 35 postdoctoral fellows in livestock genetics and breeding. Stakeholders led by the AUIBAR, and national governments should follow this good example and collaborate with appropriate institutions such as the Centre of Tropical Livestock Genetics and Research (CTLGH) to help train the next generation of livestock geneticists. Fellowships designed to enable selected graduate students to undertake advanced studies at a university or research institution could be more impactful. Well-designed fellowships will also strengthen linkages and collaboration between research institutions and or universities. In this regard, we need to encourage North-South and South-South collaborative Capacity Building Programmes which have some bias towards animal breeding training and development in Africa. Through networking opportunities, universities in countries where political or economic conditions are unfavourable can be kept operating at the highest possible standard. Models such as the European Master in Animal Breeding and Genetics (EMABG) and Erasmus Mundus MSc programme can be replicated in Africa based on international collaboration between countries. Finally, there is a need for the African Union to encourage and motivate national governments on the need to provide sustainable funding schemes for Graduate Programmes in Agriculture in general and Animal Breeding and Genetics in particular.

References

Adipala E, Blackie M, Woomer P (2010) Enhancing graduate training and agricultural research in Eastern, Central and Southern Africa: lessons and new thrusts. In: Second RUFORUM Biennial meeting 20–24 September 2010, Entebbe, Uganda

AUIBAR (2019) State of farm animal genetic resources in Africa. African Union - Interafrican Bureau for Animal Resources (AUIBAR)

Diao XEH, Kolavalli PBR, Shashidhara L, Resnick DE (eds) (2019) Ghana's economic and agricultural transformation: past performance and future prospects. Oxford University Press (OUP). https://doi.org/10.1093/oso/9780198845348.001.0001

Education, U. D. o (2011) Using Agriculture to Spur Achievement: The Walton 21st Century Rural Life Center

FAO (2007) State of the world animal genetic resources. Commission for Genetic Resources for Food and Agriculture. Food and Agriculture Organization of the United Nations, Rome, Italy

Nonnecke G, McMillan D, Kugonza DR, Masinde D (2015) Leaving the door open to new beneficiaries. In: Butler LMM (ed) Tapping philanthropy for development: lessons learned from a public-private partnership in Rural Uganda. Kumarin Press, Lynne Rienner Publishers, p 250

Ogege SO (2011) Education and the paradox of graduate unemployment: the dilemma of development in Nigeria. Afr Res Rev 5(1):253–265

Ojango JMK, Malmfors B, Mwai O, Philipsson J (2011) Training the trainers-an innovative and successful model for capacity building. In: Animal genetic resource utilization in sub-Saharan Africa and Asia. International Livestock Research Institute (ILRI) and Swedish University of Agricultural Science (SLU)

Phipps LJ, Osborne EW, Dyer JE, Ball A (2008) Handbook on agricultural education in public schools. Thomson Delmar Learning

Rege JEO, Ochieng J, Hanotte O (2021) Livestock genetics and breeding. In: Grace JMAD (ed) The impact of the International Livestock Research Institute (ILRI). ABI and the International Livestock Research Institute

Wallace I (2007) A framework for revitalisation of rural education and training systems in sub-Saharan Africa: strengthening the human resource base for food security and sustainable livelihoods. Int J Educ Dev 27:581–590. https://doi.org/10.1016/j.ijedudev.2006.08.003

Open Access This chapter is licensed under the terms of the Creative Commons Attribution 4.0 International License (http://creativecommons.org/licenses/by/4.0/), which permits use, sharing, adaptation, distribution and reproduction in any medium or format, as long as you give appropriate credit to the original author(s) and the source, provide a link to the Creative Commons license and indicate if changes were made.

The images or other third party material in this chapter are included in the chapter's Creative Commons license, unless indicated otherwise in a credit line to the material. If material is not included in the chapter's Creative Commons license and your intended use is not permitted by statutory regulation or exceeds the permitted use, you will need to obtain permission directly from the copyright holder.

Capacity Building in Livestock Breeding and Genetic Improvement in Achieving UN Sustainable Development Goals for Africa

Christian K. Tiambo, Oluyinka Opoola, Moses Okpeku, Isidore Houaga, Blaise A. Hako Touko, and Tadelle Dessie

Abstract

To achieve the Sustainable Development Goals of the United Nations (UN-SDGs) in Africa, livestock breeding efficiency is a prerequisite. Human capital in the form of professional, managerial, and technical skills, improved policies and institutional operation are the key enablers of performance for the smallholder livestock sector. Capacity building is central to guarantee African livestock efficiency along the value chains and ultimately transformed livelihoods. Currently, animal breeding and genetic improvement in Africa are constrained by a serious human and institutional capacity gap. The Livestock Development Strategy for Africa (LiDESA) seeks to transform the livestock sector for accelerated and equitable growth and for enhanced contribution to socio-economic development. Sustainable livestock development in Africa could be better realized through the development of customized capacity building strategies targeting (1) relevant participatory breeding education, hands-on training, efficient technical assistance, and collaborations / partnerships aiming to address farmers' present and future challenges, (2) enabling breeding systems for sustainable productivity, resilience of breeds, and animal seed stock management, and (3) market-led breeding solutions that are attractive for the younger and future generations.

Challenges facing capacity building in livestock breeding in Africa vary between and within stakeholders of specific value chains, and farmers lack training and suitable support programmes for their breeding activities. Beyond the lack of skilled human resources, financial capacity to enforce legislation and policies in Animal Genetic

C. K. Tiambo (✉)
Centre for Tropical Livestock Genetics and Health (CTLGH)-ILRI, Nairobi, Kenya
e-mail: c.tiambo@cgiar.org

O. Opoola · I. Houaga
Centre for Tropical Livestock Genetics and Health (CTLGH), The Roslin Institute, University of Edinburgh, Edinburgh, UK

M. Okpeku
Discipline of Genetics, School of Life Science, University of KwaZulu-Natal, Durban, South Africa

B. A. Hako Touko
Faculty of Agronomy and Agricultural Sciences, University of Dschang, Dschang, Cameroon

T. Dessie
International Livestock Research Institute (ILRI), Addis Ababa, Ethiopia

Resources (AnGR) that may lead to the design and implementation of conservation and breeding programmes is a major limiting factor in most African countries. Currently, animal breeding programmes may not thrive, unless a new cohort of breeders, infrastructures, and policies are in place and backed by long-term investment plans. The main loopholes of the desired revolution reside in the current gaps and limitations of higher education and professional trainings, in terms of human resources, infrastructure, curricular content, research, and cooperation.

The required capacity development could also be through people-centred community-based breeding programmes (CBBP), training of trainers, and continuous education with major inputs/roles by agricultural training institutions (e.g. Pan African University), the African Union InterAfrican Bureau for Animal Genetics Resources (AU-IBAR) and the African Animal Breeding Network (AABNet), in the form of collaborations for teaching, research, and outreach.

Key capacity building intervention areas needing attention are: (1) Conservation and restoration of local animal genetic diversity, their improvement and dissemination of elite locally adapted and farmers-preferred genetics, and preservation of their natural habitat; (2) Demand-led capacity development in modern livestock breeding techniques; (3) Sustainable intensification of livestock farming practices; (4) Adaptation to extreme environmental conditions and livestock emergency guidelines and standards (LEGS); (5) Breeding for animal welfare, adaptability, and biosafety; (6) Big Data and artificial intelligence for smart farming and livestock breeding programmes; and (7) Infrastructural development and policies' integration for technology uptake.

Keywords

Breeding · Livestock · Capacity building · Sustainable Development Goals · Africa

Abbreviations

AABNet	African Animal Breeding Network
ABS	Access and Benefits Sharing
AI	Artificial Intelligence
AIS	Automated Identification Systems
AMCEN	African Ministerial Conference on the Environment
AnGR	Animal Genetic Resources
APIs	Application Programming Interface
ASARECA	Association for Strengthening Agricultural Research in Eastern and Central Africa
AUC	African Union Commission
AU-IBAR	African Union-InterAfrican Bureau for Animal Genetics Resources
AWSA	Animal Welfare Strategy for Africa
CAADP	Comprehensive Africa Agriculture Development Programme
CBD	Convention on Biological Diversity
CCARDESA	Centre for Coordination of Agricultural Research and Development for Southern Africa
CGIAR	Consultative Group for International Agricultural Research
CIMMYT	International Maize and Wheat Improvement Center
COVID-19	Coronavirus Disease 2019
CRA	Collaborative Research Agreement
DA	Direct Agreement Between Firm/Farm & the Programme
DSA	Data Sharing Agreements
EPC	Electronic Product Code
FA	Farmer Assistants
FAO	Food and Agricultural Organization
FARA	Forum for Agricultural Research in Africa

GPA-AnGR	Global Plan of Action for Animal Genetic Resources	UNCSD	United Nations Conference on Sustainable Development
GPS	Global Positioning System	UNEP	United Nations Environment Programme
HLRI	Higher Learning and Research Institutions	UNSTAT	United Nations Statistics Division
ICARDA	International Center for Agricultural Research in the Dry Areas	WHO	World Health Organization
IoT	Internet of Things		
JPOI	Johannesburg Plan of Implementation		
LEGS	Livestock Emergency Guidelines and Standards		
LiDESA	Livestock Development Strategy for Africa		
ML	Machine Learning		
MOOC	Massive Open Online Course		
MSc	Master of Science		
MTA	Material Transfer Agreement		
NARI	National Agricultural Research Institute		
NCCLS	National Climate Change Learning Strategy		
NDA	Non-Disclosure Agreement.		
NEPAD	New Partnership for Africa's Development		
NRC	National Research Council		
OECD	Organization for Economic Cooperation and Development		
PAU	Pan African University		
PAUISTI	Pan African University Institute for Basic Sciences, Technology and Innovation		
PhD	Philosophy Doctor		
PRA	Data Collectors/Performance Recording Agents		
R&D	Research and Development		
RFID	Radio Frequency Identification		
RUFORUM	Regional Universities Forum for Capacity Building in Agriculture		
SADC	Southern African Development Community		
SDGs	Sustainable Development Goals		
SMART	Specific, Measurable, Achievable, Relevant, and Time-Bound		
UN	United Nations		

28.1 Introduction

Livestock is central to achieving many of the United Nation's Sustainable Development Goals (SDGs) among which the most relevant are SDGs 1, 2, 3, 5, 8, 12, 13, 15, and 17 (FAO 2015; Schneider and Tarawali 2021; Macmillan 2016). Capacity building is recognized as a crucial means of operationalizing the Johannesburg Plan of Implementation (JPOI), which is appealing for improving and fast-tracking human, institutional, and infrastructure capacity building initiatives and for supporting developing countries in building capacity to access a larger proportion of multilateral and global research and development programmes.

To achieve the UN-SDGs for Africans, livestock productivity must increase to meet the projected demand for meat, milk, and egg within a few decades, while minimizing environmental impact. This is a challenge requiring skilled manpower to lead the development in the desired direction. Few initiatives have taken place on the continent. Unfortunately, many countries and regions still suffer from a shortage of personnel with expertise in animal breeding and genetics, both at research and academic institutions and in local organizations responsible for livestock development. Sustainable livestock development for small holder livestock producers who constitute the largest producer group in Africa could be better realized through the development of customized capacity building strategies targeting: (1) provision of relevant participatory livestock breeding education, hands-on training, effective technical assistance and new collaborations/partnerships to address present and future challenges, (2) enabling breeding systems for sustainable productivity, resilience of breeds, animal seed stock management, and (3) market-led breeding solution that should be attractive for the younger and future generations.

While Chap. 27 addresses the specific issues around training and education to strengthen capacity of animal breeding practitioners—focusing on analysis of the status of animal genetic improvement education in Africa, major gaps, and recommendations of interventions to address these gaps (including re-examination of training curricula), this chapter examines capacity for livestock breeding more broadly—human, infrastructural, policy, and institutional, and roles and examples of initiatives of national, sub-regional, and continental bodies.

This chapter captures the four pathways through which livestock breeding and genetic improvement contribute to the achievement of all the 17 United Nation (UN) Sustainable Development Goal (SDG) targets and where capacity development should be strengthened in relation to the African development view from the Livestock Development Strategy for Africa (LiDeSA). It then maps the states of human resources and infrastructures to support capacity building and breeding programmes and the potential roles of African Union Commission via the pan African University (PAU) and the African Union—InterAfrican Bureau and Animal resources (AU-IBAR). Key capacity building intervention areas to contribute to achieving sustainable development goals in Africa are described, including conservation and improvement of AnGR and their environment, demand-led and market-led capacity development in modern livestock breeding techniques, sustainable intensification of livestock farming practices in Africa, adaptation to extreme environmental conditions and livestock emergency guidelines and standards (LEGS), breeding for animal welfare, adaptability, and biosafety, and Big Data and artificial intelligence for smart farming and breeding programmes.

28.2 Livestock Genetics Resources in the Context of the UN Sustainable Development Goals (SDG)

28.2.1 Contribution of Livestock Breeding and Genetic Improvement to the United Nation (UN) Sustainable Development Goal Targets

Animal-sourced foods make up 5 of 6 of the highest global value commodities (Wright 2017), and the growing demand for livestock products in developing countries, driven by population growth, higher incomes, and urbanization, makes livestock genetic resources central to achieving many of the UN Sustainable Development Goals (SDGs) and directly relevant to most of them (Macmillan 2016). This represents a huge opportunity for hundreds of millions of poor stakeholders of the various livestock value chains.

Four pathways based on livestock genetic resources directly contribute to the achievement of all the 17 SDGs, and the education system on the continent should ensure that capacity development is strengthened to preserve those assets (Fig. 28.1, FAO 2015):

- *Livestock breeding and genetic improvement for inclusive and sustainable economic growth*: The direct and indirect contributions are felt on SDGs 1, 8, and 9 through connection of poor farmers to markets. Best practices in animal breeding and genetic improvement could double the productivity of poor smallholder livestock keepers, while developing and implementing national and regional Livestock Master Plans to support effective investment will optimize livestock's contribution to continental economic growth.
- *Livestock breeding and genetic improvement for equitable livelihoods*: FAO (2014) postu-

Fig. 28.1 Contribution of animal production to the different Sustainable Development Goals (SDGs). (Source: FAO (2015))

lated that 'Poverty is a largely rural phenomenon in Africa, with the majority of (farm) households keeping livestock, ranging from Poultry, small ruminants (sheep and goats) and cattle'. The proportion of people living on less than US$ 1.25 per day which defines the extremely poor ranges from less than 5% in North African countries to over 80% in some sub-Saharan economies, such as Burundi, Chad, and Madagascar.

However, agriculture and agribusiness, including livestock, have been among the fastest growing sectors since the 1990s, and the growing demand for food and high value agricultural products, such as meat and dairy in particular, is setting the stage for continued opportunities for investment.

It is gradually increasingly acknowledged that growth of food production and agribusiness, including the farm animal industry, is fundamental to reducing poverty. A large proportion of Africa's poor comprises smallholder farmers, the majority of whom keep animals. Increased livestock productivity can improve the livelihood of smallholder producers. Also, higher production can translate into lower food prices, to the clear equitable benefit of the majority of households who are buyers of animal protein. In addition, investments in farm animal productivity can encourage the development of agroindustries and value chains that generate employment as a substitute to farming.

Livestock breeding has potential for direct and indirect impact on SDGs 1, 4, 5, 10, and 16. Within some African countries, livestock breeding is helping to achieve gender equality and to empower all women and girls (SDG 5). It is important to consider that policies will go a long way to enhance equity in investment and livelihood of both low-input livestock keepers and business-oriented ones. This is done through regulation and application of gender-transformative approaches that give women in livestock raising, processing, and trading greater access to, and control over, livestock resources. Women empowerment is both directly and indirectly impactful on the short and long term. They directly play a significant role in maintaining rural livestock-food systems while remaining the pillars of youth education in African countries. The impact of women empowerment will be also felt when developing labour-saving technologies for livestock feeding. At the regional and international levels, livestock breeding and genetic improvement could help buffer the inequality within and among countries (SDG 10). Government prioritization of the importation of commercial strains and foreign breeds used as breeding stock has proven not to be effective and sustainable for the mitigation of animal food and protein deficit in terms of quantity, quality, and economic growth. That has been the trend in sub-Saharan Africa in general and more pronounced in central African countries. Enabling policies should promote a dual track approach of addressing the gaps in the livestock food systems by mitigating short- and medium-term gaps through strategic planning of importations, while addressing long-term gaps with sustained genetic improvement of the local animal genetic resources and community-based management of more productive and resilient local stocks.

Capacity development for improved breeding techniques will safeguard pastoralists against catastrophic drought and loss of livestock in remote arid and semi-arid lands (ASALs) of Africa and develop options to reduce barriers to safe and sustainable domestic and regional trade in livestock products.

- *Livestock breeding and genetic improvement of animal-sourced foods, basic nutrition, and health*: Globally, two billion people suffer from micronutrient deficiencies, 151 million children under five suffer from stunting, and millions more have impaired cognitive development related to poor nutrition (Adegbola et al. 2019). This might be due to deficits in proteins of animal origin linked to insufficient consumption of animal-sourced foods (milk and dairy product, meat, fish, and eggs) by local or poor communities around the world. Animal-sourced foods stand as the perfect solution to the problem of malnutrition. The World Health Organization described animal-sourced foods as the best source of high-quality nutrient-rich foods for children aged 6–23 months (WHO 2014).

  Products of animal origin contain high quality protein and more bioavailable vitamin A, vitamin D3, iron, iodine, zinc, calcium, folic acid, and key essential fatty acids than plants, and consuming both animal-sourced foods and plant-based foods improves the absorption of iron in the latter (Neumann et al. 2002). Likewise, eating fish with vegetables improves the absorption of vitamin A and some fish species contain twice as high vitamin A concentration of vegetables as carrots or spinach (Kawarazuka and Béné 2010). Also, animal-sourced foods are reported to be the only natural source of vitamin B12. Vitamin B12 deficiency in individuals consuming low amounts of animal products in the developing world is linked with developmental disorders, anaemia, poorer cognitive function, and lower motor development (Stabler and Allen 2004). Within the scope of the UN SDG, livestock and food security and safe and healthy balanced diets, including animal-sourced foods, are the end goals of livestock breeding and genetic improvement which directly contribute to SDGs 2, 3, and 12: Thus, ending hunger, achieving food security and improved nutrition, and promoting sustainable agriculture (SDG2) should be part of the specific, measurable, achievable, relevant, and time-bound (SMART) objectives of livestock breeding in Africa, measurable through changing of the supply of animal-sourced foods through better feeding, breeding, and health and through the capacity of breeding techniques to reduce antimicrobial resistance through judicious use of antimicrobial drugs and antimicrobial alternatives in livestock production. This will also contribute to ensure healthy lives as well as promote the well-being of all and at all ages (SDG3), by reducing the burden of zoonotic diseases through better animal health and promotion of 'one health' approaches and improving food safety in informal markets.

- *Livestock breeding and genetic improvement for sustainable ecosystems*: According to the

EAT (a science-based global platform for food system transformation)—Lancet Commission report on food, planet, and health, animal-sourced food consumption is a global threat on sustainability and human health; this conclusion is not shared by many and judged overestimated as the tremendous variability in the environmental impact of livestock production seems to be ignored. According to Adegbola et al. (2019), the EAT-Lancet commission failed to adequately include the experience of marginalized women and children in low- and middle-income countries whose diets regularly lack the necessary nutrients.

The impact of livestock genetic improvement could easily be felt on SDGs 6, 7, 11, 13, 14, and 15. Ensuring access to water and sanitation for all (SDG 6) is achieved by improving livestock feeding regimes for efficient use of water resources and forage varieties and hence reducing and reversing land degradation and combatting desertification (SDG 15). However, it is important to structure livestock production around the world, especially in the local and poor communities where livestock intensification and the increasing adoption rate of technologies and innovations could not only be linked with income generation and improved intake of animal-sourced foods by the poor community, but also induce more carbon emission and environmental hazards. While policies and laws should be elaborated and implemented to secure the one-health approach in livestock-food systems, anticipated community programmes will certainly yield positive results.

28.2.2 African Development View from the Livestock Development Strategy for Africa (LiDeSA)

As clearly articulated in Chap. 30, Sect. 30.3.5, the Livestock Development Strategy for Africa (LiDESA) 2015–2035 seeks to transform the African livestock sector for accelerated and equitable growth and for enhanced contribution to socio-economic development. The strategy is in line with the ongoing African strategies, policy frameworks, and guidelines and is coherent with the Comprehensive Africa Agriculture Development Programme (CAADP), Frameworks and Agenda at the Continental, Regional Economic Community and Member States' levels.

28.2.3 The Challenge of Animal-Sourced Food Systems in Achieving the SDG and the Need for Genetic Improvement

The animal sector plays important roles in the context of sustainable development. These include, for instance, food security, transport, employment, financial security, and livelihoods (Keeling et al. 2019). There have been progressive advances associated with continuous growth and intensification of the animal sector to improve adaptability, resilience, and productivity. Animal-sourced foods are essential for the supply of adequate nutrients and nutrition for maintaining growth and longevity of individuals in every country, region, or continent. The SDG2 promises to ensure food security and nutrition within sustainable food systems (UN 2015; Fanzo 2019). However, achieving that goal is riddled with uncertainties as a result of the manner in which the world currently produces food. However, it is also essential to note that animal-sourced foods from animal-sourced food systems are not readily supplied in most countries due to the following: (i) poor breeding plans; (ii) low land spaces for agriculture; (iii) poor farmer knowledge on the animal-sourced food systems they practice; (iv) lack of incentives and funds to support farmers to practice animal-sourced farming systems; (v) lack of support from stakeholders involved in food systems and supply chains; and (vi) poor knowledge of individuals working with these farmers to provide support and adequate feedback to them. In addition, there are also challenges posed by the sundry environment where these animals are managed which directly affect their performance and productivity. The challenges, although not exhaustive, have been mentioned in previous studies (Rege et al. 2011; Odero-watuhui 2017; Fanzo 2019). These challenges include gaseous emissions, water and soil pollution, and ecosystem damage, issues regarding animal welfare (animal abuse and negative consequences of intensive selection and production), and animal and human health (zoonotic diseases and inappropriate use of antimicrobials and anthelmintics).

28.2.4 Challenges in Capacity Building of Livestock Breeders

Chapter 27 extensively discussed issues related to human resources' capacity and limitations in animal breeding and genetics and subsequently suggested how to turn some of the challenges into opportunities. Building capacity with trained personnel, animal experts, or professionals in the livestock sector is one of the key parameters required to enhance and optimize breeding programmes in various species. A wide range of literature have reported the challenges of capacity building and livestock breeding in Africa (Rege et al. 2011; Besbes et al. 2018; Opoola et al. 2019; Wurzinger et al. 2021). These challenges are prevalent and at varying levels mostly faced by: (i) livestock farmers (animal keepers); (ii) stakeholders involved with the farmers; and (iii) other stakeholders involved in the livestock sector such as governmental and non-governmental bodies, livestock research-based institutions, and universities. Previous reports have shown that livestock farmers lack training and suitable support programmes for breeding activities.

28.2.5 Contribution of AnGR Breeding and Genetic Improvement to SDGs

Animal Genetic Resources (AnGR) represents the pool of domesticated livestock used for food. Although the world hosts a large variety of animal types, the percentage of those used for food is comparatively low. Africa is often credited as home to the largest animal diversity of AnGRs, where local AnGRs contribute significantly to the continents' food security through production of quality animal proteins, as biological machines on farms, providing needed draught energy to drive farm machineries, supporting crop lives with their compost, and economic sustenance, particularly in poor rural agricultural systems. However, the pressure of consumption as well as uncontrolled utilization outweighs the rate of reproduction and restocking. Furthermore, the scarcity of technical know-how needed for advanced breed improvement programmes results in colossal losses and erosion of local gene pools through uncontrolled/indiscriminate crossbreeding of local AnGRs. This is exacerbated by poor managerial skills and financial support for AnGR development and sustainable utilization, challenging climatic and environmental conditions, poor or inappropriate policies, and absence of government support in some countries and regions. The multiplier effect is seen in poor production, low input/output, and over-harvesting leading to extinction of important gene pools while many others are endangered.

The Sustainable Development Goal (SDG) recognizes the immense value of local AnGRs and sides with the Global Plan of Action for Animal genetic Resources (GPA-AnGR) in the fight to curtail erosion of unsustainable utilization of AnGRs. At various fora and through different media, local and regional representatives have agreed to, and put pen to paper to, endorse this struggle to reposition African local AnGRs for more sustainable utilization. The Interlaken Declaration of 2007 particularly recognized and detailed information gaps, poor national and international inventories, lack of or/and poor characterization of local and national breeds, and poor monitoring and conservation for sustainable utilization as key limitations upon which the SDG agenda is built.

In recent times, there have been more initiatives and researches supported by local, regional, and international institutions to tackle these recognized flaws in sustainable development and utilization of the AnGRs in Africa. Genetic characterization of available AnGRs and phenomic characterization of usable metabolites from AnGRs on the continent have largely increased, bridging information gaps and providing base-line information for constructive development and improvement of indigenous AnGRs for more sustainable utilization. However, there is still a long way to go in drawing support locally and regionally to aid and implement SGD agenda of making AnGRs more sustainable. Okpeku et al. in 2019 and a host of other authorities suggested that a united private-public support is strongly needful in Africa for more sustainable solution.

28.3 Higher Learning and Research Institutions and Networking in Livestock Breeding and Genetic Improvement in Africa

According to Calestous Juma (2012), there is an urgent need to create a new generation of innovation-oriented agricultural HLRI that efficiently brings together agricultural research, training, commercialization, and extension. This requires upgrading the training, extension, and commercialization functions of existing national agricultural research institutes (NARIs). This would build on a strong research tradition of the leading Agricultural universities and colleges in Africa (Table 28.1), ongoing training efforts, connections with the private sector and farmers, and extensive international partnerships.

From Table 28.1, it appears that the capacity building may be disproportionate between regions of Africa, based on the number and quality of agricultural universities and colleges. Central Africa is the less endowed.

28.3.1 State of Human Resources and Infrastructures to Support Capacity Building and Breeding Programmes

Most countries in Africa have limited skilled human resources and financial capacity to implement legislation on policies for AnGR and to design and implement conservation and breeding programmes for AnGR (Rege et al. 2011). Moreover, animal genetic and breeding infrastructures and institutional capacities are too often inadequate or obsolete. This is in connection with the shortsighted capacity of planning and commitments from governments (Wollny 2003; FAO 2007; Boettcher and Atkin 2010). It generally appears that policy makers may not have enough knowledge about the importance of AnGR (Chagunda et al. 2015).

Very few African countries have commissioned studies on the state of their infrastructure and human resources for animal breeding and genetic improvement compared with the mapping done in the veterinary sector across the world. This constitutes one of the key issues identified in the Global Plan of Action (GPA) (FAO 2007). A recent study conducted by the African Animal Breeding Network in 38 African countries showed that lack of human capacity and infrastructures is the main challenge hindering livestock genetic evaluation and improvement in Africa (Houaga et al. 2023).

Current African animal breeding programmes will not succeed, unless a new generation of trained breeders and infrastructure such as physical facilities, operational livestock recording, and genetic evaluation systems is in place (Cardellino and Boyazoglu 2009). Skilled human resources, efficient organizations and institutions, national and regional long-term investment, and strong links between these elements are highly and urgently needed. Lack of necessary infrastructure is one of the most serious constraints in the African animal breeding sector (FAO 2011a; Philipsson et al. 2011).

28.3.2 Animal Breeding and Genetic Improvement Curricula: Gaps and Limitations of the African Higher Education Systems

Constraints are numerous for effective and efficient capacity building in Animal Breeding and Genetic improvement in Africa; these include the following:

Table 28.1 Representativity of top-50 leading agricultural universities and colleges in Africa in 2021

Africa regions	Countries (number of top Agric Universities and colleges)	Top-50 leading institutions (founding date)	Rank in Country	Rank in Africa	Rank in World
North Africa	Algeria (1)[a]	Higher National Agronomic School (1905)	38		
	Egypt (21)	Cairo University (1908)	1	21	403
		Alexandria University (1942)	2	23	434
		Assiut University (1957)	3	24	444
		Ain Shams University (1950)	4	26	470
		Zagazig University (1974)	5	42	693
		Mansoura University (1972)	6	46	716
	Tunisia (5)[a]	University of Sfax	1	68	914
Southern Africa	Botswana (1)	University of Botswana (1964)	1	40	680
	Malawi (3)	University of Malawi	1	37	649
	South Africa (21)	University of Pretoria (1908)	1	1	86
		University of Stellenbosch (1918)	2	2	111
		University of KwaZulu-Natal (2004)	3	3	127
		University of the Free State (1904)	4	5	154
		University of Cape Town (1874)	5	6	174
		University of the Witwatersrand (1922)	6	13	280
		North-West University (2004)	7	15	307
		University of Fort Hare (1916)	8	20	390
		Rhodes University (1904)	9	31	528
		Tshwane University of Technology (2003)	10	43	699
		University of South Africa (1873)	11	45	713
		University of the Western Cape	12	49	737
	Zambia (6)	University of Zambia	1	48	733
	Zimbabwe (10)	University of Zimbabwe (1955)	1	7	195
West Africa	Benin (2)	University of Abomey-Calavi (1970)	1	28	512
	Ghana (9)	University of Ghana (1948)	1	14	290
		Kwame Nkrumah University of Science and Technology (1951)	2	22	432
		University for Development Studies (1992)	3	38	652
	Niger (1)[a]	Abdou Moumouni University (1974)	1		**4947**
	Nigeria (28)	University of Ibadan (1948)	1	4	141
		Ahmadu Bello University (1962)	2	9	217
		Obafemi Awolowo University (1962)	3	10	220
		University of Nigeria—Nsukka (1962)	4	11	228
		Federal University of Agriculture, Abeokuta (1988)	5	19	381
		Federal University of Technology, Akure (1981)	6	32	554
		University of Ilorin (1975)	7	34	562
		University of Maiduguri (1975)	8	35	570
		Ladoke Akintola University of Technology (1990)	9	44	712
		University of Port Harcourt (1975)	10	47	729
		University of Benin (1970)	11	50	739
	Sierra Leone (3)[a]	Njala University (1964)	2		6146
	Togo (1)[a]	School of Agronomy of the University of Lome	1		4124

(continued)

Table 28.1 (continued)

Africa regions	Countries (number of top Agric Universities and colleges)	Top-50 leading institutions (founding date)	Rank in Country	Africa	World
East Africa	Ethiopia (12)	Haramaya University (1954)	1	16	322
		Addis Ababa University (1950)	2	18	366
		Bahir Dar University (2001)	3	33	560
		Mekelle University (1991)	4	36	575
		Hawassa University (1976)	5	39	668
	Kenya (5)	University of Nairobi (1970)	1	8	209
		Egerton University (1987)	2	27	495
		Kenyatta University (1985)	3	29	522
		Jomo Kenyatta University of Agriculture and Technology (1994)	4	41	683
	Somalia (3)[a]	SIMAD University	1	334	7567
	Sudan (9)	University of Khartoum (1956)	1	30	526
	Tanzania (1)	Sokoine University of Agriculture (1984)	1	17	343
		University of Dar es Salaam	2	25	458
	Uganda (4)	Makerere University (1922)	1	12	231
Central Africa	Cameroon (4)[a]	University of Dschang	3	54	779
		University of Buea	2	90	1153
		University of Ngaoundere	5	97	1255
	Democratic Republic of Congo (1)[a]	University of Kinshasa	1	62	3063
		University of Lubumbashi	2	102	3894
		University of Kisangani	3	117	4179

Source: compiled from Edurank (2021)
[a]No institution in the Africa's Top-50 agricultural universities from this country

- *Skilled personnel:* There are currently very few lecturers in animal breeding and genetics at agricultural institutions, and most of them have inadequate skills in biometrics and quantitative genetics, applications of molecular methods and other advanced technologies, and in the design of sustainable breeding programmes. In fact, the lack of a governmental vision in that domain and the existence of no breeding centres in most countries resulted in no applied training in teaching methods, pedagogies, and science communication in animal breeding and genetic improvement. As strongly recommended in Chap. 27 (Sects. 27.4, 27.5, 27.6, and 27.7), education in animal breeding has to be viewed in the broader context of Agriculture Education in Africa, extending from adapted curricula from primary to higher education and incentive systems to retain and grow the capacity of the trained staff.
- *Facilities:* Considering livestock in general and within the global agricultural sector, animal breeding is the poor arm in terms of equipment for training, informatics literacy, and limited access to power and internet in many African universities, which prevents access to world databases that could support up-to-date training and limit access to scientific literature and up-to-date textbooks.
- *Curricula and teaching:* there are many limiting issues in relation to curricula and teaching. Among them are the fact that most of the curriculum being taught in many African schools dealing with animal breeding and genetic improvement are obsolete and lag behind the rapid developments in knowledge of genetics, biological sciences, and bioinformatics.

Moreover, heavy teaching loads and large undergraduate class sizes contrasting with very low numbers of students at the MSc and PhD levels in animal breeding and genetic improvement and the quasi-systematic absence of experimental breeding farm in learning institutions could constitute a serious handicap to efficient animal breeding skills in Africa.
- *Philosophies and theoretical models for Africa:* There is a major gap in terms of theoretical and conceptual farming practice adapted to African context of livestock production and food systems. Academic manuals and books are shaped and modelled in the western context with no applications in the African context. The importation of foreign philosophies of research and innovation has been constant and persistent over decades. Global science and biotechnologies are driven by western priorities for extreme selection and intensification. It is evident that such context is not equitable for Africa and the majority of tropical countries where more than 60% of populations still depend on low input production systems to survive both economically and culturally. Unless this situation is addressed, the discrepancy between African livestock-food systems and both curricula and biotechnology gaps will not be bridged. An appropriate response to this may be the promotion of African scholar and academic manuals that contextualize science and technology for the sustainability goals.
- *Research and collaboration:* Most of the genetics research by local institutions in sub-Saharan Africa are limited to the typological and morphobiometrical characterizations, having no direct application in livestock improvement. Poor or no national animal performance and pedigree data recording systems exist, and the lack of data cannot support capacity development, nor provide the support needed to orient new research. Most lecturers are not sufficiently equipped to attract quality collaboration with research and development institutions within and between countries or overseas. Moreover, the separation of higher learning institutions and research institutions from producers, other key players, and sectoral ministries within countries hinders collaboration.

Despite the general picture of animal breeding on the continent being not as beautiful as in the technologically advanced countries, there are emerging opportunities for developing capacity in AnGR in Africa, such as: (1) Advances in computerization and the growing interest for bioinformatics by younger generations of African animal breeders, and the expansion of communications and information technology. (2) The existence of the African Animal Breeding Network (AABNet) as a pan African entity to interconnect the regional platforms that facilitate information sharing and resource mobilization (ASARECA, SADC, AU-IBAR, RUFORUM, FARA, and CCARDESA). (3) The Global Plan of Action (GPA) for animal genetic resources. (4) The African Union science and technology Agenda. (5) The Pan African University which could leverage on the availability of the African Diaspora for increased cooperation with advanced agricultural research institutions.

28.3.3 People-Centred Livestock Community-Based Breeding Programmes, Training of Trainers (ToT), and Continuous Education

- *Community-Oriented Research*

There is merit in developing and promoting community research schemes for many noble reasons, including: (1) It meets the immediate and future development needs of the communities; (2) It values the ancestral knowledge system; (3) It is the approach that fits the best with localized agroecosystems; (4) It provides sustainable solutions in localizing agro-food systems for food sovereignty and economic self-determination both at the local and the national levels and is therefore seen as a solution to mitigate transboundary crises like COVID-19; and (5) Community-oriented research approach is the platform for the research-community-academia nexus for sustainable development and therefore, the pipeline to develop philosophy and theoretical models for Africa.

- *The Merit of People-Centred Community-Based Breeding Programmes*

There is no doubt that the achievement of food security and poverty alleviation as a global priority targets the underprivileged and marginalized strata of the global population and Africa in particular; hence, the opinion that meeting their needs is more likely to reduce the poverty gap. It is unanimous that about 70% of the vulnerable ones live in rural communities and thus constitute the first target of SDGs. The training of trainers and continuous education within the research-community-academia nexus platform will definitely contribute to securing people-centred livestock community-based breeding programmes.

Government's Intervention and Need for a Decentralized Scheme In most African countries where the required supportive infrastructure is largely unavailable to implement well-organized breeding programmes (Kosgey et al. 2006), some governments have alternatively tried centralized breeding programmes, usually nucleus breeding units (Haile et al. 2018). These centralized systems failed to sustainably deliver the desired genetic improvements to smallholders and also failed to involve the end users in the process. Continuous training and building local capacity to design strategies that consider the farmers' needs, views, decisions, and active participation, from inception through to implementation, is key for sustainable livestock breeding programmes in the African context. The success of such initiative is based on proper understanding of farmers' breeding objectives, infrastructure, participation, and ownership (Sölkner et al. 1998; Wurzinger et al. 2011; Mueller et al. 2015; Haile et al. 2018).

28.3.4 The Potential Role of African Union Commission via the Pan African University (PAU) and AU-IBAR

For the animal breeding and genetic improvement infrastructures to develop, reshaping higher education curricula is important. Ojango et al. (2009) pointed out the lack of trained staff in animal breeding and limited higher education possibilities in sub-Saharan Africa. According to Chagunda et al. (2015), most of the few animal breeders and geneticists work at universities and research institutions where limited breeding activities are effectively carried out. It is also known that most students shy away from animal genetics justified by the lengthy and cumbersome nature of the programme.

A cost-effective way of producing a critical mass of animal breeders in Africa could be to develop and implement continental programmes within institutions of higher learning and within countries, in response to current needs and the needs of both the existing and the perceived future needs of the African livestock industry. The Pan African University Institute for basic Sciences, Technology and Innovation (PAUISTI) and the African Union Inter-African Bureau for Animal Resources could be the best Troy horse, together with National universities and agricultural research organizations, to craft the current and future generations of professional animal breeders. Sustainable animal breeding, genetic improvement, and conservation programmes in Africa need to be given greater emphasis in higher learning curricula. Despite the current role being played by PAU, the learning curricula should include aspects related to animal quantitative genetics and genomics, sustainable breeding programme design, and simulation and optimization adapted to various livestock production systems in Africa. Additionally, teaching methods require modernization and increased activities in order to attract students who often shy away from the subject.

28.3.5 The Role of the African Animal Breeding Network (AABNet)

National agricultural research scientists and university professors need opportunities to strengthen their subject knowledge, didactic skills, and applied field and modern technology aptitudes in animal breeding and genetic improvement.

The creation of AABNet (https://www.animalbreeding-africa.org/) was driven by past experiences and opportunities offered through initiatives focusing on the development and application of tools, resources, and other innovations in genetics, genomics, animal breeding, and data science to support livestock genetic improvement in Africa. AABNet is a pool of top-notch experts in industry, academia, government, international research organizations and world leading animal breeders. Its contribution consists of supporting livestock genetic improvement in Africa through four main pillars: Multi-country animal genetic evaluation, professional development in animal breeding and genetic improvement, advocacy and awareness and business development, and collaboration, networking, and partnerships. The AABNet through organization of massive open online courses (MOOCs), annual summer classes, and research placements will strengthen the capacity of the African livestock breeders' community.

28.4 Key Capacity Building Intervention Areas

28.4.1 Conservation and Improvement of the Livestock Genetic Diversity and Their Environment

Pan African commitment and awareness creation to maintain and improve the diversity at livestock species and breed levels will be of great contribution in achieving the UN sustainable development goals.

Ensuring *Genetics for future use* and their environment is instrumental for successful breeding programs (Notter 1999) to secure the unique contribution of each breed. Breeds also contribute to national or regional identity and thus capacity building for their conservation has a cultural aspect and should require substantial public investments without immediate understanding of benefits.

Conservation of livestock genetic diversity (and their environment) has been addressed at the international and regional levels, assisting policy development at the national level. Almost all African countries are parties to the Convention on Biological Diversity (CBD) and recognize the need to conserve genetic resources for food and agriculture. Target 13 of the Strategic Plan for Biodiversity 2011-2020 required that 'by 2020, the genetic diversity of farmed and domesticated animals is maintained, and strategies developed and implemented to lessen genetic erosion and preserve their genetic diversity' (CBD 2010). This obligation has not been fulfilled largely because of the absence of capacities at various levels to understand, develop, and implement national and regional livestock diversity strategies and action plans or related sectoral programmes.

The importance of building capacities to conserve the diversity of locally adapted livestock breeds is also reflected in the Sustainable Development Goals of the United Nations Agenda 2030. Target 2.5 focuses on the maintenance of genetic diversity of domestic animals (https://unstats.un.org/sdgs/metadata/?Text=&Goal=2&Target=2.5, accessed on 21 September 2021).

Conservation and preservation of Animal Genetic Resources is a too heavy and complex task to be left to local communities of farmers alone. Faced with increasing demand for niche products and meat, both nationally and internationally, many endangered local breeds will face extinction if government actions do not secure and prioritize conservation and breeding programmes for local breeds. Maintaining livestock genetic diversity with strict control of inbreeding

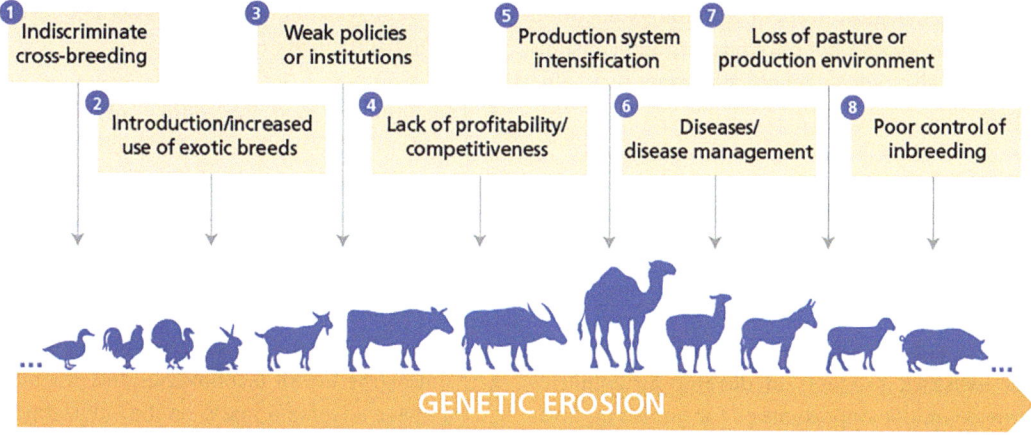

Fig. 28.2 Eight capacity building-related domains to address the causes of genetic erosion in African livestock genetic diversity. (Source: FAO (2021))

in consecutive generations under selection requires skills not only within the national breeding programme but also at the herd level. Those skills should be extended to the 8 others related sectors as proposed by FAO (2021) to address the main causes of genetic erosion (Fig. 28.2). Therefore, African breeders and livestock keepers should be trained to understand the importance of taking measures to avoid mating of related individuals and preserve genetic diversity in their herds. Such issues should be included in capacity building programmes for African farmers and breeders.

28.4.2 Demand-Led and Market-Led Capacity Development in Modern Livestock Breeding Techniques

The projection that demand for livestock products will double during the next twenty years presents the greatest opportunity for profitable increase in livestock production in Africa for the smallholders. However, this also presents emerging constraints in meeting the demands for quantity and quality preferences (Henriksen and Rota 2014). Smallholders therefore have to produce what is demanded. This also requires capacity building for animal breeding to be commercial-oriented.

To contribute to achieving the UN sustainable development goals for Africa, one of many goals of the African Animal Breeding Academy of AABNet is to maximize the benefits of African breeders' expertise for smallholder livestock keepers on the continent. For that to happen, the capacity development of animal breeders needs to be end users' demand-oriented. Generally, improved livestock breeds are a key focus of animal breeding research and development. This could highly contribute to achieving the SDGs, but technology solutions alone are not enough to bring the needed transformation. End users' demand is a critical success factor and should be catered for in the design of breeding programmes. Many capacity development programmes in Africa led by national, regional, or international livestock organizations are failing to achieve their objectives because instead of focusing on the demand from the end users of technology transfer, they have focused on the global trends in agri-biotech-innovations.

The discovery of new genes, research to understand their function, and incorporation of key beneficial traits into the best locally adapted, productive, and preferred breeds should be the building blocks of capacity building in animal breeding and genetic improvement. The emphasis which is frequently on technology should equally put customers at the heart of R&D. Capacity building has to involve end users before the scientific work or technology adoption starts, considering not only farmers and consumers, but also all stakeholders along the value chain.

28.4.3 Sustainable Intensification of Livestock Farming Practices in Africa

Present and expected future increases in demand for livestock products in developing countries provide unique opportunities for improving livelihoods and stewardship of the environment. This improvement could be a complex procedure that requires to be strengthened by enabling policies and pro-poor investments in institutional capacities and innovant technologies (McDermott et al. 2010). If animal breeding in Africa is to satisfy that demand, production must become much more productive, efficient, demand-driven, and market-driven. The 'million dollars' question is to know if intensifying production by smallholders can be sustainable. The models of smallholder dairy in east Africa and small ruminant system in western and southern Africa (Kaitibie et al. 2008; McDermott et al. 2010) illustrate successful intensification. CIMMYT (2020) defines sustainable intensification as an approach using innovations to increase productivity on existing agricultural land with positive environmental and social impacts, where both, *sustainable* and *intensification*, carry equal weight.

The capacity development for sustainable livestock intensification should encompass technical, infrastructural, and policy aspects at pan African level, to jointly produce animal-sourced food and other goods for farmers and markets, while contributing to a range of valued public goods, such as clean water, wildlife and habitats, carbon sequestration, flood protection, groundwater recharge, landscape amenity value, and leisure and tourism opportunities. In its configuration, capacity building should capitalize on the synergies and efficiencies that arise from complex ecosystems, social and economic forces (NRC 2010) focusing on:

- Development and utilization of livestock breeds with a high level of productivity using internally and externally derived inputs, while avoiding the unnecessary use of external inputs
- Harnessing agroecological processes such as nutrient cycling, biological nitrogen fixation, allelopathy, predation, and parasitism
- Minimizing use of technologies and practices that have adverse impacts on the environment and human health
- Making valuable use of human capital in the form of knowledge and capacity to livestock adaptation and innovation and of social capital to bring integrated solutions to common landscape-scale and system-wide problems
- Minimizing the impacts of livestock system management on related domains such as GHG emissions, clean water, carbon sequestration, biodiversity, and spreading of pests and pathogens

Capacity building for sustainable intensification of livestock in Africa could target specific objectives with clearly defined expected impacts (Table 28.2).

The sustainable intensification of agricultural systems should thus be seen as part of a wide range of initiatives and efforts to create greener economies. Green growth and the greener economies have become important targets for national and international organizations, including the OECD (2011), UNEP (2011), World Bank (2012), the Rio+20 conference (UNCSD 2021), and the Global Green Growth Initiative (2012).

As clearly evidenced in other chapters of this book, sustainable intensification of livestock farming practices in Africa cannot overemphasize the role and considerations of indigenous, local, and locally adapted animal genetic resources in the application of novel production technologies and marketing practices, women and youth empowerment, and development of the

Table 28.2 Objectives and expected impacts of capacity building for sustainable intensification of livestock in Africa

Objectives of the action	Expected impact
Achieve primary goals of livestock farmers	Improved yields and incomes, improved natural capital in on- and off-farm landscapes, build knowledge and social capital
Knowledge development for breed improvement	Collaborations between relevant research/learning institutions and other stakeholders as key to emergence of agroecological design; participatory research and development leads to new technologies and practices
Knowledge dissemination, policies, and incentives to adopt sustainable intensification	Conventional extension combined with participatory dissemination via peer-to-peer learning
Stewardship of conservation, integrated disease management, and ecosystem services	Greater appreciation of the contribution of multiple ecosystem services provided by agricultural landscapes and awareness of the two-way relationship between agricultural and non-agricultural components of landscapes.

appropriate capacity building and legislative and policy structures.

28.4.4 Adaptation to Extreme Environmental Conditions and Livestock Emergency Guidelines and Standards

Sub-Saharan Africa remains one of the most vulnerable regions to impacts of climate change and variability (Niang et al. 2014). Mainstreaming of climate change adaptation into policy development is vital across all sectors (Chevallier 2010; England et al. 2018). However, most of the countries in Sub-Saharan Africa are failing to adapt sufficiently to impacts of climate variability and change (Ziervogel et al. 2014). The situation is aggravated by the interaction of 'multiple stressors', including high dependence on rain-fed agriculture, widespread poverty, and weak adaptive capacity (African Union 2014). The changing climate will have adverse effects on natural ecosystems and livestock production, with consequences on Africa's capacities to realize its SDGs targets. Therefore, there is an urgent need for Africa to design robust capacities and approaches to effectively address the challenges associated with extreme environmental risks. Capacity building efforts should give emphasis on supporting the improvement of climate data acquisition, information, and services, including the adoption of the climate strategy of the New Partnership for Africa's Development (NEPAD).

In Malawi, 2011 Capacity and Training Needs Assessments identified gaps for both state and non-state actors operating in the climate change field, which informed about the development of the National Climate Change Learning Strategy (NCCLS) (Malawi Government 2011a, b). The assessment highlighted inadequate knowledge of technical personnel to conduct 'critical analysis' and capacity gaps in the understanding of climate change-related concepts among different stakeholders and has recommended strengthening skills development in human resources as one of the priority areas.

The frontline extension workers in animal production are expected to be the principal stakeholders to teach farmers how to cope with climate change. Consequently, there is a need to develop suitable capacity building packages for their training on climate change issues and its effects (Ozor 2009), based on all the adaptation strategies and practices locally available and affordable.

On the other side, the Training of Trainers in Livestock Emergency Guidelines and Standards (LEGS) will reinforce the provision of tools and guidance for participatory design and implementation of timely and suitable livestock emergency responses. Such training should consider important topics such as emergency prediction and assessment, response identification, and technical aspects including destocking, veterinary services, water, feed, shelter, and restocking.

According to Watson (2011), the LEGS Handbook and training programme have been well accepted and plans have been sketched to expand the programme to additional regions, including North and Central Africa.

28.4.5 Breeding for Animal Welfare, Adaptability, and Biosafety

AU-IBAR in partnership with other organizations led the formulation of an Animal welfare strategy for Africa (AWSA) that furthered the LIDESA strategic priorities and the OIE's standards in the region. From the AWSA, the key animal welfare issues in Africa emanate from lack of adequate education and awareness, with top issues being: (1) Inadequate stakeholder's engagement and involvement, and (2) lack of home-grown science and research data (AU-IBAR 2019). For these reasons, the first priority of the strategy is to develop the capacity of national and regional institutions to coordinate multi-country and multi-regional animal welfare improvement efforts and forging partnerships with a range of stakeholders including the private sector, farming communities, regional organizations, international organizations, and the donor community.

Breeders should be equipped to improve animal health and welfare through selection, in addition to increasing economic performance. The adoption of such goals will contribute to the well-being of animals within smallholder and commercial production systems.

From research data, the performance of local AnGR is more often compared with improved or imported livestock rather than resilience and adaptability traits. This transferred knowledge system is a setback for conservation research actions. While improved or imported breeds are important to address the increasing demand for meat and animal proteins, the local livestock genetic resources remain paramount for their resilience, adaptability in challenging environment, and breeding conditions, and contribution to nutrition, income, and cultural well-being of rural farm families; hence, the importance of promoting home-grown science and research in Africa (AU-IBAR 2019).

28.4.6 Big Data and Artificial Intelligence for Smart Farming and Livestock Breeding Programmes

Smart farming is a development that emphasizes the use of information and communication technology in the cyber-physical farm management cycle (Wolfert et al. 2017). New technologies such as the Internet of Things and Cloud Computing are expected to leverage this development and introduce more robots and artificial intelligence (AI) in livestock breeding. AI using Big Data will support evidence-based decision-making.

In the perspectives of African Animal Breeding development, capacity development for Big Data will be key to provide predictive visions in livestock production operations, drive real-time operational decisions, and redesign business processes for game-changing business models. It is advocated that Big Data will cause major moves in roles and power relations among different stakeholders in current African livestock agri-food supply chain networks. Using Fig. 28.3, Wolfert et al. (2014) summarizes the concept of smart farming along the management cycle as a cyber-physical system.

In modern livestock breeding and improvement, Big Data applications in smart farming will boost farm processes, farm management, and the data chain (Lambert and Cooper 2000), the later being the sequence of activities from data capture to decision-making and livestock data marketing (Miller and Mork 2013; Chen et al. 2014). The livestock data chain will include all activities that are needed to manage data for farm management. Figure 28.4 illustrates the main steps in data chain.

The capacity development for Big Data application could be very instrumental for the development of smart farming in the African setting. Table 28.3 provides an overview of some of the applications.

Fig. 28.3 The cyber-physical management cycle of smart farming enhanced by cloud-based event and Big Data management. (Source Wolfert et al. (2014))

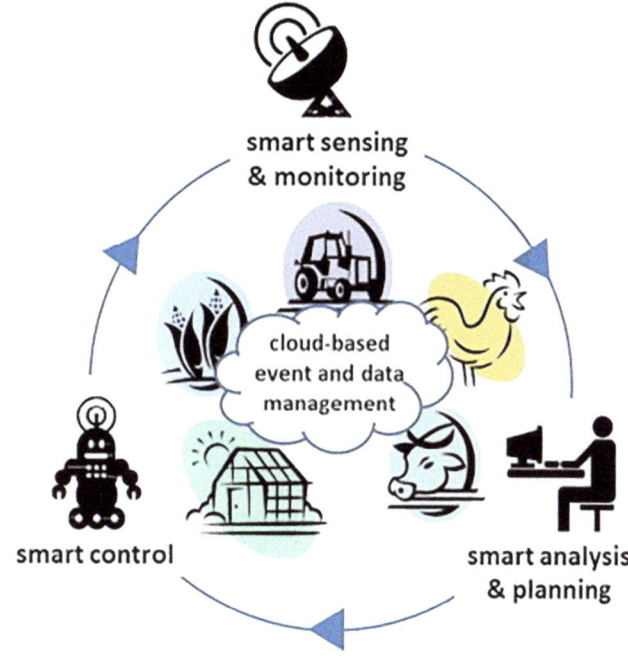

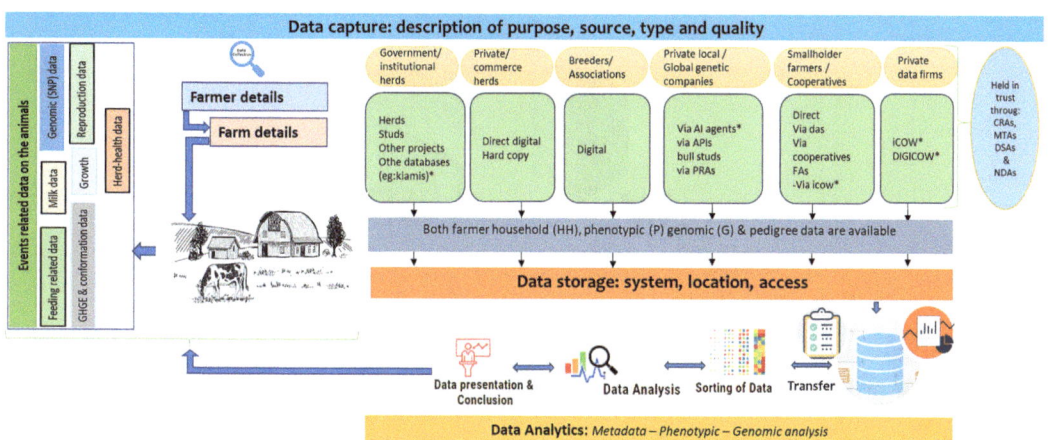

Fig. 28.4 The data chain of Big Data applications in livestock breeding. APIs Application Programming Interface, PRA Data Collectors/Performance Recording Agents, DA Direct Agreement between firm/farm & the programme, FA Farmer Assistants, MTA Material Transfer Agreement, CRA Collaborative Research Agreement, DSA Data Sharing Agreements, NDA Non-Disclosure Agreement

Nowadays, computers, sensors, cloud computing technologies, machine learning (ML), and artificial intelligence (AI) are essential tools to transform the livestock industry, by creating greater gains and more efficiencies. This supports the need for these advanced technologies to be incorporated into the academic curricula. Sensors, Big Data, and advanced AI & ML algorithms go hand in hand to provide a complete solution (Fig. 28.5).

Capacity building for sensors, Big Data, AI, and ML will constitute new possibilities to farmers who, instead of reacting to diseases after they become evident or pro-actively using the services of doctors, will provide an opportunity to constantly monitor key animal health parameters

Table 28.3 Examples of Big Data applications/aspects in livestock and fishery smart farming processes

Cycle of smart farming	Livestock	Fishery
Smart sensing and monitoring	Biometric sensing, GPS tracking (Sonka 2015)	Automated Identification Systems (AIS) (Natale et al. 2015)
Smart analysis and planning	Breeding, monitoring (Cole et al. 2012)	Surveillance, monitoring (Yan et al. 2013)
Smart control	Milk robots (Wolfert et al. 2017)	Surveillance, monitoring (Yan et al. 2013)
Big Data in the cloud	Livestock movements (Faulkner and Cebul 2014; Wamba and Wicks 2010)	Market data (Yan et al. 2013) Satellite data, (European Space Agency 2016)

such as movement, air quality, and consumption of food and fluids, detect diseases like mastitis, and estimate milk yield, reproductive performance, calving time, and breeding values among others (Neethirajan 2020) (Fig. 28.6).

28.4.7 Infrastructural Development and Policy Integration for Technology Uptake

It should be emphasized that capacity building in human resource and infrastructure to use the best breeding and genetics tools are not the only solutions to increasing livestock productivity to achieve UN SDGs in Africa. As recalled by Marshall et al. (2019), any intervention programmes should focus on strengthening the

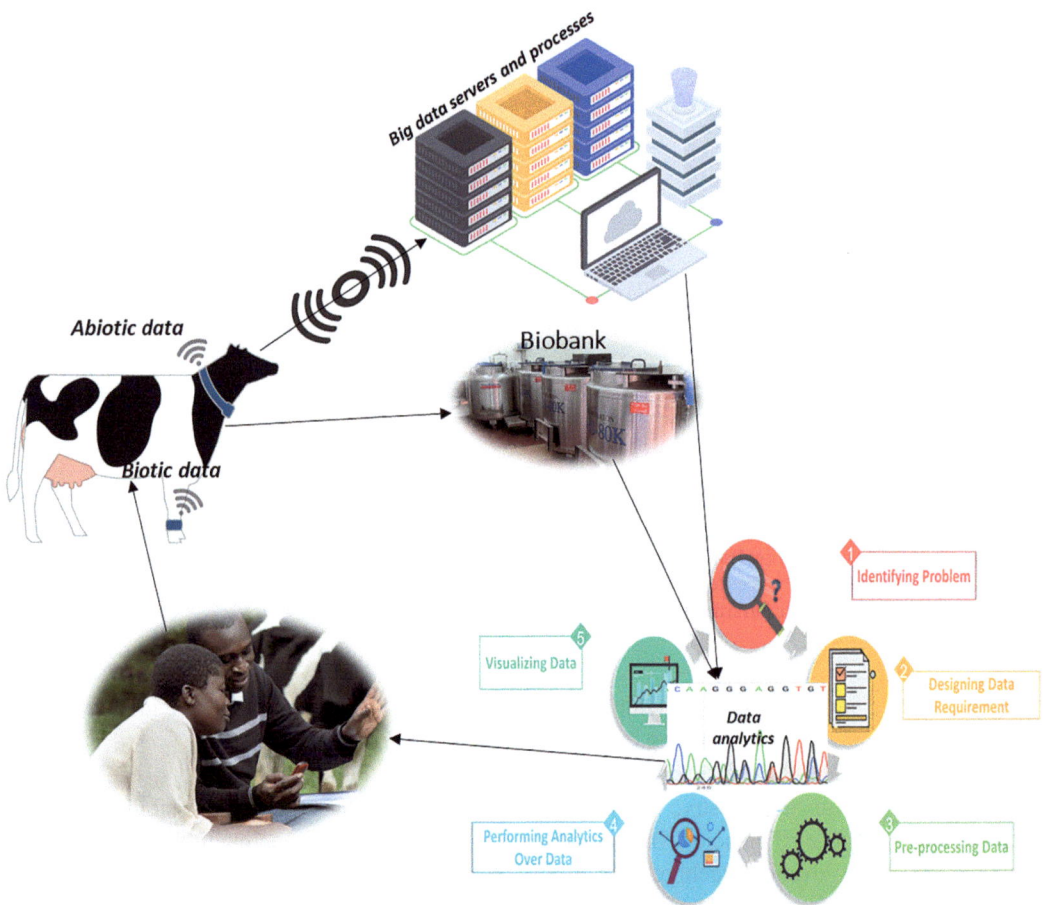

Fig. 28.5 The collection of technologies that we refer to as advanced technologies can help animal farmers create better outcomes

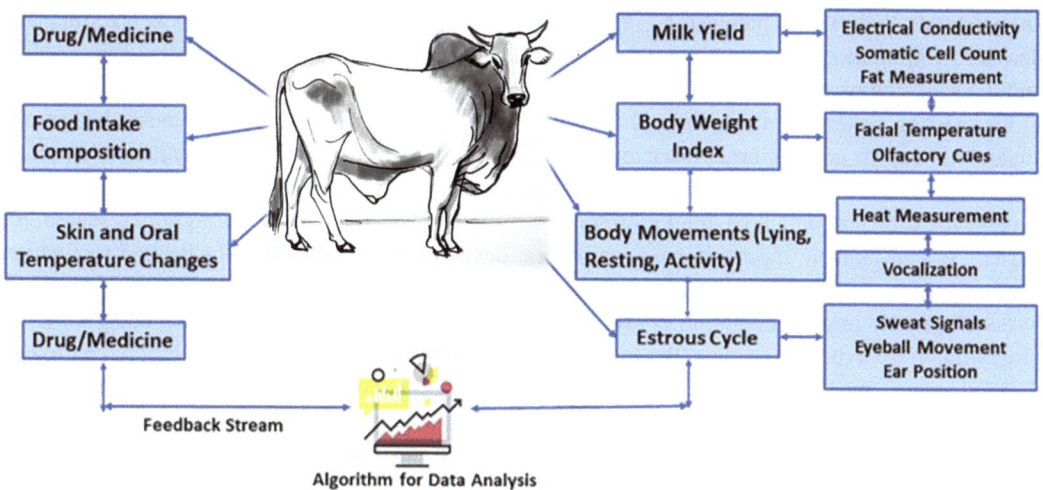

Fig. 28.6 Key domains of capacity building on machine learning algorithms to create optimal performance in livestock breeding. (Adapted from Neethirajan 2020)

capacity building of the livestock keepers and other value chain stakeholders, ensuring the availability and accessibility of inputs, provision of new technologies or customization of existing ones, support for involvement of the private and/or public sector, and most importantly, development of supportive policies.

Development of national and regional livestock strategies and policies requires in-depth understanding of various aspects of the current operation of the sector; hence, the need for capacity building for policy makers for subjects such as the legal and institutional framework, livestock production systems and their economic performance, genetic resources, available capacity, including human resources, and society expectation towards the sector (FAO 2011b).

Setting national and regional strategies should be accompanied by the development of animal breeding laws that address both the technical aspects of capacities in AnGR management (e.g. herd books, animal identification and registration, performance recording, selection and conservation programmes) and other factors that may influence the general implementation of the legislation (e.g. institutional arrangements, roles, and responsibilities of various bodies and decision making process) (Ingrassia et al. 2005). Policies should be clear on the establishment and activities of farmers' associations/breeders' societies and their role in the breeding and genetic improvement process. Policy makers should be trained to develop and enforce animal breeding laws that facilitate application of modern science-based breeding methods and adoption of balanced breeding objectives, taking into account traits supporting health and welfare and contributing to the achievement of sustainability in the sector.

On another note, capacity building and awareness creation for relevant legislations related to the implementation of the Nagoya Protocol and their implementation should target all stakeholders along the provider-user line, so as to alleviate obstacles in the regional exchange of elite animal breeds and their germinal products. Livestock research and development is often affected by the Access and Benefits Sharing (ABS) measures which can become substantial barriers to exchange when the stakeholders and policy makers are not on the same line of understanding (Martyniuk et al. 2018).

28.5 Conclusion and Future Perspectives

Livestock genetic resources play important roles in Agri-food systems and economies of African countries and the local and global demands for animal-food will keep increasing in future decades. Importation of animal-sourced foods and the said 'improved' breed for intensification of the existing systems can only be a short-term strategic action to address urgent demand gaps, but could never be a sustainable solution to address the recurrent deficit of animal protein in Africa. As a sustainable approach, the promotion of research for novel productive genes and local livestock genetic improvement and conservation programmes are expected to close the long-term demand gap and effectively secure local livestock-food systems. The emergence of the strategic partnerships for achieving the UN SDGs and the African Union-InterAfrican Bureau for Animal Resources (AU-IBAR) through Livestock Development Strategic Agenda (LiDeSA) will certainly continue to be the major driver of national and regional joint efforts of multilateral partners. The identified gaps are to be addressed by the national government and regional and international institutions for the achievement of the goals. The training of trainers and continuous education within the academia-research-community nexus platform will definitely contribute to promoting home-grown science and research in view of securing people-centred and community-based livestock breeding and improvement programmes. This will require appropriate funding and efficient governance system. The efficiency of the extension services could not be over-emphasized at the intersection of the research, academia, and farmer communities. These approaches are paramount conditions for the building of decentralized and sustainable livestock-food systems that are more resilient to global and transboundary health and socio-economic crises. However, people-centred community-based models in low-input production systems are fragile and must be protected by effective government interventions. Resources and infrastructural gaps are to be addressed in the short and medium terms. Besides, equitable policies and effective regulatory and implementation framework of actions must be taken in the short term for the preservation, conservation, improvement, and sustainable utilization of local breeds in decentralized livestock-food systems from 'one health' approach. From the current context of capacity building in livestock breeding and genetic improvement in Africa, few key strategic areas could emerge as priority capacity investments: the capacity to (a) Harness the genetic potential of local and locally adapted and performant livestock seeds and their demand-led breeding and improvement. (b) Action research for development to cope with the desired production and productivity level and application of available technologies. (c) Regional integration for availability of inputs, efficient supply chain, and service delivery. (d) Efficient livestock emergency guidelines and standard. (e) Reduce post-harvest lost through value addition, biosafety, market information and infrastructure, and international competitiveness. (f) Improve and harmonize national policies, legislations, and institutional frameworks for effective AfCFTA. To those perspectives, the Africa Animal breeding Network (AABNet), as the convener of the Animal seed working group of the Africa Seed and Biotechnology Partnership Platform, has positioned itself as the credible body to support Africa in the followin:

- Multi-country genetic evaluations to be able to rank genetic materials
- Animal seed certification systems on the continent
- Professionalizing the African animal industry through development of capacity of livestock industry practitioners, including ethical practice to eliminate 'fake seed'
- Incubation of ideas and technologies to generate new animal genetic products or approaches and best practices and their injection in industry
- Advocacy and awareness, to make a business case for livestock genetic improvement and advice countries on genetics and livestock improvement.

References

Adegbola T, Adesogan AH, Havelaar SL, McKune ME, Geoffrey ED (2019) Animal source foods: sustainability problem or malnutrition and sustainability solution? Perspective matters. Global Food Security 2020–25 - 1003252 https://www.sciencedirect.com/

science/article/pii/S2211912419300525. Accessed on 27/10/2021.

African Union (AMCEN-15-REF-11) (2014) African Climate Change Strategy. 86p. https://wedocs.unep.org/bitstream/handle/20.500.11822/20579/AMCEN_15_REF_11_Draft_African_Union_strategy_on_climate_change_English.pdf?sequence=1&isAllowed=y. Accessed 27/10/2021.

AU-IBAR, Animal Welfare Strategy for Africa (AWSA) (2019). https://rr-africa.oie.int/wp-content/uploads/2019/05/awsa_executive_summary_layout_eng_2017.pdf. Accessed 27/10/2021

Besbes B, Leroy G, Boettcher P, Baumung R (2018) Challenges for livestock breeding in developing countries in FAO reports. https://meetings.eaap.org/wp-content/uploads/2018/Session35/S35_01_Baumung.pdf. Accessed 27/10/21

Boettcher P, Atkin O (2010) Current arrangements for national and regional conservation of animal genetic resources. Anim Genet Resour 47:73–83

Cardellino RA, Boyazoglu J (2009) Research opportunities in the field of animal genetic resources. Livest Sci 120(3):166–173

CBD (2010) Decision X/2 2010 The Strategic Plan for Biodiversity 2011–2020 and the Aichi Biodiversity Targets. Available online: https://www.cbd.int/decisions/cop/?m=cop-10. Accessed on 21 Sept 2021

Chagunda MGG, Gibson JP, Dzama K, Rege JEO (2015) Options for enhancing efficiency and effectiveness of research capacity for livestock genetics in, and for, sub-Saharan Africa. Anim Genet Resour 56:145–153

Chen M, Mao S, Liu Y (2014) Big Data: a survey. Mobile Netw Appl 19:171–209. https://doi.org/10.1007/s11036-013-0489-0

Chevallier R (2010) Integrating adaptation into development strategies: the Southern African perspective. Clim Dev 2(2):191–200. https://doi.org/10.3763/cdev.2010.0039

CIMMYT (2020) Conservation agriculture key in meeting UN Sustainable Development Goals New analysis shows benefits of conservation agriculture to crop performance, water efficiency and climate action in South Asia. By Alison Doody. https://www.cimmyt.org/news/conservation-agriculture-key-in-meeting-unsustainable-development-goals/#:~:text=Jat%2C%20a%20principal%20scientist%20at,climate%20action%20and%20clean%20wate

Cole JB, Newman S, Foertter F, Aguilar I, Coffey M (2012) Breeding and genetics symposium: really Big Data: processing and analysis of very large data sets. J Anim Sci 90:723–733

Edurank (2021) Best Agricultural Science Schools in Africa. https://edurank.org/biology/agriculture/af/. Accessed 27/10/2021.

England MI, Dougill AJ, Stringer LC, Vincent KE, Pardoe J, Kalaba FK, Afionis S (2018) Climate change adaptation and cross-sectoral policy coherence in Southern Africa. Reg Environ Chang 18(7):2059–2071. https://doi.org/10.1007/s10113-018-1283-0

European Space Agency (2016) Fish farms guided by Sentinels and the cloud http://www.esa.int/Our_Activities/Observing_the_Earth/Fish_farms_guided_by_Sentinels_and_the_cloud. Consulted 14th October 2021.

Fanzo J (2019) Healthy and sustainable diets and food systems: the key to achieving sustainable development goal 2? Food Ethics 4:159–174. https://doi.org/10.1007/s41055-019-00052-6

FAO (2014) SDG Indicators Data Portal. https://www.fao.org/sustainable-development-goals-data-portal/resources/news/news-detail/fao-releases-proposedtargets-and-indicators-on-14-themes-for-sustainable-developmentgoals/en#:~:text=FAO%20releases%20proposed%20targets%20and%20indicators%20on%2014%20themes%20for%20Sustainable%20Development%20Goals,-16/06/2014&text=The%20Food%20and%20Agriculture%20Organization,FAO's%202014%20themes, viewed on 2nd June 2025.

Food and Agricultural Organisation (FAO) of the United Nations (2007) Global Plan of Action for Animal Genetic Resources and the Interlaken Declaration. Rome: Food and Agricultural Organizations of the United Nations. https://www.fao.org/policy-support/tools-and-publications/resources-details/fr/c/453629/. Consulted on 27/10/2021.

Food and Agricultural Organisation (FAO) of the United Nations (2015) FAO Synthesis—Livestock and the Sustainable Development Goals Global Agenda for Sustainable Livestock. Draft prepared by FAO-AGAL Livestock Information, Sector Analysis and Policy Branch. Available online at: http://www.livestockdialogue.org/fileadmin/templates/res_livestock/docs/2016/Panama/FAO-AGAL_synthesis_Panama_Livestock_and_SDGs.pdf. Accessed 27 Oct 2021.

Food and Agricultural Organisation (FAO) of the United Nations (2021) Animal genetics. https://www.fao.org/animal-genetics/en/. Accessed on October 27th 2021

Faulkner A, Cebul K (2014) Agriculture gets smart: the rise of data and robotics. Cleantech Agriculture Report Cleantech Group. https://www.cleantech.com/report/agriculture-gets-smart-the-rise-of-data-and-robotics/. Consulted 14th October 2021

Food and Agricultural Organisation (FAO) of the United Nations (2011a) Developing the institutional framework for the management of animal genetic resources. (FAO Animal Production and Health Guidelines, No. 6. Rome. https://www.fao.org/3/ba0054e/ba0054e00.htm. Accessed on 27/10/2021.

Food and Agricultural Organisation (FAO) of the United Nations (2011b) Guidelines for the preparation of livestock sector reviews. In: Animal production and health guidelines No. 5. FAO, Rome, Italy, pp 1–66. Available online: https://www.fao.org/3/i2294e/i2294e00.pdf. Accessed on 21 September 2021

Global Green Growth Institute (2012) Green recovery, green jobs and NetZero2050, https://gggi.org/. Consulted on 14th October 2021.

Haile A, Wurzinger M, Mueller J, Mirkena T, Duguma G, Rekik M, Mwacharo J, Mwai O, Sölkner J, Rischkowsky B (2018) Guidelines for Setting Up community-based small ruminants breeding programs in Ethiopia. ICARDA, Tools and guidelines No.1. Beirut, Lebanon. https://www.ilri.org/publications/guidelines-setting-community-based-small-ruminants-breeding-programs-second-edition. Accessed on 27/10/2021/

Henriksen J, Rota A (2014) Commercialization of livestock production; towards a profitable and market-oriented smallholder livestock production system. Livest Res Rural Dev 26:Article #90. Retrieved 21 Sept 2021, from http://www.lrrd.org/lrrd26/5/hend26090.html

Houaga I, Mrode R, Opoola O, Chagunda M, Okeyo M, Rege JEO, Olori VE, Nash O, Okeno TO, Djikeng A (2023) Livestock phenomics and genetic evaluation approaches in Africa: current state and future perspectives. Front Genet 14:1115973. https://doi.org/10.3389/fgene.2023.1115973

Ingrassia A, Manzella D, Martyniuk E (2005) The legal framework for the management of animal genetic resources; FAO legislative study 2005–89. FAO, Rome, Italy, pp 1–168. Available online: https://www.fao.org/fileadmin/user_upload/legal/docs/ls89-e.pdf. Accessed on 21 Sept 2021

Kaitibie S, Omore A, Rich K, Salasya B, Hooton N, Mwero D, Kristjanson P (2008) Influence pathways and economic impacts of policy change in the Kenyan dairy sector: the role of smallholder dairy project. Research report for the CGIAR Standing Panel on Impact Assessment, Nairobi, p 64. https://agrilinks.org/sites/default/files/resource/files/Influence%20pathways%20and%20economic%20policy%20change%20in%20Kenya%20Dairy%20Sector_ILRI%202009.pdf. Accessed 27/10/2021

Kawarazuka N, Béné C (2010) Linking small-scale fisheries and aquaculture to household nutritional security: an overview. Food Secur 2(4):343–357. https://doi.org/10.1007/s12571-010-0079-y

Keeling L, Tunón H, Olmos Antillón G, Berg C, Jones M, Stuardo L, Swanson J, Wallenbeck A, Winckler C and Blokhuis H (2019) Animal Welfare and the United Nations Sustainable Development Goals. Front Vet Sci 6:336. https://doi.org/10.3389/fvets.2019.00336

Kosgey IS, Baker RL, Udod JAM, van Arendon JAM (2006) Successes and failures of small ruminant breeding programmes in the tropics: a review. Small Rumin Res 61:13–28. https://www.sciencedirect.com/science/article/pii/S0921448805000258. Accessed on 27/10/2021

Lambert DM, Cooper MC (2000) Issues in supply chain management *Ind*. Mark Manag 29:65–83. https://drdouglaslambert.com/wp-content/uploads/2020/05/Lambert-and-Cooper-Issues-in-Supply-Chain-Management-IMM-2000.pdf. Accessed on 27/10/2021

Macmillan S (2016) Livestock and the Sustainable Development Goals, https://news.ilri.org/2016/02/29/livestock-and-the-sustainable-development-goals/. Consulted 02/07/2021.

Malawi Government (2011a) Capacity needs assessment for climate change management structure in Malawi, 2011 June. National Climate Change Programme. Ministry of Finance and Development Planning. https://reliefweb.int/sites/reliefweb.int/files/resources/NCCM-Policy-Final-06-11-2016.pdf. Accessed on 27/10/2021

Malawi Government (2011b) Training needs assessment for climate change management structures in Malawi, 2011 June. National Climate Change Programme. Ministry of Finance and Development Planning. https://reliefweb.int/sites/reliefweb.int/files/resources/NCCM-Policy-Final-06-11-2016.pdf. Accessed on 27/10/2021

Marshall K, Gibson JP, Mwai O, Mwacharo JM, Haile A, Getachew T, Mrode R, Kemp SJ (2019) Livestock genomics for developing countries–African examples in practice. Front Genet 10:1–13. Available online: https://www.frontiersin.org/articles/10.3389/fgene.2019.00297/full. Accessed on 21 Sept 2021

Martyniuk E, Berger B, Bojkovski D, Bouchel D, Hiemstra SJ, Marguerat C, Matlova V, Sæther N (2018) Possible consequences of the Nagoya Protocol for animal breeding and the worldwide exchange of animal genetic resources. Acta Agric Scand 67:96–106

McDermott JJ, Staal SJ, Freeman HA, Herrero M, Van de Steeg JA (2010) Sustaining intensification of smallholder livestock systems in the tropics. Livest Sci 130:95–109

Miller HG, Mork P (2013) From data to decisions: a value chain for Big data. IT Prof 15:57–59. https://doi.org/10.1109/MITP.2013.11

Mueller JP, Rischkowsky B, Haile A, Philipsson J, Mwai O, Besbes B, Valle Zarate A, Tibbo M, Mirkena T, Duguma G, Sölkner J, Wurzinger M (2015) Community-based livestock breeding programmes: essentials and examples. J Anim Breed Genet 132:155–168

Natale F, Gibin M, Alessandrini A, Vespe M, Paulrud A (2015) Mapping fishing effort through AIS data. PLoS One 10. https://doi.org/10.1371/journal.pone.0130746

Neethirajan S (2020) The role of sensors, big data and machine learning in modern animal farming. Sens Bio-Sens Res 29:100367

Neumann C, Harris D, Rogers L (2002) Contribution of animal source foods in improving diet quality and function in children in the developing world. Nutr Res 22(1–2):193–220

Niang I, Ruppel OC, Abdrabo MA, Essel A, Lennard C, Padgham J, Urquhart P (2014) Africa. In: Barros VR, Field CB, Dokken DJ, Mastrandrea MD, Mach KJ, Bilir TE, Chatterjee M, Ebi KL, Estrada YO, Genova RC, Girma B, Kissel ES, Levy AN, MacCracken S, Mastrandrea PR, White LL (eds) Climate change 2014: impacts, adaptation, and vulnerability. Part B: regional aspects. Contribution of working group II to the fifth assessment report of the Intergovernmental

Panel on climate change. Cambridge University Press, Cambridge, pp 1199–1265

Notter DR (1999) The importance of genetic diversity in livestock populations of the future. J Anim Sci 77:61–69

NRC (2010) Towards sustainable agricultural systems in the 21st century. National Academies Press Committee on Twenty-First Century Systems Agriculture, Washington, DC. https://www.nap.edu/catalog/12832/toward-sustainable-agricultural-systems-in-the-21st-century. Accessed on 27/10/2021

OECD (better Policies for Better lives) (2011) Towards green growth. https://www.oecd.org/env/towards-green-growth-9789264111318-en.htm. Consulted on 14th October 2021

Odero-Waitituh JA (2017) Smallholder dairy production in Kenya: A review. Ministry of Livestock Development.

Ojango JMK, Malmfors B, Okeyo AM, Philipsson J (2009) Capacity building for sustainable use of animal genetic resources in developing countries. Appl Anim Husband Rural Dev J 2:23–26. https://www.ilri.org/publications/capacity-building-sustainable-use-animal-genetic-resources-developing-countries. Accessed 27/10/2021

Okpeku M, Ogah MD, Adeleke AM (2019) A review of challenges to genetic improvement of indigenous livestock for improved food production in Nigeria. Afr J Food Agric Nutr Dev 19(1):13959–13978. https://doi.org/10.18697/ajfand.84.BLFB1021. http://ajfand.net/Volume19/No1/BLFB1021.pdf

Opoola O, Mrode R, Banos G, Ojango J, Banga C, Chagunda MGG, et al. (2019) Current situations of animal data recording, dairy improvement infrastructure, human capacity and strategic issues affecting dairy production in sub-Saharan Africa. Trop Anim Health Prod 51, 1699–1705. https://doi.org/10.1007/s11250-019-01871-9

Ozor N (2009) Implications of Climate Change for National Development: The Way Forward. Debating Policy Options for National Development; Enugu Forum Policy Paper 10; African Institute for Applied Economics (AIAE); Enugu, Nigeria: p.19-32. Available at: http://www.aiaenigeria.org/Publications/Policypaper10.pdf viewed 2nd June 2025

Philipsson J, Rege JEO, Zonabend E, Okeyo AM (2011) Sustainable breeding programmes for tropical farming systems. Animal Genetic Training Resources. version 3. Available from: http://agtr.ilri.cgiar.org/index.php?option=com_content&view=article&id=27&Itemid=267

Rege JEO, Marshall K, Notenbaert A, Ojango JMK, Okeyo AM (2011) Pro-poor animal improvement and breeding—what can science do? Livest Sci 136:15–28

Schneider F, Tarawali S (2021) Sustainable development goals and livestock systems. Rev Sci Tech Off Int Epiz 40(2):585

Sölkner J, Nakimbigwe H, Valle-Zarate A (1998) Analysis of determinants for success and failure of village breeding programs. In: Proceedings of the 6th world congress on genetics applied to livestock production, 11–16 January 1998, Armidale, NSW, Australia, pp 273–280

Sonka S (2015) Big Data: from hype to agricultural tool. Farm Policy J 12:1–9

Stabler SP, Allen RH (2004) Vitamin B12 deficiency as a worldwide problem. Ann Rev Nutr 24:299–326. https://doi.org/10.1146/annurev.nutr.24.012003.132440

UN (2015) Transforming our world: the 2030 agenda for sustainable development, https://sustainabledevelopment.un.org/content/documents/21252030%20Agenda%20for%20Sustainable%20Development%20web.pdf. Date accessed; 21.09.2021.

UNCSD (2021) United Nations Conference on Sustainable Development, Rio+20, https://sustainabledevelopment.un.org/rio20. Accessed on 14 Oct 2021.

UNEP (2011) Towards a Green Economy: Pathways to Sustainable Development and Poverty Eradication. https://sustainabledevelopment.un.org/index.php?page=view&type=400&nr=126&menu=35. Consulted on 14th October 2021

Wamba SF, Wicks A (2010) RFID deployment and use in the dairy value chain: applications, current issues and future research directions. In: 2010 IEEE international symposium on technology and society, pp 172–179. https://doi.org/10.1109/ISTAS.2010.5514642. https://ieeexplore.ieee.org/document/5514642. Accessed on 27/10/2021

Watson C (2011) Protecting livestock, protecting livelihoods: the Livestock Emergency Guidelines and Standards (LEGS). Pastoralism 1, 9. https://doi.org/10.1186/2041-7136-1-9

Wolfert J, Sørensen CG, Goense D (2014) A future internet collaboration platform for safe and healthy food from farm to fork, Global Conference (SRII), 2014 annual SRII. IEEE, San Jose, CA, USA, pp 266–273

Wolfert S, Cor Verdouw L, Bogaardt M (2017) Big data in smart farming – A review. Agricult Syst 153, 69–80. https://doi.org/10.1016/j.agsy.2017.01.023

Wollny CBA (2003) The need to conserve farm animal genetic resources in Africa: should policy makers be concerned? Ecol Econ 45(3):341–351. https://doi.org/10.1016/S0921-8009(03)00089-2

World Bank (2012) Inclusive green growth. The pathway to sustainable development. © World Bank, Washington, DC. https://openknowledge.worldbank.org/handle/10986/6058 License: CC BY 3.0 IGO

World Health Organization (2014) World Health Assembly Global Nutrition Targets 2025: Stunting Policy Brief. Retrieved from 575 http://www.who.int/nutrition/topics/globaltargets_stunting_policybrief.pdf

Wright IA (2017) The roles of livestock in achieving the Sustainable Development Goals. 25th Anniversary Conference, Ethiopian Society for Animal Production, Haramaya University 24–26 August.

Wurzinger M, Sölkner J, Iñiguez L (2011) Important aspects and limitations in considering community-based breeding programs for low-input smallholder livestock systems. Small Rumin Res 98(1–3):170–175

Wurzinger M, Gutiérrez GA, Sölkner J and Probst L (2021) Community-Based Livestock Breeding: Coordinated

Action or Relational Process? Front. Vet. Sci. 8:613505. https://doi.org/10.3389/fvets.2021.613505

Yan B, Shi P, Huang G (2013) Development of traceability system of aquatic foods supply chain based on RFID and EPC internet of things. Trans Chinese Soc Agric Eng 29:172–183

Ziervogel G, New M, Archer van Garderen E, Midgley G, Taylor A, Hamann R, Warburton M (2014) Climate change impacts and adaptation in South Africa. Wiley Interdiscip Rev Clim Chang 5(5):605–620. https://doi.org/10.1002/wcc.295

Open Access This chapter is licensed under the terms of the Creative Commons Attribution 4.0 International License (http://creativecommons.org/licenses/by/4.0/), which permits use, sharing, adaptation, distribution and reproduction in any medium or format, as long as you give appropriate credit to the original author(s) and the source, provide a link to the Creative Commons license and indicate if changes were made.

The images or other third party material in this chapter are included in the chapter's Creative Commons license, unless indicated otherwise in a credit line to the material. If material is not included in the chapter's Creative Commons license and your intended use is not permitted by statutory regulation or exceeds the permitted use, you will need to obtain permission directly from the copyright holder.

29. Harnessing Multi-Country Cooperations, Initiatives, Facilities, and Technologies for Advancing Livestock Genetic Improvement in Africa

Ntanganedzeni O. Mapholi, Cuthbert Banga, Raphael Mrode, Oluyinka Opoola, Isidore Houaga, Lucy T. Nesengani, and Eveline M. Ibeagha-Awemu

Abstract

Multi-country initiatives have facilitated the development and adoption of technologies in developed nations, but such cooperations are less evident in Africa. National breed-specific genetic evaluations have been the most commonly practiced globally; however, there has been a growing interest and opportunities have arisen for across-country collaborations, or international consortiums combining data into large-scale multi-country genetic evaluations to achieve better genetic gains. Despite the need to establish multi-country genetic evaluation schemes for livestock production suitable for the African continent, few initiatives have been put in place to achieve this. This chapter draws attention to multi-country initiatives and benefits for performance recording as well as success stories in Africa (Sect. 29.2), the benefits of data sharing (Sect. 29.3), the role and importance of core facilities, especially those associated with the technologies of genomics, proteomics, metabolomics, metagenomics, and statistical/bioinformatics management of data (Sect. 29.4), germplasm centers for the storage of genetic materials or seed stock (Sect. 29.5), and the important place and prospect of multi-country genetic evaluations and practical application in Africa (Sects. 29.6 and 29.7). Such initiatives may provide an opportunity for genetic improvement of indigenous livestock populations and the possibility to open up new markets for African germplasm as well as inter-country germplasm trade within the continent. This would have a positive impact on the continent's livestock economy.

N. O. Mapholi (✉) · L. T. Nesengani
Department of Agriculture and Animal Health, University of South Africa, Pretoria, Gauteng, South Africa
e-mail: maphon@unisa.ac.za

C. Banga
Department of Animal Sciences, Faculty of Animal and Veterinary Sciences, Botswana University of Agriculture and Natural Resources (BUAN), Gaborone, Botswana

R. Mrode
Scotland's Rural College (SRUC), Edinburgh, UK

O. Opoola · I. Houaga
Centre for Tropical Livestock Genetics and Health (CTLGH), The Roslin Institute, University of Edinburgh, Edinburgh, UK

E. M. Ibeagha-Awemu
Sherbrooke Research & Development Centre, Agriculture and Agri-Food Canada, Sherbrooke, QC, Canada

Keywords

Multi-trait across-country evaluations · Core facility · Germplasm centre · Livestock data · Genetic improvement · Africa

Abbreviations

AABNet	African Animal Breeding Network
ABI	African Biosciences Initiative
ACENET	Atlantic Computational Excellence Network
ADGG	African Dairy Genetic Gains
AfricanBP	African BioGenome
AnGR	Animal Genetic Resources
ARC	Agricultural Research Council
ARC-SA	Agricultural Research Council of South Africa
ART	Assisted Reproductive Technologies
AU	African Union
AU-PANVAC	African Union Pan African Veterinary Vaccine Centre
AWS	Amazon Web Services
BECA	Biosciences Eastern and Central Africa
BecA-ILRI Hub	Bioscience Eastern and Central Africa-International Livestock Research Institute Hub
BixCoP	Bioinformatics Community of Practice
BLUP	Best Linear Unbiased Prediction
BTP	Biotechnology Platform
CDC	Centers for Disease Control and Prevention
CES	Centre D'Expertise Et De Services Génome Québec
CGT	African Centre for Gene Technologies
ChIP-Seq	Chromatin Immunoprecipitation Sequencing
CIDA	Canadian International Development Agency
CIRDES	Centre International De Recherche En Afrique
CLSI	Clinical and Laboratory Standards Institute
CPGR	Centre for Proteomic and Genomic Research
DAR	Department of Agricultural Research
DFID	Department for International Development
DNA	Deoxyribonucleic Acid
EMBL	European Molecular Biology Laboratory
FAO	Food and Agricultural Organization
FPGAs	Field-Programmable Gate Arrays
GC	Genome Core or Centres
GMACE	Genomic Multi-Country Evaluations
GPA	Global Plan of Action
GPUs	Graphics Processing Units
H3ABioNet	Pan African Bioinformatics Network
HPC	High-Performance Computing
IBM	International Business Machines
ICAR	International Committee for Animal Recording
ILRI	International Livestock Research Institute
INTERBULL	International Bull Evaluation Services
INTERGIS	Integrated Registration and Genetic Information System
KyD	Kaonafatso Ya Dikgomo
MCI	Multi-Country Cooperations and Initiatives
MTA	Material Transfer Agreement
NAGRIC & DB	National Animal Genetic Resource Centre and Data Bank
NEPAD	New Partnership for Africa's Development

NGS	Next-Generation Sequencing
NHS	National Health Services
NRF	National Research Fund
QMS	Quality Management System
QSEs	Quality System Essentials
RMRD SA	Red Meat Research and Development South African
RN	Ribonucleic Acid
RNA-Seq	RNA Sequencing
SADC	Southern African Development Community
SAMRC	South African Medical Research Council
scRNA-Seq	Single Cell Ribonucleic Acid Sequencing
SFA	Science for Africa
SNP	Single Nucleotide Polymorphism
SOP	Standard Operating Procedures
TIA	Technology Innovation Agency
UCT	University of Cape Town
UK	United Kingdom
UniPort	University of Port Harcourt
UNISA	University of South Africa
USA	United State of America
WGBS	Whole Genome Bisulfite Sequencing
WGS	Whole Genome Sequencing

29.1 Introduction

Multi-country cooperations and initiatives (MCI) have been in the forefront of technology development and advancement in developed countries (Jit et al. 2021). They are a driving tool in strengthening and motivating the sharing of knowledge as well as building capacity within countries. This has been effective in implementing new technologies such as genomics for the benefit of human health and improving genetic material in livestock species (Opoola et al. 2020). A recent example of such initiative is the agreement and declaration for delivering across-border access to genomic information by 13 European countries. This agreement will see these countries linking genomic databases as a key factor in revolutionizing research to advance health standards. Such initiatives are, however, either limited or completely lacking in developing countries, especially in Africa. In Africa, where livestock productivity remains low, there is a greater need to improve and advance livestock genetics. Livestock genetic evaluation programmes are crucial to the establishment of sound breeding programmes and the achievement of high rates of genetic improvement (Mueller et al. 2015; Martin-Collado et al. 2021). Multi-country cooperations and initiatives can act as a driving tool to improve and implement livestock genetic evaluation programmes for the benefit of the continent at large.

Genomic evaluations for livestock selection are swiftly substituting traditional evaluation systems (Upshaw et al. 2021; Cesarani et al. 2021; Harris and Johnson 2010). This is influenced by the advancements in genotyping technologies, computational speed, and analytical approaches, as well as the growing need for higher rates of sustainable genetic change.

Implementation of across-country genetic evaluations is an evolving process in the livestock industry. Recently, there has been a rapid increase in initiatives to share livestock data and genetic material across countries, and even continents, in an effort to achieve higher rates of genetic improvement in livestock populations (Cesarani et al. 2021). For example, the benefits of across-country genetic evaluations have been reported in countries like Norway and New Zealand where de Oliveira et al. (2022) reported the feasibility of across-country genomic predictions of economic traits such as birth weight and weaning weight in a composite sheep population. The possibility to predict genomic estimated breeding values for traits not recorded in Norway using data from New Zealand was also indicated (de Oliveira et al. 2022). This is largely achieved due to improved accuracy of genetic evaluations, resulting from increased genotypic, phenotypic, and pedigree information that is made available by the participating countries (Cardoso et al. 2021). Additional benefits of such programs include increased opportunities for trade in semen, embryos, and animals among the participating countries.

Exchange of biological materials or data remains a challenge in Africa, due to inadequate infrastructure, technical capacity, and financial resources (Mulder et al. 2017). African researchers are still exploring the opportunities for initiating collaborations in data recording, sharing, and analysis across countries. In the African context, the success of such initiatives will largely depend on collaboration and sharing of not only data and biological materials, but also resources and requisite facilities such as data management and genome centers. Therefore, there is a need for multi-country strategic plans, which include the development of multi-lateral collaboration frameworks and agreements. It is also important to build capacity in the skills that are needed for the effective implementation of such plans.

29.2 Performance Recording

Essentially, performance recording entails the measurement and recording of the performance of individual animals on traits that are desirable to improve. The primary objective of recording such data is to provide reliable information to breeders and farmers to be used as a selection aid to improve herd or flock performance. A sound performance recording system is based on sustainable and continuous recording of animal data to monitor livestock performance in various management systems. Many developed countries have achieved great successes in measuring and evaluating desirable traits such as production and functional traits. Genetic evaluation of these traits is obtained from phenotypic, pedigree, and genomic data of individual animals within the population under selection. See Chap. 20 for more details on performance recording tools and genetic improvement of livestock.

29.2.1 Animal Identification and Recording Schemes

Performance recording programmes in Africa are not as robust as the specialized recording platforms in developed countries. Livestock recording schemes in Africa are hindered by factors such as poor animal identification systems and inadequate infrastructure for collecting and processing data (Ojango et al. 2017). A general lack of sustainable performance recording schemes has hampered efforts to improve Africa's livestock production systems through genetic improvement. Genetic parameter estimates for genetic evaluation within livestock populations are mostly unavailable, due to the small herd sizes and lack of data. Lack of genotyping capacity also makes it difficult to implement genomic selection in these livestock populations.

29.2.2 Success Stories in Africa

29.2.2.1 South Africa

In South Africa, the Agricultural Research Council (ARC-SA) together with the Department of Agriculture, Land Reform, and Rural Development manages the national livestock database, known as the Integrated Registration and Genetic Information System (INTERGIS) (Van der Westhuizen et al. 2005; Zuma-Netshiukhwi et al. 2022). The INTERGIS caters for animal data recording needs of the livestock industry players, including performance recording agencies. This system has provided the South African livestock industry with a professional and internationally recognized recording and genetic improvement facility. Most of the recording and identification is, however, confined to commercial herds and flocks, with limited participation of smallholder farmers like in most African countries. A beef recording scheme for smallholder farmers, called Kaonafatso ya Dikgomo (KyD) scheme, was established in 1995 and collects basic data that are recorded on the INTERGIS. More than 8000 emerging and smallholder farmers are participating in this scheme (Ngarava et al. 2020).

29.2.2.2 African Dairy Genetic Gains

The African Dairy Genetic Gains (ADGG) project has carried out systematic performance recording from over 200,000 farmers in Kenya, Ethiopia, and Tanzania using simple mobile phone technologies (Mrode et al. 2021). The aim is to enhance country collaborations through data sharing to monitor future dairy improvements and provide feedback to the farmers who are the providers of the performance data.

29.2.3 Pedigree and Phenotypic Data

Phenotypic and pedigree data are vital inputs of a genetic evaluation programme. Although performance recording mainly serves to identify opportunities to improve livestock management systems, it also provides a means to generate the information needed to achieve sustained genetic change toward a defined bio-economic development objective. In performance recording for genetic improvement, there is a great need for maintenance of detailed records of phenotypic measures of performance as well as parentage and of genetic relationships among the recorded animals. The recording of such data must be sustained over time (animal generations) and maintained in a rigorously consistent manner in order to be effective.

Phenotypic measures of animal performance that are recorded in performance recording include output, product quality, life history, and adaptation traits. Animal production levels, or outputs, are an important component of overall animal merit. In dairy production, this involves recording of milk production levels at various points during lactation and of the duration of lactation. Such a set of measurements allows estimation of total milk production. When meat production is involved, measures of body size are the main indicators of production. Body size is ideally measured in all animals at similar ages and, assuming that birth dates are known, can be further adjusted to a constant age. Generally, measures of body size may be taken at weaning, at times when animals are chosen for marketing, and at times when management or nutrition change (for example, at transitions from wet to dry seasons or when animals are brought in from summer grazing for winter). In most cases, an indicator of body size at a relatively young age is desirable because, at least in suckling mammals, such a measure provides information on both the growth potential of the offspring and the mothering ability of the dam. Attention to product quality is appropriate only if improving product quality leads to increased food security and product prices or if measures of product quality are indicators of other favourable production characteristics (e.g. lean animals usually have higher feed efficiency). The most common measures of product quality in dairy production are fat and protein content of the milk. In fibre production, fibre diameter ("fineness") is the most common measure of value, although there are other potentially important quality attributes (including scouring yield and fibre length, uniformity, colour, strength, etc.). Life history traits are those that define critical events in the animal's productive life. They are generally time-related, including such things as dates of birth, weaning, mating, parturitions, sale, medical treatments, ending of lactation, and death. Evaluation of adaptation involves identification of the most important animal stressors, including diseases, climatic variables, endo- and ectoparasites, dietary deficiencies, and seasonal feed shortages. Appropriate and measurable indicators of susceptibility to these stressors are recorded as adaptation phenotypes.

Limited recording of pedigree and phenotypic data is one of the major challenges to the development of sound genetic evaluation programmes in Africa (Opoola 2019). Most African countries do not have sustainable animal performance recording programmes, which makes it difficult to evaluate the genetic merit of animals under local conditions. South Africa is the only African country with a well-established and sustainable animal performance recording scheme and is a member of the International Committee for Animal Recording (ICAR). It is also one of the countries that has a performance recording scheme for the less developed livestock system through Kaunafatso ya Dikgomo (KYD), as explained above. A few other countries, such as Botswana, Kenya, and Zimbabwe, have had either on and off or collapsed performance recording programmes. A recent study (Opoola et al. 2020) revealed that African countries can realize large benefits from pooling together available phenotypic and pedigree data and conducting across-country genetic evaluations. A critical requirement for the successful implementation of such across-country genetic evaluations is the standardization of trait definition across countries. This could be achieved by ensuring that within country performance recording programmes, all conform to ICAR guidelines on the definition of the various traits.

29.2.4 Joint Participation in International Committee for Animal Recording

The International Committee for Animal Recording (ICAR, https://www.icar.org/) provides a platform where its members are provided with guidelines and standards for measuring economically important traits of farm animals. The are only a few ICAR member organizations from Africa, such as the Agricultural Research Council and Studbook of South Africa and the Department of Veterinary Services of Botswana. These institutions have ICAR Certificates of Quality and routinely record production, fertility, and herd health data from exotic, indigenous, and crossbred animals in accordance with ICAR guidelines (Jorjani et al. 2003). Previous African member organizations include the Department of Research and Specialist Services and Zimbabwe Dairy Services Association of Zimbabwe.

Cost of membership is a major factor limiting the participation of African countries in ICAR. It is also desirable, although not a requirement, that prospective ICAR members already have ongoing animal recording schemes. As a way of broadening its membership by increasing the participation of developing countries, ICAR instituted a Developing Countries Working Group which later became reconstituted into the Outreach Working Group. This Working Group presents a good platform for African countries to lobby or request for conditions that will promote their participation in ICAR, such as joint membership or discounted membership fees. Alternatively, African countries could form consortia or regional recording organizations that could apply for ICAR membership. Such an approach would reduce the cost of membership to individual countries.

29.3 Data Access and Benefit Sharing

Initially, the issue of data sharing is examined in terms of multi-country genetic evaluations. Subsequently, issues relating to the sharing of genetic materials are briefly discussed.

29.3.1 Sharing Data for Multi-Country Evaluations

The success of multi-country genetic evaluation relies on the consented agreement to pool data across the various countries involved. Therefore, the advantages of the systems must explicitly and implicitly be obvious to the participating countries, and well-defined protocol to ensure data security and confidentiality should be in place.

Outputs from the multi-country evaluations should reflect shared benefits to all the participating countries. These benefits include increased accuracy of genetic predictions resulting from the use of information across the countries, the ranking of animals on the basis of genetic merit tailored to the production environment of each country and, finally, identification of the most genetically superior animals available to all the countries.

Several models are available for issues relating to impartiality, data security, and confidentiality. For example, joint genetic evaluations are taking place for dairy cattle in the Scandinavian countries of Denmark, Sweden, and Finland. The system comprises of a separate unit that handles the evaluations, with each country undertaking research and implementation of the evaluation system for different traits. The separate unit ensures impartiality, with no concerns about data confidentiality as data flow freely among the countries. On a much wider scale are the multi-country evaluations undertaken by the International Bull Evaluation Services (INTERBULL) for dairy cattle, which involves about 35 countries. INTERBULL is a subcommittee of ICAR and has been involved in international genetic evaluations for dairy cattle since 1990. It is an independent unit which ensures impartiality in terms of the evaluation system. Sharing of phenotypic data among these countries, however, possesses a challenge in terms of security, confidentiality, and the huge size of the data sets. These challenges have been overcome by countries sharing only information on their bulls. Countries participating in INTERBULL multi-country genetic evaluations send the so-called de-regressed breeding values which represent the average daughter performances for these bulls in addition to the number of effective daughters. In addition, each country joining the multi-country evaluations of INTERBULL signs a code of practice (https://interbull.org/ib/cop_chap4) with INTERBULL. This code of practice outlines the responsibilities of each party and defines certain legal frameworks governing the use of the data submitted to INTERBULL for evaluation, research, or any other requirements.

Similar to the dairy cattle system, INTERBULL also undertakes across-country evaluations for beef cattle and the systems is referred as INTERBEEF. In contrast to the dairy system, participating countries send to INTERBEEF actual phenotypic records for the evaluation system. Again, INTERBEEF acts as an independent unit ensuring impartiality, security, and confidentiality. Similar to the dairy situation, each participating country signs the INTERBEEF code of practice, and in addition, ICAR, the mother organization of INTERBULL and INTERBEEF, clearly states it aims to protect personal data and commitment to privacy (https://www.icar.org/index.php/about-us-icar-facts/protection-of-personal-data-and-commitment-to-privacy/).

The situations examined so far involve the sharing and joint utilization of phenotypic and pedigree information. The picture is slightly different when it comes to sharing of genotypic data. For example, INTERBULL is unable to undertake full genotypic analysis for the Holstein breed, due to the unwillingness of participating countries to share bull genotypes. Countries only share the genomic breeding values and their associated effective number of daughters and INTERBULL undertakes the Genomic multi-country evaluations (GMACE). However, for the Brown Swiss breed, the participating countries were able to share genotypes and INTERBULL implements a genomic analysis referred to as Intergenomic (Durr and Philipsson 2012) involving genotypes and phenotypes. These activities are undergirded by the INTERBULL code of practice mentioned earlier, providing a legal framework for data sharing and utilization.

In Africa, national genetic evaluations are only conducted in a few countries such as South Africa, Kenya, Tanzania, and Ethiopia. For the latter two countries, this is through the African Dairy Genetic Gains Project (ADGG) (Mrode et al. 2021). Research on the feasibility of across-country evaluations was carried out recently (Opoola et al. 2020) and involved pooling together data from Kenya, South Africa, and Zimbabwe. Permission to use the data was obtained from the countries' respective organizations hosting the data.

Reviewing the existing models for multi-country genetic evaluation, initial steps that will promote data sharing for a joint genetic evaluation system in Africa will include: (i) encouraging countries to establish routine systems of data collection and national genetic evaluation; (ii) establishment of an independent body to undertake the joint evaluations to ensure impartiality, data security, and confidentiality; (iii) establishment of a technical committee with the know-how as part of the independent body; (iv) implementation of an efficient evaluation system that provides clear benefits to all participating countries in terms of increased accuracy, evaluations expressed on the country scale, and identification of top genetic merit animals across all countries; and (v) putting in place government systems to promote more liberal movement of animals or semen or embryos across African countries.

29.3.2 Sharing Genetic Material

The sharing of genetic material (Deoxyribonucleic acid (DNA) or samples with DNA) for the purposes of genomic characterization requires additional protocol for fairness and confidentiality as such analysis may generate data relevant for intellectual property claims. The general recommendation from FAO (Food and Agricultural Organization of the United Nations, www.fao.org) is that a formal agreement for acquisition and exchange of the material is made with the provider of the material prior to the start of the study. The standard approach is for all involved parties to sign a material transfer agreement (MTA), which stipulates the terms of the material exchange and describes how the material will be used and handled after characterization and analysis (FAO 2021). Such an MTA should provide a legal framework and certainty as to what can be done with the material and any data resulting from the study. Details of what such an MTA should include and sample MTAs are presented in the FAO (2021) document on "Genomic characterization of animal genetic resources—updated technical guidelines".

29.4 Core Facilities

29.4.1 What They Are and Their Importance

Efforts to advance livestock improvement in Africa will require multiple technologies/approaches, including genomics, proteomics, metabolomics, metagenomics, statistical/bioinformatics management of data, and involvement of multidisciplinary teams. New technologies, including genomics technologies (see Chap. 20), are expensive and their application requires significant know-how in following rigorous procedures while maintaining transparency, acquisition of the necessary infrastructure, equipment, and reagents, and the ability to keep pace with evolving processes and technologies. It will be futile for individual researchers, laboratories, or institutions to keep pace with the cost of acquisition/maintenance of expensive equipment and the ever-evolving technologies. This is particularly critical for developing countries which have less funds devoted to research and development. A smart solution would be the creation of core facilities through which resources, expertise, infrastructure, and quality services can be effectively managed for the benefit of all stakeholders.

A core facility or laboratory generally is a centralized technology-driven facility or laboratory that maintains and supports sophisticated equipment, integrates expertise, maintains highly qualified personnel, and builds networks to support optimum use in a cost-effective manner. Thus, consolidation of state-of-the-art technologies, equipment, and expertise under the umbrella of a core facility is a crucial environment for access to up-to-date technologies by all, irrespective of funding level, collaborative research, which is necessary for competitive science, and a necessary driver for breakthrough discoveries. Given the present life sciences ecosystem, core facilities are regarded as essential and an important route to providing cutting-edge technologies and expertise in a cost-effective or affordable manner (Meder et al. 2016; Gould 2015). Moreover, core facilities generate a significant and increasing portion of the research data produced by academic science and biomedical research facilities as well as serve as a central storehouse for knowledge and expertise in technology and its applications (Mische et al. 2020).

The concept of core facilities was necessitated by the fast-evolving nature of new technologies, the cost of the technologies, the expertise to maintain the technologies, and the evolving nature of research questions needing advance technologies. Starting in the early 2000s with the coming together of a large group of researchers to sequence the human genome and the need for advanced technologies to get the job done, a paradigm shift from individual-based to team-oriented research supported by diverse technologies started. Thus, several institutions in Europe and North America established core facilities which specialized in specific technologies or disciplines, such as genomics, proteomics, metabolomics, metagenomics, bioinformatics, etc. The main goals of core facilities are to (1) provide expected services using state-of-the-art technologies; (2) act as support units to research teams; (3) act as independent expert units open to all; (4) provide consultation services to users; (5) acquire and maintain state-of-the-art technologies and expertise to support research; (6) provide high quality services within reasonable time frames; (7) provide services to all interested users, both internal and external; and (8) provide cost-effective services based on a funding model which combines institutional and user fees (fee-for-service-based). For effective service delivery, institutions understand the importance of specializing in one or a limited number of technologies or disciplines, leading to the establishment of alliances to share expertise between institutions' core labs, country-wide core facilities, or across-country core facilities. Examples include (1) Life Sciences Complex Core Facilities at McGill University, Canada (https://www.mcgill.ca/lifesciencescomplex/facilities) which includes an Advanced Bioimaging Facility, an Animal Resources Centre, a Biology Platform, Cystic Fibrosis Translational Research Centre Platforms, a Flow Cytometry Core, a Histology Core, an Infection and Inflammation Core, a Mass Cytometry Core, a Metabolomics Core, Pharmacology and Therapeutics Platforms, a Structural Biology Centre, and a Transgenic Core. (2) Core For Life (https://coreforlife.eu/) established in 2012 as an Excellence Alliance of Core Facilities in Europe, bringing together about eight institutions with each having several core facilities. The member institutions for Core For Life are: The Centre for Genomic Regulation in Barcelona, Spain; EMBL (European Molecular Biology Laboratory) core facilities with headquarters in Heidelberg, Germany; The Center for Technological Resources and Research of Institut Pasteur, Paris, France; Max Planck Institute of Molecular Biology and Genetics, Dresden, Germany; Vienna Biocenter Core Facilities, Wien, Austria; Functional Genomics Center, Zurich, Switzerland; VIB Core Facilities in several locations in Belgium and CurieCoreTech Facilities of Institut Curie, Paris, France.

Core facility operational models include (Meder et al. 2016): (1) Service provider—provides services and training to users; (2) User laboratories—provides laboratory space to users as well as guide users in every stage of laboratory analysis and data management; and (3) All-inclusive service provider—provides services as well as laboratory spaces for users.

29.4.2 Genome Centres

Developments in the study of the genome were aided by advances in second- and third- (offer long read sequencing solutions) generation sequencing technologies, which over the last 20 years were accompanied by multiple generations of sequencing technologies and platforms, increased speed of data generation, and data quality. Although the genomic technologies have greatly impacted our appreciation of biology including disease and livestock production, the sheer volume of data outputted by these technologies requires a certain expertise to manage them. Consequently, developing standardized procedures, data formats, and comprehensive data quality management practices became a necessity (Endrullat et al. 2016; Hutchins et al. 2019). Moreover, the fast pace of the technologies necessitates adaptation by users, making it difficult and expensive for individual researchers, laboratories, or institutions to keep pace with the constantly evolving technologies. Therefore, Genome Core or centres (GC) were established by most institutions or countries to specialize in adapting genome technologies and in providing quality services to multiple users.

Most GC specialize in offering critical genomics-based technologies and applications in one or more genome subspecialties including DNA sequencing, next-generation sequencing (NGS), gene expression, genotyping, metagenomics, gene editing, etc. to their users (both academic and industry) and to establish, optimize, and share standard operating procedures and best practices of fast growing technologies with users and partner core facilities. For further illustration, genome and high-performance computing/data management centres are discussed in more details below.

29.4.2.1 Genome Centres: What They Offer

Most GC offer the following services (Fig. 29.1):

1. *GC house state-of-the-art equipment*: State-of-the-art equipment are used for performing genome-related applications like DNA sequencing, NGS, genotyping, protein sequencing, metabolome profiling, and sample processing services, among others. Considering NGS as an example, GC acquire and maintain state-of-the-art equipment from major providers like Illumina SOLiD, IonTorrent, PacBio, and Oxford Nanopore, etc. to support a variety of services, from sample preparation, amplicon sequencing to whole genome sequencing.
2. *Develop standard operating procedures to ensure quality and reproducible service at all times*: To render quality services, GC develop standard operating procedures (SOP) related to equipment handling and maintenance, sample handling, analysis workflows, data management, and laboratory management. SOPs are updated regularly and also adapted to suit specific environments as well as ease of understanding and adoption by personnel. According to Gargis Amy et al. (2016), NGS has contributed enormously to the enhancement of clinical diagnostics and public health surveillance efforts in the United State of America (USA). However, public health laboratories in the USA face a number of challenges when implementing NGS procedures such as: (1) difficulty in adapting to frequent updates to sequencing chemistries, platforms, softwares, and workflows; (2) difficulty setting up protocols and in troubleshooting them; (3) difficulty in implementing multistep processes with numerous potential sources of variation; (4) constantly evolving and significant personnel training requirements; and (5) difficulty in evaluating the quality of materials and data as sequencing progress through the NGS workflow (Kozyreva et al. 2017; Association of Public Health Laboratories

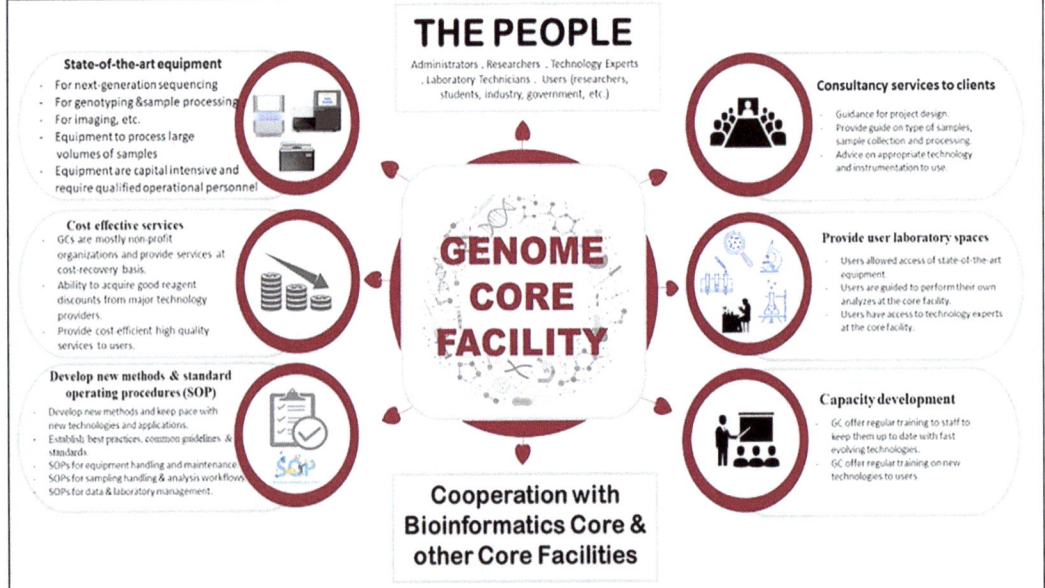

Fig. 29.1 Organization and core elements of a typical Genome Core facility

(APHL) 2016). To address these issues within Centers for Disease Control and Prevention (CDC) laboratories in the USA, a CDC NGS Quality Workgroup composed of laboratory scientists, quality managers, technicians, and bioinformaticians was formed in 2015, with the mandate to assess problems faced by member laboratories, develop and implement NGS, leverage best practices, and develop practical guidance or SOPs for laboratories conducting NGS within a quality management system (QMS) (Hutchins et al. 2019). Twelve quality system essentials (QSEs) identified by the Clinical and Laboratory Standards Institute (CLSI) as their QMS framework have been adopted by many laboratories. The 12 QSEs (facilities and safety, personnel, organization, equipment, information management, customer focus, process management, purchasing and inventory, documents and records, assessments, nonconforming events management, and continual improvement (https://clsi.org/standards-development/) (Hutchins et al. 2019)) were used by the CDC NGS Quality Workgroup as a reference point to develop a number of SOPs and information documents designed to provide a solid foundation for individual laboratories to use in developing assay-specific documentation by public health laboratories performing NGS (Table 29.1) (Hutchins et al. 2019).

3. *Qualified personnel*: Applications offered in GC are complex requiring deep understanding and commitment by a multidisciplinary staff to understand the processes and to perform them proficiently. For example, NGS equipment, reagents, and workflows are constantly evolving, requiring frequent training of personnel. Following installation of newly acquired equipment with the help of providers or with the arrival of each new process, staff have to commit time and be willing to gain new knowledge related to each technology and process. To implement NGS technology, basic understanding of system requirements is necessary and laboratory personnel must be proficient in maintaining and using them. To

Table 29.1 Documents and standard operating procedures

Document #	Title	Type	Quality system essential
1	Ion PGM sequencer competency assessment form	Form	Personnel
2	Ion PGM sequencer competency assessment SOP	SOP	Personnel
3	MiSeq competency assessment form	Form	Personnel
4	MiSeq competency assessment SOP	SOP	Personnel
5	Ion PGM sequencer training form	Form	Personnel
6	Ion PGM sequencer trainer designation form	Form	Personnel
7	Ion PGM sequencer training SOP	SOP	Personnel
8	MiSeq employee training form	Form	Personnel
9	MiSeq trainer designation form	Form	Personnel
10	MiSeq training SOP	SOP	Personnel
11	MinION training SOP	SOP	Personnel
12	MinION employee training form	Form	Personnel
13	Ion Chef preventive maintenance log	Log	Equipment
14	Ion Chef preventive maintenance SOP	SOP	Equipment
15	Ion OneTouch 2 preventive maintenance log	Log	Equipment
16	Ion OneTouch 2 preventive maintenance SOP	SOP	Equipment
17	Ion OneTouch ES preventive maintenance log	Log	Equipment
18	Ion OneTouch ES preventive maintenance SOP	SOP	Equipment

(continued)

Table 29.1 (continued)

Document #	Title	Type	Quality system essential
19	Ion PGM equipment error log	Log	Equipment
20	Ion PGM in-use equipment daily maintenance log	Log	Equipment
21	Ion PGM in-use equipment weekly maintenance log	Log	Equipment
22	Ion PGM power-off equipment maintenance log	Log	Equipment
23	Ion PGM preventive maintenance wash flowchart	Job aid	Equipment
24	Ion PGM sequencer preventive maintenance SOP	SOP	Equipment
25	MiSeq equipment error log	Log	Equipment
26	MiSeq in-use equipment maintenance log	Log	Equipment
27	MiSeq preventive maintenance SOP	SOP	Equipment
28	MiSeq preventive maintenance wash flowchart	Job aid	Equipment
29	MiSeq standby equipment maintenance log	Log	Equipment
30	Ion PGM system equipment preinstallation checklist	Job aid	Equipment
31	MiSeq equipment preinstallation checklist	Job aid	Equipment
32	Vendor-performed IQ/OQ cover sheet	Form	Equipment
33	Ion PGM sequencer software update evaluation SOP	SOP	Equipment
34	Ion PGM sequencer software update form	Form	Equipment
35	MiSeq software update evaluation SOP	SOP	Equipment
36	MiSeq software update form	Form	Equipment
37	NGS QC guidance for Illumina Workflows	SOP	Process Management
38	Bioinformatics QC workflows	SOP	Process Management
39	Sequencing QC SOP	SOP	Process Management
40	Preanalysis QC SOP	SOP	Process Management
41	Assembly QC SOP	SOP	Process Management
42	NGS QC guidance for MinION 1D workflows	SOP	Process Management
43	NGS QC guidance for MinION rapid sequencing workflows	SOP	Process Management
44	MGS methods validation SOP	SOP	Process Management
45	NGS methods validation plan template	Form	Process Management
46	NGS methods validation report template	Form	Process Management

Developed by the Center for Disease Control NGS quality Workshop (Hutchins et al. 2019)

ensure that staff perform proficiently and are well equipped to handle evolving technologies, GC develop training SOPs as well as offer regular training to staff and also to users of the platform. Therefore, to provide high quality services at all times, GC personnel must be highly qualified, possessing the necessary fundamental knowledge, and follow regular training processes.

4. *Consultancy services*: Most GC specialize in offering a number of services. To help users benefit optimally from their services, GC offer consultancy services to clients, mostly in the area of project design, sample type, sample collection/processing, and appropriate

applications for each project or research question. Thus, GC offer to clients the needed expertise, guidance from project design, and all critical aspects necessary for successful experimentation and generation of quality data.

5. *Provide cost-effective services*: Most GC are non-profit organizations obtaining most of their initial funding through government grants and donations. GC have large basin of users, granting them access to very good reagent discounts from major technology providers like Illumina, Pacific Biosciences, Qiagen, Agena, etc. Because GC are non-profit organizations, they apply the discounted products to the price of their services, which translates into competitive prices to the research community. It is quite difficult for individual researchers or university laboratories to reach such high-volume discounts from providers. Therefore, the services of GC are cost beneficial to users.

6. *Provide training*: GC offer regular training to staff and to users. Staff are trained to be proficient in handling constantly evolving technologies and processes. Staff are trained in process management which covers the lifespan of samples and associated processes from receipt of samples, processing, and data or output reporting. Users, on the other hand, are trained on the various applications offered by GC, sample requirement of each process, and expected output. GC, therefore, offer regular workshops and training sessions to staff and users.

29.4.2.2 Examples of Genome Centres and Mode of Operations in Various Countries

Canada Canada, by not participating in the *Escherichia coli*, *Saccharomyces cerevisiae*, or the human genome sequencing projects of the late 1990s and early 2000s, lagged behind in genomics research, and to address this, the federal government created Génome Canada (https://www.genomecanada.ca/) in year 2000 and six regional genome centres (Genome British Columbia, Genome Alberta, Genome Prairie, Ontario Genomics, Genome Quebec, and Genome Atlantic) starting in year 2002. The goals of Genome Canada are to (1) lead the vision in genome research in Canada and to speed up genome research through funding of large-scale genome research projects (to the tune of $10 M in 4 years); (2) support genomics researchers across Canada in the pursuit of their research objectives; and (3) support genomics researchers financially and technologically in all areas of the life sciences: human and animal health, agriculture, forestry, and environment. Genome Canada is funded by the federal government while the regional centres are funded by provincial governments. For example, Génome Québec (https://www.genomequebec.com/) is funded by the Ministère de l'Économie et de l'Innovation (MEI). Since its creation, Genome Quebec has invested and managed more than one billion dollars in genomics research in Québec. Its mission is to promote development and excellency in genomics research as well as its integration and effective utilization. Genome Quebec is the pillar of Quebec's bioeconomy, contributing to its social and sustainable development and influence.

Genome Québec is the only centre out of the six regional centres of Genome Canada to operate a technology platform: Centre d'expertise et de services Génome Québec (CES) (https://cesgq.com/). The CES offers high-quality genomic services to the scientific community, and it is a resource and networking platform for varied research activities in human health, infectious diseases, agriculture, forestry, and sustainable development. The CES offers a range of complete RNA and DNA services and processes from a few samples to thousands of samples per week. The range of services include (1) Genomic DNA and RNA extraction from a range of sample types (blood, buffy coat, hair saliva, bacteria, soil, leaves and cultured cells, etc.); (2) Next-generation sequencing, which includes library preparation and the following applications—Whole genome sequencing (WGS), Exome sequencing, Ribonucleic acid (RNA) sequencing (RNA-Seq), Single cell sequencing (scRNA-

Seq), Chromatin immunoprecipitation sequencing (ChIP-Seq), Whole genome bisulfite sequencing (WGBS), High-throughput amplicon sequencing with the Fluidigm Access ArrayTM System, and Amplicon sequencing. Its NGS services are based on the sequencing technologies of Illumina NovaSeq6000, Illumina MiSeq, and PacBio Sequel; (3) Sanger sequencing and single nucleotide polymorphism (SNP) discovery; (4) Genotyping (SNP genotyping and microsatellite analysis) services using TaqMan® genotyping, Agena® iPLEX® Gold®, Illumina® Infinium HD, and Affymetrix Axiom technologies; (5) Gene expression (RNA quality testing, gene expression arrays such as GeneChip or Clariom S, transcriptome arrays such as Clariom™ Assays, gene expression FFPE analysis, RNA and small RNA sequencing, scRNA-Seq) using ThermoFisher (Affymetrix) microarrays, Illumina NovaSeq6000, HiSeq4000, Illumina MiSeq, and PacBio Sequel II technologies; (6) Methylation (quantitative) analysis of genomic DNA using Agena® EpiTYPER®, Qiagen PyroMark® Q24, Illumina® Infinium Methylation, Methylation FFOE analysis, and Whole-genome bisulfite sequencing (WGBS) technologies; (7) Cytogenetics or copy number variation analysis services using OncoScan™ FFPE Assay, CytoScan, Illumina® Infinium HD, and TaqMan® CNV technologies; and (8) Bioinformatics services. These services are supported by solid infrastructure, state-of-the-art equipment, a web tool for sample submission and data transfer (Nanuq), and unique expertise in bioinformatics through close collaboration with the Canadian Centre for Computational Genomics (www.computationalgenomics.ca).

To support effective delivery of its services, the CES has a staff of 38 fulltime employees (directors, managers, technical personnel, informaticians/bioinformaticians, and a client management office) to support and provide access and services to a community that averages about 1000 research teams annually. Genome Quebec is a private non-profit organization and operates on a cost-recovery basis, meaning that the salaries of its employees and the price of reagents (sequencing, genotyping, etc.) are imbedded directly in the price of the services it provides to the community. Services provided by the CES generate approximately $15M annually. The operational costs of the CES are majorly covered by the services provided while 10% funding is provided by the provincial government (MEI). Other institutions across Canada like universities, hospitals, research institutions, etc. also operate core facilities in various disciplines.

United Kingdom (UK) The UK National Health Services (NHS) operates 13 Genomic Medicine Centres in 13 different locations in the Kingdom (www.england.nhs.uk/genomics/nhs-genomic-med-service/). The centres were established with main goal to support delivery of the 100,000 genomes project launched in December 2012 (Wheway et al. 2019) and to harness the power of genomic science and technologies to advance human health. Other major goals include: (1) offer whole genome sequencing as part of routine patient care; (2) transform healthcare for maximum patient benefit through knowledge generated from sequencing whole genomes of 100,000 people; (3) make accessible molecular diagnostics and routine genomic testing to patients with cancer and other conditions; (4) use genomic data to provide new treatments and diagnostic approaches; (5) develop and operate common national SOPs, specifications, and protocols; and (6) support industry and academic research and development through establishment of a national genomic knowledge base. Several institutions and universities also operate core facilities, providing services to users in the UK. For example, the Edinburgh Clinical Research Facility (www.ed.ac.uk/clinical-research-facility/core-services) operates scientific cores and specialist support services in the following areas: Nursing and Clinical; Imaging and Image Analysis; Genetics, Mass Spectrometry; Education, Epidemiology and Statistics; Information Technology and Research Support.

South Africa In South Africa, core facilities are linked to national institutions like the Agricultural

Research Council (ARC), the National Institute of Communicable Diseases, and universities. For example, the ARC operates a Biotechnology Platform (BTP) (www.arc.agric.za/Pages/BTP.aspx) which makes available high-throughput technologies and resources necessary for implementation of genomics, marker-assisted breeding, quantitative genetics, and bioinformatics know-how to the agricultural sector. The ARC-BTP is both a service provider and research-driven institution, providing an enabling environment for highly skilled personnel and users. The services offered by the platform are available to members of the ARC community, collaborators, industry, science councils, researchers, and students in South Africa. The services offered by the ACR-BTP include NGS, SNP genotyping, and high-throughput nucleic acid extraction. Another example is The African Centre for Gene Technologies (ACGT, https://acgt.co.za/) which is an initiative involving four institutions. Namely, the Council for Scientific and Industrial Research, the University of Pretoria, the University of Johannesburg, and the University of the Witwatersrand. The main goal of the ACGT is to create a collaborative network of excellence in advanced biotechnologies, with main focus on gene and genome analysis, and associated applications. In addition, individual universities also maintain platforms, but they are small in scope and mainly support the research needs of limited research groups. Presently, the genomic analysis needs of researchers at universities are outsourced to commercial companies (e.g. Inqaba biotech, https://inqababiotec.co.za/). Developing initiatives like The African BioGenome (AfricaBP) Project (https://www.earthbiogenome.org/affiliated-project-networks) may set the stage for the creation of a network of core laboratories across Africa to support successful delivery of the project.

29.4.2.3 Sequencing and Genotyping Platforms, Infrastructures, and Technologies

Sequencing Platforms Starting with developments that led to the first-generation sequencing technology, also known as Sanger sequencing in 1977, and successful sequencing of the genome of phage φX174 (Sanger et al. 1977a, b), second-generation and third-generation sequencing technologies and platforms have advanced our ability to sequence the genomes of living things at unprecedented depth and accuracy. Following the appearance of the first high-throughput platform in 2005, the Roche 454 pyrosequencer (Margulies et al. 2005), numerous NGS platforms have been developed by major providers like Illumina, Life Technologies, IonTorrent, Pacific Biosciences and Oxford Nanopore, etc., which perform a variety of services from amplicon sequencing to whole genome sequencing. The amount of data, accuracy, and reproducibility among the different platforms is influenced by many factors, including the intrinsic features of the platform and the desired analysis types and pipelines (Li et al. 2014; Shin and Park 2016; Mardis 2017). More details on available sequencing platforms, their uses, and attributes are presented in Chap. 20.

Genotyping Platforms Developments in the ability to sequence the genomes of animals and plants, and bioinformatics processing of data, led to the identification of a high number of sequence variations in the genome and knowledge of their association to traits or characters began to emerge. The sequence variations which include single nucleotide polymorphisms (SNP), insertions, deletions, copy number variations, inversions, and translocations have been used to produce genotyping arrays (microarrays) by major technology providers (Illumina, IonTorrent, etc.) and are frequently used to genotype (at a lower cost) and to document the genotype of individual animals in a population. Low-density (a few hundred markers) and high-density (thousands of markers) genotyping arrays are available for genotyping in livestock species (see Chap. 21 for details on the available livestock genotyping assays). Genotyping platforms acquire the necessary genotyping equipment and the genotyping assays in bulk and in turn offer genotyping services at cheaper rates to their clients.

29.4.3 High-Performance Computing/Data Management Centres

The information for making and implementing breeding decisions is extracted from different types of data including genotype, production, phenotype, and environmental variables among others. In particular, genomics and related areas like transcriptomics, proteomics, phenomics, metagenomics, metabolomics, etc., based on NGS and related technologies, are advancing our understanding of livestock health and production and providing the necessary information to advancing livestock health and productivity. To benefit from these technologies and data sources, the required strategies, skills, and resources must be put in place to support generation of genomics data, processing, and storage of that data. Next-generation sequencing of a few livestock genomes can generate terabytes of data requiring substantial high-performance computer resources for data processing and analysis and storage. Thus, information technology and high-performance computing (HPC)/data management capabilities are basic requirements for the effective management of big data. Therefore, it is necessary to put in place cutting-edge digital transformation technologies, infrastructure, and manpower that can cater to the compute challenges and large data presented by the NGS revolution for the advancement of livestock agriculture in Africa.

29.4.3.1 Available HPC/Data ManagementInfrastructure and Accessibility

High-performance computing is generally regarded as the ability to process big data and perform complex calculations at high speeds. HPC is achieved by using hardware and software to divide tasks into groups of discrete and independent computations allowing them to be processed in parallel and with seamless integration of output/results. A number of possible HPC solutions/areas that can be tailored to meet computational demands such as clusters, graphics processing units (GPUs), cloud computing platforms, and field-programmable gate arrays (FPGAs) (Wiewiórka et al. 2014; Lightbody et al. 2019; Mushtaq et al. 2015) are summarized in Fig. 29.2. These systems range from small numbers to several thousands of interconnected computers (compute nodes), and each approach differs in technology, performance, scalability, cost, and ease of implementation.

Information technology and HPC are capital-intensive requiring concerted efforts and pulling of resources by many institutions to acquire sufficient resources. To fulfill the conditions for effective data transformation and utilization, Stephens et al. (2015) indicated that access to adequate and affordable compute resources and efficient exploitation of that infrastructure through the use of algorithms that can reduce big data bottlenecks associated with genomics data and other data sources is necessary. This consequently led to unparalleled growth in adaptation and development of compute technologies and analytical workflows within the genomics community. Moreover, the need for adequate and easily accessible HPC propelled some institutions and national governments to set up HPC facilities. For example, to meet the computing and data management needs of Canadian researchers, the government of Canada established "compute Canada" (www.computecanada.ca), which delivers world-class and up-to-date advanced research HPC infrastructure/support to Canadian innovators and researchers. Compute Canada is composed of a nationally coordinated network of hardware and software resources, which provide a broad spectrum of advanced computing services in support of research, as well as technical support, training, application, and development. Compute Canada supports diverse disciplines such as aerospace engineering, agriculture, chemistry, computational criminology, medicine, as well as the analysis of works of fiction. The compute Canada network or federation is made up of 38 universities and four regional organizations (WestGrid, Calcu Quebec, Compute Ontario, and Atlantic Computational Excellence Network (ACENET)). Together, they play critical roles in supporting research computing needs in Canada, such as provision of essential advanced research computing infrastructure,

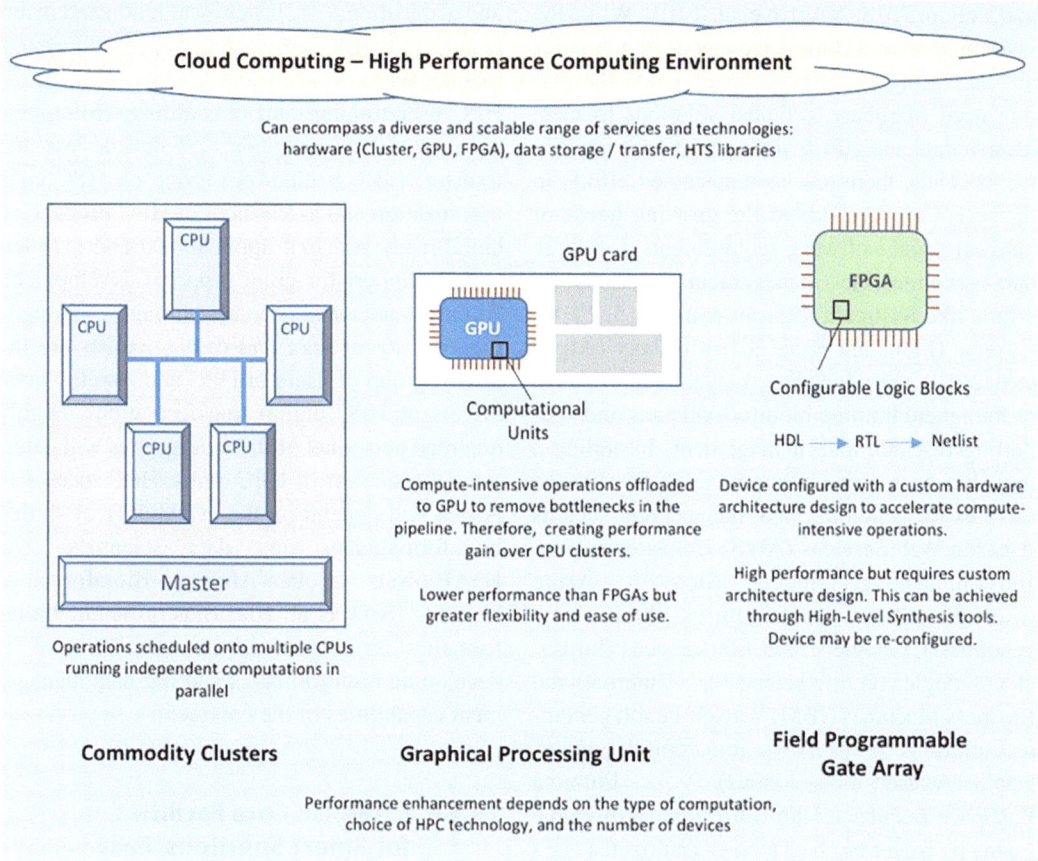

Fig. 29.2 Overview of high-performance computing options showing common clusters (e.g. GPUs, FPGAs, and cloud solutions) and highlighting differences in flexibility, performance, and options for custom design. CPU: Central Processing Units; GPU: Graphics Processing Unit; FPGA: Field Programmable Gate Arrays; HDL: Hardware Description Language; RTL: Register-Transfer Level. (From Lightbody et al. (2019), licensed under a CC BY 4:0)

software and services to supporting research needs across a wide-ranging spectrum of scope/needs, from small individual projects or initiatives or collaborations to large scale big science projects. The highly qualified staff (programmers, analysts, and computational science specialists) of Compute Canada offers a range of computational needs to researchers and users across Canada such as (1) HPC/big data and GPU computing and storage; (2) a variety of specialized softwares; (3) a portal (Globus) to share and transfer data in a fast and secure environment; (4) portals for specialized data and tools according to research disciplines; (5) ownCloud on dedicated managed servers offers data integrity and storage space across multiple devices; (6) platform to access and use genomic datasets (GenAP); (7) cloud environment development space; and (8) data storage and back-up systems with adequate stability and security. Many western countries (Australia, Germany, etc.) have set up national HPC facilities to support life science research among others (Tauch and Al-Dilaimi 2019; Schneider et al. 2019).

The growing need for adequate HPC/data management infrastructure has propelled the adaptation of cloud technologies in genomics and big data science with the need to algorithmically tackle vast amounts of data where volume, velocity, variety, and veracity are important (Langmead and Nellore 2018; Navarro et al. 2019). While the instrumentation technological needs of genomics are increasing at unprecedented rates, there is dire need of robust technical solutions to meet diverse data and HPC/data management requirements. Thus, there has been sustained efforts in the recent past to address the growing needs of hardware and software requirements for large data (genomics, etc.) management via large-scale efforts like Jetstream (Stewart et al. 2015, 2018), Cyverse (Merchant et al. 2016), Galaxy (Afgan et al. 2018), etc. which have played major roles in enabling and training bioinformaticians and supporting research data management. In addition, there has been a rise in commercial cloud-based services for genomics data management such as Amazon Web Services (AWS) (https://aws.amazon.com/health/genomics), Microsoft Azure (https://azure.microsoft.com/en-gb/services/genomics/), Google Cloud life-sciences (https://cloud.google.com/life-sciences), International Business Machines (IBM) Watson-health genomics solutions (https://www.ibm.com/uk-en/marketplace/watson-for-genomics), Illumina BasSpace Sequence Hub (https://emea.illumina.com/products/by-type/informatics products/basespace-sequence-hub.html), and 10x Genomics Cloud Analysis (https://www.10xgenomics.com/products/cloudanalysis). These providers accord customers or users the opportunity to innovate by making genomics data management more accessible and useful. In the same light, efforts to meet data management needs have seen the development of data analysis pipelines suitable for different data management applications for community use such as nf-core (Ewels et al. 2020) (https://nf-co.re/pipelines). A practical guide to processing and analyzing large-scale genomic data to extract meaningful information was provided recently (Tanjo et al. 2021).

A typical HPC core facility is made up of the computing hardware, skilled data management personnel (e.g. bioinformatician, computational biologist, data scientist, etc.), administrative personnel, specialized service units or centres, and the users (Fig. 29.3). The main functions of a typical HPC core facility are the following: (1) provide HPC services such as hardware design and acquisition; (2) cluster installation and maintenance; (3) HPC software and operating system maintenance; (4) data and system security; (5) HPC user training and onboarding; (6) storage services; (7) user support; (8) data access and transfer; (9) consultation service on HPC code optimization and assessment of HPC needs; (10) cloud-ready research applications; and (11) letters of support for grant proposal development. Bioinformaticians, computational biologist (genomics/genetics), and data scientists are the largest group of users of HPC and also the main drivers of HPC digital transformations. Highly qualified personnel of these categories will determine the success of HPC cores. HPC success in Africa will depend on the availability of skilled bioinformaticians and data scientists. The H3ABioNet, a Pan-African Bioinformatics Network (Aron et al. 2021a), is providing bioinformatics training in Africa with the goal of developing bioinformatics and big-data management capabilities of the continent.

29.4.4 Establish Core Facilities for Smart Solutions, Easy Access, and Shared Opportunities for Africa

As indicated in sections above, genomic technologies, HPC/data management infrastructure, and technologies offered by other core facilities not discussed above are capital-intensive requiring expensive equipment and technologies and highly qualified personnel to operate equipment and technical know-how in the application of the technologies. The implication is that it is futile for individual researchers or institutions to acquire the capital-intensive equipment, technologies, and the know-how to optimally benefit from these technologies. For present Africa to benefit from these advanced technologies, setting up GC, HPC/data management centres, and other core facilities at country or regional levels will leverage fund resources while providing cost-efficient high quality services through shared usage and funding. A schematic representation of a typical GC is shown in Fig. 29.2. Some

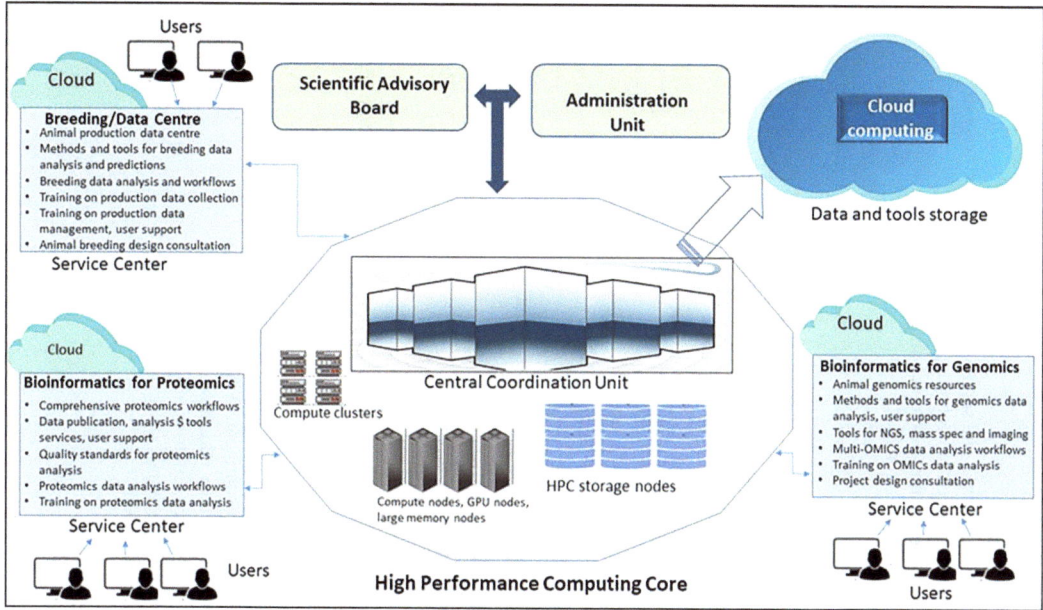

Fig. 29.3 Overview of a high-performance computing core with service centres in other locations

platforms providing genomic solutions in Africa are summarized in Table 29.2.

Therefore, establishment of a network of GC and HPC/data management centres, at least one in each region and similarly for other core facilities (metabolomics, proteomics, etc.), will be beneficial for stimulating and advancing quality research at shared cost across Africa, for the following reasons (Figs. 29.1, 29.2, and 29.3):

1. Exploits fast evolving technologies of modern biology and informatics to the fullest through pooling resources such as the necessary expertise to implement, run, improve, and adjust technologies to the needs of individual researchers, research groups, laboratories, institutions, and countries.
2. Pull resources to acquire and maintain necessary infrastructure and equipment as well as expensive technologies and make such available to all interested persons (researchers, students, laboratories, institutions and industry, etc.).
3. Pull resources to attract highly qualified personnel to operate core facilities and lead relevant technological advancements in their areas of expertise.
4. Fulfil the needs of today's research questions and projects requiring advanced technologies, large funding, and technical knowhow.
5. Fulfil the needs of interdisciplinary research teams by maintaining all the advanced technologies suitable for the needs of each project, which otherwise would be near impossible for individual laboratories, institutions, and individual countries to support.
6. Bring together an excellence alliance of highly qualified persons (core laboratories, universities, research institutions and industry, etc.) working together to solve common problems.
7. Harness and share expertise, resources, and technologies across laboratories, institutions, countries, and continents to speed up scientific discoveries and advance knowledge for the benefit of all.
8. Establish best practices, common guidelines and standards, and best operating procedures for sound science, and sharing of such with participating core facilities and users.
9. Enable optimum use of expensive technologies and equipment through shared usage.

Table 29.2 Some centres that provide genomic and high-performance computing services in Africa

Genome centre	Country	Services	Goal	Website
National Centre for Biotechnology Research (CRBt)	Algeria	Services are performed by a network of 17 laboratories: Animal biotechnology laboratory, Bacteriology and virology laboratory, Biochemistry laboratory, Bioremediation and ecotoxicology laboratory, Cell culture laboratory, Electrophoresis laboratory, Environmental analyses laboratory, Food testing and analysis laboratory, Genomics platform, GMO detection laboratory, Microscopy and genetic diagnosis platform, Molecular biology laboratory, Quality analysis laboratory, Molecular and macromolecular synthesis and characterization laboratory, Mycology laboratory, Plant in vitro culture laboratory, Immunology laboratory, Proteomics platform, and Bioinformatics unit	Coordinate research networks and support research activities in applied biotechnology in agriculture, aquaculture, animal health, human health, food, and the environment	https://www.crbt.dz/
Centre for Proteomic and Genomic Research (CPGR)	South Africa	Genomics Proteomics Transcriptomics Metabolomics	All-encompassing ecosystem that nurtures and drives Omics research and life science innovation in Africa to create a community of empowered scientists	https://www.cpgr.org.za/training/
South African Medical Research Council Genomics Centre	South Africa	Genomics Proteomics Transcriptomics Metagenomics	Capacity development in not only genomic sequencing but data science	https://www.samrc.ac.za/innovation/genomics-centre
Indalo Bio	South Africa	Genetic testing services Screening services	To bring advanced genetic services accessible to everyone on the African continent. With collaborations in the tertiary educational and research institutions, they aim to contribute greatly towards future genomic discoveries across sub-Saharan Africa.	https://www.indalobio.com/

Genome centre	Country	Services	Goal	Website
The African Centre for Gene Technologies (ACGT)	South Africa	Genomics and its subsets. Functional genomics (gene function and regulation). Structural genomics (structural analysis of gene and protein families). Proteomics (evaluation of all-expressed proteins). Transcriptomics (genes expressed through mRNA). Bioinformatics. Computational biology.	To create a collaborative network of excellence in advanced biotechnology, with specific focus on the "-omics". The ACGT addresses research priorities that focus on Africa's needs in health and agriculture and will derive value from Africa's genetic diversity.	https://acgt.co.za/about-us/
Ingaba biotech, Africa's genomic company	South Africa, Tanzania, Kenya, Nigeria, Senegal, and Ghana	MassARRAY genotyping Oligo synthesis Sanger sequencing Next-generation sequencing Bioinformatics Animal genetics qPCR analysis Fragment analysis Instrument services Workshops	To contribute to further develop science and technology in sub-Saharan Africa.	https://inqababiotec.co.za/
Biotechnology platform—Agricultural Research Council (ARC-BTP)	South Africa	Next-generation sequencing SNP genotyping High-throughput nucleic acid extraction Bioinformatics Proteomics Genomics Liquid handling robotics systems and services Laser capture micro-dissection microscope services	To create the high-throughput resources and technologies required for applications in genomics, quantitative genetics, marker-assisted breeding, and bioinformatics within the agricultural sector. The focus of the ARC-BTP is to establish itself as both a research and service-driven institution, providing an environment in which highly skilled researchers can be hosted and trained	http://www.arc.agric.za/pages/btp.aspx

(continued)

Table 29.2 (continued)

Genome centre	Country	Services	Goal	Website
African Biosciences Ltd (ABL)	Nigeria	DNA testing Agriculture and food DNA testing Next-generation DNA sequencing Sanger DNA sequencing Lab setup and management Molecular diagnostics PCR primers	ABL provides life science research services, such as reagents supply, technical training, research consultancy, molecular diagnostic/DNA testing, lab-for-rent, and contract research services to African scientists and students.	https://www.africanbio.com/
The Council for Scientific and Industrial Research (CSIR) centre for high-performance computing (CHPC)	South Africa	Introduction to Linux, scripting, and Python for scientists Parallel programming Users' induction course	The CHPC seeks to advance scientific boundaries and foster innovation through effective partnership and through the training of a new generation of computationally skilled researchers in areas underpinned by high-end computing, particularly those of national and continental strategic importance, to the benefit of basic and applied research in the public and private sectors	https://www.chpc.ac.za/
Laboratory of Applied Biotechnology in Agriculture	Tunisia	Use molecular tools to promote biological control of pests and disease. Use in vitro and molecular markers to the tolerance of cereals to abiotic stress	Develop biotechnological processes to promote biological control of pests and diseases. Develop marker-assisted selection procedures to breed abiotic stress-resistant cereal crops, among others.	http://www.inrat.agrinet.tn/index.php/en/research/laboratory-applied-biotechnology-in-agriculture.html

10. Ability to acquire large volumes of materials at cost-effective basis and to provide services at a cost-effective rate to users.
11. Accord affordability of advanced technologies to individual researchers and students.
12. Enable joint scouting, validation, and application of emerging technologies.
13. Enable establishment of a training network of personnel across core facilities.
14. Enable shared usage of equipment, capacity as well as technology investments across core facilities and countries.
15. Lower cost of establishing and maintaining core facilities through joint funding by participating countries or bodies.

29.4.4.1 Case Study 1: Biosciences Eastern and Central Africa-International Livestock Research Institute (BECA-ILRI) Hub

The BECA-ILRI Hub (https://hub.africabiosciences.org/), established as part of the AU/NEPAD (AfricanUnion/New Partnership for Africa's Development) and ABI (African Biosciences Initiative), is a shared biosciences and agricultural research platform. Headquartered in ILRI, Nairobi, Kenya, the BECA-ILRI platform provides access to state-of-the-art laboratories to African and international scientists conducting research on all aspects of African agricultural challenges. The main goal of BECA-ILRI hub is to mobilize bioscience research for Africa's development through supporting research, capacity building, and product development by African scientists and by empowering institutions in the Central/Eastern African region to leverage the power of advanced and emerging innovations for improved agricultural productivity, food/nutritional security, and incomes. In support of its goal of training, the BECA-ILRI hub offers regular training activities and workshops to researchers and students, such as training/workshops on Breeding Management System, Introduction to Molecular Biology and Bioinformatics, Animal Quantitative Genetics and Genomics Training, Basic Molecular Biology and Genomics Training, Introduction Workshop and Advanced Genomics and Bioinformatics, etc. BECA operates regional nodes in universities in Cameroon (University of Buea), Ethiopia (Ethiopian Institute of Agricultural Research), Uganda (National Agricultural Research Organization), and Kenya (University of Nairobi and Jomo Kenyatta University of Agriculture and Technology). The BECA-ILRI hub is funded by financial contributions from the Canadian Government through the Canadian International Development Agency (CIDA), AU-NEPAD, ILRI, various governments, foundations, and through various partnerships such as the Bill and Melinda Gates Foundation, Syngenta Foundation for Sustainable Agriculture, Swedish Funded Partnership on Food Security and Climate Change Adaptation, the UK Government through the Department for International Development (DFID), the United States Defense Threat Reduction Agency, the United Kingdom's Biotechnology and Biological Sciences Research Council, and the Alliance For Accelerated Crop Improvement in Africa and the European Commission—Horizon 2020.

29.4.4.2 Case Study 2: Sustainable Bioinformatics Network for Africa (H3ABioNet)

The Human Heredity and Health in Africa (H3Africa) consortium created a Pan African Bioinformatics Network (H3ABioNet) comprising 28 Nodes which are distributed among 17 countries, in which 16 are African countries (The-H3Africa-Consortium 2014). The consortium was established as an initiative to support genomics research through building bioinformatics capacity in Africa. It was developed through funding from the United States' National Institutes of Health Common Fund and the Welcome trust. The specific aims for H3ABioNet include (1) implementation of a Pan African informatics infrastructure; (2) development of an H3Africa data coordinating Centre; (3) make available high-quality informatics support to H3Africa; (4) enable enhanced innovative translational research; and (5) address outreach, development, and sustainability.

This consortium was established to build capacity in bioinformatics skills in Africa. The need was borne out of the lack of expertise in bioinformatics skills in African research institutions resulting in poor bioinformatics research capacity in Africa. Thus, the H3ABioNet provides training and opportunities that ensure the content and quality of bioinformatics capacity of African researchers. Some of the notable challenges affecting the process of delivering bioinformatics training in Africa are due to inadequate infrastructure, unstable internet connectivity, limited access to training facilities, and lack of human expertise. The H3ABioNet overcomes these challenges by making use of multi-faceted approaches through distance, face-to-face courses at various levels, webinars, hackathons, internships to mixed-model online training (Aron et al. 2021b). The learning process includes theoretical and practical sessions to allow participants to gain experience in using various tools and resources applied in bioinformatics. For example, internship programmes are offered to facilitate one-on-one interaction with experts in bioinformatics. During internship training, interns are encouraged to: (1) work on their own data; (2) improve skills and professional development; (3) perform specific research activities; (4) attend courses at the host institution; (5) gain exposure to human genetics and/or bioinformatics/high-performance computing environments; (6) gain experience in grants/project development; (7) develop educational materials, and (8) establish/further existing collaborations.

29.4.4.3 Case Study 3: Open Institute of the African BioGenome Project

A recent initiative named The African BioGenome Project (AfricaBP, https://africanbiogenome.org/) which is a coordinated pan-African effort to build capacity (and infrastructure) and to generate, analyse, and deploy genomics data for the improvement and sustainable use of biodiversity and agriculture across Africa seeks to sequence the genomes of 105,000 endemic animals, plants, fungi, protists, and other eukaryotes species on the African continent with the goal to make available genome information to facilitate the development of resilient animal and plant species for sustainability of African food systems (Ebenezer et al. 2022). The project is a continent-wide initiative which involves about 200 African scientists with expertise in genomics, molecular biology, bioinformatics, mathematics, etc. and over 50 institutions in Africa and out of Africa. This initiative births the creation of an Open Institute to facilitate the genomics and bioinformatics knowledge exchange programs for the AfricaBP (Sharaf et al. 2023). To achieve this aim, the AfricaBP has developed the 2030 Strategic Plan that builds on the position paper published in Nature in 2022 titled—Africa: Sequence 100,000 species to safeguard biodiversity (Ebenezer et al. 2022). The strategic plan is underpinned by AfricaBP's unique value proposition and strategic position as an enabler of genomics practices, integrator of discovery science, applications to deliver crop, fungi, animal, and livestock genetic innovations across Africa. With this strategic plan, AfricaBP aims to mobilize its key partners (i.e. Technology Innovation Agency (TIA), University of South Africa (UNISA), Agricultural Research Council (ARC), Centre for Proteomic and Genomic Research (CPGR), South African Medical Research Council (SAMRC), University of Cape Town (UCT), University of Port Harcourt (UniPort), and Mohammed V University). Under this case for support, AfricaBP and UNISA will focus on research, capacity building, and knowledge exchange to leverage strategic partnerships to contribute to more resilient, productive, efficient, and environmentally sustainable tropical livestock production systems, while Science for Africa (SFA) Foundation will focus on coordination and resource mobilization.

29.4.5 Funding of Core Facilities in Africa

Core facilities play a central role in facilitating research and the speed of discovery. Funding of research is capital-intensive and genomic, bioinformatics, and HPC services and supplies are even more expensive. Therefore, establishment of GC, HPC/data management centres, and other core facilities is for the public good and the initial start-up capital should come from public funds. Thereafter, core facilities can run on a cost-recovery basis whereby the cost of maintaining the facilities can be factored into the cost of the services provided. Core facilities can also be funded through grants (international development partners), endowment funds, and donations.

29.5 Germplasm Centres

Germplasms are genetic materials of germ cells, such as seeds, pollen, rootstock, semen, or tissues that are conserved for animal and plant breeding for the purpose of preservation and other research purposes. These resource materials could take the form of seed collections stored in seed banks, trees growing in nurseries, animal lines maintained in breeding programs, and genebanks. Therefore, the germplasm centres (i.e. genebanks) are where these genetic materials are stored under specified regulations for utilization in breeding and research.

Germplasm centres were set up in 2012 via FAO guidelines to ensure the cryo-conservation of Animal Genetic Resources (AnGR). The main aim is to cover fundamental issues involved in development and operation of genebanks as part of national strategies for the management of animal genetic resources for food and agriculture. The essence of germplasm centres is to ensure the goal of conservation and preservation and generally emphasize the goal of conservation to reconstitute a population that has gone extinct in vivo.

The establishment of germplasm centres is essential for Africa's food security and strategic contributions to livelihood and income of its growing population. The FAO's Global Plan of Action (GPA) for AnGR has recognized the relevance of gene banking in the conservation of AnGR. Therefore, governments and the international community have agreed on concerted and collaborative measures to ensure implementation and sustainability of genebanks at both regional and continental levels.

29.5.1 Opportunities for Setting Up Regional Biological Banks

Several initiatives have been taken up to ensure the conservation of both foreign and indigenous breeds currently in Africa. For instance, Taurus Jersey and Jersey South Africa (Nyamushamba et al. 2017) (www.jerseysa.co.za) ensures the dissemination of Jersey germplasm to countries within and outside the Southern Africa regions. The Nguni programme of South Africa was established to preserve the Nguni breed in the Eastern Province (Bester et al. 2003; Gwala 2013). This was in response to uncontrolled mating procedures reported among communal and small-scale commercial areas which threaten the existence of the indigenous breed (Obert 2012). The Inter African Bureau for Animal Resources with funding from the European Union has set up five regional animal genebanks across Africa (Fig. 29.4). However, the animal genebanks are yet to be operational due to non-availability of legal framework regulating genetic material transfer, conservation, and uses across the participating countries.

The feasibility of multi-country collaborations to support the conservation of genetic resources in Africa has been reported (Morandini et al. 1977; Rischkowsky and Pilling 2007). However, the strengthening of such collaboration through data sharing, germplasms, and capacity building is plausible and rather challenging. Most countries in Africa lack national genebanks majorly due to intricacy in governance and complexity in national policies that inhibit the establishment and sustainable operations of national genebanks. Therefore, in an attempt to mitigate these complexities in African countries, the African Union Inter-African Bureau of Animal Resources (African Union Secretariat 2020) has been able to establish five sub-regional genebanks in strategic locations in countries throughout the continent funded by an European Union project. These locations have regional representations which include Eastern Africa (National Animal Genetic Resource Centre and Data Bank (NAGRIC & DB) Entebbe, Uganda), Southern Africa (Department of Agricultural Research (DAR),

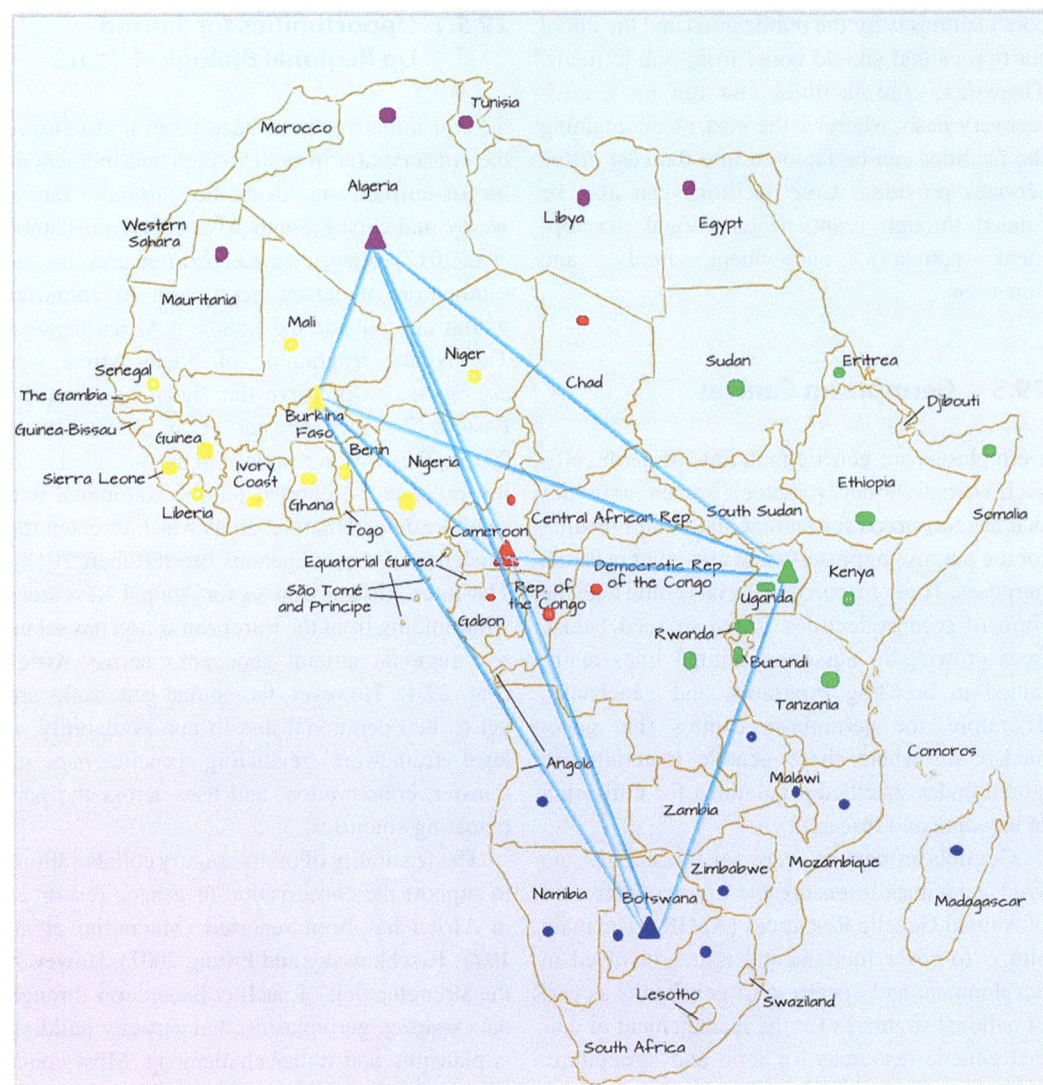

Fig. 29.4 Mapping the five African Regional Animal Genebanks. African regional Genebanks (large triangles- pink, yellow, red, blue, and green colours) showing inter- connections between them (blue lines) and countries served by each Genebank (small circles in the colours of their respective regional Genebanks)

Gaborone in Botswana), Western Africa (CIRDES (International Centre for Research and Development on Livestock in Sub-Humid Areas) in Bobo-Dioulasso, Burkina Faso), and Central Africa (Centre Régional de Recherche Agricole de Bambui-Mankon, Bambui, Cameroun). For Northern Africa, Algeria was previously selected as the potential animal germplasm centre. However, recent communication has shown that the current germplasm centre for Northern Africa is Tunisia (Banque de gènes Tunisienne). In addition, the African Union Pan African Veterinary Vaccine Centre (AU-PANVAC) is mandated to support the regional genebanks in Africa through holding regional germplasm samples and the preservation of germplasm copies available in each regional genebank to prevent accidental losses. Figure 29.4 shows the five (5) African regional genebanks with countries linked to each of the genebanks.

29.5.2 Available Infrastructure and Accessibility

The introduction of foreign germplasm has played a key role in improving desirable traits of interest for most livestock breeds. These efforts have been made through the introduction of assisted reproductive technologies (ART) to prevent undefined use of other reproductive methods such as natural mating. The ART mainly practiced in Africa include artificial insemination and embryo transfer techniques using foreign germplasm. Therefore, various breeding policies, breeder's organizations, breeding companies, and germplasm centres have been set up in some countries to monitor utilization of germplasms and ensure efficient performance recording and genetic and genomic evaluations in Africa. These include Kenya Stud Book and Kenya Livestock Breeders Association of Kenya, The BreedPlan of South Africa (Marle-Koster and Webb 2014), Limousin Cattle Breeders Society of South Africa, etc. Approaches to strengthen capacity in animal breeding in Africa have been developed. These include, for example, the Africa Bioscience Challenge Fund Programme developed by the Bioscience Eastern and Central Africa (BecA-ILRI Hub) and The African Animal Breeding Network (AABNet) (http://animalbreeding-africa.org/) (Djikeng et al. 2024). The BecA-ILRI Hub, through the ABCF programme, has trained for years students and research professionals in various aspects of animal breeding including genomics, bioinformatics, and quantitative genetics. AABNet has been recently established and hosted by the Egerton University in Kenya. Establishing robust systems for livestock genetic evaluation across multiple countries in Africa is the flagship pillar and a long-term and strategic objective of AABNet. Through the AABNet platform, animal breeders and livestock development professionals will access methodologies, tools, and technical support for multi-countries' collaboration in livestock genetic evaluation in Africa. The AABNet, through its inaugural training being organized, will support educators and students in Africa by strengthening their skills in animal quantitative genetics and genomics.

29.5.3 Dissemination of Superior Germplasm

After identification and selection of superior animal germplasms, the next step is the multiplication and dissemination of the improved strains within the country and from one country to another. The transfer of genetically improved strains will require a material transfer agreement as well as the access and benefits sharing protocol between organizations and countries involved.

29.5.4 Funding of Germplasm Centres

The value of traditional local livestock breeds to rural families in African communities needs to be acknowledged, monitored, and assisted for increasing productivity. Despite the availability of crossbred livestock in Africa, the need to conserve and maintain the indigenous local livestock breeds and their genetic resources in Africa cannot be overemphasized. The reliance on foreign commercially improved germplasm hinders efforts to develop independent breeds that could compete at international levels and threatens the long-term sustainability of local smallholder production systems and the potential to adapt to global climatic changes. Therefore, multiple well-funded and specific goal-oriented programmes are needed to characterize and preserve traditional breeds at national and regional levels.

29.6 Multi-Country Genetic Evaluations

Unlike non-ruminants, breeding activities in ruminant livestock such as dairy, beef cattle, sheep, and goats involve many individual farmers and several breeding companies. In addition, developments in reproductive technologies such as artificial insemination and semen and embryo freezing have led to widespread usage of breeds in several countries (see Chap. 20). Moreover, transboundary breeds such as the Boran imply these animals are reared in several countries.

The widespread usage of breed across countries has necessitated the developments of methods for fair comparisons of breeds across several countries. In developed countries and particularly in dairy cattle breeding, considerable research has resulted in the development of procedures for across-country genetic evaluations of bulls under the guidance of INTERBULL as mentioned in Sect. 29.3.

The current methodology used for multi-trait across-country evaluation (MACE) by INTERBULL for dairy cattle is a multi-trait Best Linear Unbiased Prediction (BLUP) procedure (Schaeffer 1994). MACE regards records on the same trait from different countries as different traits. This implies that good genetic links are required among the participating countries to estimate genetic correlations among the countries for the trait being evaluated. Usually these genetic correlations tend to be high with values varying from 0.8 to 0.95 for production traits such as milk yield but lower for reproductive and fertility traits (Simm et al. 2021). The use of different values of genetic correlations in MACE between traits in different countries results in different rankings of evaluated sires among countries. This multi-trait approach allows MACE to account for differences among countries in terms of the heritability of the trait and differences in scale used to express or measure the traits in the different countries. It also accounts for genotype-by-environment interaction, for example, milk yield in different production environments such as intensive versus grazing management systems. This highlights one of the advantages of MACE as it produces separate estimates of genetic merit for sires for each country involved in the evaluation which are expressed on the local scale and are therefore ready to be used. Second, the use of daughter information across several countries implies that MACE produces more accurate genetic evaluations for bulls which can be used across countries.

Although much of the research on across-country genetic evaluations have been developed in the evaluation of dairy sires, a similar service was recently introduced for beef cattle and it is referred to as INTERBEEF (Phocas et al. 2006; Bonifazi et al. 2020a, b). It is based on the same methodology as MACE but uses actual phenotypic records from the several countries.

In summary, some of the important factors for the implementation of across-country evaluation include (i) routine data collection systems, (ii) existence of within-country national evaluation systems validated to be efficient and accurate, and (iii) existence of genetic links among the countries through exchange of genetic materials (semen, embryos, etc.) or use of related animals.

29.6.1 Prospects for Across-Country Genetic Evaluations

The multi-country approach to genetic evaluation provides a good opportunity for transboundary breeds such as the Ankole or Boran beef breeds in Africa to be evaluated. However, in the dairy cattle situation, a lot of cross-breeding of indigenous breeds with important European dairy breeds such as the Holsteins, Friesians, Ayrshires, Jerseys, and Guernsey have taken place in many African countries. It is possible that the same bulls or related bulls have been utilized in these cross-breeding efforts in many African countries and this therefore provides an opportunity to investigate and implement across-country evaluations for dairy production traits. The critical question to address is whether there is adequate data and genetic links to enable an across-country evaluation in the Africa setting. If adequate and sufficient links exist, procedure such as MACE or an approach similar to that in INTERBEEF can be implemented.

However, when genetic links are inadequate for the estimation of the genetic correlations among countries, a simplified across-country evaluation that assumes a genetic correlation of unity can be implemented at the initial step. The disadvantage of this approach is that only a single ranking of bulls is produced for all countries involved in the system and it is unable to account for genotype by environment interaction. However, such pool of data across countries will result in more accurate genetic merits for animals which are related across several countries. Opoola et al. (2020) utilized the across-country approach assuming a genetic correlation of unity to examine several production and reproductive traits in the Holstein and Jerseys cattle breeds in South Africa, Kenya, and Zimbabwe. Results showed that the genetic variance and heritability were not always estimable within-country, but were significantly different from zero in the across-country evaluation. In all three countries, there was greater genetic gains accrued in all traits from an across-country genetic evaluation due to greater accuracy of selection compared to within-country. Collated data showed that Kenya stood to benefit most followed by Zimbabwe then South Africa from an across-country evaluation. If genetic links exist across countries, an across-country breeding programme using joint genetic evaluation is feasible and would provide a platform for accelerated genetic progress through selection and germplasm exchange between sub-Sahara African countries.

29.6.2 Genetic Evaluations Funding

Across-country evaluations are usually funded by the participating countries. For instance, in the case of MACE for dairy cattle run by INTERBULL, participating countries are charged a fee based on the number of registered cows by ICAR. In Africa, where well-established systems are lacking, initial funding could be based on the size of the data submitted for the joint evaluations by each country. In addition, the availability of data in the centre undertaking the joint evaluations implies that the centre can support research activities for post-graduate studies in several universities under well-defined protocols that do not comprise data confidentiality and security.

29.7 Perspectives and Practical Application of Across-Country Genetic Evaluation in the African Context

Practical application of across-country genetic evaluation in Africa will depend on the fairness cooperation between countries involved. This will require governments to play a role in providing adequate tools for African scientists/researchers to generate, store, analyse, and interpret genomic data by African researchers for the benefit of the African continent. Mulder et al. (2017) also emphasized that the most critical point in establishing cross-country cooperation in the African context is to create an environment of fairness and equity in data sharing. This is an important and critical factor that needs to be well negotiated between countries involved. In this context, it is of pivotal importance that adequate survey literature is available to help make informed decisions.

29.7.1 Surveys and Literature

There are few available literatures on the implantation of genomics programme for across-country genetic evaluation from the African context. This might be explained by the element that the genomics technology is still relatively new and not well implemented across Africa. There is no doubt that implementation of across-country genetic evaluation will see greater results in advancing livestock production, especially in diverse African indigenous breeds. However, such approach will require adequate planning and efforts from different bodies and disciplines. van Marle-Köster et al. (2015) emphasized that livestock scientists and their co-workers from developing countries such as Southern African Development Community (SADC) will need to integrate genomics, quantitative breeding, nutrition, physiology, and management with a careful consideration of the production system and the socioeconomic environment to ensure food security. They (van Marle-Köster et al. 2015) further suggested that the use of genomic tools will be of importance for traits associated with adaptation and disease resistance and unique product traits. This will be realized with establishment of effective cross-country genetic evaluation initiatives to increase accuracy and statistical power. The immediacy of initiating this cooperation will be driven by the need to initiate genomic programs in developing countries. Africa is largely represented by smallholder farming which can be challenging to implement genomic evaluation programmes. Muchadeyi and team (Muchadeyi et al. 2020) also highlighted difficulties associated with setting up transcriptome experiments in largely uncontrolled smallholder farming systems of developing countries. There is still a need for surveys and literature on this subject from the African context. This will be useful in making informed decisions and sourcing of funds to support the setting up of cross-country genetic evaluation programmes/initiatives.

29.7.2 Case Studies

To date, there are few case studies on across-country genetic evaluations from the African context. There has been an interest for across-country studies using genomics across different fields which can provide an important baseline to establish across-country genetic evaluation programmes. A study by Olivieri et al. (2021) investigated the complexity of ticks' microbial communities from livestock, wildlife, and pets across three countries (Egypt, Kenya, and Ethiopia). In this study, they employed metagenomics, genomics, and ecology to demonstrate the dynamics, with possible important consequences on human and animal health, economy, and on the preservation of endangered species for an in-depth analysis of the interactions among the tick-borne microorganisms. Conversely, in a classic study on multi-country genetic evaluation, Cardoso et al. (2021) pooled tick resistance phenotype reference populations from three countries (Brazil, South Africa, and Australia) from three continents. The team then jointly analysed existing tick infestation datasets to assess the possibility of improving host resistance in cattle through multi-population, multi-trait genomic selection (Cardoso et al. 2021). They indicated that a joint genomic evaluation of Angus, Hereford, Brangus, Braford, and Brahman using multivariate genomic best linear unbiased prediction (BLUP) can be readily implemented to improve tick resistance. These studies have proven and provided baseline on the potential benefits and scientific power that can be realized with implementation of across-country programmes such as multi-country genetic evaluation programmes.

29.7.3 Information Dissemination Across Countries

Information dissemination is one of the critical aspects that should be handled with care and professionalism. In addition to Sect. 29.3.1 above, this will require establishment of MTA to ensure fairness and confidentiality in the use of the shared materials between associated countries. This can be realized with involvement and commitments from governments and associated institutions.

29.7.3.1 Facilitate Access to Training and Research Facilities

The experience in recent decades in Africa is that the development of the sciences of animal breeding and genetics, and in particular genomic breeding and genetic evaluation programmes, is limited due to the shortage of professionally trained capacity. Although a number of cooperative graduate training programmes in animal genetics and breeding have been established globally, Africa still has limited expertise in genetic evaluations and in statistical/bioinformatics management of large data. Moreover, there are a limited number of established genomic centers necessary to generate the needed genomic data, and these pose challenges in initiating training facilities across the continent. These essential skills are required to achieve the ability to develop proper genetic evaluation programmes, theoretical and statistical frameworks for the analysis of large datasets suitable for cross-countries analysis, and these can only be achieved through multiple strategic partnerships with major breeding companies, research institutions, government ministries, and other industry partners. Building international professional networks among research and development staff, who have been working at a small scale in different African countries, is of vital importance. However, in some countries, the major barriers are stringent livestock breeding policies and practices, which are set by following government policies. These problems have declined in some countries as governments have reduced their activities and control of the livestock breeding practice. Currently, the implementation of livestock improvement programmes depends on the availability of appropriately trained technical staff. This applies in the areas of animal management, data management, and in the management of the infrastructure required for a modern breeding scheme.

29.7.3.2 Available Infrastructure for Knowledge Dissemination/Training and Accessibility

Currently, there are few research institutions that provide training in genetic evaluation and genomics services in Africa. However, the sharing of resources between African countries has the potential to unlock the challenges of training facilities. Recently, the Africa Biosciences Challenge Fund Bioinformatics Community of Practice (BixCoP) launched an 8-month residential training course which gave 14 early-career scientists from eight African countries the chance to gain and apply critical skills in bioinformatics data analysis, including programming (Linux and shell scripting/Python/R), NGS analyses, quantitative genetics, and phylogenetics. This training programme aimed at equipping African scientists with the skills to deploy third-generation genomics, particularly for Agri biosciences. This is one of the cross-countries initiatives that can be utilized to build training capacity in Africa. In the same context, in recent years, Africa has seen development of few genomics companies (Table 29.2) established, mostly based in South Africa. Companies such as Inqaba biotech - Africa's genomic company, Biotechnology platform, Agricultural Research Council (ARC-BTP), The African Centre for Gene Technologies (ACGT), Centre for Proteomic and Genomic Research (CPGR) and South African Medical Research Council Genomics Centre, as well as African Biosciences in Nigeria. These companies provide a wide range of genomic services to many countries in Africa (Table 29.2). Most of these companies have played an integral part in advancing genetic services and collaborations with tertiary educational and research institutes.

29.7.3.3 Formal and Informal Qualifications

As discussed above (Sect. 29.7), the limitation of expertise in key research areas such as genetic evaluation, bioinformatics, and adequate genomic facilities has been a major challenge in Africa. There have been initiatives by universities such as University of Cape Town and Stellenbosch University, etc. to provide short courses in bioinformatics in an effort to build capacity. This proved to be effective only in providing basic skills to researchers and academics. There is a greater need to establish and enhance this effort at a larger scale to build effective capacity. Conversely, on the same effort, there are companies that offer informal and formal training such as Centre for Proteomic and Genomic Research (CPGR) (www.cpgr.org.za/) outside universities that provide training and qualifications in OMICs technologies in South Africa. Despite these efforts, Africa has seen a slow growth in experts with notable bioinformatics and genomic skills and establishment of genomic centres. The slow and lack of interest into these fields may be driven by inadequate infrastructure and low support structure. Today, we have very few institutions dedicated to genomics and bioinformatics in Africa. This, in part, may explain the lack of experts in genomics and bioinformatics to enforce cross-country genomic consortium such as across-country genetic evaluations. For one to develop genetic evaluation skills, it is recommended to have a fair background on mathematics or statistics.

29.7.4 Perspectives, Practical Application, and Opportunities for Regional/Continental Training Centres

Advances in computational speed and analytical processes have resulted in remarkable progress in livestock breeding. This progress has made it possible to envisage the idea of cross-country collaboration to pull together pedigree and performance data for the purpose of cross-country genetic evaluations to achieve a better production rate. These efforts have proven to be generating better results in developed countries. To see this effort to fruition in our continent, Africa needs to establish a more vigorous initiative in terms of knowledge sharing and training capacity between researchers and academics across the continent. This can only be realized with efforts and commitments from governments, private institutions, and educational and research institutions. Institutions such as Agricultural Research Council in South Africa and Research institution of Zimbabwe had such agreement previously, which saw exchange of scientists to share research skills in genetic evaluation. Such agreements are necessary to build capacity and enforce collaboration. These efforts need to be escalated to a larger scale for it to be effective. Other funding institutions/bodies such as Erasmus and National Research Foundation of South Africa have funding programmes to motivate such initiatives and to encourage sharing of research skills and establishment of collaborations.

29.7.4.1 Case Study 1: Capacity Building and Appropriate Skilling

A key factor in establishing and enhancing capacity within African researchers is to identify required skills and build capacity (see Chaps. 27 and 28). An example is the capacity building programme in genomics recently initiated by the BECA-ILRI Hub. The training follows three phases, the first two phases being residential in ILRI (Nairobi, Kenya), and a third phase being conducted remotely at the participants' home institutions. Below is a summary of the activities envisioned as part of this programme:

Development phase: for 4 months, participants will be exposed to in-depth training modules on molecular biology and bioinformatics skills for next-generation sequencing (high molecular weight DNA extraction protocols, Unix command line, Python programming, genomics, theory, and practice of genome assembly, etc.). These training modules are comprised of formal lectures and hands-on sessions. Besides the series of lectures being delivered by internationally recognized scientists, a team of tutors (including trainees from previous BixCoP batches) provide ad-hoc supervision to trainees during individual and group exercises.

Demonstration phase: Intertwined with the Development phase and extending for 1.5 months and beyond, this phase engages trainees in project-based learning, building on the skills learnt during the Development phase, and demonstrates their use in real-life research projects in genomics for agriculture or conservation, co-designed with the training team.

Deployment phase: during this last phase of the training, the trainees return to their home institutions, where they launch a small third-generation genomics unit and implement a research project to address agricultural challenges with direct relevance to their home country or region. Wherever possible, this project will be integrated into research partnerships with ILRI or another CGIAR institute (www.cgiar.org) in their home country. Trainees benefit from the support of highly experienced mentors from the BecA-ILRI Hub, its UK-based partners, and the CGIAR consortium. For Africa to achieve success in building capacity and skills, initiatives like these are needed to effectively build capacity at larger scale.

29.7.4.2 Case Study 2: The African Animal Breeding Network (AABNet)

The African Animal Breeding Network (AABNet, http://animalbreeding-africa.org/) was formed in 2019 by bringing together some of the world's leading animal breeders and related scientific disciplines in academia and industry (Djikeng et al. 2025). The primary goal is to drive the development and dissemination of livestock-improved genetics and broader genetic improvement solutions in Africa. It comprises of different pillars, multi-country genetic evaluation which will drive the establishment of robust systems for genetic evaluation of livestock across African countries. This will serve as a long-term and strategic objective of AABNet. It will also serve as a hub to access methodologies, tools, and technical support by animal breeders to determine how different genotypes interact with environmental factors. It will ensure a comprehensive portfolio of activities and available data by adequately collected and well annotated and synthesized data on countries with ongoing and robust national evaluations. Available national data to demonstrate the value of multi-country evaluation and to build relevant systems that will support genetic evaluation focusing on key livestock productivity and resilience challenge. The second pillar is professional development which will focus on capacity building in animal breeding and associated expertise and developing new and promising talents and innovations, specifically in quantitative genetics, animal breeding, bioinformatics, genomics, data science, animal performance recording/phenotyping, reproductive technologies, and animal biotechnology. The third pillar comprises advocacy, awareness, and business development which will focus on building and making available relevant expertise to support the development of business cases for livestock improvement and credible data as well as communication that will enable governments, development partners, farmers, and other stakeholders to demonstrate compelling need for investments and other support to ensure profitable, productive, and resilient production systems in Africa that serve livestock producers. The fourth pillar of AABNet comprises collaboration, networking, and partnerships which will see operational engagements between academia, industry, farmers organizations, public sector, philanthropic, and development agencies that will be a driving factor for development and dissemination of livestock-improved genetics in Africa.

29.7.5 Funding and Collaborations for Genetic Evaluations

For across-country genetic evaluations and collaboration to be effective, these efforts will need financial support to establish proper infrastructure and maintaining logistics. Funding can be sourced from involved government and private sectors. In addition to what has been discussed in Sect. 29.6, there are few funding opportunities within countries that can provide funding for across-country initiatives. Most of these funding will depend on the countries involved. For example, in South Africa, funding bodies such as the National Research Fund (NRF) and Red Meat Research and Development South African (RMRD SA) can be useful bodies in these initiatives.

29.8 Conclusion and Future Perspectives

Multi-country cooperations and initiatives (MCI), involving the sharing of facilities, data, biological materials, and human and other resources, are key to applying efficient technologies for advancing livestock genetic improvement in Africa. Currently, such across-country cooperations and initiatives are few in Africa. African countries are generally faced with many challenges when it comes to implementing genetic evaluation and improvement schemes, as well as new technologies such as genomics, which provides strong motivation for MCIs. There have, however, been notable initiatives aimed at addressing such challenges through MCIs. These initiatives include H3ABioNet, which aims to build bioinformatics capacity, and AABNet, which focuses on development and dissemination of improved livestock and broader genetic improvement solutions in Africa. These initiatives serve as valuable successful examples of how capacity to adopt and implement technologies for achieving genetic improvement can be achieved through cooperations in Africa. Within-country facilities schemes, such as the integrated registration and genetic information system (INTERGIS) and Kaonafatso ya dikgomo (KYD), which have been used successfully to facilitate livestock improvement in commercial and smallholder farms in South Africa, can be replicated in other African countries. Alternatively, they can be shared with other

African countries, which would enable standardized performance recording protocol to promote livestock genetic improvement programs across the continent. Infrastructure and facilities for providing genomics research, such as genotyping platforms, genome sequencing, and high-performance computing/data centres, are particularly scarce across the African continent. Implementation of MCIs to advance livestock genetic improvement can go a long way towards addressing these needs. This should be underpinned by capacity building programmes using shared training facilities at higher education institutions or research councils in Africa.

References

Afgan E et al (2018) The Galaxy platform for accessible, reproducible and collaborative biomedical analyses: 2018 update. Nucleic Acids Res 46(W1):W537–W544

African Union Secretariat (2020) AU-IBAR promo video

Aron S et al (2021a) Ten simple rules for developing bioinformatics capacity at an academic institution. PLoS Comput Biol 17:e1009592

Aron S et al (2021b) The development of a sustainable bioinformatics training environment within the H3Africa Bioinformatics Network (H3ABioNet). Front Educ 6:356

Association of Public Health Laboratories (APHL) (2016) Next generation sequencing implementation guide. https://www.aphl.org/aboutAPHL/publications/Documents/ID-NGS-Implementation-Guide102016.pdf

Bester J et al (2003) The Nguni: a case study. In: Vilakati D, Morupisi C, Setshwaelo L, Wollny C, von Lossau A, Drews A (eds) Community based management of animal genetic resources. FAO, Rome, pp 45–68

Bonifazi R et al (2020a) Impact of Interbeef on national beef cattle evaluations. Acta Fytotech Zootech 23:144–155

Bonifazi R et al (2020b) International single-step genomic evaluations in beef cattle. In: Book of abstracts of the 71st annual meeting of the European Federation of Animal Science. Wageningen Academic Publishers

Cardoso FF et al (2021) Multiple country and breed genomic prediction of tick resistance in beef cattle. Front Immunol 12:620847

Cesarani A et al (2021) Genomic information allows for more accurate breeding values for milkability in dual-purpose Italian Simmental cattle. J Dairy Sci 104(5):5719–5727

de Oliveira HR et al (2022) Across-country genomic predictions in Norwegian and New Zealand Composite sheep populations with similar development history. J Anim Breed Genet 139(1):1–12

Djikeng A, Olori VE, Houaga I, Aggrey SE, Mwai O, Ibeagha-Awemu EM, Mrode R, Chagunda MGG, Tiambo CK, Rekaya R, Nash O, Nziku Z, Opoola O, Ntanganedzeni M, Ekine-Dzivenu C, Kahi A, Okeno T, Hickey JM, Enyew N, Rege EJO. The African Animal Breeding Network as a pathway towards genetic improvement of livestock. Nat Genet. 2025 Mar;57(3):498-504. doi: https://doi.org/10.1038/s41588-025-02079-4. Epub 2025 Feb 10. PMID: 39930083

Dürr J, Philipsson J (2012) International cooperation: The pathway for cattle genomics. Animal Frontiers 2(1):16–21

Ebenezer TE et al (2022) Africa: sequence 100,000 species to safeguard biodiversity. Nature 603(7901):388–392

Endrullat C et al (2016) Standardization and quality management in next-generation sequencing. Appl Transl Genom 10:2–9

Ewels PA et al (2020) The nf-core framework for community-curated bioinformatics pipelines. Nat Biotechnol 38(3):276–278

Gargis Amy S et al (2016) Assuring the quality of next-generation sequencing in clinical microbiology and public health laboratories. J Clin Microbiol 54(12):2857–2865

Gould J (2015) Core facilities: shared support. Nature 519(7544):495–496

Gwala L (2013) Effect of agricultural extension services on beneficiaries of the Nguni cattle project: the case of Ncera and Kwezana villages, Eastern Cape Province. University of Fort Hare

Harris BL, Johnson DL (2010) Genomic predictions for New Zealand dairy bulls and integration with national genetic evaluation. J Dairy Sci 93(3):1243–1252

Hutchins RJ et al (2019) Practical guidance to implementing quality management systems in public health laboratories performing next-generation sequencing: personnel, equipment, and process management (phase 1). J Clin Microbiol 57(8):e00261-19

Jit M et al (2021) Multi-country collaboration in responding to global infectious disease threats: lessons for Europe from the COVID-19 pandemic. Lancet Reg Health Eur 9:100221

Jorjani H, Klei L, Emanuelson U (2003) A simple method for weighted bending of genetic (co)variance matrices. J Dairy Sci 86(2):677–679

Kozyreva VK et al (2017) Validation and implementation of clinical laboratory improvements act-compliant whole-genome sequencing in the public health microbiology laboratory. J Clin Microbiol 55(8):2502–2520

Langmead B, Nellore A (2018) Cloud computing for genomic data analysis and collaboration. Nat Rev Genet 19(4):208–219

Li S et al (2014) Multi-platform assessment of transcriptome profiling using RNA-seq in the ABRF next-generation sequencing study. Nat Biotechnol 32(9):915–925

Lightbody G et al (2019) Review of applications of high-throughput sequencing in personalized medicine: barriers and facilitators of future progress in research and clinical application. Brief Bioinform 20(5):1795–1811

Mardis ER (2017) DNA sequencing technologies: 2006–2016. Nat Protoc 12(2):213–218

Margulies M et al (2005) Genome sequencing in microfabricated high-density picolitre reactors. Nature 437(7057):376–380

Marle-Koster V, Webb EC (2014) A perspective on the impact of reproductive technologies on food production in Africa. Adv Exp Med Biol 752:199–211

Martin-Collado D et al (2021) Measuring farmers' attitude towards breeding tools: the Livestock Breeding Attitude Scale. Animal 15(2):100062

Meder D et al (2016) Institutional core facilities: prerequisite for breakthroughs in the life sciences: core facilities play an increasingly important role in biomedical research by providing scientists access to sophisticated technology and expertise. EMBO Rep 17(8):1088–1093

Merchant N et al (2016) The iPlant collaborative: cyberinfrastructure for enabling data to discovery for the life sciences. PLoS Biol 14(1):e1002342

Mische SM et al (2020) A review of the scientific rigor, reproducibility, and transparency studies conducted by the ABRF Research Groups. J Biomol Tech 31(1):11–26

Morandini R et al (1977) FAO panel of experts on forest gene resources. Report of the fourth session. FAO

Mrode R et al (2021) Genomic prediction of crossbred dairy cattle in Tanzania: a route to productivity gains in smallholder dairy systems. J Dairy Sci 104(11):11779–11789

Muchadeyi FC et al (2020) Editorial: why livestock genomics for developing countries offers opportunities for success. Front Genet 11:626

Mueller JP et al (2015) Community-based livestock breeding programmes: essentials and examples. J Anim Breed Genet 132(2):155–168

Mulder N et al (2017) Genomic research data generation, analysis and sharing – challenges in the African setting. CODATA Data Sci J 16(49):1–15

Mushtaq S et al (2015) Long-term prognostic effect of coronary atherosclerotic burden: validation of the computed tomography-Leaman score. Circ Cardiovasc Imaging 8(2):e002332

Navarro FC et al (2019) Genomics and data science: an application within an umbrella. Genome Biol 20(1):1–11

Ngarava S, Mushunje A, Chaminuka P (2020) Qualitative benefits of livestock development programmes. Evidence from the Kaonafatso ya Dikgomo (KyD) scheme in South Africa. Eval Program Plann 78:101722

Nyamushamba G et al (2017) Conservation of indigenous cattle genetic resources in Southern Africa's smallholder areas: turning threats into opportunities—a review. Asian Australas J Anim Sci 30(5):603

Obert T (2012) Breeding objectives for in-situ conservation of indigenous Nguni cattle under low-input production systems in South Africa. University of Fort Hare

Ojango JM et al (2017) Improving smallholder dairy farming in Africa. In: Achieving sustainable production of milk, vol 2. Burleigh Dodds Science Publishing, Cambridge, pp 371–396

Olivieri E et al (2021) Multi-country investigation of the diversity and associated microorganisms isolated from tick species from domestic animals, wildlife and vegetation in selected African countries. Exp Appl Acarol 83(3):427–448

Opoola O (2019) Across-country dairy breeding strategies in sub-Saharan Africa. Thesis

Opoola O et al (2020) Joint genetic analysis for dairy cattle performance across countries in sub-Saharan Africa. S Afr J Anim Sci 50(4):507–520

Phocas F et al (2006) Genetic correlations between temperament and breeding traits in Limousin heifers. Anim Sci 82(6):805–811

Rischkowsky B, Pilling D (2007) The state of the world's animal genetic resources for food and agriculture. FAO

Sanger F et al (1977a) Nucleotide sequence of bacteriophage phi X174 DNA. Nature 265(5596):687–695

Sanger F, Nicklen S, Coulson AR (1977b) DNA sequencing with chain-terminating inhibitors. Proc Natl Acad Sci USA 74(12):5463–5467

Schaeffer L (1994) Multiple-country comparison of dairy sires. J Dairy Sci 77(9):2671–2678

Schneider L et al (2019) ClinOmicsTrailbc: a visual analytics tool for breast cancer treatment stratification. Bioinformatics 35(24):5171–5181

Sharaf A et al (2023) Bridging the gap in African biodiversity genomics and bioinformatics. Nat Biotechnol 41(9):1348–1354

Shin S, Park J (2016) Characterization of sequence-specific errors in various next-generation sequencing systems. Mol BioSyst 12(3):914–922

Simm G et al (2021) Dairy cattle breeding. In: Genetic improvement of farmed animals. CABI, Wallingford, Oxfordshire, pp 234–291

Stephens ZD et al (2015) Big data: astronomical or genomical? PLoS Biol 13(7):e1002195

Stewart CA et al (2015) Jetstream: a self-provisioned, scalable science and engineering cloud environment. In: Proceedings of the 2015 XSEDE conference: scientific advancements enabled by enhanced cyberinfrastructure. ACM, New York

Stewart CA et al (2018) Return on investment for three cyberinfrastructure facilities: a local campus supercomputer, the NSF-funded Jetstream cloud system, and XSEDE (the eXtreme Science and Engineering Discovery Environment). In: 2018 IEEE/ACM 11th international conference on utility and cloud computing (UCC). IEEE

Tanjo T et al (2021) Practical guide for managing large-scale human genome data in research. J Hum Genet 66(1):39–52

Tauch A, Al-Dilaimi A (2019) Bioinformatics in Germany: toward a national-level infrastructure. Brief Bioinform 20(2):370–374

The-H3Africa-Consortium et al (2014) Enabling the genomic revolution in Africa. Science 344(6190):1346–1348

Upshaw KW et al (2021) Utilization of genomic testing for the selection of desirable traits in cattle. In: Bovine reproduction. Wiley, Hoboken, pp 945–977

Van der Westhuizen J, Scholtz MM, Mamabolo MJ (2005) Importance of infrastructure and system for livestock recording and improvement in developing countries. In: The role of biotechnology in animal agriculture to address poverty in Africa: opportunities and challenges. TSAP/ILRI, Dar es Salaam/Nairobi, p 343

van Marle-Köster E et al (2015) Genomic technologies for food security: a review of challenges and opportunities in Southern Africa. Food Res Int 76:971–979

Wheway G et al (2019) Opportunities and challenges for molecular understanding of ciliopathies–the 100,000 Genomes Project. Front Genet 10:127

Wiewiórka MS et al (2014) SparkSeq: fast, scalable and cloud-ready tool for the interactive genomic data analysis with nucleotide precision. Bioinformatics 30(18):2652–2653

Zuma-Netshiukhwi G, Netshilema T, Nongovhela NB (2022) Breeding soundness inspection and production features of bullocks under ecological conditions in Free State region, South Africa. Eur J Agric Food Sci 4(1):39–46

Open Access This chapter is licensed under the terms of the Creative Commons Attribution 4.0 International License (http://creativecommons.org/licenses/by/4.0/), which permits use, sharing, adaptation, distribution and reproduction in any medium or format, as long as you give appropriate credit to the original author(s) and the source, provide a link to the Creative Commons license and indicate if changes were made.

The images or other third party material in this chapter are included in the chapter's Creative Commons license, unless indicated otherwise in a credit line to the material. If material is not included in the chapter's Creative Commons license and your intended use is not permitted by statutory regulation or exceeds the permitted use, you will need to obtain permission directly from the copyright holder.

Policies, Frameworks, Strategies, and Action Plans for Conservation and Sustainable Use of African Animal Genetic Resources

30

John E. O. Rege and Mure Agbonlahor

Abstract

Animal genetic resources (AnGR) refer to all animal species, breeds, and strains (and their wild relatives) that are of economic, scientific, and cultural interest to humankind in terms of food and agricultural production for the present or in the future. There are more than 40 species of animals that have been domesticated (or semi-domesticated) during the past 10,000–12,000 years which contribute directly or indirectly to food and agriculture. Common species include cattle, sheep, goats, chickens, pigs, horses, and buffalo, but many other domesticated species such as camels, donkeys, elephants, various poultry species, reindeer, rabbits, etc. are important to different cultures and regions of the world. In developing countries, AnGR also play an important role in the subsistence of many communities and the sustainability of crop–livestock systems. Within these species, breeds in certain regions and production systems are increasingly endangered and many have been lost, many more are at risk of loss, and the situation has been getting worse over the recent decades. Yet, it is the extensive genetic diversity in these breeds that allows the existence of livestock in diverse environments across the globe, providing a range of products and functions even in the most extreme production environments.

African nations are increasingly aware of the need for better understanding and safeguarding the diversity of the continent's unique AnGR. It is increasingly evident that the most significant threat to indigenous AnGR is the marginalization of traditional production systems and the associated neglect of local breeds, driven by the rapid spread of intensive livestock production, often large-scale, and utilizing a narrow range of increasingly exotic breeds. As the availability of adapted local breeds decreases, these changes are also marginalizing smallholder farmers who depend on the adaptive attributes of indigenous AnGR.

Unfortunately, policies and legal frameworks influencing the livestock sector are not always favourable to the conservation and sustainable use of indigenous AnGR. This chapter examines the global policy and strategy landscape on AnGR that is relevant for Africa, past and present initiatives to develop AnGR policies and strategies, the extent to which

J. E. O. Rege (✉)
Emerge Centre for Innovations–Africa,
Nairobi, Kenya
e-mail: ed.rege@emerge-africa.org

M. Agbonlahor
African Union Commission, Addis Ababa, Ethiopia

these are being translated into plans, and institutionalized practices by African countries, and the challenges being faced and opportunities for interventions going forward. It concludes that, while the African Union, through the Inter-African Bureau for Animal Resources (AU-IBAR), in partnership with Regional Economic Communities, has created significant awareness in Africa around global AnGR policy frameworks/instruments and actions needed at national levels, these efforts have delivered only limited change—with only a few countries having developed or in the process of developing policies and strategies, but even these largely remain on shelves, with little evidence of implementation.

Keywords

Policies · Frameworks · Strategies · Characterization · Conservation · Sustainable use · Animal genetic resources

Acronyms

AAGRIS	African Animal Genetic Resources Information System
ABGeN-NG	Animal Breeders' and Geneticists' Network of Nigeria
ADF	Agricultural Development Fund
AEC	African Economic Community
AGRIS	Agricultural Information System
AGS	Alpha-Gal Syndrome
AI	Artificial Insemination
AIA	Advance Informed Agreement
AMCEN	African Ministers Conference on the Environment
AnGR	Animal Genetic Resources
ARC	Animal Resources Committee
ARDMP	Animal Resources Data Management Platform
ARIS	Animal Resources Information System
AU	African Union
AUC	African Union Commission
AU-IBAR	African Union-Inter-African Bureau on Animal Resources
AU-PANVAC	African Union Pan-African Veterinary Vaccine Centre
BAHPA	Bulletin of Animal Health and Production in Africa
BRICS	Brazil, China, India, and South Africa
CAADP	Comprehensive Africa Agricultural Development Programme
CAIS	Central Artificial Insemination Station
CAP	Common Agricultural Policy
CBD	Convention on Biological Diversity
CEDEAO	Economic Community of West African States (Fr)
CEN–SAD	Community of Sahel–Saharan States
CIRDES	Centre International De Recherche-Développement Sur l'Élevage En Zone Sub-Humide
COMESA	Common Market for Eastern and Southern Africa
CSIR-ARI	Council for Scientific and Industrial Research—Animal Research Institute
CUT	Chinhoyi University of Technology
DAD-Nets	Domestic Animal Diversity Networks
DAFF	Department of Agriculture, forestry, and Fisheries
DAHS	Federal Department of Animal Husbandry Services
DAR	Farm Animal Genetic Resource Gene Bank, Department of Agricultural Research
EAC	East African Community
ECCAS	Economic Community of Central African States
ECDPM	European Centre for Development Policy Management
ECOWAP	Ecowas Agricultural Policy
ECOWAS	Economic Community of West African States
EU	European Union

FACT	Farm Animal Conservation Trust	LiDeSA	Livestock Development Strategy for Africa
FAO	Food and Agriculture Organization	LIMS	Livestock Information Management System
FBO	Farmer-Based Organizations	LMO	Living Modified Organisms
FMARD	Federal Ministry of Agriculture and Rural Development	LSTP	Livestock Sector Transformation Plan
GCRB	Germplasm Conservation and Reproductive Biotechnologies	MAA	Material Acquisition Agreements
GDP	Gross Domestic Product	MDG	Millennium Development Goals
GEF	Global Environment Facility		
GIIBS	Grupo Inter-Institucional Sobre Biosegurança	MOFA	Ministry of Food and Agriculture
GM	Genetically Modified	MOU	Memorandum of Understanding
GMO	Genetically Modified Organisms		
GPA	Global Plan of Action	MTA	Material Transfer Agreements
GSN	Genetics Society of Nigeria		
HIV/AIDS	Human Immunodeficiency Virus/Acquired Immunodeficiency Syndrome	NAC	National Advisory Committee
		NACGRAB&DB	National Animal Genetic Resource Centre and Data Bank
IBC	Institute of Biodiversity Conservation		
IBED	Inter-African Bureau of Epizootic Diseases	NAGRC	National Animal Genetic Resources Centre
ICPALD	IGAD Centre for Pastoral Areas and Livestock Development	NAIP	National Agriculture Investment Plans
ICT	Information and Communication Technology	NBF	National Biosafety Frameworks
IDDRSI	IGAD Drought Disaster Resilience and Sustainability Initiative	NCRST	National Commission on Research, Science, and Technology
IG	Interventional Genetics	NGO	Non-governmental Organizations
IGAD	Intergovernmental Authority on Development		
IGADD	Intergovernmental Authority on Drought and Development	NIAS	Nigeria Institute of Animal Science
ILRI	International Livestock Research Institute	NLP	National Livestock Policy
INTERGIS	Integrated Registration and Genetic Information System	NSAPs	National Strategies and Action Plans
IR	Insect-Resistant	NVRC	The National Varietal Release Committee
IR-ABS	International Regime on Access and Benefit-Sharing	PFJ	Planting for Food and Jobs
		PPP	Public-Private Partnerships
ITC	International Transhumance Certificate	PRIASAN	Programme for Agricultural Investment, Food Safety, and Nutrition
KAGRC	Kenya Animal Genetic Resources Centre		
LAP	Livestock Action Plan	RAIP	Regional Agricultural Investment Plan
LDPS	Livestock Development Policy and Strategy	RBI	Rare Breeds International

RBST	Rare Breeds Survival Trust
REC	Regional Economic Communities
RFJ	Rearing for Food and Jobs
SADC	Southern African Development Community
SADCC	Southern African Development Coordination Conference
SADC-RAP	Southern African Development Community-Regional Agricultural Policy
SME	Small and Midsize Enterprise
SoW-AnGR	State of The World's Animal Genetic Resources for Food and Agriculture
SP	Strategic Priorities
SPA	Strategic Priority Areas
S-RFP	Sub-Regional Focal Points
SSA	Sub-Saharan Africa
TLMI	Tanzania Livestock Modernization Initiative
TLMP	Tanzania Livestock Master Plan
UMA	Arab Maghreb Union
UNDP	United Nations Development Programme
UNEP	United Nations Environment Programme
VR	Virus-Resistant
WTO	World Trade Organization

30.1 Introduction

Animal genetic resources (AnGR) refer to all animal species, breeds, and strains (and their wild relatives) that are of economic, scientific, and cultural interest to humankind in terms of food and agricultural production for the present or in the future. There are more than 40 species of animals that have been domesticated (or semi-domesticated) during the past 10,000–12,000 years which contribute directly or indirectly to agricultural production. Common species include cattle, sheep, goats, chickens, pigs, horses, and buffalo, but many other domesticated species such as camels, donkeys, elephants, various poultry species, reindeer, rabbits, etc. are important to different cultures and regions of the world. These AnGR are vital to the economic development of most countries in the world. In developing countries, they also play an important role in the subsistence of many communities and the sustainability of crop–livestock systems. The world's domestic animal genetic resources represent an important resource of economic development and livelihood security. Within these species, breeds in certain regions and production systems are increasingly endangered and many have been lost, many more are at risk of loss, and the situation has been getting worse over the recent decades. Yet, it is the extensive genetic diversity in these breeds that allows the existence of livestock in diverse environments across the globe, providing a range of products and functions even in the most extreme production environments.

African policymakers are increasingly aware of the need for better understanding and safeguarding the diversity of the continent's unique animal genetic resources (AnGR). African indigenous livestock breeds have, over centuries, adapted to diverse, stressful tropical environments by undergoing natural and artificial selection. They have acquired a range of unique adaptive traits, such as tolerance to disease and heat, water scarcity, and ability to cope with poor-quality feeds. However, this unique pool of AnGR is threatened with extinction. Several causes are contributing to the loss of these genetic resources. These include the lack of supportive policies and legislation and inadequate measures to ensure the conservation and sustainable use of these resources. Policy gaps include absence of policies and strategies for conservation, genetic improvement, and sustainable use of indigenous AnGR.

Moreover, these genetic resources are also being directly undermined by some existing policies and practices. Policies and legal frameworks influencing the livestock sector are not always favourable to the conservation and sustainable use of indigenous AnGR. These include overt or hidden subsidies which promote large-scale production based on exotic breeds and crossbreds at the expense of the smallholder systems that utilize local AnGR. Moreover, development and post-disaster livestock rehabilitation programmes do not assess potential impacts on genetic diversity to ensure that the breeds used are appropriate to local smallholder production systems, and disease control strategies do not incorporate measures which are disproportionately harmful to indigenous breeds. As the availability of adapted local breeds decreases, these changes are also marginalizing smallholder farms who depend on the adaptive attributes of indigenous AnGR.

A look at livestock (production and productivity) strategies, plans, and programmes across the African continent reveals that attention is consistently given to animal health, feed production and management, and markets—traditionally considered to be the most productivity-limiting factors—while, other than their role in crossbreeding, only occasional and casual reference is made to conservation and use of indigenous animal genetic resources. The gaps include lack of accurate and up-to-date information on indigenous AnGR, trends in their population sizes, and drivers of these trends—to inform planning the conservation of these resources. Yet, these resources are key to productivity and resilience of smallholder systems today and hold promise for dealing with production challenges in the uncertain future, especially in the face of climate change. There is evidence that historical attention to animal health in Africa since pre-independence has delivered visible results, for example, the eradication of rinderpest. In the introduction to its current (2018–2023) strategy, the African Union-Inter-African Bureau on Animal Resources (AU-IBAR) recognizes that it has not done well in certain areas, including AnGR, and explicitly refers to "… *the need to build on its success in achieving important targets such as in animal disease prevention, control and eradication*".

This chapter examines the global policy and strategy landscape on AnGR that is relevant for Africa, as well as initiatives on development and implementation of AnGR policies and strategies in the continent, and the extent to which these are being translated into strategies and practices by African countries and the challenges being faced and opportunities for interventions going forward. Definitions of terms used in this chapter are in Box 30.1.

Box 30.1 Definition of Terms
1. *Animal Genetic Resources* include all species, breeds, and strains of animals that are of economic, scientific, and cultural interest to agriculture, now and in the future. Common species include cattle, sheep, goats, pigs, chickens, camels, and donkeys. There are many other domesticated animals that are important for food and agriculture in different regions and cultures of the world. These include horses, buffaloes, elephants, rabbits, and rodents, *inter-alia*.
2. *Biosafety* involves the reduction and elimination of potential risks resulting from the use of biotechnology and its products. Biosafety has also been defined as the avoidance of risk to human health and safety, and to the protection of the environment, resulting from the use of genetically engineered organisms.
3. *Characterization* of animal genetic resources is the assessment of the attributes of the animal populations and their production environments to establish their status and identify strengths that can be enhanced and weaknesses that need to be overcome, for example through genetic improvement programmes. It can also help to inform conservation strategies.
4. *Conservation* of animal genetic resources refers to measures taken to prevent the loss of genetic diversity in livestock populations, including to protect breeds from extinction. It can involve both the conservation of live populations and the cryopreservation (preservation through freezing at low temperatures) of material such as semen or embryos.
5. *Genetic improvement* (animal breeding) is based on the principle that the products (milk, meat, eggs, wool, etc.) and

(continued)

services (e.g., transport, draught power, or cultural services) provided by animals are a function of their genes and the environmental influences that they are exposed to. Improvement can be achieved by selecting genetically superior animals to be the parents of the next generation. *"Genetically superior"* means possessing gene combinations which confer superiority in terms of a particular sets of characteristics, which usually include productivity in the environmental conditions now and/or expected in the future, but should also consider traits such as fertility, disease resistance, or longevity that relate to costs of production.
6. *Policy*. A policy is a *statement of intent* and is implemented as guidelines, procedures, or protocols. Policy has also been defined as a set of ideas or plans that are used as a basis for making decisions—that is, what to do in particular situations, especially in politics, economics, or business. Policies are generally adopted by a governance body within an institution to assist in both subjective (based on the relative merits and demerits of multiple factors and changing contexts) and objective decision-making.
7. *Sustainability*. Sustainability, as used in this chapter, means practices that fulfil the needs of today's generations without compromising the needs of future generations, while also ensuring a balance between economic growth, environmental care, and social well-being. Thus, the three pillars of sustainability, "profit", planet, and people.

30.2 Global Instruments Relevant for the Management of AnGR

30.2.1 The Convention on Biological Diversity (CBD)

While concern for the environment is not new—and has featured in development discourses through the last century—there was heightened concern about human impact on the environment and loss of species and ecosystems in the 1970s. In 1972, the United Nations Conference on the Human Environment in Stockholm, Sweden, resolved to establish the United Nations Environment Programme (UNEP). Governments signed several regional and international agreements in this regard, including those targeting the protection of wetlands and regulation of international trade in endangered species and controls on toxic chemicals and pollution. However, these turned out to be stopgap actions only. It has subsequently been recognized that what is ultimately needed is for humans to learn to use biological resources in a way that minimizes their depletion. The challenge is to put in place economic policies that motivate conservation and sustainable use by creating financial incentives for those who would otherwise over-use or damage the resource.

In 1987, the World Commission on Environment and Development concluded that economic development must become less ecologically destructive: *"Humanity has the ability to make development sustainable—to ensure that it meets needs of the present without compromising the ability of future generations to meet their own needs"*.

In 1992, the largest-ever meeting of world leaders—dubbed the "Earth Summit"—took place at the United Nations Conference on Environment and Development in Rio de Janeiro, Brazil. Among the raft of instruments signed at this Summit were two binding agreements, the

Convention on Climate Change, which targets industrial and other emissions of greenhouse gases, and the Convention on Biological Diversity (CBD). The latter was the first global agreement on the conservation and sustainable use of biological diversity. This treaty gained rapid and widespread acceptance and was signed by over 150 governments at the Rio conference. To date, more than 196 countries have ratified the CBD.

The CBD has three main goals: The conservation of biodiversity; Sustainable use of the components of biodiversity; and Sharing the benefits arising from the commercial and other utilization of genetic resources in a fair and equitable way. The CBD recognizes that the conservation of biological diversity as "*a common concern of humankind*" and is integral to the development process. It covers all ecosystems, species, and genetic resources. It links traditional conservation efforts to the economic goal of using biological resources sustainably. It also covers the rapidly expanding field of biotechnology, addressing technology development and transfer, benefit-sharing, and biosafety. Even more importantly, the CBD is legally binding: countries that join it are obliged to implement its provisions. While past conservation efforts were aimed at protecting species and habitats, the CBD recognizes that ecosystems, species, and genes must be used for the benefit of humans, but in a way and at a rate that does not lead to the long-term decline of biological diversity. To date, the CBD has 193 Parties—a near universal participation among countries. Indeed, only four Member States of the United Nations—Andorra, South Sudan, United States of America, and the Holy See (the Vatican)—are not Parties to the CBD. Thus, other than South Sudan, all African countries have ratified the CBD. Although action is highly variable among countries, Member States of the African Union recognize the critical importance of biodiversity and ecosystem services—and their contribution to economic growth, sustainable development, livelihoods, and human well-being in Africa and to achieving the African Union Agenda 2063 and the 2030 Agenda for Sustainable Development.

The conservation, use, and transfer of biological diversity at national, regional, and international levels are governed by a number of international agreements, global plans of action, and other instruments, all of which are intended to facilitate the operationalization of the CBD. Three such global instruments that have direct relevance for Animal Genetic Resources are covered in this chapter. They are: The *Cartagena Protocol on Biosafety to the CBD*; the *Nagoya Protocol on Access to Genetic Resources and the Fair and Equitable Sharing of Benefits Arising from their Utilization to the CBD*; and the *Global Plan of Action for Animal Genetic Resources*. With regard to international Treaties, Conventions, and Protocols, Box 30.2 summarizes descriptors that explain level or status of engagement by countries. These descriptors are applicable in the rest of this chapter.

Box 30.2 Treaty State Descriptors—Ratification, Accession, Approval, and Acceptance

The legal incidents or implications of *ratification*, *accession*, *approval*, and *acceptance* are the same:

The instrument (e.g. treaty) becomes legally binding on the State or the regional economic integration organization. All the countries that have either *ratified, acceded to, approved, or accepted* the Treaty, Convention, or Protocol are therefore *Parties* to it.

Ratification and Accession

The primary (and traditional) distinction is only between ratification and accession. In this regard, it is only States which have signed a treaty, when it was open for signature, that can proceed to ratify it. Signature of itself does not establish consent to be bound, hence the need for the further act of ratification.

States which have not signed a treaty during the time when it is open for signa-

(continued)

ture can only *accede* to it. Therefore, the term "*accession*".

Acceptance and Approval

The terms "*acceptance*" and "*approval*" are of more recent origin and apply under the same conditions as those that apply to ratification. The legal effect is the same as ratification. The uses of these terms have to do with the diversity of legal systems across countries. Certain countries, especially some East European States, use the terms acceptance or approval for purposes of participation in treaties. The terms are also used in cases where organizations rather than States become Parties to an international treaty, for example the EU.

Succession

Succession occurs when one State is replaced by another in the responsibility for the international relations of territory. Generally, a newly independent State which makes a notification of succession is considered a party to a treaty from the date of the succession of States or from the date of entry into force of the treaty, whichever is the later date.

30.2.2 The Cartagena Protocol on Biosafety

Concerns have been raised regarding the potential adverse effects on biological diversity of the use of living modified organisms (LMOs, also known as genetically modified organisms or GMOs). The Cartagena Protocol on Biosafety seeks to address these safety concerns at the international level. The Protocol also covers potential toxic effects of insect-resistant crops on non-target organisms and potential ecological effects of gene flow from modified crops, fish, microorganisms, or insects to wild species or counterparts. The Protocol was adopted on 29 January 2000 as a supplementary agreement to the Convention on Biological Diversity and entered into force on 11 September 2003. It focuses primarily on transboundary movements and is therefore relevant to international trade. It includes provisions on import decision-making, risk assessment and management, information-sharing, documentation, capacity-building, compliance, liability and redress, public awareness and participation, and socio-economic considerations.

Key Elements of the Cartagena Protocol

The Protocol promotes biosafety by establishing rules and procedures for the safe transfer, handling, and use of LMOs/GMOs and facilitates information sharing. The five main pillars of the Protocol are: Procedures for moving LMOs across borders; risk assessment procedures; the biosafety clearing house; capacity building; and public awareness. It provides Parties the opportunity to obtain information before new biotech organisms are imported. It acknowledges each country's right to regulate bio-engineered organisms, subject to existing international obligations. It also creates the framework to help improve the capacity of developing countries to protect biodiversity. Specifically, the Protocol (CBD 2000):

(a) Establishes an online portal, *Biosafety Clearing-House*, to help Parties share scientific, technical, environmental, and legal information about living modified organisms (LMOs).

(b) Creates an advanced informed agreement (AIA) procedure that in effect requires exporting entities to seek consent from destination (importing) country before the first shipment of an LMO meant to be introduced into the environment (seeds—and by extension animal semen and ova, fish for release, or microorganisms).

(c) Requires shipments of LMOs (e.g. cereals) that are intended for direct use as food, feed, or for processing, to be accompanied by documentation stating that such shipments "may contain" living modified organisms and are "not intended for intentional introduction into the environment (i.e. use as seed)." The Protocol establishes a process for considering more detailed identification and docu-

mentation of LMO commodities in international trade.
(d) Sets out information to be included on documentation accompanying LMOs destined for contained use, including handling requirements and contact points for further information and for the receiving entity (consignee).
(e) Includes a clause which states that the agreement shall not be interpreted as implying a change in the rights and obligations of a Party under any existing international agreement, including, for example, WTO agreements. This is termed a *savings clause*.
(f) Calls on Parties to cooperate with developing countries in building their capacity for managing modern biotechnology.

With regard to an importing Party's decision-making process regarding the import of an LMO, there is a precaution with reference to Principle 15 of the Rio Declaration on Environment and Development. This precaution (hence the *precautionary approach*) states that where there are threats of serious or irreversible damage, lack of full scientific certainty shall not be used as a reason for postponing cost-effective measures to prevent environmental damage, that is:

> Lack of scientific certainty due to insufficient relevant scientific information and knowledge regarding the extent of the potential adverse effects of a living modified organism on the conservation and sustainable use of biological diversity in the Party of import, taking also into account risks to human health, shall not prevent that Party from taking a decision, as appropriate, with regard to the import of that living modified organism in order to avoid or minimize such potential adverse effects.

Thus, both the substantive content of the Protocol's precaution provisions and the preambular savings clause make clear that a Party's use of precaution in decision-making must be consistent with the Party's trade and other international obligations. Moreover, the Protocol states that the *transboundary movement of living modified organisms between Parties and non-Parties shall be consistent with the objective of the Protocol*. The implication is that, although the Protocol only requires trade between Parties and non-Parties in LMOs to be consistent with the objective of the Protocol, entities (individuals and institutions) in non-Party countries wishing to export to Parties will need to abide by domestic regulations put in place in the importing Parties for compliance with the Protocol.

Cartagena Protocol and Animal Genetic Resources in Africa

A total of 173 States and Regional Integration Organizations have ratified the Cartagena Protocol (CBD 2023a). To date, 49 African States are Parties to the Protocol (Table 30.7).

The Cartagena Protocol marked a significant milestone in how countries cooperate towards the safe transfer, handling, and use of living modified organisms that come from modern biotechnology. However, the ultimate success of this international agreement depends on the capacity of Parties to fully implement the agreement. The Global Environment Facility (GEF), as the financial mechanism to both the Convention on Biological Diversity and the Cartagena Protocol, has played an important role in building the capacity in biosafety since the adoption of the Protocol. The GEF, together with UNEP, UNDP, and the World Bank, has assisted countries in developing and implementing National Biosafety Frameworks (NBFs) and participating in the *Biosafety Clearing House* mechanism.

An NBF is a combination of policy, legal, administrative, and technical instruments put in place to address safety for the environment and human health in the context of modern biotechnology. These frameworks often focus on GMOs and have been generally driven by the crop sector, but they are meant to cover broader biotechnology research and applications. Although NBFs vary from country to country, they usually contain certain common elements. These include policy on biosafety, often as part of a broader policy on biotechnology; regulatory regime for biosafety including an act and regulations; and a system to handle notifications or requests for authorizations for certain activities, such as registration of activities ("*contained use*") and field releases of GMOs into the environment. This involves public participation and risk assessment, a mechanism for monitoring and inspections, and a system for public awareness and public information, i.e. how to inform stakeholders about the development and implementation of the NBF (UNEP 2006).

The road to the establishment of NBFs across the continent has been long, and countries remain at significantly different stages of putting in place the various components of functional NBFs. At the same time, countries are benefiting from regional integration activities being advocated for by the African Union and the Regional Economic Communities to seek common mechanisms that would complement NBFs. For example, COMESA Member States have developed their biosafety frameworks under the UNEP-GEF project during 2001–2004 (UNEP 2006). Beyond ratification and domestication of the key international agreements, political goodwill as well as human and financial capacities are important determinants of successful establishment of NBFs and their practical application in a country. Rege et al. (2022) have summarized the status of sub-Saharan Africa (SSA) countries on the development and implementation of NBFs (Table 30.1) and classified SSA countries based on the prevailing enabling environment (including policy and legislation) for biotechnology application in livestock.

In their assessment of enabling environment for biotechnology applications, Rege et al. (2022) used the following broad criteria: existence of a biosafety framework—incorporating policy and the extent to which this recognizes the potential of biotechnology and promotes improvement in the biological efficiency of livestock through such technologies; existence of acts of parliament, legislation, regulation, and responsible authority or agency mandated to implement the framework; the extent to which the framework has been operationalized; and level of investment in livestock biotechnologies. Table 30.2 summarizes the clustering of SSA countries into five categories and shows that most countries are in the "very weak" (21) and "weak" (8) categories, with only 10 in the "medium" category. Although improvements (especially in relation to investments) are still needed in all countries, Kenya and, to some extent, Nigeria ("strong") and South Africa ("very strong") currently have, overall, better enabling environment for livestock biotechnology applications than other African countries.

Implications of Developments in Transgenic Technologies

There has been an increase in research and development in Africa aimed at addressing constraints to agricultural productivity in recent years. Examples in crops include: Insect-resistant (IR) maize in Kenya; IR cotton in Ghana, Kenya, Nigeria, South Africa, and Uganda; Virus-resistant (VR) cassava in Kenya and Uganda; Fungus-resistant banana in Uganda; Drought-tolerant maize in Kenya, Mozambique, South Africa, Tanzania, and Uganda, and nutritionally enhanced sorghum in Burkina Faso, Kenya, and South Africa.

In the livestock sector, the increasing availability of GM animal feeds—e.g. soya, maize, and alfalfa—is of great relevance to Africa since feed is one of the major constraints to livestock production in the continent. Approval, local availability, and access to cheaper, more productive crops used as animal feeds will have significant implications for Africa in terms of movements into Africa from other continents and between African countries. From a biosafety standpoint, African countries have to be alive to the risks, if any, that are associated with these crops.

Globally, transgenic technologies today form the core of *"interventional genetics"*—approaches involving analysing and interfering with gene function in whole animals. In livestock, transgenic animals are being engineered to improve production and/or adaptive traits, for example, increased milk yield or milk protein, or enhanced disease resistance. With regard to transgenic meats, an example is the *GalSafe* pig (Dolgin 2021), already approved for commercial use in the US, which has been developed to be free of detectable alpha-gal sugar on its cell surfaces. People with Alpha-gal syndrome (AGS) may have allergic reactions to alpha-gal sugar found in red meat (e.g. beef, pork, and lamb). Another example is GM Salmon, the

Table 30.1 Status of National Biosafety Frameworks in SSA countries

Country	Biosafety Act	Biosafety legislation, regulation, policy	Regulatory agency
1. Botswana	Bill under review	National Biosafety policy 2013	
2. Burkina Faso	Act 2006; revised 2013	Biosafety Decree 2004; Biosafety Law 2011; Policy on Biotech	National Biosafety Authority
3. Cameroon	Biosafety Act 2003; revised 2007	Biosafety Guidelines 1995; draft	
4. Eswatini	Biosafety Act 2012	Legislation under review	
5. Ghana	Act 2011; Enacted law 2012	Regulatory framework yet to be finalized; Policy on Biotech; Regulatory Communication Strategy 2014	National Biosafety Committee
6. Kenya	Act 2009	Regulation & Guidelines 2011; National Biotechnology Policy 2006	National Biosafety Authority
7. Lesotho	National Biosafety bill 2005; amended 2014	National biosafety policy; National biosafety awareness strategy 2013; draft legislation	National Biosafety Council under Ministry of Tourism, Environment, and Culture
8. Malawi	Act 2002	Biosafety Guidelines 1995; Biosafety regulatory framework 2007; National Biotech policy 2008; draft legislation	Department of Environmental Affairs, Ministry of Environment and Climate Change Management
9. Mali	Biosafety law 2008; review of biosafety decree 2010		National Biosafety Committee
10. Mauritius	GMO Act 2004 Plant Protection Bill 2006		Ministry of Agro Industry and Food Security
11. Mozambique	Biosafety law 2007 revised 2012 Bill under review	Draft biosafety regulations/ guidelines	Grupo Inter-Institucional Sobre BioSegurança (GIIBS); NBC
12. Namibia	Biosafety Act promulgated 2006	Draft legislation Biotechnology and biosafety policy 1999	Biosafety Council of the National Commission on Research, Science, and Technology (NCRST)
13. Nigeria	Biosafety bill 2011. Bill passed second reading July 2014, referred to Senate Committees on Agriculture, Science and Technology	Biosafety guidelines 2001	Federal Ministry of Environment; National Biosafety Management Agency
14. Senegal	Biosafety law 2009		Ministry of Environment
15. South Africa	GMO Act 1997	Biosafety guidelines; National Biotechnology Policy and Strategy 2001	Directorate of Biosafety
16. Sudan	Law of Biosafety 2010	National Biosafety Framework 2008	Sudan National Biosafety Council
17. Tanzania	Environment Management Act of 2004	Biosafety regulation 2009; Biotech policy 2010; Policy under review	National Biosafety Committee
18. Togo	National Biosafety Framework 2004 biosafety Law 2009	Draft legislation	Ministry for Environment and Forestry, National Biosafety Committee

(continued)

Table 30.1 (continued)

Country	Biosafety Act	Biosafety legislation, regulation, policy	Regulatory agency
19. Uganda	National biosafety bill 2012; passed 2017, not yet assented into law	Biosafety guidelines 1995; Draft Biotech and Biosafety Policy 2013; National Biotechnology Policy 2008	Uganda National Council for Science and Technology; National Biosafety Committee
20. Zambia	Biosafety Act 2007, reviewed 2013	Nat biosafety policy 2013; Nat biosafety body 2013	National Biosafety Authority
21. Zimbabwe	National Biotech Authority Act 2000	Biosafety guidelines 1998	National Biotechnology Authority of Zimbabwe

Source: Rege et al. (2022)

Table 30.2 Classification of SSA countries on the basis of enabling environment for applications of biotechnology in livestock

Enabling environment	Countries
Very weak	Angola, Benin, Burundi, Chad, Central African Republic, Republic of Congo, Djibouti, DRC, Gambia, Equatorial Guinea, Eritrea, Eswatini, Gabon, Guinea, Guinea Bissau, Lesotho, Liberia, Niger, Sierra Leone, Togo, Somalia, South Sudan
Weak	Burkina Faso, Cameroon, Côte d'Ivoire, Madagascar, Malawi, Mozambique, Rwanda, Senegal
Medium	Botswana, Ethiopia, Ghana, Mali, Namibia, Sudan, Tanzania, Uganda, Zambia, Zimbabwe
Strong	Kenya, Nigeria
Very strong	South Africa

Source: Rege et al. (2022)

AquAdvantage Salmon (Clifford 2014), which has been genetically modified to reach a critical growth point faster and has been approved for commercial release in the USA.

The above examples of commercially released GM animals notwithstanding, to date, transgenic technology have mainly been used in basic research, in studies of gene expression, gene function, and disease processes. For example, targeting of genes encoding toxins to specific cell types in the mouse (specific cell ablation) has been used to answer questions of mammalian development and cell lineage analysis. Mutagenesis or disruption of the functioning of endogenous genes by insertion of transgenes (insertional mutagenesis) has also been used both to identify and to clone developmentally important genetic loci. Animal modelling of human disease is an area in which transgenic technology has made a huge impact. The use of large transgenic animals or "*bioreactors*" has been attempted, e.g. for the cost-effective production of high-value pharmaceuticals in the blood or milk of transgenic sheep or cows. Another promising area for the exploitation of the transgenic technology has been in the field of xenotransplantation, where transgenic pigs are engineered to express immunological characteristics that would make their organs well tolerated after transplantation into humans.

Infectious diseases adversely affect livestock production as well as animal welfare and have direct impacts upon human health and public opinion of livestock production. Using transgenic animals for the benefit of animal/human health is becoming increasingly common, driven by the combination of new technologies that facilitate the efficient production of genetically modified animals, new tools to modify gene expression, and advancement of basic understanding of causative agents and the mechanism of disease.

Clearly, African States need to be aware of the implications of these technological advances, even if their application in Africa remains relatively limited for now. Genetic engineering of animals, ethical issues, and implications for the environment have been a subject of review over recent years (e.g. Ormandy et al. (2011) and Maksimenko et al. (2013)).

The number of genetically modified organisms that can potentially harm animal diversity and the environment is staggering. This is especially so given the increasing ease of movement of genetic material across the globe. The Cartagena Protocol is, therefore, of great relevance and importance to Africa. Thus, the fact that only 21 countries have some form of biosafety laws and regulations (in the form of NBF) in place (Table 30.1), despite the fact that nearly all countries are Parties to the Protocol (Table 30.7), is worrying. Moreover, even among those countries with NBFs in place, there are those without capacity and government commitment to effectively operationalize the frameworks. Indeed, capacity building for biosafety regulators must be a key priority for African states (Obonyo et al. 2011). There is also a need to address the general lack of awareness of biosafety issues by the public and the lack of harmonized biosafety regulations at the sub-regional and continental levels. While the development of the 2007 African Model Law on Biosafety in Biotechnology is applaudable, it takes a more stringent approach than the Cartagena Protocol (Gupta et al. 2008). In any case there is need for a harmonized framework for operationalization of regulations to ensure effective biosafety enforcement without creating unnecessary hurdles to movement of genetic material.

30.2.3 The Nagoya Protocol

The *Nagoya Protocol on Access to Genetic Resources and the Fair and Equitable Sharing of Benefits Arising from their Utilization to the Convention on Biological Diversity (CBD)* (CBD 2010) is an international agreement which aims at sharing the benefits arising from the utilization of genetic resources in a fair and equitable way. It was adopted at the tenth meeting of the Conference of Parties (to the CBD) on 29 October 2010, in Nagoya, Japan, and entered into force on 12 October 2014. The Protocol is a supplementary agreement to the CBD with a focus on the implementation of one of the three objectives of the CBD, that is, the *fair and equitable sharing of benefits arising out of the utilization* of genetic resources, so as to contribute to the conservation and sustainable use of biodiversity. It sets out obligations for the providers and users of the genetic resource in question as well as other contracting parties to take measures in relation to access to genetic resources, benefit-sharing, and compliance.

The Protocol significantly advances the CBD's third objective on fairness and equity by providing a strong basis for greater legal certainty and transparency for both providers and users of genetic resources. Specific obligations to support compliance with domestic legislation or regulatory requirements of the party providing genetic resources and contractual obligations reflected in mutually agreed terms are a significant innovation of the protocol. By promoting the use of genetic resources and associated traditional knowledge, and by strengthening the opportunities for fair and equitable sharing of benefits from their use, the Protocol has potential to create incentives to conserve biological diversity, sustainably use its components, and further enhance the contribution of biological diversity to sustainable development and human well-being.

Under the Protocol, domestic-level benefit-sharing measures aim to provide for the fair and equitable sharing of benefits arising from the utilization of genetic resources with the contracting party providing genetic resources. Utilization includes research and development on the genetic or biochemical composition of genetic resources, as well as subsequent applications and commercialization. Sharing is subject to mutually agreed terms. Benefits may be monetary or non-monetary, such as royalties and the sharing of research results. In summary, the domestic-level access measures aim to:

- Create legal certainty, clarity, and transparency.
- Provide fair and non-arbitrary rules and procedures.
- Establish clear rules and procedures for prior informed consent and mutually agreed-on terms.
- Provide for issuance of a permit or equivalent when access is granted.
- Create conditions to promote and encourage research contributing to biodiversity conservation and sustainable use.

- Pay due regard to cases of present or imminent emergencies that threaten human, animal, or plant health.
- Consider the importance of crop and livestock genetic resources for food security.

The Nagoya Protocol and African Animal Genetic Resources in Africa

Working as a Group, African States were very active in the Nagoya Protocol negotiations because access and benefit-sharing are issues of major importance for Africa—the continent having rich heritage of biological diversity, genetic resources, and associated traditional knowledge. These assets have often been misappropriated or utilized without fair and equitable sharing of the accruing benefits with the countries of origin or indigenous and local communities. The African Group therefore considered that, because Africa is primarily a provider of genetic resources and associated indigenous knowledge, preventing injustices of this nature had to be a priority for the continent.

However, Africa is increasingly also a user of genetic resources, and crop varieties and livestock breeds originating from other parts of the world make (or can potentially make) major contributions to Africa's agriculture and food security. Most African countries are Parties to the International Treaty on Plant Genetic Resources for Food and Agriculture, and the continent has participated, and continues to participate, in negotiations at the FAO Commission on Genetic Resources for Food and Agriculture on Access and Benefit-Sharing measures both for plants and other groups of genetic resources, including animals, aquatic organisms, invertebrates, microorganisms, and forestry resources.

African institutions also form part of international research networks around, for example, taxonomy (requiring access to specimens), health (requiring access to human, animal, and plant pathogens), and climate change adaptation (requiring access to genetic resources adapted to different environmental conditions). Ensuring that Africa benefits from such research has substantial access and benefit-sharing implications. With increasing scientific and technological capacity, Africa has great potential to turn its genetic resources and associated indigenous knowledge into novel biotechnology products. Thus, access and benefit-sharing, if properly implemented, with appropriate training, technology transfer, and investments offer opportunities to increase Africa's ability to add value to, and benefit from, its natural and cultural resources and to help alleviate poverty, stimulate inclusive economic development while also serving as an incentive for sustainable use and conservation of biodiversity.

The adoption of the Nagoya Protocol was seen by African stakeholders as opportunity for a more coordinated approach to access and benefit-sharing that is coherent and in synergy with agreed African positions and related regional and international instruments (AUC 2015). Thus, the provisions of the Protocol were analysed against the background of *The African Common Position for the Negotiations of the International Regime on Access and Benefit-Sharing (IR-ABS)* which was adopted by the African Conference of Ministers in charge of Access and Benefit-Sharing held in March 2010 in Windhoek, Namibia. This process included a reflection on how Africa could best implement the Nagoya Protocol in synergy with relevant regional instruments such as the 2001 *African Model Law for the Protection of the Rights of the Local Communities, Farmers and Breeders and for the Regulation of Access to Biological Resources* ("the African Model Law") (AUC 2015). Building on this, the 14th meeting of the African Ministers Conference on the Environment (AMCEN), held in Arusha, Tanzania, in September 2012, resolved "*to encourage the African Union Commission to continue its ongoing work in the development of Guidelines to support the coordinated implementation of the Nagoya Protocol in Africa*".

These processes culminated into the *African Union Policy Framework and Guidelines for the Coordinated Implementation of the Nagoya Protocol in Africa* (AUC 2015), and this was adopted by the AU Assembly at its 25th Ordinary Session held in Johannesburg, South Africa, in 2015. The Guidelines for the Coordinated

Implementation of the Nagoya Protocol in Africa provide guidance for the implementation of the Protocol and for an access and benefit-sharing system at national and regional levels.

The Guidelines urge AU Member States to become Party to the Protocol and emphasize that Member States should exercise their sovereign rights over their natural resources in the implementation of the Protocol by establishing in their domestic law that Prior Informed Consent is required for access to, and utilization of, their genetic resources—including those held by indigenous and local communities and associated traditional knowledge. The Guidelines also remind Member States to take note of the definitions of *"utilisation of genetic resources"*, "biotechnology", and "derivatives" in Article 2 of the Nagoya Protocol to ensure that domestic legislative, administrative, or policy measures comprehensively cover the full scope of utilization of genetic resources, in order to trigger an effective benefit-sharing mechanism.

Objectives of the Guidelines

The stated objectives of the *African Union Policy Framework and Guidelines for the Coordinated Implementation of the Nagoya Protocol in Africa* are to:

(a) Provide practical guidance to African Union Member States on how national Access And Benefit-Sharing (ABS) systems can be implemented in a regionally coordinated manner, consistent with the provisions of the Nagoya Protocol, so as to preserve key African interests and positions while preventing a "race-to-the-bottom" scenario in which users of genetic resources and associated traditional knowledge play off African Union Member States and/or African indigenous local communities against one another.
(b) Establish a coordinated and cooperative regional approach to preventing misappropriation of African genetic resources and/or associated traditional knowledge and to punishing such misappropriation when it occurs.
(c) Encourage utilization of Africa's genetic resources and associated traditional knowledge assets in ways that support regional objectives and strategies on human resource development, technology transfer, scientific and technical capacity building, food security, and economic growth, while encouraging conservation and sustainable use of natural and human capital, including the rights of indigenous local communities.
(d) Facilitate the establishment of common African Access and Benefit-Sharing standards to inform benefit-sharing arrangements.

The Guidelines also provide guidance to Member States on institutional arrangements to regulate and manage Access and Benefit-Sharing, Prior Informed Consent, and Mutually Agreed Terms under the Protocol, including steps for establishing application and permitting processes. Suggested institutional arrangements include the establishment of National Focal Points and Competent National Authorities, including those of Indigenous Local Communities. These institutional obligations are intended, inter alia, to provide legal certainty for applicants seeking access to genetic resources and associated traditional knowledge. The Guidelines further advise that African Union Member States may wish to consider establishing mechanisms such as National Inter-Agency Access and Benefit-Sharing Committees or National Multi-Stakeholder Committees involving representatives from relevant government ministries and other authorities, indigenous local communities, and other relevant stakeholders to foster internal coordination, communication, and dialogue regarding regulation of access and benefit-sharing at the national level and streamlining institutional/administrative and decision-making arrangements and procedures.

Out of the 54 African countries, 47 are Parties to the Nagoya Protocol (CBD 2023b) (Table 30.7). Among the countries which have established National Focal Points for Access and Benefit-Sharing, some are Party States, others are not. In addition, many African governments lack regulations on bioprospecting, resulting in the unregulated exploitation of their biological and genetic resources, including livestock. Fortunately, as evidenced by their engagement in international negotiations of these instruments, African governments are increasingly recognizing the role legal frameworks play in providing controlled and legally secure access to potential users of locally available biological and genetic resources. However, as is the case with the Cartagena Protocol, operationalization of the Nagoya Protocol has been highly variable across countries. Unfortunately, no systematic quantification has been done to determine if and how countries are meeting their obligations under the CBD and the Nagoya Protocol. This is also true with regard to the extent of use of the AU Guidelines and what benefits have accrued to them so far. Also similar to the status of the Cartagena Protocol, a major concern regarding the operationalization of the Nagoya Protocol is the lack of capacity, including institutions and systems, to implement it effectively in developing regions of the world. African governments are reporting little progress and pointing out that reluctance by genetic resources user countries (mostly in the Global North) to share benefits, by paying for the true value of AnGR from the South, remains a challenge.

30.2.4 The Global Plan of Action for Animal Genetic Resources

Globally, it is estimated (FAO 2015) that more than 8800 domestic animal breeds or populations (of mammals and birds) within 38 species have been developed by farmers and pastoralists in diverse environments in the 12,000 years since the first livestock species were domesticated. These breeds and populations now represent unique combinations of genes conferring numerous characteristics, including adaptability. Today, faced with the needs of a growing population, changes in consumer demand, and the challenge posed by climate change and emerging diseases, we need to recognize and utilize these adaptive attributes and potentials to face an uncertain future.

The State of the World's Animal Genetic Resources for Food and Agriculture (SoW-AnGR) published in 2007 (FAO 2007) provided the first global assessment and analysis of the status and trends in animal genetic resources. It raised global awareness and understanding of their ongoing erosion. It also identified significant gaps in capacities of countries to manage animal genetic resources, particularly in the developing world. The Sow-AnGR report formed an important cornerstone in the creation of global awareness. As part of this process, the international community, in September 2007, adopted the first ever *Global Plan of Action (GPA) for Animal Genetic Resources*, comprising 23 *Strategic Priorities* which fall under *four Strategic Priority Areas* aimed at combating the erosion of animal genetic diversity and using animal genetic resources sustainably. It was considered that the implementation of the *Global Plan of Action* would contribute significantly to achieving the Millennium Development Goals (MDG 1, to eradicate extreme poverty and hunger; and MDG 7 to ensure environmental sustainability).

Managing AnGR involves considerations of complex interrelated issues and questions—those specific to the resources themselves (e.g. selective breeding, use, and conservation) and those which are cross-sectorial (e.g. animal health, trade standards, and environmental management). Moreover, responsibilities are invariably shared across sectors and institutions, nationally and internationally. Further, policymakers in many countries, and internationally, are seldom aware of the diverse and significant contributions of animal genetic resources to food and agriculture and of traditional rights of livestock keepers. Consequently, the sustainable use and conservation of animal genetic resources has remained a low priority in the development of agricultural, environmental, trade, and human and animal health policies. The effect has been a failure to invest adequately and appropriately in essential institutional development and associated human and infrastructural capacity.

The GPA provides a framework that countries can use to develop and operationalize policies for characterization, conservation, and use of AnGR. It aims to promote a pragmatic, systematic, and efficient overall approach, which harmoniously addresses the development of institutions, human resources, cooperative frameworks, and resource mobilization.

The aims of the GPA (FAO 2007) are to:

- Promote the sustainable use and development of animal genetic resources, for food security, sustainable agriculture, and human well-being in all countries.
- Ensure the conservation of the important animal genetic resource diversity, for present and future generations, and to halt the random loss of these crucial resources.
- Promote, as per Nagoya Protocol, a fair and equitable sharing of the benefits arising from the use of animal genetic resources for food and agriculture and recognize the role of traditional knowledge, innovations, and practices relevant to the conservation of animal genetic resources and their sustainable use, and, where appropriate, put in place effective policies and legislative measures.
- Meet the needs of pastoralists and farmers, individually and collectively, within the framework of national law, to have non-discriminatory access to genetic material, information, technologies, financial resources, research results, marketing systems, and natural resources, so that they may continue to manage and improve animal genetic resources and benefit from economic development.
- Promote agro-ecosystems approaches for the sustainable use, development, and conservation of animal genetic resources.
- Assist countries and institutions responsible for the management of animal genetic resources to establish, implement, and regularly review national priorities for the sustainable use, development, and conservation of animal genetic resources.
- Strengthen national programmes and enhance institutional capacity—in particular, in developing countries and countries with economies in transition—and develop relevant regional and international programmes. Such programmes should include education, research and training to address the characterization, inventory, monitoring, conservation, development, and sustainable use of animal genetic resources.
- Promote activities aimed at raising public awareness and bringing the needs of sustainable use and conservation of animal genetic resources to the attention of concerned governments and international organizations.

The *Strategic Priorities for Action* cover the four Strategic Priority Areas, namely:

Strategic Priority Area 1: Characterization, Inventory, and Monitoring of Trends and Associated Risks. The actions provide a consistent, efficient, and effective approach to the classification of AnGR and to assess trends in, and risks to, the AnGR.

Strategic Priority Area 2: Sustainable Use and Development. Focus is on ensuring sustainability in animal production systems, especially on food security and rural development.

Strategic Priority Area 3: Conservation. That is, steps needed to preserve genetic diversity and integrity, for the benefit of current and future generations.

Strategic Priority Area 4: Policies, Institutions, and Capacity-building. Focus is on actions which address the key questions of practical implementation, through coherent and synergistic development of the necessary institutions and capacities.

The complete list of the 23 strategic priorities of the GPA is summarized in Table 30.3 by Strategic Priority Area.

Institutional Framework for Operationalizing the Global Plan of Action (GPA)

Among instruments for operationalizing the GPA is an institutional framework (FAO 2011) which provides a basis for effective management of animal genetic resources, both nationally and internationally, and guidelines which present an overview of the components of the global network for the management of animal genetic resources and how they can be strengthened at national and regional levels. Guidance is provided on the role of National Coordinators for the

Table 30.3 Strategic Priority Areas (SPA) and Strategic Priorities (SP) of the Global Plan of Action for AnGR

Geographical target	SPA1 Characterization, inventory, and regularity of monitoring	SPA2 Sustainable use and development	SPA3 Conservation	SPA4 Policies, institutions, and capacity-building
National	*SP1*: Inventory and characterize AnGR, monitor trends and risks associated with them, and establish country-based early-warning and response systems	*SP3*: Establish and strengthen national sustainable use policies *SP4*: Establish national species and breed development strategies and programmes *SP5*: Promote agro-ecosystem approaches to the management of AnGR *SP6*: Support indigenous and local production systems and knowledge systems	*SP7*: Establish national conservation policies *SP8*: Establish or strengthen in situ conservation programmes *SP9*: Establish or strengthen ex-situ conservation programmes	*SP12*: Establish or strengthen national institutions, including national focal points *SP13*: Establish or strengthen national educational and research facilities *SP14*: Strengthen national human capacity for characterization, inventory, and monitoring of trends and associated risks *SP18*: Raise national awareness of the roles and values of AnGR *SP20*: Review and develop national policies and legal frameworks for AnGR
Regional			*SP10*: Develop and implement regional and global long-term conservation strategies	*SP17*: Establish regional focal points and strengthen international network
International	*SP2*: Develop international technical standards and protocols for characterization, inventory, and monitoring of trends and associated risks		*SP11*: Develop approaches and technical standards for conservation	*SP15*: Establish or strengthen international information sharing, research, and education *SP16*: Strengthen international cooperation to build capacities *SP19*: Raise regional and international awareness of the roles and values of AnGR *SP21*: Review and develop international policies and regulatory frameworks relevant to AnGR *SP22*: Coordinate the Commission's efforts on AnGR policy with other international forums *SP23*: Strengthen efforts to mobilize resources, including financial, for the conservation, sustainable use, and development of AnGR

Source: FAO (2007)

Management of Animal Genetic Resources and the development and operation of National Focal Points for the Management of Animal Genetic Resources supported by National Advisory Committees, working groups, and country stakeholder networks.

National Focal Points are supposed to initiate, lead, facilitate, and coordinate country activities related to the implementation of National Strategies and Action Plans (NSAPs) for AnGR and to interface with AnGR stakeholders within the country and cooperate with the *Regional Focal Point* (where available) and with the *Global Focal Point*, to plan and develop regional and global initiatives as appropriate. The Global Focal Point is located at FAO headquarters in Rome, while the Regional Focal Points are established within the respective regions (Regional Economic Communities or other Sub-Regional Organizations) as a means of facilitating regional communications, providing technical assistance and leadership in AnGR management and coordinating activities that can best be implemented at regional level or will benefit from coordination among countries within the region.

The National Coordinator is expected to work within the country's National Focal Point and lead national efforts to implement the Global Plan of Action. The National Coordinator is defined as: "*the government nominated person who coordinates national implementation of the Global Plan of Action for Animal Genetic Resources and leads the development and operation of a national network on animal genetic resources. He or she is the contact person for communication with FAO on matters relating to the implementation of the Global Plan of Action for Animal Genetic Resources and with global and regional animal genetic resources networks.*"

The Global Plan of Action and African States
Cao et al. (2021) have undertaken an analysis of progress with implementation of the GPA and observed clear differences in the implementation across countries and regions. Results from the 2019 round of the SoW-AnGR show that, in general, countries with higher GDP per capita report higher scores regarding Global Plan implementation. Countries reporting high scores but relatively low GDP per capita were almost systematically BRICS countries (specifically Brazil, China, India, and South Africa) having made significant investment in livestock production in recent years to meet rising consumer demand. By contrast, some Near and Middle East countries with relatively high GDP per capita (Qatar and Saudi Arabia) reported very low scores in Global Plan implementation, which is possibly in relation to their relatively small livestock sectors and thus lack of emphasis on livestock production in country policies.

For African countries, greater progress was reported for SPA4 (Policies, Institutions, and Capacity-Building). This was considered to be related to the implementation of donor-funded projects dedicated to the management of AnGR at regional level (mainly the interventions by AU-IBAR) or to the development of national strategies considering AnGR issues, such as the Livestock Master Plans—supported by the International Livestock Research Institute (ILRI)—developed by Ethiopia, Rwanda, and Tanzania. The lack of progress regarding SPA3 was observed in all regions except Africa where, building on existing facilities, AU-IBAR has engaged some RECs and countries to facilitate establishment of regional gene banks for AnGR as part of the support being given to countries to establish National Strategies and Action Plans. To date, 15 countries have established NSAPs and 12 are in the process of doing so (Sect. 30.5.1).

30.3 The African Union Inter-African Bureau for Animal Resources

Originally established as the Inter-African Bureau of Epizootic Diseases (IBED) in 1951 to study the epidemiological situation of, and lead the fight against, rinderpest in Africa, the organization was renamed African Union–Inter-African Bureau for Animal Resources (AU-IBAR), to reflect its broadened mandate which is to support and coordinate the improved utilization of animals (livestock, fish, and wildlife) as a resource for human well-being in the Member States of the African Union (AU) and to contribute to economic development, particularly in rural areas. An examination of the role of AU-IBAR in animal resources policy is best done by looking at the specific areas of its mandate (referred to as niche in its 2018–2023 Strategy (AU-IBAR 2018)), which are:

(a) Formulation, validation, and dissemination of animal resources continental strategic frameworks.
(b) Coordination of Africa's contribution to the development of relevant standards and regulations and enhancement of compliance by Member States.
(c) Strengthening institutional capacity and support of policy coherence and harmonization at national, regional, and continental levels.
(d) Packaging and dissemination of information and knowledge on animal resources to Member States, Regional Economic Communities, and other continental and regional institutions.
(e) Coordination of the African Voice (and common positions) in Animal resources development.
(f) Enhancing the pan-African coordination, networking, and partnerships in Animal Resources.
(g) Providing technical support and associated tools to Member States for effective implementation of their policies and strategies.

AU-IBAR's main clients are the African Union (AU) Member States and the Regional Economic Communities (RECs) to which the Member States belong. Over the years, AU-IBAR has developed a track record across domains of animal production and health, including in pastoral systems, and is recognized in the development community as providing leadership on animal resources issues at the continental level.

AU-IBAR's strategic positioning and approach to the delivery on its mandate is guided by the recognition that the productive use of animal resources is only sustainable when clear national livestock development policies, adequate institutional frameworks, relevant scientific research activities, and sustained technical and financial resourcing are in place.

30.3.1 The African Animal Genetic Resources Project (2013–2018)

Recognizing that Africa was lagging in the implementation of the Global Plan of Action (GPA) for sustainable use of AnGR, the AU-IBAR implemented a project in the period 2013–2018 (AU-IBAR 2014), funded by the European Union (EU), titled "Animal Genetic Resources Project" which sought to fast-track the implementation of the GPA in the continent. With a theme *Strengthening the Capacity of African Countries to Conserve and Sustainably Use African Animal Genetic Resources*, the project focused on building capacity for effective formulation and implementation of policies and strategies for the management of AnGR and creating awareness for its inclusion into national and regional agricultural investment priorities. The initiative also aimed to develop a suite of biodiversity indicators suitable for the African context to assess and report on progress towards the set targets. Project interventions were at two levels:

- *Regional Economic Community (REC) level*: The focus here was on harmonization of national policies and regional policies for transboundary breeds. The project envisaged to leverage on regional complementarity and pools of resources for the establishment of regional gene banks (see below) and creation of the institutional environment for the implementation of the GPA at regional levels.
- *National level*: The focus was on building capacity for formulation of national policies and strategies as well as national action plans for the implementation of the GPA, ensuring the mainstreaming of the AnGR issues in national sectoral and inter-sectoral policies, strategies, and plans.

The Project also aimed to build consensus on the methodology and tools for the characterization and inventory of AnGR. Communication of outputs and outcomes included policy briefs, specialized dissemination, and training sessions on the use of data tailored to specific target groups and use of community radios and local publications in local languages. The major deliverables of the Project are summarized in Table 30.4 by objective.

30.3.2 The State of Farm Animal Genetic Resources in Africa

An important rationale for the AU-IBAR programme was the fact that African AnGR are undervalued, and the awareness about their economic potential among policy makers and the general public is limited, and the recognition that this state of affairs had resulted in a lack of appropriate infrastructure, institutions, capacity, and technologies to harness the inherent potential of these genetic resources. It was also recognized that, since the adoption of the Global Plan of Action (GPA), only a few countries in Africa had formulated National Strategies and Action Plans (NSAPs) for implementation. Moreover, in countries where the NSAPs had been developed, low technical and financial capacity as well as a lack of information on the current population status and trends of Farm AnGR was constraining their implementation. In addition, the absence of harmonized characterization, inventory, and monitoring approaches had resulted in lack of credible evidence to inform policy making.

It was therefore imperative that the "baseline" situation be established to inform scope of work going forward. As part of its objective to catalogue information and knowledge gaps, the initiative sought to identify the most significant gaps and needs for conservation and utilization of AnGR, to provide the basis for designing strategic national, regional, and international policies to underpin the implementation of priority activities.

The assessment of the current situation leading to the publication, in 2019, of *the State of Farm Animal Genetic Resources in Africa* (AU-IBAR 2019) (in 2019) identified the state of countries in regard to livestock polices and strategies relevant for AnGR. Although livestock policies and/or strategies of many countries include at least some reference to AnGR, implementation is limited—almost in all countries. Indeed, many countries have had livestock policies over decades, but these have never been implemented.

The assessment report proposes the following priority areas for the development of Farm AnGR in Africa:

(a) Strengthening national and regional efforts in characterization and inventories of indigenous breeds.
(b) Establishment of sustainable breeding and conservation strategies for native and transboundary breeds.
(c) Establishment of regional and continental programmes to understand existing crossbred populations and creation of productive and resilient synthetic breeds.
(d) Establishment and strengthening of information systems—linked to development of monitoring tools.
(e) Creation of AnGR monitoring committees at country and regional levels.
(f) Coordination mechanisms: Mapping of key actors and development of networks, partnership arrangements, and collaboration around activities underpinned by AnGR Focal Points in each country.
(g) Harmonization of policies and legislation, and monitoring tools.
(h) Ex-situ conservation: Strengthening existing or facilitating establishment of new national and regional gene banks, and a continental backup facility to serve all regions and countries.
(i) Strengthening the capacity of value chain actors—individuals and institutions.
(j) Advocacy and awareness raising of Farm AnGR issues and needed actions.

30.3.3 Policy Framework for Sustainable Use of AnGR

As can be seen from the summary of strategic areas in Table 30.4 (Sect. 30.3.1), a major output of the AU-IBAR AnGR work has been the devel-

Table 30.4 Summary of achievements of AU-IBAR's Animal Genetic Resources Project

Result area	Objective	Achievements
1. The Status and trends of AnGR in Africa	1. Identify threatened and at-risk breeds	Forty-four Member States were supported to compile their AnGR status reports as part of the FAO-led second State of the World AnGR Report Publication of the State of Animal Genetic Resources in Africa (Sect. 30.3.2)
	2. Take stock of existing policies and regulations on the use of AnGR	Inventory of Policies and Strategies related to AnGR was completed and validated by stakeholders in East, Central, West, North, and Southern Africa. This showed that: Only a few countries had policies and legislation specific to the management of AnGR (Sect. 30.5.1) Most policies and regulations relate to control of animal movement—focusing on disease control, particularly transboundary diseases Policies and legislation in many countries are outdated There are no policies and legislation addressing crossbreeding or transboundary breed issues
	3. Assess the genetic impact of livestock production systems	The assessment of the impact of livestock mobility (nomadism) was initiated in Central and East Africa. It highlighted impacts through interbreeding and identified lessons and good practices
	4. Assess impact of breeding programs on genetic diversity and socio-economic status	An assessment was initiated to identify the key selection and breeding programmes in the regions, focusing on transboundary breeds
2. Policy frameworks for sustainable use of AnGR	1. Develop guidelines for crossbreeding policies	Issues paper on crossbreeding and conservation and management of transboundary AnGR in Africa has been developed—and will be a basis for consultations to identify key issues to be covered in the regional policy and legislative frameworks on crossbreeding
	2. Develop technical standards and protocols for exchange and use of genetic materials	The African Union Strategic Guidelines for the Coordinated Implementation of the Nagoya Protocol—and analysis of the legal environment for the exchange and movement of animal genetic material in the continent. This will culminate in the development of technical standards, legislation, or regulatory requirements for the exchange and movement of animal genetic materials

(continued)

Table 30.4 (continued)

Result area	Objective	Achievements
3. Conservation and improvement strategies and initiatives	1. Support countries to develop and implement National Strategies and Action Plans (NSAPs) for AnGR	Work has been initiated on strengthening capacity of countries to develop NSAPs and mainstream AnGR in the national ag. investment plans (NAIPs) through the CAADP country processes
	2. Support countries to develop or strengthen breeding and conservation strategies as part of their NSAPs	A total of 15 countries completed the development of their NSAPs, and 12 were underway at the close of the project (Sect. 30.5.1)
	3. Support development of regional conservation policy and strategic frameworks for trans-boundary livestock breeds at risk	Inventory of policies and legislation conducted in Southern (SADC) and Eastern Africa (EAC and IGAD) countries to be used to guide the approach for other regions
	4. Support development of regional facilities for ex-situ conservation	Regional gene banks proposed or initiated (Sect. 30.5.1)
	5. Support development and/or strengthening of Livestock Breeders' Associations	Explored different models for advisory and financial support to existing Livestock Breeders' Associations
4. Knowledge, attitude, and practice regarding contribution of livestock to economic growth, food security, and poverty reduction	1. Develop harmonized tools and protocols for characterization and inventory of AnGR	An AnGR Taxonomy Advisory Group (AnGR-TAG) has been established to develop and/or improve tools and guidelines for characterization, inventory, and monitoring
	2. Establish AnGR databases	A module on AnGR, *the African Animal Genetic Resources Information System* (AAGRIS), has been created in *Animal Resources Information System 2* (ARIS2) ARIS addresses data and information gaps to underpin assessment of capacities and to facilitate effective responses to animal health and broader animal resources challenges; the software will support better policy and decision-making by enhancing information and knowledge management to improve planning of interventions The provision of ICT equipment to 31 Member States to facilitate capture and analysis of data on AnGR
	3. Establish and strengthen systems for monitoring of trends of breeds and associated risks	Tools for characterization, inventory, and monitoring of trends and risks have been developed, validated, and piloting done in a few countries
	4. Develop regional networks for information sharing	Regional Domestic Animal Diversity Networks (DAD-Nets) have been established. To-date, DAD-Net West Africa, DAD-Net for Central Eastern Africa, Northern, Southern Africa regions are operational under the Sub-Regional Focal Points and AU-IBAR
	5. Establish or strengthen regional focal points for AnGR	Sub-Regional Focal Points (S-RFPs) for AnGR established and are at different levels of operation
	6. Document and disseminate best practices and lessons on AnGR conservation and improvement	Dissemination of AnGR information through the AU-IBAR and partners' websites, African DAD-Nets, and international workshops Publication of AnGR characterization, conservation, and use issues through the Bulletin of Animal Health and Production in Africa (BAHPA)

opment of a policy framework, consisting of a range of initiatives at national and regional (REC) levels, for the sustainable use of AnGR. The policy framework provides, inter alia, guidelines for the formulation and harmonization of crossbreeding, protocols and tools for characterization and inventory of AnGR, in situ and ex-situ conservation, technical standards, protocols for the exchange and use of genetic materials, and establishment and implementation of action plans on AnGR. As part of these efforts, AU-IBAR has organized and facilitated capacity building support to countries and RECs around sustainable use and conservation of AnGR through institutionalizing national and regional policy and legal and technical instruments. These efforts have also included catalyzing and facilitating the formation or strengthening of regional and national livestock breeders' associations.

30.3.4 AnGR Gene Banking Strategy

AU-IBAR is developing a strategy for the use of AnGR gene banks (AU-IBAR 2019, 2020). A regional approach that is driven and owned by Member States but ensures efficiency and effectiveness through sharing of facilities hosted at designated regional locations has been proposed in this regard. Potential advantages of regional gene banking include: providing access to cryopreservation facilities for countries lacking national gene banks; acting as a back-up storage for national gene banks; providing storage of material from trans-boundary breeds; and providing uniform tools and protocols for identification and evaluation of breeds to be preserved.

However, although regional animal gene banks have these advantages which make them potentially cost-effective, several issues still require attention. One of these is the fact that Member States who are signatories to the Convention on Biological Diversity need to ensure that international transfer of animal genetic resources is consistent with the terms of the Nagoya Protocol on Access and Benefit-Sharing (Sect. 30.2.3). The FAO Guidelines for cryo-conservation of animal genetic resources will inform the development of common technical standards for operation of the animal gene banks. The analysis undertaken by AU-IBAR suggests that regional animal gene banks should be located in national facilities and, to ensure they are functional, must be supported with the necessary equipment, training, and supplies. Special attention has to be given to animal health to prevent the spread of animal diseases through the movement of genetic material. This includes accurate records of the health status of sampled animals. Legal protocols must be in place to ensure that the samples in a regional animal gene bank remain in the ownership of the country of origin with provision made for appropriate access by interested parties presenting valid claims for the use of such material. Users will also be expected to replenish the gene bank, as needed, with samples (e.g. of semen or embryos) from the regenerated animals. Development and harmonization of legal instruments have been initiated, i.e. Material Acquisition Agreements (MAAs) and Material Transfer Agreements (MTA).

Regional gene banks proposed or initiated to date are:

- West Africa: Bobo Dioulasso, Burkina Faso, hosted by CIRDES
- Southern Africa: Gaborone, Botswana, Department of Agricultural Research (DAR)
- Eastern Africa: Entebbe, Uganda, National Animal Genetic Resource Centre and Data Bank (NAGRIC & DB)
- Central Africa: Bambui, Cameroun, Centre Régional de Recherche Agricole de Bambui-Mankon
- North Africa: To be determined
- Continental backup: Debre Zeit, Ethiopia, The African Union Pan-African Veterinary Vaccine Centre (AU-PANVAC)

30.3.5 The 2015–2035 Livestock Development Strategy for Africa (LiDeSA)

African Heads of State and Governments endorsed the first ever 20-year Livestock

Development Strategy for Africa (LiDeSA) (AU-IBAR 2015) in 2017. The strategy calls for an agricultural sector transformation agenda and underscores the importance of utilizing animal genetic resources to address the continent's pressing needs for food and nutritional security. It was framed as the continent blueprint for the transformation of the livestock sector through energizing its under-utilized potentials. The strategy identified the insufficient attention paid by Member States to the merits of exploiting the resilience and adaptive capacity of local animal breeds. It noted that Member States have tended to go for perceived quick fixes through importation of exotic germplasm, and in the process putting indigenous breeds at risk. The strategy identified the following six pathways for sustainable use of AnGR in African livestock transformation:

(a) Accelerate genetic improvement and access to appropriate, productive, and resilient breeds that best match the production contexts, informed by research and development, innovation (including use of ICT), and appropriate partnerships.
(b) Formulate and promote supportive, inclusive, and integrated institutional and policy frameworks for the sustainable management of animal genetic resources.
(c) Conduct inventory and characterization and innovatively utilize indigenous bio-resources (e.g. rumen and milk microbes, genes) with potential industrial and business application in African production systems
(d) Design and implement innovative and sustainable breeding and conservation programmes at national and regional levels.
(e) Develop appropriate and more effective delivery systems of appropriate and superior genetic material.
(f) Develop and support inclusive community-public-private partnerships and business models 1for generation, implementation, and delivery of appropriate genetic resources.

It was envisaged that the implementation of the LiDeSA as planned would dramatically transform the African livestock sector, towards its potential as a major contributor to the continent's socio-economic development.

30.4 Africa's Regional Economic Communities (RECs)

The Regional Economic Communities (RECs) are regional groupings of African states. Their creation was proposed by the 1980 Lagos Plan of Action for the Development of Africa and the 1991 Abuja Treaty as the basis for wider African integration. Each with their own role and structure, the RECs aim to facilitate regional economic integration between members of the region and through the wider African Economic Community (AEC), established under the Abuja Treaty. The RECs are closely integrated with work of the African Union (AU). Increasingly, RECs are involved in coordinating the interests of African Union Member States in broad areas such as peace and security, development, and governance. Figure 30.1 (map) shows which RECs each African country belongs to and indicates the various combinations of overlapping memberships to the regional economic communities in the continent. The AU recognizes eight RECs, namely the Arab Maghreb Union (UMA), the Common Market for Eastern and Southern Africa (COMESA), the Community of Sahel–Saharan States (CEN–SAD), the East African Community (EAC), the Economic Community of Central African States (ECCAS), the Economic Community of West African States (ECOWAS), the Intergovernmental Authority on Development (IGAD), and the Southern African Development Community (SADC). This section presents engagements by the different RECs in the development and operationalization of policies and strategies relevant for AnGR. The section has benefited from a 2016 review of agricultural and livestock policies undertaken by AU-IBAR (in

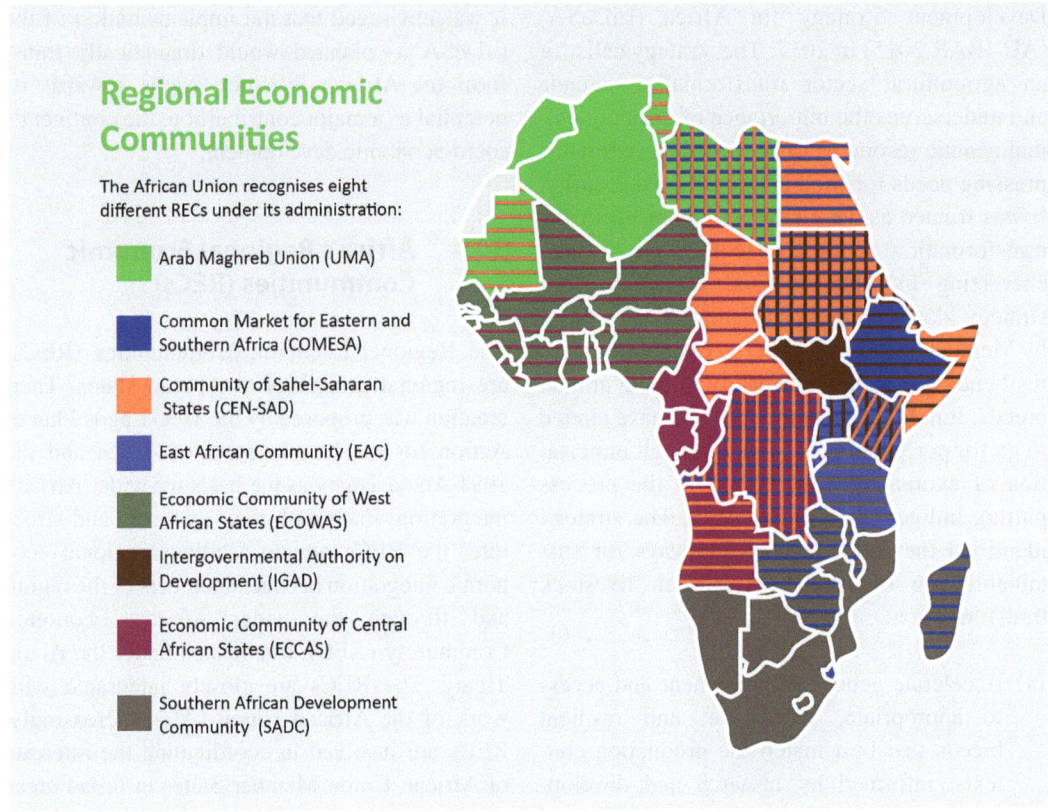

Fig. 30.1 The Regional Economic Communities in Africa

collaboration with RECs) as part of its Reinforcing Veterinary Governance in Africa (AU-IBAR 2016).

30.4.1 Common Market for Eastern and Southern Africa (COMESA)

Informed by, and significantly aligned to, the AU-IBAR's *"Livestock Development Strategy for Africa"* (LiDeSA), the COMESA Regional Livestock Policy Framework (COMESA 2015) includes *"Policies to enhance Livestock Production and Animal Health to increase Productivity and Resilience of livestock production systems."*

Specifically, policies to *improve the genetic potential and performance of animals include*:

- Support Member States to accelerate genetic improvement and access to appropriate, productive, and resilient breeds that best match the production systems/environments mediated by innovative research and development, business, public-private partnerships, and ICT.
- Formulate and promote supportive, inclusive, and integrated institutional and policy frameworks for the sustainable management of animal genetic resource.
- Support Member States to conduct inventory and characterization and innovatively utilize indigenous bio-resources (e.g. rumen and milk microbes, genes) with potential industrial and business applications in African production system.
- Support Member States to design and implement innovative and sustainable breeding and conservation programmes at national and regional levels.
- Support Member States to develop appropriate and more effective delivery systems of appropriate and superior genetic material.

- Support Member States to develop and implement inclusive community-public-private partnerships and business models for generation, implementation, and delivery of appropriate genetic resources.

While not specific to AnGR, two other relevant policy areas are: policies to enhance innovation, generation, and utilization of technologies, capacities, and entrepreneurship skills of livestock value chain actors; and policies to enhance access to markets, services, and value addition.

30.4.2 East African Community (EAC)

The EAC was established with a vision to set up a prosperous, competitive, secure, stable, and politically united East Africa and provide a platform to widen and deepen Economic, Political, Social, and Culture integration to improve the quality of life of the people of East Africa. It is a regional intergovernmental organization of eight Partner States: The Republic of Burundi, the Democratic Republic of the Congo, the Federal Republic of Somalia, the Republic of Kenya, the Republic of Rwanda, the Republic of South Sudan, the Republic of Uganda, and the United Republic of Tanzania, with its headquarters in Arusha, Tanzania. The work of the EAC is guided by its Treaty which established the Community. It was signed on 30 November 1999 and entered into force on 7 July, 2000 following its ratification by the original three Partner States—Kenya, Tanzania, and Uganda.

The EAC Council of Ministers adopted the EAC Livestock Policy in 2016. The policy (EAC 2016) envisions a *"vibrant livestock industry contributing significantly to improved living standards of EAC citizens, economic growth and sustainable natural resources management by 2025."* The six strategic pillars which explicitly cover AnGR are:

1. *Policy instruments and strategies to enhance conservation and sustainable utilization of livestock resources:*
   - Strengthen institutional capacities and engagements for strategic partnerships in implementation of the Global Plan of Action (GPA) for animal genetic resources.
   - Promote sustainable utilization of indigenous breeds with high priority to rare breeds with unique attributes through branding, incentives, and protection frameworks.
   - Support investments in regional conservation facilities.
   - Strengthen institutional capacities and engagement with strategic partners for effective and successful implementation of the GPA for animal genetic resources.
2. *Policy instruments and strategies to enhance access to quality breeding stock*:
   - Support public investment in infrastructure for animal breed improvement and multiplication of high-quality breeding stock.
   - Promote sustainable utilization and conservation of indigenous breeds with unique attributes.
   - Facilitate regional trade and investment in animal semen, embryos, and biotechnology research for production of quality breeding stock.
   - Support public-private partnerships in livestock improvement and multiplication of quality breeding stock.
   - Advice on, and support, sustainable livestock breeding programmes in Partner States.
   - Support Partner States to establish animal conservation and breeding centres
   - Domesticate and implement the GPA on AnGR conservation and utilization.
3. *Policy instruments and strategies to enhance wider application and benefits of biotechnology:*
   - Increase investment in establishment of regional advanced biotechnology research facilities and centres of excellence.
   - Prioritize biotechnology research in diagnostics product development and vaccines for diseases of economic importance.
   - Prioritize biotechnology research in genetic and product improvement technologies, animal feeds, and nutrition.
4. Policy instruments and strategies to enhance access to financial services and products:

- Advocate for implementation of fiscal and monetary policies that include livestock recording, registration, and transactions as collateral for securing loans.
- Promote strengthening and capacity building of cooperatives, village banks, and Self-Help Groups to provide financial services and products to livestock producers, traders, and SMEs.

5. Policy instruments and strategies to strengthen research in livestock subsector:
   - Advocate for Partner States commitment to allocating at least 5% of the annual research budget to livestock research.
   - Mobilize resources for establishment and administration of competitive livestock research funds.
   - Invest in capacity building, laws, and programmes strengthening intellectual property rights.
   - Prioritize research support to animal-based mechanization and integrated nutrient management research.
   - Support collaborative public-private research agendas in priority concerns to livestock value chain actors.

6. Policy instruments and strategies to enhance contribution of livestock to ecosystem productivity:
   - Support and promote implementation of payment for environmental services schemes in livestock systems.
   - Support actions for enhancing marketing of livestock-related environmental goods and services from carbon trade, biogas, and ecotourism.
   - Support actions for implementation of rules and regulations and pricing regimes to control access and use of fragile ecosystems
   - Support participatory processes and actions for implementation of community management of common pastures and water and other fragile ecosystems in each Partner state.

The EAC Livestock Policy proposes that a regional Animal Resources Committee (ARC) be formed to advise the EAC Agriculture and Food Security Council of ministers on technical matters of animal resources development. The EAC Secretariat commits to working with Member States to prepare a national master plan of action and implementation strategy of rolling out the policy in their national livestock programmes, and that the national master plans will facilitate prioritization of actions by each Member State to reflect its context, and it will detail implementation timeframe, specific actions, investments, responsible actors, and the indicators of progress for Monitoring and Evaluation towards achieving the policy objectives and targets.

30.4.3 Economic Community of Central African States (ECCAS)

The Economic Community of Central African States (ECCAS) was established in 1983 through a Constitutive Treaty, which was revised and adopted and entered into force in 2019. ECCAS is made up of 11 Member States—Angola, Burundi, Cameroon, the Central African Republic, Congo, Gabon, Equatorial Guinea, the Democratic Republic of Congo, Rwanda, Sao Tome and Principe, and Chad.

The ECCAS region prides itself in possessing significant agricultural and forestry potential. It hosts the Congo Basin Forest, Africa's largest rainforest, and also known as *"the Green Heartland of Africa"*. Relevant policy and strategy instruments in the field of Agriculture and food security include the development and adoption, in 2015, of the Common Agricultural Policy of ECCAS (CAP-ECCAS) and the Regional Programme for Agricultural Investment, Food Safety, and Nutrition (PRIASAN). In addition, Ministers of ECCAS Member States in charge of Agriculture, Livestock, and Fisheries made a Declaration in support of the transformation of Agriculture in Central Africa in accordance with the commitments of the 2014 AU Malabo Declaration.

In 2022, AU-IBAR convened a high-level technical Regional Workshop on Capacity

Building on the Guidelines to Strengthen the inclusion of Livestock in the Regional and National Agricultural Investment Plans (RAIPs and NAIPs) as part of the initial Livestock Sector Review of the ECCAS CAP and PRIASAN. AU-IBAR is partnering with ECCAS, FAO, and other partners to reformulate the ECCAS CAP and the PRIASAN. The formulation of the CAP and PRIASAN is considered as critical strategic initiative. However, to date, there is no regional livestock stand-alone policy document in the ECCAS Region, and individual Member States do not have specific national livestock development policies. At the same time, there are no national animal health/veterinary policies and strategies.

With funding from the European Union, the AU-IBAR is supporting the development of the livestock for livelihood in Africa (Live2Africa (AU-IBAR 2023)) project. Live2Africa is implementing the continental aspects of the Livestock Development Strategy (LiDeSA) (Sect. 30.3.5). One of the result areas is "To enhance the implementation of breeding and conservation programmes for sustainable utilization and management of African livestock." Part of this funding is being used to establish a regional chicken Layer-parent stock farm in ECCAS, at a location in Cameroon. The establishment of the farm will support other parts of the chicken value chain, including small and medium layers' commercial producers.

30.4.4 Economic Community of West African States (ECOWAS)

The Economic Community of West African States (ECOWAS—also known as CEDEAO in French and Portuguese) is a regional political and economic union of 15 countries of West Africa. It was established on 28 May 1975, with the signing of the Treaty of Lagos, with a mission to promote economic integration across the region. The stated goal of ECOWAS is to achieve "*collective self-sufficiency*" for its Member States by creating a single large trade bloc—building a full economic and trading union. ECOWAS aims to raise living standards and to promote economic development. A revised version of the treaty was agreed and signed on 24 July 1993 in Cotonou, Benin.

While ECOWAS does not have a livestock-specific policy document, the ECOWAS Agricultural Policy (ECOWAP) seeks broadly to ensure food sovereignty by reducing dependence on imports for food security and to reduce poverty, and the previous ECOWAP included a Livestock Action Plan (LAP 2011–2020). The goals of LAP cover: (1) promotion of livestock sector, specifically meat and dairy; (2) provision of security for cross-border movement of livestock and reduction of conflicts; (3) restructuring of the livestock production sector; and (4) creating a favourable environment for the development of livestock, meat, and dairy products (ECOWAS Commission 2010). In addition, ECOWAS has established an International Transhumance Certificate (ITC) which seeks to facilitate cross-border transhumance among ECOWAS member countries to assure free seasonal movement of livestock in search of water and pasture.

The operationalization of ECOWAP was done through action programs targeting six themes, namely (ECOWAS Commission 2009): (1) Water management; (2) Management of other shared natural resources; (3) Sustainable development of farms; (4) Markets and supply chains; (5) Prevention and management of food crises and other natural disasters; and (6) Institutional strengthening. In addition, there was a cross-cutting programme focused on the development of an Agricultural Information System (AGRIS).

While ECOWAP, through the LAP, addresses issues critical for livestock keepers in the region, it does not explicitly cover policies and strategies aimed at characterization, conservation, and sustainable use of indigenous AnGR. An examination of the policy landscape of individual ECOWAS member countries reveals that a large proportion of Member States have overarching visions and agriculture sector development policies and strategies, but more than half of them do not have specific livestock development policies or strategies. Similarly, most of the Member States have not developed animal health/veterinary policies or strategies, and the existing animal health legislations were enacted decades ago and are generally out of tune with current needs.

30.4.5 The Intergovernmental Authority on Development (IGAD)

The Intergovernmental Authority on Development (IGAD) in Eastern Africa was created in 1996. It succeeded the Intergovernmental Authority on Drought and Development (IGADD) which was founded in 1986 to mitigate the effects of the recurring droughts and other natural disasters that resulted in widespread famine, ecological degradation, and economic hardships in the region. The founding members were Djibouti, Ethiopia, Kenya, Somalia, Sudan, and Uganda. Eritrea joined in 1993 and South Sudan joined in 2011. During the launching of IGAD, members decided to enhance regional cooperation in three priority areas, namely, food security and environmental protection; economic cooperation, regional integration and social development; and peace and security.

Approximately 70% of the IGAD region is arid and semi-arid and is home to about 75% of the livestock population of the IGAD region. It is, therefore, not surprising that IGAD has developed many initiatives and programmes that address issues germane to animal resources. IGAD Drought Disaster Resilience and Sustainability Initiative (IDDRSI) was established following the severe drought that devastated the region in 2010/2011 period. IDDRSI aims at addressing the effects of drought and related shocks in the IGAD region in a sustainable and holistic manner. It has seven priority intervention areas, namely, (1) Natural Resources and Environment Management; (2) Market Access, Trade, and Financial Services; (3) Livelihood support and Basic Social services; (4) Disaster Risk Management, Preparedness, and Effective Response; (5) Research, Knowledge Management, and Technology Transfer; (6) Conflict Prevention, Resolution, and Peace Building; and (7) Coordination, Institutional Strengthening, and Partnerships.

The IGAD Centre for Pastoral Areas and Livestock Development (ICPALD) was established in 2012 as a specialized institution to "*promote and facilitate sustainable and equitable drylands and livestock development in the IGAD region*". ICPALD is expected to provide a platform for regional cooperation and coordination in drylands, and livestock development, including animal health, production and marketing, enhanced dryland agriculture, and value-added alternative livelihood products.

In 2022, IGAD developed a Strategy (2022–2037) for Sustainable and Resilient Livestock Development in View of Climate Change (IGAD 2022). The goal and purpose of the Strategy is to provide a framework to support the identification and prioritization of policies and actions to increase the resilience and sustainability of the livestock sector to climate change impact in the IGAD region. One of the priority intervention areas is on natural resource base and ecosystem services for livestock production, and one of its aims is to facilitate preservation of regional biodiversity.

IGAD has also collaborated with the AU-IBAR (IGAD and AU-IBAR 2018), to develop a model policy framework to guide IGAD Member Countries as they develop policy and legal instruments to support characterization, conservation, and sustainable use, including access to, and benefit-sharing of, animal genetic resources. The model policy framework is expected to contribute to the development of the National Action Plans on AnGR as provided for in the Global Plan of Action. The framework has five policy objectives, namely:

Policy Objective 1: Ensure the conservation of indigenous farm animal genetic resources and significantly improve their contribution to the livelihood of livestock keepers and the national economy.

Policy Objective 2: Ensure the national actions and multi-national collaboration of countries in the conservation and use of trans-boundary indigenous farm animal genetic resources.

Policy Objective 3: Develop a mechanism for sustainable utilization and development of indigenous animal genetic resources in tandem with the larger global community.

Policy Objective 4: Ensure mainstreaming of conservation and sustainable utilization of animal genetic resources in the various sectors and within and without the continent's borders.

Policy Objective 5: Ensure that global commitments are met, and agreements are kept in the implementation of indigenous farm animal genetic resources conservation, sustainable use and access, and equitable benefit-sharing.

It is envisaged that lessons from this process and its eventual rollout in IGAD Member States will inform approaches to be applied to support other African regions and countries towards development and operationalization of policies and strategies on animal genetic resources.

The policy landscape in IGAD Member States remains diverse. While some countries (e.g. Djibouti, Ethiopia, and Eritrea) have developed and operationalized national livestock policies and strategies, previously unstable countries such as South Sudan and breakaway Somalia republics have not. However, other than Ethiopia (Sect. 30.5.2), IGAD Member States, like the majority of African countries, still do not have operational policies that specifically address characterization, conservation, and sustainable use of animal resources.

30.4.6 The Southern African Development Community (SADC)

The Southern African Development Coordination Conference, SADCC, the forerunner of the Southern African Development Community (SADC), was established in April 1980 by governments of nine Southern African countries. Today's SADC comprises 16 Member States: Angola, Botswana, Comoros, Democratic Republic of Congo, Eswatini, Lesotho, Madagascar, Malawi, Mauritius, Mozambique, Namibia, Seychelles, South Africa, United Republic Tanzania, Zambia, and Zimbabwe. The mission of SADC is to promote sustainable and equitable economic growth and socio-economic development through efficient, productive systems, deeper cooperation and integration, good governance, and durable peace and security so as to be a competitive and effective player in international relations and the world economy.

The SADC has developed a Regional Agricultural Policy (SADC-RAP) which defines agreed objectives and measures to guide, promote, and support actions at regional and national levels in the agricultural sector of the SADC Member States in contribution to regional integration and the attainment of the SADC Common Agenda (SADC 2013). Observing that few farmers are able to access improved and adequate levels of supply of technologically improved plant and animal genetic material, and that this seriously impedes agricultural production and productivity, the policy commits SADC to complement national interventions by promoting and supporting measures to improve farmer access to improved plant and animal genetic material and the adoption of biotechnology in crop and livestock development. The policy also identifies the lack of capacity to regulate GMOs as a potential deterrent to embracing and adequately harnessing biotechnology. In this regard, interventions proposed include (1) Facilitating private-public sector initiatives that promote economies of scale in production and distribution of improved seeds, as well as other plant and animal genetic materials; (2) Promoting forage research and the establishment of viable forage seed industry; (3) Promoting regional value-chain partnerships in stock-feed production; (4) Facilitating agreement on a harmonized approach to the safe use of modern biotechnology and clarifying how to deal with GMOs; (5) Promoting national capacity and regional collaboration for research in biotechnology and biosafety; and (6) Facilitating the implementation of the SADC Harmonized Seed Regulatory System.

Of more direct relevance to animal genetic resources, the policy recognizes that *the SADC region is endowed with a wealth of plant and animal genetic resources and indigenous innova-*

tions with the potential to contribute significantly to economic growth and people's well-being, but that these genetic resources are rapidly being lost. To address this, *SADC shall complement national interventions by promoting and supporting measures for conservation and* sustainable use *of plant and animal genetic resources for food and agriculture.* Like in the EAC livestock policy, the SADC agricultural policy proposes specific interventions that speak to conservation and sustainable use of animal resources. These include:

- Promoting ex-situ and in situ (on-farm) conservation of genetic resources for food and agriculture.
- Strengthening national and regional capacity to conserve genetic resources.
- Promoting the development of crop varieties and animal breeds that are adaptable to climate change and variability.
- Promoting the access and benefit-sharing of plant and animal genetic resources for food and agriculture.
- Promoting collection and dissemination of information on genetic resources to enhance utilization of conserved material.

In the R&D domain, the regional policy states that SADC shall complement and support measures by Member States to promote agricultural research and development in crops, livestock, fisheries, and forestry, with specific intervention areas being: Providing policy guidance for the region's agricultural R&D; Promoting innovation for strategic and demand-driven R&D; Promoting private, public, and international partnerships in R&D; Harmonizing policy frameworks on protection of intellectual and property rights including indigenous knowledge systems; and Informing strategies and instruments to effectively promote increased adoption rates of appropriate technologies by farmers, such as through research, capacity building, and exchanges and dissemination of experiences.

In addition to the uniqueness of the SADC livestock policy in its explicit attention to animal (and plant) genetic resources, it is also unique in the way it was developed and the ownership this has created among its Member States and other crucial stakeholders. Member States acknowledge the key success factors in the implementation of the Regional Agricultural Policy as: (1) Recognition of the high potential of SADC region to be a leading agricultural producer and commitment to unleashing this potential; (2) Member States agree to focus on a few actions with the most rapid impact on agricultural growth, and within those specific objectives, focus on a few selected strategic interventions and commodities at a time; (3) Strong political will and commitment to the Policy by all Member States and key stakeholders through its alignment with, and domestication into, national policies, strategies, and programmes; (4) Establishment of an Agricultural Development Fund (ADF) as an in-built financial instrument under the Policy's Regional Agricultural Investment Plan (RAIP) that corresponds to interventions under Member States' own CAADP National Agricultural Investment Plans (NAIPs); (5) Establishment of a strong monitoring and evaluation system involving SADC Secretariat, Member States, and key stakeholders; (6) Creation, partly through adaptation of existing arrangements, of a mechanism for full engagement of private sector organizations, e.g. through public, private partnerships (PPP), in the implementation of accelerated agricultural development programmes promoted under the Policy; (7) Confidence built among financial partners regarding the effectiveness of the policy agenda and its governance; (8) Progressive elimination of barriers to trade and investment in compliance with the SADC Trade Protocol and Annexes; (9) Establishment of the requisite infrastructure to support agricultural development; and (10) Intentional mainstreaming of gender, youth, HIV/AIDS, environment, climate change and variability, and other cross-cutting issues into the Policy interventions at both regional and national levels. This Policy is implemented in 5-year reviewable cycles. This is to ensure responsiveness to change and progressivity, including recognition that capacity and systems will be developed and strengthened over time.

At the national level, the SADC livestock policy landscape is quite variable. Although Member States have demonstrated ownership of the regional policy, its translation, specifically the livestock dimensions relevant for animal genetic resources, into national policies remains limited. Indeed, only a few SADC countries (see Sect. 30.5.1) have developed livestock policies and strategies, although the situation is beginning to change. For example, Zambia has had no specific Policy to guide livestock development since independence and, until 2020, the livestock sector operated under a legal framework with a multiplicity of laws that governed specific areas of livestock development. National stakeholders recognize that the lack of a policy and legal framework has constrained effective development of the Zambian livestock sector. An example of an outcome of the AU-IBAR engagement with countries through SADC is the development of the Zambia National Livestock Development Policy (2020) (Zambia-MoFL 2020) which specifically addresses the promotion of sustainable Farm Animal Genetic Resources through, among other things, evaluation and conservation of indigenous breeds and the creation of animal genetic resources management database—a Livestock Information Management System (LIMS). Malawi has also recently developed a National Livestock Development Policy (2021–2026) (Malawi-MoA 2021). However, the policy does not articulate specific interventions that touch on indigenous animal resources, especially with regard to conservation and sustainable use.

On the positive side, all SADC Member States have elaborate animal health legislations, although some are decades old and could benefit from revision.

30.5 National Livestock Policy Instruments, Strategies, and Actions

As is clear from Sect. 31.4, progress made by the different Regional Economic Community blocks in regard to policies, strategies, and actions on AnGR differ substantially. But even more important, the differences among countries within and between these blocks are so substantial that the location of a country in a REC block is not a good indicator of status of progress. In addition, while most national livestock development policy and strategy documents recognize the importance of AnGR in transforming the livestock sector and provide recommendations for action, implementation is quite limited and varied among countries.

30.5.1 Overview of Country Situations

This section provides an overview of country situations in regard to the development and implementation of AnGR-relevant policies and strategies. As has been alluded to in earlier sections, only a small number of countries have stand-alone livestock policies, and AnGR is covered to varying extents in livestock sector policies and strategies. In most countries AnGR policy coverage is embedded in broader agricultural policies, strategies, and plans. Table 30.5 summarizes the current status of African countries in terms of policies and strategies specific to AnGR and associated implementation instruments. The latter include National Strategies and Action Plans (NSAPs) as per the Global Plan of Action (GPA) and Animal Performance Recording Schemes. Only six countries have policies which explicitly address AnGR. This compares to 15 countries which have established NSAPs and 14 which have Animal Recording Systems in place. This speaks to the fact that, while intentions expressed in policies are important, operational systems can be, and has been in some countries, put in place in the absence of policies.

Table 30.5 also indicates which countries have gene banks for AnGR, as well as countries with private sector-driven breed societies—which are crucial vehicles for sustainable management of AnGR. The institutional contexts and governance mechanisms of these gene banks are quite varied. The range includes those that were originally

Table 30.5 Summary of status of policies and implementation instruments by African countries

Policy, strategy, framework of plans	Established or planned	No. of countries
1. Specific policies and strategies for AnGR	Ghana, Kenya, Uganda, South Africa, Tanzania, Zambia	6
1.1.2. National strategies and action plans for AnGR	• *Developed and implementation underway (at differing stages and paces):* Mali, Mauritania, Guinea, Sierra Leone, Liberia, Sudan, DRC, Uganda, Kenya, Zambia, Zimbabwe, Namibia, Lesotho, Eswatini, Madagascar	15
	• *Development is in progress:* Algeria, Niger, Nigeria, Burkina Faso, Benin, Ghana, Ethiopia, Rwanda, Tanzania, Mozambique, Botswana, and South Africa	12
	• *Development is planned (and resources allocated):* Senegal, The Gambia, Cameroon, Gabon, and Burundi	5
3. National animal performance recording schemes or breeding centres	Algeria, Botswana, Burundi, Cameroon, Egypt, Ethiopia, Kenya, Lesotho, Namibia, South Africa, Swaziland, Tunisia, Zambia, Zimbabwe	14
4. Livestock breeders' associations	Algeria, Mauritania, Mali, Burkina Faso, Cote d'Ivoire, Guinea, Niger, Nigeria, Cameroon, Egypt, Uganda, Kenya, Tanzania, Rwanda, DRC, Zimbabwe, Mozambique, Namibia, South Africa	19

(continued)

Table 30.5 (continued)

Policy, strategy, framework of plans	Established or planned	No. of countries
5. Gene banks for AnGR	Botswana: *FAnGR Gene Bank*, *DAR*[a] Burkina Faso: *CIRDES*[b], *Bobo-Dioulasso* Cameroon: *The University of Dschang* Egypt: *National Gene Bank*, *Cairo* Ethiopia: *Biodiversity Institute*, *Addis Ababa* Kenya: *Kenya AnGR Centre*, *Kabete* Nigeria: *NACGRAB*[c], *Ibadan* Rwanda: *National Gene Bank*, *Huye* South Africa: *GCRB*[d], *Irene*; and *ARC*[e], *Irene* Tunisia: *Banque Nationale de Genes*, *Tunis* Uganda: *NAGRC&DB*[f], *Entebbe* Zimbabwe: *CUT Bank*[g], *Chinhoyi*	12

[a]Farm Animal Genetic Resource gene bank, Department of Agricultural Research (DAR)
[b]Centre International de Recherche-Développement sur l'Élevage en zone Sub-humide (CIRDES)
[c]National Centre for Genetic Resources and Biotechnology (NACGRAB)
[d]Germplasm Conservation and Reproductive Biotechnologies (GCRB)
[e]Agricultural Research Council (ARC) Loskop Conservation Centre
[f]National Animal Genetic Resource Centre and Data Bank (NAGRC&DB) in Entebbe, Uganda, in East Africa
[g]Chinhoyi University of Technology Semen Station and Cryobank (CUT Bank)

established as Artificial Insemination (AI) Centres under government ministries or departments responsible for livestock development (e.g. Kenya AnGR Centre came out of what used to be the Central Artificial Insemination Station, CAIS), to those established specifically to facilitate AnGR conservation (e.g. Ethiopia's

Biodiversity Institute—originally focused on crops, before expansion to cover livestock), to one initiated independently by a University (e.g. Chinhoyi University of Technology Semen Station and Cryobank, the CUT Bank, in Zimbabwe), and one established by a regional livestock centre, but now serving as a national facility (the CIRDES Gene Bank in Burkina Faso). The case of the National Animal Genetic Resources Centre and Data Bank (NAGRC and DB) in Uganda is quite unique. It was established through an Act of Parliament, the *Uganda Animal Breeding Act of 2001*, to provide for the promotion, regulation and control, marketing, import and export, and quality assurance of animal and fish genetic materials and to provide for the implementation of the national breeding policy in Uganda.

30.5.2 Selected Country Case Studies

As alluded to above, countries differ substantially in terms of where they are on policy, strategies, and action plans relevant for conservation and use of AnGR (Sect. 30.5.1). This section provides case studies of six countries (Ethiopia, Ghana, Kenya, Nigeria, South Africa, and Tanzania) to illustrate the diversity of country contexts and current situation, including specific interventions that different countries are pursuing. The case studies illustrate the many dimensions of differences: the role of the private sector, especially breed societies and Stud Book; the role of public sector beyond development of policy and legislation—e.g. actions catalyzing industry practice; the role civil society can play; and the strong links that can be leveraged among characterization, conservation, and use.

30.5.2.1 Ethiopia

Ethiopia has a well-established Biodiversity Institute (originally focused on plant genetic resources) and has fully mainstreamed and institutionalized biodiversity conservation and sustainable use. The country has developed several policies and legal frameworks in this regard. The Biodiversity Institute is mandated to undertake conservation and promote development and sustainable utilization of the country's biological resources, namely: plants, animals, and microbial genetic resources as well as associated traditional knowledge and the ecosystems. The institute has the responsibility to implement international conventions, agreements, and obligations on biodiversity to which Ethiopia is a party. These include the proclamation on Access and Benefit-Sharing, the establishment of animal genetic resources directorate within the Biodiversity Institute, and the enactment of animal breeding policy. The breeding policy emphasizes on indigenous genetic resources through selection, improvement, and conservation.

Ethiopia's National Biodiversity Conservation and Research Policy (1998) and, subsequently, the National Biodiversity Strategy and Action Plan (2005) (IBC 2005) provide guidelines for conservation, development, and sustainable use of biodiversity. The policy objective is to ensure that genetic resources and essential ecosystems of the country are conserved, developed, and sustainably used, asserting national sovereignty over genetic resources, enriching the country's biological resources through restoration, integrating biodiversity conservation with sectoral and cross-sectoral strategies and programs, recognizing and protecting traditional knowledge, ensuring that the local communities obtain shares of benefits arising from the use of genetic resources and indigenous knowledge, and promoting regional and international cooperation.

In 2012, Ethiopia developed a draft National Strategy and Plan of Action for Conservation and Sustainable Utilization of AnGR in line with the Global Plan of Action (IBC 2012). The vision of the Strategy is conserving *animal genetic resources for food and agriculture and promoting its use in support of national* food security *and sustainable development for present and future generations*. Its goals are: Promotion of sustainable use and development of animal genetic resources for food security, sustainable agriculture, and human well-being; Ensuring conservation of animal genetic resources diversity for present and future generations, and halting loss and erosion of these resources; Recognizing and promoting the role of indigenous knowledge, innovations, and practices relevant to the conservation of animal genetic resources and their sustainable use; Ensuring fair and equitable sharing of the benefits arising from access to, and use of, animal genetic resources and associated indigenous knowledge; Putting in place effective policies and legislative measures to ensure use, sustainable development, and conservation of animal genetic resources for food and agriculture; Meeting the needs of pastoralists and farmers, individually and collectively, within the framework of national law, to have non-discriminatory access to genetic material, information, technologies, financial resources, research results, marketing systems, and natural resources, so that they may continue to manage and improve animal genetic resources and benefit from economic development; Promotion of agro-ecosystems approaches for the sustainable use, development, and conservation of animal genetic resources; Assisting national regional states and institutions to establish, implement, and regularly review national and regional priorities for the sustainable use, development, and conservation of animal genetic resources; Strengthening Federal and national regional states' programs and enhancing institutional capacity, namely education, research, and training to address the characterization, inventory, monitoring, conservation, development, and sustainable use of animal genetic resources; and Promotion of activities aimed at raising public awareness and bringing the needs of sustainable use, development, and conservation of animal genetic resources to the attention of concerned stakeholders.

Other policies and strategies that address AnGR issues directly or indirectly include Ethiopia's Education Policy, Conservation Strategies of Regional States, Agricultural Growth Program, Livestock Growth Program, and regulations and control on the introduction and export of domestic animal genetic materials.

While the Strategy and Action Plan is quite ambitious in its coverage, its operationalization has been slow. A workshop held in November 2018 to evaluate implementation Strategy and Action Plan concluded that there is a need for capacity building in farm AnGR use, development, and conservation as well resourcing of implementation programmes, with breed characterization being a top priority.

30.5.2.2 Ghana

A number of policies and strategies have been put in place to improve the performance of the livestock sector in Ghana over the years. Some of these policies and strategies have been implemented in projects such as the National Livestock Services Project, Agriculture Sub-Sector Investment Project, and Livestock Development Project. In addition, several studies addressing the development of the livestock sector have been carried out since 1995. Those relevant for AnGR include: the National Animal Breeding Plans for the Republic of Ghana (1995); the State of Ghana's Animal Genetic Resources (2003); the role of Livestock in Rural Livelihoods in Ghana (2002); and Livestock Sector Review 2010 and 2012.

Although implementation has been limited, Ghana's livestock policies (those that have been stand-alone as well as those part of overall agricultural policy) have historically consistently recognized the importance of the country's wealth of indigenous AnGR and recommended interventions on their genetic improvement and conservation. The 2004 Livestock Development in Ghana—Policies and Strategies (MOFA 2004), for example, includes a policy guideline on *improvement of* indigenous breeds *through selection of local breeds to build on the existing desired traits and conservation of local breeds to maintain the desirable resistant attributes of such breeds for future exploitation.*

Ghana's Livestock Development Policy and Strategy (LDPS) (MOFA 2016) of 2016 is underpinned by a vision of a livestock sector that is modernized, competitive, and sustainable, contributing to food and nutritional security, employment generation, equity and poverty reduction while preserving the environment. The policy document recognizes that *"previous policies have largely been ineffective as a result of non-implementation or inadequate implementation of various* strategies *as well as lack of effec-*

tive monitoring and evaluation mechanisms". The stated objective of the Policy and Strategy is to develop a competitive and more efficient livestock industry that increases domestic production, reduces importation of meat and livestock products, and contributes to the improvement of the livelihoods of all livestock value chain actors and the national economy, while protecting the environment, *preserving livestock biodiversity*, and ensuring bio-security.

The specific objectives of the LDPS are (1) to support the existing livestock production systems for improving production, productivity, and income of livestock producers; (2) to strengthen overall animal health; (3) to support training, research, and development for improving livestock production, productivity, and health as well as profitability; (4) to improve the production and productivity of livestock by promoting and disseminating technologies developed by the research system; (5) to *promote conservation and genetic improvement of* indigenous breeds *of livestock*; (6) to increase availability of feed and fodder resources to meet the requirement of various species of livestock; (7) to encourage value addition of livestock products; and (8) To create an enabling environment to attract investment and finance for improving livestock infrastructure, production, processing, value addition, and marketing.

Strategic interventions covering objective 5—promotion, conservation, and genetic improvement of indigenous breeds of livestock—include the following:

(a) Conservation of indigenous genetic material through the establishment of gene banks.
(b) Improvement of existing livestock breeding stations.
(c) Establishment of open nucleus breeding schemes.

The LDPS envisages that these interventions will be driven by the Council for Scientific and Industrial Research—Animal Research Institute (CSIR-ARI), Universities and Colleges, Ministry of Food and Agriculture (Animal Production Department), Livestock Breeders, and Farmer-Based Organizations (FBOs).

The commitment by the government to implement these policy recommendations is evident through their inclusion in the "Rearing for Food and Jobs" (RFJ) which is one of the five models of the government's flagship 5-year (2019–2023) agricultural improvement programme under the theme "Planting for Food and Jobs" (PFJ). The objective of the RFJ is to develop a competitive and more efficient livestock industry that increases domestic production, reduces importation of livestock products, and contributes to employment generation and to the improvement of livelihoods of livestock value chain actors and the national economy. Priority areas under this programme include:

- Characterization, conservation, genetic improvement, and commercialization of local chicken and guinea fowl ecotypes.
- Genetic improvement of the local Ashanti Black/Dwarf pig.
- Genetic improvement of the local sheep and goat breeds.
- Genetic improvement of Ghana Shorthorn, Sanga, N'Dama cattle breeds, and their crosses.

In its 2014 State of the World's AnGR report (to FAO), Ghana reported that the country had not done an inventory of its AnGR and that the country lacked an early warning system for AnGR. In 2015/2016 period, with AU-IBAR support, the Ghana National Animal Resources Data Management Platform (ARDMP) was developed and officially launched in August 2016. In this connection, stakeholders suggested that, to facilitate implementation, a memorandum of understanding (MOU) be developed between the Veterinary Services Directorate, the Ghana Statistical Service, School of Veterinary Medicines, Ghana Private Veterinarians Association, and other partners that will be involved in the platform, and an annual workplan be developed to guide the implementation of the National ARDMP.

30.5.2.3 Kenya

Session Paper Number 2 of 2008 on National Livestock Policy (Kenya-MoLD 2008) addresses the challenges in the livestock sub-sector, covering livestock breeding, nutrition and feeding, disease control, value-addition and marketing, and research and extension. It includes an objective on improvement and conservation of animal genetic resources, with plans to:

- Undertake a survey of livestock species, breeds, and types and evaluate them to identify potential candidates for further conservation or improvement.
- Establish nucleus herds of indigenous animal genetic resources for breeding, characterization, conservation, and utilization. Linked to this, a national gene bank for storage of AnGR germplasm would be set up and livestock registration and recording schemes will be strengthened.
- Put in place appropriate regulations to guide management and exploitation and use of animal genetic resources and establish breeding programmes with long-term selection of indigenous breeds for high productivity and local adaptive traits, especially in the arid and semi-arid areas.
- Establish a central authority—a National Focal Point—to facilitate coordination of Animal Genetic Resources activities in the country and as a link to similar regional or global focal points; and that this should be done recognizing the multiplicity of institutions involved in provision of animal breeding services in Kenya (the Kenya Stud Book, the Livestock Recording Centre, the Central Artificial Insemination Station, the Kenya National Artificial Insemination Service, the Breed Associations, individual farmers, etc.).
- Develop appropriate guidelines on breeds and production systems for various ecological zones of the country.
- Set up mechanisms to strengthen the management of breeding services and regulate breeding service providers, improve delivery of extension services, and institute export and import controls of semen, embryos, and live animals, and modernization of quality control laboratories in the Central Artificial Insemination Station (CAIS) to meet required international standards. In addition, expand mandate of CAIS to serve as a gene bank for all species of livestock.

In 2018, a draft National Action Plan on Animal Genetic Resources of Kenya (Kenya – Ministry of Agriculture and Irrigation 2018), an adoption of the Global Plan of Action, was drafted and is still under discussion. Subsequently, Kenya has embarked on the formulation of policies that guide the management of AnGR in the form of the national livestock policy, dairy development policy, poultry development policy, and the feeds policy.

The Establishment of the Kenya Animal Genetic Resources Centre (Kagrc)

Established in 2011, the Kenya Animal Genetic Resources Centre (KAGRC) succeeded the Central Artificial Insemination Station (CAIS) which was established back in 1946 with the objective of controlling reproductive diseases and genetic improvement of exotic dairy cattle.

KAGRC has the mandate to produce, preserve, and conserve animal genetic materials (semen embryo, tissues, and live animals) and to produce breeding sires to produce high-quality disease-free semen to meet the national demand and for export. To meet its mandate, KAGRC works in close collaboration with other breeding organizations such as the Kenya Stud Book, the Dairy Recording Services of Kenya, and the Livestock Recording Centre. Together, these organizations implement the Contract Mating and Progeny Testing Programmes. KAGRC works closely with the breed societies as well as individual and institutional farms which provide herds for the breeding programme.

A draft National Livestock Policy (Kenya-MoLD 2019) was developed in 2019 but has not been operationalized. However, Sessional Paper Number 3 of 2020 (Kenya-MoLD 2020) on National Livestock Policy amplifies and seeks to operationalize the content of the 2019 draft policy. Both instruments emphasize intentions of

Sessional Paper Number 2 (2008). However, having been developed after the creation of devolved County governments, both the draft policy of 2019 and Sessional Paper Number 3 propose roles of national (national infrastructure, legislation, national animal identification system, import and export procedures) and sub-national (County) governments (mostly focused on characterization, animal performance recording, and conservation through use). The livestock dimensions of the 2021 Agricultural Policy (Kenya-MoALC 2021) also express commitment to biodiversity conservation, including attention to indigenous crops and livestock as critical components to ensure adaptable and resilient agriculture.

30.5.2.4 Nigeria

The Nigerian National Centre for Genetic Resources and Biotechnology (NACGRAB) is a Federal Government Parastatal institution which was established in 1987 under the Federal Ministry of Science and Technology with a mandate to collect and conserve valuable genetic resources for food and agriculture and ensure that they are used sustainably. The Federal Government of Nigeria promulgated Decree No. 33 of 1987 (Federal Government of Nigeria 2016) (now amended as Act of Parliament, 2016). The Act establishes the National Crop Varieties and Livestock Breeds Register and provides for the insertion in the Register of the names of old and new crop varieties and livestock breeds. The Act also establishes the National Crop Varieties and Livestock Breeds Registration and Release Committee (the National Varietal Release Committee, NVRC), which has responsibility for crop varieties and livestock breed validation, registration, naming, and release in the country.

One of the activities of NACGRAB is the coordination of the National Committee on Naming, Registration, and Release of new Crop Varieties and Livestock/Fisheries breeds through various Technical Sub-Committees. The Centre maintains gene banks for both Plant and Animal Genetic Resources. The stored materials are registered and subsequently evaluated for performance characteristics. The Technical Sub-Committee of Livestock/Fisheries of the NVRC has compiled descriptors for registration of both the existing and new livestock/fisheries breeds or strains into a database—the National Register or Catalogue.

The NACGRAB gene bank for crops is located in Ibadan in south-west Nigeria. The AnGR unit was established in 2004 as one of the core units of the Centre and currently has ex situ conservation activities on poultry, fishery, snails, small ruminants, cane rats, and rabbits. The functions of the unit include:

- Exploration, collection, and conservation of indigenous animal genetic resources.
- Preservation of endangered animal genetic resources.
- Genetic characterization and evaluation of animal genetic resources.
- Conduct of research in conservation and utilization of animal genetic resources, including the use biotechnology tools.
- Assisting National Varietal Release Committee (NVRC) in the naming of livestock breeds and fisheries strains, as well as registration, and release activities.

The coordination of AnGR activities in Nigeria is the remit of the Federal Department of Animal Husbandry Services (DAHS) of the Federal Ministry of Agriculture and Rural Development (FMARD) which is also the FAO representative for the AnGR management for Nigeria. One of the activities of the DAHS is to document and regularly update information on AnGR. Nigeria's National Advisory Committee (NAC) on Animal Genetic Resources was inaugurated in 2018. Nigeria has also developed a draft National Animal Breeding Policy. This document was developed by FMARD in collaboration with Animal Breeders' and Geneticists' Network of Nigeria (ABGeN-NG), which is a network of animal breeding and genetics experts drawn from the Nigeria Institute of Animal Science (NIAS), Genetics Society of Nigeria (GSN), and the Nigerian Society for Animal Production (NSAP).

Nigeria's Progress Report for the 2014–2019 period on the National Strategy and Action Plans on AnGR states that the NSAP was due to be completed in 2020. However, it has stalled "due to lack of funding". Indeed, while institutional roles for AnGR are well laid out in various documents, not much has happened in the last decade, mostly due to inadequate budget allocation. Consequently, information on status of breeds in Nigeria remains limited. Most of the figures cited on breed sizes and characteristics are from old and limited data.

30.5.2.5 South Africa

The *Animal Improvement Act* of South Africa was passed in 1998 (Act No. 62) (Republic of South Africa 1998) and its associated regulations were finalized in 2002. Its purpose is to provide for the breeding, identification, and utilization of genetically superior animals in order to improve the production and performance of animals in the interest of the Republic. The Act focuses on the implementation and documentation of breeding animals and addresses the type and scope of performance testing and recording in compulsory breeding programmes. In addition, the Act has provisions on the implementation and use of biotechnology in artificial insemination and embryo transfer, the approval of breeding associations, breeding companies, insemination centres, embryo transfer institutes and their breeding programmes, herd-book regulations (herd-book recording) and their areas of activity, conditions for import and export of animal genetic material, and the roles of breed companies and the stud book in these processes. The responsibility for the implementation of the Act lies with the Registrar of the Animal Improvement Act who is appointed by the Minister responsible for Livestock.

To support legislation and regulation of the Animal Improvement Act, and to guide implementation, an Animal Improvement Policy (Republic of South Africa 2007) was developed and gazetted in November 2007. The objectives of the policy include promoting the sustainable use of Animal Genetic Resources as major contributor to national food security and facilitating the conservation of Animal Genetic Resources for Food and Agriculture. The policy explicitly recognizes the importance of indigenous livestock breeds, the threats they face, the need for action to conserve and commercialize them as a means of ensuring their sustainable use, and the role of indigenous knowledge and the rights of owners and breeders.

In addition to the Animal Improvement Act, other instruments which underpin the Animal Improvement Policy include: the Animal Identification Act of 2002; National Animal Improvement Schemes; and a National Advisory Committee for AnGR conservation and improvement.

South Africa and the Global Plan of Action

South Africa has submitted to FAO its report on The First and Second State of the World's Animal Genetic Resources for Food and Agriculture. In addition, the AU-IBAR has engaged the Department of Agriculture, Forestry and Fisheries (DAFF) of South Africa through the Directorate of Genetic Resources to initiate the formulation of the National Strategy and Action Plan for AnGR (NSAP) as provided for in the FAO-led Global Plan of Action (GPA), and a draft of the NSAP has been prepared in form of a comprehensive report of the NSAP process (Republic of South Africa 2016). The emerging NSAP has three strategic priority areas, namely, (a) characterization and inventory of farm AnGR; (b) sustainable use and development of farm AnGR; and (c) conservation of farm AnGR.

A validation stakeholder workshop was held in 2019. However, the NSAP has not been operationalized in South Africa to date.

The South Africa Stud Book

South Africa presents one of the best examples of commercializing local breeds as a means for ensuring their sustained use. The South Africa Stud Book, an association of 63 livestock and animal breeders' societies, is a crucial instrument for development and commercialization of livestock breeds. Current breeders' societies include 26 cattle, 15 small ruminants (sheep and goats), 16 horse breeders' societies, and six breeders' societies for other species. These societies cover 77 different animal breeds. The Stud Book also handles the registrations of 29 other breeds—minor breeds that do not have breed societies.

The vision of the Stud Book (The South Africa Stud Book 2023) is to develop exceptional domestic animals and establish them as a sustainable resource and treasure for all South Africans. Its mission is to be a significant role player in the genetic improvement of animals in South Africa. Its aims are to:

- Promote the breeding, conservation, and genetic improvement of the production potential of animals under its jurisdiction
- Facilitate pedigree and performance recording of animals and issue certificates of registration for recorded animals
- Safeguard and advance the collective interests of stud breeders and their breeders' societies and act as a mouthpiece for the stud-breeding industry
- Represent the collective interest of animal breeders and their societies on various national and international bodies and forums
- Provide technical and advisory services to breeders' societies, their members, and participants in the Integrated Registration and Genetic Information System (INTERGIS)
- Promote the export of animals with pedigrees registered or recorded with the association and of semen, ova or embryos from animals thus registered or recorded.

While the SA Stud Book covers exotic breeds (dairy cattle, beef cattle, and sheep), it is actively involved with several indigenous breed societies, namely: Nguni, Afrikaner, Bonsmara, and Drakensberger cattle breeds; Dohne, Dorper, Persian, Pedi, Damara sheep breeds; and Savannah, and Boer Goat breeds. This list also demonstrates the extent to which South Africa has been involved in developing and commercializing local composite breeds—Bonsmara and Drakensberger cattle, and Dohne Merino and Dorper sheep.

The extent of commercial orientation of South Africa's livestock industry is also reflected in the number of registered breeds and those with breed societies (Table 30.6). Unfortunately, as is the case in most African countries, stud-breeding has not made much inroad into the traditional smallholder (or communal) livestock sector of South Africa. Institutionalizing the integration of stud breeding in the smallholder sector and accompanying increase in numbers of pure stock (of indigenous breeds preferred by certain communities)—to the benefit of both the communal and stud cattle breeders—can potentially significantly improve management of AnGR in this historically marginalized sector of South Africa.

In some cases, where no formal breed society exists, breeder clubs undertake these roles, especially the promotion and marketing.

Table 30.6 Number of livestock breeds and breed societies in South Africa

Species	No. of breeds	No. of breed societies
Cattle	42	25
Dog	2	2
Goat	11	3
Horse	28	17
Ostrich	1	1
Pig	5	1
Sheep	26	12
Total	*115*	*61*

Source: The South Africa Stud Book (2023)

The Farm Animal Conservation Trust (FACT)

South Africa is effectively engaging Farm Animal Conservation Trust (FACT), an NGO, in the conservation of farm AnGR. FACT is modelled on the Rare Breeds International (RBI) and the Rare Breeds Survival Trust (RBST) in the United Kingdom. FACT has been involved as a partner in a national initiative that involves public sector institutions to create awareness and promote hitherto endangered indigenous livestock breeds. The other partners involved are the Department of Agriculture, Land Reform and Rural Development, the Agricultural Research Council, some public universities, and the Department of Sport, and Arts and Culture (responsible for identification, assessment, management, protection, and promotion of heritage resources in South Africa). As part of promotion of endangered indigenous breeds, FACT activities include the preparation and dissemination of information on them and facilitating conservation through commercial use. In this regard, FACT has facilitated the publication of a booklet on South Africa's landrace breeds. In the meantime, South Africa has made significant progress in collecting, cataloguing, and storing available literature on the origins, domestication, and historical usage and evolution of its indigenous breeds and their production environments. The database is based at the Irene Animal Improvement Institute. It is considered that availability of this information will spur interest in these neglected breeds, including by smallholder farmers in traditionally marginalized communities.

30.5.2.6 Tanzania

Tanzania is an excellent case study of challenges African countries face in the *Policy-to-Practice* connection. In the last three decades, the country has had a good track record of consistently developing livestock policy frameworks and strategies—either as stand-alone documents or as part of overall agricultural policies. However, there has been limited translation of these policies to actions on the ground. Moreover, attention to indigenous animal genetic resources has been flip-flopping and, as in many African countries, there is little evidence that the intentions expressed in these past policy and strategy instruments have made a difference in improving the state of indigenous animal genetic resources. The National Livestock Policy (NLP) of 2006 (Tanzania-MoLD 2006) document states that its aim is to streamline livestock areas which were previously neglected (in the *1997 Agricultural and Livestock Policy* (Tanzania-MoAC 1997)) combined with crop-related services such as regulatory, technical, and cross-sectoral services. The 1997 policy statements on livestock emphasized crossbreeding or upgrading with high-performance exotic breeds to improve productivity. The 2006 policy expresses the intention to promote livestock breed inventory, characterization, evaluation, and genetic improvement, as well as to strengthen technical support services in animal breeding. It also recognizes the need to promote Breeders Associations to ensure sustainable conservation, breeding, and use.

The Tanzania Livestock Modernization Initiative (TLMI) (Tanzania-MoLFD 2015) of 2015 aimed to implement the National Livestock Policy (NLP) of 2006 in the context of the National Five-Year Development Plan and other national and sectoral plans and programmes. One of the strategic priorities was the improvement of the quality and performance of Tanzanian livestock populations, the preservation of the resiliency of indigenous breeds while optimizing performance gains, underpinned by conservation interventions. The TLMI calls out selection of desired characteristics of resilient and productive improved breeds, paying attention both to genetic and production environment, including feed and health, improvement. It recognizes that genetic heterogeneity of Tanzania's indigenous livestock can be exploited to develop adaptability to climate change, refers to the existence of a Livestock Identification, Registration and Traceability Act (Tanzania-MoLD 2010), and calls for *"Swift enactment of an Animal Breeding Act, that incorporates and facilitates the major tenets of the Tanzanian livestock breed improvement programme"*.

In 2017, Tanzania developed a Livestock Master Plan (TLMP) (Tanzania-MoLF 2017). The TLMP recognized that better coordination of the development and protection of animal genetic resources in Tanzania should include the establishment of reliable and sustainable germplasm delivery systems and the involvement of the private sector in animal genetic improvement. It proposed the establishment and enforcement of a legal framework, including the development of an animal breeding policy and the implementation of the animal breeding bill (under development then). It recognized that livestock selection for genetic improvement needed to focus on: (1) Ensuring effective breeding, selection, and conservation programmes are in place, including open nucleus breeding schemes and the renovation of public livestock farms and artificial insemination centres; (2) the establishment of performance recording systems for on-station and on-farm breed evaluation programmes for both locally adapted and exotic breeds and their crosses; and (3) training and support to strengthen animal breeding infrastructure, such as artificial insemination and laboratories. Unfortunately, like the previous ambitious instruments, the intentions espoused in the TLMP are not supported in the more recent Livestock Sector Transformation Plan (LSTP, 2022) (Tanzania-MoLF 2022) which, while recognizing that the country has a large population of indigenous livestock breeds, decries their low productivity, and proposes strengthening of artificial insemination to upgrade them. The LSTP focuses heavily on productivity improvements and does not make any explicit plans for improvement and/or conservation of the well-adapted indigenous animal genetic resources, which could be put at risk by the proposed emphasis on crossbreeding with exotic breeds. Of note is that Tanzania is credited with the development of a highly promising composite cattle breed, the Mpwapa (see Chap. 4). Unfortunately, lack of policy to underpin sustained investment in this initiative has led to the neglect of this breed. Today the Mpwapwa breed is disappearing fast, and will be extinct soon, unless intentional conservation and further development are undertaken urgently.

In 2019, supported by AU-IBAR, Tanzania developed a *National Compact Strategies and Action Plan to implement Global Plan of Action for Animal Genetic Resources* (Tanzania-MoLF 2019). The goal of the Action Plan was to guide conservation and sustainable utilization of AnGR and its contribution to the livelihoods of the people in Tanzania and the national economy. The stated objectives are to: Carry out a comprehensive inventory, characterization, and monitoring of trends and associated risks of livestock breeds for all species existing in the country, including their production systems; promote sustainable use of AnGR through adoption of modern technologies and indigenous knowledge; develop a comprehensive framework and ensure conservation of AnGR; and influence policy and strengthen institutional capacity in terms of technical and infrastructure. Not only has the Action Plan not been implemented, but key livestock stakeholders, including those active in AnGR work in the country, do not even know of its existence and/or status of implementation.

In an analysis of food production and supply in Tanzania during the period 1961–2001, Isinika et al. (2003) observed frequent policy changes and remarked that despite good intentions of the government, some of the policies failed to deliver intended results due to market and institutional distortions which were introduced. The government has tended to overreact to emerging problems that affect agricultural production and other development plans (and policies and strategies are often developed without accompanying implementation instruments, especially resourcing). The analysis alludes to limited capacity for policy analysis in order to determine their broad ramifications before they were adopted for implementation and that the frequent policy changes brought uncertainty and increased risk for producers and traders (investors) with consequent detrimental effects to the nation. It is estimated that during the period in question, Tanzania had a policy change or a government directive that affected the agricultural sector every 2–3 years.

30.6 Summary and Conclusions

African livestock keepers are encountering challenges in their efforts to achieve food security and improve their livelihoods. Sustainable intensification of livestock production systems is essential to successfully address these challenges, and indigenous AnGR have a critical role to play in this. While awareness about needed actions, including the role of policies, legislation, strategies, and plans, has increased in Africa substantially, especially since development of the Global Plan of Action (GPA), and many initiatives have been initiated in Africa to catalyse action, indigenous AnGR continue to decline. Promotion of a limited number of supposedly more productive breeds is a major driver of the continuing decline of local breeds, a crucial resource for low-input livestock systems of poor smallholder farmers.

Taken together, the analysis of efforts by AU-IBAR and its partnership with RECs suggest that the development of policies on conservation and sustainable use in Africa has, by and large, not translated into improvements in the state of indigenous AnGR in Africa. Only a limited number of countries have developed policies and strategies targeting indigenous AnGR. Even for these countries, a major challenge has been the absence of effective operational modalities that translate intentions articulated in the policies and strategies into practice. As stipulated in the GPA, establishment of AnGR National Focal Point and appointment of National Coordinators are critical underpinnings of effective management of AnGR at national levels. Unfortunately, where these have been established, the majority have remained ceremonial, occasionally being requested by FAO to prepare country reports on AnGR—the contents of which have generally remained static, reflecting limited progress. Indeed, the status of AnGR in most countries has deteriorated significantly over the last two decades. Establishment and resourcing of functioning National Focal Points, with dynamic Coordinators and Steering (or Advisory) Committees, is clearly a priority for the continent. In addition, it is critical that the institutional frameworks put in place under the leadership of the Focal Point institution facilitate coordination/collaboration mechanisms with agencies responsible for livestock extension, research, trade, foreign affairs, pastoral development, natural resources, and environment to ensure that these agencies work synergistically to design and implement effective programmes that deliver the desired conservation and sustainable use outcomes. In addition, the policy function of government ministries responsible for livestock should explicitly include conservation and sustainable use of indigenous AnGR.

Annex

Table 30.7 Cartagena Protocol on Biosafety and Nagoya Protocol on Access and Benefit-sharing: status of ratification by African countries and entry into force

No	Country	Cartagena Protocol				Nagoya Protocol			
			Ratified				Ratified		
		Signed	Date	Instr*	Party	Signed	Date	Instr*	Party
1.	Algeria	May 25, 2000	Aug 05, 2004	rtf	Nov 03, 2004	-	-	-	-
2.	Angola		Feb 27, 2009	acs	May 28, 2009		Feb 06, 2017	acs	May 07, 2017
3.	Benin	May 24, 2000	Mar 02, 2005	rtf	May 31, 2005	Oct 28, 2011	Jan 22, 2014	rtf	Oct 12, 2014
4.	Botswana	Jun 01, 2001	Jun 11, 2002	rtf	Sep 11, 2003		Feb 21, 2013	acs	Oct 12, 2014
5.	Burkina Faso	May 24, 2000	Aug 04, 2003	rtf	Nov 02, 2003	Sep 20, 2011	Jan 10, 2014	rtf	Oct 12, 2014
6.	Burundi		Oct 02, 2008	acs	Dec 31, 2008		Jul 03, 2014	acs	Oct 12, 2014
7.	Cameroon	Feb 09, 2001	Feb 20, 2003	rtf	Sep 11, 2003		Nov 30, 2016	acs	Feb 28, 2017
8.	Central African Republic	May 24, 2000	Nov 18, 2008	rtf	Feb 16, 2009	Apr 06, 2011	July 27, 2018	rtf	Oct 25, 2018
9.	Chad	May 24, 2000	Nov 01, 2006	rtf	Jan 30, 2007	Jan 31, 2012	Oct 11, 2017	rtf	Jan 09, 2018
10.	Comoros	-	-	-	-		May 28, 2013	acs	Oct 12, 2014
11.	Congo	Nov 21, 2000	Jul 13, 2006	rtf	Oct 11, 2006	Sep 23, 2011	May 14, 2015	rtf	Aug 12, 2015
12.	Côte d'Ivoire		Mar 12, 2015	acs	Jun 10, 2015	Jan 25, 2012	Sep 24, 2013	rtf	Oct 12, 2014
13.	Democratic Republic of the Congo		Mar 23, 2005	acs	Jun 21, 2005	Sep 21, 2011	Feb 04, 2015	rtf	May 05, 2015
14.	Djibouti		Apr 08, 2002	acs	Sep 11, 2003	Oct 19, 2011	Oct 01, 2015	rtf	Dec 30, 2015
15.	Egypt	Dec 20, 2000	Dec 23, 2003	rtf	Mar 21, 2004	Jan 25, 2012	Oct 28, 2013	rtf	Oct 12, 2014
16.	Eritrea		Mar 10, 2005	acs	Jun 08, 2005		Mar 13, 2019	acs	Jun 11, 2019
17.	Eswatini		Jan 13, 2006	acs	Apr 13, 2006		Sep 21, 2016	acs	Dec 20, 2016
18.	Ethiopia	May 24, 2000	Oct 09, 2003	rtf	Jan 07, 2004		Nov 16, 2012	acs	Oct 12, 2014
19.	Gabon		May 02, 2007	acs	Jul 31, 2007	May 13, 2011	Nov 11, 2011	acp	Oct 12, 2014
20.	Gambia	May 24, 2000	Jun 09, 2004	rtf	Sep 07, 2004		Jul 03, 2014	acs	Oct 12, 2014
21.	Ghana		May 30, 2003	acs	Sep 11, 2003	May 20, 2011	Aug 08, 2019	rtf	Nov 06, 2019
22.	Guinea	May 24, 2000	Dec 11, 2007	rtf	Mar 10, 2008	Dec 09, 2011	Oct 07, 2014	rtf	Jan 05, 2015
23.	Guinea-Bissau		May 19, 2010	acs	Aug 17, 2010	Feb 01, 2012	Sep 24, 2013	acp	Oct 12, 2014
24.	Kenya	May 15, 2000	Jan 24, 2002	rtf	Sep 11, 2003	Feb 01, 2012	Apr 07, 2014	rtf	Oct 12, 2014
25.	Lesotho		Sep 20, 2001	acs	Sep 11, 2003		Nov 12, 2014	acs	Feb 10, 2015
26.	Liberia		Feb 15, 2002	acs	Sep 11, 2003		Aug 17, 2015	acs	Nov 15, 2015
27.	Libya		Jun 14, 2005	acs	Sep 12, 2005	-	-	-	-
28.	Madagascar	Sep 14, 2000	Nov 24, 2003	rtf	Feb 22, 2004	Sep 22, 2011	Jul 03, 2014	rtf	Oct 12, 2014
29.	Malawi	May 24, 2000	Feb 27, 2009	rtf	May 28, 2009		Aug 26, 2014	acs	Nov 24, 2014
30.	Mali	Apr 04, 2001	Aug 28, 2002	rtf	Sep 11, 2003	Apr 19, 2011	Aug 31, 2016	rtf	Nov 29, 2016
31.	Mauritania		Jul 22, 2005	acs	Oct 20, 2005	May 18, 2011	Aug 18, 2015	rtf	Nov 16, 2015
32.	Mauritius		Apr 11, 2002	acs	Sep 11, 2003		Dec 17, 2012	acs	Oct 12, 2014
33.	Morocco	May 25, 2000	Apr 25, 2011	rtf	Jul 24, 2011	Dec 09, 2011	Apr 22, 2022	rtf	Jul 21, 2022
34.	Mozambique	May 24, 2000	Oct 21, 2002	rtf	Sep 11, 2003	Sep 26, 2011	Jul 07, 2014	rtf	Oct 12, 2014
35.	Namibia	May 24, 2000	Feb 10, 2005	rtf	May 11, 2005		May 15, 2014	acs	Oct 12, 2014
36.	Niger	May 24, 2000	Sep 30, 2004	rtf	Dec 29, 2004	Sep 26, 2011	Jul 02, 2014	rtf	Oct 12, 2014
37.	Nigeria	May 24, 2000	Jul 15, 2003	rtf	Oct 13, 2003	Feb 01, 2012	Jun 29, 2022	rtf	Sep 27, 2022
38.	Rwanda	May 24, 2000	Jul 22, 2004	rtf	Oct 20, 2004	Feb 28, 2011	Mar 20, 2012	rtf	Oct 12, 2014
39.	Senegal	Oct 31, 2000	Oct 08, 2003	rtf	Jan 06, 2004	Jan 26, 2012	Mar 03, 2016	rtf	Jun 01, 2016
40.	Seychelles	Jan 23, 2001	May 13, 2004	rtf	Aug 11, 2004	Apr 15, 2011	Apr 20, 2012	rtf	Oct 12, 2014
41.	Sierra Leone		Jun 15, 2020	acs	Sep 13, 2020		Nov 01, 2016	acs	Jan 30, 2017
42.	Somalia		Jul 26, 2010	acs	Oct 24, 2010	-	-	-	-
43.	South Africa		Aug 14, 2003	acs	Nov 12, 2003	May 11, 2011	Jan 10, 2013	rtf	Oct 12, 2014
44.	Sudan		Jun 13, 2005	acs	Sep 11, 2005	Apr 21, 2011	Jul 07, 2014	rtf	Oct 12, 2014
45.	Togo	May 24, 2000	Jul 02, 2004	rtf	Sep 30, 2004	Sep 27, 2011	Feb 10, 2016	rtf	May 10, 2016
46.	Tunisia	Apr 19, 2001	Jan 22, 2003	rtf	Sep 11, 2003	May 11, 2011	Aug 27, 2021	rtf	Nov 25, 2021
47.	Uganda	May 24, 2000	Nov 30, 2001	rtf	Sep 11, 2003		Jun 25, 2014	acs	Oct 12, 2014
48.	United Republic of Tanzania		Apr 24, 2003	acs	Sep 11, 2003		Jan 19, 2018	acs	Apr 19, 2018
49.	Zambia		Apr 27, 2004	acs	Jul 25, 2004		May 20, 2016	acs	Aug 18, 2016
50.	Zimbabwe	Jun 04, 2001	Feb 25, 2005	rtf	May 26, 2005		Sep 01, 2017	acs	Nov 30, 2017

[a]Instr: Instrument of ratification (rtf), or approval (apv), or acceptance (acp), or accession (acs) deposited with the Depositary. The column entitled "Party" indicates the dates when the Protocol enters into force for the country, i.e. 90 days after instrument is deposited.

References

AUC (2015) Practical guidelines for the coordinated implementation of the Nagoya Protocol in Africa. Department of Human Resources, Science and Technology, The African Union Commission (AUC), Addis Ababa, Ethiopia. https://absch.cbd.int/api/v2013/documents/ACA06BA7-2ED4-19C0-F096-883C14068E94/attachments/202597/AUPracticalGuidelinesOnABS_20150215_Druck.pdf. Accessed 17 Oct 2023

AU-IBAR (2014) African animal genetic resources project: strengthening the capacity of African countries to conserve and sustainably use African animal genetic resources. AU-IBAR, Nairobi, Kenya. https://www.au-ibar.org/sites/default/files/2020-11/doc_20150717_year_1_genetics_project_report_en.pdf. Accessed 15 June 2023

AU-IBAR (2015) Livestock Development Strategy for Africa (LiDeSA). AU-IBAR, Nairobi, Kenya. http://repository.au-ibar.org/bitstream/handle/123456789/540/2015-LiDeSA.pdf?sequence=1&isAllowed=y. Accessed 16 Nov 2023

AU-IBAR (2016) VET-GOV – livestock policy landscape in Africa: a review. African Union – InterAfrican Bureau on Animal Resources, Nairobi, Kenya. https://www.au-ibar.org/sites/default/files/2020-11/doc_20160524_livestock_policy_lanscape_africa_en.pdf. Accessed 8 Oct 2023

AU-IBAR (2018) AU-IBAR strategic plan 2018–2023. AU-IBAR, Nairobi, Kenya. https://www.au-ibar.org/sites/default/files/2020-10/sd_20190805_au-ibar_strategic_plan_2018-2023_en.pdf

AU-IBAR (2019) The state of farm animal genetic resources in Africa. AU-IBAR, Nairobi, Kenya. https://www.au-ibar.org/sites/default/files/2020-10/gi_20191107_state_farm_animal_genetic_resources_africa_full_book_en.pdf. Accessed 15 June 2023

AU-IBAR (2020) Regional Animal Genebanks for Africa: a strategy to ensure the sustainability and efficient maintenance of important animal genetic resources. AU-IBAR, Nairobi, Kenya. https://www.au-ibar.org/sites/default/files/2020-11/doc_20141127_regional_animal_genebanks_africa_en.pdf. Accessed 5 Nov 2023

AU-IBAR (2023) Sustainable development of livestock for livelihoods in Africa (Live2Africa) [2017–2023]. AU-IBAR, Nairobi, Kenya. https://www.au-ibar.org/au-ibar-projects/sustainable-development-livestock-livelihoods-africa. Accessed 9 Jan 2024

Cao J, Baumung R, Boettcher P, Scherf B, Besbes B, Leroy G (2021) Monitoring and progress in the implementation of the Global Plan of Action on Animal Genetic Resources. Sustainability 13(2):775. https://doi.org/10.3390/su13020775

CBD (2000) Text of the Cartagena Protocol on Biosafety to the Convention on Biological Diversity. https://bch.cbd.int/protocol/text/. Accessed 17 Oct 2023

CBD (2010) Text of the Nagoya Protocol on access to genetic resources and the fair and equitable sharing of benefits arising from their utilization to the Convention on Biological Diversity (CBD). https://www.cbd.int/abs/text/articles/?sec=abs-01. Accessed 17 Oct 2023

CBD (2023a) Status of ratification and entry into force of the Cartagena Protocol on Biosafety to the Convention on Biological Diversity. https://bch.cbd.int/protocol/parties/. Accessed 17 Oct 2023

CBD (2023b) Parties to the Nagoya Protocol on access to genetic resources and the fair and equitable sharing of benefits arising from their utilization to the Convention on Biological Diversity (CBD). https://www.cbd.int/abs/nagoya-protocol/signatories/. Accessed 17 Oct 2023

Clifford H (2014) AquAdvantage Salmon – a pioneering application of biotechnology in aquaculture. BMC Proc 8(Suppl 4):O31. https://doi.org/10.1186/1753-6561-8-S4-O31. PMCID: PMC4204346. https://www.ncbi.nlm.nih.gov/pmc/articles/PMC4204346/

COMESA (2015) Common Market for Eastern and Southern Africa (COMESA) regional livestock policy framework. COMESA, Lusaka, Zambia. https://www.comesa.int/wp-content/uploads/2020/10/Livestock-Policy-Framework-En.pdf. Accessed 22 Aug 2023

Dolgin E (2021) First GM pigs for allergies. Could xenotransplants be next? Nat Biotechnol 39:397–400. https://doi.org/10.1038/s41587-021-00885-9

EAC (2016) East African Community (EAC) livestock policy. Adopted by 34th Council of Ministers of the EAC on 5th September 2016. Arusha, Tanzania. https://faolex.fao.org/docs/pdf/mul204334.pdf. Accessed 21 Aug 2023

ECOWAS Commission (2009) Regional partnership compact for the implementation of ECOWAP/CAADP. Adopted during the International Conference on Financing Regional Agricultural Policy in West Africa (ECOWAP/CAADP), Abuja, Nigeria, 12 November 2009. https://www.oecd.org/swac/publications/44426979.pdf. Accessed 9 Jan 2024

ECOWAS Commission (2010) Strategic action plan for the development and transformation of livestock sector in the ECOWAS region (2011–2020). ECOWAS, Abuja, Nigeria

FAO (2007) Global Plan of Action for Animal Genetic Resources and the Interlaken Declaration. Adopted by the International Technical Conference on Animal Genetic Resources for Food and Agriculture, Interlaken, Switzerland, 3–7 September 2007. Commission on Genetic Resources for Food and Agriculture. The Food and Agriculture Organization of the United Nations, Rome. https://www.fao.org/3/a1404e/a1404e.pdf. Accessed 23 July 2023

FAO (2011) Developing the institutional framework for the management of animal genetic resources. FAO Animal Production and Health Guidelines. No. 6. FAO, Rome. https://www.fao.org/3/ba0054e/ba0054e00.pdf. Accessed 21 Sept 2023

FAO (2015) The second report on the state of the world's animal genetic resources for food and agriculture, edited by B.D. Scherf & D. Pilling. FAO Commission on Genetic Resources for Food and Agriculture Assessments. FAO, Rome. https://doi.org/10.4060/I4787E

Federal Government of Nigeria (2016) National Crop Varieties and Livestock Breeds (Registration, ETC) Act 1987. Lagos, Nigeria. https://faolex.fao.org/docs/pdf/nig120555.pdf. Accessed 12 Jan 2024

Gupta K, Karihaloo JL, Khetarpal RK (2008) Biosafety regulations of Asia-Pacific countries. Asia-Pacific Association of Agricultural Research Institutions/Asia-Pacific Consortium on Agricultural Biotechnology/Food and Agricultural Organization of the United Nations, Bangkok/New Delhi/Rome. https://www.researchgate.net/publication/274063516_Biosafety_Regulations_of_Asia-Pacific_Countries

IBC (2005) National biodiversity strategy and action plan. The Institute of Biodiversity Conservation (IBC) of the Government of the Federal Democratic Republic of Ethiopia, Addis Ababa, Ethiopia. https://www.cbd.int/doc/world/et/et-nbsap-01-en.pdf. Accessed 11 Jan 2024

IBC (2012) National strategy and plan of action for conservation, sustainable use, and development of animal genetic resources. The Institute of Biodiversity Conservation (IBC) of the Government of the Federal Democratic Republic of Ethiopia, Addis Ababa, Ethiopia. https://www.ebi.gov.et/wp-content/uploads/2021/10/Ethiopian-National-Strategy-and-Plan-of-Action-for-Conservation-Sustainable-Use-and-Development-of-Animal-Genetic-Resources-.pdf. Accessed 11 Jan 2024

IGAD (2022) IGAD strategy for sustainable and resilient livestock development in view of climate change (2022–2037). IGAD Centre for Pastoral Areas and Livestock Development (ICPALD), IGAD, Djibouti. https://igad.int/download/igad-strategy-for-sustainable-and-resilient-livestock-development-in-view-of-climate-change-2022-2037/. Accessed 4 Sept 2023

IGAD and AU-IBAR (2018) Regional model policy/legal framework for conservation, sustainable utilization, and access and benefit sharing of farm animal genetic resources. IGAD Centre for Pastoral Areas and Livestock Development (ICPALD), Intergovernmental Authority on Development and AU-IBAR, Nairobi, Kenya. https://icpald.org/wp-content/uploads/2019/07/Regional-Model-Policy-Legal-Framework.pdf. Accessed 4 Sept 2023

Isinika AC, Ashimogo GC, Mlangwa JED (2003) Tanzania macro study: final research study. Mimeo. Sokoine Agricultural University Africa in Transition Research Programme. Sokoine Agricultural University, Morogoro, Tanzania

Kenya – Ministry of Agriculture and Irrigation (2018) National strategy and action plan on management of animal genetic resources for Kenya. Ministry of Agriculture and Irrigation, Nairobi, Kenya, p 58. https://www.fao.org/3/ca4779en/ca4779en.pdf. Accessed 11 Jan 2024

Kenya-MoALC (2021) Agricultural Policy 2021. Ministry of Agriculture, Livestock and Cooperative (MoALC), Nairobi, Kenya. https://kilimo.go.ke/wp-content/uploads/2022/05/Agricultural-Policy-2021.pdf. Accessed 11 Jan 2024

Kenya-MoLD (2008) Sessional paper number 2 of 2008 on National Livestock Policy. Ministry of Livestock Development (MoLD), Nairobi, Kenya. https://faolex.fao.org/docs/pdf/ken159358.pdf. Accessed 11 Jan 2024

Kenya-MoLD (2019) Draft National Livestock Policy. Ministry of Livestock Development (MoLD), Nairobi, Kenya. https://repository.kippra.or.ke/bitstream/handle/123456789/483/Draft-reviewed-National-Livestock-Policy-February-2019.pdf?sequence=1&isAllowed=y. Accessed 11 Jan 2024

Kenya-MoLD (2020) Sessional paper number 3 of 2020 on National Livestock Policy. Ministry of Livestock Development (MoLD), Nairobi, Kenya. https://repository.kippra.or.ke/bitstream/handle/123456789/3062/Sessional%20Paper%20No.3%20of%202020.pdf?sequence=1&isAllowed=y. Accessed 11 Jan 2024

Maksimenko OG, Deykin AV, Khodarovich YM, Georgiev PG (2013) Use of transgenic animals in biotechnology: prospects and problems. Acta Nat 5(1):33–46. PMID: 23556129; PMCID: PMC3612824

Malawi-MoA (2021) National livestock development policy. Department of Animal Health and Livestock Development, Ministry of Agriculture (MoA), Lilongwe, Malawi. https://faolex.fao.org/docs/pdf/mlw214429.pdf. Accessed 5 Dec 2023

MOFA (2004) Livestock development in Ghana – policies and strategies (2004–2015). Ministry of Food and Agriculture (MOFA), Animal Production Directorate, Veterinary Services Directorate, and Livestock Planning and Information Unit, Accra, Ghana

MOFA (2016) Ghana livestock development policy and strategy. Ministry of Food and Agriculture (MOFA), Accra, Ghana

Obonyo ND, Nfor LM, Uzochukwu S, Araya-Quesada M, Farolfi F, Ripandelli D, Craig W (2011) Identified gaps in biosafety knowledge and expertise in sub-Saharan Africa. AgBioforum 14(2):71–82

Ormandy EH, Dale J, Griffin G (2011) Genetic engineering of animals: ethical issues, including welfare concerns. Can Vet J 52(5):544–550. PMID: 22043080; PMCID: PMC3078015

Rege JEO, Kiambi D, Ochieng JW (2022) Chapter 3: The state of the enabling environment for agricultural biotechnology applications in crop and livestock sectors. In: Rege JEO, Sones K (eds) Agricultural biotechnology in sub-Saharan Africa – capacity, enabling environment and applications in crops, livestock, forestry and aquaculture. Springer, Cham, pp 33–55

Republic of South Africa (1998) The Animal Improvement Act of South Africa. 28 September 1998, Gazette. Cape Town, South Africa. https://www.gov.za/sites/default/files/gcis_document/201409/a62-98.pdf. Accessed 12 Jan 2024

Republic of South Africa (2007) Animal Improvement Policy of South Africa. Republic of South Africa, Department of Agriculture, Directorate of Animal and Aquaculture Production, Pretoria, Republic of South Africa. Government Gazette, 16 November 2007. https://cer.org.za/wp-content/uploads/2003/11/Animal-Improvement-Policy-FULL.pdf. Accessed 12 Jan 2024

Republic of South Africa (2016) National plan for conservation and sustainable use of farm animal genetic resources (draft report). Department of Agriculture, Forestry and Fisheries, Republic of South Africa. https://old.dalrrd.gov.za/doaDev/sideMenu/geneticResources/docs/National%20Plan.pdf. Accessed 12 Jan 2024

SADC (2013) SADC regional agricultural policy (SADC-RAP). Southern African Development Community (SADC). SADC Secretariat, Gaborone, Botswana. https://www.inter-reseaux.org/wp-content/uploads/Regional_Agricultural_Policy_SADC.pdf. Accessed 12 Sept 2023

Tanzania-MoAC (1997) National agriculture and livestock policy. Ministry of Agriculture and Cooperatives (MoAC) of the United Republic of Tanzania, Dar es Salaam, Tanzania. http://www.tzonline.org/pdf/agricultureandlivestockpolicy.pdf. Accessed 17 Nov 2023

Tanzania-MoLD (2006) National Livestock Policy (NLP). Ministry of Livestock Development (MoLD) of the United Republic of Tanzania, Dar es Salaam, Tanzania. https://faolex.fao.org/docs/pdf/tan169547.pdf. Accessed 17 Nov 2023

Tanzania-MoLD (2010) The Livestock Identification, Registration and Traceability Act 2010. Ministry of Livestock Development (MoLD) of the United Republic of Tanzania, Dodoma, Tanzania. https://faolex.fao.org/docs/pdf/tan97294.pdf. Accessed 12 Jan 2024

Tanzania-MoLF (2017) Tanzania livestock master plan (TLMP) (2017–2022). Ministry of Livestock and Fisheries of the United Republic of Tanzania, Dodoma, Tanzania. https://www.mifugouvuvi.go.tz/uploads/projects/1553601793-TANZANIA%20LIVESTOCK%20MASTER%20PLAN.pdf. Accessed 5 Dec 2023

Tanzania-MoLF (2019) National Compact Strategies and Action Plan to implement Global Plan of Action for Animal Genetic Resources in Tanzania. Ministry of Livestock and Fisheries (MoLF) of the United Republic of Tanzania, Dodoma, Tanzania, p 46. https://www.mifugouvuvi.go.tz/uploads/publications/sw1638868464-MPANGO%20MKAKATI%20WA%20UBORESHAJI%20NA%20UENDELEZAJI%20KOSAAFU%20ZA%20MIFUGO%20NCHINI%20TANZANIA.pdf. Accessed 13 Dec 2023

Tanzania-MoLF (2022) Tanzania livestock sector transformation plan (LSTP) (2022–2027). Ministry of Livestock and Fisheries of the United Republic of Tanzania, Dodoma, Tanzania. https://www.mifugouvuvi.go.tz/uploads/publications/sw1675840376-LIVESTOCK%20SECTOR%20TRANSFORMATION%20PLAN%20(LSTP)%20202223%20-%202026 27.pdf. Accessed 18 Nov 2023

Tanzania-MoLFD (2015) The Tanzania livestock modernization initiative (TLMI) (2015–2021). Ministry of Livestock and Fisheries of the United Republic of Tanzania, Dodoma, Tanzania. https://livestocklivelihoodsandhealth.org/wp-content/uploads/2015/07/Tanzania_Livestock_Modernization_Initiative_July_2015.pdf. Accessed 20 Nov 2023

The South Africa Stud Book. https://studbook.co.za/directory.asp?CID=2. Accessed 17 Nov 2023

UNEP (2006) Building biosafety capacity: the role of UNEP and the Biosafety Unit. UNEP-GEF Biosafety Unit, Geneva, Switzerland. https://wedocs.unep.org/handle/20.500.11822/10000. Accessed 13 Dec 2023

Zambia-MoFL (2020) National livestock development policy. Ministry of Fisheries and Livestock (MoFL), Lusaka, Republic of Zambia. https://www.mfl.gov.zm/wp-content/uploads/2022/08/National-Livestock-Development-Policy.pdf. Accessed 12 Sept 2023

Open Access This chapter is licensed under the terms of the Creative Commons Attribution 4.0 International License (http://creativecommons.org/licenses/by/4.0/), which permits use, sharing, adaptation, distribution and reproduction in any medium or format, as long as you give appropriate credit to the original author(s) and the source, provide a link to the Creative Commons license and indicate if changes were made.

The images or other third party material in this chapter are included in the chapter's Creative Commons license, unless indicated otherwise in a credit line to the material. If material is not included in the chapter's Creative Commons license and your intended use is not permitted by statutory regulation or exceeds the permitted use, you will need to obtain permission directly from the copyright holder.

Resourcing and Institutional Arrangements to Deliver Sustainable Animal Genetic Improvement in Africa

31

Eveline M. Ibeagha-Awemu, Victor Olori, Ismail Muritala, Olubunmi I. Duduyemi, Mizeck G. G. Chagunda, and John E. O. Rege

Abstract

The contents of the chapters of this book attest to the fact that sustainable livestock production is imperative to meeting the food needs and the economic development of Africa. To achieve this, all stakeholders involved must understand their roles and also be willing to pull together resources through sustainable cooperations, guided by clear local, regional, and continent-wide policies to achieve the common objective of sustainable livestock improvement. This chapter presents the roles of various institutions—within countries, between countries, and multinationals engaged in supporting animal improvement in Africa. It discusses the importance of functional linkages and needed cooperations between the various stakeholders. It concludes that the crucial elements necessary for the effective utilization of the factors presented in this book to deliver resilient and sustainable animal genetic improvement for the transformation of the African livestock industry are enabling policies, functioning institutional arrangements, access to appropriate technologies, funding and information, and adequate infrastructure and trained personnel.

Keywords

Enabling policy · Stakeholders · Resourcing institutions · Institutional arrangements · Technology · Trained personnel · Sustainable livestock production

E. M. Ibeagha-Awemu (✉)
Sherbrooke Research and Development Centre,
Agriculture and Agri-Food Canada,
Sherbrooke, QC, Canada
e-mail: eveline.ibeagha-awemu@agr.gc.ca

V. Olori
Aviagen, Midlothian, UK

I. Muritala
Department of Biotechnology, Osun State University, Osogbo, Nigeria

O. I. Duduyemi
Department of Animal Science, McGill University,
Sainte-Anne-de-Bellevue, QC, Canada

M. G. G. Chagunda
Animal Breeding and Husbandry in the Tropics and Subtropics, University of Hohenheim,
Stuttgart, Germany

J. E. O. Rege
Emerge Centre for Innovations–Africa, Nairobi, Kenya

Abbreviations

AABNet	African Animal Breeders Network
AACAA	All Africa Conference on Animal Agriculture

AASAP	All Africa Society for Animal Production	ICAR	International Committee for Animal Recording
ACGG	African Chicken Genetic Gains	ICBF	Irish Cattle Breeding Federation
ADGG	African Dairy Genetics Gains	IDF	International Dairy federation
AnGR	Animal Genetic Resources	ILRI	International Livestock Research Institute
ASAN	Animal Science Association of Nigeria	INTERBEEF	International Bull for BEEF Evaluation Centre
ASARECA	Association for Strengthening Agricultural Research in Eastern and Central Africa	INTERBULL	International Bull Evaluation Centre
ASBPP	African Seed and Biotechnology Partnership Platform	IPC	International Poultry Council
		IRAD	Institut de recherche agricole pour le développement (Agricultural Research Institute for Development)
AU	African Union		
AU-IBAR	African Union–Inter-African Bureau for Animal Resources	IRZV	Institut des Recherches Zootechniques et Vètèrinaires (Institute of Animal and Veterinary Research)
BCBS	Boran Cattle Breeders' Society of Kenya		
BDCS	Bamenda Dairy Cooperative Society	KALPA	Kenya Livestock Producers' Association
CAADP	Comprehensive Africa Agriculture Development Programme	LiDeSA	Livestock Development Strategy for Africa
CCFD	Catholic Committee Against Hunger and for Development	MESRES	Ministry of Higher Education and Scientific Research
CGIAR	Consultative Group on International Agricultural Research	MINEPIA	Ministry of Livestock, Fisheries, and Animal Industries
COMESA	Common Market for Eastern and Southern Africa	MINREL	Ministry of Livestock
		MMPA	Malawi Milk Producers Association
DAD-IS	Domestic Animal Diversity Information System	NGO	Non-Governmental Organizations
DAGRIS	Domestic Animal Genetic Resources Information System	NOVID	Nederlandse Organisatie Voor Internationale Ontwikkelings samenwerking (Netherlands Organization for International Development Cooperation)
DGAK	Dairy Goats Association of Kenya		
ECOWAS	Economic community of West African States		
EU	European Union	NOWEPIFAC	Northwest Pig Farmers Cooperative
FARA	Forum for Agricultural Research in Africa	NSAP	Nigerian Society for Animal Production
GAA	German Agro Action		
GASL	Global Agenda for Sustainable Livestock	ONAREST	Office National de la Recherche Scientifique et Technique (National Office for Scientific and Technical Research (of Cameroon))
GDP	Gross Domestic Product		
HI	Heifer International		
HPI	Heifer Project International		

PRTC	Presbyterian Rural Training Center
SADC	Southern African Development Community
SAILD	Services d'Appui aux Initiatives Locales de Développement
SAJOCAH	Saint Joseph's Children and Adult Homes
SDG	Sustainable Development Goals
SF	Vétérinaires Sans Frontières (Veterinarians Without Borders)
TPGS	Tropical Poultry Genetic Solutions
UN-SDG	United Nation's Sustainable Development Goals
USA	United States of America
USAID	United States Agency for International Development
USDA	United State Department for Agriculture
VSF	Vétérinaires Sans Frontières (Veterinarians Without Borders)

31.1 Introduction

Improving the animal genetic resources (AnGR), and in particular livestock genetic resources, in Africa is an important element of feeding the growing African population, solving malnutrition, and driving sustainable development of the continent. The various components, drivers, and challenges of successfully putting in place sustainable livestock genetic improvement programs in Africa have been discussed in Chapters 1 to 30. A well-planned and implemented sustainable livestock genetic improvement process leads to increased genetic gain in traits under selection which translates to increased productivity, better health of animals, and increased profitability which benefits the producer, consumers, the livestock industry, and everyone or group or institution along the livestock-value-chain. Collectively, they form the stakeholders in the livestock-value-chain process. Therefore, stakeholders in livestock genetic improvement generally refer to those individuals or groups or organizations that have an interest in or benefit from the improvement of farm animals. In this regard, livestock will refer to animals (cattle, chicken, sheep, goats, pigs, etc., see Chapters 1 to 12) that are farmed for the provision of meat, milk, eggs, wool, and other animals' products. Generally for livestock, it involves the primary keepers of livestock animals (generally associated with small-scale livestock farmers), breeders (farmers involved in more sophisticated breeding programs, who supply the animal seed stock and could work individually or in groups/associations), regulatory bodies (those that regulate the quality and supply of animal breeding stock and the activities of the livestock industry, e.g. governments), livestock companies (those that use the improved animal seed stock to produce animals and products to meet consumer preferences), consumers (utilize animal products for food, sports, and farming), processors (those that process the primary animal products into secondary value-added products), marketers (those involved in every aspect of marketing animals and animal products), associations or cooperatives, service industries (make products for the livestock industry), funding bodies, civil society organizations, training and research institutions, and governments. Collectively, these stakeholders resource the animal genetic improvement process either directly or indirectly, requiring functioning institutional arrangements for the gainful involvement of all stakeholders.

Thus, it has been severally emphasized that the successful implementation and sustainability of AnGR management activities, including livestock genetic improvement, depend on the many key factors discussed in previous chapters of this book and the participations of the diverse stakeholders in the livestock-value-chain listed above (Wurzinger et al. 2011; Zonabend et al. 2013; Mueller et al. 2015; Leroy et al. 2017; Ibeagha-Awemu et al. 2019). The importance of clear arrangements and policy-guiding stakeholder participation is underscored by the past failed years of the lack of recognition of the common interest in sustainable improvement of livestock production. Increased livestock productivity does not benefit the primary producer only, but everyone along the livestock-value-chain, the populace and governments. Stakeholder involvement includes participation in every aspect of sustainable livestock production such as direct animal management activities (feeding, healthcare, reproduction management such as artificial insemination services, animal identification, data collection and recording programs, genetic evaluation, breeding, training, infrastructure, etc.), marketing (market access, processors, transportation, etc.), services (feed, drugs, processors, transportation, etc.), clear policies (favourable policies, implementation, provision of services, decision making, etc.), and funding (direct cash provision in the form of grants, access to loans, etc.). It should be noted that the involvement and level of participation between stakeholders may differ between regions of the same country, between countries, and continental regions as defined by the specie, the stakeholder parties, their relationships, and level of engagement in livestock production and related matters.

A sustainable and successful animal genetic improvement process requires a long-term perspective of production, inputs (funding), and market requirements, for achieving a positive genetic improvement outcome, which is cumulative over time. Thus, the collaborative effort through various functional institutional arrangements with stakeholders will strengthen partnerships within countries, between countries, continental, and multicontinental bodies/organizations engaged in resourcing animal improvement in Africa and will (1) sustainably support animal genetic improvement; (2) boost overall productivity and profitability ; (3) fulfil market demands; (4) boost livestock health, growth rate, carcass output, fleece weight, fibre diameter, and birth weight variance, among other attributes; (5) boost activities and profitability of the livestock-value-chain; (6) support economic development, and (7) boost GDP growth, among others.

The key components that drive successful (Chapteranimal genetic improvement programs include breed characterization, clearly defined breeding goals, infrastructure (genetic evaluation, data maintenance, information dissemination), human capacity/training, marketing channels, and the necessary policy guidelines. The animals whose genetic makeup and environmental influences determine the specific breeds and medium for the dissemination of genetic improvement are the most important component. Chapters 2, 3, 4, 5, 6, 7, 8, 9, 10, 11, 12, 13 and 14 of this book present a description of the breed characteristics, the production systems, and the effects of raising livestock on climate change. The livestock specie in turn determine the structure of the breeding program (Chapters 15, 16, 17, 18 and 19), which is the organization of people and the infrastructure dedicated to the genetic improvement processes (Chapters 20, 21, 22, 23 and 24), the technology, techniques, and procedures (Chapters 20, 21, 22, 23 and 24) deployed to achieve livestock genetic improvement and the standards or key regulatory roles and the institutional arrangements (Chapters 25, 26, 27, 28, 29, 30 and 31) guiding effective implementation of the livestock genetic improvement processes. A plethora of individuals, institutions, and organizations work hand in hand to resource animal improvement programs in Africa.

This chapter examines the critical stakeholders resourcing animal genetic improvement, the necessary institutional arrangements and the mitigating factors, and the key elements in achieving sustainable genetic improvement in the African continent.

31.2 Stakeholders Resourcing Animal Genetic Improvement

31.2.1 Role of the Private Sector in Animal Genetic Improvement

The private sector's contribution in African animal genetic improvement is of utmost importance and if properly harnessed, it will sustainably balance different policy objectives such as maintaining animal genetic diversity and environmental integrity, meeting the increasing demand for livestock products, responding to changing consumer requirements, ensuring food safety, and contributing to rural development and the alleviation of hunger and poverty.

31.2.1.1 Custodians of the Animals: Individual Farmers/Keepers

The most essential stakeholder group in the improvement of livestock comprises the owners of the livestock. This includes the individual farmers, farmer associations/cooperatives, and breed associations. The individual farmer, in this regard, also refers to the farm where it is a corporate entity owned by more than one individual.

The farmer is an important stakeholder because they are the primary keepers of livestock and the primary user of the end product of all genetic improvement efforts. In this regard, their demand for a specific breed or breeds is the most important motivation for investment in the development and distribution of improved animal seed stock. Furthermore, as custodians of the animals, they are also the most important source of the data required for genetic improvement of livestock. Recording the performance and pedigree of individual animals in their care is the most important contribution of the farmers to the genetic improvement of a breed. For example, in order to facilitate application of molecular methods for animal improvement, the farmers supply samples from their animals (e.g. tissue, blood, hair, etc.) from which DNA is extracted for the genotyping of individual animals.

Individual farmers play key roles in the conservation of animal genetic resources and hence the available variation in breeds used for production. They do this through the choice of the breed of animals they keep and the production practices. By keeping indigenous breeds based purely on interest or regional availability, they ensure the survival of marginalized breeds that would otherwise have gone extinct. By practicing random breeding with no directional breeding, they help maintain variation at the molecular level. This practise also, unfortunately, results in high gene flow between breeds and the dilution of breed specific characteristics. This could explain the wide array of breeds of cattle, sheep, goats, and chicken found in Africa.

There is the risk that as farmers adopt modern production practices and decide to keep only a few highly improved breeds for the main production objective, there will be a narrowing of genetic variation of the population. While this risk is real, it depends on how much other less productive breeds are neglected. This is why the Food and Agriculture Organization (FAO) of the United Nations has taken up the leadership in providing a framework for national governments to help in the conservation of AnGR (FAO 2011). The conservation of genetic variation within and across breeds and support for the creation of awareness in this key aspect of our livestock is perhaps the most important contribution of the FAO to genetic improvement of livestock. This is because future improvement depends on today's variation and today's variation also ensures we can change to alternatives in the future if the production environment or consumer preferences changes. In all of these, the choices of the individual farmers play a key role.

31.2.1.2 Farmer Organizations or Cooperatives

The impact that each individual farmer can have on overall genetic improvement is limited. This is because genetic improvement relies on expression of variation which can only be accurately estimated when information on performance is recorded on many animals across many different production environments and farms. Farmer organizations are formed on the basis of geographical location and/or the rearing of the same species. Formal or informal membership-based, farmer groups are collective action institutions with the purpose of assembling and possessing established organizational structure to support members in pursuing their individual and collective interests. One essential function is to organize relations with independent parties to mediate between members and others who act in their economic, institutional, and political environment. This definition includes farmer associations, farmer cooperatives, farmer clubs, farmer groups, producer organizations, and women's groups. These are usually farmer-based formed at regional and/or national levels.

Farmer groups assist individual farmers by consolidating the outputs of multiple producers, procure large quantities of inputs at reduced prices, and provide members with access to agricultural support services. Given their significant scale, cooperatives possess sufficient market influence to increase/determine price of products and provide a more stable and secure income for its members. Several agricultural associations also incorporate savings and loan programs for their members. These programs assist farmers in managing finances, maintaining records, and acquiring crucial financial skills to enhance their productivity. Therefore, farmer organization services are actions, strategies, or activities undertaken to help members generate more income and have better access to raw materials, inputs, training, marketing outlets, agricultural extension services, and veterinarian services, among others. Farmer organizations are found in many countries in Africa. Examples are shown in Table 31.1.

The Northwest Pig Farmers Cooperative (NOWEPIFAC) is composed of enthusiastic pig farming groups in Cameroon. The main motivation of NOWEPIFAC is to collectively fight the challenges (e.g. limited feed resources, limited market channels/access, high disease burden, and high cost of conventional drugs, among others) limiting pig farming in Cameroon. Founded in April 2004 in Bamenda (headquarters) by Mr. Atomba Titus Tegwi, the cooperative has its own building for offices and meetings in Bawock, and a slaughter house where pigs are slaughtered, inspected, and sold. NOWEPIFAC is principally made-up of different groups of mixed crop-livestock producers, pig producers (e.g. Netap Mixed Farming group, Tchi-Ncha Mixed Farming group, Baform Mixed Farming group, Fumjoh Mixed Farming group, Matter Horn Farming group, Cui-Cui Mixed Farming group, Struggling Women Mulang group, Feshang Women Group Baba 1 Ndop and Living Together Pig Farmers Nitop 1, etc.), and service providers (e.g. Mobile consultancy livestock service). These groups are composed of individual small crop-livestock farm families that have decided to work their way out of poverty through pig production.

In Cameroon, livestock including pig are kept as a way of savings and are sold for cash to cater for needs like school fees and supplies (when schools re-open in September of every year), medical bills and celebrations, among others. The farmers in the North West Region of Cameroon are principally crop farmers (maize, beans, cassava, plantains, banana, groundnuts, yams and coco-yam of different varieties, etc.) from which food for human and animal consumption are formulated. Problems like underdeveloped infrastructure (roads, housing, equipment, market, etc.) and language barrier (the North West and South West rRegions constitute the minority anglophone populations of Cameroon) hamper the sale of crops produced, so pig farming is seen as a way of solving these issues. With the main aim of fighting against vulnerability (poor crop harvest, limited market access, animal diseases, etc.) and helping cooperative members and their children sustainably support their livelihoods, membership in the cooperative is growing with members drawn from six divisions of the North West Region. NOWEPIFAC therefore seeks to achieve the following: (1) share pig farming facilities and equipment; (2) training and knowledge/wisdom sharing to develop and maintain quality pig farming, pork production, and marketing; (3) manage the pig production value-chain business (pig production, feed milling, pig slaughter, and sale of pork and other products); (3) financial independence for its members; (4) support the education of their children through the sale of pigs; (5) see beneficiaries become leaders of their communities and support the growth of their communities; (6) participate in the growth of Cameroon as a leader in the Central African region, and (7) encourage sharing and collaboration to end poverty and conflict between communities.

NOWEPIFAC receives support from many local and foreign donors which helps to grow the cooperative and keep farmers' hopes alive. The capacity of NOWEPIFAC was enhanced through its first project that started in 2006 (with the sup-

Table 31.1 Sample farmer organizations in Africa

Species	Organization name	Website	Location
Cattle	Dairy Farmers of Kenya		Kenya
	The Malawi Milk Producers Association	https://mwmilkproducers.org	Malawi
	Muturu Cattle Research Network	None	Nigeria
Goat	Dairy Goats' Association of Kenya (DGAK)	https://dgak.or.ke/	Kenya
	Meru Goat Breeders Association (MGBA)	None	Kenya
	Kitui-Mwingi Goat Breeders Association (K-MGBA)	None	Kenya
	Sheep and Goat Farmers Association of Nigeria	https://www.linkedin.com/company/shegofan/about/	Nigeria
Fish	Farmers' Organization Network in Ghana	https://fongh.org	Ghana
Poultry	Poultry Association of Nigeria	poultryassociationofnigeria.org	Nigeria
	Greater Accra Poultry Farmers Association	https://gapfaghana.org	Ghana
	South Africa Poultry Association	https://www.sapoultry.co.za/	South Africa
Pig	Northwest Pig Farmers Cooperative (NOWEPIFAC)	http://nowepifac.com	Cameroon
	Pig Farmers Association of Ghana	https://www.facebook.com/groups/5108668112530325/	Ghana
Livestock	Eastern Africa Farmers Federation (livestock, fish and crops; comprise 24 member organizations and 25 million smallholder farmers in 10 countries)	https://www.eaffu.org/	Eastern Africa)
	Association of Pastoralist Community for Change (formally Oromia Pastoralist Association) (APCfC)	https://apcfc.org.et/	Ethiopia
	Kenya Livestock Producers' Association	www.klpakenya.org	Kenya

port by Mr. Arie Peter Wingelaar and Mr. Hanneke Mertens, both of the Netherlands) with the provision of working materials (such as office supplies like cupboards, computers, microscope, tables, and chairs), feed mill, cooperative truck, a pig demonstration farm (for the training of members), training of its coordinator (international diploma in animal feed production) in the Netherlands and maintenance of vehicle and equipment, through funding by the Dutch Ministry of Development Cooperation. The second project of the cooperative came through a Dutch girl named Djouke (introduced to cooperative by Mr. Wingler) who started the project named, "A Pig Van Djouke" (which means a pig from Djouke) in 2007. As a way of alleviating poverty in Cameroon, Djouke sends regular small cash (Euros) donations to promote pig production, and by November 2014, more than 100 farm families have benefited. The project now renamed "The Pig Van Djouke POG Project" consists of donating pig feed and female piglets to farmers to breed and then pass on the gift to other deserving farmers in a process known as "Passing on the Gift: POG" (Fig. 31.1). Since the initial donation, Djouke has been able to assist hundreds of farmers, working with village communities of the North West Region through its founder, Mr Atomba Titus Tegwi, who is a veterinarian and vision bearer of NOWEPIFAC.

A slaughter house project founded by PUM (www.pum.nl) of the Netherlands enabled the cooperative to participate in the slaughter and marketing of pigs produced by members. Funding by the Foundation DIO of the Netherlands supported training in livestock farming (pig, poultry, and rabbit) of over 200 farmers in various areas of livestock production, especially pig farming. Eight training sessions over a period of 2 years were given by the local project coordinators and Dutch experts from PUM (www.pum.nl). Over the years, proper functioning of the cooperative has been achieved through collaboration with other civil society organizations (e.g. North West Farmer's Organization, North West Association of Development Organizations, etc.) and

Fig. 31.1 Recipients of piglets and feed through The Pig Van Djouke POG (Pass on the Gift) Project. (Photo by NOWEPIFAC)

non-governmental organizations (NGOs) (local and international NGOs).

Kenya Livestock Producers' Association (KALPA) is an apex body of livestock producers in Kenya. One of its objectives is strengthening the market linkages of its farmers and other members. This objective became a reality when the Djibouti Agro-Pastoral Association set out to implement a pilot project importing 45 dairy goats bred by Kenyan dairy goat farmers. The project, which was funded by the FAO and facilitated by the Eastern Africa Farmers Federation, together with other stakeholders, aimed to find out whether these dairy goats could be productive in Djibouti.

In the 1990s, Kenyan farmers came together to form groups for the purpose of breeding and marketing of dairy goats. This led to the formation of the Dairy Goats Association of Kenya (DGAK). The DGAK is an active member of KLPA. It ensures that its 16,000 members, who rear between 3 and 50 dairy goats, produce and rear the best dairy goat breeds. The association has been upgrading local breeds with bucks sourced from Germany since 1992. And because of the mix of indigenous and exotic characteristics, survival and high yields are achievable with improved animal husbandry. Government and donor funding encouraged DGAK members to rear dairy goats as a source of income for families that could not afford the land space or money to maintain a cow. In early 2000, the farmers established group nurseries of fodder shrubs for the purpose of providing quality feeds to dairy goats and cows in addition to other benefits accrued from fodder shrubs. They also ventured into production of improved fruit trees. DGAK is mainly involved in managing the breeding and marketing of improved dairy goats. The association has put in place a complex system of buck rotation between the groups as a mechanism of controlling inbreeding in their flocks. It has elaborate training programmes that precede the distribution of improved goats to new areas. The training programme has a feeding component where the use of fodder shrubs as goat feeds is highly recommended. The association has thus helped in scaling up the adoption of fodder shrubs in the East African region and its impacts are well perceived in the region. One of the challenges faced by DGAK farmers is lack of quality breeds due to inbreeding which slowed down the production of pedigree goats, leading to decline milk production and importation of semen from German and French Alpine goat breeds.

The Meru Goat Breeders Association and the Kitui-Mwingi Goat Breeders Association, both in Kenya, evolved from a Farm-Africa Goat improvement initiative (Peacock et al. 2011). These Associations have facilitated the successful development of a high yielding synthetic goat breed based on ¾ genes of an imported Toggenburg breed and ¼ genes of a local indigenous meat goat breed found to be most acceptable in terms of productivity and adaptation to local production environments (Ahuya et al. 2002). The maintenance of the pure line Toggenburgh breed as well as the planned selection and mating to produce the optimum synthetic breed is facilitated by the collaborative efforts of the members of the groups who provide the data to facilitate selection and share in the beneficial use of the identified superior bucks and does.

The Malawi Milk Producers Association (MMPA), an umbrella organization made up of three Regional dairy associations: Mpoto Dairy Farmers Association in the North, Shire Highland Milk Producers Association in the South, and the Central Region Milk Producers Association, has a membership of about 17,000 smallholder Dairy Farmers with herd sizes of one to two dairy cows each. The MMPA operates 121 Milk Collection Centres (or Bulking Groups) where milk produced by farmers is cooled and collected daily by one of five Dairy processors in Malawi for processing and sale to consumers.

The South African Beef (Becker 2016) and Dairy Genomics (Joubert 2016, 2018) programs established in 2015/16 and composed of diverse breed societies, technology innovation agencies, industry players, and international participants, has played a vital role in livestock improvement. The foremost activity of the consortia was the discovery of single nucleotide polymorphisms, establishment of DNA sequencing initiative for South African indigenous cattle breeds, and establishment of genomic selection in South African dairy cattle breeds. This led to the setting up of an accelerated long-term, sustainable programme for genomics and genetic improvement of South African breeds including establishment of a data bank for storing and managing genotypes generated, a bio-bank for biological samples, and development of a reliable methodology to derive and estimate genomic estimated breeding values for use in the beef and dairy industries.

These examples show that the collaboration of a group of farmers in recording performance information, sharing genetic resources to create genetic links across farms and production environments, and desiring to use improved stock is a strong driver for genetic improvement, or collectively collating their products for sale is a strong driver of profitable livestock farming. Facilitating the coming together of farmers for common purposes is an essential aspect in the development of sustainable livestock breeding programs.

31.2.1.3 Breed Associations

A breed association is a special kind of farmer group comprised of members with common interest in the promotion of a single breed. The main interest is usually for the promotion of the breed for commercial production or for aesthetics of the breed as often seen in companion animals. Different breed societies have been founded across the globe, addressing issues bordering on genetic improvement of livestock species. Examples include the Boran cattle societies of Kenya and South Africa, the pig breeders society of South Africa, among others (See Table 31.2 for more African examples). These associations like many others have a keen interest in the development, preservation, and promotion of their breed and its use for the production of the product for which the breed is best suitable. To do this, accurate records of animals and their location are kept, which facilitate traceability. They facilitate the registration of new births and keep meticulous record of parentage, thus allowing to have a good indication of numbers and demography of the breed. The records of births and parentage certification are very useful in generating the pedigree records on which accurate genetic evaluation relies. Members are encouraged to keep records of performance which constitutes the data for genetic evaluation and play a key role in facilitating progressive competition through the organization and participation in "shows" during

Table 31.2 Sample livestock breed societies in Africa

Breed	Organization name	Website	Locations
Cattle	Boran Breeders Society, Kenya	www.borankenya.org	Kenya and South Africa
	Nguni Cattle Breeders Society	www.nguni.co.za	Southern Africa and Zimbabwe
	Tuli Cattle Breeders Society	www.tulicattle.co.zw	Southern Africa and Zimbabwe
	Jersey South Africa	www.jerseysa.co.za	South Africa
	Bonsmara Cattle Breeders' Society	https://bonsmara.co.za/	South Africa
	Africa Droughtmaster Cattle Breeders Society	https://www.droughtmastersa.co.za/	South Africa, Namibia, and Botswana
	Senepol Cattle Breeders' Society of South Africa	https://www.senepolsa.co.za/	South Africa
Goat	Boer Goat Breeders	www.boerboksa.co.za	South Africa
	South African Mohair Growers' Association	https://www.angoras.co.za/	South Africa
Sheep	Afrino Sheep Breeders	http://www.afrino.org.za	South Africa
Pig	Pig Breeders Society	http://pigsa.co.za	South Africa
Horse	Arab Horse Society	http://arabhorse.co.za	South Africa

which animals on show are judged on key features of the breed and winners given awards and honours. This act contributes a lot to the hunger of These awards or prizes serve as motivation to individual farmers to improve their stock and hence the desire for genetic improvement. This desire is a key element in the compliance of individual members of the association to registrater the births of animals and record performance. The desire of the breed associations to promote their breed among farmers is the reason why they cooperate readily in the formation of independent genetic evaluation centres which compare different breeds. Such centres constitute a key resource in animal genetic improvement programs facilitating within- and across-countries comparison of animals and breeds.

In 1950, for example, the Kenyan Boran Cattle Breeders' Society (BCBS) was established, after the registration of the first Boran with the Kenya stud book. Thereafter, the Borans started moving out of Kenya. The late 1950s saw the arrival of the first Borons in Northern Zimbabwe, followed by heifers and bulls. Throughout the 1960s and 1970s, the Kenyan Boran breeders prospered from export of their stock to Zambia, Uganda, and Zaire. The Society sets and maintains standards, which are used to approve the registration of individual animals in the Kenya Stud Book. Animals showing constitutional defects such as excessively pendulous sheaths, lack of pigment, faulty feet and legs, wildness, and abnormalities of any kind are not registered. Borans registered in the Kenya Stud Book are eligible (under Society rules) for live exports and as parent stock for embryo and semen sales. The Society handles export protocol and holds an annual show and sale in Nairobi where the top bulls are on offer.

The BCBS was the first breed society in East Africa to create guidelines for improving indigenous cattle. In so doing, it established a strain of the East African Shorthorn Zebu as a recognized breed—the Boran. An important objective of the BCBS is to retain the efficiency and adaptation of the breed to harsh conditions. Hardiness and fertility in the Boran have made it the premier breed for crossbreeding and rangeland beef production in Kenya. The Society guards against losing these qualities.

The South African Boer Goat Breeders' Association, founded in 1959, has established the standards followed by breeders around the world. The primary aim of the society is to promote the breeding of quality and functional Milk Goats by providing relevant information to breeders on goat breeding and keeping.

31.2.1.4 Livestock Breeding Companies

Historically, private breeding companies have facilitated breed genetic improvement in Africa through the development and supply of seed stock from elite animals to livestock farmers. While notable cattle breeding companies facilitate the development of elite bulls and supply of semen through Artificial Insemination (AI) companies, poultry breeding companies supply fertile eggs, day old chicks, or parent stock which are used to generate the production population with superior performance in broiler meat and egg production. The major companies that have impacted the African livestock industry include but are not limited to the following.

ABS Global https://www.absglobal.com/: Apart from the supply of semen for AI, ABS also develops and makes available the Genetic Management System® (GMS) which is a mating program that allows farmers to make a good choice of bulls and replacement cows based on their genetic merit for various traits. This allows farmers to maximize productivity and profit in their flocks while managing inbreeding and preventing the introduction of deleterious recessive genes into their flock.

Genus plc https://www.genusplc.com/ is a global player in cattle and pig breeding. It is the parent company of ABS through which it supplies improved cattle seed stock and PIC® through which it supplies elite sows and boars as well as boar semen for pig breeding. Its impact on dairy production in Africa is mostly through Genus Breeding India and ABS South Africa. The availability of sexed semen through the cattle breeding companies allows farmers in Africa and elsewhere to breed their replacement animals from their own good cows much quickly, resulting in faster herd improvement. De Novo Genetics https://www.absglobal.com/uk/services/de-novo-genetics/ is a collaboration between Genus ABS Global and De-Su which specializes in producing superior Holstein Friesian cows for milk production. These are marketed in Africa through ABS Global.

Poultry breeding is currently undertaken by a handful of companies, most of which have market shares in Africa. These include Cobb Vantress https://cobbgenetics.com/ which focusses on broiler chicken production and Hendrix Genetics https://www.hendrix-genetics.com/en/ which is involved with multiple species breeding such as chicken, turkeys, pigs, fish, and shrimps. One poultry breeding company that is a key player in Africa supplying the well-known brands such as Ross, Arbor-Acre (AA), and Lohman India River (LIR) is Aviagen® https://aviagen.com//. Other companies that supply layer chickens include Hy-Line Genetics based in the US https://www.hyline.co.uk/about-us/history/ and Lohmann Breeders based in Germany https://lohmann-breeders.com/. In addition to these big players, there are a handful of poultry breeding organizations in Africa which specialize mostly on indigenous chicken improvement and supply locally.

Sustainable Supply

Poultry breeding companies like Aviagen play an important role in livestock genetic improvement and sustainable production in Africa. They fund research and development activities to bring about sustainable genetic improvement in traits of economic, biological, and welfare importance. However, the most important impact they have in Africa is the sustainable supply of high genetic merit poultry seed stock in the form of parent stock, fertile egg, or day old chicks on which commercial poultry production relies. The distribution network of the breeding company includes fully and co-owned grandparent stock companies located in Africa such as Aviagen East Africa based in Tanzania, Aviagen South Africa based in South Africa, and Ross Central Africa based in Zambia, as well as clients that rear parent stock for producing day old chicks. By setting up these subsidiaries closer to their clients and farmers, Aviagen ensures security of supply and minimizes the environmental footprint of the supply chain. Reducing the supply distance between hatchery and farmers also improves the welfare of day old chicks.

Technical Support

Another key contribution is the provision of technical services to farmer and hatcheries. Highly knowledgeable technical staff regularly visit farmers and hatcheries to help producers identify problems and proffer solutions to ensure they get the benefit of the full potential of their high genetic merit birds. This service, provided free of charge, supports the distributors by providing client support in the areas of nutrition and feeding, veterinary care and animal welfare, and a host of other issues. The visits to these farms also allow the technicians to hear directly from the farmers, product trends, and client requirements which are fed back to the breeding program to determine the direction of genetic improvement.

Education, Training, and Capacity Development

Another important role played by the primary breeding companies is the provision of educational materials, hands-on training of poultry production staff, and overall capacity development. For example, Aviagen runs the Aviagen production management school https://aviagen.com/en/about-us/school-overview/ which celebrated its 60 years anniversary in 2023. Its comprehensive approach to poultry training is provided through a 1-month long in-person training annually. A dedicated course for customers in Europe, Middle East, Africa, and Asia was established in 2012 offering similar experience to Africans. To date, about 1100 students from over 74 countries in these regions have benefited from the training, gaining practical skills in poultry house ventilation, hatchery management, and broiler and breeder management. In addition to the training, Aviagen also develops and maintains a repository of education materials in their "info centre" https://aviagen.com/eu/tech-center/ which can be searched to quickly access expert technical documents on a search topic. These materials which are freely available represent excellent teaching materials for academics for use in training the next generation of poultry farmers in various formal educational institutions. These examples are some of the various ways poultry primary breeders impact the availability of high genetic merit chicken and turkey birds to African poultry farmers and how best to manage them for improved productivity, bird welfare, and farmer profitability.

31.2.1.5 Animal Production Supply Chain

The animal production supply chain or livestock value chain can be regarded as the compendium of activities by different individuals involved in the transformation or processing of livestock products or engaged in different phases of the production process (e.g. live animal, chicks, egg, meat, feed, medication, fibre, manure, leather, etc.) and delivery of finished products to the consumer. The livestock value chain is also regarded as a market-focused partnership among different interested parties who produce and market value-added livestock products.

A simple livestock value chain could include the primary producers, intermediates and processors, packaging houses, wholesalers, and retailers of finished products. Individuals with similar interests along the livestock value chain and for the purpose of streamlining processes (e.g. sharing cost and knowledge sharing, etc.) form associations. Supply chain management which bothers on monitoring the flow of products and services from raw materials to finished goods (Lysons and Farrington 2012) supports the smooth function of the livestock-value-chain. In Africa, many actors are responsible for these activities. These supply chain actors and the complementary services they provide help small-scale producers upgrade their practices, raise their productivity, and subsequently improve their welfare. They include agro-dealers, processors, traders, and cooperatives who purchase livestock products from small-scale producers and process for sale to consumers.

Each participant (actor) in a value chain contributes value as the product progresses from the start of the chain to the end consumer. By contributing this value, all players are rewarded with an economic benefit—the primary advantage or motivation for engaging in a value chain. Livestock producers must possess efficient organizational abilities to thrive in an increasingly competitive market. Membership of a farmer association is an essential step for small-scale farmers aspiring to enhance their income and gain a greater share of value in the value chain. Farmer groups possess the necessary resources to effectively engage and establish connections with many stakeholders in the value chain, both at the local and international levels, which individual farmers may lack.

The poultry production supply chain starts with the primary breeder who keeps several pure lines from which current and future products are created. At the breeding nucleus, animals are kept and data are collected at various stages of the bird's life cycle for relevant traits. These data are analysed to determine the genetic merit of individuals in different cohorts of birds from which parents of future generations are selected. The bulk of the birds from the breeding nucleus are multiplied and crossed at the next stage in the supply line. The birds at this stage are the great grandparents (GGP) of the broilers that will be sold to consumers. GGP operations are managed by different companies in collaboration with the primary breeder. They add value by crossing different lines to create the parents of the desired product as well as multiplying the bird's numbers to ensure sufficient products can be supplied to the next level on demand. Their products constitute the grandparents of the desired broiler product and they sell these to the Grandparent (GP) operators. These GP operators are located in various regions including Africa and are privately owned operations that produce and distribute the parent stock (PS). They add value by crossing various male and female grandparent lines to produce the parents of the broiler product. GP operators multiply the grandparents of the broiler birds depending on the number of clients they have and their holdings. The PS operators are closer to the farmers who rear broilers, hence they are even more decentralized and more in number within most African countries. They often produce fertile eggs and run large hatcheries to produce broiler day old chicks for distribution to various broiler farmers. The broiler farmers buy the exact number of birds they can rear to market weight from the hatcheries and finish these to the desired market weight several times in a year. They add value by feeding the birds according to specification so that they can make the birds achieve the desired market weight or the weight desired by processors on time.

In many countries where the market is dominated by live bird sales, the broiler birds from the farms go directly to the market through a number of wholesaler or off takers. In other cases, the birds go to a process plant where value is added by dressing the birds with or without cutting into chicken parts. These then are transported to be sold fresh in supermarkets and butcher stalls or by cold room merchants who are thus able to store large quantity of processed birds over a longer period. More recently in Africa, the increase in number of fast food operations has necessitated further processing of dressed broilers into chicken parts like wings, thighs, and breast fillets which are sold to restaurants and specialized chicken fast food vendors. Cold room operators also stock cut-up chicken parts. The processing and cutting up of chicken into parts make the meat accessible to people of various means who otherwise could not afford to purchase whole chickens or have the willingness, skill, or resources to process live birds in their homes.

31.2.2 The Role of Training and Research Institutions

Education and training and knowledge generation are vital components of sustainable agricultural productivity and livestock improvement. These can be achieved through formal educational institutions, research institutions, professional networks, professional meetings and farmer focused meetings among others.

31.2.2.1 Academic and Research Institutions

Academic and research institutions (universities, colleges, specialized schools, research centres, research farms, etc.) play vital roles in building capacity and generating knowledge for all aspects of livestock genetic improvement. Livestock and agriculture in general are vital components of human survival and are major contributors to the economic development of all African countries.

Recognizing the important place of agriculture, national governments of African countries created public-funded agricultural centred universities, departments, colleges, and specialized institutions with the important task of capacity development to support the growth and development of the agricultural sector. The training and research activities at these institutions are the bedrock of technological developments and scientific innovations in sustaining the livestock industry and the agriculture sector in general. The academic institutions and research organizations train the students (future capacity), producers, and other value-chain players and generate and document knowledge on all aspects of the livestock-value-chain. All levels of capacity development are offered by the agricultural-centred institutions in Africa and internationally.

The top agricultural universities and colleges' building capacity in all areas of livestock production are discussed in Chapter 27, while Chapter 28 presents an overview of human capacity building in livestock genetic improvement and the limitations of current curricular as well as the opportunities for adopting training options that are suitable for the African environment. To increase effective programs delivery by these institutions, greater cooperation with primary producer groups, non-governmental organizations, and governments and other players in the livestock-value-chain will ensure the flow and sharing of knowledge, innovations, and information. International institutions (like the International Livestock Research Institute (ILRI)), national and international initiatives, and donors have played significant roles in resourcing livestock genetic improvement in Africa through funding and training.

31.2.2.2 Livestock or Breed Improvement Networks

Livestock or breed improvement networks are a grouping of researchers, farmers, and persons interested in the improvement of livestock or specific breeds. The African Goat Improvement Network (AGIN), established in 2011, is a collaborative group of scientists interested in the genetic improvement of goats in smallholder communities in Africa (Van Tassell et al. 2023). The goals of the AGIN are to: improve the productivity of indigenous goat breeds; characterize existing goat populations; facilitate germplasm preservation; apply genomic approaches to better understand adaptation; develop cost-effective strategies to implement genomic strategies to improve goat productivity without sacrificing adaptation; perform genome-wide genetic variation-enabled genetic diversity studies; facilitate improved germplasm preservation decisions; and take necessary steps to initiate large-scale genetic improvement programs. These propositions were partly executed via a sequence of community-level breeding schemes that actively involved and empowered local small-scale farmers, particularly women, to foster the sustainability of the production system. One of AGIN's goals is to facilitate concurrent, multi-faceted capacity development for researchers, students, farmers, communities, as well as local and regional government officials.

The African Animal Breeders Network (AABNet) (Djikeng et al. 2025) is another initiative that brings together animal breeding/scientist professionals and concerned stakeholders with the common goal of the sustainable development of the African livestock sector.

31.2.2.3 Professional Animal Production Societies

Professional animal production societies are a grouping of individuals (professionals, students, farmers and industry personnel) with interest to advance the improvement (genetics, health, management, reproduction, welfare, etc.) of livestock and promote viable and sustainable production systems in the face of changing environmental conditions. Membership in professional animal production societies is drawn from agricultural institutions (animal, plant, and veterinary medicine) of learning, producers, government departments, support institutions, and service industries. The societies could be national, continent-wide, or international. Sample animal science societies are listed in Table 31.3, and through their listed website addresses, specific information about each society can be seen. Their main goals are

Table 31.3 Sample professional animal production-based societies in Africa and internationally

Name of society	Acronym	Website
African Animal Breeding Network	AABNet	https://www.animalbreeding-africa.org/
All Africa Society for Animal Production	AASAP	None
American Society of Animal Science	ASAS	https://www.asas.org/
Animal Science Association of Nigeria	ASAN	https://animalscienceassociation.org.ng/
Botswana Society of Livestock Science and Production	BSLSP	None
Ethiopian Society of Animal Production	ESAP	https://landportal.org/node/62954
Ghana Society of Animal Production	GSAP	None
International Dairy Federation	IDF	https://fil-idf.org/
International Poultry Council	IPC	https://internationalpoultrycouncil.org/
Animal Production Society of Kenya	APSK	https://apsk.or.ke/
Nigerian Society for Animal Production	NSAP	http://www.nsap.org.ng/
South African Society for Animal Science	SASAS	https://www.sasas.co.za/
American Society of Animal Science	ASAS	https://www.asas.org/
American Dairy Science Association	ADSA	https://www.adsa.org/
Canadian Society for Animal Science	CSAS	https://www.asas.org/CSAS
European Association for Animal Production	EAAP	https://www.eaap.org/
World Association of Animal Production	WAAP	https://waap.it/

similar and include the following: advance all aspects of the husbandry of livestock species and the products derived from them; advance the nutrition, genetics, physiology, reproduction, health, and breeding of livestock animals; ensure the humane care of livestock; ensure that animal sourced food is produced ethically and economically; provide networking opportunities among members; share knowledge and practices through regular meetings; share knowledge on scientific innovations through publications like journals, magazines, and books, among other goals specific to each society.

The Animal Science Association of Nigeria (ASAN), for example, focuses to improve understanding of the animal sciences and the ways food is produced ethically and economically. The ASAN supports the career advancement of scientists and animal producers and serves as a platform to encourage the younger generation into pursuing animal science careers. It also advances the discovery, sharing, and application of scientific knowledge for sustainable livestock production and for the responsible use of animals to enhance human well-being and environment sustainability. The ASAN draws membership from individuals, firms, and organizations with interest in instruction, research and application, and extension services in animal science, or individuals involved in livestock production, processing, distribution, and marketing of livestock and its products. The ASAN holds regular scientific meetings and workshops for its members and other interested parties; an avenue where the latest scientific innovations in the animal sciences are shared and networking opportunities established. The ASAN publishes the Nigerian Journal of Animal Science, which covers diverse topics within the framework of sustainable livestock production including animal products, biotechnology, health, production, nutrition, management, physiology, feeds/feedstuff, genetics and breeding, reproduction, socio-economics, farming systems, crop/livestock interactions and extension, among others.

The Nigerian Society for Animal Production (NSAP) has contributed immensely to livestock improvement. Inaugurated during the First International Symposium on Animal Production in the Tropics at the University of Ibadan in Nigeria in March 1973, the NSAP mission is to promote the study and practice of all facets of animal production; provide a platform for discussions on scientific and industrial policies and educational and social issues of relevance to the development of animal production in Nigeria;

engage in the dissemination of scientific and educational information related to animal production; interact with organizations with similar ideologies within Nigeria and internationally. The membership of NSAP includes individuals and corporate bodies that are interested in all aspects of animal production such as animal scientists, livestock economists, farmers, feed millers, livestock sociologists, pasture agronomists, veterinarians, and students of animal science and related disciplines. The NSAP holds annual meetings where its members meet to share the latest scientific innovations in the animal sciences, network, and form linkages. The Nigerian Journal for Animal Production is the flagship journal of the NSAP. It has maintained consistent publication since its first issue in 1974 and has over 50 volumes to date. The Nigerian Society for Animal Production is a member of the World Association for Animal Production.

The All Africa Society for Animal Production (AASAP) brings together individuals, groups, organizations, and institutions interested in the art, science, and practice of animal sciences relevant for animal agriculture such as genetics and breeding, animal nutrition and feeding, health, welfare, reproduction, and all other aspects of animal husbandry. The AASAP main goal is to facilitate the use of technical, policy, institutional, and scientific innovations to address current and emerging challenges of animal agriculture in Africa through engagement of animal-related communities of practitioners in Africa and beyond. Every 4 years, the AASAP organizes an international meeting titled "The All Africa Conference on Animal Agriculture" (AACAA) in collaboration with other Animal Production Societies, drawing participants from the African continent and internationally. Each meeting addresses specific themes of relevance to the African continent. In 2023, the AASAP organized its 8th meeting (AACAA8) which was co-hosted by Botswana University of Agriculture and Natural Resources and in association with two Livestock Production Societies (Botswana Society of Livestock Science and Production and South African Society of Animal Science) and the National Agricultural Research and Development Institute. The AACAA8 addressed the following eight themes: (1) enhancing youth engagement in animal agriculture in Africa; (2) towards sustainable resolution of farmer-herder conflicts in Africa; (3) climate change adaptation and mitigation: the role of animal scientists; (4) the One Health Concept—managing zoonotic disease transmission through controlled contacts between domestic livestock and wildlife; (5) animal food safety and consumer health: current status in Africa and required interventions; (6) animal science, ecological goods, and rangeland services in Africa; (7) information and communication technology and data science: opportunities for utilization of innovative technologies in sustainable animal agriculture in Africa; and (8) appropriate animal genetics, feeds, and health inputs: towards solutions for access and last mile delivery. The AASAP is a member of the World Association of Animal Production (with its secretariat in Rome).

The International Poultry Council (IPC), a member-driven organization, is composed of national sector associations, poultry businesses, and other key industry stakeholders. Members of the IPC are responsible for over 88% of the world poultry meat production and 95% of the global poultry meat trade. The IPC is proactive in promoting science-based solutions and information-sharing for regulatory authorities and poultry producers worldwide and is uniquely positioned to provide a holistic view of the global poultry sector. The IPC is therefore regarded as the voice of the global poultry meat industry as it brings together global poultry industry stakeholders, represents the whole value chain (from farm to fork, and from genetics to consumption), and leverages the resources and strengths of national organizations.

International dairy federation (IDF) is a recognized international authority in the development of science-based standards for the dairy sector. IDF ensures that the right policies, standards, practices, and regulations are in place to ensure the world's dairy products are safe and sustainable.

31.2.3 The Role of National and Regional Governments

National and regional governments are vital stakeholders in resourcing animal genetic improvement because of their interest to ensure national food security, minimize carbon footprint, protect the environment, create jobs, empower youths, support rural development, and grow their Gross Domestic Product (GDP). More importantly, agriculture is a major contributor to the GDP of African countries and hence a key sector of national economies. In practical terms, government's role in resourcing animal genetic improvement involves: (a) supporting the development of the concept and setup of agencies responsible for genetic improvement; (b) support the development and availability of human capacity to man the agencies, i.e. people with the right training, knowledge, and skills to man various positions in the organization; (c) create and support education and training institutions for manpower development; (d) facilitate the operation of agencies, in terms of policy framework, regulation, standardization, and funding; and (e) provide the enabling environment for seed stock development, dissemination, and utilization which includes the support for and sensitization of farmers as to the value of utilising high genetic merit animals in farming and providing the required environments for these to thrive in order to attain sustainable improvement in livestock production.

In some African countries like South Africa, agencies dedicated to livestock genetic improvement exist and are operating routinely. In others like Nigeria, agencies such as the Centre for Genomic Research and Innovation, embedded within the National Biotechnology Development Agency https://nabda.gov.ng/, are in the development phase with a vision to set up the key infrastructure for livestock genetic improvement (Olori and Nash 2021). The task of resourcing animal genetic improvement in Africa is particularly daunting and may require initial government involvement. This is because the livestock production sector is less evolved compared to the crop production sector. Most livestock are reared under traditional production systems characterized by small and highly dispersed farms, and, for large ruminants, migratory or nomadic in nature. These make the task of animal identification and data collection difficult, requiring a new approach in breeding program design and operation. Evolving and resourcing new production systems that facilitate efficient livestock production may require government policy and intervention. These include land use, water resource and energy resource development and supply, road networks, and market development. It will also require government intervention in allocating resources to human capacity development by way of formal education as well as training of various supporting personnel and farmers. To fulfil its role in agriculture and hence livestock production, many African countries have a ministry dedicated to agriculture and livestock production. Depending on the country and government of the day, the portfolio of the ministry could include rural development and/or fisheries.

31.2.3.1 Development of Policy Framework

The primary role of national governments in livestock genetic improvement revolves around policy development, facilitation, regulation and funding. In this regard, agencies of governments have a responsibility to provide the policy framework that facilitates and promotes livestock farming, conservation of animal genetic resources, generation and dissemination of animals' seed stock, and safe utilization of products of genetic improvement for the maintenance of human health, human and animal welfare, as well as protection of the environment. Government support for livestock genetic improvement programs fulfils the multifaceted role of facilitating conservation and utilization of available animal genetic resources as well as generation and dissemination of animal seed stock to support sustainable livestock production. This indirectly reflects government interest in rural development, job creation, food security, protection of the environment, and minimization of carbon footprint of the agricultural industry. In many western countries like the UK, the government also has interest in protecting the rural landscape and hence the utilization of the available land for agriculture and especially livestock production.

Governments develop the policy framework that governs sustainable livestock production, animal welfare and health as well as the protection of human health from health risks emanating from livestock production and consumption. These policies therefore directly affect livestock producers, production environment, productivity, and hence genetic improvement. The development of policy framework implies that governments will have a clear indication of capacity requirement (including human capacity, land, and stock), principles of practice (including ethics, welfare, and health considerations), and funding requirement (level, allocation, and accounting). A good policy framework will give clarity on who can undertake livestock genetic improvement, where it can be undertaken based on environmental and health considerations, and the standard of the products that can be disseminated to farmers to improve productivity. It is pertinent to note that government regulations must be matched by government funding to enhance compliance. Such funding can sometimes be in the form of public-private partnerships or wholly private with appropriate regulation of source. For example, if for the purpose of protecting human health a farmer is asked to slaughter and appropriately dispose of all their stock, compliance will depend on how well the farmer is compensated for the loss of stock and revenue, and how well the government can monitor and supervise implementation of the directive.

31.2.3.2 Standards and Operational Regulations

One aspect of policy is the development of standards. For the purpose of protecting human health, animal health and welfare, or protecting the environment, government or their agencies sometime define the standards for the use of various biotechnologies in livestock production and genetic improvement. Examples include the use of antibiotics in livestock, the level of slurry spreading on agricultural land, or nitrogen fertilizer application to the soil where nitrogen leaching to the water table is a problem. This is another key role of government in livestock farming and hence genetic improvement. Governments also play a key role in creating bilateral agreements with different countries or group of countries to facilitate cooperation in livestock production matters. This could include the sale of livestock or animal seed stock across borders, the standard of imported inputs and bilateral export certification for biosecurity, and confidence in the health status of imported stock.

31.2.3.3 Development of Infrastructure

The most important direct role of the government is the resourcing of agencies that provide the key infrastructure that supports livestock production and genetic improvement. These include general infrastructure such as roads and other transport facilities, power/electricity, piped water supply, telecommunication facility as well as market outlets. More importantly, government mostly has the responsibility to provide infrastructure for education and training to develop the needed human capacity for livestock production and genetic improvement. A good road and transportation network is important to facilitate humane animal transportation which supports trade in animals, especially the improved genetic seed stock. For example, the transportation of day old chicks from the hatchery to the farmer's farm must happen quickly under good transportation conditions in order to minimize early mortality. In Africa where temperatures and humidity are generally high, poor transport facilities mean birds will spend more time on the road in very poor conditions which could reduce the viability of the chicks (Yerpes et al. 2021).

In terms of specific infrastructure for genetic improvement, the most important use of public funding will be in setting up an independent organization with responsibility for all aspects of livestock genetic improvement. This takes the burden from government and allows a specialized organization to take the responsibility with government support. In this regard, the government initially acts as the convener urging and facilitating all stakeholders to work with the specialist organization in all aspects of livestock improvement. For example, genetic improvement relies heavily on the presence of a unique animal identification system, the recording of livestock performance from different farms, and the condition under which they perform. The organization also needs to maintain the identity, pedigree, genomic, and performance data generated from farms and analyse the data to provide selection information to farmers.

Setting up such an organization in many countries requires national and/or regional government grants. Such expenditure can be justified for many reasons. First, such an organization can support the maintenance of genomic database of livestock and the estimation of population parameters, as well as the conservation of animal genetic resources. It can also facilitates the selection of genetically superior animals for breeding the next generation which brings about genetic improvement and hence efficient production and consequent minimization of national carbon footprint from agriculture. Furthermore, a good animal identification system is not only important for genetic improvement, but it is also critical for government to have accurate information on livestock populations and demography which is needed for accurate planning and siting of key infrastructure. The ability to track the movement of animals is critical in disease prevention and control. Identification also facilitates traceability which helps consumer confidence and is critical for livestock security. For example, the ability to trace the owner and origin of cattle will discourage large-scale cattle rustling which is a key security issue among migratory pastoral cattle farmers.

The key infrastructure required to support an efficient animal identification system is an institution with computing facilities and human capacity for safe storage and processing of animal identification records. Such institution will need to devise a process of issuing unique animal IDs as well as capturing and verifying such IDs. This requires an adequate livestock database to facilitate the capturing and storage of associated information. Many countries have a system where every animal born is immediately registered. For example, in many European Union (EU) countries, it is a government requirement for every animal born to be registered with the appropriate government department or designated organization. In the EU, the incentive for this is cash payments to qualifying farmers through various national schemes under the European common agricultural policy. An example of a national scheme is the Beef data and genomics program of the Republic of Ireland https://www.gov.ie/en/service/5b44a8-beef-data-and-genomics-programme-20152020/#.

In many countries, these roles may be consolidated in one or shared between a few organizations. For example, the Irish Cattle Breeding Federation (ICBF) (www.icbf.com), the South Africa Stud Book and Animal Improvement Association (http://studbook.co.za) and Lactanet (www.lactanet.ca) in the Republic of Ireland, South Africa, and Canada, respectively, combine the responsibility of data collection, storage, and management of farmers' data with the provision of a genetic evaluation service. The development of a central database in Ireland (Olori et al. 2005) allowed the ICBF to consolidate the functions of many different organizations, thereby avoiding duplication and minimizing costs. In other countries, these responsibilities are shared between two or more organizations, for example, in the USA, The Council on Dairy Cattle Breeding in the USA (www.uscdcb.com) is responsible for the genetic evaluation, while data collection, storage, and management are undertaken by a group of public-private state or regional Dairy Record Providers and Dairy Record Processing Centres.

31.2.3.4 Research and Livestock Development

Many countries in Africa have agricultural research councils expected to provide farmers and government with direction and leadership regarding livestock research and development. In addition to these direct agencies of governments, many Universities, other education institutions, and dedicated livestock research institutions supported with government funds play key roles in conducting research to support genetic improvement efforts. These include basic research as well as socio-cultural research to support estimation of population parameters, to improve the understanding of the biological basis of observed variation, especially with genomic information, development of livestock inputs, improvement in health care and biosecurity as well as supporting the uptake and utilization of animals of high genetic merit by farmers. Starting from the 1960s, many countries with the cooperation of universities and agricultural research departments undertook livestock improvement programs. The examples of Nigeria and Gambia are highlighted below.

Highlight of Livestock Development Programs by the Government of Nigeria

Efforts by the Nigerian government to improve livestock productivity started in the 1960s with the importation of Holstein Friesian (improved cattle breed) and distribution to farms (Agege Dairy Farm, Iwo Road Dairy, Ibadan) and universities (University of Ibadan and Ahmadu Bello University, Zaria) and again in the 1990s to develop crossbreds with the White Fulani cattle breed by the Nigerian Veterinary Research Station in Vom and use for higher milk production by West African Milk Company (Adebambo 2013). Poor management practices, lack of adequate funding, and political instability contributed to the failure of these programs.

Efforts to develop sheep and goat meat production by the government of Nigeria started in 1978 at three sites, Tuma in Katsina state (sheep), Ladanawa in Kaduna state (sheep), and Zugu in Sokoto state (goat). The objectives of the small ruminant program were to: (1) maintain pure breed herds of Sokoto goat and Balami sheep, (2) develop suitable management practices for sheep and goat production such as supplementary feeding ingredients, optimum pasture utilization, controlled mating programs, 3) establish within breed genetic selection strategies, and 4) develop sound animal health management regimes, among others. This was followed by the set up of sheep and goat multiplication centres by individual states such as at Kaltungo in Bauchi State, Marguba in Borno State, Pampegua in Kaduna State, Rano in Kano State, and Fasola in Oyo State. The main objective of these programs was to develop and distribute breeding animals to local farmers. Thus, the various livestock species were kept and managed in Breeding and Multiplication Centres in geographical zones where they are best adapted.

The federal government of Nigeria furthered efforts to genetically improve indigenous breeds of livestock in the country by establishing different research centres with specific mandates, including: (1) The National Centre for Genetic Resources (with main functions being to develop animal genetic resources - cattle, poultry, goat, sheep, etc.) and the naming, registration, and release of livestock breeds. (2) The National Animal Production Research Institute (with a mandate to genetically improve the indigenous livestock resources). (3) The Nigerian Conservation Foundation (responsible for the conservation of the animal genetic resources of Nigeria to ensure their sustainable use through research and development). All centres collaborated on sharing data on indigenous animals, especially those of economic and conservation importance.

The Federal government of Nigeria also established many departments of agriculture and veterinary medicine at major Nigerian universities as well as technical agricultural institutions for the purpose of training and research. The training and research activities on agricultural matters were tied to the National breeding policy in the implementation of the road map in Animal genetic resource development, among others. The goal of the policy is to increase the productivity of indigenous livestock and poultry and the number of officially registered livestock breeds and to popularize the use of registered species. Research activities have led to great stripes in the genetic improvement of Nigerian indigenous livestock breeds. Glaring examples are the development of improved breeds of chicken in Nigeria- FUNAAB ALPHA (sponsored by Bill and Melinda Gates Foundation) and NOILER. A recent livestock productivity and resilience project (L-PRES) of the Federal Ministry of Agriculture and rural development has as development objectives to improve livestock productivity and resilience, promote the commercialization of selected value chains, and strengthen the capacity of Nigeria to respond to crisis situations or emergencies. The L-Press project seen as a strategic step in Nigeria's long-term approach to sustainable transformation of the livestock sector seeks to (1) strengthen national livestock institutions for improved environment and service delivery; (2) enhance value chain performance; and (3) prepare for crisis prevention and management (conflict mitigation, peace building, and project coordination). The project considers modern breeding facilities, increased research efforts, and application of in situ, in vivo, and in vitro conservation methods as vital aspects for the improvement of livestock productivity and conservation. Within the goat-value-chain of the L-PRES project, clusters of goat farmers received pure-bred Boer goats in fulfilment of the program's goal to genetically improve indigenous goat productivity. The program faced many challenges such as: (1) low or non-existing animal genetic conservation; (2) more emphasis placed on crop than livestock; (3) high hybridization of indigenous livestock species; (4) very low infrastructural and personnel capacities for conservation; (5) low and inconsistent power supply to power storage and research facilities; (6) weak linkages between research institutions, universities, and the ministry of agriculture; and (7) moribund breeding and multiplication centres. As mitigation strategies, the L-PRES project (1) supported functional centres for the conservations of genetic resources along ecosystems with comparative advantages for some livestock breeds; such as Sokoto Gudali cattle, uda and sahel sheep at the North West; Rahaji cattle, Balami and Sahel sheep at the North East; and -Muturu, Ndama, Keteku cattle, West African Dwarf goat and sheep at the South West; (2) undertook concurrent conservation and crossbreeding to address food security; (3) promoted collaboration with research centres and foreign breeding companies to set up semen and embryo collection centres to serve the needs of Nigeria and the surplus for export.

To maintain pedigree details of animals, the government of Nigeria plans to implement genetic improvement programmes for which performance recording is a prerequisite, as well as animal identification and traceability recording systems, by livestock improvement organizations. The goal is to assist farmers in the management of their herds. A further goal of the Nigerial government is to establish veterinary health institutions and organizations to manage the health and vaccination of herds or individual animals (FRN 2019).

Highlight of Livestock Development Programs by the Government of Ghana
Genetic Improvement Programme in Ghana

Up until 1990, there was no livestock genetics improvement program in existence in Ghana (Okantah 2009) that addressed the multifunctional roles played by various livestock species. The Ministry of Food and Agriculture, university faculties of Agriculture, and the Animal Research Institute were the major organizations that showed interest in breed evaluation and genetic improvement of cattle in Ghana. The programs included the use of Zebu crossed with Taurine (e.g. White Fulani × West African Shorthorn), exotic × local (e. g. Friesian × Sanga), or pure-bred exotics (e.g. Friesian). According to Okantah (2009), the exotics and their crosses had better growth rates and milk yields than indigenous breeds. However, the exotic breeds also had poorer adaptation to the local environment. It was concluded that the very elaborate breed evaluation and genetic improvement experiments conducted by the universities were limited to stations, resulting in little or no impact on the livestock production systems. The experiments by the Ministry of Food and Agriculture were the most successful, as the Sanga progeny from the Zebu × Taurine crossbreeding projects was adopted by farmers in all cattle-rearing regions in the country (Okantah 2009), but the genetic improvement efforts were largely uncoordinated at the time.

In recent years, Ghana Ministry of Food and Agriculture attempted the modification of the existing livestock development policies and strategies in the country with the overall objective to develop a competitive and more efficient livestock industry with capacity to increase domestic production, reduce importation of meat and livestock products, and contribute to the improvement of the livelihoods of all livestock value chain actors and the national economy while protecting the environment, preserving livestock biodiversity, and ensuring bio-security. The strategies put in place were (1) to promote conservation and genetic improvement of indigenous breeds of livestock with a rich diversity of animal genetic resources that have over the years survived the existing harsh environments and disease challenges and (2) to improve the production and productivity of livestock by promoting and disseminating improved technologies, among others (GMFA 2016).

The necessity for all stakeholders to contribute to the genetic improvement of indigenous livestock genetic resources to ensure sustained food and nutrition security, particularly for rural households, was recognized. Within this recognition, the following categories of poultry and livestock species were areas of particular importance and focus: domestication and improvement of Grasscutter farming; characterization, conservation, genetic improvement, and commercialization of local chicken and guinea fowl ecotypes; genetic improvement of the local Ashanti Black/Dwarf pig, indigenous sheep and goat breeds, and indigenous cattle breeds and their crosses (Ghana Shorthorn, Sanga, N'Dama and their crosses).

Different livestock breeding programs that have been implemented in Ghana so far are: traditional livestock breeding system which involves the use of supposedly good males, often selected by visual assessment for upgrading flocks in the extensive (characterized by uncontrolled mating) and semi-intensive systems; and open nucleus breeding scheme which is based on individual animal performance. The open nucleus breeding scheme was undertaken by Animal Production Directorate in its genetic improvement program in conjunction with participating breeders. This scheme covered Djallonke sheep and goat, the Ashanti Black Forest pigs, and the West African Shorthorn cattle. There were on-farm selection of breeding animals (male and female) and on-station at Animal Production Directorate breeding farms. The aftermath of the selection and breeding was distribution of selected males to participating breeders for multiplication and sale of selected male offspring to other farmers for the genetic improvement of their flocks. However, the increasing use of exotic livestock and lack of conservation of indigenous breeds has been implicated in the decrease in the relative abundance of the West African shorthorn from 65% to 47% of Ghana national cattle hard in 2001 which later declined to 39.3% in 2011 (GMFA 2016).

To accomplish successful growth in the near futur, comprehensive policy and strategy instruments are required that will enable the country's livestock resources and the many actors in the livestock value chain to express their potential. A road map for the efficient application of suitable policies and strategies will need to be implemented on time, and closely monitored and evaluated.

31.3 Role of Non-governmental Organizations and Other Charities

According to the Britannica dictionary, non-governmental organizations (NGO) are a voluntary gathering of persons or organizations, not affiliated to any government, with common interests to provide a service or advocate on common topics. While majority are not-for-profit organizations, a small number of them are for-profit organizations. NGOs are involved in a wide variety of activities including trade unions, religious organizations, community groups, agriculturally based organizations, among others. The 1970s onwards saw an explosion of international NGOs in Africa focusing on different areas of development and relieve activities including livestock improvement (Thomas-Slayer 1992). In Africa, NGOs have thus been instrumental in advancing livestock improvement with the main goal of helping local communities out of poverty. Many international NGOs have contributed significantly in advancing the livestock sector in Africa including Catholic Committee against Hunger and for Development (CCFD) (French Catholic founded NGO), German Agro Action (GAA), Heifer International (HI) (USA founded NGO), OXFAM (British-founded confederation of 21 NGOs), Nederlandse organisatie voor internationale ontwikkelingssamenwerking (Netherlands Organization for International Development Cooperation) (NOVIB), Services d'Appui aux Initiatives Locales de Développement (SAILD) (local NGO in Cameroon), and Vétérinaires Sans Frontières (Veterinarians Without Borders) (VSF) (France, Belgium, and Switzerland), among others. These NGOs work with many stakeholders to drive their specific agendas in areas of focus.

Responding to achieving the United Nation's sustainable development goals (SDGs) (FAO 2015), a multistakeholder partnership of governments, organizations, civil society, research/academia, private sector, donors, NGOs, inter-governmental and multi-lateral organizations committed to the sustainable development of the livestock sector developed a common Global Agenda for Sustainable Livestock (www.livestockdialogue.org) development. The NGO cluster (The Donkey Sanctuary, HI, Kyeema Foundation, the LIFE network, and VSF) is committed to working with smallholder livestock farmers and pastorals in low- and middle-income countries to meet set targets. The NGOs are specialized in different aspects of relevance to pastoralism and smallholder livestock farming, animal health and welfare, environment, resilient livestock-based livelihoods, and integrated service delivery for poverty reduction. A recent policy brief by the NGO cluster examined smallholder and pastoral systems and highlighted the role of sustainability and innovation, and the impact that modern technologies (e.g. mobile phones), scientific research (e.g. new immunization methods), and favourable policies have on sustainable development (GASL 2019). The role of HI or Heifer Project International (HPI) is further highlighted as a case study to give better insights into the activities and functioning of an NGO in resourcing livestock genetic improvement in Africa and Cameroon in particular.

31.3.1 Case Study: Four Decades of HPI Role in Driving Livestock Genetic Improvement in Cameroon

Heifer International (HI) or Heifer Project International (HPI) (www.heifer.org) is a development not-for-profit organization (NGO) working to end hunger, poverty, and care for the earth by using sustainable practices and engaging smallholder farmers in agricultural development. Heifer International is founded on the simple belief that ending hunger begins with giving people the means to feed themselves, generate income, and achieve sustainable livelihoods. With this strong foundation, Heifer focuses her pro-poor work by creating wealth through viable value chains that harness the social capital of communities to drive market development. Heifer International works in about 21 countries across the World, and these countries are located in Africa, Asia, and the Americas, and USA. Notably, HPI has worked with over 46 million people worldwide. The activities of HPI in Cameroon will be further highlighted as an example of its achievements in many countries.

The HPI has been involved in dairy development in Cameroon since 1974. Prior to this period, Cameroon was a typical tropical developing country with relatively little or no dairy tradition apart from milking of cattle under an extensive traditional system of management by Fulani pastoralist and a small dairy farm established in Buea (Upper Farms) in the 1930s with the introduction of the German Brown Cattle by colonialists (Atekwana and Maximuangu 1981). HPI's first attempt to jumpstart the development of the dairy industry in Cameroon began in 1974 with the introduction of 22 Holstein–Friesian and Jersey (Fig. 31.2) cattle imported from the United States of America (USA). This was followed by subsequent importations of live cattle (Holstein and Jersey) and semen, dairy goats, pigs, and chicks (chicken), equipment, and supplies in sub-

Fig. 31.2 Imported Holstein (**a**) and Jersey (**b**) and Holstein (1995 batch) (**c**) at IRZV Bambui centre, as well as imported hens (**d**). (Photos by HPI, extracted from HPI Cameroon archives and provided by Emmanuel Bassam (HPI Director of Program))

sequent years. The aim of these importations was to improve the genetic make-up of indigenous cattle (Gudali, Red Mbororo (or Red Fulani), and White Fulani), sheep, goat, pig, and chicken populations through crossbreeding for improved milk, meat, and egg production. The imported animals and supplies were used for research and development purposes at the Institut des Recherches Zootechniques et Vètèrinaires -IRZ (Institute of Animal and Veterinary Research) (presently IRAD - Institut de Recherche Agricole pour le Développement (Agricultural Research Institute for Development)) centres at Bambui, Mankon, and Wakwa. Crossbreds (progenies) of exotic cattle and indigenous cattle (same for other livestock species) developed at these research centres were distributed to local farmers via extension services. This marked the official beginning of HPI agricultural development activities in Cameroon.

Following HPI's initial activities in 1974, the next 10-year period of operations (1975–1984) was regarded as the decade of implantation and experimentation. A HPI and the National Office for Scientific and Technical Research (ONAREST) of Cameroon (ONAREST—HPI) cooperation started in 1976 with objective "to make animal protein more readily available, and cheaply too, to the population of Cameroon". The specific objectives of the program were to (1) build a genetic base on which the Cameroon livestock industry will expand; (2) improve the economic and nutritional status of rural people and increase the supply of livestock products to urban centres; (3) reduce foreign exchange on the importation of animal products and by-products;

(4) provide employment opportunities to curb rural exodus; (5) provide good quality breeding stock to smallholders at subsidized prices; and (6) provide training and management opportunities to livestock farmers. The IRZV Bambui centre was focused on dairy cattle development, while the IRZV Mankon centre was focused on dairy goat development. A 1978 evaluation of the IRZV Bambui dairy program indicated that the exotic dairy breeds adapted well, reproduced, and produced normally under the prevailing climatic and environmental conditions of Bambui, which facilitated the extension of the dairy program to local farmers. To join the program, training sessions were organized and farmers who completed the training as well as demonstrated capacities (had pastures of sufficient quality and quantity) and well-constructed milking sheds were offered dairy animals.

Working jointly with the Institute of Animal Research (IAR), HPI began distributing crossbred dairy cows to smallholder farmers in 1978. By 1981, a semi-intensive dairy production system based on the South-South Workshop on Contributions of HPI to small-scale dairy development in Cameroon was instituted in the North West Province of Cameroon (Fig. 31.3a–c). HPI also engaged in pasture and forage development (Fig. 31.3d) and in using local knowledge, herbs, and practices (ethnoveterinary medicine) to treat sick animals.

The HPI-Farmer activities resulted in the establishment of the Bamenda Dairy Cooperative Society (BDCS), which functioned by distributing feed to farmers on a monthly basis, collected and sold farmers' milk on a daily basis, and held monthly appraisal meetings with farmers. Farmers needed 5–6 years of technical advice, encouragement, and support to fully establish as smallholder dairy farmers.

During this period, the productivity of the Holstein as expected was higher (3500–4000 L) compared to crossbred animals (Holstein × zebu) (2700–3300 L) and zebus (300–800 L) in a 270–300-day lactation period. HPI also initiated collaborative work with several local institutions, including the Presbyterian Rural Training Centre (PRTC) at Mfonta, the Saint Joseph's Children and Adult Homes (SAJOCAH), and the Sisterhood of Emmanuel in Bafut in 1981. HPI also worked with many smallholder farmers, gave practical training courses, and distributed animals on PASSING On the Gift (POG) basis, thus initiating the shift in focus from research to working directly with farmers. The initial decade of operation helped define HPI-specific objectives in Cameroon (Box 31.1).

> **Box 31.1 HPI-Specific Objectives in Cameroon**
>
> 1. Respond to requests for development assistance such as provide livestock and related training and technical assistance, to enable families attain self-reliance in food production and income generation on a sustainable basis.
> 2. Encourage farmers who receive in-kind loans of livestock from HPI to pay back by passing on the gift to other farmers in a manner that enhances dignity and accord an opportunity to everyone to make a difference in the fight to alleviate hunger and poverty.
> 3. Support project leadership development at the local level, farm-family unity, and cultural unification.
> 4. Work with farmers to develop appropriate techniques for year-round on-farm feed production, integrated agriculture, and better farm management, thereby ensuring reduced livestock mortality rates and increased animal and crop production.
> 5. Provide extension services to HPI-assisted farmers in collaboration with farmer leaders and collaborating agencies.
> 6. Actively support the formation and management of viable small-scale farmer cooperatives and other self-help activities, mostly those by women groups, and encourage project sustainability, farm-family self-reliance, and savings in credit unions.

7. Exploit the potentials of agroforestry in enhancing sustainable livestock production and agricultural production in general, and environmental protection.
8. Promote parity between men and women by (i) consulting development partners at the various stages of project plan implementation, accomplishment, and evaluation; (ii) accompany development partners through exchange; (iii) develop, test, and apply effective gender balance practices and also share same; (iv) gather and disseminate gender assessment tools, methodologies, and practices; and (v) collaborate with agencies that prioritize gender.
9. Conduct nutrition studies and education, public awareness programs, and market studies to promote the sale and consumption of crops and livestock.
10. Make HPI-assisted livestock and integrated projects profitable through family employment, human nutrition improvement, and income-generating activities.
11. Apply indigenous veterinary medicine practices to control and treat economically important livestock diseases and other complications.
12. Set up and operate para-veterinary associations and use well-trained and experienced Fulani grazers and other capacity building activities for staff (HPI and local), volunteers, and student interns.
13. Conduct training and other capacity-building activities for staff (HPI and local), volunteers, and student interns.
14. Collaborate with other organizations in co-financing and facilitation of technical and other networking arrangements that enhance the progress of projects.
15. Cost-effectively use human and material resources to achieve project goals.

Between 1980 and 1983, HPI extended its training initiatives by sponsoring 14 persons for short- (8 persons) and long- (6 persons) term courses in the USA, as well as provided infrastructural support to the IRZV Mankon station and purebred Holsteins (imported from USA and supported by the United State Department for Agriculture (USDA)) to IRZV Wakwa station. In 1982, HPI distributed Jerseys and Holstein Friesians to individual farmers and several institutions like Mambu Health Centre (Bafut), Emmanuel Sisterhood of Bafut, PRTC Fonta, Mbengwi Monastery, Catholic Mission Njinikom, Mbingo Hospital, Ndu Baptist College, and Shisong hospital. Following an evaluation of the productivity of animals, HPI determined that crossbred (exotic × zebu) dairy cattle had superior adaptability to the prevailing environmental conditions and local management practices and lower mortality than exotic purebreds.

A cooperation between IRZV, HPI, and USAID which started in 1976 and renewable every 5 years established a *"HPI Small Farmer Livestock and Poultry Program"*. The program ended in 1985 with an end-of-project evaluation report that strongly recommended among others that: (1) the "crossbreeding *program between Holstein Friesians and white Fulani be continued in IRZV Bambui station*; (2) *the* crossbreeding *program between Holstein Friesian and Gudali be discontinued in Bambui but continued in Wakwa*; and (3) *artificial insemination with frozen semen be encouraged at Bambui and Wakwa Stations"*.

A dairy goat program with goal to "improve the management of dairy goats" was instituted at the IRZV Mankon centre. Animals under the program performed poorly due mainly to poor management resulting from the lack of experience in goat management and training in livestock production by the staff of the Centre, most of whom were hired at entry level with no formal training in livestock management. To solve the problem, four persons were trained on different aspects of animal production in the USA to supplement the lack of experience by IRZV staff in

Fig. 31.3 Farmers receive Holstein-Friesian cattle at Nseh Mbabu village (**a**), Holstein-Friesian under semi-intensive system of management at Bamendakwe (**b, c**); and HPI staff at a pasture and forage development site at Fonta, Bambui (**d**). (Photos by HPI, culled from HPI Cameroon archives and provided by Emmanuel Bassam (HPI Director of Program))

goat production management. The goat herd which included 42 dairy goats from the HPI shipment of 1976 and 50 local goats bought with funding support by the International Foundation for Science of Sweden was not sustainable due to poor management. The management failures included (1) lack of training and supervision of livestock attendants; (2) overgrazed paddocks; (3) disease, theft, lack of adequate nutrition and disease management, and poor housing conditions; (4) inadequate grass/forage production for confinement feeding and supplemental feeding in the dry season; (5) lack of water for dry season cleaning of the goat and sheep barns and for filling dipping vats; (6) underfeeding of kids; (7) irregular feed supply; (8) incompetent veterinary personnel; (9) failure to treat sick animals; (10) lack of a breeding program; and (11) neglect of routine management practices.

In the face of many challenges of maintaining exotic animals, HPI in 1983 discontinued the importation of exotic animals and concentrated on crossbreeding (natural service and artificial insemination) programs. To ensure an adequate supply of dairy cattle to participating farmers, HPI instituted crossbreed "multiplier herd" farms in collaboration with local institutions like Mambu Health Centre (Bafut), Emmanuel Sisterhood of Bafut, PRTC Fonta, Mbengwi Monastery, Catholic Mission Njinikom, Mbingo Hospital, Ndu Baptist College and Shisong hospital, and colonel Valentine's centre in Ndop.

In the third decade (1985–1994) of operation in Cameroon, HPI focused more on direct grassroot assistance and support. While USAID and MESRES (Ministry of Higher Education and Scientific Research) in collaboration with MINEPIA were willing to participate in exten-

Fig. 31.4 A ram from HPI Project in the far North Region of Cameroon (**a**); Sheep placement in the Mbam Upper Sanaga Valley Integrated Sheep/Goat Project in the Center Region of Cameroon by HPI (**b**); and Sheep placement in the Bui Donga Small Holder Integrated Sheep & Goat Project in the Northwest Region of Cameroon by HPI (**c**). (Photos by HI, extracted from HPI Cameroon archives and provided by Emmanuel Bassam (HPI Director of Program))

sion and on-farm research with HPI, Cameroon through ONAREST laid more emphasis on research, while on-farm extension services were neglected. This disagreement in direction of focus caused a fracas in the IRZ/HPI/USAID cooperation causing HPI to downsize on its activities, withdraw most of its staff from Cameroon in 1985, and hand over most of its assets to IRZV. Now operating at a limited scale, HPI focused on its traditional on-farm/extension approach with farmer training and direct placement of animals on small-scale farms (Fig. 31.4).

Following the Lake Nyos gas disaster in Cameroon in 1986 (Kling et al. 1987), HPI responded by providing much needed relieve to livestock farmers. Under a 1987 HPI project tagged "*Lake Nyos Livestock Restocking Project*" (sponsored by HPI and Equatorial Foundation), livestock (pregnant Zebu cows, Holstein Friesians, Boran and cross-bred Boran × Holstein Friesians) (Fig. 31.5) were imported from Kenya and distributed to Fulani livestock farmers who lost their cattle during the Lake Nyos disaster for resilience building in the gas-affected communities of Nyos. Furthermore, HPI participated actively by providing relief materials and resettling the Lake Nyos gas disaster survivors living in camps out of the disaster area. The resettlement efforts by HPI were facilitated by a collaboration with the Presbyterian Church in Cameroon

Fig. 31.5 Boran animals imported from Kenya (**a, b**) and offspring of Boran × Zebu (**c**) distributed to farmers under "*Lake Nyos Livestock Restocking Project*". (Photos by HI, extracted from HPI Cameroon archives and provided by Emmanuel Bassam (HPI Director of Program))

to mutually undertake smallholder agricultural and livestock projects in the country, a cooperation that lasted until 1994. To achieve resettlement objectives, HPI worked in collaboration with several individuals and organizations such as Wum Catholic Mission with Rev Fred Ten Horn, Wum Area Development Authority, Peace Corps, and Equatorial Foundation to provide livestock and assorted crop seeds, among others, to survivors at resettlement camps in Waindo, Buabua, Essu, Kumfutu, Yemge, Kimbi, and Epalem for planting as a source of food security. These activities strengthen HPI's yearning for sustainable integrated agriculture, with staff and farmer capacity-building initiatives. In 1989 and 1990, ethno-veterinary and Women in Livestock Development (WiLD) projects, respectively, were initiated.

In the period 1990–1995, HPI funded the SAJOCAH project with Aidfund assistance provided by Mr, Jeop Duynstee. The Aidfund assistance to SAJOCAH included livestock (13 heads of dairy cattle, one pair of oxen, 60 rabbits, and 300-day-old chicks), agricultural equipment (locally made brooders and brooder guards, locally made poultry drinkers, cement troughs for cattle, rabbit cages, cement feeding troughs for pigs, corn shelter machine with spare parts and maintenance services, equipment for small scale milk processing, grinding mill, and fence), and yearly operating cost to sustain the project. During the same period, HPI in collaboration with Laboratoire National Vétérinaire (LANAVET) and Pan-African Rinderpest Campaign (PARC), initiated the Project of Organization for African Unity/Inter-African Bureau for Animal Resources in 1993 to promote the use of heat-stable Rinderpest vaccine by Community-Based Animal Healthcare agents in Cameroon. Meanwhile, a partnership established with Mieke and Joep from Holland in 1994 enabled HPI to operate a regional strategy-based umbrella project that extended its activities to the West and Southwest Regions under British High Commission sponsorship with funds from the Department for Foreign and International Development. In the same year (1994), HPI started a consolidated small scale Dairy Development Project, managed semi-intensively or by zero-grazing, with emphasis on improved productivity on minimum land and quality animal management and breeding through farmer-managed artificial insemination centres. Purebred Holstein and Friesian cattle were used in the zero-grazing production system with the aim to achieve milk production levels comparable to those of developed countries. The main forage species fed to animals were Brachiaria, Guatemala, elephant grass, Desmodium and Stylosanthes, and concentrates. Milk produced was sold mainly to SOTRAMILK, established through a loan granted by the RABO Bank of Holland. SOTRAMILK mainly served farmers in Mezam division due to proximity to the factory, while farmers in Bui and Donga/Mantung divisions were not served due to distance. However, the marketing of milk was hampered by many problems such as the dispersed location of farms and low yields causing problem of collection and transportation to the factory. Figure 31.6 shows project participants showing off their milk to government officials and veterinarians during the 40th anniversary celebration of HPI presence in Cameroon.

An evaluation of the Dairy Development Value Chain project in the Northwest and Western regions of Cameroon to improve livelihoods for the period 1995–2009 is presented in Box 31.2.

HPI continued field projects and activities in Cameroon until 2015 which culminated with its 40th-anniversary celebration (26 June 2015), field staff departure, asset disposal (August 2015), and closure of Heifer Cameroon Program (December 2015). Today, HPI continues to work in 19 countries across four continents including nine African countries (Ethiopia, Kenya, Malawi, Nigeria, Rwanda, Senegal, Tanzania, Uganda, and Zambia).

HPI's four decades work in Cameroon was facilitated by partnerships and funding from indi-

Fig. 31.6 Livestock farmers showing off their milk products to Cameroon government officials and veterinarians during Heifer International Cameroon's 40th Anniversary Celebration in Bamenda in 2015. (Photo by HPI, extracted from HPI Cameroon archives and provided by Emmanuel Bassam (HPI Director of Program))

Box 31.2

Performance indicators of HPI Dairy Development Value Chain in the Northwest and Western regions of Cameroon to improve livelihoods from 1995 to 2009.

Project location: Noun Division in the West Region and Mezam and Bui Divisions in the North West Region. The North West and West Regions make up the Western Highlands agroecological Zone.

Variables:

- Number of farm families supported: 20 (1995) to 971 (2009)
- Number of purebred dairy cattle: 20 (1995) to 633 (2009)
- Average daily milk yield/cow: 6–10 L (up to 1994) to 16–21 L (2009)
- Average calving interval: 16–18 months (up to 1994) to 12–15 months (2009)
- Objective of dairy farming: largely subsistence (up to 1994) and market-oriented (2009)
- Number of groups assisted: 1 (1995) to 22 (2009)

Successes recorded:

1. Milk production improved with increasing parity: first lactation yield ranged from 10 to 22 L with an average of 13.7 L daily; second lactation yield ranges from 10 to 25 L of milk with an average of 15.7 L of milk daily.
2. Sustainably improved household socio-economic growth/development: about 404 farm families recorded increased household income through earnings from sale of milk and livestock; increased income used to improve livelihoods through the purchase of household furniture/assets, farm tools and inputs, farmland, and motorcycles and build new homes.
3. Socio-economic empowerment of women and improved gender relations in supported household: about 167 women owned dairy cattle as direct project partners, supported their husbands by providing basic household needs, and paying children's school fees; women increasingly involved in household decision making at all levels in their households; and about 37% (61 out of 167) of participating women occupied leadership positions within dairy cooperatives/common Initiative Groups and other community groups.
4. Improved education of children in supported households: A total of 709 children (358 boys and 351 girls) from 259 farm families acquired an education (basic, secondary, vocational, high School, and university education) with income from project activities; and academic performance of children improved through regular attendance of classes and having adequate school needs.
5. Increased use of animal manure-enhanced soil nutrient and structure: (1) Majority of supported farm families (89.1%) used ecologically sustainable farming practices such as animal manure, composting, use of crop residues in feeding livestock, planting of leguminous trees, contouring, and ridging across the slopes. Moreover, a cumulative of 642 non-assisted members in assisted project communities adopted composting and use of crop residues. (2) Participating farm families (404) planted 121,200 environmentally friendly trees like Calliandra, Acacia, and Leucaena and established 303 hectares of Bracharia and Guatemala forage plots. These practices in addition to other environmentally friendly practices reduced soil erosion and improved soil fertility leading to increased crop yields. There was increased crop yields (from the same size of land) by majority of farm families (76%).

6. Mitigation of the impact of HIV and AIDS: Participating families sensitized on HIV and AIDS prevention and care; persons (565 men and 296 women) within 404 project partners adopted responsible sexual behaviours. In addition, a cumulative of 2666 non-Heifer-assisted persons within project communities adopted responsible sexual behaviours.
7. Organization of assisted farmers in dairy cooperatives and common initiative groups: Beginning with one Dairy Cooperative in 1995, HPI Cameroon supported small-scale dairy development program working with 15 dairy cooperatives and seven (7) Common Initiative Groups, organized under the umbrella of the Cameroon Union of Dairy Cooperative Societies.
8. Other project spread effects: Private Dairy Processing Plans such as Cameroon Dairy Industry and SOTRAMILK established as a result of the project; other private individuals and institutions started dairy development within and without project region; project built the capacity of students within and outside Cameroon; project inspired the initiation of a national dairy development program by the government.

vidual donations, organizations, community organizations and groups, national government, other NGOs, and international organizations. To support its activities, HPI brought in various experts into Cameroon (administrators, experts in various areas of livestock production, agronomist, economist, etc.) who worked in collaboration with IRZV (now IRAD), MINREL, MESRES, MINEPIA, USAID, local farmers, local farmer groups/associations, local institutions, Christian institutions, and other Government departments. Funding of its activities came from many sources including USAID, International Foundation for Science (Sweden), BOTHAR (Ireland), WILD GEESE (Holland), ASSIA (Holland), and Bread for the World (Germany), etc. For example, USAID invested 1.3 million USD during the period 1980–1985 for the construction of laboratory and research facilities and other infrastructures at IRZV Mankon and promoted small farmer livestock and poultry projects with the training of 14 personnel for short- and long-term courses; while between 1974 and 1985, HPI spent about *USD 1,494,625.85* as a direct investment on livestock purchase/imports, purchase of equipment and supplies and training, among others.

HPI's activities in Cameroon clearly demonstrate multi-partnership engagement in resourcing livestock improvement. Clearly, HPI positively impacted the livestock industry in Cameroon and can be regarded as the initiator of the development of the livestock sector in Cameroon using improved and pure breeds of cattle, pig, guinea pig, rabbit, poultry, and goat, and indigenous livestock breeds. Despite the success of the HPI partnership with other institutions to deliver livestock improvement in Cameroon, many obstacles were evident. For example, limited funding, knowledge gaps, and a lack of clear livestock development policy and commitment by the government of Cameroon where the major limiting factors faced by HPI during her operations in Cameroon.

This report on HPI is based on HPI annual reports (1974–2015), a report on the contribution of HPI to Small Scale Dairy Development in Cameroon and a background report of HPI in Cameroon (written by HPI Cameroon Staff to mark the celebration of her 40 years of effective presence in the development space in Cameroon). These reports were provided by Mr. Emmauel Bassam, Director of Program at HPI (2012–2015).

31.4 Regional, Continental, and International Organizations

31.4.1 International Development Institutions

International institutions and funding bodies have played significant roles in resourcing animal improvement programs in Africa such as the Consultative Group on International Agricultural Research (CGIAR), ILRI, Biosciences eastern and central Africa (BecA), the EU, USAID, Canadian International Development Agency, among many others. The international agencies focusing on livestock genetic improvement in Africa include mostly research-based institutions such as the ILRI and BECA. While these agencies undertake a wide spectrum of research in line with the mandate from their funders, a lot of their work lend itself to support livestock genetic improvement. In many cases, they have very good facilities and skilled staff which facilitate training and manpower development in highly specialized fields of genetics and livestock improvement. On many occasions, these agencies get funding to host key research projects directly related to livestock improvement.

The ILRI, as an example, has significantly impacted the livestock sector development and training in Africa (https://www.ilri.org/). ILRI, a CGIAR research centre located in Kenya and Ethiopia with offices in other African countries and Asia, focuses on improving livelihoods through livestock research and development in developing countries. In collaboration with its partners, ILRI works to (1) develop, test, adapt, and promote sustainable and scalable science-based practices to support better lives through livestock; (2) provide persuasive scientific evidence that clear and favourable policies and significant investments for livestock research can bring about significant socio-economic growth, positive health and environmental gains to farm families and developing nations; and (3) increase capacity building among key stakeholders to promote better utilization of investments in livestock and livestock science (knowledge) for enhanced livelihoods. With funding from hosting nations and international partners, ILRI's research activities in livestock have directly or indirectly impacted the African livestock sector. For example, ILRI in collaboration with multiple partners including western research institutions, national dairy genetic companies, international dairy genetic companies, farmer organizations, national livestock seed regulators, national agricultural research and extension systems planned and implemented the "The African Dairy Genetics Gains (ADGG) (https://portal.adgg.ilri.org/)" and "African Chicken Genetic Gains (ACGG) (https://africacgg.net/)" projects aimed at enhancing productivity in smallholder farming systems (ILRI 2022). The ADGG program supported by Bill and Melinda Gates Foundation and CGIAR trust fund drives improvements in national breeding schemes by analysing performance data collected in the production environments to produce and publish breeding values for enhanced productivity and adaptation. It has established centralized dairy performance recording centre's in Kenya and Tanzania with an initial 12,000 smallholder dairy herds; developed desktop, web, mobile tablet and sms-based solutions with dynamically defined user roles; and established a pilot information and communication technology-based platform for collation, evaluation, and feedback on phenotypic and genomic data from smallholder farms in Tanzania and Ethiopia (Ojango et al. 2022; Mrode et al. 2020). Following the evaluation of pedigree data collected in 2020, restricted selection indexes and a genomic prediction pipeline for animals in smallholder production environments were developed and results shared with farmers through extension messages sent digitally through the iCow system. The timely sharing of genomic evaluation and breeding management support messages helps farmers make informed decisions for improving the productivity of their animals.

The ACGG, an African-wide multistakeholder initiative led by ILRI, aimed to provide improved chicken genetics to smallholder farmers in Africa. Specifically, develop high genetic producing breeds well adapted to low-input production systems; make available breeds preferred by farmers; develop innovation platforms and solutions across the chicken-value-chain; encourage the active participation of women and promote public-private partnerships for improvement; and multiplication and delivery of improved chicken genetics to needing communities. The success of the ACGG project includes the following: (1) enhanced poultry productivity (for example, the results of on-station and on-farm tests show 200% and 300% increase in egg production and body weight, respectively, from indigenous and tropically adapted chicken breeds); (2) made available locally adapted more productive chicken breeds preferred by farmers; (3) empowered women; (4) delivered training in poultry farming; (5) developed tools that enhanced poultry farming and contributed to improving the livelihoods (nutrition and income) of participating households; (6) created linkages across the chicken production value chain in all three project countries (Ethiopia, Nigeria and Tanzania); and (7) developed and maintained notable stakeholder engagement that ensured the success of the program. The DCGG at its inception in 2014 initially focused on three countries (Ethiopia, Tanzania, and Nigeria) but has recently extended its activities to six other countries (Kenya, Ghana, Zimbabwe, Cambodia, Myanmar, and Vietnam) causing a change in appellation from ACGG to Tropical Poultry Genetic Solutions (TPGS). The TPGS intends to continue the implementation of the successfull ACGG program in all nine countries and, in addition, will address nutrition, health, finance, and policy aspects of chicken production. The ACGG/TPGS activities are facilitated with funding from the Bill and Melinda Gates foundation and CGIAR Trust fund and its multi-partners (e.g. the *Federal University of Agriculture Abeokuta, Nigeria*; National Animal Production Research Institute, Nigeria; Obafemi Awolowo University, Nigeria; Ethiopian Institute of Agricultural Research; Haramaya University, Ethiopia; Tanzania Livestock Research Institute; Sokoine University of Agriculture, Tanzania; Emerge Centre for Innovations-Africa (previously PICO-Eastern Africa); Animal Breeding and Genomics Centre, Wageningen University, Netherlands; Koepon Foundation; farmer groups; among others).

These projects (ACGG and TCGG) have demonstrated, most importantly, that the numerous challenges in bringing about livestock genetic improvement in Africa can be overcome through multistakeholder engagement. They also demonstrated that with the genomic revolution, genetic improvement in Africa can follow other paths that can be more efficient and climate-friendly. A good example from the ADGG project is that multi-county genetic evaluation with centralized regional facilities is feasible in Africa (Mrode et al. 2019). Moreover, the emergence of the African Animal Breeding Network (AABNet) with "multi-county genetic evaluation" as one of its pillars (Djikeng et al. 2025) further supports this concept.

31.4.2 Global Initiatives for Genetic Evaluations and Sustainable Livestock Improvement

While individual national or regional organizations facilitate data collection, storage, management, and utilization within their respective countries/regions, collective efforts through coming together under one umbrella organization facilitate international cooperation and collaboration. Examples of such international cooperations are the International Bull Evaluation centre (INTERBULL), the International Committee for Animal Recording (ICAR), and the Global Agenda for Sustainable Livestock (GASL). The INTERBULL is based in Uppsala, Sweden. It was set up specifically to undertake and regulate international evaluation of dairy bulls. It specializes in providing international genetic evaluation of dairy bulls across countries so that these can be ranked on the same scale across countries. This ranking of bulls on the same scale facilitates across-country comparison and trade in dairy cattle genetic resources. The initiative facilitates international trade in dairy bull semen such that farmers in Ireland can choose to buy and use semen from a bull in New Zealand and another bull in Canada in addition to bulls in Ireland. Using bull semen across countries allows rapid improvement of national stock performance in line with defined breeding objectives and facilitates the creation of genetic links across countries which make international genetic evaluation even more accurate. The INTERBULL project has been so successful that it has now been extended to beef cattle with the establishment of INTERBEEF (https://www.icar.org/index.php/technical-bodies/working-groups/interbeef-working-group/).

The ICAR, based in Rome, Italy, is an international non-governmental agency (NGO) which aims to support the development of animal identification systems and the recording of animal performance, both of which are directly relevant for genetic evaluation and livestock improvement. By providing guidelines and international standards for animal identification and recording of performance, ICAR helps to develop unique animal identification systems and standardized systems for measuring performance. For example, it certifies the equipment used for milk recording and testing so that data collected globally can be compared on the same scale and hence suitable for across-country analysis. ICAR supports individual member countries/organizations in developing unique animal identification systems and hence the traceability of individual animals. This is critical in situations where individual animals such as bulls have offspring in different countries and production systems. Unique identification allows the daughters of the bull to be tracked in the different countries and by helping to standardize recording equipment, the performance of such bull daughters in different countries and production systems can be compared on the same scale.

Organizations like INTERBULL and ICAR will be critical in the development of genetic improvement programs in Africa where across-country genetic evaluations have been advocated as a means of minimizing cost by pooling resources across countries to facilitate continent-wide genetic improvement of livestock. Both these organizations are global in outreach, hence membership is open to all countries or group of countries with a joint evaluation. Presently, South Africa through the South African Studbook is a member of both the INTERBULL and ICAR. While participating in INTERBULL international evaluations requires the country to meet some basic requirements, ICAR can accept new entrants without any requirement and can help new member nations develop their infrastructure. It is logical to first join ICAR and then subsequently join INTERBULL when facilities for national genetic evaluations are in place and functioning.

The Global Agenda for Sustainable Livestock (GASL, www.livestockdialogue.org) is an example of multi-stakeholder partnership in resourcing global livestock improvement. Founded in 2011, the GASL is a partnership that capitalizes on the strength of the private and public sectors, producers, academic and research institutions, NGOs, foundations, social movements, and community-based organizations who have pulled forces to address human health, global food and nutrition security at the animal-human-environment interface, equity and growth including viable growth in value chains, and resources and climate change mitigation. The goals of the multi-stakeholder partnership include: (1) voluntary commitment to work together towards a common goal; (2) facilitate dialogue between numerous stakeholders across a broad range of sectors; (3) derive mutual benefits from working together through win-win agreements; (4) learn from each other; (5) strive for equitable representation and inclusiveness for all relevant stakeholders; (6) agree on modalities of governance such as rules and modalities of cooperation; and (7) ensure accountability and transparency.

The GASL is playing a key role in the implementation of the UN 2030 sustainable development goals (SDGS) (FAO 2015) by catalyzing and guiding the sustainable development of the livestock sector focusing on the economic, social, and environmental outcomes. The GASL prioritizes SDGS 1, 2, 3, 5, 8, 12, 13, 15, and 17 identified as having direct links to the livestock sector. Ending 2022, the GASL had 121 official partners, including 20 governments, 24 private sector organizations, 7 multilateral and intergovernmental organizations, 27 academic and research institutions, 5 social movements, 29 NGOs, and 9 donors (Confédération Nationale de l'Élevage, France; Agriculture and Agri-Food Canada, Government of Canada; Ministry of Foreign Affairs, Government of France; Department of Agriculture, Food and the Marine, Government of Ireland; Swiss Federal Office for Agriculture, Government of Switzerland; Ministry of Economic Affairs, Government of the Netherlands; Bill and Melinda Gates Foundation; United States of America Aid, Ethiopia; and Global Dairy Platform). Funding by the donor organizations is vital for the achievement of the objectives of the GASL.

31.4.3 Continental and Regional Communities and Regional Collaboration for Livestock Development

The African Union (AU) plays a very important role in livestock genetic improvement in the continent through its agencies, the African Union–Inter-African Bureau for Animal Resources (AU-IBAR), the Forum for Agricultural Research in Africa (FARA) (https://faraafrica.org/) (See Chapter 30), and others.

The AU-IBAR was founded in 1951 with the mandate to study the epidemiological situation and fight rinderpest in Africa. Today, the AU-IBAR's mandate covers all aspects of animal resources, including livestock, fisheries, and wildlife, across the entire African continent. The objectives of the AU-IBAR include coordination of the improved utilization of livestock species in the continent through setting enabling policies and providing support to member states. It encourages regional economic communities to set common livestock policies to facilitate the movement and trade in livestock. The AU-IBAR documents the AnGR in the content, supports training and dissemination of information, and maintains a database of the livestock genetic resources in Africa. To achieve its objectives, the AU-IBAR developed a 20-year Livestock Development Strategy for Africa (LiDeSA) to address Africa's development needs and challenges in the livestock sector. The Strategy is aligned with regional strategies, policy frameworks, and guidelines and coherent with the Comprehensive Africa Agriculture Development Programme, Frameworks, and Agenda (see Chapter 30). The main goal of the strategy is to transform the livestock sector by energizing its under-utilized potential through (i) encouraging increased investments from both public and private sources for the transformation of the sector to enhance its contribution to socio-economic development and equitable growth; and (ii) accelerate reforms in the sector.

The FARA is the apex continental organization responsible for coordinating and advocating for agricultural research for development. FARA serves as the technical arm of the Africa Union Commission on matters concerning agricultural science, technology, and innovation. The FARA has the mandate to provide leadership and guidance in agricultural research and development at the continental level. They achieve these through five key mandates; Policy and advocacy, research management and leadership, knowledge management and outreach, capacity development and agri-preneurship, observation and coordination, and finally multi-stakeholder partnerships. In July 2021, FARA became the hosting organization to facilitate the activities of the African Seed and Biotechnology Partnership Platform (ASBPP). This is the regional body tasked with the responsibility of helping to coordinate policy formulation on plant and animal seed generation and utilization across the continent based on sound research evidence (African Union 2021). The animal seed working group of ASBPP will take leadership advocating and guiding the formulation of policy regarding livestock genetic resource development and hence the resourcing of genetic improvement programs in Africa. Through the ASBPP, therefore, it is hoped that FARA will start to have an impact on improvement of livestock production in line with recorded success in crop production in Africa.

The various economic communities of African states (Economic community of West African States (ECOWAS), Common Market for Eastern and Southern Africa (COMESA), Southern African Development Community (SADC), etc.) have established common livestock policies with main goal of increasing productivity and trade in livestock between member states (See Chapter 30). For example, the ECOWAS through its Agricultural Policy established a Livestock Action Plan for 2011–2020 with the goals to: promote the livestock, meat, and dairy sector; structure the livestock production sector; create a favourable environment for the development of livestock, meat, and dairy products (ECOWAS 2010). Current initiatives include facilitating pastoralism and ending livestock-crop farmer conflicts. An intergovernmental organization for Strengthening Agricultural Research in Eastern and Central Africa (ASARECA) (15 member states) has the mandate to (1) strengthen, catalyse, facilitate, and coordinate regional agricultural research for development activities; (2) improve delivery of national and regional research-driven initiatives for sustainable agricultural transformation and development outcomes; (3) enhance sustainable livelihoods for all; (4) support value addition of selected commodities; (5) develop climate smart agricultural practices; (6) enable smallholder farmers to deal with climate change issues; (7) enhance livelihoods among regional pastoral communities; (8) help member states develop agricultural development plans; (9) coordinate ASARECA Regional Agricultural Market Information System, and (10) manage Agricultural Digital Transformation for Development Hub.

31.4.4 Relevant International Agricultural Agencies

The Food and Agricultural Organization (FAO) of the United Nations (UN) sets the global agenda for the cataloguing, development, and conservation of AnGR as a means of ensuring food security and maintaining world biodiversity. The FAO maintains a database of AnGR and drives the AnGR agenda through policies, information collation, and sharing. With the adoption of the Global Plan of Action for AnGR (FAO 2007) in 2007, the world recognized the vital importance of AnGR biodiversity and in particular livestock biodiversity for agriculture, rural development, and food and nutrition security. Activities pertaining to sustainable use and conservation of biodiversity for food and agriculture are overseen by the Commission on Genetic Resources for Food and Agriculture (https://www.fao.org/cgrfa/meetings/commission/es/). Considering that livestock are central to achieving many of UN's 17 Sustainable Development Goals (SDGs) (FAO 2015; Schneider and Tarawali 2021), the FAO in collaboration with national governments, NGOs, and other support organizations is providing information and training in all areas of livestock production. The FAO actively collates and disseminates information on global livestock genetic resources. It maintains the Domestic Animal Diversity Information Systems (DAD-IS) data base (https://www.fao.org/dad-is/) . The DAD-IS is a searchable database of global livestock breeds, and breed-related information with links to other online resources on livestock diversity. This database allows the exploration of the diversity of livestock breeds at national, regional, and global levels as well as gives information on the status of breeds regarding their risk of extinction. Another database, the Domestic Animal Genetic Resources Information System (DAGRIS), is developed and administered by the ILRI. The DAGRIS facilitates the compilation, organization, and dissemination of data from research findings on the origin, distribution, diversity, use, and risk status of indigenous farm animal genetic resources. The FAO compiles and publishes on a regular basis books and reports on AnGR. Table 31.4 presents sample FAO books and reports on livestock.

31.5 The Place of Policy, Funding, and Technology in Facilitating Sustainable Livestock Production

Chapter 30 examines the global policy and strategy landscape on AnGR that is relevant for Africa, the extent to which these are being trans-

Table 31.4 Sample FAO books, reports, and databases on animal genetic resources and their management and improvement strategies

Title of book or report or database[a]	Available at	Year of publication
Animal genetic resources information		
Animal Production and Health—Annual report 2022	https://www.fao.org/documents/card/es?details=CC6433EN	2023 (Yearly reports)
Rinderpest and its eradication	https://doi.org/10.20506/9789295115606	2022
Methane emissions in livestock and rice systems—Sources, quantification, mitigation, and metrics	https://doi.org/10.4060/cc7607en	2023
A Framework for Gender-Responsive Livestock Development	https://www.fao.org/3/cc7155en/cc7155en.pdf	2023
Africa Sustainable Livestock 2050: Livestock and viral emerging infectious diseases	https://doi.org/10.4060/cc2160en	2022
Africa Sustainable Livestock 2050	https://www.fao.org/3/i7222e/i7222e.pdf	2017
Expert consultation on the sustainable management of parasites in livestock challenged by the global emergence of resistance—Part 1: Current status and management of acaricide resistance in livestock ticks. Report of the FAO Expert Consultation—9–10 November 2021	https://doi.org/10.4060/cc2981en	2022
Expert consultation on the sustainable management of parasites in livestock challenged by the global emergence of resistance—Part 2: African animal trypanosomosis and drug resistance—a challenge to progressive, sustainable disease control, 9–10 November 2021.	https://doi.org/10.4060/cc2988en	2022
The role of animal health in national climate commitments	https://doi.org/10.4060/cc0431en	2022
Peste des Petits Ruminants Global Eradication Programme II & III: Overview of the plan of action	https://www.fao.org/documents/card/en?details=cc2759en	2022
Pastoralism—Making variability work.	https://doi.org/10.4060/cb5855en	2021
Veterinary Vaccines Principles and Applications	https://www.fao.org/documents/card/es/c/cc2031en/	2021
The State of the World's Animal Genetic Resources for Food and Agriculture	https://www.fao.org/3/a1250e/a1250e.pdf	2007
The second report of the State of the World's Animal Genetic Resources for Food and Agriculture	https://doi.org/10.4060/I4787E	2015
Domestic Animal Diversity Information Systems	https://www.fao.org/dad-is/	
Domestic Animal Genetic Resources Information System	https://www.fao.org/dad-is/	
DAGRIS[b]	http://dagris.ilri.cgiar.org/	
FAOSTAT	https://www.fao.org/faostat/	

[a]The books or reports are published by the FAO (Food and Agricultural Organization) only or in collaboration with other organizations such as ILRI, IFAD (International Fund for Agricultural Development), World Bank, World Organization for Animal Health (WOAH), John Wiley & Sons
[b]A livestock database maintained by ILRI

lated into strategies, plans, and institutionalized practices by African countries, and the challenges being faced and opportunities for interventions.

The policy and institutional environments that drive characterization, improvement, conservation, and sustainable use of animal genetic resources are influenced by actions at multiple levels. At the global level, the FAO facilitates and catalyses interventions, cross-cutting technical issues, using a lens that also takes account of interactions with other aspects of natural-resources management, production-system dynamics, and general socio-economic development, and markets. A flagship contribution of FAO in this regard has been the development of the Global Plan of Action (GPA) which has provided a framework that countries can use to develop and operationalize policies for characterization, conservation, and use of AnGR. FAO has also played an important role in helping countries, especially in developing regions, including Africa, in domesticating the global instruments relevant for animal genetic resources, namely the Nagoya Protocol on Access to Genetic Resources and the Fair and Equitable Sharing of Benefits Arising from their Utilization to the Convention on Biological Diversity (CBD), and the Cartagena Protocol on Biosafety.

At the continental level, the AU-IBAR works with AU Member States and the Regional Economic Communities (RECs) to which the Member States belong to catalyse and facilitate development of policies, strategies, and plans for the livestock sector, covering the domains of animal health and animal production, including characterization, conservation, and use of animal genetic resources.

The chapter 30 concludes that, while the AU, through the AU-IBAR, in partnership with Regional Economic Communities, has created significant awareness in Africa around global AnGR policy frameworks and instruments as well as actions needed at national levels, these efforts have delivered only limited change—with only a few countries having developed or in the process of developing policies and strategies, but even these largely remain on shelves, with little evidence of implementation. The technological developments and specific examples for Africa are discussed in Chapters 20, 21, 22, 23 and 24.

31.6 Towards Sustainable Livestock Breeding Programmes

Approaches for the development of sustainable livestock breeding programmes have been discussed in other chapters (Chapters 15, 16, 17, 18 and 19). This section presents a summary of key considerations for establishing sustainable breeding programmes in Africa:

(a) Across-country genetic evaluation: This is considered a potentially powerful tool to improve genetic evaluations in Africa by increasing the number of animals involved, hence accuracy of evaluations. It entails putting data together to estimate breeding values (EBV) for individual sires and dams, followed by ranking based on their genetic merit in the different environments (e.g. countries and herds). The across-country ranking is used to account for the presence of genotype by environment interactions (G × E). This approach can be used to determine whether animals selected in one environment can perform effectively in another environment. The general increase in the value of genetic parameter estimates and accuracy of selection, especially where there are insufficient data available in individual countries, is a very attractive proposition for different livestock producing countries. Pooling data for across countries is not only important for genetic evaluations, but can also help to inform individual countries on their farm management practices. Also, the sharing of information is an important component to inform future breeding strategies and genetic gain. This approach can also bring on board countries that are currently not undertaking any animal evaluations, because of lack of capacity, or other challenges. Across-country genetic evaluation for joint genetic evaluation is feasible also through sharing already determined breeding values ranked accordingly.

(b) Establishment and maintenance of adapted indigenous breeds: Although importation of exotic breeds is a relatively quick genetic improvement pathway that can change the gene and genotypic frequency of the base population, it has been shown not to be sustainable. A more practical and sustainable approach would be through the development of new tropically adapted breeds. For example, creating a synthetic breed in conjunction with local farmers using crossbred sires selected from within the local herds. This strategy is different from previous strategies where synthetic breeds have been developed on government research farms and later distributed to farmers. This strategy allows for the creation of synthetic breeds in a process in which farmers are involved—through a systematic and continuous crossing of females with improved crossbred sires selected within a population.

(c) Harness the power of technological advances and innovations for novel phenotyping and genomics: The production environment in the tropics and specifically in Africa is confronted with disturbances that include high temperatures, heat stress, inadequate and low nutritional value feeds, seasonal and unpredictable rainfall, as well as disease and parasite burdens. These environmental disturbances compromise the ability of animals to perform optimally and also affect the welfare of the animals. Monitoring of body signals and biomarkers, such as saliva cortisol, blood lactate, heart rate variability, and temperature profiles, passively without disturbing the animal, is a great way to help the farmer make sound management decisions based on information directly coming from the animals (Guevara et al. 2022; Neethirajan 2017). These methods can detail the change and trajectory of change in the animal's physiological parameters and the biophysical conditions of the animal. In cases where the target parameters are difficult to measure, proxies and indicators can be used. These direct measurements also allow assessment of the welfare of animals, although scoring systems are also widely used. At the same time, these data can provide an excellent platform to define and capture novel phenotypes. For example, traits related to thermal stress can be measured using techniques such as thermal imagery. New ITC is helping develop farmer-centred decision-support tool/solution that integrates data from different sources. In the future, this will also include "digital twinning"—and to apply the twinning as a continuum from a static animal breeding modelling to full-fledged pairing with breeding populations including bi-directional data flows. Specific breeding populations and breeding programmes can be used to inform novel phenotypes through more in-depth qualitative data (thick data) and high-throughput quantitative data generated by ICT (big data). In the area of big data, future breeding programmes can also embrace the potential contribution of other technologies, including the Internet of Things, artificial intelligence, extended reality, and cloud storage (See Chapter 20).

Advances in genome-wide scanning technologies and computational capacity have increased the range of tools available in animal breeding. Tools to detect patterns of single nucleotide polymorphisms (SNPs) that became prevalent in genomes because of environmental selection pressure (signatures of selection) are becoming readily available. These signatures provide fundamental insights into the genome dynamics of a species and trace possible causative links to productivity and resilience traits in livestock. Genomics is also assisting in identifying the specific genes underlying difference traits. Whole genome sequences have greatly increased the ability to precisely infer population genetic parameters, demographic processes, and selection signatures (See Chap. 20). In future breeding programmes, genomics will not only present the opportunity to increase the accuracy of selection but also accelerate the genetic gains[53] (See Chapter 20). This calls for concerted efforts to catalogue the genomic variations in African livestock species and relate same with traits of interest to support the accurate estimation of genomic breeding values for the acceleration of the rate of genomic progress in livestock breeding programs. The recently initiated African BioGenome Project (AfricaBP, https://africanbiogenome.org/), which aims to sequence the genomes of 105,000 endemic animals, plants fungi, protists, and other eukaryotes species on the African continent (Ebenezer et al. 2022) (see Chap. 29), is a positive step towards the characterization of the genomic variations in African livestock.

(d) Account for farmer and consumer needs: To account for farmer, consumer, religious and cultural needs, future breeding goals should include consumer preferences (e.g. healthy fatty acid profiles, more muscle, high protein content, etc.) and attitudes towards animal-derived food products. Consumer willingness to pay for food safety and quality and animal welfare costs is increasing due to increased income per capita. There is a need to explore how consumer value-specific animal food attributes linked to food safety and nutrition for incorporation in animal breeding goals.

(e) Ecosystem perspective: On the one hand, land-use pressures, including conversion of rangelands, fencing, and fragmentation of land, pose a major challenge to some of the traditional livestock production systems such as pastoralism. On the other hand, moderate levels of livestock grazing in some parts of Africa (e.g. East Africa) have been associated with beneficial outcomes in terms of wildlife biodiversity and soil quality. Meanwhile, the expansion of livestock populations in some areas has been associated with negative effects such as wildlife population declines. Future breeding strategies should include traits (e.g. feed and water efficiency, methane emissions, disease resistance, etc.) that demonstrate that livestock are part of the ecosystem.

(f) Adaptation to climate change: Changing climatic conditions is among factors militating against livestock productivity. Rising temperatures accompanying limited rainfall, drought conditions, expanding desert lands, limited feed resources and quality, and high disease burden are the direct consequences of changing climatic conditions. To maintain/increase livestock productivity under these conditions, traits such as heat and drought resistance must be considered in future breeding strategies for resilient livestock.

(g) Gender: When characterizing the production environment, gender roles and power relations should be analysed sufficiently to gain an understanding of intra-household issues that develop with the introduction of innovative breeding technologies. This will enable the selection of traits to target women or men specifically, in order to meet their specific needs. The roles of male, female, and young farmers and their interaction with new technologies, which may create a synergistic effect in society, should be considered in breeding programmes. This calls for participatory approaches in design of breeding programmes, ensuring that the voices of men, women, and the youth are intentionally incorporated. Gender-responsive strategies aim at incorporating diverse gender roles and needs.

(h) Economic perspective: The economic aspect of livestock production is not entrenched in the livestock productivity goals of smallholder livestock farmers, whose interest is to keep a few animals for food and for sale when a need arises. To exploit the economic potential of livestock production, training in the various aspects of the livestock-value-chain should be actively promoted. A realization of the huge economic potentials of livestock production, for example, the production of livestock products to meet specific market niches may be a driving force to increased production.

These points are not exhaustive, but they provide a starting point for upscaling using a sustainable business model in an affordable way. The implementation of these programs and commercialization of animal production require the cooperation and joint program delivery by stakeholders involved in the livestock production value chain.

31.7 Conclusion and Future Perspectives

Livestock has been identified as a major pillar in the fight against hunger and poverty alleviation in Africa. The overarching message of this book is that sustainable development of the African live-

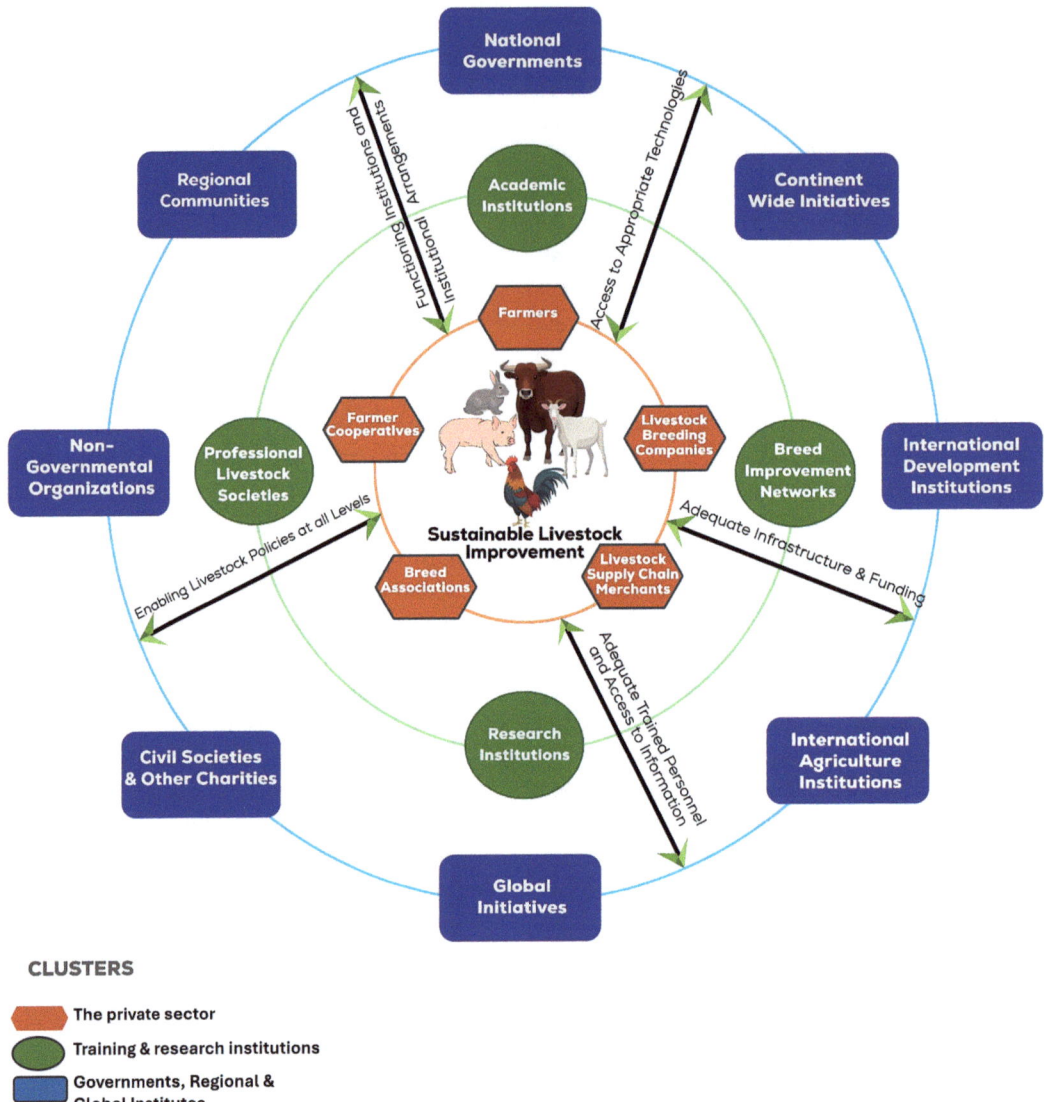

Fig. 31.7 Collaboration web for successful delivery of sustainable livestock development in Africa

stock sector requires attention to efforts towards enhancing the genetics of livestock, an area that has been historically neglected. For this to happen, there is need to pay attention to the development of enabling policies and functioning institutions that specifically address the current challenges to development and implementation of effective strategies, plans, and programmes that address animal genetic resources challenges, including conservation of indigenous breeds and sustainable genetic improvements that respond to needs of specific contexts. Access to appropriate modern technologies and adequate resourcing of priority programmes are critical pieces. Some of the items in the puzzle are already in place needing the right coordination, policies, and resourcing to drive sustainability of the sector. To be successful, functioning linkages and viable collaborations between stakeholders (discussed in previous sections) are critical (Fig. 31.7).

The recently created African Animal Breeders Network (AABNet) (Djikeng et al. 2025) (http://www.animalbreeding-africa.org) will be a key player in this regard. The AABNet proposes a bold ambition to develop into a unique platform of highly skilled and experienced animal breeders and professionals with the capability to provide highly needed support and advice on animal genetics and sustainable livestock improvement across the African continent, paying attention to human capacity development, advocacy and resource mobilization, and to foster partnerships with global and African institutions to advance livestock genetic improvement in Africa. Its main goals are to (1) coordinate livestock improvement program(s) within and across African regions, and (2) facilitate collaborations with key stakeholders, investors, development agencies, international institutions, and relevant non-governmental organizations to drive a sustainable transformative agenda of Africa's livestock industry. The AABNet's transformative agenda for the development of the African livestock industry is achievable through four core activities: (i) spearhead multi-country genetic evaluation linked to the design of breeding programs for the optimization of genetic gains; (ii) actively participate in professional development through talent and technology incubation; (iii) advocacy, create awareness, and facilitate the business development of the livestock sector; and (iv) actively create collaboration, networking, and partnerships with stakeholders. As part of its objectives of participating in professional development and creating awareness, the AABNet has already initiated training programs on various topics on sustainable livestock production and business development with courses already delivered in the following areas: (1) Practical animal breeding, (2) Data science and bioinformatics for tackling animal breeding, (3) Biothermoimaging tools for management of climate smart and precision livestock-assisted reproduction, etc. In particular, its inaugural 3-weeks training activity exposed 48 participants (early career animal breeders) from 30 countries to practical skills in animal breeding and genetics. AABNet's strategy focuses on demand-led capacity development, starting with highest priority capacity gaps. This book is one of AABNet's flagship projects to characterize the African livestock genetic resources, production systems, and breeding strategies by documenting what is already known and identify gaps that needs to be addressed.

In conclusion, the currently underutilized African livestock genetic resources can be sustainably transformed. The basic elements and requirements for the implementation of sustainable and resilient livestock improvement programmes in Africa have been presented in this volume, and the key drivers that will underpin sustainable and resilient livestock sector transformation in Africa are enabling policies, functional institutional arrangements, access to appropriate technologies, funding and information, and adequate infrastructure and trained personnel.

Acknowledgement We thank Mr. Emmanuel Bassam (HPI Director of Program) for providing the reports of HPI Cameroon and for extracting and making available photos from HPI Cameroon archives.

References

Adebambo O (2013) Introductory animal agriculture. Federal University of Agriculture, Department of Animal Breeding and Genetics

African Union (2021) African seed and biotechnology partnership platform operational guidelines. African Union. Available from https://au.int/sites/default/files/bids/39824-revised_asb_platform_guidelines_with_inputs_from_the_virtual_meeting_of_4_september_2020sn.pdf. Accessed 9 July 2021

Ahuya CO, Okeyo AM, Hendy CRC (2002) Community-based livestock improvement: a case study of FARM-Africa's goat improvement project in Meru, Kenya. Paper presented at the 28th Tanzania Society of Animal Production, Scientific Conference, Tanesco Training Institute, Morogoro, Tanzania, 7–9 August 2001

Appolinaire, Djikeng Victor E., Olori Isidore, Houaga Samuel E., Aggrey Okeyo, Mwai Eveline M., Ibeagha-Awemu Raphael, Mrode Mizeck G. G., Chagunda Christian K., Tiambo Romdhane, Rekaya Oyenkanmi, Nash Zabron, Nziku Oluyinka, Opoola Mapholi, Ntanganedzeni Chinyere, Ekine-Dzivenu Alexander, Kahi Tobias, Okeno John M., Hickey Negussie, Enyew Edward J. O., Rege (2025) The African Animal Breeding Network as a pathway towards genetic improvement of livestock. Nature Genetics 57(3):498-504. https://doi.org/10.1038/s41588-025-02079-4

Atekwana JA, Maximuangu JC (1981) The performance of imported dairy cattle breeds in Cameroon. Paper presented at sixth meeting of the African Scientific Council, Libreville, Gabon

Becker J (2016) Beef Genomics Programme: sequencing for bovine superiority: production. Red Meat/Rooivleis 7(5):58–63. https://doi.org/10.10520/EJC196215. Acccessed 20 Mar 2024

Djikeng A, Olori VE, Aggrey SE, Mwai O, Ibeagha-Awemu EM, Mrode R, Chagunda MGG, Tiambo CK, Houaga I, Rekaya R et al (2025) The African Animal Breeding Network: pathway towards genetic improvement of livestock. Nature Genetics 57(3):498–504. https://doi.org/10.1038/s41588-025-02079-4

Ebenezer TE, Muigai AWT, Nouala S, Badaoui B, Blaxter M, Buddie AG, Jarvis ED, Korlach J, Kuja JO, Lewin HA et al (2022) Africa: sequence 100,000 species to safeguard biodiversity. Nature 603(7901):388–392. https://doi.org/10.1038/d41586-022-00712-4

ECOWAS (2010) Strategic action plan for the development and transformation of livestock sector in the ECOWAS region (2011–2020). ECOWAS, Abuja

FAO (2007) Food and Agricultural Organisation (FAO) of the United Nations, Global plan of action for animal genetic resources and the Interlaken Declaration. Food and Agricultural Organizations of the United Nations, Rome. https://www.fao.org/policy-support/tools-and-publications/resources-details/fr/c/453629/2007. Consulted on 27 Oct 2021

FAO (2011) Developing the institutional framework for the management of animal genetic resources, FAO Animal Production and Health Guidelines. No. 6. FAO, Rome. Available at https://www.fao.org/publications/card/en/c/054fcb55-a353-5127-a4d9-6dfce197e16f/. Accessed April 2024

FAO (2015) Food and Agricultural Organisation (FAO) of the United Nations, FAO Synthesis – livestock and the Sustainable Development Goals Global Agenda for Sustainable Livestock. Draft prepared by FAO-AGAL Livestock Information, Sector Analysis and Policy Branch. Available at http://www.livestockdialogue.org/fileadmin/templates/res_livestock/docs/2016/Panama/FAO-AGAL_synthesis_Panama_Livestock_and_SDGs.pdf

FRN (2019) Report on a Bill for an Act to Provide for the Establishment of the National Livestock Identification and Management Bureau for the Purpose of Animal Identification, Traceability Registration, and for Other Related Matters, 2019 SB 373. Federal Republic of Nigeria National Assembly. Senate Committee on Agriculture & Rural Development Committee on Agriculture & Rural Development, 26 p

GASL (2019) Smallholder livestock systems innovations for sustainability: a policy brief from the NGO cluster of the Global Agenda for Sustainable Livestock (GASL). Available at https://www.livestockdialogue.org/fileadmin/templates/res_livestock/docs/2019_Sept_Kansas/Policy_brief_ago_2019_ok_web.pdf. Accessed on 19 Mar 2024

GMFA (2016) Ghana livestock development policy and strategy. Ghana Minstry of Food and Agriculture. Available at https://www.fao.org/faolex/results/details/en/c/LEX-FAOC169291/. Accessed 22 Nov 2023

Guevara RD, Pastor JJ, Manteca X, Tedo G, Llonch P (2022) Systematic review of animal-based indicators to measure thermal, social, and immune-related stress in pigs. PLoS One 17(5):e0266524. https://doi.org/10.1371/journal.pone.0266524

Ibeagha-Awemu EM, Peters SO, Bemji MN, Adeleke MA, Do DN (2019) Leveraging available resources and stakeholder involvement for improved productivity of African livestock in the era of genomic breeding. Front Genet 10:357. https://doi.org/10.3389/fgene.2019.00357

ILRI (2022) Overview of the Africa dairy genetic gains (ADGG) project. Presented at workshop on Training Performance Recording agents on data capture tools, Kawempe, Uganda, 18–22 July 2022. ILRI, Nairobi. https://hdl.handle.net/10568/128305

Joubert R (2016) Welcome to the DGP: technology. Dairy Mail 23(9):98–101. https://doi.org/10.10520/EJC194733. Acccessed 20 Mar 2024

Joubert R (2018) DGP update. Dairy Mail 25(2):77–79. https://doi.org/10.10520/EJC-c716f94ce. Acccessed 20 Mar 2024

Kling GW, Clark MA, Wagner GN, Compton HR, Humphrey AM, Devine JD, Evans WC, Lockwood JP, Tuttle ML, Koenigsberg EJ (1987) The 1986 lake nyos gas disaster in Cameroon, west Africa. Science 236(4798):169–175. https://doi.org/10.1126/science.236.4798.169

Leroy G, Baumung R, Notter D, Verrier E, Wurzinger M, Scherf B (2017) Stakeholder involvement and the management of animal genetic resources across the world. Livest Sci 198:120–128. https://doi.org/10.1016/j.livsci.2017.02.018

Lysons K, Farrington B (2012) Purchasing and supply chain management. Pearson Education Limited, London

Mrode R, Aliloo H, Ekine C, Ojango JD, Gibson JP, Okeyo M (2019) The application of several genomic models for the analysis of small holder dairy cattle data. Proceedings of the 2019 Interbull meeting. Interbull Bull 55:70–76. https://journal.interbull.org/index.php/ib/article/view/1486

Mrode R, Ojango JMK, Chinyere E, Mwai O (2020) The platform for Africa dairy genetics gain (ADGG) new index for dairy cattle in Tanzania and guidelines on its use. Nairobi, Kenya. https://africadgg.files.wordpress.com/2020/08/adgg-tzn-index-2020.pdf

Mueller JP, Rischkowsky B, Haile A, Philipsson J, Mwai O, Besbes B, Valle Zárate A, Tibbo M, Mirkena T, Duguma G et al (2015) Community-based livestock breeding programmes: essentials and examples. J Anim Breed Genet 132(2):155–168. https://doi.org/10.1111/jbg.12136. Acccessed 20 Mar 2024

Neethirajan S (2017) Recent advances in wearable sensors for animal health management. Sens Biosensing Res 12:15–29. https://doi.org/10.1016/j.sbsr.2016.11.004

Ojango JMK, Okeyo AM, Mrode R, Chinyere E, Gebreyohanes G, Meseret S, Mogaka D, Lyatuu E, Komwihangilo D, Msuta G et al (2022) 459. Bridging the gap in data from smallholder dairy systems; developing the Africa dairy genetic gains (ADGG) data platform. In: Proceedings of 12th World Congress on Genetics Applied to Livestock Production (WCGALP). Wageningen Academic Publishers, pp 1909–1912

Okantah SA (2009) A review of studies on breed evaluation and genetic improvement of cattle in Ghana. Ghana J Agric Sci 42:195–206. https://doi.org/10.4314/gjas.v42i1-2.60659

Olori VE, Nash O (2021) Livestock Bio-Resource Development Centre: facilitating the determination of the genetic variation and potential of individual animals in Nigeria. Unpublished: Paper presented at the 3rd Strategic Interest Research Group meeting on Genetic Improvement of Livestock. IITA, Ibadan, Nigeria, 2–3 September 2021

Olori VE, Cromie AR, Wickham B (2005) Practical aspects in setting up a cattle breeding program for Ireland. Invited Plenary presentation at the 56th annual meeting of the European Association for Animal Production. G5.2. Uppsala, Sweden

Peacock C, Ahuya CO, Ojango JMK, Okeyo AM (2011) Practical crossbreeding for improved livelihoods in developing countries: the FARM Africa goat project. Livest Sci 136(1):38–44. https://doi.org/10.1016/j.livsci.2010.09.005

Schneider F, Tarawali S (2021) Sustainable Development Goals and livestock systems. Rev Sci Tech 40(2):585. https://doi.org/10.20506/rst.40.2.3247

Thomas-Slayer BP (1992) Implementing effective local management of natural resources: new roles for NGOs in Africa. Hum Organ 15(2):136–143

Van Tassell CP, Rosen BD, Woodward-Greene MJ, Silverstein JT, Huson HJ, Sölkner J, Boettcher P, Rothschild MF, Mészáros G, Nakimbugwe HN et al (2023) The African Goat Improvement Network: a scientific group empowering smallholder farmers. Front Genet 14:1183240. https://doi.org/10.3389/fgene.2023.1183240

Wurzinger M, Sölkner J, Iñiguez L (2011) Important aspects and limitations in considering community-based breeding programs for low-input smallholder livestock systems. Small Rumin Res 98(1):170–175. https://doi.org/10.1016/j.smallrumres.2011.03.035

Yerpes M, Llonch P, Manteca X (2021) Effect of environmental conditions during transport on chick weight loss and mortality. Poult Sci 100(1):129–137. https://doi.org/10.1016/j.psj.2020.10.003

Zonabend E, Okeyo A, Ojango JM, Hoffmann I, Moyo S, Philipsson J (2013) Infrastructure for sustainable use of animal genetic resources in Southern and Eastern Africa. Anim Genet Resour 53:79–93

Open Access This chapter is licensed under the terms of the Creative Commons Attribution 4.0 International License (http://creativecommons.org/licenses/by/4.0/), which permits use, sharing, adaptation, distribution and reproduction in any medium or format, as long as you give appropriate credit to the original author(s) and the source, provide a link to the Creative Commons license and indicate if changes were made.

The images or other third party material in this chapter are included in the chapter's Creative Commons license, unless indicated otherwise in a credit line to the material. If material is not included in the chapter's Creative Commons license and your intended use is not permitted by statutory regulation or exceeds the permitted use, you will need to obtain permission directly from the copyright holder.

Index

A
Adaptation to harsh environments, 191
Adaptive traits, 31, 51, 220, 227, 303, 305, 329, 377, 389, 652, 735, 773, 804, 820–822, 843, 1096
Africa, 4, 14, 186, 240, 329, 358, 399, 452, 499, 595, 620, 679, 693, 760, 798, 820, 852
African Confederation of Equestrian Sports (ACES), 552
African indigenous horse breeds, 501, 519, 524, 532, 538, 547, 552, 574, 575, 578–579, 583, 584
African indigenous livestock breeds, 4
African indigenous pig breeds, 359, 362, 363, 380, 391, 825, 842
African indigenous poultry breeds, 346, 347
African production systems, 693–695, 707–751, 773, 820–844, 858
African taurine cattle, 66–110
Animal agriculture, 7, 14, 30, 45, 363, 390, 391, 677, 741, 828, 842
Animal genetic resources (AnGR), 3, 5, 7–10, 14, 15, 19, 23, 45, 51, 52, 55, 56, 186, 187, 189, 303, 347, 362, 373, 381, 382, 388, 389, 419, 443, 444, 581, 604, 620–666, 697, 698, 711, 761, 763, 776, 787, 823, 828, 831, 839, 842, 843, 854, 857, 1076, 1131, 1200, 1202, 1203, 1205, 1207–1209, 1215–1219, 1237
Animal health, 44, 48, 56, 382, 388, 389, 418, 419, 437, 491, 584, 683, 709, 710, 764, 790, 795, 809, 823, 856, 858, 861, 867, 875, 882, 1052
Animal identification, 304, 314, 708, 726, 727, 737, 799, 803, 855, 858
Animal nutrition science, 1041–1042
Animal production, 15, 30, 38, 44, 46, 49, 59, 255, 382, 388, 613, 680, 715, 788, 791, 798, 807, 831, 832, 853, 855, 857–867, 875, 876, 882, 1244–1248
Animal relationship Matrix (A-Matrix), 1027, 1028
Animal reproduction, 841
Arrangements, 9, 36, 502, 610, 722, 863

B
Barb horse, 500, 501, 503–506, 518, 520, 521, 523, 525, 526, 528, 530, 538, 545, 552, 556, 557, 564, 565, 570, 574–577, 584

Best Linear Unbiased Prediction (BLUP), 330, 728, 731, 736, 748, 797, 801, 1024–1025, 1027, 1028, 1176, 1178
Biodiversity conservation, 443, 666
Biotechnologies, 51, 347, 390, 432, 580, 611–613, 682, 745, 787, 791, 841, 842, 875, 883, 886
Breeding, 7, 30, 190, 242, 328, 362, 400, 453, 506, 604, 621, 676, 692, 760, 787, 820, 853
Breeding goals, 7, 50–51, 228, 242, 304–309, 381, 418, 583, 584, 692–701, 707–751, 760–779, 787–812, 820, 824–825, 827–833, 840, 843, 1273
Breeding goals for low input systems, 698
Breeding products, 740, 760, 761, 763, 766, 777, 778
Breeding program, 7, 46, 108, 121, 219, 304, 349, 362, 441, 490, 583, 604, 623, 692, 708, 760, 787, 820, 855, 914, 993, 1023, 1051, 1073, 1093, 1112, 1126, 1151, 1206, 1235
Breeding strategies in Africa, 7, 230, 452–491, 499–585, 691–701, 707–751, 760–779, 787–812, 820–844
Breeding technologies, 584, 743, 823, 856, 865, 871–875
Breeds, 4, 15, 186, 242, 358, 400, 453, 500, 596, 639, 692, 708, 760, 787, 820, 854
Breed substitution, 329, 715–726, 802–803
Busia pig, 359, 363, 370

C
Caballus, 499, 500, 503
Camel, 22, 25, 26, 30, 32, 36, 39, 360, 398–402, 408–410, 419–422, 424, 426, 432, 438, 439, 443, 444, 452, 477, 482, 491, 529, 547, 551, 574, 696, 710, 811, 871
Capacity building, 9, 51, 60, 374, 382, 386, 388, 391, 490, 584, 712, 715, 823, 827, 842–843, 857, 1063, 1201
Characterization/characterisation, 45, 46, 59, 109, 189–227, 230, 242, 269, 298–301, 313, 329, 330, 347, 362, 366–369, 371–374, 380, 389–391, 407, 439, 453, 458, 459, 477–479, 490, 512, 514, 526, 534, 553, 574–576, 579, 580, 604, 607–610, 625, 635, 646–650, 662–665, 683, 726, 739, 765, 779, 788, 823, 830, 831, 875, 1201

Climate change, 7, 23, 29, 30, 34, 45, 46, 48–49, 52, 56, 58, 142, 220, 332, 345, 382, 438, 443, 501, 621, 623, 665, 676–677, 680, 684, 685, 701, 709, 750, 751, 801, 802, 823, 827, 834, 835, 856, 865, 921, 1214

Community-based breeding program (CBBP), 308, 381, 613, 700, 736–739, 773, 836, 866, 1076

Conservation, 8, 49, 189, 242, 329, 358, 419, 453, 505, 612, 623, 726, 761, 798, 827, 870

Conservation and utilization, 336, 381

Conservation of African livestock genetic resources, 8–9

Core facilities, 1166–1172

Crossbreeding, 8, 47, 50, 51, 159, 190, 191, 216, 219, 243, 246, 258, 266, 271, 277, 278, 288, 289, 295, 303, 306–308, 314, 328, 329, 332, 349, 350, 363, 374, 375, 378, 380, 383, 386–387, 389, 404, 421, 422, 444, 461, 503, 506, 509, 516–518, 520, 523, 526, 528, 531–533, 535, 536, 540, 541, 543, 545, 555, 576, 583, 584, 643, 695–697, 699, 700, 708, 712, 714–726, 728, 730, 738–741, 744, 764, 766, 767, 769, 771, 773, 789, 796–798, 802, 803, 806, 822, 823, 828, 831, 832, 836, 854–855, 887

Crossbreeding system, 173, 789, 797, 802

Cryoconservation, 347, 1077, 1173, 1208

Cultural values, 14, 514, 621, 693–695, 709

D

Databases, 329, 439, 443, 510–512, 515–517, 519–521, 523, 524, 526, 528–530, 538, 580, 599, 610, 647, 694, 708, 760, 830, 858, 866, 874

Digital tools, 694, 708, 866, 891

Disease, 25, 68, 121, 187, 276, 329, 358, 402, 459, 500, 600, 623, 683, 693, 707, 762, 789, 820, 853, 911, 992, 1023, 1050, 1073, 1092, 1112, 1128, 1153, 1188, 1238

Disease surveillance, 388

Diversity, 4, 29, 186, 254, 328, 361, 398, 453, 522, 597, 626, 682, 700, 715, 760, 807, 830

Donkey, 4–7, 14, 15, 22, 23, 25, 29, 40, 360, 421, 423, 429, 452–491, 499, 543, 546, 554, 555, 557, 573, 583, 601, 696, 697

Dromedary, 4–7, 360, 397–444, 580, 871

E

Ecological settings, 69, 72, 75–78, 80–86, 107–108, 117, 119, 124, 125, 129–131, 153

Ecological zones, 83, 206, 230, 255, 256, 262, 634, 663, 716

Economic valuation, 9, 1095

Enabling livestock policies, 51, 55–56

Enabling policies, 10, 51, 52

Exotic cattle breeds, 68, 70, 118–177

Ex-situ conservation, 45, 303, 347, 348, 581, 612, 842, 844, 1207

F

Farmers participation, 701, 736, 746, 750

Food security, 4, 43, 51, 58, 227, 228, 314, 331, 345, 351, 358, 362, 380, 381, 389, 392, 438, 610, 665, 666, 676–677, 685, 707, 751, 812, 823, 833, 842, 844, 1227

Frameworks, 4, 9, 10, 46, 48, 51, 55, 60, 347, 443, 477, 552, 603, 697, 737, 761, 825, 1195, 1218, 1249–1250

G

Gene banks, 347, 381, 666, 828, 842, 843

Gene editing, 710, 741–743, 853, 875, 876, 878–882

Gene mapping, 600–603

Genetic characteristics, 89, 187, 191–197, 201–203, 205, 206, 208, 209, 214, 216–220, 222, 253, 255, 263, 269, 270, 272, 275–283, 286, 287, 290–295, 303, 334–344, 836

Genetic diversity, 7, 9, 19, 50, 51, 187, 189–227, 242, 254, 298, 300–304, 308, 314, 328–331, 338, 341, 346, 348, 349, 360–362, 369, 370, 389, 397, 441–443, 478–479, 500, 515, 516, 576–578, 580, 581, 584, 597, 604–610, 612, 621, 623, 634, 647–648, 664–666, 708, 728, 765, 777, 787, 831, 836, 840, 852, 855, 877

Genetic improvement, 9, 30, 51, 58, 216, 219, 224, 305–308, 311, 313, 314, 338, 350, 352, 380, 389, 390, 401, 436, 551, 580, 583–584, 600, 663, 684, 698, 699, 707, 708, 713, 715, 718–720, 722, 723, 726–729, 732, 733, 735–737, 739–750, 764, 779, 787–791, 794, 795, 797–801, 803, 805, 807–812, 820, 824, 825, 828, 830–832, 837, 839, 841, 853, 855, 858, 865, 871, 873–875, 886, 890, 892, 1059, 1126, 1128, 1131–1134, 1189, 1221, 1249, 1253, 1255–1265

Genetic parameters, 311, 728, 731, 733, 764, 765, 795, 797

Genetic resources, 3–10, 15–23, 49, 51, 186–230, 240–315, 328–352, 358–392, 397–444, 452–491, 499–585, 604, 611–613, 621, 632–635, 639–645, 697, 726, 727, 736, 737, 760, 761, 763, 766, 770, 771, 773, 776, 787, 789, 803, 812, 821–822, 830–833, 835, 842–844, 1199

Genetics, 6, 15, 187, 242, 328, 359, 397, 453, 500, 597, 625, 683, 692, 707, 762, 787, 820, 853

Genomic BLUP (GBLUP), 710, 874, 998, 1028

Genomics, 8, 9, 46, 47, 50, 51, 54, 187, 223–225, 229, 254, 299, 301, 302, 304, 306, 314, 330, 381, 398, 583, 639, 647, 648, 708–710, 713, 715, 741–743, 751, 765, 769, 778, 779, 791, 797–801, 804, 805, 807–809, 826, 828, 831, 839–841, 844, 852–854, 856, 866–886, 890, 891, 1025, 1246

Genotype-by environment interaction, 821, 835

Index

Germplasm centres, 1173–1175
Goats, 4, 14, 186, 243, 360, 398, 452, 500, 599, 676, 693, 760, 810, 854, 995
Goliath frog, 7, 621–623, 665, 666
Greenhouse gas emissions, 676–686, 714, 835, 865, 884

H
Honeybee, 7, 620, 650–665
Horse production and breeding systems, 545–550
Human resource development, 1199
Husbandry practices, 40, 44, 49, 136, 146, 193, 200, 202, 203, 206, 207, 210, 211, 216–219, 257, 271, 272, 280, 284, 285, 289, 334–345, 367, 371, 374, 378, 388, 443, 505, 506, 610–611, 730, 737, 738, 918, 938, 1063, 1085

I
Improve disease management, 490
Improved utilization and modern technologies, 8
Improve management, 866
Indigenous breeds, 7, 14, 30, 36, 39, 41, 43, 45, 50–52, 158, 195, 201–206, 267, 295, 307, 308, 335–340, 343, 344, 346, 363, 365, 366, 371–373, 378, 380, 387, 389, 525, 528, 530, 574, 584, 644, 645, 699, 700, 709, 712, 718, 719, 722, 727, 728, 737, 765–767, 769, 771, 773, 774, 778, 779, 793, 803, 829, 831, 832, 843, 855, 1096
Indigenous livestock, 34, 36, 45, 46, 50, 51, 58, 230, 697, 735, 854, 855, 870
Indigenous pig breeds, 36, 362–364, 367–374, 380, 391, 821, 822, 828, 831, 832, 836, 837, 844
Information systems, 364, 858–859
In situ conservation, 303, 347, 380–382, 581, 611, 612, 737, 828, 837, 842
Institutional, 8, 9, 23, 44, 46, 48, 51, 52, 351, 358, 362, 379, 381, 385, 387–389, 391, 712, 728, 729, 737, 744, 746, 765, 771, 774, 776, 790, 811, 812, 822, 823, 828, 841–844, 857, 1204
Institutional arrangements, 9–10, 52, 762–764

K
Kolbroek, 359, 362, 363, 366, 372–373

L
Livestock, 4, 14, 219, 240, 330, 358, 406, 452, 502, 597, 620, 676, 693, 761, 788, 820, 852
Livestock data, 855
Livestock farming, 4, 7, 9, 19, 30, 32, 33, 43, 51, 58, 680, 853, 859, 882, 912
Livestock marketing, 55, 857

M
Machine learning (ML), 864
Microbiome, 682
Mitigation, 15, 54, 56, 58, 679, 680, 682–684, 701, 709, 751, 857, 865, 866
Mobile application, 694, 858
Modern technology, 8, 315, 390, 743, 778, 794, 812, 842, 843, 852–854, 857, 862, 890–892, 973, 992–1015
Modular system, 1011
Mukota pig, 359, 363, 372
Multi-trait across-country evaluations (MACE), 1154–1156, 1176, 1177, 1181

N
Newcastle, 39, 337, 1037, 1038
Non-conventional animal, 15, 19
Nutrition, health and growth technologies, 912

O
Origins of African cattle, 66–69

P
Phenomics, 8, 852–855, 857–866
Phenotyping technology, 792, 839, 865–866
Physical and production characteristics, 7, 297
Pig production systems, 37, 358, 368, 370, 375–380, 383, 823, 825–830, 832–833, 843
Policies, 8, 27, 303, 346, 358, 418, 489, 684, 697, 761, 798, 821, 853, 1236
Poultry, 6, 14, 328, 459, 643, 676, 696, 787, 839, 854, 911, 1001, 1022, 1054, 1098, 1111, 1127, 1188, 1239
Poultry breeding, 352, 643, 787, 788, 791, 794, 796–799, 801, 802, 806, 839, 1023, 1024, 1027, 1030, 1031, 1039, 1243
Poultry farming systems, 330–332
Poultry improvement breeding, 328, 329, 349, 350
Poultry production, 23, 37–39, 327, 328, 331, 335, 337–340, 345, 349–352, 785, 794, 797, 798, 800, 802, 804, 810–812, 921, 939, 946, 1021–1023, 1033, 1035, 1037, 1042–1045, 1055, 1243–1245
Precision feeding, 54
Probiotics, 937, 938, 949
Production, 4, 14, 186, 240, 328, 358, 400, 452, 502, 596, 626, 676, 693, 707, 760, 787, 819, 852, 1127
Production efficiency, 15, 711, 712, 741, 767, 807, 823, 843, 877, 914–916

Q
Quantitative and genomic technologies, 1162

R
Rabbit, 4–7, 14–16, 19, 26, 34, 37, 360, 620, 639–650, 665, 666, 696, 747, 871, 875, 876, 881

Reproduction, 8, 40, 47, 48, 54, 122, 226–227, 298, 305–307, 309, 347, 377, 421, 434–436, 477, 488, 490, 521, 551, 552, 579–580, 604, 607, 608, 622, 631, 632, 634–637, 639, 649, 693, 701, 707, 713, 714, 728, 731, 736–738, 740, 745–748, 762–764, 788–793, 795, 796, 801, 804, 806, 810, 825, 833, 835, 840, 843, 852–854, 858, 861, 882, 887–889, 891, 1236

Reproductive and molecular technologies, 741–749, 764, 779

River buffalo, 360, 595–599, 601, 603, 610

Ruminants, 7, 14, 15, 19, 21, 24, 31, 37, 47, 49, 54, 56, 58, 187, 188, 206, 228, 314, 332, 358, 400, 424, 427, 431, 434, 436, 606, 608, 676–680, 682–684, 698, 709, 729, 735, 747, 760–779, 810, 856, 862, 865, 866, 886, 888, 889

S

Selection, 4, 6, 7, 9, 15, 46, 47, 49–51, 54, 125, 131, 172, 197, 202, 208, 212, 216, 217, 221, 224, 226, 242, 254, 279, 282, 289, 292, 298, 302, 305–311, 328, 330, 346, 381, 387, 400, 402, 418, 421–422, 431, 432, 437, 439, 444, 477, 488, 520, 532, 535, 537–539, 541, 543, 574, 579, 582, 584, 649, 664, 684, 692, 694–697, 699, 700, 707–709, 715, 717–719, 721–723, 726–743, 747, 748, 750, 762, 764–767, 773–776, 778, 787, 789–810, 820, 822, 824–826, 828, 830, 832–837, 839–841, 844, 854, 862–864, 870, 871, 873–875, 886, 1075, 1098, 1220

Sheep, 4–8, 14–17, 21–30, 32, 34, 36, 40, 41, 55, 193, 202, 206, 210, 228, 229, 240–315, 360, 421, 430, 438, 452, 454, 475, 500, 547, 557, 574, 599, 676, 679–681, 684, 693, 695, 696, 700, 734, 744, 749, 760–775, 778–779, 811, 854–856, 861, 863, 871–873, 875, 876, 878, 886, 887, 890, 1012, 1061, 1081, 1096, 1097, 1225, 1242, 1260

Snail, 6, 7, 37, 620, 624–639, 665, 666

Social and economic benefits, 9, 51, 711

Social, cultural and economic use, 711

Socio-cultural values, 711, 762, 820

Stakeholders, 5, 10, 45, 46, 50, 51, 54, 55, 59, 60, 313, 358, 373, 374, 381, 383, 386–388, 390, 391, 443, 491, 581, 582, 623, 711, 726, 727, 748, 763, 764, 779, 821, 824, 825, 828, 836, 839, 841–843, 1266

Strategies, 3–10, 15, 30, 45, 47, 48, 50–52, 54–58, 216, 227, 228, 230, 242, 302–309, 311, 314, 336, 346, 347, 352, 370, 380, 382, 385, 390, 406, 419, 421, 422, 432, 435, 436, 438, 443, 452–491, 499–585, 604, 611, 635–637, 649, 666, 678, 680, 682–684, 692–701, 707–751, 760–779, 787–812, 819–844, 853, 864, 866–868, 947, 1195, 1209, 1218

Subsistence livestock farming, 59

Sus scrofa, 358–361, 364

Sustainable Development Goals (SDGs), 19, 1125, 1127, 1129, 1267

Sustainable genetic improvement, 304–309, 313–314, 853

Sustainable livestock production, 8, 10, 46, 52, 58, 854, 856, 857, 974

Sustainable use, 9, 45, 51, 303–304, 347, 358, 362, 363, 374, 377–382, 387–391, 443, 623, 761, 766, 773, 810–812, 827, 831, 842–844, 857, 1201

T

Technological innovation, 52, 428

Technology, 5, 15, 219, 304, 347, 358, 424, 579, 604, 660, 698, 713, 762, 790, 821, 852

Total economic value (TEV), 652

Trained personnel, 10, 763

Traits of economic importance, 8, 381, 743, 788, 805, 808, 820, 823, 833–836, 864, 870, 871

Tropical climates, 95, 204

U

Uses, 7, 33, 139, 140, 155, 156, 166, 187, 201, 205, 220, 228, 242, 263, 272, 299, 362, 380, 398, 400, 427, 428, 430–434, 444, 453, 466, 480–485, 502, 511–515, 517–521, 523, 524, 526–529, 531–535, 537, 538, 550, 553–562, 573, 653–654, 807, 861, 868, 882–884, 934, 942, 966, 1092

W

Welfare and management, 919

Z

Zebu cattle, 41, 68, 118–147, 280, 709, 714, 716, 720, 721, 726, 732, 737–739, 746

Zebu-Taurine derivatives, 7

If you have any concerns about our products,
you can contact us on
ProductSafety@springernature.com

In case Publisher is established outside the EU,
the EU authorized representative is:
**Springer Nature Customer Service Center GmbH
Europaplatz 3, 69115 Heidelberg, Germany**

Printed by Libri Plureos GmbH
in Hamburg, Germany